This book contains a list of all the vascular plants found in the vast territory of the former USSR, an area extending from the Barents, Baltic, and Black seas to the Pacific Ocean and from the Arctic to the almost subtropical zone in south-central Asia. The list comprises over 22,000 species and subspecies of wild, naturalized, and adventive plants belonging to 1,945 genera and 216 families. Also included are hybrid genera, the most widely distributed cultivated plants, and 203 species that may be found within the territory, since they occur in neighboring countries.

Vascular Plants of Russia and Adjacent States is an enlarged and critically revised edition of the original Russian-language edition, *Vascular Plants of the USSR,* published in 1981. This new edition has taken into account the numerous changes in the composition, taxonomy, and nomenclature of the vascular plants of the former USSR that have been recognized since 1980.

VASCULAR PLANTS OF RUSSIA AND ADJACENT STATES
(THE FORMER USSR)

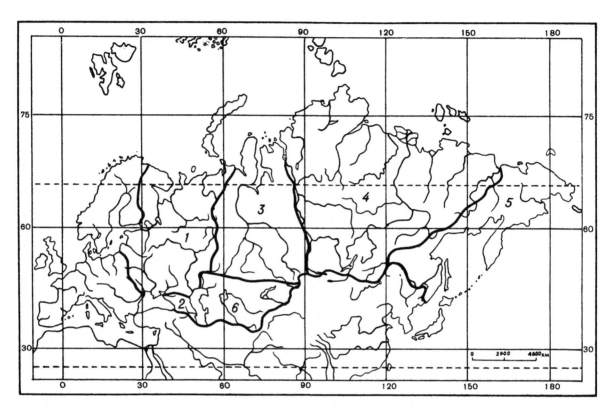

Regions of the former USSR. (1) Eastern Europe (European part of the former USSR).
(2) Caucasus. (3) Western Siberia. (4) Eastern Siberia. (5) Far East. (6) Middle Asia.

VASCULAR PLANTS OF RUSSIA AND ADJACENT STATES (THE FORMER USSR)

S. K. CZEREPANOV

V. L. Komarov Botanical Institute of the Russian Academy of Sciences

CAMBRIDGE UNIVERSITY PRESS
Cambridge, New York, Melbourne, Madrid, Cape Town, Singapore, São Paulo

Cambridge University Press
The Edinburgh Building, Cambridge CB2 8RU, UK

Published in the United States of America by Cambridge University Press, New York

www.cambridge.org
Information on this title: www.cambridge.org/9780521450065

First published 1995
This digitally printed version 2007

A catalogue record for this publication is available from the British Library

Library of Congress Cataloguing in Publication data

Czerepanov, Sergei Kirillovich.

Vascular plants of Russia and adjacent states (the former USSR) /
S. K. Czerepanov

p.   cm.

"Enlarged and critically revised edition of the original Russian-
language edition. Vascular plants of the USSR, published in 1981."

Includes index.

ISBN 0-521-45006-3 (hc)

1. Botany – Former Soviet Republics – Nomenclature. I. Czerepanov,
Sergei Kirillovich. Sosudistye rasteniia SSSR. II. Title.
QK321.C41   1995
580.947 – dc20   94-6992
CIP

ISBN 978-0-521-45006-5 hardback
ISBN 978-0-521-04483-7 paperback

CONTENTS

PREFACE

Vascular Plants of Russia and Adjacent States (the former USSR) is a new edition of the *Plantae Vasculares URSS* published in 1981. As was the earlier edition, this book is an attempt to compile a modern and to a certain extent critical list of the flora of the territory covered in *Flora URSS* (30 vols., 1934–1964), the standard work for the study of vascular plants in the territory of the former USSR. Since this monumental work was published, however, many further floristic and taxonomic studies have been undertaken and completed. The list after the Preface gives the principal floristic works that have appeared since the publication of *Plantae Vasculares URSS* in 1981.

This book covers 21,770 species and 500 subspecies belonging to a total of 1,945 genera and 216 families of both native and alien wild plants as well as cultivated plants that are in the process of becoming naturalized. In addition there are included 22 hybrid genera, 594 hybrids with binary names, and 203 species (marked in the text with a small black [■] and plain roman type) that may possibly be found in the territory of the former USSR, since they occur in neighboring countries. Also included are 533 species, in 129 genera and 14 families, of the most widely distributed cultivated plants. These are marked in the text with an asterisk (*).

As in the first edition, all the families, genera, species, subspecies, and varieties of species as well as synonyms of the taxa are arranged alphabetically by their Latin names. All accepted taxa are in roman bold type, but in the reference system and notes they appear in plain roman. Synonyms appear in italics, followed in the main list by an equals sign (=) and the accepted name in plain roman. If the name of a new taxon for the flora of the region in question is a nomenclatural combination, its basionym appears in the first place in the synonymy. New synonyms that are not cited in the *Flora URSS* are listed under accepted names. Taxa names that were recognized in the first edition as separate, but are regarded in our treatment as synonyms, are also listed.

The spelling of a considerable number of the generic names and species epithets has been checked with the original publications and, where necessary, has been altered to agree with article 73 of the *International Code of Botanical Nomenclature* (1988). Abbreviations of authors of plant names generally follow those in *Index Alphabeticus Auctorum ad Taxa Adductorum*, published by M.Kirpicznikov as an appendix to *Indices Alphabetici*, 30 vols., of *Flora URSS* (with exceptions such as Dumort., Bieb., Medik., Reichenb., Regel, and Wahlenb, instead of the *Index Alphabeticus's* Dum., M. B., Medic., Reichb.,

Rgl., and Wahl., respectively). If they do not appear in this index, the authors' names are spelled as in Appendix 1 of *Flora Europaea*. In some cases the abbreviations proposed by the authors of the taxa themselves are accepted. Original publications describing taxa are not cited.

In this book the geographical distribution of all wild and most cultivated species and subspecies of plants is indicated by reference to the six large geographical regions described in the *Flora URSS*. The numbers in the text adjacent to the species paragraphs indicate the regions where the plants can be found. These regions are as follows (see map):

1 Eastern Europe (European part of the former USSR)
2 Caucasus
3 Western Siberia
4 Eastern Siberia
5 Far East
6 Middle Asia

These regions are as they are defined in the *Flora URSS*, with two exceptions: the Arctic is not considered as a separate region and the western boundary of the Far East is the same as that stated in *Plantae Vasculares Orientalis Extremi Sovietici* (1985–1991).

The treatment of the family Apiaceae was done by M. G. Pimenov and V. N. Tikhomirov (Moscow State University). All other families are the work of the author. Because of the size and taxonomic complexity of the work the author has in many cases gladly deferred to the judgment of other taxonomic or regional specialists.

The author expresses his sincere graitude to D.V. Geltman, Yu. R. Roskov, M. L. Samutina, and G. R. Ivanova for their great assistance in preparing the book for publication. Camera-ready copy was prepared using software and hardware provided by the Missouri Botanical Garden, whose help is greatly acknowledged.

WORKS REFERRED TO IN THE PREFACE

All of the books concerning the floras of Russia and the countries in the former USSR are in Russian excepting *Flora Gruzii*, which is in Georgian. In the following list the alternative Latin titles are used, if they are present.

Conspectus florae Asiae Mediae. Tashkent: Fan.
  Vol. 6 (1981), 396 pp.
  Vol. 7 (1983), 416 pp.
  Vol. 8 (1986), 101 pp.
  Vol. 9 (1987), 400 pp.

Flora Arctica URSS. Leningrad: Nauka.
  Fasc. 8, [part 1] (1980), 334 pp.
  Fasc. 8, part 2 (1983), 52 pp.
  Fasc. 9, part 1 (1984), 334 pp.
  Fasc. 9, part 2 (1986), 188 pp.
  Fasc. 10 (1987), 411 pp.

Flora Abkhasii [Flora of Abkhasia]. A. A. Kolakovsk. Tbilisi: Metsniereba.
  Vol. 1 (1980), 212 pp.
  Vol. 2 (1982), 284 pp.
  Vol. 3 (1985), 294 pp.
  Vol. 4 (1986), 365 pp.

Flora Armenii [Flora of Armenia]. Yerevan: Izdatelstvo Akademii Nauk Arm SSR.
  Vol. 7 (1980), 292 pp.
  Vol. 8 (1987), 418 pp.

Flora Europaea in 5 volumes. E. G. Tutin et al. (eds). Cambridge: Cambridge University Press, 1968–1993.

Flora Gruzii [Flora of Georgia], 2d Ed. Tblisi: Metsniereba.
  Vol. 6 (1980), 320 pp.
  Vol. 7 (1981), 516 pp.
  Vol. 8 (1983), 378 pp.
  Vol. 9 (1984), 343 pp.
  Vol. 10 (1985), 391 pp.
  Vol. 11 (1987), 287 pp.

Flora Partis Europaeae URSS. Leningrad: Nauka
    Vol. 5 (1981), 380 pp.
    Vol. 6 (1987), 254 pp.
    Vol. 7 (in press).
    Vol. 8 (1989), 412 pp.

Flora Sibiriae. Novosibirsk: Nauka.
    [Vol. 1] (1988), Lycopodiaceae – Hydrocharitaceae. 200 pp.
    Vol. 2 (1990), Poaceae. 361 pp.
    [Vol. 3] (1990), Cyperaceae. 280 pp.
    [Vol. 4] (1987), Araceae – Orchidaceae. 248 pp.
    Vol. 5 (1992), Salicaceae – Amaranthaceae. 312 pp.
    Vol. 8 (1988), Rosaceae. 200 pp.

Flora Tadjikskoy SSR [Flora of the Tadjik SSR]. Leningrad: Nauka.
    Vol. 6 (1981), 728 pp.
    Vol. 7 (1984), 563 pp.
    Vol. 8 (1986), 519 pp.
    Vol. 9 (1988), 568 pp.
    Vol. 10 (1991), 624 pp.

International Code of Botanical Nomenclature adopted by the Fourteenth International Botanical Congress, Berlin, July–August, 1987. W. Greuter et al. (eds.). Koenigstein: Koeltz Scientific Books, 1988. 328 pp.

Opredelitel Rasteniy Sovetskogo Dalnego Vostoka [Key to plants of the Soviet Far East]. V. N. Woroschilow. Moskva: Nauka, 1982. 672 pp.

Opredelitel Rasteniy Turkmenistana [Key to plants of Turkmenistan. V. V. Nikitin and A. M. Geldykhanov. Leningrad: Nauka, 1988. 680 pp.

Opredelitel Rasteniy Tuvinskoy ASSR [Key to plants of Tuva ASSR]. Novosibirsk: Nauka, 1984. 336 pp.

Opredelitel Sosudistykh Rasteniy Kamchatskoy Oblasti [Key to vascular plants of the Kamchatka region]. Moskva: Nauka, 1981. 412 pp.

Opredelitel Vysshikh Rasteniy Moldavskoy SSR [Key to higher plants of Moldavian SSR]. 3d Ed. T. S. Heideman. Kishinev: Shtiintsa, 1986. 636 pp.

Plantae Asiae Centralis. Leningrad: Nauka.
    Fasc. 8a (1988), 125 pp
    Fasc. 9 (1989), 149 pp.

Plantae Vasculares Orientalis Extremi Sovietici. Leningrad: Nauka.
    Vol. 1 (1985), 398 pp.
    Vol. 2 (1987), 446 pp.
    Vol. 3 (1988), 421 pp.
    Vol. 4 (1989), 380 pp.
    Vol. 5 (1991), 390 pp.
    Vol. 6 (1992), 428 pp.

Plantae Vasculares URSS. S. K. Czerepanov. Leningrad: Nauka, 1981. 510 pp.

ACANTHACEAE Juss.
Acanthus L.

dioscoridis L. - 2

ACERACEAE Juss.
Acer L.

adscharicum Gatsch. - 2
assyriacum Pojark. = A. cinerascens
barbinerve Maxim. - 5
campestre L. - 1, 2
cappadocicum auct. p.p. = A. laetum
cappadocicum Gled. subsp. *divergens* (C. Koch ex Pax) E. Murr. = A. divergens
- subsp. *mayrii* (Schwer.) E. Murr. = A. mayrii
- subsp. *mono* (Maxim.) E. Murr. = A. mono
- subsp. *turkestanicum* (Pax) E. Murr. = A. turkestanicum
caudatum Wall. subsp. *ukurunduense* (Trautv. & C.A. Mey.) E. Murr. = A. ukurunduense
cinerascens Boiss.(*A. assyriacum* Pojark., *A. monspessulanum* L. subsp. *assyriacum* (Pojark.) Rech. fil., *A. monspessulanum* subsp. *cinerascens* (Boiss.) Yaltirik) - 2
■ divergens C. Koch ex Pax (*A. cappadocicum* Gled. subsp. *divergens* (C. Koch ex Pax) E. Murr.)
ginnala Maxim. - 5
heldreichii Orph. subsp. *trautvetteri* (Medw.) E. Murr. = A. trautvetteri
hyrcanum Fisch. & C.A. Mey.(*A. opalus* Mill. subsp. *hyrcanum* (Fisch. & C.A. Mey.) E. Murr.) - 2
- subsp. *stevenii* (Pojark.) E. Murr. = A. stevenii
ibericum Bieb. (*A. monspessulanum* L. subsp. *ibericum* (Bieb.) Yaltirik) - 2
japonicum Thunb. - 5
komarovii Pojark.(*A. tschonoskii* Maxim. subsp. *komarovii* (Pojark.) Urussov & Nedolushko, *A. tschonoskii* var. *rubripes* Kom.) - 5
laetum C.A. Mey. (*A. cappadocicum* auct. p.p.) - 2
mandshuricum Maxim. - 5
mayrii Schwer. (*A. cappadocicum* Gled. subsp. *mayrii* (Schwer.) E. Murr., *A. mono* Maxim. subsp. *mayrii* (Schwer.)Kitam., *A. mono* var. *mayrii* (Schwer.) Koidz. ex Nemoto, *A. truncatum* Bunge subsp. *mayrii* (Schwer.) E. Murr., *A. pictum* auct.) - 5
mono Maxim. (*A. cappadocicum* Gled. subsp. *mono* (Maxim.) E. Murr., *A. truncatum* Bunge subsp. *mono* (Maxim.) E. Murr.) - 5
- subsp. *mayrii* (Schwer.) Kitam. = A. mayrii
- var. *mayrii* (Schwer.) Koidz. ex Nemoto = A. mayrii
monspessulanum L. subsp. *assyriacum* (Pojark.) Rech. fil. = A. cinerascens
- subsp. *cinerascens* (Boiss.) Yaltirik = A. cinerascens
- subsp. *ibericum* (Bieb.) Yaltirik = A. cinerascens
- subsp. *turcomanicum* (Pojark.) E. Murr. = A. turcomanicum
- subsp. *turcomanicum* (Pojark.) Rech. fil. = A. turcomanicum
*****negundo** L. - 1, 2, 5, 6
opalus Mill. subsp. *hyrcanum* (Fisch. & C.A. Mey.) E. Murr. = A. hyrcanum
- subsp. *stevenii* (Pojark.) E.Murr. = A. stevenii
ovczinnikovii V. Zapr. = A. pubescens
*****pensylvanicum** L. - 1
pentapomicum auct. = A. pubescens
pictum auct. = A. mayrii
platanoides L. - 1, 2
pseudoplatanus L. - 1, 2
pseudosieboldianum (Pax) Kom. - 5
pubescens Franch. (*A. ovczinnikovii* V. Zapr., *A. regelii* Pax, *A. xerophilum* Butk., *A. pentapomicum* auct.) - 6
regelii Pax = A. pubescens
*****rubrum** L. - 1
*****saccharinum** L. - 1
saccharum Marsh. - 1
semenovii Regel & Herd. (*A. tataricum* L. subsp. *semenovii* (Regel & Herd.) E. Murr.) - 6
sosnowskyi Doluch. - 2
stevenii Pojark.(*A. hyrcanum* Fisch. & C.A. Mey. subsp. *stevenii* (Pojark.) E. Murr., *A. opalus* Mill. subsp. *stevenii* (Pojark.) E. Murr.) - 1
svaneticum Gatsch. - 2
tataricum L. - 1, 2, 3
- subsp. *semenovii* (Regel & Herd.) E. Murr. = A. semenovii
tegmentosum Maxim. - 5
trautvetteri Medw. (*A. heldreichii* Orph. subsp. *trautvetteri* (Medw.) E. Murr.) - 2
truncatum Bunge subsp. *mayrii* (Schwer.) E. Murr. = A. mayrii
- subsp. *mono* (Maxim.) E. Murr. = A. mono
tschonoskii Maxim. - 5
- subsp. *komarovii* (Pojark.) Urussov & Nedolushko = A. komarovii
- var. *rubripes* Kom. = A. komarovii
turcomanicum Pojark. (*A. monspessulanum* L. subsp. *turcomanicum* (Pojark.) E. Murr., *A. monspessulanum* subsp. *turcomanicum* (Pojark.) Rech. fil. comb. superfl.) - 6
turkestanicum Pax (*A. cappadocicum* Gled. subsp. *turkestanicum* (Pax) E. Murr.) - 6
ukurunduense Trautv. & C.A. Mey. (*A. caudatum* Wall. subsp. *ukurunduense* (Trautv. & C.A. Mey.) E. Murr.) - 5
velutinum Boiss. - 2
xerophilum Butk. = A. pubescens

ACTINIDIACEAE Hutch.
Actinidia Lindl.

arguta (Siebold & Zucc.) Planch. ex Miq. - 5
- var. *giraldii* (Diels) Worosch. = A. giraldii
giraldii Diels (*A. arguta* (Siebold & Zucc.) Planch. ex Miq. var. *giraldii* (Diels) Worosch.) - 5
kolomikta (Maxim.) Maxim. (*A. kolomikta* var. *sugawarana* (Koidz.) Worosch. nom. invalid., *A. sugawarana* Koidz. nom. invalid.) - 5
- var. *sugawarana* (Koidz.) Worosch. = A. kolomikta
polygama (Siebold & Zucc.) Miq. - 5
sugawarana Koidz. = A. kolomikta

ADIANTACEAE (C. Presl) Ching
Adiantum L.

capillus-veneris L. - 1, 2, 6
pedatum L. - 5

ADOXACEAE Trautv.
Adoxa L.

insularis Nepomn. (*A. moschatellina* L. var. *insularis* (Nepomn.) Shi-you Li & Zhu-hua Ning) - 5
moschatellina L. - 1, 2, 3, 4, 5, 6
- var. *insularis* (Nepomn.) Shi-you Li & Zhu-hua Ning = A. insularis
orientalis Nepomn. - 5

AGAVACEAE auct. = HOSTACEAE

AIZOACEAE Rudolphi.
See also MOLLUGINACEAE.
Aizoon L.

hispanicum - 2, 6

ALDROVANDACEAE Nakai = DROSE-RACEAE

ALISMATACEAE Vent. (*ELISMATACEAE* Nakai)
Alisma L.

bjoerkqvistii Tzvel. - 1, 3
canaliculatum A. Br. & Bouche - 5
gramineum Lej. (*A. gramineum* C.C. Gmel. nom. illegit., *A. loeselii* Gorski) - 1, 2, 3, 4, 5, 6
- subsp. *wahlenbergii* Holmb. = A. wahlenbergii
juzepczukii Tzvel. - 1
lanceolatum With. (*A. plantago-aquatica* L. subsp. *michaletii* Aschers. & Graebn. var. *stenophyllum* Aschers. & Graebn., *A. plantago-aquatica* subsp. *stenophyllum* (Aschers. & Graebn.) Holmb., *A. stenophyllum* (Aschers. & Graebn.) Sam.) - 1, 2, 6
loeselii Gorski = A. gramineum
orientale (Sam.) Juz. - 5
plantago-aquatica L. - 1, 2, 3, 4, 5, 6
- subsp. *michaletii* Aschers. & Graebn. var. *stenophyllum* Aschers. & Graebn. = A. lanceolatum
- subsp. *stenophyllum* (Aschers. & Graebn.) Holmb. = A. lanceolatum

ranunculoides L. = Baldellia ranunculoides
reniforme D.Don = Caldesia reniformis
stenophyllum (Aschers. & Graebn.) Sam. = A. lanceolatum
wahlenbergii (Holmb.) Juz. (*A. gramineum* Lej. subsp. *wahlenbergii* Holmb.) - 1

Baldellia Parl. (*Echinodorus* auct.)

ranunculoides (*Alisma ranunculoides* L., *Echinodorus ranunculoides* (L.) Engelm. ex Aschers.) - 1(?)

Caldesia Parl.

parnassifolia (L.) Parl. - 1, 2
reniformis (D. Don) Makino (*Alisma reniforme* D. Don.) - 5

Damasonium Hill

alisma Mill. (*D. alisma* subsp. *bourgaei* (Coss.) Maire, comb. superfl., *D. alisma* subsp. *bourgaei* (Cass.) K. Richt., *D. bourgaei* Coss., *D. constrictum* Juz.) - 1, 2, 3, 6
- subsp. *bourgaei* (Coss.) Maire = D.alisma
- subsp. *bourgaei* (Cass.) K.Richt. = D.alisma
bourgaei Coss. = D.alisma
constrictum Juz. = D.alisma

Echinodorus auct. = Baldellia

ranunculoides (L.) Engelm. ex Aschers. = Baldellia ranunculoides

Elisma Buchenau = Luronium

natans (L.) Buchenau = Luronium natans

Luronium Rafin. (*Elisma* Buchenau)

natans (L.) Rafin. (*Elisma natans* (L.) Buchenau) - 1

Sagittaria L.

aginashi Makino - 5
graminea auct. = S. platyphylla
graminea Michx. var. *platyphylla* Engelm. = S. platyphylla
*latifolia** Willd. - 1
natans Pall. - 1, 3, 4, 5
platyphylla (Engelm.) J.G. Smith (*S. graminea* Michx. var. *platyphylla* Engelm., *S. graminea* auct.) - 2
sagittifolia L. - 1, 2, 3, 4, 6
trifolia L. - 1, 2, 3, 4, 5, 6

ALLIACEAE J. Agardh
Allium L. (*Cepa* Hill).
See also Calloscordum and Nectaroscordum

aemulans Pavl. = A. caesium
affine Ledeb.(*A. transcaucasicum* Grossh. nom. invalid.) - 2
afghanicum Wendelbo - 6
aflatunense B. Fedtsch. - 6
akaka S.G. Gmel. ex Schult. & Schult. fil. - 2
alaicum Vved. - 6
albanum Grossh. - 1, 2, 6
alberti Regel = A. pallasii
albidum Fisch. ex Bieb. - 1, 2
- subsp. **caucasicum** (Regel) Stearn (*A. angulosum* L. var. *caucasicum* Regel) - 2

2

albiflorum Omelcz. - 1
albovianum Vved. - 2
alexandrae Vved. - 6
alexeianum Regel (*A. alexeianum* var. *hissaricum* Lipsky,
    A. nevskianum Vved. nom. illegit., *A. nevskianum*
    Wendelbo) - 6
- var. *hissaricum* Lipsky = A. alexeianum
altaicum Pall. (*A. microbulbum* Prokh.) - 3, 4, 5, 6
altissimum Regel - 6
altyncolicum Friesen - 3
amblyophyllum Kar. & Kir. (*A. platyspathum* Schrenk
    subsp. *amblyophyllum* (Kar. & Kir.) Friesen) - 3, 6
ampeloprasoides Grossh. = A. gramineum
ampeloprasum L. - 6
- subsp. *leucanthum* (C. Koch) K. Richt. = A. leucanthum
amphibolum Ledeb. - 3, 4
angulosum L. - 1, 3, 4
- var. *caucasicum* Regel = A. albidum subsp. caucasicum
anisopodium Ledeb. (*A. anisopodium* subsp. *argunense*
    Peschkova) - 3, 4, 5, 6
- subsp. *argunense* Peschkova = A. anisopodium
anisotepalum Vved. - 6
antonjanii Bordz. = A. scabriscapum
aroides M. Pop. & Vved. - 6
artvinense Miscz. ex Grossh. - 2
ascalonicum L. - 1
asperiflorum Miscz. ex Grossh. - 2
atrosanguineum Kar. & Kir.(*A. atrosanguineum* Schrenk) -
    6
atroviolaceum Boiss. - 1, 2, 6
- var. *firmotunicatum* (Fomin) Grossh. = A. firmotunicatum
aucheri Boiss. - 2
auctum Omelcz. p.p. = A. cyrillii and A. decipiens
austrosibiricum Friesen - 3, 4
baissunense Lipsky - 6
bajtulinii Bajt. & Kamenetzkaja - 6
barsczewskii Lipsky - 6
baschkyzylsaicum Krassovskaja - 6
bellulum Prokh. - 3, 4
bidentatum Fisch. ex Prokh. - 4, 5, 6
blandum Wall. (*A. carolinianum* auct.) - 6
bodeanum Regel - 6
bogdoicolum Regel - 3, 4, 6
boissieri Regel - 6
borszczowii Regel - 6
botschantzevii R. Kam. - 6
brachyodon Boiss. - 6
brachyscapum Vved. - 6
brevidens Vved. - 6
brevidentiforme Vved. - 6
bucharicum Regal - 6
bulgaricum (Janka) Prod. = Nectaroscordum bulgaricum
burjaticum Friesen - 4
caeruleum Pall. - 1, 3, 6
caesium Schrenk (*A. aemulans* Pavl., *A. renardii* Regel.) - 3,
    6
caespitosum Siev ex Bong. & C.A. Mey. (*A. caespitosum*
    Siev. ex Pall. nom. invalid.) - 6
callidictyon C.A. Mey. ex Kunth - 2
candolleanum Albov - 2
cardiostemon Fisch. & C.A. Mey. - 2
caricoides Regel = A. kokanicum
carinatum L. - 1(?)
carolinianum auct. = A. blandum
carolinianum DC. (*A. polyphyllum* Kar. & Kir.) - 6
caspium (Pall.) Bieb. - 1, 2, 6
*cepa L. - 1, 2, 3, 4, 5, 6

chamarense M. Ivanova - 4
charadzeae Tscholokaschvili - 2
charaulicum Fomin - 2
chevsuricum Tscholokaschvili - 2
chinense G. Don fil. - 6
christophii auct. = A. cristophii
cilicicum Boiss. - 2
circassicum Kolak. - 2
clathratum Ledeb. - 3, 4, 6
clausum Vved. - 6
collis-magni R. Kam. - 6
condensatum Turcz. - 4, 5
confragosum Vved. - 6
convallarioides Grossh. = A. myrianthum
costatovaginatum R. Kam. & Levichev - 6
cristophii Trautv. (*A. christophii* Trautv.) - 6
crystallinum Vved. - 6
cupuliferum Regel - 6
cyrillii Ten. (*A. auctum* Omelcz. p.p. incl. typo) - 1
daghestanicum Grossh. - 2
darwasicum Regel - 6
dasyphyllum Vved. - 6
dauricum Friesen - 4, 5
decipiens Fisch. ex Schult. & Schult. fil. (*A. auctum*
    Omelcz. p.p. excl. typo) - 1, 2, 3, 6
delicatulum Siev. ex Schult. & Schult. fil. - 1, 3, 4, 6
derderianum Regel - 2
deserticolum M. Pop. = A. teretifolium
dictyoprasum C.A. Mey. ex Kunth - 2
dictyoscordum Vved. - 6
dioscoridis auct. = Nectaroscordum bulgaricum
dodecadontum Vved. - 6
dolichomischum Vved. - 6
dolychostylum Vved. - 6
drepanophyllum Vved. - 6
drobovii Vved. - 6
dshambulicum Pavl. = A. parvulum
dshungaricum Vved. = A. globosum
eduardii Stearn (*A. fischeri* Regel, 1875, non Schult. &
    Schult. fil. 1830) - 3, 4, 6
elatum Regel (*A. lucens* E. Nikit.) - 6
elegans Drob. - 6
eremoprasum Vved. - 6
ericetorum Thore (*A. ochroleucum* Waldst. & Kit.) - 1
eriocoleum Vved. - 6
erubescens C. Koch - 1, 2
eugenii Vved. - 6
fedtschenkoanum Regel - 6
ferganicum Vved. - 6
fetisowii Regel (*A. simile* Regel, p.p.) - 6
fibrosum Regel - 6
filidens Regel - 6
filidentiforme Vved. - 6
filifolium Regel = A. kokanicum
firmotunicatum Fomin (*A. atroviolaceum* Boiss. var. *fir-*
    motunicatum (Fomin) Grossh.) - 1, 2
fischeri Regel = A. eduardii
*fistulosum L. - 1, 3, 4, 5
flavellum Vved. - 6
flavescens Bess. - 1, 3
flavidum Ledeb. - 3, 6
flavum L. subsp. *tauricum* (Bess. ex Reichenb.) Stearn =
    A. paczoskianum
- subsp. *tauricum* (Reichenb.) K. Richt. = A. paczoskia-
    num
- var. *tauricum* Bess. ex Reichenb. = A. paczoskianum
fominianum Miscz. ex Grossh. = A. gramineum

fuscoviolaceum Fomin - 2
galanthum Kar. & Kir. - 3, 6
giganteum Regel - 6
glaciale Vved. - 6
glaucum Schrad. = A. senescens subsp. glaucum
globosum Bieb. ex Redoute (*A. dshungaricum* Vved. nom. invalid.) - 1, 2, 3, 6
glomeratum Prokh. - 6
goloskokovii Vved. - 6
gracilescens Somm. & Levier - 2
gracillimum Vved. - 6
gramineum C. Koch (*A. ampeloprasoides* Grossh., *A. fominianum* Miscz. ex Grossh., *A. gramineum* var. *ampeloprasoides* (Grossh.) Grossh.) - 2
- var. *ampeloprasoides* (Grossh.) Grossh. = A. gramineum
grande Lipsky - 2
griffithianum Boiss. - 6
gubanovii R. Kam. - 4, 5
gultschense B. Fedtsch. - 6
gunibicum Miscz. ex Grossh. - 2
gusaricum Regel - 6
guttatum Stev. - 1
gypsaceum M. Pop. & Vved. - 6
gypsodictyum Vved. - 6
helicophyllum Vved. - 6
hexaceras Vved. - 6
hissaricum Vved. - 6
hymenorhizum Ledeb. (*A. kaschianum* Regel) - 3, 6
iliense Regel - 6
- subsp. nuratense R. Kam. - 6
inaequale Janka - 1, 2, 6
inconspicuum Vved. - 6
incrustatum Vved. - 6
inderiense Fisch. ex Bunge - 1, 3, 6
inops Vved. - 6
insufficiens Vved. - 6
jacquemontii auct. = A. pamiricum
jajlae Vved. (*A. scorodoprasum* L. subsp. *jajlae* (Vved.) Stearn.) -1, 2
jaxarticum Vved. - 6
jodanthum Vved. - 6
jucundum Vved. - 6
karataviense Regel - 6
karelinii Poljak. - 6
karsianum Fomin - 2
kaschianum Regel = A. hymenorhizum
kasteki M. Pop. - 6
kaufmannii Regel - 6
kirindicum Bornm. - 6
kokanicum Regel (*A. caricoides* Regel, *A. filifolium* Regel) - 6
komarovianum Vved. - 5
komarowii Lipsky - 6
kopetdagense Vved. - 6
korolkowii Regel - 6
kossoricum Fomin - 2
krylovii K. Sobol = A. mongolicum
kujukense Vved. - 6
kunthianum Vved. - 2
kurssanovii M. Pop. (*A. pseudoglobosum* M. Pop. ex Gamajun.) - 6
kysylkumi R. Kam. - 6
lacerum Freyn - 2
lasiophyllum Vved. - 6
latissimum Prokh. = A. ochotense
ledebourianum Schult. & Schult. fil. - 3
ledshanense Conrath & Freyn - 2

lehmannianum Merckl. - 6
lenkoranicum Miscz. ex Grossh. - 2, 6
leonidii Grossh. - 2
leptomorphum Vved. - 6
leucanthum C. Koch (*A. ampeloprasum* L. subsp. *leucanthum* (C. Koch) K. Richt.) - 2
leucocephalum Turcz. ex Ledeb. - 4
leucosphaerum Aitch. & Baker - 6
lineare L. - 1, 3, 6
lipskyanum Vved. - 6
litvinovii Drob. ex Vved. - 6
longicuspis Regel - 6
longiradiatum (Regel) Vved. - 6
lucens E. Nikit. = A. elatum
lutescens Vved. - 6
maackii (Maxim.) Prokh. ex Kom. - 4, 5
macrostemon Bunge - 5
majus Vved. - 6
malyschevii Friesen - 4
margaritae B. Fedtsch. - 6
margaritiferum Vved. - 6
mariae Bordz. - 2
marschallianum Vved. - 1
materculae Bordz. - 2
maximowiczii Regel (*A. schoenoprasum* L. subsp. *maximowiczii* (Regel) Bondar.) - 4, 5
meliophilum Juz. = Nectaroscordum bulgaricum
microbulbum Prokh. = A. altaicum
microdictyon Prokh. - 1, 3, 4
minutum Vved. - 6
mirzajevii Tscholokaschvili - 2
mogoltavicum Vved. - 6
monadelphum Less. ex Kunth - 4
monanthum Maxim. - 5
mongolicum Regel (*A. krylovii* K. Sobol.) - 4
monophyllum Vved. ex Czerniak. - 6
montanum F.W. Schmidt, 1794, non Schrank, 1785 (*A. senescens* L. subsp. *montanum* (F.W. Schmidt) Holub) - 1
This name must be replaced because it is a later homonym.
moschatum L. - 1, 2
motor R. Kam. & Levichev - 6
multitabulatum S. Cicina = A. proliferum
mutabile Zing. = A. scabriscapum
myrianthum Boiss. (*A. convallarioides* Grossh.) - 2, 6
neriniflorum (Herb.) Baker = Calloscordum neriniflorum
nerinifolium auct. = Calloscordum neriniflorum
nevskianum Vved. = A. alexeianum
nevskianum Wendelbo = A. alexeianum
nutans L. - 1, 3, 4, 5, 6
obliquum L. - 1, 3, 4, 6
ochotense Prokh. (*A. latissimum* Prokh., *A. victorialis* L. subsp. *platyphyllum* Hult.) - 5
ochroleucum Waldst. & Kit. = A. ericetorum
odorum L. = A. ramosum
oleraceum L. - 1, 2
oliganthum Kar. & Kir. - 3, 4, 6
ophiophyllum Vved. - 6
oreodictyum Vved. - 6
oreophiloides Regel - 6
oreophilum C.A. Mey. - 2, 6
oreoprasoides Vved. - 6
oreoprasum Schrenk - 6
oreoscordum Vved. - 6
oschaninii O. Fedtsch. - 6
otschiauriae Tscholokaschvili - 2
paczoskianum Tuzs. (*A. flavum* L. var. *tauricum* Bess. ex

Reichenb. nom. illegit., *A. flavum* subsp. *tauricum* (Bess. ex Reichenb.) Stearn, comb. superfl., *A. flavum* subsp. *tauricum* (Reichenb.) K. Richt., *A. tauricum* (Bess. ex Reichenb.) Grossh., *A. pulchellum* auct.) - 1, 2

pallasii Murr. (*A. alberti* Regel) - 3, 6

pamiricum Wendelbo (*A. jacquemontii* auct.) - 6

pangasicum Turakulov - 6

paniculatum L. - 1, 2, 3
- subsp. *rupestre* (Stev.) K. Richt. = A. rupestre
- var. *podolicum* Aschers. & Graebn. = A. podolicum

paradoxum (Bieb.) G. Don fil. - 2, 6

parvulum Vved. (*A. dshambulicum* Pavl.) - 6

paulii Vved. - 6

pervestitum Klok. - 1

petraeum Kar. & Kir. - 6

platyspathum Schrenk - 3, 6
- subsp. *amblyophyllum* (Kar. & Kir.) Friesen = A. amblyophyllum

podolicum (Aschers. & Graebn.) Blocki ex Racib. (*A. paniculatum* L. var. *podolicum* Aschers. & Graebn.) - 1

polyphyllum Kar. & Kir. = A. carolinianum

polyrhizum Turcz. ex Regel - 4, 6

ponticum Miscz. ex Grossh. - 2

popovii Vved. - 6

*****porrum** L. - 1

praemixtum Vved. - 6

praescissum Reichenb. - 1, 3

prokhanovii (Worosch.) Barkalov (*A. splendens* Willd. ex Schult. & Schult. fil. subsp. *prokhanovii* Worosch.) - 5

*****proliferum** (Moench) Schrad. ex Willd. (*Cepa prolifera* Moench, *Aliium multitabulatum* S. Cicina) - 6

prostratum Trev. - 4

pseudoampeloprasum Miscz. ex Grossh. - 2

pseudoflavum Vved. (*A. stamineum* auct.) - 2

pseudoglobosum M. Pop. ex Gamajun. = A. kurssanovii

pseudopulchellum Omelcz. - 2

pseudoseravschanicum M. Pop. & Vved. - 6

pseudostrictum Albov - 2

pskemense B. Fedtsch. - 6

pulchellum auct. = A. paczoskianum

pumilum Vved. - 3, 4

ramosum L. (*A. odorum* L.) - 3, 4, 5, 6

regelianum A. Beck. - 1, 2

regelii Trautv. - 6

renardii Regel = A. caesium

rhodanthum Vved. - 6

roborowskianum Regel - 6

robustum Kar. & Kir. - 6

rosenbachianum Regel - 6

rotundum L. (*A. scorodoprasum* L. subsp. *rotundum* (L.) Stearn) - 1, 2
- subsp. *waldsteinii* (G.Don fil.) K. Richt. = A. waldsteinii
- subsp. *waldsteinii* (G.Don fil.) Soo = A. waldsteinii

rubellum Bieb. - 2

rubens Schrad. ex Willd. - 1, 3, 4, 6

rudolfii Turakulov - 6

rupestre Stev. (*A. paniculatum* L. subsp. *rupestre* (Stev.) K. Richt.) - 1, 2

ruprechtii Boiss. = A. saxatile

sabulosum Stev. ex Bunge - 1, 6

sacculiferum Maxim. - 5

salthynicum Tscholokaschvili - 2

samurense Tscholokaschvili - 2

saposhnikovii E. Nikit. - 6

sarawschanicum Regel - 6

*****sativum** L. - 1, 2, 3, 4, 5, 6

savranicum Bess. - 1

saxatile Bieb. (*A. ruprechtii* Boiss.) - 2

scabrellum Boiss. & Buhse = A. umbilicatum

scabriscapum Boiss. & Kotschy (*A. antonjanii* Bordz., ? *A. mutabile* Zing.) - 2, 6

schachimardanicum Vved. - 6

schischkinii K. Sobol. - 4

schoenoprasoides Regel - 6

schoenoprasum L. (*A. schoenoprasum* subsp. *sibiricum* (L.) Hayek & Markgraf, *A. sibiricum* L.) - 1, 2, 3, 4, 5, 6
- subsp. *maximowiczii* (Regel) Bondar. = A. maximowiczii
- subsp. *sibiricum* (L.) Hayek & Markgraf = A. schoenoprasum

schrenkii Regel - 6

schubertii Zucc. - 6

schugnanicum Vved. - 6

scorodoprasum L. - 1
- subsp. *jajlae* (Vved.) Stearn = A. jajlae
- subsp. *rotundum* (L.) Stearn = A. rotundum
- subsp. *waldsteinii* (G. Don. fil.) Stearn = A. waldsteinii

scrobiculatum Vved. - 6

scythicum Zoz - 1

semenowii Regel - 6

senescens L. - 3, 4, 5
- subsp. **glaucum** (Schrad.) Friesen (*A. glaucum* Schrad.) - 3, 4
- subsp. *montanum* (F.W. Schmidt) Holub = A. montanum

sergii Vved. - 6

setifolium Schrenk - 6

sewerzowii Regel (*A. simile* Regel, p.p. quoad specim. Regelii) - 6

sewerzowii sensu Vved. = A. tschimganicum

sibiricum L. = A. schoenoprasum

simile Regel, p.p. = A. fetisowii and A. sewerzowii

sordidiflorum Vved. - 6

sphaerocephalon L. - 1, 2

sphaeropodum Klok. - 1

spirale Willd. ex Schlecht. - 5

splendens Willd. ex Schult. & Schult. fil. - 4, 5
- subsp. **insulare** Worosch. - 5
- subsp. *prokhanovii* Worosch. = A. prokhanovii

stamineum auct. = A. pseudoflavum

stellerianum Willd. - 4
- subsp. *tuvinicum* Friesen = A. tuvinicum

stephanophorum Vved. - 6

stipitatum Regel - 6

strictum Schrad. (*A. volhynicum* Bess.) - 1, 3, 4, 5, 6

subquinqueflorum Boiss. - 2

subtilissimum Ledeb. - 3, 6

sulphureum Vved. - 6

suworowii Regel - 6

svetlanae Vved. ex Filimonova - 6

syntamanthum C. Koch . - 2

szovitsii Regel - 2

taciturnum Vved. - 6

taeniopetalum M. Pop. & Vved. - 6

talassicum Regel - 6

talijevii Klok. - 1

talyschense Miscz. ex Grossh. - 2

tauricum (Bess. ex Reichenb.) Grossh. = A. paczoskianum

tenuicaule Regel - 6

tenuissimum L. - 4

teretifolium Regel (*A. deserticolum* M. Pop.) - 6

tianschanicum Rupr. - 6

x **tokaliense** R. Kam. & Levichev. - A. motor R. Kam. & Levichev x A. sewerzowii Regel - 6

trachyscordum Vved. - 6
transcaucasicum Grossh. = A. affine
transvestiens Vved. - 6
trautvetterianum Regel - 6
tripedale Trautv. = Nectaroscordum tripedale
tschimganicum B. Fedtsch. (*A. sewerzowii* sensu Vved.) - 6
tulipifolium Ledeb. - 1, 3, 6
turcomanicum Regel - 6
turkestanicum Regel - 6
turtschicum Regel - 6
tuvinicum (Friesen) Friesen (*A. stellerianum* Willd. subsp. *tuvinicum* Friesen) - 4
tytthanthum Vved. - 6
tytthocephalum Schult. & Schult. fil. - 3, 4
ubsicolum Regel - 3, 4
ucrainicum (Kleop. & Oxner) Bordz. = A. ursinum
udinicum Antsupova - 4
umbilicatum Boiss. (*A. scabrellum* Boiss. & Buhse) - 6
ursinum L. (*A. ucrainicum* (Kleop. & Oxner) Bordz., *A. ursinum* subsp. *ucrainicum* Kleop. & Oxner, *A. ursinum* var. *ucrainicum* (Kleop. ex Oxner) Soo) - 1, 2
- subsp. *ucrainicum* Kleop. & Oxner = A. ursinum
- var. *ucrainicum* (Kleop. ex Oxner) Soo = A. ursinum
valentinae Pavl. - 6
vavilovii M. Pop. & Vved. - 6
verticillatum (Regel) Regel - 6
victorialis L. - 1, 2
- subsp. *platyphyllum* Hult. = A. ochotense
victoris Vved. - 6
vineale L. - 1, 2
viride Grossh. - 2
viridiflorum Pobed. - 6
vodopjanovae Friesen - 3, 4
- subsp. **czemalense** Friesen - 3
volhynicum Bess. = A. strictum
vvedenskyanum Pavl. - 6
waldsteinii G. Don fil. (*A. rotundum* L. subsp. *waldsteinii* (G. Don fil.) Soo, *A. rotundum* subsp. *waldsteinii* (G. Don fil.) K. Richt., *A. scorodoprasum* subsp. *waldsteinii* (G. Don. fil.) Stearn) - 1, 2
warzobicum R. Kam. - 6
weschnjakowii Regel - 6
winklerianum Regel - 6
woronowii Miscz. ex Grossh. - 2
xiphopetalum Aitch. & Baker - 6
yatei Aitch. & Baker - 6
zaprjagajevii Kassacz - 6

Calloscordum Herb.

neriniflorum Herb. (*Allium neriniflorum* (Herb.) Baker, errore "nerinifolium", *A. nerinifolium* auct.) - 4

Cepa Hill = Allium

prolifera Moench = Allium proliferum

Nectaroscordum Lindl.

bulgaricum Janka (*Allium bulgaricum* (Janka) Prod., *A. meliophilum* Juz., *Nectaroscordum dioscoridis* (Smith) Stank. p.p. excl. typo, *N. dioscoridis* (Smith) Zahar. f. *meliophilum* (Juz.) Zahar., *N. meliophilum* Juz., *N. siculum* (Ucria) Lindl. subsp. *bulgaricum* (Janka) Stearn, *N. siculum* subsp. *dioscoridis* (Smith) Bordz. p.p. excl. typo, *Allium dioscoridis* auct.) - 1
dioscoridis (Smith) Stank. p.p. excl. typo = N. bulgaricum
dioscoridis (Smith) Zahar. f. *meliophilum* (Juz.) Zahar. = N. bulgaricum
meliophilum Juz. = N. bulgaricum
siculum (Ucria) Lindl. subsp. *bulgaricum* (Janka) Stearn = N. bulgaricum
- subsp. *dioscoridis* (Smith) Bordz. p.p. excl. typo = N. bulgaricum
tripedale (Trautv.) Grossh. (*Allium tripedale* Trautv.) - 2

AMARANTHACEAE Juss.
Achyranthes L.

bidentata auct. = A. japonica
bidentata Blume var. *japonica* Miq. = A. japonica
japonica (Miq.) Nakai (*A. bidentata* Blume var. *japonica* Miq., *A. bidentata* auct.) - 5

Acnida L. = Amaranthus

tuberculata Moq. = Amaranthus tuberculatus

Alternanthera Forssk.

sessilis (L.) DC. - 2

Amaranthus L. (*Acnida* L., *Euxolus* Rafin.)

albus L. - 1, 2, 3, 4, 5, 6
blitoides S. Wats. - 1, 2, 3, 4, 5, 6
blitum L. (*A. lividus* L.) - 1, 2, 3, 5, 6
blitum sensu Vass. = A. sylvestris
bouchonii Thell. (*A. hybridus* L. subsp. *bouchonii* (Thell.) O. Bolos & Vigo) - 5
caudatus L. (*A. leucospermus* sensu Vass.) - 1, 2, 5, 6
chlorostachys auct. = A. powellii
chlorostachys Willd. = A. hybridus
crispus (Lesp. & Thev.) N. Terr. (*Euxolus crispus* Lesp. & Thev.) - 1
cruentus L. (*A. hybridus* L. subsp. *cruentus* (L.) Thell., *A. hybridus* subsp. *paniculatus* (L.) Hejny, *A. paniculatus* L.) - 1, 2, 3, 4, 5, 6
deflexus L. - 1, 2, 6
graecizans L. - 1, 2, 6
- subsp. *sylvestris* (Vill.) Brenan = A. sylvestris
- subsp. *sylvestris* (Vill.) O. Bolos & Vigo = A. sylvestris
- subsp. *thellungianus* (Nevski) Gusev = A. thellungianus
hybridus auct. = A. powellii
hybridus L. (? *A. chlorostachys* Willd., *A. patulus* Bertol.) - 1
- subsp. *bouchonii* (Thell.) O. Bolos & Vigo = A. bouchonii
- subsp. *cruentus* (L.) Thell. = A. cruentus
- subsp. *paniculatus* (L.) Hejny = A. cruentus
hypochondriacus L. (*A. leucocarpus* S. Wats., *A. leucospermus* S. Wats.) - 1, 5
leucocarpus S. Wats. = A. hypochondriacus
leucospermus sensu Vass. = A. caudatus
leucospermus S. Wats. = A. hypochondriacus
lividus L. = A. blitum
palmeri S. Wats. - 1, 5
paniculatus L. = A. cruentus
patulus Bertol. = A. hybridus
powellii S. Wats. (*A. chlorostachys* auct., *A. hybridus* auct.) - 1
retroflexus L. - 1, 2, 3, 4, 5, 6
rudis Sauer - 1
spinosus L. - 1, 2, 5
sylvestris Vill. (*A. graecizans* L. subsp. *sylvestris* (Vill.) Brenan, *A. graecizans* subsp. *sylvestris* (Vill.) O. Bolos

& Vigo, comb. superfl., *A. blitum* sensu Vass.) - 1, 2, 6
(Nevski) Gusev) - 6
thellungianus Nevski (*A. graecizans* L. subsp. *thellungianus*
(Nevski) Gusev) - 6
tricolor L. - 6
tuberculatus (Moq.) Sauer (*Acnida tuberculata* Moq.) - 1
viridis L. - 2, 5, 6

Celosia L.

***argentea** L. - 6
- f. *cristata* (L.) Schinz = C. cristata
***cristata** L. (*C. argentea* L. f. *cristata* (L.) Schinz) - 6

Euxolus Rafin. = Amaranthus

crispus Lesp. & Thev. = Amaranthus crispus

Gomphrena L.

***globosa** L. - 6

AMARYLLIDACEAE J. St.-Hil.

Erinosma Herb. = Leucojum

carpathicum (Loud.) Herb. = Leucojum vernum var. carpathicum

Galanthus L.

alpinus Sosn. (*G. schaoricus* Kem.-Nath.) - 2
angustifolius G. Koss (*G. nivalis* L. subsp. *angustifolius* (G.
Koss) Artjushenko) - 2
bortkewitschianus G.Koss - 2
cabardensis G. Koss = G. lagodechianus
caspius (Rupr.) Grossh. = G. transcaucasicus
caucasicus (Baker) Grossh. (*G. nivalis* L. subsp. *caucasicus*
Baker) - 2
cilicicus Baker (*G. glaucescens* A. Khokhr.) - 2
elwesii Hook. fil. var. **maximus** (Velen.) G. Beck (*G.
maximus* Velen., *G. graecus* Orph. ex Boiss. var.
maximus (Velen.) Hayek, *G. graecus* f. oec. *maximus*
(Velen.) Zahar., *G. graecus* auct. p.p.) - 1
glaucescens A. Khokhr. = G. cilicicus
graecus auct. p.p. = G. elwesii var. maximus
graecus Orph. ex Boiss. var. *maximus* (Velen.) Hayek = G.
elwesii var. maximus
- f. oec. *maximus* (Velen.) Zahar. = G. elwesii var. maximus
ikariae auct. p.p. = G. woronowii
ikariae Baker subsp. *latifolius* Stern = G. platyphyllus
kemulariae Kuth. = G. lagodechianus
ketzkhovelii Kem.-Nath. = G. lagodechianus
krasnovii A. Khokhr. (*G. valentinae* Panjut. ex Grossh.
nom. invalid.) - 2
- subsp. **maculatus** A. Khokhr. - 2
lagodechianus Kem.-Nath. (*G. cabardensis* G. Koss, *G.
kemulariae* Kuth., *G. ketzkhovelii* Kem.-Nath.) - 2
latifolius Rupr. = G. platyphyllus
latifolius Salisb. = G. plicatus
maximus Velen. = G. elwesii var. maximus
nivalis L. - 2
- subsp. *angustifolius* (G. Koss) Artjushenko = G. angustifolius
- subsp. *caucasicus* Baker = G. caucasicus
- var. *caspius* Rupr. = G. transcaucasicus
platyphyllus Traub & Moldenke (*G. latifolius* Rupr. 1868
non Salisb. 1866, *G. ikariae* Baker subsp. *latifolius*

Stern) - 2
plicatus Bieb. (*G. latifolius* Salisb. 1866, non Rupr. 1868) -
1
schaoricus Kem.-Nath. = G. alpinus
transcaucasicus Fomin (*G. caspius* (Rupr.) Grossh. nom.
illegit., *G. nivalis* L. var. *caspius* Rupr.) - 2
valentinae Panjut. ex Grossh. = G. krasnovii
woronowii Losinsk. (*G. ikariae* auct. p.p.) - 2

Hermione Salisb. = Narcissus

lacticolor Haw. = Narcissus lacticolor

Leucojum (*Erinosma* Herb.)

aestivum L. - 1, 2
carpathicum (Loud.) Sweet = L. vernum var. carpathicum
vernum L. - 1
- subsp. *carpathicum* (Herb.) K. Richt. = L. vernum var.
carpathicum
- subsp. *carpathicum* (Loud.) E. Murr. = L. vernum var.
carpathicum
- subsp. *carpathicum* (Spring) O. Schwarz = L. vernum var.
carpathicum
- var. **carpathicum** Loud. (*Erionosma carpathicum* (Loud.)
Herb., *Leucojum carpathicum* (Loud.) Sweet, *L.
vernum* subsp. *carpathicum* (Loud.) E. Murr. comb.
superfl., *L. vernum* subsp. *carpathicum* (Herb.) K.
Richt., *L. vernum* subsp. *carpathicum* (Spring) O.
Schwarz, comb. superfl.) - 1

Narcissus (*Hermione* Salisb.)

angustifolius Curt. (*N. poeticus* L. subsp. *angustifolius*
(Curt.) Aschers. & Graebn., *N. poeticus* subsp. *radiiflorus* (Salisb.) Baker, *N. radiiflorus* Salisb.) - 1
caucasicus (Fomin) Gorschk. = N. lacticolor
italicus Ker-Gawl. = N. lacticolor
***jonquilla** L. - 1
***lacticolor** (Haw.) Steud. (*Hermione lacticolor* Haw., *Nar-
cissus caucasicus* (Fomin) Gorschk., *N. italicus* Ker-
Gawl., *N. tazetta* L. subsp. *italicus* (Ker-Gawl.) Baker,
N. tazetta L. subsp. *lacticolor* (Haw.) Baker) - 2, 6
***odorus** L. - 1
***poeticus** L. 1, 6
- subsp. *angustifolius* (Curt.) Aschers. & Graebn. = N.
angustifolius
- subsp. *radiiflorus* (Salisb.) Baker = N. angustifolius
***pseudonarcissus** L. - 2
radiiflorus Salisb. = N. angustifolius
***tazetta** L. - 2, 6
- subsp. *italicus* (Ker-Gawl.) Baker = N. lacticolor
- subsp. *lacticolor* (Haw.) Baker = N. lacticolor

Pancratium L.

maritimum L. - 2

Sternbergia Waldst. & Kit.

alexandrae Sosn. = S. colchiciflora var. alexandrae
colchiciflora Waldst. & Kit. - 1, 2
- var. **alexandrae** (Sosn.) Artjushenko (*S. alexandrae*
Sosn.) - 2
fischeriana (Herb.) M. Roem. - 2, 6
- subsp. **hissarica** (Kapinos) Artjushenko (*S. fischeriana* f.
hissarica Kapinos) - 6
lutea (L.) Spreng. - 2, 6

Ungernia Bunge

badghysi Botsch. - 6
ferganica Vved. ex Artjushenko - 6
minor Vved. = U. oligostroma
oligostroma M. Pop. & Vved. (*U. minor* Vved. nom. invalid.) - 6
sewerzowii (Regel) B. Fedtsch. - 6
spiralis Proskorjakov - 6
tadshicorum Vved. ex Artjushenko - 6
trisphaera Bunge - 6
victoris Vved. ex Artjushenko - 6
vvedenskyi Khamidsh. - 6

ANACARDIACEAE Lindl.

Cotinus Hill

coggygria Scop. - 1, 2

Pistacia L.

atlantica Desf. subsp. *kurdica* (Zohary) Rech. fil. = P. eurycarpa
- subsp. *mutica* (Fisch. & C.A. Mey.) Rech. fil. = P. mutica
- var. *kurdica* Zohary = P. eurycarpa
badghysi K. Pop. - 6
eurycarpa Yaltirik (*P. atlantica* Desf. subsp. *kurdica* (Zohary) Rech. fil., *P. atlantica* var. *kurdica* Zohary) - 2
mutica Fisch. & C.A. Mey. (*P. atlantica* Desf. subsp. *mutica* (Fisch. & C.A. Mey.) Rech. fil.) - 1, 2
vera L. - 6

Rhus L.

ambigua Lav. ex Dipp. = Toxicodendron orientale
coriaria L. - 1, 2, 6
hirta (L.) Sudw. = R. typhina
***javanica** L. - 1(?)
***typhina** L. (*R. hirta* (L.) Sudw. 1892, non Harv.) - 1, 6

Toxicodendron Hill

orientale Greene (?*Rhus ambigua* Lav. ex Dipp.) - 5
***radicans** (L.) O. Kuntze
trichocarpum (Miq.) O. Kuntze - 5
***vernicifluum** (Stokes) Lincz. - 1, 2, 6
***vernix** (L.) O. Kuntze - 1

*ANNONACEAE Juss.

*Asimina Adans.

***triloba** (L.) Dun. - 2

APIACEAE Lindl. (UMBELLIFERAE Juss., *HYDROCOTYLACEAE* Hyl.)

Actinolema Fenzl

eryngioides auct. = A. macrolema
macrolema Boiss. (*A. eryngioides* auct.) - 2

Aegopodium L.

See also Spuriopimpinella.
alpestre Ledeb. (*Ae. henryi* auct.) - 3, 4, 5, 6
brachycarpum (Kom.) Schischk. = Spuriopimpinella calycina

calycinum (Maxim.) Worosch. = Spuriopimpinella calycina
henryi auct. = Ae. alpestre
kashmiricum (R.R. Stewart ex Dunn) M. Pimen. (*Pimpinella kashmirica* R.R. Stewart ex Dunn) - 3, 6
latifolium Turcz. - 4
podagraria L. - 1, 2, 3, 4
tadshikorum Schischk. - 6

Aethusa L.

cynapium L. (*Ae. cynapium* subsp. *agrestis* (Wallr.) Dostal, *Ae. cynapium* subsp. *cynapioides* (Bieb.) Nym., *Ae. cynapium* subsp. *segetalis* (Boenn.) Schuebl. & Martens) - 1, 2
- subsp. *agrestis* (Wallr.) Dostal = Ae. cynapium
- subsp. *cynapioides* (Bieb.) Nym. = Ae. cynapium
- subsp. *segetalis* (Boenn.) Schuebl. & Martens = Ae. cynapium

Agasyllis Spreng. (*Siler* auct.)

latifolia (Bieb.) Boiss. (*Siler latifolia* (Bieb.) Hiroe) - 2

Albertia Regel & Schmalh. = Kozlovia

paleacea Regel & Schmalh. = Kozlovia paleacea

Albovia Schischk. = Pimpinella

tripartita (Kalen.) Schischk. = Pimpinella tripartita

Alposelinum M. Pimen. = Lomatocarpa

albomarginatum (Schrenk) M. Pimen. = Lomatocarpa albomarginata

Ammi L.

See also Visnaga.
majus L. - 1
visnaga (L.) Lam. = Visnaga daucoides

Anethum L.

***graveolens** L. - 1, 2, 6
involucratum Korov. - 6

Angelica L. (*Archangelica* N.M. Wolf, *Coelopleurum* Ledeb., *Ostericum* Hoffm., *Xanthogalum* Ave-Lall., *Athamanta* auct.)

adzharica M. Pimen. - 2
amurensis Schischk. = A. cincta
anomala Ave-Lall. (*A. jaluana* Nakai) - 4, 5
archangelica L. (*Archangelica officinalis* Hoffm., *A. norvegica* Rupr.) - 1
- subsp. *archangelica* var. *decurrens* (Ledeb.) Weinert = A. decurrens
- - var. *decurrens* f. *komarovii* (Schischk.) Weinert = A. komarovii
- - var. *decurrens* f. *tschimganica* (Korov.) Weinert = A. tschimganica
- subsp. *decurrens* (Ledeb.) Kuvajev = A. decurrens
- subsp. *litoralis* (Fries) Thell. = A. litoralis
arenaria (Waldst. & Kit.) Hiroe = Taeniopetalum arenarium
brevicaulis (Rupr.) B. Fedtsch. - 6
cincta Boissieu (*A. amurensis* Schischk.) - 5
czernaevia (Fisch. & C.A. Mey.) Kitag. (*Conioselinum*

8

czernaevia *Fisch. & C.A. Mey.) - 4, 5*
Some authors separate this species as a monotypic genus Czernaevia Turcz. under the name C. laevigata Turcz.
dahurica (Fisch. ex Hoffm.) Benth. & Hook. fil. ex Franch. & Savat. - 4, 5
decurrens (Ledeb.) B. Fedtsch. (*Archangelica decurrens* Ledeb., *Angelica archangelica* L. subsp. *archangelica* var. *decurrens* (Ledeb.) Weinert, *A. archangelica* subsp. *decurrens* (Ledeb.) Kuvajev) - 1(?), 3, 4
decursiva (Miq.) Franch. & Savat. - 5
edulis Miyabe (*A. keiskei* auct.) - 5
genuflexa Nutt. ex Torr. & Gray (*A. genuflexa* subsp. *refracta* (Fr. Schmidt) Hiroe, *A. refracta* Fr. Schmidt) - 5
- subsp. *refracta* (Fr. Schmidt) Hiroe = A. genuflexa
gmelinii (DC.) M. Pimen. (*Coelopleurum gmelinii* (DC.) Ledeb., *C. lucidum* (L.) Fern. var. *gmelinii* (DC.) Hara, *C. lucidum* subsp. *gmelinii* (DC.) A. & D. Löve) - 5
- subsp. *saxatilis* (Turcz. ex Ledeb.) Worosch. = A. saxatilis
grosseserrata Maxim. (*A. koreana* Maxim.) - 5
hultenii (Fern.) Hiroe = Ligusticum scoticum
jaluana Nakai = A. anomala
keiskei auct. = A. edulis
komarovii (Schischk.) V. Tichomirov (*Archangelica komarovii* Schischk., *Angelica archangelica* L. subsp. *archangelica* var. *decurrens* (Ledeb.) Weinert f. *komarovii* (Schischk.) Weinert, *A. komarovii* (Schischk.) Hiroe, comb. superfl.) - 6
koreana Maxim. = A. grosseserrata
litoralis Fries (*A. archangelica* L. subsp. *litoralis* (Fries) Thell., *Archangelica litoralis* (Fries) Agardh, *A. officinalis* Hoffm. subsp. *litoralis* (DC.) Dostal) - 1
maximowiczii (Fr. Schmidt) Benth. ex Maxim. (*Ostericum maximowiczii* (Fr. Schmidt) Kitag.) - 5
miqueliana Maxim. - 5
multicaulis M. Pimen. - 6
oreoselinum (L.) Hiroe = Peucedanum oreoselinum
pachyptera Ave-Lall. - 2
palustris (Boiss.) Hoffm. (*Ostericum palustre* (Bess.) Bess.) - 1, 2, 3, 6
purpurascens (Ave-Lall.) Gilli (*Xanthogalum purpurascens* Ave-Lall.) - 2
refracta Fr. Schmidt = A. genuflexa
sachalinensis Maxim. - 5
sachokiana (Karjag.) M. Pimen. & V. Tichomirov (*Xanthogalum sachokianum* Karjag.) - 2
saxatilis Turcz. ex Ledeb. (*A. gmelinii* (DC.) M. Pimen. subsp. *saxatilis* (Turcz. ex Ledeb.) Worosch., *Archangelica gmelinii* DC. subsp. *saxatilis* (Ledeb.) Weinert, *Coelopleurum alpinum* Kitag., *C. saxatile* (Ledeb.) Drude) - 4, 5
sibirica (Steph. ex Spreng.) Hiroe = Phlojodicarpus sibiricus
sylvestris L. - 1, 2, 3, 4
tatianae Bordz. (*Xanthogalum tatianae* (Bordz.) Schischk.) - 2
tenuifolia (Pall. ex Spreng.) M. Pimen. (*Athamanta tenuifolia* Pall. ex Spreng., *Peucedanum salinum* Pall. ex Spreng.) - 3, 4, 5
ternata Regel & Schmalh. - 6
tschimganica (Korov.) V. Tichomirov (*A. archangelica* L. subsp. *archangelica* var. *decurrens* (Ledeb.) Weinert f. *tschimganica* (Korov.) Weinert, *Archangelica tschimganica* (Korov.) Schischk.) - 6
ursina (Rupr.) Maxim. - 5
viridiflora (Turcz.) Benth. ex Maxim. - 4, 5

Anidrum auct. = Schrenkia

transitorium (Lipsky) K.-Pol. = Schrenkia golickeana

Anisosciadium auct. = Echinophora

tenuifolium (L.) Tamamsch. subsp. *sibthorpianum* (Guss.) Tamamsch. = Echinophora sibthorpiana

Anisum Hill = Pimpinella

vulgare Gaertn. = Pimpinella anisum

Anthriscus Pers.

aemula (Woronow) Schischk. = A. sylvestris var. nemorosa
caucalis Bieb. (*A. scandicina* Mansf., *A. vulgaris* Pers. 1805, non Bernh.1800, *Caucalis scandicina* Wigg. nom. illegit.) - 1, 2
cerefolium (L.) Hoffm. (*A. cerefolium* var. *longirostris* (Bertol.) Cannon, *A. longirostris* Bertol.) - 1, 2, 6
- var. *longirostris* (Bertol.) Cannon = A. cerefolium
glacialis Lipsky - 6
kotschyi Fenzl ex Boiss. & Bal. (*A. sosnovskyi* Schischk.) - 2
longirostris Bertol. = A. cerefolium
nemorosa (Bieb.) Spreng. = A. sylvestris var. nemorosa
nitida (Wahlenb.) Hazslinszky - 1
ruprechtii Boiss. - 2
scandicina Mansf. = A. caucalis
schmalhausenii (Albov) K.-Pol. - 2
sosnovskyi Schischk. = A. kotschyi
sylvestris (L.) Hoffm. - 1, 2, 3, 4, 5, 6
- subsp. *aemula* (Woronow) Kitag. = A. sylvestris var. nemorosa
- var. **nemorosa** (Bieb.) Trautv. (*A. aemula* (Woronow) Schischk., *A. nemorosa* (Bieb.) Spreng., *A. sylvestris* subsp. *aemula* (Woronow) Kitag.) - 1, 2, 3, 4, 5, 6
- var. **sylvestris** - 1
velutina Somm. & Levier - 2
vulgaris Pers. = A. caucalis

Aphanopleura Boiss.

capillifolia (Regel & Schmalh.) Lipsky - 6
fedtschenkoana K.-Pol. = A. leptoclada
leptoclada (Aitch. & Hemsl.) Lipsky (*A. fedtschenkoana* K.-Pol.) - 6
trachysperma Boiss. - 2
zangelanica Goghina & Matz. - 2

Apium L. (*Helosciadium* Koch)

graveolens L. - 1, 2, 6
leptophyllum (Pers.) F. Muell. ex Benth. = Ciclospermum leptophyllum
nodiflorum (L.) Lag. (*Helosciadium nodiflorum* (L.) Koch) - 6

Arafoe M. Pimen. & Lavrova

aromatica M. Pimen. & Lavrova (*Ligusticum arafoe* Albov) - 2

Archangelica N.M. Wolf = Angelica

decurrens Ledeb. = Angelica decurrens
gmelinii DC. subsp. *saxatilis* (Ledeb.) Weinert = Angelica saxatilis
komarovii Schischk. = Angelica komarovii
litoralis (Fries) Agardh = Angelica litoralis

norvegica Rupr. = A. archangelica
officinalis Hoffm. = A. archangelica
- subsp. *litoralis* (DC.) Dostal = Angelica litoralis
tschimganica (Korov.) Schischk. = Angelica tschimganica

Astoma DC. = Astomaea

galiocarpum (Korov.) M. Pimen. & Kljuykov = Astomaea galiocarpa

Astomaea Reichenb. (*Astoma* DC. 1829, non S.F. Gray, 1821, *Astomatopsis* Korov.)

galiocarpa (Korov.) M. Pimen. & Kljuykov (*Astomatopsis galiocarpa* Korov., *Astoma galiocarpum* (Korov.) M. Pimen. & Kljuykov) - 6

Astomatopsis Korov. = Astomaea

galiocarpa Korov. = Astomaea galiocarpa

Astrantia L. (*Transcaucasia* Hiroe)

biebersteinii Trautv. - 2
colchica Albov - 2
major L. - 1
maxima Pall. (*Transcaucasia armwnia* Hiroe) - 2
ossica Woronow ex Grossh. - 2
pontica Albov - 2
trifida Hoffm. - 2

Astrodaucus Drude

littoralis (Bieb.) Drude - 1, 2
orientalis (L.) Drude - 1, 2

Athamanta auct. = Angelica

tenuifolia Pall. ex Spreng. = Angelica tenuifolia

Aulacospermum Ledeb. (*Trachydium* auct.)

alaicum M. Pimen. & Kljuykov - 6
alatum (Korov.) Korov. = Lomatocarpa korovinii
anomalum (Ledeb.) Ledeb. - 3, 4, 6
coloratum Korov. = A. roseum
darwasicum (Lipsky) Schischk. (*A. stylosum* auct.) - 6
dichotomum (Korov.) Kljuykov, M. Pimen. & V. Tichomirov (*Trachydium dichotomum* Korov.) - 6
gonocaulum M. Pop. - 6
gracile M. Pimen. & Kljuykov - 6
isetense (Spreng.) Schischk. = A. multifidum
multifidum (Smith) Meinsh. (*Ligusticum multifidum* Smith, *Aulacospermum isetense* (Spreng.) Schischk.) - 1
plicatum M. Pimen. & Kljuykov - 6
popovii (Korov.) Kljuykov, M. Pimen. & V. Tichomirov (*Selinum popovii* (Korov.) Schischk.) - 6
roseum Korov. (*A. coloratum* Korov.) - 6
rupestre M. Pop. = A. simplex
simplex Rupr. (*A. rupestre* M. Pop., *Pleurospermum rupestre* (M. Pop.) K.T. Fu & Y.C. Ho, *Trachydium commutatum* (Regel & Schmalh.) Hiroe) - 6
stylosum auct. = A. darwasicum
tenuisectum Korov. - 6
tianschanicum (Korov.) C. Norm. (*Trachydium tianschanicum* Korov., *Silaum rubtzovii* (Schischk.) Czer., *S. rubtzovii* (Schischk.) Hiroe, comb. superfl., *Silaus rubtzovii* Schischk.) - 6
turkestanicum (Franch.) Schischk. - 6

vesiculoso-alatum (Rech. fil.) Kljuykov, M. Pimen. & V. Tichomirov (*Trachydium vesiculoso-alatum* Rech. fil., *T. kopetdaghense* Korov.) - 6

Autumnalia M. Pimen.

botschantzevii M. Pimen. - 6
innopinata M. Pimen. - 6

Berula Koch (*Siella* M. Pimen. nom. illegit. superfl.)

angustifolia (L.) Mert. & Koch = B. erecta
erecta (Huds.) Cov.(*B. angustifolia* (L.) Mert. & Koch, *B. orientalis* Woronow, *Siella erecta* (Huds.) M. Pimen., *Sium orientale* (Woronow) Soo) - 1, 2, 6
orientalis Woronow = B. erecta

Bifora Hoffm.

kultiassovii (Korov.) Hiroe = Schrenkia kultiassovii
radians Bieb. - 1, 2
testiculata (L.) Spreng. - 2, 6

Bilacunaria M. Pimen. & V. Tichomirov (*Hippomarathrum* auct.)

caspia (DC.) M. Pimen. & V. Tichomirov (*Hippomarathrum caspium* (DC.) Grossh.) - 2
microcarpa (Bieb.) M. Pimen. & V. Tichomirov (*Hippomarathrum longilobum* (DC.) B. Fedtsch. ex Grossh., *H. microcarpum* (Bieb.) V. Petrov) - 2

Buniella Schischk. = Bunium

chaerophylloides (Regel & Schmalh.) Schischk. = Bunium chaerophylloides

Bunium L. (*Buniella* Schischk., *Sympodium* C. Koch)

afghanicum Beauverd (*B. badghysi* (Korov.) Korov.) - 6
angreni Korov. - 6
badachschanicum R. Kam. - 6
badghysi (Korov.) Korov. = B. afghanicum
bourgaei (Boiss.) Freyn & Sint. ex Freyn = B. microcarpum
capillifolium Kar. & Kir. = B. setaceum
capusii (Franch.) Korov. (*B. gypsaceum* Korov.) - 6
chaerophylloides (Regel & Schmalh.) Drude (*Buniella chaerophylloides* (Regel & Schmalh.) Schischk.) - 6
cylindraceum Freyn var. *minor* Freyn = B. longipes subsp. minor
cylindricum (Boiss. & Hohen.) Drude - 2
elegans (Fenzl) Freyn - 2
fedtschenkoanum Korov. ex R. Kam. - 6
ferulaceum auct. = B. microcarpum
gypsaceum Korov. = B. capusii
hissaricum Korov. (*B. longilobum* Kljuykov, *B. tenuisectum* Korov.) - 6
intermedium Korov. - 6
kopetdagense Geldykhanov - 6
korovinii R. Kam. & Geldykhanov - 6
kuhitangi Nevski - 6
latilobum Korov. - 6
longilobum Kljuykov = B. hissaricum
longipes Freyn - 6
- subsp. **minor** (Freyn) Geldykhanov (*B. cylindraceum* Freyn var. *minor* Freyn) - 6

luteum Hoffm. = Elaeosticta lutea
microcarpum (Boiss.) Freyn & Sint. ex Freyn (*B. bourgaei* (Boiss.) Freyn & Sint. ex Freyn, *B. microcarpum* subsp. *bourgaei* (Boiss.) Hedge & Lamond, *B. ferulaceum* auct.) - 1, 2
- subsp. *bourgaei* (Boiss.) Hedge & Lamond = B. microcarpum
oeroilanicum (Korov.) Hiroe = Hyalolaena lipskyi
paucifolium DC. - 2
persicum (Boiss.) B. Fedtsch. - 6
salsum Korov. - 6
scabrellum Korov. - 2
seravschanicum Korov. - 6
setaceum (Schrenk) H. Wolff (*Carum setaceum* Schrenk, *Bunium capillifolium* Kar. & Kir. 1841, non Bertol. 1837, *Scaligeria setacea* (Schrenk) Korov.) - 6
simplex (C. Koch) Kljuykov (*Sympodium simplex* C. Koch) - 2
tenuisectum Korov. = B. hissaricum
transcaspicum (Korov.) Hiroe = Hyalolaena transcaspica
transitorium (Korov.) Hiroe = Elaeosticta transitoria
vaginatum Korov. - 6

Bupleurum L.

abchasicum Manden. - 2
aenigma K.-Pol. = B. veronense
affine Sadl. - 1, 2
aitchisonii (Boiss.) H. Wolff - 6
alaunicum K.-Pol. = B. multinerve
americanum Coult. & Rose = B. triradiatum
asperuloides Heldr. ex Boiss. - 1, 2
atargense Gorovoi - 5
aureum Fisch. ex Hoffm. = B. longifolium subsp. aureum
badachschanicum Lincz. - 6
bicaule Helm (*B. bicaule* var. *pusillum* (Kryl.) Gubanov, *B. pusillum* Kryl.) - 3, 4
- var. *pusillum* (Kryl.) Gubanov = B. bicaule
boissieri Post - 2
brachiatum C. Koch - 1, 2
chinense DC. var. *komarovianum* (Lincz.) Liou & Huang = B. komarovianum
commutatum auct. = B. gerardii
contractum Korov. = B. lipskyanum
czimganicum Lincz. = B. exaltatum var. czimganicum
densiflorum Rupr. - 6
diversifolium Rochel - 1(?)
euphorbioides Nakai - 5
exaltatum Bieb. (*B. kotschyanum* Boiss., *B. linearifolium* auct.) - 1(?),2, 6
- var. **czimganicum** (Lincz.) M. Pimen. & Sdobnina (*B. czimganicum* Lincz.) - 6
- var. *subnivale* Galushko = B. subnivale
falcatum L. (*B. rossicum* (K.-Pol.) Woronow) - 1, 2(?)
- subsp. *komarovianum* (Lincz.) Worosch. = B. komarovianum
- subsp. *persicum* (Boiss.) K.-Pol. = B. persicum
- subsp. *scorzonerifolium* (Willd.) K.-Pol. = B. scorzonerifolium
ferganense Lincz. - 6
fruticosum L. - 1
gerardii All. (*B. commutatum* auct.) - 2, 6
glaucum Robill. & Cast. ex DC. = B. semicompositum
gulczense O. & B. Fedtsch. - 6
isphairamicum M. Pimen. - 6
komarovianum Lincz. (*B. chinense* DC. var. *komarovianum* (Lincz.) Liou & Huang, *B. falcatum* L. subsp. *komarovianum* (Lincz.) Worosch.) - 5

koso-poljanskyi Grossh. - 2
kotschyanum Boiss. = B. exaltatum
krylovianum Schischk. - 3, 6
lancifolium Hornem. - 6
linczevskii M. Pimen. & Sdobnina - 6
linearifolium auct. = B. exaltatum
lipskyanum (K.-Pol.) Lincz. (*B. contractum* Korov.) - 6
longifolium L. subsp. **aureum** (Fisch. ex Hoffm.) Soo (*B. aureum* Fisch. ex Hoffm.) - 1, 3, 4, 6
- subsp. **vapincense** (Vill.) Todor (*B. vapincense* Vill.) - 1
longiinvolucratum Kryl. - 3, 4
longiradiatum Turcz. (*B. longiradiatum* subsp *sachalinense* (Fr. Schmidt) Kitag., *B. longiradiatum* subsp. *shikotanense* (Hiroe) Worosch., *B. sachalinense* Fr. Schmidt, *B. shikotanense* Hiroe, *B. nipponicum* auct.) - 4, 5
- subsp. *sachalinense* (Fr. Schmidt) Kitag. = B. longiradiatum
- subsp. *shikotanense* (Hiroe) Worosch. = B. longiradiatum
marschallianum C.A. Mey. (*B. tenuissimum* L. subsp. *gracile* (Bieb.) H. Wolff) - 1, 2
martjanovii Kryl. - 3, 4
multinerve DC. (*B. alaunicum* K.-Pol.) - 1, 3, 4
nipponicum auct. = B. longiradiatum
nordmannianum Ledeb. - 2
odontites L. - 1
oroboides Sosn. = B. polyphyllum
pauciradiatum Fenzl ex Boiss. - 1, 2
persicum Boiss. (*B. falcatum* L. subsp. *persicum* (Boiss.) K.-Pol.) - 2
polymorphum Albov = B. polyphyllum
polyphyllum Ledeb. (*B. oroboides* Sosn., *B. polymorphum* Albov) - 2
pusillum Kryl. = B. bicaule
ranuculoides L. - 1
- Ḅ. *arcticum* Regel = B. triradiatum
rischawii Albov - 2
rossicum (K.-Pol.) Woronow = B. falcatum
rosulare Korov. ex M. Pimen. & Sdobnina - 6
rotundifolium L. - 1, 2, 3, 5, 6
sachalinense Fr. Schmidt = B. longiradiatum
scorzonerifolium Willd. (*B. falcatum* L. subsp. *scorzonerifolium* (Willd.) K.-Pol.) - 3, 4, 5
semicompositum L. (*B. glaucum* Robill. & Cast. ex DC.) - 2, 6
shikotanense Hiroe = B. longiradiatum
sibiricum Vest - 3, 4
sosnowskyi Manden. - 2
subfalcatum Schur - 1
subnivale (Galushko) Galushko (*B. exaltatum* Bieb. var. *subnivale* Galushko) - 2
sulphureum Boiss. & Bal. - 2
tenuissimum L. - 1, 2
- subsp. *gracile* (Bieb.) H. Wolff = B. marschallianum
thianschanicum Freyn - 6
triradiatum Adams ex Hoffm. (*B. americanum* Coult. & Rose, *B. ranunculoides* Ḅ. *arcticum* Regel, *B. triradiatum* subsp. *arcticum* (Regel) Hult.) - 3, 4, 5
- subsp. *arcticum* (Regel) Hult. = B. triradiatum
vapincense Vill. = B. longifolium subsp. vapincense
veronense Turra (*B. aenigma* K.-Pol.) - 2
wittmannii Stev. - 2
woronowii Manden. - 1, 2(?)

Cachrys auct. = Prangos

acaulis DC. = Prangos acaulis
alata Bieb. = Prangos ferulacea
alpina Bieb. = Prangos trifida

arenaria (Schischk.) Hiroe = Prangos arenaria
bucharica (B. Fedtsch.) Herrnst. & Heyn = Prangos bucharica
didyma Regel = Prangos didyma
ferulacea (L.) Calest. = Prangos ferulacea
herderi Regel = Prangos herderi
latiloba (Korov.) Herrnst. & Heyn = Prangos latiloba
lophoptera (Boiss.) Takht. = Prangos lophoptera
macrocarpa Ledeb. = Prangos ledebourii
meliocarpoides (Boiss.) Herrnst. & Heyn = Prangos meliocarpoides
- var. *arcis-romanae* (Boiss. & Huet) Herrnst. & Heyn = Prangos meliocarpoides var. arcis-romanae
odontalgica Pall. = Prangos odontalgica
pabularia (Lindl.) Herrnst. & Heyn = Prangos pabularia
pubescens (Pall. ex Spreng.) Schischk. = Prangos odontalgica
uloptera (DC.) Herrnst. & Heyn = Prangos uloptera
uloptera (DC.) Takht. = Prangos uloptera

Calestania K.-Pol. = Thyselium

palustris (L.) K.-Pol. = Thyselium palustre

Caropodium Stapf & Wettst. = Grammosciadium

platycarpum (Boiss. & Hausskn.) Schischk. = Grammosciadium platycarpum

Carum L.

See also Vicatia.
alpinum (Bieb.) Benth. & Hook. fil. = Seseli alpinum
atrosanguineum Kar. & Kir. = Vicatia atrosanguinea
buriaticum Turcz. - 3, 4, 5
carvi L. - 1, 2, 3, 4, 5, 6
- subsp. **rosellum** (Woronow) Worosch. (*C. rosellum* Woronow, *C. carvi* var. *roseum* Trautv.) - 2, 3, 4, 5
- var. *roseum* Trautv. = C. carvi subsp. rosellum
caucasicum (Bieb.) Boiss. - 2
grossheimii Schischk. - 2
karatavicum (Korov.) Hiroe = Oedibasis platycarpa
komarovii Karjag. - 2
leucocoleum auct. p.p. = C. porphyrocoleon
meifolium (Bieb.) Boiss. - 2
porphyrocoleon (Freyn & Sint.) Woronow (*C. leucocoleum* auct. p.p.) - 2
rosellum Woronow = C. carvi subsp. rosellum
saxicolum Albov = Seseli saxicolum
setaceum Schrenk = Bunium setaceum

Caucalis L.

bischoffii K.-Pol. = C. platycarpos var. muricata
daucoides L. = Orlaya daucoides
- subsp. *muricata* Bisch. ex Celak. = C. platycarpos var. muricata
- var. *muricata* (Bisch. ex Celak.) Tamamsch. = C. platycarpos var. muricata
lappula (Web.) Grande = C. platycarpos
platycarpos L. (*C. lappula* (Web.) Grande) - 1, 2, 5, 6
- subsp. *bischoffii* (K.-Pol.) Soo = C. platycarpos var. muricata
- subsp. *muricata* (Bisch. ex Celak.) Holub = C. platycarpos var. muricata
- var. **muricata** (Bisch. ex Celak.) V. Tichomirov (*C. daucoides* L. subsp. *muricata* Bisch. ex Celak., *C. bischoffii*

K.-Pol., *C. daucoides* var. *muricata* (Bisch. ex Celak.) Tamamsch., *C. platycarpos* subsp. *bischoffii* (K.-Pol.) Soo, *C. platycarpos* subsp. *muricata* (Bisch. ex Celak.) Holub) - 1, 2
scandicina Wigg. = Anthriscus caucalis

Cenolophium Koch

denudatum (Hornem.) Tutin (*C. fischeri* (Spreng.) Koch) - 1, 3, 4, 6
fischeri (Spreng.) Koch = C. denudatum

Centella L.

asiatica (L.) Urban - 2

Cephalopodum Korov.

badachschanicum Korov. - 6
hissaricum M. Pimen. - 6

Cervaria Gaertn.

aegopodioides (Boiss.) M. Pimen. (*Physospermum aegopodioides* Boiss., *Peucedanum latifolium* sensu Schischk.) - 2
caucasica (Bieb.) M. Pimen. (*Peucedanum caucasicum* (Bieb.) C. Koch) - 2
cervariifolia (C.A. Mey.) M. Pimen. (*Peucedanum cervariifolium* C.A. Mey., *P. sintenisii* H. Wolff) - 2, 6
rivinii Gaertn. (*Peucedanum cervia* (L.) Lapeyr.) - 1

Chaerophyllum L. (*Golenkinianthe* K.-Pol.)

angelicifolium Bieb. - 2
aromaticum L. - 1
astrantiae Boiss. & Bal. - 2
aureum L. (*Ch. maculatum* Willd. ex DC., *Ch. temuloides* Boiss.) - 1, 2
bobrovii Schischk. = Ch. bulbosum
borodinii Albov - 2
bulbosum L. (*Ch. bobrovii* Schischk., *Ch. bulbosum* subsp. *bobrovii* (Schischk.) Soo, *Ch. caucasicum* (Hoffm.) Schischk.) - 1, 2, 6
- subsp. *bobrovii* (Schischk.) Soo = Ch. bulbosum
- subsp. *prescottii* (DC.) Nym. = Ch. prescottii
caucasicum (Hoffm.) Schischk. = Ch. bulbosum
cicutaria Vill. = Ch. hirsutum
confusum Woronow - 2
crinitum Boiss. - 2
hirsutum L. (*Ch. cicutaria* Vill.) - 1
humile Stev. (*Ch. kiapazi* Woronow ex Schischk.) - 2
khorossanicum Czerniak. ex Schischk. - 6
kiapazi Woronow ex Schischk. = Ch. humile
macrospermum (Willd. ex Spreng.) Fisch. & C.A. Mey. (*Golenkinianthe gilanica* (S.G. Gmel.) K.-Pol.) - 2, 6
maculatum Willd. ex DC. = Ch. aureum
meyeri Boiss. & Buhse - 2
millefolium DC. - 2
prescottii DC. (*Ch. bulbosum* L. subsp. *prescottii* (DC.) Nym.) - 1, 2, 3, 6
roseum Bieb. - 2
rubellum Albov - 2
temuloides Boiss. = Ch. aureum
temulum L. - 1, 2

Chamaesciadium C.A. Mey.

acaule (Bieb.) Boiss. - 2

Chrysosciadium Tamamsch. = Echinophora

sibthorpianum (Guss.) Tamamsch. = Echinophora sibthorpiana

Chymsydia Albov

agasylloides (Albov) Albov - 2
colchica (Albov) Woronow - 2

Ciclospermum Lag. (*Ptilimnium* auct.)

leptophyllum (Pers.)Sprague (*Pimpinella leptophylla* Pers., *Apium leptophyllum* (Pers.) F. Muell. ex Benth., *Ptilimnium nuttallii* auct.) - 2

Cicuta L.

virosa L. - 1, 2, 3, 4, 5, 6

Cnidiocarpa M. Pimen.

alaica M. Pimen. - 6
grossheimii (Manden.) M. Pimen. (*Cnidium grossheimii* Manden., *C. pauciradium* Somm. & Levier var. *grossheimii* (Manden.) Tamamsch.) - 2

Cnidium Cuss. ex Juss.

ajanense (Regel & Til.) Drude = Tilingia ajanensis
cnidiifolium (Turcz.) Schischk. (*Conioselinum cnidiifolium* (Turcz.) Pors.) - 4, 5
davuricum (Jacq.) Turcz. ex Fisch. & C.A. Mey. (*Selinum davuricum* (Jacq.) Leute) - 4, 5
dubium (Schkuhr) Thell. = Kadenia dubia
filisectum Nakai & Kitag. = Rupiphila tachiroei
grossheimii Manden. = Cnidiocarpa grossheimii
jeholense Nakai & Kitag. = Conioselinum jeholense
mandenovae Gagnidze = Seseli saxicolum
monnieri (L.) Cuss. ex Juss. - 4, 5
multicaule Ledeb. = Lithosciadium multicaule
olaense Gorovoi & N.S. Pavlova = Magadania olaensis
pauciradium Somm. & Levier = Seseli saxicolum
- var. *grossheimii* (Manden.) Tamamsch. = Cnidiocarpa grossheimii
salinum Turcz. = Kadenia salina
tachiroei (Franch. & Savat.) Makino = Rupiphila tachiroei
victoris (Schischk.) A. Khokhr. = Magadania victoris

Coelopleurum Ledeb. = Angelica

alpinum Kitag. = Angelica saxatilis
gmelinii (DC.) Ledeb. = Angelica gmelinii
lucidum (L.) Fern. subsp. *gmelinii* (DC.) A. & D. Löve = Angelica gmelinii
- var. *gmelinii* (DC.) Hara = Angelica gmelinii
saxatile (Ledeb.) Drude = Angelica saxatilis

Conioselinum Hoffm.

ajanense (Regel & Til.) A. Khokhr. = Tilingia ajanensis
boreale Schischk. = C. tataricum
chinense (L.) Britt., Sterns & Pogg. (*C. kamtschaticum* sensu Schischk.) - 5
- subsp. *boreale* (Schischk.) A. & D. Löve = C. tataricum
cnidiifolium (Turcz.) Pors. = Cnidium cnidiifolium
czernaevia Fisch. & C.A. Mey. = Angelica czernaevia
filicinum (H. Wolff) Hara (*Peucedanum filicinum* H. Wolff) - 5
jeholense (Nakai & Kitag.) M. Pimen. (*Cnidium jeholense* Nakai & Kitag., *Ligusticum jeholense* (Nakai & Kitag.) Nakai & Kitag., *Tilingia jeholensis* (Nakai & Kitag.) Leute) - 5
kamtschaticum Rupr. = Tilingia ajanensis
kamtschaticum sensu Schischk. = C. chinense
latifolium Rupr. = C. tataricum
longifolium Turcz. - 4
pinnatifolium (Korov.) Schischk. = Vvedenskia pinnatifolia
schugnanicum B. Fedtsch. - 6
tataricum Hoffm. (*C. boreale* Schischk., *C. chinense* (L.) Britt., Sterns & Pogg. subsp. *boreale* (Schischk.) A. & D. Löve, *C. latifolium* Rupr., *C. vaginatum* Thell.) - 1, 3, 4, 6
vaginatum Thell. = C. tataricum
victoris Schischk. = Magadania victoris

Conium L.

maculatum L. - 1, 2, 3, 4, 5, 6

Coriandrum L.

sativum L. - 1, 2, 5, 6

Crithmum L.

maritimum L. - 1, 2

Cryptodiscus Schrenk = Prangos

ammophilus Bunge = Prangos ammophila
arenarius Schischk. = Prangos arenaria
cachroides Schrenk = Prangos cachroides
didymus (Regel) Korov. = Prangos didyma

Cryptotaenia DC.

flahaultii (Woronow) K.-Pol. - 2
japonica Hassk. - 5

Cuminum L.

borsczowii (Regel & Schmalh.) K.-Pol. (*Psammogeton borsczowii* (Regel & Schmalh.) Lipsky) - 6
*****cyminum** L. - 6
setifolium (Boiss.) K.-Pol. (*Psammogeton setifolium* (Boiss.) Boiss.) - 6

Cymbocarpum DC. ex C.A. Mey.

anethoides DC. ex C.A. Mey. (*C. erythraeum* (DC.) Boiss.) - 2
erythraeum (DC.) Boiss. = C. anethoides

Cymopterus auct. = Halosciastrum

crassus (Nakai) Hiroe = Halosciastrum melanotilingia

Danaa All. = Physospermum

denaensis Schischk. = Sphaerosciadium denaense
nudicaulis (Bieb.) Grossh. = Physospermum cornubiense

Daucus L.

carota L. (*D. kotovii* Hiroe) - 1, 2, 3, 6
- subsp. *sativus* (Hoffm.) Arcang. = D. sativus
kotovii Hiroe = D. carota
*****sativus** (Hoffm.) Roehl. (*D. carota* L. subsp. *sativus* (Hoffm.) Arcang.)

Demavendia M. Pimen.

pastinacifolia (Boiss. & Hausskn.) M. Pimen. (*Peucedanum pastinacifolium* Boiss. & Hausskn.) - 6

Dimorphosciadium M. Pimen.

gayoides (Regel & Schmalh.) M. Pimen. (*Ligusticum mutellina* (L.) Crantz subsp. *gayoides* (Regel & Schmalh.) Weinert, p.p., *Pachypleurum gayoides* (Regel & Schmalh.) Schischk.) - 6

Dorema D. Don

aitchisonii Korov. ex M. Pimen. (*D. pruinosum* Korov. nom. invalid., *D. pruinosum* K. Korol. nom. invalid.) - 6
badhysi M. Pimen. (*D. gummiferum* (Jaub. & Spach) Korov. p.p. excl. typo, *D. gummiferum* (Jaub. & Spach) K. Korol. p.p. excl. typo, comb. superfl.) - 6
balchanorum M. Pimen. - 6
glabrum Fisch. & C.A. Mey. - 2
gummiferum (Jaub. & Spach) K. Korol. p.p. excl. typo = D. badhysi
gummiferum (Jaub. & Spach) Korov. p.p. excl. typo = D. badhysi
hyrcanum K.-Pol. - 6
karataviense Korov. - 6
kopetdaghense M. Pimen. - 6
microcarpum Korov. (*D. namanganicum* K. Korol. nom. invalid.) - 6
namanganicum K. Korol. = D. microcarpum
pruinosum K. Korol. = D. aitchisonii
pruinosum Korov. = D. aitchisonii
sabulosum Litv. - 6

Echinophora L. (*Anisosciadium* DC., *Chrysosciadium* Tamamsch.)

orientalis Hedge & Lamond (*E. trichophylla* auct.) - 2
sibthorpiana Guss. (*Anisosciadium tenuifolium* (L.) Tamamsch. subsp. *sibthorpianum* (Guss.) Tamamsch., *Chrysosciadium sibthorpianum* (Guss.) Tamamsch. comb. invalid., *Echinophora tenuifolia* L. subsp. *sibthorpiana* (Guss.) Tutin) - 2, 6
tenuifolia L. subsp. *sibthorpiana* (Guss.) Tutin = E. sibthorpiana
trichophylla auct. = E. orientalis

Elaeopleurum Korov. = Seseli

monococcum Korov. = Seseli ledebourii

Elaeosticta Fenzl (*Muretia* Boiss., *Scaligeria* auct.)

alaica (Lipsky) Kljuykov, M. Pimen. & V. Tichomirov (*Scaligeria alaica* (Lipsky) Korov.) - 6
allioides (Regel & Schmalh.) Kljuykov, M. Pimen. & V. Tichomirov (*Scaligeria allioides* (Regel & Schmalh.) Boiss.) - 6
bucharica (Korov.) Kljuykov, M. Pimen. & V. Tichomirov (*Scaligeria bucharica* Korov.) - 6
conica Korov. (*Scaligeria conica* (Korov.) Korov.) - 6
ferganensis (Lipsky) Kljuykov, M. Pimen. & V. Tichomirov (*Scaligeria ferganensis* Lipsky, *S. korshinskyi* (Lipsky) Korov. p.p. incl. typo) - 6
glaucescens (DC.) Boiss. (*Scaligeria glaucescens* (DC.)Boiss.) - 2
gracilis R. Kam. & M. Pimen. = Galagania gracilis

hirtula (Regel & Schmalh.) Kljuykov, M. Pimen. & V. Tichomirov (*Scaligeria hirtula* (Regel & Schmalh.) Lipsky, *S. korshinskyi* (Lipsky) Korov. p.p. excl. typo, *S. oedibasioides* R. Kam.) - 6
knorringiana (Korov.) Korov. (*Scaligeria knorringiana* Korov., *Elaeosticta knorringiana* (Korov.) Kljuykov, M. Pimen. & V. Tichomirov, comb. superfl., *Hyalolaena collina* Korov.) - 6
korovinii (Bobr.) Kljuykov, M. Pimen. & V. Tichomirov (*Scaligeria korovinii* Bobr.) - 6
kuramensis Korov. = E. transitoria
lipskyi (Korov.) Kljuykov, M. Pimen. & V. Tichomirov = Hyalolaena lipskyi
lutea (Hoffm.) Kljuykov, M. Pimen. & V. Tichomirov (*Bunium luteum* Hoffm., *Muretia lutea* (Hoffm.) Boiss.) - 1, 2, 6
paniculata (Korov.) Kljuykov & M. Pimen. (*Hyalolaena paniculata* Korov.) - 6
platyphylla (Korov.) Kljuykov, M. Pimen. & V. Tichomirov (*Scaligeria platyphylla* Korov.) - 6
polycarpa (Korov.) Kljuykov, M. Pimen. & V. Tichomirov (*Scaligeria polycarpa* Korov.) - 6
samarcandica (Korov.) Kljuykov, M. Pimen. & V. Tichomirov (*Scaligeria samarcandica* Korov.) - 6
seravschanica Kljuykov & M. Pimen.
transcaspica (Korov.) Kljuykov, M. Pimen. & V. Tichomirov (*Scaligeria transcaspica* Korov.) - 6
transitoria (Korov.) Kljuykov, M. Pimen. & V. Tichomirov (*Muretia transitoria* Korov., *Bunium transcaspicum* (Korov.) Hiroe, *Elaeosticta kuramensis* Korov.) - 6
tschimganica (Korov.) Kljuykov, M. Pimen. & V. Tichomirov (*Scaligeria tschimganica* Korov.) - 6
ugamica (Korov.) Korov. (*Scaligeria ugamica* Korov., *Elaeosticta ugamica* (Korov.) Kljuykov, M. Pimen.& V. Tichomirov, comb. superfl.) - 6
vvedenskyi (R. Kam.) Kljuykov, M. Pimen. & V. Tichomirov (*Scaligeria vvedenskyi* R. Kam.) - 6

Eleutherospermum C. Koch

cicutarium (Bieb.) Boiss. - 2
lazicum Boiss. & Bal. = Tamamschjania lazica

Eremodaucus Bunge

lehmannii Bunge - 2(alien), 6

Eriosynaphe DC.

longifolia (Fisch. ex Spreng.) DC. - 1, 2, 3(?), 6

Eryngium L.

balchanicum Bobr. = E. billardieri
biebersteinianum Nevski = E. caucasicum
billardieri Delaroche (*E. balchanicum* Bobr., *E. noeanum* Boiss., *E. nigromontanum* Boiss. & Buhse) - 2, 6
bungei Boiss. - 6
caeruleum Bieb. = E. caucasicum
campestre L. - 1, 2
caucasicum Trautv. (*E. biebersteinianum* Nevski, *E. caeruleum* Bieb. nom. illeg., *E. pskemense* Pavl.) - 2, 6
giganteum Bieb. - 2
incognitum Pavl. = E. macrocalyx
karatavicum Iljin - 6
macrocalyx Schrenk (*E. incognitum* Pavl., *E. polycephalum* auct.) - 6
maritimum L. - 1, 2
mirandum Bobr. = E. octophyllum

nigromontanum Boiss. & Buhse = E. billardieri
noeanum Boiss. = E. billardieri
octophyllum Korov. (*E. mirandum* Bobr.) - 6
planum L. - 1, 2, 3, 4, 5, 6
polycephalum auct. = E. macrocalyx
pskemense Pavl. = E. caucasicum
wanaturii Woronow - 2

Eulophus auct. = Galagania

ferganensis (Korov.) Hiroe = Galagania ferganensis
tenuisectus (Regel & Schmalh.) Hiroe = Galagania tenuisecta

Falcaria Fabr.

falcarioides (Bornm. & H. Wolff) H. Wolff = Gongylosciadium falcarioides
neglectissima Klok. = F. vulgaris
sioides (Wib.) Aschers. = F. vulgaris
vulgaris Bernh. (*Drepanophyllum sioides* Wib. nom. illegit., *Falcaria neglectissima* Klok., *F. sioides* (Wib.) Aschers.) - 1, 2, 3, 6

Fergania M. Pimen.

polyantha (Korov.) M. Pimen. (*Ferula polyantha* Korov., *Peucedanum polyanthum* (Korov.) Korov.) - 6

Ferula L. (*Merwia* B. Fedtsch., *Schumannia* O. Kuntze, *Scorodosma* Bunge, *Soranthus* Ledeb.)

aitchisonii K.-Pol. = F. karatavica
akitschkensis B. Fedtsch. ex K.-Pol. (*F. angustiloba* M. Pimen., *F. transitoria* Korov.) - 6
alaica M. Pimen. & Melibaev - 6
androssowii (B. Fedtsch.) Hiroe = F. litwinowiana
angreni Korov. - 6
angustiloba M. Pimen. = F. akitschkensis
arida (Korov.) Korov. = F. soongarica
assa-foetida auct. = F. foetida
badhysi Korov. = F. oopoda
badrakema K.-Pol. - 6
balchaschensis Bajt. = F. teterrima
botschantzevii Korov. - 6
bucharica (Lipsky) K.-Pol. = Ladyginia bucharica
calcarea M. Pimen. - 2
canescens (Ledeb.) Ledeb. - 6
caspica Bieb. (*F. gracilis* (Ledeb.) Ledeb.) - 1, 2, 3, 6
caucasica Korov. - 2
ceratophylla Regel & Schmalh. - 6
clematidifolia K.-Pol. - 6
conocaula Korov. - 6
czatkalensis M. Pimen. - 6
daghestanica (Schischk.) Hiroe = Ferulago galbanifera
decurrens Korov. - 6
dissecta (Ledeb.) Ledeb. - 6
diversivittata Regel & Schmalh. - 6
dshaudshamyr Korov. = F. dubjanskyi
dshizakensis Korov. (*F. helenae* Rakhmankulov & Melibaev) - 6
dubjanskyi Korov. ex Pavl. (*F. dshaudshamyr* Korov.) - 6
equisetacea K.-Pol. - 6
eremophila Korov. = F. karatavica
eugenii R. Kam. = F. violacea
euxina M. Pimen. - 1
fedoroviorum M. Pimen. - 6

fedtschenkoana K.-Pol. - 6
ferganensis Lipsky ex Korov. - 6
ferulaeoides (Steud.) Korov. - 6
foetida (Bunge) Regel (*Scorodosma foetidum* Bunge, *Ferula assa-foetida* auct.) - 6
foetidissima Regel & Schmalh. - 6
foliosa Lipsky ex Korov. = F. kelleri
galbanifera Mill. = Ferulago galbanifera
gigantea B. Fedtsch. (*F. inflata* Korov., *F. latifolia* Korov.) - 6
glaberrima Korov. - 6
gracilis (Ledeb.) Ledeb. - 3
grigoriewii B. Fedtsch. - 6
gummosa Boiss. - 6
gypsacea Korov. - 6
helenae Rakhmankulov & Melibaev = F. dshizakensis
hispida Friv. = Opopanax hispidus
iliensis Krasn. ex Korov. (*F. popovii* Korov.) - 6
inciso-serrata M. Pimen. & J. Baranova - 6
inflata Korov. = F. gigantea
involucrata Korov. = F. leucographa
jaeschkeana auct. = F. kuhistanica
juniperina Korov. - 6
karakalensis Korov. (*F. tersakensis* Korov.) - 6
karatavica Regel & Schmalh. (*F. aitchisonii* K.-Pol., *F. eremophila* Korov.) - 6
karataviensis (Regel & Schmalh.) Korov. - 6
karategina Lipsky ex Korov. - 6
karelinii Bunge (*Schumannia karelinii* (Bunge) Korov.) - 6
kaschkarovii Korov. = F. penninervis
kelifi Korov. (*F. primaeva* Korov.) - 6
kelleri K.-Pol. (*F. foliosa* Lipsky ex Korov.) - 6
kirialovii M. Pimen. (*F. pseudooreoselinum* (Regel & Schmalh.) K.-Pol. p.p. excl. typo) - 6
kokanica Regel & Schmalh. - 6
kopetdaghensis Korov. - 6
korshinskyi Korov. - 6
koso-poljanskyi Korov. - 6
krylovii Korov. - 6
kuhistanica Korov. (*F. jaeschkeana* auct.) - 6
kyzylkumica Korov. - 6
lapidosa Korov. - 6
latifolia Korov. = F. gigantea
latiloba Korov. - 6
latisecta auct. p.p. = F. undulata
lehmannii Boiss. - 6
leiophylla Korov. - 6
leucographa Korov. (*F. involucrata* Korov.) - 6
ligulata Korov. = F. xeromorpha
linczevskii Korov. - 6
lipskyi Korov. - 6
lithophila M. Pimen. (*Peucedanum mogoltavicum* Korov.) - 6
litwinowiana K.-Pol. (*F. androssowii* (B. Fedtsch.) Hiroe, *Merwia androssowii* B. Fedtsch.) - 6
malacophylla M. Pimen. & J. Baranova - 6
microcarpa Korov. = F. ovina
microloba Boiss. = F. szowitsiana
minkwitzae Korov. = F. tschimganica
mogoltavica Lipsky ex Korov. - 6
mollis Korov. - 6
moschata K.-Pol. = F. sumbul
nevskii Korov. - 6
nuda Spreng. - 1, 6
nuratavica M. Pimen. (*Silaus popovii* Korov., *Johrenia popovii* (Korov.) Korov., *Silaum popovii* (Korov.) Hiroe) - 6

olgae Regel & Schmalh. = F. penninervis
oopoda (Boiss. & Buhse) Boiss. (*F. badhysi* Korov.) - 6
orientalis L. - 2
ovczinnikovii M. Pimen. - 6
ovina (Boiss.) Boiss. (*F. microcarpa* Korov., *F. pachycarpa* Korov. ex Pavl., *F. stylosa* Korov.) -6
pachycarpa Korov. ex Pavl. = F. ovina
pachyphylla Korov. - 6
pallida Korov. - 6
penninervis Regel & Schmalh. (*F. kaschkarovii* Korov., *F. olgae* Regel & Schmalh.) - 6
persica Willd. - 2
petiolaris DC. = Leutea petiolaris
peucedanifolia Willd. ex Spreng. = F. sibirica
plurivittata Korov. - 6
polyantha Korov. = Fergania polyantha
popovii Korov. = F. iliensis
potaninii Korov. ex Pavl. - 6
prangifolia Korov. - 6
pratovii Khassanov & Maltzev - 6
primaeva Korov. = F. kelifi
pseudooreoselinum (Regel & Schmalh.) K.-Pol. = F. kirialovii and F. sumbul
renardii (Regel & Schmalh.) M. Pimen. (*Peucedanum renardii* Regel & Schmalh., *Ferula talassica* (Korov.) M. Pimen., *Peucedanum talassicum* Korov.)- 6
rigidula DC. - 2
rubroarenosa Korov. - 6
samarkandica Korov. - 6
schair Borszcz. = F. varia
schischkinii Hiroe = Ferulago latiloba .
schtschurowskiana Regel & Schmalh. - 6
seravschanica M. Pimen. & J. Baranova - 6
seseloides C.A. Mey. = Johreniopsis seseloides
sibirica Willd. (*F. peucedanifolia* Willd. ex Spreng., *Soranthus meyeri* Ledeb.) - 1(?), 6
sjugatensis Bajt. - 6
soongarica Pall. ex Spreng. (*F. arida* (Korov.) Korov., *F. soongarica* var. *arida* Korov.) - 6
- var. *arida* Korov. = F. soongarica
stylosa Korov. = F. ovina
subtilis Korov. - 6
subvelutina (Rech. fil.) Hiroe = Ferulago subvelutina
sumbul (Kauffm.) Hook. fil. (*F. moschata* K.-Pol. nom. nud., *F. pseudooreoselinum* (Regel & Schmalh.) K.-Pol. p.p. incl. typo, *F. urceolata* Korov.) - 6
syreitschikowii K.-Pol. - 6
szowitsiana DC. (*F. microloba* Boiss.) - 2, 6
tadshikorum M. Pimen. - 6
talassica (Korov.) M. Pimen. = F. renardii
tatarica Fisch. ex Spreng. - 1, 2, 3, 6
taucumica Bajt. - 6
taurica (Schischk.) Hiroe = Ferulago galbanifera var. brachyloba
tenuisecta Korov. - 6
tersakensis Korov. = F. karakalensis
teterrima Kar. & Kir. (*F. balchaschensis* Bajt.) - 6
transiliensis (Herd.) M. Pimen. (*Peucedanum transiliense* Herd.) - 6
transitoria Korov. = F. akitschkensis
tschimganica Lipsky ex Korov. (*F. minkwitzae* Korov.) - 6
tschuiliensis Bajt. - 6
tuberifera Korov. - 6
turcomanica (Schischk.) Hiroe = Ferulago subvelutina
turcomanica (Schischk.) M. Pimen. = Leutea petiolaris
ugamica Korov. - 6
undulata M. Pimen. & J. Baranova (*F. latisecta* auct. p.p.) - 6

urceolata Korov. = F. sumbul
varia (Schrenk) Trautv. (*F. schair* Borszcz.) - 6
vicaria Korov. - 6
violacea Korov. (*F. eugenii* R. Kam.) - 6
xeromorpha Korov. (*F. ligulata* Korov.) - 6

Ferulago Koch

campestris (Bess.) Grec. = F. galbanifera
daghestanica Schischk. = F. galbanifera
galbanifera (Mill.) Koch (*Ferula galbanifera* Mill., *F. daghestanica* (Schischk.) Hiroe, *Ferulago campestris* (Bess.) Grec., *F. daghestanica* Schischk.) - 1, 2
- var. **brachyloba** Boiss. (*F. taurica* Schischk., *Ferula taurica* (Schischk.) Hiroe) - 1, 2
▪ latiloba Schischk. (*Ferula schischkinii* Hiroe)
setifolia C. Koch - 2
subvelutina Rech. fil. (*Ferula subvelutina* (Rech. fil.) Hiroe, *F. turcomanica* (Schischk.) Hiroe, *Ferulago turcomanica* Schischk.) - 6
sylvatica (Bess.) Reichenb. - 1
taurica Schischk. = F. galbanifera var. brachyloba
turcomanica Schischk. = F. subvelutina

Ferulopsis Kitag.

hystrix (Bunge) M. Pimen. (*Peucedanum hystrix* Bunge, *Phlojodicarpus nudiusculis* Turcz. ex M. Pop. nom. invalid., *Ph. turczaninovii* Sipl.) - 3, 4

Foeniculum Hill

vulgare Mill. - 1, 2, 6

Froriepia C. Koch

subpinnata (Ledeb.) Baill. - 2

Fuernrohria C. Koch

setifolia C. Koch - 2

Galagania Lipsky (*Korovinia* Nevski & Vved., *Eulophus* auct.)

ferganensis (Korov.) M. Vassil. & M. Pimen. (*Korovinia ferganensis* Korov., *Eulophus ferganensis* (Korov.) Hiroe) - 6
fragrantissima Lipsky (*Muretia fragrantissima* (Lipsky) Korov.) - 6
gracilis (R. Kam. & M. Pimen.) R. Kam. & M. Pimen. (*Elaeosticta gracilis* R. Kam. & M. Pimen.) - 6
margiana M. Pimen. & M. Vassil. - 6
neglecta M. Vassil. & Kljuykov - 6
platypoda (Aitch. & Hemsl.) M. Vassil. & M. Pimen. (*Korovinia microcarpa* (Korov.) Korov.) - 6
tenuisecta (Regel & Schmalh.) M. Vassil. & M. Pimen. (*Eulophus tenuisectus* (Regel & Schmalh.) Hiroe, *Korovinia tenuisecta* (Regel & Schmalh.) Nevski & Vved.) - 6

Gasparrinia Bertol. = Seseli

donetzica Dubovik = Seseli peucedanoides
peucedanoides = Seseli peucedanoides

Glehnia Fr. Schmidt ex Miq.

littoralis Fr. Schmidt ex Miq. - 5

Golenkinianthe K.-Pol. = Chaerophyllum

gilanica (S.G. Gmel.) K.-Pol. = Chaerophyllum macrospermum

Gongylosciadium Rech. fil.

falcarioides (Bornm. & H. Wolff) Rech. fil. (*Falcaria falcarioides* (Bornm. & H. Wolff) H. Wolff) - 2

Grammosciadium DC. (*Caropodium* Stapf & Wettst.)

daucoides DC. - 2
platycarpum Boiss. & Hausskn. (*Caropodium platycarpum* (Boiss. & Hausskn.) Schischk.) - 2

Haloscias Fries = Ligusticum

hultenii (Fern.) Holub = Ligusticum scoticum
scoticum (L.) Fries = Ligusticum scoticum

Halosciastrum Koidz. (*Cymopterus* auct.)

melanotilingia (Boissieu) M. Pimen. & V. Tichomirov (*Selinum melanotilingia* Boissieu, *Cymopterus crassus* (Nakai) Hiroe, *Ligusticum melanotilingia* (Boissieu) Kitag., *L. purpureopetalum* Kom., *Ostericum crassum* (Nakai) Kitag., *Pimpinella* ? *crassa* Nakai) - 5

Hansenia Turcz.

mongholica Turcz. (*Ligusticum mongholicum* (Turcz.) Krylov) - 3, 4

Helosciadium Koch = Apium

nodiflorum (L.) Koch = Apium nodiflorum

Heracleum L.

aconitifolium Woronow - 2
albovii Manden. (*H. pastinacifolium* C. Koch subsp. *incanum* (Boiss. & Huet) P.H. Davis, p.p.) - 2
antasiaticum Manden. - 2
apiifolium Boiss. - 2
asperum (Hoffm.) Bieb. - 2
australe (C. Hartm.) Stank. & Maleev = H. sphondylium
barbatum Ledeb. = H. dissectum var. barbatum
- subsp. *moehlendorfii* (Hance) Hiroe = H. dissectum
calcareum Albov - 2
- var. **colchicum** (Lipsky) Satzyperova (*H. colchicum* Lipsky, *H. freynianum* Somm. & Levier) - 2
carpaticum Porc. (*H. sphondylium* L. subsp. *carpaticum* (Porc.) Soé) - 1
chorodanum (Hoffm.) DC. - 2
circassicum Manden. = H. pubescens
clausii Ledeb. = Pastinaca clausii
colchicum Lipsky = H. calcareum var. colchicum
cyclocarpum C. Koch (*H. sphondylium* L. subsp. *cyclocarpum* (C. Koch) P.H. Davis) - 2
dissectum Ledeb. (*H. barbatum* Ledeb. subsp. *moehlendorfii* (Hance) Hiroe, *H. dissectum* subsp. *moehlendorfii* (Hance) Worosch., *H. lanatum* Michx. subsp. *moehlendorfii* (Hance) Hara, *H. moehlendorfii* Hance) - 3, 4, 5, 6
- subsp. *moehlendorfii* (Hance) Worosch. = H. dissectum
- var. **barbatum** (Ledeb.) Kryl. (*H. barbatum* Ledeb.) - 3
- var. **voroschilovii** (Gorovoi) M. Pimen. (*H. voroschilovii* Gorovoi, *H. lanatum* subsp. *voroschilovii* (Gorovoi)

Worosch.) - 5
dulce Fisch. = H. lanatum
egrissicum Gagnidze - 2
freynianum Sommier & Levier = H. calcareum var. colchicum
grandiflorum Stev. ex Bieb. - 2
graveolens (Bieb.) Spreng. = Pastinaca clausii
grossheimii Manden. = H. mantegazzianum
idae Kulieva - 2
lanatum Michx. (*H. dulce* Fisch., *H. lanatum* subsp. *asiaticum* Hiroe, *H. lanatum* var. *asiaticum* (Hiroe) Hara, *H. sphondylium* L. subsp. *lanatum* (Michx.) A. & D. Löve) - 5
- subsp. *asiaticum* Hiroe = H. lanatum
- subsp. *moehlendorfii* (Hance) Hara = H. dissectum
- subsp. *voroschilovii* (Gorovoi) Worosch. = H. dissectum var. voroschilovii
- var. *asiaticum* (Hiroe) Hara = H. lanatum
lehmannianum Bunge - 6
leskovii Grossh. - 2
ligusticifolium Bieb. - 1
mandenovae Satzyperova - 2
mantegazzianum Somm. & Levier (*H. grossheimii* Manden.) - 2
moehlendorfii Hance = H. dissectum
nanum Satzyperova - 2
olgae Regel & Schmalh. = Tetrataenium olgae
ossethicum Manden. - 2
palmatum Baumg. (*H. palmatum* subsp. *transsilvanicum* (Schur) Nym., *H. sphondylium* L. subsp. *transsilvanicum* (Schur) Brummitt, *H. transsilvanicum* Schur) - 1
- subsp. *transsilvanicum* (Shur) Nym. = H. palmatum
pastinacifolium C. Koch - 2
- subsp. *incanum* (Boiss. & Huet) P.H. Davis, p.p. = H. albovii
- subsp. **schelkovnikovii** (Woronow) Satzyperova (*H. schelkovnikovii* Woronow) - 2
- subsp. **transcaucasicum** (Manden.) P.H. Davis (*H. transcaucasicum* Manden.) - 2
ponticum (Lipsky) Schischk. ex Grossh. - 2
pubescens (Hoffm.) Bieb. (*H. circassicum* Manden.) - 1, 2
roseum Stev. - 2
scabrum Albov - 2
schelkovnikovii Woronow = H. pastinacifolium subsp. schelkovnikovii
sibiricum L. (*H. sphondylium* L. subsp. *sibiricum* (L.) Simonk.) - 1, 2, 3, 5(alien)
sommieri Manden. - 2
sosnowskyi Manden. - 1(alien),2, 5(alien)
sphondylium L. (*H. australe* (C. Hartm.) Stank. & Maleev) - 1
- subsp. *carpaticum* (Porc.) Soo = H. carpaticum
- subsp. *cyclocarpum* (C. Koch) P.H. Davis = H. cyclocarpum
- subsp. *lanatum* (Michx.) A. & D. Löve = H. lanatum
- subsp. *sibiricum* (L.) Simonk. = H. sibiricum
- ▪ subsp. **trachycarpum** (Soj@k) Holub (*H. trachycarpum* Soj@k) - 1
- subsp. *transsilvanicum* (Schur) Brummitt = H. palmatum
stevenii Manden. - 1, 2
trachycarpum Sojak = H. sphondylium subsp. trachycarpum
trachyloma Fisch. & C.A. Mey. - 2
transcaucasicum Manden. = H. pastinacifolium subsp. transcaucasicum
transiliense (Regel & Herd.) O. & B. Fedtsch. = Semenovia transiliensis
transsilvanicum Schur = H. palmatum
voroschilovii Gorovoi = H. dissectum var. voroschilovii
wilhelmsii Fisch. & Ave-Lall. - 2

Hippomarathrum auct. = Bilacunaria

caspium (DC.) Grossh. = Bilacunaria caspia
insigne (Lipsky) Hiroe = Lipskya insignis
longilobum (DC.) B. Fedtsch. ex Grossh. = Bilacunaria
   microcarpa
microcarpum (Bieb.) V. Petrov = Bilacunaria microcarpa
papillare (Regel & Schmalh.) Hiroe = Schrenkia papillaris

Hohenackeria Fisch. & C.A. Mey.

exscapa (Stev.) K.-Pol. - 2

Hyalolaena Bunge (*Hymenolyma* Korov.)

bupleuroides (Schrenk) M. Pimen. & Kljuykov (*Hymeno-lyma bupleuroides* (Schrenk) Korov.) - 6
collina Korov. = Elaeosticta knorringiana
depauperata Korov. (*Selinum depauperatum* (Korov.)
   Hiroe) - 6
intermedia M. Pimen. & Kljuykov - 6
issykkulensis M. Pimen. & Kljuykov - 6
jaxartica Bunge (*H. praemontana* Korov.) - 6
lipskyi (Korov.) M. Pimen. & Kljuykov (*Scaligeria lipskyi*
   Korov., *Bunium oeroilanicum* (Korov.)Hiroe, *Elaeos-ticta lipskyi* (Korov.) Kljuykov, M. Pimen. & V.
   Tichomirov, *Muretia oeroilanica* Korov.) - 6
melanorrhiza M. Pimen. & Kljuykov - 6
paniculata Korov. = Elaeosticta paniculata
praemontana Korov. = H. jaxartica
transcaspica (Korov.) M. Pimen. & Kljuykov (*Muretia transcaspica* Korov., *Bunium transcaspicum* (Korov.)
   Hiroe) - 6
trichophylla (Schrenk) M. Pimen. & Kljuykov (*Hymenoly-ma trichophyllum* (Schrenk) Korov.) - 6
tschuiliensis (Pavl. ex Korov.) M. Pimen. & Kljuykov
   (*Seseli tschuiliense* Pavl. ex Korov.) - 6
viridiflora Kljuykov - 6

Hydrocotyle L.

natans Cyr. = H. ranunculoides
ramiflora Maxim. - 2, 5
ranunculoides L. fil. (*H. natans* Cyr.) - 2
vulgaris L. - 1, 2

Hymenolaena DC.

alpina Schischk. = Paulita alpina
badachschanica Pissjauk. - 6
nana Rupr. - 6
pimpinellifolia Rupr. - 6

Hymenolyma Korov. = Hyalolaena

bupleuroides (Schrenk) Korov. = Hyalolaena bupleuroides
trichophyllum (Schrenk) Korov. = Hyalolaena trichophylla

Johrenia DC.

paucijuga (DC.) Hoffm. - 2
platycarpa Boiss. - 6
popovii (Korov.) Korov. = Ferula nuratavica

Johreniopsis M. Pimen.

seseloides (C.A. Mey.) M. Pimen. (*Ferula seseloides* C.A.
   Mey., *Peucedanum paucifolium* Ledeb., *P. meyeri* auct.
   p. p.) - 2, 6

Kadenia Lavrova & V. Tichomirov

dubia (Schkuhr) Lavrova & V. Tichomirov (*Cnidium dubium* (Schkuhr) Thell., *Selinum dubium* (Schkuhr)
   Leute) - 1, 3, 4, 6
salina (Turcz.) Lavrova & V. Tichomirov (*Cnidium salinum*
   Turcz., *Selinum dubium* (Schkuhr) Leute subsp. *sali-num* (Turcz.) Leute, *S. salinum* (Turcz.) N. Vodopia-nova) - 4

Kafirnigania R. Kam. & Kinz.

hissarica (Korov.) R. Kam. & Kinz. (*Peucedanum hissari-cum* Korov.) - 6

Karatavia M. Pimen. & Lavrova

kultiassovii (Korov.) M. Pimen. & Lavrova (*Selinum kul-tiassovii* Korov., *Sphaenolobium kultiassovii* (Korov.)
   M. Pimen.) - 6

Kitagawia M. Pimen.

baicalensis (Redow. ex Willd.) M. Pimen. (*Peucedanum baicalense* (Redow. ex Willd.) Koch) - 3, 4
eryngiifolia (Kom.) M. Pimen. (*Peucedanum eryngiifolium*
   Kom.) - 5
komarovii M. Pimen. (*Peucedanum elegans* Kom. 1909, non
   Sweet, 1830) - 5
litoralis (Worosch. & Gorovoi) M. Pimen. (*Peucedanum litorale* Worosch. & Gorovoi, *P. japonicum* auct.) - 5
terebinthacea (Fisch. ex Spreng.) M. Pimen. (*Selinum tere-binthaceum* Fisch. ex Spreng., *Peucedanum deltoideum*
   Makino & Yabe, *P. paishanense* Nakai, *P. terebintha-ceum* (Fisch. ex Trev.) Ledeb., *P. terebinthaceum*
   subsp. *aculeolatum* Worosch., *P. terebinthaceum* subsp.
   deltoideum (Makino & Yabe) Worosch., *P. terebintha-ceum* var. *paishanense* (Nakai) Huang, *Selinum tere-binthaceum* Fisch. ex Trev. nom. nud.,*Peucedanum formosanum* auct.) - 4, 5
- subsp. **trichootheca** M. Pimen. - 5

Koelzella Hiroe = Prangos

pabularia (Lindl.) Hiroe = Prangos pabularia
uloptera (DC.) Hiroe = Prangos uloptera

Komarovia Korov.

anisosperma Korov. (sphalm. "anisospermum" in "Fl.
   URSS") - 6

Korovinia Nevski & Vved. = Galagania

ferganensis Korov. = Galagania ferganensis
goloskokovii Bajt. = Hyalolaena trichophylla
microcarpa (Korov.) Korov. = Galagania platypoda
tenuisecta (Regel & Schmalh.) Nevski & Vved. = Galagania
   tenuisecta

Korshinskya Lipsky

bupleuroides Korov. - 6
kopetdaghensis (Korov.) M. Pimen. & Kljuykov (*Physo-spermum kopetdaghense* Korov., *Scaligeria kopetda-ghensis* (Korov.) Schischk.) - 6
olgae (Regel & Schmalh.) Lipsky - 6

Kosopoljanskia Korov.

hebecarpa R. Kam & M. Pimen. - 6
turkestanica Korov. - 6

Kozlovia Lipsky (*Albertia* Regel. & Schmalh. 1878, non W. Schimp. 1837, nec *Alberta* E. Mey. 1838)

paleacea (Regel & Schmalh.) Lipsky (*Albertia paleacea* Regel & Schmalh., *Trachydium paleaceum* (Regel & Schmalh.) Hiroe) - 6

Krasnovia M. Pop. ex Schischk.

longiloba (Kar. & Kit.) M. Pop. ex Schischk. - 6

Ladyginia Lipsky

bucharica Lipsky (*Ferula bucharica* (Lipsky) K.-Pol.) - 6

Lagoecia L.

cuminoides L. - 1

Laser Borkh.

trilobum (L.) Borkh. - 1, 2

Laserpitium L.

affine Ledeb. -2
alpinum Waldst. & Kit. (*L. krapfii* auct. p.p.) - 1
hispidum Bieb. - 1, 2
krapfii auct. p. p. = L. alpinum
latifolium L. - 1
panjutinii (Manden. & Schischk.) Hiroe = Polylophium panjutinii
prutenicum L. - 1
stevenii Fisch. & Trautv. - 2

Lecokia DC.

cretica (Lam.) DC. - 2

Ledebouriella H. Wolff

divaricata (Turcz.) Hiroe = Saposhnikovia divaricata
multiflora (Ledeb.) H. Wolff (*Stenocoelium tenuifolium* Korov.) - 6
seseloides (Hoffm.) H. Wolff - 6

Leutea M. Pimen.

petiolaris (DC.) M. Pimen. (*Ferula petiolaris* DC., *F. turcomanica* (Schischk.) M. Pimen. 1981, non (Schischk.) Hiroe 1979, *Peucedanum turcomanicum* Schischk.) - 6

*Levisticum Hill

*officinale Koch

Libanotis Haller ex Zinn = Seseli

abolinii (Korov.) Korov. = Seseli abolinii
afghanica Podlech = Seseli afghanicum
amurensis Schischk. = Seseli seseloides
arctica Rupr. = Seseli condensatum
buchtormensis (Fisch. ex Spreng.) DC. = Seseli buchtormense
calycina Korov. = Seseli calycinum
condensata (L.) Crantz = Seseli condensatum
- subsp. *arctica* (Rupr.) V. Sergienko = Seseli condensatum
dolichostyla Schischk. = Seseli mucronatum

eriocarpa Schrenk = Seseli eriocarpum
fasciculata Korov. = Seseli fasciculatum
iliensis (Regel & Schmalh.) Korov. = Seseli iliense
incana (Steph. ex Willd.) B. Fedtsch. = Seseli incanum
intermedia Rupr. = Seseli libanotis
juncea Korov. = Seseli tenuisectum
korovinii (Schischk.) Korov. = Seseli korovinii
krylovii V. Tichomirov = Seseli krylovii
leiocarpa (Heuff.)Simonk. = Seseli libanotis
marginata Korov. = Seseli marginatum
merkulowiczii Korov. = Seseli merkulowiczii
michajlovae Korov. = Seseli abolinii
mironovii Korov. = Seseli mironovii
monstrosa (Willd. ex Spreng.) DC. = Sajanella monstrosa
montana Crantz = Seseli libanotis
- subsp. *bipinnata* (Celak.) Dostal = Seseli libanotis
nevskii Korov. = Seseli nevskii
petrophila Korov. = Phlojodicarpus villosus
pyrenaica (L.) Bourg. = Seseli libanotis
- subsp. *bipinnata* (Celak.) Holub = Seseli libanotis
- subsp. *montana* (Crantz) Lemke & Rothm. = Seseli libanotis
schrenkiana C.A. Mey. ex Schischk. = Seseli schrenkianum
seseloides Turcz. = Seseli seseloides
setifera (Korov. ex Pavl.) Schischk. = Seseli setiferum
sibirica (L.) C.A. Mey. = Seseli libanotis
- var. *gracilis* Kryl. = Seseli krylovii
sibirica sensu Schischk. = Seseli krylovii
songorica (Schischk.) Korov. = Seseli abolinii
talassica Korov = Pilopleura tordyloides
taurica N. Rubtz. = Seseli libanotis
tenuisecta (Regel & Schmalh.) Korov. = Seseli tenuisectum
transcaucasica Schischk. = Seseli transcaucasicum
turajgyrica Bajt. = Seseli abolinii
unicaulis Korov. = Seseli unicaule

Ligusticum L. (*Haloscias* Fries)

afghanicum Rech. fil. = Lomatocarpa afghanica
alatum (Bieb.) Spreng. = Macrosciadium alatum
arafoe Albov = Arafoe aromatica
caucasicum Somm. & Levier - 2
discolor Ledeb. = Paraligusticum discolor
fedtschenkoanum Schischk. = Seseli seravschanicum
filisectum (Nakai & Kitag.) Hiroe = Rupiphila tachiroei
hultenii Fern. = L. scoticum
jeholense (Nakai & Kitag.) Nakai & Kitag. = Conioselinum jeholense
macounii Coult. & Rose = Podistera macounii
melanotilingia (Boissieu) Kitag. = Halosciastrum melanotilingia
mongholicum (Turcz.) Kryl. = Hansenia mongholica
mucronatum (Schrenk) Leute = Seseli mucronatum
multifidum Smith = Aulacospermum multifidum
mutellina (L.) Crantz = Mutellina purpurea
- subsp. *gayoides* (Regel & Schmalh.) Weinert, p. p. = Dimorphosciadium gayoides
physospermifolium Albov = Macrosciadium physospermifolium
pumilum Korov. = Seseli talassicum
purpureopetalum Kom. = Halosciastrum melanotilingia
scoticum L. (*Angelica hultenii* (Fern.) Hiroe, *Haloscias hultenii* (Fern.) Holub, *H. scoticum* (L.) Fries, *Ligusticum hultenii* Fern., *L. scoticum* subsp. *hultenii* (Fern.) Hult.) - 1, 5
- subsp *hultenii* (Fern.) Hult. = Ligusticum scoticum
seseloides Fisch. & C.A. Mey. ex Ledeb. = Seseli seseloides
setiferum Korov. ex Pavl. = Seseli setiferum

steineri Podlech = Lomatocarpa afghanica
tachiroei (Franch. & Savat.) Hiroe & Constance = Rupiphila tachiroei
talassicum Korov. = Seseli talassicum
thomsonii Clarke = Seseli mucronatum
turkestanicum Hiroe = Seseli talassicum

Lipskya (K.-Pol.) Nevski

insignis (Lipsky) Nevski (*Schrenkia insignis* Lipsky, *Hippomarathrum insigne* (Lipsky) Hiroe) - 6

Lisaea Boiss.

armena Schischk. = L. papyracea
heterocarpa (DC.) Boiss. - 2
papyracea Boiss. (*L. armena* Schischk.) - 2

Lithosciadium Turcz.

multicaule Turcz. (*Cnidium multicaule* Ledeb., *Selinum multicaule* (Turcz.) Leute) - 4

Lomatocarpa M. Pimen. (*Alposelinum* M. Pimen., *Meum* auct., *Neogaya* auct.)

afghanica (Rech. fil.) M. Pimen. comb. nova (*Ligusticum afganicum* Rech. fil. 1963, Dan. Biol. Skr. 13, 4 : 94; *L. steineri* Podlech, *Lomatocarpa steineri* (Podlech) M. Pimen., *Pachypleurum linearilobum* Korov.) - 6
albomarginata (Schrenk) M. Pimen. & Lavrova (*Alposelinum albomarginatum* (Schrenk) M. Pimen., *Neogaya simplex* (L.) Meissn. var. *albomarginata* Schrenk, *Pachypleurum aemulans* Korov. nom. invalid.) - 6
korovinii M. Pimen. (*Aulacospermum alatum* (Korov.) Korov., *Meum alatum* Korov. 1947, non Baill. 1884) - 6
steineri (Podlech) M. Pimen. = L. afghanica

Macrosciadium V. Tichomirov & Lavrova

alatum (Bieb.) V. Tichomirov & Lavrova (*Ligusticum alatum* (Bieb.) Spreng.) - 2
physospermifolium (Albov) V. Tichomirov & Lavrova (*Ligusticum physospermifolium* Albov) - 2

Macroselinum Schur

latifolium (Bieb.) Schur (*Peucedanum latifolium* (Bieb.) DC., *P. macrophyllum* Schischk.) - 1, 2

Magadania M. Pimen. & Lavrova (*Ochotia* A. Khokhr.)

olaensis (Gorovoi & N.S. Pavlova) M. Pimen. & Lavrova (*Cnidium olaense* Gorovoi & N.S. Pavlova, *Ochotia olaensis* (Gorovoi & N.S. Pavlova) A. Khokhr.) - 5
victoris (Schischk.) M. Pimen. & Lavrova (*Conioselinum victoris* Schischk., *Cnidium victoris* (Schischk.) A. Khokhr., *Ochotia victoris* (Schischk.) A. Khokhr.) - 5

Malabaila Hoffm.

aucheri Boiss. (*M. secacul* (Mill.) Boiss. subsp. *aucheri* (Boiss.) C.C. Townsend, *Peucedanum assyriaense* Hiroe) - 2
dasyantha (C. Koch) Grossh. - 2
dasycarpa (Regel & Schmalh.) Schischk. = Semenovia dasycarpa
graveolens (Bieb.) Hoffm. = Pastinaca clausii

secacul (Mill.) Boiss. subsp. *aucheri* (Boiss.) C.C. Townsend = M. aucheri
sulcata Boiss. (*Peucedanum kochii* Hiroe) - 2

Mandenovia Alava

komarovii (Manden.) Alava (*Tordylium komarovii* Manden., *Trigonosciadium komarovii* (Manden.) Tamamsch.) - 2

Mediasia M. Pimen.

macrophylla (Regel & Schmalh.) M. Pimen. (*Seseli macrophyllum* Regel & Schmalh.) - 6

Merwia B. Fedtsch. = Ferula

androssowii B. Fedtsch. = Ferula litwinowiana

Merwiopsis Saphina = Pilopleura

goloskokovii (Korov.) Saphina = Pilopleura goloskokovii

Meum auct. = Lomatocarpa

alatum Korov. = Lomatocarpa korovinii

Meum Hill

athamanticum Jacq. - 1

Mogoltavia Korov.

narynensis M. Pimen. & Kljuykov - 6
sewerzowii (Regel) Korov. - 6

Muretia Boiss. = Elaeosticta

fragrantissima (Lipsky) Korov. = Galagania fragrantissima
lutea (Hoffm.) Boiss. = Elaeosticta lutea
oeroilanica Korov. = Hyalolaena lipskyi
transcaspica Korov. = Hyalolaena transcaspica
transitoria Korov. = Elaeosticta transitoria

Mutellina N.M. Wolf

purpurea (Poir.) Thell. (*Oenanthe purpurea* Poir., *Ligusticum mutellina* (L.) Crantz) - 1

Myrrhis Hill

odorata (L.) Scop. - 1(alien), 2(alien)

Myrrhoides Heist. ex Fabr. = Physocaulis

nodosa (L.) Cannon = Physocaulis nodosus

Neocryptodiscus Hedge & Lamond = Prangos

ammophilus (Bunge) Hedge & Lamond = Prangos ammophila
didymus (Regel) Hedge & Lamond = Prangos didyma

Neogaya auct. = Lomatocarpa

simplex (L.) Meissn. var. *albomarginata* Schrenk = Lomatocarpa albomarginata

Neopaulia M. Pimen. & Kljuykov = Paulita

alaica M. Pimen. & Kljuykov = Paulita alaica
alpina (Schischk.) M. Pimen. & Kljuykov = Paulita alpina
ovczinnikovii (Korov.) M. Pimen. & Kljuykov = Paulita ovczinnikovii

Neoplatytaenia Geldykhanov = Semenovia

dichotoma (Boiss.) Geldykhanov = Semenovia dichotoma
pamirica (Lipsky) Geldykhanov = Semenovia pamirica
pimpinelloides (Nevski) Geldykhanov = Semenovia pimpinelloides

Ochotia A. Khokhr. = Magadania

olaensis (Gorovoi & N.S. Pavlova) A. Khokhr. = Magadania olaensis
victoris (Schischk.) A. Khokhr. = Magadania victoris

Oedibasis K.-Pol.

apiculata (Kar. & Kir.) K.-Pol. - 6
- subsp. **australis** R. Kam. - 6
chaerophylloides (Regel & Schmalh.) Korov. = Oe. tamerlanii
karatavica Korov. = Oe. platycarpa
platycarpa (Lipsky) K.-Pol. (*Carum karatavicum* (Korov.)
Hiroe, *Oedibasis karatavica* Korov.) - 6
tamerlanii (Lipsky) Korov. ex Nevski (*Oe. chaerophylloides*
(Regel & Schmalh.) Korov.) - 6

Oenanthe L.

abchasica Schischk. - 2
aquatica (L.) Poir. - 1, 2, 3, 4, 6
banatica Heuff. - 1
decumbens (Thunb.) K.-Pol. p.p. = Oe. javanica
fedtschenkoana K.-Pol. - 6
fistulosa L. - 1, 2
heterococca Korov. - 6
javanica (Blume) DC. (*Sium javanicum* Blume, *Oenanthe
decumbens* (Thunb.) K.-Pol. p.p. excl. basionymo) - 5
longifoliolata Schischk. = Oe. silaifolia
media Griseb. = Oe. silaifolia
pimpinelloides L. - 1, 2
purpurea Poir. = Mutellina purpurea
silaifolia Bieb. (*Oe. longifoliolata* Schischk., *Oe. media*
Griseb., *Oe. silaifolia* subsp. *media* (Griseb.) Bertova) - 1, 2
- subsp. *media* (Griseb.) Bertova = Oe. silaifolia
sophie Schischk. - 2

Opopanax Koch

armeniacus Bordz. = O. persicus
hispidus (Friv.) Griseb. (*Ferula hispida* Friv.) - 2
persicus Boiss. (*O. armeniacus* Bordz., *Peucedanum armeniacum* (Bordz.) Hiroe) - 2

Orlaya Hoffm.

daucoides (L.) Greuter (*Caucalis daucoides* L., *Orlaya
kochii* Heywood, *O. platycarpos* (L.) Koch, p.p. excl.
basionymo) - 1, 2
grandiflora (L.) Hoffm. - 1, 2
kochii Heywood = O. daucoides
platycarpos (L.) Koch, p.p. = O. daucoides

Ormopterum Schischk.

turcomanicum (Korov.) Schischk. (*Selinum turcomanicum*
(Korov.) Hiroe) - 6

Osmorhiza Rafin.

aristata (Thunb.) Rydb. (*O. aristata* (Thunb.) Makino &
Yabe, comb. superfl.) - 2, 3, 5

Ostericum Hoffm. = Angelica

crassum (Nakai) Kitag. = Halosciastrum melanotilingia
maximowiczii (Fr. Schmidt) Kitag. = Angelica maximowiczii
palustre (Bess.) Bess. = Angelica palustris

Pachypleurum Ledeb.

aemulans Korov. = Lomatocarpa albomarginata
alpinum Ledeb. (*P. schischkinii* Serg.) - 1, 3, 4, 5, 6
altaicum Revusch. - 3
condensatum (L.) Korov. = Seseli condensatum
gayoides (Regel & Schmalh.) Schishk. = Dimorphosciadium gayoides
linearilobum Korov. = Lomatocarpa afghanica
mucronatum (Schrenk) Schischk. = Seseli mucronatum
nemorosum (Korov.) Korov. = Seseli nemorosum
schischkinii Serg. = P. alpinum
talassicum Bajt. = Seseli nemorosum

Palimbia Bess.

defoliata (Ledeb.) Korov. (*Seseli defoliatum* Ledeb.) - 3
rediviva (Pall.) Thell. = P. salsa
salsa (L. fil.) Bess. (*Sison salsum* L. fil., *Palimbia
rediviva* (Pall.) Thell.) - 1, 2, 3
turgaica Lipsky ex Woronow - 3, 6(?)

Paraligusticum V. Tichomirov

discolor (Ledeb.) V. Tichomirov (*Ligusticum discolor*
Ledeb., *Pleurospermum discolor* (Ledeb.) Hiroe) - 6

Parasilaus Leute (*Scaphospermum* Korov.)

afghanicus (Gilli) Leute = P. asiaticus
asiaticus (Korov.) M. Pimen. (*Scaphospermum asiaticum*
Korov., *Parasilaus afghanicus* (Gilli) Leute, *Silaus
afghanicus* Gilli) - 6

Pastinaca L.

armena Fisch. & C.A. Mey.
aurantiaca (Albov) Kolak. (*Peucedanum aurantiacum*
(Albov) Hiroe) - 2
biebersteinii Galushko = P. clausii
clausii (Ledeb.) M. Pimen. (*Heracleum clausii* Ledeb., *H.
graveolens* (Bieb.) Spreng., *Malabaila graveolens* (Bieb.)
Hoffm., *Pastinaca biebersteinii* Galushko, *P. graveolens*
Bieb. 1808, non Bernh. 1800, *Peucedanum biebersteinii*
Schmalh. non basionymum) - 1, 2, 6
dasycarpa Regel & Schmalh. = Semenovia dasycarpa
glacialis (Golosk.) Hiroe = Pastinacopsis glacialis
glandulosa Boiss. & Hausskn. - 2
graveolens Bieb. = P. clausii
opaca Bernh. ex Hornem. = P. umbrosa
pimpinellifolia Bieb. - 2
***sativa** L.
- subsp. *sylvestris* (Mill.) Rouy & Camus = P. sylvestris
- subsp. *umbrosa* (Stev. ex DC.) Bondar. = P. umbrosa
- subsp. *urens* (Req. ex Godr.) Celak. = P. umbrosa
sylvestris Mill. (*P. sativa* L. subsp. *sylvestris* (Mill.) Rouy &
Camus) - 1, 2, 3, 4, 5(alien)
umbrosa Stev. ex DC. (*P. opaca* Bernh. ex Hornem., *P.
sativa* L. subsp. *umbrosa* (Stev. ex DC.) Bondar., *P.
sativa* subsp. *urens* (Req. ex Godr.) Celak.) - 1, 2

Pastinacopsis Golosk.

glacialis Golosk. (*Pastinaca glacialis* (Golosk.) Hiroe) - 6

Paulia Korov. = Paulita

alpina (Schischk.) Korov. ex M. Pimen. & Kljuykov = Paulita alpina
ovczinnikovii Korov. = Paulita ovczinnikovii

Paulita Sojak (*Paulia* Korov. 1973, non Fee, *Neopaulia* M. Pimen. & Kjuykov)

alaica (M. Pimen. & Kljuykov) M. Pimen. & Kljuykov (*Neopaulia alaica* M. Pimen. & Kljuykov) - 6
alpina (Schischk.) Sojak (*Hymenolaena alpina* Schischk., *Neopaulia alpina* (Schischk.) M. Pimen. & Kljuykov, *Paulia alpina* (Schischk.) Korov. ex M. Pimen. & Kljuykov, *Pleurospermum alpinum* (Schischk.) Hiroe) - 6
ovczinnikovii (Korov.) Sojak (*Paulia ovczinnikovii* Korov., *Neopaulia ovczinnikovii* (Korov.) M. Pimen. & Kljuykov) - 6

*Petroselinum Hill

*crispum (Mill.) A.W.Hill

Peucedanum L.

adae Woronow - 2
alpestre L. = Silaum silaus
alsaticum L. = Xanthoselinum alsaticum
arenarium Waldst. & Kit. = Taeniopetalum arenarium
armeniacum (Bordz.) Hiroe = Opopanax persicus
assyriaense Hiroe = Malabaila aucheri
aurantiacum (Albov) Hiroe = Pastinaca aurantiaca
baicalense (Redow. ex Willd.) Koch = Kitagawia baicalensis
biebersteinii Schmalh. = Pastinaca clausii
borysthenicum Klok. ex Schischk. = Taeniopetalum arenarium
calcareum Albov = Peucedanum longifolium
carvifolia Vill. (*P. chabraei* (Jacq.) Reichenb. var. *podolicum* (Bess.) Todor, *P. euphimiae* Kotov, *P. podolicum* (Bess.) Eichw., *P. schottii* Bess. ex DC.) - 1
caucasicum (Bieb.) C. Koch = Cervaria caucasica
cervaria (L.) Lapeyr. = Cervaria rivinii
cervariifolium C.A. Mey. = Cervaria cervariifolia
chabraei (Jacq.) Reichenb. var. *podolicum* (Bess.) Todor = P. carvifolia
deltoideum Makino & Yabe = Kitagawia terebinthacea
elegans Kom. = Kitagawia komarovii
eryngiifolium Kom. = Kitagawia eryngiifolia
euphimiae Kotov = P. carvifolia
falcaria Turcz. - 3, 4
filicinum H. Wolff = Conioselinum filicinum
formosanum auct. = Kitagawia terebinthacea
hissaricum Korov. = Kafirnigania hissarica
hystrix Bunge = Ferulopsis hystrix
japonicum auct. = Kitagawia litoralis
kochii Hiroe = Malabaila sulcata
latifolium (Bieb.) DC. = Macroselinum latifolium
latifolium sensu Schischk. = Cervaria aegopodioides
litorale Worosch. & Gorovoi = Kitagawia litoralis
longifolium Waldst. & Kit. (*P. calcareum* Albov, *P. officinale* L. subsp. *longifolium* (Waldst. & Kit.) R. Frey) - 2
lubimenkoanum Kotov = Xanthoselinum alsaticum
luxurians Tamamsch. - 2

macrophyllum Schischk. = Macroselinum latifolium
meyeri auct. p.p. = Johreniopsis seseloides
mogoltavicum Korov. = Ferula lithophila
morisonii Bess. ex Spreng. (*P. songoricum* Schischk.) - 3, 6
officinale L. subsp. *longifolium* (Waldst. & Kit.) R. Frey = P. longifolium
oreoselinum (L.) Moench (*Angelica oreoselinum* (L.) Hiroe) - 1, 2
paishanense Nakai = Kitagawia terebinthacea
palustre (L.) Moench = Thyselium palustre
pastinacifolium Boiss. & Hausskn. = Demavendia pastinacifolia
paucifolium Ledeb. = Johreniopsis seseloides
pauciradiatum Tamamsch. = Zeravschania pauciradiata
podolicum (Bess.) Eichw. = P. carvifolia
polyanthum (Korov.) Korov. = Fergania polyantha
pschawicum Boiss. - 2
puberulum (Turcz.) Schischk. - 3, 4
renardii Regel & Schmalh. = Ferula renardii
ruthenicum Bieb. - 1, 2
salinum Pall. ex Spreng. = Angelica tenuifolia
schottii Bess. ex DC. = P. carvifolia
sintenisii H. Wolff = Cervaria cervariifolia
songoricum Schischk. = P. morisonii
talassicum Korov. = Ferula renardii
tauricum Bieb. - 1, 2
terebinthaceum (Fisch. ex Spreng.) Ledeb. = Kitagawia terebinthacea
- subsp. *aculeolatum* Worosch. = Kitagawia terebinthacea
- subsp. *deltoideum* (Makino & Yabe) Worosch. = Kitagawia terebinthacea
- var. *paishanense* (Nakai) Huang = Kitagawia terebintacea
transiliense Herd. = Ferula transiliensis
turcomanicum Schischk. = Leutea petiolaris
vaginatum Ledeb. - 3, 4
wolffianum Fedde ex H. Wolff - 5
  Dubious species. Described from the Island of Saghalin, but the herbarium material is not available.
zedelmeyerianum Manden. - 2

Phlojodicarpus Turcz. ex Ledeb.

baicalensis M. Pop. = Ph. popovii Sipl.
eudahuricus M. Pop. = Ph. sibiricus
komarovii Gorovoi (*Ph. sibiricus* (Steph. ex Spreng.) K.-Pol. subsp. *komarovii* (Gorovoi) Worosch.) - 5
nudiusculus Turcz. ex M. Pop. = Ferulopsis hystrix
popovii Sipl. (*Ph. baicalensis* M. Pop. nom. illegit., *Ph. sibiricus* (Steph. ex Spreng.) K.-Pol. var. *baicalensis* (M. Pop.) Serg. nom. illegit.) - 3, 4, 5
sibiricus (Steph. ex Spreng.) K.-Pol. (*Angelica sibirica* (Steph. ex Spreng.) Hiroe, *Phlojodicarpus eudahuricus* M. Pop. nom. illegit., *Ph. sibiricus* var. *eudahuricus* (M. Pop.) Serg. nom. illegit.) - 3, 4, 5
- subsp. *komarovii* (Gorovoi) Worosch. = Ph. komarovii
- subsp. *villosus* (Turcz. ex Fisch. & C.A. Mey.) Worosch. = Ph. villosus
- var. *baicalensis* (M. Pop.) Serg. = Ph. popovii
- var. *eudahuricus* (M. Pop.) Serg. = Ph. sibiricus
- var. *villosus* (Turcz. ex Fisch. & C. A. Mey.) Chu = Ph. villosus
turczaninovii Sipl. = Ferulopsis hystrix
villosus (Turcz. ex Fisch. & C.A. Mey.) Ledeb. (*Libanotis petrophila* Korov., *Phlojodicarpus sibiricus* (Steph. ex Spreng.) K.-Pol. subsp. *villosus* (Turcz. ex Fisch. & C.A. Mey.) Worosch., *Ph. sibiricus* var. *villosus* (Turcz. ex Fisch. & C. A. Mey.) Chu) - 1, 3, 4, 5

Physocaulis (DC.) Tausch (*Myrrhoides* Heist. ex Fabr. nom. invalid.)

nodosus (L.) Koch (*Myrrhoides nodosa* (L.) Cannon) - 1, 2, 6

Physopsermum Cuss. ex Juss. (*Danaa* All. nom. rejic.)

aegopodioides Boiss. = Cervaria aegopodioides
cornubiense (L.) DC. (*Danaa nudicaulis* (Bieb). Grossh., *Physospermum danaa* (Bieb.) Schischk. ex N. Rubtz.) - 1, 2
danaa (Bieb.) Schischk. ex N. Rubtz. = Ph. cornubiense
denaense (Schischk.) Czer. = Sphaerosciadium denaense
denaense (Schischk.) B. Fedtsch. ex Weinert = Sphaerosciadium denaense
kopetdaghense Korov. = Korshinskya kopetdaghensis

Pilopleura Schischk. (*Merwiopsis* Saphina)

goloskokovii (Korov.) M. Pimen. (*Platytaenia goloskokovii* Korov., *Merwiopsis goloskokovii* (Korov.) Saphina, *Semenovia goloskokovii* (Korov.) Bajt. nom. invalid.) - 6
kozo-poljanskii Schischk. = P. tordyloides
tordyloides (Korov.) M. Pimen. (*Zosima tordyloides* Korov. p.p., *Libanotis talassica* Korov., *Pilopleura kozo-poljanskii* Schischk.) - 6

Pimpinella L. (*Albovia* Schichk., *Anisum* Hill, *Reutera* Boiss.)

affinis Ledeb. = P. peregrina
- var. *multiradiata* Boiss. = P. peregrina
anisactis Rech. fil. (*P. litvinovii* Schischk.) - 6
*****anisum** L. (*Anisum vulgare* Gaertn.) - 1, 2, 6
anthriscoides Boiss. - 2
armena Schischk. (*P. cappadocica* auct.) - 2
aromatica Bieb. (*P. schatilensis* Otschiauri) - 2
aurea DC. (*Reutera aurea* (DC.) Boiss.) - 2, 6
bobrovii (Woronow ex Schischk.) Hiroe (*Reutera bobrovii* Woronow ex Schischk., *Pimpinella bobrovii* (Woronow ex Schischk.) E. Axenov & V. Tichomirov, comb. superfl.) - 6
calycina Maxim. = Spuriopimpinella calycina
cappadocica auct. = P. armena
confusa Woronow = P. pseudotragium
crassa Nakai = Halosciastrum melanotilingia
daghestanica Schischk. = P. tragium
dissecta Retz. = P. saxifraga
glauca L. = Trinia glauca
grossheimii Schischk. = P. tragium
idae Takht. = P. tragium
kaschmirica R.R. Stewart ex Dunn = Aegopodium kashmiricum
korovinii R. Kam. = P. peregrina
korshinskyi Schischk. = Seseli korshinskyi
leptophylla Pers. = Ciclospermum leptophyllum
lithophila Schischk. = P. tragium
litvinovii Schischk. = P. anisactis
major (L.) Huds. - 1
multiradiata (Boiss.) Korov. = P. peregrina
nigra Mill. - 1
- subsp. **arenaria** (Brylin) Weide - 1
nudicaulis auct. = P. squamosa
peregrina L. (*P. affinis* Ledeb., *P. affinis* var. *multiradiata*

Boiss., *P. korovinii* R. Kam., *P. multiradiata* (Boiss.) Korov. 1959, non Santapau, 1948, *P. taurica* (Ledeb.) Steud.) - 1, 2, 6
peucedanifolia Fisch. ex Ledeb. - 2
pseudotragium DC. (*P. confusa* Woronow, *P. tragium* Vill. subsp. *pseudotragium* (DC.) Matthews) - 2
puberula (DC.) Boiss. - 2, 6
rhodantha Boiss. - 2
saxifraga L. (*P. dissecta* Retz.) - 1, 2, 3, 4, 5(alien)
schatilensis Otschiauri = P. aromatica
squamosa Karjag. (*P. nudicaulis* auct.) - 2
taurica (Ledeb.) Steud. = P. peregrina
thellungiana H. Wolff - 4, 5
titanophila Woronow = P. tragium
tomiophylla (Woronow) Stank. = P. tragium
tragium Vill. (*P. daghestanica* Schischk., *P. grossheimii* Schischk., *P. idae* Takht., *P. lithophila* Schischk., *P. titanophila* Woronow, *P. tomiophylla* (Woronow) Stank., *P. tragium* subsp. *litophila* (Schischk.) Tutin, *P. tragium* subsp. *titanophila* (Woronow) Tutin) - 1, 2
- subsp. *litophila* (Schischk.) Tutin = P. tragium
- subsp. *pseudotragium* (DC.) Matthews = P. pseudotragium
- subsp. *titanophila* (Woronow) Tutin = P. tragium
tripartita Kalen. (*Albovia tripartita* (Kalen.) Schischk., *Scaligeria tripartita* (Kalen.) Tamamsch.) - 2
turcomanica Schischk. - 6

Platytaenia Nevski & Vved. = Semenovia

bucharica Schischk. = Semenovia bucharica
dasycarpa (Regel & Schmalh.) Korov. = Semenovia dasycarpa
depauperata Schischk. = Seseli depauperatum
dichotoma (Boiss.)Manden. = Semenovia dichotoma
goloskokovii Korov. = Pilopleura goloskokovii
heterodonta Korov. = Semenovia heterodonta
komarovii Manden. ex Schischk. = Semenovia dasycarpa
olgae (Regel & Schmalh.) Korov. = Tetrataenium olgae
pamirica (Lipsky) Nevski & Vved. = Semenovia pamirica
pimpinelloides Nevski = Semenovia pimpinelloides
rubtzovii Schischk. = Semenovia rubtzovii
tordyloides (Korov.) Korov. = Zosima korovinii
tragioides (Boiss.) Nevski & Vved. p.p. = Semenovia dichotoma

Pleurospermum Hoffm.

alpinum (Schischk.) Hiroe = Paulita alpina
austriacum (L.) Hoffm. - 1
- subsp. *uralense* (Hoffm.) Somm. = P. uralense
camtschaticum Hoffm. = P. uralense
discolor (Ledeb.) Hiroe = Paraligusticum discolor
rupestre (M. Pop.) K.T. Fu & Y.C. Ho = Aulacospermum simplex
uralense Hoffm. (*P. austriacum* (L.) Hoffm. subsp. *uralense* (Hoffm.) Somm., *P. camtschaticum* Hoffm.) - 1, 3, 4, 5

Podistera S. Wats.

macounii (Coult. & Rose) Math. & Const. (*Ligusticum macounii* Coult. & Rose) - 5

Polylophium Boiss.

panjutinii Manden. & Schischk. (*Laserpitium panjutinii* (Manden. & Schischk.) Hiroe) - 2

Prangos Lindl. (*Cryptodiscus* Schrenk, 1841, non Corda, 1838, *Koelzella* Hiroe, *Neocryptodiscus* Hedge & Lamond, *Cachrys* auct.)

acaulis (DC.) Bornm. (*Cachrys acaulis* DC.) - 2
alata (Bieb.) Grossh. = P. ferulacea
ammophila (Bunge) M. Pimen. & V. Tichomirov (*Cryptodiscus ammophilus* Bunge, *Neocryptodiscus ammophilus* (Bunge) Hedge & Lamond) - 6
arcis-romanae Boiss. & Huet = P. meliocarpoides var. arcis-romanae
arenaria (Schischk.) Pimen. & V. Tichomirov (*Cryptodiscus arenarius* Schischk., *Cachrys arenaria* (Schischk.) Hiroe) - 6
biebersteinii Karjag. = P. ferulacea
bucharica B. Fedtsch. (*Cachrys bucharica* (B. Fedtsch.) Herrnst. & Heyn) - 6
cachroides (Schrenk) M. Pimen. & V. Tichomirov (*Cryptodiscus cachroides* Schrenk) - 6
cylindrocarpa Korov. = P. pabularia subsp. cylindrocarpa
didyma (Regel) M. Pimen. & V. Tichomirov (*Cachrys didyma* Regel, *Cryptodiscus didymus* (Regel) Korov., *Neocryptodiscus didymus* (Regel) Hedge & Lamond) - 6
dzhungarica M. Pimen. - 6
equisetoides Kuzmina - 6
fedtschenkoi (Regel & Schmalh.) Korov. - 6
ferulacea (L.) Lindl. (*Cachrys alata* Bieb., *C. ferulacea* (L.) Calest., *Prangos alata* (Bieb.) Grossh. 1949, non Benth. & Hook. fil. ex Drude, 1898, *P. biebersteinii* Karjag.) - 2
gyrocarpa Kuzmina - 6
herderi (Regel) Herrnst. & Heyn (*Cachrys herderi* Regel) - 3, 6
hissarica Korov. = P. pabularia
isphairamica B. Fedtsch. = P. lipskii
lachnantha (Korov.) M. Pimen. & Kljuykov (*Schrenkia lachnantha* Korov.) - 6
lamellata Korov. = P. pabularia subsp. *lamellata*
latiloba Korov. (*Cachrys latiloba* (Korov.) Herrnst. & Heyn) - 6
ledebourii Herrnst. & Heyn (*Cachrys macrocarpa* Ledeb., non Prangos macrocarpa Boiss.) - 3, 6
lipskyi Korov. (*P. isphairamica* B. Fedtsch.) - 6
lophoptera Boiss. (*Cachrys lophoptera* (Boiss.) Takht.) - 2
meliocarpoides Boiss. (*Cachrys meliocarpoides* (Boiss.) Herrnst. & Heyn) var. **arcis-romanae** (Boiss. & Huet) Herrnst. & Heyn (*P. arcis-romanae* Boiss. & Huet, *Cachrys meliocarpoides* var. *arcis-romanae* (Boiss. & Huet) Herrnst. & Heyn) - 2
odontalgica (Pall.) Herrnst. & Heyn (*Cachrys odontalgica* Pall., *C. pubescens* (Pall. ex Spreng.) Schischk.) - 1, 2, 6
ornata Kuzmina (*P. quasiperforata* Kuzmina) - 6
pabularia Lindl. (*Cachrys pabularia* (Lindl.) Herrnst. & Heyn, *Koelzella pabularia* (Lindl.) Hiroe, *Prangos hissarica* Korov., *P. sarawschanica* (Regel & Schmalh.) Korov.) - 6
- subsp. **cylindrocarpa** (Korov.) M. Pimen. & V. Tichomirov (*P. cylindrocarpa* Korov.) - 6
- subsp. **lamellata** (Korov.) M. Pimen. & V. Tichomirov (*P. lamellata* Korov.) - 6
quasiperforata Kuzmina = P. ornata
sarawschanica (Regel & Schmalh.) Korov. = P. pabularia
trifida (Mill.) Herrnst. & Heyn (*Cachrys alpina* Bieb.) - 1
tschimganica B. Fedtsch. - 6
uloptera DC. (*Cachrys uloptera* (DC.) Herrnst. & Heyn, comb. superfl., *C. uloptera* (DC.) Takht., *Koelzella uloptera* (DC.) Hiroe) - 6

Psammogeton Edgew.

borszowii (Regel & Schmalh.) Lipsky = Cuminum borszowii
canescens (DC.) Vatke - 6
setifolium (Boiss.) Boiss. = Cuminum setifolium

Ptilimnium auct. = Ciclospermum

nuttallii auct. = Ciclospermum leptophyllum

Reutera Boiss. = Pimpinella

aurea (DC.) Boiss. = Pimpinella aurea
bobrovii Woronow ex Schischk. = Pimpinella bobrovii

Rumia Hoffm.

crithmifolia (Willd.) K.-Pol. - 1
hispida (Hoffm.) Stank. = Trinia hispida
hoffmannii (Bieb.) Stank. = Trinia hispida

Rupiphila M. Pimen. & Lavrova

tachiroei (Franch. & Savat.) M.Pimen. & Lavrova (*Seseli tachiroei* Franch. & Savat., *Cnidium filisectum* Nakai & Kitag., *C. tachiroei* (Franch. & Savat.) Makino, *Ligusticum filisectum* (Nakai & Kitag.) Hiroe, *L. tachiroei* (Franch. & Savat.) Hiroe & Constance, *Tilingia filisecta* (Nakai & Kitag.) Nakai & Kitag., *T. tachiroei* (Franch. & Savat.) Kitag.) - 5

Sajanella Sojak (*Sajania* M. Pimen. 1974, non Vologdin, 1962)

monstrosa (Willd. ex Spreng.) Sojak (*Libanotis monstrosa* (Willd. ex Spreng.) DC., *Sajania monstrosa* (Willd. ex Spreng.) M. Pimen., *Schulzia monstrosa* (Willd. ex Spreng.) Hiroe) - 3, 4

Sajania M. Pimen. = Sajanella

monstrosa (Willd. ex Spreng.) M. Pimen. = Sajanella monstrosa

Sanicula L.

chinensis Bunge (*S. kurilensis* Pobed.) - 5
europaea L. - 1, 2, 3
giraldii H. Wolff - 1, 3
kurilensis Pobed. = S. chinensis
rubriflora Fr. Schmidt ex Maxim. - 5

Saposhnikovia Schischk.

divaricata (Turcz.) Schischk. (*Ledebouriella divaricata* (Turcz.) Hiroe) - 4, 5

Scaligeria auct. = Elaeosticta

alaica (Lipsky) Korov. = Elaeosticta alaica
allioides Regel & Schmalh. = Elaeosticta allioides
bucharica Korov. = Elaeosticta bucharica
conica (Korov) Korov. = Elaeosticta conica
ferganensis Lipsky = Elaeosticta ferganensis
glaucescens (DC.)Boiss. = Elaeosticta glaucescens
hirtula (Regel & Schmalh.) Lipsky = Elaeosticta hirtula
knorringiana Korov. = Elaeosticta knorringiana

kopetdagensis (Korov.) Schischk. = Korshinskya kopetdaghensis
korovinii Bobr. = Elaeosticta korovinii
korshinskyi (Lipsky) Korov. = Elaeosticta ferganensis and
E. hirtula
lipskyi Korov. = Hyalolaena lipskyi
oedibasioides R. Kam. = Elaeosticta hirtula
platyphylla Korov. = Elaeosticta platyphylla
polycarpa Korov. = Elaeosticta polycarpa
samarkandica Korov. = Elaeosticta samarkandica
setacea (Schrenk) Korov. = Bunium setaceum
transcaspica Korov. = Elaeosticta transcaspica
tripartita (Kalen.) Tamamsch. = Pimpinella tripartita
tschimganica Korov. = Elaeosticta tschimganica
ugamica Korov. = Elaeosticta ugamica
vvedenskyi R. Kam. = Elaeosticta vvedenskyi

Scandix L.

aucheri Boiss. - 2, 6
australis L. (*S. falcata* Londes, *S. pontica* (Vierh.) Stank., *S.
taurica* Stev., *S. grandiflora* auct.) - 1, 2
falcata Londes = S. australis
fedtschenkoana K.-Pol. = S. stellata
grandiflora auct. = S. australis
iberica Bieb. - 2
macrorhyncha C.A. Mey. (*S. pecten-veneris* L. subsp.
macrorhyncha (C.A. Mey.) Rouy & Camus) - 1
pecten-veneris L. - 1, 2, 6
- subsp. *macrorhyncha* (C. A. Mey.) Rouy & Camus = S.
macrorhyncha
■ **persica** Mart.
pontica (Vierh.) Stank. = S. australis
stellata Banks & Soland. (*S. fedtschenkoana* K.-Pol.) - 1, 2,
6
taurica Stev. = S. australis

Scaphospermum Korov. = Parasilaus

asiaticum Korov. = Parasilaus asiaticus

Schrenkia Fisch. & C.A. Mey. (*Anidrum* auct.)

congesta Korov. - 6
fasciculata Korov. = Sch. pungens
golickeana (Regel & Schmalh.) B. Fedtsch. (*Anidrum transitorium* (Lipsky) K.-Pol., *Schrenkia vaginata* (Ledeb.)
Fisch. & C.A. Mey. var. *transitoria* Lipsky) - 6
insignis Lipsky = Lipskya insignis
involucrata Regel & Schmalh. - 6
kultiassovii Korov. (*Bifora kultiassovii* (Korov.) Hiroe) - 6
lachnantha Korov. = Prangos lachnantha
mogoltavica O. Politova = Sch. vaginata
papillaris Regel & Schmalh. (*Hippomarathrum papillare*
(Regel & Schmalh.) Hiroe) - 6
pulverulenta M. Pimen. - 6
pungens Regel & Schmalh. (*Sch. fasciculata* Korov., *Sch.
transitoria* R. Kam. non *Anidrum transitorium* (Lipsky)
K.-Pol.) - 6
transitoria R. Kam. = Sch. pungens
ugamica Korov. - 6
vaginata (Ledeb.) Fisch. & C.A. Mey. (*Sch. mogoltavica* O.
Politova) - 3, 6
- var. *transitoria* Lipsky = Sch. golickeana

Schtschurowskia Regel & Schmalh.

margaritae Korov. - 6
meifolia Regel & Schmalh. - 6

pentaceros (Korov.) Schischk. = Sclerotiaria pentaceros

Schulzia Spreng.

albiflora (Kar. & Kir.) M. Pop. (*Schulzia crinita* (Pall.)
Spreng. subsp. *albiflora* (Kar. & Kir.) M. Pimen.) - 6
crinita (Pall.) Spreng. - 3, 4
- subsp. *albiflora* (Kar.&Kir.) M. Pimen. = Sch. albiflora
monstrosa (Willd. ex Spreng.) Hiroe = Sajanella monstrosa
prostrata M. Pimen. & Kljuykov - 6

Schumannia O. Kuntze = Ferula

karelinii (Bunge) Korov. = Ferula karelinii

Sclerotiaria Korov.

pentaceros (Korov.) Korov. (*Schtschurowskia pentaceros*
(Korov.) Schischk.) - 6

Scorodosma Bunge = Ferula

foetidum Bunge = Ferula foetida

Selinum L.

carvifolia (L.) L. - 1, 3
coriaceum Korov. = Sphaenolobium coriaceum
davuricum (Jacq.) Leute = Cnidium davuricum
depauperatum (Korov.) Hiroe = Hyalolaena depauperata
dubium (Schkuhr) Leute = Kadenia dubia
- subsp. *salinum* (Turcz.) Leute = Kadenia salina
kultiassovii Korov. = Karatavia kultiassovii
melanotilingia Boissieu = Halosciastrum melanotilingia
multicaule (Turcz.) Leute = Lithosciadium multicaule
pauciradium (Somm. & Levier) Leute = Seseli saxicolum
popovii (Korov.) Schischk. = Aulacospermum popovii
salinum (Turcz.) N. Vodopianova = Kadenia salina
tenuisectum Korov. = Sphaenolobium tenuisectum
terebinthaceum Fisch. ex Spreng. = Kitagawia terebinthacea
terebinthaceum Fisch. ex Trev. = Kitagawia terebinthacea
tianschanicum Korov. = Sphaenolobium tianschanicum
turcomanicum (Korov.) Hiroe = Ormopterum turcomanicum

Semenovia Regel & Herd. (*Neoplatytaenia* Geldykhanov, *Platytaenia* Nevski & Vved. 1937, non Kuhn, 1882)

bucharica (Schischk.) Manden. (*Platytaenia bucharica*
Schischk., *Zosima bucharica* (Schischk.) Hiroe) - 6
dasycarpa (Regel & Schmalh.) Korov. ex Czer. (*Pastinaca
dasycarpa* Regel & Schmalh., *Malabaila dasycarpa*
(Regel & Schmalh.) Schischk., *Platytaenia dasycarpa*
(Regel & Schmalh.) Korov., *P. komarovii* Manden. ex
Schischk., *Semenovia komarovii* (Manden. ex
Schischk.) Manden., *Zosima komarovii* (Manden. ex
Schischk.) Hiroe) - 6
depauperata (Schischk.) Manden. = Seseli depauperatum
dichotoma (Boiss.) Manden. (*Zosima dichotoma* Boiss.,
Neoplatytaenia dichotoma (Boiss.) Geldykhanov, *Platytaenia dichotoma* (Boiss.) Manden., *P. tragioides*
(Boiss.) Nevski & Vved. p.p. quoad plantas, excl. typo,
Semenovia tragioides (Boiss.) M. Pimen. & V. Tichomirov, p.p. quoad plantas, excl. typo) - 6
furcata Korov. (*S. involucrata* Korov. ex Ikonn. nom. invalid.) - 6
goloskokovii (Korov.)Bajt. = Pilopleura goloskokovii

heterodonta (Korov.) Manden. (*Platytaenia heterodonta* Korov., *Zosima heterodonta* (Korov.) Hiroe) - 6

involucrata Korov. ex Ikonn. = S. furcata

komarovii (Manden. ex Schischk.) Manden. = S. dasycarpa

pamirica (Lipsky) Manden. (*Neoplatytaenia pamirica* (Lipsky) Geldykhanov, *Platytaenia pamirica* (Lipsky) Nevski & Vved.) - 6

pimpinelloides (Nevski) Manden. (*Platytaenia pimpinelloides* Nevski, *Neoplatytaenia pimpinelloides* (Nevski) Geldykhanov, *Zosima pimpinelloides* (Nevski) Hiroe) - 6

rubtzovii (Schischk.) Manden. (*Platytaenia rubtzovii* Schischk., *Zosima rubtzovii* (Schischk.) R. Kam., *Zosima rubtzovii* (Schischk.) Hiroe, comb. superfl.) - 6

tragioides (Boiss.) M. Pimen. & V. Tichomirov, p.p. = S. dichotoma

transiliensis Regel & Herd. (*Heracleum transiliense* (Regel & Herd.) O. & B. Fedtsch.) - 6

zaprjagaevii Korov. - 6

Seseli L. (*Elaeopleurum* Korov., *Gasparrinia* Bertol., *Libanotis* Haller ex Zinn, *Sphenocarpus* Korov.)

abolinii (Korov.) Schischk. (*Libanotis abolinii* (Korov.) Korov., *L. michajlovae* Korov., *L. songorica* (Schischk.) Korov., *L. turajgyrica* Bajt., *Seseli songoricum* Schischk.) - 6

aemulans M. Pop. - 6

afghanicum (Podlech) M. Pimen. (*Libanotis afghanica* Podlech, *Seseli subaphyllum* Korov.) - 6

alaicum M. Pimen. - 6

alexeenkoi Lipsky - 2

alpinum Bieb. (*Carum alpinum* (Bieb.) Benth. & Hook. fil.) - 2

annuum L. - 1

arenarium Bieb. = Seseli tortuosum

asperulum (Trautv.) Schischk. - 6

betpakdalense Bajt. - 6

buchtormense (Fisch. ex Hornem.) Koch (*Libanotis buchtormensis* (Fisch. ex Hornem.) DC.) - 3, 4, 6

calycinum (Korov.) M. Pimen. & Sdobnina (*Libanotis calycina* Korov., *Seseli calycinum* (Korov.) Hiroe, comb. superfl.) - 6

campestre Bess. = S. tortuosum

condensatum (L.) Reichenb. fil. (*Libanotis arctica* Rupr., *L. condensata* (L.) Crantz, *L. condensata* subsp. *arctica* (Rupr.) V. Sergienko, *Pachypleurum condensatum* (L.) Korov.) - 1, 3, 4, 5, 6

coronatum Ledeb. - 3

cuneifolium Bieb. - 2

defoliatum Ledeb. = Palimbia defoliata

depauperatum (Schischk.) V. Vinogradova (*Platytaenia depauperata* Schischk., *Semenovia depauperata* (Schischk.) Manden., *Zosima depauperata* (Schischk.) Hiroe) - 6

desertorum M. Pimen. - 6

dichotomum Pall. ex Bieb. - 1, 2

dolichostylum (Schischk.) Hiroe = S. mucronatum

elegans Schischk. = S. peucedanoides

eriocarpum (Schrenk) B. Fedtsch. (*Libanotis eriocarpa* Schrenk) - 3, 6

eriocephalum (Pall. ex Spreng.) Schischk. (*S. platyphyllum* (Schrenk) O. & B. Fedtsch.) - 3, 6

eryngioides (Korov.) M. Pimen. & V. Tichomirov (*Sphenocarpus eryngioides* Korov.) - 6

fasciculatum (Korov.) Korov. ex Schischk. (*Libanotis fasci-*

culata Korov.) - 6

floribundum Somm. & Levier = S. petraeum

foliosum (Somm. & Levier) Manden. - 2

giganteum Lipsky - 6

glabratum Willd. ex Spreng. - 3, 6

grandivittatum (Somm. & Levier) Schischk. (*S. varium* Trev. subsp. *grandivittatum* (Somm. & Levier) Soo - 2

gummiferum Pall. ex Smith - 1

hippomarathrum Jacq. - 1

iliense (Regel & Schmalh.) Lipsky (*Libanotis iliensis* (Regel & Schmalh.) Korov.) - 6

incanum (Steph. ex Willd.) B. Fedtsch. (*Libanotis incana* (Steph. ex Willd.) B. Fedtsch.) - 6

intermedium (Rupr.) N. Vodopianova = S. libanotis

jomuticum Schischk. - 6

karatavicum Schischk. = S. marginatum

karateginum Lipsky - 6

kaschgaricum M. Pimen. & Kljuykov - 6

korovinii Schischk. (*Libanotis korovinii* (Schischk.) Korov.) - 6

korshinskyi (Schischk.) M. Pimen. (*Pimpinella korshinskyi* Schischk.) - 6

krylovii (V. Tichomirov) M. Pimen. & Sdobnina (*Libanotis krylovii* V. Tichomirov, *L. sibirica* C.A. Mey. var. *gracilis* Kryl., *L. sibirica* sensu Schischk.) - 1, 3

ledebourii G. Don fil. (*Elaeopleurum monococcum* Korov.) - 3, 6

lehmannianum (Bunge) Boiss. - 6

lehmannii Degen - 1

leptocladum Woronow - 2

libanotis (L.) Koch (*Libanotis intermedia* Rupr., *L. leiocarpa* (Heuff.) Simonk., *L. montana* Crantz, *L. montana* subsp. *bipinnata* (Celak.) Dostal, *L. pyrenaica* (L.) Bourg., *L. pyrenaica* subsp. *bipinnata* (Celak.) Holub, *L. pyrenaica* subsp. *montana* (Crantz) Lemke & Rothm., *L. sibirica* (L.) C.A. Mey., *L. taurica* N. Rubtz., *Seseli intermedium* (Rupr.) N. Vodopianova, *S. libanotis* subsp. *intermedium* (Rupr.) P.W. Ball, *S. libanotis* var. *bipinnatum* Celak.) 1, 2, 3, 4, 6

- subsp. *intermedium* (Rupr.) P.W. Ball = S. libanotis

- var. *bipinnatum* Celak. = S. libanotis

luteolum M. Pimen. - 6

macrophyllum Regel & Schmalh. = Mediasia macrophylla

marginatum (Korov.) M. Pimen. & Sdobnina (*Libanotis marginata* Korov., *Seseli karatavicum* Schischk.) - 6

merkulowiczii (Korov.) M. Pimen. & Sdobnina (*Libanotis merkulowiczii* Korov.) - 6

mironovii (Korov.) M. Pimen. & Sdobnina (*Libanotis mironovii* Korov.) - 6

mucronatum (Schrenk) M. Pimen. & Sdobnina (*Libanotis dolichostyla* Schischk., *Ligusticum mucronatum* (Schrenk) Leute, *L. thomsonii* Clarke, *Pachypleurum mucronatum* (Schrenk) Schischk., *Seseli dolichostictum* (Schischk.) Hiroe) - 6

nemorosum (Korov.) M. Pimen. (*Pachypleurum nemorosum* (Korov.) Korov., *P. talassicum* Bajt.) - 6

nevskii (Korov.) M. Pimen. & Sdobnina (*Libanotis nevskii* Korov.) - 6

osseum Crantz - 1

pallasii Bess. = S. varium

pauciradiatum Schischk. = S. tortuosum

petraeum Bieb. (*S. floribundum* Somm. & Levier) - 2

peucedanifolium (Spreng.) Bess. = S. tortuosum

peucedanoides (Bieb.) K.-Pol. (*Gasparrinia donetzica* Dubovik, *G. peucedanoides* (Bieb.) Thell., *Seseli elegans* Schischk., *Silaum peucedanoides* (Bieb.) Hiroe) - 1, 2

platyphyllum (Schrenk) O. & B. Fedtsch. = S. eriocephalum
ponticum Lipsky - 2
rigidum Waldst. & Kit. subsp. *peucedanifolium* (Spreng.)
    Nym. = S. tortuosum
rimosum M. Pimen. - 6
rivinianum (Ledeb.) Hiroe = S. seseloides
rupicola Woronow - 2
saxicolum (Albov) M. Pimen. (*Carum saxicolum* Albov,
    Cnidium mandenovae Gagnidze, *C. paucuradium*-
    Somm. & Levier, *Selinum pauciradium* (Somm. &
    Levier) Leute) - 2
schrenkianum (C.A. Mey. ex Schischk.) M. Pimen. &
    Sdobnina (*Libanotis schrenkiana* C.A. Mey. ex
    Schischk., *Seseli schrenkianum* (C.A. Mey. ex
    Schischk.) Hiroe, comb. superfl.) - 6
sclerophyllum Korov. - 6
seravschanicum M. Pimen. & Sdobnina (*Ligusticum fedt-*
    schenkoanum Schischk. 1950, non *Seseli fedtschenkoa-*
    num Regel & Schmalh. 1881) - 6
seseloides (Turcz.) Hiroe (*Libanotis seseloides* Turcz., *L.*
    amurensis Schischk., *Ligusticum seseloides* Fisch. &
    C.A. Mey. ex Ledeb., *Seseli rivinianum* (Ledeb.)
    Hiroe) - 4, 5
sessiliflorum Schrenk (*S. squarrosum* Schischk., *S.*
    tschuense E. Nikit. nom. invalid.) - 6
setiferum M. Pimen. & Sdobnina (*Ligusticum setiferum*
    Korov. ex Pavl. nom. nud., *Libanotis setifera* (Korov. ex
    Pavl.) Schischk. nom. invalid., *Seseli setiferum*
    (Korov.ex Pavl.) Hiroe, nom. invalid. & superfl.) - 6
songoricum Schischk. = S. abolinii
squarrosum Schischk. = S. sessiliflorum
strictum Ledeb. - 1, 3, 6
subaphyllum Korov. = S. afghanicum
tachiroei Franch. & Savat. = Rupiphila tachiroei
talassicum (Korov.) M. Pimen. & Sdobnina (*Ligusticum*
    talassicum Korov., *L. turkestanicum* Hiroe, *L. pumilum*
    Korov. 1924, non *L. pumilum* Wall. ex DC. 1830, nec
    Seseli pumilum L. 1759) - 6
tenderiense Kotov = S. tortuosum
tenellum M. Pimen. - 6
tenuisectum Regel & Schmalh. (*Libanotis juncea* Korov., *L.*
    tenuisecta (Regel & Schmalh.) Korov.) - 6
tortuosum L. (*S. arenarium* Bieb., *S. campestre* Bess., *S.*
    pauciradiatum Schischk., *S. peucedanifolium* (Spreng.)
    Bess., *S. rigidum* Waldst. & Kit. subsp. *peucedanifoli-*
    um (Spreng.) Nym., *S. tenderiense* Kotov, *S. tortuosum*
    var. *pauciradiatum* (Schischk.) Tamamsch.) - 1
- var. *pauciradiatum* (Schischk.) Tamamsch. = S. tortuosum
transcaucasicum (Schischk.) M. Pimen. & Sdobnina
    (*Libanotis transcaucasica* Schischk., *Seseli transcauca-*
    sicum (Schischk.) Hiroe, comb. superfl.) - 2
tschuense E. Nikit. = S. sesseliflorum
tschuiliense Pavl. ex Korov. = Hyalolaena tschuiliensis
turbinatum Korov. - 6
unicaule (Korov.) M. Pimen. (*Libanotis unicaulis* Korov.) -
    6
valentinae M. Pop. - 6
varium Trev. (*S. pallasii* Bess. nom. nud.) - 1, 2
- subsp. *grandivittatum* (Somm. & Levier) Soo = S. grandiv-
    ittatum

Seselopsis Schischk.

pusilla M. Pimen. & Lavrova - 6
tianschanica Schischk. - 6

Siella M. Pimen. = Berula

erecta (Huds.) M. Pimen. = Berula erecta

Silaum Mill. (*Silaus* Bernh.)

alpestre (L.) Thell. = S. silaus
besseri (DC.) Galushko = S. silaus
peucedanoides (Bieb.) Hiroe = Seseli peucedanoides
popovii (Korov.) Hiroe = Ferula nuratavica
rubtzovii (Schischk.) Czer. = Aulacospermum tianscha-
    nicum
rubtzovii (Schischk.) Hiroe = Aulacospermum tianscha-
    nicum
silaus (L.) Schinz & Thell. (*Peucedanum alpestre* L., *Silaum*
    alpestre (L.) Thell., *S. besseri* (DC.) Galushko, *Silaus*
    besseri DC., *S. pratensis* (Crantz) Bess.) - 1, 2, 3, 6

Silaus Bernh. = Silaum

afghanicus Gilli = Parasilaus asiaticus
besseri DC. = Silaum silaus
popovii Korov. = Ferula nuratavica
pratensis (Crantz) Bess. = Silaum silaus
rubtzovii Schischk. = Aulacospermum tianschanicum
saxatilis Bajt. - 6
    M.S. Bajtenov in 1983 had separated this species into the genus
    Tschulaktavia Bajt., but this name was not validly published (nom.
    provis.).

Siler auct. = Agasyllis

latifolia (Bieb.) Hiroe = Agasyllis latifolia

Sison L.

amomum L. - 2

Sium L.

javanicum Blume = Oenanthe javanica
latifolium L. - 1, 2, 3, 4, 6
medium Fisch. & C.A. Mey. - 3, 6
orientale (Woronow) Soo = Berula erecta
sisaroideum DC. (*S. sisarum* L. subsp. *sisaroideum* (DC.)
    Soo) - 1, 2, 3, 6
***sisarum** L.
- subsp. *sisaroideum* (DC.) Soo = S. sisaroideum
suave Walt. - 4, 5
tenue (Kom.) Kom. - 5

Smyrniopsis Boiss.

armena Schischk. = S. aucheri
aucheri Boiss. (*S. armena* Schischk.) - 2

Smyrnium L.

cordifolium Boiss. - 6(alien)
olusatrum L. - 1(alien)
perfoliatum L. - 1, 2

Soranthus Ledeb. = Ferula

meyeri Ledeb. = Ferula sibirica

Sphaenolobium M. Pimen.

coriaceum (Korov.) M. Pimen. (*Selinum coriaceum*
    Korov.) - 6
kultiassovii (Korov.) M. Pimen. = Karatavia kultiassovii
tenuisectum (Korov.) M. Pimen. (*Selinum tenuisectum*
    Korov.) - 6
tianschanicum (Korov.) M. Pimen. (*Selinum tianschanicum*
    Korov.) - 6

Sphaerosciadium M. Pimen. & Kljuykov

denaense (Schischk.) M. Pimen. & Kljuykov (*Danaa denaensis* Schischk., *Physospermum denaense* (Schischk.) Czer., comb. superfl., *Ph. denaense* (Schischk.) B. Fedtsch. ex Weinert) - 6

Sphallerocarpus Bess. ex DC.

gracilis (Bess. ex Trev.) K.-Pol. - 3, 4, 5

Sphenocarpus Korov. = Seseli

eryngioides Korov. = Seseli eryngioides

Spuriopimpinella (Boissieu) Kitag.

calycina (Maxim.) Kitag. (*Aegopodium calycinum* (Maxim.) Worosch., *Ae. brachycarpum* (Kom.) Schischk., p.p. incl typo, *Pimpinella calycina* Maxim.) - 5

Stenocoelium Ledeb.

athamantoides (Bieb.) Ledeb. - 3, 4
tenuifolium Korov. = Ledebouriella multiflora
trichocarpum Schrenk - 6

Stenotaenia Boiss.

daralaghezica (Takht.) Schischk. = S. macrocarpa subsp. daralaghezica
iliensis Bajt. - 6

> M.S. Bajtenov in 1983 had separated this species into the genus Turaja Bajt., but this generic name was not validly pulished (nom. nud. & provis.).

macrocarpa Freyn & Sinth. ex Freyn subsp. **daralaghezica** (Takht.) Takht. (*S. daralaghezica* (Takht.) Schischk.). - 2

Symphyoloma C.A. Mey.

graveolens C.A. Mey. - 2

Sympodium C. Koch = Bunium

simplex C. Koch = Bunium simplex

Szovitsia Fisch. & C.A. Mey.

callicarpa Fisch. & C.A. Mey. - 2

Taeniopetalum Vis.

arenarium (Waldst. & Kit.) V. Tichomirov (*Angelica arenaria* (Waldst. & Kit.) Hiroe, *Peucedanum arenarium* Waldst. & Kit., *P. borysthenicum* Klok. ex Schischk.) - 1

Tamamschjania M. Pimen. & Kljuykov

lazica (Boiss. & Bal. ex Boiss.) M. Pimen. & Kljuykov (*Eleutherospermum lazicum* Boiss. & Bal. ex Boiss.) - 2

Tetrataenium (DC.) Manden.

olgae (Regel & Schmalh.) Manden. (*Heracleum olgae* Regel & Schmalh., *Platytaenia olgae* (Regel & Schmalh.) Korov.) - 6

Thyselium Rafin. (*Calestania* K.-Pol.)

palustre (L.) Rafin. (*Calestania palustris* (L.) K.Pol., *Peuce-*
danum palustre (L.) Moench) - 1, 3

Tilingia Regel

ajanensis Regel & Til. (*Cnidium ajanense* (Regel & Til.) Drude, *Conioselinum ajanense* (Regel & Til.) A. Khokhr., *C. kamtschaticum* Rupr.) - 4, 5
filisecta (Nakai & Kitag.) Nakai & Kitag. = Rupiphila tachiroei
jeholensis (Nakai & Kitag.) Leute = Conioselinum jeholense
tachiroei (Franch. & Savat.) Kitag. = Rupiphila tachiroei

Tordylium L.

komarovii Manden. = Mandenovia komarovii
maximum L. - 1, 2, 6

Torilis Adans.

arvensis (Huds.) Link - 1, 2, 6
- subsp. *heterophylla* (Guss.) Thell. = T. heterophylla
- subsp. **turcomanica** Geldykhanov - 6
heterophylla Guss. (*T. arvensis* (Huds.) Link subsp. *heterophylla* (Guss.) Thell.) - 1, 2
japonica (Houtt.) DC. - 1, 2, 5
- subsp. *ucranica* (Spreng.) Soo = T. ucranica
leptophylla (L.) Reichenb. fil. (*T. xanthotricha* (Stev.) Stank.) - 1, 2, 6
nodosa (L.) Gaertn. - 1, 2, 6
radiata Moench - 1, 2
stocksiana (Boiss.) Drude - 2(?),6
tenella (Delile) Reichenb. fil. - 2
ucranica Spreng. (*T. japonica* (Houtt.) DC. subsp. *ucranica* (Spreng.) Soo - 1
xanthotricha (Stev.) Stank. = T. leptophylla

Trachidium auct. = Aulacospermum

commutatum (Regel & Schmalh.) Hiroe = Aulacospermum simplex
dichotomum Korov. = Aulacospermum dichotomum
kopetdaghense Korov. = Aulacospermum vesiculoso-alatum
paleaceum (Regel & Schmalh.) Hiroe = Kozlovia paleacea
tianschanicum Korov. = Aulacospermum tianschanicum
vesiculoso-alatum Rech. fil. = Aulacospermum vesiculoso-alatum

*Trachyspermum Link

*ammi (L.) Sprague - 6

Transcaucasia Hiroe = Astrantia

armwnia Hiroe = Astrantia maxima

Trigonosciadium auct. = Mandenovia

komarovii (Manden.) Tamamsch. = Mandenovia komarovii

Trinia Hoffm.

biebersteinii Fedoronchuk (*T. kitaibelii* sensu Schischk.) - 1
glauca (L.) Dumort. (*Pimpinella glauca* L., *Trinia stankovii* Schischk.) - 1
hispida Hoffm. (*Rumia hispida* (Hoffm.) Stank., *R. hoffmannii* (Bieb.) Stank.) - 1, 2, 3, 6
- subsp. *leiogona* (C.A. Mey.) Fedoronchuk = T. leiogona
kitaibelii Bieb. (*T. ramosissima* (Fisch. ex Trev.) Reichenb. 1832, non Ledeb. 1829, *T. ucrainica* Schischk.) - 1

kitaibelii sensu Schischk. = T. biebersteinii
leiogona (C.A. Mey.) B. Fedtsch. (*T. hispida* Hoffm. subsp. *leiogona* (C.A. Mey.) Fedoronchuk) - 2
multicaulis (Poir.) Schischk. - 1
muricata Godet (*T. ramosissima* Ledeb. subsp. *muricata* (Godet) Fedoronchuk) - 1, 3, 6
polyclada Schischk. = T. ramosissima
ramosissima (Fisch. ex Trev.) Reichenb. = T. kitaibelii
ramosissima Ledeb. (*T. polyclada* Schischk.) - 3, 6
- subsp. *muricata* (Godet) Fedoronchuk = T. muricata
stankovii Schischk. = T. glauca
ucrainica Schischk. = T. kitaibelii

Turgenia Hoffm.

latifolia (L.) Hoffm. - 1, 2, 3, 5(alien),6

Vicatia DC.

atrosanguinea (Kar. & Kir.) P.K. Mukherjee & M. Pimen. (*Carum atrosanguineum* Kar. & Kir., *Vicatia wolffiana* (Fedde ex H. Wolff) C. Norm., *V. coniifolia* auct.) - 3, 4, 6
coniifolia auct. = V. atrosanguinea
wolffiana (Fedde ex H. Wolff) C. Norm. = V. atrosanguinea

Visnaga Mill.

daucoides Gaertn. (*Ammi visnaga* (L.) Lam.) - 2

Vvedenskia Korov.

pinnatifolia Korov. (*Conioselinum pinnatifolium* (Korov.) Schischk.) - 6

Xanthogalum Ave-Lall. = Angelica

purpurascens Ave-Lall. = Angelica purpurascens
sachokianum Karjag. = Angelica sachokiana
tatianae (Bordz.) Schischk. = Angelica tatianae

Xanthoselinum Schur

alsaticum (L.) Schur (*Peucedanum alsaticum* L., *P. lubimenkoanum* Kotov) - 1, 2, 3, 6

Zeravschania Korov.

pauciradiata (Tamamsch.) M. Pimen. (*Peucedanum pauciradiatum* Tamamsch. - 2
regeliana Korov. - 6
scabrifolia M. Pimen. - 6

Zosima Hoffm.

absinthifolia auct. = Z. orientalis
bucharica (Schischk.) Hiroe = Semenovia bucharica
depauperata (Schischk.) Hiroe = Seseli depauperatum
dichotoma Boiss. = Semenovia dichotoma
heterodonta (Korov.) Hiroe = Semenovia heterodonta
komarovii (Manden. ex Schischk.) Hiroe = Semenovia dasycarpa
korovinii M. Pimen. (*Platytaenia tordyloides* (Korov.) Korov. p.p. excl. typo, *Zosima tordyloides* sensu Schischk.) - 6
orientalis Hoffm. (*Z. absinthifolia* auct.) - 2, 6
pimpinelloides (Nevski) Hiroe = Semenovia pimpinelloides
rubtzovii (Schischk.) Hiroe = Semenovia rubtzovii
rubtzovii (Schischk.) R. Kam. = Semenovia rubtzovii

tordyloides Korov. p.p. = Pilopleura tordyloides
tordyloides sensu Schischk. = Z. korovinii

APOCYNACEAE Juss.

Apocynum auct. = Trachomitum

sarmatiense (Woodson) Wissjul. = Trachomitum sarmatiense

*Nerium L.

*****oleander** L. - 1, 2, 6

Poacynum Baill.

hendersonii (Hook. fil.) Woodson - 6
pictum (Schrenk) Baill. - 6

Trachomitum Woodson (*Apocynum* auct.)

armenum (Pobed.) Pobed. (*T. venetum* (L.) Woodson subsp. *armenum* (Pobed.) Rech. fil.) - 2
lancifolium (Russan.) Pobed. (*T. venetum* (L.) Woodson var. *lancifolium* (Russan.) Hara) - 3, 4, 6
russanovii (Pobed.) Pobed. - 1
sarmatiense Woodson (*Apocynum sarmatiense* (Woodson) Wissjul., *Trachomitum venetum* (L.) Woodson subsp. *sarmatiense* (Woodson) V. Avet.) - 1, 2
scabrum (Russan.) Pobed. (*T. venetum* (L.) Woodson subsp. *scabrum* (Russan.) Rech. fil., *T. venetum* var. *scabrum* (Beg. & Bell.) Kitam.) - 6
tauricum (Pobed.) Pobed. (*T. venetum* (L.) Woodson subsp. *tauricum* (Pobed.) Greuter & Burdet) - 1
venetum (L.) Woodson subsp. *armenum* (Pobed.) Rech. fil. = T. armenum
- subsp. *sarmatiense* (Woodson) V. Avet. = T. sarmatiense
- subsp. *scabrum* (Russan.) Rech. fil. = T. scabrum
- subsp. *tauricum* (Pobed.) Greuter & Burdet = T. tauricum
- var. *lancifolium* (Russan.) Hara = T. lancifolium
- var. *scabrum* (Beg. & Bell.) Kitam. = T. scabrum

Vinca L.

erecta Regel & Schmalh. - 6
herbacea Waldst. & Kit. (*V. semidesertorum* Ponert) - 1, 2, 6(cult.)
major L. - 1, 2, 6(cult.)
- subsp. *hirsuta* (Boiss.) Stearn = V. pubescens
- var. *hirsuta* Boiss. = V. pubescens
minor L. - 1, 2, 6(cult.)
pubescens D'Urv. (*V. major* L. var. *hirsuta* Boiss., *V. major* subsp. *hirsuta* (Boiss.) Stearn) - 2
semidesertorum Ponert = V. herbacea

AQUIFOLIACEAE Bartl.
Ilex L.

*****aquifolium** L. - 1, 2
colchica Pojark. - 2
- subsp. *imerethica* (Gagnidze) Gagnidze = I. imerethica
crenata Thunb. - 5
hyrcana Pojark. (*I. spinigera* auct. p.p.) - 2
imerethica Gagnidze (*I. colchica* Pojark. subsp. *imerethica* (Gagnidze) Gagnidze) - 2
rugosa Fr. Schmidt - 5
spinigera auct. p.p. = I. hyrcana

stenocarpa Pojark. - 2
sugerokii Maxim. - 5

ARACEAE Juss.
Acorus L.
asiaticus Nakai = A. calamus
calamus L. (*A. asiaticus* Nakai) - 1, 2, 3, 4, 5

Arisaema Mart.
amurense Maxim. - 5
- subsp. *robustum* (Engl.) H. Ohashi & J. Murata = A. robustum
- var. *robustum* Engl. = A. robustum
- var. *sachalinense* Miyabe & Kudo = A. sachalinense
- var. *serratum* Nakai = A. komarovii
japonicum Blume (*A. peninsulae* auct. fl. kuril.) - 5
japonicum sensu Kuzen. = A. peninsulae
komarovii Tzvel. (*A. amurense* Maxim. var. *serratum* Nakai, non *A. serratum* (Thunb.) Schott) - 5
ovale Nakai var. *sadoense* (Nakai) J. Murata = A. sadoense
peninsulae auct. fl. kuril. = A. japonicum
peninsulae auct. fl. sachal. = A. sachalinense
peninsulae Nakai (*A. japonicum* sensu Kuzen.) - 5
robustum (Engl.) Nakai (*A. amurense* Maxim. var. *robustum* Engl., *A. amurense* subsp. *robustum* (Engl.) H. Ohashi & J. Murata) - 5
sachalinense (Miyabe & Kudo) J. Murata (*A. amurense* Maxim. var. *sachalinense* Miyabe & Kudo, *A. peninsulae* auct. fl. sachal.) - 5
sadoense Nakai (*A. ovale* Nakai var. *sadoense* (Nakai) J. Murata) - 5

Arum L.
albispathum Stev. ex Ledeb. (*A. italicum* Mill. subsp. *albispathum* Stev. ex Ledeb.) Prime, *A. italicum* auct.) - 1, 2
alpinum Schott & Kotschy (*A. maculatum* L. subsp. *alpinum* (Schott & Kotschy) K. Richt., *A. orientale* Bieb. subsp. *alpinum* (Schott & Kotschy) H. Riedl, *A. maculatum* sensu Kuzen. p.p.) - 1
besserianum Schott (*A. maculatum* L. subsp. *besserianum* (Schott) Nym., *A. orientale* Bieb. subsp. *besserianum* (Schott) Holub, *A. immaculatum* auct., *A. maculatum* sensu Kuzen. p.p.) - 1
detruncatum C.A. Mey. ex Schott = A. elongatum subsp. detruncatum
elongatum Stev. - 1, 2
- subsp. **detruncatum** (C.A. Mey. ex Schott) H. Riedl (*A. detruncatum* C.A. Mey. ex Schott) - 2
immaculatum auct. = A. besserianum
italicum auct. = A. albispathum
italicum Mill. subsp. *albispathum* (Stev. ex Ledeb.) Prime = A. albispathum
jacquemontii Blume - 6
korolkowii Regel - 6
maculatum L. subsp. *alpinum* (Schott & Kotschy) K. Richt. = A. alpinum
- subsp. *besserianum* (Schott) Nym. = A. besserianum
maculatum sensu Kuzen. p.p. = A. alpinum and A. besserianum
orientale Bieb. - 1, 2
- subsp. *alpinum* (Schott & Kotschy) H. Riedl = A. alpinum
- subsp. *besserianum* (Schott) Holub = A. besserianum

Calla L.
palustris L. - 1, 3, 4, 5

Eminium (Blume) Schott
alberti (Regel) Engl. ex B. Fedtsch. - 6
lehmannii (Bunge) O. Kuntze (*E. regelii* Vved.) - 6
regelii Vved. = E. lehmannii

Lysichiton Schott
camtschatcense (L.) Schott - 5

Symplocarpus Salisb. ex W. Barton
foetidus auct. = S. renifolius
foetidus (L.) Nutt. f. *latissimus* Makino = S. renifolius
renifolius Shott ex Tzvel. (*S. foetidus* (L.) Nutt. f. *latissimus* Makino, *S. foetidus* auct.) - 5

ARALIACEAE Juss.
Acanthopanax (Decne. & Planch.) Miq. = Eleutherococcus
sessiliflorus (Rupr. & Maxim.) Seem. = Eleutherococcus sessiliflorus

Aralia L.
continentalis Kitag. - 5
cordata Thunb. (*A. schmidtii* Pojark.) - 5
elata (Miq.) Seem. (*A. mandshurica* Rupr. & Maxim.) - 5
mandshurica Rupr. & Maxim. = A. elata
schmidtii Pojark. = A. cordata

Echinopanax Decne. = Oplopanax
elatus Nakai = Oplopanax elatus

Eleutherococcus Maxim. (*Acanthopanax* (Decne. & Planch.) Miq.)
senticocus (Rupr. & Maxim.) Maxim. - 5
sessiliflorus (Rupr. & Maxim.) S.Y. Hu (*Acanthopanax sessiliflorus* (Rupr. & Maxim.) Seem.) - 5

*Fatsia Decne. & Planch.
*japonica (Thunb.) Decne. & Planch.

Hedera L.
caucasigena Pojark. = H. helix
chrysocarpa Walsh (*H. helix* L. subsp. *poetarum* Nym.) - 2
colchica (C. Koch) C. Koch - 2
helix L. (*H. caucasigena* Pojark., *H. helix* subsp. *caucasigena* (Pojark.) Takht. & Mulk., *H. taurica* Carriere) - 1, 2
- subsp. *caucasigena* (Pojark.) Takht. & Mulk. = H. helix
- subsp. *poetarum* Nym. = H. chrysocarpa
pastuchowii Woronow - 2
taurica Carriere = H. helix

Kalopanax Miq.
septemlobus (Thunb.) Koidz. - 5

30

Oplopanax (Torr. & Gray) Miq. (*Echinopanax* Decne. & Planch. nom. invalid.)

elatus (Nakai) Nakai (*Echinopanax elatus* Nakai, *Oplopanax horridus* (Smith) Miq. subsp. *elatus* (Nakai) Hara) - 5
horridus (Smith) Miq. subsp. *elatus* (Nakai) Hara = Oplopanax elatus

Panax L.

ginseng C.A. Mey. (*Panax schin-seng* Nees, nom. illegit.) - 5
schin-seng Nees = P. ginseng

*ARECACEAE Sch. Bip. (PALMAE Juss. nom. altern.)

*Butia (Becc.) Becc.
*capitata (Mart.) Becc. - 2

*Chamaerops L.
fortunei Hook. = Trachycarpus fortunei
*humilis L. - 2

*Jubaea H. B. K.
*chilensis (Mol.) Baill. (*J. spectabilis* H. B. K.) - 2
spectabilis H. B. K. = J. chilensis

*Phoenix L.
*canariensis Chabaud - 2
*sylvestris (L.) Roxb. - 2

Pritchardia auct. = Washingtonia
filifera Linden ex Andre = Washingtonia filifera

*Trachycarpus H. Wendl.
excelsa auct. = T. fortunei
*fortunei (Hook.) H. Wendl. (*Chamaerops fortunei* Hook., *Trachycarpus excelsa* auct.) - 1, 2

*Washingtonia H. Wendl. (*Pritchardia* auct.)
filamentosa (H. Wendl.) O. Kuntze = W. filifera
*filifera (Linden ex Andre) H. Wendl. (*Pritchardia filifera* Linden ex Andre, *Washingtonia filamentosa* (H. Wendl.) O. Kuntze) - 2
*robusta H. Wendl. - 2

ARISTOLOCHIACEAE Juss.
Aristolochia L. (*Hocquartia* Dumort., *Isotrema* Rafin.)

bottae Jaub. & Spach - 2
clematitis L. - 1, 2
contorta Bunge - 5
*durior Hill (*A. macrophylla* Lam., *Isotrema macrophyllum* (Lam.) C.F. Reed)
fimbriata Cham. - 2
iberica Fisch. & C.A. Mey. ex Boiss. - 2
macrophylla Lam. = A. durior
manshuriensis Kom. (*Hocquartia manshuriensis* (Kom.) Nakai) - 5

pontica Lam. - 1(alien),2
steupii Woronow - 2

Asarum L. (*Asiasarum* F. Maek.)
europaeum L. - 1, 3
- subsp. *caucasicum* (Duchartre) Soo = A. intermedium
- var. *caucasicum* Duchartre = A. intermedium
- var. *intermedium* C.A. Mey. = A. intermedium
heterotropoides Fr. Schmidt (*A. sieboldii* Miq. subsp. *heterotropoides* (Fr.Schmidt) Kitam.) - 5
ibericum Stev. ex Ledeb. = A. intermedium
intermedium (C.A. Mey.) Grossh. (*A. europaeum* L. var. *intermedium* C.A. Mey., *A. europaeum* var. *caucasicum* Duchartre, *A. europaeum* subsp. *caucasicum* (Duchartre) Soo, *A. ibericum* Stev. ex Ledeb. nom. invalid.) - 2
sieboldii Miq. - 5
- subsp. *heterotropoides* (Fr. Schmidt) Kitam. = A. heterotropoides

Asiasarum F. Maek. = Asarum

Hocquartia Dumort. = Aristolochia
manshuriensis (Kom.) Nakai = A. manshuriensis

Isotrema Rafin. = Aristolochia
macrophyllum (Lam.) C.F. Reed = Aristolochia macrophylla

ASCLEPIADACEAE R. Br.
Alexitoxicon St.-Lag. = Vincetoxicum
acuminatum (Decne.) Pobed. = Vincetoxicum acuminatum
albowianum (Kusn.) Pobed. = Vincetoxicum albowianum
amplexicaule (Siebold & Zucc.) Pobed. = Vincetoxicum amplexicaule
atratum (Bunge) Pobed. = Vincetoxicum atratum
cretaceum (Pobed.) Pobed. = Vincetoxicum hirundinaria
darvasicum (B. Fedtsch.) Pobed. = Vincetoxicum darvasicum
funebre (Boiss. & Kotschy) Pobed. = Vincetoxicum funebre
inamoenum (Maxim.) Pobed. = Vincetoxicum inamoenum
intermedium (Taliev) Pobed. = Vincetoxicum intermedium
jailicola (Juz.) Pobed. = Vincetoxicum jailicola
juzepczukii (Pobed.) Pobed. = Vincetoxicum juzepczukii
laxum (Bartl.) Pobed. ex Kuth. = Vincetoxicum hirundinaria
maeoticum (Kleop.) Pobed. = Vincetoxicum maeoticum
minus (C. Koch) Pobed. = Vincetoxicum fuscatum
mugodsharicum (Pobed.) Pobed. = Vincetoxicum mugodsharicum
officinale (Moench) St.-Lag. = Vincetoxicum hirundinaria
- var. *stepposum* (Pobed.) Serg. = Vincetoxicum albowianum
pumilum (Decne.) Pobed. = Vincetoxicum pumilum
raddeanum (Albov) Pobed. = Vincetoxicum raddeanum
rehmannii (Boiss.) Pobed. = Vincetoxicum rehmanni
rossicum (Kleop.) Pobed. = Vincetoxicum rossicum
scandens (Somm. & Levier) Pobed. = Vincetoxicum scandens
schmalhausenii (Kusn.) Pobed. = Vincetoxicum schmalhausenii
sibiricum (L.) Pobed. = Vincetoxicum sibiricum
stauropolitanum (Pobed.) Pobed. = Vincetoxicum hirundinaria

stepposum (Pobed.) Pobed. = Vincetoxicum albowianum
tauricum (Pobed.) Pobed. = Vincetoxicum tauricum
vincetoxicum (L.) H.P. Fuchs = Vincetoxicum hirundinaria
volubile (Maxim.) Pobed. = Vincetoxicum volubile

Antitoxicum Pobed. = Vincetoxicum

acuminatum (Decne.) Pobed. = Vincetoxicum acuminatum
albowianum (Kusn.) Pobed. = Vincetoxicum albowianum
amplexicaule (Siebold & Zucc.) Pobed. = Vincetoxicum amplexicaule
atratum (Bunge) Pobed. = Vincetoxicum atratum
cretaceum Pobed. = Vincetoxicum hirundinaria
darvasicum (B. Fedtsch.) Pobed. = Vincetoxicum darvasicum
funebre (Boiss. & Kotschy) Pobed. = Vincetoxicum funebre
inamoenum (Maxim.) Pobed. = Vincetoxicum inamoenum
intermedium (Taliev) Pobed. = Vincetoxicum intermedium
jailicola (Juz.) Pobed. = Vincetoxicum jailicola
juzepczukii Pobed. = Vincetoxicum juzepczukii
laxum (Bartl.) Pobed. = Vincetoxicum hirundinaria
maeoticum (Kleop.) Pobed. = Vincetoxicum maeoticum
minus (C. Koch) Pobed. = Vincetoxicum fuscatum
mugodsharicum (Pobed.) Pobed. = Vincetoxicum mugodsharicum
officinale (Moench) Pobed. = Vincetoxicum hirundinaria
pumilum (Decne.) Pobed. = Vincetoxicum pumilum
raddeanum (Albov) Pobed. = Vincetoxicum raddeanum
rehmannii (Boiss.) Pobed. = Vincetoxicum rehmannii
rossicum (Kleop.) Pobed. = Vincetoxicum rossicum
scandens (Somm. & Levier) Pobed. = Vincetoxicum scandens
schmalhausenii (Kusn.) Pobed. = Vincetoxicum schmalhausenii
sibiricum (L.) Pobed. = Vincetoxicum sibiricum
stauropolitanum (Pobed.) Pobed. = Vincetoxicum hirundinaria
stepposum Pobed. = Vincetoxicum albowianum
tauricum (Pobed.) Pobed. = Vincetoxicum tauricum
volubile (Maxim.) Pobed. = Vincetoxicum volubile

*Araujia Brot.

*sericifera Brot. - 2

Asclepias L.

fuscata Hornem. = Vincetoxicum fuscatum
syriaca L. - 1, 2

Cynanchum L.

acutum L. - 1, 2, 6
- subsp. *sibiricum* (Willd.) Rech. fil. = C. sibiricum
ascyrifolium (Franch. & Savat.) Matsum. = Vincetoxicum acuminatum
caudatum (Miq.) Maxim. (*C. maximoviczii* Pobed.) - 5
darvasicum (B. Fedtsch.) R. Kam. = Vincetoxicum darvasicum
jailicola (Juz.) Juz. = Vincetoxicum jailicola
juzepczukii (Pobed.) Borhidi = Vincetoxicum juzepczukii
maximoviczii Pobed. = C. caudatum
minus C. Koch = Vincetoxicum fuscatum
pamirense Tsiang & Zhang = C. sibiricum
rossicum (Kleop.) Borhidi = Vincetoxicum rossicum
sibiricum Willd. (*C. acutum* L. subsp. *sibiricum* (Willd.) Rech. fil., *C. pamirense* Tsiang & Zhang, *C. stellatum* Pobed.) - 3, 6

stellatum Pobed. = C. sibiricum
stepposum (Pobed.) Juz. = Vincetoxicum albowianum
thesioides (Freyn) Schumann = Vincetoxicum sibiricum
vincetoxicum (L.) Pers. subsp. *laxum* (Bartl.) V. Avet. = Vincetoxicum hirundinaria
wilfordii (Franch. & Savat.) Hook. fil. = Seutera wilfordii

Cynoctonum E. Mey.

purpureum (Pall.) Pobed. - 4, 5

Gomphocarpus R. Br.

fruticosus (L.) Ait. fil. - 2

Metaplexis R. Br.

japonica (Thunb.) Makino - 5

Periploca L.

graeca L. - 2, 6
sepium Bunge - 5

Pycnostelma Bunge ex Decne.

paniculata (Bunge) Schumann - 4, 5

Seutera Reichenb.

wilfordii (Franch. & Savat.) Pobed. (*Cynanchum wilfordii* (Franch. & Savat.) Hook. fil.) - 5

Vincetoxicum N.M. Wolf (*Alexitoxicon* St.-Lag., *Antitoxicum* Pobed.)

acuminatum Decne. (*Alexitoxicon acuminatum* (Decne.) Pobed., *Antitoxicum acuminatum* (Decne.) Pobed., *Cynanchum ascyrifolium* (Franch. & Savat.) Matsum., *Vincetoxicum ascyrifolium* Franch. & Savat.) - 5
albowianum (Kusn.) Pobed. (*Alexitoxicon albowianum* (Kusn.) Pobed., *A. officinale* (Moench) St.-Lag. var. *stepposum* (Pobed.) Serg., *A. stepposum* (Pobed.) Pobed., *Antitoxicum albowianum* (Kusn.) Pobed., *A. stepposum* Pobed., *Cynanchum stepposum* (Pobed.) Juz., *Vincetoxicum hirundinaria* Medik. subsp. *stepposum* (Pobed.) Markgraf, *V. stepposum* (Pobed.) A. & D. Löve, *V. stepposum* (Pobed.) Pobed. comb. superfl.) - 1, 2, 3
amplexicaule Siebold & Zucc. (*Alexitoxicon amplexicaule* (Siebold & Zucc.) Pobed., *Antitoxicum amplexicaule* (Siebold & Zucc.) Pobed.) - 5
amplifolium C. Koch = V. scandens
ascyrifolium Franch. & Savat. = V. acuminatum
atratum (Bunge) Morr. & Decne. (*Alexitoxicon atratum* (Bunge) Pobed., *Antitoxicum atratum* (Bunge) Pobed.) - 5
cretaceum (Pobed.) Wissjul. = V. hirundinaria
darvasicum B. Fedtsch. (*Alexitoxicon darvasicum* (B. Fedtsch.) Pobed., *Antitoxicum darvasicum* (B. Fedtsch.) Pobed., *Cynanchum darvasicum* (B. Fedtsch.) R. Kam.) - 6
funebre Boiss. & Kotschy (*Alexitoxicon funebre* (Boiss. & Kotschy) Pobed., *Antitoxicum funebre* (Boiss. & Kotschy) Pobed.) - 2
fuscatum auct. p.p. = V. intermedium and V. maeoticum
fuscatum (Hornem.) Endl. (*Asclepias fuscata* Hornem., *Alexitoxicon minus* (C. Koch) Pobed., *Antitoxicum minus* (C. Koch) Pobed., *Cynanchum minus* C. Koch, *Vincetoxicum minus* (C. Koch) C. Koch) - 1, 2

hirundinaria Medik. (*Alexitoxicon cretaceum* (Pobed.) Pobed., *A. laxum* (Bartl.) Pobed. ex Kuth., *A. officinale* (Moench) St.-Lag., *A. stauropolitanum* (Pobed.) Pobed., *A. vincetoxicum* (L.) H.P. Fuchs, *Antitoxicum cretaceum* Pobed., *A. laxum* (Bartl.) Pobed., *A. officinale* (Moench) Pobed., *A. stauropolitanum* (Pobed.) Pobed., *Cynanchum vincetoxicum* (L.) Pers. subsp. *laxum* (Bartl.) V. Avet., *Vincetoxicum cretaceum* (Pobed.) Wissjul., *V. hirundinaria* subsp. *cretaceum* (Probed.) Markgraf, *V. laxum* (Bartl.) Gren. & Godr., *V. stauropolitanum* Pobed., *V. tanaicense* P. Smirn.) - 1, 2
- subsp. *cretaceum* (Pobed.) Markgraf = V. hirundinaria
- subsp. *jailicola* (Juz.) Markgraf = V. jailicola
- subsp. *stepposum* (Pobed.) Markgraf = V. albowianum
inamoenum Maxim. (*Alexitoxicon inamoenum* (Maxim.) Pobed., *Antitoxicum inamoenum* (Maxim.) Pobed.) - 5
intermedium Taliev (*Alexitoxicon intermedium* (Taliev) Pobed., *Antitoxicum intermedium* (Taliev) Pobed., *Vincetoxicum fuscatum* auct. p.p.) - 1
jailicola Juz. (*Alexitoxicon jailicola* (Juz.) Pobed., *Antitoxicum jailicola* (Juz.) Pobed., *Cynanchum jailicola* (Juz.) Juz., *Vincetoxicum hirundinaria* Medik. subsp. *jailicola* (Juz.) Markgraf, *V. laxum* (Bartl.) Gren. & Godr. var. *jailicola* (Juz.) Wissjul.) - 1
juzepczukii (Pobed.) Privalova ex Wissjul. (*Antitoxicum juzepczukii* Pobed., *Alexitoxicon juzepczukii* (Pobed.) Pobed., *Cynanchum juzepczukii* (Pobed.) Borhidi, *Vincetoxicum juzepczukii* (Pobed.) Markgraf, comb. superfl.) - 1
laxum (Bartl.) Gren. & Godr. = V. hirundinaria
- var. *jailicola* (Juz.) Wissjul. = V. jailicola
maeoticum (Kleop.) Barbar. (*Alexitoxicon maeoticum* (Kleop.) Pobed., *Antitoxicum maeoticum* (Kleop.) Pobed., *Vincetoxicum fuscatum* auct. p.p.) - 1
minus (C. Koch) C. Koch = V. fuscatum
mugodsharicum Pobed. (*Alexitoxicon mugodsharicum* (Pobed.) Pobed., *Antitoxicum mugodsharicum* (Pobed.) Pobed.) - 6
officinale Moench var. *rossicum* (Kleop.) K. Grodzinska = V. rossicum
pumilum Decne. (*Alexitoxicon pumilum* (Decne.) Pobed., *Antitoxicum pumilum* (Decne.) Pobed.) - 6
■ **raddeanum** Albov (*Alexitoxicon raddeanum* (Albov) Pobed., *Antitoxicum raddeanum* (Albov) Pobed.)
rehmannii Boiss. (*Alexitoxicon rehmannii* (Boiss.) Pobed., *Antitoxicum rehmannii* (Boiss.) Pobed.) - 2
rossicum (Kleop.) Barbar. (*Alexitoxicon rossicum* (Kleop.) Pobed., *Antitoxicum rossicum* (Kleop.) Pobed., *Cynanchum rossicum* (Kleop.) Borhidi, *Vincetoxicum officinale* Moench var. *rossicum* (Kleop.) K. Grodzinska) - 1
scandens Somm. & Levier (*Alexitoxicon scandens* (Somm. & Levier) Pobed., *Antitoxicum scandens* (Somm. & Levier) Pobed., *Vincetoxicum amplifolium* C. Koch, nom. dubium) - 1, 2
schmalhausenii (Kusn.) Stank. (*Alexitoxicon schmalhausenii* (Kusn.) Pobed., *Antitoxicum schmalhausenii* (Kusn.) Pobed., *Vincetoxicum schmalhausenii* (Kusn.) Markgraf, comb. superfl.) - 1, 2
sibiricum (L.) Decne. (*Alexitoxicon sibiricum* (L.) Pobed., *Antitoxicum sibiricum* (L.) Pobed., *Cynanchum thesioides* (Freyn) Schumann, *Vincetoxicum thesioides* Freyn) - 3, 4, 5, 6
stauropolitanum Pobed. = V. hirundinaria
stepposum (Pobed.) A. & D. Löve = V. albowianum
stepposum (Pobed.) Pobed. = V. albowianum
tanaicense P. Smirn. = V. hirundinaria
tauricum Pobed. (*Alexitoxicon tauricum* (Pobed.) Pobed.,

Antitoxicum tauricum (Pobed.) Pobed.) - 1
thesioides Freyn = V. sibiricum
volubile Maxim. (*Alexitoxicon volubile* (Maxim.) Pobed., *Antitoxicum volubile* (Maxim.) Pobed.) - 5"

ASPARAGACEAE Juss.
Asparagus L.
angulofractus Iljin (*A. soongoricus* Iljin, *A. ledebourii* auct. p.p.) - 6
botschantzevii Vlassova - 6
brachyphyllus Turcz. (*A. trichophyllus* auct. p.p.) - 6
breslerianus Schult. & Schult. fil. (*A. komarovianus* Vved.) - 1, 2
bucharicus Iljin - 6
burjaticus Peschkova - 4
caspius Schult. & Schult. fil. = A. officinalis
davuricus Fisch. ex Link - 4, 5
ferganensis Vved. (*A. monoclados* Vved.) - 6
gibbus Bunge (*A. tuberculatus* Bunge ex Iljin, p.p. incl. typo) - 4
gypsaceus Vved. - 6
inderiensis Blum ex Pacz. (*A. kasakstanicus* Iljin) - 1, 6
kasakstanicus Iljin = A. inderiensis
komarovianus Vved. = A. breslerianus
ledebourii auct. p.p. = A. angulofractus
ledebourii Miscz. - 2
leptophyllus Schischk. = A. persicus
levinae Klok. = A. maritimus
litoralis Stev. - 1
maritimus (L.) Mill. (*A. officinalis* L. α . *maritimus* L., *A. levinae* Klok.) - 1
misczenkoi Iljin = A. neglectus
monoclados Vved. = A. ferganensis
neglectus Kar. & Kir. (*A. misczenkoi* Iljin) - 3, 4, 6
officinalis L. (*A. caspius* Schult. & Schult. fil., *A. polyphyllus* Stev., *A. setiformis* Kryl.) - 1, 2, 3, 4, 6
α . *maritimus* L. = A. maritimus
oligoclonos Maxim. - 5
oligophyllus Baker = A. persicus
pachyrrhizus Ivanova ex Vlassova - 6
pallasii Miscz. (*A. trichophyllus* auct. p.p.) - 1, 3, 4
persicus Baker (*A. leptophyllus* Schischk., *A. oligophyllus* Baker) - 1, 2, 6
polyphyllus Stev. = A. officinalis
popovii Iljin = A. turkestanicus
pseudoscaber Grec. - 1
schoberioides Kunth - 4, 5
setiformis Kryl. = A. officinalis
soongoricus Iljin = A. angulofractus
tamariscinus Ivanova ex Grub. - 4, 5
tenuifolius Lam. - 1
trichophyllus auct. p.p. = A. brachyphyllus and A. pallasii
tuberculatus Bunge ex Iljin, p.p. = A. gibbus
turkestanicus M. Pop. (*A. popovii* Iljin) - 6
verticillatus L.- 1, 2, 6
vvedenskyi Botsch. - 6

ASPHODELACEAE Juss.
Ammolirion Kar. & Kir. = Eremurus
comosum (O. Fedtsch.) A. Khokhr. = Eremurus comosus
inderiense (Bieb.) Regel ex A. Khohr. = Eremurus inderiensis
inderiense Bunge ex Regel = Eremurus inderiensis

Anthericum L.

calyculatum L. = Tofieldia calyculata (Melanthiaceae)
liliago L. - 1
ossifragum L. = Narthecium ossifragum (Melanthiaceae)
ramosum L. - 1, 2

Asphodeline Reichenb. (*Asphodelus* auct.)

dendroides (Hoffm.) Woronow - 2
lutea (L.) Reichenb. - 1, 2
szovitsii (C. Koch) Miscz. (*Asphodelus szovitsii* C. Koch)- 2
taurica (Pall. ex Bieb.) Endl. (*Asphodelus tauricus* Pall. ex Bieb.) - 1, 2
tenuiflora (C. Koch) Miscz. (*A. tenuior* (Bieb.) Ledeb. subsp. *tenuiflora* (C. Koch) E. Tuzlaci) - 2
tenuior (Bieb.) Ledeb. (*Asphodelus tenuior* Bieb.) - 2
- subsp. *tenuiflora* (C. Koch) E. Tuzlaci = A. tenuiflora

Asphodelus auct. = Asphodeline

szovitsii C. Koch = Asphodeline szovitsii
tauricus Pall. ex Bieb. = Asphodeline taurica
tenuior Bieb. = Asphodeline tenuior

Eremurus Bieb. (*Ammolirion* Kar. & Kir., *Henningia* Kar. & Kir., *Selonia* Regel)

aitchisonii Baker (*Henningia aitchisonii* (Baker) A. Khokhr.) - 6
alaicus Chalkuziev - 6
alberti Regel (*Henningia alberti* (Regel) A. Khokhr.) - 6
altaicus (Pall.) Stev. - 3, 6
- f. *fuscus* O. Fedtsch. = E. fuscus
ambigens Vved. (*E. stenophyllus* (Boiss. & Buhse) Baker subsp. *ambigens* (Vved.) Wendelbo, *Henningia ambigens* (Vved.) A. Khokhr., *Eremurus stenophyllus* auct. p.p.) - 6
ammophilus Vved. - 6
angustifolius Baker - 6
anisopterus (Kar. & Kir.) Regel (*Henningia anisoptera* Kar. & Kir.) - 6
aurantiacus Baker (*E. stenophyllus* (Boiss. & Buhse) Baker subsp. *aurantiacus* (Baker) Wendelbo, p.p., *Henningia stenophylla* (Boiss. & Buhse) A. Khokhr., p.p. excl. typo, *Eremurus stenophyllus* auct. p.p.) - 6
azerbajdzhanicus Charkev. - 2
baissunensis O. Fedtsch. (*Henningia baissunensis* (O. Fedtsch.) A. Khokhr.) - 6
brachystemon Vved. (*Selonia brachystemon* (Vved.) A. Khokhr.) - 6
bucharicus Regel (*Henningia bucharica* (Regel) A. Khokhr.) - 6
candidus Vved. (*Henningia candida* (Vved.) A. Khokhr.) - 6
capusii Franch. = E. luteus
chloranthus M. Pop. nom. invalid. (*Henningia chlorantha* (M. Pop.) A. Khokhr. nom. invalid.) - 6
comosus O. Fedtsch. (*Ammolirion comosum* (O. Fedtsch.) A. Khokhr.) - 6
cristatus Vved. - 6
fuscus (O. Fedtsch.) Vved. (*E. altaicus* (Pall.) Stev. f. *fuscus* O. Fedtsch.) - 6
hilariae M. Pop. & Vved. (*Henningia hilariae* (M. Pop. & Vved.) A. Khokhr.) - 6
hissaricus Vved. (*E. pectiniformis* T. Rjabova) - 6
iae Vved. - 6
inderiensis (Stev.) Regel (*Ammolirion inderiense* Bunge ex Regel, nom. invalid., *A. inderiense* (Bieb.) Regel ex A.

Khokhr. nom. invalid.) - 1, 3, 6
jungei Juz. - 1
kaufmannii Regel (*Henningia kaufmannii* Regel, nom. nud.) - 6
kopetdaghensis M. Pop. ex B. Fedtsch. - 6
korolkowii Regel (*Henningia korolkowii* (Regel) A. Khokhr.) - 6
korovinii B. Fedtsch. (*Henningia korovinii* (B. Fedtsch.) A. Khokhr.) - 6
korshinskyi O. Fedtsch. - 6
lachnostegius Vved. (*Henningia lachnostegia* (Vved.) A. Khokhr.) - 6
lactiflorus O. Fedtsch. (*Henningia lactiflora* (O. Fedtsch.) A. Khokhr.) - 6
x **ludmillae** Levichev & Priszter. - E. regelii Vved. x E. turkestanicus Regel - 6
luteus Baker (*E. capusii* Franch., *Henningia lutea* (Baker) A. Khokhr., *H. capusii* (Franch.) A. Khokhr.) - 6
micranthus Vved. - 6
mirabilis T. Rjabova = E. roseolus
nuratavicus A. Khokhr. - 6
olgae Regel (*Henningia olgae* Regel, nom. nud.) - 6
parviflorus Regel (*Henningia parviflora* (Regel) A. Khokhr.) - 6
pectiniformis T. Rjabova = E. hissaricus
pubescens Vved. (*Henningia pubescens* (Vved.) A. Khokhr.) - 6
regelii Vved. (*E. spectabilis* Bieb. subsp. *regelii* (Vved.) Wendelbo) - 6
robustus (Regel) Regel (*Henningia robusta* Regel) - 6
roseolus Vved. (*E. mirabilis* T. Rjabova, *Henningia roseola* (Vved.) A. Khokhr.) - 6
saprjagajevii B. Fedtsch. (*Henningia saprjagajevii* (B. Fedtsch.) A. Khokhr.) - 6
sogdianus (Regel) Franch. (*Selonia sogdiana* Regel) - 6
spectabilis Bieb. - 1, 2, 6
- subsp. *regelii* (Vved.) Wendelbo = E. regelii
- subsp. *subalbiflorus* (Vved.) Wendelbo = E. subalbiflorus
stenophyllus auct. p.p. = E. ambigens and E. aurantiacus
stenophyllus (Boiss. & Buhse) Baker subsp. *ambigens* (Vved.) Wendelbo = E. ambigens
- subsp. *aurantiacus* (Baker) Wendelbo, p.p. = E. aurantiacus
subalbiflorus Vved. (*E. spectabilis* Bieb. subsp. *subalbiflorus* (Vved.) Wendelbo) - 6
suworowii Regel (*Henningia suworowii* (Regel) A. Khokhr.) - 6
tadshikorum Vved. - 6
tauricus Stev. - 1, 2
thiodanthus Juz. - 1
tianschanicus Pazij & Vved. ex Golosk. (?*Henningia altissima* A. Khokhr.) - 6
turkestanicus Regel (*Selonia turkestanica* (Regel) A. Khokhr.) - 6
zenaidae Vved. - 6
zoae Vved. - 6

Henningia Kar. & Kir. = Eremurus

aitchisonii (Baker) A. Khokhr. = Eremurus aitchisonii
alberti (Regel) A. Khokhr. = Eremurus alberti
altissima A. Khokhr. = Eremurus tianschanicus
ambigens (Vved.) A. Khokhr. = Eremurus ambigens
anisoptera Kar. & Kir. = Eremurus anisopterus
baissunensis (O. Fedtsch.) A. Khokhr. = Eremurus baissunensis
bucharica (Regel) A. Khokhr. = Eremurus bucharicus
candida (Vved.) A. Khokhr. = Eremurus candidus
capusii (Franch.) A. Khokhr. = Eremurus luteus

chlorantha (M. Pop.) A. Khokhr. = Eremurus chloranthus
hilariae (M. Pop. & Vved.) A. Khokhr. = Eremurus hilariae
kaufmannii Regel = Eremurus kaufmannii
korolkowii (Regel) A. Khokhr. = Eremurus korolkowii
korovinii (B. Fedtsch.) A. Khokhr. = Eremurus korovinii
lachnostegia (Vved.) A. Khokhr. = Eremurus lachnostegius
lactiflora (O. Fedtsch.) A. Khokhr. = Eremurus lactiflorus
lutea (Baker) A. Khokhr. = Eremurus luteus
olgae Regel = Eremurus olgae
parviflora (Regel) A. Khokhr. = Eremurus parviflorus
pubescens (Vved.) A. Khokhr. = Eremurus pubescens
robusta Regel = Eremurus robustus
roseola (Vved.) A. Khokhr. = Eremurus roseolus
saprjagajevii (B. Fedtsch.) A. Khokhr. = Eremurus saprjagajevii
stenophylla (Boiss. & Buhse) A. Khokhr. p.p. = Eremurus
   aurantiacus
suworowii (Regel) A. Khokhr. = Eremurus suworowii

Selonia Regel = Eremurus

brachystemon (Vved.) A. Khokhr. = Eremurus brachystemon
sogdiana Regel = Eremurus sogdianus
turkestanica (Regel) A. Khokhr. = Eremurus turkestanicus

ASPIDIACEAE Mett. ex Frank = DRYOPTERIDACEAE

ASPLENIACEAE Newm.
Asplenium L.

adiantum-nigrum L. - 1, 2, 6
▪ x adulteriniforme Lovis, Melzen & Reichstein. - A. trichomanes L. x A. viride Huds.
altajense (Kom.) Grub. = A. pekinense
x **alternifolium** Wulf. (*A. germanicum* auct.). - A. septentrionale (L.) Hoffm. x A. trichomanes L. - 1
anagrammoides auct. = A. tenuicaule
billotii F.Schultz (*A. obovatum* auct.) - 1
ceterach L. = Ceterach officinarum
conilii Franch. & Savat. = Athyriopsis conilii (Athyriaceae)
cuneifolium Viv. - 1
daghestanicum Christ - 2
exiguum Bedd. - 3
fontanum auct. = A. pseudofontanum
fontanum (L.) Bernh. subsp. *pseudofontanum* (C.Koss.)
   Reichstein & Schneller = A. pseudofontanum
germanicum auct. = A. x alternifolium
x *germanicum* Weis subsp. *heufleri* (Reichardt) A.Bobr. =
   A. x heufleri
haussknechtii Godet & Reut. (*A. hermanni-christii* Fomin,
   A. lepidum C. Presl subsp. *haussknechtii* (Godet &
   Reut.) Brownsey, *A. lepidum* subsp. *haussknechtii* var.
   samarkandense (C. Koss.) Brownsey, *A. samarkandense* C. Koss.) - 2, 6
henryi Baker = Lunathyrium henryi (Athyriaceae)
hermanni-christii Fomin = A. haussknechtii
x **heufleri** Reichardt (*A. x germanicum* Weis subsp. *heufleri*
   (Reichardt) A.Bobr.). - A. septentrionale (L.) Hoffm.
   x A. trichomanes L. subsp. quadrivalens D.E. Mey. - 1
incisum Thunb. - 5
japonicum Thunb. = Athyriopsis japonica (Athyriaceae)
kamtschatkanum Gilbert = Lunathyrium pterorachis
   (Athyriaceae)
lepidum C. Presl subsp. *haussknechtii* (Godet & Reut.)

Brownsey = A. haussknechtii
- - var. *samarkandense* (C.Koss.) Brownsey = A. haussknechtii
melanocaulon Willd. = A. trichomanes
- subsp. *inexpectans* (Lovis) A. & D. Löve = A. trichomanes
   subsp. inexpectans
x **murbeckii** Doerfl. - A. ruta-muraria L. x A. septentrionale
   (L.) Hoffm. - 2
obovatum auct. = A. billotii
pekinense Hance (*A. altajense* (Kom.) Grub., *A. sarelii*
   Hook. var. *altajense* Kom., *A. sarelii* auct.) - 3, 4
pseudofontanum C. Koss. (*A. fontanum* (L.) Bernh. subsp.
   pseudofontanum (C. Koss.) Reichstein & Schneller, *A.
   fontanum* auct.) - 6
pseudolanceolatum Fomin = A. woronowii
ruprechtii Kurata = Camptosorus sibiricus
ruta-muraria L. - 1, 2, 3, 4, 5, 6
sajanense Gudoschn. & Krasnob. - 4
samarkandense C.Koss. = A. haussknechtii
sarelii auct. = A. pekinense
sarelii Hook. var. *altajense* Kom. = A. pekinense
scolopendrium L. = Phyllitis scolopendrium
septentrionale (L.) Hoffm. - 1, 2, 3, 6
- subsp. **caucasicum** Fraser-Jenkins & Lovis - 2
sibiricum Turcz. ex G. Kunze = Diplazium sibiricum
   (Athyriaceae)
x **souchei** Litard. - A. billotii F. Schultz x A. septentrionale
   (L.) Hoffm. - 1
tenuicaule Hayata (*A. anagrammoides* auct., *A. varians*
   auct.) - 5
trichomanes L. (*A. melanocaulon* Willd., *A. trichomanes*
   subsp. *bivalens* D.E. Mey.) - 1, 2, 6
- subsp. *bivalens* D.E. Mey. = A. trichomanes
- subsp. **inexpectans** Lovis (*A. melanocaulon* Willd. subsp.
   inexpectans (Lovis) A. & D. Löve) - 1
- subsp. **quadrivalens** D.E. Mey. - 1, 2, 6
varians auct. = A. tenuicaule
vidalii Franch. & Savat. = Athyrium vidalii (Athyriaceae)
viride Huds. - 1, 2, 3, 4, 5, 6
woronowii Christ (*A. pseudolanceolatum* Fomin) - 2

Camptosorus Link

sibiricus Rupr. (*Asplenium ruprechtii* Kurata) - 4, 5

Ceterach Willd.

officinarum Willd. (*Asplenium ceterach* L.) - 1, 2, 6

Phyllitis Hill

japonica Kom. - 5
scolopendrium (L.) Newm. (*Asplenium scolopendrium* L.) -
   1, 2, 6

ASTERACEAE Dumort. (COMPOSITAE Giseke)
Acanthocephalus Kar. & Kir.

amplexifolius Kar. & Kir. - 6
benthamianus Regel - 6

Acantholepis Less.

orientalis Less. - 2, 6

Acanthoxanthium (DC.) Fourr. = Xanthium

spinosum (L.) Fourr. = Xanthium spinosum

Achillea L.

acuminata (Ledeb.) Sch. Bip. = Ptarmica acuminata
alpina L. = Ptarmica alpina
- subsp. *alpina* var. *discoidea* (Regel) Kitam. = Ptarmica ptarmicoides
- subsp. *camtschatica* (Heimerl) Kitam. = Ptarmica camtschatica
- subsp. *japonica* (Heimerl) Kitam. = Ptarmica japonica
- var. *mongolica* (Fisch. ex Spreng.) Kitag. = Ptarmica alpina
apiculata Orlova - 1
asiatica Serg. (*A. setacea* Waldst. & Kit. subsp. *asiatica* (Serg.) Worosch.) - 1, 3, 4, 5
asplenifolia auct. = A. distans
biebersteinii Afan. - 1, 2, 6
birjuczensis Klok. - 1
biserrata Bieb. = Ptarmica biserrata
borealis Bong. - 5
bucharica C. Winkl. - 6
camtschatica (Rupr. ex Heimerl) Botsch. = Ptarmica camtschatica
carpatica Blocki ex Dubovik (*A. sudetica* auct.) - 1
cartilaginea Ledeb. ex Reichenb. = Ptarmica cartilaginea
coarctata Poir. - 1
collina J. Beck. ex Reichenb. (*A. millefolium* L. subsp. *collina* (J. Beck. ex Reichenb.) Weiss, *A. submillefolium* Klok. & Krytzka, p.max.p. excl. typo) - 1
cuneatiloba Boiss. & Buhse - 2
distans Waldst. & Kit. ex Willd. (*A. tanacetifolia* All. subsp. *distans* (Waldst. & Kit. ex Willd.) Gajic, *A. asplenifolia* auct.) - 1
- subsp. *stricta* (Schleich. ex Gremli) Janch. = A. stricta
- subsp. *tanacetifolia* (All.) Janch. = A. subtanacetifolia
euxina Klok. - 1
filipendulina Lam. - 1, 2
glaberrima Klok. - 1
griseo-virens Albov = Ptarmica griseo-virens
x **illiczevskyi** Tzvel. - A. micrantha Willd. x A. collina J. Beck. ex Reichenb. - 1
impatiens L. = Ptarmica impatiens
inundata Kondr. - 1, 3, 4(alien)
japonica Heimerl = Ptarmica japonica
kermanica Gand. = A. wilhelmsii
kitaibelliana Soo = A. ochroleuca
latiloba Ledeb. - 2
ledebourii Heimerl = Ptarmica ledebourii
leptophylla Bieb. - 1
x **leptophylloides** Tzvel. - A. leptophylla Bieb. x A. setacea Waldst. & Kit. - 1
lingulata Waldst. & Kit. = Ptarmica lingulata
macrocephala Rupr. = Ptarmica macrocephala
micrantha Willd. - 1, 2, 3, 6
micranthoides Klok. - 1
millefolium L. (*A. submillefolium* Klok. & Krytzka, p.p. incl. typo) - 1, 3, 5(alien)
- subsp. *collina* (J. Beck. ex Reichenb.) Weiss = A. collina
- subsp. *pannonica* (Scheele) Hayek = A. pannonica
- *nigrescens* E. Mey. = A. nigrescens
mongolica Fisch. ex Spreng. = Ptarmica alpina
neilreichii A. Kerner (*A. nobilis* L. subsp. *neilreichii* (A. Kerner) Form. comb. superfl., *A. nobilis* subsp. *neilreichii* (A. Kerner) Takht. comb. superfl., *A. nobilis* subsp. *neilreichii* (A. Kerner) Velen.) - 1
nigrescens (E. Mey.) Rydb. (*A. millefolium* L. *nigrescens* E. Mey.) - 1, 3, 4, 5
nobilis L. - 1, 2, 3, 6
- subsp. *neilreichii* (A. Kerner) Form. = A. neilreichii

- subsp. *neilreichii* (A. Kerner) Takht. = A. neilreichii
- subsp. *neilreichii* (A. Kerner) Velen. = A. neilreichii
ochroleuca Ehrh. (*A. kitaibelliana* Soo) - 1
oxyloba (DC.) Sch. Bip. subsp. *schurii* (Sch.Bip.) Heimerl = Ptarmica tenuifolia
pannonica Scheele (*A. millefolium* L. subsp. *pannonica* (Scheele) Hayek) - 1
ptarmica L. = Ptarmica vulgaris
- subsp. *cartilaginea* (Ledeb. ex Reichenb.) Heimerl = Ptarmica cartilaginea
ptarmicifolia (Willd.) Rupr. ex Heimerl = Ptarmica ptarmicifolia
ptarmicoides Maxim. = Ptarmica ptarmicoides
sachokiana Sosn. = Ptarmica sachokiana
salicifolia Bess. = Ptarmica salicifolia
- subsp. *septentrionalis* (Serg.) Uotila = Ptarmica septentrionalis
santolina auct. = A. wilhelmsii
schischkinii Sosn. - 2(?)
schurii Sch.Bip. = Ptarmica tenuifolia
sedelmeyeriana Sosn. = Ptarmica sedelmeyeriana
septentrionalis (Serg.) Botsch. = Ptarmica septentrionalis
setacea Waldst. & Kit. - 1
- subsp. *asiatica* (Serg.) Worosch. = A. asiatica
stepposa Klok. & Krytzka -1, 3, 6
stricta Schleich. ex Gremli (*A. distans* Waldst. & Kit. subsp. *stricta* (Schleich. ex Gremli) Janch.) - 1
x **submicrantha** Tzvel. - A. micrantha Willd. x A. setacea Waldst. & Kit. - 1
submillefolium Klok. & Krytzka = A. collina and A. millefolium
subtanacetifolia Tzvel. (*A. tanacetifolia* All. 1785, non Mill. 1768, *A. distans* Waldst. & Kit. subsp. *tanacetifolia* (All.) Janch.) - 1
x **subtaurica** Tzvel. - A. taurica Bieb. x A. setacea Waldst. & Kit. - 1
sudetica auct. = A. carpatica
tanacetifolia All. = A. subtanacetifolia
- subsp. *distans* (Waldst. & Kit. ex Willd.) Gajic = A. distans
taurica Bieb. - 1, 6
tenuifolia Lam. - 2
vermicularis Trin. - 2
wilhelmsii C. Koch (*A. kermanica* Gand., *A. santolina* auct.) - 2, 6

Achyrophorus auct. = Trommsdorfia

ciliatus (Thunb.) Sch. Bip. = Trommsdorfia ciliata
crepidioides (Miyabe & Kudo) Kitag. = Trommsdorfia crepidioides
maculatus (L.) Scop. = Trommsdorfia maculata
uniflorus (Vill.) Bluff & Fingerh. = Trommsdorfia uniflora

Acosta Adans. = Centaurea

aemulans (Klok.) Holub = Centaurea aemulans
aggregata (Fisch. & C.A. Mey.) Sojak = Centaurea aggregata
arenaria (Bieb.) Sojak = Centaurea arenaria
besseriana (DC.) Sojak = Centaurea besseriana
biebersteinii (DC.) Dostal = Centaurea biebersteinii
borysthenica (Grun.) Sojak = Centaurea borysthenica
diffusa (Lam.) Sojak = Centaurea diffusa
micranthos (S.G. Gmel.) Sojak = Centaurea biebersteinii
odessana (Prod.) Sojak = Centaurea odessana
ovina (Pall. ex Willd.) Sojak = Centaurea ovina
pseudomaculosa (Dobrocz.) Sojak = Centaurea pseudomaculosa

rhenana (Boreau) Sojak = Centaurea rhenana
savranica (Klok.) Holub = Centaurea savranica
squarrosa (Willd.) Sojak = Centaurea squarrosa
steveniana (Klok.) Holub = Centaurea steveniana

Acrocentron Cass. = Centaurea

scabiosa (L.) A. & D. Löve = Centaurea scabiosa

Acroptilon Cass.

australe Iljin (*A. repens* (L.) DC. subsp. *australe* (Iljin) Rech.fil.) - 6
repens (L.) DC. - 1, 2, 3, 6
- subsp. *australe* (Iljin) Rech. fil. = A. australe

Adenocaulon Hook.

adhaerescens Maxim. (*A. himalaicum* auct.) - 1(alien), 5
himalaicum auct. = A. adhaerescens

Adenostyles Cass. (*Pojarkovia* Asker. p.p. excl. typo)

alliariae (Gouan) A. Kerner - 1
macrophylla (Bieb.) Czer. comb. nova (*Cacalia macrophylla* Bieb. 1808, Fl. Taur.-Cauc. 2 : 286, non *Senecio macrophyllus* DC. 1838; *Adenostyles rhombifolia* (Adams) M. Pimen., *Pojarkovia macrophylla* (Bieb.) Asker., *Senecio rhombifolius*(Adams) Sch.Bip.) - 2
platyphylloides (Somm. & Levier) Czer. (*Senecio platyphylloides* Somm. & Levier, *Adenostyles rhombifolia* (Adams) M. Pimen. subsp. *platyphylloides* (Somm. & Levier) M. Pimen., *Pojarkovia platyphylloides* (Somm. & Levier) Asker.) - 2
rhombifolia (Adams) M.Pimen. = A. macrophylla
- subsp. *platyphylloides* (Somm. & Levier) M. Pimen. = A. platyphylloides
similiflora (Kolak.) Konechn. (*Senecio similiflorus* Kolak.) - 2

Adventina Rafin. = Galinsoga

ciliata Rafin. = Galinsoga ciliata

Aetheopappus Cass.

caucasicus Sosn. - 2
pulcherrimus (Willd.) Cass. - 2
vvedenskii (Sosn.) Sosn. - 2

*Ageratum L.

*houstonianum Mill. - 1, 2, 6

Ajania Poljak.

abolinii Kovalevsk. - 6
fastigiata (C. Winkl.) Poljak. - 6
fruticulosa (Ledeb.) Poljak. - 3, 4, 6
gracilis (Hook. fil. & Thoms.) Poljak. - 6
kokanica (Krasch.) Tzvel. = Tanacetum kokanicum
korovinii Kovalevsk. - 6
manshurica Poljak. (*Chrysanthemum manshuricum* (Poljak.) Kitag.) - 5
pallasiana (Fisch. ex Bess.) Poljak. (*Dendranthema pallasianum* (Fisch. ex Bess.) Worosch.) - 5
scharnhorstii (Regel & Schmalh.) Tzvel. - 6
tibetica (Hook. fil. & Thoms.) Tzvel. - 6
trilobata Poljak. - 6

Alfredia Cass.

acantholepis Kar. & Kir. - 6
cernua (L.) Cass. - 3, 6
fetissowii Iljin - 6
integrifolia (Iljin) Tuljaganova (*A. nivea* Kar. & Kir. f. *integrifolia* Iljin) - 6
nivea Kar. & Kir. - 6
- f. *integrifolia* Iljin = A. integrifolia

Amberboa (Pers.) Less.

amberboi (L.) Tzvel. - 2, 6
bucharica Iljin - 6
glauca (Willd.) Grossh. - 2
moschata (L.) DC. - 2
- subsp. *sosnovskyi* (Iljin) Takht. = A. sosnovskyi
nana (Boiss.) Iljin - 2
sosnovskyi Iljin (*A. moschata* (L.) DC. subsp. *sosnovskyi* (Iljin) Takht.) - 2
turanica Iljin - 1, 2, 3, 6

Amblyocarpum Fisch. & C.A. Mey.

inuloides Fisch. & C.A. Mey. - 2

Ambrosia L.

aptera DC. - 1, 2
artemisiifolia L. - 1, 2, 5, 6
coronopifolia auct. = A. psylostachya
psylostachya DC. (*A. coronopifolia* auct.) - 1, 2
trifida L. - 1, 2, 5

Amphoricarpos Vis.

elegans Albov - 2

Anacantha (Iljin) Sojak (*Modestia* Charadze & Tamamsch. 1956, non Modesta Rafin. 1838)

darwasica (C. Winkl.) Sojak (*Modestia darwasica* (C. Winkl.) Charadze & Tamamsch.) - 6
jucunda (C. Winkl.) Sojak (*Modestia jucunda* (C.Winkl.) Charadze & Tamamsch.) - 6
mira (Iljin) Sojak (*Modestia mira* (Iljin) Charadze & Tamamnsch.) - 6

Anacyclus L.

ciliatus Trautv. - 2
clavatus (Desv.) Pers. - 1(alien)
*officinarum Hayne - 1

Anaphalis DC.

darvasica Boriss. - 6
depauperata Boriss. - 6
garanica Boriss. - 6
latifolia Kinz. & Vainberg - 6
margaritacea (L.) A. Gray - 5
possietica Kom. = A. pterocaulon
pterocaulon (Franch. & Savat.) Maxim. (*Gnaphalium pterocaulon* Franch. & Savat., *Anaphalis possietica* Kom.) - 5
racemifera Franch. - 6
roseoalba Krasch. - 6
sarawschanica (C. Winkl.) B. Fedtsch. - 6
scopulosa Boriss. - 6
subtilis Kinz. & Vainberg - 6
tenuicaulis Boriss. - 6

velutina Krasch. - 6
virgata Thoms. - 6

Ancathia DC.

igniaria (Spreng.) DC. - 2, 3, 4, 6

Antennaria Gaertn.

alaskana Malte (*A. friesiana* (Trautv.) Ekman subsp. *alaskana* (Malte) Hult.) - 5
alpina (L.) Gaertn. - 1
- var. *compacta* (Malte) Welsh = A. compacta
angustata Greene (*A. komarovii* Juz., *A. monocephala* DC. subsp. *angustata* (Greene) Hult.) -5
atriceps Fern. - 5
carpatica (Wahlenb.) Bluff & Fingerh. - 1
caucasica Boriss. - 2
compacta Malte (*A. alpina* (L.) Gaertn. var. *compacta* (Malte) Welsh, *A. friesiana* (Trautv.) Ekman subsp. *compacta* (Malte) Hult.) - 5
dioica (L.) Gaertn. - 1, 2, 3, 4, 5, 6
dioiciformis Kom. - 5
- subsp. **paucicapitata** Ju. Kozhevn. - 5
ekmaniana A. Pors. = A. friesiana
friesiana (Trautv.) Ekman (*A. ekmaniana* A. Pors.) - 4, 5
- subsp. *alaskana* (Malte) Hult. = A. alaskana
- subsp. **beringensis** Petrovsky - 5
- subsp. *compacta* (Malte) Hult. = A. compacta
- subsp. **pseudoisolepis** Petrovsky - 5
komarovii Juz. = A. angustata
lanata (Hook.) Greene (*A. villifera* Boriss.) - 1, 3, 4
monocephala DC. - 5
- subsp. *angustata* (Greene) Hult. = A. angustata
porsildii Ekman - 1
pseudoarenicola Petrovsky - 5
rubicunda C. Koch = Helichrysum rubicundum
villifera Boriss. = A. lanata

Anthemis L. (*Cota* J. Gay)

abagensis Fed. - 2
x **adulterina** Wallr. - A. arvensis L. x A. tinctoria L.
■ albida Boiss. (*A. cretica* L. subsp. *albida* (Boiss.) Grierson)
altissima L. (*Cota altissima* (L.) J. Gay) - 1, 2, 6
anahytae Woronow ex Sosn. = A. iberica
■ anatolica Boiss. (*A. cretica* L. subsp. *anatolica* (Boiss.) Grierson)
arvensis L. - 1, 2
atropatana Iranshahr - 2
austriaca Jacq. (*Cota austriaca* (Jacq.) J. Gay) - 1, 2, 6
biebersteiniana (Adams) Boiss. var. *pectinata* Boiss. = A. caucasica
x **bollei** Sch. Bip. - A. cotula L. x A. tinctoria L. - 1
■ calcarea Sosn.
candidissima Willd. ex Spreng. - 2, 6
carpatica Waldst. & Kit. ex Willd. (*A. cretica* L. subsp. *carpatica* (Willd.) Grierson) - 1
caucasica Chandjian (*A. biebersteiniana* (Adams.) Boiss. var. *pectinata* Boiss., *A. marschalliana* Willd. subsp. *pectinata* (Boiss.) Grierson) - 2
coelopoda Boiss. - 2, 6
- var. **bourgaei** Boiss. - 2, 6
cotula L. - 1, 2
cretacea Zefir. = A. monantha
cretica L. subsp. *albida* (Boiss.) Grierson = A. albida
- subsp. *anatolica* (Boiss.) Grierson = A. anatolica

- subsp. *carpatica* (Willd.) Grierson = A. carpatica
- subsp. *iberica* (Bieb.) Grierson = A. iberica
- subsp. *saportana* (Albov) Chandjian = A. saportana
■ debilis Fed.
deserticola Krasch. & M. Pop. - 6
dubia Stev. (*Cota dubia* (Stev.) Holub) - 1
dumetorum Sosn. (*Cota dumetorum* (Sosn.) Holub) - 1, 2
emiliae Sosn. - 2
euxina Boiss. (*A. tinctoria* L. var. *euxina* (Boiss.) Grierson, *Cota euxina* (Boiss.) Holub) - 2
fruticulosa Bieb. - 2
grossheimii Sosn. = A. haussknechtii
haussknechtii Boiss. & Reut. (*A. grossheimii* Sosn., *A. karabaghensis* A.D. Mikheev) - 2
hirtella C. Winkl. - 6
iberica Bieb. (*A. anahytae* Woronow ex Sosn., *A. cretica* L. subsp. *iberica* (Bieb.) Grierson, *A. ptarmiciformis* C. Koch, *A. tempskyana* Freyn & Sint.) - 2
jailensis Zefir. (*Cota jailensis* (Zefir.) Holub) - 1
karabaghensis A.D. Mikheev = A. haussknechtii
linczevskyi Fed. - 6
> Probably, a valid name for this taxon should be **A. khorasanica** Rech. fil. (A. triumfetii (L.) All. subsp. khorasanica (Rech. fil.) Iranshahr).
lithuanica (DC.) Trautv. - 1
macroglossa Somm. & Levier - 2
maris-nigri Fed. - 2
markhotensis Fed. - 2
marschalliana Willd. (*A. marschalliana* subsp. *biebersteiniana* (Adams) Grierson) - 2
- subsp. *biebersteiniana* (Adams) Grierson = A. marschalliana
- subsp. *pectinata* (Boiss.) Grierson = A. caucasica
- subsp. *sosnovskyana* (Fed.) Grierson = A. sosnovskyana
melanoloma Trautv. (*Cota melanoloma* (Trautv.) Holub) - 2
microcephala (Schrenk) B. Fedtsch. - 6
moghanica Iranshahr - 2
monantha Willd. (*A. cretacea* Zefir., *A. parviceps* Dobrocz. & Fed. ex Klok., *Cota cretacea* (Zefir.) Holub) - 1
odontostephana Boiss. - 6
parnassica (Boiss. & Heldr.) R. Fernandes (*Cota parnassica* Boiss. & Heldr.) - 1(?)
parviceps Dobrocz. & Fed. ex Klok. = A. monantha
ptarmiciformis C. Koch = A. iberica
rigescens Willd. (*Cota rigescens* (Willd.) Holub) - 2
ruthenica Bieb. - 1, 2
saguramica Sosn. - 2
saportana Albov (*A. cretica* L. subsp. *saportana* (Albov) Chandjian) - 2
schischkiniana Fed. - 2
sosnovskyana Fed. (*A. marschalliana* Willd. subsp. *sosnovskyana* (Fed.) Grierson) - 2
sterilis Stev. - 1
subtinctoria Dobrocz. (*A. tinctoria* L. subsp. *subtinctoria* (Dobrocz.) Soo, *A. zephyrovii* Dobrocz., *Cota tinctoria* (L.) J. Gay subsp. *subtinctoria* (Dobrocz.) Holub) - 1, 2, 3, 4, 6
talyscensis Fed. - 2
tempskyana Freyn & Sint. = A. iberica
tenuifolia Schur = Ptarmica tenuifolia
tinctoria L. (*Cota tinctoria* (L.) J. Gay) - 1
- subsp. *subtinctoria* (Dobrocz.) Soo = A. subtinctoria
- var. *euxina* (Boiss.) Grierson = A. euxina
tranzscheliana Fed. - 1
trotzkiana Claus - 1, 6
wiedemanniana Fisch. & C.A. Mey. (*Cota wiedemanniana* (Fisch. & C.A. Mey.) Holub) - 2

woronowii Sosn. (*Cota woronowii* (Sosn.) Holub) - 2
zephyrovii Dobrocz. = A. subtinctoria
zyghia Woronow (*Cota zyghia* (Woronow) Holub) - 2

Anura (Juz.) Tscherneva

pallidivirens (Kult.) Tscherneva - 6

Apargia Scop. = Leontodon

pratensis Link = Leontodon autumnalis

Aposeris Cass.

foetida (L.) Less. - 1

Arctanthemum (Tzvel.) Tzvel.

arcticum (L.) Tzvel. (*Dendranthema arcticum* (L.) Tzvel.) -
    5
- subsp. *kurilense* (Tzvel.) Tzvel. = A. kurilense
- subsp. *polare* (Hult.) Tzvel. = A. hultenii
hultenii (A. & D. Löve) Tzvel. (*A. arcticum* (L.) Tzvel.
    subsp. *polare* (Hult.) Tzvel., *Dendranthema arcticum*
    (L.) Tzvel. subsp. *polare* (Hult.) Heywood, *D. hultenii*
    (A. & D. Löve) Tzvel.) - 1, 3, 4, 5
integrifolium (Richards.) Tzvel. = Hulteniella integrifolia
kurilense (Tzvel.) Tzvel. (*A. arcticum* (L.) Tzvel. subsp.
    kurilense (Tzvel.) Tzvel., *Dendranthema kurilense*
    Tzvel., *Chrysanthemum arcticum* L. subsp. *maekawa-
    num* auct.) - 5

Arctium L. (*Lappa* Scop.)

x **ambiguum** (Celak.) Nym. - A. lappa L. x A. tomentosum
    Mill. - 1
chaorum Klok. = A. lappa
x **cimbricum** (E. Krause) Hayek (*Lappa* x *cimbrica* E.
    Krause). - A. lappa L. x A. nemorosum Lej. - 1
glabrescens Klok. = A. nemorosum
lappa L. (*A. chaorum* Klok.) - 1, 2, 3, 4, 5, 6
x **leiobardanum** Juz. & C. Serg. - A. leiospermum Juz. & C.
    Serg. x A. tomentosum Mill. - 4, 6
    The name of this hybrid has not been validly published.
leiospermum Juz. & C. Serg. - 3, 4, 6
leptophyllum Klok. = A. tomentosum
x **maassii** (M. Schultze) Rouy (*Lappa* x *maassii* M.
    Schulze). - A. minus (Hill) Bernch. x A. nemorosum
    Lej. - 1
minus (Hill) Bernh. - 1, 2, 3, 5
- subsp. **pubens** (Bab.) Aren. (*A. pubens* Bab.) - 1, 2
x **mixtum** (Simonk.) Nym. - A. minus (Hill) Bernh. x A.
    tomentosum Mill. - 1
nemorosum Lej. (*A. glabrescens* Klok.) - 1, 2
x **neumannii** Rouy. - A. nemorosum Lej. x A. tomentosum
    Mill. - 1
x **nothum** (Ruhm.) Weiss. - A. lappa L. x A. minus (Hill)
    Bernh. - 1
palladinii (Marc.) Grossh. (*A. tomentosum* Mill. subsp.
    palladinii (Marc.) Takht.) - 2
platylepis (Boiss. & Bal.) Sosn. ex Grossh. - 2
pubens Bab. = A. minus subsp. pubens
radula Juz. & C. Serg. - 2
sardaimionense Rassulova & B. Scharipova - 6
tomentosum Mill. (*A. leptophyllum* Klok.) - 1, 2, 3, 4, 5, 6
- subsp. *palladinii* (Marc.) Takht. = A. palladinii

Arctogeron DC.

gramineum (L.) DC. - 3, 4

*Argyranthemum Sch. Bip.

*****frutescens** (L.) Sch. Bip. - 1

Arnica L.

alpina (L.) Olin & Ladau = A. fennoscandica
angustifolia auct. fl. URSS = A. iljinii
angustifolia Vahl subsp. *alpina* (L.) I.K. Ferguson = A.
    fennoscandica
- subsp. *iljinii* (Maguire) I.K. Ferguson = A. iljinii
attenuata auct. fl. URSS = A. iljinii
fennoscandica Jurtz. & Korobkov (*A. alpina* (L.) Olin &
    Ladau, 1799, non *A. alpina* Salisb. 1796, *A. angustifolia*
    Vahl subsp. *alpina* (L.) I.K. Ferguson) - 1
frigida C.A. Mey. ex Iljin (*A. griscomii* Fern. subsp. *frigida*
    (Iljin) S.J. Wolf, *A. lessingii* auct. fl. URSS) - 4, 5
griscomii Fern. subsp. *frigida* (Iljin) S.J. Wolf = A. frigida
iljinii (Maguire) Iljin (*A. angustifolia* Vahl subsp. *iljinii*
    (Maguire) I.K. Ferguson, *A. angustifolia* auct. fl.
    URSS, *A. attenuata* auct. fl. URSS) - 1, 3, 4
intermedia Turcz. - 4, 5
lessingii auct. fl. URSS = A. frigida
montana L. - 1
porsildiorum Boivin - 5
sachalinensis (Regel) A. Gray - 5
unalaschcensis Less. - 5

Arnoseris Gaertn.

minima (L.) Schweigg. & Koerte - 1

Artemisia L. (*Oligosporus* Cass., *Seriphidium* (Bess.) Poljak.)

abrotanum L. (*A. elatior* Klok., *A. procera* Willd.) - 1, 2, 3, 6
absinthium L. - 1, 2, 3, 4, 6
adamsii Bess. - 4
albicerata Krasch. = A. arenaria
albida auct. = A. compacta
alcockii Pamp. - 6
alpina Pall. ex Willd. = A. caucasica
altaiensis Krasch. - 3, 4
amoena Poljak. (*Seriphidium amoenum* (Poljak.) Poljak.) -
    6
anethifolia Web. - 3, 4
annua L. - 1, 2, 3, 4, 5, 6
aralensis Krasch. (*Seriphidium aralense* (Krasch.) Poljak.) -
    6
araratica Krasch. = A. marschalliana
araxina Takht. - 2
arctica Less. - 4, 5
- subsp. **ehrendorferi** Korobkov - 4, 5
- subsp. **ochotensis** (Bess.) Leonova (*A. chamissoniana*
    Bess. var. *ochotensis* Bess.) - 4, 5
arctisibiria Korobkov (*A. negleta* Leonova, 1971, non
    Spreng. 1809, *A. taimyrensis* Krasch. ex Ameljczenko,
    nom. invalid.) - 4, 5
arenaria DC. (*A. albicerata* Krasch., *Oligosporus albiceratus*
    (Krasch.) Poljak., *O. arenarius* (DC.) Poljak.) - 1, 2, 6
arenicola Krasch. - 6
argentata Klok. = A. caucasica
argyi Levl. & Vaniot - 1(alien),5
argyrophylla Ledeb. - 3, 4
armeniaca Lam. - 1, 2, 3
aschurbajewii C. Winkl. - 6
aurata Kom. - 5
austriaca Jacq. (*A. repens* Pall. ex Willd.) - 1, 2, 3, 5, 6
avarica Minatullaev - 2

badhysi Krasch. & Lincz. ex Poljak. (*Seriphidium badhysi* Krasch. & Lincz. ex Poljak.) - 6

balchanorum Krasch. (*Seriphidium balchanorum* (Krasch.) Poljak.) - 6

baldshuanica Krasch. & Zapr. (*Seriphidium baldshuanicum* (Krasch. & Zapr.) Poljak.) - 6

bargusinensis Spreng. (*A. borealis* Pall. var. *bargusinensis* (Spreng.) Worosch., *Oligosporus bargusinensis* (Spreng.) Poljak.) - 6

bejdemaniae Leonova - 4

biennis Willd. - 1(alien)

borealis Pall. (*A. campestris* L. subsp. *borealis* (Pall.) H.M. Hall & Clements, *A. remosa* Sugaw., *Oligosporus borealis* (Pall.) Poljak.) - 1, 3, 4, 5
- subsp. *bottnica* (Lundstr. ex Kindb.) Hult. = A. bottnica
- subsp. *ledebourii* (Bess.) Ameljczenko, p.p. = A. borealis subsp. mertensii
- subsp. **mertensii** (Bess.) Ameljczenko (*A. borealis* var. *mertensii* Bess., *A. borealis* var. *ledebourii* Bess., *A. borealis* subsp. *ledebourii* (Bess.) Ameljczenko,p.p.) - 5
- subsp. **richardsoniana** (Bess.) Korobkov (*A. richardsoniana* Bess., *A. henriettae* Krasch., *Oligosporus henriettae* (Krasch.) Poljak.) - 4, 5
- var. *bargusinensis* (Spreng.) Worosch. = A. bargusinensis
- var. *ledebourii* Bess. = A. borealis subsp. mertensii
- var. *mertensii* Bess. = A. borealis subsp. mertensii

borotalensis Poljak. - 6

boschniakiana (Bess.) DC. = A. santonica

bottnica Lundstr. ex Kindb. (*A. borealis* Pall. subsp. *bottnica* (Lundstr. ex Kindb.) Hult., *A. campestris* L. subsp. *bottnica* (Lundstr. ex Kindb.) Tutin) - 1

caespitosa Ledeb. - 3, 4

camelorum Krasch. (*Seriphidium camelorum* (Krasch.) Poljak.) - 6

campestris L. (*A. dniproica* Klok. p.p., *Oligosporus campestris* (L.) Cass.) - 1
- subsp. *borealis* (Pall.) H.M. Hall & Clements = A. borealis
- subsp. *bottnica* (Lundstr. ex Kindb.) Tutin = A. bottnica

capillaris Thunb. (*Oligosporus capillaris* (Thunb.) Poljak.) - 5

caucasica Willd. (*A. alpina* Pall. ex Willd., *A. argentata* Klok., *A. lanulosa* Klok., *A. pedemontana* auct.) - 1, 2

chamaemelifolia Vill. - 2

chamissoniana Bess. f. *ochotensis* Bess. = A. arctica subsp. ochotensis

cina Berg ex Poljak. (*Seriphidium cinum* (Berg ex Poljak.) Poljak.) - 6

ciniformis Krasch. & M. Pop. ex Poljak. (*Seriphidium ciniforme* (Krasch. & M. Pop. ex Poljak.) Poljak.) - 6

coarctata Forsell. = A. vulgaris

codringtonii Rech.fil. = Turaniphytum codringtonii

commutata Bess. (*A. pubescens* Ledeb., *Oligosporus commutatus* (Bess.) Poljak.) - 1, 3, 4, 5
- subsp. *dolosa* (Krasch.) Ameljczenko = A. dolosa
- var. *dolosa* (Krasch.) Poljak. = A. dolosa

compacta Fisch. ex DC. (*Seriphidium compactum* (Fisch. ex DC.) Poljak., *Artemisia albida* auct.) - 1, 3, 4, 6

cretacea Kotov = A. nutans

cuspidata Krasch. - 4

czukavinae Filat. - 6

daghestanica Krasch. & A. Poretzky (*Oligosporus daghestanicus* (Krasch. & A. Poretzky) Poljak.) - 2

demissa auct. = A. pewzowii

depauperata Krasch. - 3, 6

deserti Krasch. (*Seriphidium deserti* (Krasch.) Poljak.) - 6

desertorum Spreng. (*Artemisia japonica* Thunb. f. *manshur-ica* Kom., *A. japonica* var. *manshurica* (Kom.) Kitag., *A. manshurica* (Kom.) Kom., *Oligosporus desertorum* (Spreng.) Poljak.) - 4, 5

diffusa Krasch. ex Poljak. - 6

dimoana M. Pop. (*Oligosporus dimoanus* (M. Pop.) Poljak.) - 6

dniproica Klok. p.p. = A. campestris and A. marschalliana

dolosa Krasch. (*A. commutata* Bess. subsp. *dolosa* (Krasch.) Ameljczenko, *A. commutata* var. *dolosa* (Krasch.) Poljak.) - 3, 4

dracunculiformis Krasch. = A. dracunculus

dracunculus L. (*A. dracunculiformis* Krasch., *Oligosporus dracunculiformis* (Krasch.) Poljak., *O. dracunculus* (L.) Poljak.) - 1, 2, 3, 4, 5, 6

dubia Wall. (*A. umbrosa* (Bess.) Pamp.) - 1(alien),4,5

dubjanskyana Krasch. ex Poljak. (*Seriphidium dubjanskya-num* (Krasch. ex Poljak.) Poljak.) - 6

dudinensis Ameljczenko = A. leucophylla var. dudinensis

dumosa Poljak. (*Seriphidium dumosum* (Poljak.) Poljak., *Artemisia sieberi* auct.) - 6

dzevanovskyi Leonova - 1

elatior Klok. = A. abrotanum

elongata Filat. & Ladygina - 6

eremophila Krasch. & Butk. ex Poljak. - 6

feddei Levl. & Vaniot (*A. lavandulifolia* DC. 1838, non Salisb. 1796) - 5

fedtschenkoana Krasch. (*Seriphidium fedtschenkoanum* (Krasch.) Poljak.) - 6

ferganensis Krasch. ex Poljak. (*Seriphidium ferganense* (Krasch. ex Poljak.) Poljak.) - 6

flava Jurtz. - 5

fragrans Willd. = A. lerchiana
- subsp. *gurganica* Krasch. = A. gurganica

freyniana (Pamp.) Krasch. - 5

frigida Willd. - 1, 3, 4, 5, 6

fulvella Filat. & Ladygina - 6

furcata Bieb. (*A. trifurcata* Steph. ex Spreng.) - 4, 5
- subsp. **flavida** Worosch. & Neczaev - 5
- subsp. *insulana* (Krasch.) Worosch. = A. insulana

galinae Ikonn. - 6

gigantea Kitam. = A. montana

glabella Kar. & Kir. - 3, 6

glanduligera Krasch. ex Poljak. (*Seriphidium glanduligerum* (Krasch. ex Poljak.) Poljak.) - 6

glauca Pall. ex Willd. (*Oligosporus dracunculus* (L.) Poljak. subsp. *glaucus* (Pall. ex Willd.) A. & D. Löve, *O. glaucus* (Pall. ex Willd.) Poljak.) - 1, 3, 4, 6

glaucina Krasch. ex Poljak. - 6

globosa Krasch. - 4

globularia Bess. - 5

glomerata Ledeb. - 4, 5

gmelinii Web. - 3, 4, 5, 6
- subsp. **scheludjakoviae** Korobkov - 3, 4, 5

gnaphalodes Nutt. (*A. ludoviciana* Nutt. subsp. *gnaphalodes* (Nutt.) A. & D. Löve) - 1(alien)

gorjaevii Poljak. - 6

gracilescens Krasch. & Iljin (*Seriphidium gracilescens* (Krasch. & Iljin) Poljak.) - 1, 3, 6

graveolens Minatullaev = A. taurica

gurganica (Krasch.) Filat. (*A. fragrans* Willd. subsp. *gurganica* Krasch.) - 6

gypsacea Krasch., M. Pop. & Lincz. ex Poljak. (*Seriphidium gypsaceum* (Krasch., M. Pop. & Lincz. ex Poljak.) Poljak.) - 6

halodendron Turcz. ex Bess. (*Oligosporus halodendron* (Turcz. ex Bess.) Poljak.) - 4

halophila Krasch. (*Seriphidium halophilum* (Krasch.) Poljak.) - 6

hedinii Ostenf. - 6
henriettae Krasch. = A. borealis subsp. richardsoniana
heptapotamica Poljak. (*Seriphidium heptapotamicum* (Poljak.) Ling & Y.R. Ling) - 6
hippolyti Butk. - 6
hololeuca Bieb. ex Bess. - 1
hulteniana Worosch. - 5
hultenii Maximova = A. tilesii
incana (L.) Druce - 2
insulana Krasch. (*A. furcata* Bieb. subsp *insulana* (Krasch.) Worosch.) - 5
insularis Kitam. - 5
integrifolia L. (*A. komarovii* Poljak.) - 3, 4, 5
- f. *subulata* (Nakai) Kitag. = A. stenophylla
issykkulensis Poljak. (*Seriphidium issykkulense* (Poljak.) Poljak.) - 6
jacutica Drob. - 4
japonica Thunb. (*Oligosporus japonicus* (Thunb.) Poljak.) - 5
- var. *manshurica* (Kom.) Kitag. = A. desertorum
- f. *manshurica* Kom. = A. desertorum
juncea Kar. & Kir. (*Seriphidium junceum* (Kar. & Kir.) Poljak., *S. junceum* var. *macrosciadium* (Poljak.) Ling & Y.R. Ling) - 6
karatavica Krasch. & Abol. ex Poljak. (*Seriphidium karatavicum* (Krasch. & Abol. ex Poljak.) (Ling & Y.R Ling) - 6
karavajevii Leonova - 4
kasakorum (Krasch.) Pavl. = A. scopiformis
kaschgarica Krasch. (*Seriphidium kaschgaricum* (Krasch.) Poljak.) - 6
keiskeana Miq. - 5
kelleri Krasch. (*Oligosporus kelleri* (Krasch.) Poljak.) - 6
kemrudica Krasch. (*Seriphidium kemrudicum* (Karasch.) Poljak.) - 6
knorringiana Krasch. (*Seriphidium knorringianum* (Krasch.) Poljak.) - 6
kochiiformis Krasch. & Lincz. ex Poljak. (*Seriphidium kochiiforme* (Krasch. & Lincz. ex Poljak.) Poljak.) - 6
koidzumii Nakai - 5
komarovii Poljak. = A. integrifolia
kopetdaghensis Krasch. ex Poljak. (*Seriphidium kopetdaghense* (Krasch. ex Poljak.) Poljak.) - 6
korovinii Poljak. (*Seriphidium korovinii* (Poljak.) Poljak.) - 6
korshinskyi Krasch. ex Poljak. - 6
kruhsiana Bess. - 4, 5
- subsp. condensata Korobkov - 4
- subsp. *multisecta* (Leonova) Korobkov = A. multisecta
kulbadica Boiss. & Buhse - 6
kumykorum Minatullaev = A. santonica
kuschakewiczii C. Winkl. (*Oligosporus kuschakewiczii* (C. Winkl.) Poljak.) - 6
laciniata Willd. - 1, 3, 4, 5
laciniatiformis Kom. - 4, 5
lagocephala (Bess.) DC. - 4, 5
- subsp. lithophila Malysch. (*A. lithophila* Turcz. nom. invalid.) - 4
lagopus Fisch. ex Bess. - 4, 5
- subsp. abbreviata Krasch. ex Korobkov - 4
- subsp. jarovoi Korobkov - 4
- subsp. *triniana* (Bess.) Korobkov = A. triniana
lanulosa Klok. = A. caucasica
latifolia Ledeb. - 1, 3, 4, 5, 6
- subsp. maximowiczii (Fr. Schmidt) Worosch. = A. maximovicziana
- var. *maximowiczii* Fr. Schmidt = A. maximovicziana

lavandulifolia DC. = A. feddei
ledebouriana Bess. (*Oligosporus ledebourianus* (Bess.) Poljak.) - 4
lehmanniana Bunge (*Seriphidium lehmannianum* (Bunge) Poljak.) - 6
leontopodioides Fish. ex Bess. - 5
lerchiana Web. (*A. fragrans* Willd., *A. nachitschevanica* Rzazade, *A. pseudofragrans* Klok., *Seriphidium fragrans* (Willd.) Poljak., *S. lerchianum* (Web.) Poljak.) - 1, 2, 3, 6
lessingiana Bess. (*Seriphidium lessingianum* (Bess.) Poljak.) - 1, 3, 6
leucodes Schrenk (*Seriphidium leucodes* (Schrenk) Poljak.) - 6
leucophylla (Turcz. ex Bess.) Pamp. (*A. leucophylla* subsp. *subarctica* Ameljczenko, *A. mongolica* (Bess.) Fisch. ex Nakai var. *leucophylla* (Turcz. ex Bess.) W. Wang & H.T. Ho) - 1, 3, 4, 5
- subsp. *subarctica* Ameljczenko = A. leucophylla
- var. dudinensis (Ameljczenko) Korobkov (*A. dudinensis* Ameljczenko) - 4
- var. rotundatiloba (Jurtz.) Korobkov (*A. vulgaris* L. var. *rotundatiloba* Jurtz.) - 4
leucotricha Krash. ex Ladygina - 6
limosa Koidz. (*Oligosporus limosus* (Koidz.) Poljak.) - 5
lipskyi Poljak. (*Oligosporus lipskyi* (Poljak.) Poljak.) - 6
lithophila Turcz. = A. lagocephala subsp. lithophila
littoricola Kitam. (*Oligosporus littoricola* (Kitam.) Poljak.) - 5
lobulifolia Boiss. = A. santolina
ludoviciana Nutt. subsp. *gnaphalodes* (Nutt.) A. & D. Löve = A. gnaphalodes
macilenta (Maxim.) Krasch. (*Oligosporus macilentus* (Maxim.) Poljak.) - 5
macrantha Ledeb. - 1, 3, 4, 6
macrobotrys Ledeb. - 4
macrocephala Jacq. ex Bess. - 3, 4, 6
macrorhiza Turcz. - 4
manshurica (Kom.) Kom. = A. desertorum
maracandica Bunge - 6
maritima L. (*Seriphidium maritimum* (L.) Poljak.) subsp. humifusa (Fries ex C. Hartm.) K. Persson (*A. maritima* var. *humifusa* Fries ex C.Hartm.) - 1(alien)
- subsp. *kasakorum* Krasch. = A. scopiformis
- subsp. *monogyna* (Waldst. & Kit.) Gams = A. monogyna
- var. *humifusa* Fries ex C. Hartm. = A. maritima subsp. humifusa
marschalliana Spreng. (*A. araratica* Krasch., *A. dniproica* Klok. p.p., *A. sosnovskyi* Krasch. & Novopokr., *A. tschernieviana* Bess.) - 1, 2, 3, 4, 6
martjanovii Krasch. ex Poljak. - 4
matricarioides Less. = Lepidotheca suaveolens
maximovicziana Krasch. ex Poljak. (*A. latifolia* Ledeb. subsp. *maximowiczii* (Fr. Schmidt) Worosch., *A. latifolia* var. *maximowiczii* Fr. Schmidt) - 5
medioxima Krasch. ex Poljak. - 5
messerschmidtiana Bess. - 5
minima L. = Centipeda minima
mogoltavica Poljak. - 6
mongolica (Bess.) Fisch. ex Nakai - 4, 5
- var. *leucophylla* (Turcz. ex Bess.) W. Wang & H.T. Ho = A. leucophylla
mongolorum Krasch. (*Seriphidium mongolorum* (Krasch.) Ling & Y. R. Ling) - 3, 4
- subsp. *saissanica* Krasch. = A. saissanica
monogyna sensu Poljak. = A. santonica
■ monogyna Waldst. & Kit. (*A. maritima* L. subsp. *monogy-*

na (Waldst. & Kit.) Gams, *A. santonica* L. subsp. *monogyna* (Waldst. & Kit.) Leonova, *Seriphidium monogynum* (Waldst. & Kit.) Poljak. p.p. quoad nomen)

montana (Nakai) Pamp. (*A. gigantea* Kitam.) - 5

mucronulata Poljak. - 6

multisecta Leonova (*A. kruhsiana* Bess. subsp. *multisecta* (Leonova) Korobkov) - 4, 5

nachitschevanica Rzazade = A. lerchiana

namanganica Poljak. (*Seriphidium namanganicum* (Poljak.) Poljak.) - 6

neglecta Leonova = A. arctisibiria

nigricans Filat. & Ladygina - 6

nitrosa Web. (*Seriphidium nitrosum* (Web.) Poljak.) - 1, 3, 4, 6

norvegica Fries - 1, 3

nutans Willd. (*A. cretacea* Kotov) - 1, 3, 6

obscura Pamp. - 5

obtusiloba Ledeb. - 3, 4, 6

olchonensis Leonova - 4

olgensis (Vorobiev) Worosch. = A. pannosa

oliveriana J. Gay ex Bess. - 6

opulenta Pamp. - 5

palustris L. - 3, 4, 5

pamirica C. Winkl. (*Oligosporus pamiricus* (C. Winkl.) Poljak.) - 6

pannosa Krasch. (*A. olgensis* (Vorobiev) Worosch., *A. pannosa* var. *olgensis* Vorobiev, *Oligosporus pannosus* (Krasch.) Poljak.) - 5

- var. *olgensis* Vorobiev = A. pannosa

pauciflora Web. (*Seriphidium pauciflorum* (Web.) Poljak.) - 1, 3, 6

pedemontana auct. = A. caucasica

persica Boiss. - 6

pewzowii C. Winkl. (*A. demissa* auct.) - 6

phaeolepis Krasch. - 3, 4

polysticha Poljak. - 6

pontica L. - 1, 2, 3, 6

porrecta Krasch. ex Poljak. (*Seriphidium porrectum* (Krasch. ex Poljak.) Poljak.) - 6

prasina Krasch. ex Poljak. (*Seriphidium prasinum* (Krasch. ex Poljak.) Poljak.) - 6

praticola Klok. = A. santonica

procera Willd. = A. abrotanum

proceriformis Krasch. - 3

prolixa Krasch. ex Poljak. (*Seriphidium prolixum* (Krasch. ex Poljak.) Poljak.) - 6

pseudofragrans Klok. = A. lerchiana

pubescens Ledeb. = A. commutata

punctigera Krasch. ex Poljak. - 5

*purshiana Bess. - 1

pycnorhiza Ledeb. (*Oligosporus pycnorhizus* (Ledeb.) Poljak.) - 3, 4, 6

quinqueloba Trautv. (*Oligosporus quinquelobus* (Trautv.) Poljak.) - 6

remosa Sugaw. = A. borealis

remotiloba Krasch. ex Poljak. - 4

repens Pall. ex Willd. = A. austriaca

rhodantha Rupr. (*Seriphidium rhodanthum* (Rupr.) Poljak.) - 6

richardsoniana Bess. = A. borealis subsp. richardsoniana

rubripes Nakai - 1(alien), 5

rupestris L. - 1, 3, 4, 6

- subsp. *viridis* (Willd.) Ameljczenko = A. viridis

rutifolia Steph. ex Spreng. - 3, 4, 6

saissanica (Krasch.) Filat. (*A. mongolorum* Krasch. subsp. *saissanica* Krasch.) - 6

saitoana Kitam. - 5

salsoloides Willd. (*A. tanaitica* Klok., *Oligosporus salsoloides* (Willd.) Poljak.) - 1, 2, 3

samoiedorum Pamp. - 4

santolina Schrenk (*A. lobulifolia* Boiss., *Seriphidium lobulifolium* (Boiss.) Poljak., *Seriphidium santolinum* (Schrenk) Poljak.) - 6

santolinifolia Turcz. ex Bess. - 1, 3, 4, 6

santonica L. (*A. boschniakiana* (Bess.) DC., *A. kumykorum* Minatullaev, *A. praticola* Klok., *A. monogyna* sensu Poljak.) - 1, 2, 6

- subsp. *monogyna* (Waldst. & Kit.) Leonova = A. monogyna

saposhnikovii Krasch. ex Poljak. (*Oligosporus saposhnikovii* (Krasch. ex Poljak.) Poljak.) - 6

schischkinii Krasch. - 3

schmidtiana Maxim. - 5

schrenkiana Ledeb. (*Seriphidium schrenkianum* (Ledeb.) Poljak.) - 4, 6

scoparia Waldst. & Kit. (*Oligosporus scoparius* (Waldst. & Kit.) Less.) - 1, 2, 3, 4, 5, 6

scopiformis Ledeb. (*A. kasakorum* (Krasch.) Pavl., *A. maritima* L. subsp. *kasakorum* Krasch., *Seriphidium scopiforme* (Ledeb.) Poljak.) - 6

scotina Nevski (*Seriphidium scotinum* (Nevski) Poljak.) - 6

selengensis Turcz. ex Bess. - 1(alien),4,5

semiarida (Krasch. & Lavr.) Filat. (*A. terrae-albae* Krasch. subsp. *semiarida* Krasch. & Lavr., *Seriphidium semiaridum* (Krasch. & Lavr.) Ling & Y.R. Ling) - 3

senjavinensis Bess. - 5

sericea Web. - 1, 3, 4, 6

serotina Bunge (*Seriphidium serotinum* (Bunge) Poljak.) - 6

sieberi auct. = A. dumosa

sieversiana Willd. - 1, 3, 4, 5, 6

skorniakowii C. Winkl. - 6

sogdiana Bunge (*Seriphidium sogdianum* (Bunge) Poljak.) - 6

songarica Schrenk (*Oligosporus songaricus* (Schrenk) Poljak.) - 6

sosnovskyi Krasch. & Novopokr. = A. marschalliana

spicigera C. Koch (*Seriphidium spicigerum* (C. Koch) Poljak.) - 2

splendens Willd. - 2

stelleriana Bess. - 5

stenocephala Krasch. ex Poljak. (*Seriphidium stenocephalum* (Krasch. ex Poljak.) Poljak.) - 6

stenophylla Kitam. (*A. integrifolia* L. f. *subulata* (Nakai) Kitag., *A. subulata* Nakai) - 5

stolonifera (Maxim.) Kom. - 5

subarctica Krasch. - 4

subchrysolepis Filat. - 6

sublessingiana Krasch. ex Poljak. (*Seriphidium sublessingianum* (Krasch. ex Poljak.) Poljak.) - 3, 6

subsalsa Filat. - 6

subulata Nakai = A. stenophylla

subviscosa Turcz. ex Bess. - 4

succulenta Ledeb. - 3, 6

sylvatica Maxim. (*A. ussuriensis* Poljak.) - 5

szowitziana (Bess.) Grossh. (*Seriphidium szowitzianum* (Bess.) Poljak.) - 2

taimyrensis Krasch. ex Ameljczenco = A. arctisibirica

tanacetifolia L. - 1, 4, 5

tanaitica Klok. = A. salsoloides

taurica Willd. (*A. graveolens* Minatullaev, *A. taurica* subsp. *graveolens* (Minatullaev) Vlassova, *Seriphidium tauricum* (Willd.) Poljak.) - 1, 2

- subsp. *graveolens* (Minatullaev) Vlassova = A. taurica

tecti-mundi Podlech - 6

tenuisecta Nevski (*Seriphidium tenuisectum* (Nevski) Poljak.) - 6

terrae-albae Krasch. (*Seriphidium terrae-albae* (Krasch.) Poljak.) - 1, 6

- subsp. *semiarida* Krasch. & Lavr. = A. semiarida

tianschanica Krasch. ex Poljak. - 6

tilesii Ledeb. (*A. hultenii* Maximova, *A. tilesii* subsp. *hultenii* (Maximova) Ameljczenko) - 1, 3, 4, 5

- subsp. *hultenii* (Maximova) Ameljczenko = A. tilesii

tomentella Trautv. (*Oligosporus tomentellus* (Trautv.) Poljak.) - 4, 6

tournefortiana Reichenb. - 1(alien),2, 6

transbaicalensis Leonova - 4

transiliensis Poljak. (*Seriphidium transiliense* (Poljak.) Poljak.) - 6

trautvetteriana Bess. (*Oligosporus trautvetterianus* (Bess.) Poljak.) - 1

trifurcata Steph. ex Spreng. = A. furcata

triniana Bess. (*A. lagopus* Fisch. ex Bess. subsp. *triniana* (Bess.) Korobkov) - 4

tschernieviana Bess. = A. marschalliana

turanica Krasch. (*Seriphidium turanicum* (Krasch.) Poljak.) - 6

turcomanica Gand. (*Seriphidium turcomanicum* (Gand.) Poljak.) - 6

umbrosa (Bess.) Pamp. = A. dubia

unalaskensis Rydb. - 5

ussuriensis Poljak. = A. sylvatica

vachanica Krasch. ex Poljak. (*Seriphidium vachanicum* (Krasch. ex Poljak.) Poljak.) - 6

valida Krasch. ex Poljak. (*Seriphidium validum* (Krasch. ex Poljak.) Poljak.) - 6

verlotiorum Lamotte - 1, 2, 6

viridis Willd. (*A. rupestris* L. subsp. *viridis* (Willd.) Ameljczenko) - 3, 4, 6

vulgaris L. (*A. coarctata* Forsell., *A. vulgaris* subsp. *coarctata* (Forsell. ex Bess.) Ameljczenko, comb. invalid., *A. vulgaris* subsp. *urjanchaica* Ameljczenko, nom. invalid.) - 1, 2, 3, 4, 6

- subsp. *coarctata* (Forsell. ex Bess.) Ameljczenko = A. vulgaris

- subsp. *urjanchaica* Ameljczenko = A. vulgaris

- var. *rotundatiloba* Jurtz. = A. leucophylla var. rotundatiloba

Askellia W.A. Weber = Crepis

alaica (Krasch.) W.A. Weber = Crepis alaica

corniculata (Regel & Schmalh.) W.A. Weber = Crepis corniculata

flexuosa (Ledeb.) W.A. Weber = Crepis flexuosa

karelinii (M. Pop. & Schischk. ex Czer.) W.A. Weber = Crepis karelinii

lactea (Lipsch.) W.A. Weber = Crepis lactea

nana (Richards.) W.A. Weber = Crepis nana

sogdiana (Krasch.) W.A. Weber = Crepis sogdiana

Aster L. (*Symphyotrichum* auct., *Virgulus* auct.)

ageratoides Turcz. (*A. luxurifolius* Tamamsch., *A. pensauensis* Tamamsch., *A. see-burejensis* Tamamsch., *A. suputinicus* Tamamsch., *A. sutschanensis* Kom., *A. trinervius* D. Don subsp. *ageratoides* (Turcz.) Grierson) - 5

x **alpino-amellus** Novopokr. ex Tzvel. - A. bessarabicus Bernh. ex Reichenb. x A. serpentimontanus Tamamsch. - 1

alpinus L. (*A. alpinus* var. *fallax* (Tamamsch.) Polozh., *A.*

alpinus var. *fallax* (Tamamsch.) Y. Ling, comb. superfl., *A. fallax* Tamamsch., *A. korshinskyi* Tamamsch.) - 1, 2, 3, 4, 5, 6

- subsp. *parviceps* Novopokr. = A. serpentimontanus

- subsp. *serpentimontanus* (Tamamsch.) A. & D. Löve = A. serpentimontanus

- subsp. *tolmatschevii* (Tamamsch.) A. & D. Löve = A. tolmatschevii

- var. *fallax* (Tamamsch.) Polozh. = A. alpinus

- var. *fallax* (Tamamsch.) Y. Ling = A. alpinus

- var. *serpentimontanus* (Tamamsch.) Y. Ling = A. serpentimontanus

amelloides Bess. = A. bessarabicus

amellus L. - 1, 3

- subsp. *bessarabicus* (Bernh. ex Reichenb.) Soo = A. bessarabicus

- subsp. *ibericus* (Bieb.) V. Avet. = A. ibericus

bellidiastrum (L.) Scop. = Bellidastrum michelii

bessarabicus Bernh. ex Reichenb. (*A. amelloides* Bess. 1822, non Hoffm. 1801, *A. amellus* L. subsp. *bessarabicus* (Bernh. ex Reichenb.) Soo - 1, 2, 6

biennis Ledeb. = Heteropappus biennis

canescens (Nees) Fisjun = Heteropappus canescens

canus Waldst. & Kit. subsp. *punctatus* (Waldst. & Kit.) Soo = Galatella punctata

- subsp. *rossicus* (Novopokr.) Soo = Galatella rossica

fallax Tamamsch. = A. alpinus

fauriei Levl. & Vaniot = A. spathulifolius

fauriei sensu Tamamsch. = A. tataricus

glehnii Fr. Schmidt - 5

ibericus Bieb. (*A. amellus* L. subsp. *ibericus* (Bieb.) V. Avet.) - 1, 2

korshinskyi Tamamsch. = A. alpinus

laevis L. (*Symphyotrichum laeve* (L.) A. & D. Löve) - 2(alien)

lanceolatus Willd. - 1, 5

luxurifolius Tamamsch. = A. ageratoides

maackii Regel - 5

*****novae-angliae** L. (*Virgulus novae-angliae* (L.) Reveal & Keener) - 1, 6

*****novi-belgii** L. - 1

oleifolius (Lam.) Wagenitz = Galatella villosa

oligocephalus (Schrenk) Kitam. = Chamaegeron oligocephalus

pannonicus Jacq. = Tripolium pannonicum

parviceps (Burg.) Mackenz. & Bush - 2(alien)

pensauensis Tamamsch. = A. ageratoides

punctatus Waldst. & Kit. subsp. *rossicus* (Novopokr.) Soo = Galatella rossica

richardsonii Spreng. = A. sibiricus

*****salignus** Willd. - 1, 3

Some authors recognize this taxon as a hybrid A. lanceolatus Willd. x A. novi-belgii L.

sedifolius auct. p.p. = Galatella punctata

sedifolius L. subsp. *angustissimus* (Tausch) Merxmuller = Galatella angustissima

- subsp. *dracunculoides* (Lam.) Merxmuller = Galatella dracunculoides

- var. *dracunculoides* (Lam.) Soo = Galatella dracunculoides

see-burejensis Tamamsch. = A. ageratoides

serpentimontanus Tamamsch. (*A. alpinus* L. subsp. *parviceps* Novopokr., *A. alpinus* subsp. *serpentimontanus* (Tamamsch.) A. & D. Löve, *A. alpinus* var. *serpentimontanus* (Tamamsch.) Y. Ling) - 1, 3, 4, 6

The name of this species was not validly published.

sibiricus L. (*A. richardsonii* Spreng., *A. sibiricus* subsp. *richardsonii* (Spreng.) A. & D. Löve, *A. sibiricus* subsp.

subintegerrimus (Trautv.) A. & D. Löve, *A. subintegerrimus* (Trautv.) Ostenf. & Resv.) - 1, 3, 4, 5
- subsp. *richardsonii* (Spreng.) A. & D. Löve = A. sibiricus
- subsp. *subintegerrimus* (Trautv.) A. & D. Löve = A. sibiricus
spathulifolius Maxim. (*A. fauriei* Levl. & Vaniot, *Erigeron oharae* (Nakai) Botsch.) - 5
subintegerrimus (Trautv.) Ostenf. & Resv. = A. sibiricus
suputinicus Tamamsch. = A. ageratoides
sutschanensis Kom. = A. ageratoides
tarbagatensis (C. Koch) Merxmuller = Galatella tatarica
tataricus L. fil. (*A. fauriei* sensu Tamamsch.) - 4, 5
tolmatschevii Tamamsch. (*A. alpinus* L. subsp. *tolmatschevii* (Tamamsch.) A. & D.Löve) - 1, 3, 4, 5, 6
trinervius D. Don subsp. *ageratoides* (Turcz.) Grierson = A. ageratoides
tripolium L. subsp. *pannonicus* (Jacq.) Soo = Tripolium pannonicum
versicolor Willd. - 1
villosus (L.) Sch. Bip. = Galatella villosa
vvedenskyi Bondar. - 6
woroschilowii Zdorovjeva & Schapoval - 5

Asterothamnus Novopokr.

fruticosus (C. Winkl.) Novopokr. - 6
heteropappoides Novopokr. - 3, 4
poliifolius Novopokr. - 4
schischkinii Tamamsch. - 6

Atalanthus D. Don

acanthodes (Boiss.) Kirp. - 6

Atractylodes DC.

ovata (Thunb.) DC. - 5

*Baccharis L.

***halimifolia** L. - 1, 2, 6

Balsamita Mill. = Pyrethrum

balsamitoides (Sch. Bip.) Tzvel. = Pyrethrum balsamita

Barkhausia Moench = Crepis

kotschyana Boiss. = Crepis kotschyana
pseudoalpina Klok. = Crepis alpina

Bellidastrum Scop.

michelii Cass. (*Aster bellidiastrum* (L.) Scop., *Doronicum bellidiastrum* L.) - 1

Bellis L.

hyrcanica Woronow - 2
perennis L. - 1, 2, 5
sylvestris Cyr. - 1

Bidens L.

bipinnata L. - 2
cernua L. - 1, 2, 3, 4, 5, 6
connata Muehl. ex Willd. - 1
frondosa L. (*B. melanocarpa* Wieg.) - 1, 2, 5, 6
kamtschatica Vass. - 5
maximowicziana Oetting. - 4, 5

melanocarpa Wieg. = B. frondosa
minor (Wimm. & Grab.) Worosch. - 1, 2, 3, 4, 6
orientalis Velen. = B. tripartita
parviflora Willd. - 4, 5
radiata Thuill. - 1, 3, 4, 5, 6
taquetii Levl. & Vaniot - 5
tripartita L. (*B. orientalis* Velen.) - 1, 2, 3, 4, 5, 6

Boltonia L'Her.

integrifolia (Turcz.) Lauener = Kalimeris integrifolia
lautureana Deb. - 4, 5

Bombycilaena (DC.) Smoljian. (*Micropus* auct.)

bombycina (Lag.) Sojak = D. discolor
erecta (L.) Smoljian. - 1, 2
discolor (Pers.) Lainz (*Micropus discolor* Pers., *Bombycilaena bombycina* (Lag.) Sojak, *Micropus bombycinus* Lag.) - 1

Borkonstia Ignatov = Rhinactinidia

eremophila (Bunge) Ignatov = Rhinactinidia eremophila
limoniifolia (Less.) Ignatov = Rhinactinidia limoniifolia
novopokrovskyi (Krasch. & Iljin) Ignatov = Rhinactinidia novopokrovskyi
popovii (Botsch.) Ignatov = Rhinactinidia popovii

Brachanthemum DC.

baranovii (Krasch. & Poljak.) Krasch. - 3
fruticulosum (Ledeb.) DC. - 3, 6
kasakhorum Krasch. - 3, 6
kirghisorum Krasch. - 6
titovii Krasch. - 6

Brachyactis Ledeb.

angusta auct. = B. ciliata
ciliata (Ledeb.) Ledeb. (*B. angusta* auct.) - 1, 3, 4, 5, 6

Breea Less. = Cirsium

ochrolepidea (Juz.) Sojak = Cirsium ochrolepideum
setosa (Willd.) Sojak = Cirsium setosum

Cacalia L. (*Koyamacalia* H. Robinson & R.D. Bretell, *Parasenecio* W.W. Smith & Small)

auriculata DC. (*Koyamacalia auriculata* (DC.) H. Robinson & R.D. Brettell) - 5
- subsp. *kamtschatica* (Maxim.) Hult. = C. kamtschatica
- subsp. *praetermissa* (Pojark.) Worosch. = C. praetermissa
hastata L. (*Koyamacalia hastata* (L.) H. Robinson & R.D. Brettell) - 1, 3, 4, 5
- subsp. *komaroviana* (Pojark.) Kitag. = C. tschonoskii
kamtschatica (Maxim.) Kudo (*C. auriculata* DC. subsp. *kamtschatica* (Maxim.) Hult.) - 5
kitamurae auct. = C. tschonoskii
komaroviana (Pojark.) Pojark. = C. tschonoskii
macrophylla Bieb. = Adenostyles macrophylla
praetermissa (Pojark.) Pojark. (*C. auriculata* DC. subsp. *praetermissa* (Pojark.) Worosch.) - 5
robusta Tolm. - 5
tschonoskii Koidz. (*C. hastata* L. subsp. *komaroviana* (Pojark.) Kitag., *C. komaroviana* (Pojark.) Pojark., *C. kitamurae* auct.) - 5

Calitrapa Adans. = Centaurea

solstitialis (L.) Lam. subsp. *adamii* (Willd.) Sojak = Centaurea adamii

Calendula L.

alata Rech. fil. - 6
arvensis L. - 1, 2
gracilis DC. - 2, 6
hybrida L. = Dimorphoteca hybrida
karakalensis Vass. - 6
***officinalis** L.
persica C.A. Mey. - 2, 6
pluvialis L. = Dimorphotheca pluvialis

Callicephalus C.A. Mey.

nitens (Bieb.) C.A. Mey. - 2, 6

*Callistephus Cass.

***chinensis** (L.) Nees - 1, 2, 5, 6

Calycocorsus F.M. Schmidt (*Willemetia* Cass. 1827, non Maerklin, 1809, *Willemetia* Neck. 1790, nom. invalid.)

tuberosus (Fisch. & C.A. Mey.) Rauschert (*Willemetia tuberosa* Fisch. & C.A. Mey.) - 2

Cancrinia Kar. & Kir.

botschantzevii (Kovalevsk.) Tzvel. = Tanacetopsis botschantzevii
chrysocephala Kar. & Kir. - 6
discoidea (Ledeb.) Poljak. - 3, 6
ferganensis (Kovalevsk.) Tzvel. = Tanacetopsis ferganensis
goloskokovii (Poljak.) Tzvel. = Tanacetopsis goloskokovii
karatavica Tzvel. = Tanacetopsis karataviensis
krasnoborovii V. Khan. - 4
mucronata (Regel & Schmalh.) Tzvel. = Tanacetopsis mucronata
nevskii Tzvel. = Tanacetopsis krascheninnikovii
pamiralaica (Kovalevsk.) Tzvel. = Tanacetopsis pamiralaica
paropamisica Krasch. = Tanacetopsis paropamisica
pjataevae (Kovalevsk.) Tzvel. = Tanacetopsis pjataevae
santoana (Krasch., M. Pop. & Vved.) Poljak. = Tanacetopsis santoana
setacea (Regel & Schmalh.) Tzvel. = Tanacetopsis setacea
submarginata (Kovalevsk.) Tzvel. = Tanacetopsis submarginata
subsimilis (Rech. fil.) Tzvel. = Tanacetopsis subsimilis
tadshikorum (Kudr.) Tzvel. = Polychrysum tadshikorum
tianschanica (Krasch.) Tzvel. - 6
urgutensis (M. Pop.) Tzvel. = Tanacetopsis urgutensis

Cancriniella Tzvel.

krascheninnikovii (N. Rubtz.) Tzvel. - 6

Carduus L.

acanthocephalus C.A. Mey. - 2
acanthoides L. (*C. fortior* Klok.) - 1, 2, 3, 5(alien)
adpressus C.A. Mey. (*C. multijugus* C. Koch) - 2
albidus Bieb. = C. arabicus
arabicus Jacq. (*C. albidus* Bieb., *C. pycnocephalus* L. subsp. *albidus* (Bieb.) Kazmi) - 1, 2, 6

- subsp. *cinereus* (Bieb.) Kazmi = C. cinereus
atropatanicus Sosn. ex Grossh. - 2
attenuatus Klok. = C. thoermeri
beckerianus Tamamsch. - 2, 6
bicolorifolius Klok. (*C. personata* (L.) Jacq. subsp. *albidus* (Adamov.) Kazmi, *C. personata* var. *albidus* Adamov., *C. personata* auct.) - 1
candicans Waldst. & Kit. - 1
cinereus Bieb. (*C. arabicus* Jacq. subsp. *cinereus* (Bieb.) Kazmi, *C. pycnocephalus* L. subsp. *cinereus* (Bieb.) P.H. Davis) - 1, 2, 6
collinus Waldst. & Kit. - 1
coloratus Tamamsch. = C. nutans
crassifolius Willd. subsp. *glaucus* (Baumg.) Kazmi = C. glaucinus
crispus L. (*C. crispus* subsp. *incanus* (Klok.) Soo, *C. incanus* Klok.) -1, 2, 3, 4, 5, 6
- subsp. *dahuricus* Aren. = C. dahuricus
- subsp. *incanus* (Klok.) Soo = C. crispus
dahuricus (Aren.) Kazmi (*C. crispus* L. subsp. *dahuricus* Aren.) - 4, 5
defloratus L. subsp. *glaucus* (Baumg.) Nym. = C. glaucinus
fortior Klok. = C. acanthoides
furiosus Tamamsch. - 2
glaucinus Holub (*C. glaucus* Baumg. 1816, non Cav. 1794, *C. crassifolius* Willd. subsp. *glaucus* (Baumg.) Kazmi, *C. defloratus* L. subsp. *glaucus* (Baumg.) Nym.) - 1
glaucus Baumg. = C. glaucinus
hajastanicus Tamamsch. - 2
hamulosus Ehrh. (*C. tyraicus* Klok.) - 1, 2
- subsp. *hystrix* (C.A. Mey.) Kazmi = C. hystrix
- subsp. *pseudocollinus* (Schmalh.) Soo = C. pseudocollinus
- subsp. *sordidus* (DC.) Soo, p.p. = C. tauricus
hohenackeri Kazmi - 2
hystrix C.A. Mey. (*C. hamulosus* Ehrh. subsp. *hystrix* (C.A. Mey.) Kazmi, *C. stenocephalus* Tamamsch.) - 1, 2, 3, 6
incanus Klok. = C. crispus
kerneri Simonk. - 1
kondratjukii Gorlaczova = C. nutans
laciniatus Ledeb. (*C. multijugus* sensu Tamamsch.) - 2
leiophyllus Petrovic = C. thoermeri
multijugus C. Koch = C. adpressus
multijugus sensu Tamamsch. = C. laciniatus
nawaschinii Bordz. - 2
nervosus C. Koch - 2
nikitinii Tamamsch. - 6
nutans L. (*C. coloratus* Tamamsch., *C. kondratjukii* Gorlaczova, *C. nutans* var. *armenus* Boiss., *C. schischkinii* Tamamsch., *C. songoricus* Tamamsch., *C. thoermeri* Weinm. subsp. *armenus* (Boiss.) Kazmi) - 1, 2, 3, 5, 6
- subsp. *leiophyllus* (Petrovic) Stojan. & Stef. = C. thoermeri
- var. *armenus* Boiss. = C. nutans
onopordioides Fisch. ex Bieb. - 2
personata auct. = C. bicolorifolius
personata (L.) Jacq. subsp. *albidus* (Adamov.) Kazmi = C. bicolorifolius
- var. *albidus* Adamov. = C. bicolorifolius
poliochrus Trautv. - 2
pseudocollinus (Schmalh.) Klok. (*C. hamulosus* Ehrh. subsp. *pseudocollinus* (Schmalh.) Soo) - 1
pycnocephalus L. - 1, 2, 6
- subsp. *albidus* (Bieb.) Kazmi = C. arabicus
- subsp. *cinereus* (Bieb.) P.H. Davis = C. cinereus
schischkinii Tamamsch. = C. nutans
seminudus Bieb. - 1, 2, 6
songoricus Tamamsch. = C. nutans

stenocephalus Tamamsch. = C. hystrix
tauricus Klok. (*C. hamulosus* Ehrh. subsp. *sordidus* (DC.) Soo) - 1
thoermeri Weinm. (*C. attenuatus* Klok., *C. leiophyllus* Petrovic, *C. nutans* L. subsp. *leiophyllus* (Petrovic) Stojan. & Stef.) - 1, 2, 3, 4, 6
- subsp. *armenus* (Boiss.) Kazmi = C. nutans
tortuosus Gorlaczova & Kondratjuk - 1
transcaspicus Gand. - 6
tyraicus Klok. = C. hamulosus
uncinatus Bieb. - 1, 2, 3, 6
- subsp. **davisii** Kazmi - 1, 2

Carlina L.

acaulis L. - 1
biebersteinii Bernh. ex Hornem. (*C. stricta* (Rouy) Fritsch, *C. vulgaris* L. subsp. *stricta* (Rouy) Domin) - 1, 2, 3, 4
cirsioides Klok. - 1
intermedia Schur (*C. vulgaris* L. subsp. *intermedia* (Schur) Hayek) - 1, 2, 6
lanata L. - 2
onopordifolia Bess. ex Szaf., Kucz. & Pawl. - 1
stricta (Rouy) Fritsch = C. biebersteinii
taurica Klok. = C. vulgaris
vulgaris L. (*C. taurica* Klok.) - 1, 2
- subsp. *intermedia* (Schur) Hayek = C. intermedia
- subsp. *stricta* (Rouy) Domin = C. biebersteinii

Carpesium L.

abrotanoides L. - 2
cernuum L. - 1, 2, 5, 6
eximium C. Winkl. = C. macrocephalum
macrocephalum Franch. & Savat. (*C. eximium* C. Winkl.) - 5
triste Maxim. - 5

Carthamus L.

glaucus Bieb. - 1, 2
gypsicola Iljin - 2, 6
lanatus L. - 1, 2, 6
- subsp. *turkestanicus* Hanelt = C. x turkestanicus
oxyacanthus Bieb. - 2, 6
pterocaulos C.A. Mey. = Cousinia pterocaulos
tamamschjanae Gabr. - 2
***tinctorius** L. - 1, 2, 6
x **turkestanicus** M. Pop. (*C. lanatus* L. subsp. *turkestanicus* Hanelt). - C. lanatus L. x C. oxyacanthus Bieb. - 6
> The name C. x turkestanicus M. Pop. was not validly published.

Centaurea L. (*Acosta* Adans., *Acrocentron* Cass., *Calcitrapa* Adans., *Colymbada* Hill, *Cyanus* Hill, *Heterolophus* Cass., *Jacea* Hill, *Odontolophus* Cass., *Psephellus* Cass., *Xanthopsis* (DC.) C. Koch)

abbreviata (C. Koch) Hand.-Mazz. (*C. phrygia* L. subsp. *abbreviata* (C. Koch) Dostal, *Jacea abbreviata* (C. Koch) Sojak) - 1, 2
abchasica (Albov) Sosn. - 2
abnormis Czer. - 2
absinthifolia (Galushko) Czer. (*Psephellus absinthifolius* Galushko) - 2
acmophylla Boiss. (*Cyanus acmophyllus* (Boiss.) Sojak) - 2
adamii Willd. (*Calcitrapa solstitialis* (L.) Lam. subsp.

adamii (Willd.) Sojak, *Centaurea solstitialis* L. subsp. *adamii* (Willd.) Nym.) - 1, 2
adjarica Albov - 2
adpressa Ledeb. (*C. apiculata* Ledeb. subsp. *adpressa* (Ledeb.) Dostal, *Colymbada adpressa* (Ledeb.) Holub) - 1, 2, 3, 6
aemulans Klok. (*Acosta aemulans* (Klok.) Holub) - 1, 2
aggregata Fisch. & C.A. Mey. (*Acosta aggregata* (Fisch. & C.A. Mey.) Sojak) - 2
- subsp. *albida* (C. Koch) Bornm. = C. albida
ahverdovii Gabl. = Grossheimia ahverdovii
alaica Iljin - 6
albida C. Koch (*C. aggregata* Fisch. & C.A. Mey. subsp. *albida* (C. Koch) Bornm.) - 2
albovii Sosn. - 2
alexandri Bordz. - 2
alpestris auct. = C. ossethica
alutacea Dobrocz. (*Jacea alutacea* (Dobrocz.) Sojak) - 1, 2
amblyolepis Ledeb. - 2
andina (Galushko & Alieva) Czer. (*Psephellus andinus* Galushko & Alieva) - 2
androssovii Iljin - 6
angelescui Grint. (*C. stricta* Waldst. & Kit. var. *angelescui* (Grint.) Borza, *C. stricta* subsp. *angelescui* (Grint.) Prod., *C. triumfettii* All. subsp. *angelescui* (Grint.) Dostal, *Cyanus angelescui* (Grint.) Holub) - 1
annae (Galushko) Czer. (*Psephellus annae* Galushko) - 2
apiculata Ledeb. (*Colymbada apiculata* (Ledeb.) Holub) - 1, 2, 3
- subsp. *adpressa* (Ledeb.) Dostal = C. adpressa
appendicata Klok. (*C. margaritacea* Ten. subsp. *appendicata* (Klok.) Dostal) - 1
araxina Gabr. - 2
arenaria Bieb. (*Acosta arenaria* (Bieb.) Sojak) - 1, 2
- subsp. *borysthenica* (Grun.) Dostal = C. borysthenica
- subsp. *majorovii* (Dumb.) Dostal = C. majorovii
- subsp. *odessana* (Prod.) Dostal = C. odessana
- subsp. *sophiae* (Klok.) Dostal = C. sophiae
armena Boiss. - 2
arpensis (Czer.) Wagenitz (*C. arpensis* (Czer.) Czer. nom. superfl.) - 2
aucheri (DC.) Wagenitz = Tomanthea aucheri
- subsp. *szowitii* (Boiss.) Wagenitz = Tomanthea phaeopappa
- var. *szowitsii* (Boiss.) Wagenitz = Tomanthea phaeopappa
avarica Tzvel. - 2
bagadensis Woronow - 2
baksanica Czer. (*Psephellus lactiflorus* G. Koss ex Tschuchrukidze, 1976, non Centaurea lactiflora Halacsy, 1898) - 2
barbeyi (Albov) Sosn. - 2
behen L. - 2, 6
belangeriana (DC.) Stapf - 2, 6
bella Trautv. - 2
besseriana DC. (*Acosta besseriana* (DC.) Sojak, *Centaurea ovina* Pall. ex Willd. subsp. *besseriana* (DC.) Dostal) - 1
biebersteinii DC. (*Acosta biebersteinii* (DC.) Dostal, *A. micranthos* (S.G. Gmel.) Sojak, *Centaurea micranthos* S.G. Gmel. ex Hayek, 1901, non C. micrantha Duf. 1831) - 1, 2, 3
borysthenica Grun. (*Acosta borysthenica* (Grun.) Sojak, *Centaurea arenaria* Bieb. subsp. *borysthenica* (Grun.) Dostal) - 1
breviceps Iljin (*C. margaritacea* Ten. subsp. *breviceps* (Iljin) Dostal) - 1
buschiorum (Sosn.) Czer. comb. nova (*Psephellus buschio-*

rum Sosn. 1948, Zam. Syst. Geogr. (Tbilisi) 17 : 7, non *Centaurea buschiorum* Sosn. ex Grossh. 1949, nom. invalid.) - 2

cabardensis (G. Koss ex Tschuchrukidze) Czer. (*Psephellus cabardensis* G. Koss ex Tschuchrukidze) - 2

calcitrapa L. - 1

cana auct. = C. fuscomarginata

caprina auct. p.p. = C. steveniana

caprina Stev. (*C. koktebelica* Klok., *C. ovina* Pall. ex Willd. subsp. *koktebelica* (Klok.) Dostal, *C. steveniana* Klok. p.p. excl. typo) - 1

carbonata Klok. (*Heterolophus carbonatus* (Klok.) Sojak) - 1

carduiformis DC. (*Colymbada carduiformis* (DC.) Holub) - 2

- subsp. **orientalis** Wagenitz - 2

carpatica (Porc.) Porc. (*C. phrygia* L. subsp. *carpatica* (Porc.) Dostal, *Jacea carpatica* (Porc.) Sojak, *J. phrygia* (L.) Sojak subsp. *carpatica* (Porc.) Dostal.) - 1

carthalinica (Sosn.) Sosn. - 2

caspia Grossh. - 2

cheiranthifolia Willd. (*Cyanus cheiranthifolius* (Willd.) Sojak) - 2

- var. *purpurascens* (DC.) Wagenitz = C. fischeri

circassica (Albov) Sosn. (*Psephellus circassicus* (Albov) Galushko) - 2

ciscaucasica Sosn. (*Psephellus ciscaucasicus* (Sosn.) Galushko) - 2

colchica (Sosn.) Sosn. - 2

x **comperiana** Stev. - C. caprina Stev. x C. substituta Czer. - 1

cyanus L. - 1, 2, 3, 4, 5, 6

czerepanovii (Alieva) Czer. (*Psephellus czerepanovii* Alieva) - 2

czerkessica Dobrocz. & Kotov (*C. pseudotanaitica* Galushko, nom. invalid.) - 2

czirkejensis Gussejnov - 2

daghestanica (Lipsky) Wagenitz (*C. daghestanica* (Lipsky) Czer. nom. superfl.) - 2

dealbata Willd. (*C. tschuchrukidzei* Czer., *Psephellus sosnowskyi* Tschuchrukidze) - 2

declinata Bieb. - 1, 2

depressa Bieb. (*Cyanus depressus* (Bieb.) Sojak) - 1, 2, 6

diffusa Lam. (*Acosta diffusa* (Lam.) Sojak, *Centaurea microcalathina* A. Tarass.) - 1, 2, 4(alien)

dimitriewiae Sosn. - 2

x **dobroczaevae** Tzvel. - C. diffusa Lam. x C. pseudomaculosa Dobrocz. - 1

dominii (Dostal) Dubovik (*C. triumfettii* All. subsp. *dominii* Dostal) - 1

donetzica Klok. (*C. margaritacea* Ten. subsp. *donetzica* (Klok.) Dostal) - 1

dubjanskyi Iljin (*C. margaritacea* Ten. subsp. *dubjanskyi* (Iljin) Dostal) - 1

edmondii Sosn. - 2

erivanensis (Lipsky) Bordz. (*Xanthopsis erivanensis* (Lipsky) Sojak) - 2

- subsp. **holargyrea** (Bornm. & Woronow) Gabr. (*Psephellus holargyreus* Bornm. & Woronow) - 2

exsurgens Sosn. - 2

fajvuschii Gabr. - 2

fischeri Schlecht. (*C. cheiranthifolia* Willd. var. *purpurascens* (DC.) Wagenitz, *C. montana* L. var. *purpurascens* DC., *Cyanus fischeri* (Schlecht.) Sojak) - 2

fuscomarginata (C. Koch) Juz. (*C. cana* auct., *C. triumfettii* All. subsp. *cana* auct. p.p) - 1

galushkoi (Alieva) Czer. (*Psephellus galushkoi* Alieva) - 2

georgica Klok. - 2

gerberi Stev. (*C. margaritacea* Ten. subsp. *gerberi* (Stev.) Dostal) - 1

glehnii Trautv. (*C. pseudoscabiosa* Boiss. & Buhse subsp. *glehnii* (Trautv.) Wagenitz, *Colymbada glehnii* (Trautv.) Holub) - 2

gontscharovii Iljin - 6

grossheimii Sosn. - 2

gulissachvilii Dumb. - 2

hajastana Tzvel. - 2

■ **hedgei** Wagenitz (*C. taochia* (Sosn.) Sosn.1963, non Sosn. 1931)

holophylla (Socz. & Lipat.) Socz. & Lipat. (*Psephellus holophyllus* Socz. & Lipat.) - 2

huetii Boiss. (*Cyanus hultii* (Boiss.) Sojak) - 2

hymenolepis Trautv. - 2

x **hypanica** Pacz. - C. diffusa Lam. x C. margarita-alba Klok. - 1

hypoleuca DC. (*C. maris-nigri* Sosn.) - 2

hyrcanica Bornm. (*Jacea hyrcanica* (Bornm.) Holub) - 2

iberica Trev. ex Spreng. - 1, 2, 6

x **iljiniana** Illar. - C. caprina Stev. x C. sterilis Stev. - 1

iljinii Czerniak. - 6

indurata Janka - 1

integrifolia Tausch - 1, 3

inuloides Fisch. ex Janka = Phalacrachena inuloides

jacea L. 1, 2(alien), 3(alien), 4(alien), 5(alien)

kachetica (Rehm.) Czer. (*Psephellus kacheticus* Rehm.) - 2

karabaghensis (Sosn.) Sosn. (*C. zangezuri* (Sosn.) Sosn.) - 2

kasakorum Iljin - 1, 3, 6

kemulariae (Charadze) Czer. (*Psephellus kemulariae* Charadze) - 2

x **klokovii** Tzvel. - C. pseudomaculosa Dobrocz. x C. substituta Czer. - 1

kobstanica Tzvel. (*Odontolophus kobstanicus* (Tzvel.) Sojak) - 2

koktebelica Klok. = C. caprina

kolakovskyi Sosn. - 2

konkae Klok. (*C. margaritacea* Ten. subsp. *konkae* (Klok.) Dostal) - 1

kopetdaghensis Iljin - 6

kotschyana Heuff. (*Colymbada kotschyana* (Heuff.) Holub) - 1

kotschyi (Boiss. & Heldr.) Hayek var. *persica* (Boiss.) Wagenitz = Cheirolepis persica

kubanica Klok. - 1, 2

kultiassovii Iljin - 6

lasiopoda M. Pop. & Kult. - 6

latiloba Klok. - 2

lavrenkoana Klok. (*C. ovina* Pall. ex Willd. subsp. *lavrenkoana* (Klok.) Dostal) - 1

leucophylla Bieb. - 2

leuzeoides (Jaub. & Spach) Walp. - 2

x **livonica** Weinm. (*Jacea livonica* (Weinm.) Sojak). - C. jacea L. x C.phrygia L. - 1

x **longiaristata** Illar. - ? C.diffusa Lam. x C. sterilis Stev. - 1

majorovii Dumb. (*C. arenaria* Bieb. subsp. *majorovii* (Dumb.) Dostal) - 1, 2

maleevii (Sosn.) Sosn. - 2

margarita-alba Klok. (*C. margaritacea* Ten. subsp. *margarita-alba* (Klok.) Dostal) - 1

margaritacea Ten. (*Jacea margaritacea* (Ten.) Sojak) - 1

- subsp. *appendicata* (Klok.) Dostal = C. appendicata

- subsp. *breviceps* (Iljin) Dostal = C. breviceps

- subsp. *donetzica* (Klok.) Dostal = C. donetzica

- subsp. *dubjanskyi* (Iljin) Dostal = C. dubjanskyi

- subsp. *gerberi* (Stev.) Dostal = C. gerberi

- subsp. *konkae* (Klok.) Dostal = C. konkae
- subsp. *margarita-alba* (Klok.) Dostal = C. margarita-alba
- subsp. *paczoskii* (Kotov ex Klok.) Dostal = C. paczoskii
- subsp. *pineticola* (Iljin) Dostal = C. pineticola
- subsp. *protogerberi* (Klok.) Dostal = C. protogerberi
- subsp. *protomargaritacea* (Klok.) Dostal = C. protomargaritacea
- subsp. *pseudoleucolepis* (Kleop.) Dostal = C. pseudoleucolepis

maris-nigri Sosn. = C. hypoleuca

marmarosiensis (Jav.) Czer. (*C. mollis* Waldst. & Kit. subsp. *marmarosiensis* (Jav.) Soo, *Cyanus marmarosiensis* (Jav.) Dostal, *C. montanus* (L.) Hill subsp. *marmarosiensis* (Jav.) Sojak) - 1

marschalliana Spreng. (*Heterolophus marschallianus* (Spreng.) Sojak) - 1, 2

melanocalathia Borb. (*C. nigriceps* Dobrocz., *C. phrygia* L. subsp. *melanocalathia* (Borb.) Dostal, *C. phrygia* subsp. *nigriceps* (Dobrocz.) Dostal) - 1

meskhetica Sosn. - 2

meyeriana Tzvel. - 2

micranthos S.G. Gmel. ex Hayek = C. biebersteinii

microcalathina A. Tarass. = C. diffusa

modesti Fed. - 6

mollis Waldst. & Kit. (*Cyanus montanus* (L.) Hill subsp. *mollis* (Waldst. & Kit.) Sojak) - 1

- subsp. *marmarosiensis* (Jav.) Soo = C. marmarosiensis

montana L. var. *purpurascens* DC. = C. fischeri

napulifera Rochel subsp. *thirkei* (Sch. Bip.) Dostal = C. thirkei

nathadzeae Sosn. - 2

nigriceps Dobrocz. = C. melanocalathia

nigrofimbria (C. Koch) Sosn. (*Cyanus nigrofimbrius* (C. Koch) Sojak) - 2

x **ninae** Juz. - C. caprina Stev. x C. vankovii Klok.

nogmovii (G. Koss ex Tschuchrukidze) Czer. (*Psephellus nogmovii* G. Koss ex Tschuchrukidze) - 2

novorossica Klok. - 2

odessana Prod. (*Acosta odessana* (Prod.) Sojak, *Centaurea arenaria* Bieb. subsp. *odessana* (Prod.) Dostal) - 1

oltensis Sosn. - 2

orientalis L. (*Colymbada orientalis* (L.) Holub) - 1, 2

ossethica Sosn. (*C. alpestris* auct.) - 1, 2

ovina Pall. ex Willd. (*Acosta ovina* (Pall. ex Willd.) Sojak) - 2

- subsp. *besseriana* (DC.) Dostal = C. besseriana
- subsp. *koktebelica* (Klok.) Dostal = C. koktebelica
- subsp. *lavrenkoana* (Klok.) Dostal = C. lavrenkoana
- subsp. *steveniana* (Klok.) Dostal = C. steveniana

paczoskii Kotov ex Klok. (*C. margaritacea* Ten. subsp. *paczoskii* (Kotov ex Klok.) Dostal) - 1

pambakensis (Sosn.) Sosn. - 2

pannonica (Heuff.) Simonk. (*Jacea pannonica* (Heuff.) Sojak) - 1

- subsp. *substituta* (Czer.) Dostal = C. substituta

pauciloba Trautv. - 2

▪ **pecho** Albov

phaeopappoides Bordz. - 2

phrygia L. (*Jacea phrygia* (L.) Sojak) - 1, 3

- subsp. *abbreviata* (C. Koch) Dostal = C. abbreviata
- subsp. *carpatica* (Porc.) Dostal = C. carpatica
- subsp. *melanocalathia* (Borb.) Dostal = C. melanocalathia
- subsp. *nigriceps* (Dobrocz.) Dostal = C. melanocalathia
- subsp. *pseudophrygia* (C.A. Mey.) Gugl. = C. pseudophrygia

phyllopoda Iljin - 6

pineticola Iljin (*C. margaritacea* Ten. subsp. *pineticola*

(Iljin) Dostal) - 1

polyphylla Ledeb. = Grossheimia polyphylla

polypodiifolia Boiss. - 2

- var. *szovitsiana* (Boiss.) Wagenitz = C. szovitsiana

prokhanovii (Galushko) Czer. (*Psephellus prokhanovii* Galushko) - 2

protogerberi Klok. (*C. margaritacea* Ten. subsp. *protogerberi* (Klok.) Dostal) - 1

protomargaritacea Klok. (*C. margaritacea* Ten. subsp. *protomargaritacea* (Klok.) Dostal) - 1

pseudocoriacea Dobrocz. - 1

pseudoleucolepis Kleop. (*C. margaritacea* Ten. subsp. *pseudoleucolepis* (Kleop.) Dostal) - 1

pseudomaculosa Dobrocz. (*Acosta pseudomaculosa* (Dobrocz.) Sojak, *Centaurea rhenana* Boreau subsp. *pseudomaculosa* (Dobrocz.) Dostal) - 1, 3

pseudoovina Illar. - 1

pseudophrygia C.A. Mey. (*C. phrygia* L. subsp. *pseudophrygia* (C.A. Mey.) Gugl., *Jacea pseudophrygia* (C.A. Mey.) Holub) - 1, 3

pseudoscabiosa Boiss. & Buhse (*Colymbada pseudoscabiosa* (Boiss. & Buhse) Holub) - 2

- subsp. *glehnii* (Trautv.) Wagenitz = C. glehnii

pseudotanaitica Galushko = C. czerkessica

razdorskyi Karjag. - 2

reflexa Lam. (*Colymbada reflexa* (Lam.) Holub) - 2

rhenana Boreau (*Acosta rhenana* (Boreau) Sojak) - 1

- subsp. *pseudomaculosa* (Dobrocz.) Dostal = C. pseudomaculosa
- subsp. *savranica* (Klok.) Dostal = C. savranica

rhizantha C.A. Mey. - 2, 6

rhizanthoides Tzvel. - 2

rubriflora Illar. - 1

ruprechtii (Boiss.) Wagenitz (*C. ruprechtii* (Boiss.) Czer. comb. superfl.) - 2

ruthenica Lam. - 1, 2, 3, 6

salicifolia Bieb. (*Jacea salicifolia* (Bieb.) Sojak) - 2

salonitana Vis. (*Colymbada salonitana* (Vis.) Holub) - 1, 2

salviifolia (Boiss.) Sosn. - 2

sarandinakiae Illar. - 1

savranica Klok. (*Acosta savranica* (Klok.) Holub, *Centaurea rhenana* Boreau subsp. *savranica* (Klok.) Dostal) - 1

scabiosa L. (*Acrocentron scabiosa* (L.) A. & D. Löve, *Colymbada scabiosa* (L.) Holub, *C. scabiosa* (L.) Rauschert, comb. superfl.) - 1, 3, 4, 5(alien)

schelkovnikovii Sosn. - 2

schistosa Sosn. (*Psephellus schistosus* (Sosn.) Alieva) - 2

semijusta Juz. (*C. sterilis* Stev. subsp. *semijusta* (Juz.) Dostal) - 1

sergii Klok. (*Heterolophus sergii* (Klok.) Sojak) - 3, 6

sevanensis Sosn. - 2

sibirica L. - 1, 3

simplicicaulis Boiss. & Huet - 2(?)

sintenisiana Gand. - 6

solstitialis L. - 1, 2, 6

- subsp. *adamii* (Willd.) Nym. = C. adamii

somchetica (Sosn.) Sosn. - 2

sophiae Klok.(*C. arenaria* Bieb. subsp. *sophiae* (Klok.) Dostal) - 1

sosnowskyi Grossh. (*Colymbada sosnowskyi* (Grossh.) Holub) - 2

squarrosa Willd. (*Acosta squarrosa* (Willd.) Sojak) - 2, 6

stankovii Illar. - 1

stenolepis A. Kerner (*Jacea stenolepis* (A. Kerner) Sojak) - 1

stereophylla Bess. (*Colymbada stereophylla* (Bess.) Holub) - 1
sterilis Stev. - 1
- subsp. *semijusta* (Juz.) Dostal = C. semijusta
- subsp. *vankovii* (Klok.) Dostal = C. vankovii
steveniana Klok. (*Acosta steveniana* (Klok.) Holub, *Centaurea ovina* Pall. ex Willd. subsp. *steveniana* (Klok.) Dostal, *C. caprina* auct. p.p.) - 1
stevenii Bieb. - 2
stricta Waldst. & Kit. (*C. ternopoliensis* Dobrocz., *C. triumfettii* All. subsp. *stricta* (Waldst. & Kit.) Dostal, *Cyanus strictus* (Waldst. & Kit.) Sojak, *C. triumfettii* (All.) Dostal ex A. & D. Löve subsp. *strictus* (Waldst. & Kit.) Dostal) - 1
- subsp. *angelescui* (Grint.) Prod. = C. angelescui
- var. *angelescui* (Grint.) Borza = C. angelescui
subacaulis Ledeb. = C. urvillei subsp. stepposa
substituta Czer. (*C. pannonica* (Heuff.) Simonk. subsp. *substituta* (Czer.) Dostal, *Jacea substituta* (Czer.) Sojak) - 1, 2
sumensis Kalen. (*Heterolophus sumensis* (Kalen.) Sojak) - 1
svanetica T. Mardalejschvili - 2
szovitsiana Boiss. (*C. polypodiifolia* Boiss. var. *szovitsiana* (Boiss.) Wagenitz) - 2
takhtajanii Gabr. & Tonjan - 2
taliewii Kleop. - 1
tamanianiae Agababjan - 2
tanaitica Klok. (*C. triumfettii* All. subsp. *tanaitica* (Klok.) Dostal, *Cyanus tanaiticus* (Klok.) Sojak) - 1
taochia (Sosn.) Sosn. = C. hedgei
ternopoliensis Dobrocz. = C. stricta
thirkei Sch. Bip. (*C. napulifera* Rochel subsp. *thirkei* (Sch. Bip.) Dostal, *Cyanus thirkei* (Sch. Bip.) Holub) - 1
transcaucasica Sosn. ex Grossh. - 2
trichocephala Bieb. (*Jacea trichocephala* (Bieb.) Sojak) - 1, 2, 3
trinervia Steph. (*Odontolophus trinervius* (Steph.) Dobrocz.) - 1, 2
triumfettii All. subsp. *angelescui* (Grint.) Dostal = C. angelescui
- subsp. *cana* auct. p.p. = C. fuscomarginata
- subsp. *dominii* Dostal = C. dominii
- subsp. *stricta* (Waldst. & Kit.) Dostal = C. stricta
- subsp. *tanaitica* (Klok.) Dostal = C. tanaitica
troitzkyi (Sosn.) Sosn. - 2
tschuchrukidzei Czer. = C. dealbata
tuapsensis Sosn. - 2
turgaica Klok. - 3, 6
turkestanica Franch. - 6
x **tzebeldensis** Woronow. - C. iberica Trev. ex Spreng. x C. salicifolia Bieb. - 2
urvillei DC. subsp. **stepposa** Wagenitz (*C. subacaulis* Ledeb.) - 2
vankovii Klok. (*C. sterilis* Stev. subsp. *vankovii* (Klok.) Dostal) - 1
vavilovii Takht. & Gabr. - 2
vicina Lipsky - 2
willdenowii Czer. (*Cyanus willdenowii* (Czer.) Sojak) - 2
wolgensis DC. - 1, 2, 6
woronowii Bornm. ex Sosn. (*Cyanus woronowii* (Bornm. ex Sosn.) Sojak) - 2
xanthocephala (DC.) Sch. Bip. (*C. xanthocephala* (DC.) Sosn. comb. superfl.) - 2
- subsp. *xanthocephaloides* (Tzvel.) Gabr. = C. xanthocephaloides
xanthocephaloides Tzvel. (*C. xanthocephala* (DC.) Sch. Bip. subsp. *xanthocephaloides* (Tzvel.) Gabr., *Xanthopsis*

xanthocephaloides (Tzvel.) Sojak) - 2
zangezuri (Sosn.) Sosn. = C. karabaghensis
zuvandica (Sosn.) Sosn. - 2

Centipeda Lour.

minima (L.) A. Br. & Aschers. (*Artemisia minima* L., *Centipeda orbicularis* Lour.) - 5
orbicularis Lour. = C. minima

Cephalorrhynchus Boiss.

brassicifolius (Boiss.) Tuisl (*Lactuca brassicifolia* Boiss.) - 6
gorganicus (Rech. fil. & Esfand.) Tuisl (*Lactuca gorganica* Rech. fil. & Esfand., *Cephalorrhynchus talyschensis* Kirp.) - 2
kirpicznikovii Grossh. - 2
kossinskyi (Krasch.) Kirp. - 6
polycladus (Boss.) Kirp. - 6
soongoricus (Regel) Kovalevsk. - 6
subplumosus Kovalevsk. - 6
takhtadzhianii (Sosn.) Kirp. - 2
talyschensis Kirp. = C. gorganicus
tuberosus (Stev.) Schchian - 1, 2

Chamaegeron Schrenk

bungei (Boiss.) Botsch. - 6
oligocephalus Schrenk (*Aster oligocephalus* (Schrenk) Kitam.) - 6

Chamaemelum Mill.

nobile (L.) All. - 1

Chamomilla S.F. Gray = Matricaria

aurea (L.) J. Gay ex Coss. & Kral = Lepidotheca aurea
discoidea (DC.) J. Gay ex A. Br. = Lepidotheca suaveolens
recutita (L.) Rauschert = Matricaria recutita
suaveolens (Pursh) Rydb. = Lepidotheca suaveolens
tzvelevii (Pobed.) Rauschert = Matricaria tzvelevii

Chardinia Desf.

macrocarpa C. Koch = Ch. orientalis
orientalis (L.) O. Kuntze (*Ch. macrocarpa* C. Koch) - 2, 6

Chartolepis Boiss.

biebersteinii Jaub. & Spach - 2
glastifolia (L.) Cass. - 2
intermedia Boiss. - 1, 3, 6
pterocaula (Trautv.) Czer. - 2

Cheirolepis Boiss.

kotschyi Boiss & Heldr. subsp. *persica* (Boiss.) Takht. = Ch. persica
persica Boiss. (*Centaurea kotschyi* (Boiss. & Heldr.) Hayek var. *persica* (Boiss.) Wagenitz, *Cheirolepis kotschyi* Boiss & Heldr. subsp. *persica* (Boiss.) Takht.) - 2

Chlorocrepis Monn. = Hieracium

tristis (Willd. ex Spreng.) A. & D. Löve = Hieracium triste

Chondrilla L.

acantholepis Boiss. (*Ch. juncea* L. subsp. *acantholepis* (Boiss.) Takht.) - 1, 2, 6

ambigua Fisch. ex Kar. & Kir. - 1, 3, 6
aspera Poir. (*Prenanthes aspera* Schrad. ex Willd. iv-xii 1803, non Michx. iii 1803) - 3, 6
bosseana Iljin - 6
brevirostris Fisch. & C.A. Mey. - 1, 3, 5, 6
canescens Kar. & Kir. - 1, 3, 6
gibbirostris M. Pop. - 6
graminea Bieb. - 1
hispida (Pall.) Poir. - 1
juncea L. - 1, 2, 6
- subsp. *acantholepis* (Boiss.) Takht. = Ch. acantholepis
kusnezovii Iljin - 6
laticoronata Leonova - 3, 6
latifolia Bieb. - 1, 2, 6
lejosperma Kar. & Kir. - 6
macra Iljin - 6
macrocarpa Leonova - 6
maracandica Bunge - 6
mujunkumensis Iljin & Igolk. - 6
ornata Iljin - 6
pauciflora Ledeb. - 1, 3, 6
phaeocephala Rupr. - 6
piptocoma Fisch. & C.A. Mey. - 3, 6
rouillieri Kar. & Kir. - 6
tenuiramosa Pratov & Tagaev - 6

Chorisis DC.

repens (L.) DC. - 5

Chrysanthemum L.

arcticum L. subsp. *maekawanum* auct. = Arctanthemum kurilense
asiaticum Worosch. = Tanacetum boreale
***carinatum** Schousb. - 1
coreanum (Levl. & Vaniot) Nakai subsp. *maximowiczii* (Kom.) Worosch. = Dendranthema maximowiczii
***coronarium** L. - 1, 6
komarovii (Krasch. & N. Rubtz.) S.G. Hu = Kaschgaria komarovii
leucanthemum L. var. *margaritae* Gayer ex Jav. = Leucanthemum margaritae
manshuricum (Poljak.) Kitag. = Ajania manshurica
millefolium (L.) E.I. Nyarady = Tanacetum millefolium
polycladum Rech. fil. = Xylanthemum polycladum
popovii Gilli = Xylanthemum rupestre
roxburghii Cass. - 5
segetum L. - 1, 2
sichotense (Tzvel.) Worosch. = Dendranthema sichotense
turkestanicum (Regel & Schmalh.) Gilli = Lepidolopsis turkestanica
vulgare (L.) Bernh. subsp. *boreale* (Fisch. ex DC.) Worosch. = Tanacetum boreale
zawadskii Herbich subsp. *erubescens* (Stapf) Kitag. = Dendranthema erubescens

Chrysocoma auct. = Galatella

tatarica Less. = Galatella tatarica

Cicerbita Wallr.

alpina (L.) Wallr. - 1
azurea (Ledeb.) Beauverd - 3, 4, 6
bourgaei (Boiss.) Beauverd - 2
deltoidea (Bieb.) Beauverd - 2
kovalevskiana Kirp. - 6

macrophylla (Willd.) Wallr. (*C. sevanensis* Kirp., *Prenanthes cacaliifolia* (Bieb.) Beauverd, *Sonchus cacaliifolius* Bieb.) - 1, 2
- subsp. *uralensis* (Rouy) P.D. Sell = C. uralensis
madatapensis Gagnidze - 2
olgae Leskov - 2
petiolata (C. Koch) Gagnidze (*Mulgedium petiolatum* C. Koch, *Prenanthes cacaliifolia* sensu Kirp.) - 2
pontica (Boiss.) Grossh. (*Prenanthes pontica* (Boiss.) Leskov) - 2
prenanthoides (Bieb.) Beauverd - 2
racemosa (Willd.) Beauverd - 2
rosea (M. Pop. & Vved.) Kovalevsk. - 6
sevanensis Kirp. = C. macrophylla
thianschanica Regel & Schmalh. - 6
uralensis (Rouy) Beauverd (*C. macrophylla* (Willd.) Wallr. subsp. *uralensis* (Rouy) P.D. Sell) - 1
zeravschanica M. Pop. ex Kovalevsk. - 6

Cichorium L.

***endivia** L. - 1
- subsp. *divaricatum* (Schousb.) P.D. Sell = C. pumilum
glabratum C. Presl = C. intybus subsp. glabratum
glandulosum Boiss. & Huet - 2
intybus L. - 1, 2, 3, 4, 5(alien),6
- subsp. *divaricatum* (Schousb.) Bonnier & Layens = C. pumilum
- subsp. **glabratum** (C. Presl) Arcang. (*C. glabratum* C. Presl, *C. intybus* subsp. *glaucum* (Hoffmgg. & Link) Tzvel.) - 1, 2, 6
- subsp. *glaucum* (Hoffmgg. & Link) Tzvel. = C. intybus subsp. glabratum
pumilum Jacq. (*C. endivia* L. subsp. *divaricatum* (Schousb.) P.D. Sell, *C. intybus* L. subsp. *divaricatum* (Schousb.) Bonnier & Layens) - 2

Cineraria auct. = Tephroseris

congesta R. Br. = Tephroseris palustris
crispa Jacq. = Tephroseris crispa
frigida Richards. f. *tomentosa* Kjellm. = Tephroseris kjellmanii
- b. *taimyrensis* Herd. = Tephroseris atropurpurea
heterophylla Fisch. = Tephroseris heterophylla
pratensis Hoppe = Tephroseris integrifolia
subdentata Bunge = Tephroseris subdentata

Cirsium Hill (*Breea* Less., *Lamyropsis* (Charadze) M. Dittrich)

abkhasicum (Petrak) Grossh. - 2
acaule Scop. (*C. acaulon* (L.) Scop. nom. illegit.) - 1
- subsp. *esculentum* (Siev.) Werner = C. esculentum
acaulon (L.) Scop. = C. acaule
adjaricum Somm. & Levier - 2
aduncum Fisch. & C.A. Mey. ex DC. - 2
x **affine** Tausch. - C. heterophyllum (L.) Hill x C. oleraceum (L.) Scop. - 1
aggregatum Ledeb. - 2
alatum (S.G. Gmel.) Bobr. - 1, 3, 6
alberti Regel & Schmalh. - 6
albowianum Somm. & Levier - 2
anatolicum (Petrak) Grossh. (*C. cataonicum* auct.) - 2
apiculatum DC. = C. libanoticum
- subsp. *glaberrimum* Petrak = C. glaberrimum
arachnoideum (Bieb.) Bess. - 2

argillosum V. Petrov ex Charadze - 2
armenum DC. - 2
arvense (L.) Scop. - 1, 2
badakhschanicum Charadze - 6
balkharicum Charadze - 2
biebersteinii Charadze = C. canum
bornmuelleri Sint. ex Bornm. - 6
bracteosum DC. - 2
brevipapposum Tscherneva - 2, 6
buschianum Charadze - 2
canum (L.) All. (*C. biebersteinii* Charadze) - 1, 2, 3
caput-medusae Somm. & Levier - 2
cataonicum auct. = C. anatolicum
caucasicum (Adams) Petrak - 2
cephalotes Boiss. - 2
charadzeae (Kimeridze) Czer. comb. nova (*Lamyropsis charadzeae* Kimeridze, 1981, Zam. Syst. Geogr. (Tbilisi) 37 : 31) - 2
chlorocomos Somm. & Levier - 2
x **ciliatiforme** Petrak. - C. ciliatum (Murr.) Moench x C. kosmelii (Adams) Fisch. ex Hohen.
ciliatum (Murr.) Moench - 1, 2, 3
congestum Fisch. & C.A. Mey. ex DC. - 2, 6
- var. *sorocephalum* (Fisch. & C.A. Mey.) Petrak = C. sorocephalum
coryletorum Kom. - 5
cosmelii auct. = C. kosmelii
czerkessicum Charadze - 2
daghestanicum Charadze - 2
dealbatum Bieb. - 2
decussatum Janka - 1
depilatum Boiss. & Bal. (*C. depilatum* var. *glomeratum* Freyn & Sint., *C. pubigerum* (Desf.) DC. var. *glomeratum* (Freyn & Sint.) P.H. Davis & Parris) - 2
- var. *glomeratum* Freyn & Sint. = C. depilatum
echinus (Bieb.) Hand.-Mazz. - 2
elbrusense Somm. & Levier - 2
elodes Bieb. (*C. rhabdotolepis* Petrak, *C. subinerme* Fisch. & C.A. Mey.) - 2
erisithales (Jacq.) Scop. - 1
x **erucagineum** DC. - C. oleraceum (L.) Scop. x C. rivulare (Jacq.) All. - 1
erythrolepis C. Koch - 2
esculentum (Siev.) C.A. Mey. (*C. acaule* Scop. subsp. *esculentum* (Siev.) Werner) - 1, 3, 4, 6
euxinum Charadze - 2
fominii Petrak - 2
frickii Fisch. & C.A. Mey. - 2
gagnidzei Charadze - 2
glaberrimum (Petrak) Petrak (*C. apiculatum* DC. subsp. *glaberrimum* Petrak) - 6
glabrifolium (C. Winkl.) O. & B. Fedtsch. - 6
x **grossheimii** Petrak. - C. echinus (Bieb.) Hand.-Mazz. x C. subinerme Fisch. & C.A. Mey. - 2
helenioides (L.) Hill - 1, 3, 4, 5, 6
heterophyllum (L.) Hill - 1, 3, 4
- subsp. *angarense* M. Pop. - 4
x **hybridum** Koch ex DC. - C. oleraceum (L.) Scop. x C. palustre (L.) Scop. - 1
hydrophiloides Charadze - 2
hydrophilum Boiss. - 2
hypoleucum DC. - 2
imereticum Boiss. - 2
incanum (S.G. Gmel.) Fisch. - 1, 2, 6
x **ispolatovii** Iljin ex Tzvel. - C. roseolum Gorlaczova x C. vulgare (Savi) Ten. - 1
kamtschaticum Ledeb. - 5
- subsp. *weyrichii* (Maxim.) Worosch. = C. weyrichii

kemulariae Charadze - 2
ketzkhovelii Charadze - 2
komarovii Schischk. - 3
kosmelii (Adams) Fisch. ex Hohen. (*C. cosmelii* auct.) - 2
x **kozlowskyi** Petrak. - C. obvallatum (Bieb.) Fisch. x C. simplex C.A. Mey. - 2
kusnezowianum Somm. & Levier - 2
lamyroides Tamamsch. (*Lamyropsis macrantha* (Schrenk) M. Dittrich) - 6
laniflorum (Bieb.) Fisch. - 1
lappaceum (Bieb.) Fisch. - 2
libanoticum DC. (*C. apiculatum* DC.) - 6
x **lojkae** Somm. & Levier. - C. echinus (Bieb.) Hand.-Mazz. x C. obvallatum (Bieb.) Fisch. - 2
longiflorum Charadze - 2
maackii Maxim. - 5
macrobotrys (C. Koch) Boiss. - 2
macrocephalum C.A. Mey. - 2
megricum Charadze - 2
munitum (Bieb.) Fisch. = C. pugnax
oblongifolium C. Koch - 2
obvallatum (Bieb.) Fisch. - 2
ochrolepideum Juz. (*Breea ochrolepidea* (Juz.) Sojak) - 6
oleraceum (L.) Scop. - 1, 3
osseticum (Adam) Petrak - 2
palustre (L.) Scop. - 1, 3, 4
pannonicum (L. fil.) Link - 1
pectinellum A. Gray - 5
pendulum Fisch. ex DC. - 4, 5
x **petrakii** Kozl. & Woronow ex Grossh. - C. caucasicum (Adams) Petrak x C. ciliatum (Murr.) Moench - 2
  The name of this hybris, apparently, was not validly published.
polonicum (Petrak) Iljin - 1
polyacanthum Kar. & Kir. (*C. sieversii* (Fisch. & C.A. Mey.) Petrak) - 3, 6
x **prativagum** Petrak. - C. kuznezowianum Somm. & Levier x C. obvallatum (Bieb.) Fisch. - 2
pseudolappaceum Charadze (*C. turkestanicum* (Regel) Petrak var. *pseudolappaceum* (Charadze) Petrak) - 6
pubigerum (Desf.) DC. var. *glomeratum* (Freyn & Sint.) P.H. Davis & Parris = C. depilatum
pugnax Somm. & Levier (*C. munitum* (Bieb.) Fisch. nom. superfl. illegit., *Cnicus munitus* Bieb. nom. superfl. illegit.) - 2
rassulovae B. Scharipova - 6
x **reichenbachianum** M. Loehr. - C. arvense (L.) Scop. x C. oleraceum (L.) Scop. - 1
x **reichardtii** Juratzka. - C. palustre (L.) Scop. x C. waldsteinii Rouy - 1
rhabdotolepis Petrak = C. elodes
rhizocephalum C.A. Mey. - 2, 6
- subsp. **sinuatum** (Boiss.) P.H. Davis & Parris (*C. rhizocephalum* var. *sinuatum* Boiss.) - 2
x **rigens** (Ait.) Wallr. - C. acaule Scop. x C. oleraceum (L.) Scop. - 1
rigidum DC. - 2
rivulare (Jacq.) All. - 1
roseolum Gorlaczova - 1
sairamense (C. Winkl.) O. & B. Fedtsch. - 6
schantarense Trautv. & C.A. Mey. - 5
schelkownikowii Petrak = C. turkestanicum
schischkinii Serg. - 3, 4
semenovii Regel - 6
serratuloides (L.) Hill - 3, 4, 5
serrulatum (Bieb.) Fisch. - 1, 3
setosum (Willd.) Bess. (*Breea setosa* (Willd.) Sojak) - 1, 2, 3, 4, 5, 6

sieversii (Fisch. & C.A. Mey.) Petrak = C. polyacanthum

x **silesiakum** Sch. Bip. - C. canum (L.) All. x C. palustre (L.) Scop. - 1

simplex C.A. Mey. - 2

sinuatum (Trautv.) Boiss. (*Lamyropsis sinuata* (Trautv.) M. Dittrich) - 2

sorocephalum Fisch. & C.A. Mey. (*C. congestum* Fisch. & C.A. Mey. ex DC. var. *sorocephalum* (Fisch. & C.A. Mey.) Petrak) - 2

sosnowskyi Charadze - 2

strigosum (Bieb.) Fisch. - 2

x **subalpium** Gaudin. - C. palustre (L.) Scop. x C. rivulare (Jacq.) Scop. - 1

subinerme Fisch. & C.A. Mey. = C. elodes

sublaniflorum Sojak - 1

x **succinctum** Ledeb. - ? C. kosmelii (Adams) Fisch. ex Hohen. x C. tricholoma Fisch. & C.A. Mey. - 2

svaneticum Somm. & Levier - 2

sychnosanthum Petrak - 2

szovitsii (C. Koch) Boiss. - 2

x **tataricum** (Jacq.) All. - C. canum (L.) All. x C. oleraceum (L.) Scop. - 1

tauricum Sojak - 1

tindaicum Charadze - 2

tomentosum C.A. Mey. - 2

tricholoma Fisch. & C.A. Mey. - 2

x **trifurcum** Petrak. - C. caucasicum (Adams) Petrak x C. kosmelii (Adams) Fisch. ex Hohen. - 2

turkestanicum (Regel) Petrak (*C. schelkownikowii* Petrak) - 2, 6

- var. *pseudolappaceum* (Charadze) Petrak = C. pseudolappaceum

ukranicum Bess. - 1

uliginosum (Bieb.) Fisch. - 2

vlassovianum Fisch. - 4, 5

vulgare (Savi) Ten. - 1, 2, 3, 4, 5, 6

waldsteinii Rouy - 1

x **wankelii** H. Reichardt. - C. heterophyllum (L.) Hill x C. palustre (L.) Scop. - 1

weyrichii Maxim. (*C. kamtschaticum* Ledeb. subsp. *weyrichii* (Maxim.) Worosch.) - 5

x **woronowii** Petrak. - C. hypoleucum DC. x C. obvallatum (Bieb.) Fisch. - 2

x **xenogenum** Petrak. - C. aduncum Fisch. & C.A. Mey. ex DC. x C. ciliatum (Murr.) Moench - 2

Cladochaeta DC.

candissima (Bieb.) DC. - 2

caspica Sosn. ex Grossh. - 2

Cnicus L.

benedictus L. - 1, 2, 6

munitus Bieb. = Cirsium pugnax

Codonocephalum Fenzl

peacockianum Aitch. & Hemsl. (*Inula peacockiana* (Aitch. & Hemsl.) Korov.) - 6

Coleostephus Cass.

myconis (L.) Reichenb. fil. - 1(alien)

Colymbada Hill = Centaurea

adpressa (Ledeb.) Holub = Centaurea adpressa

apiculata (Ledeb.) Holub = Centaurea apiculata

carduiformis (DC.) Holub = Centaurea carduiformis

glehnii (Trautv.) Holub = Centaurea glehnii

kotschyana (Heuff.) Holub = Centaurea kotschyana

orientalis (L.) Holub = Centaurea orientalis

pseudoscabiosa (Boiss. & Buhse) Holub = Centaurea pseudoscabiosa

reflexa (Lam.) Holub = Centaurea reflexa

salonitana (Vis.) Holub = Centaurea salonitana

scabiosa (L.) Holub = Centaurea scabiosa

scabiosa (L.) Rauschert = Centaurea scabiosa

sosnowskyi (Grossh.) Holub = Centaurea sosnowskyi

stereophylla (Bess.) Holub = Centaurea stereophylla

Conyza Less.

bonariensis (L.) Cronq. (*Erigeron bonariensis* L.) - 2

canadensis (L.) Cronq. (*Erigeron canadensis* L.) - 1, 2, 3, 4, 5, 6

Conyzanthus Tamamsch.

graminifolius (Spreng.) Tamamsch. - 1, 2

squamatus (Spreng.) Tamamsch. - 2

*Coreopsis L.

*grandiflora Hoog ex Sweet - 1, 6

*tinctoria Nutt. - 1, 2, 6

*Cosmos Cav.

*bipinnatus Cav. - 1, 2, 4, 6

*sulphureus Cav. - 1

Cota J. Gay = Anthemis

altissima (L.) J. Gay = Anthemis altissima

austriaca (Jacq.) J. Gay = Anthemis austriaca

cretacea (Zefir.) Holub = Anthemis monantha

dubia (Stev.) Holub = Anthemis dubia

dumetorum (Sosn.) Holub = Anthemis dumetorum

euxina (Boiss.) Holub = Anthemis euxina

jailensis (Zefir.) Holub = Anthemis jailensis

melanoloma (Trautv.) Holub = Anthemis melanoloma

parnassica Boiss. & Heldr. = Anthemis parnassica

rigescens (Willd.) Holub = Anthemis rigescens

tinctoria (L.) J. Gay = Anthemis tinctoria

- subsp. *subtinctoria* (Dobrocz.) Holub = Anthemis subtinctoria

wiedemanniana (Fisch. & C.A. Mey.) Holub = Anthemis wiedemanniana

woronowii (Sosn.) Holub = Anthemis woronowii

zygia (Woronow) Holub = Anthemis zygia

Cotula L.

aurea L. = Lepidotheca aurea

coronopifolia L. - 5

Cousinia Cass.

abbreviata Tscherneva - 6

abolinii Kult. ex Tscherneva - 6

acrodroma Tscherneva - 6

adenophora Juz. - 6

affinis Schrenk - 6

agelocephala Tscherneva - 6

aitchinsonii C. Winkl. (*C. lasiosiphon* Juz.) - 6

alaica Juz. ex Tscherneva - 6

alata Schrenk - 6

alberti Regel & Schmalh. - 6
albertoregelia C. Winkl. - 6
albicaulis Boiss.& Buhse - 2
albiflora (Bornm. & Sint.) Bornm. - 6
allolepis Tscherneva & Vved. - 6
alpestris (Bornm.) Juz. - 6
alpina Bunge - 6
ambigens Juz. - 6
amoena C. Winkl. - 6
androssovii Juz. - 6
angreni Juz. - 6
angusticeps Juz. - 6
anomala Franch. - 6
antonowii C. Winkl. - 6
apiculata Tscherneva - 6
arachnoidea Fisch. & C.A. Mey. - 6
araxena Takht. - 2
arctioides Schrenk - 6
arctotidifolia Bunge - 6
armena Takht. - 2
aspera (Kult.) Karmyscheva (*C. chrysantha* Kult. var. *aspera* Kult.) - 6
astracanica (Spreng.) Tamamsch. - 1, 6
x **atripurpurea** Juz. - C. coronata Franch. x C. microcarpa Boiss. - 6
aurea C. Winkl. - 6
auriculata auct. = C. semidecurrens
badghysi Kult. - 6
baranovii Juz. ex Tscherneva - 6
batalinii C. Winkl. - 6
beckeri Trautv. - 6
bobrovii Juz. - 6
bonvalotii Franch. - 6
bornmuelleri C. Winkl. - 2
botschantzevii Juz. ex Tscherneva - 6
brachyptera DC. - 2
bungeana Regel & Schmalh. - 6
buphtalmoides Regel - 6
butkovii Tscherneva & Vved. - 6
caespitosa C. Winkl. - 6
calva Juz. - 6
campyloraphis Tscherneva - 6
x **cana** Juz. - C. ninae Juz. x C. sewerzowii Regel - 6
candicans Juz. - 6
carduncelloidea Regel & Schmalh. - 6
centauroides Fisch. & C.A. Mey. ex Bunge - 6
ceratophora Kult. - 6
chaetocephala Kult. - 6
chejrabadensis Kult. - 6
chlorantha Kult. - 6
chlorocephala C.A. Mey. - 2
chrysantha Kult. - 6
- var. *aspera* Kult. = C. aspera
coerulea Kult. ex Tscherneva - 6
congesta Bunge - 6
coronata Bunge - 6
corymbosa C. Winkl. - 6
cryptadena Juz. - 6
cynaroides (Bieb.) C.A. Mey. - 2, 6
daralaghezica Takht. - 2
darwasica C. Winkl. - 6
decurrentifolia Juz. ex Tscherneva - 6
dichotoma Bunge - 6
dichromata Kult. - 6
dimoana Kult. - 6
dissecta Kar. & Kir. - 6
dissectifolia Kult. - 6

divaricata C. Winkl. - 6
dolichoclada Juz. - 6
dolicholepsis Schrenk - 6
dolychophylla Kult. - 6
dshisakensis Kult. - 6
x **dualis** Juz. - C. pseudarctium Bornm. x C. umbrosa Bunge - 6
dubia M. Pop. - 6
egens Juz. - 6
egregia Juz. - 6
erectispina Tscherneva - 6
eriotricha Juz. - 6
erivanensis Bornm. - 2
eryngioides Boiss. - 6
eugenii Kult. - 6
fallax C. Winkl. - 6
fascicularis Juz. - 6
fedorovii Takht. - 2
fedtschenkoana Bornm. - 6
ferganensis Bornm. - 6
ferruginea Kult. - 6
fetissowii C. Winkl. - 6
finitima Juz. - 6
franchetii C. Winkl. - 6
freynii Bornm. & Sint. - 6
gabrieljaniae Takht. & Tamanjan - 2
gigantolepis Rech. fil. - 2
glabriseta Kult. - 6
glandulosa Kult. - 6
glaphyrocephala Juz. ex Tscherneva - 6
glochidiata Kult. - 6
gnezdilloi Tscherneva - 6
gomolitzkii Juz. ex Tscherneva - 6
gontscharowii Juz. - 6
grandifolia Kult. - 6
grigoriewii Juz. - 6
grisea Kult. - 6
gulczensis Kult. - 6
haesitabunda Juz. - 6
hamadae Juz. - 6
hastifolia C. Winkl. - 6
x **heterogenetos** Bornm. - C. outichaschensis Franch. x C. pulchella Bunge - 6
x **heteromorpha** Bornm. - C. alpestris (Bornm.) Juz. x C. submutica Franch. - 6
hilariae Kult. - 6
hohenackeri Fisch. & C.A. Mey. - 2
hoplophylla Tscherneva - 6
horrescens Juz. - 6
horridula Juz. - 6
hypopolia Bornm. & Sint. - 6
hystrix C.A. Mey. = C. pterocaulos
iljinii Takht. - 2
integrifolia Franch. - 6
x **iskanderi** Bornm. - C. outichaschensis Franch. x C. submutica Franch. - 6
jassyensis C. Winkl. - 6
juzepczukii Tscherneva - 6
karatavica Regel & Schmalh. - 6
kazachorum Juz. ex Tscherneva - 6
knorringiae Bornm. - 6
kokanica Regel & Schmalh. - 6
komarowii (O. Kuntze) C. Winkl. - 6
korolkowii Regel & Schmalh. - 6
korshinskyi C. Winkl. - 6
kovalevskiana Tscherneva - 6
krauseana Regel & Schmalh. - 6

kuekenthalii Bornm. - 6
laetevirens C. Winkl. - 6
lanata C. Winkl. - 6
laniceps Juz. - 6
lappacea Schrenk - 6
lasiosiphon Juz. = C. aitchinsonii
leiocephala (Regel) Juz. - 6
leptacantha (Bornm.) Juz. - 6
leptacma Tscherneva - 6
leptocampyla Bornm. - 6
leptocephala Fisch. & C.A. Mey. ex Bunge - 6
leptoclada Kult. - 6
leptocladoides Tscherneva - 6
leucantha Bornm. & Sint. - 6
linczewskii Juz. - 6
lindbergii Rech. fil. - 6
litwinowii Kult. ex Juz. - 6
lomakinii C. Winkl. -2
lyrata Bunge - 6
macilenta C. Winkl. - 6
macrocephala C.A. Mey. - 2
macroptera C.A. Mey. - 2
magnifica Juz. - 6
maracandica Juz. - 6
margaritae Kult. - 6
margiana Juz. - 6
medians Juz. - 6
meghrica Takht. - 2
microcarpa Boiss. - 6
microcephala C.A. Mey. - 2
mindschelkensis B. Fedtsch. - 6
minkwitziae Bornm. - 6
minuta Boiss. = C. prolifera
mogoltaica Tscherneva & Vved. - 6
mollis Schrenk - 6
mucida Kult. - 6
mulgediifolia Bornm. - 6
multiloba DC. - 6
murgabica Tscherneva - 6
neglecta Juz. - 6
newesskiana C. Winkl. - 6
ninae Juz. - 6
olgae Regel & Schmalh. - 6
omphalodes Tscherneva - 6
onopordioides Ledeb. - 6
oopoda Juz. - 6
oreodoxa Bornm. & Sint. - 6
oreoxerophila Kult. - 6
orientalis (Adams) C. Koch - 2
orthacantha Tscherneva - 6
ortholepis Juz. ex Tscherneva - 6
outichaschensis Franch. - 6
ovczinnikovii Tscherneva - 6
oxiana Tscherneva - 6
oxytoma Rech. fil. - 6
pannosa C. Winkl. - 6
pannosiformis Tscherneva - 6
x paraatripurpurea R. Kam. - C. microcarpa Boiss. x C.
    radians Bunge - 6
patentispina Tscherneva - 6
pauciramosa Kult. - 6
peduncularis Juz. ex Tscherneva - 6
pentacantha Regel & Schmalh. - 6
pentacanthoides Juz. ex Tscherneva - 6
perovskiensis (Bornm.) Juz. ex Tscherneva - 6
platylepis Schrenk - 6
platystegia Tscherneva - 6

podophylla Tscherneva - 6
polycephala Rupr. - 6
polytimetica Tscherneva - 6
praestans Tscherneva & Vved. - 6
princeps Franch. - 6
prolifera Jaub. & Spach (*C. minuta* Boiss.) - 6
proxima Juz. - 6
psammophila Kult. - 6
pseudarctium Bornm. - 6
pseudoaffinis Kult. - 6
pseudodshisakensis Tscherneva & Vved. - 6
pseudolanata M. Pop. ex Tscherneva - 6
pseudomollis C. Winkl. - 6
pterocaulos (C. A. Mey.) Rech. fil. (*Carthamus pterocaulos*
    C.A. Mey., *Cousinia hystrix* C.A. Mey.) - 2
pterolepida Kult. - 6
pulchella Bunge - 6
pulchra C. Winkl. - 6
pungens Juz. - 6
purpurea C.A. Mey. ex DC. - 2
pusilla C. Winkl. - 6
pycnocephala Rech. fil. - 6
pygmaea C. Winkl. - 6
qaradaghensis Rech. fil. - 2
raddeana C. Winkl. - 6
radians Bunge - 6
ramulosa Rech. fil. - 6
rava C. Winkl. - 6
refracta (Bornm.) Juz. - 6
regelii C. Winkl. - 6
resinosa Juz. - 6
rhodantha Kult. - 6
rigida Kult. - 6
rosea Kult. - 6
rotundifolia C. Winkl. - 6
rubiginosa Kult. - 6
sarawschanica C. Winkl. - 6
scabrida Juz. - 6
schepsaica Karmyscheva - 6
schischkinii Juz. - 6
schistoptera Juz. - 6
schmalhausenii C. Winkl. - 6
schtschurowskiana Regel & Schmalh. - 6
schugnanica Juz. - 6
scleracantha Kult. ex Tscherneva - 6
sclerophylla Juz. - 6
semidecurrens C. Winkl. (*C. auriculata* auct.) - 6
semilacera Juz. - 6
sewerzowii Regel - 6
simulatrix C. Winkl. - 6
smirnowii Trautv. - 6
sogdiana Bornm. - 6
sororia Juz. - 6
speciosa C. Winkl. - 6
spiridonovii Juz. - 6
splendida C. Winkl. - 6
sporadocephala Juz. - 6
spryginii Kult. - 6
stahliana Bornm. & Gauba - 6
stellaris Bornm. - 6
stenophylla Kult. - 6
stephanophora C. Winkl. - 6
stricta Tscherneva - 6
strobilocephala Tscherneva & Vved. - 6
subappendiculata Kult. - 6
subcandicans Tscherneva - 6
submutica Franch. - 6

sylvicola Bunge - 6
syrdariensis Kult. - 6
takhtajanii Tamanian - 2
talassica (Kult.) Juz. ex Tscherneva - 6
tamarae Juz. - 6
tedshenica Tscherneva - 6
tenella Fisch. & C.A. Mey. - 2, 6
tenuisecta Juz. - 6
tianschanica Kult. - 6
tomentella C. Winkl. - 6
trachyphylla Juz. - 6
transiliensis Juz. - 6
transoxana Tscherneva - 6
x triacantha Kult. - C. alberti Regel & Schmalh. x C. umbrosa Bunge - 6
triceps Kult. - 6
trichophora Kult. - 6
triflora Schrenk - 6
tscherneviae Beldyev - 6
turcomanica C. Winkl. - 6
turkestanica (Regel) Juz. - 6
turkmenorum Bornm. - 6
ugamensis Karmyscheva - 6
ulotoma Bornm. - 6
umbilicata Juz. - 6
umbrosa Bunge - 6
vavilovii Kult. - 6
verticillaris Bunge - 6
vicaria Kult. - 6
vvedenskyi Tscherneva - 6
waldheimiana Bornm. - 6
xanthina Bornm. - 6
xanthiocephala Tscherneva - 6

Cousiniopsis Nevski

atractyloides (C. Winkl.) Nevski - 6

Crepis L.   (*Askellia* W.A. Weber, *Barkhausia* Moench)

alaica Krasch. (*Askellia alaica* (Krasch.) W.A. Weber) - 6
albescens Kuvajev & Demidova - 1
alikeri Tamamsch. - 2
alpina L. (*Barkhausia pseudoalpina* Klok.) - 1, 2
altaica (Babc. & Stebb.) Roldug. = Youngia altaica
astrachanica Stev. ex Czer. - 1, 2
biennis L. - 1
bungei Ledeb. - 3, 4
burejensis Fr. Schmidt - 5
- subsp. *hokkaidoensis* (Babc.) Worosch. = C. hokkaidoensis
capillaris (L.) Wallr. - 1, 2(alien)
caucasica C.A. Mey. - 2
caucasigena Czer. - 2
chrysantha (Ledeb.) Turcz. - 1, 3, 4, 5, 6
ciliata C. Koch - 2
conyzifolia (Gouan) A. Kerner - 1
corniculata Regel & Schmalh. (*Askellia corniculata* (Regel & Schmalh.) W.A. Weber) - 6
crocea (Lam.) Babc. - 3, 4
czuensis Serg. - 3
darvazica Krasch. - 6
flexuosa (Ledeb.) Clarke (*Askellia flexuosa* (Ledeb.) W. A. Weber) - 3, 6
foetida L. - 1(alien), 2
- subsp. *rhoeadifolia* (Bieb.) Celak. = C. rhoeadifolia

foliosa Babc. - 1
glabra Boiss. - 2
gmelinii (L.) Tausch - 4
hokkaidoensis Babc. (*C. burejensis* Fr. Schmidt subsp. *hokkaidoensis* (Babc.) Worosch.) - 5
jacquinii Tausch - 1(?)
karakuschensis Czer. - 2
karelinii M. Pop. & Schischk. ex Czer. (*Askellia karelinii* (M. Pop. & Schischk. ex Czer.) W.A. Weber) - 3, 6
■ khorassanica Boiss.
kotschyana (Boiss.) Boiss. (*Barkhausia kotschyana* Boiss.) - 6
lactea Lipsch. (*Askellia lactea* (Lipsch.) W.A. Weber) - 6
lyrata (L.) Froel. - 3, 4
marschallii (C.A. Mey.) F. Schultz - 2
micrantha Czer. - 1, 2, 6
mollis (Jacq.) Aschers. (*C. planitierum* Klok.) - 1
multicaulis Ledeb. - 1, 3, 4, 6
nana Richards. (*Askellia nana* (Richards.) W.A. Weber) - 3, 4, 5, 6(?)
nigrescens Pohle (*C. tectorum* L. subsp. *nigrescens* (Pohle) A. & D. Löve, *C. tectorum* subsp. *nigrescens* (Pohle) P.D. Sell, comb. superfl.) - 1, 3
oreades Schrenk - 6
paludosa (L.) Moench - 1, 3
pannonica (Jacq.) C. Koch - 1, 2, 3
planitierum Klok. = C. mollis
polytricha (Ledeb.) Turcz. - 3, 4
pontica C.A. Mey. - 2
praemorsa (L.) Tausch - 1, 3, 4
pulchra L. - 1, 2, 6
- subsp. turkestanica Babc. - 6
ramosissima D'Urv. - 1
rhoeadifolia Bieb. (*C. foetida* L. subsp. *rhoeadifolia* (Bieb.) Celak.) - 1, 2, 6
setosa Hall. fil. - 1, 2
sibirica L. - 1, 2, 3, 4, 6
sogdiana (Krasch.) Czer. (*Askellia sogdiana* (Krasch.) W.A. Weber) - 6
sonchifolia (Bieb.) C.A. Mey. - 2
tatewakii Kudo = Hieracium tatewakii
tectorum L. - 1, 3, 4, 5, 6
- subsp. *nigrescens* (Pohle) A. & D. Löve = C. nigrescens
- subsp. *nigrescens* (Pohle) P.D. Sell = C. nigrescens
trichocephala (Krasch.) V.V. Nikit. - 6
tristis Klok. - 1
tungusica Egor. & Sipl. - 4
turcomanica Krasch. - 6
willdenowii Czer. - 2
willemetioides Boiss. - 6

Crinitaria Cass. = Galatella

fominii (Kem.-Nath.) Czer. = Galatella fominii
fominii (Kem.-Nath.) Sojak = Galatella fominii
grimmii (Regel & Schmalh.) Grierson = Pseudolinosyris grimmii
linosyris (L.) Less. = Galatella linosyris
pontica (Lipsky) Czer. = Galatella pontica
pontica (Lipsky) Sojak = Galatella pontica
tatarica (Less.) Czer. = Galatella tatarica
tatarica (Less.) Sojak = Galatella tatarica
villosa (L.) Grossh. = Galatella villosa

Crupina (Pers.) DC.

crupinastrum (G. Moris) Vis. - 2
vulgaris Cass. - 1, 2, 6

Cyanus Hill = Centaurea

acmophyllus (Boiss.) Sojak = Centaurea acmophylla
angelescui (Grint.) Holub = Centaurea angelescui
cheiranthifolius (Willd.) Sojak = Centaurea cheiranthifolia
depressus (Bieb.) Sojak = Centaurea depressa
fischeri (Schlecht.) Sojak = Centaurea fischeri
huetii (Boiss.) Sojak = Centaurea huetii
marmarosiensis (Jav.) Dostal = Centaurea marmarosiensis
montanus (L.) Hill subsp. *marmarosiensis* (Jav.) Sojak = Centaurea marmarosiensis
- subsp. *mollis* (Waldst. & Kit.) Sojak = Centaurea mollis
nigrofimbrius (C. Koch) Sojak = Centaurea nigrofimbria
strictus (Waldst. & Kit.) Sojak = Centaurea stricta
tanaiticus (Klok.) Sojak = Centaurea tanaitica
thirkei (Sch. Bip.) Holub = Centaurea thirkei
triumfettii (All.) Dostal ex A. & D. Löve subsp. *strictus* (Waldst. & Kit.) Dostal = Centaurea stricta
willdenowii (Czer.) Sojak = Centaurea willdenowii
woronowii (Bornm. ex Sosn.) Sojak = Centaurea woronowii

Cyclachaena Fresen.

xanthiifolia (Nutt.) Fresen. - 1, 2, 5

Cymbolaena Smoljian. (*Stylocline* auct.)

griffithii (A. Gray) Wagenitz (*Stylocline griffithii* A. Gray, *Cymbolaena longifolia* (Boiss. & Reut.) Smoljian.) - 2, 6
longifolia (Boiss. & Reut.) Smoljian. = Cymbolaena griffithii

*Cynara L.

*scolymus L. - 1

*Dahlia Cav.

*pinnata Cav. - 1, 6

Dendranthema (DC.) DesMoul.

arcticum (L.) Tzvel. = Arctanthemum arcticum
- subsp. *polare* (Hult.) Heywood = Arctanthemum hultenii
coreanum (Levl. & Vaniot) Worosch. (*Matricaria coreana* Levl. & Vaniot, *Dendranthema littorale* (F. Maek.) Tzvel. subsp. *coreanum* (Levl. & Vaniot) L.A. Lauener) - 5
erubescens (Stapf) Tzvel. (*Chrysanthemum zawadskii* Herbich subsp. *erubescens* (Stapf) Kitag.) - 5
hultenii (A. & D. Löve) Tzvel. = Arctanthemum hultenii
*indicum (L.) DesMoul. - 1, 2, 6
integrifolium (Richards.) Tzvel. = Hulteniella integrifolia
kurilense Tzvel. = Arctanthemum kurilense
littorale (F. Maek.) Tzvel. - 5
- subsp. *coreanum* (Levl. & Vaniot) L.A. Lauener = D. coreanum
maximowiczii (Kom.) Tzvel. (*Chrysanthemum coreanum* (Levl. & Vaniot) Nakai subsp. *maximowiczii* (Kom.) Worosch.) - 5
mongolicum (Ling) Tzvel. - 4, 5
*morifolium (Ramat.) Tzvel. - 1, 2, 6
naktongense (Nakai) Tzvel. - 5
pallasianum (Fisch. ex Bess.) Worosch. = Ajania pallasiana
sichotense Tzvel. (*Chrysanthemum sichotense* (Tzvel.) Worosch.) - 5
sinuatum (Ledeb.) Tzvel. - 3
weyrichii (Maxim.) Tzvel. - 5

xeromorphum A. Khokhr. - 5
zawadskii (Herbich) Tzvel. - 1, 3, 4, 5

Dichorocephala L'Her. (*Hippia* auct.)

bicolor (Roth) Schlecht. = D. integripolia
integripolia (L.fil.) O. Kuntze (*Hippia integrifolia* L. fil., *Dichorocephala bicolor* (Roth) Schlecht.) - 2

*Dimorphotheca Moench

*hybrida (L.) DC. (*Calendula hybrida* L.) - 1
*pluvialis (L.) Moench (*Calendula pluvialis* L.) - 1

Dipterocome Fisch. & C.A. Mey.

pusilla Fisch. & C.A. Mey. - 2, 6

Doellingeria Nees

scabra (Thunb.) Nees - 5

Dolichorrhiza (Pojark.) Galushko

caucasica (Bieb.) Galushko (*Ligularia caucasica* (Bieb.) G. Don fil.) - 2
correvoniana (Albov) Galushko (*Ligularia correvoniana* (Albov) Pojark.) - 2
renifolia (C.A. Mey.) Galushko (*Ligularia renifolia* (C.A. Mey.) DC.) - 2

Doronicum L.

altaicum Pall. (*D. bargusinense* Serg.) - 3, 4, 6
austriacum Jacq. - 1
balansae Cavill. - 2
bargusinense Serg. = D. altaicum
bellidiastrum L. = Bellidastrum michelii
carpaticum (Griseb. & Schenk) Nym. (*Doronicum columnae* auct.) - 1
clusii (All.) Tausch - 1
columnae auct. = D. carpaticum
dolichotrichum Cavill. = D. macrophyllum
hakkiaricum J.R. Edmondson = D. macrophyllum
hungaricum Reichenb. fil. (*D. longifolium* auct.) - 1
hyrcanum Widd. & Rech. fil. = D. macrophyllum
longifolium auct. = D. hungaricum
macrophyllum Fisch. ex Hornem. (*D. dolichotrichum* Cavill., *D. hakkiaricum* J.R. Edmondson, *D. hyrcanum* Widd. & Rech. fil.) - 2
oblongifolium DC. - 2, 3, 6
orientale Hoffm. - 1, 2
pardalianches L. - 1
schischkinii Serg. - 3
turkestanicum Cavill. - 3, 6

*Echinacea Moench

*purpurea (L.) Moench - 1, 6

Echinops L.

abstersibilis Iljin - 6
albicaulis Kar. & Kir. - 6
araxinus Mulk. = E. orientalis
armatus Stev. (*E. bannaticus* auct.) - 1
armenus Grossh. = E. orientalis
babatagensis Tscherneva - 6
bannaticus auct. = E. armatus
brevipenicillatus Tscherneva - 6

56

chantavicus Trautv. - 6
chloroleucus Rech. fil. - 6
chodzha-mumini Rassulova & B. Scharipova - 6
colchicus Sosn. - 2
■ **connatus** C. Koch
conrathii Freyn = E. pungens
cornigerus DC. (*E. polygraphus* Tscherneva) - 6
dagestanicus Iljin - 2
dasyanthus Regel & Schmalh. - 6
dissectus Kitag. - 5
dubjanskyi Iljin - 6
elatus auct. = E. lipskyi
erevanensis Mulk. = E. sphaerocephalus
exaltatus Schrad. - 1
fastigiatus R. Kam. & Tscherneva - 6
fedtschenkoi Iljin - 6
foliosus Somm. & Levier - 2
galaticus Freyn - 2
gmelinii Turcz. - 3
grossheimii Iljin = E. polygamus
hissaricus Rassulova & B. Scharipova - 6
humilis Bieb. - 3, 4
iljinii Mulk. = E. transcaucasicus
integrifolius Kar. & Kit. - 3, 6
kafirniganus Bobr. - 6
karabachensis Mulk. - 2
karatavicus Regel & Schmalh. - 6
kasakorum Pavl. - 6
knorringianus Iljin - 6
latifolius Tausch - 4
leiopolyceras Bornm. - 6
leucographus Bunge - 6
lipskyi Iljin (*E. elatus* auct.) - 6
macrophyllus Boiss. & Hausskn. - 6
maracandicus Bunge - 6
meyeri (DC.) Iljin (*E. ritro* L. subsp. *meyeri* (DC.) Kozuharov) - 1, 3, 6
multicaulis Nevski - 6
nanus Bunge - 3, 6
nuratavicus A. Li - 6
obliquilobus Iljin - 6
opacifolius Iljin - 2
orientalis Trautv. (*E. araxinus* Mulk., *E. armenus* Grossh.) - 2
ossicus C. Koch - 2
polyacanthus Iljin (*E. pungens* Trautv. var. *polyacanthus* (Iljin) Hedge) - 2
polygamus Bunge (*E. grossheimii* Iljin) - 2
polygraphus Tscherneva = E. cornigerus
praetermissus Nevski - 6
pseudomaracandicus Rassulova & B. Scharipova - 6
pubisquameus Ilin - 6
pungens Trautv.(*E. conrathii* Freyn, *E. szovitsii* Fisch. & C.A. Mey. ex DC.) - 2
- var. *polyacanthus* (Iljin) Hedge = E. polyacanthus
- var. *transcaucasicus* (Iljin) Hedge = E. transcaucasicus
ritro auct. = E. ruthenicus
- subsp. *meyeri* (DC.) Kozuharov = E. meyeri
ritrodes Bunge - 6
rubromontanus R. Kam. - 6
ruthenicus Bieb. (*E. ritro* auct.) - 1, 2, 3, 6
saissanicus (B. Keller) Bobr. - 3, 6
sevanensis Mulk. = E. transcaucasicus
sphaerocephalus L. (*E. erevanensis* Mulk.) - 1, 2, 3, 4, 6
spiniger Iljin ex Bobr. - 6
The name of this species was not validly published.
subglaber Schrenk

szovitsii Fisch. & C.A. Mey. ex DC. = E. pungens
talassicus Golosk. - 6
tjanschanicus Bobr. - 6
tournefortii Ledeb. - 2
transcaspicus Bornm. - 6
transcaucasicus Iljin (*E. iljinii* Mulk., *E. pungens* Trautv var. *transcaucasicus* (Iljin) Hedge, *E. sevanensis* Mulk.) - 2
transiliensis Golosk. - 6
tricholepis Schrenk - 6
tschimganicus B. Fedtsch. - 6
viridifolius Iljin - 2
wakhanicus Rech.fil. - 6

Eclipta L.
alba (L.) Hassk. - 2
prostrata (L.) L. - 2

Endocellion Turcz. ex Herd.
glaciale (Ledeb.) Toman (*Nardosmia glacialis* Ledeb., *Endocellion glaciale* (Ledeb.) A. & D. Löve, comb. superfl., *Petasites glacialis* (Ledeb.) Polun.) - 4, 5
gmelinii (Turcz. ex DC.) A. & D. Löve = E. sibiricum
sibiricum (J.F. Gmel.) Toman (*Tussilago sibirica* J.F. Gmel., *Endocellion gmelinii* (Turcz. ex DC.) A. & D. Löve, *Nardosmia gmelinii* Turcz.ex DC., *Petasites gmelinii* (Turz.ex DC.) Polun., *P. sibiricus* (J.F. Gmel.) Dingwall) - 1, 3, 4, 5

Epilasia (Bunge) Benth.
acrolasia (Bunge) Clarke - 6
hemilasia (Bunge) Clarke - 2, 6
mirabilis Lipsch. - 6

Erechtites Rafin.
hieracifolia (L.) Rafin. ex DC. (*Senecio hieracifolius* L.) - 1
valerianifolia (Wolf ex Reichenb.) DC. - 2

Erigeron L. (*Psychrogeton* Boiss.)
acer auct. = E. acris
acris L. (*E. acer* auct.) - 1, 2, 3, 4, 5, 6
- subsp. *brachycephalus* (Lindb.fil.) Hiit. = E. uralensis
- subsp. *decoloratus* (Lindl.fil.) Hiit. = E. uralensis
- subsp. *droebachiensis* (O.F. Muell.) Arcang. = E. droebachiensis
- subsp. *elongatiformis* Novopokr. = E. uralensis
- subsp. *kamtschaticus* (DC.) Hara = E. kamtschaticus
- subsp. *macrophyllus* auct. p.p. = E. podolicus
- subsp. *manshuricus* (Kom.) Worosch. = E. manshuricus
- subsp. *politus* (Fries) Lindb. fil. = E. politus
- subsp. *tilingii* (Worosch.) Worosch. = E. politus
- var. *manshuricus* Kom. = E. manshuricus
alaskanus Cronq. = E. hyperboreus
alexeenkoi (Krasch.) Botsch. (*Psychrogeton alexeenkoi* Krasch., *Erigeron karatavicus* Pavl.) - 6
allochrous Botsch. - 6
alpinus L. - 1, 2
altaicus M. Pop. - 3, 6
amorphoglossus Boiss. (*E. leucophyllus* (Bunge) Boiss., *E. mollissimus* Rech. fil. & Koeie, *Psychrogeton amorphoglossus* (Boiss.) Novopokr., *P. leucophyllus* (Bunge) Novopokr., *P. mollissimus* (Rech. fil. & Koeie) Ikonn.) - 6
andryaloides (DC.) Benth. (*Psychrogeton andryaloides*

(DC.) Novopokr. ex Krasch.) - 6

angulosus Gaudin - 1

annuus (L.) Pers. = Phalacroloma annuum

- subsp. *septentrionalis* (Fern. & Wieg.) Wagenitz = Phalacroloma septentrionale

- subsp. *strigosus* (Muehl. ex Willd.) Wagenitz = Phalacroloma strigosum

aurantiacus Regel - 6

azureus Regel ex M. Pop. - 6

badachschanicus Botsch. - 6

baicalensis Botsch. - 4

bellidiformis M. Pop. - 6

biramosus Botsch. (*Psychrogeton biramosus* (Botsch.) Grierson) - 6

bonariensis L. = Conyza bonariensis

borealis (Vierh.) Simm. - 1, 3, 4

brachycephalus Lindb. fil. = E. uralensis

brachyspermus Botsch. (*Psychrogeton brachyspermus* (Botsch.) Grierson) - 6

cabulicus (Boiss.) Botsch. (*Psychrogeton cabulicus* Boiss.) - 6

canadensis L. = Conyza canadensis

caucasicus Stev. - 2

- subsp. *venustus* (Botsch.) Grierson = E. venustus

compositus Pursh - 5

crispus Pourr. - 2, 6

decoloratus Lindb. fil. = E. uralensis

dolichostylus Botsch. - 6

droebachiensis O.F. Muell. (*E. acris* L. subsp. *droebachiensis* (O.F. Muell.) Arcang.) - 1

elongatiformis (Novopokr.) Serg. = E. uralensis

elongatus Ledeb. = E. politus

eriocalyx (Ledeb.) Vierh. (*E. korotkyi* Novopokr., *E. uniflorus* L. subsp. *eriocalyx* (Ledeb.) A. & D. Löve) - 1, 3, 4, 5

eriocephalus J. Vahl (*E. plurifolius* Botsch.) - 1, 2, 4, 5

flaccidus (Bunge) Botsch. - 3, 4

grandiflorus Hook. - 5

heterochaeta (Benth. ex Clarke) Botsch. - 6

hissaricus Botsch. - 6

humilis J. Grah. (*E. unalaschkensis* (DC.) Vierh.) - 5

hyperboreus Greene (*E. alaskanus* Cronq.) - 5

kamtschaticus DC. (*E. acris* L. subsp. *kamtschaticus* (DC.) Hara) - 5

karatavicus Pavl. = E. alexeenkoi

khorassanicus Boiss. (*Psychrogeton aucheri* (DC.) Grierson., p.p.) - 6

komarovii Botsch. (*E. komarovii* var. *koraginensis* (Kom.) Worosch., *E. koraginensis* (Kom.) Botsch. p.p. incl. typo, *E. thunbergii* A. Gray subsp. *komarovii* (Botsch.) A. & D. Löve, *E. thunbergii* subsp. *koraginensis* (Kom.) A. & D. Löve) - 1, 4, 5

- var. *koraginensis* (Kom.) Worosch. = E. komarovii

koraginensis (Kom.) Botsch. = E. komarovii

korotkyi Novopokr. = E. eriocalyx

krylovii Serg. - 3, 6

lachnocephalus Botsch. - 6

leioreades M. Pop. - 3

leucophyllus (Bunge) Boiss. = E. amorphoglossus

lonchophyllus Hook. - 3, 4, 6

macrophyllus auct. = E. podolicus

manshuricus (Kom.) Worosch. (*E. acris* L. var. *manshuricus* Kom., *E. acris* subsp. *manshuricus* (Kom.) Worosch.) - 5

mollissimus Rech. fil. & Koeie = E. amorphoglossus

nabidshonii R. Kam. (*Psychrogeton nabidshonii* (R.Kam.) Koczk. & Zhogoleva) - 6

nigromontanus Boiss. & Buhse (*Psychrogeton nigromontanus* (Boiss. & Buhse) Grierson) - 6

oharae (Nakai) Botsch. = Aster spathulifolius

olgae Regel & Schmalh.(*Psychrogeton olgae* (Regel & Schmalh.) Novopokr. ex Nevski) - 6

oreades (Schrenk) Fisch. & C.A. Mey. - 3, 4, 6

orientalis Boiss. - 1, 2, 6

pallidus M. Pop. - 6

pamiricus Botsch. & Koczk. - 6

peregrinus (Pursh) Greene - 5

petiolaris Vierh. - 3, 6

petroiketes Rech. fil. - 6

plurifolius Botsch. = E. eriocephalus

podolicus Bess. (*E. acris* L. subsp. *macrophyllus* auct. p.p., *E. macrophyllus* auct.) - 1, 2, 3, 6

politus Fries (*E. acris* L. subsp. *politus* (Fries) Lindb. fil., *E. acris* subsp. *tilingii* (Worosch.) Worosch. nom. invalid., *E. elongatus* Ledeb. 1829, non Moench, 1794, *E. tilingii* Worosch.) - 1, 3, 4, 5

polymorphus Scop. - 2

poncinsii (Franch) Botsch. (*Psychrogeton andryaloides* (DC.) Novopokr. ex Krasch. var. *poncinsii* (Franch.) Grierson, *Psychrogeton poncinsii* (Franch.) Ling & Y.L. Chen) - 6

popovii Botsch. - 6

primuloides M. Pop. (*Psychrogeton primuloides* (M. Pop.) Grierson) - 6

pseuderigeron (Bunge) M. Pop. - 6

pseuderiocephalus M. Pop. - 6

pseudoelongatus Botsch. - 2

pseudoneglectus M. Pop. - 6

pseudoseravschanicus Botsch. - 3, 6

pyrami Botsch. - 6

ramosus (Walt.) Britt., Sterns & Pogg. var. *septentrionalis* Fern. & Wieg. = Phalacroloma septentrionale

sachalinensis Botsch. - 5

schalbusi Vierh. - 2

schmalhausenii M. Pop. - 3, 6

seravschanicus M. Pop. - 6

silenifolius (Turcz.) Botsch. - 1, 3, 4

sogdianus M. Pop. - 6

*****speciosus** (Lindl.) DC. (*Stenactis speciosa* Lindl.) - 1

strigosus Muehl. ex Willd. = Phalacroloma strigosum

thunbergii A. Gray - 5

- subsp. *komarovii* (Botsch.) A. & D. Löve = E. komarovii

- subsp. *koraginensis* (Kom.) A. & D. Löve = E. komarovii

tianschanicus Botsch. - 6

tilingii Worosch. = E. politus

trimorphopsis Botsch. - 6

umbrosus (Kar. & Kir.) Boiss. - 6

unalaschkensis (DC.) Vierh. = E. humilis

uniflorus L. - 1, 2

- subsp. *eriocalyx* (Ledeb.) A. & D. Löve = E. eriocalyx

uralensis Less. (*E. acris* L. subsp. *brachycephalus* (Lindb. fil.) Hiit., *E. acris* subsp. *decoloratus* (Lindb. fil.) Hiit., *E. acris* subsp. *elongatiformis* Novopokr. nom. invalid., *E. brachycephalus* Lindb. fil., *E. decoloratus* Lindb. fil., *E. elongatiformis* (Novopokr.) Serg. nom. invalid.) - 1, 3, 4, 5

venustus Botsch. (*E. caucasicus* Stev. subsp. *venustus* (Botsch.) Grierson) - 2

vicarius Botsch. - 6

violaceus M. Pop. - 6

Eupatorium L.

cannabinum L. - 1, 2, 6

glehnii F. Schmidt ex Trautv. - 5

lindleyanum DC. - 5

Euthamia Nutt.

graminifolia (L.) Nutt. - 2

Evax Gaertn.

anatolica Boiss. & Heldr. (*Filago anatolica* (Boiss. & Heldr.) Chrtek & Holub) - 2

arenaria Smoljian. (*Filago arenaria* (Smoljian.) Chrtek & Holub) - 6

contracta Boiss. (*Filago contracta* (Boiss.) Chrtek & Holub) - 2

filaginoides Kar. & Kir. = Filago filaginoides

micropodioides (Willk.) Willk. = E. nevadensis

nevadensis Boiss. (*E. micropodioides* (Willk.) Willk. nom. illegit., *Filago iberica* Chrtek & Holub, nom. illegit.) - 2

Filaginella Opiz

baicalensis (Kirp.) Czer. comb. nova, (*Gnaphalium baicalense* Kirp. 1960, Bot. Mat. (Leningrad) 20 : 300) - 4

kasachstanica (Kirp.) Tzvel. (*Gnaphalium kasachstanicum* Kirp., *Filaginella uliginosa* (L.) Opiz subsp. *kasachstanica* (Kirp.) Holub) - 1, 3, 4, 6

mandshurica (Kirp.) Czer. comb. nova (*Gnaphalium mandshuricum* Kirp. 1960, Bot. Mat. (Leningrad) 20 : 298) - 5

pilularis (Wahlenb.) Tzvel. (*Gnaphalium pilulare* Wahlenb., *Filaginella uliginosa* (L.) Opiz subsp. *sibirica* (Kirp.) Holub, *Gnaphalium sibiricum* Kirp., *G. uliginosum* L. subsp. *pilulare* (Wahlenb.) Nym.) - 1, 3, 4, 5

rossica (Kirp.) Tzvel. (*Gnaphalium rossicum* Kirp., *Filaginella uliginosa* (L.) Opiz subsp. *rossica* (Kirp.) Holub) - 1, 2, 3, 4, 6

tranzschelii (Kirp.) Holub (*Gnaphalium tranzschelii* Kirp.) - 5

uliginosa (L.) Opiz (*Gnaphalium uliginosum* L.) - 1, 5
- subsp. *kasachstanica* (Kirp.) Holub = F. kasachstanica
- subsp. *rossica* (Kirp.) Holub = F. rossica
- subsp. *sibirica* (Kirp.) Holub = F. pilularis

Filago L. (*Gifola* Cass., *Logfia* Cass., *Oglifa* Cass.)

anatolica (Boiss. & Heldr.) Chrtek & Holub = Evax anatolica

arenaria (Smoljian.) Chrtek & Holub = Evax arenaria

arvensis L. (*Logfia arvensis* (L.) Holub, *Oglifa arvensis* (L.) Cass.) - 1, 2, 3, 6
- subsp. *lagopus* (Steph.) Nym. = F. lagopus
bornmuelleri Hausskn. ex Bornm. = F. hurdwarica
contracta (Boiss.) Chrtek & Holub = Evax contracta
desertorum Pomel - 2, 6
eriocephala Guss. (*Gifola eriocephala* (Guss.) Chrtek & Holub) - 1, 2
filaginoides (Kar. & Kir.) Wagenitz (*Evax filaginoides* Kar. & Kir.) - 6
gallica L. (*Logfia gallica* (L.) Coss. & Germ., *Oglifa gallica* (L.) Chrtek & Holub) - 2
germanica L. = F. vulgaris
germanica (L.) Huds. = F. pyramidata
hurdwarica (DC.) Wagenitz (*Gnaphalium hurdwaricum* Wall. ex DC., *Filago bornmuelleri* Hausskn. ex Bornm., *Gifola bornmuelleri* (Hausskn. ex Bornm.) Chrtek & Holub) - 6
iberica Chrtek & Holub = Evax nevadensis
lagopus (Steph.) Parl. (*F. arvensis* L. subsp. *lagopus* (Steph.) Nym., *Oglifa lagopus* (Steph.) Chrtek & Holub) - 2, 6

minima (Smith) Pers. (*Logfia minima* (Smith) Dumort., *Oglifa minima* (Smith) Reichenb. fil.) - 1

montana L. - 2, 6

paradoxa Wagenitz - 6

pyramidata L. (*F. germanica* (L.) Huds. 1762, non L. 1763, *Gifola spathulata* (C. Presl) Reichenb. fil. p.p., *Gnaphalium germanicum* L., *Filago spathulata* auct.) - 1, 2, 6

spathulata auct. = F. pyramidata

vulgaris Lam. (*F. germanica* L. 1763, non Huds. 1762, *Gifola germanica* (L.) Dumort. nom. illegit.) - 1, 2, 6

Filifolium Kitam.

sibiricum (L.) Kitam. - 4, 5

Frolovia Lipsch. = Saussurea

gilesii (Hemsl.) B. Scharipova = Saussurea gilesii

*Gaillardia Foug.

*****aristata** Pursh - 1, 6

*****pulchella** Foug. - 1, 6

Galatella Cass. (*Crinitaria* Cass., *Linosyris* Cass. 1825, non Ludw. 1757, *Chrysocoma* auct.)

altaica Tzvel. - 3

angustissima (Tausch) Novopokr. (*Aster sedifolius* L. subsp. *angustissimus* (Tausch) Merxmuller) - 1, 3, 4, 6

biflora (L.) Nees (*G. novopokrovskii* Zefir.) - 1, 3, 4, 6

chromopappus Novopokr. - 6

coriacea Novopokr. - 6

crinitoides Novopokr. - 1, 3, 6

dahurica DC. - 4, 5

divaricata (Fisch. ex Bieb.) Novopokr. - 1, 3, 6

dracunculoides (Lam.) Nees (*Aster sedifolius* L. subsp. *dracunculoides* (Lam.) Merxmuller, *Aster sedifolius* var. *dracunculoides* (Lam.) Soo) - 1, 2

eldarica Kem.-Nath. - 2

fastigiiformis Novopokr. - 6

fominii (Kem.-Nath.) Czer. comb. nova (*Linosyris fominii* Kem.-Nath. 1927, Vest. Tifl. Bot. Sada, 3-4 : 142; *Crinitaria fominii* (Kem.-Nath.) Czer. comb. superfl., *C. fominii* (Kem.-Nath.) Sojak) - 2

hauptii (Ledeb.) Lindl. - 3, 6

hissarica Novopokr. - 6

linosyris (L.) Reichenb. fil. (*Crinitaria linosyris* (L.) Less., *Linosyris vulgaris* Cass. ex Less.) - 1, 2

litvinovii Novopokr. - 6

macrosciadia Gand. - 3, 4, 6

novopokrovskii Zefir. = G. biflora

pastuchovii (Kem.-Nath.) Tzvel. - 1, 2

polygaloides Novopokr. - 6

pontica (Lipsky) Novopokr. & Bogdan (*Crinitaria pontica* (Lipsky) Czer. comb. superfl., *C. pontica* (Lipsky) Sojak, *Linosyris pontica* (Lipsky) Novopokr., *L. vulgaris* Cass. ex Less. var. *pontica* Lipsky) - 2

punctata (Waldst. & Kit.) Nees (*Aster canus* Waldst. & Kit. subsp. *punctatus* (Waldst. & Kit.) Soo, *Aster sedifolius* auct. p.p.) - 1

regelii Tzvel. - 6

rossica Novopokr. (*Aster canus* Waldst. & Kit. subsp. *rossicus* (Novopokr.) Soo, *A. punctatus* Waldst. & Kit. subsp. *rossicus* (Novopokr.) Soo) - 1, 3, 6

saxatilis Novopokr. - 6

scoparia (Kar. & Kir.) Novopokr. - 6

sosnowskyana Kem.-Nath. - 2

x **sublinosyris** Tzvel. - G. linosyris (L.) Reichenb. fil. x G. angustissima (Tausch) Novopokr. - 1

x **subtatarica** Tzvel. - G. tatarica (Less.) Novopokr. x G. villosa (L.) Reichenb. fil. - 1

x **subvillosa** Tzvel. - G. linosyris (L.) Reichenb. fil. x G. villosa (L.) Reichenb. fil. - 1

tatarica (Less.) Novopokr. (*Chrysocoma tatarica* Less., *Aster tarbagatensis* (C. Koch) Merxmuller, *Crinitaria tatarica* (Less.) Czer. comb. superfl., *C. tatarica* (Less.) Sojak, *Linosyris tarbagatensis* C. Koch, *L. tatarica* (Less.) C.A. Mey.) - 1, 3, 6

tianschanica Novopokr. - 6

trinervifolia (Less.) Novopokr. - 1, 3, 6

villosa (L.) Reichenb. fil. (*Aster oleifolius* (Lam.) Wagenitz, *A. villosus* (L.) Sch. Bip. 1855, non Thunb. 1800, *Crinitaria villosa* (L.) Grossh., *Linosyris villosa* (L.) DC.) - 1, 2, 3, 6

villosula Novopokr. - 6

Galinsoga Ruiz & Pav. (*Adventina* Rafin.)

ciliata (Rafin.) Blake (*Adventina ciliata* Rafin., *Galinsoga quadriradiata* auct.) - 1, 2, 5

parviflora Cav. - 1, 2, 5, 6

quadriradiata auct. = G. ciliata

Garhadiolus Jaub. & Spach

angulosus Jaub. & Spach. (*Rhagadiolus angulosus* (Jaub. & Spach) Kupicha) - 2, 6

papposus Boiss. & Buhse - 2, 6

Gerbera Cass.

*****jamesonii** Bolus ex Hook. fil. - 2

Geropogon L.

hybridus (L.) Sch. Bip. - 1, 2

Gifola Cass. = Filago

bornmuelleri (Hausskn. ex Bornm.) Chrtek & Holub = Filago hurdwarica

eriocephala (Guss.) Chrtek & Holub = Filago eriocephala

germanica (L.) Dumort. = Filago vulgaris

spathulata (C. Presl) Reichenb. fil. p.p. = Filago pyramidata

Gnaphalium L. (*Pseudognaphalium* auct.)

affine D. Don (*Pseudognaphalium luteo-album* (L.) O.M. Hilliard & B.L. Burtt subsp. *affine* (D. Don) O.M. Hilliard & B.L. Burtt) - 2(alien)

baicalense Kirp. = Filaginella baicalensis

caucasicum Somm. & Levier = Omalotheca caucasica

germanicum L. = Filago pyramidata

hoppeanum Koch = Omalotheca hoppeana

hurdwaricum Wall. ex DC. = Filago hurdwarica

kasachstanicum Kirp. = Filaginella kasachstanica

luteo-album L. (*Pseudognaphalium luteo-album* (L.) O.M. Hilliard & B.L. Burtt) - 1, 2, 6

mandshuricum Kirp. = Filaginella mandshurica

norvegicum Gunn. = Omalotheca norvegica

pilulare Wahlenb. = Filaginella pilularis

pterocaulon Franch. & Savat = Anaphalis pterocaulon

rossicum Kirp. = Filaginella rossica

sibiricum Kirp. = Filaginella pilularis

stewartii Clarke = Omalotheca stewartii

supinum L. = Omalotheca supina

sylvaticum L. = Omalotheca sylvatica

tranzschelii Kirp. = Filaginella tranzschelii

uliginosum L. = Filaginella uliginosa

- subsp. *pilulare* (Wahlenb.) Nym. = Filaginella pilularis

Grindelia Willd.

squarrosa (Pursh) Dun. - 1, 2, 5, 6

Grossheimia Sosn. & Takht.

ahverdovii (Gabr.) Gabr. (*Centaurea ahverdovii* Gabr.) - 2

macrocephala (Muss.-Puschk. ex Willd.) Sosn. & Takht. - 2

ossica (C. Koch) Sosn. & Takht. = G. polyphylla

polyphylla (Ledeb.) Holub (*G. ossica* (C. Koch) Sosn. & Takht., *Centaurea polyphylla* Ledeb.) - 2

*Guizotia Cass.

***abyssinica** (L.fil.) Cass. - 1

Gundelia L.

tournefortii L. - 2, 6

Gymnarrhena Desf.

micrantha Desf. - 6

Gymnaster Kitam. = Miyamayomena

savatieri (Makino) Kitam. = Miyamayomena savatieri

Gymnoclina auct. = Tanacetum

argyrophylla C. Koch = Tanacetum argyophyllum

Handelia Heimerl

trichophylla (Schrenk) Heimerl - 6

Hedypnois Hill (*Hyoseris* L. p.p.)

cretica (L.) Dum.-Cours. (*H. rhagadioloides* (L.) F.W. Schmidt subsp. *cretica* (L.) Hayek) - 1, 2

persica Bieb. = H. rhagadioloides

rhagadioloides (L.) F.W. Schmidt (*Hyoseris rhagadioloides* L., *Hedypnois persica* Bieb.) - 2

- subsp. *cretica* (L.) Hayek = H. cretica

*Helenium L.

***autumnale** L. - 6

***nudiflorum** Nutt. - 6

Helianthus L.

***annuus** L. - 1, 2, 6

- subsp. *lenticularis* (Dougl. ex Lindl.) Cockerell = H. lenticularis

***atrorubens** L. - 6

***decapetalus** L. - 1

giganteus L. - 1(alien)

***laetiflorus** Pers. - 1, 5

lenticularis Dougl. ex Lindl. (*H. annuus* L. subsp. *lenticularis* (Dougl. ex Lindl.) Cockerell) - 1, 3, 4, 5

petiolaris Nutt. - 1(alien), 5(alien)

***rigidus** (Cass.) Desf. (*Harpalium rigidum* Cass.) - 1, 5

- subsp. **subrhomboideus** (Rydb.) Heiser (*H. subrhomboideus* Rydb.) - 5

***strumosus** L. - 1

***subcanescens** (A. Gray) E.E. Wats. (*H. tuberosus* L. var. *subcanescens* A. Gray) - 1

subrhomboideus Rydb. = H. rigidus subsp. subrhomboideus
*tuberosus L. - 1, 2, 3, 4, 5, 6
- var. *subcanescens* A. Gray = H. subcanescens

Helichrysum Mill.

araxinum Takht. ex Kirp. (*H. armenium* DC. subsp. *araxinum* (Takht. ex Kirp.) Takht.) - 2
arenarium (L.) Moench - 1, 2, 3, 4, 6
- subsp. *rubicundum* (C. Koch) P.H. Davis & Kupicha = H. rubicundum
armenium DC. - 2
- subsp. *araxinum* (Takht. ex Kirp.) Takht. = H. araxinum
aurantiacum Boiss. & Huet = H. graveolens
bracteatum (Vent.) Andr. = Xerochrysum bracteatum
buschii Juz. - 1
callichrysum DC. (*H. polylepis* Bordz. ex Grossh.) - 2
corymbiforme Opperm. ex Katina - 1
graveolens (Bieb.) Sweet (*H. aurantiacum* Boiss. & Huet) - 1, 2
kopetdagense Kirp. = H. oocephalum
maracandicum M. Pop. ex Kirp. - 6
mussae Nevski - 6
nuratavicum Krasch. - 6
oocephalum Boiss. (*H. kopetdagense* Kirp.) - 6
pallasii (Spreng.) Ledeb. - 2
* **petiolare** Hillard & B. Burtt (*H. petiolatum* auct.) - 1
petiolatum auct. = H. petiolare
plicatum DC. - 2
- subsp. *polyphyllum* (Ledeb.) P.H. Davis & Kupicha = H. polyphyllum
polylepis Bordz. ex Grossh. = H. callichrysum
polyphyllum Ledeb. (*H. plicatum* DC. *polyphyllum* (Ledeb.) P.H. Davis & Kupicha) - 2
rubicundum (C. Koch) Bornm. (*Antennaria rubicunda* C. Koch, *Helichrysum arenarium* (L.) Moench subsp. *rubicundum* (C. Koch) P.H. Davis & Kupicha, *H. rubicundum* (C. Koch) Grossh. comb. superfl., *H. undulatum* Ledeb. subsp. *rubicundum* (C. Koch) Takht., comb. incorrecta, *H. undulatum* Ledeb.) - 2
sosnowskyi Grossh. - 2
tanaiticum P. Smirn. - 1, 2
thianschanicum Regel - 6
undulatum Ledeb. = H. rubicundum
- subsp. *rubicundum* (C. Koch) Takht. = H. rubicundum

*Heliopsis Pers.

helianthoides (L.) Sweet subsp. *scabra* (Dun.) Fisher = H. scabra
- var. *scabra* (Dun.) Fern. = H. scabra
*scabra Dun. (*H. helianthoides* (L.) Sweet subsp. *scabra* (Dun.) Ficher, *H. helianthoides* var. *scabra* (Dun.) Fern.) - 1, 6

Helminthia Juss. = Helminthotheca

echioides (L.) Juss. = Helminthotheca echioides

Helminthotheca Zann (*Helminthia* Juss.)

echioides (L.) Holub (*Helminthia echioides* (L.) Juss., *Helminthotheca echioides* (L.) Lack, comb. superfl.) - 1, 2, 6(alien)

Hemipappus C. Koch

canus C. Koch (*Tanacetum argenteum* (Lam.) Willd. subsp. *canum* (C. Koch) Grierson) - 2(alien)

Hemizonia DC.

pungens (Hook. & Arn.) Torr. & Gray - 5

Heteracia Fisch. & C.A. Mey.

epapposa (Regel & Schmalh.) M. Pop. = H. szovitsii
szovitsii Fisch. & C.A. Mey. (*H. epapposa* (Regel & Schmalh.) M. Pop.) - 1, 2, 6

Heteroderis (Bunge) Boiss.

leucocephala (Bunge) Leonova (*H. pusilla* (Boiss.) Boiss. var. *leucocephala* (Bunge) Rech. fil.) - 6
pusilla (Boiss.) Boiss. - 6
- var. *leucocephala* (Bunge) Rech. fil. = H. leucocephala

Heterolophus Cass. = Centaurea

carbonatus (Klok.) Sojak = Centaurea carbonata
marschallianus (Spreng.) Sojak = Centaurea marschalliana
sergii (Klok.) Sojak = Centaurea sergii
sumensis (Kalen.) Sojak = Centaurea sumensis

Heteropappus Less.

alberti (Regel) Novopokr. & Tamamsch. = H. canescens
altaicus (Willd.) Novopokr. (*H. altaicus* var. *distortus* (Turcz. ex Ave-Lall.) Gubanov, *H. distortus* (Turcz. ex Ave-Lall.) Tamamsch.) - 3, 4, 6
- var. *distortus* (Turcz. ex Ave-Lall.) Gubanov = H. altaicus
biennis (Ledeb.) Tamamsch. ex Grub. (*Aster biennis* Ledeb., *Heteropappus tataricus* (Lindl.) Tamamsch.) - 1, 3, 4, 5
canescens (Nees) Novopokr. (*Aster canescens* (Nees) Fisjun, *Heteropappus alberti* (Regel) Novopokr. & Tamamsch.) - 2, 3, 6
decipiens Maxim. = H. hispidus
- var. *elisabethinus* (Tamamsch.) Zdorovjeva = H. hispidus var. elisabethinus
distortus (Turcz. ex Ave-Lall.) Tamamsch. = H. altaicus
elisabethinus Tamamsch. = H. hispidus var. elisabethinus
hispidus (Thunb.) Less. (*H. decipiens* Maxim., *H. hispidus* var. *decipiens* (Maxim.) Worosch., *H. hispidus* var. *meyendorffii* (Regel & Maack) Zdorovjeva, *H. hispidus* var. *pinetorum* (Kom.) Zdorovjeva, *H. meyendorffii* (Regel & Maack) Kom., *H. pinetorum* Kom., *H. saxomarinus* Kom.) - 5
- var. *decipiens* (Maxim.) Worosch. = H. hispidus
- var. **elisabethinus** (Tamamsch.) Worosch. (*H. elisabethinus* Tamamsch., *H. decipiens* var. *elisabethinus* (Tamamsch.) Zdorovjeva) - 5
- var. *meyendorffii* (Regel & Maack) Zdorovjeva = H. hispidus
- var. *pinetorum* (Kom.) Zdorovjeva = H. hispidus
- var. **villosus** (Kom.) Worosch. (*H. villosus* Kom., *H. noneifolius* Tamamsch.) - 5
medius (Kryl.) Tamamsch. - 3, 4
meyendorffii (Regel & Maack) Kom. = H. hispidus
noneifolius Tamamsch. = H. hispidus var. villosus
pinetorum Kom. = H. hispidus
saxomarinus Kom. = H. hispidus
tataricus (Lindl.) Tamamsch. = H. biennis
villosus Kom. = H. hispidus var. villosus

Hieracium L. (*Chlorocrepis* Monn., *Pilosella* Hill, *Stenotheca* Monn.)

x **abakurae** Schelk. & Zahn (*Pilosella abakurae* (Schelk. & Zahn) Sojak) - 2

abastumanense Juxip - 2
abortiens Norrl. = *H.* x *zizianum*
abruptorum Schljak. - 1
accline Norrl. = *H.* x *dubium*
achalzichiense Juxip - 2
acinacifolium Schljak. - 1
acrifolium (Dahlst.) K. Joh. (*H. rigidum* C. Hartm. subsp.
  acrifolium Dahlst.) - 1(?), 3(?)
acrochlorum (Zahn) Juxip = *H.* x *prussicum*
acrocomum Naeg. & Peter = *H.* x *dubium* L.
acrogymnon (Malte) Juxip (*H. silvaticum* (L.) Gouan subsp.
  acrogymnon Malte) - 1
acroleucoides Dahlst. - 1
acroleucum Stenstr. - 1
acrosciadium (Naeg. & Peter) Juxip = *H.* x *densiflorum*
acrothyrsum Naeg. & Peter = *H.* x *melinomelas*
acrotrichum Rehm. = *H.* x *progenitum*
acroxanthum (Sosn. & Zahn) Juxip - 2
acuminatifolium (Litv. & Zahn) Juxip - 2
acuminatum Jord. - 1
acutangulum (Kozl. & Zahn) Juxip - 2
acutisquamum (Naeg. & Peter) Juxip (*H. auricula* L. var.
  acutisquamum (Naeg. & Peter) E.I. Nyarady, *Pilosella
  acutisquama* (Naeg. & Peter) Schljak., *P. lactucella*
  (Wallr.) P.D. Sell & C. West subsp. *acutisquama*
  (Naeg. & Peter) Sojak) - 1
aczelmanicum Schischk. & Serg. - 3
adakense Schljak. - 1
adelum Juxip - 1
adenoactis Juxip - 2
x adenobrachion (Litv. & Zahn) Juxip - 2
adenocladum (Rehm.) Czer. comb. nova (*H. magyaricum*
  Peter subsp. *adenocladum* Rehm. 1897, Verh. Zool.-
  Bot. Ges. Wien, 47 : 293; *Pilosella adenoclada*
  (Rehm.) Schljak.) - 1
adjarianum Peter - 2
adlerzii Almq. ex K. Joh. - 1
adspersum auct. = *H. glabriligulatum*
adunans Norrl. - 1
adunantidens Schljak. - 1
aeruginascens (Norrl.) Norrl. = *H.* x *blyttianum*
agathanthum (Rehm.) Czer. comb. nova (*H. magyaricum*
  Peter subsp. *agathanthum* Rehm. 1897, Verh. Zool.-
  Bot. Ges. Wien, 47 : 294; *Pilosella agathantha* (Rehm.)
  Schljak.) - 1
agnostum Juxip = *H. expallidiforme*
agronesaeum Juxip - 1
akinfiewii (Woronow & Zahn) Juxip (*H. verruculatum* Link
  subsp. *akinfiewii* Woronow & Zahn) - 2
akjaurense Norrl. - 1
alatavicum (Zahn) Juxip - 5
albellipes (Schelk. & Zahn) Juxip - 2
alberti Schljak. - 1
albidulum Stenstr. - 1
albipes (Dahlst.) Juxip (*H. caesium* (Fries) Fries subsp.
  albipes Dahlst.) - 1
albocinereum Rupr. = *H.* x *rothianum*
albocostatum (Norrl.) Juxip - 1
alienatum Norrl. = *H. fuliginosum*
alphostictum Dahlst. - 1
alpinum L. (*H. crispum* (Elfstr.) Elfstr., *H. gymnogenum*
  (Zahn) Juxip) - 1, 3
- subsp. *apiculatum* (Tausch) E.I. Nyarady = *H. apiculatum*
- subsp. *apiculatum* (Tausch) Zahn = *H. apiculatum*
- subsp. *eximiiforme* Dahlst. = *H. eximiiforme*
altaicum (Naeg. & Peter) Juxip (*Pilosella altaica* (Naeg. &
  Peter) Schljak.) - 1, 3

alticaule (Litv. & Zahn) Juxip - 2
altipes (Lindb. fil. ex Zahn) Juxip (*H. murorum* L. subsp.
  altipes Lindb. fil. ex Zahn) - 1
alupkanum (Zahn) Juxip = *H.* x *auriculoides*
amauranthum (Peter) Juxip = *H.* x *zizianum*
amaureilema (Naeg. & Peter) Juxip (*H. auricula* L. var.
  amaureilema (Naeg. & Peter) E.I. Nyarady, *Pilosella
  amaureilema* (Naeg. & Peter) Schljak., *P. lactucella*
  (Wallr.) P.D. Sell & C. West subsp. *amaureilema*
  (Naeg. & Peter) Sojak) - 1
amaurobasis (Litv. & Zahn) Juxip - 2
amaurochlorellum (Zahn) Juxip = *H. amaurochlorum*
amaurochlorum (Zahn) Czer. comb. nova (*H. pratense*
  Tausch subsp. *amaurochlorum* Zahn, 1911, Trav. Mus.
  Bot. Acad. Sci. Petersb. 9 : 15; *H. amaurochlorellum*
  (Zahn) Juxip, nom. illegit., *H. pratense* subsp. *amaurochlorellum* Zahn, nom. illegit., *Pilosella amaurochlora*
  (Zahn) Schljak.) - 1
ambiguum Ehrh. = *H.* x *glomeratum*
amblylobum Juxip - 1
amnoon (Naeg. & Peter) Juxip (*Pilosella amnoon* (Naeg. &
  Peter) Schljak., *P. glaucescens* (Bess.) Sojak subsp.
  amnoon (Naeg. & Peter) Sojak) - 1
amoeniceps (Zahn) Juxip = *H.* x *macrostolonum*
amoenoarduum Juxip - 4
amphileion (Pohle & Zahn) Juxip - 1
amphitephrodes (Sosn. & Zahn) Juxip - 2
anacraspedum (Rehm.) Juxip = *H.* x *flagellare*
andrzejowskii Blocki & Woloszcz. = *H. vaillantii*
anfractum (Fries) Fries (*H. lachenalii* C.C. Gmel. subsp.
  anfractum (Fries) E. I. Nyarady) - 2
- subsp. *cacuminatum* Dahlst. = *H. cacuminatum*
angustifrons Schljak. - 1
angustilobatum Schljak. - 1
angustisquamatum Schljak. - 1
anisocephalum (Rehm.) Juxip = *H.* x *flagellare*
anocladum (Naeg. & Peter) Juxip = *H.* x *mollicaule*
antecursorum Schljak. - 1(?),3
antennarioidiforme (Zahn) Juxip - 2
apatelioides (Zahn) Juxip = *H.* x *poliodermum*
x apatelium Naeg. & Peter (*H. aupaense* (Naeg. & Peter)
  Juxip, *Pilosella* x *apatelia* (Naeg. & Peter) Sojak) - 1
- nm. *piloselliflorum* (Naeg. & Peter) Lepage = *H. piloselliflorum*
apatitorum Juxip = *H. congruens*
apatorium (Naeg. & Peter) Juxip = *H.* x *arvicola*
aphanum Juxip - 1
apicifolium Fagerstrom - 1(?)
apiculatiforme Elfstr. = *H. uralense*
apiculatum Tausch (*H. alpinum* L. subsp. *apiculatum*
  (Tausch) E.I. Nyarady, comb. superfl., *H. alpinum* L.
  subsp. *apiculatum* (Tausch) Zahn) - 1
approximabile (Zahn) Juxip = *H.* x *mollicaule*
aquilonare (Naeg. & Peter) Juxip (*Pilosella aquilonaris*
  (Naeg. & Peter) Schljak., *P. obscura* (Reichenb.) Soo
  subsp. *aquilonaris* (Naeg. & Peter) Sojak) - 1
arctogenum Norrl. (*Pilosella arctogena* (Norrl.) Schljak.) - 1
arctogeton (Zahn) Juxip - 1
arctomurmanicum Schljak. - 1
arcuatidens (Zahn) Juxip - 1
argillaceoides (Litv. & Zahn) Juxip - 2
argillaceum Jord. - 1
x aridum Freyn (*H. venetianum* Naeg. & Peter, *Pilosella* x
  arida (Freyn) Schljak.) - 1
armeniacum (Naeg. & Peter) Juxip, 1960, non Arv.-Tour.
  1883 (*Pilosella armeniaca* (Naeg. & Peter) Schljak.) -
  1, 2
  This name must be replaced because it is a later homonym.

■ artabirense (Zahn) Juxip
■ artvinense (Woronow & Zahn) Juxip
arvense (Naeg. & Peter) Juxip = H. x auriculoides
x **arvicola** Naeg. & Peter (*H. apatorium* (Naeg. & Peter) Juxip, *H. assimilatum* (Norrl.) Norrl., *H. curvulum* Norrl., *H. erythrochristum* (Naeg. & Peter) Juxip, *H. floribundiforme* (Naeg. & Peter) Juxip, p.p. quoad plantas, *H. hirtulum* (Peter) Juxip, *H. leucocraspedum* (Peter) Juxip, *Pilosella x arvicola* (Naeg. & Peter) Sojak) - 1
arvorum (Naeg. & Peter) Pugsl. (*Pilosella arvorum* (Naeg. & Peter) Schljak., *P. bauhini* (Bess.) Arv.-Tour. subsp. *arvorum* (Naeg. & Peter) Sojak, *P. praealta* (Vill. ex Gochn.) F. Schuttz & Sch. Bip. subsp. *arvorum* (Naeg. & Peter) P.D. Sell & C. West) - 1, 2
aryslynense (Zahn) Juxip - 6
asiaticum (Naeg. & Peter) Juxip (*H. multifolium* (Peter) Juxip, *Pilosella asiatica* (Naeg. & Peter) Schljak.) - 1, 2, 3, 6
asperrimum Schur = H. x auriculoides
assimilatum (Norrl.) Norrl. = H. x arvicola
asterodermum (Woronow & Zahn) Juxip - 2
astibes Juxip - 1
asymmetricum Schljak. - 1
atratulum Norrl. (*H. senescentifrons* Elfstr.) - 1
atratum auct. = H. atrellum
atratum Fries subsp. *ihrowyszczense* Zahn = H. ihrowyszczense
atrellum (Zahn) Juxip (*H. atratum* auct.) - 1
atricollum Schljak. - 1
atriplicifolium Schljak. - 1
aupaense (Naeg. & Peter) Juxip = H. x apatelium
aurantiacum L. (*Pilosella aurantiaca* (L.) F. Schultz & Sch. Bip.) - 1, 5(alien)
- subsp. *carpathicola* Naeg. & Peter = H. carpathicola
- subsp. *neriodowae* Woloszcz. & Zahn = H. neriodowae
- subsp. *subaurantiacum* Naeg. & Peter = H. subaurantiacum
auratum Fries (*H. sabaundum* L. subsp. *boreale* (Fries) E.I. Nyarady var. *auratum* (Fries) E.I. Nyarady) - 1, 2
aureiceps Norr. ex Schljak. - 1
auricula auct. = H. lactucella
auricula L. subsp. *melaneilema* Peter = H. melaneilema
- var. *acutisquamum* (Naeg. & Peter) E.I. Nyarady = H. acutisquamum
- var. *amaureilema* (Naeg. & Peter) E.I. Nyarady = H. amaureilema
- var. *lithuanicum* (Naeg. & Peter) E.I. Nyarady = H. lithuanicum
- var. *magnauricula* (Naeg. & Peter) E.I. Nyarady = H. magnauricula
- var. *melaneilema* (Naeg. & Peter) E.I. Nyarady = H. melaneilema
- var. *tricheilema* (Naeg. & Peter) E.I. Nyarady = H. tricheilema
x **auriculoides** Lang (*H. alupkanum* (Zahn) Juxip, *H. arvense* (Naeg. & Peter) Juxip, *H. asperrimum* Schur, *H. auriculoides* var. *sarmentosum* (Froel.) E.I. Nyarady, *H. echiocephalum* (Naeg. & Peter) Juxip, *H. echiogenes* (Naeg. & Peter) Juxip, *H. longisetum* (Naeg. & Peter) Juxip, *H. mirum* (Naeg. & Peter) Juxip, *H. pareyssianum* (Naeg. & Peter) Juxip, *H. sarmentosum* Froel. 1838, non Salisb. 1796, *H. tanythrix* (Naeg. & Peter) Juxip, *H. umbellosum* (Naeg. & Peter) Juxip, *Pilosella x auriculoides* (Lang) F. Schultz, *P. x auriculodes* (Lang) P.D. Sell & C. West, comb. superfl., *P. auriculoides* subsp. *arvensis* (Naeg. & Peter) Sojak, *P. auriculoides* subsp. *asperrima* (Schur) Sojak, *P. auriculoides* subsp. *echiocephala* (Naeg. & Peter) Sojak, *P. auriculoides* subsp. *echiogenes* (Naeg. & Peter) Sojak, *P. auriculoides* subsp. *longiseta* (Naeg. & Peter) Sojak, *P. auriculoides* subsp. *mira* (Naeg. & Peter) Sojak, *P. auriculoides* subsb. *tanythrix* (Naeg. & Peter) Sojak, *P. auriculoides* subsp. *umbellosa* (Naeg. & Piter) Sojak) - 1, 2
- var. *sarmentosum* (Froel.) E.I. Nyarady = H. x auriculoides
aurorinii Juxip = H. caesiiflorum
aurosulum Norrl. = H. x macrostolonum
austericaule Norrl. = H. x zizianum
baenitzii (Naeg. & Peter) Juxip = H. x floribundum
bakurianense (Fomin & Zahn) Juxip - 2
balansae Boiss. = H. x macrotrichum
barbulatulum (Pohle & Zahn) Elfstr. (*H. barbulatulum* (Pohle & Zahn) Juxip, comb. superfl., *H. semisagittatum* Schljak.) - 1
basifolium auct. = H. cereolinum
basifolium (Almq.) Stenstr. subsp. *cereolinum* Norrl. = H. cereolinum
basileucum (Litv. & Zahn) Juxip - 2
bathycephalum Elfstr. var. *floccinops* Elfstr. = H. floccinops
bauhini Bess. (*H. praealtum* Vill. ex Gochn subsp. *bauhini* (Bess.) Petunn., *Pilosella bauhini* (Bess.) Arv.-Tour.) - 1
- subsp. *bauhini* var. *cymanthum* (Naeg. & Peter) E.I. Nyarady = H. cymanthum
- - var. *fastigiatum* (Naeg. & Peter) E.I. Nyarady = H. fastigiatum
- - var. *ingricum* (Naeg. & Peter) E.I. Nyarady = H. ingricum
- - var. *melachaetum* (Tausch) E.I. Nyarady = H. melachaetum
- - var. *obscuribracteatum* (Naeg. & Peter) E.I. Nyarady = H. obscuribracteatum
- - var. *plicatulum* (Zahn) E.I. Nyarady = H. plicatulum
- - var. *thaumasioides* (Naeg. & Peter) E.I. Nyarady = H. thaumasioides
- - var. *thaumasium* (Peter) E.I. Nyarady = H. thaumasium
- - var. *viscidulum* (Tausch) E.I. Nyarady = H. viscidulum
- subsp. *cymanthum* (Naeg. & Peter) Zahn var. *limenyense* Zahn = H. limenyense
- subsp. *erythrophylloides* Zahn = H. erythrophylloides
- subsp. *magyaricum* (Peter) Zahn var. *besserianum* (Spreng.) E.I. Nyarady = H. besserianum
- - var. *filiferum* (Tausch) E.I. Nyarady = H. filiferum
- - var. *heothinum* (Naeg. & Peter) E.I. Nyarady = H. heothinum
- - var. *marginale* (Naeg. & Peter) E.I. Nyarady = H. marginale
- - var. *megalomastix* (Naeg. & Peter) E.I. Nyarady = H. megalomastix
- - var. *nigrisetum* (Naeg. & Peter) E.I. Nyarady = H. nigrisetum
- - var. *pseudauriculoides* (Naeg. & Peter) E.I. Nyarady = H. pseudauriculoides
- subsp. *pseudosparsum* Zahn = H. pseudosparsum
- - b. *viscidulum* Tausch = H. viscidulum
bauhiniflorum (Naeg. & Peter) Juxip = H. mollicaule
beschtaviciforme Juxip - 2
beschtavicum (Litv. & Zahn) Juxip - 2
besserianum Spreng. (*H. bauhini* Bess. subsp. *magyaricum* (Peter) Zahn var. *besserianum* (Spreng.) E.I. Nyarady, *Pilosella besseriana* (Sprend.) Schljak, *P. glaucescens* (Bess.) Sojak subsp. *besseriana* (Spreng.) Sojak) - 1, 2

bichloricolor (Ganesch. & Zahn) Juxip - 4

x **biebersteinii** Litv. & Zahn - 2

bifidum Kit. subsp. *bifidum* var. *caesiiflorum* (Almq.) E.I. Nyarady = H. caesiiflorum

- subsp. *lonchopodum* Zahn = H. lonchopodum

x **bifurcum** Bieb. (*H. longipes* (C. Koch ex Naeg. & Peter) Juxip, 1960, non Feyn & Sint., *H. peczoryense* Juxip, *H. sterromastix* (Naeg. & Peter) Juxip, *H. vindobonae* (Zahn) Juxip, *Pilosella* x *bifurca* (Bieb.) F. Schultz & Sch. Bip., *P.* x *bifurca* subsp. *sterromastix* (Naeg. & Peter) Sojak) - 1, 2

bimanum Norrl. - 1

x **blyttianum** Fries (*H. aeruginascens* (Norrl.) Norrl., *H. clinoglossum* Norrl., *H. discoloratum* (Norrl.) Norrl., *H. elfvingii* Norrl., *H. fulveolutescens* Norrl., *H. integrilingua* Norrl., *H. parvipunctatum* Norrl., *H. pseudoblyttii* (Norrl.) Norrl., *H. pulvinatum* (Norrl.) Norrl., *H. sepositum* Norrl., *H. torquescens* Norrl., *H. vernicosum* (Norrl.) Norrl., *Pilosella* x *blyttiana* (Fries) F. Schultz & Sch. Bip., *Hieracium fuscum* auct.) - 1

bobrovii Juxip = H. congruens

boreum auct. = H. pseudoboreum

borodinianum Juxip - 1

borshomiense Juxip - 2

borzawae (Woloszcz. & Zahn) Schljak. (*H. liptoviense* Borb. subsp. *borzawae* Woloszcz. & Zahn) - 1

botniense Brenn. - 1

botrychodes (Zahn) Juxip (*Pilosella botrychodes* (Zahn) Schljak.) - 1

x **brachiatum** Bertol. ex DC. (*H. dmitrovense* (Peter) Juxip, *H. ilyassowoense* (Zahn) Juxip, *H. pseudobrachiatum* Celak., *H. tubuliflorum* (Naeg. & Peter) Juxip, *Pilosella* x *brachiata* (Bertol. ex DC.) F. Schultz & Sch. Bip., *P.* x *brachiata* subsp. *pseudobrachiata* (Celak.) Sojak.) - 1, 2

brachyacron (Rehm.) Juxip = H. x flagellare

brachycephalum (Norrl.) Norrl. = H. x fennicum

brachyschistum (Zahn) Juxip = H. x flagellare

branae (Naeg. & Peter) Juxip (*Pilosella glaucescens* (Bess.) Sojak subsp. *branae* (Naeg. & Peter) Sojak) - 2

brandisianum Zahn - 2

brittatense Juxip - 2

brotheri Norrl. - 1

buhsei (Naeg. & Peter) Juxip = H. x maschukense

bupleurifolioides (Zahn) Juxip - 1, 2

bupleurifolium Tausch - 1, 2

bupleuroides auct. = H. schenkii

bupleuroides C.C. Gmel. var. *schenkii* Griseb. = H. schenkii

buschianum Juxip - 2

cacuminatum (Dahlst.) K. Joh. (*H. anfractum* (Fries) Fries subsp. *cacuminatum* Dahlst.) - 1

caesiiflorioides Juxip - 1

caesiiflorum Almq. (*H. aurorinii* Juxip, *H. bifidum* Kit. subsp. *bifidum* var. *caesiiflorum* (Almq.) E.I. Nyarady) - 1

caesiogenum Woloszcz. & Zahn - 1

caesiomurorum Lindeb. - 1

caesitiifolium Norrl. - 1

caesium (Fries) Fries (*H. vulgatum* Fries var. *caesium* Fries) - 1

- subsp. *albipes* Dahlst. = H. albipes

- subsp. *lugiorum* Zahn = H. lugiorum

- subsp. *osiliae* Dahlst. = H. osiliae

- subsp. *ravusculum* Dahlst. = H. ravusculum

caespiticola Norrl. - 1

caespitosum Dumort. (*H. pratense* Tausch, *Pilosella caespitosa* (Dumort.) P.D. Sell & C. West) - 1, 3

- subsp. *brevipilum* (Naeg. & Peter) P. D. Sell, p.p. = H. karelicum and H. onegense

- subsp. *caespitosum* var. *colliniforme* (Peter) E.I. Nyarady = H. colliniforme

- - var. *dissolutum* (Naeg. & Peter) E.I. Nyarady = H. dissolutum

- - var. *sudetorum* (Naeg. & Peter) E.I. Nyarady = H. sudetorum

- subsp. *colliniforme* (Peter) P.D. Sell = H. colliniforme

cajanderi Norrl. - 1

calcigenum Rehm. - 1(?)

callichlorum Litv. & Zahn - 2

callicymum (Rehm.) Juxip (*Pilosella callicyma* (Rehm.) Schljak.) - 1

x **callimorphoides** Zahn (*Pilosella* x *callimorphoides* (Zahn) Sojak) - 1

callimorphopsis (Zahn) Juxip = H. x progenitum

x **callimorphum** Naeg. & Peter (*Pilosella* x *callimorpha* (Naeg. & Peter) Sojak) - 1

x **calodon** Tausch ex Peter (*H. golitzynianum* Juxip, *H. multiceps* (Naeg. & Peter) Juxip, *H. ochrophyllum* (Naeg. & Peter) Juxip, *H. pskowiense* (Zahn) Juxip, *H. strictiramum* (Naeg. & Peter) Juxip, *H. tenuiceps* (Naeg. & Peter) Juxip, *Pilosella* x *calodon* (Tausch ex Peter) Sojak, *P.* x *calodon* subsp. *multiceps* (Naeg. & Peter) Sojak, *P.* x *calodon* subsp. *tenuiceps* (Naeg. & Peter) Sojak) - 1, 2

calodontopsis (Litv. & Zahn) Kem.-Nath. (*H. calodontopsis* (Litv. & Zahn) Juxip, comb. superfl.) - 2

calolepideum (Norrl.) Juxip (*Pilosella calolepidea* (Norrl.) Schljak.) - 1

x **calomastix** Peter (*Pilosella* x *calomastix* (Peter) Sojak.) - 1

■ **caloprasinum** (Zahn) Juxip

caniceps Norrl. - 1

canitiosum (Dahlst.) Brenn. (*H. silvaticum* (L.) Gouan *[subsp.] *canitiosum* Dahlst.) - 1

canum Peter = H. x kalksburgense

carcarophyllum K. Joh. - 1

cardiobasis (Zahn) Juxip - 1

cardiophyllum Jord. ex Sudre - 2

carpathicola (Naeg. & Peter) Czer. comb. nova (*H. aurantiacum* L. subsp. *carpathicola* Naeg. & Peter 1885, Hier. Mitt. Eur. 1 : 290; *Pilosella aurantiaca* (L.) F. Schultz & Sch. Bip. subsp. *carpathicola* (Naeg. & Peter) Dostal, *P. carpathicola* (Naeg. & Peter) Schljak.) - 1

carpathicum Bess. subsp. *gracilentipes* Dahlst. ex Zahn = H. gracilentipes

- subsp. *subgracilescens* Dahlst ex Zahn = H. subgracilescens

casparyanum (Naeg. & Peter) Juxip = H. x prussicum

caucasiciforme (Litv. & Zahn) Juxip = H. x maschukense

caucasicum Naeg. & Peter - 2

caucasiense Arv.-Touv. - 2

cauri Juxip - 1

cercidotelmatodes Juxip - 1

cereolinum (Norrl.) Juxip (*H. basifolium* (Almq.) Stenstr. *[subsp.] *cereolinum* Norrl., *H. basifolium* auct.) - 1

cernuiforme (Naeg. & Peter) Zahn = H. x macrostolonum

chadatense Juxip = H. soczavae

chaetothyrsoides (Litv. & Zahn) Juxip - 2

chaetothyrsum (Litv. & Zahn) Juxip - 2

chaunocymum (Rehm.) Czer. comb. nova (*H. magyaricum* Peter subsp. *chaunocymum* Rehm. 1897, Verh. Zool.-Bot. Ges. Wien, 47 : 298; *Pilosella chaunocyma* (Rehm.) Schljak.) - 1

cheriense Jord. ex Boreau - 1

chibinense Schljak. - 1
chibinicola Schljak. - 1
chlorelliceps Norrl. ex Juxip - 1
chlorellum Norrl. - 1
chlorochromum Sosn. & Zahn - 2
chloroleucolepium (Kozl. & Zahn) Juxip - 2
chloromaurum K. Jon. (*H. lyratum* Norrl. 1889, non L. 1753, *H. tenuiglandulosum* Norrl.) - 1
chlorophaeum Norrl. = H. subchlorophaeum
chlorophilum (Kozl. & Zahn) Juxip - 2
chlorophyllum Jord. ex Boreau - 1
x **chloroprenanthes** Litv. ex Zahn - 2
x **chlorops** (Naeg. & Peter) Zahn (*Pilosella* x *chlorops* (Naeg. & Peter) Sojak) - 1
christianense Dahlst. ex Stenstr. (*H. maculosum* Dahlst. ex Stenstr.) - 1
christoglossum (Zahn) Juxip = H. tephrocephalum
chromolepium (Zahn) Kem.-Nath. (*H. chromolepium* (Zahn) Juxip, comb. superfl.) - 2
chrysophthalmum Norrl. = H. x flagellare
ciesielskii Blocki = H. x densiflorum
■ cilicicum (Naeg. & Peter) Zahn (*Pilosella cilicica* (Naeg. & Peter) Holub, *P. hoppeana* (Schult.) F. Schultz & Sch. Bip. subsp. *cilicica* (Naeg. & Peter) P.D. Sell & C.West)
cincinnatum Fries - 2
cinerellisquamum (Litv. & Zahn) Schljak. (*H. murorum* L. subsp. *cinerellisquamum* Litv. & Zahn) - 1
■ cinereostriatum (Woronow & Zahn) Juxip
cischibinense Schljak. - 1
cisuralense Schljak. (*H. frondiferum* (Elfstr.) Elfstr. f. *wologdense* Elfstr., *H. frondiferum* auct.) - 1, 3(?)
ciuriwkae (Woloszcz. & Zahn) Schljak. (*H. laevigatum* Willd. subsp. *ciuriwkae* Woloszcz. & Zahn) - 1
clinoglossum Norrk. = H. x blyttianum
cochleatum (Naeg. & Peter) Norrl. = H. x suecicum
colliniforme (Peter) Roffey, 1925, non Dalla Torre & Sarnth. 1912, (*H. collinum* Gochn. subsp. *colliniforme* Peter, *H. caespitosum* Dumort. subsp. *caespitosum* var. *colliniforme* (Peter) E.I. Nyarady, *H. caespitosum* subsp. *colliniforme* (Peter) P.D. Sell, *Pilosella caespitosa* (Dumort.) P.D. Sell & C. West subsp. *colliniformis* (Peter) P.D. Sell & C. West, *P. colliniformis* (Peter) Dostal) - 1

This name must be replaced as later homonym.

x **collinum** Gochn. (*H. durisetum* (Naeg. & Peter) Juxip, *H. fallax* Willd., *H. permense* (Zahn) Juxip, *Pilosella* x *collina* (Gochn.) Sojak, *P. collina* subsp. *fallax* (Willd.) Sojak, *P. fallax* (Willd.) Arv. Touv.) - 1, 2, 4, 6
- subsp. *colliniforme* Peter = H. colliniforme
- subsp. *dublanense* Rehm. = H. dublanense
- subsp. *glaucochroum* Naeg. & Peter = H. glaucochroum
- subsp. *sudeterum* Peter = H. sudetorum
coloratum Elfstr. - 1
comatuloides (Zahn) Schljak. (*H. nigrescens* Willd. subsp. *comatuloides* Zahn) - 1
commilitonum Juxip = H. subcompositum
comosum Elfstr. = H. glabriligulatum
concinnidens (Zahn) Juxip - 2
concoloriforme Norrl. (*H. latvaense* Norrl., *Pilosella concoloriformis* (Norrl.) Schljak.) - 1
conferciens Norrl. = H. x macranthelum
conflectens Norrl. (*Pilosella conflectens* (Norrl.) Schljak.) - 1
congruens Norrl. (*H. apatitorum* Juxip, *H. bobrovii* Juxip, *H. frigidellum* (Pohle & Zahn) Juxip, *H. ischnoadenum* Juxip, *H. kolicola* Juxip, *H. murmanicola* Juxip, 1959,

non (Zahn) Juxip, 1960, *H. subarctoum* Norrl., *H. voroniense* Juxip) - 1
coniciforme (Litv. & Zahn) Juxip - 2
conicum Arv.-Touv. - 1, 2
coniops Norrl. - 1
connatum Norrl. = H. penduliforme
consociatum Jord. ex Boreau - 1
constringensiforme Juxip = H. subramosum
contractum (Norrl.) Norrl. (*Pilosella contracta* (Norrl.) Schljak.) - 1
coreanum Nakai - 5
corymbosum Fries var. *pachycephalum* Fries ex Uechtr. = H. pachycephalum
crassifoliiforme Schljak. - 1
x *crassisetum* auct. = H. x fallaciforme
creperiforme Juxip - 1
crepidioides Norrl. - 1
cretaceum Arv.-Touv. & Gautier ex Sudre - 1
crispans Norrl. = H. crispulum
crispulum Norrl. (*H. crispans* Norrl. nom. illegit. superfl.) - 1
crispum (Elfstr.) Elfstr. = H. alpinum
crocatum Fries - 1
cruciatum Schljak. - 1
cruentiferum (Norrl. & Lindb. fil. ex Norrl.) K. Joh. (*H. rigidum* C. Hartm. *[subsp.] *cruentiferum* Norrl. & Lindb. fil. ex Norrl.) - 1
curvescens (Norrl.) Norrl. (*Pilosella curvescens* Norrl.) - 1
curvicollum Norrl. = H. x macranthelum
curvulatum (Zahn) Juxip = H. x scandinavicum
curvulum Norrl. = H. x arvicola
cuspidelliforme Juxip - 1
cuspidellum (Pohle & Zahn) Juxip - 1
cylindriceps (Naeg. & Peter) Juxip (*Pilosella cylindriceps* (Naeg. & Peter) Schljak.) - 1(?)
cymanthodes (Kozl. & Zahn) Juxip - 2
cymanthum (Naeg. & Peter) Juxip (*H. bauhini* Bess. subsp. *bauhini* var. *cymanthum* (Naeg. & Peter) E.I. Nyarady, *Pilosella bauhini* (Bess.) Atv.-Tour. subsp. *cymantha* (Naeg. & Peter) Sojak, *P. cymantha* (Naeg. & Peter) Schljak.) - 1, 2
cymiflorum Naeg. & Peter subsp. *tubulatum* Vollm. = H. x tubulatum
cymigerum Reichenb. = H. vaillantii
cymiramum (Schelk. & Zahn) Juxip - 2
cymosiforme (Naeg. & Peter) Juxip = H. x densiflorum
cymosum L. (*Pilosella cymosa* (L.) F. Schultz & Sch. Bip.) - 1
- subsp. *cymigerum* (Reichenb.) Peter = H. vaillantii
- subsp. *cymosum* var. *holmiense* Naeg. & Peter = H. holmiense
- subsp. *euryanthelum* Dahlst. = H. euryanthelum
- subsp. *leptothyrsoides* Zahn = H. x leptothyrsoides
- subsp. *samaricum* Zahn = H. samaricum
cyrtophyllum Norrl. = H. x zizianum
czaiense Schischk. & Serg. - 3
czeremoszense Woloszcz. & Zahn - 1
czeschaense Schljak. - 1
czunense Schljak. - 1
czywczynae (Woloszcz. & Zahn) Schljak. (*H. krasanii* Woloszcz. subsp. *czywczynae* Woloszcz. & Zahn) - 1
dagoense Juxip - 1
deanum (Zahn) Schljak. (*H. trebevicianum* K. Maly subsp. *deanum* Zahn) - 1
■ debilescens (Woronow & Zahn) Juxip
dechyi (Kozl. & Zahn) Juxip - 2
decipiens Tausch (*H. nigrescens* Willd. var. *decipiens*

(Tausch) E.I. Nyarady) - 1

decipientiforme (Woloszcz. & Zahn) Schljak. (*H. nigrescens* Willd. subsp. *decipientiforme* Woloszcz. & Zahn) - 1

declivium Norrl. ex Juxip - 1

decurrens Norrl. - 1

dedovii Schljak. - 1

demetrii Schljak. - 1

x **densiflorum** Tausch (*H. acrosciadium* (Naeg. & Peter) Juxip, 1960, non Froel. 1838, *H. ciesielskii* Blocki, *H. cymosiforme* (Naeg. & Peter) Juxip, *H. excellens* Blocki ex Ostenf. 1906, non J. Murr, 1901, *H. longiradiatum* (Zahn) Juxip, *H. melanolepium* (Rehm.) Juxip, *H. mnoophorum* Naeg. & Peter, *H. pseudoglomeratum* Blocki ex Woloszcz., *H. pycnomnoon* (Rehm.) Juxip, *H. subexcellens* (Zahn) Juxip, *H. tauschii* Zahn var. *umbelliferum* (Naeg. & Peter) E.I. Nyarady, *H. umbelliferum* Naeg. & Peter, *H. wjasowoense* (Zahn) Juxip, *Pilosella x densiflora* (Tausch) Sojak, *P. densiflora* subsp. *acrosciadia* (Naeg. & Peter) Sojak, *P. densiflora* subsp. *cymosiformis* (Naeg. & Peter) Sojak, *P. densiflora* subsp. *umbellifera* (Naeg. & Peter) Sojak) - 1, 3

dentatum auct. = H. sarmaticum

dentatum Hoppe subsp. *sarmaticum* Zahn = H. sarmaticum

denticuliferum (Norrl.) Norrl. (*Pilosella denticulifera* Norrl.) - 1

derivatum Norrl. (*H. naniceps* Elfstr.) - 1

detonsum (Norrl.) Norrl. = H. x glomeratum

diaphanoides Lindeb. (*H. diaphanum* auct.) - 1

- subsp. *ornatum* Dahlst. = H. ornatum

■ diaphanoidiceps (Woronow & Zahn) Juxip

diaphanum auct. = H. diaphanoides

dilutius (Rehm.) Czer. comb. nova (*H. florentinum* All. subsp. *dilutius* Rehm. 1897, Verh. Zool.-Bot. Ges. Wien, 47 :284; *Pilosella dilutior* (Rehm.) Schljak.) - 1

diminuens (Norrl.) Norrl. (*H. murorum* L. subsp. *diminuens* Norrl., *H. revocans* Juxip) - 1

x **dimorphoides** (Norrl.) Norrl. (*Pilosella x dimorphoides* Norrl., *Hieracium norrliniiforme* Pohle & Zahn, *H. pericaustum* (Norrl.) Norrl., *H. tephranthelum* (Zahn) Juxip, *Pilosella norrliniiformis* (Pohle & Zahn) Sojak) - 1

diremtum Norrl. - 1

discolor (Naeg. & Peter) Juxip = H. x mollicaule

discoloratum (Norrl.) Norrl. = H. x blyttianum

dissolutum (Naeg. & Peter) Juxip (*H. caespitosum* Dumort. subsp. *caespitosum* var. *dissolutum* (Naeg. & Peter) E.I. Nyarady, *Pilosella caespitosa* (Dumort.) P.D.Sell & C. West subsp. *dissoluta* (Naeg. & Peter) Sojak, *P. dissoluta* (Naeg. & Peter) Schljak.) - 1

distractifolium Schljak. - 1

distractum Norrl. - 1

dmitrovense (Peter) Juxip = H. x brachiatum

dobromilense Rehm. (*Pilosella dobromilensis* (Rehm.) Schljak.) - 1

dolabratum (Norrl.) Norrl. (*H. rigidum* C. Hartm. *[subsp.] *dolabratum* Norrl.) - 1

dolichanthelum Schljak. - 1

dolichotrichum Schljak. - 1

dolinense (Rehm.) Czer. comb. nova (*H. florentinum* All. subsp. *dolinense* Rehm. 1897, Verh. Zool.-Bot. Ges. Wien, 47 : 285; *Pilosella dolinensis* (Rehm.) Schljak.) - 1

donetzicum Kotov - 1

dshurdshurense Juxip - 1

x **dubium** L. (*H. accline* Norrl., *H. acrocomum* Naeg. & Peter, *H. floribundoides* (Zahn) Juxip, *H. macrastrum* (Zahn) Juxip, *H. pilipes* T. Sael., *H. violaceipes* (Zahn)

Juxip, *Pilosella x dubia* (L.) Fries, *P. dubia* (L.) Sojak, comb. superfl., *P. pilipes* (T. Sael.) Norrl., *P. pilipes* subsp. *acrocoma* (Naeg. & Peter) Sojak) - 1

dublanense (Rehm.) Czer. comb. nova (*H. collinum* Gochn. subsp. *dublanense* Rehm. 1895, Verh. Zool.-Bot. Ges. Wien, 45 : 323; *Pilosella dublanensis* (Rehm.) Schljak.) - 1

dublizkii B. Fedtsch. & Nevski - 3, 4, 6

duderhofense Juxip (*H. karelorum* (Norrl.) Juxip, *H. meinshausenianum* Juxip, *H. multiglandulosum* Juxip) - 1

x *duplex* auct. p.p. = H. x macrostolonum and H. x prussicum

durisetum (Naeg. & Peter) Juxip = H. x collinum

x *dybowskianum* (Rehm.) Zahn (*H. trigenes* Peter subsp. *dybowskianum* Rehm., *Pilosella x dybowskiana* (Rehm.) Schljak., *Hieracium trigenes* auct.) - 1

echiocephalum (Naeg. & Peter) Juxip = H. x auriculoides

echiogenes (Naeg. & Peter) Juxip = H. x auriculoides

echioides Lumn. (*H. echioides* var. *freynii* (Naeg. & Peter) E.I. Nyarady, *H. freynii* (Naeg. & Peter) Juxip, *Pilosella echioides* (Lumn.) F. Schultz & Sch. Bip., *P. echioides* subsp. *freynii* (Naeg. & Peter) Sojak) - 1, 2, 3, 4, 6

- subsp. *procerum* (Fries) P.D. Sell = H. procerum

- var. *freynii* (Naeg. & Peter) E.I. Nyarady = H. echioides

- var. *macrocymum* (Naeg. & Peter) E.I. Nyarady = H. macrocymum

eichvaldii Juxip - 1

elaeochlorum Schljak. - 1

elfvingii Norrl. = H. x blyttianum

elimense Schljak. - 1

empodistum (Naeg. & Peter) Juxip (*Pilosella empodista* (Naeg. & Peter) Schljak.) - 1(intr.?)

endaurovae Juxip - 2

ensiferum Norrl. (*Pilosella ensifera* (Norrl.) Schljak.) - 1

epichlorum (Litv. & Zahn) Juxip - 2

ericeticola Schljak. - 1

ericetorum (Naeg. & Peter) Juxip, 1960, non Freyn, 1887 (*Pilosella ericetorum* (Naeg. & Peter) Schljak.) - 1

This name must be replaced because it is a later homonym.

ermaniense Juxip - 2

erraticum Norrl. (*H. litoreum* Norrl., *Pilosella erratica* (Norrl.) Schljak.) - 1

erythrocarpoides (Litv.& Zahn) Kem.-Nath. (*H. erythrocarpoides* (Litv. & Zahn) Juxip, comb. superfl.) - 2

x **erythrocarpum** Peter - 2, 6

erythrochristum (Naeg. & Peter) Juxip = H. x arvicola

erythrophylloides (Zahn) Czer. comb. nova (*H. bauhini* Bess. subsp. *erythrophylloides* Zahn, 1911, Trav. Mus. Bot. Acad. Sci. Petersb. 9 : 38; *Pilosella erythrophylloides* (Zahn) Schljak.) - 1

erythropoides (Norrl.) Schljak. = H. x suecicum

x *euchaetium* auct. = H. x longum

euchaetium Naeg. & Peter subsp. *longum* Naeg. & Peter = H. x longum

euirriguum (Zahn) Schljak. (*H. lachenalii* C.C. Gmel. subsp. *euirriguum* Zahn, *H. irriguum* auct.) - 1

eurofinmarkicum (Norrl.) Schljak. (*H. gemellum* Elfstr. *[subsp.] *eurofinmarkicum* Norrl.) - 1

euryanthelum (Dahlst.) Dahlst. (*H. cymosum* L. subsp. *euryanthelum* Dahlst., *Pilosella euryanthela* (Dahlst.) Schljak.) - 1

eusciadium (Naeg. & Peter) Dahlst. (*Pilosella eusciadia* (Naeg.& Peter) Schljak.) - 1

excellens Blocki ex Ostenf. = H. x densiflorum

excubitum Elfstr. - 1, 3

eximiiforme (Dahlst.) Dahlst. (*H. alpinum* L. subsp. *exim-*

iiforme Dahlst.) - 1, 3

exotericum Jord. ex Boreau - 2

expallidiforme Dahlst. ex Stenstr. (*H. agnostum* Juxip) - 1

falcidentatum Juxip - 1

x **fallaciforme** Litv.& Zahn (*H. subfallaciforme* (Zahn) Juxip, *Pilosella* x *fallaciformis* (Litv. & Zahn) Schljak., *Hieracium crassisetum* auct.) - 1

x **fallacinum** F. Schultz (*Pilosella* x *fallacina* (F. Schultz) Sch. Bip.) - 1(?)

fallax Willd. = H. x collinum

farinifloccosum (Degen & Zahn) Schljak. (*H. rauzense* J. Murr subsp. *farinifloccosum* Degen & Zahn) - 1

fariniramum (Ganesch. & Zahn) Juxip - 3, 4

farinodermum (Litv. & Zahn) Juxip - 2

farreum (Norrl.) Norrl. (*Pilosella farrea* Norrl.) - 1

fastigiatum Fries - 1

fastigiatum (Tausch ex Naeg. & Peter) Juxip, 1960, non Fries, 1848 (*H. bauhini* Bess. var. *fastigiatum* (Naeg. & Peter) E.I. Nyarady, *Pilosella bauhini* (Bess.) Arv.-Tour. subsp. *fastigiata* (Tausch ex Naeg. & Peter) Sojak, *P. fastigiata* (Tausch ex Naeg. & Peter) Schljak.) - 1

This name must be replaced because it is a later homonym.

fellmanii Norrl. - 1

x **fennicum** (Norrl.) Norrl. ex Wain (*H. suecicum* Fries *[subsp.] fennicum* Norrl., *H. brachycephalum* (Norrl.) Norrl., *H. fulvescens* (Naeg. & Peter) Juxip, 1960, non Norl. 1888, *H. ladogense* (Norrl.) Norrl., *H. leopolitanum* (Zahn) Juxip, *H. limbatum* (Naeg. & Peter) Juxip, *H. longatum* (Peter) Juxip, *H. nemoricola* (Norrl.) Norrl., *H. oeneororatum* Norrl., *H. papyrodes* (Norrl.) Juxip, *H. polysarcoides* (Zahn) Juxip, *H. pubens* (Naeg. & Peter) Norrl., *H. riganum* (Syr. & Zahn) Juxip, *H. spathophyllum* Peter, *H. subpratense* (Norrl.) Norrl., *H. subpubens* (Norrl.) Norrl., *H. subswirense* (Norrl.) Juxip, *H. swirense* Norrl., *H. xanthostigma* (Norrl.) Norrl., *Pilosella* x *fennica* (Norrl.) Norrl., *P. fennica* subsp. *fulvescens* (Naeg. & Peter) Sojak, *P. fennica* subsp. *spathophylla* (Peter) Sojak, *P. ladogensis* Norrl., *P. subpratensis* Norrl.) - 1

fennoorbicans Norrl. - 1

fennoorbicantiforme Juxip - 1

ferroviae (Rehm.) Czer. comb. nova (*H. magyaricum* Peter subsp. *ferroviae* Rehm. 1897, Verh. Zool.-Bot. Ges. Wien, 47 : 291; *Pilosella ferroviae* (Rehm.) Schljak.) - 1

festinum Jord. ex Boreau - 1, 2

filiferum Tausch (*H. bauhini* Bess. subsp. *magyaricum* (Peter) Zahn var. *filiferum* (Tausch) E.I. Nyarady, *Pilosella filifera* (Tausch) Schljak., *P. glaucescens* (Bess.) Sojak subsp. *filifera* (Tausch.) Sojak) - 1, 2

filifolium Juxip (*H. umbellatum* L. subsp. *filifolium* (Juxip) Tzvel.) - 1

finmarkicum Elfstr. - 1

firmicaule (Norrl.) Norrl. (*Pilosella firmicaulis* Norrl.) - 1

firmipes (Rehm.) Juxip = H. x macrostolonum

x **flagellare** Willd. (*H. anacraspedum* (Rehm.) Juxip, *H. anisocephalum* (Rehm.) Juxip, *H. brachyacron* (Rehm.) Juxip, *H. brachyschistum* (Zahn) Juxip, *H. chrysophthalmum* Norrl., *H. flagellare* subsp. *pseuduliginosum* Zahn, nom. illegit., *H. flagellare* subsp. *tatrense* Peter, *H. homostegium* Norrl., *H. inceptans* Norrl., *H. petunnikovii* (Peter) Juxip, *H. prognatum* Norrl., *H. pseuduliginisum* Juxip, *H. tatrense* (Peter) Juxip, *H. tweriense* (Zahn) Juxip, *Pilosella* x *flagellaris* (Willd.) Arv.-Tour., *P.* x *flagellaris* (Willd.) P.D. Sell & C. West, comb. superfl., *P. flagellaris* subsp. *tatrensis* (Peter) Sojak) - 1

- subsp. *pseuduliginosum* Zahn. = H. x flagellare
- subsp. *tatrense* Peter = H. x flagellare
- nm. *moskoviticum* (Peter) Lepage = H. x macrostolonum

flagellariforme G. Schneid. = H. x progenitum

flexicaule Elfstr. = H. subflexicaule

floccicomatum (Woronow & Zahn) Juxip - 2(?)

floccinops (Elfstr.) Schljak. (*H. bathycephalum* Elfstr. var. *floccinops* Elfstr., *H. pergrandidens* (Zahn) Juxip, p.p. excl. basionymo) - 1

flocciparum (Schelk. & Zahn) Juxip - 2

floccipedunculum (Naeg. & Peter) Juxip (*Pilosella floccipeduncula* (Naeg. & Peter) Schljak.) - 1

florentinum All. subsp. *dilutius* Rehm. = H. dilutius
- subsp. *dolinense* Rehm. = H. dolinense
- subsp. *setosopetiolatum* Rehm. = H. setosopetiolatum
- subsp. *stellatum* Tausch ex Zahn = H. stellatum

floribundiforme (Naeg. & Peter) Juxip = H. x arvicola and H. x glomeratum

floribundoides (Zahn) Juxip = H. x dubium

x **floribundum** Wimm. & Grab. (*H. baenitzii* (Naeg. & Peter) Juxip, *H. regiomontanum* (Naeg. & Peter) Juxip, *H. sudavicum* (Naeg. & Peter) Juxip, *Pilosella floribunda* (Wimm. & Grab.) Arv.-Touv. comb. superfl., *P.* x *floribunda* (Wimm. & Grab.) Fries, *P. floribunda* subsp. *regiomontana* (Naeg. & Peter) Sojak) - 1
- subsp. *cochleatum* Naeg. & Peter = H. x suecicum

floridum (Naeg. & Peter) Juxip = H. x iseranum

foliolatum Schljak. - 1

folioliferum (Elfstr.) Norrl. - 1

▪ **foliosissimum** (Woronow & Zahn) Juxip

fominianum (Woronow & Zahn) Juxip - 2

freynii (Naeg. & Peter) Juxip = H. echioides

frickii (Zahn) Juxip - 2

friesii C. Hartm. - 1

frigidellum (Pohle & Zahn) Juxip = H. congruens

fritzei F. Schultz - 1

frondiferum auct. = H. cisuralense

frondiferum (Elfstr.) Elfstr. f. *wologdense* Elfstr. = H. cisuralense

frondosum (Naeg. & Peter) Juxip = H. x schultesii

fuliginosiforme Schljak. - 1

fuliginosum (Laest.) Norrl. (*H. alienatum* Norrl.) - 1

fulveolutescens Norrl. = H. x blyttianum

fulvescens (Naeg. & Peter) Juxip = H. x fennicum

fulvescens Norrl. (*H. sordidescens* Norrl. nom. illegit.) - 1

furfuraceoides (Zahn) Juxip - 1

furvescens (Dahlst.) Omang (*H. prenanthoides* Vill. var. *furvescens* Dahlst., *H. imandrense* Juxip, *H. reducatum* (Norrl.) Juxip) - 1

x **fuscoatrum** Naeg. & Peter (*H. chaetodermum* (Pohle & Zahn) Juxip, *H. rubroonegense* Norrl., *H. semionegense* Norrl., *Pilosella* x *fuscoatra* (Naeg. & Peter) Sojak) - 1

fuscocinereum auct. = H. oistophyllum

fuscocinereum Norrl. subsp. *sagittatum* (Lindeb.) Brautig. = H. oistophyllum

fuscum auct. = H. x blyttianum

galbanum (Dahlst.) K. Joh. (*H. murorum* L. subsp. *galbanum* Dahlst.) - 1

galiciense Blocki (*Pilosella galiciensis* (Blocki) Schljak.) - 1

ganeschinii Zahn - 3, 4

gemellum Elfstr. *[subsp.] eurofinmarkicum* Norrl. = H. eurofinmarkicum

geminatiforme Schljak. - 1

geminatum Norrl. - 1

gentile Jord. ex Boreau - 1, 2

georgicum Fries - 2

x **gigantellum** Litv. & Zahn - 2

giganticaule (Zahn) Juxip = H. x glomeratum

glabriligulatum Norrl. (*H. comosum* Elfstr., *H. lujaurense* Norrl. p.p. incl. typo, *H. adspersum* auct.) - 1

glaucescens Bess. (*Pilosella glaucescens* (Bess.) Sojak) - 1

glaucinoides K. Maly - 1

glaucochroum (Naeg. & Peter) Czer. comb. nova (*H. collinum* Gochn. subsp. *glaucochroum* Naeg. & Peter, 1885, Hier. Mitt.-Eur. 1 : 314; *Pilosella glaucochroa* (Naeg. & Peter) Schljak.) - 1

glehnii Juxip - 1

glomerabile (Norrl.) Czer. comb. nova (*H. pubescens* (Lindbl.) Norrl. *[subsp.] glomerabile* Norrl. 1906, in Cajander, Suomen Kasvio : 670; *Pilosella glomerabilis* (Norrl.) Schljak.) - 1

glomeratiforme (Zahn) Juxip = H. x scandinavicum

x **glomeratum** Froel. (*H. ambiguum* Ehrh. nom. invalid., *H. detonsum* (Norrl.) Norrl., *H. floribundiforme* (Naeg. & Peter) Juxip, p.p. excl. plantis, *H. giganticaule* (Zahn) Juxip, *H. x glomeratum* subsp. *ambiguum* Ehrh. ex Naeg. & Peter, *H. griseum* (Norrl.) Norrl., *H. haraldii* Norrl., *H. lamprophthalmum* Norrl., *H. luteoglandulosum* T. Sael. ex Norrl., *H. micans* Norrl., *H. neglectum* (Norrl.) Norrl., *H. prolongatum* (Naeg. & Peter) Juxip, *H. pycnothyrsum* (Peter) Juxip, *H. reflorescens* Norrl., *H. rusanum* (Zahn) Juxip, *H. subambiguum* (Naeg. & Peter) Juxip, *H. vitellinum* Norrl., *Pilosella ambigua* (Ehrh. ex Naeg. & Peter) Sojak, *P. glomerata* (Froel.) Arv.-Touv. comb. superfl., *P. x glomerata* (Froel.) Fries, *P. glomerata* subsp. *ambigua* (Ehrh. ex Naeg. & Peter) Sojak, *P. glomerata* subsp. *prolongata* (Naeg. & Peter) Sojak, *P. glomerata* subsp. *subambigua* (Naeg. & Peter) Sojak) - 1

- subsp. *ambiguum* Ehrh. ex Naeg. & Peter = H. x glomeratum

glomerellum (Zahn) Juxip - 2

gnaphalium (Naeg. & Peter) Juxip = H. x prussicum

golitzynianum Juxip = H. x calodon

gorczakovskii Schljak. - 1, 3

goriense (Kozl. & Zahn) Juxip - 2

gorodkowianum Juxip - 4

gothicifrons (Zahn) Juxip - 2

gothicum Fries - 1

gracilentipes (Dahlst. ex Zahn) Noto (*H. carpathicum* Bess. subsp. *gracilentipes* Dahlst. ex Zahn) - 1

gracilentum Backh. var. *varangerense* Elfstr. = H. varangerense

grandidens Dahlst. = H. silvularum

- subsp. *torticeps* Dahlst. = H. torticeps

granitophilum Norrl. (*Pilosella granitophila* (Norrl.) Schljak.) - 1

granvicum Juxip - 1

griseum (Norrl.) Norrl. = H. x glomeratum

grofae Woloszcz. - 1

x **gudergomiense** Juxip - 2

gudissiense Juxip - 2

guentheri Norrl. - 1

gustavianum Juxip - 2

x *guthnickianum* auct. = H. x roxolanicum

guthnickianum Hegetschw. var. *rehmannii* (Naeg. & Peter) E.I. Nyarady = H. x roxolanicum

guttenfeldense (Zahn) Juxip = H. x prussicum

gymnogenum (Zahn) Juxip = H. alpinum

gynaeconesaeum (Zahn) Juxip - 2

haematoglossum (Kozl. & Zahn) Juxip - 2

haraldii Norrl. = H. x glomeratum

heothinum (Naeg. & Peter) Juxip (*H. bauhini* Bess. subsp. *magyaricum* (Peter) Zahn var. *heothinum* (Naeg. &

Peter) E.I. Nyarady, *Pilosella glaucescens* (Bess.) Sojak subsp. *heothina* (Naeg. & Peter) Sojak, *P. heothina* (Naeg. & Peter) Schljak.) - 1

heterodontoides (Litv. & Zahn) Juxip - 2

hirsuticaule Schljak. - 1

hirtulum (Peter) Juxip = H. x arvicola

hispidissimum (Rehm. ex Naeg. & Peter) Juxip (*H. magyaricum* Peter subsp. *hispidissimum* Rehm. ex Naeg. & Peter, *Pilosella hispidissima* (Rehm. ex Naeg. & Peter) Schljak.) - 1, 2

hjeltii Norrl. - 1

hohenackeri (Naeg. & Peter) Juxip (*Pilosella hohenackeri* (Naeg. & Peter) Sojak) - 2

holmiense (Naeg. & Peter) Dahlst. (*H. cymosum* L. subsp. *cymosum* var. *holmiense* Naeg. & Peter, *Pilosella holmiensis* (Naeg. & Peter) Schljak.) - 1

hololeion Maxim. - 5

homochroum Norrl. ex Schljak. - 1

homostegium Norrl. = H. x flagellare

hopense Juxip - 2

hoppeanum Schult. (*Pilosella hoppeana* (Schult.) F. Schultz & Sch. Bip.) - 1(?),2

- subsp. *przybyslawskii* Rehm. = H. przybyslawskii
- subsp. *testimoniale* Naeg. & Peter = H. multisetum
- subsp. *vulpinum* Rehm. = H. vulpinum
- var. *multisetum* (Naeg. & Peter) E.I. Nyarady = H. multisetum

hosjense Schljak. - 1

hryniawiense Woloszcz. - 1

hryniewieckii Juxip - 2

hylocomum Juxip - 1

hylogeton (Kozl. & Zahn) Juxip - 2

x **hypeuryum** Peter (*Pilosella x hypeurya* (Peter) Sojak) - 2

- subsp. *podolicum* Rehm. = H. x podolicum

hypoglaucum (Litv. & Zahn) Juxip (*H. prenanthoides* auct.) - 2

hypopityforme Juxip - 2

hypopitys (Litv. & Zahn) Juxip - 2

hypopogon (Litv. & Zahn) Juxip - 2

igoschinae Juxip - 1

ihrowyszczense (Zahn) Schljak. (*H. atratum* Fries subsp. *ihrowyszczense* Zahn) - 1

ilyassowoense (Zahn) Juxip = H. x brachiatum

imandrense Juxip = H. furvescens

imandricola Schljak. - 1

imponens Norrl. (*Pilosella imponens* (Norrl.) Schljak.) - 1

x **incaniforme** Litv. & Zahn (*Pilosella incaniformis* (Litv. & Zahn) Sojak, *Hieracium x sintenisii* auct.) - 2

incanum Bieb. 1808, non L. 1753 - 2

This name must be replaced because it is a later homonym.

- subsp. *sosnowskyi* Zahn = H. sosnowskyi

inceptans Norrl. = H. x flagellare

incomptum Norrl. (*H. subincomptum* (Zahn) Juxip - 1

inconveniens Juxip - 1

incrassatiforme Norrl. = H. x zizianum

incurrens T. Sael. ex Norrl. - 1

ingricum (Naeg. & Peter) Juxip (*H. bauhini* Bess. subsp. *bauhini* var. *ingricum* (Naeg. & Peter) E.I. Nyarady, *H. macrum* (Naeg. & Peter) Juxip, *Pilosella ingrica* (Naeg & Peter) Schljak.) - 1

insolens Norrl. (*Pilosella insolens* (Norrl.) Schljak.) - 1

■ *insolitum* (Zahn) Juxip

insulicola Schljak. - 1

integratum Dahlst. ex Stenstr. - 1

integrilingua Norrl. = H. x blyttianum

intercessum Juxip - 1

intersitum Jord. ex Boreau - 1
inuloides Tausch subsp. *polycomatum* Zahn = H. polycomum
iremelense (Elfstr.) Juxip (*H. oncodes* Omang var. *iremelense* Elfstr.) - 1, 3
irriguiceps (Zahn) Schljak. (*H. vulgatum* Fries subsp. *irriguiceps* Zahn) - 1
irriguum auct. = H. euirriguum
ischnoadenum Juxip = H. congruens
x **iseranum** Uechtr. (*H. floridum* (Naeg. & Peter) Juxip, *H. nigriceps* Naeg. & Peter, *H. subnigriceps* (Zahn) Juxip, *Pilosella* x *iserana* (Uechtr.) Sojak, *P. iserana* subsp. *florida* (Naeg. & Peter) Sojak, *P. iserana* subsp. *subnigriceps* (Zahn) Sojak) - 1
issatschenkoi Schljak. - 1
isthmicola Norrl. = H. x suecicum
ivdelense Schljak. - 1
jablonicense Woloszcz. - 1
jackardii Zahn - 1
jailanum (Zahn) Juxip = H. tephrocephalum
jangajuense Juxip - 3
jaworowae (Zahn) Schljak. (*H. praecurrens* Vuk. subsp. *odorans* (Borb.) Zahn var. *jaworowae* Zahn, *H. praecurrens* auct.) - 1
x **juranum** Fries - 2
juratzkanum (Zahn) Schljak. (*H. vulgatum* Fries subsp. *juratzkanum* Zahn) - 1
kabanovii Juxip - 1
kaczurinii Juxip - 1
■ **kajanense** Malmgr. (*Pilosella kajanensis* (Malmgr.) Norrl.)
x **kalksburgense** Wiesb. (*H. canum* Peter, 1884, non Vuk. 1858, *H. laschii* Zahn, *H. scopulorum* Juxip, *H. spurium* Chaix ex Froel. nom. invalid., *Pilosella* x *kalksburgensis* (Wiesb.) Sojak, *P. kalksburgensis* subsp. *laschii* (Zahn) Sojak, *P. laschii* F. Schultz & Sch. Bip. nom. invalid., *P. spuria* (Chaix ex Froel.) Sojak) - 1
kandalakschae Schljak. - 1
kaninense Schljak. - 1
karelicum (Norrl.) Norrl. (*H. caespitosum* Dumort. subsp. *brevipilum* (Naeg. & Peter) P.D. Sell, p.p., *Pilosella caespitosa* (Dumort.) P.D. Sell & C. West subsp. *brevipula* (Naeg. & Peter) P.D. Sell & C. West, p.p., *P. karelica* Norrl.) - 1
karelorum (Norrl.) Juxip = H. duderhofense
karjaginii Juxip - 2
karpinskyanum (Naeg. & Peter) Juxip (*Pilosella karpinskyana* (Naeg. & Peter) Schljak.) - 1
kemulariae Juxip - 2
kiderense (Zahn) Juxip - 2
kievejense Schljak. - 1
x **kihlmanii** (Norrl.) Juxip (*Pilosella* x *kihlmanii* Norrl.) - 1
kildinense Schljak. - 1
kirghisorum Juxip - 6
knafii (Celak.) Zahn - 1
knappii Blocki (*Pilosella knappii* (Blocki) Schljak.) - 1
kochtanum (Kozl. & Zahn) Juxip - 2
■ **koenigianum** (Zahn) Juxip
x **koernickeanum** (Naeg. & Peter) Zahn (*Pilosella* x *koernickeana* (Naeg. & Peter) Sojak) - 1
kolenatii (Naeg. & Peter) Kem.-Nath. (*H. kolenatii* (Naeg. & Peter) Juxip, comb. superfl.) - 2
kolgujevense Schljak. - 1
kolicola Juxip = H. congruens
konshakovskianum Juxip = H. subramosum
korshinskyi Zahn - 3, 4, 6
- f. *sershukense* (Juxip) Serg. = H. sershukense

kosvinskiense Juxip - 1
kovdense Juxip = H. progrediens
x **kozlowskyanum** Zahn (*Pilosella kozlowskyana* (Zahn) Sojak) - 2
krasanii Woloszcz. - 1
- subsp. *czywczynae* Woloszcz. & Zahn = H. czywczynae
- subsp. *kukulense* Woloszcz. & Zahn = H. kukulense
- subsp. *pikujense* Woloszcz. & Zahn = H. pikujense
- subsp. *pseudonigritiforme* Zahn = H. pseudonigritiforme
kreczetoviczii Juxip - 2
krivanense (Woloszcz. & Zahn) Schljak. (*H. lampromegas* Zahn subsp. *krivanense* Woloszcz. & Zahn.) - 1
krylovii Nevski ex Schljak. - 1, 3, 4, 6
kubanicum (Litv. & Zahn) Juxip - 2
kubenskense Juxip - 1
kukulense (Woloszcz. & Zahn) Schljak. (*H. krasanii* Woloszcz. subsp. *kukulense* Woloszcz. & Zahn) - 1
kulkowianum (Zahn) Juxip - 1
kultukense Serg. & Juxip - 4
kumbelicum B. Fedtsch. & Nevski - 6
kupfferi Dahlst. - 1
kuroksarense Juxip - 1
kusnetzkiense Schischk. & Serg. - 3
kuusamoense Wainio - 1
kuzenevae Juxip - 1
lachenalii auct. p.p. = H. vulgatum
lachenalii C.C. Gmel. subsp. *anfractum* (Fries) E.I. Nyarady = H. anfractum
- subsp. *euirriguum* Zahn = H. euirriguum
lackschewitzii (Dahlst.) Juxip (*H. silvaticum* (L.) Gouan subsp. *lackschewitzii* Dahlst.) - 1
lactucella Wallr. (*Pilosella lactucella* (Wallr.) P.D. Sell & C. West, *Hieracium auricula* auct.) - 1
- subsp. *magnauricula* (Naeg. & Peter) P.D. Sell = H. magnauricula
ladogense (Norrl.) Norrl. = H. x fennicum
laetevirens Somm. & Levier - 2
laevicaule Jord. - 1
laevigans (Zahn) Juxip = H. laevigatum
laevigatum Willd. (*H. laevigans* (Zahn) Juxip) - 1, 2
- subsp. *ciuriwkae* Woloszcz. & Zahn = H. ciuriwkae
- subsp. *puschlachtae* Pohle & Zahn = H. puschlachtae
lailanum (Schelk. & Zahn) Juxip - 2
lamprocomoides (Woronow & Zahn) Juxip - 2
x **lamprocomum** (Naeg. & Peter) Juxip (*Pilosella hypeurya* (Peter) Sojak subsp. *lamprocoma* (Naeg. & Peter) Fuchs-Eckert, *P.* x *lamprocoma* (Naeg. & Peter) Schljak.) - 1, 2
lampromegas Zahn subsp. *krivanense* Woloszcz. & Zahn. = H. krivanense
- subsp. *orthobrachion* Woloszcz. & Zahn = H. orthobrachion
- subsp. *wysokae* Woloszcz. & Zahn = H. wysokae
lamprophthalmum Norrl. = H. x glomeratum
lanceatum Schljak. - 1
lanceolatum Vill. - 1
lancidens (Zahn) Juxip - 1, 2
lapponicifolium Schljak. - 1
lapponicum Fries - 1
largum Fries - 1
- var. **pallonianum** (Zahn) Schljak. (*H. pallonianum* Zahn.) - 1
laschii Zahn = H. x kalksburgense
lasiophorum (Naeg. & Peter) Juxip (*Pilosella auriculoides* (Lang) F. Schultz. subsp. *lasiophora* (Naeg. & Peter) Sojak) - 2
lasiothrix (Naeg. & Peter) Juxip (*Pilosella hypeurya* (Peter) Sojak subsp. *lasiothrix* (Naeg. & Peter) Fuchs-Eck-

ert) - 2

latens Juxip - 1

laterale Norrl. - 1

lateriflorum Norrl. (*H. sublateriflorum* Schljak.) - 1

laticeps (Norrl.) Norrl. (*Pilosella laticeps* Norrl., *Hieracium peleteranum* Merat subsp. *sabulosorum* Dahlst., *H. sabulosorum* (Dahlst.) Juxip) - 1

latpariense Peter - 2

latvaense Norrl. = H. concoloriforme

latypeum Norrl. - 1

laurinum Arv.-Tour. - 1

lehbertii (Zahn) Juxip - 1

lenkoranense Juxip - 2

leopolitanum (Zahn) Juxip = H. x suecicum

lepiduliforme Dahlst. - 1

lepistoides (K. Joh. ex Dahlst.) Norrl. - 1

leptadeniiforme Juxip - 3

leptadenium (Dahlst.) Dahlst. (*Pilosella leptadenia* (Dahlst.) Schljak.) - 1

leptocaulon (Naeg. & Peter) Juxip (*Pilosella caespitosa* (Dumort.) P.D. Sell & C. West subsp. *leptocaulon* (Naeg. & Peter) Sojak, *P. leptocaulon* (Naeg. & Peter) Schljak.) - 1

x **leptoclados** Peter (*Pilosella* x *leptoclados* (Peter) Sojak) - 1

x **leptogrammoides** Juxip - 2

leptopholis Schljak. - 1

leptophyes (Peter) Juxip = H. x sciadophorum

leptophyllum (Naeg. & Peter) Juxip = H. x zizianum

x **leptophytomorphum** Litv. & Zahn - 2

leptophyton Naeg. & Peter = H. x mollicaule

- var. *anocladum* (Naeg. & Peter) E.I. Nyarady = H. x mollicaule

- var. *discolor* (Naeg. & Peter) E.I. Nyarady = H. x mollicaule

x **leptoprenanthes** Litv. & Zahn - 2

x **leptothyrsoides** (Zahn) Juxip (*H. cymosum* L. subsp. *leptothyrsoides* Zahn, *Pilosella* x *leptothyrsoides* (Zahn) Schljak.) - 1

leptothyrsum Peter (*Pilosella leptothyrsa* (Peter) Schljak.) - 1

lespinassei (Kozl. & Zahn) Juxip - 2

leucocraspedum (Peter) Juxip = H. x arvicola

leucothyrsogenes (Kozl. & Zahn) Juxip - 2

leucothyrsoides (Kozl. & Zahn) Juxip - 2

leucothyrsum (Litv. & Zahn) Juxip - 2

levicaule Jord. subsp. *prilakenense* Zahn = H. prilakenense

levieri Peter (*Pilosella levieri* (Peter) Sojak) - 2

lignyotum Norrl. - 1

limbatum (Naeg. & Peter) Juxip = H. x fennicum

limenyense (Zahn) Czer. comb. nova (*H. bauhini* Bess. subsp. *cymanthum* (Nees & Peter) Zahn var. *limenyense* Zahn, 1923, in Engler, Pflanzenreich, 82 : 1426; *Pilosella limenyensis* (Zahn) Schljak.) - 1

linahamariense Juxip (*H. saxifragum* auct.) - 1

linifolium T. Sael. (*H. rigidiusculum* Schljak.) - 1

lipnickianum (Rehm.) Juxip = H. x prussicum

lippmaae Juxip - 1

lipskyanum Juxip - 2

liptoviense Borb. subsp. *borzawae* Woloszcz. & Zahn = H. borzawae

lissolepium (Zahn) Roffey (*H. lissolepium* (Zahn) Juxip, comb. superfl.) - 1, 2

lithuanicum (Naeg. & Peter) Juxip (*H. auricula* L. var. *lithuanicum* (Naeg. & Peter) E.I. Nyarady, *Pilosella lactucella* (Wallr.) P.D. Sell & C. West subsp. *lithuanica* (Naeg. & Peter) Sojak, *P. lithuanica* (Naeg. &

Peter) Schljak.) - 1

litorale Schljak. - 1

litoreum Norrl. = H. erraticum

x **litwinowianum** Zahn - 2

livescens Norrl. - 1

livescentiforme Schljak. - 1

ljapinense Juxip - 1, 3

x **lobarzewskii** Rehm. (*Pilosella* x *lobarzewskii* (Rehm.) Sojak) - 1

lomnicense Woloszcz. - 1

lonchophyllum Schljak. - 1

lonchopodum (Zahn) Schljak. (*H. bifidum* Kit. subsp. *lonchopodum* Zahn) - 1

longatum (Peter) Juxip = H. x fennicum

longipes (C. Koch ex Naeg. & Peter) Juxip = H. x bifurcum

longipubens Schljak. - 1

longiradiatum (Zahn) Juxip = H. x densiflorum

x **longiscapum** Boiss. & Kotschy ex Naeg. & Peter - 2

longisetum (Naeg. & Peter) Juxip = H. x auriculoides

x **longisquamum** Peter (*H. pachylodes* Naeg. & Peter, *Pilosella* x *longisquama* (Peter) Holub, *P. pachylodes* (Naeg. & Peter) Sojak) - 1

x **longum** (Naeg. & Peter) Juxip (*H. euchaetium* Naeg. & Peter subsp. *longum* Naeg. & Peter, *Pilosella* x *longa* (Naeg. & Peter) Schljak., *H.* x *euchaetium* auct.) - 1

loriense Juxip - 2

lovozericum Schljak. - 1

lugdunense (Rouy) Juxip - 1

lugiorum (Zahn) Schljak. (*H. caesium* (Fries) Fries subsp. *lugiorum* Zahn) - 1

lujaurense auct. p.p. = H. marjokense

lujaurense Norrl. p.p. = H. glabriligulatum

- var. *marjokense* Norrl. = H. marjokense

luteoglandulosum T. Sael. ex Norrl. = H. x glomeratum

lutnjaermense Schljak. - 1

lutulenticeps Schljak. - 1

lutulentum Norrl. - 1

lyccense (Naeg. & Peter) Juxip (*Pilosella lyccensis* (Naeg. & Peter) Schljak.) - 1

lychnaeum (Norrl.) Juxip (*Pilosella lychnaea* Norrl.) - 1

lydiae Schischk. & Steinb. - 3

lyratum Norrl. = H. chloromaurum

lyrifolium Schljak. - 1

x **macranthelum** Naeg. & Peter (*H. conferciens* Norrl., *H. curvicollum* Norrl., *H. permicum* (Zahn) Juxip, *Pilosella* x *macranthela* (Naeg. & Peter) Sojak) - 1

macrastrum (Zahn) Juxip = H. x dubium

macrochaetium (Naeg. & Peter) Juxip = H. procerum

macrochlorellum Litv. & Zahn ex Juxip - 1

macrocladum Schljak. (*H. malaxotifolium* Schljak.) - 1

macrocymum (Naeg. & Peter) Juxip (*Pilosella echioides* (Lumn.) F. Schultz & Sch. Bip. subsp. *macrocyma* (Naeg. & Peter) Sojak, *P. macrocyma* (Naeg. & Peter) Schljak.) - 1

macroglossum (Rehm.) Juxip = H. x prussicum

macrolepidiforme (Zahn) Juxip - 2

macrolepioides (Zahn) Juxip - 2

macrolepis Boiss. - 2

macrolepium (Naeg. & Peter) Juxip (*Pilosella submacrolepis* Schljak.) - 1

x **macrolygodes** (Zahn) Juxip - 2

macrophyllopodum (Zahn) Juxip - 2

macroradium (Zahn) Juxip - 2

x **macrostolonum** G. Schneid. (*H. amoeniceps* (Zahn) Juxip, *H. aurosulum* Norrl., *H. cernuiforme* (Naeg. & Peter) Zahn, *H. firmipes* (Rehm.) Juxip, *H. flagellare* nm. *moskoviticum* (Peter) Lepage, *H. macrostolonum* subsp. *cernuiforme* (Naeg. & Peter) Duvigneaud & Auquier, *H. moskoviticum* (Peter) Juxip, *H. pseudoflagellare* Blocki, *H. wilkense* (Rehm.) Juxip, *Pilosella* x *macrostolona* (G. Schneid.) Sojak, *P. macrostolona* subsp. *cernuiformis* (Naeg. & Peter) Sojak, *Hieracium duplex* auct.) - 1

- subsp. *cernuiforme* (Naeg. & Peter) Duvigneaud & Auquier = H. x macrostolonum

x **macrotrichum** Boiss. (*H. balansae* Boiss., *Pilosella* x *macrotricha* (Boiss.) F. Schultz & Sch. Bip.) - 2

macrum (Naeg. & Peter) Juxip = H. ingricum

maculatum Schrank - 1

maculosum Dahlst. ex Stenstr. = H. christianense

magnauricula (Naeg. & Peter) Juxip (*H. auricula* L. var. *magnauricula* (Naeg. & Peter) E.I. Nyarady, *H. lactucella* Wallr. subsp. *magnauricula* (Naeg. & Peter) P.D. Sell, *Pilosella lactucella* (Wallr.) P.D. Sell & C. West subsp. *magnauricula* (Naeg. & Peter) Sojak, *P. magnauricula* (Naeg. & Peter) Dostal) - 1

magocsyanum Jav. - 1

magyaricum Peter subsp. *adenocladum* Rehm. = H. adenocladum

- subsp. *agathanthum* Rehm. = H. agathanthum
- subsp. *chaunocymum* Rehm. = H. chaunocymum
- subsp. *ferroviae* Rehm. = H. ferroviae
- subsp. *hispidissimum* Rehm. ex Naeg. & Peter = H. hispidissimum
- subsp. *mnoocladum* Rehm. = H. mnoocladum
- subsp. *myriothrichum* Rehm. = H. myriothrichum
- subsp. *polyanthemum* Naeg. & Peter = H. polyanthemum
- subsp. *pseudomegalomastix* Rehm. = H. pseudomegalomastix
- subsp. *saevum* Rehm. = H. saevum
- subsp. *thaumasioides* Peter = H. thaumasioides

malacotrichum (Naeg. & Peter) Juxip (*Pilosella malacotricha* (Naeg. & Peter) Schljak.) - 1

malaxotifolium Schljak. = H. macrocladum

malmei (Dahlst.) Wiinst. = H. pycnodon

manifestum Juxip - 3

marginale (Naeg. & Peter) Juxip (*H. bauhini* Bess. subsp. *magyaricum* (Peter) Zahn var. *marginale* (Naeg. & Peter) E.I. Nyarady, *Pilosella glaucescens* (Bess.) Sojak subsp. *marginalis* (Naeg. & Peter) Sojak, *P. marginalis* (Naeg. & Peter) Schljak.) - 1, 2

marjokense (Norrl.) Schljak. (*H. lujaurense* Norrl. var. *marjokense* Norrl., *H. lujaurense* auct. p.p.) - 1

x **maschukense** Litv. & Zahn (? *H. buhsei* (Naeg. & Peter) Juxip, *H. caucasiciforme* (Litv. & Zahn) Juxip, *H. woronowianum* Zahn, *Pilosella* x *maschukensis* (Litv. & Zahn) Sojak, *P. woronowiana* (Zahn) Sojak) - 2

■ matrense (Naeg. & Peter) Juxip (*Pilosella brachiata* Bertol. ex DC. subsp. *matrensis* (Naeg. & Peter) Sojak)

maurocybe Juxip (*Pilosella maurocybe* (Juxip) Schljak.) - 1

medianiforme (Litv. & Zahn) Juxip - 2

medschedsense Zahn - 2

megalomastix (Naeg. & Peter) Juxip (*H. bauhini* subsp. *magyaricum* (Peter) Zahn var. *megalomastix* (Naeg. & Peter) E.I. Nyarady, *H. piloselloides* Vill. subsp. *megalomastix* (Naeg. & Peter) P.D. Sell, *Pilosella glaucescens* (Bess.) Sojak subsp. *megalomastix* (Naeg. & Peter) Sojak, *P. megalomastix* (Naeg. & Peter) Schljak., *P. piloselloides* (Vill.) Sojak subsp. *megalomastix* (Naeg. & Peter) P.D. Sell & C. West) - 1, 2

meinshausenianum Juxip = H. duderhofense

melachaetum Tausch (*H. bauhini* Bess. subsp. *bauhini* var. *melachaetum* (Tausch) E.I. Nyarady, *Pilosella bauhini*

(Bess.) Arv.-Tour. subsp. *melachaeta* (Tausch) Sojak, *P. melachaeta* (Tausch) Schljak.) - 1

melaneilema (Peter) Weiss (*H. auricula* L. subsp. *melaneilema* Peter, *H. auricula* var. *melaneilema* (Peter) E.I. Nyarady, *Pilosella lactucella* (Wallr.) P.D. Sell & C. West subsp. *melaneilema* (Peter) Sojak, *P. melaneilema* (Peter) Schljak.) - 1

melanocephalum Tausch - 1

melanocybe Norrl. = H. melanophaeum

melanolepium (Rehm.) Juxip = H. x densiflorum

melanophaeum (Norrl.) Czer. comb. nova (*Pilosella melanophaea* Norrl. 1895, Acta Soc. Fauna Fl. Fenn. 12, 4 : 54; *Hieracium melanocybe* Norrl. nom. illegit.) - 1

melanophilum (Rehm.) Juxip = H. x mollicaule

melanopsiforme (Zahn) Juxip = H. x piloselliflorum

x **melinomelas** Peter (*H. acrothyrsum* Naeg. & Peter, *Pilosella acrothyrsa* (Naeg. & Peter) Sojak, *P.* x *melinomelas* (Peter) Holub) - 1

membranulatum (Litv. & Zahn) Juxip - 2

mendelii (Naeg. & Peter) Juxip = H. x schultesii

meringophorum (Naeg.& Peter) Juxip = H. x stoniflorum

mestianum (Zahn) Juxip - 2

miansarofii (Kozl. & Zahn) Juxip - 2

micans Norrl. = H. x glomeratum

mickiewiczii (Rehm.) Juxip = H. x stoniflorum

micrastrum (Zahn) Juxip = H. x dubium

microplacerum Norrl. - 1, 3

microsphaericum (Zahn) Juxip = H. x piloselliflorum

microspilon (Jord. ex Sudre) Schljak. (*H. murorum* L. subsp. *gentile* (Jord. ex Boreau) Sudre b. *microspilon* Jord. ex Sudre) - 1

microtum Boiss. - 1

mikulinkae (Zahn) Schljak. (*H. praecurrens* Vuk. subsp. *mikulinkae* Zahn) - 1

mirobauhinii (Zahn) Juxip - 2

mirum (Naeg. & Peter) Juxip = H. x auriculoides

mixopolium (Dahlst.) Norrl. - 1

mnoocladum (Rehm.) Czer. comb. nova (*H. magyaricum* Peter subsp. *mnoocladum* Rehm. 1897, Verh. Zool.-Bot. Ges. Wien, 47 : 296; *Pilosella mnooclada* (Rehm.) Schljak.) - 1

mnoophorum Naeg. & Peter = H. x densiflorum

modiciforme Juxip - 1

mohburgenense (Zahn) Juxip = H. x prussicum

x **mollicaule** Vuk. (*H. anocladum* (Naeg. & Peter) Juxip, *H. approximabile* (Zahn) Juxip, *H. bauhiniflorum* (Naeg. & Peter) Juxip, *H. discolor* (Naeg. & Peter) Juxip, *H. leptophyton* Naeg. & Peter, *H. leptophyton* var. *anocladum* (Naeg. & Peter) E.I. Nyarady, *H. leptophyton* var. *discolor* (Naeg. & Peter) E.I. Nyarady, *H. melanophilum* (Rehm.) Juxip, *H. nematoclados*) Juxip, *Pilosella* x *mollicaulis* (Vuk.) Sojak., *P. mollicaulis* subsp. *anoclada* (Naeg. & Peter) Sojak., *P. mollicaulis* subsp. *bauhiniflora* (Naeg. & Peter) Sojak., *P. mollicaulis* subsp. *discolor* (Naeg. & Peter) Sojak, *P. mollicaulis* subsp. *leptophyton* (Naeg. & Peter) Sojak) - 1, 2

mollisetum (Naeg. & Peter) Dahlst. (*Pilosella molliseta* (Naeg. & Peter) Schljak.) - 1

monczecola Juxip - 1

moskoviticum (Peter) Juxip = H. x macrostolonum

mukaczevense Juxip - 1

multiceps (Naeg. & Peter) Juxip = H. x calodon

multifolium (Peter) Juxip = H. asiaticum

multifrons Brenn. - 1

multiglandulosum Juxip = H. duderhofense

multisetum (Naeg. & Peter) Rech. fil. (*H. hoppeanum* Schult. subsp. *testimoniale* Naeg. & Peter, *H. hoppea-*

num var. *multisetum* (Naeg. & Peter) E.I. Nyarady, *H. multisetum* (Naeg. & Peter) Juxip, comb. superfl., *Pilosella hoppeana* (Schult.) Sch. Bip. & F. Schutz subsp. *testimonialis* (Naeg. & Peter) P.D. Sell & C. West) - 2

munkacsense Zahn - 1

muratovoense (Zahn) Juxip = H. x scandinavicum

muricellum Fries = Picris hieracioides

murmanense Schljak. (*H. murmanicola* (Zahn) Juxip, 1960, non Juxip, 1959) - 1

murmanicola Juxip = H. congruens

murmanicola (Zahn) Juxip = H. murmanense

murmanicum Norrl. - 1

murorum L. subsp. *altipes* Lindb. fil. ex Zahn = H. altipes

- subsp. *cinerellisquamum* Litv. & Zahn = H. cinerellisquamum

- subsp. *diminuens* (Norrl.) Norrl. = H. diminuens

- subsp. *galbanum* Dahlst. = H. galbanum

- subsp. *gentile* (Jord. ex Boreau) Sudre b. *microspilon* Jord. ex Sudre = H. microspilon

mutilatum Almq. ex Elfstr. - 1

myriothrichum (Rehm.) Czer. comb. nova (*H. magyaricum* Peter subsp. *myriothrichum* Rehm. 1897, Verh. Zool.-Bot. Ges. Wien, 47 : 292; *Pilosella myriothricha* (Rehm.) Schljak.) - 1

nalczikense Juxip - 2

naniceps Elfstr. = H. derivatum

narymense Schischk. & Serg. - 3

neglectum (Norrl.) Norrl. = H. x glomeratum

nematoclados (Rehm.) Juxip. = H. x mollicaule

nemoricola (Norrl.) Norrl. = H. x fennicum

nenukovii Juxip - 1

neopinnatifidum Pugsl. - 1

neriedowae (Woloszcz. & Zahn) Czer. comb. nova (*H. aurantiacum* L. subsp. *neriedowae* Woloszcz. & Zahn, 1911, Magyar Bot. Lapok, 10 : 126; *Pilosella neriedowae* (Woloszcz. & Zahn) Schljak.) - 1

neroikense Juxip - 1, 3

nesaeum Juxip - 1

niankowiense (Rehm.) Juxip = H. x progenitum

nigrescens Willd. - 1

- subsp. *comatuloides* Zahn = H. comatuloides

- subsp. *decipientiforme* Woloszcz. & Zahn = H. decipientiforme

- subsp. *stenopiforme* Pohle & Zahn = H. stenopiforme

- subsp. *subflexicaule* Zahn = H. subflexicaule

- var. *decipiens* (Tausch) E.I. Nyarady = H. decipiens

nigriceps Naeg. & Peter = H. x iseranum

nigrisetum (Naeg. & Peter) Juxip (*H. bauhini* Bess. subsp. *magyaricum* (Peter) Zahn var. *nigrisetum* (Naeg. & Peter) E.I. Nyarady, *Pilosella glaucescens* (Bess.) Sojak subsp. *nigriseta* (Naeg. & Peter) Sojak., *P. nigriseta* (Naeg. & Peter) Schljak.) - 1

nigritum Uechtr. - 1

nigrosilvarum Juxip (*Pilosella nigrosilvarum* (Juxip) Schljak.) - 1

niphocladum (Schelk. & Zahn) Juxip - 2

nivaense Schljak. - 1

niveolimbatum Juxip - 1

norrliniiforme Pohle & Zahn = H. x dimorphoides

notense Schljak. - 1

obliquum Jord. - 1

oblongum Jord. - 1

obornyanum Naeg. & Peter = H. x polymastix

obscuribracteatum (Naeg. & Peter) Juxip (*H. bauhini* Bess. subsp. *bauhini* var. *obscuribracteatum* (Naeg. & Peter) E.I. Nyarady, *Pilosella obscuribracteata* (Naeg. &

Peter) Schljak.) - 1

obscuricaule (Litv. & Zahn) Juxip - 2

obscuriceps Dahlst. (*H. subobscuriceps* (Zahn) Juxip) - 1

obscurum Reichenb. (*H. piloselloides* Vill. subsp. *subcymiregum* (Peter) Zahn var. *obscurum* (Reichenb.) E.I. Nyarady, *Pilosella obscura* (Reichenb.) Sojak) - 1

obsistens Norrl. (*Pilosella obsistens* (Norrl.) Schljak.) - 1

ochanskiense (Zahn) Juxip ex Schljak. (*H. vulgatum* Fries subsp. *ochanskiense* Zahn) - 1

ochrophyllum (Naeg. & Peter) Juxip = H. x calodon

oeneororatum Norrl. = H. x fennicum

oioense (Dahlst.) Juxip (*H. silvaticum* (L.) Gouan subsp. *oioense* Dahlst.) - 1

oistophyllum Pugsl. (*H. fuscocinereum* Norrl. subsp. *sagittatum* (Lindeb.) Brautig., *H. sagittatum* (Lindeb.) Norrl. 1888, non Hoffmgg. & Link, 1820, *H. fuscocinereum* auct.) - 1

olympicum Boiss. - 2

omangii (Zahn) Juxip = H. spatalops

oncodes Omang var. *iremelense* Elfstr. = H. iremelense

onegense (Norrl.) Norrl. (*Pilosella onegensis* Norrl., *Hieracium caespitosum* Dumort. subsp. *brevipilum* (Naeg. & Peter) P.D. Sell, p.p., *Pilosella caespitosa* (Dumort.) P.D. Sell & C. West subsp. *brevipila* (Naeg. & Peter) P.D. Sell & C. West, p.p.) - 1, 3, 4

▪ onosmaceum (Zahn) Juxip

orbicans Almq. ex Stenstr. - 1

ornatum (Dahlst.) K. Joh. (*H. diaphanoides* Lindeb. subsp. *ornatum* Dahlst.) - 1

orthobrachion (Woloszcz. & Zahn) Schljak. (*H. lampromegas* Zahn subsp. *orthobrachion* Woloszcz. & Zahn) - 1

orthocladum Zahn - 2

orthopodum Dahlst. - 1(?)

osiliae (Dahlst.) Juxip (*H. caesium* (Fries) Fries subsp. *osiliae* Dahlst.) - 1

oswaldii (Norrl.) Juxip = H. polycomum and H. ueksipii

ovaliceps (Norrl.) Elfstr. (*H. vulgatum* Fries *[subsp.] *ovaliceps* Norrl.) - 1

▪ ovalifrons (Woronow & Zahn) Juxip

ovatifrons Dahlst. ex Noto - 1(?)

pachycephalum (Fries ex Uechtr.) Oborny (*H. corymbosum* Fries var. *pachycephalum* Fries ex Uechtr.) - 1

pachylodes Naeg. & Peter = H. x longisquamum

pahnschii Juxip - 1

paldiskiense Juxip - 1

pallidum Biv. = H. schmidtii

pallonianum Zahn. = H. largum var. pallonianum

palmenii Norrl. - 1

panaeoliforme (Pohle & Zahn) Juxip - 1

panjutinii Juxip - 2

x **pannoniciforme** Litv. & Zahn (*Pilosella x pannoniciformis* (Litv. & Zahn) Sojak) - 2

pannosum Boiss. - 2

pantepsilon (Rehm.) Juxip = H. x paragogum

papyrodes (Norrl.) Juxip = H. x fennicum

paragogiceps (Zahn) Juxip = H. x paragogum

paragogiforme Oborny = H. x paragogum

x **paragogum** Naeg. & Peter (*H. pantepsilon* (Rehm.) Juxip, *H. paragogiceps* (Zahn) Juxip, *H. paragogiforme* Oborny, 1905, non Besse & Zahn ex Kaeser, 1907, *Pilosella x paragoga* (Naeg. & Peter) Sojak) - 1

pareyssianum (Naeg. & Peter) Juxip = H. x auriculoides

parvipunctatum Norrl. = H. x blyttianum

parvistolonum (Naeg. & Peter) Juxip (*Pilosella parvistolona* (Naeg. & Peter) Schljak.) - 1

pasense Juxip - 1

peczoryense Juxip = H. x bifurcum

peleteranum Merat subsp. *sabulosorum* Dahlst. = H. laticeps

pellucidum Laest. - 1

penduliforme (Dahlst.) K. Joh. (*H. silvaticum* (L.) Gouan subsp. *penduliforme* Dahlst., *H. connatum* Norrl.) - 1

pendulum (Dahlst.) Omang - 1

perasperum (Zahn) Juxip - 2

percissum Jord. ex Boreau - 1

pereffusum Elfstr. - 1

perfoliatum Froel. - 2

perfugii Juxip = H. x zizianum

pergrandidens (Zahn) Juxip, p.p. = H. floccinops

pericaustum (Norrl.) Norrl. = H. x dimorphoides

perileucum (Schelk. & Zahn) Juxip - 2

permense (Zahn) Juxip = H. x collinum

permicum (Zahn) Juxip = H. x macranthelum

x **persicum** Boiss. - 2

persimile (Dahlst.) K. Joh. (*H. praetenerum* Almq. ex Dahlst. subsp. *persimile* Dahlst.) - 1

pervagum Jord. ex Boreau - 1, 2, 3, 5

petiolatum (Elfstr.) Norrl. - 1

petrofundii Juxip - 1

petropavlovskanum Juxip - 1

petunnikovii (Peter) Juxip = H. x flagellare

philanthrax (Stenstr.) K. Joh. & Sam. - 1

phrygicum (Zahn) Juxip = H. procerum

pikujense (Woloszcz. & Zahn) Schljak. (*H. krasanii* Woloszcz. subsp. *pikujense* Woloszcz. & Zahn) - 1

pilipes T.Sael. = H. x dubium

pilisquamum (Naeg. & Peter) Juxip (*Pilosella hoppeana* (Schult.) Sch. Bip. & F. Schultz subsp. *pilisquama* (Naeg. & Peter) P.D. Sell & C. West, *P. pilisquama* (Naeg. & Peter) Dostal) - 2

pilosella L. (*Pilosella officinarum* F. Schultz & Sch. Bip.) - 1, 2, 3

- subsp. **micradenium** Naeg. & Peter (*Pilosella officinarum* subsp. *micradenia* (Naeg. & Peter) P.D. Sell & C. West) - 2

- subsp. **tricholepium** Naeg. & Peter (*Pilosella officinarum* subsp. *tricholepia* (Naeg. & Peter) P.D. Sell & C. West) - 2

x **piloselliflorum** Naeg. & Peter (*H. apatelium* Naeg. & Peter nm. *piloselliflorum* (Naeg. & Peter) Lepage, *H. melanopsiforme* (Zahn) Juxip, *H. microsphaericum* (Zahn) Juxip, *H. stenozon* (Zahn) Juxip, *Pilosella* x *piloselliflora* (Naeg. & Peter) Sojak) - 1

piloselloides Vill. subsp. *megalomastix* (Naeg. & Peter) P.D. Sell = H. megalomastix

- subsp. *subcymigerum* (Peter) Zahn var. *obscurum* (Reichenb.) E.I. Nyarady = H. obscurum

pilosissimum Friv. - 2

pineum Schischk. & Serg. - 3

x **plaicense** Woloszcz. (*Pilosella* x *plaicensis* (Woloszcz.) Sojak) - 1

plantaginifrons Schljak. - 1

pleiophyllogenes (Zahn) Schljak. (*H. silvaticum* (L.) Gouan subsp. *pleiophyllogenes* Zahn) - 1

pleiophyllopsis (Zahn) Schljak. (*H. praecurrens* Vuk. subsp. *pleiophyllopsis* Zahn) - 1(?)

pleuroleucum (Dahlst.) Juxip (*H. serratifrons* Almq. subsp. *pleuroleucum* Dahlst.) - 1

plicatulum (Zahn) Juxip (*H. bauhini* Bess. subsp. *bauhini* var. *plicatulum* (Zahn) E.I. Nyarady, *Pilosella plicatula* (Zahn) Schljak.) - 1

pluricaule Schischk. & Serg. - 3

plurifoliosum Schischk. & Steinb. - 1, 3

poaczense Schljak. - 1

pocuticum Woloszcz. - 1

podkumokense Juxip - 2

x **podolicum** (Rehm.) Czer. comb. nova (*H. hyperyum* Peter subsp. *podolicum* Rehm. 1896, Verh. Zool.-Bot. Ges. Wien, 46 : 343; *Pilosella* x *podolica* (Rehm.) Schljak.) - 1

pohlei Pohle & Zahn - 1

pojoritense Woloszcz. - 1(?)

x **poliodermum** Dahlst. (*H. apatelioides* (Zahn) Juxip, *H. transbalticum* (Dahlst.) Juxip, *Pilosella* x *polioderma* (Dahlst.) Sojak) - 1

poliophyton Zahn ex Juxip = H. velutinum

poliudovense Juxip - 1

polyanthemum (Naeg. & Peter) Czer. comb. nova (*H. magyaricum* Peter subsp. *polyanthemum* Naeg. & Peter, 1885, Hier. Mitt.-Eur. 1 : 587; *Pilosella polyanthema* (Naeg. & Peter) Schljak.) - 1

polycomatum (Zahn) Noto = H. polycomum

polycomum Dahlst. ex Norrl. (*H. inuloides* Tausch subsp. *polycomatum* Zahn, *H. oswaldii* (Norrl.) Juxip, p.p. quoad nomen, *H. polycomatum* (Zahn) Noto, nom. illegit. superfl.) - 1

x **polymastix** Peter (*H. obornyanum* Naeg. & Peter, nom. illegit., *Pilosella obornyana* (Naeg. & Peter) Sojak, *Pilosella oborniana* subsp. *polymastix* (Peter) Sojak, *Pilosella* x *polymastix* (Peter) Holub) - 1

polymnoon (Naeg. & Peter) Norrl. (*Pilosella polymnoon* (Naeg. & Peter) Schljak.) - 1

polymorphophyllum Elfstr. - 1

polysarcoides (Zahn) Juxip = H. x fennicum

pomoricum Juxip - 1

porphyrii Schischk. & Serg. - 3

porrigens Almq. ex Loennr. - 1

- subsp. *virenticeps* Dahlst. = H. subvirenticeps

praealtum Vill. ex Gochn. (*H. siphonanthum* Juz. & Bystr., *Pilosella praealta* (Vill. ex Cochn.) F. Schultz & Sch. Bip.) - 1

- subsp. *bauhini* (Bess.) Petunn. = H. bauhini

- subsp. *thaumasium* (Peter) P.D. Sell = H. thaumasium

praecipuum Dahlst. ex Norrl. - 1

praecurrens auct. = H. jaworowae

praecurrens Vuk. subsp. *mikulinkae* Zahn = H. mikulinkae

- subsp. *odorans* (Borb.) Zahn var. *jaworowae* Zahn = H. jaworowae

- subsp. *pleiophyllopsis* Zahn = H. pleiophyllopsis

praetenerifrons Schljak. - 1

praetenerum Almq. ex Dahlst. (*H. proximum* (Norrl.) Norrl. 1895, non F.J. Hanb. 1889) - 1

- subsp. *persimile* Dahlst. = H. persimile

praetermissum Juxip - 1

praetervisum Juxip - 1

pratense Tausch = H. caespitosum

- subsp. *amaurochlorellum* Zahn = H. amaurochlorum

- subsp. *amaurochlorum* Zahn = H. amaurochlorum

- subsp. *rawaruskanum* Zahn = H. rawaruskanum

praticola Sudre - 2

prenanthoides auct. = H. hypoglaucum

prenanthoides Vill. var. *furvescens* Dahlst. = H. furvescens

prilakenense (Zahn) Schljak. (*H. levicaule* Jord. subsp. *prilakenense* Zahn) - 1

proceriforme (Naeg. & Peter) Zahn (*Pilosella proceriformis* (Naeg. & Peter) Sojak) - 1, 2

x **procerigenum** Litv. & Zahn - 2

procerum Fries (*H. echioides* Lumn. subsp. *procerum* (Fries) P.D. Sell, *H. macrochaetium* (Naeg. & Peter)

Juxip, *H. phrygicum* (Zahn) Juxip, *H. procerum* subsp. *phrygicum* Zahn, nom. illegit., *Pilosella echioides* (Lumn.) F. Schult. & Sch. Bip. subsp. *procera* (Fries) P.D. Sell & C. West, *P. procera* (Fries) F. Schultz & Sch. Bip.) - 1, 2, 3, 4, 6,
- subsp. *phrygicum* Zahn = H. procerum
x **progenitum** (Norrl.) Norrl. (*Pilosella x progenita* Norrl., *Hieracium acrotrichum* Rehm., *H. callimorphopsis* (Zahn) Juxip, *H. flagellariforme* G. Schneid., *H. niankowiense* (Rehm.) Juxip, *Pilosella flagellariformis* (G. Schneid.) Sojak) - 1
prognatum Norrl. = H. x flagellare
progrediens Norrl. (*H. kovdense* Juxip, *H. subbetuletorum* Juxip) - 1
prolatatum auct. = H. prolatescens
prolatescens K. Joh. & Sam. (*H. prolatatum* auct.) - 1
prolixiforme Norrl. - 1
prolixum Norrl. - 1
prolongatum (Naeg. & Peter) Juxip = H. x glomeratum
proximum (Norrl.) Norrl. = H. praetenerum
pruiniferum (Norrl.) Norrl. - 1
x **prussicum** Naeg. & Peter (*H. acrochlorum* (Zahn) Juxip, *H. casparyanum* (Naeg. & Peter) Juxip, *H. gnaphalium* (Naeg. & Peter) Juxip, *H. guttenfeldense* (Zahn) Juxip, *H. lipnickianum* (Rehm.) Juxip, *H. macroglossum* (Rehm.) Juxip, 1960, non Brenn. 1895, *H. mohrungenense* (Zahn) Juxip, *H. tephrantheloides* (Zahn) Juxip, *Pilosella x prussica* (Naeg. & Peter) Sojak., *Hieracium x duplex* auct. p.p.) - 1
przybyslawskii (Rehm.) Czer. comb. nova (*H. hoppeanum* Schult. subsp. *przybyslawskii* Rehm. 1896, Verh. Zool.-Bot. Ges. Wien, 46 : 330; *Pilosella przybyslawskii* (Rehm.)Schljak.) - 1
psammophilum (Naeg. & Peter) Juxip = H. x calodon
pseudarctophilum Schljak. - 1
pseudatratum Woloszcz. - 1
pseudauricula (Naeg. & Peter) Juxip = H. x suecicum
pseudauriculoides (Naeg. & Peter) Juxip (*H. bauhini* Bess. subsp. *magyaricum* (Peter) Zahn var. *pseudauriculoides* (Naeg. & Peter) E.I. Nyarady, *Pilosella glaucescens* (Bess.) Sojak subsp. *pseudauriculoides* (Naeg. & Peter) Sojak, *P. pseudauriculoides* (Naeg. & Peter) Schljak.) - 1
pseuderectum Schljak. - 1, 3
pseudobifidum Schur - 1(?)
pseudobipes Elfstr. - 1
pseudoblyttii (Norrl.) Norrl. = H. x blyttianum
pseudoboreum Schljak. (*H. boreum* auct.) - 1
pseudobrachiatum Celak. = H. x brachiatum
pseudocongruens Schljak. - 1
x **pseudoconstrictum** Zahn - 2
pseudoflagellare Blocki = H. x macrostolonum
pseudofritzei Benz & Zahn - 1(?)
pseudoglabridens Schljak. - 1
pseudoglomeratum Blocki ex Woloszcz. = H. x densiflorum
pseudograndidens Schljak. = H. silvularum
pseudohypochnoides Schljak. - 1
pseudojuranum Arv.-Touv. - 2
pseudokajanense (Schljak.) Czer. comb. nova (*Pilosella pseudokajanensis* Schljak. 1989, Flora Partis Europ. URSS,8 : 354) - 1
pseudolatypeum Schljak. - 1
pseudolepistoides Schljak. - 1
pseudomegalomastix (Rehm.) Czer. comb. nova (*H. magyaricum* Peter subsp. *pseudomegalomastix* Rehm. 1897, Verh. Zool.-Bot. Ges. Wien, 47 : 289; *Pilosella pseudomegalomastix* (Rehm.) Schljak.) - 1

pseudonigritiforme (Zahn) Schljak. (*H. krasanii* Woloszcz. subsp. *pseudonigritiforme* Zahn) - 1
pseudoomangii Schljak. - 1
pseudopalmenii Schljak. - 1
pseudophyllodes (Zahn) Juxip - 1
pseudopikujense Zahn - 1
x **pseudopiloselliflorum** Rehm. (*Pilosella x pseudopiloselliflora* (Rehm.) Sojak) - 1
pseudosparsum (Zahn) Czer. comb. nova (*H. bauhini* Bess. subsp. *pseudosparsum* Zahn, 1907, Feddes Repert. 4 : 181; *Pilosella pseudosparsa* (Zahn) Schljak.) - 1, 2
pseudostygium Woloszcz. - 1
x **pseudosvaneticum** Peter - 2
pseudothaumasium (Zahn) Juxip (*Pilosella pseudothaumasia* (Zahn) Schljak.) - 1
pseuduliginosum Juxip = H. x flagellare
psilobrachion (Woronow & Zahn) Juxip - 2
pskowiense (Zahn) Juxip = H. x calodon
pubens (Naeg. & Peter) Norrl. = H. x fennicum
pubescens (Lindbl.) Norrl. *[subsp.] *glomerabile* Norrl. = H. glomerabile
pulvinatum (Norrl.) Norrl. = H. x blyttianum
purpureibracteum (Zahn) Juxip = H. tephrocephalum
purpureovittatum (Zahn) Juxip = H. tephrocephalum
purpuristictum Juxip - 1
puschlachtae (Pohle & Zahn) Juxip (*H. laevigatum* Willd. subsp. *puschlachtae* Pohle & Zahn) - 1, 3(?)
pycnodon (Dahlst.) K. Joh. (*H. sagittatum* (Lindeb.) Norrl. subsp. *pycnodon* Dahlst., *H. malmei* (Dahlst.) Wiinst., *H. silvaticum* (L.) Gouan subsp. *malmei* Dahlst.) - 1
pycnomnoon (Rehm.) Juxip = H. x densiflorum
pycnothyrsum (Peter) Juxip = H. x glomeratum
pyrsjuense Juxip - 1
querceticola Jord. ex Boreau - 1
quinquemonticola Juxip - 2
racemosum Waldst. & Kit. ex Willd. - 1
x **raddeanum** Zahn - 2
■ radiatellum (Woronow & Zahn) Juxip
rapunculoidiforme Woloszcz. & Zahn - 1
ratluense Zahn - 2
rauzense J. Murr subsp. *farinifloccosum* Degen & Zahn = H. farinifloccosum
ravusculum (Dahlst.) Dahlst. (*H. caesium* (Fries) Fries subsp. *ravusculum* Dahlst.) - 1
rawaruskanum (Zahn) Czer. comb. nova (*H. pratense* Tausch subsp. *rawaruskanum* Zahn, 1923, in Engler, Pflanzenreich, 82 : 1271; *Pilosella rawaruskana* (Zahn) Schljak.) - 1
reducatum (Norrl.) Juxip = H. furvescens
reflorescens Norrl. = H. x glomeratum
regelianum Zahn - 2, 6
regiomontanum (Naeg. & Peter) Juxip = H. x floribundum
rehmannii (Naeg. & Peter) Juxip = H. x roxolanicum
renidescens Norrl. = H. x scandinavicum
retroversilobatum (Schelk. & Zahn) Juxip - 2
revocans Juxip = H. diminuens
revocantiforme Schljak. - 1
riganum (Syr. & Zahn) Juxip = H. x fennicum
x **rigidellum** Litv. & Zahn - 2
rigidiusculum Schljak. = H. linifolium
rigidum C. Hartm. - 1, 2
- subsp. *acrifolium* Dahlst. = H. acrifolium
- *[subsp.] *cruentiferum* Norrl. & Lindb. fil. ex Norrl. = H. cruentiferum
- *[subsp.] *dolabratum* Norrl. = H. dolabratum
riparium Juxip - 1
x **robustum** Fries - 1, 2, 3, 4, 5, 6

rohacsense Kit. ex Kanitz - 1(?)
rojowskii (Rehm.) Juxip (*Pilosella eriophylla* (Vuk.) Sojak subsp. *rojowskii* (Rehm.) Sojak., *P. rojowskii* (Rehm.) Schljak.) - 1
rossicum Schljak. - 1
x **rothianum** Wallr. (*H. albocinereum* Rupr., *Pilosella* x *rothiana* (Wallr.) F. Schultz & Sch. Bip.) - 1, 2
rotundatum auct. = H. transsilvanicum
x **roxolanicum** Rehm. (*H. guthnickianum* Hegetschw. var. *rehmannii* (Naeg. & Peter) E.I. Nyarady, *H. rehmannii* (Naeg. & Peter) Juxip, *H. rubricymigerum* (Naeg. & Peter) Juxip, *Pilosella* x *roxolanica* (Rehm.) Sojak, *P. roxolanica* subsp. *rehmannii* (Naeg. & Peter) Sojak, *P. roxolanica* subsp. *rubricymigera* (Naeg. & Peter) Sojak, *Hieracium* x *guthnickianum* auct.) - 1
rubricymigerum (Naeg. & Peter) Juxip = H. x roxolanicum
rubrobauhini (Schelk. & Zahn) Juxip - 2
rubroonegense Norrl. = H. fuscoatratum
rubropannonicum (Litv. & Zahn) Juxip - 2
rupicoloides Woloszcz. - 1
ruprechtii Boiss. = H. tephrocephalum
rusanum (Zahn) Juxip = H. x glomeratum
sabaudum L. subsp. *boreale* (Fries) E.I.Nyarady var. *auratum* (Fries) E.I. Nyarady = H. auratum
sagittatum (Lindeb.) Norrl. = H. oistophyllum
- subsp. *pycnodon* Dahlst. = H. pycnodon
sagittipotens Norrl. - 1
samaricum (Zahn) Juxip (*H. cymosum* L. subsp. *samaricum* Zahn, *Pilosella samarica* (Zahn) Schljak.) - 1
samurense (Zahn) Juxip - 2
sanii (Naeg. & Peter) Juxip (*Pilosella sanii* (Naeg. & Peter) Schljak.) - 1
sarmaticum (Zahn) Schljak. (*H. dentatum* Hoppe subsp. *sarmaticum* Zahn, *H. dentatum* auct.) - 1
sarmentosum Froel. = H. x auriculoides
■ *sarykamyschense* Juxip
savokarelicum Norrl. - 1
saxifragum auct. = H. linahamariense
sbaense Juxip - 2
scabiosum (Sudre) Juxip - 1, 2
x **scandinavicum** Dahlst. (*H. curvulatum* (Zahn) Juxip, *H. glomeratiforme* (Zahn) Juxip, *H. muratovoense* (Zahn) Juxip, *H. renidescens* Norrl., *H. subfloribundum* (Naeg. & Peter) Juxip, *Pilosella* x *scandinavica* (Dahlst.) Schljak.) - 1
x **schelkownikowii** Zahn (*Pilosella schelkownikowii* (Zahn) Sojak) - 2
schellianum Juxip - 1
schemachense Juxip - 2
schenkii (Griseb.) Schljak. (*H. bupleuroides* C.C. Gmel. var. *schenkii* Griseb., *H. bupleuroides* auct.) - 1
schennikovii Schljak. - 1
schipczinskii Juxip - 3
schischkinii Juxip - 3
schliakovii Juxip - 1
schmalhausenianum Litv. & Zahn - 2
schmidtii Tausch (*H. pallidum* Biv. 1838, nom. illegit.) - 1
x **schultesii** F. Schultz (*H. frondosum* (Naeg. & Peter)

Juxip, *H. mendelii* (Naeg. & Peter) Juxip, *H. squarrosulum* Norrl., *H. subatriceps* (Zahn) Juxip, *Pilosella* x *schultesii* (F. Schultz) F. Schultz & Sch. Bip., *P. schultesii* subsp. *mendelii* (Naeg. & Peter) Sojak) - 1
x **sciadophorum** Naeg. & Peter (*H. leptophyes* (Peter) Juxip, *Pilosella* x *sciadophora* (Naeg. & Peter) Sojak) - 1
scitulum Woloszcz. - 1
scopulorum Juxip = H. x kalksburgense
scotaiolepis Elfstr. - 1
scotodes Norrl. (*Pilosella scotodes* (Norrl.) Schljak.) - 1
sedutrix (Rehm.) Juxip (*Pilosella sedutrix* (Rehm.) Schljak.) - 1
segevoldense Syr. & Zahn - 1
semichlorellum Norrl. - 1
semicurvatum Norrl. - 1
semicymigerum (Zahn) Juxip = H. x densiflorum
semilitoreum Norrl. (*Pilosella semilitorea* (Norrl.) Schljak.) - 1
semionegense Norrl. = H. x fuscoatrum
semipraecox (Zahn) Juxip - 2
semisagittatum Schljak. = H. barbulatulum
senescentifrons Elfstr. = H. atratulum
sepositum Norrl. = H. x blyttianum
septentrionale (Norrl.) Norrl. (*Pilosella septentrionalis* Norrl.) - 1
sericicaule (Schelk. & Zahn) Juxip - 2
serratifolium Jord. ex Boreau - 1
serratifrons Almq. subsp. *pleuroleucum* Dahlst. = H. pleuroleucum
sershukense Juxip (*H. korshinskyi* Zahn f. *sershukense* (Juxip) Serg.) - 3
seticollum Norrl. - 1
setosopetiolatum (Rehm.) Czer. comb. nova (*H. florentinum* All. subsp. *setosopetiolatum* Rehm. 1897, Verh. Zool.-Bot. Ges. Wien, 47 :284; *Pilosella setosopetiolata* (Rehm.) Schljak.) - 1
sexangulare Schljak. - 1
shaparenkoi Schljak. - 1
signiferum (Norrl.) Juxip (*Pilosella signifera* Norrl.) - 1
silenii Norrl. - 1
sillamaeense Juxip - 1
silvaticum (L.) Gouan subsp. *acrogymnon* Malte = H. acrogymnon
- *[subsp.] canitiosum* Dahlst. = H. canitiosum
- subsp. *lackschewitzii* Dahlst. = H. lackschewitzii
- subsp. *malmei* Dahlst. = H. pycnodon
- subsp. *oioense* Dahlst. = H. oioense
- subsp. *penduliforme* Dahlst. = H. penduliforme
- subsp. *pleiophyllogenes* Zahn = H. pleiophyllogenes
- subsp. *sublividum* Dahlst. = H. sublividum
- subsp. *submaculosum* Dahlst. = H. submaculosum
- subsp. *triangulare* Almq. = H. triangulare
silvicomum Juxip - 1
silvularum Jord. ex Boreau (*H. grandidens* Dahlst., *H. pseudograndidens* Schljak.) - 1, 2
simplicicaule (Somm. & Levier) Peter - 2
x *sintenisii* auct. = H. incaniforme
siphonanthum Juz. & Bystr. = H. praealtum
sivorkae Juxip - 1
sobrinatum (Litv. & Zahn) Juxip - 2
soczavae Juxip (*H. chadatense* Juxip) - 1, 3
solonieviczii Schljak. - 1
solovetzkiense Juxip - 1
sordidescens Norrl. = H. fulvescens
sororians Norrl. (*Pilosella sororians* (Norrl.) Schljak.) - 1
sosnowskyi (Zahn) Kem.-Nath. (*H. incanum* Bieb. subsp.

sosnowskyi Zahn) - 2
sosvaense Schljak. - 1, 3
sparsum Friv. - 1
spatalops Omang (*H. omangii* (Zahn) Juxip) - 1
spathophyllopsis (Zahn) Juxip (*Pilosella spathophyllopsis* (Zahn) Schljak.) - 1
spathophyllum Peter = H. x fennicum
spurium Chaix ex Froel. = H. x kalksburgense
squarrosulum Norrl. = H. x schultesii
stauropolitanum Juxip - 2
steinbergianum Juxip - 1
stellatum (Tausch ex Zahn) Juxip, 1960, non Norrl. 1885 (*H. florentinum* All. subsp. *stellatum* Tausch ex Zahn, *Pilosella praealta* (Vill ex Gochn.) F. Schultz & Sch. Bip. subsp. *stellata* (Tausch ex Zahn) Sojak, *P. stellata* (Tausch ex Zahn) Schljak.) - 1
    This name must be replaced because it is a later homonym.
stenolepis Lindeb. - 1
stenomischum Omang var. *vulsum* Elfstr. = H. vulsum
stenopiforme (Pohle & Zahn) Elfstr. (*H. nigrescens* Willd. subsp. *stenopiforme* Pohle & Zahn) - 1, 3
stenozon (Zahn) Juxip = H. x piloselliflorum
sterromastix (Naeg. & Peter) Juxip = H. x bifurcum
x **stoniflorum** Waldst. & Kit. (*H. meringophorum* (Naeg.& Peter) Juxip, *H. mickiewiczii* (Rehm.) Juxip, *Pilosella* x *stoloniflora* (Waldst. & Kit.) Fries) - 1
streptotrichum Zahn - 2
strictiramum (Naeg. & Peter) Juxip = H. x calodon
strictissimum Froel. - 2, 6
■ **stupposipilum** (Woronow & Zahn) Juxip
stygium Uechtr. - 1
subambiguum (Naeg. & Peter) Juxip = H. x glomeratum
subaquilonare Juxip - 1
subarctophilum Schljak. - 1, 3, 4
subarctoum Norrl. = H. congruens
■ subartvinense Juxip
subasperellum (Zahn) Juxip - 1, 3(?)
subatriceps (Zahn) Juxip = H. x schultesii
subaurantiacum (Naeg. & Peter) Czer. comb. nova (*H. aurantiacum* L. subsp. *subaurantiacum* Naeg. & Peter, 1885, Hier. Mitt.-Eur. 1 : 287; *Pilosella subaurantiaca* (Naeg. & Peter) Schljak.) - 1
subauricula (Naeg. & Peter) Juxip = H. x suecicum
subbakurianiense Juxip - 2
subbauhiniflorum (Woronow & Zahn) Juxip - 2
subbetuletorum Juxip = H. progrediens
subchlorophaeum Schljak. (*H. chlorophaeum* Norrl. 1912, non Dahlst. ex Noto, 1910) - 1
subcinerascens Norrl. ex Schljak. - 1
subcompositum Juxip (*H. commilitonum* Juxip) - 1
subcrassifolium (Zahn) Juxip - 1
subcymigerum (Peter) Weiss (*Pilosella subcimigera* (Peter) Sojak) - 1
suberectum Schischk. & Steinb. - 1(?),3
subexcellens (Zahn) Juxip = H. x densiflorum
subfallaciforme (Zahn) Juxip = H. x fallaciforme
subfariniramum Ganesch. & Zahn - 4
subfarinosiceps Schljak. - 1
subfiliferum (Zahn) Juxip (*Pilosella subfilifera* (Zahn) Schljak.) - 1
subflexicaule (Zahn) Schljak. (*H. nigrescens* Willd. subsp. *subflexicaule* Zahn, *H. flexicaule* Elfstr. 1914, non Tausch, 1828, nec Freyn & Vand. 1895) - 1
subfloribundum (Naeg. & Peter) Juxip = H. x scandinavicum
subgalbanum (Dahlst.) Juxip - 1
subglandulosipes Schljak. - 1

subgracilescens (Dahlst. ex Zahn) Noto (*H. carpathicum* Bess. subsp. *subgracilescens* Dahlst ex Zahn) - 1
■ subhastulatum (Zahn) Juxip
subhirsutissimum Juxip - 1
subimandrae Juxip - 1
subincaniforme Kozl. & Zahn - 2
subincomptum (Zahn) Juxip = H. incomptum
sublactucaceum auct. = H. vagum
sublasiophorum (Litv. & Zahn) Juxip - 2
sublateriflorum Schljak. = H. lateriflorum
sublividum (Dahlst.) K. Joh. (*H. silvaticum* (L.) Gouan subsp. *sublividum* Dahlst.) - 1
submaculigerum Schljak. - 1
submaculosum (Dahlst.) Juxip (*H. silvaticum* (L.) Gouan subsp. *submaculosum* Dahlst.) - 1
submarginellum (Zahn) Juxip - 1
submedianum (Zahn) Juxip - 1
submelanolepis Schljak. - 1
submirum (Litv. & Zahn) Juxip - 2
subnigrescens (Fries ex Norrl.) Dahlst. - 1
subnigriceps (Zahn) Juxip = H. x iseranum
subniviferum Schljak. - 1
subobscuriceps (Zahn) Juxip = H. obscuriceps
subpellucidum (Norrl.) Norrl. (*H. vulgatum* Fries *[subsp.] *subpellucidum* Norrl.) - 1
subpleiophyllum (Zahn) Schljak. (*H. trebevicianum* K. Maly subsp. *subpleiophyllum* Zahn) - 1
subpollichii (Litv. & Zahn) Juxip - 2
subpratense (Norrl.) Norrl. = H. x fennicum
subramosum Loennr. (*H. constringensiforme* Juxip, *H. konshakovskianum* Juxip) - 1, 3
subrigidum Almq. ex Stenstr. - 1
subrubellum (Schelk. & Zahn) Juxip - 2
subsimplex Somm. & Levier - 2
substoloniferum (Peter) Juxip (*Pilosella substolonifera* (Peter) Sojak) - 2
substrictipilum Schljak. - 1
subsvaneticum (Litv. & Zahn) Juxip - 2
subswirense (Norrl.) Juxip = H. x fennicum
subtriangulatum Schljak. - 1
subumbelliforme (Zahn) Juxip - 2
subviolascentiforme (Pohle & Zahn) Juxip - 1
subvirenticeps Schljak. (*H. porrigens* Almq. ex Loennr. subsp. *virenticeps* Dahlst., *H. virenticeps* (Dahlst.) Juxip, 1960, non Norrl. 1912) - 1
subviriduliceps (Zahn) Schljak. (*H. vulgatum* Fries subsp. *subviriduliceps* Zahn) - 1
subvulgatiforme Schljak. (*H. vulgatiforme* (Dahlst.) K. Joh. 1897, non Arv.-Tour. 1876) - 1
suchonense Norrl. (*Pilosella suchonensis* (Norrl.) Schljak.) - 1
sudavicum (Naeg. & Peter) Juxip = H. x floribundum
sudeticum Sternb. - 1(?)
sudetorum (Peter) Weiss (*H. collinum* Gochn. subsp. *sudeterum* Peter, *H. caespitosum* Dumort. subsp. *caespitosum* var. *sudetorum* (Naeg. & Peter) E.I. Nyarady, *Pilosella caespitosa* (Dumort.) P.D. Sell & C. West subsp. *sudetorum* (Peter) Sojak., *P. sudetorum* (Peter) Dostal) - 1
x **suecicum** Fries (*H. cochleatum* (Naeg. & Peter) Norrl., *H. erythropoides* (Norrl.) Schljak., *H. floribundum* Wimm. & Grab. subsp. *cochleatum* Naeg. & Peter, *H. isthmicola* Norrl., *H. pseudauricula* (Naeg. & Peter) Juxip, *H. subauricula* (Naeg. & Peter) Juxip, *Pilosella cochlearis* (Norrl.) Weiss, *P. cochlearis* subsp. *pseudauricula* (Naeg. & Peter) Sojak., *P. cochlearis* subsp. *subauricula* (Naeg. & Peter) Sojak., *P. erythropoides* Norrl., *P. floribunda* (Wimm. & Grab.) Arv.-Tour. subsp. *suecica*

(Fries) Sojak, *P.* x *suecica* (Fries) F. Schultz & Sch. Bip.) - 1

-*[subsp.] *fennicum* Norrl. = H. x fennicum

sukaczewii (Zahn) Schischk. & Serg. = H. taigense

sulphurelliforme (Kozl. & Zahn) Juxip - 2

sulphurellum (Kozl. & Zahn) Juxip - 2

x **sulphureum** Doell (*Pilosella* x *sulphurea* (Doell) F. Schultz & Sch. Bip.) - 1

suomense (Norrl.) Norrl. (*Pilosella suomensis* Norrl., *P. vaillantii* (Tausch) Sojak subsp. *suomensis* (Norrl.) Sojak.) - 1

svaneticiforme (Litv. & Zahn) Kem.-Nath. - 2

swirense Norrl. = H. x fennicum

syreistschikovii (Zahn) Juxip - 2

syrjaenorum (Norrl.) Juxip (*Pilosella syrjaenorum* Norrl.) - 1

sysolskiense (Zahn) Juxip (*Pilosella sysolskiensis* (Zahn) Schljak.) - 1

szovitsii (Naeg. & Peter) Juxip - 2

tabergense (Dahlst.) Juxip (*Pilosella tabergensis* (Dahlst.) Schljak.) - 1

taigense Schischk. & Serg. (*H. sukaczewii* (Zahn) Schischk. & Serg., *H. ussense* (Pohle & Zahn) Juxip) - 1, 3, 4

tanense Elfstr. - 1

tanfiliewii Zahn ex Schljak. - 1

tanythrix (Naeg. & Peter) Juxip = H. x auriculoides

tatewakii (Kudó) Tatew. & Kitam. (*Crepis tatewakii* Kudo) - 2

tatrense (Peter) Juxip = H. x flagellare

tauschii Zahn var. *umbelliferum* (Naeg. & Peter) E.I. Nyarady = H. x densiflorum

tazense Schljak. - 3

teberdaefontis (Litv. & Zahn) Juxip - 2

x **teberdense** (Litv. & Zahn) Juxip - 2

teliforme Schljak. - 1

teligerum Norrl. - 1

teliumbellatum Schljak. - 1

tenacicaule Norrl. (*Pilosella tenacicaulis* (Norrl.) Schljak.) - 1

tenebricans (Norrl.) Norrl. (*Pilosella tenebricans* (Norrl.) Schljak.) - 1

tenuiceps (Naeg. & Peter) Juxip = H. x calodon

tenuiglandulosum Norrl. = H. chloromaurum

tephrantheloides (Zahn) Juxip = H. x prussicum

tephranthelum (Zahn) Juxip = H. x dimorphoides

tephrocephalum Vuk. (*H. christoglossum* (Zahn) Juxip, *H. jailanum* (Zahn) Juxip, *H. purpureibracteum* (Zahn) Juxip, *H. purpureovittatum* (Zahn) Juxip, *H. ruprechtii* Boiss., *H. tephropodoides* (Zahn) Juxip, *H. tephropodum* (Zahn) Juxip, *H. tuscheticum* (Zahn) Juxip, *Pilosella ruprechtii* (Boiss.) P.D. Sell & C. West, *P. ruprechtii* (Boiss.) Dostal, comb. superfl., *P. tephrocephala* (Vuk.) Sojak) - 1, 2

tephrochlorellum (Ganesch. & Zahn) Juxip - 3, 4

tephrophilum (Kozl. & Zahn) Juxip - 2

tephropodoides (Zahn) Juxip = H. tephrocephalum

tephropodum (Zahn) Juxip = H. tephrocephalum

teplouchovii Juxip - 1, 3

x **terekianum** (Litv. & Zahn) Juxip - 2

tericum Schljak. - 1

tetraodon Schljak. - 1

thaumasioides (Peter) Weiss (*H. magyaricum* Peter subsp. *thaumasioides* Peter, *H. bauhini* Bess. subsp. *bauhini* var. *thaumasioides* (Naeg. & Peter) E.I. Nyarady, *Pilosella bauhini* (Bess.) Arv.-Tour. subsp. *thaumasioides* (Bess.) Sojak, *P. thaumasioides* (Peter) Schljak.) - 1

thaumasium (Peter) Weiss (*H. bauhini* subsp. *bauhini* var. *thaumasium* (Peter) E.I. Nyarady, *H. praealtum* Vill. ex Gochn. subsp. *thaumasium* (Peter) P.D. Sell, *Pilosella bauhini* (Bess.) Arv.-Tour. subsp. *thaumasia* (Peter) Sojak, *P. thaumasia* (Peter) Dostal) - 1

thracicum (Naeg. & Peter) Juxip - 2

thyraicum Blocki - 1

tilingii Juxip - 5

timanense Schljak. - 1

tjapomense Norrl. (*Pilosella tjapomensis* (Norrl.) Schljak.) - 1

tjumentzevii Serg. & Juxip - 3

tolmatchevii Schljak. - 1

tolvaense Schljak. - 1

torquescens Norrl. = H. x blyttianum

torticeps (Dahlst.) K. Joh. (*H. grandidens* Dahlst. subsp. *torticeps* Dahlst.) - 1

transbalticum (Dahlst.) Juxip = H. x poliodermum

transnivense Schljak. - 1

transpeczoricum Schljak. - 1

transsilvanicum Heuff. (*H. rotundatum* auct.) - 1

trebevicianum K. Maly subsp. *deanum* Zahn = H. deanum

- subsp. *subpleiophyllum* Zahn = H. subpleiophyllum

triangulare (Almg.) Stenstr. (*H. silvaticum* (L.) Gouan subsp. *triangulare* Almq.) - 1

tricheilema (Naeg. & Peter) Juxip (*H. auricula* L. var. *tricheilema* (Naeg. & Peter) E.I. Nyarady, *Pilosella lactucella* (Wallr.) P.D. Sell & C. West subsp. *tricheilema* (Naeg. & Peter) Sojak, *P. tricheilema* (Naeg. & Peter) Schlajak.) - 1

trichobrachium Juxip - 1

trichocymosoides Juxip (*Pilosella trichocymosoides* (Juxip) Schljak.) - 1

trichocymosum (Zahn) Juxip (*Pilosella trichocymosa* (Zahn) Schljak.) - 1

tridentaticeps (Zahn) Juxip - 2

tridentatum (Fries) Fries (*H. vulgatum* Fries var. *tridentatum* Fries) - 1, 2

trigenes auct. = H. x dybowskianum

trigenes Peter subsp. *dybowskianum* Rehm. = H. x dybowskianum

triste Willd. ex Spreng. (*Chlorocrepis tristis* (Willd. ex Spreng.) A. & D. Löve, *Stenotheca tristis* (Willd. ex Spreng.) Schljak.) - 5

tritum Juxip - 1

trivialiforme Schljak. - 1

truncipilum (Thaisz & Zahn) Schljak. (*H. vulgatum* Fries subsp. *truncipilum* Thaisz & Zahn) - 1

tschkhubianischwilii Kem.-Nath. - 2

tuadaschense Schljak. - 1

x **tubulatum** (Vollm.) Vollm. (*H. cymiflorum* Naeg. & Peter subsp. *tubulatum* Vollm., *Pilosella* x *tubulata* (Vollm.) Sojak) - 1

tubuliflorum (Naeg. & Peter) Juxip = H. x brachiatum

tulomense Schljak. - 1

tunguskanum Ganesch. & Zahn - 4

turbidum Norrl. - 1

turcomanicum Gand. - 6

turkestanicum (Zahn) Juxip - 6

tuscheticum (Zahn) Juxip = H. tephrocephalum

tuvinicum Krasnob. & Schaulo - 4

tweriense (Zahn) Juxip = H. x flagellare

tzagwerianum (Kozl. & Zahn) Juxip - 2

uczanssuense Juxip - 1

ueksipii Schljak. (*H. oswaldii* (Norrl.) Juxip, p.p. quoad plantas) - 1

ugandiense Juxip - 1

ukierniae Woloszcz. & Zahn - 1
ulothrix Norrl. - 1
umbellaticeps (Pohle & Zahn) Juxip - 1, 3
umbellatum L. - 1, 2, 3, 4, 5, 6
- subsp. *filifolium* (Juxip) Tzvel. = H. filifolium
umbelliferum Naeg. & Peter = H. x densiflorum
umbellosum (Naeg. & Peter) Juxip = H. x auriculoides
umbrosum Jord. - 1
uralense Elfstr. (*H. apiculatiforme* Elfstr.) - 1, 3
uranopoleos Juxip - 1
ussense (Pohle & Zahn) Juxip = H. taigense
vagae Juxip - 1
vagum Jord. (*H. sublactucaceum* auct.) - 1, 2
vaidae Juxip - 1
vaillantii Tausch (*H. andrzejowskii* Blocki & Woloszcz., *H. cymigerum* Reichenb., *H. cymosum* L. subsp. *cymigerum* (Reichenb.) Peter, *Pilosella vaillantii* (Tausch) Sojak, *P. vaillantii* subsp. *cymigera* (Reichenb.) Sojak) - 1, 3
valmierense Juxip - 1
varangerense (Elfstr.) Elfstr. (*H. gracilentum* Backh. var. *varangerense* Elfstr.) - 1
varatinense Woloszcz. (*Pilosella varatinensis* (Woloszcz.) Schljak.) - 1
■ variegatisquamum (Zahn) Juxip
varsugae Schljak. - 1
vasconicum Jord. ex Zahn - 1
velutinum Hegetschw. (*H. poliophyton* Zahn ex Juxip, nom. invalid., *Pilosella velutina* (Hegetschw.) F. Schultz & Sch. Bip.) - 1
venetianum Naeg. & Peter = H. x aridum
veresczaginii Schischk. & Serg. - 3
vernicosum (Norrl.) Norrl. = H. x blyttianum
verruculatum Link (*Pilosella verruculata* (Link) Sojak) - 2
- subsp. *akinfiewii* Woronow & Zahn = H. akinfiewii
vestipes Schljak. - 1
viburgense Norrl. - 1
villosellipes (Zahn) Juxip - 2
villosipes Pax - 1
villosum Jacq. - 1
vindobonae (Zahn) Juxip = H. x bifurcum
violaceipes (Zahn) Juxip = H. x dubium
violascentiforme (Pohle & Zahn) Juxip - 1
virelliceps Norrl. - 1
virenticeps (Dahlst.) Juxip = H. subvirenticeps
virentisquamum (Naeg. & Peter) Juxip (*Pilosella hoppeana* (Schult) F. Schultz & Sch. Bip. subsp. *virentisquama* (Naeg. & Peter) Fuchs-Eckert) - 2
virgultorum Jord. - 1
■ virosiforme Woronow & Zahn
virosum Pall. - 1, 2, 3, 4, 5, 6
vischerae Juxip - 1, 3
viscidulum Tausch (*H. bauhini* Bess. b. *viscidulum* Tausch, *H. bauhini* subsp. *bauhini* var. *viscidulum* (Tausch) E.I. Nyarady, *Pilosella bauhini* (Bess.) Arv.-Tour. subsp. *viscidula* (Tausch) Sojak, *P. viscidula* (Tausch) Schljak.) - 1
vitellicolor Elfstr. - 1(?)
vitellinum Norrl. = H. x glomeratum
volhynicum (Naeg. & Peter) Juxip (*Pilosella volhynica* (Naeg. & Peter) Schljak.) - 1
voroniense Juxip = H. congruens
vulgatiforme (Dahlst.) K. Joh. = H. subvulgatiforme
vulgatum Fries (*H. lachenalii* auct. p. p.) - 1
- subsp. *irriguiceps* Zahn = H. irriguiceps
- subsp. *juratzkanum* Zahn = H. juratzkanum
- subsp. *ochanskiense* Zahn = H. ochanskiense

- *[subsp.] *ovaliceps* Norrl. = H. ovaliceps
- *[subsp.] *subpellucidum* Norrl. = H. subpellucidum
- subsp. *subviriduliceps* Zahn = H. subviriduliceps
- subsp. *truncipilum* Thaisz & Zahn = H. truncipilum
- var. *caesium* Fries = H. caesium
- var. *tridentatum* Fries = H. tridentatum
vulpinum (Rehm.) Czer. comb. nova (*H. hoppeanum* Schult. subsp. *vulpinum* Rehm. 1896, Verh. Zool.-Bot. Ges. Wien, 46 : 330; *Pilosella vulpina* (Rehm.) Schljak.) - 1
vulsum (Elfstr.) Schljak. (*H. stenomischum* Omang var. *vulsum* Elfstr.) - 1
vvedenskyi Juxip - 2
wainioi Norrl. - 1
wilkense (Rehm.) Juxip = H. x macrostolonum
wimmeri Uechtr. - 1
wjasowoense (Zahn) Juxip = H. x densiflorum
wolczankense Juxip - 1
x wolgense Zahn (*Pilosella* x *wolgensis* (Zahn) Sojak) - 1
wologdense (Pohle & Zahn) Juxip - 1
worochtae Woloszcz. & Zahn - 1
woronowianum Zahn = H. x maschukense
wysokae (Woloszcz. & Zahn) Schljak. (*H. lampromegas* Zahn subsp. *wysokae* Woloszcz. & Zahn) - 1
xanthostigma (Norrl.) Norrl. = H. x fennicum
zelencense Schljak. - 1
zinserlingianum Juxip - 1
zinserlingii Schljak. - 1
x zizianum Tausch (*H. abortiens* Norrl., *H. amauranthum* (Peter) Juxip, *H. austericaule* Norrl., *H. cyrtophyllum* Norrl., *H. incrassatiforme* Norrl., *H. leptophyllum* (Naeg. & Peter) Juxip, *H. perfugii* Juxip, *Pilosella* x *ziziana* (Tausch) F. Schultz & Sch. Bip., *P. ziziana* subsp. *leptophylla* (Naeg. & Peter) Sojak) - 1

Hippia auct. = Dichrocephala
integrifolia L. fil. = Dichorocephala integripolia

Hippolytia Poljak.

darwasica (C. Winkl.) Poljak. - 6
herderi (Regel & Schmalh.) Poljak. - 6
megacephala (Rupr.) Poljak. - 6
schugnanica (C. Winkl.) Poljak. - 6

Homogyne Cass.

alpina (L.) Cass. - 1

Hulteniella Tzvel.

integrifolia (Richards.) Tzvel. (*Arctanthemum integrifolium* (Richards.) Tzvel., *Dendranthema integrifolium* (Richards.) Tzvel.) - 5

Hyalea (DC.) Jaub. & Spach

pulchella (Ledeb.) C. Koch - 1, 2, 6
tadshicorum (Tzvel.) Sojak - 6

Hyoseris L. p.p. = Hedypnois

rhagadioloides L. = Hedypnois rhagadioloides

Hypacanthium Juz.

echinopifolium (Bornm.) Juz. - 6
evidens Tscherneva - 6

Hypochoeris L.
glabra L. - 1
radicata L. - 1, 2

Inula L.
acaulis Schott & Kotschy ex Boiss. - 2
arabica L. = Pulicaria arabica
aspera Poir. (*I. salicina* L. subsp. *aspera* (Poir.) Jav., *I. salicina* var. *aspera* (Poir.) G. Beck) - 1, 2, 3, 6
aucheriana DC. (*I. seidlitzii* Boiss.) - 2
auriculata Boiss. & Bal. - 2
britannica L. - 1, 2, 3, 4, 5, 6
- subsp. *japonica* (Thunb.) Kitam. = I. japonica
- subsp. *linariifolia* (Turcz.) Kitam. = I. linariifolia
caspica Blum ex Ledeb. - 1, 2, 3, 6
conyza DC. (*I. squarrosa* (L.) Bernh. 1800, non L. 1763, *I. vulgaris* Trevis.) - 1, 2
decurrens M. Pop. - 6
ensifolia L. - 1, 2
germanica L. - 1, 2, 3, 6
glauca C. Winkl. - 6
grandiflora Willd. - 2
grandis Schrenk = I. macrophylla
grombczewskii C. Winkl. - 6
helenium L. - 1, 2, 3, 5, 6
- subsp. **turcoracemosa** Grierson - 2
hirta L. - 1, 2, 3, 6
hissarica R. Nabiev - 6
japonica Thunb. (*I. britannica* L. subsp. *japonica* (Thunb.) Kitam.) - 5
kitamurana Tatew. ex Honda = I. salicina
linariifolia Turcz. (*I. britannica* L. subsp. *linariifolia* (Turcz.) Kitam.) - 5
macrolepis Bunge - 6
macrophylla Kar. & Kir. (*I. grandis* Schrenk) - 6
magnifica Lipsky - 2
mariae Bordz. - 2
montbretiana DC. - 2
multicaulis Kar. - 6
oculus-christi L. - 1, 2, 6
orientalis Lam. - 2
peacockiana (Aitch. & Hemsl.) Korov. = Codonocephalum peacockianum
rhizocephala Schrenk - 6
x **rigida** Doell. - I. hirta L. x I. salicina L. - 1
rubtzovii Gorschk. = I. schischkinii
sabuletorum Czern. ex Lavr. (*I. salicina* L. subsp. *sabuletorum* (Czern. ex Lavr.) Sojak) - 1, 2, 6
salicina L. (*I. kitamurana* Tatew. ex Honda) - 1, 2, 3, 4, 5, 6
- subsp. *aspera* (Poir.) Jav. = I. aspera
- subsp. *sabuletorum* (Czern. ex Lavr.) Sojak = I. sabuletorum
- var. *aspera* (Poir.) G. Beck = I. aspera
salsoloides (Turcz.) Ostenf. = Limbarda salsoloides
schischkinii Gorschk. (*I. rubtzovii* Gorschk.) - 6
schmalhausenii C. Winkl. - 6
seidlitzii Boiss. = I. aucheriana
squarrosa (L.) Bernh. = I. conyza
thapsoides (Bieb.) Spreng. - 1, 2
vulgaris Trevis. = I. conyza

Iranecio B. Nordenstam = Senecio
othonnae (Bieb.) B. Nordenstam = Senecio othonnae
paucilobus (DC.) B. Nordenstam = Senecio paucilobus

Ixeridium (A. Gray) Tzvel.
chinense (Thunb.) Tzvel. - 4, 5
- subsp. *graminifolium* (Ledeb.) Tzvel. = I. graminifolium
- subsp. *versicolor* (Fisch. ex Link) Tzvel. = I. gramineum
dentatum (Thunb.) Tzvel. - 2(alien), 5
gramineum (Fisch.) Tzvel. (*I. chinense* (Thunb.) Tzvel. subsp. *versicolor* (Fisch. ex Link) Tzvel.) - 4, 5
graminifolium (Ledeb.) Tzvel. (*I. chinense* (Thunb.) Tzvel. subsp. *graminifolium* (Ledeb.) Tzvel.) - 4
strigosum (Levl. & Vaniot) Tzvel. - 5

Jacea Hill = Centaurea
abbreviata (C. Koch) Sojak = Centaurea abbreviata
alutacea (Dobrocz.) Sojak = Centaurea alutacea
carpatica (Porc.) Sojak = Centaurea carpatica
hyrcanica (Bornm.) Holub = Centaurea hyrcanica
livonica (Weinm.) Sojak = Centaurea x livonica
margaritacea (Ten.) Sojak = Centaurea margaritacea
pannonica (Heuff.) Sojak = Centaurea pannonica
phrigia (L.) Sojak = Centaurea phrygia
- subsp. *carpatica* (Porc.) Dostal = Centaurea carpatica
pseudophrygia (C.A. Mey.) Holub = Centaurea pseudophrygia
salicifolia (Bieb.) Sojak = Centaurea salicifolia
stenolepis (A. Kerner) Sojak = Centaurea stenolepis
substituta (Czer.) Sojak = Centaurea substituta
trichocephala (Bieb.) Sojak = Centaurea trichocephala

Jurinea Cass. (*Outreya* Jaub. & Spach, *Stechmannia* DC.)
abolinii Iljin - 6
abramowii Regel & Herd. - 6
adenocarpa Schrenk - 6
akinfievii Nemirova - 2
alata (Desf.) Cass. - 2
albicaulis Bunge - 3, 6
- subsp. *laxa* (Fisch. ex Iljin) Kozuharov = J. longifolia
albovii Galushko & Nemirova - 2
algida Iljin - 6
almaatensis Iljin - 6
altaica Iljin - 3
amplexicaulis (S.G. Gmel.) Bobr. = J. polyclonos
androssovii Iljin - 6
annae Sosn. - 2
antoninae Iljin - 6
antonowii C. Winkl. - 6
apoda Iljin = J. caespitosa
arachnoidea Bunge (*J. consanguinea* DC. subsp. *arachnoidea* (Bunge) Kozuharov) - 1, 2, 6
armeniaca Sosn. = J. elegans
asperifolia Iljin - 6
atropurpurea C. Winkl. ex Iljin - 6
aucheriana DC. - 2
baissunensis Iljin - 6
baldschuanica C. Winkl. - 6
bellidioides Boiss. - 2
bipinnatifida C. Winkl. - 6
blanda (Bieb.) C.A. Mey. - 2
bobrovii Iljin - 6
botschantzevii Iljin - 6
brachycephala Klok. - 1
brachypappa Nemirova - 2
bracteata Regel & Schmalh. - 6
bucharica C. Winkl. - 6
caespitans Iljin - 6

caespitosa C. Winkl. (*J. apoda* Iljin) - 6
calcarea Klok. - 1
capusii Franch. - 6
carduiformis (Jaub. & Spach) Boiss. (*Outreya carduiformis* Jaub. & Spach) - 6
cartaliniana Boiss. - 2
centauroides Klok. - 1
cephalopoda Iljin - 6
chaetocarpa (Ledeb.) Ledeb. - 6
charcoviensis Klok. - 1
ciscaucasica (Sosn.) Iljin - 2
■ consanguinea DC. (*J. demetrii* Iljin)
- subsp. *arachnoidea* (Bunge) Kozuharov = J. arachnoidea
coronopifolia Somm. & Levier - 2
cretacea Bunge - 1
creticola Iljin - 1
cyanoides (L.) Reichenb. - 1, 2, 3
- subsp. *tenuiloba* (Bunge) Nym. = J. tenuiloba
czilikiniana Iljin - 6
darvasica Iljin - 6
demetrii Iljin = J. consanguinea
densisquamea Iljin - 6
depressa (Stev.) C.A. Mey. var. *pinnatisecta* Boiss. = Jurinella subacaulis
derderioides C. Winkl. - 6
dolomitica Galushko - 2
dshungarica (N. Rubtz.) Iljin - 6
eduardi-regelii Iljin - 6
elegans (Stev.) DC. (*J. armeniaca* Sosn.) - 2
elegantissima Iljin - 6
ewersmannii Bunge - 1, 3, 6
eximia Tek. - 6
exuberans (Trautv.) Sosn. - 2
fedtschenkoana Iljin - 6
ferganica (Iljin) Iljin - 6
filicifolia Boiss. - 2
foliosa (Iljin) Iljin = J. modesti
galushkoi Nemirova - 2
gorodkovii Iljin - 3
gracilis Iljin - 6
granitica Klok. - 1
grossheimii Sosn. - 2
grumosa Iljin - 6
hamulosa N. Rubtz. - 6
helenae Sobko - 1
helichrysifolia M. Pop. ex Iljin - 6
iljinii Grossh. - 2
iljinii Soo = J. modesti
impressinervis Iljin - 6
kamelinii Iljin - 6
kapelkinii O. Fedtsch. - 3, 6
karabugasica Iljin - 6
karatavica Iljin - 6
kasakorum Iljin - 1
kazachstanica Iljin - 6
kirghisorum Janisch. - 1, 6
knorringiana Iljin - 6
kokanica Iljin - 6
komarovii Iljin - 6
korotkovae Tarakulov & Khassanov - 6
krascheninnikovii Iljin - 6
kultiassovii Iljin - 6
kuraminensis Iljin - 6
lanipes Rupr. - 6
- subsp. *foliosa* Iljin = J. modesti
lasiopoda Trautv. - 6
laxa Fisch. ex Iljin = J. longifolia

ledebourii Bunge - 1
levieri Albov - 2
lipskyi Iljin - 6
lithophila N. Rubtz. - 6
longicorollaris (Iljin) Iljin = J. multiloba
longifolia DC. (*J. albicaulis* Bunge subsp. *laxa* (Fisch. ex Iljin) Kozuharov, *J. laxa* Fisch. ex Iljin, *J. paczoskiana* Iljin) - 1
ludmilae Iljin (*Stechmannia ludmilae* (Iljin) Sojak) - 6
lydiae Iljin (*Stechmannia lydiae* (Iljin) Sojak) - 6
macranthodia Iljin - 6
margalensis Iljin - 6
mariae Pavl. - 6
maxima C. Winkl. - 6
michelsonii Iljin - 1
modesti Czer. (*J. lanipes* Rupr. subsp. *foliosa* Iljin, *J. foliosa* (Iljin) Iljin, 1962, non Sonklar, 1870, *J. iljinii* Soo, 1969, non Grossh. 1947) - 6
mollissima Klok. - 1
monocephala Aitch. & Hemsl. subsp. *sintenisii* (Bornm.) Wagenitz = J. sintenisii
monticola Iljin - 6
mugodsharica Iljin - 6
multiceps Iljin - 6
multiflora (L.) B. Fedtsch. - 1, 2, 3, 4, 6
multiloba Iljin (*J. longicorollaris* (Iljin) Iljin) - 6
- subsp. aulieatensis (Iljin) Iljin - 6
nivea C. Winkl. - 6
olgae Regel & Schmalh. - 6
orientalis (Iljin) Iljin - 6
pachysperma Klok. - 1
paczoskiana Iljin = J. longifolia
persimilis Iljin - 6
pineticola Iljin - 3
pjatajevae Iljin = J. sangardensis
poacea Iljin - 6
polyclonos (L.) DC. (*J. amplexicaulis* (S.G. Gmel.) Bobr. nom. dubium) - 1, 3, 6
popovii Iljin (*Stechmannia popovii* (Iljin) Sojak) - 6
praetermissa Galushko & Nemirova - 2
prokhanovii Nemirova - 6
propinqua Iljin - 6
psammophila Iljin - 6
pseudocyanoides Klok. - 1
pseudoiljinii Galushko & Nemirova - 2
pseudomollis Klok. - 1
pteroclada Iljin - 6
pulchella (Fisch. ex Hornem.) DC. (*Serratula pulchella* Fisch. ex Hornem.) - 2
pumila Albov - 2
rhizomatoidea Iljin - 6
robusta Schrenk - 6
ruprechtii Boiss. - 2
salicifolia Grun. - 1
sangardensis Iljin (*J. pjatajevae* Iljin, *J. serpenticaulis* Iljin) - 6
schachimardanica Iljin - 6
schischkiniana Iljin - 1, 3
semenowii (Herd.) C. Winkl. - 6
serpenticaulis Iljin = J. sangardensis
serratuloides Iljin - 6
sintenisii Bornm. (*J. monocephala* Aitch. & Hemsl. subsp. *sintenisii* (Bornm.) Wagenitz) - 6
sordida Stev. - 1
sosnowskyi Grossh. - 2
spectabilis Fisch. & C.A. Mey. - 2
spiridonovii Iljin - 6

spissa Iljin - 6
stenophylla Iljin - 6
stoechadifolia (Bieb.) DC. - 1, 2
suffruticosa Regel - 6
suidunensis Korsh. - 6
tadshikistanica Iljin - 6
talijevii Klok. - 1
tanaitica Klok. - 1
tapetodes Iljin - 6
tenuiloba Bunge (*J. cyanoides* (L.) Reichenb. subsp. *tenuiloba* (Bunge) Nym.) - 1, 6
thianschanica Regel & Schmalh. - 6
thyrsiflora Klok. - 1
tortisquamea Iljin - 6
transhyrcanica Iljin - 6
transuralensis Iljin - 1, 3, 6
trautvetteriana Regel & Schmalh. - 6
trifurcata Iljin - 6
tyraica Klok. - 1
venusta Iljin - 2
winkleri Iljin - 6
woronowii Iljin - 2
xeranthemoides Iljin - 6
xerophytica Iljin - 6
zakirovii Iljin - 6

Jurinella Jaub. & Spach

microcephala (Boiss.) Wagenitz = Perplexia microcephala
moschus (Habl.) Bobr. - 2
- subsp. *pinnatisecta* (Boiss.) Danin & P.H. Davis = J. subacaulis
squarrosa (Fisch. & C.A. Mey.) Iljin - 2
subacaulis (Fisch. & C.A. Mey.) Iljin (*Jurinea depressa* (Stev.) C.A. Mey. var. *pinnatisecta* Boiss., *Jurinella moschus* (Habl.) Bobr. subsp. *pinnatisecta* (Boiss.) Danin & P.H. Davis) - 2

Kalimeris (Cass.) Cass.

incisa (Fisch.) DC. - 4, 5
integrifolia Turcz. (*Boltonia integrifolia* (Turcz.) Lauener) - 4, 5

Karelinia Less.

caspia (Pall.) Less. - 1, 6

Kaschgaria Poljak.

brachanthemoides (C. Winkl.) Poljak. - 6
komarovii (Krasch. & N. Rubtz.) Poljak. (*Chrysanthemum komarovii* (Krasch. & N. Rubtz.) S.Y. Hu) - 6

Kemulariella Tamamsch.

abchasica (Kem.-Nath.) Tamamsch. - 2
albovii Tamamsch. - 2
caucasica (Willd.) Tamamsch. - 2
colchica (Albov) Tamamsch. - 2
rosea (Stev. ex Bieb.) Tamamsch. - 2
tugana (Albov) Tamamsch. - 2

Kitamuraea Rauschert = Miyamayomena

savatieri (Makino) Rauschert = Miyamayomena savatieri

Kitamuraster Sojak = Miyamayomena

savatieri (Makino) Sojak = Miyamayomena savatieri

Klasea Cass. = Serratula

aphyllopoda (Iljin) Holub = Serratula aphyllopoda
cardunculus (Pall.) Holub = Serratula cardunculus
coriacea (Fisch. & C.A. Mey.) Holub = Serratula coriacea
gmelinii (Tausch) Holub = Serratula gmelinii
haussknechtii (Boiss.) Holub = Serratula haussknechtii
lycopifolia (Vill.) A. & D. Löve = Serratula lycopifolia
marginata (Tausch) Kitag. = Serratula marginata
procumbens (Regel) Holub = Serratula procumbens

Koelpinia Pall.

linearis Pall. - 1, 2, 6
macrantha C. Winkl. - 6
tenuissima Pavl. & Lipsch. - 6
turanica Vass. - 6

Koyamacalia H. Robinson & R.D. Brettell = Cacalia

auriculata (DC.) H. Robinson & R.D. Brettell = Cacalia auriculata
hastata (L.) H. Robinson & R.D. Brettell = Cacalia hastata

Krylovia Schischk. ex Tamamsch. = Rhinactinidia

eremophila (Bunge) Schischk. ex Tamamsch. = Rhinactinidia eremophila
limoniifolia (Less.) Schischk. ex Tamamsch. = Rhinactinidia limoniifolia
novopokrovskyi (Krasch. & Iljin) Tamamsch. = Rhinactinidia novopokrovskyi
popovii (Botsch.) Tamamsch. = Rhinactinidia popovii

Lachnophyllum Bunge

gossypinum Bunge - 6

Lactuca L. (*Lactucella* Nazarova, *Lagedium* Sojak, *Mulgedium* Cass., *Pterocypsela* Shih)

altaica Fisch. & C.A. Mey. - 1, 2, 3, 6
auriculata DC. = L. dissecta
blinii Levl. = Prenanthes blinii
brassicifolia Boiss. = Cephalorrhynchus brassicifolius
chaixii Vill. - 1, 2
dissecta D. Don (*L. auriculata* DC.) - 6
elata Hemsl. (*Pterocypsela elata* (Hemsl.) Shih) - 5
georgica Grossh. - 2, 6
glauciifolia Boiss. - 6
gorganica Rech. fil. & Esfand. = Cephalorrhynchus gorganicus
indica L. (*L. squarrosa* (Thunb.) Maxim., *Pterocypsela indica* (L.) Shih) - 4, 5
mira Pavl. - 6
perennis L. - 1(?)
quercina L. (*L. stricta* Waldst. & Kit., *Mulgedium quercinum* (L.) C. Jeffrey) - 1(?), 2
- subsp. *wilhelmsiana* (Fisch. & C.A. Mey. ex DC.) Ferakova = L. wilhelmsiana
raddeana Maxim. (*Pterocypsela raddeana* (Maxim.) Shih) - 5
rosularis Boiss. - 6
saligna L. - 1, 2, 3, 5, 6
***sativa** L. - 1, 2, 3(?), 4, 5, 6
serriola L. - 1, 2, 3, 4, 6

sibirica (L.) Maxim. (*Lagedium sibiricum* (L.) Sojak) - 1, 3, 4, 5
spinidens Nevski - 6
squarrosa (Thunb.) Maxim. = L. indica
stricta Waldst. & Kit. = L. quercina
taraxacifolia Chalkuziev - 6
tatarica (L.) C.A. Mey. - 1, 2, 3, 4, 5(alien), 6
triangulata Maxim. (*Pterocypsela triangulata* (Maxim.) Shih) - 5
undulata Ledeb. (*Lactucella undulata* (Ledeb.) Nazarova - 2, 3, 6
wilhelmsiana Fisch. & C.A. Mey. ex DC. (*L. quercina* L. subsp. *wilhelmsiana* (Fisch. & C.A. Mey. ex DC.) Ferakova) - 1(?), 2
winkleri Kirp. - 6

Lactucella Nazarova = Lactuca
undulata (Ledeb.) Nazarova = Lactuca undulata

Lagedium Sojak = Lactuca
sibiricum (L.) Sojak = Lactuca sibirica

Lagoseriopsis Kirp.
popovii (Krasch.) Kirp. - 6

Lagoseris Bieb.
aralensis (Bunge) Boiss. - 6
callicephala Juz. - 1
glaucescens (C. Koch) Sosn. - 2
macrantha (Bunge) Iljin - 1, 6
ovata (Boiss. & Noe) Bornm. - 6
purpurea (Willd.) Boiss. (*L. robusta* Czer.) - 1
robusta Czer. = L. purpurea
sahendi (Boiss. & Buhse) Czer. - 2
sancta (L.) K. Maly - 1, 2, 6

Lamyra (Cass.) Cass. (*Ptilostemon* auct.)
echinocephala (Willd.) Tamamsch. (*Ptilostemon echinocephalum* (Willd.) Greuter) - 1, 2

Lamyropappus Knorr. & Tamamsch.
schakaptaricus (B. Fedtsch.) Knorr. & Tamamsch. - 6

Lamyropsis (Charadze) M. Dittrich = Cirsium
charadzeae Kimeridze = Cirsium charadzeae
macrantha (Schrenk) M. Dittrich = Cirsium lamyroides
sinuata (Trautv.) M. Dittrich = Cirsium sinuatum

Lappa Scop. = Arctium
x *cimbrica* E. Krause = Arctium x cimbricum
x *maassii* M. Schulze = Arctium x maassii

Lapsana L.
aipetriensis Vass. = L. intermedia
alpina auct. = L. intermedia
communis L. - 1, 2, 3, 4, 5(alien), 6
- subsp. *alpina* (Boiss. & Bal.) P.D. Sell, p.p. = L. intermedia
- subsp. *grandiflora* (Bieb.) P.D. Sell = L. grandiflora
- subsp. *intermedia* (Bieb.) Hayek = L. intermedia
grandiflora Bieb. (*L. communis* L. subsp. *grandifora* (Bieb.)

P.D. Sell) - 2
intermedia Bieb. (*L. aipetriensis* Vass., *L. communis* subsp. *alpina* (Boiss. & Bal.) P.D. Sell, p.p., *L. communis* subsp. *intermedia* (Bieb.) Hayek, *L. alpina* auct.) - 1, 2

Lasiopogon Cass.
muscoides (Desf.) DC. - 2, 6

Launaea auct. p.p. = Paramicrorhynchus
procumbens (Roxb.) Ramayya & Rajagopal = Paramicrorhynchus procumbens

*****Layia** Hook. & Arn. ex DC.
*****elegans** Torr. & Gray - 1

Leibnitzia Cass.
anandria (L.) Turcz. - 4, 5
knorringiana (B. Fedtsch.) Pobed. - 6

Leontodon L. (*Apargia* Scop., *Oporinia* D. Don, *Scorzoneroides* Moench)
asper (Waldst. & Kit.) Poir. = L. biscutellifolius
asperrimus (Willd.) Endl. (*L. asperrimus* (Willd.) Ball, comb. superfl., *L. crispus* Vill. subsp. *asperrimus* (Willd.) Finch & P.D. Sell) - 1, 2, 6
autumnalis L. (*Apargia pratensis* Link, *Leontodon autumnalis* subsp. *pratensis* (Link) Arcang., *L. gutzulorum* V. Vassil., *L. keretinus* Nyl., *L. pratensis* (Link) Reichenb. 1831, non Lam. 1778, *Oporinia pratensis* (Link) Less., *Scorzoneroides autumnalis* (L.) Moench, *S. autumnalis* subsp. *pratensis* (Link) Holub) - 1, 2, 3, 4(alien), 5(alien)
- subsp. *pratensis* (Link) Arcang. = L. autumnalis
biscutellifolius DC. (*L. asper* (Waldst. & Kit.) Poir. 1814, non Forssk. 1775, *L. crispus* Vill. subsp. *asper* (Waldst. & Kit.) Rohlena, *L. crispus* auct.) - 1, 2
caucasicus (Bieb.) Fisch. (*L. jailae* Klok.) - 1, 2
crispus auct. = L. biscutellifolius
crispus Vill. subsp. *asper* (Waldst. & Kit.) Rohlena = L. biscutellifolius
- subsp. *asperrimus* (Willd.) Finch & P.D. Sell = L. asperrimus
croceus Haenke (*L. vagneri* Marg., *Scorzoneroides crocea* (Haenke) Holub) - 1
- subsp. *rilaensis* (Hayek) Finch & P.D. Sell = L. rilaensis
danubialis Jacq. (*L. hastilis* L. subsp. *danubialis* (Jacq.) Ball, *L. hastilis* var. *glabratus* Koch, *L. hispidus* L. subsp. *danubialis* (Jacq.) Simonk., *L. hispidus* subsp. *glabratus* (Koch) Celak., *L. hispidus* var. *glabratus* (Koch) Bisch., *L. schischkinii* V. Vassil.) - 1, 2
gutzulorum V. Vassil. = L. autumnalis
hastilis L. subsp. *danubialis* (Jacq.) Ball = L. danubialis
- var. *glabratus* Koch = L. danubialis
hispidus L. - 1, 2
- subsp. *danubialis* (Jacq.) Simonk. = L. danubialis
- subsp. *glabratus* (Koch) Celak. = L. danubialis
- var. *glabratus* (Koch) Bisch. = L. danubialis
jailae Klok. = L. caucasicus
keretinus Nyl. = L. autumnalis
kotschyi Boiss. - 2
kulczynskii M. Pop. (*L. repens* Schur, nom. illegit.) - 1
lanatus L. = Scorzonera czerepanovii
montanus Lam. subsp. *pseudotaraxaci* (Schur) Finch & P.D. Sell = L. pseudotaraxaci
pratensis (Link) Reichenb. = L. autumnalis

pseudotaraxaci Schur (*L. montanus* Lam. subsp. *pseudotaraxaci* (Schur) Finch & P.D. Sell, *Scorzoneroides pseudotaraxaci* (Schur) Holub) - 1
repens Schur = L. kulczynskii
rilaensis Hayek (*L. croceus* Haenke subsp. *rilaensis* (Hayek) Finch & P.D. Sell, *Scorzoneroides rilaensis* (Hayek) Holub) - 1(?)
saxatilis Lam. - 1
schischkinii V. Vassil. = L. danubialis
vagneri Marg. = L. croceus

Leontopodium (Pers.) R. Br. (*Gnaphalium* L. + + + *Leontopodium* Pers. stat. indefinit.)

alpinum Cass. - 1
antennarioides Socz. - 5
blagoveshczenskyi Worosch. - 5
brachyactis Gand. - 6
campestre (Ledeb.) Hand.-Mazz. = L. fedtschenkoanum
conglobatum (Turcz.) Hand.-Mazz. (*L. ochroleucum* Beauverd var. *conglobatum* (Turcz.) Grub.) - 3, 4, 5
discolor Beauverd - 5
fedtschenkoanum Beauverd (*L. campestre* (Ledeb.) Hand.-Mazz., *L. ochroleucum* Beauverd var. *campestre* (Ledeb.) Grub.) - 3, 4, 6
kamtschaticum Kom. - 5
kurilense Takeda - 5
leontopodioides (Willd.) Beauverd - 4, 5
nanum (Hook. fil. & Thoms.) Hand.-Mazz. - 6
ochroleucum Beauverd - 3, 4, 6
- var. *campestre* (Ledeb.) Grub. = L. fedtschenkoanum
- var. *conglobatum* (Turcz.) Grub. = L. conglobatum
palibinianum Beauverd - 5
stellatum A. Khokhr. - 5

Lepidolopha C. Winkl.

fedtschenkoana Knorr. - 6
filifolia Pavl. = L. komarowii
gomolitzkii Kovalevsk. & Safral - 6
karatavica Pavl. - 6
komarowii C. Winkl. (*L. filifolia* Pavl., *Tanacetum komarowii* (C.Winkl.) Muradjan, 1976, non T. komarovii Krasch. & N. Rubtz. 1946) - 6
krascheninnikovii Czil. ex Kovalevsk. & Safral - 6
mogoltavica (Krasch.) Krasch. - 6
nuratavica Krasch. (*Tanacetum nuratavicum* (Krasch.) Muradjan) - 6
talassica Kovalevsk. & Safral - 6

Lepidolopsis Poljak.

turkestanica (Regel & Schmalh.) Poljak. (*Chrysanthemum turkestanicum* (Regel & Schmalh.) Gilli) - 6

Lepidotheca Nutt. (*Santolina* auct.)

aurea (L.) Kovalevsk. (*Cotula aurea* L., *Chamomilla aurea* (L.) J. Gay ex Coss. & Kral., *Matricaria aurea* (L.) Sch. Bip.) - 2, 6
suaveolens (Pursh) Nutt. (*Santolina suaveolens* Pursh, *Artemisia matricarioides* Less., *Chamomilla discoidea* (DC.) J. Gay ex A. Br., *C. suaveolens* (Pursh) Rydb., *Matricaria discoidea* DC., *M. matricarioides* (Less.) Porter, *M. suaveolens* (Pursh) Buchenau, 1894, non L. 1755) - 1, 2, 3, 4, 5, 6

Leucanthemella Tzvel.

linearis (Matsum.) Tzvel. (*Leucanthemum lineare* (Matsum.) Worosch.) - 5
serotina (L.) Tzvel. - 1

Leucanthemopsis (Giroux) Heywood

alpina (L.) Heywood (*Pyrethrum alpinum* (L.) Schrank) - 1(?)

Leucanthemum Hill

adustum (Koch) Gremli subsp. *margaritae* (Gayer ex Jav.) Holub = L. margaritae
ircutianum (Turcz.) DC. (*L. vulgare* Lam. subsp. *ircutianum* (Turcz. ex DC.) Tzvel.) - 1, 2, 3, 4, 5, 6
kurilense (Tzvel.) Worosch. = Arctanthemum kurilense
lineare (Matsum.) Worosch. = Leucanthemella linearis
margaritae (Gayer ex Jav.) Zeleny (*Chrysanthemum leucanthemum* var. *margaritae* Gayer ex Jav., *Leucanthemum adustum* (Koch) Gremli subsp. *margaritae* (Gayer ex Jav.) Holub, *L. margaritae* (Gayer ex Jav.) Soo, nom. illegit., *L. vulgare* Lam. subsp. *margaritae* (Gayer ex Jav.) Soo) - 1
*maximum (Ramond) DC. - 1
raciborskii M. Pop. & Chrshan. = L. subalpinum
rotundifolium (Waldst. & Kit.) DC. = L. waldsteinii
subalpinum (Schur) Tzvel. (*L. raciborskii* M. Pop. & Chrshan., *L. vulgare* Lam. subsp. *subalpinum* (Schur) Soo) - 1
vulgare Lam. - 1, 2
- subsp. *ircutianum* (Turcz. ex DC.) Tzvel. = L. ircutianum
- subsp. *margaritae* (Gayer ex Jav.) Soo = L. margaritae
- subsp. **multicaule** A. Khokhr. - 2
- subsp. *subalpinum* (Schur) Soo = L. subalpinum
waldsteinii (Sch. Bip.) Pouzar (*Tanacetum waldsteinii* Sch. Bip., *Leucanthemum rotundifolium* (Waldst. & Kit.) DC. 1838, non Opiz, 1825) - 1

Leuzea auct. p.p. = Stemmacantha

altaica (Spreng.) Link = Stemmacantha serratuloides
aulieatensis (Iljin) Holub = Stemmacantha aulieatensis
carthamoides (Willd.) DC. = Stemmacantha carthamoides
- subsp. *orientalis* Serg. = Stemmacantha orientalis
integrifolia (C. Winkl.) Holub = Stemmacantha integrifolia
karatavica (Regel & Schmalh.) Holub = Stemmacantha karatavica
lyrata (Iljin) Holub = Stemmacantha lyrata
namanganica (Iljin) Holub = Stemmacantha namanganica
nana (Lipsky) Holub = Stemmacantha nana
nitida (Fisch.) Holub = Stemmacantha nitida
pulchra (Fisch. & C.A. Mey.) Holub = Stemmacantha pulchra
satzyperovii (Sosk.) Holub = Stemmacantha satzyperovii
uniflora (L.) Holub = Stemmacantha uniflora

*Liatris Gaertn. ex Schreb.

*spicata (L.) Willd.

Ligularia Cass.

abakanica Pojark. - 3, 4
alpigena Pojark. - 6
altaica DC. - 3, 6
alticola Worosch. - 5
altissima Pojark. - 6

arctica Pojark. - 1, 3
bucovinensis Nakai - 1
calthifolia Maxim. - 5
carpatica (Schott, Nym. & Kotschy) Pojark. - 1
caucasica (Bieb.) G. Don fil. = Dolichorrhiza caucasica
correvoniana (Albov) Pojark. = Dolichorrhiza correvoniana
fischeri (Ledeb.) Turcz. (*L. glabrescens* Worosch., *L. sachalinensis* Nakai) - 4, 5
glabrescens Worosch. = L. fischeri
glauca (L.) O. Hoffm. - 3
heterophylla Rupr. - 6
hodgsonii Hook. - 5
jaluensis Kom. - 5
kaialpina auct. = L. vorobievii
karataviensis (Lipsch.) Pojark. - 6
kareliniana Stschegl. - 6
knorringiana Pojark. - 6
lanipes (Worosch.) Vyschin (*L. sichotensis* Pojark. var. *lanipes* Worosch.) - 5
longipes Pojark. = L. sibirica
lydiae Minder. - 1
macrophylla (Ledeb.) DC. - 3, 6
mongolica (Turcz.) DC. - 5
narynensis (C. Winkl.) O. & B. Fedtsch. - 6
pavlovii (Lipsch.) Cretz. - 6
renifolia (C.A. Mey.) DC. = Dolichorrhiza renifolia
robusta (Ledeb.) DC. - 3
sachalinensis Nakai = L. fischeri
schischkinii Rubtz. - 6
schmidtii (Maxim.) Makino - 5
sibirica (L.) Cass. (*L. longipes* Poljak.) - 1, 3, 4, 5, 6
sichotensis Pojark. - 5
- var. *lanipes* Worosch. = L. lanipes
songarica (Fisch.) Ling - 6
splendens (Levl. & Vaniot) Nakai - 5
subsagittata Pojark. - 2
talassica Pojark. - 6
thomsonii (Clarke) Pojark. - 6
thyrsoidea (Ledeb.) DC. - 3, 6
trichocephala Pojark. - 5
vorobievii Worosch. (*L. kaialpina* auct.) - 5

Limbarda Adans.

salsoloides (Turcz.) Ikonn. (*Inula salsoloides* (Turcz.) Ostenf.) - 6

Linosyris Cass. = Galatella

fominii Kem.-Nath. = Galatella fominii
pontica (Lipsky) Novopokr. = Galatella pontica
tarbagatensis C. Koch = Galatella tatarica
tatarica (Less.) C.A. Mey. = Galatella tatarica
villosa (L.) DC. = Galatella villosa
vulgaris Cass. ex Less. = Galatella linosyris
- var. *pontica* Lipsky = Galatella pontica

Lipskyella Juz.

annua (C. Winkl.) Juz. - 6

Logfia Cass. = Filago

arvensis (L.) Holub = Filago arvensis
gallica (L.) Coss. & Germ. = Filago gallica

minima (Smith) Dumort. = Filago minima

*Madia Mol.

*sativa Mol. - 1

Matricaria auct. = Tripleurospermum

ambigua (Ledeb.) Kryl. = Tripleurospermum ambiguum
aserbaidshanica Rauschert = Tripleurospermum tzvelevii
australis (Pobed.) Rauschert = Tripleurospermum australe
breviradiata (Ledeb.) Rauschert = Tripleurospermum breviradiatum
caucasica (Willd.) Poir. = Tripleurospermum caucasicum
colchica (Manden.) Rauscheart = Tripleurospermum elongatum
coreana Levl. & Vaniot = Dendranthema coreanum
daghestanica (Rupr. ex Boiss.) Rauschert = Pyrethrum daghestanicum
decipiens (Fisch. & C.A. Mey.) C. Koch = Tripleurospermum decipiens
disciformis (C.A. Mey.) DC. = Tripleurospermum disciforme
elongata (Fisch. & C.A. Mey. ex DC.) Hand.-Mazz. = Tripleurospermum elongatum
grossheimii (Fed.) Rauschert = Tripleurospermum grossheimii
hookeri (Sch. Bip.) Czer. = Tripleurospermum hookeri
karjaginii (Manden. & Sof.) Rauschert = Tripleurospermum karjaginii
limosa (Maxim.) Kudo = Tripleurospermum limosum
maritima L. subsp. *phaeocephala* (Rupr.) Rauschert = Tripleurospermum hookeri
- subsp. *subpolaris* (Pobed.) Rauschert = Tripleurospermum subpolare Pobed.
parviflora (Willd.) Poir. = Tripleurospermum parviflorum
perforata Merat = Tripleurospermum perforatum
phaeocephala (Rupr.) Stefanss. = Tripleurospermum hookeri
rupestris (Somm. & Levier) Rauschert = Tripleurospermum rupestre
sevanensis (Manden.) Rauschert = Tripleurospermum sevanense
subnivalis (Pobed.) Rauschert = Tripleurospermum caucasicum
subpolaris (Pobed.) Holub = Tripleurospermum subpolare
szowitzii (DC.) Rauschert = Tripleurospermum szowitzii
tchihatchewii (Boiss.) Voss = Tripleurospermum tchihatchewii
tetragonosperma (Fr. Schmidt) Hara & Kitam. = Tripleurospermum tetragonospermum
transcaucasica (Manden.) Rauschert = Tripleurospermum transcaucasicum

Matricaria L. (*Chamomilla* S.F.Gray)

aurea (L.) Sch. Bip. = Lepidotheca aurea
chamomilla L. 1753 = Tripleurospermum perforatum
- subsp. *tzvelevii* (Pobed.) Soo = M. tzvelevii
- var. *recutita* (L.) Grierson = M. recutita
chamomilla sensu L. 1755 = M. recutita
discoidea DC. = Lepidotheca suaveolens
inodora L. = Tripleurospermum perforatum
maritima L. = Tripleurospermum maritimum
matricarioides (Less.) Porter = Lepidotheca suaveolens
recutita L. (*Chamomilla recutita* (L.) Rauschert, *Matricaria chamomilla* L. var. *recutita* (L.) Grierson, *M. chamomilla* sensu L. 1755) - 1, 2, 3, 4, 5, 6

suaveolens (Pursh) Buchenau = Lepidotheca suaveolens
tzvelevii Pobed. (*Chamomilla tzvelevii* (Pobed.) Rauschert, *Matricaria chamomilla* L. subsp. *tzvelevii* (Pobed.) Soo) - 1, 2

Mausolea Bunge ex Podlech
eriocarpa (Bunge) Poljak. ex Podlech - 6

Microcephala Pobed.
lamellata (Bunge) Pobed. - 6
subglobosa (Krasch.) Pobed. - 6
turcomanica (C. Winkl.) Pobed. - 6

Micropus auct. = Bombycilaena
bombycinus Lag. = Bombycilaena discolor
discolor Pers. = Bombycilaena discolor

Miyamayomena Kitam. (*Gymnaster* Kitam. 1937, non Schutt. 1891, *Kitamuraea* Rauschert, *Kitamuraster* Sojak)
savatieri (Makino) Kitam. (*Gymnaster savatieri* (Makino) Kitam., *Kitamuraea savatieri* (Makino) Rauschert, *Kitamuraster savatieri* (Makino) Sojak) - 2(alien)

Modestia Charadze & Tamamsch. = Anacantha
darwasica (C. Winkl.) Charadze & Tamamsch. = Anacantha darwasica
jucunda (C. Winkl.) Charadze & Tamamsch. = Anacantha jucunda
mira (Iljin) Charadze & Tamamsch. = Anacantha mira

Mulgedium Cass. = Lactuca
petiolatum C. Koch = Cicerbita petiolata
quercinum (L.) C. Jeffrey = Lactuca quercina

Mycelis Cass.
muralis (L.) Dumort. - 1, 2

Myriactis Less.
gmelinii (Fisch. & C.A. Mey.) DC. = M. wallichii
wallichii Less. (*M. gmelinii* (Fisch. & C.A. Mey.) DC.) - 2, 6

Nardosmia Cass. = Petasites
angulosa Cass. = Petasites frigidus
fominii (Bordz.) Kuprian. = Petasites fominii
frigida (L.) Hook. = Petasites frigidus
glacialis Ledeb. = Endocellion glaciale
gmelinii Turcz.ex DC. = Endocellion sibiricum
japonica auct. = Petasites amplus
laevigata (Willd.) DC. = Petasites radiatus
nivalis (Greene) Jurtz. p.p. = Petasites frigidus
palmata auct. = Petasites tatewakianus
radiata (J.F. Gmel.) Holub = Petasites radiatus
saxatilis Turcz. = Petasites rubellus

Neopallasia Poljak.
pectinata (Pall.) Poljak. - 4, 6

Nikitinia Iljin
leptoclada (Bornm. & Sint.) Iljin - 6

Notobasis (Cass.) Cass.
syriaca (L.) Cass. - 2

Odontolophus Cass. = Centaurea
kobstanicus (Tzvel.) Sojak = Centaurea kobstanica
trinervius (Steph.) Dobrocz. = Centaurea trinervia

Oglifa Cass. = Filago
arvensis (L.) Cass. = Filago arvensis
gallica (L.) Chrtek & Holub = Filago gallica
lagopus (Steph.) Chrtek & Holub = Filago lagopus
minima (Smith) Reichenb. fil. = Filago minima

Olgaea Iljin
baldschuanica (C. Winkl.) Iljin - 6
chodshamuminensis B. Scharipova - 6
eriocephala (C. Winkl.) Iljin - 6
▪ **lanipes** (C. Winkl.) Iljin
longifolia (C. Winkl.) Iljin - 6
nidulans (Rupr.) Iljin - 6
nivea (C. Winkl.) Iljin - 6
pectinata Iljin - 6
petri-primi B. Scharipova - 6
spinifera Iljin - 6
vvedenskyi Iljin - 6

Oligochaeta (DC.) C. Koch
divaricata (Fisch. & C.A. Mey.) C. Koch - 2
minima (Boiss.) Briq. - 6
tomentosa Czer. - 2

Oligosporus Cass. = Artemisia
albiceratus (Krasch.) Poljak. = Artemisia arenaria
arenarius (DC.) Poljak. = Artemisia arenaria
bargusinensis (Spreng.) Poljak. = Artemisia bargusinensis
borealis (Pall.) Poljak. = Artemisia borealis
campestris (L.) Cass. = Artemisia campestris
capillaris (Thunb.) Poljak. = Artemisia capillaris
commutatus (Bess.) Poljak. = Artemisia commutata
daghestanicus (Krasch. & A. Poletzky) Poljak. = Artemisia daghestanica
desertorum (Spreng.) Poljak. = Artemisia desertorum
dimoanus (M. Pop.) Poljak. = Artemisia dimoana
dracunculiformis (Krasch.) Poljak. = Artemisia dracunculus
dracunculus (L.) Poljak. = Artemisia dracunculus
- subsp. *glaucus* (Pall. ex Willd.) A. & D. Löve = Artemisia glauca
glaucus (Pall. ex Willd.) Poljak. = Artemisia glauca
halodendron (Turcz. ex Bess.) Poljak. = Artemisia halodendron
henriettae (Krasch.) Poljak. = Artemisia borealis subsp. richardsoniana
japonicus (Thunb.) Poljak. = Artemisia japonica
kelleri (Krasch.) Poljak. = Artemisia kelleri
kuschakewiczii (C. Winkl.) Poljak. = Artemisia kuschakewiczii
ledebourianus (Bess.) Poljak. = Artemisia ledebouriana
limosus (Koidz.) Poljak. = Artemisia limosa
lipskyi (Poljak.) Poljak. = Artemisia lipskyi
littoricola (Kitam.) Poljak. = Artemisia littoricola
macilentus (Maxim.) Poljak. = Artemisia macilenta
pamiricus (C. Winkl.) Poljak. = Artemisia pamirica
pannosus (Krasch.) Poljak. = Artemisia pannosa

pycnorhizus (Ledeb.) Poljak. = Artemisia pycnorhiza
quinquelobus (Trautv.) Poljak. = Artemisia quinqueloba
salsoloides (Willd.) Poljak. = Artemisia salsoloides
saposhnikovii (Krasch. ex Poljak.) Poljak. = Artemisia saposhnikovii
scoparius (Waldst. & Kit.) Less. = Artemisia scoparia
songaricus (Schrenk) Poljak. = Artemisia songarica
tomentellus (Trautv.) Poljak. = Artemisia tomentella
trautvetterianus (Bess.) Poljak. = Artemisia trautvetteriana

Omalotheca Cass. (*Synchaeta* Kirp.)

caucasica (Somm. & Levier) Czer. (*Gnaphalium caucasicum* Somm. & Levier, *Synchaeta caucasica* (Somm. & Levier) Kirp.) - 2
hoppeana (Koch) Sch. Bip. & F. Schultz (*Gnaphalium hoppeanum* Koch) - 1
norvegica (Gunn.) Sch. Bip. & F. Schultz (*Gnaphalium norvegicum* Gunn., *Synchaeta norvegica* (Gunn.) Kirp.) - 1, 3, 4, 6
stewartii (Clarke) Holub (*Gnaphalium stewartii* Clarke) - 2
supina (L.) DC. (*Gnaphalium supinum* L.) - 1, 2, 3, 6
sylvatica (L.) Sch. Bip. & F. Schultz (*Gnaphalium sylvaticum* L., *Synchaeta sylvatica* (L.) Kirp.) - 1, 2, 3, 4, 5, 6

Onopordum L.

acanthium L. - 1, 2, 3, 4, 6
armenum Grossh. (*O. frickii* Tamamsch., *O. candidum* auct.) - 2
candidum auct. = O. armenum
cinereum Grossh. - 2, 6
frickii Tamamsch. = O. armenum
heteracanthum C.A. Mey. - 2
leptolepis DC. - 6
prjachinii Tamamsch. - 6
seravschanicum Tamamsch. - 6
tauricum Willd. - 1
turcicum Danin - 2

Oporinia D. Don = Leontodon

pratensis (Link) Less. = Leontodon autumnalis

Otanthus Hoffmgg. & Link

maritimus (L.) Hoffmgg. & Link - 2

Othonna L. p. min. p. = Tephroseris

integrifolia L. = Tephroseris integrifolia
palustris. L. = Tephroseris palustris

Outreya Jaub. & Spach = Jurinea

carduiformis Jaub. & Spach = Jurinea carduiformis

Packera A. & D. Löve = Tephroseris

hyperborealis (Greenm.) A. & D. Löve = Tephroseris hyperborealis
resedifolia (Less.) A. & D. Löve = Tephroseris heterophylla

Pallenis (Cass.) Cass.

spinosa (L.) Cass. - 1, 2, 6

Paraixeris Nakai

denticulata (Houtt.) Nakai - 5

pinnatipartita (Makino) Tzvel. - 5
saxatilis (A. Baran.) Tzvel. - 5
serotina (Maxim.) Tzvel. - 5

Paramicrorhynchus Kirp. (*Launaea* auct. p.p.)

procumbens (Roxb.) Kirp. (*Launaea procumbens* (Roxb.) Ramayya & Rajagopal) - 6

Parasenecio W.W. Smith & Small = Cacalia

*Parthenium L.

*argentatum A. Gray - 1, 2, 6

Pentanema Cass. (*Vicoa* Cass. nom. cons. prop.)

albertoregelia (C. Winkl.) Gorschk. - 6
chodzhakasiani Kinz. - 6
discoideum Nabiev (*Vicoa discoidea* (Nabiev) R. Kam.) - 6
divaricatum Cass. - 6
glanduligerum (Krasch.) Gorschk. - 6
krascheninnikovii (R. Kam.) Czer. (*Vicoa krascheninnikovii* R. Kam.) - 6
parietarioides (Nevski) Gorschk. - 6
propinquum Nevski - 6
rupicola (Krasch.) Gorschk. - 6
varzobicum R. Kam. & Kinz. - 6

Perplexia Iljin

microcephala (Boiss.) Iljin (*Jurinella microcephala* (Boiss.) Wagenitz) - 6

Petasites Hill (*Nardosmia* Cass.)

albus (L.) Gaertn. (*P. glabratus* auct., *P. kablikianus* auct.) - 1, 2
amplus Kitam. (*P. japonicus* (Siebold & Zucc.) Maxim. subsp. *giganteus* Kitam., *Nardosmia japonica* auct., *Petasites japonicus* auct.) - 5
x **celakovskyi** Matouschek. - P. albus (L.) Gaertn. x P. kablikianus Tausch ex Bercht. - 1
fominii Bordz. (*Nardosmia fominii* (Bordz.) Kuprian.) - 2
frigidus (L.) Fries (*Nardosmia angulosa* Cass. nom. illegit., *N. frigida* (L.) Hook., *N. nivalis* (Greene) Jurtz. p.p. excl.basionymo, *Petasites frigidus* subsp. *sibiricus* Toman) - 1, 3, 4, 5
- subsp. *sibiricus* Toman = P. frigidus
georgicus Manden. (*P. hybridus* (L.) Gaertn. Mey. & Scherb. subsp. *georgicus* (Manden.) Toman) - 2
glabratus auct. = P. albus
glacialis (Ledeb.) Polun. = Endocellion glaciale
gmelinii (Turcz. ex DC.) Polun. = Endocellion sibiricum
hybridus (L.) Gaertn., Mey. & Scherb. - 1, 2
- subsp. *georgicus* (Manden.) Toman = P. georgicus
- subsp. *ochroleucus* (Boiss. & Huet) Sourek = P. ochroleucus
x **intercedens** Matouschek. - P. hybridus (L.) Gaertn., Mey. & Scherb. x P. kablikianus Tausch ex Bercht. - 1
japonicus auct. = P. amplus
japonicus (Siebold & Zucc.) Maxim. subsp. *giganteus* Kitam. = P. amplus
kablikianus auct. = P. albus
laevigatus (Willd.) Reichenb. = P. radiatus
ochroleucus Boiss. & Huet (*P. hybridus* (L.) Gaertn., Mey. & Scherb. subsp. *ochroleucus* (Boiss. & Huet) Sourek) - 1, 2

palmatus auct. = P. tatewakianus
radiatus (J.F. Gmel.) Toman (*Tussilago radiata* J.F. Gmel., *Nardosmia laevigata* (Willd.) DC., *N. radiata* (J.F. Gmel.) Holub, *Petasites laevigatus* (Willd.) Reichenb.) - 1, 3, 4
x **rechingeri** Hayek. - P. albus (L.) Gaertn. x P. hybridus (L.) Gaertn., Mey. & Scherb. - 1
rubellus (J. F. Gmel.) Toman (*Tussilago rubella* J.F. Gmel., *Nardosmia saxatilis* Turcz., *Petasites saxatilis* (Turcz.) Kom.) - 3, 4, 5
x **sachalinensis** Toman. - P. amplus Kitam. x P. tatewakianus Kitam. - 5
saxatilis (Turcz.) Kom. = P. rubellus
sibiricus (J. F. Gmel.) Dingwall = Endocellion sibiricum
spurius (Retz.) Reichenb. - 1, 3, 6
tatewakianus Kitam. (*Nardosmia palmata* auct., *Petasites palmatus* auct.) - 5

Phagnalon Cass.

androssovii B. Fedtsch. - 6
darvasicum Krasch. - 6

Phalacrachena Iljin

calva (Ledeb.) Iljin - 3, 6
inuloides (Fisch. ex Janka) Iljin (*Centaurea inuloides* Fisch. ex Janka) - 1

Phalacroloma Cass. (*Stenactis* Cass. p.p. excl. lectotypo)

annuum (L.) Dumort. (*Erigeron annuus* (L.) Pers., *Stenactis annua* (L.) Cass.) - 1
- subsp. *septentrionale* (Fern. & Wieg.) Adema = Ph. septentrionale
- subsp. *strigosum* (Muehl. ex Willd.) Adema = Ph. strigosum
septentrionale (Fern. & Wieg.) Tzvel. (*Erigeron ramosus* (Walt.) Britt., Sterns & Pogg. var. *septentrionalis* Fern. & Wieg., *E. annuus* (L.) Pers. subsp. *septentrionalis* (Fern. & Wieg.) Wagenitz, *Phalacroloma annuum* (L.) Dumort. subsp. *septentrionale* (Fern. & Wieg.) Adema, *Stenactis annua* (L.) Cass. subsp. *septentrionalis* (Fern. & Wieg.) A. & D. Löve, *S. septentrionalis* (Fern. & Wieg.) Holub) - 1, 2, 5
strigosum (Muehl. ex Willd.) Tzvel. (*Erigeron strigosus* Muehl. ex Willd., *E. annuus* (L.) Pers. subsp. *strigosus* (Muehl. ex Willd.) Wagenitz, *Phalacroloma annuum* (L.) Dumort. subsp. *strigosum* (Muehl. ex Willd.) Wagenitz, *Stenactis strigosa* (Muehl. ex Willd.) DC.) - 1(?), 5

Picnomon Adans.

acarna (L.) Cass. - 1, 2, 6

Picris L.

canescens (Stev.) V. Vassil. (*P. strigosa* Bieb. subsp. *canescens* (Stev.) Lack) - 2
crepidioides Miyabe & Kudo = Trommsdorfia crepidioides
davurica Fisch. - 3, 4, 6
- var. *koreana* (Kitam.) Kitag. = P. koreana
echioides L. var. *kamtschatica* (Ledeb.) B. Boivin = P. kamtschatica
hieracioides L. (*Hieracium muricellum* Fries) - 1, 2, 3, 6
- subsp. **grandiflora** (Ten.) Arcang. - 1
- subsp. *koreana* (Kitam.) Worosch. = P. koreana

japonica Thunb. - 5
- var. *koreana* Kitam. = P. koreana
kamtschatica Ledeb. (*P. echioides* L. var. *kamtschatica* (Ledeb.) B. Boivin) - 5
koreana (Kitam.) Worosch. (*P. japonica* Thunb. var. *koreana* Kitam., *P. davurica* Fisch. var. *koreana* (Kitam.) Kitag., *P. hieracioides* L. subsp. *koreana* (Kitam.) Worosch.) - 5
nuristanica Bornm. (*P. similis* V. Vassil.) - 6
pauciflora Willd. - 1, 2
rigida Ledeb. ex Spreng. - 1, 2, 3, 6
similis V. Vassil. = P. nuristanica
strigosa Bieb. (*P. strigosa* subsp. *turcomanica* Bornm. & Sint., *P. turcomanica* (Bornm. & Sint.) Sojak, nom. illegit., *P. turcomanica* Gand.) - 1, 2, 6
- subsp. *canescens* (Stev.) Lack = P. canescens
- subsp. *turcomanica* Bornm. & Sint. = P. strigosa
turcomanica (Bornm. & Sint.) Sojak = P. strigosa
turcomanica Gand. = P. strigosa

Pilosella Hill = Hieracium

abakurae (Schelk. & Zahn) Sojak = Hieracium x abakurae
acrothyrsa (Naeg. & Peter) Sojak = Hieracium x melinomelas
acutisquama (Naeg. & Peter) Schljak. = Hieracium acutisquamum
adenoclada (Rehm.) Schljak. = Hieracium adenocladum
agathantha (Rehm.) Schljak. = Hieracium agathanthum
altaica (Naeg. & Peter) Schljak. = Hieracium altaicum
amaureilema (Naeg. & Peter) Schljak. = Hieracium amaureilema
amaurochlora (Zahn) Schljak. = Hieracium amaurochlorum
ambigua (Ehrh. ex Naeg. & Peter) Sojak = Hieracium x glomeratum
amnoon (Naeg. & Peter) Schljak. = Hieracium amnoon
x *apatelia* (Naeg. & Peter) Sojak = Hieracium x apatelium
aquilonaris (Naeg. & Peter) Schljak. = Hieracium aquilonare
arctogena (Norrl.) Schljak. = Hieracium arctogenum
x *arida* (Freyn) Schljak. = Hieracium x aridum
armeniaca (Naeg. & Peter) Schljak. = Hieracium armeniacum
x *arvicola* (Naeg. & Peter) Sojak = Hieracium x arvicola
arvorum (Naeg. & Peter) Schljak. = Hieracium arvorum
asiatica (Naeg. & Peter) Schljak. = Hieracium asiaticum
aurantiaca (L.) F. Schultz & Sch. Bip. = Hieracium aurantiacum
- subsp. *carpathicola* (Naeg. & Peter) Dostal = Hieracium carpathicola
x *auriculoides* (Lang) F. Schultz = Hieracium x auriculoides
- subsp. *arvensis* (Naeg. & Peter) Sojak = Hieracium x auriculoides
- subsp. *asperrima* (Schur) Sojak = Hieracium x auriculoides
- subsp. *echiocephala* (Naeg. & Peter) Sojak = Hieracium x auriculoides
- subsp. *echiogenes* (Naeg. & Peter) Sojak = Hieracium x auriculoides
- subsp. *lasiophora* (Naeg. & Peter) Sojak = Hieracium lasiophorum
- subsp. *longiseta* (Naeg. & Peter) Sojak = Hieracium x auriculoides
- subsp. *mira* (Naeg. & Peter) Sojak = Hieracium x auriculoides
- subsp. *tanythrix* (Naeg. & Peter) Sojak = Hieracium x auriculoides
- subsp. *umbellosa* (Naeg. & Peter) Sojak = Hieracium x auriculoides Lang

x *auriculoides* (Lang) P.D. Sell & C. West = Hieracium x auriculoides

bauhini (Bess.) Arv.-Touv. = Hieracium bauhini

- subsp. *arvorum* (Naeg. & Peter) Sojak = Hieracium arvorum

- subsp. *cymantha* (Naeg. & Peter) Sojak = Hieracium cymanthum

- subsp. *fastigiata* (Tausch ex Naeg. & Peter) Sojak = Hieracium fastigiatum

- subsp. *melachaeta* (Tausch) Sojak = Hieracium melachaetum

- subsp. *obscuribracteata* (Naeg. & Peter) Sojak = Hieracium obscuribracteatum

- subsp. *thaumasia* (Peter) Sojak = Hieracium thaumasium

- subsp. *thaumasioides* (Bess.) Sojak = Hieracium thaumasioides

- subsp. *viscidula* (Tausch) Sojak = Hieracium viscidulum

bessariana (Spreng.) Sojak = Hieracium besserianum

x *bifurca* (Bieb.) F. Schultz & Sch. Bip. = Hieracium x bifurcum

- subsp. *sterromastix* (Naeg. & Peter) Sojak = Hieracium x bifurcum

x *blyttiana* (Fries) F. Schultz & Sch. Bip. = Hieracium x blyttianum

botrychodes (Zahn) Schljak. = Hieracium botrychodes

x *brachiata* (Bertol. ex DC.) F. Schultz & Sch. Bip. = Hieracium x brachiatum

- subsp. *matrensis* (Naeg. & Peter) Sojak = Hieracium matrense

- subsp. *pseudobrachiata* (Celak.) Sojak = Hieracium x brachiatum

caespitosa (Dumort.) P.D. Sell & C. West = Hieracium caespitosum

- subsp. *brevipila* (Naeg. & Peter) P.D. Sell & C. West = Hieracium karelicum and H. onegense

- subsp. *colliniformis* (Peter) P.D. Sell & C. West = Hieracium colliniforme

- subsp. *dissoluta* (Naeg. & Peter) Sojak = Hieracium dissolutum

- subsp. *leptocaulon* (Naeg. & Peter) Sojak = Hieracium leptocaulon

- subsp. *sudetorum* (Peter) Sojak = Hieracium sudetorum

callicyma (Rehm.) Schljak. = Hieracium callicymum

x *callimorpha* (Naeg. & Peter) Sojak = Hieracium x callimorphum

x *callimorphoides* (Zahn) Sojak = Hieracium x callimorphoides

x *calodon* (Tausch ex Peter) Sojak = Hieracium x calodon

- subsp. *multiceps* (Naeg. & Peter) Sojak = Hieracium x calodon

- subsp. *tenuiceps* (Naeg. & Peter) Sojak = Hieracium x calodon

calolepidea (Norrl.) Schljak. = Hieracium calolepideum

x *calomastix* (Peter) Sojak = Hieracium x calomastix

carpathicola (Naeg. & Peter) Schljak. = Hieracium carpathicola

chaunocyma (Rehm.) Schljak. = Hieracium chaunocymum

x *chlorops* (Naeg. & Peter) Sojak = Hieracium chlorops

cilicica (Naeg. & Peter) Holub = Hieracium cilicicum

cochlearis (Norrl.) Weiss = Hieracium x suecicum

- subsp. *pseudauricula* (Naeg. & Peter) Sojak = Hieracium x suecicum

- subsp. *subauricula* (Naeg. & Peter) Sojak = Hieracium x suecicum

x *collina* (Gochn.) Sojak = Hieracium x collinum

- subsp. *fallax* (Willd.) Sojak = Hieracium x collinum

colliniformis (Peter) Dostal = Hieracium colliniforme

concoloriformis (Norrl.) Schljak. = Hieracium concoloriforme

conflectens (Norrl.) Schljak. = Hieracium conflectens

contracta (Norrl.) Schljak. = Hieracium contractum

curvescens Norrl. = Hieracium curvescens

cylindriceps (Naeg. & Peter) Schljak. = Hieracium cylindriceps

cymantha (Naeg. & Peter) Schljak. = Hieracium cymanthum

cymosa (L.) F. Schultz & Sch. Bip. = Hieracium cymosum

x *densiflora* (Tausch) Sojak = Hieracium x densiflorum

- subsp. *acrosciadia* (Naeg. & Peter) Sojak = Hieracium x densiflorum

- subsp. *cymosiformis* (Naeg. & Peter) Sojak = Hieracium x densiflorum

- subsp. *umbellifera* (Naeg. & Peter) Sojak = Hieracium x densiflorum

denticulifera Norrl. = Hieracium denticuliferum

dilutior (Rehm.) Schljak. = Hieracium dilutius

x *dimorphoides* Norrl. = Hieracium x dimorphoides

dissoluta (Naeg. & Peter) Schljak. = Hieracium dissolutum

dobromilensis (Rehm.) Schljak. = Hieracium dobromilense

dolinensis (Rehm.) Schljak. = Hieracium dolinense

x *dubia* (L.) Fries = Hieracium x dubium

dubia (L.) Sojak = Hieracium x dubium

dublanensis (Rehm.) Schljak. = Hieracium dublanense

x *dybowskiana* (Rehm.) Schljak. = Hieracium x dybowskianum

echioides (Lumn.) F. Schultz & Sch. Bip. = Hieracium echioides

- subsp. *freynii* (Naeg. & Peter) Sojak = Hieracium echioides

- subsp. *macrocyma* (Naeg. & Peter) Sojak = Hieracium macrocymum

- subsp. *procera* (Fries) P.D. Sell & C. West = Hieracium procerum

empodista (Naeg. & Peter) Schljak. = Hieracium empodistum

ensifera (Norrl.) Schljak. = Hieracium ensiferum

ericetorum (Naeg. & Peter) Schljak. = Hieracium ericetorum

eriophylla (Vill.) Sojak subsp. *rojowskii* (Rehm.) Sojak = Hieracium rojowskii

erratica (Norrl.) Schljak. = Hieracium erraticum

erythrophylloides (Zahn) Schljak. = Hieracium erythrophylloides

erythropoides Norrl. = Hieracium x suecicum

euryanthela (Dahlst.) Schljak. = Hieracium euryanthelum

eusciadia (Naeg. & Peter) Schljak. = Hieracium eusciadium

x *fallaciformis* (Litv. & Zahn) Schljak. = Hieracium x fallaciforme

x *fallacina* (F. Schultz) Sch. Bip. = Hieracium x fallacinum

fallax (Willd.) Arv.-Touv. = Hieracium x collinum

farrea Norrl. = Hieracium farreum

fastigiata (Tausch ex Naeg. & Peter) Schljak. = Hieracium fastigiatum

x *fennica* (Norrl.) Norrl. = Hieracium x fennicum

- subsp. *fulvescens* (Naeg. & Peter) Sojak = Hieracium x fennicum

- subsp. *spathophylla* (Peter) Sojak = Hieracium x fennicum

ferroviae (Rehm.) Schljak. = Hieracium ferroviae

filifera (Tausch) Schljak. = Hieracium filiferum

firmicaulis Norrl. = Hieracium firmicaule

flagellariformis (G. Schneid.) Sojak = Hieracium x progenitum

x *flagellaris* (Willd.) Arv.-Touv. = Hieracium x flagellare x *flagellaris* (Willd.) P.D. Sell & C. West = Hieracium x flagellare

- subsp. *tatrensis* (Peter) Sojak = Hieracium x flagellare

floccipeduncula (Naeg. & Peter) Schljak. = Hieracium floccipedunculum

floribunda (Wimm. & Grab.) Arv.-Touv. = Hieracium x floribundum

- subsp. *regiomontana* (Naeg. & Peter) Sojak = Hieracium x floribundum

- subsp. *suecica* (Fries) Sojak = Hieracium x suecicum

x *floribunda* (Wimm. & Grab.) Fries = Hieracium x floribundum

x *fuscoatra* (Naeg. & Peter) Sojak = Hieracium x fuscoatrum

galiciensis (Blocki) Schljak. = Hieracium galiciense

glaucescens (Bess.) Sojak = Hieracium glaucescens

- subsp. *amnoon* (Naeg. & Peter) Sojak = Hieracium amnoon

- subsp. *besseriana* (Spreng.) Sojak = Hieracium besserianum

- subsp. *branae* (Naeg. & Peter) Sojak = Hieracium branae

- subsp. *filifera* (Tausch) Sojak = Hieracium filiferum

- subsp. *heothina* (Naeg. & Peter) Sojak = Hieracium heothinum

- subsp. *marginalis* (Naeg. & Peter) Sojak = Hieracium marginale

- subsp. *megalomastix* (Naeg. & Peter) Sojak = Hieracium megalomastix

- subsp. *nigriseta* (Naeg. & Peter) Sojak = Hieracium nigrisetum

- subsp. *pseudauriculoides* (Naeg. & Peter) Sojak = Hieracium pseudauriculoides

glaucochroa (Naeg. & Peter) Schljak. = Hieracium glaucochroum

glomerabilis (Norrl.) Schljak. = Hieracium glomerabile

glomerata (Froel.) Arv.-Touv. = Hieracium x glomeratum

- subsp. *ambigua* (Ehrh. ex Naeg. & Peter) Sojak = Hieracium x glomeratum

- subsp. *prolongata* (Naeg. & Peter) Sojak = Hieracium x glomeratum

- subsp. *subambigua* (Naeg. & Peter) Sojak = Hieracium x glomeratum

x *glomerata* (Froel.) Fries = Hieracium x glomeratum

granitophila (Norrl.) Schljak. = Hieracium granitophilum

heothina (Naeg. & Peter) Schljak. = Hieracium heothinum

hispidissima (Rehm. ex Naeg. & Peter) Schljak. = Hieracium hispidissimum

hohenackeri (Naeg. & Peter) Sojak = Hieracium hohenackeri

holmiensis (Naeg. & Peter) Schljak. = Hieracium holmiense

hoppeana (Schult.) F. Schultz & Sch. Bip. = Hieracium hoppeanum

- subsp. *cilicica* (Naeg. & Peter) P.D. Sell & C. West = Hieracium cilicicum

- subsp. *pilisquama* (Naeg. & Peter) P.D. Sell & C. West = Hieracium pilisquamum

- subsp. *testimonialis* (Naeg. & Peter) P.D. Sell & C. West = Hieracium multisetum

- subsp. *virentisquama* (Naeg. & Peter) Fuchs-Eckert = Hieracium virentisquamum

x *hypeurya* (Peter) Sojak = Hieracium x hypeuryum

- subsp. *lamprocoma* (Naeg. & Peter) Fuchs-Eckert = Hieracium x lamprocomum

- subsp. *lasiothrix* (Naeg. & Peter) Fuchs-Eckert = Hieracium lasiothrix

imponens (Norrl.) Schljak. = Hieracium imponens

incaniformis (Litv. & Zahn) Sojak = Hieracium x incaniforme

ingrica (Naeg. & Peter) Schljak. = Hieracium ingricum

insolens (Norrl.) Schljak. = Hieracium insolens

x *iserana* (Uechtr.) Sojak = Hieracium x iseranum

- subsp. *florida* (Naeg. & Peter) Sojak = Hieracium x iseranum

- subsp. *subnigriceps* (Zahn) Sojak = Hieracium x iseranum

kajanensis (Malmgr.) Norrl. = Hieracium kajanense

x *kalksburgensis* (Wiesb.) Sojak = Hieracium x kalksburgensis

- subsp. *laschii* (Zahn) Sojak = Hieracium x kalksburgensis

karelica Norrl. = Hieracium karelicum

karpinskyana (Naeg. & Peter) Schljak. = Hieracium karpinskyanum

kihlmanii Norrl. = Hieracium x kihlmanii

knappii (Blocki) Schljak. = Hieracium knappii

x *koernickeana* (Naeg. & Peter) Sojak = Hieracium x koernickeanum

kozlowskyana (Zahn) Sojak = Hieracium x kozlowskyanum

lactucella (Wallr.) P.D. Sell & C. West = Hieracium lactucella

- subsp. *acutisquama* (Naeg. & Peter) Sojak = Hieracium acutisquamum

- subsp. *amaureilema* (Naeg. & Peter) Sojak = Hieracium amaureilema

- subsp. *lithuanica* (Naeg. & Peter) Sojak = Hieracium lithuanicum

- subsp. *magnauricula* (Naeg. & Peter) Sojak = Hieracium magnauricula

- subsp. *melaneilema* (Peter) Sojak = Hieracium melaneilema

- subsp. *tricheilema* (Naeg. & Peter) Sojak = Hieracium tricheilema

ladogensis Norrl. = Hieracium x fennicum

x *lamprocoma* (Naeg. & Peter) Schljak. = Hieracium x lamprocomum

laschii F. Schultz & Sch. Bip. = Hieracium x kalksburgensis

laticeps Norrl. = Hieracium laticeps

leptadenia (Dahlst.) Schljak. = Hieracium leptadenium

leptocaulon (Naeg. & Peter) Schljak. = Hieracium leptocaulon

x *leptoclados* (Peter) Sojak = Hieracium x leptoclados

leptothyrsa (Peter) Schljak. = Hieracium leptothyrsum

x *leptothyrsoides* (Zahn) Schljak. = Hieracium x leptothyrsoides

levieri (Peter) Sojak = Hieracium levieri

limenyensis (Zahn) Schljak. = Hieracium limenyense

lithuanica (Naeg. & Peter) Schljak. = Hieracium lithuanicum

x *lobarzewskii* (Rehm.) Sojak = Hieracium x lobarzewskii

x *longa* (Naeg. & Peter) Schljak. = Hieracium x longum

x *longisquama* (Peter) Holub = Hieracium x longisquamum

lyccensis (Naeg. & Peter) Schljak. = Hieracium lyccense

lychnaea Norrl. = Hieracium lychnaeum

x *macranthela* (Naeg. & Peter) Sojak = Hieracium x macranthelum

macrocyma (Naeg. & Peter) Schljak. = Hieracium macrocymum

x *macrostolona* (G. Schneid.) Sojak = Hieracium x macrostolonum

- subsp. *cernuiformis* (Naeg. & Peter) Sojak = Hieracium x macrostolonum

x *macrotricha* (Boiss.) F. Schultz & Sch. Bip. = Hieracium x macrotrichum

magnauricula (Naeg. & Peter) Dostal = Hieracium magnauricula *malacotricha* (Naeg. & Peter) Schljak. = Hieracium malacotrichum

marginalis (Naeg. & Peter) Schljak. = Hieracium marginale

x *maschukensis* (Litv. & Zahn) Sojak = Hieracium x maschukense

maurocybe (Juxip) Schljak. = Hieracium maurocybe

megalomastix (Naeg. & Peter) Schljak. = Hieracium megalomastix

melachaeta (Tausch) Schljak. = Hieracium melachaetum

melaneilema (Peter) Schljak. = Hieracium melaneilema

melanophaea Norrl. = Hieracium melanophaeum

x *melinomelas* (Peter) Holub = Hieracium x melinomelas

mnooclada (Rehm.) Schljak. = Hieracium mnoocladum

x *mollicaulis* (Vuk.) Sojak = Hieracium x mollicaule

- subsp. *anoclada* (Naeg. & Peter) Sojak = Hieracium x mollicaule

- subsp. *bauhiniflora* (Naeg. & Peter) Sojak = Hieracium x mollicaule

- subsp. *discolor* (Naeg. & Peter) Sojak = Hieracium x mollicaule

- subsp. *leptophyton* (Naeg. & Peter) Sojak = Hieracium x mollicaule

molliseta (Naeg. & Peter) Schljak. = Hieracium mollisetum

myriothricha (Rehm.) Schljak. = Hieracium myriothrichum

neriodowae (Woloszcz. & Zahn) Schljak. = Hieracium neriodowae

nigriseta (Naeg. & Peter) Schljak. = Hieracium nigrisetum

nigrosilvarum (Juxip) Schljak. = Hieracium nigrosilvarum

norrliniiformis (Pohle & Zahn) Sojak = Hieracium x dimorphoides

obornyana (Naeg. & Peter) Sojak = Hieracium x polymastix

- subsp. *polymastix* (Peter) Sojak = Hieracium x polymastix Peter

obscura (Reichenb.) Sojak = Hieracium obscurum

- subsp. *aquilonaris* (Naeg. & Peter) Sojak = Hieracium aquilonare

obscuribracteata (Naeg. & Peter) Schljak. = Hieracium obscuribracteatum

obsistens (Norrl.) Schljak. = Hieracium obsistens

officinarum F. Schultz & Sch. Bip. = Hieracium pilosella

- subsp. *micradenia* (Naeg. & Peter) P.D. Sell & C. West = Hieracium pilosella subsp. micradenium

- subsp. *tricholepia* (Naeg. & Peter) P.D. Sell & C. West = Hieracium pilosella subsp. tricholepium

onegensis Norrl. = Hieracium onegense

pachylodes (Naeg. & Peter) Sojak = Hieracium x longisquamum

x *pannoniciformis* (Litv. & Zahn) Sojak = Hieracium x pannoniciforme

x *paragoga* (Naeg. & Peter) Sojak = Hieracium x paragogum

parvistolona (Naeg. & Peter) Schljak. = Hieracium parvistolonum

pilipes (T. Sael.) Norrl. = Hieracium x dubium

- subsp. *acrocoma* (Naeg. & Peter) Sojak = Hieracium x dubium

pilisquama (Naeg. & Peter) Dostal = Hieracium pilisquamum

x *piloselliflora* (Naeg. & Peter) Sojak = Hieracium x piloselliflorum

piloselloides (Vill.) Sojak subsp. *megalomastix* (Naeg. & Peter) P.D. Sell & C. West = Hieracium megalomastix

x *plaicensis* (Woloszcz.) Sojak = Hieracium x plaicense

plicatula (Zahn) Schljak. = Hieracium plicatulum

x *podolica* (Rehm.) Schljak. = Hieracium x podolicum

x *polioderma* (Dahlst.) Sojak = Hieracium x poliodermum

polyanthema (Naeg. & Peter) Schljak. = Hieracium polyanthemum

x *polymastix* (Peter) Holub = Hieracium x polymastix

polymnoon (Naeg. & Peter) Schljak. = Hieracium polymnoon

praealta (Vill. ex Gochn.) F. Schultz & Sch. Bip. = Hieracium praealtum

- subsp. *arvorum* (Naeg. & Peter) P.D. Sell & C. West = Hieracium arvorum

- subsp. *stellata* (Tausch ex Zahn) Sojak = Hieracium stellatum

procera (Fries) F. Schultz & Sch. Bip. = Hieracium procerum

proceriformis (Naeg. & Peter) Sojak = Hieracium proceriforme

x *progenita* Norrl. = Hieracium x progenitum

x *prussica* (Naeg. & Peter) Sojak = Hieracium x prussicum

przybyslawskii (Rehm.) Schljak. = Hieracium przybyslawskii

pseudauriculoides (Naeg. & Peter) Schljak. = Hieracium pseudauriculoides

pseudokajanensis Schljak. = Hieracium pseudokajanense

pseudomegalomastix (Rehm.) Schljak. = Hieracium pseudomegalomastix

x *pseudopiloselliflora* (Rehm.) Sojak = Hieracium x pseudopiloselliflorum

pseudosparsa (Zahn) Schljak. = Hieracium pseudosparsum

pseudothaumasia (Zahn) Schljak. = Hieracium pseudothaumasium

rawaruskana (Zahn) Schljak. = Hieracium rawaruskanum

rojowskii (Rehm.) Schljak. = Hieracium rojowskii

x *rothiana* (Wallr.) F. Schultz & Sch.Bip. = Hieracium x rothianum

roxolanica (Rehm.) Sojak = Hieracium x roxolanicum

- subsp. *rehmannii* (Naeg. & Peter) Sojak = Hieracium x roxolanicum

- subsp. *rubricymigera* (Naeg. & Peter) Sojak = Hieracium x roxolanicum

ruprechtii (Boiss.) Dostal = Hieracium tephrocephalum

ruprechtii (Boiss.) P.D. Sell & C. West = Hieracium tephrocephalum

saeva (Rehm.) Schljak. = Hieracium saevum

samarica (Zahn) Schljak. = Hieracium samaricum

sanii (Naeg. & Peter) Schljak. = Hieracium sanii

x *scandinavica* (Dahlst.) Schljak. = Hieracium x scandinavicum

schelkownikowii (Zahn) Sojak = Hieracium x schelkownikowii

x *schultesii* (F. Schultz) F. Schultz & Sch. Bip. = Hieracium x schultesii

- subsp. *mendelii* (Naeg. & Peter) Sojak = Hieracium x schultesii

x *sciadophora* (Naeg. & Peter) Sojak = Hieracium x sciadophorum

scotodes (Norrl.) Schljak. = Hieracium scotodes

sedutrix (Rehm.) Schljak. = Hieracium sedutrix

semilitorea (Norrl.) Schljak. = Hieracium semilitoreum

septentrionalis Norrl. = Hieracium septentrionale

setosopetiolata (Rehm.) Schljak. = Hieracium setosopetiolatum

signifera Norrl. = Hieracium signiferum

sororians (Norrl.) Schljak. = Hieracium sororians

spathophyllopsis (Zahn) Schljak. = Hieracium spathophyllopsis

spuria (Chaix ex Froel.) Sojak = Hieracium x kalksburgensis

stellata (Tausch ex Zahn) Schljak. = Hieracium stellatum

x *stoloniflora* (Waldst. & Kit.) Fries = Hieracium x stoniflorum
subaurantiaca (Naeg. & Peter) Schljak. = Hieracium
  subaurantiacum
subcymigera (Peter) Sojak = Hieracium subcymigerum
subfilifera (Zahn) Schljak. = Hieracium subfiliferum
submacrolepis Schljak. = Hieracium macrolepium
subpratensis Norrl. = Hieracium x fennicum
substolonifera (Peter) Sojak = Hieracium substoloniferum
suchonensis (Norrl.) Schljak. = Hieracium suchonense
sudetorum (Peter) Dostal = Hieracium sudetorum
x *suecica* (Fries) F. Schultz & Sch. Bip. = Hieracium x
  suecicum
x *sulphurea* (Doell) F. Schultz & Sch. Bip. = Hieracium x
  sulphureum
suomensis Norrl. = Hieracium suomense
syrjaenorum Norrl. = Hieracium syrjaenorum
sysolskiensis (Zahn) Schljak. = Hieracium sysolskiense
tabergensis (Dahlst.) Schljak. = Hieracium tabergense
tenacicaulis (Norrl.) Schljak. = Hieracium tenacicaule
tenebricans (Norrl.) Schljak. = Hieracium tenebricans
tephrocephala (Vuk.) Sojak = Hieracium tephrocephalum
thaumasia (Peter) Dostal = Hieracium thaumasium
thaumasioides (Peter) Schljak. = Hieracium thaumasioides
tjapomensis (Norrl.) Schljak. = Hieracium tjapomense
tricheilema (Naeg. & Peter) Schljak. = Hieracium tricheilema
trichocymosa (Zahn) Schljak. = Hieracium trichocymosum
trichocymosoides (Juxip) Schljak. = Hieracium trichocymosoides
x *tubulata* (Vollm.) Sojak = Hieracium x tubulatum
vaillantii (Tausch) Sojak = Hieracium vaillantii
- subsp. *cymigera* (Reichenb.) Sojak = Hieracium vaillantii
- subsp. *suomensis* (Norrl.) Sojak = Hieracium suomense
varatinensis (Woloszcz.) Schljak. = Hieracium varatinense
velutina (Hegetschw.) F. Schultz & Sch. Bip. = Hieracium
  velutinum
verruculata (Link) Sojak = Hieracium verruculatum
viscidula (Tausch) Schljak. = Hieracium viscidulum
volhynica (Naeg. & Peter) Schljak. = Hieracium volhynicum
vulpina (Rehm.) Schljak. = Hieracium vulpinum
x *wolgensis* (Zahn) Sojak = Hieracium x wolgense
woronowiana (Zahn) Sojak = Hieracium x maschukense
x *ziziana* (Tausch) F. Schultz & Sch. Bip. = Hieracium x
  zizianum
- subsp. *leptophylla* (Naeg. & Peter) Sojak = Hieracium x
  zizianum

Pilostemon Iljin

filifolius (C. Winkl.) Iljin - 6
karategini (Lipsky) Iljin - 6

Plagiobasis Schrenk

centauroides Schrenk - 6

Podospermum DC. = Scorzonera

alpigenum C. Koch = Scorzonera alpigena
bicolor (Freyn & Sint.) Kuth. = Scorzonera bicolor
grossheimii (Lipsch. & Vass.) Kuth. = Scorzonera grossheimii
kirpicznikovii (Lipsch.) Kuth. = Scorzonera kirpicznikovii
lipschitzii Kuth. = Scorzonera lipschitzii
schischkinii (Lipsch. & Vass.) Kuth. = Scorzonera schischkinii
tenuisectum Grossh. & Sosn. = Scorzonera tenuisecta

Pojarkovia Asker p.p. = Adenostyles and Senecio

macrophylla (Bieb.) Asker. = Adenostyles macrophylla
pauciloba (DC.) Asker. = Senecio paucilobus
platyphylloides (Somm. & Levier) Asker. = Adenostyles
  platyphylloides
stenocephala (Boiss.) Asker. = Senecio pojarkovae

Polychrysum (Tzvel.) Kovalevsk.

tadshikorum (Kudr.) Kovalevsk. (*Cancrinia tadshikorum*
  (Kudr.) Tzvel.) - 6

Polytaxis Bunge

lehmannii Bunge - 6
pulchella Rassulova & B. Scharipova - 6
winkleri Iljin - 6

Prenanthes L.

abietina (Boiss. & Bal.) Kirp. - 2
angustifolia Boulos - 3, 4
aspera Schrad. ex Willd. = Chondrilla aspera
blinii (Levl.) Kitag. (*Lactuca blinii* Levl., *Prenanthes
  maximowiczii* Kirp.) - 5
cacaliifolia (Bieb.) Beauverd = Cicerbita macrophylla
cacaliifolia sensu Kirp. = Cicerbita petiolata
maximowiczii Kirp. = P. blinii
pontica (Boiss.) Leskov = Cicerbita pontica
purpurea L. - 1, 2
tatarinowii Maxim. - 5

Psephellus Cass. = Centaurea

absinthifolius Galushko = Centaurea absinthifolia
andinus Galushko & Alieva = Centaurea andina
annae Galushko = Centaurea annae
buschiorum Sosn. = Centaurea buschiorum
cabardensis G. Koss ex Tschuchrukidze = Centaurea
  cabardensis
circassicus (Albov) Galushko = Centaurea circassica
ciscaucasicus (Sosn.) Galushko = Centaurea ciscaucasica
czerepanovii Alieva = Centaurea czerepanovii
galushkoi Alieva = Centaurea galushkoi
holargyreus Bornm. = Centaurea erivanensis subsp. holargyrea
holophyllus Socz. & Lipat. = Centaurea holophylla
kacheticus Rehm. = Centaurea kachetica
kemulariae Charadze = Centaurea kemulariae
lactiflorus G. Koss ex Tschuchrukidze = Centaurea baksanica
nogmovii G. Koss ex Tschuchrukidze = Centaurea nogmovii
prokhanovii Galushko = Centaurea prokhanovii
schistosus (Sosn.) Alieva = Centaurea schistosa
sosnowskyi Tschuchrukidze = Centaurea dealbata

Pseudognaphalium auct. = Gnaphalium

luteo-album (L.) O.M. Hilliard & B.L. Burtt = Gnaphalium
  luteo-album
- subsp. *affine* (D. Don) O.M. Hilliard & B.L. Burtt =
  Gnaphalium affine

Pseudohandelia Tzvel.

umbellifera (Boiss.) Tzvel. - 6

Pseudolinosyris Novopokr.

grimmii (Regel & Schmalh.) Novopokr. (*Crinitaria grimmii* (Regel & Schmalh.) Grierson) - 6
microcephala (Novopokr.) Tamamsch. - 6
sintenisii (Bornm.) Tamamsch. - 6

Pseudopodospermum (Lipsch. & Krasch.) Kuth. = Scorzonera

aragatzi (Kuth.) Kuth. = Scorzonera aragatzi
leptophyllum (DC.) Kuth. = Scorzonera leptophylla
molle (Bieb.) Kuth. = Scorzonera mollis
suberosum (C. Koch.) Kuth. = Scorzonera suberosa
szovitzii (DC.) Kuth. = Scorzonera szovitzii
turkeviczii (Krasch. & Lipsch.) Kuth. p.p. = Scorzonera mollis

Psychrogeton Boiss. = Erigeron

alexeenkoi Krasch. = Erigeron alexeenkoi
amorphoglossus (Boiss.) Novopokr. = Erigeron amorphoglossus
andryaloides (DC.) Novopokr. ex Krasch. = Erigeron andryaloides
- var. *poncinsii* (Franch.) Grierson = Erigeron poncinsii
aucheri (DC.) Grierson, p.p. = Erigeron khorassanicus
biramosus (Botsch.) Grierson = Erigeron biramosus
brachyspermus (Botsch.) Grierson = Erigeron brachyspermus
cabulicus Boiss. = Erigeron cabulicus
leucophyllus (Bunge) Novopokr. = Erigeron amorphoglossus
mollissimus (Rech. fil. & Koeie) Ikonn. = Erigeron amorphoglossus
nabidshonii (R. Kam.) Koczk. & Zhogoleva = Erigeron nabidshonii
nigromontanus (Boiss. & Buhse) Grierson = Erigeron nigromontanus
obovatus (Benth.) Grierson = Rhinactinidia popovii
olgae (Regel & Schmalh.) Novopokr. ex Nevski = Erigeron olgae
poncinsii (Franch.) Ling & Y.L. Chen = Erigeron poncinsii
primuloides (M. Pop.) Grierson = Erigeron primuloides

Ptarmica Hill

acuminata Ledeb. (*Achillea acuminata* (Ledeb.) Sch. Bip.) - 4, 5
alpina (L.) DC. (*Achillea alpina* L., *A. alpina* var. *mongolica* (Fisch. ex Spreng.) Kitag., *A. mongolica* Fisch. ex Spreng., *Ptarmica mongolica* (Fisch. ex Spreng.) DC.) - 4, 5
biserrata (Bieb.) DC. (*Achillea biserrata* Bieb.) - 2
borysthenica Klok. & Sakalo = P. salicifolia
camtschatica (Rupr. ex Heimerl) Kom. (*Achillea alpina* L. subsp. *camtschatica* (Heimerl) Kitam., *A. camtschatica* (Rupr. ex Heimerl) Botsch.) - 4, 5
cartilaginea (Ledeb. ex Reichenb.) Ledeb. (*Achillea cartilaginea* Ledeb. ex Reichenb., *A. ptarmica* L. subsp. *cartilaginea* (Ledeb. ex Reichenb.) Heimerl, *Ptarmica salicifolia* (Bess.) Serg. subsp. *cartilaginea* (Ledeb. ex Reichenb.) Tzvel.) - 1, 3, 4, 5(alien)
griseo-virens (Albov) Galushko (*Achillea griseo-virens* Albov) - 2
impatiens (L.) DC. (*Achillea impatiens* L.) - 3, 4
japonica (Heimerl) Worosch. (*Achillea japonica* Heimerl, *A. alpina* L. subsp. *japonica* (Heimerl) Kitam., *Ptarmi-*

ca japonica (Heimerl) Klok. & Krytzka, comb. superfl.) - 5
ledebourii (Heimerl) Klok. & Krytzka (*Achillea ledebourii* Heimerl) - 3
lingulata (Waldst. & Kit.) DC. (*Achillea lingulata* Waldst. & Kit.) - 1
macrocephala (Rupr.) Kom. (*Achillea macrocephala* Rupr.) - 5
mongolica (Fisch. ex Spreng.) DC. = P. alpina
ptarmicifolia (Willd.) Galushko (*Achillea ptarmicifolia* (Willd.) Rupr. ex Heimerl) - 2
ptarmicoides (Maxim.) Worosch. (*Achillea ptarmicoides* Maxim., *A. alpina* L. subsp. *alpina* var. *discoidea* (Regel) Kitam.) - 4, 5
sachokiana (Sosn.) Klok. & Krytzka (*Achillea sachokiana* Sosn.) - 2
salicifolia (Bess.) Serg. (*Achillea salicifolia* Bess., *Ptarmica borysthenica* Klok. & Sakalo, *P. salicifolia* subsp. *borysthenica* (Klok. & Sakalo) Tzvel.) - 1, 2, 3, 6
- subsp. *borysthenica* (Klok. & Sakalo) Tzvel. = P. salicifolia
- subsp. *cartilaginea* (Ledeb. ex Reichenb.) Tzvel. = P. cartilaginea
- subsp. *septentrionalis* (Serg.) Tzvel. = P. septentrionalis
- var. *septentrionalis* (Serg.) Tzvel. = P. septentrionalis
sedelmeyeriana (Sosn.) Klok. (*Achillea sedelmeyeriana* Sosn.) - 2
septentrionalis (Serg.) Klok. & Krytzka (*Achillea salicifolia* Bess. subsp. *septentrionalis* (Serg.) Uotila, *A. septentrionalis* (Serg.) Botsch., *Ptarmica salicifolia* (Bess.) Serg. var. *septentrionalis* (Serg.) Tzvel., *P. salicifolia* subsp. *septentrionalis* (Serg.) Tzvel.) - 1, 3
tenuifolia (Schur) Schur (*Anthemis tenuifolia* Schur, 1851, non *Achillea tenuifolia* Lam. 1783, *Achillea schurii* Sch. Bip., *A. oxyloba* (DC.) Sch. Bip. subsp. *schurii* (Sch. Bip.) Heimerl) - 1
vulgaris Hill (*Achillea ptarmica* L., *Ptarmica vulgaris* Blakw. ex DC.) - 1, 3(alien)

Pterocypsela Shih = Lactuca

elata (Hemsl.) Shih = Lactuca elata
indica (L.) Shih = Lactuca indica
raddeana (Maxim.) Shih = Lactuca raddeana
triangulata (Maxim.) Shih = Lactuca triangulata

Ptilostemon auct. = Lamyra

echinocephalus (Willd.) Greuter = Lamyra echinocephala

Pulicaria Gaertn.

arabica (L.) Cass. (*Inula arabica* L., *Pulicaria uliginosa* Stev. ex DC. var. *stenophylla* (Boiss.) Golubk.) - 6
dysenterica (L.) Bernh. (*P. uliginosa* Stev. ex DC.) - 1, 2, 6
gnaphalodes (Vent.) Boiss. - 6
prostrata (Gilib.) Aschers. = P. vulgaris
salviifolia Bunge - 6
uliginosa Stev. ex DC. = P. dysenterica
- var. *stenophylla* (Boiss.) Golubk. = P. arabica
vulgaris Gaertn. (*P. prostrata* (Gilib.) Aschers. nom. illegit.) - 1, 2, 3, 5, 6

Pyrethrum Zinn (*Balsamita* Mill.)

abrotanifolium Bunge ex Ledeb. - 3, 6
alatavicum (Herd.) O. & B. Fedtsch. - 3, 6
alpinum (L.) Schrank = Leucanthemopsis alpina
arassanicum (C. Winkl.) O. & B. Fedtsch. = P. djilgense
arctodzhungaricum Golosk. - 6
aromaticum Tzvel. = P. daghestanicum

■ aucherianum DC.

balsamita (L.) Willd. (*Balsamita balsamitoides* (Sch. Bip.) Tzvel., *Tanacetum balsamita* L. subsp. *balsamitoides* (Sch. Bip.) Grierson) - 2

balsamitoides (Nabel.) Tzvel. (*Tanacetum balsamitoides* (Nabel.) Chandjian) - 2

carneum Bieb. (*Tanacetum coccineum* (Willd.) Grierson subsp. *carneum* (Bieb.) Grierson, *Pyrethrum coccineum* sensu Tzvel., *P. roseum* sensu Tzvel.) - 2

chamaemelifolium (Somm. & Levier) Sosn. (*Tanacetum coccineum* (Willd.) Grierson subsp. *chamaemelifolium* (Somm. & Levier) Griereson) - 2

*****cinerariifolium** Trev. - 1, 2, 6

clusii Fisch. ex Reichenb. (*Tanacetum clusii* (Reichenb.) Sojak, *T. corymbosum* (L.) Sch. Bip. subsp. *clusii* (Fisch. ex Reichenb.) Heywood) - 1

coccineum sensu Tzvel. = P. carneum

coccineum (Willd.) Worosch. (*P. roseum* (Adams) Bieb., *Tanacetum coccineum* (Willd.) Grierson, p.p.) - 2

corymbiforme Tzvel. - 6

corymbosum (L.) Scop. - 1, 2, 3

daghestanicum (Rupr. ex Boiss.) Fler. (*Matricaria daghestanica* (Rupr. ex Boiss.) Rauschert, *Pyrethrum aromaticum* Tzvel.) - 2

demetrii Manden. - 2

djilgense (Franch.) Tzvel. (*P. arassanicum* (C. Winkl.) O. & B. Fedtsch., *Tanacetum djilgense* (Franch.) Podlech) - 6

dolomiticum Galushko - 2

fruticulosum Biehl. - 2

galae M. Pop. - 6

galushkoi Prima - 2

glanduliferum Somm. & Levier - 2

grossheimii Sosn. (*Tanacetum grossheimii* (Sosn.) Muradjan) - 2, 6

■ heldereichianum Fenzl

hissaricum Krasch. - 6

karelinii Krasch. - 6

kelleri (Kryl. & Plotn.) Krasch. (*Tanacetum kelleri* (Kryl. & Plotn.) Takht.) - 3

komarovii Sosn. (*Tanacetum zangezuricum* Chandjian) - 2

kotschyi Boiss. (*Tanacetum kotschyi* (Boiss.) Grierson) - 2

krylovianum Krasch. - 3

kubense Grossh. (*Tanacetum kubense* (Grossh.) Muradjan) - 2

lanuginosum (Sch. Bip. & Herd.) Tzvel. - 4

leontopodium (C. Winkl.) Tzvel. - 6

leptophyllum Stev. ex Bieb. - 2

*****macrophyllum** (Waldst. & Kit.) Willd. - 1, 2

*****majus** (Desf.) Tzvel. - 1, 2, 6

marionii Albov - 2

mikeschinii Tzvel. (*Tanacetum mikeschinii* (Tzvel.) Takht.) - 6

neglectum Tzvel. - 6

ordubadense Manden. - 2

■ oxylepis (Bordz.) Tzvel. (*Tanacetum oxylepis* (Bordz.) Grierson)

parthenifolium Willd. - 1, 2, 6

*****parthenium** (L.) Smith - 1, 2, 3, 4, 5, 6

peucedanifolium (Sosn.) Manden. - 2

poteriifolium Ledeb. (*Tanacetum poteriifolium* (Ledeb.) Grierson) - 2

pulchellum Turcz. ex DC. - 3, 4

pulchrum Ledeb. - 3

punctatum (Desr.) Bordz. ex Grossh. & Schischk. (*Tanacetum punctatum* (Desr.) Grierson) - 2

pyrethroides (Kar. & Kir.) B. Fedtsch. ex Krasch. (*Tanace-tum pyrethroides* (Kar. & Kir.) Muradjan, comb. superfl., *T. pyrethroides* (Kar. & Kir.) Podlech, comb. superfl., *T. pyrethroides* (Kar. & Kir.) Sch.Bip.) - 3

roseum (Adams) Bieb. = P. coccineum

roseum sensu Tzvel. = P. carneum

semenovii (Herd.) C. Winkl. ex O. & B. Fedtsch. - 6

sericeum (Adams) Bieb. - 2

sevanense Sosn. - 2, 6

silaifolium Stev. - 2

songaricum Tzvel. - 6

sorbifolium Boiss. (*Tanacetum sorbifolium* (Boiss.) Grierson) - 2

sovetkinae Kovalevsk. - 6

tianschanicum Krasch. = Xylanthemum tianschanicum

transiliense (Herd.) Regel & Schmalh. - 6

tricholobum Sosn. ex Manden. - 2

Reichardia Roth

dichotoma (Vahl) Freyn, p.p. = R. glauca

glauca Matthews (*R. dichotoma* (Vahl) Freyn, p.p. excl. basionymo) - 2

Rhabdotheca Cass.

korovinii (M. Pop.) Kirp. - 6

Rhagadiolus Hill

angulosus (Jaub. & Spach) Kupicha = Garhadiolus angulosus

edulis Gaertn. (*R. stellatus* auct.) - 1, 2

hebelaenus (DC.) Vass. - 2

stellatus auct. = R. edulis

Rhaponticum auct. = Stemmacantha

aulieatense Iljin = Stemmacantha aulieatensis

carthamoides (Willd.) Iljin = Stemmacantha carthamoides

- subsp. *orientale* (Serg.) Sosk. = Stemmacantha orientalis

chamarense Peschkova = Stemmacantha chamarensis

integrifolium C. Winkl. = Stemmacantha integrifolia

karatavicum Regel & Schmalh. = Stemmacantha karatavica

lyratum C. Winkl. ex Iljin = Stemmacantha lyrata

namanganicum Iljin = Stemmacantha namanganica

nanum Lipsky = Stemmacantha nana

nitidum Fisch. = Stemmacantha nitida

orientale (Serg.) Peschkova = Stemmacantha orientalis

pulchrum Fisch. & C.A. Mey. = Stemmacantha pulchra

satzyperovii Sosk. = Stemmacantha satzyperovii

serratuloides (Georgi) Bobr. = Stemmacantha serratuloides

uniflorum (L.) DC. = Stemmacantha uniflora

- subsp. *satzyperovii* (Sosk.) Worosch. = Stemmacantha satzyperovii

zardabii Rzazade = Stemmacantha pulchra subsp. zardabii

Rhinactina Less. = Rhinactinidia

Rhinactinidia Novopokr. (*Borkonstia* Ignatov, nom. superfl., *Krylovia* Schischk. ex Tamamsch. 1959, non Chachlov, 1939, *Rhinactina* Less. 1831, non Willd. 1807)

eremophila (Bunge) Botsch. (*Borkonstia eremophila* (Bunge) Ignatov, *Krylovia eremophila* (Bunge) Schischk. ex Tamamsch.) - 3, 6

limoniifolia (Less.) Botsch. (*Borkonstia limoniifolia* (Less.) Ignatov, *Krylovia limoniifolia* (Less.) Schischk. ex Tamamsch.) - 3, 6

novopokrovskyi (Krasch. & Iljin) Botsch. (*Borkonstia novopokrovskyi* (Krasch. & Iljin) Ignatov, *Krylovia novopokrovskyi* (Krasch. & Iljin) Tamamsch.) - 4
popovii (Botsch.) Botsch. (*Borkonstia popovii* (Botsch.) Ignatov, *Krylovia popovii* (Botsch.) Tamamsch., *Psychrogeton obovatus* (Benth.) Grierson, p.p.) - 6

*Rudbeckia L.

*hirta L. - 1, 2, 5, 6
*laciniata L. - 1, 6
*speciosa Wend. - 1

Russowia C. Winkl.

sogdiana (Bunge) B. Fedtsch. - 6

Santolina auct. = Lepidotheca

suaveolens Pursh = Lepidotheca suaveolens

*Santolina L.

*chamaecyparissus L. - 1, 2
*virens Mill. (*Santolina viridis* Willd.) - 1, 2
viridis Willd. = S. virens

*Sanvitalia Lam.

*procumbens Lam. - 1

Saussurea DC. (*Frolovia* Lipsch., *Theodorea* Cass.)

acuminata Turcz. - 4
- var. *duiensis* (Fr. Schmidt) Worosch. = S. duiensis
ajanensis (Regel) Lipsch. - 5
alata DC. (*S. crepidifolia* Turcz. p.max.p., *Theodorea crepidifolia* (Turcz.) Sojak) - 4
alberti Regel & C. Winkl. - 6
alpina (L.) DC. - 1, 3, 4, 6
- subsp. *esthonica* (Baer ex Rupr.) Kupff. = S. esthonica
amara (L.) DC. (*Theodorea glomerata* (Poir.) Sojak) - 1, 3, 4, 6
ambigua Kryl. ex Serg. = S. pseudoalpina
amurensis Turcz. - 4, 5
angustifolia (Willd.) DC. - 5
arctecapitulata Lipsch. - 4
asbukinii Iljin (*S. froloviaeformis* E. Nikit.) - 6
baicalensis (Adams) Robins. - 3, 4
blanda Schrenk - 6
caespitans Iljin - 6
cana Ledeb. - 3, 4, 6
■ **canescens** C. Winkl.
caprifolia Iljin & Zapr. - 6
carduicephala (Iljin) Iljin - 6
ceterachifolia Lipsch. - 4
chamarensis Peschkova - 4
chondrilloides C. Winkl. - 6
congesta Turcz. - 4
- subsp. *poljakowii* (Glehn) Lipsch. = S. poljakowii
- subsp. *soczavae* (Lipsch.) Worosch. = S. soczavae
controversa DC. - 1, 3, 4
coronata Schrenk - 6
crepidifolia Turcz. = S. alata and S. runcinata
czichaczevii Maneev & Krasnob. - 4
davurica Adams - 3, 4
discolor (Willd.) DC. - 1
dorogostaiskii Palib. - 4

dubia Freyn - 4, 5
duiensis Fr. Schmidt (*S. acuminata* Turcz. var. *duiensis* (Fr.Schmidt) Worosch.) - 5
elata Ledeb. - 3, 6
elegans Ledeb. - 3, 6
elongata DC. - 4
- subsp. *recurvata* (Maxim.) Worosch. = S. recurvata
esthonica Baer ex Rupr. (*S. alpina* (L.) DC. subsp. *esthonica* (Baer ex Rupr.) Kupff.) - 1
faminziniana Krasn. - 6
fauriei Franch. - 5
firma (Kitag.) Kitam. (*S. ussuriensis* Maxim. var. *firma* Kitag.) - 5
foliosa Ledeb. - 3, 4
froloviaeformis E. Nikit. = S. asbukinii
frolowii Ledeb. - 3, 4, 6
fulcrata A. Khokhr. & Worosch. - 5
gilesii Hemsl. (*Frolovia gilesii* (Hemsl.) B. Scharipova) - 6
glabriuscula A. Khokhr. = S. tilesii var. glabriusula
glacialis Herd. - 3, 6
gnaphalodes (Royle) Sch. Bip. - 6
grandifolia Maxim. - 5
grubovii Lipsch. (*S. x paradoxa* Lipsch.) - 3
hypargyrea Lipsch. & Vved. - 4
involucrata (Kar. & Kir.) Sch. Bip. - 6
ispajensis Iljin - 6
jadrinzevii Krylov - 3
kabadiana Rassulova & B. Scharipova - 6
karaartscha Saposhn. - 6
kaschgarica Rupr. - 6
kitamurana Miyabe & Tatew. - 5
kolesnikovii A. Khokhr. & Worosch. - 5
krylovii Schischk. & Serg. - 3, 4, 6(?)
kurilensis Tatew. - 5
kuschakewiczii C. Winkl. - 6
laciniata Ledeb. - 3, 6
larionowii C. Winkl. (*Theodorea larionowii* (C. Winkl.) Sojak) - 6
latifolia Ledeb. - 3, 4, 6
lenensis M. Pop. ex Lipsch. - 4
leucophylla Schrenk - 3, 4, 6
lipschitzii Filat. - 6
■ **lomatolepis** Lipsch.
manshurica Kom. - 5
masarica Lipsky - 6
maximowiczii Herd. - 5
mikeschinii Iljin - 6
neopulchella Lipsch. (*Theodorea neopulchella* (Lipsch.) Sojak) - 5
neoserrata Nakai (*S. parviflora* (Poir.) DC. subsp. *neoserrata* (Nakai) Worosch.) - 4, 5
ninae Iljin - 6
nuda Ledeb. - 5
- var. *oxyodonta* (Hult.) Worosch. = S. oxyodonta
nupuripoensis Miyabe & Miyake - 5
odontolepis (Herd.) Sch. Bip. ex Maxim. - 5
orgaadayi V. Khan. & Krasnob. - 4
ovata Benth. - 6
oxyodonta Hult. (*S. nuda* Ledeb. var. *oxyodonta* (Hult.) Worosch.) - 5
x paradoxa Lipsch. = S. grubovii
parviflora (Poir.) DC. - 1, 3, 4
- subsp. *neoserrata* (Nakai) Worosch. = S. neoserrata
- subsp. *purpurata* (Fisch. ex Herd.) Lipsch. = S. purpurata
petiolata Kom. ex Lipsch. - 4
poljakowii Glehn (*S. congesta* Turcz. subsp. *poljakowii* (Glehn) Lipsch.) - 4

porcellanea Lipsch. - 5
porcii Degen - 1
pricei auct. = S. subacaulis
pricei Simps. (*S. sumneviczii* Serg.) - 3, 4
prostrata C. Winkl. - 6
pseudoalpina Simps. (*S. ambigua* Kryl. ex Serg.) - 3
pseudoangustifolia Lipsch. - 4, 5
pseudoblanda Lipsch. ex Filat. - 6
pseudochondrilloides Rassulova & B. Scharipova - 6
pseudosquarrosa M. Pop. & Lipsch. - 4
pseudotilesii Lipsch. - 5
pulchella (Fisch.) Fisch. - 4, 5
■ **pulviniformis** C. Winkl.
purpurata (Fisch. ex Herd.) Lipsch. (*S. serrata* DC. var. *purpurata* Fisch. ex Herd., *S. parviflora* (Poir.) DC. subsp. *purpurata* (Fisch. ex Herd.) Lipsch.) - 4
recurvata (Maxim.) Lipsch. (*S. elongata* DC. subsp. *recurvata* (Maxim.) Worosch.) - 5
riederii Herd. - 5
- subsp. **yezoensis** (Maxim.) Kitam. - 5
rigida Ledeb. - 3
robusta Ledeb. - 3, 6
runcinata DC. (*S. crepidifolia* Turcz. p.min.p.) - 4
sachalinensis Fr. Schmidt - 5
sajanensis Gudoschn. - 4
salemannii C. Winkl. - 5
salicifolia (L.) DC. - 3, 4
salsa (Pall. ex Bieb.) Spreng. - 1, 2, 3, 4, 6
saxosa Lipsch. - 6
schanginiana (Wydl.) Fisch. ex Herd. - 3, 4, 6
serrata DC. var. *purpurata* Fisch. ex Herd. = S. purpurata
serratuloides Turcz. - 3
shiretokoensis Sugaw. - 5
sinuata Kom. - 5
soczavae Lipsch. (*S. congesta* Turcz. subsp. *soczavae* (Lipsch.) Worosch.) - 5
sordida Kar. & Kir. - 6
sovietica Kom. - 5
splendida Kom. - 5
squarrosa Turcz. - 4
stubendorffii Herd. - 3, 4
subacaulis (Ledeb.) Serg. (*S. pricei* auct.) - 3, 4
subtriangulata Kom. - 5
sukaczevii Lipsch. - 4
sulcata Iljin - 6
sumneviczii Serg. = S. pricei
tadshikorum Iljin & Gontsch. - 6
tilesii (Ledeb.) Ledeb. (*S. tschuktschorum* Lipsch.) - 3, 4, 5
- subsp. **putoranica** Ju. Kozhevn. - 4
- subsp. *viscida* (Hult.) Ju. Kozhevn. = S. viscida
- var. **glabriuscula** (A. Khokhr.) Korobkov (*S. glabriuscula* A. Khokhr.) - 5
tomentosa Kom. - 5
tomentosella A. Khokhr. - 5
triangulata Trautv. & C.A. Mey. - 5
tschuktschorum Lipsch. = S. tilesii
turgaiensis B. Fedtsch. - 1, 3, 6
umbrosa Kom. - 4, 5
x **uralensis** Lipsch. - S. alpina (L.) DC. x S. controversa DC. - 1
ussuriensis Maxim. - 5
- var. *firma* Kitag. = S. firma
viscida Hult. (*S. tilesii* (Ledeb.) Ledeb.) subsp. *viscida* (Hult.) Ju. Kozhevn.) - 5
vvedenskyi Lipsch. - 6

Scariola F.W. Schmidt

albertoregelia (C. Winkl.) Kirp. - 6
orientalis (Boiss.) Sojak - 2, 6
viminea (L.) F.W. Schmidt - 1, 2, 6

Schischkinia Iljin
albispina (Bunge) Iljin - 6

Schmalhausenia C. Winkl.
nidulans (Regel) Petrak - 6

Schumeria Iljin
latifolia (Boiss.) Iljin - 6
litwinowii (Iljin) Iljin - 6

Scolymus L.
hispanicus L. - 1, 2
maculatus L. - 1

Scorzonera L. (*Podospermum* DC., *Pseudopodospermum* (Lipsch. & Krasch.) Kuth.)

acanthoclada Franch. (*S. kuhistanica* M. Pop.) - 6
alaica Lipsch. - 6
albertoregelia C. Winkl. - 6
albicaulis Bunge (*S. albicaulis* subsp. *macrosperma* (Turcz. ex DC.) Worosch., *S. macrosperma* Turch. ex DC.) - 4, 5
- subsp. *macrosperma* (Turcz. ex DC.) Worosch. = S. albicaulis
alpigena (C. Koch) Grossh. (*Podospermum alpigenum* C. Koch) - 2
aragatzi Kuth. (*Pseudopodospermum aragatzi* (Kuth.) Kuth.) - 2
armeniaca (Boiss. & Huet) Boiss. - 2
austriaca Willd. - 1, 3, 4, 6
- subsp. *crispa* (Bieb.) Nym. = S. crispa
baldshuanica Lipsch. (*S. tragopogonoides* Regel & Schmalh. subsp. *baldshuanica* (Lipsch.) Tagaev, comb. invalid.) - 6
bicolor Freyn & Sint. (*Podospermum bicolor* (Freyn & Sint.) Kuth., *Scorzonera incisa* auct.) - 2
biebersteinii Lipsch. - 1(?), 2
bracteosa C. Winkl. - 6
bungei Krasch. & Lipsch. - 6
calcitrapifolia Vahl (*S. laciniata* L. subsp. *calcitrapifolia* (Vahl) Maire) - 2
cana (C.A. Mey.) O. Hoffm. (*S. meyeri* (C. Koch) Lipsch.) - 1, 2, 3, 6
chantavica Pavl. = S. inconspicua
charadzeae Papava - 2
cinerea Boiss. - 2
circumflexa Krasch. & Lipsch. - 6
- subsp. **glaberrima** (Bunge) Tagaev (*S. tuberosa* Pall. var. *glaberrima* Bunge) - 6
- subsp. **oxiana** M. Pop. ex Tagaev (*S. oxiana* M. Pop. nom. invalid.) - 6
crassifolia Krasch. & Lipsch. (*S. ovata* Trautv. subsp. *crassifolia* (Krasch. & Lipsch.) Tagaev) - 6
crispa Bieb. (*S. austriaca* Willd. subsp. *crispa* (Bieb.) Nym.) - 1, 6
curvata (Popl.) Lipsch. - 4
czerepanovii R. Kam. nom. novum (*Leontodon lanatus* L. 1759, Amoen. Acad. 4 : 287; *Scorzonera czerepanovii* R. Kam. 1986, Bot. Zhurn. 71, 12 : 1680, nom. invalid.; *S. lanata* (L.) Hoffm. 1806, non Schrank, 1789) - 2
dianthoides (Lipsch. & Krasch.) Lipsch. - 3

dzhawakhetica Sosn. ex Grossh. (*S. sosnowskyi* Lipsch.) - 2
ensifolia Bieb. - 1, 3, 6
ferganica Krasch. - 6
filifolia Boiss. (*S. lipskyi* Lipsch.) - 2
franchetii Lipsch. - 6
gageoides Boiss. - 6
glabra Rupr. (*S. ruprechtiana* Lipsch. & Krasch. ex
    Lipsch.) - 1, 3, 4
x glastifolia Willd. - S. stricta Hornem. x S. taurica Bieb. -
    1
gorovanica Nazarova - 2
gracilis Lipsch. (*S. pubescens* DC. subsp. *gracilis* (Lipsch.)
    Tagaev) - 6
grigoraschvilii (Sosn.) Lipsch. - 2
grossheimii Lipsch. & Vass. (*Podospermum grossheimii*
    (Lipsch. & Vass.) Kuth.) - 2
*hispanica L. - 1
- subsp. *taurica* (Bieb.) Tzvel. = S. taurica
hissarica C. Winkl. - 6
humilis L. - 1
idae (Sosn.) Lipsch. - 2
ikonnikovii Lipsch. & Krasch. - 4
iliensis Krasch. = S. racemosa
incisa auct. = S. bicolor
inconspicua Lipsch. ex Pavl. (*S. chantavica* Pavl.) - 3, 6
karataviensis Kult. (*S. tau-saghyz* Lipsch. & Bosse subsp.
    karataviensis (Kult.) R. Kam.) - 6
ketzkhowelii Sosn. ex Grossh. - 2
kirpicznikovii Lipsch. (*Podospermum kirpicznikovii*
    (Lipsch.) Kuth.) - 2
kozlowskyi Sosn. ex Grossh. - 2
kuhistanica M. Pop. = S. acanthoclada
lachnostegia (Woronow) Lipsch. - 2
laciniata L. - 1, 2, 6
- subsp. *calcitrapifolia* (Vahl) Maire = S. calcitrapifolia
lanata (L.) Hoffm. = S. czerepanovii
latifolia (Fisch. & C.A. Mey.) DC. - 2
leptophylla (DC.) Grossh. (*Pseudopodospermum leptophyl-*
    lum (DC.) Kuth.) - 2
lipschitzii (Kuth.) Czer. (*Podospermum lipschitzii* Kuth.) - 2
lipskyi Lipsch. = S. filifolia
litwinowii Krasch. & Lipsch. - 6
longipes Kult. (*S. tau-saghyz* Lipsch. & Bosse subsp. *lon-*
    gipes (Kult.) R. Kam.) - 6
- var. *tautaryensis* Kult. = S. tau-saghyz subsp. tautaryensis
macrosperma Turcz. ex DC. = S. albicaulis
mariae Kult. (*S. tau-saghyz* Lipsch. & Bosse subsp. *mariae*
    (Kult.) R. Kam. & Tagaev) - 6
meyeri (C. Koch) Lipsch. = S. cana
mollis Bieb. (*Pseudopodospermum molle* (Bieb.) Kuth., *P.*
    turkeviczii (Krasch. & Lipsch.) Kuth. p.p. excl. typo,
    Scorzonera turkeviczii Krasch. & Lipsch. var. *kubanica*
    Krasch. & Lipsch., *S. semicana* auct. p.p., *S. turkeviczii*
    auct.) - 1, 2
- subsp. *szovitzii* (DC.) Chamberlain = S. szovitzii
mongolica Maxim. - 6
ovata Trautv. - 6
- subsp. *crassifolia* (Krasch. & Lipsch.) Tagaev = S. crassi-
    folia
oxiana M. Pop. = S. circumflexa subsp. oxiana
papposa DC. - 2
paradoxa Fisch. & C.A. Mey. - 2
parviflora Jacq. - 1, 2, 3, 6
petrovii Lipsch. - 5
pratorum (Krasch.) Stank. - 3, 4
pseudodivaricata Lipsch. - 6
pseudolanata Grossh. = S. psychrophila

psychrophila Boiss. & Hohen. (*S. pseudolanata* Grossh.) - 2
pubescens DC. - 3, 6
- subsp. *gracilis* (Lipsch.) Tagaev = S. gracilis
pulchra Lomak. - 2
purpurea L. - 1, 3, 6
- subsp. *rosea* (Waldst. & Kit.) Nym. = S. rosea
pusilla Pall. = Takhtajaniantha pusilla
racemosa Franch. (*S. iliensis* Krasch.) - 6
raddeana C. Winkl. (*S. turcomanica* Krasch. & Lipsch.) - 6
radiata Fisch. ex Ledeb. - 3, 4, 5
ramosissima DC. - 2
rebuensis Tatew. & Kitam. ex Kitam. - 5
rigida Auch. ex DC. - 2
rindak Ovcz. (*S. tau-saghyz* Lipsch. & Bosse subsp. *rindak*
    (Ovcz.) R. Kam.) - 6
rosea Waldst. & Kit. (*S. purpurea* L. subsp. *rosea* (Waldst.
    & Kit.) Nym.) - 1
rubroviolacea Godwinski = S. sericeolanata
ruprechtiana Lipsch. & Krasch. ex Lipsch. = S. glabra
safievii Grossh. - 2
schischkinii Lipsch. & Vass. (*Podospermum schischkinii*
    (Lipsch. & Vass.) Kuth.) - 2
seidlitzii Boiss. - 2
semicana auct. p.p. = S. mollis
sericeolanata (Bunge) Krasch. & Lipsch. (*S. rubroviolacea*
    Godwinski) - 6
songorica (Kar. & Kir.) Lipsch. & Vass. - 6
sosnowskyi Lipsch. = S. dzhawakhetica
stricta Hornem. - 1, 2, 3, 6
subacaulis Lipsch. - 6
suberosa C. Koch (*Pseudopodospermum suberosum* (C.
    Koch) Kuth.) - 2
szovitzii DC. (*Pseudopodospermum szovitzii* (DC.) Kuth.,
    Scorzonera mollis Bieb. subsp. *szovitzii* (DC.) Cham-
    berlain) - 2
tadshikorum Krasch. & Lipsch. - 6
taurica Bieb. (*S. hispanica* L. subsp. *taurica* (Bieb.) Tzvel.) -
    1, 2, 3, 6
tau-saghyz Lipsch. & Bosse. - 6
- subsp. *karataviensis* (Kult.) R. Kam. = S. karataviensis
- subsp. *longipes* (Kult.) R. Kam. = S. longipes
- subsp. *mariae* (Kult.) R. Kam. & Tagaev = S. mariae
- subsp. *rindak* (Ovcz.) R. Kam. = S. rindak
- subsp. **tautaryensis** (Kult.) R. Kam. (*S. longipes* Kult. var.
    tautaryensis Kult.) - 6
- subsp. *usbekistanica* (Czevr. & Bondar.) Tagaev = S.
    usbekistanica
tenuisecta (Grossh. & Sosn.) Czer. comb. nova (*Podosper-*
    mum tenuisectum Grossh. & Sosn. 1947, Not. Syst.
    Geogr. (Tbilisi) 13 : 67) - 2
tragopogonoides Regel & Schmalh. - 6
- subsp. *baldshuanica* (Lipsch.) Tagaev = S. baldshuanica
transiliensis M. Pop. - 6
tuberosa Pall. - 1, 3, 6
- var. *glaberrima* Bunge = S. circumflexa subsp. glaberrima
turcomanica Krasch. & Lipsch. = S. raddeana
turkestanica Franch. - 6
- subsp. **susamyrica** Tagaev - 6
turkeviczii auct. = S. mollis
turkeviczii Krasch. & Lipsch. var. *kubanica* Krasch. &
    Lipsch. = S. mollis
usbekistanica Czevr. & Bondar. (*S. tau-saghyz* Lipsch. &
    Bosse subsp. *usbekistanica* (Czevr. & Bondar.) Tagaev,
    comb. invalid.) - 6
vavilovii Kult. - 6
woronowii Krasch. - 2

Scorzoneroides Moench. = Leontodon

autumnalis (L.) Moench = Leontodon autumnalis
- subsp. *pratensis* (Link) Holub = Leontodon autumnalis
crocea (Haenke) Holub = Leontodon croceus
pseudotaraxaci (Schur) Holub = Leontodon pseudotaraxaci
rilaensis (Hayek) Holub = Leontodon rilaensis

Senecio L. (*Iranecio* B. Nordenstam, *Pojarkovia* Asker. p.p. incl. typo)

abrotanifolius L. subsp. *carpathicus* (Herbich) Nym. = S. carpathicus
acrabatensis auct. = S. paulsenii
ambraceus Turcz. ex DC. - 3, 4, 5
amurensis Schischk. = Tephroseris kirilowii and T. subdentata
andrzejowskyi Tzvel. - 1
aquaticus Hill - 1
- subsp. *barbareifolius* (Wimm. & Grab.) Walters = S. erraticus
- subsp. *erraticus* (Bertol.) Matthews = S. erraticus
- var. *barbareifolius* Wimm. & Grab. = S. erraticus
aquilonaris Schischk. = Tephroseris atropurpurea
arcticus Rupr. = Tephroseris palustris
x *arctisibiricus* Jurtz. & Korobkov = Tephroseris x arctisibirica
arenarius Bieb. = S. grandidentatus
argunensis Turcz. - 4, 5
asiaticus Schischk. & Serg. = Tephroseris praticola
atropurpureus (Ledeb.) B. Fedtsch. = Tephroseris atropurpurea
- subsp. *tomentosus* (Kjellm.) Hult. = Tephroseris kjellmanii
aurantiacus (Hoppe ex Willd.) Less. = Tephroseris aurantiaca
barbareae foliis Krock. = S. erraticus
barbareifolius (Wimm. & Grab.) Reichenb. = S. erraticus
besserianus Minder. = Tephroseris besseriana
bicolor (Willd.) Tod. = S. cineraria
- subsp. *cineraria* (DC.) Chater = S. cineraria
borysthenicus (DC.) Andrz. ex Czern. (*S. divaricatus* Andrz. 1862, non L. 1753) - 1
buschianus Sosn. - 2
calvertii Boiss. - 2
campestris (Retz.) DC. = Tephroseris integrifolia
- subsp. *kirilowii* (Turcz. ex DC.) Kitag. = Tephroseris kirilowii
- var. *lenensis* (Schischk.) Hamet-Ahti = Tephroseris integrifolia
cannabifolius Less. - 4, 5
- subsp. *litvinovii* (Schischk.) Boyko = S. litvinovii
- var. *integrifolius* (Koidz.) Kitam. = S. litvinovii
capitatus (Wahlenb.) Steud. = Tephroseris capitata
carniolicus Willd. (*S. incanus* L. subsp. *carniolicus* (Willd.) Br.-Bl.) - 1
carpathicus Herbich (*S. abrotanifolius* L. subsp. *carpathicus* (Herbich) Nym.) - 1
caucasigenus Schischk. = Tephroseris caucasigena
cineraria DC. (*S. bicolor* (Willd.) Tod. 1860, non Viv. 1802, *S. bicolor* subsp. *cineraria* (DC.) Chater, *S. cineraria* subsp. *bicolor* (Willd.) Arcang.) - 1
- subsp. *bicolor* (Willd.) Arcang. = S. cineraria
cladobotrys Ledeb. = Tephroseris cladobotrys
congestus (R. Br.) DC. = Tephroseris palustris
- subsp. *palustris* (L.) Rauschert = Tephroseris palustris
crispus (Jacq.) Kitt. = Tephroseris crispa

czernjaevii Minder. = Tephroseris czernjaevii
divaricatus Andrz. = S. borysthenicus
doria auct. p.p. = S. macrophyllus Bieb. and S. schvetzovii
doria L. subsp. *kirghisicus* (DC.) Chater = S. paucifolius
- subsp. *umbrosus* (Waldst. & Kit.) Soo = S. umbrosus
dubitabilis C. Jeffrey & Y.L. Chen (*S. dubius* Ledeb. vii-xii 1833, non Beck, v-vi 1833) - 1,3,4,5,6
dubius Ledeb. = S. dubitabilis
erraticus Bertol. (*S. aquaticus* Hill var. *barbareifolius* Wimm. & Grab., *S. aquaticus* subsp. *barbareifolius* (Wimm. & Grab.) Walters, *S. aquaticus* subsp. *erraticus* (Bertol.) Matthews, *S. barbareae foliis* Krock. nom. invalid., *S. barbareifolius* (Wimm. & Grab.) Reichenb.) - 1, 2
erucifolius L. (*S. tenuifolius* Jacq. 1775, non Burm. fil.) - 1, 3, 4, 6
- subsp. *arenarius* Soo = S. grandidentatus
- subsp. *grandidentatus* (Ledeb.) V. Avet. = S. grandidentatus
- subsp. *grandidentatus* (Ledeb.) B. Nordenstam = S. grandidentatus
euxinus Minder. = S. vernalis
ferganensis Schischk. - 1, 6
flammeus Turcz. ex DC. = Tephroseris flammea
fluviatilis Wallr. (*S. sarracenicus* L. nom. ambig.) - 1, 2, 3, 4, 6
franchetii C. Winkl. - 6
frigidus (Richards.) Less. = Tephroseris frigida
fuchsii C.C. Gmel. = S. ovatus
fuscatus auct. = Tephroseris pseudoaurantiaca
grandidentatus Ledeb. (*S. arenarius* Bieb. 1816, non Thunb. 1800, *S. erucifolius* L. subsp. *arenarius* Soo, *S. erucifolius* subsp. *grandidentatus* (Ledeb.) V. Avet., *S. erucifolius* subsp. *grandidentatus* (Ledeb.) B. Nordenstam, comb. superfl.) - 1, 2, 3, 6
hieracifolius L. = Erechtites hieracifolia
hieraciformis Kom. = Tephroseris hieraciformis
hyperborealis Greenm. = Tephroseris hyperborealis
- subsp. *wrangelicus* Jurtz., Korobkov & Petrovsky = Tephroseris hyperborealis subsp. wrangelica
igoschinae Schischk. = Tephroseris igoschinae
iljinii Schischk. - 6
incanus L. subsp. *carniolicus* (Willd.) Br.-Bl. = S. carniolicus
integrifolius (L.) Clairv. = Tephroseris integrifolia
- subsp. *amurensis* (Schischk.) Worosch. = Tephroseris kirilowii
- subsp. *atropurpureus* (Ledeb.) Cuf. = Tephroseris atropurpurea
- subsp. *aurantiacus* (Hoppe ex Willd.) Briq. & Cavill. = Tephroseris aurantiaca
- subsp. *capitatus* (Wahlenb.) Cuf. = Tephroseris capitata
- subsp. *czernjaevii* (Minder.) Chater = Tephroseris czernjaevii
- subsp. *tundricola* (Tolm.) Chater = Tephroseris tundricola
- var. *lindstroemii* Ostenf. = Tephroseris pseudoaurantiaca
jacobaea L. - 1, 2, 3, 4, 6
jacquinianus Reichenb. (*S. nemorensis* L. subsp. *jacquinianus* (Reichenb.) Celak.) - 1
jacuticus Schischk. = Tephroseris jacutica
jailicola Juz. = Tephroseris jailicola
kamtschaticus Kom. = Tephroseris pseudoaurantiaca
karelinioides C. Winkl. - 6
karjaginii Sof. = Tephroseris karjaginii
kawakamii Makino = Tephroseris kawakamii
khorossanicus Rech. fil. & Aell. - 6

kirghisicus DC. = S. paucifolius
kirilowii Turcz. ex DC. = Tephroseris kirilowii
kjellmanii A. Pors. = Tephroseris kjellmanii
kolenatianus C.A. Mey. - 2
- subsp. *pseudoorientalis* (Schischk.) V. Avet. = S. pseudoorientalis
korshinskyi Krasch. - 6
krascheninnikovii Schischk. - 6
kubensis Grossh. - 2
lapsanoides DC. - 2
lenensis Schischk. = Tephroseris integrifolia
leucanthemifolius Poir. var. *vernalis* (Waldst. & Kit.) J.C.M. Alexander = S. vernalis
lindstroemii (Ostenf.) A. Pors. = Tephroseris pseudoaurantiaca
lipskyi Lomak. - 2
litvinovii Schischk. (*S. cannabifolius* Less. var. *integrifolius* (Koidz.) Kitam., *S. cannabifolius* subsp. *litvinovii* (Schischk.) Boyko, *S. palmatus* (Pall.) Ledeb. var. *integrifolius* Koidz.) - 5
lorentii Hochst. - 2
macrophyllus Bieb. (*S. doria* auct.) - 2
massagetovii Schischk. - 2
nemorensis L. - 1, 3, 4, 5, 6
- subsp. **boikoanus** Worosch. & Schlothg. - 5
- subsp. *fuchsii* (C.C. Gmel.) Celak. = S. ovatus
- subsp. *jacquinianus* (Reichenb.) Celak. = S. jacquinianus
noeanus Rupr. - 1, 2, 6
nuraniae Roldug. - 6
olgae Regel & Schmalh. - 6
othonnae Bieb. (*Iranecio othonnae* (Bieb.) B. Nordenstam) - 2
ovatus Willd. (*S. fuchsii* C.C. Gmel., *S. nemorensis* L. subsp. *fuchsii* (C.C. Gmel.) Celak., *S. sarracenicus* sensu Minder.) - 1
palmatus (Pall.) Ledeb. var. *integrifolius* Koidz. = S. litvinovii
paludosus L. (*S. riparius* Wallr.) - 1
palustris (L.) Hook. = Tephroseris palustris
pandurifolius C. Koch - 2
papposus (Reichenb.) Less. = Tephroseris papposa
paucifolius S.G. Gmel. (*S. doria* L. subsp. *kirghisicus* (DC.) Chater, *S. kirghisicus* DC.) - 1, 3, 6
paucilobus DC. (*Iranecio paucilobus* (DC.) B. Nordenstam, *Pojarkovia pauciloba* (DC.) Asker.) - 2
paulsenii O. Hoffm. (*S. acrabatensis* auct.) - 6
platyphylloides Somm. & Levier = Adenostyles platyphylloides
pojarkovae Schischk. (*Pojarkovia stenocephala* (Boiss.) Asker., *Senecio stenocephalus* Boiss. 1875, non Maxim. 1871) - 2
polycephalus Ledeb. - 2
porphyranthus Schischk. = Tephroseris porphyrantha
pratensis (Hoppe) DC. = Tephroseris integrifolia
praticola Schischk. & Serg. = Tephroseris praticola
pricei Simps. = Tephroseris pricei
propinquus Schischk. - 2
pseudoarnica Less. - 5
pseudoaurantiacus Kom. = Tephroseris pseudoaurantiaca
pseudoorientalis Schischk. (*S. kolenatianus* C.A. Mey. subsp. *pseudoorientalis* (Schischk.) V. Avet.) - 2
pyroglossus Kar. & Kir. = Tephroseris pyroglossa
quinqueligulatus C. Winkl. - 6
racemosus (Bieb.) DC. - 2
racemulifer Pavl. - 6
renardii C. Winkl. - 6
resedifolius Less. = Tephroseris heterophylla

reverdattoi K. Sobol. = Tephroseris turczaninovii
rhombifolius (Adams) Sch. Bip. = Adenostyles macrophylla
riparius Wallr. = S. paludosus
rivularis (Waldst. & Kit.) DC. = Tephroseris crispa
saposhnikovii Krasch. & Schipcz. - 6
sarracenicus L. = S. fluviatilis
sarracenicus sensu Minder. = S. ovatus
schischkinianus Sof. - 2
schistosus Charkev. = Tephroseris schistosa
schvetzovii Korsh. (*S. doria* auct. p.p.) - 1, 3, 6
sichotensis Kom. = Tephroseris sichotensis
similiflorus Kolak. = Adenostyles similiflora
sosnovskyi Sof. (*S. vernalis* Waldst. & Kit. subsp. *sosnovskyi* (Sof.) V. Avet.) - 2
stenocephalus Boiss. = S. pojarkovae
subalpinus Koch - 1
subdentatus Ledeb. - 2, 3, 4, 6
subfloccosus Schischk. = Tephroseris subfloccosa
subfrigidus Kom. = Tephroseris subfrigida
subscaposus Kom. = Tephroseris subscaposa
succisifolius Kom. = Tephroseris integrifolia
sukaczevii Schischk. = Tephroseris sukaczevii
sumneviczii Schischk. & Serg. = Tephroseris turczaninovii
sylvaticus L. - 1, 2
taimyrensis (Herd.) Gorodk. ex Tichomirov = Tephroseris atropurpurea
taraxacifolius (Bieb.) DC. - 2
tataricus Less. - 1, 3
tauricus Konechn. - 1
tenuifolius Jacq. = S. erucifolius
thianschanicus Regel & Schmalh. (*S. tianschanicus* auct.) - 6
thyrsophorus C. Koch - 2
tianschanicus auct. = S. thianschanicus
tichomirovii Schischk. = Tephroseris kjellmanii
trautvetteri Maximova = Tephroseris atropurpurea
tubicaulis Mansf. = Tephroseris palustris
tundricola Tolm. = Tephroseris tundricola
- subsp. *lindstroemii* (Ostenf.) Korobkov = Tephroseris pseudoaurantiaca
turczaninovii DC. = Tephroseris turczaninovii
- subsp. *czikoicus* Maximova = Tephroseris turczaninovii
umbrosus Waldst. & Kit. (*S. doria* L. subsp. *umbrosus* (Waldst. & Kit.) Soo) - 1
veresczaginii Schischk. & Serg. = Tephroseris veresczaginii
vernalis Waldst. & Kit. (*S. euxinus* Minder., *S. leucanthemifolius* Poir. var. *vernalis* (Waldst. & Kit.) J.C.M. Alexander) - 1, 2, 6
- subsp. *sosnovskyi* (Sof.) V. Avet. = S. sosnovskyi
viscosus L. - 1, 2, 4(alien), 5(alien)
vulgaris L. - 1, 2, 3, 4, 5, 6

Seriphidium (Cass.) Poljak. = Artemisia
amoenum (Poljak.) Poljak. = Artemisia amoena
aralense (Krasch.) Poljak. = Artemisia aralensis
badhysi (Krasch. & Lincz. ex Poljak.) Poljak. = Artemisia badhysi
balchanorum (Krasch.) Poljak. = Artemisia balchanorum
baldshuanicum (Krasch. & Zapr.) Poljak. = Artemisia baldshuanica
camelorum (Krasch.) Poljak. = Artemisia camelorum
ciniforme (Krasch. & M. Pop. ex Poljak.) Poljak. = Artemisia ciniformis
cinum (Berg ex Poljak.) Poljak. = Artemisia cina
compactum (Fisch. ex DC.) Poljak. = Artemisia compacta
deserti (Krasch.) Poljak. = Artemisia deserti

dubjanskyanum (Krasch. ex Poljak.) Poljak. = Artemisia dubjanskyana

dumosum (Poljak.) Poljak. = Artemisia dumosa

fedtschenkoanum (Krasch.) Poljak. = Artemisia fedtschenkoana

ferganense (Krasch. ex Poljak.) Poljak. = Artemisia ferganensis

fragrans (Willd.) Poljak. = Artemisia lerchiana

glanduligerum (Krasch. ex Poljak.) Poljak. = Artemisia glanduligera

gracilescens (Krasch. & Iljin) Poljak. = Artemisia gracilescens

gypsaceum (Krasch., M. Pop. & Lincz. ex Poljak.) Poljak. = Artemisia gypsacea

halophilum (Krasch.) Poljak. = Artemisia halophila

heptapotamicum (Poljak.) Ling & Y.R. Ling = Artemisia heptapotamica

issykkulense (Poljak.) Poljak. = Artemisia issykkulensis

junceum (Kar. & Kir.) Poljak. = Artemisia juncea

- var. *macrosciadium* (Poljak.) Ling & Y.R. Ling = Artemisia juncea

karatavicum (Krasch. & Abol. ex Poljak.) Ling & Y.R Ling = Artemisia karatavica

kaschgaricum (Krasch.) Poljak. = Artemisia kaschgarica

kemrudicum (Krasch.) Poljak. = Artemisia kemrudica

knorringianum (Krasch.) Poljak. = Artemisia knorringiana

kochiiforme (Krasch. & Lincz. ex Poljak.) Poljak. = Artemisia kochiiformis

kopetdaghense (Krasch. ex Poljak.) Poljak. = Artemisia kopetdaghensis

korovinii (Poljak.) Poljak. = Artemisia korovinii

lehmannianum (Bunge) Poljak. = Artemisia lehmanniana

lerchianum (Web.) Poljak. = Artemisia lerchiana

lessingianum (Bess.) Poljak. = Artemisia lessingiana

leucodes (Schrenk) Poljak. = Artemisia leucodes

lobulifolium (Boiss.) Poljak. = Artemisia santolina

maritimum (L.) Poljak. = Artemisia maritima

mongolorum (Krasch.) Ling & Y.R. Ling = Artemisia mongolorum

monogynum (Waldst. & Kit.) Poljak. = Artemisia monogyna

namanganicum (Poljak.) Poljak. = Artemisia namanganica

nitrosum (Web.) Poljak. = Artemisia nitrosa

pauciflorum (Web.) Poljak. = Artemisia pauciflora

porrectum (Krasch. ex Poljak.) Poljak. = Artemisia porrecta

prasinum (Krasch. ex Poljak.) Poljak. = Artemisia prasina

prolixum (Krasch. ex Poljak.) Poljak. = Artemisia prolixa

rhodanthum (Rupr.) Poljak. = Artemisia rhodantha

santolinum (Schrenk) Poljak. = Artemisia santolina

schrenkianum (Ledeb.) Poljak. = Artemisia schrenkiana

scopiforme (Ledeb.) Poljak. = Artemisia scopiformis

scotinum (Nevski) Poljak. = Artemisia scotina

semiaridum (Krasch. & Lavr.) Ling & Y.R. Ling = Artemisia semiarida

serotinum (Bunge) Poljak. = Artemisia serotina

sogdianum (Bunge) Poljak. = Artemisia sogdiana

spicigerum (C. Koch) Poljak. = Artemisia spicigera

stenocephalum (Krasch. ex Poljak.) Poljak. = Artemisia stenocephala

sublessingianum (Krasch. ex Poljak.) Poljak. = Artemisia sublessingiana

szowitzianum (Bess.) Poljak. = Artemisia szowitziana

tauricum (Willd.) Poljak. = Artemisia taurica

tenuisectum (Nevski) Poljak. = Artemisia tenuisecta

terrae-albae (Krasch.) Poljak. = Artemisia terrae-albae

transiliense (Poljak.) Poljak. = Artemisia transiliensis

turanicum (Krasch.) Poljak. = Artemisia turanica

turcomanicum (Gand.) Poljak. = Artemisia turcomanica

vachanicum (Krasch. ex Poljak.) Poljak. = Artemisia vachanica

validum (Krasch. ex Poljak.) Poljak. = Artemisia valida

Serratula L. (*Klasea* Cass.)

alatavica C.A. Mey. - 6

algida Iljin - 3, 6

angulata Kar. & Kir. - 6

aphyllopoda Iljin (*Klasea aphyllopoda* (Iljin) Holub) - 6

biebersteiniana (Iljin ex Grossh.) Takht. - 2

bracteifolia (Iljin ex Grossh.) Stank. = S. radiata

bulgarica Acht. & Stojan. (*S. caput-najae* Zahar.) - 1

caput-najae Zahar. = S. bulgarica

cardunculus (Pall.) Schischk. (*Klasea cardunculus* (Pall.) Holub) - 1, 3, 6

caucasica Boiss. - 2

centauroides L. - 4

chartacea C. Winkl. - 6

coriacea Fisch. & C.A. Mey. (*Klasea coriacea* (Fisch. & C.A. Mey.) Holub) - 2

coronata L. (*S. wolffii* Andrae) - 1, 2, 3, 4, 5, 6

- var. *manshurica* (Kitag.) Kitag. = S. manshurica

dissecta Ledeb. - 6

donetzica Dubovik - 1

dshungarica Iljin - 6

erucifolia (L.) Boriss. (*S. xeranthemoides* Bieb.) - 1, 2, 3, 6

gmelinii Tausch (*Klasea gmelinii* (Tausch) Holub) - 1, 3

hastifolia Kult. & Korov. - 6

haussknechtii Boiss. (*Klasea haussknechtii* (Boiss.) Holub, Serratula transcaucasica (Bornm.) Sosn. ex Grossh.) - 2

heterophylla auct. = S. lycopifolia

hungarica Klok. ex Dobrocz. = S. radiata

inermis Gilib. = S. tinctoria

kirghisorum Iljin - 3, 6

komarovii Iljin - 4, 5

lancifolia Zak. - 6

lycopifolia (Vill.) A. Kerner (*Klasea lycopifolia* (Vill.) A. & D. Löve, Serratula heterophylla auct.) - 1

lyratifolia Schrenk - 6

manshurica Kitag. (*S. coronata* L. var. *manshurica* (Kitag.) Kitag.) - 5

marginata Tausch (*Klasea marginata* (Tausch) Kitag.) - 3, 4, 6

modesti Boriss. - 6

procumbens Regel (*Klasea procumbens* (Regel) Holub) - 6

pulchella Fisch. ex Hornem. = Jurinea pulchella

quinquefolia Bieb. ex Willd. - 2

radiata (Waldst. & Kit.) Bieb. (*S. bracteifolia* (Iljin ex Grossh.) Stank., *S. hungarica* Klok. ex Dobrocz. nom. invalid.) - 1, 2

serratuloides (Fisch. & C.A. Mey.) Takht. - 2

sogdiana Bunge - 6

suffruticosa Schrenk - 6

tanaitica P. Smirn. - 1

tianschanica Saposhn. & E. Nikit. - 6

tinctoria L. (*S. inermis* Gilib. nom. invalid.) - 1

transcaucasica (Bornm.) Sosn. ex Grossh. = S. haussknechtii

wolffii Andrae = S. coronata L.

xeranthemoides Bieb. = S. erucifolia

Siebera J. Gay

pungens (Lam.) DC. - 6

Sigesbeckia L.

glabrescens Makino (*S. pubescens* (Makino) Makino var. *glabrescens* (Makino) Worosch.) - 5
orientalis L. (*S. orientalis* subsp. *caspica* (Fisch. & C.A. Mey.) Kitam.) - 1, 2, 6
- subsp. *caspica* (Fisch. & C.A. Mey.) Kitam. = S. orientalis
*pubescens (Makino) Makino - 5
- var. *glabrescens* (Makino) Worosch. = S. glabrescens

*Silphium L.

*perfoliatum L. - 1

Silybum Adans.

marianum (L.) Gaertn. - 1, 2, 3, 6

Solidago L.

alpestris Waldst. & Kit. ex Willd. (*S. virgaurea* L. subsp. *alpestris* (Waldst. & Kit. ex Willd.) Hayek) - 1
*altissima L. - 1
*arguta Ait. - 2
armena Kem.-Nath. ex Grossh. - 2
*canadensis L. - 1, 2, 6
caucasica Kem.-Nath. - 2
compacta Turcz. - 5
cuprea Juz. = S. spiraeifolia
dahurica Kitag. (*S. gebleri* Juz.) - 3, 4, 6
decurrens Lour. (*S. virgaurea* L. subsp. *leiocarpa* (Benth.) Hult.) - 5
gebleri Juz. = S. dahurica
gigantea Ait. - 5
- subsp. *serotina* (O. Kuntze) McNeill = S. serotinoides
gigantea auct. = S. serotinoides
*graminifolia (L.) Salisb. - 1
jailarum Juz. (*S. virgaurea* L. subsp. *jailarum* (Juz.) Tzvel.) - 1
*juncea Ait. - 2, 4
kuhistanica M. Pop. ex Juz. - 6
kurilensis Juz. (*S. virgaurea* L. subsp. *kurilensis* (Juz.) Worosch.) - 5
lapponica With. (*S. virgaurea* L. subsp. *lapponica* (With.) Tzvel.) - 1
- subsp. *stenophylla* G.E. Schultz = S. virgaurea subsp. stenophylla
multiradiata Ait. - 5
*neglecta Torr. & Gray - 2
*odora Ait. - 2
pacifica Juz. (*S. virgaurea* L. subsp. *coreana* (Nakai) Kitag.) - 5
*rupestris Rafin. - 2
serotina Ait. = S. serotinoides
*serotinoides A. & D. Löve (*S. serotina* Ait. 1789, non Retz. 1781, *S. gigantea* Ait. subsp. *serotina* (O. Kuntze) McNeill, *S. gigantea* auct.) - 1
spiraeifolia Fisch. ex Herd. (*S. cuprea* Juz.) - 5
taurica Juz. (*S. virgaurea* L. subsp. *taurica* (Juz.) Tzvel.) - 1
turfosa Woronow ex Grossh. - 2
virgaurea L. - 1, 2, 3
- subsp. *alpestris* (Waldst. & Kit. ex Willd.) Hayek = S. alpestris
- subsp. *coreana* (Nakai) Kitag. = S. pacifica
- subsp. *jailarum* (Juz.) Tzvel. = S. jailarum

- subsp. *kurilensis* (Juz.) Worosch. = S. kurilensis
- subsp. *lapponica* (With.) Tzvel. = S. lapponica
- subsp. *leiocarpa* (Benth.) Hult. = S. decurrens
- subsp. stenophylla (G.E. Schultz) Tzvel. (*S. lapponica* With. subsp. *stenophylla* G.E. Schultz, *S. virgaurea* b. *angustifolia* Meinsh. 1878, non Koch, 1837) - 1
- subsp. *taurica* (Juz.) Tzvel. = S. taurica
- b. *angustifolia* Meinsh. = S. virgaurea subsp. stenophylla

Sonchus L.

araraticus Nazarova & Barsegian - 2
arenicola Worosch. (*S. arvensis* L. subsp. *arenicola* (Worosch.) Worosch.) - 5
arvensis L. - 1, 2, 3, 4, 5, 6
- subsp. *arenicola* (Worosch.) Worosch. = S. arenicola
- subsp. *brachyotus* (DC.) Kitam. = S. brachyotus
- subsp. uliginosus (Bieb.) Nym. (*S. uliginosus* Bieb.) - 1, 2, 3, 4, 5, 6
- f. *brachyotus* (DC.) Kirp. = S. brachyotus
asper (L.) Hill - 1, 2, 3, 4, 5, 6
- subsp. *glaucescens* (Jord.) Ball = S. nymanii
brachyotus DC. (*S. arvensis* f. *brachyotus* (DC.) Kirp., *S. arvensis* L. subsp. *brachyotus* (DC.) Kitam.) - 3, 4, 5
cacaliifolius Bieb. = Cicerbita macrophylla
glaucescens Jord. = S. nymanii
humilis Orlova - 1
nymanii Tineo & Guss. (*S. asper* (L.) Hill subsp. *glaucescens* (Jord.) Ball, *S. glaucescens* Jord.) - 6
oleraceus L. - 1, 2, 6
palustris L. - 1, 2, 3, 6
- subsp. *sosnowskyi* (Schchian) Boulos = S. sosnowskyi
sosnowskyi Schchian (*S. palustris* L. subsp. *sosnowskyi* (Schchian) Boulos) - 2
tenerrimus L. - 2
transcaspicus Nevski - 2, 6
uliginosus Bieb. = S. arvensis subsp. uliginosus

Spathipappus Tzvel.

griffithii (Clarke) Tzvel. (*Tanacetum griffithii* (Clarke) Muradjan) - 6

Sphaeranthus L.

volgensis Tzvel. - 1

Stechmannia DC. = Jurinea

ludmilae (Iljin) Sojak = Jurinea ludmilae
lydiae (Iljin) Sojak = Jurinea lydiae
popovii (Iljin) Sojak = Jurinea popovii

Stemmacantha Cass. (*Leuzea* auct. p.p., *Rhaponticum* auct.)

aulieatensis (Iljin) M. Dittrich (*Rhaponticum aulieatense* Iljin, *Leuzea aulieatensis* (Iljin) Holub) - 6
carthamoides (Willd.) M. Dittrich (*Leuzea carthamoides* (Willd.) DC., *Rhaponticum carthamoides* (Willd.) Iljin) - 3, 4, 6
chamarensis (Peschkova) Czer. comb. nova (*Rhaponticum chamarense* Peschkova, 1977, Bot. Zhurn. 62, 2 : 227) - 4
integrifolia (C. Winkl.) M. Dittrich (*Rhaponticum integrifolium* C. Winkl., *Leuzea integrifolia* (C. Winkl.) Holub) - 6
karatavica (Regel & Schmalh.) M. Dittrich (*Rhaponticum karatavicum* Regel & Schmalh., *Leuzea karatavica* (Regel & Schmalh.) Holub) - 6

lyrata (C. Winkl. ex Iljin) M. Dittrich (*Rhaponticum lyratum* C. Winkl. ex Iljin, *Leuzea lyrata* (Iljin) Holub) - 6
namanganica (Iljin) M. Dittrich (*Rhaponticum namanganicum* Iljin, *Leuzea namanganica* (Iljin) Holub) - 6
nana (Lipsky) M. Dittrich (*Rhaponticum nanum* Lipsky, *Leuzea nana* (Lipsky) Holub) - 6
nitida (Fisch.) M. Dittrich (*Rhaponticum nitidum* Fisch., *Leuzea nitida* (Fisch.) Holub) - 6
orientalis (Serg.) Czer. comb. nova (*Leuzea carthamoides* (Willd.) DC. subsp. *orientalis* Serg. 1949, in Krylov, Fl. Sib. Occid. 11 : 2943; *Rhaponticum carthamoides* (Willd.) Iljin subsp. *orientale* (Serg.) Sosk., *R. orientale* (Serg.) Peschkova) - 3, 4
pulchra (Fisch. & C.A. Mey.) M. Dittrich (*Rhaponticum pulchrum* Fisch. & C.A. Mey., *Leuzea pulchra* (Fisch. & C.A. Mey.) Holub) - 2
- subsp. **zardabii** (Rzazade) M. Dittrich (*Rhaponticum zardabii* Rzazade) - 2
satzyperovii (Sosk.) Czer. comb. nova (*Rhaponticum satzyperovii* Sosk. 1959, Bot. Mat. (Leningrad) 19 : 400; *Leuzea satzyperovii* (Sosk.) Holub, *Rhaponticum uniflorum* (L.) DC. subsp. *satzyperovii* (Sosk.) Worosch., *Stemmacantha uniflora* (L.) M. Dittrich subsp. *satzyperovii* (Sosk.) M. Dittrich) - 5
serratuloides (Georgi) M. Dittrich (*Leuzea altaica* (Spreng.) Link, *Rhaponticum serratuloides* (Georgi) Bobr.) - 1, 3, 6
uniflora (L.) M. Dittrich (*Leuzea uniflora* (L.) Holub, *Rhaponticum uniflorum* (L.) DC.) - 4, 5
- subsp. *satzyperovii* (Sosk.) M. Dittrich = S. satzyperovii

Stenactis Cass. p.p. = Phalacroloma
annua (L.) Cass. = Phalacroloma annuum
- subsp. *septentrionalis* (Fern. & Wieg.) A. & D. Löve = Phalacroloma septentrionale
septentrionalis (Fern. & Wieg.) Holub = Phalacroloma septentrionale
speciosa Lindl. = Erigeron speciosus
strigosa (Muehl. ex Willd.) DC. = Phalacroloma strigosum

Stenotheca Monn. = Hieracium
tristis (Willd. ex Spreng.) Schljak. = Hieracium triste

Steptorhamphus Bunge
crambifolius - 6
crassifolius (Trautv.) Kirp. - 6
czerepanovii Kirp. - 2
linczevskii Kirp. - 6
persicus (Boiss.) O. & B. Fedtsch. - 2, 6
petraeus (Fisch. & C.A. Mey.) Grossh. (*S. tuberosus* (Jacq.) Grossh. subsp. *petraeus* (Fisch. & C.A. Mey.) Takht.) - 2
tuberosus (Jacq.) Grossh. - 1, 2
- subsp. *petraeus* (Fisch. & C.A. Mey.) Takht. = S. petraeus

***Stevia** Cav.
***ovata** Willd. - 2

Stizolophus Cass.
balsamita (Lam.) Cass. ex Takht. - 2, 6
coronopifolius (Lam.) Cass. - 2

Stylocline auct. = Cymbolaena
griffithii A. Gray = Cymbolaena griffithii

Symphyllocarpus Maxim.
exilis Maxim. - 5

Symphyotrichum auct. = Aster
laeve (L.) A. & D. Löve = Aster laevis

Synchaeta Kirp. = Omalotheca
caucasica (Somm. & Levier) Kirp. = Omalotheca caucasica
norvegica (Gunn.) Kirp. = Omalotheca norvegica
sylvatica (L.) Kirp. = Omalotheca sylvatica

Syneilesis Maxim.
aconitifolia (Bunge) Maxim. - 5

Synurus Iljin
deltoides (Ait.) Nakai - 4, 5

Syreitschikovia Pavl.
spinulosa (Franch.) Pavl. - 6
tenuifolia (Bong.) Pavl. = S. tenuis
tenuis (Bunge) Botsch. (*Syreitschikovia tenuifolia* (Bong.) Pavl.) - 6

Tagetes L.
***erecta** L. - 1, 2, 6
minuta L. - 2(alien)
***patula** L. - 1, 2, 6
signata Bartl. = T. tenuifolia
***tenuifolia** Cav. (*T. signata* Bartl.) - 1, 6

Taktajaniantha Nazarova
pusilla (Pall.) Nazarova (*Scorzonera pusilla* Pall.) - 1, 2, 3, 6

Tanacetopsis (Tzvel.) Kovalevsk.
botschantzevii (Kovalevsk.) Kovalevsk. (*Cancrinia botschantzevii* (Kovalevsk.) Tzvel.) - 6
czukavinae Kovalevsk. & Junuss. - 6
ferganensis (Kovalevsk.) Kovalevsk. (*Cancrinia ferganensis* (Kovalevsk.) Tzvel.) - 6
goloskokovii (Poljak.) Karmyscheva (*Cancrinia goloskokovii* (Poljak.) Tzvel.) - 6
handeliiformis Kovalevsk. - 6
kamelinii Kovalevsk. - 6
karataviensis (Kovalevsk.) Kovalevsk. (*Cancrinia karatavica* Tzvel.) - 6
kjurendaghi Kurbanov - 6
korovinii Kovalevsk. - 6
krascheninnikovii (Nevski) Kovalevsk. (*Cancrinia nevskii* Tzvel.) - 6
mucronata (Regel & Schmalh.) Kovalevsk. (*Cancrinia mucronata* (Regel & Schmalh.) Tzvel.) - 6
nepliana Tzvel. = T. santoana
pamiralaica (Kovalevsk.) Kovalevsk. (*Cancrinia pamiralaica* (Kovalevsk.) Tzvel.) - 6
paropamisica (Krasch.) Kovalevsk. (*Cancrinia paropamisica* (Krasch.) Tzvel.) - 6
pjataevae (Kovalevsk.) Karmyscheva (*Cancrinia pjataevae* (Kovalevsk.) Tzvel.) - 6
popovii R. Kam. & Kovalevsk. - 6
santoana (Krasch., M. Pop. & Vved.) Kovalevsk. (*Cancrinia santoana* (Krasch., M. Pop. & Vved.) Poljak., *Tanace-*

topsis nepliana Tzvel.) - 6

setacea (Regel & Schmalh.) Kovalevsk. (*Cancrinia setacea* (Regel & Schmalh.) Tzvel.) - 6

submarginata (Kovalevsk.) Kovalevsk. (*Cancrinia submarginata* (Kovalevsk.) Tzvel.) - 6

subsimilis (Rech. fil.) Kovalevsk. (*Cancrinia subsimilis* (Rech. fil.) Tzvel.) - 6

urgutensis (M. Pop.) Kovalevsk. (*Cancrinia urgutensis* (M. Pop.) Tzvel.) - 6

Tanacetum L. (*Gymnocline* auct.)

abrotanifolium (L.) Druce - 2

achilleifolium (Bieb.) Sch. Bip. - 1, 2, 3, 6

akinfiewii (Alexeenko) Tzvel. - 2

argenteum (Lam.) Willd. subsp. *canum* (C. Koch) Grierson = Hemipappus canus

argyrophyllum (C. Koch) Tzvel. (*Gymnocline argyrophylla* C. Koch, *Tanacetum myriophyllum* Willd. nom. illegit., *T. polycephalum* Sch. Bip. subsp. *argyrophyllum* (C. Koch) Podlech) - 2

balsamita L. subsp. *balsamitoides* (Sch. Bip.) Grierson = Pyrethrum balsamita

balsamitoides (Nabel.) Chandjian = Pyrethrum balsamitoides

bipinnatum (L.) Sch. Bip. - 1, 3, 4, 5
- subsp. **sabulosum** Karav. & Tzvel. - 4

boreale Fisch. ex DC. (*Chrysanthemum asiaticum* Worosch., *Ch. vulgare* (L.) Bernh. subsp. *boreale* (Fisch. ex DC.) Worosch., *Tanacetum vulgare* L. subsp. *boreale* (Fisch. ex DC.) A. & D. Löve, *T. vulgare* subsp. *boreale* (Fisch. ex DC.) Kuvajev, comb. superfl.) - 1(?), 3, 4, 5, 6

canescens DC. - 2

chiliophyllum (Fisch. & C.A. Mey.) Sch. Bip. - 2

clusii (Reichenb.) Sojak = Pyrethrum clusii

coccineum (Willd.) Grierson, p.p. = Pyrethrum coccineum
- subsp. *carneum* (Bieb.) Grierson = Pyrethrum carneum
- subsp. *chamaemelifolium* (Somm. & Levier) Grierson = Pyrethrum chamaemelifolium

corymbosum (L.) Sch. Bip. subsp. *clusii* (Fisch. ex Reichenb.) Heywood = Pyrethrum clusii

crassipes (Stschegl.) Tzvel. - 3, 6

djilgense (Franch.) Podlech = Pyrethrum djilgense

duderanum (Boiss.) Tzvel. (*T. polycephalum* Sch. Bip. subsp. *duderanum* (Boiss.) Podlech) - 2

flavovirens (Boiss.) Tzvel. subsp. *tamrutense* (Sosn.) Takht. = T. tamrutense

griffithii (Clarke) Muradjan = Spathipappus griffithii

grossheimii (Sosn.) Muradjan = Pyrethrum grossheimii

heterophyllum Boiss. - 6

karelinii Tzvel. - 3, 6

kelleri (Kryl. & Plotn.) Takht. = Pyrethrum kelleri

kittaryanum (C.A. Mey.) Tzvel. - 1, 3, 6
- subsp. *sclerophyllum* (Krasch.) Tzvel. = T. sclerophyllum
- subsp. *uralense* (Krasch.) Tzvel. = T. uralense

kokanicum Krasch. (*Ajania kokanica* (Krasch.) Tzvel.) - 6
It is necessary to clarify the generic identity of this species.

komarowii (C. Winkl.) Muradjan = Lepidolopha komarowii

kotschyi (Boiss.) Grierson = Pyrethrum kotschyi

kubense (Grossh.) Muradjan = Pyrethrum kubense

longipedunculatum (Sosn.) Tzvel. - 2

mikeschinii (Tzvel.) Takht. = Pyrethrum mikeschinii

millefolium (L.) Tzvel. (*Chrysanthemum millefolium* (L.) E.I. Nyarady) - 1, 2, 3, 6

mindshelkense Kovalevsk. - 6

myriophyllum Willd. = T. argyrophyllum

nuratavicum (Krasch.) Muradjan = Lepidolopha nuratavica

odessanum (Klok.) Tzvel. - 1

oligocephalum (DC.) Sch. Bip. - 2

oxylepis (Bordz.) Grierson = Pyrethrum oxylepis

paczoskii (Zefir.) Tzvel. - 1

pinnatum auct. p.p. = T. tamrutense

polycephalum Sch. Bip. subsp. *argyrophyllum* (C. Koch) Podlech = T. argyrophyllum
- subsp. *duderanum* (Boiss.) Podlech = T. duderanum

poteriifolium (Ledeb.) Grierson = Pyrethrum poteriifolium

pseudoachillea C. Winkl. - 6

punctatum (Desr.) Grierson = Pyrethrum punctatum

pyrethroides (Kar. & Kir.) Muradjan = Pyrethrum pyrethroides

pyrethroides (Kar. & Kir.) Podlech = Pyrethrum pyrethroides

pyrethroides (Kar. & Kir.) Sch. Bip. = Pyrethrum pyrethroides

santolina C. Winkl. - 1, 3, 6

saxicola (Krasch.) Tzvel. - 6

sclerophyllum (Krasch.) Tzvel. (*T. kittaryanum* (C.A. Mey.) Tzvel. subsp. *sclerophyllum* (Krasch.) Tzvel.) - 1

scopulorum (Krasch.) Tzvel. - 6

sorbifolium (Boiss.) Grierson = Pyrethrum sorbifolium

tabrisianum (Boiss.) Sosn. & Takht. - 2

tamrutense (Sosn.) Sosn. (*T. flavovirens* (Boiss.) Tzvel. subsp. *tamrutense* (Sosn.) Takht., *T. pinnatum* auct. p.p.) - 2

tanacetoides (DC.) Tzvel. - 3, 6

tenuissimum (Trautv.) Grossh. - 2

tomentosum (Decne.) Muradjan = Waldheimia tomentosa

turcomanicum (Krasch.) Tzvel. - 6

turlanicum (Pavl.) Tzvel. - 3, 6

ulutavicum Tzvel. - 6

uniflorum (Fisch. & C.A. Mey.) Sch. Bip. - 2

uralense (Krasch.) Tzvel. (*T. kittaryanum* (C.A. Mey.) subsp. *uralense* (Krasch.) Tzvel.) - 1, 3, 6

vulgare L. - 1, 2, 3, 4, 5(alien), 6
- subsp. *boreale* (Fisch. ex DC.) A. & D. Löve = T. boreale
- subsp. *boreale* (Fisch. ex DC.) Kuvajev = T. boreale

waldsteinii Sch. Bip. = Leucanthemum waldsteinii

walteri (C. Winkl.) Tzvel. - 6

zangezuricum Chandjian = Pyrethrum komarovii

Taraxacum Wigg.

acricorne Dahlst. (*T. dilutum* Dahlst.) - 4, 5

acuminatum Markl. - 1

acutangulum Markl. = T. aequilobum

adamii Claire - 1

aequilobum Dahlst. (*T. acutangulum* Markl., *T. biformatum* Lindb. fil.) - 1

afghanicum Soest (*T. lipskyi* Schischk.) - 6

ajanense Worosch. - 5

aksaicum Schischk. = T. dealbatum

alaicum Schischk. - 6

alascanum Rydb. - 5

alatavicum Schischk. - 6

alatum Lindb. fil. (*T. expallidiforme* Dahlst., *T. privum* Dahlst., *T. sundbergii* Dahlst.) - 1, 3

albescens Dahlst. - 4, 5

almaatense Schischk. - 6

alpigenum Dshanaeva - 6

alpinum auct. = T. panalpinum

altaicum Schischk. = T. ceratophorum

amgense Kuvajev - 4

anadyricum Tzvel. - 5

anchorifolium Kom. = T. perlatescens

andersonii Hagl. - 4, 5

androssovii Schischk. (*T. spinulosum* Soest subsp. *calo-*

cephaloides Soest) - 6
angustisectum Lindb. fil. - 1
angustisquameum Dahlst. ex Lindb. fil. - 1
angustissimum Lindb. fil. - 1
anzobicum Schischk. & Vainberg - 6
arasanum R. Doll - 6
arcticum (Trautv.) Dahlst. - 1, 3, 4, 5
armeniacum Schischk. - 2
armerifolium Soest - 4
aschabadense Schischk. = T. monochlamydeum
assurgens Markl. - 1, 3
atrans Schischk. (*T. pamiricum* Schischk.) - 6
atratum Schischk. = T. pseudoatratum
aurosulum Lindb. fil. - 1
bachczisaraicum Tzvel. - 1
badachschanicum Schischk. - 6
badzhalense Worosch. & Schlothg. - 5
baicalense Schischk. = T. dissectum
bakuense R. Doll - 2
balticum Dahlst. - 1
beckeri Soest - 1, 2, 3, 6
behzudicum Soest - 2
bessarabicum (Hornem.) Hand.-Mazz. - 1, 2, 3, 4, 6
bezidum R. Doll - 2
bicorne Dahlst. - 1, 3, 4, 5, 6
biformatum Lindb. fil. = T. aequilobum
borgvallii Dahlst. & Hagl. = T. mimulum
bornuuriense R. Doll - 4
botschantzevii Schischk. - 6
brachyceras Dahlst. = T. ceratophorum
brachycranum (Dahlst.) Markl. = T. falcatum
brachyglossum (Dahlst.) Dahlst. - 1
brassicifolium Kitam. - 5
brevicorne Dahlst. = T. ceratophorum
brevicorniculatum V. Korol. - 6
brevirostre Hand.-Mazz. - 6
brevum R. Doll - 2
butkovii Kovalevsk. - 6
cachkadzorum R. Doll - 2
cacuminatum Hagl. - 1
calcareum V. Korol. - 6
calocapitatum R. Doll (*T. calocephaloides* R. Doll, 1977, non A.J. Richards, 1976) - 2
calocephaloides R. Doll = T. calocapitatum
caloschistum Dahlst. = T. pectinatiforme
canoviride Lindb. fil. ex Puolanne (*T. multilobum* Dahlst. ex Puolanne) - 1
carinthiacum Soest - 1
carneocoloratum auct. = T. stepanovae
caudatulum Dahlst. = T. distantilobum
cephalum R. Doll - 2
ceratolobum Dahlst. = T. croceum
ceratophorum (Ledeb.) DC. (*T. altaicum* Schischk., *T. brachyceras* Dahlst., *T. brevicorne* Dahlst., *T. chamarense* Peschkova, *T. chamissonis* Greene, *T. evittatum* Dahlst., *T. hultenii* Dahlst., *T. kljutschevskoanum* Kom., *T. koraginense* Kom., *T. koraginicola* Kom., *T. latisquameum* Dahlst., *T. malaisei* Dahlst., *T. platyceras* Dahlst., *T. sachalinense* Kitam. 1933, non Koidz. 1934, *T. trigonolobum* Dahlst.) - 1, 3, 4, 5
- subsp. *lacerum* (Greene) A. & D. Löve = T. lacerum
chamarense Peschkova = T. ceratophorum
chamissonis Greene = T. ceratophorum
chirieanum Kitam. - 5

ciscaucasicum Schischk. - 2
collariatum Worosch. - 5
collinum DC. - 3, 4
comitans Kovalevsk. - 6
commixtiforme Soest - 4
commixtum Hagl. - 1
compactum Schischk. = T. microspermum
complicatum Kovalevsk. = T. svetlanae
confusum Schischk. - 2
conicum R. Doll - 2
constrictifrons Markl. - 1
contristans Kovalevsk. - 6
■ coreanum Nakai
crassipes Lindb. fil. - 1
crebridens Lindb. fil. - 1
crepidiforme DC. = T. stevenii
crispifolium Lindb. fil. (*T. pachylobum* Dahlst.) - 1
croceiflorum Dahlst. - 1
croceum Dahlst. (*T. ceratolobum* Dahlst., *T. officinale* Wigg. var. *lapponicum* Kihlm., *T. lapponicum* (Kihlm.) Hand.-Mazz.) - 1
- subsp. *repletum* Dahlst. = T. repletum
- var. *repletum* (Dahlst.) Tzvel. = T. repletum
curvilobatum Sahlin (*T. curvilobum* Markl. 1926, non Brenn. 1925) - 1
curvilobum Markl. = T. curvilobatum
cyanolepis Dahlst. - 1
czaunense Jurtz. & Tzvel. - 5
czuense Schischk. = T. sinicum
czukoticum Jurtz. - 5
dahlstedtii Lindb. fil. (*T. ekmanii* Dahlst., *T. huelphersianum* Dahlst., *T. lingulatum* Markl., *T. polyodon* Dahlst., *T. praecox* Dahlst. ex Puolanne, *T. retroflexum* Lindb. fil., *T. rhodopodum* Dahlst. ex M.P. Christiansen & Wiinst., *T. vastisectum* Markl. ex Puolanne) - 1, 2, 3, 4, 5(alien), 6
daralagesicum Schischk. - 2
darschajense Orazova - 6
darvasicum Vainberg - 6
dealbatum Hand.-Mazz. (*T. aksaicum* Schischk.) - 3, 4, 5, 6
decipiens Raunk. - 1
decolorans Dahlst. - 1
desertorum Schischk. - 2
dezhnevii Jurtz. = T. tamarae
dilatatum Lindb. fil. - 1
dilutum Dahlst. = T. acricorne
dissectum (Ledeb.) Ledeb. (*T. baicalense* Schischk., *T. haneltii* Soest) - 4
dissimile Dahlst. - 1
distantilobum Lindb. fil. (*T. caudatulum* Dahlst., *T. florstroemii* Markl., *T. involucratum* Dahlst., *T. leptophyllum* Lindb. fil. ex Saltin, *T. polychroum* Ekman ex Th. Lange, *T. semiglobosum* Lindb. fil., *T. subcanescens* Markl.) - 1, 2, 3, 4, 6
distinctum Lindb. fil. - 1
divulsiforme R. Doll - 2
dombaiense R. Doll - 2
duplidens Lindb. fil. = T. ostenfeldii
ecmiadzinum R. Doll - 2
ecornutum Kovalevsk. - 6
egregium Markl. - 1
ekmanii Dahlst. = T. dahlstedtii
elongatum Kovalevsk. - 6
eriobasis Kovalevsk. - 6
eriopodum (D. Don) DC. - 4, 6(?)
erostre Zak. - 5

erythrospermum Andrz. (*T. laevigatum* auct.) - 1, 2
- subsp. *gotlandicum* Dahlst. = T. gotlandicum
- subsp. *laetum* Dahlst. = T. laetum
- subsp. *marginatum* Dahlst. = T. marginatum
- subsp. *proximum* Dahlst. = T. proximum
- subsp. *tenuilobum* Dahlst. = T. tenuilobum
estonicum Markl. = T. mucronatum
euoplocarpum Markl. - 1
evittatum Dahlst. = T. ceratophorum
expallidiforme Dahlst. = T. alatum
fagerstroemii Saltin - 1
falcatum Brenn. (*T. brachycranum* (Dahlst.) Markl.) - 1
fasciatum Dahlst. - 1
fedtschenkoi Hand.-Mazz. - 6
florstroemii Markl. = T. distantilobum
flugum R. Doll - 2
fontanicola auct. = T. peralatum
fontanum auct. = T. peralatum
forellense R. Doll - 2
fulvipile Harv. - 2
fulvum Raunk. (*T. isthmicola* Lindb. fil.) - 1
galbanum Dahlst. - 1
galeatum Dahlst. - 1
geminatum Hagl. - 1(?)
georgicum Soest - 2
glabellum Schischk. - 6
glaberrimum R.Doll (*T. glabratum* R. Doll, 1977, non (Banks & Soland.) Cockayne, 1907) - 2
glabratum R. Doll = T. glaberrimum
glabrum DC. - 1, 4
glaucanthum (Ledeb.) DC. - 1, 3, 4, 6
glaucivirens Schischk. - 6
gnezdilloi Kovalevsk. - 6
goloskokovii Schischk. - 6
gorodkovii Charkev. & Tzvel. - 5
gotlandicum (Dahlst.) Dahlst. (*T. erytrospermum* Andrz. subsp. *gotlandicum* Dahlst.) - 1
graciliforme R. Doll - 2
gracilium R. Doll - 2
grandisquamatum Koidz. - 5
grossheimii Schischk. - 2
grypodiforme R. Doll - 6
guntense Dengubenko - 6
haematicum Hagl. - 1
haematopus Lindb. fil. - 1
hamatilobum Dahlst. - 1
haneltii Soest = T. dissectum
hastigerum Markl. ex Back - 1
heleonastes auct. = T. hollandicum
hellenicum Dahlst. - 1
heptapotamicum Schischk. - 6
hjeltii (Dahlst.) Dahlst. - 1
hohenackerianum Soest - 2
hollandicum Soest (*T. heleonastes* auct.) - 1
holophyllum Schischk. = T. monochlamydeum
huelphersianum Dahlst. = T. dahlstedtii
hultenii Dahlst. = T. ceratophorum
hybernum Stev. - 1
hypanicum Tzvel. - 1
hyparcticum Dahlst. - 5
idiolepium Markl. = T. litorale
ikonnikovii Schischk. = T. luridum
insigne Ekman ex Wiinst. & Jess. - 1
intercedens Markl. - 1
intumescens Hagl. = T. mimulum
involucratum Dahlst. = T. distantilobum
isophyllum Hagl. - 1

isthmicola Lindb. fil. = T. fulvum
jacuticum Tzvel. - 4
jaschilkuliense Vainberg - 6
jurtzevii Tzvel. - 5
juzepczukii Schischk. - 6
kamtschaticum Dahlst. - 5
karakoricum Soest - 6
karatavicum Pavl. ex Schischk. - 6
karelicum Lindb. fil. & Markl. - 1
kasachiforme R. Doll - 6
kasachum R. Doll - 6
ketoiense Tatew. & Kitam. - 5
kimuranum Kitam. - 5
kirghizicum Schischk. - 6
kjellmanii Dahlst. - 1
kljutschevskoanum Kom. = T. ceratophorum
klokovii Litvinenko - 1
koidzumii Nemoto (*T. sachalinense* Koidz. 1934, non Kitam. 1933) - 5
kojimae Kitam. - 5
kok-saghyz Rodin - 1(alien), 6
kolymense A. Khokhr. - 5
koraginense Kom. = T. ceratophorum
koraginicola Kom. = T. ceratophorum
korjakense Charkev. & Tzvel. - 5
korjakorum Charkev. & Tzvel. - 1, 4, 5
kotschyi Soest - 2
kovalevskiae Vainberg - 6
kuusamoense Lindb. fil. & Palmgr. - 1
kuvajevii Tzvel. - 4, 5
laceratum (Brenn.) Brenn. = T. parvuliceps
lacerum Greene (*T. ceratophorum* (Ledeb.) DC. subsp. *lacerum* (Greene) A. & D. Löve) - 5
laciniosifrons Wiinst. = T. penicilliforme
laciniosum Dahlst. (*T. pallescens* Dahlst.) - 1
lacinulatum Markl. = T. penicilliforme
lacistophyllum (Dahlst.) Raunk. - 1
laeticolor Dahlst. = T. officinale
laetum (Dahlst.) Dahlst. (*T. erytrospermum* Andrz. subsp. *laetum* Dahlst.) - 1
- subsp. *obscurans* Dahlst. = T. obscurans
laevigatum auct. = T. erythrospermum
lapponicum (Kihlm.) Hand.-Mazz. = T. croceum
lasianthum Koidz. - 5
lateritium Dahlst. - 1, 3, 4, 5
laticordatum Markl. - 1
latisectum Lindb. fil. - 1
latisquameum Dahlst. = T. ceratophorum
lehbertii Lindb. fil. - 1
lenense Tzvel. - 4, 5
lenkoranense R. Doll - 2
leptoceras Dahlst. - 5
leptophyllum Lindb. fil. ex Saltin = T. distantilobum
leucanthum (Ledeb.) Ledeb. - 3, 4, 6
leucocarpum Jurtz. & Tzvel. - 5
leucoglossum Brenn. (*T. turiense* Orlova) - 1
lilacinum Krasn. ex Schischk. - 6
limbatum auct. = T. obscurans
linczevskii Schischk. - 6
lineare Worosch. & Schaga - 5
linguatifrons Markl. - 1
linguicuspis Lindb. fil. - 1
lingulatum Markl. = T. dahlstedtii
lipskyi Schischk. = T. afghanicum
lissocarpum (Dahlst.) Dahlst. = T. lividum
litorale Raunk. (*T. idiolepium* Markl.) - 1
litwinowii Schischk. = T. stevenii

lividum (Waldst. & Kit.) Peterm. (*T. lissocarpum* (Dahlst.) Dahlst.) - 1

longicorne Dahlst. (*T. printzii* Dahlst.) - 4, 5

longipes Kom. = T. perlatescens

longipyramidatum Schischk. - 6

longirostre Schischk. - 6

longisquameum Lindb. fil. - 1

lucidum Dahlst. - 1

lugubriforme R. Doll - 6

luridum Hagl. (*T. ikonnikovii* Schischk., *T. pojarkoviae* Schischk.) - 6

lyngeanum Hagl. - 1, 5

lyratum (Ledeb.) DC. - 3

macilentum Dahlst. - 1, 3, 4, 5

macroceras Dahlst. - 1, 4, 5

macrochlamydeum Kovalevsk. - 6

macrolepium Schischk. - 2(?)

maculatum Jord. - 1

magnum Kovalevsk. - 6

majus Schischk. - 6

malaisei Dahlst. = T. ceratophorum

maracandicum Kovalevsk. (*T. pseudoalpinum* Schischk. ex Orazova) - 6

marginatum (Dahlst.) Dahlst. (*T. erythrospermum* Andrz. subsp. *marginatum* Dahlst.) - 1

marklundii Palmgr. - 1

maurostigma Markl. - 1

medeense R. Doll - 6

megricum Tzvel. - 2

melanostigma Lindb. fil. - 1

mexicanum DC. - 5

microcarpum Koidz. - 5

microcranum Markl. - 1

microlobum Markl. - 1

microspermum Schischk. (*T. compactum* Schischk.) - 3, 6

mimulum Dahlst. ex Lindb. fil. (*T. borgvallii* Dahlst. & Hagl., *T. intumescens* Hagl.) - 1

minutilobum M. Pop. ex Kovalevsk. - 6

miyakei Kitam. - 5

modestum Schischk. - 6

mongolicum Hand.-Mazz. - 4, 5

mongoliforme R. Doll - 5

monochlamydeum Hand.-Mazz. (*T. aschabadense* Schischk., *T. holophyllum* Schischk.) - 6

montanum (C.A. Mey.) DC. - 2, 6

mucronatum Lindb. fil. (*T. estonicum* Markl.) - 1, 2, 3

mujense Petroczenko - 4

multilobum Dahlst. ex Puolanne = T. canoviride

multiscaposum Schischk. - 6

multisectum Kitam. - 5

murbeckianum auct. = T. pseudomurbeckianum

murgabicum Vainberg - 6

muricatum Schischk. = T. neolobulatum

murmanicum Orlova - 1

nairoense Koidz. - 5

nanaunii Jurtz. - 5

natschikense Kom. = T. rubiginans

neokamtschaticum Worosch. - 5

neolobulatum Soest (*T. muricatum* Schischk.) - 2, 6

neosachalinense Koidz. - 5

neosivaschicum Tzvel. - 1

nevskii Juz. - 6

nigricans (Kit.) Reichenb. - 1

nigrocephalum A. Khokhr. - 5

nikitinii Schischk. - 6

nikolai Vainberg - 6

nivale Lange ex Kihlm. (*T. tundricola* Hand.-Mazz.) - 1, 3,

4

norvegicum (Dahlst.) Dahlst. - 1

novae-zemliae Holmboe - 1, 4

nudiscaposum Worosch. - 5

nuratavicum Schischk. - 6

obliquelobum Dahlst. - 1

obliquum (Fries) Dahlst. - 1

obscurans (Dahlst.) Hagl. (*T. laetum* (Dahlst.) Dahlst. subsp. *obscurans* Dahlst., *T. limbatum* auct.) - 1

- f. *psammophilum* Hagl. = T. psammophilum

ochotense Worosch. - 5

officinale Wigg. (*T. laeticolor* Dahlst.) - 1, 5

- subsp. *laceratum* Brenn. = T. parviliceps

- var. *lapponicum* Kihlm. = T. croceum

ohwianum Kitam. - 5

oliganthum Schott & Kotschy ex Hand.-Mazz. - 2

omissum Hagl. - 1

oschense Schischk. - 6

ostenfeldii Raunk. (*T. duplidens* Lindb. fil.) - 1, 2, 3, 4, 5(alien), 6

otagirianum Koidz. ex Kitam. - 5

pachylobum Dahlst. = T. crispifolium

pallescens Dahlst. = T. laciniosum

pallidulum Lindb. fil. - 1

pamiricum Schischk. = T. atrans

panalpinum Soest (*T. alpinum* auct.) - 1

pannonicum Sonck & Soest - 1

pannulatum Dahlst. (*T. piceatum* Dahlst.) - 1

papposum R. Doll - 2

parviliceps Lindb. fil. (? *T. laceratum* (Brenn.) Brenn., ? *T. officinale* Wigg. subsp. *laceratum* Brenn.) - 1

paucijugum Markl. - 1

pavlovii Orazova - 6

pawlodarskum R. Doll (*T. ustamenum* R. Doll) - 3

pectinatiforme Lindb. fil. (*T. caloschistum* Dahlst., *T. trilobatum* Palmgr.) - 1

penicilliforme Lindb. fil. (*T. laciniosifrons* Wiinst., *T. lacinulatum* Markl., *T. subintegrum* Dahlst., *T. subpenicilliforme* Lindb. fil. ex Dahlst.) - 1, 3

peralatum Soest (*T. fontanicola* auct., *T. fontanum* auct.) - 1

perfiljevii Orlova - 1, 3, 4

perlatescens Dahlst. (*T. anchorifolium* Kom., *T. longipes* Kom.) - 5

perpusillum Schischk. - 6

petri-primi Vainberg - 6

petrovskyi Tzvel. - 4, 5

phymatocarpum J. Vahl - 4, 5

piceatum Dahlst. = T. pannulatum

pineticola Klok. = T. proximum

pingue Schischk. - 6

planum Raunk. - 1, 2

platyceras Dahlst. = T. ceratophorum

platycranum Dahlst. - 5

platyglossum Raunk. - 1

platylepium Dahlst. - 1, 3, 4, 5

platypecidum Diels - 5

pobedimoviae Schischk. - 1

podkumokense Galushko - 2

pohlii Soest - 1

pojarkoviae Schischk. = T. luridum

polonicum Matecka & Soest - 1

polozhiae Kurbatsk. - 4

polychroum Ekman ex Th. Lange = T. distantilobum

polyodon Dahlst. = T. dahlstedtii

popovii Kovalevsk. ex Vainberg - 6

porphyranthum Boiss. - 2

praecox Dahlst. ex Puolanne = T. dahlstedtii
praecox Schischk. = T. taschkenticum
praestans Lindb. fil. - 1
praticola Schischk. = T. prilipkoi
prilipkoi Czer. nom. nov. (*T. praticola* Schischk. 1934, in Grossh. Fl. Cauc. 4 : 245, non Dahlst. 1911) - 2
printzii Dahlst. = T. longicorne
privum Dahlst. = T. alatum
promontoriorum Dshanaeva - 6
proximum (Dahlst.) Dahlst. (*T. erytrospermum* Andrz. subsp. *proximum* Dahlst., *T. pineticola* Klok.) - 1
pruinosum Koidz. - 5
psammophilum (Hagl.) Saarsoo (*T. obscurans* (Dahlst.) Hagl. f. *psammophilum* Hagl.) - 1
pseudoalpinum Schischk. ex Orazova = T. maracandicum
pseudoatratum Orazova (*T. atratum* Schischk. 1964, non Hagl. 1948) - 6
pseudobrevirostre Vainberg - 6
pseudocalocephalum Soest - 6
pseudodissimile Soest - 6
pseudofulvum Lindb. fil. - 1
pseudoglabrum Dahlst. = T. rubiginans
pseudogracilens R. Doll - 2
pseudolasianthum Koidz. - 5
pseudolaxum R. Doll - 2
pseudolugubre R. Doll - 6
pseudominutilobum Kovalevsk. - 6
pseudomurbeckianum Tzvel. (*T. murbeckianum* auct.) - 1
pseudonigricans Hand.-Mazz. - 2
pseudonivale Malysch. - 4
pseudophaleratum R. Doll - 2
pseudoplatylepium Jurtz. - 5
pseudoroseum Schischk. - 6
pseudosilesiacum R. Doll - 2
pseudotianschanicum R. Doll - 3
pulcherrimum Lindb. fil. - 1
pulchrifolium Markl. - 1
pycnolobum Dahlst. - 1
raikoviae Vainberg - 6
recurvum Dahlst. = T. reflexilobum
reflexilobum Lindb.fil. (*T. recurvum* Dahlst.) - 1
reinthalii Markl. - 1
repandum Pavl. - 6
repletum (Dahlst.) Dahlst. (*T. croceum* Dahlst. subsp. *repletum* Dahlst., *T. croceum* var. *repletum* (Dahlst.) Tzvel.) - 1, 3, 4
retroflexum Lindb. fil. = T. dahlstedtii
revalense Lindb. fil. - 1
rhodopodum Dahlst. ex M.P. Christiansen & Wiinst. = T. dahlstedtii
rizaense R. Doll - 2
- subsp. **jerevanum** R. Doll - 2
rubidum Schischk. - 6
rubiginans Dahlst. (*T. natschikense* Kom., *T. pseudoglabrum* Dahlst.) - 5
rubrisquameum M.P. Christiansen - 1(?)
rubtzovii Schischk. = T. songoricum
rufum Dahlst. - 5
sachalinense Kitam. = T. ceratophorum
sachalinense Koidz. = T. koidzumii
sagittifolium Lindb. fil. ex Dahlst. - 1
sagittipotens Dahlst. & Ohlsen ex Hagl. - 1(?)
sangilense Krasnob. & V. Khan. - 4
saposhnikovii Schischk. - 6
scanicum Dahlst. - 1, 5
schelkovnikovii Schischk. - 2
schirakium R. Doll - 2

schischkinii V. Korol. - 6
schugnanicum Schischk. - 6
scolopendriforme R. Doll - 2
semiglobosum Lindb. fil. = T. distantilobum
semitubulosum Jurtz. - 4
senjavinense Jurtz. & Tzvel. - 5
seravschanicum Schischk. - 6
serotinum (Waldst. & Kit.) Poir. - 1, 2, 3
shikotanense Kitam. - 5
shimushirense Tatew. & Kitam. - 5
shumushuense Kitam. - 5
sibiricum Dahlst. - 4, 5
sieheaniforme R. Doll - 2
silesiacum Dahlst. ex Hagl. - 1
simulum Brenn. - 1
sinicum Kitag. (*T. czuense* Schischk.) - 1(alien), 3, 4, 5
soczavae Tzvel. - 5
songoricum Schischk. (*T. rubtzovii* Schischk.) - 6
spinulosum Soest subsp. *calocephaloides* Soest = T. androssovii
stanjukoviczii Schischk. - 6
stellare Markl. - 1
stenolepium Hand.-Mazz. - 2
stenolobum Stschegl. - 3, 5
stepanovae Worosch. (*T. carneocoloratum* auct.) - 4, 5
stevenii DC. (*T. crepidiforme* DC., *T. litwinowii* Schischk.) - 2
steveniiforme R. Doll - 2
strizhoviae Vainberg - 6
strobilocephalum Kovalevsk. - 6
subalternilobum A. Khokhr. - 5
subcanescens Markl. = T. distantilobum
subglaciale Schischk. - 6
subintegrum Dahlst. = T. penicilliforme
sublaciniosum Dahlst. & Lindb. fil. - 1
sublaeticolor Dahlst. = T. tenebricans
subpenicilliforme Lindb. fil. ex Dahlst. = T. penicilliforme
suecicum Hagl. - 1
sugawarae Koidz. - 5
sumneviczii Schischk. - 3, 6
sundbergii Dahlst. = T. alatum
svetlanae Czer. nom. nov. (*T. complicatum* Kovalevsk. 1962, Bot. Mat. (Tashkent) 17 : 14, non Soest, 1959) - 6
syriacum Boiss. - 2, 6
syrtorum Dshanaeva - 6
tadshikorum Ovcz.ex Schischk. - 6
taeniatum Hagl. - 1
taimyrense Tzvel. - 4
tamarae Charkev. & Tzvel. (*T. dezhnevii* Jurtz. nom. invalid.) - 5
taschkenticum Orazova (*T. praecox* Schischk. 1964, non Dahlst. ex Puolanne, 1933) - 6
tatewakii Kitam. - 5
tauricum Kotov - 1
tenebricans (Dahlst.) Dahlst. (*T. sublaeticolor* Dahlst.) - 1, 2, 3, 4, 5(alien), 6
tenellisquameum Markl. - 1
tenuilobum (Dahlst.) Dahlst. (*T. erytrospermum* Andrz. subsp. *tenuilobum* Dahlst.) - 1
tenuisectum Somm. & Levier - 2
thracicum Soest - 1
tianschanicum Pavl. - 6
tinctum Markl. - 1
tolmaczevii Jurtz. - 5
tortilobum Florstrom - 1
triangulare Lindb. fil. - 1
trigonolobum Dahlst. = T. ceratophorum
trilobatum Palmgr. = T. pectinatiforme
tujuksuense Orazova - 6

tundricola Hand.-Mazz. = T. nivale
turgaicum Schischk. - 1, 3, 6
turiense Orlova = T. leucoglossum
tuvinense Krasnob. & A. Krasnikov - 4
tzvelevii Schischk. - 6
undulatiforme Dahlst. - 1(?)
urdzharense Orazova - 3
uschakovii Jurtz. - 5
ussuriense Kom. - 5
ustamenum R. Doll = T. pawlodarskum
vardenium R. Doll - 2
variegatum Kitag. - 5
varioviolaceum A. Khokhr. - 5
varsobicum Schischk. - 6
vassilczenkoi Schischk. - 6
vastisectum Markl. ex Puolanne = T. dahlstedtii
vestitum Worosch. - 5
violaceum R. Doll - 6
vitalii Orazova - 6
voronovii Schischk. - 2
vulcanorum Koidz. - 5
vulgum R. Doll - 2
wardenium R. Doll - 2
woroschilovii Gubanov - 5
wrangelicum Tzvel. - 5
xanthostigma Lindb. fil. - 1
xerophilum Markl. - 1
yamamotoi Koidz. - 5
yetrofuense Kitam. - 5
zhukovae Tzvel. - 5
zineralum R. Doll - 2
ziwaschum R. Doll - 1

Telekia Baumg.

speciosa (Schreb.) Baumg. - 1, 2

Tephroseris (Reichenb.) Reichenb. (*Othonna* L. p.min.p., *Packera* A. & D. Löve, *Cineraria* auct.)

adenolepis C. Jeffrey & Y.C. Chen - 5
aquilonaris (Schischk.) A. & D. Löve = T. atropurpurea
x **arctisibirica** (Jurtz. & Korobkov) Czer. comb. nova (*Senecio* x *arctisibiricus* Jurtz. & Korobkov, 1987, Fl. Arct. URSS, 10 : 235). - T. atropurpurea (Ledeb.) Holub x T. palustris (L.) Reichenb. - 4, 5
atropurpurea (Ledeb.) Holub (*Cineraria frigida* Richards. b. *taimyrensis* Herd., *Senecio aquilonaris* Schischk., *S. atropurpureus* (Ledeb.) B. Fedtsch., *S. integrifolius* (L.) Clairv. subsp. *atropurpureus* (Ledeb.) Cuf., *S. taimyrensis* (Herd.) Gorodk. ex Tichomirov, *S. trautvetteri* Maximova, *Tephroseris aquilonaris* (Schischk.) A. & D. Löve, *T. integrifolia* (L.) Holub subsp. *atropurpurea* (Ledeb.) B. Nordenstam) - 1, 3, 4, 5
- subsp. *frigida* (Richards.) A. & D. Löve = T. frigida
- subsp. *tomentosa* auct. p.p. = T. jacutica
aurantiaca (Hoppe ex Willd.) Griseb. & Schenk (*Senecio aurantiacus* (Hoppe ex Willd.) Less., *S. integrifolius* (L.) Clairv. subsp. *aurantiacus* (Hoppe ex Willd.) Briq. & Cavill., *Tephroseris integrifolia* (L.) Holub subsp. *aurantiaca* (Hoppe ex Willd.) B. Nordenstam) - 1
besseriana (Minder.) Czer. comb. nova (*Senecio besserianus* Minder. 1956, Ukr. Bot. Zhurn. 13, 3 : 58) - 1
campestris (Retz.) Reichenb. = T. integrifolia
capitata (Wahlenb.) Griseb. & Schenk (*Senecio capitatus* (Wahlenb.) Steud., *S. integrifolius* (L.) Clairv. subsp.

capitatus (Wahlenb.) Cuf., *Tephroseris capitata* (Wahlenb.) Holub, comb. superfl., *T. integrifolia* (L.) Holub subsp. *capitata* (Wahlenb.) B.Nordenstam) - 1
caucasigena (Schischk.) Czer. comb. nova (*Senecio caucasigenus* Schischk. 1954, Bot. Mat. (Leningrad) 16 : 427) - 2
cladobotrys (Ledeb.) Griseb. & Schenk (*Senecio cladobotrys* Ledeb.) - 2
crispa (Jacq.) Reichenb. (*Cineraria crispa* Jacq., *Senecio crispus* (Jacq.) Kitt. 1844, non Thunb. 1794, *S. rivularis* (Waldst. & Kit.) DC., *Tephroseris rivularis* (Waldst. & Kit.) Schur) - 1
czernjaevii (Minder.) Holub (*Senecio czernjaevii* Minder., *S. integrifolius* (L.) Clairv. subsp. *czernjaevii* (Minder.) Chater, *Tephroseris integrifolia* (L.) Holub subsp. *czernjaevii* (Minder.) B. Nordenstam) - 1, 3
flammea (DC.) Holub (*Senecio flammeus* Turcz. ex DC.) - 4, 5
frigida (Richards.) Holub (*Senecio frigidus* (Richards.) Less., *Tephroseris atropurpurea* (Ledeb.) Holub subsp. *frigida* (Richards.) A. & D. Löve) - 4, 5
heterophylla (Fisch.) Konechn. (*Cineraria heterophylla* Fisch. 1812, non *Senecio heterophyllus* Thunb. 1794, *Packera resedifolia* (Less.) A. & D. Löve, *Senecio resedifolius* Less.) - 1, 3, 4, 5
hieraciformis (Kom.) Czer. comb. nova (*Senecio hieraciiformis* Kom. 1926, Bot. Mat. (Leningrad) 6, 1 : 16) - 5
hyperborealis (Greenm.) Czer. comb. nova (*Senecio hyperborealis* Greenm. 1916, Ann. Missouri Bot. Gard. 3 : 98; *Packera hyperborealis* (Greenm.) A. & D. Löve) - 5
- subsp. **wrangelica** (Jurtz., Korobkov & Petrovsky) Czer. comb. nova (*Senecio hyperborealis* subsp. *wrangelicus* Jurtz., Korobkov & Petrovsky, 1987, Fl. Arct. URSS, 10 : 216) - 5
igoschinae (Schischk.) B.Nordenstam (*Senecio igoschinae* Schischk.) - 1, 3
integrifolia (L.) Holub (*Othonna integrifolia* L., *Cineraria pratensis* Hoppe, *Senecio campestris* (Retz.) DC., *S. campestris* var. *lenensis* (Schischk.) Hamet-Ahti, *S. integrifolius* (L.) Clairv., *S. lenensis* Schischk., *S. pratensis* (Hoppe) DC., *S. succisifolius* Kom., *Tephroseris campestris* (Retz.) Reichenb., *T. lenensis* (Schischk.) Holub) - 1, 3, 4, 5
- subsp. *atropurpurea* (Ledeb.) B. Nordenstam = T. atropurpurea
- subsp. *aurantiaca* (Hoppe ex Willd.) B. Nordenstam = T. aurantiaca
- subsp. *capitata* (Wahlenb.) B. Nordenstam = T. capitata
- subsp. *czernjaevii* (Minder.) B. Nordenstam = T. czernjaevii
- subsp. *kirilowii* (Turcz. ex DC.) B. Nordenstam = T. kirilowii
- subsp. *tundricola* (Tolm.) B. Nordenstam = T. tundricola
jacutica (Schischk.) Holub (*Senecio jacuticus* Schischk., *Tephroseris atropurpurea* (Ledeb.) Holub subsp. *tomentosa* auct. p.p.) - 4, 5
jailicola (Juz.) Konechn. (*Senecio jailicola* Juz.) - 1
karjaginii (Sof.) Holub (*Senecio karjaginii* Sof.) - 2
kawakamii (Makino) Holub (*Senecio kawakamii* Makino) - 5
kirilowii (Turcz. ex DC.) Holub (*Senecio kirilowii* Turcz. ex DC., *S. amurensis* Schischk. p.p. excl. syn. *S. pratensis* var. *polycephalus* Regel, *S. campestris* (Retz.) DC. subsp. *kirilowii* (Turcz. ex DC.) Kitag., *S. integrifolius* (L.) Clairv. subsp. *amurensis* (Schischk.) Worosch. comb. invalid., *Tephroseris integrifolia* (L.) Holub subsp. *kirilowii* (Turcz. ex DC.) B. Nordenstam) - 5

kjellmanii (A. Pors.) Holub (*Senecio kjellmanii* A. Pors., *Cineraria frigida* Richards. f. *tomentosa* Kjellm., *S. atropurpureus* (Ledeb.) B. Fedtsch. subsp. *tomentosus* (Kjellm.) Hult., *S. tichomirovii* Schischk., *Tephroseris tichomirovii* (Schischk.) Holub) - 5

lenensis (Schischk.) Holub = T. integrifolia

lindstroemii (Ostenf.) A. & D. Löve = T. pseudoaurantiaca

palustris (L.) Reichenb. (*Othonna palustris* L., *Cineraria congesta* R. Br., *Senecio arcticus* Rupr., *S. congestus* (R. Br.) DC., *S. congestus* subsp. *palustris* (L.) Rauschert, *S. palustris* (L.) Hook. 1834, non Vellozo, 1827, *S. tubicaulis* Mansf., *Tephroseris palustris* subsp. *congesta* (R.Br.) Holub) - 1, 3, 4, 5, 6

- subsp. *congesta* (R. Br.) Holub = T. palustris

papposa (Reichenb.) Schur (*Senecio papposus* (Reichenb.) Less., *Tephroseris papposa* (Reichenb.) Holub, comb. superfl.) - 1

porphyrantha (Schischk.) Holub (*Senecio porphyranthus* Schischk.) - 4

praticola (Schischk. & Serg.) Holub (*Senecio praticola* Schischk. & Serg., *S. asiatica* Schischk. & Serg. nom. illegit. superfl.) - 3, 4

pricei (Simps.) Holub (*Senecio pricei* Simps.) - 3, 4

pseudoaurantiaca (Kom.) Czer. comb. nova (*Senecio pseudoaurantiacus* Kom. 1930, Fl. Kamch. 3 : 168; *S. integrifolius* (L.) Clairv. var. *lindstroemii* Ostenf., *S. kamtschaticus* Kom., *S. lindstroemii* (Ostenf.) A. Pors., *S. tundricola* Tolm. subsp. *lindstroemii* (Ostenf.) Korobkov, *Tephroseris lindstroemii* (Ostenf.) A. & D. Löve, *Senecio fuscatus* auct.) - 5

pyroglossa (Kar. & Kir.) Holub (*Senecio pyroglossus* Kar. & Kir.) - 6

rivularis (Waldst. & Kit.) Schur = T. crispa

schistosa (Charkev.) Czer. comb. nova (*Senecio schistosus* Charkev. 1979, Bot. Zhurn. 64, 4 :555) - 5

sichotensis (Kom.) Holub (*Senecio sichotensis* Kom.) - 5

subdentata (Bunge) Holub (*Cineraria subdentata* Bunge, *Senecio amurensis* Schischk. p.p. quoad syn. *S. pratensis* var. *polycephalus* Regel) - 5

subfloccosa (Schischk.) Czer. comb. nova (*Senecio subfloccosus* Schischk. 1953, Bot. Mat. (Leningrad) 15 : 402) - 2

subfrigida (Kom.) Holub (*Senecio subfrigidus* Kom.) - 5

subscaposa (Kom.) Czer. comb. nova (*Senecio subscaposus* Kom. 1926, Bot. Mat. (Leningrad) 6, 1 : 16) - 5

sukaczevii (Schischk.) Holub (*Senecio sukaczevii* Schischk.) - 4

tichomirovii (Schischk.) Holub = T. kjellmanii

tundricola (Tolm.) Holub (*Senecio tundricola* Tolm., *S. integrifolius* (L.) Clairv. subsp. *tundricola* (Tolm.) Chater, *Tephroseris integrifolia* (L.) Holub subsp. *tundricola* (Tolm.) B. Nordenstam) - 1, 3, 4, 5

turczaninovii (DC.) Holub (*Senecio turczaninovii* DC., *S. turczaninovii* subsp. *czikoicus* Maximova, *S. reverdattoi* K. Sobol., *S. sumneviczii* Schischk. & Serg.) - 3, 4, 5, 6

veresczaginii (Schischk. & Serg.) Holub (*Senecio veresczaginii* Schischk. & Serg.) - 3

Theodorea Cass. = Saussurea

crepidifolia (Turcz.) Sojak = Saussurea alata
glomerata (Poir.) Sojak = Saussurea amara
larionowii (C. Winkl.) Sojak = Saussurea larionowii
neopulchella (Lipsch.) Sojak = Saussurea neopulchella

Thevenotia DC.

scabra (Boiss.) Boiss. - 6

Thrincia Roth

hispida Roth - 1

Tomanthea DC.

aucheri DC. (*Centaurea aucheri* (DC.) Wagenitz) - 2
carthamoides (DC.) Takht. - 2
daralaghezica (Fomin) Takht. - 2
phaeopappa (DC.) Takht. ex Czer. (*Centaurea aucheri* (DC.) Wagenitz subsp. *szowitsii* (Boiss.) Wagenitz, *C. aucheri* var. *szowitsii* (Boiss.) Wagenitz) - 2
spectabilis (Fisch. & C.A. Mey.) Takht. - 2

Tragopogon L.

acanthocarpus Boiss. - 2
alaicus S. Nikit. - 2
altaicus S. Nikit. & Schischk. - 3
artemczukii Klok. = T. ucrainicus
***australis** Jord. - 1
badachschanicus Boriss. - 6
bjelorussicus Artemcz. (*T. lithuanicus* (DC.) Boriss.) - 1
borysthenicus Artemcz. (*T. dolichocarpus* Klok.) - 1
brevirostris DC. (*T. floccosus* Waldst. & Kit. var. *brevirostris* (DC.) E.I. Nyarady) - 2
- var. *volgensis* S. Nikit. = T. podolicus
buphthalmoides (DC.) Boiss. - 2
capitatus S. Nikit. - 6
charadzeae Kuth. = T. filifolius
colchicus Albov - 2
collinus DC. - 2
coloratus C.A. Mey. - 2
- subsp. *nachitschevanicus* (Kuth.) T.N. Pop. = T. pterocarpus
conduplicatus S. Nikit. - 6
cretaceus S. Nikit. - 1
daghestanicus (Artemcz.) Kuth. = T. dasyrhynchus var. daghestanicus
dasyrhynchus Artemcz. - 1, 2
- var. **daghestanicus** (Artemcz.) Tzvel. (*T. daghestanicus* (Artemcz.) Kuth.) - 2
desertorum (Lindem.) Klok. (*T. major* Jacq. var. *desertorum* Lindem., *T. dubius* Scop. subsp. *desertorum* (Lindem.) Tzvel., *T. tesquicola* Klok.) - 1
dolichocarpus Klok. = T. borysthenicus
donetzicus Artemcz. = T. tanaiticus
dubius Scop. (*T. tauricus* Klok.) - 1, 2
- subsp. *desertorum* (Lindem.) Tzvel. = T. desertorum
- subsp. *major* (Jacq.) Vollm. = T. major
dubjanskyi Krasch. & S. Nikit. - 1(?), 6
elatior Stev. - 1
elongatus S. Nikit. - 3, 6
filifolius Rehm. ex Boiss. (*T. charadzeae* Kuth.) - 2
floccosus Waldst. & Kit. var. *brevirostris* (DC.) E.I. Nyarady = T. brevirostris
gaudanicus Boriss. - 6
gorskianus Reichenb. fil. - 1
gracilis D. Don - 6
graminifolius DC. - 2, 6
- subsp. *serotinus* (Sosn.) T.N. Pop. = T. serotinus
heterospermus Schweigg. - 1
idae Kuth. - 2
karelinii S. Nikit. - 6
karjaginii Kuth. - 2
kasahstanicus S. Nikit. = T. ruber
kemulariae Kuth. - 2
ketzkhovelii Kunth. - 2

kopetdaghensis Boriss. - 6
krascheninnikovii S. Nikit. (*T. longirostris* auct.) - 2, 6
kultiassovii M. Pop. ex S. Nikit. - 6
latifolius Boiss. - 2
leiorhynchus Klok. = T. podolicus
leonidae Kuth. - 2
lithuanicus (DC.) Boriss. = T. bjelorussicus
longirostris auct. = T. krascheninnikovii
macropogon C.A. Mey. - 2
major Jacq. (*T. dubius* Scop. subsp. *major* (Jacq.) Vollm.) - 1, 2
- var. *desertorum* Lindem. = T. desertorum
makaschwilii Kuth. - 2
malikus S. Nikit. - 6
marginatus Boiss. & Buhse - 2
marginifolius Pavl. - 1, 6
maturatus Boriss. - 6
melananthenus Klok. = T. orientalis
meskheticus Kuth. - 2
moldavicus Klok. = T. orientalis
montanus S. Nikit. - 6
nachitschevanicus Kuth. = T. pterocarpus
orientalis L. (*T. melananthenus* Klok., *T. moldavicus* Klok., *T. pratensis* L. subsp. *orientalis* (L.) Celak., *T. transcarpaticus* Klok., *T. xanthantherus* Klok.) - 1, 3, 4(alien), 5(alien), 6
otschiaurii Kuth. - 2
paradoxus S. Nikit. - 6
plantagineus Boiss. & Huet - 2
podolicus (DC.) S. Nikit. (*T. brevirostris* DC. var. *volgensis* S. Nikit., *T. leiorhynchus* Klok., *T. stepposus* (S. Nikit.) Stank., *T. volgensis* (S. Nikit.) S. Nikit.) - 1, 3, 6
***porrifolius** L. - 1
pratensis L. - 1
- subsp. *orientalis* (L.) Celak. = T. orientalis
pseudomajor S. Nikit. - 6
pterocarpus DC. (*T. coloratus* C.A. Mey. subsp. *nachitschevanicus* (Kuth.) T.N. Pop., *T. nachitschevanicus* Kuth.) - 2
pusillus Bieb. - 1, 2, 6
reticulatus Boiss. & Huet - 2
ruber S.G. Gmel. (*T. kasahstanicus* S. Nikit.) - 1, 3, 6
ruthenicus Bess. ex Krasch. & S. Nikit. - 1, 2, 6
- subsp. *donetzicus* (Artemcz.) I.B.K. Richardson = T. tanaiticus
sabulosus Krasch. & S. Nikit. - 3, 6
savranicus Sobko - 1
scoparius S. Nikit. - 6
segetus Kuth. - 2
serawschanicus S. Nikit. - 6
serotinus Sosn. (*T. graminifolius* DC. subsp. *serotinus* (Sosn.) T.N. Pop.) - 2
sibiricus Ganesch. - 1, 3, 4
songoricus S. Nikit. - 3, 6
sosnowskyi Kuth. - 2
stepposus (S. Nikit.) Stank. = T. podolicus
subalpinus S. Nikit. - 6
tanaiticus Artemcz. (*T. donetzicus* Artemcz., *T. ruthenicus* Bess. subsp. *donetzicus* (Artemcz.) I.B.K. Richardson) - 1
tasch-kala Kuth. - 2
tauricus Klok. = T. dubius
tesquicola Klok. = T. desertorum
tomentosulus Boriss. - 6
trachycarpus S. Nikit. - 4
transcarpaticus Klok. = T. orientalis
tuberosus C. Koch - 2

turkestanicus S. Nikit. ex Pavl. - 6
ucrainicus Artemcz. (*T. artemczukii* Klok.) - 1
undulatus Jacq. - 1
volgensis (S. Nikit.) S. Nikit. = T. podolicus
vvedenskyi M. Pop. ex Pavl. - 6
xanthantherus Klok. = T. orientalis

Trichanthemis Regel & Schmalh.

aulieatensis (B. Fedtsch.) Krasch. - 6
aurea Krasch. - 6
butkovii Kovalevsk. - 6
karataviensis Regel & Schmalh. - 6
litwinowii (Krasch.) Tzvel. (*T. simulans* (Pavl.) Karmysch.) - 6
paradoxos (C. Winkl.) Tzvel. - 6
radiata Krasch. & Vved. - 6
simulans (Pavl.) Karmysch. = T. litwinowii

Tricholepis DC.

trichocephala Lincz. - 6

Tridactylina (DC.) Sch. Bip.

kirilowii (Turcz.) Sch. Bip. - 4

Tripleurospermum Sch. Bip. (*Matricaria* auct.)

ambiguum (Ledeb.) Franch. & Savat. (*Matricaria ambigua* (Ledeb.) Kryl.) - 3, 6
australe Pobed. (*Matricaria australis* (Pobed.) Rauschert) - 2
breviradiatum (Ledeb.) Pobed. (*Matricaria breviradiata* (Ledeb.) Rauschert) - 4
caucasicum (Willd.) Hayek (*Matricaria caucasica* (Willd.) Poir., *M. subnivalis* (Pobed.) Rauschert, *Tripleurospermum subnivale* Pobed.) - 2
colchicum (Manden.) Pobed. = T. elongatum
■ decipiens (Fisch. & C.A. Mey.) Bornm. (*Matricaria decipiens* (Fisch. & C.A. Mey.) C. Koch)
disciforme (C.A. Mey.) Sch. Bip. (*Matricaria disciformis* (C.A. Mey.) DC.) - 2, 6
elongatum (Fisch. & C.A. Mey. ex DC.) Bornm. (*Matricaria colchica* (Manden.) Rauschert, *M. elongata* (Fisch. & C.A. Mey. ex DC.) Hand.-Mazz., *Tripleurospermum colchicum* (Manden.) Pobed.) - 2
■ grossheimii (Fed.) Pobed. (*Matricaria grossheimii* (Fed.) Rauschert)
hookeri Sch. Bip. (*Matricaria hookeri* (Sch. Bip.) Czer. comb. superfl., *M. hookeri* (Sch. Bip.) Hutch., *M. maritima* L. subsp. *phaeocephala* (Rupr.) Rauschert, *M. phaeocephala* (Rupr.) Stefanss., *Tripleurospermum maritimum* (L.) Koch subsp. *phaeocephalum* (Rupr.) Hamet-Ahti, *T. phaeocephalum* (Rupr.) Pobed.) - 1, 3, 4, 5
inodorum (L.) Sch. Bip. = T. perforatum
karjaginii (Manden. & Sof.) Pobed. (*Matricaria karjaginii* (Manden. & Sof.) Rauschert) - 2
limosum (Maxim.) Pobed. (*Matricaria limosa* (Maxim.) Kudo) - 5
maritimum (L.) Koch (*Matricaria maritima* L.) - 1
- subsp. *phaeocephalum* (Rupr.) Hamet-Ahti = T. hookeri
- subsp. *subpolare* (Pobed.) Hamet-Ahti = T. subpolare
oreades (Boiss.) Rech. fil. var. *tchihatchewii* (Boiss.) E. Hossain = T. tchihatchewii
parviflorum (Willd.) Pobed. (*Matricaria parviflora* (Willd.) Poir.) - 1, 2, 6

perforatum (Merat) M. Lainz (*Matricaria perforata* Merat, *M. chamomilla* L. 1753, nom. ambig., non sensu L. 1755, *M. inodora* L. nom. illegit., *Tripleurospermum inodorum* (L.) Sch. Bip. nom. illegit.) - 1, 2, 3, 4, 5, 6

phaeocephalum (Rupr.) Pobed. = T. hookeri

rupestre (Somm. & Levier) Pobed. (*Matricaria rupestris* (Somm. & Levier) Rauschert) - 2

sevanense (Manden.) Pobed. (*Matricaria sevanensis* (Manden.) Rauschert) - 2

subnivale Pobed. = T. caucasicum

subpolare Pobed. (*Matricaria maritima* L. subsp. *subpolaris* (Pobed.) Rauschert, *M. subpolaris* (Pobed.) Holub, *Tripleurospermum maritimum* (L.) Koch subsp. *subpolare* (Pobed.) Hamet-Ahti) - 1, 3, 4, 5

szowitzii (DC.) Pobed. (*Matricaria szowitzii* (DC.) Rauschert) - 2

▪ **tchihatchewii** (Boiss.) Bornm. (*Matricaria tchihatchewii* (Boiss.) Voss., *Tripleurospermum oreades* (Boiss.) Rech. fil. var. *tschihatchewii* (Boiss.) E. Hossain)

tetragonospermum (Fr. Schmidt) Pobed. (*Matricaria tetragonosperma* (Fr. Schmidt) Hara & Kitam.) - 5

transcaucasicum (Manden.) Pobed. (*Matricaria transcaucasica* (Manden.) Rauschert) - 2

tzvelevii Pobed. (*Matricaria aserbaidshanica* Rauschert) - 2

Tripolium Nees

pannonicum (Jacq.) Dobrocz. (*Aster pannonicus* Jacq., *A. tripolium* L. subsp. *pannonicus* (Jacq.) Soo) - 1, 2, 3, 4, 5, 6

- subsp. *maritimum* Holub = T. vulgare

vulgare Nees (*T. pannonicum* (Jacq.) Dobrocz. subsp. *maritimum* Holub) - 1

Trommsdorfia Bernh. (*Achyrophorus* auct.)

ciliata (Thunb.) Sojak (*Achyrophorus ciliatus* (Thunb.) Sch. Bip.) - 3, 4, 5

crepidioides (Miyabe & Kudo) Sojak (*Picris crepidioides* Miyabe & Kudo, *Achyrophorus crepidioides* (Miyabe & Kudo) Kitag.) - 5

maculata (L.) Bernh. (*Achyrophorus maculatus* (L.) Scop.) - 1, 2, 3, 4

uniflora (Vill.) Sojak (*Achyrophorus uniflorus* (Vill.) Bluff & Fingerh.) - 1

Turaniphytum Poljak

codringtonii (Rech. fil.) Podlech (*Artemisia codringtonii* Rech. fil., *Turaniphytum kopetdaghense* Poljak.) - 6

eranthemum (Bunge) Poljak. - 6

kopetdaghense Poljak. = T. codringtonii

Turczaninowia DC.

fastigiata (Fisch.) DC. - 4, 5

Tussilago L.

farfara L. - 1, 2, 3, 4, 5, 6

radiata J.F. Gmel. = Petasites radiatus

rubella J.F. Gmel. = Petasites rubellus

sibirica J. F. Gmel. = Endocellion sibiricum

Uechtritzia Freyn

▪ armena Freyn

kokanica (Regel & Schmalh.) Pobed. - 6

Ugamia Pavl.

angrenica (Krasch.) Tzvel. - 6

Urospermum Scop.

picroides (L.) Scop. ex F.W. Schmidt - 2, 6

Varthemia DC.

persica DC. - 6

Vicoa Cass. = Pentanema

discoidea (Nabiev) R. Kam. = Pentanema discoideum

krascheninnikovii R. Kam. = Pentanema krascheninnikovii

Virgulus auct. = Aster

novae-angliae (L.) Reveal & Keener = Aster novae-angliae

Waldheimia Kar. & Kir.

glabra (Decne.) Regel (*W. tridactylites* Kar. & Kir. subsp. *glabra* (Decne.) Podlech) - 6

stoliczkae (Clarke) Ostenf. - 6

tomentosa (Decne.) Regel (*Tanacetum tomentosum* (Decne.) Muradjan) - 6

transalaica Tzvel. - 6

tridactylites Kar. & Kir. - 3, 6

- subsp. *glabra* (Decne.) Podlech = W. glabra

Willemetia Cass. = Calycocorsus

Willemetia Neck. = Calycocorsus

tuberosa Fisch. & C.A. Mey. = Calycocorsus tuberosus

Xanthium L. (*Acanthoxanthium* (DC.) Fourr.)

albinum (Widd.) H. Scholz (*X. riparium* Itz. & Hertsch var. *albinum* Widd., *X. orientale* L. var. *albinum* (Widd.) T. Adema & M.T. Jansen, *X. cavanillesii* auct., *X. echinatum* auct., *X. occidentale* auct., *X. speciosum* auct.) - 1, 2, 5, 6

- subsp. *ripicola* (Holub) Dostal = X. ripicola

brasilicum Vell. - 1, 2

californicum Greene - 1, 2, 6

cavanillesii auct. = X. albinum

echinatum auct. = X. albinum

italicum Moretti (*X. strumarium* L. subsp. *italicum* (Moretti) D. Löve) - 1

japonicum auct. = X. sibiricum

occidentale auct. = X. albinum

orientale L. var. *albinum* (Widd.) T. Adema & M.T. Jansen = X. albinum

palustre Greene - 1

pensylvanicum Wallr. - 1

riparium Itz. & Hertsch var. *albinum* Widd. = X. albinum

riparium Lasch = X. ripicola

ripicola Holub (*X. albinum* (Willd.) H. Scholtz subsp. *ripicola* (Holub) Dostal, *X. riparium* Lasch, 1856 & sensu Widd. 1923, non *X. riparium* Itz. & Hertsch, 1854, nom. illegit.) - 1

sibiricum Patrin ex Widd. (*X. japonicum* auct.) - 1, 2, 3, 4, 5, 6

speciosum auct. = X. albinum

spinosum L. (*Acanthoxanthium spinosum* (L.) Fourr.) - 1, 2, 3, 5, 6

strumarium L. - 1, 2, 3, 5, 6

- subsp. *italicum* (Moretti) D. Löve = X. italicum

Xanthopsis (DC.) Koch = Centaurea

erivanensis (Lipsky) Sojak = Centaurea erivanensis
xanthocephaloides (Tzvel.) Sojak = Centaurea xanthocephaloides

Xeranthemum L. (*Xeroloma* Cass.)

annuum L. - 1, 2
bracteatum Vent. = Xerochrysum bracteatum
cylindraceum Sibth. & Smith (*Xeroloma cylindracea* (Sibth. & Smith) Holub) - 1, 2
inapertum (L.) Mill. - 2, 6(?)
longepapposum Fisch. & C.A. Mey. - 2, 6
squarrosum Boiss. - 2, 6

*Xerochrysum Tzvel.

***bracteatum** (Vent.) Tzvel. (*Xeranthemum bracteatum* Vent., *Helichrysum bracteatum* (Vent.) Andr.) - 1

Xeroloma Cass. = Xeranthemum

cylindracea (Sibth. & Smith) Holub = Xeranthemum cylindraceum

Ximenesia Cav.

encelioides Cav. - 1(alien)

Xylanthemum Tzvel.

fisherae (Aitch. & Hemsl.) Tzvel. - 6
pamiricum (O. Hoffm.) Tzvel. - 6
polycladum (Rech. fil.) Tzvel. (*Chrysanthemum polycladum* Rech. fil.) - 6
rupestre (M. Pop. ex Nevski) Tzvel. (*Chrysanthemum popovii* Gilli) - 6
tianschanicum (Krasch.) Muradjan (*Pyrethrum tianschanicum* Krasch.) - 6

Youngia Cass.

altaica (Babc. & Stebb.) Czer. (*Crepis altaica* (Babc. & Stebb.) Roldug.) - 3, 6
diversifolia (Ledeb.) Ledeb. - 3, 6
serawschanica (B. Fedtsch.) Babc. & Stebb. - 6
stenoma (Turcz.) Ledeb. - 4
tenuicaulis (Babc. & Stebb.) Czer. - 3, 4, 6
tenuifolia (Willd.) Babc. & Stebb. - 3, 4, 5

Zacintha Hill

verrucosa Gaertn. - 1

*Zinnia L.

***elegans** Jacq. 1, 2, 6

Zoegea L.

baldschuanica C. Winkl. (*Z. crinita* Boiss. subsp. *baldschuanica* (C. Winkl.) Rech. fil.) - 6
crinita Boiss. subsp. *baldschuanica* (C. Winkl.) Rech. fil. = Z. baldschuanica
purpurea Fresen. - 6

ATHYRIACEAE Alst.

Allantodia auct. = Diplazium

crenata (Sommerf.) Ching = Diplazium sibiricum

Athyriopsis Ching (*Deparia* auct. p.p.)

▪ **conilii** (Franch. & Savat.) Ching (*Asplenium conilii* Franch. & Savat., *Athyrium conilii* (Franch. & Savat.) Tagawa, *Deparia conilii* (Franch. & Savat.) M. Kato, *Lunathyrium conilii* (Franch. & Savat.) Kurata)
japonica (Thunb.) Ching (*Asplenium japonicum* Thunb., *Athyrium japonicum* (Thunb.) Copel., *Deparia japonica* (Thunb.) M. Kato, *Diplazium japonicum* (Thunb.) Bedd., *D. thunbergii* Nakai ex Momose, *Lunathyrium japonicum* (Thunb.) Kurata, *Athyrium grammitoides* auct.) - 5

Athyrium Roth (*Aspidium* Sw. p.p. nom. illegit., *Nephrodium* auct.)

acrostichoides auct. = Lunathyrium pycnosorum
alpestre (Hoppe) Clairv. var. *americanum* Butters = A. americanum
alpestre (Hoppe) Nyl. = A. distentifolium
americanum (Butters) Maxon (*A. alpestre* (Hoppe) Clairv. var. *americanum* Butters) - 5
austro-ussuriense (Kom.) Fomin = Cornopteris crenulatoserrulata
brevifrons auct. = A. submelanolepis
changpaishanense Ching ex Worosch. = A. sinense
conilii (Franch. & Savat.) Tagawa = Athyriopsis conilii
coreanum Christ = Lunathyrium henryi
crenatum (Sommerf.) Rupr. = Diplazium sibiricum
crenulatoserrulatum Makino = Cornopteris crenulatoserrulata
cyclosorum (Rupr.) Maxon (*A. filix-femina* (L.) Roth var. *cyclosorum* Rupr., *A. filix-femina* subsp. *cyclosorus* (Rupr.) C. Chr.) - 5
distentifolium Tausch ex Opiz (*A. alpestre* (Hoppe) Nyl. 1844, non Clairv. 1813) - 1, 2, 3, 4
fauriei (Chirst) Makino (*Nephrodium fauriei* Christ, *Athyrium yokoscense* (Franch. & Savat.) Christ var. *fauriei* (Christ) Tagawa) - 5
filix-femina (L.) Roth - 1, 2, 3, 4, 6
- subsp. *cyclosorum* (Rupr.) C. Chr. = A. cyclosorum
- subsp. *jacutense* (V. Petrov) Sipl. = A. monomachii
- subsp. *longipes* (Hara) Tzvel. = A. submelanolepis
- subsp. *sitchense* (Rupr.) Tzvel. = A. sitchense
- var. *cyclosorum* Rupr. = A. cyclosorum
- var. *jacutense* V. Petrov = A. monomachii
- var. *longipes* Hara = A. submelanolepis
- var. *sitchense* Rupr. = A. sitschense
goeringianum sensu Fomin = A. yokoscense
goeringianum sensu Worosch. = A. vidalii
grammitoides auct. = Athyriopsis japonica
henryi (Baker) Diels = Lunathyrium henryi
japonicum (Thunb.) Copel. = Athyriopsis japonica
lucidulum Kom. = A. yokoscense
melanolepis auct. = A. submelanolepis
monomachii (Kom.) Kom. (*A. filix-femina* (L.) Roth subsp. *jacutense* (V. Petrov) Sipl., *A. filix-femina* var. *jacutense* V. Petrov) - 4, 5
possieticum Kom. = A. yokoscense
pterorachis Christ = Lunathyrium pterorachis
pycnosorum Christ = Lunathyrium pycnosorum
- var. *vegetius* Kitag. = Lunathyrium pycnosorum var. vegetius
quelpaertense (Christ) Ching = Oreopteris quelpaertensis (Thelypteridaceae)
x **rechstenii** Schneller & Rasbach. - A. filix-femina (L.) Roth x A. distentifolium Tausch ex Opiz
rubripes (Kom.) Kom. = A. sinense

rupestre Kodama - 5

sinense Rupr. (*A. changpaishanense* Ching ex Worosch. nom. invalid., *A. rubripes* (Kom.) Kom.) - 5

- subsp. *longipes* (Hara) Worosch. = A. submelanolepis

sitchense (Rupr.) Tzvel. (*A. filix-femina* (L.) Roth var. *sitchense* Rupr., *A. filix-femina* subsp. *sitchense* (Rupr.) Tzvel.) - 5

spinulosum (Maxim.) Milde = Pseudocystopteris spinulosa

submelanolepis Tzvel. (*A. filix-femina* (L.) Roth var. *longipes* Hara, non *A. longipes* Christ, 1905, *A. filix-femina* subsp. *longipes* (Hara) Tzvel., *A. sinense* Rupr. subsp. *longipes* (Hara) Worosch., *A. brevifrons* auct., *A. melanolepis* auct.) - 5

subspinulosum (Christ) Ching ex Worosch. = A. yokoscense

vidalii (Franch. & Savat.) Nakai (*Asplenium vidalii* Franch. & Savat., *Athyrium goeringianum* sensu Worosch.) - 5

wardii (Hook.) Makino - 5

yokoscense (Franch. & Savat.) Christ (*Aspidium subspinulosum* Christ, *Athyrium lucidulum* Kom., *A. possieticum* Kom., *A. subspinulosum* (Christ) Ching ex Worosch., *A. goeringianum* sensu Fomin) - 5

- var. *fauriei* (Christ) Tagawa = A. fauriei

Carpogymnia A. & D. Löve = Gymnocarpium

continentalis (V. Petrov) A. & D. Löve = Gymnocarpium jessoense

disjuncta (Rupr.) A. & D. Löve = Gymnocarpium dryopteris

dryopteris (L.) A. & D. Löve = Gymnocarpium dryopteris

heterospora auct. = Gymnocarpium x intermedium

robertiana (Hoffm.) A. & D. Löve = Gymnocarpium robertianum

Cornopteris Nakai (*Pseudathyrium* auct.)

crenulatoserrulata (Makino) Nakai (*Athyrium crenulatoserrulatum* Makino, *A. austro-ussuriense* (Kom.) Fomin, *Pseudathyrium crenulatoserrulatum* (Makino) Nakai) - 5

pterorachis (Christ) Tardieu-Blot = Lunathyrium pterorachis

Currania Copel. = Gymnocarpium

dryopteris (L.) Wherry = Gymnocarpium dryopteris

robertiana (Hoffm.) Wherry = Gymnocarpium robertianum

Cystopteris Bernh.

almaatensis Kotuchov = C. fragilis

dickieana R. Sim (*C. fragilis* (L.) Bernh. subsp. *dickieana* (R. Sim) Hyl., *C. fragilis* var. *dickieana* (R. Sim) Moore) - 1, 2, 3, 4, 5, 6

emarginato-denticulata Fomin = C. fragilis

filix-fragilis (L.) Borb. = C. fragilis

fragilis (L.) Bernh. (*C. almaatensis* Kotuchov, *C. emarginato-denticulata* Fomin, *C. filix-fragilis* (L.) Borb., *C. fragilis* subsp. *emarginato-denticulata* Fomin, *C. thermalis* A. Khokhr.) - 1, 2, 3, 4, 5, 6

- subsp. *dickieana* (R. Sim) Hyl. = C. dickieana

- subsp. *emarginato-denticulata* Fomin = C. fragilis

- var. *dickieana* (R. Sim) Moore = C. dickieana

montana (Lam.) Desv. = Rhizomatopteris montana

regia (L.) Desv. - 1, 2

sudetica A. Br. & Milde = Rhizomatopteris sudetica

thermalis A. Khokhr. = Cystopteris fragilis

Deparia auct. p.p. = Athyriopsis and Lunathyrium

conilii (Franch. & Savat.) M. Kato = Athyriopsis conilii

henryi (Baker) M. Kato = Lunathyrium henryi

japonica (Thunb.) M. Kato = Athyriopsis japonica

pterorachis (Christ) M. Kato = Lunathyrium pterorachis

pycnosora (Christ) M. Kato = Lunathyrium pycnosorum

Diplazium Sw. (*Allantodia* auct.)

japonicum (Thunb.) Bedd. = Athyriopsis japonica

sibiricum (Turcz. ex G. Kunze) Kurata (*Asplenium sibiricum* Turcz. ex G. Kunze, *Allantodia crenata* (Sommerf.) Ching, *Athyrium crenatum* (Sommerf.) Rupr. non *Diplazium crenatum* Poir., *Diplazium sibiricum* (Turcz. ex G. Kunze) Jermy, comb. superfl., *D. sommerfeldii* A. & D. Löve) - 1, 3, 4, 5

sommerfeldii A. & D. Löve = D. sibiricum

thunbergii Nakai ex Momose = Athyriopsis japonica

Dryoathyrium Ching = Lunathyrium

coreanum (Christ) Tagawa = Lunathyrium henryi

henryi (Baker) Ching = Lunathyrium henryi

pterorachis (Christ) Ching = Lunathyrium pterorachis

Filix auct. = Gymnocarpium

pumila Gilib. = Gymnocarpium dryopteris

Gymnocarpium Newm. (*Carpogymnia* A. & D. Löve, *Currania* Copel., *Filix* auct.)

continentale (V. Petrov) Pojark. = G. jessoense

disjunctum (Rupr.) Ching = G. dryopteris

disjunctum (Rupr.) A. & D. Löve = G. dryopteris

dryopteris (L.) Newm. (*Carpogymnia disjuncta* (Rupr.) A. & D. Löve, *C. dryopteris* (L.) A. & D. Löve, *Currania dryopteris* (L.) Wherry, *Dryopteris disjuncta* (Rupr.) C.V. Morton, *D. linneana* C. Chr., *D. pumila* V. Krecz., *Filix pumila* Gilib. nom. invalid., *Gymnocarpium disjunctum* (Rupr.) Ching, *G. disjunctum* (Rupr.) A. & D. Löve, comb. superfl., *G. dryopteris* subsp. *disjunctus* (Rupr.) Sarvela, *Polypodium dryopteris* L. var. *Polypodium disjunctum* Rupr.) - 1, 2, 3, 4, 5, 6

- subsp. *disjunctus* (Rupr.) Sarvela = G. dryopteris

fedtschenkoanum Pojark. - 6

heterosporum auct. = G. x intermedium

x **intermedium** Sarvela (*Dryopteris heterospora* (Wagner) Kuvajev, p.p. excl. basionymo, *Carpogymnia heterospora* auct., *Gymnocarpium heterosporum* auct.). - G. jessoense (Koidz.) Koidz. x G. dryopteris (L.) Newm. - 1, 4

jessoense (Koidz.) Koidz. (*Dryopteris jessoensis* Koidz., *Aspidium dryopteris* (L.) Baumg. var. *longulum* Christ, *Carpogymnia continentalis* (V. Petrov) A. & D. Löve, *Dryopteris continentalis* (V. Petrov) Fomin, *Gymnocarpium continentale* (V. Petrov) Pojark., *G. jessoense* subsp. *parvulum* Sarvela, *G. longulum* (Christ) Kitag., *G. robertianum* (Hoffm.) Newm. subsp. *longulum* (Christ) Toyokuni, *G. tenuipes* Pojark., *G. remote-pinnatum* auct.) - 1, 3, 4, 5

- subsp. *parvulum* Sarvela = G. jessoense

longulum (Christ) Kitag. = G. jessoense

remote-pinnatum auct. = G. jessoense

robertianum (Hoffm.) Newm. (*Carpogymnia robertiana* (Hoffm.) A. & D. Löve, *Currania robertiana* (Hoffm.)

Wherry, *Dryopteris robertiana* (Hoffm.) C. Chr.) - 1, 2, 3, 4(?)
- subsp. *longulum* (Christ) Toyokuni = G. jessoense
tenuipes Pojark. = G. jessoense

Lunathyrium Koidz. (*Dryoathyrium* Ching, *Parathyrium* Holttum, *Deparia* auct. p.p.)

conilii (Franch. & Savat.) Kurata = Athyriopsis conilii
coreanum (Christ) Ching = L. henryi
henryi (Baker) Kurata (*Asplenium henryi* Baker, *Athyrium coreanum* Christ, *A. henryi* (Baker) Diels, *Deparia henryi* (Baker) M. Kato, *Dryoathyrium coreanum* (Christ) Tagawa, *D. henryi* (Baker) Ching, *Lunathyrium coreanum* (Christ) Ching) - 5
japonicum (Thunb.) Kurata = Athyriopsis japonica
pterorachis (Christ) Kurata (*Athyrium pterorachis* Christ, *Asplenium kamtschatkanum* Gilbert, *Cornopteris pterorachis* (Christ) Tardieu-Blot, *Deparia pterorachis* (Christ) M. Kato, *Dryoathyrium pterorachis* (Christ) Ching, *Parathyrium pterorachis* (Christ) Holttum, nom. illegit.) - 5
pycnosorum (Christ) Koidz. (*Athyrium pycnosorum* Christ, *Deparia pycnosora* (Christ) M. Kato, *Athyrium acrostichoides* auct.) - 5
- var. **vegetius** (Kitag.) Kurata (*Athyrium pycnosorum* var. *vegetius* Kitag., *Lunathyrium vegetius* (Kitag.) Ching) - 5
vegetius (Kitag.) Ching = Lunathyrium pycnosorum var. vegetius

Nephrodium auct. = Athyrium

fauriei Christ = Athyrium fauriei

Parathyrium Holttum = Lunathyrium

pterorachis (Christ) Holttum = Lunathyrium pterorachis

Pseudathyrium auct. = Cornopteris

crenulatoserrulatum (Makino) Nakai = Cornopteris crenulatoserrulata

Pseudocystopteris Ching

spinulosa (Maxim.) Ching (*Athyrium spinulosum* (Maxim.) Milde) - 5

Rhizomatopteris A. Khokhr.

montana (Lam.) A. Khokhr. (*Cystopteris montana* (Lam.) Desv.) - 1, 2, 3, 4, 5
sudetica (A. Br. & Milde) A. Khokhr. (*Cystopteris sudetica* A. Br. & Milde) - 1, 2, 3, 4, 5

*AUCUBACEAE J. Agardh
*Aucuba Thunb.

***japonica** Thunb.
    This plant is cultivated indoors only.

AZOLLACEAE Wettst.
Azolla Lam.

caroliniana Willd. - 1
filiculoides Lam. - 1

BALSAMINACEAE A. Rich.
Impatiens L.

***balsamina** L. - 2, 5, 6
brachycentra Kar. & Kir. - 6
furcellata Hemsl. - 5
glandulifera Royle (*I. roylei* Walp.) - 1, 4, 5
komarovii Pobed. = I. noli-tangere
maackii Hook. ex Kom. - 5
nevskii Pobed. = I. parviflora
noli-tangere L. (*I. komarovii* Pobed.) - 1, 2, 3, 4, 5, 6
parviflora DC. (*I. nevskii* Pobed.) - 1, 3, 6
roylei Walp. = I. glandulifera
textorii Miq. - 5
uralensis A. Skvorts. - 1

BERBERIDACEAE Juss. (*LEONTICACEAE* Bercht. & J. Presl)
Berberis L.

amurensis Maxim. - 5
bykoviana Pavl. = B. sphaerocarpa
crataegina auct. = B. iberica
densiflora Boiss. & Buhse - 2, 6
heterobotrys E. Wolf (*B. oblonga* auct. p.p.) - 6
heteropoda Schrenk = B. sphaerocarpa
iberica Stev. & Fisch. ex DC. (*B. tragacanthoides* DC., *B. crataegina* auct.) - 2, 6
iliensis M. Pop. - 6
integerrima Bunge (*B. oblonga* (Regel) Schneid.) - 6
- var. *eriwanensis* Schneid. = B. turcomanica subsp. eriwanensis
integerrima sensu B. Fedtsch. = B. nummularia
karkaralensis Kornilova & Potapov - 6
kaschgarica Rupr. - 6
multispinosa V. Zapr. - 6
nummularia Bunge (*B. integerrima* sensu B. Fedtsch.) - 6
oblonga (Regel) Schneid. = B. integerrima
orientalis Schneid. (*B. vulgaris* L. var. *orientalis* (Schneid.) Papava, *B. vulgaris* subsp. *orientalis* (Schneid.) Takht.) - 2
orthobotrys Bien. ex Aitch. - 6
■ poiretii Schneid.
sibirica Pall. - 3, 4, 6
sphaerocarpa Kar. & Kir. (*B. bykoviana* Pavl., *B. heteropoda* Schrenk) - 6
stolonifera Koehne & E. Wolf - 6
tragacanthoides DC. = B. iberica
turcomanica Kar. - 2, 6
- subsp. **eriwanensis** (Schneid.) Takht. (*B. integerrima* Bunge var. *eriwanensis* Schneid.) - 2
vulgaris L. - 1, 2
- subsp. *orientalis* (Schneid.) Takht. = B. orientalis
- var. *orientalis* (Schneid.) Papava = B. orientalis

Bongardia C.A. Mey.

chrysogonum (L.) Spach - 2, 6

Caulophyllum Michx.

robustum Maxim. (*C. thalictroides* (L.) Michx. subsp. *robustum* (Maxim.) Hiroe) - 5
thalictroides (L.) Michx. subsp. *robustum* (Maxim.) Hiroe = C. robustum

Diphylleia Michx.

cymosa Michx. subsp. *grayi* (Fr. Schmidt) Kitam. = D. grayi
grayi Fr. Schmidt (*D. cymosa* Michx. subsp. *grayi* (Fr. Schmidt) Kitam.) - 5

Epimedium L.

circinatocucullatum Sosn. (*E. pinnatum* Fisch. subsp. *circinatum* Stearn) - 2
colchicum (Boiss.) Trautv. (*E. pinnatum* Fisch. subsp. *colchicum* (Boiss.) N. Busch) - 2
grandiflorum Morr. subsp. *koreanum* (Nakai) Kitam. = E. koreanum
koreanum Nakai (*E. grandiflorum* Morr. subsp. *koreanum* (Nakai) Kitam.) - 5
macrosepalum Stearn - 5
pinnatum Fisch. - 2
- subsp. *circinatum* Stearn = E. circinatocucullatum
- subsp. *colchicum* (Boiss.) N. Busch = E. colchicum
pubigerum (DC.) Morr & Decne. - 2

Gymnospermium Spach

alberti (Regel) Takht. (*Leontice alberti* Regel) - 6
altaicum (Pall.) Spach (*Leontice altaica* Pall.) - 3
- subsp. *odessanum* (DC.) E. Mayer & Pulevic = G. odessanum
darwasicum (Regel) Takht. (*Leontice darwasica* Regel) - 6
odessanum (DC.) Takht. (*G. altaicum* (Pall.) Spach subsp. *odessanum* (DC.) E. Mayer & Pulevic, *Leontice odessana* (DC.) Fisch.) - 1
smirnowii (Trautv.) Takht. (*Leontice smirnowii* Trautv.) - 2
vitellianum M. Kral - 6

Jeffersonia auct. = Plagiorhegma

dubia (Maxim.) Benth. & Hook. fil. ex Baker & Moore = Plagiorhegma dubia

Leontice L.

alberti Regel = Gymnospermium alberti
altaica Pall. = Gymnospermium altaicum
armeniaca Belanger (*L. leontopetalum* L. subsp. *armeniaca* (Belanger) Coode, *L. minor* Boiss.) - 2, 6
darwasica Regel = Gymnospermium darwasicum
ewersmanni Bunge (*L. leontopetalum* L. subsp. *ewersmannii* (Bunge) Coode) - 6
incerta Pall. - 6
leontopetalum L. subsp. *armeniaca* (Belanger) Coode = L. armeniaca
- subsp. *ewersmannii* (Bunge) Coode = L. ewersmannii
minor Boiss. = L. armeniaca
odessana (DC.) Fisch. = Gymnospermium odessanum
smirnowii Trautv. = Gymnospermium smirnowii

Plagiorhegma Maxim. (*Jeffersonia* auct.)

dubia Maxim. (*Jeffersonia dubia* (Maxim.) Benth. & Hook. fil. ex Baker & Moore) - 5

BETULACEAE S.F. Gray (*CARPINACEAE* Kuprian. nom. superfl., *CARPINACEAE* Vest, *CORYLACEAE* Mirb.)

Alnaster Spach = Duschekia

fruticosus (Rupr.) Ledeb. = Duschekia fruticosa

glutipes Jarm. ex Czer. = Duschekia fruticosa
kamtschaticus (Regel) Czer. = Duschekia kamtschatica
manshuricus (Call.) Jarm. = Duschekia manshurica
maximowiczii (Call.) Czer. = Duschekia maximowiczii
sinuatus (Regel) Czer. = Duschekia sinuata
viridis (Chaix) Spach = Duschekia abnobetula

Alnus Hill

alnobetula (Ehrh.) C. Koch = Duschekia abnobetula
barbata C.A. Mey. (*A. glutinosa* (L.) Gaertn. subsp. *barbata* (C.A. Mey.) Yaltirik) - 2
crispa Pursh subsp. *maximowiczii* (Call.) Hult. = Duschekia maximowiczii
fruticosa Rupr. = Duschekia fruticosa
- subsp. *kamtschatica* (Regel) Ju. Kozhevn. = Duschekia kamtschatica
glutinosa (L.) Gaertn. - 1, 2, 3, 6
- subsp. *barbata* (C.A. Mey.) Yaltirik = A. barbata
glutipes (Jarm. ex Czer.) Worosch. = Duschekia fruticosa
hirsuta (Spach) Turcz. ex Rupr. (*A. incana* (L.) Moench subsp. *hirsuta* (Spach) A. & D. Löve, *A. tinctoria* Sarg.) - 4, 5
incana (L.) Moench - 1, 2, 3
- subsp. *hirsuta* (Spach) A. & D. Löve = A. hirsuta
- subsp. *kolaensis* (Orlova) A. & D. Löve = A. kolaensis
japonica (Thunb.) Steud. (*Betula japonica* Thunb., *Alnus japonica* Siebold & Zucc. nom. illegit.) - 5
kamtschatica (Regel) Kom. = Duschekia kamtschatica
kolaensis Orlova (*A. incana* (L.) Moench subsp. *kolaensis* (Orlova) A. & D. Löve) - 1
manshurica (Call.) Hand.-Mazz. = Duschekia manshurica
maximowiczii Call. = Duschekia maximowiczii
x **pubescens** Tausch. - A. glutinosa (L.) Gaertn. x A. incana (L.) Moench - 1
sibirica (Spach) Turcz. ex Kom. - 4, 5
sinuata (Regel) Rydb. = Duschekia sinuata
subcordata C.A. Mey. - 2
tinctoria Sarg. = A. hirsuta
viridis (Chaix) DC. = Duschekia abnobetula
- subsp. *maximowiczii* (Call.) D. Löve = Duschekia maximowiczii
- lusus *kamtschatica* Regel = Duschekia kamtschatica
- var. *sinuata* Regel = Duschekia sinuata

Betula L.

A detailed study of Betula species is necessary. Undoubtedly, the number of species is very overstated, because forms of infraspecific variation and hybrid have been described as species.

adasii Egor. & Sipl. - 4
aischatiae Gussejnov - 2
ajanensis Kom. = B. platyphylla
alajica Litv. (*B. pseudoalajica* V. Vassil.) - 6
alatavica Muscheg. = B. tianschanica
alba L. = B. pubescens
- subsp. *latifolia* (Tausch) Regel var. *kamtschatica* Regel = B. kamtschatica
- - var. *tauschii* Regel = B. platyphylla
- var. *carelica* Merckl. = B. pendula var. carelica
- var. *japonica* Miq. = B. kamtschatica
alnobetula Ehrh. = Duschekia abnobetula
andreji V. Vassil. - 4
x **aurata** Borkh. - B. pendula Roth x B. pubescens Ehrh.
austrosichotensis V. Vassil. & V. Baranov - 5
baicalensis Sukacz. - 4
Probably, a hybrid B. divaricata Ladeb. x B. pendula Roth.
baicalia V. Vassil. = B. x irkutensis

bornmuelleri V. Vassil. = B. seravschanica
borysthenica Klok. - 1
x **bottnica** Mela. - B. nana L. x B. pendula Roth
brachylepis V. Vassil. - 3
bucharica V. Vassil. = B. pamirica
cajanderi Sukacz. - 4, 5
- subsp. *minutifolia* Ju. Kozhevn. = B. platyphylla subsp. minutifolia
callosa Lindq. (*B. pubescens* Ehrh. subsp. *callosa* (Noto) A. & D. Löve) - 1
concinna Gunnarss. (*B. pubescens* Ehrh. subsp. *concinna* (Gunnarss.) A. & D. Löve) - 1
coriaceifolia V.Vassil. - 6
costata Trautv. - 5
crassijulis Muscheg. = B. tianschanica
czatkalensis V. Vassil. - 6
czerepanovii Orlova (*B. pubescens* Ehrh. subsp. *czerepanovii* (Orlova) Hamet-Ahti, *B. tortuosa* auct. fl. Europ. bor.) - 1, 3
darvasica Ovcz., Czuk. & Schibk. - 6
davurica Pall. - 4, 5
demetrii Ig. Vassil. (*B. grandifolia* Litv. 1905, non Ettingsh. 1866) - 4, 5
divaricata Ledeb. (*B. henriettae* Sukacz. & V. Vassil., *B. itelmenorum* V. Vassil., *B. middendorffii* Trautv. & C.A. Mey., *B. yoshimurae* Miyabe & Tatew.) - 4, 5
dolicholepis Ovcz., Czuk. & Schibk. - 6
ellipticifolia V. Vassil. - 4
ermanii Cham. (*B. komarovii* Perf. & Kolesn., *B. paraermanii* V. Vassil, *B. grossa* auct., *B. ulmifolia* auct.) - 4, 5
- subsp. *lanata* (Regel) A. Skvorts. = B. lanata
- var. *lanata* Regel = B. lanata
evenkiensis Polozh. - 4
exilis Sukacz. (*B. nana* L. subsp. *exilis* (Sukacz.) Hult., *B. nana* var. *sibirica* Ledeb., *B. sibirica* (Ledeb.) P'ei) - 4, 5
extremiorientalis Kuzen. & V. Vassil. (*B. fruticosa* Pall. subsp. *extremiorientalis* (Kuzen. & V. Vassil.) Worosch.) - 5
falcata V. Vassil. - 3
fedtschenkoana V. Vassil. - 6
x **fennica** Doerfl. - B. nana L. x B. pendula Roth - 1
ferganensis V. Vassil. - 6
fruticosa Pall. - 4, 5
- subsp. *extremiorientalis* (Kuzen. & V. Vassil.) Worosch. = B. extremiorientalis
- subsp. *fusca* (Pall. ex Georgi) M. Schemberg = B. fusca
- subsp. *gmelinii* (Bunge) Kitag. = B. fusca
- subsp. **ruprechtiana** (Trautv.) Kitag. (*B. fruticosa* var. *ruprechtiana* Trautv.) - 5
- var. *ovalifolia* (Rupr.) S.L. Tung = B. ovalifolia
- var. *ruprechtiana* Trautv. = B. fruticosa subsp. ruprechtiana
fusca Pall. ex Georgi (*B. fruticosa* Pall. subsp. *fusca* (Pall. ex Georgi) M. Schemberg, *B. fruticosa* subsp. *gmelinii* (Bunge) Kitag., *B. gmelinii* Bunge) - 4
gmelinii Bunge = B. fusca
golitsinii V. Vassil. - 1
grandifolia Litv. = B. demetrii
grossa auct. = B. ermanii
henriettae Sukacz. & V. Vassil. = B. divaricata
heptopotamica V. Vassil. - 6
hippolyti Sukacz. - 4
hissarica V. Vassil. = B. seravschanica
humilis Schrank - 1, 3, 4
insularis V. Vassil. (*B. uschkanensis* Sukacz. nom. invalid.) - 4

x **intermedia** Thomas. - B. nana L. x B. pubescens Ehrh. - 1
x **irkutensis** Sukacz. (*B. baicalia* V. Vassil.). - B. platyphylla Sukacz. x B. pubescens Ehrh. - 4
itelmenorum V. Vassil. = B. divaricata
jacutica V. Vassil. - 4
japonica auct. = B. kamtschatica
japonica Thunb. = Alnus japonica
jarmolenkoana Golosk. (*B. saposhnikovii* Sukacz. subsp. *jarmolenkoana* (Golosk.) Ovcz. & Czuk.) - 6
kamtschatica (Regel) Jansson ex V. Vassil. (*B. alba* L. subsp. *latifolia* (Tausch) Regel var. *kamtschatica* Regel, *B. alba* var. *japonica* Miq., *B. mandshurica* (Regel) Nakai var. *japonica* (Miq.) Rehd., *B. mandshurica* var. *kamtschatica* (Regel) Rehd., *B. pendula* Roth subsp. *kamtschatica* (Regel) M. Shemberg, *B. platyphylla* Sukacz. var. *japonica* (Miq.) Hara, *B. platyphylla* var. *kamtschatica* (Regel) Hara, *B. platyphylla* subsp. *kamtschatica* (Regel) Worosch., *B. japonica* auct.) - 5
karagandensis V. Vassil. - 3, 6
kelleriana Sukacz. - 3
kirghisorum Sav.-Rycz. - 6
klokovii Zaverucha - 1
komarovii Perf. & Kolesn. = B. ermanii
korshinskyi Litv. - 6
kossogolica V. Vassil. - 4
krylovii G. Kryl. - 3, 4
x **kusmisscheffii** (Regel) Sukacz. - B. nana L. x ? B. czerepanovii Orlova - 1
lanata (Regel) V. Vassil. (*B. ermanii* Cham. var. *lanata* Regel, *B. ermanii* subsp. *lanata* (Regel) A. Skvorts., *B. velutina* V. Vassil.) - 4, 5
lipskyi V. Vassil. (*B. procurva* Litv. subsp. *lipskyi* (V. Vassil.) Ovcz.) - 6
litwinowii Doluch. - 2
- var. *recurvata* Ig. Vassil. = B. recurvata
longilobata Sipl. - 4
> Probably, a hybrid B. divaricata Ledeb. x B. lanata (Regel) V. Vassil.

ludmilae V. Vassil. - 4
maarensis V. Vassil. & Gussejnov - 2
mandshurica (Regel) Nakai (*B. platyphylla* Sukacz. subsp. *mandshurica* (Regel) Kitag., *B. platyphylla* var. *mandshurica* (Regel) Hara) - 4, 5
- var. *japonica* (Miq.) Rehd. = B. kamtschatica
- var. *kamtschatica* (Regel) Rehd. = B. kamtschatica
margusarica V. Vassil. (*B. tianschanica* Rupr. subsp. *margusarica* (V.Vassil.) Ovcz.) - 6
maximowicziana Regel (*B. maximowiczii* Regel, 1865, non Rupr. 1856) - 5
maximowiczii Regel = B. maximowicziana
medwediewii Regel - 2
megrelica Sosn. - 2
microlepis Ig. Vassil. (*B. pendula* Roth var. *microlepis* (Ig.Vassil.) Doluch.) - 1, 2
microphylla Bunge - 3, 4
middendorffii Trautv. & C.A. Mey. = B. divaricata
x **montana** V. Vassil. - B. lanata (Regel) V. Vassil. x B. platyphylla Sukacz. - 4
murgabica V. Vassil. (*B. saposhnikovii* Sukacz. subsp. *murgabica* (V. Vassil.) Ovcz. & Czuk.) - 6
nana L. - 1, 3, 4
- subsp. *exilis* (Sukacz.) Hult. = B. exilis
- subsp. *perfiljevii* (V. Vassil.) A. & D. Löve = B. perfiljevii
- subsp. *rotundifolia* (Spach) Malysch. = B. rotundifolia
- subsp. *tundrarum* (Perf.) A. & D. Löve = B. tundrarum
- var. *sibirica* Ledeb. = B. exilis

obscura A. Kotula (*B. verrucosa* Ehrh. subsp. *obscura* (A. Kotula) A. & D. Löve) - 1
ovalifolia Rupr. (*B. fruticosa* Pall. var. *ovalifolia* (Rupr.) S.L. Tung) - 5
ovczinnikovii V. Vassil. - 6
oycoviensis Bess. (*B. verrucosa* Ehrh. subsp. *oycoviensis* (Bess.) A. & D. Löve) - 1
pamirica Litv. (*B. bucharica* V. Vassil., *B. stenolepis* V. Vassil.) - 6
paraermanii V. Vassil. = B. ermanii
paramushirensis Barkalov - 5
pendula Roth (*B. platyphylloides* V. Vassil., *B. talassica* Poljak., *B. verrucosa* Ehrh.) - 1, 2, 3, 4, 6
- subsp. *kamtschatica* (Regel) M. Shemberg = B. kamtschatica
- var. **carelica** (Merckl.) Hamet-Ahti (*B. alba* L. var. *carelica* Merckl.) - 1
- var. *microlepis* (Ig. Vassil.) Doluch. = B. microlepis
perfiljevii V. Vassil. (*B. nana* L. subsp. *perfiljevii* (V. Vassil.) A. & D. Löve) - 5
platyphylla Sukacz. (*B. ajanensis* Kom., *B. alba* L. subsp. *latifolia* (Tausch) Regel var. *tauschii* Regel, *B. tauschii* (Regel) Koidz.) - 4, 5
- subsp. *kamtschatica* (Regel) Worosch. = B. kamtschatica
- subsp. *mandshurica* (Regel) Kitag. = B. mandshurica
- subsp. **minutifolia** (Ju. Kozhevn.) Ju. Kozhevn. *B. cajanderi* Sukacz. subsp. *minutifolia* Ju. Kozhevn. 5
- var. *japonica* (Miq.) Hara = B. kamtschatica
- var. *kamtschatica* (Regel) Hara = B. kamtschatica
- var. *mandshurica* (Regel) Hara = B. mandshurica
platyphylloides V. Vassil. = B. pendula ˏ
potamophila V. Vassil. - 6
prochorowii Kuzen. & Litv. - 5
procurva Litv. - 6
- subsp. *lipskyi* (V. Vassil.) Ovcz. = B. lipskyi
- subsp. *schugnanica* (B. Fedtsch.) Ovcz. = B. schugnanica
psammophila V. Vassil. - 6
pseudoalajica V. Vassil. = B. alajica
x **pseudomiddendorffii** V. Vassil. - B. pubescens Ehrh. s.l. x B. rotundifolia Spach - 4
pseudopendula V. Vassil. - 4
pubentissima V. Vassil. = B. tadshikistanica
pubescens Ehrh. (*B. alba* L. nom. ambig.) - 1, 2, 3, 4
- subsp. *callosa* (Noto) A. & D. Löve = B. callosa
- subsp. *concinna* (Gunnarss.) A. & D. Löve = B. concinna
- subsp. *czerepanovii* (Orlova) Hamet-Ahti = B. czerepanovii
- subsp. *subarctica* (Orlova) A. & D. Löve = B. subarctica
- subsp. *tortuosa* (Ledeb.) Nym. = B. tortuosa
pyrolifolia V. Vassil. - 6
raddeana Trautv. - 2
recurvata (Ig. Vassil.) V. Vassil. (*B. litwinowii* Doluch. var. *recurvata* Ig. Vassil.) - 2
regeliana V. Vassil. - 6
rezniczenkoana (Litv.) Schischk. - 3
rossica Min. - 1
rotundifolia Spach (*B. nana* L. subsp. *rotundifolia* (Spach) Malysch.) - 3, 4
sajanensis V. Vassil. - 4
saksarensis Polozh. & Maltzeva - 4
saposhnikovii Sukacz. - 6
- subsp. *jarmolenkoana* (Golosk.) Ovcz. & Czuk. = B. jarmolenkoana
- subsp. *murgabica* (V. Vassil.) Ovcz. & Czuk. = B. murgabica
saviczii V. Vassil. - 3
schmidtii Regel - 5

schugnanica (B. Fedtsch.) Litv. (*B. procurva* Litv. subsp. *schugnanica* (B. Fedtsch.) Ovcz.) - 6
seravschanica V. Vassil. (*B. bornmuelleri* V. Vassil., *B. hissarica* V. Vassil.) - 6
sibirica (Ledeb.) P'ei = B. exilis
stenolepis V. Vassil. = B. pamirica
subarctica Orlova (*B. pubescens* Ehrh. subsp. *subarctica* (Orlova) A. & D. Löve) - 1
substepposa V. Vassil. - 4
tadshikistanica V. Vassil. (*B. pubentissima* V. Vassil.) - 6
talassica Poljak. = B. pendula
tauschii (Regel) Koidz. = B. platyphylla
tianschanica Rupr. (*B. alatavica* Muscheg., *B. crassijulis* Muscheg.) - 6
- subsp. *margusarica* (V.Vassil.) Ovcz. = B. margusarica
tiulinae V. Vassil. - 4
tortuosa auct. fl. Europ. bor. = B. czerepanovii
tortuosa Ledeb. (*B. pubescens* Ehrh. subsp *tortuosa* (Ledeb.) Nym.) - 3
transbaicalensis V. Vassil. - 4
tundrarum Perf. (*B. nana* L. subsp. *tundrarum* (Perf.) A. & D. Löve) - 1
turkestanica Litv. - 6
tuturinii V. Vassil. - 6
ulmifolia auct. = B. ermanii
uschkanensis Sukacz. = B. insularis
vassiljevii Dyl. - 5
velutina V. Vassil. = B. lanata
verrucosa Ehrh. = B. pendula
- subsp. *obscura* (A. Kotula) A. & D. Löve = B. obscura
- subsp. *oycoviensis* (Bess.) A. & D. Löve = B. oycoviensis
victoris Gussejnov - 2
viridis Chaix = Duschekia abnobetula
vladimirii V. Vassil. - 4
yoshimurae Miyabe & Tatew. = B. divaricata
zinserlingii V. Vassil. - 6

Carpinus L.

betulus L. (*C. caucasica* Grossh.) - 1, 2
caucasica Grossh. = C. betulus
- var. *oxycarpa* (H. Winkl.) Grossh. = C. x grosseserrata
cordata Blume - 5
geoktschaica Radde-Fomina = C. schuschaensis
x **grosseserrata** H. Winkl. (*C. caucasica* Grossh. var. *oxycarpa* (H. Winkl.) Grossh., *C. oxycarpa* H. Winkl., *C. schuschaensis* H. Winkl. var. *grosseserrata* (H. Winkl.) Grossh.). - C. betulus L. x C. schuschaensis H. Winkl. - 2
x **hybrida** H. Winkl. - C. betulus L. x C. schuschaensis H. Winkl. - 2
macrocarpa (Willk.) H. Winkl. = C. orientalis subsp. macrocarpa
orientalis Mill. - 1, 2
- subsp. **macrocarpa** (Willk.) Browicz (*C. orientalis* var. *macrocarpa* Willk., *C. macrocarpa* (Willk.) H. Winkl.) - 2
- var. *macrocarpa* Willk. = C. orientalis subsp. macrocarpa
oxycarpa H.Winkl. = C. x grosseserrata
schuschaensis H. Winkl. (? *C. geoktschaica* Radde-Fomina) - 2
- var. *grosseserrata* (H. Winkl.) Grossh. = C. x grosseserrata

Corylus L.

abchasica Kem.-Nath. - 2
avellana L. - 1, 2
brevituba Kom. = C. mandshurica

cervorum V. Petrov = C. colurna
colchica Albov - 2
colurna L. (*C. cervorum* V. Petrov, *C. iberica* Wittm. ex Kem.-Nath.) - 1(cult.), 2
egrissiensis Kem.-Nath. - 2
x fominii Kem.-Nath (x *C. gudarethica* Kem.-Nath.). - C. avellana L. x C. colurna L. - 2
x gudarethica Kem.-Nath. = C. x fominii
heterophylla Fisch. ex Trautv. - 4, 5
iberica Wittm. ex Kem.-Nath. = C. colchica
imeretica Kem.-Nath. - 2
kachetica Kem.-Nath. - 2
mandshurica Maxim. (*C. brevituba* Kom., *C. mandshurica* f. *brevituba* (Kom.) Kitag.) - 5
- f. *brevituba* (Kom.) Kitag. = C. mandshurica
***maxima** Mill. - 1, 2
pontica C. Koch - 1(cult.), 2

Duschekia Opiz (*Alnaster* Spach)

abnobetula (Ehrh.) Pouzar (*Betula alnobetula* Ehrh. 1783, *Alnaster viridis* (Chaix) Spach, *Alnus alnobetula* (Ehrh.) C. Koch, *A. viridis* (Chaix) DC., *Betula viridis* Chaix, 1786, *Duschekia viridis* (Chaix) Opiz) - 1
fruticosa (Rupr.) Pouzar (*Alnus fruticosa* Rupr., *Alnaster fruticosus* (Rupr.) Ledeb., *A. glutipes* Jarm. ex Czer., *Alnus glutipes* (Jarm. ex Czer.) Worosch., *Duschekia glutipes* (Jarm. ex Czer.) Pouzar) - 1, 3, 4, 5
glutipes (Jarm. ex Czer.) Pouzar = D. fruticosa
kamtschatica (Regel) Pouzar (*Alnus viridis* (Chaix) DC. lusus *kamtschatica* Regel, *Alnaster kamtschaticus* (Regel) Czer., *Alnus fruticosa* Rupr. subsp. *kamtschatica* (Regel) Ju. Kozhevn., *A. kamtschatica* (Regel) Kom., *Duschekia sinuata* (Regel) Pouzar subsp. *kamtschatica* (Regel) Holub) - 5
manshurica (Call.) Pouzar (*Alnaster manshuricus* (Call.) Jarm., *Alnus manshurica* (Call.) Hand.-Mazz.) - 5
maximowiczii (Call.) Pouzar (*Alnus maximowiczii* Call., *Alnaster maximowiczii* (Call.) Czer., *Alnus crispa* Pursh subsp. *maximowiczii* (Call.) Hult., *A. viridis* (Chaix) DC. subsp. *maximowiczii* (Call.) D. Löve) - 5
sinuata (Regel) Pouzar (*Alnus viridis* (Chaix) DC. var. *sinuata* Regel, *Alnaster sinuatus* (Regel) Czer., *Alnus sinuata* (Regel) Rydb.) - 5
- subsp. *kamtschatica* (Regel) Holub = D. kamtschatica
viridis (Chaix) Opiz = D. abnobetula

Ostrya Scop.

carpinifolia Scop. - 2

BIEBERSTEINIACEAE Endl.
Biebersteinia Steph.

multifida DC. - 2, 6
odora Steph. - 3, 6

BIGNONIACEAE Juss.
*Campsis Lour.

***radicans** (L.) Seem. - 1, 2, 6

*Catalpa Scop.

***bignonioides** Walt. - 1, 2, 6
***ovata** D. Don fil. - 1, 2, 6
***speciosa** (Warder ex Barney) Warder ex Engelm. - 1, 2, 6

Incarvillea Juss.

olgae Regel - 6
semiretschenskia (B. Fedtsch.) Grierson = Niedzwedzkia semiretschenskia
sinensis Lam. - 5

Niedzwedzkia B. Fedtsch.

semiretschenskia B. Fedtsch. (*Incarvillea semiretschenskia* (B. Fedtsch.) Grierson) - 6

BLECHNACEAE (C. Presl) Copel.
Blechnum L. (*Spicantopsis* Nakai, *Struthiopteris* Weis, *Lomaria* auct.)

nipponicum (G. Kunze) Makino (*Lomaria nipponica* G. Kunze, *Blechnum spicant* (L.) Roth subsp. *nipponicum* (G. Kunze) A. & D. Löve, *Spicantopsis nipponica* (G. Kunze) Nakai, *Struthiopteris nipponica* (G. Kunze) Nakai) - 5
spicant (L.) Roth (*Osmunda spicant* L.) - 1, 2
- subsp. *nipponicum* (G. Kunze) A. & D. Löve = B. nipponicum

Lomaria auct. = Blechnum

nipponica G. Kunze = Blechnum nipponicum

Spicantopsis Nakai = Blechnum

nipponica (G. Kunze) Nakai = Blechnum nipponicum

Struthiopteris Weis = Blechnum

nipponica (G. Kunze) Nakai = Blechnum nipponicum

BORAGINACEAE Juss.
Aegonychon S.F. Gray

arvense (L.) S.F. Gray subsp. *sibthorpianum* (Griseb.) Dostal = Buglossoides sibthorpiana
purpureocaeruleum (L.) Holub (*Lithospermum purpureocaeruleum* L., *Buglossoides purpureocaerulea* (L.) Johnst.) - 1, 2

Aipyanthus Stev. = Nonea

echioides (L.) Stev. = Huynhia pulchra and Nonea echioides
pulcher (Willd. ex Roem. & Schult.) V. Avet. = Huynhia pulchra
pulcher (Roem. & Schult.) Kolak. = Huynhia pulchra Schult.) Greuter & Burdet

Alkanna Tausch

cordifolia C. Koch - 2
orientalis (L.) Boiss. - 2
tinctoria auct. = A. tuberculata
tuberculata (Forssk.) Meikle (*Anchusa tuberculata* Forssk., *Alkanna tinctoria* auct.) - 1

Allocarya Greene

asiatica Kom. = A. orientalis
orientalis (L.) Brand (*A. asiatica* Kom.) - 5

Amblynotus (A. DC.) Johnst.

obovatus (Ledeb.) Johnst. = A. rupestris
rupestris (Pall. ex Georgi) M. Pop. ex Serg. (*Myosotis rupestris* Pall. ex Georgi, *Amblynotus obovatus* (Ledeb.) Johnst., *Eritrichium rupestre* (Pall. ex Georgi) Bunge, p.p. quoad typum) - 3, 4

Amsinckia Lehm.

intermedia Fisch. & C.A. Mey. - 1
menziesii auct. = A. micrantha
micrantha Suksd. (*A. menziesii* auct.) - 1(alien), 5(alien)

Anchusa L.

arvensis (L.) Bieb. = Lycopsis arvensis
- subsp. *orientalis* (L.) Nordh. = Lycopsis orientalis
azurea Mill. (*A. italica* Retz.) - 1, 2, 5(alien), 6
barrelieri (All.) Vitm. = Cynoglottis barrelieri
gmelinii Ledeb. - 1, 2
incana Ledeb. (*A. leptophylla* Roem. & Schult. subsp. *incana* (Ledeb.) Chamberlain) - 2
italica Retz. = A. azurea
leptophylla Roem. & Schult. - 1, 2
- subsp. *incana* (Ledeb.) Chamberlain = A. incana
ochroleuca Bieb. - 1, 2
- subsp. *popovii* Gusul. = A. popovii
officinalis L. - 1, 2
orientalis (L.) Reichenb. fil. = Lycopsis orientalis
ovata Lehm. = Lycopsis orientalis
popovii (Gusul.) Dobrocz. (*A. ochroleuca* Bieb. subsp. *popovii* Gusul.) - 1
procera Bess. - 1, 2
pseudoochroleuca Shost. - 1
pusilla Gusul. - 1, 2
stylosa Bieb. - 1
thessala Boiss. & Sprun. - 1, 2
tuberculata Forssk. = Alkanna tuberculata

Anoplocaryum Ledeb.

compressum (Turcz.) Ledeb. - 4
turczaninovii Krasnob. - 4

Argusia Boehm. (*Messerschmidia* Hebenstreit, *Tournefortia* L. p.p.)

sibirica (L.) Dandy (*Tournefortia sibirica* L., *Messerschmidia sibirica* (L.) L.) - 1, 2, 3, 4, 5, 6
sogdiana (Bunge) Czer. (*Heliotropium sogdianum* Bunge, *Messerschmidia sogdiana* (Bunge) H. Riedl, *Tournefortia sogdiana* (Bunge) M. Pop.) - 6

Arnebia Forrsk.

baldshuanica (Lipsky) Schischk. ex Nevski = A. coerulea
cana (Tzvel.) Czer. = Macrotomia cana
coerulea Schipcz. (*A. baldshuanica* (Lipsky) Schischk. ex Nevski, *A. grandiflora* (Trautv.) M. Pop.) - 6
decumbens (Vent.) Coss. & Kral. - 1, 2, 3, 6
densiflora (Ledeb.) Ledeb. = Macrotomia densiflora
euchroma (Royle) Johnst. = Macrotomia euchroma
grandiflora (Trautv.) M. Pop. = A. coerulea
guttata Bunge - 3, 6
linearifolia auct. = A. minima
longiflora C. Koch = Huynhia pulchra
minima Wettst. (*A. linearifolia* auct.) - 2, 6
obovata Bunge (*A. olgae* Regel) - 6

olgae Regel = A. obovata
paucisetosa A. Li - 6
pulchra (Roem. & Schult.) J.R. Edmondson = Huynhia pulchra
thomsonii Clarke = A. tibetana
tibetana Kurz (*A. thomsonii* Clarke) - 6
transcaspica M. Pop. - 6
tschimganica (B. Fedtsch.) G.L. Zhu = Ulugbekia tschimganica
ugamensis (M. Pop.) H. Riedl = Macrotomia ugamensis

Asperugo L.

procumbens L. - 1, 2, 3, 4, 5, 6

Beruniella Zak. & Nabiev

micrantha (Pall.) Zak. & Nabiev (*Heliotropium micranthos* (Pall.) Bunge) - 1, 3, 6

Bilegnum Brand = Rindera

bungei (Boiss.) Brand = Rindera bungei

Borago L.

macranthera Banks & Soland. = Caccinia macranthera
officinalis L. - 1, 2, 3, 5, 6

Bothriospermum Bunge

tenellum (Hornem.) Fisch. & C.A. Mey. - 5, 6

Brachybotrys Maxim. ex Olivier

paridiformis Maxim. ex Olivier

Brunnera Stev.

macrophylla (Adams) Johnst. (*Myosotis macrophylla* Adams) - 1(cult.), 2
sibirica Stev. - 4, 5

Buglossoides Moench (*Rhytispermum* Link)

arvensis (L.) Johnst. (*Lithospermum arvense* L., *Rhytispermum arvense* (L.) Link) - 1, 3, 4, 5, 6
- subsp. *sibthorpiana* (Griseb.) R. Fernandes = B. sibthorpiana
czernjajevii (Klok.) Czer. (*Lithospermum czernjajevii* Klok.) - 1
purpureocaerulea (L.) Johnst. = Aegonychon purpureocaeruleum
sibthorpiana (Griseb.) Czer. (*Lithospermum sibthorpianum* Griseb., *Aegonychon arvense* (L.) S.F. Gray subsp. *sibthorpianum* (Griseb.) Dostal, *Buglossoides arvensis* (L.) Johnst. subsp. *sibthorpiana* (Griseb.) R. Fernandes) - 1, 2
tenuiflora (L. fil.) Johnst. (*Lithospermum tenuiflorum* L. fil., *Rhytispermum tenuiflorum* (L. fil.) Link) - 1, 2, 6

Caccinia Savi

crassifolia (Vent.) C. Koch = C. macranthera
dubia Bunge - 6
macranthera (Banks & Soland.) Brand (*Borago macranthera* Banks & Soland., *Caccinia crassifolia* (Vent.) C. Koch, *C. turkestanica* Gand.) - 2, 6
turkestanica Gand. = C. macranthera

Cerinthe L.

alpina Kit. = C. glabra
glabra Mill. (*C. alpina* Kit.) - 2
- subsp. **caucasica** Hadac - 2
*****major** L. - 6
minor L. (*C. quinquemaculata* Wahlenb.) - 1, 2, 3
quinquemaculata Wahlenb. = C. minor

Craniospermum Lehm.

canescens DC. - 3
echioides (Schrenk) Bunge - 3
mongolicum Johnst. - 4
subfloccosum Kryl. - 3
subvillosum Lehm. - 4

Cryptantha Lehm. (*Oreocarya* auct.)

spiculifera (Piper) Payson (*Oreocarya spiculifera* Piper) - 5

Cynoglossum L.

asperrimum Nakai = Paracynoglossum asperrimum
biebersteinii (DC.) Greuter & Burdet = Solenanthus biebersteinii
capusii (Franch.) Pazij (*Paracaryum capusii* Franch., *Cynoglossum tianschanicum* M. Pop., *Lindelofia capusii* (Franch.) M. Pop. p.p., quoad nomen) - 6
circinnatum (Ledeb.) Greuter & Burdet = Solenanthus circinnatus
creticum Mill. (*C. pictum* Soland.) - 1, 2, 6
divaricatum Steph. - 3, 4, 5
germanicum Jacq. (*C. montanum* sensu M. Pop.) - 1, 2
glochidiatum Wall. ex Benth. = Paracynoglossum glochidiatum
holosericeum Stev. - 2
hungaricum Simonk. - 1(?)
imeretinum Kusn. = Paracynoglossum glochidiatum
incanum (Ledeb.) Greuter & Burdet = Paracaryum incanum
laxiflorum (Trautv.) Greuter & Burdet = Paracaryum laxiflorum
montanum L. (*C. teheranicum* H. Riedl, *C. nebrodense* auct.) - 2
montanum sensu M. Pop. = C. germanicum
nebrodense auct. = C. montanum
officinale L. - 1, 2, 3, 4, 5, 6
paracaryum Greuter & Burdet = Paracaryum strictum
pictum Soland. = C. creticum
seravschanicum (B. Fedtsch.) M. Pop. - 6
sintenisii (Bornm.) Greuter & Burdet = Paracaryum strictum
strictum (C. Koch) Greuter & Burdet = Paracaryum strictum
teheranicum H. Riedl = C. montanum
tetraspis (Pall.) Greuter & Burdet = Rindera tetraspis
tianschanicum M. Pop. = C. capusii
viridiflorum Pall. ex Lehm. - 3, 6
wallichii G. Don fil. var. *glochidiatum* (Wall. ex Benth.) Kazmi = Paracynoglossum glochidiatum

Cynoglottis (Gusul.) Vural & Kit Tan

barrelieri (All.) Vural & Kit Tan (*Anchusa barrelieri* (All.) Vitm.) - 1

Echinospermum Lehm. = Lappula

brachycentrum Ledeb. = Lappula brachycentra

caspium Fisch. & C.A. Mey. = Lappula semiglabra

Echioides auct. = Huynhia

longiflora (C. Koch) Johnst. = Huynhia pulchra

Echium L.

amoenum Fisch. & C.A. Mey. - 2
biebersteinii (Lacaita) Dobrocz. (*E. italicum* L. var. *biebersteinii* Lacaita, *E. italicum* subsp. *biebersteinii* (Lacaita) Greuter & Burdet, *E. italicum* auct.) - 1, 2, 6
creticum L. subsp. *plantagineum* (L.) R. de P. Malagarriga Heras = E. plantagineum
italicum auct. = E. biebersteinii
italicum L. subsp. *biebersteinii* (Lacaita) Greuter & Burdet = E. biebersteinii
- var. *biebersteinii* Lacaita = E. biebersteinii
lycopsis auct. = E. plantagineum
maculatum auct. = E. russicum
plantagineum L. (*E. creticum* L. subsp. *plantagineum* (L.) R. de P. Malagarriga Heras, *E. lycopsis* auct.) - 1, 2
popovii Dobrocz. - 1
rubrum Jacq. = E. russicum
russicum J.F. Gmel. (*E. rubrum* Jacq. 1778, non Forssk. 1774, *E. maculatum* auct.) - 1, 2, 6
vulgare L. - 1, 3, 4, 5, 6

Eritrichium Schrad. ex Gaudin

altaicum M. Pop. (*E. rupestre* (Pall. ex Georgi) Bunge, p.p. excl. typo) - 3
arctisibiricum (Petrovsky) A. Khokhr. (*E. sericeum* (Lehm.) DC. subsp. *arctisibiricum* Petrovsky) - 3, 4, 5
aretioides (Cham.) DC. - 5
- var. **chamissonis** (DC.) Petrovsky (*E. chamissonis* DC.) - 5
basifixum Clarke = E. villosum
canum auct. = E. subjacquemontii
caucasicum (Albov) Grossh. - 2
chamissonis DC. = E. aretioides var. chamissonis
czekanowskii Trautv. = Myosotis czekanowskii
deflexum (Wahlenb.) Y.S. Lian & J.Q. Wang = Hackelia deflexa
dubium O. Fedtsch. - 6
fetisowii Regel - 6
incanum A. DC. - 4, 5
- subsp. *sichotense* (M. Pop.) Starchenko & Gavrilenko = E. sichotense
- var. **maackii** (Maxim.) Worosch. (*E. maackii* Maxim.) - 5
jacuticum M. Pop. - 4, 5
jenisseense Turcz. ex A. DC. - 3, 4
kamtschaticum Kom. = E. sericeum
kungejense Bajt. & Kudabaeva - 6
latifolium Kar. & Kir. - 6
maackii Maxim. = E. incanum var. maackii
■ **mandshuricum** M. Pop.
■ **nanum** (All.) Schrad. ex Gaudin (*Myosotis nana* All.)
nipponicum Makino (*E. sachalinense* M. Pop.) - 5
ochotense Jurtz. & A. Khokhr. - 5
pamiricum B. Fedtsch. (*Tianschaniella wakhanica* H. Riedl) - 6
pauciflorum (Ledeb.) DC. subsp. *sajanense* (Malysch.) Malysch. = E. sajanense
- subsp. *subrupestre* (M. Pop.) Malysch. = E. subrupestre
pectinatum (Pall.) DC. - 1, 3, 4
persicum Boiss. = Lepechiniella persica
pseudolatifolium M. Pop. - 6

pseudostrictum M. Pop. - 6
pulviniforme M. Pop - 4
relictum Kudabaeva - 6
rupestre (Pall. ex Georgi) Bunge = E. altaicum and Amblynotus rupestris
- subsp. *sajanense* Malysch. = E. sajanense
sachalinense M. Pop. = E. nipponicum
sajanense (Malysch.) Sipl. (*E. rupestre* (Pall. ex Georgi) Bunge subsp. *sajanense* Malysch., *E. pauciflorum* (Ledeb.) DC. subsp. *sajanense* (Malysch.) Malysch.) - 4
sericeum (Lehm.) A. DC. (*E. kamtschaticum* Kom.) - 4, 5
- subsp. *arctisibiricum* Petrovsky = E. arctisibiricum
sichotense M. Pop. (*E. incanum* A. DC. subsp. *sichotense* (M. Pop.) Starchenko & Gavrilenko) - 5
subjacquemontii M. Pop. (*E. canum* auct.) - 6
subrupestre M. Pop. (*E. pauciflorum* (Ledeb.) DC. subsp. *subrupestre* (M. Pop.) Malysch.) - 3
thymifolium (DC.) Y.S. Lian & J.Q. Wang = Hackelia thymifolia
tianschanicum Iljin ex M. Pop = E. villosum
tschuktschorum Jurtz. & Petrovsky - 5
turkestanicum Franch. - 6
tuvinense M. Pop. - 4
uralense Serg. - 3
villosum (Ledeb.) Bunge (*E. basifixum* Clarke, *E. tianschanicum* Iljin ex M. Pop. nom. invalid.) - 1, 3, 4, 5, 6
- subsp. **pulvinatum** Petrovsky - 1, 3, 4

Gastrocotyle Bunge

hispida (Forrsk.) Bunge - 6

Hackelia Opiz

deflexa (Wahlenb.) Opiz (*Eritrichium deflexum* (Wahlenb.) Y.S. Lian & J.Q. Wang) - 1, 3, 4, 5, 6
murgabica Czuk. - 6
popovii Czuk. - 6
thymifolia (DC.) Johnst. (*Eritrichium thymifolium* (DC.) Y.S. Lian & J.Q. Wang, *Hackelia thymifolia* (DC.) M. Pop. comb. superfl.) - 3, 4

Heliotropium L. (*Piptoclaina* G. Don fil.)

acutiflorum Kar. & Kir. - 6
arguzioides Kar. & Kir. - 1, 6
biannulatiforme M. Pop. - 6
bogdanii Czuk. - 6
bucharicum B. Fedtsch. - 6
chorassanicum Bunge - 6
dasycarpum Ledeb. - 6
dolosum De Not. (*H. littorale* Stev.) - 1, 2
ellipticum Ledeb. (*H. lasiocarpum* Fisch. & C.A. Mey., *H. strictum* Ledeb.) - 1, 2, 3, 6
europaeum L. (*H. stevenianum* Andrz.) - 1, 2
- subsp. *suaveolens* (Bieb.) Litard. = H. suaveolens
fedtschenkoanum M. Pop. - 6
grande M. Pop. (*H. popovii* H. Riedl) - 6
intermedium Andrz. = H. suaveolens
kowalenskyi sensu M. Pop. = H. tzvelevii
kowalenskyi Stschegl. = H. szovitsii
lasiocarpum Fisch. & C.A. Mey. = H. ellipticum
littorale Stev. = H. dolosum
litwinowii M. Pop. - 6
magistri Raenko - 6
micranthos (Pall.) Bunge = Beruniella micrantha
olgae Bunge - 6

parvulum M. Pop. - 6
***peruvianum** L. - 1
pileiforme Czuk. - 6
popovii H. Riedl = H. grande
ramosissimum (Lehm.) Sieb. ex DC. (*H. undulatum* Vahl var. *ramosissimum* Lehm., *H. turcomanicum* M. Pop. & Korov.) - 6
seravschanicum M. Pop. - 6
sogdianum Bunge = Argusia sogdiana
stevenianum Andrz. = H. europaeum
strictum Ledeb. = H. ellipticum
styligerum Trautv. - 2
suaveolens Bieb. (*H. europaeum* L. subsp. *suaveolens* (Bieb.) Litard., *H. intermedium* Andrz.) - 1, 2
supinum L. (*Piptoclaina supina* (L.) G. Don fil.) - 2, 6
szovitsii (Stev.) Bunge (*H. kowalenskyi* Stschegl.) - 2
transoxanum Bunge - 6
turcomanicum M. Pop. & Korov. = H. ramosissimum
tzvelevii T.N. Pop. (*H. kowalenskyi* sensu M. Pop.) - 2
undulatum Vahl var. *ramosissimum* Lehm. = H. ramosissimum

Heterocaryum A. DC.

echinophorum (Pall.) Brand - 1
laevigatum (Kar. & Kir.) A. DC. - 6
macrocarpum Zak. - 2, 6
oligacanthum (Boiss.) Bornm. = H. subsessile
rigidum A. DC. - 2, 6
subsessile Vatke (*H. oligacanthum* (Boiss.) Bornm. nom. illegit., *Lappula subsessilis* (Vatke) Greuter & Burdet) - 6
szovitsianum (Fisch. & C.A. Mey.) A. DC. - 2, 6

Huynhia Greuter (*Echioides* auct.)

pulchra (Roem. & Schult.) Greuter & Burdet (*Lycopsis pulchra* Willd. ex Roem. & Schult., *Aipyanthus echioides* (L.) Stev. p.p. excl. basionymo, *A. pulcher* (Willd. ex Roem. & Schult.) V. Avet., *A. pulcher* (Roem. & Schult.) Kolak. nom. illegit. superfl., *Arnebia longiflora* C. Koch, *A. pulchra* (Roem. & Schult.) J.R. Edmondson, *Echioides longiflora* (C. Koch) Johnst., *Macrotomia echioides* (L.) Boiss. p.p. excl. basionymo) - 2

Kuschakewiczia Regel & Smirn.

turkestanica Regel & Smirn. (*Solenanthus turkestanicus* (Regel & Smirn.) Kusn.) - 6

Lappula Moench (*Echinospermum* Lehm.)

aktaviensis M. Pop. & Zak. - 6
alaica (M. Pop.) Nabiev = Lepechiniella alaica
alatavica Golosk. (*L. rupestris* (Schrenk) Guerke var. *alatavica* M. Pop. nom. invalid.) - 6
anisacantha (Turcz. ex Bunge) Guerke = L. redowskii
austrodshungarica (Golosk.) Nabiev = Lepechiniella austrodshungarica
badachschanica M. Pop. ex Ikonn. - 6
bajtenovii Kudabaeva - 6
balchaschensis M. Pop. ex Golosk. - 6
barbata (Bieb.) Guerke - 1, 2, 6
betpakdalensis Nabiev = Lepechiniella balchaschensis
brachycentra (Ledeb.) Guerke (*Echinospermum brachycentrum* Ledeb., *Lappula tuberculata* (Lipsky) Zak.) - 3, 6
brachycentroides M. Pop. - 3, 6
caspia (Fisch. & C.A. Mey.) M. Pop. ex Dobrocz. = L. semiglabra

ceratophora (M. Pop.) M. Pop. (*L. spinocarpos* (Forssk.) Aschers. subsp. *ceratophora* (M. Pop.) R. Kam. & Raenko) - 6
consanguinea (Fisch. & C.A. Mey.) Guerke - 1, 2, 3, 4, 5, 6
coronifera M. Pop. - 6
cristata (Bunge) B. Fedtsch. - 3
cymosa (Stschegl.) B. Fedtsch. = L. semiglabra
diploloma (Schrenk) Guerke - 6
drobovii (M. Pop.) M. Pop. ex Pavl. - 6
duplicicarpa Pavl. = L. semiglabra
dzharkentica M. Pop. ex Golosk. = L. semiglabra
echinata Gilib. = L. squarrosa
glabrata M. Pop. - 6
heteracantha (Ledeb.) Borb. (*L. myosotis* Moench subsp. *heteracantha* (Ledeb.) A. & D. Löve, *L. squarrosa* (Retz.) Dumort. subsp. *heteracantha* (Ledeb.) Chater) - 1, 2
heterocarpa Klok. & Artemcz. = L. semicincta
intermedia (Ledeb.) M. Pop. = L. redowskii
ketmenica Kudabaeva - 6
korshinskyi M. Pop. - 6
kulikalonica Zak. - 6
lipschitzii M. Pop. - 6
lipskyi M. Pop. - 3
macra M. Pop. ex Golosk. - 6
macrantha (Ledeb.) Guerke = L. semiglabra
marginata (Bieb.) Guerke - 1, 2
- subsp. *patula* (Lehm.) Soo = L. patula
michaelis (Golosk.) Nabiev = Lepechiniella michaelis
microcarpa (Ledeb.) Guerke - 3, 6
minuta (Lipsky) Nabiev = Lepechiniella minuta
mogoltavica M. Pop. ex Czuk. - 6
myosotis Moench = L. squarrosa
- subsp. *heteracantha* (Ledeb.) A. & D. Löve = L. heteracantha
nevskii Raenko - 6
nuratavica Nabiev & Zak. - 6
occultata M. Pop. - 6
omphaloides (Schrenk) B. Fedtsch. = Lepechiniella omphaloides
parvula Nabiev & Zak. - 6
patula (Lehm.) Menyharth (*L. marginata* (Bieb.) Guerke subsp. *patula* (Lehm.) Soo) - 1, 2, 3, 4, 6
paulsenii (Brand) M. Pop. - 6
pavlovii Golosk. - 6
petrophila Pavl. = Paracaryum bungei
physacantha Golosk. - 6
popovii Zak. - 6
redowskii (Hornem.) Greene (*L. anisacantha* (Turcz. ex Bunge) Guerke, *L. intermedia* (Ledeb.) M. Pop.) - 3, 4, 5
rupestris (Schrenk) Guerke - 6
- var. *alatavica* M. Pop. = L. alatavica
rupicola Zak. - 6
sarawschanica (Lipsky) Nabiev = Lepechiniella sarawschanica
saurica Bajt. & Kudabaeva - 6
semialata M. Pop. - 6
semicincta (Stev.) M. Pop. ex Dobrocz. (*L. heterocarpa* Klok. & Artemcz.) - 1
semiglabra (Ledeb.) Guerke (*Echinospermum caspium* Fisch. & C.A. Mey., *Lappula caspia* (Fisch. & C.A. Mey.) M. Pop. ex Dobrocz., *L. cymosa* (Stschegl.) B. Fedtsch., *L. duplicicarpa* Pavl., *L. dzharkentica* M. Pop. ex Golosk., *L. macrantha* (Ledeb.) Guerke) - 1, 3, 6
sericata M. Pop. - 6
sessiliflora (Boiss.) Guerke - 2, 6

sinaica (DC.) Aschers. ex Schweinf. - 2, 6
spinocarpos (Forssk.) Aschers. - 1, 2, 3, 6
- subsp. *ceratophora* (M. Pop.) R. Kam. & Raenko = L. ceratophora
squarrosa (Retz.) Dumort. (*Myosotis squarrosa* Retz., *Lappula echinata* Gilib. nom. invalid., *L. myosotis* Moench) - 1, 2, 3, 4, 5, 6
- subsp. *heteracantha* (Ledeb.) Chater = L. heteracantha
stricta (Ledeb.) Guerke (*L. tenuis* (Ledeb.) Guerke) - 1, 3, 6
subcaespitosa M. Pop. ex Golosk. - 6
subsessilis (Vatke) Greuter & Burdet = Heterocaryum subsessile
tadshikorum M. Pop. - 6
tenuis (Ledeb.) Guerke = L. stricta
tianschanica M. Pop. & Zak. - 6
transalaica (M. Pop.) Nabiev = Lepechiniella transalaica
tuberculata (Lipsky) Zak. = L. brachycentra
ulacholica M. Pop. - 6

Lepechiniella M. Pop.

alaica M. Pop. (*Lappula alaica* (M. Pop.) Nabiev) - 6
arassanica (B. Fedtsch.) M. Pop. - 6
austrodshungarica Golosk. (*Lappula austrodshungarica* (Golosk.) Nabiev) - 6
balchaschensis M. Pop. (*Lappula betpakdalensis* Nabiev) - 6
ferganensis M. Pop. - 6
korshinskyi M. Pop. - 6
michaelis Golosk. (*Lappula michaelis* (Golosk.) Nabiev) - 6
minuta (Lipsky) M. Pop. (*Lappula minuta* (Lipsky) Nabiev) - 6
omphaloides (Schrenk) M. Pop. (*Lappula omphaloides* (Schrenk) B. Fedtsch.) - 6
persica (Boiss.) H. Riedl (*Eritrichium persicum* Boiss.) - 6
- subsp. **kopetdaghensis** R. Kam. & Raenko - 6
sarawschanica (Lipsky) M. Pop. (*Lappula sarawschanica* (Lipsky) Nabiev) - 6
transalaica M. Pop (*Lappula transalaica* (M. Pop.) Nabiev) - 6

Lindelofia Lehm.

anchusoides auct. = L. macrostyla
anchusoides (Lindl.) Lehm. subsp. *macrostyla* (Bunge) R. Kam. = L. macrostyla
angustifolia (Schrenk) Brand = L. stylosa
capusii (Franch.) M. Pop. = L. tschimganica and Cynoglossum capusii
macrostyla (Bunge) M. Pop. (*L. anchusoides* (Lindl.) Lehm. subsp. *macrostyla* (Bunge) R. Kam., *L. anchusoides* auct.) - 6
olgae (Regel & Smirn.) Brand (*Mertensia lindelofioides* Rech. fil. & H. Riedl, *Pseudomertensia lindelofioides* (Rech. fil. & H. Riedl) H. Riedl) - 6
pterocarpa (Rupr.) M. Pop. = L. stylosa var. pterocarpa
stylosa (Kar. & Kir.) Brand (*L. angustifolia* (Schrenk) Brand) - 6
- var. **pterocarpa** (Rupr.) M. Pop. ex Czuk. (*L. pterocarpa* (Rupr.) M. Pop.) - 6
tschimganica (Lipsky) M. Pop. ex Pazij (*Solenanthus olgae* Regel & Smirn. γ. *tschimganicus* Lipsky, *Lindelofia capusii* (Franch.) M. Pop. p.p. quoad plantas) - 6

Lithospermum L.

apulum (L.) Vahl = Neatostema apulum
arvense L. = Buglossoides arvense

czernjajevii Klok. = Buglossoides czernjajevii
erythrorhizon Siebold & Zucc. - 4, 5
officinale L. - 1, 2, 3, 4, 6
purpureocaeruleum L. = Aegonychon purpureocaeruleum
sibthorpianum Griseb. = Buglossoides sibthorpiana
tenuiflorum L. fil. = Buglossoides tenuiflora
tschimganicum B. Fedtsch. = Ulugbekia tschimganica

Lycopsis L.

arvensis L. (*Anchusa arvensis* (L.) Bieb.) - 1, 5(alien)
echioides L. = Nonea echioides
orientalis L. (*Anchusa arvensis* (L.) Bieb. subsp. *orientalis*
   (L.) Nordh., *A. orientalis* (L.) Reichenb. fil. 1858, non
   L. 1753, *A. ovata* Lehm.) - 1, 2, 6
pulchra Willd. ex Roem. & Schult. = Huynhia pulchra
pulla L. = Nonea pulla

Macrotomia DC.

cana Tzvel. (*Arnebia cana* (Tzvel.) Czer.) - 6
densiflora (Ledeb.) Macbr. (*Arnebia densiflora* (Ledeb.)
   Ledeb.) - 2
echioides (L.) Boiss. = Huynhia pulchra and Nonea
   echioides
euchroma (Royle) Pauls. (*Arnebia euchroma* (Royle)
   Johnst.) - 6
ugamensis M. Pop. (*Arnebia ugamensis* (M. Pop.) H.
   Riedl) - 6

Mattiastrum (Boiss.) Brand = Paracaryum

bungei (Boiss.) Rech. fil. & H. Riedl = Paracaryum bungei
crista-galli Rech. fil. & H. Riedl = Paracaryum crista-galli
emiri (M. Pop.) Czer. = Paracaryum himalayense
gracile (Czerniak.) Czer. = Paracaryum gracile
himalayense (Klotzsch) Brand = Paracaryum himalayense
- subsp. *fallax* Rech. fil. & H. Riedl = Paracaryum himalayense
incanum (Ledeb.) Brand = Paracaryum incanum
karataviense (Pavl. ex M. Pop.) Czer. = Paracaryum karataviense
laxiflorum (Trautv.) Czer. = Paracaryum laxiflorum
turcomanicum (Bornm. & Sint.) Brand = Paracaryum
   turcomanicum

Mertensia Roth

asiatica (Takeda) Macbr. = M. maritima
davurica (Pall. ex Sims) G. Don fil. - 3, 4
dshagastanica Regel - 6
jenissejensis M. Pop. = M. sibirica
kamczatica (Turcz.) DC. = M. pubescens
lindelofioides Rech. fil. & H. Riedl = Lindelofia olgae
maritima (L.) S.F. Gray (*M. asiatica* (Takeda) Macbr., *M.
   maritima* subsp. *asiatica* Takeda, *M. maritima* var.
   asiatica (Takeda) Kitag., *M. simplicissima* (Ledeb.) G.
   Don fil., *Pulmonaria simplicissima* Ledeb.) - 1, 5
- subsp. *asiatica* Takeda = M. maritima
- var. *asiatica* (Takeda) Kitag. = M. maritima
pallasii (Ledeb.) G. Don fil. - 3
pilosa auct. = M. pubescens
popovii Rubtz. - 6
pterocarpa (Turcz.) Tatew. & Ohwi - 5
pubescens (Roem. & Schult.) DC. (*M. kamczatica* (Turcz.)
   DC., *M. pilosa* auct.) - 5
rivularis (Turcz.) DC. - 5
serrulata (Turcz.) DC. - 4

sibirica (L.) G. Don fil. (*M. jenissejensis* M. Pop., *M. sibirica* var. *jenissejensis* (M. Pop.) Polozh.) - 4
- var. *jenissejensis* (M. Pop.) Polozh. = M. sibirica
simplicissima (Ledeb.) G. Don fil. = M. maritima
stylosa (Fisch.) DC. - 4
tarbagataica B. Fedtsch. - 6

Messerschmidia Hebenstreit = Argusia

sibirica (L.) L. = Argusia sibirica
sogdiana (Bunge) H. Riedl = Argusia sogdiana

Microparacaryum (M. Pop. ex H. Riedl) Hilger & Podlech = Paracaryum

intermedium (Fresen.) Hilger & Podlech = Paracaryum
   intermedium

Moltkia Lehm.

coerulea (Willd.) Lehm. - 1, 2

Myosotis L. (*Strophiostoma* Turcz.)

alpestris F.W. Schmidt - 1, 2
amoena (Rupr.) Boiss. - 2
arvensis (L.) Hill - 1, 2, 3, 4, 6
- var. *versicolor* Pers. = M. discolor
asiatica (Vestergren) Schischk. & Serg. (*M. suaveolens*
   Waldst. & Kit. subsp. *asiatica* (Vestergner) Ju. Kozhevn.) - 1, 3, 4, 5, 6
baltica Sam. ex Lindm. (*M. laxa* Lehm. subsp. *baltica* (Sam.
   ex Lindm.) Hyl. ex Nordh., *M. laxa* auct.) - 1
brevisetacea (R. Schuster) Holub (*M. nemorosa* Bess.
   subsp. *brevisetacea* R. Schuster) - 1
cespitosa K.F. Schultz (*M. laxa* Lehm. subsp. *cespitosa*
   (K.F. Schultz) Hyl. ex Nordh.) - 1, 2, 3, 4, 5, 6
collina auct. = M. ramosissima
czekanowskii (Trautv.) R. Kam. & V. Tichomirov (*Eritrichium czekanowskii* Trautv.) - 4
daralaghezica T.N. Pop. - 2
decumbens Host (*M. frigida* (Vestergren) Czernov, comb.
   superfl., *M. frigida* (Vestergren) A. & D. Löve, *M.
   sylvatica* Hoffm. subsp. *frigida* Vestergren) - 1
densiflora C. Koch (*M. sylvatica* Ehrh. ex Hoffm. subsp.
   cyanea Vestergren) - 2(?)
discolor Pers. (*M. arvensis* (L.) Hill var. *versicolor* Pers., *M.
   versicolor* (Pers.) Smith) - 1, 2
frigida (Vestergren) Czernov = M. decumbens
frigida (Vestergren) A. & D. Löve = M. decumbens
heteropoda Trautv. - 2
hispida Schlecht. = M. ramosissima
idaea Boiss. & Heldr. = M. incrassata
imitata Serg. (*M. suaveolens* auct. p.p.) - 3, 4, 5, 6
incrassata Guss. (*M. idaea* Boiss. & Heldr.) - 1, 2
krylovii Serg. - 3, 4, 6
laxa auct. = M. baltica
laxa Lehm. subsp. *baltica* (Sam. ex Lindm.) Hyl. ex
   Nordh. = M. baltica
- subsp. *cespitosa* (K.F. Schultz) Hyl. ex Nordh. = M. cespitosa
laxiflora Reichenb. (*M. palustris* (L.) L. subsp. *laxiflora*
   (Reichenb.) Soo, *M. palustris* subsp. *laxiflora* (Reichenb.) Sychowa, comb. superfl.) - 1
lazica M. Pop. - 2
lithospermifolia (Willd.) Hornem. (*M. sylvatica* Ehrh. ex
   Hoffm. var. *lithospermifolia* (Willd.) Cincovic & Kojic,
   M. suaveolens auct. p.p.) - 1, 2

lithuanica (Schmalh.) Bess. ex Dobrocz. (*M. palustris* (L.) L. b. *lithuanica* Schmalh.) - 1
litoralis Stev. ex Bieb. - 1
ludomilae Zaverucha - 1
macrophylla Adams = Brunnera macrophylla
magnifera A. Khokhr. - 2
micrantha Pall. ex Lehm. (*M. stricta* Link ex Roem. & Schult.) - 1, 2, 3, 4, 6
nana All. = Eritrichium nanum
nemorosa Bess. (*M. palustris* (L.) L. subsp. *nemorosa* (Bess.) C.C. Berg & Kaastra, comb. superfl., *M. palustris* subsp. *nemorosa* (Bess.) Jav., *M. palustris* subsp. *nemorosa* (Bess.) Soo, comb. superfl.) - 1, 3, 4
- subsp. *brevisetacea* R. Schuster = M. brevisetacea
palustris (L.) L. (*M. scorpioides* L. p.p. nom. ambig.) - 1, 2
- subsp. *laxiflora* (Reichenb.) Soo = M. laxiflora
- subsp. *laxiflora* (Reichenb.) Sychowa = M. laxiflora
- subsp. *nemorosa* (Bess.) C.C. Berg & Kaastra = M. nemorosa
- subsp. *nemorosa* (Bess.) Jav. = M. nemorosa
- subsp. *nemorosa* (Bess.) Soo = M. nemorosa
- subsp. *strigulosa* (Reichenb.) Arcang. = M. strigulosa
- var. *strigulosa* (Reichenb.) Mert. & Koch = M. strigulosa
- b. *lithuanica* Schmalh. = M. lithuanica
pineticola Klok. & Shost. - 1
popovii Dobrocz. (*M. suaveolens* auct. p.p.) - 1
propinqua (Turcz.) A. DC. - 2, 6
pseudopropinqua M. Pop. = M. sparsiflora
pseudovariabilis M. Pop. (*M. sylvatica* Ehrh. ex Hoffm. subsp. *pseudovariabilis* (M. Pop.) Petrovsky) - 4
radix-palaris A. Khokhr. - 2
ramosissima Rochel ex Schult. (*M. hispida* Schlecht., *M. collina* auct.) - 1, 2
refracta Boiss. - 1, 6
- subsp. **chitralica** Kazmi - 6
rupestris Pall. ex Georgi = Amblynotus rupestris
sachalinensis M. Pop. = M. sylvatica
scorpioides L. = M. palustris
■ *sicula* Guss.
sparsiflora Pohl (*M. pseudopropinqua* M. Pop.) - 1, 2, 3, 4, 6
squarrosa Retz. = Lappula squarrosa
stenophylla Knaf (*M. suaveolens* auct. p.p.) - 1
stricta Link ex Roem. & Schult. = M. micrantha
strigulosa Reichenb. (*M. palustris* (L.) L. subsp. *strigulosa* (Reichenb.) Arcang., *M. palustris* var. *strigulosa* (Reichenb.) Mert. & Koch) - 1
suaveolens auct. = M. imitata, M. lithospermifolia, M. popovii and M. stenophylla
suaveolens Waldst. & Kit. - 1
- subsp. *asiatica* (Vestergren) Ju. Kozhevn. = A. asiatica
superalpina A. Khokhr. - 2
sylvatica Ehrh. ex Hoffm. (*M. sachalinensis* M. Pop., *M. sylvatica* var. *sachalinensis* (M. Pop.) Tolm.) - 1, 2, 5
- subsp. *cyanea* Vestergren = M. densiflora
- subsp. *frigida* Vestergren = M. decumbens
- subsp. *pseudovariabilis* (M. Pop.) Petrovsky = M. pseudovariabilis
- subsp. **rivularis** Vestergren - 2
- var. *lithospermifolia* (Willd.) Cincovic & Kojic = M. lithospermifolia
- var. *sachalinensis* (M. Pop.) Tolm. = M. sylvatica
ucrainica Czern. - 1, 6
versicolor (Pers.) Smith = M. discolor

Neatostema Johnst.

apulum (L.) Johnst. (*Lithospermum apulum* (L.) Vahl) - 1

Nonea Medik. (*Aipyanthus* Stev.)

alpestris (Stev.) G. Don fil. - 2
armeniaca (Kusn.) Grossh. (*N. pulla* DC. var. *armeniaca* Kusn., *N. pulla* subsp. *armeniaca* (Kusn.) Soo) - 2
calceolaris Pazij - 6
candaharensis auct. = N. caspica
caspica (Willd.) G. Don fil. (*N. candaharensis* auct.) - 1, 2, 6
- subsp. *melanocarpa* (Boiss.) H. Riedl = N. melanocarpa
- subsp. **schmidii** H. Riedl - 6
daghestanica Kusn. - 2
decurrens (C.A. Mey.) G. Don fil. - 2
echioides (L.) Roem. & Schult. (*Lycopsis echioides* L., *Aipyanthus echioides* (L.) Stev. p.p. quoad basionymum, *Macrotomia echioides* (L.) Boiss. p.p. quoad basionymum, *Nonea ventricosa* (Smith) Griseb.) - 1, 2
flavescens (C.A. Mey.) Fisch. & C.A. Mey. - 2, 6
intermedia Ledeb. - 2
lutea (Desr.) DC. - 1, 2
macropoda M. Pop. - 6
melanocarpa Boiss. (*N. caspica* (Willd.) G. Don fil. subsp. *melanocarpa* (Boiss.) H. Riedl) - 2, 6
ovczinnikovii Czuk. - 6
pallens Petrovic - 1
persica Boiss. (*N. pulla* DC. var. *persica* (Boiss.) M. Pop.) - 2
picta (Bieb.) Fisch. & C.A. Mey. - 1, 2
pulla DC. (*Lycopsis pulla* L. 1759, non Loefl. 1758) - 1
- subsp. *armeniaca* (Kusn.) Soo = N. armeniaca
- subsp. *rossica* (Stev.) Soo = N. rossica
- subsp. *taurica* (Ledeb.) Soo = N. taurica
- var. *armeniaca* Kusn. = N. armeniaca
- var. *persica* (Boiss.) M. Pop. = N. persica
- var. *rossica* (Stev.) M. Pop. = N. rossica
- var. *taurica* (Ledeb.) Kusn. = N. taurica
rosea (Bieb.) Link - 2
rossica Stev. (*N. pulla* DC. subsp. *rossica* (Stev.) Soo, *N. pulla* var. *rossica* (Stev.) M. Pop.) - 1, 2, 3, 4, 5, 6
setosa (Lehm.) Roem. & Schult. - 2
taurica (Ledeb.) Ledeb. (*N. pulla* DC. subsp. *taurica* (Ledeb.) Soo, *N. pulla* var. *taurica* (Ledeb.) Kusn.) - 1, 2
turcomanica M. Pop. - 6
ventricosa (Smith) Griseb. = N. echioides
versicolor (Stev.) Sweet - 2

Omphalodes Hill

cappadocica (Willd.) DC. - 2
glochidiata Bunge = Paracaryum intermedium
kusnetzovii Kolak. - 2
linifolia (L.) Moench - 1, 2
lojkae Somm. & Levier - 2
physodes Bunge = Paracaryum intermedium
rupestris Rupr. ex Boiss. - 2
scorpioides (Haenke) Schrank - 1, 2
verna Moench - 1

Onosma L.

albicaulis M. Pop. - 6
araratica H. Riedl - 2
arenaria auct. = O. borysthenica
armeniaca Klok. ex M. Pop. = O. setosa
atrocyanea Franch. - 6
aucheriana DC. = O. montana
azurea Schipcz. - 6
baldshuanica Lipsky - 6
barsczewskii Lipsky - 6

borysthenica Klok. (*O. arenaria* auct.) - 1
bourgaei Boiss. - 2
brevipilosa Schischk. ex M. Pop. - 6
calycina Stev. = O. visianii
caucasica Levin ex M. Pop. - 2
- subsp. oligotricha T.N. Pop. - 2
dichroantha Boiss. - 1, 2
ferganensis M. Pop. - 6
gehardica T.N. Pop. - 2
gmelinii Ledeb. - 3, 4, 6
gracilis Trautv. - 2
graniticola Klok. - 1
guberlinensis Dobrocz. & V. Vinogradova - 1
iricolor Klok. (*O. mugodzharica* Klok.) - 1, 3, 6
irritans M. Pop. ex Pavl. - 6
isaurica Boiss. & Heldr. - 2
leucocarpa M. Pop. - 6
levinii T.N. Pop. - 2
lipskyi Klok. - 1
liwanowii M. Pop. - 6
longiloba Bunge - 6
macrochaeta Klok. & Dobrocz. = O. visianii
macrorhiza M. Pop. - 6
maracandica Zak. - 6
microcarpa Stev. ex DC. - 2
montana Smith (*O. aucheriana* DC.) - 2
mugodzharica Klok. = O. iricolor
polychroma Klok. ex M. Pop. - 1, 2
polyphylla Ledeb. - 1, 2
pseudotinctoria Klok. - 1
rigida Ledeb. - 1, 2
rupestris Bieb. = O. tenuiflora
sabalanica Ponert - 2
samarica Klok. - 1
sericea Willd. - 6
setosa Ledeb. (*O. armeniaca* Klok. ex M. Pop.) - 1, 2
simplicissima L. - 1, 3, 4, 6
staminea Ledeb. - 6
subtinctoria Klok. - 1
tanaitica Klok. - 1
taurica Pall. ex Willd. - 1, 2
tenuiflora Willd. (*O. rupestris* Bieb.) - 2
tinctoria Bieb. - 1, 2
trachycarpa Levin - 6
transrhymnensis Klok. ex M. Pop. - 1, 3, 4
visianii Clementi (*O. calycina* Stev. 1851, non Ave-Lall. 1842, *O. macrochaeta* Klok. & Dobrocz.) - 1, 2(?)
volgensis Dobrocz. - 1
zangezura T.N. Pop. - 2
zerizamina Lipsky - 6

Oreocarya auct. = Cryptantha

spiculifera Piper = Cryptantha spiculifera

Paracaryum (DC.) Boiss. (*Mattiastrum* (Boiss.) Brand, *Microparacaryum* (M. Pop. ex H. Riedl) Hilger & Podlech)

bungei (Boiss.) Brand (*Lappula petrophila* Pavl. nom. invalid., *Mattiastrum bungei* (Boiss.) Rech. fil. & H. Riedl, *Paracaryum petrophilum* Golosk.) - 6
capusii Franch. = Cynoglossum capusii
crista-galli (Rech. fil. & H. Riedl) R. Kam. & Raenko (*Mattiastrum crista-galli* Rech. fil. & H. Riedl) - 6
emiri M. Pop. = P. himalayense
glochidiatum (Bunge) H. Riedl = P. intermedium

gracile Czerniak. (*Mattiastrum gracile* (Czerniak.) Czer.) - 6
himalayense (Klotzsch) Clarke (*Mattiastrum emiri* (M. Pop.) Czer., *M. himalayense* (Klotzsch) Brand, ? *M. himalayense* subsp. *fallax* Rech. fil. & H. Riedl, *Paracaryum emiri* M. Pop.) - 6
hirsutum (DC.) Boiss. = P. strictum
incanum (Ledeb.) Boiss. (*Cynoglossum incanum* (Ledeb.) Greuter & Burdet, *Mattiastrum incanum* (Ledeb.) Brand) - 2
integerrimum Myrzakulov - 6
intermedium (Fresen.) Lipsky (*Microparacaryum intermedium* (Fresen.) Hilger & Podlech, *Omphalodes glochidiata* Bunge, *O. physodes* Bunge, *Paracaryum glochidiatum* (Bunge) H. Riedl, *P. physodes* (Bunge) H. Rield) - 6
- subsp. papillosum (A. DC.) R. Kam. & Raenko - 6
karataviense Pavl. ex M. Pop. (*Mattiastrum karataviense* (Pavl. ex M. Pop.) Czer.) - 6
laxiflorum Trautv. (*Cynoglossum laxiflorum* (Trautv.) Greuter & Burdet, *Mattiastrum laxiflorum* (Trautv.) Czer.) - 2
petrophilum Golosk. = P. bungei
physodes (Bunge) H. Rield = P. intermedium
salsum Boiss. - 6
sintenisii Hausskn. ex Bornm. = P. strictum
strictum (C. Koch) Boiss. (*Cynoglossum paracaryum* Greuter & Burdet, *C. sintenisii* (Bornm.) Greuter & Burdet, *C. strictum* (C. Koch) Greuter & Burdet, *Paracaryum hirsutum* (DC.) Boiss. 1849, non *Cynoglossum hirsutum* Thunb. 1794, *P. sintenisii* Hausskn. ex Bornm.) - 2
turcomanicum Bornm. & Sint. (*Mattiastrum turcomanicum* (Bornm. & Sint.) Brand - 6

Paracynoglossum M. Pop.

asperrimum (Nakai) M. Pop. (*Cynoglossum asperrimum* Nakai) - 5(?)
denticulatum (A. DC.) M. Pop. = P. glochidiatum
glochidiatum (Wall. ex Benth.) M. Pop. ex Czuk. (*Cynoglossum glochidiatum* Wall. ex Benth., *C. imeretinum* Kusn., *C. wallichii* G. Don fil. var. *glochidiatum* (Wall. ex Benth.) Kazmi, *Paracynoglossum denticulatum* (A. DC.) M. Pop., *P. imeretinum* (Kusn.) M. Pop.) - 2, 6
imeretinum (Kusn.) M. Pop. = P. glochidiatum

Phyllocara Gusul.

aucheri (A. DC.) Gusul. - 2

Piptoclaina G. Don fil. = Heliotropium

supina (L.) G. Don fil. = Heliotropium supinum

Pseudomertensia H. Riedl (*Scapicephalus* Ovcz. & Czuk.)

lindelofioides (Rech. fil. & H. Riedl) H. Riedl = Lindelofia olgae
rosulata (Ovcz. & Czuk.) Ovcz. & Czuk. (*Scapicephalus rosulatus* Ovcz. & Czuk.) - 6

Pulmonaria L.

angustifolia L. - 1
dacica (Simonk.) Simonk. = P. mollis
filarszkyana Jav. (*P. rubra* Schott subsp. *filarszkyana* (Jav.) Domin) - 1
mollis Wulf. ex Hornem. (*P. dacica* (Simonk.) Simonk., *P. mollis* subsp. *mollissima* (A. Kerner) A. & D. Löve, *P.*

mollissima A. Kerner, *P. rubra* Schott var. *dacica* Simonk.) - 1, 2, 3, 4, 6
- subsp. *mollissima* (A. Kerner) A. & D. Löve = P. mollis
mollissima A. Kerner = P. mollis
murinii J. Majovsky subsp. **continentalis** J. Majovsky - 4
obscura Dumort. - 1, 3
■ **officinalis** L.
rubra Schott. - 1
- subsp. *filarszkyana* (Jav.) Domin = P. filarszkyana
- var. *dacica* Simonk. = P. mollis
simplicissima Ledeb. = Mertensia maritima

Rhytispermum Link = Buglossoides

arvense (L.) Link = Buglossoides arvensis
tenuiflorum (L. fil.) Link = Buglossoides tenuiflora

Rindera Pall. (*Bilegnum* Brand)

austroechinata M. Pop. - 6
baldshuanica Kusn. = R. tetraspis
bungei (Boiss.) Guerke (*Bilegnum bungei* (Boiss.) Brand) - 6
coechinata M. Pop. - 6
cristulata Lipsky (*R. turkestanica* Kusn.) - 6
cyclodonta Bunge = R. tetraspis
echinata Regel - 6
ferganica M. Pop. - 6
fornicata Pazij - 6
glabrata Pazij - 6
holochiton M. Pop. - 6
korshinskyi (Lipsky) Brand - 6
kuramensis Turakulov - 6
lanata (Lam.) Bunge - 2
oblongifolia M. Pop. - 6
ochroleuca Kar. & Kir. - 6
oschensis M. Pop. - 6
tetraspis Pall. (*Cynoglossum tetraspis* (Pall.) Greuter & Burdet, *Rindera baldshuanica* Kusn., *R. cyclodonta* Bunge) - 1, 2, 3, 6
tianschanica M. Pop. - 6
tschotkalensis M. Pop. - 6
turkestanica Kusn. = R. cristulata
umbellata (Waldst. & Kit.) Bunge - 1

Rochelia Reichenb.

bungei Trautv. - 6
campanulata M. Pop. & Zak. - 6
cardiosepala Bunge - 2, 6
claviculata M. Pop. & Zak. - 6
disperma (L. fil.) C. Koch - 2, 6
- subsp. *retorta* (Pall.) Kotejowa = R. retorta
- var. *microcalycina* (Bornm.) J.R. Edmondson = R. microcalycina
dispermoides Bondar. & Zak. = R. retorta
jackabaghi (Lipsky) Pavl. - 6
karsensis M. Pop. = R. microcalycina
leiocarpa Ledeb. - 3, 6
leiosperma Golosk. - 6
macrocalyx Bunge = R. rectipes
■ **microcalycina** Bornm. (*R. disperma* (L. fil.) C. Koch var. *microcalycina* (Bornm.) J.R. Edmondson, *R. karsensis* M. Pop.)
pamirica Dengubenko - 6
peduncularis Boiss. - 6
persica Bunge - 2, 3, 6
rectipes Stocks (*R. macrocalyx* Bunge) - 6

retorta (Pall.) Lipsky (*R. disperma* (L. fil.) C. Koch subsp. *retorta* (Pall.) Kotejowa, *R. dispermoides* Bondar. & Zak.) - 1, 2, 6

Scapicephalus Ovcz. & Czuk. = Pseudomertesia

rosulatus Ovcz. & Czuk. = Pseudomertensia rosulata

Solenanthus Ledeb.

albiflorus Czuk. & Meling - 6
biebersteinii DC. (*Cynoglossum biebersteinii* (DC.) Greuter & Burdet) - 1, 2
brachystemon Fisch. & C.A. Mey. - 2
circinnatus Ledeb. (*Cynoglossum circinnatum* (Ledeb.) Greuter & Burdet, *Solenanthus coronatus* Regel) - 2, 3, 6
coronatus Regel = S. circinnatus
hirsutus Regel - 6
karateginus Lipsky - 6
kokanicus Regel - 6
olgae Regel & Smirn. var. *tschimganicus* Lipsky = Lindelofia tschimganica
plantaginifolius Lipsky - 6
stamineus (Desf.) Wettst. - 2, 6
turkestanicus (Regel & Smirn.) Kusn. = Kuschakewiczia turkestanica

Stenosolenium Turcz.

saxatile (Pall.) Turcz. - 4

Stephanocaryum M. Pop.

olgae (B. Fedtsch.) M. Pop. - 6
popovii R. Kam. - 6

Strophiostoma Turcz. = Myosotis

Suchtelenia Kar. ex Meissn.

calycina (C.A. Mey.) A. DC. (? *S. eriophora* Bornm.) - 2, 6
eriophora Bornm. = S. calycina

Symphytum L.

abchasicum Trautv. = S. grandiflorum
armeniacum Bucknall = S. asperum
asperum Lepech. (*S. armeniacum* Bucknall, *S. asperum* var. *armeniacum* (Bucknall) Kurtto) - 1, 2
- var. *armeniacum* (Bucknall) Kurtto = S. asperum
besseri Zaverucha = S. microcalyx
bohemicum F.W. Schmidt - 1
carpaticum Frolov - 1
caucasicum Bieb. - 1, 2
ciscaucasicum Gviniaschvili = S. grandiflorum
cordatum Waldst. & Kit. ex Willd. - 1
grandiflorum DC. (*S. abchasicum* Trautv., *S. ciscaucasicum* Gviniaschvili, *S. ibericum* Stev., *S. ibericum* var. *abchasicum* (Trautv.) Gviniaschvili) - 2
hajastanum Gviniaschvili - 2
ibericum Stev. = S. grandiflorum
- var. *abchasicum* (Trautv.) Gviniaschvili = S. grandiflorum
incarnatum S.G. Gmel. - 1
microcalyx Opiz (*S. besseri* Zaverucha, *S. tuberosum* auct. p.p.) - 1
officinale L. - 1, 2, 3, 5(alien), 6
orientale auct. = S. tauricum
peregrinum Ledeb. - 1(cult.), 2, 5(alien)

podcumicum Frolov - 2
popovii Dobrocz. (*S. tuberosum* auct. p.p.) - 1
tanaicense Stev. - 1
tauricum Willd. (*S. orientale* auct.) - 1, 2
tuberosum auct. = S. microcalyx and S. popovii

Tianschaniella B. Fedtsch. ex M. Pop.

umbellulifera B. Fedtsch. ex M. Pop. - 6
wakhanica H. Riedl = Eritrichium pamiricum

Tournefortia L. p.p. = Argusia

sibirica L. = Argusia sibirica
sogdiana (Bunge) M. Pop. = Argusia sogdiana

Trachelanthus G. Kunze

hissaricus Lipsky - 6
korolkowii Lipsky - 6

Trachystemon D. Don.

orientalis (L.) G. Don fil. - 2

Trichodesma R. Br.

incanum (Bunge) A. DC. - 6

Trigonocaryum Trautv.

involucratum (Stev.) Kusn. - 2

Trigonotis Stev.

koreana Nakai = T. radicans
myosotidea (Maxim.) Maxim. - 4, 5
nakai Hara - 5
peduncularis (Trev.) Benth. ex Baker & S. Moore - 1, 2, 3, 4, 5, 6
radicans (Turcz.) Stev. (*T. koreana* Nakai) - 4, 5

Ulugbekia Zak.

tschimganica (B. Fedtsch.) Zak. (*Lithospermum tschimganicum* B. Fedtsch., *Arnebia tschimganica* (B. Fedtsch.) G.L. Zhu) - 6

BOTRYCHIACEAE Horan.

Botrychium Sw. (*Botrypus* Michx., *Japanobotrychium* Masamune, *Osmundopteris* (Milde) Small, *Sceptridium* Lyon)

boreale Milde (*B. boreale* subsp. *obtusilobum* (Rupr.) Clausen, *B. crassinervium* Rupr. var. *obtusilobum* Rupr.) - 1, 3, 4, 5
- subsp. *obtusilobum* (Rupr.) Clausen = B. boreale
charcoviense Portenschl. = B. virginianum
crassinervium Rupr. var. *obtusilobum* Rupr. = B. boreale
lanceolatum (S.G. Gmel.) Angstr. (*Osmunda lanceolata* S.G. Gmel.) - 1, 3, 4, 5
lunaria (L.) Sw. - 1, 2, 3, 4, 5, 6
matricariifolium A. Br. ex Koch (*B. ramosum* (Roth) Aschers. p.p. excl. typo, 1864, non Sailer, 1844, *Osmunda lunaria* L. var. *matricariifolia* Retz.) - 1
multifidum (S.G. Gmel.) Rupr. (*Sceptridium multifidum* (S.G. Gmel.) Tagawa) - 1, 2, 3, 4, 5
- subsp. *robustum* (Rupr.) Clausen = B. robustum
ramosum (Roth) Aschers. p.p. = B. matricariifolium

robustum (Rupr.) Underw. (*B. multifidum* (S.G. Gmel.) Rupr. subsp. *robustum* (Rupr.) Clausen, *Sceptridium multifidum* (S.G. Gmel.) Tagawa var. *robustum* (Rupr.) Nishida & Tagawa) - 5
simplex E. Hitchc. - 1
strictum Underw. (*Botrypus strictus* (Underw.) Holub, *Osmundopteris stricta* (Underw.) Nishida) - 5
virginianum (L.) Sw. (*B. charcoviense* Portenschl., *B. virginianum* var. *europaeum* Angstr., *B. virginianum* subsp. *europaeum* (Angstr.) Clausen, comb. superfl., *B. virginianum* subsp. *europaeum* (Angstr.) Jav., *Botrypus virginianus* (L.) Michx, *B. virginianus* (L.) Holub, comb. superfl., *B. virginianus* subsp. *europaeus* (Angstr.) Holub, *Japanobotrychium virginianum* (L.) Nishida & Tagawa, *Osmundopteris virginiana* (L.) Small) - 1, 3, 4, 5
- subsp. *europaeum* (Angstr.) Clausen = B. virginianum
- subsp. *europaeum* (Angstr.) Jav. = B. virginianum
- var. *europaeum* Angstr. = B. virginianum

Botrypus Michx. = Botrychium

strictus (Underw.) Holub = Botrychium strictum
virginianus (L.) Michx. = Botrychium virginianum
virginianus (L.) Holub = B. virginianum
- subsp. *europaeus* (Angstr.) Holub = Botrychium virginianum

Japanobotrychium Masamune = Botrychium

virginianum (L.) Nishida & Tagawa = Botrychium virginianum

Osmundopteris (Milde) Small = Botrychium

stricta (Underw.) Nishida = Botrychium strictum
virginiana (L.) Small = Botrychium virginianum

Sceptridium Lyon = Botrychium

multifidum (S.G. Gmel.) Tagawa = Botrychium multifidum
- var. *robustum* (Rupr.) Nishida & Tagawa = Botrychium robustum

BRASSICACEAE Burnett (CRUCIFERAE Juss. nom. altern.)

Acachmena H.P. Fuchs = Erysimum

cuspidata (Bieb.) H.P. Fuchs = Erysimum cuspidatum
krynkensis (Lavr.) H.P. Fuchs = Erysimum krynkense

Achoriphragma Sojak (*Neuroloma* Andrz. 1824, non Nevroloma Rafin. 1819, *Parrya* auct. p.p.)

ajanense (N. Busch) Sojak = A. nudicaule
albidum (M. Pop.) Sojak (*Parrya albida* M. Pop., *Neuroloma albidum* (M. Pop.) Botsch.) - 6
album (E. Nikit.) Czer. comb. nova (*Parrya alba* E. Nikit. 1967, Fl. Kirg. SSR, Suppl. 1 : 113. 76; *Neuroloma album* (E. Nikit.) Pachom.) - 6
angrenicum (Botsch. & Vved.) Sojak (*Parrya angrenica* Botsch. & Vved., *Neuroloma angrenicum* (Botsch. & Vved.) Botsch.) - 6
asperrimum (B. Fedtsch.) Sojak (*Parrya nudicaulis* (L.) Regel var. *asperrima* B. Fedtsch., *Neuroloma asperrimum* (B. Fedtsch.) Botsch., *Parrya asperrima* (B.

Fedtsch.) M. Pop., *P. golenkinii* Lipsch. & Pavl.) - 6

australe (Pavl.) Czer. comb. nova (*Parrya australis* Pavl. 1949, Vest. AN Kaz. SSR, 1 : 28; *Neuroloma australe* (Pavl.) Botsch.) - 6

beketovii (Krasn.) Sojak (*Parrya beketovii* Krasn., *Neuroloma beketovii* (Krasn.) Botsch., *Parrya michaelis* A. Vassil.) - 6

botschantzevii (Pachom.) Sojak (*Neuroloma botschantzevii* Pachom.) - 6

darvazicum (Botsch. & Vved.) Sojak (*Parrya darvazica* Botsch. & Vved., *Neuroloma darvazicum* (Botsch. & Vved.) Botsch.) - 6

fruticulosum (Regel & Schmalh.) Sojak (*Parrya fruticulosa* Regel & Schmalh., *Neuroloma fruticulosum* (Regel & Schmalh.) Botsch.) - 6

korovinii (A.Vassil.) Botsch. (*Parrya korovinii* A. Vassil., *Neuroloma korovinii* (A. Vassil.) Botsch.) - 6

kuramense (Botsch.) Sojak (*Parrya kuramensis* Botsch., *Neuroloma kuramense* (Botsch.) Botsch.) - 6

lancifolium (M. Pop.) Sojak (*Parrya lancifolia* M. Pop., *Neuroloma lancifolium* (M. Pop.) Botsch.) - 6

longicarpum (Krasn.) Sojak (*Parrya longicarpa* Krasn., *Neuroloma longicarpum* (Krasn.) Botsch., *Parrya siliquosa* Krasn.) - 6

maidantalicum (M. Pop. & P. Baran.) Sojak (*Parrya maidantalica* M. Pop. & P. Baran., *Christolea maidantalica* (M. Pop. & P. Baran.) N. Busch, *Neuroloma maidantalicum* (M. Pop. & P. Baran.) Botsch.) - 6

minutum (Botsch. & Vved.) Czer. comb. nova (*Parrya minuta* Botsch. & Vved. 1948, Bot. Mat. (Tashkent) 12 : 9; *Neuroloma minutum* (Botsch. & Vved.) Botsch.) - 6

nudicaule (L.) Sojak (*A. ajanense* (N. Busch) Sojak, *Neuroloma ajanense* (N. Busch) Botsch., *N. nudicaule* (L.) DC., *Parrya ajanensis* N. Busch, *P. nudicaulis* (L.) Regel, *P. nudicaulis* subsp. *septentrionalis* Hult., *P. arctica* auct.) - 1, 3, 4, 5

nuratense (Botsch. & Vved.) Sojak (*Parrya nuratensis* Botsch. & Vved., *Neuroloma nuratense* (Botsch. & Vved.) Botsch.) - 6

pavlovii (A.Vassil.) Sojak (*Parrya pavlovii* A. Vassil., *Neuroloma pavlovii* (A. Vassil.) Botsch., *Parrya linearifolia* Pavl. 1949, non W.W. Smith, 1919) - 6

pazijae (Pachom.) Sojak (*Neuroloma pazijae* Pachom.) - 6

pinnatifidum (Kar. & Kir.) Sojak (*Parrya pinnatifida* Kar. & Kir., *Neuroloma pinnatifidum* (Kar. & Kir.) Botsch.) - 6

pjataevae (Pachom.) Sojak (*Neuroloma pjataevae* Pachom.) - 6

popovii (Botsch.) Sojak (*Parrya popovii* Botsch., *Neuroloma popovii* (Botsch.) Botsch.) - 6

pulvinatum (M. Pop.) Sojak (*Parrya pulvinata* M. Pop., *Neuroloma pulvinatum* (M. Pop.) Botsch.) - 6

runcinatum (Regel & Schmalh.) Sojak (*Neuroloma runcinatum* (Regel & Schmalh.) Botsch., *Parrya runcinata* (Regel & Schmalh.) N. Busch) - 6

saposhnikovii (A.Vassil.) Sojak (*Parrya saposhnikovii* A. Vassil., *Neuroloma saposhnikovii* (A. Vassil.) Botsch.) - 6

sauricum (Pachom.) Sojak (*Neuroloma sauricum* Pachom.) - 6

saxifragum (Botsch. & Vved.) Sojak (*Parrya saxifraga* Botsch. & Vved., *Neuroloma saxifragum* (Botsch. & Vved.) Botsch.) - 6

schugnanum (Lipsch.) Sojak (*Parrya schugnana* Lipsch., *Neuroloma schugnanum* (Lipsch.) Botsch.) - 6

simulatrix (E. Nikit.) Sojak (*Parrya simulatrix* E. Nikit.,

Neuroloma simulatrix (E. Nikit.) Botsch.) - 6

stenocarpum (Kar. & Kir.) Sojak (*Parrya stenocarpa* Kar. & Kir., *Neuroloma stenocarpum* (Kar. & Kir.) Botsch.) - 3, 6

stenophyllum (M. Pop.) Czer. comb. nova (*Parrya stenophylla* M. Pop. 1939, Fl. URSS, 8 : 646, 266; *Neuroloma stenophyllum* (M. Pop.) Botsch.) - 6

subsiliquosum (M. Pop.) Sojak (*Parrya subsiliquosa* M. Pop., *Neuroloma subsiliquosum* (M. Pop.) Botsch.) - 6

tianschanicum (E. Nikit.) Sojak (*Parrya tianschanica* E. Nikit., *Neuroloma tianschanicum* (E. Nikit.) Botsch.) - 6

turkestanicum (Korsh.) Sojak (*Neuroloma turkestanica* (Korsh.) Botsch., *Parrya nudicaulis* (L.) Regel subsp. *turkestanica* (Korsh.) Hult., *P. turkestanica* (Korsh.) N. Busch) - 4, 6

villosulum (Botsch. & Vved.) Czer. comb. nova (*Parrya villosula* Botsch. & Vved. 1941, Bot. Mat. (Tashkent) 3 : 16; *Neuroloma villosulum* (Botsch. & Vved.) Botsch.) - 6

Aethionema R. Br.

arabicum (L.) Lipsky (*Ae. arabicum* (L.) Rothm. comb. superfl.) - 2

armenum Boiss. - 2

caespitosum (Boiss.) Boiss. = Iberidella caespitosa

cardiophyllum Boiss. & Heldr. (*Ae. koenigii* Woronow) - 2

carneum (Banks & Soland) B. Fedtsch. = Campyloptera carnea

cordatum (Desf.) Boiss. (*Thlaspi cordatum* Desf.) - 2

diastrophis Bunge - 2

edentulum N. Busch (*Ae. membranaceum* auct.) - 2

elongatum auct. = Ae. szowitsii

fimbriatum Boiss. - 2

grandiflorum Boiss. & Hohen. - 2

heterophyllum (Boiss. & Buhse) Boiss. = Iberidella heterophylla

koenigii Woronow = Ae. cardiophyllum

kopetdaghi Lipsky - 6

levandovskyi N. Busch - 2

lipskyi N. Busch = Ae. szowitsii

membranaceum auct. = Ae. edentulum

oppositifolium (Pers.) Hedge (*Iberis oppositifolia* Pers.) - 2

pulchellum Boiss. & Huet - 2

sagittatum (Boiss.) Boiss. = Iberidella sagittata

salmasium Boiss. = Iberidella trinervia

schistosum Boiss. & Kotschy - 2(?)

spinosum (Boiss.) Bornm. = Moriera spinosa

szowitsii Boiss. (*Ae. lipskyi* N. Busch, *Ae. woronowii* Schischk., *Ae. elongatum* auct., *Ae. virgatum* auct. p.p. quoad pl. cauc.) - 2

transhyrcanum (Czerniak.) N. Busch (*Moriera transhyrcana* Czerniak.) - 6

trinervium (DC.) Boiss. = Iberidella trinervia

virgatum auct. p.p. = Ae. szowitsii

woronowii Schischk. = Ae. szowitsii

Alaida Dvorak = Dimorphostemon

glandulosa (Kar. & Kir.) Dvorak = Dimorphostemon glandulosus

pectinata (DC.) Dvorak = Dimorphostemon pectinatus

Alliaria Heist. ex Fabr.

alliacea (Salisb.) Britten & Rendle = A. petiolata

brachycarpa Bieb. - 2

officinalis Andrz. ex Bieb. = A. petiolata

petiolata (Bieb.) Cavara & Grande (*Arabis petiolata* Bieb., *Alliaria alliacea* (Salisb.) Britten & Rendle, nom. illegit., *A. officinalis* Andrz. ex Bieb., *Sisymbrium alliaceum* Salisb. nom. illegit.) - 1, 2, 5(alien), 6

Alyssoides Hill (*Vesicaria* Adans.)

graeca (Reut.) Jav. = A. utriculata

utriculata (L.) Medik. (*Alyssum utriculatum* L., *Alyssoides graeca* (Reut.) Jav., *Vesicaria graeca* Reut., *V. utriculata* (L.) Lam.) - 2

Alyssopsis Boiss.

kotschyi Boiss. = Dielsiocharis kotschyi

mollis (Jacq.) O. Schulz (*Sisymbrium molle* Jacq., *Arabis secunda* N. Busch) - 2

trinervis Botsch. & Sejfulin - 6

Alyssum L. (*Odontarrhena* C.A. Mey., *Psilonema* C.A. Mey.)

alpestre L. var. *orbiculare* Regel = A. orbiculare

alyssoides (L.) L. = A. calycinum

americanum Greene = A. obovatum

andinum Rupr. (*A. schemachense* N. Busch) - 2

arcticum Wormsk. ex Hornem. = Lesquerella arctica

■ armenum Boiss.

artvinense N. Busch = A. ochroleucum

baicalicum E.I. Nyarady (*A. obovatum* (C.A. Mey.) Turcz. subsp. *baicalicum* (E.I. Nyarady) Peschkova) - 4

betpakdalense N. Rubtz. = A. heterotrichum

biovulatum N. Busch = A. obovatum

borzaeanum E.I. Nyarady (*A. tenderiense* Kotov, *A. tortuosum* Waldst. & Kit. subsp. *borzaeanum* (E.I. Nyarady) Stojan. & Stef.) - 1

- subsp. aequiacutum E.I. Nyarady - 1

bracteatum auct. = A. tortuosum

buschianum Grossh. = A. stapfii

caliacrae E.I. Nyarady (*A. tortuosum* Waldst. & Kit. ex Willd. subsp. *caliacrae* (E.I. Nyarady) Stojan. & Stef.) - 1

calycinum L. (*A. alyssoides* (L.) L., *A. campestre* (L.) L., *Clypeola alyssoides* L. nom. ambig., *Psilonema alyssoides* (L.) Heideman, *P. calycinum* (L.) C.A. Mey.) - 1, 2, 5(alien)

calycocarpum Rupr. - 1

campestre auct. = A. parviflorum and A. strigosum

campestre (L.) L. = A. calycinum

canescens DC. = Ptilotrichum canescens

caucasicum (E.I. Nyarady) E.I. Nyarady = A. murale

cretaceum (Kotov) Kotov = A. gymnopodum

cuneipetalum E.I. Nyarady = A. tortuosum

cupreum Freyn & Sint. = Meniocus linifolius

czernjakowskae Rech. fil. = A. lanceolatum

daghestanicum Rupr. - 2

dasycarpum Steph. (*Psilonema dasycarpum* (Steph.) C.A. Mey.) - 1, 2, 3, 6

dechyanum E.I. Nyarady = A. gehamense

desertorum Stapf = A. turkestanicum var. desertorum

diversicaule P. Smirn. = A. tortuosum

fallax E.I. Nyarady

Described from Siberia. The additional investigation is necessary.

fedtschenkoanum N. Busch - 3, 6

fischerianum DC. = A. lenense

gehamense Fed. (*A. dechyanum* E.I. Nyarady) - 2

globosum Grossh. = Takhtajaniella globosa

gmelinii Jord. (*A. montanum* L. subsp. *gmelinii* (Jord.) Hegi & E. Schmid, *A. montanum* auct.) - 1

x grintescui E.I. Nyarady. - *A. obtusifolium* Stev. ex DC. x *A. tortuosum* Waldst. & Kit. ex Willd. subsp. *heterophyllum* E.I. Nyarady - 1

gymnopodum P. Smirn. (*A. cretaceum* (Kotov) Kotov, *A. tortuosum* Waldst. & Kit. ex Willd. subsp. *cretaceum* Kotov) - 1

hajastanum V. Avet. - 2

hakaszkii E.I. Nyarady = A. microphyllum

halacsyi E.I. Nyarady f. densum E.I. Nyarady - 1

heterotrichum Boiss. (*A. betpakdalense* N. Rubtz.) - 6

hirsutum Bieb. (*A. minus* (L.) Rothm. subsp. *hirsutum* (Bieb.) Stojan. & Stef.) - 1, 2

inflatum E.I. Nyarady - 6

iranicum Czerniak. = A. lanceolatum

kotovii A. Iljinskaja - 1

lanceolatum Baumgartner (*A. czernjakowskae* Rech. fil., *A. iranicum* Czerniak. 1924, non Haußkn. ex Baumgartner, 1907, *A. shahrudum* Parsa) - 6

lenense Adams (*A. fischerianum* N. Busch) - 1, 3, 4, 5, 6

lepidulum E.I. Nyarady - 1, 2

- subsp. lepidulum (*A. lepidulum* subsp. *genuinum* E.I. Nyarady) - 1

- subsp. congreatum E.I. Nyarady - 2

linifolium Steph. var. *cupreum* (Freyn & Sint.) T.R. Dudley = Meniocus linifolius

longistylum (Somm. & Levier) Grossh. = A. tortuosum

macrostylum Boiss. & Huet = A. xanthocarpum

marginatum Steud. ex Boiss. = A. szovitsianum

medium A. Iljinskaja - 1

micranthum C.A. Mey. = A. strigosum

micropetalum auct. = A. strigosum

micropetalum Fisch. ex DC. = A. parviflorum

micropetalum Fisch. ex Hornem. = A. parviflorum

microphylliforme E.I. Nyarady

Described from Siberia and Mongolia. Additional investigation is necessary.

microphyllum (C.A. Mey.) Steud. (*Odontarrhena microphylla* C.A. Mey., *Alyssum hakaszkii* E.I. Nyarady) - 3, 4

minus (L.) Rothm. = A. parviflorum

- subsp. *hirsutum* (Bieb.) Stojan. & Stef. = A. hirsutum

- subsp. *micranthum* (C.A. Mey.) Breistroffer = A. strigosum

- var. *micranthum* (C.A. Mey.) T.R. Dudley = A. strigosum

- var. *strigosum* (Banks & Soland.) Zohary = A. strigosum

minutum Schlecht. ex DC. - 1, 2

montanum auct. = A. gmelinii

montanum L. subsp. *gmelinii* (Jord.) Hegi & E. Schmid = A. gmelinii

muelleri Boiss. & Buhse = A. persicum

murale Waldst. & Kit. (*A. caucasicum* (E.I. Nyarady) E.I. Nyarady, *A. murale* subsp. *caucasicum* E. I. Nyarady) - 1, 2

- subsp. *caucasicum* E.I. Nyarady = A. murale

mutabile Vent. = Berteroa mutabilis

obovatum (C.A. Mey.) Turcz. (*Odontarrhena obovata* C.A. Mey., *Alyssum biovulatum* N. Busch, *A. sibiricum* auct.) - 1, 3, 4, 5, 6

- subsp. *baicalicum* (E.I. Nyarady) Peschkova = A. baicalicum

- subsp. *orbiculare* (Regel) Peschkova = A. orbiculare

obtusifolium Stev. ex DC. - 1, 2

- subsp. cordatocarpum E.I. Nyarady - 1, 2

■ ochroleucum Boiss. & Huet (*A. artvinense* N. Busch)

odessanum E.I. Nyarady = A. tortuosum

orbiculare (Regel) E.I. Nyarady (*A. alpestre* L. var. *orbiculare* Regel, *A. obovatum* (C.A. Mey.) Turcz. subsp. *orbiculare* (Regel) Peschkova) - 4

parviflorum Fisch. ex Bieb. (*A. micropetalum* Fisch. ex Hornem. 1819, *A. micropetalum* Fisch. ex DC. 1821, nom. superfl., *A. minus* (L.) Rothm. comb. invalid., *A. rothmalleri* Galushko, nom. invalid., *Clypeola minor* L. nom. nud., *A. campestre* auct. p.p.) - 1, 2, 3, 6

peltarioides Boiss. - 2

persicum Boiss. (*A. muelleri* Boiss. & Buhse) - 2

podolicum Bess. = Schivereckia podolica

repens Baumg. - 1(?)

- *trichostachyum* (Rupr.) Hayek = A. trichostachyum

rostratum Stev. - 1, 2

rothmalleri Galushko = A. parviflorum

savranicum Andrz. (*A. tortuosum* Waldst. & Kit. ex Willd. subsp. *savranicum* (Andrz.) E.I. Nyarady) - 1

saxatile L. = Aurinia saxatilis

schemachense N. Busch = A. andinum

sergievskajae Krasnob. = Stevenia sergievskajae

serpyllifolium Desf. var. *longistylum* Somm. & Levier = A. tortuosum

shahrudum Parsa = A. lanceolatum

sibiricum auct. = A. obovatum

smyrnaeum C.A. Mey. - 1

stapfii Vierh. (*A. buschianum* Grossh.) - 2, 6

stenostachyum Botsch. & Vved. - 6

strictum Willd. - 2

strigosum Banks & Soland. (*A. micranthum* C.A. Mey., *A. minus* (L.) Rothm. subsp. *micranthum* (C.A. Mey.) Breistroffer, *A. minus* var. *micranthum* (C.A. Mey.) T.R. Dudley, *A. minus* var. *strigosum* (Banks & Soland.) Zohary, *A. campestre* auct. p.p., *A. micropetalum* auct.) - 2, 6

subbaicalicum E.I. Nyarady - 4

szarabiacum E.I. Nyarady - 6

szovitsianum Fisch. & C.A. Mey. (*A. marginatum* Steud. ex Boiss.) - 2, 6

tenderiense Kotov = A. borzaeanum

tenuifolium Steph. = Ptilotrichum tenuifolium

tortuosum Waldst. & Kit. ex Willd. (*A. cuneipetalum* E.I. Nyarady, *A. diversicaule* P. Smirn., *A. longistylum* (Somm. & Levier) Grossh., *A. odessanum* E.I. Nyarady, *A. serpyllifolium* Desf. var. *longistylum* Somm. & Levier, 1900, non Boiss. 1842, *A. tortuosum* var. *longistylum* (Somm. & Levier) N. Busch, *A. transiens* E.I. Nyarady, *A. bracteatum* auct.) - 1, 2, 6

- subsp. *borzaeanum* (E.I. Nyarady) Stojan. & Stef. = A. borzaeanum
- subsp. *caliacrae* (E.I. Nyarady) Stojan. & Stef. = A. caliacrae
- subsp. *cretaceum* Kotov = A. gymnopodum
- subsp. cuspidatum E.I. Nyarady - 2
- subsp. heterophyllum E.I. Nyarady - 1
- subsp. *savranicum* (Andrz.) E.I. Nyarady = A. savranicum
- var. *longistylum* (Somm. & Levier) N. Busch = A. tortuosum

transiens E.I. Nyarady = A. tortuosum

trichostachyum Rupr. (*A. repens* Baumg. subsp. *trichostachyum* (Rupr.) Hayek) - 1, 2

turkestanicum Regel & Schmalh. - 1, 2, 3, 4, 6

- var. desertorum (Stapf) Botsch. (*A. desertorum* Stapf) - 1, 2, 3, 4, 6

umbellatum Desv. - 1

utriculatum L. = Alyssoides utriculata

wierzbickii Heuff. - 1

xanthocarpum Boiss. (*A. macrostylum* Boiss. & Huet) - 2

Anchonium DC.

brachycarpum (Trautv.) Vass. = A. elichrysifolium

elichrysifolium (DC.) Boiss. (*A. brachycarpum* (Trautv.) Vass.) - 2

ramosissimum O.E. Schulz = Sterigmostemum ramosissimum

sterigmoides Lipsky ex Vass. = Sterigmostemum ramosissimum

Andreoskia DC. = Dontostemon

dentata Bunge = Dontostemon dentatus

Andrzeiowskia Reichenb.

cardamine (*A. cardaminifolia* auct.) - 2

cardaminifolia auct. = A. cardamine

Aphragmus Andrz.

involucratus (Bunge) O.E. Schulz - 3

oxycarpus (Hook. fil. & Thoms.) Jafri (*Braya oxycarpa* Hook. fil. & Thoms.) - 6

Apterigia Galushko = Atropatenia and Noccaea

pumila (Stev.) Galushko = Noccaea pumila

rostrata (N. Busch) Galushko = Atropatenia rostrata

Arabidopsis (DC.) Heynh.

arenosa (L.) Lawalree = Cardaminopsis arenosa

bactriana Ovcz. & Junussov - 6

bursifolia (DC.) Botsch. (*Nasturtium bursifolium* DC., *Arabidopsis trichopoda* (Turcz.) Botsch., *Arabis bursifolia* (DC.) A. Khokhr., *A. trichopoda* Turcz., *Arabidopsis mollis* auct.) - 4, 5

griffithiana (Boiss.) N. Busch = A. pumila

korshinskyi Botsch. - 6

mollis auct. = A. bursifolia

mollissima (C.A. Mey.) N. Busch - 3, 4, 5, 6

ovczinnikovii Botsch. - 6

parvula (Schrenk) O.E. Schulz - 1, 2, 6

pumila (Steph.) N. Busch (*A. griffithiana* (Boiss.) N. Busch, *A. pumila* var. *griffithiana* (Boiss.) Jafri, *Microsisymbrium griffithianum* (Boiss.) O.E. Schulz., *Sisymbrium griffithianum* Boiss., *Microsisymbrium minutiflorum* sensu Vass.) - 1, 2, 6

- var. *griffithiana* (Boiss.) Jafri = A. pumila

suecica (Fries) Norrl. = Hylandra suecica

taraxacifoia (T. Anders.) Jafri = A. wallichii

thaliana (L.) Heynh. - 1, 2, 3, 6

toxophylla (Bieb.) N. Busch - 1, 3, 6

trichopoda (Turcz.) Botsch. = A. bursifolia

tschuktschorum (Jurtz.) Jurtz. (*Arabis tschuktschorum* Jurtz.) - 5

verna (C. Koch) N. Busch = Drabopsis nuda

wallichii (Hook. fil. & Thoms.) N. Busch (*Sisymbrium wallichii* Hook. fil. & Thoms., *Arabidopsis taraxacifoia* (T. Anders.) Jafri, *Arabis bucharica* (Lipsky) N. Busch, comb. superfl., *A. bucharica* (Lipsky) Nevski, *A. taraxacifoia* T. Anders.) - 6

Arabis L.

allionii auct. p.p. = A. sudetica

allionii DC. subsp. *sudetica* (Tausch) Sojak = A. sudetica

alpina L. - 1, 3, 4

- var. *japonica* A. Gray = A. japonica
amurensis N. Busch = Cardaminopsis petraea
arenosa (L.) Scop. = Cardaminopsis arenosa
armena N. Busch = A. carduchorum
aucheri Boiss. - 6
auriculata auct. = A. recta
billardieri DC.
> This species had been recorded for the Caucasus in error.

boissieuana auct. = A. glauca
borealis Andrz. - 1, 3, 4
brachycarpa Rupr. - 2
bucharica (Lipsky) N. Busch = Arabidopsis wallichii
bucharica (Lipsky) Nevski = Arabidopsis wallichii
bursifolia (DC.) A. Khokhr. = Arabidopsis bursifolia
carduchorum Boiss. (*A. armena* N. Busch) - 2
caucasica Schlecht. (*A. alpina* L. subsp. *caucasica*
(Schlecht.) Briq., *A. caucasica* var. *dolichothrix* (N.
Busch) Grossh., *A. dolichothrix* (N. Busch) N. Busch,
A. flaviflora Bunge, *A. ionocalyx* auct.) - 1, 2
- var. *dolichothrix* (N. Busch) Grossh. = A. caucasica
- var. *farinacea* (Rupr.) Grossh. = A. farinacea
christianii N. Busch = A. mollis
colchica Kolak. - 2
coronata Nakai = Cardaminopsis gemmifera
dolichothrix (N. Busch) N. Busch = A. caucasica
erysimoides Kar. & Kir. = Prionotrichon erysimoides
excelsa Prokh. - 2
farinacea Rupr. (*A. caucasica* Schlecht. var. *farinacea*
(Rupr.) Grossh.) - 2
flaviflora Bunge = A. caucasica
fruticulosa C.A. Mey. - 3, 4, 6
gegamica Mtzchvetadze - 2
gemmifera (Matsum.) Makino = Cardaminopsis gemmifera
gerardii (Bess.) Koch (*Turritis gerardii* Bess., *T. nemorensis*
Hoffm., *Arabis hirsuta* (L.) Scop. subsp. *gerardii*
(Bess.) C. Hartm. fil., *A. nemorensis* (Hoffm.) Reichenb. 1832, non C.A. Mey. 1831, *A. planisiliqua*
(Pers.) Reichenb. subsp. *nemorensis* (Hoffm.) Sojak,
A. planisiliqua auct.) - 1, 2
> Under the name A. planisiliqua (Pers.) Reichenb. a West
> European species had been described.

glabra (L.) Benth. = Turritis glabra
glandulosa Kar. & Kir. = Dimorphostemon glandulosus
glauca Boissieu (*A. serrata* Franch. & Savat. var. *glauca*
(Boissieu) Ohwi, *A. boissieuana* auct.) - 5
- subsp. *pseudoauriculata* (Boissieu) Worosch. = A. pseudoauriculata
halleri L. = Cardaminopsis halleri
hirsuta auct. = A. sagittata
hirsuta (L.) Scop. subsp. *gerardii* (Bess.) C. Hartm. fil. = A.
gerardii
- subsp. *hornungiana* (Schur) Sojak = A. hornungiana
- subsp. *sudetica* (Tausch) Oborny = A. sudetica
hispida Mygind = Cardaminopsis petraea
hornungiana Schur (*A. hirsuta* (L.) Scop. subsp. *hornungiana* (Schur) Sojak) - 1
ionocalyx auct. = A. caucasica
japonica (A. Gray) A. Gray (*A. alpina* var. *japonica* A.
Gray, *A. stelleri* DC. var. *japonica* (A. Gray) Fr.
Schmidt, *A. stelleri* subsp. *japonica* (A. Gray) Worosch.) - 5
kamelinii Botsch. - 6
kamtschatica (Fisch.) Ledeb. = Cardaminopsis lyrata
karatavica (Lipsch. & Pavl.) Botsch. & Vved. = Botschantzevia karatavica
karategina Lipsky - 6
kazbegi Mtzchvetadze - 2

kokanica Regel & Schmalh. - 6
laxa Sibth. & Smith = Turritis laxa
lyrata L. = Cardaminopsis lyrata
- subsp. *kamtschatica* (Fisch.) Hult. = Cardaminopsis lyrata
- var. *A. kamtschatica* Fisch. = Cardaminopsis lyrata
magna (N. Busch) Grossh. = A. sachokiana
maximowiczii N. Busch = Cardaminopsis gemmifera
media N. Busch = Cardaminopsis petraea
mindshilkensis Bajt. & Myrzakulov - 6
mollis Stev. (*A. christianii* N. Busch, *A. nepetifolia* auct.) - 2
montbretiana Boiss. = A. recta
multijuga Borb. = Cardaminopsis multijuga
neglecta Schult. = Cardaminopsis neglecta
nemorensis (Hoffm.) Reichenb. = A. gerardii
nepetifolia auct. = A. mollis
■ nepetifolia Boiss.
nordmanniana Rupr. - 2
nova auct. = A. recta
nuda Belanger = Drabopsis nuda
ovirensis Wulf. = Cardaminopsis ovirensis
pachyrhiza Kar. & Kir. = Rhammatophyllum pachyrhizum
pendula L. - 1, 3, 4, 5, 6
petiolata Bieb. = Alliaria petiolata
petraea (L.) Lam. = Cardaminopsis petraea
- subsp. *septentrionalis* (N. Busch) Tolm. = Cardaminopsis
petraea
- subsp. *umbrosa* (Turcz.) Tolm. = Cardaminopsis petraea
petrogena A. Kerner = Cardaminopsis petrogena
planisiliqua auct. = A. gerardii
planisiliqua (Pers.) Reichenb. subsp. *nemorensis* (Hoffm.)
Sojak = A. gerardii
popovii Botsch. & Vved. - 6
pseudoauriculata Boissieu (*A. glauca* Boissieu subsp.
pseudoauriculata (Boissieu) Worosch.) - 5
recta Vill. (*A. montbretiana* Boiss., *A. auriculata* auct., *A.
nova* auct.) - 1, 2, 6
rupicola Kryl. - 3
sachokiana (N. Busch) N. Busch (*Cardamine sachokiana* N.
Busch, *Arabis magna* (N. Busch) Grossh., *Draba
magna* (N. Busch) Tolm.) - 2
sagittata (Bertol.) DC. (*Turritis sagittata* Bertol., *Arabis
hirsuta* auct.) - 1, 2, 3, 4, 5, 6
secunda N. Busch = Alyssopsis mollis
septentrionalis (N. Busch) O.E. Schulz = Cardaminopsis
petraea
serrata Franch. & Savat. var. *glauca* (Boissieu) Ohwi = A.
glauca
sinuata Turcz. = Neotorularia humilis
stelleri DC. - 4, 5
- subsp. *japonica* (A. Gray) Worosch. = A. japonica
- var. *japonica* (A. Gray) Fr. Schmidt = A. japonica
sudetica Tausch (*A. allionii* DC. subsp. *sudetica* (Tausch)
Sojak, *A. hirsuta* (L.) Scop. subsp. *sudetica* (Tausch)
Oborny, *A. allionii* auct. p.p.) - 1
suecica Fries = Hylandra suecica
taraxacifolia T. Anders. = Arabidopsis wallichii
tianschanica Pavl. - 6
tibetica Hook. fil. & Thoms. - 6
tilesii (Ledeb.) Karavaev = Cardaminopsis lyrata
trichopoda Turcz. = Arabidopsis bursifolia
tschuktschorum Jurtz. = Arabidopsis tschuktschorum
turczaninowii Ledeb. - 4, 5
turrita L. - 1, 2
umbrosa Turcz. = Cardaminopsis petraea
verna (L.) R. Br. (*Hesperis verna* L.) - 1

Armoracia Gaertn., Mey. & Scherb.

*macrocarpa (Waldst. & Kit.) Baumg. (*Cochlearia macrocarpa* Waldst. & Kit.) - 1
rusticana Gaertn., C.A. Mey. & Scherb. (*Cochlearia rusticana* Lam. nom. illegit.) - 1, 2, 3, 5, 6
sisymbrioides (DC.) Cajand. - 3, 4, 5

Asperuginoides Rauschert (*Buchingera* Boiss. & Hohen. 1849, non F. Schultz, 1847)

axillaris (Boiss. & Hohen.) Rauschert (*Buchingera axillaris* Boiss. & Hohen.) - 2, 6

Asterotricha V. Boczantzeva = Pterigostemon

spathulata (Kar. & Kir.) V. Boczantzeva = Pterygostemon spathulatus

Atelanthera Hook. fil. & Thoms.

perpusilla Hook. fil. & Thoms. - 6

Atropatenia F.K. Mey. (*Apterigia* Galushko, p.p.)

rostrata (N. Busch) F.K. Mey. (*Thlaspi rostratum* N. Busch, *Apterigia rostrata* (N. Busch) Galushko) - 2
zangezura (Tzvel.) F.K. Mey. (*Thlaspi zangezurum* Tzvel.) - 2

Aurinia Desv.

cyclocarpa (Boiss.) Czer. (*Ptilotrichum cyclocarpum* Boiss., *Aurinia rupestris* (Ten.) Cullen & T.R. Dudley subsp. *cyclocarpa* (Boiss.) Cullen & T.R. Dudley) - 2
rupestris (Ten.) Cullen & T.R. Dudley subsp. *cyclocarpa* (Boiss.) Cullen & T.R. Dudley = A. cyclocarpa
saxatilis (L.) Desv. (*Alyssum saxatile* L.) - 1

Barbarea R. Br. (*Campe* Dulac)

arcuata (Opiz ex J. & C. Presl) Reichenb. = B. vulgaris
grandiflora N. Busch - 2
inregrifolia DC. - 2
ketzkhovelii T. Mardalejschvili - 2
minor C. Koch - 2, 6
orthoceras Ledeb. - 4, 5
plantaginea DC. - 2
stricta Andrz. (*Campe stricta* (Andrz.) W. Wight ex Piper) - 1, 2, 3, 4, 6
taurica DC. = B. vulgaris
verna (Mill.) Aschers. (*Erysimum vernum* Mill., *Campe verna* (Mill.) Heller) - 1
vulgaris R. Br. (*B. arcuata* (Opiz ex J. & C. Presl) Reichenb., *B. taurica* DC., *B. vulgaris* subsp. *arcuata* (Opiz ex J. & C. Presl) Celak., *B. vulgaris* var. *arcuata* (Opiz ex J. & C. Presl.) Fries, *Campe barbarea* (L.) W. Wight ex Piper, *C. barbarea* subsp. *arcuata* (Opiz ex J. & C. Presl) Rauschert, *Erysimum arcuatum* Opiz ex J. & C. Presl) - 1, 2, 3, 4, 5(alien), 6
- subsp. *arcuata* (Opiz ex J. & C. Presl) Celak. = B. vulgaris
- var. *arcuata* (Opiz ex J. & C. Presl) Fries = B. vulgaris

Berteroa DC.

ascendens C. Koch = B. mutabilis
incana (L.) DC. - 1, 2, 3, 4, 5, 6

mutabilis (Vent.) DC. (*Alyssum mutabile* Vent., *Berteroa ascendens* C. Koch) - 2
spathulata (Steph.) C.A. Mey. = Galitzkya spathulata

Biscutella L.

laevigata - 1

Borodinia N. Busch

baicalensis N. Busch = B. tilingii
tilingii (Regel) Berkutenko (*Braya tilingii* Regel, *Borodinia baicalensis* N. Busch) - 4, 5

Botschantzevia Nabiev (*Koeiea* auct p.p., *Parrya* auct. p.p.)

karatavica (Lipsch. & Pavl.) Nabiev (*Parrya karatavica* Lipsch., *Arabis karatavica* (Lipsch.) Botsch. & Vved., *Koeiea karatavica* (Lipsch.) Rech. fil.) - 6

Brachilobus Desv. = Rorippa

hispidus Desv. = Rorippa hispida

Brassica L.

armoracioides Czern. ex Turcz. = Erucastrum armoracioides
campestris L. (*B. rapa* L. subsp. *campestris* (L.) Clapham) - 1, 2, 3, 4, 5, 6
cretica Lam. - 1
elongata auct. = Erucastrum armoracioides
elongata Ehrh. subsp. *armoracioides* (Czern. ex Turcz.) Aschers. & Graebn. = Erucastrum armoracioides
- subsp. *integrifolia* (Boiss.) Breistroffer = Erucastrum armoracioides
- subsp. *pinnatifida* (Schmalh.) Greuter & Burdet = Erucastrum cretaceum
- var. *pinnatifida* Schmalh. = Erucastrum cretaceum
juncea (L.) Czern. - 1, 2, 3, 4, 5, 6
napus L. (*B. rapa* L. subsp. *napus* (L.) Schuebl. & Martens) - 1, 2, 3, 4, 5, 6
nigra (L.) Koch (*Sinapis nigra* L.) - 1, 2, 5, 6
*oleracea L. - 1, 2, 3, 4, 5, 6
*rapa L. - 1, 2, 3, 4, 5, 6
- subsp. *campestris* (L.) Clapham = B. campestris
- subsp. *napus* (L.) Schuebl. & Martens = B. napus
sisymbrioides (Fisch.) Grossh. (*B. tournefortii* Gouan var. *sisymbrioides* Fisch., *B. tournefortii* auct.) - 2, 6
subhastata Willd. = Sisymbrium subhastatum
subspinescens Fisch. & C.A. Mey. = Sisymbrium subspinescens
sylvestris (L.) Mill. subsp. *taurica* Tzvel. = B. taurica
taurica (Tzvel.) Tzvel. (*B. sylvestris* (L.) Mill. subsp. *taurica* Tzvel.) - 1
tournefortii auct. = B. sisymbrioides
tournefortii Gouan var. *sisymbrioides* Fisch. = B. sisymbrioides

Braya Sternb. & Hoppe

aenea Bunge (*B. rosea* Bunge var. *aenea* (Bunge) Malysch.) - 3, 4, 5
- subsp. **pseudoaenea** Petrovsky - 5
angustifolia (N. Busch) Vass. - 4
brachycarpa Vass. = B. rosea

humilis (C.A. Mey.) Robins. = Neotorularia humilis
- subsp. *arctica* (Bocher) Rollins = Neotorularia humilis
oxycarpa Hook. fil. & Thoms. = Aphragmus oxycarpus
pamirica (Korsh.) O. Fedtsch. - 6
pilosa Hook. (*B. purpurascens* (R. Br.) Bunge subsp. *pilosa* (Hook.) Hult.) - 4, 5
- subsp. *thorild-wulffii* (Ostenf.) Petrovsky = B. thorild-wulffii
purpurascens (R. Br.) Bunge - 1, 3, 4, 5
- subsp. *pilosa* (Hook.) Hult. = B. pilosa
- subsp. *thorild-wulffii* (Ostenf.) Hult. = B. thorild-wulffii
rosea Bunge (*B. brachycarpa* Vass., *B. rosea* var. *brachycarpa* (Vass.) Malysch.) - 3, 4, 6
- var. *aenea* (Bunge) Malysch. = B. aenea
- var. *brachycarpa* (Vass.) Malysch. = B. rosea
scharnhorstii Regel & Schmalh. - 6
siliquosa Bunge - 3, 4
thorild-wulffii Ostenf. (*B. pilosa* Hook. subsp. *thorild-wulffii* (Ostenf.) Petrovsky, *B. purpurascens* (R. Br.) Bunge subsp. *thorild-wulffii* (Ostenf.) Hult.) - 5
tilingii Regel = Borodinia tilingii

Buchingera Boiss. & Hohen. = Asperuginoides
axillaris Boiss. & Hohen. = Asperuginoides axillaris

Bunias L.

arvensis Jord. - 1
cochlearioides Murr. - 1, 3, 4, 6
orientalis L. - 1, 2, 3, 5(alien)

Cakile Mill.

arctica Pobed. (*C. edentula* (Bigel.) Hook. subsp. *islandica* auct. p.p., *C. edentula* auct. p.p.) - 1
baltica Jord. ex Pobed. (*C. maritima* Scop. f. *baltica* Jord. ex Rouy & Fouc., *C. maritima* subsp. *baltica* (Jord. ex Rouy & Fouc.) Hyl. ex P.W. Ball, *C. maritima* sensu N. Busch, p.p.) - 1
edentula auct. p.p. = C. arctica and C. lapponica
edentula (Bigel.) Hook. subsp. *islandica* auct. p.p. = C. arctica and C. lapponica
euxina Pobed. (*C. maritima* Scop. subsp. *euxina* (Pobed.) E.I. Nyarady, *C. maritima* sensu N. Busch, p.p.) - 1, 2
lapponica Pobed. (*C. edentula* (Bigel.) Hook. subsp. *islandica* auct. p.p., *C. edentula* auct. p.p.) - 1
maritima Scop. subsp. *baltica* (Jord. ex Rouy & Fouc.) Hyl. ex P.W. Ball = C. baltica
- subsp. *euxina* (Pobed.) E.I. Nyarady = C. euxina
- subsp. *maritima*, p.p. = C. monosperma
- f. *baltica* Jord. ex Rouy & Fouc. = C. baltica
maritima sensu N. Busch = C. baltica and C. euxina
monosperma Lange (*C. maritima* subsp. *maritima*, p.p.) - 1(?)

Calepina Adans.

irregularis (Asso) Thell. - 1, 2, 6

Callothlaspi F.K. Mey.

abchasicum - 2

Calymmatium O.E. Schulz

draboides (Korsh.) O.E. Schulz (*Capsella draboides* Korsh.) - 6

Camelina Crantz

albiflora (Boiss.) N. Busch = C. rumelica
alyssum (Mill.) Thell. (*Myagrum alyssum* Mill., *Camelina linicola* Schimp. & Spenn., *C. sativa* (L.) Crantz subsp. *linicola* (Schimp. & Spenn.) N. Zing.) - 1
- subsp. *integerrima* (Celak.) Smejkal = C. macrocarpa
- subsp. *macrocarpa* (Wierzb. ex Reichenb.) Hiit. = C. macrocarpa
barbareifolia DC. = Rorippa barbareifolia
caucasica (Sinsk.) Vass. = C. sativa
foetida Fries a[= subsp.] *integerrima* (Celak.) Celak. = C. macrocarpa
- var. *integerrima* Celak. = C. macrocarpa
glabrata (DC.) Fritsch = C. sativa
laxa C.A. Mey. - 2
linicola Schimp. & Spenn. = C. alyssum
macrocarpa Wierzb. ex Reichenb. (*C. alyssum* (Mill.) Thell. subsp. *integerrima* (Celak.) Smejkal, *C. alyssum* subsp. *macrocarpa* (Wierzb. ex Reichenb.) Hiit., *C. foetida* var. *integerrima* Celak., *C. foetida* a[= subsp.] *integerrima* (Celak.) Celak., *C. sativa* (L.) Crantz subsp. *macrocarpa* (Wierzb. ex Reichenb.) Soo) - 1
microcarpa Andrz. - 1, 2, 3, 4, 5, 6
- subsp. *pilosa* (DC.) Jav. = C. pilosa
- subsp. *pilosa* (DC.) Soo = C. pilosa
- subsp. *sylvestris* (Wallr.) Hiit. = C. sylvestris
pilosa (DC.) N. Zing. (*C. microcarpa* Andrz. subsp. *pilosa* (DC.) Jav., *C. microcarpa* subsp. *pilosa* (DC.) Soo, comb. superfl., *C. sativa* (L.) Crantz subsp. *pilosa* (DC.) N. Zing.) - 1, 2
rumelica Velen. (*C. albiflora* (Boiss.) N. Busch, *C. sativa* (L.) Crantz subsp. *rumelica* (Velen.) O. Bolos & Vigo) - 1, 2, 5(alien), 6
- subsp. *transcaspica* (Fritsch) Hedge = C. transcaspica
sativa (L.) Crantz (*Myagrum sativum* L., *Camelina caucasica* (Sinsk.) Vass., *C. glabrata* (DC.) Fritsch, *C. sativa* subsp. *glabrata* (DC.) N. Zing.) - 1, 2, 3, 4, 5, 6
- subsp. *glabrata* (DC.) N. Zing. = C. sativa
- subsp. *linicola* (Schimp. & Spenn.) N. Zing. = C. alyssum
- subsp. *macrocarpa* (Wierzb. ex Reichenb.) Soo = C. macrocarpa
- subsp. *pilosa* (DC.) N. Zing. = C. pilosa
- subsp. *rumelica* (Velen.) O. Bolos & Vigo = C. rumelica
sylvestris Wallr. (*C. microcarpa* Andrz. subsp. *sylvestris* (Wallr.) Hiit.) - 1, 2, 6
transcaspica Fritsch (*C. rumelica* Velen. subsp. *transcaspica* (Fritsch) Hedge) - 6

Campe Dulac = Barbarea
barbarea (L.) W. Wight ex Piper = Barbarea vulgaris
- subsp. *arcuata* (Opiz ex J. & C. Presl) Rauschert = Barbarea vulgaris
stricta (Andrz.) W. Wight ex Piper = Barbarea stricta
verna (Mill.) Heller = Barbarea verna

Campyloptera Boiss.

carnea (Banks & Soland.) Botsch. & Vved. (*Aethionema carneum* (Banks & Soland.) B. Fedtsch.) - 2, 6

Capsella Medik.

bursa-pastoris (L.) Medik. (*C. hyrcana* Grossh.) - 1, 2, 3, 4, 5, 6
draboides Korsh. = Calymmatium draboides
hyrcana Grossh. = C. bursa-pastoris
orientalis Klok. - 1
rubella Reut. - 1

Cardamine L. (*Sphaerotorrhiza* (O.E. Schulz) A. Khokhr.)

acris auct. = C. seidlitziana
amara L. - 1, 3
- subsp. *opizii* (J. & C. Presl) Celak. = C. opizii
bellidifolia L. - 1, 3, 4, 5
blaisdellii Eastw. (*C. hyperborea* O.E. Schulz, nom. superfl., p.p. excl. var. *oxyphylla* (Trautv.) O.E. Schulz & syn. *C. digitata* Richards., *C. microphylla* auct. p.p.) - 4, 5
> The question of the correct name of this species has not been solved yet. It is necessary to study type material of C. blaisdellii Eastw. The name C. hyperborea must be replaced however, since it has always been superfluous.

chiriensis Miyabe & Tatew. = C. victoris
conferta Jurtz. - 4, 5
crassifolia auct. = C. rivularis
dasycarpa Bieb. = C. impatiens
densiflora Gontsch. - 6
dentata Schult. (*C. hayneana* Welw. ex Reichenb. nom. invalid., *C. palustris* (Wimm. & Grab.) Peterm., *C. pratensis* L. subsp. *dentata* (Schult.) Celak., *C. pratensis* var. *palustris* Wimm. & Grab., *C. matthioli* auct.) - 1, 3, 4
digitata Richards. (*C. hyperborea* O.E. Schulz, nom. superfl., p.p. quoad var. *oxyphylla* (Trautv.) O.E. Schulz & syn. *C. digitata* Richards. non *Dentaria digitata* Lam., *C. richardsonii* Hult. nom. superfl.) - 5
x **dubia** Zapal. - C. opizii J. & C. Presl x C. pratensis L. - 1
fauriei Franch. - 5
flexuosa With. (*C. scutata* Thunb. subsp. *flexuosa* (With.) Hara) - 1
fontana Lam. = Nasturtium officinale
gemmifera Matsum. = Cardaminopsis gemmifera
glanduligera O. Schwarz = Dentaria glandulosa
- subsp. *sibirica* (O.E. Schulz) O. Schwarz = Dentaria sibirica
graeca L. - 1
hayneana Welw. ex Reichenb. = C. dentata
- var. *iliciana* Fritsch = C. iliciana
hirsuta L. - 1, 2, 6
hyperborea O.E. Schulz = C. blaisdellii and C. digitata
iliciana (Fritsch) N. Busch (*C. hayneana* Welw. ex Reichenb. var. *iliciana* Fritsch, *C. pratensis* L. subsp. *iliciana* (Fritsch) O.E. Schulz) - 2
impatiens L. (*C. dasycarpa* Bieb.) - 1, 2, 3, 4, 5, 6
- var. *pectinata* (Pall. ex DC.) Trautv. = C. pectinata
lazica Boiss. & Bal. - 2
leucantha (Tausch) O.E. Schulz - 4, 5
- subsp. **tomentella** Worosch. - 5
lyrata Bunge - 4, 5
macrophilla Willd. (*C. sachalinensis* Miyabe & Miyake) - 1, 3, 4, 5
manshurica (Kom.) Nakai (*C. parviflora* L. var. *manshurica* Kom.) - 5
matthioli auct. = C. dentata
microphylla Adams (*C. minuta* Willd. ex DC.) - 4, 5
microphylla auct. p.p. = C. blaisdellii
minuta Willd. ex DC. = C. microphylla
nymanii Gand. - 1, 3, 4, 5
opizii J. & C. Presl (*C. amara* L. subsp. *opizii* (J. & C. Presl) Celak.) - 1
palustris (Wimm. & Grab.) Peterm. = C. dentata
parviflora L. - 1, 2, 3, 4, 5, 6
- var. *manshurica* Kom. = C. manshurica
pectinata Pall. ex DC. (*C. impatiens* L. var. *pectinata* (Pall. ex DC.) Trautv.) - 1, 2
pedata Regel & Til. - 5

petraea L. = Cardaminopsis petraea
pratensis L. - 1, 3, 4, 5
- subsp. **angustifolia** (Hook.) O.E. Schulz. - 1, 3, 4, 5
- subsp. *dentata* (Schult.) Celak. = C. dentata
- subsp. *iliciana* (Fritsch) O.E. Schulz = C. iliciana
- subsp. *rivularis* (Schur) Jav. = C. rivularis
- subsp. *rivularis* (Schur) Simonk. = C. rivularis
- var. *palustris* Wimm. & Grab. = C. dentata
prorepens Fisch. - 4, 5
- subsp. **roseiflora** (Worosch.) Worosch. (*C. yezoensis* Maxim. subsp. *roseiflora* Worosch.) - 5
purpurea Cham. & Schlecht. - 5
regeliana Miq. (*C. scutata* Thunb. subsp. *regeliana* (Miq.) Hara, *C. scutata* auct.) - 5
richardsonii Hult. = C. digitata
rivularis Schur (*C. pratensis* L. subsp. *rivularis* (Schur) Jav., comb. superfl., *C. pratensis* subsp. *rivularis* (Schur) Simonk., *C. crassifolia* auct.) - 1
sachalinensis Miyabe & Miyake = C. macrophylla
sachokiana N. Busch = Arabis sachokiana
schinziana O.E. Schulz - 5
scutata auct. = C. regeliana
scutata Thunb. subsp. *flexuosa* (With.) Hara = C. flexuosa
- subsp. *regeliana* (Miq.) Hara = C. regeliana
seidlitziana Albov (*C. acris* auct.) - 2
seravschanica Botsch. - 6
sphenophylla Jurtz. - 5
tenera S.G. Gmel. ex C.A. Mey. - 1, 2
tenuifolia (Ledeb.) Turcz. = C. trifida
trifida (Poir.) B.M.G. Jones (*Dentaria trifida* Poir., *Cardamine tenuifolia* (Ledeb.) Turcz., *Dentaria alaunica* Golits., *D. tenuifolia* Ledeb., *Sphaerotorrhiza trifida* (Poir.) A. Khokhr.) - 1, 3, 4, 5
trifolia L. - 1
x **uliginosa** Bieb. - C. amara L. x C. dentata Schult. - 2
umbellata Greene - 5
victoris N. Busch (? *C. chiriensis* Miyabe & Tatew.) - 5
yezoensis Maxim. - 5
- subsp. *roseiflora* Worosch. = C. prorepens subsp. roseiflora

Cardaminopsis (C.A. Mey.) Hayek

amurensis (N. Busch) O.E. Schulz = C. petraea
arenosa (L.) Hayek. (*Arabidopsis arenosa* (L.) Lawalree, *Arabis arenosa* (L.) Scop.) - 1
- subsp. *borbasii* (Zapal.) Pawl. = C. multijuga
- var. *multijuga* (Borb.) Kotov = C. multijuga
dependens Soo = C. multijuga
gemmifera (Matsum.) Berkutenko (*Cardamine gemmifera* Matsum., *Arabis coronata* Nakai, *A. gemmifera* (Matsum.) Makino, *A. maximowiczii* N. Busch, *Cardaminopsis maximowiczii* (N. Busch) O.E. Schulz) - 4, 5
halleri (L.) Hayek (*Arabis halleri* L.) - 1
- subsp. *ovirensis* (Wulf.) Hegi & E. Schmid = C. ovirensis
hispida (Mygind) Hayek = C. petraea
kamtschatica (Fisch.) O.E. Schulz = C. lyrata
lyrata (L.) Hiit. (*Arabis lyrata* L., *A. kamtschatica* (Fisch.) Ledeb., *A. lyrata* var. *A. kamtschatica* (Fisch.) Hult., *A. tilesii* (Ledeb.) Karavaev, *Cardaminopsis kamtschatica* (Fisch.) O.E. Schulz, *Sisymbrium tilesii* Ledeb.) - 4, 5
maximowiczii (N. Busch) O.E. Schulz = C. gemmifera
media (N. Busch) O.E. Schulz = C. petraea
multijuga (Borb.) Czer. (*Arabis multijuga* Borb., *Cardaminopsis arenosa* (L.) Hayek subsp. *borbasii* (Zapal.) Pawl., *C. arenosa* var. *multijuga* (Borb.) Kotov, *C. dependens* Soo) - 1

neglecta (Schult.) Hayek (*Arabis neglecta* Schult.) - 1
ovirensis (Wulf.) Thell. ex Jav. (*Arabis ovirensis* Wulf., *Cardaminopsis halleri* (L.) Hayek subsp. *ovirensis* (Wulf.) Hegi & E. Schmid, *C. ovirensis* (Wulf.) O. Schwarz, comb. superfl.) - 1
petraea (L.) Hiit. (*Cardamine petraea* L., *Arabis amurensis* N. Busch, *A. hispida* Mygind, *A. media* N. Busch, *A. petraea* (L.) Lam., *A. petraea* subsp. *septentrionalis* (N. Busch) Tolm., *A. petraea* subsp. *umbrosa* (Turcz.) Tolm., *A. septentrionalis* (N. Busch) O.E. Schulz, *A. umbrosa* Turcz., *Cardaminopsis amurensis* (N. Busch) O.E. Schulz, *C. hispida* (Mygind) Hayek, *C. media* (N. Busch) O.E. Schulz, *C. petraea* subsp. *media* (N. Busch) Hamet-Ahti, *C. petraea* subsp. *umbrosa* (Turcz.) Peschkova, *C. septentrionalis* (N. Busch) O.E. Schulz, *C. umbrosa* (Turcz.) Czer.) - 1, 3, 4, 5
- subsp. *media* (N. Busch) Hamet-Ahti = C. petraea
- subsp. *umbrosa* (Turcz.) Peschkova = C. petraea
petrogena (A. Kerner) Mesicek (*Arabis petrogena* A. Kerner) - 1
septentrionalis (N. Busch) O.E. Schulz = C. petraea
sinuata (Turcz.) O.E. Schulz = Neotorularia humilis
suecica (Fries) Hiit. = Hylandra suecica
umbrosa (Turcz.) Czer. = C. petraea

Cardaria Desv. (*Hymenophysa* C.A. Mey.)

boissieri (N. Busch) Soo (*Lepidium boisseri* N. Busch) - 2
draba (L.) Desv. (*Lepidium draba* L.) - 1, 2, 3, 4, 5(alien), 6
- subsp. *chalepensis* auct. p.p. = C. propinqua and C. repens
macrocarpa (Franch.) Rollins (*Hymenophysa macrocarpa* Franch.) - 6
propinqua (Fisch. & C.A. Mey.) N. Busch (*Lepidium propinquum* Fisch. & C.A. Mey., *Cardaria propinqua* (Fisch. & C.A. Mey.) Soo, comb. superfl., *C. draba* (L.) Desv. subsp. *chalepensis* auct. p.p.) - 2
pubescens (C.A. Mey.) Jarm. (*Hymenophysa pubescens* C.A. Mey., *Cardaria pubescens* (C.A. Mey.) Rollins, comb. superfl.) - 1(alien), 3, 6
repens (Schrenk) Jarm. (*C. repens* (Schrenk) Soo, comb. superfl., *Lepidium repens* (Schrenk) Boiss., *Cardaria draba* (L.) Desv. subsp. *chalepensis* auct. p.p.) - 6

Carpoceras (DC.) Link = Thlaspi

brevistylum N. Busch = Noccaea tatianae
ceratocarpum (Pall.) N. Busch = Thlaspi ceratocarpum
hastulatum (DC.) Boiss. = Noccidium hastulatum
stenocarpum Boiss. = Kotschyella stenocarpa
tatianae (Bordz.) Grossh. = Noccaea tatianae

Carpoceras sensu N. Busch, p.p. = Kotschyella, Noccaea and Noccidium

Catenularia Botsch. = Catenulina

hedysaroides Botsch. = Catenulina hedysaroides

Catenulina Sojak (*Catenularia* Botsch. 1957, non Grove, 1886, *Chodsha-Kasiana* Rauschert)

hedysaroides (Botsch.) Sojak (*Catenularia hedysaroides* Botsch., *Chodsha-Kasiana hedysaroides* (Botsch.) Rauschert) - 6

Chalcanthus Boiss.

renifolius (Boiss.) Boiss. (*Hesperis renifolia* Boiss., *Chalcan-*

thus tuberosus (Kom.) Kom.) - 6
tuberosus (Kom.) Kom. = Ch. renifolius

Chartoloma Bunge

platycarpum (Bunge) Bunge - 6

*Cheiranthus L.

annuus L. = Matthiola annua
armeniacum Sims = Erysimum armeniacum
bicornis Sibth. & Smith = Matthiola bicornis
canus Pill. & Mitt. = Syrenia cana
***cheiri** L. - 1
himalayensis Cambess. = Oreoblastus himalayensis
incanus L. = Matthiola incana
longipetalus Vent. = Matthiola longipetala
montanus Pall. = Syrenia montana
odoratissimus Pall. ex Bieb. = Matthiola odoratissima
sulphureus Banks & Soland. = Sterigmostemum sulphureum
taraxacifolius Balb. = Strigosella africana
torulosus Bieb. = Sterigmostemum incanum

Chodsha-Kasiana Rauschert = Catenulina

hedysaroides (Botsch.) Rauschert = Catenulina hedysaroides

Chorispora R. Br. ex DC.

bungeana Fisch. & C.A. Mey. - 3, 6
elegans Cambess. - 6
greigii Regel - 6
iberica (Bieb.) DC. - 2
insignis Pachom. - 6
macropoda Trautv. - 6
pamirica Pachom. - 6
sibirica (L.) DC. - 3, 4, 6
songarica Schrenk - 6
tenella (Pall.) DC. - 1, 2, 3, 5(alien), 6

Christolea Cambess.

crassifolia Cambess. (*Ermania crassifolia* (Cambess.) Ovcz. & Junussov) - 6
flabellata (Regel) N. Busch = Oreoblastus flabellatus
incana Ovcz. = Oreoblastus incanus
linearis N. Busch = Oreoblastus linearis
maidantalica (M. Pop. & P. Baran.) N. Busch = Achoriphragma maidantalicum
pamirica Korsh. (*Ermania pamirica* (Korsh.) Ovcz. & Junussov) - 6
parryoides (Cham.) N. Busch = Ermania parryoides
susloviana Jafri = Desideria pamirica

Chrysanthemopsis Rech. fil. = Smelowskia

koelzii Rech. fil. = Smelowskia calycina

Chrysochamela (Fenzl) Boiss.

draboides Woronow = Ch. elliptica
elliptica (Boiss.) Boiss. (*Nasturtium ellipticum* Boiss., *Chrysochamela draboides* Woronow) - 1

Cithareloma Bunge

lehmannii Bunge - 6
vernum Bunge - 6

Clausia Korn.-Tr.

aprica (Steph.) Korn.-Tr. - 1, 2, 3, 4, 5, 6
gracillima M. Pop. ex Botsch. & Vved. = Pseudoclausia
   gracillima
hispida (Regel) Lipsky = Pseudoclausia hispida
kasakhorum Pavl. - 6
mollissima Lipsky = Pseudoclausia mollissima
olgae (Regel & Schmalh.) Lipsky = Pseudoclausia olgae
papillosa Vass. = Pseudoclausia papillosa
robusta Pachom. - 6
sarawschanica (Regel & Schmalh.) Lipsky = Pseudoclausia
   sarawschanica
tschimganica M. Pop. ex Botsch. & Vved. = Pseudoclausia
   tschimganica
turkestanica Lipsky = Pseudoclausia turkestanica

Clypeola L.

alyssoides L. = Alyssum calycinum
aspera (Grauer) Turrill (*Peltaria aspera* Grauer, *Clypeola
   echinata* DC., *C. lasiocarpa* Juss. ex Pers.) - 6
campestris L. = Alyssum calycinum
dichotoma Boiss. = Pseudanastatica dichotoma
echinata DC. = C. aspera
■ elegans Boiss. & Huet
jonthlaspi L. - 1, 2, 6
- subsp. *microcarpa* (G. Moris) Arcang. = C. microcarpa
lasiocarpa Juss. ex Pers. = C. aspera
maritima L. = Lobularia maritima
microcarpa G. Moris (*C. jonthlaspi* L. subsp. *microcarpa*
   (G. Moris) Arcang.) - 1, 2
minor L. = Alyssum parviflorum
■ raddeana Albov

Cochlearia L. (*Cochleariopsis* A. & D. Löve)

anglica L. - 1
arctica Schlecht. ex DC. (*C. lenensis* Adams ex Fisch., *C.
   officinalis* L. subsp. *arctica* (Schlecht. ex DC.) Hult.,
   Cochleariopsis groenlandica (L.) A. & D. Löve subsp.
   arctica (Schlecht. ex DC.) A. & D. Löve, *Cochlearia
   fenestrata* auct.) - 1, 3, 4, 5
- subsp. *oblongifolia* (DC.) Petrovsky = C. oblongifolia
aucheri Boiss. = Pseudosempervivum aucheri
danica L. - 1
fenestrata auct. = C. arctica
fenestrata R. Br. ex DC. = C. groenlandica
groenlandica L. (*C. fenestrata* R. Br. ex DC., *C. officinalis*
   L. subsp. *groenlandica* (L.) A. Pors., *C. polaris* Pobed.,
   Cochleariopsis groenlandica (L.) A. & D. Löve) - 1, 3,
   4, 5
lenensis Adams ex Fisch. = C. arctica
macrocarpa Waldst. & Kit. = Armoracia macrocarpa
oblongifolia DC. (*C. arctica* Schlecht. ex DC. subsp.
   oblongifolia (DC.) Petrovsky, *C. officinalis* L. subsp.
   oblongifolia (DC.) Hult., *Cochleariopsis groenlandica*
   (L.) A. & D. Löve subsp. *oblongifolia* (DC.) A. & D.
   Löve) - 5
officinalis L. - 1
- subsp. *arctica* (Schlecht. ex DC.) Hult. = C. arctica
- subsp. *groenlandica* (L.) A. Pors. = C. groenlandica
- subsp. *oblongifolia* (DC.) Hult. = C. oblongifolia
planisiliqua Boiss. = Peltariopsis planisiliqua
polaris Pobed. = C. groenlandica
polonica auct. = C. pyrenaica
pyrenaica L. (*C. polonica* auct.) - 1
rusticana Lam. = Armoracia rusticana

sibirica Willd. = Sobolewskia sibirica
wasabii Siebold = Eutrema japonicum

Cochleariopsis A. & D. Löve = Cochlearia

groenlandica (L.) A. & D. Löve = Cochlearia groenlandica
- subsp. *arctica* (Schlecht. ex DC.) A. & D. Löve = Cochlearia arctica
- subsp. *oblongifolia* (DC.) A. & D. Löve = Cochlearia
   oblongifolia

Coluteocarpus Boiss.

vesicaria (L.) Holmboe - 2

Conringia Adans.

austriaca (Jacq.) Sweet - 2
clavata Boiss. (*C. perfoliata* (C.A. Mey.) N.
   Busch, 1939, non Link, 1822) - 2, 6
orientalis (L.) Dumort. - 1, 2, 3, 5, 6
perfoliata (C.A. Mey.) N. Busch = C. clavata
persica Boiss. - 2, 6
planisiliqua Fisch. & C.A. Mey. - 2, 6

Coronopus Zinn

didymus (L.) Smith - 2
procumbens Gilib. = C. squamatus
squamatus (Forssk.) Aschers (*Lepidium squamatum*
   Forssk., *Coronopus procumbens* Gilib. nom. invalid.) -
   1, 2, 6

Crambe L.

aculeolata (N. Busch.) Czerniak. - 2
amabilis Butk. & Majlun = C. orientalis
armena N. Busch - 2
aspera Bieb. (*C. buschii* (O.E. Schulz) Stank. p.p. quoad
   plantas) - 1
buschii (O.E. Schulz) Stank. = C. aspera and C. gibberosa
caspica Habl. = Physoptychis caspica
cordifolia Stev. - 2
- subsp. *kotschyana* (Boiss.) Jafri = C. kotschyana
cretacea (Czerniak.) Czerniak. = C. edentula
edentula Fisch. & C.A. Mey. ex Korsh. (*C. cretacea* (Czerniak.) Czerniak., *C. edentula* var. *cretacea* Czerniak.) -
   6
- var. *cretacea* Czerniak. = C. edentula
gibberosa Rupr. (*C. buschii* (O.E. Schulz) Stank. p.p. quoad
   nomen, *C. tataria* Sebeok var. *buschii* O.E. Schulz) - 2
gordjaginii Spryg. & M. Pop. - 6
grandiflora DC. - 1, 2
grossheimii I. Khalilov - 2
juncea Bieb. - 2
koktebelica (Junge) N. Busch - 1, 2
kotschyana Boiss. (*C. cordifolia* Stev. subsp. *kotschyana*
   (Boisss.) Jafri, *C. palmatifida* Regel & Schmalh., *C.
   sewerzowii* Regel) - 6
litwinowii K. Gross - 1
maritima L. - 1
- subsp. *pinnatifida* (R. Br.) Schmalh. = C. pinnatifida
maritima sensu Czerniak. = C. pontica
mitridatis Juz. - 1
orientalis L. (*C. amabilis* Butk. & Majlun) - 2, 6
palmatifida Regel & Schmalh. = C. kotschyana
pinnatifida R. Br. (*C. maritima* L. subsp. *pinnatifida* (R.
   Br.) Schmalh.) - 1, 2
pontica Stev. ex Rupr. (*C. maritima* sensu Czerniak.) - 1
schugnana Korsh. - 6

sewerzowii Regel = C. kotschyana
steveniana Rupr. - 1, 2
tataria Sebeok - 1, 2, 3
- var. *buschii* O.E. Schulz = C. gibberosa

Cryptospora Kar. & Kir.

dentata Freyn & Sint. = Neotorularia dentata
falcata Kar. & Kir. - 6
omissa Botsch. - 6
trichocarpa Botsch. - 6

Cuspidaria auct. = Erysimum

cuspidata (Bieb.) Takht. = Erysimum cuspidatum

Cymatocarpus O.E. Schultz

grossheimii N. Busch - 2
heterophyllus (M. Pop.) N. Busch = C. popovii
pilosissimus (Trautv.) O.E. Schulz - 6
popovii Botsch. & Vved. (*C. heterophyllus* (M. Pop.) N. Busch, nom. illegit., *Sisymbrium heterophyllum* M. Pop. 1923, non Forst. 1785) - 6

Deilosma Andrz. = Hesperis

suaveolens Andrz. = Hesperis suaveolens

Dentaria L.

alaunica Golits. = Cardamine trifida
bipinnata C.A. Mey. - 2
bulbifera L. - 1, 2
glandulosa Waldst. & Kit. (*Cardamine glanduligera* O. Schwarz) - 1
microphylla Willd. - 2
quinquefolia Bieb. - 1, 2
sibirica (O.E. Schulz) N. Busch (*Cardamine glanduligera* O. Schwarz subsp. *sibirica* (O.E. Schulz) O. Schwarz) - 3, 4
tenuifolia Ledeb. = Cardamine trifida
trifida Poir. = Cardamine trifida

Descurainia Webb & Berth.

▪ **kochii** (Petri) O.E. Schulz
richardsonii (Sweet) O.E. Schulz - 1
sophia (L.) Webb ex Prantl - 1, 2, 3, 4, 5, 6
sophioides (Fisch. ex Hook.) O.E. Schulz - 3, 4, 5

Desideria Pamp.

mirabilis auct. = D. pamirica
pamirica Suslova (*Christolea susloviana* Jafri, *Desideria mirabilis* auct.) - 6

Dichasianthus Ovcz. & Junussov

brachycarpus (Vass.) Sojak = Neotorularia brachycarpa
brevipes (Kar. & Kir.) Sojak = Neotorularia brevipes
contortuplicatus (Steph.) Sojak = Neotorularia contortuplicata
dentatus (Freyn & Sint.) Sojak = Neotorularia dentata
eldaricus (Grossh.) Sojak = Neotorularia eldarica
humilis (C.A. Mey.) Sojak = Neotorularia humilis
korolkowii (Regel & Schmalh.) Sojak = Neotorularia korolkowii
rossicus (O.E. Schulz) Sojak = Neotorularia rossica
sergievskianus (Polozh.) Sojak = Neotorularia sergievskiana

subtilissimus (M. Pop.) Ovcz. & Junussov (*Sisymbrium subtilissimum* M. Pop., *Torularia subtilissima* (M. Pop.) Botsch.) - 6
torulosus (Desf.) Sojak = Neotorularia torulosa

Didymophysa Boiss.

aucheri Boiss. - 2, 6
fedtschenkoana Regel - 6
- subsp. **incisa** Junussov & R. Kam. - 6

Dielsiocharis O.E. Schulz (*Alyssopsis* Boiss. p.p.)

kotschyi (Boiss.) O.E. Schulz (*Alyssopsis kotschyi* Boiss.) - 6

Dilophia Thoms.

salsa Thoms. - 6

Dimorphostemon Kitag. (*Alaida* Dvorak, nom. superfl.)

asper (Pall.) Kitag. = D. glandulosus
glandulosus (Kar. & Kir.) Golubk. (*Arabis glandulosa* Kar. & Kir., *Alaida glandulosa* (Kar. & Kir.) Dvorak, *Torularia glandulosa* (Kar. & Kir.) Vass.) - 6
pectinatus (DC.) Golubk. (*Sisymbrium pectinatum* DC., *Alaida pectinata* (DC.) Dvorak, *Dimorphostemon asper* (Pall.) Kitag. nom. illegit., *Dontostemon asper* (Pall.) Schischk. nom. illegit., *D. pectinatus* (DC.) Ledeb., *Sisymbrium an asperum* ? Pall. non L., *Torularia pectinata* (DC.) Ovcz. & Junussov) - 4, 5

Diplotaxis DC.

cretacea Kotov - 1
muralis (L.) DC. - 1, 2
x **tanaitica** Schtscherbina. - D. cretacea Kotov x D. muralis (L.) DC. - 1
tenuifolia (L.) DC. - 1
viminea (L.) DC. - 1

Diptychocarpus Trautv.

strictus (Fisch. ex Bieb.) Trautv. - 1, 2, 6

Dontostemon Andrz. ex C.A. Mey. (*Andreoskia* DC. 1824, non Reichenb. 1823)

asper (Pall.) Schischk. = Dimorphostemon pectinatus
brevipes Bunge = Strigosella brevipes
crassifolius Bunge - 3
dentatus (Bunge) Ledeb. (*Andreoskia dentata* Bunge) - 4, 5
hispidus Maxim. - 5
integrifolius (L.) C.A. Mey. - 3, 4, 5
intermedius Worosch. - 5
micranthus C.A. Mey. - 3, 4, 5
pectinatus (DC.) Ledeb. = Dimorphostemon pectinatus
perennis C.A. Mey. - 3, 4
senilis Maxim. - 4

Draba L. (*Holargidium* Turcz. ex Ledeb., *Nesodraba* Greene)

adamsii Ledeb. = D. pauciflora
aizoides L. - 1
ajanensis N. Busch = D. cinerea
alajica Litv. - 6
alberti Regel & Schmalh. - 6

aleutica Ekman (*D. behringii* Tolm., *D. tschuktschorum*
auct. p.p.) - 5
- subsp. *arctoberingensis* Jurtz. & Petrovsky = D. alpina
algida Adams var. *brachycarpa* Bunge = D. oreades
alpina L. (*D. aleutica* Ekman subsp. *arctoberingensis* Jurtz.
& Petrovsky) - 1, 3, 4, 5, 6
- subsp. *brachycarpa* (Bunge) Malysch. = D. oreades
altaica (C.A. Mey.) Bunge - 3, 6
alticola Kom. - 6
aradani N. Busch = D. fladnizensis
araratica Rupr. - 2
arctica auct. = D. groenlandica
arctica J. Vahl subsp. *groenlandica* (Ekman) Bocher = D.
groenlandica
arctogena (Ekman) Ekman = D. groenlandica
arseniewii (B. Fedtsch.) Gilg ex Tolm. - 6
aucheri Boiss. = D. rosularis
baicalensis Tolm. - 4
barbata Pohle = D. macrocarpa
behringii Tolm. = D. aleutica
borealis DC. (*D. kurilensis* (Turcz.) Fr. Schmidt) - 4, 5
- var. *sachlalinensis* Fr. Schmidt = D. sachalinensis
bruniifolia Stev. (*D. diversifolia* Boiss. & Huet, *D. globifera*
Ledeb.) - 2
- subsp. *heterocoma* (Fenzl) Coode & Cullen = D. heterocoma
- subsp. **kurdica** Coode & Cullen - 2(?)
- subsp. *olympica* (Sibth. ex DC.) Coode & Cullen = D.
olympica
bryoides DC. - 2
caesia auct. = D. palanderiana
camtschatica (Cham. & Schlecht.) Andrz. ex Tolm. = D.
kamtschatica
cana Rydb. (*D. incurvata* A. Vassil. & Golosk., *D. lanceolata* auct.) - 1(?), 3, 4, 5, 6
cardaminiflora Kom. - 5
carinthiaca Hoppe (*D. siliquosa* Bieb. subsp. *carinthiaca*
(Hoppe) O. Bolos & Vigo) - 1
chamissonis G. Don fil. - 5
cinerea Adams (*D. ajanensis* N. Busch, *D. parvisiliquosa*
Tolm.) - 1, 3, 4, 5
corymbosa auct. = D. macrocarpa
crassifolia Graham. - 5
cuspidata Bieb. - 1
darwasica Lipsky - 6
daurica DC. = D. hirta
densifolia auct. = D. stenopetala
diversifolia Boiss. & Huet = D. bruniifolia
elisabethae N. Busch - 2
eriopoda Turcz. ex Ledeb. - 3, 4
eschscholtzii Pohle ex N. Busch = D. macrocarpa
fedtschenkoi (Pohle) Gilg ex Tolm. - 6
fladnizensis Wulf. (*D. aradani* N. Busch, *D. tschuktschorum*
Trautv.) - 1, 3, 4, 5, 6
frigida Saut. var. *kamtschatica* Ledeb. = D. kamtschatica
glabella Pursh = D. hirta
glacialis Adams - 1, 3, 4
globifera Ledeb. = D. bruniifolia
grandiflora C.A. Mey. = Pachyneurum grandiflorum
groenlandica Ekman (*D. arctica* J. Vahl subsp. *groenlandica*
(Ekman) Bocher, *D. arctogena* (Ekman) Ekman, *D.*
groenlandica var. *arctogena* Ekman, *D. arctica* auct.) -
4, 5
- var. *arctogena* Ekman = D. groenlandica
heterocoma Fenzl (*D. bruniifolia* Stev. subsp. *heterocoma*
(Fenzl) Coode & Cullen) - 2
hirta L. (*D. daurica* DC., *D. glabella* Pursh) - 1, 3, 4, 5

- var. *leiocarpa* Regel f. *parviflora* Regel = D. parviflora
hispida Willd. - 2
hissarica Lipsky - 6
huetii Boiss. - 2, 6
hyperborea (L.) Desv. (? *D. hyperborea* subsp. *platytricha*
Hult., *Nesodraba hyperborea* (L.) Jurtz.) - 5
- subsp. *platytricha* Hult. = D. hyperborea
imeretica (Rupr.) Rupr. - 2
incana L. - 1
incompta Stev. - 2
incurvata A. Vassil. & Golosk. = D. cana
insularis Pissjauk. (*D. valida* Pissjauk. 1956, non Goodding,
1904) - 1
juvenilis Kom. (*D. longipes* Raup) - 4, 5
kamtschatica (Ledeb.) N. Busch (*D. frigida* var. *kamtschatica* Ledeb., *D. camtschatica* (Cham. & Schlecht.)
Andrz. ex Tolm., *D. lonchocarpa* Rydb. subsp. *kamtschatica* (Ledeb.) Calder & Taylor) - 4, 5
kjellmanii Lid ex Ekman - 1
koeiei Rech. fil. - 6
korshinskyi (O. Fedtsch.) Pohle - 6
kuramensis Junussov - 6
kurilensis (Turcz.) Fr. Schmidt = D. borealis
kusnetzowii (Turcz. ex Ledeb.) Hayek (*Holargidium kusnetzowii* Turcz. ex Ledeb.) - 3, 4, 5
lactea Adams (*D. pseudopilosa* Pohle) - 1, 3, 4, 5
lanceolata auct. = D. cana
lasiophylla Royle - 6
lipskyi Tolm. - 6
lonchocarpa Rydb. - 4, 5
- subsp. *kamtschatica* (Ledeb.) Calder & Taylor = D.
kamtschatica
longipes Raup = D. juvenilis
longisiliqua Schmalh. - 2
macrocarpa Adams (*D. barbata* Pohle, *D. eschscholtzii*
Pohle ex N. Busch, *D. corymbosa* auct.) - 1, 3, 4, 5
magadanensis Berkutenko & A. Khokhr. - 5
magna (N. Busch) Tolm. = Arabis sachokiana
majae Berkutenko & A. Khokhr. - 5
melanopus Kom. - 6
meskhetica Chinth. - 2
microcarpella A. Vassil. & Golosk. - 6
micropetala sensu Tolm. = D. pauciflora
mingrelica Schischk. - 2
mollissima Stev. - 2
mongolica Turcz. - 4, 5
muralis L. - 1, 2
nemorosa L. - 1, 2, 3, 4, 5, 6
nivalis Liljebl. - 1, 3, 4, 5
norvegica Gunn. - 1
oblongata auct. = D. pauciflora
oblongata R. Br. subsp. *minuta* Petrovsky = D. pauciflora
subsp. minuta
ochroleuca Bunge - 3, 4, 5, 6
odudiana Lipsky - 6
olgae Regel & Schmalh. - 6
olympica Sibth. ex DC. (*D. bruniifolia* Stev. subsp. *olympica*
(Sibth. ex DC.) Coode & Cullen) - 2(?)
oreades Schrenk (*D. algida* Adams var. *brachycarpa* Bunge,
D. alpina L. subsp. *brachycarpa* (Bunge) Malysch.) - 3,
4, 6
ossetica (Rupr.) Somm. & Levier - 2
palanderiana Kjellm. (*D. caesia* auct.) - 4, 5
pamirica (O. Fedtsch.) Pohle - 6
parryoides Cham. = Ermania parryoides
parviflora (Regel) O.E. Schulz (*D. hirta* L. var. *leiocarpa*
Regel f. *parviflora* Regel) - 6

parvisiliquosa Tolm. = D. cinerea

pauciflora R. Br. (*D. adamsii* Ledeb., *D. micropetala* sensu Tolm., *D. oblongata* auct.) - 1, 3, 4, 5

- subsp. **minuta** (Petrovsky) Czer. comb. nova (*D. oblongata* R. Br. subsp. *minuta* Petrovsky, 1981, Bot. Zhurn. 66, 3 : 385) - 5

physocarpa Kom. - 6

pilosa DC. - 4, 5

pohlei Tolm. - 4

polytricha Ledeb. - 2

primuloides Turcz. ex N. Busch - 4

prozorowskii Tolm. - 4

x **pseudonivalis** N. Busch. - D. fladnizensis Wulf. x D. nivalis Liljebl.

pseudopilosa Pohle = D. lactea

pygmaea Turcz. ex N. Busch - 4

rosularis Boiss. (*D. aucheri* Boiss.) - 6

sachalinensis (Fr. Schmidt) Trautv. (*D. borealis* DC. var. *sachalinensis* Fr. Schmidt) - 5

sambukii Tolm. - 4

scabra C.A. Mey. - 2

sibirica (Pall.) Thell. - 1, 2, 3, 4, 5, 6

siliquosa Bieb. (*D. subglabra* (Rupr.) Tolm.) - 2

- subsp. **carinthiaca** (Hoppe) O. Bolos & Vigo = D. carinthiaca

stenocarpa Hook. fil. & Thoms. - 6

stenopetala Trautv. (*D. densifolia* auct.) - 5

stylaris J. Gay ex Koch - 2

subamplexicaulis C.A. Mey. - 3, 4, 6

subcapitata Simm. - 1, 3, 4, 5

subfladnizensis Kuvajev - 4

subglabra (Rupr.) Tolm. = D. siliquosa

subsecunda Somm. & Levier - 2

supranivalis Rupr. - 2

supravillosa A. Khokhr. - 5

taimyrensis Tolm. - 4

talassica Pohle - 6

tibetica Hook. fil. & Thoms. - 6

tichomirovii Ju. Kozhevn. - 5

tschuktschorum auct. p.p. = D. aleutica

turczaninowii Pohle & N. Busch - 3, 4, 5, 6

ussuriensis Pohle (*D. villosula* Tolm.) - 5

valida Pissjauk. = D. insularis

villosula Tolm. = D. ussuriensis

vvedenskyi Kovalevsk. - 6

yunussovii Tolm. - 6

Drabopsis C. Koch

nuda (Belanger) Stapf (*Arabis nuda* Belanger, *Arabidopsis verna* (C. Koch) N. Busch, *Drabopsis verna* C. Koch) - 2, 6

verna C. Koch = D. nuda

Eremoblastus Botsch.

caspicus Botsch. - 6

Ermania Cham. ex Botsch.

crassifolia (Cambess.) Ovcz. & Junussov = Christolea crassifolia

flabellata (Regel) O.E. Schulz = Oreoblastus flabellatus

incana (Ovcz.) Botsch. = Oreoblastus incanus

linearis (N. Busch) Botsch. = Oreoblastus linearis

microcarpa (Ledeb.) Dvorak = Pachyneurum grandiflorum

pamirica (Korsh.) Ovcz. & Junussov = Christolea pamirica

parryoides (Cham.) Botsch. (*Draba ? parryoides* Cham.,

Christolea parryoides (Cham.) N. Busch, *Smelowskia parryoides* (Cham.) Polun.) - 5

saposhnikovii A. Vassil. = Oreoblastus saposhnikovii

Erophila DC.

krockeri Andrz. (*E. verna* (L.) Chevall. var. *krockeri* (Andrz.) Aschers. & Graebn., *E. verna* var. *krockeri* (Andrz.) Diklic, comb. superfl., *E. verna* var. *krockeri* (Andrz.) Stojan. & Stef. comb. superfl.) - 1

minima C.A. Mey. - 2, 6

praecox (Stev.) DC. (*E. verna* (L.) Chevall. var. *praecox* (Stev.) Diklic, *E. verna* subsp. *praecox* (Stev.) Walters) - 1, 2, 6

spathulata Lang (*E. verna* (L.) Chevall. subsp. *spathulata* (Lang) Walters) - 1

verna (L.) Bess. (*E. verna* (L.) Chevall. comb. superfl.) - 1, 2, 6

verna (L.) Chevall. = E. verna (L.) Bess.

- subsp. *praecox* (Stev.) Walters = E. praecox

- subsp. *spathulata* (Lang) Walters = E. spathulata

- var. *krockeri* (Andrz.) Aschers. & Graebn. = E. krockeri

- var. *krockeri* (Andrz.) Diklic = E. krockeri

- var. *krockeri* (Andrz.) Stojan. & Stef. = E. krockeri

- var. *praecox* (Stev.) Diklic = E. praecox

Eruca Hill

sativa Mill. (*Eruca vesicaria* (L.) Cav. subsp. *sativa* (Mill.) Thell.) - 1, 2, 3, 5, 6

vesicaria (L.) Cav. subsp. *sativa* (Mill.) Thell. = E. sativa

Erucastrum C. Presl

armoracioides (Czern. ex Turcz.) Cruchet (*Brassica armoracioides* Czern. ex Turcz., *B. elongata* Ehrh. subsp. *armoracioides* (Czern ex Turcz.) Aschers. & Graebn., *B. elongata* subsp. *intergrifolia* (Boiss.) Breistoffer, *B. elongata* auct.) - 1, 2, 3, 6

cretaceum Kotov (*Brassica elongata* Ehrh. var. *pinnatifida* Schmalh., *B. elongata* subsp. *pinnatifida* (Schmalh.) Greuter & Burdet) - 1

gallicum (Willd.) O.E. Schulz (*Sisymbrium gallicum* Willd.) - 1, 5(alien)

nasturtiifolium (Poir.) O.E. Schulz (*Sinapis nasturtiifolia* Poir.) - 1

Erysimastrum (DC.) Rupr. = Erysimum

lazistanicum Rupr. = Erysimum lazistanicum

Erysimum L. (*Acachmaena* H.P. Fuchs, *Erysimastrum* (DC.) Rupr., *Syreniopsis* H.P. Fuchs, 1959, non Jaub. & Spach, 1842, *Cuspidaria* auct.)

aksaricum Pavl. - 6

alaicum Novopokr. ex E. Nikit. - 6

altaicum C.A. Mey. = E. flavum

amurense Kitag. (*E. aurantiacum* (Bunge) Maxim, 1889, non Leyb. 1855, *E. aurantiacum* var. *amurense* (Kitag.) Y.L. Chang) - 4, 5

andrzejovskianum Bess. = E. canescens

arcuatum Opiz ex J. & C. Presl = Barbarea vulgaris

argyrocarpum N. Busch - 2

armeniacum (Sims) J. Gay (*Cheiranthus armeniacum* Sims) - 2

■ **artvinense** N. Busch

aurantiacum (Bunge) Maxim. = E. amurense
- var. *amurense* (Kitag.) Y.L. Chang = E. amurense
aureum Bieb. (*E. sylvaticum* Bieb.) - 1, 2
azerbaidzanicum M. Kassumov = E. chazarjurti
babadagense Prima - 2
babataghi Korsh. - 6
badghysi (Korsh.) Lipsky - 6
boreale C.A. Mey. - 5
brachycarpum Boiss. - 2
brevistylum Somm. & Levier - 2
buschii M. Kassumov = E. subulatum
callicarpum Lipsky - 2
canescens Roth (*E. andrzejovskianum* Bess., *E. diffusum* auct.) - 1, 2, 3, 4, 5(?), 6
canum (Pill. & Mill.) Polatschek = Syrenia cana
caspicum N. Busch - 2
caucasicum Trautv. - 2
chazarjurti N. Busch (*E. azerbaidzanicum* M. Kassumov, *E. iljinii* M. Kassumov, *E. nachyczevanicum* M. Kassumov) - 2
cheiranthoides L. (*E. cheiranthoides* subsp. *altum* Ahti) - 1, 2(alien), 3, 4, 5, 6
- subsp. *altum* Ahti = E. cheiranthoides
chrysanthum Botsch. & Vved. - 6
clausioides Botsch. & Vved. - 6
collinum (Bieb.) Andrz. - 2
contractum Somm. & Levier - 2
crassipes Fisch. & C.A. Mey. (*E. transcaucasicum* M. Kassumov) - 2
cretaceum (Rupr.) Schmalh. - 1
croceum M. Pop. - 6
cuspidatum (Bieb.) DC. (*Acachmena cuspidata* (Bieb.) H.P. Fuchs, *Cuspidaria cuspidata* (Bieb.) Takht., *Syrenia cuspidata* (Bieb.) Reichenb., *Syreniopsis cuspidata* (Bieb.) H.P. Fuchs) - 1, 2
cyaneum M. Pop. - 6
czernjajevii N. Busch - 3, 6
diffusum auct. = E. canescens
durum J. & C. Presl = E. hieracifolium
epikeimenum N. Busch = E. samarkandicum
exaltatum Andrz. (*E. hieracifolium* L. subsp. *exaltatum* (Andrz.) Borza) - 1
feodorovii M. Kassumov = E. leptophyllum
ferganicum Botsch. & Vved. (*E. microtrichon* Botsch. & Vved.) - 6
flavum (Georgi) Bobr. (*Hesperis flava* Georgi, *Erysimum altaicum* C.A. Mey.) - 3, 4, 6
franchetii N. Busch = E. samarkandicum
gaudanense Litv. = Prionotrichon gaudanense
gelidum Bunge - 2
- subsp. *krynitzkii* (Bordz.) V.I. Dorof. = E. krynitzkii
grubovii Botsch. - 3
gypsaceum Botsch. & Vved. - 6
hajastanum Wissjul. & Bordz. = E. leptophyllum
hieracifolium L. (*E. durum* J. & C. Presl, *E. hieracifolium* subsp. *durum* (J. & C. Presl) Celak., *E. marschallianum* Andrz., *E. strictum* Gaertn., Mey. & Scherb.) - 1, 3, 4, 5, 6
- subsp. *durum* (J. & C. Presl) Celak. = E. hieracifolium
- subsp. *exaltatum* (Andrz.) Borza = E. exaltatum
humillimum (C.A. Mey.) N. Busch - 3, 6
hungaricum Zapal. - 1
ibericum (Adams) DC. - 2
iljinii M. Kassumov = E. chazarjurti
inense N. Busch - 3
ischnostylum Freyn & Sint. (*E. persepolitanum* auct.) - 6
jodonyx Botsch. & Vved. - 6

kazachstanicum Botsch. - 6
kerbabaevii Kurbanov & Gudkova - 6
komarovii M. Kassumov = E. subulatum
krynitzkii Bordz. (*E. gelidum* Bunge subsp. *krynitzkii* (Bordz.) V.I. Dorof., *E. subnivale* Prima) - 2
krynkense Lavr. (*Acachmena krynkensis* (Lavr.) H.P. Fuchs, *Syreniopsis krynkensis* (Lavr.) H.P. Fuchs) - 1
lazistanicum (Rupr.) Lipsky (*Erysimastrum lazistanicum* Rupr., *Erysimum pulchellum* (Willd.) J. Gay var. *grandiflorum* (Trautv.) Boiss., *Sisymbrium pulchellum* (Willd.) Trautv. var. *grandiflorum* Trautv.) - 2
leptophyllum (Bieb.) Andrz. (*E. feodorovii* M. Kassumov, *E. hajastanum* Wissjul. & Bordz.) - 2
leptostylum DC. - 1, 2
leucanthemum (Steph.) B. Fedtsch. (*E. passgalense* Boiss.) - 2
lilacinum Steinb. - 2
marschallianum Andrz. = E. hieracifolium
meyerianum (Rupr.) N. Busch (*E. substrigosum* (Rupr.) N. Busch subsp. *meyerianum* (Rupr.) V.I. Dorof.) - 2
michaelis Adylov (*E. popovii* Botsch. & Vved. 1948, non Rothm. 1940) - 6
microtrichon Botsch. & Vved. = E. ferganicum
nabievii Adylov - 6
nachyczevanicum M. Kassumov = E. chazarjurti
nuratense M. Pop. ex Botsch. & Vved. - 6
odoratum Ehrh. = E. pannonicum
pallasii (Pursh) Fern. - 1, 3, 4, 5
pannonicum Crantz (*E. odoratum* Ehrh.) - 1
passgalense Boiss. = E. leucanthemum
persepolitanum auct. = E. ischnostylum
persicum Boiss. = E. subulatum
popovii Botsch. & Vved. = E. meyerianum
pulchellum (Willd.) J. Gay - 2
- var. *grandiflorum* (Trautv.) Boiss. = E. lazistanicum
repandum L. - 1, 2, 5, 6
samarkandicum M. Pop. (*E. epikeimenum* N. Busch, *E. franchetii* N. Busch) - 6
sisymbrioides C.A. Mey. - 1, 2, 3, 6
strictisiliquum N. Busch - 2
strictum Gaertn., Mey. & Scherb. = E. hieracifolium
subnivale Prima = E. krynitzkii
substrigosum (Rupr.) N. Busch - 2
- subsp. *meyerianum* (Rupr.) V.I. Dorof. = E. meyerianum
subulatum J. Gay (*E. buschii* M. Kassumov, *E. komarovii* M. Kassumov, *E. persicum* Boiss.) - 2
sylvaticum Bieb. = E. aureum
szowitsianum Boiss. - 2
transcaucasicum M. Kassumov = E. crassipes
transiliense M. Pop. - 6
transsilvanicum Schur = E. witmannii
ucranicum J. Gay - 1, 2
vernum Mill. = Barbarea verna
versicolor (Bieb.) Andrz. - 1, 2, 3, 6
violascens M. Pop. - 6
vitellinum M. Pop. - 6
wagifii M. Kassumov - 2
witmannii Zawadzki (*E. transsilvanicum* Schur, *E. witmannii* subsp. *transsilvanicum* (Schur) P.W. Ball) - 1
- subsp. *transsilvanicum* (Schur) P.W. Ball = E. witmannii

Euclidium R. Br.

syriacum (L.) R. Br. - 1, 2, 3, 5(alien), 6
tenuissimum (Pall.) B. Fedtsch. = Litwinowia tenuissima

Eunomia DC.

rotundifolia C.A. Mey. - 2

Eutrema R. Br.

compactum O.E. Schulz = E. edwardsii
cordifolium Turcz. ex Ledeb. - 4
edwardsii R. Br. (*E. compactum* O.E. Schulz) - 1, 3, 4, 5, 6
- var. *parviflorum* (Turcz. ex Ledeb.) N. Busch = E. parviflorum
- var. *septigerum* (Bunge) N. Busch = Thlaspi septigerum
integrifolium (DC.) Bunge - 3, 6
japonicum (Miq.) Koidz. (*Lunaria* ? *japonica* Miq., *Cochlearia* ? *wasabii* Siebold, nom. nud., *Eutrema wasabii* Maxim.) - 5
parviflorum Turcz. ex Ledeb. (*E. edwardsii* R. Br. var. *parviflorum* (Turcz. ex Ledeb.) N. Busch) - 4
pseudocordifolium M. Pop. - 6
septigerum Bunge = Thlaspi septigerum
wasabii Maxim. = E. japonicum

Farsetia auct. = Fibigia and Pterygostemon

eriocarpa DC. = Fibigia eriocarpa
macrocarpa Boiss. = Fibigia macrocarpa
macroptera Kotschy & Boiss. ex Fourn. = Fibigia macrocarpa
spathulata Kar. & Kir. = Pterygostemon spathulatus

Fedtschenkoa Regel & Schmalh. = Strigosella

africana (L.) Dvorak = Strigosella africana
brevipes (Bunge) Dvorak = Strigosella brevipes
brevipes (Kar. & Kir.) Dvorak, p.p. = Strigosella brevipes
bucharica (Vass.) Dvorak = Strigosella grandiflora
circinata (Bunge) Dvorak = Strigosella circinata
grandiflora (Bunge) Dvorak = Strigosella grandiflora
hispida (Litv.) Dvorak = Strigosella hispida
malacotricha (Botsch. & Vved.) Dvorak = Strigosella malacotricha
multisiliqua (Vass.) Dvorak = Strigosella scorpioides
pamirica (Botsch. & Vved.) Dvorak = Strigosella strigosa
scorpioides (Bunge) Dvorak = Strigosella scorpioides
spryginoides (Botsch. & Vved.) Dvorak = Strigosella spryginioides
stenopetala (Bernh. ex Fisch. & C.A. Mey.) Dvorak = Strigosella stenopetala
taraxacifolia (Balb.) Dvorak = Strigosella africana and S. intermedia
turkestanica Regel & Schmalh. = Strigosella grandiflora

Fibigia Medik. (*Farsetia* auct.)

clypeata (L.) Medik. - 1, 2
eriocarpa (DC.) Boiss. (*Farsetia eriocarpa* DC.) - 2
macrocarpa (Boiss.) Boiss. (*Farsetia macrocarpa* Boiss., *F. macroptera* Kotschy & Boiss. ex Fourn., *Fibigia macroptera* (Kotschy & Boiss. ex Fourn.) Boiss.) - 2
macroptera (Kotschy & Boiss. ex Fourn.) Boiss. = F. macrocarpa
suffruticosa (Vent.) Sweet - 2, 6

Gagria M. Kral

lobata M. Kral - 2

Galitzkya V. Boczantzeva (*Hormathophylla* Cullen & T. R. Dudley, p.p.)

spathulata (Steph.) V. Boczantzeva (*Berteroa spathulata* (Steph.) C.A. Mey., *Hormathophylla spathulata* (Steph.) Cullen & T.R. Dudley) - 3, 6

Goldbachia DC.

ikonnikovii Vass. = G. laevigata
laevigata (Bieb.) DC. (*G. ikonnikovii* Vass., *G. laevigata* var. *ikonnikovii* (Vass.) Kuan & Y.C. Ma, *G. reticulata* (O. Kuntze) Vass.) - 1, 3, 6
- var. *ikonnikovii* (Vass.) Kuan & Y.C. Ma = G. laevigata
papillosa Vass. = G. tetragona
pendula Botsch. - 1, 2, 6
reticulata (O. Kuntze) Vass. = G. laevigata
tetragona Ledeb. (*G. papillosa* Vass.) - 2, 6
torulosa DC. - 2, 6
verrucosa Kom. - 6

Gorodkovia Botsch. & Karav.

jacutica Botsch. & Karav. - 4, 5

Graellsia Boiss. (*Physalidium* Fenzl)

graellsiifolia (Lipsky) Poulter (*Physalidium graellsiifolium* Lipsky) - 6
hissarica Junussov - 6
saxifragifolia (DC.) Boiss. - 6

Guepinia Bast = Teesdalia

coronopifolia (J.P. Bergeret) Sojak = Teesdalia coronopifolia
nudicaulis (L.) Bast = Teesdalia nudicaulis

Guillenia auct. = Microsisymbrium

minutiflora (Hook. fil. & Thoms.) S.S.R. Bennet = Microsisymbrium minutiflorum

Hedinia Ostenf. (*Hediniopsis* Botsch. & Petrovsky)

altaica Pobed. (*Smelowskia altaica* (Pobed) Botsch.) - 3
czukotica (Botsch. & Petrovsky) Jurtz., Korobk. & Balandin (*Hediniopsis czukotica* Botsch. & Petrovsky) - 5
■ mongolica (Kom.) Veliczkin (*Sophiopsis mongolica* (Kom.) N.Busch)
tibetica (Thoms.) Ostenf. - 6

Hediniopsis Botsch. & Petrovsky = Hedinia

czukotica Botsch. & Petrovsky = Hedinia czukotica

Heldreichia Boiss. = Stroganowia

longifolia Boiss. = Stroganowia longifolia
silaifolia Hook. fil. & Thoms. = Winklera silaifolia

Hesperis L. (*Deilosma* Andrz.)

adenosepala Borb. = H. sibirica
adzharica Tzvel. (*H. matronalis* L. subsp. *adzharica* (Tzvel.) Cullen) - 2
alyssifolia DC. = Matthiola alyssifolia
armena auct. = H. buschiana, H. transcaucasica and H. voronovii
■ bicuspidata (Willd.) Poir. (*H. violacea* Boiss.)
■ buschiana Tzvel. (*H. armena* auct. p.p.)
candida Kit. ex Muggenb., Kanitz & Knapp (*H. matronalis* L. subsp. *candida* (Kit. ex Muggenb., Kanitz & Knapp) Hegi & E. Schmid, ? *H. nivea* sensu Kotov) - 1
caucasica Rupr. = H. matronalis and H. voronovii
cinerea (Dvorak) Dvorak (*H. steveniana* DC. subsp. *cinerea*

Dvorak) - 2

crenulata C.A. Mey. = Zuvanda meyeri

elata Hornem. = H. sibirica

flava Georgi = Erysimum flavum

hirsutissima (N. Busch) Tzvel. (*H. matronalis* L. var. *hirsutissima* N. Busch., *H. matronalis* subsp. *hirsutissima* (N. Busch) V. Avet.) - 2

hyrcana Bornm. & Gauba (*H. unguicularis* Boiss. subsp. *hyrcana* (Bornm. & Gauba) Dvorak) - 2(?)

karsiana N. Busch = H. persica

matronalis L. (*H. caucasica* Rupr. p.p.) - 1, 2, 5(alien)

- subsp. *adzharica* (Tzvel.) Cullen = H. adzharica

- subsp. *candida* (Kit. ex Muggenb., Kanitz & Knapp) Hegi & E. Schmid = H. candida

- subsp. *hirsutissima* (N. Busch) V. Avet. = H. hirsutissima

- subsp. *sibirica* (L.) Hayek = H. sibirica

- subsp. *voronovii* (N. Busch) P.W. Ball = H. voronovii

- var. *hirsutissima* N. Busch. = H. hirsutissima

- var. *robusta* N. Busch = H. robusta

meyeriana (Trautv.) N. Busch = H. voronovii

nivea Baumg. - 1(?)

nivea sensu Kotov = H. candida

persica Boiss. (*H. karsiana* N. Busch) - 2, 6

pontica Zapal. = H. sibirica

pseudonivea Tzvel. - 3, 4, 6

pycnotricha Borb. & Degen - 1, 2, 3

renifolia Boiss. = Chalcanthus renifolius

∎ robusta (N. Busch) Tzvel. (*H. matronalis* L. var. *robusta* N. Busch)

sibirica L. (*H. adenisepala* Borb., *H. elata* Hornem., *H. matronalis* L. subsp. *sibirica* (L.) Hayek, *H. pontica* Zapal., *H. sylvestris* auct. p.p.) - 1, 3, 4, 6

steveniana DC. - 1

- subsp. *cinerea* Dvorak = H. cinerea

suaveolens (Andrz.) Steud. (*Deilosma suaveolens* Andrz., *Hesperis sylvestris* auct. p.p.) - 1

sylvestris auct. = H. sibirica, H. suaveolens and H. voronovii

∎ sylvestris Crantz

transcaucasica Tzvel. (*H. armena* auct. p.p., *H. unguicularis* auct. p.p.) - 2

tristis L. - 1, 2

unguicularis auct. p.p. = H. transcaucasica

unguicularis Boiss. subsp. *hyrcana* (Bornm. & Gauba) Dvorak = H. hyrcana

verna L. = Arabis verna

violacea Boiss. = H. bicuspidata

voronovii N. Busch (*H. caucasica* Rupr. p.p., *H. matronalis* L. subsp. *voronovii* (N. Busch) P.W. Ball, *H. meyeriana* (Trautv.) N. Busch, *H. armena* auct. p.p., *H. sylvestris* auct. p.p.) - 1, 2

Hirschfeldia Moench

incana (L.) Lagr.-Foss. - 1, 2, 5(alien), 6

- subsp. **geniculata** (Desf.) Tzvel. (*Sinapis geniculata* Desf.) - 2

- subsp. **leptocarpa** Tzvel. - 1, 2

Holargidium Turcz. ex Ledeb. = Draba

kusnetzowii Turcz. ex Ledeb. = Draba kusnetzowii

Hormathophylla Cullen & T.R. Dudley, p.p. = Galitzkya

spathulata (Steph.) Cullen & T.R. Dudley = Galitzkya spathulata

Hornungia Reichenb. (*Hutchinsia* R. Br. p.p. nom. illegit.)

petraea (L.) Reichenb. (*Hutchinsia petraea* (L.) R. Br.) - 1

Hutchinsia R. Br. p.p. = Hornungia and Noccaea

bifurcata Ledeb. = Smelowskia bifurcata

hastulata DC. = Noccidium hastulatum

pectinata Bunge = Smelowskia pectinata

petraea (L.) R. Br. = Hornungia petraea

Hylandra A. Löve

x **suecica** (Fries) A. Löve (*Arabis suecica* Fries, *Arabidopsis suecica* (Fries) Norrl., *Cardaminopsis suecica* (Fries) Hiit.). - Arabidopsis thaliana (L.) Heynh. x Cardaminopsis arenosa (L.) Hayek - 1

Hymenolobus Nutt. ex Torr. & Gray

austriaca (Jacq.) Sweet - 2

procumbens (L.) Fourr. - 1, 2, 3, 6

puberulus (Rupr.) N.Busch - 2

Hymenophysa C.A. Mey. = Cardaria

macrocarpa Franch. = Cardaria macrocarpa

pubescens C.A. Mey. = Cardaria pubescens

Iberidella Boiss.

caespitosa Boiss. (*Aethionema caespitosum* (Boiss.) Boiss.) - 2

heterophylla Boiss. & Buhse (*Aethionema heterophyllum* (Boiss. & Buhse) Boiss.) - 2

sagittata Boiss. (*Aethionema sagittatum* (Boiss.) Boiss.) - 2, 6

trinervia (DC.) Boiss. (*Aethionema salmasium* Boiss., *Ae. trinervium* (DC.) Boiss.) - 2

Iberis L.

amara L. - 1, 2

oppositifolia Pers. = Aethionema oppositifolium

oschtenica Charkev. = I. taurica

pinnata L. - 1

pumila Stev. = Noccaea pumila

saxatilis L. - 1

simplex auct. p.p. = I. taurica

taurica DC. (*I. oschtenica* Charkev., *I. simplex* auct. p.p.) - 1, 2

*****umbellata** L. - 1

Isatis L.

aleppica Scop. = I. lusitanica

anceps N. Busch = I. steveniana

apscheronica N. Busch - 2

araratica Rupr. = I. kozlowskyi

arnoldiana N. Busch - 2, 6

atropatana Grossh. = I. steveniana

besseri sensu N. Busch = I. steveniana

boissierana Reichenb. fil. (*Sameraria boissieriana* (Reichenb. fil.) Nabiev) - 6

brachycarpa C.A. Mey. (*I. somchetica* Bordz.) - 2

brevipes (Bunge) Jafri = Pachypterygium brevipes
bullata Aitch. & Hemsl. (*Sameraria aitchisonii* (Korsh.) B. Fedtsch., *S. bullata* (Aitch. & Hemsl.) B. Fedtsch., *S. bullata* var. *aitchisonii* (Korsh.) Botsch.) - 6
bungeana Seidl. - 2
buschiana Karjag. = I. karjaginii
buschiana Schischk. (*I. collina* auct., *I. glauca* auct.) - 2
buschiorum Hovh. - 2
campestris Stev. ex DC. (*I. maeotica* DC.) - 1, 2
canaliculata (Vass.) V. Boczantzeva (*Sameraria canaliculata* Vass.) - 6
canescens auct. = I. tomentella
cappadocica Desv. subsp. *steveniana* (Trautv.) P.H. Davis = I. steveniana
- subsp. *subradiata* (Rupr.) P.H. Davis = I. subradiata
caucasica (Rupr.) N. Busch - 2
collina auct. = I. buschiana
costata C.A. Mey. (*I. lasiocarpa* Ledeb.) - 1, 3, 4, 6
deserti (N. Busch) V. Boczantzeva (*Sameraria deserti* N. Busch) - 6
emarginata Kar. & Kir. - 6
frutescens Kar. & Kir. - 3
funebris Fed. - 6
glauca auct. = I. buschiana
grossheimii N. Busch = I. tomentella
hirtocalyx Franch. (*Sameraria hirtocalyx* (Franch.) Nabiev) - 6
iberica Stev. - 2
jacutensis (N. Busch) N. Busch - 4
japonica auct. = I. tinctoria
karjaginii Schischk. (*I. buschiana* Karjag. 1938, non Schischk. 1928) - 2
kozlowskyi Grossh. (? *I. araratica* Rupr. nom. dubium) - 2
laevigata Trautv. - 1, 3, 6
lasiocarpa Ledeb. = I. costata
latisiliqua Stev. - 2
leuconeura Boiss. & Buhse - 6
littoralis Stev. ex DC. - 1
lusitanica L. (*I. allepica* Scop.) - 6
maeotica DC. = I. campestris
maxima Pavl. - 6
minima Bunge - 6
multicaulis (Kar. & Kir.) Jafri = Pachypterygium multicaule
nummularia sensu Grossh. = I. steveniana
nummularia Trautv. (*I. psilocarpa* sensu N. Busch) - 2
oblongata DC. - 4
- var. *yezoensis* (Ohwi) Y.L. Chang = I. tinctoria var. yezoensis
ornithorhynchus N. Busch - 2
pavlovii Jarm. ex Pavl. = I. violascens
praecox Kit. ex Tratt. - 1, 2, 3, 6
psilocarpa Ledeb. = I. steveniana
psilocarpa sensu N. Busch = I. nummularia
reticulata C.A. Mey. - 2
sabulosa Stev. ex Ledeb. - 1, 2
sevangensis N. Busch - 2
somchetica Bordz. = I. brachycarpa
steveniana Trautv. (*I. anceps* N. Busch, *I. atropatana* Grossh., *I. cappadocica* Desv. subsp. *steveniana* (Trautv.) P.H. Davis, *I. psilocarpa* Ledeb. non sensu N. Busch, *Sameraria litvinovii* N. Busch, *Isatis besseri* sensu N. Busch, *I. nummularia* sensu Grossh.) - 2
■ subdidyma (N. Busch) V. Avet. (*Stubendorffia subdidyma* N. Busch)
subradiata Rupr. (*I. cappadocica* Desv. subsp. *subradiata* (Rupr.) P.H. Davis, *I. tphilisiensis* Grossh.) - 2
takhtadjanii V. Avet. - 2

taurica Bieb. - 1, 2
tinctoria L. (*I. japonica* auct.) - 1, 2, 5, 6
- subsp. *tomentella* (Boiss. & Bal.) P.H. Davis = I. tomentella
- var. **yesoensis** (Ohwi) Ohwi (*I. yezoensis* Ohwi, *I. oblongata* DC. var. *yezoensis* (Ohwi) Y.L. Chang) - 5
tomentella Boiss. & Bal. (*I. grossheimii* N. Busch, *I. tinctoria* L. subsp. *tomentella* (Boiss. & Bal.) P.H. Davis, *I. canescens* auct.) - 1, 2
tphilisiensis Grossh. = I. subradiata
trachycarpa Trautv. - 6
turcomanica Korsh. (*Sameraria turcomanica* (Korsh.) B. Fedtsch.) - 6
villarsii Gaudin - 1
violascens Bunge (*I. pavlovii* Jarm. ex Pavl.) - 6
yezoensis Ohwi = I. tinctoria var. yesoensis

Iskandera N. Busch

alaica (Korsh.) Botsch. & Vved. (*Matthiola albicaulis* Boiss. var. *alaica* Korsh.) - 6
hissarica N. Busch - 6

Koeiea auct. p.p. = Botschantzevia

karatavica (Lipsch.) Rech. fil. = Botschantzevia karatavica

Konig Adans. = Lobularia

Koniga auct. = Lobularia

maritima (L.) R. Br. = Lobularia maritima

Kotschyella F.K. Mey. (*Carposeras* sensu N. Busch, p.p.)

stenocarpa (Boiss.) F.K. Mey. (*Carpoceras stenocarpum* Boiss., *Thlaspi stenocarpum* (Boiss.) Hedge) - 6

Lachnoloma Bunge

lehmannii Bunge - 6

Leavenworthia Torr.

torulosa Gray - 1

Leiospora (C.A. Mey) Dvorak (*Parrya* auct. p.p.)

bellidifolia (Danguy) Botsch. & Pachom. (*Parrya bellidifolia* Danguy) - 6
crassifolia (Botsch. & Vved.) A. Vassil. (*Parrya crassifolia* Botsch. & Vved.) - 6
eriocalyx (Regel & Schmalh.) Dvorak (*Parrya eriocalyx* Regel & Schmalh.) - 6
exscapa (C.A. Mey.) Dvorak (*Parrya exscapa* C.A. Mey., *Leiospora exscapa* (C.A. Mey.) A. Vassil. comb. superfl.) - 3, 6
pamirica (Botsch. & Vved.) Botsch. & Pachom. (*Parrya pamirica* Botsch. & Vved.) - 6
subscapigera (Botsch. & Vved.) Botsch. & Pachom. (*Parrya subscapigera* Botsch. & Vved., *P. seravschanica* A. Vassil.) - 6

Lepidium L.

affine Ledeb. (*L. sibiricum* Schweigg. 1812, non Pall. 1776) - 1(alien), 3, 6
amplexicaule Willd. - 3, 6

apetalum auct. = L. densiflorum
aucheri Boiss. (*L. borsczovii* N. Busch) - 6
boisseri N. Busch = Cardaria boissieri
bonariense L. - 1
borsczovii N. Busch = L. aucheri
borysthenicum Kleop. = L. crassifolium
bupleuroides Rech. fil. = Stroganowia bupleuroides
campestre (L.) R. Br. - 1, 2
cartilagineum (J. Mayer) Thell. (*Thlaspi cartilagineum* J. Mayer) - 1, 2, 3, 6
- subsp. *crassifolium* (Waldst. & Kit.) Thell. = L. crassifolium
- subsp. *pumilum* (Boiss. & Bal.) Hedge = L. pumilum
cordatum Willd. ex DC. - 3, 5(alien), 6
coronopifolium Fisch. ex Ledeb. (*L. lyratum* L. subsp. *coronopifolium* (Fisch. ex Ledeb.) Thell.) - 1, 3, 6
crassifolium Waldst. & Kit. (*L. borysthenicum* Kleop., *L. cartilagineum* (J. Mayer) Thell. subsp. *crassifolium* (Waldst. & Kit.) Thell.) - 1, 2, 3, 6
densiflorum Schrad. (*L. apetalum* auct.) - 1, 3, 4, 5, 6
deserti Pavl. - 6
draba L. = Cardaria draba
eremophilum Schrenk - 3
ferganense Korsh. - 6
graminifolium L. - 1
jarmolenkoi R. Vinogradova - 6
karataviense Regel & Schmalh. - 6
lacerum C.A. Mey. (*L. lyratum* L. subsp. *lacerum* (C.A. Mey.) Thell., *L. persicum* auct.) - 3, 6
latifolium L. - 1, 2, 3, 5(?), 6
lyratum L. - 2
- subsp. *coronopifolium* (Fisch. ex Ledeb.) Thell. = L. coronopifolium
- subsp. *lacerum* (C.A. Mey.) Thell. = L. lacerum
meyeri Claus - 1
neglectum Thell. - 1
obtusum Basin. - 6
perfoliatum L. - 1, 2, 3, 5(alien), 6
persicum auct. = L. lacerum
pinnatifidum Ledeb. - 1, 2, 6
propinquum Fisch. & C.A. Mey. = Cardaria propinqua
pumilum Boiss. & Bal. (*L. cartilagineum* (J. Mayer) Thell. subsp. *pumilum* (Boiss. & Bal.) Hedge) - 1
ramosissimum Nels. - 1
repens (Schrenk) Boiss. = Cardaria repens
rubtzovii Vass. - 6
ruderale L. - 1, 2, 3, 4, 5, 6
*****sativum** L. - 1, 2, 5, 6
seravschanicum Ovcz. & Junussov - 6
sibiricum Schweigg. = L. affine
songaricum Schrenk - 3, 6
squamatum Forssk. = Coronopus squamatus
stylosum Pers. = Rorippa pyrenaica
subcordatum Botsch. & Vved. - 6
syvaschicum Kleop. - 1
texanum Buckl. - 2
turczaninowii Lipsky - 1
vesicarium L. - 2
virginicum L. - 1, 2, 5

Leptaleum DC.

filifolium (Willd.) DC. - 1, 2, 6

Lesquerella S. Wats.

arctica (Wormsk. ex Hornem.) S. Wats. (*Alyssum arcticum* Wormsk. ex Hornem., *Vesicaria arctica* (Wormsk. ex Hornem.) Richards., *V. leiocarpa* (Trautv.) N. Busch) - 4, 5

Litwinowia Woronow

tenuissima (Pall.) Woronow ex Pavl. (*Euclidium tenuissimum* (Pall.) B. Fedtsch., *Litwinowia tenuissima* (Pall.) N. Busch ex Jarm. comb. superfl., *L. tenuissima* (Pall.) N. Busch. ex Vass. comb. superfl.) - 1, 3, 6

Lobularia Desv. (*Konig* Adans., *Koniga* auct.)

maritima (L.) Desv. (*Clypeola maritima* L., *Koniga maritima* (L.) R. Br.) - 1, 2

Lunaria L.

*****annua** L. - 1, 2
japonica Miq. = Eutrema japonicum
rediviva L. - 1

Macropodium R. Br.

nivale (Pall.) R. Br. - 3, 4, 6
pterospermum Fr. Schmidt - 5

Malcolmia auct. p.p. = Strigosella and Zuvanda

africana (L.) R. Br. = Strigosella africana
- var. *divaricata* Fisch. = Strigosella stenopetala
- var. *stenopetala* Bernh. ex Fisch. & C.A. Mey. = Strigosella stenopetala
boissierana Jafri = Strigosella circinata
bucharica Vass. = Strigosella grandiflora
circinata (Bunge) Boiss. = Strigosella circinata
crenulata (C.A. Mey.) Vass. = Zuvanda meyeri
divaricata (Fisch.) Fisch. = Strigosella stenopetala
grandiflora (Bunge) O. Kuntze = Strigosella grandiflora
hispida Litv. = Strigosella hispida
hyrcanica Freyn & Sint. = Strigosella hyrcanica
intermedia C.A. Mey. = Strigosella intermedia
karelinii Lipsky = Strigosella brevipes
komarovii Vass. = Strigosella grandiflora
laxa (Lam.) DC. = Strigosella africana var. laxa
malacotricha Botsch. & Vved. = Strigosella malacotricha
meyeri Boiss. = Zuvanda meyeri
multisiliqua Vass. = Strigosella scorpioides
pamirica Botsch. & Vved. = Strigosella strigosa
scorpioides (Bunge) Boiss. = Strigosella scorpioides
spryginoides Botsch. & Vved. = Strigosella spryginioides
stenopetala (Bernh. ex Fisch. & C.A. Mey.) Ledeb. = Strigosella stenopetala
strigosa Boiss. = Strigosella strigosa
tadshikistanica Vass. = Strigosella tadshikistanica
taraxacifolia auct. = Strigosella intermedia
taraxacifolia (Balb.) DC. = Strigosella africana
tenuissima Botsch. = Strigosella tenuissima
trichocarpa Boiss. & Buhse = Strigosella trichocarpa
turkestanica Litv. = Strigosella turkestanica

Maresia Pom.

meyeri (Boiss.) Dvorak = Zuvanda meyeri
nana (DC.) Batt. - 2

Matthiola R. Br.

afghanica Rech. fil. & Koie - 6
albicaulis Boiss. = M. alyssifolia
- var. *alaica* Korsh. = Iskandera alaica

alyssifolia (DC.) Bornm. (*Hesperis alyssifolia* DC., *Matthiola albicaulis* Boiss.) - 6

*annua (L.) Sweet (*Cheiranthus annuus* L.) - 1

*bicornis (Sibth. & Smith) DC. (*Cheiranthus bicornis* Sibth. & Smith, *Matthiola longipetala* (Vent.) DC. subsp. *bicornis* (Sibth. & Smith) P.W. Ball) - 1

boissieri Grossh. (*M. odoratissima* (Bieb.) R. Br. subsp. *boissieri* (Grossh.) Takht.) - 2

bucharica Czerniak. - 6

caspica (N. Busch) Grossh. (*M. odoratissima* (Bieb.) R. Br. var. *caspica* N. Busch) - 2

chenopodiifolia Fisch. & C.A. Mey. - 6

czerniakowskae Botsch. & Vved. - 6

daghestanica (Conti) N. Busch - 2

farinosa Bunge. - 6

fragrans Bunge - 1, 3, 6

*incana (L.) R. Br. (*Cheiranthus incanus* L.) - 1

integrifolia Kom. - 6

*longipetala (Vent.) DC. (*Cheiranthus longipetalus* Vent., *Matthiola longipetala* var. *oxyceras* (DC.) Zohary, *M. oxyceras* DC.) - 1

- subsp. *bicornis* (Sibth. & Smith) P.W. Ball = M. bicornis

- var. *oxyceras* (DC.) Zohary = M. longipetala

obovata Bunge (*M. runcinata* Regel) - 6

odoratissima (Bieb.) R. Br. (*Cheiranthus odoratissimus* Pall. ex Bieb., *Matthiola odoratissima* subsp. *taurica* (Conti) Stank., *M. odoratissima* var. *taurica* Conti, *M. taurica* (Conti) Grossh.) - 1, 2

- subsp. *boissieri* (Grossh.) Takht. = M. boissieri

- subsp. *taurica* (Conti) Stank. = M. odoratissima

- var. *caspica* N. Busch = M. caspica

- var. *taurica* Conti = M. odoratissima

oxyceras DC. = M. longipetala

robusta Bunge - 2, 6

runcinata Regel = M. obovata

stoddartii Bunge - 6

superba Conti - 3, 4, 6

tatarica (Pall.) DC. - 1, 3, 6

taurica (Conti) Grossh. = M. odoratissima

tianschanica Sarkisova - 6

Megacarpaea DC.

gigantea Regel - 6

gracilis Lipsky - 6

iliensis Golosk. & Vass. - 6

megalocarpa (Fisch. ex DC.) B. Fedtsch. - 1, 3, 6

mugodzharica Golosk. & Vass. - 6

orbiculata B. Fedtsch. - 6

schugnanica B. Fedtsch. - 6

Megadenia Maxim.

bardunovii M. Pop. - 4

speluncarum Vorobiev, Worosch. & Gorovoi - 5

Meniocus Desv.

linifolius (Steph.) DC. (*Alyssum cupreum* Freyn & Sint., *A. linifolium* Steph. var. *cupreum* (Freyn & Sint.) T.R. Dudley) - 1, 2, 3, 6

Microsisymbrium O.E. Schulz (*Guillenia* auct.)

griffithianum (Boiss.) O.E. Schulz = Arabidopsis pumila

minutiflorum (Hook. fil. & Thoms.) O.E. Schulz (*Sisymbrium minutiflorum* Hook. fil. & Thoms., *Guillenia minutiflora* (Hook. fil. & Thoms.) S.S.R. Bennet) - 6

minutiflorum sensu Vass. = Arabidopsis pumila

murgabicum Ikonn. - 6

Microstigma Trautv.

deflexum (Bunge) Juz. - 3

Microthlaspi F.K. Mey.

perfoliatum (L.) F.K. Mey. (*Thlaspi perfoliatum* L.) - 1, 2, 3, 6

umbellatum (Stev.) F.K. Mey. (*Thlaspi umbellatum* Stev.) - 2

Moriera Boiss.

spinosa Boiss. (*Aethionema spinosum* (Boiss.) Bornm.) - 6

transhyrcana Czerniak. = Aethionema transhyrcanum

Murbeckiella Rothm. (*Phryne* auct.)

huetii (Boiss.) Rothm. (*Phryne huetii* (Boiss.)) O.E. Schulz) - 2

Myagrum L.

alyssum Mill. = Camelina alyssum

perfoliatum L. - 1, 2

prostratum J.P. Bergeret = Rorippa x anceps

pyrenaicum Lam. = Rorippa pyrenaica

sativum L. = Camelina sativa

Nasturtium R. Br.

armoracioides Tausch = Rorippa x armoracioides

astylon Reichenb. = Rorippa x astyla

barbareioides Tausch = Rorippa x astyla

bursifolium DC. = Arabidopsis bursifolia

camelinae Fisch. & C.A. Mey. = Rorippa camelinae

ellipticum Boiss. = Chrysochamela elliptica

fontanum (Lam.) Aschers. = N. officinale

microphyllum Boenn. ex Reichenb. (*N. officinale* R. Br. var. *olgae* (Regel & Schmalh.) N. Busch, *Rorippa microphylla* (Boenn. ex Reichenb.) Hyl. ex A. & D. Löve, *R. nasturtium-aquaticum* (L.) Hayek subsp. *microphylla* (Boenn. ex Reichenb.) O. Bolos & Vigo) - 1, 2, 6

natans DC. = Rorippa amphibia

officinale R. Br. (*Cardamine fontana* Lam. nom. illegit., *Nasturtium fontanum* (Lam.) Aschers. nom. illegit., *Rorippa nasturtium-aquaticum* (L.) Hayek, *R. officinalis* (R. Br.) P. van Royen, *Sisymbrium nasturtium-aquaticum* L.) - 1, 2, 6

- subsp. rotundifolium A. Khokhr. - 2

- var. *olgae* (Regel & Schmalh.) N. Busch = N. microphyllum

proliferum Heuff. = Rorippa prolifera

riparium Wallr. = Rorippa amphibia

turczaninowii Czern. ex Turcz. = Rorippa x armoracioides

Neotorularia Hedge & J. Leonard (*Torularia* (Coss.) O.E. Schulz, 1924, non Bonnemaison, 1828)

aculeolata (Boiss.) Hedge & J. Leonard (*Sisymbrium aculeolatum* Boiss., *Torularia aculeolata* (Boiss.) O.E. Schulz) - 6

brachycarpa (Vass.) Hedge & J. Leonard (*Torularia brachycarpa* Vass., *Dichasianthus brachycarpus* (Vass.) Sojak) - 6

brevipes (Kar. & Kir.) Hedge & J. Leonard (*Torularia brevipes* (Kar. & Kir.) O.E. Schulz, *Dichasianthus brevipes* (Kar. & Kir.) Sojak) - 6

contortuplicata (Steph.) Hedge & J. Leonard (*Dichasianthus contortuplicatus* (Steph.) Sojak, *Torularia contortuplicata* (Steph.) O.E. Schulz, *T. ledebourii* (Boiss.) Grossh.) - 1, 2, 6

dentata (Freyn & Sint.) Hedge & J. Leonard (*Cryptospora dentata* Freyn & Sint., *Dichasianthus dentatus* (Freyn & Sint.) Sojak, *Sisymbrium adpressum* Trautv. 1884, non Turcz. 1854, *Torularia adpressa* (Trautv.) O.E. Schulz, nom. illegit., *T. dentata* (Freyn & Sint.) Kitam.) - 6

eldarica (Grossh.) V. Avet. (*Torularia eldarica* Grossh., *Dichasianthus eldaricus* (Grossh.) Sojak) - 2

humilis (C.A. Mey.) Hedge & J. Leonard (*Arabis sinuata* Turcz., *Braya humilis* (C.A. Mey.) Robins., *B. humilis* subsp. *arctica* (Bocher) Rollins, *Cardaminopsis sinuata* (Turcz.) O.E. Schulz, *Dichasianthus humilis* (C.A. Mey.) Sojak, *Torularia humilis* (C.A. Mey.) O.E. Schulz, *T. humilis* subsp. *arctica* Bocher) - 3, 4, 5, 6

karatavica (Myrzakulov & Bajt.) Czer. comb. nova (*Torularia karatavica* Myrzakulov & Bajt. 1979, Bot. Mat. (Alma-Ata) 11 : 54) - 6

korolkowii (Regel & Schmalh.) Hedge & J. Leonard (*Dichasianthus korolkowii* (Regel & Schmalh.) Sojak, *Torularia korolkowii* (Regel & Schmalh.) O.E. Schulz) - 6

rossica (O.E. Schulz) Hedge & J. Leonard (*Torularia rossica* O.E. Schulz, *Dichasianthus rossicus* (O.E. Schulz) Sojak) - 1, 6

sergievskiana (Polozh.) Czer. comb. nova (*Torularia sergievskiana* Polozh. 1974, Nov. Syst. Pl. Vasc. 11 : 210; *Dichasianthus sergievskianus* (Polozh.) Sojak) - 4

sulphurea (Korsh.) Ikonn. (*Torularia sulphurea* (Korsh.) O.E. Schulz) - 6

torulosa (Desf.) Hedge & J. Leonard (*Dichasianthus torulosus* (Desf.) Sojak, *Torularia torulosa* (Desf.) O.E. Schulz) - 1, 2, 6

Neslia Desv.

apiculata Fisch. & C.A. Mey. (*N. paniculata* (L.) Desv. var. *apiculata* (Fisch. & C.A. Mey.) Boivin, *N. paniculata* subsp. *apiculata* (Fisch. & C.A. Mey.) Maire & Weiller, *N. paniculata* subsp. *thracica* (Velen.) Bornm., *N. thracica* Velen.) - 2, 6

paniculata (L.) Desv. - 1, 2, 3, 4, 5, 6
- subsp. *apiculata* (Fisch. & C.A. Mey.) Maire & Weiller = N. apiculata
- subsp. *thracica* (Velen.) Bornm. = N. apiculata
- var. *apiculata* (Fisch. & C.A. Mey.) Boivin = N. apiculata
thracica Velen. = N. apiculata

Nesodraba Green = Draba

hyperborea (L.) Jurtz. = Draba hyperborea

Neuroloma Andrz. = Achoriphragma

ajanense (N. Busch) Botsch. = Achoriphragma nudicaule
albidum (M. Pop.) Botsch. = Achoriphragma albidum
album (E. Nikit.) Pachom. = Achoriphragma album
angrenicum (Botsch. & Vved.) Botsch. = Achoriphragma angrenicum
asperrimum (B. Fedtsch.) Botsch. = Achoriphragma asperrimum
australe (Pavl.) Botsch. = Achoriphragma australe
beketovii (Krasn.) Botsch. = Achoriphragma beketovii
botschantzevii Pachom. = Achoriphragma botschantzevii

darvazicum (Botsch. & Vved.) Botsch. = Achoriphragma darvazicum
fruticulosum (Regel & Schmalh.) Botsch. = Achoriphragma fruticulosum
korovinii (A. Vassil.) Botsch. = Achoriphragma korovinii
kuramense (Botsch.) Botsch. = Achoriphragma kuramense
lancifolium (M. Pop.) Botsch. = Achoriphragma lancifolium
longicarpum (Krasn.) Botsch. = Achoriphragma longicarpum
maidantalicum (M. Pop. & P. Baran.) Botsch. = Achoriphragma maidantalicum
minutum (Botsch. & Vved.) Botsch. = Achoriphragma minutum
nudicaule (L.) DC. = Achoriphragma nudicaule
nuratense (Botsch. & Vved.) Botsch. = Achoriphragma nuratense
pavlovii (A. Vassil.) Botsch. = Achoriphragma pavlovii
pazijae Pachom. = Achoriphragma pazijae
pinnatifidum (Kar. & Kir.) Botsch. = Achoriphragma pinnatifidum
pjataevae Pachom. = Achoriphragma pjataevae
popovii (Botsch.) Botsch. = Achoriphragma popovii
pulvinatum (M. Pop.) Botsch. = Achoriphragma pulvinatum
runcinatum (Regel & Schmalh.) Botsch. = Achoriphragma runcinatum
saposhnikovii (A. Vassil.) Botsch. = Achoriphragma saposhnikovii
sauricum Pachom. = Achoriphragma sauricum
saxifragum (Botsch. & Vved.) Botsch. = Achoriphragma saxifragum
schugnanum (Lipsch.) Botsch. = Achoriphragma schugnanum
simulatrix (E. Nikit.) Botsch. = Achoriphragma simulatrix
stenocarpum (Kar. & Kir.) Botsch. = Achoriphragma stenocarpum
stenophyllum (M. Pop.) Botsch. = Achoriphragma stenophyllum
subsiliquosum (M. Pop.) Botsch. = Achoriphragma subsiliquosum
tianschanicum (E. Nikit.) Botsch. = Achoriphragma tianschanicum
turkestanicum (Korsh.) Botsch. = Achoriphragma turkestanicum
villosulum (Botsch. & Vved.) Botsch. = Achoriphragma villosulum

Neurotropis (DC.) F.K. Mey.

armena (N. Busch) Czer. (*Thlaspi armenum* N. Busch) - 2
kotschyana (Boiss. & Hohen.) Czer. (*Thlaspi kotschyanum* Boiss. & Hohen.) - 6
orbiculata (Stev.) F.K. Mey. (*Thlaspi orbiculatum* Stev.) - 2
platycarpa (Fisch. & C.A. Mey.) F.K. Mey. (*Thlaspi platycarpum* Fisch. & C.A. Mey.) - 2
szowitsiana (Boiss.) F.K. Mey. (*Thlaspi szowitsianum* Boiss.) - 2

Noccaea Moench (*Apterigia* Galushko, p.p. incl. typo, *Hutchinsia* R. Br. p.p. nom. illegit., *Carpoceras* sensu N. Busch, p.p.)

annua (C. Koch) F.K. Mey. (*Thlaspi annuum* Koch) - 2
cochleariformis (DC.) A. & D. Löve (*Thlaspi cochleariforme* DC., *Noccaea exauriculata* (Kom.) Czer., *Thlaspi exauriculatum* Kom.) - 3, 4, 5, 6

dacica (Heuff.) F.K. Mey. (*Thlaspi dacicum* Heuff.) - 1
exauriculata (Kom.) Czer. = N. cochleariformis
ferganensis (N. Busch) Czer. (*Thlaspi ferganense* N. Busch) - 6
freynii (N. Busch) Czer. (*Thlaspi freynii* N. Busch, *T. macranthum* N. Busch var. *somkheticum* Bordz., *T. somkheticum* (Bordz.) Bordz.) - 2
kamtschatica (Karav.) Czer. (*Thlaspi kamtschaticum* Karav.) - 4, 5
kovatsii (Heuff.) F.K. Mey. (*Thlaspi kovatsii* Heuff., *T. avalanum* Panc.) - 1
macrantha (Lipsky) F.K. Mey. (*Thlaspi praecox* Wulf. var. *macranthum* Lipsky, *T. macranthum* (Lipsky) N. Busch) - 1, 2
praecox (Wulf.) F.K. Mey. (*Thlaspi praecox* Wulf.) - 1
pumila (Stev.) Steud. (*Iberis pumila* Stev., *Apterigia pumila* (Stev.) Galushko, *Thlaspi pumilum* (Stev.) Ledeb.) - 2
sarmatica F.K. Mey. - 1
tatianae (Bordz.) F.K. Mey. (*Thlaspi tatianae* Bordz., *Carpoceras brevistylum* N. Busch, *C. tatianae* (Bordz.) Grossh.) - 2

Noccidium F.K. Mey. (*Carpoceras* sensu N. Busch, p.p.)

hastulatum (DC.) F.K. Mey. (*Hutchinsia hastulata* DC., *Carpoceras hastulatum* (DC.) Boiss.) - 2

Octoceras Bunge

lehmannianum Bunge - 6

Odontarrhena C.A. Mey. = Alyssum

microphylla C.A. Mey. = Alyssum microphyllum
obovata C.A. Mey. = Alyssum obovatum

Oreoblastus Suslova

flabellatus (Regel) Suslova (*Christolea flabellata* (Regel) N. Busch, *Ermania flabellata* (Regel) O.E. Schulz, nom. invalid.) - 6
himalayensis (Cambess.) Suslova (*Cheiranthus himalayensis* Cambess.) - 6
incanus (Ovcz.) Suslova (*Christolea incana* Ovcz., *Ermania incana* (Ovcz.) Botsch. nom. invalid.) - 6
linearis (N. Busch) Suslova (*Christolea linearis* N. Busch, *Ermania linearis* (N. Busch) Botsch. nom. invalid.) - 6
saposhnikovii (A. Vassil.) Czer. (*Ermania saposhnikovii* A. Vassil.) - 6

Pachyneurum Bunge (*Parrya* auct. p.p)

grandiflorum (C.A. Mey.) Bunge (*Draba grandiflora* C.A. Mey., *Ermania microcarpa* (Ledeb.) Dvorak, *Parrya grandiflora* (C.A. Mey.) Schischk., *P. microcarpa* Ledeb.) - 3, 4

Pachyphragma (DC.) Reichenb.

macrophyllum (Hoffm.) N. Busch - 2

Pachypterygium Bunge

brevipes Bunge (*Isatis brevipes* (Bunge) Jafri) - 6
densiflorum Bunge - 6
echinatum Jarm. ex Junussov = P. multicaule
multicaule (Kar. & Kir.) Bunge (*Isatis multicaulis* (Kar. & Kir.) Jafri, *Pachypterygium echinatum* Jarm. ex Junus-sov, *P. praemontanum* Jarm.) - 6
praemontanum Jarm. = P. multicaule

Parrya auct. = Achoriphragma, Botschantzevia, Leiospora, Pachyneurum and Phaeonychum

ajanense N. Busch = Achoriphragma nudicaule
alba E. Nikit. = Achoriphragma album
albida M. Pop. = Achoriphragma albidum
angrenica Botsch. & Vved. = Achoriphragma angrenicum
arctica auct. = Achoriphragma nudicaule
asperrima (B. Fedtsch.) M. Pop. = Achoriphragma asperrimum
australis Pavl. = Achoriphragma australe
beketovii Krasn. = Achoriphragma beketovii
bellidifolia Danguy = Leiospora bellidifolia
crassifolia Botsch. & Vved. = Leiospora crassifolia
darvazica Botsch. & Vved. = Achoriphragma darvazicum
eriocalyx Regel & Schmalh. = Leiospora eriocalyx
exscapa C.A. Mey. = Leiospora exscapa
fruticulosa Regel & Schmalh. = Achoriphragma fruticulosum
golenkinii Lipsch. & Pavl. = Achoriphragma asperrimum
grandiflora (C.A. Mey.) Schischk. = Pachyneurum grandiflorum
karatavica Lipsch. & Pavl. = Botschantzevia karatavica
korovinii A. Vassil. = Achoriphragma korovinii
kuramensis Botsch. = Achoriphragma kuramense
lancifolia M. Pop. = Achoriphragma lancifolium
linearifolia Pavl. = Achoriphragma pavlovii
longicarpa Krasn. = Achoriphragma longicarpum
maidantalica M. Pop. & P. Baran. = Achoriphragma maidantalicum
michaelis A. Vassil. = Achoriphragma beketovii
microcarpa Ledeb. = Pachyneurum grandiflorum
minuta Botsch. & Vved. = Achoriphragma minutum
nudicaulis (L.) Regel = Achoriphragma nudicaule
- subsp. *septentrionalis* Hult. = Achoriphragma nudicaule
- subsp. *turkestanica* (Korsh.) Hult. = Achoriphragma turkestanicum
- var. *asperrima* B. Fedtsch. = Achoriphragma asperrimum
nuratensis Botsch. & Vved. = Achoriphragma nuratense
pamirica Botsch. & Vved. = Leiospora pamirica
pavlovii A. Vassil. = Achoriphragma pavlovii
pinnatifida Kar. & Kir. = Achoriphragma pinnatifidum
popovii Botsch. = Achoriphragma popovii
pulvinata M. Pop. = Achoriphragma pulvinatum
runcinata (Regel & Schmalh.) N. Busch = Achoriphragma runcinatum
saposhnikovii A. Vassil. = Achoriphragma saposhnikovii
saxifraga Botsch. & Vved. = Achoriphragma saxifragum
schugnana Lipsch. = Achoriphragma schugnanum
seravschanica A. Vassil. = Leiospora subscapigera
siliquosa Krasn. = Achoriphragma longicarpum
simulatrix E. Nikit. = Achoriphragma simulatrix
stenocarpa Kar. & Kir. = Achoriphragma stenocarpum
stenophylla M. Pop. = Achoriphragma stenophyllum
subscapigera Botsch. & Vved. = Leiospora subscapigera
subsiliquosa M. Pop. = Achoriphragma subsiliquosum
surculosa N. Busch = Phaeonychium surculosum
tianschanica E. Nikit. = Achoriphragma tianschanicum
turkestanica (Korsh.) N. Busch = Achoriphragma turkestanicum
villosula Botsch. & Vved. = Achoriphragma villosulum

Peltaria Jacq. (*Ricotia* auct.)

aspera Grauer = Clypeola aspera

■ aucheri Boiss. (*P. woronowii* N. Busch, *Ricotia aucheri* (Boiss.) B.L. Burtt)
turkmena Lipsky - 6
woronowii N. Busch = P. aucheri

Peltariopsis N. Busch (*Pseudocamelina* (Boiss.) N. Busch, p.p.)

grossheimii N. Busch - 2
planisiliqua (Boiss.) N. Busch (*Cochlearia planisiliqua* Boiss., *Pseudocamelina szowitsii* (Boiss.) N. Busch) - 2, 6

Phaeonychium O.E. Schulz (*Parrya* auct. p.p.)

abalakovii Junussov - 6
surculosum (N. Busch) Botsch. (*Parrya surculosa* N. Busch) - 6

Phryne auct. = Murbeckiella

huetii (Boiss.) O.E. Schulz = Murbeckiella huetii

Physalidium Fenzl = Graellsia

graellsiifolium Lipsky = Graellsia graellsiifolia

Physoptychis Boiss.

caspica (Habl.) V. Boczantzeva (*Crambe caspica* Habl., *Physoptychis gnaphalodes* (DC.) Boiss.) - 2
gnaphalodes (DC.) Boiss. = Ph. caspica

Prionotrichon Botsch. & Vved.

erysimoides (Kar. & Kir.) Botsch. & Vved. (*Arabis erysimoides* Kar. & Kir.) - 6
gaudanense (Litv.) Botsch. (*Erysimum gaudanense* Litv.) - 6
pseudoparrya Botsch. & Vved. - 6

Pseudanastatica (Boiss.) Grossh.

dichotoma (Boiss.) Grossh. (*Clypeola dichotoma* Boiss.) - 2

Pseudocamelina (Boiss.) N. Busch, p.p. = Peltariopsis

szowitsii (Boiss.) N. Busch = Peltariopsis planisiliqua

Pseudoclausia M. Pop.

gracillima (M. Pop. ex Botsch. & Vved.) A. Vassil. (*Clausia gracillima* M. Pop. ex Botsch. & Vved.) - 6
hispida (Regel) M. Pop. ex A. Vassil. (*Clausia hispida* (Regel) Lipsky) - 6
kuramensis Ovcz. & Junussov - 6
mollissima (Lipsky) A. Vassil. (*Clausia mollissima* Lipsky) - 6
olgae (Regel & Schmalh.) Botsch. (*Clausia olgae* (Regel & Schmalh.) Lipsky) - 6
papillosa (Vass.) A. Vassil. (*Clausia papillosa* Vass.) - 6
sarawschanica (Regel & Schmalh.) Botsch. (*Clausia sarawschanica* (Regel & Schmalh.) Lipsky) - 6
tschimganica (M. Pop. ex Botsch. & Vved.) A. Vassil. (*Clausia tschimganica* M. Pop. ex Botsch. & Vved.) - 6
turkestanica (Lipsky) A. Vassil. (*Clausia turkestanica* Lipsky) - 6
vvedenskyi Pachom. - 6

■ Pseudosempervivum (Boiss.) Grossh.
■ aucheri (Boiss.) Pobed. (*Cochlearia aucheri* Boiss., *Pseudosempervivum karsianum* (N.Busch) Grossh.)
karsianum (N. Busch) Grossh. = P. aucheri

Pseudovesicaria (Boiss.) Rupr.

digitata (C.A. Mey.) Rupr. - 2

Psilonema C.A. Mey. = Alyssum

alyssoides (L.) Heideman = Alyssum calycinum
calycinum (L.) C.A. Mey. = Alyssum calycinum
dasycarpum (Steph.) C.A. Mey. = Alyssum dasycarpum

Pterygostemon V. Boczantzeva (*Asterotrica* V. Boczantzeva, 1976, non DC. 1829, *Farsetia* auct.)

spathulatus (Kar. & Kir.) V. Boczantzeva (*Farsetia spathulata* Kar. & Kir., *Asterotricha spathulata* (Kar. & Kir.) V. Boczantzeva) - 6

Ptilotrichum C.A. Mey.

canescens (DC.) C.A. Mey. (*Alyssum canescens* DC.) - 3, 6
- subsp. *tenuifolium* (Steph.) Hanelt & Davamzac = P. tenuifolium
cyclocarpum Boiss. = Aurinia cyclocarpa
dahuricum Peschkova - 4
elongatum (DC.) C.A. Mey. = P. tenuifolium
tenuifolium (Steph.) C.A. Mey. (*Alyssum tenuifolium* Steph., *Ptilotrichum canescens* (DC.) C.A. Mey. subsp. *tenuifolium* (Steph.) Hanelt & Davamzac, *P. elongatum* (DC.) C.A. Mey.) - 3, 4

Pugionum Gaertn.

pterocarpum Kom. - 4

Raphanistrum Hill = Raphanus

odessanum Andrz. = Raphanus maritimus

Raphanus L. (*Raphanistrum* Hill)

candidus Worosch. - 1
maritimus Smith (*Raphanistrum odessanum* Andrz., *Raphanus odessanus* (Andrz.) Spreng., *R. raphanistrum* L. subsp. *maritimus* (Smith) Thell., *R. raphanistrum* subsp. *odessanus* (Andrz.) Schmalh.) - 1, 2
*****niger** Mill. - 1
odessanus (Andrz.) Spreng. = R. maritimus
*****raphanistroides** (Makino) Sinsk. - 2, 5
raphanistrum L. - 1, 2, 3, 4, 5, 6
- subsp. *maritimus* (Smith) Thell. = R. maritimus
- subsp. *odessanus* (Andrz.) Schmalh. = R. maritimus
- subsp. *rostratus* (DC.) Thell. = R. rostratus
- subsp. *sativus* (L.) Schmalh. = R. sativus
rostratus DC. (*R. raphanistrum* L. subsp. *rostratus* (DC.) Thell.) - 2
*****sativus** L. (*R. raphanistrum* L. subsp. *sativus* (L.) Schmalh.) - 1, 2, 3, 5, 6

Rapistrum Crantz

orientale auct. = R. rugosum
perenne (L.) All. - 1
rugosum (L.) All. (*R. orientale* auct.) - 1, 2, 5(alien), 6

Redowskia Cham. & Schlecht.

sophiifolia Cham. & Schlecht. - 4

Rhammatophyllum O.E. Schulz

frutex Botsch. & Vved. - 6
krascheninnikovii A. Vassil. = R. pachyrhizum
pachyrhizum (Kar. & Kir.) O.E. Schulz (*Arabis pachyrhiza*
Kar. & Kir., *Rhammatophyllum krascheninnikovii* A.
Vassil.) - 6

Ricotia auct. = Peltaria and Rorippa

aucheri (Boiss.) B.L. Burtt = Peltaria aucheri
cantoniensis Lour. = Rorippa cantoniensis

Rorippa Scop. (*Brachilobus* Desv., *Tetrapoma*
Turcz., *Ricotia* auct.)

amphibia (L.) Bess. (*Nasturtium natans* DC., *N. riparium*
Wallr., *Rorippa amphibia* var. *natans* (DC.) Kotov, *R.
amphibia* var. *riparia* (Wallr.) Kotov) - 1, 2, 3, 4, 6
- var. *natans* (DC.) Kotov = R. amphibia
- var. *riparia* (Wallr.) Kotov = R. amphibia
x **anceps** (Wahlenb.) Reichenb. (*Myagrum prostratum* J.P.
Bergeret, *Rorippa anceps* (Wahlenb.) Grossh. comb.
superfl., *R.* x *prostrata* (J.P. Bergeret) Schinz & Thell.
nom. ambig.). - R. amphibia (L.) Bess. x R. sylvestris
(L.) Bess. - 1, 2, 3
x **armoracioides** (Tausch) Fuss (*Nasturtium armoracioides*
Tausch, *N. turczaninowii* Czern. ex Turcz., *Rorippa* x
turczaninowii (Czern. ex Turcz.) Simonk.). - R. austriaca (Crantz) Bess. x R. sylvestris (L.) Bess. - 1(?)
x **astyla** (Reichenb.) Reichenb. (*Nasturtium astylon* Reichenb., *N. barbareioides* Tausch, *Rorippa barbareioides*
(Tausch) Celak.). - R. palustris (L.) Bess. x R. sylvestris (L.) Bess. - 1(?)
austriaca (Crantz) Bess. - 1, 2, 5(alien), 6
barbareifolia (DC.) Kitag. (*Camelina barbareifolia* DC.,
Rorippa hispida (Desv.) Britt. var. *barbareifolia*
(DC.) Hult., *R. islandica* (Oed. ex Murr.) Borb. subsp.
barbareifolia (DC.) Scoggan, *Tetrapoma kruhsianum*
Fisch. & C.A. Mey.) - 5
barbareioides (Tausch) Celak. = R. x astyla
brachycarpa (C.A. Mey.) Hayek (*R. brachycarpa* (C.A. Mey.)
Woronow, comb. superfl., *R. hybrida* Klok.) - 1, 2, 3, 6
camelinae (Fisch. & C.A. Mey.) Spach (*Nasturtium camelinae* Fisch. & C.A. Mey.) - 5
cantoniensis (Lour.) Ohwi (*Ricotia ? cantoniensis* Lour.,
Rorippa microsperma (DC.) Bailey, *R. microsperma*
(DC.) Vass. comb. superfl.) - 5
dogadovae Tzvel. (*R. islandica* (Oed. ex Murr.) Borb.
subsp. *dogadovae* (Tzvel.) Jonsell) - 1(?), 3, 4
globosa (Turcz. ex Fisch. & C.A. Mey.) Hayek (*R. globosa*
(Turcz. ex Fisch. & C.A. Mey.) Vass. comb. superfl.) -
4, 5
hispida (Desv.) Britt. (*Brachilobus hispidus* Desv., *Rorippa
islandica* (Oed. ex Murr.) Borb. var. *hispida* (Desv.)
Butters & Abbe) - 4, 5
- var. *barbareifolia* (DC.) Hult. = R. barbareifolia
hybrida Klok. = R. brachycarpa
islandica auct. = R. palustris
islandica (Oed. ex Murr.) Borb. subsp. *barbareifolia* (DC.)
Scoggan = R. barbareifolia
- subsp. *dogadovae* (Tzvel.) Jonsell = R. dogadovae
- var. *hispida* (Desv.) Butters & Abbe = R. hispida
microphylla (Boenn. ex Reichenb.) Hyl. ex A. & D. Löve =

Nasturtium microphyllum
microsperma (DC.) Bailey = R. cantoniensis
microsperma (DC.) Vass. = R. cantoniensis
nasturtium-aquaticum (L.) Hayek = Nasturtium officinale
- subsp. *microphylla* (Boenn. ex Reichenb.) O. Bolos &
Vigo = Nasturtium microphyllum
officinalis (R.Br.) P. van Royen = Nasturtium officinale
palustris (L.) Bess. (*Sisymbrium amphibium* L. var. *palustre* L., *Rorippa islandica* auct.) - 1, 2, 3, 4, 5, 6
x **podolica** Zapal. - R. amphibia (L.) Bess. x R. austriaca
(Crantz) Bess. - 1(?)
prolifera (Heuff.) Neilr. (*Nasturtium proliferum* Heuff.) - 1
x *prostrata* (J.P. Bergeret) Schinz & Thell. = R. x anceps
pyrenaica (Lam.) Reichenb. (*Myagrum pyrenaicum* Lam.,
Lepidium stylosum Pers., *Rorippa stylosa* (Pers.)
Mansf. & Rothm., *Sisymbrium pyrenaicum* L. 1759,
non L. ex Loerf. 1758) - 1
x **sodalis** Zapal. - R. amphibia (L.) Bess. x R. sylvestris (L.)
Bess. - 1(?)
x **stenophylla** Borb. - R. pyrenaica (Lam.) Reichenb. x R.
sylvestris (L.) Bess. - 1(?)
stylosa (Pers.) Mansf. & Rothm. = R. pyrenaica
sylvestris (L.) Bess. - 1, 2
x *turczaninowii* (Czern. ex Turcz.) Simonk. = R. x armoracioides
viaria V.I. Dorof. - 1

Sameraria Desv.

aitchisonii (Korsh.) B. Fedtsch. = Isatis bullata
armena (L.) Desv. (*S. bidentata* Botsch.) - 2, 6
armena sensu N. Busch & Vass. = S. cardiocarpa
bidentata Botsch. = S. armena
boissierana (Reichenb. fil.) Nabiev = Isatis boissieriana
bullata (Aitch. & Hemsl.) B. Fedtsch. = Isatis bullata
- var. *aitchisonii* (Korsh.) Botsch. = Isatis bullata
canaliculata Vass. = Isatis canaliculata
cardiocarpa Trautv. (*S. armena* sensu N. Busch & Vass.) - 2
deserti N. Busch = Isatis deserti
glastifolia (Fisch. & C.A. Mey.) Boiss. (*S. sclerocarpa*
Bordz.) - 2
hirtocalyx (Franch.) Nabiev = Isatis hirtocalyx
litvinovii N. Busch = Isatis steveniana
odontogera Bordz. (*S. odontophora* Bordz.) - 2
odontophora Bordz. = S. odontogera
sclerocarpa Bordz. = S. glastifolia
turcomanica (Korsh.) B. Fedtsch. = Isatis turcomanica

Schivereckia Andrz. ex DC.

berteroides Fisch. ex M. Alexeenko = Sch. podolica
kusnezovii M. Alexeenko = Sch. podolica
monticola M. Alexeenko = Sch. podolica
- subsp. *mutabilis* M. Alexeenko = Sch. podolica
- subsp. *stenocarpa* M. Alexeenko = Sch. podolica
mutabilis (M. Alexeenko) M. Alexeenko = Sch. podolica
podolica (Bess.) Andrz. ex DC. (*Alyssum podolicum* Bess.,
Schivereckia berteroides Fisch. ex M. Alexeenko, *Sch.
kusnezovii* M. Alexeenko, *Sch. monticola* M. Alexeenko, *Sch. monticola* subsp. *mutabilis* M. Alexeenko, *Sch.
monticola* subsp. *stenocarpa* M. Alexeenko, *Sch.
mutabilis* (M. Alexeenko) M. Alexeenko) - 1

Sinapis L.

alba L. - 1, 2, 3, 4, 5, 6
- subsp. *dissecta* (Lag.) Bonnier = S. dissecta
- subsp. *dissecta* (Lag.) Simk. = S. dissecta

arvensis L. - 1, 2, 3, 4, 5, 6
dissecta Lag. (*S. alba* L. subsp. *dissecta* (Lag.) Bonnier, comb. superfl., *S. alba* subsp. *dissecta* (Lag.) Simk.) - 1
geniculata Desf. = Hirschfeldia incana subsp. geniculata
nasturtiifolia Poir. = Erucastrum nasturtiifolium
nigra L. = Brassica nigra

Sisymbriopsis Botsch. & Tzvel.

mollipila (Maxim.) Botsch. (*Torularia mollipila* (Maxim.) O.E. Schulz) - 6
schugnana Botsch. & Tzvel. - 6

Sisymbrium L.

aculeolatum Boiss. = Neotorularia aculeolata
adpressum Trautv. = Neotorularia dentata
alliaceum Salisb. = Alliaria petiolata
altissimum L. - 1, 2, 3, 5(alien), 6
amphibium L. var. *palustre* L. = Rorippa palustris
an asperum ? Pall. = Dimorphostemon pectinatus
austriacum Jacq. - 1
bilobum (C. Koch) Grossh. = S. septulatum
brachycarpum (N. Busch) Vass. = S. erucastrifolium
brassiciforme C.A. Mey. - 3, 6
confertum Stev. ex Turcz. - 1
daghestanicum Vass. = S. orientale
dahuricum Turcz. ex Fourn. = S. heteromallum
decipiens Bunge = S. loeselii
elatum C. Koch - 2
erucastrifolium (Rupr.) Trautv. (*S. brachycarpum* (N. Busch) Vass.) - 2
gallicum Willd. = Erucastrum gallicum
griffithianum Boiss. = Arabidopsis pumila
heteromallum C.A. Mey. (*S. dahuricum* Turcz. ex Fourn.) - 3, 4
heterophyllum M. Pop. = Cymatocarpus popovii
hirsutum Lag. ex DC. = S. runcinatum
irio L. - 1, 2, 6
isfarense Vass. (*S. subspinescens* Bunge subsp. *isfarense* (Vass.) R. Kam.) - 6
junceum Bieb. - 1, 3
lasiocalyx Prokh. - 2
lipskyi N. Busch - 2
loeselii L. (*S. decipiens* Bunge, *S. turcomanicum* Litv.) - 1, 2, 3, 4, 5, 6
luteum (Maxim.) O.E. Schulz - 5
minutiflorum Hook. fil. & Thoms. = Microsisymbrium minutiflorum
molle Jacq. = Alyssopsis mollis
nasturtium-aquaticum L. = Nasturtium officinale
officinale (L.) Scop. - 1, 2, 3, 4, 5, 6
orientale L. (*S. daghestanicum* Vass.) - 1, 2, 5(alien), 6
- var. *subhastatum* (Willd.) Thell. = S. subhastatum
pectinatum DC. = Dimorphostemon pectinatus
polymorphum (Murr.) Roth - 1, 2, 3, 4, 5, 6
praetermissum T. Mardalejschvili - 2
pulchellum (Willd.) Trautv. var. *grandiflorum* Trautv. = Erysimum lazistanicum
pyrenaicum L. = Rorippa pyrenaica
runcinatum Lag. ex DC. (*S. hirsutum* Lag. ex DC., *Thellungiella runcinata* (Lag. ex DC.) Bondar.) - 2, 6
septulatum DC. (*S. bilobum* (C. Koch) Grossh.) - 2, 6
strictissimum L. - 1
subhastatum (Willd.) Hornem. (*Brassica subhastata* Willd., *Sisymbrium orientale* L. var. *subhastatum* (Willd.) Thell.) - 1

subspinescens Bunge (*Brassica subspinescens* Fisch. & C.A. Mey. nom. nud.) - 6
- subsp. *isfarense* (Vass.) R. Kam. = S. isfarense
subtilissimum M. Pop. = Dichasianthus subtilissimus
supinum L. - 1
thellungii O.E. Schulz - 1, 5
tilesii Ledeb. = Cardaminopsis lyrata
turcomanicum Litv. = S. loeselii
wallichii Hook. fil. & Thoms. = Arabidopsis wallichii
wolgense Bieb. ex Fourn. - 1, 5(alien)

Smelowskia C.A. Mey. (*Chrysanthemopsis* Rech. fil.)

alba (Pall.) Regel - 3, 4, 5
- var. *tilingii* Regel = S. tilingii
altaica (Pobed.) Botsch. = Hedinia altaica
asplenifolia Turcz. = S. bifurcata
bifurcata (Ledeb.) Botsch. (*Hutchinsia bifurcata* Ledeb., *Smelowskia asplenifolia* Turcz.) - 4
calycina (Steph.) C.A. Mey. (*Chrysanthemopsis koelzii* Rech. fil., *Smelowskia koelzii* (Rech. fil.) Rech. fil.) - 3, 4, 6
- var. *porsildii* Drury & Rollins = S. porsildii
inopinata (Kom) Kom. - 5
- subsp. **pseudoalba** Worosch. - 5
jurtzevii Veliczkin = S. porsildii
koelzii (Rech. fil.) Rech. fil. = S. calycina
parryoides (Cham.) Polun. = Ermania parryoides
pectinata (Bunge) Veliczkin (*Hutchinsia pectinata* Bunge) - 1
porsildii (Drury & Rollins) Jurtz. (*S. calycina* (Steph.) C.A. Mey. var. *porsildii* Drury & Rollins, *S. jurtzevii* Veliczkin) - 5

> The priority name for this species is probably S. integrifolia Seeman.

tianschanica Veliczkin - 6
tilingii (Regel) Worosch. (*S. alba* (Pall.) Regel var. *tilingii* Regel) - 5

Sobolewskia Bieb.

caucasica (Rupr.) N. Busch (*S. lithophila* Bieb. var. *caucasica* Rupr.) - 2
clavata (Boiss.) Fenzl - 2
lithophila Bieb. = S. sibirica
- var. *caucasica* Rupr. = S. caucasica
sibirica (Willd.) P.W. Ball (*Cochlearia sibirica* Willd., *Sobolewskia lithophila* Bieb.) - 1
truncata N. Busch - 2

Sophiopsis O.E. Schulz

annua (Rupr.) O.E. Schulz - 6
flavissima (Kar. & Kir.) O.E. Schulz - 6
micrantha Botsch. & Vved. - 6
mongolica (Kom.) N. Busch = Hedinia mongolica
sisymbrioides (Regel & Herd.) O.E. Schulz - 6

Sphaerotorrhiza (O.E. Schulz) A. Khokhr. = Cardamine

trifida (Poir.) A. Khokhr. = Cardamine trifida

Spirorhynchus Kar. & Kir.

sabulosus Kar. & Kir. - 6

Spryginia M. Pop.

crassifolia (Botsch.) Botsch. (*S. winkleri* (Regel) M. Pop. subsp. *crassifolia* Botsch.) - 6
falcata Botsch. - 6
gracilis Botsch. - 6
pilosa Botsch. - 6
undulata Botsch. - 6
winkleri (Regel) M. Pop. (*S. winkleri* subsp. *araneosa* Botsch.) - 6
- subsp. *araneosa* Botsch. = S. winkleri
- subsp. *crassifolia* Botsch. = S. crassifolia

Sterigmostemum Bieb.

acanthocarpum Fisch. & C.A. Mey. - 2
incanum Bieb. (*Cheiranthus torulosus* Bieb. 1808, non Thunb. 1800, *Sterigmostemum torulosum* (Bieb.) Stapf, nom. illegit.) - 2
ramosissimum (O.E. Schulz) Rech. fil. (*Anchonium ramosissimum* O.E. Schulz, *A. sterigmoides* Lipsky ex Vass. nom. invalid.) - 6
sulphureum (Banks & Soland.) Bornm. (*Cheiranthus sulphureus* Banks & Soland.) - 2
tomentosum (Willd.) Bieb. - 1, 2, 3, 6
torulosum (Bieb.) Stapf = S. incanum

Stevenia Adams & Fisch.

alyssoides Adams & Fisch. (*S. zinaidae* Malysch.) - 4
■ **axillaris** (Kom.) N.Busch
cheiranthoides DC. - 3, 4, 5
sergievskajae (Krasnob.) R. Kam & Gubanov (*Alyssum sergievskajae* Krasnob.) - 4
zinaidae Malysch. = S. alyssoides

Streptoloma Bunge

desertorum Bunge - 6
sumbarense (Lipsky) Botsch. (*Torularia sumbarensis* (Lipsky) O.E. Schulz) - 6

Strigosella Boiss. (*Fedtschenkoa* Regel & Schmalh., *Malcolmia* auct. p.max.p.)

africana (L.) Botsch. (*Cheiranthus taraxacifolius* Balb. 1814, non Steph. 1800, *Fedtschenkoa africana* (L.) Dvorak, *F. taraxacifolia* (Balb.) Dvorak, p.p. nom. illegit., *Malcomia africana* (L.) R. Br., *M. taraxacifolia* (Balb.) DC. nom. illegit.) - 1, 2, 3, 6
- var. **laxa** (Lam.) Botsch. (*Malcolmia laxa* (Lam.) DC.)
brevipes (Bunge) Botsch. (*Dontostemon brevipes* Bunge, *Fedtschenkoa brevipes* (Bunge) Dvorak, nom. illegit., *F. brevipes* (Kar. & Kir.) Dvorak, p.p., *Malcolmia karelinii* Lipsky) - 6
circinata (Bunge) Botsch. (*Fedtschenkoa circinata* (Bunge) Dvorak, *Malcolmia boissieriana* Jafri, *M. circinata* (Bunge) Boiss. 1867, non Hook. fil. & Thoms. 1861) - 6
grandiflora (Bunge) Botsch. (*Fedtschenkoa bucharica* (Vass.) Dvorak, *F. grandiflora* (Bunge) Dvorak, *F. turkestanica* Regel & Schmalh., *Malcolmia bucharica* Vass., *M. grandiflora* (Bunge) O. Kuntze, *M. komarovii* Vass.) - 6
hispida (Litv.) Botsch. (*Malcolmia hispida* Litv., *Fedtschenkoa hispida* (Litv.) Dvorak) - 6
hyrcanica (Freyn & Sint.) Botsch. (*Malcolmia hyrcanica* Freyn & Sint.) - 6
intermedia (C.A. Mey.) Botsch. (*Malcolmia intermedia* C.A. Mey., *Fedtschenkoa taraxacifolia* (Balb.) Dvorak. p.p., *Malcolmia taraxacifolia* auct.) - 1, 2, 6
latifolia Bondar. & Botsch. - 6
leptopoda Bondar. & Botsch. - 6
malacotricha (Botsch. & Vved.) Botsch. (*Malcolmia malacotricha* Botsch. & Vved., *Fedtschenkoa malacotricha* (Botsch. & Vved.) Dvorak) - 6
multisiliqua (Vass.) Ikonn. = S. scorpioides
myrzakulovii Bajt. - 6
scorpioides (Bunge) Botsch. (*Fedtschenkoa multisiliqua* (Vass.) Dvorak, *F. scorpioides* (Bunge) Dvorak, *Malcolmia multisiliqua* Vass., *M. scorpioides* (Bunge) Boiss., *Strigosella multisiliqua* (Vass.) Ikonn.) - 6
spryginioides (Botsch. & Vved.) Botsch. (*Malcolmia spryginioides* Botsch. & Vved., *Fedtschenkoa spryginioides* (Botsch. & Vved.) Dvorak) - 6
stenopetala (Bernh. ex Fisch. & C.A. Mey.) Botsch. (*Malcomia africana* (L.) R. Br. var. *stenopetala* Bernh. ex Fisch. & C.A. Mey., *Fedtschenkoa stenopetala* (Bernh. ex Fisch. & C.A. Mey.) Dvorak, *Malcolmia africana* var. *divaricata* Fisch., *M. divaricata* (Fisch.) Fisch., *M. stenopetala* (Bernh. ex Fisch. & C.A. Mey.) Ledeb.) - 1, 2, 3, 6
strigosa (Boiss.) Botsch. (*Malcolmia strigosa* Boiss., *Fedtschenkoa pamirica* (Botsch. & Vved.) Dvorak. *Malcolmia pamirica* Botsch.& Vved.) - 6
tadshikistanica (Vass.) Botsch. (*Malcolmia tadshikistanica* Vass.) - 6
tenuissima (Botsch.) Botsch. (*Malcolmia tenuissima* Botsch.) - 6
trichocarpa (Boiss. & Buhse) Botsch. (*Malcolmia trichocarpa* Boiss. & Buhse) - 6
turkestanica (Litv.) Botsch. (*Malcolmia turkestanica* Litv.) - 6
vvedenskyi Bondar. & Botsch. - 6

Stroganowia Kar. & Kir. (*Heldreichia* Boiss.)

afghana (Boiss.) Pavl. p.p. = S. bupleuroides
angustifolia Botsch. & Vved. - 6
brachyota Kar. & Kir. (*S. desertorum* (Schrenk) Botsch.) - 6
bupleuroides (Rech. fil.) Botsch. (*Lepidium bupleuroides* Rech. fil., *Stroganowia afghana* (Boiss.) Pavl. p.p. quoad pl. turcom.) - 6
cardiophylla Pavl. - 6
desertorum (Schrenk) Botsch. = S. brachyota
gracilis Pavl. = Stubendorffia gracilis
intermedia Kar. & Kir. - 6
litwinowii Lipsky - 6
longifolia (Boiss.) Botsch. & Vved. (*Heldreichia longifolia* Boiss.) - 6
minor Botsch. & Vved. - 6
paniculata Regel & Schmalh. - 6
■ **persica** N. Busch
robusta Pavl. - 6
rubtzovii Botsch. - 6
sagittata Kar. & Kir. - 6
saravschanica Bulgakova - 6
subalpina (Kom.) Thell. ex Pavl. - 6
tianschanica Botsch. & Vved. - 6
tolmaczovii Junussov - 6
trautvetteri Botsch. - 6

Stubendorffia Schrenk

aptera Lipsky - 6
botschantzevii R. Vinogradova - 6
curvinervia Botsch. & Vved. - 6

gracilis (Pavl.) Botsch. & Vved. (*Stroganowia gracilis* Pavl.) - 6
lipskyi N. Busch - 6
olgae R. Vinogradova - 6
orientalis Schrenk - 6
pterocarpa Botsch. & Vved. - 6
subdidyma N. Busch = Isatis subdidyma

Subularia L.
aquatica L. - 1, 3, 4, 5

Syrenia Andrz.
angustifolia (Ehrh.) Reichenb. = S. cana
aucta Klok. = S. montana
cana (Pill. & Mitt.) Neilr. (*Cheiranthus canus* Pill. & Mitt., *Erysimum canum* (Pill. & Mitt.) Polatschek, *Syrenia angustifolia* (Ehrh.) Reichenb., *S. ucrainica* Klok.) - 1, 6
cuspidata (Bieb.) Reichenb. = Erysimum cuspidatum
dolichostylos Klok. = S. montana
macrocarpa Vass. - 3, 6
montana (Pall.) Klok. (*Cheiranthus montanus* Pall., *Syrenia aucta* Klok., *S. dolichostylos* Klok., *S. sessiliflora* (DC.) Ledeb.) - 1, 2, 3, 6
pedunculata Klok. - 2
praevisa Klok. - 1
sessiliflora (DC.) Ledeb. = S. montana
siliculosa (Bieb.) Andrz. - 1, 2, 3, 6
talijevii Klok. - 1
ucrainica Klok. = S. cana

Syreniopsis H.P. Fuchs = Erysimum
cuspidata (Bieb.) H.P. Fuchs = Erysimum cuspidatum
krynkensis (Lavr.) H.P. Fuchs = Erysimum krynkensis

Takhtajaniella V. Avet.
globosa (Grossh.) V. Avet. (*Alyssum globosum* Grossh.) - 2

Taphrospermum C.A. Mey.
altaicum C.A. Mey. - 3, 6
platypetalum Schrenk - 6

Tauscheria Fisch. ex DC.
lasiocarpa Fisch. ex DC. (*T. oblonga* Vass.) - 1, 3, 6
oblonga Vass. = T. lasiocarpa

Teesdalia R. Br. (*Guepinia* Bast. 1812, non *Guepinia* Fries, nom. cons.)
coronopifolia (J.P. Bergeret) Thell. (*Thlaspi coronopifolium* J.P. Bergeret, *Guepinia coronopifolia* (J.P. Bergeret) Sojak) - 1
nudicaulis (L.) R. Br. (*Guepinia nudicaulis* (L.) Bast.) - 1

Tetracme Bunge
bucharica (Korsh.) O.E. Schulz = Tetracmidion bucharicum
glochidiata (Botsch. & Vved.) Pachom. = Tetracmidion glochidiatum
leptopoda Pachom. - 6
pamirica Vass. - 6
quadricornis (Steph.) Bunge - 1, 6
recurvata Bunge - 6

Tetracmidion Korsh.
bucharicum Korsh. (*Tetracme bucharica* (Korsh.) O.E. Schulz) - 6
glochidiatum Botsch. & Vved. (*Tetracme glochidiata* (Botsch. & Vved.) Pachom.) - 6

Tetrapoma Turcz. = Rorippa
kruhsianum Fisch. & C.A. Mey. = Rorippa barbareifolia

Thellungiella O.E. Schulz
halophila (C.A. Mey.) O.E. Schulz - 3, 6
runcinata (Lag. ex DC.) Bondar. = Sisymbrium runcinatum
salsuginea (Pall.) O.E. Schulz - 1, 3, 4, 6

Thlaspi L.
alliaceum L. - 1(alien)
alpestre L. - 1
annuum C. Koch = Noccaea annua
armenum N. Busch = Neurotropis armena
arvense L. - 1, 2, 3, 4, 5, 6
avalanum Panc. = Noccaea kovatsii
cartilagineum J. Mayer = Lepidium cartilagineum
ceratocarpum (Pall.) N. Busch (*Carpoceras ceratocarpum* (Pall.) N. Busch) - 3, 6
cochleariforme DC. = Noccaea cochleariformis
cordatum Desf. = Aethionema cordatum
coronopifolium J.P. Bergeret = Teesdalia coronopifolia
dacicum Heuff. = Noccaea dacica
exauriculatum Kom. = Noccaea cochleariformis
ferganense N. Busch = Noccaea ferganensis
freynii N. Busch = Noccaea freynii
huetii Boiss. - 2
kamtschaticum Karav. = Noccaea kamtschatica
kotschyanum Boiss. & Hohen. = Neurotropis kotschyana
kovatsii Heuff. = Noccaea kovatsii
macranthum N. Busch = Noccaea macrantha
- var. *somkheticum* Bordz. = Noccaea freynii
orbiculatum Stev. = Neurotropis orbiculata
perfoliatum L. = Microthlaspi perfoliatum
platycarpum Fisch. & C.A. Mey. = Neurotropis platycarpa
praecox Wulf. = Noccaea praecox
- var. *macranthum* Lipsky = Noccaea macrantha
pumilum (Stev.) Ledeb. = Noccaea pumila
rostratum N. Busch = Atropatenia rostrata
septigerum (Bunge) Jafri (*Eutrema ? septigerum* Bunge, *E. edwardsii* R. Br. var. *septigerum* (Bunge) N. Busch) - 3
somkheticum (Bordz.) Bordz. = Noccaea freynii
stenocarpum (Boiss.) Hedge = Kotschyella stenocarpa
szowitsianum Boiss. = Neurotropis szowitsiana
tatianae Bordz. = Noccaea tatianae
umbellatum Stev. = Microthlaspi umbellatum
zangezurum Tzvel. = Atropatenia zangezura

Torularia (Coss.) O.E. Schulz = Neotorularia
aculeolata (Boiss.) O.E. Schulz = Neotorularia aculeolata
adpressa (Trautv.) O.E. Schulz = Neotorularia dentata
brachycarpa Vass. = Neotorularia brachycarpa
brevipes (Kar. & Kir.) O.E. Schulz = Neotorularia brevipes
contortuplicata (Steph.) O.E. Schulz = Neotorularia contortuplicata
dentata (Freyn & Sint.) Kitam. = Neotorularia dentata
eldarica Grossh. = Neotorularia eldarica

glandulosa (Kar. & Kir.) Vass. = Dimorphostemon glandulosus
humilis (C.A. Mey.) O.E. Schulz = Neotorularia humilis
- subsp. *arctica* Bocher = Neotorularia humilis
karatavica Myrzakulov & Bajt. = Neotorularia karatavica
korolkowii (Regel & Schmalh.) O.E. Schulz = Neotorularia
korolkowii
ledebourii (Boiss.) Grossh. = Neotorularia contortuplicata
mollipila (Maxim.) O.E. Schulz = Sisymbriopsis mollipila
pectinata (DC.) Ovcz. & Junussov = Dimorphostemon
pectinatus
rossica O.E. Schulz = Neotorularia rossica
sergievskiana Polozh. = Neotorularia sergievskiana
subtilissima (M. Pop.) Botsch. = Dichasianthus subtilissimus
sulphurea (Korsh.) O.E. Schulz = Neotorularia sulphurea
sumbarensis (Lipsky) O.E. Schulz = Streptoloma sumbarense
torulosa (Desf.) O.E. Schulz = Neotorularia torulosa

Trichochiton Kom.

inconspicuum Kom. - 6
- var. *umbrosum* Kom. = T. umbrosum
umbrosum (Kom.) Botsch. & Vved. (*T. inconspicuum*
Kom. var. *umbrosum* Kom.) - 6

Turritis L.

gerardii Bess. = Arabis gerardii
glabra L. (*Arabis glabra* (L.) Bernh.) - 1, 2, 3, 4, 5, 6
laxa (Sibth. & Smith) Hayek (*Arabis laxa* Sibth. & Smith) -
2
nemorensis Hoffm. = Arabis gerardii
sagittata Bertol. = Arabis sagittata

Uranodactylus Gilli = Winklera

patrinioides (Regel) Gilli = Winklera patrinioides
silaifolius (Hook. fil. & Thoms.) Gilli = Winklera silaifolia
silaifolius (Hook. fil. & Thoms.) Jafri = Winklera silaifolia

Vesicaria Adans. = Alyssoides

arctica (Wormsk. ex Hornem.) Richards. = Lesquerella
arctica
graeca Reut. = Alyssoides utriculata
leiocarpa (Trautv.) N. Busch = Lesquerella arctica
utriculata (L.) Lam. = Alyssoides utriculata

Winklera Regel (*Uranodactylus* Gilli)

patrinioides Regel (*Uranodactylus patrinioides* (Regel)
Gilli) - 6
silaifolia (Hook. fil. & Thoms.) Korsh. (*Heldreichia silaifo-
lia* Hook. fil. & Thoms., *Uranodactylus silaifolius*
(Hook fil. & Thoms.) Jafri, *U. silaifolius* (Hook. fil. &
Thoms.) Gilli, comb. superfl.) - 6

Zuvanda (Dvorak) Askerova   (*Malcolmia*
auct. p.p.)

meyeri (Boiss.) Askerova (*Malcolmia meyeri* Boiss., *Hesperis
crenulata* C.A. Mey. 1831, non DC. 1821, *Malcolmia
crenulata* (C.A. Mey.) Vass. 1939, non (DC.) Boiss.
1867, *Maresia meyeri* (Boiss.) Dvorak) - 2

*BUDDLEJACEAE Wilhelm
*Buddleja L.

*davidii Franch. - 2

BUTOMACEAE Rich.
Butomus L.

junceus Turcz. - 4, 6
umbellatus L. - 1, 2, 3, 4, 5, 6

BUXACEAE Dumort.
Buxus L.

colchica Pojark. - 2
hyrcana Pojark. (*B. sempervirens* L. subsp. *hyrcana*
(Pojark.) Takht.) - 2
*****sempervirens** L. - 1, 2, 6
- subsp. *hyrcana* (Pojark.) Takht. = B. hyrcana

Pachysandra Michx.

terminalis Siebold & Zucc. - 5(?)

CABOMBACEAE A. Rich.
Brasenia Schreb.

schreberi J.F. Gmel. - 5

CAESALPINIACEAE R. Br. = FABACEAE

CALLIGONACEAE Chalkuziev = POLYGO-
NACEAE

CALLITRICHACEAE Link
Callitriche L.

anceps Fern. subsp. *subanceps* (V. Petrov) A. & D. Löve =
C. palustris
autumnalis L. = C. hermaphroditica
brutia Petagna (*C. intermedia* Hoffm. subsp. *pedunculata*
(DC.) Clapham, *C. pedunculata* DC., *C. intermedia*
auct.) - 2
- subsp. *hamulata* (Kutz. ex Koch) O. Bolos & Vigo = C.
hamulata
cophocarpa Sendtner (*C. polymorpha* Loennr., *C. platycarpa* auct.) - 1, 2
elegans V. Petrov = C. palustris
fallax V. Petrov = C. palustris
fimbriata (Schotsman) Tzvel. (*C. truncata* Guss. subsp.
fimbriata Schotsman) - 1
hamulata Kutz. ex Koch (*C. brutia* Petagna subsp. *hamulata*
(Kutz. ex Koch) O. Bolos & Vigo) - 1
     Probably, the priority name of this species should be C. inter-
media Hoffm.
hermaphroditica L. (*C. autumnalis* L.) - 1, 3, 4, 5, 6
intermedia auct. = C. brutia
intermedia Hoffm. subsp. *pedunculata* (DC.) Clapham = C.
brutia
palustris L. (*C. anceps* Fern. subsp. *subanceps* (V. Petrov)
A. & D. Löve, *C. elegans* V. Petrov, *C. fallax* V. Petrov,
C. palustris var. *elegans* (V. Petrov) Y.L. Chang, *C.
subanceps* V. Petrov, *C. verna* L., *C. verna* var. *elegans*
(V. Petrov) Kitag.) - 1, 2, 3, 4, 5, 6
- var. *elegans* (V. Petrov) Y.L. Chang = C. palustris
pedunculata DC. = C. brutia
platycarpa auct. = C. cophocarpa

polymorpha Loennr. = C. cophocarpa
stagnalis Scop. - 1, 2
subanceps V. Petrov = C. palustris
transvolgensis Tzvel. - 1
truncata Guss. subsp. *fimbriata* Schotsman = C. fimbriata
verna L. = C. palustris
- var. *elegans* (V. Petrov) Kitag. = C. palustris

CAMPANULACEAE Juss.
Adenophora Fisch.

collina Kitag. - 5
coronopifolia Fisch. - 4, 5
crispata (Korsh.) Kitag. - 4
divaricata Franch. & Savat. - 5
gmelinii (Spreng.) Fisch. - 4, 5
golubinzevaeana Reverd. - 3
himalayana Feer - 6
jacutica Fed. - 4, 5
kurilensis Nakai = A. triphylla
lamarckii Fisch. - 3, 4, 6
lilifolia (L.) A. DC. - 1, 3, 6
onoei Tatew. & Kitam. = A. pereskiifolia
pereskiifolia (Fisch. ex Schult.) G. Don fil. (*A. onoei* Tatew. & Kitam., *A. polyantha* auct.) - 4, 5
- subsp. *sublata* (Kom.) Worosch. = A. sublata
polyantha auct. = A. pereskiifolia
remotiflora (Siebold & Zucc.) Miq. (*Campanula remotiflora* Siebold & Zucc., *Adenophora trachelioides* auct.) - 5
rupestris Reverd. - 4
stenanthina (Ledeb.) Kitag. - 3, 4, 5
sublata Kom. (*A. pereskiifolia* (Fisch. ex Schult.) G. Don fil. subsp. *sublata* (Kom.) Worosch.) - 4, 5
taurica (Sukacz.) Juz. - 1
tetraphylla auct. p.p. = A. verticillata
tetraphylla (Thunb.) Fisch. = A. triphylla
thunbergiana Kudo = A. triphylla
trachelioides auct. = A. remotiflora
tricuspidata (Fisch. ex Schult.) A. DC. - 4, 5
triphylla auct. p.p. = A. verticillata
triphylla (Thunb.) A. DC. (*A. kurilensis* Nakai, *A. tetraphylla* (Thunb.) Fisch., *A. thunbergiana* Kudo, *A. triphylla* var. *kurilensis* (Nakai) Kitam.) - 5
- var. *kurilensis* (Nakai) Kitam. = A. triphylla
verticillata Fisch. (*A. tetraphylla* auct. p.p., *A. triphylla* auct. p.p.) - 4, 5

Annaea Kolak. = Campanula

hieracioides (Kolak.) Kolak. = Campanula hieracioides

Astrocodon Fed.

expansus (J. Rudolph) Fed. (*A. kruhseanus* (Fisch. ex Regel & Til.) Fed.) - 5
kruhseanus (Fisch. ex Regel & Til.) Fed. = A. expansus

Asyneuma Griseb. & Schenk

amplexicaule (Willd.) Hand.-Mazz. - 2
argutum (Regel) Bornm. - 6
- subsp. *baldshuanicum* (O. Fedtsch.) Damboldt = A. baldshuanicum
- subsp. **pavlovii** Damboldt - 6
attenuatum (Franch.) Bornm. - 6
baldshuanicum (O. Fedtsch.) Fed. (*A. argutum* (Regel) Bornm. subsp. *baldshuanicum* (O. Fedtsch.) Dam-

boldt) - 6
campanuloides (Bieb. ex Sims) Bornm. (*Phyteuma campanuloides* Bieb. ex Sims) - 2
canescens (Waldst. & Kit.) Griseb. & Schenk - 1
- subsp. *salicifolium* (Kit.) Soo = A. salignum
▪ cichoriforme (Boiss.) Bornm. (*A. virgatum* (Labill.) Bornm. subsp. *cichoriforme* (Bornm.) Damboldt)
debile Fed. - 6
japonicum (Miq.) Briq. - 5
lanceolatum (Willd.) Hand.-Mazz. = A. rigidum
▪ leianthum (Trautv.) Bornm. (*A. virgatum* auct. p.p.)
limonifolium auct. p.p. = A. otites
▪ lobelioides (Willd.) Hand.-Mazz.
▪ otites (Boiss.) Bornm. (*A. limonifolium* auct. p.p.)
pulchellum (Fisch. & C.A. Mey.) Bornm. - 6
ramosum Pavl. - 6
rigidum (Willd.) Grossh. (*A. lanceolatum* (Willd.) Hand.-Mazz., *Phyteuma lanceolatum* Willd. 1789, non Vill. 1787) - 2
salignum (Waldst. & Kit. ex Bess.) Fed. (*A. canescens* (Waldst. & Kit.) Griseb. & Schenk subsp. *salicifolium* (Kit.) Soo) - 1, 2
talyschense Fed. - 2
thomsonii (Hook. fil.) Bornm. (*Campanula thomsonii* Hook. fil.) - 6
trautvetteri (B. Fedtsch.) Bornm. - 6
▪ urceolatum (Fomin) Fed.
virgatum auct. p.p. = A. leianthum
virgatum (Labill.) Bornm. subsp. *cichoriforme* (Bornm.) Dambolt = A. cichoriforme
▪ woronowii (Fomin) Bornm.

Brachycodon Fed. = Brachycadonia

fastigiatus (Duf. ex A. DC.) Fed. = Brachycodonia fastigiata

Brachycodonia Fed. (*Brachycodon* Fed. 1957, non Progel, 1865)

fastigiata (Duf. ex A. DC.) Fed. (*Brachycodon fastigiatus* (Duf. ex A. DC.) Fed.) - 2, 6

Campanula L. (*Annaea* Kolak., *Fedorovia* Kolak. 1980, non Yakovl. 1971, *Hemisphaera* Kolak., *Hyssaria* Kolak., *Mzymtella* Kolak., *Neocodon* Kolak. & Serdjukova, *Pseudocampanula* Kolak., *Roucela* Dumort., *Sachokiella* Kolak., *Symphyandra* A. DC. p.p. excl. typo, *Theodorovia* Kolak., *Thesium* auct.)

abietina Griseb. & Schenk (*C. patula* L. subsp. *abietina* (Griseb. & Schenk) Simonk., *C. patula* subsp. *abietina* var. *vajdae* (Penzes) Soo, *C. vajdae* Penzes, *Neocodon abietinus* (Griseb. & Schenk) Kolak. & Serdjukova) - 1
achverdovii Charadze - 2
akuschensis Gussejnov - 2
alberti Trautv. (*Neocodon alberti* (Trautv.) Kolak. & Serdjukova) - 6
albovii Kolak. - 2
aldanensis Fed. & Karav. - 4
alliariifolia Willd. (*C. ochroleuca* Kem.-Nath.) - 2
alpigena C. Koch - 2
alpina Jacq. - 1
altaica Ledeb. (*C. stevenii* Bieb. subsp. *altaica* (Ledeb.) Fed., *Neocodon altaicus* (Ledeb.) Kolak. & Serdjukova) - 1, 3, 6
andina Rupr. - 2

annae Kolak. = C. collina
anomala Fomin - 2
antiqua (Kolak.) Kolak. & Serdjukova (*Symphyandra antiqua* Kolak.) - 2
ardonensis Rupr. - 2
argunensis Rupr. - 2
armazica Charadze - 2
armena Stev. (*Symphyandra armena* (Stev.) A. DC.) - 2
aucheri A. DC. - 2
autraniana Albov - 2
bayerniana Rupr. (*C. elegantissima* Grossh., *C. takhtadzhianii* Fed.) - 2
- subsp. *choziatowskyi* (Fomin) Oganesian = C. choziatowskyi
beauverdiana Fomin (*C. stevenii* Bieb. subsp. *beauverdiana* (Fomin) Rech. fil. & Schiman-Czeika, *Neocodon beauverdianus* (Fomin) Kolak. & Serdjukova) - 2
bellidifolia Adams - 2
besenginica Fomin - 2
■ betulifolia C. Koch
biebersteiniana Schult. (*C. tridens* Rupr., *C. tridentata* Schreb. subsp. *biebersteiniana* (Schult.) Oganesian, *Hemisphaera tridens* (Rupr.) Kolak.) - 2
bononiensis L. - 1, 2, 3, 6
brassicifolia Somm. & Levier = C. imeretina
brotheri Somm. & Levier = C. raddeana
bzybica Jabr.-Kolak. - 2
calcarata Somm. & Levier - 2
calcarea (Albov) Charadze - 2
capusii (Franch.) Fed. - 6
carpatica Jacq. (*Neocodon carpaticus* (Jacq.) Kolak. & Serdjukova) - 1
cashmeriana Royle (*C. evolvulacea* Royle ex A. DC.) - 6
caucasica Bieb. - 2
cephalotes Nakai (*C. glomerata* L. subsp. *canescens* (Maxim.) Beauverd) - 4, 5
cervicaria L. - 1, 3, 4
- subsp. *macrostachya* (Waldst. & Kit. ex Willd.) Tacik = C. macrostachya
chamissonis Fed. (*C. dasyantha* Bieb. var. *chamissonis* (Fed.) Toyokuni & Nosaka) - 5
charadzeae Grossh. - 2
charkeviczii Fed. (*C. sibirica* L. subsp. *charkeviczii* (Fed.) Fed.) - 1
choziatowskyi Fomin (*C. bayerniana* Rupr. subsp. *choziatowskyi* (Fomin) Oganesian) - 2
chysnysuensis Czer. nom. nov. (*C. foliosa* Galushko, 1971, Nov. Syst. Pl. Vasc. 8 : 266, non Ten. 1815) - 2
ciliata Stev. - 2
circassica Fomin (*Hemisphaera circassica* (Fomin) Kolak. & Serdjukova) - 2
collina Sims (*C. collina* Bieb., *C. annae* Kolak.) - 2
- subsp. **gapschimensis** Gussejnov - 2
cordifolia C. Koch (*C. rapunculoides* L. subsp. *cordifolia* (C. Koch) Damboldt) - 2
coriacea P.H. Davis (*C. radula* auct.) - 2
crispa Lam. - 2
czerepanovii Fed. - 2
daghestanica Fomin - 2
daralaghezica (Grossh.) Kolak. & Serdjukova (*Symphyandra daralaghezica* Grossh., *S. armena* (Stev.) A. DC. subsp. *daralaghezica* (Grossh.) Fed.) - 2
darialica Charadze = C. hohenackeri var. darialica
dasyantha Bieb. - 3, 4, 5
- var. *chamissonis* (Fed.) Toyokuni & Nosaka = C. chamissonis
dolomitica E. Busch - 2

doluchanovii Charadze - 2
dzaaku Albov (*Pseudocampanula dzaaku* (Albov) Kolak.) - 2
dzyschrica Kolak. - 2
elatior (Fomin) Grossh. = C. praealta
elegantissima Grossh. = C. bayerniana
elliptica Kit. (*C. glomerata* L. subsp. *pulchra* (Wissjul.) Soo, *C. pulchra* Wissjul.) - 1
engurensis Charadze - 2
erinus L. (*C. nanella* P. Smirn.) - 1, 2
eugeniae Fed. - 6
evolvulacea Royle ex A. DC. = C. cashmeriana
farinosa Andz. (*C. glomerata* L. subsp. *farinosa* (Andrz.) Kirschl.) - 1
fedorovii Charadze - 2
fedtschenkiana Trautv. = C. incanescens
■ finitima Fomin
foliosa Galushko = C. chysnysuensis
fominii Grossh. - 2
fonderwisii Albov - 2
galushkoi Prima - 2
giesekiana Vest ex Schult. = C. rotundifolia
glomerata L. (*C. glomerata* subsp. *farinosa* (Rochel ex Bess.) Jav. f. *polessica* (Wissjul.) Soo, *C. polessica* Wissjul.) - 1, 2, 3, 4, 6
- subsp. *canescens* (Maxim.) Beauverd = C. cephalotes
- subsp. *elliptica* (Kit.) Jav. f. *subcapitata* (M. Pop.) Soo = C. subcapitata
- subsp. *farinosa* (Andrz.) Kirschl. = C. farinosa
- subsp. *farinosa* (Rochel ex Bess.) Jav. f. *polessica* (Wissjul.) Soo = C. glomerata
- subsp. *oblongifolia* (C. Koch) Fed. = C. oblongifolia
- subsp. *pulchra* (Wissjul.) Soo = C. elliptica
- subsp. **semgensis** Gussejnov - 2
- subsp. *subcapitata* (M. Pop.) Fed. = C. subcapitata
groenlandica Berl. = C. rotundifolia
grossheimii Charadze = C. rapunculoides
hemschinica C. Koch (*Neocodon hemschinicus* (C. Koch) Kolak. & Serdjukova, *Campanula olympica* auct. p.p.) - 2
hieracioides Kolak. (*Annaea hieracioides* (Kolak.) Kolak.) - 2
hissarica R. Kam. - 6
hohenackeri Fisch. & C.A. Mey. (*C. sibirica* L. subsp. *hohenackeri* (Fisch. & C.A. Mey.) Damboldt) - 2
- var. **darialica** (Charadze) Serdjukova (*C. darialica* Charadze) - 2
hypopolia Trautv. - 2
imeretina Rupr. (*C. brassicifolia* Somm. & Levier) - 2
incanescens Boiss. (*C. fedtschenkiana* Trautv.) - 6
irinae Kuth. - 2
jadvigae Kolak. - 2
kachetica Kantsch. - 2
kadargavanica Amirchanov & Komzha - 2
kantschavelii Zagareli - 2
karakuschensis Grossh. (*C. minsteriana* Grossh., *Fedorovia karakuschensis* (Grossh.) Kolak., *Theodorovia karakuschensis* (Grossh.) Kolak. comb. invalid.) - 2
kemulariae Fomin - 2
ketzkhovelii Sosn. = C. massalskyi
kirpicznikovii Fed. - 2
kladniana auct. = C. polymorpha
kladniana (Schur) Witas. subsp. *polymorpha* Witas. = C. polymorpha
kluchorica Kolak. - 2
kolakovskyi Charadze - 2
kolenatiana C.A. Mey. ex Rupr. - 2

komarovii Maleev - 2
kryophila Rupr. - 2
lactiflora Bieb. = Gadellia lactiflora
lambertiana A. DC. (*C. rapunculus* L. subsp. *lambertiana*
    (A. DC.) Rech. fil., *Neocodon lambertianus* (A.DC.)
    Kolak. & Serdjukova) - 2
langsdorffiana Fisch. ex Trautv. & C.A. Mey = C. rotundifo-
    lia
lasiocarpa Cham. - 5
- subsp. latisepala (Hult.) Hult. (*C. lasiocarpa* var. *latisepa-
    la* Hult.) - 5
latifolia L. - 1, 2, 3
■ lazica (Boiss. & Bal.) Charadze (*Symphyandra lazica*
    Boiss. & Bal.)
■ ledebouriana Trautv.
lehmanniana Bunge (*Hyssaria lehmanniana* Kolak.) - 6
- subsp. integerrima R. Kam. - 6
- subsp. pseudohissarica R. Kam. - 6
leskovii Fed. - 2
letschchumensis Kem.-Nath. - 2
lezgina (Alexeenko ex Lipsky) Kolak. & Serdjukova
    (*Symphyandra lezgina* Alexeenko ex Lipsky) - 2
longistyla Fomin - 2
lyrata Lam. - 2
macrochlamys Boiss. & Huet (*Sachokiella macrochlamys*
    (Boiss. & Huet) Kolak.) - 2
macrostachya Waldst. & Kit. ex Willd. (*C. cervicaria* L.
    subsp. *macrostachya* (Waldst. & Kit. ex Willd.)
    Tacik) - 1
makaschvilii E. Busch - 2
maleevii Fed. - 2
massalskyi Fomin (*C. ketzkhovelii* Sosn.) - 2
*medium L. - 1
megrelica Manden. & Kuth. - 2
meyeriana Rupr. - 2
minsteriana Grossh. = C. karakuschensis
mirabilis Albov - 2
nanella P. Smirn. = C. erinus
napuligera Schur = C. serrata
nefedovii Galushko - 2
oblongifolia (C. Koch) Charadze (*C. glomerata* L. subsp.
    oblongifolia (C. Koch) Fed.) - 2
oblongifolioides Galushko - 2
ochroleuca Kem.-Nath. = C. alliariifolia
odontosepala Boiss. (*Symphyandra odontosepala* (Boiss.)
    Esfandiari) - 2
olympica auct. p.p. = C. hemschinica
ossetica Bieb. - 2
panjutinii Kolak. - 2
paradoxa Kolak. - 2
patula L. (*Neocodon patulus* (L.) Kolak. & Serdjukova) - 1,
    3(?)
- subsp. *abietina* (Griseb. & Schenk) Simonk. = C. abietina
- - var. *vajdae* (Penzes) Soo = C. abietina
pendula Bieb. (*Symphyandra pendula* (Bieb.) A. DC.) - 2
persicifolia L. (*Neocodon persicifolius* (L.) Kolak. & Ser-
    djukova) - 1, 2, 3
petrophila Rupr. - 2
polessica Wissjul. = C. glomerata
polymorpha Witas. (*C. kladniana* (Schur) Witas. subsp.
    polymorpha Witas., *C. rotundifolia* L. subsp. *polymor-
    pha* (Witas.) Tacik, *C. kladniana* auct., *C. polymorpha*
    auct., *C. tatrae* auct.) - 1
pontica Albov (*Neocodon ponticus* (Albov) Kolak. & Serd-
    jukova) - 2
praealta Galushko (*C. elatior* (Fomin) Grossh. 1949, non
    Hoffmgg & Link, *C. sibirica* L. subsp. *elatior* (Fomin)

Fed.) - 1, 2
propinqua Fisch. & C.A. Mey. (*Roucela propinqua* (Fisch.
    & C.A. Mey.) Charadze) - 2
pulchra Wissjul. = C. elliptica
punctata Lam. - 4, 5
radchensis Charadze - 2
raddeana Trautv. (*C. brotheri* Somm. & Levier) - 2
radula auct. = C. coriacea
rapunculoides L. (*C. grossheimii* Charadze) - 1, 2, 3, 6
- subsp. *cordifolia* (C. Koch) Damboldt = C. cordifolia
rapunculus L. (*Neocodon ranunculus* (L.) Kolak. & Sedju-
    kova) - 1, 2
- subsp. *lambertiana* (A. DC.) Rech. fil. = C. lambertiana
remotiflora Siebold & Zucc. = Adenophora remotiflora
repens (Karpissonova) Czer. comb. nova (*Symphyandra
    repens* Karpissonova, 1979, Bull. Soc. Nat. Moscou, 84,
    6 : 123) - 2
rotundifolia L. (*C. giesekiana* Vest ex Schult., *C. groenlandi-
    ca* Berl., *C. langsdorffiana* Fisch. ex Trautv. & C.A.
    Mey., *C. rotundifolia* subsp. *langsdorffiana* (Fisch. ex
    Trautv. & C.A. Mey.) Vodopianova) - 1, 3, 4, 5
- subsp. *langsdorffiana* (Fisch. ex Trautv. & C.A. Mey.)
    Vodopianova = C. rotundifolia
- subsp. *polymorpha* (Witas.) Tacik = C. polymorpha
ruprechtii auct. = C. tridentata
sarmatica Ker-Gawl. - 2
saxifraga Bieb. (*Hemisphaera saxifraga* (Bieb.) Kolak.) - 2
scapifoliosa A. Khokhr. - 2
schelkownikowii Grossh. ex Fed. - 2
schischkinii Kolak. & Sachok. - 2
schistosa Kolak. - 2
sclerophylla (Kolak.) Czer. comb. nova (*Mzymtella sclero-
    phylla* Kolak. 1981, Proc. Acad. Sci. Georg. 103, 1 :
    149-152) - 2
sclerotricha Boiss. - 2
serrata (Schult.) Hendrych (*Thesium serratum* Schult.,
    Campanula napuligera Schur) - 1
sibirica L. - 1, 3, 4, 6
- subsp. *charkeviczii* (Fed.) Fed. = C. charkeviczii
- subsp. *elatior* (Fomin) Fed. = C. praealta
- subsp. *hohenackeri* (Fisch. & C.A. Mey.) Damboldt = C.
    hohenackeri
- subsp. *talievii* (Juz.) Fed. = C. talievii
- subsp. *taurica* (Juz.) Fed. = C. taurica
siegizmundi Fed. - 2
sommieri Charadze - 2
songutica Amirchanov - 2
sosnowskyi Charadze - 2
sphaerocarpa Kolak. - 2
stevenii Bieb. (*Neocodon stevenii* (Bieb.) Kolak. & Serdju-
    kova) - 2
- subsp. *altaica* (Ledeb.) Fed. = C. altaica
- subsp. *beauverdiana* (Fomin) Rech. fil. & Schiman-
    Czeika = C. beauverdiana
- subsp. *wolgensis* (P. Smirn.) Fed. = C. wolgensis
■ stricta L.
suanetica Rupr. - 2
subcapitata M. Pop. (*C. glomerata* L. subsp. *elliptica* (Kit.)
    Jav. f. *subcapitata* (M. Pop.) Soo, *C. glomerata* subsp.
    subcapitata (M. Pop.) Fed.) - 1
symphytifolia (Albov) Kolak. - 2
takhtadzhianii Fed. = C. bayerniana
talievii Juz. (*C. sibirica* L. subsp. *talievii* (Juz.) Fed.) - 1
tatrae auct. = C. polymorpha
taurica Juz. (*C. sibirica* L. subsp. *taurica* (Juz.) Fed.) - 1, 2
thomsonii Hook. fil. = Asyneuma thomsonii
trachelium L. - 1, 2, 3

transcaucasica (Somm. & Levier) Kolak. & Serdjukova (*Symphyandra transcaucasica* (Somm. & Levier) Grossh.) - 2

trautvetteri Grossh. ex Fed. - 2

tridens Rupr. = C. biebersteiniana

tridentata Schreb. (*C. ruprechtii* auct.) - 2

- subsp. *biebersteiniana* (Schult.) Oganesian = C. biebersteiniana

tschuktschorum Jurtz. & Fed. - 5

turczaninovii Fed. (*Neocodon turczaninovii* (Fed.) Kolak. & Serdjukova) - 3, 4, 5

uniflora L. - 1, 3, 4, 5

vajdae Penzes = C. abietina

wolgensis P. Smirn. (*C. stevenii* Bieb. subsp. *wolgensis* (P. Smirn.) Fed., *Neocodon wolgensis* (P. Smirn.) Kolak. & Serdjukova) - 1, 3, 6

woronowii Charadze - 2

zangezura (Lipsky) Kolak. & Serdjukova (*Symphyandra zangezura* Lipsky) - 2

zeyensis Amirchanov & Tavasiev - 2

Codonopsis Wall.

clematidea (Schrenk) Clarke - 6

lanceolata (Siebold & Zucc.) Benth. & Hook. fil. - 5

pilosula (Franch.) Nannf. - 5

ussuriensis (Rupr. & Maxim.) Hemsl. - 5

Cryptocodon Fed.

monocephalus (Trautv.) Fed. - 5

Cylindrocarpa Regel

sewerzowii (Regel) Regel - 6

Edraianthus auct. = Muehlenbergella

owerinianus Rupr. = Muehlenbergella oweriniana

Fedorovia Kolak. = Campanula

karakuschensis (Grossh.) Kolak. = Campanula karakuschensis

Gadellia Schulkina

lactiflora (Bieb.) Schulkina (*Campanula lactiflora* Bieb.) - 2

Hemisphaera Kolak. = Campanula

circassica (Fomin) Kolak. & Serdjukova = Campanula circassica

saxifraga (Bieb.) Kolak. = Campanula saxifraga

tridens (Rupr.) Kolak. = Campanula biebersteiniana

Hyssaria Kolak. = Campanula

lehmanniana (Bunge) Kolak. = Campanula lehmanniana

Jasione L.

■ heldreichii Boiss. & Orph.

montana L. - 1

Legousia Durande

falcata (Ten.) Fritsch - 2, 6

hybrida (L.) Delarb. - 1, 2

pentagonia (L.) Druce - 2

skvortsovii G. Proskuriakova - 6

speculum-veneris (L.) Chaux - 2

Michauxia L'Her. nom. conserv. (*Mindium* Adans.)

laevigata Vent. (*Mindium laevigatum* (Vent.) Rech. fil. & Schiman-Czeika) - 2

Mindium Adans. = Michauxia

laevigatum (Vent.) Rech. fil. & Schiman-Czeika = Michauxia laevigata

Muehlenbergella Feer (*Edraianthus* auct.)

oweriniana (Rupr.) Feer (*Edraianthus owerinianus* Rupr.) - 2

Mzymtella Kolak. = Campanula

sclerophylla Kolak. = Campanula sclerophylla

Neocodon Kolak. & Serdjukova = Campanula

abietinus (Griseb. & Schenk) Kolak. & Serdjukova = Campanula abietina

alberti (Trautv.) Kolak. & Serdjukova = Campanula alberti

altaicus (Ledeb.) Kolak. & Serdjukova = Campanula altaica

beauverdianus (Fomin) Kolak. & Serdjukova = Campanula beauverdiana

carpaticus (Jacq.) Kolak. & Serdjukova = Campanula carpatica

hemschinicus (C. Koch) Kolak. & Serdjukova = Campanula hemschinica

lambertianus (A. DC.) Kolak. & Serdjukova = Campanula lambertiana

patulus (L.) Kolak. & Serdjukova = Campanula patula

persicifolius (L.) Kolak. & Serdjukova = Campanula persicifolia

ponticus (Albov) Kolak. & Serdjukova = Campanula pontica

ranunculus (L.) Kolak. & Sedjukova = Campanula rapunculus

stevenii (Bieb.) Kolak. & Serdjukova = Campanula stevenii

turczaninovii (Fed.) Kolak. & Serdjukova = Campanula turczaninovii

wolgensis (P. Smirn.) Kolak. & Serdjukova = Campanula wolgensis

Ostrowskia Regel

magnifica Regel - 6

Peracarpa Hook. fil. & Thoms.

circaeoides (Fr. Schmidt) Feer - 5

Phyteuma L.

campanuloides Bieb. ex Sims = Asyneuma campanuloides

lanceolatum Willd. = Asyneuma rigidum

nigrum F.W. Schmidt - 1

orbiculare L. - 1

spicatum L. - 1

tetramerum Schur - 1

vagneri A. Kerner - 1

Platycodon A. DC.

grandiflorus (Jacq.) A. DC. - 4, 5

Popoviocodonia Fed.

stenocarpa (Trautv. & C.A. Mey.) Fed. - 5
uyemurae (Kudo) Fed. - 5

Pseudocampanula Kolak. = Campanula

dzaaku (Albov) Kolak. = Campanula dzaaku

Roucela Dumort. = Campanula

propinqua (Fisch. & C.A. Mey.) Charadze = Campanula propinqua

Sachokiella Kolak. = Campanula

macrochlamys (Boiss. & Huet) Kolak. = Campanula macrochlamys

Sergia Fed.

regelii (Trautv.) Fed. - 6
sewerzowii (Regel) Fed. - 6

Symphyandra A. DC. p.p. = Campanula

antiqua Kolak. = Campanula antiqua
armena (Stev.) A. DC. = Campanula armena
- subsp. *daralaghezica* (Grossh.) Fed. = Campanula daralaghezica
daralaghezica Grossh. = Campanula daralaghezica
lazica Boiss. & Bal. = Campanula lazica
lezgina Alexeenko ex Lipsky = Campanula lezgina
odontosepala (Boiss.) Esfandiari = Campanula odontosepala
pendula (Bieb.) A. DC. = Campanula pendula
repens Karpissonova = Campanula repens
transcaucasica (Somm. & Levier) Grossh. = Campanula transcaucasica
zangezura Lipsky = Campanula zangezura

Theodorovia Kolak. = Campanula

karakuschensis (Grossh.) Kolak. = Campanula karakuschensis

Thesium auct. = Campanula

serratum Kit. = Campanula serrata

CANNABACEAE Endl.

Antidesma auct. = Humulopsis

scandens Lour. = Humulopsis scandens

Cannabis L.

x **intersita** Sojak (*C. sativa* L. subsp. x *intersita* (Sojak) Sojak) - 1
ruderalis Janisch. - 1, 2, 3, 4, 6
sativa L. - 1, 2, 3, 4, 5, 6
- subsp. x *intersita* (Sojak) Sojak = C. x intersita

Humulopsis Grudz. (*Antidesma* auct.)

scandens (Lour.) Grudz. (*Antidesma scandens* Lour., *Humulus japonicus* Siebold & Zucc., *Humulus scandens* (Lour.) Merr.) - 5

Humulus L.

cordifolius Miq. (*Humulus lupulus* L. var. *cordifolius* (Miq.) Maxim.) - 5
japonicus Siebold & Zucc. = Humulopsis scandens
lupulus L. - 1, 2, 3, 4, 5, 6
- var. *cordifolius* (Miq.) Maxim. = H. cordifolius
scandens (Lour.) Merr. = Humulopsis scandens

CAPPARACEAE Juss. (*CLEOMACEAE* Horan., *CLEOMACEAE* (Pax) Airy Shaw, nom. superfl.)

Buhsea Bunge

coluteoides (Boiss.) Bunge (*Cleome coluteoides* Boiss.) - 6
raddeana (Trautv.) Boriss. (*Cleome raddeana* Trautv.) - 6

Capparis L.

herbacea Willd. (*C. ovata* Desf. var. *herbacea* (Willd.) Zohary, *C. spinosa* auct.) - 1, 2, 6
ovata Desf. var. *herbacea* (Willd.) Zohary = C. herbacea
rosanowiana B. Fedtsch. - 6
spinosa auct. = C. herbacea

Cleome L.

ariana Hedge & Lamond (*C. griffithiana* auct., *C. ornithopodioides* auct. p.p.) - 6
canescens Stev. ex DC. (*C. ornithopodioides* L. subsp. *canescens* (Stev. ex DC.) Tzvel., *C. ornithopodioides* auct. p.p.) - 1
circassica Tzvel. - 2
coluteoides Boiss. = Buhsea coluteoides
daghestanica (Rupr.) Tzvel. (*C. ornithopodioides* L. f. *daghestanica* Rupr.) - 2
donetzica Tzvel. (*C. ornithopodioides* L. subsp. *donetzica* (Tzvel.) Tzvel.) - 1
fimbriata Vicary (*C. griffithiana* Rech. fil., *C. noeana* Boiss., *C. quinquenervia* DC. var. *noeana* (Boiss.) Parsa) - 6
gordjaginii M. Pop. - 6
griffithiana auct. = C. ariana
griffithiana Rech. fil. = C. fimbriata
iberica DC. (*C. ornithopodioides* auct. p.p.) - 2
khorassanica Bunge & Bien. - 6
lipskyi M. Pop. - 6
noeana Boiss. = C. fimbriata
ornithopodioides auct. = C. ariana, C. canescens, C. iberica and C. steveniana
ornithopodioides L. subsp. *canescens* (Stev. ex DC.) Tzvel. = C. canescens
- subsp. *donetzica* (Tzvel.) Tzvel. = C. donetzica
- f. *daghestanica* Rupr. = C. daghestanica
quinquenervia DC. - 6
- var. *noeana* (Boiss.) Parsa = C. fimbriata
raddeana Trautv. = Buhsea raddeana
rostrata Bobr. - 6
steveniana Schult. & Schult. fil. (*C. ornithopodioides* auct. p.p.) - 2
tomentella M. Pop. - 6
turkmena Bobr. - 6

Polanisia Rafin.

dodecandra (L.) DC. subsp. **dodecandra** var. **trachysperma**

(Torr. & Gray) Iltis (*P. trachysperma* Torr. & Gray) - 5

trachysperma Torr. & Gray = P. dodecandra subsp. *dodecandra* var. trachysperma

CAPRIFOLIACEAE Juss.
Abelia R. Br.
coreana Nakai - 5
corymbosa Regel & Schmalh. - 6

Linnaea L.
borealis L. - 1, 2, 3, 4, 5

Lonicera L. (*Metalonicera* M. Wang & A.G. Gu)
alberti Regel - 6
alpigena L. subsp. *glehnii* (Fr. Schmidt) Hara = L. glehnii
- subsp. *glehnii* (Fr. Schmidt) Nedoluzhko = L. glehnii
altaica Pall. (*L. caerulea* L. subsp. *altaica* (Pall.) Gladkova) - 1, 3, 4
altmannii Regel & Schmalh. - 6
*x amoena Zab. - L. korolkowii Stapf x L. tatarica L.
anisotricha Bondar. - 6
asperifolia (Decne.) Hook. fil. & Thoms. - 6
baltica Pojark. - 1
*x **bella** Zab. - L. morrowii A. Gray x L. tatarica L.
bracteolaris Boiss. & Buhse - 2, 6
buschiorum Pojark. - 2
caerulea L. (*L. caerulea* var. *hirsuta* Regel, *L. caerulea* subsp. *hirsuta* (Regel) Kuvajev, *L. caerulea* subsp. *kamtschatica* (Sevast.) Gladkova, *L. caerulea* subsp. *venulosa* (Maxim.) Worosch., *L. edulis* Turcz. ex Freyn var. *turczaninowii* (Pojark.) Kitag., *L. emphyllocalyx* Maxim., *L. kamtschatica* (Sevast.) Pojark., *L. regeliana* Boczkarn. 1975, non Petzold & Kirchner, 1864, *L. turczaninowii* Pojark., *L. venulosa* Maxim.) - 1, 2, 3, 4, 5, 6
- subsp. *altaica* (Pall.) Gladkova = L. altaica
- subsp. *edulis* (Turcz. ex Freyn) Hult. = L. edulis
- subsp. *hirsuta* (Regel) Kuvajev = L. caerulea
- subsp. *kamtschatica* (Sevast.) Gladkova = L. caerulea
- subsp. *pallasii* (Ledeb.) Browicz = L. pallasii
- subsp. *venulosa* (Maxim.) Worosch. = L. caerulea
- var. *hirsuta* Regel = L. caerulea
caprifolium L. - 1(cult.), 2, 6(cult.)
caucasica Pall. = L. orientalis
chamissoi Bunge ex P. Kir. - 5
chrysantha Turcz. ex Ledeb.(*L. chrysantha* subsp. *gibbiflora* (Rupr.) Kitag., *L. gibbiflora* (Rupr.) Dipp., *L. regeliana* Petzold & Kirchner, 1864, non Boczkarn. 1975) - 4, 5
- subsp. *gibbiflora* (Rupr.) Kitag. = L. chamissoi
cinerea Pojark. = L. zaravschanica
edulis Turcz. ex Freyn (*L. caerulea* L. subsp. *edulis* (Turcz. ex Freyn) Hult., *Metalonicera edulis* (Turcz. ex Freyn) M. Wang & A.G. Gu) - 4, 5
- var. *turczaninowii* (Pojark.) Kitag. = L. caerulea
emphyllocalyx Maxim. = L. caerulea
etrusca Santi - 1, 2
floribunda Boiss. & Buhse - 2, 6
gibbiflora (Rupr.) Dipp. = L. chamissoi
glehnii Fr. Schmidt (*L. alpigena* L. subsp. *glehnii* (Fr. Schmidt) Hara, *L. alpigena* subsp. *glehnii* (Fr. Schmidt) Nedoluzhko, comb. superfl.) - 5
heterophylla auct. = L. karelinii
heterotricha Pojark. & Zak. - 6

hispida Pall. ex Schult. - 3, 6
humilis Kar. & Kir. (*L. popovii* Golosk.) - 6
iberica Bieb. - 2
iliensis Pojark. - 6
***japonica** Thunb. - 1, 2, 6
kamtschatica (Sevast.) Pojark. = L. caerulea
karataviensis Pavl. = L. tatarica
karelinii Bunge ex P. Kir. (*L. heterophylla* auct.) - 6
korolkowii Stapf (*L. lanata* Pojark.) - 6
lanata Pojark. = L. korolkowii
maackii (Rupr.) Herd. - 5
maximowiczii (Rupr.) Regel - 5
- subsp. *sachalinensis* (Rupr.) Nedoluzhko = L. sachalinensis
micrantha (Trautv.) Regel = L. tatarica
*x **micranthoides** Zab. - L. nigra L. x L. tatarica L.
microphylla Willd. ex Schult. - 3, 6
monantha Nakai (*L. subhispida* Nakai) - 5
nigra L. - 1
nummariifolia Jaub. & Spach - 6
olgae Regel & Schmalh. - 6
orientalis Lam. (*L. caucasica* Pall.) - 2
pallasii Ledeb. (*L. caerulea* L. subsp. *pallasii* (Ledeb.) Browicz) - 1, 3, 4
pamirica Pojark. - 6
paradoxa Pojark. - 6
***periclymenum** L. - 1
popovii Golosk. = L. humilis
praeflorens Batal. - 5
regeliana Boczkarn. = L. caerulea
regeliana Petzold & Kirchner = L. chrysantha
ruprechtiana Regel - 5
sachalinensis (Fr. Schmidt) Nakai (*L. maximowiczii* (Rupr.) Regel subsp. *sachalinensis* (Rupr.) Nedoluzhko) - 5
semenovii Regel - 6
simulatrix Pojark. - 6
sovetkinae V. Tkaczenko - 6
stenantha Pojark. - 3, 6
steveniana Fisch. ex Pojark. - 2
*x **subarctica** Pojark. - L. altaica Pall. x L. pallasii Ledeb. - 1, 3, 4
subhispida Nakai = L. monantha
tatarica L. (*L. karataviensis* Pavl., *L. micrantha* (Trautv.) Regel) - 1, 3, 6
tianschanica Pojark. - 6
tolmatchevii Pojark. - 5
turczaninowii Pojark. = L. caerulea
venulosa Maxim. = L. caerulea
* **xylosteoides** Tausch. - L. tatarica L. x L. xylosteum L.
xylosteum L. - 1, 3
zaravschanica (Rehd.) Pojark. (*L. cinerea* Pojark.) - 6

Metalonicera M. Wang & A.G. Gu = Lonicera
edulis (Turcz. ex Freyn) M. Wang & A.G. Gu = Lonicera edulis

*Symphoricarpos Duham.
***rivularis** Suksdorf - 1

Triosteum L.
sinuatum Maxim. - 5

Weigela Thunb.
middendorffiana (Carr.) C. Koch - 5

praecox (Lemoine) Bailey - 5
suavis (Kom.) Bailey - 5

CARPINACEAE Kuprian. = BETULACEAE

CARPINACEAE Vest = BETULACEAE

CARYOPHYLLACEAE Juss.
Acanthophyllum C.A. Mey.

■ acerosum Sosn.
aculeatum Schischk. - 6
adenophorum Freyn - 6
albidum Schischk. - 6
borsczowii Litv. - 6
brevibracteatum Lipsky - 6
brevicalycale Sosk. = A. pungens
bungei (Boiss.) Trautv. = Allochrusa bungei
coloratum Schischk. - 6
crassinodum Jukhananov & J. Edmondson - 6
cyrtostegium Vved. - 6
elatius Bunge - 6
elongatum Preobr. ex Schischk. = A. gracile
glandulosum Bunge - 6
gracile Boiss. (*A. elongatum* Preobr. ex Schischk.) - 6
gypsophiloides Regel = Allochrusa gypsophiloides
knorringianum Schischk. = Kuhitangia knorringiana
korolkowii Regel & Schmalh. - 6
korshinskyi Schischk. - 6
krascheninnikovii Schischk. - 6
laxiusculum Schiman-Czeika - 6
leiostegium Vved. = A. subglabrum
lilacinum Schischk. - 6
microcephalum Boiss. - 6
mikeschinianum Jukhananov & Kuvajev - 6
mucronatum C.A. Mey. - 2
pachystegium Rech. fil. - 6
paniculatum Regel = Allochrusa paniculata
popovii (Preobr.) Barkoudah = Kuhitangia popovii
pulchrum Schischk. - 6
pungens (Bunge) Boiss. (*A. brevicalycale* Sosk., *A. ruptile* Vved., *A. squarrosum* auct.) - 2, 6
ruptile Vved. = A. pungens
sarawschanicum Golenk. - 6
schugnanicum Schischk. - 6
sordidum Bunge - 6
speciosum Rech. fil. & Schiman-Czeika - 6
squarrosum auct. = A. pungens
stenostegium Freyn - 6
subglabrum Schischk. (*A. leiostegium* Vved., *A. turcomanicum* Schischk.) - 6
tadshikistanicum Schischk. = Allochrusa tadshikistanica
tenuifolium Schischk. - 6
transhyrcanum Preobr. = Diaphanoptera transhyrcana
turcomanicum Schischk. = A. subglabrum
versicolor Fisch. & C.A. Mey. = Allochrusa versicolor

Agrostemma L.

githago L. (*A. githago* var. *linicola* (Terech.) Hammer, *A. githago* var. *microspermum* (Levina) Hammer, *A. linicola* Terech., *A. macrospermum* Levina) - 1, 2, 3, 4, 5, 6
- var. *linicola* (Terech.) Hammer = A. githago
- var. *microspermum* (Levina) Hammer = A. githago

involucratum (Cham. & Schlecht.) G. Don fil. = Gastrolychnis involucrata
linicola Terech. = A. githago
macrospermum Levina = A. githago

Allochrusa Bunge

bungei Boiss. (*Acanthophyllum bungei* (Boiss.) Trautv.) - 2
gypsophiloides (Regel) Schischk. (*Acanthophyllum gypsophiloides* Regel, *Allochrusa gypsophiloides* (Regel) Ovcz. & Czuk. comb. superfl.) - 6
paniculata (Regel) Ovcz. & Czuk. (*Acanthophyllum paniculatum* Regel, *Allochrusa paniculata* subsp. *ferganensis* Jusupov ex Jukhananov) - 6
- subsp. *ferganensis* Jusupov ex Jukhananov = A. paniculata
tadshikistanica Schischk. (*Acanthophyllum tadshikistanicum* Schischk. nom. altern.) - 6
takhtajanii Gabr. & Dittr. - 2
transhyrcana (Preobr.) Czer. = Diaphanoptera transhyrcana
versicolor (Fisch. & C.A. Mey.) Boiss. (*Acanthophyllum versicolor* Fisch. & C.A. Mey.) - 2

Alsinanthe (Fenzl) Reichenb. = Minuartia

elegans (Cham. & Schlecht.) A. & D. Löve = Minuartia elegans
rossii (R. Br.) A. & D.Löve = Minuartia rossii

Alsine L. p.p. = Spergularia

circassica Albov = Minuartia circassica
glandulosa Boiss. & Huet = Minuartia glandulosa
gracilipes Kom. = Minuartia gracilipes
juniperina (L.) Wahlenb. var. *lineata* Boiss. = Minuartia lineata
neglecta (Weihe) A. & D. Löve = Stellaria neglecta
oxypetala Woloszcz. = Minuartia oxypetala
pinifolia Fenzl var. *eglandulosa* Fenzl = Minuartia eglandulosa
segetalis L. = Spergularia segetalis
tschuktschorum (Regel) Steffen = Eremogone tschuktschorum
verna (L.) Wahlenb. var. *glacialis* Fenzl = Minuartia rubella
zarecznyi Zapal. = Minuartia zarecznyi

Alsinopsis auct. = Minuartia

obtusiloba Rydb. = Minuartia obtusiloba

Alsinula Dostal = Stellaria

media (L.) Dostal = Stellaria media
neglecta (Weihe) Dostal = Stellaria neglecta
pallida (Dumort.) Dostal = Stellaria pallida

Ammodenia J.G. Gmel. ex S.G. Gmel. = Honckenya

major (Hook.) Heller = Honckenya oblongifolia
peploides (L.) Rupr. = Honckenya peploides

Arenaria L.

anomala (Waldst. & Kit.) Shinners = Dichodon viscidum
armena Schischk. = Minuartia glandulosa
asiatica Schischk. = Eremogone asiatica
biebersteinii Schlecht. = Eremogone biebersteinii

biflora L. - 1
blepharophylla Boiss. subsp. *steveniana* (Boiss.) T.N. Pop. = Eremogone steveniana
- var. *parviflora* (Fenzl) McNeill = Eremogone steveniana
brachypetala (Grossh.) T.N. Pop. = Eremogone brachypetala
brevifolia Gilib. = A. leptoclados
brotheriana Trautv. = Minuartia trautvetteriana
bungei Barkoudah = Dichoglottis alsinoides
caespitosa (Cambess.) Ju. Kozhevn. = Thylacospermum caespitosum
capillaris Poir. = Eremogone capillaris
caricifolia Boiss. = Eremogone caricifolia
cephalotes Bieb. = Eremogone cephalotes
ciliata auct. p.p. = A. pseudofrigida
cucubaloides Smith = Eremogone cucubaloides
dianthoides Smith = Eremogone dianthoides
dicranoides (Cham. & Schlecht.) Hult. = Stellaria dicranoides
ferganica Schischk. = Eremogone ferganica
formosa Fisch. ex Ser. = Eremogone formosa
glaucescens H. Winkl. = Eremogone glaucescens
graminea C.A. Mey. = Eremogone graminea
- var. *brachypetala* Grossh. = Eremogone brachypetala
graminifolia Schrad. = Eremogone saxatilis
graminifolia sensu Schischk. & Knorr. = Eremogone biebersteinii, E. micradenia and E. saxatilis
griffithii Boiss. = Eremogone griffithii
gypsophiloides L. = Eremogone gypsophiloides
holostea Bieb. = Eremogone holostea
- subsp. *macrantha* (Schischk.) McNeill = Eremogone macrantha
humifusa Wahlenb. - 1
insignis Litv. = Eremogone insignis
jacutorum A. Khokhr. - 4
juncea Bieb. = Eremogone juncea
juniperina L. = Minuartia juniperina
koriniana Fisch. ex Fenzl = Eremogone koriniana
ledebouriana Fenzl = Eremogone ledebouriana
leptoclados (Reichenb.) Guss. (*A. serpyllifolia* L. var. *leptoclados* Reichenb., *A. brevifolia* Gilib. nom. invalid., *A. serpyllifolia* subsp. *leptoclados* (Reichenb.) Celak.) - 1, 2, 6
litwinowii Schischk. = Eremogone litwinowii
longifolia Bieb. = Eremogone longifolia
longipedunculata Hult. - 5
lychnidea Bieb. = Eremogone lychnidea
macrantha Schischk. = Eremogone macrantha
marginata DC. = Spergularia maritima
marina (L.) All. = Spergularia salina
maritima All. = Spergularia maritima
media auct. = Spergularia maritima
merckioides Maxim. - 5
meyeri Fenzl = Eremogone meyeri
micradenia P. Smirn. = Eremogone micradenia
mongolica Schischk. = Eremogone mongolica
multicaulis L. subsp. *pseudofrigida* (Ostenf. & Dahl) A. & D. Löve = A. pseudofrigida
oosepala Bordz. = Eremogone oosepala
orbicularis Vis. - 1(alien)
paulsenii H. Winkl. = Eremogone paulsenii
peploides L. var. *diffusa* Hornem. = Honckenya oblongifolia
- var. *major* Hook. = Honckenya oblongifolia
pineticola Klok. = Eremogone pineticola
polaris Schischk. = Eremogone polaris
potaninii Schischk. - 3, 6

procera auct. = Eremogone biebersteinii
procera Spreng. = Eremogone saxatilis
- subsp. *glabra* (F. Williams) Holub = Eremogone micradenia and E. saxatilis
- subsp. *pubescens* (Fenzl) Jalas = Eremogone biebersteinii
pseudofrigida (Ostenf. & Dahl) Juz. ex Schischk. (*A. multicaulis* L. subsp. *pseudofrigida* (Ostenf. & Dahl) A. & D. Löve, *A. ciliata* auct. p.p.) - 1
redowskii Cham. & Schlecht. - 4, 5
- subsp. **ajanensis** Worosch. - 5
rigida Bieb. = Eremogone rigida
rossii R. Br. = Minuartia rossii
- subsp. *elegans* (Cham. & Schlecht.) Maguire = Minuartia elegans
rotundifolia M. Bieb. (*A. turkestanica* Schischk.) - 2, 6
rubra L. = Spergularia rubra
saxatilis L. = Eremogone saxatilis
serpyllifolia L. - 1, 2, 3, 5(alien), 6
- subsp. *leptoclados* (Reichenb.) Celak. = A. leptoclados
- var. *leptoclados* Reichenb. = A. leptoclados
stenophylla Ledeb. = Eremogone saxatilis
- subsp. *polaris* (Schischk.) E. Selivan. = Eremogone polaris
steveniana Boiss. = Eremogone steveniana
syreistschikowii P. Smirn. = Eremogone saxatilis
szowitsii Boiss. = Eremogone szowitsii
talassica Adylov = Eremogone talassica
trigyna (Vill.) Shinners = Dichodon ceratoides
tschuktschorum Regel = Eremogone tschuktschorum
turkestanica Schischk. = A. rotundifolia
turlanica Bajt. = Eremogone turlanica
ucranica Spreng. ex Steud. = Eremogone micradenia
uralensis Pall. ex Spreng. - 1
zozii Kleop. - 1

Behen Moench = Oberna

cserei (Baumg.) Gusul. = Oberna cserei

Behenantha (Otth) Schur = Oberna
behen (L.) Ikonn. = Oberna behen
commutata (Guss.) Ikonn. = Oberna commutata
crispata (Stev.) Ikonn. = Oberna crispata
cserei (Baumg.) Schur = Oberna cserei
lacera (Stev.) Ikonn. = Oberna lacera
littoralis (Rupr.) Ikonn. = Oberna littoralis
multifida (Adams) Ikonn. = Oberna multifida
procumbens (Murr.) Ikonn. = Oberna procumbens
uniflora (Roth) Ikonn. = Oberna uniflora
wallichiana (Klotzsch) Ikonn. = Oberna wallichiana

Bolbosaponaria Bondar.
babatagi (Ovcz.) Bondar. (*Saponaria babatagi* Ovcz.) - 6
bucharica (B. Fedtsch.) Bondar. (*Gypsophila bucharica* B. Fedtsch.) - 6
fedtschenkoana (Schischk.) Bondar. (*Gypsophila fedtschenkoana* Schischk.) - 6
intricata (Franch.) Bondar. (*Gypsophila intricata* Franch.) - 6
kafirniganica V.A. Shulz - 6
sewerzowii (Regel & Schmalh.) Bondar. (*Saponaria sewerzowii* Regel & Schmalh., *Gypsophila sewerzowii* (Regel & Schmalh.) R.Kam.) - 6
villosa (Barkoudah) Bondar. (*Gypsophila villosa* Barkoudah) - 6

Bufonia L.

macrocarpa auct. = B. sintenisii
oliveriana Ser. - 6

parviflora Griseb. (*B. tenuifolia* auct.) - 1, 2
sintenisii Freyn (*B. macrocarpa* auct.) - 6
tenuifolia auct. = B. parviflora

Cerastium L.

aleuticum Hult. - 5
alexeenkoanum Schischk. - 6
alpinum L. - 1
- subsp. *glabratum* ("C. Hartm.") A. & D. Löve = C. glabratum
- subsp. *lanatum* (Lam.) Aschers. & Graebn. = C. lanatum
- subsp. *lanatum* (Lam.) Simonk. = C. lanatum
- var. *caespitosum* Malmgr. = C. regelii subsp. caespitosum
- var. *glabratum* Wahlenb. = C. glabratum
amurense (Regel) Ohwi = C. pauciflorum
annae I. Sokolova - 2
anomalum Waldst. & Kit. = Dichodon viscidum
araraticum Rupr. - 2
arcticum Lange (*C. arcticum* subsp. *hyperboreum* (Tolm.) W. Bocher, *C. hyperboreum* Tolm.) - 1
- subsp. *hyperboreum* (Tolm.) W. Bocher = C. arcticum
argenteum Bieb. - 2
▪ **armeniacum** Gren.
arvense L. - 1, 2, 3, 4, 5, 6
atriusculum Klok. - 1
balearicum F. Herm. (*C. dentatum* Moschl, *C. semidecandrum* L. subsp. *balearicum* (F. Herm.) R. Litardiere, *C. semidecandrum* var. *dentatum* (Moschl) Charadze, *C. semidecandrum* subsp. *dentatum* (Moschl) Maire & Weiller) - 1, 2, 6
banaticum auct. p.p. = C. sosnowskyi
beeringianum Cham. & Schlecht. (*C. sugawarae* Koidz. & Ohwi) - 1, 4, 5
- subsp. *bialynickii* (Tolm.) Tolm. = C. bialynickii
- subsp. **continentale** Peschkova (*C. furcatum* auct. p.p.) - 4
- subsp. *jenisejense* (Hult.) Worosch. = C. jenisejense
beeringianum sensu Schischk. = C. jenisejense
bialynickii Tolm. (*C. beeringianum* Cham. & Schlecht. subsp. *bialynickii* (Tolm.) Tolm.) - 3, 4, 5
biebersteinii DC. (*C. biebersteinii* subsp. *transcaucasicum* Buschmann) - 1, 2
- subsp. *transcaucasicum* Buschmann = C. biebersteinii
boreale Takeda = C. fischerianum
borisii Zak. - 6
brachypetalum Desp. ex Pers. (*C. brachypetalum* subsp. *tauricum* (Spreng.) Murb., *C. brachypetalum* f. *tauricum* (Spreng.) Prod., *C. tauricum* Spreng.) - 1, 2
- subsp. *tauricum* (Spreng.) Murb. = C. brachypetalum
- f. *tauricum* (Spreng.) Prod. = C. brachypetalum
bungeanum Vved. (*C. falcatum* auct.) - 3, 6
caespitosum Gilib. = C. holosteoides
- subsp. *triviale* (Spenn.) Hiit. = C. holosteoides
capillatum I. Sokolova - 2
cerastoides (L.) Britt. = Dichodon cerastoides
▪ **chloriifolium** Fisch. & C.A. Mey. (*Dichodon chloriifolium* (Fisch. & C.A. Mey.) A. & D.Löve)
ciliatum Turcz. = C. subciliatum
crassiusculum Klok. - 1
daghestanicum Schischk. - 2
davuricum Fisch. ex Spreng. (*Dichodon davuricum* (Fisch. ex Spreng.) A. & D. Löve) - 1, 2, 3, 4, 6
dentatum Moschl = C. balearicum

dichotomum L. - 2, 6
- subsp. *inflatum* (Link) Cullen = C. inflatum
diffusum Pers. - 1
dubium (Bast.) Guepin = Dichodon viscidum
dubium (Bast.) O. Schwarz = Dichodon viscidum
elbrusense Boiss. (*C. purpurascens* Adams var. *elbrusense* (Boiss.) Moschl) - 2
falcatum auct. = C. bungeanum
fischerianum Ser. (*C. boreale* Takeda, *C. schmidtianum* Takeda, *C. tatewakii* Miyabe, *C. unalaschkense* Takeda) - 5
- subsp. **albimarginatum** (Worosch.) Worosch. (*C. fischerianum* var. *albimarginatum* Worosch.) - 5
- var. *albimarginatum* Worosch. = C. fischerianum subsp. albimarginatum
flavescens H. Gartner (*C. rigidum* Ledeb. 1815, non Vitm. 1789, *C. furcatum* auct. p.p., *C. rubescens* auct. p.p.) - 4
    The name C. furcatum Cham. & Schlect. is recognized as nomen dubium.
- subsp. **jablonense** H. Gartner - 4
fontanum Baumg. (*C. fontanum* subsp. *macrocarpum* (Schur) Jalas, *C. lucorum* (Schur) Moschl subsp. *macrocarpum* (Schur) Lainz, *C. macrocarpum* Schur, *C. vulgatum* L. subsp. *macrocarpum* (Schur) B. Kotula) - 1, 2
- subsp. *holosteoides* (Fries) Salman, van Ommering & de Voogd = C. holosteoides
- subsp. *lucorum* (Schur) Soo = C. lucorum
- subsp. *macrocarpum* (Schur) Jalas = C. fontanum
- subsp. *scandicum* H. Gartner = C. scandicum
- subsp. *triviale* (Spenn.) Jalas = C. holosteoides
- subsp. **turcicum** H. Gartner - 2(?)
- subsp. *vulgare* (C. Hartm.) Greuter & Burdet = C. holosteoides
fragillimum Boiss. - 2
furcatum auct. = C. beeringianum, C. flavescens and C. subciliatum
glabratum (Wahlenb.) C. Hartm. (*C. alpinum* L. var. *glabratum* Wahlenb., *C. alpinum* subsp. *glabratum* ("C. Hartm.") A. & D. Löve) - 1
glanduliferum Schur var. *lucorum* Schur = C. lucorum
glomeratum Thuill. - 1, 2, 6
glutinosum Fries (*C. pallens* (F. Schultz) F. Schultz, *C. pumilum* Curt. subsp. *glutinosum* (Fries) Jalas, *C. pumilum* subsp. *pallens* (F. Schultz) Schinz & Thell.) - 1
gnaphalodes Fenzl - 2
- subsp. *album* Buschmann = C. szowitsii var. album
gorodkovianum Schischk. - 1
gracile auct. p.p. = C. schmalhausenii and C. pseudobulgaricum
hemschinicum Schischk. - 2
heterotrichum Klok. - 1
holosteoides Fries (*C. caespitosum* Gilib. nom. invalid., *C. caespitosum* subsp. *triviale* (Spenn.) Hiit., *C. fontanum* Baumg. subsp. *holosteoides* (Fries) Salman, van Ommering & de Voogd, *C. fontanum* subsp. *triviale* (Spenn.) Jalas, *C. fontanum* subsp. *vulgare* (C. Hartm.) Greuter & Burdet, *C. holosteoides* subsp. *triviale* (Spenn.) Moschl, ? *C. rigidulum* Takeda, *C. triviale* Link, nom. illegit., *C. vulgare* C. Hartm., *C. vulgatum* auct.) - 1, 2, 3, 4, 5, 6
- subsp. *triviale* (Spenn.) Moschl = C. holosteoides
holosteum Fisch. ex Hornem. - 2
- var. **meyerianum** (Rupr.) I. Sokolova, comb. nova (*C. meyerianum* Rupr. 1869, Fl. Cauc. 1 : 221) - 2
hyperboreum Tolm. = C. arcticum
igoschiniae Pobed. - 1

inflatum Gren. (*C. dichotomum* L. subsp. *inflatum* (Link) Cullen, nom. invalid.) - 2, 6

jenisejense Hult. (*C. beeringianum* Cham. & Schlecht. subsp. *jenisejense* (Hult.) Worosch., *C. beeringianum* sensu Schischk.) - 1, 3, 4, 5

kasbek Parrot - 2

kioviense Klok. - 1

krylovii Schischk. & Gorczak. - 1

lanatum Lam. (*C. alpinum* L. subsp. *lanatum* (Lam.) Aschers. & Graebn. comb. superfl., *C. alpinum* subsp. *lanatum* (Lam.) Simonk.) - 1

lithospermifolium Fisch. - 3, 4, 6

longifolium Willd. - 2

lucorum (Schur) Moschl (*C. glanduliferum* Schur var. *lucorum* Schur, *C. fontanum* Baumg. subsp. *lucorum* (Schur) Soo, *C. macrocarpum* Schur subsp. *lucorum* (Schur) H. Gartner) - 1

- subsp. *macrocarpum* (Schur) Lainz = C. fontanum

luridum Guss. - 2

macrocalyx Buschmann - 2

Dubious species.

macrocarpum Schur = C. fontanum

- subsp. *lucorum* (Schur) H. Gartner = C. lucorum

maximum L. (*Dichodon maximum* (L.) A. & D. Löve) - 1, 3, 4, 5

meyerianum Rupr. = C. holosteum var. meyerianum

microspermum C.A. Mey. - 2

multiflorum C.A. Mey. - 2

nemorale Bieb. - 1, 2

odessanum Klok. - 1

oreades Schischk. - 2

pallens (F. Schultz) F. Schultz = C. glutinosum

pauciflorum Stev. ex Ser. (*C. amurense* (Regel) Ohwi, *C. pauciflorum* var. *amurense* (Regel) Mizushima, *C. pilosum* Ledeb. var. *amurense* Regel) - 1, 3, 4, 5, 6

- var. *amurense* (Regel) Mizushima = C. pauciflorum

pentandrum L. = C. semidecandrum

perfoliatum L. (*Dichodon perfoliatum* (L.) A. & D. Löve) - 1, 2, 6

pilosum Ledeb. var. *amurense* Regel = C. pauciflorum

polymorphum Rupr. - 2

ponticum Albov - 2

porphyrii Schischk. - 3

pseudobulgaricum Klok. (*C. gracile* auct. p.p.) - 1

pseudokasbek Vysokoostr. - 2

pumilum Curt. - 1, 2

- subsp. *glutinosum* (Fries) Jalas = C. glutinosum

- subsp. *pallens* (F. Schultz) Schinz & Thell. = C. glutinosum

purpurascens Adams - 2

- subsp. **parviflorum** (Trautv.) I. Sokolova, comb. nova (*C. purpurascens* var. *parviflorum* Trautv. 1873, Acta Horti Petropol. 2, 2 : 514) - 2

- var. *elbrusense* (Boiss.) Moschl = C. elbrusense

- var. *parviflorum* Trautv. = C. purpurascens subsp. parviflorum

pusillum sensu Schischk. p.p. = C. subciliatum

pusillum Ser. - 3, 6

Further investigation is needed for the Altai and Middle Asian plants, which we have recorded under this name. The name C. pusillum Ser. is recognized as nomen dubium.

regelii Ostenf. - 1, 3, 4, 5

- subsp. **caespitosum** (Malmgr.) Tolm. (*C. alpinum* L. var. *caespitosum* Malmgr.) - 1, 3

- subsp. **regelii** - 3, 4, 5

rigidulum Takeda = C. holosteoides

rigidum Ledeb. = C. flavescens

rotundatum Schur = C. semidecandrum

rubescens auct. p.p. = C. flavescens and C. subciliatum

ruderale Bieb. - 2

saccardoanum Diratzouyan - 2

scammaniae Polun. - 5(?)

An additional study of this taxon is necessary.

scandicum (H. Gartner) Kuzen. (*C. fontanum* Baumg. subsp. *scandicum* H. Gartner) - 1

schischkinii Grossh. = Dichodon schischkinii

schmalhausenii Pacz. (*C. gracile* auct. p.p.) - 1

schmidtianum Takeda = C. fischerianum

semidecandrum L. (*C. pentandrum* L., *C. rotundatum* Schur) - 1, 2

- subsp. *balearicum* (F. Herm.) R. Litardiere = C. balearicum

- subsp. *dentatum* (Moschl) Maire & Weiller = C. balearicum

- var. *dentatum* (Moschl) Charadze = C. balearicum

sosnowskyi Schischk. (*C. banaticum* auct. p.p.) - 2

stevenii Schischk. - 1

subciliatum H. Gartner (*C. ciliatum* Turcz. 1842, non Waldst. & Kit. 1805, *C. furcatum* auct. p.p., *C. pusillum* sensu Schischk. p.p., *C. rubescens* auct. p.p.) - 4

- subsp. **sajanense** H. Gartner - 4

sugawarae Koidz. & Ohwi = C. beeringianum

svanicum Charadze - 2

sylvaticum Waldst. & Kit. - 1

syvaschicum Kleop. - 1, 2

szowitsii Boiss. - 2

- var. **album** (Buschmann) I. Sokolova, comb. nova (*C. gnaphalodes* Fenzl subsp. *album* Buschmann, 1939, Feddes Repert. 46 : 34) - 2

taschkendicum Adylov & Vved. - 6

tatewakii Miyabe = C. fischerianum

tauricum Spreng. = C. brachypetalum

tenoreanum Ser. - 1

tianschanicum Schischk. (*C. vulgatum* L. var. *tianschanicum* (Schischk.) Ju. Kozhevn.) - 6

trigynum Vill. = Dichodon cerastoides

triviale Link = C. holosteoides

ucrainicum Pacz. ex Klok. - 1

unalaschkense Takeda = C. fischerianum

undulatifolium Somm. & Levier - 2

uralense Grub. - 1

vulgare C. Hartm. = C. holosteoides

vulgatum auct. = C. holosteoides

vulgatum L. subsp. *macrocarpum* (Schur) B. Kotula = C. fontanum

- var. *tianschanicum* (Schischk.) Ju. Kozhevn. = C. tianschanicum

zhiguliense S. Saksonov - 1

Charesia N. Busch = Silene

akinfievii (Schmalh.) E. Busch = Silene akinfievii

Cherleria auct. = Stellaria

dicranoides Cham. & Schlecht. = Stellaria dicranoides

Coccyganthe (Reichenb.) Reichenb.

flos-cuculi (L.) Fourr. (*Lychnis flos-cuculi* L., *Coronaria flos-cuculi* (L.) R. Br., *Silene flos-cuculi* (L.) Greuter & Burdet) - 1, 2, 3, 4, 5(alien)

Conosilene (Rohrb.) = Pleconax

conica (L.) Fourr. = Pleconax conica

- subsp. *conica* var. *subconica* (Friv.) A. Löve & Kjellqvist = Pleconax subconica

- subsp. *coniflora* (Nees ex Otth) A. Löve & Kjellqvist =

Pleconax coniflora
- subsp. *conoidea* (L.) A. Löve & Kjellqvist = Pleconax conoidea

Coronaria Hill

coriacea (Moench) Schischk. & Gorschk. (*Lychnis coronaria* (L.) Desr.) - 1, 2, 5(alien), 6
flos-cuculi (L.) R.Br. = Coccyganthe flos-cuculi

Cucubalus L.

baccifer L. - 1, 2, 3, 6
- var. *japonicus* Miq. = C. japonicus
japonicus (Miq.) Worosch. (*C. baccifer* L. var. *japonicus* Miq.) - 5
multiflorus Ehrh. = Silene multiflora
spergulifolius Desf. = Silene spergulifolia
spergulifolius Willd. = Silene spergulifolia
venosus Gilib. = Oberna behen

Dianthus L.

abchasicus Gvinianidze - 2
acantholimonoides Schischk. - 2
acicularis Fisch. ex Ledeb. - 1, 3, 6
alatavicus M. Pop. = D. turkestanicus
alpinus L. subsp. *repens* (Regel) Ju. Kozhevn. = D. repens
amurensis Jacques = D. chinensis
■ **andronakii** Woronow ex Schischk.
andrzejowskianus (Zapal.) Kulcz. (*D. capitatus* Balb. ex DC. subsp. *andrzejowskianus* Zapal.)
angrenicus Vved. - 6
arenarius L. (*D. krylovianus* Juz.) - 1, 2
- subsp. *borussicus* Vierh. = D. borussicus
- subsp. *pseudoserotinus* (Blocki) Tutin = D. pseudoserotinus
- subsp. *pseudosquarrosus* (Novak) Kleop. = D. pseudosquarrosus
armeria L. - 1, 2, 5(alien)
attenuatus V. Pavl. - 6
awaricus Charadze - 2
azkurensis Sosn. - 2
baldshuanicus Lincz. (*D. crinitus* Smith subsp. *baldshuanicus* (Lincz.) Rech. fil.) - 6
*****barbatus** L. - 1, 2, 3, 5, 6
- subsp. *compactus* (Kit.) Heuff. = D. compactus
- subsp. *compactus* (Kit.) Nym. = D. compactus
- subsp. *compactus* (Kit.) Stojan. = D. compactus
bessarabicus Klok. (*D. polymorphus* Bieb. subsp. *bessarabicus* Kleop. non basionymum, *D. polymorphus* var. *bessarabicus* (Klok.) Sanda) - 1
bicolor Adams (*D. preobrashenskii* Klok.) - 2
borbasii Vandas - 1, 2, 3, 6
- subsp. *capitellatus* (Klok.) Tutin = D. capitellatus
borussicus (Vierh.) Juz. (*D. arenarius* L. subsp. *borussicus* Vierh.) - 1
brachyodontus Boiss. & Huet - 2
brevipetalus Vved. - 6
bucoviensis (Zapal.) Klok. (*D. carthusianorum* L. var. *bucoviensis* Zapal.) - 1
caesius Smith = D. gratianopolitanus
calocephalus Boiss. - 2
campestris Bieb. (*D. campestris* subsp. *arenarius* Sirj., *D. campestris* subsp. *steppaceus* Sirj., *D. pseudoversicolor* Klok.) - 1, 3, 6
- subsp. *arenarius* Sirj. = D. campestris
- subsp. **laevigatus** (Grun.) Klok. (*D. campestris* var. *laevi-*

gatus Grun.) - 1
- subsp. *steppaceus* Sirj. = D. campestris
- var. *laevigatus* Grun. = D. campestris subsp. laevigatus
canescens C. Koch - 2
capitatus Balb. ex DC. - 1, 2
- subsp. *andrzejowskianus* Zapal. = D. andrzejowskianus
capitellatus Klok. (*D. borbasii* Vandas subsp. *capitellatus* (Klok.) Tutin) - 1
carbonatus Klok. - 1
carpaticus Woloszcz. (*D. tenuifolius* auct.) - 1
carthusianorum L. - 1
- subsp. *polonicus* (Zapal.) M. Kovanda = D. polonicus
- var. *bucoviensis* Zapal. = D. bucoviensis
- var. *commutatus* Zapal. = D. commutatus
*****caryophyllus** L.
caucaseus Smith (*D. discolor* Smith, *D. seguieri* auct. p.p.) - 2
charadzeae Gagnidze & Gviniaschvili - 2
chinensis L. (*D. amurensis* Jacques, *D. chinensis* var. *amurensis* (Jacques) Kitag.) - 5, in 1, 3, 6 cultivated only.
- subsp. **paracampestris** Worosch. - 5
- subsp. **reflexus** Worosch. - 5
- subsp. *repens* (Willd.) Worosch. = D. repens
- subsp. *versicolor* (Fisch. ex Link) Worosch. = D. versicolor
- var. *amurensis* (Jacques) Kitag. = D. chinensis
- var. *versicolor* (Fisch. ex Link) Y.C. Ma = D. versicolor
collinus Waldst. & Kit. - 1
- subsp. *glabriusculus* (Kit.) Soo = D. glabriusculus
- subsp. *glabriusculus* (Kit.) Thaisz = D. glabriusculus
commutatus (Zapal.) Klok. (*D. carthusianorum* L. var. *commutatus* Zapal.) - 1
compactus Kit. (*D. barbatus* L. subsp. *compactus* (Kit.) Heuff., *D. barbatus* subsp. *compactus* (Kit.) Nym. comb. superfl., *D. barbatus* subsp. *compactus* (Kit.) Stojan. comb. superfl.) - 1
x **courtoisii** Reichenb. - D. barbarus L. x D. superbus L.
cretaceus Adams (*D. liboschitzianus* Ser., *D. petraeus* Bieb.) - 2
crinitus Smith - 1
- subsp. *baldshuanicus* (Lincz.) Rech. fil. = D. baldshuanicus
- subsp. *soongoricus* (Schischk.) Ju. Kozhevn. = D. soongoricus
- subsp. *tetralepis* (Nevski) Rech. fil. = D. tetralepis
- subsp. *turcomanicus* (Schischk.) Rech. fil. = D. turcomanicus
crossopetalus (Boiss. ex Fenzl) Grossh. - 2
cyri Fisch. & C.A. Mey. - 2, 6
dagestanicus Charadze - 2
darvazicus Lincz. - 6
deltoides L. - 1, 3, 5(alien)
discolor Smith = D. caucaseus
elatus Ledeb. - 3, 6
elbrusense Charadze - 2
eugeniae Kleop. (*D. tesquicola* Klok.) - 1
euponticus Zapal. = D. pseudobarbatus
fischeri Spreng. - 1
floribundus Boiss. (*D. schischkinii* Grossh.) - 2
fragrans Adams - 2
glabriusculus (Kit.) Borb. (*D. collinus* Waldst. & Kit. subsp. *glabriusculus* (Kit.) Soo, *D. collinus* subsp. *glabriusculus* (Kit.) Thaisz) - 1
gratianopolitanus Vill. (*D. caesius* Smith) - 1
grossheimii Schischk. (*D. seguieri* auct. p.p.) - 2
guttatus Bieb. (*D. mariae* Klok. nom. invalid.) - 1

helenae Vved. - 6

x **helwigii** Aschers. - A. armeria L. x D. deltoides L.

hoeltzeri C. Winkl. = D. superbus

humilis Willd. ex Ledeb. - 1

hypanicus Andrz. - 1

imereticus (Rupr.) Schischk. - 2

inamoenus Schischk. (*D. pallens* Smith subsp. *inamoenus* (Schischk.) Sanda) - 2

jacobsii Rech. fil. - 2

x **jaczonis** Aschers. - D. deltoides L. x D. superbus L.

jaroslavii Galushko - 2

karataviensi Pavl. ex Schischk. - 6

ketzkhovelii Makaschvili - 2

kirghizicus Schischk. - 6

krylovianus Juz. = D. arenarius

kubanensis Schischk. - 2

kuschakewiczii Regel & Schmalh. - 6

kusnezovii Marc. - 2

lanceolatus Stev. ex Reichbenb. = D. pallens

lenkoranicus Charadze - 2

leptopetalus Willd. - 1, 3, 6

libanotis Labill. - 2

liboschitzianus Ser. = D. cretaceus

macronyx Fenzl ex Schischk. (*D. orientalis* Adams subsp. *stenocalyx* (Boiss.) Rech. fil.) - 6

> Probably. the priority name of this taxa should be D. pulverulentus Stapf.

maeoticus Klok. - 1

mariae Klok. = D. guttatus

marschallii Schischk. (*D. pallens* Smith subsp. *marschallii* (Schischk.) Sanda) - 1

membranaceus Borb. - 1

monadelphus subsp. *pallens* (Smith) Greuter & Burdet = D. pallens

multicaulis Bouss. & Huet - 2

multisquameus Bondar. & R. Vinogradova - 6

orientalis Adams - 2, 6

- subsp. *stenocalyx* (Boiss.) Rech. fil. = D. macronyx

oschtenicus Galushko - 2

pallens Sibth. & Sm. (*D. lanceolatus* Stev. ex Reichb., *D. monadelphus* subsp. *pallens* (Smith) Greuter & Burdet) - 1, 2, 6

- subsp. *inamoenus* (Schischk.) Sanda = D. inamoenus

- subsp. *marschallii* (Schischk.) Sanda = D. marschallii

pallidiflorus Ser. - 1, 2

pamiralaicus Lincz. - 6

parviflorus Boiss. - 2

patentisquameus Bondar. & R. Vinogradova - 6

petraeus Bieb. = D. cretaceus

platyodon Klok. (*D. polymorphus* Bieb. var. *platyodon* (Klok.) Sanda) - 1

polonicus Zapal. (*D. carthusianorum* L. subsp. *polonicus* (Zapal.) M. Kovanda) - 1

polylepis Bien. ex Boiss. - 6

polymorphus Bieb. - 1, 2

- subsp. *bessarabicus* Kleop. = D. bessarabicus

- var. *bessarabicus* (Klok.) Sanda = D. bessarabicus

- var. *platyodon* (Klok.) Sanda = D. platyodon

pontederae A. Kerner - 1

pratensis Bieb. - 1

preobrashenskii Klok. = D. bicolor

pseudoarmeria Bieb. - 1, 2

pseudobarbatus Bess. ex Ledeb. (*D. euponticus* Zapal., *D. trifasciculatus* Kit. subsp. *euponticus* (Zapal.) Kleop., *D. trifasciculatus* subsp. *pseudobarbatus* (Schmalh.) Jalas) - 1

pseudoserotinus Blocki (*D. arenarius* L. subsp. *pseudose-*

rotinus (Blocki) Tutin) - 1

pseudosquarrosus (Novak) Klok. (*D. arenarius* L. subsp. *pseudosquarrosus* (Novak) Kleop.) - 1

pseudoversicolor Klok. = D. campestris

raddeanus Vierh. - 2

ramosissimus Pall. ex Poir. - 3, 6

■ recognitus Schischk.

rehmannii Blocki - 2

repens Willd. (*D. alpinus* L. subsp. *repens* (Regel) Ju. Kozhevn., *D. chinensis* L. subsp. *repens* (Willd.) Worosch.) - 1, 3, 4, 5

rigidus Bieb. - 1, 3, 6

rogowiczii Kleop. - 1

ruprechtii Schischk. - 2

schemachensis Schischk. - 2

schischkinii Grossh. = D. floribundus

seguieri auct. p.p. = D. caucaseus and D. grossheimii

semenovii (Regel & Herd.) Vierh. - 6

seravschanicus Schischk. - 6

serotinus Waldst. & Kit. - 1

soongoricus Schischk. (*D. crinitus* Smith subsp. *soongoricus* (Schischk.) Ju. Kozhevn.) - 6

speciosus Reichenb. (*D. superbus* L. subsp. *alpestris* Kablik ex Celak., *D. superbus* subsp. *speciosus* (Reichenb.) Hayek) - 1

spiculifolius Schur - 1

squarrosus Bieb. - 1, 3, 6

stenocalyx Juz. - 1

subscabridus Lincz. - 6

subulosus Freyn & Conrath - 2

superbus L. (*D. hoeltzeri* C. Winkl.) - 1, 3, 4, 5, 6

- subsp. *alpestris* Kablik ex Celak. = D. speciosus

- subsp. *speciosus* (Reichenb.) Hayek = D. speciosus

tabrisianus Bien. ex Boiss. - 2

talyschensis Boiss. & Buhse - 2

tenuifolius auct. = D. carpaticus

tesquicola Klok. = D. eugeniae

tetralepis Nevski (*D. crinitus* Smith subsp. *tetralepis* (Nevski) Rech. fil.) - 6

tianschanicus Schischk. - 6

tlaratensis Gussejnov - 2

transcaucasicus Schischk. - 2

trifasciculatus Kit. subsp. *euponticus* (Zapal.) Kleop. = D. pseudobarbatus

- subsp. *pseudobarbatus* (Schmalh.) Jalas = D. pseudobarbatus

turcomanicus Schischk. (*D. crinitus* Smith subsp. *turcomanicus* (Schischk.) Rech. fil.) - 6

turkestanicus Preobr. (*D. alatavicus* M. Pop., *D. versicolor* Fisch. ex Link subsp. *turkesnicus* (Preobr.) Ju. Kozhevn.) - 6

ugamicus Vved. - 6

uralensis Korsh. - 1, 3

uzbekistanicus Lincz. - 6

velutinus Guss. = Kohlrauschia velutina

versicolor Fisch. ex Link (*D. chinensis* L. subsp. *versicolor* (Fisch. ex Link) Worosch., *D. chinensis* var. *versicolor* (Fisch. ex Link) Y.C. Ma) - 1, 3, 4, 5, 6

- subsp. *turkestanicus* (Preobr.) Ju. Kozhevn. = D. turkestanicus

vladimiri Galushko - 2

volgicus Juz. - 1

Diaphanoptera Rech. fil.

transhyrcana (Preobr.) Rech. fil. & Schiman-Czeika (*Acanthophyllum transhyrcanum* Preobr., *Allochrusa transhyrcana* (Preobr.) Czer.) - 6

Dichodon (Bartl.) Reichenb. (*Provencheria* B. Boiv.)

cerastoides (L.) Reichenb. (*Arenaria trigyna* (Vill.) Shinners, *Cerastium cerastoides* (L.) Britt., *C. trigynum* Vill., *Provencheria cerastoides* (L.) B. Boiv.) - 1, 2, 3, 4, 6

chlorifolium (Fisch. & C.A. Mey) A. & D. Löve = Cerastium chlorifolium

davuricum (Fisch. ex Spreng.) A. & D. Löve = Cerastium davuricum

dubium (Bast.) Ikonn. = D. viscidum

maximum (L.) A. & D. Löve = Cerastium maximum

perfoliatum (L.) A. & D. Löve = Cerastium perfoliatum

schischkinii (Grossh.) Ikonn. (*Cerastium schischkinii* Grossh.) - 2

viscidum (Bieb.) Holub (*Stellaria viscida* Bieb. nom. legit., *Arenaria anomala* (Waldst. & Kit.) Shinners, nom. illegit., *Cerastium anomalum* Waldst. & Kit. 1799, non Schrank, 1795, *C. dubium* (Bast.) Guepin, *C. dubium* (Bast.) O. Schwarz, comb. superfl., *Dichodon dubium* (Bast.) Ikonn., *Provencheria dubia* (Bast.) B. Boiv., *Stellaria dubia* Bast.) - 1, 2

Dichoglottis Fisch. & C.A. Mey.

alsinoides (Bunge) Walp. (*Gypsophila alsinoides* Bunge, *Arenaria bungei* Barkoudah, *Stellaria blatteri* Mattf.) - 6

linearifolia Fisch. & C.A. Mey. (*Gypsophila linearifolia* (Fisch. & C.A. Mey.) Boiss.) - 1, 3, 6

Elisanthe (Fenzl) Reichenb. = Silene

adenophora (Schischk.) A. Devjatov & V. Tickomirov = Silene adenophora

akinfievii (Schmalh.) Grossh. = Silene akinfievii

aprica (Turcz. ex Fisch. & C.A. Mey.) Peschkova = Silene aprica

erubescens (Schischk.) A. Devjatov & V. Tichomirov = Silene erubescens

ferganica (Preobr.) A. Devjatov & V. Tichomirov = Silene ferganica

firma (Siebold & Zucc.) A. Devjatov & V. Tichomirov = Silene firma

noctiflora (L.) Rupr. = Silene noctiflora

olgae (Maxim.) A. Devjatov & V. Tichomirov = Silene olgae

ovalifolia (Regel & Schmalh.) A. Devjatov & V. Tichomirov = Silene ovalifolia

quadriloba (Turcz. ex Kar. & Kir.) Ikonn. = Silene quadriloba

ruinarum (M. Pop.) A. Devjatov & V. Tichomirov = Silene ruinarum

suaveolens (Kar. & Kir.) A. Devjatov & V. Tichomirov = Silene suaveolens

zawadskii (Herbich) Fuss = Silene zawadskii

zawadskii (Herbich) Klok. = Silene zawadskii

Eremogone Fenzl

asiatica (Schischk.) Ikonn. (*Arenaria asiatica* Schischk.) - 3

biebersteinii (Schlecht.) Holub (*Arenaria biebersteinii* Schlecht., *A. procera* Spreng. subsp. *pubescens* (Fenzl) Jalas, *A. graminifolia* sensu Schischk. & Knorr., *A. procera* auct., *Eremogone procera* sensu Ikonn. p.p.) - 1, 3, 6

brachypetala (Grossh.) Czer. (*Arenaria graminea* C.A. Mey. var. *brachypetala* Grossh., *A. brachypetala* (Grossh.)

T.N. Pop.) - 2

capillaris (Poir.) Fenzl (*Arenaria capillaris* Poir.) - 4, 5

caricifolia (Boiss.) Ikonn. (*Arenaria caricifolia* Boiss.) - 2

cephalotes (Bieb.) Fenzl (*Arenaria cephalotes* Bieb.) - 1

cucubaloides (Smith) Fenzl (*Arenaria cucubaloides* Smith) - 2

dianthoides (Smith) Ikonn. (*Arenaria dianthoides* Smith) - 2

ferganica (Schischk.) Ikonn. (*Arenaria ferganica* Schischk.) - 6

formosa (Fisch. ex Ser.) Fenzl (*Arenaria formosa* Fisch. ex Ser.) - 3, 4, 5

glaucescens (H. Winkl.) Ikonn. (*Arenaria glaucescens* H. Winkl., *A. griffithii* auct. p.p.) - 6

graminea (C.A. Mey.) C.A. Mey. (*Arenaria graminea* C.A. Mey.) - 2

graminifolia (Schrad.) Fenzl = E. saxatilis

griffithii (Boiss.) Ikonn. (*Arenaria griffithii* Boiss.) - 6

gypsophiloides (L.) Fenzl (*Arenaria gypsophiloides* L.) - 2

holostea (Bieb.) Rupr. (*Arenaria holostea* Bieb.) - 2

insignis (Litv.) Ikonn. (*Arenaria insignis* Litv.) - 6

juncea (Bieb.) Fenzl (*Arenaria juncea* Bieb.) - 4, 5

koriniana (Fisch. ex Fenzl) Ikonn. (*Arenaria koriniana* Fisch. ex Fenzl) - 1, 3, 6

ladyginae Ikonn. - 6

▪ **ledebouriana** (Fenzl) Ikonn. (*Arenaria ledebouriana* Fenzl)

litwinowii (Schischk.) Ikonn. (*Arenaria litwinowii* Schischk.) - 6

longifolia (Bieb.) Fenzl (*Arenaria longifolia* Bieb.) - 1, 2, 3, 6

lychnidea (Bieb.) Rupr. (*Arenaria lychnidea* Bieb.) - 2

macrantha (Schischk.) Ikonn. (*Arenaria macrantha* Schischk., *A. holostea* Bieb. subsp. *macrantha* (Schischk.) McNeill) - 2

meyeri (Fenzl) Ikonn. (*Arenaria meyeri* Fenzl) - 3, 6

micradenia (P. Smirn.) Ikonn. (*Arenaria micradenia* P. Smirn., *A. procera* Spreng. subsp. *glabra* (F. Williams) Holub, p.p., *A. ucranica* Spreng. ex Steud. nom. nud., *A. graminifolia* sensu Schischk. & Knorr. p.p.) - 1

mongolica (Schischk.) Ikonn. (*Arenaria mongolica* Schischk.) - 3

oosepala (Bordz.) Czer. (*Arenaria oosepala* Bordz.) - 2

paulsenii (H. Winkl.) Ikonn. (*Arenaria paulsenii* H. Winkl.) - 6

pineticola (Klok.) Klok. (*Arenaria pineticola* Klok.) - 1

polaris (Schischk.) Ikonn. (*Arenaria polaris* Schischk., *A. stenophylla* Ledeb. subsp. *polaris* (Schischk.) E. Selivan.) - 1, 3, 4

procera sensu Ikonn. p.p. = E. biebersteinii

procera (Spreng.) Reichenb. = E. saxatilis

rigida (Bieb.) Fenzl (*Arenaria rigida* Bieb.) - 1

saxatilis (L.) Ikonn. (*Arenaria saxatilis* L., *A. graminifolia* Schrad. 1809, non Ard. 1759, *A. procera* Spreng., *A. procera* subsp. *glabra* (F. Williams) Holub, p.p., *A. stenophylla* Ledeb., *A. syreistschikowii* P. Smirn., *Eremogone graminifolia* (Schrad.) Fenzl, nom. illegit., *E. procera* (Spreng.) Reichenb., *E. stenophylla* (Ledeb.) Fisch. & C.A. Mey., *Arenaria graminifolia* sensu Schischk. & Knorr. p.p.) - 1, 2, 3, 4

stenophylla (Ledeb.) Fisch. & C.A. Mey. = E. saxatilis

steveniana (Boiss.) Ikonn. (*Arenaria steveniana* Boiss., *A. blepharophylla* Boiss. subsp. *steveniana* (Boiss.) T.N. Pop., *A. blepharophylla* var. *parviflora* (Fenzl) McNeill) - 2

▪ **szowitsii** (Boiss.) Ikonn. (*Arenaria szowitsii* Boiss.)

talassica (Adylov) Czer. (*Arenaria talassica* Adylov) - 6

tschuktschorum (Regel) Ikonn. (*Arenaria tschuktschorum* Regel, *Alsine tschuktschorum* (Regel) Steffen) - 5

turlanica (Bajt.) Czer. (*Arenaria turlanica* Bajt.) - 6

Fiedleria Reichenb. = Petrorhagia

alpina (Habl.) Ovcz. = Petrorhagia alpina
cretica (L.) Ovcz. = Petrorhagia cretica

Fimbripetalum (Turcz.) Ikonn.

radians (L.) Ikonn. (*Stellaria radians* L.) - 4, 5

Gastrolychnis (Fenzl) Reichenb. (*Physolychnis* (Benth.) Rupr.

affinis (J. Vahl ex Fries) Tolm. & Kozhanczikov = G. involucrata

angustiflora Rupr. (*Melandrium angustiflorum* (Rupr.) Walp., *M. furcatum* (Rafin.) Hult. subsp. *angustiflorum* (Rupr.) Hult., *Silene furcata* Rafin. subsp. *angustiflora* (Rupr.) Walters, *S. involucrata* (Cham. & Schlecht.) Bocquet subsp. *angustiflora* (Rupr.) Hult.) - 1, 3
- subsp. *tenella* (Tolm.) Tolm. & Kozhanczikov = G. taimyrensis

apetala (L.) Tolm. & Kozhanczikov (*Lychnis apetala* L., *Gastrolychnis apetala* subsp. *uralensis* (Rupr.) A. & D. Löve, *G. uralensis* Rupr., *Melandrium apetalum* (L.) Fenzl, *Silene uralensis* (Rupr.) Bocquet, *S. uralensis* subsp. *apetala* (L.) Bocquet, *S. wahlbergella* Chowdhuri) - 1, 3, 4, 5, 6
- subsp. **arctica** (Fries) A. & D. Löve - 5
- subsp. *attenuata* (Farr) Petrovsky = G. attenuata
- subsp. *uralensis* (Rupr.) A. & D. Löve = G. apetala

attenuata (Farr) Czer. (*Lychnis attenuata* Farr, *Gastrolychnis apetala* (L.) Tolm. & Kozhanczikov subsp. *attenuata* (Farr) Petrovsky, *Lychnis apetala* L. subsp. *attenuata* (Farr) Maguire, *Melandrium apetalum* (L.) Fenzl subsp. *attenuatum* (Farr) Hara, *Silene uralensis* (Rupr.) Bocquet subsp. *attenuata* (Farr) McNeill, *S. wahlbergella* Chowdhuri subsp. *attenuata* (Farr) Hult.) - 5

brachypetala (Hornem.) Tolm. & Kozhanczikov (*Lychnis brachypetala* Hornem., *Gastrolychnis brachypetala* var. *baicalensis* (Tolm.) Peschkova, *Melandrium baicalense* Tolm., *M. brachypetalum* (Hornem.) Fenzl, *M. brachypetalum* var. *baicalense* (Tolm.) Sukacz., *M. songaricum* Fisch., C.A. Mey. & Ave-Lall., *Silene duthiei* Majumdar, *S. songarica* (Fisch., C.A. Mey. & Ave-Lall.) Bocquet, *S. songarica* (Fisch., C.A. Mey. & Ave-Lall.) Worosch. comb. superfl.) - 3, 4, 5, 6
- var. *baicalensis* (Tolm.) Peschkova = G. brachypetala

gonosperma (Rupr.) Czer. comb. nova (*Physolychnis gonosperma* Rupr. 1869, Mem. Acad. Sci. Petersb. (Sci. Phys. Math.), ser. 7, 14, 4 : 41; *Silene gonosperma* (Rupr.) Bocquet) - 6

gracilis (Tolm.) Czer. (*Melandrium gracile* Tolm.) - 4

involucrata (Cham. & Schlecht.) A. & D. Löve (*Lychnis apetala* L. var. *involucrata* Cham. & Schlecht., *Agrostemma involucratum* (Cham. & Schlecht.) G. Don fil., *Gastrolychnis affinis* (J. Vahl ex Fries) Tolm. & Kozhanczikov, *Lychnis affinis* J. Vahl ex Fries, *Melandrium affine* (J. Vahl ex Fries) J. Vahl, *M. furcatum* (Rafin.) Hadac, *M. furcatum* (Rafin.) Hult. comb. superfl., *M. furcatum* (Rafin.) Hyl. comb. superfl., *Silene furcata* Rafin. nom. dubium, *S. involucrata* (Cham. & Schlecht.) Bocquet) - 1, 3, 4, 5
- subsp. **elatior** (Regel) A. & D. Löve (*Lychnis apetala* var. *elatior* Regel, *Silene involucrata* subsp. *elatior* (Regel)

Bocquet)
- subsp. *tenella* (Tolm.) A. & D. Löve = G. taimyrensis

longicarpophora (Kom.) Czer. (*Melandrium longicarpophorum* Kom., *Silene longicarpophora* (Kom.) Bocquet) - 6

macrosperma (A. Pors.) Tolm. & Kozhanczikov (*Melandrium macrospermum* A. Pors., *Silene macrosperma* (A. Pors.) Hult.) - 5

ostenfeldii (A. Pors.) Petrovsky (*Melandrium ostenfeldii* A. Pors., *Gastrolychnis triflora* (R. Br.) Tolm. & Kozhanczikov subsp. *dawsonii* (Robins.) A. & D. Löve, p.p. excl. syn. *Melandrium taimyrense* Tolm., *Lychnis triflora* R. Br. var. *dawsonii* Robins., *Melandrium dawsonii* (Robins.) Hult.) - 5

popovii Peschkova - 4

saxatilis (Turcz. ex Fisch. & C.A. Mey.) Peschkova (*Lychnis saxatilis* Turcz. ex Fisch. & C.A. Mey., *Melandrium saxatile* (Turcz. ex Fisch. & C.A. Mey.) A. Br., *Silene tolmatchevii* Bocquet) - 4, 5

soczaviana (Schischk.) Tolm. & Kozhanczikov (*Melandrium soczavianum* Schischk., *Lychnis soczaviana* (Schischk.) J.P. Anders., *Silene soczaviana* (Schischk.) Bocquet) - 5

sordida (Kar. & Kir.) Czer. (*Lychnis sordida* Kar. & Kir., *Melandrium sordidum* (Kar. & Kir.) Rohrb., *Silene karekirii* Bocquet) - 6

taimyrensis (Tolm.) Czer. (*Melandrium taimyrense* Tolm., *Gastrolychnis angustiflora* Rupr. subsp. *tenella* (Tolm.) Tolm. & Kozhanczikov, *G. involucrata* (Cham. & Schlecht.) A. & D. Löve subsp. *tenella* (Tolm.) A. & D. Löve, *G. tenella* (Tolm.) Kuvajev, *Lychnis taimyrensis* (Tolm.) Polun., *Melandrium affine* (J. Vahl ex Fries) J. Vahl subsp. *tenellum* Tolm., *M. angustiflorum* (Rupr.) Walp. subsp. *tenellum* (Tolm.) Ju. Kozhevn., *M. tenellum* (Tolm.) Tolm., *Silene involucrata* (Cham. & Schlecht.) Bocquet subsp. *tenella* (Tolm.) Bocquet, *S. taimyrensis* (Tolm.) Bocquet) - 4

tenella (Tolm.) Kuvajev = G. taimyrensis

triflora (R. Br.) Tolm. & Kozhanczikov (*Lychnis triflora* R. Br., *Melandrium triflorum* (R. Br.) J. Vahl) - 4, 5
> Probably, a species of hybrid origin (G. apetala (L.) Tom. & Kozhanczikov x G. involucrata (Cham. & Schlecht.) A. & D. Löve).
- subsp. *dawsonii* (Robins.) A. & D. Löve, p.p. = G. ostenfeldii

tristis (Bunge) Czer. (*Lychnis tristis* Bunge, *Melandrium triste* (Bunge) Fenzl, *Silene bungei* Bocquet) - 3, 4, 6

uralensis Rupr. = G. apetala

violascens Tolm. (*Lychnis violascens* (Tolm.) A. Khokhr., *Melandrium violascens* (Tolm.) A.Khokhr.) - 4, 5

Gypsophila L.

acutifolia Fisch. ex Spreng. - 1, 2
albida Schischk. - 2
alpina Habl. = Petrorhagia alpina
alsinoides Bunge = Dichoglottis alsinoides
altissima L. (*G. tianschanica* M. Pop. & Schischk.) - 1, 2, 3, 4, 6
anatolica Boiss. & Heldr. = G. perfoliata
antoninae Schischk. - 6
aretioides Boiss. - 2, 6
aulieatensis B. Fedtsch. - 6
belorossica Barkoudah - 1
bicolor (Freyn & Sint.) Grossh. - 2, 6
■ brachypetala Trautv.
bucharica B. Fedtsch. = Bolbosaponaria bucharica
capitata Bieb. - 2
capituliflora Rupr. (*G. dshungarica* Czerniak.) - 6
cappadocica Boiss. & Bal. = G. laricina

cephalotes (Schrenk) Kom. - 3, 6
collina Stev. ex Ser. (*G. dichotoma* sensu Schischk.) - 1
davurica Turcz. ex Fenzl (*G. patrinii* Ser. subsp. *davurica*
(Fenzl) Ju. Kozhevn.) - 4, 5
desertorum (Bunge) Fenzl - 3
dichotoma Bess. = G. fastigiata
dichotoma sensu Schischk. = G. collina
diffusa Fisch. & C.A. Mey. ex Rupr. - 2, 6
dshungarica Czerniak. = G. capituliflora
elegans Bieb. - 1, 2, 5(alien), 6
▪ **eriocalyx** Boiss.
fastigiata L. (? *G. dichotoma* Bess., *G. ucrainica* Kleop.) - 1
fedtschenkoana Schischk. = Bolbosaponaria fedtschenkoana
ferganica Vved. = G. preobrashenskyi
filipes (Boiss.) Schischk. = Psammophiliella filipes
floribunda (Kar. & Kir.) Turcz. ex Fenzl = Psammophiliella
floribunda
glandulosa (Boiss.) Walp. - 2
glauca Stev. ex Ser. - 2
globulosa Stev. ex Boiss. = G. glomerata
glomerata Pall. ex Adams (*G. globulosa* Stev. ex Boiss.) - 1,
2
glomerata Pall. ex Bieb. = G. pallasii
gypsacea (Vved.) Bondar. = Saponaria gypsacea
herniarioides Boiss. - 6
heteropoda Freyn & Sint. - 2, 6
hispida Boiss. - 2
imbricata Rupr. - 2
intricata Franch. = Bolbosaponaria intricata
juzepczukii Ikonn. - 1
knorringiana (Schischk.) Vved. = Kuhitangia knorringiana
krascheninnikovii Schischk. - 6
▪ **laricina** Schreb. (*G. cappadocica* Boiss. & Bal., *G. sphaerocephala* Fenzl)
licentiana Hand.-Mazz. - 6
linearifolia (Fisch. & C.A. Mey.) Boiss. = Dichoglottis
linearifolia
lipskyi Schischk. (*G. nabelekii* Schischk. var. *lipskyi*
(Schischk.) Barkoudah) - 2
litwinowii K.-Pol. - 1
meyeri Rupr. - 2
microphylla (Schrenk) Fenzl - 6
muralis L. = Psammophiliella muralis
- subsp. *stepposa* (Klok.) Soo = Psammophiliella stepposa
nabelekii Schischk. var. *lipskyi* (Schischk.) Barkoudah = G.
lipskyi
oligosperma A. Krasnova - 1
orientalis (L.) Bondar. = Saponaria orientalis
pacifica Kom. - 5
pallasii Ikonn. (*G. glomerata* Pall. ex Bieb. 1808, non Pall.
ex Adams, 1805) - 1, 2
paniculata L. - 1, 2, 3, 6
patrinii Ser. - 1, 3, 4, 6
- subsp. *davurica* (Fenzl) Ju. Kozhevn. = G. davurica
paulii Klok. = G. perfoliata
perfoliata L. (*G. anatolica* Boiss. & Heldr., *G. paulii* Klok.,
G. perfoliata var. *anatolica* (Boiss. & Heldr.) Barkoudah, *G. trichotoma* Wend.) - 1, 2, 3, 6
- var. *anatolica* (Boiss. & Heldr.) Barkoudah = G. perfoliata
picta (Boiss.) Boiss. = Psammophiliella picta
pilosa Huds. = Pseudosaponaria pilosa
pinegensis Perf. = G. uralensis
popovii Preobr. = Kuhitangia popovii
porrigens (L.) Boiss. = Pseudosaponaria pilosa
preobrashenskyi Czerniak. (*G. ferganica* Vved.) - 6

pulvinaria Rech. fil. - 2
robusta Grossh. (*G. yorae* Woronow) - 2
rupestris A. Kuprian. - 3, 6
sambukii Schischk. - 4
scorzonerifolia Ser. - 1, 2
sericea (Ser.) Kryl. - 3, 4
serotina Hayne ex Willd. = Psammophiliella muralis
sewerzowii (Regel & Schmalh.) R. Kam. = Bolbosaponaria
sewerzowii
silenoides Rupr. - 2
▪ **simulatrix** Bornm. & Woronow
spathulifolia Fenzl = Saponaria spathulifolia
sphaerocephala Fenzl = G. laricina
stepposa (Klok.) Ikonn. = Psammophiliella stepposa
steupii Schischk. - 2
stevenii Fisch. ex Schrank - 2
szovitsii Fisch. & C.A. Mey. ex Fenzl - 2
tadzhikistanica Botsch. = Saponaria tadzhikistanica
takhtadzhanii Schischk. ex Ikonn. - 2
tenuifolia Bieb. - 2
thyraica A. Krasnova - 1
tianschanica M. Pop. & Schischk. = G. altissima
transalaica Ikonn. - 6
transcaucasica Barkoudah - 2
trichotoma Wend. = G. perfoliata
turkestanica Schischk. - 6
ucrainica Kleop. = G. fastigiata
uralensis Less. (*G. pinegensis* Perf.) - 1
vedeneevae Lepesch. - 6
villosa Barkoudah = Bolbosaponaria villosa
vinogradovii Safonov - 2
violacea (Ledeb.) Fenzl - 5
virgata Boiss. - 2
volgensis A. Krasnova - 1
yorae Woronow = G. robusta
zhegulensis A. Krasnova - 1

Heliosperma (Reichenb.) Reichenb. = Ixoca

alpestre auct. = Ixoca arcana
arcanum Zapal. = Ixoca arcana
carpaticum (Zapal.) Klok. = Ixoca carpatica
quadrifidum auct. = Ixoca carpatica
quadrifidum (L.) Reichenb. subsp. *carpaticum* Zapal. =
Ixoca carpatica

Herniaria L.

besseri Fisch. ex Hornem. - 1, 2
caucasica Rupr. - 2, 3, 6
cinerea DC. = H. hirsuta
euxina Klok. - 1
glabra L. (*H. suavis* Klok.) - 1, 2, 3, 6
hirsuta L. (*H. cinerea* DC.) - 2, 6
incana Lam. - 1, 2, 6
kotovii Klok. - 1
polygama J. Gay - 1, 3, 6
suavis Klok. = H. glabra

Holosteum L.

glutinosum (Bieb.) Fisch. & C.A. Mey. (*H. umbellatum* L.
subsp. *glutinosum* (Bieb.) Nym.) - 1, 2, 6
marginatum Fisch. & C.A. Mey. - 2
polygamum C. Koch - 1, 2, 6
sperguloides Lehm. = Spergularia sperguloides
subglutinosum Klok. - 1
syvaschicum Kleop. = H. umbellatum

umbellatum L. (*H. syvaschicum* Kleop.) - 1, 2, 6
- subsp. *glutinosum* (Bieb.) Nym. = H. glutinosum

Honckenya Ehrh. (*Ammodenia* J.G. Gmel ex S.G. Gmel. nom. invalid.)

diffusa (Hornem.) A. & D. Löve = H. oblongifolia
frigida Pobed. = H. oblongifolia
oblongifolia Torr. & Gray (*Ammodenia major* (Hook.) Heller, *Arenaria peploides* L. var. *diffusa* Hornem., *A. peploides* var. *major* Hook., *Honckenya diffusa* (Hornem.) A. & D. Löve, *H. frigida* Pobed., *H. peploides* (L.) Ehrh. subsp. *diffusa* (Hornem.) A. Löve, *H. peploides* subsp. *major* (Hook.) Hult.) - 1, 3, 4, 5
peploides (L.) Ehrh. (*Ammodenia peploies* (L.) Rupr.) - 1
- subsp. *diffusa* (Hornem.) A. Löve = H. oblongifolia
- subsp. *major* (Hook.) Hult. = H. oblongifolia

Ixoca Rafin. (*Heliosperma* (Reichenb.) Reichenb.)

arcana (Zapal.) Ikonn. (*Heliosperma arcanum* Zapal., *H. alpestre* auct., *Ixoca quadrifida* auct.) - 1
carpatica (Zapal.) Ikonn. (*Heliosperma quadrifidum* (L.) Reichenb. subsp. *carpaticum* Zapal., *H. carpaticum* (Zapal.) Klok., *H. quadrifidum* auct., *Ixoca pusilla* auct., *Silene pusilla* auct. p.p.) - 1
pusilla auct. = I. carpatica
quadrifida auct. = I. arcana

Kohlrauschia Kunth

prolifera (L.) Kunth (*Petrorhagia prolifera* (L.) P.W.Ball & Heywood) - 1, 2
saxifraga (L.) Dandy = Petrorhagia saxifraga
velutina (Guss.) Reichenb. (*Dianthus velutinus* Guss., *Petrorhagia velutina* (Guss.) P.W. Ball & Heywood) - 2

Krascheninnikowia Turcz. ex Fenzl = Pseudostellaria

borodinii Kryl. = Pseudostellaria borodinii
davidii Franch. = Pseudostellaria davidii
heterantha Maxim. = Pseudostellaria heterantha
heterophylla Miq. = Pseudostellaria heterophylla
japonica Korsh. = Pseudostellaria japonica
maximowicziana Franch. & Savat. = Pseudostellaria heterantha
rigida Kom. = Pseudostellaria rigida
rupestris Turcz. = Pseudostellaria rupestris
sylvatica Maxim. = Pseudostellaria sylvatica

Kuhitangia Ovcz.

knorringiana (Schischk.) Bondar. (*Acanthophyllum knorringianum* Schischk., *Gypsophila knorringiana* (Schischk.) Vved.) - 6
popovii (Preobr.) Ovcz. (*Gypsophila popovii* Preobr., *Acanthophyllum popovii* (Preobr.) Barkoudah) - 6

Lepigonum Wahlb.

microspermum Kindb. = Spergularia microsperma

Lepyrodiclis Fenzl

holosteoides (C.A. Mey.) Fenzl (*L. holosteoides* (C.A. Mey.) Fenzl ex Fisch. & C.A. Mey. comb. invalid.) - 2, 6
- var. *stellarioides* (Schrenk) Ju. Kozhevn. = L. stellarioides

stellarioides Schrenk (*L. holosteoides* (C.A. Mey.) Fenzl var. *stellarioides* (Schrenk) Ju. Kozhevn.) - 2, 6

Lidia A. & D. Löve = Minuartia

arctica (Stev. ex Ser.) A. & D. Löve = Minuartia arctica
biflora (L.) A. & D. Löve = Minuartia biflora
obtusiloba (Rydb.) A. & D. Löve = Minuartia obtusiloba

Lychnis L.

affinis J. Vahl ex Fries = Gastrolychnis involucrata
ajanensis (Regel & Til.) Regel (*Silene ajanensis* (Regel & Til.) Worosch.) - 5
alba Mill. = Melandrium album
alpina L. = Steris alpina
apetala L. = Gastrolychnis apetala
- subsp. *attenuata* (Farr) Maguire = Gastrolychnis attenuata
- var. *elatior* Regel = Gasrolychnis involucrata subsp. elatior
- var. *involucrata* Cham. & Schlecht. = Gastrolychnis involucrata
attenuata Farr = Gasrolychnis attenuata
brachypetala Hornem. = Gasrolychnis brachypetala
chalcedonica L. - 1, 3, 4, 6
cognata Maxim. - 5
coronaria (L.) Desr. = Coronaria coriacea
dioica L. p.p. = Melandrium dioicum
- var. *rubra* Weig. = Melandrium dioicum
divaricata Reichenb. = Melandrium latifolium
flos-cuculi L. = Coccyganthe flos-cuculi
fulgens Fisch. ex Curt. - 4, 5
pratensis Rafn = Melandrium album
samojedorum (Sambuk) Perf. (*L. sibirica* L. subsp. *samojedorum* Sambuk, *L. samojedorum* (Sambuk) Gorschk. comb. superfl., *L. sibirica* subsp. *jakutensis* Sambuk, *L. sibirica* subsp. *samojedorum* var. *jakutensis* (Sambuk) Jurtz., *Silene sibirica* (L.) Pers. subsp. *jakutensis* (Sambuk) Hamet-Ahti) - 1, 3, 4, 5
saxatilis Turcz. ex Fisch. & C.A. Mey. = Gasrolychnis saxatilis
sibirica L. (*Silene linnaeana* Worosch.) - 3, 4, 5
- subsp. *jakutensis* Sambuk = L. samojedorum
- subsp. *samojedorum* Sambuk = L. samojedorum
- - var. *jakutensis* (Sambuk) Jurtz. = L. samojedorum
- subsp. *villosula* (Trautv.) Tolm. = L. villosula
soczaviana (Schischk.) J.P. Anders. = Gastrolychnis soczaviana
sordida Kar. & Kir. = Gastrolychnis sordida
suecica Lodd. = Steris alpina
taimyrensis (Tolm.) Polun. = Gastrolychnis taimyrensis
triflora R. Br. = Gastrolychnis triflora
- var. *dawsonii* Robins. = Gastrolychnis ostenfeldii
tristis Bunge = Gastrolychnis tristis
villosula (Trautv.) Gorschk. (*L. sibirica* L. subsp. *villosula* (Trautv.) Tolm., *Silene ajanensis* (Regel & Til.) Worosch. subsp. *villosula* (Trautv.) Worosch.) - 4, 5
violascens (Tolm.) A. Khokhr. = Gastrolychnis violascens
viscaria L. = Steris viscaria
- subsp. *atropurpurea* (Griseb.) Chater = Steris atropurpurea
viscosa Scop. = Steris viscaria
wilfordii (Regel) Maxim. - 5

Malachium Fries = Myosoton

aquaticum (L.) Fries = Myosoton aquaticum

Melandrium Roehl.

adenophorum Schischk. = Silene adenophora
affine (J. Vahl ex Fries) J. Vahl = Gastrolychnis involucrata
- subsp. *tenellum* Tolm. = Gastrolychnis taimyrensis
akinfievii (Schmalh.) Schischk. = Silene akinfievii
album (Mill.) Garcke (*Lychnis alba* Mill., *L. pratensis* Rafn, *Melandrium dioicum* (L.) Coss. & Germ. subsp. *album* (Mill.) D. Löve, *M. pratense* (Rafn) Roehl., *Silene alba* (Mill.) E. Krause, 1901, non Britt. 1893, *S. latifolia* Poir. subsp. *alba* (Mill.) Greuter & Burdet, *S. pratense* (Rafn) Godr.) - 1, 2, 3, 4, 5, 6
- subsp. *eriocalycinum* (Boiss.) Hayek = M. eriocalycinum
angustiflorum (Rupr.) Walp. = Gastrolychnis angustiflora
- subsp. *tenellum* (Tolm.) Ju. Kozhevn. = Gastrolychnis taimyrensis
apetalum (L.) Fenzl = Gastrolychnis apetala
- subsp. *attenuatum* (Farr) Hara = Gastrolychnis attenuata
apricum (Turcz. ex Fisch. & C.A. Mey.) Rohrb. = Silene aprica
astrachanicum Pacz. (*Silene astrachanica* (Pacz.) Takht.) - 1
baicalense Tolm. = Gastrolychnis brachypetala
balansae Boiss. = M. latifolium
boissieri Schischk. = M. latifolium
brachypetalum (Hornem.) Fenzl = Gastrolychnis brachypetala
- var. *baicalense* (Tolm.) Sukacz. = Gastrolychnis brachypetala
dawsonii (Robins.) Hult. = Gastrolychnis ostenfeldii
dioicum (L.) Coss. & Germ. (*Lychnis dioica* L. p.p., *L. dioica* var. *rubra* Weig., *Melandrium dioicum* subsp. *rubrum* (Weig.) D. Löve, *M. rubrum* (Weig.) Garcke, *M. sylvestre* (Schkuhr) Roehl. nom. illegit., *Silene dioica* (L.) Clairv.) - 1, 2, 3, 6
- subsp. *album* (Mill.) D. Löve = M. album
- subsp. *divaricatum* (Reichenb.) A. & D. Löve = M. latifolium
- subsp. *rubrum* (Weig.) D. Löve = M. dioicum
divaricatum (Reichenb.) Fenzl = M. latifolium
eriocalycinum Boiss. (*M. album* (Mill.) Garcke subsp. *eriocalycinum* (Boiss.) Hayek, *Silene alba* (L.) E. Krause subsp. *eriocalycina* (Boiss.) Walters, *S. latifolia* Poir. subsp. *eriocalycina* (Boiss.) Greuter & Burdet, *S. pratensis* (Rafn) Godr. subsp. *eriocalycina* (Boiss.) McNeill & Prentice) - 1, 6
- var. *persicum* Boiss. & Buhse = M. persicum
erubescens Schischk. = Silene erubescens
fedtschenkoanum (Preobr.) Schischk. = Silene fedtschenkoana
ferganicum (Preobr.) Schischk. = Silene ferganica
firmum (Siebold & Zucc.) Rohrb. = Silene firma
furcatum (Rafin.) Hadac = Gastrolychnis involucrata
furcatum (Rafin.) Hult. = Gastrolychnis involucrata
- subsp. *angustiflorum* (Rupr.) Hult. = Gastrolychnis angustiflora
glaberrimum Bondar. & Vved. - 6
gracile Tolm. = Gastrolychnis gracilis
lapponicum (Simm.) Kuzen. (*M. rubrum* (Weig.) Garcke subsp. *lapponicum* Simm., *Silene dioica* (L.) Clairv. subsp. *lapponica* (Simm.) Tolm. & Kozhanczikov) - 1
latifolium (Poir.) Maire (*Silene latifolia* Poir., *Lychnis divaricata* Reichenb., *Melandrium balansae* Boiss., *M. boissieri* Schischk., *M. divaricatum* (Reichenb.) Fenzl, *M. dioicum* (L.) Coss. & Germ. subsp. *divaricatum* (Reichenb.) A. & D. Löve, *Silene alba* (Mill.) E. Krause subsp. *divaricata* (Reichenb.) Walters, *S. pratensis* (Rafn) Godr. subsp. *divaricata* (Reichenb.)

McNeill & Prentice) - 1, 2
longicarpophorum Kom. = Gastrolychnis longicarpophora
macrospermum A. Pors. = Gastrolychnis macrosperma
noctiflorum (L.) Fries = Silene noctiflora
olgae Maxim. = Silene olgae
ostenfeldii A. Pors. = Gastrolychnis ostenfeldii
ovalifolium Regel & Schmalh. = Silene ovalifolia
persicum (Boiss.& Buhse) Bohnm. (*M. eriocalycinum* Boiss. var. *persicum* Boiss. & Buhse, *Silene latifolia* Poir. subsp. *persica* (Boiss. & Buhse) Melzheimer) - 2, 6
pratense (Rafn) Roehl. = M. album
quadrilobum (Turcz. ex Kar. & Kir.) Schischk. = Silene quadriloba
rubrum (Weig.) Garcke = M. dioicum
- subsp. *lapponicum* Simm. = M. lapponicum
ruinarum M. Pop. = Silene ruinarum
sachalinense (Fr. Schmidt) Kudo = Silene sachalinensis
sachalinense (Fr. Schmidt) Schischk. = Silene sachalinensis
saxatile (Turcz. ex Fisch. & C.A. Mey.) A. Br. = Gastrolychnis saxatilis
saxosum A. Khokhr. - 2
soczavianum Schischk. = Gastrolychnis soczaviana
songaricum Fisch., C.A. Mey. & Ave-Lall. = Gastrolychnis brachypetala
sordidum (Kar. & Kir.) Rohrb. = Gastrolychnis sordida
suaveolens (Kar. & Kir.) Schischk. = Silene suaveolens
sylvestre (Schkuhr) Roehl. = M. dioicum
taimyrense Tolm. = Gastrolychnis taimyrensis
tenellum (Tolm.) Tolm. = Gastrolychnis taimyrensis
triflorum (R. Br.) J. Vahl = Gastrolychnis triflora
triste (Bunge) Fenzl = Gastrolychnis tristis
turkestanicum (Regel) Vved. = Silene turkestanica
violascens (Tolm.) A. Khokhr. = Gastrolychnis violascens
viscosum (L.) Celak. = Silene viscosa
zawadskii (Herbich) A. Br. = Silene zawadskii

Merckia Fisch. ex Cham. & Schlecht. = Wilhelmsia

physodes (Ser.) Fisch. ex Cham. & Schlecht. = Wilhelmsia physodes

Mesostemma Vved.

alexeenkoanum (Schischk.) Ikonn. (*Stellaria alexeenkoana* Schischk.) - 6
karatavicum (Schischk.) Vved. (*Stellaria karatavica* Schischk.) - 6
kotschyanum (Fenzl ex Boiss.) Vved. (*Stellaria kotschyana* Fenzl ex Boiss.) - 6
- subsp. **afghanicum** Rech. fil. - 6
martjanovii (Kryl.) Ikonn. (*Stellaria martjanovii* Kryl.) - 3
schugnanicum (Schischk.) Ikonn. (*Stellaria schugnanica* Schischk.) - 6

Minuartia L. (*Alsinanthe* (Fenzl) Reichenb., *Lidia* A. & D. Löve, *Sabulina* Reichenb., *Somerauera* Hoppe, *Tryphane* (Fenzl) Reichenb., *Wierzbickia* Reichenb., *Alsinopsis* auct.)

abchasica Schischk. - 2
adenotricha Schischk. - 1
aizoides (Boiss.) Bornm. - 2
akinfiewii (Schmalh.) Woronow - 2
anatolica (Boiss.) Woronow - 2

arctica (Stev. ex Ser.) Graebn. (*Lidia arctica* (Stev. ex Ser.) A. & D. Löve) - 1, 3, 4, 5, 6
aucta Klok. (*M. setacea* auct. p.p.) - 1
biebersteinii (Rupr.) Schischk. - 2
biflora (L.) Schinz & Thell. (*Lidia biflora* (L.) A. & D. Löve) - 1, 3, 4, 5, 6
bilykiana Klok. - 1
birjuczensis Klok. - 1
brotheriana sensu Schischk. p.p. = M. trautvetteriana
brotheriana (Trautv.) Woronow (*Stellaria brotheriana* Trautv.) - 2
buschiana Schischk. - 2
caucasica (Adams ex Rupr.) Mattf. = M. circassica
circassica (Albov) Woronow (*Alsine circassica* Albov, *Minuartia caucasica* (Adams ex Rupr.) Mattf. nom. illegit.) - 2
colchica Charadze - 2
dianthifolia (Boiss.) Hand.-Mazz. (*Somerauera dianthifolia* (Boiss.) A. & D. Löve) - 2
eglandulosa (Fenzl) Klok. (*Alsine pinifolia* Fenzl var. *eglandulosa* Fenzl) - 1
elegans (Cham. & Schlecht.) Schischk. (*Alsinanthe elegans* (Cham. & Schlecht.) A. & D. Löve, *Arenaria rossii* R. Br. subsp. *elegans* (Cham. & Schlecht.) Maguire, *Minuartia rossii* (R. Br.) Graebn. subsp. *elegans* (Cham. & Schlecht.) Rebr., *M. rossii* var. *elegans* (Cham. & Schlecht.) Hult.) - 5
euxina Klok. (*M. setacea* auct. p.p.) - 1
gerardii (Willd.) Fritsch = M. verna
glandulosa (Boiss. & Huet) Bornm. (*Alsine glandulosa* Boiss. & Huet, *Arenaria armena* Schischk. nom. invalid.) - 2
glomerata (Bieb.) Degen - 1, 2
gracilipes (Kom.) Kom. (*Alsine gracilipes* Kom., *Minuartia stricta* (Sw.) Hiern subsp. *gracilipes* (Kom.) Ju. Kozhevn.) - 5
granuliflora (Fenzl) Grossh. - 2
hamata (Hausskn. & Bornm.) Mattf. = Queria hispanica
helmii (Fisch. ex Ser.) Schischk. - 1
hirsuta (Bieb.) Hand.-Mazz. - 1
hybrida (Vill.) Schischk. (*M. tenuifolia* (L.) Hiern subsp. *hybrida* (Vill.) Mattf., *Sabulina hybrida* (Vill.) Fourr.) - 1, 2, 6
- subsp. *turcica* McNeill - 2
hypanica Klok. - 1
imbricata (Bieb.) Woronow - 2
inamoena (C.A. Mey.) Woronow - 2
intermedia (Boiss.) Hand.-Mazz. - 2
jacutica Schischk. - 4
juniperina (L.) Maire & Petitm. (*Arenaria juniperina* L.) - 2
krascheninnikovii Schischk. - 1
kryloviana Schischk. - 3, 6
laricina (L.) Mattf. - 4, 5
leiosperma Klok. (*M. setacea* auct. p.p.) - 1
lineata (C.A. Mey.) Bornm. (*Alsine juniperina* (L.) Wahlenb. var. *lineata* Boiss.) - 2
litwinowii Schischk. - 6
macrocarpa (Pursh) Ostenf. (*Wierzbickia macrocarpa* (Pursh) Reinchenb.). - 1, 3, 4, 5
- subsp. *minutiflora* (Hult.) Worosch. = M. minutiflora
- var. *minutiflora* Hult. = M. minutiflora
mesogitana (Boiss.) Hand.-Mazz. subsp. *lydia* (Boiss.) McNeill var. *turcomanica* (Schischk.) McNeill = M. turcomanica
- subsp. *turcomanica* (Schischk.) McNeill = M. turcomanica
meyeri (Boiss.) Bornm. - 2, 6
micrantha Schischk. - 2

minutiflora (Hult.) Worosch. (*M. macrocarpa* (Pursh) Ostenf. var. *minutiflora* Hult., *M. macrocarpa* subsp. *minutiflora* (Hult.) Worosch.) - 5
montana auct. p.p. = M. wiesneri
montana L. subsp. *wiesneri* (Stapf) McNeill = M. wiesneri
obtusiloba (Rydb.) House (*Alsinopsis obtusiloba* Rydb., *Lidia obtusiloba* (Rydb.) A. & D. Löve) - 5(?)
oreina (Mattf.) Schischk. (*M. recurva* (All.) Schinz & Thell. subsp. *oreina* (Mattf.) McNeill) - 2
orthotrichoides Schischk. = M. rubella
oxypetala (Woloszcz.) Kulcz. (*Alsine oxypetala* Woloszcz., *Minuartia verna* (L.) Hiern subsp. *oxypetala* (Woloszcz.) Halliday) - 1
palyzanica G. Proskuriakova - 1, 6
piskunovii Klok. = M. viscosa
pseudohybrida Klok. - 1
pusilla Schischk. = M. schischkinii
recurva (All.) Schinz & Thell. subsp. *oreina* (Mattf.) McNeill = M. oreina
regeliana (Trautv.) Mattf. - 1, 2, 3, 6
rhodocalyx (Albov) Woronow - 2
rossii (R.Br.) Graebn. (*Arenaria rossii* R. Br., *Alsinanthe rossii* (R. Br.) A. & D. Löve) - 5
- subsp. *elegans* (Cham. & Schlecht.) Rebr. = M. elegans
- var. *elegans* (Cham. & Schlecht.) Hult. = M. elegans
- var. *orthotrichoides* (Schischk.) Hult. = M. rubella
rubella (Wahlenb.) Hiern (*Alsine verna* (L.) Wahlenb. var. *glacialis* Fenzl, *Minuartia rossii* (R. Br.) Gaertn. var. *orthotrichoides* (Schischk.) Hult., *M. orthotrichoides* Schischk., *M. rubella* var. *glabrata* (Cham. & Schlecht.) Peschkova, *M. verna* (L.) Hiern subsp. *glacialis* (Fenzl) Kuvajev, *Tryphane rubella* (Wahlenb.) Reichenb.) - 1, 3, 4, 5
- var. *glabrata* (Cham. & Schlecht.) Peschkova = M. rubella
ruprechtiana Charadze - 2
schischkinii Adylov (*M. pusilla* Schischk. 1936, non Mattf. 1929) - 6
sclerantha (Fisch. & C.A. Mey.) Thell. - 2, 6
setacea auct. p.p. = M. aucta, M. euxina and M. thyraica
stricta (Sw.) Hiern - 1, 3, 4, 5
- subsp. *gracilipes* (Kom.) Ju. Kozhevn. = M. gracilipes
subuniflora (Albov) Woronow - 2
taurica (Stev.) Graebn. - 1
tenuifolia (L.) Hiern subsp. *hybrida* (Vill.) Mattf. = M. hybrida
thyraica Klok. (*M. setacea* auct. p.p.) - 1
trautvetteriana Sosn. & Charadze (*Arenaria brotheriana* Trautv., *Minuartia brotheriana* sensu Schischk. p.p.) - 2
tricostata A. Khokhr. - 5
turcomanica Schischk. (*M. mesogitana* (Boiss.) Hand.-Mazz. subsp. *lydia* (Boiss.) McNeill var. *turcomanica* (Schischk.) McNeill, *M. mesogitana* subsp. *turcomanica* (Schischk.) McNeill) - 6
verna (L.) Hiern (*M. gerardii* (Willd.) Fritsch, *Tryphane verna* (L.) Reichenb.) - 1, 2, 3, 4, 5, 6
- subsp. *glacialis* (Fenzl) Kuvajev = M. rubella
- subsp. *oxypetala* (Woloszcz.) Halliday = M. oxypetala
viscosa (Schreb.) Schinz & Thell. (*M. piskunovii* Klok.) - 1, 2
wiesneri (Stapf) Schischk. (*M. montana* L. subsp. *wiesneri* (Stapf) McNeill, *M. montana* auct. p.p.) - 1, 2
woronowii Schischk. - 2
yukonensis Hult. - 5
zarecznyi (Zapal.) Klok. (*Alsine zarecznyi* Zapal.) - 1

Moehringia L.

elongata Schischk. = M. lateriflora

hypanica Grynj & Klok. - 1
lateriflora (L.) Fenzl (*M. elongata* Schischk., *M. lateriflora* var. *elongata* (Schischk.) Worosch.) - 1, 3, 4, 5
- var. *elongata* (Schischk.) Worosch. = M. lateriflora
muscosa L. - 1
trinervia (L.) Clairv. - 1, 2, 3, 4, 6
umbrosa (Bunge) Fenzl - 3, 6

Myosoton Moench (*Malachium* Fries)

aquaticum (L.) Moench (*Malachium aquaticum* (L.) Fries) - 1, 2, 3, 4, 5, 6

Oberna Adans. (*Behen* Moench, 1794, non Hill, 1762, *Behenantha* (Otth) Schur, nom. superfl.)

behen (L.) Ikonn. (*Behenantha behen* (L.) Ikonn., *Cucubalus venosus* Gilib. nom. invalid., *Silene cucubalus* Wib., *S. latifolia* (Mill.) Britt. & Rendle, 1907, non Poir. 1789, *S. venosa* Aschers, *S. vulgaris* (Moench) Garcke) - 1, 2, 3, 4, 5, 6
carpatica (Zapal.) Czer. (*Silene venosa* Aschers. var. *carpatica* Zapal., *S. carpatica* (Zapa.) Czopik) - 1
commutata (Guss.) Ikonn. (*Silene commutata* Guss., *Behenantha commutata* (Guss.) Ikonn., *Silene cucubalus* Wib. subsp. *commutata* (Guss.) Rech. fil., *S. vulgaris* (Moench) Garcke subsp. *commutata* (Guss.) Hayek, *S. vulgaris* var. *commutata* (Guss.) Coode & Cullen) - 1, 2
crispata (Stev.) Ikonn. (*Silene crispata* Stev., *Behenantha crispata* (Stev.) Ikonn., *Silene fabaria* auct. p.p.) - 1, 2
cserei (Baumg.) Ikonn. (*Silene cserei* Baumg., *Behen cserei* (Baumg.) Gusul., *Behenantha cserei* (Baumg.) Schur, *Silene coringiifolia* Andrz., *S. saponariifolia* Schott ex Ledeb., *S. fabaria* auct. p.p.) - 1
lacera (Stev.) Ikonn. (*Behenantha lacera* (Stev.) Ikonn., *Silene lacera* (Stev.) Sims) - 2
littoralis (Rupr.) Ikonn. (*Silene inflata* (Salisb.) Smith var. *littoralis* Rupr., *Behenantha littoralis* (Rupr.) Ikonn., *Silene vulgaris* (Moench) Garcke var. *littoralis* (Rupr.) Jalas) - 1
multifida (Adams) Ikonn. (*Behenantha multifida* (Adams) Ikonn., *Silene fimbriata* Sims., *S. multifida* (Adams) Rohrb. 1868, non Edgew. 1846) - 2
procumbens (Murr.) Ikonn. (*Silene procumbens* Murr., *Behenantha procumbens* (Murr.) Ikonn.) - 1, 2, 3, 6
uniflora (Roth) Ikonn. (*Silene uniflora* Roth, *Behenantha uniflora* (Roth) Ikonn., *Silene maritima* With., *S. vulgaris* (Moench) Garcke subsp. *maritima* (With.) A. & D. Löve) - 1
wallichiana (Klotzsch) Ikonn. (*Silene wallichiana* Klotzsch, *Behenantha wallichiana* (Klotzsch) Ikonn.) - 6

Otites Adans. = Silene

artemisetorum Klok. = Silene artemisetorum
baschkirorum (Janisch.) Holub = Silene baschkirorum
baschkirorum (Janisch.) Stank. = Silene baschkirorum
borysthenica (Grun.) Klok. = Silene borysthenica
- subsp. *parviflora* (Hornem.) Holub = Silene borysthenica
chersonensis (Zapa.) Klok. = Silene exaltata
cyri (Schischk.) Grossh. = Silene cyri
densiflora (D'Urv.) Grossh. = Silene densiflora
dolichocarpa Klok. = Silene dolichocarpa
donetzica (Kleop.) Klok. = Silene donetzica
eugeniae (Kleop.) Klok. = Silene eugeniae
exaltata (Friv.) Holub = Silene exaltata

graniticola Klok. = Silene graniticola
hellmannii (Claus) Klok. = Silene hellmannii
jenissensis Klok. = Silene wolgensis
krymensis (Kleop.) Holub = Silene krymensis
krymensis (Kleop.) Klok. = Silene krymensis
maeotica Klok. = Silene maeotica
media (Litv.) Klok. = Silene media
moldavica Klok. = Silene moldavica
orae-syvaschicae Klok. = Silene wolgensis
parviflora (Ehrh.) Grossh. = Silene borysthenica
polaris (Kleop.) Holub = Silene polaris
polaris (Kleop.) Stank. = Silene polaris
pseudotites (Bess. ex Reichenb.) Klok. = Silene pseudotites
wolgensis (Hornem.) Grossh. = Silene wolgensis

Paronychia Hill

argentea Lam. - 2
azerbaijanica Chaudhri - 2
cephalotes (Bieb.) Bess. (*P. taurica* Borhidi & Sikura) - 1, 2
- subsp. *pontica* Borhidi = P. pontica
kurdica Boiss. - 2
pontica (Borhidi) Chaudhri (*P. cephalotes* (Bieb.) Bess. subsp. *pontica* Borhidi) - 1
splendens Stev. - 2
taurica Borhidi & Sikura = P. cephalotes

Petrocoma Rupr.

hoefftiana (Fisch.) Rupr. - 2

Petrorhagia (Ser. ex DC.) Link (*Fiedleria* Reichenb., *Tunica* auct.)

alpina (Habl.) P.W. Ball & Heywood (*Gypsophila alpina* Habl., *Fiedleria alpina* (Habl.) Ovcz., *Tunica alpina* (Habl.) Bobr., *T. stricta* (Bunge) Fisch. & C.A. Mey.) - 2, 3, 4, 6
cretica (L.) P.W. Ball & Heywood (*Saponaria cretica* L., *Fiedleria cretica* (L.) Ovcz., *Tunica pachygona* Fisch. & C.A. Mey.) - 2, 6
prolifera (L.) P.W. Ball & Heywood = Kohlrauschia prolifera
saxifraga (L.) Link (*Kohlrauschia saxifraga* (L.) Dandy, *Tunica saxifraga* (L.) Scop.) - 2
velutina (Guss.) P.W. Ball & Heywood = Kohlrauschia velutina

Physolychnis (Benth.) Rupr. = Gastrolychnis

gonosperma Rupr. = Gastrolychnis gonosperma

Pleconax Rafin. (*Conosilene* (Rohrb.) Fourr.)

conica (L.) Sourkova (*Silene conica* L., *Conosilene conica* (L.) Fourr., *Pleconax conica* (L.) A. Löve & Kjellqvist, comb. superfl.) - 1, 2, 6
- subsp. *conica* var. *subconica* (Friv.) A. Löve & Kjellqvist = P. subconica
- subsp. *coniflora* (Ness & Otth) A. Löve & Kjellqvist = P. coniflora
- subsp. *conoidea* (L.) A. Löve & Kjellqvist = P. conoidea
coniflora (Nees ex Otth) Sourkova (*Silene coniflora* Nees ex Otth, *Conosilene conica* (L.) Fourr. subsp. *coniflora* (Nees ex Otth) A. Löve & Kjellqvist, *Pleconax conica* (L.) A. Löve & Kjellqvist subsp. *coniflora* (Nees ex Otth) A. Löve & Kjellqvist) - 2, 6
conoidea (L.) Sourkova (*Silene conoidea* L., *Conosilene conica* (L.) Fourr. subsp. *conoidea* (L.) A. Löve &

Kjellqvist, *Pleconax conica* (L.) A. Löve & Kjellqvist subsp. *conoidea* (L.) A. Löve & Kjiellqvist) - 2, 6

subconica (Friv.) Sourkova (*Silene subconica* Friv., *Conosilene conica* (L.) Fourr. subsp. *conica* var. *subconica* (Friv.) A. Löve & Kjellqvist, *Pleconax conica* (L.) A. Löve & Kjellqvist subsp. *conica* var. *subconica* (Friv.) A. Löve & Kjellqvist, *Silene conica* subsp. *subconica* (Friv.) Garioli) - 2

Pleioneura Rech. fil. = Saponaria

griffithiana (Boiss.) Rech. fil. = Saponaria griffithiana

Polycarpon Loefl. ex L.

tetraphyllum (L.) L. - 2

Provencheria B. Boiv. = Dichodon

cerastoides (L.) B. Boiv. = Dichodon cerastoides
dubia (Bast.) B. Boiv. = Dichodon viscidum

Psammophila Fourr. = Psammophiliella

filipes (Boiss.) Ikonn. = Psammophiliella filipes
floribunda (Kar. & Kir.) Ikonn. = Psammophiliella floribunda
muralis (L.) Fourr. = Psammophiliella muralis
picta (Boiss.) Ikonn. = Psammophiliella picta
stepposa (Klok.) Ikonn. = Psammophiliella stepposa

Psammophiliella Ikonn. (*Psammophila* Fourr. 1868, non Schult. 1822)

filipes (Boiss.) Ikonn. (*Saponaria filipes* Boiss., *Gypsophila filipes* (Boiss.) Schischk., *Psammophila filipes* (Boiss.) Ikonn.) - 6

floribunda (Kar. & Kir.) Ikonn. (*Gypsophila floribunda* (Kar. & Kir.) Turcz. ex Fenzl, *Psammophila floribunda* (Kar. & Kir.) Ikonn., *Saponaria floribunda* (Kar. & Kir.) Boiss.) - 3, 6

muralis (L.) Ikonn. (*Gypsophila muralis* L., *G. serotina* Hayne ex Willd., *Psammophila muralis* (L.) Fourr.) - 1, 2, 3, 4, 5, 6

■ **picta** (Boiss.) Ikonn. (*Gypsophila picta* (Boiss.) Boiss., *Psammophila picta* (Boiss.) Ikonn.)

stepposa (Klok.) Ikonn. (*Gypsophila stepposa* (Klok.) Ikonn., *G. muralis* L. subsp. *stepposa* (Klok.) Soo, *Psammophila stepposa* (Klok.) Ikonn.) - 1, 3, 6

Pseudosaponaria (F. Williams) Ikonn.

pilosa (Huds.) Ikonn. (*Gypsophila pilosa* Huds., *G. porrigens* (L.) Boiss.) - 2, 6

Pseudostellaria Pax (*Krascheninnikowia* Turcz. ex Fenzl, 1840, non Gueldenst. 1772)

borodinii (Kryl.) Pax (*Krascheninnikowia borodinii* Kryl.) - 3

davidii (Franch.) Pax (*Krascheninnikowia davidii* Franch.) - 5

heterantha (Maxim.) Pax (*Krascheninnikowia heterantha* Maxim., *K. maximowicziana* Franch. & Savat.) - 5

heterophylla (Miq.) Pax (*Krascheninnikowia heterophylla* Miq.) - 5

japonica (Korsh.) Pax (*Krascheninnikowia japonica* Korsh.) - 5

rigida (Kom.) Pax (*Krascheninnikowia rigida* Kom.) - 5

rupestris (Turcz.) Pax (*Krascheninnikowia rupestris* Turcz.) - 4, 5

sylvatica (Maxim.) Pax (*Krascheninnikowia sylvatica* Maxim.) - 5

Pteranthus Forssk.

dichotomus Forssk. - 2

Queria L.

hispanica L. (*Minuartia hamata* (Hausskn. & Bornm.) Mattf., *Scleranthus hamatus* Hausskn. & Bornm.) - 1, 2, 6

Sabulina Reichenb. = Minuartia

hybrida (Vill.) Fourr. = Minuartia hybrida

Sagina L. (*Spergella* Reichenb.)

apetala Ard. (*S. ciliata* Fries) - 1, 2
- subsp. *erecta* (Hornem.) F. Herm. = S. micropetala
ciliata Fries = S. apetala
crassicaulis S. Wats. = S. maxima
intermedia Fenzl ex Ledeb. (*Spergella intermedia* (Fenzl) A. & D. Löve, *Sagina nivalis* auct.) - 1, 3, 4, 5, 6
japonica (Sw.) Ohwi (*Spergula japonica* Sw.) - 5
litoralis Hult. = S. maxima
maritima G. Don - 1
maxima A. Gray (*S. crassicaulis* S. Wats., *S. litoralis* Hult., *S. maxima* subsp. *crassicaulis* (S. Wats.) G.E. Crow, *S. maxima* f. *crassicaulis* (S. Wats.) Mizushima, *S. maxima* var. *crassicaulis* (S. Wats.) Worosch.) - 5
- subsp. *crassicaulis* (S. Wats.) G.E. Crow = S. maxima
- var. *crassicaulis* (S. Wats.) Worosch. = S. maxima
- f. *crassicaulis* (S. Wats.) Mizushima = S. maxima
micrantha Bunge = S. saginoides
micropetala Rauschert (*A. apetala* Ard. subsp. *erecta* (Hornem.) F. Herm.) -
nivalis auct. = S. intermedia
nodosa (L.) Fenzl. (*Spergella nodosa* (L.) Reichenb.) - 1, 2, 3, 4
x **normaniana** Lagerh. - S. procumbens L. x S. saginoides (L.) Karst.
oxysepala Boiss. - 2
procumbens L. - 1, 2, 3, 4, 5(alien)
saginoides (L.) Karst. (*S. micrantha* Bunge, *S. saginoides*(L.) Dalla Torre, comb. superfl.) - 1, 2, 3, 4, 5, 6
subulata (Sw.) C. Presl - 1

Saponaria L. (*Pleioneura* Rech. fil.)

atocioides Boiss. var. *calvertii* Boiss. = Spanizium prostratum
babatagi Ovcz. = Bolbosaponaria babatagi
cerastoides Fisch. ex C.A. Mey. - 2
cretica L. = Petrorhagia cretica
filipes Boiss. = Psammophiliella filipes
floribunda (Kar. & Kir.) Boiss. = Psammophiliella floribunda
glutinosa M. Bieb. - 1, 2
griffithiana Boiss. (*Pleioneura griffithiana* (Boiss.) Rech. fil.) - 6
gypsacea Vved. (*Gypsophila gypsacea* (Vved.) Bondar.) - 6
hispanica Mill. = Vaccaria hispanica
officinalis L. - 1, 3, 4, 5(alien)
orientalis L. (*Gypsophila orientalis* (L.) Bondar.) - 2, 6

parvula Bunge = S. spathulifolia
prostrata Willd. = Spanizium prostratum
- subsp. *calvertii* (Boiss.) Hedge = Spanizium prostratum
segetalis Neck. = Vaccaria hispanica
sewerzowii Regel & Schmalh. = Bolbosaponaria sewerzowii
spathulifolia (Fenzl) Vved. (*Gypsophila spathulifolia* Fenzl, *Saponaria parvula* Bunge) - 6
tadzhikistanica (Botsch.) V.A. Shultz (*Gypsophila tadzhikistanica* Botsch.) - 6
vaccaria L. var. *grandiflora* Fisch. ex DC. = Vaccaria hispanica subsp. grandiflora
viscosa C.A. Mey. - 2, 6

Scleranthus L.

annuus L. - 1, 2, 5(alien)
- subsp. *polycarpos* (L.) Thell. = S. polycarpos
hamatus Hausskn. & Bornm. = Queria hispanica
orientalis Rossler - 2
perennis L. - 1, 2
polycarpos L. (*S. annuus* L. subsp. *polycarpos* (L.) Thell.) - 2
syvaschicus Kleop. - 1
tauricus Knaf - 1
uncinatus Schur - 1, 2

Sclerocephalus Boiss.

arabicus Boiss. - 6

Silene L. (*Charesia* E. Busch, *Elisanthe* (Fenzl) Reichenb., *Otites* Adans., *Melandrium* sensu Schischk. p.p.)

acaulis (L.) Jacq. (*S. acaulis* subsp. *arctica* A. & D. Löve) - 1, 3, 5
- subsp. *arctica* A. & D. Löve = S. acaulis
acutidentata Bondar. & Vved. - 6
adenopetala Raik. - 6
adenophora (Schischk.) Czer. comb. nova (*Melandrium adenophorum* Schischk. 1936, Fl. URSS, 6 : 890, 705; *Elisanthe adenophora* (Schischk.) A. Devjatov & V. Tichomirov) - 6
ajanensis (Regel & Til.) Worosch. = Lychnis ajanensis
- subsp. *villosula* (Trautv.) Worosch. = Lychnis villosula
akinfievii Schmalh. (*Charesia akinfievii* (Schmalh.) E. Busch, *Elisanthe akinfievii* (Schmalh.) Grossh., *Melandrium akinfievii* (Schmalh.) Schischk.) - 2
alba (Mill.) E. Krause = Melandrium album
- subsp. *divaricata* (Reichenb.) Walters = Melandrium latifolium
- subsp. *eriocalycina* (Boiss.) Walters = Melandrium eriocalycinum
alexandrae B. Keller - 6
alexeji Kolak. - 2
alpicola Schischk. - 2
altaica Pers. - 1, 3, 4, 6
anglica L. = S. gallica
anisoloba Schrenk - 3, 6
apetala Willd. - 6
apiculata Ovcz. p.p. = S. guntensis and S. kuschakewiczii
aprica Turcz. ex Fisch. & C.A. Mey. (*Elisanthe aprica* (Turcz. ex Fisch. & C.A. Mey.) Peschkova, *Melandrium apricum* (Turcz. ex Fisch. & C.A. Mey.) Rohrb.) - 3, 4, 5
araratica Schischk. - 2
arenosa C. Koch - 2
argentea Ledeb. (*S. cappadocica* Boiss. & Heldr.) - 2

arguta auct. = S. sisianica
▪ armena Boiss.
armeria L. - 1, 4, 5
artemisetorum (Klok.) Czer. (*Otites artemisetorum* Klok., *Silene chersonensis* (Zapal.) Kleop. subsp. *littoralis* Kleop., *S. otites* auct. p.p.) - 1
artwinensis Schischk. = S. odontopetala
astrachanica (Pacz.) Takht. = Melandrium astrachanicum
aucheriana Boiss. - 2, 6
badachschanica Ovcz. - 6
balchaschensis Schischk. - 6
baldshuanica B. Fedtsch. - 6
baranovii Ovcz. & Z. Kurbanbekov - 6
baschkirorum Janisch. (*Otites baschkirorum* (Janisch.) Holub, *O. baschkirorum* (Janisch.) Stank. comb. invalid.) - 1, 3
betpakdalensis Bajt. - 6
bobrovii Schischk. - 6
bogdanii Ovcz. = S. nevskii
boissieri Panjut. (*S. panjutinii* Kolak.) - 2
bornmuelleriana Freyn - 2
borysthenica (Grun.) Walters (*S. otites* (L.) Wib. var. *borysthenica* Grun., *Otites borysthenica* (Grun.) Klok., *O. borysthenica* subsp. *parviflora* (Hornem.) Holub, *O. parviflora* (Ehrh.) Grossh., *Silene borysthenica* subsp. *parviflora* (Hornem.) Sourkova, *S. ehrhartiana* Soo, *S. parviflora* (Ehrh.) Pers. 1805, non Moench, 1794, *Viscago parviflora* Hornem.) - 1, 2, 3, 4, 6
- subsp. *parviflora* (Hornem.) Sourkova = S. borysthenica
brachypetala Robill. & Cast. ex DC. = S. nocturna
brahuica Boiss. - 6
- subsp. **megacalyx** R.Kam. - 6
brotheriana Somm. & Levier - 2
bucharica M. Pop. - 6
bungei Bocquet = Gastrolychnis tristis
bupleuroides L. (*S. longiflora* Ehrh.) - 2, 6
- subsp. **ramosa** Chowdhuri - 6
caespitosa Stev. - 2
▪ capitellata Boiss.
cappadocica Boiss. & Heldr. = S. argentea
carpatica (Zapal.) Czopik = Oberna carpatica
caucasica (Bunge) Boiss. - 2
caudata Ovcz. - 6
cephalantha Boiss. - 2
chaetodonta Boiss. - 6
chamarensis Turcz. (*S. tenuis* Willd. subsp. *chamarensis* (Turcz.) Ju. Kozhevn.) - 3, 4
- subsp. *paucifolia* (Ledeb.) Kuvajev = S. paucifolia
chersonensis (Zapal.) Kleop. subsp. *littoralis* Kleop. = S. artemisetorum
chlorantha (Willd.) Ehrh. - 1, 2, 3, 4, 5, 6
chlorifolia Smith - 2
chloropetala Rupr. = S. longipetala
claviformis Litv. (*S. heptapotamica* Schischk.) - 6
commelinifolia Boiss. - 2
commutata Guss. = Oberna commutata
compacta Fisch. ex Hornem. - 1, 2, 5(alien)
conformifolia (Preobr.) Preobr. ex Schischk. (*S. samarkandensis* Preobr. subsp. *conformifolia* Preobr.) - 6
conica L. = Pleconax conica
- subsp. *subconica* (Friv.) Garioli = Pleconax subconica
coniflora Nees ex Otth = Pleconax coniflora
conoidea L. = Pleconax conoidea
coringiifolia Andrz. = Oberna cserei
cretacea Fisch. ex Spreng. - 1, 3
crispans Litv. - 6
crispata Stev. = Oberna crispata
cserei Baumg. = Oberna cserei

cucubalus Wib. = Oberna behen
- subsp. *commutata* (Guss.) Rech. fil. = Oberna commutata
cyri Schischk. (*Otites cyri* (Schischk.) Grossh., *Silene turcomanica* Kleop.) - 1, 2, 6
czopandagensis Bondar. (*S. turcomanica* Schischk. 1936, non Kleop. 1936) - 6
daghestanica Rupr. - 2
densiflora D'Urv. (*Otites densiflora* (D'Urv.) Grossh., *Silene otites* (L.) Wib. subsp. *densiflora* (D'Urv.) Aschers. & Graebn.) - 1, 2, 6
- subsp. *wolgensis* (Hornem.) Slavic = S. wolgensis
- var. *wolgensis* (Hornem.) Jordanov & Panov = S. wolgensis
depressa Bieb. - 2
dianthoides Pers. - 2
dichotoma Ehrh. - 1, 2, 3
- subsp. *euxina* (Rupr.) Coode & Cullen = S. euxina
- subsp. *racemosa* (Otth) Graebn. & Graebn. fil. = S. racemosa
- var. *euxina* Rupr. = S. euxina
dioica (L.) Clairv. = Melandrium dioicum
- subsp. *lapponica* (Simm.) Tolm. & Kozhanczikov = Melandrium lapponicum
dolichocarpa (Klok.) Czer. (*Otites dolichocarpa* Klok.) - 1
donetzica Kleop. (*Otites donetzica* (Kleop.) Klok.) - 1
dubia Herbich (*S. nutans* L. subsp. *dubia* (Herbich) Zapal.) - 1
duthiei Majumdar = Gastrolychnis brachypetala
ehrhartiana Soo = S. borysthenica
eremitica Boiss. - 2
erubescens (Schischk.) Czer. (*Melandrium erubescens* Schischk., *Elisanthe erubescens* (Schischk.) A. Devjatov & V. Tichomirov) - 6
eugeniae Kleop. (*Otites eugeniae* (Kleop.) Klok.) - 1
euxina (Rupr.) Hand.-Mazz. (*S. dichotoma* Ehrh. var. euxina Rupr., *S. dichotoma* subsp. *euxina* (Rupr.) Coode & Cullen) - 2
eviscosa Bondar. & Vved. - 6
exaltata Friv. (*Otites chersonensis* (Zapal.) Klok., *O. exaltata* (Friv.) Holub) - 1
excedens Bondar. & Vved. - 6
fabaria auct. = Oberna crispata and O. cserei
fedtschenkoana Preobr. (*Melandrium fedtschenkoanum* (Preobr.) Schischk.) - 6
fedtschenkoi Bondar. & Vved. - 6
ferganica Preobr. (*Elisanthe ferganica* (Preobr.) A. Devjatov & V. Tichomirov, *Melandrium ferganicum* (Preobr.) Schischk.) - 6
fetissovii Lazkov - 6
fimbriata Sims = Oberna multifida
firma Siebold & Zucc. (*Elisanthe firma* (Siebold & Zucc.) A. Devjatov & V. Tichomirov, *Melandrium firmum* (Siebold & Zucc.) Rohrb.) - 5
flexuosa Ovcz. = S. tachtensis
flos-cuculi (L.) Greuter & Burdet = Coccyganthe flos-cuculi
foliosa Maxim. (*S. foliosa* var. *macrostyla* (Maxim.) Worosch., *S. macrostyla* Maxim.) - 5
- var. *macrostyla* (Maxim.) Worosch. = S. foliosa
furcata Rafin. = Gastrolychnis involucrata
- subsp. *angustiflora* (Rupr.) Walters = Gastrolychnis angustiflora
gallica L. (*S. anglica* L., *S. gallica* subsp. *anglica* (L.) A. & D. Löve) - 1, 2
- subsp. *anglica* (L.) A. & D. Löve = S. gallica
gasimailikensis B. Fedtsch. - 6
gawrilowii (Krasn.) M. Pop. - 6

gebleriana Schrenk - 6
glaucescens Schischk. - 6
gonosperma (Rupr.) Bocquet = Gastrolychnis gonosperma
graminifolia Otth (*S. tenuis* Willd. nom. dubium) - 3, 4, 6
graniticola (Klok.) Sourkova (*Otites graniticola* Klok.) - 1
grossheimii Schischk. - 2
guntensis B. Fedtsch. (*S. apiculata* Ovcz. p.p. incl. typo, *S. incanescens* Ovcz., *S. kuhistanica* Ovcz. p.p.) - 6
gynodioica Ghazanfar subsp. **glandulosa** Melzheimer - 6
hellmannii Claus (*Otites hellmannii* (Claus) Klok.) - 1, 6
heptapotamica Schischk. = S. claviformis
hispidula Ovcz. = S. kuschakewiczii
hissarica M. Pop. = S. longicalycina and S. tachtensis
holopetala Bunge (*S. komarovii* Schischk. p.p.) - 6
humilis C.A. Mey. - 2
hypanica Klok. - 1
iberica Bieb. - 2
incanescens Ovcz. = S. guntensis
incurvifolia Kar. & Kir. - 3, 6
indeprensa Schischk. - 6
inflata (Salisb.) Smith var. *littoralis* Rupr. = Oberna littoralis
involucrata (Cham. & Schlecht.) Bocquet = Gastrolychnis involucrata
- subsp. *angustiflora* (Rupr.) Hult. = Gastrolychnis angustiflora
- subsp. *elatior* (Regel) Bocquet = Gastrolychnis involucrata subsp. elatior
- subsp. *tenella* (Tolm.) Bocquet = Gastrolychnis taimyrensis
▪ *ispirensis* Boiss. & Huet
italica (L.) Pers. - 1, 2, 6
- subsp. *nemoralis* (Waldst. & Kit.) Nym. = S. nemoralis
jailensis N. Rubtz. - 1
jaxartica Pavl. - 6
jenisseensis Willd. (*S. tuvinica* K. Sobol.) - 3, 4, 5
- subsp. *popovii* V. Zuev - 4
jucunda Pavl. = S. longicalycina
jundzillii Zapal. - 1
karaczukuri B. Fedtsch. - 6
karekirii Bocquet = Gastrolychnis sordida
karkaralensis A. Dmitr. & M. Pop. - 6
kawashimae Miyabe & Tatew. = S. sachalinensis
kirgisensis Bajt. & Nelina - 6
komarovii Schischk. = S. holopetala and S. scabrifolia
koreana Kom. - 5
korshinskyi Schischk. - 6
krymensis Kleop. (*Otites krymensis* (Kleop.) Klok., *O. krymensis* (Kleop.) Holub, comb. superfl.) - 1
kubanensis Somm. & Levier - 2
kudrjaschevii Schischk. - 6
kuhistanica Ovcz. = S. guntensis and S. kuschakewiczii
kulabensis B. Fedtsch. - 6
▪ **kungessana** B.Fedtsch.
kuschakewiczii Regel & Schmalh. (*S. apiculata* Ovcz. p.p., *S. hispidula* Ovcz., *S. kuhistanica* Ovcz. p.p. incl. typo) - 6
lacera (Stev.) Sims = Oberna lacera
ladyginae Lazkov - 6
lasiantha C. Koch (*S. olympica* Boiss. subsp. *lasiantha* (C. Koch) Chowdhuri) - 2
latifolia (Mill.) Britt. & Rendle = Oberna behen
latifolia Poir. = Melandrium latifolium
- subsp. *alba* (Mill.) Greuter & Burdet = Melandrium album
- subsp. *eriocalycina* (Boiss.) Greuter & Burdet = Melandrium eriocalycinum
- subsp. *persica* (Boiss. & Buhse) Melzheimer = Melandri-

um persicum
lazica Boiss. - 2
lepidifera Ovcz. - 6
leptocaulis Schischk. = S. plurifolia
linearifolia Otth - 2
linnaeana Worosch. = Lychnis sibirica
lithophila Kar. & Kir. - 6
lithuanica Zapal. - 1
litwinowii Schischk. - 6
longicalycina Kom. (*S. hissarica* M. Pop., *S. jucunda* Pavl. 1950, non Jord. & Fourr. 1866) - 6
longicarpophora (Kom.) Bocquet = Gastrolychnis longicarpophora
longidens Schischk. - 2
longiflora Ehrh. = S. bupleuroides
longipetala Vent. (*S. chloropetala* Rupr.) - 2
lychnidea C.A. Mey. - 2
macrosperma (A. Pors.) Hult. = Gastrolychnis macrosperma
macrostyla Maxim. = S. foliosa
maeotica (Klok.) Czer. (*Otites maeotica* Klok.) - 1
marcowiczii Schischk. - 2
mariae Klok. - 1
maritima With. = Oberna uniflora
marschallii C.A. Mey. - 2
media (Litv.) Kleop. (*Otites media* (Litv.) Klok.) - 1, 3, 6
megalantha Bondar. & Vved. - 6
meyeri Fenzl ex Boiss. & Buhse - 2
michelsonii Preobr. - 6
microphylla Boiss. (*S. pamirensis* (H. Winkl.) Preobr. ex Schischk.) - 6
moldavica (Klok.) Sourkova (*Otites moldavica* Klok.) - 1
monantha Bondar. & Vved. - 6
montbretiana Boiss. - 2
morganae Freyn - 2
multifida (Adams) Rohrb. = Oberna multifida
multiflora (Ehrh.) Pers. (*Cucubalus multiflorus* Ehrh.) - 1, 3, 6
muslimii Pavl. - 6
nana Kar. & Kir. - 6
nemoralis Waldst. & Kit. (*S. italica* (L.) Pers. subsp. *nemoralis* (Waldst. & Kit.) Nym.) - 1
nevskii Schischk. (*S. bogdanii* Ovcz.) - 6
noctiflora L. (*Elisanthe noctiflora* (L.) Rupr., *Melandrium noctiflorum* (L.) Fries) - 1, 2, 3, 5(alein), 6
nocturna L. (*S. brachypetala* Robill. & Cast. ex DC.) - 2
nuratavica R. Kam. - 6
nutans L. - 1, 2, 3, 4
- subsp. *dubia* (Herbich) Zapa. = S. dubia
obovata Schischk. - 6
obscura Worosch. - 5
obtusidentata B. Fedtsch. & M. Pop. - 6
odessana Klok. - 1
odontopetala Fenzl (*S. artwinensis* Schischk., *S. raddeana* Trautv.) - 2
- subsp. *physocalyx* (Ledeb.) Bornm. & Gauba = S. physocalyx
odoratissima Bunge (*S. odoratissima* var. *olgiana* (B. Fedtsch.) Ju. Kozhevn., *S. olgiana* B. Fedtsch.) - 6
- var. *olgiana* (B. Fedtsch.) Ju. Kozhevn. = S. odoratissima
oldhamiana Miq. - 5
olgae (Maxim.) Rohrb. (*Melandrium olgae* Maxim., *Elisanthe olgae* (Maxim.) A. Devjatov & V. Tichomirov) - 5
olgiana B. Fedtsch. = S. odoratissima
olympica Boiss. subsp. *lasiantha* (C. Koch) Chowdhuri = S. lasiantha

orae-syvaschicae (Klok.) Czer. = S. wolgensis
oreina Schischk. - 6
otites auct. p.p. = S. artemisetorum, S. polaris and S. pseudotites
otites (L.) Wib. subsp. *densiflora* (D'Urv.) Aschers. & Graebn. = S. densiflora
- subsp. *polaris* (Kleop.) A. & D. Löve = S. polaris
- var. *borysthenica* Grun. = S. borysthenica
ovalifolia (Regel & Schmalh.) Melzheimer (*Melandrium ovalifolium* Regel & Schmalh., *Elisanthe ovalifolia* (Regel & Schmalh.) A. Devjatov & V. Tichomirov) - 6
pachyneura Schischk. = S. sisianica
pamirensis (H. Winkl.) Preobr. ex Schischk. = S. microphylla
panjutinii Kolak. = S. boissieri
paranadena Bondar. & Vved. - 6
parviflora (Ehrh.) Pers. = S. borysthenica
paucifolia Ledeb. (*S. chamarensis* Turcz. subsp. *paucifolia* (Ledeb.) Kuvajev, *S. tenuis* Willd. subsp. *paucifolia* (Ledeb.) Ju. Kozhevn.) - 1, 3, 4
peduncularis Boiss. - 2
pendula L. - 1, 2
physocalyx Ledeb. (*S. odontopetala* Fenzl subsp. *physocalyx* (Ledeb.) Bornm. & Gauba) - 2(?)
plurifolia Schischk. (*S. leptocaulis* Schischk.) - 6
polaris Kleop. (*Otites polaris* (Kleop.) Holub, *Otites polaris* (Kleop.) Stank. nom. invalid., *Silene otites* (L.) Web.subsp. *polaris* (Kleop.) A. & D. Löve, *S. otites* auct. p.p.) - 1, 3, 4
popovii Schischk. - 6
praelonga Ovcz. - 6
praemixta M. Pop. - 6
praestans Schischk. - 2
pratense (Rafn) Godr. = Melandrium album
- subsp. *eriocalycina* (Boiss.) McNeill & Prentice = Melandrium eriocalycinum
- subsp. *divaricata* (Reichenb.) McNeill & Prentice = Melandrium latifolium
prilipkoana Schischk. - 6
procumbens Murr. = Oberna procumbens
propinqua Schischk. - 2
■ **pruinosa** Boiss. (*S. supina* subsp. *pruinosa* (Boiss.) Chowdhuri)
pseudotenuis Schischk. - 6
pseudotites Bess. ex Reichenb. (*Otites pseudotites* (Bess. ex Reichenb.) Klok., *Silene otites* auct. p.p.) - 1
pubicalyx Bondar. & Vved. - 6
pugionifolia M. Pop. - 6
pusilla auct. p.p. = Ixoca carpatica
pygmaea Adams - 2
quadriloba Turcz. ex Kar. & Kir. (*Elisanthe quadriloba* (Turcz. ex Kar. & Kir.) Ikonn., *Melandrium quadrilobum* (Turcz. ex Kar. & Kir.) Schischk., *Silene viscosa* (L.) Pers. subsp. *quadriloba* (Trautv.) Melzheimer, *S. viscosa* var. *quadriloba* Trautv.) - 3, 4, 6
racemosa Otth (*S. dichotoma* Ehrh. subsp. *racemosa* (Otth) Graebn. & Graebn. fil., *S. thirkeana* C. Koch) - 2
raddeana Trautv. = S. odontopetala
repens Patrin - 1, 3, 4, 5, 6
- subsp. **alpina** (Kom.) Ju. Kozhevn. (*S. repens* f. *alpina* Kom.) - 5
- f. *alpina* Kom. = S. repens subsp. alpina
ruinarum M. Pop. (*Elisanthe ruinarum* (M. Pop.) A. Devjatov & V. Tichomirov, *Melandrium ruinarum* M. Pop. nom. altern.) - 6
rupestris L. - 1

ruprechtii Schischk. (*S. saxatilis* Bieb. 1808, non Sims, 1803) - 2

sachalinensis Fr. Schmidt (*Melandrium sachalinense* (Fr. Schmidt) Kudo, *M. sachalinense* (Fr. Schmidt) Schischk. comb. superfl., *Silene kawashimae* Miyabe & Tatew.) - 5

samarkandensis Preobr. - 6

- subsp. *conformifolia* Preobr. = S. conformifolia

saponariifolia Schott ex Ledeb. = Oberna cserei

sarawschanica Regel & Schmalh. - 6

saxatilis Bieb. = S. ruprechtii

scabrifolia Kom. (*S. komarovii* Schischk. p.p. incl. typo) - 6

schafta S.G. Gmel. ex Hohen. - 2

schischkinii K. Sobol. = S. sobolevskajae

schischkinii (M. Pop.) Vved. (*S. trajectorum* Kom. var. *schischkinii* M. Pop) - 6

schugnanica B. Fedtsch. - 6

semenovii Regel & Herd. - 6

sibirica (L.) Pers. - 1, 3, 6

- subsp. *jakutensis* (Sambuk) Hamet-Ahti = Lychnis samojedorum

sisianica Boiss. & Buhse (*S. pachyneura* Schischk., *S. arguta* auct.) - 2

sobolevskajae Czer. (*S. schischkinii* K. Sobol. 1953, non Vved. 1953) - 4

soczaviana (Schischk.) Bocquet = Gastrolychnis soczaviana

solenantha Trautv. - 2

songarica (Fisch., C.A. Mey. & Ave-Lall.) Bocquet = Gastrolychnis brachypetala

songarica (Fisch., C.A. Mey. & Ave-Lall.) Worosch. = Gastrolychnis brachypetala

spergulifolia (Willd.) Bieb. (*Cucubalus spergulifolius* Willd. 1799, *C. spergulifolius* Desf. 1804) - 2

stenantha Ovcz. - 6

stenophylla Ledeb. - 4, 5

steppicola Kleop. (*S. steppicola* subsp. *glabra* Kleop., *S. steppicola* subsp. *pubescens* Kleop.) - 1

- subsp. *glabra* Kleop. = S. steppicola
- subsp. *pubescens* Kleop. = S. steppicola

suaveolens Kar. & Kir. (*Elisanthe suaveolens* (Kar. & Kir.) A. Devjatov & V. Tichomirov, *Melandrium suaveolens* (Kar. & Kir.) Schischk.) - 6

subadenophora Ovcz. - 6

subconica Friv. = Pleconax subconica

suecica (Lodd.) Greuter & Burdet = Steris alpina

suffrutescens Bieb. - 1, 2, 3, 6

supina Bieb. - 1, 2

- subsp. *pruinosa* (Boiss.) Chowdhuri = S. pruinosa

swertiifolia Boiss. - 6

syreitschikowii P. Smirn. - 1

syvaschica Kleop. - 1

tachtensis Franch. (*S. flexuosa* Ovcz. nom. invalid., *S. hissarica* M. Pop. p.p. incl. typo, *S. tachtensis* subsp. *flexuosa* (Ovcz.) Ovcz. nom. invalid., *S. tachtensis* subsp. *hissarica* (M. Pop.) Ovcz., *S. tachtensis* subsp. *nemoralis* Ovcz. nom. invalid., *S. tachtensis* subsp. *varzobica* Ovcz. nom. invalid.) - 6

- subsp. *flexuosa* (Ovcz.) Ovcz. = S. tachtensis
- subsp. *hissarica* (M. Pop.) Ovcz. = S. tachtensis
- subsp. *nemoralis* Ovcz. = S. tachtensis
- subsp. *varzobica* Ovcz. = S. tachtensis

taimyrensis (Tolm.) Bocquet = Gastrolychnis taimyrensis

taliewii Kleop. - 1

talyschensis Schischk. - 2

tatarica (L.) Pers. - 1, 2, 3

tatjanae Schischk. - 2

tenella C.A. Mey. - 2

tenuis Willd. = S. graminifolia

- subsp. *chamarensis* (Turcz.) Ju. Kozhevn. = S. chamarensis
- subsp. *paucifolia* (Ledeb.) Ju. Kozhevn. = S. paucifolia

thirkeana C. Koch = S. racemosa

tianschanica Schischk. - 6

tolmatchevii Bocquet = Gastrolychnis saxatilis

tomentella Schischk. - 6

trajectorum Kom. - 6

- var. *schischkinii* M. Pop = S. schischkinii

turcomanica Kleop. = S. cyri

turcomanica Schischk. = S. czopandagensis

turgida Bieb. ex Bunge - 3, 4

turkestanica Regel (*Melandrium turkestanicum* (Regel) Vved.) - 6

tuvinica K. Sobol. = S. jeniseensis

ucrainica Klok. - 1

uniflora Roth = Oberna uniflora

uralensis (Rupr.) Bocquet = Gastrolychnis apetala

- subsp. *apetala* (L.) Bocquet = Gastrolychnis apetala
- subsp. *attenuata* (Farr) McNeill = Gastrolychnis attenuata

vachschii Ovcz. - 6

venosa Aschers = Oberna behen

- var. *carpatica* Zapal. = Oberna carpatica

viridiflora L. - 1

viscaria (L.) Jess. = Steris viscaria

viscosa (L.) Pers. (*Melandrium viscosum* (L.) Celak.) - 1, 2, 3, 4, 6

- subsp. *quadriloba* (Trautv.) Melzheimer = S. quadriloba
- var. *quadriloba* Trautv. = S. quadriloba

vulgaris (Moench) Garcke = Oberna behen

- subsp. *commutata* (Guss.) Hayek = Oberna commutata
- subsp. *maritima* (With.) A. & D. Löve = Oberna uniflora
- var. *commutata* (Guss.) Coode & Cullen = Oberna commutata
- var. *littoralis* (Rupr.) Jalas = Oberna littoralis

wahlbergella Chowdhuri = Gastrolychnis apetala

- subsp. *attenuata* (Farr) Hult. = Gastrolychnis attenuata

wallichiana Klotzsch = Oberna wallichiana

wolgensis (Hornem.) Bess. ex Spreng. (?*Otites jenissensis* Klok., *O. orae-syvaschicae* Klok., *O. wolgensis* (Hornem.) Grossh., *Silene densiflora* D'Urv. var. *wolgensis* (Hornem.) Jordanov & Panov, *S. densiflora* subsp. *wolgensis* (Hornem.) Slavic, *S. orae-syvaschicae* (Klok.) Czer.) - 1, 2, 3, 4, 6

zangezura A. Jelen. - 2

zawadskii Herbich (*Elisanthe zawadskii* (Herbich) Fuss, *E. zawadskii* (Herbich) Klok. nom. illegit. superfl., *Melandrium zawadskii* (Herbich) A. Br.) - 1

Somerauera Hoppe = Minuartia

dianthifolia (Boiss.) A. & D. Löve = Minuartia dianthifolia

Spanizium Griseb

prostratum (Willd.) V.A. Shultz (*Saponaria prostrata* Willd., *S. atocioides* Boiss. var. *calvertii* Boiss., *S. prostrata* subsp. *calvertii* (Boiss.) Hedge) - 2

Spergella Reichenb. = Sagina

intermedia (Fenzl) A. & D. Löve = Sagina intermedia

nodosa (L.) Reichenb. = Sagina nodosa

Spergula L.

arvensis L. (*S. arvensis* subsp. *sativa* (Boenn.) Celak., *S. arvensis* subsp. *vulgaris* (Boenn.) Celak., *S. arvensis* subsp. *vulgaris* (Boenn.) O. Schwarz, comb. superfl., *S. sativa* Boenn. nom. illegit., *S. vulgaris* Boenn.) - 1, 2, 3, 4, 5
- subsp. *maxima* (Weihe) O. Schwarz = S. maxima
- subsp. *sativa* (Boenn.) Celak. = S. arvensis
- subsp. *vulgaris* (Boenn.) Celak. = S. arvensis
- subsp. *vulgaris* (Boenn.) O. Schwarz = S. arvensis
japonica Sw. = Sagina japonica
linicola Boreau (*S. sativa* Boenn. subsp. *linicola* (Boreau) O. Schwarz) - 1
marina (L.) Bartl. & Wendl. = Spergularia salina
maritima (All.) T.M. Pedersen = Spergularia maritima
maxima Weihe (*S. arvensis* L. subsp. *maxima* (Weihe) O. Schwarz) - 1, 3, 5
morisonii Boreau (*S. vernalis* Willd. nom. illegit.) - 1
pentandra L. - 1(?)
sativa Boenn. = S. arvensis
- subsp. *linicola* (Boreau) O. Schwarz = S. linicola
vernalis Willd. = S. morisonii
vulgaris Boenn. = S. arvensis

Spergularia (Pers.) J. & C. Presl (*Alsine* L. p.p.)

adenophora Vved. = S. diandra
campestris (L.) Aschers. = S. rubra
diandra (Guss.) Boiss. (*S. adenophora* Vved., *S. nematopoda* Vved.) - 1, 2, 3, 6
- var. *microspermoides* (Vved.) Tackholm & Boulos = S. microsperma
glaucophylla Vved. = S. microsperma
marginata (DC.) Kitt. = S. maritima
marina (L.) Griseb. = S. salina
maritima (All.) Chiov. (*Arenaria maritima* All., *A. marginata* DC. nom. illegit., *Spergula maritima* (All.) T.M. Pedersen, *Spergularia marginata* (DC.) Kitt., *S. media* (L.) C. Presl, p.p. excl. basionymo, *Arenaria media* auct.) - 1, 2, 3, 6
media (L.) C. Presl, p.p. = S. maritima
microsperma (Kindb.) Aschers. (*Lepigonum microspermum* Kindb., *Spergularia diandra* (Guss.) Boiss. var. *microspermoides* (Vved.) Tackholm & Boulos, *S. glaucophylla* Vved., *S. microspermoides* Vved.) - 6
microspermoides Vved. = S. microsperma
nematopoda Vved. = S. diandra
rubra (L.) J. & C. Presl (*Arenaria rubra* L. p.p., *Spergularia campestris* (L.) Aschers.) - 1, 2, 3, 4, 5, 6
salina J. & C. Presl (*Arenaria marina* (L.) All., *Spergula marina* (L.) Bartl. & Wendl., *Spergularia marina* (L.) Griseb.) - 1, 2, 3, 4, 5, 6
segetalis (L.) G. Don fil. (*Alsine segetalis* L.) - 1, 6
sperguloides (Lehm.) Heynh. (*Holosteum sperguloides* Lehm.) - 6

Stellaria L. (*Alsinula* Dostal, nom. invalid., *Cherleria* auct.)

alatavica M. Pop. (*S. brachypetala* Bunge var. *alatavica* (M. Pop.) Ju. Kozhevn.) - 6
alexeenkoana Schischk. = Mesostemma alexeenkoanum
alsine Grimm (*S. uliginosa* Murr.) - 1, 3, 5(alien)
- subsp. *inundata* (Worosch.) Worosch. = S. inundata
- subsp. *undulata* (Thunb.) Worosch. = S. undulata
- var. *undulata* (Thunb.) Ohwi = S. undulata
alsinoides Boiss. & Buhse = Tytthostemma alsinoides

amblyosepala Schrenk - 3, 6
anagalloides C.A. Mey. ex Rupr. - 2
angarae M. Pop. - 4
arctica Schischk. = S. laeta
arenicola Raup - 5
barthiana Schur - 1
blatteri Mattf. = Dichoglottis alsinoides
brachypetala Bunge - 3, 6
- var. *alatavica* (M. Pop.) Ju. Kozhevn. = S. alatavica
brotheriana Trautv. = Minuartia brotheriana
bungeana Fenzl - 1, 3, 4, 5
calycantha (Ledeb.) Bong. - 1, 3
- subsp. *pilosella* Ju. Kozhevn.
cherleriae (Fisch. ex Ser.) F. Williams - 4, 5
ciliatosepala Trautv. - 1, 3, 4, 5
- subsp. *laxmannii* (Fisch. ex Ser.) Ju. Kozhevn. = S. laxmannii
- var. *arctica* (Schischk.) Hult. = S. laeta
crassifolia Ehrh. - 1, 2, 3, 4, 5, 6
crassipes Hult. - 1, 4, 5
crispa Cham. & Schlecht. - 5
dahurica Willd. ex Schlecht. - 4
darvasievii R. Kam. - 6
dichotoma L. - 3, 4, 5
dicranoides (Cham. & Schlecht.) Fenzl (*Cherleria dicranoides* Cham. & Schlecht., *Arenaria dicranoides* (Cham. & Schlecht.) Hult.) - 5
diffusa Willd. ex Schlecht. = S. longifolia
discolor Turcz. - 4, 5
dubia Bast. = Dichodon viscidum
ebracteata Kom. - 5
edwardsii R. Br. - 1, 3, 4, 5
eschscholtziana Fenzl - 4, 5
fennica (Murb.) Perf. (*S. palustris* Retz. var. *fennica* Murb.) - 1
filicaulis Makino (*S. filicaulis* f. *jaluana* (Nakai) Kitag., *S. filipes* M. Pop., *S. jaluana* Nakai) - 5
- f. *jaluana* (Nakai) Kitag. = S. filicaulis
filipes M. Pop. = S. filicaulis
fischeriana Ser. (*S. florida* Fisch.) - 4, 5
- subsp. *viridifolia* A.Khokhr. = S. viridifolia
florida Fisch. = S. fischeriana
fontana M. Pop. - 6
fragilis Klok. - 1
glandulifera N. Zolot. - 3
graminea L. - 1, 2, 3, 4, 5, 6
- var. *hippoctona* Czern. = S. hippoctona
gypsophiloides Fenzl - 5
hebecalyx Fenzl - 1, 3
hippoctona (Czern.) Klok. (*S. graminea* L. var. *hippoctona* Czern.) - 1
holostea L. - 1, 2, 3
humifusa Rottb. - 1, 3, 4, 5
imbricata Bunge - 3
inundata Worosch. (*S. alsine* Grimm subsp. *inundata* (Worosch.) Worosch.) - 5
irrigua Bunge (*S. umbellata* Turcz. ex Kar. & Kir.) - 3, 4, 5, 6
jacutica Schischk. - 4, 5
jaluana Nakai = S. filicaulis
karatavica Schischk. = Mesostemma karatavicum
kolymensis A. Khokhr. - 5
kotschyana Fenzl ex Boiss. = Mesostemma kotschyanum
laeta Richards. (*S. arctica* Schischk., *S. ciliatosepala* Trautv. var. *arctica* (Schischk.) Hult.) - 5
laxmannii Fisch. ex Ser. (*S. ciliatosepala* Trautv. subsp. *laxmannii* (Fisch. ex Ser.) Ju. Kozhevn., *S. palustris* Retz. var. *laxmannii* (Fisch. ex Ser.) Simonk.) - 4

longifolia Muehl. ex Willd. (*S. diffusa* Willd. ex Schlecht.) - 1, 3, 4, 5

longipes Goldie - 5

- subsp. *monantha* (Hult.) W.A. Weber = S. monantha

martjanovii Kryl. = Mesostemma martjanovii

media (L.) Vill. (*Alsinula media* (L.) Dostal, comb. invalid.) - 1, 2, 3, 4, 5, 6

- subsp. *neglecta* (Weihe) Murb. = S. neglecta
- subsp. *pallida* (Dumort.) Aschers. & Graebn. = S. pallida

monantha Hult. (*S. longipes* Goldie subsp. *monantha* (Hult.) W.A. Weber) - 5

montana auct. = S. nemorum

neglecta Weihe (*Alsine neglecta* (Weihe) A. & D. Löve, *Alsinula neglecta* (Weihe) Dostal, *Stellaria media* (L.) Vill. subsp. *neglecta* (Weihe) Murb.) - 1, 2, 6

nemorum L. (*S. montana* auct.) - 1, 2

pallida (Dumortier) Pire (*Alsinula pallida* (Dumort.) Dostal, comb. invalid., *Stellaria media* (L.) Vill. subsp. *pallida* (Dumort.) Aschers. & Graebn.) - 1, 2, 6

palustris Retz. - 1, 2, 3, 4, 5, 6

- var. *fennica* Murb. = S. fennica

peduncularis Bunge - 1, 3, 4, 5

persica Boiss. - 2

petraea Bunge - 3, 4, 6

radians L. = Fimbripetalum radians

ruscifolia Pall. ex Schlect. - 4(?), 5

schischkinii Peschkova - 4

schugnanica Schischk. = Mesostemma schugnanicum

sibirica (Regel & Til.) Schischk. - 5

soongorica Roshev. - 6

turkestanica Schischk. - 6

uliginosa Murr. = S. alsine

umbellata Turcz. ex Kar. & Kir. = S. irrigua

undulata Thunb. (*S. alsine* Grimm var. *undulata* (Thunb.) Ohwi, *S. alsine* subsp. *undulata* (Thunb.) Worosch.) - 5

viridifolia (A. Khokhr.) A. Khokhr. & V. Pavl. (*S. fischeriana* Ser. subsp. *viridifolia* A. Khokhr.) - 4

viscida Bieb. = Dichodon viscidum

winkleri (Briq.) Schischk. - 6

Steris Adans. (*Viscaria* Bernh.)

alpina (L.) Sourkova (*Lychnis alpina* L., *L. suecica* Lodd., *Silene suecica* (Lodd.) Greuter & Burdet, *Viscaria alpina* (L.) G. Don fil.) - 1

atropurpurea (Griseb.) Holub (*Viscaria atropurpurea* Griseb., *Lychnis viscaria* L. subsp. *atropurpurea* (Griseb.) Chater, *Viscaria vulgaris* Bernh. subsp. *atropurpurea* (Griseb.) Stojan.) - 1(?)

viscaria (L.) Rafin. (*Lychnis viscaria* L., *L. viscosa* Scop., *Silene viscaria* (L.) Jess., *Viscaria viscosa* (Scop.) Aschers., *V. vulgaris* Bernh.) - 1, 2, 3, 6

Telephium L.

imperati L. subsp. *orientale* (Boiss.) Nym. = T. orientale

- subsp. *orientale* (Boiss.) Rech.fil. = T. orientale

oligospermum Steud. ex Boiss. - 2

orientale Boiss. (*T. imperati* L. subsp. *orientale* (Boiss.) Nym., *T. imperati* subsp. *orientale* (Boiss.) Rech. fil. comb. superfl.) - 2, 6

Thylacospermum Fenzl

caespitosum (Cambess.) Schischk. (*Arenaria caespitosa* (Cambess.) Ju. Kozhevn.) - 6

Tryphane (Fenzl) Reichenb. = Minuartia

rubella (Wahlenb.) Reichenb. = Minuartia rubella

verna (L.) Reichenb. = Minuartia verna

Tunica auct. = Petrorhagia

alpina (Habl.) Bobr. = Petrorhagia alpina

pachygona Fisch. & C.A. Mey. = Petrorhagia cretica

saxifraga (L.) Scop. = Petrorhagia saxifraga

stricta (Bunge) Fisch. & C.A. Mey. = Petrorhagia alpina

Tytthostemma Nevski

alsinoides (Boiss. & Buhse) Nevski (*Stellaria alsinoides* Boiss. & Buhse) - 6

Vaccaria N.M. Wolf

grandiflora (Fisch. ex DC.) Jaub. & Spach = V. hispanica subsp. grandiflora

hispanica (Mill.) Rauschert (*Saponaria hispanica* Mill., *S. segetalis* Neck. nom. illegit., *Vaccaria pyramidata* Medik., *V. segetalis* Garcke) subsp. **grandiflora** (DC.) Holub (*Saponaria vaccaria* var. *grandiflora* Fisch. ex DC., *Vaccaria grandiflora* (Fisch. ex DC.) Jaub. & Spach, *V. hispanica* var. *grandiflora* (Fisch. ex DC.) Gvinianidze, *V. hispanica* var. *grandiflora* (Fisch. ex DC.) Leonard, comb. superfl., *V. pyramidata* Medik. subsp. *grandiflora* (Fisch. ex DC.) Hayek, *V. pyramidata* var. *grandiflora* (Fisch. ex DC.) Cullen) - 1, 2, 3, 4, 5, 6

- subsp. *oxyodonta* (Boiss.) Greuter & Burdet = V. oxyodonta
- var. *grandiflora* (Fisch. ex DC.) Gvinianidze = V. hispanica subsp. grandiflora
- var. *grandiflora* (Fisch. ex DC.) Leonard = V. hispanica subsp. grandiflora
- var. *oxyodonta* (Boiss.) Leonard = V. oxydonta

oxyodonta Boiss. (*V. hispanica* (Mill.) Rauschert subsp. *oxyodonta* (Boiss.) Greuter & Burdet, *V. hispanica* var. *oxyodonta* (Boiss.) Leonard, *V. pyramidata* Medik. var. *oxyodonta* (Boiss.) Zohary) - 6

pyramidata Medik. = V. hispanica

- subsp. *grandiflora* (Fisch. ex DC.) Hayek = V. hispanica subsp. grandiflora
- var. *grandiflora* (Fisch. ex DC.) Cullen = V. hispanica subsp. grandiflora
- var. *oxyodonta* (Boiss.) Zohary = V. oxyodonta

segetalis Garcke = V. hispanica

Velezia L.

rigida L. - 1, 2, 6

Viscago Zinn

parviflora Hornem. = Silene borysthenica

Viscaria Bernh. = Steris

alpina (L.) G. Don fil. = Steris alpina

atropurpurea Griseb. = Steris atropurpurea

viscosa (Scop.) Aschers. = Steris viscaria

vulgaris Bernh. = Steris viscaria

- subsp. *atropurpurea* (Griseb.) Stojan. = Steris atropurpurea

Wierzbickia Reichenb. = Minuartia

macrocarpa (Pursh) Reinchenb. = Minuartia macrocarpa

Wilhelmsia Reichenb. (*Merckia* Fisch. ex Cham. & Schlecht. 1826, non *Merkia* Borkh. 1792)

physodes (Ser.) McNeill (*Merckia physodes* (Ser.) Fisch. ex Cham. & Schlecht.) - 4, 5

CELASTRACEAE R. Br.
Celastrus L.

alata Thunb. = Euonymus alata
flagellaris Rupr. - 5
orbiculata Thunb. - 4(alien), 5
- var. *strigillosa* (Nakai) Hara = C. strigillosa
striata Thunb. = Euonymus alata
strigillosa Nakai (*C. orbiculata* Thunb. var. *strigillosa* (Nakai) Hara) - 5

Euonymus L. (*Kalonymus* (G. Beck) Prokh., *Masakia* (Nakai) Nakai,*Turibana* (Nakai) Nakai)

alata (Thunb.) Siebold (*Celastrus alata* Thunb., *C. striata* Thunb., *Euonymus alata* f. *striata* (Thunb.) Kitag., *E. striata* (Thunb.) Loes.) - 5
- subsp. *sacrosancta* (Koidz.) Worosch. = E. sacrosancta
- f. *striata* (Thunb.) Kitag. = E. alata
armasica Gatsch. = E. leiophloea
bulgarica Velen. = E. europaea
czernjaevii Klok. (*E. europaea* L. subsp. *moldavica* (Klok.) Grosset, *E. europaea* subsp. *pontica* Savul. & Rayss, *E. europaea* subsp. *subvelutina* (Savul. & Rayss) Grosset, *E. europaea* var. *involuta* Lindem., *E. europaea* var. *suberoso-alata* Lindem., *E. europaea* var. *subvelutina* Savul. & Rayss, *E. moldavica* Klok., *E. odessana* Klok., *E. pubescens* Stev. nom. provis.) - 1
europaea L. (*E. bulgarica* Velen., *E. europaea* f. *bulgarica* (Velen.) I. Ganchev, *E. floribunda* Stev., *E. medirossica* Klok., *E. suberosa* Klok.) - 1, 2
- subsp. *moldavica* (Klok.) Grosset = E. czernjaevii
- subsp. *pontica* Savul. & Rayss = E. czernjaevii
- subsp. *subvelutina* (Savul. & Rayss) Grosset = E. czernjaevii
- var. *involuta* Lindem. = E. czernjaevii
- var. *latifolia* L. = E. latifolia
- var. *suberoso-alata* Lindem. = E. czernjaevii
- var. *subvelutina* Savul. & Rayss = E. czernjaevii
- f. *bulgarica* (Velen.) I. Ganchev = E. europaea
floribunda Stev. = E. europaea
hamiltoniana Wall. ex Roxb. subsp. *sieboldiana* (Blume) Hara = E. sieboldiana
*japonica Thunb. (*Masakia japonica* (Thunb.) Nakai) - 1, 2, 6
ketzhovelii Gatsch. = E. leiophloea
koopmannii Lauche - 6
latifolia (L.) Mill. (*E. europaea* var. *latifolia* L., *Kalonymus latifolia* (L.) Prokh.) - 1, 2
- var. *planipes* Koehne = E. planipes
leiophloea Stev. (*E. armasica* Gatsch., *E. ketzhovelii* Gatsch., *E. leiophloea* var. *armasica* (Gatsch.) Gagnidze, *Kalonymus leiophloea* (Stev.) Prokh.) - 2
- var. *armasica* (Gatsch.) Gagnidze = E. leiophloea
maackii Rupr. - 4, 5
macroptera Rupr. (*Kalonymus macroptera* (Rupr.) Prokh., *Turibana macroptera* (Rupr.) Nakai) - 5
- var. *miniata* (Tolm.) Worosch. = E. miniata
maximowicziana Prokh. (*Kalonymus maximowicziana*

Prokh., *Euonymus maximowicziana* (Prokh.) Worosch. comb. superfl.) - 5
medirossica Klok. = E. europaea
miniata Tolm. (*E. macroptera* Rupr. var. *miniata* (Tolm.) Worosch., *Kalonymus* x *miniata* (Tolm.) Prokh.) - 5
moldavica Klok. = E. czernjaevii
nana Bieb. - 1, 2
odessana Klok. = E. czernjaevii
oxyphylla auct. = E. planipes
pauciflora Maxim. - 5
planipes (Koehne) Koehne (*E. latifolia* (L.) Mill. var. *planipes* Koehne, *Kalonymus oxyphylla* (Miq.) Prokh. p.p. excl. basionymo, *K. yezoensis* (Koidz.) Prokh. p.p. excl. basionymo, *Turibana planipes* (Koehne) Nakai, *Euonymus oxyphylla* auct., *E. yezoensis* auct.) - 5
pubescens Stev. = E. czernjaevii
sachalinensis (Fr. Schmidt) Maxim. (*E. tricarpa* Koidz., *Kalonymus sachalinensis* (Fr. Schmidt) Prokh., *Turibana sachalinensis* (Fr. Schmidt) Nakai, *T. tricarpa* (Koidz.) Nakai) - 5
sacrosancta Koidz. (*E. alata* (Thunb.) Siebold subsp. *sacrosancta* (Koidz) Worosch.) - 4, 5
semenovii Regel & Herd. - 6
sieboldiana Blume (*E. hamiltoniana* Wall. ex Roxb. subsp. *sieboldiana* (Blume) Hara) - 5
striata (Thunb.) Loes. = E. alata
suberosa Klok. = E. europaea
tricarpa Koidz. = E. sachalinensis
velutina Fisch. & C.A. Mey. - 2, 6
verrucosa Scop. - 1, 2
yezoensis auct. = E. planipes

Kalonymus (G. Beck) Prokh. = Euonymus

latifolia (L.) Prokh. = Euonymus latifolia
leiophloea (Stev.) Prokh. = Euonymus leiophloea
macroptera (Rupr.) Prokh. = Euonymus macroptera
maximowicziana Prokh. = Euonymus maximowicziana
x *miniata* (Tolm.) Prokh. = Euonymus miniata
oxyphylla (Miq.) Prokh. p.p. = Euonymus planipes
sachalinensis (Fr. Schmidt) Prokh. = Euonymus sachalinensis
yesoensis (Koidz.) Prokh. p.p. = Euonymus planipes

Masakia (Nakai) Nakai = Euonymus

japonica (Thunb.) Nakai = Euonymus japonica

Turibana (Nakai) Nakai = Euonymus

macroptera (Rupr.) Nakai = Euonymus macroptera
planipes (Koehne) Nakai = Euonymus planipes
sachalinensis (Fr. Schmidt) Nakai = Euonymus sachalinensis
tricarpa (Koidz.) Nakai = Euonymus sachalinensis

CELTIDACEAE Link
Celtis L.

aspera (Ledeb.) Stev. = C. tournefortii
australis L. - 2
- subsp. *caucasica* (Willd.) C.C. Townsend = C. caucasica
caucasica Willd. (*C. australis* L. subsp. *caucasica* (Willd.) C.C. Townsend) - 2, 6
- subsp. *caudata* (Planch.) Grudz. = C. tupalangi
- var. *caudata* Planch. = C. tupalangi

glabrata Stev. ex Planch. - 1, 2
tournefortii Lam. (*C. aspera* (Ledeb.) Stev.) - 2
tupalangi Vass. (*C. caucasica* Willd. var. *caudata* Planch., *C. caucasica* subsp. *caudata* (Planch.) Grudz.) - 6

CERATOPHYLLACEAE S.F. Gray

Ceratophyllum L.

affine Z. Troitz. - 3
demersum L. (*C. oxyacanthum* Cham., *C. tuberculatum* Cham.) - 1, 2, 3, 4, 5, 6
- subsp. *pentacanthum* (Haynald) Soo = C. pentacanthum
komarovii Kuzen. = C. pentacanthum
kossinskyi Kuzen. (*C. muricatum* subsp. *kossinskyi* (Kuzen.) Les Donald) - 1
muricatum subsp. *kossinskyi* (Kuzen.) Les Donald = C. kossinskyi
oryzetorum Kom. - 4, 5, 6
oxyacanthum Cham. = C. demersum
pentacanthum Haynald (*C. komarovii* Kuzen., *C. demersum* L. subsp. *pentacanthum* (Haynald) Soo, *C. platyacanthum* auct.) - 1
platyacanthum auct. = C. pentacanthum
submersum L. - 1, 2, 3, 6
tanaiticum Sapeg. - 1, 6
tuberculatum Cham. = C. demersum

CHEILANTHACEAE Nayar = SINOPTE-RIDACEAE

CHELIDONIACEAE Nakai = PAPAVE-RACEAE

CHENOPODIACEAE Vent.

Aellenia Ulbr. = Halothamnus

auricula (Moq.) Ulbr. = Halothamnus auriculus
glauca (Bieb.) Aell. = Halothamnus glaucus
- subsp. *euglauca* Aell. = Halothamnus glaucus
- subsp. *hispidula* (Bunge) Aell. = Halothamnus hispidulus
- subsp. *lancifolia* (Boiss.) Aell. = Halothamnus acutifolius
hispidula (Bunge) Botsch. = Halothamnus hispidulus
iliensis (Lipsky) Aell. = Halothamnus iliensis
lancifolia (Boiss.) Aell. = Halothamnus acutifolius
lancifolia (Boiss.) Ulbr. = Halothamnus acutifolius
subaphylla (C.A. Mey.) Aell. = Halothamnus subaphyllus
- subsp. *eusubaphylla* Aell. = Halothamnus subaphyllus
- subsp. *turcomanica* Aell. = Halothamnus subaphyllus
turcomanica (Aell.) Czer. = Halothamnus subaphyllus

Agriophyllum Bieb. ex C.A. Mey.

arenarium Bieb. ex C.A. Mey. = A. squarrosum
lateriflorum (Lam.) Moq. - 2, 6
latifolium Fisch. & C.A. Mey. - 6
minus Fisch. & C.A. Mey. - 6
paletzkianum Litv. - 6
pungens (Vahl) Link = A. aquarrosum
squarrosum (L.) Moq. (*Coryspermum squarrosum* L., *Agriophyllum arenarium* Bieb. ex C.A. Mey., *A. pungens* (Vahl) Link, *Coryspermum pungens* Vahl) - 1, 2, 3, 4, 6

Alexandra Bunge

lehmannii Bunge - 6

Anabasis L.

abolinii Iljin = A. brevifolia
affinis Fisch. & C.A. Mey. = A. brevifolia
annua Bunge (*A. micradena* Iljin) - 6
aphylla L. - 1, 2, 6
- subsp. **australis** Iljin - 2
- subsp. **rubra** Iljin - 2
balchaschensis Iljin = Arthrophytum balchaschense
brachiata Fisch. & C.A. Mey. ex Kar. & Kir. (*A. undulata* Iljin) - 2, 6
brevifolia C.A. Mey. (*A. abolinii* Iljin, *A. affinis* Fisch. & C.A. Mey.) - 3, 4, 6
cretacea Pall. - 1, 3, 6
ebracteolata Korov. ex Botsch. - 6
elatior (C.A. Mey.) Schischk. - 6
eriopoda (Schrenk) Benth. ex Volkens - 6
eugeniae Iljin - 2
ferganica Drob. - 6
gontscharowii Iljin = A. turkestanica
gypsicola Iljin - 6
hispidula (Bunge) Benth. ex Volkens = A. jaxartica
jaxartica (Bunge) Benth. ex Volkens (*A. hispidula* (Bunge) Benth. ex Volkens) - 6
korovinii Iljin cx A. Vassil. - 6
macroptera auct. = A. tianschanica
micradena Iljin = A. annua
pauciflora M. Pop. ex Iljin = Arthrophytum balchaschense
pelliotii Danguy - 6
pulcherrima Iljin = A. truncata
ramosissima Minkw. = A. salsa
salsa (C.A. Mey.) Benth. ex Volkens (*A. ramosissima* Minkw.) - 1, 2, 3, 6
tianschanica Botsch. (*A. macroptera* auct.) - 6
truncata (Schrenk) Bunge (*A. pulcherrima* Iljin) - 6
turgaica Iljin & Krasch. - 6
turkestanica Korov. & Iljin (*A. gontscharowii* Iljin) - 6
undulata Iljin = A. brachiata

Anthochlamys Fenzl

polygaloides (Fisch. & C.A. Mey.) Fenzl - 2, 6
tjanschanica Iljin ex Aell. - 6
turcomanica Iljin - 6

Arthrocnemon auct. = Microcnemum

coralloides Loscos & Pardo = Microcnemum coralloides

Arthrocnemum Moq. p.p. = Halostachys

berangerianum Moq. = Halostachys belangeriana

Arthrophytum Schrenk

affine Korov. & Miron. = A. betpakdalense
balchaschense (Iljin) Botsch. (*Anabasis balchaschensis* Iljin, *A. pauciflora* M. Pop. ex Iljin) - 6
betpakdalense Korov. & Miron. (*A. affine* Korov. & Miron.) - 6
iliense Iljin - 6
korovinii Botsch. - 6
lehmannianum Bunge (*A. litwinowii* Korov.) - 6
leptocladum M. Pop. ex Iljin = Hammada leptoclada
litwinowii Korov. = A. lehmannianum

longibracteatum Korov. - 6
pulvinatum Litv. - 6
subulifolium Schrenk - 6
wakhanicum (Pauls.) Iljin = Hammada wakhanica

Atriplex L. (*Obione* Gaertn.)

acuminata Waldst. & Kit. = A. sagittata
amblyostegia Turcz. = A. aucheri
arazdajanica Kapell. = A. tatarica
aucheri Moq. (*A. amblyostegia* Turcz., *A. desertorum* (Iljin)
   Sosn., *A. hortensis* L. subsp. *desertorum* (Iljin) Aell., *A.*
   nitens Schkuhr subsp. *desertorum* Iljin, *A. nitens* subsp.
   aucheri (Moq.) Takht.) - 1, 2, 6
babingtonii J. Woods = A. glabriuscula
bucharica Iljin = A. moneta
calotheca (Rafn) Fries (*A. hastata* L. nom. ambig., *A. pros-*
   trata Boucher ex DC. subsp. *calotheca* (Rafn) M.
   Gust.) - 1
cana C.A. Mey. - 1, 2, 3, 6
centralasiatica Iljin (*A. sibirica* L. var. *centralasiatica* (Iljin)
   Grub., *Obione centralasiatica* (Iljin) Kitag.) - 6
- var. *megalotheca* (M. Pop.) G.L. Chu = A. megalotheca
crassifolia C.A. Mey. - 1, 3, 4, 6
deltoidea Bab. = A. prostrata subsp. deltoidea
desertorum (Iljin) Sosn. = A. aucheri
dimorphostegia Kar. & Kir. - 6
fera (L.) Bunge - 3, 4
flabellum Bunge - 6
fominii Iljin - 2, 6
glabriuscula Edmondston (*A. babingtonii* J. Woods) - 1
gmelinii C.A. Mey. - 5
- subsp. *dilatata* (Franch. & Savat.) Kitam. = A. subcordata
hastata L. = A. calotheca
- subsp. *polonica* (Zapal.) Aell. = A. polonica
hastata sensu Iljin = A. prostrata
heterosperma Bunge = A. micrantha
hortensis L. - 1, 2, 3, 5(alien), 6
- subsp. *desertorum* (Iljin) Aell. = A. aucheri
- subsp. *heterosperma* (Bunge) Meijden = A. micrantha
iljinii Aell. - 3, 6
kuzenevae N. Semen. - 1
laevis C.A. Mey. - 1, 3, 4, 6
- var. *patens* (Litv.) Grub. = A. patens
lapponica Pojark. = A. nudicaulis
lasiantha Boiss. - 6
latifolia Wahlenb. = A. prostrata subsp. latifolia
leucoclada Boiss. subsp. *turcomanica* (Moq.) Aell. = A.
   turcomanica
- var. *turcomanica* (Moq.) Zohary = A. turcomanica
littoralis L. - 1, 2, 3
- var. *dilatata* Franch. & Savat. = A. subcordata
longipes auct. p.p. = A. nudicaulis
longipes Drej. - 1
- subsp. *lapponica* (Pojark.) A. & D. Löve = A. nudicaulis
- subsp. *praecox* (Hulph.) Turesson = A. praecox
megalotheca M. Pop. (*A. centralasiatica* Iljin var. *megaloth-*
   eca (M. Pop.) G.L. Chu) - 6
micrantha C.A. Mey. (*A. heterosperma* Bunge, *A. hortensis*
   L. subsp. *heterosperma* (Bunge) Meijden) - 1, 2, 3,
   5(alien), 6
moneta Bunge (*A. bucharica* Iljin, 1937, non Gand. 1919) -
   6
multicolora Aell. = A. tatarica
nitens Schkuhr = A. sagittata
- subsp. *aucheri* (Moq.) Takht. = A. aucheri
- subsp. *desertorum* Iljin = A. aucheri
nudicaulis Bogusl. (*A. lapponica* Pojark., *A. longipes* Drej.

subsp. *lapponica* (Pojark.) A. & D. Löve, *A. longipes*
   auct. p.p.) - 1
oblongifolia Waldst. & Kit. - 1, 2, 6
olivieri Moq. - 2
ornata Iljin - 6
pamirica Iljin - 6
patens (Litv.) Iljin (*A. laevis* C.A. Mey. var. *patens* (Litv.)
   Grub.) - 1, 2, 3, 4, 5(alien), 6
patula L. - 1, 2, 3, 4, 5(alien), 6
pedunculata L. = Halimione pedunculata
polonica Zapal. (*A. hastata* L. subsp. *polonica* (Zapal.)
   Aell., *A. prostrata* Boucher ex DC. subsp. *polonica*
   (Zapal.) Uotila) - 1
praecox Hulph. (*A. longipes* Drej. subsp. *praecox* (Hulph.)
   Turesson) - 1
procumbens Less. - 3
prostrata Boucher ex DC. (*A. hastata* sensu Iljin) - 1, 2, 3,
   4, 5(alien), 6
- subsp. *calotheca* (Rafn) M. Gust. = A. calotheca
- subsp. **deltoidea** (Bab.) Rauschert (*A. deltoidea* Bab.)
- subsp. **latifolia** (Wahlenb.) Rauschert (*A. latifolia* Wahlenb.)
- subsp. *polonica* (Zapal.) Uotila = A. polonica
- subsp. **triangularis** (Willd.) Rauschert (*A. triangularis*
   Willd.)
pungens Trautv. - 6
rosea L. - 1, 4(alien)
sagittata Borkh. (*A. acuminata* Waldst. & Kit., *A. nitens*
   Schkuhr, nom. illegit.) - 1, 2, 3, 4, 6
schugnanica Iljin - 6
sibirica L. (*Obione sibirica* (L.) Fisch.) - 1(alien), 3, 4,
   5(alien), 6
- var. *centralasiatica* (Iljin) Grub. = A. centralasiatica
sphaeromorpha Iljin - 1, 2, 6
subcordata Kitag. (*A. gmelinii* C.A. Mey. subsp. *dilatata*
   (Franch. & Savat.) Kitam., *A. littoralis* L. var. *dilatata*
   Franch. & Savat.) - 5
tatarica L. (*A. arazdajanica* Kapell., *A. multicolora* Aell.) -
   1, 2, 3, 4, 5(alien), 6
thunbergiifolia (Boiss. & Noe) Boiss. - 6
tianschanica Pratov - 6
triangularis Willd. = A. prostrata subsp. triangularis
turcomanica (Moq.) Boiss. (*A. leucoclada* Boiss. subsp.
   turcomanica (Moq.) Aell., *A. leucoclada* var. *turco-*
   manica (Moq.) Zohary) - 2, 6
verrucifera Bieb. = Halimione verrucifera

Axyris L.

amaranthoides L. - 1, 3, 4, 5, 6
caucasica (Somm. & Levier) Lipsky - 2
hybrida L. - 3, 4, 5, 6
prostrata L. - 3, 4, 6
sphaerosperma Fisch. & C.A. Mey. - 3, 4, 5(?)

Bassia All. (*Echinopsilon* Moq.)

dasyphylla (Fisch. & C.A. Mey.) O. Kuntze (*Kochia dasy-*
   phylla Fisch. & C.A. Mey., *Bassia fiedleri* Aell., *Echi-*
   nopsilon dasyphyllum (Fisch. & C.A. Mey.) Moq., *E.*
   divaricatum Kar. & Kir. 1841, non *Bassia divaricata* F.
   Muell, 1882) - 4, 6
eriantha (Fisch. & C.A. Mey.) O. Kuntze = Londesia eriantha
fiedleri Aell. = B. dasyphylla
hirsuta (L.) Aschers. (*Echinopsilon hirsutum* (L.) Moq.) -
   1, 2, 3, 6
hyssopifolia (Pall.) O. Kuntze (*Echinopsilon hyssopifolium*
   (Pall.) Moq.) - 1, 2, 3, 4, 6

krylovii (Litv.) A.J. Scott = Kochia krylovii
laniflora (S.G. Gmel.) A.J. Scott = Kochia laniflora
melanoptera (Bunge) A.J. Scott = Kochia melanoptera
prostrata (L.) A.J. Scott = Kochia prostrata
scoparia (L.) A.J. Scott = Kochia scoparia
sedoides (Pall.) Aschers. (*Echinopsilon sedoides* (Pall.) Moq.) - 1, 2, 3, 6

Beta L.

cicla L. = B. vulgaris subsp. cicla
corolliflora Zosimovic ex Buttler - 2
esculenta Salisb. = B. vulgaris subsp. esculenta
lomatogona Fisch. & C.A. Mey. - 2
macrorhiza Stev. - 2
maritima L. (*B. perennis* (L.) Freyn, *B. vulgaris* L. subsp. *maritima* (L.) Arcang., *B. vulgaris* subsp. *perennis* (L.) Aell.) - 1(?), 2, 6
- subsp. **orientalis** (Roth) Burenin (*B. orientalis* Roth, *B. vulgaris* subsp. *orientalis* (Roth) Aell.) - 2(?)
orientalis Roth = B. maritima subsp. orientalis
perennis (L.) Freyn = B. maritima
trigyna Waldst. & Kit. - 1, 2
*****vulgaris** L. - 1, 2, 5, 6
- subsp. **cicla** (L.) Schuebl. & Martens (*B. cicla* L.)
- subsp. **esculenta** Guerke (*B. esculenta* Salisb. nom. illegit., *B. vulgaris* subsp. *esculenta* Coutinho)
- subsp. *esculenta* Coutinho = B. vulgaris subsp. esculenta
- subsp. *maritima* (L.) Arcang. = B. maritima
- subsp. *orientalis* (Roth) Aell. = B. maritima subsp. orientalis
- subsp. *perennis* (L.) Aell. = B. maritima

Bienertia Bunge

cycloptera Bunge - 1, 2, 6

Borsczowia Bunge

aralocaspica Bunge - 6

Camphorosma L.

annua Pall. - 1
lessingii Litv. (*C. monspeliaca* L. subsp. *lessingii* (Litv.) Aell.) - 1, 2, 3, 6
monspeliaca L. - 1, 2, 3, 6
- subsp. *lessingii* (Litv.) Aell. = C. lessingii
songorica Bunge - 1, 2, 3, 6

Caroxylon Moq. = Halothamnus

acutifolium (Moq.) Moq. = Halothamnus acutifolius
lancifolium Boiss. = Halothamnus acutifolius

Caspia Galushko = Salsola

foliosa (L.) Galushko = S. foliosa

Ceratocarpus L.

arenarius L. - 1, 2, 3, 4, 6
- subsp. *utriculosus* (Bluk.) Takht. = C. utriculosus
turkestanicus Sav.-Rycz. ex Iljin = C. utriculosus
utriculosus Bluk. (*C. arenarius* L. subsp. *utriculosus* (Bluk.) Takht., *C. turkestanicus* Sav.-Rycz. ex Iljin) - 2, 6

Ceratoides Gagnebin, p.p. = Krasheninnikovia

arborescens (Losinsk.) Tsien & C.G. Ma = Krascheninniko-via arborescens
ewersmanniana (Stschegl.ex Losinsk.) Botsch. & Ikonn. = Krascheninnikovia ewersmanniana
fruticulosa (Pazij) Czer. = Krascheninnikovia fruticulosa
latens (J.F. Gmel.) Reveal & Holmgren = Krascheninniko-via ceratoides
lenensis (Kumin.) Jurtz. & R. Kam. = Krascheninnikovia lenesis
papposa Botsch. & Ikonn. = Krascheninnikovia ceratoides
pungens (M. Pop.) Czer. = Krascheninnikovia pungens

Ceratospermum Pers. = Krascheninnikovia

papposum Pers. = Krascheninnikovia ceratoides

Chenopodium L. (*Neobotrydium* Mold.)

acerifolium Andrz. (*Ch. album* L. var. *hastatum* Klinggr., *Ch. album* subsp. *hastatum* (Klinggr.) J. Murr, *Ch. klinggraeffii* (Abrom.) Aell.) - 1, 2, 3, 4
acuminatum Willd. - 1, 3, 4, 5(alien), 6
album L. (*Ch. album* subsp. *virgatum* (Thunb.) Blom, *Ch. centrorubrum* (Makino) Nakai, *Ch. virgatum* Thunb., *Ch. giganteum* auct., *Ch. viride* auct.) - 1, 2, 3, 4, 5, 6
- subsp. *hastatum* (Klinggr.) J. Murr = Ch. acerifolium
- subsp. *reticulatum* (Aell.) Beauge ex Greuter & Burdet = Ch. reticulatum
- subsp. *striatum* (Krasan) J. Murr = Ch. strictum
- subsp. *virgatum* (Thunb.) Blom = Ch. album
- var. *hastatum* Klinggr. = Ch. acerifolium
ambrosioides L. (*Teloxys ambrosioides* (L.) W.A. Weber) - 1, 2
amurense Ignatov - 5
anthelminticum L. - 2
aristatum L. = Teloxys aristata
badachschanicum Tzvel. - 6
berlandieri Moq. (*Ch. zschackei* J. Murr) - 1
betaceum Andrz. = Ch. strictum
bonus-henricus L. - 1
botryodes Sm. = Ch. chenopodioides
botrys L. (*Neobotrydium botrys* (L.) Mold., *Teloxys botrys* (L.) W.A. Weber) - 1, 2, 3, 5(alien), 6
bryoniifolium Bunge - 4, 5
capitatum (L.) Aschers. - 1, 4
centrorubrum (Makino) Nakai = Ch. album
chenopodioides (L.) Aell. (*Ch. botryodes* Smith, *Ch. rubrum* L. subsp. *botryodes* (Smith) Slavnic, comb. invalid.) - 1, 2, 3, 4, 6
ficifolium Smith (*Ch. serotinum* auct.) - 1, 3, 4, 5, 6
foetidum Schrad. = Ch. schraderianum
foliosum Aschers. (*Ch. korshinskyi* Litv., *Monolepis litwi-nowii* Pauls.) - 1, 2, 3, 4, 6
frutescens C.A. Mey. - 3, 4
x **fursajevii** Aell. & Iljin. - Ch. album L. x Ch. suecicum J. Murr - In places of common occurrence.
giganteum auct. = Ch. album
gigantospermum Aell. = Ch. hybridum subsp. giganto-spermum
glaucophyllum Aell. (*Ch. strictum* Roth subsp. *glaucophyl-lum* (Aell.) Aell.) - 1
glaucum L. (*Ch. glaucum* subsp. *orientale* Worosch., *Ch. wolfii* Simonk.) - 1, 2, 3, 4, 5, 6
- subsp. *orientale* Worosch. = Ch. glaucum
x **gruellii** Aell. - Ch. ficifolium Smith x Ch. suecicum J. Murr - 1
hircinum Schrad. - 1(alien)
x **humuliforme** (J. Murr) Dvorak. - Ch. album L. x Ch. ficifolium Smith x Ch. strictum Roth - 1

hybridum L. - 1, 2, 3, 4, 5, 6
- subsp. **gigantospermum** (Aell.) Hult. (*Ch. gigantospermum* Aell.) - 4
iljinii Golosk. - 3, 4, 6
x **jedlicae** Dvorak. - Ch. album L. x Ch. ficifolium Smith - 1
jenissejense Aell. & Iljin - 1, 3, 4
karoi (J. Murr) Aell. = Ch. prostratum subsp. karoi
klinggraeffii (Abrom.) Aell. = Ch. acerifolium
korshinskyi Litv. = Ch. foliosum
missouriense Aell. - 1
murale L. - 1, 2, 6
nidorosum Otschiauri - 2
opulifolium Schrad. - 1, 2, 3, 5(alien), 6
pamiricum Iljin - 6
pedunculare Bertol. - 1
pratericola Rydb. - 1
x **preissmannii** J. Murr. - Ch. album. x Ch. opulifolium Schrad. - In places of common occurrence.
probstii Aell. - 1
prostratum Bunge - 1, 3, 4, 5, 6
- subsp. **karoi** (J. Murr) Lomonosova (*Ch. karoi* (J. Murr) Aell.) - 1, 3, 4, 5, 6
- subsp. **prostratum** - 4
pumilio R. Br. - 1, 5
*****quinoa** Willd. - 1
reticulatum Aell. (*Ch. album* L. subsp. *reticulatum* (Aell.) Beauge ex Greuter & Burdet) - 1
rubrum L. - 1, 2, 3, 4, 5, 6
- subsp. *botryodes* (Smith) Slavnic = Ch. chenopodioides
schraderianum Schult. (*Ch. foetidum* Schrad. 1808, non Lam. 1778, *Teloxys schraderiana* (Schult.) W.A. Weber) - 1, 2
x **schulzeanum** J. Murr. - Ch. glaucum L. x Ch. rubrum L. - In places of common occurrence.
serotinum auct. = Ch. ficifolium
x **smardae** Dvorak. - Ch. album L. x Ch. ficifolium Smith x Ch. suecicum J. Murr - 1
sosnowskyi Kapell. - 2
striatiforme J. Murr (*Ch. strictum* Roth subsp. *striatiforme* (J. Murr) Uotila) - 1, 2
striatum (Krasan) J. Murr = Ch. strictum
strictum Roth (*Ch. album* L. subsp. *striatum* (Krasan) J. Murr, *Ch. betaceum* Andrz., *Ch. striatum* (Krasan) J. Murr, *Ch. strictum* subsp. *striatum* (Krasan) Aell. & Iljin) - 1, 2, 3, 4, 5, 6
- subsp. *glaucophyllum* (Aell.) Aell. = Ch. glaucophyllum
- subsp. *striatiforme* (J. Murr) Uotila = Ch. striatiforme
- subsp. *striatum* (Krasan) Aell. & Iljin = Ch. strictum
suecicum J. Murr - 1, 2, 3, 4, 5
x **thellungii** J. Murr. - Ch. opulifolium Schrad. x Ch. suecicum J. Murr - In places of common occurence.
urbicum L. - 1, 2, 3, 4, 5, 6
vachelii Hook. & Arn. - 1(alien), 5
virgatum Thunb. = Ch. album
viride auct. = Ch. album
vulvaria L. - 1, 2, 3, 4, 5, 6
wolffii Simonk. = Ch. glaucum
zerovii Iljin - 1
zschackei J. Murr = Ch. berlandieri

Climacoptera Botsch.

affinis (C.A. Mey.) Botsch. (*Salsola affinis* C.A. Mey.) - 1, 6
amblyostegia (Botsch.) Botsch. (*Salsola amblyostegia* Botsch.) - 6
aralensis (Iljin) Botsch. (*Salsola aralensis* Iljin) - 1, 6
brachiata (Pall.) Botsch. (*Salsola brachiata* Pall.) - 1, 2, 3, 6
bucharica (Iljin) Botsch. (*Salsola bucharica* Iljin) - 6

crassa (Bieb.) Botsch. (*Salsola crassa* Bieb.) - 1, 2, 6
czelekenica Pratov - 6
ferganica (Drob.) Botsch. (*Salsola ferganica* Drob.) - 6
glaberrima Botsch. - 6
intricata (Iljin) Botsch. (*Salsola intricata* Iljin) - 6
kasakorum (Iljin) Botsch. (*Salsola kasakorum* Iljin) - 6
korshinskyi (Drob.) Botsch. (*Salsola korshinskyi* Drob.) - 6
lachnophylla (Iljin) Botsch. (*Salsola lachnophylla* Iljin) - 6
lanata (Pall.) Botsch. (*Salsola lanata* Pall.) - 1, 6
longistylosa (Iljin) Botsch. (*Salsola longistylosa* Iljin) - 6
malyginii (Korov. ex Botsch.) Botsch. (*Salsola malyginii* Korov. ex Botsch.) - 6
merkulowiczii (Zak.) Botsch. (*Salsola merkulowiczii* Zak.) - 6
minkwitziae (Korov.) Botsch. (*Salsola minkwitziae* Korov.) - 6
narynensis Pratov - 6
obtusifolia (Schrenk) Botsch. (*Halimocnemis obtusifolia* Schrenk, *Salsola heptapotamica* Iljin) - 6
olgae (Iljin) Botsch. (*Salsola olgae* Iljin) - 6
oxyphylla Pratov - 6
pjataevae Pratov - 6
ptiloptera Pratov - 6
subcrassa (M. Pop.) Botsch. (*Salsola subcrassa* M. Pop.) - 6
sukaczevii Botsch. (*Salsola sukaczevii* (Botsch.) A.J. Li) - 6
sysamyrica Pratov - 6
transoxana (Iljin) Botsch. (*Salsola transoxana* Iljin) - 1, 6
turcomanica (Litv.) Botsch. (*Salsola turcomanica* Litv., *Halanthium lipskyi* Pauls.) - 1, 6
turgaica (Iljin) Botsch. (*Salsola turgaica* Iljin) - 6
tyshchenkoi Pratov - 6
ustjurtensis Pratov - 6
vachschi Kinzikaeva & Pratov - 6

Corispermum L.

algidum Iljin - 1
altaicum Iljin - 3, 4
aralo-caspicum Iljin - 1, 6
- subsp. *caucasicum* Iljin = C. caucasicum
bardunovii M. Pop ex Lomonosova - 4
bjelorussicum Klok. & A. Krasnova - 1
borysthenicum Andrz. - 1
calvo-borysthenicum Klok. - 2
    Probably, a hybrid C. borysthenicum Andrz. x C. nitidum Schult.
calvum Klok. - 1
canescens Schult. - 1
caucasicum (Iljin) Iljin (*C. aralo-caspicum* Iljin subsp. *caucasicum* Iljin) - 1, 2, 6
chinganicum Iljin - 4, 6
confertum Bunge = C. elongatum
crassifolium Turcz. - 4, 5
czernjaevii Klok. - 1
declinatum Steph. ex Iljin - 1, 3, 4, 5, 6(?)
dutreuilii Iljin = C. hilariae
elongatum Bunge (*C. confertum* Bunge, *C. stauntonii* Moq. subsp. *elongatum* (Bunge) Worosch.) - 1(alien), 5
erosum Iljin - 3
falcatum Iljin - 4
filifolium C.A. Mey. ex A. Beck. - 1
gelidum Iljin - 6
glabratum Klok. - 1
heptapotamicum Iljin - 6
hilariae Iljin (*C. dutreuilii* Iljin) - 6
hybridum Bess. ex Andrz. (*C. hyssopifolium* L. var. *leptopterum* Aschers., *C. leptopterum* (Aschers.) Iljin) - 1
hyssopifolium L. - 1, 3, 6

- var. *leptopterum* Aschers. = C. hybridum
insulare Klok. - 1
intermedium Schweigg. - 1(?)
komarovii Iljin - 4
korovinii Iljin - 6
krylovii Iljin - 3, 4
laxiflorum Schrenk - 6
lehmannianum Bunge - 6
leptopterum (Aschers.) Iljin = C. hybridum
macrocarpum Bunge - 4, 5
marschallii Stev. (*C. volgicum* Klok.) - 1, 6
maynense Ignatov - 5
mongolicum Iljin - 3, 4
nitidulum Klok. - 1
nitidum Schult. - 1, 2, 6
ochotense Ignatov - 5
orientale Lam. - 1, 2, 3, 4, 6
pamiricum Iljin - 6
papillosum (O. Kuntze) Iljin - 6
piliferum Iljin - 4, 6
pungens Vahl = Agriophyllum squarrosum
redowskii Fisch. ex Fenzl - 1(alien), 4
sibiricum Iljin (*C. stauntonii* Moq. subsp. *sibiricum* (Iljin) Worosch.) - 3, 4
squarrosum L. = Agriophyllum squarrosum
- subsp. *monticola* Iljin = C. uralense subsp. monticola
- subsp. *silvicola* Iljin = C. uralense subsp. silvicola
- subsp. *uralense* Iljin = C. uralense
squarrosum sensu Iljin = C. uralense
stauntonii Moq. - 5
- subsp. *elongatum* (Bunge) Worosch. = C. elongatum
- subsp. *sibiricum* (Iljin) Worosch. = C. sibiricum
stenopterum Klok. - 1
tibeticum Iljin - 6
ucrainicum Iljin - 1
ulopterum Fenzl - 4
uralense (Iljin) Aell. (*C. squarrosum* L. subsp. *uralense* Iljin, *C. squarrosum* sensu Iljin) - 1, 3. 4
- subsp. **monticola** (Iljin) Aell. (*C. squarrosum* subsp. *monticola* Iljin) - 4
- subsp. **silvicola** (Iljin) Aell. (*C. squarrosum* subsp. *silvicola* Iljin) - 3, 4
volgicum Klok. = C. marschallii

Cornulaca Delile
korshinskyi Litv. - 6

Echinopsilon Moq. = Bassia
dasyphyllum (Fisch. & C.A. Mey.) Moq. = Bassia dasyphylla
divaricatum Kar. & Kir. = Bassia dasyphylla
hirsutum (L.) Moq. = Bassia hirsuta
hyssopifolium (Pall.) Moq. = Bassia hyssopifolia
sedoides (Pall.) Moq. = Bassia sedoides

Eurotia Adans. = Krascheninnikovia
arborescens Losinsk. = Krascheninnikovia arborescens
ceratoides (L.) C.A. Mey. = Krascheninnikovia ceratoides
- var. *pungens* M. Pop. = Krascheninnikovia pungens
- f. *deserticola* Losinsk. = Krascheninnikovia fruticulosa
- f. *tragacanthoides* Losinsk. = Krascheninnikovia pungens
ewersmanniana Stschegl. ex Losinsk. = Krascheninnikovia ewersmanniana
fruticulosa Pazij. = Krascheninnikovia fruticulosa
lenensis Kumin. = Krascheninnikovia lenensis

pungens (M. Pop.) Pazij = Krascheninnikovia pungens

Gamanthus Bunge
commixtus Bunge - 6
ferganicus Iljin - 6
gamocarpus (Moq.) Bunge (*G. kelifi* Korov.) - 6
kelifi Korov. = G. gamocarpus
leucophysus Botsch. - 6
pilosus (Pall.) Bunge - 2

Girgensohnia Bunge
diptera Bunge - 6
minima Korov. - 6
oppositiflora (Pall.) Fenzl - 6

Hablitzia Bieb.
tamnoides Bieb. - 2

Halanthium C. Koch
kulpianum (C. Koch) Bunge - 2
lipskyi Pauls. = Climacoptera turcomanica
rarifolium C. Koch (*H. rarifolium* subsp. *lanatum* Iljin, *H. roseum* (Trautv.) Iljin) - 2
- subsp. *lanatum* Iljin = H. rarifolium
roseum (Trautv.) Iljin = H. rarifolium

Halimione Aell.
pedunculata (L.) Aell. (*Atriplex pedunculata* L., *Obione pedunculata* (L.) Moq.) - 1, 2, 3, 6
verrucifera (Bieb.) Aell. (*Atriplex verrucifera* Bieb., *Obione verrucifera* (Bieb.) Moq.) - 1, 2, 3, 6

Halimocnemis C.A. Mey.
aequipila Iljin - 6
beresinii Iljin - 6
fragilis Tarass. = Halotis pilifera
glaberrima Iljin - 6
karelinii Moq. - 1, 6
lasiantha Iljin - 6
latifolia Iljin - 6
longifolia Bunge - 6
macrantha Bunge (*H. macranthera* Bunge) - 6
macranthera Bunge = H. macrantha
mironovii Botsch. - 6
molissima Bunge - 6
obtusifolia Schrenk = Climacoptera obtusifolia
pilifera Moq. = Halotis pilifera
sclerosperma (Pall.) C.A. Mey. - 1, 2, 6
smirnowii Bunge - 6
tomentosa Moq. = Salsola tomentosa
villosa Kar. & Kir. - 6

Halocharis Moq.
gossypina Korov. & Kinzikaeva - 6
hispida (Schrenk) Bunge - 6
lachnantha Korov. - 6
turcomanica Iljin - 6

Halocnemum Bieb.
caspicum (Pall.) Bieb. = Halostachys belangeriana
strobilaceum (Pall.) Bieb. - 1, 2, 3, 4, 6

Halogeton C.A. Mey.

acutifolius Bunge = Salsola mutica
arachnoideus Moq. = Micropeplis arachnoidea
georgicus Moq. = Salsola ericoides
glomeratus C.A. Mey. - 1, 3, 4, 6
tibeticus Bunge - 6

Halopeplis Bunge ex Ung.-Sternb.

pygmaea (Pall.) Bunge ex Ung.-Sternb. - 1, 2, 6

Halostachys C.A. Mey. (*Arthrocnemum* Moq. p.p.)

belangeriana (Moq.) Botsch. (*Arthrocnemum belangeria-num* Moq., *Halocnemum caspicum* (Pall.) Bieb. nom. illegit., *Halostachys caspia* C.A. Mey. nom. nud., *Salicornia caspica* Pall. 1771, non L. 1753) - 1, 2, 6
caspia C.A. Mey. = H. belangeriana

Halothamnus Jaub. & Spach (*Aellenia* Ulbr., *Caroxylon* Moq.)

acutifolius (Moq.) Botsch. (*Salsola auricula* Moq. var. *acutifolia* Moq., *Aellenia glauca* (Bieb.) Aell. subsp. *lancifolia* (Boiss.) Aell., *A. lancifolia* (Boiss.) Aell. comb. superfl., *A. lancifolia* (Boiss.) Ulbr., *Caroxylon acutifolium* (Moq.) Moq., *C. lancifolium* Boiss., *Salsola acutifolia* (Moq.) Bunge, *S. lancifolia* (Boiss.) Boiss.) - 6
auriculus (Moq.) Botsch. (*Aellenia auricula* (Moq.) Ulbr.) - 6
babatagi Botsch. - 6
ferganensis Botsch. - 6
glaucus (Bieb.) Botsch. (*Salsola glauca* Bieb., *Aellenia glauca* (Bieb.) Aell., *A. glauca* subsp. *euglauca* Aell.) - 2, 6
heptapotamicus Botsch. - 6
hispidulus (Bunge) Botsch. (*Aellenia glauca* (Bieb.) Aell. subsp. *hispidula* (Bunge) Aell., *A. hispidula* (Bunge) Botsch., *Salsola hispidula* (Bunge) Bunge) - 6
iliensis (Lipsky) Botsch. (*Salsola iliensis* Lipsky, *Aellenia iliensis* (Lipsky) Aell.) - 6
moquinianus (Jaub. & Spach) Botsch. (*Salsola moquiniana* Jaub. & Spach) - 6
oxianus Botsch. - 6
psammophilus Botsch. - 6
schurobi Botsch. - 6
seravschanicus Botsch. - 6
subaphyllus (C.A. Mey.) Botsch. (*Salsola subaphylla* C.A. Mey., *Aellenia subaphylla* (C.A. Mey.) Aell., *A. subaphylla* subsp. *eusubaphylla* Aell. nom. illegit., *A. subaphylla* subsp. *turcomanica* Aell., *A. turcomanica* (Aell.) Czer., *Salsola subaphylla* var. *arenaria* Drob., *S. subaphylla* subsp. *arenaria* (Drob.) Iljin, *S. subaphylla* var. *typica* Drob. nom. illegit., *S. subaphylla* subsp. *typica* (Drob.) Iljin, nom. illegit.) - 6
tianschanicus Botsch. - 6
turcomanicus Botsch. - 6

Halotis Bunge

pilifera (Moq.) Botsch. (*Halimocnemis pilifera* Moq., *H. fragilis* Tarass., *Halotis pilosa* (Moq.) Iljin) - 6
pilosa (Moq.) Iljin = H. pilifera

Haloxylon Bunge

ammodendron (C.A. Mey.) Bunge - 6

aphyllum (Minkw.) Iljin - 6
persicum Bunge ex Boiss. & Buhse - 6

Hammada Iljin

eriantha Botsch. - 6
leptoclada (M. Pop. ex Iljin) Iljin (*Arthrophytum leptocladum* M. Pop. ex Iljin) - 6
wakhanica (Pauls.) Iljin (*Arthrophytum wakhanicum* (Pauls.) Iljin) - 6

Horaninovia Fisch. & C.A. Mey.

anomala (C.A. Mey.) Moq. - 6
excellens Iljin - 6
minor Schrenk - 1, 6
sogdiana Blag. = H. ulicina
ulicina Fisch. & C.A. Mey. (*H. sogdiana* Blag.) - 1, 6

Iljinia Korov.

regelii (Bunge) Korov. - 6

Kalidium Moq.

arabicum (L.) Moq. var. *cuspidatum* Ung.-Sternb. = K. cuspidatum
caspicum (L.) Ung.-Sternb. - 1, 2, 3, 6
cuspidatum (Ung.-Sternb.) Grub. (*K. arabicum* (L.) Moq. var. *cuspidatum* Ung.-Sternb.) - 6
foliatum (Pall.) Moq. - 1, 2, 3, 4, 6
schrenkianum Bunge ex Ung.-Sternb. - 6

Kirilowia Bunge

eriantha Bunge - 6

Kochia Roth

angustifolia (Turcz.) Peschkova (*K. scoparia* (L.) Schrad. var. *angustifolia* Turcz.) - 4
dasyphylla Fisch. & C.A. Mey. = Bassia dasyphylla
densiflora (Moq.) Aell. (*K. scoparia* (L.) Schrad. var. *densiflora* Moq., *K. scoparia* subsp. *densiflora* (Moq.) Aell., *K. sieversina* auct.) - 1(alien), 3, 4, 6
iranica Bornm. - 2, 4, 6
krylovii Litv. (*Bassia krylovii* (Litv.) A.J. Scott) - 3
laniflora (S.G. Gmel.) Borb. (*Bassia laniflora* (S.G. Gmel.) A.J. Scott) - 1, 2, 3, 4, 6
melanoptera Bunge (*Bassia melanoptera* (Bunge) A.J. Scott) - 3, 6
odontoptera Schrenk (*K. schrenkiana* (Moq.) Iljin) - 6
prostrata (L.) Schrad. (*Bassia prostrata* (L.) A.J. Scott, *Kochia prostrata* var. *villosissima* Bong. & C.A. Mey., *K. villosissima* (Bong. & C.A. Mey.) Serg.) - 1, 2, 3, 4, 6
- subsp. **grisea** Pratov - 6
- subsp. **virescens** (Fenzl) Pratov (*K. prostrata* var. *virescens* Fenzl) - 6
- var. *villosissima* Bong. & C.A. Mey. = K. prostrata
schrenkiana (Moq.) Iljin = K. odontoptera
scoparia (L.) Schrad. (*Bassia scoparia* (L.) A.J. Scott) - 1, 2, 3, 4, 5, 6
- subsp. *densiflora* (Moq.) Aell. = K. densiflora
- var. *angustifolia* Turcz. = K. angustifolia
- var. *densiflora* Moq. = K. densiflora
sieversina auct. = K. densiflora
tianschanica Pavl. - 6
villosissima (Bong. & C.A. Mey.) Serg. = K. prostrata

Krascheninnikovia Gueldenst. (*Ceratoides* Gagnebin, p.p., nom. illegit., *Ceratospermum* Pers., *Eurotia* Adans. nom. illegit.)

arborescens (Losinsk.) Czer. comb. nova (*Eurotia arborescens* Losinsk. 1930, Bull. Acad. Sci. URSS, ser. 7, 9 : 990; *Ceratoides arborescens* (Losinsk.) Tsien & C.G. Ma) - 6

ceratoides (L.) Gueldenst. (*Ceratoides latens* (J.F. Gmel.) Reveal & Holmgren, *C. papposa* Botsch. & Ikonn., *Ceratospermum papposum* Pers. nom. illegit. superfl., *Eurotia ceratoides* (L.) C.A. Mey., *Krascheninnikovia latens* J.F. Gmel. nom. illegit. superfl.) - 1, 2, 3, 4, 6
- subsp. *deserticola* (Losinsk.) Ovcz. & Kinzikaeva = Krascheninnikovia fruticulosa
- subsp. *tragacanthoides* (Losinsk.) Ovcz. & Kinzikaeva = Krascheninnikovia pungens

ewersmanniana (Stschegl. ex Losinsk.) Grub. (*Eurotia ewersmanniana* Stschegl. ex Losinsk., *Ceratoides ewersmanniana* (Stschegl. ex Losinsk.) Botsch. & Ikonn.) - 6

fruticulosa (Pazij) Czer. comb. nova (*Eurotia fruticulosa* Pazij, 1941, Bot. Mat. (Taschkent) 3 : 31; *Ceratoides fruticulosa* (Pazij) Czer., *Eurotia ceratoides* (L.) C.A. Mey. f. *deserticola* Losinsk., *Krascheninnikovia ceratoides* (L.) Gueldenst. subsp. *deserticola* (Losinsk.) Ovcz. & Kinzikaeva) - 6

latens J.F. Gmel. = Krascheninnikovia ceratoides

lenensis (Kumin.) Tzvel. (*Eurotia lenensis* Kumin., *Ceratoides lenensis* (Kumin.) Jurtz. & R. Kam.) - 1, 3, 4

pungens (M. Pop.) Czer. comb. nova (*Eurotia ceratoides* (L.) C.A. Mey. var. *pungens* M. Pop. 1937, in M. Pop. & Androsov, Rastit. Zapovedn. Guralash i Zaamin. Lesn. Dachi, ed. 2 : 26; *Ceratoides pungens* (M. Pop.) Czer., *Eurotia ceratoides* f. *tragacanthoides* Losinsk., *E. pungens* (M. Pop.) Pazij, *Krascheninnikovia ceratoides* (L.) Gueldenst. subsp. *tragacanthoides* (Losinsk.) Ovcz. & Kinzikaeva, *K. pungens* (Pazij) Podlech, nom. invalid.) - 6

Londesia Fisch. & C.A. Mey.

eriantha Fisch. & C.A. Mey. (*Bassia eriantha* (Fisch. & C.A. Mey.) O. Kuntze) - 6

Microcnemum Ung.-Sternb. (*Arthrocnemon* auct.)

coralloides (Loscos & Pardo) Buen (*Arthrocnemon coralloides* Loscos & Pardo, *Microcnemum coralloides* (Loscos & Pardo) Font Quer, comb. superfl.) - 2
- subsp. **anatolicum** Wagenitz - 2

Microgynoecium Hook. fil.

tibeticum Hook. fil. - 6

Micropeplis Bunge

arachnoidea (Moq.) Bunge (*Halogeton arachnoideus* Moq.) - 3, 4, 6

Monolepis Schrad.

asiatica Fisch. & C.A. Mey. - 4, 5
litwinowii Pauls. = Chenopodium foliosum

Nanophyton Less.

botschantzevii Pratov - 6

erinaceum (Pall.) Bunge - 1, 3, 6
- subsp. **karataviense** Pratov - 6
- subsp. **subulifolium** Pratov - 6
grubovii Pratov - 4
iliense Pratov - 6
narynense Pratov - 6
pulvinatum Pratov - 6
saxatile Botsch. - 6

Neobotrydium Mold. = Chenopodium

botrys (L.) Mold. = Chenopodium botrys

Noaea Moq.

leptoclada (Woronow) Iljin - 2
minuta Boiss. & Bal. - 2
mucronata (Forssk.) Aschers. & Schweinf. - 2, 6

Obione Gaertn. = Atriplex

centralasiatica (Iljin) Kitag. = Atriplex centralasiatica
pedunculata (L.) Moq. = Halimione pedunculata
sibirica (L.) Fisch. = Atriplex sibirica
verrucifera (Bieb.) Moq. = Halimione verrucifera

Ofaiston Rafin.

monandrum (Pall.) Moq. - 1, 3, 6

Panderia Fisch. & C.A. Mey.

pilosa Fisch. & C.A. Mey. - 2
turkestanica Iljin - 2, 6

Petrosimonia Bunge

brachiata (Pall.) Bunge - 1, 2, 3, 6
brachyphylla (Bunge) Iljin - 1, 6
crassifolia (Pall.) Bunge = P. oppositifolia
glauca (Pall.) Bunge - 2, 6
glaucescens (Bunge) Iljin - 1, 2, 6
hirsutissima (Bunge) Iljin - 6
litwinowii Korsh. - 1, 3, 4, 6
monandra (Pall.) Bunge - 1, 3, 6
oppositifolia (Pall.) Litv. (*Polycnemum oppositifolium* Pall., *Petrosimonia crassifolia* (Pall.) Bunge) - 1, 2, 3, 6
sibirica (Pall.) Bunge - 3, 6
squarrosa (Schrenk) Bunge - 2, 6
triandra (Pall.) Simonk. - 1, 2, 3, 4, 6

Physandra Botsch.

halimocnemis (Botsch.) Botsch. (*Salsola halimocnemis* Botsch.) - 6

Piptoptera Bunge

turkestana Bunge - 6

Polycnemum L.

arvense L. (*P. minus* Kitt.) - 1, 2, 3, 4, 6
heuffelii Lang - 1
majus A. Br. - 1, 2, 3, 6
minus Kitt. = P. arvense
oppositifolium Pall. = Petrosimonia oppositifolia
perenne Litv. - 6
verrucosum Lang - 1

Rhaphidophyton Iljin

regelii (Bunge) Iljin - 6

Salicornia L.

borysthenica Tzvel. - 1
caspica Pall. = Halostachys belangeriana
dolichostachya auct. p.p. = S. pojarkovae
dolichostachya Moss subsp. *pojarkovae* (N. Semen.) Piirainen = S. pojarkovae
europaea L. (*S. herbacea* (L.) L.) - 1
herbacea (L.) L. = S. europaea
- subsp. *pojarkovae* (N. Semen.) V. Sergienko = S. pojarkovae
perennans Willd. (*S. prostrata* Pall.) - 1, 2, 3, 4, 5, 6
pojarkovae N. Semen. (*S. dolichostachya* Moss subsp. *pojarkovae* (N. Semen.) Piirainen, *S. herbacea* (L.) L. subsp. *pojarkovae* (N. Semen.) V. Sergienko, *S. dolichostachya* auct. p.p.) - 1
prostrata Pall. = S. perennans

Salsola L. (*Caspia* Galushko)

abrotanoides Bunge - 4
acutifolia (Bunge) Botsch. = S. mutica
acutifolia (Moq.) Bunge = Halothamnus acutifolius
affinis C.A. Mey. = Climacoptera affinis
amblyostegia Botsch. = Climacoptera amblyostegia
androssowii Litv. - 6
angusta Botsch. - 6
aperta Pauls. - 6
aralensis Iljin = Climacoptera aralensis
arborescens L. var. *richteri* Moq. = S. richteri
arbuscula Pall. - 1, 6
arbusculiformis Drob. - 6
aucheri (Moq.) Bunge ex Iljin - 6
auricula Moq. var. *acutifolia* Moq. = Halothamnus acutifolius
australis R. Br. (*S. iberica* (Sennen & Pau) Botsch., *S. kali* L. subsp. *ruthenica* (Iljin) Soo, *S. pestifer* Nels., *S. ruthenica* Iljin, *S. tragus* L. subsp. *iberica* Sennen & Pau) - 1, 2, 3, 4, 5, 6
badghysi Botsch. - 6
baranovii Iljin - 6
botschantzevii Kurbanov - 6
brachiata Pall. = Climacoptera brachiata
bucharica Iljin = Climacoptera bucharica
bungeana (Botsch.) Botsch. (*S. tomentosa* (Moq.) Spach subsp. *bungeana* Botsch.) - 6
camphorosma Iljin (*S. camphorosmoides* Iljin, 1938, non Desf. 1789) - 2
camphorosmoides Iljin = S. camphorosma
cana C. Koch - 2
carinata C.A. Mey. = S. leptoclada
- subsp. *carinatiformis* Kinzikaeva = S. leptoclada
carinatiformis Kinzikaeva = S. leptoclada
chiwensis M. Pop. - 6
collina Pall. (*S. kali* L. subsp. *collina* (Pall.) O. Bolos) - 1, 3, 4, 5, 6
crassa Bieb. = Climacoptera crassa
daghestanica (Turcz.) Turcz. - 2
dendroides Pall. - 1, 2, 6
deserticola Iljin - 6
drobovii Botsch. - 6
dschungarica Iljin - 6
ericoides Bieb. (*Halogeton ? georgicus* Moq., *Salsola ericoides* subsp. *sulphurea* Iljin, *S. georgica* (Moq.) Bunge) - 2
- subsp. *sulphurea* Iljin = S. ericoides
euryphylla Botsch. - 6
ferganica Drob. = Climacoptera ferganica

flavovirens Iljin = S. tomentosa
flexuosa Botsch. - 6
foliosa (L.) Schrad. (*Caspia foliosa* (L.) Galushko) - 1, 2, 3, 6
forcipitata Iljin - 6
futilis Iljin - 2
gemmascens Pall. - 4, 6
- subsp. *nodulosa* (Moq.) Botsch. = S. nodulosa
- subsp. **oreina** Botsch. - 4, 6
- subsp. **subglabra** Botsch. - 4
georgica (Moq.) Bunge = S. ericoides
glabella Botsch. - 6
glauca Bieb. = Halothamnus glaucus
gossypina Bunge - 6
halimocnemis Botsch. = Physandra halimocnemis
heptapotamica Iljin = Climacoptera obtusifolia
hispidula (Bunge) Bunge = Halothamnus hispidulus
iberica (Sennen & Pau) Botsch. = S. australis
iliensis Lipsky = Halothamnus iliensis
iljinii Botsch. - 6
implicata Botsch. - 6
incanescens C.A. Mey. - 2, 6
intricata Iljin = Climacoptera intricata
kali L. - 1
- subsp. *collina* (Pall.) O. Bolos = S. collina
- subsp. *ruthenica* (Iljin) Soo = S. australis
- subsp. *tragus* (L.) Celak. = S. tragus
- var. *pontica* Pall. = S. pontica
kasakorum Iljin = Climacoptera kasakorum
komarovii Iljin - 5
kopetdaghensis (Botsch.) Botsch. (*S. tomentosa* (Moq.) Spach subsp. *kopetdaghensis* Botsch.) - 6
korshinskyi Drob. = Climacoptera korshinskyi
kurbanovii Botsch. - 6
lachnophylla Iljin = Climacoptera lachnophylla
lanata Pall. = Climacoptera lanata
lancifolia (Boiss.) Boiss. = Halothamnus acutifolius
laricifolia Turcz. & Litv. - 6
laricina Pall. - 1, 2, 3, 6
leptoclada Gand. (*S. carinata* C.A. Mey. 1831, non Spreng. 1825, *S. carinata* subsp. *carinatiformis* Kinzikaeva, nom. invalid., *S. carinatiformis* Kinzikaeva) - 6
lipschitzii Botsch. - 6
longistylosa Iljin = Climacoptera longistylosa
macera Litv. = S. nitraria
malyginii Korov. ex Botsch. = Climacoptera malyginii
maracandica Iljin = S. montana
merkulowiczii Zak. = Climacoptera merkulowiczii
micranthera Botsch. - 6
minkwitziae Korov. = Climacoptera minkwitziae
monoptera Bunge - 3, 4
montana Litv. (*S. maracandica* Iljin) - 6
moquiniana Jaub. & Spach = Halothamnus moquinianus
mutica C.A. Mey. (*Halogeton acutifolius* Bunge, *Salsola acutifolia* (Bunge) Botsch. 1963, non Bunge, 1860) - 1, 3, 6
nitraria Pall. (*S. macera* Litv.) - 1, 2, 6
nodulosa (Moq.) Iljin (*S. gemmascens* Pall. subsp. *nodulosa* (Moq.) Botsch.) - 2
olgae Iljin = Climacoptera olgae
oreophila Botsch. - 6
orientalis S.G. Gmel. (*S. rigida* Pall.) - 1, 2, 6
pachyphylla Botsch. - 6
paletzkiana Litv. - 6
passerina Bunge - 4, 6
paulsenii Litv. (*S. pellucida* Litv.) - 1, 2, 6
pellucida Litv. = S. paulsenii
pestifer Nels. = S. australis

pontica (Pall.) Degen (*S. kali* L. var. *pontica* Pall.) - 1, 2
praecox Litv. - 1, 2, 6
pulvinata Botsch. - 6
richteri (Moq.) Kar. ex Litv. (*S. arborescens* L. var. *richteri* Moq.) - 6
rigida Pall. = S. orientalis
rosacea L. - 3, 4, 6
roshevitzii Iljin - 6
ruthenica Iljin = S. australis
sclerantha C.A. Mey. - 6
soda L. - 1, 2, 3, 6
sogdiana Bunge - 6
stellulata Korov. (*S. tomentosa* (Moq.) Spach subsp. *stellulata* (Korov.) Botsch.) - 6
subaphylla C.A. Mey. = Halothamnus subaphyllus
- subsp. *arenaria* (Drob.) Iljin = Halothamnus subaphyllus
- subsp. *typica* (Drob.) Iljin = Halothamnus subaphyllus
- var. *arenaria* Drob. = Halothamnus subaphyllus
- var. *typica* Drob. = Halothamnus subaphyllus
subcrassa M. Pop. = Climacoptera subcrassa
sukaczevii (Botsch.) A.J. Li = Climacoptera sukaczevii
takhtadshjanii Iljin = S. tomentosa
tamamschjanae Iljin - 2
tamariscina Pall. - 1, 2, 3, 6
tianschanica Botsch. - 6
titovii Botsch. (*S. titovii* subsp. *canescens* Botsch.) - 6
- subsp. *canescens* Botsch. = S. titovii
tomentosa (Moq.) Spach (*Halimocnemis tomentosa* Moq., *Salsola flavovirens* Iljin, *S. takhtadshjanii* Iljin) - 2, 6
- subsp. *bungeana* Botsch. = S. bungeana
- subsp. *kopetdaghensis* Botsch. = S. kopetdaghensis
- subsp. *stellulata* (Korov.) Botsch. = S. stellulata
tragus L. (*S. kali* L. subsp. *tragus* (L.) Celak.) - 1, 2, 6
- subsp. *iberica* Sennen & Pau = S. australis
transhyrcanica Iljin - 6
transoxana Iljin = Climacoptera transoxana
turcomanica Litv. = Climacoptera turcomanica
turgaica Iljin = Climacoptera turgaica
turkestanica Litv. - 6
vvedenskyi Iljin & M. Pop. - 6

Seidlitzia Bunge

florida (Bieb.) Bunge - 2
rosmarinus Bunge - 6

Spinacia L.

*oleracea L. - 1, 2, 3, 4, 5, 6
tetrandra Stev. - 2
turkestanica Iljin - 6

Suaeda Forssk. ex Scop.

acuminata (C.A. Mey.) Moq. (*S. pterantha* (Kar. & Kir.) Bunge, *S. pygmaea* (Kar. & Kir.) Iljin, *S. transoxana* (Bunge) Boiss.) - 3, 4, 6
altissima (L.) Pall. - 1, 2, 3, 6
arctica Jurtz. & Petrovsky - 5
arcuata Bunge (*S. lipskyi* Litv.) - 6
baccifera Pall. - 1, 2
balchaschensis Pavl. = S. kossinskyi
confusa Iljin - 1, 2, 3, 6
corniculata (C.A. Mey.) Bunge - 1, 3, 4, 6
- var. *olufsenii* (Pauls.) G.L. Chu = S. olufsenii
crassifolia Pall. (*S. drepanophylla* Litv. nom. provis.) - 1, 2, 6
dendroides (C.A. Mey.) Moq. - 1, 2, 6

drepanophylla Litv. = S. crassifolia
eltonica Iljin - 1, 6
glauca (Bunge) Bunge - 4, 5
heterophylla (Kar. & Kir.) Bunge - 1, 2, 6
heteroptera Kitag. (*S. ussuriensis* Iljin) - 5
kossinskyi Iljin (*S. balchaschensis* Pavl.) - 1, 3, 4, 6
linifolia Pall. - 1, 3, 4, 6
lipskyi Litv. = S. arcuata
maritima (L.) Dumort. - 1, 6
- subsp. *prostrata* (Pall.) Soo = S. prostrata
- subsp. *salsa* (L.) Jav. = S. salsa
- subsp. *salsa* (L.) Soo = S. salsa
microphylla Pall. - 1, 2, 6
microsperma (C.A. Mey.) Fenzl - 6
olufsenii Pauls. (*S. corniculata* (C.A. Mey.) Bunge var. *olufsenii* (Pauls.) G.L. Chu) - 6
pannonica G. Beck - 1
paradoxa (Bunge) Bunge - 6
physophora Pall. - 1, 2, 3, 6
prostrata Pall. (*S. maritima* (L.) Dumort. subsp. *prostrata* (Pall.) Soo) - 1, 2, 3, 4, 6
pterantha (Kar. & Kir.) Bunge = S. acuminata
pygmaea (Kar. & Kir.) Iljin = S. acuminata
salsa (L.) Pall. (*S. maritima* (L.) Dumort. subsp. *salsa* (L.) Jav., *S. maritima* subsp. *salsa* (L.) Soo, comb. superfl.) - 1, 2, 3, 4, 6
transoxana (Bunge) Boiss. = S. acuminata
turkestanica Litv. - 6
ussuriensis Iljin = S. heteroptera

Sympegma Bunge

regelii Bunge - 6

Teloxys Moq.

ambrosioides (L.) W.A. Weber = Chenopodium ambrosioides
aristata (L.) Moq. (*Chenopodium aristatum* L.) - 1, 3, 4, 5, 6
botrys (L.) W.A. Weber = Chenopodium botrys
schraderiana (Schult.) W.A. Weber = Chenopodium schraderianum

CHLORANTHACEAE R.Br. ex Lindl.
Chloranthus Sw. (*Nigrina* Thunb.)

japonicus Siebold - 5
serratus (Thunb.) Roem. & Schult. (*Nigrina serrata* Thunb.) - 5

Nigrina Thunb. = Chloranthus

serrata Thunb. = Chloranthus serratus

CISTACEAE Juss.
Cistus L.

creticus auct. p.p. = C. ponticus and C. tauricus
creticus L. subsp. *eriocephalus* (Viv.) Greuter & Burdet = C. tauricus
eriocephalus Viv. = C. tauricus
hirsutus Thuill. = Helianthemum ovatum
incanus auct. p.p. = C. tauricus
ovatus Viv. = Helianthemum ovatum
ponticus Juz. (*C. creticus* auct. p.p.) - 2
salviifolius L. - 2

tauricus C. Presl (*C. creticus* L. subsp. *eriocephalus* (Viv.) Greuter & Burdet, *C. eriocephalus* Viv., *C. creticus* auct. p.p., *C. incanus* auct. p.p.) - 1
thymifolius L. = Fumana thymifolia

Fumana (Dun.) Spach

arabica auct. = F. viscidula
procumbens (Dun.) Gren. & Godr. - 1, 2, 6
thymifolia (L.) Spach ex Webb (*Cistus thymifolius* L.) - 1(?)
viscidula (Stev. ex Palib.) Juz. (*F. arabica* auct.) - 1, 2

Helianthemum Hill (*Rhodax* Spach)

alpestre (Jacq.) DC. (*H. oelandicum* (L.) DC. subsp. *alpestre* (Jacq.) Breistroffer, *Rhodax italicus* (L.) Holub subsp. *alpestiris* (Jacq.) A. & D. Löve) - 1(?)
arcticum (Grosser) Janch. - 1
buschii (Palib.) Juz. & Pozd. (*H. canum* (L.) Baumg. subsp. *buschii* (Palib.) Kupatadze, *H. italicum* (L.) Pers. var. *buschii* (Palib.) Stank.) - 2
canum (L.) Baumg. = H. canum (L.) Hornem.
- subsp. *buschii* (Palib.) Kupatadze = H. buschii
- subsp. *stevenii* (Rupr. ex Juz. & Pozd.) M.C.F. Proctor = H. stevenii
canum (L.) Hornem. (*H. canum* (L.) Baumg. comb. superfl., *Rhodax canus* (L.) Fuss) - 1
- subsp. **baschkirorum** Juz. ex Kupatadze - 1
chamaecistus Mill. subsp. *obscurum* Celak. = H. ovatum
- subsp. *ovatum* (Viv.) Petersen = H. ovatum
ciscaucasicum Juz. & Pozd. (*Rhodax canus* (L.) Fuss subsp. *ciscaucasicus* (Juz. & Pozd.) Holub) - 2
cretaceum (Rupr.) Juz. ex Dobrocz. (*H. italicum* (L.) Pers. var. *cretaceum* Rupr.) - 1
creticola Klok. & Dobrocz. - 1
cretophilum Klok. & Dobrocz. - 1
dagestanicum Rupr. - 2
georgicum Juz. & Pozd. - 2
grandiflorum (Scop.) DC. (*H. nummularium* (L.) Mill. subsp. *grandiflorum* (Scop.) Schinz & Thell.) - 1, 2
- subsp. *glabrum* (Koch) Holub = H. nitidum
- subsp. **glaucescens** (Murb.) Holub - 1
- subsp. *obscurum* (Wahlenb.) Holub = H. ovatum
hirsutum (Thuill.) Merat = H. ovatum
italicum auct. = H. orientale
italicum (L.) Pers. var. *buschii* (Palib.) Stank. = H. buschii
- var. *cretaceum* Rupr. = H. cretaceum
lasiocarpum Jacques & Herincq (*H. lasiocarpum* Willk. nom. illegit., *H. ledifolium* (L.) Mill. subsp. *lasiocarpum* (Jacques & Herincq) Nym., *H. ledifolium* subsp. *lasiocarpum* (Willk.) Bornm. nom. illegit.) - 2
ledifolium (L.) Mill. - 2(?), 6
- subsp. *lasiocarpum* (Jacques & Herincq) Nym. = H. lasiocarpum
- subsp. *lasiocarpum* (Willk.) Bornm. = H. lasiocarpum
marginale C.A. Mey. - 1(?)
nitidum Clementi (*H. grandiflorum* (Scop.) DC. subsp. *glabrum* (Koch) Holub, *H. nummularium* (L.) Mill. subsp. *glabrum* (Koch) Wilczek, *H. vulgare* Gaertn. var. *glabrum* Koch) - 1, 2
nummularium (L.) Mill. - 1, 2
- subsp. *glabrum* (Koch) Wilczek = H. nitidum
- subsp. *grandiflorum* (Scop.) Schinz & Thell. = H. grandiflorum
- subsp. *obscurum* (Celak.) Holub = H. ovatum
- subsp. *ovatum* (Viv.) Schinz & Thell. = H. ovatum
- subsp. *tomentosum* (Scop.) Schinz & Thell. = H. tomentosum

obscurum Pers. = H. ovatum
oelandicum (L.) DC. subsp. *alpestre* (Jacq.) Breistroffer = H. alpestre
- subsp. *orientale* (Grosser) M.C.F. Proctor = H. orientale
- subsp. *rupifragum* (A. Kerner) Breistroffer = H. rupifragum
- subsp. *stevenii* (Juz. & Pozd.) Greuter & Burdet = H. stevenii
orientale (Grosser) Juz. & Pozd. (*H. oelandicum* (L.) DC. subsp. *orientale* (Grosser) M.C.F. Proctor, *Rhodax orientalis* (Grosser) Holub, *Helianthemum italicum* auct.) - 1, 2
ovatum (Viv.) Dun. (*Cistus ovatus* Viv., *C. hirsutus* Thuill. 1800, non Lam. 1778, *Helianthemum chamaecistus* Mill. subsp. *obscurum* Celak., *H. chamaecistus* subsp. *ovatum* (Viv.) Petersen, *H. grandiflorum* (Scop.) DC. subsp. *obscurum* (Wahlenb.) Holub, *H. hirsutum* (Thuill.) Merat, nom. illegit., *H. nummularium* (L.) Mill. subsp. *obscurum* (Celak.) Holub, *H. nummularium* subsp. *ovatum* (Viv.) Schinz & Thell., *H. obscurum* Pers. nom. illegit., *H. vulgare* Gaertn. var. *obscurum* Wahlenb.) - 1, 2
rupifragum A. Kerner (*H. oelandicum* (L.) DC. subsp. *rupifragum* (A. Kerner) Breistroffer, *Rhodax italicus* (L.) Holub subsp. *rupifragus* (A. Kerner) A. & D. Löve, *R. rupifragus* (A. Kerner) Holub) - 1
salicifolium (L.) Mill. - 1, 2, 6
songaricum Schrenk - 6
stevenii Rupr. ex Juz. & Pozd. (*H. canum* (L.) Baumg. subsp. *stevenii* (Rupr. ex Juz. & Pozd.) M.C.F. Proctor, *H. oelandicum* (L.) DC. subsp. *stevenii* (Juz. & Pozd.) Greuter & Burdet, *Rhodax stevenii* (Juz. & Pozd.) Holub) - 1
tomentosum (Scop.) S.F. Gray (*H. nummularium* (L.) Mill. subsp. *tomentosum* (Scop.) Schinz & Thell.) - 2
vulgare Gaertn. var. *glabrum* Koch = H. nitidum
- var. *obscurum* Wahlenb. = H. ovatum

Rhodax Spach = Helianthemum

canus (L.) Fuess = Helianthemum canum
- subsp. *ciscaucasicus* (Juz. & Pozd.) Holub = Helianthemum ciscaucasicum
italicus (L.) Holub subsp. *alpestiris* (Jacq.) A. & D. Löve = Helianthemum alpestre
- subsp. *rupifragus* (A. Kerner) A. & D. Löve = Helianthemum rupifragum
orientalis (Grosser) Holub = Helianthemum orientale
rupifragus (A. Kerner) Holub = Helianthemum rupifragum
stevenii (Juz. & Pozd.) Holub = Helianthemum stevenii

CLEOMACEAE Horan. = CAPPARACEAE

CLEOMACEAE (Pax) Airy Shaw = CAPPARACEAE

COLCHICACEAE DC. = MELANTHIACEAE

COMMELINACEAE R. Br.

Aneilema auct. = Mardannia

japonicum auct. = Mardannia keisak
keisak Hassk. = Mardannia keisak

Commelina L.
benghalensis L. - 2
communis L. - 1, 2, 3, 4, 5

Mardannia Royle (*Aneilema* auct.)
keisak (Hassk.) Hand.-Mazz. (*Aneilema keisak* Hassk., *A. japonicum* auct.) - 5

Streptolirion Edweg.
volubile Edweg. - 5

COMPOSITAE Giseke = ASTERACEAE

CONVALLARIACEAE Horan.
Clintonia Rafin.
udensis Trautv. & C.A. Mey. - 5

Convallaria L.
keiskei Miq. (*C. majalis* L. subsp. *manshurica* (Kom.) Bordz.) - 4, 5
latifolia Mill. = C. majalis
majalis L. (*C. latifolia* Mill.) - 1
- subsp. *transcaucasica* (Utkin ex Grossh.) Bordz. = C. transcaucasica
odorata Mill. = Polygonatum odoratum
transcaucasica Utkin ex Grossh. (*C. majalis* L. subsp. *transcaucasica* (Utkin ex Grossh.) Bordz.) - 2
trifolia L. = Smilacina trifolia

Disporum Salisb. ex D. Don (*Uvularia* auct.)
sessile D. Don (*Uvularia sessilis* Thunb. nom. superfl.) - 5
smilacinum A. Gray (*D. viridescens* (Maxim.) B. Fedtsch. comb. superfl., *D. viridescens* (Maxim.) Nakai) - 5
viridescens (Maxim.) B. Fedtsch. = D. smilacinum
viridescens (Maxim.) Nakai = D. smilacinum

Maianthemum Wigg.
bifolium (L.) F. W. Schmidt - 1, 2, 3, 4, 5
dilatatum (Wood) Nels. & Macbr. - 5
intermedium Worosch. - 5

Polygonatum Hill
acuminatifolium Kom. - 5
buschianum Tzvel. - 1
chinense Kunth - 4
desoulavyi Kom. - 5
glaberrimum C. Koch - 2
hirtum (Bosc. ex Poir.) Pursh (*Convallaria hirta* Bosc. ex Poir., *C. latifolia* Jacq. 1775, non Mill. 1768, *Polygonatum latifolium* (Jacq.) Desf.) - 1
humile Fisch. ex Maxim. - 3, 4, 5
inflatum Kom. - 5
involucratum (Franch. & Savat.) Maxim. - 5
latifolium (Jacq.) Desf. = P. hirtum
maximowiczii Fr. Schmidt (*P. odoratum* (Mill.) Druce var. *maximowiczii* (Fr. Schmidt) Koidz.) - 5
multiflorum (L.) All. - 1, 2
obtusifolium (C. Koch) Miscz. ex Grossh. - 2
odoratum (Mill.) Druce (*Convallaria odorata* Mill., *Polygonatum officinale* All.) - 1, 2, 3, 4, 5

- var. *maximowiczii* (Fr. Schmidt) Koidz. = P. maximowiczii
officinale All. = P. odoratum
orientale Desf. (*P. polyanthemum* (Bieb.) A. Dietr. comb. superfl., *P. polyanthemum* (Bieb.) Link) - 1, 2
ovatum Miscz. ex Knorr. nom. invalid. - 2
polyanthemum (Bieb.) A. Dietr. = P. orientale
polyanthemum (Bieb.) Link = P. orientale
roseum (Ledeb.) Kunth - 3, 6
sewerzowii Regel - 6
sibiricum Delaroche - 4
stenophyllum Maxim. - 5
verticillatum (L.) All. - 2

Smilacina Desf.
davurica Fisch. & C.A. Mey. - 4, 5
hirta Maxim. (*S. japonica* A. Gray subsp. *hirta* (Maxim.) Worosch., *S. japonica* auct.) - 5
japonica auct. = S. hirta
japonica A. Gray subsp. *hirta* (Maxim.) Worosch. = S. hirta
trifolia (L.) Desf. (*Convallaria trifolia* L.) - 3, 4, 5

Streptopus Michx.
amplexifolius (L.) DC. - 1, 4, 5
- subsp. amplexifolius - 1
- subsp. papillatus (Ohwi) A. & D. Löve (*S. amplexifolius* var. *papillatus* Ohwi) - 4, 5
streptopoides (Ledeb.) Frye & Rigg - 4, 5

Uvularia auct. = Disporum
sessilis Thunb. = Disporum sessile

CONVOLVULACEAE Juss.
Calystegia R. Br.
americana (Sims) Daniels = C. inflata
amurensis Probat. - 5
dahurica (Herb.) Choisy (*Convolvulus dahuricus* Herb., *Calystegia pellita* (Ledeb.) G. Don fil., *C. pellita* subsp. *stricta* Brummitt) - 3, 4, 5
hederacea Wall. - 3, 5, 6
inflata Sweet (*C. americana* (Sims) Daniels, *C. sepium* (L.) R. Br. subsp. *americana* (Sims) Brummitt, *Convolvulus americanus* (Sims) Greene, *C. sepium* L. var. *americanus* Sims) - 1, 4, 5
japonica Choisy - 5
x lucana (Ten.) G. Don fil. (*Convolvulus lucanus* Ten.). - C. sepium (L.) R. Br. x C. silvatica (Kit.) Griseb. - 2
pellita (Ledeb.) G. Don fil. = C. dahurica
- subsp. *stricta* Brummitt = C. dahurica
sepium (L.) R. Br. - 1, 2, 3, 4, 5, 6
- subsp. *americana* (Sims) Brummitt = C. inflata
- subsp. baltica Rothm. - 1
silvatica (Kit.) Griseb. (*Convolvulus silvaticus* Kit., *Calystegia sylvestris* (Willd.) Roem. & Schult.) - 1, 2
soldanella (L.) R. Br. - 1, 2, 5
subvolubilis (Ledeb.) D. Don fil. (*Convolvulus subvolubilis* Ledeb.) - 3, 4, 5
> It is likely to be a hybrid C. dahurica (Herb.) Choisy x C. sepium (L.) R. Br.
sylvestris (Willd.) Roem. & Schult. = C. silvatica

Convolvulus L.
americanus (Sims) Greene = Calystegia inflata
ammanii Desr. - 3, 4, 6
arvensis L. - 1, 2, 3, 4, 5, 6

askabadensis Bornm. & Sint. ex Bornm. - 6
betonicifolius Mill. (*C. hirsutus* Bieb.) - 1, 2
bracteosus Juz. (*C. calvertii* Boiss. subsp. *bracteosus* (Juz.) Smoljian.) - 1
calvertii Boiss. - 2, 6
- subsp. *bracteosus* (Juz.) Smoljian. = C. bracteosus
- subsp. *tauricus* (Bornm.) Smoljian. = C. tauricus
campanulatus Zapr. - 6
cantabrica L. - 1, 2
chinensis Ker-Gawl. (*C. fischerianus* V. Petrov) - 3, 4, 5
commutatus Boiss. - 2
dahuricus Herb. = Calystegia dahurica
divaricatus Regel & Schmalh. (*C. michelsonii* V. Petrov, p.p.) - 6
dorycnium L. subsp. *subhirsutus* (Regel & Schmalh.) Sa'ad = C. subhirsutus
eremophilus Boiss. & Buhse - 2
erinaceus Ledeb. - 6
fischerianus V. Petrov = C. chinensis
fruticosus Pall. - 3, 6
georgicus Sa'ad - 2
gortschakovii Schrenk - 3, 6
hamadae (Vved.) V. Petrov - 6
hederaceus L. = Ipomoea hederacea
hirsutus Bieb. = C. betonicifolius
holosericeus Bieb. - 1, 2
korolkowii Regel & Schmalh. (*C. michelsonii* V. Petrov, p.p., *C. turcomanicus* (O. Kuntze) V. Petrov) - 6
krauseanus Regel & Schmalh. - 6
leiocalycinus Boiss. (*C. lycioides* Boiss.) - 6
lineatus L. - 1, 2, 3, 6
lucanus Ten. = Calystegia x lucana
lycioides Boiss. = C. leiocalycinus
michelsonii V. Petrov = C. divaricatus and C. korolkowii
olgae Regel & Schmalh. - 6
pennatus Desr. = Quamoclit pennata
persicus L. - 2, 6
pilosellifolius Desr. - 2, 6
pseudocantabrica Schrenk - 6
■ **pseudoscammonia** C. Koch (*C. scammonia* L. var. *pseudoscammonia* (C. Koch) Sa'ad)
ruprechtii Boiss. - 2
scammonia L. - 1
- var. *pseudoscammonia* (C. Koch) Sa'ad = C. pseudoscammonia
sepium L. var. *americanus* Sims = Calystegia inflata
sericocephalus Juz. (*C. tauricus* (Bornm.) Juz. var. *sericocephalus* (Juz.) Wissjul.) - 1
silvaticus Kit. = Calystegia silvatica
spinifer M. Pop. - 6
stocksii Boiss. - 6
subhirsutus Regel & Schmalh. (*C. dorycnium* L. subsp. *subhirsutus* (Regel & Schmalh.) Sa'ad) - 6
subsericeus Schrenk - 6
tauricus (Bornm.) Juz. (*C. calvertii* Boiss. subsp. *tauricus* (Bornm.) Smoljian.) - 1
- var. *sericocephalus* (Juz.) Wissjul. = C. sericocephalus
tragacanthoides Turcz. - 6
tschimganicus M. Pop. & Vved. - 6
tshegemensis Galushko - 2
tujuntauensis Kinzikaeva - 6
turcomanicus (O. Kuntze) V. Petrov = C. korolkowii

Cressa L.

cretica - 2, 6

Ipomoea L.

angulata Roem. & Schult. = Quamoclit angulata
***batatas** (L.) Lam. - 1, 2, 6
coccinea L. = Quamoclit coccinea
hederacea (L.) Jacq. (*Convolvulus hederaceaus* L.) - 1(cult.), 5(alien)
hispida (Vahl) Roem. & Schult. - 6
lacunosa L. - 2(alien), 5(alien)
***purpurea** (L.) Roth - 1, 2, 5(alien), 6
sibirica (L.) Pers. - 4, 5

Quamoclit Hill

angulata (Roem. & Schult.) Boj. (*Ipomoea angulata* Roem. & Schult.) - 5(alien)
coccinea (L.) Moench (*Ipomoea coccinea* L.) - 5(alien)
pennata (Desr.) Boj. (*Convolvulus pennatus* Desr.) - 2(alien), 5(alien)

COPTACEAE (Gregory) A. & D. Löve = RANUNCULACEAE

CORNACEAE Dumort.
Bothrocaryum (Koehne) Pojark.

controversum (Hemsl. ex Prain) Pojark. (*Cornus controversa* Hemsl. ex Prain, *Swida controversa* (Hemsl. ex Prain) Sojak) - 5

Chamaepericlymenum Hill

canadense (L.) Aschers. & Graebn. - 5
suecicum (L.) Aschers. & Graebn. - 1, 5
x unalaschkense (Ledeb.) Rydb. - Ch. canadense (L.) Aschers. & Graebn. x Ch. suecicum (L.) Aschers. & Graebn. - 5

Cornus L.

controversa Hemsl. ex Prain = Bothrocaryum controversum
darvasica (Pojark.) Pilipenko = Swida darvasica
hungarica Karpati = Swida sanguinea subsp. hungarica
mas L. - 1, 2
meyeri (Pojark.) Pilipenko = Swida meyeri
sanguinea L. subsp. *australis* (C.A. Mey.) Jav. = Swida australis
- subsp. *czerniaewii* Grosset = Swida sanguinea subsp. czerniaewii
- subsp. *hungarica* (Karpati) Soo = Swida sanguinea subsp. hungarica
- var. *hungarica* (Karpati) Soo = Swida sanguinea subsp. hungarica
sericea L. = Swida sericea
subumbellata Komatsu = Swida alba

*Cynoxylon (Rafin.) Small

***capaitata** (Wall. ex Roxb.) Nakai - 1, 2
***florida** (L.) Britt. & Shaf. - 1, 2

Swida Opiz (*Thelycrania* (Dumort.) Fourr.)

alba (L.) Opiz (*Cornus subumbellata* Komatsu, *Swida subumbellata* (Komatsu) Holub, *Thelycrania alba* (L.) Pojark.) - 1, 3, 4, 5
australis (C.A. Mey.) Pojark. ex Grossh. (*Cornus sanguinea*

L. subsp. *australis* (C.A. Mey.) Jav., *Swida sanguinea* (L.) Opiz subsp. *australis* (C.A. Mey.) Kubat, *S. sanguinea* subsp. *australis* (C.A. Mey.) Sojak, comb. superfl., *S. sanguinea* subsp. *australis* (C.A. Mey.) Takht. comb. superfl., *Thelycrania australis* (C.A. Mey.) Sanadze) - 1, 2

brachypoda (C.A. Mey.) Sojak (*Thelycrania brachypoda* (C.A. Mey.) Pojark.) - 5

controversa (Hemsl. ex Prain) Sojak = Bothrocaryum controversum

czerniaewii (Grosset) Sojak = S. sanguinea subsp. czerniaewii

darvasica (Pojark.) Sojak (*Thelycrania darvasica* Pojark., *Cornus darvasica* (Pojark.) Pilipenko) - 6

hungarica (Karpati) Sojak = S. sanguinea subsp. hungarica

iberica (Woronow) Pojark. ex Grossh. (*Thelycrania iberica* (Woronow) Pojark.) - 2

koenigii (Schneid.) Pojark. ex Grossh. (*Thelycrania koenigii* (Schneid.) Sanadze) - 2

meyeri (Pojark.) Sojak (*Thelycrania meyeri* Pojark., *Cornus meyeri* (Pojark.) Pilipenko) - 2, 6

sanguinea (L.) Opiz (*Thelycrania sanguinea* (L.) Fourr.) - 1
- subsp. *australis* (C.A. Mey.) Kubat = S. australis
- subsp. *australis* (C.A. Mey.) Sojak = S. australis
- subsp. *australis* (C.A. Mey.) Takht. = S. australis
- subsp. **czerniaewii** (Grosset) Gubanov & V. Tichomirov (*Cornus sanguinea* subsp. *czerniaewii* Grosset, *Swida czerniaewii* (Grosset) Sojak) - 1
- subsp. **hungarica** (Karpati) Kubat (*Cornus hungarica* Karpati, *C. sanguinea* var. *hungarica* (Karpati) Soo, *C. sanguinea* subsp. *hungarica* (Karpati) Soo, *Swida hungarica* (Karpati) Sojak, *S. sanguinea* subsp. *hungarica* (Karpati) Gubanov & V. Tichomirov, comb. superfl.) - 1

*****sericea** (L.) Holub (*Cornus sericea* L., *Swida stolonifera* (Michx.) Rydb., *Thelycrania stolonifera* (Michx.) Pojark.) - 1, 3, 6

stolonifera (Michx.) Rydb. = S. sericea

subumbellata (Komatsu) Holub = S. alba

Thelycrania (Dumort.) Fourr. = Swida

alba (L.) Pojark. = Swida alba
australis (C.A. Mey.) Sanadze = Swida australis
brachypoda (C.A. Mey.) Pojark. = Swida brachypoda
darvasica Pojark. = Swida darvasica
iberica (Woronow) Pojark. = Swida iberica
koenigii (Schneid.) Sanadze = Swida koenigii
meyeri Pojark. = Swida meyeri
sanguinea (L.) Fourr. = Swida sanguinea
stolonifera (Michx.) Pojark. = Swida sericea

CORYLACEAE Mirb. = BETULACEAE

CRASSULACEAE DC.

Aizopsis Grulich = Sedum

aizoon (L.) Grulich = Sedum aizoon
hybrida (L.) Grulich = Sedum hybridum
kamtschatica (Fisch.) Grulich = Sedum kamtschaticum
middendorfiana (Maxim.) Grulich = Sedum middendorfianum
selskiana (Regel & Maack) Grulich = Sedum selskianum

Anacampseros Hill = Hylotelephium

triphylla Haw. = Hylotelephium triphyllum

Asterosedum Grulich = Sedum

obtusifolium (C.A. Mey.) Grulich = Sedum spurium
spurium (Bieb.) Grulich = Sedum spurium
stevenianum (Rouy & Camus) Grulich = Sedum stevenianum

Chiastophyllum (Ledeb.) Stapf ex Berger (*Umbilicus* auct.)

oppositifolium (Ledeb.) Berger (*Umbilicus oppositifolius* (Ledeb.) Ledeb.) - 2

Clementsia Rose

semenovii (Regel & Herd.) Boriss. (*C. semenovii* (Regel & Herd.) W.A. Weber, comb. superfl., *Rhodiola semenovii* (Regel & Herd.) Boriss.) - 6

Crassula auct. = Sedum and Tillaea

alata (Viv.) Berger = Tillaea alata
aquatica (L.) Schoenl. = Tillaea aquatica
caespitosa Cav. = Sedum caespitosum
vaillantii (Willd.) Roth = Tillaea vaillantii

Diopogon Jord. & Fourr. = Jovibarba

hirtus (L.) H.P. Fuchs ex H. Huber subsp. *borealis* H. Huber = Jovibarba sobolifera

Etiosedum A. & D. Löve = Sedum

annuum (L.) A. & D. Löve = Sedum annuum

Hjaltalinia A. & D. Löve = Sedum

villosa (L.) A. & D. Löve = Sedum villosum

Hylotelephium H. Ohba (*Anacampseros* Hill, 1753, non L. 1758, nom. cons., *Telephium* Hill, 1756, non L. 1753)

argutum (Haw.) Holub subsp. *carpaticum* (G. Reuss) Dostal = H. carpaticum

carpaticum (G. Reuss) Sojak (*Sedum carpaticum* G. Reuss, *Hylotelephium argutum* (Haw.) Holub subsp. *carpaticum* (G. Reuss) Dostal, *Sedum fabaria* auct., *S. vulgare* auct.) - 1

caucasicum (Grossh.) H. Ohba (*Sedum caucasicum* (Grossh.) Boriss., *S. telephium* L. subsp. *caucasicum* (Grossh.) Takht.) - 2

cyaneum (J. Rudolph) H. Ohba (*Sedum cyaneum* J. Rudolph) - 4, 5

erythrostictum (Miq.) H. Ohba (*Sedum erythrostictum* Miq., *S. alboroseum* Baker) - 5

eupatorioides (Kom.) H. Ohba (*Sedum eupatorioides* (Kom.) Kom.) - 5

ewersii (Ledeb.) H. Ohba (*Sedum ewersii* Ledeb.) - 3. 4. 6

maximum (L.) Holub (*H. telephium* (L.) H. Ohba subsp. *maximum* (L.) H. Ohba, *Sedum maximum* (L.) Hoffm., *S. telephium* L. subsp. *maximum* (L.) Krock.) - 1
- subsp. *ruprechtii* (Jalas) Dostal = H. ruprechtii

mugodsharicum (Boriss.) Grulich (*Sedum mugodsharicum* Boriss.) - 6

pallescens (Freyn) H. Ohba (*Sedum pallescens* Freyn) - 4, 5

parvistamineum (V. Petrov) Czer. comb. nova (*Sedum*

parvistamineum V. Petrov, 1927, Bull. Jard. Bot. Princip. URSS, 26, 2 : 183) - 5(?)
pluricaule (Kudo) H. Ohba (*Sedum pluricaule* Kudo) - 5
populifolium (Pall.) H. Ohba (*Sedum populifolium* Pall.) - 3, 4
purpureum (L.) Holub = H. triphyllum
purpureum (L.) S.H. Fu = H. triphyllum
ruprechtii (Jalas) Tzvel. (*Sedum telephium* L. subsp. *ruprechtii* Jalas, 1954, *Hylotelephium maximum* (L.) Holub subsp. *ruprechtii* (Jalas) Dostal, *H. telephium* (L.) H. Ohba subsp. *ruprechtii* (Jalas) H. Ohba, *Sedum maximum* (L.) Hoffm. subsp. *ruprechtii* (Jalas) Soo, *S. ruprechtii* (Jalas) Omelcz., *S. telephium* sensu Boriss. p.p.) - 1
  Probably, the priority name should be H. polonicum (Blocki) Holub (*Sedum polonicum* Blocki).
stepposum (Boriss.) Tzvel. (*Sedum stepposum* Boriss., *S. telephium* sensu Boriss. p.p.) - 1, 3
telephium (L.) H. Ohba, p.p. = H. triphyllum
- subsp. *maximum* (L.) H. Ohba = H. maximum
- subsp. *ruprechtii* (Jalas) H. Ohba = H. ruprechtii
triphyllum (Haw.) Holub (*Anacampseros triphylla* Haw., *Hylotelephium purpureum* (L.) Holub, *H. purpureum* (L.) S.H. Fu, comb. superfl., *H. telephium* (L.) H.Ohba, p.p., *Sedum purpureum* (L.) Schult., *S. telephium* sensu Czer.) - 1, 3, 4, 5
ussuriense (Kom.) H. Ohba (*Sedum ussuriense* Kom.) - 5
verticillatum (L.) H. Ohba (*Sedum verticillatum* L.) - 5
viviparum (Maxim.) H. Ohba (*Sedum viviparum* Maxim.) - 5

Jovibarba Opiz (*Diopogon* Jord. & Fourr.)

heuffelii auct. = J. preissiana
hirta auct. = J. preissiana
hirta (L.) Opiz subsp. *borealis* (H. Huber) Soo = J. sobolifera
- subsp. *glabrescens* (Sabr.) Soo & Jav. = J. preissiana
preissiana (Domin) Omelcz. & Czopik (*Sempervivum preissianum* Domin, *Jovibarba hirta* (L.) Opiz subsp. *glabrescens* (Sabr.) Soo & Jav., *Sempervivum hirtum* L. subsp. *glabrescens* (Sabr.) Jav., *S. hirtum* subsp. *preissianum* (Domin) Dostal, *S. hirtum* f. *glabrescens* Sabr., *S. soboliferum* Sims subsp. *glabrescens* (Sabr.) Soo, *S. soboliferum* subsp. *preissianum* (Domin) S. Pawl., *?Jovibarba heuffelii* auct., *J. hirta* auct., *Sempervivum hirtum* auct. p.p.) - 1
sobolifera (Sims) Opiz (*Sempervivum soboliferum* Sims., *Diopogon hirtus* (L.) H.P. Fuchs ex H. Huber subsp. *borealis* H. Huber, *Jovibarba hirta* (L.) Opiz subsp. *borealis* (H. Huber) Soo) - 1

Kirpicznikovia A. & D. Löve = Rhodiola

quadrifida (Pall.) A. & D. Löve = Rhodiola quadrifida

Macrosepalum Regel & Schmalh.

aetnense (Tineo) Palanov (*Sedum aetnense* Tineo) - 1
tetramerum (Trautv.) Palanov (*Sedum tetramerum* Trautv., *S. aetnense* Tineo subsp. *tetramerum* (Trautv.) Breistroffer) - 2, 6

Oreosedum Grulich = Sedum

album (L.) Grulich = Sedum album
nanum (Boiss.) Grulich = Sedum nanum
subulatum (C.A. Mey.) Grulich = Sedum subulatum
tenellum (Bieb.) Grulich = Sedum tenellum

villosum (L.) Grulich = Sedum villosum

Orostachys Fisch. (*Umbilicus* auct.)

aggregata auct. = O. furusei
cartilaginea Boriss. = O. erubescens
erubescens (Maxim.) Ohwi (*Umbilicus erubescens* Maxim., *Orostachys cartilaginea* Boriss., *Sedum erubescens* (Maxim.) Ohwi, 1965, non Sennen 1927, *Orostachys japonica* auct.) - 5
fimbriata (Turcz.) Berger - 4
furusei Ohwi (*Sedum furusei* (Ohwi) Ohwi, *Orostachys aggregata* auct.) - 5
iwarenge auct. = O. malacophylla
japonica auct. = O. erubescens
malacophylla (Pall.) Fisch. (*O. iwarenge* auct.) - 4, 5
paradoxa (A. Khokhr. & Worosch.) Czer. comb. nova (*Sedum paradoxum* A. Khokhr. & Worosch. 1975, Bull. Gl. Bot. Sada, 75 : 42) - 5
spinosa (L.) C.A. Mey. - 1, 3, 4, 5, 6
thyrsiflora Fisch. - 1, 3, 4, 6

Petrosedum Grulich = Sedum

reflexum (L.) Grulich = Sedum reflexum

Prometheum (Berger) H. Ohba (*Pseudorosularia* Gurgenidze)

pilosum (Bieb.) H. Ohba (*Pseudorosularia pilosa* (Bieb.) Gurgenidze, *Rosularia pilosa* (Bieb.) Boriss.) - 2
sempervivoides (Fisch. ex Bieb.) H. Ohba (*Pseudorosularia sempervivoides* (Fisch. ex Bieb.) Gurgenidze, *Rosularia sempervivoides* (Fisch. ex Bieb.) Boriss.) - 2

Pseudorosularia Gurgenidze = Prometheum

pilosa (Bieb.) Gurgenidze = Prometheum pilosum
sempervivoides (Fisch. ex Bieb.) Gurgenidze = Prometheum sempervivoides

Pseudosedum (Boiss.) Berger

bucharicum Boriss. - 6
campanuliflorum Boriss. - 6
condensatum Boriss. - 6
fedtschenkoanum Boriss. - 6
ferganense Boriss. - 6
kamelinii Palanov - 6
karatavicum Boriss. - 6
kuramense Boriss. = P. longidentatum
lievenii (Ledeb.) Berger - 6
longidentatum Boriss. (*P. kuramense* Boriss.) - 6
multicaule (Boiss. & Buhse) Boriss. - 6

Rhodiola L. (*Kirpicznikovia* A. & D. Löve, *Tolmachevia* A. & D. Löve)

algida (Ledeb.) Fisch. & C.A. Mey. - 3, 4
arctica Boriss. (*R. rosea* L. subsp. *arctica* (Boriss.) A. & D. Löve, *Sedum roseum* (L.) Scop. subsp. *arcticum* (Boriss.) Ju. Kozhevn.) - 1
atropurpurea (Turcz.) Trautv. & C.A. Mey. = R. integrifolia
borealis Boriss. = R. rosea
coccinea (Royle) Boriss. (*Sedum quadrifidum* Pall. subsp. *coccineum* (Royle) Ju. Kozhevn.) - 4, 6
crassipes (Wall. ex Hook. fil. & Thoms.) Boriss. var. *step-

hanii (Cham.) Jacobsen = R. stephanii
fastigiata (Hook. fil. & Thoms.) Fu var. *gelida* (Schrenk) Jacobsen = R. gelida
gelida Schrenk (*R. fastigiata* (Hook. fil. & Thoms.) Fu var. *gelida* (Schrenk) Jacobsen) - 6
heterodonta (Hook. fil. & Thoms.) Boriss. (*R. viridula* Boriss.) - 6
himalensis (D. Don) Fu var. *ishidae* (Miyabe & Kudo) Jacobsen = R. ishidae
integrifolia Rafin. (*R. atropurpurea* (Turcz.) Trautv. & C.A. Mey., *R. rosea* L. subsp. *atropurpurea* (Turcz.) Jacobsen, *R. rosea* subsp. *integrifolia* (Rafin.) Ju. Kozhevn., *R. rosea* subsp. *integrifolia* (Rafin.) E. Murray, comb. superfl., *R. rosea* subsp. *integrifolia* (Rafin.) Petrovsky, comb. superfl., *Sedum roseum* (L.) Scop. subsp *integrifolium* (Rafin.) Hult., *Tolmachevia atropurpurea* (Turcz.) A. & D. Löve, *T. integrifolia* (Rafin.) A. & D. Löve) - 5
iremelica Boriss. - 1
ishidae (Miyabe & Kudo) Hara (*Sedum ishidae* Miyabe & Kudo, *Rhodiola himalensis* (D. Don) Fu var. *ishidae* (Miyabe & Kudo) Jacobsen) - 5
kaschgarica Boriss. - 6
kirilowii (Regel) Maxim. - 6
komarovii Boriss. - 5
krivochizhinii Sipl. = R. rosea
krylovii Polozh. & Revjakina - 3
linearifolia Boriss. - 6
litwinowii Boriss. - 6
pamiroalaica Boriss. (*Sedum pamiroalaicum* (Boriss.) Jansson) - 6
pinnatifida Boriss. (*Sedum pinnatifidum* (Boriss.) Ju. Kozhevn.) - 4
- subsp. **subpinnata** Krasnob. - 4
quadrifida (Pall.) Fisch. & C.A. Mey. (*Kirpicznikovia quadrifida* (Pall.) A. & D. Löve) - 1, 3, 4, 5
recticaulis Boriss. (*Sedum recticaule* (Boriss.) Wendelbo) - 6
rosea L. (*R. borealis* Boriss., *R. krivochizhinii* Sipl., *R. rosea* subsp. *elongata* (Ledeb.) Jacobsen, *Tolmachevia krivochizhinii* (Sipl.) A. & D. Löve) - 1, 3, 4, 5, 6
- subsp. *arctica* (Boriss.) A. & D. Löve = R. arctica
- subsp. *atropurpurea* (Turcz.) Jacobsen = R. integrifolia
- subsp. *elongata* (Ledeb.) Jacobsen = R. rosea
- subsp. *integrifolia* (Rafin.) Ju. Kozhevn. = R. integrifolia
- subsp. *integrifolia* (Rafin.) E. Murray = R. integrifolia
- subsp. *integrifolia* (Rafin.) Petrovsky = R. integrifolia
sachalinensis Boriss. (*Sedum sachalinense* (Boriss.) Worosch.) - 5
semenovii (Regel & Herd.) Boriss. = Clementsia semenovii
stephanii (Cham.) Trautv. & C.A. Mey. (*R. crassipes* (Wall. ex Hook. fil. & Thoms.) Boriss. var. *stephanii* (Cham.) Jacobsen) - 5
viridula Boriss. = R. heterodonta

Rosularia (DC.) Stapf (*Umbilicus* auct.)

aizoon (Fenzl) Berger (*Umbilicus aizoon* Fenzl, *Rosularia chrysantha* (Boiss.) Takht. p.p. excl. typo) - 2
alpestris (Kar. & Kir.) Boriss. - 6
borissovae Pratov - 6
chrysantha (Boiss.) Takht. p.p. = R. aizoon
elymaitica (Boiss. & Hausskn.) Berger - 2, 6
glabra (Regel & C. Winkl.) Berger - 6
hissarica Boriss. - 6
kokanica (Regel & Schmalh.) Boriss. - 6
lipskyi Boriss. - 2
lutea Boriss. - 6

paniculata (Regel & Schmalh.) Berger - 6
persica (Boiss.) Berger - 2
pilosa (Bieb.) Boriss. = Prometheum pilosum
platyphylla (Schrenk) Berger - 6
radiciflora Steud. ex Boriss. - 2
schischkinii Boriss. - 6
sempervivoides (Fisch. ex Bieb.) Boriss. = Prometheum sempervivoides
sempervivum (Bieb.) Berger - 2
subspicata (Freyn & Sint.) Boriss. - 6
tadzhikistana Boriss. - 6
turkestanica (Regel & C. Winkl.) Berger - 6
vvedenskyi Pratov - 6

Sedum L. (*Aizopsis* Grulich, *Asterosedum* Grulich, *Etiosedum* A. & D. Löve, *Hjaltalinia* A. & D. Löve, *Oreosedum* Grulich, *Petrosedum* Grulich, *Spathulata* (Boriss.) A. & D. Löve, *Crassula* auct.)

abchasicum Kolak. - 2
acre L. - 1, 2, 3, 4, 5(alien)
- subsp. *sexangulare* (L.) O. Schwarz = S. sexangulare
aetnense Tineo = Macrosepalum aetnense
- subsp. *tetramerum* (Trautv.) Breistroffer = Macrosepalum tetramerum
aizoon L. (*Aizopsis aizoon* (L.) Grulich, *Sedum hyperaizoon* Kom., *S. litorale* Kom.) - 3, 4, 5
- subsp. *maximowiczii* (Regel) Worosch. = S. maximowiczii
alberti Regel - 3, 6
alboroseum Baker = Hylotelephium erythrostictum
album L. (*Oreosedum album* (L.) Grulich) - 1, 2
alpestre Vill. - 1
annuum L. (*Etiosedum annuum* (L.) A. & D.Löve) - 2
antiquum Omelcz. & Zaverucha - 1
argunense Galushko - 2
atratum L. - 1
berunii Pratov - 6
borissovae Balk. - 1
bucharicum Boriss. - 6
caespitosum (Cav.) DC. (*Crassula caespitosa* Cav., *Sedum rubrum* (L.) Thell. 1912, non Royle ex Edgew. 1846) - 1, 2
carpaticum G. Reuss = Hylotelephium carpaticum
caucasicum (Grossh.) Boriss. = Hylotelephium caucasicum
corymbosum Grossh. - 2
cyaneum J. Rudolph = Hylotelephium cyaneum
ellacombianum Praeger = S. sikokianum
erubescens (Maxim.) Ohwi = Orostachys erubescens
erythrostictum Miq. = Hylotelephium erythrostictum
eupatorioides (Kom.) Kom. = Hylotelephium eupatorioides
ewersii Ledeb. = Hylotelephium ewersii
fabaria auct. = Hylotelephium carpaticum
furusei (Ohwi) Ohwi = Orostachys furusei
gracile C.A. Mey. - 2
hillebrandtii Fenzl (*S. sartorianum* Boiss. subsp. *hillebrandtii* (Fenzl) D.A. Webb) - 1
hispanicum L. - 1, 2
hybridum L. (*Aizopsis hybrida* (L.) Grulich) - 1, 3, 4, 6
hyperaizoon Kom. = S. aizoon
involucratum Bieb. - 2
ishidae Miyabe & Kudo = Rhodiola ishidae
kamtschaticum Fisch. (*Aizopsis kamtschatica* (Fisch.) Grulich) - 5
- var. *ellacombianum* (Praeger) T.B. Clausen = S. sikokianum

- var. *middendorfianum* (Maxim.) T.B. Clausen = S. middendorfianum
kurilense Worosch. (*S. sikokianum* Maxim. subsp. *kurilense* (Worosch.) Worosch.) - 5
lenkoranicum Grossh. - 2
listoniae Vis. = S. obtusifolium
litorale Kom. = S. aizoon
▪ **lydium** Boiss.
maximowiczii Regel (*S. aizoon* L. subsp. *maximowiczii* (Regel) Worosch.) - 5
maximum (L.) Hoffm. = Hylotelephium maximum
- subsp. *ruprechtii* (Jalas) Soo = Hylotelephium ruprechtii
middendorfianum Maxim. (*Aizopsis middendorfiana* (Maxim.) Grulich, *Sedum kamtschaticum* Fisch. var. *middendorfianum* (Maxim.) T.B. Clausen) - 4, 5
- subsp. **arcuatum** Worosch. & Schlothg. - 5
- subsp. *sichotense* (Worosch.) Worosch. = S. sichotense
mugodsharicum Boriss. = Hylotelephium mugodsharicum
▪ **nanum** Boiss. (*Oreosedum nanum* (Boiss.) Grulich)
obtusifolium C.A. Mey. (*Asterosedum obtusifolium* (C.A. Mey.) Grulich, *Sedum listoniae* Vis.) - 2
oppositifolium Sims - 2
pallescens Freyn = Hylotelephium mugodsharicum
pallidum Bieb. - 1, 2
pamiroalaicum (Boriss.) Jansson = Rhodiola pamiroalaica
paradoxum A. Khokhr. & Worosch. = Orostachys paradoxa
parvistamineum V. Petrov = Hylotelephium parvistamineum
pentapetalum Boriss. - 2, 6
pinnatifidum (Boriss.) Ju. Kozhevn. = Rhodiola pinnatifida
pluricaule Kudo = Hylotelephium pluricaule
polytrichoides Hemsl. - 5
populifolium Pall. = Hylotelephium populifolium
pseudohybridum Worosch. & Schloth. - 5
purpureum (L.) Schult. = Hylotelephium triphyllum
quadrifidum Pall. subsp. *coccineum* (Royle) Ju. Kozhevn. = Rhodiola coccinea
recticaule (Boriss.) Wendelbo = Rhodiola recticaulis
reflexum L. (*Petrosedum reflexum* (L.) Grulich, *Sedum rupestre* L. subsp. *reflexum* (L.) Hegi & E. Schmid) - 1, 2
roseum (L.) Scop. subsp. *arcticum* (Boriss.) Ju.Kozhevn. = Rhodiola arctica
- subsp *integrifolium* (Rafin.) Hult. = Rhodiola integrifolia
rubens L. - 1(?)
rubrum (L.) Thell. = S. caespitosum
rupestre L. subsp. *reflexum* (L.) Hegi & E. Schmid = S. reflexum
ruprechtii (Jalas) Omelcz. = Hylotelephium ruprechtii
sachalinense (Boriss.) Worosch. = Rhodiola sachalinensis
sartorianum Boiss. subsp. *hillebrandtii* (Fenzl) D.A. Webb = S. hillebrandtii
selskianum Regel & Maack (*Aizopsis selskiana* (Regel & Maack) Grulich) - 5
sexangulare L. (*S. acre* L. subsp. *sexangulare* (L.) O. Schwarz) - 1
sichotense Worosch. (*S. middendorfianum* Maxim. subsp. *sichotense* (Worosch.) Worosch.) - 5
sikokianum Maxim. (*S. ellacombianum* Praeger, *S. kamtschaticum* Fisch. var. *ellacombianum* (Praeger) T.B. Clausen) - 5
- subsp. *kurilense* (Worosch.) Worosch. = S. kurilense
spurium Bieb. (*Asterosedum spurium* (Bieb.) Grulich, *Spathulata spuria* (Bieb.) A. & D. Löve) - 2
stepposum Boriss. = Hylotelephium stepposum
stevenianum Rouy & Camus (*Asterosedum stevenianum* (Rouy & Camus) Grulich) - 2

stoloniferum S.G. Gmel. - 2
subulatum (C.A. Mey.) Boiss. (*Oreosedum subulatum* (C.A. Mey.) Grulich) - 1, 2
sukaczevii Maximova - 4
telephium L. subsp. *caucasicum* (Grossh.) Takht. = Hylotelephium caucasicum
- subsp. *maximum* (L.) Krock. = Hylotelephium maximum
- subsp. *ruprechtii* Jalas = Hylotelephium ruprechtii
telephium sensu Czer. = Hylotelephium triphyllum
telephium sensu Boriss. = Hylotelephium ruprechtii and H. stepposum
tenellum Bieb. (*Oreosedum tenellum* (Bieb.) Grulich) - 2
tetramerum Trautv. = Macrosepalum tetramerum
ussuriense Kom. = Hylotelephium ussuriense
verticillatum L. = Hylotelephium verticillatum
villosum L. (*Hjaltalinia villosa* (L.) A. & D. Löve, *Oreosedum villosum* (L.) Grulich) - 2(?)
viviparum Maxim. = Hylotelephium viviparum
vulgare auct. = Hylotelephium carpaticum

Sempervivum L.

altum Turrill = S. pumilum
annae Gurgenidze - 2
armenum Boiss. & Huet (*S. glabrifolium* Boriss.) - 2
artvinense Muirhead - 2
borissovae Wale = S. caucasicum
caucasicum Rupr. ex Boiss. (*S. borissovae* Wale, *S. caucasicum* var. *borissovae* (Wale) Gurgenidze) - 2
- var. *borissovae* (Wale) Gurgenidze = S. caucasicum
charadzeae Gurgenidze - 2
dominii R. Konop & K. Konopova - 2
dzhavachischvilii Gurgenidze - 2
ermanicum Gurgenidze - 2
georgicum Gurgenidze = S. transcaucasicum
glabrifolium Boriss. = S. armenum
globiferum auct. p.p. = S. transcaucasicum
hirtum auct. p.p. = Jovibarba preissiana
hirtum L. subsp. *glabrescens* (Sabr.) Jav. = Jovibarba preissiana
- subsp. *preissianum* (Domin) Dostal = Jovibarba preissiana
- f. *glabrescens* Sabr. = Jovibarba preissiana
ingwersenii Wale - 2
marmoreum Griseb. (*S. schlehanii* Schott) - 1(?)
montanum L. subsp. **carpaticum** Wettst. ex Hayek - 1
ossetiense Wale - 2
preissianum Domin = Jovibarba preissiana
pumilum Bieb. (*S. altum* Turrill) - 2
ruthenicum Schnittsp. & C.B. Lehm. - 1
schlehanii Schott = S. marmoreum
soboliferum Sims = Jovibarba sobolifera
- subsp. *glabrescens* (Sabr.) Soo = Jovibarba preissiana
- subsp. *preissianum* (Domin) S. Pawl. = Jovibarba preissiana
sosnowskyi Ter-Chatsch. - 2
*****tectorum** L. - 1
transcaucasicum Muirhead (*S. georgicum* Gurgenidze, *S. globiferum* auct. p.p.) - 2

Spathulata (Boriss.) A. & D. Löve = Sedum

spuria (Bieb.) A. & D. Löve = Sedum spurium

Telephium Hill = Hylotelephium

Tillaea L. (*Crassula* auct.)

alata Viv. (*Crassula alata* (Viv.) Berger) - 2

aquatica L. (*Crassula aquatica* (L.) Schoenl.) - 1, 3, 4, 5, 6
vaillantii Willd. (*Crassula vaillantii* (Willd.) Roth) - 1, 6

Tolmachevia A. & D. Löve = Rhodiola
atropurpurea (Turcz.) A. & D. Löve = Rhodiola integrifolia
integrifolia (Rafin.) A. & D. Löve = Rhodiola integrifolia
krivochizhinii (Sipl.) A. & D. Löve) = Rhodiola rosea

Umbilicus auct. =   Chiastophyllum, Orostachys
and Rosularia
aizoon Fenzl = Rosularia aizoon
erubescens Maxim. = Orostachys erubescens
oppositifolius (Ledeb.) Ledeb. = Chiastophyllum oppositifolium

CRUCIFERAE Juss. = BRASSICACEAE

CRYPTOGRAMMACEAE Pichi Sermolli
Allosorus auct. = Cryptogramma
minutus Turcz. ex Trautv. = Cryptogramma stelleri
raddeanus (Fomin) Ching = Cryptogramma raddeana

Cryptogramma R. Br. (*Allosorus* auct.)
acrostichoides R. Br. (*C. crispa* (L.) R. Br. subsp. *acrosti-*
choides (R. Br.) Hult.) - 5
crispa (L.) R.Br. - 1, 2, 5
- subsp. *acrostichoides* (R.Br.) Hult. = C. acrostichoides
- subsp. *raddeana* (Fomin) Hult. = C. raddeana
raddeana Fomin (*Allosorus raddeanus* (Fomin) Ching,
Cryptogramma crispa (L.) R. Br. subsp. *raddeana*
(Fomin) Hult.) - 4, 5
stelleri (S.G. Gmel.) Prantl (*Allosorus minutus* Turcz. ex
Trautv.) - 1, 3, 4, 5, 6

CUCURBITACEAE Juss.
Actinostemma Griff.
lobatum (Maxim.) Maxim. ex Franch. & Savat. - 5

Bryonia L.
alba L. - 1, 2, 6
aspera Stev. ex Ledeb. - 2
cretica L. subsp. *dioica* (Jacq.) Tutin = B. dioica
dioica Jacq. (*B. cretica* L. subsp. *dioica* (Jacq.) Tutin, *B.*
transoxana Vass.) - 1, 2, 6
lappifolia Vass. - 6
melanocarpa Nabiev - 6
monoica Aitch. & Hemsl. - 6
transoxana Vass. = B. dioica

Citrullus Schrad.
colocynthis (L.) Schrad. - 6
***lanatus** (Thunb.) Matsum. & Nakai (*Momordica lanata*
Thunb., *Citrullus lanatus* (Thunb.) Mansf. comb.
superfl.)
- subsp. **vulgaris** (Schrad.) Fursa (*C. vulgaris* Schrad.)
vulgaris Schrad. = C. lanatus subsp. vulgaris

***Cucumis** L.
agrestis (Naud.) Grebensc. = Melo agrestis

flexuosus L. = Melo flexuosus
***myriocarpus** Naud.
orientalis Kudr. = Melo orientalis
***sativus** L.

***Cucurbita** L.
***maxima** Duch.
- subsp. **turbaniformis** (M. Roem.) Vass.
***moschata** (Duch.) Poir. (*Cucurbita pepo* L. A. *moschata*
Duch.)
***pepo** L.
- A. *moschata* Duch. = C. moschata
siceraria Mol. = Lagenaria siceraria

Ecballium A. Rich.
elaterium (L.) A. Rich. - 1, 2, 6

Echinocystis Torr. & Gray
echinata (Muehl. ex Willd.) Britt., Sterns & Pogg. = E.
lobata
echinata (Muehl.) Vass. = E. lobata
lobata (Michx.) Torr. & Gray (*E. echinata* (Muehl. ex
Willd.) Britt., Sterns & Pogg., *E. echinata* (Muehl.)
Vass. comb. superfl. invalid.) - 1, 3, 4, 5

Gynostemma Blume (*Vitis* auct.)
pentaphyllum (Thunb.) Makino (*Vitis pentaphylla* Thunb.) -
5

***Lagenaria** Ser.
***siceraria** (Mol.) Standl. (*Cucurbita siceraria* Mol., *Lage-*
naria vulgaris Ser.)
vulgaris Ser. = L. siceraria

***Luffa** Mill.
***acutangula** (L.) Roxb.
***cylindrica** (L.) M. Roem.

***Melo** Hill
***agrestis** (Naud.) Pang. (*Cucumis agrestis* (Naud.) Gre-
bensc.)
***dudaim** (L.) Sager.
***flexuosus** (L.) Sager. ex M. Roem. (*Cucumis flexuosus* L.)
***orientalis** (Kudr.) Nabiev (*Cucumis orientalis* Kudr.)
***sativus** Sager. ex M. Roem.

***Momordica** L.
balsamina L. - 1, 6
lanata Thunb. = Citrullus lanatus

Schizopepon Maxim.
bryoniifolius Maxim. - 5

Sicyos L.
angulatus L. - 1

Thladiantha Bunge
dubia Bunge - 1, 5

Vitis auct. = Gynostemma
pentaphylla Thunb. = Gynostemma pentaphyllum

CUPRESSACEAE Rich. ex Bartl.

Biota (D. Don) Endl. = Platycladus

orientalis (L.) Endl. = Platycladus orientalis

*Chamaecyparis Spach

*lawsoniana (A. Murr.) Parl. (*Cupressus lawsoniana* A. Murr.) - 1
*pisifera Siebold & Zucc. - 1

*Cupressus L.

lawsoniana A. Murr. = Chamaecyparis lawsoniana
*sempervirens L. - 1, 2

Juniperus L. (*Sabina* Mill.)

alpina (Suter) S.F. Gray = J. sibirica
communis L. - 1, 2, 3, 4, 6
- subsp. *alpina* (Suter) Celak. = J. sibirica
- subsp. *hemisphaerica* (C. Presl) Nym. = J. hemisphaerica
- subsp. *nana* (Willd.) Syme = J. sibirica
- subsp. *oblonga* (Bieb.) Galushko = J. oblonga
- subsp. *pygmaea* (C. Koch) Imch. = J. pygmaea
- var. *alpina* Suter = J. sibirica
conferta Parl. (*J. litoralis* Maxim.) - 5
davurica Pall. - 4, 5
- subsp. **litoralis** Urussov - 5
- subsp. **maritima** Urussov - 5
depressa Stev. = J. hemisphaerica
drobovii Sumn. = J. semiglobosa
excelsa Bieb. - 1, 2
- subsp. *polycarpos* (C. Koch) Takht. = J. polycarpos
- subsp. *seravschanica* (Kom.) R. Kam. ex Imch. = J. seravschanica
- subsp. *turcomanica* (B. Fedtsch.) Imch. = J. turcomanica
- var. *polycarpos* (C. Koch) J. Silba = J. polycarpos
foetidissima Willd. - 1, 2
hemisphaerica auct. p.p. = J. pygmaea
hemisphaerica C. Presl (*J. communis* L. subsp. *hemisphaerica* (C. Presl) Nym., *J. depressa* Stev. 1857, non Rafin. 1818, *J. pygmaea* auct. p.p.) - 1, 2
intermedia Drob. = J. x media
isophyllos C. Koch (*J. excelsa* auct. p.p.) - 1, 2
kokbulakensis Dmitr. = J. x media
kulsaica Dmitr. = J. seravschanica
litoralis Maxim. = J. conferta
x **media** Dmitr. (*J. intermedia* Drob. 1941, non Schur, 1851, *J. kokbulakensis* Dmitr.). - J. semiglobosa Regel x J. turcomanica Kom. - 6
nana Willd. = J. sibirica
oblonga Bieb. (*J. communis* L. subsp. *oblonga* (Bieb.) Galushko) - 2
oxycedrus L. (*J. oxycedrus* subsp. *rufescens* Deb., *J. rufescens* Link, nom. illegit.) - 1, 2
- subsp. *rufescens* Deb. = J. oxycedrus
polycarpos C. Koch (*J. excelsa* Bieb. subsp. *polycarpos* (C. Koch) Takht., *J. excelsa* var. *polycarpos* (C. Koch) J. Silba, *J. excelsa* auct. p.p.) - 6
- var. *seravschanica* (Kom.) Kitam. = J. seravschanica
polysperma Dmitr. = J. seravschanica
pseudosabina Fisch. & C.A. Mey. (*Sabina pseudosabina* (Fisch. & C.A. Mey.) W.C. Cheng & W.T. Wang) - 3, 4, 6
- var. *turcomanica* (Kom.) J. Silba = J. turkestanica
pygmaea auct. p.p. = J. hemisphaerica
pygmaea C. Koch (*J. communis* L. subsp. *pygmaea* (C.

Koch) Imch., *J. hemisphaerica* auct. p.p., *J. sibirica* auct. p.p.) - 2
rigida Siebold & Zucc. - 5
- subsp. **litoralis** Urussov - 5
rufescens Link = J. oxycedrus
sabina L. - 1, 2, 3, 6
sargentii (A. Henry) Takeda ex Koidz. (*Sabina chinensis* (L.) Ant. var. *sargentii* (A. Henry) W.C. Cheng & L.K. Fu, *S. sargentii* (A. Henry) Miyabe & Tatew.) - 5
schugnanica Kom. = J. semiglobosa
semiglobosa Regel (*J. drobovii* Sumn., *J. schugnanica* Kom., *J. tianschanica* Sumn.) - 6
seravschanica Kom. (*J. excelsa* Bieb. subsp. *seravschanica* (Kom.) R. Kam. ex Imch., *J. kulsaica* Dmitr., *J. polycarpos* C. Koch var. *seravschanica* (Kom.) Kitam. nom. invalid., *J. polysperma* Dmitr., *J. zaaminica* Dmitr., *Sabina seravschanica* (Kom.) Nevski, *Juniperus excelsa* auct. p.p.) - 6
sibirica Burgsd. (*J. alpina* (Suter) S.F. Gray, *J. communis* L. var. *alpina* Suter, *J. communis* subsp. *alpina* (Suter) Celak., *J. communis* subsp. *nana* (Willd.) Syme, *J. nana* Willd.) - 1, 2, 3, 4, 5, 6
x **talassica** Lipsky. - J. sabina L. x J. semiglobosa Regel - 6
tianschanica Sumn. = J. semiglobosa Regel
turcomanica B. Fedtsch. (*J. excelsa* Bieb. subsp. *turcomanica* (B. Fedtsch.) Imch.) - 6
turkestanica Kom. (*J. pseudosabina* Fisch. & C.A. Mey. var. *turkestanica* (Kom.) J. Silba, *Sabina pseudosabina* (Fisch. & C.A. Mey.) W.C. Cheng & W.T. Wang var. *turkestanica* (Kom.) C.Y. Yang) - 6
*virginiana L. - 1, 6
zaaminica Dmitr. = J. seravschanica

Microbiota Kom.

decussata Kom. - 5

*Platycladus Spach (*Biota* (D. Don) Endl.)

*orientalis (L.) Franco (*Thuja orientalis* L., *Biota orientalis* (L.) Endl.) - 1, 2, 6

Sabina Mill. = Juniperus

chinensis (L.) Ant. var. *sargentii* (A. Henry) W.C. Cheng & L.K. Fu = Juniperus sargentii
pseudosabina (Fisch. & C.A. Mey.) W.C. Cheng & W.T. Wang = Juniperus pseudosabina
- var. *turkestanica* (Kom.) C.Y. Yang = Juniperus turkestanica
sargentii (A. Henry) Miyabe & Tatew. = Juniperus sargentii
seravschanica (Kom.) Nevski = Juniperus seravschanica

*Thuja L.

*occidentalis L. - 1, 2, 3, 5
orientalis L. = Platycladus orientalis

CUSCUTACEAE Dumort.

Cuscuta L. (*Grammica* Lour., *Kadurias* Rafin., *Monogynella* DesMoul.)

alba C. Presl - 1, 2
approximata Bab. (*C. cupulata* Engelm.) - 1, 6
araratica Butk. (*C. epithymum* (L.) L. subsp. *araratica* (Butk.) R. Kam.) - 2
australis auct. = C. cesatiana and C. tinei

australis R. Br. subsp. *cesatiana* (Bertol.) Feinbrun = C. cesatiana

- subsp. *cesatiana* (Bertol.) O. Schwarz = C. cesatiana
- subsp. *tinei* (Insenga) Feinbrun = C. tinei

babylonica Auch. ex Choisy - 6

balansae Boiss. & Reut. ex Yunck. = C. palaestina subsp. balansae

basarabica Buia = C. cesatiana

breviflora sensu Zefir. = C. cesatiana

breviflora Vis. = C. tinei

brevistyla auct. p.p. = C. palaestina and C. planiflora

bucharica Palib. (*Monogynella bucharica* (Palib.) Hadac & Chrtek) - 6

callinema Butk. - 6

campestris Yunck. (*Grammica campestris* (Yunck.) Hadac & Chrtek) - 1, 4, 6

capitata auct. = C. stenocalycina

cesatiana Bertol. (*C. australis* R. Br. subsp. *cesatiana* (Bertol.) Feinbrun, comb. superfl., *C. australis* subsp. *cesatiana* (Bertol.) O. Schwarz, *C. basarabica* Buia, *C. scandens* Brot. subsp. *cesatiana* (Bertol.) Greuter & Burdet, comb. superfl., *C. scandens* subsp. *cesatiana* (Bertol.) Soo, *Grammica basarabica* (Buia) Hadac & Chrtek, *Cuscuta australis* auct. p.p., *C. breviflora* sensu Zefir.) - 1, 2, 3, 6

chinensis Lam. (*Grammica chinensis* (Lam.) Hadac & Chrtek) - 5, 6

colorans auct. = C. japonica

convallariiflora Pavl. (*Kadurias convallariiflora* (Pavl.) Hadac & Chrtek) - 6

cupulata Engelm. = C. approximata

cupulata sensu Butk. = C. planiflora

elpassiana Pavl. - 6

engelmannii Korsh. - 6

epilinum Weihe - 1, 2, 3, 4, 5, 6

epithymum (L.) L. (*C. europaea* L. var. *epithymum* L., *C. trifolii* Bab.) - 1, 2, 3, 6

- subsp. *araratica* (Butk.) R. Kam. = C. araratica
- subsp. *kotschyi* (DesMoul.) Arcang. = C. kotschyi

europaea L. (*C. viciae* Koch, Schnizl. & Schoenh.) - 1, 2, 3, 4, 5, 6

- var. *epithymum* L. = C. epithymum

ferganensis Butk. - 6

gigantea Griff. (*C. reflexa* auct.) - 6

globularis Bertol. = C. palaestina

gronovii Willd. ex Schult. (*Grammica gronovii* (Willd. ex Schult.) Hadac & Chrtek) - 1

halophyta Fries - 1

indica (Engelm.) V. Petrov = C. pellucida

indica (Engelm.) V. Petrov ex Schischk. = C. pellucida

japonica Choisy (*Monogynella japonica* (Choisy) Hadac & Chrtek, *Cuscuta colorans* auct.) - 4, 5

karatavica Pavl. - 6

kotschyana Boiss. - 2, 6

kotschyi DesMoul. (*C. epithymum* (L.) L. subsp. *kotschyi* (DesMoul.) Arcang.) - 1, 2, 6

lehmanniana Bunge (*Monogynella lehmanniana* (Bunge) Hadac & Chrtek) - 1(alien), 2(alien), 6

lophosepala Butk. (*Monogynella lophosepala* (Butk.) Hadac & Chrtek) - 6

lupuliformis Krock. (*Monogynella lupuliformis* (Krock.) Hadac & Chrtek) - 1, 2, 3, 4, 6

monogyna Vahl (*Monogynella monogyna* (Vahl) Hadac & Chrtek) - 1, 2, 3, 6

nipponica Franch. & Savat. - 5(?)

palaestina Boiss. (*C. globularis* Bertol., *C. brevistyla* auct. p.p.) - 2

- subsp. **balansae** (Yunck.) Plitm. (*C. balansae* Boiss. & Reut. ex Yunck.) - 2(?)

pamirica Butk. (*Kadurias pamirica* (Butk.) Hadac & Chrtek) - 6

pedicellata Ledeb. - 1, 2, 3, 6

pellucida Butk. (*C. indica* (Engelm.) V. Petrov, nom. invalid., *C. indica* (Engelm.) V. Petrov ex Schischk. nom. illegit.) - 1, 2, 3, 6

planiflora Ten. (*C. brevistyla* auct. p.p., *C. cupulata* sensu Butk.) - 1, 2, 3, 6

pulchella Engelm. - 6

reflexa auct. = C. gigantea

ruschanica Junussov - 6

scandens Brot. subsp. *cesatiana* (Bertol.) Greuter & Burdet = C. cesatiana

- subsp. *cesatiana* (Bertol.) Soo = C. cesatiana

stapfiana Palib. - 6

stenocalycina Palib. (*C. capitata* auct.) - 6

suaveolens Ser. - 1, 6

syrtorum Arbajeva - 6

tianschanica Palib. (*Monogynella tianschanica* (Palib.) Hadac & Chrtek) - 6

tinei Insenga (*C. australis* R. Br. subsp. *tinei* (Insenga) Feinbrun, *C. breviflora* Viz., *Grammica australis* (R. Br.) Hadac & Chrtek subsp. *tinei* (Insenga) Dostal, *Cuscuta australis* auct. p.p.) - 1

trifolii Bab. = C. epithymum

viciae Koch, Schnizl. & Schoenh. = C. europaea

Grammica Lour. = Cuscuta

australis (R. Br.) Hadac & Chrtek subsp. *tinei* (Insenga) Dostal = Cuscuta tinei

basarabica (Buia) Hadac & Chrtek = Cuscuta cesatiana

campestris (Yunck.) Hadac & Chrtek = Cuscuta campestris

chinensis (Lam.) Hadac & Chrtek = Cuscuta chinensis

gronovii (Willd. ex Schult.) Hadac & Chrtek = Cuscuta gronovii

Kadurias Rafin. = Cuscuta

convallariiflora (Pavl.) Hadac & Chrtek = Cuscuta convallariiflora

pamirica (Butk.) Hadac & Chrtek = Cuscuta pamirica

Monogynella DesMoul. = Cuscuta

bucharica (Palib.) Hadac & Chrtek = Cuscuta bucharica

japonica (Choisy) Hadac & Chrtek = Cuscuta japonica

lehmanniana (Bunge) Hadac & Chrtek = Cuscuta lehmanniana

lophosepala (Butk.) Hadac & Chrtek = Cuscuta lophosepala

lupuliformis (Krock.) Hadac & Chrtek = Cuscuta lupuliformis

monogyna (Vahl) Hadac & Chrtek = Cuscuta monogyna

tianschanica (Palib.) Hadac & Chrtek = Cuscuta tianschanica

*CYCADACEAE Pers.

*Cycas L.

revoluta Thunb. - 2

CYNOMORIACEAE Lindl.
Cynomorium L.

coccineum L. subsp. *songaricum* (Rupr.) J. Leonard = C. songaricum

songaricum Rupr. (*C. coccineum* L. subsp. *songaricum* (Rupr.) J. Leonard) - 6

CYPERACEAE Juss. (*KOBRESIACEAE* Gilly)

Abildgaardia auct. p.p. = Bulbostylis

densa (Wall.) Lye = Bulbostylis densa

Acorellus Palla = Juncellus

distachyos (All.) Palla = Juncellus distachyos
laevigatus auct. = Juncellus distachyos
laevigatus (L.) Palla subsp. *distachyos* (All.) Holub = Juncellus distachyos
pannonicus (Jacq.) Palla = Juncellus pannonicus

Baeothryon A. Dietr. (*Eriophorella* Holub, *Trichophorum* Pers. p.p. quoad *T. alpinum* (L.) Pers.)

alpinum (L.) Egor. (*Eriophorella alpina* (L.) Holub, *Eriophorum hudsonianum* Michx., *Scirpus hudsonianus* (Michx.) Fern., *Trichophorum alpinum* (L.) Pers.) - 1, 3, 4, 5
cespitosum (L.) A. Dietr. (*Trichophorum bracteatum* (Bigel.) V. Krecz. ex Czernov, nom. invalid., *T. cespitosum* (L.) C. Hartm.) - 1, 3, 4, 5
dolichocarpum (Zak.) Egor. (*Trichophorum dolichocarpum* Zak.) - 6
pumilum (Vahl) A. & D. Löve (*B. pumilum* (Vahl) T. Koyama, comb. superfl., *Eriophorella pumila* (Vahl) Kit Tan, *Trichophorum pumilum* (Vahl) Schinz & Thell.) - 2, 3, 4, 6
uniflorum (Trautv.) Egor. (*Scirpus uniflorus* Trautv., *Trichophorum uniflorum* (Trautv.) Karav.) - 4, 5

Blysmopsis Oteng-Yeboah = Blysmus

rufa (Huds.) Oteng-Yeboah = Blysmus rufus

Blysmus Panz. ex Schult. (*Blysmopsis* Oteng-Yeboah, *Nomochloa* Beauv.)

compressus (L.) Panz. ex Link (*Nomochloa compressa* (L.) Beetle) - 1, 2, 6
exilis (Printz) Ivanova = B. rufus
rufus (Huds.) Link (*Blysmopsis rufa* (Huds.) Oteng-Yeboah, *Blysmus exillis* (Printz) Ivanova, *B. rufus* subsp. *exilis* Printz, *Nomochloa rufa* (Huds.) Beetle) - 1, 3, 4, 6
- subsp. *exilis* Printz = B. rufus
sinocompressus Tang & Wang - 4, 6

Bolboschoenus (Aschers.) Palla

affinis (Roth) Drob. p.p. = B. popovii
compactus (Hoffm.) Drob. = B. maritimus var. compactus
desoulavii (Drob.) A.E. Kozhevnikov (*B. maritimus* (L.) Palla var. *desoulavii* Drob.) - 4, 5
koshewnikowii (Litv.) A.E. Kozhevnikov (*Scirpus koshewnikowii* Litv.) - 1, 3, 5, 6
macrostachys (Willd.) Grossh. = B. maritimus var. macrostachys
maritimus (L.) Palla - 1, 2, 3, 4, 6
- var. **compactus** (Hoffm.) Egor. (*B. compactus* (Hoffm.) Drob.)
- var. *desoulavii* Drob. = B. desoulavii
- var. **macrostachys** (Willd.) Egor. (*B. macrostachys*

(Willd.) Grossh.)
planiculmis (Fr. Schmidt) Egor. (*Scirpus planiculmis* Fr. Schmidt, *S. biconcavus* Ohwi) - 5
popovii Egor. (*B. affinis* (Roth) Drob. p.p. excl. typo, *B. strobilinus* (Roxb.) V. Krecz. p.p. excl. typo) - 1, 2, 3, 6
strobilinus (Roxb.) V. Krecz. p.p. = B. popovii
yagara (Ohwi) A.E. Kozhevnikov (*Scirpus yagara* Ohwi) - 5

Bulbostylis Kunth (*Abildgaardia* auct. p.p.)

capillaris auct. = B. densa
densa (Wall.) Hand.-Mazz. (*Scirpus densus* Wall., *Abildgaardia densa* (Wall.) Lye, *Bulbostylis capillaris* auct.) - 5
tenerrima Fisch. & C.A. Mey. ex Grossh. - 2
woronowii Palla = Fimbristylis hispidula

Carex L. (*Vignea* Beauv.)

aa Kom. = C. heleonastes
x **abortiva** Holmb. - C. brunnescens (Pers.) Poir. x C. cinerea Poll.
accrescens Ohwi = C. pallida
acrifolia V. Krecz. (*C. enervis* C.A. Mey. subsp. *acrifolia* (V. Krecz.) Egor., *C. pseudofoetida* Kuk. subsp. *acrifolia* (V. Krecz.) Kukkonen) - 2
acuta L. (*C. acuta* f. *prolixa* (Fries) Sylven, *C. dichroandra* V. Krecz., *C. fuscovaginata* Kuk., *C. graciliformis* V. Krecz. 1933, non Ohwi, 1931, *C. gracilis* Curt., *C. prolixa* Fries, *C. sareptana* V. Krecz.) - 1, 2, 3, 4, 6
- f. *prolixa* (Fries) Sylven = C. acuta
acutiformis Ehrh. - 1, 2, 3, 4, 6
adelostoma V. Krecz. (*C. buxbaumii* Wahlenb. subsp. *alpina* (C. Hartm.) Liro, *C. buxbaumii* subsp. *mutica* (C. Hartm.) Isoviita, *C. buxbaumii* var. *alpina* C. Hartm., *C. buxbaumii* var. *mutica* C. Hartm., *C. polygama* Schkuhr subsp. *alpina* (C. Hartm.) Cajand., *C. polygama* subsp. *mutica* (C. Hartm.) Cajand.) - 1, 3, 4
aequivoca V. Krecz. = C. medwedewii
ajanensis Worosch. = C. alba
alajica Litv. - 6
alba Scop. (*C. ajanensis* Worosch.) - 1, 2, 3, 4, 5, 6
- var. *ussuriensis* (Kom.) T. Koyama = C. ussuriensis
albata Boott ex Franch. & Savat. - 5
x **albidula** Holmb. - C. cinerea Poll. x C. loliacea L.
alexeenkoana Litv. (*C. kenkolensis* Litv.) - 6
algida Turcz. ex V. Krecz. = C. quasivaginata
x **almii** Holmb. - C. dioica L. x C. tenuiflora Wahlenb.
alpina Sw. = C. norvegica
- var. *inferalpina* Wahlenb. = C. media
x **alsatica** Zahn. - C. flava L. x C. serotina Merat
altaica (Gorodk.) V. Krecz. (*C. bigelowii* Torr. ex Schwein. subsp. *altaica* (Gorodk.) Malysch., *C. ensifolia* Turcz. ex V. Krecz. subsp. *altaica* (Gorodk.) Malysch., *C. orbicularis* Boott subsp. *altaica* (Gorodk.) Egor., *C. rigida* Good. subsp. *altaica* Gorodk.) - 3, 4
alticola Popl. ex Sukacz. (*C. argunensis* Turcz. ex Trev. subsp. *alticola* (Popl. ex Sukacz.) Malysch., *C. rupestris* All. subsp. *alticola* (Popl. ex Sukacz.) Worosch.) - 4, 5
amblyorhyncha V. Krecz. = C. marina
amgunensis Fr. Schmidt - 1, 3, 4, 5
amurensis Kuk. = C. sordida
aneurocarpa V. Krecz. - 3, 6
angarae Steud. = C. media
- subsp. *brachylepis* Kalela = C. media
angustior Mackenz. (*C. basilata* Ohwi, *Vignea angustior* (Mackenz.) Sojak) - 5
anisoneura V. Krecz. - 6

anthoxanthea C. Presl - 5

aomorensis Franch. (*C. capillacea* Boott subsp. *aomorensis* (Franch.) Egor., *C. ontakensis* auct.) - 5

aperta auct. = C. schmidtii

aphanolepis Franch. & Savat. - 5

appendiculata (Trautv. & C.A. Mey.) Kuk. (*C. gaudichaudiana* Kuk. subsp. *appendiculata* (Trautv. & C.A. Mey.) A. & D. Löve, *C. gaudichaudiana* var. *appendiculata* (Trautv. & C.A. Mey.) T. Koyama, *C. koidzumiana* Ohwi, *C. spongiifolia* A.E. Kozhevnikov, *C. thunbergii* Steud. var. *appendiculata* (Trautv. & C.A. Mey.) Ohwi) - 4, 5

appropinquata Schum. (*Vignea appropinquata* (Schum.) Sojak) - 1, 2, 3, 4

approximata All. = C. ericetorum

aquatilis Wahlenb. - 1, 3, 4
- subsp. *hyperborea* (Drej.) K. Richt. = C. bigelowii
- subsp. *stans* (Drej.) Hult. = C. concolor

arakamensis Clarke = C. nesophila

arcatica Meinsh. = C. orbicularis
- f. *taldycola* (Meinsh.) Ovcz. & Czuk. = C. orbicularis

arctisibirica (Jurtz.) Czer. (*C. ensifolia* Turcz. ex V. Krecz. subsp. *arctisibirica* Jurtz., *C. bigelowii* Torr. ex Schwein. subsp. *arctisibirica* (Jurtz.) A. & D. Löve, *C. bigelowii* subsp. *ensifolia* (Gorodk.) Holub var. *arctisibirica* (Jurtz.) Malysch.) - 1, 3, 4, 5

arctogena H. Smith (*C. capitata* L. subsp. *arctogena* (H. Smith) Hiit., *C. capitata* subsp. *arctogena* (H. Smith) Bocher, comb. superfl., *C. capitata* f. *arctogena* (H. Smith) Raymond) - 1

arenaria L. - 1

arenicola Fr. Schmidt - 5

argunensis Turcz. ex Trev. (*C. rupestris* All. subsp. *argunensis* (Turcz. ex Trev.) Worosch.) - 4, 5
- subsp. *alticola* (Popl. ex Sukacz.) Malysch. = C. alticola

argyroglochin Hornem. = C. ovalis

aristata R. Br. = C. atherodes
- subsp. *eriophylla* (Kuk.) Worosch. = C. raddei
- subsp. *orthostachys* (C.A. Mey.) Kuk. = C. atherodes
- subsp. *raddei* (Kuk.) Kuk. = C. raddei

arnellii Christ (*C. subconcolor* Kitag.) - 1, 3, 4, 5

arrhyncha Franch. = C. tenuiflora

x **arthuriana** Beckm. & Figert. - C. cinerea Poll. x C. remota L.

aspratilis V. Krecz. (*C. distans* L. subsp. *aspratilis* (V. Krecz.) Egor., *C. karelinii* Meinsh. var. *aspratilis* (V. Krecz.) Serg.) - 1, 2, 3, 4, 6

aterrima Hoppe (*C. atrata* L. subsp. *aterrima* (Hoppe) Celak., *C. atrata* subsp. *perfusca* (V. Krecz.) T. Koyama, *C. caucasica* Stev. subsp. *perfusca* (V. Krecz.) T. Koyama, *C. perfusca* V. Krecz.) - 1, 3, 4, 5, 6
- subsp. *medwedewii* (Leskov) Egor. = C. medwedewii

atherodes Spreng. (*C. aristata* R. Br. 1823, non Honck. 1792, *C. aristata* subsp. *orthostachys* (C.A. Mey.) Kuk., *C. atherodes* var. *orthostachys* (C.A. Mey.) A.E. Kozhevnikov, *C. orthostachys* C.A. Mey., *C. pergrandis* V. Krecz. & Lucznik, *C. siegertiana* Uechtr.) - 1, 2, 3, 4, 5, 6
- var. *orthostachys* (C.A. Mey.) A.E. Kozhevnikov = C. atherodes

atrata L. - 1, 3
- subsp. *aterrima* (Hoppe) Celak. = C. aterrima
- subsp. *perfusca* (V. Krecz.) T. Koyama = C. aterrima

atrofusca Schkuhr (*C. oxyleuca* V. Krecz., *C. sisukensis* Akiyama, *C. stilbophaea* V. Krecz., *C. taskanensis* A. Khokhr.) - 1, 3, 4, 5, 6

augustinowiczii Meinsh. - 5

- subsp. *soyaeensis* (Kuk.) Egor. = C. soyaeensis

auriculata Franch. = C. campylorhina

austroussuriensis A.E. Kozhevnikov - 5

basilata Ohwi = C. angustior

behringensis Clarke = C. podocarpa

bergrothii Palmgr. (*C. viridula* Michx. var. *bergrothii* (Palmgr.) B. Schmid) - 1

x **bicharica** Simonk. - C. cinerea Poll. x C. echinata Murr.

bicolor All. - 1, 3, 4, 5

bigelowii Torr. ex Schwein. (*C. aquatilis* Wahlenb. subsp. *hyperborea* (Drej.) K. Richt. comb. incorrecta, *C. bigelowii* subsp. *hyperborea* (Drej.) Bocher, *C. bigelowii* subsp. *rigida* (Good.) W. Schultze-Motel, p.p. quoad nomen, *C. hyperborea* Drej., *C. rigida* Good. 1794, non Schrank, 1789, *C. rigida* subsp. *inferalpina* (Laest.) Gorodk. p.p.) - 1
- subsp. *altaica* (Gorodk.) Malysch. = C. altaica
- subsp. *arctisibirica* (Jurtz.) A. & D. Löve = C. arctisibirica
- subsp. *dacica* (Heuff.) Egor. = C. dacica
- subsp. *ensifolia* (Gorodk.) Holub = C. ensifolia
- - var. *arctisibirica* (Jurtz.) Malysch. = C. arctisibirica
- subsp. *hyperborea* (Drej.) Bocher = C. bigelowii
- subsp. *lugens* (H.T. Holm) Egor. = C. lugens
- subsp. *nardeticola* Holub = C. dacica
- subsp. *paishanensis* (Nakai) Worosch. = C. paishanensis
- subsp. *rigida* (Good.) W. Schultze-Motel = C. bigelowii and C. dacica
- subsp. *rigidioides* (Gorodk.) Egor. = C. rigidioides

bipartita sensu Malysch. = C. lachenalii

biwensis auct. = C. jankowskii

biwensis Franch. subsp. *jankowskii* (Gorodk.) Egor. = C. jankowskii

blepharicarpa Franch. - 5

x **bogstadensis** Kuk. - C. rhynchophysa C.A. Mey. x C. vesicaria L.

bohemica Schreb. (*C. cyperoides* Murr., *Vignea bohemica* (Schreb.) Sojak) - 1, 2, 3, 4, 5, 6

bonanzensis Britt. (*Vignea bonanzensis* (Britt.) Sojak) - 3, 4, 5

bordzilowskii V. Krecz. = C. schkuhrii

bostrychostigma Maxim. - 5

brachylepis Turcz. ex Bess. = C. media

brevicollis DC. - 1, 2

brizoides L. - 1

brunnescens (Pers.) Poir. (*C. brunnescens* subsp. *vitilis* (Fries) Kalela, *C. vitilis* Fries, *Vignea brunnescens* (Pers.) Sojak, *V. brunnescens* subsp. *vitilis* (Fries) Sojak) - 1, 2(?), 3, 4, 5, 6
- subsp. **pacifica** Kalela (*Vignea brunnescens* subsp. *pacifica* (Kalela) Sojak) - 5
- subsp. *vitilis* (Fries) Kalela = C. brunnescens

bucculenta V. Krecz. = C. maritima

bucharica Kuk. (*Kobresia bucharica* (Kuk.) Ivanova) - 6

buschiorum V. Krecz. = C. humilis

buekii Wimm. - 1, 2

buxbaumii Wahlenb. - 1, 2, 3, 4, 6
- subsp. *alpina* (C. Hartm.) Liro = C. adelostoma
- subsp. *hartmanii* (Cajand.) Domin = C. hartmanii
- subsp. *mutica* (C. Hartm.) Isoviita = C. adelostoma
- var. *alpina* C. Hartm. = C. adelostoma
- var. *mutica* C. Hartm. = C. adelostoma

caespitosa auct. = C. cespitosa

callitrichos V. Krecz. (*C. nanella* Ohwi var. *callitrichos* (V. Krecz.) Worosch.) - 4, 5

camptotropa V. Krecz. = C. maritima

campylorhina V. Krecz. (*C. auriculata* Franch. 1895, non Bailey, 1889) - 5

canescens auct. = C. cinerea

canescens L. subsp. *macilenta* (Fries) K. Richt. = C. x macilenta

capillacea auct. = C. jankowskii

capillacea Boott subsp. *aomorensis* (Franch.) Egor. = C. aomorensis

- var. *sachlinensis* (Fr.Schmidt) Ohwi = C. sachalinensis

capillaris L. (*C. capillaris* subsp. *chlorostachys* (Stev.) A. Löve, D. Löve & Raymond, p.p., *C. capillaris* var. *chlorostachys* (Stev.) Grossh., *C. chlorostachys* Stev., *C. karoi* (Freyn) Freyn) - 1, 2, 3, 4, 5, 6

- subsp. *chlorostachys* (Stev.) A. Löve, D. Löve & Raymond, p.p. = C. capillaris

- subsp. *fuscidula* (V. Krecz. ex Egor.) A. & D. Löve = C. fuscidula

- subsp. *ledebouriana* (C.A. Mey. ex Trev.) Worosch. = C. ledebouriana

- var. *chlorostachys* (Stev.) Grossh. = C. capillaris

capitata L. (*Vignea capitata* (L.) Sojak) - 1, 2, 3, 4, 5

- subsp. *arctogena* (H. Smith) Bocher = C. arctogena

- subsp. *arctogena* (H. Smith) Hiit. = C. arctogena

- f. *arctogena* (H. Smith) Raymond = C. arctogena

capitellata Boiss. & Bal. - 2

capituliformis Meinsh. ex Maxim. (*C. onoei* Franch. & Savat. subsp. *capituliformis* (Meinsh. ex Maxim.) Egor., *C. onoei* var. *capituliformis* (Meinsh. ex Maxim.) Kitag., *C. onoei* auct.) - 5

capricornis Meinsh. ex Maxim. - 4, 5

caryophyllea Latourr. (*C. ruthenica* V. Krecz., *C. scabricuspis* V. Krecz., *C. verna* Chaix) - 1, 2, 3, 4, 6

- subsp. *conspissata* (V. Krecz.) Hamet-Ahti = C. conspissata

- var. *microtricha* (Franch.) Kuk. = C. microtricha

caucasica Stev. (*C. karacolica* Polozh., *C. urbis-malorum* M. Pop.) - 1, 2, 3, 6

- subsp. *perfusca* (V. Krecz.) T. Koyama = C. aterrima

cenantha A.E. Kozhevnikov - 5

cespitosa L. (*C. cespitosa* var. *inumbrata* (V. Krecz.) K. Eichw., *C. cespitosa* f. *retorta* (Fries) Sylven, *C. inumbrata* V. Krecz., *C. retorta* (Fries) V. Krecz., *C. rubra* Levl. & Vaniot, *C. caespitosa* auct.) - 1, 2, 3, 4, 5, 6

- subsp. *dacica* (Hauff.) K. Richt. = C. dacica

- subsp. *minuta* (Franch.) Worosch. = C. minuta

- var. *inumbrata* (V. Krecz.) K. Eichw. = C. cespitosa

- f. *retorta* (Fries) Sylven = C. cespitosa

chalcodeta V. Krecz. = C. pallescens

chamarensis Egor. = C. heterolepis

chamissonis Meinsh. - 5

charkeviczii A.E. Kozhevnikov - 5

chinganensis Litv. = C. chloroleuca

chishimana Ohwi = C. glareosa

chloroleuca Meinsh. (*C. chinganensis* Litv.) - 4, 5

chlorostachys Stev. = C. capillaris

chordorrhiza Ehrh. - 1, 2, 3, 4, 5, 6

chosenica Ohwi (*C. dahurica* Kuk. subsp. *chosenica* (Ohwi) Egor.) - 5

cilicica Boiss. - 2

cinerascens Kuk. (*C. micrantha* Kuk.) - 4, 5

cinerea Poll. (*C. curta* Good., *C. hylaea* V. Krecz., *Vignea cinerea* (Poll.) Dostal, *C. canescens* auct.) - 1, 2, 3, 4, 5, 6

▪ circinata C.A. Mey.

x **clausa** Holmb. - C. cinerea Poll. x C. lapponica O. Lang

coarcta Boott = C. divisa

colchica J. Gay (*Vignea colchica* (J. Gay) Sojak) - 1, 2, 6

- subsp. *ligerica* (J. Gay) Egor. = C. ligerica

compacta Lam. = C. vulpina

compacta sensu V. Krecz. = C. otrubae

concolor R. Br. (*C. aquatilis* Wahlenb. subsp. *stans* (Drej.) Hult., *C. stans* Drej., *C. uzoni* Kom.) - 1, 3, 4, 5

x **connectens** Holmb. - C. limosa L. x C. paupercula Michx.

consimilis H.T. Holm = C. lugens

conspissata V. Krecz. (*C. caryophyllea* Latourr. subsp. *conspissata* (V. Krecz.) Hamet-Ahti) - 4, 5

contigua Hoppe (*C. lumnitzeri* Rouy, *C. muricata* L. subsp. *lumnitzeri* (Rouy) Soo, *C. pairaei* F. Schultz subsp. *lumnitzeri* (Rouy) Soo, ? *C. spicata* Huds. nom. ambig., *C. spicata* subsp. *lumnitzeri* (Rouy) Soo, *Vignea spicata* (Huds.) Sojak) - 1, 2, 3, 4, 5(alien?)

coriophora Fisch. & C.A. Mey. - 3, 4

cryptocarpa C.A. Mey. (*C. lyngbyei* Hornem. subsp. *crypto-carpa* (C.A. Mey.) Hult., *C. pedunculifera* Kom., *C. riabushinskii* Kom., *C. lyngbyei* auct., *C. lyngbyei* sensu Worosch. p.p.) - 5

x **csomadensis** Simonk. - C. riparia Curt. x C. vesicaria L.

cuprina (Sandor ex Heuff.) Th. Nendtv. ex A. Kerner = C. muricata

curaica Kunth (*Vignea curaica* (Kunth) Sojak) - 3, 4

- subsp. *pycnostachya* (Kar. & Kir.) Egor. = C. pycnostachya

curta Good. = C. cinerea

curvula All. - 1

cuspidata Host (*C. flacca* Schreb. subsp. *cuspidata* (Host) Vicioso, *C. flacca* subsp. *serrulata* (Biv.) Greuter, *C. glauca* Scop. subsp. *serrulata* (Biv.) Arcang. comb. superfl., *C. glauca* subsp. *serrulata* (Biv.) K. Richt., *C. serrulata* Biv.) - 1, 2

cyperoides Murr. = C. bohemica

czarwakensis Litv. = C. diluta

dacica Heuff. (*C. bigelowii* Torr. ex Schwein. subsp. *dacica* (Heuff.) Egor., *C. bigelowii* subsp. *nardeticola* Holub, *C. bigelowii* subsp. *rigida* (Good.) W. Schultze-Motel, p.p. excl. typo, *C. cespitosa* L. subsp. *dacica* (Heuff.) K. Richt., *C. fusca* All. subsp. *dacica* (Heuff.) Grint., *C. nigra* (L.) Reichard subsp. *dacica* (Heuff.) Soo) - 1

dahurica Kuk. - 4, 5

- subsp. *chosenica* (Ohwi) Egor. = C. chosenica

davalliana Smith (*C. scabra* Hoppe) - 1

decaulescens V. Krecz. = C. popovii

delicata Clarke (*C. karoi* sensu V. Krecz.) - 3, 4, 5(?)

- subsp. *selengensis* (Ivanova) Egor. = C. selengensis

demissa Hornem. - 1

depauperata Curt. ex With. - 1, 2

depressa auct. = C. transsilvanica

depressa Link subsp. *transsilvanica* (Schur) Egor. = C. transsilvanica

- subsp. *transsilvanica* (Schur) K. Richt. = C. transsilvanica

x **descendens** Kuk. - C. apppendiculata (Trautv. & C.A. Mey.) Kuk. x C. juncella (Fries) Th. Fries

diandra Schrank (*Vignea diandra* (Schrank) Sojak) - 1, 2, 3, 4, 5, 6

diastena V. Krecz. (*C. dominii* auct.) - 5

dichroa (Freyn) V. Krecz. (*C. pamirensis* Clarke ex B. Fedtsch. subsp. *dichroa* (Freyn) Malysch.) - 2, 3, 4, 6

- subsp. *pamirensis* (Clarke ex B. Fedtsch.) Egor. = C. pamirensis

dichroandra V. Krecz. = C. acuta

digitata L. - 1, 2, 3

- var. *pallens* Fristedt = C. pallens

diluta Bieb. (*C. czarwakensis* Litv., *C. karelinii* Meinsh.) - 1, 2, 3, 4, 6

dimorphotheca Stschegl. (*C. duriusculiformis* V. Krecz., *C. rigescens* (Franch.) V. Krecz. p.p. quoad nomen, *C. stenophylla* Wahlenb. subsp. *stenophylloides* (V.

Krecz.) Egor., *C. stenophylloides* V. Krecz.) - 2, 6
dioica L. - 1, 3, 4
- subsp. *gynocrates* (Wormsk.) Hult. = C. gynocrates
diplasiocarpa V. Krecz. - 4, 5
discolor Nyl. = C. salina
dispalata Boott - 5
disperma Dew. (*C. dominii* Levl. & Vaniot, *Vignea disperma* (Dew.) Sojak) - 1, 3, 4, 5
dissitiflora Franch. - 5
distans L. - 1, 2
- subsp. *aspratilis* (V. Krecz.) Egor. = C. aspratilis
disticha Huds. (*C. disticha* subsp. *grossheimii* (V. Krecz.) Egor., *C. grossheimii* V. Krecz.) - 1, 2, 3, 4, 6
- subsp. *grossheimii* (V. Krecz.) Egor. = C. disticha
- subsp. *lithophila* (Turcz.) Hamet-Ahti = C. lithophila
diversicolor auct. = C. flacca
divisa Huds. (*C. coarcta* Boott, *C. divisa* var. *coarcta* (Boott) T. Koyama) - 1, 2, 6
- var. *coarcta* (Boott) T. Koyama = C. divisa
divulsa Stokes (*Vignea divulsa* (Stokes) Reichenb.) - 1, 2, 6
- subsp. *leersii* (Kneuck.) Walo Koch = C. polyphylla
doenitzii Boeck. - 5
dolichocarpa C.A. Mey. ex V. Krecz. - 5
dominii auct. = C. diastena
dominii Levl. & Vaniot = C. disperma
doniana Spreng. - 5
drymophila Turcz. ex Steud. - 4, 5
duriuscula C.A. Mey. (*C. eleocharis* Bailey, *C. rigescens* (Franch.) V. Krecz. p.p. excl. typo, *C. stenophylla* Wahlenb. subsp. *eleocharis* (Bailey) Hult., *Vignea duriuscula* (C.A. Mey.) Sojak) - 3, 4, 5
duriusculiformis V. Krecz. = C. dimorphotheca
echinata Murr. (*C. stellulata* Good., *C. muricata* sensu V. Krecz.) - 1, 2
egena Levl. & Vaniot (*C. filipes* Franch. & Savat. subsp. *oligostachys* (Maxim.) T. Koyama, *C. oligostachys* Meinsh. ex Maxim. 1886, non *C. oligostachya* Nees, 1834) - 5
elata All. - 1
- subsp. *omskiana* (Meinsh.) Jalas = C. omskiana
eleocharis Bailey = C. duriuscula
eleusinoides Turcz. ex Kunth - 3, 4, 5
elongata L. - 1, 2, 3, 4
emasculata V. Krecz. = C. hartmanii
enervis C.A. Mey. (*C. similigena* V. Krecz., *Vignea enervis* (C.A. Mey.) Sojak) - 3, 4, 6
- subsp. *acrifolia* (V. Krecz.) Egor. = C. acrifolia
ensifolia Turcz. ex V. Krecz (*C. bigelowii* Torr. ex Schwein. subsp. *ensifolia* (Gorodk.) Holub, *C. rigida* Good. subsp. *ensifolia* Gorodk.) - 1, 3, 4
- subsp. *altaica* (Gorodk.) Malysch. = C. altaica
- subsp. *arctisibirica* Jurtz. = C. arctisibirica
- subsp. *lugens* (H.T. Holm) Worosch. = C. lugens
- subsp. *soczavaeana* (Gorodk.) Worosch. = C. soczavaeana
erawinensis Korotk. - 4
eremopyroides V. Krecz. - 4
ericetorum Poll. (*C. approximata* All., *C. ericetorum* subsp. *approximata* (All.) K. Richt.) - 1, 2, 3, 4
- subsp. *approximata* (All.) K. Richt. = C. ericetorum
eriophylla (Kuk.) Kom. = C. raddei
erythrobasis Levl. & Vaniot - 5
euxina (Woronow & Marc.) V. Krecz. = C. transsilvanica
extensa Good. - 1, 2
falcata Turcz. - 4, 5
fedia Nees - 6
fedtschenkoana Kuk. = C. songorica
ferruginea Scop. - 1

filipes Franch. & Savat. subsp. *oligostachys* (Maxim.) T. Koyama = C. egena
firma Host. - 1
flacca Schreb. (*C. glauca* Scop., *C. diversicolor* auct.) - 1
- subsp. *cuspidata* (Host) Vicioso = C. cuspidata
- subsp. *serrulata* (Biv.) Greuter = C. cuspidata
flava L. (*C. flavella* V. Krecz., *C. nevadensis* Boiss. & Reut. subsp. *flavella* (V. Krecz.) Podlech) - 1, 2, 4
flavella V. Krecz. = C. flava
flavocuspis Franch. & Savat. (*C. spectabilis* Dew. subsp. *flavocuspis* (Franch. & Savat.) T. Koyama) - 5(?)
- subsp. *krascheninnikovii* (Kom. ex V. Krecz.) Egor. = C. krascheninnikovii
foliabunda A.E. Kozhevnikov - 5
foliosissima Fr. Schmidt - 5
forficula Franch. & Savat. - 5
x **fridtzi** Holmb. - C. brunnescens (Pers.) Poir. x C. parallela (Laest.) Sommerf.
x **friesii** Blytt. - C. rhynchophysa C.A. Mey. x C. rostrata Stokes
fujitae Kudo = C. livida
fuliginosa Schkuhr - 1
- subsp. *misandra* (R. Br.) W. Dietr. = C. misandra
- subsp. *misandra* (R. Br.) Nym. = C. misandra
fusca All. subsp. *dacica* (Heuff.) Grint. = C. dacica
fuscidula V. Krecz. ex Egor. (*C. capillaris* L. subsp. *fuscidula* (V. Krecz. ex Egor.) A. & D. Löve, *C. capillaris* auct. p.p., *C. lenaensis* sensu V. Krecz. p.p.) - 1, 3, 4, 5
fusco-cuprea (Kuk.) V. Krecz. = C. limosa
fuscovaginata Kuk. = C. acuta
x **fussii** Simonk. - C. elongata L. x C. paniculata L.
gaudichaudiana Kuk. subsp. *appendiculata* (Trautv. & C.A. Mey.) A. & D. Löve = C. appendiculata
- var. *appendiculata* (Trautv. & C.A. Mey.) T. Koyama = C. appendiculata
x **gaudiniana** Guthm. - C. dioica L. x C. C. echinata Murr.
geantha Ohwi = C. jacens
x **gerhardtii** Figert. - C. echinata Murr. x C. remota L.
glabrescens Ohwi - 5
glacialis Mackenz. - 1, 3, 4, 5
glareosa Wahlenb. (*C. chishimana* Ohwi, *C. glareosa* subsp. *chishimana* (Ohwi) Worosch., *C. soriofkensis* Levl. & Vaniot, *C. ushishirensis* Ohwi, *C. marina* sensu V. Krecz.) - 1, 4, 5
- subsp. *chishimana* (Ohwi) Worosch. = C. glareosa
- subsp. *marina* (Dew.) A. & D. Löve = C. marina
- subsp. *pribylovensis* (J.M. Macoun) Halliday & Chater = C. pribylovensis
glauca Scop. = C. flacca
- subsp. *serrulata* (Biv.) Arcang. = C. cuspidata
- subsp. *serrulata* (Biv.) K. Richt. = C. cuspidata
glauciformis Meinsh. - 4, 5
glehnii Fr. Schmidt = C. parciflora
globularis L. - 1, 3, 4, 5
gmelinii Hook. & Arn. - 5
gorodkovii V. Krecz. = C. ktausipali
gotoi Ohwi (*C. songorica* Kar. & Kir. subsp. *gotoi* (Ohwi) M. Pop., *C. sukaczovii* V. Krecz., *C. haematostachys* auct.) - 4, 5
graciliformis V. Krecz. = C. acuta
graciliformis Ohwi = C. kirganica
gracilis Curt. = C. acuta
x **grahamii** Boott - C. saxatilis L. x C vesicaria L.
griffithii Boott (*C. oliveri* Boeck., *C. nivalis* Boott var. *griffithii* (Boott) T. Koyama) - 6
griffithii sensu V. Krecz. = C. tianschanica
grioletii Roem. - 2

grossheimii V. Krecz. = C. disticha
x **grossii** Fiek. - C. hirta L. x C. vesicaria L.
gynocrates Wormsk. (*C. dioica* L. subsp. *gynocrates* (Wormsk.) Hult., *Vignea gynocrates* (Wormsk.) Sojak) - 4, 5
haematostachys auct. = C. gotoi
hakkodensis Franch. (? *C. rhizopoda* Maxim.) - 5
halleri Gunn. p.p. = C. norvegica
halleriana Asso - 1, 2
halophila Nyl. = C. recta
hancockiana Maxim. - 3, 4
hartmanii Cajand. (*C. buxbaumii* Wahlenb. subsp. *hartmanii* (Cajand.) Domin, *C. emasculata* V. Krecz.) - 1, 2, 3, 6
heleonastes Ehrh. (*C. aa* Kom.) - 1, 2, 3, 4, 5
x **helvola** Blytt. - C. cinerea Poll. x C. tripartita All.
hepburnii Boott (*C. nardina* Fries subsp. *hepburnii* (Boott) A. Löve, D. Löve & Kapoor, *Kobresia hepburnii* (Boott) Ivanova, *Vignea hepburnii* (Boott) Sojak, *Carex nardina* auct.) - 5
heterolepis Bunge (*C. chamarensis* Egor.) - 4, 5
x **heterophyta** Holmb. - C. cinerea Poll. x C. heleonastes L.
hindsii Clarke (*C. kelloggii* auct.) - 5
hirta L. - 1, 2
holostoma Drej. - 1, 4, 5
holotricha Ohwi - 5
hordeistichos Vill. - 1, 2
hostiana DC. - 1, 2
hudsonii auct. = C. omskiana
huetiana Boiss. (*C. umbrosa* Host subsp. *huetiana* (Boiss.) Soo, *C. umbrosa* subsp. *huetiana* (Boiss.) Egor. comb. superfl.) - 2
humilis Leyss. (*C. buschiorum* V. Krecz.) - 1, 2, 3, 4
- var. *scirrobasis* (Kitag.) Y.L. Chang & Y.L. Yang = C. scirrobasis
hylaea V. Krecz. = C. cinerea
hypaneura V. Krecz. - 2
hyperborea Drej. = C. bigelowii
hypochlora auct. = C. pseudosabynensis
hypochlora Freyn = C. leucochlora
ikonnikovii Egor. (*C. montis-everesti* auct.) - 6
iljinii V. Krecz. - 3, 4
incisa Boott - 5
inflata Huds. = C. vesicaria
inflata sensu V. Krecz. = C. rostrata
infuscata Nees (*C. trispiculata* auct.) - 6
insaniae Koidz. - 5
inumbrata V. Krecz. = C. cespitosa
irrigua (Wahlenb.) Smith ex Hoppe = C. paupercula
jacens Clarke (*C. geantha* Ohwi) - 5
jacutica V. Krecz. - 4, 5
jaluensis Kom. - 5
jankowskii Gorodk. (*C. biwensis* Franch. subsp. *jankowskii* (Gorodk.) Egor., *C. biwensis* auct., *C. capillacea* auct.) - 5
japonica Thunb. - 5
jemtlandica (Palmgr.) Palmgr. (*C. lepidocarpa* Tausch subsp. *jemtlandica* Palmgr.) - 1
jucunda V. Krecz. = C. maritima
juncella (Fries) Th. Fries (*C. vulgaris* Fries * *juncella* Fries, *C. juncella* subsp. *wiluica* (Meinsh.) Egor., *C. nigra* (L.) Reichard subsp. *juncella* (Fries) Lemke, *C. nigra* subsp. *juncella* (Fries) A. & D. Löve, comb. superfl., *C. nigra* subsp. *wiluica* (Meinsh.) A. & D. Löve, *C. stolonifera* Hoppe subsp. *juncella* (Fries) O. Schwarz, *C. wiluica* Meinsh.) - 1, 3, 4, 5
- subsp. *wiluica* (Meinsh.) Egor. = C. juncella

kabanovii V. Krecz. - 5
kamtschatica (Gorodk.) V. Krecz. = C. soczavaeana
karacolica Polozh. = C. caucasica
karafutoana Ohwi - 5
karelinii Meinsh. = C. diluta
- var. *aspratilis* (V. Krecz.) Serg. = C. aspratilis
karoi (Freyn) Freyn = C. capillaris
- subsp. *selengensis* (Ivanova) Egor. = C. selengensis
- var. *selengensis* (Ivanova) Serg. = C. selengensis
karoi sensu V. Krecz. = C. delicata
kattegatensis Fries ex Lindm. = C. recta
kelloggii auct. = C. hindsii
kenkolensis Litv. = C. alexeenkoana
kirganica Kom. (*C. graciliformis* Ohwi) - 4, 5
kirilowii Turcz. = C. pediformis
knorringiae Kuk. ex V. Krecz. - 6
kobomugi Ohwi (*Vignea kobomugi* (Ohwi) Sojak) - 5
koidzumiana Ohwi = C. appendiculata
koidzumii Honda (*C. lasiocarpa* Ehrh. subsp. *occultans* (Franch.) Hult., *C. occultans* (Franch.) V. Krecz.) - 5
komarovii Koidz. = C. tenuiformis
koraginensis Meinsh. = C. riishirensis
koreana Kom. = C. tenuiformis
korkischkoae A.E. Kozhevnikov - 5
korshinskyi Kom. (*C. supina* Wahlenb. subsp. *korshinskyi* (Kom.) Hult.) - 3, 4, 5
koshewnikowii Litv. - 6
kotschyana Boiss. & Hohen. (*C. orbicularis* Boott subsp. *kotschyana* (Boiss. & Hohen.) Egor. comb. superfl., *C. orbicularis* subsp. *kotschyana* (Boiss. & Hohen.) Kukkonen) - 2
krascheninnikovii Kom. ex V. Krecz. (*C. flavocuspis* Franch. & Savat. subsp. *krascheninnikovii* (Kom. ex V. Krecz.) Egor., *C. melanostoma* Fisch. ex V. Krecz. p.p. excl. typo) - 5
krausei Boeck. (*C. lenaensis* sensu V. Krecz. p.p.) - 1, 3, 4, 5
kreczetoviczii Egor. (*C. laeviculmis* auct.) - 4, 5
ktausipali Meinsh. (*C. gorodkovii* V. Krecz., *stenantha* auct.) - 5
x **kurilensis** Ohwi. - C. cespitosa L. x C. cryptocarpa C.A. Mey.
x **kyyhkynenii** Hiden. - C. cinerea Poll. x C. tenuiflora Wahlenb.
lachenalii Schkuhr (*C. lagopina* Wahlenb., *C. leporina* L. nom. ambig., *Vignea lachenalii* (Schkuhr) Sojak, *Carex bipartita* sensu Malysch., *C. tripartita* auct.) - 1, 3, 4, 5
laeviculmis auct. = C. kreczetoviczii
laevissima Nakai - 5
lagopina Wahlenb. = C. lachenalii
lanceata Dew. = C. salina
lanceolata Boott (*C. lanceolata* var. *pseudolanceolata* (V. Krecz.) Worosch., *C. prevernalis* Kitag., *C. pseudolanceolata* V. Krecz., *C. subpediformis* (Kuk.) Suto & Suzuki) - 4, 5
- var. *pseudolanceolata* (V. Krecz.) Worosch. = C. lanceolata
lancibracteata A.E. Kozhevnikov - 5
lapponica O. Lang - 1, 3, 4, 5
lasiocarpa Ehrh. - 1, 2, 3, 4, 5, 6
- subsp. *occultans* (Franch.) Hult. = C. koidzumii
latifrons V. Krecz. (*C. sylvatica* Huds. subsp. *latifrons* (V. Krecz.) O. Nilsson) - 2
latisquamea Kom. - 5
laxa Wahlenb. - 1, 3, 4, 5
ledebouriana C.A. Mey. ex Trev. (*C. capillaris* L. subsp. *ledebouriana* (C.A. Mey. ex Trev.) Worosch., *C. le-*

naensis Kuk., *C. seiskoensis* Freyn) - 1, 3, 4, 5, 6
- subsp. **substepposa** Malysch. - 3
- subsp. *tenuiformis* (Levl. & Vaniot) Egor. = C. tenuiformis
- subsp. **transbaicalensis** Malysch. - 4, 5
leersiana Rauschert = C. polyphylla
leersii F. Schultz = C. polyphylla
leiogona Franch. (*C. semiplena* Kuk.) - 5
leiorhyncha C.A. Mey. (*Vignea leiorhyncha* (C.A. Mey.)
   Sojak) - 4, 5
lenaensis Kuk. = C. ledebouriana
lenaensis sensu V. Krecz. = C. fuscidula and C. krausei
lepidocarpa Tausch (*C. viridula* Michx. var. *lepidocarpa*
   (Tausch) B. Schmid) - 1
- subsp. *jemtlandica* Palmgr. = C. jemtlandica
leporina L. = C. lachenalii
leporina sensu V. Krecz. = C. ovalis
x **leptoblasta** Holmb. - C. brunnescens (Pers.) Poir. x C.
   disperma Dew.
leucochlora Bunge (*C. hypochlora* Freyn, *C. umbrosa* Host
   subsp. *hypochlora* (Freyn) Egor. p.p. quoad nomen) - 5
x **lidii** Flatb. - C. cinerea Poll. x C. chordorrhiza Ehrh.
ligerica J. Gay (*C. colchica* J. Gay subsp. *ligerica* (J. Gay)
   Egor., *C. pseudoarenaria* Reichenb., *Vignea ligerica* (J.
   Gay) Sojak) - 1
limosa L. (*C. fusco-cuprea* (Kuk.) V. Krecz.) - 1, 2, 3, 4, 5
- var. *irrigua* Wahlenb. = C. paupercula
- var. *livida* Wahlenb. = C. livida
- var. *rariflora* Wahlenb. = C. rariflora
lineolata Cham. ex V. Krecz. = C. schmidtii
liparocarpos Gaudin (*C. nitida* Host, 1801, non Hoppe,
   1800) - 1
- subsp. *bordzilowskii* (V. Krecz.) Egor. = C. schkuhrii
lithophila Turcz. (*C. disticha* Huds. subsp. *lithophila*
   (Turcz.) Hamet-Ahti, *Vignea lithophila* (Turcz.)
   Sojak) - 4, 5
litwinowii Kuk. - 6
livida (Wahlenb.) Willd. (*C. limosa* var. *livida* Wahlenb., *C.
   fujitae* Kudo) - 1, 3, 4, 5
loliacea L. - 1, 3, 4, 5, 6
longirostrata C.A. Mey. - 5
lucidula auct. = C. subebracteata
lugens H.T. Holm (*C. bigelowii* Torr. ex Schwein. subsp.
   lugens (H.T. Holm) Egor., *C. consimilis* H.T. Holm, *C.
   ensifolia* Turcz. ex V. Krecz. subsp. *lugens* (H.T.
   Holm) Worosch.) - 4, 5
lumnitzeri Rouy = C. contigua
lyngbyei auct. = C. cryptocarpa
lyngbyei Hornem. subsp. *cryptocarpa* (C.A. Mey.) Hult. = C.
   cryptocarpa
- var. *prionocarpa* (Franch.) Kuk. = C. prionocarpa
lyngbyei sensu Worosch. = C. cryptocarpa and C. prionocarpa
maackii Maxim. - 5
x **macilenta** Nyl. (*C. canescens* L. subsp. *macilenta* (Fries)
   K. Richt.). - C. brunnescens (Pers.) Poir. x C. loliacea
   L.
mackenziei V. Krecz. (*C. pribylovensis* sensu V. Krecz.) - 1,
   4, 5
macloviana auct. = C. pyrophila
macloviana D'Urv. subsp. *pachystachya* (Cham. ex Steud.)
   Hult. p.p. = C. pyrophila
macrocephala Willd. ex Spreng. (*Vignea macrocephala*
   (Willd. ex Spreng.) Sojak) - 5
macrochaeta C.A. Mey. - 5
macrogyna Turcz. ex Steud. - 3, 4, 5
macrostigmatica Kuk. - 4

macroura Meinsh. (*C. pediformis* C.A. Mey. subsp. *macroura* (Meinsh.) Podp., *C. pediformis* var. *macroura*
   (Meinsh.) Worosch.) - 1, 3, 4, 5, 6
- subsp. *kirilowii* (Turcz.) Malysch. = C. pediformis
magellanica auct. = C. paupercula
magellanica Lam. subsp. *irrigua* (Wahlenb.) Hiit. = C.
   paupercula
- subsp. *irrigua* (Wahlenb.) Hult. = C. paupercula
majae A. Khokhr. = C. paupercula
malyschevii Egor. - 4, 5(?)
mandshurica Meinsh. - 5
marina Dew. (*C. amblyorhyncha* V. Krecz., *C. glareosa*
   Wahlenb. subsp. *marina* (Dew.) A. & D. Löve, p.p.
   quoad nomen) - 3, 4
marina sensu V. Krecz. = C. glareosa
maritima Gunn. (*C. bucculenta* V. Krecz., *C. camptotropa*
   V. Krecz., *C. jucunda* V. Krecz., *C. orthocaula* V.
   Krecz., *C. psychroluta* V. Krecz., *C. transmarina* V.
   Krecz., *Vignea maritima* (Gunn.) Reichenb.) - 1, 3, 4
- subsp. *setina* (Christ) Egor. = C. setina
x **marshalii** A. Benn. - C. flava L. x C. saxatilis L.
martynenkoi Zolot. - 3
maximowiczii Miq. - 5
- subsp. *suifunensis* (Kom.) Worosch. = C. suifunensis
media R. Br. in Richards (*C. alpina* Sw. var. *inferalpina*
   Wahlenb., *C. angarae* Steud., *C. angarae* subsp. *brachylepis* Kalela, *C. brachylepis* Turcz. ex Bess. nom. nud.,
   C. norvegica Retz. subsp. *inferalpina* (Wahlenb.)
   Hult.) - 1, 3, 4, 5, 6
medwedewii Leskov (*C. aequivoca* V. Krecz., *C. aterrima*
   Hoppe subsp. *medwedewii* (Leskov) Egor.) - 2
meinshauseniana V. Krecz. - 2
melanantha C.A. Mey. - 3, 4, 6
melananthiformis Litv. - 2, 3, 4, 6
melanocarpa Cham. ex Trautv. - 1, 3, 4, 5
melanocephala Turcz. - 3, 4, 6
melanolepis Boeck. = C. orbicularis
melanostachya Bieb. ex Willd. (*C. ripariiformis* Litv.) - 1, 2,
   3, 4, 6
melanostoma Fisch. ex V. Krecz. = C. krascheninnikovii
   and C. nesophila
membranacea Hook. - 5
meyeriana Kunth - 3, 4, 5
michelii Host - 1, 2
micrantha Kuk. = C. cinerascens
microchaeta H.T. Holm subsp. *nesophila* (H.T. Holm) D.F.
   Murray = C. nesophila
microglochin Wahlenb. - 1, 2, 3, 4, 6
micropoda C.A. Mey. (*C. pyrenaica* Wahlenb. subsp. *micropoda* (C.A. Mey.) Hult., *C. pyrenaica* var. *micropoda*
   (C.A. Mey.) Boivin) - 5
micropodioides V. Krecz. (*C. pyrenaica* auct.) - 2
x **microstachya** Ehrh. - C. cinerea Poll. x C. dioica L.
x **microstachyoides** Montell. - C. dioica L. x C. heleonastes
   Ehrh.
x **microstyla** J. Gay. - C. brunnescens (Pers.) Poir. x. C.
   foetida All.
microtricha Franch. (*C. caryophyllea* Latourr. var. *microtricha* (Franch.) Kuk., *C. verna* Chaix var. *microtricha*
   (Franch.) Ohwi, *C. sikokiana* sensu Worosch.) - 5
middendorfii Fr. Schmidt - 5
mimula V. Krecz. = C. norvegica
mingrelica Kuk. - 2
minuta Franch. (*C. cespitosa* L. subsp. *minuta* (Franch.)
   Worosch.) - 4, 5
minutiscabra Kuk. ex V. Krecz. - 6
misandra R. Br. (*C. fuliginosa* Schkuhr subsp. *misandra* (R.

Br.) Nym., *C. fuliginosa* subsp. *misandra* (R. Br.) W. Dietr. comb. superfl.) - 1, 3, 4, 5

x **mithala** Callme. - C. cinerea Poll. x C. loliacea L.

mollicula Boott - 5

mollissima Christ - 1, 3, 4, 5

monile Tuckerm. - 5

montana L. - 1, 2(?), 3

montanensis Bailey = C. podocarpa

montano-altaica Zotov - 3

montis-everesti auct. = C. ikonnikovii

mucronata All. - 2

x **mucronulata** Holmb. - C. heleonastes Ehrh. x C. tenuiflora Wahlenb.

muricata L. (*C. cuprina* (Sandor ex Heuff.) Th. Nendtv. ex A. Kerner, *Vignea cuprina* (Sandor ex Heuff.) Sojak, *V. muricata* (L.) Reichenb.) - 1, 2, 3, 4

- subsp. *leersii* (Kneuck.) Aschers. & Graebn. = C. polyphylla

- subsp. *lumnitzeri* (Rouy) Soo = C. contigua

- var. *leersii* (F. Schultz) Kneuck. = C. polyphylla

muricata sensu V. Krecz. = C. echinata

musartiana Kuk. ex V. Krecz. = C. podocarpa

myosuroides Vill. = Kobresia myosuroides

nanella Ohwi (*C. scirrobasis* auct.) - 4, 5

- var. *callitrichos* (V. Krecz.) Worosch. = C. callitrichos

nanelliformis A.E. Kozhevnikov - 5

nardina auct. = C. hepburnii

nardina Fries subsp. *hepburnii* (Boott) A. Löve, D. Löve & Kapoor = C. hepburnii

nemurensis Franch. (*Vignea nemurensis* (Franch.) Sojak) - 5

neosachalinensis A.E. Kozhevnikov - 5

nervata Franch. & Savat. - 5

nesophila H.T. Holm (*C. arakamensis* Clarke, *C. melanostoma* Fisch. ex V. Krecz. p.p. incl. typo, *C. microchaeta* H.T. Holm subsp. *nesophila* (H.T. Holm) D.F. Murray, *C. spectabilis* auct., *C. tolmiei* auct.) - 5

neurocarpa Maxim. - 5

nevadensis Boiss. & Reut. subsp. *flavella* (V. Krecz.) Podlech = C. flava

nigra (L.) Reichard (*C. acuta* L. p.p. quoad var. *nigra* L., *C. stolonifera* Hoppe, *C. vulgaris* Fries) - 1, 3, 4

- subsp. *dacica* (Heuff.) Soo = C. dacica

- subsp. *juncella* (Fries) Lemke = C. juncella

- subsp. *juncella* (Fries) A. & D. Löve = C. juncella

- subsp. *wiluica* (Meinsh.) A. & D. Löve = C. juncella

nigricans C.A. Mey. - 5(?)

nigrita Fisch. ex Worosch. = C. pluriflora

nikolskensis Kom. (*C. papulosa* auct.) - 5

A.E. Kozhevnikov (1988, Pl. Vasc. Orient. Extr. Sov. 3 : 333) has accepted the name C. papulosa Boott. for this species.

nitida Host = C. liparocarpos

nivalis Boot var. *griffithii* (Boott) T. Koyama = C. griffithii

norvegica Retz. (*C. alpina* Sw. 1798, non Schrank, 1789, *C. halleri* Gunn. p.p. excl. typo, nom. ambig., *C. mimula* V. Krecz.) - 1, 3, 4, 5

- subsp. **conicorostrata** Kalela - 5

- subsp. *inferalpina* (Wahlenb.) Hult. = C. media

novograblenovii Kom. = C. williamsii

obtusata Liljebl. - 1, 2, 3, 4, 5, 6

occultans (Franch.) V. Krecz. = C. koidzumii

oederi auct. = C. serotina

oederi Retz. subsp. *fennica* Palmgr. = C. serotina

- subsp. *pulchella* Loennr. = C. scandinavica

- subsp. *viridula* (Michx.) Hult. = C. viridula

- var. *philocrena* (V. Krecz.) T. Koyama = C. serotina

oligantha Steud. - 2

oligosperma auct. = C. tsuishikarensis

oligosperma Michx. var. *tsuishikarensis* (Koidz. & Ohwi) Boivin = C. tsuishikarensis

oligostachys Meinsh. ex Maxim. = C. egena

oliveri Boeck. = C. griffithii

omiana Franch. & Savat. (*Vignea omiana* (Franch. & Savat.) Sojak) - 5

omskiana Meinsh. (*C. elata* All. subsp. *omskiana* (Meinsh.) Jalas, *C. hudsonii* auct.) - 1, 2, 3, 4

onoei auct. = C. capituliformis

onoei Franch. & Savat. subsp. *capituliformis* (Meinsh. ex Maxim.) Egor. = C. capituliformis

- var. *capituliformis* (Meinsh. ex Maxim.) Kitag. = C. capituliformis

ontakensis auct. = C. aomorensis

orbicularis Boott (*C. arcatica* Meinsh., *C. arcatica* f. *taldycola* (Meinsh.) Ovcz. & Czuk., ? *C. melanolepis* Boeck., *C. taldycola* Meinsh., *C. transalaica* Tzvel.) - 3, 6

- subsp. *altaica* (Gorodk.) Egor. = C. altaica

- subsp. *kotschyana* (Boiss. & Hohen.) Egor. = C. kotschyana

- subsp. *kotschyana* (Boiss. & Hohen.) Kukkonen = C. kotschyana

oreophila C.A. Mey. (*Vignea oreophila* (C.A. Mey.) Sojak) - 2

ornithopoda Willd. (*C. pedata* sensu V. Krecz. vix L. 1763, nom. dubium) - 1

orthocaula V. Krecz. = C. maritima

orthostachys C.A. Mey. = C. atherodes

otrubae Podp. (*Vignea otrubae* (Podp.) Sojak, *Carex compacta* sensu V. Krecz.) - 1, 2, 6

ovalis Good. (*C. argyroglochin* Hornem., *C. ovalis* var. *argyroglochin* (Hornem.) De Langhe & Lambinon, *Vignea ovalis* (Good.) Sojak, *Carex leporina* sensu V. Krecz.) - 1, 2, 3, 4, 5(alien)

- var. *argyroglochin* (Hornem.) De Langhe & Lambinon = C. ovalis

oxyandra (Franch. & Savat.) Kudo - 5

- var. **pauzhetica** (A.E. Kozhevnikov) A.E. Kozhevnikov (*C. pauzhetica* A.E. Kozhevnikov) - 5

oxyleuca V. Krecz. = C. atrofusca

pachystachya Cham. ex Steud. - 5

pachystachya sensu V. Krecz. = C. pyrophila

pachystylis J. Gay (*Vignea pachystylis* (J. Gay) Sojak) - 2, 6

pairaei F. Schultz subsp. *lumnitzeri* (Rouy) Soo = C. contigua

paishanensis Nakai (*C. bigelowii* Torr. ex Schwein. subsp. *paishanensis* (Nakai) Worosch.) - 5

paleacea Wahlenb. (*C. paralia* V. Krecz.) - 1

pallens (Fristedt) Harmaja (*C. digitata* L. var. *pallens* Fristedt.) - 1

pallescens L. (*C. chalcodeta* V. Krecz., *C. pallescens* var. *chalcodeta* (V. Krecz.) O.Nilsson) - 1, 2, 3, 4, 6

- var. *chalcodeta* (V. Krecz.) O. Nilsson = C. pallescens

pallida C.A. Mey. (*C. accrescens* Ohwi) - 4, 5

pamirensis Clarke ex B. Fedtsch. (*C. dichroa* (Freyn) V. Krecz. subsp. *pamirensis* (Clarke ex B. Fedtsch.) Egor.) - 6

- subsp. *dichroa* (Freyn) Malysch. = C. dichroa

panicea L. - 1, 2, 3, 4, 6

paniculata L. (*Vignea paniculata* (L.) Reichenb.) - 1, 2

- subsp. *szovitsii* (V. Krecz.) O. Nilsson = C. szovitsii

x **pannewitziana** Figert. - C. rostrata Stokes x C. vesicaria L.

papulosa auct. = C. nikolskensis

paralia V. Krecz. = C. paleacea

parallela (Laest.) Sommerf. (*Vignea parallela* (Laest.) Sojak) - 1, 3

- subsp. *redowskiana* (C.A. Mey.) Egor. = C. redowskiana
parciflora Boott (*C. glehnii* Fr. Schmidt) - 5
parva Nees - 6
pauciflora Lightf. - 1, 3, 4, 5, 6
paupercula Michx. (*C. irrigua* (Wahlenb.) Smith ex Hoppe, *C. limosa* L. var. *irrigua* Wahlenb., *C. magellanica* Lam. subsp. *irrigua* (Wahlenb.) Hiit., *C. magellanica* subsp. *irrigua* (Wahlenb.) Hult. comb. superfl., *C. majae* A. Khokhr., *C. paupercula* subsp. *irrigua* (Wahlenb.) A. & D. Löve, *C. magellanica* auct.) - 1, 2, 3, 4, 5
- subsp. *irrigua* (Wahlenb.) A. & D. Löve = C. paupercula
pauxilla V. Krecz. = C. podocarpa
pauzhetica A.E. Kozhevnikov = C. oxyandra var. pauzhetica
pedata sensu V. Krecz. = C. ornithopoda
pediformis C.A. Mey. (? *C. kirilowii* Turcz., *C. macroura* Meinsh. subsp. *kirilowii* (Turcz.) Malysch., *C. sutschanensis* Kom.) - 1, 3, 4, 5
- subsp. *macroura* (Meinsh.) Podp. = C. macroura
- subsp. *reventa* (V. Krecz.) Malysch. = C. reventa
- subsp. *rhizodes* (Blytt ex Meinsh.) Lindb. fil. = C. rhizina
- var. *macroura* (Meinsh.) Worosch. = C. macroura
- var. *reventa* (V. Krecz.) Worosch. = C. reventa
pedunculifera Kom. = C. cryptocarpa
peiktusani Kom. - 5
pendula Huds. - 1, 2
perfusca V. Krecz. = C. aterrima
pergrandis V. Krecz. & Lucznik = C. atherodes
peshemskyi Malysch. = C. williamsii
petricosa Dew. - 5
> A.E. Kozhevnikov (1988, Pl. Vasc. Orient. Extr. Sov. 3 : 298) suggested that this North American species does not occur on the investigated territory.
philocrena V. Krecz. = C. serotina
phyllostachys C.A. Mey. - 2
physocarpa C. Presl - 5
physodes Bieb. - 1, 6
- subsp. *subphysodes* (M. Pop. ex V. Krecz.) Kukkonen = C. subphysodes
x **pieperana** P. Jungl. - C. flava L. x C. lepidocarpa Tausch
pilosa Scop. - 1
x **pilosiuscula** Gobi. - C. hirta L. x C. rhynchophysa C.A. Mey.
pilulifera L. - 1
planiculmis Kom. - 5
platyrhyncha Franch. & Savat. = C. pumila
x **ploetteriana** Beyer. - C. elongata L. x C. remota L.
pluriflora Hult. (*C. rariflora* (Wahlenb.) Smith var. *pluriflora* (Hult.) Boivin, *C. rariflora* subsp. *pluriflora* (Hult.) Egor., *C. rariflora* var. *pluriflora* (Hult.) Worosch. comb. superfl., *C. nigrita* Fisch. ex Worosch. nom. invalid., *C. stygia* sensu V. Krecz. p.p.) - 5
podocarpa R. Br. (*C. behringensis* Clarke, *C. montanensis* Bailey, *C. musartiana* Kuk. ex V. Krecz., *C. pauxilla* V. Krecz.) - 4, 5
- var. *koraginensis* (Meinsh.) Worosch. = C. riishirensis
polygama Schkuhr subsp. *alpina* (C. Hartm.) Cajand. = C. adelostoma
- subsp. *mutica* (C. Hartm.) Cajand. = C. adelostoma
polyphylla Kar. & Kir. (*C. divulsa* Stokes subsp. *leersii* (Kneuck.) Walo Koch, *C. leersiana* Rauschert, *C. leersii* F. Schultz, 1870, non Willd. 1787, *C. muricata* L. var. *leersii* (F. Schultz) Kneuck., *C. muricata* subsp. *leersii* (Kneuck.) Aschers. & Graebn., *Vignea divulsa* (Stokes) Reichenb. subsp. *leersiana* (Rauschert) Dostal, *V. polyphylla* (Kar. & Kir.) Sojak) - 1, 2, 3, 4, 6
pontica Albov - 2
popovii V. Krecz. (*C. decaulescens* V. Krecz.) - 6

praecox Schreb. (*Vignea praecox* (Schreb.) Sojak) - 1, 2, 3, 4, 6
prevernalis Kitag. = C. lanceolata
pribylovensis J.M. Macoun (*C. glareosa* Wahlenb. subsp. *pribylovensis* (J.M. Macoun) Halliday & Chater, *Vignea pribylovensis* (J.M. Macoun) Sojak) - 5
pribylovensis sensu V. Krecz. = C. mackenziei
prionocarpa Franch. (*C. lyngbyei* Hornem. var. *prionocarpa* (Franch.) Kuk., *C. lyngbyei* sensu Worosch. p.p.) - 5
procerula V. Krecz. = C. saxatilis subsp. laxa
prolixa Fries = C. acuta
pseudoarenaria Reichenb. = C. ligerica
pseudobrizoides Chavand (*C. reichenbachii* Bonnet) - 1
pseudocuraica Fr. Schmidt - 4, 5
pseudocyperus L. - 1, 2, 3, 4, 6
pseudodahurica A. Khokhr. - 5
pseudofoetida Kuk. (*C. slobodovii* V. Krecz., *C. vulpinaris* auct.) - 3, 4, 6
- subsp. *acrifolia* (V. Krecz.) Kukkonen = C. acrifolia
x **pseudohelvola** Kihlm. - C. cinerea Poll. x C. mackenziei V. Krecz.
pseudolanceolata V. Krecz. = C. lanceolata
pseudololiacea Fr. Schmidt - 5
pseudosabynensis (Egor.) A.E. Kozhevnikov (*C. umbrosa* Host subsp. *pseudosabynensis* Egor., *C. umbrosa* subsp. *hypochlora* (Freyn) Egor. p.p. quoad plantas, *C. hypochlora* auct., *C. recticulmis* auct.) - 5
psychroluta V. Krecz. = C. maritima
pulchella (Loennr.) Lindm. = C. scandinavica
pulchrifolia A.E. Kozhevnikov - 5
pulla Good. var. *laxa* Trautv. = C. saxatilis subsp. laxa
pulicaris L. - 1
pumila Thunb. (*C. platyrhyncha* Franch. & Savat.) - 5
x **putjatini** Kom. - C. meyeriana Kunth x C. schmidtii Meinsh.
pycnostachya Kar. & Kir. (*C. curaica* Kunth subsp. *pycnostachya* (Kar. & Kir.) Egor.) - 3, 4, 6
pyrenaica auct. = C. micropodioides
pyrenaica Wahlenb. subsp. *micropoda* (C.A. Mey.) Hult. = C. micropoda
- var. *micropoda* (C.A. Mey.) Boivin = C. micropoda
pyrophila Gand. (*C. macloviana* D'Urv. subsp. *pachystachya* (Cham. ex Steud.) Hult. p.p. quoad pl. kamtsch., *C. macloviana* auct., *C. pachystachya* sensu V. Krecz.) - 5
quadriflora (Kuk.) Ohwi - 5
quasivaginata Clarke (*C. algida* Turcz. ex V. Krecz., *C. vaginata* Tausch subsp. *quasivaginata* (Clarke) Malysch.) - 1, 3, 4, 5
raddei Kuk. (*C. aristata* R. Br. subsp. *eriophylla* (Kuk.) Worosch. comb. invalid., *C. aristata* subsp. *raddei* (Kuk.) Kuk., *C. eriophylla* (Kuk.) Kom.) - 5
ramenskii Kom. (*C. salina* Wahlenb. subsp. *ramenskii* (Kom.) Egor., *C. subspathacea* Wormsk. ex Hornem. subsp. *ramenskii* (Kom.) Egor.) - 5
rariflora (Wahlenb.) Smith (*C. limosa* L. var. *rariflora* Wahlenb.) - 1, 3, 4, 5
- subsp. *pluriflora* (Hult.) Egor. = C. pluriflora
- subsp. *stygia* (Fries) Anderss. = C. x stygia
- subsp. *stygia* (Fries) A. & D. Löve = C. x stygia
- subsp. *stygia* (Fries) Nym. = C. x stygia
- var. *pluriflora* (Hult.) Boivin = C. pluriflora
- var. *pluriflora* (Hult.) Worosch. = C. pluriflora
recta Boott (*C. halophila* Nyl., ? *C. kattegatensis* Fries ex Lindm., ? *C. vacillans* Drej.) - 1
recticulmis auct. = C. pseudosabynensis
redowskiana C.A. Mey. (*C. parallela* (Laest.) Sommerf. subsp. *redowskiana* (C.A. Mey.) Egor., *Vignea redow-*

skiana (C.A. Mey.) Sojak) - 1, 3, 4
regeliana (Kuk.) Litv. - 6
reichenbachii Bonnet = C. pseudobrizoides
relaxa V. Krecz. - 4
remota L. - 1, 2
remotiuscula Wahlenb. (*Vignea remotiuscula* (Wahlenb.) Sojak) - 5
reptabunda (Trautv.) V. Krecz. (*Vignea reptabunda* (Trautv.) Sojak) - 3, 4
retorta (Fries) V. Krecz. = C. cespitosa
reventa V. Krecz. (*C. pediformis* C.A. Mey. subsp. *reventa* (V. Krecz.) Malysch., *C. pediformis* var. *reventa* (V. Krecz.) Worosch., *C. rhizina* Blytt ex Lindbl. subsp. *reventa* (V. Krecz.) Egor.) - 5
rhizina Blytt ex Lindbl. (*C. pediformis* C.A. Mey. subsp. *rhizodes* (Blytt ex Meinsh.) Lindb. fil., *C. rhizodes* Blytt ex Meinsh.) - 1, 2, 3
- subsp. *reventa* (V. Krecz.) Egor. = C. reventa
rhizodes Blytt ex Meinsh. = C. rhizina
rhizopoda Maxim. = C. hakkodensis
rhynchophysa C.A. Mey. - 1, 3, 4, 5, 6
riabushinskii Kom. = C. cryptocarpa
rigescens (Franch.) V. Krecz. = C. dimorphotheca and C. duriuscula
rigida Good. = C. bigelowii
- subsp. *altaica* Gorodk. = C. altaica
- subsp. *ensifolia* Gorodk. = C. ensifolia
- subsp. *inferalpina* (Laest.) Gorodk. p.p. = C. bigelowii
rigidioides (Gorodk.) V. Krecz. (*C. bigelowii* Torr. ex Schwein. subsp. *rigidioides* (Gorodk.) Egor.) - 4, 5
riishirensis Franch. (*C. koraginensis* Meinsh., *C. podocarpa* R. Br. var. *koraginensis* (Meinsh.) Worosch.) - 5
riparia Curt. - 1, 2, 3, 4, 6
ripariiformis Litv. = C. melanostachya
rochebrunii Franch. & Savat. - 5(alien)
rostrata Stokes (*C. inflata* sensu V. Krecz., *C. utriculata* auct.) - 1, 2, 3, 4, 5, 6
rotundata Wahlenb. (*C. ruesanensis* Kudo) - 1, 3, 4, 5
rubra Levl. & Vaniot = C. cespitosa
ruesanensis Kudo = C. rotundata
rugulosa Kuk. (*C. smirnovii* V. Krecz.) - 3, 4, 5
rupestris All. - 1, 2, 3, 4, 5
- subsp. *alticola* (Popl. ex Sukacz.) Worosch. = C. alticola
- subsp. *argunensis* (Turcz. ex Trev.) Worosch. = C. argunensis
ruthenica V. Krecz. = C. caryophyllea
sabulosa Turcz. ex C.A. Mey. - 4, 6
sabynensis Less. ex Kunth (*C. recticulmis* Freyn, *C. umbrosa* Host subsp. *sabynensis* (Less. ex Kunth) Kuk.) - 1, 3, 4, 5
sachalinensis Fr. Schmidt (*C. capillacea* Boott var. *sachalinensis* (Fr. Schmidt) Ohwi) - 5
sadoensis Franch. - 5
sajanensis V. Krecz. - 4
salina Wahlenb. (*C. discolor* Nyl., ? *C. lanceata* Dew.) - 1
- subsp. *ramenskii* (Kom.) Egor. = C. ramenskii
sareptana V. Krecz. = C. acuta
saxatilis L. - 1, 3, 4, 5
- subsp. **laxa** (Trautv.) Kalela (*C. pulla* Good. var. *laxa* Trautv., *C. procerula* V. Krecz., *C. saxatilis* var. *laxa* (Trautv.) Hyl. comb. superfl., *C. saxatilis* var. *laxa* (Trautv.) Ohwi) - 1, 3, 4, 5
- subsp. **saxatilis** - 1
- var. *laxa* (Trautv.) Hyl. = C. saxatilis subsp. laxa
- var. *laxa* (Trautv.) Ohwi = C. saxatilis subsp. laxa
scabra Hoppe = C. davalliana
scabricuspis V. Krecz. = C. caryophyllea

scabrifolia Steud. - 5
scabrinervia Franch. (*C. scita* Maxim. var. *scabrinervia* (Franch.) Kuk.) - 5
scandinavica E.W. Davies (*C. oederi* Retz. subsp. *pulchella* Loennr., *C. pulchella* (Loennr.) Lindm. 1918, non Berggr. 1878, *C. serotina* Merat subsp. *pulchella* (Loennr.) Ooststr., *C. viridula* Michx. subsp. *pulchella* (Loennr.) Malysch., *C. viridula* var. *pulchella* (Loennr.) B. Schmid) - 1
schkuhrii Willd. (*C. bordzilowskii* V. Krecz., *C. liparocarpos* Gaudin subsp. *bordzilowskii* (V. Krecz.) Egor.) - 1, 2
schmidtii Meinsh. (*C. lineolata* Cham. ex V. Krecz., *C. aperta* auct.) - 4, 5
x **schuetzeana** Figert (*Vignea* x *schuetzeana* (Figert) Dostal). - C. appropinquata Schum. x C. cinerea Poll.
scirpoidea Michx. - 4, 5
scirrobasis auct. = C. nanella
scirrobasis Kitag. (*C. humilis* Leyss. var. *scirrobasis* (Kitag.) Y.L. Chang & Y.L. Yang) - 5
scita auct. = C. tenuiseta
scita Maxim. var. *scabrinervia* (Franch.) Kuk. = C. scabrinervia
secalina Willd. ex Wahlenb. - 1, 2, 3, 4, 5(alien), 6
sedakowii C.A. Mey. ex Meinsh. - 3, 4, 5
seiskoensis Freyn = C. ledebouriana
selengensis Ivanova (*C. delicata* Clarke subsp. *selengensis* (Ivanova) Egor. comb. invalid., *C. karoi* (Freyn) Freyn subsp. *selengensis* (Ivanova) Egor., *C. karoi* var. *selengensis* (Ivanova) Serg.) - 4
semiplena Kuk. = C. leiogona
sempervirens Vill. (*C. sempervirens* var. *pseudotristis* Domin, *C. sempervirens* subsp. *pseudotristis* (Domin) Pawl.) - 1
- subsp. *pseudotristis* (Domin) Pawl. = C. sempervirens
- var. *pseudotristis* Domin = C. sempervirens
serotina Merat (*C. oederi* Retz. subsp. *fennica* Palmgr., *C. oederi* var. *philocrena* (V. Krecz.) T. Koyama, *C. philocrena* V. Krecz., *C. serotina* subsp. *fennica* (Palmgr.) A. & D.Löve, *C. serotina* subsp. *philocrena* (V. Krecz.) Kukkonen, *C. viridula* Michx. subsp. *serotina* (Merat) Malysch., *C. oederi* auct.) - 1, 2, 3, 4, 6
- subsp. *fennica* (Palmgr.) A. & D. Löve = C. serotina
- subsp. *philocrena* (V. Krecz.) Kukkonen = C. serotina
- subsp. *pulchella* (Loennr.) Ooststr. = C. scandinavica
serrulata Biv. = C. cuspidata
setina (Christ) V. Krecz. (*C. maritima* Gunn. subsp. *setina* (Christ) Egor., *Vignea maritima* (Gunn.) Reichenb. subsp. *setina* (Christ) Sojak) - 1, 4, 5
shimidzensis Franch. - 5
shiriyajirensis Akiyama - 5
siderosticta Hance - 5
siegertiana Uechtr. = C. atherodes
sikokiana sensu Worosch. = C. microtricha
x **silesiaca** Figert. - C. cinerea Poll. x C. paniculata L.
similigena V. Krecz. = C. enervis
sisukensis Akiyama = C. atrofusca
slobodovii V. Krecz. = C. pseudofoetida
smirnovii V. Krecz. = C. rugulosa
soczavaeana Gorodk. (*C. ensifolia* Turcz. ex V. Krecz. subsp. *soczavaeana* (Gorodk.) Worosch., *C. kamtschatica* (Gorodk.) V. Krecz.) - 4, 5
songorica Kar. & Kir. (*C. fedtschenkoana* Kuk.) - 2, 3, 4, 6
- subsp. *gotoi* (Ohwi) M. Pop. = C. gotoi
sordida Heurck & Muell. Arg. (*C. amurensis* Kuk.) - 4, 5
soriofkensis Levl. & Vaniot = C. glareosa
soyaeensis Kuk. (*C. augustinowiczii* Miensh. subsp. *soyaeensis* (Kuk.) Egor.) - 5

spaniocarpa Steud. (*C. supina* Wahlenb. subsp. *spaniocarpa* (Steud.) Hult.) - 4, 5
spectabilis auct. = C. nesophila
spectabilis Dew. subsp. *flavocuspis* (Franch. & Savat.) T. Koyama = C. flavocuspis
spicata Huds. = C. contigua
- subsp. *lumnitzeri* (Rouy) Soo = C. contigua
spongiifolia A.E. Kozhevnikov = C. appendiculata
x **squamigera** V. Krecz. & Lucznik. - C. apppendiculata (Trautv. & C.A. Mey.) Kuk. x C. schmidtii Meinsh.
stans Drej. = C. concolor
stellulata Good. = C. echinata
stenantha auct. = C. ktausipali
stenocarpa Turcz. ex V. Krecz. - 3, 4, 6
stenolepis Less. - 1
stenophylla Wahlenb. (*C. uralensis* Clarke) - 1, 3, 6
- subsp. *eleocharis* (Bailey) Hult. = C. duriuscula
- subsp. *stenophylloides* (V. Krecz.) Egor. = C. dimorphotheca
stenophylloides V. Krecz. = C. dimorphotheca
stilbophaea V. Krecz. = C. atrofusca
stipata Muehl. ex Willd. - 5
stolonifera Hoppe = C. nigra
- subsp. *juncella* (Fries) O. Schwarz = C. juncella
x **stricticulmis** Holmb. - C. dioica L. x C. glareosa Wahlenb.
strigosa Huds. - 1, 2
x **stygia** Fries (*C. rariflora* (Wahlenb.) Smith subsp. *stygia* (Fries) Anderss. comb. superfl., *C. rariflora* subsp. *stygia* (Fries) A. & D. Löve, comb. superfl., *C. rariflora* subsp. *stygia* (Fries) Nym.). - C. paupercula Michx. x C. rariflora (Wahlenb.) Smith
stygia sensu V. Krecz. p.p. = C. pluriflora
stylosa C.A. Mey. - 4, 5
subconcolor Kitag. = C. arnellii
x **subcostata** Holmb. - C. dioica L. x C. loliacea L.
subebracteata (Kuk.) Ohwi (*C. lucidula* auct.) - 4, 5
x **subpatula** Holmb. - C. brunnescens (Pers.) Poir. x C. lapponica O. Lang
subpediformis (Kuk.) Suto & Suzuki = C. lanceolata
subphysodes M. Pop. ex V. Krecz. (*C. physodes* Bieb. subsp. *subphysodes* (M. Pop. ex V. Krecz.) Kukkonen) - 6
subspathacea Wormsk. ex Hornem. - 1, 3, 4, 5
- subsp. *ramenskii* (Kom.) Egor. = C. ramenskii
subumbellata Meinsh. - 5
suifunensis Kom. (*C. maximowiczii* Miq. subsp. *suifunensis* (Kom.) Worosch.) - 5
sukaczovii V. Krecz. = C. gotoi
supermascula V. Krecz. (*C. sutschanensis* sensu V.Krecz.) - 4, 5
supina Willd. ex Wahlenb. - 1, 2, 3, 4, 6
- subsp. *korshinskyi* (Kom.) Hult. = C. korshinskyi
- subsp. *spaniocarpa* (Steud.) Hult. = C. spaniocarpa
sutschanensis Kom. = C. pediformis
sutschanensis sensu V. Krecz. = C. supermascula
sylvatica Huds. - 1, 2, 3
- subsp. *latifrons* (V. Krecz.) O. Nilsson = C. latifrons
x **sylvenii** Holmb. - C. dioica L. x C. mackenziei V. Krecz.
szovitsii V. Krecz. (*C. paniculata* L. subsp. *szovitsii* (V. Krecz.) O. Nilsson) - 2
taldycola Meinsh. = C. orbicularis
tarumensis Franch. - 5
taskanensis A. Khokhr. = C. atrofusca
tasorum Kom. = C. tenuiformis
tatjanae Malysch. - 4
tegulata Levl. & Vaniot - 5
x **tenebricans** Holmb. - C. brunnescens (Pers.) Poir. x C.

dioica L.
x **tenelliformis** Holmb. - C. cinerea Poll. x C. disperma Dew.
tenuiflora Wahlenb. (*C. arrhyncha* Franch., *Vignea tenuiflora* (Wahlenb.) Sojak) - 1, 2, 3, 4, 5
tenuiformis Levl. & Vaniot (*C. komarovii* Koidz., *C. koreana* Kom. 1901 non *C. coreana* Bailey, 1889, *C. ledebouriana* C.A. Mey. ex Trev. subsp. *tenuiformis* (Levl. & Vaniot) Egor., *C. tasorum* Kom.) - 5
tenuiseta Franch. (*C. scita* auct.) - 5
thunbergii Steud. - 5
- var. *appendiculata* (Trautv. & C.A. Mey.) Ohwi = C. appendiculata
tianschanica Egor. (*C. griffithii* sensu V. Krecz.) - 6
titovii V. Krecz. - 6
x **toezensis** Simonk. - C. melanostachya Bieb. ex Willd. x C. riparia Curt.
tolmiei auct. = C. nesophila
tomentosa L. - 1, 2, 3, 4, 6
traiziscana Fr. Schmidt - 5
transalaica Tzvel. = C. orbicularis
transcaucasica Egor. - 2
transmarina V. Krecz. = C. maritima
transsilvanica Schur (*C. depressa* Link subsp. *transsilvanica* (Schur) Egor. comb. superfl., *C. depressa* subsp. *transsilvanica* (Schur) K. Richt., *C. euxina* (Woronow & Marc.) V. Krecz., *C. depressa* auct.) - 1, 2
trautvetteriana Kom. - 4, 5
tripartita auct. = C. lachenalii
trispiculata auct. = C. infuscata
tristis Bieb. - 2
tsuishikarensis Koidz. & Ohwi (*C. oligosperma* Michx. var. *tsuishikarensis* (Koidz. & Ohwi) Boivin, *C. oligosperma* auct.) - 5
tuminensis Kom. - 5
turkestanica Regel (*C. verae* Ovcz. & Czuk.) - 3, 6
- subsp. **beleensis** Zolot. - 3
uda Maxim. - 5
ulobasis V. Krecz. - 5
umbrosa Host - 1
- subsp. *huetiana* (Boiss.) Egor. = C. huetiana
- subsp. *huetiana* (Boiss.) Soo = C. huetiana
- subsp. *hypochlora* (Freyn) Egor. = C. leucochlora and C. pseudosabynensis
- subsp. *pseudosabynensis* Egor. = C. pseudosabynensis
- subsp. *sabynensis* (Less. ex Kunth) Kuk. = C. sabynensis
ungurensis Litv. - 6
uralensis Clarke = C. stenophylla
urbis-malorum M. Pop. = C. caucasica
urostachys Franch. - 5
ursina Dew. (*Vignea ursina* (Dew.) Sojak) - 1, 3, 4, 5
ushishirensis Ohwi = C. glareosa
ussuriensis Kom. (*C. alba* Scop. var. *ussuriensis* (Kom.) T. Koyama) - 5
utriculata auct. = C. rostrata
uzoni Kom. = C. concolor
vacillans Drej. = C. recta
vaginata Tausch - 1, 2, 3, 4
- subsp. *quasivaginata* (Clarke) Malysch. = C. quasivaginata
vanheurckii Muell. Arg. - 4, 5
- subsp. **crassispiculata** (Małysch.) Malysch. (*C. vanheurckii* var. *crassispiculata* Małysch.) - 4
verae Ovcz. & Czuk. = C. turkestanica
verna Chaix = C. caryophyllea
- var. *microtricha* (Franch.) Ohwi = C. microtricha
vesicaria L. (? *C. inflata* Huds.) - 1, 2, 3, 4, 6
- subsp. *vesicata* (Meinsh.) Egor. = C. vesicata

vesicata Meinsh. (*C. vesicaria* L. subsp. *vesicata* (Meinsh.) Egor.) - 4, 5

x **vierhapperi** Beck. - C. echinata Murr. x C. remota L.

viridula Michx. (*C. oederi* Retz. subsp. *viridula* (Michx.) Hult.) - 5

- subsp. *pulchella* (Loennr.) Malysch. = C. scandinavica
- subsp. *serotina* (Merat) Malysch. = C. serotina
- var. *bergrothii* (Palmgr.) B. Schmid = C. bergrothii
- var. *lepidocarpa* (Tausch) B. Schmid = C. lepidocarpa
- var. *pulchella* (Loennr.) B. Schmid = C. scandinavica

vitilis Fries = C. brunnescens

vorobjevii A.E. Kozhevnikov - 5

vulgaris Fries = C. nigra

- *juncella* Fries = C. juncella

vulpina L. (*C. compacta* Lam.) - 1, 2, 3, 4, 6

vulpinaris auct. = C. pseudofoetida

vulpinoidea Michx. (*Vignea vulpinoidea* (Michx.) Sojak) - 1(alien)

wendelboi Nelmes (*Vignea wendelboi* (Nelmes) Sojak) - 6

williamsii Britt. (*C. novograblenovii* Kom., *C. peshemskyi* Malysch.) - 1, 3, 4, 5

wiluica Meinsh. = C. juncella

x **wolteri** Gross. - C. pseudocyperus L. x C. vesicaria L.

woroschilovii A.E. Kozhevnikov - 5

x **xanthocarpa** Degl. - C. flava L. x C. hostiana DC.

xyphium Kom. - 5

x **zahnii** Kneuck. - C. brunnescens (Pers.) Poir. x C. lachenalii Schkuhr

Chlorocyperus Rikki = Cyperus

longus (L.) Palla subsp. *badius* (Desf.) Soo = Cyperus badius

Cladium P. Br.

grossheimii Pobed. = C. martii

mariscus (L.) Pohl - 1, 2

- subsp. *martii* (Roem. & Schult.) Egor. = C. martii
- subsp. *martii* (Roem. & Schult.) Soo = C. martii

martii (Roem. & Schult.) K. Richt. (*Isolepis martii* Roem. & Schult., *Cladium grossheimii* Pobed., *C. mariscus* (L.) Pohl subsp. *martii* (Roem. & Schult.) Egor., *C. mariscus* subsp. *martii* (Roem. & Schult.) Soo, comb. invalid.) - 1, 2, 6

Cobresia auct. = Kobresia

Cyperus L. (*Chlorocyperus* Rikki)

amuricus Maxim. - 5

aureus Ten. = C. esculentus

badius Desf. (*Chlorocyperus longus* (L.) Palla subsp. *badius* (Desf.) Soo, *Cyperus longus* L. subsp. *badius* (Desf.) Murb.) - 1(?), 2

capitatus Vand. - 2

congestus Vahl = Mariscus congestus

diaphanus auct. = Pycreus setiformis

difformis L. - 1, 2, 5, 6

eragrostis Vahl = Pycreus sanguinolentus

*****esculentus** L. (*C. aureus* Ten.) - 2

ferax Rich. subsp. *transcaucasicus* Kuk. = Torulinium caucasicum

flavidus Retz. = Pycreus flavidus

flavidus sensu Schischk. = C. tenuispicus

fuscus L. (*C. haspan* auct.) - 1, 2, 3, 4, 5(alien), 6

glaber L. - 1, 2, 6

globosus All. = Pycreus flavidus

glomeratus L. - 1, 2, 3, 5, 6

haspan auct. = C. fuscus

iria L. - 6

korshinskyi Meinsh. = Pycreus sanguinolentus

laevigatus L. subsp. *distachyos* (All.) Maire & Weiller = Juncellus distachyos

latespicatus Boeck. var. *setiformis* (Korsh.) T. Koyama = Pycreus setiformis

litwinowii Meinsh. = Juncellus serotinus

longus L. - 1, 2, 6

- subsp. *badius* (Desf.) Murb. = C. badius

nipponicus Franch. & Savat. = Dichostylis nipponica

orthostachyus Franch. & Savat. (*C. truncatus* Turcz. ex Ledeb. 1852, non A. Rich. 1850) - 4, 5

polystachyos Rottb. = Pycreus polystachyos

rotundus L. - 2, 6

sanguinolentus Vahl = Pycreus sanguinolentus

- var. *korshinskyi* (Meinsh.) Kuk. = Pycreus sanguinolentus
- var. *pratorum* (Korotk.) Kuk. = Pycreus pratorum

soongoricus Kar. & Kir. - 3

tenuispicus Steud. (*C. flavidus* sensu Schischk.) - 6

truncatus Turcz. ex Ledeb. = C. orthostachyus

Dichostylis Beauv. ex Lestib.

hamulosa (Bieb.) Nees = Mariscus hamulosus

limosa (Maxim.) A.E. Kozhevnikov (*Pycreus limosus* (Maxim.) Schischk.) - 5

micheliana (L.) Nees (*D. wolgensis* A. Tarass.) - 1, 2, 3, 4, 5, 6

nipponica (Franch. & Savat.) Palla (*Cyperus nipponicus* Franch. & Savat.) - 5

pygmaea (Rottb.) Nees - 2, 6

wolgensis A. Tarass. = D. micheliana

Eleocharis R. Br. (*Heleocharis* R. Br. corr. Lestib.)

acicularis (L.) Roem. & Schult. - 1, 2, 3, 4, 5, 6

- subsp. *yokoscensis* (Franch. & Savat.) Egor. = E. yokoscensis

afflata Steud. = E. pellucida

argyrolepidoides Zinserl. = E. mitracarpa

argyrolepis Kier. (*E. crassa* sensu Zinserl. 1935, p.p. quoad pl. ex As. Med.) - 6

atropurpurea (Retz.) C. Presl - 6

attenuata (Franch. & Savat.) Palla (*Scirpus attenuatus* Franch. & Savat.) - 5

austriaca Hayek (*E. leptostylopodiata* Zinserl. p.p. incl. typo, *E. mamillata* Lindb. fil. subsp. *austriaca* (Hayek) Strandhede) - 1, 2, 3

carinata Sakalo = E. uniglumis

carniolica Koch - 1

congesta D. Don var. *japonica* (Miq.) T. Koyama = E. pellucida

- var. *thermalis* (Hult.) T. Koyama = E. thermalis

crassa Fisch. & C.A. Mey. ex Zinserl. = E. palustris

crassa sensu Zinserl. 1935 = E. argyrolepis, E. mitracarpa and E. palustris

czernjajevii Zoz = E. quinqueflora

ecarinata Zinserl. = E. palustris

equisetiformis (Meinsh.) B. Fedtsch. = E. mitracarpa

eupalustris Lindb. fil. = E. palustris

euuniglumis Zinserl. = E. uniglumis

fennica Palla (*E. uniglumis* (Link) Schult. subsp. *fennica* (Palla) Egor., *E. uniglumis* subsp. *uniglumis* var. *fennica* (Palla) Hyl.) - 1

globularis Zinserl. = E. palustris

intersita Zinserl. = E. palustris and E. vulgaris
japonica Miq. = E. pellucida
kamtschatica (C.A. Mey.) Kom. (*E. kamtschatica* var. *komarovii* (Zinserl.) Worosch., *E. komarovii* Zinserl., *E. sachalinensis* (Meinsh.) Kom.) - 5
- var. *komarovii* (Zinserl.) Worosch. = E. kamtschatica
kasakstanica Zinserl. = E. palustris
klingei (Meinsh.) B. Fedtsch. (*E. korshinskyana* Zinserl., *E. scythica* Zinserl.) - 1, 4, 6
komarovii Zinserl. = E. kamtschatica
korshinskyana Zinserl. = E. klingei
lehmannii Kier. = E. mitracarpa
leptostylopodiata Zinserl. = E. austriaca and E. ussuriensis
levinae Zoz = E. palustris
macrocarpa Zoz = E. uniglumis
mamillata Lindb. fil. (*E. palustris* (L.) Roem. & Schult. var. *mamillata* (Lindb. fil.) E.I. Nyarady, *Scirpus (Heleocharis) mamillata* Lindb. fil.) - 1, 3, 4, 5, 6
- subsp. *austriaca* (Hayek) Strandhede = E. austriaca
- subsp. *ussuriensis* (Zinserl.) Egor. = E. ussuriensis
- f. *ussuriensis* (Zinserl.) Y.L. Chang = E. ussuriensis
margaritacea (Hult.) Miyabe & Kudo - 5
maximoviczii Zinserl. (*E. pellucida* C. Presl var. *maximoviczii* (Zinserl.) Ohwi) - 5
meridionalis Zinserl. (*E. quinqueflora* (F.X. Hartm.) O. Schwarz subsp. *meridionalis* (Zinserl.) Egor., *E. quinqueflora* var. *meridionalis* (Zinserl.) Raymond) - 6
mitracarpa Steud. (*E. argyrolepidoides* Zinserl., *E. equisetiformis* (Meinsh.) B. Fedtsch., ? *E. lehmannii* Kier., *E. turcomanica* Zinserl., *E. crassa* sensu Zinserl. p.p.) - 1, 2, 6
multicaulis (Smith) Desv. (*Scirpus multicaulis* Smith) - 1
multiseta Zinserl. = E. uniglumis
nipponica Makino - 5
ovata (Roth) Roem. & Schult. (*Scirpus ovatus* Roth, 1793, non Gilib. 1792, nom. invalid., *Eleocharis soloniensis* (Dubois) Hara, *Scirpus soloniensis* Dubois) - 1, 2, 3, 4, 5, 6
oxylepis (Meinsh.) B. Fedtsch. - 1, 6
oxystachys Sakalo = E. palustris
palustris (L.) Roem. & Schult. (*Scirpus palustris* L., *Eleocharis crassa* Fisch. & C.A. Mey. ex Zinserl. 1929, *E. ecarinata* Zinserl., *E. eupalustris* Lindb. fil., *E. globularis* Zinserl., *E. intersita* Zinserl. p.p. incl. typo, *E. kasakstanica* Zinserl., *E. levinae* Zoz, *E. oxystachys* Sakalo, *E. palustris* subsp. *globularis* (Zinserl.) Egor., *E. palustris* var. *globularis* (Zinserl.) A.E. Kozhevnikov, *E. palustris* subsp. *microcarpa* Walters, *E. crassa* sensu Zinserl. 1935, p.p.) - 1, 2, 3, 4, 5, 6
- subsp. *globularis* (Zinserl.) Egor. = E. palustris
- subsp. *microcarpa* Walters = E. palustris
- subsp. *vulgaris* Walters = E. vulgaris
- var. *globularis* (Zinserl.) A.E. Kozhevnikov = E. palustris
- var. *mamillata* (Lindb. fil.) E.I. Nyarady = E. mamillata
parvula (Roem. & Schult.) Bluff, Nees & Schauer (*E. parvula* subsp. *oppermannii* Zoz) - 1, 2, 4, 5, 6
- subsp. *oppermannii* Zoz = E. parvula
paucidentata Zinserl. = E. uniglumis
pauciflora (Lightf.) Link = E. quinqueflora
pellucida C. Presl (*E. afflata* Steud., *E. congesta* D. Don var. *japonica* (Miq.) T. Koyama, *E. japonica* Miq.) - 5
- var. *maximoviczii* (Zinserl.) Ohwi = E. maximoviczii
- var. *thermalis* (Hult.) Hara = E. thermalis
petasata (Maxim.) Zinserl. = E. wichurae
quinqueflora (F.X. Hartm.) O. Schwarz (*Scirpus quinqueflorus* F.X. Hartm., *Eleocharis czernjajevii* Zoz, *E. pauciflora* (Lightf.) Link, *E. meridionalis* auct. fl.

cauc.) - 1, 2, 3, 4, 5
- subsp. *meridionalis* (Zinserl.) Egor. = E. meridionalis
- var. *meridionalis* (Zinserl.) Raymond = E. meridionalis
sachalinensis (Meinsh.) Kom. = E. kamtschatica
sareptana Zinserl. = E. uniglumis
scythica Zinserl. = E. klingei
septentrionalis Zinserl. (*E. uniglumis* (Link) Schult. subsp. *septentrionalis* (Zinserl.) Egor., *E. uniglumis* subsp. *uniglumis* var. *septentrionalis* (Zinserl.) Strandhede) - 1
soloniensis (Dubois) Hara = E. ovata
svensonii Zinserl. = E. yokoscensis
tetraquetra auct. = E. wichurae
tetraquetra Nees - 5
thermalis (Hult.) Egor. (*Scirpus japonicus* (Miq.) Franch. & Savat. var. *thermalis* Hult., *Eleocharis congesta* D. Don var. *thermalis* (Hult.) T. Koyama, *E. pellucida* C. Presl var. *thermalis* (Hult.) Hara) - 5
transcaucasica Zinserl. = E. uniglumis
turcomanica Zinserl. = E. mitracarpa
tuvinica Bubnova - 4, 6
uniglumis (Link) Schult. (*Scirpus uniglumis* Link, ? *Eleocharis carinata* Sakalo, *E. euuniglumis* Zinserl., *E. macrocarpa* Zoz, *E. multiseta* Zinserl., *E. paucidentata* Zinserl., *E. sareptana* Zinserl., *E. transcaucasica* Zinserl., *E. uniglumis* var. *transcaucasica* (Zinserl.) T. Koyama, *E. zinserlingii* Zoz) - 1, 2, 3, 4, 5, 6
- subsp. *fennica* (Palla) Egor. = E. fennica
- subsp. *septentrionalis* (Zinserl.) Egor. = E. septentrionalis
- subsp. *uniglumis* var. *fennica* (Palla) Hyl. = E. fennica
- - var. *septentrionalis* (Zinserl.) Strandhede = E. septentrionalis
- var. *transcaucasica* (Zinserl.) T. Koyama = E. uniglumis
ussuriensis Zinserl. (*E. leptostylopodiata* Zinserl. p.p., *E. mamillata* Lindb. fil. subsp. *ussuriensis* (Zinserl.) Egor., *E. mamillata* f. *ussuriensis* (Zinserl.) Y.L. Chang) - 4, 5
vulgaris (Walters) A. & D. Löve (*E. palustris* (L.) Roem. & Schult. subsp. *vulgaris* Walters, *E. intersita* Zinserl. p.min.p. excl. typo) - 1
wichurae Boeck. (*E. petasata* (Maxim.) Zinserl., *E. wichurae* f. *petasata* (Maxim.) Hara, *E. tetraquetra* auct. fl. kuril.) - 4, 5
- f. *petasata* (Maxim.) Hara = E. wichurae
yokoscensis (Franch. & Savat.) Tang & Wang (*Scirpus yokoscensis* Franch. & Savat, *Eleocharis acicularis* (L.) Roem. & Schult. subsp. *yokoscensis* (Franch. & Savat.) Egor., *E. svensonii* Zinserl.) - 5
zinserlingii Zoz = E. uniglumis

Elyna Schrad. = Kobresia

myosuroides (Vill.) Fritsch = Kobresia myosuroides
sibirica Turcz. ex Ledeb. = Kobresia sibirica

Eriophorella Holub = Baeothryon

alpina (L.) Holub = Baeothryon alpinum
pumila (Vahl) Kit Tan = Baeothryon pumilum

Eriophorum L.

altaicum Meinsh. (*E. scheuchzeri* Hoppe subsp. *altaicum* (Meinsh.) N. Bondareva) - 3, 6
angustifolium Honck. = E. polystachion
- subsp. *komarovii* (V. Vassil.) Worosch. = E. komarovii
- subsp. *subarcticum* (V. Vassil.) Hult. = E. polystachion
- subsp. *triste* (Th. Fries) Hult. = E. polystachion
- var. *triste* Th. Fries = E. polystachion

asiaticum V. Vassil. = E. gracile
brachyantherum Trautv. & C.A. Mey. (*E. opacum* (Bjornstr.) Fern., *Scirpus brachyantherus* (Trautv. & C.A. Mey.) T. Koyama) - 1, 3, 4, 5
callitrix Cham. ex C.A. Mey. - 4, 5
chamissonis C.A. Mey. = E. medium and E. russeolum
- subsp. *mandschuricum* (Meinsh.) Worosch. = E. russeolum
coreanum Palla = E. gracile
eximium V. Vassil. = E. russeolum
fauriei E.G. Camus = E. vaginatum
gracile Koch (*E. asiaticum* V. Vassil., *E. coreanum* Palla, *E. gracile* subsp. *asiaticum* (V. Vassil.) Tolm. nom. provis., *E. gracile* subsp. *coreanum* (Palla) Tolm., *E. gracile* var. *coreanum* (Palla) Ohwi, *Scirpus ardea* T. Koyama, *S. ardea* var. *coreanus* (Palla) T. Koyama) - 1, 3, 4, 5
- subsp. *asiaticum* (V. Vassil.) Tolm. = E. gracile
- subsp. *coreanum* (Palla) Hult. = E. gracile
- var. *coreanum* (Palla) Ohwi = E. gracile
hudsonianum Michx. = Baeothryon alpinum
humile Turcz. ex Steud. - 3, 4, 5, 6
japonicum Maxim. = Scirpus maximowiczii
komarovii V. Vassil. (*E. angustifolium* Honck. subsp. *komarovii* (V. Vassil.) Worosch.) - 4, 5
latifolium Hoppe (*Scirpus angustifolius* (Honck.) T. Koyama subsp. *latifolius* (Hoppe) T. Koyama) - 1, 2
leucocephalum Boeck. = E. scheuchzeri
mandschuricum Meinsh. = E. russeolum
maximowiczii (Clarke) Beetle = Scirpus maximowiczii
medium Anderss. (*E. chamissonis* C.A. Mey. p.min.p. excl. lectotypo, nom. confusum) - 1, 3, 4, 5
opacum (Bjornstr.) Fern. = E. brachyantherum
polystachion L. (*E. angustifolium* Honck., *E. angustifolium* subsp. *subarcticum* (V. Vassil.) Hult., *E. angustifolium* var. *triste* Th. Fries, *E. angustifolium* subsp. *triste* (Th. Fries) Hult., *E. subarcticum* V. Vassil., *E. triste* (Th. Fries) Hadac & A. Löve, *Scirpus angustifolius* (Honck.) T. Koyama) - 1, 2, 3, 4, 5, 6
russeolum Fries (*E. chamissonis* C.A. Mey. p.p. incl. lectotypo, nom. confusum, *E. chamissonis* subsp. *mandschuricum* (Meinsh.) Worosch. comb. invalid., ? *E. eximium* V. Vassil., *E. mandschuricum* Meinsh., *E. strigosum* Miyabe & Kudo, *Scirpus chamissonis* (C.A. Mey.) T. Koyama, *S. russeolus* (Fries) T. Koyama) - 1, 3, 4, 5
scheuchzeri Hoppe (*E. leucocephalum* Boeck., *Scirpus leucocephalus* (Boeck.) T. Koyama) - 1, 3, 4, 5, 6
- subsp. *altaicum* (Meinsh.) N. Bondareva = E. altaicum
strigosum Miyabe & Kudo = E. russeolum
subarcticum V. Vassil. = E. polystachion
triste (Th. Fries) Hadac & A. Löve = E. polystachion
vaginatum L. (*E. fauriei* E.G. Camus, *E. vaginatum* subsp. *fauriei* (E.G. Camus) A. & D. Löve, *E. vaginatum* var. *fauriei* (E.G. Gamus) Kitag., *Scirpus fauriei* (E.G. Camus) T. Koyama, *S. fauriei* subsp. *vaginatus* (L.) T. Koyama) - 1, 2, 3, 4, 5, 6
- subsp. *fauriei* (E.G. Camus) A. & D. Löve ≠ E. vaginatum
- var. *fauriei* (E.G. Gamus) Kitag. = E. vaginatum

Fimbristylis Vahl

aestivalis (Retz.) Vahl (*Scirpus aestivalis* Retz., *Fimbristylis leiocarpa* Maxim.) - 5
annua (All.) Roem. & Schult. (*F. dichotama* (L.) Vahl f. *annua* (All.) Ohwi) - 2
bisumbellata (Forssk.) Bub. (*Scirpus bisumbellatus* Forssk., *Fimbristylis dichotama* auct.) - 1, 2, 6

dichotoma auct. = F. bisumbellata
dichotoma (L.) Vahl f. *annua* (All.) Ohwi = F. annua
ferruginea auct. = F. turkestanica
hispidula (Vahl) Kunth (*Scirpus hispidulus* Vahl, *Bulbostylis woronowii* Palla) - 2
leiocarpa Maxim. = F. aestivalis
makinoana Ohwi = F. velata
maracandica Zak. - 6
ochotensis (Meinsh.) Kom. - 5
quinquangularis (Vahl) Kunth - 6
schischkinii Pobed. = F. sieberiana
sieberiana Kunth (*F. schischkinii* Pobed.) - 2
squarrosa Vahl - 2, 5
subbispitata Nees & Meyen - 5
turkestanica (Regel) B. Fedtsch. (*F. ferruginea* auct.) - 6
velata R.Br. (*F. makinoana* Ohwi) - 5
verrucifera (Maxim.) Makino - 5

Heleocharis R. Br. corr. Lestib. = Eleocharis

Holoschoenus Link = Scirpoides

romanus (L.) Fritsch = Scirpoides holoschoenus
- subsp. *holoschoenus* (L.) Greuter = Scirpoides holoschoenus
vulgaris Link = Scirpoides holoschoenus

Hymenochaeta Beauv. ex Lest. = Scirpus

juncoides (Roxb.) Nakai = Scirpus juncoides
lacustris (L.) Nakai = Scirpus lacustris
tabernaemontani (C. C. Gmel.) Nakai = Scirpus tabernaemontani
triquetra (L.) Nakai = Scirpus triqueter

Isolepis R. Br. = Scirpus

martii Roem. & Schult. = Cladium martii
oryzetorum Steud. = Scirpus lateriflorus
roylei Nees = Scirpus roylei
setacea (L.) R. Br. = Scirpus setaceus

Juncellus (Griseb.) Clarke (*Acorellus* Palla)

alopecuroides auct. = J. serotinus
distachyos (All.) Turrill (*Acorellus distachyos* (All.) Palla, *Cyperus laevigatus* L. subsp. *distachyos* (All.) Maire & Weiller, *Juncellus distachyos* (All.) Egor. comb. superfl., *J. laevigatus* (L.) Clarke subsp. *distachyos* (All.) P.H. Davis, *Acorellus laevigatus* auct.) - 6
laevigatus (L.) Clarke subsp. *distachyos* (All.) P.H. Davis = J. distachyos
pannonicus (Jacq.) Clarke (*Acorellus pannonicus* (Jacq.) Palla) - 1, 2, 3, 6
serotinus (Rottb.) Clarke (*Cyperus litwinowii* Meinsh., *Juncellus alopecuroides* auct.) - 1, 2, 5, 6

Kobresia Willd. (*Elyna* Schrad., *Cobresia* auct.)

bellardii (All.) Degl. = K. myosuroides
bucharica (Kuk.) Ivanova = Carex bucharica
capillifolia auct. = K. capilliformis, K. macrolepis and K. ovczinnikovii
capillifolia (Decne.) Clarke subsp. *capilliformis* (Ivanova) Ovcz. = K. capilliformis
- subsp. *pamirica* Ovcz. = K. ovczinnikovii
capilliformis Ivanova (*K. capillifolia* (Decne.) Clarke subsp.

capilliformis (Ivanova) Ovcz., *K. capillifolia* auct. p.p.) - 3, 4, 6
filifolia (Turcz.) Clarke - 3, 4, 5
- subsp. *subfilifolia* Egor., Jurtz. & Petrovsky = K. simpliciuscula subsp. subfilifolia
hepburnii (Boott) Ivanova = Carex hepburnii
hissarica (Pissjauk.) Sojak = Schoenoxiphium hissaricum
humilis (C.A. Mey. ex Trautv.) Serg. - 6
macrolepis Meinsh. (*K. capillifolia* auct. p.p.) - 2
myosuroides (Vill.) Fiori (*Carex myosuroides* Vill., *Elyna myosuroides* (Vill.) Fritsch, *Kobresia bellardii* (All.) Degl.) - 1, 3, 4, 5, 6
ovczinnikovii Egor. (*K. capillifolia* (Decne.) Clarke subsp. *pamirica* Ovcz. nom. invalid., *K. capillifolia* auct. p.p.) - 6
pamiroalaica Ivanova - 6
paniculata Meinsh. = K. stenocarpa
persica Kuk. & Bornm. (*K. humilis* (C.A. Mey. ex Trautv.) Serg. p.p.) - 2, 6
royleana auct. = K. stenocarpa
schoenoides (C.A. Mey.) Steud. - 2
sibirica (Turcz. ex Ledeb.) Boeck. (*Elyna sibirica* Turcz. ex Ledeb.) - 1, 3, 4, 5
simpliciuscula (Wahlenb.) Mackenz. subsp. **subfilifolia** (Egor., Jurtz. & Petrovsky) Egor. (*K. filifolia* (Turcz.) Clarke subsp. *subfilifolia* Egor., Jurtz. & Petrovsky) - 3, 4, 5
- subsp. **subholarctica** Egor. (*K. simpliciuscula* var. *subholarctica* (Egor.) A.E. Kozhevnikov) - 1, 3, 4, 5
- var. *subholarctica* (Egor.) A.E. Kozhevnikov = K. simpliciuscula subsp. subholarctica
smirnovii Ivanova - 3, 4, 6
stenocarpa (Kar. & Kir.) Steud. (*K. paniculata* Meinsh., *K. royleana* auct.) - 4, 6

Kyllinga Rottb.

brevifolia Rottb. - 2, 5
- subsp. *kamtschatica* (Meinsh.) Worosch. = K. kamtschatica
- var. *leiolepis* (Franch. & Savat.) Hara = K. gracillima
gracillima Miq. (*K. brevifolia* Rottb. var. *leiolepis* (Franch. & Savat.) Hara) - 2, 5
kamtschatica Meinsh. (*K. brevifolia* Rottb. subsp. *kamtschatica* (Meinsh.) Worosch.) - 5
metzii Hochst. ex Steud. = K. squamulata
nemoralis (Forst. & Forst. fil.) Dandy ex Hutch. (*Thryocephalon nemorale* Forst. & Forst. fil.) - 2
squamulata Vahl (*K. metzii* Hochst. ex Steud.) - 2

Mariscus Vahl

congestus (Vahl) Clarke (*Cyperus congestus* Vahl, *Mariscus cyri* Grossh.) - 2
cyri Grossh. = M. congestus
hamulosus (Bieb.) Hooper (*Dichostylis hamulosa* (Bieb.) Nees, *Mariscus hamulosus* (Bieb.) Egor. comb. superfl.) - 1, 3,6

Maximoviczia A. Khokhr. = Scirpus

japonica (Maxim.) A. Khokhr. = Scirpus maximowiczii

Nomochloa Beauv. = Blysmus

compressa (L.) Beetle = Blysmus compressus
rufa (Huds.) Beetle = Blysmus rufus

Pycreus Beauv.

colchicus (C. Koch) Schischk. (*P. tremulus* auct.) - 2
eragrostis (Vahl) Palla = P. sanguinolentus
flavescens (L.) Beauv. ex Reichenb. - 1, 2, 6
flavidus (Retz.) T. Koyama (*Cyperus flavidus* Retz., *C. globosus* All. 1789, non Forrsk. 1775, *Pycreus globosus* (All.) Reichenb. nom. illegit.) - 2, 6
globosus (All.) Reichenb. = P. flavidus
korshinskyi (Meinsh.) V. Krecz. = P. sanguinolentus
limosus (Maxim.) Schischk. = Dichostylis limosa
nilagiricus (Hochst. ex Steud.) E.G. Camus - 4, 5, 6
polystachyos (Rottb.) Beauv. (*Cyperus polystachyos* Rottb.) - 5
pratorum (Korotk.) Schischk. (*Cyperus sanguinolentus* Vahl var. *pratorum* (Korotk.) Kuk.) - 5
rehmannii (Boiss.) Palla ex Grossh. = P. sanguinolentus
sanguinolentus (Vahl) Nees (*Cyperus sanguinolentus* Vahl, *C. eragrostis* Vahl, 1806, non Lam. 1791, *C. korshinskyi* Meinsh., *C. sanguinolentus* var. *korshinskyi* (Meinsh.) Kuk., *Pycreus eragrostis* (Vahl) Palla, *P. korshinskyi* (Meinsh.) V. Krecz., *P. rehmannii* (Boiss.) Palla ex Grossh.) - 2, 5, 6
setiformis (Korsh.) Nakai (*Cyperus latespicatus* Boeck. var. *setiformis* (Korsh.) T. Koyama, *Pycreus setiformis* (Korsh.) Schischk., *Cyperus diaphanus* auct.) - 5
tremulus auct. = P. colchicus

Rhynchospora Vahl

alba (L.) Vahl - 1, 2, 3, 4, 5
caucasica Palla - 2
faberi Clarke - 5
faberi sensu Worosch. = R. fujiiana
fauriei auct. = R. fujiiana
fujiiana Makino (*R. faberi* sensu Worosch., *R. fauriei* auct.) - 5
fusca (L.) Ait. fil. - 1

Schoenoplectus Palla = Scirpus

americanus auct. = Scirpus pungens
bucharicus (Roshev.) Grossh. = Scirpus bucharicus
ehrenbergii (Boeck.) Sojak = Scirpus ehrenbergii
grossheimii Pobed. = Scirpus hippolyti
hippolyti (V. Krecz.) V. Krecz. = Scirpus hippolyti
juncoides (Roxb.) Palla = Scirpus juncoides
kalmussii (Aschers., Abrom. & Graebn.) Palla = Scirpus kalmussii
komarovii (Roshev.) Ohwi = Scirpus komarovii
komarovii (Roshev.) Sojak = Scirpus komarovii
lacustris (L.) Palla = Scirpus lacustris
- subsp. *tabernaemontani* (C.C. Gmel.) A. & D. Löve = Scirpus tabernaemontani
lateriflorus (J.F. Gmel.) Lye = Scirpus lateriflorus
litoralis (Schrad.) Palla = Scirpus litoralis
- subsp. *kasachstanicus* (Dobroch.) Sojak = Scirpus kasachstanicus
lupulinus (Nees) V. Krecz. = Scirpus roylei
melanospermus (C.A. Mey.) Grossh. = Scirpus melanospermus
mucronatus (L.) Palla = Scirpus mucronatus
nipponicus (Makino) Ohwi = Scirpus nipponicus
nipponicus (Makino) Sojak = Scirpus nipponicus
oryzetorum (Steud.) Ohwi = Scirpus lateriflorus
oryzetorum (Steud.) V. Krecz. = Scirpus lateriflorus
pungens (Vahl) Palla = Scirpus pungens
roylei (Nees) Lye = Scirpus roylei

roylei (Nees) Ovcz. & Czuk. = Scirpus roylei
setaceus (L.) Palla = Scirpus setaceus
supinus (L.) Palla = Scirpus supinus
- subsp. *lateriflorus* (J.F. Gmel.) T. Koyama = Scirpus lateriflorus
- subsp. *lateriflorus* (J.F. Gmel.) Sojak = Scirpus lateriflorus
tabernaemontani (C.C. Gmel.) Palla = Scirpus tabernaemontani
triqueter (L.) Palla = Scirpus triqueter
triquetriformis V. Krecz. = Scirpus triquetriformis
validus (Vahl) Ovcz. & Czuk. p.p. = Scirpus hippolyti

Schoenoxiphium Nees

hissaricum Pissjauk. (*Kobresia hissarica* (Pissjauk.) Sojak) - 6

Schoenus L.

ferrugineus L. - 1
nigricans L. - 1, 2, 6

Scirpoides Seguier (*Holoschoenus* Link)

holoschoenus (L.) Sojak (*Holoschoenus romanus* (L.) Fritsch, *H. romanus* subsp. *holoschoenus* (L.) Greuter, *H. vulgaris* Link, *Scirpoides romana* (L.) Sojak) - 1, 2, 3, 6
romana (L.) Sojak = S. holoschoenus

Scirpus L. (*Hymenochaeta* Beauv. ex Lest., *Isolepis* R. Br., *Maximoviczia* A. Khokhr., *Schoenoplectus* Palla, *Trichophorum* Pers. p.p. excl. *T. alpino* (L.) Pers.)

aestivalis Retz. = Fimbristylis aestivalis
americanus auct. = S. pungens
angustifolius (Honck.) T. Koyama = Eriophorum polystachion
- subsp. *latifolius* (Hoppe) T. Koyama = Eriophorum latifolium
ardea T. Koyama = Eriophorum gracile
- var. *coreanus* (Palla) T. Koyama = Eriophorum gracile
asiaticus Beetle (*S. wichurae* sensu Roshev.) - 5
attenuatus Franch. & Savat. = Eleocharis attenuata
avatschensis Kom. = S. microcarpus
biconcavus Ohwi = Bolboschoenus planiculmis
bisumbellatus Forssk. = Fimbristylis bisumbellata
brachyantherus (Trautv. & C.A. Mey.) T. Koyama = Eriophorum brachyantherum
bucharicus Roshev. (*Schoenoplectus bucharicus* (Roshev.) Grossh.) - 2, 6
chamissonis (C.A. Mey.) T. Koyama = Eriophorum russeolum
colchicus Kimeridze - 2
densus Wall. = Bulbostylis densa
depauperatus Kom. = S. nipponicus
ehrenbergii Boeck. (*Schoenoplectus ehrenbergii* (Boeck.) Sojak) - 1, 3, 6
erectus sensu Roshev. = S. hotarui and S. juncoides
etuberculatus (Steud.) O. Kuntze subsp. *nipponicus* (Makino) T. Koyama = S. nipponicus
fauriei (E.G. Camus) T. Koyama = Eriophorum vaginatum
- subsp. *vaginatus* (L.) T. Koyama = Eriophorum vaginatum
grossheimii (Pobed.) Czer. = S. hippolyti
hippolyti V. Krecz. (*Schoenoplectus grossheimii* Pobed., *Sch. hippolyti* (V. Krecz.) V. Krecz., *Sch. validus*

(Vahl) Ovcz. & Czuk. p.p. excl. typo, *Scirpus grossheimii* (Pobed.) Czer., *S. validus* auct.) - 1, 2, 3, 4, 5, 6
hispidulus Vahl = Fimbristylis hispidula
hotarui Ohwi (*S. erectus* sensu Roshev. p.p., *S. juncoides* sensu Worosch.) - 5
hudsonianus (Michx.) Fern = Baeothryon alpinum
japonicus (Maxim.) Fern. = S. maximowiczii
japonicus (Miq.) Franch. & Savat. var. *thermalis* Hult. = Eleocharis thermalis
juncoides Roxb. (*Hymenochaeta juncoides* (Roxb.) Nakai, *Schoenoplectus juncoides* (Roxb.) Palla, *Scirpus erectus* sensu Roshev. p.p.) - 2, 6
juncoides sensu Worosch. = S. hotarui
kalmussii Aschers., Abrom. & Graebn. (*Schoenoplectus kalmussii* (Aschers., Abrom. & Graebn.) Palla) - 1
kasachstanicus Dobroch. (*Schoenoplectus litoralis* (Schrad.) Palla subsp. *kasachstanicus* (Dobroch.) Sojak) - 6
komarovii Roshev. (*Schoenoplectus komarovii* (Roshev.) Ohwi, *Sch. komarovii* (Roshev.) Sojak, comb. superfl.) - 5
koshewnikowii Litv. = Bolboschoenus koshewnikowii
lacustris L. (*Hymenochaeta lacustris* (L.) Nakai, *Schoenoplectus lacustris* (L.) Palla) - 1, 2, 3, 4, 6
- subsp. *tabernaemontani* (C.C. Gmel.) Syme = S. tabernaemontani
lateriflorus J.F. Gmel. (*Isolepis oryzetorum* Steud., *Schoenoplectus lateriflorus* (J.F. Gmel.) Lye, *Sch. oryzetorum* (Steud.) Ohwi, nom. altern., *Sch. oryzetorum* (Steud.) V. Krecz. comb. superfl., *Sch. supinus* (L.) Palla subsp. *lateriflorus* (J.F. Gmel.) T. Koyama, comb. superfl., *Sch. supinus* subsp. *lateriflorus* (J.F. Gmel.) Sojak, *Scirpus oryzetorum* (Steud.) Ohwi, *S. supinus* L. var. *lateriflorus* (J.F. Gmel.) T. Koyama) - 3, 6
leucocephalus (Boeck.) T. Koyama = Eriophorum scheuchzeri
lineatus Michx. subsp. *wichurae* (Boeck.) T. Koyama = S. wichurae
lineolatus Franch. & Savat. - 5
litoralis Schrad. (*Schoenoplectus litoralis* (Schrad.) Palla) - 1, 2, 6
lupulinus (Nees) Roshev. = S. roylei
mamillata Lindb. fil. = Eleocharis mamillata
maximowiczii Clarke (*Eriophorum japonicum* Maxim., *E. maximowiczii* (Clarke) Beetle, *Maximoviczia japonica* (Maxim.) A. Khokhr., *Scirpus japonicus* (Maxim.) Fern. 1905, non Franch. & Savat. 1877) - 4, 5
melanospermus C.A. Mey. (*Schoenoplectus melanospermus* (C.A. Mey.) Grossh.) - 1, 2, 3, 6
microcarpus C. Presl (*S. avatschensis* Kom.) - 5
mucronatus L. (*Schoenoplectus mucronatus* (L.) Palla) - 1, 2, 5, 6
multicaulis Smith = Eleocharis multicaulis
nipponicus Makino (*Schoenoplectus nipponicus* (Makino) Ohwi, *Sch. nipponicus* (Makino) Sojak, comb. superfl., *Scirpus depauperatus* Kom. 1901, non Poir. 1804, *S. etuberculatus* (Steud.) O. Kuntze subsp. *nipponicus* (Makino) T. Koyama) - 5
oligosetus A.E. Kozhevnikov - 5
orientalis Ohwi (*S. sylvaticus* L. subsp. *orientalis* (Ohwi) Worosch.) - 5
oryzetorum (Steud.) Ohwi = S. lateriflorus
ovatus Roth = Eleocharis ovata
palustris L. = Eleocharis palustris
planiculmis Fr. Schmidt = Bolboschoenus planiculmis
pungens Vahl (*Schoenoplectus pungens* (Vahl) Palla, *Sch. americanus* auct., *Scirpus americanus* auct.) - 1(?)

quinqueflorus F.X. Hartm. = Eleocharis quinqueflora
radicans Schkuhr - 1, 3, 4, 5
roylei (Nees) Parker (*Isolepis roylei* Nees, *Schoenoplectus lupulinus* (Nees) V. Krecz., *Sch. roylei* (Nees) Lye, comb. superfl., *Sch. roylei* (Nees) Ovcz. & Czuk., *S. lupulinus* (Nees) Roshev. 1935, non Spreng. 1807) - 6
russeolus (Fries) T. Koyama = Eriophorum russeolum
setaceus L. (*Isolepis setacea* (L.) R. Br., *Schoenoplectus setaceus* (L.) Palla) - 1, 2, 3, 6
soloniensis Dubois = Eleocharis ovata
supinus L. (*Schoenoplectus supinus* (L.) Palla) - 1, 2, 3, 4, 5(alien), 6
- var. *lateriflorus* (J.F. Gmel.) T. Koyama = S. lateriflorus
sylvaticus L. - 1, 2, 3, 4, 6
- subsp.*orientalis* (Ohwi) Worosch. = S. orientalis
tabernaemontani C.C. Gmel. (*Hymenochaeta tabernaemontani* (C.C. Gmel.) Nakai, *Schoenoplectus lacustris* (L.) Palla subsp. *tabernaemontani* (C.C. Gmel.) A. & D. Löve, *Sch. tabernaemontani* (C.C. Gmel.) Palla, *Scirpus lacustris* L. subsp. *tabernaemontani* (C.C. Gmel.) Syme) - 1, 2, 3, 4, 5, 6
triangulatus Roxb. - 5
triqueter L. (*Hymenochaeta triquetra* (L.) Nakai, *Schoenoplectus triqueter* (L.) Palla) - 1, 2, 3, 5, 6
triquetriformis (V. Krecz.) Egor. (*Schoenoplectus triquetriformis* V. Krecz.) - 6
uniflorus Trautv. = Baeothryon uniflorum
uniglumis Link = Eleocharis uniglumis
validus auct. = S. hippolyti
wichurae Boeck. (*S. lineatus* Michx. subsp. *wichurae* (Boeck.) T. Koyama) - 5
wichurae sensu Roshev. = S. asiaticus
yagara Ohwi = Bolboschoenus yagara
yokoscensis Franch. & Savat = Eleocharis yokoscensis

Torulinium Desv. ex Hamilt.

caucasicum Palla (*Cyperus ferax* Rich. subsp. *transcaucasicus* Kuk., *Torulinium ferax* (Rich.) Desv. ex Hamilt. subsp. *caucasicum* (Palla) Hadac, *T. ferax* auct., *T. odoratum* auct.) - 2
ferax auct. = T. caucasicum
ferax (Rich.) Desv. ex Hamilt. subsp. *caucasicum* (Palla) Hadac = T. caucasicum
odoratum auct. = T. caucasicum

Trichophorum Pers. = Baeothryon and Scirpus

alpinum (L.) Pers. = Baeothryon alpinum
bracteatum (Bigel.) V. Krecz. ex Czernov = Baeothryon cespitosum
cespitosum (L.) C. Hartm. = Baeothryon cespitosum
dolichocarpum Zak. = Baeothryon dolichocarpum
pumilum (Vahl) Schinz & Thell. = Baeothryon pumilum
uniflorum (Trautv.) Karav. = Baeothryon uniflorum

Vignea Beauv. = Carex

angustior (Mackenz.) Sojak = Carex angustior
appropinquata (Schum.) Sojak = Carex appropinquata
bohemica (Schreb.) Sojak = Carex bohemica
bonanzensis (Britt.) Sojak = Carex bonanzensis
brunnescens (Pers.) Sojak = Carex brunnescens
- subsp. *pacifica* (Kalela) Sojak = Carex brunnescens subsp. pacifica
- subsp. *vitilis* (Fries) Sojak = Carex brunnescens
capitata (L.) Sojak = Carex capitata
cinerea (Poll.) Dostal = Carex cinerea

colchica (J. Gay) Sojak = Carex colchica
cuprina (Sandor ex Heuff.) Sojak = Carex muricata
curaica (Kunth) Sojak = Carex curaica
diandra (Schrank) Sojak = Carex diandra
disperma (Dew.) Sojak = Carex disperma
divulsa (Stokes) Reichenb. = Carex divulsa
- subsp. *leersiana* (Rauschert) Dostal = Carex polyphylla
duriuscula (C.A. Mey.) Sojak = Carex duriuscula
enervis (C.A. Mey.) Sojak = Carex enervis
gynocrates (Wormsk.) Sojak = Carex gynocrates
hepburnii (Boott) Sojak = Carex hepburnii
kobomugi (Ohwi) Sojak = Carex kobomugi
lachenalii (Schkuhr) Sojak = Carex lachenalii
leiorhyncha (C.A. Mey.) Sojak = Carex leiorhyncha
ligerica (J. Gay) Sojak = Carex ligerica
lithophila (Turcz.) Sojak = Carex lithophila
macrocephala (Willd. ex Spreng.) Sojak = Carex macrocephala
maritima (Gunn.) Reichenb. = Carex maritima
- subsp. *setina* (Christ) Sojak = Carex setina
muricata (L.) Reichenb. = Carex muricata
nemurensis (Franch.) Sojak = Carex nemurensis
omiana (Franch. & Savat.) Sojak = Carex omiana
oreophila (C.A. Mey.) Sojak = Carex oreophila
otrubae (Podp.) Sojak = Carex otrubae
ovalis (Good.) Sojak = Carex ovalis
pachystylis (J. Gay) Sojak = Carex pachystylis
paniculata (L.) Reichenb. = Carex paniculata
parallela (Laest.) Sojak = Carex parallela
polyphylla (Kar. & Kir.) Sojak = Carex polyphylla
praecox (Schreb.) Sojak = Carex praecox
pribylovensis (J.M. Macoun) Sojak = Carex pribylovensis
redowskiana (C.A. Mey.) Sojak = Carex redowskiana
remotiuscula (Wahlenb.) Sojak = Carex remotiuscula
reptabunda (Trautv.) Sojak = Carex reptabunda
x *schuetzeana* (Figert) Dostal = Carex x schuetzeana
spicata (Huds.) Sojak = Carex contigua
tenuiflora (Wahlenb.) Sojak = Carex tenuiflora
ursina (Dew.) Sojak = Carex ursina
vulpinoidea (Michx.) Sojak = Carex vulpinoidea
wendelboi (Nelmes) Sojak = Carex wendelboi

CYTINACEAE Brongn. (*RAFFLESIACEAE* auct.)
Cytinus L.
ruber (Fourr.) Kom. - 2

DAPHNIPHYLLACEAE Muell. Arg.
Daphniphyllum Blume
humile Maxim. ex Franch. & Savat. - 5

DATISCACEAE Lindl.
Datisca L.
cannabina L. - 2, 6

DENNSTAEDTIACEAE Lotsy
Coptidipteris Nakai & Momose = Dennstaedtia
wilfordii (Moore) Nakai & Momose = Dennstaedtia wilfordii

Davallia auct. = Dennstaedtia

hirsuta Sw. = Dennstaedtia hirsuta
pilosella Hook. = Dennstaedtia hirsuta

Dennstaedtia Bernh. (*Coptidipteris* Nakai & Momose, *Fuziifilix* Nakai & Momose, *Davallia* auct., *Microlepia* auct.)

hirsuta (Sw.) Mett. (*Davallia hirsuta* Sw., *D. pilosella* Hook., *Dennstaedtia pilosella* (Hook.) Ching, *Fuziifilix pilosella* (Hook.) Nakai & Momose, *Microlepia hirsuta* (Sw.) Sw., *M. pilosella* (Hook.) Moore) - 5
pilosella (Hook.) Ching = D. hirsuta
wilfordii (Moore) Christ (*Microlepia wilfordii* Moore, *Coptidipteris wilfordii* (Moore) Nakai & Momose, *Dennstaedtia wilfordii* (Moore) Koidz. ex Tagawa, comb. superfl.) - 5

Fuziifilix Nakai & Momose = Dennstaedtia

pilosella (Hook.) Nakai & Momose = Dennstaedtia hirsuta

Microlepia auct. = Dennstaedtia

hirsuta (Sw.) Sw. = Dennstaedtia hirsuta
pilosella (Hook.) Moore = Dennstaedtia hirsuta
wilfordii Moore = Dennstaedtia wilfordii

DIAPENSIACEAE Lindl.
Diapensia L.

lapponica L. - 1, 3
- subsp. *obovata* (Fr. Schmidt) Hult. = D. obovata
obovata (Fr. Schmidt) Nakai (*D. lapponica* L. subsp *obovata* (Fr. Schmidt) Hult.) - 4, 5

DIOSCOREACEAE R. Br.
Dioscorea L.

batatas Decne. (*D. tenuipes* auct.) - 5
caucasica Lipsky - 2
nipponica Makino (*D. polystachya* auct.) - 5
polystachya auct. = D. nipponica
tenuipes auct. = D. batatas

Tamus L.

communis L. - 1, 2

DIPSACACEAE Juss.
Cephalaria Schrad. ex Roem. & Schult.

aristata C. Koch (*C. armena* Grossh.) - 2
armena Grossh. = C. aristata
armeniaca Bordz. - 2
    C. sparsipilosa Mattews (<u>C. pilosa</u> Boiss. & Huet, 1856, non Gren. & Godr. 1850) does not occur on the investigated territory.
balkharica E. Busch (*C. svanetica* Kolak.) - 2
brevipalea (Somm. & Levier) Litv. = C. calcarea
calcarea Albov (*C. brevipalea* (Somm. & Levier) Litv.) - 2
charadzeae Schchian - 2
coriacea (Willd.) Steud. - 1, 2
dagestanica Bobr. - 2
demetrii Bobr. - 1
dipsacoides Kar. & Kir. = Dipsacus dipsacoides
gigantea (Ledeb.) Bobr. - 2

grossheimii Bobr. = C. kotschyi
kotschyi Boiss. & Hohen. (*C. grossheimii* Bobr.) - 2
litvinovii Bobr. - 1
media Litv. - 2
microdonta Bobr. - 2
nachiczevanica Bobr. - 2
pilosa auct. = C. armeniaca
procera Fisch. & Ave-Lall. - 2
sosnowskyi Kolak. - 2
svanetica Kolak. = C. balkharica
syriaca (L.) Schrad. ex Roem. & Schult. (*C. syriaca* subsp. *turanica* Bobr.) - 2, 6
- subsp. *transcaucasica* Bobr. = C. transcaucasica
- subsp. *turanica* Bobr. = C. syriaca
tchihatchewii Boiss. - 2
transcaucasica (Bobr.) Galushko (*C. syriaca* (L.) Schrad. ex Roem. & Schult. subsp. *transcaucasica* Bobr.) - 2
transsylvanica (L.) Schrad. ex Roem. & Schult. - 1, 2
uralensis (Murr.) Schrad. ex Roem. & Schult. - 1, 2, 3
velutina Bobr. - 2

Dipsacella Opiz = Dipsacus

pilosa (L.) Sojak = Dipsacus pilosus

Dipsacus L. (*Dipsacella* Opiz, *Virga* Hill)

azureus Schrenk = D. dipsacoides
dipsacoides (Kar. & Kir.) Botsch. (*Cephalaria dipsacoides* Kar. & Kir., *Dipsacus azureus* Schrenk) - 6
fullonum L. p.p. = D. sylvestris
gmelinii Bieb. - 1, 2, 3, 6
laciniatus L. - 1, 2, 6
pilosus L. (*Dipsacella pilosa* (L.) Sojak, *Virga pilosa* (L.) Hill) - 1, 2, 3(alien)
sativus (L.) Honck. - 1, 2, 6
strigosus Willd. ex Roem. & Schult. (*Virga strigosa* (Willd. ex Roem. & Schult.) Holub) - 1, 2, 6
sylvestris Huds. (*D. fullonum* L. p.p. nom. ambig.) - 1, 2

Knautia L. (*Trichera* Schrad.)

arvensis (L.) Coult. (*Trichera arvensis* (L.) Schrad.) - 1, 2, 3(alien), 4(alien), 5(alien), 6
dipsacifolia Kreutzer (*Scabiosa dipsacifolia* Host, 1827, non Schrank, 1824, *Trichera dipsacifolia* (Kreutzer) Nym.) subsp. **pocutica** (Szabo) Ehrend. (*K. sylvatica* (L.) Duby var. *pocutica* Szabo, *K. lancifolia* (Heuff.) Fuss subsp. *pocutica* (Szabo) Holub, *K. sylvatica* subsp. *pocutica* (Szabo) Soo, *Trichera dipsacifolia* subsp. *pocutica* (Szabo) Sojak) - 1
involucrata Somm. & Levier - 2
kitaibelii (Schult.) Borb. (*Scabiosa kitaibelii* Schult.) - 1
lancifolia (Heuff.) Fuss subsp. *pocutica* (Szabo) Holub = K. dipsacifolia subsp. pocutica
longifolia auct. = Scabiosa opaca
montana (Bieb.) DC. - 2
orientalis L. - 2
sylvatica (L.) Duby subsp. *pocutica* (Szabo) Soo = K. dipsacifolia subsp. pocutica
- var. *pocutica* Szabo = K. dipsacifolia subsp. pocutica
tatarica (L.) Szabo (*Trichera tatarica* (L.) Sojak) - 1

Lomelosia Rafin. = Scabiosa

argentea (L.) Greuter & Burdet = Scabiosa argentea
caucasica (Bieb.) Greuter & Burdet = Scabiosa caucasica

micrantha (Desf.) Greuter & Burdet = Scabiosa micrantha
olivieri (Coult.) Greuter & Burdet = Scabiosa olivieri
persica (Boiss.) Greuter & Burdet = Scabiosa persica
rotata (Bieb.) Greuter & Burdet = Scabiosa rotata

Pterocephalus Adans.

afghanicus (Aitch. & Hemsl.) Boiss. - 6
fruticulosus Korov. - 6
■ khorassanicus Czerniak.
plumosus (L.) Coult. - 1, 2, 6
szovitsii Boiss. - 2

Scabiosa L. (*Lomelosia* Rafin. med. 1838, *Trochocephalus* (Mert. & Koch) Opiz, Aug.-Sept. 1838)

adzharica Schchian - 2
alexeenkoana Sulak. - 2
alpestris Kar. & Kir. (*Trochocephalus alpestris* (Kar. & Kir.) A. & D. Löve) - 6
amoena Jacq. fil. - 2
argentea L. (*Lomelosia argentea* (L.) Greuter & Burdet, *Scabiosa taurica* Kotov, *Trochocephalus argenteus* (L.) A. & D. Löve) - 1, 2
austroaltaica Bobr. (*S. isetensis* L. var. *austroaltaica* (Bobr.) Serg., *Trochocephalus austroaltaicus* (Bobr.) A. & D. Löve) - 3
bipinnata C. Koch - 2
caucasica Bieb. (*Lomelosia caucasica* (Bieb.) Greuter & Burdet, *Trochocephalus caucasicus* (Bieb.) A. & D. Löve) - 2
colchica Stev. - 2
columbaria L. - 1, 2, 6
comosa Fisch. ex Roem. & Schult. (*Trochocephalus comosus* (Fisch. ex Roem. & Schult.) A. & D. Löve) - 4
correvoniana Somm. & Levier - 2
deserticola Rech. fil. - 6
dipsacifolia Host = Knautia dipsacifolia
flavida Boiss. & Hausskn. - 2, 6
georgica Sulak. - 2
gumbetica Boiss. (*Trochocephalus gumbeticus* (Boiss.) A. & D. Löve) - 2
hyrcanica Stev. - 2
imeretica (Somm. & Levier) Sulak. - 2
isetensis L. (*Trochocephalus isetensis* (L.) A. & D. Löve) - 1, 2, 3, 6
- var. *austroaltaica* (Bobr.) Serg. = S. austroaltaica
kitaibelii Schult. = Knautia kitaibelii
lachnophylla Kitag. (*Trochocephalus lachnophyllus* (Kitag.) A. & D. Löve) - 4, 5
lucida auct. = S. opaca
meskhetica Schchian - 2
micrantha Desf. (*Lomelosia micrantha* (Desf.) Greuter & Burdet, *Trochocephalus micranthus* (Desf.) A. & D. Löve) - 1, 2, 6
ochroleuca L. - 1, 2, 3, 4, 6
olgae Albov (*Trochocephalus olgae* (Albov) A. & D. Löve) - 2
olivieri Coult. (*Lomelosia olivieri* (Coult.) Greuter & Burdet, *Trochocephalus olivieri* (Coult.) A. & D. Löve) - 2, 6
opaca Klok. (*Knautia longifolia* auct., *Scabiosa lucida* auct.) - 1
owerinii Boiss. - 2
persica Boiss. (*Lomelosia persica* (Boiss.) Greuter & Burdet, *Scabiosa setulosa* Fisch. & C.A. Mey., *Trocho-*

cephalus persicus (Boiss.) A. & D. Löve) - 2, 6
praemontana Privalova - 1
rhodantha Kar. & Kir. - 2, 6
rotata Bieb. (*Lomelosia rotata* (Bieb.) Greuter & Burdet, *Trochocephalus rotatus* (Bieb.) A. & D. Löve) - 1, 2, 6
setulosa Fisch. & C.A. Mey. = S. persica
songarica Schrenk (*Trochocephalus songaricus* (Schrenk) A. & D. Löve) - 6
sosnowskyi Sulak. - 2
taurica Kotov = S. argentea
transcaspica Rech. fil. - 2, 6
tschiliensis Grunn. - 5
ucranica L. (*Trochocephalus ucranicus* (L.) A. & D. Löve) - 1, 2
ulugbekii Zak. - 6
velenovskiana Bobr. - 2

Succisa Hall.

pratensis Moench - 1, 2, 3, 4

Succisella G. Beck

inflexa (Kluk) G. Beck - 1, 2

Trichera Schrad. = Knautia

arvensis (L.) Schrad. = Knautia arvensis
dipsacifolia (Kreutzer) Nym. = Knautia dipsacifolia
- subsp. *pocutica* (Szabo) Sojak = Knautia dipsacifolia subsp. pocutica
tatarica (L.) Sojak = Knautia tatarica

Trochocephalus (Mert. & Koch) Opiz = Scabiosa

alpestris (Kar. & Kir.) A. & D. Löve = Scabiosa alpestris
argenteus (L.) A. & D. Löve = Scabiosa argentea
austroaltaicus (Bobr.) A. & D. Löve = Scabiosa austroaltaica
caucasicus (Bieb.) A. & D. Löve = Scabiosa caucasica
comosus (Fisch. ex Roem. & Schult.) A. & D. Löve = Scabiosa comosa
gumbeticus (Boiss.) A. & D. Löve = Scabiosa gumbetica
isetensis (L.) A. & D. Löve = Scabiosa isetensis
lachnophyllus (Kitag.) A. & D. Löve = Scabiosa lachnophylla
micranthus (Desf.) A. & D. Löve = Scabiosa micrantha
olgae (Albov) A. & D. Löve = Scabiosa olgae
olivieri (Coult.) A. & D. Löve = Scabiosa olivieri
persicus (Boiss.) A. & D. Löve = Scabiosa persica
rotatus (Bieb.) A. & D. Löve = Scabiosa rotata
songaricus (Schrenk) A. & D. Löve = Scabiosa songarica
ucranicus (L.) A. & D. Löve = Scabiosa ucranica

Virga Hill = Dipsacus

pilosa (L.) Hill = Dipsacus pilosus
strigosa (Willd. ex Roem. & Schult.) Holub = Dipsacus strigosus

DROSERACEAE Salisb. (*ALDROVANDACEAE* Nakai)

Aldrovanda L.

vesiculosa L. - 1, 2, 5, 6

Drosera L.

anglica Huds. - 1, 3, 4, 5
intermedia Hayne - 1
x obovata Mert. & Koch. - D. anglica Huds. x D. rotundifolia L.
rotundifolia L. - 1, 2, 3, 4, 5

DRYOPTERIDACEAE Ching (*ASPIDIACEAE* Mett. ex Frank, nom. illegit.)

Arachniodes Blume (*Polystichopsis* auct. p.p., *Rumohra* auct. p.p.)

miqueliana (Maxim. ex Franch. & Savat.) Ohwi = Leptorumohra miqueliana
mutica (Franch. & Savat.) Ohwi (*Aspidium muticum* Franch. & Savat., *Dryopteris mutica* (Franch. & Savat.) C. Chr., *Leptorumohra mutica* (Franch. & Savat.) Czer., *Polystichopsis mutica* (Franch. & Savat.) Tagawa, nom. illegit., *Rumohra mutica* (Franch. & Savat.) Ching) - 5

Aspidium Sw. = Dryopteris, Polystichum, and also Athyrium (Athyriaceae) and Thelypteris (Thelypteridaceae)

affine Fisch. & C.A. Mey. = Dryopteris caucasica
x *bicknellii* Christ = Polystichum x bicknellii
braunii Spenn. = Polystichum braunii
caucasicum A. Br. = Dryopteris caucasica
dryopteris (L.) Baumg. var. *longulum* Christ = Gymnocarpium jessoense (Athyriaceae)
goeringianum G. Kunze = Dryopteris goeringiana
x *illiricum* Borb. = Polystichum x illiricum
luerssenii Doerfl. = Polystichum x luerssenii
microchlamys Christ = Polystichum microchlamys
miquelianum Maxim. ex Franch. & Savat. = Leptorumohra miqueliana
muticum Franch. & Savat. = Arachnioides mutica
nipponicum Franch. & Savat. = Parathelypteris nipponica (Thelypteridaceae)
rigidum Hoffm. ex Sw. var. *remotum* A. Br. ex Doll = Dryopteris remotum
spinulosum (O.F. Muel.) Sw. = Dryopteris carthusiana
- b. *uliginosum* A.Br. ex Doll = Dryopteris x uliginosa
subspinulosum Christ = Athyrium yokoscense (Athyriaceae)

Dryopteris Adans. (*Aspidium* Sw. p.p.nom. illegit., *Lastrea* auct., *Nephrodium* auct.)

abbreviata auct. = D. oreades
abbreviata (DC.) Newm. ex Manton, p.p. = D. affinis and D. oreades
aemula (Ait.) O. Kuntze (*D. liliana* Golits.) - 2
affinis (Lowe) Fraser-Jenkins (*Nephrodium affine* Lowe, *Dryopteris abbreviata* (DC.) Newm. ex Manton, 1950, p.p. incl. typo, non *D. abbrreviata* (Schrad.) O. Kuntze, 1891, *D. filix-mas* (L.) Schott subsp. *abbreviata* (DC.) Schidlay, p.p. incl. typo, *D. filix-mas* var. *abbreviata* (DC.) Newm. p.p. incl. typo, *D. mediterranea* Fomin, p.p. excl. typo, *Polystichum abbreviatum* DC.) - 1, 2
- subsp. *borreri* (Newm.) Fraser-Jenkins = D. pseudomas
- subsp. **cambrensis** Fraser-Jenkins - 2
- subsp. *robusta* Oberholzer & Tavel ex Fraser-Jenkins = D. pseudomas subsp. robusta

alexeenkoana Fomin = D. dilatata
x **ambroseae** Fraser-Jenkins & Jermy. - D. dilatata (Hoffm.) A. Gray x D. assimilis S. Walker
amurensis Christ = Leptorumohra amurensis
amurensis (Milde) Takeda = Leptorumohra amurensis
assimilis S. Walker (*D. expansa* (C. Presl) Fraser-Jenkins & Jermy subsp. *assimilis* (S. Walker) Tzvel., *D. spinulosa* (O.F. Muel.) Watt subsp. *assimilis* (S. Walker) Schidlay) - 1, 2, 3, 4, 5
austriaca (Jacq.) Woynar, p.p. = D. dilatata and Pteridium aquilinum
barbellata Fomin = D. sichotensis
barbigera (Hook.) O. Kuntze (*Nephrodium barbigerum* Hook., *Dryopteris barbigera* subsp. *komarovii* (C. Koss.) Fraser-Jenkins, *D. komarovii* C. Koss.) - 6
- subsp. *komarovii* (C. Koss.) Fraser-Jenkins = D. barbigera
borreri (Newm.) V. Krecz. = D. pseudomas
borreri (Newm.) Oberholzer & Tavel = D. pseudomas
x **brathaica** Fraser-Jenkins & Reichstein. - D. carthusiana (Vill.) H.P. Fuchs x D. filix-mas (L.) Schott
buschiana Fomin = D. crassirhizoma
carthusiana (Vill.) H.P. Fuchs (*Polypodium carthusianum* Vill., *Aspidium spinulosum* (O.F. Muell.) Sw., *Dryopteris lanceolatocristata* (Hoffm.) Alst., *D. spinulosa* (O.F. Muell.) O. Kuntze, comb. superfl., *D. spinulosa* (O.F. Muell.) Watt, *Polypodium lanceolatocristatum* Hoffm. p.p., *P. spinulosum* O.F. Muell. 1777, non Burm. fil. 1768) - 1, 2, 3, 4
caucasica (A.Br.) Fraser-Jenkins & Corley (*Aspidium caucasicum* A. Br., *A. affine* Fisch. & C.A. Mey. 1838, non Blume 1828) - 2
chinensis (Baker) Koidz. (*Nephrodium chinense* Baker, *Dryopteris subtripinnata* auct.) - 5
continentalis (V. Petrov) Fomin = Gymnocarpium jessoense (Athyriaceae)
coreano-montana Nakai = D. sichotensis
crassirhizoma Nakai (*D. buschiana* Fomin) - 5
cristata (L.) A. Gray - 1, 3
x **deweveri** (Jansen) Jansen & Wachter (*Aspidium deweveri* Jansen). - Dryopteris carthusiana (Vill.) H.P. Fuchs x D. dilatata (Hoffm.) A. Gray
dilatata (Hoffm.) A. Gray (*Polypodium dilatatum* Hoffm., *Dryopteris alexeenkoana* Fomin, *D. austriaca* (Jacq.) Woynar, p.p. solum excl. basionymo (*Polypodium austriacum* Jacq. = Pteridium aquilinum (L.) Kuhn), *Polypodium lanceolatocristatum* Hoffm. p.p.) - 1, 2, 3, 4
disjuncta (Rupr.) C.V. Morton = Gymnocarpium dryopteris (Athyriaceae)
x **doluchanovii** A. Askerov. - D. dilatata (Hoffm.) A. Gray x D. pseudomas (Wollaston) Holub & Pouzar - 2
x **euxinensis** Fraser-Jenkins & Corley. - D. caucasica (A. Br.) Fraser-Jenkins & Corley x D. filix-mas (L.) Schott - 2
expansa (C. Presl) Fraser-Jenkins & Jermy (*Nephrodium expansum* C. Presl, *Dryopteris extremiorientalis* V. Vassil. nom. invalid.) - 5
- subsp. *assimilis* (S. Walker) Tzvel. = D. assimilis
extremiorientalis V. Vassil. = D. expansa
filix-mas (L.) Schott - 1, 2, 3, 4, 6
- subsp. *abbreviata* (DC.) Schidlay, p.p. = D. affinis
- subsp. *borreri* (Newm.) Becherer & Tavel = D. pseudomas
- subsp. x *tavelii* (Rothm.) Schidlay = D. x tavelii
- var. *abbreviata* (DC.) Newm. p.p. = D. affinis and D. oreades

- var. *borreri* Newm. = D. pseudomas
fragrans (L.) Schott - 1, 3, 4, 5
fragrantiformis Tzvel. - 5
goeringiana (G. Kunze) Koidz. (*Aspidium goeringianum* G. Kunze, *Dryopteris laeta* (Kom.) C. Chr., *D. wladiwostokensis* (B. Fedtsch.) Kom.) - 5
goldiana (Hook.) A. Gray subsp. *monticola* (Makino) Fraser-Jenkins = D. monticola
heterospora (Wagner) Kuvajev, p.p. = Gymnocarpium x intermedium (Athyriaceae)
x **initialis** Fraser-Jenkins & Corley. - D. caucasica (A. Br.) Fraser-Jenkins & Corley x D. oreades Fomin
jessoensis Koidz. = Gymnocarpium jessoense (Athyriaceae)
kamtschatica Kom. = Oreopteris quelpaertensis (Thelypteridaceae)
kemulariae Mikh. = D. remota
komarovii C. Koss. = D. barbigera
laeta (Kom.) C. Chr. = D. goeringiana
lanceolatocristata (Hoffm.) Alst. = D. carthusiana
liliana Golits. = D. aemula
limbosperma (All.) Becherer = Oreopteris limbosperma (Thelypteridaceae)
linneana C. Chr. = Gymnocarpium dryopteris (Athyriaceae)
x **mantoniae** Fraser-Jenkins & Corley. - D. filix-mas (L.) Schott x D. oreades Fomin - 2
mediterranea Fomin, p.p. = D. affinis and D. pseudomas
mindshelkensis Pavl. (*D. villarii* (Bell.) Woynar ex Schinz & Thell. subsp. *mindshelkensis* (Pavl.) Fraser-Jenkins) - 6
miqueliana (Maxim. & Franch. & Savat.) C. Chr. = Leptorumohra miqueliana
monticola (Makino) C. Chr. (*Nephrodium monticola* Makino, *Dryopteris goldiana* (Hook.) A. Gray subsp. *monticola* (Makino) Fraser-Jenkins) - 5
mutica (Franch. & Savat.) C. Chr. = Arachnioides mutica
nipponica (Franch. & Savat.) C. Chr. = Parathelypteris nipponica (Thelypteridaceae)
oreades Fomin (*D. abbreviata* (DC.) Newm. ex Manton (1950), p.p. excl. typo, non *D. abbreviata* (Schrad.) O. Kuntze (1891), nec *Polystichum abbreviatum* DC. (1805), *D. filix-mas* (L.) Schott var. *abbreviata* (DC.) Newm. p.p. excl. typo, *D. abbreviata* auct.) - 2
oreopteris (Ehrh.) Maxon = Oreopteris limbosperma (Thelypteridaceae)
paleacea (Sw.) C. Chr. var. *borreri* (Newm.) H. Wolff = D. pseudomas
pallida (Bory) Fomin subsp. *raddeana* (Fomin) Fraser-Jenkins = D. raddeana
- subsp. *raddeana* (Fomin) Nardi = D. raddeana
phegopteris (L.) C. Chr. = Phegopteris connectilis (Thelypteridaceae)
pseudomas (Wollaston) Holub & Pouzar (*Lastrea pseudomas* Wollaston, *Dryopteris affinis* (Lowe) Fraser-Jenkins subsp. *borreri* (Newm.) Fraser-Jenkins, *D. borreri* (Newm.) V. Krecz. comb. superfl., *D. borreri* (Newm.) Oberholzer & Tavel, *D. filix-mas* (L.) Schott var. *borreri* Newm., *D. filix-mas* subsp. *borreri* (Newm.) Becherer & Tavel, *D. mediterranea* Fomin, p.p. incl. typo, *D. paleacea* (Sw.) C. Chr. var. *borreri* (Newm.) H. Wolff) - 2
- subsp. **robusta** (Fraser-Jenkins) Holub (*D. affinis* (Lowe) Fraser-Jenkins subsp. *robusta* Oberholzer & Tavel ex Fraser-Jenkins) - 2(?)
pumila V. Krecz. = Gymnocarpium dryopteris (Athyriaceae)
quelpaertensis Christ = Oreopteris quelpaertensis (Thelypteridaceae)

raddeana (Fomin) Fomin (*D. pallida* (Bory) Fomin subsp. *raddeana* (Fomin) Fraser-Jenkins, *D. pallida* subsp. *raddeana* (Fomin) Nardi, comb. superfl.) - 2
remota (A. Br. ex Doll) Druce (*Aspidium rigidum* Hoffm. ex Sw. var. *remotum* A. Br. ex Doll, *Dryopteris kemulariae* Mikh.) - 1, 2
rigida auct. = D. submontana
robertiana (Hoffm.) C. Chr. = Gymnocarpium robertianum (Athyriaceae)
x **sarvelae** Fraser-Jenkins & Jermy. - D. carthusiana (Vill.) H.P. Fuchs x D. assimilis S. Walker
x **shorapanensis** A. Askerov. - D. caucasica (A. Br.) Fraser-Jenkins & Corley x D. pseudomas (Wollaston) Holub & Pouzar - 2
sichotensis Kom. (*D. barbellata* Fomin, *D. coreano-montana* Nakai, *D. sichotensis* subsp. *coreano-montana* (Nakai) Worosch.) - 5
- subsp. *coreano-montana* (Nakai) Worosch. = D. sichotensis
spinulosa (O.F. Muell.) O. Kuntze = D. carthusiana
spinulosa (O.F. Muell.) Watt = D. carthusiana
- subsp. *assimilis* (S. Walker) Schidlay = D. assimilis
submontana (Fraser-Jenkins & Jermy) Fraser-Jenkins (*D. villarii* (Bell.) Woynar ex Schinz & Thell. subsp. *submontana* Fraser-Jenkins & Jermy, *D. rigida* auct., *D. villarii* auct.) - 2
subtripinnata auct. = D. chinensis
x **tavelii** Rothm. (*D. filix-mas* (L.) Schott subsp. x *tavelii* (Rothm.) Schidlay). - D. affinis (Lowe) Fraser-Jenkins x D. filix-mas (L.) Schott - 2
thelypteris (L.) A. Gray = Thelypteris palustris (Thelypteridaceae)
x **uliginosa** (A. Br. ex Doll) Druce (*Aspidium spinulosum* (O.F. Muell.) Sw. b. *uliginosum* A. Br. ex Doll). - D. carthusiana (Vill.) H.P. Fuchs x D. cristata (L.) A. Gray
villarii auct. = D. submontana
villarii (Bell.) Woynar ex Schinz & Thell. subsp. *mindshelkensis* (Pavl.) Fraser-Jenkins = D. mindshelkensis
- subsp. *submontana* Fraser-Jenkins & Jermy = D. submontana
wladiwostokensis (B. Fedtsch.) Kom. = D. goeringiana

Lastrea auct. = Dryopteris
pseudomas Wollaston = Dryopteris pseudomas

Leptorumohra (H. Ito) H. Ito (*Leptorumohra* (H. Ito) Serizawa, comb. superfl., *Polystichopsis* auct. p.p., *Rumohra* auct. p.p.)
amurensis (Christ) Tzvel. (*Dryopteris amurensis* Christ, *D. amurensis* (Milde) Takeda) - 5
miqueliana (Maxim. ex Franch. & Savat.) H. Ito (*Aspidium miquelianum* Maxim. ex Franch. & Savat., *Arachniodes miqueliana* (Maxim. ex Franch. & Savat.) Ohwi, *Dryopteris miqueliana* (Maxim. ex Franch. & Savat.) C. Chr., *Polystichopsis miqueliana* (Maxim. ex Franch. & Savat.) Tagawa, nom. illegit., *Rumohra miqueliana* (Maxim. ex Franch. & Savat.) Ching) - 5
mutica (Franch. & Savat.) Czer. = Arachniodes mutica

Nephrodium auct. = Dryopteris
affine Lowe = Dryopteris affinis
barbigerum Hook. = Dryopteris barbigera
chinense Baker = Dryopteris chinensis

expansum C. Presl = Dryopteris expansa
monticola Makino = Dryopteris monticola

Polystichopsis auct. p.p. = Arachniodes and Leptorumohra

miqueliana (Maxim. ex Franch. & Savat.) Tagawa = Leptorumohra miqueliana
mutica (Franch. & Savat.) Tagawa = Arachniodes mutica

Polystichum Roth (*Aspidium* Sw. p.p. nom. illegit.)

abbreviatum DC. = Dryopteris affinis
aculeatum (L.) Roth (*Polypodium aculeatum* L., *P. lobatum* Huds., *Polystichum lobatum* (Huds.) Bast.) - 1, 2, 6
- var. *retroso-paleaceum* Kodama = P. retroso-paleaceum
angulare (Kit. ex Willd.) C. Presl. = P. setiferum
x **bicknellii** (Christ) Hahne (*Aspidium* x *bicknellii* Christ). - P. aculeatum (L.) Roth x P. setiferum (Forssk.) Moore ex Woynar
braunii (Spenn.) Fee (*Aspidium braunii* Spenn.) - 1, 2, 3, 4, 5
- subsp. *kamtschaticum* (C. Chr. & Hult.) A. & D. Löve = P. microchlamys
- var. *kamtschaticum* C. Chr. & Hult. = P. microchlamys
craspedosorum (Maxim.) Diels - 4, 5
x **dmitrievae** A. Askerov - 2
x **fominii** A. Askerov & A. Bobr. - P. aculeatum (L.) Roth x P. woronowii Fomin - 2
x **illiricum** (Borb.) Hahne (*Aspidium* x *illiricum* Borb.). - P. aculeatum (L.) Roth x P. lonchitis (L.) Roth
kadyrovii A. Askerov & A. Bobr. - 2
kamtschaticum (C. Chr. & Hult.) Fomin = P. microchlamys
lobatum (Huds.) Bast. = P. aculeatum
lonchitis (L.) Roth - 1, 2, 3, 4, 5, 6
x **luerssenii** (Doerfl.) Hahne (*Aspidium luerssenii* Doerfl.). - P. aculeatum (L.) Roth x P. braunii (Spenn.) Fee
x **meyeri** Sleep & Reichstein. - P. braunii (Spenn.) Fee x P. lonchitis (L.) Roth
microchlamys (Christ) Matsum. (*Aspidium microchlamys* Christ, *Polystichum braunii* (Spenn.) Fee var. *kamtschaticum* C. Chr. & Hult., *P. braunii* subsp. *kamtschaticum* (C. Chr. & Hult.) A. & D. Löve, *P. kamtschaticum* (C. Chr. & Hult.) Fomin) - 5
retroso-paleaceum (Kodama) Tagawa (*P. aculeatum* (L.) Roth var. *retroso-paleaceum* Kodama) - 5(?)
x **safarovii** A. Askerov & A. Bobr. - P. braunii (Spenn.) Fee x P. woronowii Fomin - 2
setiferum (Forssk.) Moore ex Woynar (*Polypodium setiferum* Forssk., *Polystichum angulare* (Kit. ex Willd.) C. Presl) - 1, 2
subtripteron Tzvel. - 5
tripteron (G. Kunze) C. Presl - 5
x **wirtgenii** Hahne. - P. braunii (Spenn.) Fee x P. setiferum (Forssk.) Moore ex Woynar
woronowii Fomin - 2

Rumohra auct. p.p. = Arachniodes and Leptorumohra

miqueliana (Maxim. ex Franch. & Savat.) Ching = Leptorumohra miqueliana
mutica (Franch. & Savat.) Ching = Arachniodes mutica

EBENACEAE Guerke
Diospyros L.

*****kaki** Thunb. - 1, 2, 6
lotus L. - 1(cult.), 2, 6
*****virginiana** L. - 2, 6

ELAEAGNACEAE Juss.
Elaeagnus L.

*****angustifolia** L. - 1
- subsp. *orientalis* (L.) Sojak = E. orientalis
- var. *caspica* Sosn. = E. caspica
- var. *iliensis* Musheg. = E. iliensis
*****argentea** Pursh - 1, 6
caspica (Sosn.) Grossh. (*E. angustifolia* L. var. *caspica* Sosn.) - 2
hortensis Bieb. var. *songarica* Bernh. ex Schlecht. = E. songarica
iliensis (Musheg.) Musheg. (*E. angustifolia* L. var. *iliensis* Musheg.) - 6
*****multiflora** Thunb. (*E. umbellata* auct.) - 5
orientalis L. (*E. angustifolia* L. subsp. *orientalis* (L.) Sojak) - 2, 6
oxycarpa Schlecht. - 3, 6
songarica (Bernh. ex Schlecht.) Schlecht. (*E. hortensis* Bieb. var. *songarica* Bernh. ex Schlecht.) - 6
turcomanica N. Kozl. - 6
umbellata auct. = E. multiflora

Hippophae L.

rhamnoides L. (*H. rhamnoides* subsp. *carpatica* Rousi, *H. rhamnoides* subsp. *caucasica* Rousi, *H. rhamnoides* subsp. *mongolica* Rousi, *H. rhamnoides* subsp. *pamiroalaica* Avdeev, ?*H. rhamnoides* subsp. *salicifolia* (D. Don) Servettaz, *H. rhamnoides* subsp. *turkestanica* Rousi) - 1, 2, 3, 4, 6
- subsp. *carpatica* Rousi = H. rhamnoides
- subsp. *caucasica* Rousi = H. rhamnoides
- subsp. *mongolica* Rousi = H. rhamnoides
- subsp. *pamiroalaica* Avdeev = H. rhamnoides
- subsp. *salicifolia* (D. Don) Servettaz = H. rhamnoides
- subsp. *turkestanica* Rousi = H. rhamnoides

ELATINACEAE Dumort.
Bergia L.

ammannioides Heyne ex Roth - 6
aquatica Roxb. - 2
koganii V.V. Nikit. ex Tuljaganova - 6

Elatine L. (*Peplis* auct.)

alsinastrum L. - 1, 2, 3, 6
ambigua Wight - 3
americana (Pursh) Arn. (*Peplis americana* Pursh, *Elatine triandra* Schkuhr subsp. *americana* (Pursh) A. & D. Löve) - 5
callitrichoides (W. Nyl.) Kauffm. = E. triandra
gyrosperma Dueben = E. hydropiper
■ **hexandra** (Lapierre) DC.
hungarica Moesz (*E. macropoda* Guss. f. *hungarica* (Moesz) Soo) - 1, 3
hydropiper L. (*E. gyrosperma* Dueben) - 1, 2
- subsp. *orthosperma* (Dueben) F. Herm. = E. orthosperma

macropoda Guss. f. *hungarica* (Moesz) Soo = E. hungarica
orthosperma Dueben (*E. hydropiper* L. subsp. *orthosperma* (Dueben) F. Herm., ? *E. spathulata* Gorski) - 1, 2(alien), 5
> According N.N. Tzvelev E. spathulata Gorski is likely a synonym of E. hydropiper L.

spathulata Gorski = E. orthosperma
triandra Schkuhr (*E. callitrichoides* (W. Nyl.) Kauffm.) - 1, 5, 6
- subsp. *americana* (Pursh) A. & D. Löve = E. americana

Peplis auct. = Elatine
americana Pursh = Elatine americana

ELISMATACEAE Nakai = ALISMATACEAE

EMPETRACEAE S.F. Gray
Empetrum L.
albidum V. Vassil. - 5
androgynum V. Vassil. - 4, 5
- var. *caucasicum* V. Vassil. = E. caucasicum
arcticum V. Vassil. p.p. = E. subholarcticum
asiaticum (Nakai) Nakai ex V. Vassil. = E. sibiricum
caucasicum Juz. (*E. androgynum* V. Vassil. var. *caucasicum* V. Vassil. nom. nud.) - 2
eamesii Fern. & Wieg. subsp. *hermaphroditum* (Lange) D. Löve = E. hermaphroditum
hermaphroditum Hagerup (*E. eamesii* Fern. & Wieg. subsp. *hermaphroditum* (Lange) D. Löve, *E. nigrum* L. subsp *hermaphroditum* (Lange) Bocher, *E. nigrum* f. *hermaphroditum* Lange, nom. nud.) - 1, 3
kardakovii V. Vassil. (*E. purpureum* auct.) - 5
kurilense V. Vassil. = E. sibiricum var. japonicum
nigrum L. - 1, 3
- subsp. *hermaphroditum* (Lange) Bocher = E. hermaphroditum
- subsp. *japonicum* (R. Good) Hult. = E. sibiricum var. japonicum
- var. *asiaticum* Nakai = E. sibiricum
- var. *japonicum* Siebold & Zucc. ex C. Koch = E. sibiricum var. japonicum
- f. *hermaphroditum* Lange = E. hermaphroditum
- f. *japonicum* R. Good = E. sibiricum var. japonicum
polare V.Vassil. = E. subholarcticum
pubescens V. Vassil. p.max.p. = E. sibiricum
purpureum auct. = E. kardakovii
sibiricum V. Vassil. (*E. asiaticum* (Nakai) Nakai ex V. Vassil., *E. nigrum* L. var. *asiaticum* Nakai, *E. pubescens* V. Vassil. p.max.p. incl. typo, *E. subholarcticum* V. Vassil. var. *pubescens* (V. Vassil.) V. Vassil. p.max.p.) - 4, 5
- var. **japonicum** (Siebold & Zucc. ex C. Koch) Tzvel. (*E. nigrum* var. *japonicum* Siebold & Zucc. ex C. Koch, *E. kurilense* V. Vassil., *E. nigrum* f. *japonicum* R. Good, *E. nigrum* subsp. *japonicum* (R. Good) Hult.) - 5
stenopetalum V. Vassil. (*E. subholarcticum* V. Vassil. f. *stenopetalum* (V. Vassil.) V. Vassil.) - 4, 5
subholarcticum V. Vassil. (*E. arcticum* V. Vassil. p.p. quoad pl. asiat., *E. polare* V. Vassil.) - 1, 3, 4, 5
- var. *pubescens* (V. Vassil.) V. Vassil. p.max.p. = E. sibiricum

- f. *stenopetalum* (V. Vassil.) V. Vassil. = E. stenopetalum

EPHEDRACEAE Dumort.
Ephedra L.
aitchisonii (Stapf) V. Nikit. p.p. = E. kokanica
arborea Lag. - 1
aurantiaca Takht. & Pachom. - 2, 6
botschantzevii Pachom. - 6
ciliata auct. = E. kokanica
dahurica Turcz. (*E. pseudodistachya* Pachom., *E. sinica* auct.) - 4, 6
distachya L. - 1, 2, 3, 6
eleutherolepis V. Nikit. - 6
> Some authors recognize this species as a hybrid E. intermedia Schrenk & C.A. Mey x E. strobilacea Bunge.

equisetina Bunge (*E. nebrodensis* Guss. subsp. *equisetina* (Bunge) Greuter & Burdet) - 2, 3, 4, 6
fedtschenkoae Pauls. - 6
ferganensis V. Nikit. = E. intermedia
gerardiana Wall. ex Stapf - 6
glauca Regel (*E. heterosperma* V. Nikit.) - 6
glauca sensu V. Nikit. = E. intermedia
heterosperma V. Nikit. = E. glauca
intermedia Schrenk & C.A. Mey. (*E. ferganensis* V. Nikit., *E. intermedia* var. *persica* Stapf, ? *E. microsperma* V. Nikit., *E. persica* (Stapf) V. Nikit., *E. tesquorum* V. Nikit., *E. tibetica* (Stapf) V. Nikit. p.p. quoad plantas, *E. valida* V. Nikit., *E. glauca* sensu V. Nikit.) - 3, 6
- var. *persica* Stapf = E. intermedia
kaschgarica B. Fedtsch. & Bobr. = E. przewalskii
kokanica Regel (*E. aitchisonii* (Stapf) V. Nikit. p.p. quoad pl. ex Asia Media, *E. ciliata* auct.) - 6
lomatolepis Schrenk - 6
major Host subsp. *procera* (Fisch. & C.A. Mey.) Bornm. = E. procera
- subsp. *procera* (Fisch. & C.A. Mey.) Markgraf = E. procera
microsperma V. Nikit. = E. intermedia
minuta sensu V. Nikit. = E. regeliana
monosperma C.A. Mey. - 3, 4, 5
nebrodensis Guss. subsp. *equisetina* (Bunge) Greuter & Burdet = E. equisetina
- subsp. *procera* (Fisch. & C.A. Mey.) K. Richt. = E. procera
persica (Stapf) V. Nikit. = E. intermedia
procera Fisch. & C.A. Mey. (*E. major* Host subsp. *procera* (Fisch. & C.A. Mey.) Bornm., *E. major* subsp. *procera* (Fisch. & C.A. Mey.) Markgraf, comb. superfl., *E. nebrodensis* Guss. subsp. *procera* (Fisch. & C.A. Mey.) K. Richt.) - 2
przewalskii Stapf (*E. kaschgarica* B. Fedtsch. & Bobr., *E. przewalskii* var. *kaschgarica* (B. Fedtsch. & Bobr.) C.Y. Cheng) - 6
- var. *kaschgarica* (B. Fedtsch. & Bobr.) C.Y. Cheng = E. przewalskii
pseudodistachya Pachom. = E. dahurica
pulvinaris V. Nikit. = E. regeliana
regeliana Florin (*E. pulvinaris* V. Nikit., *E. minuta* sensu V. Nikit.) - 6
sinica auct. = E. dahurica
strobilacea Bunge - 6
tesquorum V. Nikit. = E. intermedia
tibetica (Stapf) V. Nikit. = E. intermedia
valida V. Nikit. = E. intermedia
vvedenskyi Pachom. - 6

EQUISETACEAE Rich. ex DC.

Equisetum L. (*Hippochaete* Milde)

arvense L. (*E. arvense* subsp. *boreale* (Bong.) Tolm., *E. boreale* Bong.) - 1, 2, 3, 4, 5, 6
- subsp. *boreale* (Bong.) Tolm. = E. arvense
boreale Bong. = E. arvense
fluviatile L. (*E. heleocharis* Ehrh., *E. limosum* L.) - 1, 2, 3, 4, 5, 6
heleocharis Ehrh. = E. fluviatile
x **hybridum** Huter. - E. arvense L. x E. variegatum Schleich. ex Web. & Mohr
hyemale L. (*E. komarovii* Iljin, *Hippochaete hyemalis* (L.) Bruhin) - 1, 2, 3, 4, 5, 6
- subsp. *moorei* (Newm.) Syme & N.E. Br. = E. x moorei
- subsp. *trachyodon* A. Br. = E. x trachyodon
- subsp. *variegatum* (Schleich. ex Web. & Mohr) A. Br. = E. variegatum
- var. *moorei* (Newm.) Hook. & Arn. = E. x moorei
komarovii Iljin = E. hyemale
limosum L. = E. fluviatile
x **litorale** Kuhl. ex Rupr. - E. arvense L. x E. fluviatile L. - 1, 3, 4
majus Gars. = E. telmateia
maximum auct. = E. telmateia
x **meridionale** (Milde) Chiov. (*E. variegatum* Schleich. ex Web. & Mohr var. *meridionale* Milde, *Hippochaete* x *meridionalis* (Milde) Holub). - E. ramosissimum Desf. x E. variegatum Schleich. ex Web. & Mohr
x **mildeanum** Rothm. - E. pratense L. x E. sylvaticum L.
x **moorei** Newm. (*E. hyemale* L. subsp. *moorei* (Newm.) Syme & N.E. Br., *E. hyemale* var. *moorei* (Newm.) Hook. & Arn., *E. ramosissimum* Desf. subsp. *moorei* (Newm.) Maire, *Hippochaete* x *moorei* (Newm.) H.P. Fuchs). - E. hyemale L. x E. ramosissimum Desf.
palustre L. - 1, 2, 3, 4, 5, 6
pratense Ehrh. - 1, 2, 3, 4, 5, 6
ramosissimum Desf. (*Hippochaete ramosissima* (Desf.) Boern.) - 1, 2, 3, 6
- subsp. *moorei* (Newm.) Maire = E. x moorei
x *rothmaleri* Page = E. x torgesianum
scirpoides Michx. (*Hippochaete scirpoides* (Michx.) Farw., *H. scirpoides* (Michx.) Rothm. comb. superfl.) - 1, 3, 4, 5
sylvaticum L. - 1, 2, 3, 4, 5, 6
telmateia Ehrh. (*E. majus* Gars., *E. maximum* auct.) - 1, 2
x **torgesianum** Rothm. (*E.x rothmaleri* Page). - A. arvense L. x E. palustre L.
x **trachyodon** A. Br. (*E. hyemale* L. subsp. *trachyodon* A. Br., *Hippochaete* x *trachyodon* (A. Br.) Boern.). - E. hyemale L. x E. variegatum Schleich. ex Web. & Wohr - 1, 2
variegatum Schleich. ex Web. & Mohr (*E. hyemale* L. subsp. *variegatum* (Schleich. ex Web. & Mohr) A. Br., *Hippochaete variegata* (Schleich. ex Web. & Mohr) Bruhin) - 1, 2, 3, 4, 5
- var. *meridionale* Milde = E. x meridionale

Hippochaete Milde = Equisetum

hyemalis (L.) Bruhin = Equisetum hyemale
x *meridionalis* (Milde) Holub = Equisetum x meridionale
x *moorei* (Newm.) H.P. Fuchs = Equisetum x moorei
ramosissima (Desf.) Boern. = Equisetum ramosissimum
scirpoides (Michx.) Farw. = Equisetum scirpoides
scirpoides (Michx.) Rothm. = Equisetum scirpoides
x *trachyodon* (A. Br.) Boern. = Equisetum x trachyodon
variegata (Schleich. ex Web. & Mohr) Bruhin = Equisetum variegatum

ERICACEAE Juss. (*VACCINIACEAE* S.F. Gray)

Andromeda L.

polifolia L. - 1, 3, 4, 5
- subsp. **pumila** V. Vinogradova - 5

Arbutus L.

andrachne L. - 1, 2
***unedo** L. - 1, 2

Arcterica Cov.

nana (Maxim.) Makino - 5

Arctostaphylos Adans.

caucasica Lipsch. - 2
uva-ursi (L.) Spreng. - 1, 3, 4, 5

Arctous (A. Gray) Niedenzu

alpina (L.) Niedenzu - 1, 3, 4, 5, 6
- subsp. *erythrocarpa* (Small) M. Ivanova = A. erythrocarpa
- subsp. *japonica* (Nakai) Tatew. = A. japonica
erythrocarpa Small (*A. alpina* (L.) Niedenzu subsp. *erythrocarpa* (Small) M. Ivanova) - 4, 5
japonica Nakai (*A. alpina* (L.) Niedenzu subsp. *japonica* (Nakai) Tatew.) - 5

Botryostege Stapf

bracteata (Maxim.) Stapf - 5

■ Bruckenthalia (Salisb.) Reichenb.

■ **spiculifolia** (Salisb.) Reichenb.

Bryanthus S.G. Gemel.

gmelinii D. Don - 5

Calluna Salisb.

vulgaris (L.) Hull - 1, 3, 4

Cassiope D. Don

x **anadyrensis** Jurtz. - C. ericoides (Pall.) D. Don x C. tetragona (L.) D. Don - 5
ericoides (Pall.) D. Don - 4, 5
lycopodioides (Pall.) D. Don - 5
redowskii (Cham. & Schlecht.) G. Don fil. - 5
tetragona (L.) D. Don - 1, 3, 4, 5

Chamaedaphne Moench

calyculata (L.) Moench - 1, 3, 4, 5

Epigaea L.

gaultherioides (Boiss. & Bal.) Takht. - 2

Erica L.

arborea L. - 2
tetralix L. - 1

Eubotryoides (Nakai) Hara

grayana (Maxim.) Hara - 5

Gaultheria L.
miqueliana Takeda - 5

Harrimanella Cov.
hypnoides (L.) Cov. - 5
stelleriana (Pall.) Cov. - 5

Ledum L.
decumbens (Ait.) Lodd. ex Steud. (*L. palustre* L. subsp. *decumbens* (Ait.) Hult., *Rhododendron subarcticum* Harmaja) - 1, 3, 4, 5
hypoleucum Kom. (*L. macrophyllum* Tolm. p.p. quoad typum, *L. palustre* L. var. *diversipilosum* Nakai, *L. palustre* subsp. *diversipilosum* (Nakai) Hara, *L. palustre* var. *macrophyllum* (Tolm.) Kitag., *Rhododendron hypoleucum* (Kom.) Harmaja, *R. tolmachevii* Harmaja) - 5
macrophyllum Tolm. p.p. = L. hypoleucum
maximum (Nakai) A. Khokhr. & Mazurenko (*L. palustre* L. var. *maximum* Nakai) - 5
palustre L. (*L. tomentosum* Stokes, *Rhododendron palustre* (L.) Kron & Judd, 1990, non *R. palustre* Turcz. ex DC. 1839, *R. tomentosum* (Stokes) Harmaja) - 1, 3, 4, 5
- subsp. *angustissimum* Worosch. = L. subulatum
- subsp. *decumbens* (Ait.) Hult. = L. decumbens
- subsp. *diversipilosum* (Nakai) Hara = L. hypoleucum
- var. *diversipilosum* Nakai = L. hypoleucum
- var. *macrophyllum* (Tolm.) Kitag. = L. hypoleucum
- var. *maximum* Nakai = L. maximum
- var. *subulatum* Nakai = L. subulatum
palustriforme A. Khokhor. & Mazurenko - 5
subulatum A. Khokhr. & Mazurenko (*L. palustre* L. var. *subulatum* Nakai, *L. palustre* subsp. *angustissimum* Worosch.) - 3, 4, 5
tomentosum Stokes = L. palustre

Loiseleuria Desv.
procumbens (L.) Desv. - 1, 3, 4, 5

Menziesia Smith
pentandra Maxim. - 5

Oxycoccus Hill
hagerupii A. & D. Löve = O. palustris subsp. hagerupii
microcarpus Turcz. ex Rupr. (*Vaccinium microcarpum* (Turcz. ex Rupr.) Hook. fil., *V. oxycoccos* L. subsp. *microcarpum* (Turcz. ex Rupr.) A. Blytt, *V. oxycoccos* subsp. *microcarpum* (Turcz. ex Rupr.) Kitam. comb. superfl.) - 1, 3, 4, 5
palustris Pers. (*O. quadripetalus* Gilib., *Vaccinium oxycoccos* L.) - 1, 3, 4, 5
- subsp. **hagerupii** (A. & D. Löve) A. Khokhor. & Mazurenko (*O. hagerupii* A. & D. Löve) - 1, 3, 4, 5
quadripetalus Galib. = O. palustris

Phyllodoce Salisb.
aleutica (Spreng.) Heller - 5
caerulea (L.) Bab. - 1, 3, 4, 5

x *Rhodocinium* Avror. = Vaccinium
intermedium (Ruthe) Avror. = Vaccinium x intermedium

Rhodococcum (Rupr.) Avror. = Vaccinium
x *hybridum* Avror. = Vaccinium x hybridum
minus (Lodd.) Avror. = Vaccinium minus
vitis-idaea (L.) Avror. = Vaccinium vitis-idaea

Rhododendron L.
adamsii Rehd. - 4, 5
aureum Georgi - 3, 4, 5
- subsp. *hypopitys* (Pojark.) Worosch. = R. hypopitys
- var. *hypopitys* (Pojark.) Chamberlain = R. hypopitys
brachycarpum D. Don - 5
x **burjaticum** Malysch. - R. adamsii Rehd. x R. parvifolium Adams - 4
camtschaticum Pall. - 5
- subsp. **glandulosum** (Standl.) Hult. - 5
caucasicum Pall. - 2
x **charadzeae** A. Khokhr. & Mazurenko. - R. ungernii Trautv. x R. smirnowii Trautv. - 2
dauricum L. - 4, 5
- subsp. *mucronulatum* (Turcz.) Worosch. = R. mucronulatum
- subsp. *sichotense* (Pojark.) M.S. Alexandr. & P. Schmidt = R. sichotense
- subsp. *sichotense* (Pojark.) Worosch. = R. sichotense
fauriei Franch. - 5
hypoleucum (Kom.) Harmaja = Ledum hypoleucum
hypopitys Pojark. (*R. aureum* Georgi var. *hypopitys* (Pojark.) Chamberlain, *R. aureum* subsp. *hypopitys* (Pojark.) Worosch.) - 5
kotschyi Simonk. = R. myrtifolium
lapponicum (L.) Wahlenb. subsp. **alpinum** (Glehn) A. Khokhr. (*R. parvifolium* Adams var. *alpinum* Glehn) - 4, 5
ledebourii Pojark. - 3, 4
luteum Sweet - 1, 2
mucronulatum Turcz. (*R. dauricum* L. subsp. *mucronulatum* (Turcz.) Worosch.) - 5
- subsp. *sichotense* (Pojark.) A. Khokhr. = R. sichotense
myrtifolium Schott & Kotschy (*R. kotschyi* Simonk.) - 1
palustre (L.) Kron & Judd = Ledum palustre
parvifolium Adams - 4, 5
- var. *alpinum* Glehn = R. lapponicum subsp. alpinum
ponticum L. - 2
redowskianum Maxim. - 4, 5
schlippenbachii Maxim. - 5
sichotense Pojark. (*R. dauricum* L. subsp. *sichotense* (Pojark.) M.S. Alexandr. & P. Schmidt, *R. dauricum* subsp. *sichotense* (Pojark.) Worosch. comb. invalid., *R. mucronulatum* Turcz. subsp. *sichotense* (Pojark.) A. Khokhr.) - 5
smirnowii Trautv. - 2
x **sochadzeae** Charadze & Davlianidze. - R. caucasicum Pall. x R. ponticum L. - 2
subarcticum Harmaja = Ledum decumbens
tolmachevii Harmaja = Ledum hypoleucum
tomentosum (Stokes) Harmaja = Ledum palustre
tschonoskii Maxim. - 5
ungernii Trautv. - 2

Vaccinium L. (x *Rhodocinium* Avror., *Rhodococcum* (Rupr.) Avror.)
arctostaphylos L. - 2
axillare Nakai - 5
buergeri Miq. = V. smallii
gaultherioides Bigel. = V. uliginosum subsp. microphyllum

hirtum auct. = V. smallii

x **hybridum** (Avror.) Czer. (*Rhodococcum* x *hybridum* Avror.). - V. minus (Lodd.) Worosch. x V. vitis-idaea L. - 1

x **intermedium** Ruthe (x *Rhodocinium intermedium* (Ruthe) Avror.). - V. mirtillus L. x V. vitis-idaea L. - 1

microcarpum (Turcz. ex Rupr.) Hook. fil. = Oxycoccus microcarpus

microphyllum (Lange) Hagerup. = V. uliginosum subsp. microphyllum

minus (Lodd.) Worosch. (*V. vitis-idaea* L. *minus* Lodd., *Rhodococcum minus* (Lodd.) Avror., *Vaccinium minus* (Lodd.) Maximova, comb. superfl., *V. vitis-idaea* subsp *minus* (Lodd.) Hult.) - 1, 3, 4, 5

myrtillus L. - 1, 2, 3, 4, 5

ovalifolium Smith - 5

oxycoccos L. = Oxycoccus palustris

- subsp. *microcarpum* (Turcz. ex Rupr.) A. Blytt = Oxycoccus microcarpus

- subsp. *microcarpum* (Turcz. ex Rupr.) Kitam. = Oxycoccus microcarpus

pedris (Harshberger) Holub = V. uliginosum subsp. pedris

praestans Lamb. - 5

pubescens Wormsk. ex Hornem. = V. uliginosum subsp. alpinum

smallii A. Gray (*V. buergeri* Miq., *V. hirtum* auct.) - 5

uliginosum L. - 1, 2, 3, 4, 5

- subsp. **alpinum** (Bigel.) Hult. (*V. uliginosum* var. *alpinum* Bigel., *V. pubescens* Wormsk. ex Hornem., *V. uliginosum* subsp. *pubescens* (Wormsk. ex Hornem.) S.B. Young)

- subsp. *gaultherioides* (Bigel.) S.B. Young = V. uliginosum subsp. microphyllum

- subsp. **khokhrjakovii** Mazurenko - 2

- subsp. **microphyllum** Lange (*V. gaultherioides* Bigel., *V. microphyllum* (Lange) Hagerup, non Reinw ex Blume, 1878, nec Rydb. 1897, *V. uliginosum* subsp. *gaultherioides* (Bigel.) S.B. Young) - 1, 3, 4, 5

- subsp. **pedris** (Harshberger) S.B. Young (*V. uliginosum* var. *pedris* Harshberger, *V. pedris* (Harshberger) Holub, *V. uliginosum* f. *pedris* (Harshberger) Lepage) - 5

- subsp. *pubescens* (Wormsk. ex Hornem.) S.B. Young = V. uliginosum subsp. alpinum

- var. *alpinum* Bigel. = V. uliginosum subsp. alpinum

- var. *pedris* Harshberger = V. uliginosum subsp. pedris

- var. *vulcanorum* (Kom.) Jurtz. = V. vulkanorum

- f. *pedris* (Harshberger) Lepage = V. uliginosum subsp. pedris

vitis-idaea L. (*Rhodococcum vitis-idaea* (L.) Avror.) - 1, 2, 3, 4, 5

- subsp. *minus* (Lodd.) Hult. = V. minus

- *minus* Lodd. = V. minus

vulkanorum Kom. (*V. uliginosum* L. var. *vulcanorum* (Kom.) Jurtz.) - 5

yatabei Makino - 5

ERIOCAULACEAE Desv.
Eriocaulon L.

atrum Nakai - 5

chinorossicum Kom. - 5

cinereum R. Br. (*E. sieboldianum* Siebold & Zucc. ex Steud., *E. cinereum* var. *sieboldianum* (Siebold & Zucc. ex Steud.) Murata, *E. cinereum* var. *sieboldianum* (Siebold & Zucc. ex Steud.) T. Koyama) - 6

- var. *sieboldianum* (Siebold & Zucc. ex Steud.) Murata = E. cinereum

- var. *sieboldianum* (Siebold & Zucc. ex Steud.) T. Koyama = E. cinereum

decemflorum Maxim. - 5

desulavii Tzvel. - 5

komarovii Tzvel. (*E. robustius* auct.) - 5

robustius auct. = E. komarovii

sachalinense Miyabe & Nakai - 5

schischkinii Tzvel. - 5

sieboldianum Siebold & Zucc. ex Steud. = E. cinereum

ussuriense Koern. ex Regel - 5

EUCOMMIACEAE Engl.
*Eucommia Oliv.

*ulmoides Oliv. - 2

EUPHORBIACEAE Juss.
Acalypha L.

australis L. - 1, 2, 5

indica L. - 2(?)

*Aleurites Forst. & Forst. fil.

*fordii Hemsley - 2

Andrachne L.

asperula Nevski - 6

buschiana Pojark. (*A. fruticulosa* auct. p.p.) - 2

fedtschenkoi C. Koss. - 6

filiformis Pojark. (*A. fruticulosa* auct. p.p.) - 2

fruticulosa auct. = A. buschiana and A. filiformis

pusilla Pojark. - 6

pygmaea C. Koss. - 6

rotundifolia C.A. Mey. (*A. virga-tenuis* Nevski) - 2, 6

rupestris Pazij - 6

stenophylla C. Koss. - 6

telephioides L. - 1, 2

virga-tenuis Nevski = A. rotundifolia

vvedenskyi Pazij - 6

Arachne Neck. = Leptopus

colchica (Fisch. & C.A. Mey. ex Boiss.) Pojark. = Leptopus colchicus

Chamaesyce S.F. Gray = Euphorbia

granulata (Forssk.) Sojak var. *turcomanica* (Boiss.) V.S. Raju & P.N. Rao = Euphorbia turcomanica

massiliensis (DC.) Galushko = Euphorbia massiliensis

Chrozophora Adr. Juss.

cordifolia Pazij - 6

gracilis Fisch. & C.A. Mey. ex Ledeb. - 6

hierosolymitana Spreng. - 2, 6

lepidocarpa Pazij - 6

mujukumi Nassimova - 6

obliqua (Vahl) Adr. Juss. ex Spreng. - 6

sabulosa Kar. & Kir. - 6

tinctoria (L.) Adr. Juss. - 1, 2, 6

Cystidospermum Prokh. = Euphorbia

cheirolepis (Fisch. & C.A. Mey. ex Ledeb.) Prokh. = Euphorbia cheirolepis

petiolatum (Banks & Soland.) Nassimova = Euphorbia petiolata

Dematra Rafin. = Euphorbia

cheirolepis (Fisch. & C.A. Mey. ex Ledeb.) Sojak = Euphorbia cheirolepis

petiolata (Banks & Soland.) Sojak = Euphorbia petiolata

Euphorbia L. (*Chamaesyce* S.F. Gray, *Cystidospermum* Prokh., *Dematra* Rafin., *Galarhoeus* Haw., *Tithymalus* Gaertn.)

acuminata Lam. = E. falcata
agraria Bieb. - 1, 2
alaica Prokh. - 6
alatavica Boiss. - 6
aleppica L. - 1, 2
alpina C.A. Mey. - 3, 4
altaica C.A. Mey. - 3, 4
amygdaloides L. (*Tithymalus amygdaloides* (L.) Gars.) - 1, 2
andrachnoides Schrenk - 3, 6
angulata Jacq. - 1
anisopetala Prokh. - 6
ardonensis Galushko (*Tithymalus ardonensis* (Galushko) Galushko, *T. ardonensis* (Galushko) Sojak, comb. superfl.) - 2
aristata Schmalh. (*E. soongarica* Boiss. subsp. *aristata* (Schmalh.) Prokh. & Kuzm., *Tithymalus soongaricus* (Boiss.) Prokh. subsp. *aristatus* (Schmalh.) Sojak) - 2
armena Prokh. (*Tithymalus marschallianus* (Boiss.) Klotzsch & Garcke subsp. *armenus* (Prokh.) Sojak) - 2
arvalis Boiss. & Heldr. - 2
aserbajdzhanica Bordz. - 2, 6
astrachanica C.A. Mey. ex Claus = E. praecox
astrachanica C.A. Mey. ex Prokh. = E. praecox
aucheri Boiss. - 6
aulacosperma Boiss. - 2
austriaca A. Kerner subsp. *sojakii* Chrtek & Krisa = E. sojakii
- subsp. *tauricola* (Prokh.) Chrtek & Krisa = E. tauricola
balkhanica Tarass. = E. oidorhiza
baxanica Galushko = E. subtilis
bessarabica Klok. = E. klokoviana
biglandulosa Desf. = E. rigida
blepharophylla C.A. Mey. - 6
boissieriana (Woronow) Prokh. (*E. virgata* Waldst. & Kit. subsp. *boissieriana* (Woronow) Prokh.) - 2, 6
borodinii Sambuk - 1
borszczowii Prokh. - 1, 3
buchtormensis C.A. Mey. - 3, 6
buhsei Boiss. - 6
bungei Boiss. - 6
buschiana Grossh. (*Tithymalus buschianus* (Grossh.) Sojak) - 2
canescens L. (*E. chamaesyce* L. subsp. *canescens* (L.) Prokh.) - 2, 6
carniolica Jacq. - 1
carpatica Woloszcz. (*Tithymalus carpaticus* (Woloszcz.) A. & D. Löve) - 1
caucasica Dubovik - 2
chamaesyce L. - 1, 2
- subsp. *canescens* (L.) Prokh. = E. canescens
- subsp. *massiliensis* (DC.) Thell. = E. massiliensis

chankoana Worosch. (*E. esula* L. subsp. *chankoana* (Worosch.) Worosch., *Tithymalus chankoanus* (Worosch.) Sojak) - 5
cheirolepis Fisch. & C.A. Mey. ex Ledeb. (*Cystidospermum cheirolepis* (Fisch. & C.A. Mey. ex Ledeb.) Prokh., *Dematra cheirolepis* (Fisch. & C.A. Mey. ex Ledeb.) Sojak) - 6
chimaera Lipsky (*Tithymalus chimaerus* (Lipsky) Galushko) - 2
colchica (Litv.) Geltm. comb. nova, hoc loco (*E. petrophila* C.A. Mey. var. *colchica* Litv. 1898, Acta Horti Petropol. 14, 2 : 299) - 2
condylocarpa Bieb. - 2
coniosperma Boiss. & Buhse - 2
consanguinea Schrenk - 6
cretophila Klok. - 1
croizatii (Hurusawa) Kitag. = E. leoncroizatii
cyparissias L. - 1
cyrtophylla Prokh. - 6
dahurica Peschkova - 4
deltobracteata Prokh. - 6
densa Schrenk - 6
- subsp. **badghysi** Botsch. - 6
densiuscula M. Pop. - 6
densiusculiformis (Pazij) Botsch. (*Tithymalus densiusculiformis* (Pazij) Pazij) - 6
dentata Michx. - 1(alien), 2(alien)
■ **denticulata** Lam.
discolor Ledeb. - 3, 4, 5
- subsp. *karoi* (Freyn) Basargin = E. maackii
djimilensis Boiss. - 2
dubovikiae Qudejans (*E. pinetorum* Dubovik, 1977, non (Small) Webster, 1967) - 2
dulcis L. (*E. purpurata* auct.) - 1
epithymoides auct. p.p. = E. lingulata
eriophora Boiss. - 2
esula L. - 1
- subsp. *chankoana* (Worosch.) Worosch. = E. chankoana
eugeniae Prokh. - 2
exigua L. - 1, 2
falcata L. (*E. acuminata* Lam., *E. falcata* subsp. *acuminata* (Lam.) Simonk., *E. tatianae* Al.Theod., *Tithymalus falcatus* (L.) Klotzsch & Garcke, *T. falcatus* subsp. *acuminatus* (Lam.) Sojak) - 1, 2, 5(alien), 6
- subsp. *acuminata* (Lam.) Simonk. = E. falcata
ferganensis B. Fedtsch. - 6
fischeriana Steud. (*E. pallasii* Turcz. ex Ledeb., *Tithymalus fischerianus* (Steud.) Sojak) - 4
- var. *komaroviana* (Prokh.) Chu = E. komaroviana
flerowii Woronow & Fler. - 2
forskalii J. Gay - 2
franchetii B. Fedtsch. - 6
gerardiana Jacq. var. *hohenackeri* Boiss. = E. seguieriana subsp. hohenackeri
glaberrima C. Koch - 2
glareosa Pall. ex Bieb. (*E. maleevii* Tamamsch., *E. nicaeensis* All. subsp. *glareosa* (Pall. ex Bieb.) A. Radcliffe-Smith, *E. nicaeensis* subsp. *maleevii* (Tamamsch.) Qudejans) - 1, 2
- subsp. *stepposa* (Zoz) Kuzm. = E. stepposa
glomerulans Prokh. - 2
gmelinii Steud. - 1, 3
  Probably, a hybrid E. discolor Ledeb. x E. microcarpa Prokh.
goldei Prokh. (*E. nicaeensis* All.subsp. *goldei* (Prokh.) Greuter & Burdet) - 1
graeca Boiss. & Sprun. = E. taurinensis
graminifolia Vill. subsp. *zhiguliensis* (Prokh.) Qudejans = E. zhiguliensis

granulata Forssk. var. *turcomanica* (Boiss.) Hadidi = E. turcomanica

grossheimii Prokh. - 2

guntensis Prokh. - 6

helioscopia L. - 1, 2, 6

- subsp. *helioscopioides* (Loscos & Pardo) O. Bolos = E. helioscopioides

- subsp. *hiemalis* A. Khokhr. = E. helioscopioides

helioscopioides Loscos & Pardo (*E. helioscopia* L. subsp. *helioscopioides* (Loscos & Pardo) O. Bolos, *E. helioscopia* subsp. *hiemalis* A. Khokhr.) - 1, 2

heptapotamica Golosk. - 6

heteradena Jaub. & Spach (*E. ispahanica* Boiss.) - 2

hirsuta L. (*E. pubescens* Vahl) - 2

humifusa Schlecht. - 1, 2, 3, 4, 5, 6

humilis C.A. Mey. - 3, 6

hyrcana Grossh. - 2

iberica Boiss. (*E. vedica* Ter-Chatsch.) - 2, 5(alien)

inderiensis Less. ex Kar. & Kir. - 6

indica Lam. - 2

irgisensis Litv. - 6

ispahanica Boiss. = E. heteradena

jasiewiczii (Chrtek & Krisa) Dubovik (*Tithymalus jasiewiczii* Chrtek & Krisa, *Euphorbia jasiewiczii* (Chrtek & Krisa) A. Radcliffe-Smith, comb. superfl.) - 1

jaxartica Prokh. (*E. virgata* Waldst. & Kit. subsp. *jaxartica* (Prokh.) Prokh., *E. waldsteinii* (Sojak) A. Radcliffe-Smith subsp. *jaxartica* (Prokh.) Qudejans, *Tithymalus graminifolius* (Vill.) Sojak subsp. *jaxarticus* (Prokh.) Sojak) - 6

kaleniczenkoi Czern. - 1

karoi Freyn = E. maackii

kemulariae Ter-Chatsch. - 2

kitaibelii Klok. & Dubovik = E. virgata

klokoviana Railjan (*E. bessarabica* Klok. 1955, non E. bessarabica Prod. 1930, *Tithymalus klokovianus* (Railjan) Holub) - 1

klokovii Dubovik (*E. pseudovillosa* Klok. 1955, non Prod. 1953) - 1

komaroviana Prokh. (*E. fischeriana* Steud. var. *komaroviana* (Prokh.) Chu) - 5

kopetdaghi Prokh. - 6

kotovii Klok. - 1

kudrjaschevii (Pazij) Prokh. - 6

lamprocarpa Prokh. (*E. soongarica* Boiss. subsp. *lamprocarpa* (Prokh.) Prokh., *Tithymalus soongaricus* (Boiss.) Prokh. subsp. *lamprocarpus* (Prokh.) Sojak) - 6

lanata Sieber ex Spreng. = E. petiolata

lathyris L. - 2

latifolia C.A. Mey. - 3, 6

ledebourii Boiss. - 1, 2

leoncroizatii Qudejans (*E. croizatii* (Hurusawa) Kitag. 1956, non Leandri, 1946, *Galarhoeus croizatii* Hurusawa) - 5

leptocaula Boiss. - 1, 2

- var. *praecox* Fisch. ex Boiss. = E. praecox

lingulata Heuff. (*E. polychroma* A. Kerner subsp. *lingulata* (Heuff.) Soo, *E. epithymoides* auct. p.p., *E. polychroma* auct.) - 1

lipskyi Prokh. - 6

lucida Waldst. & Kit. - 1, 2, 3

lucorum Rupr. ex Maxim. - 5

maackii Meinsh. (*E. discolor* Ledeb. subsp. *karoi* (Freyn) Basargin, *E. karoi* Freyn, *Galarhoeus esula* (L.) Rydb. subsp. *maackii* (Meinsh.) Hurusawa, p.p.) - 4

macrocarpa Boiss. & Buhse (*Tithymalus notabilis* Sojak) - 2

macroceras Fisch. & C.A. Mey. - 2

macroclada Boiss. - 2

macrorhiza C.A. Mey. - 3, 6

maculata L. - 1(alien), 2(alien), 5(alien)

maleevii Tamamsch. = E. glareosa

mandshurica Maxim. (*Tithymalus mandshuricus* (Maxim.) Sojak) - 5

marginata Pursh (*Tithymalus marginatus* (Pursh) Cockerell ex F.P. Daniels) - 2(alien)

marschalliana Boiss. - 2

- subsp. *woronowii* (Grossh.) Prokh. = E. woronowii

massiliensis DC. (*Chamaesyce massiliensis* (DC.) Galushko, *Euphorbia chamaesyce* L. subsp. *massiliensis* (DC.) Thell.) - 1

meyeriana Galushko = E. subtilis

micrantha Steph. = E. stricta

microcarpa Prokh. - 3, 6

microsphaera Boiss. - 2, 6

mongolica Prokh. - 4

monocyathium Prokh. - 6

monostyla Prokh. - 6

mucronulata Prokh. - 6

myrsinites L. - 1

nicaeensis All. subsp. *glareosa* (Pall. ex Bieb.) A. Radcliffe-Smith = E. glareosa

- subsp. *goldei* (Prokh.) Greuter & Burdet = E. goldei

- subsp. *maleevii* (Tamamsch.) Qudejans = E. glareosa

- subsp. *stepposa* (Zoz) Greuter & Burdet = E. stepposa

normannii Schmalh. ex Lipsky - 2

novorossica Dubovik - 2

nutans Lag. - 2

oblongifolia (C. Koch) C. Koch - 1, 2

oidorhiza Pojark. (*E. balkhanica* Tarass., *Tithymalus oidorhizus* (Pojark.) Sojak) - 6

orientalis L. - 2

oschtenica Galushko (*Tithymalus oschtenicus* (Galushko) Galushko, *T. oschtenicus* (Galushko) Sojak, comb. superfl.) - 2

pachyrhiza Kar. & Kir. - 6

pallasii Turcz. ex Ledeb. = E. fischeriana

palustris L. - 1, 2, 3

pamirica Prokh. - 6

panjutinii Grossh. (*Tithymalus panjutinii* (Grossh.) Galushko, comb. superfl., *T. panjutinii* (Grossh.) Sojak) - 2

paralias L. - 1, 2

peplis L. - 1, 2

peploides Gouan (*Tithymalus peplus* (L.) Gaertn. subsp. *peploides* (Gouan) Sojak) - 6

peplus L. - 1, 2

petiolata Banks & Soland. (*Cystidospermum petiolatum* (Banks & Soland.) Nassimova, *Dematra petiolata* (Banks & Soland.) Sojak, *Euphorbia lanata* Sieber ex Spreng., *Tithymalus petiolatus* (Banks & Soland.) Sojak) - 6

petrophila C.A. Mey. (*E. subhastifolia* Klok., *Tithymalus petrophilus* (C.A. Mey.) Sojak) - 1, 2

- var. *colchica* Litv. = E. colchica

pilosa L. - 3, 4, 6

pinetorum Dubovik = E. dubovikiae

platyphyllos L. - 1, 2

poecilophylla Prokh. - 6

polychroma auct. = E. lingulata

polychroma A. Kerner subsp. *lingulata* (Heuff.) Soo = E. lingulata

polytimetica Prokh. - 6

pontica Prokh. - 2

popovii Rotschild (*Tithymalus popovii* (Rotschild) Sojak) - 6

Probably, a hybrid E. consanquinea Schrenk x E. sororia Schrenk.

praecox (Fisch. ex Boiss.) B. Fedtsch. & Fler. (*E. leptocaula* var. *praecox* Fisch. ex Boiss., *E. astrachanica* C.A. Mey. ex Claus, nom. invalid., *E. astrachanica* C.A. Mey. ex Prokh. nom. illegit. superfl.) - 1

procera Bieb. - 2

prokhanovii M. Pop. = E. tianschanica

pseudagraria P. Smirn. - 1, 3

pseudoglareosa Klok. - 1

pseudovillosa Klok. = E. klokovii

pubescens Vahl = E. hirsuta

purpurata auct. = E. dulcis

rapulum Kar. & Kir. - 6

rhabdotosperma A. Radcliffe-Smith - 2

rigida Bieb. (*E. biglandulosa* Desf., *Tithymalus rigidus* (Bieb.) Sojak) - 1, 2

roschanica (Ikonn.) Czer. (*Tithymalus roschanicus* Ikonn., *Euphorbia roschanica* (Ikonn.) Qudejans, comb. superfl.) - 6

rosularis Al. Theod. (*Tithymalus rosularis* (Al. Theod.) Pazij) - 6

rupestris C.A. Mey. - 3

salicifolia Host - 1

sarawschanica Regel - 6

sareptana A. Beck. - 1, 2

savaryi Kiss - 5

schizoceras Boiss. & Hohen. - 2

schugnanica B. Fedtsch. - 6

sclerocyathium Korov. & M. Pop. - 6

scripta Somm. & Levier - 2

seguieriana Neck. - 1, 2, 3, 6

- subsp. **hohenackeri** (Boiss.) Rech. fil. (*E. gerardiana* Jacq. var. *hohenackeri* Boiss.) - 2

semivillosa Prokh. - 1, 3

serrulata Thuill. = E. stricta

sewerzowii Herd. ex Prokh. - 6

sieboldiana Morr. & Decne. (*Tithymalus sieboldianus* (Morr. & Decne.) Prokh. nom. altern., *T. sieboldianus* (Morr. & Decne.) Hara, comb. superfl.) - 5

sogdiana M. Pop. - 6

sojakii (Chrtek & Krisa) Dubovik (*E. austriaca* A. Kerner subsp. *sojakii* Chrtek & Krisa, *Tithymalus sojakii* (Chrtek & Krisa) Chrtek & Krisa, *T. sojakii* (Chrtek & Krisa) Holub, comb. superfl.) - 1

soongarica Boiss. - 1, 3, 6

- subsp. *aristata* (Schmalh.) Prokh. & Kuzm. = E. aristata

- subsp. *lamprocarpa* (Prokh.) Prokh. = E. lamprocarpa

sororia Schrenk - 6

spinidens Bornm. ex Prokh. - 6

squamosa Willd. - 2

stepposa Zoz (*E. glareosa* Pall. ex Bieb. subsp. *stepposa* (Zoz) Kuzm., *E. nicaeensis* All. subsp. *stepposa* (Zoz) Greuter & Burdet) - 1, 2

stocksiana Boiss. (*Tithymalus stocksianus* (Boiss.) Sojak) - 6

stricta L. (*E. micrantha* Steph., *E. serrulata* Thuill., *Tithymalus micranthus* (Steph.) Sojak, *T. serrulatus* (Thuill.) Holub) - 1, 2, 6

subcordata C.A. Mey. (*Galarhoeus caesius* (C.A. Mey.) Hurusawa, *G. esula* (L.) Rydb. subsp. *maackii* (Meinsh.) Hurusava var. *subcordatus* (C.A. Mey.) Hurusawa) - 1, 3, 6

subhastifolia Klok. = E. petrophila

subtilis Prokh. (*E. baxanica* Galushko, *E. meyeriana* Galushko, *Tithymalus baxanicus* (Galushko) Galushko, *T. meyerianus* (Galushko) Galushko) - 1, 2, 3(?)

szovitsii Fisch. & C.A. Mey. - 2, 6

talastavica Prokh. - 6

tanaitica Pacz. (*Tithymalus tanaiticus* (Pacz.) Galushko) - 1

tatianae Al. Theod. = E. falcata

tauricola Prokh. (*E. austriaca* A. Kerner. subsp. *tauricola* (Prokh.) Chrtek & Krisa, *Tithymalus tauricola* Prokh. nom. altern., *T. tauricola* (Prokh.) Holub, comb. superfl.) - 1

taurinensis All. (*E. graeca* Boiss. & Sprun.) - 1, 2

terracina L. - 2

teskensuensis Orazova - 6

tianschanica Prokh. (*E. prokhanovii* M. Pop.) - 6

tibetica Boiss. - 6

transoxana Prokh. - 6

tranzschelii Prokh. - 6

triodonta Prokh. - 6

tristis Bess. ex Bieb. - 1

tshuiensis (Prokh.) Serg. - 3

turcomanica Boiss. (*Chamaesyce granulata* (Forrsk.) Sojak var. *turcomanica* (Boiss.) V.S. Raju & P.N. Rao, *Euphorbia granulata* Forrsk. var. *turcomanica* (Boiss.) Hadidi) - 2, 6

turczaninowii Kar. & Kir. - 6

turkestanica Regel - 6

tyraica Klok. & Artemcz. - 1

undulata Bieb. - 1, 2, 3, 6

uralensis Fisch. ex Link - 1, 3, 6

valdevillosocarpa Arvat & E.I. Nyarady (*Tithymalus valdevillosocarpus* (Arvat & E.I. Nyarady) Chrtek & Krisa) - 1

vedica Ter-Chatsch. = E. iberica

villosa Waldst. & Kit. - 1, 2

virgata Waldst. & Kit. (*E. kitaibelii* Klok. & Dubovik, nom. invalid., *E. waldsteinii* (Sojak) Czer., *E. waldsteinii* (Sojak) A. Radcliffe-Smith, comb. superfl., *Tithymalus graminifolius* (Vill.) Sojak subsp. *waldsteinii* (Sojak) Sojak, *T. tommasinianus* (Bertol.) Sojak subsp. *waldsteinii* (Sojak) Sojak, *T. waldsteinii* Sojak) - 1, 2, 3, 4, 5

- subsp. *boissieriana* (Woronow) Prokh. = E. boissieriana

- subsp. *jaxartica* (Prokh.) Prokh. = E. jaxartica

- subsp. *zhiguliensis* (Prokh.) Prokh. = E. zhiguliensis

virgultosa Klok. (*Tithymalus tommasinianus* (Bertol.) Sojak subsp. *virgultosus* (Klok.) Sojak, *T. virgultosus* (Klok.) Holub) - 1

volgensis Krysht. - 1

volhynica Bess. ex Racib. (*Tithymalus volhynicus* (Bess. ex Racib.) Holub) - 1

waldsteinii (Sojak) Czer. = E. virgata

waldsteinii (Sojak) A. Radcliffe-Smith = E. virgata

- subsp. *jaxartica* (Prokh.) Qudejans = E. jaxartica

wittmannii Boiss. - 2

woronowii Grossh. (*E. marschalliana* Boiss. subsp. *woronowii* (Grossh.) Prokh., *Tithymalus marschallianus* (Boiss.) Klotzsch & Garcke subsp. *woronowii* (Grossh.) Sojak) - 1

yaroslavii Poljak. (*Tithymalus yaroslavii* (Poljak.) Sojak) - 6

zhiguliensis Prokh. (*E. graminifolia* Vill. subsp. *zhiguliensis* (Prokh.) Qudejans, *E. virgata* Waldst. & Kit. subsp. *zhiguliensis* (Prokh.) Prokh., *Tithymalus graminifolius* (Vill.) Sojak subsp. *zhiguliensis* (Prokh.) Sojak) - 1

Fluggea Willd.

ussuriensis Pojark. - 5

Galarhoeus Haw. = Euphorbia

caesius (C. A. Mey.) Hurusawa = Euphorbia subcordata

croizatii Hurusawa = Euphorbia leoncroizatii
esula (L.) Rydb. subsp. *maackii* (Meinsh.) Hurusawa, p.p. = Euphorbia maackii
- subsp. *maackii* (Meinsh.) Hurusawa var. *subcordatus* (C.A. Mey.) Hurusawa = Euphorbia subcordata

Leptopus Decne. (*Arachne* Neck. nom. specif. uninom.)

colchicus (Fisch. & C.A. Mey. ex Boiss.) Pojark. (*Arachne colchica* (Fisch. & C.A. Mey. ex Boiss.) Pojark.) - 2

Mercurialis L.

annua L. - 1, 2
ovata Sternb. & Hoppe - 1, 2
perennis L. - 1, 2
taurica Juz. - 1, 2

Phyllanthus L.

ussuriensis Rupr. & Maxim. - 5

*Ricinus L.

***communis** L. - 1, 2, 6

Securinega Comm. ex Juss.

suffruticosa (Pall.) Rehd. - 4, 5

Tithymalus Gaertn. = Euphorbia

amygdaloides (L.) Gars. = Euphorbia amygdaloides
ardonensis (Galushko) Galushko = Euphorbia ardonensis
ardonensis (Galushko) Sojak = Euphorbia ardonensis
baxanicus (Galushko) Galushko = Euphorbia subtilis
buschianus (Grossh.) Sojak = Euphorbia buschiana
carpaticus (Woloszcz.) A. & D. Löve = Euphorbia carpatica
chankoanus (Worosch.) Sojak = Euphorbia chankoana
chimaerus (Lipsky) Galushko = Euphorbia chimaera
densiusculiformis (Pazij) Pazij = Euphorbia densiusculiformis
falcatus (L.) Klotzsch & Garcke = Euphorbia falcata
- subsp. *acuminatus* (Lam.) Sojak = Euphorbia falcata
fischerianus (Steud.) Sojak = Euphorbia fischeriana
graminifolius (Vill.) Sojak subsp. *jaxarticus* (Prokh.) Sojak = Euphorbia jaxartica
- subsp. *waldsteinii* (Sojak) Sojak = Euphorbia virgata
- subsp. *zhiguliensis* (Prokh.) Sojak = Euphorbia zhiguliensis
jasiewiczii Chrtek & Krisa = Euphorbia jasiewiczii
klokovianus (Railjan) Holub = Euphorbia klokoviana
mandshuricus (Maxim.) Sojak = Euphorbia mandshurica
marginatus (Pursh) Cockerell ex F.P. Daniels = Euphorbia marginata
marschallianus (Boiss.) Klotzsch & Garcke subsp. *armenus* (Prokh.) Sojak = Euphorbia armena
- subsp. *woronowii* (Grossh.) Sojak = Euphorbia woronowii
meyerianus (Galushko) Galushko = Euphorbia subtilis
micranthus (Steph.) Sojak = Euphorbia stricta
notabilis Sojak = Euphorbia macrocarpa
oidorhizus (Pojark.) Sojak = Euphorbia oidorhiza
oschtenicus (Galushko) Galushko = Euphorbia oschtenica
oschtenicus (Galushko) Sojak = Euphorbia oschtenica
panjutinii (Grossh.) Galushko = Euphorbia panjutinii
panjutinii (Grossh.) Sojak = Euphorbia panjutinii
peplus (L.) Gaertn. subsp. *peploides* (Gouan) Sojak =

Euphorbia peploides
petiolatus (Banks & Soland.) Sojak = Euphorbia petiolata
petrophilus (C.A. Mey.) Sojak = Euphorbia petrophila
popovii (Rotschild) Sojak = Euphorbia popovii
rigidus (Bieb.) Sojak = Euphorbia rigida
roschanicus Ikonn. = Euphorbia roschanica
rosularis (Al. Theod.) Pazij = Euphorbia rosularis
serrulatus (Thuill.) Holub = Euphorbia stricta
sieboldianus (Morr. & Decne.) Hara = Euphorbia sieboldiana
sieboldianus (Morr. & Decne.) Prokh. = Euphorbia sieboldiana
sojakii (Chrtek & Krisa) Chrtek & Krisa = Euphorbia sojakii
sojakii (Chrtek & Krisa) Holub = Euphorbia sojakii
soongaricus (Boiss.) Prokh. subsp. *aristatus* (Schmalh.) Sojak = Euphorbia aristata
- subsp. *lamprocarpus* (Prokh.) Sojak = Euphorbia lamprocarpa
stocksianus (Boiss.) Sojak = Euphorbia stocksiana
tanaiticus (Pacz.) Galushko = Euphorbia tanaitica
tauricola Prokh. = Euphorbia tauricola
tauricola (Prokh.) Holub = Euphorbia tauricola
tommasinianus (Bertol.) Sojak subsp. *virgultosus* (Klok.) Sojak = Euphorbia virgultosa
- subsp. *waldsteinii* (Sojak) Sojak = Euphorbia virgata
valdevillosocarpus (Arvat & E.I. Nyarady) Chrtek & Krisa = Euphorbia valdevillosocarpa
virgultosus (Klok.) Holub = Euphorbia virgultosa
volhynicus (Bess. ex Racib.) Holub = Euphorbia volhynica
waldsteinii Sojak = Euphorbia virgata
yaroslavii (Poljak.) Sojak = Euphorbia yaroslavii

FABACEAE Lindl. (*CAESALPINIACEAE* R. Br., *MIMOSACEAE* R. Br.)[LEGUMINOSAE Juss., PAPILIONACEAE Giseke]

*Acacia Hill

***dealbata** Link - 2
***melanoxylon** R. Br. - 2
***retinodes** Schlecht. - 2

Albizia Durazz.

julibrissin Durazz. - 1(cult.), 2, 6(cult.)

Alhagi Hill

canescens (Regel) Shap. (*A. maurorum* Medik. subsp. *canescens* (Regel) Yakovl.) - 6
kirghisorum Schrenk (*A. maurorum* Medik. subsp. *kirghisorum* (Schrenk) Yakovl.) - 3, 6
maurorum auct. p.p. = A. persarum and A. pseudalhagi
maurorum Medik. subsp. *canescens* (Regel) Yakovl. = A. canescens
- subsp. *kirghisorum* (Schrenk) Yakovl. = A. kirghisorum
persarum Boiss. & Buhse (*A. pseudalhagi* (Bieb.) Fisch. subsp. *persarum* (Boiss. & Buhse) Takht., *A. maurorum* auct. p.p.) - 2, 6
pseudalhagi (Bieb.) Fisch. (*A. maurorum* auct. p.p.) - 1, 2, 3, 6
- subsp. *persarum* (Boiss. & Buhse) Takht. = A. persarum Buhse
sparsifolia (Shap.) Shap. - 6

Alophotropis (jaub. & Spach) Grossh. = Vavilovia

aucheri (Jaub. & Spach) Grossh. = Vavilovia formosa
formosa (Stev.) Grossh. = Vavilovia formosa

Amarenus C. Presl = Chrysaspis

spadiceus (L.) C. Presl = Chrysaspis spadicea

Amblyotropis Kitag. = Gueldenstaedtia

monophylla (Fisch.) C.Y. Wu = Gueldenstaedtia monophylla
verna (Georgi) Kitag. = Gueldenstaedtia verna

Ammodendron Fisch. (*Podalyria* Willd. p.p.)

argenteum (Pall.) Kryl. = A. bifolium
argenteum (Siev.) O. Kuntze = A. bifolium
bifolium (Pall.) Yakovl. (*Sophora bifolia* Pall., *Ammodendron argenteum* (Pall.) Kryl., *A. argenteum* (Siev.) O. Kuntze, comb. invalid., *A. floribundum* Zing., *A. lehmannii* Bunge, *Podalyria argentea* (Pall.) Willd., *Robinia argentea* Siev. nom. nud., *Sophora argentea* Pall. 1797, non Salisb. 1796) - 3, 6
conollyi Bunge - 6
eichwaldii Ledeb. - 6
floribundum Zing. = A. bifolium
karelinii Fisch. & C.A. Mey. (*A. longiracemosum* Raik.) - 6
lehmannii Bunge = A. bifolium
longiracemosum Raik. = A. karelinii

Ammopiptanthus Cheng fil. (*Piptanthus* auct., *Podalyria* auct.)

nanus (M. Pop.) Cheng fil. (*Piptanthus nanus* M. Pop., *Podalyria nana* (M. Pop.) M. Pop.) - 6

Ammothamnus Bunge

lehmannii Bunge (*Sophora lehmannii* (Bunge) Yakovl.) - 6
songoricus (Schrenk) Lipsky ex Pavl. (*Sophora songorica* Schrenk) - 6

Amoria C. Presl (*Galearia* C. Presl, *Mistyllus* C. Presl)

ambigua (Bieb.) Sojak (*Trifolium ambiguum* Bieb.) - 1, 2
angulata (Waldst. & Kit.) C. Presl (*Trifolium angulatum* Waldst. & Kit.) - 1, 2
bobrovii (Chalilov) Roskov (*Trifolium bobrovii* Chalilov) - 2
bonannii (C. Presl) Roskov (*Trifolium bonannii* C.Presl, *Galearia fragifera* (L.) C. Presl subsp. *bonannii* (C.Presl) Sojak, *Trifolium fragiferum* L. subsp. *bonannii* (C. Presl) Sojak, *T. neglectum* C.A. Mey.) - 1, 2, 6
bordzilowskyi (Grossh.) Roskov (*Trifolium bordzilowskyi* Grossh.) - 2
egrissica (A.D. Mikheev & Magulaev) Roskov (*Trifolium egrissicum* A.D. Mikheev & Magulaev) - 2
elizabethae (Grossh.) Roskov (*Trifolium elizabethae* Grossh.) - 2
fragifera (L.) Roskov (*Trifolium fragiferum* L., *Galearia fragifera* (L.) Bobr. comb. superfl., *G. fragifera* (L.) C. Presl) - 1, 2, 3, 6
glomerata (L.) Sojak (*Trifolium glomeratum* L.) - 2
hybrida (L.) C. Presl (*Trifolium hybridum* L., *Amoria hybrida* subsp. *elegans* (Savi) Sojak, *Trifolium elegans* Savi,

T. hybridum subsp. *elegans* (Savi) Aschers. & Graebn., *T. hybridum* var. *elegans* (Savi) Boiss.) - 1, 2, 3, 4, 5, 6
- subsp. *elegans* (Savi) Sojak = A. hybrida
montana (L.) Sojak (*Trifolium montanum* L.) - 1, 2, 3, 4(alien), 5(alien), 6
nigrescens (Viv.) Fourr. (*Trifolium nigrescens* Viv.) - 2
pallescens (Schreb.) C. Presl (*Trifolium pallescens* Schreb., *T. glareosum* Schleich.) - 1
physodes (Stev. ex Bieb.) Roskov (*Trifolium physodes* Stev. ex Bieb., *Galearia physodes* (Stev.) Sojak) - 2
raddeana (Trautv.) Roskov (*Trifolium raddeanum* Trautv., *Galearia raddeana* (Trautv.) Bobr., *G. raddeana* (Trautv.) Sojak, comb. superfl.) - 2
repens (L.) C. Presl (*Trifolium repens* L., *Amoria repens* subsp. *orbelica* (Velen.) Sojak, *Trifolium orbelicum* Velen., *T. repens* subsp. *orbelicum* (Velen.) Pawl.) - 1, 2, 3, 4, 5, 6
- subsp. *orbelica* (Velen.) Sojak = A. repens
resupinata (L.) Roskov (*Trifolium resupinatum* L.) - 1(alien), 2, 6(cult.)
retusa (L.) Dostal (*Trifolium retusum* L., *T. parviflorum* Ehrh.) - 1, 2
ruprechtii (Tamamsch. & Fed.) Roskov (*Trifolium ruprechtii* Tamamsch. & Fed.) - 2
spumosa (L.) Roskov (*Trifolium spumosum* L., *Mistyllus spumosus* (L.) Bobr.) - 2
suffocata (L.) Roskov (*Trifolium suffocatum* L.) - 2
talyschensis (Chalilov) Roskov (*Trifolium talyschense* Chalilov, *T. tumens* Stev. ex Bieb. subsp. *talyschense* (Chalilov) Zohary) - 2
tomentosa (L.) Roskov (*Trifolium tomentosum* L.) - 2
tumens (Stev. ex Bieb.) Roskov (*Trifolium tumens* Stev. ex Bieb.) - 2
vesiculosa (Savi) Roskov (*Trifolium vesiculosum* Savi) - 1, 2

*Amorpha L.
*fruticosa L. - 1, 5, 6

Amphicarpaea Ell. (*Falcata* auct.)

japonica (Oliv.) B. Fedtsch. (*Falcata japonica* (Oliv.) Kom.) - 5

Anthyllis L.

affinis auct. = A. carpatica
alpestris (Kit. ex Schult.) Reichenb. (*A. vulneraria* L. var. *alpestris* Kit. ex Schult.) - 1
arenaria (Rupr.) Juz. - 1
- var. *fennica* (Jalas) Kloczkova = A. fennica
x **baltica** Juz. ex Kloczkova. - A. maritima Schweigg. x A. vulneraria L. - 1
biebersteiniana Popl. (*A. daghestanica* Chinth., *A. vulneraria* L. var. *biebersteiniana* Taliev, nom. provis., *A. vulneraria* subsp. *pulchella* auct. p.p.) - 1, 2
boissieri (Sagor.) Grossh. p.p. = A. lachnophora and A. variegata
carpatica Pant. (*A. vulneraria* L. subsp. *carpatica* (Pant.) Nym., *A. affinis* auct.) - 1
caucasica (Grossh.) Juz. = A. variegata
coccinea (L.) G.Beck - 1
colorata Juz. - 1
daghestanica Chinth. = A. biebersteiniana
fennica (Jalas) Akulova (*A. vulneraria* L. subsp. *fennica* Jalas, *A. arenaria* (Rupr.) Juz. var. *fennica* (Jalas) Kloczkova, *A. vulneraria* subsp. *lapponica* (Hyl.) Jalas var. *fennica* (Jalas) Cullen) - 1

grossheimii Chinth. = A. macrocephala and A. variegata
incisa Willd. = Cicer incisum
irenae Juz. = A. variegata
kuzenevae Juz. (*A. vulneraria* L. subsp. *lapponica* (Hyl.)
    Jalas, *A. vulneraria* var. *lapponica* Hyl.) - 1
lachnophora Juz. (*A. boissieri* (Sagor.) Grossh. p.p. excl.
    basionymo, *A. vulneraria* L. subsp. *boissieri* auct. p.p.) -
    2
linnaei (Sagor.) Juz. = A. vulneraria
macrocephala Wend. (*A. grossheimii* Chinth. p.p. excl. typo,
    A. polyphylla Kit. ex Loud., *A. vulneraria* L. subsp.
    polyphylla (Ser.) Nym. p.p.) - 1, 2
- var. *schiwereckii* (Ser.) Soo = A. schiwereckii
maritima Schweigg. (*A. vulneraria* L. subsp. *maritima*
    (Schweigg.) Corb.) - 1
polyphylla Kit. ex Loud. = A. macrocephala
x **polyphylloides** Juz. - A. arenaria (Rupr.) Juz. x A. macro-
    cephala Wend. - 1
schiwereckii (Ser.) Blocki (*A. macrocephala* Wend. var.
    schiwereckii (Ser.) Soo) - 1
taurica Juz. (*A. vulneraria* L. subsp. *boissieri* auct. p.p.) - 1
variegata Boiss. ex Grossh. (*A. boissieri* (Sagor.) Grossh.
    p.p. excl. basionymo, *A. caucasica* (Grossh.) Juz., *A.
    grossheimii* Chinth. p.p. quoad typum, *A. irenae* Juz.,
    A. vulneraria L. subsp. *hispidissima* (Sagor.) Cullen,
    p.p. excl. basionymo, *A. vulneraria* subsp. *subscaposa*
    Cullen) - 2
vulneraria L. (*A. linnaei* (Sagor.) Juz., *A. vulneraria* subsp.
    linnaei Sagor. p.p. incl. typo) - 1
- subsp. *boissieri* auct. p.p. = A. lachnophora and A. taurica
- subsp. *carpatica* (Pant.) Nym. = A. carpatica
- subsp. *fennica* Jalas = A. fennica
- subsp. *hispidissima* (Sagor.) Cullen, p.p. = A. variegata
- subsp. *lapponica* (Hyl.) Jalas = A. kuzenevae
- - var. *fennica* (Jalas) Cullen = A. fennica
- subsp. *linnaei* Sagor. p.p. = A. vulneraria
- subsp. *maritima* (Schweigg.) Corb. = A. maritima
- subsp. *polyphylla* (Ser.) Nym. p.p. = A. macrocephala
- subsp. *pulchella* auct. p.p. = A. biebersteiniana
- subsp. *subscaposa* Cullen = A. variegata
- var. *alpestris* Kit. ex Schult. = A. alpestris
- var. *biebersteiniana* Taliev = A. biebersteiniana
- var. *lapponica* Hyl. = A. kuzenevae

*Arachis L.
*hypogaea L. - 1, 2

Argyrolobium Eckl. & Zeyh.
biebersteinii P.W. Ball (*Cytisus calycinus* Bieb. nom. su-
    perfl., *Argyrolobium calycinum* (Bieb.) Jaub. & Spach,
    nom. superfl.) - 1, 2
calycinum (Bieb.) Jaub. & Spach = A. biebersteinii
lotoides Bunge = A. trigonelloides
prilipkoanum Grossh. = A. trigonelloides
trigonelloides Jaub. & Spach (*A. lotoides* Bunge, *A. prilip-
    koanum* Grossh.) - 2

Asphalthium auct. = Psoralea
acaule (Stev.) Hutch. = Psoralea acaulis

Astracantha Podlech (*Astragalus* L. subgen.
    Tragacantha Bunge, non *Tragacantha* Mill.,
    Tragacantha auct.)
▪ acmophylloides (Grossh.) Podlech (*Astragalus acmophyl-*

loides Grossh., *A. sommieri* Freyn subsp. *acmophyl-
    loides* (Grossh.) Ponert, *A. transcaucasicus* Boriss.
    nom. altern., *Tragacantha transcaucasica* Boriss.)
alexeenkoana (B. Fedtsch. & Ivanova) Podlech (*Astragalus
    alexeenkoanus* B. Fedtsch. & Ivanova) - 6
andreji (Rzazade) Czer. comb. nova (*Astragalus andreji*
    Rzazade, 1953, Dokl. AN AzSSR, 9, 7 : 407) - 2
arnacantha (Bieb.) Podlech (*Astragalus arnacantha* Bieb.) -
    1
- subsp. *arnacanthoides* (Boriss.) Reer & Podlech = A.
    arnacanthoides
arnacanthoides (Boriss.) Podlech (*Astragalus arnacan-
    thoides* Boriss. nom. altern., *Astracantha arnacantha*
    (Bieb.) Podlech subsp. *arnacanthoides* (Boriss.) Reer
    & Podlech, *Tragacantha arnacanthoides* Boriss.) - 2
asaphes (Bunge) Podlech (*Astragalus asaphes* Bunge) - 6
atenica (Ivanischvili) Podlech (*Astragalus atenicus* Iva-
    nischvili) - 2
aurea (Willd.) Podlech (*Astragalus aureus* Willd., *A. mac-
    ropodius* E. Fisch.) - 2
bactriana (Fisch.) Podlech (*Astragalus bactrianus* Fisch.) - 6
barba-caprina (Al. Theod., Fed. & Rzazade) Podlech
    (*Astragalus barba-caprina* Al. Theod., Fed. & Rza-
    zade) - 2
carthlica (Al. Theod., Fed. & Rzazade) Podlech = A.
    microcephala
caspica (Bieb.) Podlech (*Astragalus caspicus* Bieb.) - 2
caucasica (Pall.) Podlech (*Astragalus caucasicus* Pall.) - 2
cerasocrena (Bunge) Podlech (*Astragalus cerasocrenus*
    Bunge) - 6
chionocalyx (Nevski) Czer. comb. nova (*Tragacantha
    chionocalyx* Nevski, 1937, Fl. Tadzhik. 5 : 681, 492;
    Astragalus chionocalyx (Nevski) Boriss.) - 6
chodsha-bakirganica (B. Kom.) Czer. comb. nova (*Astra-
    galus chodsha-bakirganicus* B. Kom. 1967, Opred.
    Rast. Sev. Tadzhik. : 275; *Tragacantha chodsha-bakir-
    ganica* (B. Kom.) Rassulova) - 6
coarctata (Trautv.) Greuter (*Astragalus coarctatus*
    Trautv.) - 2
condensata (Ledeb.) Podlech (*Astragalus condensatus*
    Ledeb.) - 2
consentanea (Boriss.) Czer. comb. nova (*Tragacantha
    consentanea* Boriss. 1947, Bot. Mat. (Leningrad), 10 :
    61; *Astragalus consentaneus* Boriss.) - 6
densissima (Boriss.) Czer. comb. nova (*Tragacantha den-
    sissima* Boriss. 1936, Tr. Bot. Inst. AN SSSR, ser. 1, 3 :
    222; *Astragalus densissimus* (Boriss.) Boriss. comb.
    superfl., *A. densissimus* (Boriss.) Sirj.) - 6
denudata (Stev.) Podlech (*Astragalus denudatus* Stev.,
    Astracantha marschalliana (Fisch.) Podlech, *A. terek-
    ensis* (Al. Theod., Fed. & Rzazade) Podlech, *Astraga-
    lus marschallianus* Fisch., *A. terekensis* Al. Theod.,
    Fed. & Rzazade) - 2, 6
devia (Boriss.) Czer. comb. nova (*Tragacantha devia* Bo-
    riss. 1947, Bot. Mat. (Leningrad), 10 : 62; *Astragalus
    devius* Boriss.) - 6
dissecta (B.Fedtsch. & Ivanova) Podlech (*Astragalus dissec-
    tus* B. Fedtsch. & Ivanova, *A. proximus* (Boriss.) Bo-
    riss., *Tragacantha dissecta* (B. Fedtsch. & Ivanova)
    Boriss.) - 6
dolona (Rassulova & B. Scharipova) Czer. comb. nova
    (*Tragacantha dolona* Rassulova & B. Scharipova, 1978,
    Fl. Tadzh. SSR 5 : 630, 420; *Astragalus dolonus* (Rassu-
    lova & B. Scharipova) Czer.) - 6
erinacea (Fisch. & C.A. Mey.) Podlech = A. microcephala
▪ **ferox** (Boriss.) Czer. comb. nova (*Tragacantha ferox*
    Boriss. 1947, Bot. Mat. (Leningrad), 10 : 77; *Astragalus*

ferox (Boriss) Boriss.)

flavirubens (Al. Theod., Fed. & Rzazade) Podlech (*Astragalus flavirubens* Al. Theod., Fed. & Rzazade) - 2

flexilispina (Boriss.) Podlech (*Tragacantha flexilispina* Boriss., *Astragalus flexilispinus* (Boriss.) Boriss.) - 6

gudrathi (Al. Theod., Fed. & Rzazade) Podlech (*Astragalus gudrathi* Al. Theod., Fed. & Rzazade) - 2

hilariae (Boriss.) Czer. comb. nova (*Tragacantha hilariae* Boriss. 1937, Fl. Tadzhik. SSR 5 : 683, 493; *Astragalus hilariae* (Boriss.) Boriss. comb. superfl., *A. hilariae* (Boriss.) Sirj.) - 6

▪ **imbricata** (Boriss.) Podlech (*Tragacantha imbricata* Boriss., *Astragalus imbricatus* Boriss.)

insidiosa (Boriss.) Podlech (*Tragacantha insidiosa* Boriss., *Astragalus insidiosus* Boriss.) - 2

intermixta (Boriss.) Czer. comb. nova (*Tragacantha intermixta* Boriss. 1936, Tr. Bot. Inst. AN SSSR, ser. 1, 3 : 222; *Astragalus intermixtus* (Boriss.) Boriss. comb. superfl., *A. intermixtus* (Boriss.) Sirj.) - 6

jucunda (Al. Theod., Fed. & Rzazade) Czer. comb. nova (*Astragalus jucundus* Al. Theod., Fed. & Rzazade, 1954, Bot. Mat. (Leningrad), 16 :230) - 2

karabaghensis (Bunge) Podlech (*Astragalus karabaghensis* Bunge) - 2

- subsp. *vedica* (Takht.) Podlech = A. vedica

karakalensis (Freyen & Sint.) Podlech (*Astragalus karakalensis* Freyn & Sint.) - 6

karjaginii (Boriss.) Podlech (*Tragacantha karjaginii* Boriss., *Astragalus karjaginii* Boriss.) - 2

kuhitangi (Nevski) Podlech (*Astragalus kuhitangi* (Nevski) Boriss. comb. superfl., *A. kuhitangi* (Nevski) Sirj.) - 6

kuramensis (Boriss.) Podlech (*Tragacantha kuramensis* Boriss., *Astragalus kuramensis* (Boriss.) Czer.) - 6

▪ **lagowskyi** (Trautv.) Podlech (*Astragalus lagowskyi* Trautv.)

lasiostyla (Fisch.) Podlech (*Astragalus lasiostylus* Fisch.) - 6

longiramosa (Boriss.) Czer. comb. nova (*Tragacantha longiramosa* Boriss. 1936, Tr. Bot. Inst. AN SSSR, ser. 1, 3 : 223; *Astragalus longiramosus* (Boriss.) Boriss. comb. superfl., *A. longiramosus* (Boriss.) Sirj.) - 6

macrantha (Boriss.) Podlech (*Tragacantha macrantha* Boriss., non *Astragalus macranthus* Willd., *Astracantha vladimiri* (Sirj.) Podlech, *Astragalus macranthoides* Boriss., *A. vladimiri* Sirj.) - 6

marschalliana (Fisch.) Podlech = A. denudata

meana (Boriss.) Podlech (*Tragacantha meana* Boriss., *Astragalus meanus* Boriss.) - 6

meraca (Boriss.) Podlech (*Tragacantha meraca* Boriss., *Astragalus meracus* Boriss.) - 6

meschchedensis (Bunge) Podlech (*Astragalus meschchedensis* Bunge) - 6

meyeri (Boriss.) Podlech (*Astragalus meyeri* Boriss.) - 2

microcephala (Willd.) Podlech (*Astragalus microcephalus* Willd., *Astracantha carthlica* (Al. Theod., Fed. & Rzazade) Podlech, *A. erinacea* (Fisch. & C.A. Mey.) Podlech, *Astragalus carthlicus* Al. Theod., Fed. & Rzazade, *A. erinaceus* Fisch. & C.A. Mey., *A. pycnophyllus* Stev.) - 2, 6

multifoliolata (Boriss.) Podlech (*Astragalus multifoliolatus* (Boriss.) Boriss. comb. superfl., *A. multifoliolatus* (Boriss.) Sirj.) - 6

nuratavica (Boriss.) Czer. comb. nova (*Tragacantha nuratavica* Boriss. 1947, Bot. Mat. (Leningrad), 10 : 70; *Astragalus nuratavicus* Boriss.) - 6

oleifolia (DC.) Podlech (*Astragalus oleifolius* DC.) - 2

▪ **oltensis** (Grossh.) Podlech (*Astragalus oltensis* Grossh.)

paliurus (Boriss.) Czer. comb. nova (*Tragacantha paliurus* Boriss. 1947, Bot. Mat. (Leningrad), 10 : 64; *Astragalus*

paliurus Boriss.) - 6

piletoclada (Freyn & Sint.) Czer. comb. nova (*Astragalus piletocladus* Freyn & Sint. 1904, Bull. Herb. Boiss., ser. 2, 4 : 1108) - 6

plumata (Boriss.) Podlech (*Tragacantha plumata* Boriss., *Astragalus plumatus* Boriss.) - 6

prominens (Boriss.) Podlech (*Tragacantha prominens* Boriss., *Astragalus prominens* Boriss.) - 6

pterocephala (Bunge) Podlech (*Astragalus pterocephalus* Bunge) - 6

pulvinata (Bunge) Podlech (*Astragalus pulvinatus* Bunge, *A. wrangelii* Sirj., *A. wrangelii* subsp. *memorabilis* Sirj.) - 6

pycnantha (Boriss.) Podlech (*Astragalus pycnanthus* (Boriss.) Boriss. comb. superfl., *A. pycnanthus* (Boriss.) Sirj.) - 6

rubens (B. Fedtsch. & Ivanova) Podlech (*Astragalus rubens* B. Fedtsch. & Ivanova) - 6

sommieri (Freyn) Podlech (*Astragalus sommieri* Freyn, *Tragacantha sommieri* (Freyn) Boriss.) - 2

stenonychioides (Freyn & Bornm.) Podlech (*Astragalus stenonychioides* Freyn & Bornm., *Astracantha strictifolia* (Boiss.) Greuter, *Astragalus strictifolius* Boiss. 1849, non Willd. 1794) - 2

stipulosa (Boriss.) Podlech (*Tragacantha stipulosa* Boriss., *Astragalus stipulosus* Boriss.) - 6

strictifolia (Boiss.) Greuter = A. stenonychioides

tenuispina (Boriss.) Czer. comb. nova (*Tragacantha tenuispina* Boriss. 1937, Fl. Tadzhik. 5 : 682, 492, non *Astragalus tenuispinus* Bunge; *Astragalus munitus* Boriss.) - 6

terekensis (Al. Theod., Fed. & Rzazade) Podlech = A. denudata

theodoriana (Fed. & Rzazade) Czer. comb. nova (*Astragalus theodorianus* Fed. & Rzazade, 1953, Dokl. AN AzSSR, 9, 10 : 605) - 6

transoxana (Fisch.) Podlech (*Astragalus transoxanus* Fisch.) - 6

turkmenorum (Boriss.) Podlech (*Astragalus turkmenorum* (Boriss.) Boriss. comb. superfl., *A. turkmenorum* (Boriss.) Sirj.) - 6

unguiculata (Boriss.) Czer. comb. nova (*Tragacantha unguiculata* Boriss. 1947, Bot. Mat. (Leningrad), 10 : 57;*Astragalus unguiculatus* Boriss.) - 6

vedica (Takht.) Czer. comb. nova (*Astragalus vedicus* Takht. 1940, Not. Syst. Geogr. (Tbilisi), 9 : 15; *Astracantha karabaghensis* (Bunge) Podlech subsp. *vedica* (Takht.) Podlech, *Astragalus karabaghensis* Bunge subsp. *vedicus* (Takht.) Takht.) - 2

vladimiri (Sirj.) Podlech = A. macrantha

▪ **voronoviana** (Boriss.) Podlech (*Tragacantha voronoviana* Boriss., *Astragalus voronovianus* Boriss.)

Astragalus L. (*Didymopelta* Regel & Schmalh., *Dipelta* Regel & Schmalh. 1878, non Maxim. 1878, *Ophiocarpus* (Bunge) Ikonn., *Sewerzowia* Regel & Schmalh., *Thlaspidium* (Lipsky) Rassulova)

subgen. *Tragacantha* Bunge = Astracantha

abbreviatus Kar. & Kir. - 6

abolinii M. Pop. - 6

aboriginum auct. = A. tugarinovii

abramovii Gontsch. (*A. pseudobrachytropis* Gontsch.) - 6

abruptus Krytzka = A. pseudotataricus

acanthocarpus Boriss. - 6

achtalensis Conrath & Freyn - 2

achundovii Grossh. ex Fed. - 2
ackerbergensis Freyn - 6
acmophylloides Grossh. = Astracantha acmophylloides
acormosus Basil. - 6
adpressepilosus Gontsch. - 6
adsurgens Pall. - 4, 5
- subsp. *oreogenus* (Jurtz.) Worosch. = A. inopinatus subsp.
   oreogenus
aduncus Willd. - 2
adzharicus M. Pop. - 2
aegobromus Boiss. & Hohen (*A. torrentum* Bunge) - 2
aemulans (Nevski) Gontsch. - 6
aflatunensis B. Fedtsch. - 6
agameticus Lipsky - 2
agassii Manden. - 2
agrestis Dougl. - 5
aiwadzhi B. Fedtsch. = A. unifoliolatus
aksaicus Schischk. - 6
aksaricus Pavl. - 3
aksuensis Bunge - 6
aktauensis Gontsch. - 6
- subsp. chodshamastonicus Rassulova - 6
alaarczensis Vass. - 6
alabugensis B. Fedtsch. - 6
alaicus Freyn - 6
alatavicus Kar. & Kir. - 6
albanicus Grossh. - 2
alberti Bunge - 6
albertoregelia C. Winkl. & B. Fedtsch. - 6
albescens Boriss. = A. ninae
albicans Bong. - 6
albicaulis DC. - 1, 2, 3
albidus Waldst. & Kit. (*A. vesicarius* auct.) - 1
aleschii Bondar. (*A. jaxarticus* Bondar. 1961, non Pavl.
   1935) - 6
alexandri Charadze, 1942, non Sirj. 1939 - 2
   This name must be replaced because it is a later homonym.
alexeenkoanus B. Fedtsch. & Ivanova = Astracantha alexeenkoana
alexeenkoi Gontsch. - 6
alexeji Gontsch. - 6
alitschuri B. Fedtsch. - 6
allotricholobus Nabiev - 6
alopecias Pall. - 3, 6
alopecurus Pall. - 1, 3, 4, 6
alpinus L. (*A. astragalinus* (Hook.) A. & D. Löve, *A. grossheimianus* Sosn., *A. salicetorum* Kom.) - 2, 3, 4, 5, 6
- subsp. alaskanus Hult. - 4, 5
- subsp. *arcticus* (Bunge) Hult. = A. subpolaris
- subsp. *borealis* Ju. Kozhevn. = A. subpolaris
altaicus Bunge - 3, 6
▪ alyssoides Lam.
amabilis M. Pop. - 6
amarus Pall. - 1, 6
ambigens M. Pop. - 6
ammodendron Bunge (*A. transcaspicus* Freyn) - 1, 6
ammodytes Pall. - 3, 6
ammophilus Kar. & Kir. (*A. vachanicus* Boriss. & A.
   Korol., *A. biovulatus* auct.) - 2, 6
ammotrophus Bunge - 6
amygdalinus Bunge - 6
andarabicus Podlech - 6
andaulgensis B. Fedtsch. - 6
andreji Rzazade = Astracantha andreji
androssovianus Gontsch. - 6
angarensis Turcz. ex Bunge - 4
- subsp. ozjorensis Peschkova - 4

angreni Lipsky - 6
angustidens Freyn & Sint. - 6
angustiflorus C. Koch - 2
angustissimus Bunge - 6
anisomerus Bunge - 6
ankylotus Fisch. & C.A. Mey. - 1, 6
anrachaicus Golosk. = A. lanuginosus
antoninae Grig. - 6
aphanassjevii Gontsch. - 6
apiculatus Gontsch. - 6
applicatus Boriss. = A. kochianus
arbuscula Pall. - 3, 6
arcuatus Kar. & Kir. - 1, 3, 6
arenarius L. - 1
arganaticus Bunge ex Regel & Herd. - 6
argillosus Manden. - 2
arguricus Bunge (*A. kozlovskyi* Grossh.) - 2
argutensis Bunge - 3
argyroides G. Beck ex Stapf - 2
arianus Gontsch. - 6
aridicolus Sosn. = A. humilis
aridus Freyn = A. lasiosemius
arkalycensis Bunge - 3, 4, 6
armeniacus auct. = A. humilis
arnacantha Bieb. = Astracantha arnacantha
arnacanthoides Boriss. = Astracantha arnacanthoides
arpilobus auct. = A. harpilobus
artemisiiformis Rassulova - 6
artvinensis M. Pop. = A. taochius
arvatensis Gontsch. - 6
asaphes Bunge = Astracantha asaphes
aschuturi B. Fedtsch. - 6
asper Jacq. - 1, 2
aspindzicus Manden. & Chinth. (*A. sevangensis* Grossh.
   subsp. *aspindzicus* (Manden.& Chinth.) Sytin) - 2
asterias Stev. ex Ledeb. (*A. cruciatus* auct.) - 2, 6
astragalinus (Hook.) A. & D. Löve = A. alpinus
atenicus Ivanischvili = Astracantha atenica
atlasovii Kom. = A. polaris
atraphaxifolius Rassulova - 6
▪ atropurpureus Boiss. (*A. campylosema* Boiss. subsp.
   atropurpureus (Boiss.) Chamberlain)
atrovinosus M. Pop. - 6
aulieatensis M. Pop. - 6
auratus Gontsch. - 6
aureus Willd. = Astracantha aurea
australis auct. p.p. = A. gorczakovii and A. kaufmannii
australis (L.) Lam. subsp. *krajinae* Domin-Kostrakiewicz =
   A. krajinae
austriacus Jacq. - 1, 2, 3, 6
austrodshungaricus Golosk. - 6
austroferganicus R. Kam. & R. Vinogradova - 6
austrosachalinensis N.S. Pavlova - 5
austrosibiricus Schischk. - 3, 4, 6
austrotadzhikistanicus Czer. = A. modesti
aznabjurticus Grossh. - 2
babatagi M. Pop. - 6
- subsp. indurescens Rassulova - 6
bachardeni R. Kam. & Kovalevsk. - 6
bachmarensis Grossh. - 2
bactrianus Fisch. = Astracantha bactriana
badachschanicus Boriss. - 6
badamensis M. Pop. - 6
badamliensis Chalilov - 2
badghysi M. Pop. - 6
baissunensis Lipsky - 6
bakaliensis Bunge - 6

bakuensis Bunge - 2
balchanensis Boriss. - 6
balchaschensis Sumn. - 6
baldshuanicus M. Pop. - 6
baranovii M. Pop. - 6
barba-caprina Al. Theod., Fed. & Rzazade = Astracantha
    barba-caprina
barnassari Grossh. - 2
barrowianus Aitch. & Baker - 6
- subsp. *ochranthus* (Gontsch.) R. Kam. = A. ochranthus
basilii R. Kam. & Kovalevsk. - 6
basineri Trautv. - 6
beckerianus Trautv. - 2
beketowii (Krasn.) B. Fedtsch. - 6
bibracteatus Ovcz. & Rassulova - 6
▪ bicolor Lam.
- subsp. *owerinii* (Bunge) Ponert = A. onobrychioides
biebersteinii Bunge - 2
bifidus Turcz. - 4
biovulatus auct. = A. ammophilus
bischkendicus Gontsch. - 6
bobrovii B. Fedtsch. - 6
bogensis Rassulova - 6
bor-bulakensis Rassulova - 6
boreomarinus A. Khokhr. (*A. marinus* Boriss. *boreomarinus*
    (A. Khokhr.) N.S. Pavlova) - 5
borissianus Gontsch. - 6
borissovae Grossh. = A. bungeanus
bornmuellerianus B. Fedtsch. (*A. bornmuellerianus* Sirj.) -
    6
bornmuellerianus Sirj. = A. bornmuellerianus
borodinii Krasn. (*A. projecturus* Sumn.) - 6
borysthenicus Klok. = A. onobrychis
bosbutooensis E. Nikit. & Sudn. - 6
bossuensis M. Pop. - 6
botschantzevii R. Kam. & Rassulova - 6
brachybotrys Bunge - 3
brachycarpus Bieb. - 2
brachyceras Ledeb. = A. hamosus
brachylobus Fisch. - 1, 2, 3, 6
brachymorphus Nikif. = A. stalinskyi
brachypetalus Trautv. - 2, 6
brachypodus Boiss. (*A. karsianus* Bunge) - 2
brachypus Schrenk - 6
brachyrachis M. Pop. - 6
brachytropis (Stev.) C.A. Mey. (*A. norvegicus* Grauer var.
    brachytropis (Stev.) Hashimov) - 2
brevicarinus Bajt. - 6
brevidens Freyn & Sint. - 6
brevifolius Ledeb. - 3
breviscapus B. Fedtsch. - 6
brotherusii Podlech - 6
bryophilus Greene = Oxytropis bryophila
bucharicus Regel - 6
buchtormensis Pall. - 3, 6
bungeanus Boiss. (*A. borissovae* Grossh., *A. kukurttavicus*
    Prokh., *A. perembelicus* Grossh.) - 2
bungei C. Winkl. & B. Fedtsch. = A. camptoceras
buschiorum Galushko - 2
butkovii M. Pop. - 6
bylowae A. Jelen. - 2
caespitosulus Gontsch. - 6
calycinus Bieb. - 1, 2
camptoceras Bunge (*A. bungei* C. Winkl. & B. Fedtsch.) - 2,
    6
campylorrhynchus Fisch. & C.A. Mey - 2, 6
▪ campylosema Boiss.

- subsp. *atropurpureus* (Boiss.) Chamberlain = A. atropur-
    pureus
campylotrichus Bunge - 6
cancellatus Bunge (*A. perrarus* Boriss., *A. pseudocancellatus*
    Grossh.) - 2
candidissimus Ledeb. - 6
candolleanus Boiss. (*A. latifolius* Lam. subsp. *candolleanus*
    (Boiss.) Ponert) - 2
canescens auct. = A. sevangensis
canoflavus M. Pop - 6
captiosus Boriss. (*A. interpositus* Boriss., *A. klukhoricus*
    Sosn.) - 2
- var. *ketzkhovelii* (Charadze) Sytin (*A. ketzkhovelii* Cha-
    radze) - 2
caraganae Fisch. & C.A. Mey. - 2
carthlicus Al. Theod., Fed. & Rzazade = Astracantha
    microcephala
cartilagineus Gontsch. - 6
caspicus Bieb. = Astracantha caspica
caucasicus Pall. = Astracantha caucasica
caudicosus Galkina & Nabiev - 6
caulescens (Gontsch.) Abduss. (*Oxytropis caulescens*
    Gontsch.) - 6
centralis Sheld. (*A. kisylkumi* Boriss.) - 6
cerasocrenus Bunge = Astracantha cerasocrena
ceratoides Bieb. - 3, 6
cernuiflorus Gontsch. - 6
chadjanensis Franch. - 6
chaetodon Bunge - 6
chaetolobus Bunge - 6
chaeturus M. Pop. - 3
chakassiensis Polozh. - 4
chalilovii Grossh. ex Fed. - 2
charadzeae Grossh. - 2
charguschanus Freyn - 5
chinensis L. fil. - 5
chingoanus R. Kam. - 6
chionanthus M. Pop. - 6
chionocalyx (Nevski) Boriss. = Astracantha chionocalyx
chiwensis Bunge - 6
chlorodontus Bunge - 6
chodsha-bakirganicus B. Kom. = Astracantha chodsha-
    bakirganica
chodshamastonicus Pachom. - 6
chodshenticus B. Fedtsch. (*A. thaumatothamnus* Vved.) - 6
choicus Bunge (*A. heteromorphus* Boriss., *A. latifolius*
    auct.) - 2
chomutowii B. Fedtsch. - 6
chordorrhizus Fisch. ex Bunge - 2
chorgossicus Lipsky - 6
chorinensis Bunge - 4
chorizanthus auct. = A. marguzaricus
chrysomallus Bunge - 6
chrysostachys Boiss. - 6
cicer L. - 1, 2
ciceroides Sosn. - 2
cinerascens M. Pop. = A. remanens
cinereus Willd. - 2
circassicus Grossh. - 1, 2
cisdarvasicus Gontsch. = A. kabadianus
citoinflatus Bondar. - 6
clerceanus Iljin & Krasch. - 1, 3
coarctatus Trautv. = Astracantha coarctata
coelestis Boiss. - 2
cognatus Schrenk - 6
coluteocarpus Boiss. - 6
commixtus Bunge - 1, 2, 6

compactus Willd. 1794, non Lam. 1783 - 2

This name must be replaced as later homonym.

compositus Pavl. (*Sewerzowia composita* (Pavl.) Rassulova) - 6

compressus Ledeb. - 3

concavus Boriss. - 1

condensatus Ledeb. = Astracantha condensata

confiniorum Boriss. - 6

conrathii Freyn = A. ornithopodioides

consanguineus Bong. & C.A. Mey. - 6

consentaneus Boriss. = Astracantha consentanea

conspicuus Boriss. - 2

contortuplicatus L. - 1, 2, 3, 6

corniculatus Bieb. - 1, 2

cornubovis Lipsky - 6

cornutus Pall. (*A. cretophilus* Klok., *A. odessanus* Bess.) - 1, 2, 3, 6

corrugatus Bertol. - 2, 6

corydalinus Bunge - 6

cottonianus Aitch. & Baker - 6

cretophilus Klok. = A. cornutus

cristophii Trautv. - 6

cruciatus auct. = A. asterias

curvipes Trautv. - 6

cuscutae Bunge - 2

cyri Fomin (*A. glaucophylloides* auct.) - 2

cyrtobasis Bunge ex Boiss. - 6

cysticalyx Ledeb. - 6

cysticarpus Boriss. - 6

cytisoides Bunge - 6

czapdarinus Ovcz. & Rassulova (*A. karaculensis* Ovcz. & Rassulova, 1976, non Bunge, 1851)

czilduchtaroni R. Kam. - 6

▪ **czorochensis** Charadze

czuiliensis Golosk. = A. globiceps

daghestanicus Grossh. - 2

danicus Retz. - 1, 2, 3, 4, 5, 6

- subsp. *dasyglottis* (Fisch.) Worosch. = A. dasyglottis

darwasicus Basil. - 6

dasyanthus Pall. - 1

dasyglottis Fisch. (*A. danicus* Retz. subsp. *dasyglottis* (Fisch.) Worosch.) - 3, 4, 6

davuricus (Pall.) DC. - 3, 4, 5

dealbatus Pall. = A. glaucus

debilis Ovcz. & Rassulova = A. sardaimionensis

declinatus Willd. = A. pinetorum

degilmonus Rassulova - 6

demetrii Charadze - 2

dendroides Kar. & Kir. - 6

densiflorus Kar. & Kir. - 6

densissimus (Boriss.) Boriss. = Astracantha densissima

densissimus (Boriss.) Sirj. = Astracantha densissima

densus M. Pop. - 6

denudatus Stev. = Astracantha denudata

depauperatus Ledeb. - 3, 4

derbenticus Bunge - 2

deserti M. Pop. = A. turcomanicus

devestitus Pazij & Vved. - 6

devius Boriss. = Astracantha devia

dianthoides Boriss. - 6

dianthus Bunge - 6

dictamnoides Gontsch. (*A. stenophysus* Vved. & Zak.) - 6

dignus Boriss. - 6

dilutus Bunge - 6

dipelta Bunge (*Didymopelta turkestanica* (Regel & Schmalh.) Regel & Schmalh., *Dipelta turkestanica* Regel & Schmalh.) - 6

discessiflorus Gontsch. - 6

dissectus B. Fedtsch. & Ivanova = Astracantha dissecta

distentus Boriss. - 6

djilgensis Franch. (*A. tranzschelii* Boriss.) - 6

dolichocarpus M. Pop. - 6

dolichophyllus Pall. - 1, 2, 3, 6

dolichopodus Freyn - 6

dolonus (Rassulova & B. Scharipova) Czer. = Astracantha dolona

doluchanovii Manden. = A. viciifolius

drobovii M. Pop. & Vved. - 6

dschangartensis Sumn. - 6

dsharfi B. Fedtsch. - 6

dsharkenticus M. Pop. - 6

dshimensis Gontsch. - 6

duanensis Saposhn. ex Sumn. - 6

dubius Krasn. - 6

dzhebrailicus Grossh. (*A. schuschensis* Grossh.) - 2

elbrusensis Boiss. (*A. schischkinii* Grossh.) - 2, 6

elegans Bunge - 2

ellipsoideus Ledeb. (*A. transiliensis* Gontsch.) - 3, 6

▪ **elongatus** Willd.

ephemeretorum Gontsch. - 6

- subsp. **bilobulatus** Rassulova - 6

eremobius M. Pop. = A. petunnikowii

eremospartoides Regel - 6

erinaceus Fisch. & C.A. Mey. = Astracantha microcephala

eriocarpus DC. (*A. protractus* Boriss., *A. suluklensis* Freyn & Sint.) - 6

erioceras Fisch. & C.A. Mey. ex Ledeb. - 6

erioceratiformis M. Pop. = A. jaxarticus

eriolobus Bunge - 3

erivanensis Bornm. & Woronow - 2

eucosmus auct. = A. sealei

eucosmus Robbins subsp. *sealei* (Lepage) Hult. = A. sealei

eugenii Grossh. - 2

euoplus Trautv. - 2

eupeplus Barneby (*A. holosericeus* M. Pop. 1937, non Jones, 1895) - 6

exasperatus Basil. - 6

excedens M. Pop. & Kult. - 6

excelsior M. Pop. - 6

exilis A. Korol. - 6

eximius Bunge - 6

exscapus L. - 1

- subsp. *pubiflorus* (DC.) Soo = A. pubiflorus

fabaceus Bieb. - 2

falcatus Lam. - 1, 2, 3

falcigerus M. Pop. - 6

farctissimus Lipsky - 6

farctus Bunge - 6

fedorovii Takht. - 2

fedtschenkoanus Lipsky - 6

ferganensis (M. Pop.) B. Fedtsch. ex A. Korol. - 6

ferox Boriss. = Astracantha ferox

fetissowii B. Fedtsch. - 6

filicaulis Fisch. & C.A. Mey. ex Kar. & Kir. (*A. rytilobus* Bunge) - 6

finitimus Bunge (*A. macrocephalus* Willd. subsp. *finitimus* (Bunge) Chamberlain) - 2

fissuralis Alexeenko - 2

flaccidus Bieb. = A. viciifolius

flavidus M. Pop. - 6

flavirubens Al. Theod., Fed. & Rzazade = Astracantha flavirubens

flexicaulis Sosn. - 2

flexilispinus Boriss. = Astracantha flexilispina

flexus Fisch. - 6
floccosifolius Sumn. - 6
▪ fodinarum Boiss. & Noe ex Bunge
follicularis Pall. - 3, 6
fragrans auct. = A. resupinatus
fraxinifolius auct. = A. glycyphylloides
freynii Albov - 2
frickii Bunge (*A. salareticus* Boriss.) - 2
frigidus (L.) A. Gray (*A. frigidus* subsp. *grigorjewii* (B. Fedtsch.) Chater, *A. grigorjewi* B. Fedtsch., *A. kolaensis* Kuzen.) - 1, 3, 4, 5
- subsp. **frigidus** var. **grigorjewii** (B. Fedtsch.) Jurtz. - 1
- subsp. *grigorjewii* (B. Fedtsch.) Chater = A. frigidus
- subsp. *minutulus* Kuvajev = A. frigidus subsp. parviflorus
- subsp. **parviflorus** (Turcz.) Hult. (*Phaca parviflora* Turcz., *Astragalus frigidus* subsp. *minutulus* Kuvajev) - 4, 5
- subsp. *secundus* (DC.) Worsch. = A. secundus
fruticosus Pall. = A. suffruticosus
fuhsii Freyn & Sint. - 6
fuliginosus G. Beck ex Stapf - 2
▪ fumosus Boriss.
galactites Pall. - 4
galegiformis L. - 1, 2, 6
gardanikaphtharicus Rassulova - 6
gaudanensis B. Fedtsch. - 6
gawrilowii Krasn. - 6
gebleri Fisch. ex Bong. & C.A. Mey. - 6
georgii Gontsch. - 6
gezeldarensis Grossh. - 2
gjunaicus Grossh. (*A. ketzkhovelianus* Manden., *A. mukusiensis* Rech. fil., *A. psoraloides* auct.) - 2
glabrescens Gontsch. - 6
glaucophylloides auct. = A. cyri
glaucus Bieb. (*A. dealbatus* Pall. nom. illegit.) - 1
globiceps Bunge (*A. czuiliensis* Golosk.) - 6
glochideus Boriss. (*A. stevenianus* DC. subsp. *glochideus* (Boriss.) Sytin) - 2
glomeratus Ledeb. - 3
glycyphylloides DC. (*A. glycyphyllos* L. subsp. *glycyphylloides* (DC.) Matthews, *A. fraxinifolius* auct.) - 1, 2
glycyphyllos L. - 1, 2, 3
- subsp. *glycyphylloides* (DC.) Matthews = A. glycyphylloides
goktschaicus Grossh. (*A. kosmaljanicus* Rzazade, *A. onobrychis* L. subsp. *goktschaicus* (Grossh.) Ponert) - 2
gontsharovii Vass. - 6
gorczakovii L. Vassil. (*A. uralensis* Litv. 1892, non L. 1753, *A. australis* auct. p.p.) - 1, 3
gorodkovii Jurtz. - 3
gracilipes Benth. ex Bunge (*A. ninae* Gontsch. 1947, non Pavl. 1934, *A. schurae* Pavl.) - 6
grammocalyx Boiss. & Hohen. - 2
grandiflorus Bunge = A. korovinianus
grigorjewii B. Fedtsch. = A. frigidus
grossheimianus Sosn. = A. alpinus
gudrathi Al. Theod., Fed. & Rzazade = Astracantha gudrathi
haesitabundus Lipsky - 2
hajastanus Grossh. - 2
▪ halicacabus Lam.
hamosus L. (*A. brachyceras* Ledeb.) - 1, 2, 6
harpilobus Kar. & Kir. (*A. arpilobus* auct.) - 1, 6
harpocarpus Meff. - 6
helmii Fisch. - 1, 3
hemiphaca Kar. & Kir. (*A. tauczilikensis* Gontsch.) - 6
henningii (Stev.) Klok. (*A. novoascanicus* Klok.) - 1

heptapotamicus Sumn. (*A. poljakovii* M. Pop.) - 6
heterodontus Boriss. - 6
heteromorphus Boriss. = A. choicus
heterotrichus Gontsch. - 6
hilariae (Boriss.) Boriss. = Astracantha hilariae
hilariae (Boriss.) Sirj. = Astracantha hilariae
hirtulus Ledeb. (*A. viciifolius* DC. subsp. *hirtulus* (Ledeb.) Sytin) - 2
hissaricus Lipsky - 6
hohenackeri Boiss. - 2
holargyreus Bunge - 6
holdichianus Aitch. & Baker
    This species has been recorded for the investigated territory by mistake.
holophyllus Boriss. - 2
holopterus Turcz. ex Bunge - 4
holosericeus M.Pop. = A. eupeplus
humilis Bieb. (*A. aridicolus* Sosn., *A. humilis* subsp. *theodori* (Grossh.) Hashimov, *A. kikodzeanus* Sosn., *A. theodorii* Grossh., *A. armeniacus* auct.) - 2
- subsp. *theodori* (Grossh.) Hashimov = A. humilis
husseinovii Rzazade - 2
hyalolepis Bunge - 2
hybridus S.G. Gmel. = A. onobrychis
hypanicus Krytzka - 1
hypogaeus Ledeb. - 3
hypoglottoides Baker = Oxytropis hypoglottoides
hyrcanus Pall. - 2
igniarius M. Pop. - 2
igoschinae R. Kam. & Jurtz. - 1
ikonnikovii Podlech - 6
iliensis Bunge - 6
iljinii R. Kam. = A. modesti
iljinii Rzazade - 2
imbricatus Boriss. = Astracantha imbricata
imetensis Boriss. - 6
inaequalifolius Basil. - 6
incertus Ledeb. - 2
indurescens Gontsch. - 6
inflatus DC. - 3
infractus Sumn. - 6
innominatus Boriss. - 6
inopinatus Boriss. - 4, 5
- subsp. *oreogenus* Jurtz. (*A. adsurgens* Pall. subsp. *oreogenus* (Jurtz.) Worosch.) - 4
- subsp. *pseudoadsurgens* (Jurtz.) N.S. Pavlova = A. pseudoadsurgens
insidiosus Boriss = Astracantha insidiosa
insignis Gontsch. - 6
intarrensis Franch. - 6
intermedius Kar. & Kir. - 6
intermixtus (Boriss.) Boriss. = Astracantha intermixta
intermixtus (Boriss.) Sirj. = Astracantha intermixta
interpositus Boriss. = A. captiosus
involutivus Sumn. - 6
ionae Palib. - 4
irinae B. Fedtsch. - 6
irisuensis Boriss. - 6
ischnocarpus Gontsch. = A. kudrjaschovii
ishigensis Maxim. ex Kom. - 5
ishkamishensis Podlech - 6
iskanderi Lipsky - 6
isophysus Nabiev = A. melanocomus
ispahanicus Boiss. - 6
isphairamicus B. Fedtsch. - 6
▪ ispirensis Boiss.
jagnobicus Lipsky - 6

janischewskii M. Pop. - 6
japonicus Boissieu (*A. kurilensis* Matsum.) - 5
jarmolenkoi Gontsch. - 6
jaxarticus Bondar. = A. aleschii
jaxarticus Pavl. (*A. erioceratiformis* M. Pop.) - 6
■ jodostachys Boiss. & Buhse (*A. stevenianus* DC. subsp. *jodostachys* (Boiss. & Bunge) Ponert)
johannis Rzazade - 2
jolderensis B. Fedtsch. - 6
jucundus Al. Theod., Fed. & Rzazade = Astracantha jucunda
juniperetorum Gontsch. - 6
junussovii Rassulova - 6
juratzkanus Freyn & Sint. (*A. neilreichianus* Freyn & Sint., *A. xanthoxiphidium* Freyn & Sint.) - 6
juzepczukii (Galushko) Galushko (*A. levieri* Freyn ex Somm. & Levier var. *juzepczukii* Galushko) - 2
kabadianus Lipsky (*A. cisdarvasicus* Gontsch.) - 6
kabristanicus Grossh. - 2
kadschorensis Bunge - 2
kaghysmani Gontsch. - 2
kahiricus DC. (*A. subkahiricus* Gontsch.) - 6
kamtschaticus (Kom.) Gontsch. = A. schelichowii
karabaghensis Bunge = Astracantha karabaghensis
- subsp. *vedicus* (Takht.) Takht. = Astracantha vedica
karabilicus M. Pop. - 6
karaculensis Ovcz. & Rassulova = A. czapdarinus
karakalensis Freyn & Sint. = Astracantha karakalensis
karakugensis Bunge - 6
karakuschensis Gontsch. - 2
karataviensis Pavl. - 6
karategini Gontsch. - 6
karatjubeki Golosk. - 6
karelinianus M. Pop. - 1, 3
karjaginii Boriss. = Astracantha karjaginii
karkarensis M. Pop. - 6
karsianus Bunge = A. brachypodus
kasachstanicus Golosk. - 6
kaschkadarjensis Gontsch. - 6
kaufmannii Kryl. (*A. ausralis* auct. p.p.) - 3, 4
kawakamii Matsum. - 5
kazbeki Charadze - 2
kazymbeticus Saposhn. ex Sumn. - 6
kelifi Lipsky - 6
kelleri M. Pop. - 6
keminensis K. Isakov - 6
kemulariae Grossh. = A. raddeanus
kendyrlyki M. Pop. - 6
kenkolensis B. Fedtsch. - 6
kessleri Trautv. - 6
ketzkhovelianus Manden. = A. gjunaicus
ketzkhovelii Charadze = A. captiosus var. ketzkhovelii
kikodzeanus Sosn. = A. humilis
kirghisorum Gontsch. - 6
kirpicznikovii Grossh. - 2
kisylkumi Boriss. = A. centralis
kjurendaghi V.V. Nikit. - 6
klopotovskyi Sosn. = A. vavilovii
klukhoricus Sosn. = A. captiosus
knorringianus Boriss. - 6
kochianus Sosn. (*A. applicatus* Boriss., *A. stevenianus* DC. subsp. *kochianus* (Sosn.) Takht., *A. stevenianus* var. *kochianus* (Sosn.) Chamberlain) - 2
koikitaensis Rassulova - 6
kokandensis Bunge - 6
kolaensis Kuzen. = A. frigidus
kolymensis Jurtz. (*A. tugarinowii* Basil. subsp. *kolymensis*

(Jurtz.) Ju. Kozhevn.) - 4, 5
komarovii Lipsky - 6
kopalensis Lipsky ex R. Kam. - 6
kopetdaghi Boriss. - 6
korolkowii Bunge - 6
korotkovae R. Kam. & Kovalevsk. - 6
korovinianus Barneby (*A. grandiflorus* Bunge, 1880, non L. 1753) - 6
kosmaljanicus Rzazade = A. goktschaicus
kozlovskyi Grossh. = A. arguricus
krajinae Domin (? *A. australis* (L.) Lam. subsp. *krajinae* Domin-Kostrakiewicz) - 1
krascheninnikovii R. Kam. - 6
krasnovii M. Pop. - 6
krassnovianus Gontsch. = A. pseudocytisoides
krauseanus Regel - 6
kronenburgii B. Fedtsch. ex Kneuck. - 6
kschtutensis Rassulova - 6
kubensis Grossh. - 2
kucanensis Rech. fil. - 6
kudrjaschovii A. Korol. (*A. ischnocarpus* Gontsch.) - 6
kugartensis Boriss. - 6
kuhitangi (Nevski) Boriss. = Astracantha kuhitangi
kuhitangi (Nevski) Sirj. = Astracantha kuhitangi
kukurttavicus Prokh. = A. bungeanus
kulabensis Lipsky - 6
kungurensis Boriss. - 1
kuramensis (Boriss.) Czer. = Astracantha kuramensis
kurdaicus Saposhn. ex Sumn. - 6
kurgankolensis Ovcz. & Rassulova (*A. pamirensis* Ovcz. & Rassulova, 1976, non Franch. 1896) - 6
kurilensis Matsum. = A. japonicus
kurtschumensis Bunge - 3
kuschakewiczi B. Fedtsch. - 6
kuschkensis Boriss. - 6
kushmasarensis Vass. - 6
kusnetzovii M. Pop. ex Kovalevsk. - 6
kustanaicus M. Pop. - 6
laceratus Lipsky - 6
lachnolobus Kovalevsk. & Vved. - 6
lactiflorus Ledeb. (*A. poliotes* Bunge) - 3, 6
lagopoides Lam. var. *persicus* DC. = A. persicus
lagowskyi Trautv. = Astracantha lagowskyi
laguroides Pall. - 3, 4
lagurus Willd. - 2
- subsp. *persicus* (DC.) Ponert = A. persicus
lancifolius Gontsch. - 6
lanuginosus Kar. & Kir. (*A. anrachaicus* Golosk.) - 6
lasiocalyx Gontsch. - 6
lasioglottis Stev. ex Bieb. - 2
lasiopetalus Bunge - 6
lasiophyllus Ledeb. - 1, 6
lasiosemius Boiss. (*A. aridus* Freyn, *A. latistylus* Freyn) - 6
lasiostylus Fisch. = Astracantha lasiostyla
latifolius auct. = A. choicus
latifolius Lam. subsp. *candolleanus* (Boiss.) Ponert = A. candolleanus
latistylus Freyn = A. lasiosemius
lavrenkoi R. Kam. - 6
laxmannii Jacq. - 6
lehmannianus Bunge - 1, 2, 6
leiophysa Bunge - 6
leiosemius (Lipsky) M. Pop. - 6
lentilobus R. Kam. & Kovalevsk. (*A. mediomontanus* R. Kam. & Kovalevsk. nom. invalid.) - 6
- subsp. **bilobatoalatus** Rassulova - 6
leonidae Manden. = A. sevangensis

lepagei Hult. = A. tugarinovii
lepsensis Bunge - 6
leptocaulis Ledeb. - 3
leptophysus Vved. - 6
leptopoides M. Pop. = A. michaelis
leptostachys Pall. (*A. macropterus* DC.) - 3, 4, 6
leucocalyx M. Pop. - 6
leucocladus Bunge - 6
levieri Freyn ex Somm. & Levier - 2
- var. *juzepczukii* Galushko = A. juzepczukii
linczevskyi Gontsch. - 5
linearifolius Ovcz. & Rassulova = A. madruschkendicus
lipschitzii Pavl. - 6
lipskyi M. Pop. - 6
lithophilus Kar. & Kir. - 6
litwinowianus Gontsch. - 6
litwinowii Lipsky = A. maximowiczii
longicuspis Bunge - 2
longiflorus Pall. = A. longipetalus
longipes Kar. & Kir. - 6
longipetalus Chater (*A. longiflorus* Pall. nom. illegit. superfl.) - 1, 2, 3, 6
longipetiolatus M. Pop. - 6
longiramosus (Boriss.) Boriss. = Astracantha longiramosa
longiramosus (Boriss.) Sirj. = Astracantha longiramosa
longisepalus Rassulova - 6
longistipitatus Boriss. (*A. ovczinnikovii* Boriss.) - 6
lorinseranus Freyn - 6
lunatus Pall. - 2
lupulinus Pall. - 4
lussiae Rzazade - 2
luxurians Bunge - 3
mackewiczii Gontsch. - 6
macranthoides Boriss. = Astracantha macrantha
macrobotrys Bunge = A. squarrosus
macrocephalus Willd. subsp. *finitimus* (Bunge) Chamberlain = A. finitimus
macroceras C.A. Mey. - 3
macrocladus Bunge - 6
macronyx Bunge - 6
macropetalus Schrenk - 6
macrophysus Somm. & Levier - 2
macropodium Lipsky - 6
macropodius E. Fisch. = Astracantha aurea
macropterus DC. = A. leptostachys
macropus Bunge (*A. olgianus* Krytzka) - 1, 3, 6
macrostachys DC. - 2
macrotropis Bunge - 6
macrourus Fisch. & C.A. Mey. - 2
maculatus Bunge = A. strictilobus
madruschkendicus Ovcz. & Rassulova (*A. linearifolius* Ovcz. & Rassulova, 1976, non Pers. 1807) - 6
magnificus Kolak. - 2
mailiensis B. Fedtsch. - 6
majevskianus Kryl. - 3
managildensis B. Fedtsch. - 6
maraziensis Rzazade - 2
marguzaricus Lipsky (*A. chorizanthus* auct.) - 6
marinus Boriss. - 5
- subsp. *boreomarinus* (A. Khokhr.) N.S. Pavlova = A. boreomarinus
marschallianus Fisch. = Astracantha denudata
massagetowii B. Fedtsch. = A. stipitatus
massalskyi Grossh. ex Fed. - 2
maverranagri M. Pop. - 6
maximowiczii Trautv. (*A. litwinowii* Lipsky) - 2(?), 6
maximus Willd. - 2

meanus (Boriss.) Boriss. = Astracantha meana
medius Schrenk - 1, 3, 6
megalanthus DC. - 3
megalomerus Bunge - 6
megalotropis C.A. Mey. ex Bunge - 2
megricus Grossh. - 2
melanocladus Lipsky - 6
melanocomus M. Pop. (*A. isophysus* Nabiev) - 6
melanostachys Benth. ex Bunge - 6
melilotoides Pall. - 3, 4
membranaceus (Fisch.) Bunge - 4, 5
- f. *propinquus* (Schischk.) Kitag. = A. propinquus
mendax Freyn = A. taldycensis
meracus Boriss. = Astracantha meraca
merkensis R. Kam. & Kovalevsk. - 6
meschchedensis Bunge = Astracantha meschchedensis
mesites Boiss. & Buhse - 2
meskheticus Manden. (*A. stevenianus* DC. subsp. *meskheticus* (Manden.) Sytin) - 2
meyeri Boriss. = Astracantha meyeri
michaelis Boriss. (*A. leptopoides* M. Pop.) - 6
micracme Boiss. - 2
microcephalus Willd. = Astracantha microcephala
miklaschewskii Basil. (*A. palibinii* Polozh.) - 4
miniatus Bunge - 4
mirabilis Lipsky - 6
- subsp. **chodshamastonicus** Rassulova - 6
- subsp. **czargicus** Rassulova - 6
mironovii Pachom. & Rassulova - 6
modesti R. Kam. & Kovalevsk. (*A. iljinii* R. Kam. 1966, non Rzazade, 1954, *A. austrotadzhikistanicus* Czer.) - 6
mogoltavicus M. Pop. - 6
mokeevae M. Pop. - 6
mollis Bieb. - 2
mongholicus Bunge - 3, 4
mongutensis Lipsky - 6
monophyllus Bunge - 4
monspessulanus L. - 1
montis-aquilis Grossh. - 2
mucidus Bunge - 6
mugodsharicus Bunge - 3, 6
mukusiensis Rech. fil. = A. gjunaicus
multicaulis Ledeb. - 3, 4
multifoliolatus (Boriss.) Boriss. = Astracantha multifoliolata
multifoliolatus (Boriss.) Sirj. = Astracantha multifoliolata
multijugus (Trautv.) Grossh. = A. ornithopodioides
munitus Boriss. = Astracantha tenuispina
muschketowii B. Fedtsch. - 6
myriophyllus Bunge = A. pamirensis
nachitschevanicus Rzazade - 2
namanganicus M. Pop. - 6
nathaliae Meff. = A. nivalis
neilreichianus Freyn & Sint. = A. juratzkanus
nematodes Bunge - 6
neo-lipskyanus M. Pop. - 6
neo-popovii Golosk. - 6
nephtonensis Freyn - 6
neurophyllus Franch. - 6
nevskii Gontsch. - 6
nicolai Boriss. - 6
nigrescens M. Pop. = A. nigricans
nigricans Barneby (*A. nigrescens* M. Pop., 1947, non Pall. 1800, nec Nutt. 1848) - 6
nigriceps M. Pop. - 6
nigrivestitus Podlech - 6
nigrocalyx Slob. - 6

nigrocarpus Chasanov & I. Maltsev - 6
nigromontanus M. Pop. - 6
nikitinae B. Fedtsch. (*A. promontoriorum* Gontsch.) - 6
ninae Gontsch. = A. gracilipes
ninae Pavl. (*A. albescens* Boriss. nom. invalid.) - 6
nitens Boiss. & Heldr. subsp. *viridis* (Bunge) Ponert = A. viridis
nivalis Kar. & Kir. (*A. nathaliae* Meff., *A. orthanthoides* Boriss.) - 6
nobilis Bunge ex B. Fedtsch. - 6
norvegicus Grauer (*A. oroboides* Hornem.) - 1, 3, 4, 5
- var. *brachytropis* (Stev.) Hashimov = A. brachytropis
novoascanicus Klok. = A. henningii
nuciferus Bunge - 6
nucleosus M. Pop. - 6
nuratavicus Boriss. = Astracantha nuratavica
nuratensis M. Pop. - 6
ochotensis A. Khokhr. - 5
ochranthus Gontsch. (*A. barrowianus* Aitch. & Baker subsp. *ochranthus* (Gontsch.) R. Kam.) - 6
odessanus Bess. = A. cornutus
odoratus Lam. - 2, 6
oeroilanicus M. Pop. = A. rubromarginatus
olchonensis Gontsch. - 4
oldenburgii B. Fedtsch. - 6
oleifolius DC. = Astracantha oleifolia
olgae Bunge (*A. peltatus* auct.) - 6
olgianus Krytzka = A. macropus
oltensis Grossh. = Astracantha oltensis
omissus Pachom. - 6
onobrychioides Bieb. (*A. bicolor* Lam. subsp. *owerinii* (Bunge) Ponert, *A. owerinii* Bunge, *A. ruprechtii* Bunge) - 2
onobrychis L. (*A. borysthenicus* Klok., *A. hybridus* S.G. Gmel., *A. troizkii* Grossh.) - 1, 2, 3, 6
- subsp. *goktschaicus* (Grossh.) Ponert = A. goktschaicus
ophiocarpus Benth. ex Boiss. (*Ophiocarpus paulsenii* (Freyn) Ikonn.) - 6
orbiculatus Ledeb. - 3, 6
ordubadensis Grossh. - 2
oreades C.A. Mey. - 2
ornithopodioides Lam. (*A. conrathii* Freyn, *A. multijugus* (Trautv.) Grossh., *A. stevenianus* DC. var. *multijugus* Trautv.) - 2
ornithorrhynchus M. Pop. - 6
oroboides Hornem. = A. norvegicus
orthanthoides Boriss. = A. nivalis
ortholobiformis Sumn. - 6
ortholobus Bunge - 3
ovatus DC. - 2
ovczinnikovii Boriss. = A. longistipitatus
owerinii Bunge = A. onobrychioides
oxyglottis Stev. ex Bieb. - 1, 2, 6
oxypterus Boriss. - 6
pachyrhizus M. Pop. - 6
palibinii Polozh. = A. miklaschewskii
paliurus Boriss. = Astracantha paliurus
pallescens Bieb. - 1
pamirensis Franch. (*A. myriophyllus* Bunge, 1880, non Pall. 1800) - 6
pamirensis Ovcz. & Rassulova = A. kurgankolensis
paradoxus Bunge - 2
paraglycyphyllus Boissieu = A. schelichowii
patentivillosus Gontsch. - 6
paucijugus Schrenk - 6
pauper Bunge - 6
pauperiformis B. Fedtsch. - 6

- subsp. flavescens Rassulova - 6
pavlovianus Gamajun. - 6
peduncularis Royle - 6
pedunculosus M. Pop. - 6
peltatus auct. = A. olgae
penduliflorus auct. = A. propinquus
pendulinus M. Pop. & B. Fedtsch. - 6
perembelicus Grossh. = A. bungeanus
perrarus Boriss. = A. cancellatus
persicus Fisch. & C.A. Mey. ex Bunge (*A. lagopoides* Lam. var. *persicus* DC., *A. lagurus* Willd. subsp. *persicus* (DC.) Ponert) - 2
petkoffii B. Fedtsch. - 6
petraeus Kar. & Kir. - 6
petri-primi Rassulova & Strizhova - 6
petropolitanus Sheld. (*A. trichocalyx* Trautv. 1876, non Nutt. ex Torr. & Gray, 1838) - 2
petropylensis Bunge - 3, 6
petunnikowii Litv. (*A. eremobius* M. Pop.) - 6
physocarpus Ledeb. - 3, 6
physodes L. - 1
piletocladus Freyn & Sint. = Astracantha piletoclada
pinetorum Boiss. (*A. declinatus* Willd. 1800, non Salisb. 1796, *A. talyschensis* Bunge) - 2
pischtovensis Gontsch. - 6
platyphyllus Kar. & Kir. - 6
plumatus Boriss. = Astracantha plumata
plumbeus (Nevski) Gontsch. - 6
podocarpus C.A. Mey. - 2
podolobus Boiss. & Hohen. - 6
polaris Benth. ex Hook. (*A. atlasovii* Kom.) - 5
poliotes Bunge = A. lactiflorus
politovii Kryl. - 3
poljakovii M. Pop. = A. heptapotamicus
polozhiae Timochina - 4
polyceras Kar. & Kir. - 6
polygala Pall. - 2
polyphyllus Bunge - 2
polytimeticus M. Pop. - 6
polyzygus M. Pop. - 6
ponticus Pall. - 1, 2
popovii Pavl. - 6
prilipkoanus Grossh. - 2
projecturus Sumn. = A. borodinii
prominens Boriss. = Astracantha prominens
promontoriorum Gontsch. = A. nikitinae
propinquus Schischk. (*A. membranaceus* Bunge f. *propinquus* (Bunge) Kitag., *A. penduliflorus* auct.) - 3, 4, 6
protractus Boriss. = A. eriocarpus
proximus (Boriss.) Boriss. = Astracantha dissecta
psammophilus Golosk. - 6
pschuknurensis Golosk. - 6
pseudanthylloides Gontsch. - 6
pseudoadsurgens Jurtz. (*A. inopinatus* Boriss. subsp. *pseudoadsurgens* (Jurtz.) N.S. Pavlova) - 4, 5
pseudoamygdalinus M. Pop. - 6
pseudoaustralis Fisch. & C.A. Mey. - 3
pseudobabatagi Pachom. & Rassulova - 6
pseudobrachytropis Gontsch. = A. abramovii
pseudocancellatus Grossh. = A. cancellatus
pseudocytisoides M. Pop. (*A. krassnovianus* Gontsch.) - 6
pseudodianthus Nabiev - 6
pseudoeremophysa M. Pop. - 6
pseudoglaucus Klok. - 1
pseudomacropterus Karmyscheva - 6
pseudomegalomerus M. Pop. - 6
pseudonobilis M. Pop. - 6

pseudonobrychis Andrz. - 1
Dubious species.
pseudopendulinus R. Kam. - 6
pseudorhacodes Gontsch. - 6
pseudoscoparius Gontsch. - 6
pseudotataricus Boriss. (*A. abruptus* Krytzka) - 1, 2
pseudotetrastichus Abdullaeva - 6
pseudoutriger Grossh. - 2
psiloglottis Stev. ex DC. - 2
psilolobus Putschkova - 6
psilophus Schrenk - 6
pskemensis M. Pop. - 6
psoraloides auct. = A. gjunaicus
pterocephalus Bunge = Astracantha pterocephala
puberulus Ledeb. - 3
pubiflorus DC. (*A. exscapus* L. subsp. *pubiflorus* (DC.)
Soo) - 1
pulcher Korov. - 6
pulposus M. Pop. - 6
pulvinatus Bunge = Astracantha pulvinata
■ **punctatus** Bunge
pycnanthus (Boriss.) Boriss. = Astracantha pycnantha
pycnanthus (Boriss.) Sirj. = Astracantha pycnantha
pycnolobus Bunge - 3
pycnophyllus Stev. = Astracantha microcephala
pygmaeus Pall. = Oxytropis gorodkovii
quisqualis Bunge - 6
- subsp. **congregatus** Rassulova - 6
- subsp. **kuhifrushensis** Rassulova - 6
- subsp. **longiscapus** Rassulova - 6
raddeanus Regel (*A. kemulariae* Grossh., *A. viciifolius* DC.
subsp. *raddeanus* (Regel) Sytin) - 2
raddei Basil. - 6
ramitensis Rassulova - 6
rariflorus Ledeb. - 3, 6
rarissimus M. Pop. - 6
rawlinsianus Aitch. & Baker - 6
reduncus Pall. - 1, 2
refractus C.A. Mey. - 2
regelii Trautv. - 2
remanens Nabiev (*A. cinerascens* M. Pop. 1947, non DC.
1840) - 6
resupinatus Bieb. (*A. fragrans* auct.) - 2
retamocarpus Boiss. & Hohen. - 6
reticulatus Bieb. - 1, 6
reverdattoanus Sumn. - 6
rhacodes Bunge - 6
richardsonii auct. = A. tolmaczevii
robustus Bunge - 2
- subsp. *subrobustus* (Boriss.) Ponert = A. subrobustus
■ **rollovii** Grossh.
roschanicus B. Fedtsch. - 6
roseus Ledeb. - 3, 6
rostratus C.A. Mey. - 2
rotundus Gontsch. - 6
rubellus Gontsch. - 6
rubens B. Fedtsch. & Ivanova = Astracantha rubens
rubescens Kovalevsk. & Vved. - 6
rubrifolius V.V. Nikit. ex Kovalevsk. - 6
rubrigalli M. Pop. - 6
rubrivenosus Gontsch. - 6
rubromarginatus Czerniak. (*A. oeroilanicus* M. Pop.) - 6
rubtzovii Boriss. - 6
rumpens Meff. - 6
rupifragiformis M. Pop. - 6
rupifragus Pall. (*A. sareptanus* A. Beck.) - 1, 3, 6
ruprechtii Bunge = A. onobrychioides

rytidocarpus Ledeb. - 3, 4
rytilobus Bunge = A. filicaulis
rzaevii Grossh. - 2
sabuletorum Ledeb. - 3, 6
saccocalyx Schrenk (*A. suidunensis* Bunge) - 6
sachalinensis Bunge - 5
sachokianus Grossh. - 2
saganlugensis Trautv. - 2
sahendi Buhse - 2
salareticus Boriss = A. frickii
salatavicus Bunge - 2
salicetorum Kom. = A. alpinus
salsugineus Kar. & Kir. - 6
sandalaschensis E. Nikit. - 6
sangesuricus Boriss. - 2
sangonensis Sirj. & Rech. fil. = A. squarrosus
sanguinolentus Bieb. - 2
saralensis Gontsch. - 3, 4
saratagius Bunge - 6
- subsp. **artschamajdani** Rassulova - 6
- subsp. **sarimensis** Rassulova - 6
sarbasnensis B. Fedtsch. - 6
sarchanensis Gontsch. - 6
sardaimionensis Ovcz. & Rassulova (*A. debilis* Ovcz. &
Rassulova, 1977, non A. Gray, 1864) - 6
sareptanus A. Beck. = A. rupifragus
sarygorensis Rassulova - 6
sarypulensis B. Fedtsch. - 6
sarytavicus M. Pop - 6
satteotoichus Gontsch. - 6
scaberrimus Bunge - 4
scabrisetus Bong. - 6
schachbuzensis Rzazade - 2
schachdarinus Lipsky - 6
schachimardanus Basil. - 6
schahrudensis Bunge - 6
schanginianus Pall. - 3, 6
- subsp. **neo-schanginianus** Golosk. - 6
schelichowii Turcz. (*A. kamtschaticus* (Kom.) Gontsch., *A.
paraglycyphyllus* Boissieu) - 4, 5
schemachensis Karjag. - 2
scheremetewianus B. Fedtsch. - 6
schischkinianus Gontsch. = A. takhtadzhjani
schischkinii Grossh. = A. elbrusensis
■ **schizopterus** Boiss.
schmalhausenii Bunge - 6
schrenkianus Fisch. & C.A. Mey. - 6
schugnanicus B. Fedtsch. - 6
schumilovae Polozh. - 4
schurae Pavl. = A. gracilipes
schuschensis Grossh. = A. dzhebrailicus
schutensis Gontsch. - 6
scleropodius Ledeb. - 3, 6
scleroxylon Bunge - 6
scoparius Schrenk - 6
scopiformis Ledeb. = A. tenuifolius
sealei Lepage (*A. eucosmus* Robbins subsp. *sealei* (Lepage)
Hult., *A. eucosmus* auct.) - 5
secundiflorus Rassulova - 6
secundus DC. (*A. frigidus* (L.) A. Gray subsp. *secundus*
(DC.) Worosch.) - 4, 5
semenovii Bunge - 6
semideserti Gontsch. - 6
sericeocanus Gontsch. - 4
sericeopuberulus Boriss. - 6
sericopetalus Trautv. - 6
sesamoides Boiss. - 6

setiferus Ovcz. & Rassulova = A. strizhovae
setosulus Gontsch. (*A. sphaeranthus* auct.) - 1
sevangensis Grossh. (*A. leonidae* Manden., *A. canescens* auct.) - 2
- subsp. *aspindzicus* (Manden. & Chinth.) Sytin = A. aspindzicus
sewerzowii Bunge - 6
shagalensis Grossh. - 2
shelkovnikovii Grossh. - 2
shinanensis Ohwi - 5
sieversianus Pall. - 6
similis Boriss. - 1
sinaicus Boiss. - 1
sisyrodites Bunge - 6
skorniakowii B. Fedtsch. - 6
sobolevskiae Polozh. - 4
sogdianus Bunge - 6
sogotensis Lipsky - 6
sommieri Freyn = Astracantha sommieri
- subsp. *acmophylloides* (Grossh.) Ponert = Astracantha acmophylloides
sosnowskyi auct. = A. tanae
spartioides Kar. & Kir. - 6
speciosissimus Pavl. (*A. tumescens* M. Pop.) - 6
sphaeranthus auct. = A. setosulus
sphaerocalyx Ledeb. - 2
sphaerocephalus Stev. - 2
sphaerocystis Bunge - 6
sphaerophysa Kar. & Kir. - 6
spinescens Bunge - 6
spongocarpus Meff. - 6
spryginii M. Pop. - 6
squarrosus Bunge (*A. macrobotrys* Bunge, *A. sangonensis* Sirj. & Rech. fil., *A. squarrosus* var. *sangonensis* (Sirj. & Rech. fil.) Ali) - 6
- var. *sangonensis* (Sirj. & Rech. fil.) Ali = A. squarrosus
stalinskyi Sirj. (*A. brachymorphus* Nikif.) - 1, 6
steinbergianus Sumn. - 6
stenanthus Bunge - 6
stenocarpus Gontsch. - 6
stenoceras C.A. Mey. - 3, 4, 6
stenoceroides Boriss. - 6
stenocystis Bunge - 6
stenonychioides Freyn & Bornm. = Astracantha stenonychioides
stenophysus Vved. & Zak. = A. dictamnoides
stephenianus Aitch. & Baker - 6
stevenianus DC. - 2
- subsp. *glochideus* (Boriss.) Sytin = A. glochideus
- subsp. *jodostachys* (Boiss. & Bunge) Ponert = A. jodostachys
- subsp. *kochianus* (Sosn.) Takht. = A. kochianus
- subsp. *meskheticus* (Manden.) Sytin = A. meskheticus
- var. *kochianus* (Sosn.) Chamberlain = A. kochianus
- var. *multijugus* Trautv. = A. ornithopodioides
stipitatus Benth. ex Bunge (*A. massagetowii* B. Fedtsch.) - 6
stipulosus Boriss. = Astracantha stipulosa
striatellus Pall. ex Bieb. - 1, 2, 6
strictifolius Boiss. = Astracantha stenonychioides
strictilobus Barneby (*A. maculatus* Bunge, 1868, non Lam. 1783) - 6
strizhovae Ovcz. & Rassulova (*A. setiferus* Ovcz. & Rassulova, 1976, non DC. 1825) -
subalabugensis M. Pop. - 6
subangustidens V.V. Nikit. - 6
subarcuatus M. Pop. - 6
subauriculatus Gontsch. - 6

subbarbellatus Bunge - 6
subbijugus Ledeb. - 6
subdjenarensis L. Vassil. - 6
subexcedens Gontsch. - 6
subinduratus Gontsch. - 6
subkahiricus Gontsch. = A. kahiricus
submaculatus Boriss. - 6
subpolaris Boriss. & Schischk. (*A. alpinus* L. subsp. *arcticus* Lindm., *A. alpinus* subsp. *borealis* Ju. Kozhevn.) - 1, 3, 4, 5
subrobustus Boriss. (*A. robustus* Bunge subsp. *subrobustus* (Boriss.) Ponert) - 2
subrosularis Gontsch. - 6
subscaposus M. Pop. ex Boriss. - 6
subschachimardanus M. Pop. - 6
subspinescens M. Pop. - 6
subspongocarpus Ovcz. & Rassulova - 6
substenoceras Boriss. - 6
substipitatus Gontsch. - 6
subternatus Pavl. - 6
subtrijugus M. Pop. - 6
subulatus Pall. = A. subuliformis
subuliformis DC. (*A. subulatus* Pall. 1800, non Desf. 1799) - 1, 2, 3
subverticillatus Gontsch. - 6
suffruticosus DC. (*A. fruticosus* Pall. 1800, non Forssk. 1775) - 3, 4, 5
suidunensis Bunge = A. saccocalyx
sukaczevii Derviz & A. Jelen. - 2
sulcatus L. - 1, 3, 4, 6
suluklensis Freyn & Sint. = A. eriocarpus
sumbari M. Pop. - 6
sumneviczii Pavl. - 6
supinus Bunge - 2
supralanatus Freyn - 6
suprapilosus Gontsch. - 1
surchanensis M. Pop. = A. unifoliolatus
surchobi Gontsch. - 6
syreitschikovii Pavl. - 6
szovitsii Fisch. & C.A. Mey. - 2
takhtadzhjanii Grossh. (*A. schischkinianus* Gontsch.) - 2
talassicus M. Pop. - 6
taldycensis Franch. (*A. mendax* Freyn) - 6
talievii Sirj. = A. tibetanus
talyschensis Bunge = A. pinetorum
tanae Sosn. (*A. sosnowskyi* auct.) - 2
tanaiticus C. Koch - 1
tanchasi Gontsch. - 6
▪ **taochius** Woronov (*A. artvinensis* M. Pop.)
tarchankuticus Boriss. - 1
taschkendicus Bunge - 6
tashkutanus V. Nikit. - 6
tatjanae Lincz. - 6
tauczilikensis Gontsch. = A. hemiphaca
tauricus Pall. - 1
- subsp. *scopiformis* (Ledeb.) L. Vassil. = A. tenuifolius
teberdensis Grossh. - 2
tecti-mundi Freyn - 6
▪ **teheranicus** Boiss.
tekessicus Bajt. - 6
tekutjevii Gontsch. - 6
temirensis M. Pop. - 1, 6
tenuifolius L. (*A. scopiformis* Ledeb., *A. tauricus* Pall. subsp. *scopiformis* (Ledeb.) L. Vassil.) - 1, 3, 6
tenuis Turcz. - 4
tephrolobus Bunge - 3
terekensis Al. Theod., Fed. & Rzazade = Astracantha denudata

terekliensis Gontsch. - 6
terektensis Fisjun - 6
terrae-rubrae Butk. - 6
testiculatus Pall. - 1, 2, 3, 4, 6
tetrastichus Bunge - 6
thaumatothamnus Vved. = A. chodshenticus
theodorianus Fed. & Rzazade = Astracantha theodoriana
theodorii Grossh. = A. humilis
thlaspi Lipsky (*Thlaspidium thlaspi* (Lipsky) Rassulova) - 6
tianschanicus Bunge - 6
tibetanus Benth. ex Bunge (*A. talievii* Sirj.) - 3, 6
titovii Gontsch. - 6
tolmaczevii Jurtz. (*A. richardsonii* auct.) - 3, 4, 5
torrentum Bunge = A. aegobromus
trachycarpus Gontsch. - 6
transcaspicus Freyn = A. ammodendron
transcaucasicus Boriss. = Astracantha acmophylloides
transhyrcanus M. Pop. = A. xiphidioides
transiliensis Gontsch. = A. ellipsoideus
transnominatus Abdullaeva (*A. trifoliolatus* Pavl. 1949, non
   Boiss. 1843) - 6
transoxanus Fisch. = Astracantha transoxana
tranzschelii Boriss. = A. djilgensis
trautvetteri Bunge - 3, 6
tribuloides Delile - 1, 2, 6
trichanthus Golosk. - 6
trichocalyx Trautv. = A. petropolitanus
trifoliolatus Pavl. = A. transnominatus
trigonelloides Boiss. - 2(?)
trigonocarpus (Turcz.) Bunge - 4
troizkii Grossh. = A. onobrychis
tscharynensis M. Pop. - 6
tschimganicus M. Pop. - 6
tschuensis Bunge - 3
tshegemensis Galushko - 2
tugarinovii Basil. (*A. lepagei* Hult., *A. aboriginum* auct.) - 4,
   5
- subsp. *kolymensis* (Jurtz.) Ju. Kozhevn. = A. kolymensis
tulinovii B. Fedtsch. - 6
tumescens M. Pop. = A. speciosissimus
tumninensis N.S. Pavlova & Bassargin - 5
tupalangi Gontsch. - 6
turajgyricus Golosk. - 6
turbinatus Bunge - 6
turcomanicus Bunge (*A. deserti* M. Pop.) - 6
turczaninowii Kar. & Kir. - 6
turkestanus Bunge - 6
turkmenorum (Boriss.) Boriss. = Astracantha turkmenorum
turkmenorum (Boriss.) Sirj. = Astracantha turkmenorum
turlanicus Bajt. & Myrzakulov - 6
tuvinicus Timochina - 4
tyttocarpus Gontsch. (*A. woldemarii* Juz.) - 6
ucrainicus M. Pop. & Klok. - 1
ufraensis Freyn & Sint. - 6
ugamicus M. Pop. - 6
ujalensis Gontsch. - 6
uliginosus L. - 3, 4, 5
umbellatus Bunge - 1, 3, 4, 5
unguiculatus Boriss. = Astracantha unguiculata
unifoliolatus Bunge (*A. aiwadzhi* B. Fedtsch., *A. surchanensis* M. Pop.) - 6
unijugus Bunge - 6
unilateralis Kar. & Kir. - 3, 6
unilocularis R. Kam. & Pachom. - 6
uninodus M. Pop. & Vved. - 6
uralensis Litv. = A. gorczakovii
uraniolimneus Boiss. - 2

urgutinus Lipsky - 6
urmiensis Bunge - 6
ustiurtensis Bunge - 6
utriger Pall. - 1, 2
vachanicus Boriss. & A. Korol. = A. ammophilus
vaginatus Pall. - 1, 3, 4
vallicola Gontsch. - 4
vallicoloides A. Khokhr. - 5
vardziae Charadze & Chinth. - 2
variegatus Franch. - 6
variocarinus A. Khokhr. - 2, 6
varius S.G. Gmel. (*A. virgatus* Pall.) - 1, 2, 3, 6
varzobicus Gontsch. - 6
vassilczenkoanus Golosk. - 6
vassilczenkoi Berdyev - 6
vavilovii Tamamsch. & Fed. (*A. klopotovskyi* Sosn.) - 2
vedicus Takht. = Astracantha vedica
vegetior Gontsch. - 6
velatus Trautv. - 6
veresczaginii Kryl. & Sumn. - 3
vernus Georgi = Gueldenstaedtia verna
versicolor Pall. - 4
vesicarius auct. = A. albidus
vicarius Lipsky (*Sewerzowia vicaria* (Lipsky) Rassulova) - 6
viciifolius DC. (*A. doluchanovii* Manden., *A. flaccidus*
   Bieb.) - 2
- subsp. **abchasicus** Sytin - 2
- subsp. *hirtulus* (Ledeb.) Sytin = A. hirtulus
- subsp. *raddeanus* (Regel) Sytin = A. raddeanus
villosissimus Bunge (*A. violaceus* Basil.) - 6
violaceus Basil. = A. villosissimus
virens Pavl. - 6
virgatus Pall. = A. varius
▪ virgeus Boriss.
viridiflorus Boriss. - 6
viridis Bunge (*A. nitens* Boiss. & Heldr. subsp. *viridis*
   (Bunge) Ponert) - 2
vladimiri Sirj. = Astracantha macrantha
voronovianus Boriss. = Astracantha voronoviana
vulpinus Willd. - 1, 3, 6
vvedenskyi M. Pop. - 6
wachschi B. Fedtsch. - 6
willisii M. Pop. - 6
winkleri Trautv. - 6
woldemarii Juz. = A. tyttocarpus
wolgensis Bunge - 1, 3
woronowii Bornm. - 2(?)
wrangelii Sirj. = Astracantha pulvinata
- subsp. *memorabilis* Sirj. = Astracantha pulvinata
xanthomeloides Korov. & M. Pop. - 6
xanthotrichus Ledeb. - 3
xanthoxiphidium Freyn & Sint. = A. juratzkanus
xerophilus Ledeb. - 2
xiphidioides Freyn & Sint. (*A. transhyrcanus* M. Pop.) - 6
xiphidium Bunge - 2
xipholobus M. Pop. - 6
zaissanensis Sumn. - 3, 6
zangelanus Grossh. - 2
zaprjagaevii Gontsch. - 6
zarokoensis Rassulova - 6
zerabulaki M. Pop. - 6
zingeri Korsh. - 1
zuvanticus Grossh. - 2

Bona Medik. = Vicia

narbonensis (L.) Medik. = Vicia narbonensis
serratifolia (Jacq.) Stankevicz = Vicia serratifolia

Botryolotus Jaub. & Spach = Melilotoides

adscendens Nevski = Melilotoides adscendens

*Caesalpinia L.

*gilliesii (Hook.) Dietr. - 1, 2, 6

*Cajanus DC.

*cajan (L.) Millspaugh (*C. indicus* Spreng.)
indicus Spreng. = C. cajan

Calispepla Vved.

aegacanthoides Vved. - 6

Calophaca Fisch. ex DC.

grandiflora Regel - 6
hovenii Schrenk = C. soongorica
kultiassovii S. Korov. = C. soongorica var. tianschanica
pskemica Gorbunova - 6
reticulata Sumn. - 6
sericea B. Fedtsch. ex Boriss. - 6
soongorica Kar. & Kir. (*C. hovenii* Schrenk) - 6
- var. **tianschanica** (B. Fedtsch.) Gorbunova (*C. kultiassovii* S. Korov., *C. tianschanica* (B. Fedtsch.) Boriss.) - 6
tianschanica (B. Fedtsch.) Boriss. = C. soongorica var. tianschanica
wolgarica (L. fil.) DC. - 1

Calycomorphum C. Presl = Trifolium

subterraneum (L.) C. Presl = Trifolium subterraneum

*Canavalia DC.

*gladiata (Jacq.) DC.

Caragana Fabr.

acanthophylla Kom. - 6
alaica Pojark. - 6
alexeenkoi R. Kam. - 6
altaica (Kom.) Pojark. = C. pygmaea
andassaica Bajt. = C. balchaschensis
arborescens Lam. (*C. fruticosa* (Pall.) Kom. p.p. incl. typo) - 3, 4, 5(cult.), 6
aurantiaca Koehne - 6
balchaschensis (Kom.) Pojark. (*C. andassaica* Bajt.) - 6
bongardiana (Fisch. & C.A. Mey.) Pojark. - 6
bungei Ledeb. - 3, 4
buriatica Peschkova (*C. microphylla* Lam. subsp. *buriatica* (Peschkova) Yakovl.) - 4
camilli-schneideri Kom. - 3, 6
■ dasyphylla Pojark.
frutescens (Pall.) DC. = C. frutex
frutex (L.) C. Koch (*C. frutescens* (Pall.) DC.) - 1, 2, 3, 4, 6
fruticosa (Pall.) Bess. = C. arborescens and C. manshurica
grandiflora (Bieb.) DC. - 2, 6
hololeuca Bunge ex Kom. = C. tragacanthoides
jubata (Pall.) Poir. - 4, 5, 6
kirghisorum Pojark. - 6
komarovii Schischk. = C. pygmaea
laeta Kom. (*C. turfanensis* (Krasn.) Kom.) - 6
laetevirens Pojark. = C. leucospina
leucophloea Pojark. (*C. pygmaea* (L.) DC. subsp. *leucophloea* (Pojark.) Polovinko) - 6
leucospina Kom. (*C. laetevirens* Pojark.) - 6

manshurica (Kom.) Kom. (*C. microphylla* Lam. f. *manshurica* Kom., *C. fruticosa* (Pall.) Kom. p.p. quoad pl.) - 5
microphylla Lam. - 4
- subsp. *buriatica* (Peschkova) Yakovl. = C. buriatica
- f. *manshurica* Kom. = C. manshurica
mollis (Bieb.) Bess. - 1, 2
pleiophylla (Regel) Pojark. - 6
prainii auct. = C. turkestanica
pruinosa Kom. - 6
pumila Pojark. - 3, 6
pygmaea (L.) DC. (*C. altaica* (Kom.) Pojark., *C. komarovii* Schischk. nom. invalid., *C. splendens* Schischk. ex K. Sobol.) - 3, 4
- subsp. **austrotuvinica** Bondareva - 4
- subsp. *leucophloea* (Pojark.) Polovinko = C. leucophloea
scythica (Kom.) Pojark. - 1
spinosa (L.) Vahl ex Hornem. - 3, 4
splendens Schischk. ex K. Sobol. = C. pygmaea
stenophylla Pojark. - 4
tragacanthoides (Pall.) Poir. (*C. hololeuca* Bunge ex Kom.) - 3, 6
turfanensis (Krasn.) Kom. = C. laeta
turkestanica Kom. (*C. prainii* auct.) - 6
ussuriensis (Regel) Pojark. - 5

Cassia L.

*marylandica L. - 2
nictitans L. - 2
nomame (Siebold) Honda - 5
occidentalis L. - 5
tora L. - 1(alien), 5(alien)

*Ceratonia L.

*siliqua L. - 2

Cercis L.

griffithii Boiss. - 6
siliquastrum L. - 1, 2, 6

Chamaecytisus Link (*Cytisus* L. p.p.)

aggregatus (Schur) Czer. = Ch. supinus
albus (Hacq.) Rothm. (*Cytisus albus* Hacq.) - 1
austriacus (L.) Link (*Cytisus austriacus* L.) - 1, 2
blockianus (Pawl.) Klaskova (*Cytisus blockianus* Pawl., *C. blockii* V. Krecz.) - 1
borysthenicus (Grun.) Klaskova (*Cytisus borysthenicus* Grun.) - 1, 2
caucasicus (Grossh.) Holub = Ch. ruthenicus
*glaber (L. fil.) Rothm. (*Cytisus glaber* L. fil.) - 1
- var. **elongatus** (Waldst. & Kit.) Tzvel. (*Cytisus elongatus* Waldst. & Kit., *C. hirsutus* L. subsp. *elongatus* (Waldst. & Kit.) Briq., *C. ratisbonensis* Schaeff. subsp. *elongatus* (Waldst. & Kit.) Gams) - 1
graniticus (Rehm.) Rothm. (*Cytisus graniticus* Rehm.) - 1
hirsutissimus (C. Koch) Czer. (*Cytisus hirsutissimus* C.Koch, *Chamaecytisus hirsutus* (L.) Link subsp. *hirsutissimus* (C. Koch) Ponert) - 2
hirsutus (L.) Link (*Cytisus hirsutus* L., *Chamaecytisus hirsutus* subsp. *leucotrichus* (Schur) Ponert, *Ch. leucotrichus* (Schur) Czer., *Cytisus hirsutus* var. *leucotrichus* Schur, *C. leucotrichus* (Schur) Schur) - 1
- subsp. *hirsutissimus* (C. Koch) Ponert = Ch. hirsutissimus
- subsp. *leucotrichus* (Schur) Ponert = Ch. hirsutus

- subsp. *polytrichus* (Bieb.) Ponert = Ch. polytrichus
kreczetoviczii (Wissjul.) Holub (*Cytisus kreczetoviczii* Wissjul.) - 1
leiocarpus (A. Kerner) Rothm. (*Cytisus leiocarpus* A. Kerner) - 1
leucotrichus (Schur) Czer. = Ch. hirsutus
lindemannii (V. Krecz.) Kraskova (*Cytisus lindemannii* V. Krecz.) - 1, 2
litwinowii (V. Krecz.) Klaskova (*Cytisus litwinowii* V. Krecz.) - 1
paczoskii (V. Krecz.) Klaskova (*Cytisus paczoskii* V. Krecz.) - 1
podolicus (Blocki) Klaskova (*Cytisus podolicus* Blocki) - 1
polytrichus (Bieb.) Rothm. (*Cytisus polytrichus* Bieb., *Chamaecytisus hirsutus* (L.) Link subsp. *polytrichus* (Bieb.) Ponert) - 1
ponomarjovii (Seredin) Czer. (*Cytisus ponomarjovii* Seredin) - 2
***purpureus** (Scop.) Link (*Cytisus purpureus* Scop.) - 1
ratisbonensis (Schaeff.) Rothm. (*Cytisus ratisbonensis* Schaeff.) - 1
- subsp. *ruthenicus* (Fisch. ex Woloszcz.) Zielinski = Ch. ruthenicus
rochelii (Wierzb.) Rothm. (*Cytisus rochelii* Wierzb.) - 1
ruthenicus (Fisch. ex Woloszcz.) Klaskova (*Cytisus ruthenicus* Fisch. ex Woloszcz., *Chamaecytisus caucasicus* (Grossh.) Holub, *Ch. ratisbonensis* (Schaeff.) Rothm. subsp. *ruthenicus* (Fisch. ex Woloszcz.) Zielinski, *Cytisus caucasicus* Grossh.) - 1, 2, 3
- var. *syreiszczikowii* (V. Krecz.) Tzvel. (*Cytisus syreiszczikowii* V. Krecz.) - 1
skrobiszewskii (Pacz.) Klaskova (*Cytisus skrobiszewskii* Pacz.) - 1
supinus (L.) Link (*Cytisus supinus* L., *C. aggregatus* Schur, *Chamaecytisus aggregatus* (Schur) Czer., *Ch. supinus* subsp. *aggregatus* (Schur) A. & D. Löve) - 1(?)
- subsp. *aggregatus* (Schur) A. & D. Löve = Ch. supinus
wulffii (V. Krecz.) Klaskova (*Cytisus wulffii* V. Krecz.) - 1
zingeri (Nenuk. ex Litv.) Klaskova (*Cytisus zingeri* (Nenuk. ex Litv.) V. Krecz.) - 1

Chamaespartium Adans. = Genista and Genistella

sagittale (L.) P. Gibbs = Genistella sagittalis

Chesneya Lindl. (*Chesniellla* Boriss.)

antoninae Rassulova & B. Scharipova - 6
astragalina Jaub. & Spach - 6
badachschanica Boriss. - 6
borissovae Pavl. - 6
botschantzevii R. Vinogradova - 6
crassipes Boriss. - 6
darvasica Boriss. - 6
dshungarica Golosk. - 6
■ elegans Fomin
ferganensis Korsh. (*Chesniella ferganensis* (Korsh.) Boriss.) - 6
gracilis (Boriss.) R. Kam. (*Chesniella gracilis* Boriss.) - 6
hissarica Boriss. - 6
isfarensis Tarakulov - 6
karatavica R. Kam. - 6
kopetdaghensis Boriss. - 6
kschtutica Rassulova & B. Scharipova - 6
latefoliolata Rassulova & B. Scharipova - 6
linczevskyi Boriss. - 6
neplii Boriss. - 6

quinata Fed. - 6
tadzhikistana Boriss. - 6
ternata (Korsh.) M. Pop. - 6
tribuloides Nevski (*Chesniella tribuloides* (Nevski) Boriss.) - 6
trijuga Boriss. - 6
turkestanica Franch. - 6
villosa (Boriss.) R. Kam. & R. Vinogradova (*Chesniella villosa* Boriss.) - 6

Chesniella Boriss. = Chesneya

ferganensis (Korsh.) Boriss = Chesneya ferganensis
gracilis Boriss = Chesneya gracilis
tribuloides (Nevski) Boriss. = Chesneya tribuloides
villosa Boriss. = Chesneya villosa

Chrysaspis Desv. (*Amarenus* C. Presl)

aurea (Poll.) Greene (*Trifolium aureum* Poll., *T. agrarium* L. p.p. nom. ambig., *T. strepens* Crantz, nom. illegit.) - 1, 2, 3, 5(alien)
badia (Schreb.) Greene (*Trifolium badium* Schreb.) - 1
- subsp. *rytidosemia* (Boiss. & Hohen.) Hendrych = Ch. rytidosemia
campestris (Schreb.) Desv. (*Trifolium campestre* Schreb., *T. agrarium* L. p.p. nom. ambig.) - 1, 2, 5(alien), 6
dubia (Sibth.) Desv. (*Trifolium dubium* Sibth.) - 1, 2
grandiflora (Schreb.) Hendrych (*Trifolium grandiflorum* Schreb., *T. speciosum* Willd.) - 1, 2
karatavica (Pavl.) Roskov (*Trifolium karatavicum* Pavl.) - 6
micrantha (Viv.) Hendrych (*Trifolium micranthum* Viv.) - 2
patens (Schreb.) Holub (*Trifolium patens* Schreb.) - 1
rytidosemia (Boiss. & Hohen.) Roskov (*Trifolium rytidosemium* Boiss. & Hohen., *Chrysaspis badia* (Schreb.) Greene subsp. *rytidosemia* (Boiss. & Hohen.) Hendrych, *Trifolium badium* Schreb. subsp. *rytidosemium* (Boiss. & Hohen.) Hossain) - 2
sebastianii (Savi) Hendrych (*Trifolium sebastianii* Savi) - 2
sintenisii (Freyn) Hendrych (*Trifolium sintenisii* Freyn, *T. stipitatum* Boiss. & Bal. 1872, non Clos. 1847) - 2
spadicea (L.) Greene (*Trifolium spadiceum* L., *Amarenus spadiceus* (L.) C. Presl) - 1, 2, 3, 4(alien), 5(alien)

Cicer L.

acanthophyllum Boriss (*C. garanicum* Boriss.) - 6
anatolicum Alef. - 2
***arietinum** L. - 1, 2, 6
balcaricum Galushko - 2
baldshuanicum (M. Pop.) Lincz. - 6
caucasicum Bornm. = C. incisum
chorassanicum (Bunge) M. Pop. (*Ononis chorassanica* Bunge) - 6
ervoides Brign. = Lens ervoides
ervoides (Sieb.) Fenzl = C. incisum
fedtschenkoi Lincz. - 6
flexuosum Lipsky - 6
- subsp. *grande* M. Pop. = C. grande
garanicum Boriss. = C. acanthophyllum
grande (M. Pop.) Korotk. (*C. flexuosum* Lipsky subsp. *grande* M. Pop.) - 6
incanum Korotk. - 6
incisum (Willd.) K. Maly (*Anthyllis incisa* Willd., *Cicer caucasicum* Bornm., *C. ervoides* (Sieb.) Fenzl, 1843, non Brign. 1810) - 2
jacquemontii auct. = C. microphyllum

kopetdaghense Lincz. (*C. tragacanthoides* Jaub. & Spach var. *turcomanicum* M. Pop.) - 6
> Probably, the priority name for this taxon should be C. strausii Bornm.

korshinskyi Lincz. - 6
laetum Rassulova & B. Scharipova - 6
luteum Rassulova & B. Scharipova - 6
macracanthum M. Pop. - 6
microphyllum Royle (*C. jacquemontii* auct., *C. multijugum* Maesen, 1972, non Rassulova & B. Scharipova, 1978) - 6
minutum Boiss. & Hohen. (*C. pimpinellifolium* Jaub. & Spach subsp. *minutum* (Boiss. & Hohen.) Ponert) - 2
mogoltavicum (M. Pop.) A. Korol. - 6
multijugum Maesen = C. microphyllum
multijugum Rassulova & B. Scharipova = C. rassuloviae
paucijugum Nevski (*C. songaricum* Steph. ex DC. var. *paucijugum* M. Pop. nom. nud.) - 6
pimpinellifolium Jaub. & Spach subsp. *minutum* (Boiss. & Hohen.) Ponert = C. minutum
pungens Boiss. (*C. spinosum* M. Pop.) - 6
rassuloviae Lincz. (*C. multijugum* Rassulova & B. Scharipova, 1978, non Maesen, 1972) - 6
songaricum Steph. ex DC. - 6
- var. *paucijugum* M. Pop. = C. paucijugum
spinosum M. Pop. = C. pungens
tragacanthoides Jaub. & Spach var. *turcomanicum* M. Pop. = C. kopetdaghense

*Cladrastis Rafin. (*Virgilia* auct.)

***kentukea** (Dum.-Cours.) Rudd (*Sophora kentukea* Dum.-Cours., *Cladrastis lutea* (Michx. fil.) C. Koch, *Virgilia lutea* Michx. fil) - 1, 2, 6
lutea (Michx. fil.) C. Koch = C. kentukea

Colutea L.

acutifolia Shap. - 2
arborescens L. - 1, 6
- subsp. *cilicica* (Boiss. & Bal.) Ponert = C. cilicica
armena Boiss. & Huet - 2
atabaevii B. Fedtsch. (*C. kopetdaghensis* B. Fedtsch.) - 6
brachyptera Sumn. - 6
buhsei (Boiss.) Shap. - 6
canescens Shap. = C. paulsenii
cilicica Boiss. & Bal. (*C. arborescens* L. subsp. *cilicica* (Boiss. & Bal.) Ponert) - 1, 2
gracilis Freyn & Sint. - 6
guntensis Rassulova & B. Scharipova - 6
hybrida Shap. = C. orbiculata
jarmolenkoi Shap. - 6
komarovii Takht. - 2
kopetdaghensis B. Fedtsch. = C. atabaevii
x **media** Willd. - C. arborescens L. x C. orientalis Mill. - 2, 6
mesantha Shap. ex Ali = C. paulsenii
orbiculata Sumn. (*C. hybrida* Shap.) - 1, 2
orientalis Mill. - 1, 2
paulsenii Freyn & Sint. (*C. canescens* Shap., *C. mesantha* Shap. ex Ali, *C. paulsenii* var. *canescens* (Shap.) Browicz, *C. paulsenii* subsp. *mesantha* (Shap. ex Ali) Ali, *C. paulsenii* var. *mesantha* (Shap. ex Ali) Browicz, *C. paulsenii* f. *mesantha* (Shap. ex Ali) Rassulova & B. Scharipova, *C. rostrata* Sumn.) - 6
- subsp. *mesantha* (Shap. ex Ali) Ali = C. paulsenii
- var. *canescens* (Shap.) Browicz = C. paulsenii
- var. *mesantha* (Shap. ex Ali) Browicz = C. paulsenii
- f. *mesantha* (Shap. ex Ali) Rassulova & B. Scharipova =

C. paulsenii
rostrata Sumn. = C. paulsenii
x **variabilis** Browicz. - C. cilicica Boiss. & Bal. x C. orientalis Mill. - 2

Coronilla L.

balansae (Boiss.) Grossh. = Securigera balanse
cappadocica Willd. var. *balansae* Boiss. = Securigera balanse
charadzeae Chinth. & Tschuchrukidze = Securigera charadzeae
coronata L. - 1, 2
cretica L. = Securigera cretica
elegans Panc. = Securigera elegans
emeroides Boiss. & Sprun. = Hippocrepis emeroides
emerus L. subsp. *emeroides* (Boiss. & Sprun.) Hayek = Hippocrepis emeroides
- subsp. *emeroides* (Boiss. & Sprun.) Holmboe = Hippocrepis emeroides
hyrcana Prilipko = Securigera hyrcana
latifolia (Hazsl.) Jav. = Securigera elegans
orientalis Mill. = Securigera orientalis
parviflora Willd. = Securigera parviflora
rostrata Boiss. & Sprun. = Securigera parviflora
scorpioides (L.) Koch - 1, 2
varia L. = Securigera varia
- subsp. *hirta* (Bunge ex Boiss.) Rech. fil. = Securigera varia subsp. hirta
- subsp. *orientalis* (Uhrova) Jahn = Securigera varia subsp. orientalis
- var. *hirta* Bunge ex Boiss. = Securigera varia subsp. hirta
- var. *orientalis* Uhrova = Securigera varia subsp. orientalis

Corothamnus (Koch) C. Presl

procumbens (Waldst. & Kit. ex Willd.) C. Presl (*Genista procumbens* Waldst. & Kit. ex Willd., *Cytisus procumbens* (Waldst. & Kit. ex Willd.) Spreng.) - 1(?)

Crimaea Vass. = Melilotoides

cretacea (Bieb.) Vass. = Melilotoides cretacea

*Crotalaria L.

***juncea** L. - 2, 6

*Cytisus L.

aggregatus Schur = Chamaecytisus supinus
albus Hacq. = Chamaecytisus albus
austriacus L. = Chamaecytisus austriacus
blockianus Pawl. = Chamaecytisus blockianus
blockii V. Krecz. = Chamaecytisus blockianus
borysthenicus Grun. = Chamaecytisus borysthenicus
calycinus Bieb. = Argyrolobium biebersteinii
caucasicus Grossh. = Chamaecytisus ruthenicus
elongatus Waldst. & Kit. = Chamaecytisus glaber var. elongatus
glaber L. fil. = Chamaecytisus glaber
graniticus Rehm. = Chamaecytisus graniticus
hirsutissimus C. Koch = Chamaecytisus hirsutissimus
hirsutus L. = Chamaecytisus hirsutus
- subsp. *elongatus* (Waldst. & Kit.) Briq. = Chamaecytisus glaber var. elongatus
- var. *leucotrichus* Schur = Chamaecytisus hirsutus
kreczetoviczii Wissjul. = Chamaecytisus kreczetoviczii
leiocarpus A. Kerner = Chamaecytisus leiocarpus

leucotrichus (Schur) Schur = Chamaecytisus hirsutus
lindemanni V. Krecz. = Chamaecytisus lindemannii
litwinowii V. Krecz. = Chamaecytisus litwinowii
paczoskii V. Krecz. = Chamaecytisus paczoskii
podolicus Blocki = Chamaecytisus podolicus
polytrichus Bieb. = Chamaecytisus polytrichus
ponomarjovii Seredin = Chamaecytisus ponomarjovii
procumbens (Waldst. & Kit. ex Willd.) Spreng. = Corot-
    hamnus procumbens
purpureus Scop. = Chamaecytisus purpureus
ratisbonensis Schaeff. = Chamaecytisus ratisbonensis
- subsp. *elongatus* (Waldst. & Kit.) Gams = Chamaecytisus
    glaber var. elongatus
rochelii Wierzb. = Chamaecytisus rochelii
ruthenicus Fisch. ex Woloszcz. = Chamaecytisus ruthenicus
sessilifolius L.
skrobiszewskii Paczosky = Chamaecytisus skrobiszewskii
supinus L. = Chamaecytisus supinus
syreiszczikowii V. Krecz. = Chamaecytisus ruthenicus var.
    syreiszczikowii
wulffii V. Krecz. = Chamaecytisus wulffii
zingeri (Nenuk. ex Litv.) V. Krecz. = Chamaecytisus zingeri

Dendrobrychis (DC.) Galushko = Onobrychis

cornuta (L.) Galushko = Onobrychis cornuta

Desmodium auct. = Podocarpium

fallax auct. = Podocarpium mandshuricum
mandshuricum (Maxim.) Schindl. = Podocarpium mand-
    shuricum
oldhamii Oliv. = Podocarpium oldhamii
oxyphyllum DC. var. *mandshuricum* (Maxim.) Ohashi =
    Podocarpium mandshuricum
podocarpum DC. var. *mandshuricum* Maxim. = Podocarpi-
    um mandshuricum
racemosum (Thunb.) DC. var. *mandshuricum* (Maxim.)
    Ohwi = Podocarpium mandshuricum

Didymopelta Regel & Schmalh. = Astragalus

turkestanica (Regel & Schmalh.) Regel & Schmalh =
    Astragalus dipelta

Dipelta Regel & Schmalh = Astragalus

turkestanica Regel & Schmalh. = Astragalus dipelta

Dolichos auct. = Pueraria

lobatus Willd. = Pueraria lobata

Dolichos L. p.p. = Glycine and Vigna

sinensis L. = Vigna unguiculata
soja L. = Glycine max
unguiculatus L. = Vigna unguiculata

Dorycnium Mill.

graecum (L.) Ser. - 1, 2
herbaceum Vill. (*D. pentaphyllum* Scop. subsp. *herbaceum*
    (Vill.) Rouy) - 1, 2
intermedium Ledeb. - 1, 2
pentaphyllum Scop. (*D. suffruticosum* auct.) - 1
- subsp. *herbaceum* (Vill.) Rouy = D. herbaceum
strictum (Fisch. & C.A. Mey.) Lassen = Lotus strictus
suffruticosum auct. = D. pentaphyllum

Edwardsia auct. = Keyserlingia

griffithii (Stocks) Pilipenko = Keyserlingia griffithii
hortensis Boiss. & Buhse = Keyserlingia hortensis

Eremosparton Fisch. & C.A. Mey.

aphyllum (Pall.) Fisch. & C.A. Mey. - 1, 2, 6
flaccidum Litv. - 6
songoricum (Litv.) Vass. - 6

Ervilia Link = Vicia

sativa Link = Vicia ervilia

Ervum auct. = Lens

cyaneum Boiss. & Hohen. = Lens cyanea

Ervum L. = Vicia

hirsutum L. = Vicia hirsuta
lathyroides (L.) Stankevicz = Vicia lathyroides
tsydenii (Malysch.) Stankevicz = Vicia tsydenii

Ewersmannia Bunge

botschantzevii Sarkisova - 6
sarytavica Sarkisova - 6
sogdiana Ovcz. - 6
subspinosa (Fisch. ex DC.) B. Fedtsch. - 6

Faba Hill = Vicia

bona Medik. = Vicia faba
vulgaris Moench = Vicia faba

Falcata auct. = Amphicarpaea

japonica (Oliv.) Kom. = Amphicarpaea japonica

Galearia C. Presl = Amoria

fragifera (L.) Bobr. = Amoria fragifera
fragifera (L.) C. Presl = Amoria fragifera
- subsp. *bonannii* (C. Presl) Sojak = Amoria bonannii
physodes (Stev.) Sojak = Amoria physodes
raddeana (Trautv.) Bobr. = Amoria raddeana
raddeana (Trautv.) Sojak = Amoria raddeana

Galega L.

officinalis L. - 1, 2
orientalis Lam. - 2

Genista L. (*Chamaespartium* Adans. p.p. emend.
    Spach)

abchasica Sachok. - 2
adzharica M. Pop. = G. suanica
albida Willd. - 1
- var. *pontica* Zelen. = G. millii
angustifolia Schischk. - 2
armeniaca Spach - 2
■ **artwinensis** Schischk.
borysthenica Kotov = G. sibirica
compacta Schischk. - 2
depressa Bieb. (*G. tinctoria* L. var. *depressa* (Bieb.) P.
    Gibbs, comb. superfl., *G. tinctoria* var. *depressa* (Bieb.)
    Schmalh.) - 1

donetzica Kotov = G. tinctoria var. donetzica
dracunculoides Spach (*G. transcaucasica* Schischk.) - 2
elata (Moench) Wend. = G. tinctoria
flagellaris Somm. & Levier - 2
germanica L. - 1
glaberrima Novopokr. = G. suanica
godetii Spach - 1
humifusa L. (*G. lipskyi* Novopokr. & Schischk., *G. sachoki-ana* A. Kuth.) - 2
juzepczukii Tzvel. - 1, 2
kolakowskyi Sachok. - 2
lipskyi Novopokr. & Schischk. = G. humifusa
marginata Bess. = G. tinctoria var. pubescens
millii Heldr. ex Boiss. (*G. albida* Willd. var. *pontica* Zelen., *G. pontica* (Zelen.) Juz.) - 1
mingrelica Albov - 2
oligosperma (Andrae) Simonk. = G. rupestris
ovata Waldst. & Kit. - 1
patula Bieb. - 2
pilosa L. - 1(?)
pontica (Zelen.) Juz. = G. millii
procumbens Waldst. & Kit. ex Willd. = Corothamnus procumbens
pubescens Lang = G. tinctoria var. pubescens
rupestris Schur (? *G. alpicola* Schur, *G. oligosperma* (Andrae) Simonk. nom. illegit., *G. tinctoria* L. var. *oligosperma* Andrae, *G. tinctoria* subsp. *oligosperma* (Andrae) Jav.) - 1
sachokiana A. Kuth. = G. humifusa
sagittalis L. = Genistella sagittalis
scythica Pacz. - 1
sibirica L. (*G. borysthenica* Kotov) - 1
suanica Schischk. (*G. adzharica* M. Pop., *G. glaberrima* Novopokr.) - 2
tanaitica P. Smirn. - 1
taurica Dubovik - 1
tetragona Bess. - 1
tinctoria L. (*G. elata* (Moench) Wend., *G. elatior* Koch, *G. tinctoria* subsp. *elata* (Moench) Aschers. & Graebn. ex Morariu, *G. virgata* Willd.) - 1, 2, 3
- subsp. *elata* (Moench) Aschers. & Graebn. ex Morariu = G. tinctoria
- subsp. *oligosperma* (Andrae) Jav. = G. rupestris
- var. *depressa* (Bieb.) P. Gibbs = G. depressa
- var. *depressa* (Bieb.) Schmalh. = G. depressa
- var. **donetzica** (Kotov) Wissjul. (*G. donetzica* Kotov) - 1
- var. *oligosperma* Andrae = G. rupestris
- var. **pubescens** (Lang) Heuff. (*G. pubescens* Lang, *G. marginata* Bess.) - 1
transcaucasica Schischk. = G. dracunculoides
verae Juz. - 1
virgata Willd. = G. tinctoria

Genistella Ort. (*Chamaespartium* Adans. p.p. emend. Dandy)

sagittalis (L.) Gams (*Genista sagittalis* L., *Chamaespartium sagittale* (L.) P. Gibbs) - 1

Gleditsia L.

caspia Desf. (*G. horrida* (Thunb.) Makino subsp. *caspia* (Desf.) Paclt) - 2
horrida (Thunb.) Makino subsp. *caspia* (Desf.) Paclt = G. caspia
triacanthos L. - 1, 2, 6

Glycine Willd. (*Dolichos* L. p.p.)

gracilis Skvorts. = G. soja
hispida (Moench) Maxim. = G. max
*****max** (L.) Merr. (*Phaseolus max* L., *Dolichos soja* L., *Glycine hispida* (Moench) Maxim.) - 1, 2, 5
- subsp. *soja* (Siebold & Zucc.) Ohashi = G. soja
soja Siebold & Zucc. (*G. gracilis* Skvorts., *G. max* (L.) Merr. subsp. *soja* (Siebold & Zucc.) Ohashi, *G. soja* var. *gracilis* (Skvorts.) L.Z. Wang, *G. ussuriensis* Regel & Maack) - 5
- var. *gracilis* (Skvorts.) L.Z. Wang = G. soja
ussuriensis Regel & Maack = G. soja

Glycyrrhiza L.

aspera Pall. (*G. laxissima* Vass., *G. zaissanica* Serg.) - 1, 2, 3, 6
bucharica Regel = Meristotropis bucharica
echinata L. - 1, 2, 3, 6
erythrocarpa (Vass.) Abdullaeva = Meristotropis triphylla
foetidissima Tausch (*G. macedonica* Boiss. & Orph.) - 1, 2, 6
glabra L. (*G. glabra* subsp. *glandulifera* (Waldst. & Kit.) Ponert, *G. glandulifera* Waldst. & Kit., *G. hirsuta* Pall.) - 1, 2, 3, 6
- subsp. *glandulifera* (Waldst. & Kit.) Ponert = G. glabra
glandulifera Waldst. & Kit. = G. glabra
gontscharovii Masl. (*Meristotropis bucharica* (Regel) Kruganova f. *gontscharovii* (Masl.) Maltzeva) - 6
hirsuta Pall. = G. glabra
korshinskyi Grig. - 1, 3, 6
kulabensis Masl. = Meristotropis bucharica
laxissima Vass. = G. aspera
macedonica Boiss. & Orph. = G. foetidissima
pallidiflora Maxim. - 5
uralensis Fisch. - 3, 4, 6
zaissanica Serg. = G. aspera

Goebelia Bunge (*Vexibia* auct. p.p.)

alopecuroides (L.) Bunge = Pseudosophora alopecuroides
- var. *tomentosa* Bunge = Pseudosophora alopecuroides
pachycarpa (C.A. Mey.) Bunge (*Sophora pachycarpa* C.A. Mey., *Vexibia pachycarpa* (C.A. Mey.) Yakovl.) - 6
prodanii (E. Anders.) Grossh. = Sophora jaubertii

Gueldenstaedtia Fisch.(*Amblyotropis* Kitag.)

monophylla Fisch. (*Amblyotropis monophylla* (Fisch.) C.Y. Wu) - 3
pauciflora (Pall.) Fisch. = G. verna
verna (Georgi) Boriss. (*Astragalus vernus* Georgi, *Amblyotropis verna* (Georgi) Kitag., *Gueldenstaedtia pauciflora* (Pall.) Fisch.) - 3, 4, 5

Guilandina L. p.p. = Gymnocladus

dioica L. = Gymnocladus dioicus

*Gymnocladus Lam. (*Guilandina* L. p.p.)

*****dioicus** (L.) C. Koch (*Guilandina dioica* L.) - 1

Halimodendron Fisch. ex DC.

halodendron (Pall.) Voss - 1, 2, 3, 6

Hedysarum L.

aculeatum Golosk. = H. krylovii
acutifolium Bajt. - 6
alabukense E. Nikit. = H. turkestanicum
alaicum B. Fedtsch. - 6
alpinum L. - 1, 3, 4, 5
- subsp. **boreo-europaeum** Jurtz. - 1, 3
amankutanicum B. Fedtsch. - 6
americanum (Michx.) Britt. (*H. auriculatum* Eastw.) - 5
angrenicum Korotk. - 6
arcticum B. Fedtsch. (*H. hedysaroides* (L.) Schinz & Thell. subsp. *arcticum* (B. Fedtsch.) P.W. Ball) - 1, 3, 4, 5
arenarium Kit. = Onobrychis arenaria
argenteum Bieb. = H. biebersteinii
argyrophyllum Ledeb. - 2
armenum Boiss. & Tchih. - 2
atropatanum Bunge ex Boiss. - 2
auriculatum Eastw. = H. americanum
auriculatum sensu B. Fedtsch. = H. truncatum
austrokurilense (N.S. Pavlova) N.S. Pavlova (*H. sachalinense* B. Fedtsch. subsp. *austrokurilense* N.S. Pavlova, *H. komarovii* auct.) - 5
austrosibiricum B. Fedtsch. (*H. hedysaroides* (L.) Schinz & Thell. subsp. *austrosibiricum* (B. Fedtsch.) Jurtz.) - 3, 4, 6
babatagicum Korotk. = H. iomuticum
baicalense B. Fedtsch. = H. setigerum
balchanense Boriss. = H. macranthum
baldshuanicum B. Fedtsch. (*H. vvedenskyi* Korotk.) - 6
bectauatavicum Bajt. - 6
biebersteinii Zertova (*H. argenteum* Bieb. 1808, non L. 1774) - 1, 2
bordzilovskyi Grossh. - 2
boreale Nutt. subsp. *mackenzii* (Richards.) A. & D. Löve = H. mackenzii
- subsp. *mackenzii* (Richards.) Welsh = H. mackenzii
- var. *mackenzii* (Richards.) C.L. Hitchc. = H. mackenzii
branthii Trautv. & C.A. Mey. - 4, 5
bucharicum B. Fedtsch. - 6
candidum Bieb. - 1, 2
caucasicum Bieb. (*H. kemulariae* Sachok. & Ghinth.) - 2
cephalotes Franch. = H. minjanense
chaitocarpum Regel & Schmalh. - 6
chaiyrakanicum Kurbatsky - 4
chantavicum M. Pop. ex Bajt. - 6
cisbaicalense Malysch. - 4
cisdarvasicum R. Kam. & Karimova - 6
confertum (N.S. Pavlova) N.S. Pavlova (*H. sachalinense* B. Fedtsch. subsp. *confertum* N.S. Pavlova) - 5
connatum (B. Fedtsch.) B. Fedtsch. = H. inundatum
consanguineum DC. - 3, 4
cretaceum Fisch. - 1
cumuschtanicum B. Sultanova - 6
czatkalense B. Sultanova = H. dmitrievae
daghestanicum Rupr. ex Boiss. - 2
dahuricum Turcz. ex B. Fedtsch. = H. gmelinii
daraut-kurganicum B. Sultanova - 6
dasycarpum Turcz. - 4, 5
denticulatum Regel & Schmalh. - 6
dmitrievae Bajt. (*H. czatkalense* B. Sultanova) - 6
drobovii Korotk. - 6
dshambulicum Pavl. = H. plumosum
elegans Boiss. & Huet - 2
enaffae B. Sultanova - 6
fedtschenkoanum Regel = H. omissum and H. plumosum
ferganense Korsh. (*H. pumilum* (Ledeb.) B. Fedtsch. nom. illegit., *H. schischkinii* Sumn.) - 3, 6

flavescens Regel & Schmalh. - 6
flavum Rupr. = H. semenovii
formosum Fisch. & C.A. Mey. ex Basin. - 2
fruticosum Pall. - 4
gmelinii Ledeb. (*H. dahuricum* Turcz. ex B. Fedtsch.) - 1, 3, 4, 6
grandiflorum Pall. - 1
gypsaceum Korotk. - 6
hedysaroides (L.) Schinz & Thell. - 1
- subsp. *arcticum* (B. Fedtsch.) P.W. Ball = H. arcticum
- subsp. *austrosibiricum* (B. Fedtsch.) Jurtz. = H. austrosibiricum
- subsp. *tschuktschorum* Jurtz. = H. truncatum
hemithamnoides Korotk. - 6
ibericum Bieb. - 2
iliense B. Fedtsch. (*H. kandyktassicum* Bajt.) - 6
inundatum Turcz. (*H. connatum* (B. Fedtsch.) B. Fedtsch.) - 4, 5
iomuticum B. Fedtsch. (*H. babatagicum* Korotk.) - 6
issykkulense E. Nikit. = H. songoricum
jaxarticum M. Pop. - 6
junceum L. fil. = Lespedeza juncea
kamcziraki Karimova - 6
kandyktassicum Bajt. = H. iliense
karataviense B. Fedtsch. - 6
kasteki Bajt. = H. plumosum
kemulariae Sachok. & Chinth. = H. caucasicum
kirghisorum B. Fedtsch. - 6
komarovii auct. = H. austrokurilense
kopetdaghi Boriss. - 6
korshinskyanum B. Fedtsch. - 6
krasnovii B. Fedtsch.
> According to G.P. Yakovlev, this species does not occur on the investigated territory.
krylovii Sumn. (*H. aculeatum* Golosk., *H. linczevskyi* Bajt.) - 6
kudrjaschevii Korotk. = H. nuratense
kuhitangi Boriss. = H. plumosum
kungeicum Bajt. = H. semenovii
latibracteatum N.S. Pavlova - 5
lehmannianum Bunge - 6
linczevskyi Bajt. = H. krylovii
lipskianum L. Vassil. - 6
lipskyi B. Fedtsch. = Onobrychis laxiflora
mackenzii Richards. (*H. boreale* Nutt. var. *mackenzii* (Richards.) C.L. Hitchc., *H. boreale* subsp. *mackenzii* (Richards.) A. & D. Löve, comb. superfl., *H. boreale* subsp. *mackenzii* (Richards.) Welsh) - 5
macranthum (Freyn & Sint.) Bornm. & Gauba (*H. micropterum* Bunge subsp. *macranthum* Freyn & Sint., *H. balchanense* Boriss.) - 6
macrocarpum Korotk. ex Kovalevsk. - 6
magnificum Kudr. - 6
microphyllum Turcz. = H. turczaninovii
micropterum auct. = H. wrightianum
micropterum Bunge subsp. *macranthum* Freyn & Sint. = H. macranthum
mindshilkense Bajt. - 6
minjanense Rech. fil. (*H. cephalotes* Franch. 1883, non Roxb. 1832) - 6
minussinense B. Fedtsch. - 4
mogianicum (B. Fedtsch.) B. Fedtsch. - 6
monophyllum Boriss. - 6
montanum (B. Fedtsch.) B. Fedtsch. = H. songoricum
narynense E. Nikit. - 6
neglectum Ledeb. - 3, 4, 6
nikolai Kovalevsk. (*H. villosum* Pavl. 1954, non Mill. 1768) - 6

nuratense M. Pop. (*H. kudrjaschevii* Korotk.) - 6
olgae B. Fedtsch. - 6
omissum Korotk. ex Kovalevsk. (*H. fedtschenkoanum* Regel, p.p. excl. lectotypo) - 6
ovczinnikovii Karimova - 6
pallidiflorum Pavl. - 6
parviflorum Bajt. = H. plumosum
parvum B. Sultanova - 6
pavlovii Bajt. (*H. ulcunburulicum* Bajt.) - 6
plumosum Boiss. & Hausskn. (*H. dshambulicum* Pavl., *H. fedtschenkoanum* Regel, p.p. incl. lectotypo, *H. kasteki* Bajt., *H. kuhitangi* Boriss., *H. parviflorum* Bajt.) - 6
poncinsii Franch. - 6
popovii Korotk. - 6
pskemense M. Pop. ex B. Fedtsch. - 6
pulchrum E. Nikit. - 6
pumilum (Ledeb.) B. Fedtsch. = H. ferganense
razoumovianum Fisch. & Helm - 1
roseum Sims - 4
sachalinense B. Fedtsch. - 5
- subsp. *austrokurilense* N.S. Pavlova = H. austrokurilense
- subsp. *confertum* N.S. Pavlova = H. confertum
sangilense Krasnob. & Timochina - 4
santalaschi B. Fedtsch. - 6
schischkinii Sumn. = H. ferganense
scoparium Fisch. & C.A. Mey. - 6
semenovii Regel & Herd. (*H. flavum* Rupr., *H. kungeicum* Bajt.) - 6
sericeum Bieb. - 2
setigerum Turcz. ex Fisch. & C.A. Mey. (*H. baicalense* B. Fedtsch.) - 4, 5
setosum Vved. (*H. sussamyrense* E. Nikit.) - 6
severzovii Bunge - 6
songoricum Bong. (*H. issykkulense* E. Nikit., *H. montanum* (B. Fedtsch.) B. Fedtsch., *H. subglabrum* (Kar. & Kir.) B. Fedtsch., *H. tenuifolium* (B. Fedtsch.) B. Fedsch.) - 6
splendens Fisch. - 3, 6
subglabrum (Kar. & Kir.) B. Fedtsch. = H. songoricum
sussamyrense E. Nikit. = H. setosum
syriacum Boiss. (*H. varium* Willd. subsp. *syriacum* (Boiss.) Townsend) - 2(?)
talassicum E. Nikit. & B. Sultanova - 6
taschkendicum M. Pop. - 6
tauricum Pall. ex Willd. - 1, 2
tenuifolium (B. Fedtsch.) B. Fedsch. = H. songoricum
theinum Krasnob. - 3
truncatum Eastw. (*H. hedysaroides* (L.) Schinz & Thell. subsp. *tschuktschorum* Jurtz., *H. auriculatum* sensu B. Fedtsch.) - 5
turczaninovii Peschkova (*H. microphyllum* Turcz. 1842, non Thunb. 1784) - 4
turkestanicum Regel & Schmalh. (*H. alabukense* E. Nikit.) - 6
turkewiczii B. Fedtsch. - 2
ucrainicum Kaschm. - 1
ulcunburulicum Bajt. = H. pavlovii
ussuriense I. Schischk. & Kom. - 5
varium Willd. - 2
- subsp. *syriacum* (Boiss.) Townsend = H. syriacum
vegetius (Trautv.) B. Fedtsch. = H. wrightianum
vicioides Turcz. - 4, 5
villosum Pavl. = H. nikolai
vvedenskyi Korotk. = H. baldshuanicum
wrightianum Aitch. & Baker (*H. vegetius* (Trautv.) B. Fedtsch., *H. micropterum* auct.) - 2, 6
zundukii Peschkova - 4

Hippocrepis L.

biflora Spreng. - 2, 6
bisiliqua Forssk. (*H. unisiliquosa* L. subsp. *bisiliqua* (Forssk.) Bornm.) - 6
ciliata Willd. (*H. multisiliquosa* auct.) - 1
comosa L. - 1
emeroides (Boiss. & Sprun.) Czer. comb. nova (*Coronilla emeroides* Boiss. & Sprun. 1843, in Boiss., Diagn. Pl. Or., ser. 1, 2 : 100; *C. emerus* L. subsp. *emeroides* (Boiss. & Sprun.) Hayek, comb. superfl., *C. emerus* subsp. *emeroides* (Boiss. & Sprun.) Holmboe, *Hippocrepis emerus* (L.) Lassen subsp. *emeroides* (Boiss. & Sprun.) Lassen) - 1, 2
emerus (L.) Lassen subsp. *emeroides* (Boiss. & Sprun.) Lassen = H. emeroides
multisiliquosa auct. = H. ciliata
unisiliquosa L. - 1, 2
- subsp. **armena** Lassen - 2
- subsp. *bisiliqua* (Forssk.) Bornm. = H. bisiliqua
- subsp. **unisiliquosa** - 1

*Indigofera L.

***tinctoria** L. - 2

Keyserlingia Bunge (*Edwardsia* auct.)

griffithii (Stocks) Bunge (*Sophora griffithii* Stocks, *Edwardsia griffithii* (Stocks) Pilipenko, *Sophora mollis* (Royle) Baker subsp. *griffithii* (Stocks) Ali) - 6
hortensis (Boiss. & Buhse) Yakovl. (*Edwardsia hortensis* Boiss. & Buhse, *Keyserlingia mollis* (Royle) Boiss. p.p. excl. typo, *Sophora griffithii* Stocks subsp. *hortensis* (Boiss. & Buhse) Yakovl., *S. hortensis* (Boiss. & Buhse) Rech. fil., *S. mollis* Grah. ex Baker, p.p.) - 6
    Probably, this species should be named K. mollis (Edwardsia mollis Royle).
mollis (Royle) Boiss. p.p. = K. hortensis

Kummerowia Schindl.

stipulacea (Maxim.) Makino - 5
striata (Thunb.) Schindl. - 5

*Laburnum Medik.

***anagyroides** Medik. - 1, 2, 6

Lagonychium Bieb. (*Proposis* auct.)

farctum (Banks & Soland.) Bobr. (*Prosopis farcta* (Banks & Soland.) Macbr.) - 2, 6

Lathyrus L. (*Orobus* L.)

alatus (Maxim.) Kom. = L. komarovii
aleuticus (Greene) Pobed. (*L. maritimus* Bigel. *aleuticus* Greene, *L. japonicus* Willd. var. *aleuticus* (Greene) Fern., *L. japonicus* subsp. *japonicus* f. *pubescens* (C. Hartm.) Ohashi & Tateishi, *L. japonicus* subsp. *pubescens* Korobkov, *Pisum maritimum* L. var. *pubescens* C. Hartm.) - 1, 5
angulatus auct. = L. leptophyllus
annuus L. (*L. colchicus* Lipsky) - 2, 6
aphaca L. - 1, 2, 6
asiaticus (Zalk.) Kudr. = L. sativus
atropatanus (Grossh.) Sirj. (*L. nivalis* Hand.-Mazz. subsp. *atropatanus* (Grossh.) Ponert) - 2

aureus (Stev.) Brandza - 1, 2
austriacus (Crantz) Wissjul. = L. pannonicus
chloranthus Boiss. - 2
cicera L. - 1, 2, 6
ciliatidentatus Czefr. (*Orobus ciliatidentatus* (Czefr.)
    Avazneli) - 2
colchicus Lipsky = L. annuus
cyaneus (Stev.) C. Koch - 2
- subsp. *digitatus* (Bieb.) Ponert = L. digitatus
davidii Hance - 5
digitatus (Bieb.) Fiori (*L. cyaneus* (Stev.) C. Koch subsp.
    digitatus (Bieb.) Ponert) - 1
dominianus Litv. - 6
frolovii Rupr. - 3, 4
gmelinii Fritsch - 1, 3, 4, 6
hirsutus L. - 1, 2, 6
humilis (Ser.) Spreng. - 1, 3, 4, 5, 6
inconspicuus L. - 2, 6
incurvus (Roth) Roth - 1, 2, 6
japonicus Willd. - 4
- subsp. *japonicus* f. *pubescens* (C. Hartm.) Ohashi & Ta-
    teishi = L. aleuticus
- subsp. *maritimus* (L.) P.W. Ball = L. maritimus
- subsp. *pubescens* Korobkov = L. aleuticus
- var. *aleuticus* (Greene) Fern. = L. aleuticus
ketzkhovelii Avazneli - 2
komarovii Ohwi (*L. alatus* (Maxim.) Kom. 1904, non Ten.
    1811, nec Sibth. & Smith, 1813, *Orobus alatus* Maxim.
    1859, non A. Br. 1853) - 4, 5
krylovii Serg. (*L. laevigatus* (Waldst. & Kit.) Gren. subsp.
    krylovii (Serg.) Hendrych) - 3
lacaitae Czefr. (*Orobus hispanicus* Lacaita, 1928, non *Lathy-
    rus hispanicus* Rouy, 1899, *L. pannonicus* (Jacq.)
    Garcke subsp. *hispanicus* (Lacaita) Bassler) - 1
lacteus (Bieb.) Wissjul. (*Orobus lacteus* Bieb., *Lathyrus
    versicolor* auct.) - 1, 2
laevigatus (Waldst. & Kit.) Gren. - 1
- subsp. *krylovii* (Serg.) Hendrych = L. krylovii
- subsp. *transsylvanicus* (Spreng.) Breistroffer = L. transsyl-
    vanicus
- subsp. *transsylvanicus* (Spreng.) Soo = L. transsylvanicus
latidentatus A. Jelen. - 2
latifolius L. (*L. megalanthus* Steud., *L. sylvestris* L. subsp.
    latifolius (L.) Ponert) - 1
laxiflorus (Desf.) O. Kuntze - 1, 2
ledebourii Trautv. (*L. pannonicus* (Jacq.) Garcke subsp.
    ledebourii (Trautv.) Bassler, *Orobus ledebourii*
    (Trautv.) Roldug.) - 1, 3, 4
leptophyllus Bieb. (*L. angulatus* auct.) - 2
linifolius (Reichard) Bassler (*Orobus linifolius* Reichard) -
    1
- var. *montanus* (Bernh.) Bassler (*L. montanus* Bernh.) - 1
litvinovii Iljin - 1, 3
luteus (L.) Peterm. subsp. *transsylvanicus* (Spreng.)
    Dostal = L. transsylvanicus
maritimus Bigel. (*L. japonicus* Willd. subsp. *maritimus* (L.)
    P.W. Ball, *L. maritimus* (L.) Fries, nom. illegit., *Pisum
    maritimum* L.) - 1
- *aleuticus* Greene = L. aleuticus
megalanthus Steud. = L. latifolius
miniatus Bieb. ex Stev. (*L. rotundifolius* Willd. subsp.
    miniatus (Bieb. ex Stev.) P.H. Davis) - 2
miyabei Matsum. = L. pilosus var. miyabei
montanus Bernh. = L. linifolius var. montanus
mulkak Lipsky - 6
multijugus (Ledeb.) Czefr. (*Orobus lacteus* Bieb. var. *multi-
    jugus* Ledeb., *Lathyrus pannonicus* (Jacq.) Garcke

    subsp. *multijugus* (Ledeb.) Bassler) - 3, 4
niger (L.) Bernh. - 1, 2
nissolia L. - 1, 2
nivalis Hand.-Mazz. subsp. *atropatanus* (Grossh.) Ponert =
    L. atropatanus
ochrus (L.) DC. - 1
*odoratus L. - 1
pallescens (Bieb.) C. Koch (*Orobus canescens* auct., *O.
    filiformis* auct.) - 1, 2
palustris L. - 1, 2, 3, 4, 6
- subsp. *pilosus* (Cham.) Hult. = L. pilosus
- var. *pilosus* (Cham.) Ledeb. = L. pilosus
pannonicus (Jacq.) Garcke (*Orobus pannonicus* Jacq.,
    Lathyrus austriacus (Crantz) Wissjul., *Orobus austria-
    cus* Crantz) - 1
- subsp. *hispanicus* (Lacaita) Bassler = L. lacaitae
- subsp. *ledebourii* (Trautv.) Bassler = L. ledebourii
- subsp. *multijugus* (Ledeb.) Bassler = L. multijugus
pilosus Cham. (*L. palustris* L. subsp. *pilosus* (Cham.) Hult.,
    L. palustris var. *pilosus* (Cham.) Ledeb.) - 1, 3, 4, 5, 6
- var. *miyabei* (Matsum.) Hara (*L. miyabei* Matsum.) - 5
pisiformis L. - 1, 2, 3, 4, 6
platyphyllos Retz. (*L. sylvestris* L. subsp. *platyphyllos*
    (Retz.) Celak.) - 1
pratensis L. - 1, 2, 3, 4, 5, 6
quinquenervius (Miq.) Litv. ex Kom. - 4, 5
roseus Stev. - 1, 2
rotundifolius Willd. - 1
- subsp. *miniatus* (Bieb. ex Stev.) P.H. Davis = L. miniatus
sativus L. (*L. asiaticus* (Zalk.) Kudr. comb. invalid., *L.
    sativus* subsp. *asiaticus* Zalk. nom. invalid.) - 1, 2, 4, 6
- subsp. *asiaticus* Zalk. = L. sativus
saxatilis (Vent.) Vis. (*Vicia saxatilis* (Vent.) Tropea) - 1
setifolius L. - 1, 2
sphaericus Retz. - 1, 2, 6
subalpinus (Herb.) G. Beck (*Orobus subalpinus* Herb.) - 1
subrotundus Maxim. = Vicia subrotunda
sylvestris L. - 1, 2
- subsp. *latifolius* (L.) Ponert = L. latifolius
- subsp. *platyphyllos* (Retz.) Celak. = L. platyphyllos
transsylvanicus (Spreng.) Reichenb. fil. (*Orobus transsyl-
    vanicus* Spreng., *Lathyrus laevigatus* (Waldst. & Kit.)
    Gren. subsp. *transsylvanicus* (Spreng.) Breistroffer, *L.
    laevigatus* subsp. *transsylvanicus* (Spreng.) Soo, comb.
    superfl., *L. luteus* (L.) Peterm. subsp. *transsylvanicus*
    (Spreng.) Dostal) - 1
tuberosus L. - 1, 2, 3, 5(alien), 6
undulatus Boiss. - 1
venetus (Mill.) Wohlf. - 1
vernus (L.) Bernh. - 1, 2, 3, 4
versicolor auct. = L. lacteus
vinealis Boiss. & Noe - 2
woronowii Bornm. - 2

Lembotropis Griseb.

nigricans (L.) Griseb. - 1

Lens Mill. (*Ervum* auct.)

*culinaris Medik. - 1, 2, 6
- subsp. **macrosperma** (Baumg.) Czefr. (*L. esculenta
    Moench var. *macrosperma* Baumg.)
- subsp. *orientalis* (Boiss.) Ponert = L. orientalis
cyanea (Boiss. & Hohen.) Alef. (*Ervum cyaneum* Boiss. &
    Hohen.) - 6
ervoides (Brign.) Grande (*Cicer ervoides* Brign., *Lens lenti-
    cula* (Schreb.) Webb & Berth.) - 1, 2

esculenta Moench var. *macrosperma* Baumg. = Lens culinaris subsp. macrosperma
lenticula (Schreb.) Webb & Berth. = L. ervoides
nigricans (Bieb.) Webb & Berth. - 1, 2
orientalis (Boiss.) Schmalh. (*L. culinaris* Medik. subsp. *orientalis* (Boiss.) Ponert) - 1, 2, 6
pygmaea Grossh. = Vicia ervilia

Lespedeza Michx.

bicolor Turcz. - 4, 5
cyrtobotrya Miq. - 5
davurica (Laxm.) Schindl. - 4, 5
hedysaroides (Pall.) Kitag. = L. juncea
juncea (L. fil.) Pers. (*Hedysarum junceum* L. fil., *Lespedeza hedysaroides* (Pall.) Kitag.) - 2(alien), 4, 5
sericea Miq. - 2(alien)
tomentosa (Thunb.) Maxim. - 5

Lotononis (DC.) Exkl. & Zeyh.

genistoides (Fenzl) Benth. - 2

Lotus L.

aleppicus Boiss. (*L. corniculatus* L. var. *hirsutissimus* Ledeb., *L. gebelia* Vent. subsp. *hirsutissimus* (Ledeb.) Ponert) - 2
alpicola (G. Beck) Min., Ulle & Kritzk. (*L. corniculatus* L. var. *alpicola* G. Beck, *L. alpinus* auct.) - 1
alpinus auct. = L. alpicola
ambiguus Bess. ex Spreng. - 1
angustissimus L. - 1, 2
- subsp. *palustris* (Willd.) Ponert = L. palustris
arvensis Pers. - 1
balticus Min. - 1
burttii Sz.-Borsos - 6
callunetorum (Juxip) Min. (*L. corniculatus* L. f. *callunetorum* Juxip) - 1
caucasicus Kuprian. ex Juz. - 1, 2
confusus Serg. = L. krylovii
corniculatus L. - 1, 5(alien)
- subsp. *frondosus* Freyn = L. frondosus
- subsp. *tenuis* (Waldst. & Kit. ex Willd.) Briq. ex Rech. fil. = L. tenuis
- var. *alpicola* G. Beck = L. alpicola
- var. *hirsutissimus* Ledeb. = L. aleppicus
- var. *norvegicus* Chrtkova-Zertova = L. norvegicus
- f. *callunetorum* Juxip = L. callunetorum
dvinensis Min. & Ulle - 1
elisabethae Opperm. ex Wissjul. - 1, 2
frondosus (Freyn) Kuprian. s. restr. (*L. corniculatus* L. subsp. *frondosus* Freyn) - 1, 2, 6
gebelia Vent. - 2
- subsp. *hirsutissimus* (Ledeb.) Ponert = L. aleppicus
- subsp. **trdatii** (Tamamsch.) T.N. Pop. & Takht. (*L. gebelia* var. *trdatii* Tamamsch.) - 2
komarovii Min. - 1
krylovii Schischk. & Serg. (*L. confusus* Serg., *L. frondosus* (Freyn) Kuprian. p.p. excl. typo) - 3, 6
maritimus L. = Tetragonolobus maritimus
michauxianus Ser. - 2(?)
norvegicus (Chrtkova-Zertova) Min. (*L. corniculatus* L. var. *norvegicus* Chrtkova-Zertova) - 1
olgae Klok. - 1
ornithopodioides L. - 1, 2
palustris Willd. (*L. angustissimus* L. subsp. *palustris* (Willd.) Ponert) - 2

peczoricus Min. & Ulle - 1, 4(alien?)
praetermissus Kuprian. - 1, 2, 3
ruprechtii Min. - 1
sergievskiae R. Kam. & Kovalevsk. - 3, 6
strictus Fisch. & C.A. Mey. (*Dorycnium strictum* (Fisch. & C.A. Mey.) Lassen) - 2, 3, 6
tauricus Juz. - 1
tenuis Waldst. & Kit. ex Willd. (*L. corniculatus* L. subsp. *tenuis* (Waldst. & Kit. ex Willd.) Briq. ex Rech. fil.) - 1, 2, 6
ucrainicus Klok. - 1, 2, 3, 6
uliginosus Schkuhr - 1, 2(?)
zhegulensis Klok. - 1

Lupinaster Fabr. (*Ursia* Vass.)

albus Link (*Trifolium ciswolgense* Spryg. ex Iljin & Truchaleva, nom. invalid., *T. spryginii* Belaeva & Sipl.) - 1, 3, 6
baicalensis (Belaeva & Sipl.) Roskov (*Trifolium baicalense* Belaeva & Sipl.) - 4
eximius (Steph. ex Ser.) C. Presl (*Trifolium eximium* Stev. ex DC., *Lupinaster eximius* (Steph. ex DC.) Bobr. comb. superfl.) - 3, 4, 5, 6
gordejevii (Kom.) Roskov (*Melilotoides gordejevii* (Kom.) Sojak, *Melissitus gordejevii* (Kom.) Latsch., *Trifolium gordejevii* (Kom.) Z. Wei, *T. gordejevii* (Kom.) N.S. Pavlova, comb. superfl., *Trigonella gordejevii* (Kom.) Grossh., *Ursia gordejevii* (Kom.) Vass.) - 5
litwinowii (Iljin) Roskov (*Trifolium litwinowii* Iljin) - 1
pacificus (Bobr.) Latsch. (*Trifolium pacificum* Bobr.) - 5
pentaphyllus Moench (*L. pentaphyllus* subsp. *angustifolius* (Litv.) Sojak, *Trifolium lupinaster* L., *T. lupinaster* subsp. *angustifolium* (Litv.) Bobr.) - 3, 4, 5, 6
- subsp. *angustifolius* (Litv.) Sojak = Lupinaster pentaphyllus
polyphyllus (C.A. Mey.) Latsch. (*Trifolium polyphyllum* C.A. Mey., *Lupinaster polyphyllus* (C.A. Mey.) Sojak, comb. superfl.) - 2
popovii Roskov - 4

*Lupinus L.

albus L. - 1, 2
angustifolius L. - 1
hirtus auct. = L. micranthus
luteus L. - 1
micranthus Guss. (*L. hirtus* auct.) - 1
nootkatensis Donn - 5
perennis L. - 1
pilosus L. - 1
polyphyllus Lindl. - 1, 4, 5
varius L. - 1

Maackia Maxim. &Rupr.

amurensis Maxim. & Rupr. - 5

Medicago L.

afghanica (Bord.) Vass. = M. sativa
agrestis Ten. = M. rigidula
agropyretorum Vass. (*M. rivularis* Vass.) - 6
alatavica Vass. nom. invalid. - 6
arabica (L.) Huds. (*M. polymorpha* L. var. *arabica* L.) - 1, 2
borealis Grossh. = M. falcata
caerulea Less. ex Ledeb. (*M. sativa* L. subsp. *caerulea* (Less. ex Ledeb.) Schmalh.) - 1, 2, 6
- subsp. **caspica** Sinsk. - 2

- subsp. **infradaghestanica** Sinsk. - 2, 6
- subsp. **pilifera** (Urb.) Vass. (*M. sativa* subsp. *microcarpa* Urb. c. *pilifera* Urb.) - 1
- subsp. **semicoerulea** Sinsk. - 2
cancellata Bieb. - 1, 2
caucasica Vass. (*M. sativa* L. subsp. *caucasica* (Vass.) Lubenetz) - 2
constricta Durieu (*M. globosa* auct.) - 2
coronata (L.) Bartalini - 2, 6
daghestanica Rupr. ex Boiss. - 2
denticulata Willd. (*M. nigra* (L.) Krock. var. *denticulata* (Willd.) O. Bolos & Vigo) - 1, 2, 6
difalcata Sinsk. - 2, 6
- subsp. **kazachstanica** Sinsk. - 6
- subsp. **orientalicaucasica** Sinsk. - 2
dzhawakhetica Bordz. = M. papillosa
erecta Kotov = M. romanica
falcata L. (*M. borealis* Grossh., *M. glutinosa* Bieb. subsp. *praefalcata* Sinsk., *M. procumbens* Bess., *M. quasifalcata* Sinsk., *M. sativa* L. subsp. *falcata* (L.) Arcang.) - 1, 2, 3, 4, 5, 6
- subsp. **adenocarpa** Sojak - 1
- subsp. *romanica* (Prod.) Schwarz & Klinkovski = M. romanica
- subsp. *tenderiensis* (Opperm. ex Klok.) Vass. = M. romanica
- var. *ambigua* Trautv. = M. trautvetteri
- var. *revoluta* Sumn. = M. romanica subsp. revoluta
- var. *subdicycla* Trautv. = M. trautvetteri
glandulosa Davidov - 1, 2
globosa auct. = M. constricta
glutinosa Bieb. (*M. sativa* L. subsp. *glomerata* (Balb.) Rouy, *M. sativa* subsp. *glomerata* (Balb.) Tutin, p.p. comb. superfl., *M. virescens* Grossh.) - 2
- subsp. *papillosa* (Boiss.) Ponert = M. papillosa
- subsp. *polychroa* (Grossh.) Sinsk. = M. polychroa
- subsp. *praefalcata* Sinsk. = M. falcata
grandiflora (Grossh.) Vass. (*M. sativa* L. var. *grandiflora* Grossh.) - 2
grossheimii Vass. - 6
gunibica Vass. - 2
hemicycla Grossh. (*M. sativa* L. subsp. *hemicycla* (Grossh.) C.R. Gunn) - 2
- subsp. **loriensis** Sinsk. - 2
- subsp. **medidaghestanica** Sinsk. - 2
hispida Gaertn. = M. nigra
karatschaica Latsch. - 2
komarovii Vass. - 1, 6
kopetdaghi Vass. = M. transoxana
kotovii Wissjul. = M. romanica
laciniata (L.) Mill. - 1
ladak Vass. = M. sativa
lanigera C. Winkl. & B. Fedtsch. - 6
lavrenkoi Vass. = M. tianschanica
littoralis Rohde ex Loisel. (*M. truncatula* Gaertn. subsp. *littoralis* (Rohde ex Loisel.) Ponert) - 2, 6
lupulina L. - 1, 2, 3, 4, 5(alien), 6
marina L. - 1, 2
mesopotamica Vass. = M. sativa
meyeri Grun. (*M. minima* (L.) Bartalini var. *meyeri* (Grun.) Heyn) - 1, 2, 6
minima (L.) Bartalini (*M. polymorpha* L. var. *minima* L.) - 1, 2, 5(alien), 6
- var. *meyeri* (Grun.) Heyn = M. meyeri
monantha (C.A. Mey.) Trautv. subsp. *noeana* (Boiss.) Greuter & Burdet = Trigonella noeana
nigra (L.) Krock. (*M. polymorpha* L. var. *nigra* L., *M. hispi-*

da Gaertn. nom. illegit., *M. polymorpha* L. p.p. nom. ambig.) - 1(?)
- var. *denticulata* (Willd.) O. Bolos & Vigo = M. denticulata
ochroleuca Kult. = M. tianschanica
orbicularis (L.) Bartalini (*M. polymorpha* L. var. *orbicularis* L.) - 1, 2, 6
orientalis Vass. = M. sativa
papillosa Boiss. (*M. dzhawakhetica* Bordz., *M. glutinosa* Bieb. subsp. *papillosa* (Boiss.) Ponert) - 2
polia (Brand) Vass. = M. sativa
polychroa Grossh. (*M. glutinosa* Bieb. subsp. *polychroa* (Grossh.) Sinsk.) - 2
polymorpha L. p.p. = M. nigra
- var. *arabica* L. = M. arabica
- var. *minima* L. = M. minima
- var. *nigra* L. = M. nigra
- var. *orbicularis* L. = M. orbicularis
- var. *rigidula* L. = M. rigidula
- var. *scutellata* L. = M. scutellata
praecox DC. - 1
praesativa Sinsk. = M. sativa
- subsp. *spontanea* Sinsk. = M. sativa
procumbens Bess. = M. falcata
quasifalcata Sinsk. = M. falcata
radiata L. = Radiata glabra
- var. *dasycarpa* Ser. = Radiata dasycarpa
rigidula (L.) All. (*M. polymorpha* L. var. *rigidula* L., *M. agrestis* Ten., *M. rigidula* var. *agrestis* (Ten.) Burn.) - 1, 2, 6
- var. *agrestis* (Ten.) Burn. = M. rigidula
rivularis Vass. = M. agropyretorum
romanica Prod. (*M. erecta* Kotov, 1940, non Winterl., *M. falcata* L. subsp. *romanica* (Prod.) Schwarz & Klinkovski, *M. falcata* subsp. *tenderiensis* (Opperm. ex Klok.) Vass., *M. kotovii* Wissjul., *M. tenderiensis* Opperm. ex Klok.) - 1, 2, 3, 4, 6
- subsp. **revoluta** (Sumn.) Vass. (*M. falcata* var. *revoluta* Sumn.) - 6
rupestris Bieb. - 1, 2
sativa L. (*M. afghanica* (Bord.) Vass., *M. ladak* Vass., *M. mesopotamica* Vass., *M. orientalis* Vass., *M. polia* (Brand) Vass., *M. praesativa* Sinsk., *M. praesativa* subsp. *spontanea* Sinsk., *M. sativa* var. *polia* Brand, *M. sativa* grex *afganica* Bord., *M. sogdiana* Vass.) - 1, 2, 3, 5, 6
- subsp. *ambigua* (Trautv.) Tutin = M. trautvetteri
- subsp. *caerulea* (Less. ex Ledeb.) Schmalh. = M. caerulea
- subsp. *caucasica* (Vass.) Lubenetz = M. caucasica
- subsp. *falcata* (L.) Arcang. = M. falcata
- subsp. *glomerata* (Balb.) Rouy = M. glutinosa
- subsp. *glomerata* (Balb.) Tutin, p.p. = M. glutinosa
- subsp. *hemicycla* (Grossh.) C.R. Gunn = M. hemicycla
- subsp. *microcarpa* Urb. c. *pilifera* Urb. = M. caerulea subsp. pilifera
- subsp. *transoxana* (Vass.) Lubenetz = M. transoxana
- subsp. x *varia* (T. Martyn) Arcang. = M. x varia
- subsp. x *varia* (T. Martyn) O. Bolos & Vigo = M. x varia
- var. *grandiflora* Grossh. = M. grandiflora
- var. *polia* Brand = M. sativa
- grex *afganica* Bord. = M. sativa
saxatilis Bieb. - 1
schischkinii Sumn. - 6
scutellata (L.) Mill. (*M. polymorpha* L. var. *scutellata* L.) - 1
sinskiae Uljanova - 6
sogdiana Vass. = M. sativa
soleirolii Duby - 1(?)

subdicycla (Trautv.) Vass. = M. trautvetteri
tadzhicorum Vass. = M. transoxana
talyschensis Latsch. - 2
tenderiensis Opperm. ex Klok. = M. romanica
tianschanica Vass. (*M. lavrenkoi* Vass., *M. ochroleuca* Kult., *M. tianschanica* var. *lavrenkoi* (Vass.) Khassan., *M. tianschanica* var. *ochroleuca* (Kult.) Khassan.) - 6
- var. *lavrenkoi* (Vass.) Khassan. = M. tianschanica
- var. *ochroleuca* (Kult.) Khassan. = M. tianschanica
transoxana Vass. (*M. kopetdaghi* Vass., *M. sativa* L. subsp. *transoxana* (Vass.) Lubenetz, *M. tadzhicorum* Vass., *M. transoxana* var. *kopetdaghi* (Vass.) Khassan., *M. transoxana* var. *tadzhicorum* (Vass.) Khassan.) - 6
- var. *kopetdaghi* (Vass.) Khassan. = M. transoxana
- var. *tadzhicorum* (Vass.) Khassan. = M. transoxana
trautvetteri Sumn. (*M. falcata* L. var. *ambigua* Trautv., *M. falcata* var. *subdicycla* Trautv., *M. sativa* L. subsp. *ambigua* (Trautv.) Tutin, *M. subdicycla* (Trautv.) Vass., *M. trautvetteri* var. *subdicycla* (Trautv.) Khassan.) - 3, 6
- var. *subdicycla* (Trautv.) Khassan. = M. trautvetteri
tribuloides Desr. = M. truncatula
truncatula Gaertn. (*M. tribuloides* Desr.) - 1, 2, 6
- subsp. *littoralis* (Rohde ex Loisel.) Ponert = M. littoralis
vardanis Vass. - 2
x **varia** T. Martyn (*M. sativa* L. subsp. x *varia* (T. Martyn) Arcang., *M. sativa* subsp. x *varia* (T. Martyn) O. Bolos & Vigo, comb. superfl.). - M. falcata L. x M. sativa L. - 1
vassilczenkoi Worosch. = Melilotoides schischkinii
virescens Grossh. = M. glutinosa

Melilotoides Heist. ex Fabr. (*Botryolotus* Jaub. & Spach, *Crimea* Vass., *Melissitus* Medik., *Pocockia* Ser., *Turukhania* Vass.)

adscendens (Nevski) Sojak (*Botryolotus adscendens* Nevski, *Melissitus adscendens* (Nevski) Ikonn., *Trigonella adscendens* (Nevski) Afan. & Gontsch.) - 6
aristata (Vass.) Sojak (*Trigonella aristata* Vass., *Melissitus aristatus* (Vass.) Latsch.) - 6
badachschanica (Afan.) Sojak (*Trigonella badachschanica* Afan., *Melissitus badachschanicus* (Afan.) Ikonn., *M. badachschanicus* (Afan.) Latsch. comb. superfl.) - 6
biflora (Griseb.) Czer. comb. nova (*Trigonella biflora* Griseb. 1843, Spicil. Fl. Rumel. 1 : 46; *Melissitus biflorus* (Griseb.) Latsch., *Trigonella lanata* Boiss.) - 2
brachycarpa (Fisch.) Sojak (*Melissitus brachycarpus* (Fisch.) Latsch., *Trigonella brachycarpa* (Fisch.) G. Moris) - 1, 2
cretacea (Bieb.) Sojak (*Crimaea cretacea* (Bieb.) Vass., *Melissitus cretaceus* (Bieb.) Latsch., *Trigonella cretacea* (Bieb.) Taliev) - 1, 2
gontscharovii (Vass.) Sojak (*Trigonella gontscharovii* Vass., *Melissitus gontscharovii* (Vass.) Latsch., *Trigonella griffithii* auct. p.p.) - 6
gordejevii (Kom.) Sojak = Lupinaster gordejevii
iskanderi (Vass.) Sojak (*Trigonella iskanderi* Vass., *Melissitus iskanderi* (Vass.) Latsch.) - 6
kafirniganica (Vass.) Sojak (*Trigonella kafirniganica* Vass., *Melissitus kafirniganicus* (Vass.) Latsch.) - 6
karkarensis (Semen. ex Vass.) Sojak (*Trigonella karkarensis* Semen. ex Vass., *Melilotoides platycarpos* (L.) Sojak var. *karkarensis* (Semen. ex Vass.) Yakovl., *Melissitus karkarensis* (Semen. ex Vass.) Golosk.) - 6
korovinii (Vass.) Sojak (*Trigonella korovinii* Vass., *Melissitus korovinii* (Vass.) Latsch.) - 6

korshinskyi (Grossh.) Czer. comb. nova (*Trigonella korshinskyi* Grossh. 1945, Fl. URSS, 11 : 127; *Melissitus korshinskyi* (Grossh.) Latsch.) - 5
laxiflora (Aitch. & Baker) Sojak (*Trigonella laxiflora* Aitch. & Baker, *Melissitus laxiflorus* (Aitch. & Baker) Latsch.) - 6
linczevskii (Vass.) Sojak (*Trigonella linczevskii* Vass., *Melissitus linczevskii* (Vass.) Latsch.) - 6
lipskyi (Sirj.) Sojak (*Trigonella lipskyi* Sirj., *Melissitus lipskyi* (Sirj.) Latsch.) - 6
pamirica (Boriss.) Sojak (*Trigonella pamirica* Boriss., *Melissitus pamiricus* (Boriss.) Golosk., *Trigonella griffithii* auct. p.p.) - 6
platycarpos (L.) Sojak (*Trigonella platycarpos* L., *Melissitus platycarpos* (L.) Golosk., *Turukhania platycarpos* (L.) Vass.) - 1, 3, 4, 5, 6
- var. *karkarensis* (Semen. ex Vass.) Yakovl. = M. karkarensis
popovii (Korov.) Sojak (*Trigonella popovii* Korov., *Melissitus popovii* (Korov.) Golosk.) - 6
ruthenica (L.) Sojak (*Trigonella ruthenica* L., *Melissitus ruthenicus* (L.) Latsch., *M. ruthenicus* (L.) Peschkova, comb. superfl., *Pocockia ruthenica* (L.) P.Y. Fu & Y.A. Chen, *Turukhania ruthenica* (L.) N.S. Pavlova) - 3, 4, 5
schachimardanica (Vass.) Sojak (*Trigonella schachimardanica* Vass., *Melissitus schachimardanicus* (Vass.) Latsch.) - 6
schischkinii (Vass.) Sojak (*Trigonella schischkinii* Vass., *Medicago vassilczenkoi* Worosch., *Melissitus schischkinii* (Vass.) Bondar. comb. superfl., *M. schischkinii* (Vass.) Latsch., *Turukhania schischkinii* (Vass.) N.S. Pavlova) - 5
siunica (Vass.) Sojak (*Trigonella siunica* Vass., *Melissitus siunicus* (Vass.) Latsch.) - 6
squarrosa (Vass.) Czer. comb. nova (*Trigonella squarrosa* Vass. 1951, Bot. Mat. (Leningrad) 14 : 229; *Melissitus squarrosus* (Vass.) Latsch.) - 6
tianschanica (Vass.) Sojak (*Trigonella tianschanica* Vass., *Melissitus tianschanicus* (Vass.) Golosk.) - 6
zapgjagaevii (Afan. & Gontsch.) Sojak (*Trigonella zaprjagaevii* Afan. & Gontsch., *Melissitus zaprjagaevii* (Afan. & Gontsch.) Latsch., *Trigonella griffithii* auct. p.p.) - 6

Melilotus Hill.

albus Medik. (*M. officinalis* (L.) Pall. var. *albus* (Medik.) Ohashi & Tateishi) - 1, 2, 3, 4, 5, 6
altissimus Thuill. - 1, 3(?)
arenarius Grec. - 1
caspius Grun. = M. polonicus
dentatus (Waldst. & Kit.) Pers. - 1, 2, 3, 4, 6
hirsutus Lipsky - 2
indicus (L.) All. - 1, 2, 6
neapolitanus Ten. - 1, 2
officinalis (L.) Pall. - 1, 2, 3, 4, 5, 6
- var. *albus* (Medik.) Ohashi & Tateishi = M. albus
- f. *suaveolens* (Ledeb.) Ohashi & Tateishi = M. suaveolens
polonicus (L.) Pall. (*M. caspius* Grun.) - 1, 2, 6
x **scythicus** O.E. Schulz. - A. officinalis (L.) Pall. x M. polonicus (L.) Pall.
suaveolens Ledeb. (*M. officinalis* (L.) Pall. f. *suaveolens* (Ledeb.) Ohashi & Tateishi) - 4, 5, 6
tauricus (Bieb.) Ser. - 1
wolgicus Poir. - 1, 2, 3, 6

Melissitus Mediik. = Melilotoides

adscendens (Nevski) Ikonn. = Melilotoides adscendens

aristatus (Vass.) Latsch. = Melilotoides aristata
badachschanicus (Afan.) Ikonn. = Melilotoides badachschanica
badachschanicus (Afan.) Latsch. = Melilotoides badachschanica
biflorus (Griseb.) Latsch. = Melilotoides biflora
brachycarpus (Fisch.) Latsch. = Melilotoides brachycarpa
cretaceus (Bieb.) Latsch. = Melilotoides cretacea
dasycarpus (Ser.) Latsch. ex Golosk. = Radiata dasycarpa
gontscharovii (Vass.) Latsch. = Melilotoides gontscharovii
gordejevii (Kom.) Latsch. = Lupinaster gordejevii
iskanderi (Vass.) Latsch. = Melilotoides iskanderi
kafirniganicus (Vass.) Latsch. = Melilotoides kafirniganica
karkarensis (Semen. ex Vass.) Golosk. = Melilotoides karkarensis
korovinii (Vass.) Latsch. = Melilotoides korovinii
korshinskyi (Grossh.) Latsch. = Melilotoides korshinskyi
laxiflorus (Aitch. & Baker) Latsch. = Melilotoides laxiflora
linczevskii (Vass.) Latsch. = Melilotoides linczevskii
lipskyi (Sirj.) Latsch. = Melilotoides lipskyi
pamiricus (Boriss.) Golosk. = Melilotoides pamirica
platycarpos (L.) Golosk. = Melilotoides platycarpos
popovii (Korov.) Golosk. = Melilotoides popovii
radiatus (L.) Latsch. = Radiata glabra
ruthenicus (L.) Latsch. = Melilotoides ruthenica
ruthenicus (L.) Peschkova = Melilotoides ruthenica
schachimardanicus (Vass.) Latsch. = Melilotoides schachimardanica
schischkinii (Vass.) Bondar. = Melilotoides schischkinii
schischkinii (Vass.) Latsch. = Melilotoides schischkinii
siunicus (Vass.) Latsch. = Melilotoides siunica
squarrosus (Vass.) Latsch. = Melilotoides squarrosa
tianschanicus (Vass.) Golosk. = Melilotoides tianschanica
zaprjagaevii (Afan. & Gontsch.) Latsch. = Melilotoides zapgjagaevii

Meristotropis Fisch. & C.A. Mey.

bucharica (Regel) Kruganova (*Glycyrrhiza bucharica* Regel, *G. kulabensis* Masl.) - 6
- f. *gontscharovii* (Masl.) Maltzeva = Glycyrrhiza gontscharovii
erythrocarpa Vass. = M. triphylla
triphylla (Fisch. & C.A. Mey.) Fisch. & C.A. Mey. (*Glycyrrhiza erythrocarpa* (Vass.) Abdullaeva, *Meristotropis erythrocarpa* Vass., *M. xanthioides* Vass.) - 6
xanthioides Vass. = M. triphylla

Mistyllus C. Presl = Amoria

spumosus (L.) Bobr. = Amoria spumosa

Nephromedia Kostel. = Radiata

dasycarpa (Ser.) Czer. = Radiata dasycarpa
radiata (L.) Kostel. = Radiata glabra

Onobrychis Hill (*Dendrobrychis* (DC.) Galushko, *Xanthobrychis* Galushko)

alatavica Bajt. - 6
altissima Grossh. (*O. viciifolia* Scop. subsp. *altissima* (Grossh.) Ponert) - 2
amoena M. Pop. & Vved. (*Xanthobrychis amoena* (M. Pop. & Vved.) Galushko) - 6
angustifolia Chinth. (*O. petraea* (Bieb. ex Willd.) Fisch. subsp. *angustifolia* (Chinth.) Fed.) - 2
arenaria (Kit.) DC. (*Hedysarum arenarium* Kit., *Onobrychis*

ferganica (Sirj.) Grossh., *O. tanaitica* Spreng., *O. tesquicola* Krytzka, *O. transcaspica* V.V. Nikit.) - 1, 2, 3, 4, 6
- subsp. *miniata* (Stev.) P.W. Ball = O. miniata
- subsp. *sibirica* (Sirj.) P.W. Ball = O. sibirica
- var. *sibirica* Sirj. = O. sibirica
■ argyrea Boiss.
■ armena Boiss. & Huet (*O. oxyodonta* Boiss. subsp. *armena* (Boiss. & Huet) Ponert)
atropatana Boiss. (*O. ornata* (Willd.) Desv. subsp. *atropatana* (Boiss.) Ponert) - 2
aucheri Boiss. subsp. *teheranica* (Bornm.) Rech. fil. = O. teheranica
balansae Boiss. var. *major* Boiss. = O. major
baldshuanica Sirj. - 6
biebersteinii Sirj. - 2
bobrovii Grossh. (*Xanthobrychis bobrovii* (Grossh.) Galushko) - 2
borysthenica (Sirj.) Klok. (*O. gracilis* Bess. var. *longiaculeata* Pacz., *O. longiaculeata* (Pacz.) Wissjul. 1954, non Pau, *O. paczoskiana* Krytzka) - 1
bracteolata Proskurjakova = O. nikitinii
buhseana Bunge (*Xanthobrychis buhseana* (Bunge) Galushko) - 2
bungei Boiss. - 2
cadmea Boiss. (*O. montana* DC. subsp. *cadmea* (Boiss.) P.W. Ball) - 2
caput-galli (L.) Lam. - 2
chorassanica Bunge (*Xanthobrychis chorassanica* (Bunge) Galushko) - 6
cornuta (L.) Desv. (*Dendrobrychis cornuta* (L.) Galushko) - 2, 6
cyri Grossh. - 2
daghestanica Grossh. - 2
darwasica Vass. = O. echidna
dielsii (Sirj.) Vass. - 2
echidna Lipsky (*O. darwasica* Vass.) - 6
ferganica (Sirj.) Grossh. = O. arenaria
gontscharovii Vass. - 6
gracilis Bess. - 1
- var. *longiaculeata* Pacz. = O. borysthenica
grandis Lipsky - 6
grossheimii Kolak. ex Fed. - 2
hajastana Grossh. - 2
hamata Vass. - 2
heterophylla C.A. Mey. (*Xanthobrychis heterophylla* (C.A. Mey.) Galushko) - 2
hohenackeriana C.A. Mey. (*Xanthobrychis hohenackeriana* (C.A. Mey.) Galushko) - 2
iberica Grossh. - 2
inermis Stev. - 1, 2
jailae Czernova - 1
kachetica Boiss. & Buhse (*Xanthobrychis kachetica* (Boiss. & Buhse) Galushko) - 2
kemulariae Chinth. - 2
kluchorica Chinth. - 2
komarovii Grossh. - 2
laxiflora Baker (*Hedysarum lipskyi* B. Fedtsch., *Onobrychis laxiflora* subsp. *schugnanica* (B. Fedtsch.) Ali, *O. laxiflora* var. *schugnanica* (B. Fedtsch.) Ali, *O. laxiflora* f. *schugnanica* (B. Fedtsch.) Sirj. ex Rech. fil., *O. schugnanica* B. Fedtsch.) - 6
- subsp. *schugnanica* (B. Fedtsch.) Ali = O. laxiflora
- var. *schugnanica* (B. Fedtsch.) Ali = O. laxiflora
- f. *schugnanica* (B. Fedtsch.) Sirj. ex Rech. fil. = O. laxiflora
lipskyi Korov. = O. verae
longiaculeata (Pacz.) Wissjul. = O. borysthenica

- major (Boiss.) Hand.-Mazz. (*O. balansae* Boiss. var. *major* Boiss.)
majorovii Grossh. (*Xanthobrychis majorovii* (Grossh.) Galushko) - 2
megalobotrys Aitch. & Hemsl. - 6
megaloptera Kovalevsk. - 6
meschetica Grossh. - 2
michauxii DC. - 2
micrantha Schrenk - 6
miniata Stev. (*O. arenaria* (Kit.) DC. subsp. *miniata* (Stev.) P.W. Ball) - 1, 2
montana DC. subsp. *cadmea* (Boiss.) P.W. Ball = O. cadmea
nemecii Sirj. - 2
nikitinii Orazmuchommedov (*O. bracteolata* Proskurjakova) - 6
novopokrovskii Vass. - 2
ornata (Willd.) Desv. subsp. *atropatana* (Boiss.) Ponert = O. atropatana
oxyodonta Boiss. & Huet - 2
- subsp. *armena* (Boiss. & Huet) Ponert = O. armena
oxytropoides Bunge - 2
paczoskiana Krytzka = O. borysthenica
pallasii (Willd.) Bieb. (*Xanthobrychis pallasii* (Willd.) Galushko) - 1
pallida Boiss. & Kotschy (*O. sulphurea* Boiss. & Bal. subsp. *pallida* (Boiss. & Kotschy) Ponert) - 2
petraea (Bieb. ex Willd.) Fisch. - 2
- subsp. *angustifolia* (Chinth.) Fed. = O. angustifolia
- subsp. *ruprechtii* (Grossh.) Fed. = O. ruprechtii
- subsp. *sosnowskyi* (Grossh.) Fed. = O. sosnowskyi
pulchella Schrenk - 6
radiata (Desf.) Bieb. (*Xanthobrychis radiata* (Desf.) Galushko) - 6
ruprechtii Grossh. (*O. petraea* (Fisch. ex Willd.) Fisch. subsp. *ruprechtii* (Grossh.) Fed.) - 2
saravschanica B. Fedtsch. - 6
schugnanica B. Fedtsch. = O. laxiflora
schuschajensis Agaeva - 2
sibirica (Sirj.) Turcz. ex Grossh. (*O. arenaria* (Kit.) DC. var. *sibirica* Sirj., *O. arenaria* subsp. *sibirica* (Sirj.) P.W. Ball, *O. tanaitica* Spreng. var. *sibirica* (Sirj.) Polozh.) - 1, 3, 4, 5(alien), 6
sintenisii Bornm. - 6
- **sosnowskyi** Grossh. (*O. petraea* (Bieb. ex Willd.) Fisch. subsp. *sosnowskyi* (Grossh.) Fed., *O. stenostachya* Freyn subsp. *sosnowskyi* (Grossh.) Hedge)
stenostachya Freyn subsp. *sosnowskyi* (Grossh.) Hedge = O. sosnowskyi
subacaulis Boiss. - 2
sulphurea Boiss. & Bal. subsp. *pallida* (Boiss. & Kotschy) Ponert = O. pallida
tanaitica Spreng. = O. arenaria
- var. *sibirica* (Sirj.) Polozh. = O. sibirica
tavernierifolia Stocks ex Boiss. - 6
teheranica (Bornm.) Grossh. (*O. aucheri* Boiss. subsp. *teheranica* (Bornm.) Rech. fil.) - 6
tesquicola Krytzka = O. arenaria
tournefortii (Willd.) Desv. (*Xanthobrychis tournefortii* (Willd.) Galushko) - 2
transcaspica V.V. Nikit. = O. arenaria
transcaucasica Grossh. - 1, 2
vaginalis C.A. Mey. - 2
vassilczenkoi Grossh. (*Xanthobrychis vassilczenkoi* (Grossh.) Galushko) - 1, 2
verae Sirj. (*O. lipskyi* Korov.) - 6
viciifolia Scop. - 1, 5(alien)

- subsp. *altissima* (Grossh.) Ponert = O. altissima

Ononis L.
afghanica Sirj. & Rech. fil. - 6
antiquorum L. (*O. repens* L. subsp. *antiquorum* (L.) Greuter, *O. spinosa* L. subsp. *antiquorum* (L.) Briq., *O. spinosa* var. *antiquorum* (L.) Arcang.) - 2, 6
arvensis L. (*O. repens* L. subsp. *arvensis* (L.) Greuter, *O. spinosa* L. 1753, nom. rejic., *O. spinosa* subsp. *arvensis* (L.) Greuter & Burdet) - 1, 3, 4
campestris Koch (*O. repens* L. subsp. *spinosa* Greuter, *O. spinosa* L. 1759, non L. 1753) - 1
chorassanica Bunge = Cicer chorassanicum
intermedia C.A. Mey. ex Rouy - 1, 2
leiosperma Boiss. (*O. repens* L. subsp. *leiosperma* (Boiss.) Greuter, *O. spinosa* L. subsp. *leiosperma* (Boiss.) Sirj.) - 1
procurrens Wallr. = O. repens
pusilla L. - 1, 2
repens L. (*O. procurrens* Wallr.) - 1
- subsp. *antiquorum* (L.) Greuter = O. antiquorum
- subsp. *arvensis* (L.) Greuter = O. arvensis
- subsp. *leiosperma* (Boiss.) Greuter = O. leiosperma
- subsp. *spinosa* Greuter = O. campestris
spinosa L. 1753 = O. arvensis
- subsp. *antiquorum* (L.) Briq. = O. antiquorum
- subsp. *arvensis* (L.) Greuter & Burdet = O. arvensis
- subsp. *leiosperma* (Boiss.) Sirj. = O. leiosperma
- var. *antiquorum* (L.) Arcang. = O. antiquorum
spinosa L. 1759 = O. campestris

Ophiocarpus (Bunge) Ikonn. = Astragalus
paulsenii (Freyn) Ikonn. = Astragalus ophiocarpus

Ornithopus L.
compressus L. - 2
perpusillus L. - 1
sativus Brot. - 1

Orobus L. = Lathyrus
alatus Maxim. = Lathyrus komarovii
austriacus Crantz = Lathyrus pannonicus
canescens auct. = Lathyrus pallescens
ciliatidentatus (Czefr.) Avazneli = Lathyrus ciliatidentatus
filiformis auct. = Lathyrus pallescens
hispanicus Lacaita = Lathyrus lacaitae
lacteus Bieb. = Lathyrus lacteus
- var. *multijugus* Ledeb. = Lathyrus multijugus
ledebourii (Trautv.) Roldug. = Lathyrus ledebourii
linifolius Reichard = Lathyrus linifolius
pannonicus Jacq. = Lathyrus pannonicus
subalpinus Herb. = Lathyrus subalpinus
transsylvanicus Spreng. = Lathyrus transsylvanicus
versicolor auct. = Lathyrus lacteus

Oxytropis DC. (*Phaca* auct.)
acanthacea Jurtz. - 4
aciphylla Ledeb. - 3
aculeata Korsh. - 6
adamsiana (Trautv.) Jurtz. (*O. strobilacea* Bunge var. *adamsiana* Trautv.) - 4
- subsp. *janensis* Jurtz. - 4
- subsp. *subnutans* Jurtz. = O. subnutans
adamsiana Vass. = O. vassilczenkoi

adenophylla M. Pop. - 4
adscendens Gontsch. - 6
aequipetala Bunge - 6
ajanensis (Regel & Til.) Bunge - 5
- *semiglobosa* (Jurtz.) N.S. Pavlova = O. semiglobosa
alajica Drob. - 6
albana Stev. - 2
alberti-regelii Vass. - 6
albiflora Bunge - 4
albovillosa B. Fedtsch. - 6
almaatensis Bajt. - 6
alpestris Schischk. - 3
alpicola Turcz. - 4
alpina Bunge - 3, 4
altaica (Pall.) Pers. - 3, 4
ambigua (Pall.) DC. - 1, 3, 4
ammophila Turcz. - 4
ampullata (Pall.) Pers. - 3, 4, 6
anadyrensis Vass. (*O. middendorffii* Trautv. subsp. *anady-rensis* (Vass.) Ju. Kozhevn. comb. superfl., *O. midden-dorffii* subsp. *anadyrensis* (Vass.) Jurtz.) - 5
anaulgensis Pavl. - 6
anjuica Jurtz. = O. vasskovskyi
approximata Less. - 1
arassanica Gontsch. - 6
arbaeviae Vass. - 6
arctica auct. = O. karga
arctica R. Br. subsp. *taimyrensis* Jurtz. = O. karga
argentata (Pall.) Pers. - 3, 4
armeniaca Sosn. ex Mulk. - 2
arystangalievii Bajt. - 6
aspera Gontsch. - 6
assiensis Vass. - 6
astragaloides Boriss. - 6
atbaschi Saposhn. - 6
aulieatensis Vved. (*O. nauvalensis* Pavl.) - 6
aurea Vass. - 6
austrosachalinensis Vass. ex N.S. Pavlova (*O. megalantha* auct.) - 5
avis Saposhn. - 6
babataghi Abduss. - 6
baicalia (Pall.) Pers. - 4
baissunensis Vass. - 6
bajtulinii Kotuchov - 6
baldshuanica B. Fedtsch. - 6
bargusinensis Peschkova - 4
bella B. Fedtsch. - 6
bellii auct. = O. wrangelii
beringensis Jurtz. - 5
biloba Saposhn. - 2
bobrovii B. Fedtsch. - 6
boguschi B. Fedtsch. - 6
borealis DC. (*O. leucantha* sensu Vass.) - 5
borissovae Polozh. - 4
bosculensis Golosk. - 6
brachycarpa Vass. (*O. transversa* Vass.) - 6
bracteata Basil. - 4
bracteolata Vass. - 5
brevicaulis Ledeb. - 3, 6
bryophila (Greene) Jurtz. (*Astragalus bryophilus* Greene, *Oxytropis nigrescens* (Pall.) Fisch. subsp. *bryophila* (Greene) Hult.) - 5
caespitosa (Pall.) Pers. - 4, 5
caespitosula Gontsch. - 6
calcareorum N.S. Pavlova (*O. rukutamensis* Sugaw. nom. invalid.) - 5
calva Malysch. - 4

campanulata Vass. - 3, 4
campestris (L.) DC. - 1
- subsp. *sordida* (Willd.) C. Hartm. fil. = O. sordida
cana Bunge - 6
candicans (Pall.) DC. - 4
canopatula Vass. - 6
capusii Franch. - 6
carpathica Uechtr. - 1
caucasica Regel - 2
caulescens Gontsch. = Astragalus caulescens
chakassiensis Polozh. - 4
chankaensis Jurtz. = O. hailarensis
chantengriensis Vass. - 6
charkeviczii Vyschin - 5
chesneyoides Gontsch. - 6
chiliophylla Royle - 6
chionobia Bunge - 6
chionophylla Schrenk - 6
chorgossica Vass. - 6
cobresietorum Vass. = O. crassiuscula
coelestis Abduss. - 6
coerulea (Pall.) DC. - 4
- subsp. **monticola** (Malysch.) Peschkova (*O. filiformis* DC. var. *monticola* Malysch.) - 4
columbina Vass. - 6
coluteoides Vass. = O. nigrescens
confusa Bunge - 3
crassiuscula Boriss. (*O. cobresietorum* Vass.) - 6
cretacea Basil. - 1
cuspidata Bunge - 6
cyanea Bieb. - 2
czapandaghi B. Fedtsch. - 6
czekanowskii Jurtz. - 4
czerepanovii Charkev. = O. exserta
czerskii Jurtz. - 4
czukotica Jurtz. (*O. tschuktschorum* Jurtz. p.p. excl. typo) - 4, 5
darpirensis Jurtz. & A. Khokhr. - 4, 5
dasypoda Rupr. ex Boiss. - 2
deflexa (Pall.) DC. - 3, 4, 5
- subsp. **dezhnevii** (Jurtz.) Jurtz. (*O. deflexa* var. *dezhnevii* Jurtz.) - 5
dichroantha Schrenk - 6
didymophysa Bunge - 6
dorogostajskyi Kuzen. - 4
dubia Turcz. - 4
echidna Vved. - 6
erecta Kom. - 5
eriocarpa Bunge - 3
ervicarpa Vved. ex Filimonova - 6
evenorum Jurtz. & A. Khokhr. - 5
exserta Jurtz. (*O. longipes* Fisch. ex Bunge, 1874, non Bunge, 1848, *O. czerepanovii* Charkev.) - 5
falcata Bunge (*O. hedinii* Ulbr.) - 6
fedtschenkoana Vass. - 6
ferganensis Vass. - 6
fetisowii Bunge - 6
filiformis DC. - 4
- var. *monticola* Malysch. = O. coerulea subsp. monticola
fischeriana Vass. - 3
floribunda (Pall.) DC. - 1, 3, 6
fominii Grossh. - 2
frigida Kar. & Kir. - 3, 6
fruticulosa Bunge - 6
gebleri Fisch. ex Bunge - 3
gebleriana Schrenk (*O. mugodsharica* Bunge) - 3, 6
glabra (Lam.) DC. (*O. salina* Vass.) - 1, 3, 4, 6

- subsp. *riparia* (Litv.) M. Pop. = O. riparia
glandulosa Turcz. - 4
glareosa Vass. - 4, 6
globiflora Bunge (*O. pagobia* Bunge) - 6
gmelinii Fisch. ex Boriss. - 1
goloskokovii Bajt. = O. gorbunovii
 gorbunovii Boriss. (*O. goloskokovii* Bajt.) - 6
gorodkovii Jurtz. (*Astragalus pygmaeus* Pall., *Oxytropis nigrescens* (Pall.) Fisch. subsp. *pygmaea* (Pall.) Hult., *O. pygmaea* (Pall.) Fern. 1928, non Tausch ex G. Beck, 1901, *O. tschuktschorum* Jurtz. p.p. quoad typum) - 5
grandiflora (Pall.) DC. - 4
guntensis B. Fedtsch. - 6
gymnogyne Bunge - 6
hailarensis Kitag. (*O. chankaensis* Jurtz., *O. hailarensis* f. *chankaensis* (Jurtz.) Kitag., *O. hailarensis* var. *chankaensis* (Jurtz.) Kitag.) - 5
- var. *chankaensis* (Jurtz.) Kitag. = O. hailarensis
- f. *chankaensis* (Jurtz.) Kitag. = O. hailarensis
halleri Bunge ex Koch (*O. sericea* (Lam.) Simonk. 1886, non Nutt. ex Torr. & Gray, 1838) - 1
hedinii Ulbr. = O. falcata
helenae N.S. Pavlova (*O. kawasimaensis* Sugaw. nom. invalid.) - 5
heterophylla Gontsch. = O. nikolai
heteropoda Bunge - 6
heterotricha Turcz. - 4
hidakamontana Miyabe & Tatew. - 5
hippolyti Boriss. - 1
hirsuta Bunge - 3, 6
hirsutiuscula Freyn - 6
humifusa Kar. & Kir. - 6
hyperborea A. Pors. - 5
hypoglottoides (Baker) Ali (*Astragalus hypoglottoides* Baker) - 6
hystrix Schrenk - 3, 6
immersa (Baker ex Aitch.) Bunge ex B. Fedtsch. - 6
- subsp. **kussavliensis** Abduss. - 6
inaria (Pall.) DC. - 3
incana Jurtz. - 4
incanescens Freyn - 6
includens Basil. - 4
inopinata Jurtz. - 4
integripetala Bunge - 6
intermedia Bunge - 3
interposita Sipl. - 4
ircutensis M. Pop. - 4
iskanderica B. Fedtsch. - 6
itoana Tatew. - 5
jordalii auct. = O. leucantha subsp. tschukotcensis
jucunda Vved. - 6
jurtzevii Malysch. - 4
kamelinii Vass. - 6
kamtschatica Hult. (*O. sublongipes* Jurtz.) - 5
karataviensis Pavl. - 6
karavaevii Jurtz. - 4
karga Saposhn. ex Polozh. (*O. arctica* R. Br. subsp. *taimyrensis* Jurtz., *O. taimyrensis* (Jurtz.) A. & D. Löve, *O. arctica* auct., *O. roaldii* auct.) - 4
karjaginii Grossh. - 2
kasbeki Bunge - 2
kaspensis Krasnob. & Pschen. - 3
katangensis Basil. (*O. schischkinii* Vass.) - 4
kateninii Jurtz. - 5
kawasimaensis Sugaw. = O. helenae
ketmenica Saposhn. - 6
kirgisensis Abdulina - 6

kodarensis Jurtz. & Malysch. - 4
komarovii Vass. - 4
kopetdagensis Gontsch. - 6
kossinskyi B. Fedtsch. & Basil. - 3
krylovii Schipcz. - 3
kubanensis Leskov - 2
kuhistanica Abduss. - 6
kunashiriensis Kitam. - 5
kuramensis Abduss. - 6
kusnetzovii Kryl. & Steinb. - 4, 5
kyziltalensis Vass. - 6
ladyginii Kryl. - 3
lanata (Pall.) DC. (*Phaca tomentosa* Georgi) - 4
lanuginosa Kom. - 4
lapponica (Wahlenb.) J. Gay - 2, 3, 6
lasiocarpa Gontsch. - 6
lasiopoda Bunge - 4
lazica Boiss. (*O. meyeri* Bunge, *O. owerinii* Bunge var. *meyeri* (Bunge) Grossh.) - 2
lehmannii Bunge - 6
leptophylla (Pall.) DC. - 4
leptophysa Bunge - 6
leucantha (Pall.) Bunge - 4, 5
- subsp. *jordalii* (A. Pors.) Jurtz. = O. leucantha subsp. tschukotcensis
- subsp. **subarctica** Jurtz. - 4
- subsp. **tschukotcensis** Jurtz. (*O. leucantha* subsp. *jordalii* (A. Pors.) Jurtz. p.p. excl. typo, *O. jordalii* auct.) - 5
leucantha sensu Vass. = O. borealis
leucocyanea Bunge - 6
leucotricha Turcz. - 4
linczevskii Gontsch. - 6
lipskyi Gontsch. - 6
lithophila Vass. - 6
litoralis Kom. - 5
litwinowii B. Fedtsch. - 6
longibracteata Kar. & Kir. - 3, 4, 6
longipes Fisch. ex Bunge = O. exserta
longirostra DC. - 3, 4
lupinoides Grossh. ex Fed. - 2
macrobotrys Bunge - 3
macrocarpa Kar. & Kir. (*O. robusta* M. Pop.) - 6
macrodonta Gontsch. - 6
macrosema Bunge - 3
maidantalensis B. Fedtsch. - 6
malacophylla Bunge - 3
mandchurica Bunge - 5(?)
marina Vass. - 5
martjanovii Kryl. - 3
masarensis Vass. - 6
maydelliana Trautv. - 5
megalantha auct. = O. austrosachalinensis
megalorrhyncha Nevski - 6
meinshausenii Schrenk - 6
melaleuca Bunge - 3
melanotricha Bunge - 6
merkensis Bunge - 6
mertensiana Turcz. - 1, 3, 4, 5
meyeri Bunge = O. lazica
michelsonii B. Fedtsch. - 6
microcarpa Gontsch. - 6
microphylla (Pall.) DC. - 4
microsphaera Bunge - 6
middendorffii Trautv. - 4, 5
- subsp. **albida** Jurtz. - 4
- subsp. *anadyrensis* (Vass.) Ju. Kozhevn. = O. anadyrensis
- subsp. *anadyrensis* (Vass.) Jurtz. = O. anadyrensis

- subsp. **coerulescens** Jurtz. & Petrovsky - 5
- subsp. **jarovoji** Jurtz. - 4
- subsp. **orulganica** Jurtz. - 4
- subsp. *schmidtii* (Meinsh.) Jurtz. = O. schmidtii
- subsp. **submiddendorfii** Jurtz. - 5
- subsp. *trautvetteri* (Meinsh.) Jurtz. = O. trautvetteri
mixotriche Bunge - 4
mongolica Kom. - 4
mugodsharica Bunge = O. gebleriana
mumynabadensis B. Fedtsch. - 6
muricata (Pall.) DC. - 3, 4, 5
myriophylla (Pall.) DC. - 4, 5(alien)
nauvalensis Pavl. = O. aulieatensis
niedzweckiana M. Pop. - 6
nigrescens (Pall.) Fisch. (*O. coluteoides* Vass.) - 4
- subsp. *bryophila* (Greene) Hult. = O. bryophila
- subsp. *pygmaea* (Pall.) Hult. = O. gorodkovii
nikolai Filimonova & Abduss. (*O. heterophylla* Gontsch.
  1940, non Bunge, 1880) - 6
nitens Turcz. - 4
nivea Bunge - 3
norinii Ju. Kozhevn. = O. putoranica
nuda Basil. - 4
nutans Bunge - 6
ochotensis Bunge - 4, 5
ochroleuca Bunge - 6
oligantha Bunge - 3, 6
ornata Vass. - 6
ovczinnikovii Abduss. - 6
owerinii Bunge - 2
- var. *meyeri* (Bunge) Grossh. = O. lazica
oxyphylla (Pall.) DC. - 4
oxyphylloides M. Pop. - 4
pagobia Bunge = O. globiflora
pallasii Pers. - 1, 2
pamiroalaica Abduss. - 6
pauciflora Bunge - 3
pellita Bunge - 6
penduliflora Gontsch. - 6
peschkovae M. Pop. - 4
physocarpa Ledeb. - 3
piceetorum Vass. - 6
pilosa (L.) DC. - 1, 2, 3, 4
pilosissima Vved. - 6
platonychia Bunge - 6
platysema Schrenk - 6
podoloba Kar. & Kir. - 6
politovii Sumn. - 3
polyphylla Ledeb. - 3, 4
poncinsii Franch. = O. stracheana
popoviana Peschkova (*O. popovii* Peschkova, 1970, non
  Vass. 1970) - 4
popovii Peschkova = O. popoviana
prostrata (Pall.) DC. - 4
protopopovii Kom. - 5
proxima Boriss. - 6
pseudofrigida Saposhn. - 6
pseudoleptophysa Boriss. - 6
pseudorosea Filimonova - 6
puberula Boriss. - 6
pulvinata Saposhn. = O. tianschanica
pulvinoides Vass. - 6
pumila Fisch. ex DC. - 3
pumilio (Pall.) Ledeb. - 5
putoranica M. Ivanova (*O. norinii* Ju. Kozhevn.) - 4
pygmaea (Pall.) Fern. = O. gorodkovii
recognita Bunge - 3, 6

regelii Vass. - 6(?)
retusa Matsum. - 5
reverdattoi Jurtz. - 4
revoluta Ledeb. - 5
rhynchophysa Schrenk - 3, 6
riparia Litv. (*O. glabra* (Lam.) DC. subsp. *riparia* (Litv.) M.
  Pop.) - 6
rishiriensis Matsum. - 5
roaldii auct. = O. karga
robusta M. Pop. = O. macrocarpa
rosea Bunge - 6
roseiformis B. Fedtsch. - 6
roseililacina Vass. - 2
rostrata Vass. - 6
rubriargillosa Vass. - 6
rubricaudex Hult. - 5
ruebsaamenii B. Fedtsch. - 6
rukutamensis Sugaw. = O. calcareorum
rupifraga Bunge - 6
ruthenica Vass. - 5
sachalinensis Miyabe & Tatew. - 5
sajanensis Jurtz. - 4
salina Vass. = O. glabra
samurensis Bunge - 2
saposhnikovii Kryl. - 3
sarkandensis Vass. - 6
satpaevii Bajt. - 3, 6
saurica Saposhn. - 6
savellanica Bunge - 3, 6
scabrida Gontsch. - 6
schachimardanica Filimonova - 6
scheludjakovae Karav. & Jurtz. - 4, 5
schischkinii Vass. = O. katangensis
schmidtii Meinsh. (*O. middendorffii* Trautv. subsp.
  schmidtii (Meinsh.) Jurtz.) - 4
schmorgunoviae Jurtz. - 5
schrenkii Trautv. - 6
selengensis Bunge - 4
semenowii Bunge - 6
semiglobosa Jurtz. (*O. ajanensis* (Regel & Til.) Bunge
  subsp. *semiglobosa* (Jurtz.) N.S. Pavlova) - 5
seravschanica Gontsch. - 6
sericea (Lam.) Simonk. = O. halleri
setosa (Pall.) DC. - 3
sewerzowii Bunge (*O. tujuksuensis* Bajt.) - 6
siomensis Abduss. - 6
songorica (Pall.) DC. - 3, 4, 6
sordida (Willd.) Pers. (*O. campestris* (L.) DC. subsp. *sordida* (Willd.) C. Hartm. fil.) - 1, 3, 4, 5
- subsp. **arctolenensis** Jurtz. - 4
- subsp. **schamurinii** Jurtz. - 4, 5
spicata (Pall.) O. & B. Fedtsch. - 1, 3
spinifer Vass. - 6
squamulosa DC. - 3, 4
stenofoliola Polozh. - 4
stenophylla Bunge - 3
stracheana Benth. ex Baker (*O. poncinsii* Franch.) - 6
stracheyana Bunge - 6
strobilacea Bunge - 3, 4, 5
- var. *adamsiana* Trautv. = O. adamsiana
stukovii Palib. - 4
suavis Boriss. - 6
subcapitata Gontsch. - 6
sublongipes Jurtz. = O. kamtschatica
submutica Bunge (*O. valichanovii* Bajt.) - 6
subnutans (Jurtz.) Jurtz. (*O. adamsiana* (Trautv.) Jurtz.
  subsp. *subnutans* Jurtz.) - 4

subverticillaris Ledeb. - 6
sulphurea (Fisch. ex DC.) Ledeb. - 3
sumneviczii Kryl. - 3
suprajenissejensis Kuvajev & Sonnikova - 4
susamyrensis B. Fedtsch. - 6
susumanica Jurtz. - 5
sverdrupii Lynge - 5
sylvatica (Pall.) DC. - 4
tachtensis Franch. - 6
taimyrensis (Jurtz.) A. & D. Löve = O. karga
talassica Gontsch. - 6
talgarica M. Pop. - 6
tenuirostris Boriss. - 6
tenuissima Vass. - 6
terekensis B. Fedtsch. - 6
teres (Lam.) DC. - 3
tianschanica Bunge (*O. pulvinata* Saposhn.) - 6
tichomirovii Jurtz. - 4
tilingii Bunge - 5
todomoshiriensis Miyabe & Miyake - 5
tomentosa Gontsch. - 6
tompudae M. Pop. - 4
tragacanthoides Fisch. - 4
trajectorum B. Fedtsch. - 6
transalaica Vass. - 6
transversa Vass. = O. brachycarpa
trautvetteri Meinsh. (*O. middendorffii* Trautv. subsp. *trautvetteri* (Meinsh.) Jurtz.) - 5
trichocalycina Bunge - 6
trichophysa Bunge - 3
trichosphaera Freyn - 6
triphylla (Pall.) Pers. - 4
tschatkalensis L. Vassil. - 6
tschimganica Gontsch. - 6
tschujae Bunge - 3
tschuktschorum Jurtz. = O. czukotica and O. gorodkovii
tujuksuensis Bajt. = O. sewerzowii
turczaninovii Jurtz. - 4
tyttantha Gontsch. - 6
ugamensis Vass. - 6
ugamica Gontsch. - 6
uniflora Jurtz. - 5
uralensis (L.) DC. - 1
uschakovii Jurtz. - 5
valichanovii Bajt. = O. submutica
varlakovii Serg. - 4
vassilczenkoi Jurtz. (*O. adamsiana* Vass. nom. invalid.) - 4, 5
- subsp. **substepposa** Jurtz. - 4, 5
vassilevii Jurtz. - 5
vasskovskyi Jurtz. (*O. anjuica* Jurtz. nom. invalid.) - 4, 5
vermicularis Freyn - 6
vvedenskyi Filimonova - 6
wrangelii Jurtz. (*O. bellii* auct.) - 5
zaprjagaevae Abduss. - 6

Phaca auct. = Astragalus and Oxytropis

parviflora Turcz. = Astragalus frigidus subsp. parviflorus
tomentosa Georgi = Oxytropis lanata

*Phaseolus L.

acutifolius A. Gray - 1, 3
angularis (Willd.) W. Wight = Vigna angularis
aureus Roxb. = Vigna radiata
coccineus L. - 1, 2, 5, 6
max L. = Glycine max

radiatus L. = Vigna radiata
vulgaris L. - 1, 2, 3, 5, 6

Piptanthus auct. = Ammopiptanthus

nanus M. Pop. = Ammopiptanthus nanus

Pisum L.

arvense L. - 1, 2, 6
aucheri Jaub. & Spach = Vavilovia formosa
elatius Bieb. (*P. humile* Boiss. & Noe, 1856, non Mill. 1768, *P. sativum* L. subsp. *elatius* (Bieb.) Aschers. & Graebn.) - 1, 2, 6
formosum (Stev.) Alef. = Vavilovia formosa
humile Boiss. & Noe = P. elatius
maritimum L. = Lathyrus maritimus
- var. *pubescens* C. Hartm. = Lathyrus aleuticus
sativum L. - 1, 2, 3, 4, 5
- subsp. *elatius* (Bieb.) Aschers. & Graebn. = P. elatius

Pocockia Ser. = Melilotoides

ruthenica (L.) P.Y. Fu & Y.A. Chen = Melilotoides ruthenica

Podalyria auct. = Ammopiptanthus

nana (M. Pop.) M. Pop. = Ammopiptanthus nanus

Podalyria Willd. p.p. = Ammodendron

argentea (Pall.) Willd. = Ammodendron bifolium

Podocarpium (Benth.) Y.C. Yang & P.H. Huang (*Desmodium* auct.)

mandshuricum (Maxim.) Czer. comb. nova (*Desmodium podocarpum* var. *mandshuricum* Maxim. 1886, Mel. Biol. Acad. Sci. Petersb. 12 : 440; *D. mandshuricum* (Maxim.) Schindl., *D. oxyphyllum* DC. var. *mandshuricum* (Maxim.) Ohashi, *D. racemosum* (Thunb.) DC. var. *mandshuricum* (Maxim.) Ohwi, *D. fallax* auct.) - 5
oldhamii (Oliv.) Y.C. Yang & P.H. Huang (*Desmodium oldhamii* Oliv.) - 5

Prosopis auct. = Lagonychium

farcta (Banks & Soland.) Macbr. = Lagonychium farctum

Pseudomelissitus Ovcz., Rassulova & Kinzikaeva = Radiata

dasycarpus (Ser.) Ovcz., Rassulova & Kinzikaeva = Radiata dasycarpa
radiatus (L.) Ovcz., Rassulova & Kinzikaeva = Radiata glabra

Pseudosophora (DC.) Sweet (*Goebelia* Bunge, p.p., *Vexibia* auct. p.p.)

alopecuroides (L.) Sweet (*Sophora alopecuroides* L., *Goebelia alopecuroides* (L.) Bunge, *G. alopecuroides* var. *tomentosa* Bunge, *Pseudosophora alopecuroides* (L.) Galushko, comb. superfl., *P. alopecuroides* (L.) Ikonn., comb. superfl., *Sophora alopecuroides* var. *tomentosa* (Bunge) Chamberlain, *S. alopecuroides* subsp. *tomentosa* (Bunge) Ponert, comb. superfl., *S. alopecuroides* subsp. *tomentosa* (Bunge) Yakovl.,

Vexibia alopecuroides (L.) Yakovl., *V. alopecuroides*
subsp. *tomentosa* (Bunge) Yakovl.) - 1, 2, 3, 6

Psoralea L. (*Asphalthium* auct.)

acaulis Stev. (*Asphalthium acaule* (Stev.) Hutch.) - 2
bituminosa L. - 1, 2
drupacea Bunge - 6

Pueraria DC. (*Dolichos* auct.)

hirsuta (Thunb.) Matsum. = P. lobata
lobata (Willd.) Ohwi (*Dolichos lobatus* Willd., *Pueraria
hirsuta* (Thunb.) Matsum. 1902, non Kurz, 1873) -
1(cult.), 2(cult.), 5, 6(cult.)

Radiata Medik. (*Nephromeria* Kostel. nom. invalid., *Pseudomelissitus* Ovcz., Rassulova & Kinzikaeva)

dasycarpa (Ser.) Ovcz., Rassulova & Kinzikaeva (*Medicago
radiata* L. var. *dasycarpa* Ser., *Melissitus dasycarpus*
(Ser.) Latsch. ex Golosk., *Nephromedia dasycarpa*
(Ser.) Czer., *Pseudomellisitus dasycarpus* (Ser.) Ovcz.,
Rassulova & Kinzikaeva, *Radiata dasycarpa* (Ser.)
Sojak, comb. superfl., *Trigonella dasycarpa* (Ser.)
Vass.) - 6
glabra Ovcz., Rassulova & Kinzikaeva (*Medicago radiata* L.,
Melissitus radiatus (L.) Latsch., *Nephromedia radiata*
(L.) Kostel., *Pseudomellisitus radiatus* (L.) Ovcz.,
Rassulova & Kinzikaeva, *Radiata leiocarpa* Sojak,
Trigonella radiata (L.) Boiss.) - 2
leiocarpa Sojak = R. glabra

*Robinia L.

argentea Siev. = Ammodendron bifolium
***hispida** L. - 1, 2, 6
***neomexicana** A. Gray - 1, 2, 6
***pseudoacacia** L. - 1, 2, 5, 6
***viscosa** Vent. - 1, 2, 6

Rudua F. Maek. = Vigna

aurea (Roxb.) F. Maek. = Vigna radiata

Sarothamnus Wimm.

scoparius (L.) Koch - 1

Scorpiurus L.

minimus Losinsk. = S. muricatus
muricatus L. (*S. minimus* Losinsk., *S. subvillosus* L.) - 1, 2
subvillosus L. = S. muricatus

Securigera DC.

balanse (Boiss.) Czer. comb. nova (*Coronilla cappadocica*
Willd. var. *balansae* Boiss. 1872, Fl. Or. 2 : 181; *C.
balansae* (Boiss.) Grossh.) - 2
charadzeae (Chinth. & Tschuchrukidze) Czer. comb. nova
(*Coronilla charadzeae* Chinth. & Tschuchrukidze,
1980, Not. Syst. Geogr. (Tbilisi) 36 : 39) - 2
cretica (L.) Lassen (*Coronilla cretica* L.) - 1, 2
elegans (Panc.) Lassen (*Coronilla elegans* Panc., *C. latifolia*
(Hazsl.) Jav.) - 1
hyrcana (Prilipko) Czer. comb. nova (*Coronilla hyrcana*
Prilipko, 1953, Dokl. AN AzSSR, 9 : 12 : 737) - 2

orientalis (Mill.) Lassen (*Coronilla orientalis* Mill.) - 2
parviflora (Desv.) Lassen (*Artrolobium parviflorum* Desv.,
Coronilla parviflora Willd. 1802, non Moench, 1794, *C.
rostrata* Boiss. & Sprun.) - 1, 2
securidaca (L.) Degen & Doerfl. - 1, 2
varia (L.) Lassen (*Coronilla varia* L.) - 1, 2, 3, 5(alien), 6
- subsp. **hirta** (Bunge ex Boiss.) Czer. comb. nova (*Coronilla varia* var. *hirta* Bunge ex Boiss. 1872, Fl. Or. 2 : 182;
C. varia subsp. *hirta* (Bunge ex Boiss.) Rech.fil.) - 6
- subsp. **orientalis** (Uhrova) Czer. comb. nova (*Coronilla
varia* var. *orientalis* Uhrova, 1935, Beih. Bot. Centralbl.
53, B : 134; *C. varia* subsp. *orientalis* (Uhrova) Jahn) -
1, 2

Sesbania Scop.

exaltata (Rafin.) Cory (*Darwinia exaltata* Rafin.) - 5

Sewerzowia Regel & Schmalh. = Astragalus

composita (Pavl.) Rassulova = Astragalus compositus
vicaria (Lipsky) Rassulova = Astragalus vicarius

Smirnowia Bunge

turkestana Bunge - 6

Sophora L. (*Vexibia* auct. p.p.)

alopecuroides L. = Pseudosophora alopecuroides
- subsp. *jaubertii* (Spach) Borza = S. jaubertii
- subsp. *prodanii* (E. Anders.) Yakovl. = S. jaubertii
- subsp. *tomentosa* (Bunge) Ponert = Pseudosophora
alopecuroides
- subsp. *tomentosa* (Bunge) Yakovl. = Pseudosophora
alopecuroides
- var. *tomentosa* (Bunge) Chamberlain = Pseudosophora
alopecuroides
angustifolia Siebold & Zucc. = S. flavescens
argentea Pall. = Ammodendron bifolium
bifolia Pall. = Ammodendron bifolium
flavescens Soland. (*S. angustifolia* Siebold & Zucc., *S. flavescens* subsp. *angustifolia* (Siebold. & Zucc.) Yakovl.) -
4, 5
- subsp. *angustifolia* (Siebold. & Zucc.) Yakovl. = S. flavescens
griffithii Stocks = Keyserlingia griffithii
- subsp. *hortensis* (Boiss. & Buhse) Yakovl. = Keyserlingia
hortensis
- subsp. *korolkowii* (Koehne) Yakovl. = Styphnolobium
japonicum
hortensis (Boiss. & Buhse) Rech. fil. = Keyserlingia hortensis
japonica L. = Styphnolobium japonicum
jaubertii Spach (*Goebelia prodanii* (E. Anders.) Grossh.,
Sophora alopecuroides L. subsp. *jaubertii* (Spach)
Borza, *S. alopecuroides* subsp. *prodanii* (E. Anders.)
Yakovl., *S. prodanii* E. Anders., *Vexibia alopecuroides*
(L.) Yakovl. subsp. *jaubertii* (Spach) Yakovl.) - 1, 2
A special study on generic identity of this species is necessary.
kentukea Dum.-Cours. = Cladrastis kentukea
korolkowi Diecks = Styphnolobium japonicum
korolkowi Diecks ex Koehne = Styphnolobium japonicum
lehmannii (Bunge) Yakovl. = Ammothamnus lehmannii
mollis Grah. ex Baker, p.p. = Keyserlingia hortensis
mollis (Royle) Baker subsp. *griffithii* (Stocks) Ali = Keyserlingia griffithii
pachycarpa C.A. Mey. = Goebelia pachycarpa
prodanii E. Anders. = S. jaubertii
songorica Schrenk = Ammothamnus songoricus

Spartium L.
junceum L. - 1, 2, 6(cult.)

Sphaerophysa DC.
salsula (Pall.) DC. - 2, 3, 4, 6

*Stizolobium R. Br.
*utile (Wall.) Ditm.
Can be cultivated in Caucasus region.

*Styphnolobium Schott
japonicum (L.) Schott (*Sophora japonica* L., ? *S. griffithii* Stocks subsp. *korolkowii* (Koehne) Yakovl., *S. korolkowi* Diecks, ? *S. korolkowi* Diecks ex Koehne) - 1, 2, 6

Teline Medik.
monspessulana (L.) C. Koch - 2

Tetragonolobus Scop.
maritimus (L.) Roth (*Lotus maritimus* L., *Tetragonolobus maritimus* var. *siliquosus* (L.) O. Bolos & Vigo, *T. siliquosus* (L.) Roth, nom. illegit.) - 1, 2
- var. *siliquosus* (L.) O. Bolos & Vigo = T. maritimus
purpureus Moench - 1, 2
siliquosus (L.) Roth = T. maritimus

Thermopsis R. Br.
alpestris Czefr. = T. alpina
alpina (Pall.) Ledeb. (*T. alpestris* Czefr.) - 3, 4, 6
alterniflora Regel & Schmalh. - 6
bargusinensis Czefr. - 4
dahurica Czefr. = T. lanceolata
dolichocarpa V. Nikit. - 6
fabacea (Pall.) DC. = T. lupinoides
glabra Czefr. = T. lanceolata
jacutica Czefr. (*T. lanceolata* R. Br. subsp. *jacutica* (Czefr.) Schreter) - 4
lanceolata R. Br. (*T. dahurica* Czefr., *T. glabra* Czefr., *T. lanceolata* subsp. *lanceolata* var. *glabra* (Czefr.) Yakovl., *T. sibirica* Czefr.) - 1, 3, 4, 5
- subsp. *jacutica* (Czefr.) Schreter = T. jacutica
- subsp. *lanceolata* var. *glabra* (Czefr.) Yakovl. = T. lanceolata
- subsp. *turkestanica* (Gand.) Gubanov = T. turkestanica
- var. *turkestanica* (Gand.) C.Y. Yang = T. turkestanica
lupinoides (L.) Link (*T. fabacea* (Pall.) DC.) - 5
mongolica Czefr. (*T. mongolica* subsp. *schischkinii* (Czefr.) Yakovl., *T. schischkinii* Czefr.) - 6
- subsp. *schischkinii* (Czefr.) Yakovl. = T. mongolica
schischkinii Czefr. = T. mongolica
sibirica Czefr. = T. lanceolata
turkestanica Gand. (*T. lanceolata* R. Br. subsp. *turkestanica* (Gand.) Gubanov, *T. lanceolata* var. *turkestanica* (Gand.) C.Y. Yang) - 6

Thlaspidium (Lipsky) Rassulova = Astragalus
thlaspi (Lipsky) Rassulova = Astragalus thlaspi

Tragacantha auct. = Astracantha
arnacanthoides Boriss. = Astracantha arnacanthoides
chionocalyx Nevski = Astracantha chionocalyx

chodsha-bakirganica (B. Kom.) Rassulova = Astracantha chodsha-bakirganica
consentanea Boriss. = Astracantha consentanea
densissima Boriss. = Astracantha densissima
devia Boriss. = Astracantha devia
dissecta (B. Fedtsch. & Ivanova) Boriss. = Astracantha dissecta
dolona Rassulova & B. Scharipova = Astracantha dolona
ferox Boriss. = Astracantha ferox
flexilispina Boriss. = Astracantha flexilispina
imbricata Boriss. = Astracantha imbricata
insidiosa Boriss. = Astracantha insidiosa
intermixta Boriss. = Astracantha intermixta
karjaginii Boriss. = Astracantha karjaginii
kuramensis Boriss. = Astracantha kuramensis
longiramosa Boriss. = Astracantha longiramosa
macrantha Boriss. = Astracantha macrantha
meana Boriss. = Astracantha meana
meraca Boriss. = Astracantha meraca
nuratavica Boriss. = Astracantha nuratavica
paliurus Boriss. = Astracantha paliurus
plumata Boriss. = Astracantha plumata
prominens Boriss. = Astracantha prominens
sommieri (Freyn) Boriss. = Astracantha sommieri
stipulosa Boriss. = Astracantha stipulosa
tenuispina Boriss. = Astracantha tenuispina
transcaucasica Boriss. = Astracantha acmophylloides
unguiculata Boriss. = Astracantha unguiculata
voronoviana Boriss. = Astracantha voronoviana

Trifolium L. (*Calycomorphum* C. Presl)
agrarium L. = Chrysaspis aurea and Ch. campestris
alexandrinum L. - 2
alpestre L. - 1, 2
ambiguum Bieb. = Amoria ambigua
angulatum Waldst. & Kit. = Amoria angulata
angustifolium L. - 1, 2
apertum Bobr. - 2
arvense L. - 1, 2, 3, 4(alien), 5(alien), 6
aureum Poll. = Chrysaspis aurea
badium Schreb. = Chrysaspis badia
- subsp. *rytidosemium* (Boiss. & Hohen.) Hossain = Chrysaspis rytidosemia
baicalense Belaeva & Sipl. = Lupinaster baicalensis
biebersteinii Chalilov - 2
bithynicum Boiss. (*T. grossheimii* Chalilov) - 1, 2
bobrovii Chalilov = Amoria bobrovii
bonannii C. Presl = Amoria bonannii
bordzilowskyi Grossh. = Amoria bordzilowskyi
borysthenicum Grun. - 1, 3, 6
caeruleum L. = Trigonella caerulea
campestre Schreb. = Chrysaspis campestris
canescens Willd. - 2
caucasicum Tausch (*T. ochroleucon* Huds. subsp. *caucasicum* (Tausch) Kozhukharov, *T. topczibaschovii* Chalilov, *T. ochroleucon* auct. p.p., *T. panormitanum* auct.) - 1, 2
ciswolgense Spryg. ex Iljin & Truchaleva = Lupinaster albus
diffusum Ehrh. (*T. pallidum* auct. fl. cauc.) - 1, 2
dubium Sibth. = Chrysaspis dubia
echinatum Bieb. - 1, 2
egrissicum A.D. Mikheev & Magulaev = Amoria egrissica
elegans Savi = Amoria hybrida
elizabethae Grossh. = Amoria elizabethae
eximium Stev. ex DC. = Lupinaster eximius
expansum Waldst. & Kit. (*T. pratense* L. var. *americanum* Harz, *T. pratense* subsp. *americanum* (Harz) Sojak, *T.*

pratense subsp. *expansum* (Waldst. & Kit.) Jav. comb. superfl., *T. pratense* subsp. *expansum* (Waldst. & Kit.) Ponert, comb. superfl., *T. pratense* subsp. *expansum* (Waldst. & Kit.) Simonk.) - 1, 2, 3
foliosum (Brand) Min. = T. sativum
fontanum Bobr. - 2
fragiferum L. = Amoria fragifera
- subsp. *bonannii* (C. Presl) Sojak = Amoria bonannii
glareosum Schleich. = Amoria pallescens
glomeratum L. = Amoria glomerata
gordejevii (Kom.) N.S. Pavlova = Lupinaster gordejevii
gordejevii (Kom.) Z. Wei = Lupinaster gordejevii
grandiflorum Schreb. = Chrysaspis grandiflora
grossheimii Chalilov = T. bithynicum
hirtum All. - 1, 2
hybridum L. = Amoria hybrida
- subsp. *elegans* (Savi) Aschers. & Graebn. = Amoria hybrida
- var. *elegans* (Savi) Boiss. = Amoria hybrida
incarnatum L. - 1, 2, 5(alien)
- subsp. *molineri* (Balb. ex Hornem.) Syme = T. molineri
issajevii Chalilov - 2
karatavicum Pavl. = Chrysaspis karatavica
lappaceum L. - 1, 2, 6
lenkoranicum (Grossh.) Roskov (*T. pratense* L. var. *lenkoranicum* Grossh.) - 2
leucanthum Bieb. (*T. sachokianum* Grossh.) - 1, 2
litwinowii Iljin = Lupinaster litwinowii
lucanicum Gasp. ex Guss. - 2
lupinaster L. = Lupinaster pentaphyllus
- subsp. *angustifolium* (Litv.) Bobr. = Lupinaster pentaphyllus
maritimum Huds. = T. squamosum
medium L. - 1, 3, 4(alien), 5(alien)
micranthum Viv. = Chrysaspis micrantha
molineri Balb. ex Hornem. (*T. incarnatum* L. subsp. *molineri* (Balb. ex Hornem.) Syme) - 1
montanum L. = Amoria montana
neglectum C.A. Mey. = Amoria bonannii
nigrescens Viv. = Amoria nigrescens
ochroleucon auct. p.p. = T. caucasicum
ochroleucon Huds. - 1
- subsp. *caucasicum* (Tausch) Kozhukharov = T. caucasicum
orbelicum Velen. = Amoria repens
pacificum Bobr. = Lupinaster pacificus
pallescens Schreb. = Amoria pallescens
pallidum auct. fl. cauc. = T. diffusum
pannonicum Jacq. - 1
panormitanum auct. = T. caucasicum
parviflorum Ehrh. = Amoria retusa
patens Schreb. = Chrysaspis patens
phleodes Pourr. - 1, 2
physodes Stev. ex Bieb. = Amoria physodes
polyphyllum C.A. Mey. = Lupinaster polyphyllus
pratense L. (*T. ucrainicum* Opperm. ex Wissjul.) - 1, 2, 3, 4, 5(alien), 6
- subsp. *americanum* (Harz) Sojak = T. expansum
- subsp. *expansum* (Waldst. & Kit.) Jav. = T. expansum
- subsp. *expansum* (Waldst. & Kit.) Ponert = T. expansum
- subsp. *expansum* (Waldst. & Kit.) Simonk. = T. expansum
- subsp. *sativum* (Schreb.) Ponert = T. sativum
- subsp. *sativum* (Schreb.) Schuebl. & Mart. = T. sativum
- var. *americanum* Harz = T. expansum
- var. *foliosum* Brand = T. sativum
- var. *lenkoranicum* Grossh. = T. lenkoranicum
- var. *sativum* Schreb. = T. sativum

raddeanum Trautv. = Amoria raddeana
repens L. = Amoria repens
- subsp. *orbelicum* (Vel.) Pawl. = Amoria repens
resupinatum L. = Amoria resupinata
retusum L. = Amoria retusa
rubens L. - 1
ruprechtii Tamamsch. & Fed. = Amoria ruprechtii
rytidosemium Boiss. & Hohen. = Chrysaspis rytidosemia
sachokianum Grossh. = T. leucanthum
sarosiense Hazsl. - 2
sativum (Schreb.) Crome (*T. pratense* L. var. *sativum* Schreb., *T. foliosum* (Brand) Min., *T. pratense* var. *foliosum* Brand, *T. pratense* subsp. *sativum* (Schreb.) Ponert, comb. superfl., *T. pratense* subsp. *sativum* (Schreb.) Schuebl. & Mart., *T. sativum* subsp. *praecox* (Witte) Bobr., *T. sativum* subsp. *serotinum* (Witte) Bobr.) - 1, 2, 3, 4, 5, 6
- subsp. *praecox* (Witte) Bobr. = T. sativum
- subsp. *serotinum* (Witte) Bobr. = T. sativum
scabrum L. - 1, 2, 6
sebastianii Savi = Chrysaspis sebastianii
seravschanicum Ovcz. ex Bobr. - 6
sintenisii Freyn = Chrysaspis sintenisii
spadiceum L. = Chrysaspis spadicea
speciosum Willd. = Chrysaspis grandiflora
spryginii Belaeva & Sipl. = Lupinaster albus
spumosum L. = Amoria spumosa
squamosum L. (*T. maritimum* Huds.) - 1, 2
squarrosum L. - 1(alien)
stellatum L. - 2
stipitatum Boiss. & Bal. = Chrysaspis sintenisii
strepens Crantz = Chrysaspis aurea
striatum L. - 1, 2
subterraneum L. (*Calycomorphum subterraneum* (L.) C. Presl) - 1, 2
- subsp. **brachycalycinum** Katznelson & Morley - 2
suffocatum L. = Amoria suffocata
talyschense Chalilov = Amoria talyschensis
tomentosum L. = Amoria tomentosa
topczibaschovii Chalilov = T. caucasicum
trichocephalum Bieb. (*T. trichocephalum* subsp. *armenum* Chalilov, *T. trichocephalum* subsp. *sarikamischense* Chalilov) - 2
- subsp. *armenum* Chalilov = T. trichocephalum
- subsp. *sarikamischense* Chalilov = T. trichocephalum
tumens Stev. ex Bieb. = Amoria tumens
- subsp. *talyschense* (Chalilov) Zohary = Amoria talyschensis
ucrainicum Opperm. ex Wissjul. = T. pratense
vesiculosum Savi = Amoria vesiculosa

Trigonella L.

adscendens (Nevski) Afan. & Gontsch. = Melilotoides adscendens
arcuata C.A. Mey. - 1, 2, 3, 6
aristata Vass. = Melilotoides aristata
astroides Fisch. & C.A. Mey. - 2, 6
badachschanica Afan. = Melilotoides badachschanica
besseriana Ser. = T. procumbens
biflora Griseb. = Melilotoides biflora
brachycarpa (Fisch.) G. Moris = Melilotoides brachycarpa
brahuica Boiss. - 6
caerulea (L.) Ser. (*Trifolium caeruleum* L.) - 1, 2, 5(alien)
- subsp. *procumbens* (Bess.) Vass. = T. procumbens
calliceras Fisch. - 2, 6
cancellata Desf. - 1, 2

capitata Boiss. - 2
coerulescens (Bieb.) Halacsy - 1, 2, 6
cretacea (Bieb.) Taliev = Melilotoides cretacea
dasycarpa (Ser.) Vass. = Radiata dasycarpa
fischeriana Ser. - 1, 2
foenum-graecum L. - 1, 2, 6
geminiflora Bunge (*T. monantha* C.A. Mey. subsp. *gemini-flora* (Bunge) Rech. fil.) - 6
gladiata Stev. ex Bieb. - 1, 2
gontscharovii Vass. = Melilotoides gontscharovii
gordejevii (Kom.) Grossh. = Lupinaster gordejevii
grandiflora Bunge - 1(alien), 5(alien), 6
griffithii auct. p.p. = Melilotoides gontscharovii, M. pamirica and M. zaprjagaevii
incisa Benth. (*T. monantha* C.A. Mey. subsp. *incisa* (Benth.) Ali) - 6
iskanderi Vass. = Melilotoides iskanderi
kafirniganica Vass. = Melilotoides kafirniganica
karkarensis Semen. ex Vass. = Melilotoides karkarensis
korovinii Vass. = Melilotoides korovinii
korshinskyi Grossh. = Melilotoides korshinskyi
laxiflora Aitch. & Baker = Melilotoides laxiflora
linczevskii Vass. = Melilotoides linczevskii
lipskyi Sirj. = Melilotoides lipskyi
lunata Boiss. = Melilotoides biflora
monantha C.A. Mey. - 2, 6
- subsp. *geminiflora* (Bunge) Rech. fil. = T. geminiflora
- subsp. *incisa* (Benth.) Ali = T. incisa
- subsp. *noeana* (Boiss.) Huber-Morath = T. noeana
monspeliaca L. - 1, 2, 6
noeana Boiss. (*Medicago monantha* (C.A. Mey.) Trautv. subsp. *noeana* (Boiss.) Greuter & Burdet, *Trigonella monantha* C.A. Mey. subsp. *noeana* (Boiss.) Huber-Morath) - 2, 6
orthoceras Kar. & Kir. - 1, 2, 6
pamirica Boriss. = Melilotoides pamirica
platycarpos L. = Melilotoides platycarpos
popovii Korov. = Melilotoides popovii
procumbens (Bess.) Reichenb. (*T. besseriana* Ser., *T. caeru-lea* (L.) Ser. subsp. *procumbens* (Bess.) Vass.) - 1, 2
radiata (L.) Boiss. = Radiata glabra
ruthenica L. = Melilotoides ruthenica
schachimardanica Vass. = Melilotoides schachimardanica
schischkinii Vass. = Melilotoides schischkinii
siunica Vass. = Melilotoides siunica
smyrnea Boiss. - 1
spicata Sibth. & Smith - 1, 2
spruneriana Boiss. (*T. torulosa* Griseb.) - 2, 6
squarrosa Vass. = Melilotoides squarrosa
stellata Forssk. - 6
strangulata Boiss. - 2
striata L. fil. (*T. tenuis* Fisch. ex Bieb.) - 1, 2
tenuis Fisch. ex Bieb. = T. striata
tianschanica Vass. = Melilotoides tianschanica
torulosa Griseb. = T. spruneriana
turkmena M. Pop. - 6
verae Sirj. - 6
zaprjagaevii Afan. & Gontsch. = Melilotoides zaprjagaevii

Turukhania Vass. = Melilotoides

platycarpos (L.) Vass. = Melilotoides platycarpos
ruthenica (L.) N.S. Pavlova = Melilotoides ruthenica
schischkinii (Vass.) N.S. Pavlova = Melilotoides schischkinii

*Ulex L.

*europaea L. - 2

Ursia Vass. = Lupinaster

gordejevii (Kom.) Vass. = Lupinaster gordejevii

Vavilovia Fed. (*Alophotropis* (Jaub. & Spach) Grossh. nom. illegit. superfl.)

aucheri (Jaub. & Spach) Fed. = V. formosa
formosa (Stev.) Fed. (*Alophotropis aucheri* (Jaub. & Spach) Grossh., *A. formosa* (Stev.) Grossh., *Pisum aucheri* Jaub. & Spach, *P. formosum* (Stev.) Alef., *Vavilovia aucheri* (Jaub. & Spach) Fed.) - 2

Vexibia auct. p.p. = Goebelia, Pseudosophora and Sophora

alopecuroides (L.) Yakovl. = Pseudosophora alopecuroides
- subsp. *jaubertii* (Spach) Yakovl. = Sophora jaubertii
- subsp. *tomentosa* (Bunge) Yakovl. = Pseudosophora alopecuroides
pachycarpa (C.A. Mey.) Yakovl. = Goebelia pachycarpa

Vicia L. (*Bona* Medik., *Ervilia* Link, *Ervum* L., *Faba* Hill)

abbreviata Fisch. ex Spreng. (*V. truncatula* Fisch. ex Bieb.) - 2
- subsp. *balansae* (Boiss.) Stankevicz = V. balansae
- subsp. *montenegrina* (Rohlena) Kuzm. = V. montenegrina
abbreviata sensu B. Fedtsch. = V. montenegrina
akhmaganica Kazar. - 2
alpestris Stev. (*V. purpurea* Stev.) - 2
amoena Fisch. - 3, 4, 5
amphicarpa Lam. (*V. sativa* L. subsp. *amphicarpa* (Lam.) Aschers. & Graebn.) - 1, 2
amurensis Oett. (*V. japonica* A. Gray subsp. *amurensis* (Oett.) Kitam.) - 4, 5
- var. *pallida* (Turcz.) Kitag. = V. woroschilovii
anatolica Turrill (*V. hajastana* Grossh.) - 1, 2, 6
angustifolia Reichard (*V. angustifolia* var. *bobartii* (E. Forst.) Koch, *V. bobartii* E. Forst., *V. sativa* L. subsp. *nigra* (L.) Ehrh.) - 1, 2, 6
- subsp. *segetalis* (Thull.) Arcang. = V. segetalis
- var. *bobartii* (E. Forst.) Koch = V. angustifolia
- var. *segetalis* (Thuill.) Ser. = V. segetalis
antiqua Grossh. - 2
armena auct. = V. nissoliana subsp. sojuchica
articulata Hornem. (*V. monanthos* (L.) Desf. 1799, non Retz. 1783) - 1
asiatica (Freyn) Grossh. = V. elegans
baicalensis (Turcz.) B. Fedtsch. (*V. ramuliflora* (Maxim.) Ohwi f. *baicalensis* (Turcz.) P.Y. Fu & Y.A. Chen) - 4
balansae Boiss. (*V. abbreviata* Fisch. ex Spreng. subsp. *balansae* (Boiss.) Stankevicz) - 2
biebersteinii Bess. ex Bieb. (*V. grandiflora* Scop. var. *bieber-steinii* (Bess. ex Bieb.) Griseb., *V. grandiflora* subsp. *biebersteinii* (Bess. ex Bieb.) Dostal) - 1
biennis L. (*V. picta* Fisch. & C.A. Mey.) - 1, 2, 3, 6
biennis sensu B. Fedtsch. = V. lilacina
bithynica (L.) L. - 1, 2
bobartii E. Forst. = V. angustifolia
boissieri Freyn (*V. tenuifolia* Roth subsp. *boissieri* (Freyn) Radzhi) - 1, 2
calcarata auct. = V. cinerea
canescens Labill. subsp. *variegata* (Willd.) P.H. Davis = V. nissoliana
cappadocica Boiss. & Bal. (*V. paucijuga* (Trautv.) B. Fedtsch.) - 2

cassubica L. - 1, 2
- var. *denticulata* Akinf. = V. montenegrina
caucasica Ekvtim. (*V. semiglabra* Rupr. ex Boiss. subsp.
    caucasica (Ekvtim.) Radzhi) - 2
ciceroidea Boiss. (*V. costata* Ledeb. subsp. *ciceroidea*
    (Boiss.) Stankevicz, *V. rafigae* Tamamsch.) - 2
ciliatula Lipsky - 1, 2
cinerea Bieb. (*V. calcarata* auct.) - 2, 6
cordata Wulf. ex Hoppe (*V. sativa* L. subsp. *cordata* (Wulf.
    ex Hoppe) Aschers. & Graebn., *V. sativa* subsp. *incisa*
    (Bieb.) Arcang. var. *cordata* (Wulf. ex Hoppe)
    Arcang.) - 1, 2
costata Ledeb. - 3, 6
- subsp. *ciceroidea* (Boiss.) Stankevicz = V. ciceroidea
- subsp. *venulosa* (Boiss. & Hohen.) Stankevicz = V. venu-
    losa
cracca L. (*V. macrophylla* (Maxim.) B. Fedtsch., *V. lilacina*
    sensu B. Fedtsch.) - 1, 2, 3, 4, 5, 6
- subsp. *kitaibeliana* (Reichenb.) Soo = V. sordida
- subsp. *tenuifolia* (Roth) Gaudin = V. tenuifolia
crocea (Desf.) B. Fedtsch. = V. crocea
crocea (Desf.) Fritsch (*V. crocea* (Desf.) B. Fedtsch. comb.
    superfl.) - 2
dadianorum Somm. & Levier = V. orobus
dalmatica auct. = V. elegans
dasicarpa auct. = V. varia
dumetorum L. - 1
ecirrhosa Rupr. ex Boiss. = V. semiglabra
elegans Guss. (*V. asiatica* (Freyn) Grossh., *V. elegans* var.
    asiatica Freyn, *V. heracleotica* Juz., *V. dalmatica*
    auct.) - 1, 2
- var. *asiatica* Freyn = V. elegans
ervilia (L.) Willd. (*Ervilia sativa* Link, *Lens pygmaea*
    Grossh.) - 1, 2, 6
*faba L. (*Faba bona* Medik., *F. vulgaris* Moench) - 1, 2, 3, 4,
    5, 6
fedtschenkoana V.V. Nikit. - 6
geminiflora Trautv. - 4
gracilior (M. Pop.) M. Pop. - 6
gracilis Loisel. = V. tenuissima
grandiflora Scop. - 1, 2
- subsp. *biebersteinii* (Bess. ex Bieb.) Dostal = V. bieber-
    steinii
- subsp. *sordida* (Waldst. & Kit.) Dostal = V. sordida
- var. *biebersteinii* (Bess. ex Bieb.) Griseb. = V. bieber-
    steinii
- var. *kitaibeliana* Koch = V. sordida
- var. *sordida* (Waldst. & Kit.) Griseb. = V. sordida
gregaria Boiss. & Heldr. - 2(?)
grossheimii Ekvtim. - 2
hajastana Grossh. = V. anatolica
heracleotica Juz. = V. elegans
heterophylla Worosch. = V. japonica
hirsuta (L.) S.F. Gray (*Ervum hirsutum* L.) - 1, 2, 3, 4, 5, 6
hololasia Woronow - 2
hybrida L. - 1, 2, 6(alien)
hyrcanica Fisch. & C.A. Mey. - 2, 6
iberica Grossh. (*V. tenuifolia* Roth subsp. *iberica* (Grossh.)
    Radzhi) - 2
incana Gouan - 1
incisa Bieb. (*V. sativa* L. subsp. *incisa* (Bieb.) Arcang.) - 1
iranica auct. = V. subvillosa
iranica Boiss. - 6
janeae T. Mardalejschvili - 2
japonica A. Gray (*V. heterophylla* Worosch. 1978, non Phil.
    1894) - 4, 5
- subsp. *amurensis* (Oett.) Kitam. = V. amurensis

- subsp. *pallida* (Turcz.) Worosch. = V. woroschilovii
- var. *pallida* (Turcz.) Hara = V. woroschilovii
johannis Tamamsch. = V. narbonensis
kitaibeliana (Koch) Stank. = V. sordida
kokanica Regel & Schmalh. - 6
larissae Prima - 2
lathyroides L. (*Ervum lathyroides* (L.) Stankevicz) - 1
- subsp. *olbiensis* (Reut. & Shuttl. ex Rouy) Smejkal = V.
    olbiensis
laxiflora auct. = V. tenuissima
lilacina Ledeb. (*V. neglecta* Hanelt & Mettin, *V. biennis*
    sensu B. Fedtsch.) - 3
lilacina sensu B. Fedtsch. = V. cracca
litvinovii Boriss. = V. loiseleurii
loiseleurii (Bieb.) Litv. (*V. litvinovii* Boriss., *V. meyeri*
    Boiss., *V. pubescens* auct.) - 1, 2
lutea L. - 1, 2
macrantha Jurtz. (*V. multicaulis* Ledeb. subsp. *macrantha*
    (Jurtz.) Stankevicz) - 4, 5
- subsp. *olchonensis* Peschkova = V. olchonensis
macrophylla (Maxim.) B. Fedtsch. = V. cracca
megalosperma Bieb. = V. peregrina
megalotropis Ledeb. - 3, 4, 6
meyeri Boiss. = V. loiseleurii
michauxii Spreng. (*V. peregrina* L. subsp. *michauxii*
    (Spreng.) Ponert) - 6
monanthos (L.) Desf. = V. articulata
montenegrina Rohlena (*V. abbreviata* Fisch. ex Spreng.
    subsp. *montenegrina* (Rohlena) Kuzm., *V. abbreviata*
    sensu B. Fedtsch.) - 2
multicaulis Ledeb. - 1, 3, 4, 5
- subsp. *macrantha* (Jurtz.) Stankevicz = V. macrantha
narbonensis L. (*Bona narbonensis* (L.) Medik., *Vicia
    johannis* Tamamsch., *V. narbonensis* var. *ecirrhosa* M.
    Pop., *V. turkestanica* Vassilkovsk.) - 1, 2, 6
- subsp. *serratifolia* (Jacq.) Arcang. = V. serratifolia
- var. *ecirrhosa* M. Pop. = V. narbonensis
- var. *serratifolia* (Jacq.) Ser. = V. serratifolia
neglecta Hanelt & Mettin = V. lilacina
nervata Sipl. - 4
nipponica Matsum. - 5
nissoliana L. (*V. canescens* Labill. subsp. *variegata* (Willd.)
    P.H. Davis, *V. variegata* Willd.) - 2
- subsp. **sojuchica** (Radzhi) Czer. comb. nova (*V. variegata*
    subsp. *sojuchica* Radzhi, 1965, Botany, Plant Physiolo-
    gy and Plant Industry, Daghest. Univ. : 58; *V. armena*
    auct.) - 2
noeana Reut. ex Boiss. - 2(?)
ohwiana Hokokama - 5
olbiensis Reut. ex Timb.-Lagr. (*V. lathyroides* L. subsp.
    olbiensis (Reut. & Shuttl. ex Rouy) Smejkal) - 1, 2
olchonensis (Peschkova) Nikiforova (*V. macrantha* Jurtz.
    subsp. *olchonensis* Peschkova) - 4
orobus DC. (*V. dadianorum* Somm. & Levier) - 2
pallida Turcz. = V. woroschilovii
pannonica Crantz - 1, 2
- subsp. *purpurascens* (DC.) Arcang. = V. striata
- subsp. *striata* (Bieb.) Nym. = V. striata
- subsp. *striata* (Bieb.) Ponert = V. striata
- var. *purpurascens* (DC.) Ser. = V. striata
paucijuga (Trautv.) B. Fedtsch. = V. cappadocica
peregrina L. (*V. megalosperma* Bieb., *V. peregrina* subsp.
    megalosperma (Bieb.) Ponert) - 1, 2, 6
- subsp. *megalosperma* (Bieb.) Ponert = V. peregrina
- subsp. *michauxii* (Spreng.) Ponert = V. michauxii
persica Boiss. (*V. variegata* Willd. subsp. *bornmuelleri*
    Radzhi) - 2

picta Fisch. & C.A. Mey. = V. biennis
pilosa Bieb. - 1, 2
pisiformis L. - 1, 2
popovii Nikiforova - 4, 5
pseudorobus Fisch. & C.A. Mey. - 4, 5
pubescens auct. = V. loiseleurii
purpurea Stev. = V. alpestris
rafigae Tamamsch. = V. ciceroidea
ramuliflora (Maxim.) Ohwi - 4, 5
- f. *baicalensis* (Turcz.) P.Y. Fu & Y.A. Chen = V. baicalensis
sativa L. - 1, 2, 3, 4, 5, 6
- subsp. *amphicarpa* (Lam.) Aschers. & Graebn. = V.
  amphicarpa
- subsp. *cordata* (Wulf. ex Hoppe) Aschers. & Graebn. = V.
  cordata
- subsp. *incisa* (Bieb.) Arcang. = V. incisa
- - var. *cordata* (Wulf. ex Hoppe) Arcang. = V. cordata
- subsp. **linearifolia** Stankevicz - 2
- subsp. *nigra* (L.) Ehrh. = V. angustifolia
- subsp. *segetalis* (Thuill.) Dostal = V. segetalis
saxatilis (Vent.) Tropea = Lathyrus saxatilis
segetalis Thuill. (*V. angustifolia* Reichard subsp. *segetalis*
  (Thull.) Arcang., *V. angustifolia* var. *segetalis* (Thuill.)
  Ser., *V. sativa* L. subsp. *segetalis* (Thuill.) Dostal) - 1, 3,
  4, 5
semenovii (Regel & Herd.) B. Fedtsch. - 6
semiglabra Rupr. ex Boiss. (*V. ecirrhosa* Rupr. ex Boiss.) - 2
- subsp. *caucasica* (Ekvtim.) Radzhi = V. caucasica
- subsp. *sosnowskyi* (Ekvtim.) Radzhi = V. sosnowskyi
sepium L. - 1, 2, 3, 4, 5, 6
- subsp. **montana** (Koch) Hamet-Ahti (*V. sepium* var.
  montana Koch) - 1
serratifolia Jacq. (*Bona serratifolia* (Jacq.) Stankevicz, *Vicia
  narbonensis* L. subsp. *serratifolia* (Jacq.) Arcang., *V.
  narbonensis* var. *serratifolia* (Jacq.) Ser.) - 2
sordida Waldst. & Kit. (*V. cracca* L. subsp. *kitaibeliana*
  (Reichenb.) Soo, *V. grandiflora* Scop. var. *kitaibeliana*
  Koch, *V. grandiflora* subsp. *sordida* (Waldst. & Kit.)
  Dostal, *V. grandiflora* var. *sordida* (Waldst. & Kit.)
  Griseb., *V. kitaibeliana* (Koch) Stank.) - 1
sosnowskyi Ekvtim. (*V. semiglabra* Rupr. ex Boiss. subsp.
  sosnowskyi (Ekvtim.) Radzhi) - 2
sparsiflora Ten. - 1(?)
striata Bieb. (*V. pannonica* Crantz subsp. *purpurascens*
  (DC.) Arcang., *V. pannonica* subsp. *striata* (Bieb.)
  Nym., *V. pannonica* subsp. *striata* (Bieb.) Ponert,
  comb. superfl., *V. pannonica* var. *purpurascens* (DC.)
  Ser.) - 1
subrotunda (Maxim.) Czefr. (*Lathyrus subrotundus*
  Maxim.) - 5
subvillosa (Ledeb.) Boiss. (*V. iranica* auct.) - 6
sylvatica L. - 1, 3, 4
tenuifolia Roth (*V. cracca* L. subsp. *tenuifolia* (Roth)
  Gaudin) - 1, 2, 3, 4, 6
- subsp. *boissieri* (Freyn) Radzhi = V. boissieri
- subsp. *iberica* (Grossh.) Radzhi = V. iberica
tenuissima (Bieb.) Schinz & Thell. (*V. gracilis* Loisel. 1807,
  non Banks & Soland, 1794, *V. laxiflora* auct.) - 1
tetrasperma (L.) Schreb. - 1, 2, 3, 4, 5(alien), 6
truncatula Fisch. ex Bieb. = V. abbreviata
tsydenii Malysch. (*Ervum tsydenii* (Malysch.) Stankevicz) - 4
turkestanica Vassilkovsk. = V. narbonensis
unijuga A. Br. - 3, 4, 5
varia Host (*V. villosa* Roth subsp. *varia* (Host) Corb., *V.
  dasicarpa* auct.) - 1, 2
variabilis Freyn & Sint. - 2

variegata Willd. = V. nissoliana
- subsp. *bornmuelleri* Radzhi = V. persica
- subsp. *sojuchica* Radzhi = V. nissoliana subsp. sojuchica
venosa (Willd. ex Link) Maxim. - 4, 5
venulosa Boiss. & Hohen. (*V. costata* Ledeb. subsp. *venulosa* (Boiss. & Hohen.) Stankevicz) - 6
villosa Roth - 1, 2, 5(alien), 6
- subsp. *varia* (Host) Corb. = V. varia
woroschilovii N.S. Pavlova (*V. pallida* Turcz. 1842, non
  Hook. & Arn. 1833, *V. amurensis* Oett. var. *pallida*
  (Turcz.) Kitag., *V. japonica* A. Gray var. *pallida*
  (Turcz.) Hara, *V. japonica* subsp. *pallida* (Turcz.)
  Worosch.) - 4, 5

Vigna Savi (*Dolichos* L. p.p., *Rudua* F. Maek.)

angularis (Willd.) Ohwi & Ohashi (*Phaseolus angularis*
  (Willd.) W. Wight) - 5
radiata (L.) Wilczek (*Phaseolus radiatus* L., *Ph. aureus*
  Roxb., *Rudua aurea* (Roxb.) F. Maek.) - 1, 2, 5, 6
sinensis (L.) Savi ex Hassk. = V. unguiculata
unguiculata (L.) Walp. (*Dolichos unguiculatus* L., *D.
  sinensis* L., *Vigna sinensis* (L.) Savi ex Hassk.) - 2, 6

Virgilia auct. = Cladrastis

lutea Michx. = Cladrastis kentukea

Wisteria Nutt.

sinensis (Sims) Sweet - 1, 2, 6

Xanthobrychis Galushko = Onobrychis

amoena (M. Pop. & Vved.) Galushko = Onobrychis
  amoena
bobrovii (Grossh.) Galushko = Onobrychis bobrovii
buhseana (Bunge) Galushko = Onobrychis buhseana
chorassanica (Bunge) Galushko = Onobrychis chorassanica
heterophylla (C.A. Mey.) Galushko = Onobrychis heterophylla
hohenackeriana (C.A. Mey.) Galushko = Onobrychis
  hohenackeriana
kachetica (Boiss. & Buhse) Galushko = Onobrychis kachetica
majorovii (Grossh.) Galushko = Onobrychis majorovii
pallasii (Willd.) Galushko = Onobrychis pallasii
radiata (Desf.) Galushko = Onobrychis radiata
tournefortii (Willd.) Galushko = Onobrychis tournefortii
vassilczenkoi (Grossh.) Galushko = Onobrychis vassilczenkoi

FAGACEAE Dumort.

Castanea Hill
sativa Mill. - 1(cult.), 2

Fagus L.

orientalis Lipsky (*F. sylvatica* L. subsp. *orientalis* (Lipsky)
  Greuter & Burdet) - 1, 2
sylvatica L. - 1
- subsp. *orientalis* (Lipsky) Greuter & Burdet = F. orientalis

Quercus L.

anatolica Sosn. ex Bandin, p.p. = Q. pubescens
araxina (Trautv.) Grossh. = Q. boissieri

austriaca Willd. (*Q. cerris* auct.) - 1
boissieri Reut. (*Q. araxina* (Trautv.) Grossh., *Q. infectoria*
   Oliv. subsp. *boissieri* (Reut.) O. Schwarz) - 2
calcarea Troitz. = Q. dalechampii
castaneifolia C.A. Mey. - 2, 6(cult.)
cerris auct. = Q. austriaca
***cerris** L. - 1, 2, 6
colchica Kotschy ex Czeczott, p.p. = Q. dshorochensis
crispata Stev. = Q. pubescens
crispula Blume (*Q.* x *crispulimongolica* Nakai, *Q. grosseserrata* Blume, *Q. mongolica* Turcz. ex Ledeb. subsp.
   crispula (Blume) Menitsky, *Q. mongolica* subsp. *crispula* (Blume) Worosch. comb. superfl., *Q. mongolica*
   subsp. *grosseserrata* (Blume) Worosch.nom. invalid.) -
   5
x *crispulimongolica* Nakai = Q. crispula
dalechampii Ten. (*Q. calcarea* Troitz. 1931, non Gand.
   1890, *Q. lanuginosa* Lam. subsp. *medwediewii* A.
   Camus, *Q. petraea* L. ex Liebl. subsp. *medwediewii* (A.
   Camus) Menitsky) - 1, 2
dentata Thunb. - 5
dshorochensis C. Koch (*Q. colchica* Kotschy ex Czeczott,
   p.p. quoad typum, *Q. kochiana* O.Schwarz, *Q. komarowii* A. Camus, *Q. longifolia* C. Koch, 1849, non
   Rafin. 1838, *Q. petraea* L. ex Liebl. subsp. *dshorochensis* (C. Koch) Menitsky) - 2
erucifolia Stev. = Q. pedunculiflora
- subsp. *longipes* (Stev.) O. Schwarz = Q. pedunculiflora
- - var. *imeretina* (Stev. ex Woronow) O. Schwarz = Q.
   imeretina
gagriana Rossinsky - 2
grosseserrata Blume = Q. crispula
grossheimii Sachok. = Q. iberica
hartwissiana Stev. - 1(cult.), 2, 6(cult.)
hypochrysa Stev. = Q. iberica
iberica Stev. (*Q. grossheimii* Sachok., *Q. hypochrysa* Stev.,
   Q. iberica var. *kozlowskyi* (Woronow ex Grossh.)
   Gagnidze, *Q. kozlowskyi* Woronow ex Grossh., *Q.
   macrocarpa* Grossh. 1945, nom. invalid., *Q. petraea* L.
   ex Liebl. subsp. *iberica* (Stev.) Krassiln.) - 1(cult.), 2,
   6(cult.)
- var. *kozlowskyi* (Woronow ex Grossh.) Gagnidze = Q.
   iberica
- var. *polycarpa* (Schur) Zohary = Q. petraea
***ilex** L. - 1, 2
imeretina Stev. ex Woronow (*Q. erucifolia* Stev. var. *imeretina* (Stev. ex Woronow) O. Schwarz, *Q. robur* L.
   subsp. *imeretina* (Stev. ex Woronow) Menitsky) - 2
infectoria Oliv. subsp. *boissieri* (Reut.) O. Schwarz = Q.
   boissieri
kochiana O. Schwarz = Q. dshorochensis
komarowii A. Camus = Q. dshorochensis
kozlowskyi Woronow ex Grossh. = Q. iberica
lanuginosa Lam. subsp. *crispata* (Stev.) A. Camus = Q.
   pubescens
- subsp. *medwediewii* A. Camus = Q. dalechampii
longifolia C. Koch = Q. dshorochensis
longipes Stev. = Q. pedunculiflora
macranthera Fisch. & C.A. Mey. ex Hohen. - 1(cult.), 2,
   6(cult.)
macrocarpa Grossh. = Q. iberica
mongolica Fisch. ex Ledeb. - 4, 5
- subsp. *crispula* (Blume) Menitsky = Q. crispula
- subsp. *crispula* (Blume) Worosch. = Q. crispula
- subsp. *grosseserrata* (Blume) Worosch. = Q. crispula
***occidentalis** J. Gay - 1, 2
pedunculata Ehrh. = Q. robur

- var. *pedunculiflora* (C. Koch) Stojan. & Stef. = Q. pedunculiflora
pedunculiflora C. Koch (*Q. erucifolia* Stev., *Q. erucifolia*
   subsp. *longipes* (Stev.) O. Schwarz, *Q. longipes* Stev.,
   Q. pedunculata Ehrh. var. *pedunculiflora* (C. Koch)
   Stojan. & Stef., *Q. pedunculiflora* var. *erucifolia* (Stev.)
   Gagnidze, *Q. robur* L. subsp. *pedunculiflora* (C. Koch)
   Menitsky) - 2
- var. *erucifolia* (Stev.) Gagnidze = Q. pedunculiflora
petraea L. ex Liebl. (*Q. iberica* Stev. var. *polycarpa* (Schur)
   Zohary, *Q. polycarpa* Schur, *Q. robur* L. Spielart *Q.
   petrea* Mattuschka, non basionymum) - 1, 2
- subsp. *dshorochensis* (C. Koch) Menitsky = Q. dshorochensis
- subsp. *iberica* (Stev.) Krassiln. = Q. iberica
- subsp. *medwediewii* (A. Camus) Menitsky = Q. dalechampii
- subsp. *pinnatiloba* (C. Koch) Menitsky = Q. pinnatiloba
pinnatiloba C. Koch (*Q. petraea* L. ex Liebl. subsp. *pinnatiloba* (C. Koch) Menitsky) - 2
polycarpa Schur = Q. petraea
pontica C. Koch - 2
pubescens Willd. (*Q. anatolica* Sosn. ex Bandin, p.p. quoad
   plantas, nom. invalid., *Q. crispata* Stev., *Q. lanuginosa*
   Lam. subsp. *crispata* (Stev.) A. Camus, *Q. pubescens*
   subsp. *crispata* (Stev.) Greuter & Burdet) - 1, 2
- subsp. *crispata* (Stev.) Greuter & Burdet = Q. pubescens
robur L. (*Q. pedunculata* Ehrh.) - 1, 2, 3(cult.), 6(cult.)
- subsp. *imeretina* (Stev. ex Woronow) Menitsky = Q.
   imeretina
- subsp. *pedunculiflora* (C. Koch) Menitsky = Q. pedunculiflora
- Spielart *Q. petrea* Mattuschka = Q. petraea
***rubra** L. - 1
***suber** L. - 1, 2
woronowii Maleev = Q. dshorochensis

FRANKENIACEAE S.F. Gray
Frankenia L.

bucharica Basil. - 6
- subsp. *mironovii* (Botsch.) Chrtek = F. mironovii
- subsp. *transkaratavica* (Botsch.) Chrtek = F. transkaratavica
- subsp. *vvedenskyi* (Botsch.) Chrtek = F. vvedenskyi
hirsuta L. (*F. hispida* DC., *F. intermedia* DC.) - 1, 2, 3, 6
hispida DC. = F. hirsuta
intermedia DC. = F. hirsuta
laevis L. - 2
mironovii Botsch. (*F. bucharica* Basil. subsp. *mironovii*
   (Botsch.) Chrtek) - 6
pulverulenta L. - 1, 2, 3, 6
transkaratavica Botsch. (*F. bucharica* Basil. subsp. *transkaratavica* (Botsch.) Chtrek) - 6
tuvinica Lomonosova - 4
vvedenskyi Botsch. (*F. bucharica* Basil. subsp. *vvedenskyi*
   (Botsch.) Chrtek) - 6

FUMARIACEAE DC.
Adlumia Rafin. ex DC.

asiatica Ohwi - 5

Capnoides Hill = Corydalis

popovii (Nevski ex M. Pop.) Nevski = Corydalis popovii

Corydalis DC. nom. cons. (*Capnoides* Hill, *Pistolochia* Bernh.)

aitchisonii M. Pop. (*Pistolochia aitchisonii* (M. Pop.) Sojak) - 6

alba (Mill.) Mansf. = C. capnoides

alexeenkoana N. Busch (*Pistolochia alexeenkoana* (N. Busch) Holub) - 2

alpestris C.A. Mey. (*Pistolochia alpestris* (C.A. Mey.) Sojak) - 2

ambigua Cham. & Schlecht. (*Pistolochia ambigua* (Cham. & Schlecht.) Sojak) - 5

- subsp. **amurensis** (Maxim.) Worosch. (*C. ambigua* var. *amurensis* Maxim.) - 5

- var. *amurensis* Maxim. = C. ambigua subsp. amurensis

- f. *fumariola* (Maxim.) Kitag. = C. fumariola

angustifolia (Bieb.) DC. (*Pistolochia angustifolia* (Bieb.) Holub) - 1

arctica M. Pop. (*C. pauciflora* (Steph.) Pers. subsp. *arctica* (M. Pop.) Worosch., *Pistolochia arctica* (M. Pop.) Sojak) - 4, 5

bracteata (Steph.) Pers. (*Pistolochia bracteata* (Steph.) Sojak) - 3, 4

bucharica M. Pop. - 6

bulbosa (L.) DC. (*Fumaria bulbosa* L. var. *solida* L., *Corydalis halleri* (Willd.) Willd., *C. solida* (L.) Clairv., *Fumaria halleri* Willd., *F. solida* (L.) Mill., *Pistolochia bulbosa* (L.) Sojak, *P. solida* (L.) Bernh.) - 1, 3, 6

- subsp. *marschalliana* (Pers.) Chater = C. marschalliana

bulbosa (L.) Pers. = C. cava

bungeana Turcz. - 5

buschii Nakai (*Pistolochia buschii* (Nakai) Sojak) - 5

capnoides (L.) Pers. (*C. alba* (Mill.) Mansf., *Fumaria alba* Mill.) - 1, 3, 4, 6

caucasica DC. (*C. solida* (L.) Clairv. var. *caucasica* (DC.) Jordanov & Kozuharov) - 2

cava (L.) Schweigg. & Koerte (*Fumaria bulbosa* L. var. *cava* L., *Corydalis bulbosa* (L.) Pers. 1807, non DC. 1805, *C. tuberosa* DC., *Fumaria cava* (L.) Mill., *Pistolochia cava* (L.) Bernh.) - 1, 2

- subsp. *marschalliana* (Pers.) Hayek = C. marschalliana

chionophila Czerniak. (*Pistolochia chionophila* (Czerniak.) Sojak) - 6

conorhiza Ledeb. (*Pistolochia conorhiza* (Ledeb.) Sojak) - 2

darwasica Regel ex Prain (*Pistolochia darwasica* (Regel ex Prain) Sojak) - 6

emanuelii C.A. Mey. (*C. pallidiflora* (Rupr.) N. Busch, *Pistolochia emanuelii* (C.A. Mey.) Sojak, *P. pallidiflora* (Rupr.) Sojak) - 2

erdelii Zucc. (*C. rutifolia* (Sibth. & Smith) subsp. *erdelii* (Zucc.) Cullen & P.H. Davis, *Pistolochia erdelii* (Zucc.) Sojak) - 2

fabacea (Retz.) Pers. = C. intermedia

feddeana Poellnitz (*Pistolochia feddeana* (Poellnitz) Holub) - 1

fedtschenkoana Regel = Cysticorydalis fedtschenkoana

fimbrillifera Korsh. - 6

fumariifolia Maxim. (*C. ambigua* Cham. & Schlecht. f. *fumariifolia* (Maxim.) Kitag., *C. lineariloba* Siebold & Zucc. var. *fumariifolia* (Maxim.) Kitag., *C. turtschaninowii* Bess. f. *fumariifolia* (Maxim.) Y.H. Chou, *Pistolochia fumariifolia* (Maxim.) Sojak) - 5

x **gigantea** Trautv. & C.A. Mey. - 5

- subsp. *multiflora* (Mikhailova) Worosch. = C. multiflora

glaucescens Regel (*Pistolochia glaucescens* (Regel) Sojak) - 6

- subsp. **pamiroalaica** Sosk. - 6

gorinensis Van - 5

gorodkovii Karav. (*Pistolochia gorodkovii* (Karav.) Sojak) - 4, 5

- subsp. *magadanica* (A. Khokhr.) Worosch. = C. magadanica

gortschakovii Schrenk - 6

- subsp. *onobrychis* (Fedde) Wendelbo = C. onobrychis

gypsophila Mikhailova - 6

halleri (Willd.) Willd. = C. bulbosa

heterophylla Mikhailova - 6

x **hybrida** Mikhailova. - C. bracteata (Steph.) Pers. x C. bulbosa (L.) DC. - 1

impatiens (Pall.) Fisch. ex DC. - 4

inconspicua Bunge (*C. tenella* Kar. & Kir.) - 3, 4, 6

intermedia (L.) Merat (*C. fabacea* (Retz.) Pers., *Fumaria intermedia* (L.) Ehrh., *Pistolochia intermedia* (L.) Bernh.) - 1

kamelinii Kurbanov - 6

kaschgarica Rupr. - 6(?)

    M.A. Mikhailova suggests that this species does not occur on the investigated territory.

krasnovii Mikhailova - 6

kusnetzovii A. Khokhr. - 2

ledebouriana Kar. & Kir. (*Pistolochia ledebouriana* (Kar. & Kir.) Sojak) - 6

lineariloba auct. = C. ussuriensis

lineariloba Siebold & Zucc. var. *fumariifolia* (Maxim.) Kitag. = C. fumariifolia

macrantha (Regel) M. Pop. - 5

macrocalyx Litv. - 6

macrocentra Regel (*Pistolochia macrocentra* (Regel) Sojak) - 6

magadanica A. Khokhr. (*C. gorodkovii* Karav. subsp. *magadanica* (A. Khokhr.) Worosch.) - 5

malkensis Galushko - 2

maracandica Mikhailova - 6

marschalliana (Pall. ex Willd.) Pers. (*Fumaria marschalliana* Pall. ex Willd. 1802, *F. marschalliana* Pall. 1797, nom. nud., *Corydalis bulbosa* (L.) DC. subsp. *marschalliana* (Pers.) Chater, comb. invalid., *C. cava* (L.) Schweigg. & Koerte subsp. *marschalliana* (Pall.) Hayek, comb. invalid., *Pistolochia bulbosa* (L.) Sojak subsp. *marschalliana* (Pall.) Sojak, comb. invalid., *Pistolochia marschalliana* (Pall.) Holub, comb. invalid.) - 1, 2

microphylla Mikhailova - 6

multiflora Mikhailova (*C. giganthea* Trautv. & C.A. Mey. subsp. *multiflora* (Mikhailova) Worosch.) - 5

nariniana Fed. = C. persica

nevskii M. Pop. (*Pistolochia nevskii* (M. Pop.) Sojak) - 6

nobilis (L.) Pers. - 3, 6

nudicaulis Regel (*Pistolochia nudicaulis* (Regel) Sojak) - 6

ochotensis Turcz. - 5

- var. *raddeana* (Regel) Nakai = C. raddeana

onobrychis Fedde (*C. gortschakovii* Schrenk subsp. *onobrychis* (Fedde) Wendelbo) - 6

paczoskii N. Busch (*Pistolochia paczoskii* (N. Busch) Sojak) - 1

paeoniifolia (Steph.) Pers. (*C. redowskii* Fedde) - 4, 5

pallida auct. p.p. = C. speciosa

pallida Pers. - 5

pallidiflora (Rupr.) N. Busch = C. emanuelii

paniculigera Regel & Schmalh. - 6

pauciflora (Steph.) Pers. (*Pistolochia pauciflora* (Steph.) Sojak) - 3

- subsp. *arctica* (M. Pop.) Worosch. = C. arctica

persica Cham. & Schlecht. (*C. nariniana* Fed., *Pistolochia* *nariniana* (Fed.) Holub, *P. persica* (Cham. & Schlecht.) Sojak) - 2

popovii Nevski ex M. Pop. (*Capnoides popovii* (Nevski ex M. Pop.) Nevski, *Pistolochia popovii* (Nevski ex M. Pop.) Sojak) - 6

pseudoadunca M. Pop. - 6

pseudoalpestris M. Pop. (*Pistolochia pseudoalpestris* (M. Pop.) Sojak) - 6

pseudostricta M. Pop. - 6

raddeana Regel (*C. ochotensis* Turcz. var. *raddeana* (Regel) Nakai) - 5

rarissima Mikhailova - 6

redowskii Fedde = C. paeoniifolia

remota Fisch. ex Maxim. (*Pistolochia remota* (Fisch. ex Maxim.) Sojak) - 4, 5

> Probably, the priority name of this species should be C. turtschaninowii Bess. (1834, Flora (Regensb.) 17, Biebl. 1 : 6)

repens Mandl & Muehld. (*Pistolochia repens* (Mandl & Muehld.) Sojak) - 5

rupestris Kotschy ex Boiss. - 6(?)

rutifolia (Sibth. & Smith) DC. subsp. *erdelii* (Zucc.) Cullen & P.H. Davis = C. erdelii

sajanensis Peschkova - 4

schanginii (Pall.) B. Fedtsch. (*Pistolochia schanginii* (Pall.) Sojak) - 3, 6

schelesnowiana Regel & Schmalh. - 6

semenovii Regel - 6

sempervirens (L.) Pers. - 3

sewerzowii Regel (*Pistolochia sewerzowii* (Regel) Sojak) - 6

sibirica (L. fil.) Pers. - 1, 3, 4, 5

solida (L.) Clairv. = C. bulbosa

- var. *caucasica* (DC.) Jordanov & Kozuharov = C. caucasica

speciosa Maxim. (*C. pallida* auct. p.p.) - 5

stricta Steph. - 3, 6

- subsp. *pamirica* Mikhailova - 6

tarkiensis Prokh. (*Pistolochia tarkiensis* (Prokh.) Holub) - 2

teberdensis A. Khokhr. - 2

tenella Kar. & Kir. = C. inconspicua

transalaica M. Pop. - 6

tuberosa DC. = C. cava

turtschaninowii Bess. f. *fumariifolia* Cham. & Schlecht. = C. fumariifolia

udokanica Peschkova - 4

ussuriensis Aparina (*Pistolochia ussuriensis* (Aparina) Sojak, *Corydalis lineariloba* auct.) - 5

vittae Kolak. - 2

zeaensis Mikhailova - 5

zeravschanica Mikhailova - 6

Cysticorydalis Fedde ex Ikonn.

fedtschenkoana (Regel) Fedde ex Ikonn. (*Corydalis fedtschenkoana* Regel) - 6

Dicentra Bernh.

peregrina (J. Rudolph) Makino - 5

Fumaria L.

alba Mill. = Corydalis capnoides

asepala Boiss. - 2, 6

bella P.D. Sell - 1(?)

bulbosa L. var. *cava* L. = Corydalis cava

- var. *solida* L. = Corydalis bulbosa

capreolata L. - 2

cava (L.) Mill. = Corydalis cava

daghestanica Mikhailova - 2

densiflora auct. p.p. = F. micrantha

densiflora DC. subsp. *micrantha* (Lag.) Maire & Weiller = F. micrantha

x *gagrica* Mikhailova. - F. capreolata L. x F. officinalis L. - 2

halleri Willd. = Corydalis bulbosa

indica (Hausskn.) Pugsl. (*F. vaillantii* var. *indica* Hausskn.) - 6

intermedia (L.) Ehrh. = Corydalis intermedia

kralikii Jord. - 1

marschalliana Pall. = Corydalis marschalliana

marschalliana Pall. ex Willd. = Corydalis marschalliana

micrantha Lag. (*F. densiflora* DC. subsp. *micrantha* (Lag.) Maire & Weiller, *F. densiflora* auct. p.p.) - 2, 6

microcarpa Boiss. ex Hausskn. - 2

officinalis L. - 1, 2, 3, 4, 5

- subsp. *schrammii* Aschers. = F. schrammii

petteri Reichenb. subsp. *thuretii* (Boiss.) Pugsl. = F. thuretii

pikermiana Boiss. & Heldr. = F. thuretii

rostellata Knaf - 1

schleicheri Soy.-Willem. - 1, 2, 3, 6

schrammii (Aschers.) Velen. (*F. parviflora* Lam. subsp. *schrammii* Aschers., *F. vaillantii* Loisel. subsp. *schrammii* (Aschers.) Bordz.) - 1

solida (L.) Mill. = Corydalis bulbosa

thuretii Boiss. (*F. petteri* Reichenb. subsp. *thuretii* (Boiss.) Pugsl., *F. pikermiana* Boiss. & Heldr.) - 1

vaillantii Loisel. - 1, 2, 6

- subsp. *schrammii* (Aschers.) Bordz. = F. schrammii

- var. *indica* Hausskn. = F. indica

Fumariola Korsh.

turkestanica Korsh. - 6

Pistolochia Bernh. = Corydalis

aitchisonii (M. Pop.) Sojak = Corydalis aitchisonii

alexeenkoana (N. Busch) Holub = Corydalis alexeenkoana

alpestris (C.A. Mey.) Sojak = Corydalis alpestris

ambigua (Cham. & Schlecht.) Sojak = Corydalis ambigua

angustifolia (Bieb.) Holub = Corydalis angustifolia

arctica (M. Pop.) Sojak = Corydalis arctica

bracteata (Steph.) Sojak = Corydalis bracteata

bulbosa (L.) Sojak = Corydalis bulbosa

- subsp. *marschalliana* (Pall.) Sojak = Corydalis marschalliana

buschii (Nakai) Sojak = Corydalis buschii

cava (L.) Bernh. = Corydalis cava

chionophila (Czerniak.) Sojak = Corydalis chionophila

conorhiza (Ledeb.) Sojak = Corydalis conorhiza

darwasica (Regel ex Prain) Sojak = Corydalis darwasica

emanuelii (C.A. Mey.) Sojak = Corydalis emanuelii

erdelii (Zucc.) Sojak = Corydalis erdelii

feddeana (Poellnitz) Holub = Corydalis feddeana

fumariifolia (Maxim.) Sojak = Corydalis fumariifolia

glaucescens (Regel) Sojak = Corydalis glaucescens

gorodkovii (Karav.) Sojak = Corydalis gorodkovii

intermedia (L.) Bernh. = Corydalis intermedia

ledebouriana (Kar. & Kir.) Sojak = Corydalis ledebouriana

macrocentra (Regel) Sojak = Corydalis macrocentra

marschalliana (Pall.) Holub = Corydalis marschalliana

nariniana (Fed.) Holub = Corydalis persica

nevskii (M. Pop.) Sojak = Corydalis nevskii

nudicaulis (Regel) Sojak = Corydalis nudicaulis

paczoskii (N. Busch) Sojak = Corydalis paczoskii

pallidiflora (Rupr.) Sojak = Corydalis emanuelii

pauciflora (Steph.) Sojak = Corydalis pauciflora
persica (Cham. & Schlecht.) Sojak = Corydalis persica
popovii (Nevski ex M. Pop.) Sojak = Corydalis popovii
pseudoalpestris (M. Pop.) Sojak = Corydalis pseudoalpestris
remota (Fisch. ex Maxim.) Sojak = Corydalis remota
repens (Mandl & Muehld.) Sojak = Corydalis repens
schanginii (Pall.) Sojak = Corydalis schanginii
sewerzowii (Regel) Sojak = Corydalis sewerzowii
solida (L.) Bernh. = Corydalis bulbosa
tarkiensis (Prokh.) Holub = Corydalis tarkiensis
ussuriensis (Aparina) Sojak = Corydalis ussuriensis

Roborowskia Batal.

mira Batal. - 6

FUNKIACEAE Horan. = HOSTACEAE

GENTIANACEAE Juss.
Aloitis Rafin. = Gentianella
aurea (L.) A. & D. Löve = Gentianella aurea
propinqua (Richards.) A. & D.Löve = Gentianella propinqua

Anagallidium Griseb.

dichotomum (L.) Griseb. - 3, 4, 6

Arctogenia A. Löve = Gentianella
aurea (L.) A. Löve = Gentianella aurea
propinqua (Richards.) A. & D. Löve = Gentianella propinqua

Blackstonia Huds. (*Chlora* L.)
acuminata (Koch & Ziz) Domin (*Chlora acuminata* Koch & Ziz, *Blackstonia perfoliata* (L.) Huds. subsp. *serotina* (Koch ex Reichenb.) Vollm., *B. serotina* (Koch ex Reichenb.) G. Beck, *Chlora serotina* Koch ex Reichenb.) - 1
perfoliata (L.) Huds. - 1, 2
- subsp. *serotina* (Koch ex Reichenb.) Vollm. = B. acuminata
serotina (Koch ex Reichenb.) G. Beck = B. acuminata

Calathiana Delarb. = Gentiana
angulosa (Bieb.) Holub = Gentiana angulosa
nivalis (L.) Delarb. = Gentiana nivalis
oschtenica (Kusn.) Holub = Gentiana oschtenica
pontica (Soltok.) Holub = Gentiana pontica
uniflora (Georgi) Holub = Gentiana uniflora
utriculosa (L.) Holub = Gentiana utriculosa
verna (L.) Holub = Gentiana verna

Centaurium Hill (*Erythraea* Borkh., *Chironia* auct.)
anatolicum (C. Koch) Tzvel. (*Erythraea anatolica* C. Koch, *Centaurium erythraea* Rafn subsp. *turcicum* (Velen.) Melderis, *C. minus* Moench subsp. *turcicum* (Velen.) Soo, *C. turcicum* (Velen.) Bornm., *C. turcicum* (Velen.) Ronn. comb. superfl., *Erythraea turcica*

Velen.) - 1, 2
candelabrum Lindb. fil. = C. pulchellum
caspicum (Fisch. ex Griseb.) Tzvel. (*Erythraea caspica* Fisch. ex Griseb., *Centaurium pulchellum* (Sw.) Druce var. *caspicum* (Fisch. ex Griseb.) R.R. Stewart) - 2
erythraea Rafn (? *C. erythraea* subsp. *austriacum* (Ronn.) Holub, *C. minus* Moench, p.p. nom. ambig., ? *C. minus* subsp. *austriacum* (Ronn.) O. Schwarz, *C. umbellatum* Gilib. nom. invalid., ? *C. umbellatum* subsp. *austriacum* Ronn.) - 1, 2, 3, 6
- subsp. *austriacum* (Ronn.) Holub = C. erythraea
- subsp. *turcicum* (Velen.) Melderis = C. anatolicum
littorale (D. Turner) Gilmour (*Chironia littoralis* D. Turner, *Centaurium minus* Moench, p.p. nom. ambig., *C. vulgare* Rafn, nom. superfl.) - 1
- subsp. *uliginosum* (Waldst. & Kit.) Melderis = C. uliginosum
meyeri (Bunge) Druce (*C. pulchellum* (Sw.) Druce subsp. *meyeri* (Bunge) Tzvel.) - 1, 2, 3, 6
minus Moench = C. erythraea and C. littorale
- subsp. *austriacum* (Ronn.) O. Schwarz = C. erythraea
- subsp. *turcicum* (Velen.) Soo = C. anatolicum
pulchellum (Sw.) Druce (*C. candelabrum* Lindb. fil.) - 1, 2, 5(alien), 6
- subsp. *meyeri* (Bunge) Tzvel. = C. meyeri
- var. *caspicum* (Fisch. ex Griseb.) R.R. Stewart = C. caspicum
spicatum (L.) Fritsch - 1, 2, 6
tenuiflorum (Hoffmgg. & Link) Fritsch - 1, 2
turcicum (Velen.) Bornm. = C. anatolicum
turcicum (Velen.) Ronn. = C. anatolicum
uliginosum (Waldst. & Kit.) G. Beck ex Ronn. (*C. littorale* (D. Turner) Gilmour subsp. *uliginosum* (Waldst. & Kit.) Melderis, *C. vulgare* Rafn subsp. *uliginosum* (Waldst. & Kit.) Soo) - 1, 2, 6
umbellatum Gilib. = C. erythraea
- subsp. *austriacum* Ronn. = C. erythraea
vulgare Rafn = C. littorale
- subsp. *uliginosum* (Waldst. & Kit.) Soo = C. uliginosum

Chironia auct. = Centaurium
littoralis D. Turner = Centaurium littorale

Chlora L. = Blackstonia
acuminata Koch & Ziz = Blackstonia acuminata
serotina Koch ex Reichenb. = Blackstonia acuminata

Chondrophylla Nels. = Gentiana
aquatica (L.) W.A. Weber = Gentiana aquatica
nutans (Bunge) W.A. Weber = Gentiana prostrata

Ciminalis Adans. = Gentiana
acaulis (L.) Moench = Gentiana acaulis
aquatica (L.) V. Zuev = Gentiana aquatica
clusii (Perrier & Song.) Holub = Gentiana clusii
grandiflora (Laxm.) V. Zuev = Gentiana grandiflora
leucomelaena (Maxim.) V. Zuev = Gentiana leucomelaena
prostrata (Haenke) A. & D. Löve = Gentiana prostrata
pseudoaquatica (Kusn.) V. Zuev = Gentiana pseudoaquatica
riparia (Kar. & Kir.) V. Zuev = Gentiana riparia
squarrosa (Ledeb.) V. Zuev = Gentiana squarrosa
variegata V. Zuev = Gentiana variegata

Comastoma (Wettst.) Toyokuni

azureum (Bunge) V. Zuev = Gentianella azurea
dechyanum (Somm. & Levier) Holub (*Gentiana dechyana* Somm. & Levier) - 2
dichotomum (Pall.) Holub = C. tenellum
falcatum (Turcz.) Toyokuni (*Gentiana falcata* Turcz., *Gentianella falcata* (Turcz.) H. Smith) - 3, 4, 6
irinae (Pachom.) Czer. comb. nova (*Gentiana irinae* Pachom. 1986, Opred. Rast. Sr. Asii, 8 :169, 49) - 6
malyschevii (V. Zuev) V. Zuev (*Gentianella malyschevii* V. Zuev, *Gentiana malyschevii* (V. Zuev) Gubanov & R. Kam.) - 4
pulmonarium (Turcz. ex Ledeb.) Toyokuni (*Gentiana pulmonaria* Turcz. ex Ledeb., *Gentianella pulmonaria* (Turcz. ex Ledeb.) V. Zuev) - 4
tenellum (Rottb.) Toyokuni (*Gentiana tenella* Rottb., *Comastoma dichotomum* (Pall.) Holub, *Gentiana dichotoma* Pall., *Gentianella dichotoma* (Pall.) H. Smith, *G. tenella* (Rottb.) Boern., *Lomatogonium tenellum* (Rottb.) A. & D. Löve) - 1, 3, 4, 5, 6

Crawfurdia Wall.

japonica Siebold & Zucc. - 5
volubilis (Maxim.) Makino - 5

Dasystephana Adans. = Gentiana

asclepiadea (L.) Borkh. = Gentiana asclepiadea
axillariflora (Levl. & Vaniot) Sojak = Gentiana axillariflora
fischeri (P. Smirn.) Sojak = Gentiana fischeri
gelida (Bieb.) Sojak = Gentiana gelida
grossheimii (Doluch.) Sojak = Gentiana grossheimii
kolakovskyi (Doluch.) Sojak = Gentiana kolakovskyi
lagodechiana (Kusn.) Sojak = Gentiana lagodechiana
makinoi (Kusn.) Sojak = Gentiana axillariflora
owerinii (Kusn.) Sojak = Gentiana owerinii
paradoxa (Albov) Sojak = Gentiana paradoxa
pneumonanthe (L.) Sojak = Gentiana pneumonanthe
sangilenica V. Zuev = Gentiana sangilenica
scabra (Bunge) Sojak = Gentiana scabra
schistocalyx (C. Koch) Sojak = Gentiana schistocalyx
septemfida (Pall.) Sojak = Gentiana septemfida

Ericoila Borkh. = Gentiana

acaulis (L.) A. & D. Löve = Gentiana acaulis
clusii (Perrier & Song.) A. & D. Löve = Gentiana clusii

Erythraea Borkh. = Centaurium

anatolica C. Koch = Centaurium anatolicum
caspica Fisch. ex Griseb. = Centaurium caspicum
turcica Velen. = Centaurium anatolicum

Frasera auct. = Ophelia

diluta (Turcz.) Toyokuni = Ophelia diluta
tetrapetala (Pall.) Toyokuni = Ophelia tetrapetala

Gentiana L. (*Calathiana* Delarb., *Chondrophylla* Nels., *Ciminalis* Adans., *Dasystephana* Adans., *Ericoila* Borkh., *Gentianodes* A. & D. Löve, *Pneumonanthe* Gled., *Tretorhiza* Adans., *Varasia* Phil.)

acaulis L. (*Ciminalis acaulis* (L.) Moench, *Ericoila acaulis*

(L.) A. & D. Löve, *Gentiana excisa* C. Presl) - 1
acuta Michx. = Gentianella acuta
- var. *plebeja* (Cham. & Schlecht.) Worosch. = Gentianella plebeja
algida Pall. (*Gentianodes algida* (Pall.) A. & D. Löve, *Pneumonanthe algida* (Pall.) F.W. Schmidt) - 3, 4, 5, 6
amarella L. = Gentianella amarella
- subsp. *acuta* (Michx.) Hult. = Gentianella acuta
- - var. *plebeja* (Cham. & Schlecht.) Hult. = Gentianella plebeja
- subsp. *axillaris* (F.W. Schmidt) Murb. = Gentianella amarella
angulosa Bieb. (*Calathiana angulosa* (Bieb.) Holub, *Gentiana verna* L. subsp. *angulosa* (Bieb.) V. Avet.) - 2
- var. *pontica* (Soltok.) Gagnidze = G. pontica
aquatica L. (*Chondrophylla aquatica* (L.) W.A. Weber, *Ciminalis aquatica* (L.) V. Zuev, *Varasia aquatica* (L.) Sojak) - 2, 3, 4, 5, 6
- var. *pseudoaquatica* (Kusn.) S. Agrawal = G. pseudoaquatica
arctica Grossh. = G. verna
arctophila Griseb. = Gentianella arctophila
asclepiadea L. (*Dasystephana asclepiadea* (L.) Borkh.) - 1
- subsp. *schistocalyx* (C. Koch) J. Zachar. = G. schistocalyx
- var. *schistocalyx* C. Koch = G. schistocalyx
atrata Bunge = Gentianella atrata
aurea L. = Gentianella aurea
auriculata Pall. = Gentianella auriculata
axillariflora Levl. & Vaniot (*Dasystephana axillariflora* (Levl. & Vaniot) Sojak, *D. makinoi* (Kusn.) Sojak, *Gentiana makinoi* Kusn., *G. rigescens* Franch. var. *japonica* Kusn., *G. triflora* Pall. var. *japonica* (Kusn.) Hara, *G. triflora* subsp. *japonica* (Kusn.) Worosch.) - 5
axillaris (F.W. Schmidt) Reichenb. = Gentianella amarella
- var. *lingulata* (Agardh) Topa = Gentianella lingulata
- var. *livonica* (Eschsch.) Topa = Gentianella livonica
azurea Bunge = Gentianella azurea
baltica Murb. = Gentianella baltica
barbata Froel. = Gentianopsis barbata
biebersteinii Bunge = Gentianella biebersteinii
blepharophora Bordz. = Gentianopsis blepharophora
bzybica (Doluch.) Kolak. (*G. kolakovskyi* Doluch var. *bzybica* Doluch.) - 2
campestris L. = Gentianella campestris
- var. *germanica* Froel. = Gentianella campestris
- var. *suecica* Froel. = Gentianella campestris subsp. suecica
carpatica Wettst. = Gentianella carpatica
carpaticola Borb. = Gentianella carpatica
caucasea Lodd. ex Sims = Gentianella caucasea
caucasica Bieb. = Gentianella caucasea
ciliata L. = Gentianopsis ciliata
- subsp. *blepharophora* (Bordz.) Greuter = Gentianopsis blepharophora
clarkei auct. = G. prostrata
clusii Perrier & Song. (*Ciminalis clusii* (Perrier & Song.) Holub, *Ericoila clusii* (Perrier & Song.) A. & D. Löve) - 1
cordifolia C. Koch - 2
cruciata L. (*Pneumonanthe cruciata* (L.) V. Zuev, *Tretorhiza cruciata* (L.) Delarb., *T. cruciata* (L.) Opiz, comb. superfl.) - 1, 2, 3, 6
- subsp. *phlogifolia* (Schott & Kotschy) Tutin = G. phlogifolia
dahurica Fisch. (*Pneumonanthe dahurica* (Fisch.) V. Zuev, *Tretorhiza dahurica* (Fisch.) Sojak) - 4
dechyana Somm. & Levier = Comastoma dechyanum

decumbens L. fil. (*Tretorhiza decumbens* (L. fil.) Sojak) - 1, 3, 4, 6
detonsa Rottb. = Gentianopsis detonsa
dichotoma Pall. = Comastoma tenellum
djimilensis auct. = G. dshimilensis
x **doerfleri** Ronn. - G. lutea L. x G. punctata L. - 1
doluchanovii Grossh. = Gentianopsis doluchanovii
dschungarica Regel - 6
dshimilensis C. Koch (*Varasia dshimilensis* (C. Koch) Sojak, *Gentiana djimilensis* auct., *G. pyrenaica* auct. p.p.) - 2
excisa C. Presl = G. acaulis
falcata Turcz. = Comastoma falcatum
fetisowii Regel & Winkl. (*Pneumonanthe fetisowii* (Regel & Winkl.) V. Zuev, *Tretorhiza fetisowii* (Regel & Winkl.) Sojak) - 3, 6
fischeri P. Smirn. (*Dasystephana fischeri* (P. Smirn.) Sojak, *Pneumonanthe fischeri* (P. Smirn.) V. Zuev) - 3, 6
frigida Haenke (*Gentianodes frigida* (Haenke) A. & D. Löve, *Pneumonanthe frigida* (Haenke) F.W. Schmidt) - 1
gelida Bieb. (*Dasystephana gelida* (Bieb.) Sojak) - 2
germanica subsp. *carpatica* Hayek = Gentianella carpatica
glauca Pall. (*Gentianodes glauca* (Pall.) A. & D. Löve, *Pneumonanthe glauca* (Pall.) F.W. Schmidt) - 4, 5
grandiflora Laxm. (*Ciminalis grandiflora* (Laxm.) V. Zuev, *Varasia grandiflora* (Laxm.) Sojak) - 3, 4, 6
grossheimi Doluch. (*Dasystephana grossheimii* (Doluch.) Sojak) - 2
irinae Pachom. = Comastoma irinae
jamesii Hemsl. (*Varasia jamesii* (Hemsl.) Sojak, *Gentiana nipponica* sensu Grossh.) - 5
junussovii Pachom. & Taidshan. = Gentianella junussovii
karelinii Griseb. (*Varasia karelinii* (Griseb.) Sojak) - 3, 6
kaufmanniana Regel & Schmalh. (*Tretorhiza kaufmanniana* (Regel & Schmalh.) Sojak) - 3, 6
kawakamii Makino = G. nipponica
kirilowii Turcz. (*G. tianschanica* Rupr., *Tretorhiza tianschanica* (Rupr.) Sojak) - 6
kolakovskyi Doluch. (*Dasystephana kolakovskyi* (Doluch.) Sojak) - 2
- var. *bzybica* Doluch. = G. bzybica
komarovii Grossh. = Gentianopsis komarovii
krylovii Grossh. = G. uniflora
kurilensis Grossh. = G. nipponica
laciniata Kit. ex Kanitz - 1
lagodechiana (Kusn.) Grossh. (*Dasystephana lagodechiana* (Kusn.) Sojak) - 2
leucomelaena Maxim. (*Ciminalis leucomelaena* (Maxim.) V. Zuev, *Varasia leucomelaena* (Maxim.) Sojak) - 3, 4, 6
lingulata Agardh = Gentianella lingulata
lipskyi Kusn. = Gentianella lipskyi
livonica Eschsch. = Gentianella livonica
lutea L. - 1
lutescens Velen. = Gentianella lutescens
macrophylla Pall. (*Pneumonanthe macrophylla* (Pall.) V. Zuev, *Tretorhiza macrophylla* (Pall.) Sojak) - 3, 4, 5
makinoi Kusn. = G. axillariflora
malyschevii (V. Zuev) Gubanov & R. Kam. = Comastoma malyschevii
marcowiczii Kusn. = Gentianella caucasea
nipponica Maxim. (*G. kawakamii* Makino, *G. kurilensis* Grossh., *Varasia nipponica* (Maxim.) Sojak) - 5
nipponica sensu Grossh. = G. jamesii
nivalis L. (*Calathiana nivalis* (L.) Delarb.) - 1, 2
nutans Bunge = G. prostrata

olgae Regel & Schmalh. (*Tretorhiza olgae* (Regel & Schmalh.) Sojak) - 6
olivieri Griseb. (*Tretorhiza olivieri* (Griseb.) Sojak) - 2, 6
oschtenica (Kusn.) Woronow (*Calathiana oschtenica* (Kusn.) Holub) - 2
owerinii (Kusn.) Grossh. (*Dasystephana owerinii* (Kusn.) Sojak) - 2
paludicola Koidz. ex Sugaw. = Gentianella sugawarae
pamirica Grossh. = Gentianella sibirica
paradoxa Albov (*Dasystephana paradoxa* (Albov) Sojak) - 2
■ **phlogifolia** Schott & Kotschy (*G. cruciata* L. subsp. *phlogifolia* (Schott & Kotschy) Tutin, *Tretorhiza phlogifolia* (Schott & Kotschy) Sojak)
plebeja Cham. ex Bunge = Gentianella plebeja
plebeja Cham. & Schlecht. = Gentianella plebeja
pneumonanthe L. (*Dasystephana pneumonanthe* (L.) Sojak, *Pneumonanthe vulgaris* F.W. Schmidt) - 1, 2, 3, 4
pontica Soltok. (*Calathiana pontica* (Soltok.) Holub, *Gentiana angulosa* Bieb. var. *pontica* (Soltok.) Gagnidze, *G. verna* L. subsp. *pontica* (Soltok.) Hayek) - 2
praecox auct. = Gentianella lutescens
praeflorens Wettst. = Gentianella lutescens var. praeflorens
promethea Juz. = Gentianella promethea
propinqua Richards. = Gentianella propinqua
- subsp. *arctophila* (Griseb.) Hult. = Gentianella arctophila
prostrata Haenke (*Chondrophylla nutans* (Bunge) W.A. Weber, *Ciminalis prostrata* (Haenke) A. & D. Löve, *Gentiana nutans* Bunge, *Varasia nutans* (Bunge) Sojak, *V. prostrata* (Haenke) Sojak, *Gentiana clarkei* auct.) - 1, 2, 3, 4, 5, 6
pseudoaquatica Kusn. (*Ciminalis pseudoaquatica* (Kusn.) V. Zuev, *Gentiana aquatica* L. var. *pseudoaquatica* (Kusn.) S. Agrawal, *Varasia pseudoaquatica* (Kusn.) Sojak) - 3, 4
pulmonaria Turcz. ex Ledeb. = Comastoma pulmonarium
punctata L. - 1
pygmaea Regel & Schmalh. = Gentianella pygmaea
pyrenaica auct. p.p. = G. dshimilensis
rhodocalyx Kolak. - 2
rigescens Franch. var. *japonica* Kusn. = G. axillariflora
riparia Kar. & Kir. (*Ciminalis riparia* (Kar. & Kir.) V. Zuev, *Varasia riparia* (Kar. & Kir.) Sojak) - 1(?), 3, 4, 6
romanzowii Ledeb. ex Bunge - 5
sangilenica (V. Zuev) Czer. comb. nova (*Dasystephana sangilenica* V. Zuev, 1992, Bot. Zhurn. 77, 2 :96) - 4
saposhnikovii Pachom. = Gentianella saposhnikovii
scabra Bunge (*Dasystephana scabra* (Bunge) Sojak) - 4, 5
schistocalyx (C. Koch) C. Koch (*G. asclepiadea* L. var. *schistocalyx* C. Koch, *Dasystephana schistocalyx* (C. Koch) Sojak, *Gentiana asclepiadea* subsp. *schistocalyx* (C. Koch) J. Zachar.) - 2
septemfida Pall. (*Dasystephana septemfida* (Pall.) Sojak) - 1, 2
sibirica (Kusn.) Grossh. = Gentianella sibirica
squarrosa Ledeb. (*Ciminalis squarrosa* (Ledeb.) V. Zuev, *Varasia squarrosa* (Ledeb.) Sojak) - 3, 4, 5, 6
stoliczkae auct. = Gentianella turkestanorum
stricta Klotzsch = Gentianopsis stricta
sugawarae Hara = Gentianella sugawarae
susamyrensis Pachom. - 6
tenella Rottb. = Comastoma tenellum
tianschanica Rupr. = G. kirilowii
transalaica Pachom. & Tajdshan. = Gentianella transalaica
triflora Pall. (*Pneumonanthe triflora* (Pall.) F.W. Schmidt) - 4, 5
- subsp. *japonica* (Kusn.) Worosch. = G. axillariflora

- var. *japonica* (Kusn.) Hara = G. axillariflora
turkestanorum Gand. = Gentianella turkestanorum
uliginosa Willd. = Gentianella uliginosa
umbellata Bieb. = Gentianella umbellata
- var. *sibirica* (Kusn.) Serg. = Gentianella sibirica
uniflora Georgi (*Calathiana uniflora* (Georgi) Holub, *Gentiana krylovii* Grossh.) - 3, 4, 6
utriculosa L. (*Calathiana utriculosa* (L.) Holub) - 1
variegata (V. Zuev) Czer. comb. nova (*Ciminalis variegata* V. Zuev, Bot. Zhurn. 71, 10 : 1406) - 4
verna L. (*Calathiana verna* (L.) Holub, *Gentiana arctica* Grossh., *G. verna* var. *arctica* (Grossh.) Tolm.) - 1
- subsp. *angulosa* (Bieb.) V. Avet. = G. angulosa
- subsp. *pontica* (Soltok.) Hayek = G. pontica
- var. *arctica* (Grossh.) Tolm. = G. verna
vittae Kolak. - 2
vvedenskyi Grossh. = Gentianopsis vvedenskyi
walujewii Regel & Schmalh. (*Tretorhiza walujewii* (Regel & Schmalh.) Sojak) - 6
zollingeri Fawc. (*Varasia zollingeri* (Fawc.) Sojak) - 5

Gentianella Moench (*Aloitis* Rafin., *Arctogenia* A. Löve)

acuta (Michx.) Hiit. (*Gentiana acuta* Michx., *G. amarella* L. subsp. *acuta* (Michx.) Hult., *Gentianella amarella* (L.) Boern. subsp. *acuta* (Michx.) Gillett) - 4, 5
- subsp. *plebeja* (Cham. ex Bunge) Holub = G. plebeja
amarella (L.) Boern. (*Gentiana amarella* L., *G. amarella* subsp. *axillaris* (F.W. Schmidt) Murb., *G. axillaris* (F.W. Schmidt) Reichenb., *Gentianella amarella* (L.) H. Smith, comb. superfl., *G. axillaris* (F.W. Schmidt) A. & D. Löve, *G. axillaris* (F.W. Schmidt) Jovanovic-Dunjic, comb. superfl.) - 1, 2, 3, 4, 6
- subsp. *acuta* (Michx.) Gillett = G. acuta
- subsp. *lingulata* (Agardh) Holub = G. lingulata
- subsp. *livonica* (Ledeb.) Dostal = G. livonica
- subsp. *uliginosa* (Willd.) Tzvel. = G. uliginosa
arctophila (Griseb.) A. Khokhr. (*Gentiana arctophila* Griseb., *G. propinqua* Richards. subsp. *arctophila* (Griseb.) Hult., *Gentianella propinqua* (Richards.) Gillett subsp. *arctophila* (Griseb.) Tzvel.) - 5
atrata (Bunge) Holub (*Gentiana atrata* Bunge) - 3
aurea (L.) H. Smith (*Gentiana aurea* L., *Aloitis aurea* (L.) A. & D. Löve, *Arctogenia aurea* (L.) A. Löve) - 1
auriculata (Pall.) Gillett (*Gentiana auriculata* Pall.) - 4, 5
axillaris (F.W. Schmidt) A. & D. Löve = G. amarella
axillaris (F.W. Schmidt) Jovanovic-Dunjic = G. amarella
azurea (Bunge) Holub (*Gentiana azurea* Bunge, *Comastoma azureum* (Bunge) V. Zuev) - 3, 4, 6
baltica (Murb.) Boern. (*Gentiana baltica* Murb., *Gentianella baltica* (Murb.) H. Smith, comb. superfl., *G. campestris* (L.) Boern. subsp. *baltica* (Murb.) A. & D. Löve, *G. campestris* subsp. *baltica* (Murb.) Tutin, comb. superfl.) - 1
barbata (Froel.) Bercht. & J. Presl = Gentianopsis barbata
biebersteinii (Bunge) Holub (*Gentiana biebersteinii* Bunge) - 2
campestris (L.) Boern. (*Gentiana campestris* L., *G. campestris* var. *germanica* Froel., *Gentianella campestris* (L.) H. Smith, comb. superfl., *G. campestris* subsp. *germanica* (Froel.) A. & D. Löve) - 1
- subsp. *baltica* (Murb.) A. & D. Löve = G. baltica
- subsp. *baltica* (Murb.) Tutin = G. baltica
- subsp. *germanica* (Froel.) A. & D. Löve = G. campestris
- subsp. **suecica** (Froel.) Tzvel. (*Gentiana campestris* var. *suecica* Froel., *Gentianella campestris* subsp. *campes-*

tris var. *suecica* (Froel.) Dostal) - 1
carpatica Boern. (*Gentiana carpatica* Wettst. 1892, non Kit. 1814, *G. carpaticola* Borb., *G. germanica* subsp. *carpatica* Hayek, *Gentianella lutescens* (Velen.) Holub subsp. *carpatica* Holub) - 1
caucasea (Lodd. ex Sims) Holub (*Gentiana caucasea* Lodd. ex Sims, *G. caucasica* Bieb., *G. marcowiczii* Kusn., *Gentianella caucasica* (Bieb.) Czer.) - 2
caucasica (Bieb.) Czer. = G. caucasea
ciliata (L.) Borkh. = Gentianopsis ciliata
- subsp. *blepharophora* (Bordz.) Pritchard = Gentianopsis blepharophora
- subsp. *doluchanovii* (Grossh.) Pritchard = Gentianopsis doluchanovii
detonsa (Rottb.) G. Don fil. = Gentianopsis detonsa
dichotoma (Pall.) H.Smith = Comastoma tenellum
falcata (Turcz.) H. Smith = Comastoma falcatum
junussovii (Pachom. & Taidshan.) Czer. comb. nova (*Gentiana junussovii* Pachom. & Taidshan. 1986, Opred. Rast. Sr. Asii, 8 : 169, 47) - 6
lingulata (Agardh) Pritchard (*Gentiana lingulata* Agardh, *G. axillaris* (F.W. Schmidt) Reichenb. var. *lingulata* (Agardh) Topa, *Gentianella amarella* (L.) Boern. subsp. *lingulata* (Agardh) Holub) - 1, 2, 3, 4, 6
lipskyi (Kusn.) Holub (*Gentiana lipskyi* Kusn.) - 2
livonica (Eschsch.) Soo (*Gentiana livonica* Eschsch., *G. axillaris* (F.W. Schmidt) Reichenb. var. *livonica* (Eschsch.) Topa, *Gentianella amarella* (L.) Boern. subsp. *livonica* (Ledeb.) Dostal) - 1
lutescens (Velen.) Holub (*Gentiana lutescens* Velen., *Gentiana praecox* auct.) - 1
- subsp. *carpatica* Holub = G. carpatica
- var. **praeflorens** (Wettst.) Tzvel. (*Gentiana praeflorens* Wettst.) - 1
malyschevii V. Zuev = Comastoma malyschevii
pamirica (Grossh.) Holub = G. sibirica
plebeja (Cham. ex Bunge) Czer. (*Gentiana plebeja* Cham. ex Bunge, *G. plebeja* Cham. & Schlecht., *G. acuta* Michx. var. *plebeja* (Cham. & Schlecht.) Worosch., *G. amarella* subsp. *acuta* (Michx.) Hult. var. *plebeja* (Cham. & Schlecht.) Hult., *Gentianella acuta* (Michx.) Hiit. subsp. *plebeja* (Cham. ex Bunge) Holub) - 4, 5
poretzkyi Tzvel. - 2
promethea (Juz.) Holub (*Gentiana promethea* Juz.) - 2
propinqua (Richards.) Gillett (*Gentiana propinqua* Richards., *Aloitis propinqua* (Richards.) A. & D. Löve, *Arctogenia propinqua* (Richards.) A. & D. Löve) - 5
- subsp. *arctophila* (Griseb.) Tzvel. = G. arctophila
pulmonaria (Turcz. ex Ledeb.) V. Zuev = Comastoma pulmonarium
pygmaea (Regel & Schmalh.) H. Smith (*Gentiana pygmaea* Regel & Schmalh., *Gentianella pygmaea* (Regel & Schmalh.) Ikonn. comb. superfl.) - 6
saposhnikovii (Pachom.) Czer. comb. nova (*Gentiana saposhnikovii* Pachom. 1986, Opred. Rast. Sr. Asii, 8 : 170, 49) - 6
sibirica (Kusn.) Holub (*Gentiana pamirica* Grossh., *G. sibirica* (Kusn.) Grossh., *G. umbellata* Bieb. var. *sibirica* (Kusn.) Serg., *Gentianella pamirica* (Grossh.) Holub) - 3, 6
sugawarae (Hara) Czer. (*Gentiana sugawarae* Hara, *G. paludicola* Koidz. ex Sugaw. 1937, non Gilg, 1906) - 5
tenella (Rottb.) Boern. = Comastoma tenellum
transalaica (Pachom. & Tajdshan.) Czer. comb. nova (*Gentiana transalaica* Pachom. & Tajdshan. 1986, Opred. Rast. Sr. Asii, 8 : 171, 46) - 6
turkestanorum (Gand.) Holub (*Gentiana turkestanorum*

Gand., *Gentiana stoliczkae* auct.) - 3, 6

uliginosa (Willd.) Boern. (*Gentiana uliginosa* Willd., *Gentianella amarella* (L.) Boern. subsp. *uliginosa* (Willd.) Tzvel., *G. uliginosa* (Willd.) H. Smith, comb. superfl.) - 1

umbellata (Bieb.) Holub (*Gentiana umbellata* Bieb.) - 2

vvedenskyi (Grossh.) H. Smith = Gentianopsis vvedenskyi

Gentianodes A. & D. Löve = Gentiana

algida (Pall.) A. & D. Löve = Gentiana algida
frigida (Haenke) A. & D. Löve = Gentiana frigida
glauca (Pall.) A. & D. Löve = Gentiana glauca

Gentianopsis Ma

barbata (Froel.) Ma (*Gentiana barbata* Froel., *Gentianella barbata* (Froel.) Bercht. & J. Presl) - 1, 3, 4, 5, 6

blepharophora (Bordz.) Galushko (*Gentiana blepharophora* Bordz., *G. ciliata* L. subsp. *blepharophora* (Bordz.) Greuter, *Gentianella ciliata* (L.) Borkh. subsp. *blepharophora* (Bordz.) Pritchard, *Gentianopsis ciliata* (L.) Ma subsp. *blepharophora* (Bordz.) Holub) - 2

ciliata (L.) Ma (*Gentiana ciliata* L., *Gentianella ciliata* (L.) Borkh.) - 1

- subsp. *blepharophora* (Bordz.) Holub = G. blepharophora

detonsa (Rottb.) Ma (*Gentiana detonsa* Rottb., *Gentianella detonsa* (Rottb.) G. Don fil.) - 1

doluchanovii (Grossh.) Tzvel. (*Gentiana doluchanovii* Grossh., *Gentianella ciliata* (L.) Borkh. subsp. *doluchanovii* (Grossh.) Pritchard) - 1, 3, 4

komarovii (Grossh.) Czer. (*Gentiana komarovii* Grossh.) - 5

stricta (Klotzsch) Ikonn. (*Gentiana stricta* Klotzsch, *Gentianopsis stricta* (Klotzsch) Holub, comb. superfl.) - 6

vvedenskyi (Grossh.) Pissjauk. (*Gentiana vvedenskyi* Grossh., *Gentianella vvedenskyi* (Grossh.) H. Smith) - 6

Halenia Borkh.

corniculata (L.) Cornaz - 1, 3, 4, 5
elliptica D. Don - 6

Lomatogonium A. Br.

carinthiacum (Wulf.) Reichenb. - 2, 3, 4, 5, 6
rotatum (L.) Fries - 1, 3, 4, 5, 6
tenellum (Rottb.) A. & D. Löve = Comastoma tenellum
thomsonii (Clarke) Fern. = Pleurogynella thomsonii

Ophelia D. Don (*Frasera* auct.)

chinensis Bunge ex Griseb. = O. diluta

diluta (Turcz.) Ledeb. (*Frasera diluta* (Turcz.) Toyokuni, *Ophelia chinensis* Bunge ex Griseb., *Swertia chinensis* (Bunge ex Griseb.) Franch., *S. diluta* (Turcz.) Benth. & Hook. fil.) - 3, 4, 5

pseudochinensis (Hara) Czer. comb. nova (*Swertia pseudochinensis* Hara, 1950, Journ. Jap. Bot. 25 : 89) - 5

tetrapetala (Pall.) Grossh. (*Frasera tetrapetala* (Pall.) Toyokuni) - 5

tosaensis (Makino) Czer. comb. nova (*Swertia tosaensis* Makino, 1903, Bot. Mag. Tokyo, 17 : 54; *Ophelia tscherskyi* (Kom.) Grossh.) - 5

tscherskyi (Kom.) Grossh. = O. tosaensis

wilfordii A. Kerner - 5

Pleurogyne auct. = Pleurogynella

brachyanthera Clarke = Pleurogynella brachyanthera
thomsonii Clarke = Pleurogynella thomsonii

Pleurogynella Ikonn. (*Pleurogyne* auct.)

brachyanthera (Clarke) Ikonn. (*Pleurogyne brachyanthera* Clarke) - 6

thomsonii (Clarke) Ikonn. (*Pleurogyne thomsonii* Clarke, *Lomatogonium thomsonii* (Clarke) Fern.) - 6

Pneumonanthe Gled. = Gentiana

algida (Pall.) F.W. Schmidt = Gentiana algida
cruciata (L.) V. Zuev = Gentiana cruciata
dahurica (Fisch.) V. Zuev = Gentiana dahurica
fetisowii (Regel & Winkl.) V. Zuev = Gentiana fetisowii
fischeri (P. Smirn.) V. Zuev = Gentiana fischeri
frigida (Haenke) F.W. Schmidt = Gentiana frigida
glauca (Pall.) F.W. Schmidt = Gentiana glauca
macrophylla (Pall.) V. Zuev = Gentiana macrophylla
triflora (Pall.) F.W. Schmidt = Gentiana triflora
vulgaris F.W. Schmidt = Gentiana pneumonanthe

Swertia L.

alpestris Baumg. (*S. perennis* L. subsp. *alpestris* (Baumg.) Domin & Podp. comb. superfl., *S. perennis* subsp. *alpestris* (Baumg.) Simonk., *S. perennis* var. *alpestris* (Baumg.) Kozhukharov & A.V. Petrova) - 1

aucheri Boiss. = S. longifolia

baicalensis M. Pop. ex Pissjauk. - 4

chinensis (Bunge ex Griseb.) Franch. = Ophelia diluta

connata Schrenk - 6

diluta (Turcz.) Benth. & Hook. fil. = Ophelia diluta

erosula Gontsch. = S. gonczaroviana

fedtschenkoana Pissjauk. - 6

gonczaroviana Pissjauk. (*S. erosula* Gontsch. 1933, non N.E. Br. 1904) - 6

graciliflora Gontsch. - 6

haussknechtii Gilg ex Pissjauk. (*S. iberica* Fisch. ex C.A. Mey. subsp. *haussknechtii* (Gilg ex Pissjauk.) V. Avet.) - 2

iberica Fisch. & C.A. Mey. - 2

- subsp. *haussknechtii* (Gilg ex Pissjauk.) V. Avet. = S. haussknechtii

juzepczukii Pissjauk. - 6

komarovii Pissjauk. - 4

lactea Bunge - 6

longifolia Boiss. (*S. aucheri* Boiss.) - 2

marginata Schrenk - 4, 6

obtusa Ledeb. (*S. perennis* L. subsp. *obtusa* (Ledeb.) Hara) - 1, 3, 4, 5, 6

- var. *stenopetala* Regel & Til. = S. stenopetala

perennis L. - 1

- subsp. *alpestris* (Baumg.) Domin & Podp. = S. alpestris
- subsp. *alpestris* (Baumg.) Simonk. = S. alpestris
- subsp. *obtusa* (Ledeb.) Hara = S. obtusa
- subsp. *stenopetala* (Regel & Til.) Worosch. = S. stenopetala
- var. *alpestris* (Baumg.) Kozhukharov & A.V. Petrova = S. alpestris

pseudochinensis Hara = Ophelia pseudochinensis

pseudopetiolata Pissjauk. - 6

punctata Baumg. - 1

schugnanica Pissjauk. - 6

stenopetala (Regel & Til.) Pissjauk. (*S. obtusa* Ledeb. var.

stenopetala Regel & Til., *S. perennis* L. subsp. *stenopetala* (Regel & Til.) Worosch.) - 5
tosaensis Makino = Ophelia tosaensis
variabilis Pissjauk. - 6
veratroides Maxim. ex Kom. - 5

Tretorhiza Maxim. ex Kom. = Gentiana

cruciata (L.) Delarb. = Gentiana cruciata
cruciata (L.) Opiz = Gentiana cruciata
dahurica (Fisch.) Sojak = Gentiana dahurica
decumbens (L. fil.) Sojak = Gentiana decumbens
fetisowii (Regel & Winkl.) Sojak = Gentiana fetisowii
kaufmanniana (Regel & Schmalh.) Sojak = Gentiana kaufmanniana
macrophylla (Pall.) Sojak = Gentiana macrophylla
olgae (Regel & Schmalh.) Sojak = Gentiana olgae
olivieri (Griseb.) Sojak = Gentiana olivieri
phlogifolia (Schott & Kotschy) Sojak = Gentiana phlogifolia
tianschanica (Rupr.) Sojak = Gentiana kirilowii
walujewii (Regel & Schmalh.) Sojak = Gentiana walujewii

Varasia Phil. = Gentiana

aquatica (L.) Sojak = Gentiana aquatica
dshimilensis (C. Koch) Sojak = Gentiana dshimilensis
grandiflora (Laxm.) Sojak = Gentiana grandiflora
jamesii (Hemsl.) Sojak = Gentiana jamesii
karelinii (Griseb.) Sojak = Gentiana karelinii
leucomelaena (Maxim.) Sojak = Gentiana leucomelaena
nipponica (Maxim.) Sojak = Gentiana nipponica
nutans (Bunge) Sojak = Gentiana prostrata
prostrata (Haenke) Sojak = Gentiana prostrata
pseudoaquatica (Kusn.) Sojak = Gentiana pseudoaquatica
riparia (Kar. & Kir.) Sojak = Gentiana riparia
squarrosa (Ledeb.) Sojak = Gentiana squarrosa
zollingeri (Fawc.) Sojak = Gentiana zollingeri

GERANIACEAE Juss.
Erodium L'Her.

absinthoides Willd. subsp. *armenum* (Trautv.) P.H. Davis = E. armenum
anthemidifolium Bieb. - 2
armenum (Trautv.) Woronow (*E. absinthoides* Willd. subsp. *armenum* (Trautv.) P.H. Davis) - 2
beketowii Schmalh. - 1
chium (L.) Willd. (*Geranium chium* L.) - 2(?)
ciconium (L.) L'Her. - 1, 2, 6
- var. *turcmenum* Litv. = E. turcmenum
cicutarium (L.) L'Her. - 1, 2, 3, 4, 5, 6
fumarioides Stev. - 2
hoefftianum C.A. Mey. - 1, 2, 6
laciniatum auct. = E. strigosum
litwinowii Woronow - 6
malacoides (L.) L'Her. - 1, 2, 6
oxyrhynchum Bieb. - 2, 6
ruthenicum Bieb. - 1
schemachense Grossh. - 2
sosnowskianum Fed. - 2
stephanianum Willd. - 3, 4, 5, 6
stevenii Bieb. - 2
strigosum Kar. ex Ledeb. (*E. laciniatum* auct.) - 2, 6
tataricum Willd. - 4
tibetanum Edgew. - 6
turcmenum (Litv.) Grossh. (*E. ciconium* (L.) L'Her. var.

turcmenum Litv.) - 2, 6

Geranium L.

affine Ledeb. - 3, 6
albanum Bieb. - 2
albiflorum Ledeb. - 3, 4, 5, 6
alpestre Schur - 1
asphodeloides auct. = G. pallens and G. tauricum
baschkyzylsaicum Nabiev - 6
bifolium Patrin - 3, 4
bohemicum L. - 1, 2
charlesii (Aitch. & Hemsl.) Vved. ex Nevski (*G. kotschyi* Boiss. subsp. *charlesii* (Aitch. & Hemsl.) P.H. Davis) - 6
chium L. = Erodium chium
cinereum Cav. subsp. *subcaulescens* (L'Her. ex DC.) Hayek = G. subcaulescens
collinum Steph. (*G. collinum* var. *wakhanicum* Pauls., *G. minutum* Ikonn., *G. wakhanicum* (Pauls.) Ikonn.) - 1, 2, 3, 6
- var. *wakhanicum* Pauls. = G. collinum
columbinum L. - 1, 2
davuricum DC. - 4, 5
depilatum (Somm. & Levier) Grossh. - 2
dissectum L. - 1, 2, 6
divaricatum Ehrh. - 1, 2, 3, 6
elatum (Maxim.) Knuth (*G. erianthum* DC. var. *elatum* Maxim.) - 5
erianthum DC. - 5
- var. *elatum* Maxim. = G. elatum
eriostemon Fisch. - 4, 5
- var. *orientale* Maxim. = G. orientale
ferganense Bobr. - 6
finitimum Woronow (*G. pratense* L. subsp. *finitimum* (Woronow) Knuth) - 2
gracile Ledeb. - 2
gymnocaulon DC. - 2
himalayense Klotzsch (*G. meeboldii* Briq.) - 6
ibericum Cav. - 2
kemulariae Charadze - 2
kotschyi Boiss. - 6
- subsp. *charlesii* (Aitch. & Hemsl.) P.H. Davis = G. charlesii
krameri auct. = G. sieboldii
linearilobum DC. (*G. tuberosum* L. subsp. *linearilobum* (DC.) Malagarriga) - 1, 2
- subsp. *transversale* (Kar. & Kir.) P.H. Davis = G. transversale
lucidum L. - 1, 2, 6
***macrorrhizum** L. - 1
maximowiczii Regel & Maack - 5
meeboldii Briq. = G. himalayense
minutum Ikonn. = G. collinum
molle L. - 1, 2
montanum Habl. ex Pall. - 2
nepalense auct. = G. thunbergii
orientale (Maxim.) Freyn (*G. eriostemon* Fisch. var. *orientale* Maxim.) - 5
pallens Bieb. (*G. asphodeloides* auct. p.p.) - 2
palustre L. - 1, 2
- subsp. *transbaicalicum* (Serg.) Gubanov = G. transbaicalicum
pamiricum Ikonn. - 6
persicum Schonbeck-Temesy - 6(?)
phaeum L. - 1
platypetalum Fisch. & C.A. Mey. - 2

pratense L. - 1, 2, 3, 4, 5, 6
- subsp. *finitimum* (Woronow) Knuth = G. finitimum
pseudosibiricum J. Mayer (*G. sylvaticum* L. subsp. *pseudosibiricum* (J. Mayer) D.A. Webb & Ferguson) - 1, 3, 4, 6
psilostemon Ledeb. - 2
purpureum Vill. - 2
pusillum L. - 1, 2, 6
pyrenaicum Burm. fil. - 1
rectum Trautv. - 6
regelii Nevski - 6
renardii Trautv. - 2
robertianum L. - 1, 2, 3, 6
rotundifolium L. - 1, 2, 5(alien), 6
ruprechtii (Woronow) Grossh. - 2
sanguineum L. - 1, 2
saxatile Kar. & Kir. - 6
schrenkianum Trautv. ex Pavl. - 1, 3, 6
sibiricum L. - 1, 2, 3, 4, 5, 6
sieboldii Maxim. (*G. krameri* auct.) - 5
soboliferum Kom. - 5
sophiae Fed. - 6
stepporum P.H. Davis (*G. tuberosum* L. subsp. *linearifolium* (Boiss.) P.H. Davis, *G. tuberosum* var. *linearifolium* Boiss.) - 2, 6
subcaulescens L'Her. ex DC. (*G. cinereum* Cav. subsp. *subcaulescens* (L'Her. ex DC.) Hayek) - 2
subumbelliforme Knuth - 5
sylvaticum L. - 1, 2, 3, 4, 6
- subsp. *pseudosibiricum* (J. Mayer) D.A. Webb & Ferguson = G. pseudosibiricum
tauricum Rupr. (*G. asphodeloides* auct. p.p.) - 1
thunbergii Siebold & Zucc. ex Lindl. & Paxt. (*G. nepalense* auct.) - 5
transbaicalicum Serg. (*G. palustre* L. subsp. *transbaicalicum* (Serg.) Gubanov) - 4
transversale (Kar. & Kir.) Vved. (*G. linearilobum* DC. subsp. *transversale* (Kar. & Kir.) P.H. Davis) - 3, 6
tuberosum L. - 1, 2
- subsp. *linearifolium* (Boiss.) P.H. Davis = G. stepporum
- subsp. *linearilobum* (DC.) Malagarriga = G. linearilobum
- var. *linearifolium* Boiss. = G. stepporum
uralense Kuvajev - 1, 3
wakhanicum (Pauls.) Ikonn. = G. collinum
wilfordii Maxim. - 5
wlassowianum Fisch. ex Link - 4, 5
yesoense Franch. & Savat. - 5

Pelargonium L'Her.

endlicherianum Fenzl - 2

GLOBULARIACEAE DC.
Globularia L.

aphyllanthes Crantz = G. punctata
elongata Hegetschw. = G. punctata
punctata Lapeyr. (*G. elongata* Hegetschw., *G. aphyllanthes* Crantz, p.p. nom. illegit.) - 1, 2
trichosantha Fisch. & C.A. Mey. - 1, 2
vulgaris L. - 1(?)

GRAMINEAE Juss. = POACEAE

GROSSULARIACEAE DC.
Grossularia Hill

acicularis (Smith) Spach - 3, 4, 6
burejensis (Fr. Schmidt) Berger - 5
reclinata (L.) Mill. (*G. uva-crispa* (L.) Mill. subsp. *reclinata* (L.) Dostal, *Ribes uva-crispa* L. subsp. *reclinatum* (L.) Reichenb., *R. uva-crispa* subsp. *reclinatum* (L.) O. Schwarz, comb. superfl.) - 1, 2, 5
*****uva-crispa** (L.) Mill. - 1, 2
    Occasionally can be found in the wild as a naturalized species.
- subsp. *reclinata* (L.) Dostal = G. reclinata

Ribes L.

achurjani Mulk. - 2
acidum Turcz. ex Pojark. = R. glabellum
alpinum L. - 1, 2
- subsp. *lucidum* (Kit.) Pawl. = R. lucidum
- subsp. *lucidum* (Kit.) Sojak = R. lucidum
altissimum Turcz. ex Pojark. - 3, 4
armenum Pojark. - 2
atropurpureum C.A. Mey. - 3, 4, 6
*****aureum** Pursh
biebersteinii Berl. ex DC. - 2
carpaticum Schult. (*R. petraeum* auct. p.p.) - 1
diacantha Pall. - 4, 5
dikuscha Fisch. ex Turcz. - 4, 5
fontaneum Boczkarnikova (*R. japonicum* auct.) - 5
fragrans Pall. - 4, 5
glabellum (Trautv. & C.A. Mey.) Hedl. (*R. rubrum* L. var. *glabellum* Trautv. & C.A. Mey., *R. acidum* Turcz. ex Pojark., *R. rubrum* sensu Pojark.) - 1, 3, 4, 5
graveolens Bunge - 3, 4
heterotrichum C.A. Mey. - 3, 6
hispidulum (Jancz.) Pojark. (*R. spicatum* Robson subsp. *hispidulum* (Jancz.) Hamet-Ahti) - 1, 3, 4, 6
horridum Rupr. - 5
janczewskii Pojark. - 6
japonicum auct. = R. fontaneum
kolymense Turcz. ex Pojark. = R. pauciflorum
komarovii Pojark. - 5
latifolium Jancz. - 5
lucidum Kit. (*R. alpinum* L. subsp. *lucidum* (Kit.) Pawl., *R. alpinum* subsp. *lucidum* (Kit.) Sojak, comb. superfl.) - 1, 2
malvifolium Pojark. - 6
mandshuricum (Maxim.) Kom. - 5
maximoviczianum Kom. - 5
melananthum Boiss. & Hohen. - 6
meyeri Maxim. - 3, 6
- subsp. **dshabari** Ovcz. - 6
nigrum L. - 1, 2, 3, 4, 6
*****odoratum** Wendl. - 1
orientale Desf. - 2
palczewskii (Jancz.) Pojark. (*R. spicatum* Robson subsp. *palczewskii* (Jancz.) Malysch.) - 4, 5
pallidiflorum Pojark. - 5
pauciflorum Turcz. ex Pojark. (*R. kolymense* Turcz. ex Pojark.) - 4, 5
petraeum auct. p.p. = R. carpaticum
procumbens Pall. - 3, 4, 5
pubescens (C. Hartm.) Hedl. = R. spicatum
pulchellum Turcz. - 4
*****rubrum** L. (*R. sativum* Syme, *R. sylvestre* Mert. & Koch, *R. vulgare* Lam. nom. illegit.)
- var. *glabellum* Trautv. & C.A. Mey. = R. glabellum
- var. *pubescens* Sw. ex C. Hartm. = R. spicatum

rubrum sensu Pojark. = R. glabellum
sachalinense (Fr. Schmidt) Nakai - 5
*sanguineum Pursh
sativum Syme = R. rubrum
saxatile Pall. - 3, 6
scandicum Hedl. (*R. spicatum* Robson subsp. *scandicum*
(Hedl.) Hyl.) - 1
spicatum Robson (*R. rubrum* L. var. *pubescens* Sw. ex C.
Hartm., *R. pubescens* (C. Hartm.) Hedl., *R. spicatum*
subsp. *pubescens* (C. Hartm.) Hyl.) - 1
- subsp. *hispidulum* (Jancz.) Hamet-Ahti = R. hispidulum
- subsp. *palczewskii* (Jancz.) Malysch. = R. palczewskii
- subsp. *pubescens* (C. Hartm.) Hyl. = R. spicatum
- subsp. *scandicum* (Hedl.) Hyl. = R. scandicum
sylvestre Mert. & Koch = R. rubrum
triste Pall. - 4, 5
turbinatum Pojark. - 3, 6
ussuriense Jancz. - 5
uva-crispa L. subsp. *reclinatum* (L.) Reichenb. = Grossularia reclinata
- subsp. *reclinatum* (L.) O. Schwarz = Grossularia reclinata
villosum Wall. ex Roxb. - 6
vulgare Lam. = R. rubrum

GUTTIFERAE auct. = HYPERICACEAE

HALORAGACEAE R. Br.
Myriophyllum L.

alterniflorum DC. - 1, 5
*aquaticum (Velloso) Verdc. (*Enydria aquatica* Velloso) - 2
isoetophyllum Kom. - 5
sibiricum Kom. - 1, 3, 4, 5
spicatum L. - 1, 2, 3, 4, 5, 6
ussuriense (Regel) Maxim. - 5
verticillatum L. - 1, 2, 3, 4, 5, 6

HAMAMELIDACEAE R. Br.
Parrotia C.A. Mey.

persica (DC.) C.A. Mey. - 2

HEMEROCALLIDACEAE R. Br.
Hemerocallis L.

citrina auct. = H. yezoensis
coreana Nakai (*H. dumortieri* auct. p.p.) - 5
dumortieri auct. p.p. = H. coreana and H. middendorfii
dumortieri Morr. subsp. *middendorfii* (Trautv. & C.A. Mey.)
Worosch. = H. middendorfii
- var. *esculenta* (Koidz.) Kitam. = H. esculenta
- var. *middendorfii* (Trautv. & C.A. Mey.) Kitam. = H.
middendorfii
esculenta Koidz. (*H. dumortieri* Morr. var. *esculenta*
(Koidz.) Kitam., *H. middendorfii* Trautv. & C.A. Mey.
var. *esculenta* (Koidz.) Ohwi, *H. pedicellata* Nakai) - 5
flava (L.) L. = H. lilio-asphodelus
- var. *minor* (Mill.) M. Hotta = H. minor
fulva (L.) L. (*H. lilio-asphodelus* L. var. *fulva* L.) - 1(cult.), 2
lilio-asphodelus L. (*H. flava* (L.) L., *H. lilio-asphodelus* var.
flava L., *H. vespertina* auct.) - 1(cult.), 3, 4, 5
- var. *flava* L. = H. lilio-asphodelus
- var. *fulva* L. = H. fulva
middendorfii Trautv. & C.A. Mey. (*H. dumortieri* Morr.
var. *middendorfii* (Trautv. & C.A. Mey.) Kitam., *H.*

dumortieri subsp. *middendorfii* (Trautv. & C.A. Mey.)
Worosch.) - 1(cult.), 5
- var. *esculenta* (Koidz.) Ohwi = H. esculenta
minor Mill. (*H. flava* (L.) L. var. *minor* (Mill.) M. Hotta) -
1(cult.), 3, 4, 5
pedicellata Nakai = H. esculenta
thunbergii auct. = H. yezoensis
vespertina auct. = H. lilio-asphodelus
yezoensis Hara (*H. citrina* auct., *H. thunbergii* auct.) - 5

HEMIONITIDACEAE Pichi Sermolli
Anogramma Link

leptophylla (L.) Link - 1, 2, 6

Coniogramme Fee

fraxinea auct. = C. intermedia
intermedia Hieron. (*C. fraxinea* auct.) - 5

*HIPPOCASTANACEAE DC.
*Aesculus L.

*hippocastanum L. - 1, 2, 6

*HIPPURIDACEAE Link
Hippuris L.

x **lanceolata** Retz. - H. tetraphylla L. x H. vulgaris L. - 1, 3,
4, 5
melanocarpa N. Semen. - 1, 5
montana Ledeb. - 5
tetraphylla L. - 1, 3, 4, 5
vulgaris L. - 1, 2, 3, 4, 5, 6

HOSTACEAE Mathew (*FUNKIACEAE* Horan.
nom. illegit., *AGAVACEAE* auct.)

Aletris auct. = Hosta

japonica Thunb. p.p. = Hosta albomarginata

Bryocles Salisb. = Hosta

ventricosa Salisb. = Hosta ventricosa

Funkia Spreng. = Hosta

albomarginata Hook. = Hosta albomarginata
lancifolia (Thunb.) Spreng. = Hosta albomarginata
ovata sensu Czerniak. = Hosta rectifolia
ovata Spreng. = Hosta ventricosa

Hemerocallis auct. = Hosta

caerulea Andr. = Hosta ventricosa
japonica (Thunb.) Thunb. p.p. = Hosta albomarginata
lancifolia Thunb. = Hosta albomarginata
plantaginea Lam. = Hosta plantaginea

Hosta Tratt. (*Bryocles* Salisb., *Funkia* Spreng.,
Aletris auct., *Hemerocallis* auct.)

albomarginata (Hook.) Ohwi (*Funkia albomarginata*
Hook., *Aletris japonica* Thunb. p.p., *Funkia lancifolia*
(Thunb.) Spreng. nom. illegit., *Hemerocallis japonica*
(Thunb.) Thunb. p.p., *H. lancifolia* Thunb. nom.
illegit., *Hosta japonica* (Thunb.) Aschers. 1863, non

Tratt. 1812, *H. lancifolia* (Thunb.) Engl. nom. illegit., *H. clausa* auct.) - 5

caerulea (Andr.) Tratt. = H. ventricosa

clausa auct. = H. albomarginata

japonica (Thunb.) Aschers. = H. albomarginata

japonica Tratt. = H. plantaginea

lancifolia (Thunb.) Engl. = H. albomarginata

***plantaginea** (Lam.) Aschers. (*Hemerocallis plantaginea* Lam., *Hosta japonica* Tratt.)

rectifolia Nakai (*H. rectifolia* var. *sachalinensis* (Koidz.) F. Maek., *H. sachalinensis* Koidz., *Funkia ovata* sensu Czerniak.) - 5

- var. *sachalinensis* (Koidz.) F. Maek. = H. rectifolia

sachalinensis Koidz. = H. rectifolia

***ventricosa** Stearn (*Bryocles ventricosa* Salisb. nom. nud., *Funkia ovata* Spreng. nom. illegit., *Hemerocallis caerulea* Andr., *Hosta caerulea* (Andr.) Tratt. 1814, non Jacq. 1797)

HUPERZIACEAE Rothm. (*UROSTACHYACEAE* Rothm.)

Huperzia Bernh. (*Mirmau* Adans. p.p. nom. illegit., *Urostacys* Herter)

appressa auct. p.p. = H. arctica and H. petrovii

arctica (Tolm.) Sipl. (*Lycopodium selago* L. subsp. *arcticum* Tolm., *Huperzia selago* (L.) Bernh. ex Schrank & C. Mart. subsp. *arctica* (Tolm.) A. & D. Löve, *Lycopodium appressum* (Desv.) V. Petrov, p.p. excl. basionymo & pl. non arct., non Lloyd & Underw., *L. arcticum* Grossh. nom. invalid., *L. selago* L. subsp. *appressum* (Desv.) Hult. p.p., *Huperzia appressa* auct. p.p.) - 4, 5

chinensis (Christ) Czer. (*Lycopodium chinense* Christ, *Huperzia selago* (L.) Bernh. ex Schrank & C. Mart. subsp. *chinensis* (Christ) A. & D. Löve, *H. selago* var. *miyoshiana* (Makino) Taylor & MacBryde, *Lycopodium miyoshianum* Makino, *L. selago* subsp. *miyoshianum* (Makino) Calder & Taylor) - 5

laxa (Desv.) A. Khokhr. (*Lycopodium selago* L. subsp. *laxum* Desv.) - 4, 5

petrovii Sipl. (*Lycopodium appressum* (Desv.) V. Petrov, p.p. excl. basionymo & pl. arct., non Lloyd & Underw., *L. selago* L. subsp. *appressum* (Desv.) Hult. p.p., *Huperzia appressa* auct. p.p.) - 1, 3, 4, 5

selago (L.) Bernh. ex Schrank & C. Mart. (*Lycopodium selago* L., *Mirmau selago* (L.) H.P. Fuchs, *Urostachys selago* (L.) Herter, nom. illegit.) - 1, 2, 3, 4, 5, 6

- subsp. *arctica* (Tolm.) A. & D. Löve = Huperzia arctica

- subsp. *chinensis* (Christ) A. & D. Löve = Huperzia chinensis

- subsp. *serrata* (Thunb.) A. & D. Löve = Huperzia serrata

- var. *miyoshiana* (Makino) Taylor & MacBryde = Huperzia chinensis

serrata (Thunb.) Rothm. (*Lycopodium serratum* Thunb., *Huperzia selago* (L.) Bernh. ex Schrank & C. Mart. subsp. *serrata* (Thunb.) A. & D. Löve) - 5

Mirmau Adans. p.p. = Huperzia

selago (L.) H.P. Fuchs = Huperzia selago

Urostachys Herter = Huperzia

selago (L.) Herter = Huperzia selago

HYACINTHACEAE Batsch

Barnardia Lindl. = Scilla

scilloides Lindl. = Scilla scilloides

Bellevalia Lapeyr.

acutifolia (Boiss.) Deloney = Pseudomuscari paradoxum

albana Woronow - 2

aperta (Freyn & Conrath) Grossh. = Pseudomuscari apertum

araxina Woronow = B. longistyla

atroviolacea Regel (*B. turkestanica* Franch.) - 6

aucheri (Baker) Losinsk. p.p. = B. fominii

coelestis (Fomin) Grossh. = Pseudomuscari coeleste

coerulea (Losinsk.) Schchian = Pseudomuscari coeruleum

fominii Woronow (*B. aucheri* (Baker) Losinsk. p.p. quoad plantas) - 2

forniculata (Fomin) Deloney = Pseudomuscari forniculatum

glauca auct. = B. wilhelmsii

inconspicua Vved. - 6

lipskyi (Miscz.) E. Wulf - 1

longistyla (Miscz.) Grossh. (*B. araxina* Woronow) - 2

lutea Bordz. - 2

macrobotrys auct. p.p. = B. zygomorpha

makuensis Woronow ex Grossh. - 2

montana (C. Koch) Boiss. - 2

paradoxa (Fisch. & C.A. Mey.) Boiss. = Pseudomuscari paradoxum

pycnantha (C. Koch) Losinsk. - 2

sarmatica (Georgi) Woronow (*Hyacinthus sarmaticus* Pall. ex Georgi) - 1, 2

saviczii Woronow - 6

speciosa Woronow - 2

turkestanica Franch. = B. atroviolacea

turkewiczii (Woronow) Grossh. = Pseudomuscari turkewiczii

wilhelmsii (Stev.) Woronow (*B. glauca* auct.) - 2

zygomorpha Woronow (*B. macrobotrys* auct. p.p.) - 2

Camassia Lindl. (*Phalangium* auct.)

quamash (Pursh) Greene (*Phalangium quamash* Pursh) - 2
Occurs in the wild as a naturalized species.

Dipcadi Medik.

turkestanicum Vved. - 6

Honorius S.F. Gray = Ornithogalum

boucheanus (Kunth) Holub = Ornithogalum boucheanum

nutans (L.) S.F. Gray = Ornithogalum nutans

Hyacinthella Schur

atropatana (Grossh.) H. Mordak & Zakharyeva (*Scilla atropatana* Grossh.) - 2

leucophaea (C. Koch) Schur - 1, 2(?)

litwinowii (Czerniak.) M. Baranova (*Hyacinthus litwinowii* Czerniak.) - 6

pallasiana (Stev.) Losinsk. - 1

transcaspica (Litv.) Chouard (*Hyacinthus transcaspicus* Litv., *H. kopetdaghi* Czerniak., *H. transcaspicus* subsp. *kopetdaghi* (Czerniak.) Beljanina & Proskurjakova, comb. invalid.) - 6

*Hyacinthus L.

kopetdaghi Czerniak. = Hyacinthella transcaspica
litwinowii Czerniak. = Hyacinthella litwinowii
***orientalis** L. - 1, 2
pallens Bieb. = Pseudomuscari pallens
sarmaticus Pall. ex Georgi = Bellevalia sarmatica
transcaspicus Litv. = Hyacinthella transcaspica
- subsp. *kopetdaghi* (Czerniak.) Beljanina & Proskurjakova = Hyacinthella transcaspica

Leopoldia Parl.

caucasica (Griseb.) Losinsk. (*Muscari caucasicum* (Griseb.) Baker) - 1, 2
comosa (L.) Parl. (*L. tubiflora* (Stev.) Juz., *Muscari comosum* (L.) Mill.) - 1
longipes (Boiss.) Losinsk. (*Muscari longipes* Boiss.) - 2
tenuiflora (Tausch) Heldr. (*Muscari tenuiflorum* Tausch) - 1, 2
tubiflora (Stev.) Juz. = L. comosa

Moscharia Salisb. = Muscari

Muscari Hill (*Moscharia* Salisb. 1866, non Forssk. 1775, nec Ruiz & Pav. 1794, nom. cons., *Muscarimia* Kostel. ex Losinsk.)

alexandrae A. Khokhr. - 2
alpanicum Schchian - 2
apertum Freyn & Conrath = Pseudomuscari apertum
armeniacum Leichtl. ex Baker (*M. colchicum* Grossh.) - 2
botryoides (L.) Mill. (*M. carpaticum* Racib., *M. pocuticum* Zapal.) - 1
- var. *bucharicum* Regel = M. leucostomum
botryoides sensu Losinsk. = Pseudomuscari coeruleum
bucharicum Regel = M. leucostomum
carpaticum Racib. = M. botryoides
caucasicum (Griseb.) Baker = Leopoldia caucasica
coeleste Fomin = Pseudomuscari coeleste
coeruleum Losinsk. = Pseudomuscari coeruleum
colchicum Grossh. = M. armeniacum
commutatum auct. = M. inconstrictum
comosum (L.) Mill. = Leopoldia comosa
dolichanthum Woronow & Tron (*M. steupii* Woronow & Tron) - 2
dolioliforme Sobko = M. neglectum
elegantulum Schchian - 2
forniculatum Fomin = Pseudomuscari forniculatum
grossheimii Schchian - 2
inconstrictum Rech. fil. (*M. commutatum* auct.) - 2
leucostomum Woronow ex Czerniak. (*M. botryoides* (L.) Mill. var. *bucharicum* Regel, *M. bucharicum* Regel, ? nom. provis., *M. woronowii* Losinsk. & Tron) - 1, 2, 6
longipes Boiss. = Leopoldia longipes
moschatum Willd. = M. muscarimi
muscarimi Medik. (*M. moschatum* Willd., *M. racemosum* Mill. nom. confus., *Muscarimia muscari* (L.) Losinsk.) - 2
neglectum Guss. (*M. dolioliforme* Sobko, *M. racemosum* (L.) DC. 1805, non Mill. 1768) - 1, 2, 5
neglectum sensu Losinsk. = M. szovitsianum
pallens (Bieb.) Fisch. = Pseudomuscari pallens
paradoxum (Fisch. & C.A. Mey.) C. Koch = Pseudomuscari paradoxum
pendulum Trautv. - 2
pocuticum Zapal. = M. botryoides

polyanthum Boiss. - 2
racemosum (L.) DC. = M. neglectum
racemosum Mill. = M. muscarimi
sosnowskyi Schchian - 2
steupii Woronow & Tron = M. dolichanthum
stoloniferum Charkev. = Pseudomuscari pallens
szovitsianum Baker (*M. neglectum* sensu Losinsk.) - 2
tenuiflorum Tausch = Leopoldia tenuiflora
turkewiczii (Woronow) Losinsk. = Pseudomuscari turkewiczii
woronowii Losinsk. & Tron = M. leucostomum

Muscarimia Kostel. ex Losinsk. = Muscari

muscari (L.) Losinsk. = M. muscarimi

Ornithogalum L. (*Honorius* S.F. Gray)

amblyocarpum Zahar. - 2
amphibolum Zahar. - 1
arcuatum Stev. - 2
arianum Lipsky ex Vved. - 6
balansae Boiss. (*O. oligophyllum* auct. p.p.) - 2
boucheanum (Kunth) Aschers. (*Honorius boucheanus* (Kunth) Holub, *Ornithogalum nutans* L. subsp. *boucheanum* (Kunth) Hayek, comb. superfl., *O. nutans* subsp. *boucheanum* (Kunth) K. Richt.) - 1
brachystachys C. Koch (*O. narbonense* L. subsp. *brachystachys* (C. Koch) Feinbrun) - 2
cuspidatum Bertol. (*O. platyphyllum* Boiss., *O. tempskyanum* Freyn & Sint., *O. montanum* auct.) - 2
fimbriatum Willd. - 1
fischerianum Krasch. - 1, 3, 6
flavescens Lam. (*O. melancholicum* Klok. ex A. Krasnova, *O. pyrenaicum* L. subsp. *flavescens* (Lam.) K. Richt.) - 1
georgicum Agapova = O. ponticum
graciliflorum C. Koch - 2
gussonei auct. = O. kochii
gussonei Ten. subsp. *kochii* (Parl.) Holub = O. kochii
hajastanum Agapova (*O. obconicum* (Zahar.) Agapova, nom. superfl., *O. ponticum* Zahar. subsp. *obconicum* Zahar.) - 2
hyrcanum Grossh. - 2
imereticum Sosn. (*O. woronowii* Krasch. var. *imereticum* (Sosn.) Grossh.) - 2
kochii Parl. (*O. gussonei* Ten. subsp. *kochii* (Parl.) Holub, *O. orthophyllum* Ten. subsp. *kochii* (Parl.) Maire & Weiller, *O. orthophyllum* subsp. *kochii* (Parl.) Zahar. comb. superfl., *O. gussonei* auct., *O. tenuifolium* auct.) - 1, 2
magnum Krasch. & Schischk. - 2
melancholicum Klok. ex A. Krasnova = O. flavescens
montanum auct. = O. cuspidatum
nanum auct. p.p. = O. sigmoideum
narbonense L. subsp. *brachystachys* (C. Koch) Feinbrun = O. brachystachys
nutans L. (*Honorius nutans* (L.) S.F. Gray) - 1
- subsp. *boucheanum* (Kunth) Hayek = O. boucheanum
- subsp. *boucheanum* (Kunth) K. Richt. = O. boucheanum
obconicum (Zahar.) Agapova = O. hajastanum
oligophyllum auct. p.p. = O. balansae
oreoides Zahar. - 1
orthophyllum Ten. subsp. *kochii* (Parl.) Maire & Weiller = O. kochii
- subsp. *kochii* (Parl.) Zahar. = O. kochii
platyphyllum Boiss. = O. cuspidatum
ponticum Zahar. (*O. georgicum* Agapova, ?*O. ponticum*

subsp. *cyclogynum* Zahar., *O. pyrenaicum* auct.) - 1, 2, 6
- subsp. *cyclogynum* Zahar. = O. ponticum
- subsp. *obconicum* Zahar. = O. hajastanum
pyrenaicum auct. = O. ponticum
pyrenaicum L. subsp. *flavescens* (Lam.) K. Richt. = O. flavescens
refractum Schlecht. - 1
schelkownikowii auct. = O. shelkovnikovii
schischkinii Krasch. = O. sintenisii
schmalhausenii Albov - 2
shelkovnikovii Grossh. (*O. schelkownikowii* auct.) - 2
sigmoideum Freyn & Sint. (*O. nanum* auct. p.p.) - 2
sintenisii Freyn (*O. schischkinii* Krasch.) - 2
tempskyanum Freyn & Sint. = O. cuspidatum
tenuifolium auct. = O. kochii
transcaucasicum Miscz. ex Grossh. - 2
umbellatum L. - 1
woronowii Krasch. - 1, 2
- var. *imereticum* (Sosn.) Grossh. = O. imereticum

Phalangium auct. = Camassia

quamash Pursh = Camassia quamash

Pseudomuscari Garbari & Greuter

acutifolium (Boiss.) Garbari = P. paradoxum
apertum (Freyn & Conrath) Garbari (*Muscari apertum* Freyn & Conrath, *Bellevalia aperta* (Freyn & Conrath) Grossh.) - 2
■ **coeleste** (Fomin) Garbari (*Muscari coeleste* Fomin, *Bellevalia coelestis* (Fomin) Grossh.)
coeruleum (Losinsk.) Garbari (*Muscari coeruleum* Losinsk., *Bellevalia coerulea* (Losinsk.) Schchian, *Muscari botryoides* sensu Losinsk.) - 2
■ **forniculatum** (Fomin) Garbari (*Muscari forniculatum* Fomin, *Bellevalia forniculata* (Fomin) Deloney)
pallens (Bieb.) Garbari (*Hyacinthus pallens* Bieb., *Muscari pallens* (Bieb.) Fisch., *M. stoloniferum* Charkev.) - 2
paradoxum (Fisch. & C.A. Mey.) Garbari (*Bellevalia acutifolia* (Boiss.) Deloney, *B. paradoxa* (Fisch. & C.A. Mey.) Boiss., *Muscari paradoxum* (Fisch. & C.A. Mey.) C. Koch, *Pseudomuscari acutifolium* (Boiss.) Garbari) - 2
■ **turkewiczii** (Woronow) Garbari (*Bellevalia turkewiczii* (Woronow) Grossh., *Muscari turkewiczii* (Woronow) Losinsk.)

Puschkinia Adams

hyacinthoides Baker - 2
scilloides Adams - 2

Scilla L. (*Barnardia* Lindl.)

armena Grossh. (*S. siberica* Haw. subsp. *armena* (Grossh.) H. Mordak) - 2
atropatana Grossh. = Hyacinthella atropatana
autumnalis L. (*S. scythica* Kleop.) - 1, 2
bifolia L. (*S. bifolia* subsp. *nivalis* (Boiss.) Domin, comb. superfl., *S. bifolia* subsp. *nivalis* (Boiss.) K. Richt., *S. bifolia* var. *taurica* Regel, *S. nivalis* Boiss., *S. taurica* (Regel) Fuss) - 1, 2
- subsp. *nivalis* (Boiss.) Domin = S. bifolia
- subsp. *nivalis* (Boiss.) K. Richt. = S. bifolia
- var. *taurica* Regel = S. bifolia
bucharica Desjat. ex B. Fedtsch. = S. puschkinioides
caucasica Miscz. (*S. siberica* Haw. subsp. *caucasica* (Miscz.)

H. Mordak) - 2
diziensis Grossh. = S. mischtschenkoana
grossheimii Sosn. = S. mischtschenkoana
hohenackeri Fisch. & C.A. Mey. - 2
japonica (Thunb.) Baker = S. scilloides
khorassanica Meikle - 6
kladnii Schur - 1
mischtschenkoana Grossh. (*S. diziensis* Grossh., *S. grossheimii* Sosn., *S. zangezura* Grossh. nom. invalid.) - 2
monanthos C. Koch - 2
mordakiae Speta = S. sibirica
nivalis Boiss. = S. bifolia
otschiauriae H. Mordak (*S. siberica* Haw. subsp. *otschiauriae* (H. Mordak) H. Mordak) - 2
puschkinioides Regel (*S. bucharica* Desjat. ex B. Fedtsch. nom. invalid.) - 6
raewskiana Regel - 6
rosenii C. Koch - 2
scilloides (Lindl.) Druce (*Barnardia scilloides* Lindl., *Scilla japonica* (Thunb.) Baker, 1873, non Thunb., *S. sinensis* auct.) - 5
scythica Kleop. = S. autumnalis
siberica Haw. (*S. mordakiae* Speta) - 1, 2
- subsp. *armena* (Grossh.) H. Mordak = S. armena
- subsp. *caucasica* (Miscz.) H. Mordak = S. caucasica
- subsp. *otschiauriae* (H. Mordak) H. Mordak = S. otschiauriae
sinensis auct. = S. scilloides
taurica (Regel) Fuss = S. bifolia
vvedenskyi Pazij - 6
winogradowii Sosn. - 2
zangezura Grossh. = S. mischtschenkoana

HYDRANGEACEAE Dumort.

Calyptranthe (Maxim.) Nakai = Hydrangea

petiolaris (Siebold & Zucc.) Nakai = Hydrangea petiolaris

Deutzia Thunb.

amurensis (Regel) Airy Shaw = D. parviflora
glabrata Kom. - 5
parviflora Bunge (*D. amurensis* (Regel) Airy Shaw) - 5

Hydrangea L. (*Calyptranthe* (Maxim.) Nakai)

anomala D. Don subsp. *petiolaris* (Siebold & Zucc.) McClintock = H. petiolaris
paniculata Siebold - 5
petiolaris Siebold & Zucc. (*Calyptranthe petiolaris* (Siebold & Zucc.) Nakai, *Hydrangea anomala* D. Don subsp. *petiolaris* (Siebold & Zucc.) McClintock) - 5

Philadelphus L.

caucasicus Koehne - 2, 6(cult.)
*latifolius Schrad. ex DC. - 1, 6
schrenkii Rupr. & Maxim. = Ph. tenuifolius
tenuifolius Rupr. & Maxim. (*Ph. schrenkii* Rupr. & Maxim., *Ph. tenuifolius* var. *schrenkii* (Rupr. & Maxim.) Ja. Vassil.) - 5
- var. *schrenkii* (Rupr. & Maxim.) Ja. Vassil. = Ph. tenuifolius

Schizophragma Siebold & Zucc.

hydrangeoides Siebold & Zucc. - 5

HYDROCARYACEAE Raimann = TRA-
PACEAE

HYDROCHARITACEAE Juss.
Elodea Michx.
canadensis Michx. - 1, 3
***densa** (Planch.) Caspari - 2

Hydrilla Rich.
lithuanica (Bess.) Dandy = H. verticillata
verticillata (L. fil.) Royle (*H. lithuanica* (Bess.) Dandy) - 1,
3, 4, 5

Hydrocharis L. (*Pontederia* auct.)
asiatica Miq. = H. dubia
dubia (Blume) Backer (*Pontederia dubia* Blume, *Hydrocharis asiatica* Miq.) - 5
morsus-ranae L. - 1, 2, 3, 4, 6

Ottelia Pers.
alismoides (L.) Pers. - 5

Pontederia auct.
dubia Blume = Hydrocharis dubia

Stratiotes L.
aloides L. - 1, 2, 3

Vallisneria L.
asiatica Miki - 5
spiralis L. - 1, 2, 6

HYDROCOTYLACEAE Hyl. = APIACEAE

HYDROPHYLLACEAE R. Br.
Phacelia Juss.
tanacetifolia Benth. - 1, 2, 5(alien)

HYMENOPHYLLACEAE Link (*TRICHOMA-
NACEAE* Burmeist.)
Gonocormus Bosch (*Trichomanes* auct.)
minutus (Blume) Bosch (*Trichomanes minutum* Blume, *T. parvulum* auct.) - 5

Hymenophyllum Smith
tunbrigense (L.) Smith - 2
wrightii Bosch = Mecodium wrightii

Lacosteopsis (Prantl) Nakaike (*Trichomanes* auct., *Vandenboschia* auct.)
orientalis (C. Chr.) Nakaike (*Trichomanes orientale* C. Chr., *T. radicans* Sw. var. *orientale* (C. Chr.) Lellinger, *Vandenboschia orientalis* (C. Chr.) Ching, *V. radicans* (Sw.) Copel. var. *orientalis* (C. Chr.) H. Ito, *Trichomanes radicans* auct., *Vandenboschia radicans* auct.) - 5

- var. **angustata** (Christ) Nakaike (*Trichomanes japonicum* Franch. & Savat. var. *angustatum* Christ, ? *T. orientale* C. Chr. var. *abbreviatum* Miyabe & Kudo) - 5

Mecodium C. Presl ex Copel.
wrightii (Bosch) Copel. (*Hymenophyllum wrightii* Bosch) - 5

Trichomanes auct. = Gonocormus and Lacosteopsis
japonicum Franch. & Savat. var. *angustatum* Christ = Lacosteopsis orientalis var. angustata
minutum Blume = Gonocormus minutus
orientale C. Chr. = Lacosteopsis orientalis
- var. *abbreviatum* Miyabe & Kudo = Lacosteopsis orientalis var. angustata
parvulum auct. = Gonocormus minutus
radicans auct. = Lacosteopsis orientalis
radicans Sw. var. *orientale* (C. Chr.) Lellinger = Lacosteopsis orientalis (C. Chr.) Nakaike

Vandeboschia auct. = Lacosteopsis
orientalis (C. Chr.) Ching = Lacosteopsis orientalis
radicans auct. = Lacosteopsis orientalis
radicans (Sw.) Copel. var. *orientalis* (C. Chr.) H. Ito = Lacosteopsis orientalis

HYPECOACEAE (Dumort.) Willk.
Hypecoum L.
erectum L. - 3, 4, 6
grandiflorum Benth. = H. imberbe
imberbe Smith (*H. grandiflorum* Benth., *H. procumbens* L. subsp. *imberbe* (Smith) R. de Malagarriga) - 2
lactiflorum (Kar. & Kir.) Pazij - 6
leptocarpum Hook. fil. & Thoms. - 6
parviflorum Kar. & Kir. (*H. pendulum* L. var. *parviflorum* (Kar. & Kir.) Kryl., *H. pendulum* var. *parviflorum* (Kar. & Kir.) Cullen, comb. superfl.) - 3, 6
pendulum L. - 1, 2, 6
- var. *parviflorum* (Kar. & Kir.) Cullen = H. parviflorum
- var. *parviflorum* (Kar. & Kir.) Kryl. = H. parviflorum
- var. *trilobum* (Trautv.) Cullen = H. trilobum
procumbens L. subsp. *imberbe* (Smith) R. de Malagarriga = H. imberbe
trilobum Trautv. (*H. pendulum* L. var. *trilobum* (Trautv.) Cullen) - 6

HYPERICACEAE Juss. (*GUTTIFERAE* auct.)
Androsaemum Hill = Hypericum
xylosteifolium Spach = Hypericum xylosteifolium

Hypericum L. (*Androsaemum* Hill)
acutum Moench = H. quadrangulum
aemulans Koidz. = H. erectum
alpestre Stev. = H. linarioides
- subsp. *polygonifolium* (Rupr.) V. Avet. & Takht. = H. linarioides
alpigenum Kit. (*H. grisebachii* Boiss., *H. richeri* Vill. subsp. *grisebachii* (Boiss.) Nym.) - 1
androsaemum L. - 1, 2
antasiaticum Grossh. p.p. = H. elongatum
apricum auct. fl. cauc. = H. elongatum
apricum Kar. & Kir. - 3, 6

ardasenowii R. Keller & Albov = H. bithynicum
armenum Jaub. & Spach - 2
ascyron L. - 3, 4, 5
asperuloides Czern. ex Turcz. - 2
atropatanum Rzazade - 2
attenuatum Choisy - 4, 5
aurantiacum Kolak. = H. nummularioides
bithynicum Boiss. (*H. ardasenowii* R. Keller & Albov, *H. caucasicum* (Woronow) Gorschk., *H. nordmannii* Boiss., *H. montbretii* auct.) - 2
bupleuroides Griseb. - 2
buschianum Woronow = H. orientale
calycinum L. - 1, 2
caucasicum (Woronow) Gorschk. = H. bithynicum
chrysothyrsum (Woronow) Grossh. = H. lydium
crassifolium Nakai - 5(?)
crispum L. = H. triquetrifolium
davisii N. Robson = H. elongatum
elegans Steph. - 1, 2, 3, 4
eleonorae A. Jelen. = H. nummularioides
elongatum Ledeb. (*H. antasiaticum* Grossh. p.p. quoad lectotypum, *H. davisii* N. Robson, *H. elongatum* subsp. *apiculatum* N. Robson, *H. hyssopifolium* Chaix subsp. *elongatum* (Ledeb.) Woronow, *H. karjaginii* Rzazade, *H. apricum* auct. fl. cauc., *H. hyssopifolium* auct. p.p., *H. lydium* auct.) - 1, 2, 5, 6
- subsp. *apiculatum* N. Robson = H. elongatum
- subsp. **racemosum** (O. Kuntze) N. Robson (*H. hyssopifolium* var. *racemosum* O. Kuntze) - 6
erectum Thunb. (*H. aemulans* Koidz., *H. erectum* subsp. *vaniotii* Levl., *H. porphyrandrum* Levl. & Vaniot, *H. sachalinense* Levl., *H. vaniotii* Levl. nom. invalid.) - 5
- subsp. *vaniotii* Levl. = H. erectum
fissurale Woronow - 2
formosissimum Takht. - 2
gebleri Ledeb. - 3, 4, 5, 6
grisebachii Boiss. = H. alpigenum
grossheimii Kem.-Nath. ex Fed. = H. pruinatum
helianthemoides (Spach) Boiss. - 2, 6
hirsutum L. - 1, 2, 3, 4
humifusum L. - 1
hyssopifolium auct. p.p. = H. elongatum and H. lydium
hyssopifolium Chaix subsp. *chrysothyrsum* Woronow = H. lydium
- subsp. *elongatum* (Ledeb.) Woronow = H. elongatum
- var. *racemosum* O. Kuntze = H. elongatum subsp. racemosum
inodorum Willd. = H. xylosteifolium
kamtschaticum Ledeb. (*H. paramushirense* Kudo) - 5
karjaginii Rzazade = H. elongatum
karsianum (Woronow) Grossh. (*H. perplexum* Woronow subsp. *karsianum* Woronow) - 2
komarovii Gorschk. = H. perforatum
linarioides Bosse (*H. alpestre* Stev. nom. provis., *H. alpestre* subsp. *polygonifolium* (Rupr.) V. Avet. & Takht., *H. polygonifolium* Rupr.) - 1, 2
lydium auct. = H. elongatum
lydium Boiss. (*H. chrysothyrsum* (Woronow) Grossh., *H. hyssopifolium* Chaix subsp. *chrysothyrsum* Woronow, *H. ponticum* Lipsky, *H. hyssopifolium* auct. p.p.) - 2
maculatum Crantz (*H. quadrangulum* auct.) - 1, 3, 4
▪ marginatum Woronow
montanum L. - 1, 2
montbretii auct. = H. bithynicum
mutilum L. - 2
nachitschevanicum Grossh. = H. perforatum
nordmannii Boiss. = H. bithynicum

nummularioides Trautv. (*H. aurantiacum* Kolak., *H. eleonorae* A. Jelen., *H. nummularium* L. subsp. *nummularioides* (Trautv.) V. Avet. & Takht.) - 2
nummularium L. subsp. *nummularioides* (Trautv.) V. Avet. & Takht. = H. nummularioides
▪ olympicum L.
orientale L. (*H. buschianum* Woronow, *H. ptarmicifolium* Spach) - 2
origanifolium Willd. - 2
paramushirense Kudo = H. kamtschaticum
perforatum L. (*H. komarovii* Gorschk., *H. nachitschevanicum* Grossh.) - 1, 2, 3, 4, 6
- subsp. *veronense* (Schrank) Froehl. = H. veronense
perplexum Woronow subsp. *karsianum* Woronow = H. karsianum
polygonifolium Rupr. = H. linarioides
ponticum Lipsky = H. lydium
porphyrandrum Levl. & Vaniot = H. erectum
pruinatum Boiss. & Bal. (*H. grossheimii* Kem.-Nath. ex Fed.) - 2
ptarmicifolium Spach = H. orientale
quadrangulum auct. = H. maculatum
quadrangulum L. (*H. acutum* Moench, nom. illegit., *H. tetrapterum* Fries) - 1, 2
richeri Vill. subsp. *grisebachii* (Boiss.) Nym. = H. alpigenum
sachalinense Levl. = H. erectum
scabrum L. - 2, 3, 6
senanense Maxim. - 5(?)
strictum Maleev - 2
tetrapterum Fries = H. quadrangulum
theodorii Woronow - 2
▪ triquetrifolium Turra (*H. crispum* L. nom. illegit.)
vaniotii Levl. = H. erectum
venustum Fenzl - 2
veronense Schrank (*H. perforatum* L. subsp. *veronense* (Schrank) Froehl.) - 1
xylosteifolium (Spach) N. Robson (*Androsaemum xylosteifolium* Spach, *H. inodorum* Willd. 1802, non Mill. 1768) - 2
yezoense Maxim. - 5

Triadenum Rafin.

japonicum (Blume) Makino - 5

HYPOLEPIDACEAE Pichi Sermolli
Allosorus auct. = Pteridium
tauricus C. Presl = Pteridium tauricum

Pteridium Gled. ex Scop. (*Allosorus* auct.)

aquilinum (L.) Kuhn (*Dryopteris austriaca* (Jacq.) Woynar, p.p. quoad solum basionymum, *Polypodium austriacum* Jacq.) - 1, 2, 3, 4, 5, 6
- subsp. *brevipes* (Tausch) E. Wulf = P. tauricum
tauricum V. Krecz. (*Allosorus tauricus* C. Presl, nom. nud., ? *Pteridium aquilinum* (L.) Kunh subsp. *brevipes* (Tausch) E. Wulf, *Pteris aquilina* L. f. *transcaucasica* Rupr.) - 1, 2

IRIDACEAE Juss.
Belamcanda Adans.

chinensis (L.) DC. - 5

Crocus L.

adamii J. Gay (*C. biflorus* Mill. subsp. *adamii* (J. Gay) K. Richt., *C. biflorus* subsp. *adamii* (J. Gay) Mathew, comb. superfl., *C. biflorus* auct. p.p.) - 2
alatavicus Regel & Semen. - 6
albiflorus Kit. = C. vernus
angustifolius Weston (*C. susianus* Ker-Gawl.) - 1
artvinensis (Philippow) Grossh. (*C. biflorus* Mill. subsp. *artvinensis* (Philippow) Mathew) - 2
aureus Smith = C. flavus
autranii Albov - 2
banaticus J. Gay - 1
biflorus auct. p.p. = C. adamii
biflorus Mill. subsp. *adamii* (J. Gay) Mathew = C. adamii
- subsp. *adamii* (J. Gay) K. Richt. = C. adamii
- subsp. *artvinensis* (Philippow) Mathew = C. artvinensis
caspius Fisch. & C.A. Mey. - 2
*****chrysanthus** (Herb.) Herb. - 1
flavus Weston (*C. aureus* Smith, *C. maesiacus* Ker-Gawl.) - 1
geghartii Sosn. - 2
heuffelianus Herb. - 1
▪ **karsianus** Fomin
korolkowii Regel ex Maw - 6
kotschyanus C. Koch subsp. *suworowianus* (C. Koch) Mathew = C. suworowianus
maesiacus Ker-Gawl. = C. flavus
michelsonii B. Fedtsch. - 6
pallasii Goldb. - 1
polyanthus Grossh. - 2
reticulatus Stev. ex Adams (*C. variegatus* Hoppe & Hornsch.) - 1, 2
▪ **roopiae** Woronow
*****sativus** L. - 1
- var. *vernus* L. = C. vernus
scharojanii Rupr. - 2
speciosus Bieb. - 1, 2
susianus Ker-Gawl. = C. angustifolius
suworowianus C. Koch (*C. kotschyanus* C. Koch subsp. *suworowianus* (C. Koch) Mathew) - 2
tauricus (Trautv.) Puring - 1, 2
vallicola Herb. - 2
variegatus Hoppe & Hornsch. = C. reticulatus
vernus (L.) Hill (*C. sativus* L. var. *vernus* L., *C. albiflorus* Kit., *C. vernus* subsp. *albiflorus* (Kit.) Aschers. & Graebn. comb. superfl., *C. vernus* subsp. *albiflorus* (Kit.) K. Richt.) - 1
- subsp. *albiflorus* (Kit.) Aschers. & Graebn. = C. vernus
- subsp. *albiflorus* (Kit.) K. Richt. = C. vernus

Cryptobasis Nevski = Iris

loczyi (Kanitz) Ikonn. = Iris loczyi
tianschanica (Maxim.) Nevski = Iris loczyi

Gladiolus L.

apterus Klok. = G. imbricatus
atroviolaceus Boiss. - 2, 6
cuacasicus Herb. - 2
communis L. - 1, 2
dzavakheticus Eristavi - 2
halophilus Boiss. & Heldr. - 2
imbricatus L. (*G. apterus* Klok.) - 1, 2, 3
italicus Mill. (*G. segetum* Ker-Gawl., *G. tenuiflorus* C. Koch, *G. turkmenorum* Czerniak.) - 1, 2, 6
kotschyanus Boiss. (*G. tenuiflorus* auct.) - 2

palustris Gaudin - 1
segetum Ker-Gawl. = G. italicus
tenuflorus auct. = G. kotschyanus
tenuiflorus C. Koch = G. italicus
tenuis Bieb. (*G. apterus* Klok.) - 1, 2(?)
turkmenorum Czerniak. = G. italicus

Gynandriris Pall.

maricoides (Regel) Nevski (*Iris maricoides* Regel, *Gynandriris maricoides* (Regel) Rodionenko, comb. superfl., *G. maricoides* (Regel) Vved. comb. superfl.) - 6

Iridodictyum Rodionenko (*Xiphion* auct.)

hyrcanum (Woronow ex Grossh.) Rodionenko (*Iris hyrcana* Woronow ex Grossh.) - 2
kolpakowskianum (Regel) Rodionenko (*Iris kolpakowskiana* Regel, *Xiphion kolpakowskianum* (Regel) Baker) - 6
reticulatum (Bieb.) Rodionenko (*Iris reticulata* Bieb.) - 2
winkleri (Regel) Rodionenko (*Iris winkleri* Regel, *Xiphion winkleri* (Regel) Vved.) - 6
winogradowii (Fomin) Rodionenko (*Iris winogradowii* Fomin) - 2

Iris L. (*Cryptobasis* Nevski, *Oncocyclus* Siemss., *Sclerosiphon* Nevski, *Xiridion* (Tausch) Fourr.)

acutiloba C.A. Mey. (*I. fominii* Woronow ex Grossh.) - 2
- subsp. *lineolata* (Trautv.) Mathew & Wendelbo = I. helena
- var. *lineolata* Trautv. = I. helena
aequiloba Ledeb. = I. pumila
alberti Regel - 6
alexeenkoi Grossh. - 2
annae Grossh. - 2
aphylla L. (*I. aphylla* subsp. *polonica* (Aschers. & Graebn.) Soo, *I. aphylla* A. *typica* c. *polonica* Aschers. & Graebn., *I. furcata* Bieb., *I. polonica* (Aschers. & Graebn.) Fomin & Bordz.) - 1, 2
- subsp. *dacica* (Beldie) Soo = I. hungarica
- subsp. *hungarica* (Waldst. & Kit.) Hegi = I. hungarica
- subsp. *polonica* (Aschers. & Graebn.) Soo = I. aphylla
- A. *typica* c. *polonica* Aschers. & Graebn. = I. aphylla
arenaria Waldst. & Kit. subsp. *orientalis* (Ugr.) Lavr. = I. pineticola
astrachanica Rodionenko = I. scariosa
atropatana Grossh. = Juno atropatana
babadagica Rzazade & Golneva = I. hungarica
baldshuanica B. Fedtsch. = Juno baldshuanica
biglumis Vahl (*I. ensata* sensu B. Fedtsch. p.p., *I. lactea* sensu Grub. p.p.) - 4
bloudowii Ledeb. - 3, 4, 5, 6
brandzae Prod. (*I. sintenisii* Janka subsp. *brandzae* (Prod.) Prod., *I. sintenisii* subsp. *brandzae* (Prod.) D.A. Webb & Chater, comb. superfl.) - 1
brevituba (Maxim.) Vved. ex E. Nikit. (*I. ruthenica* Ker-Gawl. var. *brevituba* Maxim., *I. ruthenica* subsp. *brevituba* (Maxim.) V. Doronkin) - 3, 4, 6(?)
x **brzhezitzkyi** Grossh. (*I. x tatianae* Grossh.). - I. paradoxa Stev. x I. acutiloba C.A. Mey. - 2
bucharica M. Foster = Juno bucharica
x **caeciliae** Grossh. - I. paradoxa Stev. x I. lycotis Woronow - 2
camillae Grossh. - 2

carthaliniae Fomin (*I. spuria* L. subsp. *carthaliniae* (Fomin) Mathew) - 2
caucasica Hoffm. = Juno caucasica
coerulea B. Fedtsch. = Juno coerulea
colchica Kem.-Nath. - 2
dacica Beldie = I. hungarica
darwasica Regel - 6
demetrii Achverd. & Mirzoeva = I. prilipkoana
diantha C. Koch = I. hungarica
dichotoma Pall. = Pardanthopsis dichotoma
drepanophylla Aitch. & Baker = Juno drepanophylla
elegantissima Sosn. (*I. iberica* Hoffm. subsp. *elegantissima* (Sosn.) Fed. & Takht.) - 2
ensata sensu B. Fedtsch. p.p. = I. biglumis, I. lactea and I. pallasii
ensata Thunb. (*I. kaempferi* Siebold ex Lem.) - 3, 4, 5, 6
ewbankiana M. Foster - 6
- var. *lineolata* (Trautv.) Gawrilenko = I. helena
falcifolia Bunge - 6
flavissima Pall. = I. humilis
- subsp. *stolonifera* f. *orientalis* Ugr. = I. pineticola
- subsp. *transuralensis* Ugr. = I. humilis
***florentina** L. - 1
foetidissima auct. = I. musulmanica
fominii Woronow ex Grossh. = I. acutiloba
fosteriana Aitch. & Baker = Juno fosteriana
fragrans Lindl. (*I. moorcroftiana* Wall. ex D. Don) - 6
furcata Bieb. = I. aphylla
- var. *diantha* (C. Koch) Grossh. = I. hungarica
germanica L. - 1
glaucescens Bunge (*I. scariosa* sensu B. Fedtsch.) - 3, 6
graberana Sealy - 6
graminea L. - 1
grossheimii Woronow ex Grossh. - 2
haematophylla Fisch. ex Link = I. pallasii
halophila Pall. (*I. spuria* L. subsp. *halophila* (Pall.) Mathew & Wendelbo, *I. spuria* subsp. *halophila* (Pall.) D.A. Webb & Chater, comb. superfl.) - 1, 2, 3, 6
- var. *sogdiana* (Bunge) Grub. = I. sogdiana
helena (C. Koch) C. Koch (*Oncocyclus helena* C. Koch, *Iris acutiloba* C.A. Mey. var. *lineolata* Trautv., *I. acutiloba* subsp. *lineolata* (Trautv.) Mathew & Wendelbo, *I. ewbankiana* M. Foster var. *lineolata* (Trautv.) Gawrilenko, *I. lineolata* (Trautv.) Grossh.) - 2
hippolyti (Vved.) R. Kam. = Juno hippolyti
hoogiana Dykes - 6
humilis Bieb. = I. pontica
- var. *pontica* (Zapal.) Prod. = I. pontica
humilis Georgi (*I. flavissima* Pall., *I. flavissima* subsp. *transuralensis* Ugr., *I. mandshurica* Maxim.) - 1, 3, 4, 5
- subsp. *orientalis* (Ugr.) Soo = I. pineticola
hungarica Waldst. & Kit. (*I. aphylla* L. subsp. *dacica* (Beldie) Soo, *I. aphylla* subsp. *hungarica* (Waldst. & Kit.) Hegi, *I. babadagica* Rzazade & Golneva, *I. dacica* Beldie, *I. diantha* C. Koch, *I. furcata* Bieb. var. *diantha* (C. Koch) Grossh., *I. hungarica* subsp. *dacica* (Beldie) Prod.) - 1, 2
- subsp. *dacica* (Beldie) Prod. = I. hungarica
hyrcana Woronow ex Grossh. = Iridodicryum hyrcanum
iberica Hoffm. - 2
- subsp. *elegantissima* (Sosn.) Fed. & Takht. = I. elegantissima
- subsp. *lycotis* (Woronow) Takht. = I. lycotis
iliensis Poljak. = I. pallasii
imbricata Lindl. (*I. sulphurea* C. Koch) - 2
interior (E. Anders.) Czer. = I. setosa
issica Siehe = Juno issica

ivanovae V. Doronkin - 4
kaempferi Siebold ex Lem. = I. ensata
karategina B. Fedtsch. = I. lineata
x karjaginii Grossh. - I. paradoxa Stev. x I. medwedewii Fomin - 2
klattii Kem.-Nath. = I. musulmanica
koenigii Sosn. - 2
kolpakowskiana Regel = Iridodictyum kolpakowskianum
kopetdagensis (Vved.) Mathew & Wendelbo = Juno kopet-dagensis
korolkowii Regel - 6
kuschakewiczii B. Fedtsch. = Juno kuschakewiczii
lactea Pall. (*I. ensata* sensu B. Fedtsch. p.p.) - 4
lactea sensu Grub. = I. biglumis and I. pallasii
laevigata Fisch. & C.A. Mey. - 4, 5
lazica Albov = Siphonostylis lazica
leptorhiza Vved. = Juno leptorhiza
lineata M. Foster ex Regel (*I. karategina* B. Fedtsch. nom. invalid.) - 6
lineolata (Trautv.) Grossh. = I. helena
linifolia (Regel) O. Fedtsch. = Juno linifolia
loczyi Kanitz (*Cryptobasis loczyi* (Kanitz) Ikonn., *C. tianschanica* (Maxim.) Nevski, *Iris tianschanica* (Maxim.) Vved. ex Woronow) - 6
longiscapa Ledeb. - 6
ludwigii Maxim. - 3
lycotis Woronow (*I. iberica* Hoffm. subsp. *lycotis* (Woronow) Takht.) - 2
magnifica Vved. = Juno magnifica
mandshurica Maxim. = I. humilis
maracandica (Vved.) Wendelbo = Juno maracandica
maricoides Regel = Gynandriris maricoides
marschalliana Bobr. = I. pontica
medwedewii Fomin - 2
moorcroftiana Wall. ex D. Don = I. fragrans
musulmanica Fomin (*I. klattii* Kem.-Nath., *I. spuria* L. subsp. *musulmanica* (Fomin) Takht., *I. violacea* Klatt, 1867, non Savi, 1815, *I. foetidissima* auct.) - 2
narbutii O. Fedtsch. = Juno narbutii
narynensis O. Fedtsch. = Juno narynensis
nicolai Vved. = Juno nicolai
notha Bieb. (*I. spuria* L. subsp. *notha* (Bieb.) Aschers. & Graebn., *I. spuria* L. var. *notha* (Bieb.) R.R. Stewart) - 2
orchioides Carr. = Juno orchioides
orientalis Thunb. = I. sanguinea
oxypetala Bunge - 5
pallasii Fisch. ex Trev. (*I. haematophylla* Fisch. ex Link, *I. iliensis* Poljak., *I. ensata* sensu B. Fedtsch. p.p., *I. lactea* sensu Grub. p.p.) - 3, 6
***pallida** Lam. - 1
paradoxa Stev. - 2
parvula Vved. = Juno parvula
pineticola Klok. (*I. arenaria* Waldst. & Kit. subsp. *orientalis* (Ugr.) Lavr., *I. flavissima* Pall. subsp. *stolonifera* f. *orientalis* Ugr., *I. humilis* Georgi subsp. *orientalis* (Ugr.) Soo) - 1
polonica (Aschers. & Graebn.) Fomin & Bordz. = I. aphylla
pontica Zapal. (*I. humilis* Bieb. 1808, non Georgi, 1775, *I. humilis* var. *pontica* (Zapal.) Prod., *I. marschalliana* Bobr.) - 1, 2
popovii Vved. = Juno popovii
potaninii Maxim. - 3, 4
prilipkoana Kem.-Nath. (*I. demetrii* Achverd. & Mirzoeva, nom. nud., *I. spuria* L. subsp. *demetrii* (Achverd. & Mirzoeva) Mathew) - 2

pseudacorus L. - 1, 2, 3, 5
pseudocaucasica Grossh. = Juno pseudocaucasica
pseudocyperus Schur - 1
pseudonotha Galushko - 2
pumila L. (*I. aequiloba* Ledeb., *I. pumila* subsp. *aequiloba* (Ledeb.) Baker, *I. taurica* Lodd.) - 1, 2
- subsp. *aequiloba* (Ledeb.) Baker = I. pumila
reticulata Bieb. = Iridodictyum reticulatum
rosenbachiana Regel = Juno rosenbachiana
ruthenica Ker = Gawl. - 3, 4, 5, 6
- subsp. *brevituba* (Maxim.) V. Doronkin = I. brevituba
- var. *brevituba* Maxim. = I. brevituba
*sambucina L. - 1
sanguinea Donn (*I. orientalis* Thunb. 1794, non Mill. 1768) - 4, 5
scariosa sensu B. Fedtsch. = I. glaucescens
scariosa Willd. ex Link (*I. astrachanica* Rodionenko) - 1
schelkownikowii (Fomin) Fomin - 2
schischkinii Grossh. = Juno schischkinii
setosa Pall. ex Link (*I. interior* (E. Anders.) Czer., *I. setosa* var *interior* E. Anders., *I. setosa* subsp. *interior* (E. Anders.) Hult.) - 4, 5
- subsp. *interior* (E. Anders.) Hult. = I. setosa
- var. *interior* E. Anders. = I. setosa
sibirica L. - 1, 2, 3, 4
x sinistra Sosn. - 2
sintenisii Janka subsp. *brandzae* (Prod.) Prod. = I. brandzae
- subsp. *brandzae* (Prod.) D.A. Webb & Chater = I. brandzae
sogdiana Bunge (*I. halophila* Pall. var. *sogdiana* (Bunge) Grub., *Xyridion sogdianum* (Bunge) Nevski) - 6
songarica Schrenk (*Sclerosiphon songaricum* (Schrenk) Nevski) - 6
spuria L. subsp. *carthaliniae* (Fomin) Mathew = I. carthaliniae
- subsp. *demetrii* (Achverd. & Mirzoeva) Mathew = I. prilipkoana
- subsp. *halophila* (Pall.) D.A. Webb & Chater = I. halophila
- subsp. *musulmanica* (Fomin) Takht. = I. musulmanica
- subsp. *notha* (Bieb.) Aschers. & Graebn. = I. notha
- var. *notha* (Bieb.) R.R. Stewart = I. notha
*squalens L. - 1
stolonifera Maxim. - 6
suaveolens Boiss. & Reut. - 1(?)
subdecolorata Vved. = Juno subdecolorata
sulphurea C. Koch = I. imbricata
tadshikorum Vved. = Juno tadshikorum
x *tatianae* Grossh. = I. x brzhezitzkyi
taurica Lodd. = I. pumila
tenuifolia Pall. - 1, 3, 4, 6
tianschanica (Maxim.) Vved. ex Woronow = I. loczyi
tigridia Bunge - 3, 4
timofejewii Woronow - 2
tubergeniana M. Foster = Juno tubergeniana
uniflora Pall. ex Link - 4, 5
variegata L. - 1
ventricosa Pall. - 4, 5
vicaria Vved. = Juno vicaria
violacea Klatt = I. musulmanica
vorobievii N.S. Pavlova - 5
vvedenskyi Nevski ex Woronow = Juno vvedenskyi
warleyensis M. Foster = Juno warleyensis
willmottiana M. Foster = Juno willmottiana
winkleri Regel = Iridodictyum winkleri
winogradowii Fomin = Iridodictyum winogradowii

Juno Tratt.

almaatensis Pavl. - 6
atropatana (Grossh.) Czer. (*Iris atropatana* Grossh., *Juno atropatana* (Grossh.) Rodionenko, comb. invalid.) - 2
baldshuanica (B. Fedtsch.) Vved. (*I. baldshuanica* B. Fedtsch.) - 2
bucharica (M. Foster) Vved. (*I. bucharica* M. Foster) - 6
capnoides Vved. - 6
caucasica (Hoffm.) Klatt (*I. caucasica* Hoffm.) - 2
coerulea Poljak. (*Iris coerulea* B. Fedtsch. 1904, non Spach, 1846) - 6
drepanophylla (Aitch. & Baker) Rodionenko (*Iris drepanophylla* Aitch. & Baker) - 6
fosteriana (Aitch. & Baker) Rodionenko (*Iris fosteriana* Aitch. & Baker) - 6
hippolyti Vved. (*Iris hippolyti* (Vved.) R. Kam.) - 6
inconspicua Vved. - 6
issica (Siehe) R. Kam. (*Iris issica* Siehe) - 2
kopetdagensis Vved. (*Iris kopetdagensis* (Vved.) Mathew & Wendelbo) - 6
kuschakewiczii (B. Fedtsch.) Poljak. (*Iris kuschakewiczii* B. Fedtsch.) - 6
leptorhiza Vved. (*Iris leptorhiza* Vved.) - 6
linifolia (Regel) Vved. (*Iris linifolia* (Regel) O. Fedtsch.) - 6
linifoliiformis Chalkuziev - 6
magnifica (Vved.) Vved. (*Iris magnifica* Vved.) - 6
maracandica Vved. (*Iris maracandica* (Vved.) Wendelbo) - 6
narbutii (O. Fedtsch.) Vved. (*Iris narbutii* O. Fedtsch.) - 6
narynensis (O. Fedtsch.) Vved. (*Iris narynensis* O. Fedtsch.) - 6
nicolai Vved. (*Iris nicolai* Vved.) - 6
orchioides (Carr.) Vved. (*Iris orchioides* Carr.) - 6
parvula Vved. (*Iris parvula* Vved. nom. invalid.) - 6
popovii Vved. (*Iris popovii* Vved.) - 6
pseudocaucasica (Grossh.) Rodionenko (*Iris pseudocaucasica* Grossh.) - 2
rosenbachiana (Regel) Vved. (*Iris rosenbachiana* Regel) - 6
schischkinii (Grossh.) Czer. (*Iris schischkinii* Grossh., *Juno schischkinii* (Grossh.) Rodionenko, comb. invalid.) - 2
subdecolorata (Vved.) Vved. (*Iris subdecolorata* Vved.) - 6
svetlanae Vved. - 6
tadshikorum Vved. (*Iris tadshikorum* Vved.) - 6
tubergeniana (M. Foster) Vved. (*Iris tubergeniana* M. Foster) - 6
vicaria Vved. (*Iris vicaria* Vved. nom. invalid.) - 6
vvedenskyi (Nevski ex Woronow) Nevski (*Iris vvedenskyi* Nevski ex Woronow) - 6
warleyensis (M. Foster) Vved. (*Iris warleyensis* M. Foster) - 6
willmottiana (M. Foster) Vved. (*Iris willmottiana* M. Foster) - 6
zaprjagajevii N. Abramov - 6
zenaidae Vved. - 6

Oncocyclus Siemss. = Iris

helena C. Koch = Iris helena

Pardanthopsis (Hance) Lenz

dichotoma (Pall.) Lenz (*Iris dichotoma* Pall.) - 4, 5

Sclerosiphon Nevski = Iris

songaricum (Schrenk) Nevski = Iris songarica

Siphonostylis W. Schulze
lazica (Albov) W. Schulze (*Iris lazica* Albov) - 2

Sisyrinchium L.
angustifolium auct. = S. septentrionale
bermudiana auct. = S. septentrionale
montanum auct. = S. septentrionale
septentrionale Bicknell (*S. angustifolium* auct., *S. bermudiana* auct., *S. montanum* auct.) - 1(alien), 4(alien), 5(alien)

Xiphion auct. = Iridodictyum
kolpakowskianum (Regel) Baker = Iridodictyum kolpakowskianum
winkleri (Regel) Vved. = Iridodictyum winkleri

Xiridion (Tausch) Fourr. = Iris
sogdianum (Bunge) Nevski = Iris sogdiana

ISOETACEAE Reichenb.
Isoetes L.
asiatica (Makino) Makino (*I. echinospora* Durieu subsp. *asiatica* (Makino) A. Löve, *I. setacea* Lam. subsp. *asiatica* (Makino) Holub) - 5
beringensis Kom. = I. maritima
echinospora Durieu = I. setacea
- subsp. *asiatica* (Makino) A. Löve = I. asiatica
- subsp. *maritima* (Underw.) A. Löve = I. maritima
lacustris L. (*I. rossica* Gand.) - 1, 3, 4
maritima Underw. (*I. beringensis* Kom., *I. echinospora* Durieu subsp. *maritima* (Underw.) A. Löve, *I. muricata* Durieu subsp. *maritima* (Underw.) Hult., *I. setacea* Lam. subsp. *maritima* (Underw.) Holub) - 5
muricata Durieu subsp. *maritima* (Underw.) Hult. = I. maritima
petropolitana Gand. = I. setacea
rossica Gand. = I. lacustris
setacea Durieu (*I. echinospora* Durieu, *I. petropolitana* Gand., *I. tenella* Leman ex Desv.) - 1, 3, 4
- subsp. *asiatica* (Makino) Holub = I. asiatica
- subsp. *maritima* (Underw.) Holub = I. maritima
tenella Leman ex Desv. = I. setacea

IXIOLIRIACEAE Nakai
Ixiolirion Herb. (*Kolpakowskia* Regel)
ferganicum Kovalevsk. & Vved. - 6
karateginum Lipsky (*Kolpakowskia karategina* (Lipsky) Traub) - 6
ledebourii Fisch. & C.A. Mey. (*I. tataricum* (Pall.) Schult. & Schult. fil. subsp. *ledebourii* (Fisch. & C.A. Mey.) Kryl., *I. tataricum* subsp. *ledebourii* (Fisch. & C.A. Mey.) R.R. Mill, comb. superfl.) - 3
montanum (Labill.) Schult. & Schult. fil. (*I. tataricum* (Pall.) Schult. & Schult. fil. subsp. *montanum* (Labill.) Takht.) - 2
tataricum (Pall.) Schult. & Schult. fil. - 3, 6
- subsp. *ledebourii* (Fisch. & C.A. Mey.) Kryl. = I. ledebourii
- subsp. *ledebourii* (Fisch. & C.A. Mey.) R.R. Mill = I. ledebourii
- subsp. *montanum* (Labill.) Takht. = I. montanum

Kolpakowskia Regel = Ixiolirion
karategina (Lipsky) Traub = Ixiolirion karateginum

JUGLANDACEAE A. Rich. ex Kunth
Juglans L.
ailanthifolia Carr. (*J. sachalinensis* Komatsu, *J.sieboldiana* Maxim. 1872, non Gopert.) - 5
biflorens Tujcz. - 6
fallax Dode (*J. regia* L. subsp. *fallax* (Dode) M. Pop.) - 6
mandshurica Maxim. - 5
pistaciformis Tujcz. - 6
pterocarpa Michx. = Pterocarya pterocarpa
racemiformis Tujcz. - 6
regia L. - 2, 6
- subsp. *fallax* (Dode) M. Pop. = J. fallax
sachalinensis Komatsu = J. ailanthifolia
sieboldiana Maxim. = J. ailanthifolia

Pterocarya Kunth
fraxinifolia (Poir.) Spach = P. pterocarpa
pterocarpa (Michx.) Kunth ex I. Iljinsk. (*Juglans pterocarpa* Michx., *Pterocarya fraxinifolia* (Poir.) Spach, nom. ambig.) - 2

JUNCACEAE Juss.
Juncoides Seguier = Luzula
piperi Cov. = Luzula piperi

Juncus L.
acutiflorus Ehrh. ex Hoffm. - 1
acutus L. - 2
- subsp. *littoralis* (C.A. Mey.) Feinbrun = J. littoralis
albidus Hoffm. = Luzula luzuloides
alpigenus C. Koch - 2
alpino-articulatus Chaix (*J. alpino-articulatus* subsp. *alpestris* (C. Hartm.) Hamet-Ahti, *J. alpino-articulatus* subsp. *alpinus* (Vill.) O. Schwarz, *J. alpino-articulatus* subsp. *arthrophyllus* (Brenn.) Reichgelt, *J. alpino-articulatus* subsp. *fuscoater* (Schreb.) O. Schwarz, *J. alpinus* Vill. nom. illegit., *J. alpinus* subsp. *arthrophyllus* (Brenn.) Hyl., *J. alpinus* subsp. *fuscoater* (Schreb.) Lindb. fil., *J. fuscoater* Schreb., *J. fuscoater* [subsp.]* *J. arthrophyllus* Brenn., *J. geniculatus* Schrank subsp. *arthrophyllus* (Brenn.) Soo, *J. geniculatus* auct.) - 1, 2, 3, 4, 5
- subsp. *alpestris* (C. Hartm.) Hamet-Ahti = J. alpino-articulatus
- subsp. *alpinus* (Vill.) O. Schwarz = J. alpino-articulatus
- subsp. **americanus** (Farm.) Hamet-Ahti (*J. alpinus* var. *americanus* Farm.) - 5
- subsp. *arthrophyllus* (Brenn.) Reichgelt = J. alpino-articulatus
- subsp. **fischerianus** (Turcz. ex V. Krecz.) Hamet-Ahti (*J. fischerianus* Turcz. ex V. Krecz.,*J. alpino-articulatus* var. *macrocephalus* (Hyl.) V. Novik., *J. alpinus* subsp. *arthrophyllus* var. *macrocephalus* Hyl.) - 1, 3, 4, 5
- subsp. *fuscoater* (Schreb.) O. Schwarz = J. alpino-articulatus
- subsp. *nodulosus* (Wahlenb.) Hamet-Ahti = J. nodulosus
- subsp. *rariflorus* (C. Hartm.) Breistroffer = J. nodulosus
- subsp. *rariflorus* (C. Hartm.) Holub = J. nodulosus
- var. *macrocephalus* (Hyl.) V.Novik. = J. alpino-articulatus subsp. fischerianus
alpino-pilosus Chaix = Luzula alpino-pilosa
alpinus Vill. = J. alpino-articulatus
- subsp. *arthrophyllus* (Brenn.) Hyl. = J. alpino-articulatus
- - var. *macrocephalus* Hyl. = J. alpino-articulatus subsp.

fischerianus
- subsp. *fuscoater* (Schreb.) Lindb. fil. = J. alpino-articulatus
- subsp. *nodulosus* (Wahlenb.) Lindm. = J. nodulosus
- subsp. *rariflorus* (C. Hartm.) K. Richt. = J. nodulosus
- subsp. *turczaninowii* (Buchenau) Worosch. = J. turczaninowii
- var. *americanus* Farm. = J. alpino-articulatus subsp. americanus
ambiguus Guss. (*J. bufonius* L. subsp. *juzepczukii* (V. Krecz. & Gontsch.) Soo, *J. bufonius* var. *juzepczukii* (V. Krecz. & Gontsch.) Worosch., *J. juzepczukii* V. Krecz. & Gontsch., *J. ranarius* Song. & Perrier ex Billot) - 1, 2, 3, 4, 5, 6
- subsp. *turkestanicus* (V. Krecz. & Gontsch.) V. Novik. = J. turkestanicus
amuricus (Maxim.) V. Krecz. & Gontsch. - 5
anceps Laharpe - 1(?)
arabicus (Aschers. & Buchenau) Adamson (*J. maritimus* Lam. var. *arabicus* Aschers. & Buchenau, *J. rigidus* auct.) - 6
arcticus Willd. (*J. arcticus* subsp. *alaskanus* Hult., *J. arcticus* var. *alaskanus* (Hult.) V. Novik.) - 1, 3, 4, 5
- subsp. *alaskanus* Hult. = J. arcticus
- subsp. *balticus* (Willd.) Hyl. = J. balticus
- subsp. *sitchensis* Engelm. = J. haenkei
- var. *alaskanus* (Hult.) V. Novik. = J. arcticus
- var. *muelleri* (Trautv.) Kovtonjuk = J. muelleri
arcuatus Wahlenb. = Luzula arcuata
arianus V. Krecz. = J. kotschyi
articulatus L. (*J. geniculatus* Schrank, *J. lampocarpus* Ehrh. ex Hoffm., *J. subarticulatus* Zak. & Novopokr.) - 1, 2, 3, 4, 5, 6
- subsp. *limosus* (Worosch.) Worosch. = J. limosus
- subsp. *tatewakii* (Satake) Worosch. = J. tatewakii
- subsp. *turczaninowii* (Buchenau) Worosch. = J. turczaninowii
atratus Krock. - 1, 2, 3, 4, 6
atrofuscus Rupr. (*J. gerardii* Loisel. subsp. *atrofuscus* (Rupr.) Printz, *J. gerardii* subsp. *atrofuscus* (Rupr.) Tolm. comb. superfl.) - 1, 3, 4
balticus Willd. (*J. arcticus* Vill. subsp. *balticus* (Willd.) Hyl.) - 1
- subsp. *inundatus* (Drej.) A. & D. Löve = J. x inundatus
- subsp. *inundatus* (Drej.) K. Richt. = J. x inundatus
- subsp. *sitchensis* (Engelm.) Hult. = J. haenkei
beringensis Buchenau (*J. fauriei* Levl. & Vaniot, p.p.) - 5
biglumis L. - 1, 3, 4, 5
brachyspathus Maxim. - 1, 3, 4, 5
brachytepalus V. Krecz. & Gontsch. (*J. inflexus* L. var. *brachytepalus* (V. Krecz. & Gontsch.) Kitam., *J. inflexus* subsp. *brachytepalus* (V. Krecz. & Gontsch.) V. Novik.) - 6
bufonius L. (*J. erythropodus* V. Krecz., *J. hybridus* auct.) - 1, 2, 3, 4, 5, 6
- subsp. *juzepczukii* (V. Krecz. & Gontsch.) Soo = J. ambiguus
- subsp. *minutulus* (V. Krecz. & Gontsch.) Soo = J. minutulus
- subsp. *nastanthus* (V. Krecz. & Gontsch.) Soo = J. nastanthus
- subsp. *turkestanicus* (V. Krecz. & Gontsch.) Soo = J. turkestanicus
- var. *juzepczukii* (V. Krecz. & Gontsch.) Worosch. = J. ambiguus
- f. *minutulus* Albert & Jahand. = J. minutulus
bulbosus L. - 1

campestris L. = Luzula campestris
- var. *multiflorus* Ehrh. = Luzula multiflora
capitatus Weig. (*J. globiceps* Bajt.) - 1, 6
castaneus Smith - 1, 3, 4, 5
- subsp. *leucochlamys* (Zing. ex V. Krecz.) Hult. = J. leucochlamys
- subsp. *triceps* (Rostk.) V. Novik. = J. triceps
compressus Jacq. (*J. compressus* var. *subcompressus* (Zak. & Novopokr.) V. Novik., *J. subcompressus* Zak. & Novopokr.) - 1, 2, 3, 4, 5, 6
- subsp. *gracillimus* (Buchenau) Worosch. = J. gracillimus
- subsp. *jaxarticus* (V. Krecz. & Gontsch.) Jafri = J. jaxarticus
- var. *subcompressus* (Zak. & Novopokr.) V. Novik. = J. compressus
conglomeratus L. (*J. effusus* L. f. *conglomeratus* (L.) Reichgelt, *J. leersii* Marss., *J. subuliflorus* Drej. f. *leersii* (Marss.) Reichgelt) - 1, 3
curvatus Buchenau - 5
decipiens (Buchenau) Nakai - 5
effusus L. - 1, 2, 3
- f. *conglomeratus* (L.) Reichgelt = J. conglomeratus
elbrusicus Galushko - 2
ensifolius Wikstr. (*J. oligocephalus* Satake & Ohwi) - 5
equisetinus Proskurjakova - 6
erythropodus V. Krecz. = J. bufonius
falcatus E. Mey. subsp. *prominens* (Buchenau) V. Novik. = J. prominens
- subsp. *sitchensis* (Buchenau) Hult. = J. prominens
- var. *sitchensis* Buchenau = J. prominens
fauriei Levl. & Vaniot, p.p. = J. beringensis and J. yokoscensis
fauriensis Buchenau - 5
- subsp. *kamtschatcensis* (Buchenau) V. Novik. = J. kamtschatcensis
filiformis L. - 1, 2, 3, 4, 5, 6
fischerianus Turcz. ex V. Krecz. = J. alpino-articulatus subsp. fischerianus
flavescens Host = Luzula luzulina
fominii Zoz - 1
fontanesii J. Gay ex Laharpe - 2
- subsp. *kotschyi* (Boiss.) Snogerup = J. kotschyi
fuscoater Schreb. = J. alpino-articulatus
- [subsp.]*J. arthrophyllus* Brenn. = J. alpino-articulatus
geniculatus auct. = J. alpino-articulatus
geniculatus Schrank = J. articulatus
- subsp. *arthrophyllus* (Brenn.) Soo = J. alpino-articulatus
gerardii Loisel. (*J. gerardii* subsp. *macrocarpus* A. Tarass., *J. gerardii* subsp. *microcarpus* A. Tarass., *J. intermedius* A. Tarass. 1970, non Thuill. 1799) - 1, 2, 3, 4, 5, 6
- subsp. *atrofuscus* (Rupr.) Printz = J. atrofuscus
- subsp. *atrofuscus* (Rupr.) Tolm. = J. atrofuscus
- subsp. *macrocarpus* A. Tarass. = J. gerardii
- subsp. *microcarpus* A. Tarass. = J. gerardii
- subsp. *soranthus* (Schrenk) K. Richt. = J. soranthus
- subsp. *soranthus* (Schrenk) Snogerup = J. soranthus
- subsp. *vvedenskyi* (V. Krecz.) V. Novik. = J. vvedenskyi
glaucus Ehrh. var. *yokoscensis* Franch. & Savat. = J. yokoscensis
globiceps Bajt. = J. capitatus
gontscharovii Sumn. = J. triglumis
gracillimus (Buchenau) V. Krecz. & Gontsch. (*J. compressus* Jacq. subsp. *gracillimus* (Buchenau) Worosch.) - 4, 5
grubovii V. Novik. (*J. muelleri* Trautv. subsp. *grubovii* (V. Novik.) V. Novik.) - 4
gubanovii V. Novik. - 6

haenkei E. Mey. (*J. arcticus* Willd. subsp. *sitchensis* Engelm., *J. balticus* Willd. subsp. *sitchensis* (Engelm.) Hult.) - 5

heldreichianus Marss. ex Parl. (*J. heldreichianus* subsp. *orientalis* Snogerup, *J. heldreichianus* var. *orientalis* (Snogerup) V. Novik., *J. orientalis* (Snogerup) Filimonova) - 2, 6

- subsp. *orientalis* Snogerup = J. heldreichianus
- var. *orientalis* (Snogerup) V. Novik. = J. heldreichianus

heptopotamicus V. Krecz. & Gonsch. - 6

himalensis Klotzsch - 6

hybridus auct. = J. bufonius

hybridus Brot. subsp. *nastanthus* (V. Krecz. & Gontsch.) V. Novik. = J. nastanthus

inflexus L. (*J. paniculatus* Hoppe ex Mert. & Koch) - 1, 2, 6

- subsp. *brachytepalus* (V. Krecz. & Gontsch.) V. Novik. = J. brachytepalus
- var. *brachytepalus* (V. Krecz. & Gontsch.) Kitam. = J. brachytepalus

intermedius A.Tarass. = J. gerardii

x **inundatus** Drej. (*J. balticus* Willd. subsp. *inundatus* (Drej.) A. & D. Löve, comb. superfl., *J. balticus* subsp. *inundatus* (Drej.) K. Richt.). - J. balticus Willd. x J. filiformis L. - 1

jaxarticus V. Krecz. & Gontsch. (*J. compressus* Jacq. subsp. *jaxarticus* (V. Krecz. & Gontsch.) Jafri) - 6

juzepczukii V. Krecz. & Gontsch. = J. ambiguus

kamtschatcensis (Buchenau) Kudo (*J. fauriensis* Buchenau subsp. *kamtschatcensis* (Buchenau) V. Novik.) - 5

kotschyi Boiss. (*J. arianus* V. Krecz., *J. fontanesii* J. Gay ex Laharpe subsp. *kotschyi* (Boiss.) Snogerup, *J. kotschyi* var. *arianus* (V. Krecz.) V. Novik.) - 6

- var. *arianus* (V. Krecz.) V. Novik. = J. kotschyi

krameri Franch. & Savat. - 5

lampocarpus Ehrh. ex Hoffm. = J. articulatus

leersii Marss. = J. conglomeratus

leschenaultii J. Gay ex Laharpe (*J. wallichianus* Laharpe) - 5

leucochlamys Zing. ex V. Krecz. (*J. castaneus* Smith subsp. *leucochlamys* (Zing. ex V. Krecz.) Hult.) - 3, 4, 5

- subsp. **borealis** (Tolm.) V. Novik. (*J. leucochlamys* var. *borealis* Tolm.) - 4, 5

limosus Worosch. (*J. articulatus* L. subsp. *limosus* (Worosch.) Worosch.) - 5

littoralis C.A. Mey. (*J. acutus* L. subsp. *littoralis* (C.A. Mey.) Feinbrun) - 2, 6

- subsp. *tyraicus* (Pacz.) V. Novik. = J. tyraicus

longirostris Kuvajev - 4

luzulinus Vill. = Luzula luzulina

luzuloides Lam. = Luzula luzuloides

macer S.F. Gray = J. tenuis

macrantherus V. Krecz. & Gontsch. - 6

maritimus Lam. - 1, 2, 6(?)

- var. *arabicus* Aschers. & Buchenau = J. arabicus

membranaceus Royle - 6

minutulus V. Krecz. & Gontsch. (*J. bufonius* L. f. *minutulus* Albert & Jahand. non basionymum, *J. bufonius* subsp. *minutulus* (V. Krecz. & Gontsch.) Soo) - 1, 3, 6

x **montelii** Vierh. - J. arcticus Willd. x J. filiformis L. - 1

muelleri Trautv. (*J. arcticus* Willd. var. *muelleri* (Trautv.) Kovtonjuk) - 4

- subsp. *grubovii* (V. Novik.) V. Novik. = J. grubovii

multiflorus Retz. = Luzula multiflora

nastanthus V. Krecz. & Gontsch. (*J. bufonius* L. subsp. *nastanthus* (V. Krecz. & Gontsch.) Soo, *J. hybridus* Brot. subsp. *nastanthus* (V. Krecz. & Gontsch.) V. Novik.) - 1, 2, 3, 4, 6

nevskii V. Krecz. & Gontsch. - 6

nipponensis Buchenau = J. papillosus

nodulosus Wahlenb. (*J. alpino-articulatus* Chaix subsp. *nodulosus* (Wahlenb.) Hamet-Ahti, *J. alpino-articulatus* subsp. *rariflorus* (C. Hartm.) Breistroffer, *J. alpino-articulatus* subsp. *rariflorus* (C. Hartm.) Holub, comb. superfl., *J. alpinus* Vill. subsp. *nodulosus* (Wahlenb.) Lindm., *J. alpinus* subsp. *rariflorus* (C. Hartm.) K. Richt., *J. rariflorus* C.Hartm.) - 1, 3(?), 4, 5

obtusiflorus Ehrh. ex Hoffm. = J. subnodulosus

oligocephalus Satake & Ohwi = J. ensifolius

orchonicus V. Novik. - 3, 4

orientalis (Snogerup) Filimonova = J. heldreichianus

pallescens Wahlenb. = Luzula multiflora

paniculatus Hoppe ex Mert. & Koch = J. inflexus

papillosus Franch. & Savat. (*J. nipponensis* Buchenau) - 5

- var. *virens* (Buchenau) Worosch. = J. virens

persicus Boiss. subsp. *vvedenskyi* (V. Krecz.) V. Novik = J. vvedenskyi

potaninii Buchenau subsp. *woroschilovii* (A.A. Neczajev & V. Novik.) V. Novik. = J. woroschilovii

prominens (Buchenau) Miyabe & Kudo (*J. falcatus* E. Mey. subsp. *prominens* (Buchenau) V. Novik., *J. falcatus* var. *sitchensis* Buchenau, *J. falcatus* subsp. *sitchensis* (Buchenau) Hult.) - 5

x **raeperi** Aschers. & Graebn. - J. alpino-articulatus Chaix x J. articulatus L. - 1

ranarius Song. & Perrier ex Billot = J. ambiguus

- subsp. *turkestanicus* (V. Krecz. & Gontsch.) V. Novik. = J. turkestanicus

rariflorus C. Hartm. = J. nodulosus

rechingeri Snogerup - 6

rigidus auct. = J. arabicus

rochelianus Schult. & Schult. fil. = J. thomasii

salsuginosus Turcz. ex E. Mey. - 3, 4

- subsp. **tuvinicus** Kovtonjuk (*J. salsuginosus* var. *tuvinicus* (Kovtonjuk) V. Novik.) - 4

schischkinii Kryl. & Sumn. = J. triglumis

secundus Beauv. - 5(alien)

soranthus Schrenk (*J. gerardii* Loisel. subsp. *soranthus* (Schrenk) K. Richt., *J. gerardii* subsp. *soranthus* (Schrenk) Snogerup, comb. superfl.) - 1, 2, 3, 4, 6

sosnowskyi V. Novik. - 2

spadiceus All. = Luzula alpino-pilosa

sphaerocarpus Nees - 1, 2, 3, 4, 6

squarrosus L. - 1

stygius L. - 1, 3, 4, 5

- subsp. **americanus** (Buchenau) Hult. (*J. stygius* var. *americanus* Buchenau) - 5

subarticulatus Zak. & Novopokr. = J. articulatus

subcompressus Zak. & Novopokr. = J. compressus

subnodulosus Schrank (*J. obtusiflorus* Ehrh. ex Hoffm.) - 1

subulatus Forssk. - 6

subuliflorus Drej. f. *leersii* (Marss.) Reichgelt = J. conglomeratus

tatewakii Satake (*J. articulatus* L. subsp. *tatewakii* (Satake) Worosch.) - 5

tenageia Ehrh. ex L. fil. - 1, 2

tenuis Willd. (*J. macer* S.F. Gray) - 1, 2, 4, 5

thomasii Ten. (*J. rochelianus* Schult. & Schult. fil.) - 1

thomsonii Buchenau - 6

transsilvanicus Schur - 1

triceps Rostk. (*J. castaneus* Smith subsp. *triceps* (Rostk.) V. Novik.) - 3, 4, 5, 6

trichodes Steud. - 1(?)

trifidus L. - 1, 3, 4

triglumis L. (*J. albescens* (Lange) Fern., *J. gontscharovii*

Sumn., *J. schischkinii* Kryl. & Sumn., *J. triglumis* var. *albescens* Lange, *J. triglumis* subsp *albescens* (Lange) Hult., *J. triglumis* var. *schischkinii* (Kryl. & Sumn.) V. Novik.) - 1, 2, 3, 4, 5, 6
- subsp. *albescens* (Lange) Hult. = J. triglumis
- var. *albescens* Lange = J. triglumis
- var. *schischkinii* (Kryl. & Sumn.) V. Novik. = J. triglumis
turczaninowii (Buchenau) Freyn (*J. alpinus* Vill. subsp. *turczaninowii* (Buchenau) Worosch., *J. articulatus* L. subsp. *turczaninowii* (Buchenau) Worosch. comb. invalid.) - 4, 5
turkestanicus V. Krecz. & Gontsch. (*J. ambiguus* Guss. subsp. *turkestanicus* (V. Krecz. & Gontsch.) V. Novik., *J. bufonius* L. subsp. *turkestanicus* (V. Krecz. & Gontsch.) Soo, *J. ranarius* Song. & Perrier ex Billot subsp. *turkestanicus* (V. Krecz. & Gontsch.) V. Novik.) - 2, 4(?), 6
tyraicus (Pacz.) V. Krecz. & Gontsch. (*J. littoralis* C.A. Mey. subsp. *tyraicus* (Pacz.) V. Novik.) - 1
virens Buchenau (*J. papillosus* Franch. & Savat. var. *virens* (Buchenau) Worosch.) - 5
vvedenskyi V. Krecz. (*J. gerardii* Loisel. subsp. *vvedenskyi* (V. Krecz.) V. Novik., *J. persicus* Boiss. subsp. *vvedenskyi* (V. Krecz.) V. Novik.) - 6
wallichianus Laharpe = J. leschenaultii
woroschilovii A.A. Neczajev & V. Novik. (*J. potaninii* Buchenau subsp. *woroschilovii* (A.A. Neczajev & V. Novik.) V. Novik.) - 6
yokoscensis (Franch. & Savat.) Satave (*J. glaucus* Ehrh. var. *yokoscensis* Franch.& Savat., *J. fauriei* Levl. & Vaniot, p.p.) - 5

Luzula DC. (*Juncoides* Seguier)

abchasica V. Novik. - 2
acuminata Rafin. - 5
albida (Hoffm.) DC. = L. luzuloides
alpino-pilosa (Chaix) Breistroffer (*Juncus alpino-pilosus* Chaix, *J. spadiceus* All. 1785, non Vill. 1779, *Luzula spadicea* (All.) DC.) subsp. **obscura** Frohner (*L. obscura* (Frohner) V. Novik.) - 1
arctica auct. = L. nivalis
arctica Blytt subsp. *latifolia* (Kjellm.) A. Pors. = L. camtschadalorum
arcuata (Wahlenb.) Sw. (*Juncus arcuatus* Wahlenb.) - 1
- subsp. *unalaschkensis* (Buchenau) Hult. = L. unalaschkensis
- var. *unalaschkensis* Buchenau = L. unalaschkensis
- f. *latifolia* Kjellm. = L. camtschadalorum
beringensis Tolm. - 5
berringiana Gjaerevoll = L. tundricola
campestris (L.) DC. (*Juncus campestris* L., *Luzula subpilosa* Gilib.) - 1
- subsp. *taurica* V. Krecz. = L. taurica
- var. *nivalis* Laest. = L. nivalis
camtschadalorum (Sam.) Gorodk. ex Kryl. (*L. arctica* Blytt subsp. *latifolia* (Kjellm.) A. Pors., *L. arcuata* (Wahlenb.) Sw. f. *latifolia* Kjellm., *L. unalaschkensis* (Buchenau) Satake subsp. *camtschadalorum* (Sam.) Tolm. non rite publ.) - 5
capitata (Miq.) Kom. - 5
caspica Rupr. ex Bordz. = L. forsteri subsp. caspica
confusa Lindeb. - 1, 3, 4, 5
divulgata Kirschner - 1
fastigiata E. Mey. (*L. parviflora* (Ehrh.) Desv. subsp. *fastigata* (E. Mey.) Hamet-Ahti) - 5
flavescens (Host) Gaudin = L. luzulina
forsteri (Smith) DC. - 1, 2

- subsp. *caspica* (Rupr. ex Bordz.) V. Novik. (*L. caspica* Rupr. ex Bordz.) - 2
frigida (Buchenau) Sam. (*L. kjellmaniana* Miyabe & Kudo subsp. *frigida* (Buchenau) Schljak., *L. multiflora* (Retz.) Lej. subsp. *frigida* (Buchenau) V. Krecz.) - 1, 3, 4
japonica Buchenau = L. plumosa
jimboi Miyabe & Kudo = L. rostrata
kjellmaniana Miyabe & Kudo (*L. multiflora* (Retz.) Lej. subsp. *kjellmaniana* (Miyabe & Kudo) Tolm., *L. sudetica* (Willd.) Schult. var. *kjellmaniana* (Miyabe & Kudo) T. Shimizu) - 5
- subsp. *frigida* (Buchenau) Schljak. = L. frigida
- subsp. *sibirica* (V. Krecz.) Schljak. = L. sibirica
kobayasii Satake - 5
luzulina (Vill.) Dalla Torre & Sarnth. (*Juncus luzulinus* Vill., *J. flavescens* Host, *Luzula flavescens* (Host) Gaudin) - 1
luzuloides (Lam.) Dandy & Wilmott (*Juncus luzuloides* Lam., *J. albidus* Hoffm., *Luzula albida* (Hoffm.) DC., *L. nemorosa* (Poll.) E. Mey. 1848, non Hornem. 1815) - 1
- subsp. *cuprina* (Roch. ex Aschers.) Chrtek & Krisa - 1
macrocarpa (Buchenau) Nakai = L. rufescens
melanocarpa (Michx.) Desv. (*L. parviflora* (Ehrh.) Desv. subsp. *melanocarpa* (Michx.) Tolm., *L. spadicea* (All.) DC. f. *melanocarpa* (Michx.) Grint. nom. invalid.) - 5
multiflora (Ehrh.) Lej. (*Juncus campestris* L. var. *multiflorus* Ehrh., *J. multiflorus* Retz., *J. pallescens* Wahlenb. 1812, non Lam. 1789, *Luzula multiflora* subsp. *pallescens* (Sw.) Reichgelt, *L. pallescens* Sw.) - 1, 2, 3, 5(alien), 6
- subsp. *frigida* (Buchenau) V. Krecz. = L. frigida
- subsp. *kjellmaniana* (Miyabe & Kudo) Tolm. = L. kjellmaniana
- subsp. *pallescens* (Sw.) Reichgelt = L. multiflora
- subsp. *sibirica* V. Krecz. = L. sibirica
- subsp. *taurica* (V. Krecz.) V. Novik. = L. taurica
nemorosa (Poll.) E. Mey. = L. luzuloides
nivalis (Laest.) Spreng. (*L. campestris* (L.) DC. var. *nivalis* Laest., *L. arctica* auct.) - 1, 3, 4, 5
obscura (Frohner) V. Novik. = L. alpino-pilosa subsp. obscura
oligantha Sam. - 5
pallescens auct. = L. pallidula
pallescens Sw. = L. multiflora
pallidula Kirschner (*L. pallescens* auct.) - 1, 2, 3, 4, 5, 6
parviflora (Ehrh.) Desv. (*L. spadicea* (All.) DC. f. *parviflora* (Ehrh.) Grint.) - 1, 3, 4, 5
- subsp. *fastigata* (E. Mey.) Hamet-Ahti = L. fastigiata
- subsp. *melanocarpa* (Michx.) Tolm. = L. melanocarpa
pilosa (L.) Willd. - 1, 2, 3, 4, 6
- var. *macrocarpa* (Buchenau) B. Boivin = L. rufescens
piperi (Cov.) Jones (*Juncoides piperi* Cov., *Luzula wahlenbergii* Rupr. subsp. *piperi* (Cov.) Hult.) - 5
plumosa E. Mey. (*L. japonica* Buchenau) - 5
pseudosudetica (V. Krecz.) V. Krecz. - 2
rostrata Buchenau (*L. jimboi* Miyabe & Kudo) - 5
rufescens Fisch. ex E. Mey. (*L. macrocarpa* (Buchenau) Nakai, *L. pilosa* (L.) Willd. var. *macrocarpa* (Buchenau) B. Boivin, *L. rufescens* var. *macrocarpa* Buchenau) - 4, 5
- var. *macrocarpa* Buchenau = L. rufescens
sibirica V. Krecz. (*L. kjellmaniana* Miyabe & Kudo subsp. *sibirica* (V. Krecz.) Schljak., *L. multiflora* (Retz.) Lej. subsp. *sibirica* V. Krecz., *L. sibirica* var. *vinogradovii* (Sipl.) M. Ivanova, *L. vinogradovii* Sipl.) - 1, 3, 4, 5, 6

- var. *vinogradovii* (Sipl.) M. Ivanova = L. sibirica
spadicea (All.) DC. = L. alpino-pilosa
- f. *melanocarpa* (Michx.) Grint. = L. melanocarpa
- f. *parviflora* (Ehrh.) Grint. = L. parviflora
spicata (L.) DC. - 1, 2, 3, 4, 6
- subsp. **mongolica** V. Novik. - 3, 4, 6
- subsp. **mutabilis** Chrtek & Krisa - 1
stilbocarpa Kirschner & Krisa - 2
subpilosa Gilib. = L. campestris
sudetica (Willd.) Schult. - 1, 2
- var. *kjellmaniana* (Miyabe & Kudo) T. Shimizu = L. kjellmaniana
sylvatica (Huds.) Gaudin - 1, 2
taurica (V. Krecz.) V. Novik. (*L. campestris* (L.) DC. subsp. *taurica* V. Krecz., *L. multiflora* (Retz.) Lej. subsp. *taurica* (V. Krecz.) V. Novik.) - 1, 2
tundricola Gorodk. ex V. Vassil. (*L. berringiana* Gjaerevoll) - 3, 4, 5
unalaschkensis (Buchenau) Satake (*L. arcuata* (Wahlenb.) Sw. var. *unalaschkensis* Buchenau, *L. arcuata* subsp. *unalaschkensis* (Buchenau) Hult., *L. unalaschkensis* (Buchenau) V. Vassil. comb. superfl.) - 5
- subsp. *camtschadalorum* (Sam.) Tolm. = L. camtschadalorum
vinogradovii Sipl. = L. sibirica
wahlenbergii Rupr. - 1, 3, 4, 5
- subsp. *piperi* (Cov.) Hult. = L. piperi

JUNCAGINACEAE Rich.
Triglochin L.

asiaticum (Kitag.) A. & D.Löve (*T. maritimum* L. subsp. *asiaticum* Kitag.) - 5
komarovii Lipsch. & Pavl. = T. palustre
maritimum L. - 1, 2, 3, 4, 5, 6
- subsp. *asiaticum* Kitag. = T. asiaticum
palustre L. (*T. komarovii* Lipsch. & Pavl., *T. palustre* var. *komarovii* (Lipsch. & Pavl.) Tzvel.) - 1, 2, 3, 4, 5, 6
- var. *komarovii* (Lipsch. & Pavl.) Tzvel. = T. palustre

KOBRESIACEAE Gilly = CYPERACEAE

LABIATAE Juss. = LAMIACEAE

LAMIACEAE Lindl. (LABIATAE Juss.)
Acinos Mill.

alpinus (L.) Moench (*A. alpinus* subsp. *baumgartenii* (Simonk.) Pawl., *A. baumgartenii* (Simonk.) Klok., *Melissa baumgartenii* Simonk.) - 1
- subsp. *baumgartenii* (Simonk.) Pawl. = A. alpinus
arvensis (Lam.) Dandy (*A. schizodontus* Klok., *A. thymoides* Moench) - 1, 2, 5(alien)
- subsp. *eglandulosus* (Klok.) Tzvel. = A. villosus
- subsp. *villosus* (Pers.) Sojak = A. villosus
baumgartenii (Simonk.) Klok. = A. alpinus
eglandulosus Klok. = A. villosus
fominii Shost. = A. rotundifolius
graveolens (Bieb.) Link = A. rotundifolius
infectus Klok. = A. rotundifolius
rotundifolius Pers. (*A. fominii* Shost., *A. graveolens* (Bieb.) Link, *A. infectus* Klok., *A. subcrispus* Klok.) - 1, 2, 6
schizodontus Klok. = A. arvensis
subcrispus Klok. = A. rotundifolius

thymoides Moench = A. arvensis
villosus Pers. (*A. arvensis* (Lam.) Dandy subsp. *eglandulosus* (Klok.) Tzvel., *A. arvensis* subsp. *villosus* (Pers.) Sojak, *A. eglandulosus* Klok.) - 1

Agastache Clayt. ex Gronov.

rugosa (Fisch. & C.A. Mey.) O. Kuntze - 5

Ajuga L. (*Chamaepitys* Hill, *Phleboanthe* (Boiss.) Tausch)

chamaecistus Ging. ex Benth. - 2
- subsp. *turkestanica* (Regel) Rech. fil. = A. turkestanica
chamaepitys (L.) Schreb. (*Chamaepitys trifida* Dumort.) - 1
- subsp. *chia* (Schreb.) Arcang. = A. chia
- subsp. *chia* (Schreb.) Murb. = A. chia
- - var. *ciliata* Briq. = A. glabra
- subsp. *ciliata* (Briq.) Smejkal = A. glabra
chia Schreb. (*A. chamaepitys* (L.) Schreb. subsp. *chia* (Schreb.) Arcang., *A. chamaepitys* subsp. *chia* (Schreb.) Murb. comb. superfl., *Chamaepitys chia* (Schreb.) Holub) - 1, 2
comata auct. p.p. = A. glabra
genevensis L. - 1, 2
glabra C. Presl (? *A. chamaepitys* (L.) Schreb. subsp. *chia* (Schreb.) Murb. var. *ciliata* Briq., ? *A. chamaepitys* subsp. *ciliata* (Briq.) Smejkal, *A. pseudochia* Shost., *Chamaepitys glabra* (C.Presl) Holub, *Ajuga comata* auct. p.p.) - 1, 2, 6
laxmannii (L.) Benth. (*Phleboanthe laxmannii* (L.) Tausch) - 1, 2
mollis Gladkova (*Chamaepitys mollis* (Gladkova) Holub) - 1
multiflora Bunge - 4, 5
oblongata Bieb. - 2
orientalis L. - 1, 2
pseudochia Shost. = A. glabra
pyramidalis L. - 1
reptans L. - 1, 2
salicifolia (L.) Schreb. - 1
shikotanensis Miyabe & Tatew. - 5
turkestanica (Regel) Briq. (*A. chamaecistus* Ging. ex Benth. subsp. *turkestanica* (Regel) Rech. fil.) - 6
yezoensis Maxim. - 5

Alajja Ikonn. (*Erianthera* Benth. 1833, non Nees, 1832, *Susilkumara* Bennet, *Eriophyton* auct.)

afghanica (Rech. fil.) Ikonn. (*Eriophyton afghanicum* Rech. fil., *Erianthera afghanica* (Rech. fil.) R. Kam.) - 6
anomala (Juz.) Ikonn. (*Erianthera anomala* Juz., *Susilkumara anomala* (Juz.) Bennet) - 6
rhomboidea (Benth.) Ikonn. (*Erianthera rhomboidea* Benth., *Susilkumara rhomboidea* (Benth.) Bennet) - 6

Amaracus Gled.

rotundifolius (Boiss.) Briq. - 2

Amethystea L.

caerulea L. - 3, 4, 5, 6

Antonina Vved.

debilis (Bunge) Vved. (*Calamintha debilis* (Bunge) Benth.) - 2, 3, 6

Arischrada Pobed. = Salvia

bucharica (M. Pop.) Pobed. = Salvia bucharica
dracocephaloides (Boiss.) Pobed. = Salvia hydrangea
korolkowii (Regel & Schmalh.) Pobed. = Salvia korolkowii
multicaulis (Vahl) Pobed. = Salvia multicaulis

Ballota L.

borealis Schweigg. = B. nigra
foetida Lam. = B. nigra
grisea Pojark. - 2
longicalyx Klok. = B. nigra
nigra L. (*B. borealis* Schweigg., *B. foetida* Lam., *B. longicalyx* Klok., *B. nigra* subsp. *foetida* (Lam.) Hayek) - 1, 2
- subsp. *foetida* (Lam.) Hayek = B. nigra
- subsp. **kurdica** P.H. Davis - 2

Betonica L. = Stachys

abchasica (Bornm.) Chinth. = Stachys abchasica
bjelorussica Kossko ex Klok. = Stachys officinalis
brachyodonta Klok. = Stachys officinalis
foliosa Rupr. = Stachys betoniciflora
fusca Klok. = Stachys officinalis
grandiflora Willd. = Stachys macrantha
macrantha C. Koch = Stachys macrantha
macrostachya Wend. = Stachys macrostachya
nivea Stev. = Stachys discolor
- subsp. *abchasica* Bornm. = Stachys abchasica
- subsp. *ossetica* Bornm. = Stachys ossetica
- var. *abchasica* N. Pop. ex Grossh. = Stachys abchasica
officinalis L. = Stachys officinalis
orientalis L. = Stachys macrostachya
ossetica (Bornm.) Chinth. = Stachys ossetica
peraucta Klok. = Stachys officinalis

Calamintha Hill

ascendens Jord. = C. menthifolia
debilis (Bunge) Benth. = Antonina debilis
glandulosa (Requien) Benth. = C. parviflora
grandiflora (L.) Moench - 1, 2
largiflora Klok. = C. nepeta
macra Klok. = C. parviflora
menthifolia Host (*C. ascendens* Jord., *C. sylvatica* Bromf. subsp. *ascendens* (Jord.) P.W. Ball, *C. officinalis* sensu Boriss. p.p.) - 1, 2
- subsp. *sylvatica* (Bromf.) Menitsky = C. sylvatica
nepeta (L.) Savi (*C. largiflora* Klok., *C. officinalis* Moench) - 1, 2
- subsp. *glandulosa* (Requien) P.W. Ball = C. parviflora
officinalis Moench = C. nepeta
officinalis sensu Boriss. = C. menthifolia and C. sylvatica
parviflora Lam. (*C. glandulosa* (Requien) Benth., *C. macra* Klok., *C. nepeta* (L.) Savi subsp.*glandulosa* (Requien) P.W. Ball, *Thymus glandulosus* Requien) - 1
sylvatica Bromf. (*C. menthifolia* Host subsp. *sylvatica* (Bromf.) Menitsky, *C. officinalis* sensu Boriss. p.p.) - 1
- subsp. *ascendens* (Jord.) P.W. Ball = C. menthifolia

Chaiturus Willd.

marrubiastrum (L.) Reichenb. - 1, 2, 3, 6

Chamaepitys Hill = Ajuga

chia (Schreb.) Holub = Ajuga chia
glabra (C. Presl) Holub = Ajuga glabra

mollis (Gladkova) Holub = Ajuga mollis
trifida Dumort. = Ajuga chamaepitys

Chamaesphacos Schrenk

ilicifolius Schrenk - 6

Clinopodium L.

chinense (Benth.) O. Kuntze - 5
gracile (Benth.) O. Kuntze var. *sachalinense* (Fr. Schmidt) Ohwi = C. sachalinense
integerrimum Boriss. - 6
sachalinense (Fr. Schmidt) Koidz. (*C. gracile* (Benth.) O. Kuntze var. *sachalinense* (Fr. Schmidt) Ohwi) - 5
umbrosum (Bieb.) C. Koch (*Satureja umbrosa* (Bieb.) Greuter & Burdet) - 2
vulgare L. - 1, 2, 3, 4
- subsp. **arundanum** (Boiss.) Nym. (*Melissa arundana* Boiss.) - 2

Dalanum Dostal = Galeopsis

angustifolium (Ehrh. ex Hoffm.) Dostal = Galeopsis angustifolia
ladanum (L.) Dostal = Galeopsis ladanum

Dracocephalum L.

altaiense Laxm. = D. grandiflorum
argunense Fisch. ex Link - 4, 5
austriacum L. - 1, 2
bipinnatum Rupr. - 6
botryoides Stev. - 2
bungeanum Schischk. & Serg. (*D. origanoides* Steph. subsp. *bungeanum* (Schischk. & Serg.) A. Budantz.) - 3
butkovii Krassovskaja = D. komarovii
discolor Bunge - 3, 4
diversifolium Rupr. - 6
foetidum Bunge - 3, 4
formosum Gontsch. - 6
fragile Turcz. ex Benth. - 3, 4
fruticulosum Steph. - 4
goloskokovii Roldug. = D. integrifolium
grandiflorum L. (*D. altaiense* Laxm.) - 3, 4, 6
heterophyllum Benth. - 4, 6
- subsp. **ovalifolium** A. Budantz. - 4
imberbe Bunge - 3, 4, 6
integrifolium Bunge (*D. goloskokovii* Roldug.) - 3, 6
karataviense Pavl. & Roldug. - 6
komarovii Lipsky (*D. butkovii* Krassovskaja) - 6
krylovii Lipsky - 3
moldavica L. - 1, 2, 3, 4, 5, 6
multicaule Montbr. & Auch. ex Benth. - 2
multicolor Kom. - 5
nodulosum Rupr. - 6
nuratavicum Adylov - 6
nutans L. (*D. nutans* subsp. *subarcticum* Kuvajev) - 1, 3, 4, 5, 6
oblongifolium Regel - 6
origanoides Steph. - 3, 4, 6
- subsp. *bungeanum* (Schischk. & Serg.) A. Budantz. = D. bungeanum
palmatum Steph. - 5
paulsenii Briq. - 4, 6
pavlovii Roldug. - 6
peregrinum L. - 3, 4, 6
pinnatum L. - 4

popovii Egor. & Sipl. - 4
ruyschiana L. - 1, 2, 3, 4, 6
schischkinii Strizhova - 6
scrobiculatum Regel - 6
spinulosum M. Pop. - 6
stamineum Kar. & Kir. - 6
stellerianum Hiltebr. - 4, 5
subcapitatum (O. Kuntze) Lipsky - 6
thymiflorum L. - 1, 2, 3, 4, 5(alien), 6

Drepanocaryum Pojark.

sewerzowii (Regel) Pojark. - 6

Dysophylla Blume (*Pogostemon* auct.)

yatabeana Makino (*Pogostemon yatabeanus* (Makino) J.R. Press) - 5

Elsholtzia Willd.

ciliata (Thunb.) Hyl. (*Sideritis ciliata* Thunb., *Elsholtzia patrinii* (Lepech.) Garcke) - 1, 3, 4, 5
densa Benth. - 6
patrinii (Lepech.) Garcke = E. ciliata
pseudocristata Levl. & Vaniot - 5
serotina Kom. - 5

Eremostachys Bunge

acaulis Rech. fil. = Phlomoides speciosa
affinis Schrenk - 3, 6
alberti Regel = Phlomoides alberti
ambigua M. Pop. ex Pazij & Vved. = Phlomoides ambigua
angreni M. Pop. = Phlomoides angreni
anisochila Pazij & Vved. = Paraeremostachys anisochila
aralensis Bunge = Paraeremostachys aralensis
arctiifolia M. Pop. = Phlomoides arctiifolia
baburii Adyl. = Phlomoides baburii
baissunensis M. Pop. = Phlomoides baissunensis
baldschuanica Regel = Phlomoides baldschuanica
beckeri Regel = Phlomoides beckeri
boissieriana Regel = Phlomoides boissieriana
botschantzevii Adyl. = Phlomoides botschantzevii
canescens M. Pop. = Phlomoides mihaelis
cephalariifolia M. Pop. = Phlomoides cephalariifolia
codonocalyx Pazij & Vved. (*E. desertorum* Regel subsp. *ferganensis* M. Pop.) - 6
cordifolia Regel = Phlomoides cordifolia
czuiliensis Golosk. = Phlomoides czuiliensis
desertorum Regel = Paraeremostachys desertorum
- subsp. *ferganensis* M. Pop. = Eremostachys codonocalyx
dshungarica (M. Pop.) Golosk. = Paraeremostachys dshungarica
dzhambulensis Knorr. = Phlomoides speciosa
ebracteolata M. Pop. = Phlomoides ebracteolata
edelbergii Rech. fil. = Phlomoides speciosa
eriocalyx Regel = Phlomoides eriocalyx
- var. *taschkentica* M. Pop. = Phlomoides impressa
eriolarynx Pazij & Vved. - 6
ferganensis Ubuk. = Phlomoides kirghisorum
fetisowii Regel = Phlomoides fetisowii
fulgens Bunge = Phlomoides fulgens
glabra Boiss. ex Benth. = Phlomoides glabra
glanduligera M. Pop. = Phlomoides baissunensis
gymnocalyx Schrenk = Phlomoides gymnocalyx
gypsacea M. Pop. = Phlomoides gypsacea
hajastanica Sosn. ex Grossh. = Phlomoides laciniata
hissarica Regel = Phlomoides hissarica

iberica Vis. = Phlomoides laciniata
iliensis Regel = Phlomoides iliensis
impressa Pazij & Vved. = Phlomoides impressa
integior Pazij & Vved. = Phlomoides integior
isochila Pazij & Vved. - 6
jakkabaghi M. Pop. = Phlomoides baissunensis
karatavica Pavl. = Paraeremostachys karatavica
kaufmanniana Regel = Phlomoides kaufmanniana
korovinii M. Pop. = Phlomoides korovinii
labiosa Bunge = Phlomoides labiosa
labiosiformis (M. Pop.) Knorr. = Phlomoides labiosiformis
labiosissima Pazij & Vved. = Phlomoides labiosissima
laciniata (L.) Bunge = Phlomoides laciniata
lehmanniana Bunge = Phlomoides lehmanniana
leiocalyx Pazij & Vved. = Phlomoides leiocalyx
macrophylla Montbr. & Auch. ex Benth. (*E. moluccelloides* Bunge subsp. *macrophylla* (Montbr. & Auch. ex Benth.) Takht.) - 2
mogianica M. Pop. - 6
moluccelloides Bunge - 3, 6
- subsp. *macrophylla* (Montbr. & Auch. ex Benth.) Takht. = E. macrophylla
napuligera Franch. = Phlomoides napuligera
nuda Regel = Phlomoides nuda
- var. *septentrionalis* M. Pop. = Phlomoides septentrionalis
- f. *simplex* M. Pop. = Phlomoides integior
paniculata Regel = Paraeremostachys paniculata
pauciflora O. Kuntze = Phlomoides pauciflora
pectinata M. Pop. = Phlomoides pectinata
persimilis Aitch. & Hemsl. - 6
phlomoides Bunge = Paraeremostachys phlomoides
- f. *dshungarica* M. Pop. = Paraeremostachys dshungarica
popovii Gontsch. = Phlomoides popovii
pulchra M. Pop. = Phlomoides pulchra
regeliana Aitch. & Hemsl. = Phlomoides regeliana
rotata Schrenk ex Fisch. & C.A. Mey. - 6
sanglechensis Rech. fil. = Phlomoides sanglechensis
sarawschanica Regel = Phlomoides sarawschanica
- f. *lyrata* M. Pop. = Phlomoides lyrata
schugnanica (M. Pop.) Knorr. = Phlomoides schugnanica
septentrionalis (M. Pop.) Golosk. = Phlomoides septentrionalis
sogdiana Pazij & Vved. = Paraeremostachys sogdiana
speciosa Rupr. = Phlomoides speciosa
- var. *bipinnatifida* (Regel) M. Pop. = Phlomoides speciosa
- var. *brevicaulis* (Regel) M. Pop. = Phlomoides speciosa
subspicata M. Pop. = Phlomoides subspicata
tadschikistanica B. Fedtsch. = Phlomoides tadschikistanica
taschkentica (M. Pop.) Golosk. = Phlomoides impressa
tianschanica M. Pop. = Phlomoides tianschanica
transoxana Bunge = Paraeremostachys transoxana
tuberosa (Pall.) Bunge - 1, 3, 6
uniflora Regel = Phlomoides uniflora
zenaidae M. Pop. = Phlomoides zenaidae

Erianthera Benth. = Alajja

afghanica (Rech. fil.) R. Kam. = Alajja afghanica
anomala Juz. = Alajja anomala
rhomboidea Benth. = Alajja rhomboidea

Eriophyton auct. = Alajja

afghanicum Rech. fil. = Alajja afghanica

Galeobdolon Adans. (*Lamiastrum* Heist. ex Fabr. nom. specif. uninom., *Pollichia* auct.)

caucasicum A. Khokhr. - 2

luteum Huds. (*Lamiastrum galeobdolon* (L.) Ehrend. &
Polatschek) - 1, 2
- subsp. *montanum* (Pers.) Dvorakova = G. montanum
- subsp. *montanum* (Pers.) R. Mill = G. montanum
montanum (Pers.) Reichenb. (*Pollichia montana* Pers.,
Galeobdolon luteum Huds. subsp. *montanum* (Pers.)
Dvorakova, *G. luteum* subsp. *montanum* (Pers.) R.
Mill, *Lamiastrum galeobdolon* (L.) Ehrend. & Polat-
schek subsp. *montanum* (Pers.) Ehrend. & Polatschek,
L. montanum (Pers.) Ehrend.) - 1

Galeopsis L. (*Dalanum* Dostal, *Ladanum* Gilib.)

angustifolia Ehrh. ex Hoffm. (*Dalanum angustifolium*
(Ehrh. ex Hoffm.) Dostal) - 1
bifida Boenn. (*G. tetrahit* L. subsp. *bifida* (Boenn.) Fries) -
1, 2, 3, 4, 5, 6
ladanum L. (*Dalanum ladanum* (L.) Dostal, *Ladanum
intermedium* (Vill.) Slavikova) - 1, 2, 3, 4, 5, 6
nana Otschiauri - 2
pubescens Bess. - 1
speciosa Mill. - 1, 3, 4, 5
tetrahit L. - 1, 5(alien)
- subsp. *bifida* (Boenn.) Fries = G. bifida

Glechoma L.

hederacea L. - 1, 2, 3, 4, 5, 6
- subsp. **grandis** (A. Gray) Hara - 5
hirsuta Waldst. & Kit. - 1
longituba (Nakai) Kuprian. - 5
pannonica Borb. - 1

Gontscharovia Boriss.

popowii (B. Fedtsch. & Gontsch.) Boriss. (*Micromeria
gontscharovii* Vved. nom. illegit., *M. popowii* (B.
Fedtsch. & Gontsch.) Vved.) - 6

Hesiodia Moench = Sideritis

montana (L.) Dumort. = Sideritis montana
- subsp. *comosa* (Rochel ex Benth.) Sojak = Sideritis
comosa

Hymenocrater Fisch. & C.A. Mey.

bituminosus Fisch. & C.A. Mey. (*H. calycinus* auct., *H.
elegans* auct.) - 2, 6
calycinus auct. = H. bituminosus
elegans auct. = H. bituminosus
incisidentatus Boriss. - 6

Hypogomphia Bunge

bucharica Vved. - 6
elatior (Regel) Vass. = H. turkestana
purpurea (Regel) Vved. ex Koczk. (*H. turkestanica* var.
purpurea Regel) - 6
turkestana Bunge (*H. elatior* (Regel) Vass.) - 6
- var. *purpurea* Regel = H. purpurea

Hyssopus L.

ambiguus (Trautv.) Iljin - 3, 6
angustifolius Bieb. - 2
cretaceus Dubjan. - 1
cuspidatus Boriss. - 3, 6

ferganensis Boriss. = H. seravschanicus
macranthus Boriss. - 3, 6
majae A. Khokhr. = Satureja spicigera
officinalis L. - 1, 4
seravschanicus (Dubjan.) Pazij (*H. ferganensis* Boriss., *H.
tianschanicus* Boriss.) - 6
tianschanicus Boriss. = H. seravschanicus

Kudrjaschevia Pojark.

allotricha Pojark. - 6
grubovii Koczk. - 6
jacubi (Lipsky) Pojark. - 6
korshinskyi (Lipsky) Pojark. - 6
nadinae (Lipsky) Pojark. - 6
pojarkoviae Ikonn. - 6

Ladanum Gilib. = Galeopsis

intermedium (Vill.) Slavikova = Galeopsis ladanum

Lagochilopsis Knorr. = Lagochilus

acutiloba (Ledeb.) Knorr. = Lagochilus acutilobus
bungei (Benth.) Knorr. = Lagochilus bungei
hirta (Fisch. & C.A. Mey.) Knorr. = Lagochilus hirtus
pungens (Schrenk) Knorr. = Lagochilus pungens
subhispida (Knorr.) Knorr. = Lagochilus subhispidus

Lagochilus Bunge (*Lagochilopsis* Knorr.)

acutilobus (Ledeb.) Fisch. & C.A. Mey. (*Lagochilopsis
acutiloba* (Ledeb.) Knorr.) - 6
altaicus Wu & Hsuan = L. bungei
androssowii Knorr. - 6
balchanicus Czerniak. - 6
botschantzevii R. Kam. & Zuckerwanik - 6
bungei Benth. (*Lagochilopsis bungei* (Benth.) Knorr.,
Lagochilus altaicus Wu & Hsuan) - 3, 6
cabulicus Benth. - 2, 6
diacanthophyllus (Pall.) Benth. (*L. diacanthophyllus* subsp.
karataviensis Knorr. ex Zuckerwanik) - 6
- subsp. *karataviensis* Knorr. ex Zuckerwanik = L. diacanthophyllus
drobovii R. Kam. & Zuckerwanik - 6
ferganensis Ikramov = L. pubescens
gypsaceus Vved. (*L. intermedius* Vved.) - 6
hirsutissimus Vved. - 6
hirtus Fisch. & C.A. Mey. (*Lagochilopsis hirta* (Fisch. &
C.A. Mey.) Knorr.) - 6
ilicifolius Bunge - 4
iliensis Wu & Hsuan = L. platyacanthus
inebrians Bunge - 6
intermedius Vved. = L. gypsaceus
kaschgaricus Rupr. - 6
keminsis K. Isakov = L. platyacanthus
knorringianus Pavl. - 6
kschtutensis Knorr. (*L. kschtutensis* subsp. *pubescens* R.
Kam. & Zuckerwanik) - 6
- subsp. *pubescens* R. Kam. & Zuckerwanik = L. kschtutensis
leiacanthus Fisch. & C.A. Mey. - 6
longidentatus Knorr. - 6
macrodontus Knorr. = L. platyacanthus
nevskii Knorr. - 6
occultiflorus Rupr. (*L. tianschanicus* Pavl.) - 6
- subsp. **linearilobus** Zuckerwanik - 6
- subsp. **usunachmaticus** (Knorr.) Zuckerwanik (*L. usunachmaticus* Knorr.) - 6

olgae R. Kam. - 6
paulsenii Briq. - 6
platyacanthus Rupr. (*L. iliensis* Wu & Hsuan, *L. keminsis*
K. Isakov, *L. macrodontus* Knorr.) - 6
platycalyx Schrenk - 6
proskorjakovii Ikramov - 6
pubescens Vved. (*L. ferganensis* Ikramov) - 6
pulcher Knorr. - 6
pungens Schrenk (*Lagochilopsis pungens* (Schrenk)
Knorr.) - 6
schugnanicus Knorr. = L. seravschanicus
seravschanicus Knorr. (*L. schugnanicus* Knorr., *L. serav-*
schanicus subsp. *schugnanicus* (Knorr.) Zuckerwanik) - 6
- subsp. *schugnanicus* (Knorr.) Zuckerwanik = L. seravschanicus
setulosus Vved. - 6
subhispidus Knorr. (*Lagochilopsis subhispida* (Knorr.)
Knorr.) - 6
taukumensis Zuckerwanik - 6
tianschanicus Pavl. = L. occultiflorus
turkestanicus Knorr. - 6
usunachmaticus Knorr. = L. occultiflorus subsp. usunachmaticus
vvedenskyi R. Kam. & Zuckerwanik - 6

Lagopsis Bunge

eriostachya (Benth.) Ik.-Gal. ex Knorr. - 4
flava Kar. & Kir. - 6
marrubiastrum (Steph.) Ik.-Gal. - 3
supina (Steph.) Ik.-Gal. ex Knorr. - 4

Lallemantia Fisch. & C.A. Mey.

baldshuanica Gontsch. - 6
canescens (L.) Fisch. & C.A. Mey. - 1(naturalized), 2
iberica (Bieb.) Fisch. & C.A. Mey. - 1(alien), 2, 6
kopetdaghensis Boriss. - 6
peltata (L.) Fisch. & C.A. Mey. - 2, 6
royleana (Benth.) Benth. - 1, 2, 3, 6

Lamiastrum Heist. ex Fabr. = Galeobdolon

galeobdolon (L.) Ehrend. & Polatschek = Galeobdolon
luteum
- subsp. *montanum* (Pers.) Ehrend. & Polatschek = Galeobdolon montanum
montanum (Pers.) Ehrend. = Galeobdolon montanum

Lamium L.

album L. (*L. dumeticola* Klok.) - 1, 2, 3, 4, 5, 6
- subsp. *crinitum* (Montbr. & Auch. ex Benth.) Mennema =
L. crinitum
- subsp. *hyrcanicum* (A. Khokhr.) Menitsky = L. hyrcanicum
- subsp. **orientale** R. Kam. & A. Budantz. - 4, 5
- subsp. *sempervirens* (A. Khokhr.) Menitsky = L. sempervirens
- subsp. *transcaucasicum* (A. Khokhr.) Menistky = L. transcaucasicum
- subsp. *turkestanicum* (Kuprian.) R. Kam. & A. Budantz. =
L. turkestanicum
- var. *maculatum* L. = L. maculatum
amplexicaule L. (*L. stepposum* Kossko ex Klok.) - 1, 2, 3, 4,
5, 6
armenum Boiss. - 2
barbatum Siebold & Zucc. - 5

caucasicum Grossh. (*L. purpureum* L. subsp. *caucasicum*
(Grossh.) Menitsky) - 2
confertum Fries (*L. moluccellifolium* Fries) - 1
crinitum Montbr. & Auch. ex Benth. (*L. album* L. subsp.
crinitum (Montbr. & Auch. ex Benth.) Mennema) - 2,
6
cupreum Schott (*L. maculatum* (L.) L. subsp. *cupreum*
(Schott) Hadac) - 1
dumeticola Klok. = L. album
garganicum L. subsp. *reniforme* (Montbr. & Auch. ex
Benth.) R. Mill = L. reniforme
glaberrimum (C. Koch) Taliev - 1
gundelsheimeri C. Koch - 2
hybridum Vill. - 1
hyrcanicum A. Khokhr. (*L. album* L. subsp. *hyrcanicum* (A.
Khokhr.) Menitsky) - 2
laevigatum auct. = L. maculatum
macrodon Boiss. & Huet (*L. ordubadicum* Grossh.) - 2
maculatum (L.) L. (*L. album* L. var. *maculatum* L., *L.*
laevigatum auct.) - 1, 2
- subsp. *cupreum* (Schott) Hadac = L. cupreum
moluccellifolium Fries = L. confertum
ordubadicum Grossh. = L. macrodon
orvala L. - 1(?)
paczoskianum Worosch. - 1, 6
purpureum L. - 1, 2, 3, 4
- subsp. *caucasicum* (Grossh.) Menitsky = L. caucasicum
reniforme Montbr. & Auch. ex Benth. (*L. garganicum* L.
subsp. *reniforme* (Montbr. & Auch. ex Benth.) R.
Mill) - 2
sempervirens A. Khokhr. (*L. album* L. subsp. *sempervirens*
(A. Khokhr.) Menitsky) - 2
stepposum Kossko ex Klok. = L. amplexicaule
tomentosum Willd. - 2
transcaucasicum A. Khokhr. (*L. album* L. subsp. *transcau-*
casicum (A. Khokhr.) Menistky) - 2
turkestanicum Kuprian. (*L. album* L. subsp. *turkestanicum*
(Kuprian.) R. Kam. & A. Budantz.) - 6

Lavandula L.

***angustifolia** Mill. (*L. spica* L. nom. ambig.) - 1, 2, 6
spica L. = L. angustifolia

Leonuroides Rauschert = Panzerina

argyracea (Kuprian.) Rauschert = Panzerina lanata
lanata (L.) Rauschert = Panzerina lanata

Leonurus L.

artemisiae (Lour.) S.Y. Hu = L. japonicus
cardiaca L. - 1
- subsp. *glaucescens* (Bunge) Schmalh. = L. glaucescens
- subsp. *turkestanicus* (V. Krecz. & Kuprian.) Rech. fil. = L.
turkestanicus
- subsp. *villosus* (Desf. ex D'Urv.) Hyl. = L. quinquelobatus
deminutus V. Krecz. - 4
glaucescens Bunge (*L. cardiaca* L. subsp. *glaucescens*
(Bunge) Schmalh.) - 1, 2, 3, 6
heterophyllus Sweet = L. japonicus
incanus V. Krecz. & Kuprian. - 6
japonicus Houtt. (*L. artemisiae* (Lour.) S.Y. Hu, *L. hetero-*
phyllus Sweet, *Stachys artemisiae* Lour.) - 3(?), 4, 5
kudrjaschevii R. Kam. & Tulaganova = L. turkestanicus var.
songoricus
macranthus Maxim. - 5
mongolicus V. Krecz. & Kuprian. - 4

oreades Pavl. = L. panzerioides
panzerioides M. Pop. (*L. oreades* Pavl.) - 6
persicus Boiss. - 2(?)
quinquelobatus Gilib. (*L. cardiaca* L. subsp. *villosus* (Desf. ex D'Urv.) Hyl., *L. villosus* Desf. ex D'Urv.) - 1, 2, 3, 5(alien)
sibiricus L. - 3(?), 4, 5
tataricus L. - 3, 4
turkestanicus V. Krecz. & Kuprian. (*L. cardiaca* L. subsp. *turkestanicus* (V. Krecz. & Kuprian.) Rech. fil.) - 6
- var. **songoricus** Krestovsk. (*L. kudrjaschevii* R. Kam. & Tulaganova) - 6
villosus Desf. ex D'Urv. = L. quinquelobatus

Lophanthus Adans.

chinensis Benth. - 4, 5
elegans (Lipsky) Levin - 6
kryloii Lipsky - 3, 6
lipskyanus Ik.-Gal. & Nevski = L. ouroumitanensis
ouroumitanensis (Franch.) Koczk. (*Nepeta ouroumitanensis* Franch., *Lophanthus lipskyanus* Ik.-Gal. & Nevski) - 6
schrenkii Levin - 6
schtschurowskianus (Regel) Lipsky - 6
subnivalis Lipsky - 6
- var. *virescens* Lipsky = L. virescens
tschimganicus Lipsky - 6
varzobicus Koczk. - 6
virescens (Lipsky) Koczk. (*L. subnivalis* Lipsky var. *virescens* Lipsky) - 6

Lycopus L.

angustus Makino = L. maackianus
diantherus Buch.-Ham. ex Roxb. = Mosla dianthera
europaeus L. - 1, 2, 3, 4, 6
exaltatus L. fil. - 1, 2, 3, 4, 6
hirtellus Kom. - 5
lucidus Turcz. ex Benth. - 4, 5
maackianus (Maxim.) Makino (*L. angustus* Makino) - 5
parviflorus Maxim. = L. uniflorus
uniflorus Michx. (*L. parviflorus* Maxim.) - 5

*Majorana Hill

*hortensis Moench - 1, 6

Marrubium L.

alternidens Rech. fil. = M. anisodon
anisodon C. Koch (*M. alternidens* Rech. fil., *M. kusnezowii* N. Pop.) - 2, 6
astracanicum Jacq. (*M. goktschaicum* N. Pop., *M. purpureum* Bunge, *M. turkeviczii* Knorr.) - 2, 6
catariifolium Desr. - 2
civice Klok. = M. peregrinum
goktschaicum N. Pop. = M. astracanicum
kusnezowii N. Pop. = M. anisodon
leonuroides Desr. - 1, 2
nanum Knorr. - 2
parviflorum Fisch. & C.A. Mey. - 2
peregrinum L. (*M. civice* Klok.) - 1, 2
persicum C.A. Mey. - 2
pestalozzae auct. = M. praecox
plumosum C.A. Mey. - 2
praecox Janka (*M. pestalozzae* auct.) - 1, 2
propinquum Fisch. & C.A. Mey. - 2
purpureum Bunge = M. astracanicum
turkeviczii Knorr. = M. astracanicum

vulgare L. - 1, 2, 6
woronowii N. Pop. - 2

Meehania Britt.

urticifolia (Miq.) Makino - 5

Melissa L.

altissima Smith (*M. officinalis* L. subsp. *altissima* (Smith) Arcang.) - 2
arundana Boiss. = Clinopodium vulgare subsp. arundanum
baumgartenii Simonk. = Acinos alpinus
bicornis Klok. = M. officinalis
officinalis L. (*M. bicornis* Klok.) - 1, 2, 6
- subsp. *altissima* (Smith) Arcang. = M. altissima

Melittis L.

carpatica Klok. (*M. melissophyllum* L. subsp. *carpatica* (Klok.) P.W. Ball, *M. melissophyllum* var. *carpatica* (Klok.) Soo & Borsos, *M. melissophyllum* auct. p.p.) - 1
- var. *sarmatica* (Klok.) Soo = M. sarmatica
melissophyllum auct. = M. carpatica and M. sarmatica
melissophyllum L. subsp. *carpatica* (Klok.) P.W. Ball = M. carpatica
- subsp. *sarmatica* (Klok.) Gladkova = M. sarmatica
- var. *carpatica* (Klok.) Soo & Borsos = M. carpatica
- var. *sarmatica* (Klok.) Soo & Borsos = M. sarmatica
sarmatica Klok. (*M. carpatica* Klok. var. *sarmatica* (Klok.) Soo, *M. melissophyllum* L. subsp. *sarmatica* (Klok.) Gladkova, *M. melissophyllum* var. *sarmatica* (Klok.) Soo & Borsos, *M. melissophyllum* auct. p.p.) - 1

Mentha L. (*Pulegium* Mill.)

alaica Boriss. - 6
aquatica L. - 1, 2
arvensis L. (*M. arvensis* subsp. *parietariifolia* (J. Beck.) Briq., *M. lapponica* Wahlenb., *M. parietariifolia* J. Beck.) - 1, 2, 3, 4, 6
- subsp. *parietariifolia* (J. Beck.) Briq. = M. arvensis
- subsp. *piperascens* (Malinvaud ex Holmes) Hara = M. canadensis
- var. *piperascens* Malinvaud ex Holmes = M. canadensis
asiatica Boriss. (*M. kopetdaghensis* Boriss., *M. longifolia* (L.) Huds. var. *asiatica* (Boriss.) Rech. fil., *M. vagans* Boriss.) - 2, 3, 6
canadensis L. (*M. arvensis* L. var. *piperascens* Malinvaud ex Holmes, *M. arvensis* subsp. *piperascens* (Malinvaud ex Holmes) Hara, *M. haplocalyx* Briq., *M. sachalinensis* (Briq.) Kudo) - 3, 4, 5
x **carinthiaca** Host. - M. arvensis L. x M. rotundifolia (L.) Huds. - 1
> M. rotundifolia (L.) Huds. itself probably does not occur in this territory.

caucasica Gand. (*M. longifolia* (L.) Huds. subsp. *caucasica* (Gand.) Briq.) - 2
crispa L. p.p. = M. x piperita and M. spicata
crispata Schrad. ex Willd. = M. spicata
daghestanica Boriss. = M. pulegium
dahurica Benth. - 4, 5
x **dalmatica** Tausch. - M. arvensis L. x M. longifolia (L.) Huds. - 1
darvasica Boriss. - 6
x **dumetorum** Schult. - M. aquatica L. x M. longifolia (L.) Huds. - 1

292

x **gentilis** L. - M. arvensis L. x M. spicata L. - 1
haplocalyx Briq. = M. canadensis
x **interrupta** Boriss. 1954, non H. Br. 1890. - M. arvensis L.
   x M. asiatica Boriss. - 1
kopetdaghensis Boriss. = M. asiatica
lapponica Wahlenb. = M. arvensis
longifolia (L.) Huds. (*M. spicata* L. subsp. *longifolia* (L.)
   Tacik) - 1, 2, 3
- subsp. *caucasica* (Gand.) Briq. = M. caucasica
- var. *asiatica* (Boriss.) Rech. fil. = M. asiatica
micrantha (Fisch. ex Benth.) Litv. - 1, 6
pamiroalaica Boriss. - 6
parietariifolia J. Beck. = M. arvensis
***x piperita** L. (*M. crispa* L. p.p.). - M. aquatica L. x M.
   spicata L. - 1, 2, 6
pulegium L. (*M. daghestanica* Boriss., *Pulegium daghestani-*
   cum (Boriss.) Holub) - 1, 2, 6
rotundifolia sensu Boriss. = M. suaveolens
sachalinensis (Briq.) Kudo = M. canadensis
***spicata** L. (*M. crispa* L. p.p. quoad lectotypum, *M. crispata*
   Schrad. ex Willd.) - 1, 2, 6
- subsp. *longifolia* (L.) Tacik = M. longifolia
suaveolens Ehrh. (*M. rotundifolia* sensu Boriss.) - 1, 3, 6
vagans Boriss. = M. asiatica
x **verticillata** L. - M. aquatica L. x M. arvensis L. - 1

Metastachydium Airy Shaw ex C.Y. Wu & H.W. Li (*Metastachys* Knorr. 1954, non Tiegh. 1895)

sagittatum (Regel) C.Y. Wu & H.W. Li (*Metastachys sagit-*
   tata (Regel) Knorr.) - 6

Metastachys Knorr. = Megastachydium

sagittata (Regel) Knorr. = Metastachydium sagittatum

Micromeria Benth.

▪ elegans Boriss.
▪ elliptica C. Koch (*Satureja kochii* N.Pop. ex Greuter &
   Burdet)
fruticosa (L.) Druce   subsp. *serpyllifolia* (Bieb.) P.H.
   Davis = M. serpyllifolia
gontscharovii Vved. = Gontscharovia popowii
▪ marifolia (Willd.) Benth.
popowii (B. Fedtsch. & Gontsch.) Vved. = Gontscharovia
   popowii
serpyllifolia (Bieb.) Boiss. (*M. fruticosa* (L.) Druce subsp.
   serpyllifolia (Bieb.) P.H. Davis) - 1

Moluccella L.

laevis L. - 1, 2, 6

Mosla Buch.-Ham. ex Maxim. (*Orthodon* Benth. & Oliv. 1865, non R. Br. 1820)

dianthera (Roxb.) Maxim. (*Lycopus diantherus* Buch.-Ham.
   ex Roxb., *Orthodon grosseserratum* (Maxim.) Kudo) -
   2, 5

Navicularia Heist. ex Fabr. = Sideritis

syriaca (L.) Sojak subsp. *catillaris* (Juz.) Sojak = Sideritis
   catillaris
- subsp. *euxina* (Juz.) Sojak = Sideritis euxina
- subsp. *taurica* (Steph.) Sojak = Sideritis taurica

Nepeta L.

alaghezi Pojark. - 2
alatavica Lipsky - 6
amoena Stapf - 2
badachschanica Kudr. - 6
betonicifolia C.A. Mey. - 2
- subsp. *somkhetica* (Kapell.) Menitsky = N. somkhetica
- subsp. *strictifolia* (Pojark.) Menitsky = N. strictifolia
biebersteiniana (Trautv.) Pojark. - 2
botschantzevii Tscherneva - 6
brevifolia C.A. Mey. = N. lamiifolia
bracteata Benth. - 6
bucharica Lipsky - 6
buhsei Pojark. - 2
buschii Sosn. & Manden. (*N. supina* Stev. subsp. *buschii*
   (Sosn. & Manden.) Menitsky) - 2
cataria L. - 1, 2, 3, 5(alien), 6
consanguinea Pojark. - 6
cyanea Stev. - 2
czegemensis Pojark. - 2
daenensis Boiss. - 6
daghestanica Pojark. (*N. teucriifolia* Willd. subsp. *daghes-*
   tanica (Pojark.) A. Budantz.) - 2
densiflora Kar. & Kir. - 3, 6(?)
erivanensis Grossh. - 2
fedtschenkoi Pojark. = N. pungens
fissa C.A. Mey. = N. teucriifolia
floccosa auct. p.p. = N. vakhanica
floccosa Benth. - 6
formosa Kudr. - 6
globifera Bunge - 6
glutinosa Benth. - 6
gontscharovii Kudr. - 6
grandiflora Bieb. - 1(naturalized), 2
grossheimii Pojark. = N. strictifolia
hajastana Grossh. = N. mussinii
iberica Pojark. (*N. teucriifolia* Willd. subsp. *iberica* (Pojark.)
   A. Budantz.) - 2
ispahanica Boiss. - 6
knorringiana Pojark. = N. subhastata
kochii Czer. = N. meyeri
kokamirica Regel - 6
kokanica Regel - 6
komarovii E. Busch - 2
kopetdaghensis Pojark. (*N. ucranica* L. subsp. *kopetdaghen-*
   sis (Pojark.) Rech. fil.) - 6
kubanica Pojark. - 2
ladanolens Lipsky - 6
lamiifolia Willd. (*N. brevifolia* C.A. Mey.) - 2
lipskyi Kudr. - 6
longibracteata Benth. - 6
longituba Pojark. (*N. sosnovskyi* Asker.) - 2
manchuriensis S. Moore - 5
mariae Regel (*N. pulchella* Pojark.) - 6
maussarifi Lipsky - 6
meyeri Benth. (*N. kochii* Czer., *N. pallida* C. Koch, 1848,
   non Salisb. 1796) - 2
micrantha Bunge - 3, 6
microcephala Pojark. = N. pungens
mussinii Spreng. (*N. hajastana* Grossh., *N. reichenbachiana*
   Fisch. & C.A. Mey., *N. transcaucasica* Grossh.) - 2
noraschenica Grossh. - 2
nuda L. = N. pannonica
- subsp. *albiflora* (Boiss.) Gams = N. sulphurea
- var. *albiflora* Boiss. = N. sulphurea
odorifera Lipsky - 6
olgae Regel - 6

ouroumitanensis Franch. = Lophanthus ouroumitanensis
pallida C. Koch = N. meyeri
pamirensis Franch. - 6
pannonica L. (*N. nuda* L.) - 1, 2, 3, 4, 6
- subsp. *albiflora* (Boiss.) Soo = N. sulphurea
parviflora Bieb. (*N. ucranica* L. subsp. *parviflora* (Bieb.) M. Masclans de Bolos) - 1, 2
podostachys Benth. - 6
- subsp. **darwasica** A. Budantz. - 6
pseudofloccosa Pojark. = N. vakhanica
pseudokokanica Pojark. - 6
pulchella Pojark. = N. mariae
pungens (Bunge) Benth. (*N. fedtschenkoi* Pojark., *N. microcephala* Pojark.) - 6
reichenbachiana Fisch. & C.A. Mey. = N. mussinii
saccharata Bunge - 6
santoana M. Pop. - 6
saturejoides Boiss. - 6
schischkinii Pojark. (*N. ucranica* L. subsp. *schischkinii* (Pojark.) Rech. fil.) - 2
schugnanica Lipsky - 6
sibirica L. - 3, 4, 6
sintenisii Bornm. - 6
somkhetica Kapell. (*N. betonicifolia* C.A. Mey. subsp. *somkhetica* (Kapell.) Menitsky) - 2
sosnovskyi Asker. = N. longituba
spathulifera Benth. - 6
strictifolia Pojark. (*N. betonicifolia* C.A. Mey. subsp. *strictifolia* (Pojark.) Menitsky, *N. grossheimii* Pojark.) - 2
subhastata Regel (*N. knorringiana* Pojark.) - 6
subincisa Benth. - 6
subsessilis Maxim. (*N. yezoensis* Franch. & Savat.) - 5
sulphurea C. Koch (*N. nuda* L. var. *albiflora* Boiss., *N. nuda* subsp. *albiflora* (Boiss.) Gams, *N. pannonica* L. subsp. *albiflora* (Boiss.) Soo) - 2
supina Stev. - 2
- subsp. *buschii* (Sosn. & Manden.) Menitsky = N. buschii
teucriifolia Willd. (*N. fissa* C.A. Mey.) - 2
- subsp. *daghestanica* (Pojark.) A. Budantz. = N. daghestanica
- subsp. *iberica* (Pojark.) A. Budantz. = N. iberica
transcaucasica Grossh. = N. mussinii
transiliensis Pojark. - 6
trautvetteri Boiss. & Buhse (*N. velutina* Pojark.) - 2, 6(?)
troitzkii Sosn. - 2
tytthantha Pojark. - 6
ucranica L. - 1, 3, 6
- subsp. *kopetdaghensis* (Pojark.) Rech. fil. = N. kopetdaghensis
- subsp. *parviflora* (Bieb.) M. Masclans de Bolos = N. parviflora
- subsp. *schischkinii* (Pojark.) Rech. fil. = N. schischkinii
vakhanica Pojark. (*N. pseudofloccosa* Pojark., *N. floccosa* auct. p.p.) - 6
velutina Pojark. = N. trautvetteri
yezoensis Franch. & Savat. = N. subsessilis
zangezura Grossh. - 2

Neustruevia Juz. = Pseudomarrubium

karatavica Juz. = Pseudomarrubium eremostachydioides

Ocimum L.

basilicum L. - 1, 2, 5, 6

Origanum L.

dilatatum Klok. = O. vulgare
kopetdaghense Boriss. - 6
puberulum (G. Beck) Klok. = O. vulgare
tytthanthum Gontsch. - 6
vulgare L. (*O. dilatatum* Klok., *O. puberulum* (G. Beck) Klok., *O. vulgare* var. *puberulum* G. Beck) - 1, 2, 3, 4, 5(alien), 6
- var. *puberulum* G. Beck = O. vulgare

Orthodon Benth. & Oliv. = Mosla

grosseserratum (Maxim.) Kudo = Mosla dianthera

Orvala L.

lamioides DC. - 1

Otostegia Benth.

bucharica B. Fedtsch. - 6
fedtschenkoana Kudr. - 6
glabricalyx Vved. - 6
megastegia Vved. - 6
nikitinae V. Scharaschova - 6
olgae (Regel) Korsh. - 6
schennikovii V. Scharaschova - 6
sogdiana Kudr. - 6

Panzeria Moench = Panzerina

argyracea Kuprian. = Panzerina lanata
canescens Bunge = Panzerina canescens
lanata (L.) Bunge = Panzerina lanata

Panzerina Sojak (*Leonuroides* Rauschert, *Panzeria* Moench, 1794, non *Panzera* Cothenius, 1790)

canescens (Bunge) Sojak (*Panzeria canescens* Bunge) - 3
lanata (L.) Sojak (*Leonuroides argyracea* (Kuprian.) Rauschert, *L. lanata* (L.) Rauschert, *Panzeria argyracea* Kuprian., *P. lanata* (L.) Bunge) - 3, 4

Paraeremostachys Adyl., R. Kam. & Machmedov

anisochila (Pazij & Vved.) Adyl., R.Kam. & Machmedov (*Eremostachys anisochila* Pazij & Vved.) - 6
aralensis (Bunge) Adyl., R. Kam. & Machmedov (*Eremostachys aralensis* Bunge) - 6
desertorum (Regel) Adyl., R. Kam. & Machmedov (*Eremostachys desertorum* Regel) - 6
dshungarica (M. Pop.) Adyl., R. Kam. & Machmedov (*Eremostachys phlomoides* Bunge f. *dshungarica* M. Pop., *E. dshungarica* (M. Pop.) Golosk.) - 6
karatavica (Pavl.) Adyl., R. Kam. & Machmedov (*Eremostachys karatavica* Pavl.) - 6
paniculata (Regel) Adyl., R. Kam. & Machmedov (*Eremostachys paniculata* Regel) - 6
phlomoides (Bunge) Adyl., R. Kam. & Machmedov (*Eremostachys phlomoides* Bunge) - 6
sogdiana (Pazij & Vved.) Adyl., R. Kam. & Machmedov (*Eremostachys sogdiana* Pazij & Vved.) - 6
transoxana (Bunge) Adyl., R. Kam. & Machmedov (*Eremostachys transoxana* Bunge) - 6

Perilla L.

*frutescens (L.) Britt. (*P. ocimoides* L.) - 1, 2, 5
nankinensis (Lour.) Decne. - 2
ocimoides L. = P. frutescens

Perovskia Kar.

abrotanoides Kar. - 6
angustifolia Kudr. - 6
botschantzevii Kovalevsk. & Koczk. - 6
kudrjaschevii Gorschk. & Pjat. - 6
linczevskii Kudr. - 6
scrophulariifolia Bunge - 6
virgata Kudr. - 6

Phleboanthe (Boiss.) Tausch = Ajuga

laxmannii (L.) Tausch = Ajuga laxmannii

Phlomidoschema (Benth.) Vved.

parviflorum (Benth.) Vved. - 6

Phlomis L.

agraria Bunge = Phlomoides agraria
alaica Knorr. = Phlomoides alaica
alpina Pall. = Phlomoides alpina
angrenica Knorr. = Ph. sewerzowii
betonicifolia Regel - 6
brachystegia Bunge = Phlomoides brachystegia
bucharica Regel - 6
cancellata Bunge - 2, 6
canescens Regel = Phlomoides canescens
cashmeriana auct. = Ph. fruticetorum
caucasica Rech. fil. = Ph. orientalis
cyclodon Knorr. - 6
desertorum P. Smirn. = Phlomoides tuberosa
drobovii M. Pop. & Vved. - 6
dszumrutensis Afan. = Phlomoides dszumrutensis
ferganensis M. Pop. = Phlomoides ferganensis
fruticetorum Gontsch. (*Ph. cashmeriana* auct.) - 6
fruticosa L. - 1, 2
glandulifera Klok. = Phlomoides tuberosa
herba-venti L. subsp. *kopetdaghensis* (Knorr.) Rech. fil. = Ph. kopetdaghensis
- subsp. *lenkoranica* (Knorr.) Rech. fil. = Ph. lenkoranica
- subsp. *pungens* (Willd.) Maire ex DeFilipps = Ph. pungens
hybrida Zelen. = Phlomoides hybrida
hypanica Shost. = Phlomoides hybrida
hypoleuca Vved. - 6
jailicola Klok. = Phlomoides tuberosa
knorringiana M. Pop. = Phlomoides knorringiana
kopetdaghensis Knorr. (*Ph. herba-venti* L. subsp. *kopetdaghensis* (Knorr.) Rech. fil.) - 6
koraiensis auct. = Phlomoides alpina
lenkoranica Knorr. (*Ph. herba-venti* L. subsp. *lenkoranica* (Knorr.) Rech. fil.) - 2
linearifolia Zak. - 6
maeotica Shost. = Phlomoides hybrida
majkopensis (Novopokr.) Grossh. - 2
maximowiczii Regel = Phlomoides maximowiczii
nubilans Zak. - 6
oblongata Schrenk var. *canescens* Regel = Stachyopsis canescens
olgae Regel (*Ph. tomentosa* Regel) - 6
oreophila Kar. & Kir. = Phlomoides oreophila
orientalis Mill. (*Ph. caucasica* Rech. fil.) - 2

ostrowskiana Regel = Phlomoides ostrowskiana
piskunovii Klok. = Phlomoides tuberosa
pratensis Kar. & Kir. = Phlomoides pratensis
pseudopungens Knorr. = Ph. pungens
puberula Kryl. & Serg. = Phlomoides puberula
pungens Willd. (*Ph. herba-venti* L. subsp. *pungens* (Willd.) Maire ex DeFilipps, *Ph. pseudopungens* Knorr.) - 1, 2, 6
regelii M. Pop. (*Ph. sewerzowii* sensu Knorr.) - 6
salicifolia Regel - 6
scythica Klok. & Shost. = Phlomoides scythica
sewerzowii Regel (*Ph. angrenica* Knorr., *Ph. tenuis* Knorr., *Ph. zenaidae* Knorr.) - 6
sewerzowii sensu Knorr. = Ph. regelii
spinidens Nevski - 6
stepposa Klok. = Phlomoides tuberosa
taurica Hartwiss ex Bunge - 1
tenuis Knorr. = Ph. sewerzowii
thapsoides Bunge - 6
tomentosa Regel = Ph. olgae
tschimganica Vved. = Phlomoides tschimganica
tuberosa L. = Phlomoides tuberosa
tuvinica A. Schroeter = Phlomoides tuvinica
tytthaster Vved. = Phlomoides tytthaster
urodonta M. Pop. = Phlomoides urodonta
vavilovii M. Pop. = Phlomoides vavilovii
woroschilovii Makarov = Phlomoides woroschilovii
zenaidae Knorr. = Ph. sewerzowii

Phlomoides Moench

agraria (Bunge) Adyl., R. Kam. & Machmedov (*Phlomis agraria* Bunge) - 3, 6
alaica (Knorr.) Adyl., R. Kam. & Machmedov (*Phlomis alaica* Knorr.) - 6
alberti (Regel) Adyl., R. Kam. & Machmedov (*Eremostachys alberti* Regel. - 6
alpina (Pall.) Adyl., R. Kam. & Machmedov (*Phlomis alpina* Pall., *Ph. koraiensis* auct.) - 3, 5, 6
ambigua (M. Pop. ex Pazij & Vved.) Adyl., R. Kam. & Machmedov (*Eremostachys ambigua* M. Pop. ex Pazij & Vved., *Phlomoides kamelinii* Machmedov, nom. invalid. & superfl.) - 6
angreni (M. Pop.) Adyl., R. Kam. & Machmedov (*Eremostachys angreni* M. Pop.) - 6
arctiifolia (M. Pop.) Adyl., R. Kam. & Machmedov (*Eremostachys arctiifolia* M. Pop.) - 6
baburii (Adyl.) Adyl. (*Eremostachys baburii* Adyl.) - 6
baissunensis (M. Pop.) Adyl., R. Kam. & Machmedov (*Eremostachys baissunensis* M. Pop., *E. glanduligera* M. Pop., *E. jakkabaghi* M. Pop.) - 6
baldschuanica (Regel) Adyl., R. Kam. & Machmedov (*Eremostachys baldschuanica* Regel) - 6
beckeri (Regel) Adyl., R. Kam. & Machmedov (*Eremostachys beckeri* Regel) - 6
boissieriana (Regel) Adyl., R. Kam. & Machmedov (*Eremostachys boissieriana* Regel) - 6
botschantzevii (Adyl.) Adyl. (*Eremostachys botschantzevii* Adyl.) - 6
brachystegia (Bunge) Adyl., R. Kam. & Machmedov (*Phlomis brachystegia* Bunge) - 6
canescens (Regel) Adyl., R. Kam. & Machmedov (*Phlomis canescens* Regel) - 6
cephalariifolia (M. Pop.) Adyl., R. Kam. & Machmedov (*Eremostachys cephalariifolia* M. Pop.) - 6
cordifolia (Regel) Adyl., R. Kam. & Machmedov (*Eremostachys cordifolia* Regel) - 6

czuiliensis (Golosk.) Adyl., R. Kam. & Machmedov
(*Eremostachys czuiliensis* Golosk.) - 6

dszumrutensis (Afan.) Adyl., R. Kam. & Machmedov
(*Phlomis dszumrutensis* Afan.) - 6

ebracteolata (M. Pop.) Adyl., R. Kam. & Machmedov
(*Eremostachys ebracteolata* M. Pop.) - 6

eriocalyx (Regel) Adyl., R. Kam. & Machmedov
(*Eremostachys eriocalyx* Regel) - 6

ferganensis (M. Pop.) Adyl., R. Kam. & Machmedov
(*Phlomis ferganensis* M. Pop.) - 6

fetisowii (Regel) Adyl., R. Kam. & Machmedov
(*Eremostachys fetisowii* Regel) - 6

fulgens (Bunge) Adyl., R. Kam. & Machmedov
(*Eremostachys fulgens* Bunge) - 6

glabra (Boiss. ex Benth.) R. Kam. & Machmedov
(*Eremostachys glabra* Boiss. ex Benth.) - 2

gymnocalyx (Schrenk) Adyl., R. Kam. & Machmedov
(*Eremostachys gymnocalyx* Schrenk) - 6

gypsacea (M.Pop.) Adyl., R. Kam. & Machmedov
(*Eremostachys gypsacea* M. Pop.) - 6

hajastanica (Sosn. ex Grossh.) R. Kam. & Machmedov =
Ph. laciniata

hissarica (Regel) Adyl., R. Kam. & Machmedov
(*Eremostachys hissarica* Regel) - 6

hybrida (Zelen.) R. Kam. & Machmedov (*Phlomis hybrida*
Zelen., *Ph. hypanica* Shost., *Ph. maeotica* Shost.) - 1

iliensis (Regel) Adyl., R. Kam. & Machmedov
(*Eremostachys iliensis* Regel) - 6

impressa (Pazij & Vved.) Adyl., R. Kam. & Machmedov
(*Eremostachys impressa* Pazij & Vved., *E. eriocalyx*
Regel var. *taschkentica* M. Pop., *E. taschkentica* (M.
Pop.) Golosk. nom. illegit. superfl.) - 6

integior (Pazij & Vved.) Adyl., R. Kam. & Machmedov
(*Eremostachys integior* Pazij & Vved., *E. nuda* Regel f.
simplex M. Pop.) - 6

kamelinii Machmedov = Ph. ambigua

kaufmanniana (Regel) Adyl., R. Kam. & Machmedov
(*Eremostachys kaufmanniana* Regel) - 6

kirghisorum Adyl., R. Kam. & Machmedov (*Eremostachys
ferganensis* Ubuk. nom. invalid.) - 6

knorringiana (M. Pop.) Adyl., R. Kam. & Machmedov
(*Phlomis knorringiana* M. Pop.) - 6

korovinii (M. Pop.) Adyl., R. Kam. & Machmedov
(*Eremostachys korovinii* M. Pop.) - 6

labiosa (Bunge) Adyl., R. Kam. & Machmedov
(*Eremostachys labiosa* Bunge) - 6

labiosiformis (M. Pop.) Adyl., R. Kam. & Machmedov
(*Eremostachys labiosiformis* (M. Pop.) Knorr.) - 6

labiosissima (Pazij & Vved.) Adyl., R. Kam. & Machmedov
(*Eremostachys labiosissima* Pazij & Vved.) - 6

laciniata (L.) R. Kam. & Machmedov (*Eremostachys laciniata* (L.) Bunge, *E. hajastanica* Sosn. ex Grossh., *E.
iberica* Vis., *Phlomoides hajastanica* (Sosn. ex Grossh.)
R. Kam. & Machmedov) - 2

lanatifolia Machmedov - 6

lehmanniana (Bunge) Adyl., R. Kam. & Machmedov
(*Eremostachys lehmanniana* Bunge) - 6

leiocalyx (Pazij & Vved.) Adyl., R. Kam. & Machmedov
(*Eremostachys leiocalyx* Pazij & Vved.) - 6

lyrata (M. Pop.) Adyl. (*Eremostachys sarawschanica* Regel
f. *lyrata* M.Pop.) - 6

maximowiczii (Regel) R. Kam. & Machmedov (*Phlomis
maximowiczii* Regel) - 5

mihaelis Adyl., R. Kam. & Machmedov (*Eremostachys
canescens* M. Pop. non *Phlomoides canescens* (Regel)
Adyl., R. Kam. & Machmedov) - 6

napuligera (Franch.) Adyl., R. Kam. & Machmedov

(*Eremostachys napuligera* Franch.) - 6

nuda (Regel) Adyl., R. Kam. & Machmedov (*Eremostachys
nuda* Regel) - 6

oreophila (Kar. & Kir.) Adyl., R. Kam. & Machmedov
(*Phlomis oreophila* Kar. & Kir.) - 3, 6

ostrowskiana (Regel) Adyl., R. Kam. & Machmedov
(*Phlomis ostrowskiana* Regel) - 6

pauciflora (O. Kuntze) Adyl., R. Kam. & Machmedov
(*Eremostachys pauciflora* O. Kuntze) - 6

pectinata (M. Pop.) Adyl., R. Kam. & Machmedov
(*Eremostachys pectinata* M. Pop.) - 6

popovii (Gontsch.) Adyl., R. Kam. & Machmedov
(*Eremostachys popovii* Gontsch.) - 6

pratensis (Kar. & Kir.) Adyl., R. Kam. & Machmedov
(*Phlomis pratensis* Kar. & Kir.) - 6

puberula (Kryl. & Serg.) Adyl., R. Kam. & Machmedov
(*Phlomis puberula* Kryl. & Serg.) - 3, 6

pulchra (M. Pop.) Adyl., R. Kam. & Machmedov
(*Eremostachys pulchra* M. Pop.) - 6

regeliana (Aitch. & Hemsl.) Adyl., R. Kam. & Machmedov
(*Eremostachys regeliana* Aitch. & Hemsl.) - 6

sanglechensis (Rech. fil.) R. Kam. & Machmedov
(*Eremostachys sanglechensis* Rech. fil.) - 6

sarawschanica (Regel.) Adyl., R. Kam. & Machmedov
(*Eremostachys sarawschanica* Regel) - 6

schugnanica (M. Pop.) Adyl., R. Kam. & Machmedov
(*Eremostachys schugnanica* (M. Pop.) Knorr.) - 6

scythica (Klok. & Shost.) Czer. comb. nova (*Phlomis scythica* Klok. & Shost. 1938, Bot. Mat. (Leningrad), 8, 3 :
31) - 1

septentrionalis (M. Pop.) Adyl., R. Kam. & Machmedov
(*Eremostachys nuda* Regel var. *septentrionalis* M. Pop.,
E. septentrionalis (M. Pop.) Golosk.) - 6

speciosa (Rupr.) Adyl., R. Kam. & Machmedov
(*Eremostachys speciosa* Rupr., *E. acaulis* Rech. fil., *E.
dzhambulensis* Knorr., *E. edelbergii* Rech. fil., *E. speciosa* var. *bipinnatifida* (Regel) M. Pop., *E. speciosa* var.
brevicaulis (Regel) M. Pop.) - 6

subspicata (M. Pop.) Adyl., R. Kam. & Machmedov
(*Eremostachys subspicata* M. Pop.) - 6

tadschikistanica (B. Fedtsch.) Adyl., R. Kam. & Machmedov (*Eremostachys tadschikistanica* B. Fedtsch.) - 6

tianschanica (M. Pop.) Adyl., R. Kam. & Machmedov
(*Eremostachys tianschanica* M. Pop.) - 6

tschimganica (Vved.) Adyl., R. Kam. & Machmedov
(*Phlomis tschimganica* Vved.) - 6

tuberosa (L.) Moench (*Phlomis tuberosa* L., *Ph. desertorum*
P. Smirn., *Ph. glandulifera* Klok., *Ph. jailicola* Klok.,
Ph. piskunovii Klok., *Ph. stepposa* Klok.) - 1, 2, 3, 4, 5,
6

tuvinica (A. Schroeter) R. Kam. & Machmedov (*Phlomis
tuvinica* A. Schroeter) - 4

tytthaster (Vved.) Adyl., R. Kam. & Machmedov (*Phlomis
tytthaster* Vved.) - 6

uniflora (Regel) Adyl., R. Kam. & Machmedov
(*Eremostachys uniflora* Regel) - 6

urodonta (M. Pop.) Adyl., R. Kam. & Machmedov (*Phlomis urodonta* M. Pop.) - 6

vavilovii (M. Pop.) Adyl., R. Kam. & Machmedov (*Phlomis
vavilovii* M. Pop.) - 6

woroschilovii (Makarov) Czer. comb. nova (*Phlomis
woroschilovii* Makarov, 1982, Bull. Glavn. Bot. Sada
AN SSSR, 123 : 44) - 4

zenaidae (M. Pop.) Adyl., R. Kam. & Machmedov
(*Eremostachys zenaidae* M. Pop.) - 6

Plectranthus auct. = Rabdosia

excisus Maxim. = Rabdosia excisa
glaucocalyx Maxim. = Rabdosia japonica var. glaucocalyx
serra Maxim. = Rabdosia serra

Pogostemon auct. = Dasophylla

yatabeanus (Makino) J.R. Press = Dysophylla yatabeana

Pollichia auct. = Galeobdolon

montana Pers. = Galeobdolon montanum

Prunella L.

x **bicolor** G. Beck. - P. grandiflora (L.) Scholl. x P. laciniata
   (L.) L. - 1
grandiflora (L.) Scholl. - 1, 2
x **intermedia** Link. - P. laciniata (L.) L. x P. vulgaris L. - 1
japonica Makino (*P. vulgaris* L. var. *aleutica* Fern., *P. vulgaris* subsp. *asiatica* (Nakai) Hara) - 5
laciniata (L.) L. - 1, 2
vulgaris L. - 1, 2, 3, 4, 5, 6
- subsp. *asiatica* (Nakai) Hara = P. japonica
- subsp. **lanceolata** (Bart.) Hult. - 5
- var. *aleutica* Fern. = P. japonica

Pseuderemostachys M. Pop.

sewerzowii (Herd.) M. Pop. - 6

Pseudomarrubium M. Pop. (*Neustruevia* Juz.)

eremostachydioides M. Pop. (*Neustruevia karatavica* Juz.) -
   6

Pulegium Mill. = Mentha

daghestanicum (Boriss.) Holub = Mentha pulegium

Rabdosia (Blume) Hassk. (*Plectranthus* auct.)

excisa (Maxim.) Hara (*Plectranthus excisus* Maxim.) - 5
japonica (Burm. fil.) Hara (*Scutellaria japonica* Burm. fil.) -
   5
- var. **glaucocalyx** (Maxim.) Hara (*Plectranthus glaucocalyx*
   Maxim.) - 5
serra (Maxim.) Hara (*Plectranthus serra* Maxim.) - 5

*Rosmarinus L.

*officinalis L. - 1, 2

Salvia L. (*Arischrada* Pobed., *Schraderia* Medik. 1791, non *Schradera* Vahl, 1796, nom. cons.)

acetabulosa sensu Vahl = S. multicaulis
adenostachya Juz. = S. scabiosifolia
aequidens Botsch. - 6
aethiopis L. - 1, 2, 6
alexandri Pobed. = S. suffruticosa
alexeenkoi Pobed. - 2
amasiaca Freyn & Sint. (*S. verticillata* L. subsp. *amasiaca*
   (Freyn & Sint.) Bornm.) - 2
x **andreji** Pobed. - S. nemorosa L. x S. virgata Jacq. - 2
ariana Hedge - 6
armeniaca (Bordz.) Grossh. (*S. staminea* auct. p.p.) - 2
atropatana auct. p.p. = S. kopetdaghensis and S. linczevskii
austriaca Jacq. - 1

baldshuanica Lipsky - 6
beckeri Trautv. - 2
betonicifolia Etl. - 1
brachyantha (Bordz.) Pobed. = S. modesta
brachystemon Klok. = S. tomentosa
bucharica M. Pop. (*Arischrada bucharica* (M. Pop.) Pobed.,
   Schraderia bucharica (M. Pop.) Nevski) - 6
bzybica Kolak. - 2
campestris Bieb. = S. sibthorpii
campylodonta Botsch. - 6
canescens C.A. Mey. - 2
ceratophylla L. - 2
chloroleuca Rech. fil. & Aell. - 6
compar Trautv. ex Grossh. - 2
cremenecensis Bess. - 1
   Probably, S. nutans L. x S. sp.?
daghestanica Sosn. - 2
deserta Schang. = S. nemorosa
■ **divaricata** Montbr. & Auch. ex Benth. (*S. trigonocalyx*
   Woronow)
drobovii Botsch. - 6
dumetorum Andrz. - 1
elenevskyi Pobed. = S. virgata
extersa Klok. = S. sibthorpii
fominii Grossh. = S. limbata
forskahlei L. - 2
fugax Pobed. - 2
garedji Troitzk. - 2
glabricaulis Pobed. - 6
glutinosa L. - 1, 2
golneviana Rzazade - 2
gontscharovii Kudr. - 6
grandiflora Etl. = S. tomentosa
grossheimii Sosn. (*S. hajastana* Pobed.) - 2
hablitziana Pall. ex Willd. = S. scabiosifolia
hajastana Pobed. = S. grossheimii
horminum L. = S. viridis
hydrangea DC. ex Benth. (*Arischrada dracocephaloides*
   (Boiss.) Pobed., *Schraderia dracocephaloies* (Boiss.)
   Pobed.) - 2
illuminata Klok. (*S. nemorosa* L. subsp. *illuminata* (Klok.)
   Soo) - 1
insignis Kudr. - 6
intercedens Pobed. - 6
jailicola Klok. = S. nemorosa
kamelinii Machmedov - 6
karabachensis Pobed. = S. verbascifolia
komarovii Pobed. - 6
kopetdaghensis Kudr. (*S. atropatana* auct. p.p.) - 6
korolkowii Regel & Schmalh. (*Arischrada korolkowii* (Regel
   & Schmalh.) Pobed., *Schraderia korolkowii* (Regel &
   Schmalh.) Pobed.) - 6
kuznetzovii Sosn. - 2
lilacinocoerulea Nevski - 6
limbata C.A. Mey. (*S. fominii* Grossh., *S. prilipkoana*
   Grossh. & Sosn.) - 2
linczevskii Kudr. (*S. atropatana* auct. p.p.) - 6
lipskyi Pobed. = S. trautvetteri
macrosiphon Boiss. - 6
margaritae Botsch. - 6
modesta Boiss. (*S. brachyantha* (Bordz.) Pobed.) - 2
moldavica Klok. = S. nemorosa
■ **multicaulis** Vahl (*Arischrada multicaulis* (Vahl) Pobed.,
   Schraderia acetabulosa (Vahl) Pobed. nom. illegit.,
   Salvia acetabulosa sensu Vahl, non L.)
nachiczevanica Pobed. = S. reuteriana

nemorosa L. (*S. deserta* Schang., *S. jailicola* Klok., *S. moldavica* Klok., *S. nemorosa* subsp. *jailicola* (Klok.) Soo, *S. nemorosa* subsp. *moldavica* (Klok.) Soo) - 1, 2, 3, 6
- subsp. *illuminata* (Klok.) Soo = S. illuminata
- subsp. *jailicola* (Klok.) Soo = S. nemorosa
- subsp. *moldavica* (Klok.) Soo = S. nemorosa
- subsp. *praemontana* (Klok.) Soo = S. tesquicola
- subsp. *tesquicola* (Klok. & Pobed.) Soo = S. tesquicola
nutans L. - 1, 2
officinalis L. - 1, 2
pachystachya Trautv. - 2
plebeja R. Br. - 5
praemontana Klok. = S. tesquicola
pratensis L. - 1
prilipkoana Grossh. & Sosn. = S. limbata
reflexa Hornem. - 1
reuteriana Boiss. (*S. nachiczevanica* Pobed.) - 2
rhodantha Zefir. = S. verbenaca
ringens Smith - 2
▪ rosifolia Smith
sarawschanica Regel & Schmalh. - 6
scabiosifolia Lam. (*S. adenostachya* Juz., *S. hablitziana* Pall. ex Willd., *S. scabiosifolia* var. *adenostachya* (Juz.) Zefir.) - 1
- var. *adenostachya* (Juz.) Zefir. = S. scabiosifolia
schmalhausenii Regel - 6
sclarea L. - 1, 2, 6
semilanata Czerniak. - 6
sibthorpii Smith (*S. campestris* Bieb., *S. extersa* Klok.) - 1, 2
spinosa L. - 2, 6
***splendens** Ker-Gawl.
staminea auct. p.p. = S. armeniaca
stepposa Shost. - 1, 3, 6
submutica Botsch. & Vved. - 6
suffruticosa Montbr. & Auch. ex Benth. (*S. ɩ andri* Pobed.) - 2
sylvestris L. subsp. *tesquicola* (Klok. & Pobed.) So k = .. tesquicola
syriaca L. - 2
tesquicola Klok. & Pobed. (*S. nemorosa* L. subsp. *praemontana* (Klok.) Soo, *S. nemorosa* subsp. *tesquicola* (Klok. & Pobed.) Soo, *S. praemontana* Klok., *S. sylvestris* L. subsp. *tesquicola* (Klok. & Pobed.) Sojak) - 1, 2, 3, 6
tianschanica Machmedov - 6
tomentosa Mill. (*S. brachystemon* Klok., *S. grandiflora* Etl.) - 1, 2
transcaucasica Pobed. - 2
trautvetteri Regel (*S. lipskyi* Pobed.) - 6
trigonocalyx Woronow = S. divaricata
turcomanica Pobed. - 6
verbascifolia Bieb. (*S. karabachensis* Pobed.) - 2
verbenaca L. (*S. rhodantha* Zefir.) - 1, 2
vergeduzica Rzazade - 2
verticillata L. - 1, 2, 3, 5(alien), 6
virgata Jacq. (*S. elenevskyi* Pobed.) - 1, 2, 6
viridis L. (*S. horminum* L.) - 1, 2
vvedenskyi E. Nikit. - 6
xanthocheila Boiss. ex Benth. - 2

Satureja L.

altaica Boriss. - 3
borissovae Zeinalova - 2
bzybica Woronow - 2
confinis Boriss. - 2
densiflora Zeinalova = S. zuvandica
hortensis L. (*S. laxiflora* C. Koch, *S. pachyphylla* C. Koch) -

1, 2, 5(alien), 6
intermedia C.A. Mey. - 2
kochii N. Pop. ex Greuter & Burdet = Micromeria elliptica
laxiflora C. Koch = S. hortensis
macrantha C.A. Mey. - 2
***montana** L. - 6
- subsp. *taurica* (Velen.) P.W. Ball = S. taurica
mutica Fisch. & C.A. Mey. - 2, 6
pachyphylla C. Koch = S. hortensis
spicigera (C. Koch) Boiss. (*Hyssopus majae* A. Khokhr.) - 2
subdentata Boiss. - 2
taurica Velen. (*S. montana* L. subsp. *taurica* (Velen.) P.W. Ball) - 1
umbrosa (Bieb.) Greuter & Burdet = Clinopodium umbrosum
zuvandica D. Kapanadze (*S. densiflora* Zeinalova, 1969, non Briq.) - 2

Schizonepeta (Benth.) Briq.

annua (Pall.) Schischk. - 3, 4
multifida (L.) Briq. - 3, 4, 5

Schraderia Medik. = Salvia

acetabulosa (Vahl) Pobed. = Salvia multicaulis
bucharica (M.Pop.) Nevski = Salvia bucharica
dracocephaloides (Boiss.) Pobed. = Salvia hydrangea
korolkowii (Regel & Schmalh.) Pobed. = Salvia korolkowii

Scutellaria L.

adenostegia Briq. (*S. bucharica* Juz.) - 6
adsurgens M. Pop. - 6
alberti Juz. - 6
albida L. (*S. pallida* Bieb., *S. subalbida* Klok.) - 1, 2
- subsp. *colchica* (Rech. fil.) Edmondson = S. woronowii
alexeenkoi Juz. - 6
alpina L. subsp. *supina* (L.) I.B.K. Richardson = S. supina
altaica Fisch. ex Sweet - 3, 4
altissima L. - 1, 2
amphichlora Juz. (*S. orientalis* L. subsp. *amphichlora* (Juz.) Fed.) - 2
andina Charadze - 2
andrachnoides Vved. - 6
androssovii Juz. - 6
angrenica Juz. & Vved. = S. pycnoclada
anitae Juz. = S. physocalyx
▪ araratica Grossh. (*S. orientalis* L. subsp. *araratica* (Grossh.) Fed.)
araxensis Grossh. (*S. orientalis* L.subsp. *araxensis* (Grossh.) Fed.) - 2
▪ artvinensis Grossh. (*S. orientalis* L. subsp. *artvinensis* (Grossh.) Fed.)
baicalensis Georgi - 4, 5
baldshuanica Nevski ex Juz. - 6
botschantzevii Abdullajeva - 6
bucharica Juz. = S. adenostegia
catharinae Juz. = S. sieversii
chenopodiifolia Juz. = S. navicularis
chitrovoi Juz. - 1
chodshakasiani R. Kam. - 6
cisvolgensis Juz. - 1
colpodea Nevski - 6
comosa Juz. - 6
cordifrons Juz. - 6
creticola Juz. - 1
cristata M. Pop. - 6

daghestanica Grossh. (*S. orientalis* L. subsp. *daghestanica* (Grossh.) Fed.) - 2
darriensis Grossh. (*S. orientalis* L. subsp. *darriensis* (Grossh.) Fed.) - 2
darvasica Juz. = S. jodudiana
dentata Levl. (*S. ussuriensis* (Regel) Kudo) - 5
dependens Maxim. - 4, 5
dubia Taliev & Sirj. - 1, 3, 6
fedtschenkoi Bornm. - 6
filicaulis Regel - 6
flabellulata Juz. - 6
galericulata L. - 1, 2, 3, 4, 5, 6
glabrata Vved. - 6
gontscharovii Juz. - 6
grandiflora Sims - 3, 4
granulosa Juz. (*S. orientalis* L. subsp. *granulosa* (Juz.) Fed.) - 2
grossheimiana Juz. (*S. orientalis* L. subsp. *grossheimiana* (Juz.) Fed.) - 2
guttata Nevski ex Juz. - 6
haematochlora Juz. (*S. tschimganica* Juz.) - 6
haesitabunda Juz. ex Koczk. - 6
hastifolia L. - 1, 2, 3
helenae Albov - 2
heterochroa Juz. (*S. orientalis* L. subsp. *heterochroa* (Juz.) Fed.) - 1
heterotricha Juz. & Vved. - 6
hirtella Juz. (*S. orientalis* L. subsp. *hirtella* (Juz.) Fed.) - 1
hissarica B. Fedtsch. - 6
holosericea Gontsch. ex Juz. - 6
hypopolia Juz. (*S. orientalis* L. subsp. *hypopolia* (Juz.) Fed.) - 1
ikonnikovii Juz. (*S. regeliana* Nakai var. *ikonnikovii* (Juz.) C.Y. Wu & H.W. Li) - 4, 5
immaculata Nevski ex Juz. - 6
intermedia M. Pop. - 6
irregularis Juz. - 6
iskanderi Juz. - 6
japonica Burm. fil. = Rabdosia japonica
jodudiana B. Fedtsch. (*S. darvasica* Juz.) - 6
juzepczukii Gontsch. - 6
karatavica Juz. - 6
karatschaica Charadze - 6
karjaginii Grossh. (*S. orientalis* L. subsp. *karjaginii* (Grossh.) Fed.) - 2
karkaralensis Juz. = S. turgaica
knorringiae Juz. - 6
krasevii Kom. & I. Schischk. ex Juz. - 5
krylovii Juz. - 3, 6
kugarti Juz. - 6
kurssanovii Pavl. - 6
lanipes Juz. - 6
leptosiphon Nevski - 6
leptostegia Juz. (*S. orientalis* L. subsp. *leptostegia* (Juz.) Fed.) - 2
linczewskii Juz. - 6
lipskyi Juz. - 6
litwinowii Bornm. & Sint. ex Bornm. - 6
luteo-coerulea Bornm. & Sint. ex Bornm. - 6
macrodonta Nevski ex Juz. = S. megalodonta
megalodonta Juz. (*S. macrodonta* Nevski ex Juz. 1954, nom. invalid., non Hand.-Mazz. 1936) - 6
mesostegia Juz. - 6
microdasys Juz. - 6
microphysa Juz. - 6
x **minkwitziae** Juz. - S. microdasys Juz. x S. tschimganica Juz.

mongolica K. Sobol. - 4
moniliorrhiza Kom. - 5
navicularis Juz. (*S. chenopodiifolia* Juz.) - 6
nepetoides M. Pop. ex Juz. - 6
nevskii Juz. & Vved. - 6
novorossica Juz. (*S. orientalis* L. subsp. *novorossica* (Juz.) Fed.) - 2
ocellata Juz. - 6
oligodonta Juz. - 6
orbicularis Bunge - 6
oreophila Grossh. (*S. orientalis* L. subsp. *oreophila* (Grossh.) Fed.) - 2
orientalis L. - 2
- subsp. *amphichlora* (Juz.) Fed. = S. amphichlora
- subsp. *araratica* (Grossh.) Fed. = S. araratica
- subsp. *araxensis* (Grossh.) Fed. = S. araxensis
- subsp. *artvinensis* (Grossh.) Fed. = S. artvinensis
- subsp. *daghestanica* (Grossh.) Fed. = S. daghestanica
- subsp. *darriensis* (Grossh.) Fed. = S. darriensis
- subsp. *granulosa* (Juz.) Fed. = S. granulosa
- subsp. *grossheimiana* (Juz.) Fed. = S. grossheimiana
- subsp. *heterochroa* (Juz.) Fed. = S. heterochroa
- subsp. *hirtella* (Juz.) Fed. = S. hirtella
- subsp. *hypopolia* (Juz.) Fed. = S. hypopolia
- subsp. *karjaginii* (Grossh.) Fed. = S. karjaginii
- subsp. *leptostegia* (Juz.) Fed. = S. leptostegia
- subsp. *novorossica* (Juz.) Fed. = S. novorossica
- subsp. *oreophila* (Grossh.) Fed. = S. oreophila
- subsp. *oschtenica* (Juz.) Fed. = S. oschtenica
- subsp. *platystegia* (Juz.) Fed. = S. platystegia
- subsp. *polyodon* (Juz.) Fed. = S. polyodon
- subsp. *prilipkoana* (Grossh.) Fed. = S. prilipkoana
- subsp. *raddeana* (Juz.) Fed. = S. raddeana
- subsp. *rhomboidalis* (Grossh.) Fed. = S. rhomboidalis
- subsp. *sedelmeyerae* (Juz.) Fed. = S. sedelmeyerae
- subsp. *sevanensis* (Sosn. ex Grossh.) Fed. = S. sevanensis
- subsp. *sosnowskyi* (Takht.) Fed. = S. sosnowskyi
- subsp. *stevenii* (Juz.) Fed. = S. stevenii
- subsp. *tatianae* (Juz.) Fed. = S. tatianae
- subsp. *taurica* (Juz.) Fed. = S. taurica
oschtenica Juz. (*S. orientalis* L. subsp. *oschtenica* (Juz.) Fed.) - 2
ossethica Charadze - 2
oxyphylla Juz. = S. supina
oxystegia Juz. - 6
pacifica Juz. = S. transitra
pallida Bieb. = S. albida
pamirica Juz. - 6
paradoxa Galushko - 2
paulsenii Briq. - 6
phyllostachya Juz. - 6
physocalyx Regel & Schmalh. (*S. anitae* Juz.) - 6
picta Juz. - 6
platystegia Juz. (*S. orientalis* L. subsp. *platystegia* (Juz.) Fed.) - 2
poecilantha Nevski ex Juz. - 6
polyodon Juz. (*S. orientalis* L. subsp. *polyodon* (Juz.) Fed.) - 2
polyphylla Juz. (*S. scordiifolia* Fisch. ex Schrank var. *polyphylla* (Juz.) Worosch.) - 5
polytricha Juz. - 6
pontica C. Koch - 2
popovii Vved. - 6
prilipkoana Grossh. (*S. orientalis* L. subsp. *prilipkoana* (Grossh.) Fed.) - 2
przewalskii Juz. - 6
pycnoclada Juz. (*S. angrenica* Juz. & Vved.) - 6

raddeana Juz. (*S. orientalis* L. subsp. *raddeana* (Juz.) Fed.) - 2
ramosissima M. Pop. - 6
regeliana Nakai - 5
- var. *ikonnikovii* (Juz.) C.Y. Wu & H.W. Li = S. ikonnikovii
rhomboidalis Grossh. (*S. orientalis* L. subsp. *rhomboidalis* (Grossh.) Fed.) - 2
rubromaculata Juz. & Vved. - 6
schachristanica Juz. - 6
schugnanica B. Fedtsch. - 6
scordiifolia Fisch. ex Schrank - 3, 4, 5
- var. *polyphylla* (Juz.) Worosch. = S. polyphylla
sedelmeyerae Juz. (*S. orientalis* L. subsp. *sedelmeyerae* (Juz.) Fed.) - 2
sevanensis Sosn. ex Grossh. (*S. orientalis* L. subsp. *sevanensis* (Sosn. ex Grossh.) Fed.) - 2
shikokiana Makino - 5
sieversii Bunge (*S. catharinae* Juz.) - 3, 4, 6
soongorica Juz. - 6
sosnowskyi Takht. (*S. orientalis* L. subsp. *sosnowskyi* (Takht.) Fed.) - 2
squarrosa Nevski - 6
stevenii Juz. (*S. orientalis* L. subsp. *stevenii* (Juz.) Fed.) - 1
striatella Gontsch. - 6
strigillosa Hemsl. (*S. taquetii* Levl. & Vaniot) - 5
- var. *yezoensis* (Kudo) Kitam. = S. yezoensis
subalbida Klok. = S. albida
subcaespitosa Pavl. - 6
subcordata Juz. - 6
supina L. (*S. alpina* L. subsp. *supina* (L.) I.B.K. Richardson, *S. oxyphylla* Juz.) - 1, 3, 4, 6
talassica Juz. - 6
taquetii Levl. & Vaniot = S. strigillosa
tatianae Juz. (*S. orientalis* L. subsp. *tatianae* (Juz.) Fed.) - 2
taurica Juz. (*S. orientalis* L. subsp. *taurica* (Juz.) Fed.) - 1
titovii Juz. - 6
toguztoraviensis Juz. - 6
tournefortii Benth. - 2
transiliensis Juz. - 6
transitra Makino (*S. pacifica* Juz., *S. transitra* var. *pacifica* (Juz.) Kitag.) - 5
- var. *pacifica* (Juz.) Kitag. = S. transitra
tschimganica Juz. = S. haematochlora
tuminensis Nakai - 5
turgaica Juz. (*S.karkaralensis* Juz.) - 3, 6
tuvensis Juz. - 4
urticifolia Juz. & Vved. - 6
ussuriensis (Regel) Kudo = S. dentata
vacillans Rech. fil. subsp. *colchica* Rech. fil. = S. woronowii
velutina Juz. & Vved. - 6
verna Bess. - 1
villosissima Gontsch. ex Juz. - 6
woronowii Juz. (*S. albida* L. subsp. *colchica* (Rech. fil.) Edmondson, *S. vacillans* Rech. fil. subsp. *colchica* Rech. fil.) - 1(?), 2
xanthosiphon Juz. - 5
yezoensis Kudo (*S. strigillosa* Hemsl. var. *yezoensis* (Kudo) Kitam.) - 5
zaprjagaevii Koczk. & Zhogoleva - 6
zivari Ahmed-zade - 2

Sideritis L. (*Hesiodia* Moench, *Navicularia* Heist. ex Fabr.)

ajpetriana Klok. (*S. syriaca* auct. p.p.) - 1
atrinervia Juz. (*S. syriaca* auct. p.p.) - 1

balansae Boiss. (*Stachys woronowii* (Schischk. ex Grossh.) R. Mill) - 2
catillaris Juz. (*Navicularia syriaca* (l.) Sojak subsp. *catillaris* (Juz.) Sojak, *Sideritis conferta* Juz., *S. imbrex* Juz., *S. syriaca* L. subsp. *catillaris* (Juz.) Gladkova) - 1
- var. *chlorostegia* (Juz.) Zefir. = S. chlorostegia
chlorostegia Juz. (*S. catillaris* Juz. var. *chlorostegia* (Juz.) Zefir.) - 1
ciliata Trunb. = Elsholtzia ciliata
comosa (Rochel ex Benth.) Stank. (*Hesiodia montana* (L.) Dumort. subsp. *comosa* (Rochel ex Benth.) Sojak, *Sideritis montana* L. subsp. *comosa* (Rochel ex Benth.) Soo) - 1, 2
conferta Juz. = S. catillaris
euxina Juz. (*Navicularia syriaca* (L.) Sojak subsp. *euxina* (Juz.) Sojak) - 2
imbrex Juz. = S. catillaris
marschalliana Juz. - 1
montana L. (*Hesiodia montana* (L.) Dumort.) - 1, 2, 5(alien), 6
- subsp. *comosa* (Rochel ex Benth.) Soo = S. comosa
scythica Juz. = S. taurica
syriaca auct. p.p. = S. ajpetriana and S. atrinervia
syriaca L. subsp. *catillaris* (Juz.) Gladkova = S. catillaris
- subsp. *taurica* (Steph.) Gladkova = S. taurica
taurica Steph. (*Navicularia syriaca* (L.) Sojak subsp. *taurica* (Steph.) Sojak, *Sideritis scythica* Juz., *S. syriaca* L. subsp. *taurica* (Steph.) Gladkova) - 1

Stachyopsis M. Pop. & Vved.

canescens (Regel) Adyl. & Tuljaganova (*Phlomis oblongata* Schrenk var. *canescens* Regel) - 6
lamiiflora (Rupr.) M. Pop. & Vved. - 6
marrubioides (Regel) Ik.-Gal. - 6
oblongata (Schrenk) M. Pop. & Vved. (*S. ovata* Djugaeva) - 6
ovata Djugaeva = S. oblongata

Stachys L. (*Betonica* L.)

abchasica (N. Pop. ex Grossh.) Czer. comb. nova (*Betonica nivea* Stev. var. *abchasica* N. Pop. ex Grossh. 1932, Fl. Cauc. 3 : 309; *B. abchasica* (Bornm.) Chinth., *B. nivea* subsp. *abchasica* Bornm.) - 2
acanthodonta Klok. = S. atherocalyx
alpina L. - 1
- subsp. *macrophylla* (Albov) Bhattacharjee = S. macrophylla
angustifolia Bieb. - 1
annua (L.) L. (*S. neglecta* Klok. ex Kossko, nom. invalid.) - 1, 2, 3, 5(alien)
araxina Kapell. - 2
artemisiae Lour. = Leonurus japonicus
arvensis (L.) L. - 1
aspera Michx. (*S. aspera* var. *japonica* (Miq.) Worosch., *S. baicalensis* Fisch. ex Benth., *S. chinensis* Bunge ex Benth., *S. japonica* Miq., *S. palustris* L. subsp. *aspera* (Michx.) Derviz-Sokolova, *S. riederi* Cham.) - 1, 4, 5
- var. *japonica* (Miq.) Worosch. = S. aspera
atherocalyx C. Koch (*S. acanthodonta* Klok., *S. recta* L. subsp. *atherocalyx* (C. Koch) Derviz-Sokolova, *S. patula* auct. p.p.) - 1, 2
baicalensis Fisch. ex Benth. = S. aspera
balansae Boiss. & Kotschy (*S. germanica* L. subsp. *balansae* (Boiss. & Kotschy) Derviz-Sokolova, *S. terekensis* Knorr.) - 1, 2

betoniciflora Rupr. (*Betonica foliosa* Rupr. 1869, non C. Presl, 1826) - 6
bithynica Boiss. = S. germanica subsp. bithynica
boissieri Kapell. = S. lavandulifolia
byzantina C. Koch (*S. lanata* Jacq. 1781, non Crantz, 1769, *S. taurica* Zefir.) - 1, 2
chinensis Bunge ex Benth. = S. aspera
cordata Klok. = S. germanica
cretica auct. = S. velata
cretica L. subsp. *velata* (Klok.) Greuter & Burdet = S. velata
czernjaevii Shost. = S. recta
discolor Benth. (*Betonica nivea* Stev., *Stachys nivea* (Stev.) Briq. 1897, non Labill. 1809) - 2
fominii Sosn. - 2
    Dubious species. According T.G. Derviz-Sokolova is likely a hybrid S. inflata Benth. x S. lavandulifolia Vahl.
fruticulosa Bieb. (*S. grossheimii* Kapell.) - 2
germanica L. (*S. cordata* Klok., *S. germanica* subsp. *cordata* (Klok.) Greuter & Burdet, *S. heterodonta* Zefir., *S. lanata* Crantz) - 1, 2
- subsp. *balansae* (Boiss. & Kotschy) Derviz-Sokolova = S. balansa
- subsp. **bithynica** (Boiss.) Bhattacharjee (*S. bithynica* Boiss.) - 1
- subsp. *cordata* (Klok.) Greuter & Burdet = S. germanica
grossheimii Kapell. = S. fruticulosa
heterodonta Zefir. = S. germanica
hissarica Regel (*S. hissarica* subsp. *turkestanica* (Regel) Derviz-Sokolova, *S. pseudofloccosa* Knorr., *S. tschatkalensis* Knorr., *S. turkestanica* (Regel) M. Pop. ex Knorr.) - 6
- subsp. *turkestanica* (Regel) Derviz-Sokolova = S. hissarica
iberica Bieb. (*S. memorabilis* Klok.) - 1, 2
- subsp. **georgica** Rech. fil. - 2
inflata Benth. - 2
- subsp. *caucasica* (Stschegl.) Takht. = S. stschegleewii
- var. *caucasica* Stschegl. = S. stschegleewii
iranica Rech. fil. (*S. setifera* C.A. Mey. subsp. *iranica* (Rech. fil.) Rech. fil.) - 6
japonica Miq. = S. aspera
komarovii Knorr. - 5
krynkensis Kotov (*S. transsilvanica* auct.) - 1
lanata Crantz = S. germanica
lanata Jacq. = S. byzantina
lavandulifolia Vahl (*S. boissieri* Kapell., *S. zuvandica* Rzazade) - 2, 6
macrantha (C. Koch) Stearn (*Betonica macrantha* C. Koch, *B. grandiflora* Willd. 1800, non Thuill. 1799, *Stachys macrantha* (C. Koch) Jalas, comb. superfl.) - 2
macrophylla Albov (*S. alpina* L. subsp. *macrophylla* (Albov) Bhattacharjee) - 2
macrostachya (Wend.) Briq. (*Betonica macrostachya* Wend., *B. orientalis* L. non *Stachys orientalis* L.) - 2
maeotica Postr. = S. palustris
maritima Gouan - 2
memorabilis Klok. = S. iberica
neglecta Klok. ex Kossko = S. annua
nivea (Stev.) Briq. = S. discolor
odontophylla Freyn - 2
officinalis (L.) Trevis. (*Betonica officinalis* L., *B. bjelorussica* Kossko ex Klok., *B. brachyodonta* Klok., *B. fusca* Klok., *B. peraucta* Klok.) - 1, 2, 3
ossetica (Bornm.) Czer. comb. nova (*Betonica nivea* Stev. subsp. *ossetica* Bornm. 1936, Feddes Repert. 15 : 370; *B. ossetica* (Bornm.) Chinth.) - 2
palustris L. (*S. maeotica* Postr.) - 1, 2, 3, 4, 5, 6

- subsp. *aspera* (Michx.) Derviz-Sokolova = S. aspera
patula auct. p.p. = S. atherocalyx
patula Griseb. = S. recta
paulii Grossh. - 2
persica S.G. Gmel. ex C.A. Mey. - 2
pseudofloccosa Knorr. = S. hissarica
pubescens Ten. - 1, 2
recta L. (*S. czernjaevii* Shost., *S. patula* Griseb.) - 1, 2
- subsp. *atherocalyx* (C. Koch) Derviz-Sokolova = S. atherocalyx
riederi Cham. = S. aspera
setifera C.A. Mey. - 2
- subsp. *iranica* (Rech. fil.) Rech. fil. = S. iranica
x **sintenisii** Rech. fil. - S. lavandulifolia Vahl x S. turcomanica Trautv. - 6
sosnowskyi Kapell. - 2(?)
spectabiliformis Kapell. = S. spectabilis
spectabilis Choisy ex DC. (*S. spectabiliformis* Kapell.) - 2
stschegleewii Sosn. (*S. inflata* Benth. var. *caucasica* Stschegl., *S. inflata* subsp. *caucasica* (Stschegl.) Takht.) - 2
sylvatica L. (*S. trapezuntea* Boiss.) - 1, 2, 3, 6
talyschensis Kapell. - 2
taurica Zefir. = S. byzantina
terekensis Knorr. = S. balansae
transsilvanica auct. = S. krynkensis
trapezuntea Boiss. = S. sylvatica
trinervis Aitch. & Hemsl. - 6
tschatkalensis Knorr. = S. hissarica
turcomanica Trautv. - 6
turkestanica (Regel) M. Pop. ex Knorr. = S. hissarica
velata Klok. (*S. cretica* L. subsp. *velata* (Klok.) Greuter & Burdet, *S. cretica* auct.) - 1, 2
wolgensis Wilensky - 1
woronowii (Schischk. ex Grossh.) R. Mill = Sideritis balansae
zuvandica Rzazade = S. lavandulifolia

Susilkumara Bennet = Alajja

anomala (Juz.) Bennet = Alajja anomala
rhomboidea (Benth.) Bennet = Alajja rhomboidea

Teucrium L.

x **alexeenkoanum** Juz. - T. canum Fisch. & C.A. Mey. x T. nuchense C. Koch - 2
botrys L. - 1
canum Fisch. & C.A. Mey. - 2
chamaedrys L. (*T. pulchrius* Juz., *T. stevenianum* Klok.) - 1, 2
- subsp. *syspirense* (C. Koch) Rech. fil. = T. syspirense
cordioides Schreb. - 1
cretaceum Kotov = T. jailae
excelsum Juz. - 6
fagetorum Klok. = T. krymense
fischeri Juz. = T. krymense
hircanicum L. - 2
jailae Juz. (*T. cretaceum* Kotov, *T. montanum* L. subsp. *jailae* (Juz.) Soo) - 1
japonicum Houtt. - 5
krymense Juz. (*T. fagetorum* Klok., *T. fischeri* Juz.) - 1, 2
miquelianum (Maxim.) Kudo (*T. stoloniferum* Roxb. var. *miquelianum* Maxim.) - 5
montanum L. (*T. montanum* subsp. *praemontanum* (Klok.) S. Pawl., *T. praemontanum* Klok.) - 1
- subsp. *jailae* (Juz.) Soo = T. jailae
- subsp. *pannonicum* (A. Kerner) Domin = T. pannonicum

- subsp. *pannonicum* (A. Kerner) S. Pawl. = T. pannonicum
- subsp. *praemontanum* (Klok.) S. Pawl. = T. montanum
multinodum (Bordz.) Juz. - 2
nuchense C. Koch - 2
orientale L. - 2
pannonicum A. Kerner (*T. montanum* L. subsp. *pannonicum* (A. Kerner) Domin, *T. montanum* subsp. *pannonicum* (A. Kerner) S. Pawl. comb. superfl.) - 1
parviflorum Schreb. - 2
polium L. - 1, 2, 6
- subsp. **capitatum** (L.) Arcang. - 1, 2, 6
praemontanum Klok. = T. montanum
pulchrius Juz. = T. chamaedrys
reuticum Bogoutdinova - 1
scordioides Schreb. (*T. scordium* L. subsp. *scordioides* (Schreb.) Maire & Petitm.) - 1, 2, 6
scordium L. - 1, 3, 6
- subsp. *scordioides* (Schreb.) Maire & Petitm. = T. scordioides
stevenianum Klok. = T. chamaedrys
stoloniferum Roxb. var. *miquelianum* Maxim. = T. miquelianum
syspirense C. Koch (*T. chamaedrys* L. subsp. *syspirense* (C. Koch) Rech. fil.) - 6
taylorii Boiss. - 2
- subsp. **muticum** Menitsky - 2
trapezunticum (Rech. fil.) Juz. - 2
ussuriense Kom. - 5
veronicoides Maxim. - 5

Thuspeinantha Durand

persica (Boiss.) Briq. - 6

Thymus L.

afghanicus Ronn. - 6
alatauensis (Klok. & Shost.) Klok. - 3
alpestris Tausch ex A. Kerner (*T. subalpestris* Klok.) - 1
altaicus Klok. & Shost. - 3, 4
alternans Klok. = T. roegneri
amictus Klok. = T. marschallianus
amurensisKlok. - 5
angustifolius Pers. var. *pycnotrichus* Uechtr. = T. serpyllum
ararati-minoris Klok. & Shost. = T. fedtschenkoi
armeniacus Klok. & Shost. = T. collinus
arsenijevii Klok. - 5
aschurbajevii Klok. = T. seravschanicus
aserbaidshanicus Klok. = T. transcaucasicus
asiaticus Serg. = T. mongolicus (Ronn.) Ronn.
attenuatus Klok. = T. roegneri
baicalensis Serg. - 4
bakurianicus Klok. = T. transcaucasicus
bashkiriensis Klok. & Shost. - 1
binervulatus Klok. & Shost. = T. talijevii
bituminosus Klok. = T. pavlovii
borysthenicus Klok. & Shost. - 1
brachyodon Borb. = T. pallasianus
bucharicus Klok. = T. seravschanicus
bulgaricus auct. p.p. = T. elisabethae and T. karamarjanicus
buschianus Klok. & Shost. = T. nummularius
calcareus Klok. & Shost. (*T. cretaceus* Klok. & Shost., *T. graniticus* Klok. & Shost., *T. kaljmijussicus* Klok. & Shost., *T. pseudocretaceus* Klok.) - 1
callieri Borb. ex Velen. = T. roegneri
- subsp. *urumovii* Velen. = T. roegneri
caucasicus Willd. ex Ronn. (*T. caucasicus* var. *medwedewii* Ronn., *T. praecox* Opiz subsp. *caucasicus* (Ronn.)

Jalas, *T. praecox* subsp. *grossheimii* (Ronn.) Jalas var. *medwedewii* (Ronn.) Jalas) - 2
- subsp. *grossheimii* (Ronn.) Jalas = T. grossheimii
- var. *medwedewii* Ronn. = T. caucasicus
chamaedrys Fries = T. ovatus
chankoanus Klok. = T. przewalskii
cherlerioides auct. p.p. = T. helendzhicus and T. tauricus
ciliatissimus Klok. = T. pallasianus
cimicinus Blum ex Ledeb. (*T. dubjanskii* Klok. & Shost., *T. zheguliensis* Klok. & Shost. p.p. incl. typo) - 1
circumcinctus Klok. = T. pulcherrimus
clandestinus Schur (*T. enervius* Klok., *T. montanus* Waldst. & Kit. 1801, non Crantz, 1769, *T. serpyllum* L. subsp. *montanus* Archang., *T. pulegioides* L. subsp. *montanus* (Arcang.) Ronn.) - 1
collinus Bieb. (*T. armeniacus* Klok. & Shost., *T. deflexus* Klok., *T. perplexus* Klok., *T. terekensis* Klok.) - 2
coriifolius Ronn. (*T. sosnowskyi* Grossh.) - 2
cosacorum Klok. = T. kirgisorum
crebrifolius Klok. - 2
crenulatus Klok. - 4
cretaceus Klok. & Schost. = T. carcareus
creticola (Klok. & Shost.) Stank. = T. kirgisorum
cuneatus Klok. = T. seravschanicus
curtus Klok. - 5
"*czernjaevii*" Klok. & Shost. = T. x tschernjajevii
daghestanicus Klok. & Shost. (*T. hadzhievii* Grossh., *T. lipskyi* Klok. & Shost., *T. mashukensis* Klok., *T. shemachensis* Klok.) - 2
dahuricus Serg. (*T. quinquecostatus* auct.) - 4, 5
deflexus Klok. = T. collinus
desjatovae Ronn. - 2
didukhii Ostapko - 1
diminutus Klok. - 6
x **dimorphus** Klok. & Shost. - T. calcareus Klok. & Shost. x T. marschallianus Willd.
disjunctus Klok. - 5
diversifolius Klok. - 5
dmitrievae Gamajun. - 6
dubjanskii Klok. & Shost. = T. cimicinus
dzevanovskyi Klok. & Shost. - 1
elegans Serg. - 3, 4
elisabethae Klok. & Shost. (*T. pseudobulgaricus* Klok. p.p., *T. teberdensis* Klok., *T. bulgaricus* auct. p.p.) - 2
eltonicus Klok. & Shost. = T. kirgisorum
enervius Klok. = T. clandestinus
eravinensis Serg. - 4
eremita Klok. - 6
eriophorus Ronn. = T. kotschyanus
eubajcalensis Klok. = T. pavlovii
eupatoriensis Klok. & Shost. = T. moldavicus
evenkiensis Byczennikova = T. reverdattoanus
extremus Klok. - 4
fedtschenkoi Ronn. (*T. ararati-minoris* Klok. & Shost., *T. kjapazi* Grossh.) - 2
flexilis Klok. = T. ochotensis
fominii Klok. & Shost. = T. transcaucasicus
x **georgicus** Ronn. - T. collinus Bieb. x T. transcaucasicus Ronn. - 2
glaber Mill. - 1
glabrescens Willd. subsp. *urumovii* (Velen.) Jalas = T. roegneri
glabricaulis Klok. - 1
glacialis Klok. = T. ochotensis
glandulosus Requien = Calamintha parviflora
gobicus Tscherneva - 4
graniticus Klok. & Shost. = T. calcareus

grossheimii Ronn. (*T. caucasicus* Willd. ex Ronn. subsp. *grossheimii* (Ronn.) Jalas, *T. praecox* Opiz subsp. *caucasicus* (Ronn.) Jalas var. *grossheimii* (Ronn.) Jalas, *T. praecox* subsp. *grossheimii* (Ronn.) Jalas) - 2

guberlinensis Iljin - 3

hadzhievii Grossh. = T. daghestanicus

helendzhicus Klok. & Shost. (*T. cherlerioides* auct. p.p.) - 2

hirsutus Bieb. p.p. = T. roegneri

hirtellus Klok. = T. roegneri

hirticaulis Klok. = T. talijevii

hohenackeri Klok. - 2

iljinii Klok. & Shost. - 4

inaequalis Klok. = T. japonicus

incertus Klok. - 6

indigirkensis Karav. - 4

irtyschensis Klok. - 3

jailae (Klok. & Shost.) Stank. = T. roegneri

japonicus (Hara) Kitag. (*T. quinquecostatus* Celak. var. *japonicus* Hara, *T. inaequalis* Klok., *T. nervulosus* Klok.) - 4, 5

jenisseensis Iljin - 4

kaljmijussicus Klok. & Shost. = T. calcareus

karamarjanicus Klok. & Shost. (*T. pseudobulgaricus* Klok. p.p., *T. bulgaricus* auct. p.p.) - 2

karatavicus A. Dmitr. ex Gamajun. - 6

karjaginii Grossh. - 2

kasakstanicus Klok. & Shost. = T. kirgisorum

kirgisorum Dubjan. (*T. cosacorum* Klok., *T. creticola* (Klok. & Shost.) Stank., *T. eltonicus* Klok. & Shost., *T. kasakstanicus* Klok. & Shost., *T. lanulosus* Klok. & Shost., *T. pallasianus* H. Br. subsp. *brachyodon* (Borb.) Jalas, p.p. excl. typo, nom. illegit.) - 1, 3, 6

kjapazi Grossh. = T. fedtschenkoi

klokovii (Ronn.) Shost. = T. tiflisiensis

komarovii Serg. (*T. semiglaber* Klok.) - 5

kondratjukii Ostapko - 1

kotschyanus Boiss. & Hohen. (*T. eriophorus* Ronn., *T. kotschyanus* var. *eriophorus* (Ronn.) Jalas) - 2

- subsp. *migricus* (Klok. & Shost.) Menitsky = T. migricus

- subsp. **pseudocollinus** Menitsky - 2

- var. *eriophorus* (Ronn.) Jalas = T. kotschyanus

- var. *migricus* (Klok. & Shost.) Ronn. ex Rech. fil. = T. migricus

krylovii Byczennikova - 4

kuminovianus (Serg.) Peschkova (*T. serpyllum* L. var. *kuminovianus* Serg.) - 4

kytlymiensis Klok. = T. paucifolius

ladjanuricus Kem.-Nath. - 2

lanulosus Klok. & Shost. = T. kirgisorum

latifolius (Bess.) Andrz. = T. marschallianus

latissimus Klok. = T. pastoralis

lavrenkoanus Klok. - 3

liaculatus Klok. - 1

lipskyi Klok. & Shost. = T. daghestanicus

x **littoralis** Klok. & Shost. - T. marschallianus Willd. x T. moldavicus Klok. & Shost. - 1

loevyanus auct. = T. marschallianus

magnificus Klok. - 6

majkopensis Klok. & Shost. - 2

mandschuricus auct. = T. przewalskii

markhotensis Maleev - 2

marschallianus Willd. (*T. amictus* Klok., *T. latifolius* (Bess.) Andrz., *T. pannonicus* All. subsp. *marschallianus* (Willd.) Soo, *T. platyphyllus* Klok., *T. pseudopannonicus* Klok., *T. stepposus* Klok. & Shost., *T. loevyanus* auct., *T. pannonicus* auct.) - 1, 2, 3, 5(alien), 6

mashukensis Klok. = T. daghestanicus

michaelis R. Kam. & A. Budantz. (*T. mongolicus* Klok. 1954, non (Ronn.) Ronn. 1934) - 4(?)

migricus Klok. & Shost. (*T. kotschyanus* Boiss. & Hohen. subsp. *migricus* (Klok. & Shost.) Menitsky, *T. kotschyanus* var. *migricus* (Klok. & Shost.) Ronn. ex Rech. fil.) - 2

minussinensis Serg. - 3, 4

moldavicus Klok. & Shost. (*T. eupatoriensis* Klok. & Shost., *T. zygioides* auct. p.p.) - 1

mongolicus Klok. = T. michaelis

mongolicus (Ronn.) Ronn. (*T. serpyllum* L. var. *mongolicus* Ronn., *T. asiaticus* Serg.) - 3, 6

montanus Waldst. & Kit. = T. clandestinus

mugodzharicus Klok. & Shost. - 1, 3, 6

muscosus Zaverucha = T. serpyllum

narymensis Serg. - 3, 4

nerczensis Klok. - 4

nervulosus Klok. = T. japonicus

nummularius Bieb. (*T. buschianus* Klok. & Shost., *T. pseudonummularius* Klok. & Shost.) - 2

x **oblongifolius** Opiz (*T. podolicus* Klok. & Shost. p.max.p., *T.* x *polessicus* Klok. nom. nud.). - T.ovatus Mill. x T. serpyllum L. - 1

ochotensis Klok. (*T. flexilis* Klok., *T. glacialis* Klok.) - 4, 5

ovatus Mill. (*T. chamaedrys* Fries, *T. pulegioides* L. subsp. *chamaedrys* (Fries) Gusul., *T. serpyllum* L. subsp. *chamaedrys* (Fries) Celak., *T. ucrainicus* (Klok. & Shost.) Klok., *T. pulegioides* auct.) - 1

oxyodontus Klok. - 4

pallasianus H. Br. (*T. brachyodon* Borb., *T. ciliatissimus* Klok., *T. pallasianus* subsp. *brachyodon* (Borb.) Jalas, p.p. quoad typum, nom. illegit.) - 1, 2

- subsp. *brachyodon* (Borb.) Jalas = T. kirgisorum and T. pallasianus

pannonicus auct. = T. marschallianus

pannonicus All. subsp. *marschallianus* (Willd.) Soo = T. marschallianus

paradoxus Klok. = T. paucifolius

pastoralis Iljin ex Klok. (*T. latissimus* Klok.) - 2

paucifolius Klok. (? *T. kytlymiensis* Klok., ?*T. paradoxus* Klok., *T. pseudalternans* Klok., ?*T. purpurellus* Klok., *T. talijevii* Klok. & Shost. subsp. *paucifolius* (Klok.) P. Schmidt) - 1

pavlovii Serg. (*T. bituminosus* Klok., *T. eubajcalensis* Klok.) - 4

perplexus Klok. = T. collinus

petraeus Serg. - 3, 6

phyllopodus Klok. - 4

x **pilisiensis** Borb. - T. marschallianus Willd. x T. ovatus Mill. - 1

platyphyllus Klok. = T. marschallianus

podolicus Klok. & Shost. p.max.p. = T. x oblongifolius

x *polessicus* Klok. = T. x oblongifolius

praecox Opiz subsp. *caucasicus* (Ronn.) Jalas = T. caucasicus

- - var. *grossheimii* (Ronn.) Jalas = T. grossheimii

- subsp. *grossheimii* (Ronn.) Jalas = T. grossheimii

- - var. *medwedewii* (Ronn.) Jalas = T. caucasicus

proximus Serg. - 3, 6

przewalskii (Kom.) Nakai (*T. chankoanus* Klok., *T. mandschuricus* auct.) - 5

pseudalternans Klok. = T. paucifolius

pseudobulgaricus Klok. = T. elisabethae and T. karamarjanicus

pseudocretaceus Klok. = T. calcareus

x **pseudograniticus** Klok. & Shost. - 1

pseudohumillimus Klok. & Shost. = T. tauricus

pseudonummularius Klok. & Shost. = T. nummularius
pseudopannonicus Klok. = T. marschallianus
pseudopulegioides Klok. & Shost. - 2
pseudostepposus Klok. - 1
pulchellus C.A. Mey. - 2
pulcherrimus Schur (*T. circumcinctus* Klok.) - 1
pulegioides auct. = T. ovatus
pulegioides L. subsp. *chamaedrys* (Fries) Gusul. = T. ovatus
- subsp. *montanus* (Arcang.) Ronn. = T. clandestinus
punctulosus Klok. - 1
purpurellus Klok. = T. paucifolius
purpureo-violaceus Byczennikova & Kuvajev - 4
putoranicus Byczennikova & Kuvajev - 4
pycnotrichus (Uechtr.) Ronn. = T. serpyllum
quinquecostatus auct. = T. dahuricus
quinquecostatus Celak. var. *japonicus* Hara = T. japonicus
rariflorus C. Koch - 2
rasitatus Klok. - 3, 6
reverdattoanus Serg. (*T. evenkiensis* Byczennikova, *T. reverdattoanus* var. *evenkiensis* (Byczennikova) Kuvajev) - 3, 4, 5
- var. *evenkiensis* (Byczennikova) Kuvajev = T. reverdattoanus
roegneri C. Koch (*T. alternans* Klok., *T. attenuatus* Klok., *T. callieri* Borb. ex Velen., *T. callieri* subsp. *urumovii* Velen., *T. glabrescens* Willd. subsp. *urumovii* (Velen.) Jalas, *T. hirsutus* Bieb. p.p. quoad plantas, nom. illegit., *T. hirtellus* Klok., *T. jailae* (Klok. & Shost.) Stank., *T. zelenetzkyi* Klok. & Shost.) - 1
roseus Schipcz. - 3, 6
rotundatus Klok. = T. talijevii
schischkinii Serg. - 3, 4
semiglaber Klok. = T. komarovii
seravschanicus Klok. (*T. aschurbajevii* Klok., *T. bucharicus* Klok., *T. cuneatus* Klok.) - 6
sergievskajae Karav. - 4
serpyllum L. (*T. angustifolius* Pers. var. *pycnotrichus* Uechtr., *T. muscosus* Zaverucha, *T. pycnotrichus* (Uechtr.) Ronn., *T. serpyllum* subsp. *pycnotrichus* (Uechtr.) Pawl.) - 1, 3, 4
- subsp. *chamaedrys* (Fries) Celak. = T. ovatus
- subsp. *montanus* Archang. = T. clandestinus
- subsp. *pycnotrichus* (Uechtrl) Pawl. = T. serpyllum
- subsp. *tanaensis* (Hyl.) Jalas = T. subarcticus
- var. *kuminovianus* Serg. = T. kuminovianus
- var. *mongolicus* Ronn. = T. mongolicus
- var. *tanaensis* Hyl. = T. subarcticus
sessilifolius Klok. - 2
shemachensis Klok. = T. daghestanicus
sibiricus (Serg.) Klok. & Shost. - 3, 4
sokolovii Klok. - 5
sosnowskyi Grossh. = T. coriifolius
stepposus Klok. & Shost. = T. marschallianus
subalpestris Klok. = T. alpestris
subarcticus Klok. & Shost. (*T. serpyllum* L. subsp. *tanaensis* (Hyl.) Jalas, *T. serpyllum* var. *tanaensis* Hyl.) - 1
subnervosus Vved., Nabiev & Tuljaganova - 6
superbus Ronn. = T. transcaucasicus
talijevii Klok. & Shost. (*T. binervulatus* Klok. & Shost., *T. hirticaulis* Klok., *T. rotundatus* Klok., *T. talijevii* f. *hirticaulis* (Klok.) P. Schmidt, *T. uralensis* Klok.) - 1
- subsp. *paucifolius* (Klok.) P. Schmidt = T. paucifolius
- f. *hirticaulis* (Klok.) P. Schmidt = T. talijevii
tauricus Klok. & Shost. (*T. pseudohumillimus* Klok. & Shost., *T. cherlerioides* auct. p.p.) - 1
teberdensis Klok. = T. elisabethae
terekensis Klok. = T. collinus

terskeicus Klok. - 6
tiflisiensis Klok. & Shost. (*T. klokovii* (Ronn.) Shost.) - 2
tonsilis Klok. - 4
transcaspicus Klok. - 6
transcaucasicus Ronn. (? *T. aserbaidshanicus* Klok., *T. bakurianicus* Klok., *T. fominii* Klok. & Shost., *T. superbus* Ronn., *T. ziaratinus* Klok. & Shost.) - 2
trautvetteri Klok. & Shost. - 2
x tschernjajevii Klok. & Shost. (*T.* "*czernjaevii*" Klok. & Shost.). - T. marschallianus Willd. x T. pallasianus H. Br. - 1
turczaninovii Serg. - 4
ucrainicus (Klok. & Shost.) Klok. = T. ovatus
uralensis Klok. = T. talijevii
ussuriensis Klok. - 5
x zedelmeyeri Ronn. - T. rariflorus C. Koch x T. transcaucasicus Ronn. - 2
zelenetzkyi Klok. & Shost. = T. roegneri
zheguliensis Klok. & Shost. p.p. = T. cimicinus
ziaratinus Klok. & Shost. = T. transcaucasicus
zygioides auct. p.p. = T. moldavicus

Wiedemannia Fisch. & C.A. Mey.

multifida (L.) Benth. - 2

Ziziphora L.

afghanica Rech. fil. = Z. clinopodioides
biebersteiniana (Grossh.) Grossh. (*Z. clinopodioides* Lam. subsp. *pseudodasyantha* (Rech. fil.) Rech. fil., *Z. clinopodioides* subsp. *szowitsii* (Rech. fil.) Rech. fil., *Z. pseudodasyantha* Rech. fil., *Z. szowitsii* Rech. fil.) - 2
borzhomica Juz. = Z. clinopodioides
brantii C. Koch - 2
brevicalyx Juz. = Z. clinopodioides
bungeana Juz. = Z. clinopodioides
capitata L. (*Z. capitata* subsp. *orietalis* Sam. ex Rech. fil., *Z. capitellata* Juz.) - 1, 2, 6
- subsp. *orietalis* Sam. ex Rech. fil. = Z. capitata
capitellata Juz. = Z. capitata
clinopodioides Lam. (*Z. afghanica* Rech. fil., *Z. borzhomica* Juz., *Z. brevicalyx* Juz., *Z. bungeana* Juz., *Z. clinopodioides* subsp. *afghanica* (Rech. fil.) Rech. fil., *Z. clinopodioides* subsp. *bungeana* (Juz.) Rech. fil., *Z. denticulata* Juz., *Z. dzhavakhishvilii* Juz., *Z. turcomanica* Juz.) - 2, 3, 4, 6
- subsp. *afghanica* (Rech. fil.) Rech. fil. = Z. clinopodioides
- subsp. *bungeana* (Juz.) Rech. fil. = Z. clinopodioides
- subsp. *pseudodasyantha* (Rech. fil.) Rech. fil. p.p. = Z. biebersteiniana
- subsp. *rigida* (Boiss.) Rech. fil. = Z. rigida
- subsp. *szowitsii* (Rech. fil.) Rech. fil. = T. biebersteiniana
denticulata Juz. = Z. clinopodioides
dzhavakhishvilii Juz. = Z. clinopodioides
fasciculata C. Koch ex Grossh. = Z. rigida
fasciculata C. Koch ex Rech. fil. = Z. serpyllacea
galinae Juz. - 6
interrupta Juz. - 6
karjaginii Ter-Chatsch. = Z. puschkinii
magakjanii Ter-Chatsch. = Z. raddei
pamiroalaica Juz. (*Z. pulchella* Pavl., *Z. tomentosa* Juz.) - 6
pedicellata Pazij & Vved. - 6
persica Bunge - 1, 2, 6
pseudodasyantha Rech. fil. = Z. biebersteiniana
pulchella Pavl. = Z. pamiroalaica
puschkinii Adams (*T. karjaginii* Ter-Chatsch.) - 2
raddei Juz. (*Z. magakjanii* Ter-Chatsch.) - 2

rigida (Boiss.) Stapf (*Z. clinopodioides* Lam. subsp. *rigida* (Boiss.) Rech. fil., *Z. fasciculata* C. Koch ex Grossh. 1932, non C. Koch ex Rech. fil. 1951) - 2

serpyllacea Bieb. (*Z. fasciculata* C. Koch ex Rech. fil. 1951, non C. Koch ex Grossh. 1932) - 2

suffruticosa Pazij & Vved. - 6

szowitsii Rech. fil. = Z. biebersteiniana

taurica Bieb. - 1

tenuior L. - 1, 2, 3, 6

tomentosa Juz. = Z. pamiroalaica

turcomanica Juz. = Z. clinopodioides

vichodceviana V. Tkatsch. ex Tuljaganova - 6

woronowii Maleev - 2

LAURACEAE Juss.
Laurus L.
nobilis L. - 2

LEGUMINOSAE Juss. = FABACEAE

LEMNACEAE S.F. Gray
Lemna L.
aequinoctialis Welw. - 5

gibba L. - 1, 2

japonica Landolt - 5

minor L. - 1, 2, 3, 4, 5, 6

minuscula Herter = L. minuta

minuta Humb., Bonpl. & Kunth (*L. minuscula* Herter, nom. illegit.) - 1

trisulca L. - 1, 2, 3, 4, 5, 6

turionifera Landolt - 1, 2, 3, 4, 6

Spirodela Schleid.
polyrhiza (L.) Schleid. - 1, 2, 3, 4, 5, 6

Wolffia Horkel ex Schleid.
arrhiza (L.) Horkel ex Wimm. - 1, 2

LENTIBULARIACEAE Rich.
Pinguicula L.
algida Malysch. - 4

alpina L. - 1, 3, 4

bicolor Woloszcz. (*P. vulgaris* L. subsp. *bicolor* (Woloszcz.) A. & D. Löve) - 1

glandulosa Trautv. & C.A. Mey. = P. spathulata

macroceras Link (*P. vulgaris* L. subsp. *macroceras* (Link) Calder & Taylor) - 5

spathulata Ledeb. (*P. glandulosa* Trautv. & C.A. Mey., *P. variegata* Turcz.) - 4, 5

variegata Turcz. = P. spathulata

villosa L. - 1, 3, 4, 5

vulgaris L. - 1, 2, 3, 4

- subsp. *bicolor* (Woloszcz.) A. & D. Löve = P. bicolor

- subsp. *macroceras* (Link) Calder & Taylor = P. macroceras

Utricularia L.
australis R. Br. (*U. major* auct.) - 1, 2

bremii Heer - 1

intermedia Hayne - 1, 2, 3, 4, 5, 6

japonica Makino = U. macrorhiza

macrorhiza Le Conte (*U. japonica* Makino, *U. vulgaris* L. subsp. *macrorhiza* (Le Conte) Clausen) - 5

major auct. = U. australis

minor L. - 1, 2, 3, 4, 5, 6

- var. *multispinosa* Miki = U. multispinosa

multispinosa (Miki) Miki (*U. minor* L. var. *multispinosa* Miki) - 5

ochroleuca R. Hartm. - 1

vulgaris L. - 1, 2, 3, 4, 5, 6

- subsp. *macrorhiza* (Le Conte) Clausen = U. macrorhiza

LEONTICACEAE Bercht. & J. Presl = BERBERIDACEAE

LILIACEAE Juss.
Cardiocrinum (Endl.) Lindl. (*Hemerocallis* auct.)
cordatum (Thunb.) Makino (*Hemerocallis cordata* Thunb., *Cardiocrinum cordatum* var. *glehnii* (Fr. Schmidt) Hara, *C. glehnii* (Fr. Schmidt) Makino, *Lilium cordifolium* Thunb., *L. glehnii* Fr. Schmidt) - 5

- var. *glehnii* (Fr. Schmidt) Hara = C. cordatum

glehnii (Fr. Schmidt) Makino = C. cordatum

Eduardoregelia M. Pop. = Tulipa
heterophylla (Regel) M. Pop. = Tulipa heterophylla

Erythronium L.
caucasicum Woronow - 2

dens-canis L. - 1

japonicum Decne. - 5

sibiricum (Fisch. & C.A. Mey.) Kryl. - 3, 4

Fritillaria L. (*Petilium* Ludw.)
ariana (Losinsk. & Vved.) E.M. Rix = Rhinopetalum gibbosum

armena Boiss. (*F. caucasica* Adams var. *armena* (Boiss.) Grossh.) - 2

biebersteiniana Charkev. = F. ophioglossifolia

camschatcensis (L.) Ker-Gawl. - 5

caucasica Adams - 2

- var. *armena* (Boiss.) Grossh. = F. armena

collina Adams = F. latifolia

crassifolia Boiss. & Huet subsp. *kurdica* (Boiss. & Noe) E.M. Rix = F. grossheimiana and F. kurdica

dagana Turcz. ex Trautv. - 4

dzhabavae A. Khokhr. - 2

eduardii Regel (*Petilium eduardii* (Regel) Vved.) - 6

ferganensis Losinsk. = F. walujewii

grandiflora Grossh. (*F. kotschyana* Herb. subsp. *grandiflora* (Grossh.) E.M. Rix) - 2

■ grossheimiana Losinsk. (*F. crassifolia* Boiss. & Huet subsp. *kurdica* (Boiss. & Noe) E.M. Rix, p.p.)

*imperialis L. (*Petilium imperiale* (L.) Jaume) - 1

kotschyana Herb. - 2

- subsp. *grandiflora* (Grossh.) E.M. Rix = F. grandiflora

kurdica Boiss. & Noe (*F. crassifolia* Boiss. & Huet subsp. *kurdica* (Boiss. & Noe) E.M. Rix, p.p.) - 2

lagodechiana Charkev. = F. ophioglossifolia

latifolia Willd. (*F. collina* Adams, *F. lutea* Mill. subsp. *latifolia* (Willd.) Artjushenko) - 2

lutea Bieb. = F. ophioglossifolia
lutea Mill. subsp. *latifolia* (Willd.) Artjushenko = F. latifolia
maximowiczii Freyn - 4, 5
meleagris L. - 1
meleagroides Patrin ex Schult. & Schult. fil. - 1, 3, 6
■ michailovskyi Fomin
montana Hoppe - 1
olgae Vved. - 6
- subsp. *regelii* (Losinsk.) R. Kam. = F. regelii
ophioglossifolia Freyn & Sint. (*F. biebersteiniana* Charkev., *F. lagodechiana* Charkev., *F. lutea* Bieb. 1808, non Mill. 1768) - 2
orientalis Adams - 2
pallidiflora Schrenk - 6
raddeana Regel (*Petilium raddeanum* (Regel) Vved. ex Pazij) - 6
regelii Losinsk. (*F. olgae* Regel subsp. *regelii* (Losinsk.) R. Kam.) - 6
ruthenica Wikstr. - 1, 2, 3, 6
ussuriensis Maxim. - 5
verticillata Willd. - 3, 6
walujewii Regel (*F. ferganensis* Losinsk.) - 6

Gagea Salisb. (*Plecostigma* Turcz. ex Trautv., *Ornithogalum* auct.)

x **absurda** Levichev. - G. chomutovae (Pasch.) Pasch. x G. substilis Vved. - 6
afghanica Terr. (*G. obvoluta* Pavl.) - 6
alberti Regel - 6
alexeenkoana Miscz. - 2
altaica Schischk. & Sumn. - 3, 4
amblyopetala Boiss. & Heldr. subsp. *heldreichii* Terr. = G. chrysantha
angrenica Levichev - 6
anisanthos C. Koch - 2
anisopoda M. Pop. - 6
artemczukii A. Krasnova - 1
arvensis Dumort. = G. villosa
- subsp. *granatellii* (Parl.) Aschers. & Graebn. = G. granatellii
- subsp. *granatellii* (Parl.) K. Richt. = G. granatellii
azutavica Kotuchov - 3
baschkyzylsaica Levichev - 6
bergii Litv. - 6
bohemica (Zauschn.) Schult. & Schult. fil. (*Ornithogalum bohemicum* Zauschn., *Gagea szovitsii* (Lang) Bess. ex Schult. & Schult. fil.) - 1
boroldaica Levichev ex Czer. = G. pseudominutiflora
brevistolonifera Levichev - 6
bulbifera (Pall.) Salisb. - 1, 2, 3, 6
caelestis Levichev - 6
calantha Levichev - 6
callieri Pasch. - 1
calyptrifolia Levichev - 6
capillifolia Vved. - 6
capusii sensu Grossh. = G. praemixta and G. turkestanica
capusii Terr. (*G. triquetra* Vved.) - 6
caroli-kochii Grossh. - 2
caucasica Stapf = G. chlorantha
chanae Grossh. - 2
charadzeae Davlianidze - 2
chitralensis Dasggupta & Deb - 6
chlorantha (Bieb.) Schult. & Schult. fil. (*G. caucasica* Stapf) - 2, 6
chomutovae (Pasch.) Pasch. (*Lloydia rubroviridis* auct.) - 6
chrysantha (Jan) Schult. & Schult. fil. (*Ornithogalum chry-*

santhum Jan, *Gagea amblyopetala* Boiss. & Heldr. subsp. *heldreichii* Terr., *G. heldreichii* (Terr.) Lojac.) - 1
circumplexa Vved. - 6
commutata C. Koch = G. fibrosa
confusa Terr. - 2, 6
coreanica Koidz. - 5
czatkalica Levichev - 6
davlianidzeae Levichev - 6
delicatula Vved. - 6
divaricata Regel - 6
dschungarica Regel - 6
dubia Terr. (*G.littoralis* Artemcz.) - 1, 2, 6
emarginata Kar. & Kir. - 3, 6
erubescens (Bess.) Schult. & Schult. fil. - 1
exilis Vved. - 6
fedtschenkoana Pasch. - 3, 6
ferganica Levichev - 6
fibrosa (Desf.) Schult. & Schult. fil. (*Ornithogalum fibrosum* Desf., *Gagea commutata* C. Koch, *G. reticulata* (Pall.) Schult. & Schult. fil. subsp. *rigida* (Boiss. & Sprun.) K. Richt., *G. reticulata* var. *rigida* (Boiss. & Sprun.) Pasch., *G. rigida* Boiss. & Sprun.) - 2, 6
filiformis (Ledeb.) Kar. & Kir. (*G. filiformis* (Ledeb.) Kunth, comb. superfl., *G. minuta* Grossh., *G. pseudoerubescens* Pasch., *G. sacculifera* Regel) - 3, 6
fistulosa Ker-Gawl. (*Ornithogalum fistulosum* Ramond ex DC. nom. illegit. superfl.) - 1
gageoides (Zucc.) Vved. - 2, 6
germainae Grossh. - 1, 2
glacialis C. Koch - 2
- var. *joannis* (Grossh.) Grossh. = G. luteoides
glaucescens Levichev - 6
graminifolia Vved. - 6
granatellii (Parl.) Parl. (*G. arvensis* Dumort. subsp. *granatellii* (Parl.) Aschers. & Graebn. comb. superfl., *G. arvensis* subsp. *granatellii* (Parl.) K. Richt.) - 1
granulosa Turcz. - 1, 3, 4, 6
gymnopoda Vved. - 6
gypsacea Levichev - 6
heldreichii (Terr.) Lojac. = G. chrysantha
helenae Grossh. - 2
hiensis Pasch. - 4, 5
hissarica Lipsky - 6
holochiton M. Pop. & Czug. nom. invalid. - 6
humicola Levichev - 6
hypanica Sobko - 1
ignota Levichev - 6
iliensis M. Pop. = G. setifolia
improvisa Grossh. - 2
incrustata Vved. - 6
jaeschkei Pasch. = G. olgae
joannis Grossh. = G. luteoides
kamelinii Levichev - 6
kopetdagensis Vved. - 6
korshinskyi Grossh. = G. vvedenskyi
kuraminica Levichev - 6
lasczinskyi N. Zolotuchin - 3
leucantha M. Pop. & Czug. nom. invalid. - 6
littoralis Artemcz. = G. dubia
longiscapa Grossh. - 3, 4
ludmilae Levichev - 6
lutea (L.) Ker-Gawl. - 1, 2, 3, 4, 5
luteoides Stapf (*G. glacialis* C. Koch var. *joannis* (Grossh.) Grossh., *G. joannis* Grossh.) - 2
maeotica Artemcz. - 1
michaelis Golosk. - 6

minima (L.) Ker-Gawl. - 1
minuta Grossh. = G. filiformis
minutiflora Regel - 6
minutissima Vved. - 6
mirabilis Grossh. - 1, 3, 6
nabievii Levichev - 6
nakaiana Kitag. - 5
neo-popovii Golosk. (*G. vaginata* M. Pop. ex Golosk. 1955, non Pasch. 1906, *G. subalpina* L.Z. Shue) - 6
novoascanica Klok. - 1
obvoluta Pavl. = G. afghanica
olgae Regel (*G. jaeschkei* Pasch.) - 6
ova Stapf (*G. stipitata* Merckl. ex Bunge var. *ova* (Stapf) Pasch.) - 6
paczoskii (Zapal.) Grossh. (*G. pratensis* (Pers.) Dumort. var. *paczoskii* (Zapal.) Zahar.) - 1
paedophila Vved. - 6
pamirica Grossh. - 6
paniculata Levichev - 6
parva Vved. ex Grossh. = G. turkestanica
pauciflora Turcz. ex Ledeb. (*Plecostigma pauciflorum* Turcz. ex Trautv. non basionymum) - 4, 5
pedata Levichev - 6
pineticola Klok. - 1
podolica Schult. & Schult. fil. (*G. scythica* Artemcz.) - 1
popovii Vved. - 6
praeciosa Klok. - 1
praemixta Vved. (*G. capusii* sensu Grossh. p.p.) - 6
pratensis (Pers.) Dumort. - 1
- var. *paczoskii* (Zapal.) Zahar. = G. paczoskii
provisa Pasch. - 4
pseudoerubescens Pasch. = G. filiformis
pseudominutiflora Levichev (*G. boroldaica* Levichev ex Czer. nom. invalid.) - 6
pseudoreticulata Vved. - 6
pusilla (F.W. Schmidt) Schult. & Schult. fil. - 1, 2
- subsp. *tesquicola* (A. Krasnova) Davlianidze = G. tesquicola
reticularis Salisb. = G. reticulata
reticulata (Pall.) Schult. & Schult. fil. (*G. reticularis* Salisb.) - 1, 2, 6
- subsp. *rigida* (Boiss. & Sprun.) K. Richt. = G. fibrosa
- var. *rigida* (Boiss. & Sprun.) Pasch. = G. fibrosa
rigida Boiss. & Sprun. = G. fibrosa
rufidula Levichev - 6
rupicola Levichev - 6
sacculifera Regel = G. filiformis
samojedorum Grossh. - 1
scythica Artemcz. = G. podolica
setifolia Baker (*G. iliensis* M. Pop.) - 6
spathacea (Hayne) Salisb. - 1
stipitata Merckl. ex Bunge - 2, 6
- var. *ova* (Stapf) Pasch. = G. ova
subalpina L.Z. Shue = G. neo-popovii
subtilis Vved. - 6
sulfurea Miscz. - 2
szovitsii (Lang) Bess. ex Schult. & Schult. fil. = G. bohemica
talassica Levichev - 6
taschkentica Levichev - 6
taurica Stev. - 1, 2
- var. *ucrainica* (Klok.) Zahar. = G. ucrainica
tenera Pasch. - 2, 6
tenuifolia (Boiss.) Fomin - 2, 6
■ tenuissima Miscz.
tesquicola A. Krasnova (*G. pusilla* (F.W. Schmidt) Schult. & Schult. fil. subsp. *tesquicola* (A. Krasnova) Davlia-

nidze) - 1
toktogulii Levichev - 6
transversalis Stev. - 1
triquetra Vved. = G. capusii
x turanica Levichev. - G. ova Stapf x G. stipitata Merckl. ex Bunge - 6
turkestanica Pasch. (*G. parva* Vved. ex Grossh. nom. invalid., *G. capusii* sensu Grossh. p.p.) - 6
ucrainica Klok. (*G. taurica* Stev. var. *ucrainica* (Klok.) Zahar.) - 1
ugamica Pavl. - 6
ulazsaica Levichev - 6
x vaga Levichev. - G. calantha Levichev x G. rufidula Levichev - 6
vaginata M. Pop. ex Golosk. = G. neo-popovii
vaginata Pasch. - 5
vegeta Vved. - 6
villosa (Bieb.) Duby (*Ornithogalum villosum* Bieb., *O. arvense* Pers., *Gagea arvensis* Dumort.) - 1, 2
villosula Vved. - 6
vvedenskyi Grossh. (*G. korshinskyi* Grossh.) - 6

Hemerocallis auct. = Cardiocrinum

cordata Thunb. = Cardiocrinum cordatum

Korolkowia Regel

sewerzowii Regel - 6

Lilium L.

armenum (Miscz. ex Grossh.) Manden. (*L. szovitsianum* Fisch. & Ave-Lall. var. *armenum* Miscz. ex Grossh., *L. monadelphum* Bieb. subsp. *armenum* (Miscz. ex Grossh.) G. Kudrjaschova, *L. monadelphum* var. *armenum* (Miscz. ex Grossh.) P.H. Davis & Henderson) - 2
avenaceum Fisch. ex Maxim. = L. debile and L. distichum
bulbiferum L. - 1
buschianum Lodd. (*L. pulchellum* Fisch., *L. concolor* auct. p.p.) - 4, 5
callosum Siebold & Zucc. - 5
candidum L. - 1
carniolicum Bernh. ex Koch subsp. *ponticum* (C. Koch) P.H. Davis & Henserson = L. ponticum
caucasicum (Miscz. ex Grossh.) Grossh. (*L. martagon* L. subsp. *caucasicum* Miscz. ex Grossh.) - 2
cernuum Kom. - 5
concolor auct. p.p. = L. buschianum
cordifolium Thunb. = Cardiocrinum cordatum
dauricum Ker-Gawl. = L. pensylvanicum
debile Kittlitz (*L. avenaceum* Fisch. ex Maxim. p.p. incl. typo, *L. medeoloides* auct. p.p.) - 5
distichum Nakai (*L. avenaceum* Fisch. ex Maxim. p.p., *L. hansonii* auct., *L. medeoloides* auct. p.p.) - 5
georgicum Manden. = L. monadelphum
glehnii Fr. Schmidt = Cardiocrinum cordatum
hansonii auct. = L. distichum
kesselringianum Miscz. (*L. kosii* Orekhov & Eremin) - 2
kosii Orekhov & Eremin = L. kesselringianum
lancifolium Thunb. (*L. tigrinum* Ker-Gawl.) - 5
ledebourii (Baker) Boiss. - 2
leichtlinii Hook. fil. var. *maximowiczii* (Regel) Baker = L. pseudotigrinum
- f. *pseudotigrinum* (Carr.) Hara & Kitam. = L. pseudotigrinum

maculatum Thunb. subsp. *dauricum* (Ker-Gawl.) Hara = L. pensylvanicum

martagon L. - 1, 3, 4
- subsp. *caucasicum* Miscz. ex Grossh. = L. caucasicum
- subsp. *pilosiusculum* (Freyn) Miscz. ex Iljin = L. pilosiusculum
- subsp. *sooianum* Priszter = L. pilosiusculum
- var. *pilosiusculum* Freyn = L. pilosiusculum
maximowiczii Regel = L. pseudotigrinum
medeoloides auct. = L. debile and L. distichum

monadelphum Bieb. (*L. georgicum* Manden.) - 2
- subsp. *armenum* (Miscz. ex Grossh.) G. Kudrjaschova = L. armenum
- var. *armenum* (Miscz. ex Grossh.) P.H. Davis & Henderson = L. armenum
- var. *szovitsianum* (Fisch. & Ave-Lall.) Elwes = L. szovitsianum

pensylvanicum Ker-Gawl. (*L. dauricum* Ker-Gawl., *L. maculatum* Thunb. subsp. *dauricum* (Ker-Gawl.) Hara, *L. pensylvanicum* subsp. *sachalinense* (Vrishcz) M. Baranova, comb. invalid., *L. sachalinense* Vrishcz) - 4, 5
- subsp. *sachalinense* (Vrishcz) M. Baranova = L. pensylvanicum

pilosiusculum (Freyn) Miscz. (*L. martagon* L. var. *pilosiusculum* Freyn, *L. martagon* subsp. *pilosiusculum* (Freyn) Miscz. ex Iljin, *L. martagon* ssp. *sooianum* Priszter) - 1, 3, 4

ponticum C. Koch (*L. carniolicum* Bernh. ex Koch subsp. *ponticum* (C. Koch) P.H. Davis & Henserson, *L. pyrenaicum* Gouan subsp. *ponticum* (C. Koch) Matthews, *L. szovitsianum* Fisch. & Ave-Lall. subsp. *ponticum* (C. Koch) G. Kudrjaschova) - 2

pseudotigrinum Carr. (*L. leichtlinii* Hook. fil. var. *maximowiczii* (Regel) Baker, *L. leichtlinii* f. *pseudotigrinum* (Carr.) Hara & Kitam., *L. maximowiczii* Regel) - 5
pulchellum Fisch. = L. buschianum

pumilum Delile (*L. tenuifolium* Fisch. ex Schrank) - 4, 5
pyrenaicum Gouan subsp. *ponticum* (C. Koch) Matthews = L. ponticum
sachalinense Vrishcz = L. pensylvanicum

szovitsianum Fisch. & Ave-Lall. (*L. monadelphum* Bieb. var. *szovitsianum* (Fisch. & Ave-Lall.) Elwes) - 2
- subsp. *ponticum* (C. Koch) G. Kudrjaschova = L. ponticum
- var. *armenum* Miscz. ex Grossh. = L. armenum
tenuifolium Fisch. ex Schrank = L. pumilum
tigrinum Ker-Gawl. = L. lancifolium

Lloydia Reichenb.

rubroviridis auct. = Gagea chomutovae
serotina (L.) Reichenb. - 1, 2, 3, 4, 5, 6
triflora (Ledeb.) Baker - 5

Ornithogalum auct. = Gagea

arvense Pers. = Gagea villosa
bohemicum Zauschn. = Gagea bohemica
chrysanthum Jan = Gagea chrysantha
fibrosum Desf. = Gagea fibrosa
fistulosum Ramond ex DC. = Gagea fistulosa
villosum Bieb. = Gagea villosa

Petilium Ludw. = Fritillaria

eduardii (Regel) Vved. = Fritillaria eduardii
imperiale (L.) Jaume = Fritillaria imperialis

raddeanum (Regel) Vved. ex Pazij = Fritillaria raddeana

Plecostigma Turcz. ex Trautv. = Gagea

pauciflorum Turcz. ex Trautv. = Gagea pauciflora

Rhinopetalum Fisch. ex Alexand.

arianum Losinsk. & Vved. = R. gibbosum
bucharicum (Regel) Losinsk. - 6
gibbosum (Boiss.) Losinsk. & Vved. (*Fritillaria ariana* (Losinsk. & Vved.) E.M. Rix, *Rhinopetalum arianum* Losinsk. & Vved.) - 2, 6
karelinii Fisch. ex D. Don - 6
stenantherum Regel - 6

Tulipa L. (*Eduardoregelia* M. Pop.)

affinis Z. Botsch. - 6
alberti Regel - 6
altaica Pall. ex Spreng. - 3, 6
anadroma Z. Botsch. - 6
anisophylla Vved. - 6
armena Boiss. - 2
behmiana Regel - 6
biebersteiniana Schult. & Schult. fil. (*T. graniticola* (Klok. & Zoz) Klok., *T. hypanica* Klok. & Zoz, *T. ophiophylla* Klok. & Zoz, *T. ophiophylla* subsp. *bestashica* Klok. & Zoz, *T. ophiophylla* subsp. *donetzica* Klok. & Zoz, *T. ophiophylla* subsp. *graniticola* Klok. & Zoz, *T. quercetorum* Klok. & Zoz, *T. scythica* Klok. & Zoz) - 1, 2, 3, 6
biflora Pall. (*T. callieri* Halacsy & Levier, *T. koktebelica* Junge, *T. polychroma* Stapf) - 1, 2, 3, 6
bifloriformis Vved. - 6
binutans Vved. - 6
boettgeri Regel - 6
borszczowii Regel - 6
botschantzevae S. Abramova & Zakaljabina - 6
brachystemon Regel - 6
buhseana Boiss. - 6
butkovii Z. Botsch. - 6
callieri Halacsy & Levier = T. biflora
carinata Vved. - 6
caucasica Lipsky = T. lipskyi
confusa Gabr. (*T. karabachensis* Grossh. p.p. excl. typo) - 2
dasystemon (Regel) Regel - 6
dasystemonoides Vved. - 6
dubia Vved. - 6
eichleri Regel - 2
ferganica Vved. - 6
florenskyi Woronow (*T. karabachensis* Grossh. p.p. incl. typo) - 2
fosteriana Irving - 6
gesneriana L. (*T. schrenkii* Regel) - 1, 2, 3, 6
graniticola (Klok. & Zoz) Klok. = T. biebersteiniana
greigii Regel - 6
heteropetala Ledeb. = T. uniflora
heterophylla (Regel) Baker (*Eduardoregelia heterophylla* (Regel) M. Pop.) - 6
hissarica M. Pop. & Vved. - 6
hoogiana B. Fedtsch. (*T. sovietica* A.D. Hall) - 6
hypanica Klok. & Zoz = T. biebersteiniana
iliensis Regel - 6
ingens Th. Hoog - 6
julia C. Koch (*T. montana* auct.) - 2
karabachensis Grossh. = T. confusa and T. florenskyi
kaufmanniana Regel - 6

koktebelica Junge = T. biflora
kolpakowskiana Regel - 6
kopetdaghensis B. Fedtsch. - 6
     Dubious species.
korolkowii Regel (*T. nitida* Th. Hoog) - 6
korshinskyi Vved. - 6
krauseana Regel - 6
kuschkensis B. Fedtsch. - 6
lanata Regel - 6
lehmanniana Merckl. - 6
linifolia Regel - 6
lipskyi Grossh. (*T. caucasica* Lipsky, 1902, non Otto ex
    Steud. 1841, nec Nym. 1882) - 2
maximowiczii Regel - 6
micheliana Th. Hoog - 6
mogoltavica M. Pop. & Vved. - 6
montana auct. = T. julia
neustruevae Pobed. - 6
nitida Th. Hoog = T. korolkowii
ophiophylla Klok. & Zoz = T. biebersteiniana
- subsp. *bestashica* Klok. & Zoz = T. biebersteiniana
- subsp. *donetzica* Klok. & Zoz = T. biebersteiniana
- subsp. *graniticola* Klok. & Zoz = T. biebersteiniana
orithyioides Vved. - 6
orthopoda Vved. - 6
ostrowskiana Regel - 6
patens Agardh ex Schult. & Schult. fil. - 1, 3
platystemon Vved. - 6
polychroma Stapf = T. biflora
praestans Th. Hoog - 6
prolongata Vved. - 6
quercetorum Klok. & Zoz = T. biebersteiniana
regelii Krasn. - 6
rosea Vved. - 6
schmidtii Fomin - 6
schrenkii Regel = T. gesneriana
scythica Klok. & Zoz = T. biebersteiniana
sogdiana Bunge - 6
sosnowskyi Achverd. & Mirzoeva - 2
sovietica A.D. Hall = T. hoogiana
subbiflora Vved. - 6
subpraestans Vved. - 6
subquinquefolia Vved. - 6
sylvestris L. - 1
tarda Stapf - 6
tetraphylla Regel - 6
thianschanica Regel - 6
tschimganica Z. Botsch. - 6
tubergeniana Th. Hoog - 6
turcomanica B. Fedtsch. - 6
turkestanica (Regel) Regel - 6
uniflora (L.) Bess. ex Baker (*T. heteropetala* Ledeb.) - 3, 4,
    6
uzbekistanica Z. Botsch. & Scharipov - 6
victoris Vved. ex Z. Botsch. - 6
■ violacea Boiss. & Buhse
vvedenskyi Z. Botsch. - 6
wilsoniana Th. Hoog - 6
zenaidae Vved. - 6

LIMONIACEAE Ser. (*LIMONIACEAE* Lincz.
   nom. superfl.)

Acantholimon Boiss. (*Statice* L. p.p.)

    See also Neogontscharovia.
acerosum (Willd.) Boiss. (*Statice acerosa* Willd.) - 2

afanassievii Lincz. - 6
alaicum Czerniak. ex Lincz. - 6
alatavicum Bunge - 6
alberti Regel - 6
alexandri Fed. - 6
alexeenkoanum Czerniak. ex Ikonn. - 6
annae Lincz. - 6
anzobicum Lincz. - 6
araxanum Bunge - 2
armenum Boiss. & Huet - 2
auganum Bunge (*A. munroanum* Aitch. & Hemsl.) - 6
aulieatense Czerniak. - 6
avenaceum Bunge - 6
balchanicum Korov. - 6
blandum Czerniak. - 6
borodinii Krasn. - 6
bracteatum (Girard) Boiss. - 2
butkovii Lincz. - 6
calvertii Boiss. - 2
caryophyllaceum Boiss. - 2
compactum Korov. - 6
diapensioides Boiss. - 6
echinus (L.) Boiss. (*Statice echinus* L.) - 2
ekatherinae (B. Fedtsch.) Czerniak. ex Lincz. - 6
erinaceum (Jaub. & Spach) Lincz. - 6
erythraeum Bunge - 6
fedorovii Tamamsch. & Mirzoeva - 2
fetisowii Regel - 6
fominii Kusn. - 2
gabrieljanii Mirzoeva - 2
gaudanense Czerniak. - 6
glumaceum (Jaub. & Spach) Boiss. - 2
gontscharovii Czerniak. - 6
hedinii Ostenf. - 6
hilariae Ikonn. - 6
hissaricum Lincz. - 6
hohenackeri (Jaub. & Spach) Boiss. - 2
karadarjense Lincz. - 6
karatavicum Pavl. - 6
karelinii (Stschegl.) Bunge - 2
katrantavicum Lincz. - 6
khorassanicum Czerniak. - 6
kjurendaghi Mestscherjakov - 6
knorringianum Lincz. - 6
kokandense Bunge - 6
komarovii Czerniak. ex Lincz. - 6
korolkowii (Regel) Korov. ex Lincz. - 6
korovinii Czerniak. - 6
kuramense Lincz. - 6
langaricum O. & B. Fedtsch. - 6
laxiusculum Khassanov & Maltzev - 6
laxum Czerniak. - 6
lepturoides (Jaub. & Spach) Boiss. - 2
linczevskii Pavl. - 6
litvinovii Lincz. - 6
lycopodioides (Girard) Boiss. - 6
majewianum Regel - 6
margaritae Korov. ex Lincz. - 6
mikeschinii Lincz. - 6
minshelkense Pavl. - 6
mirandum Lincz. = Neogontscharovia miranda
mirum Lincz. = Neogontscharovia mira
muchamedshanovii Lincz. - 6
munroanum Aitch. & Hemsl. = A. auganum
nabievii Lincz. - 6
nikitinii Lincz. - 6
nuratavicum Zak. ex Lincz. - 6

pamiricum Czerniak. ex Lincz. - 6
parviflorum Regel - 6
pavlovii Lincz. - 6
procumbens Czerniak. - 6
pskemense Lincz. - 6
pterostegium Bunge - 6
puberulum Boiss. & Bal. - 2
pulchellum Korov. - 6
purpureum Korov. - 6
quinquelobum Bunge - 2
raddeanum Czerniak. - 6
raicoviae Czerniak. ex Lincz. - 6
ruprechtii Bunge - 6
sackenii Bunge - 6
sahendicum Boiss. & Buhse - 2
sarawschanicum Regel - 6
sarytavicum Lincz. - 6
schachimardanicum Lincz. - 6
schemachense Grossh. - 2
squarrosum Pavl. - 6
strictum Czerniak. ex Lincz. - 6
subavenaceum Lincz. - 6
tarbagataicum Gamajun. - 6
taschkurganicum Lincz. & Akshig. - 6
tataricum Boiss. - 6
tenuiflorum Boiss. - 2
tianschanicum Czerniak. - 6
titovii Lincz. - 6
varivtzevae Czerniak. ex Lincz. - 6
vedicum Mirzoeva - 2
velutinum Czerniak. ex Lincz. - 6
virens Czerniak. ex Lincz. - 6
vvedenskyi Lincz. - 6
zaprjagaevii Lincz. - 6

Armeria Willd. (*Statice* L. p.p.)

arctica (Cham.) Wallr. = A. scabra
elongata (Hoffm.) Koch = A. vulgaris
x **intermedia** (Marss.) Szaf. (*A. vulgaris* Willd. var. *intermedia* Marss.). - A. maritima (Mill.) Willd. x A. vulgaris Willd. - 1
labradorica Wallr. (*A. maritima* (Mill.) Willd. subsp. *labradorica* (Wallr.) Hult., *A. maritima* subsp. *sibirica* (Boiss.) Hamet-Ahti, comb. superfl., *A. maritima* subsp. *sibirica* (Turcz. ex Boiss.) Nym., *A. sibirica* Turcz. ex Boiss.) - 1, 3, 4
maritima (Mill.) Willd. (*Statice maritima* Mill.) - 1, 3, 4
- subsp. *arctica* (Cham.) Hult. = A. scabra
- subsp. *elongata* (Hoffm.) Bonnier = A. vulgaris
- subsp. *labradorica* (Wallr.) Hult. = A. labradorica
- subsp. *sibirica* (Boiss.) Hamet-Ahti = A. labradorica
- subsp. *sibirica* (Turcz. ex Boiss.) Nym. = A. labradorica
pocutica Pawl. - 1
scabra Pall. ex Schult. (*A. arctica* (Cham.) Wallr., *A. maritima* (Mill.) Willd. subsp. *arctica* (Cham.) Hult., *A. scabra* subsp. *arctica* (Cham.) Ivers., *A. vulgaris* Willd. subsp. *arctica* (Cham.) Hult. comb. superfl., *A. vulgaris* subsp. *arctica* (Cham.) Nym.) - 1, 3, 4, 5
- subsp. *arctica* (Cham.) Ivers. = A. scabra
sibirica Turcz. ex Boiss. = A. labradorica
vulgaris Willd. (*A. elongata* (Hoffm.) Koch, *A. maritima* (Mill.) Willd. subsp. *elongata* (Hoffm.) Bonnier) - 1
- subsp. *arctica* (Cham.) Hult. = A. scabra
- subsp. *arctica* (Cham.) Nym. = A. scabra
- var. *intermedia* Marss. = A. x intermedia

Cephalorhizum M. Pop. & Korov.

micranthum Lincz. - 6
oopodum M. Pop. & Korov. - 6
popovii Lincz. - 6
turcomanicum M. Pop. ex Lincz. = Popoviolimon turcomanicum

Chaetolimon (Bunge) Lincz.

limbatum Lincz. - 6
setiferum (Bunge) Lincz. - 6
sogdianum Lincz. = Vassilczenkoa sogdiana

Eremolimon Lincz.

botschantzevii Lincz. - 6
drepanostachyum (Ik.-Gal.) Lincz. (*Limonium drepanostachyum* Ik.-Gal.) - 6
fajzievii (Zak. ex Lincz.) Lincz. (*Limonium fajzievii* Zak. ex Lincz.) - 6
jarmolenkoi Lincz. - 6
kurgantjubense Lincz. - 6
piptopodum (Nevski) Lincz. (*Limonium piptopodum* Nevski) - 6
sogdianum (Ik.-Gal.) Lincz. (*Limonium sogdianum* Ik.-Gal.) - 6

Goniolimon Boiss.

besserianum (Schult.) Kusn. - 1
callicomum (C.A. Mey.) Boiss. - 3, 6
caucasicum Klok. - 2
cuspidatum Gamajun. - 6
desertorum (Trautv.) Klok. = G. graminifolium
dschungaricum (Regel) O. & B. Fedtsch. - 6
elatum (Fisch. ex Spreng.) Boiss. - 1, 3, 6
eximium (Schrenk) Boiss. - 6
glaberrimum Klok. - 2
graminifolium (Ait.) Boiss. (*G. desertorum* (Trautv.) Klok.) - 1
orae-syvashicae Klok. = G. rubellum
orthocladum Rupr. - 6
platypterum Klok. - 2
rubellum (S.G. Gmel.) Klok. (*G. orae-syvashicae* Klok.) - 1, 3, 6
sewerzowii Herd. - 6
speciosum (L.) Boiss. - 1, 3, 4, 6
tataricum (L.) Boiss. - 1, 2
tauricum Klok. - 1

Ikonnikovia Lincz.

kaufmanniana (Regel) Lincz. - 6

Limoniopsis Lincz.

owerinii (Boiss.) Lincz. - 2

Limonium Mill. (*Statice* L. p.p.)

alutaceum (Stev.) O. Kuntze = L. tomentellum
aureum (L.) Hill ex O. Kuntze - 4
bellidifolium auct. p.p. = L. caspium
bungei (Claus) Gamajun. (*L. membranaceum* (Czern.) Klok.) - 1, 3(?)
carnosum (Boiss.) O. Kuntze - 2
caspium (Willd.) Gams (*L. bellidifolium* auct. p.p.) - 1, 2, 3, 6

chodshamumynense Lincz. & Czuk. - 6
chrysocomum (Kar. & Kir.) O. Kuntze - 3, 6
congestum (Ledeb.) O. Kuntze - 3, 4
coralloides (Tausch) Lincz. - 3, 4, 6
cretaceum Tscherkasova - 3
czurjukiense (Klok.) Lavr. ex Klok. = L. tomentellum
danubiale Klok. - 1
dichroanthum (Rupr.) Ik.-Gal. - 6
donetzicum Klok. - 1
drepanostachyum Ik.-Gal. = Eremolimon drepanostachyum
dubium Gamajun. ex Klok. - 1
fajzievii Zak. ex Lincz. = Eremolimon fajzievii
ferganense Ik.-Gal. - 6
fischeri (Trautv.) Lincz. - 2
flexuosum (L.) O. Kuntze - 4
gmelinii (Willd.) O. Kuntze - 1, 3, 4, 6
- subsp. *hypanicum* (Klok.) Soo = L. hypanicum
- var. *hypanicum* (Klok.) Pawl. = L. hypanicum
hoeltzeri (Regel) Ik.-Gal. - 6
hypanicum Klok. (*L. gmelinii* (Willd.) O. Kuntze subsp. *hypanicum* (Klok.) Soo, *L. gmelinii* var. *hypanicum* (Klok.) Pawl.) - 1
kaschgaricum (Rupr.) Ik.-Gal. - 6
komarovii Ik.-Gal. ex Lincz. & Czuk. - 6
latifolium (Smith) O. Kuntze = L. platyphyllum
leptolobum (Regel) O. Kuntze - 6
leptophyllum (Schrenk) O. Kuntze - 6
lessingianum Lincz. - 6
macrorhizon (Ledeb.) O. Kuntze - 3, 4, 6
membranaceum (Czern.) Klok. = L. bungei
meyeri (Boiss.) O. Kuntze - 1, 2 6
michelsonii Lincz. - 6
myrianthum (Schrenk) O. Kuntze - 6
narynense Lincz. - 6
neoscoparium Klok. - 2
otolepis (Schrenk) O. Kuntze - 6
ovczinnikovii Lincz. & Czuk. - 6
piptopodum Nevski = Eremolimon piptopodum
platyphyllum Lincz. (*L. latifolium* (Smith) O. Kuntze, 1891, non Moench, 1794) - 1, 2
popovii Kubansk. - 6
reniforme (Girard) Lincz. - 6
rezniczenkoanum Lincz. - 6
sareptanum (A. Beck.) Gams - 1, 3, 6
semenovii (Herd.) O. Kuntze - 6
sinuatum (L.) Mill. - 2
sogdianum Ik.-Gal. = Eremolimon sogdianum
suffruticosum (L.) O. Kuntze - 1, 2, 3, 6
tianschanicum Lincz. - 6
tomentellum (Boiss.) O. Kuntze (*L. alutaceum* (Stev.) O. Kuntze, *L.* "*czurjukiense*" (Klok.) Lavr. ex Klok., *L. tschurjukiense* (Klok.) Lavr. ex Klok., *Statice tschurjukiensis* Klok.) - 1
tschurjukiense (Klok.) Lavr. ex Klok. = L. tomentellum

Neogontscharovia Lincz.

mira (Lincz.) Lincz. (*Acantholimon mirum* Lincz.) - 6
miranda (Lincz.) Lincz. (*Acantholimon mirandum* Lincz.) - 6

Popoviolimon Lincz.

turcomanicum (M. Pop. ex Lincz.) Lincz. (*Cephalorhizium turcomanicum* M. Pop. ex Lincz.) - 6

Psylliostachys (Jaub. & Spach) Nevski

anceps (Regel) Roshk. - 6

x **androssovii** Roshk. - P. anceps (Regel) Roshk. x P. leptostachya (Boiss.) Roshk. - 6
leptostachya (Boiss.) Roshk. - 6
x **myosuroides** (Regel) Roshk. - P. leptostachya (Boiss.) Roshk. x P. suworowii (Regel) Roshk. - 6
spicata (Willd.) Nevski - 1, 2, 6
suworowii (Regel) Roshk. - 6

Statice L. = Acantholimon, Armeria and Limonium

acerosa Willd. = Acantholimon acerosum
echinus L. = Acantholimon echinus
maritima Mill. = Armeria maritima
tschurjukiensis Klok. = Limonium tomentellum

Vassilczenkoa Lincz.

sogdiana (Lincz.) Lincz. (*Chaetolimon sogdianum* Lincz.) - 6

LINACEAE DC. ex S.F. Gray
Linum L.

alexeenkoanum E. Wulf (*L. armenum* (Bordz.) Galushko, *L. mucronatum* Bertol. subsp. *armenum* (Bordz.) P.H. Davis, *L. orientale* (Boiss. & Heldr.) Boiss. subsp. *armenum* Bordz., *L. orientale* auct.) - 2
altaicum Ledeb. ex Juz. - 3, 6
amurense Alef. - 5
anatolicum Boiss. - 2
angustifolium Huds. = L. bienne
armenum (Bordz.) Galushko = L. alexeenkoanum
atricalyx Juz. - 6
aucheri Planch. - 1
austriacum L. (*L. perenne* L. subsp. *austriacum* (L.) O. Bolos & Vigo, *L. squamulosum* Rudolphi) - 1, 2, 3
- subsp. *euxinum* (Juz.) Ockendon = L. euxinum
- subsp. *marschallianum* (Juz.) Greuter & Burdet = L. marschallianum
baicalense Juz. - 4
bassarabicum (Savul. & Rayss) Klok. ex Juz. - 1
*****bienne** Mill. (*L. angustifolium* Huds., *L. usitatissimum* L. subsp. *bienne* (Mill.) Stankevicz) - 2
boreale Juz. (*L. perenne* L. var. *boreale* (Juz.) Serg.) - 1, 3
brevisepalum Juz. - 4
catharticum L. - 1, 2
corymbulosum Reichenb. (*L. strictum* L. subsp. *corymbulosum* (Reichenb.) Rouy, *L. liburnicum* auct.) - 1, 2, 6
*****crepitans** (Boenn.) Dumort. (*L. usitatissimum* L. subsp. *crepitans* (Boenn.) Elladi) - 1, 5
czerniaevii Klok. - 1
euxinum Juz. (*L. austriacum* L. subsp. *euxinum* (Juz.) Ockendon) - 1, 2
extraaxillare Kit. (*L. perenne* L. subsp. *extraaxillare* (Kit.) Nym.) - 1
flavum L. - 1, 2
gallicum L. = L. trigynum
grandiflorum Desf. - 5
heterosepalum Regel - 6
hirsutum L. - 1
humile Mill. (*L. usitatissimum* L. subsp. *humile* (Mill.) Czernom.) - 1, 2, 3, 6
hypericifolium Salisb. - 2
jailicola Juz. - 1, 2
komarovii Juz. - 4, 5
lanuginosum Juz. - 1, 2

liburnicum auct. = L. corymbulosum
linearifolium (Lindem.) Jav. (*L. tauricum* Willd.) subsp.
    linearifolium (Lindem.) A. Petrova) - 1
luteolum Bieb. = L. nodiflorum
macrorhizum Juz. (*L. mesostylum* Juz.) - 6
marschallianum Juz. (*L. austriacum* L. subsp. *marschallia-*
    num (Juz.) Greuter & Burdet) - 1
mesostylum Juz. = L. macrorhizum
mucronatum Bertol. subsp. *armenum* (Bordz.) P.H. Davis =
    L. alexeenkoanum
nervosum Waldst. & Kit. (*L. nervosum* subsp. *glabratum*
    (DC.) P.H. Davis) - 1, 2
- subsp. *glabratum* (DC.) P.H. Davis = L. nervosum
nodiflorum L. (*L. luteolum* Bieb.) - 1, 2
olgae Juz. - 6
orientale auct. = L. alexeenkoanum
orientale (Boiss. & Heldr.) Boiss. subsp. *armenum* Bordz. =
    L. alexeenkoanum
pallasianum Schult. - 1
pallescens Bunge - 3, 6
perenne L. - 1, 3
- subsp. *austriacum* (L.) O. Bolos & Vigo = L. austriacum
- subsp. *extraaxillare* (Kit.) Nym. = L. extraaxillare
- var. *boreale* (Juz.) Serg. = L. boreale
seljukorum P.H. Davis - 1
spicatum (Lam.) Pers. - 2(?)
squamulosum Rudolphi = L. austriacum
stelleroides Planch. - 4, 5
strictum L. - 1, 2
- subsp. *corymbulosum* (Reichenb.) Rouy = L. corymbu-
    losum
subbiflorum Juz. - 2
tauricum Willd. - 1, 2
- subsp. *linearifolium* (Lindem.) A. Petrova = L. linearifoli-
    um
tenuifolium L. - 1, 2
trigynum L. (*L. gallicum* L. nom. illegit.) - 2
turcomanicum Juz. - 6
ucranicum Czern. - 1
uralense Juz. - 1, 2, 3, 4, 5, 6
*usitatissimum L. - 1
- subsp. *bienne* (Mill.) Stankevicz = L. bienne
- subsp. *crepitans* (Boenn.) Elladi = L. crepitans
- subsp. *humile* (Mill.) Czernom. = L. humile
- subsp. **intermedium** Czernom.
- subsp. **latifolium** (L.) Stankevicz
violascens Bunge - 3

Radiola Hill

linoides Roth - 1

LIRIODENDRACEAE F.A. Barkley = MAGNOLIACEAE

LOBELIACEAE R. Br.
Lobelia L.

dortmanna L. - 1
*erinus L. - 1
sessilifolia Lamb. - 4, 5

LORANTHACEAE Juss.
Loranthus Jacq.

europaeus Jacq. - 1

LYCOPODIACEAE Beauv. ex Mirb.
Diphasiastrum Holub (*Diphasium* C. Presl, p.p. excl. typo)

alpinum (L.) Holub (*Lycopodium alpinum* L., *Diphasium
    alpinum* (L.) Rothm.) - 1, 2, 3, 4, 5, 6
complanatum (L.) Holub (*Lycopodium complanatum* L.,
    Diphasium anceps (Wallr.) A. & D. Löve, nom. illegit.,
    D. complanatum (L.) Rothm., *D. complanatum* subsp.
    anceps (Wallr.) Dostal, *D. wallrothii* H.P. Fuchs,
    Lycopodium anceps Wallr. 1841, non C. Presl, 1830, *L.
    complanatum* subsp. *anceps* (Wallr.) Aschers. &
    Graebn.) - 1, 3, 4, 5
- subsp. *chamaecyparissus* (A. Br. ex Mutel) Kukkonen =
    D. tristachyum
- subsp. *montellii* (Kukkonen) Kukkonen = D. montellii
- subsp. *zeileri* (Rouy) Kukkonen = D. x zeileri
- f. *montellii* (Kukkonen) J. Feilberg = D. montellii
x issleri (Rouy) Holub (*Lycopodium alpinum* L. race *issleri*
    Rouy, *Diphasium complanatum* (L.) Rothm. subsp.
    issleri (Rouy) Dostal, *D. hastulatum* Sipl., *D. issleri*
    (Rouy) Holub, *Lycopodium alpinum* subsp. *issleri*
    (Rouy) Chassagne, *L. complanatum* L. subsp. *issleri*
    (Rouy) Domin, *L. issleri* (Rouy) Domin, *L. issleri*
    (Rouy) Lawalree, comb. superfl.). - D. alpinum (L.)
    Holub x D. complanatum (L.) Holub - 1, 4
montellii (Kukkonen) Min. & Ivanenko (*Diphasium com-
    planatum* (L.) Rothm. subsp. *montellii* Kukkonen,
    Diphasiastrum complanatum (L.) Holub subsp. *mon-
    tellii* (Kukkonen) Kukkonen, *D. complanatum* f.
    montellii (Kukkonen) J. Feilberg, *Diphasium compla-
    natum* subsp. *complanatum* var. *montellii* Kukkonen,
    nom. invalid.) - 1
sitchense (Rupr.) Holub (*Lycopodium sitchense* Rupr.,
    Diphasium sitchense (Rupr.) A. & D. Löve) - 5
tristachyum (Pursh) Holub (*Lycopodium tristachyum*
    Pursh, *Diphasiastrum complanatum* (L.) Holub. subsp.
    chamaecyparissus (A. Br.ex Mutel) Kukkonen, *Dipha-
    sium chamaecyparissus* (A. Br. ex Mutel) A. & D.
    Löve, *D. complanatum* (L.) Rothm. subsp. *chamaecy-
    parissus* (A. Br. ex Mutel) Dostal, nom. invalid., *D.
    complanatum* subsp. *chamaecyparissus* (A. Br. ex
    Mutel) Kukkonen, *D. tristachyum* (Pursh) Rothm.,
    Lycopodium chamaecyparissus A. Br. ex Mutel, *L.
    complanatum* L. var. *chamaecyparissus* (A. Br. ex
    Mutel) Doll, *L. complanatum* subsp. *chamaecyparissus*
    (A. Br. ex Mutel) C. Hartm., *L. complanatum* subsp.
    chamaecyparissus (A. Br. ex Mutel) Milde, *L. compla-
    natum* var. *tristachyum* (Pursh) Domin, *L. complana-
    tum* subsp. *tristachyum* (Pursh) Dostal) - 1, 2, 3
x zeileri (Rouy) Holub (*Lycopodium alpinum* L. race *zeileri*
    Rouy, *Diphasiastrum complanatum* (L.) Holub subsp.
    zeileri (Rouy) Kukkonen, *Diphasium complanatum*
    (L.) Rothm. subsp. *zeileri* (Rouy) Pacyna, *D. zeileri*
    (Rouy) Damboldt, *Lycopodium zeileri* (Rouy) Greuter
    & Burdet). - D. complanatum (L.) Holub x D.
    tristachyum (Pursh) Holub - 1

Diphasium C. Presl, p.p. = Diphasiastrum

alpinum (L.) Rothm. = Diphasiastrum alpinum
anceps (Wallr.) A. & D. Löve = Diphasiastrum
    complanatum
chamaecyparissus (A. Br. ex Mutel) A. & D. Löve = Dipha-
    siastrum tristachyum
complanatum (L.) Rothm. = Diphasiastrum complanatum
- subsp. *anceps* (Wallr.) Dostal = Diphasiastrum compla-
    natum

- subsp. *chamaecyparissus* (A. Br. ex Mutel) Dostal = Diphasiastrum tristachyum
- subsp. *chamaecyparissus* (A. Br. ex Mutel) Kukkonen = Diphasiastrum tristachyum
- subsp. *complanatum* var. *montellii* Kukkonen = Diphasiastrum montellii
- subsp. *issleri* (Rouy) Dostal = Diphasiastrum x issleri
- subsp. *montellii* Kukkonen = Diphasiastrum montellii
- subsp. *zeileri* (Rouy) Pacyna = Diphasiastrum x zeileri
hastulatum Sipl. = Diphasiastrum x issleri
issleri (Rouy) Holub = Diphasiastrum x issleri
sitchense (Rupr.) A. & D. Löve = Diphasiastrum sitchense
tristachyum (Pursh) Rothm. = Diphasiastrum tristachyum
wallrothii H.P. Fuchs = Diphasiastrum complanatum
zeileri (Rouy) Damboldt = Diphasiastrum x zeileri

Lepidotis Beauv. ex Mirb. = Lycopodiella

inundata (L.) Boern. = Lycopodiella inundata

Lycopodiella Holub (*Lepidotis* Beauv. ex Mirb. p.p. nom. illegit.)

inundata (L.) Holub (*Lycopodium inundatum* L., *Lepidotis inundata* (L.) Boern.) - 1, 2, 3, 4

Lycopodium L.

alpinum L. = Diphasiastrum alpinum
- subsp. *issleri* (Rouy) Chassagne = Diphasiastrum x issleri
- race *issleri* Rouy = Diphasiastrum x issleri
- race *zeileri* Rouy = Diphasiastrum x zeileri
anceps Wallr. = Diphasiastrum complanatum
annotinum L. - 1, 2, 3, 4, 5
- subsp. *alpestre* (C. Hartm.) A. & D. Löve = L. dubium
- subsp. *dubium* (Zoega) Kallio, Laine & Makinen = L. dubium
- subsp. *pungens* (Desv.) Hult. = L. dubium
- var. *alpestre* C. Hartm. = L. dubium
- var. *pungens* Desv. = L. dubium
appressum (Desv.) V. Petrov, p.p. = Huperzia arctica and H. petrovii (Huperziaceae)
arcticum Grossh. = Huperzia arctica (Huperziaceae)
chamaecyparissus A. Br. ex Mutel = Diphasiastrum tristachyum
chinense Christ = Huperzia chinensis (Huperziaceae)
clavatum L. - 1, 2, 3, 4, 5, 6
- subsp. *lagopus* (Laest.) Dostal = L. lagopus
- subsp. *monostachyon* (Grev. & Hook.) Selander = L. lagopus
- var. *monostachyon* Grev. & Hook. = L. lagopus
- f. *lagopus* Laest. = L. lagopus
complanatum L. = Diphasiastrum complanatum
- subsp. *anceps* (Wallr.) Aschers. & Graebn. = Diphasiastrum complanatum
- subsp. *chamaecyparissus* (A. Br. ex Mutel) C. Hartm. = Diphasiastrum tristachyum
- subsp. *chamaecyparissus* (A. Br. ex Mutel) Milde = Diphasiastrum tristachyum
- subsp. *issleri* (Rouy) Domin = Diphasiastrum x issleri
- subsp. *tristachyum* (Pursh) Dostal = Diphasiastrum tristachyum
- var. *chamaecyparissus* (A. Br. ex Mutel) Doll = Diphasiastrum tristachyum
- var. *tristachyum* (Pursh) Domin = Diphasiastrum tristachyum
dubium Zoega (*L. annotinum* L. var. *alpestre* C. Hartm., *L. annotinum* subsp. *alpestre* (C. Hartm.) A. & D. Löve,

L. annotinum subsp. *dubium* (Zoega) Kallio, Laine & Makinen, ? nom. superfl., *L. annotinum* var. *pungens* Desv., *L. annotinum* subsp. *pungens* (Desv.) Hult., *L. pungens* (Desv.) La Pyl. ex Iljin) - 1, 3, 4, 5
inundatum L. = Lycopodiella inundata
issleri (Rouy) Domin = Diphasiastrum x issleri
issleri (Rouy) Lawalree = Diphasiastrum x issleri
juniperoideum Sw. - 4, 5
lagopus (Laest.) Zinserl. ex Kuzen. (*L. clavatum* L. f. *lagopus* Laest., *L. clavatum* subsp. *lagopus* (Laest.) Dostal, *L. clavatum* var. *monostachyon* Grev. & Hook., *L. clavatum* subsp. *monostachyon* (Grev. & Hook.) Selander) - 1, 3, 4, 5
miyoshianum Makino = Huperzia chinensis (Huperziaceae)
obscurum L. - 5
pungens (Desv.) La Pyl. ex Iljin = L. dubium
rupestre L. = Selaginella rupestris (Selaginellaceae)
selago L. = Huperzia selago (Huperziaceae)
- subsp. *appressum* (Desv.) Hult. p.p. = Huperzia arctica and H. petrovii (Huperziaceae)
- subsp. *arcticum* Tolm. = Huperzia arctica (Huperziaceae)
- subsp. *miyoshianum* (Makino) Calder & Taylor = Huperzia chinensis (Huperziaceae)
- var. *laxum* Desv. = Huperzia laxa (Huperziaceae)
serratum Thunb. = Huperzia serrata (Huperziaceae)
sitchense Rupr. = Diphasiastrum sitchense
subarcticum V. Vassil. - 1, 5
tristachyum Pursh = Diphasiastrum tristachyum
zeileri (Rouy) Greuter & Burdet = Diphasiastrum x zeileri

LYTHRACEAE J. St.-Hil.
Ammannia L.

aegyptica Willd. = A. baccifera
arenaria Kunth = A. auriculata
auriculata Willd. (*A. arenaria* Kunth) - 2, 6
baccifera L. (*A. aegyptica* Willd., *A. viridis* Hornem.) - 1, 2, 6
multiflora Roxb. - 6
pubiflora (Koehne) Sosn. - 2
verticillata (Ard.) Lam. - 1, 2, 6
viridis Hornem. = A. baccifera

Lythrum L.

borysthenicum (Bieb. ex Schrank) Litv. = Middendorfia borysthenica
hybridum Klok. - 1
hyssopifolia L. - 1, 2, 3, 5, 6
intermedium Ledeb. (*L. salicaria* L. subsp. *intermedium* (Ledeb.) Hara) - 1, 2, 3, 4, 5, 6
komarovii Murav. - 3
linifolium Kar. & Kir. (*L. thesioides* Bieb. subsp. *linifolium* (Kar. & Kir.) Koehne) - 3, 6
melanospermum Savul. & Zahar - 1
nanum Kar. & Kir. - 3, 6
portula (L.) D.A. Webb = Peplis portula
salicaria L. - 1, 2, 3, 4, 5
- subsp. *intermedium* (Ledeb.) Hara = L. intermedium
schelkovnikovii Sosn. - 2
silenoides Boiss. & Noe - 6
sophiae Klok. - 1
theodorii Sosn. - 2, 6
thesioides Bieb. - 1, 2
- subsp. *linifolium* (Kar. & Kir.) Koehne = L. linifolium
thymifolia L. - 1, 2, 3, 6

tribracteatum Salzm. ex Spreng. - 1, 2, 3, 6
virgatum L. - 1, 2, 3, 4, 6
volgense D.A. Webb = Peplis alternifolia

Middendorfia Trautv.

borysthenica (Bieb. ex Schrank) Trautv. (*Peplis borystheni-ca* Bieb. ex Schrank, *Lythrum borysthenicum* (Bieb. ex Schrank) Litv.) - 1, 3

Peplis L.

alternifolia Bieb. (*Lythrum volgense* D.A. Webb) - 1, 2, 3, 6
borysthenica Bieb. ex Schrank = Middendorfia borysthenica
hyrcanica Sosn. - 2
portula L. (*Lythrum portula* (L.) D.A. Webb) - 1, 3

Rotala L.

densiflora (Roth) Koehne - 6
indica (Willd.) Koehne - 2, 6

MAGNOLIACEAE Juss. (*LIRIODENDRAC-EAE* F.A. Barkley)

*Liriodendron L.

***tulipiferum** L. - 1, 2

Magnolia L.

***grandiflora** L. - 1, 2, 6
hypoleuca Siebold. & Zucc. (*M. obovata* Thunb. p.p.) - 5
obovata Thunb. p.p. = M. hypoleuca

MALVACEAE Juss.

Abutilon Hill

theophrasti Medik. - 1, 2, 5, 6

Alcea L.

abchasica Iljin - 2
angulata (Freyn) Freyn & Sint. ex Iljin - 6
antoninae Iljin - 6
baldschuanica (Bornm.) Iljin - 6
calvertii (Boiss.) Boiss. (*Althaea calvertii* Boiss.) - 2(?)
excubita Iljin - 2
flavovirens (Boiss. & Buhse) Iljin - 2
- var. *tabrisiana* (Boiss. & Buhse) Zohary = A. tabrisiana
freyniana Iljin - 6
froloviana (Litv.) Iljin (*Althaea nudiflora* Lindl. var. *frolovi-ana* Litv.) - 3, 6
grossheimii Iljin - 2
heldreichii (Boiss.) Boiss. - 1
hyrcana (Grossh.) Grossh. - 2
karakalensis Freyn - 6
karsiana (Bordz.) Litv. - 2
kopetdaghensis Iljin - 6
kusariensis (Iljin) Iljin - 2
lenkoranica Iljin - 2
litwinowii (Iljin) Iljin - 6
nikitinii Iljin - 6
novopokrovskii Iljin - 1
nudiflora (Lindl.) Boiss. - 3, 6
pallida (Waldst. & Kit. ex Willd.) Waldst. & Kit. - 1
popovii Iljin - 6
rhyticarpa (Trautv.) Iljin - 6

***rosea** L.
rugosa Alef. - 1, 2
sachsachanica Iljin - 2
sophiae Iljin - 2
sosnovskyi Iljin - 2
sycophylla Iljin & V.V. Nikit. - 6
tabrisiana (Boiss. & Buhse) Iljin (*A. flavovirens* (Boiss. & Buhse) Iljin var. *tabrisiana* (Boiss. & Buhse) Zohary) - 2
talassica Iljin - 6
taurica Iljin - 1
tiliacea (Bornm.) Zohary (*Althaea tiliacea* Bornm.) - 6
transcaucasica (Iljin) Iljin - 2
turcomanica Iljin - 6
turkeviczii Iljin - 2
▪ **woronowii** (Iljin) Iljin

Althaea L.

armeniaca Ten. - 1, 2, 6
broussonetiifolia Iljin - 1, 6
calvertii Boiss. = Alcea calvertii
cannabina L. - 1, 2, 6
hirsuta L. - 1, 2, 6
ludwigii L. - 6
narbonensis Pourr. - 1
nudiflora Lindl. var. *froloviana* Litv. = Alcea froloviana
officinalis L. - 1, 2, 3, 4, 6
taurinensis DC. - 2
tiliacea Bornm. = Alcea tiliacea
transcaucasica (Sosn.) R. Kam. = Malvalthaea transcau-casica

Anoda Cav.

cristata (L.) Schlecht. (*Sida cristata* L.) - 5(alien)

*Gossypium L.

***albescens** Rafin. - 2, 6
barbadense auct. = G. peruvianum
***figarei** Tod. - 2, 6
***frutescens** Lastey. - 2, 6
***herbaceum** L. - 2, 6
***hirsutum** L. - 1, 2, 6
***jumelianum** (Tod.) Prokh. - 2, 6
***peruvianum** Cav. (*G. barbadense* auct.)
***zaitzevii** Prokh. - 6

Hibiscus L.

***cannabinus** L. - 2, 5, 6
***esculentus** L. - 1, 2, 6
***planifolius** Sweet
ponticus Rupr. - 2
***rosa-sinensis** L. - 2, 6
***sabdariffa** L. - 2, 6
***syriacus** L. - 1(cult.), 2(naturalized & cult.), 6(cult.)
***tiliaceus** L.
trionum L. - 1, 2, 4, 5, 6

Kitaibelia Willd.

vitifolia Willd. - 1, 6

Kosteletzkya C. Presl

pentacarpa (L.) Ledeb. - 1, 2

Lavatera L.
cashemiriana Cambess. - 6
punctata All. - 2, 6
thuringiaca L. - 1, 2, 3, 4, 6
*trimestris L.

Malope L.
malacoides L. - 2(?)
*trifida Cav.

Malva L.
alcea L. - 1
- subsp. *excisa* (Reichenb.) Holub = M. excisa
ambigua Guss. - 1, 2
armeniaca Iljin - 2
bucharica Iljin - 6
crispa (L.) L. = M. verticillata
erecta C. Presl (*M. sylvestris* L. subsp. *erecta* (C. Presl) Slavik) - 1, 2, 6
excisa Reichenb. (*M. alcea* L. subsp. *excisa* (Reichenb.) Holub) - 1
grossheimii Iljin = M. sylvestris
iljinii I. Riedl (*M. leiocarpa* Iljin, 1923, non *M. liocarpa* Philippi, 1893) - 2, 6
leiocarpa Iljin = M. iljinii
mauritiana L. = M. sylvestris
meluca Graebn. ex P. Medw. = M. verticillata
mohileviensis Downar = M. verticillata
moschata L. - 1, 5
neglecta Wallr. - 1, 2, 5(alien), 6
nicaeensis All. - 1, 2, 6
pamiroalaica Iljin - 6
parviflora L. - 2, 5(alien), 6
pusilla Smith (*M. rotundifolia* auct.) - 1, 2, 3, 4, 5, 6
rotundifolia auct. = M. pusilla
sylvestris L. (*M. grossheimii* Iljin, *M. mauritiana* L.) - 1, 2, 3, 4, 5, 6
- subsp. *erecta* (C. Presl) Slavik = M. erecta
verticillata L. (*M. crispa* (L.) L., *M. meluca* Graebn. ex P. Medw., *M. mohileviensis* Downar) - 1, 3, 4, 5, 6

Malvalthaea Iljin
transcaucasica (Sosn.) Iljin (*Althaea transcaucasica* (Sosn.) R. Kam.) - 2, 6

Malvella Jaub. & Spach
sherardiana (L.) Jaub. & Spach - 1, 2

Sida L.
cristata L. = Anoda cristata
*hermaphrodita Rusby - 1
rhombifolia L. - 1, 2
spinosa L. - 1, 2, 5

MARSILEACEAE Mirb. (*PILULARIACEAE* Bercht. & J. Presl)
Marsilea L.
aegyptiaca Willd. - 1, 6
quadrifolia L. - 1, 2, 3, 6
strigosa Willd. - 1, 2, 3, 6

Pilularia L.
globulifera L. - 1

MARTYNIACEAE Stapf
Proboscidea Schmidel
louisiana (Mill.) Thell. - 1, 2, 6

MELANTHIACEAE Batsch (*COLCHI-CACEAE* DC.)
Acelidanthus Trautv. & C.A. Mey.
anticleoides Trautv. & C.A. Mey. - 5

Bulbocodium L.
vernum L. subsp. *versicolor* (Ker-Gawl.) K. Richt. = B. versicolor
versicolor (Ker-Gawl.) Spreng. (*B. vernum* L. subsp. *versicolor* (Ker-Gawl.) K. Richt.) - 1, 2

Colchicum L.
ancyrense B.L. Burtt = C. triphyllum
autumnale L. - 1
biebersteinii Rouy = C. triphyllum
bifolium Freyn & Sint. = C. szovitsii
falcifolium Stapf (*C. serpentinum* Woronow ex Miscz.) - 2
fominii Bordz. - 2
kesselringii Regel (*C. regelii* Stef.) - 6
laetum Stev. - 1, 2
luteum Baker - 6
ninae Sosn. - 2
nivale (Boiss. & Huet) Stapf = C. szovitsii
regelii Stef. = C. kesselringii
serpentinum Woronow ex Miscz. = C. falcifolium
speciosum Stev. - 2
szovitsii Fisch. & C.A. Mey. (*C. bifolium* Freyn & Sint., *C. nivale* (Boiss. & Huet) Stapf) - 2, 6
triphyllum G. Kunze (*C. ancyrense* B.L. Burtt, *C. biebersteinii* Rouy, nom. illegit.) - 1
umbrosum Stev. - 1, 2
- subsp. **amphibolum** Zahar. & Artjushenko - 2
woronowii Bokeria - 2
zangezurum Grossh. - 2

Heloniopsis A. Gray (*Scilla* auct.)
orientalis (Thunb.) C. Tanaka (*Scilla orientalis* Thunb.) - 5

Merendera Ramond
candidissima Miscz. ex Grossh. - 2
eichleri (Regel) Boiss. = M. trigyna
ghalghana Otschiauri - 2
hissarica Regel - 6
jolantae Czerniak. - 6
mirzoevae Gabr. - 2
raddeana Regel - 2
robusta Bunge - 6
sobolifera C.A. Mey. - 2, 6
trigyna (Stev. ex Adams) Stapf (*M. eichleri* (Regel) Boiss.) - 2

Metanarthecium Maxim.
luteo-viride Maxim. - 5

Narthecium Huds.

balansae Briq. (*N. caucasicum* Miscz.) - 2
caucasicum Miscz. = N. balansae
ossifragum (L.) Huds. (*Anthericum ossifragum* L.) - 1
pusillum Michx. = Tofieldia pusilla

Scilla auct. = Heloniopsis

orientalis Thunb. = Heloniopsis orientalis

Stenanthium (A. Gray) Kunth

sachalinense Fr. Schmidt - 5

Tofieldia Huds.

calyculata (L.) Wahlenb. (*Anthericum calyculatum* L.) - 1
cernua Smith - 4, 5
coccinea Richards. (? *T. fusca* Miyabe & Kudo, *T. nutans*
    Willd. ex Schult. & Schult. fil.) - 1, 3, 4, 5
- subsp. **sphaerocephala** A. Khokhr. - 5
fusca Miyabe & Kudo = T. coccinea
nutans Willd. ex Schult. & Schult. fil. = T. coccinea
okuboi Makino - 5
palustris Huds. = T. pusilla
pusilla (Michx.) Pers. (*Narthecium pusillum* Michx., *Tofiel-
    dia palustris* Huds. p.p. nom. illegit.) - 1, 3, 4, 5

Veratrum L.

albiflorum Tolm. - 5
album L. - 1
- subsp. *lobelianum* (Bernh.) K. Richt. = V. lobelianum
- subsp. *lobelianum* (Bernh.) Schuebl. & Martens = V.
    lobelianum
- subsp. *misae* (Sirj.) Tzvel. = V. misae
- subsp. *virescens* (Gaudin) Jav. & Soo = V. lobelianum
- var. *grandiflorum* Maxim. ex Baker = V. grandiflorum
- var. *lobelianum* (Bernh.) Koch = V. lobelianum
- var. *virescens* Gaudin = V. lobelianum
alpestre Nakai - 5
calycinum Kom. = V. dolichopetalum
dahuricum (Turcz.) Loes. fil. - 4, 5
dolichopetalum Loes. fil. (*V. calycinum* Kom.) - 5
grandiflorum (Maxim. ex Baker) Loes. fil. (*V. album* L. var.
    grandiflorum Maxim. ex Baker) - 5
lobelianum Bernh. (*V. album* L. var. *lobelianum* (Bernh.)
    Koch, *V. album* subsp. *lobelianum* (Bernh.) K. Richt.
    comb. superfl., *V. album* subsp. *lobelianum* (Bernh.)
    Schuebl. & Martens, *V. album* var. *virescens* Gaudin,
    V. album subsp. *virescens* (Gaudin) Jav. & Soo) - 1, 2,
    3, 4, 5, 6
maackii Regel (*V. nigrum* L. subsp. *maackii* (Regel)
    Kitam.) - 5
misae (Sirj.) Loes. fil. (*V. album* L. subsp. *misae* (Sirj.)
    Tzvel.) - 1, 3, 4, 5
nigrum L. - 1, 3, 4, 6
- subsp. *maackii* (Regel) Kitam. = V. maackii
- subsp. *ussuriense* (Loes. fil.) Worosch. = V. ussuriense
- var. *ussuriense* Loes. fil. = V. ussuriense
oxysepalum Turcz. - 4, 5
patulum Loes. fil. - 5
ussuriense (Loes. fil.) Nakai (*V. nigrum* L. var. *ussuriense*
    Loes. fil., *V. nigrum* subsp. *ussuriense* (Loes. fil.)
    Worosch.) - 5

Zigadenus Michx.

sibiricus (L.) A. Gray - 1, 3, 4, 5

*MELIACEAE Juss.
*Melia L.

*azedarach L. - 1, 2, 6

MENISPERMACEAE Juss.
Menispermum L.

dauricum DC. - 4, 5

MENYANTHACEAE Dumort.
Fauria Franch.

crista-galli (Menz.) Makino - 5

Limnanthemum S.G. Gmel. = Nymphoides

coreanum Levl. = Nymphoides coreana

Menyanthes L.

trifoliata L. - 1, 2, 3, 4, 5, 6

Nymphoides Hill (*Limnanthemum* S.G. Gmel.)

coreana (Levl.) Hara (*Limnanthemum coreanum* Levl.) - 5
peltata (S.G. Gmel.) O. Kuntze - 1, 2, 3, 4, 5, 6

MIMOSACEAE R. Br. = FABACEAE

MOLLUGINACEAE Hutch.
Glinus L.

lotoides L. - 2

Mollugo L.

cerviana (L.) Ser. - 1, 3, 6

MONOTROPACEAE Nutt.
Hypopitys Hill

hypophegea (Wallr.) G. Don fil. (*Monotropa hypophegea*
    Wallr.) - 1
monotropa Crantz - 1, 2, 3, 4, 5, 6

Monotropa L.

hypophegea Wallr. = Hypopitys hypophegea
uniflora L. - 5

Monotropastrum Andres

globosum Andres ex Hara - 5

MORACEAE Link
*Broussonetia L'Her ex Vent.

*papyrifera (L.) Vent. (*Morus papyrifera* L.) - 2, 6

Ficus L.

afghanica (M. Pop.) Drob. = F. afghanistanica
afghanistanica Warb. (*F. afghanica* (M. Pop.) Drob., *F.
    carica* L. var. *afghanica* M. Pop., *F. johannis* Boiss.

subsp. *afghanistanica* (Warb.) Browicz) - 6
carica L. (*F. colchica* Grossh., *F. hyrcana* Grossh., *F. kopet-dagensis* Pachom.) - 1, 2, 6
- var. *afghanica* M. Pop. = F. afghanistanica
colchica Grossh. = F. carica
hyrcana Grossh. = F. carica
johannis Boiss. subsp. *afghanistanica* (Warb.) Browicz = F. afghanistanica
kopetdagensis Pachom. = F. carica

Joxylon auct. = Maclura
pomiferum Rafin. = Maclura pomifera

*Maclura Nutt. (*Joxylon* auct.)
aurantiaca Nutt. = M. pomifera
*pomifera (Rafin.) Schneid. (*Joxylon pomiferum* Rafin., *Maclura aurantiaca* Nutt.) - 1, 2, 6

Morus L.
alba L. - 1, 2, 5, 6
bombycis Koidz. - 5
nigra L. - 1, 2, 6
papyrifera L. = Broussonetia papyrifera

MORINACEAE Rafin.
Morina L.
coulteriana Royle (*M. lehmanniana* Bunge) - 6
kokanica Regel - 6
lehmanniana Bunge = M. coulteriana
parviflora Kar. & Kir. - 6
persica L. - 6

MYRICACEAE Blume
Myrica L.
gale L. - 1
- subsp. *tomentosa* (DC.) E. Murr. = M. tomentosa
tomentosa (DC.) Aschers. & Graebn. (*M. gale* L. subsp. *tomentosa* (DC.) E. Murr.) - 5

*MYRTACEAE Juss.
*Eucalyptus L'Her.
*camaldulensis Dehnh. - 2
*cinerea F. Muell. ex Benth. - 2
*citriodora Hook. - 2
*globulus Labill. - 2
*gummifera (Gaertn.) Hochr. - 2
*gunnii Hook. fil. - 2
*macarthurii H. Deane & Maid. - 2
*pauciflora Sieb. ex Spreng. - 2
*salicifolia (Soland. ex Gaertn.) Cav. - 2
*urnigera Hook. fil. - 2
*viminalis Labill. - 2

NAJADACEAE Juss.
Caulinia Willd.
amurensis (Tzvel.) Tzvel. = C. japonica
flexilis Willd. (*Najas flexilis* (Willd.) Rostk. & W.L. Schmidt) - 1, 3, 4, 5
foveolata auct. = C. orientalis

graminea (Delile) Tzvel. (*Najas graminea* Delile) - 6
japonica (Nakai) Nakai (*Najas japonica* Nakai, *Caulinia amurensis* (Tzvel.) Tzvel., *C. tenuissima* (A. Br. ex Magnus) Tzvel. subsp. *amurensis* Tzvel.) - 5
minor (All.) Coss. & Germ. (*Najas minor* All.) - 1, 2, 3, 4, 6
orientalis (Triest & Uotila) Tzvel. (*Najas orientalis* Triest & Uotila, *Caulinia foveolata* auct., *Najas foveolata* auct.) - 5
tenuissima (A. Br. ex Magnus) Tzvel. (*Najas minor* All. subsp. *tenuissima* (A. Br. ex Magnus) K. Richt., *N. tenuissima* A. Br. ex Magnus) - 1, 3, 5
- subsp. *amurensis* Tzvel. = C. japonica

Najas L.
flexilis (Willd.) Rostk. & W.L. Schmidt = Caulinia flexilis
foveolata auct. = Caulinia orientalis
graminea Delile = Caulinia graminea
japonica Nakai = Caulinia japonica
major All. (*N. marina* L. subsp. *major* (All.) Viinikka) - 1, 3, 5, 6
- var. **polonica** (Zalewsky) Tzvel. (*Najas polonica* Zalewsky) - 1
marina L. - 1, 2, 3, 6
- subsp. **aculeolata** Tzvel. - 1, 2, 6
- subsp. **brachycarpa** (Trautv.) Tzvel. - 6
- subsp. *major* (All.) Viinikka = N. major
minor All. = Caulinia minor
- subsp. *tenuissima* (A. Br. ex Magnus) K. Richt. = Caulinia tenuissima
orientalis Triest & Uotila = Caulinia orientalis
polonica Zalewsky = Najas major var. polonica
tenuissima A. Br. ex Magnus = Caulinia tenuissima

NELUMBONACEAE Dumort.
Nelumbium Juss. = Nelumbo

Nelumbo Hill (*Nelumbium* Juss.)
caspica (DC.) Fisch. (*N. nucifera* auct. p.p.) - 1, 2
komarovii Grossh. (*N. nucifera* auct. p.p.) - 5
nucifera auct. p.p. = N. caspica

NITRARIACEAE Bercht. & J. Presl.
Nitraria L.
komarovii Iljin & Lava ex Bobr. - 2, 6
pamirica L. Vassil. - 6
schoberi L. - 1, 2, 3, 6
sibirica Pall. - 3, 4, 6

NUPHARACEAE Nakai = NYMPHAE-ACEAE

NYCTAGINACEAE Juss.
Mirabilis auct. = Oxybaphus
nyctaginea (Michx.) MacMill. = Oxybaphus nyctagineus

*Mirabilis L.
*jalapa L. - 1

Oxybaphus L' Her. ex Willd. (*Mirabilis* auct.)

nyctagineus (Michx.) Sweet (*Mirabilis nyctaginea* (Michx.) MacMill.) - 1, 6(cult.)

NYMPHAEACEAE Salisb. (*NUPHARACEAE* Nakai)

Euryale Salisb.

ferox Salisb. - 5

Nuphar Smith

advena (Soland.) R. Br. - 1
intermedia Ledeb. = N. x spenneriana
japonica DC. - 5
lutea (L.) Smith - 1, 2, 3, 4, 6
pumila (Timm) DC. (*Nymphaea lutea* L. var. *pumila* Timm, *Nuphar subpumila* Miki, *N. tenella* Reichenb.) - 1, 3, 4, 5
x **spenneriana** Gaudin (*N. intermedia* Ledeb.). - N. lutea (L.) Smith x N. pumila (Timm) DC. - 1, 3
subpumila Miki = N. pumila
tenella Reichenb. = N. pumila

Nymphaea L.

alba L. (*N. minoriflora* (Simonk.) Wissjul.) - 1, 2
x **borealis** E. Camus. - N. alba L. x N. candida J. Presl - 1
candida J. Presl - 1, 3, 4, 6
colchica (Woronow ex Grossh.) Kem.-Nath. - 2
lutea L. var. *pumila* Timm = Nuphar pumila
minoriflora (Simonk.) Wissjul. = N. alba
tetragona Georgi (*N. tetragona* var. *wenzelii* (Maack) Worosch., *N. wenzelii* Maack) - 1, 3, 4, 5
- var. *wenzelii* (Maack) Worosch. = N. tetragona
wenzelii Maack = N. tetragona

OLEACEAE Hoffmgg. & Link
Fraxinus L.

*****americana** L.
angustifolia Vahl - 1, 2
- subsp. *oxycarpa* (Willd.) Franko & Rocha Afonso = F. oxycarpa
- subsp. *pannonica* Soo & Simon = F. ptacovskyi
- subsp. *syriaca* (Boiss.) Yaltirik = F. syriaca
- var. *pojarkoviana* (V. Vassil.) Karpati = F. ptacovskyi
chinensis subsp. *rhynchophylla* (Hance) E. Murr. = F. rhynchophylla
coriariifolia Scheele (*F. excelsior* L. subsp. *coriariifolia* (Scheele) E. Murr.) - 1, 2
densata Nakai - 5
excelsior L. - 1, 2
- subsp. *coriariifolia* (Scheele) E. Murr. = F. coriariifolia
*****lanceolata** Borkh.
lanuginosa Koidz. (*F. longicuspis* auct., *F. sieboldiana* auct.) - 5
longicuspis auct. = F. lanuginosa
mandshurica Rupr. (*F. nigra* subsp. *mandshurica* (Rupr.) S.S. Sun) - 5
nigra subsp. *mandshurica* (Rupr.) S.S. Sun = F. mandshurica
ornus L. - 1, 2
oxycarpa Willd. (*F. angustifolia* Vahl subsp. *oxycarpa* (Willd.) Franko & Rocha Afonso, *F. rotundifolia* Mill. subsp. *oxycarpa* (Willd.) P.S. Green, comb. superfl., *F.*

rotundifolia subsp. *oxycarpa* (Willd.) Yaltirik, *F. rotundifolia* auct. p.p.) - 1, 2
- subsp. *syriaca* (Boiss.) Yaltirik = F. syriaca
pallisae Wilmott - 1
*****pennsylvanica** Marsh.
pojarkoviana V. Vassil. = F. ptacovskyi
potamophila Herd. = F. sogdiana
ptacovskyi Domin (*F. angustifolia* Vahl subsp. *pannonica* Soo & Simon, *F. angustifolia* var. *pojarkoviana* (V. Vassil.) Karpati, *F. pojarkoviana* V. Vassil.) - 1
raibocarpa Regel - 6
rhynchophylla Hance (*F. chinensis* subsp. *rhynchophylla* (Hance) E. Murr.) - 5
rotundifolia auct. = F. oxycarpa and F. syriaca
rotundifolia Mill. subsp. *oxycarpa* (Willd.) Yaltirik = F. oxycarpa
- subsp. *syriaca* (Boiss.) Yaltirik = F. syriaca
sieboldiana auct. = F. lanuginosa
sogdiana Bunge (*F. potamophila* Herd.) - 6
syriaca Boiss. (*F. angustifolia* Vahl subsp. *syriaca* (Boiss.) Yaltirik, *F. oxycarpa* Willd. subsp. *syriaca* (Boiss.) Yaltirik, *F. rotundifolia* Mill. subsp. *syriaca* (Boiss.) Yaltirik, *F. rotundifolia* auct. p.p.) - 6

Jasminum L.

fruticans L. - 1, 2, 6
humile L. (*J. humile* var. *revolutum* (Sims) Stockes, *J. revolutum* Sims) - 6
- var. *revolutum* (Sims) Stockes = J. humile
*****nudiflorum** Lindl. - 6
officinale L. - 2
revolutum Sims = J. humile

Ligustrina Rupr. = Syringa

amurensis Rupr. = Syringa amurensis
japonica (Maxim.) V. Vassil. = Syringa reticulata

Ligustrum L.

*****japonicum** Thunb. - 1, 2
*****lucidum** Ait. fil. - 1, 2
reticulatum Blume = Syringa reticulata
tschnoskii auct. = L. yezoense
vulgare L. - 1, 2, 6(cult.)
yezoense Nakai (*L. tschonoskyi* auct.) - 5

Olea L.

*****europaea** L. - 1, 2, 6

Osmanthus Lour.

decorus (Boiss. & Bal.) Kasapligil (*Phillyrea decora* Boiss. & Bal., *Ph. medwedewii* Sred.) - 2

Phillyrea L.

*****angustifolia** L. - 1
decora Boiss. & Bal. = Osmanthus decorus
*****latifolia** L. - 1
medwedewii Sred. = Osmanthus decorus

Syringa L. (*Ligustrina* Rupr.)

amurensis Rupr. (*Ligustrina amurensis* Rupr., *Syringa reticulata* (Blume) Hara var. *amurensis* (Rupr.) J.S. Pringle) - 5
*****chinensis** Willd. - 1
josikaea Jacq. fil. - 1

persica L. - 1(cult.), 2, 6(cult.)
reticulata (Blume) Hara (*Ligustrum reticulatum* Blume, *Ligustrina japonica* (Maxim.) V. Vassil.) - 5
- var. *amurensis* (Rupr.) J.S. Pringle = S. amurensis
robusta Nakai = S. wolfii
*villosa Vahl - 1
vulgaris L. - 1, 5(cult.), 6(cult.)
wolfii Schneid. (*Syringa robusta* Nakai) - 5

ONAGRACEAE Juss.
Chamaenerion Hill (*Chamerion* (Rafin.) Rafin.)

angustifolium (L.) Scop. (*Chamerion angustifolium* (L.) Holub) - 1, 2, 3, 4, 5, 6
- subsp. *circumvagum* (Mosquin) Mold. = Ch. danielsii
- subsp. macrophyllum (Hausskn.) Czer. comb. nova (*Epilobium angustifolium* L. f. *macrophyllum* Hausskn. 1884, Monogr. Epilob. : 38; *E. angustifolium* subsp. *macrophyllum* (Haussk.) Hult.) - ?
- var. *platyphyllum* Daniels = Ch. danielsii
angustissimum (Web.) Grossh. = Ch. dodonaei
caucasicum (Hausskn.) Sosn. ex Grossh. (*Chamerion caucasicum* (Hausskn.) Galushko) - 2
colchicum (Albov) Steinb. (*Chamerion colchicum* (Albov) Holub) - 2
danielsii (D. Löve) Czer. comb. nova (*Epilobium danielsii* D. Löve, 1968, Taxon, 17, 1 : 89; *Chamaenerion angustifolium* (L.) Scop. subsp. *circumvagum* (Mosquin) Mold., *Ch. angustifolium* var. *platyphyllum* Daniels, *Chamerion danielsii* (D. Löve) Czer., *Ch. platyphyllum* (Daniels) A. & D. Löve, nom. illegit. superfl., *Epilobium angustifolium* L. subsp. *circumvagum* Mosquin, *E. platyphyllum* (Daniels) A. & D. Löve, 1966, non Rydb. 1930) - 5
dodonaei (Vill.) Kost. (*Epilobium dodonaei* Vill., *Chamaenerion angustissimum* (Web.) Grossh., *Chamerion dodonaei* (Vill.) Holub, *Epilobium angustissimum* Grauer, p.p.) - 1, 2
halimifolium Salisb. = Ch. latifolium
latifolium (L.) Th. Fries & Lange (*Ch. halimifolium* Salisb., *Ch. subdentatum* Rydb., *Chamerion latifolium* (L.) Holub, *Ch. subdentatum* (Rydb.) A. & D. Löve) - 1, 3, 4, 5, 6
stevenii (Boiss.) Sosn. ex Grossh. (*Chamerion stevenii* (Boiss.) Holub) - 2
subdentatum Rydb. = Ch. latifolium

Chamerion (Rafin.) Rafin. = Chamaenerion

angustifolium (L.) Holub = Chamaenerion angustifolium
caucasicum (Hausskn.) Galushko = Chamaenerion caucasicum
colchicum (Albov) Holub = Chamaenerion colchicum
danielsii (D. Löve) Czer. = Chamaenerion danielsii
dodonaei (Vill.) Holub = Chamaenerion dodonaei
latifolium (L.) Holub = Chamaenerion latifolium
platyphyllum (Daniels) A. & D. Löve = Chamaenerion danielsii
stevenii (Boiss.) Holub = Chamaenerion stevenii
subdentatum (Rydb.) A. & D. Löve = Chamaenerion latifolium

Circaea L.

alpina L. - 1, 2, 3, 4, 5
- subsp. *caulescens* (Kom.) Tatew. = C. caulescens
- var. *caulescens* Kom. = C. caulescens

caucasica A. Skvorts. - 2
caulescens (Kom.) Nakai (*C. alpina* L. var. *caulescens* Kom., *C. alpina* subsp. *caulescens* (Kom.) Tatew.) - 2, 3, 4, 5
cordata Royle - 5
x intermedia Ehrh. - C. alpina L. x C. lutetiana L. - 1, 2, 5
lutetiana L. (*C. quadrisulcata* (Maxim.) Franch. & Savat. p.p. quoad basionymum) - 1, 2, 3, 4, 5, 6
- subsp. quadrisulcata (Maxim.) Aschers. & Magnus - 5
mollis Siebold & Zucc. (*C. quadrisulcata* (Maxim.) Franch. & Savat. p.p. excl. basionymo) - 5
quadrisulcata (Maxim.) Franch. & Savat. = C. lutetiana and C. mollis
- var. *skvortsovii* (Boufford) Worosch. = C. x skvortsovii
x skvortsovii Boufford (*C. quadrisulcata* (Maxim.) Franch. & Savat. var. *skvortsovii* (Boufford) Worosch. comb. invalid.). - C. cordata Royle x C. lutetiana L. - 5

Epilobium L.

acradenum Pazij & Vved. = E. confusum
adenocaulon Hausskn. = E. ciliatum
- subsp. *rubescens* (Rydb.) Hiit. = E. ciliatum
adnatum Griseb. = E. tetragonum
affine Bong. = E. glandulosum
affine Maxim. = E. maximowiczii
algidum Bieb. - 2
almaatense Steinb. = E. subnivale
alpestre (Jacq.) Krock. - 1, 2
alpinum L. (*E. anagallidifolium* Lam., *E. dielsii* Levl.) - 1, 2, 3, 4, 5, 6
alsinifolium Vill. - 1
americanum Hausskn. = E. ciliatum
amurense Hausskn. (? *E. gansuense* Levl., *E. miyabei* Levl., *E. ovale* Takeda, *E. shikotanense* Takeda, *E. tenue* Kom.) - 5
anagallidifolium Lam. = E. alpinum
anatolicum Hausskn. - 2
angulatum Kom. = E. cephalostigma
angustifolium L. subsp *circumvagum* Mosquin = Chamaenerion danielsii
- subsp. *macrophyllum* (Haussk.) Hult. = Chamaenerion angustifolium subsp. macrophyllum
- f. *macrophyllum* Hausskn. = Chamaenerion angustifolium subsp. macrophyllum
angustissimum Grauer, p.p. = Chamaenerion dodonaei
arcticum Sam. = E. davuricum
baicalense M. Pop. = E. fastigiato-ramosum
behringianum Haussk. = E. sertulatum
bifarium Kom. = E. glandulosum
bongardii Hausskn. = E. sertulatum
calycinum Hausskn. = E. maximowiczii
cephalostigma Hausskn. (*E. angulatum* Kom., *E. cephalostigma* var. *angulatum* (Kom.) Worosch., *E. cephalostigma* var. *nudicarpum* (Kom.) Hara, *E. coreanum* Levl., *E. nudicarpum* Kom., *E. sugaharai* Koidz.) - 5
- var. *angulatum* (Kom.) Worosch. = E. cephalostigma
- var. *nudicarpum* (Kom.) Hara = E. cephalostigma
ciliatum Rafin. (*E. adenocaulon* Hausskn., *E. adenocaulon* subsp. *rubescens* (Rydb.) Hiit., *E. americanum* Hausskn., *E. dominii* M. Pop., *E. rubescens* Rydb.) - 1
- subsp. *glandulosum* (Lehm.) P.C. Hoch & P.H. Raven = E. glandulosum
collinum C.C. Gmel. - 1
confusum Hausskn. (*E. acradenum* Pazij & Vved.) - 2, 6
coreanum Levl. = E. cephalostigma

cylindricum D. Don (*E. tianschanicum* Pavl.) - 6
cylindrostigma Kom. = E. maximowiczii
danielsii D. Löve = Chamaenerion danielsii
davuricum Fisch. ex Hornem. (*E. arcticum* Sam., *E. davuricum* var. *arcticum* (Sam.) Polun., *E. davuricum* subsp. *arcticum* (Sam.) P.H. Raven) - 1, 3, 4, 5
- subsp. *arcticum* (Sam.) P.H. Raven = E. davuricum
- var. *arcticum* (Sam.) Polun. = E. davuricum
dielsii Levl. = E. alpinum
dodonaei Vill. = Chamaenerion dodonaei
dominii M. Pop. = E. ciliatum
fastigiato-ramosum Nakai (*E. baicalense* M. Pop.) - 3, 4, 5
fauriei Levl. - 5
foucaudianum Levl. = E. hornemannii
frigidum Hausskn. - 2
gansuense Levl. = E. amurense
gemmascens C.A. Mey. - 2
glanduligerum Pachom. - 6
glandulosum Lehm. (*E. affine* Bong. 1833, non Maxim. 1869, *E. bifarium* Kom., *E. ciliatum* Rafin. subsp. *glandulosum* (Lehm.) P.C. Hoch & P.H. Raven) - 5
- subsp. *maximowiczii* (Hausskn.) Worosch. = E. maximowiczii
x **haynaldianum** Hausskn. - E. alsinifolium Vill. x E. palustre L. - 1
hirsutum L. - 1, 2, 3, 6
hornemannii Reichenb. (*E. foucaudianum* Levl., *E. hornemannii* var. *foucaudianum* (Levl.) Hara, *E. uralense* Rupr.) - 1, 5
- subsp. *behringianum* (Hausskn.) P.C. Hoch & P.H. Raven = E. sertulatum
- var. *foucaudianum* (Levl.) Hara = E. hornemannii
- var. *lactiflorum* (Hausskn.) D. Löve = E. lactiflorum
japonicum Hausskn. = E. pyrricholophum
komarovii Ovcz. - 6
korshinskyi Morozova - 6
kurilense Nakai = E. maximowiczii
lactiflorum Hausskn. (*E. hornemannii* Reichenb. var. *lactiflorum* (Hausskn.) D. Löve) - 1, 3, 4, 5
lamyi F. Schultz (*E. tetragonum* L. subsp. *lamyi* (F. Schultz) Nym.) - 1, 2
lanceolatum Seb. & Mauri - 1, 2
laxum auct. p.p. = E. subnivale
lipschitzii Pachom. - 6
maximowiczii Hausskn. (*E. affine* Maxim. 1869, non Bong. 1833, *E. calycinum* Hausskn., *E. cylindrostigma* Kom., *E. glandulosum* Lehm. subsp. *maximowiczii* (Hausskn.) Worosch., ? *E. kurilense* Nakai, *E. punctatum* Levl.) - 5
minutiflorum Hausskn. - 2, 6
miyabei Levl. = E. amurense
modestum Hausskn. (*E. rupicola* Pavl.) - 6
montanum L. - 1, 2, 3, 4, 5(alien)
nervosum Boiss. & Buhse (*E. roseum* Schreb. subsp. *subsessile* (Boiss.) P.H. Raven, *E. roseum* var. *subsessile* Boiss.) - 1, 2, 6
nudicarpum Kom. = E. cephalostigma
nutans F.W. Schmidt - 1
obscurum Schreb. - 1
oligodontum Hausskn. = E. pyrricholophum
ovale Takeda = E. amurense
palustre L. (*E. tundrarum* Sam.) - 1, 2, 3, 4, 5, 6
parviflorum Schreb. - 1, 2, 6
x **persicinum** Reichenb. - E. parviflorum Schreb. x E. roseum Schreb. - 1
platyphyllum (Daniels) A. & D. Löve = Chamaenerion danielsii

ponticum Hausskn. - 2
prionophyllum Hausskn. - 2
punctatum Levl. = E. maximowiczii
x **purpureum** Fries. - E. palustre L. x E. roseum Schreb. - 1
pyrricholophum Franch. & Savat. (*E. japonicum* Hausskn., *E. oligodontum* Hausskn.) - 5
x **rivulare** Wahlenb. - A. palustre L. x E. parviflorum Schreb. - 1
roseum Schreb. - 1
- subsp. *consimile* (Hausskn.) P.H. Raven = E. consimile
- subsp. *subsessile* (Boiss.) P.H. Raven = E. nervosum
- var. *subsessile* Boiss. = E. nervosum
rubescens Rydb. = E. ciliatum
x **rubneri** Domin. - E. collinum C.C. Gmel. x E. roseum Schreb. - 1
rupicola Pavl. = E. modestum
sertulatum Hausskn. (*E. behringianum* Hausskn., *E. bongardii* Hausskn., *E. hornemannii* Reichenb. subsp. *behringianum* (Hausskn.) P.C. Hoch & P.H. Raven) - 5
shikotanense Takeda = E. amurense
x **simulatum** Hausskn. - E. nutans F.W. Schmidt x E. palustre L. - 1
subalgidum Hausskn. - 2, 6
subnivale M. Pop. ex Pavl. (*E. almaatense* Steinb., *E. laxum* auct. p.p.) - 6
sugaharai Koidz. = E. cephalostigma
tenue Kom. = E. amurense
tetragonum L. (*E. adnatum* Griseb.) - 1, 2, 3, 6
- subsp. *lamyi* (F. Schultz) Nym. = E. lamyi
thermophilum Pauls. - 6
tianschanicum Pavl. = E. cylindricum
tundrarum Sam. = E. palustre
turkestanicum Pazij & Vved. - 6
uralense Rupr. = E. hornemannii
velutinum Nevski - 6

Ludwigia L.

epilobioides Maxim. (*L. prostrata* auct.) - 5
palustris (L.) Ell. - 2
parviflora Roxb. = L. perennis
perennis L. (*L. parviflora* Roxb.) - 6
prostrata auct. = L. epilobioides

Oenothera L. (*Onagra* Hill)

albipercurva Renner ex Hudziok - 1
ammophila Focke - 1
biennis L. (*Onagra biennis* (L.) Scop.) - 1, 2, 3, 4, 5, 6
- subsp. *rubricaulis* (Klebahn) Stomps = O. rubricaulis
depressa Greene = O. salicifolia
erythrosepala Borb. - 1, 2
hoelscheri Renner ex Rostanski - 1
laciniata Hill = Raimannia laciniata
muricata auct. = O. strigosa
muricata L. = O. rubricaulis
odorata Jacq. - 2
parviflora L. - 1
perangusta Gates - 1
renneri H. Scholz - 1, 5(?)
rubricaulis Klebahn (*O. biennis* L. subsp. *rubricaulis* (Klebahn) Stomps, *O. muricata* L. nom. confus., *Onagra muricata* (L.) Moench) - 1, 4, 5
salicifolia Desf. ex D. Don fil. (*O. depressa* Greene, *O. strigosa* (Rydb.) Mackenz. & Bush var. *depressa* (Greene) Gates, *O. strigosa* sensu Kotov) - 1
stricta Ledeb. ex Link - 1(?)

strigosa (Rydb.) Mackenz. & Bush (*O. muricata* auct., *Onagra muricata* sensu Steinb.) - 5
- var. *depressa* (Greene) Gates = O. salicifolia
strigosa sensu Kotov = O. salicifolia
suaveolens Desf. ex Pers. - 1, 2

Onagra Hill = Oenothera
biennis (L.) Scop. = Oenothera biennis
muricata (L.) Moench = Oenothera rubricaulis
muricata sensu Steinb. = Oenothera strigosa

Raimannia Rose ex Britt. & Br.
laciniata (Hill) Rose ex Britt. & Br. (*Oenothera laciniata* Hill) - 1

ONOCLEACEAE Pichi Sermolli
Matteuccia Tod. (*Pteretis* Rafin., *Struthiopteris* Hall.)
orientalis (Hook.) Trev. (*Struthiopteris orientalis* Hook., *Pteretis orientalis* (Hook.) Ching) - 5
struthiopteris (L.) Tod. (*Struthiopteris filicastrum* All.) - 1, 2, 3, 4, 5

Onoclea L.
sensibilis L. - 5

Pteretis Rafin. = Matteuccia
orientalis (Hook.) Ching = Matteuccia orientalis

Struthiopteris Hall. = Matteuccia
filicastrum All. = Matteuccia struthiopteris
orientalis Hook. = Matteuccia orientalis

OPHIOGLOSSACEAE (R. Br.) Agardh
Ophioglossum L.
alascanum E. Britt. - 5
bucharicum (O. & B. Fedtsch.) O. & B. Fedtsch. (*O. vulgatum* L. var. *bucharicum* O. & B. Fedtsch., *O. thermale* Kom. var. *bucharicum* (O. & B. Fedtsch.) H.P. Fuchs) - 6
lusitanicum L. - 2
mironovii Sumn. = O. vulgatum
nipponicum Miyabe & Kudo - 5
thermale Kom. - 5
- var. *bucharicum* (O. & B. Fedtsch.) H.P. Fuchs = O. bucharicum
vulgatum L. (*O. mironovii* Sumn.) - 1, 2, 3, 6
- var. *bucharicum* O. & B. Fedtsch. = O. bucharicum

ORCHIDACEAE Juss.
Aceras R. Br.
anthropophorum (L.) Ait. fil. (*Ophrys anthropophora* L.) - 1

Amitostigma Schlechter
hisamatsui Miyabe & Tatew. = A. kinoshitae
kinoshitae (Makino) Schlechter (*Gymnodenia kinoshitae* Makino, *Amitostigma hisamatsui* Miyabe & Tatew.) - 5
It is likely that plants from the Kuril Islands represent a diffe-

rent species A. hisamatsui Miyabe & Tatew.

Anacamptis Rich.
pyramidalis (L.) Rich. - 1, 2, 6

Arethusa auct. = Eleorchis
japonica A. Gray = Eleorchis japonica

Bletilla auct. = Eleorchis
japonica (A. Gray) Sachlechter = Eleorchis japonica

Calypso Salisb.
bulbosa (L.) Oakes - 1, 3, 4, 5

Cephalanthera Rich.
caucasica Kraenzl. (*C. damasonium* (Mill.) Druce subsp. *caucasicum* (Kraenzl.) H. Sundermann) - 2
cucullata Boiss. & Heldr. subsp. *floribunda* (Woronow) H. Sundermann = C. floribunda
- subsp. *kurdica* (Bornm.ex Kraenzl.) H. Sundermann = C. kurdica
damasonium (Mill.) Druce (*Serapias damasonium* Mill., *S. grandiflora* L. nom. illegit., *S. lonchophyllum* L. fil., *Cephalanthera grandiflora* S.F. Gray, *C. lonchophyllum* (L. fil.) Reichenb. fil., *C. lonchophyllum* (L. fil.) Mansf. comb. superfl.) - 1, 2
- subsp. *caucasicum* (Kraenzl.) H. Sundermann = C. caucasica
epipactoides auct. = C. floribunda
floribunda Woronow (*C. cucullata* Boiss. & Heldr. subsp. *floribunda* (Woronow) H. Sundermann, *C. kurdica* Bornm. ex Kraenzl. subsp. *floribunda* (Woronow) Soo, *C. epipactoides* auct.) - 2
grandiflora S.F. Gray = C. damasonium
kurdica Bornm.ex Kraenzl. (*C. cucullata* Boiss. & Heldr. subsp. *kurdica* (Bornm. ex Kraenzl.) H. Sundermann) - 2
- subsp. *floribunda* (Woronow) Soo = C. floribunda
lonchophyllum (L. fil.) Mansf. = C. damasonium
lonchophyllum (L. fil.) Reichenb. fil. = C. damasonium
longibracteata Blume - 5
longifolia (L.) Fritsch - 1, 2, 6
rubra (L.) Rich. - 1, 2, 6

Chamaeorchis auct. = Chamorchis

Chamorchis Rich. (*Chamaeorchis* auct.)
alpina (L.) Rich. - 1

Chusua Nevski = Ponerorchis
pauciflora (Lindl.) P.F. Hunt = Ponerorchis pauciflora
secunda Nevski = Ponerorchis pauciflora

x *Coeloglossorchis* Guetrot = x Dactyloglossum

Coeloglossum C. Hartm.
bracteatum (Willd.) Schlechter = C. viride
viride (L.) C. Hartm. (*C. bracteatum* (Willd.) Schlechter, *C. viride* subsp. *bracteatum* (Willd.) Hult. comb. superfl.,

C. viride subsp. *bracteatum* (Willd.) K. Richt., *C. viride*
subsp. *bracteatum* (Willd.) Soo, comb. superfl.) - 1, 2,
3, 4, 5, 6
- subsp. *bracteatum* (Willd.) Hult. = C. viride
- subsp. *bracteatum* (Willd.) K. Richt. = C. viride
- subsp. *bracteatum* (Willd.) Soo = C. viride

Comperia C. Koch

comperiana (Stev.) Aschers. & Graebn. (*Orchis comperiana*
Stev., *Comperia taurica* C. Koch) - 1
taurica C. Koch = C. comperiana

Corallorhiza Chatel. = Corallorrhiza

Corallorrhiza Rupp. ex Gagnebin (*Corallorhiza* Chatel.)

trifida Chatel. - 1, 2, 3, 4, 5, 6

Cremastra Lindl. (*Hyacinthorchis* auct.)

variabilis (Blume) Nakai (*Hyacinthorchis variabilis*
Blume) - 5

Cypripedium L.

calceolus L. - 1, 2, 3, 4, 5
freynii Karo = C. ventricosum
guttatum Sw. - 1, 3, 4, 5
- subsp. *yatabeanum* (Makino) Hult. = C. yatabeanum
- subsp. *yatabeanum* (Makino) Murata = C. yatabeanum
- subsp. *yatabeanum* (Makino) Soo = C. yatabeanum
macranthon Sw. - 1, 3, 4, 5
- subsp. *ventricosum* (Sw.) Soo = C. ventricosum
ventricosum Sw. (*C. freynii* Karo, *C. macranthon* Sw. subsp.
ventricosum (Sw.) Soo) - 1, 3, 4, 5
yatabeanum Makino (*C. guttatum* Sw. subsp. *yatabeanum*
(Makino) Hult. comb. superfl., *C. guttatum* subsp.
yatabeanum (Makino) Murata, *C. guttatum* subsp.
yatabeanum (Makino) Soo, comb. superfl.) - 5

xDactyloglossum P.F. Hunt & Summerhayes (x*Coeloglossorchis* Guetrot, x *Haberiorcis* Rolfe, x *Orchicoeloglossum* Aschers. & Graebn.). - Dactylorhiza Nevski x Coeloglossum C. Hartm.

turcestanicum (G. Keller & Soo) Soo (x *Orchicoeloglossum
turkestanicum* G. Keller & Soo). - Dactylorhiza
umbrosa (Kar. & Kir.) Nevski x Coeloglossum viride
(L.) C. Hartm. - 6

Dactylorchis (Klinge) Vermeulen = Dactylorhiza

alpestris (Pugsl.) Vermeulen = Dactylorhiza alpestris
amblyoloba (Nevski) Soo = Dactylorhiza amblyoloba
baltica (Klinge) Vermeulen = Dactylorhiza longifolia
caucasica (Klinge) Soo = Dactylorhiza euxina
cordigera (Fries) Vermeulen = Dactylorhiza cordigera
cruenta (O.F. Muell.) Vermeulen = Dactylorhiza cruenta
curvifolia (Nyl.) Vermeulen = Dactylorhiza curvifolia
elodes (Griseb.) Vermeulen = Dactylorhiza elodes
euxina (Nevski) Soo = Dactylorhiza euxina
flavescens (C. Koch) Vermeulen = Dactylorhiza flavescens
fuchsii (Druce) Vermeulen = Dactylorhiza fuchsii

- subsp. *hebridensis* (Wilmott) H.-Harrison fil. = Dactylorhiza hebridensis
- subsp. *psychrophila* (Schlechter) Vermeulen = Dactylorhiza psychrophila
- var. *meyeri* (Reichenb. fil.) Vermeulen = Dactylorhiza
hebridensis
iberica (Bieb. ex Willd.) Vermeulen = Dactylorhiza iberica
incarnata (L.) Vermeulen = Dactylorhiza incarnata
- subsp. *ochroleuca* (Wustn. ex Boll.) H.-Harrison = Dactylorhiza ochroleuca
kotschyi (Reichenb. fil.) Soo = Dactylorhiza kotschyi
lancibracteata (C. Koch) Soo = Dactylorhiza urvilleana
lapponica (Laest.) Vermeulen = Dactylorhiza lapponica
latifolia (L.) Rothm. = Dactylorhiza majalis
- subsp. *alpestris* (Pugsl.) Soo = Dactylorhiza alpestris
- subsp. *baltica* (Klinge) Soo = Dactylorhiza longifolia
longifolia (L. Neum.) Vermeulen = Dactylorhiza longifolia
maculata (L.) Vermeulen = Dactylorhiza maculata
- subsp. *cartaliniae* (Klinge) Soo, p.p. = Dactylorhiza
amblyoloba and D. urvilleana
- subsp. *elodes* (Griseb.) Vermeulen = Dactylorhiza elodes
- - var. *sudetica* (Poch ex Reichenb. fil.) Vermeulen =
Dactylorhiza sudetica
- subsp. *ericetorum* (Linton) Vermeulen = Dactylorhiza
ericetorum
- subsp. *hebridensis* (Wilmott) Vermeulen = Dactylorhiza
hebridensis
- subsp. *lancibracteata* (C. Koch) Soo = Dactylorhiza urvilleana
- var. *schurii* (Klinge) Borsos & Soo = Dactylorhiza schurii
megapolitana Bisse = Dactylorhiza x megapolitana
osmanica (Klinge) Soo = Dactylorhiza osmanica
romana (Seb.) Vermeulen = Dactylorhiza romana
- subsp. *georgica* (Klinge) Soo = Dactylorhiza flavescens
russowii (Klinge) A. & D. Löve = Dactylorhiza russowii
salina (Turcz. ex Lindl.) Vermeulen = Dactylorhiza salina
sambucina (L.) Vermeulen = Dactylorhiza sambucina
sanasunitensis (Fleischm.) Soo = Dactylorhiza sanasunitensis
traunsteineri (Saut.) Vermeulen = Dactylorhiza traunsteineri
- subsp. *curvifolia* (Nyl.) Vermeulen = Dactylorhiza curvifolia
- subsp. *russowii* (Klinge) Vermeulen = Dactylorhiza
russowii
umbrosa (Kar. & Kir.) Wendelbo = Dactylorhiza umbrosa

Dactylorhiza Nevski (*Dactylorchis* (Klinge) Vermeulen)

affinis (C. Koch) Aver. (*Orchis affinis* C. Koch) - 2
alpestris (Pugsl.) Aver. (*Orchis alpestris* Pugsl., *Dactylorchis
alpestris* (Pugsl.) Vermeulen, *D. latifolia* (L.) Rothm.
subsp. *alpestris* (Pugsl.) Soo, *Dactylorhiza latifolia* (L.)
Soo subsp. *alpestris* (Pugsl.) Soo, *Orchis latifolia* L.
subsp. *alpestris* (Pugsl.) Janch., *O. majalis* Reichenb.
subsp. *alpestris* (Pugsl.) Hellmayr) - 1
x **ambigua** (A. Kerner) H. Sundermann (*Orchis* x *ambigua*
A. Kerner, *Dactylorhiza* x *maculatiformis* (Rouy)
Borsos & Soo, *Orchis* x *maculatiformis* Rouy). - D.
incarnta (L.) Soo x D. maculata (L.) Soo
amblyoloba (Nevski) Aver. (*Orchis amblyoloba* Nevski,
Dactylorchis amblyoloba (Nevski) Soo, nom. invalid.,
D. maculata (L.) Vermeulen subsp. *cartaliniae*
(Klinge) Soo, p.p. nom. invalid., *Dactylorhiza maculata*
(L.) Soo subsp. *triphylla* (C. Koch) H. Sundermann, *D.
saccifera* (Brongn.) Soo subsp. *cartaliniae* (Klinge)

Soo, p.p., *D. triphylla* (C. Koch) Czer., *Orchis basilica* L. subsp. *cartaliniae* Klinge, p.p., *O. cartaliniae* (Klinge) Lipsky, p.p., *O. triphylla* C. Koch, 1849, non Spreng. 1826) - 2

aristata (Fisch. ex Lindl.) Soo (*Orchis aristata* Fisch. ex Lindl.) - 5

x **aschersoniana** (Hausskn.) Borsos & Soo (*Orchis aschersoniana* Hausskn.). - D. incarnata (L.) Soo x D. majalis (Reichenb.) P.F. Hunt & Summerhayes

x **baicalica** Aver. - D. cruenta (O.F. Muell) Soo x D. salina (Turcz. ex Lindl.) Soo - 4

baldshuanica Czerniak. ex Aver. - 6

baltica (Klinge) Nevski = D. longifolia

baltica (Klinge) Orlova = D. longifolia

x **braunii** (Halacsy) Borsos & Soo (*Orchis braunii* Halacsy). - D. fuchsii (Druce) Soo x D. majalis (Reichenb.) P.F. Hunt & Summerhayes

x **carnea** (E.G. Camus) Soo. - D. elodes (Griseb.) Aver. x D. incarnata (L.) Soo

cataonica (Fleischm.) Holub (*Orchis cataonica* Fleischm., *Dactylorhiza euxina* (Nevski) Soo var. *cataonica* (Fleischm.) Soo, nom. invalid.) - 2

- subsp. *caucasica* (Klinge) Soo = D. euxina

caucasica (Klinge) Soo = D. euxina

chuhensis Renz & Taubenheim (*D. renzii* Aver. 1983, non H. Baumann & Kunkele, 1981) - 2

x **claudiopolitana** (Simk.) Soo. - D. incarnata (L.) Soo x D. schurii (Klinge) Aver.

cordigera (Fries) Soo (*Orchis cordigera* Fries, *Dactylorhiza cordigera* (Fries) Vermeulen, *Dactylorhiza majalis* (Reichhenb.) P.F. Hunt & Summerhayes subsp. *cordigera* (Fries) H. Sundermann) - 1

cruenta (O.F. Muel.) Soo (*Orchis cruenta* O.F. Muell., *Dactylorhis cruenta* (O.F. Muell.) Vermeulen, *Dactylorhiza incarnata* (L.) Soo subsp. *cruenta* (O.F. Muell.) P.D. Sell, *D. incarnata* var. *cruenta* (O.F. Muell.) Hyl. nom. invalid.) - 1, 3, 4

- subsp. *lapponica* (Laest.) E. Nels. = D. lapponica

- subsp. *salina* (Turcz. ex Lindl.) E. Nels. = D. salina

curvifolia (Nyl.) Czer. (*Orchis curvifolia* Nyl., *Dactylorhis curvifolia* (Nyl.) Vermeulen, *D. traunsteineri* (Saut.) Vermeulen subsp. *curvifolia* (Nyl.) Vermeulen, nom. altern., *Dactylorhiza traunsteineri* (Saut.) Soo subsp. *curvifolia* (Nyl.) Soo) - 1

czerniakowskae Aver. - 6

x **dufftiana** (M. Schulze) Soo (*Orchis dufftiana* M. Schulze). - D. majalis (Reichenb.) P.F. Hunt & Summerhayes x D. traunsteineri (Saut.) Soo

elodes (Griseb.) Aver. (*Orchis elodes* Griseb., *Dactylorhis elodes* (Griseb.) Vermeulen, *D. maculata* (L.) Vermeulen subsp. *elodes* (Griseb.) Vermeulen, *Dactylorhiza maculata* (L.) Soo subsp. *elodes* (Griseb.) Soo) - 1

■ ericetorum (Linton) Aver. (*Orchis maculata* L. subsp. *ericetorum* Linton, *Dactylorhis maculata* (L.) Vermeulen subsp. *ericetorum* (Linton) Vermeulen, *Dactylorhiza maculata* (L.) Soo subsp. *ericetorum* (Linton) P.F. Hunt & Summerhays, *D. maculata* var. *ericetorum* (Linton) Soo, comb. invalid., *Orchis ericetorum* (Linton) A. Benn.)

euxina (Nevski) Czer. (*Orchis euxina* Nevski, *Dactylorhis caucasica* (Klinge) Soo, nom. invalid. illegit., *D. euxina* (Nevski) Soo, nom. invalid., *Dactylorhiza cataonica* (Fleischm.) Holub subsp. *caucasica* (Klinge) Soo, *D. caucasica* (Klinge) Soo, nom. illegit., *D. euxina* (Nevski) H. Baumann & Kunkele, comb. superfl., *D. euxina* (Nevski) Soo, nom. invalid., *D. majalis* (Reichenb.) P.F. Hunt & Summerhayes subsp. *caucasica* (Klinge)

H. Sundermann, *Orchis caucasica* (Klinge) Lipsky, 1899, non Regel, 1869, *O. monticola* Klinge subsp. *caucasica* Klinge) - 2

- var. *cataonica* (Fleischm.) Soo = D. cataonica

- var. *markowitschii* (Soo) Renz & Taubenheim = D. markowitschii

fistulosa (Moench) H. Baumann & Kunkele = D. majalis

flavescens (C. Koch) Holub (*Orchis flavescens* C. Koch, *Dactylorchis flavescens* (C. Koch) Vermeulen, *D. romana* (Seb.) Vermeulen subsp. *georgica* (Klinge) Soo, nom. invalid., *Dactylorhiza flavescens* (C. Koch) H. Baumann & Kunkele, comb. superfl., *D. flavescens* (C. Koch) Seifulin, comb. superfl., *D. romana* (Seb.) Soo subsp. *georgica* (Klinge) Soo), *D. sambucina* (L.) Soo subsp. *georgica* (Klinge) H. Sundermann, *Orchis mediterranea* Klinge subsp. *georgica* Klinge, *O. georgica* (Klinge) Czerniak. comb. superfl., *O. georgica* (Klinge) Lipsky) - 2, 6

fuchsii (Druce) Soo (*Orchis fuchsii* Druce, *Dactylorchis fuchsii* (Druce) Vermeulen, *Dactylorhiza longebracteata* (F.W. Schmidt) Holub, p.p. quoad pl., *D. maculata* (L.) Soo subsp. *fuchsii* (Druce) Hyl. nom. invalid., *Orchis maculata* L. subsp. *fuchsii* (Druce) Jorgens.) - 1, 3, 4

- subsp. *fuchsii* var. *meyeri* (Reichenb. fil.) Soo = D. hebridensis

- - var. *psychrophila* (Schlechter) Soo = D. psychrophila

- subsp. *hebridensis* (Wilmott) Soo = D. hebridensis

- subsp. *psychrophila* (Schlechter) Holub = D. psychrophila

- subsp. *transsilvanica* (Schur) Frohner = D. transsilvanica

hebridensis (Wilmott) Aver. (*Orchis hebridensis* Wilmott, *Dactylorchis fuchsii* (Druce) Vermeulen subsp. *hebridensis* (Wilmott) H.-Harrison fil., *D. fuchsii* var. *meyeri* (Reichenb. fil.) Vermeulen, *D. maculata* (L.) Vermeulen subsp. *hebridensis* (Wilmott) Vermeulen, *Dactylorhiza fuchsii* (Druce) Soo subsp. *fuchsii* var. *meyeri* (Reichenb. fil.) Soo, *D. fuchsii* subsp. *hebridensis* (Wilmott) Soo, *D. maculata* (L.) Soo subsp. *meyeri* (Reichenb. fil.) Tournay, *D. meyeri* (Reichenb. fil.) Aver., *Orchis fuchsii* Druce subsp. *hebridensis* (Wilmott) Clapham, *O. maculata* L. var. *meyeri* Reichenb. fil., *O. maculata* subsp. *meyeri* (Reichenb. fil.) E.G. Camus & A. Camus, comb. superfl., *O. maculata* subsp. *meyeri* (Reichenb. fil.) K. Richt.) - 1, 3, 4

iberica (Bieb. ex Willd.) Soo (*Orchis iberica* Bieb. ex Willd., *Dactylorchis iberica* (Bieb. ex Willd.) Vermeulen) - 1, 2

incarnata (L.) Soo (*Orchis incarnata* L., *Dactylorchis incarnata* (L.) Vermeulen, *Dactylorhiza strictifolia* (Opiz) Rauschert, *Orchis strictifolia* Opiz, *O. latifolia* sensu Nevski, p.p.) - 1, 2, 3, 4, 6

- subsp. *cruenta* (O.F. Muell.) P.D. Sell = D. cruenta

- subsp. *ochroleuca* (Wustn. ex Boll.) P.F. Hunt & Summerhayes = D. ochroleuca

- subsp. *turcestanica* (Klinge) H. Sundermann = D. umbrosa

- var. *cruenta* (O.F. Muell.) Hyl. = D. cruenta

- var. *ochroleuca* (Wustn. ex Boll.) Hyl. = D. ochroleuca

x **ishorica** Aver. - D. incarnata (L.) Soo x D. longifolia (L. Neum.) Aver. - 1

x **kelleriana** (Ciferri & Giacomi) Soo. - D. fuchsii (Druce) Soo x D. traunsteineri (Saut.) Soo

x **kerneriorum** (Soo) Soo (*Orchis kerneriorum* Soo). - D. fuchsii (Druce) Soo x D. incarnata (L.) Soo

knorringiana (Kraenzl.) Ikonn. = D. kotschyi

■ kolaensis (Montell) Aver. (*Orchis maculata* L. var. *kolaensis* Montell)

x **komiensis** Aver. - D. hebridensis (Wilmott) Aver. x D.

maculata (L.) Soo - 1

kotschyi (Reichenb. fil.) Soo (*Orchis incarnata* L. var. *kotschyi* Reichenb. fil., *Dactylorchis kotschyi* (Reichenb. fil.) Soo, *Dactylorhiza knorringiana* (Kraenzl.) Ikonn., *D. umbrosa* (Kar. & Kir.) Nevski subsp. *knorringiana* (Kraenzl.) Soo, *D. umbrosa* var. *knorringiana* (Kraenzl.) Soo, *Orchis incarnata* var. *knorringiana* Kraenzl., *O. knorringiana* (Kraenzl.) Czerniak. ex E. Nikit., *O. kotschyi* (Reichenb. fil.) Schlechter) - 4, 6

x **krylowii** (Soo) Soo (*Orchis krylowii* Soo). - D. cruenta (O.F. Muell.) Soo x D. incarnata (L.) Soo

kulikalonica Czerniak. ex Aver. - 6

lancibracteata (C. Koch) Renz = D. urvilleana

■ lapponica (Laest.) Soo (*Orchis angustifolia* Krock. var. *lapponica* Laest., *Dactylorchis lapponica* (Laest.) Vermeulen, *Dactylorhiza cruenta* (O.F. Muell.) Soo subsp. *lapponica* (Laest.) E. Nels. comb. invalid, *D. majalis* (Reichenb.) P.F. Hunt & Summerhayes subsp. *lapponica* (Laest.) H. Sundermann, *D. traunsteineri* (Saut.) Soo subsp. *lapponica* (Laest.) Soo)

latifolia (L.) H. Baumann & Kunkele = D. majalis

latifolia (L.) Soo = D. majalis

- subsp. *alpestris* (Pugsl.) Soo = D. alpestris
- subsp. *baltica* (Klinge) Soo = D. longifolia
- - var. *longifolia* (L. Neum.) Soo = D. longifolia

x **lehmannii** (Klinge) Soo (*Orchis lehmannii* Klinge). - D. incarnata (L.) Soo x D. russowii (Klinge) Holub

longebracteata (F.W. Schmidt) Holub, p.p. = D. fuchsii

- subsp. *sudetica* (Poch ex Reichenb. fil.) Holub = D. sudetica

longifolia (L. Neum.) Aver. (*Orchis longifolia* L. Neum., *Dactylorchis baltica* (Klinge) Vermeulen, *D. latifolia* (L.) Rothm. subsp. *baltica* (Klinge) Soo, nom. invalid., *D. longifolia* (L. Neum.) Vermeulen, *Dactylorhiza baltica* (Klinge) Nevski, nom. invalid., *D. baltica* (Klinge) Orlova, *D. latifolia* (L.) Soo subsp. *baltica* (Klinge) Soo, *D. latifolia* subsp. *baltica* var. *longifolia* (L. Neum.) Soo, *D. majalis* (Reichenb.) P.F. Hunt & Summerhayes subsp. *baltica* (Klinge) Senghas, *D. majalis* subsp. *baltica* (Klinge) H. Sundermann, comb. superfl., *Orchis baltica* (Klinge) Nevski, *O. latifolia* L. subsp. *baltica* Klinge) - 1, 3, 4, 6

maculata (L.) Soo (*Orchis maculata* L., *Dactylorchis maculata* (L.) Vermeulen) - 1, 3, 4

- subsp. *elodes* (Griseb.) Soo = D. elodes
- - var. *schurii* (Klinge) Soo = D. schurii
- - var. *sudetica* (Poch ex Reichenb. fil.) Soo = D. sudetica
- subsp. *ericetorum* (Linton) P.F. Hunt & Summerhays = D. ericetorum
- subsp. *fuchsii* (Druce) Hyl. = D. fuchsii
- subsp. *meyeri* (Reichenb. fil.) Tournay = D. hebridensis
- subsp. *schurii* (Klinge) Soo = D. schurii
- subsp. *sudetica* (Poch ex Reichenb. fil.) Voth = D. sudetica
- subsp. *transsilvanica* (Schur) Soo = D. transsilvanica
- subsp. *triphylla* (C. Koch) H. Sundermann = D. amblyoloba
- var. *ericetorum* (Linton) Soo = D. ericetorum

x *maculatiformis* (Rouy) Borsos & Soo = D. x ambigua

magna (Czerniak.) Ikonn. (*Orchis magna* Czerniak., *Dactylorhiza umbrosa* (Kar. & Kir.) Nevski subsp. *magna* (Czerniak.) Soo) - 6

majalis (Reichenb.) P.F. Hunt & Summerhayes (*Orchis majalis* Reichenb., *Dactylorchis latifolia* (L.) Rothm. nom. invalid., *Dactylorhiza fistulosa* (Moench) H. Baumann & Kunkele, *D. latifolia* (L.) H. Baumann & Kunkele, comb. superfl., *D. latifolia* (L.) Soo, *Orchis*

fistulosa Moench, nom. illegit., *O. latifolia* L. p.p. nom. confus., *O. latifolia* sensu Nevski, p.p.) - 1

- subsp. *baltica* (Klinge) Senghas = D. longifolia
- subsp. *baltica* (Klinge) H. Sundermann = D. longifolia
- subsp. *caucasica* (Klinge) H. Sundermann = D. euxina
- subsp. *cordigera* (Fries) H. Sundermann = D. cordigera
- subsp. *lapponica* (Laest.) H. Sundermann = D. lapponica
- subsp. *russowii* (Klinge) H. Sundermann = D. russowii
- subsp. *traunsteineri* (Saut.) H. Sundermann = D. traunsteineri

markowitschii (Soo) Aver. (*Orchis caucasica* (Klinge) Lipsky var. *markowitschii* Soo, *Dactylorhiza euxina* (Nevski) Czer. var. *markowitschii* (Soo) Renz & Taubenheim) - 2

x **megapolitana** (Bisse) Soo (*Dactylorchis megapolitana* Bisse). - D. fuchsii (Druce) Soo x R. russowii (Klinge) Holub

merovensis (Grossh.) Aver. (*Orchis merovensis* Grossh.) - 2

meyeri (Reichenb. fil.) Aver. = D. hebridensis

ochroleuca (Wustn. ex Boll.) Holub (*Orchis incarnata* L. var. *ochroleuca* Wustn. ex Boll., *Dactylorchis incarnata* (L.) Vermeulen subsp. *ochroleuca* (Wustn. ex Boll.) H.-Harrison, *Dactylorhiza incarnata* (L.) Soo subsp. *ochroleuca* (Wustn. ex Boll.) P.F. Hunt & Summerhayes, *D. incarnata* var. *ochroleuca* (Wustn. ex Boll.) Hyl., *D. ochroleuca* (Wustn. ex Boll.) Aver. comb. superfl., *Orchis incarnata* subsp. *ochroleuca* (Wustn. ex Boll.) O. Schwarz, *O. latifolia* L. var. *ochroleuca* (Wustn. ex Boll.) Pugsl., *O. strictifolia* Opiz var. *ochroleuca* (Wustn. ex Boll.) Hyl.) - 1

■ osmanica (Klinge) Soo (*Orchis orientalis* Klinge subsp. *osmanica* Klinge, *Dactylorchis osmanica* (Klinge) Soo, *Orchis osmanica* (Klinge) Lipsky)

x **predaensis** (Gsell) Soo. - D. cruenta (O.F. Muell.) Soo x D. majalis (Reichenb.) P.F. Hunt & Summerhayes

psychrophila (Schlechter) Aver. (*Orchis maculata* L. var. *psychrophila* Schlechter, *Dactylorchis fuchsii* (Druce) Vermeulen subsp. *psychrophila* (Schlechter) Vermeulen, *Dactylorhiza fuchsii* (Druce) Soo var. *psychrophila* (Schlechter) Soo, *D. fuchsii* subsp. *psychrophila* (Schlechter) Holub) - 1, 3, 4

■ pycnantha (L. Neum.) Aver. (*Orchis angustifolia* Loes. ex Reichenb. subsp. *pycnantha* L. Neum.)

renzii Aver. = D. chuhensis

romana (Seb.) Soo (*Orchis romana* Seb., *Dactylorchis romana* (Seb.) Vermeulen) - 1

- subsp. *georgica* (Klinge) Soo = D. flavescens

ruprechtii Aver. - 2

russowii (Klinge) Holub (*Orchis angustifolia* Loes. ex Reichenb. var. *russowii* Klinge, *Dactylorchis russowii* (Klinge) A. & D. Löve, *D. traunsteineri* (Saut.) Vermeulen subsp. *russowii* (Klinge) Vermeulen, *Dactylorhiza majalis* (Reichenb.) P.F. Hunt & Summerhayes subsp. *russowii* (Klinge) H. Sundermann, *D. traunsteineri* (Saut.) Soo subsp. *russowii* (Klinge) Soo, *Orchis angustifolia* subsp. *russowii* (Klinge) Klinge, *O. russowii* (Klinge) Schlechter) - 1, 3, 4

saccifera (Brongn.) Soo subsp. *cartaliniae* (Klinge) Soo, p.p. = D. amblyoloba and D. urvilleana

- subsp. *lancibracteata* (C. Koch) Soo = D. urvilleana

salina (Turcz. ex Lindl.) Soo (*Orchis salina* Turcz. ex Lindl., *Dactylorchis salina* (Turcz. ex Lindl.) Vermeulen, *Dactylorhiza cruenta* (O.F. Muell.) Soo subsp. *salina* (Turcz. ex Lindl.) E. Nels., *D. sanasunitensis* sensu Czer., *Orchis sanasunitensis* sensu Nevski) - 2, 3, 4, 6

sambucina (L.) Soo (*Orchis sambucina* L., *Dactylorchis sambucina* (L.) Vermeulen) - 1

- subsp. *georgica* (Klinge) H. Sundermann = D. flavescens
x **samnaunensis** (Gsell) Soo. (*Orchis samnaunensis* Gsell). - D. cruenta (O.F. Muell.) Soo x D. maculata (L.) Soo
■ sanasunitensis (Fleischm.) Soo (*Orchis sanasunitensis* Fleischm., *Dactylorchis sanasunitensis* (Fleischm.) Soo, nom. invalid.)
sanasunitensis sensu Czer. = D. salina
schurii (Klinge) Aver. (*Orchis angustifolia* Loes. ex Reichenb. var. *recurva* Klinge f. *schurii* Klinge, *Dactylorchis maculata* (L.) Vermeulen var. *schurii* (Klinge) Borsos & Soo, *Dactylorhiza maculata* (L.) Soo subsp. *elodes* (Griseb.) Soo var. *schurii* (Klinge) Soo, *D. maculata* subsp. *schurii* (Klinge) Soo) - 1
x **stenostachys** (J. Murr) Rauschert (*Orchis* x *stenostachys* J. Murr, *Dactylorhiza* x *thellungiana* (Br.-Bl.) Soo, *Orchis* x *thellungiana* Br.-Bl.). - D. incarnata (L.) Soo x D. traunsteineri (Saut.) Soo
strictifolia (Opiz) Rauschert = D. incarnata
sudetica (Poch ex Reichenb.fil.) Aver. (*Orchis maculata* L. var. *sudetica* Poch ex Reichenb. fil., *Dactylorchis maculata* (L.) Vermeulen subsp. *elodes* (Griseb.) Vermeulen var. *sudetica* (Poch ex Reichenb. fil.) Vermeulen, *Dactylorhiza longebracteata* (F.W. Schmidt) Holub subsp. *sudetica* (Poch ex Reichenb. fil.) Holub, *D. maculata* (L.) Soo subsp. *elodes* (Griseb.) Soo var. *sudetica* (Poch ex Reichenb. fil.) Soo, *D. maculata* subsp. *sudetica* (Poch ex Reichenb. fil.) Voth) - 1, 3
x *thellungiana* (Br.-Bl.) Soo = D. x stenostachys
x **transiens** (Druce) Soo. - D. fuchsii (Druce) Soo x D. maculata (L.) Soo
transsilvanica (Schur) Aver. (*Orchis transsilvanica* Schur, *Dactylorchis maculata* (L.) Vermeulen subsp. *transsilvanica* (Schur) Vermeulen, *Dactylorhiza fuchsii* (Druce) Soo subsp. *transsilvanica* (Schur) Frohner, *D. maculata* (L.) Soo subsp. *transsilvanica* (Schur) Soo, *Orchis maculata* L. subsp. *transsilvanica* (Schur) Domin) - 1
traunsteineri (Saut.) Soo (*Orchis traunsteineri* Saut., *Dactylorchis traunsteineri* (Saut.) Vermeulen, *Dactylorhiza majalis* (Reichenb.) P.F. Hunt & Summerhayes subsp. *traunsteineri* (Saut.) H. Sundermann) - 1
- subsp. *curvifolia* (Nyl.) Soo = D. curvifolia
- subsp. *lapponica* (Laest.) Soo = D. lapponica
- subsp. *russowii* (Klinge) Soo = D. russowii
triphylla (C. Koch) Czer. = D. amblyoloba
umbrosa (Kar. & Kir.) Nevski (*Orchis umbrosa* Kar. & Kir., *Dactylorchis umbrosa* (Kar. & Kir.) Wendelbo, *Dactylorhiza incarnata* (L.) Soo subsp. *turcestanica* (Klinge) H. Sundermann, *Orchis orientalis* Klinge subsp. *turcestanica* Klinge, *O. turcestanica* (Klinge) O. Fedtsch.) - 3, 4, 6
- subsp. *knorringiana* (Kraenzl.) Soo = D. kotschyi
- subsp. *magna* (Czerniak.) Soo = D. magna
- var. *knorringiana* (Kraenzl.) Soo = D. kotschyi
urvilleana (Steud.) H. Baumann & Kunkele (*Orchis urvilleana* Steud., *Dactylorchis lancibracteata* (C. Koch) Soo, nom. invalid., *D. maculata* (L.) Vermeulen subsp. *cartaliniae* (Klinge) Soo, p.p. nom. invalid., *D. maculata* subsp. *lancibracteata* (C. Koch) Soo, nom. invalid., *Dactylorhiza lancibracteata* (C. Koch) Renz, *D. saccifera* (Brongn.) Soo subsp. *cartaliniae* (Klinge) Soo, p.p., *D. saccifera* subsp. *lancibracteata* (C. Koch) Soo, *Orchis basilica* L. subsp. *cartaliniae* Klinge, p.p., *O. cartaliniae* (Klinge) Lipsky, p.p., *O. lancibracteata* C. Koch, *O. triphylla* sensu Nevski) - 2

Dactylostalyx Reichenb. fil.
ringens Reichenb. fil. - 5

Eleorchis F. Maek. (*Arethusa* auct., *Bletilla* auct.)
japonica (A. Gray) F. Maek. (*Arethusa japonica* A. Gray, *Bletilla japonica* (A. Gray) Schlechter) - 5

Ephippianthus Reichenb. fil
sachalinensis Reichenb. fil. (*E. schmidtii* Reichenb. fil.) - 5
schmidtii Reichenb. fil. = E. sachalinensis

Epipactis Zinn
atropurpurea Rafin. = E. atrorubens
atrorubens (Hoffm. ex Bernh.) Bess. (*Serapias atrorubens* Hoffm. ex Bernh., *Epipactis atropurpurea* Rafin., *E. helleborine* (L.) Crantz a. *E. rubiginosa* Crantz, *E. rubiginosa* (Crantz) Gaudin ex Koch, *Serapias latifolia* (L.) Huds. * *S. atrorubens* Hoffm. nom. invalid.) - 1, 2, 3, 6
convallarioides Sw. = Listera convallarioides
helleborine (L.) Crantz (*Serapias helleborine* L. p.p. quoad var.*latifolia* L., *Epipactis latifolia* (L.) All.) - 1, 2, 3, 4, 6
- a. *E. rubiginosa* Crantz = E. atrorubens
- subsp. **transcaucasica** A. Khokhr. - 2
latifolia (L.) All. = E. helleborine
- var. *violacea* Dur. Duq. = E. purpurata
microphylla (Ehrh.) Sw. - 1, 2
palustris (L.) Crantz - 1, 2, 3, 4, 6
papillosa Franch. & Savat. - 5
persica (Soo) Nannf. - 2
purpurata Smith (*E. latifolia* (L.) All. var. *violacea* Dur. Duq., *E. sessilifolia* Peterm., *E. violacea* (Dur. Duq.) Boreau) - 1
royleana Lindl. - 6
rubiginosa (Crantz) Gaudin ex Koch = E. atrorubens
thunbergii A. Gray - 5
turcomanica K. Pop. & Neshataeva - 6
veratrifolia Boiss. & Hohen. - 2, 6
violacea (Dur. Duq.) Boreau = E. purpurata

Epipogium J.G. Gmel. ex Borkh. (*Epipogon* Patze & al.)
aphyllum Sw. (*Orchis aphylla* F.W. Schmidt, 1797, non Forssk. 1775) - 1, 2, 3, 4, 5

Epipogon Patze & al. = Epipogium

Eulophia R. Br. ex Lindl.
turkestanica (Litv.) Schlechter - 6

Galearis Rafin. (*Galeorchis* Rydb.)
cyclochila (Franch. & Savat.) Soo (*Galeorchis cyclochila* (Franch. & Savat.) Nevski, *Orchis cyclochila* (Franch. & Savat.) Maxim.) - 5

Galeorchis Rydb. = Galearis
cyclochila (Franch. & Savat.) Nevski = Galearis cyclochila

Gastrodia R. Br.
elata Blume - 5

Goodyera R. Br.

maximowicziana Makino - 5
repens (L.) R. Br. - 1, 2, 3, 4, 5, 6
schlechtendaliana Reichenb. fil. - 5

Gymnadenia R. Br.

albida (L.) Rich. subsp. *straminea* (Fern.) B. Ljtnant = Leucorchis straminea
alpina (Turcz. ex Reichenb. fil.) Czer. (*G. conopsea* (L.) R. Br. var. *alpina* Turcz. ex Reichenb. fil., *G. conopsea* subsp. *alpina* (Turcz. ex Reichenb. fil.) Janch. ex Soo) - 1
camtschatica (Cham. & Schlecht.) Miyabe & Kudo = Platanthera camtschatica
conopsea (L.) R. Br. - 1, 2, 3, 4, 5
- subsp. *alpina* (Turcz. ex Reichenb. fil.) Janch. ex Soo = G. alpina
- subsp. *densiflora* (Wahlenb.) K. Richt. = G. densiflora
- var. *alpina* Turcz. ex Reichenb. fil. = G. alpina
cucullata (L.) Rich. = Neottianthe cucullata
densiflora (Wahlenb.) A. Dietr. (*Orchis densiflora* Wahlenb., *Gymnadenia conopsea* (L.) R. Br. subsp. *densiflora* (Wahlenb.) K. Richt.) - 1, 2
x **intermedia** Peterm. - G. conopsea (L.) R. Br. x G. odoratissima (L.) Rich. - 1
kinoshitae Makino = Amitostigma kinoshitae
odoratissima (L.) Rich. - 1
pauciflora Lindl. = Ponerorchis pauciflora

Habenaria Willd.

dianthoides Nevski = H. radiata
linearifolia Maxim. - 5
radiata (Thunb.) Spreng. (*H. dianthoides* Nevski) - 5
straminea Fern. = Leucorchis straminea
yezoensis Hara - 5

x *Haberiorchis* Rolfe = x Dactyloglossum

Hammarbya O. Kuntze (*Malaxis* sensu Nevski)

paludosa (L.) O. Kuntze (*Malaxis paludosa* (L.) Sw.) - 1, 3, 4, 5

Herminium Hill

monorchis (L.) R. Br. - 1, 2, 3, 4, 5, 6

Himantoglossum Koch

caprinum (Bieb.) C. Koch (*H. hircinum* (L.) Koch subsp. *caprinum* (Bieb.) K. Richt., *H. hircinum* subsp. *caprinum* (Bieb.) H. Sundermann, comb. superfl.) - 1, 2
formosum (Stev.) C. Koch - 2
hircinum (L.) Koch subsp. *caprinum* (Bieb.) K. Richt. = H. caprinum
- subsp. *caprinum* (Bieb.) H. Sundermann = H. caprinum

Holopogon Kom. & Nevski = Neottia

ussuriensis Kom. & Nevski = Neottia ussuriensis

Hyacinthorchis auct. = Clemastra

variabilis Blume = Cremastra variabilis

Leucorchis E. Mey. (*Pseudorchis* Seguier, nom. specif. uninom.)

albida (L.) E. Mey. (*Pseudorchis albida* (L.) A. & D. Löve) - 1, 3
- subsp. *straminea* (Fern.) A. Löve = L. straminea
straminea (Fern.) A. Löve (*Habenaria straminea* Fern., *Gymnadenia albida* (L.) Rich. subsp. *straminea* (Fern.) B. Ljtnant, *Leucorchis albida* (L.) E. Mey. subsp. *straminea* (Fern.) A. Löve, *Pseudorchis albida* (L.) A. & D. Löve subsp. *straminea* (Fern.) A. & D. Löve) - 1

Limnorchis Rydb. = Platanthera

convallariifolia (Fisch. ex Lindl.) Rydb. = Platanthera convallariifolia
dilatata (Pursh) Rydb. = Platanthera dilatata
hologlottis (Maxim.) Nevski = Platanthera hologlottis

Limodorum Boehm.

abortivum (L.) Sw. - 1, 2

Liparis Rich.

auriculata auct. = L. kumokuri
japonica (Miq.) Maxim. - 5
krameri Franch. & Savat. - 5
kumokuri F. Maek. (*L. auriculata* auct.) - 5
loeselii (L.) Rich. - 1, 3, 6
makinoana Schlechter - 5
sachalinensis Nakai - 5

Listera R. Br.

brevidens Nevski = L. nipponica
convallarioides (Sw.) Torr. (*Epipactis convallarioides* Sw.) - 5
cordata (L.) R. Br. - 1, 2, 3, 4, 5
- var. *nipponica* (Makino) Hiroe = L. nipponica
major Nakai = L. pinetorum
nipponica Makino (*L. brevidens* Nevski, *L. cordata* (L.) R. Br. var. *nipponica* (Makino) Hiroe) - 5
ovata (L.) R. Br. - 1, 2, 3, 4, 6
pinetorum Lindl. (*L. major* Nakai, *L. savatieri* Maxim. ex Kom., *L. yatabei* Makino) - 4, 5
savatieri Maxim. ex Kom. = L. pinetorum
yatabei Makino = L. pinetorum

Lysiella Rydb.

nevskii Aver. - 6
obtusata (Pursh) Britt & Rydb. subsp. *oligantha* (Turcz.) Tolm. = L. oligantha
oligantha (Turcz.) Nevski (*L. obtusata* (Pursh) Britt. & Rydb. subsp. *oligantha* (Turcz.) Tolm., *Platanthera obtusata* (Pursh) Lindl. subsp. *oligantha* (Turcz.) Hult., *P. oligantha* Turcz.) - 4, 5

Malaxis sensu Nevski = Hammarbya

paludosa (L.) Sw. = Hammarbya paludosa

Malaxis Soland. ex Sw. (*Microstylis* auct.)

monophyllos (L.) Sw. (*Microstylis monophyllos* (L.) Lindl.) - 1, 3, 4, 5

Microstylis auct. = Malaxis

monophyllos (L.) Lindl. = Malaxis monophyllos

Myrmechis Blume (*Rhamphidia* auct.)

japonica (Reichenb. fil.) Rolfe (*Rhamphidia japonica* Reichenb. fil.) - 5

Neolindleya Kraenzl. = Platanthera

camtschatica (Cham. & Schlecht.) Nevski = Platanthera camtschatica

Neottia Guett. (*Holopogon* Kom. & Nevski)

asiatica Ohwi - 5
camtschatea (L.) Reichenb. fil. - 3, 4, 5
nidus-avis (L.) Rich. - 1, 2, 3
papilligera Schlechter - 5
ussuriensis (Kom. & Nevski) Soo (*Holopogon ussuriensis* Kom. & Nevski) - 5

Neottianthe (Reichenb.) Schlechter

cucullata (L.) Schlechter (*Gymnadenia cucullata* (L.) Rich.) - 1, 3, 4, 5

Nigritella Rich. (*Satyrium* L. p.p.)

nigra (L.) Reichenb. (*Satyrium nigrum* L.) - 1(?)

Ophrys L.

anthropophora L. = Aceras anthropophorum
apifera Huds. - 1, 2
aranifera Huds. subsp. *transhyrcana* (Czerniak.) Soo = O. transhyrcana
caucasica Woronow ex Grossh. (*O. mammosa* Desf. subsp. *caucasica* (Woronow ex Grossh.) Soo, *O. sphegodes* Mill. subsp. *caucasica* (Woronow ex Grossh.) Soo) - 2
cornuta Stev. (*O. oestrifera* Bieb. subsp. *cornuta* (Stev.) K. Richt., *O. oestrifera* subsp. *cornuta* (Stev.) Soo, comb. superfl., *O. scolopax* Cav. subsp. *cornuta* (Stev.) E.G. Camus, *O. scolopax* auct. p.p.) - 1, 2
insectifera L. (*O. muscifera* Huds.) - 1
kopetdaghensis K. Pop. & Neshataeva - 6
    Probably, the priority name of this species is O. turcomanica Renz (described from Iran).
mammosa Desf. subsp. *caucasica* (Woronow ex Grossh.) Soo = O. caucasica
- subsp. *taurica* (Agg.) Soo = O. taurica
- subsp. *transhyrcana* (Czerniak.) Buttler = O. transhyrcana
muscifera Huds. = O. insectifera
oestrifera Bieb. (*O. scolopax* Cav. subsp. *oestrifera* (Bieb.) Soo, *O. scolopax* auct. p.p.) - 1, 2
- subsp. *cornuta* (Stev.) K. Richt. = O. cornuta
scolopax auct. p.p. = O. cornuta and O. oestrifera
scolopax Cav. subsp. *cornuta* (Stev.) E.G. Camus = O. cornuta
- subsp. *oestrifera* (Bieb.) Soo = O. oestrifera
sphegodes Mill. subsp. *caucasica* (Woronow ex Grossh.) Soo = O. caucasica
- subsp. *taurica* (Agg.) Soo = O. taurica
- subsp. *transhyrcana* (Czerniak.) Soo = O. transhyrcana
spiralis L. = Spiranthes spiralis
taurica (Agg.) Nevski (*O. mammosa* Desf. subsp. *taurica* (Agg.) Soo, *O. sphegodes* Mill. subsp. *taurica* (Agg.) Soo) - 1, 2
transhyrcana Czerniak. (*O. aranifera* Huds. subsp. *transhyr-* *cana* (Czerniak.) Soo, *O. mammosa* Desf. subsp. *transhyrcana* (Czerniak.) Buttler, *O. sphegodes* subsp. *transhyrcana* (Czerniak.) Soo) - 6

x *Orchicoeloglossum* Aschers. & Graebn. = x Dactyloglossum

turcestanicum G. Keller & Soo = x Dactyloglossum turcestanicum

Orchis L. (*Vermeulenia* A. & D. Löve)

affinis C.Koch = Dactylorhiza affinis
alpestris Pugsl. = Dactylorhiza alpestris
x *ambigua* A. Kerner = Dactylorhiza x ambigua
amblyoloba Nevski = Dactylorhiza amblyoloba
x **angusticruris** Franch. - O. purpurea Huds. x O. simia Lam. - 1
angustifolia Krock. var. *lapponica* Laest. = Dactylorhiza lapponica
angustifolia Loes. ex Reichenb. subsp. *russowii* (Klinge) Klinge = Dactylorhiza russowii
- var. *recurva* Klinge f. *schurii* Klinge = Dactylorhiza schurii
- var. *russowii* Klinge = Dactylorhiza russowii
aphylla F.W. Schmidt = Epipogium aphyllum
aristata Fisch. ex Lindl. = Dactylorhiza aristata
aschersoniana Hausskn. = Dactylorhiza x aschersoniana
baltica (Klinge) Nevski = Dactylorhiza longifolia
basilica L. subsp. *cartaliniae* Klinge, p.p. = Dactylorhiza amblyoloba and D. urvilleana
beyrichii A. Kerner = O. stevenii
braunii Halacsy = Dactylorhiza x braunii
cartaliniae (Klinge) Lipsky, p.p. = Dactylorhiza amblyoloba and D. urvilleana
caspia Trautv. (*Vermeulenia caspia* (Trautv.) A. & D. Löve) - 2
cassidea Bieb. = O. fragrans
cataonica Fleischm. = Dactylorhiza cataonica
caucasica (Klinge) Lipsky = Dactylorhiza euxina
- var. *markowitschii* Soo = Dactylorhiza markowitschii
chlorotica Woronow (*Vermeulenia chlorotica* (Woronow) A. & D. Löve) - 2
collina auct. p.p. = O. fedtschenkoi
comperiana Stev. = Comperia comperiana
cordigera Fries = Dactylorhiza cordigera
coriophora L. - 1, 2
- subsp. *fragrans* (Pollini) E.G. Camus = O. fragrans
- subsp. *fragrans* (Pollini) K. Richt. = O. fragrans
- subsp. *fragrans* (Pollini) Sudre = O. fragrans
- subsp. *nervulosa* (Sakalo) Soo = O. nervulosa
- var. *cassidea* (Bieb.) Nevski = O. fragrans
- var. *fragrans* (Pollini) Boiss. = O. fragrans
- var. *nervulosa* (Sakalo) Bordz. = O. nervulosa
cruenta O.F. Muell. = Dactylorhiza cruenta
curvifolia Nyl. = Dactylorhiza curvifolia
cyclochila (Franch. & Savat.) Maxim. = Galearis cyclochila
x **darcisii** Murr. - O. fragrans Pollini x O. picta Loisel.
densiflora Wahlenb. = Gymnadenia densiflora
dufftiana M. Schulze = Dactylorhiza x dufftiana
elegans Heuff. = O. palustris
elodes Griseb. = Dactylorhiza elodes
ericetorum (Linton) A. Benn. = Dactylorhiza ericetorum
euxina Nevski = Dactylorhiza euxina
fedtschenkoi Czerniak. (*O. saccata* Ten. var. *fedtschenkoi* (Czerniak.) L. Hautzinger, *Vermeulenia fedtschenkoi* (Czerniak.) A. & D. Löve, *Orchis collina* auct. p.p.) - 6
fistulosa Moench = Dactylorhiza majalis
flavescens C. Koch = Dactylorhiza flavescens

fragrans Pollini (*O. cassidea* Bieb., *O. coriophora* L. var. *cassidea* (Bieb.) Nevski, *O. coriophora* subsp. *fragrans* (Pollini) E.G. Camus, comb. superfl., *O. coriophora* subsp. *fragrans* (Pollini) K. Richt., *O. coriophora* subsp. *fragrans* (Pollini) Sudre, comb. superfl., *O. coriophora* var. *fragrans* (Pollini) Boiss.) - 1, 2

fuchsii Druce = Dactylorhiza fuchsii

- subsp. *hebridensis* (Wilmott) Clapham = Dactylorhiza hebridensis

fuscescens L. = Tulotis fuscescens

georgica (Klinge) Czerniak. = Dactylorhiza flavescens

georgica (Klinge) Lipsky = Dactylorhiza flavescens

hebridensis Wilmott = Dactylorhiza hebridensis

x **hybrida** Boenn. - O. militaris L. x O. purpurea Huds. - 1

iberica Bieb. ex Willd. = Dactylorhiza iberica

incarnata L. = Dactylorhiza incarnata

- subsp. *ochroleuca* (Wustn. ex Boll.) O. Schwarz = Dactylorhiza ochroleuca

- var. *knorringiana* Kraenzl. = Dactylorhiza kotschyi

- var. *kotschyi* Reichenb. fil. = Dactylorhiza kotschyi

- var. *ochroleuca* Wustn. ex Boll. = Dactylorhiza ochroleuca

x **jailae** Soo. - O. mascula (L.) L. subsp. signifera (Vest) Soo x O. provincialis Balb. ex DC. - 1

x **kelleriana** Ugr. - O. coriophora L. x O. palustris Jacq. - 1

kerneriorum Soo = Dactylorhiza x kerneriorum

x **kisslingii** G. Beck. - O. mascula (L.) L. subsp. signifera (Vest) Soo x O. pallens L. - 1

knorringiana (Kraenzl.) Czerniak. ex E. Nikit. = Dactylorhiza kotschyi

kotschyi (Reichenb. fil.) Schlechter = Dactylorhiza kotschyi

krylowii Soo = Dactylorhiza x krylowii

lancibracteata C. Koch = Dactylorhiza urvilleana

latifolia L. p.p. = Dactylorhiza majalis

- subsp. *alpestris* (Pugsl.) Janch. = Dactylorhiza alpestris

- subsp. *baltica* Klinge = Dactylorhiza longifolia

- var. *ochroleuca* (Wustn. ex Boll.) Pugsl. = Dactylorhiza ochroleuca

latifolia sensu Nevski = Dactylorhiza incarnata

laxiflora Lam. - 1, 2, 6

- subsp. *dielsiana* Soo = O. pseudolaxiflora

- subsp. *elegans* (Heuff.) Soo = O. palustris

- subsp. *palustris* (Jacq.) Aschers. & Grebn. = O. palustris

- subsp. *palustris* (Jacq.) Bonnier & Layens = O. palustris

lehmannii Klinge = Dactylorhiza x lehmannii

longifolia L. Neum. = Dactylorhiza longifolia

maculata L. = Dactylorhiza maculata

- subsp. *ericetorum* Linton = Dactylorhiza ericetorum

- subsp. *fuchsii* (Druce) Jorgens. = Dactylorhiza fuchsii

- subsp. *meyeri* (Reichenb. fil.) E.G. Camus & A. Camus = Dactylorhiza hebridensis

- subsp. *meyeri* (Reichenb. fil.) K.Richt. = Dactylorhiza hebridensis

- subsp. *transsilvanica* (Schur) Domin = Dactylorhiza transsilvanica

- var. *kolaensis* Montell = Dactylorhiza kolaensis

- var. *meyeri* Reichenb. fil. = Dactylorhiza hebridensis

- var. *psychrophila* Schlechter = Dactylorhiza psychrophila

- var. *sudetica* Poch ex Reichenb. fil. = Dactylorhiza sudetica

x *maculatiformis* Rouy = Dactylorhiza x ambigua

magna Czerniak. = Dactylorhiza magna

majalis Reichenb. = Dactylorhiza majalis

- subsp. *alpestris* (Pugsl.) Hellmayr = Dactylorhiza alpestris

mascula (L.) L. (*O. morio* L. var. *mascula* L.) - 1, 2

- subsp. **pinetorum** (Boiss. & Kotschy) E.G. Camus (*O. pinetorum* Boiss. & Kotschy) - 1, 2

- subsp. **signifera** (Vest) Soo (*O. signifera* Vest, *O. speciosa* Host) - 1

- subsp. **wanjkowii** (E. Wulf) Soo (*O. wanjkowii* E. Wulf) - 1

maxima C. Koch - 2

mediterranea Klinge subsp. *georgica* Klinge = Dactylorhiza flavescens

merovensis Grossh. = Dactylorhiza merovensis

militaris L. - 1, 2, 3, 4

monticola Klinge subsp. *caucasica* Klinge = Dactylorhiza euxina

morio L. - 1

- subsp. *picta* (Loisel.) Arcang. = O. picta

- subsp. *picta* (Loisel.) Aschers. & Graebn. = O. picta

- subsp. *picta* (Loisel.) K. Richt. = O. picta

- var. *mascula* L. = O. mascula

nervulosa Sakalo (*O. coriophora* L. subsp. *nervulosa* (Sakalo) Soo, *O. coriophora* var. *nervulosa* (Sakalo) Bordz.) - 1

orientalis Klinge subsp. *osmanica* Klinge = Dactylorhiza osmanica

- subsp. *turcestanica* Klinge = Dactylorhiza umbrosa

osmanica (Klinge) Lipsky = Dactylorhiza osmanica

pallens L. - 1, 2

palustris Jacq. (*O. elegans* Heuff., *O. laxiflora* Lam. subsp. *elegans* (Heuff.) Soo, *O. laxiflora* subsp. *palustris* (Jacq.) Aschers. & Graebn. comb. superfl., *O. laxiflora* subsp. *palustris* (Jacq.) Bonnier & Layens) - 1, 2, 6

patens Desf. subsp. *viridifusca* (Albov) Soo = O. viridifusca

x **permixta** Soo. - O. pallens L. x O. provincialis Balb. ex DC. x O. mascula (L.) L. subsp. signifera (Vest) Soo

picta Loisel. (*O. morio* L. subsp. *picta* (Loisel.) Arcang. comb. superfl., *O. morio* subsp. *picta* (Loisel.) Aschers. & Graebn. comb. superfl., *O. morio* subsp. *picta* (Loisel.) K. Richt.) - 1, 2

pinetorum Boiss. & Kotschy = O. mascula subsp. pinetorum

x **plessidiaca** Renz. - O. pallens L. x O. provincialis Balb. ex DC. - 1

provincialis Balb. ex DC. - 1, 2

pseudolaxiflora Czerniak. (*O. laxiflora* Lam. subsp. *dielsiana* Soo) - 6

punctulata Stev. ex Lindl. - 1, 2

- subsp. *schelkownikowii* (Woronow) Soo = O. schelkownikowii

- subsp. *stevenii* (Reichenb. fil.) H. Sundermann = O. stevenii

purpurea Huds. - 1, 2

x **reinhardii** Ugr. ex E.G. Camus. - O. coriophora L. x O. palustris Jacq. s.l.

romana Seb. = Dactylorhiza romana

russowii (Klinge) Schlechter = Dactylorhiza russowii

saccata Ten. var. *fedtschenkoi* (Czerniak.) L. Hautzinger = O. fedtschenkoi

salina Turcz. ex Lindl. = Dactylorhiza salina

sambucina L. = Dactylorhiza sambucina

samnaunensis Gsell = Dactylorhiza x samnaunensis

sanasunitensis Fleischm. = Dactylorhiza sanasunitensis

sanasunitensis sensu Nevski = Dactylorhiza salina

schelkownikowii Woronow (*O. punctulata* Stev. ex Lindl. subsp. *schelkownikowii* (Woronow) Soo) - 2

secunda (Nevski) Worosch. = Ponerorchis pauciflora

signifera Vest = O. mascula subsp. signifera

simia Lam. - 1, 2, 6

speciosa Host = O. mascula subsp. signifera

x *stenostachys* J. Murr = Dactylorhiza x stenostachys

stevenii Reichenb. fil. (*O. beyrichii* A. Kerner, *O. punctulata* Stev. ex Lindl. subsp. *stevenii* (Reichenb. fil.) H. Sundermann) - 1, 2

strictifolia Opiz = Dactylorhiza incarnata

- var. *ochroleuca* (Wustn. ex Boll.) Hyl. = Dactylorhiza ochroleuca
x **suckowii** Kumpel. - O. maxima C. Koch x O. punctulata Stev. ex Lindl. - 2
x *thellungiana* Br.-Bl. = Dactylorhiza x stenostachys
transsilvanica Schur = Dactylorhiza transsilvanica
traunsteineri Saut. = Dactylorhiza traunsteineri
tridentata Scop. - 1, 2
triphylla C. Koch = Dactylorhiza amblyoloba
triphylla sensu Nevski = Dactylorhiza urvilleana
turcestanica (Klinge) O. Fedtsch. = Dactylorhiza umbrosa
x **ugrinskyana** Soo. - O. fragrans Pollini x O. palustris Jacq. s.l. - 1
umbrosa Kar. & Kir. = Dactylorhiza umbrosa
urvilleana Steud. = Dactylorhiza urvilleana
ustulata L. - 1, 2, 3
viridifusca Albov (*O. patens* Desf. subsp. *viridifusca* (Albov) Soo) - 2
vomeracea Burm. fil. = Serapias vomeracea
wanjkowii E. Wulf = O. mascula subsp. wanjkowii
x **wulffiana** Soo. - O. punctulata Stev. ex Lindl. x O. purpurea Huds. - 1
yooiokiana Makino = Ponerorchis pauciflora

Oreorchis Lindl.

patens (Lindl.) Lindl. - 5

Perularia Lindl. = Tulotis

fuscescens (L.) Lindl. = Tulotis fuscescens
ussuriensis (Regel & Maack) Schlechter = Tulotis ussuriensis

Platanthera Rich. (*Limnorchis* Rydb., *Neolindleya* Kraenzl., *Pseudodiphryllum* Nevski)

bifolia (L.) Rich. - 1, 2, 3, 4
- subsp. *extremiorientalis* (Nevski) Soo = P. extremiorientalis
camtschatica (Cham. & Schlecht.) Makino (*Gymnadenia camtschatica* (Cham. & Schlecht.) Miyabe & Kudo, *Neolindleya camtschatica* (Cham. & Schlecht.) Nevski, *Platanthera camtschatica* (Cham. & Schlecht.) Soo, comb. superfl.) - 5
chlorantha (Cust.) Reichenb. - 1, 2
- subsp. *orientalis* (Schlechter) Soo = P. freynii
- var. *orientalis* Schlechter = P. freynii
chorisiana auct. = P. ditmariana
chorisiana (Cham.) Reichenb. fil. (*Pseudodiphryllum chorisianum* (Cham.) Nevski, p.p. quoad typum) - 5
convallariifolia Fisch. ex Lindl. (*Limnorchis convallariifolia* (Fisch. ex Lindl.) Rydb., *Platanthera dilatata* (Pursh) Lindl. ex G. Beck subsp. *convallariifolia* (Fisch. ex Lindl.) Soo, *P. hyperborea* auct.) - 5
cornu-bovis Nevski = P. maximowicziana
dilatata (Pursh) Lindl. ex G. Beck (*Limnorchis dilatata* (Pursh) Rydb.) - 5
- subsp. *convallariifolia* (Fisch. ex Lindl.) Soo = P. convallariifolia
ditmariana Kom. (*Pseudodiphryllum chorisianum* (Cham.) Nevski, p.p. quoad pl., *Plathanthera chorisiana* auct.) - 5
extremiorientalis Nevski (*P. bifolia* (L.) Rich. subsp. *extremiorientalis* (Nevski) Soo, *P. metabifolia* F. Maek.) - 5
freynii Kraenzl. (*P. chlorantha* (Cust.) Reichenb. var. *orientalis* Schlechter, *P. chlorantha* subsp. *orientalis* (Schlechter) Soo) - 4, 5

hologlottis Maxim. (*Limnorchis hologlottis* (Maxim.) Nevski) - 4, 5
hyperborea auct. = P. convallariifolia
mandarinorum Reichenb. fil. subsp. *maximowicziana* (Schlechter) K. Inoue = P. maximowicziana
- subsp. *ophrydioides* (Fr. Schmidt) K. Inoue = P. ophrydioides
- var. *cornu-bovis* (Nevski) K. Inoue = P. maximowicziana
- var. *maximowicziana* (Schlechter) Ohwi = P. maximowicziana
- f. *maximowicziana* (Schlechter) Hiroe = P. maximowicziana
maximowicziana Schlechter (*P. cornu-bovis* Nevski, *P. mandarinorum* Reichenb. fil. var. *cornu-bovis* (Nevski) K. Inoue, *P. mandarinorum* f. *maximowicziana* (Schlechter) Hiroe, *P. mandarinorum* subsp. *maximowicziana* (Schlechter) K. Inoue, *P. mandarinorum* var. *maximowicziana* (Schlecter) Ohwi) - 5
metabifolia F. Maek. = P. extremiorientalis
obtusata (Pursh) Lindl. subsp. *oligantha* (Turcz.) Hult. = Lysiella oligantha
oligantha Turcz. = Lysiella oligantha
ophrydioides Fr. Schmidt (*P. mandarinorum* Reichenb. fil. subsp. *ophrydioides* (Fr. Schmidt) K. Inoue) - 5
sachalinensis Fr. Schmidt - 5
tipuloides (L. fil.) Lindl. - 4, 5
- var. *ussuriensis* Regel & Maack = Tulotis ussuriensis

Pogonia Juss.

japonica Reichenb. fil. - 5

Ponerorchis Reichenb. fil. (*Chusua* Nevski)

pauciflora (Lindl.) Ohwi (*Gymnadenia pauciflora* Lindl., *Chusua pauciflora* (Lindl.) P.F. Hunt, *Ch. secunda* Nevski, *Orchis secunda* (Nevski) Worosch., *O. yooiokiana* Makino) - 4, 5

Pseudodiphryllum Nevski = Plananthera

chorisianum (Cham.) Nevski = Platanthera chorisiana and P. ditmariana

Pseudorchis Seguier = Leucorchis

albida (L.) A. & D. Löve = Leucorchis albida
- subsp. *straminea* (Fern.) A. & D. Löve = Leucorchis straminea

Rhamphidia auct. = Myrmechis

japonica Reichenb. fil. = Myrmechis japonica

Satyrium L. p.p. = Nigritella

nigrum L. = Nigritella nigra

Serapias L.

atrorubens Hoffm. ex Bernh. = Epipactis atrorubens
cordigera subsp. *vomeracea* (Burm. fil.) H. Sundermann = S. vomeracea
damasonium Mill. = Cephalanthera damasonium
grandiflora L. = Cephalanthera damasonium
helleborine L. = Epipactis helleborine
latifolia (L.) Huds. *S. atrorubens* Hoffm. = Epipactis atrorubens
lonchophyllum L. fil. = Cephalanthera damasonium

vomeracea (Burm. fil.) Briq. (*Orchis vomeracea* Burm. fil., *Serapias cordigera* subsp. *vomeracea* (Burm. fil.) H. Sundermann) - 2
- subsp. **orientalis** Greuter - 2

Spiranthes Rich.

amoena (Bieb.) Spreng. (*S. sinensis* (Pers.) Ames var. *amoena* (Bieb.) Hara, *S. australis* auct.) - 1, 3, 4, 5
australis auct. = S. amoena
autumnalis (Balb.) Rich. = S. spiralis
sinensis (Pers.) Ames - 5
- var. *amoena* (Bieb.) Hara = S. amoena
spiralis (L.) Chevall. (*Ophrys spiralis* L. p.p., *Spiranthes autumnalis* (Balb.) Rich.) - 1, 2

Steveniella Schlechter

satyrioides (Stev.) Schlechter - 1, 2

Traunsteinera Reichenb.

globosa (L.) Reichenb. - 1, 2
- subsp. *sphaerica* (Bieb.) Soo = T. sphaerica
sphaerica (Bieb.) Schlechter (*T. globosa* (L.) Reichenb. subsp. *sphaerica* (Bieb.) Soo) - 2

Tulotis Rafin. (*Perularia* Lindl.)

asiatica Hara = T. fuscescens
fuscescens (L.) Czer. (*Orchis fuscescens* L., *Perularia fuscescens* (L.) Lindl., *Tulotis asiatica* Hara, nom. illegit.) - 3, 4, 5
ussuriensis (Regel & Maack) Hara (*Platanthera tipuloides* (L. fil.) Lindl. var. *ussuriensis* Regel & Maack, *Perularia ussuriensis* (Regel & Maack) Schlechter) - 5

Vermeulenia A. & D. Löve = Orchis

caspia (Trautv.) A. & D. Löve = Orchis caspia
chlorotica (Woronow) A. & D. Löve = Orchis chlorotica
fedtschenkoi (Czerniak.) A. & D. Löve = Orchis fedtschenkoi

Zeuxine Lindl.

strateumatica (L.) Schlechter - 6

OROBANCHACEAE Vent

Anoplon Reichenb. p.p. = Diphelypaea

coccineum (Bieb.) H. Riedl & Schiman-Czeika = Diphelypaea coccinea

Boschniakia C.A. Mey.

rossica (Cham. & Schlecht.) B. Fedtsch. - 1, 3, 4, 5

Cistanche Hoffmgg. & Link

ambigua (Bunge) G. Beck - 2, 6
fissa (C.A. Mey.) G. Beck - 2, 6
flava (C.A. Mey.) Korsh. - 2, 6
jodostoma Butk. & Vved. - 6
mongolica G. Beck (*C. tubulosa* auct.) - 6
ridgewayana Aitch. & Hemsl. - 6
salsa (C.A. Mey.) G. Beck - 1, 2, 6
speciosa Butk. - 6
stenostachya Butk. - 6

trivalvis (Trautv.) Korsh. - 6
tubulosa auct. = C. mongolica

Diphelypaea Nicolson (*Phelypaea* L. 1758, non *Phaelypea* P. Br. 1756, *Anoplon* Reichenb. p.p. nom. illegit)

coccinea (Bieb.) Nicolson (*Anoplon coccineum* (Bieb.) H. Riedl & Schiman-Czeika, *Phelypaea coccinea* (Bieb.) Poir.) - 1, 2
helenae (Popl.) Tzvel. (*Phelypaea helenae* Popl.) - 1
tournefortii (Desf.) Nicolson (*Phelypaea tournefortii* Desf.) - 2

Mannagettaea H. Smith

hummelii H. Smith (*M. ircutensis* M. Pop.) - 4
ircutensis M. Pop. = M. hummelii

Orobanche L.

aegyptiaca Pers. = Phelipanche aegyptiaca
alba Steph. - 1, 2, 6
- var. *bidentata* G. Beck f. *ingens* G. Beck = O. ingens
alsatica Kirschl. - 1, 2, 3, 4, 6
- subsp. *libanotidis* (Rupr.) Tzvel. = O. bartlingii
amethystea Thuill. - 2
amoena C.A. Mey. - 1, 3, 6
amurensis (G. Beck) Kom. - 5
anatolica auct. = O. colorata
androssovii Novopokr. = Phelipanche androssovii
arenaria Borkh. = Phelipanche laevis
ariana Gontsch. - 6
■ **armena** Tzvel.
artemisiae-campestris Vaucher ex Gaudin (*O. loricata* Reichenb.) - 2(?)
badchysensis Novopokr. & V.V. Nikit. - 6
bartlingii Griseb. (*O. alsatica* Kirschl. subsp. *libanotidis* (Rupr.) Tzvel., *O. libanotidis* Rupr.) - 1, 2, 3, 4, 6
borissovae Novopokr. = Phelipanche borissovae
brachypoda Novopokr. = Phelipanche brachypoda
brassicae (Novopokr.) Novopokr. = Phelipanche brassicae
brevidens Novopokr. - 6
bungeana G. Beck = Phelipanche bungeana
caesia Reichenb. = Phelipanche lanuginosa
camptolepis Boiss. & Reut. - 6
caryophyllacea Smith (*O. vulgaris* Poir.) - 1, 2, 6
cernua Loefl. - 1, 2, 6
- subsp. *cumana* (Wallr.) Soo = O. cumana
cilicica G. Beck = Phelipanche cilicica
clarkei Hook. fil. - 6
coelestis (Reut.) Boiss. & Reut. ex G. Beck = Phelipanche coelestis
coerulescens Steph. (*O. coerulescens* f. *korshinskyi* (Novopokr.) Y.C. Ma, *O. korshinskyi* Novopokr.) - 1, 2, 3, 4, 5, 6
- f. *korshinskyi* (Novopokr.) Y.C. Ma = O. coerulescens
colorata C. Koch (*O. anatolica* auct.) - 2
connata C. Koch - 2
crenata Forssk. (*O. owerinii* (G. Beck) G. Beck) - 1, 2
cumana Wallr. (*O. cernua* Loefl. subsp. *cumana* (Wallr.) Soo, *O. sarmatica* Kotov) - 1, 2, 3, 4, 6
dalmatica (G. Beck) Tzvel. = Phelipanche dalmatica
elatior Sutt. (*O. major* L. p.p. nom. ambig.) - 1, 2, 3, 6
flava C. Mart. ex F. Schultz - 1(?), 2
gamosepala Reut. - 2
gigantea (G. Beck) Gontsch. - 6

glabricaulis Tzvel. - 2
glaucantha Trautv. - 4
gracilis Smith - 1, 2
grigorjevii Novopokr. - 6
grossheimii Novopokr. - 2
hansii A. Kerner - 6
hederae Duby - 1, 2
heldreichii (Reut.) G. Beck (*Phelypaea heldreichii* Reut.) - 2
hians Stev. = O. lutea
hirtiflora (Reut.) Burkill = Phelipanche hirtiflora
hirtiflora (Reut.) Tzvel. = Phelipanche hirtiflora
hohenackeri (Reut.) Tzvel. = Phelipanche hohenackeri
hymenocalyx Reut. - 2
iberica (G. Beck) Tzvel. = Phelipanche iberica
ingens (G. Beck) Tzvel. (*O. alba* Steph. var. *bidentata* G. Beck f. *ingens* G. Beck) - 2
inulae Novopokr. & Abramov - 2
karatavica Pavl. = Phelipanche karatavica
kelleri Novopokr. = Phelipanche kelleri
korshinskyi Novopokr. = O. coerulescens
kotschyi auct. = O. spectabilis
kotschyi Reut. (*O. solmsii* auct.) - 6
krylowii G. Beck - 1, 3, 4, 6
kurdica Boiss. & Hausskn. - 2
laevis L. = Phelipanche laevis
lanuginosa (C.A. Mey.) Greuter & Burdet = Phelipanche lanuginosa
libanotidis Rupr. = O. bartlingii
linczevskyi Novopokr. = O. lutea
longibracteata Schiman-Czeika - 2
loricata Reichenb. = O. artemisiae-campestris
lucorum A. Br. - 1
lutea Baumg. (*O. hians* Stev., *O. linczevskyi* Novopokr.) - 1, 2, 6
major L. p.p. = O. elatior
minor Smith - 1, 2
mutelii F. Schultz = Phelipanche mutelii
nana (Reut.) Noe ex G. Beck = Phelipanche oxyloba
orientalis G. Beck = Phelipanche orientalis
owerinii (G. Beck) G. Beck = O. crenata
oxyloba (Reut.) G. Beck = Phelipanche oxyloba
pachypoda Butk. - 6
pallidiflora Wimm. & Grab. (*O. reticulata* Wallr. subsp. *pallidiflora* (Wimm. & Grab.) Hayek, *O. reticulata* subsp. *pallidiflora* (Wimm. & Grab.) Tzvel. comb. superfl.) - 1, 2, 3
picridis F. Schultz - 1, 2
pogonanthera Reut. - 6
pubescens D'Urv. (*O. versicolor* F. Schultz) - 1, 2
pulchella (C.A. Mey.) Novopokr. = Phelipanche pulchella
purpurea Jacq. = Phelipanche purpurea
pycnostachya Hance - 4, 5
raddeana G. Beck - 2
ramosa L. = Phelipanche ramosa
- subsp. *mutelii* (F. Schultz) Arcang. = Phelipanche mutelii
- subsp. *nana* (Reut.) Coutinho = Phelipanche oxyloba
reticulata Wallr. - 1, 2
- subsp. *pallidiflora* (Wimm. & Grab.) Hayek = O. pallidiflora
- subsp. *pallidiflora* (Wimm. & Grab.) Tzvel. = O. pallidiflora
rosea Tzvel. - 2, 6
sarmatica Kotov = O. cumana
schelkovnikovii Tzvel. - 2
septemloba (G. Beck) Tzvel. = Phelipanche septemloba
serratocalyx G. Beck = Phelipanche serratocalyx

■ sintenisii G. Beck
sogdiana Novopokr. = Phelipanche sogdiana
solenanthi Novopokr. & Pissjauk. - 6
solmsii auct. = O. kotschyi
sordida C.A. Mey. - 3, 6
spectabilis Reut. (*O. kotschyi* auct. fl. URSS) - 6
subalpina Herb. - 1
sulphurea Gontsch. - 6
teucrii Holandre - 1
transcaucasica Tzvel. - 2
uralensis G. Beck = Phelipanche uralensis
versicolor F. Schultz = O. pubescens
vitellina Novopokr. - 2
vulgaris Poir. = O. caryophyllacea

Phacellanthus Siebold & Zucc.
tubiflorus Siebold & Zucc. - 5

Phelipanche Pomel
aegyptiaca (Pers.) Pomel (*Orobanche aegyptiaca* Pers.) - 1, 2, 6
androssovii (Novopokr.) Sojak (*Orobanche androssovii* Novopokr.) - 6
arenaria (Borkh.) Pomel = Ph. laevis
borissovae (Novopokr.) Sojak (*Orobanche borissovae* Novopokr.) - 6
brachypoda (Novopokr.) Sojak (*Orobanche brachypoda* Novopokr.) - 6
brassicae (Novopokr.) Sojak (*Orobanche brassicae* (Novopokr.) Novopokr.) - 1
bungeana (G. Beck) Sojak (*Orobanche bungeana* G. Beck) - 2
caesia (Reichenb.) Sojak = Ph. lanuginosa
cilicica (G. Beck) Sojak (*Orobanche cilicica* G. Beck) - 2
coelestis (Reut.) Sojak (*Orobanche coelestis* (Reut.) Boiss. & Reut. ex G. Beck) - 2, 6
dalmatica (G. Beck) Sojak (*Orobanche dalmatica* (G. Beck) Tzvel.) - 2
hirtiflora (Reut.) Sojak (*Orobanche hirtiflora* (Reut.) Burkill, *O. hirtiflora* (Reut.) Tzvel. comb. superfl.) - 2
hohenackeri (Reut.) Sojak (*Orobanche hohenackeri* (Reut.) Tzvel.) - 2
iberica (G. Beck) Sojak (*Orobanche iberica* (G. Beck) Tzvel.) - 2
karatavica (Pavl.) Sojak (*Orobanche karatavica* Pavl.) - 6
kelleri (Novopokr.) Sojak (*Orobanche kelleri* Novopokr.) - 2, 3, 6
laevis (L.) Holub (*Orobanche laevis* L., *O. arenaria* Borkh., *Phelipanche arenaria* (Borkh.) Pomel) - 1, 2, 3, 6
lanuginosa (C.A.Mey.) Holub (*Phelypaea lanuginosa* C.A. Mey., *Orobanche caesia* Reichenb., *O. lanuginosa* (C.A. Mey.) Greuter & Burdet, *Phelipanche caesia* (Reichenb.) Sojak) - 1, 2, 3, 4, 6
mutelii (F. Schultz) Czer. comb. nova (*Orobanche mutelii* F. Schultz, 1835, in Mutel, Fl. Fr. 2 : 353; *O. ramosa* L. subsp. *mutelii* (F. Schultz) Arcang.) - 1, 2
nana (Reut.) Sojak = Ph. oxyloba
nikitae Teryokhin - 6
orientalis (G. Beck) Sojak (*Orobanche orientalis* G. Beck) - 2, 6
oxyloba (Reut.) Sojak (*Orobanche nana* (Reut.) Noe ex G. Beck, *O. oxyloba* (Reut.) G. Beck, *O. ramosa* L. subsp. *nana* (Reut.) Coutinho, *Phelipanche nana* (Reut.) Sojak, *Phelypaea nana* (Reut.) Reichenb. fil.) - 1, 2, 6
pulchella (C.A. Mey.) Sojak (*Orobanche pulchella* (C.A. Mey.) Novopokr.) - 2

purpurea (Jacq.) Sojak (*Orobanche purpurea* Jacq.) - 1, 2
ramosa (L.) Pomel (*Orobanche ramosa* L.) - 1, 2
septemloba (G. Beck) Sojak (*Orobanche septemloba* (G. Beck) Tzvel.) - 2
■ *serratocalyx* (G. Beck) Sojak (*Orobanche serratocalyx* G. Beck)
sogdiana (Novopokr.) Sojak (*Orobanche sogdiana* Novopokr.) - 6
uralensis (G. Beck) Czer. comb. nova (*Orobanche uralensis* G. Beck, 1890, Monogr. Orob. :132) - 1, 2, 6

Phelypaea L. = Diphelypaea
coccinea (Bieb.) Poir. = Diphelypaea coccinea
heldreichii Reut. = Orobanche heldreichii
helenae Popl. = Diphelypaea helenae
lanuginosa C.A. Mey. = Phelipanche lanuginosa
nana (Reut.) Reichenb. fil. = Phelipanche oxyloba
tournefortii Desf. = Diphelypaea tournefortii

OSMUNDACEAE Bercht. & J. Presl
Osmunda L.
asiatica (Fern.) Ohwi = Osmundastrum asiaticum
cinnamomea auct. = Osmundastrum asiaticum
cinnamomea L. var. *asiatica* Fern. = Osmundastrum asiaticum
claytoniana L. = Osmundastrum claytonianum
japonica Thunb. (*O. regalis* L. subsp. *japonica* (Thunb.) A. & D. Löve) - 5
lanceolata S.G. Gmel. = Botrychium lanceolatum (Botrychiaceae)
longifolia (C. Presl) A. Bobr. = Osmunda regalis
lunaria L. var. *matricariifolia* Retz. = Botrychium matricariifolium (Botrychiaceae)
pilosa Wall. ex Grev. & Hook. = Osmundastrum claytonianum subsp. pilosum
regalis L. (*O. longifolia* (C. Presl) A. Bobr., *O. regalis* var. *longifolia* C. Presl) - 2
- subsp. *japonica* (Thunb.) A. & D. Löve = O. japonica
- var. *longifolia* C. Presl = O. regalis
spicant L. = Blechnum spicant (Blechnaceae)

Osmundastrum C. Presl
asiaticum (Fern.) Tagawa (*Osmunda cinnamomea* L. var. *asiatica* Fern., *O. asiatica* (Fern.) Ohwi, *O. cinnamomea* auct., *Osmundastrum cinnamomeum* auct.) - 5
cinnamomeum auct. = O. asiaticum
claytonianum (L.) Tagawa (*Osmunda claytoniana* L.) subsp. **pilosum** (Wall. ex Grev. & Hook.) Tzvel. (*Osmunda pilosa* Wall. ex Grev. & Hook.) - 5

OXALIDACEAE R. Br.
Jonoxalis Small = Oxalis
pes-caprae (L.) Small = Oxalis pes-caprae
violacea (L.) Small = Oxalis violacea

Oxalis L. (*Jonoxalis* Small)
acetosella L. - 1, 2, 3, 4, 5
ambigua Salisb. = Xanthoxalis fontana
corniculata L. = Xanthoxalis corniculata
dillenii Jacq. = Xanthoxalis stricta
europaea Jord. = Xanthoxalis fontana
- f. *villicaulis* Wiegand = Xanthoxalis fontana subsp. villicaulis
fontana Bunge = Xanthoxalis fontana
grenadensis Urb. = Xanthoxalis grenadensis
latifolia Kunth - 1
obtriangulata Maxim. - 5
pes-caprae L. (*Jonoxalis pes-caprae* (L.) Small) - 2
stricta L. = Xanthoxalis stricta
stricta sensu Gorschk. = Xanthoxalis fontana
violacea L. (*Jonoxalis violacea* (L.) Small) - 2

Xanthoxalis Small
corniculata (L.) Small (*Oxalis corniculata* L.) - 1, 2, 5, 6(alien)
dillenii (Jacq.) Holub = X. stricta
europaea (Jord.) Mold. = X. fontana
fontana (Bunge) Holub (*Oxalis fontana* Bunge, *O. ambigua* Salisb. 1794, non Jacq. 1794, *O. europaea* Jord., *Xanthoxalis europaea* (Jord.) Mold., *Oxalis stricta* sensu Gorschk.) - 1, 2, 3, 5
- subsp. **villicaulis** (Wiegand) Tzvel. (*Oxalis europaea* f. *villicaulis* Wiegand) - 1
grenadensis (Urb.) Tzvel. (*Oxalis grenadensis* Urb.) - 1, 2, 5(alien)
stricta (L.) Small (*Oxalis stricta* L., *O. dillenii* Jacq., *Xanthoxalis dillenii* (Jacq.) Holub) - 1

PAEONIACEAE Rudolphi
Paeonia L.
abchasica Miscz. = P. wittmanniana Hartwiss ex Lindl.
albiflora Pall. = P. lactiflora
anomala L. - 1, 3, 4, 6
- var. *intermedia* (C.A. Mey.) O. & B. Fedtsch. = P. intermedia
biebersteiniana Rupr. (*P. tenuifolia* L. subsp. *biebersteiniana* (Rupr.) Takht.) - 2
carthalinica Ketzch. - 2
caucasica (Schipcz.) Schipcz. (*P. kavachensis* auct.) - 2
    Some authors accept the name P. kavachensis Aznav. for this species.
daurica Andr. (*P. triternata* Pall. ex DC., *P. taurica* auct.) - 1, 2
hybrida Pall. - 3, 6
intermedia C.A. Mey. (*P. anomala* L. var. *intermedia* (C.A. Mey.) O. & B. Fedtsch.) - 3, 6
- subsp. **pamiroalaica** Ovcz. - 6
japonica (Makino) Miyabe & Takeda = P. obovata
kavachensis auct. = P. caucasica
lactiflora Pall. (*P. albiflora* Pall.) - 4, 5
lagodechiana Kem.-Nath. - 2
lithophila Kotov - 1
macrophylla (Albov) Lomak. - 2
majko Ketzch. - 2
mlokosewitschii Lomak. - 2
obovata Maxim. (*P. japonica* (Makino) Miyabe & Takeda) - 5
oreogeton S. Moore (*P. vernalis* Mandl) - 5
peregrina Mill. - 1
ruprechtiana Kem.-Nath. - 2
x saundersii Stebb. - P. dahurica Andr. (*P. triternata* Pall. ex DC.) x P. tenuifolia L.
steveniana Kem.-Nath. (*P. wittmanniana* Stev. 1848 & sensu Schipcz. non Hartwiss ex Lindl. 1846) - 2
taurica auct. = P. daurica
tenuifolia L. - 1, 2
- subsp. *biebersteiniana* (Rupr.) Takht. = P. biebersteiniana

tomentosa (Lomak.) N. Busch - 2
triternata Pall. ex DC. = P. daurica
vernalis Mandl = P. oreogeton
wittmanniana Hartwiss ex Lindl. (*P. abchasica* Miscz.) - 2
wittmanniana Stev. = P. steveniana

PALMAE Juss. = ARECACEAE

PAPAVERACEAE Juss. (*CHELIDONIACEAE* Nakai)

Argemone L.

armeniaca L. = Papaver armeniacum
mexicana L. - 2

Chelidonium L.

asiaticum (Hara) Krachulkova (*Ch. majus* L. subsp. *asiaticum* Hara, *Ch. majus* var. *asiaticum* (Hara) Ohwi) - 5
dodecandrum Forssk. = Roemeria hybrida
majus L. - 1, 2, 3, 4, 6
- subsp. *asiaticum* Hara = Ch. asiaticum
- var. *asiaticum* (Hara) Ohwi = Ch. asiaticum
vernale (Maxim.) Ohwi = Hylomecon vernalis

Closterandra auct. = Papaver

minor Boiv. = Papaver minus

Eschscholzia Cham.

californica Cham. - 1(alien), 5(alien)

Glaucium Hill

bracteatum M. Pop. (*G. elegans* Fisch. & C.A. Mey. subsp. *bracteatum* (M. Pop.) B. Mory) - 6
corniculatum (L.) J. Rudolph - 1, 2, 5(alien), 6
elegans Fisch. & C.A. Mey. - 2, 6
- subsp. *bracteatum* (M. Pop.) B. Mory = G. bracteatum
fimbrilligerum (Trautv.) Boiss. (*G. luteum* Scop. var. *fimbrilligerum* Trautv.) - 6
flavum Crantz - 1, 2
grandiflorum Boiss. & Huet - 2
insigne M. Pop. - 6
integrifolium A. Li & Nabiev = Roemeria hybrida
■ **leiocarpum** Boiss.
luteum Scop. var. *fimbrilligerum* Trautv. = G. fimbrilligerum
oxylobum Boiss. & Buhse (*G. paucilobatum* Freyn) - 6
paucilobatum Freyn = G. oxylobum
pulchrum Stapf - 6
squamigerum Kar. & Kir. - 6

Hylomecon Maxim.

vernalis Maxim. (*Chelidonium vernale* (Maxim.) Ohwi) - 5

Papaver L. (*Closterandra* auct.)

ajanense M. Pop. = P. stubendorfii
alascanum Hult. (*P. microcarpum* DC. subsp. *alascanum* (Hult.) Tolm.) - 5
alberti A.D. Mikheev - 2
albiflorum (Bess.) Pacz. (*P. dubium* L. subsp. *albiflorum* (Bess.) Dostal) - 1
alboroseum Hult. - 5
alpinum L. var. *xanthopetalum* Trautv. = P. nudicaule

ambiguum M. Pop. (*P. dubium* L. f. *ambiguum* (M. Pop.) Grossh.) - 2
amurense (N. Busch) Tolm. (*P. nudicaule* L. subsp. *amurense* N. Busch, *P. anomalum* sensu M. Pop.) - 4, 5
anadyrense Petrovsky - 5
angrenicum Pazij = P. croceum
angustifolium Tolm. - 3, 4
anjuicum Tolm. - 5
anomalum Fedde (*P. nudicaule* L. subsp. *anomalum* (Fedde) Worosch.) - 5
anomalum sensu M. Pop. = P. amurense
arenarium Bieb. - 1, 2
argemone L. - 1
- subsp. *belangeri* (Boiss.) Takht. = P. minus
- subsp. *minus* (Boiv.) Kadereit = P. minus
armeniacum (L.) DC. (*Argemone armeniaca* L., *Papaver roopianum* (Bordz.) Sosn., *P. urbanianum* auct.) - 2
atrovirens Petrovsky - 5
belangeri Boiss. = P. minus
bipinnatum C.A. Mey. - 2
bracteatum Lindl. - 2
- var. *pseudoorientale* Fedde = P. pseudoorientale
calcareum Petrovsky - 5
canescens Tolm. - 6
caucasicum Bieb. = P. fugax
chelidoniifolium Boiss. & Buhse - 2
chibinense N. Semen. (*P. radicatum* Rottb. subsp. *lapponicum* Tolm. var. *chibinense* (N. Semen.) A. Löve) - 1
chionophilum Petrovsky - 5
commutatum Fisch. & C.A. Mey. - 2
croceum Ledeb. (*P. angrenicum* Pazij, *P. croceum* subsp. *altaicum* Serg., *P. croceum* subsp. *corydalifolium* (Fedde) Tolm., *P. croceum* subsp. *subcorydalifolium* (Fedde) Tolm. nom. provis., *P. nudicaule* L. var. *croceum* (Ledeb.) Kitag.) - 3, 4, 6
- subsp. *altaicum* Serg. = P. croceum
- subsp. *corydalifolium* (Fedde) Tolm. = P. croceum
- subsp. *subcorydalifolium* (Fedde) Tolm. = P. croceum
czekanowskii Tolm. (*P. microcarpum* DC. subsp. *czekanowskii* (Tolm.) Tolm.) - 4
dahlianum auct. p.p. = P. lapponicum subsp. jugorum and P. polare
dahlianum Nordh. subsp. *dahlianum* var. *lujaurense* (N. Semen.) A. Löve = P. lujaurense
detritophilum Petrovsky - 5
dubium L. - 1, 2
- f. *ambiguum* (M. Pop.) Grossh. = P. ambiguum
fugax Poir. (*P. caucasicum* Bieb.) - 2
gorodkovii Tolm. & Petrovsky - 5
hybridum L. - 1, 2, 6
hyrsipetes Petrovsky - 5
indigirkense Jurtz. - 4
intermedium DC. = P. pseudoorientale
involucratum M. Pop. - 6
- subsp. **nigrescente-hirsutum** Tolm. - 6
jugoricum (Tolm.) Stank. = P. lapponicum subsp. jugoricum
keelei A. Pors. (*P. macounii* auct.) - 5
lacerum M. Pop. - 2
laevigatum Bieb. - 1, 2
langeanum (Lundstr.) Tolm. (*P. rubro-aurantiacum* (Fisch. ex DC.) Lindstr. subsp. *langeanum* Lundstr.) - 4(?)
lapponicum (Tolm.) Nordh. (*P. radicatum* Rottb. subsp. *lapponicum* Tolm., *P. radicatum* auct. p.p.) - 1, 3, 4, 5
- subsp. **dasycarpum** Tolm. - 1
- subsp. **jugoricum** (Tolm.) Tolm. (*P. radicatum* subsp. *jugoricum* Tolm., *P. jugoricum* (Tolm.) Stank., *P.*

dahlianum auct. p.p., *P. radicatum* auct. p.p.) - 1, 3, 4
- subsp. *occidentale* (Lundst.) Knab. p.p. = P. radicatum subsp. occidentale
- subsp. **orientale** Tolm. - 3, 4, 5
- subsp. **porsildii** Knab. - 5
lasiothrix Fedde (*P. orientale* L. var. *lasiothrix* (Fedde) Grossh.) - 2
ledebourianum Lundstr. = P. rubro-aurantiacum
leiocarpum (Turcz.) M. Pop. (*P. nudicaule* L. j. *leiocarpum* Turcz.) - 4
leucotrichum Tolm. - 4
lisae N. Busch - 2
litwinowii Fedde ex Bornm. - 6
lujaurense N. Semen. (*P. dahlianum* Nordh. subsp. *dahlianum var. lujaurense* (N. Semen.) A. Löve) - 1
macounii auct. = P. keelei
macrostomum Boiss. & Huet - 1, 2, 6
maeoticum Klok. - 1
microcarpum DC. - 4, 5
- subsp. *alascanum* (Hult.) Tolm. = P. alascanum
- subsp. *czekanowskii* (Tolm.) Tolm. = P. czekanowskii
- subsp. *ochotense* (Tolm.) Tolm. = P. ochotense
minus (Boiv.) Meikle (*Closterandra minor* Boiv., *Papaver argemone* L. subsp. *belangeri* (Boiss.) Takht., *P. argemone* subsp. *minus* (Boiv.) Kadereit, *P. belangeri* Boiss.) - 2
minutiflorum Tolm. - 4, 5
miyabeanum Tatew. - 5
monanthum Trautv. - 2
multiradiatum Petrovsky - 5
nivale Tolm. - 4, 5
nothum sensu Klok. = P. stevenianum
nothum Stev. = P. strigosum
novokschonovii M. Pop. = P. popovii
nudicaule L. (*P. alpinum* L. var. *xanthopetalum* Trautv., *P. nudicaule* L. subsp. *xanthopetalum* (Trautv.) Fedde) - 3, 4, 5, 6
- subsp. *amurense* N. Busch = P. amurense
- subsp. *anomalum* (Fedde) Worosch. = P. anomalum
- subsp. **baicalense** Tolm. - 4
- subsp. **gracile** Tolm. - 4
- subsp. **insulare** Petrovsky - 5
- subsp. *rubro-aurantiacum* (Fisch. ex DC.) Fedde = P. rubro-aurantiacum
- subsp. *xanthopetalum* (Trautv.) Fedde = P. nudicaule
- var. *croceum* (Ledeb.) Kitag. = P. croceum
- var. *leiocarpum* Turcz. = P. leiocarpum
- var. *rubro-aurantiacum* Fisch. ex DC. = P. rubro-aurantiacum
- f. *lededourianum* (Lundstr.) Kitag. = P. rubro-aurantiacum
ocellatum Woronow (*P. pavoninum* Schrenk subsp. *ocellatum* (Woronow) Kadereit) - 2, 6
ochotense Tolm. (*P. microcarpum* DC. subsp. *ochotense* (Tolm.) Tolm.) - 5
oreophilum Rupr. - 2
orientale L. - 2
- var. *intermedium* (DC.) Grossh. = P. pseudoorientale
- var. *lasiothrix* (Fedde) Grossh. = P. lasiothrix
paczoskii A.D. Mikheev - 2
paucifoliatum (Trautv.) Fedde - 2
paucistaminum Tolm. & Petrovsky - 5
pavoninum Schrenk - 4
- subsp. *ocellatum* (Woronow) Kadereit = P. ocellatum
persicum Lindl. - 2
polare (Tolm.) Perf. (*P. radicatum* Rottb. subsp. *polare* Tolm., *P. dahlianum* auct. p.p., *P. radicatum* auct.

p.p.) - 1, 3, 4, 5
popovii Sipl. (*P. novokschonovii* M. Pop. nom. nud.) - 4
pseudocanescens M. Pop. - 3, 4
- subsp. **udocanicum** Peschkova - 4
pseudoorientale (Fedde) Medw. (*P. bracteatum* Lindl. var. *pseudoorientale* Fedde, *P. intermedium* DC. 1836, non Becker, 1828, *P. orientale* L. var. *intermedium* (DC.) Grossh. nom. illegit.) - 2
pseudostubendorfii M. Pop. - 4
pulvinatum Tolm. - 4, 5
- subsp. **interius** Petrovsky - 5
- subsp. *lenaense* Tolm. nom. invalid. - 4
- subsp. *tschuktschorum* Tolm. nom. invalid. - 5
radicatum Rottb. subsp. *occidentale* Lundstr. (*P. lapponicum* (Tolm.) Nordh. subsp. *occidentale* (Lundst.) G. Knab. p.p. quoad nomen, *P. radicatum* auct. p.p.) - 1, 3, 4, 5
- subsp. *jugoricum* Tolm. = P. lapponicum subsp. jugorum
- subsp. *lapponicum* Tolm. = P. lapponicum
- - var. *chibinense* (N. Semen.) A. Löve = P. chibinense
- - var. *tolmachevii* (N. Semen.) A. Löve = P. tolmachevii
- subsp. *polare* Tolm. = P. polare
refractum (DC.) K.F. Gunther = Roemeria refracta
rhoeas L. - 1, 2, 3
- subsp. *strigosum* (Boenn.) Simonk. = P. strigosum
- subsp. *strigosum* (Boenn.) Soo = P. strigosum
roopianum (Bordz.) Sosn. = P. armeniacum
rubro-aurantiacum (Fisch. ex DC.) Lundstr. (*P. nudicaule* L. var. *rubro-aurantiacum* Fisch. ex DC., *P. ledebourianum* Lundstr. nom. illegit., *P. nudicaule* f. *ledebourianum* (Lundstr.) Kitag., *P. nudicaule* subsp. *rubro-aurantiacum* (Fisch. ex DC.) Fedde) - 4, 5
- subsp. *langeanum* Lundstr. = P. langeanum
- subsp. **longiscapum** Randel. - 5
- subsp. *setosum* Tolm. = P. setosum
schamurinii Petrovsky - 5
schelkownikowii N. Busch - 2
■ **setigerum** DC. (*P. somniferum* L. subsp. *setigerum* (DC.) Corb.)
setosum (Tolm.) Pesckhova (*P. rubro-aurantiacum* (Fisch. ex DC.) Lundstr. subsp. *setosum* Tolm.) - 4
smirnovii Peschkova - 4
somniferum L. - 1, 2, 3, 4, 5, 6
- subsp. *setigerum* (DC.) Corb. = P. setigerum
stevenianum A.D. Mikheev (*P. nothum* sensu Klok.) - 1, 2
strigosum (Boenn.) Schur (*P. nothum* Stev., *P. rhoeas* L. subsp. *strigosum* (Boenn.) Simonk., *P. rhoeas* subsp. *strigosum* (Boenn.) Soo, comb. superfl.) - 1, 2
stubendorfii Tolm. (*P. ajanense* M. Pop.) - 4, 5
talyschense Grossh. - 2
tenellum Tolm. - 3
tianschanicum M. Pop. - 3, 6
tichomirovii (Ju. Kozhevn.) A. Khokhr. = P. uschakovii subsp. tichomirovii
tichomirovii A.D. Mikheev - 1, 2
tolmachevii N. Semen. (*P. radicatum* Rottb. var. *tolmachevii* (N. Semen.) A. Löve) - 1
tumidulum Klok. - 1
urbanianum auct. = P. armeniacum
uschakovii Tolm. & Petrovsky - 5
- subsp. **tichomirovii** Ju. Kozhevn. (*P. tichomirovii* (Ju. Kozhevn.) A. Khokhr. 1985, non A.D. Mikheev, 1981) - 5
variegatum Tolm. - 4
walpolei A. Pors. - 5
zangezuricum A.D. Mikheev - 2

Roemeria Medik.

dodecandra (Forssk.) Stapf = R. hybrida
hybrida (L.) DC. (*Chelidonium dodecandrum* Forssk.,
      Glaucium integrifolium A. Li & Nabiev, *Roemeria*
      dodecandra (Forssk.) Stapf, *R. hybrida* subsp. *dode-*
      candra (Forssk.) Maire, *R. orientalis* Boiss.) - 1, 2, 6
- subsp. *dodecandra* (Forssk.) Maire = R. hybrida
orientalis Boiss. = R. hybrida
refracta DC. (*Papaver refractum* (DC.) K.F. Gunther) - 1, 2,
      6

PAPILIONACEAE Giseke = FABACEAE

PARNASSIACEAE S.F. Gray
Parnassia L.

bifolia Nekr. - 6
kotzebuei Cham. & Schlecht. - 4, 5
laxmannii Pall. ex Schult. - 3, 4, 6
obtusiflora Rupr. = P. palustris
palustris L. (*P. obtusiflora* Rupr., *P. palustris* subsp. *obtusi-*
      flora (Rupr.) D.A. Webb, p.p.) - 1, 2, 3, 4, 5, 6
- subsp. **multiseta** (Ledeb.) Worosch. (*P. palustris* var.
      multiseta Ledeb.) - 5
- subsp. **neogaea** (Fern.) Hult. (*P. palustris* var. *neogaea*
      Fern., *P. palustris* var. *tenuis* Wahlenb., *P. tenuis*
      (Wahlenb.) A. Khokhr. & V. Pavl.) - 1, 3, 4, 5
- subsp. *obtusiflora* (Rupr.) D.A. Webb, p.p. = P. palustris
- subsp. **pseudoalpicola** Worosch. & Makarov - 5
- var. *multiseta* Ledeb. = P. palustris subsp. multiseta
- var. *neogaea* Fern. = P. palustris subsp. neogaea
- var. *tenuis* Wahlenb. = P. palustris subsp. neogaea
tenuis (Wahlenb.) A. Khokhr. & V. Pavl. = P. palustris
      subsp. neogaea

***PEDALIACEAE** R. Br.
***Sesamum** L.

***indicum** L. (*S. orientale* L.)
orientale L. = S. indicum

PEGANACEAE (Engl.) Tiegh. ex Takht.
Malacocarpus Fisch. & C.A. Mey.

crithmifolius (Retz.) C.A. Mey. - 6

Peganum L.

harmala L. - 1, 2, 3, 6
nigellastrum Bunge - 4

PENTHORACEAE Rydb. ex Britt.
Penthorum L.

chinense Pursh (*P. humile* Regel & Maack) - 5
humile Regel & Maack = P. chinense

PHRYMACEAE Schauer
Phryma L.

asiatica (Hara) O. & I. Degener (*Ph. leptostachya* L. var.
      asiatica Hara, *Ph. asiatica* (Hara) Probat. comb.
      superfl., *Ph. leptostachya* subsp. *asiatica* (Hara)
      Kitam., *Ph. leptostachya* auct.) - 5

leptostachya auct. = Ph. asiatica
leptostachya L. subsp. *asiatica* (Hara) Kitam. = Ph. asiatica
- var. *asiatica* Hara = Ph. asiatica

PHYTOLACCACEAE R. Br.
Phytolacca L.

americana L. - 1, 2

PILULARIACEAE Wettst. = MARSILE-
ACEAE

PINACEAE Lindl.
Abies Hill

ajanensis Lindl. & Gord. = Picea ajanensis
alba Mill. - 1
***balsamea** (L.) Mill. (*Pinus balsamea* L.)
- subsp. *fraseri* (Pursh) E. Murr. = A. fraseri
excelsa (Lam.) Poir. subsp. *alpestris* Bruegg. = Picea obova-
      ta
***fraseri** (Pursh) Poir. (*A. balsamea* (L.) Mill. subsp. *fraseri*
      (Pursh) E. Murr.)
gmelinii Rupr. = Larix gmelinii
gracilis Kom. = A. nephrolepis
holophylla Maxim. - 5
ledebourii Rupr. = Larix sibirica
mayriana (Miyabe & Kudo) Mayabe & Kudo (*A. sachali-*
      nensis Fr. Schmidt var. *mayriana* Miyabe & Kudo) - 5
nephrolepis (Trautv.) Maxim. (*A. gracilis* Kom.) - 5
nordmanniana (Stev.) Spach - 2
sachalinensis Fr. Schmidt (*A. wilsonii* Miyabe & Kudo) - 5
- var. *mayriana* Miyabe & Kudo = A. mayriana
semenovii B. Fedtsch. (*A. sibirica* Ledeb. var. *semenovii* (B.
      Fedtsch.) Liu) - 6
sibirica Ledeb. - 1, 3, 4, 6
- var. *semenovii* (B. Fedtsch.) Liu = A. semenovii
wilsonii Miyabe & Kudo = A. sachalinensis

***Cedrus** Trew

***deodara** (Roxb.) G. Don fil. - 1, 2

Larix Hill

altaica Fisch. ex Parl. = L. sibirica
amurensis Beissn. = L. gmelinii
amurensis Kolesn. = L. x maritima
cajanderi Mayr (*L. gmelinii* (Rupr.) Rupr. subsp. *cajanderi*
      (Mayr) Ju. Kozhevn.) - 4, 5
x czekanowskii Szaf. (*L. uszkanensis* Sukacz. & Lamak.
      nom. nud.). - L. gmelinii (Rupr.) Rupr. x L. sibirica
      Ledeb. - 4
dahurica Turcz. ex Trautv. = L. gmelinii
***decidua** Mill.
- subsp. *polonica* (Racib.) Domin = L. x polonica
- subsp. *sibirica* (Ledeb.) Domin = L. sibirica
- var. *polonica* (Racib.) Ostenf. & Larsen = L. x polonica
gmelinii (Rupr.) Rupr. (*Abies gmelinii* Rupr., *Larix amu-*
      remsis Beissn., *L. dahurica* Turcz. ex Trautv., *L.*
      pumila Doctur. & Fler.) - 4, 5
- subsp. *cajanderi* (Mayr) Ju. Kozhevn. = L. cajanderi
- subsp. *olgensis* (A. Henry) Worosch. = L. olgensis
- subsp. *polonica* (Racib.) E. Murr. = L. x polonica
kamtschatica (Rupr.) Carr. (*L. kurilensis* Mayr, *L. midden-*
      dorffii Kolesn. p.p. , *L. ochotensis* Kolesn. p.p.) - 5

komarovii Kolesn. = L. x lubarskii and L. olgensis
kurilensis Mayr = L. kamtschatica
ledebourii (Rupr.) Cinovskis = L. sibirica
***leptolepis** (Siebold & Zucc.) Gord. - 5
x **lubarskii** Sukacz. (*L. komarovii* Kolesn. p.p., *L. principis-rupprechtii* auct.). - L. gmelinii (Rupr.) Rupr. x L. kamtschatica (Rupr.) Carr. x L. olgensis A. Henry - 5
x **maritima** Sukacz. (*L. amurensis* Kolesn. nom. nud. 1946, non Beissn. 1891, *L. middendorffii* Kolesn. p.p., *L. ochotensis* Kolesn. p.p.). - L. gmelinii (Rupr.) Rupr. x L. kamtschatica (Rupr.) Carr. - 5
middendorffii Kolesn. = L. kamtschatica and L. x maritima
ochotensis Kolesn. = L. kamtschatica and L. x maritima
olgensis A. Henry (*L. gmelinii* (Rupr.) Rupr. subsp. *olgensis* (A. Henry) Worosch., *L. komarovii* Kolesn. p.p.) - 5
x **polonica** Racib. (*L. decidua* Mill. subsp. *polonica* (Racib.) Domin, *L. decidua* var. *polonica* (Racib.) Ostenf. & Larsen, *L. gmelinii* (Rupr.) Rupr. subsp. *polonica* (Racib.) E. Murr., *L. sibirica* Ledeb. subsp. *polonica* (Racib.) Sukacz.). - L. decidua Mill. x L. sibirica Ledeb. - 1
principis-rupprechtii auct. = L. x lubarskii
pumila Doctur. & Fler. = L. gmelinii
russica (Endl.) Sabine ex Trautv. = L. sibirica
sibirica Ledeb. (*Abies ledebourii* Rupr., *Larix altaica* Fisch. ex Parl., *L. decidua* Mill. subsp. *sibirica* (Ledeb.) Domin, *L. ledebourii* (Rupr.) Cinovskis, *L. russica* (Endl.) Sabine ex Trautv., *L. sukaczewii* Dyl., *Pinus larix* L. var. *russica* Endl., *P. ledebourii* (Rupr.) Endl.) - 1, 3, 4, 6
- subsp. *polonica* (Racib.) Sukacz. = L. x polonica
sukaczewii Dyl. = L. sibirica
uszkanensis Sukacz. & Lamak. = L. x czekanowskii

Picea A. Dietr.

abies (L.) Karst. (*Pinus abies* L., *Picea abies* subsp. *acuminata* (G. Beck) Parfenov, *P. abies* subsp. *europaea* (Tepl.) Hyl., *P. excelsa* (Lam.) Link, *P. excelsa* var. *acuminata* G. Beck, *P. montana* Schur, *P. vulgaris* Link var. *europaea* Tepl., *Pinus excelsa* Lam.) - 1
- subsp. *acuminata* (G. Beck) Parfenov = P. abies
- subsp. *alpestris* (Bruegg.) Domin = P. obovata
- subsp. *europaea* (Tepl.) Hyl. = P. abies
- subsp. *fennica* (Regel) Parfenov = P. x fennica
- subsp. *obovata* (Ledeb.) Domin = P. obovata
- var. *alpestris* (Bruegg.) P. Schmidt = P. obovata
ajanensis (Lindl. & Gord.) Fisch. ex Carr. (*Abies ajanensis* Lindl. & Gord., *Picea jezoensis* (Siebold & Zucc.) Carr. var. *ajanensis* (Fisch. ex Trautv. & C.A. Mey.) W.C. Cheng & L.K. Fu, *P. jezoensis* var. *komarovii* (V. Vassil.) W.C. Cheng & L.K. Fu, *P. kamtschatkensis* Lacass., *P. komarovii* V. Vassil., *P. microsperma* Carr. p.p., *P. jezoensis* auct.) - 5
alpestris (Bruegg.) Stein = P. obovata
canadensis (Mill.) Britt., Sterns & Pogg. = P. glauca
***engelmannii** Engelm. - 1
excelsa (Lam.) Link = P. abies
- var. *acuminata* G. Beck = P. abies
x **fennica** (Regel) Kom. (*P. abies* (L.) Karst. subsp. *fennica* (Regel) Parfenov). - P. abies (L.) Karst. x P. obovata Ledeb. - 1
- nothosubsp. **uralensis** (Tepl.) P. Schmidt (*P. vulgaris* Link var. *uralensis* Tepl.) - 1
***glauca** (Moench) Voss (*Pinus glauca* Moench, *Picea canadensis* (Mill.) Britt., Sterns & Pogg. 1888, non Link, 1841)
glehnii (Fr. Schmidt) Mast. - 5

jezoensis auct. = P. ajanensis
jezoensis (Siebold & Zucc.) Carr. var. *ajanensis* (Fisch. ex Trautv. & C.A. Mey.) W.C. Cheng & L.K. Fu = P. ajanensis
- var. *komarovii* (V. Vassil.) W.C. Cheng & L.K. Fu = P. ajanensis
kamtschatkensis Lacass. = P. ajanensis
komarovii V. Vassil. = P. ajanensis
koraiensis Nakai - 5
microsperma Carr. p.p. = P. ajanensis
montana Schur = P. abies
morinda Link subsp. *tianschanica* (Rupr.) Berezin = P. schrenkiana
obovata Ledeb. (*Abies excelsa* (Lam.) Poir. subsp. *alpestris* Bruegg., *Picea abies* (L.) Karst. subsp. *alpestris* (Bruegg.) Domin, *P. abies* var. *alpestris* (Bruegg.) P. Schmidt, *P. abies* subsp. *obovata* (Ledeb.) Domin, *P. alpestris* (Bruegg.) Stein, *P. obovata* subsp. *petchorica* Govor., *P. petchorica* Govor. nom. altern.) - 1, 3, 4, 5
- subsp. *petchorica* Govor. = P. obovata
orientalis (L.) Link - 2
petchorica Govor. = P. obovata
prostrata K. Isakov = P. schrenkiana
***pungens** Engelm.
robertii P. Vipper = P. schrenkiana
schrenkiana Fisch. & C.A. Mey. (*P. morinda* Link subsp. *tianschanica* (Rupr.) Berezin, *P. prostrata* K. Isakov, *P. robertii* P. Vipper, *P. tianschanica* Rupr., *P. schrenkiana* subsp. *tianschanica* (Rupr.) Bykov) - 6
- subsp. *tianschanica* (Rupr.) Bykov = P. schrenkiana
tianschanica Rupr. = P. schrenkiana
vulgaris Link var. *europaea* Tepl. = P. abies
- var. *uralensis* Tepl. = P. x fennica nothosubsp. uralensis

Pinus L.

abies L. = Picea abies
armena C. Koch = P. kochiana
balsamea L. = Abies balsamea
brutia auct. p.p. = P. eldarica and P. pityusa
brutia Ten. subsp. *eldarica* (Medw.) Nahal = P. eldarica
- subsp. *eldarica* (Medw.) R. Kam. = P. eldarica
- subsp. *pityusa* (Stev.) Nahal = P. pityusa
- subsp. *stankewiczii* (Sukacz.) Nahal = P. pityusa
- var. *eldarica* (Medw.) J. Silba = P. eldarica
- var. *pityusa* (Stev.) J. Silba = P. pityusa
cembra L. - 1
- subsp. *pumila* (Pall.) E. Murr. = P. pumila
- subsp. *sibirica* (Du Tour) E. Murr. = P. sibirica
densiflora Siebold & Zucc. (*P. sylvestris* L. subsp. *densiflora* (Siebold & Zucc.) Worosch.) - 5
eldarica Medw. (*P. brutia* Ten. subsp. *eldarica* (Medw.) Nahal, *P. brutia* subsp. *eldarica* (Medw.) R. Kam. comb. superfl., *P. brutia* var. *eldarica* (Medw.) J. Silba, *P. brutia* auct. p.p.) - 2
excelsa Lam. = Picea abies
fominii Kondr. = P. sylvestris
friesiana Wichura (*P. lapponica* (Fries ex C. Hartm.) Mayr, *P. sylvestris* L. var. *lapponica* Fries ex C. Hartm., *P. sylvestris* subsp. *lapponica* (Fries ex C. Hartm.) Holmb.) - 1
x **funebris** Kom. - P. densiflora Siebold & Zucc. x P. sylvestris L. - 5
glauca Moench = Picea glauca
halepensis subsp. *pityusa* (Stev.) E. Murr. = P. pityusa
- subsp. *stankewiczii* (Sukacz.) E. Murr. = P. pityusa
hamata (Stev.) Sosn. = P. kochiana
kochiana Klotzsch ex C. Koch (*P. armena* C. Koch, *P.

hamata (Stev.) Sosn. 1925, non Roezl, 1857, *P. sosnowskyi* Nakai, *P. sylvestris* L. subsp. *hamata* (Stev.) Fomin, *P. sylvestris* subsp. *kochiana* (Klotzsch ex C. Koch) Elicin, *P. sylvestris* auct. fl. cauc. & taur.) - 1, 2

koraiensis Siebold & Zucc. - 5

krylovii Serg. & Kondr. = P. sylvestris

lapponica (Fries ex C. Hartm.) Mayr = P. friesiana

larix L. var. *russica* Endl. = Larix sibirica

ledebourii (Rupr.) Endl. = Larix sibirica

mughus Scop. = P. mugo

mugo Turra (*P. mughus* Scop.) - 1

nigra Arnold subsp. *pallasiana* (D. Don) Holmboe = P. pallasiana

pallasiana D. Doń (*P. nigra* Arnold subsp. *pallasiana* (D. Don) Holmboe) - 1, 2

*pinea L. - 1, 2

pityusa Stev. (*P. brutia* Ten. subsp. *pityusa* (Stev.) Nahal, *P. brutia* var. *pityusa* (Stev.) J. Silba, *P. brutia* subsp. *stankewiczii* (Sukacz.) Nahal, *P. halepensis* subsp. *pityusa* (Stev.) E. Murr., *P. halepensis* subsp. *stankewiczii* (Sukacz.) E. Murr., *P. stankewiczii* (Sukacz.) Fomin, *P. brutia* auct. p.p.) - 1, 2

pumila (Pall.) Regel (*P. cembra* L. subsp. *pumila* (Pall.) E. Murr.) - 4, 5

sibirica Du Tour (*P. cembra* L. subsp. *sibirica* (Du Tour) E. Murr., *P. sibirica* (Loud.) Mayr) - 1, 3, 4

sosnowskyi Nakai = P. kochiana

stankewiczii (Sukacz.) Fomin = P. pityusa

*strobus L. - 1

sylvestris auct. = P. kochiana

sylvestris L. (*P. fominii* Kondr., *P. krylovii* Serg. & Kondr.) - 1, 3, 4, 5

- subsp. *densiflora* (Siebold & Zucc.) Worosch. = P. densiflora

- subsp. *hamata* (Stev.) Fomin = P. kochiana

- subsp. *kochiana* (Klotzsch ex C. Koch) Elicin = P. kochiana

- subsp. *lapponica* (Fries ex C. Hartm.) Holmb. = P. friesiana

- var. *lapponica* Fries ex C. Hartm. = P. friesiana

PITTOSPORACEAE R. Br.

*Pittosporum Banks ex Soland.

*tobira (Thunb.) Ait. - 1, 2

PLAGIOGYRIACEAE Bower

Lomaria auct. = Plagiogyria

matsumurana Makino = Plagiogyria matsumurana

Plagiogyria (G. Kunze) Mett. (*Lomaria* auct.)

matsumurana (Makino) Makino (*Lomaria matsumurana* Makino, *Plagiogyria semicordata* subsp. *matsumurana* (Makino) Nakaike) - 5

semicordata subsp. *matsumurana* (Makino) Nakaike = P. matsumurana

PLANTAGINACEAE Juss.

Littorella Berg.

uniflora (L.) Aschers. - 1

Plantago L. (*Psyllium* Hill)

afra L. = P. squalida

alpina auct. = P. neumannii

altissima L. - 1

arachnoidea Schrenk - 6

arenaria Waldst. & Kit. (*P. indica* L. nom. illegit., *P. scabra* Moench, nom. illegit. superfl., *Psyllium arenarium* (Waldst. & Kit.) Mirb., *P. indicum* (L.) Dum.-Cours. nom. illegit., *P. scabrum* (Moench) Holub, nom. illegit.) - 1, 2, 3, 6

- subsp. **orientalis** (Soo) Greuter & Burdet (*P. indica* subsp. *orientalis* Soo, *P. arenaria* subsp. *orientalis* (Soo) Tzvel. comb. superfl., *P. arenaria* f. *rossica* Tuzs., *P. arenaria* var. *rossica* (Tuzs.) J. Lewalle, *P. indica* var. *rossica* (Tuzs.) Pilg., *P. latifolia* Wissjul. 1960, non Salisb. 1796, *P. scabra* subsp. *orientalis* (Soo) Tzvel., *Psyllium arenarium* subsp. *orientale* (Soo) Sojak, *P. indicum* subsp. *orientale* (Soo) Sojak, *P. scabrum* subsp. *orientale* (Soo) Holub) - 1, 2

- var. *rossica* (Tuzs.) J. Lewalle = P. arenaria subsp. orientalis

- f. *rossica* Tuzs. = P. arenaria subsp. orientalis

aristata Michx. - 1(alien)

asiatica auct. = P. cornuti

atrata Hoppe subsp. **carpathica** (Pilg.) Soo (*P. montana* Lam. var. *carpathica* Pilg., *P. atrata* var. *carpathica* (Pilg.) Pilg.) - 1

- subsp. **circassica** Tzvel. - 2

- var. *carpathica* (Pilg.) Pilg. = P. atrata subsp. carpathica

- var. *saxatilis* (Bieb.) V. Avet. = P. saxatilis

biebersteinii Opiz = P. intermedia

borealis Lange = P. schrenkii

borysthenica (Rogow.) Wissjul. = P. major

camtschatica Link - 5

canescens Adams - 4, 5

- subsp. **jurtzevii** Tzvel. (*P. septata* auct.) - 5

- subsp. **tolmatschevii** Tzvel. - 4

- subsp. **trautvetteri** Tzvel. (*P. canescens* var. *glabrata* Trautv.) - 4

- var. *glabrata* Trautv. = P. canescens subsp. trautvetteri

caucasica Papava = P. saxatilis

cornuti Gouan (*P. major* L. subsp. *cornuti* (Gouan) R. de P. Malagarriga, *P. asiatica* auct.) - 1, 2, 3, 4, 5, 6

coronopus L. - 1, 2, 6

depressa Schlecht. - 3, 4, 5, 6

dubia L. (*P. eriophora* Hoffmgg. & Link, *P. glabriflora* Sakalo, *P. lanceolata* L. var. *lanuginosa* Bast., *P. lanceolata* subsp. *lanuginosa* (Bast.) Arcang., *P. lanuginosa* (Bast.) Karnauch) - 1, 2

eriophora Hoffmgg. & Link = P. dubia

gentianoides auct. = P. griffithii

gentianoides Sibth. & Smith subsp. *griffithii* (Decne.) Rech. fil. = P. griffithii

glabriflora Sakalo = P. dubia

griffithii Decne. (*P. gentianoides* Sibth. & Smith subsp. *griffithii* (Decne.) Rech. fil., *P. gentianoides* auct.) - 6

indica L. = P. arenaria

- subsp. *orientalis* Soo = P. arenaria subsp. orientalis

- var. *rossica* (Tuzs.) Pilg. = P. arenaria subsp. orientalis

intermedia DC. (*P. biebersteinii* Opiz, *P. major* L. subsp. *intermedia* (DC.) Arcang. comb. superfl., *P. major* subsp. *intermedia* (DC.) Lange, *P. major* subsp. *plejosperma* Pilg., *P. major* f. *scopulorum* Fries, *P. scopulorum* (Fries) N.M. Pavl., ? *P. uliginosa* F. W. Schmidt) - 1, 2, 3, 6

japonica Franch. & Savat. - 5

komarovii Pavl. - 6

krascheninnikovii C. Serg. - 1

lachnantha Bunge - 6

lagocephala Bunge - 6
lagopus L. - 1
lanceolata L. - 1, 2, 3, 4(alien), 5(alien), 6
- subsp. *lanuginosa* (Bast.) Arcang. = P. dubia
- var. *lanuginosa* Bast. = P. dubia
lanuginosa (Bast.) Karnauch = P. dubia
latifolia Wissjul. = P. arenaria subsp. orientalis
lessingii Fisch. & C.A. Mey. (*P. minuta* Pall. subsp. *lessingii* (Fisch. & C.A. Mey.) Tzvel.) - 1(?), 3, 6
loeflingii L. - 2
macrocarpa Cham. & Schlecht. - 5
major L. (*P. borysthenica* (Rogow.) Wissjul., *P. major* var. *borysthenica* Rogow.) - 1, 2, 3, 4, 5, 6
- subsp. *cornuti* (Gouan) R. de P. Malagarriga = P. cornuti
- subsp. *intermedia* (DC.) Arcang. = P. intermedia
- subsp. *intermedia* (DC.) Lange = P. intermedia
- subsp. *plejosperma* Pilg. = P. intermedia
- subsp. *winteri* (Wirtg.) W. Ludw. = P. winteri
- var. *borysthenica* Rogow. = P. major
- f. *scopulorum* Fries = P. intermedia
maritima L. - 1
- subsp *borealis* (Lange) Blytt & Dahl = P. schrenkii
- subsp. *ciliata* Printz = P. salsa
- subsp. *neumannii* (Opiz) Tzvel. = P. neumannii
- subsp. *salsa* (Pall.) Hult. = P. salsa
- subsp. *salsa* (Pall.) Rech. fil. = P. salsa
- subsp. *salsa* (Pall.) Sojak = P. salsa
- subsp. *salsa* (Pall.) Soo = P. salsa
- subsp. *subpolaris* (Andrejev) Tzvel. = P. subpolaris
- var. *salsa* (Pall.) Pilg. = P. salsa
- var. *salsa* (Pall.) Serg. = P. salsa
maxima Juss. ex Jacq. - 1, 2, 3, 4, 6
media L. - 1, 3, 4, 5, 6
- subsp. *longifolia* (G. Mey.) Witte = P. urvillei
- subsp. *stepposa* (Kuprian.) Soo = P. urvillei
- subsp. *urvilleana* (Rapin) Hult. = P. urvillei
- var. *urvilleana* Rapin = P. urvillei
- Spielart *longifolia* G. Mey. = P. urvillei
minuta Pall. - 1, 2, 6
- subsp. *lessingii* (Fisch. & C.A. Mey.) Tzvel. = P. lessingii
montana Lam. var. *carpathica* Pilg. = P. atrata subsp. carpathica
neumannii Opiz (*P. maritima* L. subsp. *neumannii* (Opiz) Tzvel., *P. alpina* auct.) - 1
notata Lag. - 2
ovata Forssk. - 2, 6
polysperma Kar. & Kir. - 1, 3, 6
psyllium sensu L. 1762 = P. squalida
pusilla Bunge - 4
   Dubious species.
salsa Pall. (*P. maritima* L. subsp. *ciliata* Printz, *P. maritima* subsp. *salsa* (Pall.) Hult. comb. superfl., *P. maritima* var. *salsa* (Pall.) Pilg., *P. maritima* subsp. *salsa* (Pall.) Rech. fil. comb. superfl., *P. maritima* var. *salsa* (Pall.) Serg. comb. superfl., *P. maritima* subsp. *salsa* (Pall.) Sojak, *P. maritima* subsp. *salsa* (Pall.) Soo, comb. superfl.) - 1, 2, 3, 4, 6
samojedorum Gand. = P. schrenkii
saxatilis Bieb. (*P. atrata* Hoppe var. *saxatilis* (Bieb.) V. Avet., *P. caucasica* Papava) - 2
scabra Moench = P. arenaria
- subsp. *orientalis* (Soo) Tzvel. = P. arenaria subsp. orientalis
schrenkii C. Koch (*P. borealis* Lange, *P. maritima* L. subsp. *borealis* (Lange) Blytt & Dahl, *P. samojedorum* Gand.) - 1
schwarzenbergiana Schur - 1

scopulorum N.M. Pavl. = P. intermedia
septata auct. = P. canescens subsp. jurtzevii
sphaerostachya (Mert. & Koch) A. Kerner - 1
squalida Salisb. (*P. afra* L. nom. ambig., *P. psyllium* sensu L. 1762, non L. 1753, *Psyllium squalidum* (Salisb.) Sojak) - 2, 6
stepposa Kuprian. = P. urvillei
subpolaris Andrejev (*P. maritima* L. subsp. *subpolaris* (Andrejev) Tzvel.) - 1
tenuiflora Waldst. & Kit. - 1, 2, 3, 6
togashii Miyabe & Tatew. - 5
uliginosa F.W. Schmidt = P. intermedia
urvillei Opiz (? *P. media* L. Spielart *longifolia* G. Mey., ? *P. media* subsp. *longifolia* (G. Mey.) Witte, *P. media* subsp. *stepposa* (Kuprian.) Soo, *P. media* var. *urvilleana* Rapin, *P. media* subsp. *urvilleana* (Rapin) Hult., *P. stepposa* Kuprian.) - 1, 2, 3, 4, 6
winteri Wirtg. (*P. major* L. subsp. *winteri* (Wirtg.) W. Ludw.) - 1

Psyllium Hill = Plantago

arenarium (Waldst. & Kit.) Mirb. = Plantago arenaria
- subsp. *orientale* (Soo) Sojak = Plantago arenaria subsp. orientalis
indicum (L.) Dum.-Cours. = Plantago arenaria
- subsp. *orientale* (Soo) Sojak = Plantago arenaria subsp. orientalis
scabrum (Moench) Holub = Plantago arenaria
- subsp. *orientale* (Soo) Holub = Plantago arenaria subsp. orientalis
squalidum (Salisb.) Sojak = Plantago squalida

PLATANACEAE Dumort.
Platanus L.

cuneata Willd. = P. orientalis
digitata Gord. = P. orientalis
digitifolia Palib. = P. orientalis
*occidentalis L. - 1
orientalior Dode = P. orientalis
orientalis L. (*P. cuneata* Willd., *P. digitata* Gord. 1872, non Ung. 1850, *P. digitifolia* Palib., *P. orientalior* Dode) - 1, 2, 6

PLEUROSORIOPSIDACEAE Kurata & Ikebe ex Ching
Pleurosoriopsis Fomin

makinoi (Maxim.) Fomin - 5

PLUMBAGINACEAE Juss.
Plumbagella Spach

micrantha (Ledeb.) Spach - 3, 4, 6

Plumbago L.

europaea L. - 2, 6

POACEAE Barnhart (GRAMINEAE Juss. nom. altern.)

Achnatherum Beauv. (*Aristella* (Trin.) Bertol., *Lasiagrostis* Link, *Timouria* Roshev.)

botschantzevii Tzvel. - 6
bromoides (L.) P. Beauv. (*Aristella bromoides* (L.) Bertol., *Lasiagrostis bromoides* (L.) Nevski & Roshev., *Stipa bromoides* (L.) Doerfl.) - 1, 2, 6
caragana (Trin.) Nevski (*Stipa caragana* Trin., *Lasiagrostis caragana* (Trin.) Trin. & Rupr.) - 2, 3, 6
confusum (Litv.) Tzvel. (*Stipa confusa* Litv.) - 3, 4, 5
effusum (Maxim.) Chang = A. extremiorientale
extremiorientale (Hara) Keng (*Stipa extremiorientalis* Hara, *Achnatherum effusum* (Maxim.) Chang, *A. extremiorientalis* (Hara) Keng ex Tzvel. comb. superfl., *Stipa effusa* (Maxim.) Nakai ex Honda, 1926, non Mez, 1921, *S. komarovii* P. Smirn.) - 4, 5
longiaristatum (Boiss. & Hausskn.) Nevski, p.p. = A. turcomanicum
mongholicum (Turcz. ex Trin.) Ohwi = Ptilagrostis mongholica
roshevitzii Musajev - 2
saposhnikovii (Roshev.) Nevski (*Timouria saposhnikovi* Roshev.) - 6
sibiricum (L.) Keng ex Tzvel. (*Stipa sibirica* (L.) Lam.) - 2, 3, 4, 5, 6
splendens (Trin.) Nevski (*Stipa splendens* Trin., *Lasiagrostis splendens* (Trin.) Kunth) - 1, 3, 4, 6
turcomanicum (Roshev.) Tzvel. (*Oryzopsis turcomanica* Roshev., *Achnatherum longiaristatum* (Boiss. & Hausskn.) Nevski, p.p. quoad pl. excl. typo, *Lasiagrostis longearistata* (Boiss. & Hausskn.) Roshev. & Nevski, p.p. quoad pl. excl. typo, *Stipa litwinowiana* P. Smirn., *Stipa kurdistanica* Bor, p.p. excl. typo) - 6

Aegilemma A. Löve = Aegilops
kotschyi (Boiss.) A. Löve = Aegilops kotschyi

Aegilonearum A. Löve = Aegilops
juvenale (Thell.) A. Löve = Aegilops juvenalis

Aegilopodes A. Löve = Aegilops
triuncialis (L.) A. Löve = Aegilops triuncialis
- subsp. *persica* (Boiss.) A. Löve = Aegilops triuncialis subsp. persica

Aegilops L. (*Aegilemma* A. Löve, *Aegilonearum* A. Löve, *Aegilopodes* A. Löve, *Chennapyrum* A. Löve, *Comopyrum* (Jaub. & Spach) A. Löve, *Cylindropyrum* (Jaub. & Spach) A. Löve, *Gastropyrum* (Jaub. & Spach) A. Löve, *Kiharapyrum* A. Löve, *Patropyrum* A. Löve, *Sitopsis* (Jaub. & Spach) A. Löve)

aucheri Boiss. - 2(alien)
biuncialis Vis. (*A. lorentii* Hochst., *A. macrochaeta* Shuttl. & Huet, *A. ovata* L. subsp. *biuncialis* (Vis.) Anghel & Beldie, *Triticum biunciale* (Vis.) K. Richt. 1890, non Vill. 1787, *T. lorentii* (Hochst.) Zeven, *T. macrochaetum* (Shuttl. & Huet) K. Richt.) - 1, 2, 6
columnaris Zhuk. (*Triticum columnare* (Zhuk.) Morris & Sears) - 2

comosa Sibth. & Smith (*Comopyrum comosum* (Sibth. & Smith) A. Löve) - 2(alien)
crassa Boiss. (*A. crassa* var. *macranthera* Boiss., *A. crassa* subsp. *macranthera* (Boiss.) Zhuk., *Triticum crassum* (Boiss.) Aitch. & Hemsl., *Gastropyrum crassum* (Boiss.) A. Löve) - 2, 6
- subsp. *macranthera* (Boiss.) Zhuk. = A. crassa
- var. *macranthera* Boiss. = A. crassa
cylindrica Host (*A. cylindrica* subsp. *aristulata* Zhuk., *A. cylindrica* var. *aristulata* (Zhuk.) Tzvel., *A. cylindrica* subsp. *pauciaristata* (Eig) Chennav., *Cylindropyrum cylindricum* (Host) A. Löve, *Triticum cylindricum* (Host) Cesati) - 1, 2, 5(alien)
- subsp. *aristulata* Zhuk. = A. cylindrica
- subsp. *pauciaristata* (Eig) Chennav. = A. cylindrica
- var. *aristulata* (Zhuk.) Tzvel. = A. cylindrica
geniculata Roth (*A. vagans* Jord. & Fourr., *Triticum vagans* (Jord. & Fourr.) Greuter, *Aegilops ovata* sensu Nevski) - 1, 2
glabriglumis Gand. = A. kotschyi
juvenalis (Thell.) Eig (*Aegilonearum juvenale* (Thell.) A. Löve) - 6
kotschyi Boiss. (*Aegilemma kotschyi* (Boiss.) A. Löve, *Aegilops glabriglumis* Gand., *A. triuncialis* L. subsp. *kotschyi* (Boiss.) Zhuk., *Triticum kotschyi* (Boiss.) Bowden) - 2
loliacea Jaub. & Spach = Amblyopyrum muticum
lorentii Hochst. = A. biuncialis
macrochaeta Shuttl. & Huet = A. biuncialis
mutica Boiss. = Amblyopyrum muticum
- subsp. *loliacea* (Jaub. & Spach) Zhuk. = Amblyopyrum muticum
neglecta Req. ex Bertol. (*A. ovata* L. emend. Roth, nom. rej., ? *A. triaristata* Willd. subsp. *contorta* Zhuk., ? *A. triaristata* subsp. *intermixta* Zhuk., *Triticum neglectum* (Bertol.) Greuter, *T. ovatum* (L.) Raspail, p.p., *Aegilops triaristata* sensu Nevski) - 1, 2, 6
- subsp. *recta* (Zhuk.) Hammer = A. recta
ovata L. = A. neglecta
- subsp. *biuncialis* (Vis.) Anghel & Beldie = A. biuncialis
- subsp. *triaristata* (Willd.) Jav. = A. triuncialis
ovata sensu Nevski = A. geniculata
persica Boiss. = A. triuncialis subsp. persica
■ **recta** (Zhuk.) Chennav. (*A. triaristata* Willd. subsp. *recta* Zhuk., *A. neglecta* Req. ex Bertol. subsp. *recta* (Zhuk.) Hammer, *Triticum rectum* (Zhuk.) Bowden)
speltoides Tausch (*Sitopsis speltoides* (Tausch) A. Löve) - 2(alien)
squarrosa L. = A. triuncialis
- subsp. *meyeri* (Griseb.) Zhuk. = A. tauschii
- subsp. *salina* Zhuk. = A. tauschii
- subsp. *strangulata* Eig = A. tauschii subsp. strangulata
- var. *meyeri* Griseb. = A. tauschii
squarrosa sensu Nevski = A. tauschii
tauschii Coss. (*A. squarrosa* L. b. *meyeri* Griseb., *A. squarrosa* subsp. *meyeri* (Griseb.) Zhuk., *A. squarrosa* subsp. *salina* Zhuk., *Patropyrum tauschii* (Coss.) A. Löve, *P. tauschii* subsp. *salinum* (Zhuk.) A. Löve, *Triticum tauschii* (Coss.) Schmalh., *Aegilops squarrosa* sensu Nevski) - 1, 2, 6
- subsp. *strangulata* (Eig) Tzvel. (*A. squarrosa* subsp. *strangulata* Eig, *Patropyrum tauschii* subsp. *strangulatum* (Eig) A. Löve) - 2, 6
triaristata sensu Nevski = A. neglecta
triaristata Willd. = A. triuncialis
- subsp. *contorta* Zhuk. = A. neglecta
- subsp. *intermixta* Zhuk. = A. neglecta
- subsp. *recta* Zhuk. = A. recta

triuncialis L. (*Aegilopodes triuncialis* (L.) A. Löve, *Aegilops ovata* L. subsp. *triaristata* (Willd.) Jav., *A. squarrosa* L., *A. triaristata* Willd., *Triticum triunciale* (L.) Raspail) - 1, 2, 6
- subsp. *kotschyi* (Boiss.) Zhuk. = A. kotschi
- subsp. *orientalis* Eig = A. triuncialis subsp. persica
- subsp. **persica** (Boiss.) Zhuk. (*A. persica* Boiss., *Aegilopodes triuncialis* subsp. *persica* (Boiss.) A. Löve, *Aegilops triuncialis* subsp. *orientalis* Eig, *Triticum persicum* (Boiss.) Aitch. & Hemsl.) - 2, 6

umbellata Zhuk. (*Kiharapyrum umbellatum* (Zhuk.) A. Löve) - 2
- subsp. **transcaucasica** V. Dorof. & Migusch. (*Kiharapyrum umbellatum* subsp. *transcaucasicum* (V. Dorof. & Migusch.) A. Löve) - 2

uniaristata Vis. (*Chennapyrum uniaristatum* (Vis.) A. Löve) - 2(alien)

vagans Jord. & Fourr. = A. geniculata

ventricosa Tausch (*Gastropyrum ventricosum* (Tausch) A. Löve) - 2(alien)

Aeluropus Trin.

badghyzi Tzvel. - 6

intermedius Regel (*A. littoralis* (Gouan.) Parl. subsp. *intermedius* (Regel) Tzvel.) - 1, 3, 6

korshinskyi Tzvel. (*A. littoralis* (Gouan) Parl. subsp. *korshinskyi* (Tzvel.) Tzvel.) - 6

lagopoides (L.) Trin. ex Thwaites subsp. *repens* (Desf.) Tzvel. = A. repens

littoralis (Gouan) Parl. (*A. littoralis* subsp. *kuschkensis* Tzvel.) - 1, 2, 6
- subsp. *intermedius* (Regel) Tzvel. = A. intermedius
- subsp. *korshinskyi* (Tzvel.) Tzvel. = A. korshinskyi
- subsp. *kuschkensis* Tzvel. = A. littoralis
- subsp. *pungens* (Bieb.) Tzvel. = A. pungens

mesopotamicus Nabel. = A. repens

pungens (Bieb.) C. Koch (*Poa pungens* Bieb., *Aeluropus littoralis* (Gouan) Parl. subsp. *pungens* (Bieb.) Tzvel.) - 1, 2, 3, 6

repens (Desf.) Parl. (*A. lagopoides* (L.) Trin. ex Thwaites subsp. *repens* (Desf.) Tzvel., *A. mesopotamicus* Nabel.) - 2, 6

x Agrohordeum auct. = x Elyhordeum

jordalii Melderis = x Elyhordeum jordalii

x Agropogon Fourn.

littoralis (Smith) C.E. Hubb. (*Polypogon littoralis* Smith). - Agrostis stolonifera L. x Polypogon monspeliensis (L.) Desf. - 6

Agropyron Gaertn. (*Kratzmannia* Opiz)

aegilopoides Drob. = Elytrigia gmelinii
aemulans (Nevski) N.M. Kuznetzov = Leymus aemulans
aitchisonii (Boiss.) P. Candargy = Elymus longe-aristatus subsp. canaliculatus
alaicum Drob. = Elytrigia alaica
alascanum Scribn. & Merr. = Elymus alascanus
alatavicum Drob. = Elytrigia alatavica
ambigens (Hausskn. ex Halacsy) Roshev. = Elytrigia pulcherrima
amgunense Nevski = Elytrigia amgunensis
amurense Drob. = Elymus amurensis
androssovii Roshev. = x Agrotrigia androssovii
angarense Peschkova - 4

angulare Nevski = Elytrigia caespitosa
angustifolium (Link) Schult. = A. fragile
angustiglume Nevski = Elymus mutabilis
- subsp. *irendykense* Nevski = Elymus mutabilis
x *apiculatum* Tscherning = Elytrigia x mucronata
arcuatum Golosk. = Elymus arcuatus
argenteum (Nevski) Pavl. = Elytrigia batalinii
armenum Nevski = Elytrigia armena
atbassaricum Golosk. = Elymus novae-angliae
attenuatiglume Nevski = Elytrigia attenuatiglumis
aucheri Boiss. = Elytrigia trichophora
badamense Drob. (*A. cristatum* (L.) Beauv. subsp. *badamense* (Drob.) A. Löve) - 6
barbulatum Schur = Elytrigia trichophora
batalinii (Krasn.) Roshev. = Elytrigia batalinii
x *bazargiciense* Prod. ex Burduja = Elytrigia x mucronata
bessarabicum Savul. & Rayss = Elytrigia bessarabica
boreale (Turcz.) Drob. subsp. *hyperarcticum* (Polun.) Melderis = Elymus hyperarcticus
brachypodioides (Nevski) Czerepnin = Elymus brachypodioides
brachypodioides (Nevski) Serg. = Elymus brachypodioides
breviaristatum Grossh. = Elytrigia x tesquicola
caesium J. & C. Presl = Elytrigia repens
caespitosum C. Koch = Elytrigia caespitosa
canaliculatum Nevski = Elymus longe-aristatus subsp. canaliculatus
caucasicum (C. Koch) Grossh. = Elymus caucasicus
chinense (Trin.) Ohwi = Leymus chinensis
chino-rossicum Ohwi = Leymus dasystachys
ciliare (Trin.) Franch. = Elymus ciliaris
ciliolatum Nevski = Elytrigia lolioides
cimmericum Nevski (*A. cristatum* (L.) Beauv. subsp. *birjutczense* (Lavr.) A. Löve, *A. dasyanthum* Ledeb. var. *birjutczense* Lavr., *A. dasyanthum* subsp. *birjutczense* (Lavr.) Lavr.) - 1
cognatum Hack. = Elytrigia cognata
confusum Roshev. var. *pubiflorum* Roshev. = Elymus pubiflorus
cretaceum Klok. & Prokud. = Elytrigia stipifolia
cristatum (L.) Beauv. (*Kratzmannia cristata* (L.) Jir. & Skalicky) - 1, 3, 4, 5(alien), 6
- subsp. *badamense* (Drob.) A. Löve = A. badamense
- subsp. *baicalense* Egor. & Sipl. = A. distichum
- subsp. *birjutczense* (Lavr.) A. Löve = A. cimmericum
- subsp. *dasyanthum* (Ledeb.) A. Löve = A. dasyanthum
- subsp. *desertorum* (Fisch. ex Link) A. Löve = A. desertorum
- subsp. *fragile* (Roth) A. Löve = A. fragile
- subsp. *imbricatum* (Bieb.) A. Löve = A. pectinatum
- subsp. *kasachstanicum* Tzvel. = A. kazachstanicum
- subsp. *michnoi* (Roshev.) A. Löve = A. michnoi
- subsp. *nathaliae* (Sipl.) A. Löve = A. nathaliae
- subsp. *pectinatum* (Bieb.) Tzvel. = A. pectinatum
- subsp. *pinifolium* (Nevski) Bondar. = A. pinifolium
- subsp. *ponticum* (Nevski) Tzvel. = A. ponticum
- subsp. *puberulum* (Boiss. ex Steud.) Tzvel. = A. puberulum
- subsp. *pumilum* (Steud.) A. Löve = A. pumilum
- subsp. *sabulosum* Lavr. = A. lavrenkoanum
- subsp. *sclerophyllum* Novopokr. = A. pinifolium
- subsp. *sibiricum* (Willd.) A. Löve = A. fragile
- subsp. *stepposum* (Dubovik) A. Löve = A. stepposum
- subsp. *tarbagataicum* (N. Plotnikov) Tzvel. = A. tarbagataicum
- var. *erickssonii* Melderis = A. erickssonii
- var. *pectiniforme* (Roem. & Schult.) Matveev = A. pectinatum

curvatiforme Nevski = Elymus curvatiformis
czilikense Drob. = Elymus x czilikensis
dagnae Grossh. = A. pectinatum
dasyanthum Ledeb. (*A. cristatum* (L.) Beauv. subsp. *dasyanthum* (Ledeb.) A. Löve) - 1
- subsp. *birjutczense* (Lavr.) Lavr. = A. cimmericum
- var. *birjutczense* Lavr. = A. cimmericum
desertorum (Fisch. ex Link) Schult. (*A. cristatum* (L.) Beauv. subsp. *desertorum* (Fisch. ex Link) A. Löve) - 1, 2, 3, 6
distichum (Georgi) Peschkova (*Bromus distichus* Georgi, *Agropyron cristatum* (L.) Beauv. subsp. *baicalense* Egor. & Sipl., *A. pectiniforme* Roem. & Schult. subsp. *baicalense* (Egor. & Sipl.) A. Löve) - 4
divaricatum auct. = Elytrigia attenuatiglumis
dolicholepis Melderis = Elymus sclerophyllus
dshungaricum (Nevski) Nevski = Elytrigia cognata
elongatiforme Drob. = Elytrigia elongatiformis
elongatum (Host) Beauv. = Elytrigia elongata
- subsp. *ruthenicum* (Griseb.) Anghel & Morariu = Elytrigia trichophora
erickssonii (Melderis) Peschkova (*A. cristatum* (L.) Beauv. var. *erickssonii* Melderis) - 3, 4
ferganense Drob. = Elytrigia cognata
fibrosum (Schrenk) P. Candargy = Elymus fibrosus
firmiculme Nevski = Elytrigia caespitosa
flexuosissimum Nevski = Elymus longe-aristatus subsp. flexuosissimus
fragile (Roth) P. Candargy (*A. angustifolium* (Link) Schult., *A. cristatum* (L.) Beauv. subsp. *fragile* (Roth) A. Löve, *A. cristatum* subsp. *sibiricum* (Willd.) A. Löve, *A. fragile* (Roth) Nevski, comb. superfl., *A. fragile* subsp. *sibiricum* (Willd.) Melderis, *A. sibiricum* (Willd.) Beauv., *Triticum angustifolium* Link) - 1, 2, 3, 6
- subsp. *sibiricum* (Willd.) Melderis = A. fragile
geniculatum (Trin.) C. Koch = Elytrigia geniculata
glaucissimum M. Pop. = Elymus glaucissimus
glaucum (Desf. ex DC.) Roem. & Schult. subsp. *barbulatum* (Schur) K. Richt. = Elytrigia trichophora
gmelinii (Ledeb.) Scribn. & J.G. Smith = Elymus gmelinii
gmelinii (Trin.) P. Candargy = Elytrigia gmelinii
- var. *reflexiaristatum* (Nevski) Serg. = Elytrigia reflexiaristata
- var. *roshevitzii* (Nevski) Malysch. = Elytrigia gmelinii
gracillimum Nevski = Elytrigia gracillima
hajastanicum Tzvel. = x Agrotrigia hajastanica
himalayanum (Nevski) Melderis = Elymus himalayanus
hispidum Opiz = Elytrigia intermedia
ilmense Roshev. ex Nevski = Elymus mutabilis
imbricatum Roem. & Schult. = A. pectinatum
intermedium (Host) Beauv. = Elytrigia intermedia
- subsp. *trichophorum* (Link) Reichenb. ex Hegi = Elytrigia trichophora
- subsp. *trichophorum* (Link) Volkart = Elytrigia trichophora
- var. *ambigens* Hausskn. ex Halacsy = Elytrigia pulcherrima
jacutorum Nevski = Elytrigia jacutorum
junceiforme A. & D. Löve = Elytrigia junceiformis
junceum (L.) Beauv. subsp. *boreo-atlanticum* Simon. & Guinoch. = Elytrigia junceiformis
- var. *bessarabicum* (Savul. & Rayss) Anghel & Morariu = Elytrigia bessarabica
junceum sensu Nevski = Elytrigia bessarabica
karadaghense Kotov = A. pinifolium
karataviense Pavl. = A. pectinatum
karawaewii P. Smirn. = Elytrigia villosa

karkaralense Roshev. = Elymus viridiglumis
kasteki M. Pop. = Elytrigia kasteki
kazachstanicum (Tzvel.) Peschkova (*A. cristatum* (L.) Beauv. subsp. *kasachstanicum* Tzvel.) - 3, 4, 6
kronokense Kom. = Elymus kronokensis
krylovianum Schischk. (*Elytrigia kryloviana* (Schischk.) Nevski) - 3, 4, 6
lachnophyllum (Ovcz. & Sidor.) Bondar. = Elymus lachnophyllus
latiglume (Scribn. & J.G. Smith) Rydb. subsp. *eurasiaticum* Hult. p.p. = Elymus kronokensis subsp. subalpinus
- subsp. *subalpinum* (L. Neum.) Vestergren, p.p. = Elymus kronokensis subsp. subalpinus
lavrenkoanum Prokud. (*A. cristatum* (L.) Beauv. subsp. *sabulosum* Lavr., *A. pectiniforme* Roem. & Schult. subsp. *sabulosum* (Lavr.) A. Löve) - 1
laxum (Fries) Fries = Elytrigia x littorea
lenense M. Pop. = Elymus lenensis
leptourum (Nevski) Grossh. = Elymus transhyrcanus
leptourum (Nevski) Pavl. = Elymus transhyrcanus
littorale Dumort. = Elytrigia pungens
litvinovii Prokud. = A. pectinatum
lolioides (Kar. & Kir.) P. Candargy = Elytrigia lolioides
lolioides (Kar. & Kir.) Roshev. = Elytrigia lolioides
longe-aristatum (Boiss.) Boiss. = Elymus longe-aristatus
- var. *aitchisonii* Boiss. = Elymus longe-aristatus subsp. canaliculatus
macrochaetum (Nevski) Bondar. = Elymus macrochaetus
macrochaetum (Nevski) Melderis = Elymus macrochaetus
macrolepis Drob. = Elymus x macrolepis
maeoticum Prokud. = Elytrigia maeotica
michnoi Roshev. (*A. cristatum* (L.) Beauv. subsp. *michnoi* (Roshev.) A. Löve) - 4
- subsp. *nathaliae* (Sipl.) Tzvel. = A. nathaliae
mucronatum Opiz ex Bercht. = Elytrigia x mucronata
mutabile Drob. = Elymus mutabilis
nathaliae Sipl. (*A. cristatum* (L.) Beauv. subsp. *nathaliae* (Sipl.) A. Löve, *A. michnoi* Roshev. subsp. *nathaliae* (Sipl.) Tzvel.) - 4
nodosum Nevski = Elytrigia nodosa
nomokonovii M. Pop. = Elymus macrourus
novae-angliae Scribn. = Elymus novae-angliae
orientale (L.) Roem. & Schult. subsp. *lanuginosum* (Griseb.) K. Richt. = Eremopyrum distans
- var. *sublanginosum* Drob. = Eremopyrum bonaepartis
oschense Roshev. = Elymus mutabilis
pallidissimum M. Pop. = Elymus transbaicalensis
panormitanum Parl. = Elymus panormitanus
pauciflorum auct. = Elymus novae-angliae
pauciflorum (Schwein.) Hitchc. = Elymus trachycaulus
- subsp. *novae-angliae* (Scribn.) Melderis = Elymus novae-angliae
pavlovii Nevski = x Elyhordeum pavlovii
pectinatum (Bieb.) Beauv. (*Triticum pectinatum* Bieb., *Agropyron cristatum* (L.) Beauv. subsp. *imbricatum* (Bieb.) A. Löve, *A. cristatum* subsp. *pectinatum* (Bieb.) Tzvel., *A. cristatum* var. *pectiniforme* (Roem. & Schult.) Matveev, *A. dagnae* Grossh., *A. imbricatum* Roem. & Schult., *A. karataviense* Pavl., *A. litvinovii* Prokud., *A. pectiniforme* Roem. & Schult., *Elymus pectinatus* (Bieb.) Lainz, *Kratzmannia pectinata* (Bieb.) Jir. & Skalicky, *Triticum caucasicum* Spreng. 1807, non Agropyron caucasicum (C. Koch) Grossh., *T. imbricatum* Bieb. 1808, non Lam. 1791, *T. muricatum* Link) - 1, 2, 3, 4, 5, 6
pectiniforme Roem. & Schult. = A. pectinatum
- subsp. *baicalense* (Egor. & Sipl.) A. Löve = A. distichum

- subsp. *sabulosum* (Lavr.) A. Löve = A. lavrenkoanum
pendulinum (Nevski) Worosch. = Elymus pendulinus
pertenue (C.A. Mey.) Nevski = Elytrigia pertenuis
peschkovae M. Pop. = Elymus novae-angliae
pinifolium Nevski (*A. cristatum* (L.) Beauv. subsp. *pinifolium* (Nevski) Bondar., *A. cristatum* subsp. *sclerophyllum* Novopokr. nom. altern., *A. karadaghense* Kotov, *A. sclerophyllum* Novopokr.) - 1, 2
ponticum Nevski (*A. cristatum* (L.) Beauv. subsp. *ponticum* (Nevski) Tzvel.) - 1
popovii Drob. = Elytrigia pulcherrima
praecaespitosum Nevski = Elymus praecaespitosus
propinquum Nevski = Elytrigia gmelinii
pruiniferum (Nevski) Nevski = Elytrigia pruinifera
pseudoagropyrum (Trin. ex Griseb.) Franch. = Leymus chinensis
pseudocaesium (Pacz.) Zoz = Elytrigia pseudocaesia
pseudostrigosum P. Candargy = Elymus schrenkianus
puberulum (Boiss. ex Steud.) Grossh. (*Triticum puberulum* Boiss. ex Steud., *Agropyron cristatum* (L.) Beauv. subsp. *puberulum* (Boiss. ex Steud.) Tzvel., *Eremopyrum puberulum* (Boiss. ex Steud.) Grossh. ex A. Löve) - 2
pubescens Schischk. = Elymus jacutensis
pulcherrimum Grossh. = Elytrigia pulcherrima
pumilum P. Candargy (*Triticum pumilum* Steud. 1854, non L. fil. 1781, *Agropyron cristatum* (L.) Beauv. subsp. *pumilum* (Steud.) A. Löve, *A. pumilum* (Steud.) Nevski, comb. illegit., *Elytrigia praetermissa* Nevski) - 4
pungens (Pers.) Roem. & Schult. = Elytrigia pungens
ramosum (Trin.) K. Richt. = Leymus ramosus
reflexiaristatum Nevski = Elytrigia reflexiaristata
repens (L.) P. Beauv. = Elytrigia repens
- subsp. *elongatiforme* (Drob.) D.R. Dewey = Elytrigia elongatiformis
- subsp. *pseudocaesium* (Pacz.) Lavr. = Elytrigia pseudocaesia
- var. *pseudocaesium* Pacz. = Elytrigia pseudocaesia
roshevitzii Nevski = Elytrigia gmelinii
ruthenicum (Griseb.) Prokud. = Elytrigia elongata
sachalinense Honda = Elytrigia repens
sajanense (Nevski) Grub. = Elymus sajanensis
schrenkianum (Fisch. & C.A. Mey.) P. Candargy = Elymus schrenkianus
sclerophyllum Novopokr. = A. pinifolium
scythicum Nevski = Elytrigia scythica
setuliferum (Nevski) Nevski = Elytrigia setulifera
sibiricum (Willd.) Beauv. = A. fragile
sinuatum Nevski = Elytrigia sinuata
sosnovskyi Hack. = Elytrigia sosnovskyi
stenophyllum Nevski = Elytrigia gmelinii
stepposum Dubovik (*A. cristatum* (L.) Beauv. subsp. *stepposum* (Dubovik) A. Löve) - 1, 2
stipifolium Czern. ex Nevski = Elytrigia stipifolia
strictum (Dethard.) Reichenb. = x Leymotrigia stricta
strigosum (Bieb.) Boiss. = Elytrigia strigosa
- subsp. *aegilopoides* (Drob.) Tzvel. = Elytrigia gmelinii
- subsp. *amgunense* (Nevski) Tzvel. = Elytrigia amgunensis
- subsp. *jacutorum* (Nevski) Tzvel. = Elytrigia jacutorum
- subsp. *reflexiaristatum* (Nevski) Tzvel. = Elytrigia reflexiaristata
subalpinum Golosk. = Elytrigia kasteki
tanaiticum Nevski - 1
tarbagataicum N. Plotnikov (*A. cristatum* (L.) Beauv. subsp. *tarbagataicum* (N. Plotnikov) Tzvel.) - 3, 6
tenerum auct. = Elymus novae-angliae
tenerum Vasey = Elymus trachycaulus

tesquicola Prokud. = Elytrigia x tesquicola
tianschanicum Drob.. = Elymus tianschanigenus
trachycaulon (Link) Steud. = Elymus trachycaulus
transhyrcanum (Nevski) Bondar. = Elymus transhyrcanus
transiliense M. Pop. = Elymus mutabilis
transnominatum Bondar. = Elymus sclerophyllus
trichophorum (Link) K. Richt. = Elytrigia trichophora
troctolepis (Nevski) Melderis = Elymus troctolepis
truncatum (Wallr.) Fuss subsp. *trichophorum* (Link) Soo = Elytrigia trichophora
tschimganicum Drob. = Elymus tschimganicus
tugarinovii Reverd. = Elymus jacutensis
turcomanicum Czopanov = x Agrotrigia androssovii
turkestanicum Gand. = Eremopyrum bonaepartis
turuchanense Reverd. = Elymus turuchanensis
uninerve P. Candargy = Leymus chinensis
vernicosum Nevski ex Grub. = Elymus brachypodioides
violaceum (Hornem.) Vasey var. *hyperarcticum* Polun. = Elymus hyperarcticus
wiluicum Drob. = x Leymotrigia wiluica

Agrostis L. (*Pentatherum* Nabel.)

abakanensis Less. ex Trin. = A. clavata
aenea (Trin.) Trin. = A. alascana
agrostidiformis (Roshev.) Bor (*Calamagrostis agrostidiformis* Roshev., *Agrostis olympica* (Boiss.) Bor subsp. *agrostidiformis* (Roshev.) Tzvel., *Pentatherum agrostidiforme* (Roshev.) Nevski) - 6
alascana Hult. (*A. aenea* (Trin.) Trin. 1841, non Spreng. 1827, *A. canina* var. *aenea* Trin.) - 5
alba auct. = A. gigantea
alba L. subsp. *gigantea* (Roth) Jir. = A. gigantea
- subsp. *stolonifera* (L.) Jir. = A. stolonifera
- subsp. *stolonizans* (Bess. ex Schult. & Schult. fil.) Lavr. = A. stolonifera
albida Trin. (*A. stolonifera* L. subsp. *albida* (Trin.) Tzvel.) - 1, 3
alpina Scop. - 1
x **amurensis** Probat. - A. scabra Willd. x A. trinii Turcz. - 5
anadyrensis Socz. - 4, 5
angrenica (Butk.) Tzvel. (*Pentatherum angrenicum* Butk.) - 6
balansae (Boiss.) Tzvel. (*Calamagrostis balansae* Boiss., *Pentatherum balansae* (Boiss.) Nevski) - 2
    Even if Pentatherum Nabel. is recognized as a separate genus, this species should belong to the genus Agrostis.
biebersteiniana Claus = Zingeria biebersteiniana
x **bjoerkmanii** Widen. - A. giganthea Roth x A. tenuis Sibth. - 1
bodaibensis Peschkova - 4
borealis C. Hartm. (*A. mertensii* Trin. subsp. *borealis* (C. Hartm.) Tzvel.) - 1
- subsp. *viridissima* (Kom.) Tzvel. = A. mertensii
- var. *flaccida* (Hack.) T. Koyama ex T. Shimizu = A. flaccida
canina L. (*A. sudavica* Natk.-Ivanausk.) - 1
- subsp. *montana* (C. Hartm.) C. Hartm. = A. vinealis
- subsp. *pusilla* (Dumort.) Malagarriga = A. vinealis
- subsp. *trinii* (Turcz.) Hult. = A. trinii
- var. *aenea* Trin. = A. alascana
- var. *montana* C. Hartm. = A. vinealis
capillaris auct. = A. tenuis
x **castriferrei** Waisb. - A. canina L. x A. stolonifera L.
clavata Trin. (*A. abakanensis* Less. ex Trin., *A. teberdensis* Litv.) - 1, 2, 3
- subsp. *matsumurae* (Hack. ex Honda) Tateoka = A. macrothyrsa

x **clavatiformis** Probat. - A. clavata Trin. x A. flaccida Hack. - 5

coarctata Ehrh. ex Hoffm. = A. vinealis
- subsp. *hyperborea* (Laest.) H. Scholz = A. vinealis
- subsp. *syreistschikowii* (P. Smirn.) H. Scholz = A. vinealis
- subsp. *trinii* (Turcz.) H. Scholz = A. trinii
densiflora auct. = A. exarata
diffusa Host = A. gigantea
divaricatissima Mez (*A. mongholica* Roshev.) - 3, 4, 5(alien)
elegans Thore = Neoschischkinia elegans
exarata Trin. (? *A. densiflora* auct.) - 5
fertilis Steud. = Sporobolus fertilis
flaccida Hack. (*A. borealis* C. Hartm. var. *flaccida* (Hack.) T. Koyama ex T. Shimizu) - 5
x **fouilladei** Fourn. - A. canina L. x A. tenuis Sibth.
geminata Trin. - 5
gigantea Roth (*A. alba* L. subsp. *gigantea* (Roth) Jir. 1950, non Spenn. 1825, *A. diffusa* Host, *A. graniticola* Klok., *A. praticola* Klok., *A. stolonifera* L. subsp. *gigantea* (Roth) Beldie, comb. superfl., *A. stolonifera* subsp. *gigantea* (Roth) Maire & Weiller, *Cinna karataviensis* Pavl., *Agrostis alba* auct.) - 1, 2, 3, 4, 5, 6
- subsp. *maeotica* (Klok.) Tzvel. = A. maeotica
graniticola Klok. = A. gigantea
hiemalis auct. = A. scabra
hissarica Roshev. (*A. tianschanica* Pavl., *Polypogon hissaricus* (Roshev.) Bor) - 6
- subsp. *pamirica* (Ovcz.) Tzvel. = A. pamirica
hyperborea Laest. = A. vinealis
jacutica Schischk. = A. stolonifera
x **kamtschatica** Probat. - A. kudoi Honda x A. scabra Willd. - 5
karsensis Litv. = A. stolonifera
kolymensis Kuvajev & A. Khokhr. - 5
korczaginii Senjan.-Korcz. - 1
kronokensis Probat. - 5
kudoi Honda (*A. trinii* Turcz. subsp. *kudoi* (Honda) Worosch., *A. vinealis* Schreb. subsp. *kudoi* (Honda) Tzvel.) - 4, 5
x **lapponica** Montell. - A. borealis C. Hartm. x A. tenuis Sibth. - 1
lazica Balansa - 2
lithuanica Bess. ex Schult. & Schult. fil. = A. tenuis
macrantha Schischk. = A. stolonifera
macrothyrsa Hack. (*A. clavata* Trin. subsp. *matsumurae* (Hack. ex Honda) Tateoka, *A. matsumurae* Hack. ex Honda, *A. perennans* auct.) - 5
maeotica Klok. (*A. gigantea* Roth subsp. *maeotica* (Klok.) Tzvel.) - 1
maritima auct. = A. straminea
marschalliana Seredin. = A. vinealis
matsumurae Hack. ex Honda = A. macrothyrsa
mertensii Trin. (*A. borealis* C. Hartm. subsp. *viridissima* (Kom.) Tzvel., *A. viridissima* Kom.) - 5
- subsp. *borealis* (C. Hartm.) Tzvel. = A. borealis
mongholica Roshev. = A. divaricatissima
x **murbeckii** Fouill. - A. stolonofera L. x A. tenuis Sibth.
nebulosa Boiss. & Reut. = Neoschischkinia nebulosa
nevskii Tzvel. (*Calamagrostis hissarica* Nevski, *Pentatherum hissaricum* (Nevski) Nevski) - 6

> Even if Pentatherum Nabel. is recognized as a separate genus, this species should belong to the genus Agrostis.

olympica (Boiss.) Bor (*Calamagrostis olympica* Boiss., *C. ruprechtii* Nevski, *Pentatherum olympicum* (Boiss.) Nabel., *P. ruprechtii* (Nevski) Nevski) - 2
- subsp. *agrostidiformis* (Roshev.) Tzvel. = A. agrostidi-formis

palustris Huds. = A. stolonifera
pamirica Ovcz. (*A. hissarica* Roshev. subsp. *pamirica* (Ovcz.) Tzvel.) - 6
x **paramushirensis** Probat. - A. alascana Hult. x A. flaccida Hack. - 5
paulsenii Hack. - 6
pauzhetkica Probat. - 5
perennans auct. = A. macrothyrsa
pisidica Boiss. = Zingeria pisidica
planifolia C. Koch (*A. vinealis* Schreb. subsp. *planifolia* (C. Koch) Tzvel.) - 2
praticola Klok. = A. gigantea
prorepens (Koch) Golub. = A. stolonifera
pseudoalba Klok. = A. stolonifera
pusilla Dumort. = A. vinealis
rupestris All. - 1
sabulicola Klok. - 1
salsa Korsh. - 1, 3
scabra Willd. (*A. hiemalis* auct.) - 5
semiverticillata (Forssk.) C. Chr. = Polypogon viridis
sibirica V. Petrov = A. stolonifera
sichotensis Provat. - 5
sokolovskajae Probat. (*A. stolonifera* L. subsp. *sokolovskajae* (Probat.) Worosch.) - 5
stolonifera L. (*A. alba* L. subsp. *stolonifera* (L.) Jir., *A. alba* subsp. *stolonizans* (Bess. ex Schult. & Schult. fil.) Lavr., *A. jacutica* Schischk., *A. karsensis* Litv., *A. macrantha* Schischk., *A. palustris* Huds., *A. prorepens* (Koch) Golub. nom. illegit., *A. pseudoalba* Klok., *A. sibirica* V. Petrov, *A. stolonifera* subsp. *palustris* (Huds.) Tzvel., *A. stolonifera* subsp. *stolonizans* (Bess. ex Schult. & Schult. fil.) Soo, *A. stolonizans* Bess. ex Schult. & Schult. fil., *A. zerovii* Klok.) - 1, 2, 3, 4, 5, 6
- subsp. *albida* (Trin.) Tzvel. = A. albida
- subsp. *gigantea* (Roth) Beldie = A. gigantea
- subsp. *gigantea* (Roth) Maire & Weiller = A. gigantea
- subsp. *palustris* (Huds.) Tzvel. = A. stolonifera
- subsp. *sokolovskajae* (Probat.) Worosch. = A. sokolovskajae
- subsp. *stolonizans* (Bess. ex Schult. & Schult. fil.) Soo = A. stolonifera
- subsp. *straminea* (C. Hartm.) Tzvel. = A. straminea
- subsp. *transcaspica* (Litv.) Tzvel. = A. transcaspica
stolonizans Bess. ex Schult. & Schult. fil. = A. stolonifera
straminea C. Hartm. (*A. stolonifera* L. subsp. *straminea* (C. Hartm.) Tzvel., *A. maritima* auct.) - 1, 5(alien)
stricta J.F. Gmel. = A. vinealis
- subsp. *syreistchikowii* (P. Smirn.) Soo = A. vinealis
subaristata Aitch. & Hemsl. - 6
x **subclavata** Probat. - A. clavata Trin. x A. kudoi Honda - 5
sudavica Natk.-Ivanausk. = A. canina
syreistschikowii P. Smirn. = A. vinealis
teberdensis Litv. = A. clavata
tenuifolia Bieb. = A. vinealis
tenuis Sibth. (*A. lithuanica* Bess. ex Schult. & Schult. fil., *A. vulgaris* With., *A. capillaris* auct.) - 1, 2, 3, 4, 5, 6
tianschanica Pavl. = A. hissarica
transcaspica Litv. (*A. stolonifera* L. subsp. *transcaspica* (Litv.) Tzvel.) - 2, 6
trichantha (Schischk.) Tzvel. (*Calamagrostis trichantha* Schischk., *Pentatherum trichanthum* (Schischk.) Nevski) - 2
trinii Turcz. (*A. canina* L. subsp. *trinii* (Turcz.) Hult., *A. coarctata* Ehrh. ex Hoffm. subsp. *trinii* (Turcz.) H. Scholz, *A. vinealis* Schreb. subsp. *trinii* (Turcz.) Tzvel.) - 3, 4, 5

- subsp. *kudoi* (Honda) Worosch. = A. kudoi
turkestanica Drob. (*A. vinealis* Schreb. subsp. *turkestanica* (Drob.) Tzvel.) - 6

> Probably, the priority names of this species should be A. wightii Nees ex Steud.

tuvinica Peschkova - 3, 4
x **ussuriensis** Probat. - A. stolonifera L. x A. trinii Turcz. - 5
ventricosa Gouan = Gastridium ventricosum
verticillata Vill. = Polypogon viridis
villosa Chaix = Calamagrostis villosa
vinealis Schreb. (*A. canina* L. var. *montana* C. Hartm., *A. canina* subsp. *montana* (C. Hartm.) C. Hartm., *A. canina* subsp. *pusilla* (Dumort.) Malagarriga, *A. coarctata* Ehrh. ex Hoffm., *A. coarctata* subsp. *hyperborea* (Laest.) H. Scholz, *A. coarctata* subsp. *syreistschikowii* (P. Smirn.) H. Scholz, *A. hyperborea* Laest., *A. marschalliana* Seredin, *A. pusilla* Dumort., *A. stricta* J.F. Gmel., *A. stricta* subsp. *syreistchikowii* (P. Smirn.) Soo, *A. syreistschikowii* P. Smirn., *A. tenuifolia* Bieb. 1808, non Curt. 1787, *A. vinealis* subsp. *syreitschikowii* (P. Smirn.) Soo) - 1, 2, 3, 4
- subsp. *kudoi* (Honda) Tzvel. = A. kudoi
- subsp. *planifolia* (C. Koch) Tzvel. = A. planifolia
- subsp. *syreitschikowii* (P. Smirn.) Soo = A. vinealis
- subsp. *trinii* (Turcz.) Tzvel. = A. trinii
- subsp. *turkestanica* (Drob.) Tzvel. = A. turkestanica
viridis Gouan = Polypogon viridis
viridissima Kom. = A. mertensii
vulgaris With. = A. tenuis
zerovii Klok. = A. stolonifera

x **Agrotrigia** Tzvel.

androssovii (Roshev.) Tzvel. (*Agropyron androssovii* Roshev., *A. turcomanicum* Czopanov). - Agropyron pectinatum (Bieb.) Beauv. x Elytrigia trichophora (Link) Nevski - 6
hajastanica (Tzvel.) Tzvel. (*Agropyron hajastanicum* Tzvel.). - Agropyron ? pectinatum (Bieb.) Beauv. x Elytrigia ? repens (L.) Nevski, s.l. - 2
kotovii Tzvel. - Agropyron ? pinifolium Nevski x Elytrigia ? repens (L.) Nevski, s.l. - 1

Aira L.

alpina auct. = Hierochloe alpina
ambigua De Not. = A. elegantissima
canescens L. = Corynephorus canescens
capillaris Host = A. elegans
- subsp. *ambigua* Arcang. = A. elegantissima
- var. *ambigua* (Arcang.) Aschers. & Graebn. = A. elegantissima
caryophyllea L. - 1, 2(?)
cristata L. = Koeleria cristata
divaricata Pourr. = Corynephorus divaricatus
elegans Willd. ex Gaudin (*A. capillaris* Host, 1809, non Savi, 1798, nec Lag. 1805, *A. elegantissima* Schur subsp. *hosteana* Holub) - 1, 2, 6
- subsp. *ambigua* (Arcang.) Holub = A. elegantissima
- subsp. *notarisiana* (Steud.) Sojak = A. elegantissima
elegantissima Schur (*A. ambigua* De Not. 1844, non Michx. 1803, *A. capillaris* Host subsp. *ambigua* Arcang., *A. capillaris* var. *ambigua* (Arcang.) Aschers. & Graebn., *A. elegans* Willd. ex Gaudin subsp. *ambigua* (Arcang.) Holub, *A. elegans* subsp. *notarisiana* (Steud.) Sojak, *A. notarisiana* Steud., *A. elegantissima* subsp. *ambigua* (Arcang.) M. Dogan) - 2
- subsp. *ambigua* (Arcang.) M. Dogan = A. elegantissima

- subsp. *hosteana* Holub = A. elegans
glauca Spreng. = Koeleria glauca
laevigata Smith = Deschampsia alpina
laevis Brot. = Molineriella laevis
macrantha Ledeb. = Koeleria macrantha
montana L. = Avenella flexuosa subsp. montana
notarisiana Steud. = A. elegantissima
parviflora Thuill. = Deschampsia cespitosa subsp. parviflora
praecox L. - 1, 2
pulchella Willd. = Molineriella laevis
setacea Huds. = Deschampsia setacea
varia Jacq. p.p. = Sesleria albicans
wilhelmsii Steud. = Deschampsia cespitosa subsp. wilhelmsii

Alopecurus L.

aequalis Sobol. (*A. aequalis* subsp. *amurensis* (Kom.) Hult., *A. aequalis* subsp. *aristulatus* (Michx.) Tzvel., *A. amurensis* Kom., *A. aristulatus* Michx., *A. fulvus* Smith, *A. geniculatus* L. subsp. *fulvus* (Smith) C. Hartm.) - 1, 2, 3, 4, 5, 6
- subsp. *amurensis* (Kom.) Hult. = A. aequalis
- subsp. *aristulatus* (Michx.) Tzvel. = A. aequalis
albovii Tzvel. (? *A. gerardii* auct.) - 2
alpestris (Wahlenb.) Czer. (*A. pratensis* L. var. *alpestris* Wahlenb., *A. pratensis* subsp. *alpestris* (Wahlenb.) K. Richt., *A. pratensis* subsp. *alpestris* (Wahlenb.) Seland. comb. superfl.) - 1
alpinus Smith (*A. alpinus* subsp. *borealis* (Trin.) Jurtz., *A. alpinus* var. *altaicus* Griseb., *A. altaicus* (Griseb.) V. Petrov, *A. borealis* Trin.) - 1, 3, 4, 5
- subsp. *borealis* (Trin.) Jurtz. = A. alpinus
- subsp. *glaucus* (Less.) Hult. = A. glaucus
- subsp. *pseudobrachystachyus* (Ovcz.) Tzvel. = A. pseudobrachystachyus
- subsp. *stejnegeri* (Vasey) Hult. = A. stejnegeri
- var. *altaicus* Griseb. = A. alpinus
- var. *songaricus* Schrenk = A. pratensis
altaicus (Griseb.) V. Petrov = A. alpinus
amurensis Kom. = A. aequalis
apiatus Ovcz. - 6
aristulatus Michx. = A. aequalis
armenus (C. Koch) Grossh. (*A. pratensis* L. var. *armenus* C. Koch, *A. arundinaceus* Poir. subsp. *armenus* (C. Koch) Tzvel.) - 2
arundinaceus Poir. (*A. exaltatus* Less., *A. muticus* Kar. & Kir., *A. repens* Bieb. nom. illegit., *A. ventricosus* Pers. 1805, non Huds. 1778) - 1, 2, 3, 4, 5, 6
- subsp. *armenus* (C. Koch) Tzvel. = A. armenus
aucheri Boiss. (*A. vaginatus* (Willd.) Pall. ex Kunth subsp. *aucheri* (Boiss.) Westb.) - 2
borealis Trin. = A. alpinus
borii Tzvel. - 6
brachystachyus Bieb. (*A. pratensis* L. subsp. *brachystachyus* (Bieb.) Trab.) - 4, 5
x **brachystylus** Peterm. - A. geniculatus L. x A. pratensis L.
brevifolius Grossh. = A. glacialis and A. vaginatus
caucasicus Seredin = A. ponticus
dasyanthus Trautv. - 2
exaltatus Less. = A. arundinaceus
fulvus Smith = A. aequalis
geniculatus L. - 1, 5(alien)
- subsp. *fulvus* (Smith) C. Hartm. = A. aequalis
gerardii auct. = A. albovii
glacialis C. Koch (*A. brevifolius* Grossh. p.p. incl. typo, *A. vaginatus* (Willd.) Pall. ex Kunth subsp. *glacialis* (C. Koch) Westb.) - 2

- subsp. *tuscheticus* (Trautv.) Tzvel. = A. tuscheticus
glaucus Less. (*A. alpinus* Smith subsp. *glaucus* (Less.) Hult., *A. roshevitzianus* Ovcz., *A. tenuis* Kom.) - 1, 3(?), 4, 5, 6
x **haussknechtianus** Aschers. & Graebn. - A. geniculatus L. x A. aequalis Sobol.
himalaicus Hook. fil. - 6
laguriformis Schur (*A. pratensis* L. subsp. *laguriformis* (Schur) Tzvel.) - 1
laxiflorus Ovcz. = A. pratensis
longiaristatus Maxim. - 5
longifolius Kolak. = A. vaginatus
x **marssonii** Hausskn. ex Schennik. - A. geniculatus L. x A. arundinaceus Poir.
mucronatus Hack. - 6
muticus Kar. & Kir. = A. arundinaceus
myosuroides Huds. - 1, 2, 5(alien), 6
nepalensis Trin. ex Steud. - 6
ponticus C. Koch (*A. caucasicus* Seredin, *A. sericeus* Albov) - 2
pratensis L. (*A. alpinus* Smith var. *songaricus* Schrenk, *A. laxiflorus* Ovcz., *A. seravschanicus* Ovcz., *A. songaricus* (Schrenk) V. Petrov) - 1, 2, 3, 4, 5, 6
- subsp. *alpestris* (Wahlenb.) K. Richt. = A. alpestris
- subsp. *alpestris* (Wahlenb.) Seland. = A. alpestris
- subsp. *brachystachyus* (Bieb.) Trab. = A. brachystachyus
- subsp. *laguriformis* (Schur) Tzvel. = A. laguriformis
- var. *alpestris* Wahlenb. = A. alpestris
- var. *armenus* C. Koch = A. armenus
pseudobrachystachyus Ovcz. (*A. alpinus* Smith subsp. *pseudobrachystachyus* (Ovcz.) Tzvel.) - 4, 5
repens Bieb. = A. arundinaceus
roshevitzianus Ovcz. = A. glaucus
seravschanicus Ovcz. = A. pratensis
sericeus Albov = A. ponticus
songaricus (Schrenk) V. Petrov = A. pratensis
stejnegeri Vasey (*A. alpinus* Smith subsp. *stejnegeri* (Vasey) Hult.) - 5
tenuis Kom. = A. glaucus
textilis Boiss. - 2
- subsp. *tiflisiensis* (Westb.) Tzvel. = A. tiflisiensis
tiflisiensis (Westb.) Grossh. (*A. textilis* Boiss. subsp. *tiflisiensis* (Westb.) Tzvel.) - 2, 6
turczaninovii Nikiforova - 4
tuscheticus Trautv. (*A. glacialis* C. Koch subsp. *tuscheticus* (Trautv.) Tzvel.) - 2
typhoides Burm. fil. = Pennisetum americanum convar. typhoides
vaginatus (Willd.) Pall. ex Kunth (*Polypogon vaginatus* Willd., *Alopecurus brevifolius* Grossh. p.p. excl. typo, *A. longifolius* Kolak.) - 1, 2
- subsp. *aucheri* (Boiss.) Westb. = A. aucheri
- subsp. *glacialis* (C. Koch) Westb. = A. glacialis
ventricosus Pers. = A. arundinaceus
x **winklerianus** Aschers. & Graebn. - A. aequalis Sobol. x A. pratensis L.

Amblyopyrum (Jaub. & Spach) Eig

muticum (Boiss.) Eig (*Aegilops mutica* Boiss., *A. loliacea* Jaub. & Spach, *A. mutica* subsp. *loliacea* (Jaub. & Spach) Zhuk., *Amblyopyrum muticum* subsp. *loliaceum* (Jaub. & Spach) A. Löve) - 2
- subsp. *loliaceum* (Jaub. & Spach) A. Löve = A. muticum

x *Ammocalamagrostis* P. Fourn. = x Calammophila

baltica (Flugge ex Schrad.) P. Fourn. = x Calammophila baltica

Ammochloa Boiss.

palaestina Boiss. - 2

Ammophila Host

arenaria (L.) Link - 1
baltica (Flugge ex Schrad.) Link = x Calammophila baltica

Amphilopsis Nash = Bothriochloa

ischaemum (L.) Nash = Bothriochloa ischaemum

Anachortus Jir. & Chrtek = Corynephorus

articulatus (Desf.) Jir. & Chrtek = Corynephorus divaricatus
divaricatus (Pourr.) Lainz = Corynephorus divaricatus

Andropogon L.

amplexifolius Trin. = Arthraxon langsdorffii
bladhii Retz. = Bothriochloa bladhii
caucasicus Trin. = Bothriochloa bladhii
citratus DC. ex Nees = Cymbopogon citratus
halepensis (L.) Brot. var. *sudanensis* (Piper) Suss. = Sorghum sudanense
ischaemum L. = Bothriochloa ischaemum
polydactylon L. = Chloris polydactyla
ravennae L. = Erianthus ravennae
vimineus Trin. = Microstegium vimineum
virginicus L. - 2

Aneurolepidium Nevski = Leymus

aemulans Nevski = Leymus aemulans
akmolinense (Drob.) Nevski = Leymus akmolinensis
alaicum (Korsh.) Nevski = Leymus alaicus
angustum (Trin.) Nevski = Leymus angustus
baldshuanicum (Roshev.) Nevski = Leymus tianschanicus
chinense (Trin.) Kitag. = Leymus chinensis
dasystachys (Trin.) Nevski, p.p. = Leymus dasystachys
divaricatum (Drob.) Nevski = Leymus divaricatus
fasciculatum (Roshev.) Nevski = Leymus fasciculatus
flexile Nevski = Leymus flexilis
karataviense (Roshev.) Nevski = Leymus karataviensis
karelinii (Turcz.) Nevski = Leymus karelinii and L. kopetdaghensis
kopetdaghense Nevski = Leymus kopetdaghensis
multicaule (Kar. & Kir.) Nevski = Leymus multicaulis
ovatum (Trin.) Nevski = Leymus ovatus
paboanum (Claus) Nevski = Leymus paboanus
petraeum Nevski = Leymus petraeus
pseudoagropyrum (Trin. ex Griseb.) Nevski = Leymus chinensis
ramosum (Trin.) Nevski = Leymus ramosus
regelii (Roshev.) Nevski = Leymus divaricatus
secalinum (Georgi) Kitag. = Leymus secalinus
tianschanicum (Drob.) Nevski = Leymus tianschanicus
ugamicum (Drob.) Nevski = Leymus alaicus

Anisantha C. Koch

diandra (Roth) Tutin (*Bromus diandrus* Roth, *Anisantha gussonii* (Parl.) Nevski, *A. rigens* (L.) Nevski, p.p.,

345

quoad plantas, *Bromus gussonii* Parl., *B. rigidus* Roth subsp. *gussonii* (Roth) Maire, *Zerna gussonii* (Parl.) Grossh., *Bromus rigens* auct.) - 1, 2, 5(alien), 6(alien)

gussonii (Parl.) Nevski = A. diandra

madritensis (L.) Nevski (*Bromus madritensis* L., *Zerna madritensis* (L.) S.F. Gray, *Z. madritensis* (L.) Grossh. comb. superfl.) - 1

rigens (L.) Nevski, p.p. = A. diandra

rubens (L.) Nevski (*Bromus rubens* L., *Zerna rubens* (L.) Grossh.) - 1(alien), 2, 6

sericea (Drob.) Nevski (*Bromus sericeus* Drob.) - 6

sterilis (L.) Nevski (*Bromus sterilis* L., *Zerna sterilis* (L.) Panz.) - 1, 2, 5(alien), 6

tectorum (L.) Nevski (*Bromus tectorum* L., *Zerna tectorum* (L.) Lindm.) - 1, 2, 3, 5(alien), 6

Anthosachne Steud. = Elymus

jacquemontii (Hook. fil.) Nevski, p.p. = Elymus longe-aristatus subsp. canaliculatus

longe-aristata (Boiss.) Nevski = Elymus longe-aristatus

Anthoxanthum L.

alpinum A. & D. Löve (*A. odoratum* L. subsp. *alpinum* (A. & D. Löve) B.M.G. Jones & Melderis) - 1, 2, 3, 4, 6

amarum Brot. (*A. odoratum* L. subsp. *amarum* (Brot.) K. Richt.) - 2

arcticum J.F. Veldkamp = Hierochloe pauciflora

aristatum Boiss. - 1

australe (Schrad.) J.F. Veldkamp = Hierochloe australis

glabrum (Trin.) J.F. Veldkamp = Hierochloe glabra

hirtum (Schrank) Y. Schouten & J.F. Veldkamp = Hierochloe hirta

nipponicum Honda (*A. odoratum* L. subsp. *nipponicum* (Honda) Tzvel.) - 5

nipponicum sensu Worosch. = A. odoratum

odoratum L. (*A. nipponicum* sensu Worosch.) - 1, 2, 3, 4, 5(alien)

- subsp. *alpinum* (A. & D. Löve) B.M.G. Jones & Melderis = A. alpinum

- subsp. *amarum* (Brot.) K. Richt. = A. amarum

- subsp. *nipponicum* (Honda) Tzvel. = A. nipponicum

repens (Host) J.F. Veldkamp = Hierochloe repens

stepporum (P. Smirn.) J.F. Veldkamp = Hierochloe repens

Apera Adans.

intermedia Hack. - 2, 5(alien)

interrupta (L.) Beauv. - 1, 2, 6

longiseta Klok. = A. spica-venti

maritima Klok. (*A. spica-venti* (L.) Beauv. subsp. *maritima* (Klok.) Tzvel.) - 1, 6

spica-venti (L.) Beauv. (*A. longiseta* Klok.) - 1, 2, 3, 4, 5, 6

- subsp. *maritima* (Klok.) Tzvel. = A. maritima

Apluda L.

inermis Regel = A. mutica

mutica L. (*A. inermis* Regel) - 6

Arctagrostis Griseb.

anadyrensis V. Vassil. = A. latifolia

arundinacea (Trin.) Beal (*A. caespitans* V. Vassil., *A. calamagrostidiformis* V. Vassil., *A. festucacea* V. Petrov, *A. latifolia* (R. Br.) Griseb. subsp. *arundinacea* (Trin.) Tzvel., *A. tenuis* V. Vassil., *A. ursorum* (Kom.) Roshev., *A. viridula* V. Vassil.) - 3, 4, 5

caespitans V. Vassil. = A. arundinacea

calamagrostidiformis V. Vassil. = A. arundinacea

festucacea V. Petrov = A. arundinacea

latifolia (R. Br.) Griseb. (*A. anadyrensis* V. Vassil., *A. latifolia* subsp. *gigantea* Tzvel.) - 1, 3, 4, 5

- subsp. *arundinacea* (Trin.) Tzvel. = A. arundinacea

- subsp. *gigantea* Tzvel. = A. latifolia

tenuis V. Vassil. = A. arundinacea

ursorum (Kom.) Roshev. = A. arundinacea

viridula V. Vassil. = A. arundinacea

x **Arctodupontia** Tzvel.

scleroclada (Rupr.) Tzvel. (*Poa scleroclada* Rupr.). - Arctophila fulva (Trin.) Anderss. x Dupontia psilantha Rupr. - 1, 5

Arctophila (Rupr.) Anderss.

effusa Lange = A. fulva

fulva (Trin.) Anderss. (*A. effusa* Lange, *A. fulva* subsp. *similis* (Rupr.) Tzvel., *Poa laestadii* Rupr., *P. latiflora* Rupr., *P. poecilantha* Rupr., *P. remotiflora* Rupr., *P. similis* Rupr., *P. trichoclada* Rupr.) - 1, 3, 4, 5

- subsp. *similis* (Rupr.) Tzvel. = A. fulva

Arctopoa (Griseb.) Probat.

eminens (C. Presl) Probat. (*Poa eminens* C. Presl, *P. alexeji* Sofeikova & Worosch., *P. eminens* var. *alexeji* (Sofeikova & Worosch.) Worosch., *P. kurilensis* Hack.) - 5

schischkinii (Tzvel.) Probat. (*Poa schischkinii* Tzvel.) - 3

subfastigiata (Trin.) Probat. (*Poa subfastigiata* Trin.) - 3, 4, 5

tibetica (Munro ex Stapf) Probat. (*Poa tibetica* Munro ex Stapf, *P. ciliatiflora* Roshev., *P. fedtschenkoi* Roshev., *P. princei* Simps.) - 3, 4, 6

trautvetteri (Tzvel.) Probat. (*Poa trautvetteri* Tzvel., *P. glumaris* Trin. var. *laevigata* Trautv.) - 4

Aristella (Trin.) Bertol. = Achnatherum

bromoides (L.) Bertol. = Achnatherum bromoides

Aristida L.

adscensionis auct. = A. heymannii

adscensionis L. subsp. *heymannii* (Regel) Tzvel. = A. heymannii

arachnoidea Litv. = Stipagrostis arachnoidea

heymannii Regel (*A. adscensionis* L. subsp. *heymannii* (Regel) Tzvel., *A. adscensionis* auct.) - 2, 6

karelinii (Trin. & Rupr.) Roshev. = Stipagrostis karelinii

longespica Poir. - 2(alien)

pennata Trin. = Stipagrostis pennata

- var. *minor* Litv. = Stipagrostis pennata subsp. minor

plumosa L. = Stipagrostis plumosa

- subsp. *kyzylkumica* Tzvel. = Stipagrostis plumosa subsp. kyzylkumica

- subsp. *szovitsiana* (Trin. & Rupr.) Tzvel. = Stipagrostis plumosa subsp. szovitsiana

- var. *szovitsiana* Trin. & Rupr. = Stipagrostis plumosa subsp. szovitsiana

Arrhenatherum Beauv.

bulbosum (Willd.) C. Presl = A. elatius subsp. nodosum

compactum (Boiss. & Heldr.) Potztal = Danthoniastrum compactum

elatius (L.) J. & C. Presl - 1, 2, 3, 5(alien), 6
- subsp. **nodosum** (Parl.) Arcang. (*A. bulbosum* (Willd.) C. Presl) - 2
fedtschenkoi (Hack.) Potztal = Helictotrichon fedtschenkoi
hookeri (Scribn.) Potztal = Helictotrichon hookeri
kotschyi Boiss. - 2
mongolicum (Roshev.) Potztal = Helictotrichon mongolicum

Arthratherum auct. = Stipagrostis

arachnoideum (Litv.) Tzvel. = Stipagrostis arachnoidea
karelinii (Trin. & Rupr.) Tzvel. = Stipagrostis karelinii
pennatum (Trin.) Tzvel. = Stipagrostis pennata
- subsp. *minus* (Litv.) Tzvel. = Stipagrostis pennata subsp. minor
plumosum (L.) Nees = Stipagrostis plumosa
- subsp. *kyzylkumicum* (Tzvel.) Tzvel. = Stipagrostis plumosa subsp. kyzylkumica
- subsp. *szovitsianum* (Trin. & Rupr.) Tzvel. = Stipagrostis plumosa subsp. szovitsiana

Arthraxon Beauv. (*Pleuroplitis* Trin.)

caucasicus (Rupr. ex Regel) Tzvel. (*Pleuroplitis langsdorffii* Trin. var. *caucasica* Rupr. ex Regel, *Arthraxon hispidus* (Thunb.) Makino subsp. *caucasicus* (Rupr. ex Regel) Tzvel., *Pleuroplitis caucasica* (Rupr. ex Regel) Trautv.) - 2
centrasiaticus (Griseb.) Gamajun (*Pleuroplitis centrasiatica* Griseb., *Arthraxon hispidus* (Thunb.) Makino subsp. *centrasiaticus* (Griseb.) Tzvel.) - 2, 5(alien), 6
hispidus (Thunb.) Makino (*Phalaris hispida* Thunb.) - 2, 5
- subsp. *caucasicus* (Rupr. ex Regel) Tzvel. = A. caucasicus
- subsp. *centrasiaticus* (Griseb.) Tzvel. = A. centrasiaticus
- subsp. *langsdorffii* (Trin.) Tzvel. = A. langsdorffii
langsdorffianus Hochst. = A. langsdorffii
langsdorffii (Trin.) Roshev. (*Andropogon amplexifolius* Trin., *Arthraxon hispidus* (Thunb.) Makino subsp. *langsdorffii* (Trin.) Tzvel., *A. langsdorffianus* Hochst. nom. illegit.) - 2, 5, 6
nodosus Kom. = Microstegium nodosum

Arundinaria auct. = Pleioblastus and Sinarundinaria

hindsii Munro = Pleioblastus hindsii
humilis Mitf. = Pleioblastus humilis
nitida Mitf. ex Stapf = Sinarundinaria nitida
pumila Mitf. = Pleioblastus pumilus
simonii (Carr.) A. & C. Riviere = Pleioblastus simonii

Arundinella Raddi

anomala Steud. (*A. hirta* (Thunb.) Tanaka subsp. *anomala* (Steud.) Tzvel.) - 4, 5
hirta (Thunb.) Tanaka (*Poa hirta* Thunb.) - 5
- subsp. *anomala* (Steud.) Tzvel. = A. anomala

Arundo L.

altissima Benth. = Phragmites altissimus
australis Cav. = Phragmites australis
baltica Flugge ex Schrad. = x Calammophila baltica
canescens Web. = Calamagrostis canescens
confinis Willd. = Calamagrostis lapponica
dioica Spreng. = Cortaderia selloana
donax L. - 1, 2, 6
isiaca Delile = Phragmites altissimus

maxima Forssk. = Phragmites altissimus
selloana Schult. & Schult. fil. = Cortaderia selloana
stricta Timm = Calamagrostis stricta
strigosa Wahlenb. = Calamagrostis x strigosa
varia Schrad. = Calamagrostis varia

Asperella Humb. = Hystrix

ajanensis V. Vassil. = x Leymotrix ajanensis
coreana (Honda) Nevski = Hystrix coreana
komarovii Roshev. = Hystrix komarovii
sibirica Trautv. = Hystrix sibirica

Asthenatherum Nevski

forsskalii (Vahl) Nevski (*Danthonia forsskalii* (Vahl) R. Br.) - 6

Atropis Rupr. = Puccinellia

alascana (Scribn. & Merr.) V. Krecz. = Puccinellia alascana and P. kurilensis
angustata (R. Br.) Griseb. = Puccinellia angustata
angustata (R. Br.) V. Krecz. = Puccinellia angustata
anisoclada V. Krecz. = Puccinellia gigantea
bulbosa Grossh. = Puccinellia bulbosa
chilochloa V. Krecz. = Puccinellia gigantea and P. poecilantha
choresmica V. Krecz. = Puccinellia choresmica
convoluta (Hornem.) Griseb. p.p. = Puccinellia gigantea
diffusa V. Krecz. = Puccinellia diffusa
distans (Jacq.) Griseb. = Puccinellia distans
- var. *limosa* Schur = Puccinellia limosa
dolicholepis V. Krecz. = Puccinellia dolicholepis
x *elata* Holmb. = Puccinellia x elata
geniculata V. Krecz. = Puccinellia geniculata
gigantea Grossh. = Puccinellia gigantea
glauca (Regel) V. Krecz. = Puccinellia glauca
grossheimiana V. Krecz. = Puccinellia grossheimiana
hackeliana V. Krecz. = Puccinellia hackeliana
hauptiana V. Krecz. = Puccinellia hauptiana
humilis Litv. ex V. Krecz. = Puccinellia humilis
iliensis V. Krecz. = Puccinellia hauptiana
jenisseiensis Roshev. = Puccinellia jenisseiensis
kamtschatica (Holmb.) V. Krecz. = Puccinellia kamtschatica
kurilensis Takeda = Puccinellia kurilensis
laeviuscula V. Krecz. = Puccinellia tenella
macranthera V. Krecz. = Puccinellia macranthera
macropus V. Krecz. = Puccinellia macropus
maritima (Huds.) Griseb. = Puccinellia maritima
maritima (Huds.) K. Richt. = Puccinellia maritima
montana Drob. = Puccinellia hackeliana
pamirica (Roshev.) V. Krecz. = Puccinellia pamirica
pauciramea (Hack.) V. Krecz. = Puccinellia pauciramea
paupercula (Holm) V. Krecz. = Puccinellia kurilensis and P. langeana
paupercula (Holm) Steffen = Puccinellia langeana
phryganodes (Trin.) V. Krecz. = Puccinellia phryganodes
phryganodes (Trin.) Steffen = Puccinellia phryganodes
poecilantha (C. Koch) V. Krecz. = Puccinellia poecilantha
pulvinata (Fries) V. Krecz. = Puccinellia pulvinata
roshevitsiana Schischk. = Puccinellia roshevitsiana
sclerodes V. Krecz. = Puccinellia gigantea
sevangensis (Grossh.) V. Krecz. = Puccinellia sevangensis
sibirica (Holmb.) V. Krecz. = Puccinellia sibirica
subspicata V. Krecz. = Puccinellia subspicata

suecica Holmb. = Puccinellia capillaris
tenella (Lange) V. Krecz. = Puccinellia coarctata and P. tenella
tenella (Lange) K. Richt. = Puccinellia tenella
tenuiflora Griseb. = Puccinellia tenuiflora
tenuissima Litv. ex V. Krecz. = Puccinellia tenuissima
vilfoidea (Anderss.) K. Richt. = Puccinellia vilfoidea

Avena L.

adsurgens Schur ex Simonk. = Helictotrichon praeustum
aemulans Nevski - 1
alpestris Host = Trisetum alpestre
barbata Pott ex Link (*A. strigosa* Schreb. subsp. *barbata* (Pott ex Link) Thell., *A. hirtula* auct.) - 1, 2, 6
- subsp. *wiestii* (Steud.) Mansf. = A. wiestii
- subsp. *wiestii* (Steud.) Tzvel. = A. wiestii
besseri Griseb. = Helictotrichon desertorum
bruhnsiana Grun. = A. ventricosa
byzantina C. Koch (*A. sterilis* L. subsp. *byzantina* (Koch) Thell.) - 2, 6
*****chinensis** (Fisch. ex Roem. & Schult.) Metzg. (*A. nuda* L. var. *chinensis* Fisch. ex Roem. & Schult.) - 1, 2, 3, 4, 5, 6
ciliaris Kit. = Trisetum ciliare
clauda Durieu - 2, 6
compressa Heuff. = Helictotrichon compressum
cultiformis (Malz.) Malz. (*A. fatua* L. subsp. *cultiformis* Malz.) - 1
eriantha Durieu (*Trisetum pilosum* Roem. & Schult., *Avena pilosa* (Roem. & Schult.) Bieb. 1819, non Scop. 1772) - 1, 2, 6
fatua L. (*A. septentrionalis* Malz.) - 1, 2, 3, 4, 5, 6
- subsp. *cultiformis* Malz. = A. cultiformis
- subsp. *meridionalis* Malz. = A. meridionalis
- subsp. *praegravis* (Krause) Malz. = A. georgica
- subsp. *praegravis* (Krause) Malz. prol. *grandiuscula* Malz. = A. grandis
flavescens L. var. *virescens* Regel = Trisetum seravschanicum
fusca Kit. = Trisetum ciliare
georgica Zuccagni (*A. fatua* L. subsp. *praegravis* (Krause) Malz., *A. praegravis* (Krause) Roshev., *A. sativa* L. subsp. *praegravis* (Krause) Mordv.) - 1, 2, 3, 4, 5, 6
grandis Nevski (*A. fatua* L. subsp. *praegravis* (Krause) Malz. prol. *grandiuscula* Malz.) - 6
hirtula auct. = A. barbata
hookeri Scribn. = Helictotrichon hookeri
laevigata Schur = Helictotrichon pubescens subsp. laevigatum
ludoviciana Durieu = A. persica
*****macrantha** (Hack.) Nevski (*A. sativa* L. subsp. *macrantha* (Hack.) Mordv., *A. sativa* subsp. *macrantha* (Hack.) Rocha Afonso, comb. superfl.) - 2, 6
meridionalis (Malz.) Roshev. (*A. fatua* L. subsp. *meridionalis* Malz.) - 2, 6
mollis Michx. = Trisetum molle
nodipilosa (Malz.) Malz. = A. sativa
*****nuda** L. - 1, 3
- subsp. *strigosa* (Schreb.) Mansf. = A. strigosa
- subsp. *wiestii* (Steud.) A. & D. Löve = A. wiestii
- var. *chinensis* Fisch. ex Roem. & Schult. = A. chinensis
*****orientalis** Schreb. (*A. sativa* L. var. *contracta* Neilr., *A. sativa* subsp. *contracta* (Neilr.) Celak., *A. sativa* subsp. *orientalis* (Schreb.) Jess.) - 1, 2, 3, 4, 5, 6
persica Steud. (*A. ludoviciana* Durieu, *A. sterilis* L. subsp. *ludoviciana* (Durieu) Gillet & Magne) - 1, 2, 5(alien), 6

pilosa (Roem. & Schult.) Bieb. = A. eriantha
planiculmis Schrad. = Helictotrichon planiculme
praegravis (Krause) Roshev. = A. georgica
praeusta Reichenb. = Helictotrichon praeustum
riabuschinskii Kom. = Danthonia riabuschinskii
*****sativa** L. (*A. nodipilosa* (Malz.) Malz., *A. sativa* subsp. *nodipilosa* (Malz.) Mordv.) - 1, 2, 3, 4, 5, 6
- subsp. *contracta* (Neilr.) Celak. = A. orientalis
- subsp. *macrantha* (Hack.) Mordv. = A. macrantha
- subsp. *macrantha* (Hack.) Rocha Afonso = A. macrantha
- subsp. *nodipilosa* (Malz.) Mordv. = A. sativa
- subsp. *orientalis* (Schreb.) Jess. = A. orientalis
- subsp. *praegravis* (Krause) Mordv. = A. georgica
- var. *contracta* Neilr. = A. orientalis
- var. *volgensis* Vav. = A. volgensis
septentrionalis Malz. = A. fatua
sterilis L. - 1, 2, 6
- subsp. *byzantina* (Koch) Thell. = A. byzantina
- subsp. *ludoviciana* (Durieu) Gillet & Magne = A. persica
- subsp. *trichophylla* (C. Koch) Malz. = A. trichophylla
striata Michx. subsp. *capillipes* Kom. = Schizachne komarovii
strigosa Schreb. (*A. nuda* L. subsp. *strigosa* (Schreb.) Mansf.) - 1
- subsp. *barbata* (Pott ex Link) Thell. = A. barbata
- subsp. *wiestii* (Steud.) Thell. = A. wiestii
subspicata L. var. *agrostidea* Laest. = Trisetum agrostideum
tianschanica (Roshev.) Worosch. = Helictotrichon tianschanicum
trichophylla C. Koch (*A. sterilis* L. subsp. *trichophylla* (C. Koch) Malz.) - 1, 2, 6
ventricosa Bal. ex Coss. (*A. bruhnsiana* Grun.) - 2
versicolor Vill. = Helictotrichon versicolor
volgensis (Vav.) Nevski (*A. sativa* L. var. *volgensis* Vav.) - 1
wiestii Steud. (*A. barbata* Pott ex Link subsp. *wiestii* (Steud.) Mansf., *A. barbata* subsp. *wiestii* (Steud.) Tzvel. comb. superfl., *A. nuda* L. subsp. *wiestii* (Steud.) A. & D. Löve, *A. strigosa* Schreb. subsp. *wiestii* (Steud.) Thell.) - 2, 6

Avenastrum Jess. = Helictotrichon

Avenastrum Opiz = Helictotrichon

adzharicum (Albov) Roshev. = Helictotrichon adzharicum
alpinum (Smith) Fritsch, p.p. = Helictotrichon praeustum
armeniacum (Schischk.) Roshev. = Helictotrichon armeniacum
asiaticum Roshev. = Helictotrichon adzharicum and H. hookeri
basalticum Podp. = Helictotrichon desertorum
besseri (Griseb.) Koczwara = Helictotrichon desertorum
compactum (Boiss. & Heldr.) Roshev. = Danthoniastrum compactum
dahuricum (Kom.) Roshev. = Helictotrichon dahuricum
desertorum (Less.) Podp. = Helictotrichon desertorum
fedtschenkoi (Hack.) Roshev. = Helictotrichon fedtschenkoi
hissaricum Roshev. = Helictotrichon hissaricum
krylovii Pavl. = Helictotrichon krylovii
laevigatum (Schur) Domin = Helictotrichon pubescens subsp. laevigatum
mongolicum (Roshev.) Roshev. = Helictotrichon mongolicum
planiculme (Schrad.) Opiz = Helictotrichon planiculme
pratense (L.) Opiz = Helictotrichon pratense
pubescens (Huds.) Opiz = Helictotrichon pubescens

schellianum (Hack.) Roshev. = Helictotrichon schellianum
tianschanicum Roshev. = Helictotrichon tianschanicum
versicolor (Vill.) Fritsch = Helictotrichon versicolor

Avenella Drej. (*Lerchenfeldia* Schur)

flexuosa (L.) Drej. (*Deschampsia flexuosa* (L.) Nees, *Lerchenfeldia flexuosa* (L.) Schur) - 1, 2, 3, 4, 5 (subsp. **flexuosa** - 1, 5)
- subsp. **montana** (L.) A. & D. Löve (*Aira montana* L., *Deschampsia montana* (L.) G. Don fil., *Lerchenfeldia flexuosa* (L.) Schur subsp. *montana* (L.) Tzvel.) - 1, 2, 3, 4

Avenochloa Holub = Helictotrichon

adsurgens (Simonk.) Holub = Helictotrichon praeustum
adzharica (Albov) Holub = Helictotrichon adzharicum
armeniaca (Schischk.) Holub = Helictotrichon armeniacum
asiatica (Roshev.) Holub = Helictotrichon hookeri
compressa (Heuff.) Holub = Helictotrichon compressum
dahurica (Kom.) Holub = Helictotrichon dahuricum
hookeri (Scribn.) Holub = Helictotrichon hookeri
- subsp. *schelliana* (Hack.) Sojak = Helictotrichon schellianum
planiculmis (Schrad.) Holub = Helictotrichon planiculme
praeusta (Reichenb.) Sojak = Helictotrichon praeustum
pratensis (L.) Holub = Helictotrichon pratense
pubescens (Huds.) Holub = Helictotrichon pubescens
- subsp. *laevigata* (Schur) Soo = Helictotrichon pubescens subsp. laevigatum
schelliana (Hack.) Holub = Helictotrichon schellianum
taurica (Prokud.) Holub = Helictotrichon compressum
versicolor (Vill.) Holub = Helictotrichon versicolor
- subsp. *caucasica* (Holub) Holub = Helictotrichon adzharicum

Avenula (Dumort.) Dumort. = Helictotrichon

adzharica (Albov) Holub = Helictotrichon adzharicum
armeniaca (Schischk.) Holub = Helictotrichon armeniacum
compressa (Heuff.) Holub = Helictotrichon compressum
compressa (Heuff.) Sauer & Chmelitschek = Helictotrichon compressum
dahurica (Kom.) Holub = Helictotrichon dahuricum
dahurica (Kom.) Sauer & Chmelitschek = Helictotrichon dahuricum
hookeri (Scribn.) Holub = Helictotrichon hookeri
- subsp. *schelliana* (Hack.) Lomonosova = Helictotrichon schellianum
planiculmis (Schrad.) Holub = Helictotrichon planiculme
planiculmis (Schrad.) Sauer & Chmelitschek = Helictotrichon planiculme
- subsp. *angustior* Holub = Helictotrichon planiculme subsp. angustius
praeusta (Reichenb.) Holub = Helictotrichon praeustum
pratensis (L.) Dumort. = Helictotrichon pratense
pubescens (Huds.) Dumort. = Helictotrichon pubescens
- subsp. *laevigata* (Schur) Holub = Helictotrichon pubescens subsp. laevigatum
schelliana (Hack.) Holub = Helictotrichon schellianum
schelliana (Hack.) Sauer & Chmelitschek = Helictotrichon schellianum
versicolor (Vill.) Lainz = Helictotrichon versicolor

Bambusa Schreb. (*Ludolfia* auct.)

disticha Mitf. = Pleioblastus distichus
fastuosa Mitf. = Semiarundinaria fastuosa
***glaucescens** (Willd.) Siebold ex Munro (*Ludolfia glauce-*

scens Willd., *Bambusa multiplex* auct.) - 2
kumasasa Zoll. = Shibataea kumasasa
marmorea Mitf. = Chimonobambusa marmorea
multiplex auct. = B. glaucescens
nigra Lodd. = Phyllostachys nigra
palmata Marl. ex Burb. = Sasa palmata
quadrangularis Fenzi = Chimonobambusa quadrangularis
senanensis Franch. & Savat. = Sasa senanensis
simonii Carr. = Pleioblastus simonii
tessellata Munro = Indocalamus tessellatus
variegata Miq. = Pleioblastus variegatus
veitchii Carr. = Sasa veitchii
viridi-glaucescens Carr. = Phyllostachys viridi-glaucescens
viridi-striata Siebold ex Andre = Pleioblastus viridi-striatus

Beckmannia Host

borealis (Tzvel.) Probat. (*B. eruciformis* (L.) Host subsp. *borealis* Tzvel.) - 1, 3, 4, 5
eruciformis (L.) Host - 1, 2, 3, 4, 6
- subsp. *baicalensis* (V. Kusn.) Hult. = B. syzigachne
- subsp. *borealis* Tzvel. = B. borealis
- subsp. *syzigachne* (Steud.) Breitung = B. syzigachne
- var. *baicalensis* V. Kusn. = B. syzigachne
hirsutiflora (Roshev.) Probat. (*B. syzigachne* (Steud.) Fern. var. *hirsutiflora* Roshev., *B. syzigachne* subsp. *hirsutiflora* (Roshev.) Tzvel.) - 4, 5
syzigachne (Steud.) Fern. (*B. eruciformis* (L.) Host subsp. *baicalensis* (V. Kusn.) Hult., *B. eruciformis* var. *baicalensis* V. Kusn., *B. eruciformis* subsp. *syzigachne* (Steud.) Breitung, *B. syzigachne* subsp. *baicalensis* (V. Kusn.) T. Koyama & Kawano) - 1, 3, 4, 5, 6
- subsp. *baicalensis* (V. Kusn.) T. Koyama & Kawano = B. syzigachne
- subsp. *hirsutiflora* (Roshev.) Tzvel. = B. hirsutiflora
- var. *hirsutiflora* Roshev. = B. hirsutiflora

Bellardiochloa Chiov.

polychroa (Trautv.) Roshev. (*Poa polychroa* (Trautv.) Grossh.) - 2
violacea (Bell.) Chiov. (*Poa violacea* Bell.) - 1

Boissiera Hochst. ex Steud. (*Euraphis* (Trin.) Lindl.)

pumilio (Trin.) Hack. = B. squarrosa
squarrosa (Banks & Soland.) Nevski (*Pappophorum squarrosum* Banks & Soland., *Boissiera pumilio* (Trin.) Hack., *Bromus pumilio* (Trin.) P. Smith, *Euraphis squarrosa* (Banks & Soland.) Sojak) - 2, 6

Boriskellera Terekhov = Eragrostis

arundinacea (L.) Terekhov = Eragrostis collina

Bothriochloa O. Kuntze (*Amphilopsis* Nash, *Andropogon* L. p.p., *Dichanthium* auct.)

bladhii (Retz.) S.T. Blake (*Andropogon bladhii* Retz., *A. caucasicus* Trin., *Bothriochloa caucasica* (Trin.) C.E. Hubb., *Dichanthium bladhii* (Retz.) Clayton, *D. caucasicum* (Trin.) Jain & Deshpande) - 2, 6
caucasica (Trin.) C.E. Hubb. = B. bladhii
ischaemum (L.) Keng (*Andropogon ischaemum* L., *Amphilopsis ischaemum* (L.) Nash, *Bothriochloa ischaemum* (L.) Henrard, comb. superfl., *B. ischaemum* (L.) Mansf. comb. superfl., *Dichanthium ischaemum* (L.) Roberty) - 2, 3, 6

Brachiaria (Trin.) Griseb.

eruciformis (Smith) Griseb. - 2, 6

Brachypodium Beauv. (*Brevipodium* A. & D. Löve)

caespitosum (Host) Roem. & Schult. = B. rupestre subsp. caespitosum

distachyon (L.) Beauv. = Trachynia distachya

geniculatum C. Koch = Trachynia distachya

gracile (Leyss.) Beauv. See B. rupestre.

- var. *pubescens* Peterm. = B. sylvaticum subsp. pubescens

kurilense (Probat.) Probat. (*B. sylvaticum* (Huds.) Beauv. subsp. *kurilense* Probat., *B. miserum* Koidz. subsp. *kurilense* (Probat.) Worosch., *B. miserum* auct.) - 5

miserum auct. = B. kurilense

miserum Koidz. subsp. *kurilense* (Probat.) Worosch. = B. kurilense

pinnatum (L.) Beauv. - 1, 2, 3, 4, 6

- subsp. *caespitosum* (Host) Hack. = B. rupestre subsp. caespitosum

- subsp. *gracile* (Leyss.) Soo. See B. rupestre.

- subsp. *rupestre* (Host) Schubl. & Martens = B. rupestre

- subsp. *rupestre* (Host) Tzvel. = B. rupestre

- var. *rupestre* (Host) Reichenb. = B. rupestre

pubescens (Peterm.) Musajev = B. sylvaticum subsp. pubescens

rupestre (Host) Roem. & Schult. (*B. pinnatum* (L.) Beauv. subsp. *rupestre* (Host) Schubl. & Martens, *B. pinnatum* subsp. *rupestre* (Host) Tzvel., *B. pinnatum* var. *rupestre* (Host) Reichenb.) - 1, 2

> Probably, the priority name of this species is B. gracile (Leyss.) Beauv. (Bromus gracilis Leyss., Brachypodium pinnatum subsp. gracile (Leyss.) Soo).

- subsp. **caespitosum** (Host) H. Scholz (*Bromus caespitosus* Host, *Brachypodium caespitosum* (Host) Roem. & Schult., *B. pinnatum* subsp. *caespitosum* (Host) Hack.) - 1(?), 2

sylvaticum (Huds.) Beauv. (*Brevipodium sylvaticum* (Huds.) A. & D. Löve) - 1, 2, 3, 4, 5, 6

- subsp. *dumosum* (Vill.) Tzvel. = B. sylvaticum subsp. pubescens

- subsp. *kurilense* Probat. = B. kurilense

- subsp. **pubescens** (Peterm.) Tzvel. (*B. gracile* (Leyss.) Beauv. var. *pubescens* Peterm., *B. pubescens* (Peterm.) Musajev, *B. sylvaticum* subsp. *dumosum* (Vill.) Tzvel., *Brevipodium sylvaticum* var. *dumosum* (Vill.) Pinto da Silva & Teles, *Bromus dumosus* Vill. p.p. nom. illegit.) - 1, 2(alien)

villosum Drob. = Elytrigia villosa

Brevipodium A. & D. Löve = Brachypodium

sylvaticum (Huds.) A. & D. Löve = Brachypodium sylvaticum

- var. *dumosum* (Vill.) Pinto da Silva & Teles = Brachypodium sylvaticum subsp. pubescens

Briza L.

australis Prokud. = B. elatior

elatior Sibth. & Smith (*B. australis* Prokud., *B. media* L. subsp. *elatior* (Sibth. & Smith) Rohlena) - 1, 2

humilis Bieb. = Brizochloa humilis

marcowiczii Woronow - 2

maxima L. - 1, 2, 5

media L. - 1

- subsp. *elatior* (Sibth. & Smith) Rohlena = B. elatior

minor L. - 2

spicata Sibth. & Smith = Brizochloa humilis

Brizochloa Jir. & Chrtek

humilis (Bieb.) Chrtek & Hadac (*Briza humilis* Bieb., *B. spicata* Sibth. & Smith, 1806, non Burm. fil. 1768, *Brizochloa spicata* (Sibth. & Smith) Jir. & Chrtek, nom. illegit.) - 1, 2

spicata (Sibth. & Smith) Jir. & Chrtek = B. humilis

Bromopsis Fourr. (*Zerna* Panz. p.p., nom. illegit.)

alexeenkoi (Tzvel.) Czer. (*Zerna erecta* (Huds.) S.F. Gray subsp. *alexeenkoi* Tzvel., *Bromopsis erecta* (Huds.) Fourr. subsp. *alexeenkoi* (Tzvel.) Tzvel.) - 2

alpina (Malysch.) Peschkova = B. arctica

altaica Peschkova - 3, 4

angrenica (Drob.) Holub (*Bromus angrenicus* Drob., *Bromopsis paulsenii* (Hack.) Holub subsp. *angrenica* (Drob.) Tzvel., *Zerna angrenica* (Drob.) Nevski, *Z. paulsenii* (Hack.) Nevski subsp. *angrenica* (Drob.) Tzvel.) - 6

arctica (Shear) Holub (*Bromus arcticus* Shear, *Bromopsis alpina* (Malysch.) Peschkova, *B. arctica* var. *alpina* (Malysch.) Peschkova, *B. pumpelliana* (Scribn.) Holub subsp. *artica* (Shear) A. & D. Löve, *B. pumpelliana* subsp. *arctica* (Shear) Tzvel. comb. superfl., *B. pumpelliana* var. *arctica* (Shear) Ju. Kozhevn. nom. invalid., *Bromus inermis* Leyss. subsp. *pumpellianus* (Scribn.) Wagnon var. *arcticus* (Shear) Wagnon, ? *B. paramushirensis* Kudo, *B. pumpellianus* Scribn. var. *arcticus* (Shear) Pors., *B. sibiricus* Drob. f. *alpina* Malysch., *Zerna arctica* (Shear) Tzvel., *Z. pumpelliana* (Scribn.) Tzvel. subsp. *arctica* (Shear) Tzvel.) - 4, 5

- var. *alpina* (Malysch.) Peschkova = B. arctica

aristata (C. Koch) Holub (*Festuca aristata* C. Koch, *Bromus aristatus* (C. Koch) Steud., *Zerna aristata* (C. Koch) Tzvel.) - 2

aspera (Murr.) Fourr. = B. ramosa

austrosibirica Peschkova - 4

benekenii (Lange) Holub (*B. ramosa* (Huds.) Holub subsp. *benekenii* (Lange) Tzvel., *Bromus benekenii* (Lange) Trimen, *B. ramosus* Huds. subsp. *benekenii* (Lange) Lindb. fil., *Zerna benekenii* (Lange) Lindm., *Z. ramosa* (Huds.) Lindm. subsp. *benekenii* (Lange) Tzvel.) - 1, 2, 3, 6

biebersteinii (Roem. & Schult.) Holub (*Bromus biebersteinii* Roem. & Schult., *Zerna biebersteinii* (Roem. & Schult.) Nevski) - 2

calcarea Klok. - 1

canadensis (Michx.) Holub (*Bromus canadensis* Michx., *B. canadensis* subsp. *yezoensis* (Ohwi) Worosch., *B. yezoensis* Ohwi, *Zerna canadensis* (Michx.) Tzvel., *Z. yezoensis* (Ohwi) Sugim., *Bromus ciliatus* auct., *B. richardsonii* auct. p.p.) - 5

cappadocica (Boiss. & Bal.) Holub (*Bromus cappadocicus* Boiss. & Bal., *Bromopsis tomentella* (Boiss.) Holub subsp. *cappadocica* (Boiss. & Bal.) Tzvel., *Zerna cappadocica* (Boiss. & Bal.) Nevski, p.p. quoad nomen, *Z. tomentella* (Boiss.) Nevski subsp. *cappadocica* (Boiss. & Bal.) Tzvel.) - 2

cimmerica Klok. - 1

divaricata (Tzvel.) Czer. (*Zerna riparia* (Rehm.) Nevski subsp. *divaricata* Tzvel., *Bromopsis riparia* (Rehm.) Holub subsp. *divaricata* (Tzvel.) Tzvel.) - 2

dolichophylla Klok. - 2

erecta (Huds.) Fourr. (*Bromus erectus* Huds., *Zerna erecta* (Huds.) S.F. Gray) - 1
- subsp. *alexeenkoi* (Tzvel.) Tzvel. = B. alexeenkoi
- subsp. *gordjaginii* (Tzvel.) Tzvel. = B. gordjaginii
fedtschenkoi (Tzvel.) Czer. (*Zerna ramosa* (Huds.) Lindm. subsp. *fedtschenkoi* Tzvel., *Bromopsis ramosa* (Huds.) Holub subsp. *fedtschenkoi* (Tzvel.) Tzvel.) - 6
glabrata Klok. = B. x taurica
gordjaginii (Tzvel.) Galushko (*Zerna erecta* (Huds.) S.F. Gray subsp. *gordjaginii* Tzvel., *Bromopsis erecta* (Huds.) Fourr. subsp. *gordjaginii* (Tzvel.) Tzvel.) - 2
heterophylla (Klok.) Holub (*Zerna heterophylla* Klok., *Bromopsis riparia* (Rehm.) Holub subsp. *heterophylla* (Klok.) Tzvel., *Zerna cappadocica* (Boiss. & Bal.) Nevski, p.p. quoad pl., *Z. riparia* (Rehm.) Nevski subsp. *heterophylla* (Klok.) Tzvel., *Bromus cappadocicus* sensu Nevski & Socz.) - 1
indurata (Hausskn. & Bornm.) Holub (*Bromus induratus* Hausskn. & Bornm., *Zerna indurata* (Hausskn. & Bornm.) Tzvel.) - 2
- subsp. *nachiczevanica* (Tzvel.) Tzvel. = B. nachiczevanica
inermis (Leyss.) Holub (*Bromus inermis* Leyss., *B. littoreus* Georgi, *B. pskemensis* Pavl., *Zerna inermis* (Leyss.) Lindm.) - 1, 2, 3, 4, 5, 6
- subsp. *pumpelliana* (Scribn.) W.A. Weber = B. pumpelliana
ircutensis (Kom.) A. & D. Löve (*Bromus ircutensis* Kom., *Bromopsis pumpelliana* (Scribn.) Holub subsp. *korotkiji* (Drob.) Tzvel. var. *ircutensis* (Kom.) Tzvel., *Zerna ircutensis* (Kom.) Nevski) - 4
karavajevii (Tzvel.) Czer. (*Zerna pumpelliana* (Scribn.) Tzvel. subsp. *karavajevii* Tzvel., *Bromopsis pumpelliana* (Scribn.) Holub subsp. *karavajevii* (Tzvel.) Tzvel.) - 4
kopetdagensis (Drob.) Holub (*Bromus kopetdagensis* Drob., *Zerna kopetdagensis* (Drob.) Nevski) - 6
korotkiji (Drob.) Holub (*Bromus korotkiji* Drob., *Bromopsis pumpelliana* (Scribn.) Holub subsp. *korotkiji* (Drob.) Tzvel., *Zerna korotkiji* (Drob.) Nevski, *Z. pumpelliana* (Scribn.) Tzvel. subsp. *korotkiji* (Drob.) Tzvel.) - 4
nachiczevanica (Tzvel.) Holub (*Zerna indurata* (Hausskn. & Bornm.) Tzvel. subsp. *nachiczevanica* Tzvel., *Bromopsis indurata* (Hausskn. & Bornm.) Holub subsp. *nachiczevanica* (Tzvel.) Tzvel.) - 2
ornans (Kom.) Holub (*Bromus ornans* Kom., *Bromopsis pumpelliana* (Scribn.) Holub subsp. *ornans* (Kom.) Tzvel., *Zerna ornans* (Kom.) Nevski, *Z. pumpelliana* (Scribn.) Tzvel. subsp. *ornans* (Kom.) Tzvel.) - 5
pamirica (Drob.) Holub (*Bromus pamiricus* Drob., *Bromopsis paulsenii* (Hack.) Holub subsp. *pamirica* (Drob.) Tzvel., *Zerna pamirica* (Drob.) Nevski, *Z. paulsenii* (Hack.) Nevski subsp. *pamirica* (Drob.) Tzvel.) - 6
paulsenii (Hack.) Holub (*Bromus paulsenii* Hack., *Zerna paulsenii* (Hack.) Nevski) - 6
- subsp. *angrenica* (Drob.) Tzvel. = B. angrenica
- subsp. *pamirica* (Drob.) Tzvel. = B. pamirica
- subsp. *turkestanica* (Drob.) Tzvel. = B. turkestanica
pavlovii (Roshev.) Peschkova (*Bromus pavlovii* Roshev.) - 4
pseudocappadocica Klok. (*B. riparia* (Rehm.) Holub subsp. *fibrosa* (Hack.) Tzvel. p.p. excl. typo, *Bromus pseudocappodocicus* Stank. nom. invalid., *Zerna pseudocappadocica* Klok. nom. invalid.) - 1
pumpelliana (Scribn.) Holub (*Bromus pumpellianus* Scribn., *Bromopsis inermis* (Leyss.) Holub subsp. *pumpelliana* (Scribn.) W.A. Weber, *B. sibirica* (Drob.)

Peschkova, nom. invalid., *Bromus inermis* Leyss. subsp. *pumpellianus* (Scribn.) Wagnon, *B. occidentalis* (Nevski) Pavl., *B. sibiricus* Drob., *B. uralensis* Govor., *Zerna occidentalis* Nevski, *Z. pumpelliana* (Scribn.) Tzvel., *Z. richardsonii* (Link) Nevski, p.p. quoad pl., *Bromus richardsonii* auct. p.p.) - 1, 3, 4, 5
- subsp. *artica* (Shear) A. & D. Löve = B. arctica
- subsp. *arctica* (Shear) Tzvel. = B. arctica
- subsp. *flexuosa* (Drob.) Probat. (*Bromus sibiricus* Drob. var. *flexuosus* Drob.) - 5
- subsp. *karavajevii* (Tzvel.) Tzvel. = B. karavajevii
- subsp. *korotkiji* (Drob.) Tzvel. = B. korotkiji
- - var. *ircutensis* (Kom.) Tzvel. = B. ircutensis
- subsp. *ornans* (Kom.) Tzvel. = B. ornans
- subsp. *pumpelliana* var. *taimyrensis* (Roshev.) Tzvel. = B. taimyrensis
- subsp. *vogulica* (Socz.) Tzvel. = B. vogulica
- var. *arctica* (Shear) Ju. Kozhevn. = B. arctica
ramosa (Huds.) Holub (*Bromus ramosus* Huds., *Bromopsis aspera* (Murr.) Fourr., *Bromus asper* Murr., *Zerna aspera* (Murr.) S.F. Gray, *Z. ramosa* (Huds.) Lindm.) - 1, 2
- subsp. *benekenii* (Lange) Tzvel. = B. benekenii
- subsp. *fedtschenkoi* (Tzvel.) Tzvel. = B. fedtschenkoi
riparia (Rehm.) Holub (*Bromus riparius* Rehm., *Zerna riparia* (Rehm.) Nevski) - 1, 2, 5(alien), 6(alien)
- subsp. *divaricata* (Tzvel.) Tzvel. = B. divaricata
- subsp. *fibrosa* (Hack.) Tzvel. p.p. = B. pseudocappadocica and B. x taurica
- subsp. *heterophylla* (Klok.) Tzvel. = B. heterophylla
shelkovnikovii (Tzvel.) Holub (*Zerna tomentella* (Boiss.) Nevski subsp. *shelkovnikovii* Tzvel., *Bromopsis tomentella* (Boiss.) Holub subsp. *shelkovnikovii* (Tzvel.) Tzvel.) - 2
sibirica (Drob.) Peschkova = B. pumpelliana
taimyrensis (Roshev.) Peschkova (*Bromus sibiricus* Drob. var. *taimyrensis* Roshev., *Bromopsis pumpelliana* (Scribn.) Holub subsp. *pumpeliana* var. *taimyrensis* (Roshev.) Tzvel.) - 4
x taurica Sljussarenko (*B. glabrata* Klok., *B. riparia* (Rehm.) Holub subsp. *fibrosa* (Hack.) Tzvel. p.p. excl. typo, ? *Zerna fibrosa* (Hack.) Grossh. p.p. quoad pl., *Z. riparia* (Rehm.) Nevski subsp. *fibrosa* (Hack.) Tzvel. p.p. excl. typo). - B. heterophylla (Klok.) Holub x B. riparia (Rehm.) Holub - 1, 2
tomentella (Boiss.) Holub (*Bromus tomentellus* Boiss., *Zerna tomentella* (Boiss.) Nevski) - 2
- subsp. *cappadocica* (Boiss. & Bal.) Tzvel. = B. cappadocica
- subsp. *shelkovnikovii* (Tzvel.) Tzvel. = B. shelkovnikovii
- subsp. *woronowii* (Tzvel.) Tzvel. = B. woronowii
triniana (Schult.) Holub (*Bromus trinianus* Schult., *B. tomentosus* Trin. 1818, non Rohde, 1809, *Zerna tomentosa* (Trin.) Nevski, nom. illegit., *Z. triniana* (Schult.) Tzvel.) - 2
turkestanica (Drob.) Holub (*Bromus turkestanicus* Drob., *Bromopsis paulsenii* (Hack.) Holub subsp. *turkestanica* (Drob.) Tzvel., *Zerna paulsenii* (Hack.) Nevski subsp. *turkestanica* (Drob.) Tzvel., *Z. turkestanica* (Drob.) Nevski) - 6
tyttholepis (Nevski) Holub (*Zerna tyttholepis* Nevski, *Bromus tyttholepis* (Nevski) Nevski) - 6
variegata (Bieb.) Holub (*Bromus variegatus* Bieb., *Zerna variegata* (Bieb.) Nevski) - 2
- subsp. *villosula* (Steud.) Tzvel. = B. villosula
villosula (Steud.) Holub (*Bromus villosulus* Steud., *Bromopsis variegata* (Bieb.) Holub subsp. *villosula*

(Steud.) Tzvel., *Bromus adjaricus* Somm. & Levier, *B. variegatus* Bieb. subsp. *villosulus* (Steud.) P.M. Smith, *Zerna adjarica* (Somm. & Levier) Nevski, *Z. variegata* (Bieb.) Nevski subsp. *villosula* (Steud.) Tzvel.) - 2

vogulica (Socz.) Holub (*Bromus vogulicus* Socz., *Bromopsis pumpelliana* (Scribn.) Holub subsp. *vogulica* (Socz.) Tzvel., *Bromus julii* Govor., *Zerna pumpelliana* (Scribn.) Tzvel. subsp. *vogulica* (Socz.) Tzvel., *Z. vogulica* (Socz.) Nevski) - 1, 3

woronowii (Tzvel.) Czer. (*Zerna tomentella* (Boiss.) Nevski subsp. *woronowii* Tzvel., *Bromopsis tomentella* (Boiss.) Holub subsp. *woronowii* (Tzvel.) Tzvel.) - 2

Bromus L.

adjaricus Somm. & Levier = Bromopsis villosula

anatolicus Boiss. & Heldr. (*B. japonicus* Thunb. subsp. *anatolicus* (Boiss. & Heldr.) Penzes) - 1, 2, 6

angrenicus Drob. = Bromopsis angrenica

arcticus Shear = Bromopsis arctica

aristatus (C. Koch) Steud. = Bromopsis aristata

arvensis L. - 1, 2, 3(alien), 4(alien), 5(alien), 6(alien)

asper Murr. = Bromopsis ramosa

benekenii (Lange) Trimen = Bromopsis benekenii

biebersteinii Roem. & Schult. = Bromopsis biebersteinii

briziformis Fisch. & C.A. Mey. - 1, 2, 6

caespitosus Host = Brachypodium rupestre subsp. caespitosum

canadensis Michx. = Bromopsis canadensis

- subsp. *yezoensis* (Ohwi) Worosch. = Bromopsis canadensis

cappadocicus Boiss. & Bal. = Bromopsis cappadocica

cappadocicus sensu Nevski & Socz. = Bromopsis heterophylla

carinatus Hook. & Arn. = Ceratochloa carinata

catharticus Vahl = Ceratochloa cathartica

ciliatus auct. = Bromopsis canadensis

commutatus Schrad. (*B. racemosus* L. subsp. *commutatus* (Schrad.) Maire & Weiller) - 1, 2, 5(alien), 6

danthoniae Trin. - 1(alien), 2, 6

diandrus Roth = Anisantha diandra

distachyos L. = Trachynia distachya

distichus Georgi = Agropyron distichum

dumosus Vill. p.p. = Brachypodium sylvaticum subsp. pubescens

erectus Huds. = Bromopsis erecta

gedrosianus Penzes (? *B. pectinatus* Thunb.) - 6

gracilis Leyss. See Brachypodium rupestre.

gracilis M. Pop. = B. tytthanthus

gracillimus Bunge = Nevskiella gracillima

gussonii Parl. = Anisantha diandra

hookerianus Thurb. = Ceratochloa carinata var. hookeriana

hordeaceus L. = B. mollis

- subsp. *mollis* (L.) Hyl. = B. mollis

- subsp. *thominii* (Hardouin) Hyl. = B. thominii

hordeaceus sensu Tzvel. = B. thominii

induratus Hausskn. & Bornm. = Bromopsis indurata

inermis Leyss. = Bromopsis inermis

- subsp. *pumpellianus* (Scribn.) Wagnon = Bromopsis pumpelliana

- - var. *arcticus* (Shear) Wagnon = Bromopsis arctica

ircutensis Kom. = Bromopsis ircutensis

japonicus Thunb. (*B. japonicus* var. *typicus* Hack. nom. illegit., *B. japonicus* subsp. *typicus* (Hack.) Penzes, nom. illegit.) - 1, 2, 3, 4(alien), 5(alien), 6

- subsp. *anatolicus* (Boiss. & Heldr.) Penzes = B. anatolicus

- subsp. **phrygius** (Boiss.) Penzes (*B. phrygius* Boiss.) - 2

- subsp. **sooi** Penzes (*B. japonicus* var. *sooi* (Penzes) Soo) - 1

- subsp. **subsquarrosus** (Borb.) Penzes (*B. patulus* Mert. & Koch var. *subsquarrosus* Borb.) - 1

- subsp. *typicus* (Hack.) Penzes = B. japonicus

- var. *sooi* (Penzes) Soo = B. japonicus subsp. sooi

- var. *typicus* Hack. = B. japonicus

julii Govor. = Bromopsis vogulica

kopetdagensis Drob. = Bromopsis kopetdagensis

korotkiji Drob. = Bromopsis korotkiji

lanceolatus Roth (*B. lanceolatus* subsp. *macrostachys* (Desf.) Maire, *B. lanuginosus* Poir., *B. macrostachys* Desf.) - 2, 6

- subsp. *macrostachys* (Desf.) Maire = B. lanceolatus

- subsp. *oxyodon* (Schrenk) Tzvel. = B. oxydon

lanuginosus Poir. = B. lanceolatus

lepidus Holmb. - 1

littoreus Georgi = Bromopsis inermis

x **litvinovii** Roshev. ex Nevski. - B. japonicus Thunb. x B. racemosus L.

macrostachys Desf. = B. lanceolatus

madritensis L. = Anisantha madritensis

mollis L. (*B. hordeaceus* L. nom. dubium & ambig., *B. hordeaceus* subsp. *mollis* (L.) Hyl.) - 1, 2, 5(alien)

noeanus Boiss. = B. squarrosus subsp. noeanus

occidentalis (Nevski) Pavl. = Bromopsis pumpelliana

ornans Kom. = Bromopsis ornans

oxydon Schrenk (*B. lanceolatus* Roth subsp. *oxyodon* (Schrenk) Tzvel.) - 1(alien), 5(alien), 6

pamiricus Drob. = Bromopsis pamirica

paramushirensis Kudo = Bromopsis arctica

patulus Mert. & Koch var. *subsquarrosus* Borb. = B. japonicus subsp. subsquarrosus

paulsenii Hack. = Bromopsis paulsenii

pavlovii Roshev. = Bromopsis pavlovii

pectinatus Thunb. = B. gedrosianus

pentastachyos Ten. = Trachynia pentastachya

phrygius Boiss. = B. japonicus subsp. phrygius

popovii Drob. = B. racemosus

pseudocappodocicus Stank. = Bromopsis pseudocappadocica

pseudodanthoniae Drob. - 2, 6

pskemensis Pavl. = Bromopsis inermis

pumilio (Trin.) P. Smith = Boissiera squarrosa

pumpellianus Scribn. = Bromopsis pumpelliana

- var. *arcticus* (Shear) Pors. = Bromopsis arctica

racemosus L. (*B. popovii* Drob., *B. tuzsonii* Penzes) - 1, 2, 5(alien), 6

- subsp. *commutatus* (Schrad.) Maire & Weiller = B. commutatus

ramosus Huds. = Bromopsis ramosa

- subsp. *benekenii* (Lange) Lindb. fil. = Bromopsis benekenii

richardsonii auct. = Bromopsis canadensis and B. pumpelliana

rigens auct. = Anisantha diandra

rigidus Roth subsp. *gussonii* (Roth) Maire = Anisantha diandra

riparius Rehm. = Bromopsis riparia

rubens L. = Anisantha rubens

scoparius L. - 1, 2, 6

secalinus L. - 1, 2, 3(alien), 5(alien)

sericeus Drob. = Anisantha sericea

sewerzowii Regel - 6

sibiricus Drob. = Bromopsis pumpelliana

- var. *flexuosus* Drob. = Bromopsis pumpelliana subsp. flexuosa

- var. *taimyrensis* Roshev. = Bromopsis taimyrensis

- f. *alpina* Malysch. = Bromopsis arctica

squarrosus L. (*B. squarrosus* subsp. *typicus* Penzes, nom. illegit.) - 1, 2, 3, 5(alien), 6
- subsp. **danubialis** Penzes - ?
- subsp. **noeanus** (Boiss.) Penzes (*B. noeanus* Boiss.) - 2
- subsp. *typicus* Penzes = B. squarrosus
sterilis L. = Anisantha sterilis
subulatus Griseb. = Melica subulata
tectorum L. = Anisantha tectorum
thominii Hardouin (*B. hordeaceus* L. subsp. *thominii* (Hardouin) Hyl., *B. hordeaceus* sensu Tzvel.) - 1
tomentellus Boiss. = Bromopsis tomentella
tomentosus Trin. = Bromopsis triniana
trinianus Schult. = Bromopsis triniana
turkestanicus Drob. = Bromopsis turkestanica
tuzsonii Penzes = B. racemosus
tytthanthus Nevski (*B. gracilis* M. Pop. 1915, non Leyss. 1761, nec Weig. 1772) - 6
tyttholepis (Nevski) Nevski = Bromopsis tyttholepis
tzvelevii Musajev - 2
uralensis Govor. = Bromopsis pumpelliana
variegatus Bieb. = Bromopsis variegata
- subsp. *villosulus* (Steud.) P.M. Smith = Bromopsis villosula
villosulus Steud. = Bromopsis villosula
vogulicus Socz. = Bromopsis vogulica
willdenowii Kunth = Ceratochloa unioloides
wolgensis Fisch. ex Jacq. fil. - 1, 2, 3, 6
yezoensis Ohwi = Bromopsis canadensis

Brylkinia Fr. Schmidt

caudata (Munro) Fr. Schmidt (*B. schmidtii* Ohwi, nom. illegit.) - 5
schmidtii Ohwi = B. caudata

Calamagrostis Adans. (*Deyeuxia* Clarion, *Stilpnophleum* Nevski)

aculeolata (Hack.) Ohwi = C. inexpansa
x **acutiflora** (Schrad.) Reichenb. (*C. trinii* Rupr.). - C. arundinacea (L.) Roth, s.l. x C. epigeios (L.) Roth, s.l. - 1, 3, 4
adylssuensis Seredin = C. subchalybaea
agrostidiformis Roshev. = Agrostis agrostidiformis
agrostioides Matuszk. = C. obtusata
ajanensis Charkev. & Probat. - 5
alaica Litv. (*C. schugnanica* Litv.) - 6
aleutica Trin. = C. nutkaensis
alexeenkoana Litv. ex Roshev. = C. turkestanica
alopecuroides Roshev. = C. turkestanica
amurensis Probat. - 5
x **andrejewii** Litv. - C. arundinacea (L.) Roth, s.l. x C. obtusata Trin. - 1, 3, 4
angustifolia Kom. (*C. lanceolatiformis* V. Vassil., *C. purpurea* (Trin.) Trin. subsp. *angustifolia* (Kom.) Worosch., *Deyeuxia angustifolia* (Kom.) Chang) - 5
- subsp. *tenuis* (V. Vassil.) Tzvel. = C. tenuis
- var. *tenuis* (V. Vassil.) Kitag. = C. tenuis
anthoxanthoides (Munro) Regel (*C. phalaroides* Regel, *Stilpnophleum anthoxanthoides* (Munro) Nevski) - 6
- subsp. *laguroides* (Regel) Tzvel. = C. laguroides
arctica Vasey (*C. purpurascens* R. Br. subsp. *arctica* (Vasey) Hult.) - 4, 5
x **aristata** Ohwi. - C. hakonensis Franch. & Savat. x C. purpurea (Trin.) Trin. s.l. - 5
arundinacea (L.) Roth (*C. parviflora* Rupr., *Deyeuxia arundinacea* (L.) Beauv.) - 1, 2, 3, 4
- subsp. *brachytricha* (Steud.) Tzvel. = C. brachytricha

- subsp. *distantiflora* (Lucznik) Tzvel. = C. distantiflora
- subsp. *monticola* (V. Petrov ex Kom.) Tzvel. = C. monticola
- subsp. *sugawarae* (Ohwi) Tzvel. = C. sugawarae
- var. *pamirensis* Hack. = C. fedtschenkoana
x **badzhalensis** Probat. - C. arctica Vasey x C. korotkyi Litv. - 5
baicalensis Litv. = C. langsdorffii
balansae Boiss. = Agrostis balansae
balkharica P. Smirn. - 2
barbata V. Vassil. (*C. purpurea* (Trin.) Trin. subsp. *barbata* (V. Vassil.) Tzvel.) - 4, 5
borealis Laest. = C. groenlandica
brachytricha Steud. (*C. arundinacea* (L.) Roth subsp. *brachytricha* (Steud.) Tzvel., *Deyeuxia brachytricha* (Steud.) Chang) - 4, 5
bracteolata V. Vassil. = C. deschampsioides
bungeana V. Petrov = C. holmii
caespitosa V. Vassil. = C. purpurascens
canadensis (Michx.) Beauv. subsp. *langsdorffii* (Link) Hult. = C. langsdorffii
canescens (Web.) Roth (*Arundo canescens* Web., *Calamagrostis lanceolata* Roth, *C. lithuanica* Bess.) - 1, 2, 3, 4
- subsp. *vilnensis* (Bess.) H. Scholz = C. x vilnensis
caucasica Trin. - 2
- subsp. *iberica* (Griseb.) Tzvel. = C. iberica
chalybaea (Laest.) Fries - 1, 3
chassanensis Probat. - 5
compacta (Munro ex Hook. fil.) Hack. (*Deyeuxia compacta* Munro ex Hook. fil.) - 6
confinis (Willd.) Beauv. = C. lapponica
confusa V. Vassil. = C. langsdorffii
czerepanovii Gussejnov - 2
czukczorum Socz. See C. tenuis.
deschampsioides Trin. (*C. bracteolata* V. Vassil., *C. deschampsioides* subsp. *macrantha* Piper ex Scribn. & Merr., *C. festuciformis* V. Vassil., *C. macrantha* (Piper ex Scribn. & Merr.) V. Vassil., *C. miyabei* Honda) - 1, 3, 4, 5
- subsp. *macrantha* Piper ex Scribn. & Merr. = C. deschampsioides
distantiflora Lucznik (*C. arundinacea* (L.) Roth subsp. *distantiflora* (Lucznik) Tzvel.) - 5
dubia Bunge (*C. pseudophragmites* (Hall. fil.) Koel. subsp. *dubia* (Bunge) Tzvel.) - 1, 2, 6
elata Blytt = C. phragmitoides
emodensis Griseb. var. *breviseta* Hack. = C. hedinii
epigeios (L.) Roth (? *C. lenkoranensis* Steud.) - 1, 2, 3, 4, 5, 6
- subsp. *extremiorientalis* Tzvel. = C. extremiorientalis
- subsp. *glomerata* (Boiss. & Buhse) Tzvel. = C. glomerata
- subsp. *macrolepis* (Litv.) Tzvel. = C. macrolepis
- subsp. *meinshausenii* Tzvel. = C. meinshausenii
- var. *extremiorientalis* (Tzvel.) Kitag. = C. extremiorientalis
- var. *laevis* Meinsh. = C. meinshausenii
evenkiensis Reverd. = C. holmii
extremiorientalis (Tzvel.) Probat. (*C. epigeios* (L.) Roth subsp. *extremiorientalis* Tzvel., *C. epigeios* var. *extremiorientalis* (Tzvel.) Kitag.) - 4, 5
fedtschenkoana (Tzvel.) Ikonn. (*C. tianschanica* Rupr. subsp. *fedtschenkoana* Tzvel., *C. arundinacea* (L.) Roth var. *pamirensis* Hack.) - 4, 5
festuciformis V. Vassil. = C. deschampsioides
flexuosa Rupr. = C. phragmitoides
fusca Kom. = C. purpurea
georgica C. Koch = C. glomerata
gigantea Roshev. = C. macrolepis

glauca (Bieb.) Trin. = C. pseudophragmites
glomerata Boiss. & Buhse (*C. epigeios* (L.) Roth subsp. *glomerata* (Boiss. & Buhse) Tzvel., ? *C. georgica* C. Koch, *C. koibalensis* Reverd.) - 1, 2, 3, 4, 5
gorodkovii V. Vassil. = C. lapponica
gracilis (Litv.) V. Vassil. = C. purpurea
grandiflora Hack. ex Roshev. = C. pamirica
groenlandica (Schrank) Kunth (*C. borealis* Laest., *C. neglecta* (Ehrh.) Gaertn., Mey. & Scherb. subsp. *borealis* (Laest.) Seland., *C. neglecta* subsp. *groenlandica* (Schrank) Matuszk., *C. stricta* (Timm) Koel. subsp. *groenlandica* (Schrank) A. Löve) - 1, 3, 4, 5
hakonensis Franch. & Savat. - 5
x **hartmanniana** Fries. - A. arundinacea (L.) Roth x C. canescens (Web.) Roth - 1, 3
hedinii Pilg. (*C. emodensis* Griseb. var. *breviseta* Hack., *C. littorea* (Schrad.) DC. var. *tartarica* Hook. fil., *C. pseudophragmites* (Hall. fil.) Koel. subsp. *tartarica* (Hook. fil.) Tzvel.) - 6
hirsuta V. Vassil. = C. tenuis
hissarica Nevski = Agrostis nevskii
holciformis Jaub. & Spach (*Deyeuxia holciformis* (Jaub. & Spach) Bor) - 6
- subsp. *pamirica* (Litv.) Tzvel. = C. pamirica
holmii Lange (*C. bungeana* V. Petrov, *C. evenkiensis* Reverd., *C. neglecta* (Ehrh.) Gaertn., Mey. & Scherb. subsp. *holmii* (Lange) Worosch., ? *C. steinbergii* Roshev.) - 1, 3, 4, 5
hymenoglossa Ohwi = C. monticola
hyperborea auct. = C. neglecta
iberica (Griseb.) Litv. (*C. caucasica* Trin. subsp. *iberica* (Griseb.) Tzvel.) - 2
x **indagata** Torges & Hausskn. - C. arundinacea (L.) Roth x C. villosa (Chaix) J.F. Gmel. - 1
inexpansa A. Gray (*C. aculeolata* (Hack.) Ohwi, *C. neglecta* (Ehrh.) Gaertn., Mey. & Scherb. var. *aculeolata* (Hack.) Miyabe & Kudo, *C. neglecta* subsp. *inexpansa* (A. Gray) Tzvel., *C. stricta* (Timm) Koel. var. *aculeolata* Hack.) - 5
interrupta Seredin - 2
    It is unlikely that this species belongs to the genus Calamagrostis.
jacutensis V. Petrov = C. micrantha
kalarica Tzvel. - 4
karataviensis P. Smirn. = C. macrolepis
koeleriiformis Roshev. = Trisetum litwinowii
koibalensis Reverd. = C. glomerata
kolgujewensis Gand. = C. neglecta
x **kolymensis** Kom. - C. deschampsioides Trin. x C. holmii Lange - 4, 5
korotkyi Litw. - 4, 5
- subsp. *latissima* (Worosch.) Tzvel. = C. latissima
- subsp. *monticola* (V. Petrov ex Kom.) Worosch. = C. monticola
- subsp. *turczaninowii* (Litv.) Tzvel. = C. turczaninowii
korshinskyi Litv. - 6
x **kotulae** Zapal. - C. canescens (Web.) Roth x C. villosa (Chaix) J.F. Gmel. - 1
krylovii Reverd. = C. pavlovii
x **kuznetzovii** Tzvel. - C. epigeios (L.) Roth, s.l. x C. obtusata Trin. - 3
laguroides Regel (*C. anthoxanthoides* (Munro) Regel subsp. *laguroides* (Regel) Tzvel., *Stilpnophleum laguroies* (Regel) Nevski) - 6
lancea Ohwi = C. lapponica
lanceolata Roth = C. canescens
lanceolatiformis V. Vassil. = C. angustifolia

langsdorffii (Link) Trin. (*C. baicalensis* Litv., *C. canadensis* (Michx.) Beauv. subsp. *langsdorffii* (Link) Hult., *C. confusa* V. Vassil., *C. purpurea* (Trin.) Trin. subsp. *langsdorffii* (Link) Tzvel., *C. unilateralis* V. Petrov, *C. yendoana* Honda, *Deyeuxia langsdorffii* (Link) Keng) - 1, 3, 4, 5
lapponica (Wahlenb.) C. Hartm. (? *Arundo confinis* Willd., ? *Calamagrostis confinis* (Willd.) Beauv., *C. gorodkovii* V. Vassil., *C. lancea* Ohwi, *C. lapponica* subsp. *sibirica* (V. Petrov) Tzvel., *C. pseudolapponica* V. Vassil., *C. sibirica* V. Petrov) - 1, 3, 4, 5
- subsp. *sibirica* (V. Petrov) Tzvel. = C. lapponica
latissima (Worosch.) Probat. (*C. turczaninowii* Litv. subsp. *latissima* Worosch., *C. korotkyi* Litv. subsp. *latissima* (Worosch.) Tzvel.) - 5
lenkoranensis Steud. = C. epigeios
lithuanica Bess. = C. canescens
littorea (Schrad.) DC. var. *tartarica* Hook. fil. = C. hedinii
litwinowii Kom. (*C. sachalinensis* Fr. Schmidt subsp. *litwinowii* (Kom.) Probat.) - 5
macilenta (Griseb.) Litv. (*Deyeuxia macilenta* (Griseb.) Keng) - 3, 4
macrantha (Piper ex Scribn. & Merr.) V. Vassil. = C. deschampsioides
macrolepis Litv. (*C. epigeios* (L.) Roth subsp. *macrolepis* (Litv.) Tzvel., *C. gigantea* Roshev. 1932, non Nutt. 1837, *C. karataviensis* P. Smirn.) - 1, 2, 3, 4, 5(alien), 6
magadanica V. Vassil. = C. purpurea
manarovii Gussejnov - 2
meinshausenii (Tzvel.) Viljasoo (*C. epigeios* (L.) Roth subsp. *meinshausenii* Tzvel., *C. epigeios* var. *laevis* Meinsh., *C. meinshausenii* (Tzvel.) Min. comb. superfl., *C. paralias* (Fries) Juz. ex Pobed. p.p. quoad pl.) - 1
micrantha Kearney (*C. jacutensis* V. Petrov, *C. neglecta* (Ehrh.) Gaertn., Mey. & Scherb. subsp. *micrantha* (Kearney) Tzvel., *C. praerupta* V. Vassil., *C. reverdattoi* Golub) - 4, 5
mirabilis V. Vassil. = C. tenuis
miyabei Honda = C. deschampsioides
monticola V. Petrov ex Kom. (*C. arundinacea* (L.) Roth subsp. *monticola* (V. Petrov ex Kom.) Tzvel., *C. hymenoglossa* Ohwi, *C. korotkyi* Litv. subsp. *monticola* (V. Petrov ex Kom.) Worosch.) - 5
mutabilis V. Vassil. = C. tenuis
neglecta (Ehrh.) Gaertn., Mey. & Scherb. (*C. kolgujewensis* Gand., *C. ochotensis* V. Vassil., *C. hyperborea* auct.) - 1, 2, 3, 4, 5, 6
- subsp. *borealis* (Laest.) Seland. = C. groenlandica
- subsp. *groenlandica* (Schrank) Matuszk. = C. groenlandica
- subsp. *holmii* (Lange) Worosch. = C. holmii
- subsp. *inexpansa* (A. Gray) Tzvel. = C. inexpansa
- subsp. *micrantha* (Kearney) Tzvel. = C. micrantha
- subsp. *stricta* (Timm) Tzvel. = C. stricta
- var. *aculeolata* (Hack.) Miyabe & Kudo = C. inexpansa
nivalis V. Vassil. = C. x vassiljevii
notabilis Litv. = C. purpurea
nutkaensis (C. Presl) Steud. (*Deyeuxia nutkaensis* C. Presl, *Calamagrostis aleutica* Trin.) - 5(?)
obscura Downar = C. x vilnensis
obtusata Trin. (*C. agrostioides* Matuszk. 1948, non C. agrostoides Pursh ex Spreng. 1828) - 1, 3, 4, 5
ochotensis V. Vassil. = C. neglecta
olympica Boiss. = Agrostis olympica
pamirica Litv. (*C. grandiflora* Hack. ex Roshev., *C. holciformis* Jaub. & Spach subsp. *pamirica* (Litv.) Tzvel.) - 6

x **paradoxa** Lipsky. - C. arundinacea (L.) Roth x ? C. balkharica P. Smirn. - 2
paralias (Fries) Juz. ex Pobed. p.p. = C. meinshausenii
parviflora Rupr. = C. arundinacea
pavlovii Roshev. (*C. krylovii* Reverd.) - 3, 4, 5, 6
persica Boiss. (*C. pseudophragmites* (Hall. fil.) Koel. subsp.
persica (Boiss.) Tzvel.) - 2, 6
phalaroides Regel = C. anthoxanthoides
phragmitoides C. Hartm. (*C. elata* Blytt, *C. flexuosa* Rupr.,
C. phragmitoides subsp. *flexuosa* (Rupr.) K. Richt., *C.
purpurea* (Trin.) Trin. subsp. *flexuosa* (Rupr.) Sojak, *C.
purpurea* subsp. *phragmitoides* (C. Hartm.) Tzvel.) - 1,
2, 3
- subsp. *flexuosa* (Rupr.) K. Richt. = C. phragmitoides
poaeoides V.Vassil. = C. purpurea
x **ponojensis** Montell. - C. lapponica (Wahlenb.) C. Hartm.
x C. neglecta (Ehrh.) Gaertn., Mey. & Scherb. s.l. - 1
poplawskae Roshev. = C. purpurea
praerupta V. Vassil. = C. micrantha
x **pseudodeschampsioides** Tzvel. - C. deschampsioides Trin.
x C. neglecta (Ehrh.) Gaertn., Mey. & Scherb. s.l. - 1
pseudolapponica V. Vassil. = C. lapponica
pseudophragmites (Hall. fil.) Koel. (*C. glauca* (Bieb.) Trin.
1837, non Reichenb. 1830) - 1, 2, 3, 4, 5, 6
- subsp. *dubia* (Bunge) Tzvel. = C. dubia
- subsp. *persica* (Boiss.) Tzvel. = C. persica
- subsp. *tartarica* (Hook. fil.) Tzvel. = C. hedinii
purpurascens R. Br. (*C. caespitosa* V. Vassil. 1950, non
Steud. 1854, *C. wiluica* Litv. ex V. Petrov) - 4, 5
- subsp. *arctica* (Vasey) Hult. = C. arctica
- subsp. *sesquiflora* (Trin.) Worosch. = C. sesquiflora
- subsp. *urelytra* (Hack.) Worosch. = C. urelytra
purpurea (Trin.) Trin. (*C. fusca* Kom., *C. gracilis* (Litv.) V.
Vassil. comb. invalid., *C. magadanica* V. Vassil., *C.
notabilis* Litv., *C. poaeoides* V. Vassil. non Steud., *C.
poplawskae* Roshev.) - 1, 3, 4, 5, 6
- subsp. *angustifolia* (Kom.) Worosch. = C. angustifolia
- subsp. *barbata* (V. Vassil.) Tzvel. = C. barbata
- subsp. *flexuosa* (Rupr.) Sojak = C. phragmitoides
- subsp. *langsdorffii* (Link) Tzvel. = C. langsdorffii
- subsp. *phragmitoides* (C. Hartm.) Tzvel. = C. phragmitoides
reverdattoi Golub = C. micrantha
x **rigens** Lindgr. - C. canescens (Web.) Roth x C. epigeios
(L.) Roth, s.l.
ruprechtii Nevski = Agrostis olympica
sachalinensis Fr. Schmidt - 5
- subsp. *litwinowii* (Kom.) Probat. = C. litwinowii
sajanensis Malysch. - 4
salina Tzvel. - 4
schugnanica Litv. = C. alaica
sesquiflora (Trin.) Tzvel. (*Trisetum sesquiflorum* Trin.,
Calamagrostis purpurascens R. Br. subsp. *sesquiflora*
(Trin.) Worosch.) - 5
- subsp. *urelytra* (Hack.) Probat. = C. urelytra
sibirica V. Petrov = C. lapponica
steinbergii Roshev. = C. holmii
stricta (Timm) Koel. (*Arundo stricta* Timm, *Calamagrostis
neglecta* (Ehrh.) Gaertn., Mey. & Scherb. subsp. *stricta*
(Timm) Tzvel.) - 1, 5
- subsp. *groenlandica* (Schrank) A. Löve = C. groenlandica
- var. *aculeolata* Hack. = C. inexpansa
x **strigosa** (Wahlenb.) C. Hartm. (*Arundo strigosa*
Wahlenb.). - C. epigeios (L.) Roth, s.l. x C. neglecta
(Ehrh.) Gaertn., Mey. & Scherb. s.l.
subchalybaea Tzvel. (*C. adylssuensis* Seredin) - 2
x **subepigeios** Tzvel. - C. epigeios (L.) Roth, s.l. x C. purpurea (Trin.) Trin. s.l.
x **submonticola** Probat. - C. epigeios (L.) Roth, s.l. x C.
monticola V. Petrov ex Kom. - 5
x **subneglecta** Tzvel. - C. neglecta (Ehrh.) Gaertn., Mey. &
Scherb. s.l. x C. purpurea (Trin.) Trin. s.l.
sugawarae Ohwi (*C. arundinacea* (L.) Roth subsp. *sugawarae* (Ohwi) Tzvel.) - 5
teberdensis Litv. - 2, 4, 6
tenuis V. Vassil. (*C. angustifolia* Kom. var. *tenuis* (V.
Vassil.) Kitag., *C. angustifolia* subsp. *tenuis* (V. Vassil.)
Tzvel., ? *C. hirsuta* V. Vassil., *C. mirabilis* V. Vassil., *C.
mutabilis* V. Vassil.) - 5
  Probably, the priority name of this species is C. czukczorum
  Socz.
x **thyrsoidea** C. Koch. - C. epigeios (L.) Roth, s.l. x C.
pseudophragmites (Hall. fil.) Koel. s.l.
tianschanica Rupr. (*Deyeuxia tianschanica* (Rupr.) Bor) - 6
- subsp. *fedtschenkoana* Tzvel. = C. fedtschenkoana
trichantha Schischk. = Agrostis trichantha
trinii Rupr. = C. x acutiflora
turczaninowii Litv. (*C. korotkyi* Litv. subsp. *turczaninowii*
(Litv.) Tzvel., *Deyeuxia turczaninowii* (Litv.) Chang) -
4, 5
- subsp. *latissima* Worosch. = C. latissima
turkestanica Hack. (? *C. alexeenkoana* Litv. ex Roshev., *C.
alopecuroides* Roshev.) - 3(?), 6
tzvelevii Gussejnov - 2
unilateralis V. Petrov = C. langsdorffii
uralensis Litv. - 1
urelytra Hack. (*C. purpurascens* R. Br. subsp. *urelytra*
(Hack.) Worosch., *C. sesquiflora* (Trin.) Tzvel. subsp.
urelytra (Hack.) Probat.) - 5
x **ussuriensis** Tzvel. - C. angustifolia Kom. s.l. x C. epigeios
(L.) Roth, s.l.
varia (Schrad.) Host (*Arundo varia* Schrad.) - 1
x **vassiljevii** Tzvel. (*C. nivalis* V. Vassil. 1963, non Hack.
1910). - C. angustifolia Kom. s.l. x C. deschampsioides
Trin.
veresczaginii N. Zolotuchin - 3
villosa (Chaix) J.F. Gmel. (*Agrostis villosa* Chaix) - 1
x **vilnensis** Bess. (*C. canescens* (Web.) Roth subsp. *vilnensis*
(Bess.) H. Scholz, *C. obscura* Downar). - C. canescens
(Web.) Roth x C. neglecta (Ehrh.) Gaertn., Mey. &
Scherb. s.l.
wiluica Litv. ex V. Petrov = C. purpurascens
x **yatabei** Maxim. - C. brachytricha Steud. x C. epigeios (L.)
Roth, s.l.
yendoana Honda = C. langsdorffii

x **Calammophila** Brand (x *Ammocalamagrostis* P. Fourn., x *Calamophila* O. Schwarz)

baltica (Flugge ex Schrad.) Brand (*Arundo baltica* Flugge
ex Schrad., x *Ammocalamagrostis baltica* (Flugge ex
Schrad.) P.Fourn., *Ammophila baltica* (Flugge ex
Schrad.) Link, x *Calamophila baltica* (Flugge ex
Schrad.) O. Schwarz) - 1

x *Calamophila* O. Schwarz = x Calammophila

baltica (Flugge ex Schrad.) O. Schwarz = x Calammophila
baltica

Campeiostachys Drob. = Elymus

schrenkiana (Fisch. & C.A. Mey.) Drob. = Elymus schrenkianus

Catabrosa Beauv.

aquatica (L.) Beauv. - 1, 2, 3, 4, 6
- subsp. *capusii* (Franch.) Tzvel. = C. capusii
- subsp. *pseudairoides* (Herrm.) Tzvel. = C. pseudairoides
capusii Franch. (*C. aquatica* (L.) Beauv. subsp. *capusii* (Franch.) Tzvel.) - 6
concinna Th. Fries subsp. *vacillans* Th. Fries = x Pucciphippsia vacillans
pseudairoides (Herrm.) Tzvel. (*Poa pseudairoides* Herrm., *Catabrosa aquatica* (L.) Beauv. subsp. *pseudairoides* (Herrm.) Tzvel.) - 1, 2, 6
vilfoidea Anderss. = Puccinellia vilfoidea

Catabrosella (Tzvel.) Tzvel.

araratica (Lipsky) Tzvel. (*Colpodium araraticum* (Lipsky) Woronow) - 2
calvertii (Boiss.) Czer. (*Colpodium calvertii* Boiss., *Catabrosella humilis* (Bieb.) Tzvel. subsp. *calvertii* (Boiss.) Tzvel., *C. parviflora* (Boiss. & Buhse) Alexeev ex R. Mill subsp. *calvertii* (Boiss.) Alexeev ex R. Mill) - 2
fibrosa (Trautv.) Tzvel. (*Colpodium fibrosum* Trautv.) - 2
humilis (Bieb.) Tzvel. (*Colpodium humile* (Bieb.) Griseb.) - 1, 2, 3, 6
- subsp. *calvertii* (Boiss.) Tzvel. = C. calvertii
- subsp. *ornata* (Nevski) Tzvel. = C. ornata
- subsp. *parviflora* (Boiss. & Buhse) Tzvel. = C. parviflora
- subsp. *songorica* Tzvel. = C. songorica
leiantha (Hack.) Czer. (*Colpodium leianthum* Hack., *Catabrosella variegata* (Boiss.) Tzvel. var. *leiantha* (Hack.) Alexeev ex R. Mill, *C. variegata* subsp. *leiantha* (Hack.) Tzvel., *Colpodium variegatum* (Boiss.) Griseb. var. *leianthum* (Hack.) Grossh.) - 2
ornata (Nevski) Czopanov (*Colpodium ornatum* Nevski, *Catabrosella humilis* (Bieb.) Tzvel. subsp. *ornata* (Nevski) Tzvel.) - 6
parviflora (Boiss. & Buhse) Czopanov (*Colpodium parviflorum* Boiss. & Buhse, *Catabrosella humilis* (Bieb.) Tzvel. subsp. *parviflora* (Boiss. & Buhse) Tzvel., *C. parviflora* (Boiss. & Buhse) Alexeev ex R. Mill, comb. superfl.) - 2, 6
parviflora (Boiss. & Buhse) Alexeev ex R. Mill subsp. *calvertii* (Boiss.) Alexeev ex R. Mill = C. calvertii
songorica (Tzvel.) Czer. (*C. humilis* (Bieb.) Tzvel. subsp. *songorica* Tzvel.) - 6
variegata (Boiss.) Tzvel. (*Colpodium chrysanthum* Woronow, *C. variegatum* (Boiss.) Griseb.) - 2
- subsp. *leiantha* (Hack.) Tzvel. = C. leiantha
- var. *leiantha* (Hack.) Alexeev ex R. Mill = C. leiantha

Catapodium auct. = Cutandia and Scleropoa

rigescens (Grossh.) Bor = Cutandia rigescens
rigidum (L.) C.E. Hubb. = Scleropoa rigida

Cenchrus L.

incertus Curt. = C. pauciflorus
pauciflorus Benth. (? *C. incertus* Curt., *C. tribuloides* auct.) - 1, 2
tribuloides auct. = C. pauciflorus

Ceratochloa Beauv.

carinata (Hokk. & Arn.) Tutin (*Bromus carinatus* Hook. & Arn.) var. **hookeriana** (Thurb.) Tzvel. (*Bromus hookerianus* Thurb., *Ceratochloa grandiflora* Hook.) - 1
cathartica (Vahl) Herter (*Bromus catharticus* Vahl, *Cerato-*

chloa haenkeana C. Presl) - 1, 2, 5, 6
grandiflora Hook. = Ceratochloa carinata var. hookeriana
haenkeana C. Presl = C. cathartica
polyantha (Scribn.) Tzvel. (*Bromus polyanthus* Scribn.) - 1, 6
***unioloides** (Willd.) Beauv. (*Bromus willldenowii* Kunth, *Ceratochloa willldenowii* (Kunth) W.A. Weber)
willldenowii (Kunth) W.A. Weber = C. unioloides

Chennapyrum A. Löve = Aegilops

uniaristatum (Vis.) A. Löve = Aegilops uniaristata

*Chimonobambusa Makino

***marmorea** (Mitf.) Makino (*Bambusa marmorea* Mitf.) - 2
***quadrangularis** (Fenzi) Makino (*Bambusa quadrangularis* Fenzi) - 1, 2

Chloris Sw.

***gayana** Kunth
***polydactyla** (L.) Sw. (*Andropogon polydactylon* L.)
***uliginosa** Hack.
villosa (Desf.) Pers. = Tetrapogon villosus
virgata Sw. - 2, 5, 6

Chrysopogon Trin.

gryllus (L.) Trin. - 1, 2

Cinna L.

karataviensis Pavl. = Agrostis gigantea
latifolia (Trev.) Griseb. - 1, 2, 3, 4, 5

Cleistogenes Keng (*Kengia* Packer, *Diplachne* auct.)

andropogonoides Honda = C. squarrosa
bulgarica (Bornm.) Keng (*C. maeotica* Klok. & Zoz, *C. serotina* (L.) Keng subsp. *bulgarica* (Bornm.) Tutin, *Diplachne bulgarica* (Bornm.) Bornm., *D. bulgarica* (Bornm.) Roshev. comb. superfl., *D. maeotica* Klok. & Zoz, nom. invalid., *Kengia bulgarica* (Bornm.) Packer, *K. maeotica* (Klok. & Zoz) Packer) - 1, 2
chinensis auct. = C. kitagawae
hancei Keng (*Diplachne sinensis* Hance, 1870, non Cleistogenes chinensis (Maxim.) Keng, *Kengia hancei* (Keng) Packer) - 5
kitagawae Honda (*Kengia kitagawae* (Honda) Packer, *Cleistogenes chinensis* auct., *Diplachne sinensis* auct.) - 3, 4, 5, 6
maeotica Klok. & Zoz = C. bulgarica
nakaii (Keng) Honda - 5
serotina (L.) Keng (*Diplachne serotina* (L.) Link, *Kengia serotina* (L.) Packer) - 1, 2, 6
- subsp. *bulgarica* (Bornm.) Tutin = C. bulgarica
songorica (Roshev.) Ohwi (*Diplachne songorica* Roshev., *Cleistogenes thoroldii* (Stapf) Roshev. p.p. excl. typo, *Kengia songorica* (Roshev.) Packer) - 3, 4, 6
squarrosa (Trin.) Keng (*C. andropogonoides* Honda, *C. squarrosa* subsp. *andropogonoides* (Honda) Worosch., *Diplachne squarrosa* (Trin.) Maxim., *Kengia squarrosa* (Trin.) Packer) - 1, 2, 3, 4, 5, 6
- subsp. *andropogonoides* (Honda) Worosch. = C. squarrosa
thoroldii (Stapf) Roshev. p.p. = C. songorica

356

Clinelymus (Griseb.) Nevski = Elymus

cylindricus (Franch.) Honda = Elymus dahuricus
dahuricus (Turcz. ex Griseb.) Nevski = Elymus dahuricus
excelsus (Turcz. ex Griseb.) Nevski = Elymus excelsus
nutans (Griseb.) Nevski = Elymus nutans
sibiricus (L.) Nevski = Elymus sibiricus

*Coix L.

*lacryma-jobi L. - 6

Coleanthus Seidel

subtilis (Tratt.) Seidel - 1, 3, 5

Colpodium Trin.

altaicum Trin. = Paracolpodium altaicum
araraticum (Lipsky) Woronow = Catabrosella araratica
calvertii Boiss. = Catabrosella calvertii
chrysanthum Woronow = Catabrosella variegata
colchicum (Albov) Woronow = Paracolpodium colchicum
fibrosum Trautv. = Catabrosella fibrosa
humile (Bieb.) Griseb. = Catabrosella humilis
ivanoviae Malysch. = Hyalopoa ivanoviae
lakium Woronow = Hyalopoa lakia
lanatiflorum (Roshev.) Tzvel. = Hyalopoa lanatiflora
- subsp. *momicum* Tzvel. = Hyalopoa momica
leianthum Hack. = Catabrosella leiantha
leucolepis Nevski = Paracolpodium leucolepis
ornatum Nevski = Catabrosella ornata
parviflorum Boiss. & Buhse = Catabrosella parviflora
ponticum (Bal.) Woronow = Hyalopoa pontica
schelkownikowii Grossh. = C. versicolor
vacillans (Th. Fries) Polun. = x Pucciphippsia vacillans
vahlianum (Liebm.) Nevski = Puccinellia vahliana
variegatum (Boiss.) Griseb. = Catabrosella variegata
- var. *leianthum* (Hack.) Grossh = Catabrosella leiantha
versicolor (Stev.) Schmalh. (*C. schelkownikowii* Grossh.) - 2
wrightii Scribn. & Merr. = Puccinellia wrightii

Comopyrum (Jaub. & Spach) A. Löve = Aegilops

comosum (Sibth. & Smith) A. Löve = Aegilops comosa

*Cortaderia Stapf

dioica (Spreng.) Speg. = C. selloana
*selloana (Schult. & Schult. fil.) Aschers. & Graebn. (*Arundo selloana* Schult. & Schult. fil., *A. dioica* Spreng. 1825, non Lour. 1793, *Cortaderia dioica* (Spreng.) Speg.) - 1, 2, 6

Corynephorus Beauv. (*Anachortus* Jir. & Chrtek)

articulatus (Desf.) Beauv. = C. divaricatus
canescens (L.) Beauv. (*Aira canescens* L.) - 1
divaricatus (Pourr.) Breistroffer (*Aira divaricata* Pourr., *Anachortus articulatus* (Desf.) Jir. & Chrtek, *A. divaricatus* (Pourr.) Lainz, *Corynephorus articulatus* (Desf.) Beauv., *C. divaricatus* subsp. *articulatus* (Desf.) Lainz) - 1(?), 2
- subsp. *articulatus* (Desf.) Lainz = C. divaricatus

Critesion Rafin. = Hordeum

bogdanii (Wilensky) A. Löve = Hordeum bogdanii
brevisubulatum (Trin.) A. Löve = Hordeum brevisubulatum
- subsp. *nevskianum* (Bowden) A. Löve = Hordeum nevskianum
- subsp. *turkestanicum* (Nevski) A. Löve = Hordeum turkestanicum
bulbosum (L.) A. Löve = Hordeum bulbosum
californicum (Cavas & Stebb.) A. Löve subsp. *sibiricum* (Roshev.) A. Löve = Hordeum roshevitzii
geniculatum (All.) A. Löve = Hordeum geniculatum
glaucum (Steud.) A. Löve = Hordeum glaucum
hystrix (Roth) A. Löve = Hordeum geniculatum
jubatum (L.) Nevski = Hordeum jubatum
- subsp. *breviaristatum* (Bowden) A. & D. Löve = Hordeum brachyantherum
marinum (Huds.) A. Löve = Hordeum marinum
murinum (L.) A. Löve = Hordeum murinum
- subsp. *glaucum* (Steud.) W.A. Weber = Hordeum glaucum
- subsp. *leporinum* (Link) A. Löve = Hordeum leporinum
secalinum (Schreb.) A. Löve = Hordeum secalinum
violaceum (Boiss. & Huet) A.Löve = Hordeum violaceum

Crithodium Link = Triticum

aegilopoides Link = Triticum boeoticum
jerevani (Thum.) A. Löve = Triticum boeoticum
monococcum (L.) A. Löve = Triticum monococcum
- subsp. *aegilopoides* (Link) A. Löve = Triticum boeoticum
urartu (Thum.) A. Löve = Triticum urartu

Crypsis Ait. (*Heleochloa* Host ex Roem.)

aculeata (L.) Ait. - 1, 2, 3, 6
acuminata Trin. (*C. acuminata* subsp. *borszczowii* (Regel) Kit Tan, *C. borszczowii* Regel) - 1, 6
- subsp. *borszczowii* (Regel) Kit Tan = C. acuminata
alopecuroides (Pill. & Mitt.) Schrad. (*C. phalaroides* Bieb., *Heleochloa alopecuroides* (Pill. & Mitt.) Host ex Roem.) - 1, 2, 3, 6
borszczowii Regel = C. acuminata
faktorovskyi Eig (*Heleochloa faktorovskyi* (Eig) Pilg.) - 6
phalaroides Bieb. = C. alopecuroides
schoenoides (L.) Lam. (*Heleochloa schoenoides* (L.) Host ex Roem.) - 1, 2, 3, 6
turkestanica Eig - 1, 2, 3, 6

Cutandia Willk. (*Catapodium* auct., *Desmazeria* auct. p.p.)

dichotoma (Forssk.) Trab. (*Festuca dichotoma* Forssk.) - 2
- var. *memphitica* (Spreng.) Maire & Weiller = C. memphitica
memphitica (Spreng.) Benth. (*C. dichotoma* (Forrsk.) Trab. var. *memphitica* (Spreng.) Maire & Weiller) - 2, 6
rigescens (Grossh.) Tzvel. (*Scleropoa rigescens* Grossh., *Catapodium rigescens* (Grossh.) Bor, *Desmazeria rigescens* (Grossh.) Ovcz. nom. invalid.) - 2, 6

Cuviera Koel. = Hordelymus

europaea (L.) Koel. = Hordelymus europaeus

Cylindropyrum (Jaub. & Spach) A. Löve = Aegilops

cylindricum (Host) A. Löve = Aegilops cylindrica

*Cymbopogon Spreng.

*citratus (DC. ex Nees) Stapf (*Andropogon citratus* DC. ex Nees)

Cynodon Rich.

dactylon (L.) Pers. - 1, 2, 3, 6

Cynosurus L.

aureus L. = Lamarckia aurea
coracanus L. = Eleusine coracana
cristatus L. - 1, 2, 4(alien)
echinatus L. - 1, 2
elegans auct. = C. turcomanicus
tristachyus Lam. = Eleusine tristachya
turcomanicus Proskurjakova (*C. elegans* auct.) - 6

Dactylis L.

altaica Bess. (*D. glomerata* L. subsp. *altaica* (Bess.) Domin) - 3
aschersoniana Graebn. = D. polygama
glomerata L. - 1, 2, 3, 4, 5, 6
- subsp. *altaica* (Bess.) Domin = D. altaica
- subsp. *aschersoniana* (Graebn.) Thell. = D. polygama
- subsp. **himalayensis** Domin - 3, 6
- subsp. *hispanica* (Roth) Nym. = D. hispanica
- subsp. *hyrcana* Tzvel. = D. hyrcana
- subsp. *lobata* (Drej.) Lindb. fil. = D. polygama
- subsp. *polygama* (Horvat.) Domin = D. polygama
- subsp. *slovenica* (Domin) Domin = D. slovenica
- subsp. *woronowii* (Ovcz.) Stebb. & Zohary = D. woronowii
- var. *lobata* Drej. = D. polygama
hispanica Roth (*D. glomerata* L. subsp. *hispanica* (Roth) Nym.) - 1, 2
hyrcana (Tzvel.) Musajev (*D. glomerata* L. subsp. *hyrcana* Tzvel.) - 2
lobata Bieb. = Koeleria lobata
polygama Horvat. (*D. aschersoniana* Graebn., *D. glomerata* L. subsp. *aschersoniana* (Graebn.) Thell., *D. glomerata* subsp. *lobata* (Drej.) Lindb. fil., *D. glomerata* subsp. *polygama* (Horvat.) Domin, *D. glomerata* var. *lobata* Drej.) - 1, 2
slovenica Domin (*D. glomerata* L. subsp. *slovenica* (Domin) Domin) - 1
woronowii Ovcz. (*D. glomerata* L. subsp. *woronowii* (Ovcz.) Stebb. & Zohary) - 2, 6

Danthonia DC.

alpina Vest (*D. calycina* (Vill.) Reichenb. 1829, non Roem. & Schult. 1817, *D. provincialis* DC. nom. illegit.) - 1, 2
calycina (Vill.) Reichenb. = D. alpina
compacta (Boiss. & Heldr.) Grossh. = Danthoniastrum compactum
decipiens (O. Schwarz & Bassl.) A. & D. Löve = Sieglingia decumbens subsp. decipiens
decumbens (L.) DC. subsp. *decipiens* O. Schwarz & Bassl. = Sieglingia decumbens subsp. decipiens
forsskalii (Vahl) R. Br. = Asthenatherum forsskalii
intermedia auct. = D. riabuschinskii
intermedia Vasey subsp. *riabuschinskii* (Kom.) Tzvel. = D. riabuschinskii
provincialis DC. = D. alpina
riabuschinskii (Kom.) Kom. (*Avena riabuschinskii* Kom., *Danthonia intermedia* Vasey subsp. *riabuschinskii*

(Kom.) Tzvel., *D. intermedia* auct.) - 5

Danthoniastrum (Holub) Holub

compactum (Boiss. & Heldr.) Holub (*Arrhenatherum compactum* (Boiss. & Heldr.) Potztal, *Avenastrum compactum* (Boiss. & Heldr.) Roshev., *Danthonia compacta* (Boiss. & Heldr.) Grossh., *Helictotrichon compactum* (Boiss. & Heldr.) Henrard) - 2

Dasypyrum (Coss. & Durieu) T. Durand (*Haynaldia* Schur, VI 1866, non S. Schulzer, V 1866)

villosum (L.) P. Candargy (*Haynaldia villosa* (L.) Schur) - 1, 2

Deschampsia Beauv.

alpina (L.) Roem. & Schult. (*Aira laevigata* Smith, *Deschampsia cespitosa* (L.) Beauv. subsp. *alpina* (L.) Tzvel., *D. laevigata* (Smith) Roem. & Schult.) - 1
altaica (Schischk.) Nikiforova (*D. cespitosa* (L.) Beauv. var. *altaica* Schischk.) - 3, 4
anadyrensis V. Vassil. = D. glauca
arctica sensu Roshev. = D. glauca
arctica (Spreng.) Merr. = D. brevifolia
arctica (Spreng.) Schischk. = D. brevifolia
beringensis Hult. (*D. cespitosa* (L.) Beauv. subsp. *beringensis* (Hult.) W. Lawr.) - 5
biebersteiniana Schult. = D. cespitosa
borealis (Trautv.) Roshev. (*D. cespitosa* (L.) Beauv. subsp. *borealis* (Trautv.) A. & D. Löve, *D. cespitosa* var. *minor* Kom., *D. sukatschewii* (Popl.) Roshev. subsp. *minor* (Kom.) Tzvel.) - 1, 3, 4, 5
bottnica (Wahlenb.) Trin. (*D. cespitosa* (L.) Beauv. subsp. *bottnica* (Wahlenb.) G.C.S. Clarke, comb. superfl., *D. cespitosa* subsp. *bottnica* (Wahlenb.) Tzvel.) - 1(?)
    It is likely that this species had been recorded for the investigated territory by mistake.
brevifolia auct. p. max. p. = D. glauca
brevifolia R. Br. (*D. arctica* (Spreng.) Merr., *D. arctica* (Spreng.) Schischk. comb. superfl., *D. cespitosa* (L.) Beauv. subsp. *brevifolia* (R. Br.) Tzvel.) - 3(?)
- var. *pumila* Griseb. = D. paramushirensis
cespitosa (L.) Beauv. (*D. biebersteiniana* Schult.) - 1, 2, 3, 4, 5, 6
- subsp. *alpicola* Chrtek & Jir. = D. cespitosa subsp. gaudinii
- subsp. *alpina* (L.) Tzvel. = D. alpina
- subsp. *anadyrensis* (V. Vassil.) A. & D. Löve = D. glauca
- subsp. *beringensis* (Hult.) W. Lawr. = D. beringensis
- subsp. *borealis* (Trautv.) A. & D. Löve = D. borealis
- subsp. *bottnica* (Wahlenb.) G.C.S. Clarke = D. bottnica
- subsp. *bottnica* (Wahlenb.) Tzvel. = D. bottnica
- subsp. *brevifolia* (R. Br.) Tzvel. = D. brevifolia
- subsp. **gaudinii** K. Richt. (*D. cespitosa* subsp. *alpicola* Chrtek & Jir.) - 1
- subsp. *glauca* (C. Hartm.) C. Hartm. = D. glauca
- subsp. *koelerioies* (Regel) Tzvel. = D. koelerioides
- subsp. *macrothyrsa* (Tatew. & Ohwi) Tzvel. = D. macrothyrsa
- subsp. *mezensis* (Senjan.-Korcz. & Korcz.) Tzvel. = D. mezensis
- subsp. *obensis* (Roshev.) Tzvel. = D. obensis
- subsp. *orientalis* Hult. = D. sukatschewii
- subsp. *pamirica* (Roshev.) Tzvel. = D. pamirica

- subsp. *paramushirensis* (Honda) Tzvel. = D. paramushirensis
- subsp. **parviflora** (Thuill.) K. Richt. (*Aira parviflora* Thuill., *Deschampsia cespitosa* subsp. *parviflora* (Thuill.) Chrtek & Jir. comb. superfl., *D. cespitosa* subsp. *parviflora* (Thuill.) Soo & Jav.comb. superfl.) - 1, 2, 3
- subsp. *turczaninowii* (Litv.) Tzvel. = D. turczaninowii
- subsp. **wilhelmsii** (Steud.) Tzvel. (*Aira wilhelmsii* Steud.) - 2
- var. *altaica* Schischk. = D. altaica
- var. *macrothyrsa* Tatew. & Ohwi = D. macrothyrsa
- var. *minor* Kom. = D. borealis
flexuosa (L.) Nees = Avenella flexuosa
glauca C. Hartm. (*D. anadyrensis* V. Vassil., *D. cespitosa* (L.) Beauv. subsp. *anadyrensis* (V. Vassil.) A. & D. Löve, *D. cespitosa* subsp. *glauca* (C. Hartm.) C. Hartm., *D. arctica* sensu Roshev., *D. brevifolia* auct. p. max. p.) - 1, 3, 4, 5
koelerioides Regel (*D. cespitosa* (L.) Beauv. subsp. *koelerioies* (Regel) Tzvel.) - 3, 4, 6
komarovii V. Vassil. = D. paramushirensis
laevigata (Smith) Roem. & Schult. = D. alpina
macrothyrsa (Tatew. & Ohwi) Kawano (*D. cespitosa* (L.) Beauv. var. *macrothyrsa* Tatew. & Ohwi, *D. cespitosa* subsp. *macrothyrsa* (Tatew. & Ohwi) Tzvel.) - 5
media (Gouan) Roem. & Schult. - 2
mezensis Senjan.-Korcz. & Korcz. (*D. cespitosa* (L.) Beauv. subsp. *mezensis* (Senjan.-Korcz. & Korcz.) Tzvel.) - 1
montana (L.) G. Don fil. = Avenella flexuosa subsp. montana
obensis Roshev. (*D. cespitosa* (L.) Beauv. subsp. *obensis* (Roshev.) Tzvel.) - 1, 3, 4
pacifica Tatew. & Ohwi = Vahlodea flexuosa
pamirica Roshev. (*D. cespitosa* (L.) Beauv. subsp. *pamirica* (Roshev.) Tzvel.) - 6
paramushirensis Honda (*D. brevifolia* R. Br. var. *pumila* Griseb., *D. cespitosa* (L.) Beauv. subsp. *paramushirensis* (Honda) Tzvel., *D. komarovii* V. Vassil., *D. pumila* (Griseb.) Ostenf. 1923, non Woronow, 1909) - 5
pulchella (Willd.) Trin. = Molineriella laevis
pumila (Griseb.) Ostenf. = D. paramushirensis
submutica (Trautv.) Nikiforova = D. sukatschewii
sukatschewii (Popl.) Roshev. (*D. cespitosa* (L.) Beauv. subsp. *orientalis* Hult., *D. submutica* (Trautv.) Nikiforova, *D. sukatschewii* subsp. *orientalis* (Hult.) Tzvel., *D. sukatschewii* subsp. *submutica* (Trautv.) Tzvel.) - 1, 3, 4, 5
- subsp. *minor* (Kom.) Tzvel. = D. borealis
- subsp. *orientalis* (Hult.) Tzvel. = D. sukatschewii
- subsp. *submutica* (Trautv.) Tzvel. = D. sukatschewii
turczaninowii Litv. (*D. cespitosa* (L.) Beauv. subsp. *turczaninowii* (Litv.) Tzvel.) - 4
tzvelevii Probat. - 4
vodopjanoviae Nikiforova - 4

Desmazeria auct. = Cutandia, Sclerocloa and Scleropoa

compressa Ovcz. & Schibk. = Sclerochloa woronowii
rigescens (Grossh.) Ovcz. = Cutandia rigescens
rigida (L.) Tutin = Scleropoa rigida
woronowii (Hack.) Ovcz. = Sclerochloa woronowii

Deyeuxia Clarion = Calamagrostis

angustifolia (Kom.) Chang = Calamagrostis angustifolia
arundinacea (L.) Beauv. = Calamagrostis arundinacea

brachytricha (Steud.) Chang = Calamagrostis brachytricha
compacta Munro ex Hook. fil. = Calamagrostis x compacta
holciformis (Jaub. & Spach) Bor = Calamagrostis holciformis
langsdorffii (Link) Keng = Calamagrostis langsdorffii
macilenta (Griseb.) Keng = Calamagrostis macilenta
nutkaensis C. Presl = Calamagrostis nutkaensis
tianschanica (Rupr.) Bor = Calamagrostis tianschanica
turczaninowii (Litv.) Chang = Calamagrostis turczaninowii

Diandrochloa de Winter (*Roshevitzia* Tzvel.)

diarrhena (Schult. & Schult. fil) A.N. Henry (*Poa diarrhena* Schult. & Schult. fil., *Eragrostis diarrhena* (Schult. & Schult. fil.) Steud., *E. kossinskyi* Roshev., *Roshevitzia diarrhena* (Schult. & Schult. fil.) Tzvel.) - 1

Diarrhena auct. = Neomolinia

fauriei (Hack.) Ohwi = Neomolinia fauriei
japonica Franch. & Savat. = Neomolinia japonica
koryoensis Honda = Neomolinia koryoensis
mandshurica Maxim. = Neomolinia mandshurica

Dichanthelium (Hitchc. & Chase) Gould = Panicum

lindheimeri (Nash) Gould = Panicum lindheimeri

Dichanthium auct. = Bothriochloa

bladhii (Retz.) Clayton = Bothriochloa bladhii
caucasicum (Trin.) Jain & Deshpande = Bothriochloa bladhii
ischaemum (L.) Roberty = Bothriochloa ischaemum

Digitaria Hall.

adscendens (Kunth) Henrard = D. ciliaris
aegyptiaca (Retz.) Willd. (*Panicum aegyptiacum* Retz., *Digitaria sanguinalis* (L.) Scop. subsp. *aegyptiaca* (Retz.) Henrard, *Panicum sanguinale* L. subsp. *aegypticum* (Retz.) K. Richt.) - 1, 2, 6
- subsp. *caucasica* (Henrard) Tzvel. = D. caucasica
asiatica Tzvel. (*D. ischaemum* (Schreb.) Muehl. subsp. *asiatica* (Tzvel.) Tzvel.) - 1, 2, 5(alien), 6
caucasica Henrard (*D. aegyptiaca* (Retz.) Willd. subsp. *caucasica* (Henrard) Tzvel.) - 2
chinensis (Nees) A. Camus = D. violascens
ciliaris (Retz.) Koel. (*Panicum ciliare* Retz., *Digitaria adscendens* (Kunth) Henrard, *D. sanguinalis* (L.) Scop. subsp. *ciliaris* (Retz.) Domin, p.p. quoad nomen, *Panicum adscendens* Kunth) - 2, 5(alien)
horizontalis Willd. - 2
ischaemum (Schreb.) Muehl (*Panicum ischaemum* Schreb., *Digitaria linearis* (L.) Crep. p.p. excl. typo, *Panicum arenarium* Bieb. 1808, non Brot. 1804) - 1, 2, 3, 4, 5, 6
- subsp. *asiatica* (Tzvel.) Tzvel. = D. asiatica
linearis (L.) Crep. p.p. = D. ischaemum
longiflora auct. = D. violascens
paspaloides Michx. = Paspalum paspaloides
pectiniformis (Henrard) Tzvel. (*D. sanguinalis* (L.) Scop. subsp. *pectiniformis* Henrard, *D. ciliaris* (Retz.) Koel. p.p. excl. typo) - 1, 2, 6(alien)
sabulosa Tzvel. (*D. sanguinalis* (L.) Scop. subsp. *sabulosa* (Tzvel.) Tzvel.) - 2
sanguinalis (L.) Scop. (*D. sanguinalis* subsp. *vulgaris* (Schrad.) Henrard, *D. vulgaris* (Schrad.) Bess., *Synthe-*

risma vulgaris Schrad.) - 1, 2, 6
- subsp. *aegyptiaca* (Retz.) Henrard = D. aegyptiaca
- subsp. *ciliaris* (Retz.) Domin = D. ciliaris
- subsp. *pectiniformis* Henrard = D. pectiniformis
- subsp. *sabulosa* (Tzvel.) Tzvel. = D. sabulosa
- subsp. *vulgaris* (Schrad.) Henrard = D. sanguinalis
violascens Link (*D. chinensis* (Retz.) A. Camus, 1823, non Hornem. 1819, *Paspalum chinense* Nees, *Digitaria longiflora* auct.) - 2
vulgaris (Schrad.) Bess. = D. sanguinalis

Digraphis Trin. = Phalaroides

arundinacea (L.) Trin. = Phalaroides arundinacea

Dimeria R. Br.

neglecta Tzvel. - 5

Diplachne auct. = Cleistogenes

bulgarica (Bornm.) Bornm. = Cleistogenes bulgarica
bulgarica (Bornm.) Roshev. = Cleistogenes bulgarica
maeotica Klok. & Zoz = Cleistogenes bulgarica
serotina (L.) Link = Cleistogenes serotina
sinensis Hance = Cleistogenes hancei
songorica Roshev. = Cleistogenes songorica
squarrosa (Trin.) Maxim. = Cleistogenes squarrosa

Drymochloa Holub = Festuca

drymeja (Mert. & Koch) Holub = Festuca drymeja
sylvatica (Poll.) Holub = Festuca altissima

Dupontia R. Br.

fisheri R. Br. - 1, 4, 5
- subsp. *pelligera* (Rupr.) Tzvel. = D. pelligera
- subsp. *psilosantha* (Rupr.) Hult. = D. psilosantha
pelligera (Rupr.) A. Löve & Ritchie (*Poa pelligera* Rupr., *Dupontia fisheri* R. Br. subsp. *pelligera* (Rupr.) Tzvel.) - 1, 3, 4, 5
psilosantha Rupr. (*D. fisheri* R. Br. subsp. *psilosantha* (Rupr.) Hult.) - 1, 3, 4, 5

x **Dupontopoa** Probat.

dezhnevii Probat. - Arctopoa eminens (C. Presl) Probat. x Dupontia psilosantha Rupr. - 5

Echinaria Desf.

capitata (L.) Desf. - 1, 2, 6

Echinochloa Beauv.

caudata Roshev. (*E. crusgalli* (L.) Beauv. subsp. *caudata* (Roshev.) Tzvel., *E. crusgalli* var. *caudata* (Roshev.) Kitag.) - 4, 5
coarctata Kossenko = E. oryzoides
colonum auct. = E. crusgalli
cf. **colonum** (L.) Link - 1(alien)
crusgalli (L.) Beauv. (*E. colonum* auct.) - 1, 2, 3, 4, 5, 6
- subsp. *caudata* (Roshev.) Tzvel. = E. caudata
- subsp. *spiralis* (Vasing.) Tzvel. = E. occidentalis
- var. *caudata* (Roshev.) Kitag. = E. caudata
- var. *macrocarpa* (Vasing.) Ohwi = E. oryzoides
- var. *oryzicola* (Vasing.) Ohwi = E. phyllopogon
- var. *utilis* (Ohwi & Yabuno) Kitam. = E. utilis
frumentacea Link (*Panicum frumentaceum* Roxb. 1820,

non Salisb. 1796) - 1, 6
- subsp. *utilis* (Ohwi & Yabuno) Tzvel. = E. utilis
hostii (Bieb.) Holub = E. oryzoides
hostii (Bieb.) Stev. = E. oryzoides
macrocarpa Vasing. = E. oryzoides
microstachya (Wiegand) Rydb. (*E. muricata* (Beauv.) Fern. var. *microstachya* Wiegand) - 1(alien)
muricata (Beauv.) Fern. var. *microstachya* Wiegand = E. microstachya
- var. *occidentalis* Wiegand = E. occidentalis
occidentalis (Wiegand) Rydb. (*E. muricata* (Beauv.) Fern. var. *occidentalis* Wiegand, *E. crusgalli* (L.) Beauv. subsp. *spiralis* (Vasing.) Tzvel., *E. spiralis* Vasing.) - 1, 2, 3, 4, 5, 6
oryzicola (Vasing.) Vasing. = E. phyllopogon
oryzoides (Ard.) Fritsch (*Panicum oryzoides* Ard., *Echinochloa coarctata* Kossenko, nom. illegit., *E. crusgalli* (L.) Beauv. var. *macrocarpa* (Vasing.) Ohwi, *E. hostii* (Bieb.) Holub, comb. superfl., *E. hostii* (Bieb.) Stev., *E. macrocarpa* Vasing., *Panicum coarctatum* Stev. ex Trin. nom. invalid., *P. crusgalli* L. subsp. *hostii* (Bieb.) K. Richt., *P. hostii* Bieb.) - 1, 2, 5, 6
- subsp. *phyllopogon* (Stapf) Tzvel. = E. phyllopogon
phyllopogon (Stapf) Kossenko (*Panicum phyllopogon* Stapf, *Echinochloa crusgalli* (L.) Beauv. var. *oryzicola* (Vasing.) Ohwi, *E. oryzicola* (Vasing.) Vasing., *E. oryzoides* (Ard.) Fritsch subsp. *phyllopogon* (Stapf) Tzvel., *E. phyllopogon* subsp. *oryzicola* (Vasing.) Kossenko, *E. phyllopogon* subsp. *stapfiana* Kossenko, nom. illegit.) - 1, 2, 5, 6
- subsp. *oryzicola* (Vasing.) Kossenko = E. phyllopogon
- subsp. *stapfiana* Kossenko = E. phyllopogon
pungens (Poir.) Rydb. var. *wiegandii* Fassett = E. wiegandii
spiralis Vasing. = E. occidentalis
utilis Ohwi & Yabuno (*E. crusgalli* (L.) Beauv. var. *utilis* (Ohwi & Yabuno) Kitam., *E. frumentacea* Link subsp. *utilis* (Ohwi & Yabuno) Tzvel.) - 1, 2, 3(alien), 5, 6
wiegandii (Fassett) McNeill & Dore (*E. pungens* (Poir.) Rydb. var. *wiegandii* Fassett) - 1(alien)

Eleusine Gaertn.

*****coracana** (L.) Gaertn. (*Cynosurus coracanus* L.) - 2, 6
indica (L.) Gaertn. - 2, 6
triaristata (Lam.) Kunth - 2
tristachya (Lam.) Lam. (*Cynosurus tristachyus* Lam.) - 2

x **Elyhordeum** Mansf. ex Cziczin & Petr. (x *Elymordeum* Lepage, x *Horderoegneria* Tzvel., x *Agrohordeum* auct.)

arachleicum Peschkova. - Elymus sibiricus L. x Hordeum brevisubulatum (Trin.) Link - 4
arcuatum W. Mitch. & Hodgs. - Elymus sibiricus L. x Hordeum jubatum L. - 5
bowes-lyonii (Melderis) Melderis (x *Elymordeum bowes-lyonii* Melderis). - Elymus nutans Griseb. x Hordeum brevisubulatum (Trin.) Link - 6
chatangense (Roshev.) Tzvel. (*Elymus chatangensis* Roshev., x *Horderoegneria chatangensis* (Roshev.) Tzvel.). - Elymus macrourus (Turcz.) Tzvel. s.l. x Hordeum jubatum L. - 4
jordalii (Melderis) Tzvel. (x *Agrohordeum jordalii* Melderis). - Elymus macrourus (Turcz.) Tzvel. s.l. x Hordeum brachyantherum Nevski - 5
kolymense Probat. - Elymus confusus (Roshev.) Tzvel. subsp. pilosifolius A. Khokhr. x Hordeum jubatum L. - 5

pavlovii (Nevski) Tzvel. (*Agropyron pavlovii* Nevski, *Roegneria pavlovii* (Nevski) Filat. non valide publ.). - Elymus praecaespitosus (Nevski) Tzvel. x Hordeum turkestanicum Nevski - 6

schmidii (Melderis) Melderis (x *Elymordeum schmidii* Melderis). - Elymus nutans Griseb. x Hordeum turkestanicum Nevski - 6

x *Elymordeum* Lepage = x Elyhordeum

bowes-lyonii Melderis = x Elyhordeum bowes-lyonii
schmidii Melderis = x Elyhordeum schmidii

x **Elymostachys** Tzvel.

badachschanica Tzvel. - Elymus longe-aristatus (Boiss.) Tzvel. s.l. x Psathyrostachys kronenburgii (Hack.) Nevski - 6

x **Elymotrigia** Hyl

austroaltaica Kotuchov. - Elymus gmelinii (Ledeb.) Tzvel. x Elytrigia gmelinii (Trin.) Nevski - 3
azutavica Kotuchov. - Elytrigia geniculata (Trin.) Nevski x Elymus mutabilis (Drob.) Tzvel.
bergrothii (Lindb. fil.) Hyl. = x Leymotrigia bergrothii
kalbica Kotuchov. - Elymus mutabilis (Drob.) Tzvel. x Elytrigia geniculata (Trin.) Nevski - 3
karakabinica Kotuchov. - Elymus fedtschenkoi Tzvel. x Elytrigia repens (L.) Nevski - 6
kirghizica Tzvel. - Elymus schrenkianus (Fisch. & C.A. Mey.) Tzvel. x Elytrigia batalinii (Krasn.) Nevski - 6
kurtczumica Kotuchov. - Elytrigia gmelinii (Trin.) Nevski x Elymus fedtschenkoi Tzvel. - 3
leninogorica Kotuchov. - Elymus sibiricus L. x Elytrigia geniculata (Trin.) Nevski - 3
stricta (Dethard.) Hyl. = x Leymotrigia stricta

Elymus L. (*Anthosachne* Steud., *Campeiostachys* Drob., *Clinelymus* (Griseb.) Nevski, *Goulardia* Husn., *Roegneria* C. Koch, *Semeiostachys* Drob.)

abolinii (Drob.) Tzvel. (*Goulardia abolinii* (Drob.) Ikonn., *Roegneria abolinii* (Drob.) Nevski, *Semeiostachys abolini* (Drob.) Drob.) - 6
aegilopoides (Drob.) Worosch. = Elytrigia gmelinii
aemulans (Nevski) Nikif. = Leymus aemulans
ajanensis Roshev. ex Worosch. = x Leymotrix ajanensis
alascanus (Scribn. & Merr.) A. Löve (*Agropyron alascanum* Scribn. & Merr., *Roegneria alascana* (Scribn. & Merr.) Jurtz. & Petrovsky) - 5
- subsp. *borealis* (Turcz.) A. & D. Löve = E. kronokensis subsp. subalpinus
- subsp. *borealis* (Turcz.) Melderis = E. kronokensis subsp. subalpinus
- subsp. *hyperarcticus* (Polun.) A. & D. Löve = E. hyperarcticus
- subsp. *kronokensis* (Kom.) A. & D. Löve = E. kronokensis
- subsp. *sajanensis* (Nevski) A. Löve = E. sajanensis
- subsp. *scandicus* (Nevski) Melderis = E. kronokensis subsp. subalpinus
- subsp. *subalpinus* (L. Neum.) A. & D. Löve = E. kronokensis subsp. subalpinus
- subsp. *subalpinus* (L. Neum.) Melderis = E. kronokensis subsp. subalpinus
- subsp. *villosus* (V. Vassil.) A. & D. Löve = E. vassiljevii

alatavicus (Drob.) A. Löve = Elytrigia alatavica
amurensis (Drob.) Czer. (*Agropyron amurense* Drob., *Elymus ciliaris* (Trin.) Tzvel. subsp. *amurensis* (Drob.) Tzvel., *Roegneria amurensis* (Drob.) Nevski) - 5
angustiformis Drob. = Leymus alaicus and L. latiglumis
angustiformis Pavl. = Leymus karelinii
arcuatus (Golosk.) Tzvel. (*Agropyron arcuatum* Golosk., *Roegneria arcuata* (Golosk.) Golosk. nom. invalid.) - 6
arenarius L. = Leymus arenarius
- subsp. *mollis* (Trin.) Hult. = Leymus mollis
- - var. *villosissimus* (Scribn.) Hult. = Leymus villosissimus
- subsp. *sabulosus* (Bieb.) Beldie = Leymus racemosus subsp. sabulosus
- var. *sabulosus* (Bieb.) Schmalh. = Leymus racemosus subsp. sabulosus
- var. *villosissimus* (Scribn.) Polun. = Leymus villosissimus
asiaticus A. Löve = Hystrix sibirica
attenuatus (Griseb.) K. Richt. = Leymus racemosus
badachschanicus (Tzvel.) Ikonn. (*E. longe-aristatus* (Boiss.) Tzvel. subsp. *badachschanicus* Tzvel.) - 6
batalinii (Krasn.) A. Löve = Elytrigia batalinii
- subsp. *alaicus* (Drob.) A. Löve = Elytrigia alaica
boreoochotensis A. Khokhr. - 5
brachypodioides (Nevski) Peschkova (*Roegneria brachypodioides* Nevski, *Agropyron brachypodioides* (Nevski) Czerepnin, *A. brachypodioides* (Nevski) Serg. comb. superfl., *A. vernicosum* Nevski ex Grub., *Elymus pendulinus* (Nevski) Tzvel. subsp. *brachypodioides* (Nevski) Tzvel., *E. pendulinus* var. *brachypodioides* (Nevski) Probat., *E. vernicosus* (Nevski ex Grub.) Tzvel.) - 3, 4, 5
buchtarmensis Kotuchov - 3
bungeanus (Trin.) Melderis = Elytrigia geniculata
- subsp. *pruiniferus* (Nevski) Melderis = Elytrigia pruinifera
- subsp. *scythicus* (Nevski) Melderis = Elytrigia scythica
buschianus (Roshev.) Tzvel. (*Roegneria buschiana* (Roshev.) Nevski) - 2
caespitosus Sukacz. = Psathyrostachys caespitosa
canadensis L. - 3
canaliculatus (Nevski) Tzvel. = E. longe-aristatus subsp. canaliculatus
caninus (L.) L. (*Elytrigia canina* (L.) Drob., *Goulardia canina* (L.) Husn., *Roegneria canina* (L.) Nevski, *R. tuskaulensis* Vass.) - 1, 2, 3, 4, 6
caput-medusae L. subsp. *crinitus* (Schreb.) Maire & Weiller = Taeniatherum crinitum
caucasicus (C. Koch) Tzvel. (*Roegneria caucasica* C.Koch, *Agropyron caucasicum* (C. Koch) Grossh., *Roegneria linczevskii* Czopanov) - 2, 6
charkeviczii Probat. - 5
chatangensis Roshev. = x Elyhordeum chatangense
chinensis (Trin.) Keng = Leymus chinensis
ciliaris (Trin.) Tzvel. (*Agropyron ciliare* (Trin.) Franch., *Roegneria ciliaris* (Trin.) Nevski) - 5
- subsp. *amurensis* (Drob.) Tzvel. = E. amurensis
cladostachys Turcz. = Leymus mollis
confusus (Roshev.) Tzvel. (*E. sibiricus* L. var. *confusus* (Roshev.) Worosch. comb. invalid., *Roegneria confusa* (Roshev.) Nevski) - 4, 5
- subsp. **pilosifolius** A. Khokhr. (*E. sibiricus* L. var. *pilosifolius* (A. Khokhr.) Worosch.) - 5
- var. *pubiflorus* (Roshev.) Tzvel. = E. pubiflorus
coreanus Honda = Hystrix coreana
curvatiformis (Nevski) Tzvel. (*Agropyron curvatiforme* Nevski, *Elymus curvatiformis* (Nevski) A. Löve, comb. superfl., *Goulardia curvatiformis* (Nevski) Ikonn.) - 6
cylindricus (Franch.) Honda = E. dahuricus

x **czilikensis** (Drob.) Tzvel. (*Agropyron czilikense* Drob.). - E. dahuricus Turcz. ex Griseb. x E. tianschanigenus Czer. - 6

"*czimganicus*" (Drob.) Tzvel. = E. tschimganicus

dahuricus Turcz. ex Griseb. (*Clinelymus cylindricus* (Franch.) Honda, *C. dahuricus* (Turcz. ex Griseb.) Nevski, *Elymus cylindricus* (Franch.) Honda, 1930, non Pohl, 1810, *E. dahuricus* var. *cylindricus* Franch., *E. franchetii* Kitag.) - 1(alien), 3, 4, 5, 6

- subsp. *excelsus* (Turcz. ex Griseb.) Tzvel. = E. excelsus
- subsp. *pacificus* Probat. = E. woroschilowii
- var. *cylindricus* Franch. = E. dahuricus

dasyphyllus (A. Khokhr.) A. Khokhr. = E. kronokensis

dasystachys Trin. p.p. = Leymus dasystachys
- var. *littoralis* Griseb. = Leymus littoralis
- var. *pubescens* O. Fedtsch. = Leymus pubescens
- var. *salsuginosus* Griseb. = Leymus paboanus

dentatus (Hook. fil.) Tzvel. subsp. *lachnophyllus* (Ovcz. & Sidor.) Tzvel. = E. lachnophyllus
- subsp. *ugamicus* (Drob.) Tzvel. = E. nevskii

drobovii (Nevski) Tzvel. (*Roegneria drobovii* (Nevski) Nevski, *Semeiostachys drobovii* (Nevski) Drob.) - 6

elongatus (Host) Greuter = Elytrigia elongata

elongatus (Host) Runemark = Elytrigia elongata
- subsp. *ponticus* (Podp.) Melderis = Elytrigia elongata
- subsp. *turcicus* (McGuire) Melderis = Elytrigia turcica

excelsus Turcz. ex Griseb. (*Clinelymus excelsus* (Turcz. ex Griseb.) Nevski, *Elymus dahuricus* Turcz. ex Griseb. subsp. *excelsus* (Turcz. ex Griseb.) Tzvel.) - 4, 5

farctus (Viv.) Runemark ex Melderis subsp. *bessarabicus* (Savul. & Rayss) Melderis = Elytrigia bessarabica
- subsp. *boreo-atlanticus* (Simon. & Guinoch.) Melderis = Elytrigia junceiformis

fedtschenkoi Tzvel. (*E. macrolepis* (Drob.) Tzvel. p.p. quoad. pl., *Roegneria curvata* (Nevski) Nevski, non *Elymus curvatus* Piper, 1903, *R. striata* (Steud.) Nevski, p.p. quoad pl. ex Asia Media) - 3, 6

fibrosus (Schrenk) Tzvel. (*Agropyron fibrosum* (Schrenk) P. Candargy, *Roegneria fibrosa* (Schrenk) Nevski) - 1, 3, 4, 5(alien)
- subsp. *subfibrosus* (Tzvel.) Tzvel. = E. subfibrosus

flexilis (Nevski) N.M. Kuznetzov = Leymus flexilis

franchetii Kitag. = E. dahuricus
- subsp. *pacificus* (Probat.) Peschkova = E. woroschilowii

giganteus Vahl = Leymus racemosus
- var. *attenuatus* Griseb. = Leymus racemosus
- var. *crassinervius* Kar. & Kir. = Leymus racemosus subsp. crassinervius

glaucissimus (M. Pop.) Tzvel. (*Agropyron glaucissimum* M. Pop., *Roegneria glaucissima* (M. Pop.) Filat. nom. invalid.) - 6

glaucus Regel, p.p. = Leymus dasystachys

gmelinii (Ledeb.) Tzvel. (*Triticum caninum* L. var. *gmelinii* Ledeb., *Agropyron gmelinii* (Ledeb.) Scribn. & J.G. Smith, p.p. quoad nomen, *Roegneria gmelinii* (Ledeb.) Kitag., *R. turczaninovii* (Drob.) Nevski, *Semeiostachys turczaninovii* (Drob.) Drob., *Triticum rupestre* Turcz. ex Ganesch. 1915, non Link, 1821) - 3, 4, 5, 6
- subsp. *ugamicus* (Drob.) A. Löve = E. nevskii

himalayanus (Nevski) Tzvel. (*Roegneria himalayana* Nevski, *Agropyron himalayanum* (Nevski) Melderis) - 6

hispidus (Opiz) Melderis = Elytrigia intermedia
- subsp. *barbulatus* (Schur) Melderis = Elytrigia trichophora
- subsp. *pulcherrimus* (Grossh.) Melderis = Elytrigia pulcherrima

hyalanthus Rupr. = Psathyrostachys hyalantha

hyperarcticus (Polun.) Tzvel. (*Agropyron violaceum* (Hornem.) Vasey var. *hyperarcticum* Polun., *A. boreale* (Turcz.) Drob. subsp. *hyperarcticum* (Polun.) Melderis, *Elymus alascanus* (Scribn. & Merr.) A. Löve subsp. *hyperarcticus* (Polun.) A. & D. Löve, *E. sajanensis* (Nevski) Tzvel. subsp. *hyperarcticus* (Polun.) Tzvel., *Roegneria hyperarctica* (Polun.) Tzvel.) - 4, 5
- subsp. *villosus* (V. Vassil.) Worosch. = E. vassiljevii

interior Hult. = Leymus interior

ircutensis Peschkova - 3, 4

jacutensis (Drob.) Tzvel. (*Agropyron pubescens* Schischk. 1928, non *Elymus pubescens* Davy, 1901, *A. tugarinovii* Reverd., *Roegneria jacutensis* (Drob.) Nevski, *R. pubescens* (Schischk.) Nevski, *R. trinii* Nevski, *Triticum pubescens* Trin. 1835, non Bieb. 1800, nec Hornem. 1813) - 3, 4, 5

x *jenisseiensis* Turcz. = Leymus x jenisseiensis

kamczadalorum (Nevski) Tzvel. (*Roegneria kamczadalorum* Nevski, *Elymus trachycaulus* (Link) Gould & Shinners subsp. *kamczadalorum* (Nevski) Tzvel.) - 4, 5

karakabinicus Kotuchov - 6

karelinii Turcz. = Leymus karelinii

kirghisorum Drob. = Leymus karelinii

komarovii (Nevski) Tzvel. 1968, non Ohwi, 1933 (*E. uralensis* (Nevski) Tzvel. subsp. *komarovii* (Nevski) Tzvel., *Roegneria komarovii* (Nevski) Nevski) - 3, 4, 6

　　This name should probably be replaced since there exists an earlier homonym - Elymus komarovii (Roshev.) Ohwi.

komarovii (Roshev.) Ohwi = Hystrix komarovii

kopetdaghensis Roshev. = Leymus kopetdaghensis

kronenburgii (Hack.) Nikif. = Psathyrostachys kronenburgii

kronokensis (Kom.) Tzvel. (*Agropyron kronokense* Kom., *Elymus alascanus* (Scribn. & Merr.) A. Löve subsp. *kronokensis* (Kom.) A. & D. Löve, *E. dasyphyllus* (A. Khokhr.) A. Khokhr., *E. kronokensis* subsp. *dasyphyllus* A. Khokhr., *Roegneria kronokensis* (Kom.) Tzvel.) - 1, 3, 4, 5, 6
- subsp. *borealis* (Turcz.) Tzvel. = E. kronokensis subsp. subalpinus
- subsp. *dasyphyllus* A. Khokhr. = E. kronokensis
- subsp. *kronokensis* - 4, 5
- subsp. **subalpinus** (L. Neum.) Tzvel. (*Triticum violaceum* Hornem. f. *subalpinum* L. Neum., *Agropyron latiglume* (Scribn. & J.G. Smith) Rydb. subsp. *eurasiaticum* Hult. p.p. incl. typo, *A. latiglume* subsp. *subalpinum* (L. Neum.) Vestergren, p.p., *Elymus alascanus* subsp. *borealis* (Turcz.) A. & D. Löve, *E. alascanus* subsp. *borealis* (Turcz.) Melderis, comb. superfl., *E. alascanus* subsp. *scandicus* (Nevski) Melderis, *E. alascanus* subsp. *subalpinus* (L. Neum.) A. & D. Löve, *E. alascanus* subsp. *subalpinus* (L. Neum.) Melderis, comb. superfl., *E. kronokensis* subsp. *borealis* (Turcz.) Tzvel., *E. kronokensis* var. *borealis* (Turcz.) Tzvel., *E. kronokensis* var. *scandicus* (Nevski) Tzvel., *E. neoborealis* A. Khokhr., *E. scandicus* (Nevski) A. Khokhr., *Roegneria borealis* (Turcz.) Nevski, non *Elymus borealis* Scribn. 1910, *R. scandica* Nevski) - 1, 3, 4, 5, 6
- var. *borealis* (Turcz.) Tzvel. = E. kronokensis subsp. subalpinus
- var. *scandicus* (Nevski) Tzvel. = E. kronokensis subsp. subalpinus

kugalensis E. Nikit. = Leymus karelinii

kurilensis Probat. - 5

kuznetzovii Pavl. = Leymus karelinii

lachnophyllus (Ovcz. & Sidor.) Tzvel. (*Roegneria lachnophylla* Ovcz. & Sidor., *Agropyron lachnophyllum* (Ovcz. & Sidor.) Bondar., *Elymus dentatus* (Hook. fil.) Tzvel.

subsp. *lachnophyllus* (Ovcz. & Sidor.) Tzvel.) - 6

latiglumis Nikif. = Leymus latiglumis

lazicus subsp. *attenuatiglumis* (Nevski) Melderis = Elytrigia attenuatiglumis

lenensis (M. Pop.) Tzvel. (*Agropyron lenense* M. Pop.) - 4

littoralis (Griseb.) Turcz. ex Steud. = Leymus littoralis

lolioides (Kar. & Kir.) Melderis = Elytrigia lolioides

longe-aristatus (Boiss.) Tzvel. (*Agropyron longe-aristatum* (Boiss.) Boiss., *Anthosachne longe-aristata* (Boiss.) Nevski, *Roegneria longe-aristata* (Boiss.) Drob.) - 6

- subsp. *badachschanicus* Tzvel. = E. badachschanicus

- subsp. **canaliculatus** (Nevski) Tzvel. (*Agropyron canaliculatum* Nevski, *A. aitchisonii* (Boiss.) P. Candargy, *A. longe-aristatum* var. *aitchisonii* Boiss., *Anthosachne jacquemontii* (Hook. fil.) Nevski, p.p. quoad pl., *Elymus canaliculatus* (Nevski) Tzvel., *Roegneria canaliculata* (Nevski) Ohwi, *R. jacquemontii* (Hook. fil.) Ovcz. & Sidor. p.p. quoad pl.) - 6

- subsp. **flexuosissimus** (Nevski) Tzvel. (*Agropyron flexuosissimum* Nevski) - 6

- subsp. **litvinovii** Tzvel. - 6

macrochaetus (Nevski) Tzvel. (*Roegneria macrochaeta* Nevski, *Agropyron macrochaetum* (Nevski) Bondar., *A. macrochaetum* (Nevski) Melderis, comb. superfl., *Semeiostachys macrochaeta* (Nevski) Drob.) - 6

x **macrolepis** (Drob.) Tzvel. (*Agropyron macrolepis* Drob. p.p., *Semeiostachys macrolepis* (Drob.) Drob.). - E. fedtschenkoi Tzvel. x E. praeruptus Tzvel. - 6

macrostachys Spreng. = Leymus racemosus

macrourus (Turcz.) Tzvel. (*Agropyron nomokonovii* M. Pop., *Roegneria macroura* (Turcz.) Nevski) - 1, 3, 4, 5

- subsp. *neplianus* (V. Vassil.) Tzvel. = E. neplianus

- subsp. **pilosivaginatus** (Jurtz.) Czer. comb. nova (*Roegneria macroura* subsp. *pilosivaginata* Jurtz. 1981, Bot. Zhurn. 66, 7 :1042) - 5

- subsp. *turuchanensis* (Reverd.) Tzvel. = E. turuchanensis

magadanensis A. Khokhr. - 5

marmoreus Kotuchov - 3

mollis Trin. = Leymus mollis

- subsp. *interior* (Hult.) Bowden = Leymus interior

- subsp. *villosissimus* (Scribn.) A. Löve = Leymus villosissimus

mutabilis (Drob.) Tzvel. (*Agropyron mutabile* Drob. p.p., *A. angustiglume* Nevski, nom. illegit., *A. angustiglume* subsp. *irendykense* Nevski, *A. ilmense* Roshev. ex Nevski, nom. nud., *A. oschense* Roshev., *A. transiliense* M. Pop., *Goulardia mutabilis* (Drob.) Ikonn., *Roegneria angustiglumis* (Nevski) Nevski, *R. mutabilis* (Drob.) Hyl., *R. mutabilis* var. *varsugensis* Melderis, *R. mutabilis* subsp. *varsugensis* (Melderis) A. & D. Löve, *R. oschensis* Nevski, *R. transiliensis* (M. Pop.) Filat. nom. invalid.) - 1, 2 (subsp. **barbulatus**), 3, 4, 5, 6

- subsp. **barbulatus** Nevski ex Tzvel. - 2

- subsp. *praecaespitosus* (Nevski) Tzvel. = E. praecaespitosus

- subsp. *transbaicalensis* (Nevski) Tzvel. = E. transbaicalensis

neoborealis A. Khokhr. = E. kronokensis subsp. subalpinus

neplianus (V. Vassil.) Czer. (*Roegneria nepliana* V. Vassil., *Elymus macrourus* (Turcz.) Tzvel. subsp. *neplianus* (V. Vassil.) Tzvel.) - 4, 5

nevskii Tzvel. (*E. dentatus* (Hook. fil.) Tzvel. subsp. *ugamicus* (Drob.) Tzvel., *E. gmelinii* (Ledeb.) Tzvel. subsp. *ugamicus* (Drob.) A. Löve, *Goulardia ugamica* (Drob.) Ikonn., *Roegneria ugamica* (Drob.) Nevski, non *Elymus ugamicus* Drob. 1925, *Semeiostachys ugamica* (Drob.) Drob.) - 6

nikitinii Czopanov = Leymus nikitinii

nodosus (Nevski) Melderis = Elytrigia nodosa

- subsp. *caespitosus* (C. Koch) Melderis = Elytrigia caespitosa

- subsp. *sinuatus* (Nevski) Melderis = Elytrigia sinuata

novae-angliae (Scribn.) Tzvel. (*Agropyron novae-angliae* Scribn., *A. atbassaricum* Golosk., *A. pauciflorum* (Schwein.) Hitchc. subsp. *novae-angliae* (Scribn.) Melderis, *A. peschkovae* M. Pop. nom. invalid., *Elymus trachycaulus* (Link) Gould & Shinners subsp. *novae-angliae* (Scribn.) Tzvel., *Roegneria novae-angliae* (Scribn.) Jurtz. & Petrovsky, *R. pauciflora* (Schwein.) Hyl. p.p. quoad pl., *R. trachycaulon* (Link) Nevski, p.p. quoad pl., *Agropyron pauciflorum* auct., *A. tenerum* auct.) - 1, 3, 4, 5, 6

nutans Griseb. (*Clinelymus nutans* (Griseb.) Nevski) - 6

occidentali-altaicus Kotuchov - 3

pallidissimus (M. Pop.) Peschkova = E. transbaicalensis

pamiricus Tzvel. (*E. schrenkianus* (Fisch. & C.A. Mey.) Tzvel. subsp. *pamiricus* (Tzvel.) Tzvel.) - 6

panormitanus (Parl.) Tzvel. (*Agropyron panormitanum* Parl., *Roegneria panormitana* (Parl.) Nevski, *Semeiostachys panormitana* (Parl.) Drob., *Triticum panormitanum* (Parl.) Bertol.) - 1

pauciflorus (Schwein.) Gould = E. trachycaulus

pectinatus (Bieb.) Lainz = Agropyron pectinatum

pendulinus (Nevski) Tzvel. (*Roegneria pendulina* Nevski, *Agropyron pendulinum* (Nevski) Worosch.) - 3, 5

- subsp. *brachypodioides* (Nevski) Tzvel. = E. brachypodioides

- var. *brachypodioides* (Nevski) Probat. = E. brachypodioides

petraeus (Nevski) Pavl. = Leymus petraeus

pilifer Banks & Soland. = Heteranthelium piliferum

praecaespitosus (Nevski) Tzvel. (*Agropyron praecaespitosum* Nevski, *Elymus mutabilis* (Drob.) Tzvel. subsp. *praecaespitosus* (Nevski) Tzvel., *Goulardia praecaespitosa* (Nevski) Ikonn., *Roegneria praecaespitosa* (Nevski) Nevski) - 6

praeruptus Tzvel. (*Roegneria interrupta* (Nevski) Nevski, non *Elymus interruptus* Buckl. 1862, *Semeiostachys interrupta* (Nevski) Drob.) - 6

prokudinii (Seredin) Tzvel. (*Roegneria prokudinii* Seredin, *Elymus uralensis* (Nevski) Tzvel. subsp. *prokudinii* (Seredin) Tzvel.) - 2

pseudoagropyrum (Trin. ex Griseb.) Turcz. = Leymus chinensis

pubiflorus (Roshev.) Peschkova (*Agropyron confusum* Roshev. var. *pubiflorum* Roshev., *Elymus confusus* (Roshev.) Tzvel. var. *pubiflorus* (Roshev.) Tzvel.) - 4, 5

pungens (Pers.) Melderis = Elytrigia pungens

racemosus Lam. = Leymus racemosus

- var. *sabulosus* (Bieb.) Bowden = Leymus racemosus subsp. sabulosus

ramosus (Trin.) Filat. = Leymus ramosus

reflexiaristatus (Nevski) Melderis = Elytrigia reflexiaristata

- subsp. *strigosus* (Bieb.) Melderis = Elytrigia strigosa

repens (L.) Gould = Elytrigia repens

- subsp. *elongatiformis* (Drob.) Melderis = Elytrigia elongatiformis

- subsp. *pseudocaesius* (Pacz.) Melderis = Elytrigia pseudocaesia

sabulosus Bieb. = Leymus racemosus subsp. sabulosus

sajanensis (Nevski) Tzvel. (*Roegneria sajanensis* Nevski, *Agropyron sajanense* (Nevski) Grub. p.p. quoad nomen, *Elymus alascanus* (Scribn. & Merr.) A. Löve subsp. *sajanensis* (Nevski) A. Löve) - 3, 4

- subsp. *hyperarcticus* (Polun.) Tzvel. = E. hyperarcticus
- subsp. *villosus* (V. Vassil.) Tzvel. = E. vassiljevii
salsuginosus (Griseb.) Turcz. ex Steud. = Leymus paboanus
scandicus (Nevski) A. Khokhr. = E. kronokensis subsp. subalpinus
schrenkianus (Fisch. & C.A. Mey.) Tzvel. (*Agropyron pseudostrigosum* P. Candargy, *A. schrenkianum* (Fisch. & C.A. Mey.) P. Candargy, *Campeiostachys schrenkiana* (Fisch. & C.A. Mey.) Drob., *Roegneria schrenkiana* (Fisch. & C.A. Mey.) Nevski, *Triticum strigosum* (Bieb.) Less. var. *planifolium* Regel) - 3, 4, 6
- subsp. *pamiricus* (Tzvel.) Tzvel. = E. pamiricus
schugnanicus (Nevski) Tzvel. (*Roegneria schugnanica* (Nevski) Nevski) - 6
sclerophyllus (Nevski) Tzvel. (*Roegneria sclerophylla* Nevski, *Agropyron dolicholepis* Melderis, *A. transnominatum* Bondar., *Goulardia sclerophylla* (Nevski) Ikonn., *Semeiostachys sclerophylla* (Nevski) Drob.) - 6
secalinus (Georgi) Bobr. = Leymus secalinus
sibinicus Kotuchov - 3
sibiricus L. (*Clinelymus sibiricus* (L.) Nevski, *Elymus tener* L. fil., *Triticum arctasianum* F. Herm.) - 1, 3, 4, 5, 6
- var. *confusus* (Roshev.) Worosch. = E. confusus
- var. *pilosifolius* (A. Khokhr.) Worosch. = E. confusus subsp. pilosifolius
sosnovskyi (Hack.) Melderis = Elytrigia sosnovskyi
stipifolius (Czern. ex Nevski) Melderis = Elytrigia stipifolia
subfibrosus (Tzvel.) Tzvel. (*Roegneria subfibrosa* Tzvel., *Elymus fibrosus* (Schrenk) Tzvel. subsp. *subfibrosus* (Tzvel.) Tzvel.) - 3, 4, 5
tauri (Boiss. & Bal.) Melderis subsp. *pertenuis* (C.A. Mey.) Melderis = Elytrigia pertenuis
tener L. fil. = E. sibiricus
tianschanigenus Czer. (*Agropyron tianschanicum* Drob. non *Elymus tianschanicus* Drob. 1925; *Elymus czilikensis* (Drob.) Tzvel. p.p. quoad pl., *E. uralensis* (Nevski) Tzvel. subsp. *tianschanicus* (Drob.) Tzvel., *Roegneria tianschanica* (Drob.) Nevski, *Semeiostachys tianschanica* (Drob.) Drob.) - 6
trachycaulus (Link) Gould & Shinners (*Triticum trachycaulon* Link, *Agropyron pauciflorum* (Schwein.) Hitchc., *A. tenerum* Vasey, *A. trachycaulon* (Link) Steud., *Elymus pauciflorus* (Schwein.) Gould, 1947, non Lam. 1791, *Roegneria pauciflora* (Schwein.) Hyl. p.p. quoad nomen, *Roegneria trachycaulon* (Link) Nevski, p.p. quoad nomen, *Triticum pauciflorum* Schwein.) - 1(alien), 4, 5
- subsp. *kamczadalorum* (Nevski) Tzvel. = E. kamczadalorum
- subsp. **major** (Vasey) Tzvel. - 4, 5
- subsp. *novae-angliae* (Scribn.) Tzvel. = E. novae-angliae
transbaicalensis (Nevski) Tzvel. (*Agropyron pallidissimum* M. Pop., *Elymus mutabilis* (Drob.) Tzvel. subsp. *transbaicalensis* (Nevski) Tzvel., *E. pallidissimus* (M. Pop.) Peschkova, *Roegneria burjatica* Sipl., *R. transbaicalensis* (Nevski) Nevski) - 3, 4
transhyrcanus (Nevski) Tzvel. (*Roegneria transhyrcana* Nevski, *Agropyron leptourum* (Nevski) Grossh., *A. leptourum* (Nevski) Pavl. comb. superfl., *A. transhyrcanum* (Nevski) Bondar., *Elytrigia vvedenskyi* Drob., *Goulardia transhyrcana* (Nevski) Ikonn., *Roegneria leptoura* Nevski, *Semeiostachys leptoura* (Nevski) Drob.) - 2, 6
trinii Melderis = Leymus ramosus
troctolepis (Nevski) Tzvel. (*Roegneria troctolepis* Nevski, *Agropyron troctolepis* (Nevski) Melderis) - 2
tschimganicus (Drob.) Tzvel. (*Agropyron tschimganicum*

Drob., *Elymus "czimganicus"* (Drob.) Tzvel., *Roegneria "czimganica"* (Drob.) Nevski, *R. tschimganica* (Drob.) Nevski) - 6
turgaicus Roshev. = Leymus karelinii
turuchanensis (Reverd.) Czer. (*Agropyron turuchanense* Reverd., *Elymus macrourus* (Turcz.) Tzvel. subsp. *turuchanensis* (Reverd.) Tzvel., *Roegneria turuchanensis* (Reverd.) Nevski) - 1, 4
tzvelevii Kotuchov - 3
uralensis (Nevski) Tzvel. (*Roegneria uralensis* (Nevski) Nevski) - 1
- subsp. *komarovii* (Nevski) Tzvel. = E. komarovii
- subsp. *prokudinii* (Seredin) Tzvel. = E. prokudinii
- subsp. *tianschanicus* (Drob.) Tzvel. = E. tianschanigenus
- subsp. *viridiglumis* (Nevski) Tzvel. = E. viridiglumis
vassiljevii Czer. (*Roegneria villosa* V. Vassil. non *Elymus villosus* Muhl. ex Willd. 1909, *Elymus alascanus* (Scribn. & Merr.) A. Löve subsp. *villosus* (V. Vassil.) A. & D. Löve, *E. hyperarcticus* (Polun.) Tzvel. subsp. *villosus* (V. Vassil.) Worosch., *E. sajanensis* (Nevski) Tzvel. subsp. *villosus* (V. Vassil.) Tzvel.) - 4, 5
- subsp. **coeruleus** (Jurtz.) Czer. comb. nova (*Roegneria villosa* subsp. *coerulea* Jurtz. 1989, Bot. Zhurn. 74, 1 : 113) - 5
- subsp. **laxe-pilosus** (Jurtz.) Czer. comb. nova (*Roegneria villosa* subsp. *laxe-pilosa* Jurtz. 1981, Bot. Zhurn. 66, 7 : 1041) - 5
vernicosus (Nevski ex Grub.) Tzvel. = E. brachypodioides
versicolor A. Khokhr. - 5
villosissimus Scribn. = Leymus villosissimus
viridiglumis (Nevski) Czer. (*Roegneria viridiglumis* Nevski, *Agropyron karkaralense* Roshev., *Elymus uralensis* (Nevski) Tzvel. subsp. *viridiglumis* (Nevski) Tzvel., *Roegneria karkaralensis* (Roshev.) Filat. nom. invalid., *R. taigae* Nevski) - 1, 3
woroschilowii Probat. (*E. dahuricus* Turcz. ex Griseb. subsp. *pacificus* Probat., *E. franchetii* Kitag. subsp. *pacificus* (Probat.) Peschkova) - 5
zejensis Probat. - 5

Elymus sensu Nevski = Leymus

Elytrigia Desv. (*Lophopyrum* A. Löve, *Pseudoroegneria* (Nevski) A. Löve, *Thinopyrum* A. Löve)

aegilopoides (Drob.) Peschkova = E. gmelinii
alaica (Drob.) Nevski (*Agropyron alaicum* Drob., *Elymus batalinii* (Krasn.) A. Löve subsp. *alaicus* (Drob.) A. Löve, *Elytrigia alaica* subsp. *pamirica* Tzvel., *E. batalinii* (Krasn.) Nevski subsp.. *alaica* (Drob.) Tzvel., *Roegneria carinata* Ovcz. & Sidor.) - 6
- subsp. *pamirica* Tzvel. = E. alaica
alatavica (Drob.) Nevski (*Agropyron alatavicum* Drob., *Elymus alatavicus* (Drob.) A. Löve) - 6
amgunensis (Nevski) Nevski (*Agropyron amgunense* Nevski, *A. strigosum* (Bieb.) Boiss. subsp. *amgunense* (Nevski) Tzvel., *Elytrigia strigosa* (Bieb.) Nevski subsp. *amgunensis* (Nevski) Tzvel., *Pseudoroegneria strigosa* (Bieb.) A. Löve subsp. *amgunensis* (Nevski) A. Löve) - 5
angularis (Nevski) Nevski = E. caespitosa
x *apiculata* (Tscherning) Jir. = E. x mucronata
argentea Nevski = E. batalinii
armena (Nevski) Nevski (*Agropyron armenum* Nevski, *Elytrigia stipifolia* (Czern. ex Nevski) Nevski subsp. *armena* (Nevski) Tzvel., *Pseudoroegneria armena* (Nevski) A. Löve) - 2

attenuatiglumis (Nevski) Nevski (*Agropyron attenuatiglume* Nevski, *Elymus lazicus* subsp. *attenuatiglumis* (Nevski) Melderis, *Elytrigia divaricata* (Boiss. & Bal.) Nevski subsp. *attenuatiglumis* (Nevski) Tzvel., *Pseudoroegneria divaricata* (Boiss. & Bal.) A. Löve subsp. *attenuatiglumis* (Nevski) A. Löve, *Agropyron divaricatum* auct., *Elytrigia divaricata* auct.) - 2

aucheri (Boiss.) Nevski = E. trichophora

batalinii (Krasn.) Nevski (*Agropyron argenteum* (Nevski) Pavl., *A. batalinii* (Krasn.) Roshev., *Elymus batalinii* (Krasn.) A. Löve, *Elytrigia argentea* Nevski) - 6

- subsp. *alaica* (Drob.) Tzvel. = E. alaica

bessarabica (Savul & Rayss) Prokud. (*Agropyron bessarabicum* Savul. & Rayss, *A. junceum* (L.) Beauv. var. *bessarabicum* (Savul. & Rayss) Anghel & Morariu, *Elymus farctus* (Viv.) Runemark ex Melderis subsp. *bessarabicus* (Savul. & Rayss) Melderis, *Elytrigia bessarabica* (Savul. & Rayss) Dubovik, comb. superfl., *E. bessarabica* (Savul. & Rayss) Holub, comb. superfl., *E. juncea* (L.) Nevski, p.p. excl. basionymo, *E. juncea* subsp. *bessarabica* (Savul. & Rayss) Tzvel., *Thinopyrum bessarabicum* (Savul. & Rayss) A. Löve, *Agropyron junceum* sensu Nevski) - 1, 2

caespitosa (C. Koch) Nevski (*Agropyron caespitosum* C. Koch, *A. angulare* Nevski, *A. firmiculme* Nevski, *Elymus nodosus* (Nevski) Melderis subsp. *caespitosus* (C. Koch) Melderis, *Elytrigia angularis* (Nevski) Nevski, *E. firmiculmis* (Nevski) Nevski, *Lophopyrum caespitosum* (C. Koch) A. Löve) - 2, 6

- subsp. *nodosa* (Nevski) Tzvel. = E. nodosa
- subsp. *sinuata* (Nevski) Tzvel. = E. sinuata

canina (L.) Drob. = Elymus caninus

ciliolata (Nevski) Nevski = E. lolioides

cognata (Hack.) Holub (*Agropyron cognatum* Hack., *A. dshungaricum* (Nevski) Nevski, *A. ferganense* Drob., *Elytrigia dshungarica* Nevski, *E. ferganensis* (Drob.) Nevski, *E. geniculata* (Trin.) Nevski subsp. *ferganensis* (Drob.) Tzvel., *Pseudoroegneria cognata* (Hack.) A. Löve) - 6

cretacea (Klok. & Prokud.) Klok. = E. stipifolia

divaricata auct. = E. attenuatiglumis

divaricata (Boiss. & Bal.) Nevski subsp. *attenuatiglumis* (Nevski) Tzvel. = E. attenuatiglumis

dshinalica Sablina (*Pseudoroegneria dshinalica* (Sablina) A. Löve) - 2

dshungarica Nevski = E. cognata

elongata (Host) Nevski (*Agropyron elongatum* (Host) Beauv., *A. ruthenicum* (Griseb.) Prokud. p.p. excl. basionymo, *Elymus elongatus* (Host) Greuter, comb. superfl., *E. elongatus* (Host) Runemark, *Elymus elongatus* subsp. *ponticus* (Podp.) Melderis, *Elytrigia prokudinii* Druleva, *E. ruthenica* (Griseb.) Prokud. p.p. excl. basionymo, *Lophopyrum elongatum* (Host) A. Löve, *Triticum ponticum* Podp., *T. rigidum* Schrad.) - 1, 2, 6

elongatiformis (Drob.) Nevski (*Agropyron elongatiforme* Drob., *A. repens* (L.) Beauv. subsp. *elongatiforme* (Drob.) D.R. Dewey, *Elymus repens* (L.) Gould subsp. *elongatiformis* (Drob.) Melderis, *Elytrigia quercetorum* Prokud., *E. repens* (L.) Nevski subsp. *elongatiformis* (Drob.) Tzvel. p.p.) - 1, 2, 6

ferganensis (Drob.) Nevski = E. cognata

firmiculmis (Nevski) Nevski = E. caespitosa

geniculata (Trin.) Nevski (*Agropyron geniculatum* (Trin.) C. Koch, *Elymus bungeanus* (Trin.) Melderis, *Pseudoroegneria geniculata* (Trin.) A. Löve, *Triticum bungeanum* Trin.) - 3, 4

- subsp. *ferganensis* (Drob.) Tzvel. = E. cognata
- subsp. *pruinifera* (Nevski) Tzvel. = E. pruinifera
- subsp. *scythica* (Nevski) Tzvel. = E. scythica

gmelinii (Trin.) Nevski (*Triticum gmelinii* Trin., *Agropyron aegilopoides* Drob., *A. gmelinii* (Trin.) P. Candargy, 1901, non Scribn. & J.G. Smith, 1897, *A. gmelinii* var. *roshevitzii* (Nevski) Malysch., *A. propinquum* Nevski, *A. roshevitzii* Nevski, *A. stenophyllum* Nevski, *A. strigosum* (Bieb.) Boiss. subsp. *aegilopoides* (Drob.) Tzvel., *Elymus aegilopoides* (Drob.) Worosch., *Elytrigia aegilopoides* (Drob.) Peschkova, *E. propinqua* (Nevski) Nevski, *E. stenophylla* (Nevski) Nevski, *E. stenophylla* subsp. *aegilopoides* (Drob.) Tzvel., *E. strigosa* (Bieb.) Nevski subsp. *aegilopoides* (Drob.) Tzvel., *Pseudoroegneria strigosa* (Bieb.) A. Löve subsp. *aegilopoides* (Drob.) A. Löve) - 3, 4, 6

gracillima (Nevski) Nevski (*Agropyron gracillimum* Nevski, *Pseudoroegneria gracillima* (Nevski) A. Löve) - 2

heidemaniae Tzvel. (*Pseudoroegneria heidemaniae* (Tzvel.) A. Löve) - 2

intermedia (Host) Nevski (*Agropyron hispidum* Opiz, *A. intermedium* (Host) Beauv., *Elymus hispidus* (Opiz) Melderis) - 1, 2, 6

- subsp. *barbulata* (Schur) A. Löve = E. trichophora
- subsp. *barbulata* (Schur) Sojak = E. trichophora
- subsp. *pulcherrima* (Grossh.) Tzvel. = E. pulcherrima
- subsp. *trichophora* (Link) A. & D. Löve = E. trichophora
- subsp. *trichophora* (Link) Tzvel. = E. trichophora

jacutorum (Nevski) Nevski (*Agropyron jacutorum* Nevski, *A. strigosum* (Bieb.) Boiss. subsp. *jacutorum* (Nevski) Tzvel., *Elytrigia strigosa* (Bieb.) Nevski subsp. *jacutorum* (Nevski) Tzvel., *Pseudoroegneria strigosa* (Bieb.) A. Löve subsp. *jacutorum* (Nevski) A. Löve) - 4, 5

juncea (L.) Nevski, p.p. = E. bessarabica

- subsp. *bessarabica* (Savul. & Rayss) Tzvel. = E. bessarabica
- subsp. *boreo-atlantica* (Simon. & Guinoch.) Hyl. = E. junceiformis

junceiformis A. & D. Löve (*Agropyron junceiforme* A. & D. Löve, nom. altern., *A. junceum* (L.) Beauv. subsp. *boreo-atlanticum* Simon. & Guinoch., *Elymus farctus* (Viv.) Runemark ex Melderis subsp. *boreo-atlanticus* (Simon. & Guinoch.) Melderis, *Elytrigia juncea* (L.) Nevski subsp. *boreo-atlantica* (Simon. & Guinoch.) Hyl., *Thinopyrum junceiforme* (A. & D. Löve) A. Löve) - 1

kaachemica Lomonosova & Krasnob. - 4

kasteki (M. Pop.) Tzvel. (*Agropyron kasteki* M. Pop., *A. subalpinum* Golosk.) - 6

kotovii Dubovik = E. stipifolia

kryloviana (Schischk.) Nevski = Agropyron krylovianum

levadica Kuvajev - 2

x littorea (Schum.) Hyl. (*Triticum littoreum* Schum., *Agropyron laxum* (Fries) Fries, *Triticum laxum* Fries). - E. junceiformis A. & D. Löve x E. repens (L.) Nevski - 1

lolioides (Kar. & Kir.) Nevski (*Agropyron ciliolatum* Nevski, *A. lolioides* (Kar. & Kir.) P. Candargy, *A. lolioides* (Kar. & Kir.) Roshev. comb. superfl., *Elymus lolioides* (Kar. & Kir.) Melderis, *Elytrigia ciliolata* (Nevski) Nevski, *E. pachyneura* Prokud.) - 1, 3, 4

maeotica (Prokud.) Prokud. (*Agropyron maeoticum* Prod.) - 1

x mucronata (Opiz ex Bercht.) Prokud. (*Agropyron mucronatum* Opiz ex Bercht., *A.* x *apiculatum* Tscherning, *A.* x *bazargiciense* Prod. ex Burdnja, *Elytrigia* x *apiculata* (Tscherning) Jir.). - E. intermedia (Host) Nevski x E. repens (L.) Nevski - 1

ninae Dubovik = E. stipifolia

nodosa (Nevski) Nevski (*Agropyron nodosum* Nevski, *Elymus nodosus* (Nevski) Melderis, *Elytrigia caespitosa* (C. Koch) Nevski subsp. *nodosa* (Nevski) Tzvel., *Triticum nodosum* Stev. ex Bieb. nom. invalid., *Lophopyrum nodosum* (Nevski) A. Löve) - 1

pachyneura Prokud. = E. lolioides

pertenuis (C.A. Mey.) Nevski (*Agropyron pertenue* (C.A. Mey.) Nevski, *Elymus tauri* (Boiss. & Bal.) Melderis subsp. *pertenuis* (C.A. Mey.) Melderis, *Elytrigia tauri* (Boiss. & Bal.) Tzvel. subsp. *pertenuis* (C.A. Mey.) Tzvel., *Pseudoroegneria pertenuis* (C.A. Mey.) A. Löve) - 2

praetermissa Nevski = Agropyron pumilum

prokudinii Druleva = E. elongata

propinqua (Nevski) Nevski = E. gmelinii

pruinifera Nevski (*Agropyron pruiniferum* (Nevski) Nevski, *Elymus bungeanus* (Trin.) Melderis subsp. *pruiniferus* (Nevski) Melderis, *Elytrigia geniculata* (Trin.) Nevski subsp. *pruinifera* (Nevski) Tzvel., *Pseudoroegneria geniculata* (Trin.) A. Löve subsp. *pruinifera* (Nevski) A. Löve) - 1, 3

pseudocaesia (Pacz.) Prokud. (*Agropyron repens* (L.) Beauv. var. *pseudocaesium* Pacz., *Agropyron pseudocaesium* (Pacz.) Zoz, *Agropyron repens* subsp. *pseudocaesium* (Pacz.) Lavr., *Elymus repens* (L.) Gould subsp. *pseudocaesius* (Pacz.) Melderis, *Elytrigia repens* (L.) Nevski subsp. *pseudocaesia* (Pacz.) Tzvel.) - 1, 2, 3, 6

pulcherrima (Grossh.) Nevski (*Agropyron pulcherrimum* Grossh., *A. ambigens* (Hausskn. ex Halacsy) Roshev., *A. intermedium* (Host) Beauv. var. *ambigens* Hausskn. ex Halacsy, *A. popovii* Drob., *Elymus hispidus* (Opiz) Melderis subsp. *pulcherrimus* (Grossh.) Melderis, *Elytrigia intermedia* (Host) Nevski subsp. *pulcherrima* (Grossh.) Tzvel.) - 2, 6

pungens (Pers.) Tutin (*Triticum pungens* Pers., *Agropyron littorale* Dumort., *A. pungens* (Pers.) Roem. & Schult., *Elymus pungens* (Pers.) Melderis, *Triticum littorale* Host, 1809, non Pall. 1776) - 1(alien)

quercetorum Prokud. = E. elongatiformis

reflexiaristata (Nevski) Nevski (*Agropyron reflexiaristatum* Nevski, *A. gmelinii* (Trin.) P. Candargy var. *reflexiaristatum* (Nevski) Serg. nom. invalid., *A. strigosum* (Bieb.) Boiss. subsp. *reflexiaristatum* (Nevski) Tzvel., *Elymus reflexiaristatus* (Nevski) Melderis, *Elytrigia strigosa* (Bieb.) Nevski subsp. *reflexiaristata* (Nevski) Tzvel., *Pseudoroegneria reflexiaristata* (Nevski) A. Lavrenko, *P. strigosa* (Bieb.) A. Löve subsp. *reflexiaristata* (Nevski) A. Löve) - 1

repens (L.) Nevski (*Agropyron caesium* J. & C. Presl, *A. repens* (L.) Beauv., *A. sachalinense* Honda, *Elymus repens* (L.) Gould) - 1, 2, 3, 4, 5, 6

- subsp. *elongatiformis* (Drob.) Tzvel. = E. elongatiformis

- subsp. *pseudocaesia* (Pacz.) Tzvel. = E. pseudocaesia

ruthenica (Griseb.) Prokud. = E. elongata and E. trichophora

scythica (Nevski) Nevski (*Agropyron scythicum* Nevski, *Elymus bungeanus* (Trin.) Melderis subsp. *scythicus* (Nevski) Melderis, *Elytrigia geniculata* (Trin.) Nevski subsp. *scythica* (Nevski) Tzvel., *Pseudoroegneria geniculata* (Trin.) A. Löve subsp. *scythica* (Nevski) A. Löve) - 1

setulifera Nevski (*Agropyron setuliferum* (Nevski) Nevski, *Pseudoroegneria setulifera* (Nevski) A. Löve) - 6

sinuata (Nevski) Nevski (*Agropyron sinuatum* Nevski, *Elymus nodosus* (Nevski) Melderis subsp. *sinuatus* (Nevski) Melderis, *Elytrigia caespitosa* (C. Koch) Nevski subsp. *sinuata* (Nevski) Tzvel., *Lophopyrum*

sinuatum (Nevski) A. Löve) - 2

■ sosnovskyi (Hack.) Nevski (*Agropyron sosnovskyi* Hack., *Elymus sosnovskyi* (Hack.) Melderis, *Pseudoroegneria sosnovskyi* (Hack.) A. Löve)

stenophylla (Nevski) Nevski = E. gmelinii

stipifolia (Czern. ex Nevski) Nevski (*Agropyron stipifolium* Czern. ex Nevski, *A. cretaceum* Klok. & Prokud., *Elymus stipifolius* (Czern. ex Nevski) Melderis, *Elytrigia cretacea* (Klok. & Prokud.) Klok., *E. kotovii* Dubovik, *E. ninae* Dubovik, *E. stipifolia* subsp. *stipifolia* var. *cretacea* (Klok. & Prokud.) Tzvel., *Pseudoroegneria cretacea* (Klok. & Prokud.) A. Löve, *P. kotovii* (Dubovik) A. Löve, *P. ninae* (Dubovik) A. Löve, *P. stipifolia* (Czern. ex Nevski) A. Löve, *Triticum intermedium* Host var. *stipifolium* Sirj. & Lavr. non basionymum) - 1, 2

- subsp. *armena* (Nevski) Tzvel. = E. armena

- subsp. *stipifolia* var. *cretacea* (Klok. & Prokud.) Tzvel. = E. stipifolia

strigosa (Bieb.) Nevski (*Agropyron strigosum* (Bieb.) Boiss., *Elymus reflexiaristatus* (Nevski) Melderis subsp. *strigosus* (Bieb.) Melderis, non rite publ., *Pseudoroegneria strigosa* (Bieb.) A. Löve) - 1

- subsp. *aegilopoides* (Drob.) Tzvel. = E. gmelinii

- subsp. *amgunensis* (Nevski) Tzvel. = E. amgunensis

- subsp. *jacutorum* (Nevski) Tzvel. = E. jacutorum

- subsp. *reflexiaristata* (Nevski) Tzvel. = E. reflexiaristata

tauri (Boiss. & Bal.) Tzvel. subsp. *pertenuis* (C.A. Mey.) Tzvel. = E. pertenuis

x tesquicola (Prokud.) Klok. (*Agropyron tesquicola* Prokud., *A. breviaristatum* Grossh. nom. invalid.). - E. repens (L.) Nevski x E. trichophora (Link) Nevski - 1

trichophora (Link) Nevski (*Agropyron aucheri* Boiss., *A. barbulatum* Schur, *A. elongatum* (Host) Beauv. subsp. *ruthenicum* (Griseb.) Anghel & Morariu, *A. glaucum* (Desf. ex DC.) Roem. & Schult. subsp. *barbulatum* (Schur) K. Richt., *A. intermedium* (Host) Beauv. subsp. *trichophorum* (Link) Volkart, *A. intermedium* subsp. *trichophorum* (Link) Reichenb. ex Hegi, comb. superfl., *A. ruthenicum* (Griseb.) Prokud. p.p. quoad nomen, *A. trichophorum* (Link) K. Richt., *A. truncatum* (Wallr.) Fuss subsp. *trichophorum* (Link) Soo, *Elymus hispidus* (Opiz) Melderis subsp. *barbulatus* (Schur) Melderis, *Elytrigia aucheri* (Boiss.) Nevski, *E. intermedia* (Host) Nevski subsp. *barbulata* (Schur) A. Löve, *E. intermedia* subsp. *barbulata* (Schur) Sojak, comb. superfl., *E. intermedia* subsp. *trichophora* (Link) A. & D. Löve, *E. intermedia* subsp. *trichophora* (Schur) Tzvel. comb. superfl., *E. ruthenica* (Griseb.) Prokud. p.p. quoad nomen, *Triticum rigidum* Schrad. var. *ruthenicum* Griseb.) - 1, 2, 6

turcica McGuire (*Elymus elongatus* (Host) Runemark subsp. *turcicus* (McGuire) Melderis) - 2(?)

villosa (Drob.) Tzvel. (*Brachypodium villosum* Drob. 1914, non *Agropyron villosum* Link, 1827, nec *Roegneria villosa* V. Vassil. 1954, *Agropyron karawaewii* P. Smirn., *Roegneria karawaewii* (P. Smirn.) Karav.) - 4

vvedenskyi Drob. = Elymus transhyrcanus

Enneapogon Desv. ex Beauv. (*Pappophorum* auct.)

borealis (Griseb.) Honda (*Pappophorum boreale* Griseb., *Enneapogon desvauxii* Beauv. subsp. *borealis* (Griseb.) Tzvel.) - 4, 6

desvauxii Beauv. subsp. *borealis* (Griseb.) Tzvel. = E. borealis

persicus Boiss. (*Pappophorum persicum* (Boiss.) Steud.) - 2, 6

Eragrostis N.M. Wolf (*Boriskellera* Terekhov, *Psilantha* (C. Koch) Tzvel. nom. illegit.)

aegyptiaca (Willd.) Delile - 1
amurensis Probat. (*E. pilosa* (L.) Beauv. var. *amurensis* (Probat.) Worosch.) - 5
arundinacea (L.) Roshev. = E. collina
barrelieri Daveau - 2, 6(?)
borysthenica Klok. (*E. suaveolens* A. Beck. ex Claus var. *borysthenica* Schmalh., *E. suaveolens* subsp. *borysthenica* (Schmalh.) Tzvel.) - 1
cilianensis (All.) Vign.-Lut. ex Janch. (*Poa cilianensis* All., *Eragrostis cilianensis* (All.) Vign.-Lut. 1904, nom. invalid., *E. megastachya* (Koel.) Link) - 1, 2, 5(alien), 6
- subsp. *starosselskyi* (Grossh.) Tzvel. = E. starosselskyi
collina Trin. (*Boriskellera arundinacea* (L.) Terekhov, *Eragrostis arundinacea* (L.) Roshev. 1934, non Jedwabnick, 1924, *E. tatarica* (Fisch. ex Griseb.) Nevski, *Poa tatarica* Fisch. ex Griseb., *Psilantha arundinacea* (L.) Tzvel.) - 1, 2, 3, 6
curvula (Schrad.) Nees (*Poa curvula* Schrad.) - 2, 6
diarrhena (Schult. & Schult. fil.) Steud. = Diandrochloa diarrhena
imberbis (Franch.) Probat. (*E. pilosa* (L.) Beauv. var. *imberbis* Franch., *E. pilosa* subsp. *imberbis* (Franch.) Tzvel.) - 4, 5
kossinskyi Roshev. = Diandrochloa diarrhena
megastachya (Koel.) Link = E. cilianensis
minor Host (*E. poaeoides* Beauv.) - 1, 2, 3, 4, 5(alien), 6
multicaulis Steud. (*E. pilosa* (L.) Beauv. subsp. *multicaulis* (Steud.) Tzvel.) - 1(alien), 2(alien), 5
nigra Nees ex Steud. - 2
pilosa (L.) Beauv. - 1, 2, 3, 4, 5, 6
- subsp. *imberbis* (Franch.) Tzvel. = E. imberbis
- subsp. *multicaulis* (Steud.) Tzvel. = E. multicaulis
- var. *amurensis* (Probat.) Worosch. = E. amurensis
- var. *imberbis* Franch. = E. imberbis
poaeoides Beauv. = E. minor
starosselskyi Grossh. (*E. cilianensis* (All.) Vign.-Lut. subsp. *starosselskyi* (Grossh.) Tzvel.) - 2, 6
suaveolens A. Beck. ex Claus - 1, 5(alien), 6
- subsp. *borysthenica* (Schmalh.) Tzvel. = E. borysthenica
- var. *borysthenica* Schmalh. = E. borysthenica
tatarica (Fisch. ex Griseb.) Nevski = E. collina
***tef** (Zuccagni) Trotter (*Poa tef* Zuccagni)

Eremopoa Roshev.

altaica (Trin.) Roshev. (*E. bellula* (Regel) Roshev. p.p. excl. typo) - 3, 6
- subsp. *oxyglumis* (Boiss.) Tzvel. = E. oxyglumis
- subsp. *songarica* (Schrenk) Tzvel. = E. songarica
bellula (Regel) Roshev. = E. altaica and E. songarica
glareosa Gamajun. = E. songarica
multiradiata (Trautv.) Roshev. (*E. persica* (Trin.) Roshev. subsp. *multiradiata* (Trautv.) Tzvel.) - 2
oxyglumis (Boiss.) Roshev. (*E. altaica* (Trin.) Roshev. subsp. *oxyglumis* (Boiss.) Tzvel., *E. persica* (Trin.) Roshev. var. *oxyglumis* (Boiss.) Grossh.) - 2, 6
persica (Trin.) Roshev. -
- subsp. *multiradiata* (Trautv.) Tzvel. = E. multiradiata
- var. *oxyglumis* (Boiss.) Grossh. = E. oxyglumis
- var. *songarica* (Schrenk) Bor = E. songarica
songarica (Schrenk) Roshev. (*E. altaica* (Trin.) Roshev. subsp. *songarica* (Schrenk) Tzvel., *E. bellula* (Regel)

Roshev. p.p. quoad nomen, *E. glareosa* Gamajun., *E. persica* (Trin.) Roshev. var. *songarica* (Schrenk) Bor, *Poa paradoxa* Kar. & Kir., *P. subtilis* Kar. & Kir.) - 2, 3, 6

Eremopyrum (Ledeb.) Jaub. & Spach

bonaepartis (Spreng.) Nevski (*Agropyron orientale* (L.) Roem. & Schult. var. *sublanuginosum* Drob., *A. turkestanicum* Gand., *Eremopyrum bonaepartis* subsp. *hirsutum* (Bertol.) Melderis, *E. bonaepartis* subsp. *sublanuginosum* (Drob.) A. Löve, *E. hirsutum* (Bertol.) Nevski) - 2, 6
- subsp. *hirsutum* (Bertol.) Melderis = E. bonaepartis
- subsp. *sublanuginosum* (Drob.) A. Löve = E. bonaepartis
confusum Melderis - 6(?)
distans (C. Koch) Nevski (*Agropyron orientale* (L.) Roem & Schult. subsp. *lanuginosum* (Griseb.) K. Richt.) - 1, 2, 6
hirsutum (Bertol.) Nevski = E. bonaepartis
orientale (L.) Jaub. & Spach - 1, 2, 3, 6
puberulum (Boiss. ex Steud.) Grossh. ex A. Löve = Agropyron puberulum
triticeum (Gaertn.) Nevski - 1, 2, 3, 6

Erianthus Michx.

purpurascens auct. = E. ravennae
ravennae (L.) Beauv. (*Andropogon ravennae* L., *Saccharum ravennae* (L.) Murr., *Erianthus purpurascens* auct.) - 2, 6

Erioblastus Honda = Vahlodea

flexuosus Honda = Vahlodea flexuosa

Eriochloa Kunth

succincta (Trin.) Kunth - 1, 2, 6
villosa (Thunb.) Kunth - 2, 3, 5

*Euchlaena Schrad.

***mexicana** Schrad.

Euraphis (Trin.) Lindl. = Boissiera

squarrosa (Banks & Soland.) Sojak = Boissiera squarrosa

Festuca L. (*Drymochloa* Holub, *Leiopoa* Ohwi, *Leucopoa* Griseb., *Xantochloa* auct.)

airoides Lam. (*F. ovina* L. subsp. *supina* (Schur) Schinz & R. Keller, *F. supina* Schur) - 1, 2
alaica Drob. (*F. bornmuelleri* (Hack.) V. Krecz. & Bobr. p.p. quoad nomen, *F. kotschyi* (Hack.) Grossh. p.p. quoad nomen, *F. ovina* L. subsp. *kotschyi* Hack.) - 6
- subsp. *pamirica* (Tzvel.) Tzvel. = F. pamirica
alatavica (St.-Yves) Roshev. (*F. tianschanica* Roshev.) - 6
albida (Turcz. ex Trin.) Malysch. = F. sibirica
albifolia Reverd. = F. lenensis
- var. *tschujensis* (Reverd.) Serg. = F. tschujensis
alexeenkoi E. Alexeev (*F. laevis* auct. p.p.) - 2
altaica Trin. - 3, 4, 5, 6
altissima All. (*Drymochloa sylvatica* (Poll.) Holub, *Festuca sylvatica* (Poll.) Vill. 1787, non Huds. 1762) - 1, 2, 3, 4, 6
amblyodes V. Krecz. & Bobr. (*F. amblyodes* subsp. *erecti-*

flora (Pavl.) Tzvel., *F. erectiflora* Pavl., *F. kashmiriana* auct. p.p.) - 6
- subsp. *erectiflora* (Pavl.) Tzvel. = F. amblyodes
amethystina auct. fl. carpat. = F. inarmata
amethystina L. subsp. *inarmata* (Schur) Krajina = F. inarmata
- subsp. *orientalis* Krajina = F. inarmata
amethystina sensu V. Krecz. & Bobr. = F. woronowii
amurensis E. Alexeev - 5
apennina De Not. (*F. australis* Schur, *F. elatior* L. subsp. *apennina* (De Not.) Hayek, *F. pratensis* Huds. subsp. *apennina* (De Not.) Hegi, *F. pratensis* subsp. *apennina* (De Not.) Beldie, comb. superfl.) - 1
arenaria Osbeck (*F. rubra* L. subsp. *arenaria* (Osbeck) Syme, *F. rubra* subsp. *arenaria* (Osbeck) O. Schwarz, comb. superfl., *F. villosa* Schweigg.) - 1
arietina Klok. = F. wolgensis
aristata C. Koch = Bromopsis aristata
artvinensis Markgraf-Dannenberg - 2
arundinacea Schreb. (*F. elatior* L. p.p. nom. ambig., *F. elatior* subsp. *arundinacea* (Schreb.) Celak.) - 1
- subsp. *fenas* (Lag.) Arcang. = F. interrupta
- subsp. *interrupta* (Desf.) Tzvel. = F. interrupta
- subsp. *orientalis* (Hack.) K. Richt. = F. regeliana
- subsp. *orientalis* (Hack.) Tzvel. = F. regeliana
x **aschersoniana** Doerfl. - F. arundinacea Schreb. x F. pratensis Huds.
asperrima Link = Poa sterilis
aucta V. Krecz. & Bobr. = F. rubra
auriculata Drob. - 1, 4, 5
- subsp. *chionobia* (Egor. & Sipl.) Tzvel. = F. chionobia
- subsp. *mollissima* (V. Krecz. & Bobr.) Tzvel. = F. mollissima
australis Schur = F. apennina
azgarica E. Alexeev - 2
baffinensis Polun. - 1, 4, 5
baicalensis (Griseb.) V. Krecz. & Bobr. (*F. rubra* L. subsp. *baicalensis* (Griseb.) Tzvel.) - 4
barbata L. = Schismus barbatus
bargusinensis Malysch. - 4
beckeri (Hack.) Trautv. (*F. laeviuscula* Klok.) - 1, 2, 3, 4, 6
- subsp. *polesica* (Zapal.) Tzvel. = F. polesica
- subsp. *sabulosa* (Anderss.) Tzvel. = F. sabulosa
blepharogyna (Ohwi) Ohwi (*Leiopoa blepharogyna* Ohwi, *Festuca sichotensis* Krivot.) - 5
bonassorum Bornm. = F. gigantea
borissii Reverd. - 3, 6
bornmuelleri (Hack.) V. Krecz. & Bobr. = F. alaica, F. transcaucasica and F. pseudodalmatica subsp. asperrima
brachyphylla Schult. & Schult. fil. (*F. brevifolia* R. Br. 1824, non Muhl. 1817, *F. ovina* L. subsp. *saximontana* (Rydb.) St.-Yves var. *purpusiana* St.-Yves, p.p., *F. purpusiana* (St.-Yves) Tzvel. p.p. quoad pl., *F. saximontana* Rydb. subsp. *purpusiana* (St.-Yves) Tzvel. p.p. quoad pl.) - 1, 3, 4, 5, 6
- var. *brevissima* (Jurtz.) Ju. Kozhevn. = F. brevissima
brachystachys (Hack.) K. Richt. = F. versicolor
brevifolia R. Br. = F. brachyphylla
brevissima Jurtz. (*F. brachyphylla* Schult. & Schult. fil. var. *brevissima* (Jurtz.) Ju. Kozhevn., *F. brevissima* subsp. *contracta* Jurtz., *F. ovina* L. subsp. *alaskana* Holmen) - 5
- subsp. *contracta* Jurtz. = F. brevissima
bromoides L. = Vulpia bromoides
brunnescens (Tzvel.) Galushko (*F. rupicola* Heuff. subsp. *brunnescens* Tzvel., *F. brunnescens* (Tzvel.) Markgraf-

Dannenberg, comb. superfl., *F. valesiaca* Gaudin subsp.*brunnescens* (Tzvel.) E. Alexeev) - 2
buschiana (St.-Yves) Tzvel. (*F. rubra* L. subsp. *buschiana* St.-Yves, *F. frigida* (Hack.) Grossh. p.p. quoad pl., nom. illegit.) - 2
calceolaris Somm. & Levier = F. djimilensis
callieri (Hack.) Markgraf (*F. ovina* L. subsp. *sulcata* var. *callieri* Hack., *F. taurica* sensu V. Krecz. & Bobr.) - 1, 2
- subsp. *uralensis* Tzvel. = F. uralensis
capillaris Liljebl. = Puccinellia capillaris
capillata Lam. = F. filiformis
carpatica F. Dietr. (*F. dimorpha* auct.) - 1
caucasica (Boiss.) Hack. ex Trautv. (*F. sibirica* Hack. ex Boiss. var. *caucasica* Boiss., *Leucopoa caucasica* (Boiss.) V. Krecz. & Bobr.) - 2
chalcophaea V. Krecz. & Bobr. (*F. violacea* auct. fl. cauc.) - 2
chionobia Egor. & Sipl. (*F. auriculata* Drob. subsp. *chionobia* (Egor. & Sipl.) Tzvel.) - 4, 5
ciliata Danth. = Vulpia ciliata
cinerea Vill. subsp. *pallens* (Host) Stohr = F. pallens
- subsp. *psammophila* (Hack. ex Celak.) Stohr = F. psammophila
coelestis (St.-Yves) V. Krecz. & Bobr. (*F. santyvesii* Pazij, nom. illegit.) - 6
cretacea T. Pop & Proskorjakov (*F. cretacea* (Lavr.) V. Krecz. & Bobr. nom. illegit., *F. issatchenkoi* St.-Yves) - 1
cristata L. = Rostraria cristata
cryophila V. Krecz. & Bobr. = F. richardsonii
x **czarnochorensis** Zapal. - F. apennina De Not x F. gigantea (L.) Vill. - 1
daghestanica (Tzvel.) E. Alexeev (*F. rubra* L. subsp. *daghestanica* Tzvel.) - 2
dahurica (St.-Yves) V. Krecz. & Bobr. - 4
dalmatica (Hack.) K. Richt. subsp. *dalmatica* var. *pseudodalmatica* (Krajina) Beldie = F. pseudodalmatica
- subsp. *pseudodalmatica* (Krajina) Soo = F. pseudodalmatica
dichotoma Forssk. = Cutandia dichotoma
diffusa Dumort. (*F. multiflora* Hoffm. 1800, non Walt. 1788, *F. rubra* L. subsp. *multiflora* (Hoffm.) Jir. nom. illegit., *F. rubra* var. *multiflora* Steud., *F. rubra* subsp. *multiflora* (Steud.) Piper) - 1
diffusa V.Vassil. = F. ovina
dimorpha auct. = F. carpatica
djimilensis Boiss. & Bal. (*F. calceolaris* Somm. & Levier) - 2
drymeja Mert. & Koch (*Drymochloa drymeja* (Mert. & Koch) Holub, *Festuca montana* Bieb. 1819, non Savi, 1798, *Poa bakuensis* Litv. ex Roshev.) - 1, 2
duriuscula L. = F. trachyphylla
egena V. Krecz. & Bobr. = F. rubra
elatior L. = F. arundinacea and F. pratensis
- subsp. *apennina* (De Not.) Hayek = F. apennina
- subsp. *arundinacea* (Schreb.) Celak. = F. arundinacea
elbrusica E. Alexeev - 2
erectiflora Pavl. = F. amblyodes
eriantha Honda & Tatew. = F. rubra
extremiorientalis Ohwi (*F. subulata* Trin. ex Bong. var. *japonica* Hack., *F. subulata* subsp. *japonica* (Hack.) T. Koyama & Kawano) - 3, 4, 5
fallax Thuill. (*F. pseudorubra* Schur, *F. rubra* L. subsp. *fallax* (Thuill.) Nym.) - 1
fasciculata Forssk. = Vulpia fasciculata
fenas Lag. = F. interrupta

filiformis Pourr. (*F. capillata* Lam. nom. illegit., *F. ovina* L. subsp. *tenuifolia* (Sibth.) Celak., *F. tenuifolia* Sibth.) - 1, 2

x **flischeri** Rohlena. - F. arundinacea Schreb. x F. gigantea (L.) Vill.

frigida (Hack.) Grossh. p.p. = F. buschiana

ganeschinii Drob. = F. rupicola

gigantea (L.) Vill. (? *F. bonassorum* Bornm., *F. pseudogigantea* Ovcz. & Schibk.) - 1, 2, 3, 4, 6

glauca auct. = F. pallens

glauca Lam. subsp. *pallens* (Host) K. Richt. = F. pallens

- subsp. *pallens* (Host) O. Schwarz = F. pallens

- subsp. *psammophila* Hack. ex Celak. = F. psammophila

- subsp. *psammophila* (Hack.) O. Schwarz = F. psammophila

- var. *psammophila* (Celak.) Stohr = F. psammophila

goloskokovii E. Alexeev - 6

guestphalica Boenn. ex Reichenb. = F. ovina

heterophylla Lam. - 1, 2

hirsuta Host = F. rupicola

hondoensis (Ohwi) Ohwi (*F. rubra* L. var. *hondoensis* Ohwi, *F. jacutica* Drob. subsp. *pobedimoviae* Tzvel.) - 5

hubsugulica Krivot. (*F. sumneviczii* Serg.) - 4

hyperborea Holmen ex Frederiksen - 1, 5

igoschiniae Tzvel. - 1

inarmata Schur (*F. amethystina* L. subsp. *inarmata* (Schur) Krajina, *F. amethystina* subsp. *orientalis* Krajina, *F. amethystina* auct. fl. carpat.) - 1

inguschetica E. Alexeev - 2

insularis M. Pop. = F. popovii

interrupta Desf. (*F. arundinacea* Schreb. subsp. *fenas* (Lag.) Arcang., *F. arundinacea* subsp. *interrupta* (Desf.) Tzvel., *F. fenas* Lag.) - 1, 2, 6

irtyshensis E. Alexeev - 3

issatchenkoi St.-Yves = F. cretacea

jacutica Drob. - 4, 5

- subsp. *nutans* (Malysch.) Tzvel. = F. malyschevii

- subsp. *pobedimoviae* Tzvel. = F. hondoensis

- var. *nutans* Malysch. = F. malyschevii

jenissejensis Reverd. = F. pseudosulcata

kamtschatica (St.-Yves) Tzvel. (*F. ovina* L. subsp. *laevis* var. *kamtschatica* St.-Yves, *F. ovina* subsp. *kamtschatica* (St.-Yves) Worosch., *F. pseudosulcata* Drob. subsp. *kamtschatica* (St.-Yves) Tzvel.) - 5

karabaghensis Musajev (*F. varia* subsp. *eu-varia* var. *caucasica* St.-Yves, *F. woronowii* Hack. subsp. *caucasica* (St.-Yves) E. Alexeev, *F. woronowii* subsp. *caucasica* (St.-Yves) Markgraf-Dannenberg, comb. superfl.) - 2

karadagensis Hadac & Chrtek = F. pseudodalmatica subsp. asperrima

karatavica (Bunge) B. Fedtsch. (*Leucopoa karatavica* (Bunge) V. Krecz. & Bobr., *Xanthochloa karatavica* (Bunge) R. Kam.) - 6

karavaevii E. Alexeev - 4

karsiana E. Alexeev - 2

kashmiriana auct. p.p. = F. amblyodes

kirelowii Steud. = F. richardsonii

kirghisorum (Katsch. ex Tzvel.) E. Alexeev (*F. rupicola* Heuff. subsp. *kirghisorum* Katsch. ex Tzvel., *F. valesiaca* Gaudin subsp. *kirghisorum* (Katsch. ex Tzvel.) Tzvel.) - 6

x **kolesnikovii** Tzvel. - F. molissima V. Krecz. & Bobr. x F. vorobievii Probat. - 5

kolymensis Drob. - 4, 5

komarovii Krivot. - 4

kotschyi (Hack.) Grossh. = F. alaica and F. transcaucasica

krivotulenkoae E. Alexeev - 2

kryloviana Reverd. - 3, 6

- var. *musbelica* Reverd. = F. musbelica

kurtschumica E. Alexeev - 3

laevis auct. = F. alexeenkoi and F. saxatilis

laeviuscula Klok. = F. beckeri

lenensis Drob. (*F. albifolia* Reverd., *F. lenensis* subsp. *albifolia* (Reverd.) Tzvel., *F. lenensis* var. *albifolia* (Reverd.) Tzvel.) - 3, 4, 5

- subsp. *albifolia* (Reverd.) Tzvel. = F. lenensis

- var. *albifolia* (Reverd.) Tzvel. = F. lenensis

litvinovii (Tzvel.) E. Alexeev (*F. pseudosulcata* Drob. var. *litvinovii* Tzvel., *F. pseudosulcata* subsp. *litvinovii* (Tzvel.) Tzvel.) - 4, 5

loliacea Huds. = x Festulolium loliaceum

longearistata (Hack.) Somm. & Levier = F. sommieri

longifolia auct. = F. trachyphylla

macutrensis Zapal. (*F. ovina* L. subsp. *macutrensis* (Zapal.) A. Kozlowska, *F. valesiaca* Gaudin subsp. *valesiaca* var. *macutrensis* (Zapal.) Tzvel.) - 1

malyschevii E. Alexeev (*F. jacutica* Drob. subsp. *nutans* (Malysch.) Tzvel., *F. jacutica* var. *nutans* Malysch.) - 4

malzewii (Litv.) Reverd. = F. ovina

megalura Nutt. = Vulpia megalura

mollissima V. Krecz. & Bobr. (*F. auriculata* Drob. subsp. *molissima* (V. Krecz. & Bobr.) Tzvel.) - 5

montana Bieb. = F. drymeja

multiflora Hoffm. = F. diffusa

musbelica (Reverd.) Ikonn. (*F. kryloviana* Reverd. var. *musbelica* Reverd., *F. oreophila* Markgraf-Dannenberg, *F. ovina* L. subsp. *sulcata* var. *hypsophila* St.-Yves, *F. valesiaca* Gaudin subsp. *hypsophila* (St.-Yves) Tzvel., *F. valesiaca* subsp. *musbelica* (Reverd.) Tzvel.) - 2, 3, 4, 6

olchonensis E. Alexeev - 4

olgae (Regel) Krivot. (*Leucopoa olgae* (Regel) V. Krecz. & Bobr.) - 6

oreophila Markgraf-Dannenberg = F. musbelica

orientalis (Hack.) V. Krecz. & Bobr. = F. regeliana

ovina L. (*F. diffusa* V. Vassil. 1940, non Dumort. 1823, *F. guestphalica* Boenn. ex Reichenb., *F. malzewii* (Litv.) Reverd., *F. ovina* var. *elata* Drob., *F. ovina* subsp. *elata* (Drob.) Tzvel., *F. ovina* subsp. *guestphalica* (Boenn. ex Reichenb.) K. Richt., *F. ovina* var. *malzewii* Litv.) - 1, 2, 3, 4, 5

- subsp. *alaskana* Holmen = F. brevissima

- subsp. *elata* (Drob.) Tzvel. = F. ovina

- subsp. *eu-ovina* var. *duriuscula* subvar. *trachyphylla* Hack. = F. trachyphylla

- - var. *glauca* subvar. *psammophila* Hack. = F. psammophila

- - var. *yaroschenkoi* St.-Yves = F. yaroschenkoi

- subsp. **firmulacea** (Markgraf-Dannenberg) Probat. (*F. trachyphylla* (Hack.) Krajina f. *firmulacea* Markgraf-Dannenberg) - 1

- *glauca* var. *sabulosa* Anderss. = F. sabulosa

- subsp. *guestphalica* (Boenn. ex Reichenb.) K. Richt. = F. ovina

- subsp. *kamtschatica* (St.-Yves) Worosch. = F. kamtschatica

- subsp. *kotschyi* Hack. = F. alaica

- subsp. *laevis* var. *kamtschatica* St.-Yves = F. kamtschatica

- - var. *marginata* subvar. *transcaucasica* St. Yves = F. transcaucasica

- subsp. *litoralis* (Tzvel.) E. Alexeev = F. vorobievii

- subsp. *macutrensis* (Zapal.) A. Kozlowska = F. macutrensis

- subsp. *ruprechtii* (Boiss.) Tzvel. = F. ruprechtii

- subsp. *saximontana* (Rydb.) St.-Yves var. *purpusiana* St.-Yves, p.p. = F. brachyphylla
- subsp. *sphagnicola* (B. Keller) Tzvel. = F. sphagnicola
- subsp. *sulcata* var. *callieri* Hack. = P. callieri
- - var. *hypsophila* St.-Yves = F. musbelica
- subsp. *supina* (Schur) Schinz & R. Keller = F. airoides
- subsp. *tenuifolia* (Sibth.) Celak. = F. filiformis
- subsp. **vylzaniae** E. Alexeev - 4
- var. *elata* Drob. = F. ovina
- var. *hirsuta* Link, p.p. = F. rupicola
- var. *litoralis* Tzvel. = F. vorobievii
- var. *malzewii* Litv. = F. ovina
- var. *sulcata* subvar. *asperrima* Hack. = F. pseudodalmatica subsp. asperrima
- a. *polita* Halacsy = F. polita
- var. *vivipara* L. = F. vivipara

pallens Host (*F. cinerea* Vill. subsp. *pallens* (Host) Stohr, *F. glauca* Lam. subsp. *pallens* (Host) K. Richt., *F. glauca* subsp. *pallens* (Host) O. Schwarz, comb. superfl., *F. glauca* auct., *F. vaginata* auct.) - 1
- subsp. *psammophila* (Hack. ex Celak.) Tzvel. = F. psammophila

pamirica Tzvel. (*F. alaica* Drob. subsp. *pamirica* (Tzvel.) Tzvel.) - 6

picta Kit. (*F. violacea* Gaudin subsp. *picta* (Kit.) Hegi) - 1

x **pocutica** Zapal. - F. picta Kit. x F. porcii Hack. - 1

pohleana E. Alexccv - 1

polesica Zapal. (*F. beckeri* (Hack.) Trautv. subsp. *polesica* (Zapal.) Tzvel., *F. querceto-pinetorum* Klok. nom. invalid.) - 1, 3

polita (Halacsy) Tzvel. (*F. ovina* L. a. *polita* Halacsy) - 2

polonica Zapal. - 1

popovii E. Alexeev (*F. insularis* M. Pop. 1957, non Steud. 1855) - 4

porcii Hack. - 1

pratensis Huds. (*F. elatior* L. p.p. nom. ambig.) - 1, 2, 3, 4, 5, 6
- subsp. *apennina* (De Not.) Beldie = F. apennina
- subsp. *apennina* (De Not.) Hegi = F. apennina

primae E. Alexeev - 2

probatoviae E. Alexeev - 5

prolifera (Piper) Fern. - 5(?)

psammophila (Hack. ex Celak.) Fritsch (*F. glauca* Lam. subsp. *psammophila* Hack. ex Celak., *F. cinerea* Vill. subsp. *psammophila* (Hack. ex Celak.) Stohr, *F. glauca* subsp. *psammophila* (Hack.) O. Schwarz, comb. superfl., *F. glauca* var. *psammophila* (Celak.) Stohr, *F. ovina* L. subsp. *eu-ovina* var. *glauca* subvar. *psammophila* Hack., *F. pallens* Host subsp. *psammophila* (Hack. ex Celak.) Tzvel.) - 1

pseudodalmatica Krajina (*F. dalmatica* (Hack.) K. Richt. subsp. *dalmatica* var. *pseudodalmatica* (Krajina) Beldie, *F. dalmatica* subsp. *pseudodalmatica* (Krajina) Soo, *F. valesiaca* Gaudin subsp. *pseudodalmatica* (Krajina) Soo, *F. valesiaca* var. *pseudodalmatica* (Krajina) E.I. Nyarady, *F. valesiaca* var. *wagneri* auct.) - 1, 2, 6
- subsp. **asperrima** (Hack.) E. Alexeev (*F. ovina* var. *sulcata* subvar. *asperrima* Hack., *F. bornmuelleri* (Hack.) V.Krecz. & Bobr. p.p. excl. basionymo, *F. karadagensis* Hadac & Chrtek, *F. sulcata* (Hack.) Nym. var. *assperrima* (Hack.) Hack., *F. valesiaca* var. *asperrima* (Hack.) E. Alexeev) - 2, 6

pseudogigantea Ovcz. & Schibk. = F. gigantea

pseudorubra Schur = F. fallax

pseudosulcata Drob. (? *F. jenissejensis* Reverd.) - 4, 5
- subsp. *kamtschatica* (St.-Yves) Tzvel. = F. kamtschatica

- subsp. *litvinovii* (Tzvel.) Tzvel. = F. litvinovii
- var. *litvinovii* Tzvel. = F. litvinovii

pseudovina Hack. ex Wiesb. (? *F. pulchra* Schur, nom. dubium, *F. valesiaca* Gaudin subsp. *pseudovina* (Hack. ex Wiesb.) Hegi, *F. valesiaca* var. *pseudovina* Hack. ex Wiesb.) Schinz & R. Keller) - 1, 2, 3, 6

pulchra Schur = F. pseudovina

purpusiana (St.-Yves) Tzvel. p.p. = F. brachyphylla

querceto-pinetorum Klok. = F. polesica

recognita Reverd. = F. rupicola

regeliana Pavl. (*F. arundinacea* Schreb. subsp. *orientalis* (Hack.) K. Richt., *F. arundinacea* subsp. *orientalis* (Hack.) Tzvel. comb. superfl., *F. orientalis* (Hack.) V. Krecz. & Bobr. 1934, non B. Fedtsch. 1915) - 1, 2, 3, 6

richardsonii Hook. (*F. cryophila* V. Krecz. & Bobr., *F. kirelowii* Steud., *F. richardsonii* subsp. *cryophila* (V. Krecz. & Bobr.) A. & D. Löve, *F. rubra* L. subsp. *arctica* (Hack.) Govor., *F. rubra* subsp. *cryophila* (V. Krecz. & Bobr.) Hult., *F. rubra* var. *cryophila* (V. Krecz. & Bobr.) Reverd., *F. rubra* subsp. *eu-rubra* var. *arenaria* f. *arctica* Hack., *F. rubra* subsp. *kirelowii* (Steud.) Tzvel.) - 1, 3, 4, 5, 6
- subsp. *cryophila* (V. Krecz. & Bobr.) A. & D. Löve = F. richardsonii

rubra L. (*F. aucta* V. Krecz. & Bobr., *F. egena* V. Krecz. & Bobr., *F. eriantha* Honda & Tatew., *F. rubra* subsp. *aucta* (V. Krecz. & Bobr.) Hult.) - 1, 2, 3, 4, 5, 6
- subsp. *arctica* (Hack.) Govor. = F. richardsonii
- subsp. *arenaria* (Osbeck) O. Schwarz = F. arenaria
- subsp. *arenaria* (Osbeck) Syme = F. arenaria
- subsp. *aucta* (V. Krecz. & Bobr.) Hult. = F. rubra
- subsp. *baicalensis* (Griseb.) Tzvel. = F. baicalensis
- subsp. *buschiana* St.-Yves = F. buschiana
- subsp. *cryophila* (V. Krecz. & Bobr.) Hult. = F. richardsonii
- subsp. *daghestanica* Tzvel. = F. daghestanica
- subsp. *eu-rubra* var. *arenaria* f. *arctica* Hack. = F. richardsonii
- subsp. *fallax* (Thuill.) Nym. = F. fallax
- subsp. *kirelowii* (Steud.) Tzvel. = F. richardsonii
- subsp. **limosa** E. Alexeev - 5
- subsp. *multiflora* (Hoffm.) Jir. = F. diffusa
- subsp. *multiflora* (Steud.) Piper = F. diffusa
- var. *cryophila* (V. Krecz. & Bobr.) Reverd. = F. richardsonii
- var. *hondoensis* Ohwi = F. hondoensis
- var. *multiflora* Steud. = F. diffusa

rupicola Heuff. (*F. ganeschinii* Drob., *F. hirsuta* Host, 1802, non Moench, 1802, *F. ovina* L. var. *hirsuta* Link, p.p., *F. recognita* Reverd., *F. sulcata* (Hack.) Nym. p.p. nom. illegit., *F. valesiaca* Gaudin subsp. *sulcata* (Hack.) Schinz & R. Keller, *F. valesiaca* subsp. *valesiaca* var. *ganeschinii* (Drob.) Tzvel., *F. valesiaca* subsp. *valesiaca* var. *hirsuta* (Link) E. Alexeev, *F. valesiaca* var. *wagneri* auct.) - 1, 2, 3, 6
- subsp. *brunnescens* Tzvel. = F. brunnescens
- subsp. *kirghisorum* Katsch. ex Tzvel. = F. kirghisorum
- subsp. *saxatilis* (Schur) Beldie = F. saxatilis
- subsp. *saxatilis* (Schur) Rauschert = F. saxatilis
- subsp. *saxatilis* (Schur) Tzvel. = F. saxatilis
- var. *saxatilis* (Schur) A. Nyarady = F. saxatilis
- var. *saxatilis* (Schur) Soo = F. saxatilis

ruprechtii (Boiss.) V. Krecz. & Bobr. (*F. ovina* L. subsp. *ruprechtii* (Boiss.) Tzvel.) - 1, 2, 4, 5

sabulosa (Anderss.) Lindb. fil. (*F. ovina glauca* var. *sabulosa* Anderss., *F. beckeri* (Hack.) Trautv. subsp. *sabulosa* (Anderss.) Tzvel.) - 1

sajanensis Roshev. = F. tristis
santyvesii Pazij = F. coelestis
saurica E. Alexeev - 6
saxatilis Schur (*F. rupicola* Heuff. subsp. *saxatilis* (Schur)
    Beldie, comb. superfl., *F. rupicola* subsp. *saxatilis*
    (Schur) Rauschert, *F. rupicola* subsp. *saxatilis* (Schur)
    Tzvel. comb. superfl., *F. rupicola* var. *saxatilis* (Schur)
    A. Nyarady, *F. rupicola* var. *saxatilis* (Schur) Soo,
    comb. superfl., *F. sulcata* (Hack.) Nym. subsp. *saxatilis*
    (Schur) K. Richt., *F. sulcata* var. *saxatilis* (Schur) E.I.
    Nyarady, *F. valesiaca* Gaudin subsp. *saxatilis* (Schur)
    E. Alexeev, *F. laevis* auct. p.p.) - 1, 2
saximontana Rydb. subsp. *purpusiana* (St.-Yves) Tzvel.
    p.p. = F. brachyphylla
schischkinii Krivot. - 6
x **schlickumii** Grantzow. - F. gigantea (L.) Vill. x F. praten-
    sis Huds.
sclerophylla Boiss. ex Bischoff (*F. sclerophylla* Boiss. &
    Hohen., *Leucopoa sclerophylla* (Boiss. & Hohen.) V.
    Krecz. & Bobr.) - 2
sibirica Hack. ex Boiss. (*F. albida* (Turcz. ex Trin.) Ma-
    lysch. 1965, p.p. non Lowe, 1831, *Leucopoa albida*
    (Turcz. ex Trin.) V. Krecz. & Bobr., *L. kreczetoviczii*
    K. Sobol., *L. sibirica* Griseb. nom. illegit., *Poa albida*
    Turcz. ex Trin.) - 4, 5
- var. *caucasica* Boiss. = F. caucasica
sichotensis Krivot. = F. blepharogyna
skrjabinii E. Alexeev - 4
skvortsovii E. Alexeev - 2, 6
sommieri Litardiere (*F. longearistata* (Hack.) Somm. &
    Levier, 1897, non Walp. 1849) - 2
sphagnicola B. Keller (*F. ovina* L. subsp. *sphagnicola* (B.
    Keller) Tzvel.) - 3, 4, 6
squamulosa Ovcz. & Schibk. = F. valesiaca
subulata Trin. ex Bong. subsp. *japonica* (Hack.) T. Koyama
    & Kawano = F. extremiorientalis
- var. *japonica* Hack. = F. extremiorientalis
sulcata (Hack.) Nym. p.p. = F. rupicola and F. valesiaca
- subsp. *saxatilis* (Schur) K. Richt. = F. saxatilis
- var. *assperrima* (Hack.) Hack. = F. pseudodalmatica
    subsp. asperrima
- var. *saxatilis* (Schur) E.I. Nyarady = F. saxatilis
sumneviczii Serg. = F. hubsugulica
supina Schur = F. airoides
sylvatica (Poll.) Vill. = F. altissima
taurica (Hack.) A. Kerner ex Trautv. - 1
taurica sensu V. Krecz. & Bobr. = F. callieri
tenuiflora Schrad. p.p. = Nardurus krausei
- var. *aristata* Koch = Nardurus krausei
tenuifolia Sibth. = F. filiformis
tianschanica Roshev. = F. alatavica
trachyphylla (Hack.) Krajina (*F. ovina* L. subsp. *eu-ovina*
    var. *duriuscula* subvar. *trachyphylla* Hack., *F. duriuscu-*
    la L. nom. ambig., *F. longifolia* auct.) - 1
- f. *firmulacea* Markgraf-Dannenberg = F. ovina subsp.
    firmulacea
transcaucasica (St.-Yves) Tzvel. (*F. bornmuelleri* (Hack.)
    V. Krecz. & Bobr. p.p. quoad pl. cauc., *F. ovina* L.
    subsp. *laevis* var. *marginata* subvar. *transcaucasica* St.-
    Yves, *F. kotschyi* (Hack.) Grossh. p.p. quoad pl.) - 2
tristis Kryl. & Ivanitzk. (*F. sajanensis* Roshev., *F. tristis*
    subsp. *sajanensis* (Roshev.) Tzvel.) - 3, 4, 6
- subsp. *sajanensis* (Roshev.) Tzvel. = F. tristis
tschatkalica E. Alexeev - 6
tschujensis Reverd. (*F. albifolia* Reverd. var. *tschujensis*
    (Reverd.) Serg.) - 3, 4
tzvelevii E. Alexeev - 2

uniglumis Soland. = Vulpia fasciculata
uralensis (Tzvel.) E. Alexeev (*F. callieri* (Hack.) Markgraf
    subsp. *uralensis* Tzvel.) - 1
vaginata auct. = F. pallens
vagravarica E. Alexeev - 2
valesiaca Gaudin (*F. sulcata* (Hack.) Nym. p.p., *F. squamu-*
    losa Ovcz. & Schibk.) - 1, 2, 3, 4, 5(alien), 6
- subsp. *brunnescens* (Tzvel.) E. Alexeev = F. brunnescens
- subsp. *hypsophila* (St.-Yves) Tzvel. = F. musbelica
- subsp. *kirghisorum* (Katsch. ex Tzvel.) Tzvel. = F. kirghi-
    sorum
- subsp. *musbelica* (Reverd.) Tzvel. = F. musbelica
- subsp. *pseudodalmatica* (Krajina) Soo = F. pseudodal-
    matica
- subsp. *pseudovina* (Hack. ex Wiesb.) Hegi = F. pseudovi-
    na
- subsp. *saxatilis* (Schur) E. Alexeev = F. saxatilis
- subsp. *sulcata* (Hack.) Schinz & R. Keller = F. rupicola
- subsp. *valesiaca* var. *ganeschinii* (Drob.) Tzvel. = F.
    rupicola
- - var. *hirsuta* (Link) E. Alexeev = F. rupicola
- - var. *macutrensis* (Zapal.) Tzvel. = F. macutrensis
- var. *asperrima* (Hack.) E. Alexeev = F. pseudodalmatica
    subsp. asperrima
- var. *pseudodalmatica* (Krajina) E.I. Nyarady = F. pseudo-
    dalmatica
- var. *pseudovina* (Hack. ex Wiesb.) Schinz & R. Keller =
    F. pseudovina
- var. *wagneri* auct. = F. pseudodalmatica and F. rupicola
varia auct. fl. carpat. = F. versicolor
varia Haenke subsp. *brachystachys* (Hack.) Hegi = F. versi-
    color
- subsp. *eu-varia* var. *brachystachys* Hack. = F. versicolor
- - var. *caucasica* St.-Yves = F. karabaghensis
- subsp. *woronowii* (Hack.) Tzvel. = F. woronowii
varia sensu V. Krecz. & Bobr. = F. woronowii
venusta St.-Yves - 4
versicolor Tausch (*F. brachystachys* (Hack.) K. Richt., *F.*
    varia Haenke subsp. *brachystachys* (Hack.) Hegi, *F.*
    varia subsp. *eu-varia* var. *brachystachys* Hack., *F. varia*
    auct. fl. carpat.) - 1
villosa Schweigg. = F. arenaria
violacea auct. fl. cauc. = F. chalcophaea
violacea Gaudin subsp. *picta* (Kit.) Hegi = F. picta
vivipara (L.) Smith (*F. ovina* L. var. *vivipara* L.) - 1, 3, 4, 5
vorobievii Probat. (*F. ovina* L. var. *litoralis* Tzvel., *F. ovina*
    subsp. *litoralis* (Tzvel.) E. Alexeev) - 5
wolgensis P. Smirn. (*F. arietina* Klok., *F. wolgensis* subsp.
    arietina (Klok.) Tzvel.) - 1, 3
- subsp. *arietina* (Klok.) Tzvel. = F. wolgensis
woronowii Hack. (*F. varia* Haenke subsp. *woronowii*
    (Hack.) Tzvel., *F. amethystina* sensu V. Krecz. &
    Bobr., *F. varia* sensu V. Krecz. & Bobr.) - 2
- subsp. *caucasica* (St.-Yves) E. Alexeev = F. karabaghensis
- subsp. *caucasica* (St.-Yves) Markgraf-Dannenberg = F.
    karabaghensis
yaroschenkoi (St.-Yves) E. Alexeev (*F. ovina* L. subsp. *eu-*
    ovina var. *yaroschenkoi* St.-Yves) - 2

x Festulolium Aschers. & Graebn.

loliaceum (Huds.) P. Fourn. (*Festuca loliacea* Huds.). -
    Festuca pratensis Huds. x Lolium perenne L.

Gastridium Beauv. (*Lachnagrostis* auct.)

lendigerum (L.) Desv. = G. ventricosum
phleoides (Nees & Meyen) C.E. Hubb. (*Lachnagrostis*

phleoides Nees & Meyen, *Gastridium ventricosum* (Gouan) Schinz & Thell. subsp. *phleoides* (Nees & Meyen) Tzvel.) - 2

ventricosum (Gouan) Schinz & Thell. (*Agrostis ventricosa* Gouan, *Gastridium lendigerum* (L.) Desv.) - 2

- subsp. *phleoides* (Nees & Meyen) Tzvel. = G. phleoides

Gastropyrum (Jaub. & Spach) A. Löve = Aegilops

crassum (Boiss.) A. Löve = Aegilops crassa
ventricosum (Tausch) A. Löve = Aegilops ventricosa

Gaudinia Beauv.

fragilis (L.) Beauv. - 1, 2

Gaudinopsis Eig

macra (Stev.) Eig. - 1, 2, 6

Gigachilon Seidl = Triticum

polonicum (L.) Seidl = Triticum polonicum
- subsp. *carthlicum* (Nevski) A. Löve = Triticum carthlicum
- subsp. *dicoccon* (Schrank) A. Löve = Triticum dicoccon
- subsp. *durum* (Desf.) A. Löve = Triticum durum
- subsp. *paleocolchicum* (A. & D. Löve) A. Löve = Triticum karamyschevii
- subsp. *turanicum* (Jakubz.) A. Löve = Triticum turanicum
- subsp. *turgidum* (L.) A. Löve = Triticum turgidum
timopheevii (Zhuk.) A. Löve = Triticum timopheevii
- subsp. *armeniacum* (Jakubz.) A. Löve = Triticum araraticum

Glyceria R. Br.

alnasteretum Kom. - 5
x **amurensis** Probat. - G. leptorhiza (Maxim.) Kom. x G. triflora (Korsh.) Kom. - 5
aquatica (L.) Wahlenb. = G. maxima
- var. *debilior* Trin. ex Fr. Schmidt = G. lithuanica
arundinacea Kunth (*G. maxima* (C. Hartm.) Holmb. subsp. *arundinacea* (Kunth) Hult., *Poa arundinacea* Bieb. 1808, non Moench, 1794) - 1, 2
- subsp. *triflora* (Korsh.) Tzvel. = G. triflora
caspia Trin. - 2
debilior auct. = G. triflora
debilior (Fr. Schmidt) Kudo, p.p. = G. lithuanica
declinata Breb. (*G. fluitans* (L.) R. Br. var. *declinata* (Breb.) Ghisa) - 1
depauperata Ohwi - 5
fluitans (L.) R. Br. - 1, 2, 3(alien), 4(alien)
- var. *declinata* (Breb.) Ghisa = G. declinata
ischyroneura Steud. (*G. tonglensis* auct.) - 5
kamtschatica Kom. = G. triflora
karafutensis Ohwi = G. triflora
langeana Berlin = Puccinellia langeana
leptolepis Ohwi (*G. ussuriensis* Kom.) - 5
leptorhiza (Maxim.) Kom. - 4, 5
lithuanica (Gorski) Gorski (*G. aquatica* (L.) Wahlenb. var. *debilior* Trin. ex Fr. Schmidt, *G. debilior* (Fr. Schmidt) Kudo, p.p.) - 1, 2, 3, 4, 5
maxima (C. Hartm.) Holmb. (*Molinia maxima* C. Hartm., *Glyceria aquatica* (L.) Wahlenb. 1820, non J. & C. Presl, 1819) - 1, 3
- subsp. *arundinacea* (Kunth) Hult. = G. arundinacea
- subsp. *triflora* (Korsh.) Hult. = G. triflora
natans Kom. = Torreyochloa natans

nemoralis (Uechtr.) Uechtr. & Koern. (*G. plicata* (Fries) Fries var. *nemoralis* Uechtr.) - 1, 2
nervata (Willd.) Trin. = G. striata
notata Chevall. (*G. plicata* (Fries) Fries, *G. turcomanica* Kom.) - 1, 2, 3, 5, 6
x **orientalis** Kom. - G. lithuanica (Gorski) Gorski x G. alnasteretum Kom. - 5
pallida auct. = Torreyochloa natans
paupercula Holm = Puccinellia langeana
x **pedicellata** Towns. - G. fluitans (L.) R. Br. x G. notata Chevall.
plicata (Fries) Fries = G. notata
- var. *nemoralis* Uechtr. = G. nemoralis
pumila Vasey = Puccinellia kurilensis
spiculosa (Fr. Schmidt) Roshev. - 4, 5
striata (Lam.) Hitchc. (*Poa striata* Lam., *Glyceria nervata* (Willd.) Trin.) - 1(alien)
tonglensis auct. = G. ischyroneura
triflora (Korsh.) Kom. (*G. arundinacea* Kunth subsp. *triflora* (Korsh.) Tzvel., *G. kamtschatica* Kom., *G. karafutensis* Ohwi, *G. maxima* (C. Hartm.) Holmb. subsp. *triflora* (Korsh.) Hult., *G. debilior* auct.) - 1, 3, 4, 5
turcomanica Kom. = G. notata
ussuriensis Kom. = G. leptolepis
vaginata Lange = Puccinellia vaginata
- var. *contracta* Lange = Puccinellia angustata

Goulardia Husn. = Elymus

abolinii (Drob.) Ikonn. = Elymus abolinii
canina (L.) Husn. = Elymus caninus
curvatiformis (Nevski) Ikonn. = Elymus curvatiformis
mutabilis (Drob.) Ikonn. = Elymus mutabilis
praecaespitosa (Nevski) Ikonn. = Elymus praecaespitosus
sclerophylla (Nevski) Ikonn. = Elymus sclerophyllus
transhyrcana (Nevski) Ikonn. = Elymus transhyrcanus
ugamica (Drob.) Ikonn. = Elymus nevskii

Hainardia Greuter = Monerma

cylindrica (Willd.) Greuter = Monerma cylindrica

Haynaldia Schur = Dasypyrum

villosa (L.) Schur = Dasypyrum villosum

Heleochloa Host ex Roem. = Crypsis

alopecuroides (Pill. & Mitt.) Host ex Roem. = Crypsis alopecuroides
faktorovskyi (Eig) Pilg. = Crypsis faktorovskyi
schoenoides (L.) Host ex Roem. = Crypsis schoenoides

Helictotrichon Bess. (*Avenastrum* Jess. nom. illegit., *Avenastrum* Opiz, *Avenochloa* Holub, nom. illegit., *Avenula* (Dumort.) Dumort.)

adzharicum (Albov) Grossh. (*Avenastrum adzharicum* (Albov) Roshev., *A. asiaticum* Roshev. p.p. quoad pl. cauc., *Avenochloa adzharica* (Albov) Holub, *A. versicolor* (Vill.) Holub subsp. *caucasica* (Holub) Holub, nom. invalid., *Avenula adzharica* (Albov) Holub, *Helictotrichon asiaticum* (Roshev.) Grossh. p.p. quoad pl., *H. versicolor* (Vill.) Pilg. proles *caucasicum* Holub) - 2
alpinum auct. = H. praeustum
altaicum Tzvel. (*H. desertorum* (Less.) Nevski subsp. *altaicum* (Tzvel.) Holub, *H. desertorum* subsp. *centroasiaticum* Holub) - 3, 4, 6

armeniacum (Schischk.) Grossh. (*Avenastrum armeniacum* (Schischk.) Roshev., *Avenochloa armeniaca* (Schischk.) Holub, *Avenula armeniaca* (Schischk.) Holub) - 2

asiaticum (Roshev.) Grossh. = H. adzharicum and H. hookeri

besseri (Griseb.) Klok. = H. desertorum

compactum (Boiss. & Heldr.) Henrard = Danthoniastrum compactum

compressum (Heuff.) Henrard (*Avena compressa* Heuff., *Avenochloa compressa* (Heuff.) Holub, *A. taurica* (Prokud.) Holub, *Avenula compressa* (Heuff.) Holub, 12 XI 1976, comb. superfl., *A. compressa* (Heuff.) Sauer & Chmelitschek, 16 XI 1976, *Helictotrichon tauricum* Prokud.) - 1

dahuricum (Kom.) Kitag. (*Avenastrum dahuricum* (Kom.) Roshev., *Avenochloa dahurica* (Kom.) Holub, *Avenula dahurica* (Kom.) Holub, 12 XI 1976, comb. superfl., *A. dahurica* (Kom.) Sauer & Chmelitschek, 16 X 1976) - 4, 5

desertorum (Less.) Nevski (*Avena besseri* Griseb., *Avenastrum basalticum* Podp., *A. besseri* (Griseb.) Koczwara, *A. desertorum* (Less.) Podp., *Helictotrichon besseri* (Griseb.) Klok., *H. desertorum* subsp. *basalticum* (Podp.) Holub) - 1, 3, 4, 6
- subsp. *altaicum* (Tzvel.) Holub = H. altaicum
- subsp. *basalticum* (Podp.) Holub = H. desertorum
- subsp. *centroasiaticum* Holub = H. altaicum

fedtschenkoi (Hack.) Henrard (*Arrhenatherum fedtschenkoi* (Hack.) Potztal, *Avenastrum fedtschenkoi* (Hack.) Roshev.) - 6

hissaricum (Roshev.) Henrard (*Avenastrum hissaricum* Roshev.) - 6

hookeri (Scribn.) Henrard (*Avena hookeri* Scribn., *Arrhenatherum hookeri* (Scribn.) Potztal, *Avenastrum asiaticum* Roshev. p.p. excl. pl. cauc., *Avenochloa asiatica* (Roshev.) Holub, *A. hookeri* (Scribn.) Holub, *Avenula hookeri* (Scribn.) Holub, *Helictotrichon asiaticum* (Roshev.) Grossh. p.p. quoad nomen) - 3, 4, 6
- subsp. *schellianum* (Hack.) Tzvel. = H. schellianum

krylovii (Pavl.) Henrard (*Avenastrum krylovii* Pavl.) - 4, 5

mongolicum (Roshev.) Henrard (*Arrhenatherum mongolicum* (Roshev.) Potztal, *Avenastrum mongolicum* (Roshev.) Roshev.) - 3, 4, 6
- subsp. **sajanense** Lomonosova - 4

planiculme (Schrad.) Pilg. (*Avena planiculmis* Schrad., *Avenastrum planiculme* (Schrad.) Opiz, *Avenochloa planiculmis* (Schrad.) Holub, *Avenula planiculmis* (Schrad.) Holub, 12 XI 1976, comb. superfl., *A. planiculmis* (Schrad.) Sauer & Chmelitschek, 16 X 1976) - 1
- subsp. **angustius** (Holub) Dostal (*Avenula planiculmis* subsp. *angustior* Holub) - 1

praeustum (Reichenb.) Tzvel. (*Avena praeusta* Reichenb., *A. adsurgens* Schur ex Simonk., *Avenastrum alpinum* (Smith) Fritsch, p.p. quoad pl., *Avenochloa adsurgens* (Simonk.) Holub, *A. praeusta* (Reichenb.) Sojak, *Avenula praeusta* (Reichenb.) Holub, *Helictotrichon alpinum* auct.) - 1

pratense (L.) Bess. (*Avenastrum pratense* (L.) Opiz, *Avenochloa pratensis* (L.) Holub, *Avenula pratensis* (L.) Dumort.) - 1

pubescens (Huds.) Pilg. (*Avenastrum pubescens* (Huds.) Opiz, *Avenochloa pubescens* (Huds.) Holub, *Avenula pubescens* (Huds.) Dumort.) - 1, 2, 3, 4, 5, 6
- subsp. **laevigatum** (Schur) Soo (*Avena laevigata* Schur, *Avenastrum laevigatum* (Schur) Domin, *Avenochloa*

pubescens subsp. *laevigata* (Schur) Soo, *Avenula pubescens* subsp. *laevigata* (Schur) Holub) - 1

sangilense Krasnob. - 4

schellianum (Hack.) Kitag. (*Avenastrum schellianum* (Hack.) Roshev., *Avenochloa hookeri* (Scribn.) Holub subsp. *schelliana* (Hack.) Sojak, *A. schelliana* (Hack.) Holub, *Avenula hookeri* (Scribn.) Holub subsp. *schelliana* (Hack.) Lomonosova, *A. schelliana* (Hack.) Holub, 12 XI 1976, comb. superfl., *A. schelliana* (Hack.) Sauer & Chmelitschek, 16 X 1976, *Helictotrichon hookeri* (Scribn.) Henrard subsp. *schellianum* (Hack.) Tzvel.) - 1, 3, 4, 5, 6

tauricum Prokud. = H. compressum

tianschanicum (Roshev.) Henrard (*Avenastrum tianschanicum* Roshev., *Avena tianschanica* (Roshev.) Worosch.) - 6

turcomanicum Czopanov - 6

versicolor (Vill.) Pilg. (*Avena versicolor* Vill., *Avenastrum versicolor* (Vill.) Fritsch, *Avenochloa versicolor* (Vill.) Holub, *Avenula versicolor* (Vill.) Lainz) - 1
- proles *caucasicum* Holub = H. adzharicum

Hemarthria R. Br. (*Rottboellia* auct.)

altissima (Poir.) Stapf & C.E. Hubb. (*Rottboellia altissima* Poir., *Hemarthria compressa* (L. fil.) R. Br. subsp. *altissima* (Poir.) Maire, *H. fasciculata* (Lam.) Kunth) - 1

compressa (L. fil.) R. Br. subsp. *altissima* (Poir.) Maire = H. altissima

fasciculata (Lam.) Kunth = H. altissima

japonica (Hack.) Roshev. = H. sibirica

sibirica (Gand.) Ohwi (*Rottboellia sibirica* Gand., *Hemarthria japonica* (Hack.) Roshev.) - 5

Henrardia C.E. Hubb. (*Rottboellia* auct.)

glabriglumis (Nevski) Ovcz. (*Pholiurus glabriglumis* Nevski, *Henrardia persica* (Boiss.) C.E. Hubb subsp. *glabriglumis* (Nevski) Czopanov, *H. persica* var. *glaberrima* (Hausskn. ex Bornm.) C.E. Hubb., *Lepturus persicus* Boiss. var. *glaberrimus* Hausskn. ex Bornm.) - 6

hirtella Nikif. = H. pubescens

persica (Boiss.) C.E. Hubb. (*Lepturus persicus* Boiss., *L. erectus* (Griseb.) Szov. ex Roshev. p.p. incl. typo, *Pholiurus persicus* (Boiss.) A. Camus) - 1, 2, 6
- subsp. *glabriglumis* (Nevski) Czopanov = H. glabriglumis
- var. *glaberrima* (Hausskn. ex Bornm.) C.E. Hubb. = H. glabriglumis

pubescens (Bertol.) C.E. Hubb. (*Rottboellia pubescens* Bertol., *Henrardia hirtella* Nikif., *H. pubescens* subsp. *hirtella* (Nikif.) A. Löve) - 6
- subsp. *hirtella* (Nikif.) A. Löve = H. pubescens

Heteranthelium Hochst.

piliferum (Banks & Soland.) Hochst. (*Elymus pilifer* Banks & Soland.) - 2, 6

Hierochloe R. Br. (*Savastana* Schrank)

alpina (Sw.) Roem. & Schult. (*Holcus alpinus* Sw., *Aira alpina* auct.) - 1, 3, 4, 5
- subsp. *orthantha* (Sorensen) G. Weimarck = H. orthantha

annulata V. Petrov (*H. odorata* (L.) Beauv. subsp. *kolymensis* Probat.) - 4, 5

arctica C. Presl (*H. hirta* (Schrank) Borb. subsp. *arctica* (C. Presl) G. Weimarck, *H. odorata* (L.) Beauv. subsp. *arctica* (C. Presl) Tzvel.) - 1, 2, 3, 4, 5

australis (Schrad.) Roem. & Schult. (*Anthoxanthum australe* (Schrad.) J.F. Veldkamp) - 1

baltica (G. Weimarck) Czer. (*H. odorata* (L.) Beauv. subsp. *baltica* G. Weimarck) - 1

bungeana Trin. = H. glabra

glabra Trin. (*Anthoxanthum glabrum* (Trin.) J.F. Veldkamp, *Hierochloe bungeana* Trin., *H. glabra* subsp. *bungeana* (Trin.) Peschkova, *H. glabra* subsp. *chakassica* Peschkova, *H. grandiflora* (Litv.) V. Petrov, *H. odorata* (L.) Beauv. subsp. *glabra* (Trin.) Tzvel., *H. odorata* subsp. *pubescens* (Kryl.) Hara, *H. odorata* f. *pubescens* Kryl.) - 3, 4, 5

- subsp. *bungeana* (Trin.) Peschkova = H. glabra
- subsp. *chakassica* Peschkova = H. glabra
- subsp. *kamtschatica* Probat. = H. kamtschatica
- subsp. *sachalinensis* (Printz) Tzvel. = H. sachalinensis
- subsp. *sibirica* (Tzvel.) Tzvel. = H. sibirica

grandiflora (Litv.) V. Petrov = H. glabra

hirta (Schrank) Borb. (*Savastana hirta* Schrank, *Anthoxanthum hirtum* (Schrank) Y. Schouten & J.F. Veldkamp, *Hierochloe odorata* (L.) Beauv. subsp. *hirta* (Schrank) Tzvel.) - 1

- subsp. *arctica* (C. Presl) G. Weimarck = H. arctica

kamtschatica (Probat.) Probat. (*H. glabra* Trin. subsp. *kamtschatica* Probat.) - 5

ochotensis Probat. - 5

odorata (L.) Beauv. - 1, 2, 4, 6

- subsp. *arctica* (C. Presl) Tzvel. = H. arctica
- subsp. *baltica* G. Weimarck = H. baltica
- subsp. *glabra* (Trin.) Tzvel. = H. glabra
- subsp. *hirta* (Schrank) Tzvel. = H. hirta
- subsp. *kolymensis* Probat. = H. annulata
- subsp. *pannonica* Chrtek & Jir. = H. repens
- subsp. *pubescens* (Kryl.) Hara = H. glabra
- subsp. *sachalinensis* (Printz) Tzvel. = H. sachalinensis
- subsp. *sibirica* Tzvel. = H. sibirica
- var. *sachalinensis* Printz = H. sachalinensis
- f. *pubescens* Kryl. = H. glabra

orientalis Fries ex Heuff. = H. repens

orthantha Sorensen (*H. alpina* (Sw.) Roem. & Schult. subsp. *orthantha* (Sorensen) G. Weimarck) - 5(?)

pauciflora R. Br. (*Anthoxanthum arcticum* J.F. Veldkamp) - 1, 3, 4, 5

repens (Host) Beauv. (*Holcus repens* Host, *Anthoxanthum repens* (Host) J.F. Veldkamp, *A. stepporum* (P. Smirn.) J.F. Veldkamp, *Hierochloe odorata* (L.) Beauv. subsp. *pannonica* Chrtek & Jir., *H. orientalis* Fries ex Heuff., *H. stepporum* P. Smirn.) - 1, 2, 3, 6

sachalinensis (Printz) Worosch. (*H. odorata* (L.) Beauv. var. *sachalinensis* Printz, *H. glabra* Trin. subsp. *sachalinensis* (Printz) Tzvel., *H. odorata* subsp. *sachalinensis* (Printz) Tzvel.) - 5

sibirica (Tzvel.) Czer. (*H. odorata* (L.) Beauv. subsp. *sibirica* Tzvel., *H. glabra* Trin. subsp. *sibirica* (Tzvel.) Tzvel.) - 3, 4, 5, 6

stepporum P. Smirn. = H. repens

wrangelica Jurtz. & Probat. - 5

x **zinserlingii** Tzvel. - H. alpina (Sw.) Roem. & Schult. x H. hirta (Schrank) Borb. - 1

Holcus L.

alpinus Swartz ex Willd. = Hierochloe alpina

annuus Salzm. ex C.A. Mey. (*H. setosus* Trin. nom. illegit.) - 2

cernuus Ard. = Sorghum cernuum

dochna Forssk. = Sorghum saccharatum

durra Forssk. = Sorghum durra

lanatus L. - 1, 2, 5(alien)

mollis L. - 1

repens Host = Hierochloe repens

setosus Trin. = H. annuus

spicatus L. = Pennisetum americanum convar. spicatum

Hordelymus (Jess.) Harz (*Cuviera* Koel. 1802, non DC. 1807, nom. conserv.)

asper (Simonk.) Savul. & E.I. Nyarady = Taeniatherum asperum

caput-medusae (L.) Pignatti subsp. *crinitus* (Schreb.) Pignatti = Taeniatherum crinitum

europaeus (L.) Harz (*Cuviera europaea* (L.) Koel.) - 1, 2

x Horderoegneria Tzvel. = x Elyhordeum

chatangensis (Roshev.) Tzvel. = x Elyhordeum chatangense

Hordeum L. (*Critesion* Rafin.)

aegiceras Nees ex Royle (*H. coeleste* (L.) Beauv. var. *trifurcatum* Schlecht., *H. trifurcatum* (Schlecht.) Wender, *H. vulgare* L. subsp. *aegiceras* (Nees ex Royle) A. Löve)

bogdanii Wilensky (*Critesion bogdanii* (Wilensky) A. Löve) - 1, 3, 6

boreale Scribn. & J.G. Smith = H. brachyantherum

brachyantherum Nevski (*Critesion jubatum* (L.) Nevski subsp. *breviaristatum* (Bowden) A. & D. Löve, *Hordeum boreale* Scribn. & J.G. Smith, 1897, non Gand. 1881, *H. jubatum* L. subsp. *brachyantherum* (Nevski) Bondar., *H. jubatum* subsp. *breviaristatum* Bowden) - 5

brevisubulatum sensu Nevski = H. nevskianum

brevisubulatum (Trin.) Link (*Critesion brevisubulatum* (Trin.) A. Löve, ? *Hordeum macilentum* Steud.) - 1(alien), 3, 4, 5, 6

- subsp. *nevskianum* (Bowden) Tzvel. = H. nevskianum
- subsp. *turkestanicum* (Nevski) Tzvel. = H. turkestanicum
- subsp. *violaceum* (Boiss. & Huet) Tzvel. = H. violaceum
- var. *nevskianum* (Bowden) Tzvel. = H. nevskianum

bulbosum L. (*Critesion bulbosum* (L.) A. Löve) - 1, 2, 6

coeleste (L.) Beauv. var. *trifurcatum* Schlecht. = H. aegiceras

distichon L. (*H. distichon* subsp. *nudum* (L.) Rothm. nom. invalid., *H. distichon* subsp. *zeocrithon* (L.) Celak., *H. nudum* (L.) Ard., *H. vulgare* L. subsp. *distichon* (L.) Koern., *H. zeocrithon* L.) - 1, 2, 3, 4, 5, 6

- subsp. *nudum* (L.) Rothm. = H. distichon
- subsp. *zeocrithon* (L.) Celak. = H. distichon

geniculatum All. (*Critesion geniculatum* (All.) A. Löve, 1980, non Rafin. 1819, *C. hystrix* (Roth) A. Löve, *Hordeum gussoneanum* Parl., *H. hystrix* Roth, *H. marinum* Huds. subsp. *gussoneanum* (Parl.) Thell., *H. maritimum* With. subsp. *gussoneanum* (Parl.) Aschers. & Graebn. comb. superfl., *H. maritimum* subsp. *gussoneanum* (Parl.) K. Richt., *H. maritimum* subsp. *hystrix* (Roth) Jir.) - 1, 2, 6

glaucum Steud. (*Critesion glaucum* (Steud.) A. Löve, *C. murinum* (L.) A. Löve subsp. *glaucum* (Steud.) W.A. Weber, *Hordeum leporinum* Link subsp. *glaucum* (Steud.) T. Booth & Richards, *H. murinum* L. subsp. *glaucum* (Steud.) Tzvel., *H. stebbinsii* Covas) - 1, 2, 6

gussoneanum Parl. = H. geniculatum

hexastichon L. = H. vulgare
hrasdanicum Gandiljan (*H. murinum* L. subsp. *hrasdanicum* (Gandiljan) A. Trof.) - 2
hystrix Roth = H. geniculatum
x **intermedium** (Koern.) Carlet. (*H. vulgare* L. subsp. *intermedium* Koern., *H. vulgare* convar. *intermedium* (Koern.) Mansf.). - H. distichon L. x H. vulgare L.
ischnatherum (Coss.) Koern. ex Schweinf. = H. spontaneum
ithaburense Boiss. = H. spontaneum
- var. *ischnatherum* Coss. = H. spontaneum
jubatum L. (*Critesion jubatum* (L.) Nevski) - 1, 2, 3, 4, 5, 6
- subsp. *brachyantherum* (Nevski) Bondar. = H. brachyantherum
- subsp. *breviaristatum* Bowden = H. brachyantherum
x **lagunculiforme** (Bacht.) Bacht. ex Nikif. (*H. spontaneum* C. Koch var. *lagunculiforme* Bacht.). - *H. spontaneum* C. Koch x H. vulgare L.
leporinum Link (*Critesion murinum* (L.) A. Löve subsp. *leporinum* (Link) A. Löve, *Hordeum murinum* L. subsp. *leporinum* (Link) Arcang.) - 1, 2, 6
- subsp. *glaucum* (Steud.) T. Booth & Richards = H. glaucum
macilentum Steud. = H. brevisubulatum
marinum Huds. (*Critesion marinum* (Huds.) A. Löve) - 1, 2
- subsp. *gussoneanum* (Parl.) Thell. = H. geniculatum
maritimum With. subsp. *gussoneanum* (Parl.) Aschers. & Graebn. = H. geniculatum
- subsp. *gussoneanum* (Parl.) K. Richt. = H. geniculatum
- subsp. *hystrix* (Roth) Jir. = H. geniculatum
murinum L. (*Critesion murinum* (L.) A. Löve) - 1, 2(alien)
- subsp. *glaucum* (Steud.) Tzvel. = H. glaucum
- subsp. *hrasdanicum* (Gandiljan) A. Trof. = H. hrasdanicum
- subsp. *leporinum* (Link) Arcang. = H. leporinum
nevskianum Bowden (*Critesion brevisubulatum* (Trin.) A. Löve subsp. *nevskianum* (Bowden) A. Löve, *Hordeum brevisubulatum* (Trin.) Link subsp. *nevskianum* (Bowden) Tzvel., *H. brevisubulatum* var. *nevskianum* (Bowden) Tzvel., *H. brevisubulatum* sensu Nevski) - 1, 3, 6
nodosum auct. = H. secalinum
nudum (L.) Ard. = H. distichon
roshevitzii Bowden (*Critesion californicum* (Covas & Stebb.) A. Löve subsp. *sibiricum* (Roshev.) A. Löve, *Hordeum sibiricum* Roshev. 1929, non Schenk, 1908) - 3, 4, 5, 6
secalinum Schreb. (*Critesion secalinum* (Schreb.) A. Löve, *Hordeum nodosum* auct.) - 1
sibiricum Roshev. = H. roshevitzii
spontaneum C.Koch (*H. ischnatherum* (Coss.) Koern. ex Schweinf., *H. ithaburense* Boiss., *H. ithaburense* var. *ischnatherum* Coss., *H. vulgare* L. subsp. *spontaneum* (C. Koch) Aschers. & Graebn.) - 2, 6
- var. *lagunculiforme* Bacht. = H. x lagunculiforme
stebbinsii Covas = H. glaucum
trifurcatum (Schlecht.) Wender = H. aegiceras
turkestanicum Nevski (*Critesion brevisubulatum* (Trin.) A. Löve subsp. *turkestanicum* (Nevski) A. Löve, *Hordeum brevisubulatum* (Trin.) Link subsp. *turkestanicum* (Nevski) Tzvel.) - 1(alien), 3, 5(alien), 6
violaceum Boiss. & Huet (*Critesion violaceum* (Boiss. & Huet) A. Löve, *Hordeum brevisubulatum* (Trin.) Link subsp. *violaceum* (Boiss. & Huet) Tzvel.) - 2
*****vulgare** L. (*H. hexastichon* L., *H. vulgare* subsp. *hexastichon* (L.) Celak.) - 1, 2, 3, 4, 5, 6
- subsp. *aegiceras* (Nees ex Royle) A. Löve = H. aegiceras
- subsp. *distichon* (L.) Koern. = H. distichon

- subsp. *hexastichon* (L.) Celak. = H. vulgare
- subsp. *intermedium* Koern. = H. x intermedium
- subsp. *spontaneum* (C. Koch) Aschers. & Graebn. = H. spontaneum
- convar. *intermedium* (Koern.) Mansf. = H. x intermedium
zeocrithon L. = H. distichon

Hyalopoa (Tzvel.) Tzvel.

czirahica Gussejnov - 2
ivanoviae (Malysch.) Czer. (*Colpodium ivanoviae* Malysch., *Hyalopoa lanatiflora* (Roshev.) Tzvel. subsp. *ivanoviae* (Malysch.) Tzvel.) - 4
lakia (Woronow) Tzvel. (*Colpodium lakium* Woronow) - 2
lanatiflora (Roshev.) Tzvel. (*Poa lanatiflora* Roshev., *Colpodium lanatiflorum* (Roshev.) Tzvel.) - 4
- subsp. *ivanoviae* (Malysch.) Tzvel. = H. ivanoviae
- subsp. *momica* (Tzvel.) Tzvel. = H. momica
momica (Tzvel.) Czer. (*Colpodium lanatiflorum* (Roshev.) Tzvel. subsp. *momicum* Tzvel., *Hyalopoa lanatiflora* (Roshev.) Tzvel. subsp. *momica* (Tzvel.) Tzvel.) - 4
pontica (Bal.) Tzvel. (*Colpodium ponticum* (Bal.) Woronow) - 2

Hydropyrum Link = Zizania

latifolium Griseb. = Zizania latifolia

Hystrix Moench (*Asperella* Humb. 1790, non Asprella Schreb. 1789)

coreana (Honda) Ohwi (*Elymus coreanus* Honda, *Asperella coreana* (Honda) Nevski) - 5
komarovii (Roshev.) Ohwi (*Asperella komarovii* Roshev., *Elymus komarovii* (Roshev.) Ohwi, 1933, non Tzvel. 1968, *Hystrix sachalinensis* Ohwi) - 5
sachalinensis Ohwi = H. komarovii
sibirica (Trautv.) O. Kuntze (*Asperella sibirica* Trautv., *Elymus asiaticus* A. Löve) - 4, 5

Imperata Cyr.

cylindrica (L.) Raeusch. - 2, 6
- subsp. *koenigii* (Retz.) Tzvel. = I. koenigii
koenigii (Retz.) Beauv. (*Saccharum koenigii* Retz., *Imperata cylindrica* (L.) Raeusch. subsp. *koenigii* (Retz.) Tzvel.) - 6

*Indocalamus Nakai

*****tessellatus** (Munro) Keng fil. (*Bambusa tessellata* Munro, *Sasa tessellata* (Munro) Makino & Shibata)

Kengia Packer = Cleistogenes

bulgarica (Bornm.) Packer = Cleistogenes bulgarica
hancei (Keng) Packer = Cleistogenes hancei
kitagawae (Honda) Packer = Cleistogenes kitagawae
maeotica (Klok. & Zoz) Packer = Cleistogenes bulgarica
serotina (L.) Packer = Cleistogenes serotina
songorica (Roshev.) Packer = Cleistogenes songorica
squarrosa (Trin.) Packer = Cleistogenes squarrosa

Kiharapyrum A. Löve = Aegilops

umbellatum (Zhuk.) A. Löve = Aegilops umbellata
- subsp. *transcaucasicum* (V. Dorof. & Migusch.) A. Löve = Aegilops umbellata subsp. transcaucasica

Koeleria Pers.

albovii Domin (*K. buschiana* (Domin) Gontsch., *K. caucasica* (Domin) B. Fedtsch. p.p. quoad nomen, *K. eriostachya* Panc. subsp. *caucasica* Domin, *K. eriostachya* subsp. *fominii* Domin, *K. fominii* (Domin) Gontsch., *K. eriostachya* auct. p.p.) - 2

altaica (Domin) Kryl. (*K. eriostachya* Panc. subsp. *caucasica* var. *altaica* Domin) - 3, 4, 6

x **aschersoniana** Domin (*K.* x *soongarica* Domin). - K. cristata (L.) Pers. x K. glauca (Spreng.) DC.

ascoldensis Roshev. = K. tokiensis

asiatica Domin (*K. mariae* V. Vassil.) - 1, 3, 4, 5
- subsp. *atroviolacea* (Domin) Tzvel. = K. atroviolacea
- subsp. *ledebourii* (Domin) Tzvel. = K. ledebourii

atroviolacea Domin (*K. asiatica* Domin subsp. *atroviolacea* (Domin) Tzvel., *K. geniculata* Domin) - 3, 4

barabensis (Domin) Gontsch. = K. delavignei

besseri Ujhlyi - 1

biebersteinii M. Kaleniczenko - 1

borysthenica Klok. = K. sabuletorum

brevis Stev. (*K. degenii* Domin, *K. splendens* C. Presl var. *degenii* (Domin) Stojan. & Stef.) - 1

buschiana (Domin) Gontsch. = K. albovii

calarashica M. Kaleniczenko - 1

callieri (Domin) Ujhelyi = K. lobata

caucasica (Domin) B. Fedtsch. = K. albovii and K. ledebourii

chakassica Reverd. (*K. cristata* (L.) Pers. subsp. *chakassica* (Reverd.) Tzvel.) - 4

cristata (L.) Pers. (*Aira cristata* L. p. max. p., *Koeleria gracilis* Pers. nom. illegit.) - 1, 2, 3, 4, 5, 6
- subsp. *ascoldensis* (Roshev.) Worosch. = K. tokiensis
- subsp. *chakassica* (Reverd.) Tzvel. = K. chakassica
- subsp. *krylovii* (Reverd.) Tzvel. = K. krylovii
- subsp. *mongolica* (Domin) Tzvel. = K. macrantha
- subsp. *seminuda* (Trautv.) Tzvel. = K. seminuda
- subsp. *transiliensis* (Reverd. ex Gamajun.) Tzvel. = K. transiliensis
- subsp. *transsilvanica* (Schur) K. Richt. = K. transsilvanica
- var. *tokiensis* (Domin) Hara = K. tokiensis
- c. *robusta* Pacz. ex Schmalh. = K. lobata

degenii Domin = K. brevis

delavignei Czern. ex Domin (*K. barabensis* (Domin) Gontsch., *K. delavignei* subsp. *barabensis* Domin, *K. incerta* Domin) - 1, 3, 4, 5(alien)
- subsp. *barabensis* Domin = K. delavignei
- subsp. **veresczaginii** Vlasova - 3

elata M. Kaliniczenko - 1

eriostachya auct. p.p. = K. albovii

eriostachya Panc. subsp. *caucasica* Domin = K. albovii
- - var. *altaica* Domin = K. altaica
- subsp. *fominii* Domin = K. albovii

fominii (Domin) Gontsch. = K. albovii

geniculata Domin = K. atroviolacea

glauca (Spreng.) DC. (*Aira glauca* Spreng., *Poa glauca* Schkuhr, 1799, non Vahl, 1790) - 1, 3, 4
- subsp. *pohleana* (Domin) Tzvel. = K. pohleana
- subsp. *sabuletorum* Domin = K. sabuletorum

glaucovirens Domin = K. kurdica

gorodkowii Roshev. = x Trisetokoeleria gorodkowii

gracilis Pers. = K. cristata
- subsp. *luerssenii* Domin = K. luerssenii
- subsp. *seminuda* (Trautv.) Domin = K. seminuda
- subsp. *sibirica* Domin = K. seminuda
- subsp. *transsilvanica* (Schur) Fodor = K. transsilvanica
- var. *mukdenensis* (Domin) Kitag. = K. macrantha

grandis Bess. ex Gorski (*K. polonica* Domin) - 1

grossheimiana (Tzvel.) Galushko (*K. luerssenii* (Domin) Domin subsp. *grossheimiana* Tzvel., *K. nitidula* auct.) - 2

incerta Domin = K. delavignei

karavajevii Govor. - 4

krylovii Reverd. (*K. cristata* (L.) Pers. subsp. *krylovii* (Reverd.) Tzvel.) - 3, 4

kurdica Ujhelyi (*K. glaucovirens* Domin, nom. illegit.) - 2

ledebourii Domin (*K. asiatica* Domin subsp. *ledebourii* (Domin) Tzvel., *K. caucasica* (Domin) B. Fedtsch. p.p. quoad pl.) - 1, 3, 6

litwinowii Domin = Trisetum litwinowii

lobata (Bieb.) Roem. & Schult. (*Dactylis lobata* Bieb., *Koeleria callieri* (Domin) Ujhelyi, *K. cristata* (L.) Pers. c. *robusta* Pacz. ex Schmalh., *K. robusta* (Pacz. ex Schmalh.) Janata, *K. splendens* C. Presl, *K. splendens* var. *callieri* Domin) - 1

luerssenii (Domin) Domin (*K. gracilis* Pers. subsp. *luerssenii* Domin, *K. monantha* Domin) - 2
- subsp. *grossheimiana* Tzvel. = K. grossheimiana

macrantha (Ledeb.) Schult. (*Aira macrantha* Ledeb., *Koeleria cristata* (L.) Pers. subsp. *mongolica* (Domin) Tzvel., *K. gracilis* Pers. var. *mukdenensis* (Domin) Kitag., *K. mukdenensis* Domin, *K. poaeformis* Domin, *K. tokiensis* Domin subsp. *mongolica* Domin) - 4, 5(?)
- subsp. *transsilvanica* (Schur) A. Nyarady = K. transsilvanica

mariae V. Vassil. = K. asiatica

moldavica M. Alexeenko - 1

monantha Domin = K. luerssenii

mukdenensis Domin = K. macrantha

nitidula auct. = K. grossheimiana

phleoides (Vill.) Pers. var. *glabriflora* Trautv. = Rostraria glabriflora

poaeformis Domin = K. macrantha

pohleana (Domin) Gontsch. (*K. glauca* (Spreng.) DC. subsp. *pohleana* (Domin) Tzvel.) - 1

polonica Domin = K. grandis

pyramidata (Lam.) Beauv. (*Poa pyramidata* Lam., *Koeleria cristata* (L.) Pers. p.p. quoad pl., excl. typo) - 1

robusta (Pacz. ex Schmalh.) Janata = K. lobata

sabuletorum (Domin) Klok. (*K. glauca* (Spreng.) DC. subsp. *sabuletorum* Domin, *K. borysthenica* Klok.) - 1, 3, 4, 6

sclerophylla P. Smirn. - 1, 3, 6
- subsp. **sclerophylla** - 1, 3
- subsp. **theodoriana** Klok. ex Tzvel. (*K. theodoriana* Klok. nom. invalid.) - 1, 6

seminuda (Trautv.) Gontsch. (*K. cristata* (L.) Pers. subsp. *seminuda* (Trautv.) Tzvel., *K. gracilis* Pers. subsp. *seminuda* (Trautv.) Domin, *K. gracilis* subsp. *sibirica* Domin, *K. sibirica* (Domin) Gontsch.) - 4

sibirica (Domin) Gontsch. = K. seminuda

skrjabinii Karav. & Tzvel. - 4

x *soongarica* Domin = K. x aschersoniana

splendens C. Presl = K. lobata
- var. *callieri* Domin = K. lobata
- var. *degenii* (Domin) Stojan. & Stef. = K. brevis

talievii Lavr. - 1

taurica M. Kaleniczenko - 1

theodoriana Klok. = K. sclerophylla subsp. theodoriana

thonii Domin - 4

tokiensis Domin (*K. ascoldensis* Roshev., *K. cristata* (L.) Pers. subsp. *ascoldensis* (Roshev.) Worosch., *K. cristata* var. *tokiensis* (Domin) Hara) - 5
- subsp. *mongolica* Domin = K. macrantha

transiliensis Reverd. ex Gamajun. (*K. cristata* (L.) Pers. subsp. *transiliensis* (Reverd. ex Gamajun.) Tzvel.) - 6

transsilvanica Schur (*K. cristata* (L.) Pers. subsp. *transsilvanica* (Schur) K. Richt., *K. gracilis* Pers. subsp. *transsilvanica* (Schur) Fodor, comb. invalid., *K. macrantha* (Ledeb.) Schult. subsp. *transsilvanica* (Schur) A. Nyarady) - 1
tzvelevii Vlasova - 4

Kratzmannia Opiz = Agropyron

cristata (L.) Jir. & Skalicky = Agropyron cristatum
pectinata (Bieb.) Jir. & Skalicky = Agropyron pectinatum

Lachnagrostis auct. = Gastridium

phleoides Nees & Meyen = Gastridium phleoides

Lagurus L.

ovatus L. - 1, 2, 5(alien)

*Lamarckia Moench

*aurea (L.) Moench (*Cynosurus aureus* L.)

Lasiagrostis Link = Achnatherum

alpina Fr. Schmidt = Ptilagrostis alpina
bromoides (L.) Nevski & Roshev. = Achnatherum bromoides
caragana (Trin.) Trin. & Rupr. = Achnatherum caragana
longearistata (Boiss. & Hausskn.) Roshev. & Nevski, p.p. = Achnatherum turcomanicum
splendens (Trin.) Kunth = Achnatherum splendens

Leersia Sw.

oryzoides (L.) Sw. - 1, 2, 3, 5, 6

Leiopoa Ohwi = Festuca

blepharogyna Ohwi = Festuca blepharogyna

Lepidurus Janch. = Parapholis

incurvus (L.) Janch. = Parapholis incurva

Lepturus Trin. = Parapholis

erectus (Griseb.) Szov. ex Roshev. p.p. = Henrardia persica
incurvatus (L.) Trin. var. *longiflorus* Grossh. = Parapholis longiflora
incurvus (L.) Druce = Parapholis incurva
persicus Boiss. = Henrardia persica
- var. *glaberrimus* Hausskn. ex Bornm. = Henrardia glabriglumis

Lerchenfeldia Schur = Avenella

flexuosa (L.) Schur = Avenella flexuosa
- subsp. *montana* (L.) Tzvel. = Avenella flexuosa subsp. montana

Leucopoa Griseb. = Festuca

albida (Turcz. ex Trin.) V. Krecz. & Bobr. = Festuca sibirica
caucasica (Boiss.) V. Krecz. & Bobr. = Festuca caucasica
karatavica (Bunge) V. Krecz. & Bobr. = Festuca karatavica
kreczetoviczii K. Sobol. = Festuca sibirica
olgae (Regel) V. Krecz. & Bobr. = Festuca olgae
sclerophylla (Boiss. & Hohen.) V. Krecz. & Bobr. = Festuca sclerophylla
sibirica Griseb. = Festuca sibirica

x *Leymopyron* Tzvel. = x Leymotrigia
bergrothii (Lindb. fil.) Tzvel. = x Leymotrigia bergrothii

x **Leymostachys** Tzvel.

korovinii Tzvel. - Leymus ? racemosus (Lam.) Tzvel. s.l. x Psathyrostachys juncea (Fisch.) Nevski, s.l. - 6

x **Leymotrigia** Tzvel. (x *Leymopyron* Tzvel., x *Tritordeum* auct.)

bergrothii (Lindb. fil.) Tzvel. (x *Tritordeum bergrothii* Lindb. fil., x *Elymotrigia bergrothii* (Lindb. fil.) Hyl., x *Leymopyron bergrothii* (Lindb. fil.) Tzvel.). - Leymus arenarium (L.) Hochst. x Elytrigia repens (L.) Nevski, s.l. - 1
pacifica Probat. - Leymus mollis (Trin.) Hara x Elytrigia repens (L.) Nevski - 5
roshevitzii Tzvel. - Leymus ? paboanus (Claus) Pilg. x Elytrigia repens (L.) Nevski - 6
stricta (Dethard.) Tzvel. (*Triticum strictum* Dethard., *Agropyron strictum* (Dethard.) Reichenb., x *Elymotrigia stricta* (Dethard.) Hyl.). - Leymus arenarius (L.) Hochst. x Elytrigia junceiformis A. & D. Löve - 1
wiluica (Drob.) Tzvel. (*Agropyron wiluicum* Drob.). - Leymus secalinus (Georgi) Tzvel. x Elytrigia repens (L.) Nevski, s.l. - 4
zarubinii Peschkova. - Leymus chinensis (Trin.) Tzvel. x Elytrigia repens (L.) Nevski - 4

x **Leymotrix** Charkev. & Probat.

ajanensis (V. Vassil.) Charkev. & Probat. (*Asperella ajanensis* V. Vassil., *Elymus ajanensis* Roshev. ex Worosch. nom. invalid., *Leymus ajanensis* (V. Vassil.) Tzvel.). - Leymus interior (Hult.) Tzvel. x Hystrix sibirica (Trautv.) O. Kuntze - 5

Leymus Hochst. (*Aneurolepium* Nevski, *Malacurus* Nevski, *Elymus* sensu Nevski)

aemulans (Nevski) Tzvel. (*Aneurolepidium aemulans* Nevski, *Agropyron aemulans* (Nevski) N.M. Kuznetzov, *Elymus aemulans* (Nevski) Nikif.) - 6
ajanensis (V. Vassil.) Tzvel. = x Leymotrix ajanensis
akmolinensis (Drob.) Tzvel. (*Aneurolepidium akmolinense* (Drob.) Nevski, *Leymus paboanus* (Claus) Pilg. subsp. *akmolinensis* (Drob.) Tzvel., *L. paboanus* subsp. *korshinskyi* Tzvel.) - 1, 3, 4, 6
alaicus (Korsh.) Tzvel. (*Aneurolepidium alaicum* (Korsh.) Nevski, *A. ugamicum* (Drob.) Nevski, *Elymus angustiformis* Drob. p.p. quoad typum, *Leymus angustiformis* (Drob.) Tzvel., *L. ugamicus* (Drob.) Tzvel.) - 6
- subsp. *karataviensis* (Roshev.) Tzvel. = L. karataviensis
- subsp. *petraeus* (Nevski) Tzvel. = L. petraeus
angustiformis (Drob.) Tzvel. = L. alaicus
angustus (Trin.) Pilg. (*Aneurolepidium angustum* (Trin.) Nevski, p.p., *Triticum angustum* (Trin.) F. Herm.) - 3, 4, 6
arenarius (L.) Hochst. (*Elymus arenarius* L., *Triticum arenarium* (L.) F. Herm.) - 1
- subsp. *mollis* (Trin.) Tzvel. = L. mollis
baldshuanicus (Roshev.) Tzvel. = L. tianschanicus
buriaticus Peschkova - 4

chakassicus Peschkova - 4

chinensis (Trin.) Tzvel. (*Triticum chinense* Trin., *Agropyron chinense* (Trin.) Ohwi, *A. pseudoagropyrum* (Trin. ex Griseb.) Franch., *A. uninerve* P. Candargy, *Aneurolepidium chinense* (Trin.) Kitag., *A. pseudoagropyrum* (Trin. ex Griseb.) Nevski, *Elymus chinensis* (Trin.) Keng, *E. pseudoagropyrum* (Trin. ex Griseb.) Turcz., *Leymus pseudoagropyrum* (Trin. ex Griseb.) Tzvel., *Triticum pseudoagropyrum* Trin. ex Griseb.) - 3, 4, 5

dasystachys (Trin.) Pilg. (*Elymus dasystachys* Trin. p.p., *Agropyron chino-rossicum* Ohwi, *Aneurolepidium dasystachys* (Trin.) Nevski, p.p., *Elymus glaucus* Regel, p.p., *Triticum dasystachys* (Trin.) F. Herm.) - 3, 4, 6

divaricatus (Drob.) Tzvel. (*Aneurolepidium divaricatum* (Drob.) Nevski, *A. regelii* (Roshev.) Nevski, *Leymus regelii* (Roshev.) Tzvel.) - 6

- subsp. *fasciculatus* (Roshev.) Tzvel. = L. fasciculatus

fasciculatus (Roshev.) Tzvel. (*Aneurolepidium fasciculatum* (Roshev.) Nevski, *Leymus divaricatus* (Drob.) Tzvel. subsp. *fasciculatus* (Roshev.) Tzvel.) - 6

x **fedtschenkoi** Tzvel. - L. alaicus (Korsh.) Tzvel. x L. lanatus (Korsh.) Tzvel. - 6

flexilis (Nevski) Tzvel. (*Aneurolepidium flexile* Nevski, *Elymus flexilis* (Nevski) N.M. Kuznetzov) - 6

giganteus (Vahl) Pilg. = L. racemosus

interior (Hult.) Tzvel. (*Elymus interior* Hult., *E. mollis* Trin. subsp. *interior* (Hult.) Bowden, *Leymus mollis* (Trin.) Pilg. subsp. *interior* (Hult.) A. & D. Löve) - 4, 5

x **jenisseiensis** (Turcz.) Tzvel. (*Elymus x jenisseiensis* Turcz.). - L. racemosus (Lam.) Tzvel. x L. dasystachys (Trin.) Pilg. - 4

karataviensis (Roshev.) Tzvel. (*Aneurolepidium karataviense* (Roshev.) Nevski, *Leymus alaicus* (Korsh.) Tzvel. subsp. *karataviensis* (Roshev.) Tzvel.) - 6

karelinii (Turcz.) Tzvel. (*Elymus karelinii* Turcz., *Aneurolepidium karelinii* (Turcz.) Nevski, p.p quoad nomen, *Elymus angustiformis* Pavl. 1952, non Drob. 1941, *E. kirghisorum* Drob., *E. kugalensis* E. Nikit. nom. invalid., *E. kuznetzovii* Pavl., *E. turgaicus* Roshev., *Leymus karelinii* (Turcz.) Czopanov, comb. superfl., *L. kugalensis* (E. Nikit.) Tzvel. nom. invalid., *L. kuznetzovii* (Pavl.) Tzvel.) - 1, 3, 6

kopetdaghensis (Roshev.) Tzvel. (*Elymus kopetdaghensis* Roshev., *Aneurolepidium karelinii* (Turcz.) Nevski, p.p. quoad pl., *A. kopetdaghense* Nevski) - 6

kugalensis (E. Nikit.) Tzvel. = L. karelinii

kuznetzovii (Pavl.) Tzvel. = L. karelinii

lanatus (Korsh.) Tzvel. (*Malacurus lanatus* (Korsh.) Nevski) - 6

latiglumis Tzvel. (*Elymus latiglumis* Nikif. 1968, non Phil. 1864-1865, *E. angustiformis* Drob. p.p. excl. typo) - 6

littoralis (Griseb.) Peschkova (*Elymus dasystachys* Trin. var. *littoralis* Griseb., *E. littoralis* (Griseb.) Turcz. ex Steud.) - 4

mollis (Trin.) Pilg. (*Elymus mollis* Trin., *E. arenarius* L. subsp. *mollis* (Trin.) Hult., *E. cladostachys* Turcz., *Leymus arenarius* (L.) Hochst. subsp. *mollis* (Trin.) Tzvel., *L. mollis* (Trin.) Hara, 1938, nom. invalid., *Triticum molle* (Trin.) F. Herm.) - 5

- subsp. *interior* (Hult.) A. & D. Löve = L. interior

- subsp. *villosissimus* (Scribn.) A. & D. Löve = L. villosissimus

multicaulis (Kar. & Kir.) Tzvel. (*Aneurolepidium multicaule* (Kar. & Kir.) Nevski, *Triticum aralense* (Regel) F. Herm.) - 1, 3, 6

nikitinii (Czopanov) Tzvel. (*Elymus nikitinii* Czopanov,

Leymus nikitinii (Czopanov) Czopanov, comb. superfl.) - 6

ordensis Peschkova - 4

ovatus (Trin.) Tzvel. (*Aneurolepidium ovatum* (Trin.) Nevski, *Leymus secalinus* (Georgi) Tzvel. subsp. *ovatus* (Trin.) Tzvel.) - 3, 4, 6

paboanus (Claus) Pilg. (*Aneurolepidium paboanum* (Claus) Nevski, *Elymus dasystachys* Trin. var. *salsuginosus* Griseb., *E. salsuginosus* (Griseb.) Turcz. ex Steud.) - 1, 3, 4, 6

- subsp. *akmolinensis* (Drob.) Tzvel. = L. akmolinensis

- subsp. *korshinskyi* Tzvel. = L. akmolinensis

petraeus (Nevski) Tzvel. (*Aneurolepidium petraeum* Nevski, *Elymus petraeus* (Nevski) Pavl., *Leymus alaicus* (Korsh.) Tzvel. subsp. *petraeus* (Nevski) Tzvel.) - 6

pseudoagropyrum (Trin. ex Griseb.) Tzvel. = L. chinensis

pubescens (O. Fedtsch.) Ikonn. (*Elymus dasystachys* Trin. var. *pubescens* O. Fedtsch., *Leymus secalinus* (Georgi) Tzvel. subsp. *pubescens* (O. Fedtsch.) Tzvel., *L. secalinus* var. *pubescens* (O. Fedtsch.) Tzvel.) - 6

racemosus (Lam.) Tzvel. (*Elymus racemosus* Lam., *E. attenuatus* (Griseb.) K. Richt., *E. giganteus* Vahl, *E. giganteus* var. *attenuatus* Griseb., *E. macrostachys* Spreng., *Leymus giganteus* (Vahl) Pilg.) - 1, 2, 3, 4, 6

- subsp. **crassinervius** (Kar. & Kir.) Tzvel. (*Elymus giganteus* var. *crassinervius* Kar. & Kir.) - 3, 4, 6

- subsp. **klokovii** Tzvel. - 1, 3, 4, 6

- subsp. **sabulosus** (Bieb.) Tzvel. (*Elymus sabulosus* Bieb., *E. arenarius* L. subsp. *sabulosus* (Bieb.) Beldie, *E. arenarius* var. *sabulosus* (Bieb.) Schmalh., *E. racemosus* var. *sabulosus* (Bieb.) Bowden, *Leymus racemosus* var. *sabulosus* (Bieb.) Tzvel., *L. sabulosus* (Bieb.) Tzvel., *Triticum sabulosum* (Bieb.) F. Herm.) - 1, 2, 6

- var. *sabulosus* (Bieb.) Tzvel. = L. racemosus subsp. sabulosus

x **ramosoides** Kolmakov ex Tzvel. - L. paboanus (Claus) Pilg. x L. ramosus (Trin.) Tzvel. - 1

ramosus (Trin.) Tzvel. (*Agropyron ramosum* (Trin.) K. Richt., *Aneurolepidium ramosum* (Trin.) Nevski, *Elymus ramosus* (Trin.) Filat. 1969, nom. invalid., non Desf. 1829, *E. trinii* Melderis) - 1, 3, 4, 6

regelii (Roshev.) Tzvel. = L. divaricatus

sabulosus (Bieb.) Tzvel. = L. racemosus subsp. sabulosus

secalinus (Georgi) Tzvel. (*Triticum secalinum* Georgi, *Aneurolepidium secalinum* (Georgi) Kitam., *Elymus secalinus* (Georgi) Bobr., *Triticum littorale* Pall.) - 4

- subsp. *ovatus* (Trin.) Tzvel. = L. ovatus

- subsp. *pubescens* (O. Fedtsch.) Tzvel. = L. pubescens

- var. *pubescens* (O. Fedtsch.) Tzvel. = L. pubescens

sphacelatus Peschkova - 4

tianschanicus (Drob.) Tzvel. (*Aneurolepidium baldshuanicum* (Roshev.) Nevski, *A. tianschanicum* (Drob.) Nevski, *Leymus baldshuanicus* (Roshev.) Tzvel.) - 6

tuvinicus Peschkova - 4

ugamicus (Drob.) Tzvel. = L. alaicus

villosissimus (Scribn.) Tzvel. (*Elymus villosissimus* Scribn., *E. arenarius* L. var. *villosissimus* (Scribn.) Polun., *E. arenarius* subsp. *mollis* (Trin.) Hult. var. *villosissimus* (Scribn.) Hult. comb. superfl., *E. mollis* Trin. subsp. *villosissimus* (Scribn.) A. Löve, *Leymus mollis* (Trin.) Pilg. subsp. *villosissimus* (Scribn.) A. & D. Löve) - 4, 5

Limnas Trin.

malyschevii Nikiforova - 4, 5

stelleri Trin. - 4, 5

veresczaginii Kryl. & Schischk. - 3

Littledalea Hemsl.

alaica (Korsh.) V. Petrov ex Nevski - 6

Loliolum V. Krecz. & Bobr.

orientale (Boiss.) V. Krecz. & Bobr. = L. subulatum
subulatum (Banks & Soland.) Eig (*Triticum subulatum* Banks & Soland., *Loliolum orientale* (Boiss.) V. Krecz. & Bobr., *Nardurus subulatus* (Banks & Soland.) Bor) - 2, 6

Lolium L.

arvense With. (*L. speciosum* Bieb., *L. temulentum* L. subsp. *arvense* (With.) Tzvel., *L. temulentum* subsp. *speciosum* (Bieb.) Arcang.) - 1, 2
cuneatum Nevski (*L. temulentum* L. subsp. *cuneatum* (Nevski) Tzvel.) - 6
x hybridum Hausskn. - L. multiflorum Lam. x L. perenne L.
italicum A. Br. = L. multiflorum
loliaceum (Bory & Chaub.) Hand.-Mazz. - 1, 2, 6
marschallii Stev. = L. perenne
multiflorum Lam. (*L. italicum* A. Br.) - 1, 2, 5, 6,
perenne L. (*L. marschallii* Stev.) - 1, 2, 3, 5(alien), 6
- subsp. *remotum* (Schrank) A. & D. Löve = L. remotum
- subsp. *rigidum* (Gaudin) A. & D. Löve = L. rigidum
persicum Boiss. & Hohen. - 1(alien), 2, 5(alien), 6
remotum Schrank (*L. perenne* L. subsp. *remotum* (Schrank) A. & D. Löve) - 1, 4, 5
rigidum Gaudin (*L. perenne* L. subsp. *rigidum* (Gaudin) A. & D. Löve) - 1, 2, 5(alien), 6
speciosum Bieb. = L. arvense
temulentum L. - 1, 2, 3, 5, 6
- subsp. *arvense* (With.) Tzvel. = L. arvense
- subsp. *cuneatum* (Nevski) Tzvel. = L. cuneatum
- subsp. *speciosum* (Bieb.) Arcang. = L. arvense

Lophochloa Reichenb. = Rostraria

cavanillesii (Forssk.) Bor = Trisetaria cavanillesii
cristata (L.) Hyl. = Rostraria cristata
obtusiflora (Boiss.) Gontsch. = Rostraria obtusiflora
phleoides (Vill.) Reichenb. = Rostraria cristata

Lophopyrum A. Löve = Elytrigia

caespitosum (C. Koch) A. Löve = Elytrigia caespitosa
elongatum (Host) A. Löve = Elytrigia elongata
nodosum (Nevski) A. Löve = Elytrigia nodosa
sinuatum (Nevski) A. Löve = Elytrigia sinuata

Ludolfia auct. = Bambusa

glaucescens Willd. = Bambusa glaucescens

Malacurus Nevski = Leymus

lanatus (Korsh.) Nevski = Leymus lanatus

Melica L.

altissima L. - 1, 2, 3, 4, 6
x aschersonii M. Schulze. - M. nutans L. x M. picta C. Koch
atropatana Schischk. (*M. persica* Kunth subsp. *atropatana* (Schischk.) Tzvel.) - 2, 6
breviflora Boiss. = M. jacquemontii
canescens (Regel) Lavr. (*M. jacquemontii* Decne. subsp. *canescens* (Regel) Bor, *M. persica* Kunth subsp. *canescens* (Regel) P.H. Davis) - 6

chrysolepis Klok. = M. flavescens
ciliata L. (*M. glauca* F. Schultz, *M. simulans* Klok.) - 1
- subsp. *micrantha* (Boiss. & Hohen.) Sojak = M. taurica
- subsp. *monticola* (Prokud.) Tzvel. = M. monticola
- subsp. *taurica* (C. Koch) Tzvel. = M. taurica
- subsp. *transsilvanica* (Schur) Celak. = M. transsilvanica
flavescens (Schur) Simonk. (*M. chrysolepis* Klok., *M. transsilvanica* Schur subsp. *klokovii* Tzvel.) - 1
glauca F. Schultz = M. ciliata
hohenackeri Boiss. (*M. jacquemontii* Decne. subsp. *hohenackeri* (Boiss.) Bor) - 2, 6
inaequiglumis Boiss. (*M. persica* Kunth subsp. *inaequiglumis* (Boiss.) Bor) - 2, 6
jacquemontii Decne. (*M. breviflora* Boiss., *M. persica* Kunth subsp. *jacquemontii* (Decne.) P.H. Davis) - 2, 6
- subsp. *canescens* (Regel) Bor = M. canescens
- subsp. *hohenackeri* (Boiss.) Bor = M. hohenackeri
komarovii Lucznik - 5
micrantha Boiss. & Hohen. = M. taurica
minor Hack. - 2
monticola Prokud. (*M. ciliata* L. subsp. *monticola* (Prokud.) Tzvel., *M. taurica* C. Koch subsp. *monticola* (Prokud.) Prokud.) - 1
nutans L. - 1, 2, 3, 4, 5, 6
- subsp. **amurensis** Probat. - 5
■ penicillaris Boiss. & Bal.
persica Kunth - 6
- subsp. *atropatana* (Schischk.) Tzvel. = M. atropatana
- subsp. *canescens* (Regel) P.H. Davis = M. canescens
- subsp. *inaequiglumis* (Boiss.) Bor = M. inaequiglumis
- subsp. *jacquemontii* (Decne.) P.H. Davis = M. jacquemontii
picta C. Koch - 1, 2
schafkatii Bondar. - 6
schischkinii I. Iljinsk. - 2
secunda Regel - 6
simulans Klok. = M. ciliata
subulata (Griseb.) Scribn. (*Bromus subulatus* Griseb.) - 5
taurica C. Koch (*M. ciliata* L. subsp. *micrantha* (Boiss. & Hohen.) Sojak, *M. ciliata* subsp. *taurica* (C. Koch) Tzvel., *M. micrantha* Boiss. & Hohen., *M. transsilvanica* Schur subsp. *micrantha* (Boiss. & Hohen.) K. Richt.) - 1, 2, 6
- subsp. *monticola* (Prokud.) Prokud. = M. monticola
x thuringiaca Rauschert. - M. ciliata L. x M. trassilvanica Schur
transsilvanica Schur (*M. ciliata* L. subsp. *transsilvanica* (Schur) Celak.) - 1, 2, 3, 4, 6
- subsp. *klokovii* Tzvel. = M. flavescens
- subsp. *micrantha* (Boiss. & Hohen.) K. Richt. = M. taurica
turczaninowiana Ohwi - 4, 5
uniflora Retz. - 1, 2
virgata Turcz. ex Trin. - 4
x weinii Hempel. - M. nutans L. x M. uniflora Retz.

Mibora Adans.

minima (L.) Desv. - 1

Microstegium Nees (*Pollinia* Trin. 1832, non Spreng. 1815)

imberbe (Nees) Tzvel. = M. vimineum
japonicum (Miq.) Koidz. (*Pollinia japonica* Miq., *Microstegium nudum* (Trin.) A. Camus subsp. *japonicum* (Miq.) Tzvel., *M. nudum* auct., *Pollinia nuda* auct.) - 2

nodosum (Kom.) Tzvel. (*Arthraxon nodosus* Kom., *Microstegium vimineum* (Trin.) A. Camus subsp. *nodosum* (Kom.) Tzvel., *Pollinia imberbis* sensu Roshev. p.p.) - 5

nudum auct. = M. japonicum

nudum (Trin.) A. Camus subsp. *japonicum* (Miq.) Tzvel. = M. japonicum

vimineum (Trin.) A. Camus (*Andropogon vimineus* Trin., *Microstegium imberbe* (Nees) Tzvel., *Pollinia imberbis* Nees) - 2

- subsp. *nodosum* (Kom.) Tzvel. = M. nodosum

Milium L.

caerulescens Desf. var. *angustifolium* Regel = Piptatherum angustifolium

effusum L. - 1, 2, 3, 4, 5, 6

- subsp. *schmidtianum* (C. Koch) Tzvel. = M. schmidtianum

kochii Mez = Zingeria kochii

laterale Regel = Piptatherum laterale

schmidtianum C. Koch (*M. effusum* L. subsp. *schmidtianum* (C. Koch) Tzvel.) - 2

transcaucasicum Tzvel. - 2

trichopodum Boiss. = Zingeria trichopoda

vernale Bieb. - 1, 2, 6

- subsp. **alexeenkoi** (Tzvel.) Probat. (*M. vernale* var. *alexeenkoi* Tzvel.) - 2

- subsp. **tzvelevii** Probat. - 6

Miscanthus Anderss. (*Triarrhena* (Maxim.) Nakai)

purpurascens Anderss. (*M. sinensis* Anderss. subsp. *purpurascens* (Anderss.) Tzvel., *M. sinensis* var. *purpurascens* (Anderss.) Matsum., *M. sinensis* f. *purpurascens* (Anderss.) Nakai) - 2(alien), 5

sacchariflorus (Maxim.) Benth. (*Triarrhena sacchariflora* (Maxim.) Nakai) - 5

sinensis Anderss. - 2(alien), 5

- subsp. *purpurascens* (Anderss.) Tzvel. = M. purpurascens

- var. *purpurascens* (Anderss.) Matsum. = M. purpurascens

- f. *purpurascens* (Anderss.) Nakai = M. purpurascens

Molineria Parl. = Molineriella

laevis (Brot.) Hack. = Molineriella laevis

Molineriella Rouy (*Molineria* Parl. 1850, non Colla, 1826)

laevis (Brot.) Rouy (*Aira laevis* Brot., *A. pulchella* Willd., *Deschampsia pulchella* (Willd.) Trin., *Molineria laevis* (Brot.) Hack.) - 1

Molinia Schrank (*Moliniopsis* Hayata)

caerulea (L.) Moench (*M. euxina* Pobed.) - 1, 2, 3, 6

- subsp. *litoralis* (Host) Paul = M. litoralis

euxina Pobed. = M. caerulea

fauriei Hack. = Neomolinia fauriei

japonica Hack. (*Moliniopsis japonica* (Hack.) Hayata) - 5

litoralis Host (*M. caerulea* (L.) Moench subsp. *litoralis* (Host) Paul) - 1, 2

maxima C. Hartm. = Glyceria maxima

Moliniopsis Hayata = Molinia

japonica (Hack.) Hayata = Molinia japonica

Monerma Beauv. (*Hainardia* Greuter, nom. illegit., *Rottboellia* auct.)

cylindrica (Willd.) Coss. & Durieu (*Rottboellia cylindrica* Willd., *Hainardia cylindrica* (Willd.) Greuter) - 1, 2

Muhlenbergia Schreb.

curviaristata (Ohwi) Ohwi (*M. ramosa* (Hack.) Makino var. *curviaristata* Ohwi, *M. tenuiflora* (Willd.) Britt. Sterns & Pogg. subsp. *curviaristata* (Ohwi) T. Koyama & Kawano, *M. ramosa* auct.) - 5

huegelii Trin. (*M. longistolon* Ohwi) - 5, 6

japonica Steud. - 5

longistolon Ohwi = M. huegelii

ramosa auct. = M. curviaristata

ramosa (Hack.) Makino var. *curviaristata* Ohwi = M. curviaristata

schreberi J.F. Gmel. - 2

tenuiflora (Willd.) Britt., Sterns & Pogg. subsp. *curviaristata* (Ohwi) T. Koyama & Kawano = M. curviaristata

Nardurus Reichenb.

elegans Drob. = N. krausei

krausei (Regel) V. Krecz. & Bobr. (*Festuca tenuifolia* Schrad. p.p. excl. typo, *F. tenuiflora* var. *aristata* Koch, *Nardurus elegans* Drob., *N. maritimus* (L.) Murb. subsp. *aristatus* (Koch) Tzvel., *N. maritimus* subsp. *krausei* (Regel) Tzvel., *N. tenellus* subsp. *aristatus* (Koch) Arcang., *N. tenuiflorus* (Schrad.) Boiss. p.p., *N. maritimus* auct.) - 1, 2, 6

maritimus auct. = N. krausei

maritimus (L.) Murb. subsp. *aristatus* (Koch) Tzvel. = N. krausei

- subsp. *krausei* (Regel) Tzvel. = N. krausei

subulatus (Banks & Soland.) Bor = Loliolum subulatum

tenellus subsp. *aristatus* (Koch) Arcang. = N. krausei

tenuiflorus (Schrad.) Boiss. p.p. = N. krausei

Nardus L.

glabriculmis Sakalo = N. stricta

incurva Gouan = Psilurus incurvus

stricta L. (*N. glabriculmis* Sakalo) - 1, 2, 4

Neomolinia Honda (*Diarrhena* auct.)

fauriei (Hack.) Honda (*Molinia fauriei* Hack., *Diarrhena fauriei* (Hack.) Ohwi) - 5

japonica (Franch. & Savat.) Probat. (*Diarrhena japonica* Franch. & Savat.) - 5

koryoensis (Honda) Nakai (*Diarrhena koryoensis* Honda) - 5

mandshurica (Maxim.) Honda (*Diarrhena mandshurica* Maxim.) - 5

*Neoschischkinia Tzvel.

***elegans** (Beauv.) Tzvel. (*Trichodium elegans* Leers ex Beauv., *Agrostis elegans* Thore, 1809, non Salisb. 1796)

***nebulosa** (Boiss. & Reut.) Tzvel. (*Agrostis nebulosa* Boiss. & Reut.)

Nevskiella V. Krecz. & Vved.

gracillima (Bunge) V. Krecz. & Vved. (*Bromus gracillimus* Bunge) - 6

Nowodworskya C. Presl = Polypogon

fugax (Nees ex Steud.) Nevski = Polypogon fugax
semiverticillata (Forssk.) Nevski = Polypogon viridis
verticillata (Vill.) Nevski = Polypogon viridis

Oplismenus Beauv.

burmannii (Retz.) Beauv. - 2
compositus (L.) Beauv. (*Panicum compositum* L.) - 2
undulatifolius (Ard.) Beauv. - 2

Oreochloa Link

disticha (Wulf.) Link (*Poa disticha* Wulf., *Sesleria disticha* (Wulf.) Pers.) - 1

*Oryza L.

***sativa** L. - 1, 2, 5, 6

Oryzopsis auct. = Piptatherum

aequiglumis Hook. fil. = Piptatherum aequiglume
alpestris Grig. = Piptatherum alpestre
angustifolia (Regel) Kitam. = Piptatherum angustifolium
caerulescens auct. = Piptatherum vicarium
ferganensis Litv. = Piptatherum ferganense
holciformis (Bieb.) Hack. = Piptatherum holciforme
kokanica (Regel) Roshev. = Piptatherum kokanicum
kopetdagensis Roshev. = Piptatherum holciforme
lateralis (Regel) Stapf = Piptatherum laterale
latifolia Roshev. = Piptatherum latifolium
molinioides (Boiss.) Hack. = Piptatherum molinioides
pamiroalaica Grig. = Piptatherum pamiroalaicum
platyantha (Nevski) Grig. = Piptatherum platyanthum
purpurascens Hack. = Piptatherum purpurascens
sogdiana Grig. = Piptatherum sogdianum
songarica (Trin. & Rupr.) B. Fedtsch. = Piptatherum songaricum
tianschanica Drob. & Vved. = Piptatherum kokanicum
turcomanica Roshev. = Achnatherum turcomanicum
vicaria Grig. = Piptatherum vicarium
virescens (Trin.) G. Beck = Piptatherum virescens

Panicum L. (*Dichanthelium* (Hitchc. & Chase) Gould)

acroanthum Steud. = P. bisulcatum
acuminatum subsp. *implicatum* (Scribn.) A.A. Beetle = P. implicatum
- var. *lindheimeri* (Nash) A.A. Beetle = P. lindheimeri
adhaerens Forrsk. = Setaria adhaerens
adscendens Kunth = Digitaria ciliaris
aegyptiacum Retz. = Digitaria aegyptiaca
alopecuroides L. = Pennisetum alopecuroides
americanum L. = Pennisetum americanum
***antidotale** Retz. - 6
arenarium Bieb. = Digitaria ischaemum
barbipulvinatum Nash (*P. capillare* L. subsp. *barbipulvinatum* (Nash) Tzvel.) - 1, 2, 5
bisulcatum Thunb. (*P. acroanthum* Steud.) - 5
capillare L. - 1, 2, 5
- subsp. *barbipulvinatum* (Nash) Tzvel. = P. barbipulvinatum
ciliare Retz. = Digitaria ciliaris
coarctatum Stev. ex Trin. = Echinochloa oryzoides
compositum L. = Oplismenus compositus
crusgalli L. subsp. *hostii* (Bieb.) K. Richt. = Echinochloa

oryzoides
dichotomiflorum Michx. - 1, 2, 5
frumentaceum Roxb. = Echinochloa frumentacea
hostii Bieb. = Echinochloa oryzoides
huachucae Ashe - 1
implicatum Scribn. (*P. acuminatum* subsp. *implicatum* (Scribn.) A.A. Beetle) - 1
intermedium (Roem. & Schult.) Roth = Setaria intermedia
ischaemum Schreb. = Digitaria ischaemum
lanuginosum Ell. var. *lindheimeri* (Nash) Fern. = P. lindheimeri
lindheimeri Nash (*Dichanthelium lindheimeri* (Nash) Gould, *Panicum acuminatum* var. *lindheimeri* (Nash) A.A. Beetle, *P. lanuginosum* Ell. var. *lindheimeri* (Nash) Fern.) - 2
lutescens Weig. = Setaria pumila
miliaceum L. - 1, 2, 3, 4, 5, 6
- subsp. **ruderale** (Kitag.) Tzvel. (*P. miliaceaum* var. *ruderale* Kitag.) - 3, 4, 5
miliare auct. = P. sumatrense
oryzoides Ard. = Echinochloa oryzoides
pachystachys Franch. & Savat. = Setaria pachystachys
phyllopogon Stapf = Echinochloa phyllopogon
pycnocomum Steud. = Setaria pycnocoma
sanguinale L. subsp. *aegyptiacum* (Retz.) K. Richt. = Digitaria aegyptiaca
sumatrense Roth ex Roem. & Schult. (*Panicum miliare* auct.) - 2
tomentosum Roxb. = Setaria intermedia
***virgatum** L. - 1, 6
viride L. var. *giganteum* Franch. & Savat. = Setaria pycnocoma

Pappagrostis Roshev. (*Stephanachne* Keng, p.p. excl. typo)

pappophorea (Hack.) Roshev. (*Stephanachne pappophorea* (Hack.) Keng) - 6

Pappophorum auct. = Enneapogon

boreale Griseb. = Enneapogon borealis
persicum (Boiss.) Steud. = Enneapogon persicus
squarrosum Banks & Soland. = Boissiera squarrosa

Paracolpodium (Tzvel.) Tzvel.

altaicum (Trin.) Tzvel. (*Colpodium altaicum* Trin.) - 3, 4, 6
- subsp. *leucolepis* (Nevski) Tzvel. = P. leucolepis
colchicum (Albov) Tzvel. (*Colpodium colchicum* (Albov) Woronow) - 2
leucolepis (Nevski) Tzvel. (*Colpodium leucolepis* Nevski, *Paracolpodium altaicum* (Trin.) Tzvel. subsp. *leucolepis* (Nevski) Tzvel.) - 6

Parapholis C.E. Hubb. (*Lepidurus* Janch. nom. invalid., *Lepturus* Trin. 1820, non R. Br. 1810)

incurva (L.) C.E. Hubb. (*Lepidurus incurvus* (L.) Janch. nom. invalid., *Lepturus incurvus* (L.) Druce, *Pholiurus incurvus* (L.) Schinz & Thell.) - 1, 2, 6
- subsp. *longiflora* (Grossh.) Tzvel. = P. longiflora
longiflora (Grossh.) Musajev (*Lepturus incurvatus* (L.) Trin. var. *longiflorus* Grossh., *Parapholis incurva* (L.) C.E. Hubb. subsp. *longiflora* (Grossh.) Tzvel.) - 2, 6

Paspalum L.

chinense Nees = Digitaria violascens
digitaria Poir. = P. paspalodes
dilatatum Poir. - 2, 6
distichum auct. = P. paspalodes
paspalodes (Michx.) Scribn. (*Digitaria paspalodes* Michx., *Paspalum digitaria* Poir., *P. distichum* auct.) - 1, 2, 6
scrobiculatum auct. = P. thunbergii
setaceum Michx. - 2
thunbergii Kunth ex Steud. (*P. scrobiculatum* auct.) - 2, 5(alien)

Patropyrum A. Löve = Aegilops

tauschii (Coss.) A. Löve = Aegilops tauschii
- subsp. *salinum* (Zhuk.) A. Löve = Aegilops tauschii
- subsp. *strangulatum* (Eig) A.Löve = Aegilops tauschii subsp. strangulata

Pennisetum Rich.

alopecuroides (L.) Spreng. (*Panicum alopecuroides* L., *Pennisetum villosum* auct.) - 2
americanum (L.) Leeke (*Panicum americanum* L., *Pennisetum americanum* (L.) Schumann, nom. invalid., *P. glaucum* (L.) R. Br. p.p. quoad pl.) - 1, 2, 6
- convar. **spicatum** (L.) Tzvel. (*Holcus spicatus* L., *Pennisetum americanum* subsp. *spicatum* (L.) Maire & Weiller, *Pennisetum spicatum* (L.) Koern.)
- convar. **typhoides** (Burm. fil.) Tzvel. (*Alopecurus typhoides* Burm. fil., *Pennisetum americanum* subsp. *typhoideum* (Rich.) Maire & Weiller, *P. typhoides* (Burm. fil.) Stapf & C.E. Hubb., *P. typhoideum* Rich.)
- subsp. *spicatum* (L.) Maire & Weiller = P. americanum convar. spicatum
- subsp. *typhoideum* (Rich.) Maire & Weiller = P. americanum convar. typhoides
centrasiaticum Tzvel. - 6
flaccidum Griseb. - 6
glaucum (L.) R. Br. p.p. = P. americanum
orientale Rich. (*P. setaceum* (Forrsk.) Chiov. subsp. *orientale* (Rich.) Maire) - 2, 6
setaceum (Forrsk.) Chiov. subsp. *orientale* (Rich.) Maire = P. orientale
spicatum (L.) Koern. = P. americanum convar. spicatum
typhoides (Burm. fil.) Stapf & C.E. Hubb. = P. americanum convar. typhoides
typhoideum Rich = P. americanum convar. typhoides
villosum auct. = P. alopecuroides

Pentatherum Nabel. = Agrostis

agrostidiforme (Roshev.) Nevski = Agrostis agrostidiformis
angrenicum Butk. = Agrostis angrenica
balansae (Boiss.) Nevski = Agrostis balansae
hissaricum (Nevski) Nevski = Agrostis nevskii
olympicum (Boiss.) Nabel. = Agrostis olympica
ruprechtii (Nevski) Nevski = Agrostis olympica
trichanthum (Schischk.) Nevski = Agrostis trichantha

Phalaris L.

aquatica L. (*Ph. tuberosa* L., *Ph. bulbosa* auct., *Ph. truncata* auct.) - 2
brachystachys Link - 2
bulbosa auct. = Ph. aquatica
canariensis L. - 1, 2, 3, 4, 5
hispida Thunb. = Arthraxon hispidus

japonica Steud. = Phalaroides japonica
minor Retz - 1(?), 2, 5(alien), 6
paradoxa L. - 1(alien), 2, 5(alien), 6
semiverticillata Forssk. = Polypogon viridis
subulata Savi = Phleum subulatum
tenuis Host = Phleum subulatum
truncata auct. = Ph. aquatica
tuberosa L. = Ph. aquatica

Phalaroides N.M. Wolf (*Digraphis* Trin., *Typhoides* Moench)

arundinacea (L.) Rauschert (*Digraphis arundinacea* (L.) Trin., *Typhoides arundinacea* (L.) Moench) - 1, 2, 3, 4, 5, 6
- subsp. *japonica* (Steud.) Tzvel. = Ph. japonica
japonica (Steud.) Czer. (*Phalaris japonica* Steud., *Phalaroides arundinacea* (L.) Rauschert subsp. *japonica* (Steud.) Tzvel., *Typhoides arundinacea* (L.) Moench subsp. *japonica* (Steud.) Tzvel.) - 4, 5

Phippsia (Trin.) R. Br.

algida (Soland.) R. Br. (*Ph. foliosa* V. Vassil.) - 1, 3, 4, 5
- subsp. *algidiformis* (H. Smith) A. & D. Löve = Ph. x algidiformis
- subsp. *concinna* (Th. Fries) A. & D. Löve = Ph. concinna
- subsp. *concinna* (Th. Fries) K. Richt. = Ph. concinna
x **algidiformis** (H. Smith) Tzvel. (*Ph. concinna* (Th. Fries) Lindeb. subsp. *algidiformis* H. Smith, *Ph. algida* (Soland.) R. Br. subsp. *algidiformis* (H. Smith) A. & D. Löve). - Ph. algida (Soland.) R. Br. x Ph. concinna (Th. Fries) Lindeb.
angustata (R. Br.) A. & D. Löve = Puccinellia angustata
- subsp. *palibinii* (Sorensen) A. & D. Löve = Puccinellia palibinii
borealis (Swall.) A. & D. Löve = Puccinellia borealis
- subsp. *neglecta* (Tzvel.) A. & D. Löve = Puccinella neglecta
capillaris (Liljebl.) A. & D. Löve = Puccinellia capillaris
- subsp. *pulvinata* (Fries) A. & D. Löve = Puccinellia pulvinata
coarctata (Fern. & Weath.) A. Löve = Puccinellia coarctata
concinna (Th. Fries) Lindeb. (*Ph. algida* (Soland.) R. Br. subsp. *concinna* (Th. Fries) A. & D. Löve, comb. superfl., *Ph. algida* subsp. *concinna* (Th. Fries) K. Richt.) - 1, 3, 4, 5
- subsp. *algidiformis* H. Smith = Ph. x algidiformis
distans (Jacq.) A. & D. Löve = Puccinellia distans
foliosa V. Vassil. = Ph. algida
fragiliflora (Sorensen) A. & D. Löve = Puccinellia palibinii
gorodkovii (Tzvel.) A. & D. Löve = Puccinellia gorodkovii
hauptiana (V. Krecz.) A. & D. Löve = Puccinellia hauptiana
langeana (Berlin) A. & D. Löve = Puccinellia langeana
- subsp. *alascana* (Scribn. & Merr.) A. & D. Löve = Puccinellia alascana
- subsp. *asiatica* (Sorensen) A. & D. Löve = Puccinellia tenella
lenensis (Holmb.) A. & D. Löve = Puccinellia lenensis
limosa (Schur) A. & D. Löve = Puccinellia limosa
maritima (Huds.) A. Löve = Puccinellia maritima
nutkaensis (C. Presl) A. & D. Löve subsp. *borealis* (Holmb.) A. & D. Löve = Puccinellia coarctata
phryganodes (Trin.) A. & D. Löve = Puccinellia phryganodes
sibirica (Holmb.) A. & D. Löve = Puccinellia sibirica

tenella (Lange) A. & D. Löve = Puccinellia tenella
vaginata (Lange) A. & D. Löve = Puccinellia vaginata
vahliana (Liebm.) A. & D. Löve = Puccinellia vahliana
- subsp. *byrrangensis* (Tzvel.) A. & D. Löve = Puccinellia byrrangensis
- subsp. *colpodioides* (Tzvel.) A. & D. Löve = Puccinellia colpodioides
- subsp. *jenisseiensis* (Roshev.) A. & D. Löve = Puccinellia jenisseiensis
vilfoidea (Anderss.) A. & D. Löve = Puccinellia vilfoidea
- subsp. *beringensis* A. & D. Löve = Puccinellia vilfoidea subsp. beringensis
wrightii (Scribn. & Merr.) A. & D. Löve = Puccinellia wrightii

Phleum L.

alpinum L. (*Ph. alpinum* subsp. *commutatum* (Gaudin) Hult. comb. superfl., *Ph. alpinum* subsp. *commutatum* (Gaudin) Malagarriga, comb. superfl., *Ph. alpinum* subsp. *commutatum* (Gaudin) K. Richt., *Ph. commutatum* Gaudin) - 1, 2, 3, 4, 5, 6
- subsp. *commutatum* (Gaudin) Hult. = Ph. alpinum
- subsp. *commutatum* (Gaudin) Malagarriga = Ph. alpinum
- subsp. *commutatum* (Gaudin) K. Richt. = Ph. alpinum
ambiguum Ten. (*Ph. hirsutum* Honck. subsp. *ambiguum* (Ten.) Cif. & Giac., *Ph. hirsutum* subsp. *ambiguum* (Ten.) Tzvel. comb. superfl., *Ph. michelii* All. subsp. *ambiguum* (Ten.) Arcang., *Ph. michelii* sensu Ovcz.) - 2
annuum Bieb. = Ph. paniculatum
arenarium L. - 1
asperum Jacq. subsp. *annuum* (Bieb.) K. Richt. = Ph. paniculatum
- var. *ciliatum* Boiss. = Ph. paniculatum
bertolonii DC. = Ph. nodosum
boissieri auct. = Ph. paniculatum
commutatum Gaudin = Ph. alpinum
echinatum Host - 1(?)
graecum auct. = Ph. himalaicum
himalaicum Mez (*Ph. graecum* auct.) - 6
hirsutum Honck. (*Ph. michelii* All.) - 1
- subsp. *ambiguum* (Ten.) Cif. & Giac. = Ph. ambiguum
- subsp. *ambiguum* (Ten.) Tzvel. = Ph. ambiguum
laeve Bieb. = Ph. phleoides
michelii All. = Ph. hirsutum
- subsp. *ambiguum* (Ten.) Arcang. = Ph. ambiguum
michelii sensu Ovcz. = Ph. ambiguum
montanum C. Koch (*Ph. phleoides* (L.) Karst. subsp. *montanum* (C. Koch) Tzvel.) - 1, 2
nodosum L. (*Ph. bertolonii* DC., *Ph. pratense* L. subsp. *bertolonii* (DC.) Bornm., *Ph. pratense* subsp. *bertolinii* (DC.) I. Serbanescu & E.I. Nyarady, comb. superfl., *Ph. pratense* subsp. *nodosum* (L.) Arcang.) - 1, 2, 3, 4, 6
paniculatum Huds. (*Ph. annuum* Bieb., *Ph. asperum* Jacq. subsp. *annuum* (Bieb.) K. Richt., *Ph. asperum* var. *ciliatum* Boiss., *Ph. paniculatum* subsp. *annuum* (Bieb.) Holub., *Ph. paniculatum* subsp. *ciliatum* (Boiss.) M. Dogan, *Ph. boissieri* auct.) - 1, 2, 6
- subsp. *annuum* (Bieb.) Holub. = Ph. paniculatum
- subsp. *ciliatum* (Boiss.) M. Dogan = Ph. paniculatum
phleoides (L.) Karst. (*Ph. laeve* Bieb. nom. illegit.) - 1, 2, 3, 5(alien), 6
- subsp. *montanum* (C. Koch) Tzvel. = Ph. montanum
pratense L. - 1, 2, 3, 4, 5, 6
- subsp. *bertolonii* (DC.) Bornm. = Ph. nodosum
- subsp. *bertolinii* (DC.) I. Serbanescu & E.I. Nyarady = Ph. nodosum

- subsp. *nodosum* (L.) Arcang. = Ph. nodosum
- subsp. *roshevitzii* (Pavl.) Tzvel. = Ph. roshevitzii
roshevitzii Pavl. (*Ph. pratense* L. subsp. *roshevitzii* (Pavl.) Tzvel.) - 6
subulatum (Savi) Aschers. & Graebn. (*Phalaris subulata* Savi, *Ph. tenuis* Host, *Phleum tenue* (Host) Schrad.) - 1
tenue (Host) Schrad. = Ph. subulatum
tzvelevii Dubovik - 2

Pholiurus Trin.

glabriglumis Nevski = Henrardia glabriglumis
incurvus (L.) Schinz & Thell. = Parapholis incurva
pannonicus (Host) Trin. - 1, 2, 3, 6
persicus (Boiss.) A. Camus = Henrardia persica

Phragmites Adans.

altissimus (Benth.) Nabille (*Arundo altissima* Benth., *A. isiaca* Delile, nom. illegit., ? *A. maxima* Forssk. nom. dubium, *Phragmites australis* (Cav.) Trin. ex Steud. subsp. *altissimus* (Benth.) W. Clayton, *Ph. australis* subsp. *pseudodonax* (Rabenh.) Rauschert, ? *Ph. communis* Trin. subsp. *maximus* (Forssk.) W. Clayton, *Ph. communis* var. *pseudodonax* Rabenh., *Ph. communis* subsp. *pseudodonax* (Rabenh.) Rothm. ex Tzvel., *Ph. isiacus* (Delile) Kunth, nom. illegit., ? *Ph. maximus* (Forssk.) Chiov. p.p.) - 1, 2, 4, 5, 6
australis (Cav.) Trin. ex Steud. (*Arundo australis* Cav., *Phragmites communis* Trin.) - 1, 2, 3, 4, 5, 6
- subsp. *altissimus* (Benth.) W. Clayton = Ph. altissimus
- subsp. *pseudodonax* (Rabenh.) Rauschert = Ph. altissimus
communis Trin. = Ph. australis
- subsp. *maximus* (Forssk.) W. Clayton = Ph. altissimus
- subsp. *pseudodonax* (Rabenh.) Rothm. ex Tzvel. = Ph. altissimus
- var. *pseudodonax* Rabenh. = Ph. altissimus
isiacus (Delile) Kunth = Ph. altissimus
japonicus Steud. (*Ph. serotinus* Kom.) - 5
maximus (Forssk.) Chiov. = Ph. altissimus
serotinus Kom. = Ph. japonicus

*Phyllostachys Siebold & Zucc.

*****aurea** A. & C. Riviere (*Ph. meyeri* McClure var. *aurea* (A. & C. Riviere) Pilipenko) - 2
*****bambusoides** Siebold & Zucc. (*Ph. castillonis* Marliac ex Carr., *Ph. marliacea* Mitf., *Ph. sulfurea* A. & C. Riviere) - 1, 2
castillonis Marliac ex Carr. = Ph. bambusoides
edulis Houz. = Ph. pubescens
*****flexuosa** A. & C. Riviere - 2
henonis Mitf. = Ph. nigra
heterocycla (Carr.) Mitf. = Ph. pubescens
marliacea Mitf. = Ph. bambusoides
meyeri McClure var. *aurea* (A. & C. Riviere) Pilipenko = Ph. aurea
mitis (Carr.) A. & C. Riviere, p.p. = Ph. viridis
*****nigra** (Lodd.) Munro (*Bambusa nigra* Lodd., *Phyllostachys henonis* Mitf., *Ph. puberula* (Miq.) Munro) - 2
puberula (Miq.) Munro = Ph. nigra
*****pubescens** Mazel ex Houz. (*Ph. edulis* Houz., *Ph. heterocycla* (Carr.) Mitf.) - 2
sulfurea A. & C. Riviere = Ph. bambusoides
- var. *viridis* Young = Ph. viridis
*****viridi-glaucescens** (Carr.) A. & C. Riviere (*Bambusa viridi-glaucescens* Carr.) - 2

*viridis (Young) McClure (*Ph. sulfurea* A. & C. Riviere var. *viridis* Young, *Ph. mitis* (Carr.) A. & C. Riviere, p.p. quoad pl.) - 1, 2

Piptatherum Beauv. (*Oryzopsis* auct.)

aequiglume (Hook. fil.) Roshev. (*Oryzopsis aequiglumis* Hook. fil.) - 6

alpestre (Grig.) Roshev. (*Oryzopsis alpestris* Grig., *Piptatherum laterale* (Regel) Munro ex Nevski subsp. *alpestre* (Grig.) Freitag) - 6

angustifolium (Regel) Munro ex Boiss. (*Milium caerulescens* Desf. var. *angustifolium* Regel, *Oryzopsis angustifolia* (Regel) Kitam.) - 6

badachshanicum (Tzvel.) Ikonn. (*P. hilariae* Pazij subsp. *badachschanicum* Tzvel.) - 6

binabium Pazij = P. hilariae

caerulescens auct. = P. vicarium

fedtschenkoi Roshev. (*P. sogdianum* (Grig.) Roshev. subsp. *fedtschenkoi* (Roshev.) Tzvel.) - 6

ferganense (Litv.) Roshev. ex E. Nikit. (*Oryzopsis ferganensis* Litv.) - 6

grigorjevii Tzvel. - 6

hilariae Pazij (*P. binabium* Pazij, *P. schugnanicum* Roshev., *P. tremuloides* Ovcz. & Czuk.) - 6

- subsp. *badachschanicum* Tzvel. = P. badachschanicum

hissaricum (Tzvel.) Czer. (*P. sogdianum* (Grig.) Roshev. subsp. *hissaricum* Tzvel.) - 6

holciforme (Bieb.) Roem. & Schult. (*Oryzopsis holciformis* (Bieb.) Hack., *O. kopetdagensis* Roshev., *Piptatherum karataviense* Roshev.) - 1, 2, 6

karataviense Roshev. = P. holciforme

kokanicum (Regel) Nevski (*Oryzopsis kokanica* (Regel) Roshev., *O. tianschanica* Drob. & Vved., *Piptatherum songaricum* (Trin. & Rupr.) Roshev. ex E. Nikit. var. *kokanicum* (Regel) Roshev., *P. songaricum* subsp. *tianschanicum* (Drob. & Vved.) Tzvel., *P. tianschanicum* (Drob. & Vved.) Roshev. ex E. Nikit.) - 6

laterale (Regel) Munro ex Nevski (*Milium laterale* Regel, *Oryzopsis lateralis* (Regel) Stapf, *Piptatherum laterale* (Regel) Roshev. comb. superfl.) - 6

- subsp. *alpestre* (Grig.) Freitag = P. alpestre

latifolium (Roshev.) Nevski (*Oryzopsis latifolia* Roshev.) - 6

molinioides Boiss. (*Oryzopsis molinioides* (Boiss.) Hack.) - 2

pamiroalaicum (Grig.) Roshev. ex E. Nikit. (*Oryzopsis pamiroalaica* Grig.) - 6

platyanthum Nevski (*Oryzopsis platyantha* (Nevski) Grig.) - 6

purpurascens (Hack.) Roshev. (*Oryzopsis purpurascens* Hack.) - 6

roshevitsianum Tzvel. - 6

schugnanicum Roshev. = P. hilariae

sogdianum (Grig.) Roshev. (*Oryzopsis sogdiana* Grig.) - 6

- subsp. *fedtschenkoi* (Roshev.) Tzvel. = P. fedtschenkoi

- subsp. *hissaricum* Tzvel. = P. hissaricum

songaricum (Trin. & Rupr.) Roshev. ex E. Nikit. (*Oryzopsis songarica* (Trin. & Rupr.) B. Fedtsch.) - 3, 6

- subsp. *tianschanicum* (Drob. & Vved.) Tzvel. = P. kokanicum

- var. *kokanicum* (Regel) Roshev. = P. kokanicum

tianschanicum (Drob. & Vved.) Roshev. ex E. Nikit. = P. kokanicum

tremuloides Ovcz. & Czuk. = P. hilariae

vicarium (Grig.) Roshev. ex E. Nikit. (*Oryzopsis vicaria* Grig., *O. caerulescens* auct., *Piptatherum caerulescens* auct.) - 6

virescens (Trin.) Boiss. (*Oryzopsis virescens* (Trin.) G. Beck) - 1, 2

*Pleioblastus Nakai (*Arundinaria* auct.)

*distichus (Mitf.) Nakai (*Bambusa disticha* Mitf.) - 2

*hindsii (Munro) Nakai (*Arundinaria hindsii* Munro) - 2

*humilis (Mitf.) Nakai (*Arundinaria humilis* Mitf.) - 2

*pumilus (Mitf.) Nakai (*Arundinaria pumila* Mitf.) - 2

*simonii (Carr.) Nakai (*Bambusa simonii* Carr., *Arundinaria simonii* (Carr.) A. & C. Riviere) - 1, 2

*variegatus (Miq.) Makino (*Bambusa variegata* Miq.) - 2

*viridi-striatus (Siebold ex Andre) Makino (*Bambusa viridi-striata* Siebold ex Andre) - 2

Pleuroplitis Trin. = Arthraxon

caucasica (Rupr. ex Regel) Trautv. = Arthraxon caucasicus

centrasiatica Griseb. = Arthraxon centrasiaticus

langsdorffii Trin. var. *caucasica* Rupr. ex Regel = Arthraxon caucasicus

Pleuropogon R. Br.

sabinii R. Br. - 1, 3, 4, 5

Poa L.

abbreviata R. Br. - 1, 4, 5

- subsp. *jordalii* (A. Pors.) Hult. = P. jordalii

- subsp. *jordalii* (A. Pors.) Tzvel. = P. jordalii

- var. *jordalii* (A. Pors.) Probat. = P. jordalii

acroleuca sensu Worosch. = P. ussuriensis

acroleuca Steud. - 5

acuminata Ovcz. = P. fragilis

acuticaulis Ovcz. & Czuk. = P. bucharica

aksuensis (Tzvel.) Czer. (*P. bucharica* Roshev. subsp. *aksuensis* Tzvel., *P. aksuensis* Roshev. nom. invalid.) - 6

alberti Regel = P. attenuata

albida Turcz. ex Trin. = Festuca sibirica

alexeenkoi (Tzvel.) Czer. (*P. nemoralis* L. subsp. *alexeenkoi* Tzvel.) - 2

alexeji Sofeikova & Worosch. = Arctopoa eminens

almasovii Golub - 5

alpigena (Blytt) Lindm. (*P. pratensis* L. var. *alpigena* Blytt, *P. angustifolia* L. var. *alpigena* (Blytt) Worosch., *P. pratensis* subsp. *alpigena* (Blytt) Hiit.) - 1, 3, 4, 5

- subsp. **colpodea** (Th. Fries) Jurtz. & Petrovsky (*P. stricta* Lindeb. subsp. *colpodea* Th. Fries, *P. pratensis* subsp. *colpodea* (Th. Fries) Tzvel., *P. rigens* C. Hartm. subsp. *colpodea* (Th. Fries) D. Löve) - 1, 3, 4, 5

alpina auct. = P. badensis

alpina L. (*P. elbrussica* Timpko) - 1, 2, 3, 4, 5, 6

- subsp. *badensis* (Haenke) Arcang. = P. badensis

- subsp. *vivipara* (L.) Arcang. = P. vivipara

- subsp. *vivipara* (L.) Tzvel. = P. vivipara

- var. *vivipara* L. = P. vivipara

altaica Trin. (*P. tristis* Trin.) - 3, 4, 6

anadyrica Roshev. = P. glauca

angustifolia L. (*P. pratensis* L. subsp. *angustifofia* (L.) Arcang., *P. pratensis* subsp. *angustifolia* (L.) Lindb. fil. comb. superfl., *P. setacea* Hoffm., *P. strigosa* Hoffm., *P. viridula* Palib.) - 1, 2, 3, 4, 5, 6

- subsp. *humilis* (Ehrh. ex Smith) K. Richt. = P. subcaerulea

- var. *alpigena* (Blytt) Worosch. = P. alpigena

- var. *angustiglumis* (Roshev.) Worosch. = P. pratensis

- var. *sublanata* (Reverd.) Worosch. = P. sublanata

angustiglumis Roshev. = P. pratensis

annua L. - 1, 2, 3, 5, 6

araratica Trautv. (*P. pseudosterilis* Galkin, *P. versicolor* Bess. subsp. *araratica* (Trautv.) Tzvel.) - 2, 6

arctica R. Br. (? *P. debilis* V. Vassil. 1940, non Thuill. 1799, *P. petschorica* Roshev.) - 1, 3, 4, 5
- subsp. *caespitans* Nannf. = P. tolmatchewii
- subsp. **longiculmis** Hult. - 5
- subsp. *smirnowii* (Roshev.) Malysch. = P. smirnowii
- subsp. *stricta* Nannf. = P. lindebergii
- subsp. **williamsii** (Nash) Hult. (*P. williamsii* Nash) - 5
arctostepporum Jurtz.& Probat. - 5
argunensis Roshev. (*P. attenuata* Trin. subsp. *argunensis* (Roshev.) Tzvel.) - 4
articulata Ovcz. = P. pratensis
arundinacea Bieb. = Glyceria arundinacea
attenuata Trin. (*P. alberti* Regel, *P. dahurica* Trin., *P. juldusicola* Regel, *Sesleria pavlovii* Litv.) - 3, 4, 6
- subsp. *argunensis* (Roshev.) Tzvel. = P. argunensis
- subsp. *botryoides* (Trin. ex Griseb.) Tzvel. = P. botryoides
- var. *botryoides* (Trin. ex Griseb.) Tzvel. = P. botryoides
bactriana Roshev. - 6
- subsp. *drobovii* Tzvel. = P. drobovii
- subsp. *glabriflora* (Roshev. ex Ovcz.) Tzvel. = P. glabriflora
- subsp. *zaprjagajevii* (Ovcz.) Tzvel. = P. zapjagajevii
badachschanica Ikonn. = P. litvinoviana
badensis Haenke (*P. alpina* L. subsp. *badensis* (Haenke) Arcang., *P. imeretica* Somm. & Levier, *P. alpina* auct.) - 2
bakuensis Litv. ex Roshev. = Festuca drymeja
balfourii auct. = P. nemoralis subsp. carpatica
barguzinensis M. Pop. = P. paucispicula
bedeliensis Litv. = P. lipskyi
beringiana Probat. (*P. macrocalyx* Trautv. & C.A. Mey. var. *beringiana* (Probat.) Worosch.) - 5
biebersteinii H. Pojark. (*P. sterilis* Bieb. subsp. *biebersteinii* (H. Pojark.) Tzvel.) - 1
bifida Frohner = P. supina
botryoides (Trin. ex Griseb.) Kom. (*P. attenuata* Trin. subsp. *botryoides* (Trin. ex Griseb.) Tzvel., *P. attenuata* var. *botryoides* (Trin. ex Griseb.) Tzvel.) - 3, 4, 5
brachyanthera Hult. = P. pseudoabbreviata
bracteosa Kom. = P. malacantha
- var. *penicillata* (Kom.) Worosch. = P. platyantha
- var. *plathyantha* (Kom.) Worosch. = P. platyantha
bryophila Trin. = P. glauca
bucharica Roshev. (*P. acuticaulis* Ovcz. & Czuk.) - 6
- subsp. *aksuensis* Tzvel. = P. aksuensis
- subsp. *karateginensis* (Roshev. ex Ovcz.) Tzvel. = P. karateginensis
bulbosa L. (*P. bulbosa* subsp. *pseudoconcinna* (Schur) Jir., *P. pseudoconcinna* Schur) - 2, 3, 6
- subsp. *crispa* (Thuill.) Tzvel. = P. crispa
- subsp. *delicatula* (Tzvel.) Tzvel. = P. delicatula
- subsp. *nevskii* (Roshev. ex Ovcz.) Tzvel. = P. nevskii
- subsp. *pseudoconcinna* (Schur) Jir. = P. bulbosa
- subsp. *sinaica* (Steud.) Tzvel. = P. sinaica
- subsp. *vivipara* (Koel.) Arcang. = P. crispa
- var. *vivipara* Koel. = P. crispa
caesia Smith = P. glauca
calliopsis Litv. ex Ovcz. (? *P. phariana* Bor) - 6
caucasica Trin. (*P. naltchikensis* Roshev.) - 2
cenisia auct. = P. deylii
chaixii Vill. - 1, 2
cilianensis All. = Eragrostis cilianensis
ciliatiflora Roshev. = Arctopoa tibetica
compressa L. - 1, 2, 3(alien), 4(alien), 5(alien)
contracta Ovcz. & Czuk. = P. lipskyi
crispa Thuill. (*P. bulbosa* L. var. *vivipara* Koel., *P. bulbosa* subsp. *crispa* (Thuill.) Tzvel., *P. bulbosa* subsp. *vivipara*

(Koel.) Arcang.) - 1, 2, 3, 6
curvula Schrad. = Eragrostis curvula
czernjajevii Prokud. = P. subcaerulea
dahurica Trin. = P. attenuata
debilis V. Vassil. = P. arctica
delicatula Wilhelms ex Tzvel. (*P. bulbosa* L. subsp. *delicatula* (Tzvel.) Tzvel.) - 2
densa Troitzk. - 2, 6
densissima Roshev. ex Ovcz. = P. litvinoviana
deylii Chrtek & Jir. (*P. cenisia* auct., *P. granitica* auct.) - 1
diarrhena Schult. & Schult. fil. = Diandrochloa diarrhena
disjecta Ovcz. (*P. lubrica* Ovcz.) - 6
dispansa Ovcz. = P. lipskyi
distans Jacq. = Puccinellia distans
disticha Wulf. = Oreochloa disticha
drobovii (Tzvel.) Czer. (*P. bactriana* Roshev. subsp. *drobovii* Tzvel., ? *P. spicata* Drob. 1941, non L. 1767) - 6
dschungarica Roshev. (*P. lipskyi* Roshev. subsp. *dschungarica* (Roshev.) Tzvel.) - 6
dshilgensis Roshev. = P. timoleontis var. dshilgensis
eduardii Golub = P. platyantha
elbrussica Timpko = P. alpina
eligulata Pavl. = P. korshunensis
eminens J. Presl = Arctopoa eminens
- var. *alexeji* (Sofeikova & Worosch.) Worosch. = Arctopoa eminens
erythropoda Klok. (*P. versicolor* Bess. subsp. *erythropoda* (Klok.) Tzvel.) - 1
evenkiensis Reverd. = P. glauca
extremiorientalis Ohwi (*P. glauca* Vahl subsp. *extremiorientalis* (Ohwi) Worosch., *P. pseudoattenuata* Probat.) - 4(?), 5
fagetorum P. Smirn. (*P. longifolia* Trin. subsp. *fagetorum* (P. Smirn.) Tzvel.) - 1
fedtschenkoi Roshev. = Arctopoa tibetica
x **figertii** Gerh. - P. nemoralis L. x P. compressa L.
filiculmis Roshev. - 4, 5
filipes Lange = P. tolmatchewii
flavidula Kom. = P. leptocoma
fragilis Ovcz. (*P. acuminata* Ovcz. 1933, non Scribn. 1896) - 6
ganeschinii Roshev. = P. glauca
garanica Ikonn. = P. pratensis
glabriflora Roshev. ex Ovcz. (*P. bactriana* Roshev. subsp. *glabriflora* (Roshev. ex Ovcz.) Tzvel.) - 6
glauca Schkuhr = Koeleria glauca
glauca Vahl (*P. anadyrica* Roshev., *P. bryophila* Trin., *P. caesia* Smith, *P. evenkiensis* Reverd., *P. ganeschinii* Roshev., *P. soczawae* Roshev., *P. turczaninovii* Serg., *P. udensis* Trautv. & C.A. Mey., *P. kenteica* auct.) - 1, 2, 3, 4, 5, 6
- subsp. *extremiorientalis* (Ohwi) Worosch. = P. extremiorientalis
- subsp. *litvinoviana* (Ovcz.) Tzvel. = P. litvinoviana
- subsp. *reverdattoi* (Roshev.) Tzvel. = P. reverdattoi
- var. *pekulnejensis* (Jurtz. & Tzvel.) Probat. = P. pekulnejensis
glauciculmis Ovcz. = P. litvinoviana
glumaris Trin. var. *laevigata* Trautv. = Arctopoa trautvetteri
gorbunovii Ovcz. = Puccinellia subspicata
granitica auct. = P. deylii
grisea Korotky = P. tianschanica
hartzii Gand. - 5
- var. *vrangelica* (Tzvel.) Probat. = P. vrangelica
x **herjedalica** H. Smith. - P. alpina L. x P. pratensis L.
hirta Thunb. = Arundinella hirta
hispidula Vasey = P. macrocalyx

hissarica Roshev. ex Ovcz. (*P. laudanensis* Roshev. ex Ovcz.) - 6
hybrida Gaudin - 1
hybrida sensu Roshev. = P. iberica
hypanica Prokud. (*P. nemoralis* L. subsp. *hypanica* (Prokud.) Tzvel.) - 1
iberica Fisch. & C.A. Mey. (*P. hybrida* sensu Roshev.) - 2
imeretica Somm. & Levier = P. badensis
infirma Kunth - 1(alien), 2, 6
insignis Litv. ex Roshev. (*P. sibirica* Roshev. subsp. *uralensis* Tzvel., *P. sibirica* var. *insignis* (Litv. ex Roshev.) Serg.) - 1, 3, 4, 6
x **intricata** Wein. - P. nemoralis L. x P. palustris L.
ircutica Roshev. - 4
irrigata Lindm. = P. subcaerulea
janczewskii Zapal. = P. palustris
jordalii A. Pors. (*P. abbreviata* R. Br. subsp. *jordalii* (A. Pors.) Hult., *P. abbreviata* subsp. *jordalii* (A. Pors.) Tzvel. comb. superfl., *P. abbreviata* var. *jordalii* (A. Pors.) Probat.) - 5
juldusicola Regel = P. attenuata
kamczatensis Probat. - 5
karateginensis Roshev. ex Ovcz. (*P. bucharica* Roshev. subsp. *karateginensis* (Roshev. ex Ovcz.) Tzvel.) - 6
kenteica auct. = P. glauca
ketoiensis Tatew. & Ohwi = P. macrocalyx
koelzii auct. = P. rangkulensis
koksuensis Golosk. - 6
kolymensis Tzvel. - 5
komarovii Roshev. = P. malacantha
korshunensis Golosk. (*P. eligulata* Pavl. 1949, non Hack. 1902, *P. nemoralis* L. subsp. *korshunensis* (Golosk.) Tzvel.) - 3, 6
krylovii Reverd. - 3, 4
kungeica Golosk. = P. lipskyi
kurilensis Hack. = Arctopoa eminens
laestadii Rupr. = Arctophila fulva
lanata Scribn. & Merr. (*P. petraea* Trin. ex Kom.) - 5
- var. *trivialiformis* (Kom.) Worosch. = P. trivialiformis
lanatiflora Roshev. = Hyalopoa lanatiflora
lapidosa Drob. = P. litvinoviana
lapponica Prokud. (*P. nemoralis* L. subsp. *lapponica* (Prokud.) Tzvel.) - 1, 2
latiflora Rupr. = Arctophila fulva
laudanensis Roshev. ex Ovcz. = P. hissarica
laxa auct. = P. media
laxiflora Buckley - 5(?)
leptocoma Trin. (*P. flavidula* Kom., *P. nivicola* Kom. var. *flavidula* (Kom.) Roshev.) - 5
- subsp. *paucispicula* (Scribn. & Merr.) Tzvel. = P. paucispicula
lindebergii Tzvel. (*P. stricta* Lindeb. 1855, non Roth, 1821, nec D. Don, 1821, *P. arctica* R. Br. subsp. *stricta* (Lindeb.) Nannf., *P. tolmatchewii* Roshev. var. *stricta* (Nannf.) Tzvel.) - 4
lipskyi Roshev. (*P. bedeliensis* Litv., *P. contracta* Ovcz. & Czuk. 1957, non Retz. 1783, ? *P. dispansa* Ovcz., *P. kungeica* Golosk., *P. ovczinnikovii* Ikonn., *P. pseudodisiecta* Ovcz., *P. taldyksuensis* Roshev. nom. invalid.) - 6
- subsp. *dschungarica* (Roshev.) Tzvel. = P. dschungarica
litvinoviana Ovcz. (*P. badachschanica* Ikonn., *P. densissima* Roshev. ex Ovcz., *P. glauca* Vahl subsp. *litvinoviana* (Ovcz.) Tzvel., *P. glauciculmis* Ovcz., *P. lapidosa* Drob., *P. marginata* Ovcz. 1933, non Schlecht. 1814, *P. pseudotremuloides* Ovcz. & Czuk., *P. roshevitzii* Golosk., *P. tremuloides* Litv. ex Ovcz.) - 3, 6
longifolia Trin. - 2

- subsp. *fagetorum* (P. Smirn.) Tzvel. = P. fagetorum
- subsp. *meyeri* (Trin. ex Roshev.) Tzvel. = P. meyeri
lubrica Ovcz. = P. disjecta
macrocalyx Trautv. & C.A. Mey. (*P. hispidula* Vasey, *P. ketoiensis* Tatew. & Ohwi, *P. scabriflora* Hack., ? *P. tschegolevii* V. Vassil.) - 5
- subsp. **koriakiensis** (Worosch.) Worosch. (*P. macrocalyx* var. *koriakiensis* Worosch.) - 5
- subsp. *sugawarae* (Ohwi) Worosch. = P. sugawarae
- subsp. *tatewakiana* (Ohwi) Worosch. = P. tatewakiana
- var. *beringiana* (Probat.) Worosch. = P. beringiana
- var. *koriakiensis* Worosch. = P. macrocalyx subsp. koriakiensis
- var. *sachalinensis* Koidz. = P. sachalinensis
- var. *sugawarae* (Ohwi) Worosch. = P. sugawarae
- var. *tatewakiana* (Ohwi) Ohwi = P. tatewakiana
- b. *tianschanica* Regel = P. tianschanica
x **magadanensis** Probat. - P. glauca Vahl x P. malacantha Kom. - 5
magadanica Kuvajev - 5
malacantha Kom. (*P. bracteosa* Kom., *P. komarovii* Roshev.) - 5
marginata Ovcz. = P. litvinoviana
mariae Reverd. (*P. smirnowii* Roshev. subsp. *mariae* (Reverd.) Tzvel.) - 3, 4
masenderana Freyn & Sint. - 2
maydelii Roshev. = P. pratensis
media Schur (*P. media* subsp. *ursina* (Velen.) Diklic & Nikolic, *P. ursina* Velen., *P. laxa* auct.) - 2
- subsp. *ursina* (Velen.) Diklic & Nikolic = P. media
meyeri Trin. ex Roshev. (*P. longifolia* Trin. subsp. *meyeri* (Trin. ex Roshev.) Tzvel.) - 2
naltchikensis Roshev. = P. caucasica
x **nannfeldtii** Jir. - P. annua L. x P. supina Schrad.
nemoraliformis Roshev. = P. urssulensis
nemoralis L. - 1, 2, 3, 4, 5, 6
- subsp. *alexeenkoi* Tzvel. = P. alexeenkoi
- subsp. **carpatica** Jir. (*P. balfourii* auct.) - 1
- subsp. *hypanica* (Prokud.) Tzvel. = P. hypanica
- subsp. *korshunensis* (Golosk.) Tzvel. = P. korshunensis
- subsp. *lapponica* (Prokud.) Tzvel. = P. lapponica
- subsp. *rehmannii* Aschers. & Graebn. = P. rehmannii
- subsp. *sichotensis* (Probat.) Bondar. = P. sichotensis
- subsp. *tanfiljewii* (Roshev.) Kuvajev = P. tanfiljewii
- var. *rhomboidea* (Roshev.) Grossh. = P. tanfiljewii
- var. *sichotensis* (Probat.) Worosch. = P. sichotensis
neosachalinensis Probat. - 5
nevskii Roshev. ex Ovcz. (*P. bulbosa* L. subsp. *nevskii* (Roshev. ex Ovcz.) Tzvel.) - 6
nivicola Kom. = P. paucispicula and P. shumushuensis
- var. *flavidula* (Kom.) Roshev. = P. leptocoma
nudiflora Hack. = Puccinellia nudiflora
nuttaliana Schult. = Puccinellia nuttaliana
ochotensis Trin. (*P. sphondylodes* Trin., *P. versicolor* Bess. subsp. *ochotensis* (Trin.) Tzvel.) - 4, 5
- subsp. *stepposa* (Kryl.) Tzvel. = P. transbaicalica
ovczinnikovii Ikonn. = P. lipskyi
palustris L. (? *P. janczewskii* Zapal., *P. rotundata* Trin., *P. serotina* Ehrh. ex Gaudin, ? *P. strictula* Steud.) - 1, 2, 3, 4, 5, 6
- subsp. *tanfiljewii* (Roshev.) Tzvel. = P. tanfiljewii
- subsp. *volhynensis* (Klok.) Tzvel. = P. turfosa
pamirica Roshev. ex Ovcz. = P. tianschanica
paradoxa Kar. & Kir. = Eremopoa songarica
paucispicula Scribn. & Merr. (*P. barguzinensis* M. Pop., *P. leptocoma* Trin. subsp. *paucispicula* (Scribn. & Merr.) Tzvel., *P. nivicola* Kom. p.max.p., *P. taimyrensis*

Roshev.) - 4, 5

x **pawlowskii** Jir. - P. chaixii Vill. x P. remota Forsell.

pekulnejensis Jurtz. & Tzvel. (*P. glauca* Vahl var. *pekulnejensis* (Jurtz. & Tzvel.) Probat.) - 5

pelligera Rupr. = Dupontia pelligera

penicillata Kom. = P. platyantha

petraea Trin. ex Kom. = P. lanata

petschorica Roshev. = P. arctica

phariana Bor = P. calliopsis

pinegensis Roshev. = P. pratensis

pinetorum Klok. = P. turfosa

platyantha Kom. (*P. bracteosa* Kom. var. *penicillata* (Kom.) Worosch., *P. bracteosa* var. *plathyantha* (Kom.) Worosch., *P. eduardii* Golub, *P. penicillata* Kom.) - 5

podolica (Aschers. & Graebn.) Blocki ex Zapal. = P. versicolor

poecilantha Rupr. = Arctophila fulva

polonica Blocki = P. versicolor

polychroa (Trautv.) Grossh. = Bellardiochloa polychroa

pratensis L. (*P. angustifolia* L. var. *angustiglumis* (Roshev.) Worosch., *P. angustiglumis* Roshev., *P. articulata* Ovcz. 1956, non Schrank, 1824, *P. garanica* Ikonn., *P. maydelii* Roshev., *P. pinegensis* Roshev., *P. pratensis* subsp. *angustiglumis* (Roshev.) Tzvel., *P. pratensis* subsp. *pratensis* var. *angustiglumis* (Roshev.) Bondar., *P. subglabriflora* Roshev., *P. urjanchaica* Roshev.) - 1, 2, 3, 4, 5, 6

- subsp. *alpigena* (Blytt) Hiit. = P. alpigena
- subsp. *angustifofia* (L.) Arcang. = P. angustifolia
- subsp. *angustifolia* (L.) Lindb. fil. = P. angustifolia
- subsp. *angustiglumis* (Roshev.) Tzvel. = P. pratensis
- subsp. *colpodea* (Th.Fries) Tzvel. = P. alpigena subsp. colpodea
- subsp. *irrigata* (Lindm.) Lindb. fil. = P. subcaerulea
- subsp. *pratensis* var. *angustiglumis* (Roshev.) Bondar. = P. pratensis
- - var. *turfosa* (Litv.) Bondar. = P. turfosa Litv.
- subsp. **rigens** (C. Hartm.) Tzvel. (*P. rigens* C. Hartm.) - 1
- subsp. *sabulosa* (Roshev.) Tzvel. = P. sabulosa
- subsp. *sergievskajae* (Probat.) Tzvel. = P. sergievskajae
- subsp. **skrjabinii** Tzvel. - 4
- subsp. *sobolevskiana* (Gudoschn.) Tzvel. = P. sobolevskiana
- subsp. *subcaerulea* (Smith) Hiit. = P. subcaerulea
- subsp. *subcaerulea* (Smith) Tutin = P. subcaerulea
- subsp. *turfosa* (Litv.) Worosch. = P. turfosa
- subsp. **zhukoviae** Jurtz. & Tzvel. - 5
- var. *alpigena* Blytt = P. alpigena

primae Tzvel. - 2

princei Simps. = Arctopoa tibetica

pruinosa Korotky = P. tianschanica

pseudairoides Herrm. = Catabrosa pseudairoides

pseudoabbreviata Roshev. (*P. brachyanthera* Hult.) - 4, 5

pseudoattenuata Probat. = P. extremiorientalis

pseudoconcinna Schur = P. bulbosa

pseudodisiecta Ovcz. = P. lipskyi

pseudonemoralis Skvorts. = P. pseudopalustris

pseudopalustris Keng (*P. pseudonemoralis* Skvorts. 1954, non Schur, 1866, *P. skvortzovii* Probat.) - 4, 5

pseudosterilis Galkin = P. araratica

pseudotremuloides Ovcz. & Czuk. = P. litvinoviana

pungens Bieb. = Aeluropus pungens

pyramidata Lam. = Koeleria pyramidata

radula Franch. & Savat. - 5

- subsp. *ussuriensis* (Roshev.) Worosch. = P. ussuriensis

raduliformis Probat. (*P. remota* Forsell. subsp. *raduliformis* (Probat.) Worosch.) - 4, 5

rangkulensis Ovcz. & Czuk. (*P. koelzii* auct.) - 6

rehmannii (Aschers. & Graebn.) Woloszcz. (*P. nemoralis* L. subsp. *rehmannii* Aschers. & Graebn.) - 1

relaxa Ovcz. (*P. urgutina* Drob., *P. versicolor* Bess. subsp. *relaxa* (Ovcz.) Tzvel.) - 6

remota Forsell. - 1, 2, 3, 4, 6

- subsp. *raduliformis* (Probat.) Worosch. = P. raduliformis

remotiflora Rupr. = Arctophila fulva

reverdattoi Roshev. (*P. glauca* Vahl subsp. *reverdattoi* (Roshev.) Tzvel.) - 3, 4

rhomboidea Roshev. = P. tanfiljewii

rigens C. Hartm. = P. pratensis subsp. rigens

- subsp. *colpodea* (Th. Fries) D. Löve = P. alpigena subsp. colpodea

roshevitzii Golosk. = P. litvinoviana

rotundata Trin. = P. palustris

sabulosa (Roshev.) Roshev. (*P. pratensis* L. subsp. *sabulosa* (Roshev.) Tzvel.) - 4

sachalinensis (Koidz.) Honda (*P. macrocalyx* Trautv. & C.A. Mey. var. *sachalinensis* Koidz., *P. sugawarana* Koidz. ex Sugaw. 1937, nom. invalid., non P. sugawarae Ohwi, 1935) - 5

sajanensis Roshev. = P. tianschanica

scabriflora Hack. = P. macrocalyx

schischkinii Tzvel. = Arctopoa schischkinii

scleroclada Rupr. = x Arctodupontia scleroclada

seredinii Galkin - 2

sergievskajae Probat. (*P. pratensis* L. subsp. *sergievskajae* (Probat.) Tzvel.) - 4, 5

serotina Ehrh. ex Gaudin = P. palustris

setacea Hoffm. = P. angustifolia

shumushuensis Ohwi (*P. nivicola* Kom. 1924, p.p. incl. typo, non Ridley, 1916) - 5

sibirica Roshev. - 1, 3, 4, 5, 6

- subsp. *uralensis* Tzvel. = P. insignis
- var. *insignis* (Litv. ex Roshev.) Serg. = P. insignis

sichotensis Probat. (*P. nemoralis* L. subsp. *sichotensis* (Probat.) Bondar., *P. nemoralis* var. *sichotensis* (Probat.) Worosch.) - 5

similis Rupr. = Arctophila fulva

sinaica Steud. (*P. bulbosa* L. subsp. *sinaica* (Steud.) Tzvel.) - 2, 6

skvortzovii Probat. = P. pseudopalustris

smirnowii Roshev. (*P. arctica* R. Br. subsp. *smirnowii* (Roshev.) Malysch.) - 5

- subsp. *mariae* (Reverd.) Tzvel. = P. mariae

sobolevskiana Gudoschn. (*P. pratensis* L. subsp. *sobolevskiana* (Gudoschn.) Tzvel.) - 4

soczawae Roshev. = P. glauca

sphondylodes Trin. = P. ochotensis

spicata Drob. = P. drobovii

stenantha auct. = P. trivialis

stepposa (Kryl.) Roshev. = P. transbaicalica

sterilis Bieb. (*Festuca asperrima* Link) - 1, 2

- subsp. *biebersteinii* (H. Pojark.) Tzvel. = P. biebersteinii
- subsp. *versicolor* (Bess.) K. Richt. = P. versicolor

striata Lam. = Glyceria striata

stricta Lindeb. = P. lindebergii

- subsp. *colpodea* Th. Fries = P. alpigena subsp. colpodea

strictula Steud. = P. palustris

strigosa Hoffm. = P. angustifolia

subcaerulea Smith (*P. angustifolia* L. subsp. *humilis* (Ehrh. ex Smith) K. Richt., *P. czernjajevii* Prokud., *P. humilis* Ehrh. ex Smith, *P. irrigata* Lindm., *P. pratensis* L. subsp. *irrigata* (Lindm.) Lindb. fil., *P. pratensis* subsp. *subcaerulea* (Smith) Hiit., *P. pratensis* subsp. *subcaerulea* (Smith) Tutin, comb. superfl.) - 1, 5(alien)

subfastigiata Trin. = Arctopoa subfastigiata
subglabriflora Roshev. = P. pratensis
sublanata Reverd. (*P. angustifolia* L. var. *sublanata* (Reverd.) Worosch.) - 3, 4, 5
subpolaris Kuvajev - 1, 3, 4
subtilis Kar. & Kir. = Eremopoa songarica
sugawarae Ohwi (*P. macrocalyx* Trautv. & C.A. Mey. var. *sugawarae* (Ohwi) Worosch., *P. macrocalyx* subsp. *sugawarae* (Ohwi) Worosch.) - 5
sugawarana Koidz. ex Sugaw. = P. sachalinensis
supina Schrad. (*P. bifida* Frohner, *P. supina* subsp. *ustulata* (Frohner) A. & D. Löve, *P. ustulata* Frohner) - 1, 3, 4, 5(alien), 6
- subsp. *ustulata* (Frohner) A. & D. Löve = P. supina
sylvicola Guss. (*P. trivialis* L. subsp. *sylvicola* (Guss.) Lindb. fil.) - 1, 2, 6
taimyrensis Roshev. = P. paucispicula
taldyksuensis Roshev. = P. lipskyi
tanfiljewii Roshev. (*P. nemoralis* L. var. *rhomboidea* (Roshev.) Grossh., *P. nemoralis* subsp. *tanfiljewii* (Roshev.) Kuvajev, *P. palustris* L. subsp. *tanfiljewii* (Roshev.) Tzvel., *P. rhomboidea* Roshev.) - 1, 2, 3, 4, 6
tatarica Fisch. ex Griseb. = Eragrostis collina
tatewakiana Ohwi (*P. macrocalyx* Trautv. & C.A. Mey. var. *tatewakiana* (Ohwi) Ohwi, *P. macrocalyx* subsp. *tatewakiana* (Ohwi) Worosch.) - 5
taurica H. Pojark. - 1
tef Zuccagni = Eragrostis tef
tianschanica (Regel) Hack. ex O. Fedtsch. (*P. macrocalyx* Trautv. & C.A. Mey. var. *tianschanica* Regel, ? *P. grisea* Korotky, *P. pamirica* Roshev. ex Ovcz., *P. pruinosa* Korotky, *P. sajanensis* Roshev.) - 3, 4, 6
tibetica Munro ex Stapf = Arctopoa tibetica
timoleontis Heldr. ex Boiss. var. **dshilgensis** (Roshev.) Tzvel. (*P. dshilgensis* Roshev.) - 6
tolmatchewii Roshev. (*P. arctica* R. Br. subsp. *caespitans* Nannf., ? *P. filipes* Lange) - 1, 3, 5
- var. *stricta* (Nannf.) Tzvel. = P. lindebergii
transbaicalica Roshev. (*P. ochotensis* Trin. subsp. *stepposa* (Kryl.) Tzvel., *P. stepposa* (Kryl.) Roshev., *P. versicolor* Bess. subsp. *stepposa* (Kryl.) Tzvel.) - 1, 2, 3, 4, 5, 6
trautvetteri Tzvel. = Arctopoa trautvetteri
tremuloides Litv. ex Ovcz. = P. litvinoviana
trichoclada Rupr. = Arctophila fulva
tristis Trin. = P. altaica
trivialiformis Kom. (*P. lanata* Scribn. & Merr. var. *trivialiformis* (Kom.) Worosch.) - 5
trivialis L. (*P. trivialis* var. *woronowii* (Roshev.) Grossh., *P. woronowii* Roshev., *P. stenantha* auct.) - 1, 2, 3, 4, 5, 6
- subsp. *sylvicola* (Guss.) Lindb. fil. = P. sylvicola
- var. *woronowii* (Roshev.) Grossh. = P. trivialis
tschegolevii V. Vassil. = P. macrocalyx
turczaninovii Serg. = P. glauca
turfosa Litv. (*P. palustris* L. subsp. *volhynensis* (Klok.) Tzvel., *P. pinetorum* Klok. nom. invalid., *P. pratensis* L. subsp. *pratensis* var. *turfosa* (Litv.) Bondar., *P. pratensis* subsp. *turfosa* (Litv.) Worosch., *P. volhynensis* Klok.) - 1
turneri Scribn. - 5
tzvelevii Probat. - 5
udensis Trautv. & C.A. Mey. = P. glauca
urgutina Drob. = P. relaxa
urjanchaica Roshev. = P. pratensis
ursina Velen. = P. media
urssulensis Trin. (? *P. nemoraliformis* Roshev.) - 1, 3, 4, 5, 6
ussuriensis Roshev. (*P. radula* Franch. & Savat. subsp.

ussuriensis (Roshev.) Worosch., *P. acroleuca* sensu Worosch.) - 5
ustulata Frohner = P. supina
veresczaginii Tzvel. - 3
versicolor Bess. (*P. podolica* (Aschers. & Graebn.) Blocki ex Zapal., *P. polonica* Blocki, *P. sterilis* Bieb. subsp. *versicolor* (Bess.) K. Richt.) - 1
- subsp. *araratica* (Trautv.) Tzvel. = P. araratica
- subsp. *erythropoda* (Klok.) Tzvel. = P. erythropoda
- subsp. *ochotensis* (Trin.) Tzvel. = P. ochotensis
- subsp. *relaxa* (Ovcz.) Tzvel. = P. relaxa
- subsp. *stepposa* (Kryl.) Tzvel. = P. transbaicalica
violacea Bell. = Bellardiochloa violacea
viridula Palib. = P. angustifolia
vivipara (L.) Willd. (*P. alpina* L. var. *vivipara* L., *P. alpina* subsp. *vivipara* (L.) Arcang., *P. alpina* subsp. *vivipara* (L.) Tzvel. comb. superfl.) - 1
volhynensis Klok. = P. turfosa
vorobievii Probat. - 5
vrangelica Tzvel. (*P. hartzii* Gand. var. *vrangelica* (Tzvel.) Probat.) - 5
vvedenskyi Drob. - 6
williamsii Nash = P. arctica subsp. williamsii
woronowii Roshev. = P. trivialis
zaprjagajevii Ovcz. (*P. bactriana* Roshev. subsp. *zaprjagajevii* (Ovcz.) Tzvel.) - 6

Pollinia Trin. = Microstegium

imberbis Nees = Microstegium vimineum
imberbis sensu Roshev. p.p. = Microstegium nodosum
japonica Miq. = Microstegium japonicum
nuda auct. = Microstegium japonicum

Polypogon Desf. (*Nowodworskya* C. Presl)

demissus Steud. = P. fugax
fugax Nees ex Steud. (*Nowodworskya fugax* (Nees ex Steud.) Nevski, *Polypogon demissus* Steud.) - 2, 5(alien), 6
hissaricus (Roshev.) Bor = Agrostis hissarica
littoralis Smith = x Agropogon littoralis
maritimus Willd. - 1, 2, 3, 6
monspeliensis (L.) Desf. - 1, 2, 3, 5(alien), 6
semiverticillatus (Forssk.) Hyl. = P. viridis
vaginatus Willd. = Alopecurus vaginatus
viridis (Gouan) Breistr. (*Agrostis viridis* Gouan, *A. semiverticillata* (Forssk.) C. Chr., *A. verticillata* Vill., *Nowodworskya semiverticillata* (Forssk.) Nevski, *N. verticillata* (Vill.) Nevski, *Phalaris semiverticillata* Forssk., *Polypogon semiverticillatus* (Forssk.) Hyl.) - 1, 2, 6

Psathyrostachys Nevski

caespitosa (Sukacz.) Peschkova (*Elymus caespitosus* Sukacz.) - 4
daghestanica (Alexeenko) Nevski (*P. rupestris* (Alexeenko) Nevski subsp. *daghestanica* (Alexeenko) A. Löve) - 2
fragilis (Boiss.) Nevski subsp. **secaliformis** Tzvel. - 2
hyalantha (Rupr.) Tzvel. (*Elymus hyalanthus* Rupr., *Psathyrostachys juncea* (Fisch.) Nevski subsp. *hyalantha* (Rupr.) Tzvel.) - 3, 4, 6
juncea (Fisch.) Nevski - 1, 3, 4, 6
- subsp. *hyalantha* (Rupr.) Tzvel. = P. hyalantha
kronenburgii (Hack.) Nevski (*Elymus kronenburgii* (Hack.) Nikif.) - 6
lanuginosa (Trin.) Nevski - 3, 6
rupestris (Alexeenko) Nevski - 2

- subsp. *daghestanica* (Alexeenko) A. Löve = P. daghestanica

Pseudoroegneria (Nevski) A. Löve = Elytrigia

armena (Nevski) A. Löve = Elytrigia armena
cognata (Hack.) Holub = Elytrigia cognata
cretacea (Klok. & Prokud.) A. Löve = Elytrigia stipifolia
divaricata (Boiss. & Bal.) A. Löve subsp. *attenuatiglumis* (Nevski) A. Löve = Elytrigia attenuatiglumis
dshinalica (Sablina) A. Löve = Elytrigia dshinalica
geniculata (Trin.) A. Löve = Elytrigia geniculata
- subsp. *pruinifera* (Nevski) A. Löve = Elytrigia pruinifera
- subsp. *scythica* (Nevski) A. Löve = Elytrigia scythica
gracillima (Nevski) A. Löve = Elytrigia gracillima
heidemaniae (Tzvel.) A. Löve = Elytrigia heidemaniae
kotovii (Dubovik) A. Löve = Elytrigia stipifolia
ninae (Dubovik) A. Löve = Elytrigia stipifolia
pertenuis (C.A. Mey.) A. Löve = Elytrigia pertenuis
reflexiaristata (Nevski) A. Lavrenko = Elytrigia reflexiaristata
setulifera (Nevski) A. Löve = Elytrigia setulifera
sosnovskyi (Hack.) A. Löve = Elytrigia sosnovskyi
stipifolia (Czern. ex Nevski) A. Löve = Elytrigia stipifolia
strigosa (Bieb.) A. Löve = Elytrigia strigosa
- subsp. *aegilopoides* (Drob.) A. Löve = Elytrigia gmelinii
- subsp. *amgunensis* (Nevski) A. Löve = Elytrigia amgunensis
- subsp. *jacutorum* (Nevski) A. Löve = Elytrigia jacutorum
- subsp. *reflexiaristata* (Nevski) A. Löve = Elytrigia reflexiaristata

*Pseudosasa Makino ex Nakai

***japonica** (Siebold & Zucc. ex Steud.) Makino ex Nakai (*Sasa japonica* (Siebold & Zucc. ex Steud.) Makino) - 1

Psilantha (C. Koch) Tzvel. = Eragrostis

arundinacea (L.) Tzvel. = Eragrostis collina

Psilurus Trin.

aristatus (L.) Lange = P. incurvus
incurvus (Gouan) Schinz & Thell. (*Nardus incurva* Gouan, *Psilurus aristatus* (L.) Lange) - 1, 2, 6

Ptilagrostis Griseb.

alpina (Fr. Schmidt) Sipl. (*Lasiagrostis alpina* Fr. Schmidt) - 4, 5
concinna auct. = P. schischkinii
concinna (Hook. fil.) Roshev. subsp. *schischkinii* Tzvel. = P. schischkinii
czekanowskii (V. Petrov) Sipl. = P. mongholica
junatovii Grub. - 3, 4
malyschevii Tzvel. - 6
minutiflora (Titov ex Roshev.) Czer. (*Stipa mongholica* Turcz. ex Trin. var. *minutiflora* Titov ex Roshev., *Ptilagrostis mongholica* (Turcz. ex Trin.) Griseb. var. *minutiflora* (Titov ex Roshev.) Roshev., *P. mongholica* subsp. *minutiflora* (Titov ex Roshev.) Tzvel.) - 4
mongholica (Turcz. ex Trin.) Griseb. (*Achnatherum mongholicum* (Turcz. ex Trin.) Ohwi, *Ptilagrostis czekanowskii* (V. Petrov) Sipl., *Stipa czekanowskii* V. Petrov) - 3, 4, 6
- subsp. *minutiflora* (Titov ex Roshev.) Tzvel. = P. minutiflora

- var. *minutiflora* (Titov ex Roshev.) Roshev. = P. minutiflora
purpurea (Griseb.) Roshev. = Stipa purpurea
schischkinii (Tzvel.) Czer. (*P. concinna* (Hook. fil.) Roshev. subsp. *schischkinii* Tzvel., *P. concinna* auct.) - 6
subsessiliflora (Rupr.) Roshev. = Stipa subsessiliflora

Puccinellia Parl. (*Atropis* Rupr.)

akbaitalensis Ovcz. & Czuk. = P. pamirica
alascana Scribn. & Merr. (*Atropis alascana* (Scribn. & Merr.) V. Krecz. p.p. quoad nomen, *Phippsia langeana* (Berlin) A. & D. Löve subsp. *alascana* (Scribn. & Merr.) A. & D. Löve, *Puccinellia langeana* (Berlin) Sorensen subsp. *alascana* (Scribn. & Merr.) Sorensen, *P. tenella* (Lange) Holmb. subsp. *alascana* (Scribn. & Merr.) Tzvel.) - 5
altaica Tzvel. - 3
angustata (R. Br.) Rand & Redf. (*Atropis angustata* (R. Br.) Griseb. p.p. quoad nomen, *A. angustata* (R. Br.) V. Krecz. comb. superfl., *Glyceria vaginata* Lange var. *contracta* Lange, *Phippsia angustata* (R. Br.) A. & D. Löve, *Puccinellia contracta* (Lange) Sorensen, *P. taimyrensis* Roshev.) - 1, 3, 4, 5
- subsp. *palibinii* (Sorensen) Tzvel. = P. palibinii
anisoclada V. Krecz. = P. gigantea
asiatica (Hadac & A. Löve) Czer. (*P. vilfoidea* (Anderss.) A. & D. Löve subsp. *asiatica* Hadac & A. Löve, *P. phryganodes* (Trin.) Scribn. & Merr. subsp. *asiatica* (Hadac & A. Löve) Tzvel.) - 1, 3, 4, 5
x **beckii** Holmb. - P. distans (Jacq.) Parl. x P. limosa (Schur) Holmb.
beringensis Tzvel. - 5
bilykiana Klok. - 1
borealis Swall. (*Phippsia borealis* (Swall.) A. & D. Löve) - 4, 5
- subsp. *neglecta* Tzvel. = P. neglecta
brachylepis Klok. = P. gigantea
bulbosa (Grossh.) Grossh. (*Atropis bulbosa* Grossh., *Puccinellia gigantea* (Grossh.) Grossh. subsp. *bulbosa* (Grossh.) Tzvel.) - 2
byrrangensis Tzvel. (*Phippsia vahliana* (Liebm.) A. & D. Löve subsp. *byrrangensis* (Tzvel.) A. & D. Löve) - 4
capillaris (Liljebl.) Jansen (*Festuca capillaris* Liljebl., *Atropis suecica* Holmb., *Phippsia capillaris* (Liljebl.) A. & D. Löve, *Puccinellia capillaris* (Liljebl.) Tzvel. comb. superfl., *P. suecica* (Holmb.) Holmb.) - 1
- subsp. *pulvinata* (Fries) Tzvel. = P. pulvinata
- var. *vaginata* (Lange) Tzvel. = P. coarctata
chilochloa V. Krecz. = P. gigantea and P. poecilantha
chinampoensis auct. = P. jacutica
choresmica V. Krecz. (*Atropis choresmica* V. Krecz.) - 1, 2, 6
coarctata Fern. & Weath. (*Atropis tenella* (Lange) V. Krecz. 1934, p.p. quoad pl., non K. Richt. 1890, nec Simm. 1913, *Phippsia coarctata* (Fern. & Weath.) A. Löve, *Ph. nutkaensis* (C. Presl) A. & D. Löve subsp. *borealis* (Holmb.) A. & D. Löve, *Puccinellia capillaris* (Liljebl.) Tzvel. var. *vaginata* (Lange) Tzvel. p.p. quoad pl., *P. distans* (Jacq.) Parl. subsp. *borealis* (Holmb.) W.E. Hughes, *P. retroflexa* (Curt.) Holmb. subsp. *borealis* Holmb.) - 1
colpodioides Tzvel. (*Phippsia vahliana* (Liebm.) A. & D. Löve subsp. *colpodioides* (Tzvel.) A. & D. Löve, *Puccinellia wrightii* (Scribn. & Merr.) Tzvel. subsp. *colpodioies* (Tzvel.) Tzvel.) - 5
contracta (Lange) Sorensen = P. angustata
diffusa V. Krecz. (*Atropis diffusa* Krecz.) - 6

distans (Jacq.) Parl. (*Poa distans* Jacq., *Atropis distans* (Jacq.) Griseb., *Phippsia distans* (Jacq.) A. & D. Löve, *Puccinellia pseudoconvoluta* Klok., *P. sachalinensis* Ohwi) - 1, 2, 3, 4(alien), 5(alien), 6
- subsp. *borealis* (Holmb.) W.E. Hughes = P. coarctata
- subsp. *glauca* (Regel) Tzvel. = P. glauca
- subsp. *hauptiana* (V. Krecz.) W.E. Hughes = P. hauptiana
- subsp. *limosa* (Schur) Soo & Jav. = P. limosa
- subsp. *sevangensis* (Grossh.) Tzvel. = P. sevangensis
dolicholepis V. Krecz. (*Atropis dolicholepis* V. Krecz., *Puccinellia dolichnolepis* (V. Krecz.) Pavl. comb. superfl.) - 1, 2, 3, 6
- subsp. **aksaica** Tzvel. - 6
- subsp. *fominii* (Bilyk) Tzvel. = P. fominii
x **elata** (Holmb.) Holmb. (*Atropis x elata* Holmb.). - P. capillaris (Liljebl.) Jansen x P. distans (Jacq.) Parl.
festuciformis auct. = P. poecilantha
filifolia (Trin.) Tzvel. - 4
filiformis V. Vassil. = P. hauptiana
fominii Bilyk (*P. dolicholepis* V. Krecz. subsp. *fominii* (Bilyk) Tzvel.) - 1, 2, 6
fragiliflora Sorensen = P. lenensis and P. palibinii
geniculata V. Krecz. (*Atropis geniculata* V. Krecz., *Puccinellia geniculata* (V. Krecz.) Hult. comb. superfl., *P. phryganodes* (Trin.) Scribn. & Merr. subsp. *geniculata* (V. Krecz.) Tzvel.) - 5
gigantea (Grossh.) Grossh. (*Atropis gigantea* Grossh., *A. anisoclada* V. Krecz., *A. chilochloa* V. Krecz. p.p., *A. convoluta* (Hornem.) Griseb., *A. sclerodes* V. Krecz., *Puccinellia anisoclada* V. Krecz., *P. brachylepis* Klok., *P. chilochloa* V. Krecz. p.p., *P. sclerodes* V. Krecz.) - 1, 2, 3, 5(alien), 6
- subsp. *bulbosa* (Grossh.) Tzvel. = P. bulbosa
glauca (Regel) V. Krecz. (*Atropis glauca* (Regel) V. Krecz., *Puccinellia distans* (Jacq.) Parl. subsp. *glauca* (Regel) Tzvel., *P. glauca* (Regel) Persson, comb. superfl.) - 6
gorodkovii Tzvel. (*Phippsia gorodkovii* (Tzvel.) A. & D. Löve) - 4
grossheimiana V. Krecz. (*Atropis grossheimiana* V. Krecz.) - 2
hackeliana V. Krecz. (*Atropis hackeliana* V. Krecz., *A. montana* Drob., *Puccinellia hackeliana* (V. Krecz.) Persson, comb. superfl.) - 3, 4, 6
- subsp. *humilis* (Litv. ex V. Krecz.) Tzvel. = P. humilis
hauptiana V. Krecz. (*Atropis hauptiana* V. Krecz., *A. iliensis* V. Krecz., *Phippsia hauptiana* (V. Krecz.) A. & D. Löve, *Puccinellia distans* (Jacq.) Parl. subsp. *hauptiana* (V. Krecz.) W.E. Hughes, *P. filiformis* V. Vassil. 1949, non Keng, 1938, *P. hauptiana* (V. Krecz.) Kitag. comb. superfl., *P. hauptiana* (V. Krecz.) Tzvel. comb. superfl., *P. iliensis* V. Krecz., *P. iliensis* (V. Krecz.) Serg. comb. superfl.) - 1, 3, 4, 5, 6
humilis Litv. ex V. Krecz. (*Atropis humilis* Litv. ex V. Krecz., *Puccinellia hackeliana* V. Krecz. subsp. *humilis* (Litv. ex V. Krecz.) Tzvel., *P. humilis* (Litv. ex V. Krecz.) Roshev. comb. superfl.) - 6
x **hybrida** Holmb. - P. distans (Jacq.) Parl. x P. maritima (Huds.) Parl.
iliensis V. Krecz. = P. hauptiana
iliensis (V. Krecz.) Serg. = P. hauptiana
interior Sorensen - 4, 5
jacutica Bubnova (*P. chinampoensis* auct.) - 4
jenisseiensis (Roshev.) Tzvel. (*Atropis jenisseiensis* Roshev., *Phippsia vahliana* (Liebm.) A. & D. Löve subsp. *jenisseiensis* (Roshev.) A. & D. Löve) - 4
kalininiae Bubnova - 3
kamtschatica Holmb. (*Atropis kamtschatica* (Holmb.) V. Krecz.) - 4, 5
- var. *sublaevis* Holmb. = P. sublaevis
kreczetoviczii Bubnova - 4
kulundensis Serg. - 3
kurilensis (Takeda) Honda (*Atropis kurilensis* Takeda, *A. alascana* (Scribn. & Merr.) V. Krecz. p.p. quoad pl., *A. paupercula* (Holm) V. Krecz. p.p. quoad pl., ? *Glyceria pumila* Vasey, *Puccinellia pumila* (Vasey) Hitchc.) - 5
laeviuscula V. Krecz. = P. tenella
langeana (Berlin) Sorensen (*Glyceria langeana* Berlin, *Atropis paupercula* (Holm) V. Krecz. p.p. quoad nomen, *A. paupercula* (Holm) Steffen, comb. superfl., *Glyceria paupercula* Holm, *Phippsia langeana* (Berlin) A. & D. Löve, *Puccinellia langeana* subsp. *typica* Sorensen, *P. paupercula* (Holm) Fern. & Weath. p.p., *P. tenella* (Lange) Holmb. subsp. *langeana* (Berlin) Tzvel.) - 5
- subsp. *alascana* (Scribn. & Merr.) Sorensen = P. alascana
- subsp. *asiatica* Sorensen = P. tenella
- subsp. *typica* Sorensen = P. langeana
lenensis (Holmb.) Tzvel. (*P. sibirica* Holmb. var. *lenensis* Holmb., *Phippsia lenensis* (Holmb.) A. & D. Löve, *Puccinellia fragiliflora* Sorensen, p.p. excl. typo) - 4, 5
limosa (Schur) Holmb. (*Atropis distans* (Jacq.) Griseb. a. *limosa* Schur, *Phippsia limosa* (Schur) A. & D. Löve, *Puccinellia distans* (Jacq.) Parl. subsp. *limosa* (Schur) Soo & Jav.) - 1, 2
macranthera V. Krecz. (*Atropis macranthera* V.Krecz., *Puccinellia macranthera* (V. Krecz.) T. Norlindh, comb. superfl.) - 4
macropus V. Krecz. (*Atropis macropus* V. Krecz., *Puccinellia macropus* (V. Krecz.) Pavl. comb. superfl.) - 6
maritima (Huds.) Parl. (*Atropis maritima* (Huds.) Griseb. p.p., *A. maritima* (Huds.) K. Richt. comb. superfl., *Phippsia maritima* (Huds.) A. Löve) - 1
x **mixta** Holmb. - P. capillaris (Liljebl.) Jansen x P. maritima (Huds.) Parl.
mongolica (T. Norlindh) Bubnova (*P. tenuiflora* (Griseb.) Scribn. & Merr. var. *mongolica* T. Norlindh) - 4
neglecta (Tzvel.) Bubnova (*P. borealis* Swall. subsp. *neglecta* Tzvel., *Phippsia borealis* (Swall.) A. & D. Löve subsp. *neglecta* (Tzvel.) A. & D. Löve) - 4, 5
nipponica Ohwi - 5
nudiflora (Hack.) Tzvel. (*Poa nudiflora* Hack.) - 6
nuttaliana (Schult.) Hitchc. (*Poa nuttaliana* Schult.) - 1(alien), 5
palibinii Sorensen (*Phippsia angustata* (R. Br.) A. & D. Löve subsp. *palibinii* (Sorensen) A. & D. Löve, *Ph. fragiliflora* (Sorensen) A. & D. Löve, *Puccinellia angustata* (R. Br.) Rand & Redf. subsp. *palibinii* (Sorensen) Tzvel., *P. fragiliflora* Sorensen, p.p. quoad typum) - 1
pamirica (Roshev.) V. Krecz. (*Atropis pamirica* (Roshev.) V. Krecz., *Puccinellia akbaitalensis* Ovcz. & Czuk., *P. pamirica* (Roshev.) V. Krecz. ex Roshev. comb. superfl.) - 6
- subsp. *vachanica* (Ovcz. & Czuk.) Tzvel. = P. vachanica
pauciramea (Hack.) V. Krecz. (*Atropis pauciramea* (Hack.) V. Krecz., *Puccinellia pauciramea* (Hack.) V. Krecz. ex Roshev. comb. superfl.) - 6
paupercula (Holm) Fern. & Weath. p.p. = P. langeana
phryganodes (Trin.) Scribn. & Merr. (*Atropis phryganodes* (Trin.) V. Krecz. comb. superfl., *A. phryganodes* (Trin.) Steffen, *Phippsia phryganodes* (Trin.) A. & D. Löve) - 5
- subsp. *asiatica* (Hadac & A. Löve) Tzvel. = P. asiatica
- subsp. *geniculata* (V. Krecz.) Tzvel. = P. geniculata
- subsp. *vilfoidea* (Anderss.) Tzvel. = P. vilfoidea

poecilantha (C. Koch) Grossh. (*Atropis chilochloa* V.Krecz. p.p., *A. poecilantha* (C. Koch) V. Krecz., *Puccinellia chilochloa* V. Krecz. p.p., *P. festuciformis* auct.) - 1, 2, 6

pseudoconvoluta Klok. = P. distans

pulvinata (Fries) V. Krecz. (*Atropis pulvinata* (Fries) V. Krecz., *Phippsia capillaris* (Liljebl.) A. & D. Löve subsp. *pulvinata* (Fries) A. & D. Löve, *Puccinellia capillaris* (Liljebl.) Jansen. subsp. *pulvinata* (Fries) Tzvel., *P. pulvinata* (Fries) Tzvel. comb. superfl.) - 1

pumila (Vasey) Hitchc. = P. kurilensis

retroflexa (Curt.) Holmb. subsp. *borealis* Holmb. = P. coarctata

roshevitsiana (Schischk.) Tzvel. (*Atropis roshevitsiana* Schischk.) - 3, 6

sachalinensis Ohwi = P. distans

schischkinii Tzvel. - 3, 4, 6

sclerodes V. Krecz. = P. gigantea

sevangensis Grossh. (*Atropis sevangensis* (Grossh.) V. Krecz., *Puccinellia distans* (Jacq.) Parl. subsp. *sevangensis* (Grossh.) Tzvel.) - 2

sibirica Holmb. (*Atropis sibirica* (Holmb.) V. Krecz., *Phippsia sibirica* (Holmb.) A. & D. Löve) - 1, 3, 4, 5

- var. *lenensis* Holmb. = P. lenensis

sublaevis (Holmb.) Tzvel. (*P. kamtschatica* Holmb. var. *sublaevis* Holmb.) - 5

subspicata V. Krecz. (*Atropis subspicata* V. Krecz., *Poa gorbunovii* Ovcz., *Puccinellia subspicata* (V. Krecz.) Pavl. comb. superfl.) - 6

suecica (Holmb.) Holmb. = P. capillaris

syvaschica Bilyk - 1

taimyrensis Roshev. = P. angustata

tenella (Lange) Holmb. (*Atropis laeviuscula* V. Krecz., *A. tenella* (Lange) V. Krecz. p.p. quoad nomen, comb. superfl., *A. tenella* (Lange) K. Richt., *Phippsia langeana* (Berlin) A. & D. Löve subsp. *asiatica* (Sorensen) A. & D. Löve, *Ph. tenella* (Lange) A. & D. Löve, *Puccinellia laeviuscula* V. Krecz., *P. langeana* (Berlin) Sorensen subsp. *asiatica* Sorensen) - 1, 4, 5

- subsp. *alascana* (Scribn. & Merr.) Tzvel. = P. alascana

- subsp. *langeana* (Berlin) Tzvel. = P. langeana

tenuiflora (Griseb.) Scribn. & Merr. (*Atropis tenuiflora* Griseb.) - 4, 5, 6(?)

- subsp. *tianschanica* Tzvel. = P. tianschanica

- var. *mongolica* T. Norlindh = P. mongolica

tenuissima Litv. ex V. Krecz. (*Atropis tenuissima* Litv. ex V. Krecz.) - 1, 3, 4, 5(alien), 6

tianschanica (Tzvel.) Ikonn. (*P. tenuiflora* (Griseb.) Scribn. & Merr. subsp. *tianschanica* Tzvel.) - 6

vachanica Ovcz. & Czuk. (*P. pamirica* (Roshev.) V. Krecz. subsp. *vachanica* (Ovcz. & Czuk.) Tzvel.) - 6

vacillans (Th. Fries) Scholand. = x Pucciphippsia vacillans

vaginata (Lange) Fern. & Weath. (*Glyceria vaginata* Lange, *Phippsia vaginata* (Lange) A. & D. Löve, *Puccinellia capillaris* (Liljebl.) Tzvel. var. *vaginata* (Lange) Tzvel. p.p. quoad nomen) - 4, 5

vahliana (Liebm.) Scribn. & Merr. (*Colpodium vahlianum* (Liebm.) Nevski, *Phippsia vahliana* (Liebm.) A. & D. Löve) - 1

vilfoidea (Anderss.) A. & D. Löve (*Catabrosa vilfoidea* Anderss., *Atropis vilfoidea* (Anderss.) K. Richt., *Phippsia vilfoidea* (Anderss.) A. & D. Löve, *Puccinellia phryganodes* (Trin.) Scribn. & Merr. subsp. *vilfoidea* (Anderss.) Tzvel.) - 1

- subsp. *asiatica* Hadac & A. Löve = P. asiatica

- subsp. **beringensis** (A. & D. Löve) Czer. comb. nova (*Phippsia vilfoidea* subsp. *beringensis* A. & D. Löve,

1975, Bot. Not. (Lund) 128, 4 : 501) - 5

waginiae Bubnova - 3

wrightii (Scribn. & Merr.) Tzvel. (*Colpodium wrightii* Scribn. & Merr., *Phippsia wrightii* (Scribn. & Merr.) A. & D. Löve) - 5

- subsp. *colpodioies* (Tzvel.) Tzvel. = P. colpodioides

x **Pucciphippsia** Tzvel.

czukczorum Tzvel. - Phippsia algida (Soland.) R. Br. x Puccinellia wrightii (Scribn. & Merr.) Tzvel. - 5

vacillans (Th. Fries) Tzvel. (*Catabrosa concinna* Th. Fries subsp. *vacillans* Th. Fries, *Colpodium vacillans* (Th. Fries) Polun., *Puccinellia vacillans* (Th. Fries) Scholand.). - Phippsia algida (Soland.) R. Br. x Puccinellia vahliana (Liebm.) Scribn. & Merr. - 1

Rhizocephalus Boiss.

orientalis Boiss. (*R. turkestanicus* (Litv.) Roshev.) - 2, 6

turkestanicus (Litv.) Roshev. = R. orientalis

*****Rhynchelytrum** Nees

*****repens** (Willd.) C.E. Hubb. (*Saccharum repens* Willd., *Rhynchelytrum roseum* (Nees) Stapf & C.E. Hubb., *Tricholaena rosea* Nees)

roseum (Nees) Stapf & C.E. Hubb. = R. repens

Roegneria C. Koch = Elymus

abolinii (Drob.) Nevski = Elymus abolinii

alascana (Scribn. & Merr.) Jurtz. & Petrovsky = Elymus alascanus

amurensis (Drob.) Nevski = Elymus amurensis

angustiglumis (Nevski) Nevski = Elymus mutabilis

arcuata (Golosk.) Golosk. = Elymus arcuatus

borealis (Turcz.) Nevski = Elymus kronokensis subsp. subalpinus

brachypodioides Nevski = Elymus brachypodioides

burjatica Sipl. = Elymus transbaicalensis

buschiana (Roshev.) Nevski = Elymus buschianus

canaliculata (Nevski) Ohwi = Elymus longe-aristatus subsp. canaliculatus

canina (L.) Nevski = Elymus caninus

carinata Ovcz. & Sidor. = Elytrigia alaica

caucasica C. Koch = Elymus caucasicus

ciliaris (Trin.) Nevski = Elymus ciliaris

confusa (Roshev.) Nevski = Elymus confusus

curvata (Nevski) Nevski = Elymus fedtschenkoi

"czimganica" (Drob.) Nevski = Elymus tschimganicus

drobovii (Nevski) Nevski = Elymus drobovii

fibrosa (Schrenk) Nevski = Elymus fibrosus

glaucissima (M. Pop.) Filat. = Elymus glaucissimus

gmelinii (Ledeb.) Kitag. = Elymus gmelinii

himalayana Nevski = Elymus himalayanus

hyperarctica (Polun.) Tzvel. = Elymus hyperarcticus

interrupta (Nevski) Nevski = Elymus praeruptus

jacquemontii (Hook. fil.) Ovcz. & Sidor. p.p. = Elymus longe-aristatus subsp. canaliculatus

jacutensis (Drob.) Nevski = Elymus jacutensis

kamczadalorum Nevski = Elymus kamczadalorum

karawaewii (P. Smirn.) Karav. = Elytrigia villosa

karkaralensis (Roshev.) Filat. = Elymus viridiglumis

komarovii (Nevski) Nevski = Elymus komarovii

kronokensis (Kom.) Tzvel. = Elymus kronokensis

lachnophylla Ovcz. & Sidor. = Elymus lachnophyllus

leptoura Nevski = Elymus transhyrcanus

linczevskii Czopanov = Elymus caucasicus
longe-aristata (Boiss.) Drob. = Elymus longe-aristatus
macrochaeta Nevski = Elymus macrochaetus
macroura (Turcz.) Nevski = Elymus macrourus
- subsp. *pilosivaginata* Jurtz. = Elymus macrourus subsp. pilosivaginatus
mutabilis (Drob.) Hyl. = Elymus mutabilis
- subsp. *varsugensis* (Melderis) A. & D. Löve = Elymus mutabilis
- var. *varsugensis* Melderis = Elymus mutabilis
nepliana V. Vassil. = Elymus neplianus
novae-angliae (Scribn.) Jurtz. & Petrovsky = Elymus novae-angliae
oschensis Nevski = Elymus mutabilis
panormitana (Parl.) Nevski = Elymus panormitanus
pauciflora (Schwein.) Hyl. = Elymus trachycaulus
pavlovii (Nevski) Filat. = x Elyhordeum pavlovii
pendulina Nevski = Elymus pendulinus
praecaespitosa (Nevski) Nevski = Elymus praecaespitosus
prokudinii Seredin = Elymus prokudinii
pubescens (Schischk.) Nevski = Elymus jacutensis
sajanensis Nevski = Elymus sajanensis
scandica Nevski = Elymus kronokensis subsp. subalpinus
schrenkiana (Fisch. & C.A. Mey.) Nevski = Elymus schrenkianus
schugnanica (Nevski) Nevski = Elymus schugnanicus
sclerophylla Nevski = Elymus sclerophyllus
striata (Steud.) Nevski, p.p. = Elymus fedtschenkoi
subfibrosa Tzvel. = Elymus subfibrosus
taigae Nevski = Elymus viridiglumis
tianschanica (Drob.) Nevski = Elymus tianschanigenus
trachycaulon (Link) Nevski = Elymus trachycaulus
transbaicalensis (Nevski) Nevski = Elymus transbaicalensis
transhyrcana Nevski = Elymus transhyrcanus
transiliensis (M. Pop.) Filat. = Elymus mutabilis
trinii Nevski = Elymus jacutensis
troctolepis Nevski = Elymus troctolepis
tschimganica (Drob.) Nevski = Elymus tschimganicus
turczaninovii (Drob.) Nevski = Elymus gmelinii
turuchanensis (Reverd.) Nevski = Elymus turuchanensis
tuskaulensis Vass. = Elymus caninus
ugamica (Drob.) Nevski = Elymus nevskii
uralensis (Nevski) Nevski = Elymus uralensis
villosa V. Vassil. = Elymus vassiljevii
- subsp. *coerulea* Jurtz. = Elymus vassiljevii subsp. coeruleus
- subsp. *laxe-pilosa* Jurtz. = Elymus vassiljevii subsp. laxe-pilosus
viridiglumis Nevski = Elymus viridiglumis

Roshevitzia Tzvel. = Diandrochloa

diarrhena (Schult. & Schult. fil.) Tzvel. = Diandrochloa diarrhena

Rostraria Trin. (*Lophochloa* Reichenb.)

cristata (L.) Tzvel. (*Festuca cristata* L., *Lophochloa cristata* (L.) Hyl., *L. phleoides* (Vill.) Reichenb., *Trisetaria cristata* (L.) Kerguelen, *T. phleoides* (Vill.) Nevski, *Trisetum cristatum* (L.) Potztal) - 1, 2, 6
- subsp. *glabriflora* (Trautv.) Tzvel. = R. glabriflora
- subsp. *obtusiflora* (Boiss.) Tzvel. = R. obtusiflora
- var. *glabriflora* (Trautv.) M. Dogan = R. glabriflora
glabriflora (Trautv.) Czer. (*Koeleria phleoides* (Vill.) Pers. var. *glabriflora* Trautv., *Rostraria cristata* (L.) Trautv. var. *glabriflora* (Trautv.) M. Dogan, *R. cristata* subsp. *glabriflora* (Trautv.) Tzvel.) - 1(?), 2, 6

obtusiflora (Boiss.) Holub (*Lophochloa obtusiflora* (Boiss.) Gontsch., *Rostraria cristata* (L.) Tzvel. subsp. *obtusiflora* (Boiss.) Tzvel., *Trisetum obtusiflorum* (Boiss.) Potztal) - 6

Rottboellia auct. = Hemarthria, Henrardia and Monerma

altissima Poir. = Hemarthria altissima
cylindrica Willd. = Monerma cylindrica
pubescens Bertol. = Henrardia pubescens
sibirica Gand. = Hemarthria sibirica

Saccharum L.

koenigii Retz. = Imperata koenigii
***officinarum** L.
ravennae (L.) Murr. = Erianthus ravennae
repens Willd. = Rhynchelytrum repens
sibiricum (Trin.) Roberty = Spodiopogon sibiricus
spontaneum L. - 6

Sasa Makino & Shibata

amphitricha Koidz. = S. sendaica
amplissima Koidz. = S. palmata
austrokurilensis Koidz. = S. niijimae
blepharodes Koidz. = S. megalophylla
cernua Makino = S. spiculosa
chimakisasa Koidz. = S. palmata
confusa Nakai = S. matsudae
depauperata (Takeda) Nakai (*S. nipponica* Makino & Shibata var. *depauperata* Takeda, *S. dilacerata* Koidz., *S. meakensis* Nakai, *S. yahikoensis* Makino var. *depauperata* (Takeda) Suzuki) - 5
dilacerata Koidz. = S. depauperata
fortis Koidz. = S. rivularis
harae Nakai = S. rivularis
hirta (Koidz.) Tzvel. (*S. pseudocernua* Koidz. var. *hirta* Koidz., *S. kurilensis* (Rupr.) Makino & Shibata var. *hirta* (Koidz.) Suzuki, *S. spiculosa* (Fr. Schmidt) Makino var. *hirta* (Koidz.) Tzvel., *S. naigoensis* auct.) - 5
intercedens Koidz. = S. megalophylla
japonica (Siebold & Zucc. ex Steud.) Makino = Pseudosasa japonica
kosakensis Nakai = S. suzukii
koshinaiana Koidz. = S. palmata
kozasa Nakai = S. sendaica
kurilensis (Rupr.) Makino & Shibata - 5
- var. *hirta* (Koidz.) Suzuki = S. hirta
laevissima Koidz. = S. palmata
lasionodosa Koidz. = S. makinoi
lingulata Koidz. = S. palmata
makinoi Nakai (*S. lasionodosa* Koidz.) - 5
maokateiensis Koidz. = S. septentrionalis
matsudae Nakai (*S. confusa* Nakai, *S. pseudocernua* Koidz. var. *psilonodosa* Koidz., *S. spiculosa* (Fr. Schmidt) Makino var. *psilonodosa* (Koidz.) Tzvel.) - 5
meakensis Nakai = S. depauperata
megalophylla Makino & Uchida (*S. blepharodes* Koidz., *S. intercedens* Koidz., *S. okuyezoensis* Koidz., ? *S. sugawarae* Nakai) - 5
naigoensis auct. = S. hirta
nakasiretokoensis Koidz. = S. palmata
nambuana Koidz. = S. spiculosa
niijimae Tatew. ex Nakai (*S. austrokurilensis* Koidz., *S. palmata* (Marl. ex Burb.) E.G. Camus var. *hiijimae*

(Tatew. ex Nakai) Suzuki) - 5

nipponica Makino & Shibata var. *depauperata* Takeda = S. depauperata

okuyezoensis Koidz. = S. megalophylla

oseana (Makino) Uchida (*S. paniculata* (Fr. Schmidt) Makino & Shibata var. *oseana* Makino, *S. yahikoensis* Makino var. *oseana* (Makino) Suzuki) - 5

palmata (Marl. ex Burb.) E.G. Camus (*Bambusa palmata* Marl. ex Burb., *Sasa amplissima* Koidz., *S. chimakisasa* Koidz., *S. koshinaiana* Koidz., *S. laevissima* Koidz., *S. lingulata* Koidz., *S. nakasiretokoensis* Koidz., *S. pseudobrachyphylla* Nakai, *S. shikotanensis* Nakai var. *pseudobrachyphylla* (Nakai) Koidz., *S. soyensis* Nakai) - 5

- var. *hiijimae* (Tatew. ex Nakai) Suzuki = S. niijimae

paniculata (Fr. Schmidt) Makino & Shibata = S. senanensis

- var. *harae* (Nakai) Suzuki = S. rivularis
- var. *oseana* Makino = S. oseana
- var. *paniculata* f. *rivularis* (Nakai) Suzuki = S. rivularis

pilosa Nakai = S. senanensis

pseudobrachyphylla Nakai = S. palmata

pseudocernua Koidz. var. *hirta* Koidz. = S. hirta

- var. *psilonodosa* Koidz. = S. matsudae

pseudonipponica Tatew. ex Nakai = S. senanensis

rivularis Nakai (*S. fortis* Koidz., *S. harae* Nakai, *S. paniculata* (Fr. Schmidt) Makino & Shibata var. *harae* (Nakai) Suzuki, *S. paniculata* var. *paniculata* f. *rivularis* (Nakai) Suzuki) - 5

sachalinensis Makino & Nakai = S. tyuhgokensis

sattosasa Koidz. = S. tyuhgokensis

senanensis (Franch. & Savat.) Rehd. (*Bambusa senanensis* Franch. & Savat., *Sasa paniculata* (Fr. Schmidt) Makino & Shibata, *S. pilosa* Nakai, *S. pseudonipponica* Tatew. ex Nakai, *S. stripitans* Koidz., *S. tesioensis* Tatew., *S. uyetsuensis* Koidz.) - 5

sendaica Makino (*S. amphitricha* Koidz., *S. kozasa* Nakai) - 5

septentrionalis Makino (*S. maokateiensis* Koidz.) - 5

shikotanensis Nakai - 5

- var. *pseudobrachyphylla* (Nakai) Koidz. = S. palmata

sorstitialis Koidz. = S. spiculosa

soyensis Nakai = S. palmata

spiculosa (Fr. Schmidt) Makino (*S. cernua* Makino, *S. nambuana* Koidz., *S. sorstitialis* Koidz.) - 5

- var. *hirta* (Koidz.) Tzvel. = S. hirta
- var. *psilonodosa* (Koidz.) Tzvel. = S. matsudae

stripitans Koidz. = S. senanensis

sugawarae Nakai = S. megalophylla

suzukii Nakai (*S. kosakensis* Nakai) - 5(?)

tatewakiana Makino - 5

tesioensis Tatew. = S. senanensis

tessellata (Munro) Makino & Shibata = Indocalamu ssellatus

tyuhgokensis Makino (*S. sachalinensis* Makinc & N ai, *S. sattosasa* Koidz.) - 5

uyetsuensis Koidz. = S. senanensis

*****veitchii** (Carr.) Rehd. (*Bambusa veitchii* Carr.)

yahikoensis Makino var. *depauperata* (Takeda) Suzuki = S. depauperata

- var. *oseana* (Makino) Suzuki = S. oseana

Savastana Schrank = Hierochloe

hirta Schrank = Hierochloe hirta

Schismus Beauv.

arabicus Nees (*Sch. barbatus* (L.) Thell. subsp. *arabicus*

(Nees) Maire & Weiller) - 1, 2, 6

barbatus (L.) Thell. (*Festuca barbata* L., *Schismus calycinus* (L.) C. Koch, *Sch. minutus* (Hoffm.) Roem. & Schult.) - 2, 6

- subsp. *arabicus* (Nees) Maire & Weiller = Sch. arabicus

calycinus (L.) C. Koch = Sch. barbatus

minutus (Hoffm.) Roem. & Schult. = Sch. barbatus

Schizachne Hack.

callosa (Turcz. ex Griseb.) Ohwi (*Sch. purpurascens* (Torr.) Swall. var. *callosa* (Turcz. ex Griseb.) Kitag., *Sch. purpurascens* subsp. *callosa* (Turcz. ex Griseb.) T. Koyama & Kawano) - 1, 3, 4, 5

komarovii Roshev. (*Avena striata* Michx. subsp. *capillipes* Kom., *Schizachne purpurascens* (Torr.) Swall. subsp. *capillipes* (Kom.) Tzvel., *Sch. purpurascens* auct.) - 5

purpurascens auct. = Sch. komarovii

purpurascens (Torr.) Swall. subsp. *callosa* (Turcz. ex Griseb.) T. Koyama & Kawano = Sch. callosa

- subsp. *capillipes* (Kom.) Tzvel. = Sch. komarovii
- var. *callosa* (Turcz. ex Griseb.) Kitag. = Sch. callosa

Sclerochloa Beauv. (*Desmazeria* auct. p.p.)

dura (L.) Beauv. - 1, 2, 6

woronowii (Hack.) Tzvel. ex Bor (*Sclerochloa woronowii* Hack., *Desmazeria compressa* Ovcz. & Schibk., *D. woronowii* (Hack.) Ovcz. nom. invalid., *Scleropoa compressa* (Ovcz. & Schibk.) Bondar.) - 2, 6

Scleropoa Griseb. (*Catapodium* auct., *Desmazeria* auct.)

compressa (Ovcz. & Schibk.) Bondar. = Sclerochloa woronowii

rigescens Grossh. = Cutandia rigescens

rigida (L.) Griseb. (*Catapodium rigidum* (L.) C.E. Hubb., *Desmazeria rigida* (L.) Tutin) - 1, 2

woronowii Hack. = Sclerochloa woronowii

Scolochloa Link

festucacea (Willd.) Link - 1, 2, 3, 4, 6

Secale L.

afghanicum (Vav.) Roshev. = S. segetale

anatolicum Boiss. (? *S. daralagesi* Thum. nom. invalid., *S. montanum* Guss. subsp. *anatolicum* (Boiss.) Tzvel.) - 2

ancestrale Zhuk. var. *afghanicum* (Vav.) A. Ivanov & G.V. Yakovlev = S. segetale

- var. *dighoricum* (Vav.) A. Ivanov & G.V. Yakovlev = S. dighoricum

*****cereale** L. (*S. turkestanicum* Bensin) - 1, 2, 3, 4, 5, 6

- subsp. *derzhavinii* (Tzvel.) Kobyl. = S. x derzhavinii
- subsp. *dighoricum* Vav. = S. dighoricum
- subsp. *rigidum* V. & V. Antrop.
- s. *segetale* Zhuk. = S. segetale
- subsp. **tetraploidum** Kobyl.
- subsp. **tsitsinii** Kobyl.
- subsp. *vavilovii* (Grossh.) Kobyl. = S. vavilovii
- var. *afghanicum* Vav. = S. segetale
- var. *dighoricum* (Vav.) Yaaska = S. dighoricum

chaldicum Fed. (*S. kuprijanovii* Grossh. subsp. *transcaucasicum* A. Ivanov & G.V. Yakovlev, nom. invalid., *S. montanum* Guss. subsp. *chaldicum* (Fed.) Tzvel.) - 2

daralagesi Thum. = S. anatolicum

x **derzhavinii** Tzvel. (*S. cereale* L. subsp. *derzhavinii* (Tzvel.) Kobyl.). - S. cereale L. x S. montanum Guss. s.l. - 1

dighoricum (Vav.) Roshev. (*S. cereale* L. subsp. *dighoricum* Vav., *S. ancestrale* Zhuk. var. *dighoricum* (Vav.) A. Ivanov & G.V. Yakovlev, *S. cereale* var. *dighoricum* (Vav.) Yaaska, nom. invalid., *S. segetale* (Zhuk.) Roshev. subsp. *dighoricum* (Vav.) Tzvel.) - 1, 2

kuprijanovii Grossh. (*S. kuprijanovii* subsp. *ciscaucasicum* A. Ivanov & G.V. Yakovlev, nom. illegit., *S. montanum* Guss. subsp. *kuprijanovii* (Grossh.) Tzvel.) - 2
- subsp. *ciscaucasicum* A. Ivanov & G.V. Yakovlev = S. kuprijanovii
- subsp. *transcaucasicum* A. Ivanov & G.V. Yakovlev = S. chaldicum

montanum Guss. subsp. *anatolicum* (Boiss.) Tzvel. = S. anatolicum
- subsp. *chaldicum* (Fed.) Tzvel. = S. chaldicum
- subsp. *kuprijanovii* (Grossh.) Tzvel. = S. kuprijanovii

segetale (Zhuk.) Roshev. (*S. cereale* L. subsp. *segetale* Zhuk., *S. afghanicum* (Vav.) Roshev., *S. ancestrale* Zhuk. var. *afghanicum* (Vav.) A. Ivanov & G.V. Yakovlev, *S. cereale* var. *afghanicum* Vav., *S. segetale* subsp. *afghanicum* (Vav.) Bondar.) - 2, 3, 6
- subsp. *afghanicum* (Vav.) Bondar. = S. segetale
- subsp. *dighoricum* (Vav.) Tzvel. = S. dighoricum

sylvestre Host - 1, 2, 3, 6
transcaucasicum Grossh. = S. vavilovii
turkestanicum Bensin = S. cereale

vavilovii Grossh. (*S. cereale* L. subsp. *vavilovii* (Grossh.) Kobyl., *S. transcaucasicum* Grossh. nom. invalid.) - 2

Semeiostachys Drob. = Elymus

abolinii (Drob.) Drob. = Elymus abolinii
drobovii (Nevski) Drob. = Elymus drobovii
interrupta (Nevski) Drob. = Elymus praeruptus
leptoura (Nevski) Drob. = Elymus transhyrcanus
macrochaeta (Nevski) Drob. = Elymus macrochaetus
macrolepis (Drob.) Drob. = Elymus x macrolepis
panormitana (Parl.) Drob. = Elymus panormitanus
sclerophylla (Nevski) Drob. = Elymus sclerophyllus
tianschanica (Drob.) Drob. = Elymus tianschanigenus
turczaninovii (Drob.) Drob. = Elymus gmelinii
ugamica (Drob.) Drob. = Elymus nevskii

*Semiarundinaria Makino ex Nakai

*fastuosa (Mitf.) Makino ex Nakai (*Bambusa fastuosa* Mitf.) - 2

Sesleria Scop.

alba Smith (*S. anatolica* Deyl, *S. autumnalis* (Scop.) F. Schultz subsp. *anatolica* (Deyl) Tzvel., *S. autumnalis* auct.) - 1(?), 2
albicans Kit. (*Aira varia* Jacq. p.p. excl. typo, *Sesleria caerulea* (L.) Ard. subsp. *calcaria* (Opiz) Hegi, *S. calcaria* Opiz, *S. deyliana* A. & D. Löve, *S. varia* (Jacq.) Wettst. p.p. quoad pl.) - 1(?)
anatolica Deyl = S. alba
autumnalis auct. = S. alba
autumnalis (Scop.) F. Schultz subsp. *anatolica* (Deyl) Tzvel. = S. alba
bielzii Schur = S. coerulans
caerulea (L.) Ard. (*S. caerulea* subsp. *uliginosa* (Opiz) Celak., *S. uliginosa* Opiz, *S. varia* (Jacq.) Wettst. p.p. quoad nomen) - 1
- subsp. *calcaria* (Opiz) Hegi = S. albicans

- subsp. *uliginosa* (Opiz) Celak. = S. caerulea
calcaria Opiz = S. albicans
coerulans Friv. (*S. bielzii* Schur, *S. coerulans* subsp. *bielzii* (Schur) Gergely & Beldie) - 1
- subsp. *bielzii* (Schur) Gergely & Beldie = S. coerulans
deyliana A. & D. Löve = S. albicans
disticha (Wulf.) Pers. = Oreochloa disticha
heufleriana Schur - 1, 2
pavlovii Litv. = Poa attenuata
phleoies Stev. ex Roem. & Schult. - 2
uliginosa Opiz = S. caerulea
varia (Jacq.) Wettst. = S. albicans and S. caerulea

Setaria Beauv.

adhaerens (Forssk.) Chiov. (*Panicum adhaerens* Forssk.) - 1(alien), 2
ambigua (Guss.) Guss. = S. verticilliformis
decipiens Schimp. ex F.W. Schultz = S. verticilliformis
faberi Herrm. (*S. macrocarpa* Lucznik) - 1, 2, 5
gigantea (Franch. & Savat.) Makino = S. pycnocoma
glareosa V. Petrov = S. viridis subsp. glareosa
glauca auct. = S. pumila
gussonei Kerguelen = S. verticilliformis
intermedia (Roth.) Roem. & Schult. (*Panicum intermedium* (Roem. & Schult.) Roth, 1821, non Vahl ex Hornem. 1813, *P. tomentosum* Roxb., *Setaria tomentosa* (Roxb.) Kunth) - 2
italica (L.) Beauv. - 1, 2, 3, 4, 5, 6
- subsp. *pycnocoma* (Steud.) De Wet = S. pycnocoma
ketzchovelii Menabde & Eritsian = S. pycnocoma
lutescens (Weig.) F.T. Hubb. = S. pumila
macrocarpa Lucznik = S. faberi
pachystachys (Franch. & Savat.) Matsum. (*Panicum pachystachys* Franch. & Savat., *Setaria viridis* (L.) Beauv. subsp. *pachystachys* (Franch. & Savat.) Masam. & Yanag.) - 2, 3, 5
pumila (Poir.) Schult. (*Panicum lutescens* Weig., *Setaria lutescens* (Weig.) F.T. Hubb. nom. illegit., *S. glauca* auct.) - 1, 2, 3, 4, 5, 6
pycnocoma (Steud.) Henrard ex Nakai (*Panicum pycnocomum* Steud., *P. viride* L. var. *giganteum* Franch. & Savat., *Setaria gigantea* (Franch. & Savat.) Makino, *S. italica* (L.) Beauv. subsp. *pycnocoma* (Steud.) De Wet, *S. ketzchovelii* Menabde & Eritsian, *S. viridis* (L.) Beauv. subsp. *pycnocoma* (Steud.) Tzvel.) - 1, 2, 3, 4, 5, 6
tomentosa (Roxb.) Kunth = S. intermedia
verticillata (L.) Beauv. - 1, 2, 5, 6
verticilliformis Dumort. (*S. ambigua* (Guss.) Guss. 1842, non Merat, 1836, *S. decipiens* Schimp. ex F.W. Schultz, *S. gussonei* Kerguelen) - 1, 2, 6
viridis (L.) Beauv. (*S. weinmannii* Roem.& Schult.) - 1, 2, 3, 4, 5, 6
- subsp. **glareosa** (V. Petrov) Peschkova (*S. glareosa* V. Petrov) - 4
- subsp. *pachystachys* (Franch. & Savat.) Masam. & Yanag. = S. pachystachys
- subsp. **purpurascens** (Maxim.) Peschkova (*S. viridis* var. *purpurascens* Maxim.) - 4, 5
- subsp. *pycnocoma* (Steud.) Tzvel. = S. pycnocoma
- var. *purpurascens* Maxim. = S. viridis subsp. purpurascens
weinmannii Roem. & Schult. = S. viridis

*Shibataea Makino ex Nakai

*kumasaca (Zoll.) Makino ex Nakai (*Bambusa kumasasa* Zoll.) - 5

Sieglingia Bernh.

decumbens (L.) Bernh. - 1
- subsp. **decipiens** (O. Schwarz & Bassl.) Tzvel. (*Danthonia decumbens* (L.) DC. subsp. *decipiens* O. Schwarz & Bassl., *D. decipiens* (O. Schwarz & Bassl.) A. & D. Löve, *Sieglingia decumbens* subsp. *decipiens* (O.Schwarz & Bassl.) Soo, comb. superfl.) - 1, 2

*Sinarundinaria Nakai (*Arundinaria* auct.)

*****nitida** (Mitf.ex Stapf) Nakai (*Arundinaria nitida* Mitf. ex Stapf) - 1, 2

Sitopsis (Jaub. & Spach) A. Löve = Aegilops

speltoides (Tausch) A. Löve = Aegilops speltoides

*Sorghum Moench

* x **almum** Parodi. - S. halepense (L.) Pers. x S. sudanense (Piper) Stapf - 6
*****bicolor** (L.) Moench (*S. japonicum* (Hack.) Roshev. p.p. quoad nomen, *S. saccharatum* (L.) Moench var. *bicolor* (L.) Kerguelen, *S. vulgare* sensu Roshev.) - 1, 2, 5, 6
- var. *cernuum* (Ard.) Ghisa = S. cernuum
- var. *saccharatum* (L.) Mohlenbrock = S. saccharatum
*****cernuum** (Ard.) Host (*Holcus cernuus* Ard., *Sorghum bicolor* (L.) Moench var. *cernuum* (Ard.) Ghisa, *S. vulgare* Pers.) - 1, 6
* x **derzhavinii** Tzvel. - S. halepense (L.) Pers. x S. saccharatum (L.) Moench
dochna (Forssk.) Snowd. = S. saccharatum
- var. *technicum* (Koern.) Snowd. = S. saccharatum
*****durra** (Forssk.) Stapf (*Holcus durra* Forssk.) - 6
*****halepense** (L.) Pers. - 1, 2, 5, 6
- var. *sudanense* (Piper) Soo = S. sudanense
japonicum (Hack.) Roshev. = S. bicolor and S. nervosum
*****nervosum** Bess. (*S. japonicum* (Hack.) Roshev. p.p. quoad pl.) - 4, 5
*****saccharatum** (L.) Moench (*Holcus dochna* Forssk., *Sorghum bicolor* (L.) Moench var. *saccharatum* (L.) Mohlenbrock, *S. dochna* (Forssk.) Snowd., *S. dochna* var. *technicum* (Koern.) Snowd., *S. technicum* (Koern.) Batt. & Trab.) - 1, 2, 3, 4, 5, 6
- var. *bicolor* (L.) Kerguelen = S. bicolor
- var. *sudanense* (Piper) Kerguelen = S. sudanense
*****sudanense** (Piper) Stapf (*Andropogon halepensis* (L.) Brot. var. *sudanensis* (Piper) Suss., *Sorghum halepense* (L.) Pers. var. *sudanense* (Piper) Soo, *S. saccharatum* (L.) Moench var. *sudanense* (Piper) Kerguelen) - 1, 2, 3, 6
technicum (Koern.) Batt. & Trab. = S. saccharatum
vulgare Pers. = S. cernuum
vulgare sensu Roshev. = S. bicolor

Sphenopus Trin.

divaricatus (Gouan) Reichenb. - 2, 6

Spodiopogon Trin.

sibiricus Trin. (*Saccharum sibiricum* (Trin.) Roberty) - 4, 5

Sporobolus R. Br.

fertilis (Steud.) W. Clayton (*Agrostis fertilis* Steud., *Sporobolus indicus* auct.) - 2
indicus auct. = S. fertilis

*Stenotaphrum Trin.

*****secundum** (Walt.) O. Kuntze (*Ischaemum secundum* Walt.)

Stephanachne Keng, p.p. = Pappagrostis

pappophorea (Hack.) Keng = Pappagrostis pappophorea

Stilpnophleum Nevski = Calamagrostis

anthoxanthoides (Munro) Nevski = Calamagrostis anthoxanthoides
laguroides (Regel) Nevski = Calamagrostis laguroides

Stipa L.

adoxa Klok. & Ossycznjuk - 1
aktauensis Roshev. - 6
alaica Pazij - 6
anisotricha P. Smirn. (*S. sareptana* A. Beck. subsp. *anisotricha* (P. Smirn.) Tzvel.) - 2
anomala P. Smirn. (*S. borysthenica* Klok. ex Prokud., *S. joannis* Celak. subsp. *sabulosa* (Pacz.) Lavr., *S. pennata* L. subsp. *joannis* (Celak.) Pacz. f. *sabulosa* Pacz., *S. pennata* subsp. *sabulosa* (Pacz.) Tzvel., *S. pennata* subsp. *sabulosa* var. *anomala* (P. Smirn.) Tzvel., *S. sabulosa* (Pacz.) Sljussarenko) - 1, 2, 3, 4, 6
arabica Trin. & Rupr. (*S. arabica* subsp. *arabica* var. *meyeriana* (Trin. & Rupr.) Tzvel., *S. arabica* var. *S. meyeriana* Trin. & Rupr., *S. meyeriana* (Trin. & Rupr.) Grossh., *S. szovitsiana* (Trin.) Griseb. var. *meyeriana* (Trin. & Rupr.) Roshev.) - 2, 6
- subsp. *arabica* var. *meyeriana* (Trin. & Rupr.) Tzvel. = S. arabica
- subsp. *caspia* (C. Koch) Tzvel. = S. caspia
- subsp. *prilipkoana* (Grossh.) Tzvel. = S. prilipkoana
- var. *pamirica* (Roshev.) Freitag = S. pamirica
- var. *prilipkoana* (Grossh.) Freitag = S. prilipkoana
- var. *S. meyerana* Trin. & Rupr. = S. arabica
- var. *S. szovitsiana* Trin. = S. caesia
araxensis Grossh. (? *S. lasiopoda* P. Smirn., *S. pennata* L. subsp. *pulcherrima* (C. Koch) Freitag var. *araxensis* (Grossh.) Freitag, *S. pulcherrima* C. Koch subsp. *araxensis* (Grossh.) Tzvel.) - 2
armeniaca P. Smirn. = S. ehrenbergiana
asperella Klok. & Ossycznjuk - 1
attenuata P. Smirn. = S. baicalensis
austroaltaica Kotuchov - 3
badachschanica Roshev. - 6
- subsp. *pamirica* (Roshev.) Tzvel. = S. pamirica
baicalensis Roshev. (*S. attenuata* P. Smirn.) - 3, 4, 5
barbata Desf. - 2
barchanica Lomonosova - 4
bella Drob. = S. drobovii
borysthenica Klok. ex Prokud. = S. anomala
brachyptera Klok. = S. syreistschikowii
brauneri (Pacz.) Klok. (*S. lessingiana* Trin. & Rupr. subsp. *brauneri* Pacz., *S. lessingiana* var. *brauneari* (Pacz.) Roshev.) - 1, 2
breviflora Griseb. - 6
bromoides (L.) Doerfl. = Achnatherum bromoides
bungeana Trin. - 6
canescens P. Smirn. (*S. zalesskii* Wilensky subsp. *canescens* (P. Smirn.) Tzvel.) - 2
capensis Thunb. (*S. retorta* Cav., *S. tortilis* Desf.) - 2, 6
capillata L. - 1, 2, 3, 4, 6
caragana Trin. = Achnatherum caragana

caspia C. Koch (*S. arabica* Trin. & Rupr. subsp. *caspia* (C. Koch) Tzvel., *S. arabica* var. *S. szovitsiana* Trin., *S. koenigii* Woronow, *S. szovitsiana* (Trin.) Griseb., *S. turgaica* Roshev.) - 1, 2, 6

caucasica Schmalh. - 2, 6
- subsp. *desertorum* (Roshev.) Tzvel. = S. desertorum
- subsp. *drobovii* Tzvel. = S. drobovii
- subsp. *glareosa* (P. Smirn.) Tzvel. = S. glareosa
- subsp. *iskanderkulica* Tzvel. = S. iskanderkulica
- var. *desertorum* (Roshev.) Tzvel. = S. desertorum
- f. *desertorum* Roshev. = S. desertorum

cerariorum Panc. = S. tirsa
confusa Litv. = Achnatherum confusum

consanguinea Trin. & Rupr. - 3

crassiculmis P. Smirn. (*S. pulcherrima* C. Koch subsp. *crassiculmis* (P. Smirn.) Tzvel.) - 1, 6
- subsp. **euroanatolica** Martinovsky - 1

cretacea P. Smirn. - 1

czekanowskii V. Petrov = Ptilagrostis mongholica

daghestanica Grossh. - 2

dasyphylla (Lindem.) Trautv. (*S. pennata* L. var. *dasyphylla* Lindem.) - 1, 2, 3

decipiens P. Smirn. = S. krylovii
densa P. Smirn. = S. krylovii
densiflora P. Smirn. = S. krylovii

desertorum (Roshev.) Ikonn. (*S. caucasica* Schmalh. f. *desertorum* Roshev., *S. caucasica* subsp. *desertorum* (Roshev.) Tzvel., *S. caucasica* var. *desertorum* (Roshev.) Tzvel.) - 3, 6

disjuncta Klok. = S. pennata

drobovii (Tzvel.) Czer. (*S. caucasica* Schmalh. subsp. *drobovii* Tzvel., *S. bella* Drob. 1925, non Phil. 1870) - 2, 6

effusa (Maxim.) Nakai ex Honda = Achnatherum extremiorientale

ehrenbergiana Trin. & Rupr. (*S. armeniaca* P. Smirn.) - 2

epilosa Martinovsky (*S. pulcherrima* C. Koch subsp. *epilosa* (Martinovsky) Tzvel.) - 2, 6

eriocaulis Borb. subsp. *lithophila* (P. Smirn.) Tzvel. = S. lithophila

extremiorientalis Hara = Achnatherum extremiorientale

fallacina Klok. & Ossycznjuk - 1

gaubae Bor (*S. nachiczevanica* Musajev & Sadychov) - 2

gegarkunii P. Smirn. - 2

gigantea auct. = S. pellita
gigantea Lag. var. *S. pellita* Trin. & Rupr. = S. pellita
glabrata P. Smirn. = S. zalesskii

glabrinoda Klok. - 1

glareosa P. Smirn. (*S. caucasica* Schmalh. subsp. *glareosa* (P. Smirn.) Tzvel.) - 3, 4, 6

gnezdilloi Pazij - 6

gobica Roshev. - 3(?)
- var. *klemenzii* (Roshev.) Norlindh = S. klemenzii
gobica sensu M. Pop. = S. klemenzii

gracilis Roshev. - 6

grafiana Stev. = S. pulcherrima

grandis P. Smirn. - 4

graniticola Klok. - 1

heptapotamica Golosk. - 6

heterophylla Klok. - 1

hohenackeriana Trin. & Rupr. (*S. hohenackeriana* subsp. *grossheimii* Tzvel.) - 2, 3, 6
- subsp. *grossheimii* Tzvel. = S. hohenackeriana
- subsp. **nachiczevanica** Tzvel. (*S. hohenackeriana* subsp. *ordubadica* Tzvel.) - 2, 6
- subsp. *ordubadica* Tzvel. = S. hohenackeriana subsp. nachiczevanica

holosericea Trin. & Rupr. - 2
- subsp. *transcaucasica* (Grossh.) Tzvel. = S. transcaucasica

ikonnikovii Tzvel. - 6

iljinii Roshev. = S. zalesskii

iskanderkulica (Tzvel.) Czer. (*S. caucasica* Schmalh. subsp. *iskanderkulica* Tzvel.) - 6

issaevii Musajev & Sadychov - 2

jagnobica Ovcz. & Czuk. = S. kuhitangi

joannis Celak. = S. pennata
- subsp. *penicillifera* (Pacz.) Lavr. = S. pennata
- subsp. *sabulosa* (Pacz.) Lavr. = S. anomala

karataviensis Roshev. - 6

karjaginii Musajev & Sadychov - 2

kirghisorum P. Smirn. (*S. pennata* L. subsp. *kirghisorum* (P.Smirn.) Freitag) - 3, 6

klemenzii Roshev. (*S. gobica* Roshev. var. *klemenzii* (Roshev.) Norlindh, *S. gobica* sensu M. Pop.) - 4

koenigii Woronow = S. caspia

komarovii P. Smirn. = Achnatherum extremiorientale

kopetdaghensis Czopanov - 6

korshinskyi Roshev. - 1, 3, 6

krascheninnikowii Roshev. = S. ucrainica

krylovii Roshev. (*S. decipiens* P. Smirn., *S. densa* P. Smirn., *S. densiflora* P. Smirn. 1929, non Hughes, 1921) - 3, 4, 6

kuhitangi Drob. (*S. jagnobica* Ovcz. & Czuk., *S. richteriana* Kar. & Kir. subsp. *jagnobica* (Ovcz. & Czuk.) Tzvel.) - 6

kungeica Golosk. - 6

kurdistanica Bor, p.p. = Achnatherum turcomanicum

lagascae auct. = S. pellita

lasiopoda P. Smirn. = S. araxensis

lejophylla P. Smirn. (*S. pennata* L. subsp. *lejophylla* (P. Smirn.) Tzvel.) - 2

lessingiana Trin. & Rupr. - 1, 2, 3, 6
- subsp. *brauneri* Pacz. = S. brauneri
- var. *brauneari* (Pacz.) Roshev. = S. brauneri

lingua Junge - 6

lipskyi Roshev. - 6

lithophila P. Smirn. (*S. eriocaulis* Borb. subsp. *lithophila* (P. Smirn.) Tzvel., *S. pennata* L. subsp. *lithophila* (P. Smirn.) Martinovsky) - 1

litwinowiana P. Smirn. = Achnatherum turcomanicum

longifolia Borb. = S. tirsa

longiplumosa Roshev. - 6

macroglossa P. Smirn. - 6

maeotica Klok. & Ossycznjuk - 1

magnifica Junge - 6

majalis Klok. - 1

manrakica Kotuchov - 3

margelanica P. Smirn. - 6

martinovskyi Klok. (*S. rubens* P. Smirn. subsp. *sublaevis* Martinovsky) - 1

meyeriana (Trin. & Rupr.) Grossh. = S. arabica

mongholica Turcz. ex Trin. var. *minutiflora* Titov ex Roshev. = Ptilagrostis minutiflora

nachiczevanica Musajev & Sadychov = S. gaubae

okmirii Dengubenko - 6

oreades Klok. - 1

orientalis Trin. - 1, 3, 4, 6

ovczinnikovii Roshev. - 6

pamirica Roshev. (*S. arabica* Trin. & Rupr. var. *pamirica* (Roshev.) Freitag, *S. badachschanica* Roshev. subsp. *pamirica* (Roshev.) Tzvel.) - 6

paradoxa (Junge) P. Smirn. = S. syreistschikowii

pellita (Trin. & Rupr.) Tzvel. (*S. gigantea* Lag. var. *S. pellita* Trin. & Rupr., *S. gigantea* auct., *S. lagascae* auct.) - 2

pennata L. (*S. disjuncta* Klok., *S. joannis* Celak., *S. joannis* subsp. *penicillifera* (Pacz.) Lavr., *S. pennata* subsp. *joannis* (Celak.) Hyl. comb. superfl., *S. pennata* subsp. *joannis* (Celak.) Pacz., *S. pennata* subsp. *joannis* f. *penicillifera* Pacz.) - 1, 2, 3, 4, 6
- subsp. *grafiana* (Stev.) K. Richt. = S. pulcherrima
- subsp. *joannis* (Celak.) Hyl. = S. pennata
- subsp. *joannis* (Celak.) Pacz. = S. pennata
- - f. *penicillifera* Pacz. = S. pennata
- - f. *sabulosa* Pacz. = S. anomala
- subsp. *kirghisorum* (P.Smirn.) Freitag = S. kirghisorum
- subsp. *lejophylla* (P. Smirn.) Tzvel. = S. lejophylla
- subsp. *pulcherrima* (C. Koch) Freitag = S. pulcherrima
- - var. *araxensis* (Grossh.) Freitag = S. araxensis
- subsp. *pulcherrima* (C. Koch) A. & D. Löve = S. pulcherrima
- subsp. *sabulosa* (Pacz.) Tzvel. = S. anomala
- - var. *anomala* (P. Smirn.) Tzvel. = S. anomala
- subsp. *zalesskii* (Wilensky) Freitag = S. zalesskii
- var. *dasyphylla* Lindem. = S. dasyphylla
- var. *stenophylla* Lindem. = S. tirsa
poetica Klok. - 1
pontica P. Smirn. (*S. zalesskii* Wilensky subsp. *pontica* (P. Smirn.) Tzvel.) - 1, 2
praecapillata Alech. (*S. sareptana* A. Beck. subsp. *praecapillata* (Alech.) Tzvel.) - 1
prilipkoana Grossh. (*S. arabica* Trin. & Rupr. var. *prilipkoana* (Grossh.) Freitag, *S. arabica* subsp. *prilipkoana* (Grossh.) Tzvel.) - 2, 6
pseudocapillata Roshev. (*S. spiridonovii* Roshev.) - 6
pulcherrima C. Koch (*S. grafiana* Stev., *S. pennata* L. subsp. *grafiana* (Stev.) K. Richt., *S. pennata* subsp. *pulcherrima* (C. Koch) Freitag, comb. superfl., *S. pennata* subsp. *pulcherrima* (C. Koch) A. & D. Löve, *S. pulcherrima* subsp. *grafiana* (Stev.) Pacz.) - 1, 2, 3, 6
- subsp. *araxensis* (Grossh.) Tzvel. = S. araxensis
- subsp. *crassiculmis* (P. Smirn.) Tzvel. = S. crassiculmis
- subsp. *epilosa* (Martinovsky) Tzvel. = S. epilosa
- subsp. *grafiana* (Stev.) Pacz. = S. pulcherrima
purpurea Griseb. (*Ptilagrostis purpurea* (Griseb.) Roshev.) - 6
regeliana Hack. - 6
retorta Cav. = S. capensis
richteriana Kar. & Kir. - 3, 6
- subsp. *jagnobica* (Ovcz. & Czuk.) Tzvel. = S. kuhitangi
rubens P. Smirn. = S. zalesskii
- subsp. *sublaevis* Martinovsky = S. martinovskyi
rubentiformis P. Smirn. = S. zalesskii
sabulosa (Pacz.) Sljussarenko = S. anomala
sareptana A. Beck. - 1, 2, 3, 4, 6
- subsp. *anisotricha* (P. Smirn.) Tzvel. = S. anisotricha
- subsp. *praecapillata* (Alech.) Tzvel. = S. praecapillata
sczerbakovii Kotuchov - 3
setulosissima Klok. - 1
sibirica (L.) Lam. = Achnatherum sibiricum
sosnowskyi Seredin - 2
spiridonovii Roshev. = S. pseudocapillata
splendens Trin. = Achnatherum splendens
stenophylla (Lindem.) Trautv. = S. tirsa
subsessiliflora (Rupr.) Roshev. (*Ptilagrostis subsessiliflora* (Rupr.) Roshev.) - 3, 6
syreistschikowii P. Smirn. (*S. brachyptera* Klok., *S. paradoxa* (Junge) P. Smirn. 1927, non Rasp. 1825) - 1, 2
szovitsiana (Trin.) Griseb. = S. caspia
- var. *meyeriana* (Trin. & Rupr.) Roshev. = S. arabica
talassica Pazij - 6
tianschanica Roshev. - 6

tirsa Stev. (*S. cerariorum* Panc., *S. longifolia* Borb., *S. pennata* L. var. *stenophylla* Lindem., *S. stenophylla* (Lindem.) Trautv.) - 1, 2, 3, 6
tortilis Desf. = S. capensis
transcarpatica Klok. - 1
transcaucasica Grossh. (*S. holosericea* Trin. subsp. *transcaucasica* (Grossh.) Tzvel.) - 2, 6
trichoides P. Smirn. (*S. turkestanica* Hack. subsp. *trichoides* (P. Smirn.) Tzvel.) - 6
turcomanica P. Smirn. (*S. zalesskii* Wilensky subsp. *turcomanica* (P. Smirn.) Tzvel.) - 6
turgaica Roshev. = S. caspia
turkestanica Hack. - 6
- subsp. *trichoides* (P. Smirn.) Tzvel. = S. trichoides
tzvelevii Ikonn. - 6
ucrainica P. Smirn. (*S. krascheninnikowii* Roshev., *S. zalesskii* Wilensky subsp. *ucrainica* (P. Smirn.) Tzvel.) - 1, 2, 3, 6
zaissanica Kotuchov - 3
zalesskii Wilensky (*S. glabrata* P. Smirn., *S. iljinii* Roshev., *S. pennata* L. subsp. *zalesskii* (Wilensky) Freitag, *S. rubens* P. Smirn., *S. rubentiformis* P. Smirn.) - 1, 2, 3, 4, 6
- subsp. *canescens* (P. Smirn.) Tzvel. = S. canescens
- subsp. *pontica* (P. Smirn.) Tzvel. = S. pontica
- subsp. *turcomanica* (P. Smirn.) Tzvel. = S. turcomanica
- subsp. *ucrainica* (P. Smirn.) Tzvel. = S. ucrainica
zuvantica Tzvel. - 2

Stipagrostis Nees (*Arthratherum* auct.)

arachnoidea (Litv.) de Winter (*Aristida arachnoidea* Litv., *Arthratherum arachnoideum* (Litv.) Tzvel.) - 6
karelinii (Trin. & Rupr.) Tzvel. (*Aristida karelinii* (Trin. & Rupr.) Roshev., *Arthratherum karelinii* (Trin. & Rupr.) Tzvel.) - 1, 6
pennata (Trin.) de Winter (*Aristida pennata* Trin., *Arthratherum pennatum* (Trin.) Tzvel.) - 1, 3, 6
- subsp. **minor** (Litv.) Tzvel. (*Aristida pennata* var. *minor* Litv., *Arthratherum pennatum* subsp. *minus* (Litv.) Tzvel.) - 6
plumosa (L.) Munro ex T. Anders. (*Aristida plumosa* L., *Arthratherum plumosum* (L.) Nees) - 2, 6
- subsp. **kyzylkumica** (Tzvel.) Tzvel. (*Aristida plumosa* subsp. *kyzylkumica* Tzvel., *Arthratherum plumosum* subsp. *kyzylkumicum* (Tzvel.) Tzvel.) - 6
- subsp. **szovitsiana** (Trin. & Rupr.) Tzvel. (*Aristida plumosa* var. *szovitsiana* Trin. & Rupr., *A. plumosa* subsp. *szovitsiana* (Trin. & Rupr.) Tzvel., *Arthratherum plumosum* subsp. *szovitsianum* (Trin. & Rupr.) Tzvel., *Stipagrostis szovitsiana* (Trin. & Rupr.) Musajev) - 2, 6
szovitsiana (Trin. & Rupr.) Musajev = S. plumosa subsp. szovitsiana

Syntherisma Walt.

vulgaris Schrad. = Digitaria sanguinalis

Taeniatherum Nevski

asperum (Simonk.) Nevski (*Hordelymus asper* (Simonk.) Savul. & E.I. Nyarady, *Taeniatherum caput-medusae* (L.) Nevski subsp. *asperum* (Simonk.) Melderis) - 1, 2, 6
caput-medusae (L.) Nevski subsp. *asperum* (Simonk.) Melderis = T. asperum
- subsp. *crinitum* (Schreb.) Melderis = T. crinitum
- var. *crinitum* (Schreb.) C.J. Humphries = T. crinitum

crinitum (Schreb.) Nevski (*Elymus caput-medusae* L. subsp. *crinitus* (Schreb.) Maire & Weiller, *Hordelymus caput-medusae* (L.) Pignatti subsp. *crinitus* (Schreb.) Pignatti, *Taeniatherum caput-medusae* (L.) Nevski var. *crinitum* (Schreb.) C.J. Humphries, *T. caput-medusae* subsp. *crinitum* (Schreb.) Melderis) - 1, 2, 6

Tetrapogon Desf.

villosus Desf. (*Chloris villosa* (Desf.) Pers.) - 6

Thinopyrum A. Löve = Elytrigia

bessarabicum (Savul. & Rayss) A. Löve = Elytrigia bessarabica

junceiforme (A. & D. Löve) A. Löve = Elytrigia junceiformis

Timouria Roshev. = Achnatherum

saposhnikovi Roshev. = Achnatherum saposhnikovii

Torreyochloa Church

natans (Kom.) Church (*Glyceria natans* Kom., *Torreyochloa pallida* (Torr.) Church var. *natans* (Kom.) B. Boivin, *T. pallida* subsp. *natans* (Kom.) T. Kayama & Kawano, *Glyceria pallida* auct.) - 5

pallida (Torr.) Church subsp. *natans* (Kom.) T. Kayama & Kawano = T. natans

- var. *natans* (Kom.) B. Boivin = T. natans

Trachynia Link

distachya (L.) Link (*Bromus distachyos* L., *Brachypodium distachyon* (L.) Beauv., ? *B. geniculatum* C. Koch) - 1, 2, 6

pentastachya (Ten.) Tzvel. (*Bromus pentastachyos* Ten.) - 2

Tragus Hall.

racemosus (L.) All. - 1, 2, 6

Triarrhena (Maxim.) Nakai = Miscanthus

sacchariflora (Maxim.) Nakai = Miscanthus sacchariflorus

Tripogon Roem. & Schult.

chinensis (Franch.) Hack. - 4, 5

Trisetaria Forssk.

cavanillesii (Trin.) Maire (*Trisetum cavanillesii* Trin., *Lophochloa cavanillesii* (Trin.) Bor) - 1, 2, 6
- subsp. **sabulosa** Tzvel. - 2, 6

cristata (L.) Kerguelen = Rostraria cristata

linearis Forssk. (*Trisetum lineare* (Forssk.) Boiss.) - 2

phleoides (Vill.) Nevski = Rostraria cristata

x **Trisetokoeleria** Tzvel.

gorodkowii (Roshev.) Tzvel. (*Koeleria gorodkowii* Roshev.). - Koeleria asiatica Domin x Trisetum litorale (Rupr. ex Roshev.) Czer. - 3

jurtzevii Probat. - Trisetum spicatum (L.) K. Richt. x Koeleria asiatica Domin - 5

taimyrica Tzvel. - Koeleria asiatica Domin x Trisetum agrostideum (Laest.) Fries - 4, 5

Trisetum Pers.

agrostideum (Laest.) Fries (*Avena subspicata* L. var. *agrostidea* Laest., *Trisetum subalpestre* (C. Hartm.) L. Neum.) - 4, 5

alascanum Nash (*T. molle* Kunth subsp. *alascanum* (Nash) Rebr., *T. spicatum* (L.) K. Richt. subsp. *alascanum* (Nash) Hult.) - 5

alpestre (Host) Beauv. (*Avena alpestris* Host) - 1
- subsp. **glabrescens** (Schur) Tzvel. (*T. alpestre* var. *glabrescens* Schur) - 1

altaicum Roshev. - 3, 4, 6

buschianum Seredin - 2

carpaticum auct. = T. ciliare

cavanillesii Trin. = Trisetaria cavanillesii

ciliare (Kit.) Domin (*Avena ciliaris* Kit., *A. fusca* Kit. 1814, non Ard. 1789, *T. fuscum* Roem. & Schult., *T. carpaticum* auct.) - 1

cristatum (L.) Potztal = Rostraria cristata

distichophyllum auct. = T. transcaucasicum

fedtschenkoi Henrard = T. seravschanicum

flavescens (L.) Beauv. (*T. pratense* Pers.) - 1, 2, 6
- subsp. *parvispiculatum* Tzvel. = T. parvispiculatum
- subsp. **tatricum** Chrtek - 1

fuscum Roem. & Schult. = T. ciliare

lineare (Forssk.) Boiss. = Trisetaria linearis

litorale (Rupr. ex Roshev.) A. Khokhr. (*T. sibiricum* Rupr. subsp. *litorale* Rupr. ex Roshev., *T. litorale* (Rupr. ex Roshev.) Czer. comb. superfl.) - 1, 3, 4, 5

litwinowii (Domin) Nevski (*Koeleria litwinowii* Domin) - 6

macrotrichum Hack. - 1

molle Kunth (*Avena mollis* Michx. 1803, non Salisb. 1796, *Trisetum spicatum* (L.) K. Richt. subsp. *molle* (Kunth) Hult., *T. triflorum* (Bigel.) A. & D. Löve subsp. *molle* (Kunth) A. & D. Löve) - 4, 5
- subsp. *alascanum* (Nash) Rebr. = T. alascanum

mongolicum (Hult.) Peschkova (*T. spicatum* (L.) K. Richt. subsp. *mongolicum* Hult.) - 3, 4, 6

obtusiflorum (Boiss.) Potztal = Rostraria obtusiflora

ovatipaniculatum (Hult.) Galushko (*T. spicatum* (L.) K. Richt. subsp. *ovatipaniculatum* Hult.) - 2

parvispiculatum (Tzvel.) Probat. (*T. flavescens* (L.) Beauv. subsp. *parvispiculatum* Tzvel.) - 2, 6

pilosum Roem. & Schult. = Avena eriantha

pratense Pers. = T. flavescens

rigidum (Bieb.) Roem. & Schult. - 1(?), 2
- subsp. *teberdense* (Litv.) Tzvel. = T. teberdense
- var. *teberdense* Litv. = T. teberdense

seravschanicum Roshev. (*Avena flavescens* L. var. *virescens* Regel, *Trisetum fedtschenkoi* Henrard, *T. spicatum* (L.) K. Richt. subsp. *virescens* (Regel) Tzvel., *T. virescens* (Regel) Roshev. 1914, non Nees ex Steud. 1854) - 5, 6

sesquiflorum Trin. = Calamagrostis sesquiflora

sibiricum Rupr. - 1, 3, 4, 5, 6
- subsp. *litorale* Rupr. ex Roshev. = T. litorale
- subsp. *umbratile* (Kitag.) Tzvel. = T. umbratile
- var. *umbratile* Kitag. = T. umbratile

spicatum (L.) K. Richt. - 1, 3, 4, 5, 6
- subsp. *alascanum* (Nash) Hult. = T. alascanum
- subsp. *molle* (Kunth) Hult. = T. molle
- subsp. *mongolicum* Hult. = T. mongolicum
- subsp. *ovatipaniculatum* Hult. = T. ovatipaniculatum
- subsp. *virescens* (Regel) Tzvel. = T. seravschanicum
- subsp. *wrangelense* Petrovsky = T. wrangelense

subalpestre (C. Hartm.) L. Neum. = T. agrostideum

teberdense (Litv.) Charadze (*T. rigidum* (Bieb.) Roem. &

Schult. var. *teberdense* Litv., *T. rigidum* subsp. *teberdense* (Litv.) Tzvel.) - 2

transcaucasicum Seredin (*T. distichophyllum* auct.) - 2

triflorum (Bigel.) A. & D. Löve subsp. *molle* (Kunth.) A. & D. Löve = T. molle

turcicum Chrtek - 2

umbratile (Kitag.) Kitag. (*T. sibiricum* Rupr. var. *umbratile* Kitag., *T. sibiricum* subsp. *umbratile* (Kitag.) Tzvel.) - 5

virescens (Regel) Roshev. = T. seravschanicum

wrangelense (Petrovsky) Probat. (*T. spicatum* (L.) K. Richt. subsp. *wrangelense* Petrovsky) - 5

*x **Triticale** Muntz. (x *Triticosecale* Wittm.)

***rimpaui** (Wittm.) Muntz. (x *Triticosecale rimpaui* Wittm.). - Secale cereale L. x Triticum aestivum L.

x *Triticosecale* Wittm. = x Triticale

rimpaui Wittm. = x Triticale rimpaui

Triticum L. (*Critholidium* Link, *Gigachilon* Seidl)

aegilopoides (Link) Bal. ex Koern. = T. boeoticum

*****aestivum** L. - 1, 2, 3, 4, 5, 6

- subsp. *compactum* (Host) Thell. = T. compactum
- subsp. *durum* (Desf.) Thell. = T. durum
- subsp. *hadropyrum* (Flaksb.) Tzvel. = T. asiaticum
- subsp. *inflatum* (Kudr.) Tzvel. = T. inflatum
- subsp. *macha* (Dekapr. & Menabde) Mac Key = T. macha
- subsp. *spelta* (L.) Thell. = T. spelta
- subsp. *transcaucasicum* V. Dorof. & Laptev = T. spelta
- subsp. *turgidum* (L.) Domin = T. turgidum
- subsp. *vavilovii* (Jakubz.) A. Löve = T. vavilovii
- subsp. *vavilovii* (Jakubz.) Sears = T. vavilovii

afghanicum Kudr. (*T. compactum* Host subsp. *irano-asiaticum* Flaksb.) - 6

angustifolium Link = Agropyron fragile

angustum (Trin.) F. Herm. = Leymus angustus

aralense (Regel) F. Herm. = Leymus multicaulis

araraticum Jakubz. (*Gigachilon timopheevii* (Zhuk.) A. Löve subsp. *armeniacum* (Jakubz.) A. Löve, *Triticum armeniacum* (Jakubz.) Makuschina, 1938, non Nevski, 1934, *T. chaldicum* Menabde, *T. dicoccoides* (Koern. ex Aschers. & Graebn.) Schweinf. subsp. *armeniacum* Jakubz., *T. montanum* Makuschina, nom. illegit., *T. nikolai* Fed. & Takht. ex Zhuk. nom. nud., *T. timopheevi* (Zhuk.) Zhuk. subsp. *araraticum* (Jakubz.) Mac Key, nom. illegit., *T. timopheevii* var. *araraticum* (Jakubz.) C. Yen, *T. turgidum* L. subsp. *armeniacum* (Jakubz.) A. & D. Löve, *T. dicoccoides* auct.) - 2

arctasianum F. Herm. = Elymus sibiricus

arenarium (L.) F. Herm. = Leymus arenarius

armeniacum (Jakubz.) Makuschina = T. araraticum

armeniacum (Stolet.) Nevski (*T. diccocon* (Schrank) Schuebl. subsp. *asiaticum* Vav. p.p. incl. typo, *T. dicoccon* subsp. *euroum* Flaksb. p.p. incl. typo) - 2

asiaticum Kudr. (*T. aestivum* L. subsp. *hadropyrum* (Flaksb.) Tzvel., *T. vulgare* Vill. subsp. *hadropyrum* Flaksb., *T. vulgare* subsp. *irano-asiaticum* Flaksb. p.p. incl. typo) - 1, 6

- subsp. **eligulatum** (Vav.) Kudr. (*T. vulgare* grex *eligulatum* Vav.) - 6
- subsp. **muticorigidum** Kudr. - 6
- subsp. **rigidum** (Vav.) Kudr. (*T. vulgare* grex *rigium* Vav.) - 6

- subsp. **semiaristatum** Kudr. - 6
- subsp. **speltiforme** (Vav.) Kudr. (*T. vulgare* grex *speltiforme* Vav.) - 6

biunciale (Vis.) K. Richt. = Aegilops biuncialis

boeoticum Boiss. (*Crithodium aegilopoides* Link, *C. jerevani* (Thum.) A. Löve, nom. invalid., *C. monococcum* (L.) A. Löve subsp. *aegilopoides* (Link) A. Löve, *Triticum aegilopoides* (Link) Bal. ex Koern. 1885, non Forssk. 1775, *T. boeoticum* subsp. *aegilopoides* (Link) Grossh., *T. boeoticum* subsp. *thaoudar* (Reut. ex Hausskn.) Grossh., *T. boeoticum* subsp. *thaoudar* (Reut. ex Hausskn.) Schiemann, comb. superfl., *T. jerevani* Thum. nom. invalid., *T. monococcum* L. subsp. *boeoticum* (Boiss.) Hayek, *T. monococcum* subsp. *boeoticum* (Boiss.) A. & D. Löve, comb. superfl., *T. monococcum* subsp. *boeoticum* (Boiss.) C. Yen, comb. superfl., *T. monococcum* subsp. *thaoudar* (Reut. ex Hausskn.) Zhuk. comb. invalid., *T. spontaneum* Flaksb. nom. illegit., *T. spontaneum* subsp. *aegilopoides* (Link) Flaksb. nom. illegit., *T. spontaneum* subsp. *thaoudar* (Reut. ex Hausskn.) Flaksb. nom. illegit., *T. thaoudar* Reut. ex Hausskn.) - 1, 2

- subsp. *aegilopoides* (Link) Grossh. = T. boeoticum
- subsp. *thaoudar* (Reut. ex Hausskn.) Grossh. = T. boeoticum
- subsp. *thaoudar* (Reut. ex Hausskn.) Schiemann = T. boeoticum
- subsp. *urartu* (Thum.) V. Dorof. = T. urartu

x **borisovii** Zhebrak. - T. aestivum L. x T. timopheevii (Zhuk.) Zhuk.

bungeanum Trin. = Elytrigia geniculata

caninum L. var. *gmelinii* Ledeb. = Elymus gmelinii

carthlicum Nevski (*Gigachilon polonicum* (L.) Seidl subsp. *carthlicum* (Nevski) A. Löve, *Triticum ibericum* Menabde, nom. illegit., *T. paradoxum* Parodi, *T. persicum* Vav. ex Zhuk. 1923, non Aitch. & Hemsl. 1888, *T. turgidum* L. subsp. *carthlicum* (Nevski) A. & D. Löve) - 2

caucasicum Spreng. = Agropyron pectinatum

chaldicum Menabde = T. araraticum

chinense Trin. = Leymus chinensis

columnare (Zhuk.) Morris & Sears = Aegilops columnaris

compactum Host (*T. aestivum* L. subsp. *compactum* (Host) Thell.) - 1(?), 2, 3, 4, 5, 6

- subsp. *irano-asiaticum* Flaksb. = T. afghanicum
- subsp. **muticocompactum** Kudr. - 6
- subsp. **rigidocompactum** Kudr. - 6
- subsp. **semiaristocompactum** Kudr. - 6

compositum L. = T. turgidum

crassum (Boiss.) Aitch. & Hemsl. = Aegilops crassa

cylindricum (Host) Cesati = Aegilops cylindrica

dasystachys (Trin.) F. Herm. = Leymus dasystachys

dicoccoides auct. = T. araraticum

dicoccoides (Koern. ex Aschers. & Graebn.) Schweinf. subsp. *armeniacum* Jakubz. = T. araraticum

*****dicoccon** (Schrank) Schuebl. (*T. spelta* L. var. *dicoccon* Schrank, *Gigachilon polonicum* (L.) Seidl subsp. *dicoccon* (Schrank) A. Löve, *Triticum dicoccon* subsp. *europaeum* Vav. p.p. incl. typo) - 1

- subsp. *asiaticum* Vav. p.p. = T. armeniacum
- subsp. *europaeum* Vav. p.p. = T. dicoccon
- subsp. *euroum* Flaksb. p.p. = T. armeniacum
- subsp. *georgicum* (Dekapr. & Menabde) Flaksb. = T. karamyschevii
- subsp. *subspontaneum* Tzvel. = T. subspontaneum
- subsp. *volgense* (Flaksb.) Tzvel. = T. volgense
- grex *georgicum* Dekapr. & Menabde = T. karamyschevii

- var. *chwamlicum* Supat. = T. karamyschevii
*durum Desf. (*Gigachilon polonicum* (L.) Seidl subsp. *durum* (Desf.) A. Löve, *Triticum aestivum* L. subsp. *durum* (Desf.) Thell., *T. turgidum* L. convar. *durum* (Desf.) Bowden) - 1, 2, 3, 4, 6
- subsp. **caucasicum** V. Dorof. - 2
- subsp. *turgidum* (L.) V. Dorof. = T. turgidum
x **edwardii** Zhebrak. - T. durum Desf. x T. monococcum L.
x **fungicidum** Zhuk. - T. carthlicum Nevski x T. timopheevii (Zhuk.) Zhuk.
georgicum (Dekapr. & Menabde) Dekapr. = T. karamyschevii
gmelinii Trin. = Elytrigia gmelinii
ibericum Menabde = T. carthlicum
imbricatum Bieb. = Agropyron pectinatum
imereticum Dekapr. = T. macha
inflatum Kudr. (*T. aestivum* L. subsp. *inflatum* (Kudr.) Tzvel., *T. inflatum* subsp. *mutico-inflatum* Kudr.) - 6
- subsp. *mutico-inflatum* Kudr. = T. inflatum
intermedium Host var. *stipifolium* Sirj. & Lavr. = Elytrigia stipifolia
*jakubzineri (Udachin & Shakhmedov) Udachin & Shakhmedov (*T. turgidum* L. subsp. *jakubzineri* Udachin & Shakhmedov)
jerevani Thum. = T. boeoticum
karamyschevii Nevski (*Gigachilon polonicum* (L.) Seidl subsp. *paleocolchicum* (A. & D. Löve) A. Löve, *Triticum dicoccon* (Schrank) Schuebl. var. *chwamlicum* Supat., *T. dicoccon* grex *georgicum* Dekapr. & Menabde, *T. dicoccon* subsp. *georgicum* (Dekapr. & Menabde) Flaksb., *T. georgicum* (Dekapr. & Menabde) Dekapr. nom. illegit., *T. paleocolchicum* Menabde, nom. illegit., *T. turgidum* L. subsp. *paleocolchicum* (Menabde) A. & D. Löve, nom. illegit.) - 2
kotschyi (Boiss.) Bowden = Aegilops kotschyi
laxum Fries = Elytrigia x littorea
littorale Host = Elytrigia pungens
littorale Pall. = Leymus secalinus
littoreum Schum. = Elytrigia x littorea
lorentii (Hochst.) Zeven = Aegilops biuncialis
macha Dekapre & Menabde (*T. aestivum* L. subsp. *macha* (Dekapr. & Menabde) Mac Key, *T. imereticum* Dekapr. nom. invalid., *T. spelta* L. subsp. *macha* (Dekapr. & Menabde) V. Dorof. nom. invalid., *T. tubalicum* Dekapr. nom. invalid.) - 2
macrochaetum (Shuttl. & Huet) K. Richt. = Aegilops biuncialis
michaelii Zhuk. = T. urartu
militinae Zhuk. & Migush. - 2
molle (Trin.) F. Herm. = Leymus mollis
monococcum L. (*Crithodium monococcum* (L.) A. Löve) - 1, 2
- subsp. *boeoticum* (Boiss.) Hayek = T. boeoticum
- subsp. *boeoticum* (Boiss.) A. & D. Löve = T. boeoticum
- subsp. *boeoticum* (Boiss.) C.Yen = T. boeoticum
- subsp. *michaelii* Fed. & Takht. ex Zhuk. = T. urartu
- subsp. *thaoudar* (Reut. ex Hausskn.) Zhuk. = T. boeoticum
- subsp. *urartu* (Thum.) A. & D. Löve = T. urartu
montanum Makuschina = T. araraticum
muricatum Link = Agropyron pectinatum
neglectum (Bertol.) Greuter = Aegilops neglecta
nikolai Fed. & Takht. ex Zhuk. = T. araraticum
nodosum Stev. ex Bieb. = Elytrigia nodosa
orientale Perciv. = T. turanicum
ovatum (L.) Raspail, p.p. = Aegilops neglecta
paleocolchicum Menabde = T. karamyschevii

panormitanum (Parl.) Bertol. = Elymus panormitanus
paradoxum Parodi = T. carthlicum
pauciflorum Schwein. = Elymus trachycaulus
pectinatum Bieb. = Agropyron pectinatum
percivalianum Parodi = T. turanicum
percivalii C.E. Hubb. ex Schiemann = T. turanicum
persicum (Boiss.) Aitch. & Hemsl. = Aegilops triuncialis subsp. persica
persicum Vav. ex Zhuk. = T. carthlicum
*polonicum L. (*Gigachilon polonicum* (L.) Seidl, *Triticum turgidum* L. subsp. *polonicum* (L.) A. & D. Löve, *T. turgidum* subsp. *turgidum* convar. *polonicum* (L.) Mac Key) - 1, 2, 3, 6
ponticum Podp. = Elytrigia elongata
pseudoagropyrum Trin. ex Griseb. = Leymus chinensis
puberulum Boiss. ex Steud. = Agropyron puberulum
pubescens Trin. = Elymus jacutensis
pumilum L. fil. = Agropyron cristatum
pumilum Steud. = Agropyron pumilum
pungens Pers. = Elytrigia pungens
rectum (Zhuk.) Bowden = Aegilops recta
rigidum Schrad. = Elytrigia elongata
- var. *ruthenicum* Griseb. = Elytrigia trichophora
rupestre Turcz. ex Ganesch. = Elymus gmelinii
sabulosum (Bieb.) F. Herm. = Leymus racemosus subsp. sabulosus
secalinum Georgi = Leymus secalinus
x **soveticum** Zhebrak. - T. durum Desf. x T. timopheevii (Zhuk.) Zhuk.
*spelta L. (*T. aestivum* L. subsp. *spelta* (L.) Thell., ? *T. aestivum* subsp. *transcaucasicum* V. Dorof. & Laptev, nom. invalid., *T. spelta* subsp. *kuckuckianum* Gokgol) - 1, 2, 6
- subsp. *kuckuckianum* Gokgol = T. spelta
- subsp. *macha* (Dekapr. & Menabde) V. Dorof. = T. macha
- subsp. *vavilovii* (Jakubz.) V. Dorof. = T. vavilovii
- var. *dicoccon* Schrank = T. dicoccon
sphaerococcum Perciv. - 2
spontaneum Flaksb. = T. boeoticum
- subsp. *aegilopoides* (Link) Flaksb. = T. boeoticum
- subsp. *thaoudar* (Reut. ex Hausskn.) Flaksb. = T. boeoticum
strictum Dethard. = x Leymotrigia stricta
strigosum (Bieb.) Less. var. *planifolium* Regel = Elymus schrenkianus
subspontaneum (Tzvel.) Czer. (*T. dicoccon* (Schrank) Schuebl. subsp. *subspontaneum* Tzvel.) - 2
subulatum Banks & Soland. = Loliolum subulatum
tauschii (Coss.) Schmalh. = Aegilops tauschii
thaoudar Reut. ex Hausskn. = T. boeoticum
timopheevi (Zhuk.) Zhuk. (*Gigachilon timopheevii* (Zhuk.) A. Löve, *Triticum turgidum* L. subsp. *timopheevii* (Zhuk.) A. & D. Löve) - 2
- subsp. *araraticum* (Jakubz.) Mac Key = T. araraticum
- var. *araraticum* (Jakubz.) C. Yen = T. araraticum
trachycaulon Link = Elymus trachycaulus
triunciale (L.) Raspail = Aegilops triuncialis
tubalicum Dekapr. = T. macha
turanicum Jakubz. (*Gigachilon polonicum* (L.) Seidl subsp. *turanicum* (Jakubz.) A. Löve, *Triticum orientale* Perciv. 1921, non Bieb. 1808, *T. percivalianum* Parodi, *T. percivalii* C.E. Hubb. ex Schiemann, *T. turgidum* L. subsp. *turanicum* (Jakubz.) A. & D. Löve, *T. turgidum* subsp. *turgidum* convar. *turanicum* (Jakubz.) Mac Key) - 2, 6
*turgidum L. (*Gigachilon polonicum* (L.) Seidl subsp. *turgi-*

dum (L.) A. Löve, *Triticum aestivum* L. subsp. *turgidum*
(L.) Domin, *T. compositum* L., *T. durum* Desf. subsp.
turgidum (L.) V. Dorof. nom. illegit., nom. invalid.) - 1,
2, 3
- subsp. *armeniacum* (Jakubz.) A. & D. Löve = T. araraticum
- subsp. *carthlicum* (Nevski) A. & D. Löve = T. carthlicum
- subsp. *jakubzineri* Udachin & Shakhmedov = T. jakubzineri
- subsp. *paleocolchicum* (Menabde) A. & D. Löve = T.
karamyschevii
- subsp. *polonicum* (L.) A. & D. Löve = T. polonicum
- subsp. *timopheevii* (Zhuk.) A. & D. Löve = T. timopheevii
- subsp. *turanicum* (Jakubz.) A. & D. Löve = T. turanicum
- subsp. *turgidum* convar. *polonicum* (L.) Mac Key = T.
polonicum
- - convar. *turanicum* (Jakubz.) Mac Key = T. turanicum
- subsp. *volgense* (Flaksb.) A. & D. Löve = T. volgense
- convar. *durum* (Desf.) Bowden = T. durum
urartu Thum. ex Gandiljan (*Crithodium urartu* (Thum.) A.
Löve, comb. invalid., *Triticum boeoticum* Boiss. subsp.
urartu (Thum.) V. Dorof. comb. invalid., *T. michaelii*
Zhuk. nom. nud., *T. monococcum* L. subsp. *michaelii*
Fed. & Takht. ex Zhuk. nom. nud., *T. monococcum*
subsp. *urartu* (Thum.) A. & D. Löve, nom. invalid.) - 2
vagans (Jord. & Fourr.) Greuter = Aegilops geniculata
vavilovii Jakubz. (*T. aestivum* L. subsp. *vavilovii* (Jakubz.)
A. Löve, comb. superfl., *T. aestivum* subsp. *vavilovii*
(Jakubz.) Sears, *T. spelta* L. subsp. *vavilovii* (Jakubz.)
V. Dorof. comb. invalid.) - 2
violaceum Hornem. f. *subalpinum* L. Neum. = Elymus
kronokensis subsp. subalpinus
volgense (Flaksb.) Nevski (*T. dicoccon* (Schrank) Schuebl.
subsp. *volgense* (Flaksb.) Tzvel., *T. turgidum* L. subsp.
volgense (Flaksb.) A. & D. Löve) - 1
vulgare Vill. subsp. *hadropyrum* Flaksb. = T. asiaticum
- subsp. *irano-asiaticum* Flaksb. = T. asiaticum
- grex *eligulatum* Vav. = T. asiaticum subsp. eligulatum
- grex *rigium* Vav. = T. asiaticum subsp. rigidum
- grex *speltiforme* Vav. = T. asiaticum subsp. speltiforme
zhukovskyi Menabde & Ericzjan - 2

x **Trititrigia** Tzvel.

cziczinii Tzvel. (*Triticum x agropyrotritium* Cziczin, nom.
invalid.). - Elytrigia intermedia (Host) Nevski x Triticum aestivum L.

x *Tritordeum* auct. = x Leymotrigia
bergrothii Lindb. fil. = x Leymotrigia bergrothii

Typhoides Moench = Phalaroides
arundinacea (L.) Moench = Phalaroides arundinacea
- subsp. *japonica* (Steud.) Tzvel. = Phalaroides japonica

Urochloa Beauv.
panicoides Beauv. - 5

Vahlodea Fries (*Erioblastus* Honda)
atropurpurea (Wahlenb.) Fries - 1
- subsp. *paramushirensis* (Kudo) Hult. = V. flexuosa
flexuosa (Honda) Ohwi (*Erioblastus flexuosus* Honda,
Deschampsia pacifica Tatew. & Ohwi, *Vahlodea
atropurpurea* (Wahlenb.) Fries subsp. *paramushirensis*
(Kudo) Hult., *V. paramushirensis* (Kudo) Roshev.) - 5

paramushirensis (Kudo) Roshev. = V. flexuosa

Ventenata Koel.
dubia (Leers) Coss. - 1, 2

Vulpia C.C. Gmel.
bromoides (L.) S.F. Gray (*Festuca bromoides* L., *Vulpia
dertonensis* (All.) Gola) - 1, 2
ciliata Dumort. (*Festuca ciliata* Danth. 1805, non Gouan,
1762) - 1, 2, 6
dertonensis (All.) Gola = V. bromoides
fasciculata (Forssk.) Fritsch (*Festuca fasciculata* Forssk., *F.
uniglumis* Soland., *Vulpia fasciculata* (Forssk.) Sampaio, comb. superfl., *V. uniglumis* (Soland.) Dumort.,
V. membranacea auct.) - 2
hirtiglumis Boiss. & Hausskn. (*V. villosa* T. Massl.) - 2, 6
megalura (Nutt.) Rydb. (*Festuca megalura* Nutt., *Vulpia
myuros* (L.) C.C. Gmel. subsp. *megalura* (Nutt.) Sojak,
V. myuros f. *megalura* (Nutt.) C.A. Stace & R. Cotton) - 1(alien), 5(alien)
membranacea auct. = V. fasciculata
myuros (L.) C.C. Gmel. - 1, 2, 5(alien), 6
- subsp. *megalura* (Nutt.) Sojak = V. megalura
- f. *megalura* (Nutt.) C.A. Stace & R. Cotton = V. megalura
persica (Boiss. & Buhse) V. Krecz. & Bobr. - 2, 6
uniglumis (Soland.) Dumort. = V. fasciculata
villosa T. Massl. = V. hirtiglumis

Xanthochloa auct. = Festuca
karatavica (Bunge) R. Kam. = Festuca karatavica

Zea L.
mays L. - 1, 2, 3, 4, 6

Zerna Panz. = Bromopsis
adjarica (Somm. & Levier) Nevski = Bromopsis villosula
angrenica (Drob.) Nevski = Bromopsis angrenica
arctica (Shear) Tzvel. = Bromopsis arctica
aristata (C. Koch) Tzvel. = Bromopsis aristata
aspera (Murr.) Fourr. = Bromopsis ramosa
benekenii (Lange) Lindm. = Bromopsis benekenii
biebersteinii (Roem. & Schult.) Nevski = Bromopsis biebersteinii
canadensis (Michx.) Tzvel. = Bromopsis canadensis
cappadocica (Boiss. & Bal.) Nevski = Bromopsis cappadocica and B. heterophylla
erecta (Huds.) S.F. Gray = Bromopsis erecta
- subsp. *alexeenkoi* Tzvel. = Bromopsis alexeenkoi
- subsp. *gordjaginii* Tzvel. = Bromopsis gordjaginii
fibrosa (Hack.) Grossh. p.p. = Bromopsis x taurica
gussonii (Parl.) Grossh. = Anisantha diandra
heterophylla Klok. = Bromopsis heterophylla
indurata (Hausskn. & Bornm.) Tzvel. = Bromopsis indurata
- subsp. *nachiczevanica* Tzvel. = Bromopsis nachiczevanica
inermis (Leyss.) Lindm. = Bromopsis inermis
ircutensis (Kom.) Nevski = Bromopsis ircutensis
kopetdagensis (Drob.) Nevski = Bromopsis kopetdagensis
korotkiji (Drob.) Nevski = Bromopsis korotkiji
madritensis (L.) Grossh. = Anisantha madritensis
madritensis (L.) S.F. Gray = Anisantha madritensis
occidentalis Nevski = Bromopsis pumpelliana
ornans (Kom.) Nevski = Bromopsis ornans
pamirica (Drob.) Nevski = Bromopsis pamirica

paulsenii (Hack.) Nevski = Bromopsis paulsenii
- subsp. *angrenica* (Drob.) Tzvel. = Bromopsis angrenica
- subsp. *pamirica* (Drob.) Tzvel. = Bromopsis pamirica
- subsp. *turkestanica* (Drob.) Tzvel. = Bromopsis turkestanica
pseudocappadocica Klok. = Bromopsis pseudocappadocica
pumpelliana (Scribn.) Tzvel. = Bromopsis pumpelliana
- subsp. *arctica* (Shear) Tzvel. = Bromopsis arctica
- subsp. *karavajevii* Tzvel. = Bromopsis karavajevii
- subsp. *korotkiji* (Drob.) Tzvel. = Bromopsis korotkiji
- subsp. *ornans* (Kom.) Tzvel. = Bromopsis ornans
- subsp. *vogulica* (Socz.) Tzvel. = Bromopsis vogulica
ramosa (Hudson) Lindm. = Bromopsis ramosa
- subsp. *benekenii* (Lange) Tzvel. = Bromopsis benekenii
- subsp. *fedtschenkoi* Tzvel. = Bromopsis fedtschenkoi
richardsonii (Link) Nevski, p.p. = Bromopsis pumpelliana
riparia (Rehm.) Nevski = Bromopsis riparia
- subsp. *divaricata* Tzvel. = Bromopsis divaricata
- subsp. *fibrosa* (Hack.) Tzvel. p.p. = Bromopsis x taurica
- subsp. *heterophylla* (Klok.) Tzvel. = Bromopsis heterophylla
rubens (L.) Grossh. = Anisantha rubens
sterilis (L.) Panz. = Anisantha sterilis
tectorum (L.) Lindm. = Anisantha tectorum
tomentella (Boiss.) Nevski = Bromopsis tomentella
- subsp. *cappadocica* (Boiss. & Ball.) Tzvel. = Bromopsis cappadocica
- subsp. *shelkovnikovii* Tzvel. = Bromopsis shelkovnikovii
- subsp. *woronowii* Tzvel. = Bromopsis woronowii
tomentosa (Trin.) Nevski = Bromopsis triniana
triniana (Schult.) Tzvel. = Bromopsis triniana
turkestanica (Drob.) Nevski = Bromopsis turkestanica
tyttholepis Nevski = Bromopsis tyttholepis
variegata (Bieb.) Nevski = Bromopsis variegata
- subsp. *villosula* (Steud.) Tzvel. = Bromopsis villosula
vogulica (Socz.) Nevski = Bromopsis vogulica
yezoensis (Ohwi) Sugim. = Bromopsis canadensis

Zingeria P. Smirn.

biebersteiniana (Claus) P. Smirn. (*Agrostis biebersteiniana* Claus, *Zingeria trichopoda* (Boiss.) P. Smirn. subsp. *biebersteiniana* (Claus) M. Dogan, non rite publ.) - 1, 2(?)
- subsp. *trichopoda* (Boiss.) R. Mill = Z. trichopoda
kochii (Mez) Tzvel. (*Milium kochii* Mez) - 2
pisidica (Boiss.) Tutin (*Agrostis pisidica* Boiss.) - 2(?)
trichopoda (Boiss.) P. Smirn. (*Milium trichopodum* Boiss., *Zingeria biebersteiniana* (Claus) P. Smirn. subsp. *trichopoda* (Boiss.) R. Mill) - 2
- subsp. *biebersteiniana* (Claus) M. Dogan = Z. biebersteiniana

*Zizania L. (*Hydropyrum* Link)

*aquatica** L. subsp. **angustifolia** (Hitchc.) Tzvel. (*Z. aquatica* var. *angustifolia* Hitchc., *Z. palustris* L.) - 1(alien), 2
dahurica Turcz. ex Steud. = Z. latifolia
latifolia (Griseb.) Stapf (*Hydropyrum latifolium* Griseb., *Zizania dahurica* Turcz. ex Steud.) - 1(alien), 4, 5
palustris L. = Z. aquatica subsp. angustifolia

Zoysia Willd.

japonica Steud. - 5

POLEMONIACEAE Juss.
Collomia Nutt.
linearis Nutt. - 1, 5, 6

Phlox L.

alaskensis Jordal (*Ph. borealis* Wherry, *Ph. richardsonii* Hook. subsp. *alaskensis* (Jordal) Wherry, *Ph. sibirica* L. subsp. *alaskensis* (Jordal) A. & D. Löve, *Ph. sibirica* subsp. *borealis* (Wherry) Shetler) - 5
borealis Wherry = Ph. alaskensis
*drummondii** Hook. - 1
*paniculata** L. - 1
richardsonii Hook. subsp. *alaskensis* (Jordal) Wherry = Ph. alaskensis
sibirica L. - 1, 3, 4, 5
- subsp. *alaskensis* (Jordal) A. & D. Löve = Ph. alaskensis
- subsp. *borealis* (Wherry) Shetler = Ph. alaskensis

Polemonium L.

acutiflorum Willd. ex Roem. & Schult. (*P. acutiflorum* subsp. *diminutum* (Klok.) R. Kam., *P. diminutum* Klok., *P. diminutum* var. *foliolatum* (Klok.) Serg., *P. foliolatum* Klok. p.p. incl. typo) - 1, 3, 4, 5
- subsp. *diminutum* (Klok.) R. Kam. = P. acutiflorum
boreale Adams (*P. boreale* subsp. *humile* (Willd. ex Roem. & Schult.) A. & D. Löve, *P. boreale* subsp. *macranthum* (Cham.) Hult., *P. boreale* subsp. *nudipedum* (Klok.) R. Kam., *P. boreale* var. *pseudopulchellum* (V. Vassil.) M. Pop., *P. hultenii* Hara, *P. humile* Willd. ex Roem. & Schult. 1819, non Sasisb. 1796, *P. humile macranthum* Cham., *P. hyperboreum* Tolm., *P. lapponicum* Klok., *P. macranthum* (Cham.) Klok., *P. nudipedum* Klok., *P. onegense* Klok., *P. pseudopulchellum* V. Vassil., *P. pulchellum* Bunge, *P. pulchellum* subsp. *pseudopulchellum* (V. Vassil.) Egor. & Sipl., *P. pulcherrimum* Hook. subsp. *hyperboreum* (Tolm.) A. & D. Löve, *P. villosum* J. Rudolph ex Georgi) - 1, 3, 4, 5
- subsp. *humile* (Willd. ex Roem. & Schult.) A. & D. Löve = P. boreale
- subsp. *macranthum* (Cham.) Hult. = P. boreale
- subsp. *nudipedum* (Klok.) R. Kam. = P. boreale
- var. *pseudopulchellum* (V. Vassil.) M. Pop. = P. boreale
caeruleum L. - 1, 3, 4
- subsp. *campanulatum* Th. Fries = P. campanulatum
- subsp. *caucasicum* (N. Busch) V. Avet. = P. caucasicum
- subsp. *liniflorum* (V. Vassil.) Tolm. = P. chinense
- subsp. *schmidtii* (Klok.) Tolm. = P. schmidtii
- a. *vulgare* Regel lusus *laxiflorum* Regel = P. laxiflorum
campanulatum (Th. Fries) Lindb. fil. (*P. caeruleum* L. subsp. *campanulatum* Th. Fries, *P. foliolatum* Klok. p.p. excl. typo, *P. pacificum* V. Vassil., *P. schizanthum* Klok., *P. villosum* J. Rudolphi ex Georgi subsp. *pacificum* (V. Vassil.) A. & D. Löve, *P. villosum* sensu V. Vassil.) - 1, 3, 4, 5
caucasicum N. Bush. (*P. caeruleum* L. subsp. *caucasicum* (N. Busch) V. Avet.) - 2
chinense (Brand) Brand (*P. caeruleum* L. subsp. *liniflorum* (V. Vassil.) Tolm., *P. liniflorum* V. Vassil., *P. racemosum* auct.) - 4
diminutum Klok. = P. acutiflorum
- var. *foliolatum* (Klok.) Serg. = P. acutiflorum
foliolatum Klok. = P. acutiflorum and P. campanulatum
hultenii Hara = P. boreale
humile Willd. ex Roem. & Schult. = P. boreale
- *macranthum* Cham. = P. boreale

hyperboreum Tolm. = P. boreale
lapponicum Klok. = P. boreale
laxiflorum (Regel) Kitam. (*P. caeruleum* L. a. *vulgare* Regel lusus *laxiflorum* Regel) - 5
liniflorum V. Vassil. = P. chinense
macranthum (Cham.) Klok. = P. boreale
majus Tolm. - 5
nudipedum Klok. = P. boreale
onegense Klok. = P. boreale
pacificum V. Vassil. = P. campanulatum
parviflorum Tolm. = P. pulcherrimum
pseudopulchellum V. Vassil. = P. boreale
pulchellum Bunge = P. boreale
- subsp. *pseudopulchellum* (V. Vassil.) Egor. & Sipl. = P. boreale
pulcherrimum Hook. (*P. parviflorum* Tolm.) - 4, 5
- subsp. *hyperboreum* (Tolm.) A. & D. Löve = P. boreale
racemosum auct. = P. chinense
sachalinense Worosch. - 5
schizanthum Klok. = P. campanulatum
schmidtii Klok. (*P. caeruleum* L. subsp. *schmidtii* (Klok.) Tolm.) - 4
sibiricum D. Don
x *vajgaczense* Klok. = P. x victoris
x **victoris** Klok. (*P.* x *vajgaczense* Klok.). - P. acutiflorum Willd. ex Roem. & Schult. x P. boreale Adams - 1
villosum J. Rudolph ex Georgi = P. boreale
- subsp. *pacificum* (V. Vassil.) A. & D. Löve = P. campanulatum
villosum sensu V. Vassil. = P. campanulatum

POLYGALACEAE R. Br.
Polygala L.

alata (Tamamsch.) Galushko (*P. major* Jacq. var. *alata* Tamamsch.) - 2
albowii Kem.-Nath. (*P. colchica* Tamamsch.) - 2
alpestris Reichenb. - 1, 2(?)
alpicola Rupr. - 2
amara auct. = P. amarella
amara L. subsp. *brachyptera* (Chodat) Hayek = P. subamara
amarella Crantz (*P. amara* auct.) - 1
▪ amoenissima Tamamsch.
anatolica Boiss. & Heldr. - 1, 2
andrachnoides Willd. - 1, 2
austriaca auct. = P. decipiens
caucasica Rupr. - 2
colchica Tamamsch. = P. albowii
comosa Schkuhr (*P. comosa* var. *hybrida* (DC.) Petelin, *P. hybrida* DC.) - 1, 2, 3, 4, 5, 6
- var. *hybrida* (DC.) Petelin = P. comosa
cretacea Kotov (? *P. nicaeensis* Risso ex Koch subsp. *mediterranea* Chodat) - 1
decipiens Bess. (*P. austriaca* auct.) - 1
grossheimii Kem.-Nath. - 2
hohenackeriana Fisch. & C.A. Mey. - 2
hybrida DC. = P. comosa
japonica Houtt. - 5
kemulariae Tamamsch. = P. makaschwilii
leucothyrsa Woronow - 2
major Jacq. - 1, 2
- var. *alata* Tamamsch. = P. alata
makaschwilii Kem.-Nath. (*P. kemulariae* Tamamsch.) - 2
mariamae Tamamsch. - 2
moldavica Kotov - 1
nathadzeae A. Kuth. - 2

nicaeensis Risso ex Koch subsp. *mediterranea* Chodat = P. cretacea
oxyptera Reichenb. (*P. vulgaris* L. subsp. *oxyptera* (Reichenb.) Lange) - 1
paludosa St.-Hil. - 2
▪ papilionacea Boiss.
podolica DC. - 1
pruinosa Boiss. - 2
pseudohospita (Tamamsch.) Tamamsch. - 2
rossica Kem.-Nath. - 1
schirvanica Grossh. - 2
sibirica L. - 1, 2, 3, 4, 5
sophiae Kem.-Nath. - 2
sosnowskyi Kem.-Nath. - 2
stocksiana Boiss. - 2
suanica Tamamsch. - 2
subamara Fritsch (*P. amara* L. subsp. *brachyptera* (Chodat) Hayek) - 1
▪ supina Schreb.
tatarinowii Regel - 5
tenuifolia Willd. - 3, 4, 5
transcaucasica Tamamsch. - 2
urartu Tamamsch. - 2
vulgaris L. - 1
- subsp. *oxyptera* (Reichenb.) Lange = P. oxyptera
wolfgangiana Bess. ex Ledeb. - 1

POLYGONACEAE Juss. (*CALLIGONACEAE* Chalkuziev)
Acetosa Hill

acetoselloides (Bal) Holub = Rumex acetoselloides
alpestris (Jacq.) A. Löve = Rumex alpestris
- subsp. *carpatica* (Zapal.) Dostal = Rumex rugosus
- subsp. *lapponica* (Hiit.) A. Löve = Rumex lapponicus
arifolia (All.) Schur = Rumex alpestris
fontano-paludosa (Kalela) Holub = Rumex acetosa
lapponica (Hiit.) Holub = Rumex lapponicus
oblongifolia (Tolm.) A. & D. Löve = Rumex oblongifolius
pratensis Mill. = Rumex acetosa
- subsp. *alpestris* (Jacq.) A. Löve = Rumex alpestris
- - var. *lapponica* (Hiit.) A. Löve = Rumex lapponicus
- subsp. *pseudoxyria* (Tolm.) A. Löve & Kapoor = Rumex pseudoxyria
pseudoxyria (Tolm.) Tzvel. = Rumex pseudoxyria
rugosa (Campd.) Holub = Rumex rugosus
scutata (L.) Mill. = Rumex scutatus
- subsp. *hastifolia* (Bieb.) A. Löve & Kapoor = Rumex hastifolius
thyrsiflora (Fingerh.) A. & D. Löve = Rumex thyrsiflorus
tuberosa (L.) Chaz. subsp. *turcomanica* (Rech. fil.) A. Löve & Kapoor = Rumex turcomanicus

Acetosella (Meissn.) Fourr. = Rumex

angiocarpa (Murb.) A. Löve. = Rumex angiocarpus
aureostigmatica (Kom.) Tzvel. = Rumex aureostigmaticus
beringensis (Jurtz. & Petrovsky) A. & D. Löve = Rumex beringensis
graminifolia (Lamb.) A. Löve = Rumex graminifolius
krausei (Jurtz. & Petrovsky) A. & D. Löve = Rumex krausei
multifida (L.) A. Löve = Rumex acetosella
tenuifolia (Wallr.) A. Löve = Rumex acetosella
vulgaris (Koch) Fourr. = Rumex acetosella
- subsp. *angiocarpa* (Murb.) Hadac & Hasek = Rumex

angiocarpus
- f. *multifida* (L.) Dostal = Rumex acetosella

Aconogonon (Meissn.) Reichenb. (*Pleuropteropyrum* H. Gross)

ajanense (Regel & Til.) Hara (*Polygonum ajanense* (Regel & Til.) Grig.) - 4, 5

alaskanum (Small) Sojak (*Polygonum alpinum* All. subsp *alaskanum* Small, *P. alaskanum* (Small) Wight ex Hult.) - 5

alpinum (All.) Schur (*Polygonum alpinum* All., *Aconogonon dshawachischwilii* (Charkev.) Holub, comb. superfl., *A. dshawachischwilii* (Charkev.) Sojak, *Pleuropteropyrum alpinum* (All.) Kitag., *P. alpinum* (All.) Nakai, *P. undulatum* (Murr.) A. & D. Löve, nom. illegit., *Polygonum dshawachischwilii* Charkev., *P. undulatum* Murr. 1775, non Berg. 1767) - 1, 2, 3, 4, 5, 6

amgense (V. Michaleva & V. Perfiljeva) Tzvel. (*Polygonum amgense* V. Michaleva & V. Perfiljeva, *Pleuropteropyrum amgense* (V. Michaleva & V. Perfiljeva) Sojak) - 4, 5

angustifolium (Pall.) Hara (*Polygonum angustifolium* Pall., *Pleuropteropyrum angustifolium* (Pall.) Kitag.) - 4, 5

baicalense (Sipl.) Sojak = A. ochreatum

bargusinense (Peschkova) Sojak (*Polygonum bargusinense* Peschkova) - 4

bucharicum (Grig.) Holub (*Polygonum bucharicum* Grig., *Aconogonon coriarium* (Grig.) Sojak subsp. *bucharicum* (Grig.) Sojak, *Pleuropteropyrum bucharicum* (Grig.) Nevski) - 6

chaneyi (B. Fedtsch. ex R.R. Stewart) Hara (*Polygonum chaneyi* B. Fedtsch. ex R.R. Stewart) - 4

chlorochryseum (M. Ivanova) Sojak (*Polygonum chlorochryseum* M. Ivanova) - 4

coriarium (Grig.) Sojak (*Polygonum coriarium* Grig.) - 6
- subsp. *bucharicum* (Grig.) Sojak = A. bucharicum

diffusum (Willd. ex Spreng.) Tzvel. (*Polygonum diffusum* Willd. ex Spreng.) - 1, 3

divaricatum (L.) Nakai ex Mori (*Polygonum divaricatum* L.) - 1(alien), 4, 5

dshawachischwilii (Charkev.) Holub = A. alpinum
dshawachischwilii (Charkev.) Sojak = A. alpinum

hissaricum (M. Pop.) Sojak (*Polygonum hissaricum* M. Pop.) - 6

jurii (A. Skvorts.) Holub (*Polygonum jurii* A. Skvorts., *P. luxurians* Grig. 1936, non Danser, *P. platyphyllum* auct.) - 5

laxmannii (Lepech.) Holub = A. ochreatum var. laxmannii
laxmannii (Lepch.) A. & D. Löve = A. ochreatum var. laxmannii

limosum (Kom.) Hara (*Polygonum divaricatum* L. var. *limosum* Kom., *Pleuropteropyrum limosum* (Kom.) Kitag., *Polygonum limosum* (Kom.) Kom.) - 5

middendorfii (Kongar) Holub (*Polygonum middendorfii* Kongar, *P. ajanense* (Regel & Til.) Grig. subsp. *middendorfii* (Kongar) Worosch.) - 5

ochreatum (L.) Hara (*Polygonum ochreatum* L., *Aconogonon baicalense* (Sipl.) Sojak, *Polygonum baicalense* Sipl.) - 1, 3, 4, 5
- var. **laxmannii** (Lepech.) Tzvel. (*Polygonum laxmannii* Lepech., *Aconogonon laxmannii* (Lepech.) Holub, comb. superfl., *A. laxmannii* (Lepech.) A. & D. Löve, *Pleuropteropyrum laxmannii* (Lepech.) Kitag., *Polygonum ochreatum* var. *laxmannii* (Lepech.) M. Pop.) - 5
- var. **riparium** (Georgi) Tzvel. (*Polygonum riparium* Georgi, *Aconogonon riparium* (Georgi) Hara, *Pleu-*

ropteropyrum riparium (Georgi) Kitag.) - 4

pamiricum (Korsh.) Hara = Knorringia pamirica

panjutinii (Charkev.) Sojak (*Polygonum panjutinii* Charkev., *Aconogonon panjutinii* (Charkev.) Holub, comb. superfl.) - 2

pseudoajanense Barkalov & Vyschin - 5

relictum (Kom.) Sojak (*Polygonum relictum* Kom.) - 5

riparium (Georgi) Hara = A. ochreatum var. riparium

savatieri (Nakai) Tzvel. (*Polygonum savatieri* Nakai) - 5
- var. **iturupense** (Mischurov) Tzvel. (*Polygonum iturupense* Mischurov, *Pleuropteropyrum iturupense* (Mischurov) Sojak) - 5

sericeum (Pall. ex Georgi) Hara (*Polygonum sericeum* Pall. ex Georgi) - 4

sibiricum (Laxm.) Hara = Knorringia sibirica

songaricum (Schrenk) Hara (*Polygonum songaricum* Schrenk) - 6

x **subsericeum** (M. Pop.) Sojak (*Polygonum subsericeum* M. Pop.). - A. ajanense (Regel & Til.) Hara x A. sericeum (Pall. ex Georgi) Hara - 4

tripterocarpum (A. Gray) Hara (*Polygonum tripterocarpum* A. Gray, *Pleuropteropyrum tripterocarpum* (A. Gray) H. Gross) - 4, 5

tzvelevii Barkalov & Vyschin - 5

valerii (A. Skvorts.) Sojak (*Polygonum valerii* A. Skvorts.) - 4, 5

weyrichii (Fr. Schmidt) Hara (*Polygonum weyrichii* Fr. Schmidt, *Pleuropteropyrum weyrichii* (Fr. Schmidt) H. Gross) - 5

zaravschanicum (Zak.) Sojak (*Polygonum zaravschanicum* Zak.) - 6

Ampelygonum auct. = Chylocalyx

perfoliatum (L.) Roberty & Vautier = Chylocalyx perfoliatus

Antenoron Rafin.

filiforme (Thunb.) Roberty & Vautier (*Polygonum filiforme* Thunb.) - 5

Atraphaxis L.

angustifolia Jaub. & Spach - 2

avenia Botsch. - 6

badhysi Kult. - 6

billardieri Jaub. & Spach subsp. *daghestanica* O. Lovelius = A. daghestanica

canescens Bunge - 3, 6

caucasica (Hoffm.) Pavl. - 2

compacta Ledeb. - 3, 6

daghestanica (O. Lovelius) O. Lovelius (*A. billardieri* Jaub. & Spach subsp. *daghestanica* O. Lovelius, *A. tournefortii* auct. p.p.) - 2

decipiens Jaub. & Spach - 3, 6

frutescens (L.) C. Koch - 1, 3, 4, 6

karataviensis Lipsch. & Pavl. (*A. pulcherrima* Vass.) - 6

kopetdagensis Kovalevsk. - 6

laetevirens (Ledeb.) Jaub. & Spach - 3, 6

muschketowii Krasn. - 6

pulcherrima Vass. = A. karataviensis

pungens (Bieb.) Jaub. & Spach - 3, 4

pyrifolia Bunge - 6

replicata Lam. - 1, 2, 3, 6

rodinii Botsch. - 6

seravschanica Pavl. - 6

spinosa L. - 1, 2, 3, 6

teretifolia (M. Pop.) Kom. - 6
tournefortii auct. p.p. = A. daghestanica
virgata (Regel) Krasn. - 3, 6

Bilderdykia Dumort. = Fallopia

baldschuanica (Regel) D.A. Webb = Fallopia baldschuanica
convolvulus (L.) Dumort. = Fallopia convolvulus
dentato-alata (Fr. Schmidt) Kitag. = Fallopia dentato-alata
dumetorum (L.) Dumort. = Fallopia dumetorum
multiflora (Thunb.) Roberty & Vautier = Fallopia multiflora

Bistorta Hill

abbreviata Kom. (*B. major* S.F. Gray subsp. *cordifolia*
  (Turcz.) Sojak, *Polygonum abbreviatum* Kom., *P. bistorta* L. var. *cordifolium* Turcz., *P. bistorta* subsp. *cordifolium* (Turcz.) Malysch., *P. cordifolium* (Turcz.)
  Losinsk. nom. illegit.) - 4, 5
alopecuroides (Turcz. ex Meissn.) Kom. (*Polygonum alopecuroides* Turcz. ex Meissn.) - 3, 4, 5
- var. **zeaensis** (Kom.) Tzvel. (*Polygonum zeaense* Kom.) - 5
carnea (C. Koch) Kom. (*Polygonum carneum* C. Koch,
  Bistorta major S.F. Gray subsp. *carnea* (C. Koch)
  Sojak, *Persicaria bistorta* (L.) Sampaio subsp. *carnea*
  (C. Koch) Greuter & Burdet, *Polygonum bistorta* L.
  subsp. *carneum* (C. Koch) Coode & Cullen) - 2
elliptica (Willd. ex Spreng.) Kom. (*Polygonum ellipticum*
  Willd. ex Spreng., *Bistorta major* S.F. Gray subsp. *elliptica* (Willd. ex Spreng.) A. & D. Löve, comb. superfl.,
  B. major subsp. *elliptica* (Willd. ex Spreng.) Sojak, *B.
  major* subsp. *nitens* (Fisch. & C.A. Mey.) Sojak, *Poly-
  gonum alopecuroides* Turcz. ex Meissn. var. *attenua-
  tum* (V. Petrov ex Kom.) Serg., *P. attenuatum* V.
  Petrov ex Kom. 1936, non R. Br. 1810, *P. bistorta* L.
  subsp. *attenuatum* (V. Petrov ex Kom.) V. Perfiljeva,
  P. bistorta subsp. *ellipticum* (Willd. & Spreng.) Petrovsky, *P. bistorta* subsp. *ochotense* (V. Petrov ex Kom.)
  Worosch., *P. intercedens* V. Petrov ex Kom. 1936, non
  G. Beck, 1906, *P. kolymense* A. Khokhr., *P. nitens*
  (Fisch. & C.A. Mey.) V. Petrov ex Kom., *P. ochotense*
  V. Petrov ex Kom., *P. schugnanicum* Kom., *P. vladimiri* Czer.) - 1, 3, 4, 5, 6
ensigera (Juz.) Tzvel. (*Polygonum ensigerum* Juz., *Bistorta
  major* S.F. Gray subsp. *ensigera* (Juz.) Sojak) - 1
insularis (Maximova) Sojak = B. vivipara
krascheninnikovii Ivanova (*Polygonum krascheninnikovii*
  (Ivanova) Czer.) - 1
major S.F. Gray (*Polygonum bistorta* L.) - 1, 3, 4, 5
- subsp. **carnea** (C. Koch) Sojak = B. carnea
- subsp. **cordifolia** (Turcz.) Sojak = B. abbreviata
- subsp. **elliptica** (Willd. ex Spreng.) A. & D. Löve = B.
  elliptica
- subsp. **elliptica** (Willd. ex Spreng.) Sojak = B. elliptica
- subsp. **ensigera** (Juz.) Sojak = B. ensigera
- subsp. **nitens** (Fisch. & C.A. Mey.) Sojak = B. elliptica
manshuriensis Kom. (*Polygonum manshuriense* V. Petrov
  ex Kom.) - 5
pacifica (V. Petrov ex Kom.) Kom. (*Polygonum pacificum*
  V. Petrov ex Kom., *Bistorta pacifica* (V. Petrov ex
  Kom.) Kom. ex Kitag. comb. superfl., *Polygonum
  bistorta* L. subsp. *pacificum* (V.Petrov ex Kom.)
  Worosch., *P. regelianum* Kom., *P. ussuriense* (Regel)
  V. Petrov ex Kom. 1926, non Nakai, 1922) - 5
plumosa (Small) D. Löve (*Polygonum plumosum* Small) - 4,
  5

subauriculata Kom. (*Polygonum subauriculatum* V. Petrov
  ex Kom., *P. alopecuroides* Turcz. ex Meissn. subsp.
  subauriculatum (V. Petrov ex Kom.) Worosch., *P.
  bistorta* L. subsp. *subauriculatum* (V. Petrov ex Kom.)
  Gubanov) - 5
vivipara (L.) S.F. Gray (*Polygonum viviparum* L., *Bistorta
  insularis* (Maximova) Sojak, *Polygonum insulare*
  Maximova) - 1, 2, 3, 4, 5, 6
- subsp. **macounii** (Small) Sojak (*Polygonum macounii*
  Small) - 5

Calligonum L. (*Calliphysa* Fisch. & C.A. Mey., *Pterococcus* Pall.)

acanthopterum Borszcz. (*C. repetekense* Rotov, p.p., *C.
  rotovii* Godwinski & Nardina, p.p., *C. smirnovii*
  Drob.) - 6
aciferum Godwinski & Nardina = C. setosum
aculeatum (Litv.) Mattei = C. aphyllum
aequilaterale Godwinski = C. rubicundum
affine T. Pop. = C. rubicundum
alatiforme Pavl. = C. leucocladum
- subsp. *roseum* (Drob.) Sosk. = C. leucocladum
alatum Litv. = C. aphyllum
androssowii Litv. = C. leucocladum
aphyllum (Pall.) Guerke (*C. aculeatum* (Litv.) Mattei, *C.
  alatum* Litv., *C. aphyllum* var. *commune* Litv., *C. aphyllum* var. *crispatum* Litv., *C. aphyllum* var. *lamellatum*
  Litv., *C. borszczowii* Litv., *C. cartilagineum* Pavl., *C.
  commune* (Litv.) Mattei, *C. crispatum* (Litv.) Mattei,
  C. eugenii-korovinii Pavl., *C. humile* Litv., *C. lamellatum* (Litv.) Mattei, *C. membranaceum* (Borszcz.) Litv.,
  C. oxicum Drob., *C. palibinii* Mattei, *C. pseudohumile*
  Drob., *C. rigidum* Litv., *C. rigidum* var. *aculeatum* Litv.,
  C. tenue Pavl., *C. tortile* Drob., *C. undulatum* Litv., *C.
  ustjurtense* Drob.) - 1, 2, 3, 6
- subsp. **heptapotamicum** Sosk. - 6
- var. *commune* Litv. = C. aphyllum
- var. *crispatum* Litv. = C. aphyllum
- var. *lamellatum* Litv. = C. aphyllum
aralense Borszcz. = C. leucocladum
arborescens Litv. - 6
babakianum Godwinski = C. rubicundum
bakuense Litv. (*C. petunnikowii* Litv.) - 2
x **barsukiense** Sosk. (*C.* x *pseudotetrapterum* Sosk., *C. tetrapterum* auct.). - *C. aphyllum* (Pall.) Guerke x C. murex
  Bunge - 6
batiola Litv. = C. leucocladum
borszczowii Litv. = C. aphyllum
x **bubyrii** B. Fedtsch. ex Pavl. (*C. cristatum* Pavl. 1933, non
  Bunge, 1852, *C.* x *leucanthum* Sosk.). - *C. acanthopterum* Borszcz. x C. leucocladum (Schrenk) Bunge - 6
bykovii Godwinski = C. junceum
x **calcareum** Pavl. - *C. elegans* Drob. x C. santoanum
  Korov. - 6
cancellatum Mattei - 6
caput-medusae Schrenk - 6
cartilagineum Pavl. = C. aphyllum
colubrinum Borszcz. = C. x macrocarpum
commune (Litv.) Mattei = C. aphyllum
connivens Godwinski = C. rubicundum
cordatum Korov. ex Pavl. - 6
cordiforme Godwinski = C. rubicundum
cordipterum Drob. = C. elegans
coriaceum Pavl. = C. rubicundum
crispatum (Litv.) Mattei = C. aphyllum
crispum Bunge (*C. pavlovii* Godwinski, *C. spinosissimum*

Pavl.) - 3, 6
cristatum Pavl. = C. x bubyrii
densiforme Godwinski & Nardina = C. x paletzkianum
densum Borszcz. - 6
dissectum M. Pop. = C. rubicundum
diversiforme Godwinski = C. rubicundum
drobovii Bondar. - 6
dubianskyi Pavl. - 6
durum Godwinski = C. rubicundum
elatum Litv. = C. microcarpum
elegans Drob. (*C. cordipterum* Drob., *C. integrum* Drob., *C. leucocladum* (Schrenk) Bunge subsp. *elegans* (Drob.) Sosk., *C. margelanicum* Drob., *C. parvulum* Drob.) - 6
erinaceum Borszcz. = C. x macrocarpum
eriopodum Bunge - 6
- subsp. **turkmenorum** Sosk. & Astan. - 6
eugenii-korovinii Pavl. = C. aphyllum
falcilobum Godwinski = C. rubicundum
ferganense Pavl. = C. litwinowii
flavidum Bunge = C. rubicundum
fragile Godwinski & Nardina = C. setosum
golbeckii Drob. = C. leucocladum
gracile Litv. = C. leucocladum
griseum Korov. ex Pavl. - 6
gypsaceum Drob. = C. leucocladum
humile Litv. = C. aphyllum
inaequale Godwinski = C. rubicundum
integrum Drob. = C. elegans
involutum Pavl. = C. rubicundum
irtyschense Godwinski = C. rubicundum
josephii Godwinski = C. rubicundum
junceum (Fisch. & C.A. Mey.) Litv. (*C. bykovii* Godwinski, *Calliphysa bykovii* (Godwinski) Chakuziev) - 3, 6
- subsp. **ludmilae** Sosk. - 6
karakalpakense Drob. = C. leucocladum
koslovii Losinsk. - 6
kurotschkinae Godwinski = C. rubicundum
kzyl-kumi Pavl. = C. setosum
lamellatum (Litv.) Mattei = C. aphyllum
lanciculatum Pavl. = C. leucocladum
x *leucanthum* Sosk. = C. x bubyrii
leucocladiforme Drob. = C. leucocladum
leucocladum (Schrenk) Bunge (*C. alatiforme* Pavl., *C. alatiforme* subsp. *roseum* (Drob.) Sosk., *C. androssowii* Litv., *C. aralense* Borszcz., *C. batiola* Litv., *C. golbeckii* Drob., *C. gracile* Litv., *C. gypsaceum* Drob., *C. karakalpakense* Drob., *C. lanciculatum* Pavl., *C. leucocladiforme* Drob., *C. lipskyi* Litv., *C. obtusum* Litv., *C. orthocarpum* Drob., *C. physopterum* Pavl., *C. plicatum* Pavl., *C. quadripterum* Korov. ex Pavl., *C. roseum* Drob., *C. turbineum* Pavl., *C. uzunachmatense* V. Tkatschenko) - 3, 6
- subsp. *elegans* (Drob.) Sosk. = C. elegans
- subsp. **persicum** (Boiss.& Buhse) Sosk. (*Pterococcus persicus* Boiss. & Buhse) - 6
lipskyi Litv. = C. leucocladum
litwinowii Drob. (*C. ferganense* Pavl.) - 6
x **macrocarpum** Borszcz. (*C. colubrinum* Borszcz., *C. erinaceum* Borszcz.). - C. acanthopterum Borszcz. x C. caput-medusae Schrenk - 6
margelanicum Drob. = C. elegans
matteianum Drob. - 6
membranaceum (Borszcz.) Litv. = C. aphyllum
microcarpum Borszcz. (*C. elatum* Litv., *C. minimum* Lipsky) - 6
minimum Lipsky = C. microcarpum
molle Litv. = C. setosum

muravljanskyi Pavl. - 6
murex Bunge - 6
nardinae Godwinski = C. setosum
nudatum Godwinski = C. rubicundum
obtusum Litv. = C. leucocladum
orthocarpum Drob. = C. leucocladum
orthotrichum Pavl. = C. rubescens
oxicum Drob. = C. aphyllum
ozunraicum Godwinski = C. rubicundum
x **paletzkianum** Litv. (*C. densiforme* Godwinski & Nardina). - C. rubescens Mattei x C. setosum (Litv.) Litv. - 6
palibinii Mattei = C. aphyllum
pappii Godwinski = C. rubicundum
parvulum Drob. = C. elegans
parvum Godwinski = C. rubicundum
patens Litv. - 6
pavlovii Godwinski = C. crispum
pellucidum Pavl. = C. rubescens
petunnikowii Litv. = C. bakuense
physopterum Pavl. = C. leucocladum
platyacanthum Borszcz. - 6
plicatum Pavl. = C. leucocladum
polygonoides L. - 2
przewalskii Losinsk. - 6
x *pseudodubjanskyi* Sosk. = C. x spinulosum
pseudohumile Drob. = C. aphyllum
x *pseudomuravljanskyi* Sosk. = C. x spinulosum
pseudotenue Godwinski = C. rubicundum
x *pseudotetrapterum* Sosk. = C. x barsukiense
pulcherrimum Korov. ex Pavl. = C. setosum
quadripterum Korov. ex Pavl. = C. leucocladum
repetekense Rotov, p.p. = C. acanthopterum
rigidum Litv. = C. aphyllum
- var. *aculeatum* Litv. = C. aphyllum
roseum Drob. = C. leucocladum
rotovii Godwinski & Nardina, p.p. = C. acanthopterum
rotula Borszcz. - 6
rubens auct. = C. rubescens
rubescens Mattei (*C. orthotrichum* Pavl., *C. pellucidum* Pavl., *C. turkestanicum* (Korov.) Pavl., *C. rubens* auct.) - 6
rubicundum Bunge (*C. aequilaterale* Godwinski, *C. affine* T. Pop., *C. babakianum* Godwinski, *C. connivens* Godwinski, *C. cordiforme* Godwinski, *C. coriaceum* Pavl., *C. dissectum* M. Pop., *C. diversiforme* Godwinski, *C. durum* Godwinski, *C. falcilobum* Godwinski, *C. flavidum* Bunge, *C. inaequale* Godwinski, *C. involutum* Pavl., *C. irtyschense* Godwinski, *C. josephii* Godwinski, *C. kurotschkinae* Godwinski, *C. nudatum* Godwinski, *C. ozunraicum* Godwinski, *C. pappii* Godwinski, *C. parvum* Godwinski, *C. pseudotenue* Godwinski, *C. rubidum* Godwinski, *C. russanovii* Pavl., *C. sinuato-aculeolatum* Godwinski, *C. subcomplanatum* Godwinski) - 3, 6
rubidum Godwinski = C. rubicundum
russanovii Pavl. = C. rubicundum
santoanum Korov. - 6
setosum (Litv.) Litv. (*C. aciferum* Godwinski & Nardina, *C. fragile* Godwinski & Nardina, *C. kzyl-kumi* Pavl., *C. molle* Litv., *C. nardinae* Godwinski, *C. pulcherrimum* Korov. ex Pavl.) - 6
sinuato-aculeolatum Godwinski = C. rubicundum
smirnovii Drob. = C. acanthopterum
spinosissimum Pavl. = C. crispum
x **spinulosum** Drob. (*C. x pseudodubjanskyi* Sosk., *C. x pseudomuravljanskyi* Sosk.). - C. acanthopterum Borszcz. x C. aphyllum (Pall.) Guerke - 6

squarrosum Pavl. - 6

subcomplanatum Godwinski = C. rubicundum

tenue Pavl. = C. aphyllum

tetrapterum auct. = C. x barsukiense

tortile Drob. = C. aphyllum

triste Litv. - 6

turbineum Pavl. = C. leucocladum

turkestanicum (Korov.) Pavl. = C. rubescens

undulatum Litv. = C. aphyllum

ustjurtense Drob. = C. aphyllum

uzunachmatense V. Tkatschenko = C. leucocladum

x **zaissano-muravljanskyi** Sosk. - C. crispum Bunge x C. rubicundum Bunge - 3, 6

zakirovii (Chalkuziev) Czer. comb. nova (*Calliphysa zakirovii* Chalkuziev, 1990, O Rodstvennykh Svyasyakh Nekotrykh Semeystv Rasteniy Pustynnykh Oblastey : 85) - 6

Calliphysa Fisch. & C.A. Mey. = Calligonum

bykovii (Godwinski) Chakuziev = Calligonum junceum

zakirovii Chalkuziev = Calligonum zakirovii

Cephalophilon (Meissn.) Spach

nepalense (Meissn.) Tzvel. (*Polygonum nepalense* Meissn., *P. alatum* Buch.-Ham. ex Spreng.) - 2(alien), 5, 6

runcinatum (Buch.-Ham. ex D. Don) Tzvel. (*Polygonum runcinatum* Buch.-Ham. ex D. Don) - 2(alien)

Chylocalyx Hassk. ex Miq. (*Ampelygonum* auct.)

perfoliatus (L.) Hassk. ex Miq. (*Polygonum perfoliatum* L., *Ampelygonum perfoliatum* (L.) Roberty & Vautier, *Truellum perfoliatum* (L.) Sojak) - 2(alien), 5

Coccoloba auct. = Muehlenbeckia

sagittifolia Ortega = Muehlenbeckia sagittifolia

Fagopyrum Mill.

baldschuanicum (Regel) H. Gross = Fallopia baldschuanica

convolvulus (L.) H. Gross = Fallopia convolvulus

dumetorum (L.) Schreb. = Fallopia dumetorum

*****esculentum** Moench (*F. sagittatum* Gilib. nom. invalid.)

multiflorum (Thunb.) I. Grint. = Fallopia miltiflora

sagittatum Gilib. = F. esculentum

suffruticosum Fr. Schmidt = F. tataricum

tataricum (L.) Gaertn. (*F. suffruticosum* Fr. Schmidt) - 1, 2, 3, 4, 5, 6

Fallopia Adans. (*Bilderdykia* Dumort.)

baldschuanica (Regel) Holub (*Polygonum baldschuanicum* Regel, *Bilderdykia baldschuanica* (Regel) D.A. Webb, *Fagopyrum baldschuanicum* (Regel) H. Gross, *Reynoutria baldschuanica* (Regel) Mold.) - 6

convolvulus (L.) A. Löve (*Polygonum convolvulus* L., *Bilderdykia convolvulus* (L.) Dumort., *Fagopyrum convolvulus* (L.) H. Gross, *Reynoutria convolvulus* (L.) Shinners) - 1, 2, 3, 4, 5, 6

dentato-alata (Fr. Schmit) Holub (*Polygonum dentato-alatum* Fr. Schmidt, *Bilderdykia dentato-alata* (Fr. Schmidt) Kitag.) - 5

dumetorum (L.) Holub (*Polygonum dumetorum* L., *Bilderdykia dumetorum* (L.) Dumort., *Fagopyrum dumetorum* (L.) Schreb., *Reynoutria scandens* (L.) Shinners

subsp. *dumetorum* (L.) Shinners) - 1, 2, 3, 4, 5, 6

*****multiflora** (Thunb.) K. Haraldson (*Polygonum multiflorum* Thunb., *Bilderdykia multiflora* (Thunb.) Roberty & Vautier, *Fagopyrum multiflorum* (Thunb.) I. Grint., *Fallopia multiflora* (Thunb.) Czer. comb. superfl.) - 2

pauciflora (Maxim.) Kitag. (*Polygonum pauciflorum* Maxim., *P. convolvulus* L. var. *pauciflorum* (Maxim.) Worosch.) - 5

schischkinii Tzvel. - 5

Knorringia (Czuk.) Tzvel.

pamirica (Korsh.) Tzvel. (*Polygonum pamiricum* Korsh., *Aconogonon pamiricum* (Korsh.) Hara) - 6

sibirica (Laxm.) Tzvel. (*Polygonum sibiricum* Laxm., *Aconogonon sibiricum* (Laxm.) Hara, *Pleuropteropyrum sibiricum* (Laxm.) Kitag.) - 3, 4, 5(alien), 6

Koenigia L.

islandica L. - 1, 3, 4, 5, 6

Lapathum Hill = Rumex

alpestre (Jacq.) Scop. = Rumex alpestris

sylvestre Lam. = Rumex sylvestris

Muehlenbeckia Meissn. (*Coccoloba* auct.)

sagittifolia (Ortega) Meissn. (*Coccoloba sagittifolia* Ortega, *Polygonum acetosifolium* Vent.) - 1

Oxyria Hill

digyna (L.) Hill (*O. elatior* R. Br. ex Meissn.) - 1, 2, 3, 4, 5, 6

elatior R. Br. ex Meissn. = O. digyna

Persicaria Hill

amphibia (L.) S.F. Gray (*Polygonum amphibium* L., *Persicaria amphibia* var. *amurensis* (Korsh.) Hara, *P. amphibia* subsp. *amurensis* (Korsh.) Sojak, *P. amurensis* Nieuwland, *Polygonum amphibium* var. *amurense* Korsh., *P. amphibium* subsp. *amerense* (Korsh.) Hult., *P. amurense* (Korsh.) Worosch.) - 1, 2, 3, 4, 5, 6

- subsp. *amurensis* (Korsh.) Sojak = P. amphibia

- var. *amurensis* (Korsh.) Hara = P. amphibia

amurensis Nieuwland = Persicaria amphibia

belophylla (Litv.) Kitag. = Truellum sieboldii

bistorta (L.) Sampaio subsp. *carnea* (C. Koch) Greuter & Burdet = Bistorta carnea

brittingeri (Opiz) Opiz (*Polygonum brittingeri* Opiz, *Persicaria lapathifolia* (L.) S.F. Gray subsp. *brittingeri* (Opiz) Sojak, *Polygonum lapathifolium* L. subsp. *brittingeri* (Opiz) Rech. fil.) - 1, 2, 5(alien)

bungeana (Turcz.) Nakai ex Mori (*Polygonum bungeanum* Turcz.) - 5

chrtekii Sojak = P. sungareensis

erecto-minor auct. = P. sungareensis

erecto-minor (Makino) Nakai var. *koreensis* (Nakai) Ito = P. koreensis

- var. *roseoviridis* (Kitag.) Ito = P. roseoviridis

- var. *sungareensis* (Kitag.) Kitag. = P. sungareensis

extremiorientalis (Worosch.) Tzvel. (*Polygonum extremiorientale* Worosch.) - 5

foliosa (Lindb. fil.) Kitag. (*Polygonum foliosum* Lindb. fil., *Persicaria foliosa* var. *paludicola* (Makino) Hara ex Ito, *P. paludicola* (Makino) Nakai, *Polygonum mariae*

V. Vassil., *P. foliosum* var. *paludicola* (Makino) Kitam., *P. paludicola* Makino) - 1, 3, 5
- var. *paludicola* (Makino) Hara ex Ito = P. foliosa
hastato-triloba (Meissn.) Okuyama = Truellum thunbergii
hydropiper (L.) Spach (*Polygonum hydropiper* L.) - 1, 2, 3, 4, 5, 6
- subsp. *mitis* (Schrank) Majeed Kak = P. mitis
hydropiperoides (Michx.) Small (*Polygonum hydropiperoides* Michx.) - 2(alien), 5(alien)
hypanica (Klok.) Tzvel. (*Polygonum hypanicum* Klok., *Persicaria lapathifolia* (L.) S.F. Gray subsp. *hypanica* (Klok.) Sojak) - 1
imeretina (Kom.) Sojak (*Polygonum imeretinum* Kom.) - 2
komarovii (Levl.) Sojak = P. lapathifolia
koreensis (Nakai) Nakai (*Polygonum koreense* Nakai, *Persicaria erecto-minor* (Makino) Nakai var. *koreensis* (Nakai) Ito) - 5
korshinskiana (Nakai) Nakai = Truellum hastatosagittatum
lanata (Roxb.) Tzvel. (*Polygonum lanatum* Roxb., *Persicaria lapathifolia* (L.) S.F. Gray subsp. *lanata* (Roxb.) Sojak) - 1(alien), 2(alien)
lapathifolia (L.) S.F. Gray (*Polygonum lapathifolium* L., *Persicaria komarovii* (Levl.) Sojak, *P. lapathifolia* subsp. *klokovii* Sojak, *P. lapathifolia* var. *tomentosa* (Schrank) Tzvel., *P. nodosa* (Pers.) Opiz, *Polygonum komarovii* Levl., *P. lapathifolium* subsp. *nodosum* (Pers.) Kitam., *P. nodosum* Pers., *P. paniculatum* Andrz. 1862, non Blume, 1826, *P. tomentosum* Schrank) - 1, 2, 3, 4, 5, 6
- subsp. **andrzejowskiana** (Klok.) Sojak (*Polygonum andrzejowskianum* Klok.) - 1
- subsp. *brittingeri* (Opiz) Sojak = P. brittingeri
- subsp. *hypanica* (Klok.) Sojak = P. hypanica
- subsp. *klokovii* Sojak = P. lapathifolia
- subsp. *lanata* (Roxb.) Sojak = P. lanata
- subsp. *pallida* (With.) A. Löve = P. scabra
- subsp. *saporoviensis* (Klok.) Sojak = P. saporoviensis
- subsp. **turgida** (Thuill.) Sojak (*Polygonum turgidum* Thuill.) - 1
- var. *tomentosa* (Schrank) Tzvel. = P. lapathifolia
linicola (Sutul.) Nenjuk. (*Polygonum linicola* Sutul.) - 1, 2, 3, 5(alien)
longiseta (De Bruyn) Kitag. (*Polygonum longisetum* De Bruyn, *P. blumei* Meissn. nom. illegit., *P. posumbu* sensu Kom.) - 2(alien), 5
maculata (Rafin.) A. & D. Löve (*Polygonum maculatum* Rafin., *P. persicaria* L.) - 1, 2, 3, 4, 5, 6
minor (Huds.) Opiz (*Polygonum minus* Huds.) - 1, 2, 3, 4, 5, 6
mitis (Schrank) Opiz ex Assenov (*Polygonum mite* Schrank, *Persicaria hydropiper* (L.) Spach subsp. *mitis* (Schrank) Majeed Kak, *P. mitis* (Schrank) Holub, comb. superfl.) - 1
nipponensis (Makino) H. Gross = Truellum nipponense
nodosa (Pers.) Opiz = P. lapathifolia
*orientalis (L.) Spach (*Polygonum orientale* L., *Periscaria orientalis* (L.) Assenov, comb. superfl.) - 1, 2, 5, 6
pilosa (Roxb.) Kitag. (*Polygonum pilosum* Roxb.) - 5
roseoviridis Kitag. (*P. erecto-minor* (Makino) Nakai var. *roseoviridis* (Kitag.) Ito, *Polygonum roseoviride* (Kitag.) Li & Chang) - 5
saporoviensis (Klok.) Tzvel. (*Polygonum saporoviense* Klok., *Persicaria lapathifolia* (L.) S.F. Gray subsp. *saporoviensis* (Klok.) Sojak) - 1
scabra (Moench) Mold. (*Polygonum scabrum* Moench, *Persicaria lapathifolia* (L.) S.F. Gray subsp. *pallida* (With.) A. Löve, *P. scabra* var. *incana* (F.W. Schmidt)

Tzvel., *Polygonum incanum* F.W. Schmidt, *P. lapathifolium* L. subsp. *pallidum* (With.) Fries, *P. pallidum* With.) - 1, 2, 3, 4, 5, 6
- var. *incana* (F.W. Schmidt) Tzvel. = P. scabra
sungareensis Kitag. (*P. chrtekii* Sojak, *P. erecto-minor* (Makino) Nakai var. *sungareensis* (Kitag.) Kitag., *Polygonum chrtekii* (Sojak) Czer., *P. intricatum* Kom. 1936, non Tod. ex Lojac. 1904, *P. sungareense* Kitag. nom. altern., *Persicaria erecto-minor* auct., *Polygonum erecto-minus* auct.) - 3, 4, 5
*tinctoria (Ait.) Spach (*Polygonum tinctorium* Ait.)
  Previosly had been cutivated in 1, 2, 5
trigonocarpa (Makino) Nakai (*Polygonum minus* Huds. f. *trigonocarpum* Makino, *P. trigonocarpum* (Makino) Kudo & Masam.) - 5
ussuriensis (Regel & Maack) Nakai ex Mori = Truellum hastatosagittatum
viscofera (Makino) H. Gross ex Nakai (*Polygonum viscoferum* Makino, *P. excurrens* auct., *P. makinoi* auct.) - 5
viscosa (Buch.-Ham. ex D. Don) H. Gross ex Nakai (*Polygonum viscosum* Buch.-Ham. ex D. Don) - 5
yokusaiana (Makino) Nakai (*Polygonum yokusaianum* Makino, *P. caespitosum* auct., *P. posumbu* sensu Worosch.) - 2(alien), 5

Pleuropteropyrum H. Gross = Aconogonon

alpinum (All.) Kitag. = Aconogonon alpinum
alpinum (All.) Nakai = Aconogonon alpinum
amgense (V. Michaleva & V. Perfiljeva) Sojak = Aconogonon amgense
angustifolium (Pall.) Kitag. = Aconogonon angustifolium
bucharicum (Grig.) Nevski = Aconogonon bucharicum
iturupense (Mischurov) Sojak = Aconogonon savatieri var. iturupense
laxmannii (Lepech.) Kitag. = Aconogonon ochreatum var. laxmannii
limosum (Kom.) Kitag. = Aconogonon limosum
riparium (Georgi) Kitag. = Aconogonon ochreatum var. riparium
sibiricum (Laxm.) Kitag. = Knorringia sibirica
tripterocarpum (A. Gray) H. Gross = Aconogonon tripterocarpum
undulatum (Murr.) A. & D. Löve = Aconogonon alpinum
weyrichii (Fr. Schmidt) H. Gross = Aconogonon weyrichii

Polygonum L.

abbreviatum Kom. = Bistorta abbreviata
acerosum Ledeb. ex Meissn. - 6
acetosellum Klok. = P. calcatum
acetosifolium Vent. = Muehlenbeckia sagittifolia
acetosum Bieb. (*P. jaxarticum* Sumn.) - 1, 3, 6
aequale Lindm. = P. arenastrum
agreste Sumn. = P. arenastrum
agrestinum Jord. ex Boreau = P. neglectum
ajanense (Regel & Til.) Grig. = Aconogonon ajanense
- subsp. *middendorfii* (Kongar) Worosch. = Aconogonon middendorfii
alaskanum (Small) Wight ex Hult. = Aconogonon alaskanum
alatum Buch.-Ham. ex Spreng. = Cephalophilon nepalense
alopecuroides Turcz. ex Meissn. = Bistorta alopecuroides
- subsp. *subauriculatum* (V. Petrov ex Kom.) Worosch. = Bistorta subauriculata
- var. *attenuatum* (V. Petrov ex Kom.) Serg. = Bistorta elliptica

alpestre C.A. Mey. (*P. alpestre* var. *ammanioides* (Jaub. & Spach) Boiss., *P. ammanioides* Jaub. & Spach) - 2
- var. *ammanioides* (Jaub. & Spach) Boiss. = P. alpestre
alpinum All. = Aconogonon alpinum
- *alaskanum* Small = Aconogonon alaskanum
amgense V. Michaleva & V. Perfiljeva = Aconogonon amgense
ammanioides Jaub. & Spach = P. alpestre
amphibium L. = Persicaria amphibia
- subsp. *amurense* (Korsh.) Hult. = Persicaria amphibia
- var. *amurense* Korsh. = Persicaria amphibia
amurense (Korsh.) Worosch. = Persicaria amphibia
andrzejowskianum Klok. = Persicaria lapathifolia subsp. andrzejowskiana
angustifolium Pall. = Aconogonon angustifolium
aphyllum Krock. = P. arenastrum
araraticum Kom. = P. arenastrum var. caspicum
arenarium Waldst. & Kit. - 1
- subsp. *janatae* (Klok.) Soo = P. pulchellum
- subsp. *pseudoarenarium* (Klok.) Soo = P. pseudoarenarium
- subsp. *pulchellum* (Loisel.) D.A. Webb & Chater = P. pulchellum
arenastrum Boreau (*P. aequale* Lindm., *P. agreste* Sumn., ? *P. aphyllum* Krock., *P. aviculare* L. subsp. *aequale* (Lindm.) Aschers. & Graebn., *P. retinerve* Worosch. nom. invalid., *P. sparsiflorum* Worosch., *P. suifunense* Worosch. & N.S. Pavlova, *P. aviculare* sensu Kom.) - 1, 2, 3, 4, 5, 6
- var. **caspicum** (Kom.) Tzvel. (*P. caspicum* Kom., *P. araraticum* Kom., *P. lencoranicum* Kom., *P. littorale* Meissn. 1856, non Link, 1821) - 2
argyrocoleon Steud. ex G. Kunze (*P. deciduum* Boiss. & Noe ex Meissn.) - 1, 2, 5(alien), 6
arianum Grig. - 6
aschersonianum H. Gross (*P. samarense* H. Gross, *P. scythicum* Klok.) - 1, 2
atraphaxiforme Botsch. - 6
attenuatum V. Petrov ex Kom. = Bistorta elliptica
aviculare L. (*P. aviculare* subsp. *heterophyllum* (Lindm.) Aschers. & Graebn. nom. illegit., *P. aviculare* subsp. *monspeliense* (Pers.) Chrtek, *P. heterophyllum* Lindm. nom. illegit., *P. monspeliense* Thieb. ex Pers.) - 1, 2, 3, 4, 5, 6
- subsp. *aequale* (Lindm.) Aschers. & Graebn. = P. arenastrum
- - var. *calcatum* (Lindm.) Hyl. = P. calcatum
- subsp. *calcatum* (Lindm.) Thell. = P. calcatum
- subsp. *heterophyllum* (Lindm.) Aschers. & Graebn. = P. aviculare
- subsp. *monspeliense* (Pers.) Chrtek = P. aviculare
- subsp. *rectum* Chrtek = P. neglectum
- subsp. *rurivagum* (Jord. ex Boreau) O. Bolos & Vigo = P. rurivagum
- var. *boreale* Lange = P. boreale
- var. *minutiflorum* Franch. = P. plebejum
- var. *rigidum* (Skvorts.) Fu = P. rigidum
- var. *virgatum* Peterm. = P. neglectum
aviculare sensu Kom. = P. arenastrum
baicalense Sipl. = Aconogonon ochreatum
baldschuanicum Regel = Fallopia baldschuanica
bargusinense Peschkova = Aconogonon bargusinense
bellardii All. (*P. kitaibelianum* Sadl., *P. patulum* Bieb. subsp. *kitaibelianum* (Sadl.) Aschers. & Graebn., *P. tiflisiense* Kom.) - 1, 2, 5(alien)
bellardii sensu Tzvel. = P. neglectum
belophyllum Litv. = Truellum sieboldii

biaristatum Aitch. & Hemsl. - 6
bistorta L. = Bistorta major
- subsp. *attenuatum* (V. Petrov ex Kom.) V. Perfiljeva = Bistorta elliptica
- subsp. *carneum* (C. Koch) Coode & Cullen = Bistorta carnea
- subsp. *cordifolium* (Turcz.) Malysch. = Bistorta abbreviata
- subsp. *ellipticum* (Willd. & Spreng.) Petrovsky = Bistorta elliptica
- subsp. *ochotense* (V. Petrov ex Kom.) Worosch. = Bistorta elliptica
- subsp. *pacificum* (V. Petrov ex Kom.) Worosch. = Bistorta pacifica
- subsp. *subauriculatum* (V. Petrov ex Kom.) Gubanov = Bistorta subauriculata
- var. *cordifolium* Turcz. = Bistorta abbreviata
blumei Meissn. = Persicaria longiseta
bordzilowskii Klok. (*P. gracilius* (Ledeb.) Klok. p.p. quoad pl., *P. patulum* Bieb. subsp. *bordzilowskii* (Klok.) Soo) - 1
boreale (Lange) Small (*P. aviculare* L. var. *boreale* Lange, *P. caducifolium* Worosch. p.p. quoad pl., *P. heterophyllum* Lindm. subsp. *boreale* (Lange) A. & D. Löve, non rite publ., *P. buxifolium* auct.) - 1, 5
borgoicum Tupitzina - 4
bornmuelleri Litv. (*P. fragile* Sumn.) - 6
brittingeri Opiz = Persicaria brittingeri
bucharicum Grig. = Aconogonon bucharicum
bungeanum Turcz. = Persicaria bungeana
buxifolium auct. = P. boreale
caducifolium Worosch. p.p. = P. boreale
caespitosum auct. = Persicaria yokusaiana
calcatum Lindm. (*P. acetosellum* Klok., *P. aviculare* L. subsp. *calcatum* (Lindm.) Thell., *P. aviculare* subsp. *aequale* (Lindm.) Aschers. & Graebn. var. *calcatum* (Lindm.) Hyl.) - 1, 2(alien), 3, 4, 5
carneum C. Koch = Bistorta carnea
caspicum Kom. = P. arenastrum var. caspicum
caurianum Robins. = P. humifusum
chaneyi B. Fedtsch. ex R.R. Stewart = Aconogonon chaneyi
changii Kitag. = P. plebejum
chlorochryseum M. Ivanova = Aconogonon chlorochryseum
chrtekii (Sojak) Czer. = Persicaria sungareensis
cognatum Meissn. (*P. rupestre* Kar. & Kir.) - 3, 6
convolvulus L. = Fallopia convolvulus
- var. *pauciflorum* (Maxim.) Worosch. = Fallopia pauciflora
cordifolium (Turcz.) Losinsk. = Bistorta abbreviata
coriarium Grig. = Aconogonon coriarium
corrigioloies Jaub. & Spach - 2, 6
cretaceum Kom. (*P. novoascanicum* Klok. subsp. *cretaceum* (Kom.) Tzvel.) - 1
cuspidatum Siebold & Zucc. = Reynoutria japonica
czukavinae Sojak (*P. paradoxum* Czuk. 1966, non Levl. 1909, nec T.C.E. Fries, 1924) - 6
deciduum Boiss. & Noe ex Meissn. = P. argyrocoleon
dentato-alatum Fr. Schmidt = Fallopia dentato-alata
diffusum Willd. ex Spreng. = Aconogonon diffusum
dissitiflorum Hemsl. = Truellum dissitiflorum
divaricatum L. = Aconogonon divaricatum
- var. *limosum* Kom. = Aconogonon limosum
dshawachischwilii Charkev. = Aconogonon alpinum
dumetorum L. = Fallopia dumetorum
ellipticum Willd. ex Spreng. = Bistorta elliptica
ensigerum Juz. = Bistorta ensigera
equisetiforme auct. = P. hyrcanicum
erecto-minus auct. = Persicaria sungareensis

euxinum Chrtek (*P. robertii* auct.) - 1, 2
excurrens auct. = Persicaria viscofera
extremiorientale Worosch. = Persicaria extremiorientalis
fibrilliferum Kom. - 6
filiforme Thunb. = Antenoron filiforme
floribundum Schlecht. ex Spreng. - 1, 3, 6
foliosum Lindb. fil. = Persicaria foliosa
- var. *paludicola* (Makino) Kitam. = Persicaria foliosa
fragile Sumn. = P. bornmuelleri
franchetii Worosch. = P. plebejum
fusco-ochreatum Kom. - 5
glaucescens Ivanova ex Tupitzina - 4
gracilius (Ledeb.) Klok. = P. bordzilowskii and P. patulum
hastatosagittatum Makino = Truellum hastatosagittatum
hastato-trilobum Meissn. = Truellum thunbergii
heterophyllum Lindm. = P. aviculare
- subsp. *boreale* (Lange) A. & D. Löve = P. boreale
- subsp. *rurivagum* (Jord. ex Boreau) Lindm. = P. rurivagum
hissaricum M. Pop. = Aconogonon hissaricum
humifusum Merk ex C. Koch (*P. caurianum* Robins.) - 1, 3, 4, 5
hydropiper L. = Persicaria hydropiper
hydropiperoides Michx. = Persicaria hydropiperoides
hypanicum Klok. = Persicaria hypanica
hyrcanicum Rech. fil. (*P. equisetiforme* auct.) - 2, 6
imeretinum Kom. = Persicaria imeretina
incanum F.W. Schmidt = Persicaria scabra
inflexum Kom. (*P. oxanum* Kom.) - 6
insulare Maximova = Bistorta vivipara
intercedens V. Petrov ex Kom. = Bistorta elliptica
intricatum Kom. = Persicaria sungareensis
iturupense Mischurov = Aconogonon savatieri var. iturupense
janatae Klok. = P. pulchellum
jaxarticum Sumn. = P. acetosum
junceum Ledeb. = P. pseudoarenarium
jurii A. Skvorts. = Aconogonon jurii
kitaibelianum Sadl. = P. bellardii
kolymense A. Khokhr. = Bistorta elliptica
komarovii Levl. = Persicaria lapathifolia
korotkovae Sumn. = P. vvedenskyi
korshinskianum Nakai = Truellum hastatosagittatum
kotovii Klok. = P. patulum
krascheninnikovii (Ivanova) Czer. = Bistorta krascheninnikovii
kudrjaschevii Vassilkovsk. = P. serpyllaceum
lanatum Roxb. = Persicaria lanata
lapathifolium L. = Persicaria lapathifolia
- subsp. *brittingeri* (Opiz) Rech. fil. = Persicaria brittingeri
- subsp. *nodosum* (Pers.) Kitam. = Persicaria lapathifolia
- subsp. *pallidum* (With.) Fries = Persicaria scabra
laxmannii Lepech. = Aconogonon ochreatum var. laxmannii
lencoranicum Kom. = P. arenastrum var. caspicum
liaotungense Kitag. - 5
limosum (Kom.) Kom. = Aconogonon limosum
linicola Sutul. = Persicaria linicola
littorale Meissn. = P. arenastrum var. caspicum
longipedicellatum Zak. = P. myrtillifolium
longisetum De Bruyn = Persicaria longiseta
luxurians Grig. = Aconogonon jurii
luzuloides Jaub. & Spach (*P. setosum* Jacq. subsp. *luzuloides* (Jaub. & Spach) Leblebici) - 2
maackianum Regel = Truellum maackianum
macounii Small = Bistorta vivipara subsp. macounii
maculatum Rafin. = Persicaria maculata

makinoi auct. = Persicaria viscofera
manshuriense V. Petrov ex Kom. = Bistorta manshuriensis
mariae V. Vassil. = Persicaria foliosa
maritimum L. - 1, 2
mesembricum Chrtek - 1, 2
mezianum H. Gross - 6
middendorfii Kongar = Aconogonon middendorfii
minus Huds. = Persicaria minor
- f. *trigonocarpum* Makino = Persicaria trigonocarpa
mite Schrank = Persicaria mitis
molliiforme Boiss. - 6
monspeliense Thieb. ex Pers. = P. aviculare
multiflorum Thunb. = Fallopia multiflora
myrtillifolium Kom. (*P. longipedicellatum* Zak.) - 6
neglectum Bess. (*P. agrestinum* Jord. ex Boreau, *P. aviculare* L. subsp. *rectum* Chrtek, *P. aviculare* var. *virgatum* Peterm., ? *P. nervosum* Wallr., *P. procumbens* Gilib. nom. invalid., *P. rectum* (Chrtek) H. Scholz, *P. rectum* subsp. *virgatum* (Peterm.) H. Scholz, *P. bellardii* sensu Tzvel.) - 1, 2, 3, 4, 5, 6
neglectum sensu Kom. p.p. = P. rurivagum
nepalense Meissn. = Cephalophilon nepalense
nervosum Wallr. = P. neglectum
nipponense Makino = Truellum nipponense
nitens (Fisch. & C.A. Mey.) V. Petrov ex Kom. = Bistorta elliptica
nodosum Pers. = Persicaria lapathifolia
norvegicum (Sam.) Lid (*P. raii* Bab. subsp. *norvegicum* Sam., *P. raii* auct.) - 1
novoascanicum Klok. (? *P. spectabile* auct.) - 1
- subsp. *cretaceum* (Kom.) Tzvel. = P. cretaceum
- subsp. *psammophilum* Bordz. ex Tzvel. = P. psammophilum
ochotense V. Petrov ex Kom. = Bistorta elliptica
ochreatum L. = Aconogonon ochreatum
- var. *laxmannii* (Lepech.) M. Pop. = Aconogonon ochreatum var. laxmannii
orientale L. = Persicaria orientalis
ovczinnikovii Czuk. - 6
oxanum Kom. = P. inflexum
oxyspermum C.A. Mey. & Bunge - 1
pacificum V. Petrov ex Kom. = Bistorta pacifica
pallidum With. = Persicaria scabra
paludicola Makino = Persicaria foliosa
paludosum (Kom.) Kom. = Truellum sieboldii
pamiricum Korsh. = Knorringia pamirica
pamiroalaicum Kom. = P. serpyllaceum
paniculatum Andrz. = Persicaria lapathifolia
panjutinii Charkev. = Aconogonon panjutinii
paradoxum Czuk. = P. czukavinae
paronychioides C.A. Mey. - 2, 6
patuliforme Worosch. - 1, 2
patulum Bieb. (*P. gracilius* (Ledeb.) Klok. p.p. quoad basionymum, *P. kotovii* Klok., *P. patulum* subsp. *kotovii* (Klok.) Soo, *P. patulum* var. *patulum* f. *gracilius* (Ledeb.) I. Grint.) - 1, 2, 3, 4, 5, 6
- subsp. *bordzilowskii* (Klok.) Soo = P. bordzilowskii
- subsp. *kitaibelianum* (Sadl.) Aschers. & Graebn. = P. bellardii
- subsp. *kotovii* (Klok.) Soo = P. patulum
- subsp. *pulchellum* (Loisel.) Leblebici = P. pulchellum
- var. *patulum* f. *gracilius* (Ledeb.) I. Grint. = P. patulum
pauciflorum Maxim. = Fallopia pauciflora
perfoliatum L. = Chylocalyx perfoliatus
persicaria L. = Persicaria maculata
pilosum Roxb. = Persicaria pilosa
platyphyllum auct. = Aconogonon jurii

plebejum R. Br. (*P. aviculare* L. var. *minutiflorum* Franch., *P. changii* Kitag., *P. franchetii* Worosch., *P. plebejum* subsp. *changii* (Kitag.) Worosch.) - 5
- subsp. *changii* (Kitag.) Worosch. = P. plebejum
plumosum Small = Bistorta plumosa
polycnemoides Jaub. & Spach - 2, 6
polyneuron Franch. & Savat. - 5
posumbu sensu Kom. = Persicaria longiseta
posumbu sensu Worosch. = Persicaria yokusaiana
procumbens Gilib. = P. neglectum
propinquum Ledeb. - 1, 2, 3, 4, 5, 6
psammophilum (Bordz. ex Tzvel.) Tzvel. (*P. novoascanicum* Klok. subsp. *psammophilum* Bordz. ex Tzvel.) - 1
pseudoarenarium Klok. (*P. arenarium* Waldst. & Kit. subsp. *pseudoarenarium* (Klok.) Soo, *P. junceum* Ledeb. 1850, non A. Cunn. ex Lindl. 1848) - 1, 2, 3, 6
- subsp. *janatae* (Klok.) E. Wulf = P. pulchellum
pseudoincanum Klok. - 1
pulchellum Loisel. (*P. arenarium* Waldst. & Kit. subsp. *janatae* (Klok.) Soo, *P. arenarium* subsp. *pulchellum* (Loisel.) D.A. Webb & Chater, *P. janatae* Klok., *P. patulum* Bieb. subsp. *pulchellum* (Loisel.) Leblebici, *P. pseudoarenarium* Klok. subsp. *janatae* (Klok.) E. Wulf) - 1, 2, 6
pulvinatum Kom. - 3, 6
raii auct. = P. norvegicum
raii Bab. subsp. *norvegicum* Sam. = P. norvegicum
rectum (Chrtek) H. Scholz = P. neglectum
- subsp. *virgatum* (Peterm.) H. Scholz = P. neglectum
regelianum Kom. = Bistorta pacifica
relictum Kom. = Aconogonon relictum
retinerve Worosch. = P. arenastrum
rigidum Skvorts. (*P. aviculare* L. var. *rigidum* (Skvorts.) Fu) - 5
riparium Georgi = Aconogonon ochreatum var. riparium
robertii auct. = P. euxinum
roseoviride (Kitag.) Li & Chang = Persicaria roseoviridis
rottboellioies Jaub. & Spach (*P. tubulosum* Boiss.) - 6
runcinatum Buch.-Ham. ex D. Don = Cephalophilon runcinatum
rupestre Kar. & Kir. = P. cognatum
rurivagum Jord. ex Boreau (*P. aviculare* L. subsp. *rurivagum* (Jord. ex Boreau) O. Bolos & Vigo, *P. heterophyllum* Lindm. subsp. *rurivagum* (Jord. ex Boreau) Lindm. non rite publ., *P. neglectum* sensu Kom. p.p.) - 1, 2, 5(alien)
sabulosum Worosch. - 5
> Probably, the priority name of this species is P. sylvaticum Skvorts.
sachalinense Fr. Schmidt = Reynoutria sachalinensis
sagittatum L. = Truellum sagittatum
- subsp. *sieboldii* (Meissn.) Worosch. = Truellum sieboldii
- var. *ussuriense* Regel & Maack = Truellum hastatosagittatum
salsugineum Bieb. - 1, 2, 3, 6
samarense H. Gross = P. aschersonianum
saporoviensis Klok. = Persicaria saporoviensis
savatieri Nakai = Aconogonon savatieri
scabrum Moench = Persicaria scabra
schistosum Czuk. - 6
schugnanicum Kom. = Bistorta elliptica
scythicum Klok. = P. aschersonianum
senticosum (Meissn.) Franch. & Savat. = Truellum japonicum
sericeum Pall. ex Georgi = Aconogonon sericeum
serpyllaceum Jaub. & Spach (*P. kudrjaschevii* Vassilkovsk., *P. pamiroalaicum* Kom.) - 6

setosum Jacq. - 2
- subsp. *luzuloides* (Jaub. & Spach) Leblebici = P. luzuloides
sibiricum (Laxm.) Hara = Knorringia sibirica
sieboldii Meissn. = Truellum sieboldii
songaricum Schrenk = Aconogonon songaricum
sparsiflorum Worosch. = P. arenastrum
spectabile auct. = P. novoascanicum
strigosum auct. = Truellum nipponense
subaphyllum Sumn. - 6
subauriculatum V. Petrov ex Kom. = Bistorta subauriculata
subsericeum M. Pop. = Aconogonon x subsericeum
suifunense Worosch. & N.S. Pavlova = P. arenastrum
sungareense Kitag. = Persicaria sungareensis
sylvaticum Skvorts. See P. sabulosum
tenuissimum A. Baran. & Skvorts. ex Worosch. - 2(alien), 5
thunbergii Siebold & Zucc. = Truellum thunbergii
thymifolium Jaub. & Spach - 6
tiflisiense Kom. = P. bellardii
tinctorium Ait. = Persicaria tinctoria
tomentosum Schrank = Persicaria lapathifolia
trigonocarpum (Makino) Kudo & Masam. = Persicaria trigonocarpa
tripterocarpum A. Gray = Aconogonon tripterocarpum
tubulosum Boiss. = P. rottboellioies
turgidum Thuill. = Persicaria lapathifolia subsp. turgida
turkestanicum Sumn. - 6
undulatum Murr. = Aconogonon alpinum
ussuriense (Regel) V. Petrov ex Kom. = Bistorta pacifica
ussuriense (Regel & Maak) Nakai = Truellum hastatosagittatum
valerii A. Skvorts. = Aconogonon valerii
viscoferum Makino = Persicaria viscofera
viscosum Buch.-Ham. ex D. Don = Persicaria viscosa
viviparum L. = Bistorta vivipara
vladimiri Czer. = Bistorta elliptica
volchovense Tzvel. - 1, 3
vvedenskyi Sumn. (*P. korotkovae* Sumn.) - 6
weyrichii Fr. Schmidt = Aconogonon weyrichii
yokusaianum Makino = Persicaria yokusaiana
zakirovii Czevrenidi - 6
zaravschanicum Zak. = Aconogonon zaravschanicum
zeaense Kom. = Bistorta alopecuroides var. zeaensis

Pterococcus Pall. = Calligonum

persicus Boiss. & Buhse = Calligonum leucocladum subsp. persicum

Pteropyrum Jaub. & Spach

aucheri Jaub. & Spach - 6

Reynoutria Houtt. (*Tiniaria* (Meissn.) Reichenb.)

baldschuanica (Regel) Mold. = Fallopia baldschuanica
convolvulus (L.) Shinners = Fallopia convolvulus
japonica Houtt. (*P. cuspidatum* Siebold & Zucc., *Tiniaria japonica* (Houtt.) Hedberg) - 2(cult.), 5
sachalinensis (Fr. Schmidt) Nakai (*Polygonum sachalinense* Fr. Schmidt, *Tiniaria sachalinesis* (Fr. Schmidt) Janch.) - 5
scandens (L.) Shinners subsp. *dumetorum* (L.) Shinners = Fallopia dumetorum

Rheum L.

altaicum Losinsk. = R. compactum
compactum L. (*R. altaicum* Losinsk.) - 3, 4, 5
- var. **orientale** (Losinsk.) Tzvel. (*R. orientale* Losinsk.) - 4, 5
cordatum Losinsk. - 6
darwasicum Titov ex Losinsk. - 6
fedtschenkoi Maxim. ex Regel - 6
ferganense Titov = R. macrocarpum
hissaricum Losinsk. - 6
korshinskyi Titov ex Losinsk. = R. lucidum
lobatum Litv. ex Losinsk. = R. macrocarpum
lucidum Losinsk. (*R. korshinskyi* Titov ex Losinsk.) - 6
macrocarpum Losinsk. (*R. ferganense* Titov, *R. lobatum* Litv. ex Losinsk., *R. nuratavicum* Titov, *R. plicatum* Losinsk., *R. vvedenskyi* Sumn., *R. zergericum* Titov) - 6
maximowiczii Losinsk. - 6
megalophyllum Sumn. = R. turkestanicum
moorcroftianum Royle = R. spiciforme
nanum Siev. - 3, 6
nuratavicum Titov = R. macrocarpum
orientale Losinsk. = R. compactum var. orientale
*palmatum L. - 1
plicatum Losinsk. = R. macrocarpum
renifolium Sumn. = R. turkestanicum
reticulatum Losinsk. = R. spiciforme
rhabarbarum L. (*R. undulatum* L.) - 4
*rhaponticum L. - 1
rhizostachyum Schrenk = R. spiciforme
ribes L. - 2
rupestre Litv. ex Losinsk. = R. turkestanicum
spiciforme Royle (*R. moorcroftianum* Royle, *R. reticulatum* Losinsk., *R. rhizostachyum* Schrenk) - 6
x **svetlanae** Krassovskaja. - R. cordatum Losinsk. x R. maximowiczii Losinsk. - 6
tataricum L. fil. - 1(cult.), 6
turanicum Titov = R. turkestanicum
turkestanicum Janisch. (*R. megalophyllum* Sumn., *R. renifolium* Sumn., *R. rupestre* Litv. ex Losinsk., *R. turanicum* Titov) - 6
undulatum L. = R. rhabarbarum
vvedenskyi Sumn. = R. macrocarpum
wittrockii Lundstr. - 6
zergericum Titov = R. macrocarpum

Rumex L. (*Acetosa* Mill., *Acetosella* (Meissn.) Fourr., *Lapathum* Hill)

x **abortivus** Ruhmer. - R. conglomeratus Murr. x R. obtusifolius L.
acetosa L. (*Acetosa fontano-paludosa* (Kalela) Holub, *A. pratensis* Mill., *Rumex acetosa* subsp. *fontano-paludosus* (Kalela) Hyl., *R. fontano-paludosus* Kalela) - 1, 2, 3, 4, 5, 6
- subsp. *alpestris* (Jacq.) A. Löve, p.p. = R. alpestris
- subsp. *auriculatus* (Wallr.) Blytt & Dahl = R. thyrsiflorus
- subsp. *fontano-paludosus* (Kalela) Hyl. = R. acetosa
- subsp. *lapponicus* Hiit. = R. lapponicus
- subsp. *pseudoxyria* Tolm. = R. pseudoxyria
- subsp. *thyrsiflorus* (Fingerh.) Hayek = R. thyrsiflorus
- subsp. *thyrsiflorus* (Fingerh.) Stojan. & Stef. = R. thyrsiflorus
- var. *auriculatus* Wallr. = R. thyrsiflorus
acetosella L. (*Acetosella multifida* (L.) A. Löve, *A. tenuifolia* (Wallr.) A. & D. Löve, *A. vulgaris* (Koch) Fourr., *A. vulgaris* f. *multifida* (L.) Dostal, *Rumex acetosella* var. *tenuifolius* Wallr., *R.acetosella* subsp. *tenuifolius*

(Wallr.) O. Schwarz, *R. mulitfidus* L., *R. tenuifolius* (Wallr.) A. Löve) - 1, 2, 3, 4, 5, 6
- subsp. *angiocarpus* (Murb.) Murb. = R. angiocarpus
- subsp. *tenuifolius* (Wallr.) O. Schwarz = R.acetosella
- var. *tenuifolius* Wallr. = R. acetosella
acetoselloides Bal. (*Acetosa acetoselloides* (Bal.) Holub, *Rumex fascilobus* Klok.) - 1, 2
alpestris Jacq. (*Acetosa alpestris* (Jacq.) A. Löve, *A. arifolia* (All.) Schur, *A. pratensis* Mill. subsp. *alpestris* (Jacq.) A. Löve, *Lapathum alpestre* (Jacq.) Scop., *Rumex acetosa* L. subsp. *alpestris* (Jacq.) A. Löve, p.p., *R. arifolius* All., *R. montanus* Desf.) - 1, 2
- subsp. *lapponicus* (Hiit.) Jalas = R. lapponicus
alpinus L. - 1, 2
alveolatus Losinsk. = R. drobovii
x **ambigens** Hausskn. - R. aquaticus L. x R. conglomeratus Murr.
amurensis Fr. Schmidt ex Maxim. - 5
angiocarpus Murb. (*Acetosella angiocarpa* (Murb.) A. Löve, *A. vulgaris* (Koch) Fourr. subsp. *angiocarpa* (Murb.) Hadac & Hasek, *Rumex acetosella* L. subsp. *angiocarpus* (Murb.) Murb.) - 5
angrenii Sumn. = R. rectinervis
angustifolius Campd. - 2
x **anisotyloides** Sumn. - R. crispus L. x R. halacsyi Rech.
aquaticus L. (*R. aquaticus* subsp. *euaquaticus* Rech. fil. nom. illegit., *R. aquaticus* subsp. *protractus* (Rech. fil.) Rech. fil., *R. protractus* Rech. fil., *R. fenestratus* auct.) - 1, 2, 3, 4, 5, 6
- subsp. *euaquaticus* Rech. fil. = R. aquaticus
- subsp. *insularis* Tolm. = R. insularis
- subsp. *lipschitzii* Rech. fil. = R. popovii
- subsp. *protractus* (Rech. fil.) Rech. fil. = R.aquaticus
- subsp. *schischkinii* (Losinsk.) Rech. fil. = R. schischkinii
arcticus Trautv. (*R. kamtschadalus* Kom., *R.ursinus* Maximova) - 1, 3, 4, 5
arifolius All. = R. alpestris
armenus C. Koch - 2
x **armoraciifolius** L. Neum. - R. aquaticus L. x R. longifolius DC.
x **arnottii** Druce. - R. longifolius DC. x R. obtusifolius L.
x **arpadianus** Bihari. - R. halacsyi Rech. x R. stenophyllus Ledeb.
aschabadensis Losinsk. = R. pamiricus
aureostigmaticus Kom. (*Acetosella aureostigmatica* (Kom.) Tzvel.) - 1, 4, 5
auriculatus sensu Leskov, p. max. p. = R. lapponicus
auriculatus (Wallr.) Murb. = R. thyrsiflorus
beringensis Jurtz. & Petrovsky (*Acetosella beringensis* (Jurtz. & Petrovsky) A. & D. Löve) - 5
x **callianthemus** Danser. - R. maritimus L. x R. obtusifolius L.
callosus (Fr. Schmidt ex Maxim.) Rech. fil. = R. patientia
carpaticus Zapal. = R. rugosus
caucasicus Rech. fil. - 2
chalepensis Mill. (*R. dictyocarpus* Boiss. & Buhse, *R. syriacus* Meissn., *R. syriacus* sensu Losinsk. p.p.) - 2, 6
confertus Willd. - 1, 2, 3, 5(alien), 6
conglomeratus Murr. - 1, 2, 6
x **conspersus** C. Hartm. - R. aquaticus L. x R. crispus L.
crispus L. - 1, 2, 3, 4, 5, 6
- subsp. *japonicus* (Houtt.) Kitam. = R. japonicus
dentatus L. subsp. *halacsyi* (Rech.) Rech. fil. = R. halacsyi
- subsp. *nipponicus* (Franch. & Savat.) Rech. fil. = R. nipponicus
dictyocarpus Boiss. & Buhse = R. chalepensis
x **digeneus** G. Beck - R. conglomeratus Murr. x R. hydrolapathum Huds.

divaricatus L. (*R. pulcher* L. subsp. *divaricatus* (L.) Arcang., *R. pulcher* subsp. *divaricatus* (L.) Murb. comb. superfl.) - 1, 2

x **dolosus** Valta. - R. aquaticus L. x R. confertus Willd.

domesticus C. Hartm. = R. longifolius

- var. *pseudonatronatus* Borb. = R. pseudonatronatus

drobovii Korov. (*R. alveolatus* Losinsk., *R. foveolatus* Losinsk. 1936, non Hochst. 1845, *R. losinskajae* Rech. fil., *R. syriacus* sensu Losinsk. p.p.) - 6

x **erubescens** Simonk. - R. obtusifolius L. x R. patientia L.

euxinus Klok. (*R. tuberosus* L. subsp. *euxinus* (Klok.) Borod., *R. tuberosus* auct.) - 1

evenkiensis Elisarjeva - 4

x **fallacinus** Hausskn. - R. crispus L. x R. maritimus L.

fascilobus Klok. = R. acetoselloides

fauriei Rech. fil. - 5

fenestratus auct. = R. aquaticus

fennicus (Murb.) Murb. = R. pseudonatronatus

fischeri Reichenb. - 6

fontano-paludosa Kalela = R. acetosa

foveolatus Losinsk. = R. drobovii

gmelinii Turcz. ex Ledeb. - 4, 5

gracilescens Rech. fil. - 2

graminifolius Lamb. (*Acetosella graminifolia* (Lamb.) A. Löve) - 1, 3, 4, 5

x **griffithii** Rech. fil. - R. crispus L. x ? R. chalepensis Mill.

hadroocarpus Rech. fil. = R. japonicus

haematinus Kihlm. (*R. thyrsiflorus* Fingerh. subsp. *haematinus* (Kihlm.) Borod.) - 1

halacsyi Rech. (*R. dentatus* L. subsp. *halacsyi* (Rech.) Rech. fil., *R. reticulatus* sensu Losinsk.) - 1, 2, 3, 6

haplorhizus Czern. ex Turcz. = R. thyrsiflorus

hastifolius Bieb. (*Acetosa scutata* (L.) Mill. subsp. *hastifolia* (Bieb.) A. Löve & Kapoor, *Rumex scutatus* L. subsp. *glaucus* (Jacq.) Gaudin & E. Wulf, p.p. quoad pl., *R. scutatus* L. subsp. *hastifolius* (Bieb.) Borod., *R. scutatus* sensu Losinsk.) - 1, 2

x **heterophyllus** C.F. Schultz (R. x maximus Schreb. 1811, non C.C. Gmel. 1806). - R. aquaticus L. x R. hydrolapathum Huds.

horizontalis C. Koch (*R. tuberosus* L. subsp. *horizontalis* (C. Koch) Rech. fil.) - 2

hultenii Tzvel. - 4, 5

hydrolapathum Huds. - 1, 2, 3, 5

insularis (Tolm.) Czer. (*R. aquaticus* L. subsp. *insularis* Tolm.) - 1

x **interjectus** Rech. fil. - R. confertus Willd. x R. nepalensis Spreng.

jacutensis Kom. - 4, 5

japonicus Houtt. (*R. crispus* L. subsp. *japonicus* (Houtt.) Kitam., *R. japonicus* Meissn., *R. hadroocarpus* Rech. fil.) - 5

kamtschadalus Kom. = R. arcticus

x **knafii** Celak. - R. conglomeratus Murr. x R. maritimus L.

x **knekii** Rech. - R. crispus L. x R. obtusifolius L.

komarovii Schischk. & Serg. - 3

krausei Jurtz. & Petrovsky (*Acetosella krausei* (Jurtz. & Petrovsky) A. & D. Löve) - 5

lapponicus (Hiit.) Czernov (*R. acetosa* L. subsp. *lapponicus* Hiit., *Acetosa alpestris* (Jacq.) A. Löve subsp. *lapponica* (Hiit.) A. Löve, *A. lapponica* (Hiit.) Holub, *A. pratensis* Mill. subsp. *alpestris* (Jacq.) A. Löve var. *lapponica* (Hiit.) A. Löve, *Rumex alpestris* Jacq. subsp. *lapponicus* (Hiit.) Jalas, *R. auriculatus* sensu Leskov, p. max.p.) - 1, 3, 4, 5

x **larinii** Borod. - R. maritimus x R. marschallianus Reichenb. - 1

lonaczevskii Klok. (*R. patientia* L. subsp. *orientalis* Danser, *R. orientalis* Bernh. 1830, non Campd. 1819) - 1, 2

longifolius DC. (*R. domesticus* C. Hartm.) - 1, 2, 3, 4, 5, 6

losinskajae Rech. fil. = R. drobovii

madajo Makino - 5

maritimus L. - 1, 2, 3, 4, 5, 6

- subsp. *rossicus* (Murb.) Kryl. = R. rossicus

- var. *ochotskius* (Rech. fil.) Kitag. = R. ochotskius

marschallianus Reichenb. - 1, 3, 6

x *maximus* Schreb. = R. x heterophyllus

montanus Desf. = R. alpestris

multifidus L. = R. acetosella

nepalensis Spreng. - 6

nipponicus Franch & Savat. (*R. dentatus* L. subsp. *nipponicus* (Franch. & Savat.) Rech. fil.) - 5

oblongifolius Tolm. (*Acetosa oblongifolia* (Tolm.) A. & D. Löve) - 4, 5

obtusifolius L. - 1, 2, 3, 4, 5

- subsp. **subalpinus** (Schur) Celak. - 2

- subsp. **sylvestris** (Lam.) Celak. = R. sylvestris

- subsp. **transiens** (Simonk.) Rech. fil. - 2

ochotskius Rech. fil. (*R. maritimus* L. var. *ochotskius* (Rech. fil.) Kitag.) - 5

orientalis Bernh. = R. lonaczevskii

palustris Smith - 1, 2

pamiricus Rech. fil. (*R. aschabadensis* Losinsk., *R. patientia* L. subsp. *pamiricus* (Rech. fil.) Rech. fil., *R.rechingerianus* Losinsk.) - 6

patientia L. (*R. callosus* (Fr. Schmidt ex Maxim.) Rech. fil., *R. patientia* var. *callosus* Fr. Schmidt ex Maxim., *R. patientia* subsp. *callosus* (Fr. Schmidt ex Maxim.) Rech. fil.) - 1, 2, 3, 5, 6

- subsp. *callosus* (Fr. Schmidt ex Maxim.) Rech. fil. = R. patientia

- subsp. *orientalis* Danser = R. lonaczevskii

- subsp. *pamiricus* (Rech. fil.) Rech. fil. = R. pamiricus

- var. *callosus* Fr. Schmidt ex Maxim. = R. patientia

paulsenianus Rech. fil. - 6

x **platyphyllos** Aresch. - R. aquaticus L. x R. obtusifolius L.

ponticus E. Krause - 2

popovii Pachom. (*R. aquaticus* L. subsp. *lipschitzii* Rech. fil.) - 6

x **propinquus** Aresch. - R. crispus L. x R. longifolius DC.

protractus Rech. fil. = R. aquaticus

pseudonatronatus (Borb.) Borb. ex Murb. (*R. domesticus* C. Hartm. var. *pseudonatronatus* Borb., *R. fennicus* (Murb.) Murb., *R. pseudonatronatus* subsp. *fennicus* Murb.) - 1, 2, 3, 4, 5, 6

- subsp. *fennicus* Murb. = R. pseudonatronatus

pseudoxyria (Tolm.) A. Khokhr. (*R. acetosa* L. subsp. *pseudoxyria* Tolm., *Acetosa pratensis* Mill. subsp. *pseudoxyria* (Tolm.) A. Löve & Kapoor, *A. pseudoxyria* (Tolm.) Tzvel. - 4, 5

pulcher L. - 1, 2

- subsp. **anodontus** (Hausskn.) Rech. fil. (*R. pulcher* var. *anodontus* Hausskn.) - 2

- subsp. *divaricatus* (L.) Arcang. = R. divaricatus

- subsp. *divaricatus* (L.) Murb. = R. divaricatus

rechingerianus Losinsk. = R. pamiricus

rectinervis Rech. fil. (*R. angreni* Sumn.) - 6

regelii Fr. Schmidt - 5

reticulatus sensu Losinsk. = R. halacsyi

rossicus Murb. (*R. maritimus* L. subsp. *rossicus* (Murb.) Kryl.) - 1, 3, 4, 5(alien)

rugosus Campd. (*Acetosa alpestris* (Jacq.) A. Löve subsp. *carpatica* (Zapal.) Dostal, *A. rugosa* (Campd.) Holub, *Rumex carpaticus* Zapal.) - 1

salicifolius auct. = R. triangulivalvis

salicifolius Weinm. subsp. *triangulivalvis* Danser = R. triangulivalvis

sanguineus L. - 1, 2

schischkinii Losinsk. (*R. aquaticus* L. subsp. *schischkinii* (Losinsk.) Rech. fil.) - 3

x **schreberi** Hausskn. - R. crispus L. x R. hydrolapathum Huds.

x **schultzei** Hausskn. - R. conglomeratus Murr. x R. crispus L.

scutatus L. (Acetosa scutata (L.) Mill.) - 1

- subsp. *glaucus* (Jacq.) Gaudin ex E. Wulf, p.p. = R. hastifolius

- subsp. *hastifolius* (Bieb.) Borod. = R. hastifolius

scutatus sensu Losinsk. = R. hastifolius

sibiricus Hult. - 3, 4, 5(alien)

simulans Rech. fil. - 1, 3, 6

x **skofitzii** Blocki. - R. confertus Willd. x R. crispus L.

songaricus Schrenk - 6

stenophyllus Ledeb. (*R. stenophyllus* var. *ussuriensis* (Losinsk.) Kitag., *R. ussuriensis* Losinsk.) - 1, 2, 3, 4, 5, 6

- var. *ussuriensis* (Losinsk.) Kitag. = R. stenophyllus

sylvestris (Lam.) Wallr. (*Lapathum sylvestre* Lam., *Rumex obtusifolius* L. subsp. *sylvestris* (Lam.) Celak.) - 1, 2

syriacus Meissn. = R. chalepensis

syriacus sensu Losinsk. p.p. = R. chalepensis and R. drobovii

tenuifolius (Wallr.) A. Löve = R. acetosella

thyrsiflorus Fingerh. (*Acetosa thyrsiflora* (Fingerh.) A. & D. Löve, *Rumex acetosa* L. var. *auriculatus* Wallr., *R. acetosa* subsp. *auriculatus* (Wallr.) Blytt & Dahl, *R. acetosa* subsp. *thyrsiflorus* (Fingerh.) Hayek, *R. acetosa* subsp. *thyrsiflorus* Stojan. & Stef. comb. superfl., *R. auriculatus* (Wallr.) Murb., *R. haplorhizus* Czern. ex Turcz.) - 1, 2, 3, 4, 5, 6

- subsp. *haematinus* (Kihlm.) Borod. = R. haematinus

tianschanicus Losinsk. - 6

x **transbaicalicus** Rech. fil. - R. gmelinii Turcz. ex Ledeb. x R. aquaticus L.

triangulivalvis (Danser) Rech. fil. (*R. salicifolius* Weinm. subsp. *triangulivalvis* Danser, *R. salicifolius* auct.) - 1(alien)

tuberosus auct. = R. euxinus

tuberosus L. subsp. *euxinus* (Klok.) Borod. = R. euxinus

- subsp. *horizontalis* (C. Koch) Rech. fil. = R. horizontalis

- subsp. *turcomanicus* (Rech. fil.) Rech. fil. = R. turcomanicus

- var. *turcomanicus* Rech. fil. = R. turcomanicus

turcomanicus (Rech. fil.) Czer. (*R. tuberosus* L. var. *turcomanicus* Rech. fil., *Acetosa tuberosa* (L.) Chaz. subsp. *turcomanica* (Rech. fil.) A. Löve & Kapoor, *Rumex tuberosus* subsp. *turcomanicus* (Rech. fil.) Rech. fil.) - 6

ucranicus Fisch. ex Spreng. - 1, 3, 4, 6

ujskensis Rech. fil. - 4, 5

ursinus Maximova = R. arcticus

ussuriensis Losinsk. = R. stenophyllus

x **weberi** Fischer-Benzon. - R. hydrolapathum Huds. x R. obtusifolius L.

Tiniaria (Meissn.) Reichenb. = Reynoutria

japonica (Houtt.) Hedberg = Reynoutria japonica

sachalinensis (Fr. Schmidt) Janch. = Reynoutria sachalinensis

Truellum Houtt.

dissitiflorum (Hemsl.) Tzvel. (*Polygonum dissitiflorum* Hemsl.) - 5

hastatosagittatum (Makino) Sojak (*Polygonum hastatosagittatum* Makino, *Persicaria korshinskiana* (Nakai) Nakai, *P. ussuriensis* (Regel & Maack) Nakai ex Mori, *Polygonum korshinskianum* Nakai, *P. sagittatum* L. var. *ussuriense* Regel & Maack, *P. ussuriense* (Regel & Maack) Nakai), *Truellum korshinskianum* (Nakai) Sojak) - 5

japonicum Houtt. (*Polygonum senticosum* (Meissn.) Franch. & Savat.) - 5

korshinskianum (Nakai) Sojak = T. hastatosagittatum

maackianum (Regel) Sojak (*Polygonum maackianum* Regel) - 5

nipponense (Makino) Sojak (*Polygonum nipponense* Makino, *Persicaria nipponensis* (Makino) H. Gross, *Polygonum strigosum* auct.) - 5

paludosum (Kom.) Sojak = T. sieboldii

perfoliatum (L.) Sojak = Chylocalyx perfoliatus

sagittatum (L.) Sojak (*Polygonum sagittatum* L.) - 5

sibiricum (Meissn.) Sojak = T. sieboldii

sieboldii (Meissn.) Sojak (*Polygonum sieboldii* Meissn., *Persicaria belophylla* (Litv.) Kitag., *Polygonum belophyllum* Litv., *P. paludosum* (Kom.) Kom., *P. sagittatum* L. subsp. *sieboldii* (Meissn.) Worosch., *Truellum paludosum* (Kom.) Sojak, *T. sibiricum* (Meissn.) Sojak) - 3, 4, 5

thunbergii (Siebold & Zucc.) Sojak (*Polygonum thunbergii* Siebold & Zucc., *Persicaria hastato-triloba* (Meissn.) Okuyama, *Polygonum hastato-trilobum* Meissn.) - 2(alien), 5

POLYPODIACEAE Bercht. & J. Presl

See also ADIANTACEAE, ASPLENIACEAE, ATHYRIACEAE, BLECHNACEAE, CRYPTOGRAMMACEAE, DENNSTAEDTIACEAE, DRYOPTERIDACEAE, HEMIONITIDACEAE, HYLOLEPIDACEAE, ONOCLEACEAE, PLAGIOGYRIACEAE, PLEUROSORIOPSIDACEAE, PTERIDACEAE, SINOPTERIDACEAE, THELYPTERIDACEAE and WOODSIACEAE.

Cyclophorus Desv. = Pyrrosia

lingua auct. = Pyrrosia petiolosa

Lepisorus (J. Smith) Ching = Pleopeltis

clathratus (Clarke) Ching = Pleopeltis clathrata

distans (Makino) Ching = Pleopeltis distans

linearis (Thunb.) Ching = Pleopeltis thunbergiana

thunbergianus (Kaulf.) Ching = Pleopeltis thunbergiana

ussuriensis (Regel & Maack) Ching = Pleopeltis ussuriensis

- var. *distans* (Makino) Tagawa = Pleopeltis distans

Pleopeltis Humb. & Bonpl. ex Willd. (*Lepisorus* (J. Smith) Ching)

clathrata (Clarke) Czer. comb. nova (*Polypodium clathratum* Clarke, 1880, Trans. Linn. Soc. London (Bot.), Ser.2, 1 : 559, pl. 82, fig. i; *P. alberti* Regel, *Lepisorus clathratus* (Clarke) Ching) - 3, 6

distans (Makino) Worosch. (*Polypodium lineare* Thunb. var. *distans* Makino, *Lepisorus distans* (Makino) Ching, *L. ussuriensis* (Regel & Maack) Ching var. *distans* (Makino) Tagawa, *Pleopeltis ussuriensis* Regel & Maack var. *distans* (Makino) Okuyama) - 5

kolesnikovii Tzvel. - 5
thunbergiana Kaulf. (*Lepisorus linearis* (Thunb.) Ching, nom. illegit., *L. thunbergianus* (Kaulf.) Ching, *Polypodium lineare* Thunb. 1784, non Burm. fil. 1768, nec Houtt. 1783, *P. thunbergianum* (Kaulf.) C. Chr.) - 5
ussuriensis Regel & Maack (*Lepisorus ussuriensis* (Regel & Maack) Ching, *Polypodium ussuriense* (Regel & Maack) Regel) - 5
- var. *distans* (Makino) Okuyama = P. distans

Polypodium L.

aculeatum L. = Polystichum aculeatum (Dryopteridaceae)
alberti Regel = Pleopeltis clathrata
australe Fee (*P. cambricum* L. subsp. *australe* (Fee) Greuter & Burdet, *P. serratum* (Willd.) Saut. 1882, non Aubl. 1775, *P. vulgare* L. subsp. *serrulatum* Arcang., *P. cambricum* auct.) - 1, 2
austriacum Jacq. = Pteridium aquilinum (Hylolepidaceae)
cambricum auct. = P. australe
cambricum L. subsp. *australe* (Fee) Greuter & Burdet = P. australe
carthusianum Vill. = Dryopteris carthusiana (Dryopteridaceae)
clathratum Clarke = Pleopeltis clathrata
connectile Michx. = Phegopteris connectilis (Thelypteridaceae)
dilatatum Hoffm. = Dryopteris dilatata (Dryopteridaceae)
dryopteris L. var. *Polypodium disjunctum* Rupr. = Gymnocarpium dryopteris (Athyriaceae)
fauriei Christ (*P. japonicum* (Franch.& Savat.) Maxon, 1902, non Houtt. 1783, *P. vulgare* L. var. *japonicum* Franch. & Savat.) - 5
interjectum Shivas (*P. vulgare* L. subsp. *prionodes* (Aschers.) Rothm., *P. vulgare* var. *attenuatum* Milde b. *prionodes* Aschers.) - 1
japonicum (Franch. & Savat.) Maxon = P. fauriei
lanceolatocristatum Hoffm. p.p. = Dryopteris carthusiana and D. dilatata (Dryopteridaceae)
limbospermum All. = Oreopteris limbosperma (Thelypteridaceae)
lineare Thunb. = Pleopeltis thunbergiana
- var. *distans* Makino = Pleopeltis distans
lobatum Huds. = Polystichum aculeatum (Dryopteridaceae)
nipponicum auct. = P. vulgare
oreopteris Ehrh. = Oreopteris limbosperma (Thelypteridaceae)
palustre Salisb. = Thelypteris palustris (Thelypteridaceae)
petiolosum Christ & Baroni = Pyrrosia petiolosa
serratum (Willd.) Saut. = P. australe
setiferum Forssk. = Polystichum setiferum (Dryopteridaceae)
sibiricum Sipl. (*P. virginianum* auct.) - 5
someyae auct. = P. vulgare
spinulosum O.F. Muell. = Dryopteris carthusiana (Dryopteridaceae)
thunbergianum (Kaulf.) C. Chr. = Pleopeltis thunbergiana
ussuriense (Regel & Maack) Regel = Pleopeltis ussuriensis
virginianum auct. = P. sibiricum
vulgare L. (*P. vulgare* subsp. *issaevii* A. Askerov & A. Bobr., *P. nipponicum* auct., *P. someyae* auct.) - 1, 2, 3, 5, 6
- subsp. *issaevii* A. Askerov & A. Bobr. = P. vulgare
- subsp. *prionodes* (Aschers.) Rothm. = P. interjectum
- subsp. *serrulatum* Arcang. = P. australe

- var. *attenuatum* Milde b. *prionodes* Aschers. = P. interjectum
- var. *japonicum* Franch. & Savat. = P. fauriei

Pyrrosia Mirb. (*Cyclophorus* Desv.)

lingua auct. = P. petiolosa
petiolosa (Christ & Baroni) Ching (*Polypodium petiolosum* Christ & Baroni, *Cyclophorus lingua* auct., *Pyrrosia lingua* auct.) - 5

PONTEDERIACEAE Kunth
Monochoria C. Presl

korsakowii Regel & Maack - 5
plantaginea (Roxb.) Kunth (*M. vaginalis* auct.) - 5
vaginalis auct. = M. plantaginea

PORTULACACEAE Juss.
Claytonia L.

acutifolia Pall. ex Schult. - 4, 5
- subsp. *graminifolia* Hult. = C. eschscholtzii
arctica Adams - 4, 5
czukczorum Volkova (*C. tuberosa* Pall. ex Schult. var. *czukczorum* (Volkova) Hult.) - 5
eschscholtzii Cham. (*C. acutifolia* Pall. ex Schult. subsp *graminifolia* Hult.) - 4, 5
joanneana Schult. - 3, 4, 6
sarmentosa C.A. Mey. - 5
sibirica L. - 5
soczaviana Jurtz. - 5
tuberosa Pall. ex Schult. - 4, 5
- var. *czukczorum* (Volkova) Hult. = C. czukczorum
udokanica V. Zuev - 4
vassilievii Kuzen. = Claytoniella vassilievii

Claytoniella Jurtz. (*Montiastrum* auct. p.p.)

vassilievii (Kuzen.) Jurtz. (*Claytonia vassilievii* Kuzen., *Montia vassilievii* (Kuzen.) NcNeill, *Montiastrum vassilievii* (Kuzen.) O. Nilsson) - 5
- subsp. **petrovskii** Jurtz. & M. Griczuk, nom. invalid. - 5

Montia L.

fontana L. (*M. fontana* subsp *variabilis* Walters, *M. lamprosperma* Cham., *M. rivularis* C.C. Gmel.) - 1, 5
- subsp. *chondrosperma* (Fenzl) Walters = M. minor
- subsp. *minor* (C.C. Gmel.) Celak. = M. minor
- subsp. *variabilis* Walters = M. fontana
lamprosperma Cham. = M. fontana
minor C.C. Gmel. (*M. fontana* L. subsp. *chondrosperma* (Fenzl) Walters, *M. fontana* subsp. *minor* (C.C. Gmel.) Celak.) - 2
rivularis C.C. Gmel. = M. fontana
vassilievii (Kuzen.) McNeill = Claytoniella vassilievii

Montiastrum auct. p.p. = Claytoniella

vassilievii (Kuzen.) O. Nilsson = Claytoniella vassilievii

Portulaca L.

oleracea L. - 1, 2, 5, 6

POTAMOGETONACEAE Dumort.

Coleogeton (Reichenb.) Raunkiaer = Potamogeton

pectinatus (L.) Dostal = Potamogeton pectinatus

Groenlandia J. Gay

densa (L.) Fourr. (*Potamogeton densus* L.) - 1, 2

Potamogeton L. (*Coleogeton* (Reichenb.) Raunkiaer, *Stuckenia* Borner)

acutifolius Link - 1, 2
- subsp. *manchuriensis* A. Benn. = P. manschuriensis
alpinus Balb. - 1, 2, 3, 4, 6
- subsp. *tenuifolius* (Rafin.) Hult. = P. tenuifolius
amblyophyllus C.A. Mey. - 2
anadyrensis V. Vassil. = P. sibiricus
anguillanus auct. = P. richardsonii
asiaticus A. Benn. = P. octandrus
austro-sibiricus Kaschina = P. rostratus
x **babingtonii** A. Benn. - P. lucens L. x P. praelongus Wulf. - 1
berchtoldii Fieb. (*P. pusillus* sensu Juz.) - 1, 2, 3, 4, 5, 6
biformis Hagstr. - 3, 6
borealis Rafin. - 1, 3, 4, 5
x **bottnicus** Hagstr. - P. pectinatus L. x P. vaginatus Turcz. - 1
carinatus Kupff. - 1
chakassiensis (Kaschina) Volobaev (*P. pectinatus* L. subsp. *chakassiensis* Kaschina) - 3
x **cognatus** Aschers. & Graebn. - P. perfoliatus L. x P. praelongus Wulf. - 1
coloratus Hornem. - 2
compressus L. (*P. zosterifolius* Schum.) - 1, 3, 4, 5, 6
■ x **cooperi** (Fryer) Fryer. - P. crispus L. x P. perfoliatus L.
crispus L. - 1, 2, 3, 4, 5, 6
cristatus Regel & Maack - 5
x **decipiens** Nolte. - P. lucens L. x P. perfoliatus L. - 1
densus L. = Groenlandia densa
digynus Wall. ex Hook. - 4(?), 5
distinctus A. Benn. (*P. franchetii* Baagoe ex A. Benn.) - 5
filiformis Pers. - 1, 2(?), 3, 4, 6
x **fluitans** Roth (*P.* x *sterilis* Hagstr.). - P. lucens L. x P. natans L. - 1
franchetii Baagoe ex A. Benn. = P. distinctus
friesii Rupr. - 1, 3, 4, 5
fryeri A. Benn. - 5
gramineus L. (*P. gramineus* subsp. *heterophyllus* (Schreb.) Schinz & R. Keller, *P. heterophyllus* Schreb., *P. wolfgangii* Kihlm.) - 1, 2, 3, 4, 5, 6
- subsp. *heterophyllus* (Schreb.) Schinz & R. Keller = P. gramineus
henningii A. Benn. - 1(?), 2(?)
Mysterious species. The critical investigation is necessary.
heterophyllus Schreb. = P. gramineus
interruptus auct. = P. pectinatus
intramongolicus Ma - 2
javanicus Hassk. = P. octandrus
juzepczukii P. Dorof. & Tzvel. - 4, 5
limosellifolius Maxim. ex Korsh. = P. octandrus
lucens L. - 1, 2, 3, 4, 5, 6
maackianus A. Benn. - 4, 5
macrocarpus Dobroch. - 3
malainus Miq. - 3, 4, 5
manchuriensis (A. Benn.) A. Benn. (*P. acutifolius* Link

subsp. *manchuriensis* A. Benn.) - 5
x **meinshausenii** Juz. - ? P. filiformis Pers. x P. vaginatus Turcz. - 1
miduhikimo Makino = P. octandrus
morongii A. Benn. = P. natans
natans L. (*P. morongii* A. Benn.) - 1, 2, 3, 4, 5, 6
x **nerbiger** Wolfg. - P. alpinus Balb. x P. lucens L. - 1
x **nitens** Web. (*P.* x *salicifolius* Wolfg.). - P. gramineus L. x P. perfoliatus L. - 1
nodosus Poir. - 1, 2, 3, 6
obtusifolius Mert. & Koch - 1, 3, 4, 5, 6
octandrus Poir. (*P. asiaticus* A. Benn., *P. javanicus* Hassk., *P. limosellifolius* Maxim. ex Korsh., *P. miduhikimo* Makino, *P. octandrus* subsp. *limosellifolius* (Maxim. ex Korsh.) Worosch.) - 5
- subsp. *limosellifolius* (Maxim. ex Korsh.) Worosch. = P. octandrus
oxyphyllus Miq. - 5
pamiricus Baagoe - 6
panormitanus Biv.-Bern. = P. pusillus
pectinatus L. (*Coleogeton pectinatus* (L.) Dostal, *Potamogeton interruptus* auct.) - 1, 2, 3, 4, 5, 6
- subsp. *chakassiensis* Kaschina = P. chakassiensis
- subsp. **mongolicus** (A. Benn.) Volobaev (*P. pectinatus* var. *mongolicus* A. Benn.) - 3
- var. *mongolicus* A. Benn. = P. pectinatus subsp. mongolicus
perfoliatus L. - 1, 2, 3, 4, 5, 6
- var. *richardsonii* A. Benn. = P. richardsonii
- var. *sachalinensis* Levl. = P. richardsonii
polygonifolius Pourr. - 1(?)
praelongus Wulf. - 1, 2, 3, 4, 5
pusillus L. (*P. panormitanus* Biv.-Bern.) - 1, 2, 4, 5, 6
pusillus sensu Juz. = P. berchtoldii
richardsonii (A. Benn.) Rydb. (*P. perfoliatus* L. var. *richardsonii* A. Benn., *P. perfoliatus* var. *sachalinensis* Levl., *P. sachalinensis* (Levl.) Levl., *P. anguillanus* auct.) - 5
rostratus Hagstr. (*P. austro-sibiricus* Kaschina) - 3, 4
rutilus Wolfg. - 1
sachalinensis (Levl.) Levl. = P. richardsonii
x *salicifolius* Wolfg. = P. x nitens
sarmaticus Maemets - 1, 3, 6
sibiricus A. Benn. (*P. anadyrensis* V. Vassil., *P. subsibiricus* Hagstr.) - 1, 4, 5
x **sparganiifolius** Laest. ex Beurl. - P. gramineus L. x P. natans L. - 1, 4
x *sterilis* Hagstr. = P. x fluitans
subretusus Hagstr. - 1, 3, 4, 5
subsibiricus Hagstr. = P. sibiricus
x **suecicus** K. Richt. - P. filiformis Pers. x P. pectinatus L.
tenuifolius Rafin. (*P. alpinus* Balb. subsp. *tenuifolius* (Rafin.) Hult.) - 4, 5
trichoides Cham. & Schlecht. - 1, 2, 3, 4, 6
tubulatus Hagstr. - 6
x **undulatus** Wolfg. - P. crispus L. x P. praelongus Wulf. - 1
vaginatus Turcz. (*Stuckenia vaginata* (Turcz.) Holub) - 4
x **vilnensis** Galinis. - P. gramineus L. x P. praelongus Wulf. - 1
wolfgangii Kihlm. = P. gramineus
x **zizii** Mert. & Koch. - P. gramineus L. x P. lucens L. - 1, 3
zosterifolius Schum. = P. compressus

Stuckenia Borner = Potamogeton

vaginata (Turcz.) Holub = Potamogeton vaginatus

PRIMULACEAE Vent.

Aleuritia (Duby) Opiz = Primula

algida (Adams) Sojak = Primula algida
arctica (Koidz.) Sojak = Primula arctica
auriculata (Lam.) Sojak = Primula auriculata
baldshuanica (B. Fedtsch.) Sojak = Primula baldshuanica
bayernii (Rupr.) Sojak = Primula bayernii
borealis (Duby) Sojak = Primula borealis
capitellata (Boiss.) Sojak = Primula capitellata
darialica (Rupr.) Sojak = Primula darialica
eximia (Greene) Sojak = Primula eximia
farinifolia (Rupr.) Sojak = Primula farinifolia
farinosa (L.) Opiz = Primula farinosa
fedtschenkoi (Regel) Sojak = Primula fedtschenkoi
fistulosa (Turkev.) Sojak = Primula fistulosa
flexuosa (Turkev.) Sojak = Primula flexuosa
gigantea (Jacq.) Sojak = Primula serrata
halleri (J.F. Gmel.) Sojak = Primula halleri
iljinskii (Fed.) Sojak = Primula iljinskii
japonica (A. Gray) Sojak = Primula japonica
knorringiana (Fed.) Sojak = Primula knorringiana
longipes (Freyn & Sint.) Sojak = Primula longipes
longiscapa (Ledeb.) Sojak = Primula longiscapa
luteola (Rupr.) Sojak = Primula luteola
macrophylla (D. Don) Sojak = Primula macrophylla
matsumurae (Petitm.) Sojak = Primula matsumurae
nivalis (Pall.) Sojak = Primula nivalis
nutans (Georgi) Sojak = Primula nutans
- subsp. *finmarchica* (Jacq.) Sojak = Primula finmarchica
ossetica (Kusn.) Sojak = Primula ossetica
pamirica (Fed.) Sojak = Primula pamirica
pinnata (M. Pop. & Fed.) Sojak = Primula pinnata
pulverea (Fed.) Sojak = Primula macrophylla
sachalinensis (Nakai) Sojak = Primula sachalinensis
scandinavica (Bruun) Sojak = Primula scandinavica
stricta (Hornem.) Sojak = Primula stricta
tschuktschorum (Kjellm.) Sojak = Primula tschuktschorum
warshenewskiana (B. Fedtsch.) Sojak = Primula warshenewskiana
xanthobasis (Fed.) Sojak = Primula xanthobasis

Anagallis L.

arvensis L. - 1, 2, 6
caerulea Schreb. = A. foemina
foemina Mill. (*A. caerulea* Schreb.) - 1, 2, 3, 5(alien), 6

Androsace L.

acrolasia Ovcz. & Vved. = A. akbaitalensis
acrolasia sensu Schischk. & Bobr. = A. dasyphylla
aflatunense Ovcz. - 6
akbaitalensis Derg. (*A. acrolasia* Ovcz. & Vved.) - 6
alaica Ovcz. & Astan. - 6
albana Stev. - 2
amurensis Probat. (*A. lactiflora* Fisch. ex Duby, 1844, non Kar. & Kir. 1841, *A. lactiflora* Pall. 1776, nom. nud.) - 3(?), 4, 5
angrenica Ovcz. = A. dasyphylla
arctica Cham. & Schlecht. = Douglasia ochotensis
arctica sensu Schischk. & Bobr. = Douglasia gormanii
arctisibirica (Korobkov) Probat. (*A. chamaejasme* Wulf. subsp. *arctisibirica* Korobkov) - 1, 3, 4, 5
armeniaca Duby - 2
barbulata Ovcz. = A. kozo-poljanskii
bidentata C. Koch - 2
bryomorpha Lipsky - 6

bungeana Schischk. & Bobr. = A. lehmanniana
caduca Ovcz. - 6
capitata Willd. ex Roem. & Schult. (*A. chamaejasme* Wulf. subsp. *capitata* (Willd. ex Roem. & Schult.) Korobkov) - 5
chamaejasme auct. p.p. = A. lehmanniana
chamaejasme Wulf. subsp. *arctisibirica* Korobkov = A. arctisibirica
- subsp. *capitata* (Willd. ex Roem. & Schult.) Korobkov = A. capitata
- subsp *lehmanniana* (Spreng.) Hult. = A. lehmanniana
constancei Wendelbo = Douglasia gormanii
darvasica Ovcz. - 6
dasyphylla Bunge (*A. angrenica* Ovcz., *A. villosa* L. var. *dasyphylla* (Bunge) Kar. & Kir., *A. acrolasia* sensu Schischk. & Bobr.) - 3, 4, 6
elongata L. - 1, 2, 3(?), 4
exscapa (Turcz.) Maximova (*A. septentrionalis* L. var. *exscapa* Turcz.) - 4
fedtschenkoi Ovcz. - 3, 6
filiformis Retz. - 1, 2, 3, 4, 6
gmelinii (Gaertn.) Roem. & Schult. - 3, 4, 5
gorodkovii Ovcz. & Karav. - 4, 5
- subsp. *semiperennis* (Jurtz.) Ju. Kozhevn. = A. semiperennis
incana Lam. - 3, 4, 5
intermedia Ledeb. - 2
koso-poljanskii Ovcz. (*A. barbulata* Ovcz., *A. villosa* L. subsp. *koso-poljanskii* (Ovcz.) Fed., *A. villosa* auct. p.p.) - 1
lactiflora Fisch. ex Duby = A. amurensis
lactiflora Pall. = A. amurensis
lehmanniana Spreng. (*A. bungeana* Schischk. & Bobr., *A. chamaejasme* Wulf. subsp *lehmanniana* (Spreng.) Hult., *A. olgae* Ovz., *A. chamaejasme* auct. p.p.) - 1, 2, 3, 4, 5, 6
▪ macrantha Boiss. & Huet
maxima L. (*A. maxima* subsp. *turczaninowii* (Freyn) Fed., *A. turczaninowii* Freyn) - 1, 2, 3, 4, 6
- subsp. **caucasica** (Kusn.) Fed. (*A. maxima* var. *caucasica* Kusn.) - 2
- subsp. *turczaninowii* (Freyn) Fed. = A. maxima
- var. *caucasica* Kusn. = A. maxima subsp. caucasica
ochotensis Willd. ex Roem. & Schult. = Douglasia ochotensis
olgae Ovcz. = A. lehmanniana
ovczinnikovii Schischk. & Bobr. - 3, 6
pavlovskyi Ovcz. - 6
raddeana Somm. & Levier - 2
semiperennis Jurtz. (*A. gorodkovii* Ovcz. & Karav. subsp. *semiperennis* (Jurtz.) Ju. Kozhevn.) - 5
septentrionalis L. - 1, 2, 3, 4, 5, 6
- var. *exscapa* Turcz. = A. exscapa
sericea Ovcz. - 6
taurica Ovcz. (*A. villosa* L. subsp. *taurica* (Ovcz.) Fed., *A. villosa* auct. p.p.) - 1
triflora Adams - 1, 3, 4
turczaninowii Freyn = A. maxima
umbellata (Lour.) Merr. - 5
villosa auct. p.p. = A. koso-poljanskii
villosa L. subsp. *koso-poljanskii* (Ovcz.) Fed. = A. koso-poljanskii
- subsp. *taurica* (Ovcz.) Fed. = A. taurica
- var. *dasyphylla* (Bunge) Kar. & Kir. = A. dasyphylla

Asterolinon Hoffmgg. & Link

linum-stellatum (L.) Duby - 1, 2

Auganthus Link = Primula

cortusoides (L.) Sojak = Primula cortusoides
eugeniae (Fed.) Sojak = Primula eugeniae
jesoanus (Miq.) Sojak = Primula jesoana
kaufmannianus (Regel) Sojak = Primula kaufmanniana
lactiflorus (Turkev.) Sojak = Primula lactiflora
minkwitziae (W.W. Smith) Sojak = Primula minkwitziae

Auricula-ursi Seguier = Primula

minima (L.) Sojak = Primula minima

Centunculus L.

minimus L. - 1, 5

Cortusa L.

altaica Losinsk. (*C. matthioli* L. subsp. *altaica* (Losinsk.) Korobkov) - 1, 3, 4
amurensis Fed. (*C. sachalinensis* Losinsk. var. *amurensis* (Fed.) Worosch.) - 5
brotheri Pax ex Lipsky - 6
coreana (Nakai) Nakai (*Primula coreana* Nakai) - 5(?)
discolor Worosch. & Gorovoi - 5
matthioli L. - 1
- subsp. *altaica* (Losinsk.) Korobkov = C. altaica
- subsp. *sibirica* (Andrz.) E.I. Nyarady = C. sibirica
- subsp. *turkestanica* (Losinsk.) Iranshahr & Wendelbo = C. turkestanica
pekinensis auct. = C. sachalinensis
sachalinensis Losinsk. (*C. pekinensis* auct.) - 5
sibirica Andrz. (*C. matthioli* L. subsp. *sibirica* (Andrz.) E.I. Nyarady) - 4, 5
turkestanica Losinsk. (*C. matthioli* L. subsp. *turkestanica* (Losinsk.) Iranshahr & Wendelbo) - 6

Cyclamen L.

abchasicum (Medw. ex Kusn.) Kolak. (*C. coum* Mill. subsp. *caucasicum* (C. Koch) O. Schwarz, p.p.) - 2
adzharicum Pobed. (*C. coum* Mill. subsp. *caucasicum* (C. Koch) O. Schwarz, p.p.) - 2
caucasicum Willd. ex Stev. = C. vernum
circassicum Pobed. (*C. coum* Mill. subsp. *caucasicum* (C. Koch) O. Schwarz, p.p.) - 2
colchicum (Albov) Albov (*C. ponticum* (Albov) Pobed.) - 2
coum Mill. (*C. coum* subsp. *hiemale* (Hildebr.) O. Schwarz, *C. hiemale* Hildebr., *C. hyemale* Salisb., *C. kuznetzovii* Kotov & Czernova) - 1, 2
- subsp. *alpinum* (Sprenger) O. Schwarz, p.p. = C. parviflorum
- subsp. *caucasicum* (C. Koch) O. Schwarz = C. abchasicum, C. adzharicum, C, circassicum, C. elegans and C. vernum
- subsp. *hiemale* (Hildebr.) O. Schwarz = C. coum
elegans Boiss. & Buhse (*C. coum* Mill. subsp. *caucasicum* (C. Koch) O. Schwarz, p.p.) - 2
europaeum L. var. *caucasicum* C. Koch = C. vernum
hiemale Hildebr. = C. coum
hyemale Salisb. = C. coum
kuznetzovii Kotov & Czernova = C. coum
■ **parviflorum** Pobed. (*C. coum* Mill. subsp. *alpinum* (Sprenger) O. Schwarz, p.p.)
ponticum (Albov) Pobed. = C. colchicum
purpurascens Mill. - 1(?)
vernum Sweet (? *C. caucasicum* Willd. ex Stev., *C. coum* Mill. subsp. *caucasicum* (C. Koch) O. Schwarz, p.p.,

? *C. europaeum* L. var. *caucasicum* C. Koch) - 2

Dionysia Fenzl

gandzhinae R. Kam. - 6
hissarica Lipsky - 6
involucrata Zapr. - 6
kossinskyi Czerniak. - 6
tapetodes Bunge - 6

Dodecatheon L.

frigidum Cham. & Schlecht. - 5

Douglasia Lindl.

gormanii Constance (*Androsace constancei* Wendelbo, ? *A. arctica* sensu Schischk. & Bobr.) - 5(?)
ochotensis (Willd. ex Roem. & Schult.) Hult. (*Androsace ochotensis* Willd. ex Roem. & Schult., *A. arctica* Cham. & Schlecht., *Douglasia ochotensis* subsp. *arctica* (Cham. & Schlecht.) A. & D. Löve) - 4, 5
- subsp. *actica* (Cham. & Schlecht.) A. & D. Löve = D. ochotensis

Glaux L.

maritima L. - 1, 2, 3, 4, 5, 6

Hottonia L.

palustris L. - 1

Kaufmannia Regel

brachyanthera Losinsk. = K. semenovii
semenovii (Herd.) Regel (*K. brachyanthera* Losinsk.) - 6

Lysimachia L.

barystachys Bunge - 5
clethroides Duby - 5
davurica Ledeb. (*L. vulgaris* L. subsp. *davurica* (Ledeb.) Tatew.) - 4, 5
dubia Soland. - 1(?), 2, 6
fortunei Maxim. - 2(alien)
japonica Thunb. - 2(alien)
nemorum L. - 1
nummularia L. - 1, 2
punctata L. - 1
verticillaris Spreng. - 1, 2
vulgaris L. - 1, 2, 3, 4, 6
- subsp. *davurica* (Ledeb.) Tatew. = L. davurica

Naumburgia Moench

thyrsiflora (L.) Reichenb. - 1, 3, 4, 5, 6

Primula L. (*Aleuritia* (Duby) Opiz, *Auganthus* Link, *Auricula-ursi* Seguier)

abchasica Sosn. (*P. vulgaris* Huds. subsp. *abchasica* (Sosn.) Soo) - 2
acaulis (L.) L. (*P. vulgaris* Huds.) - 1, 2
- subsp. *alpina* (M. Canak & M. Gajic) Czer. comb. nova (*P. vulgaris* subsp. *alpina* M. Canak & M. Gajic, 1979, Glasn. Prir. Muz. Beogradu, B, 33 : 74, 77) - 2
algida Adams (*Aleuritia algida* (Adams) Sojak) - 2
- subsp. *bungeana* (C.A. Mey.) Fed. = P. bungeana
amoena Bieb. (*P. elatior* (L.) Hill subsp. *amoena* (Bieb.)

Greuter & Burdet) - 2

arctica Koidz. (*Aleuritia arctica* (Koidz.) Sojak, *Primula tschuktschorum* Kjellm. subsp. *arctica* (Koidz.) A. & D. Löve, *P. tschuktschorum* var. *arctica* (Koidz.) Fern.) - 5

auriculata Lam. (*Aleuritia auriculata* (Lam.) Sojak) - 2

baldshuanica B. Fedtsch. (*Aleuritia baldshuanica* (B. Fedtsch.) Sojak) - 2

bayernii Rupr. (*Aleuritia bayernii* (Rupr.) Sojak) - 2

beringensis (A. Pors.) Jurtz. (*P. tschuktschorum* Kjellm. var. *beringensis* A. Pors., *P. tschuktschorum* subsp. *beringensis* (A. Pors.) Jurtz. & Ju. Kozhevn.) - 5

borealis Duby (*Aleuritia borealis* (Duby) Sojak) - 4, 5

botschantzevii Czuk. & Kovalevsk. - 6

bungeana C.A. Mey. (*P. algida* Adams subsp. *bungeana* (C.A. Mey.) Fed.) - 3, 6

capitellata Boiss. (*Aleuritia capitellata* (Boiss.) Sojak) - 6

carpathica (Griseb. & Schenk) Fuss (*P. elatior* (L.) Hill subsp. *carpathica* (Griseb. & Schenk) Nym., *P. elatior* var. *carpathica* (Griseb. & Schenk) Nikolic) - 1

cordifolia Rupr. - 2

coreana Nakai = Cortusa coreana

cortusoides L. (*Auganthus cortusoides* (L.) Sojak) - 3, 4

cuneifolia Ledeb. (*P. cuneifolia* subsp. *hakusanensis* (Franch.) Worosch. p.p. quoad pl., *P. saxifragifolia* Lehm., *P. hakusanensis* auct., *P. heterodonta* auct.) - 4, 5

- subsp. *hakusanensis* (Franch.) Worosch. p.p. = P. cuneifolia

darialica Rupr. (*Aleuritia darialica* (Rupr.) Sojak) - 2

drosocalyx Poljak. & Lincz. = P. minkwitziae

egaliksensis Wormsk. - 5

elatior (L.) Hill - 1

- subsp. *amoena* (Bieb.) Greuter & Burdet = P. amoena

- subsp. *carpathica* (Griseb. & Schenk) Nym. = P. carpathica

- subsp. *meyeri* (Rupr.) Valentine & Lamond = P. meyeri

- subsp. *poloninensis* (Domin) Dostal = P. poloninensis

- var. *carpathica* (Griseb. & Schenk) Nikolic = P. carpathica

eugeniae Fed. (*Auganthus eugeniae* (Fed.) Sojak) - 6

eximia Greene (*Aleuritia eximia* (Greene) Sojak) - 5

farinifolia Rupr. (*Aleuritia farinifolia* (Rupr.) Sojak) - 2

farinosa L. (*Aleuritia farinosa* (L.) Opiz) - 1, 3, 4, 5

- var. *pinnata* (M. Pop. & Fed.) M. Ivanova = P. pinnata

fauriei auct. = P. matsumurae

fedtschenkoi Regel (*Aleuritia fedtschenkoi* (Regel) Sojak) - 6

finmarchica Jacq. (*Aleuritia nutans* (Georgi) Sojak subsp. *finmarchica* (Jacq.) Sojak, *Primula nutans* Georgi subsp. *finmarchica* (Jacq.) A. & D. Löve) - 1

fistulosa Turkev. (*Aleuritia fistulosa* (Turkev.) Sojak) - 5

flexuosa Turkev. (*Aleuritia flexuosa* (Turkev.) Sojak) - 6

geranophylla Kovalevsk. - 6

gigantea Jacq. = P. serrata

hakusanensis auct. = P. cuneifolia

halleri J.F. Gmel. (*Aleuritia halleri* (J.F. Gmel.) Sojak) - 1

heterochroma Stapf - 2

heterodonta auct. = P. cuneifolia

iljinskii Fed. (*Aleuritia iljinskii* (Fed.) Sojak) - 6

inflata Lehm. subsp. *macrocalyx* (Bunge) O. Schwarz = P. macrocalyx

japonica A. Gray (*Aleuritia japonica* (A. Gray) Sojak) - 5

jesoana Miq. (*Auganthus jesoanus* (Miq.) Sojak) - 5

juliae Kusn. - 2

kaufmanniana Regel (*Auganthus kaufmannianus* (Regel) Sojak) - 6

kawasimae Hara - 5

knorringiana Fed. (*Aleuritia knorringiana* (Fed.) Sojak) - 6

komarovii Losinsk. (*P. vulgaris* Huds. subsp. *komarovii* (Losinsk.) Fed. comb. superfl., *P. vulgaris* subsp. *komarovii* (Losinsk.) Soo) - 1, 2

kusnetzovii Fed. - 2

lactiflora Turkev. (*Auganthus lactiflorus* (Turkev.) Sojak) - 6

leskeniensis G. Koss ex Smoljian. - 2

■ **longipes** Freyn & Sint. (*Aleuritia longipes* (Freyn & Sint.) Sojak)

longiscapa Ledeb. (*Aleuritia longiscapa* (Ledeb.) Sojak) - 1, 3, 4, 6

luteola Rupr. (*Aleuritia luteola* (Rupr.) Sojak) - 2

macrocalyx Bunge (*P. inflata* Lehm. subsp. *macrocalyx* (Bunge) O. Schwarz) - 1, 2, 3, 4, 5

macrophylla D.Don (*Aleuritia macrophylla* (D. Don) Sojak, *A. pulverea* (Fed.) Sojak, *Primula pulverea* Fed.) - 6

matsumurae Petitm. (*Aleuritia matsumurae* (Petitm.) Sojak, *Primula fauriei* auct.) - 5

mazurenkoae A. Khokhr. - 5

megaseifolia Boiss. & Bal. ex Boiss. - 2

meyeri Rupr. (*P. elatior* (L.) Hill subsp. *meyeri* (Rupr.) Valentine & Lamond) - 2

minima L. (*Auricula-ursi minima* (L.) Sojak) - 1

minkwitziae W.W. Smith (*Auganthus minkwitziae* (W.W. Smith) Sojak, *Primula drosocalyx* Poljak. & Lincz.) - 6

moorcroftiana Wall. ex Klatt - 6

nivalis Pall. (*Aleuritia nivalis* (Pall.) Sojak) - 3, 4, 6

- subsp. *subintegerrima* (Regel) Worosch. = P. xanthobasis

- var. *subintegerrima* Regel = P. xanthobasis

nutans Georgi (*Aleuritia nutans* (Georgi) Sojak, *Primula sibirica* Jacq.) - 3, 4, 5

- subsp. *finmarchica* (Jacq.) A. & D. Löve = P. finmarchica

olgae Regel - 6

ossetica Kusn. (*Aleuritia ossetica* (Kusn.) Sojak) - 2

pallasii Lehm. - 1, 2, 3, 4, 6

pamirica Fed. (*Aleuritia pamirica* (Fed.) Sojak) - 6

pannonica A. Kerner - 1(?)

patens (Turcz.) E. Busch (*P. sieboldii* E. Morr. var. *patens* (Turcz.) Kitag., *P. sieboldii* auct.) - 4, 5

pinnata M. Pop. & Fed. (*Aleuritia pinnata* (M. Pop. & Fed.) Sojak, *Primula farinosa* L. var. *pinnata* (M. Pop. & Fed.) M. Ivanova) - 4

poloninensis (Domin) Fed. (*P. elatior* (L.) Hill subsp. *poloninensis* (Domin) Dostal) - 1

pseudoelatior Kusn. - 2

pulverea Fed. = P. macrophylla

renifolia Volgun. - 2

repentina O. Schwarz - 2(?)

ruprechtii Kusn. - 2

sachalinensis Nakai (*Aleuritia sachalinensis* (Nakai) Sojak) - 5

saguramica Gavr. - 2

saxifragifolia Lehm. = P. cuneifolia

scandinavica Bruun (*Aleuritia scandinavica* (Bruun) Sojak) - 1

serrata Georgi (*Aleuritia gigantea* (Jacq.) Sojak, *Primula gigantea* Jacq.) - 4

sibirica Jacq. = P. nutans

sibthorpii Hoffmgg. - 2

sieboldii auct. = P. patens

sieboldii E. Morr. var. *patens* (Turcz.) Kitag. = P. patens

stricta Hornem. (*Aleuritia stricta* (Hornem.) Sojak) - 1

tournefortii Rupr. - 2

tschuktschorum Kjellm. (*Aleuritia tschuktschorum* (Kjellm.) Sojak) - 5

- subsp. *arctica* (Koidz.) A. & D. Löve = P. arctica

- subsp. *beringensis* (A. Pors.) Jurtz. & Ju. Kozhevn. = P. beringensis
- var. *arctica* (Koidz.) Fern. = P. arctica
- var. *beringensis* A. Pors. = P. beringensis
turkestanica (Haage & Schmidt) E.A. White - 6
valentinae Fed. - 6
veris L. - 1
vulgaris Huds. = P. acaulis
- subsp. *abchasica* (Sosn.) Soo = P. abchasica
- subsp. *alpina* M. Canak & M. Gajic = P. acaulis subsp. alpina
- subsp. *komarovii* (Losinsk.) Fed. = P. komarovii
- subsp. *komarovii* (Losinsk.) Soo = P. komarovii
- subsp. *woronowii* (Losinsk.) Fed. = P. woronowii
- subsp. *woronowii* (Losinsk.) Soo = P. woronowii
warshenewskiana B. Fedtsch. (*Aleuritia warshenewskiana* (B. Fedtsch.) Sojak) - 6
woronowii Losinsk. (*P. vulgaris* Huds. subsp. *woronowii* (Losinsk.) Fed. comb. superfl., *P. vulgaris* subsp. *woronowii* (Losinsk.) Soo) - 2
xanthobasis Fed. (*Aleuritia xanthobasis* (Fed.) Sojak, *P. nivalis* Pall. var. *subintegerrima* Regel, *P. nivalis* subsp. *subintegerrima* (Regel) Worosch.) - 4, 5
zeylamica Charadze & Kapell. - 2

Samolus L.

valerandi L. - 1, 2, 6

Soldanella L.

alpina L. a. *major* Neilr. = S. hungarica subsp. major
hungarica Simonk. (*S. marmarossiensis* Klast., *S. montana* Willd. subsp. *hungarica* (Simonk.) Ludi) - 1
- subsp. **major** (Neilr.) S. Pawl. (*S. alpina* L. a. *major* Neilr.) - 1
marmarossiensis Klast. = S. hungarica
montana Willd. - 1
- subsp. *hungarica* (Simonk.) Ludi = S. hungarica

Sredinskya (Stein) Fed.

grandis (Trautv.) Fed. - 2

Trientalis L.

arctica Fisch. ex Hook. - 4, 5
europaea L. - 1, 3, 4, 5

PTERIDACEAE Reichenb.
Pteris L.

cretica L. - 2
vittata L. - 2

PUNICACEAE Horan.
Punica L.

granatum L. - 1(cult.), 2, 6

PYROLACEAE Dumort.
Chimaphila Pursh

japonica Miq. - 5
umbellata (L.) W. Barton - 1, 3, 4, 5

Moneses Salisb.

uniflora (L.) A. Gray - 1, 2, 3, 4, 5, 6

Orthilia Rafin. (*Ramischia* Opiz ex Garcke)

kareliniana (A. Skvorts.) Holub (*Ramischia kareliniana* A. Skorts., *Orthilia kareliniana* (A. Skvorts.) A. Skvorts. comb. superfl.) - 6
obtusata (Turcz.) Hara (*O. obtusata* (Turcz.) Jurtz. comb. superfl., *O. secunda* (L.) House subsp *obtusata* (Turcz.) Bocher, *O. secunda* var. *obtusata* (Turcz.) House, *Pyrola secunda* L. subsp. *obtusata* (Turcz.) Hult., *Ramischia obtusata* (Turcz.) Freyn) - 1, 3, 4, 5, 6
secunda (L.) House (*Ramischia secunda* (L.) Garcke) - 1, 2, 3, 4, 5
- subsp. *obtusata* (Turcz.) Bocher = O. obtusata
- var. *obtusata* (Turcz.) House = O. obtusata

Pyrola L.

alpina Andres - 5
asarifolia auct. = P. incarnata
asarifolia Michx. subsp. *incarnata* (DC.) Haber & Takahashi = P. incarnata
- subsp. *incarnata* (DC.) E. Murr. = P. incarnata
californica auct. = P. rotundifolia
carpatica Holub & Krisa - 1
chlorantha Sw. (*P. virens* Schweigg., *P. virescens* auct.) - 1, 2, 3, 4, 5
conferta Fisch. ex Cham. & Schlecht. = P. minor
dahurica (Andres) Kom. (*P. rotundifolia* L. c. *dahurica* Andres, *P. incarnata* (DC.) Freyn subsp. *dahurica* (Andres) Krisa, *P. rotundifolia* L. subsp. *dahurica* (Andres) Andres) - 4, 5
faurieana Andres (*P. minor* L. subsp. *faurieana* (Andres) Worosch. ex A. Khokhr. & Mazurenko) - 5
grandiflora Radius (*P. rotundifolia* L. subsp. *rotundifolia* var. *grandiflora* (Radius) Fern.) - 1, 3, 4, 5
- subsp. *norvegica* (Knab.) A. & D. Löve = P. norvegica
incarnata (DC.) Freyn (*P. rotundifolia* L. var. *incarnata* DC., *P. asarifolia* Michx. subsp. *incarnata* (DC.) Haber & Takahashi, comb. superfl., *P. asarifolia* subsp. *incarnata* (DC.) E. Murr., *P. rotundifolia* subsp. *rotundifolia* var. *incarnata* (DC.) A. Khokhr., *P. asarifolia* auct.) - 1, 3, 4, 5
- subsp. *dahurica* (Andres) Krisa = P. dahurica
intermedia auct. p.p. = P. norvegica
japonica Klenze ex Alef. (*P. japonica* var. *subaphylla* (Maxim.) Andres, *P. subaphylla* Maxim.) - 5
- var. *subaphylla* (Maxim.) Andres = P. japonica
media Sw. - 1, 2, 3, 4
minor L. (*P. conferta* Fisch. ex Cham. & Schlecht., *P. minor* subsp. *minor* var. *conferta* (Fisch. ex Cham. & Schlecht.) A. Khokhr.) - 1, 2, 3, 4, 5, 6
- subsp. *faurieana* (Andres) Worosch. ex A. Khokhr. & Mazurenko = P. faurieana
nephrophylla (Andres) Andres (*P. rotundifolia* L. var. *nephrophylla* Andres) - 5(?)
norvegica Knab. (*P. grandiflora* Radius subsp. *norvegica* (Knab.) A. & D. Löve, *P. rotundifolia* L. subsp. *norvegica* (Knab.) Hamet-Ahti, *P. intermedia* auct. p.p.) - 1
renifolia Maxim. - 5
rotundifolia L. (*P. tianschanica* Poljak., *P. californica* auct.) - 1, 2, 3, 4, 6
- subsp. *dahurica* (Andres) Andres = P. dahurica
- subsp. *norvegica* (Knab.) Hamet-Ahti = P. norvegica
- subsp. *rotundifolia* var. *grandiflora* (Radius) Fern. = P. grandiflora

- - var. *incarnata* (DC.) A. Khokhr. = P. incarnata
- c. *dahurica* Andres = P. dahurica
- var. *incarnata* DC. = P. incarnata
- var. *nephrophylla* Andres = P. nephrophylla
secunda L. subsp. *obtusata* (Turcz.) Hult. = Orthilia obtusata
subaphylla Maxim. = P. japonica
tianschanica Poljak. = P. rotundifolia
virens Schweigg. = P. chlorantha
virescens auct. = P. chlorantha

Ramischia Opiz ex Garcke = Orthilia

kareliniana A. Skvorts. = Orthilia kareliniana
obtusata (Turcz.) Freyn = Orthilia obtusata
secunda (L.) Garcke = Orthilia secunda

RAFFLESIACEAE auct. = CYTINACEAE

RANUNCULACEAE Juss. (*COPTACEAE* (Gregory) A. &. D. Löve)

Aconitella Spach (*Aconitopsis* Kem.-Nath.)

barbata (Bunge) Sojak (*Delphinium barbatum* Bunge, *Aconitopsis barbata* (Bunge) Kem.-Nath., *Consolida barbata* (Bunge) Schroding.) - 6
hohenackeri (Boiss.) Sojak (*Delphinium hohenackeri* Boiss., *Aconitopsis hohenackeri* (Boiss.) Kem.-Nath., *Consolida hohenackeri* (Boiss.) Grossh.) - 2

Aconitopsis Kem.-Nath. = Aconitella

barbata (Bunge) Kem.-Nath. = Aconitella barbata
hohenackeri (Boiss.) Kem.-Nath. = Aconitella hohenackeri

Aconitum L. (*Lycoctonum* (DC.) Fourr.)

ajanense Steinb. (*A. ranunculoides* Turcz. ex Ledeb. subsp. *ajanense* (Steinb.) Worosch., *Lycoctonum ajanense* (Steinb.) Nakai) - 5
alatavicum Worosch. = A. soongaricum
albo-violaceum Kom. (*Lycoctonum albo-violaceum* (Kom.) Nakai) - 5
altaicum Steinb. - 3, 4
ambiguum Reichenb. - 4, 5
- subsp. **jacuticum** Worosch. - 4
amurense Nakai = A. volubile
angusticassidatum Steinb. (*A. chasmanthum* auct.) - 6
anthora L. - 1
- subsp. *confertiflorum* (DC.) Worosch. = A. confertiflorum
- subsp. *nemorosum* (Bieb. ex Reichenb.) Worosch. = A. nemorosum
- var. *confertiflorum* DC. = A. confertiflorum
anthoroideum DC. - 3, 4, 6
apetalum (Huth) B. Fedtsch. (*A. monticola* Steinb.) - 3, 6
arcuatum Maxim. = A. sczukinii
- subsp. *axilliflorum* (Worosch.) Kadota = A. axilliflorum
axilliflorum Worosch. (*A. arcuatum* Maxim. subsp. *axilliflorum* (Worosch.) Kadota, *A. vorobievii* Worosch.) - 5
baburinii (Worosch.) Schlothgauer (*A. karafutense* Miyabe & Nakai var. *baburinii* Worosch., *A. karafutense* subsp. *baburinii* (Worosch.) Worosch.) - 5
baicalense Turcz. ex Steinb. = A. turczaninowii
baicalense Turcz. ex Rapaics (*A. contractum* Worosch., *A. contractum* subsp. *tukuringraense* Worosch., *A. czekanovskyi* Steinb., *A. jamalicum* Govor.) - 3, 4, 5

barbatum Pers. (*Lycoctonum barbatum* (Pers.) Nakai, *L. ochranthum* (C.A. Mey.) Nakai, *L. sibiricum* (Poir.) Nakai) - 3, 4, 5
besserianum Andrz. - 1
biflorum Fisch. ex DC. - 3, 4
birobidshanicum Worosch. = A. kusnezoffii subsp. birobidshanicum
brachynasum Kem.-Nath. = A. nasutum
bucovinense Zapal. (*A. callibotryon* Reichenb. subsp. *bucovinense* (Zapal.) Grint., *A. firmum* Reichenb. subsp. *bucovinense* (Zapal.) Aschers. & Graebn.) - 1
callibotryon Reichenb. subsp. *bucovinense* (Zapal.) Grint. = A. bucovinense
x **cammarum** L. (*A.* x *stoerkianum* Reichenb.) - A. napellus L. x A. variegatum L.
- var. *gracile* Reichenb. = A. gracile
caucasicum N. Busch subsp. *pubiceps* (Rupr.) N. Busch var. *tuscheticum* N. Busch = A. nasutum
chamissonis Reichenb. = A. delphinifolium
charkeviczii Worosch. - 5
chasmanthum auct. = A. angusticassidatum
chasmanthum Stapf - 4
cochleare Worosch. (*A. nasutum* Fisch. ex Reichenb. subsp. *cochleare* (Worosch.) Worosch., *A. tuscheticum* (N. Busch) N. Busch, p.p. quoad pl. karabach.) - 2
confertiflorum (DC.) Gayer (*A. anthora* L. var. *confertiflorum* DC., *A. anthora* subsp. *confertiflorum* (DC.) Worosch., *A. confertiflorum* (DC.) Worosch. comb. superfl.) - 1, 2
consanguineum Worosch. - 5
contractum Worosch. = A. baicalense
- subsp. *tukuringraense* Worosch. = A. baicalense
coreanum (Levl.) Rapaics (*A. delavayi* Franch. var. *coreanum* Levl., *A. komarovii* Steinb.) - 5
crassifolium Steinb. - 5
curvirostre (Kryl.) Serg. = A. decipiens
cymbulatum (Schmalh.) Lipsky (*A. nasutum* Fisch. ex Reichenb. subsp. *cymbulatum* (Schmalh.) Worosch.) - 2
czekanovskyi Steinb. = A. baicalense
decipiens Worosch. & Anfalov (*A. curvirostre* (Kryl.) Serg., *A. napellus* L. var. *curvirostre* Kryl.) - 3, 4
degenii Gayer - 1
delavayi Franch. var. *coreanum* Levl. = A. coreanum
delphinifolium DC. (*A. chamissonis* Reichenb., *A. delphinifolium* subsp. *chamissonis* (Reichenb.) Hult.) - 5
- subsp. *anadyrense* Worosch. = A. productum
- subsp. *chamissonis* (Reichenb.) Hult. = A. delphinifolium
- subsp. *delphinifolium* var. *paradoxum* (Reichenb.) Jurtz. = A. paradoxum
- subsp. *kuzenevae* (Worosch.) Worosch. = A. kuzenevae
- subsp. *paradoxum* (Reichenb.) Hult. = A. paradoxum
- subsp. *pavlovae* (Worosch.) Worosch. = A. pavlovae
- subsp. *pseudokusnezowii* (Worosch.) Worosch. = A. pseudokusnezowii
- subsp. *subglandulosum* (A. Khokhr.) A. Luferov = A. subglandulosum
desoulavyi Kom. - 5
eulophum Reichenb. - 1
excelsum Reichenb. = A. septentrionale
firmum Reichenb. (*A. napellus* L. subsp. *firmum* (Reichenb.) Gayer, *A. napellus* subsp. *hians* (Reichenb.) Gayer, *A. napellus* subsp. *skerisorae* (Gayer) Seitz, *A. skerisorae* Gayer) - 1
- subsp. *bucovinense* (Zapal.) Aschers. & Graebn. = A. bucovinense
- subsp. *fissurae* E.I. Nyarady = A. flerovii
fischeri Reichenb. (*A. lubarskyi* Reichenb.) - 5

fissurae E.I. Nyarady = A. flerovii

flagellare (Fr. Schmidt) Steinb. = A. karafutense

flerovii Steinb. (*A. fissurae* E.I. Nyarady, *A. firmum* Reichenb. subsp. *fissurae* E.I. Nyarady, nom. altern., *A. napellus* L. subsp. *fissurae* (E.I. Nyarady) Seitz) - 1

gigas Levl. & Vaniot (*A. ranunculoides* Turcz. ex Ledeb. subsp. *gigas* (Levl. & Vaniot) Worosch.) - 5

glandulosum A. Khokhr. = A. subglandulosum

gracile (Reichenb.) Gayer (*A.* x *cammarum* L. var. *gracile* Reichenb.) - 1

helenae Worosch. - 5

hosteanum Schur - 1

jacquinii Reichenb. - 1

jaluense auct. = A. saxatile

jamalicum Govor. = A. baicalense

jenisseense Polozh. - 4

karafutense Miyabe & Nakai (*A. flagellare* (Fr. Schmidt) Steinb.) - 5

- subsp. *baburinii* (Worosch.) Worosch. = A. baburinii
- var. *baburinii* Worosch. = A. baburinii

karakolicum Rapaics - 6

kirinense Nakai (*Lycoctonum kirinense* (Nakai) Nakai) - 5

komarovii Steinb. = A. coreanum

krylovii Steinb. - 3, 4

kunasilense Nakai (*A. maximum* Pall. ex DC. var. *kunasilense* (Nakai) Tamura & Namba, *A. maximum* subsp. *kunasilense* (Nakai) Tamura & Namba) - 5

kurilense Takeda (*A. maximum* Pall. ex DC. subsp. *kurilense* (Takeda) Kadota) - 5

kusnezoffii Reichenb. (*A. pulcherrimum* Nakai) - 4, 5

- subsp. **birobidshanicum** (Worosch.) A. Luferov (*A. birobidshanicum* Worosch., *A. pulcherrimum* Nakai subsp. *birobidshanicum* (Worosch.) Worosch., *A. tokii* Nakai subsp. *birobidshanicum* (Worosch.) Worosch. nom. invalid.) - 5

kuzenevae Worosch. (*A. delphinifolium* DC. subsp. *kuzenevae* (Worosch.) Worosch.) - 4, 5

lasiocarpum Gayer - 1

lasiostomum Reichenb. - 1

leucostomum Worosch. - 3, 4, 6

longiracemosum Worosch. - 4

lubarskyi Reichenb. = A. fischeri

luxurians (Reichenb.) Nakai = A. maximum

lycoctonum L. subsp. *moldavicum* (Hacq.) Jalas = A. moldavicum

- subsp. *septentrionale* (Koelle) Mela & Cajand. = A. septentrionale

macrorhynchum Turcz. ex Ledeb. - 4, 5

maximum Pall. ex DC. (*A. luxurians* (Reichenb.) Nakai) - 5

- subsp. *kunasilense* (Nakai) Tamura & Namba = A. kunasilense
- subsp. *kurilense* (Takeda) Kadota = A. kurilense
- var. *kunasilense* (Nakai) Tamura & Namba = A. kunasilense

miyabei Nakai (*A. neosachalinense* Levl. subsp. *miyabei* (Nakai) Worosch.) - 5

moldavicum Hacq. (*A. lycoctonum* L. subsp. *moldavicum* (Hacq.) Jalas) - 1

montibaicalense Worosch. - 4

monticola Steinb. = A. apetalum

nanum (Baumg.) Simonk. (*A. napellus* L. var. *nanum* Baumg., *A. tauricum* Wulf. subsp. *nanum* (Baumg.) Grint.) - 1

*****napellus** L. - 1

- subsp. *firmum* (Reichenb.) Gayer = A. firmum
- subsp. *fissurae* (E.I. Nyarady) Seitz = A. flerovii
- subsp. *hians* (Reichenb.) Gayer = A. firmum

- subsp. *skerisorae* (Gayer) Seitz = A. firmum
- var. *curvirostre* Kryl. = A. decipiens
- var. *nanum* Baumg. = A. nanum

nasutum Fisch. ex Reichenb. (*A. brachynasum* Kem.-Nath., *A. caucasicum* N. Busch subsp. *pubiceps* (Rupr.) N. Busch var. *tuscheticum* N. Busch, *A. nasutum* subsp. *anfalovii* Worosch., *A. pubiceps* (Rupr.) Trautv., *A. tuscheticum* (N. Busch) N. Busch, p.p. incl. typo, *A. variegatum* L. subsp. *nasutum* (Fisch. ex Reichenb.) E. Goetz) - 2

- subsp. *anfalovii* Worosch. = A. nasutum
- subsp. *cochleare* (Worosch.) Worosch. = A. cochleare
- subsp. *cymbulatum* (Schmalh.) Worosch. = A. cymbulatum

nemorosum Bieb. ex Reichenb. (*A. anthora* L. subsp. *nemorosum* (Bieb. ex Reichenb.) Worosch.) - 1

nemorum M. Pop. (*A. saposhnikovii* B. Fedtsch., *A. tranzschelii* Steinb.) - 6

neokurilense Worosch. = A. sachalinense subsp. neokurilense

neosachalinense Levl. (*A. suginomei* Nakai) - 5

- subsp. *miyabei* (Nakai) Worosch. = A. miyabei

ochotense Reichenb. - 5

odontandrum Wissjul. - 1

orientale Mill. - 2

paniculatum Lam. - 1

- var. *podolicum* Zapal. = A. podolicum

paradoxum Reichenb. (*A. delphinifolium* DC. subsp. *delphinifolium* var. *paradoxum* (Reichenb.) Jurtz., *A. delphinifolium* subsp. *paradoxum* (Reichenb.) Hult.) - 5

pascoi Worosch. - 4

- subsp. **arcto-alpinum** Worosch. - 4

pavlovae Worosch. (*A. delphinifolium* DC. subsp. *pavlovae* (Worosch.) Worosch.) - 5

phragmitincola Kom. ex Worosch. = A. volubile

podolicum (Zapal.) Worosch. (*A. paniculatum* Lam. var. *podolicum* Zapal.) - 1

popovii Steinb. & Schischk. ex Sipl. = A. tanguticum

possieticum Worosch. = A. volubile

productum Reichenb. (*A. delphinifolium* DC. subsp. *anadyrense* Worosch.) - 4, 5

pseudanthora Blocki ex Pacz. - 1

pseudokusnezowii Worosch. (*A. delphinifolium* DC. subsp. *pseudokusnezowii* (Worosch.) Worosch.) - 5

pubiceps (Rupr.) Trautv. = A. nasutum

puchonroenicum Uyeki & Sataka (*A. ranunculoides* Turcz. ex Ledeb. subsp. *puchonroenicum* (Uyeki & Sataka) Worosch.) - 5

pulcherrimum Nakai = A. kusnezoffii

- subsp. *birobidshanicum* (Worosch.) Worosch. = A. kusnezoffii subsp. birobidshanicum

raddeanum Regel - 5

ranunculoides Turcz. ex Ledeb. (*Lycoctonum ranunculoides* (Turcz. ex Ledeb.) Nakai) - 4, 5

- subsp. *ajanense* (Steinb.) Worosch. = A. ajanense
- subsp. *gigas* (Levl. & Vaniot) Worosch. = A. gigas
- subsp. *puchonroenicum* (Uyeki & Sataka) Worosch. = A. puchonroenicum
- subsp. *umbrosum* (Korsh.) Worosch. = A. umbrosum

rogoviczii Wissjul. - 1

romanicum Woloszcz. - 1

rotundifolium Kar. & Kir. - 6

- subsp. *iliense* Worosch. - 6

rubicundum Fisch. - 4

sachalinense Fr. Schmidt - 5

- subsp. **neokurilense** (Worosch.) A. Luferov (*A. neokuri-*

lense Worosch.) - 5

sajanense Kumin. - 4

saposhnikovii B. Fedtsch. = A. nemorum

saxatile Worosch. & Vorobiev (*A. jaluense* auct.) - 5

sczukinii Turcz. (*A. arcuatum* Maxim.) - 5

- subsp. *subalpinum* (Baran.) Worosch. = A. subalpinum

selemdshense Worosch. = A. subvillosum

septentrionale Koelle (*A. excelsum* Reichenb., *A. lycoctonum* L. subsp. *septentrionale* (Koelle) Mela & Cajand., *Lycoctonum excelsum* (Reichenb.) Nakai) - 1, 3, 4

seravschanicum Steinb. - 6

sichotense Kom. - 5

skerisorae Gayer = A. firmum

smirnovii Steinb. - 4

soongaricum Stapf (*A. alatavicum* Worosch.) - 3, 6

x *stoerkianum* Reichenb. = A. x cammarum

stoloniferum Worosch. - 5

stubendorffii Worosch. - 4

subalpinum Baran. (*A. sczukinii* Turcz. subsp. *subalpinum* (Baran.) Worosch.) - 5

subglandulosum A. Khokhr. (*A. glandulosum* A. Khokhr. 1985, non Rapais, *A. delphinifolium* DC. subsp. *subglandulosum* (A. Khokhr.) A. Luferov) - 5

subvillosum Worosch. (*A. selemdshense* Worosch.) - 5

suginomei Nakai = A. neosachalinense

sukaczevii Steinb. - 4

taigicola Worosch. (*A. tokii* auct.) - 4, 5

talassicum M. Pop. - 6

tanguticum (Maxim.) Stapf (*A. popovii* Steinb. & Schischk. ex Sipl.) - 4

tauricum Wulf. subsp. *nanum* (Baumg.) Grint. = A. nanum

tokii auct. = A. taigicola

tokii Nakai subsp. *birobidshanicum* (Worosch.) Worosch. = A. kusnezoffii subsp. birobidshanicum

tranzschelii Steinb. = A. nemorum

turczaninowii Worosch. (*A. baicalense* Turcz. ex Steinb. 1937, non Turcz. ex Rapaics, 1907) - 4

tuscheticum (N. Busch) N. Busch = A. cochleare and A. nusutum

umbrosum (Korsh.) Kom. (*A. ranunculoides* Turcz. ex Ledeb. subsp. *umbrosum* (Korsh.) Worosch., *Lycoctonum umbrosum* (Korsh.) Nakai) - 5

variegatum L. - 1

- subsp. *nasutum* (Fisch. ex Reichenb.) E. Goetz = A. nasutum

villosum Reichenb. = A. volubile

- var. *amurense* (Nakai) S.H. Li & Y.H. Huang = A. volubile

volubile Pall. ex Koelle (*A. amurense* Nakai, *A. phragmitincola* Kom. ex Worosch., *A. possieticum* Worosch., *A. villosum* Reichenb., *A. villosum* var. *amurense* (Nakai) S.H. Li & Y.H. Huang) - 1, 3, 4, 5

vorobievii Worosch. = A. axilliflorum

woroschilovii A. Luferov - 5

Actaea L.

acuminata auct. = A. asiatica

asiatica Hara (*A. spicata* L. var. *asiatica* (Hara) S.H. Li & Y.H. Huang, *A. acuminata* auct.) - 5

densiflora M. Kral - 2

erythrocarpa Fisch. (*A. spicata* L. subsp. *erythrocarpa* (Fisch.) Hult.) - 1, 3, 4, 5

spicata L. - 1, 2, 3

- subsp. *erythrocarpa* (Fisch.) Hult. = A. erythrocarpa

- var. *asiatica* (Hara) S.H. Li & Y.H. Huang = A. asiatica

Adonanthe Spach = Adonis

amurensis (Regel & Radde) Chrtek & Slavikova = Adonis amurensis

leiosepala (Butk.) Chrtek & Slavikova = Adonis leiosepala

ramosa (Franch.) Chrtek & Slavikova = Adonis ramosa

sibirica (Ledeb.) Spach = Adonis sibirica

tianschanica (Adolf) Chrtek & Slavikova = Adonis tianschanica

turkestanica (Korsh.) Chrtek & Slavikova = Adonis turkestanica

vernalis (L.) Spach = Adonis vernalis

villosa (Ledeb.) Chrtek & Slavikova = Adonis villosa

wolgensis (Stev.) Chrtek & Slavikova = Adonis wolgensis

Adonis L. (*Adonanthe* Spach, *Chrysocyathus* Falc.)

aestivalis L. - 1, 2, 3, 6

- subsp. *parviflora* (Fisch. ex DC.) N. Busch = A. parviflora

amurensis Regel & Radde (*Adonanthe amurensis* (Regel & Radde) Chrtek & Slavikova) - 5

annua L. (*A. annua* var. *atrorubens* L., *A. annua* subsp. *autumnalis* (L.) Maire & Weiller, *A. atrorubens* (L.) L., *A. autumnalis* L.) - 1

- subsp. *autumnalis* (L.) Maire & Weiller = A. annua

- var. *atrorubens* L. = A. annua

apennina auct. = A. sibirica

atrorubens (L.) L. = A. annua

autumnalis L. = A. annua

bienertii Butk. - 2

chrysocyathus Hook. fil. & Thoms. (*Chrysocyathus falconeri* Chrtek & Slavikova) - 6

dentata auct. p.p. = A. persica

dentata Delile subsp. *persica* (Boiss.) H. Riedl = A. persica

eriocalycina Boiss. - 2(?)

flammea Jacq. - 1, 2

- subsp. **cortiana** C. Steinberg - 2

leiosepala Butk. (*Adonanthe leiosepala* (Butk.) Chrtek & Slavikova) - 6

parviflora Fisch. ex DC. (*A. aestivalis* L. subsp. *parviflora* (Fisch. ex DC.) N. Busch) - 2, 6

persica Boiss. (*A. dentata* Delile subsp. *persica* (Boiss.) H. Riedl, *A. dentata* auct. p.p.) - 6

ramosa Franch. (*Adonanthe ramosa* (Franch.) Chrtek & Slavikova) - 5

scrobiculata Boiss. - 2

sibirica Patrin ex Ledeb. (*Adonanthe sibirica* (Ledeb.) Spach, *Adonis apennina* auct.) - 1, 3, 4, 6

tianschanica (Adolf) Lipsch. (*Adonanthe tianschanica* (Adolf) Chrtek & Slavikova) - 6

turkestanica (Korsh.) Adolf (*Adonanthe turkestanica* (Korsh.) Chrtek & Slavikova) - 6

vernalis L. (*Adonanthe vernalis* (L.) Spach) - 1, 2, 3, 4

villosa Ledeb. (*Adonanthe villosa* (Ledeb.) Chrtek & Slavikova) -

wolgensis Stev. (*Adonanthe wolgensis* (Stev.) Chrtek & Slavikova) - 1, 2, 3, 6

Anemonastrum Holub

biarmiense (Juz.) Holub (*Anemone biarmiensis* Juz., *A. narcissiflora* L. subsp. *biarmiensis* (Juz.) Jalas) - 1, 3

brevipedunculatum (Juz.) Holub (*Anemone brevipedunculata* Juz., *A. narcissiflora* L. var. *brevipedunculata* (Juz.) Tamura) - 5

calvum (Juz.) Holub (*Anemone calva* Juz., *Anemonastrum narcissiflorum* (L.) Holub subsp. *calvum* (Juz.) A. & D. Löve) - 4, 5

crinitum (Juz.) Holub (*Anemone crinita* Juz., *A. narcissiflora* L. subsp. *crinita* (Juz.) Kitag., *A. narcissiflora* var. *crinita* (Juz.) Tamura) - 3, 4

fasciculatum (L.) Holub (*Anemone fasciculata* L., *A. narcissiflora* L. f. *fasciculata* (L.) A. Nyarady) - 2

impexum (Juz.) Holub (*Anemone impexa* Juz., *A. narcissiflora* L. var. *willdenowii* Boiss., *A. narcissiflora* subsp. *willdenowii* (Boiss.) P.H. Davis) - 2

narcissiflorum (L.) Holub (*Anemone narcissiflora* L., *A. laxa* (Ulbr.) Juz.) - 1
- subsp. *calvum* (Juz.) A. & D. Löve = A. calvum
- subsp. *sibiricum* (L.) A. & D. Löve = A. sibiricum
- subsp. *villosissimum* (DC.) A. & D. Löve = A. villosissimum

protractum (Ulbr.) Holub (*Anemone protracta* (Ulbr.) Juz.) - 6

sachalinense (Juz.) Starodub. (*Anemone sachalinensis* Juz.) - 5

schrenkianum (Juz.) Holub (*Anemone schrenkiana* Juz.) - 6

sibiricum (L.) Holub (*Anemone sibirica* L., *Anemonastrum narcissiflorum* (L.) Holub subsp. *sibiricum* (L.) A. & D. Löve, *Anemone narcissiflora* L. subsp. *sibirica* (L.) Hult., *A. narcissiflora* var. *sibirica* (L.) Tamura) - 4, 5

speciosum (Adams ex G. Pritz.) Galushko (*Anemone speciosa* Adams ex G. Pritz.) - 2

villosissimum (DC.) Holub (*A. narcissiflorum* (L.) Holub subsp. *villosissimum* (DC.) A. & D. Löve, *Anemone narcissiflora* L. subsp. *villosissima* (DC.) Hult., *A. villosissima* (DC.) Juz.) - 5

Anemone L.

albana Stev. subsp. *armena* (Boiss.) Smirn. = Pulsatilla armena

almaatensis Juz. - 6

altaica Fisch. ex C.A. Mey. = Anemonoides altaica

ambigua Turcz. ex Hayek = Pulsatilla ambigua

amurensis (Korsh.) Kom. = Anemonoides amurensis
- subsp. *kamtschatica* (Kom.) Starodub. = Anemonoides amurensis subsp. kamtschatica

baikalensis Turcz. ex Ledeb. = Arsenjevia baikalensis
- subsp. *glabrata* (Maxim.) Kitag. = Arsenjevia glabrata
- subsp. *rossii* (S. Moore) Starodub. = Arsenjevia rossii

baissunensis Juz. ex Scharipova - 6

biarmiensis Juz. = Anemonastrum biarmiense

blanda Schott & Kotschy = Anemonoides blanda

brevipedunculata Juz. = Anemonastrum brevipedunculatum

bucharica (Regel) Fin. & Gagnep. - 6

caerulea DC. = Anemonoides caerulea

calva Juz. = Anemonastrum calvum

caucasica Willd. ex Rupr. = Anemonoides caucasica

crinita Juz. = Anemonastrum crinitum

debilis Fisch. ex Turcz. = Anemonoides debilis

dichotoma L. = Anemonidium dichotomum

drummondii auct. = A. multiceps

eranthioides Regel - 6

fasciculata L. = Anemonastrum fasciculatum

fischeriana DC. = Anemonoides caerulea

flaccida Fr. Schmidt = Arsenjevia flaccida

glabrata (Maxim.) Juz. = Arsenjevia glabrata

gortschakowii Kar. & Kir. - 6

impexa Juz. = Anemonastrum impexum

jenisseensis (Korsh.) Kryl. = Anemonoides jenisseensis

juzepczukii Starodub. = Anemonoides juzepczukii

kusnetzowii Woronow ex Grossh. - 2

laxa (Ulbr.) Juz. = Anemonastrum narcissiflorum

litoralis (Litv.) Juz. = Arsenjevia rossii

multiceps (Greene) Standl. (*Pulsatilla multiceps* Greene,

Anemone drummondii auct.) - 5

narcissiflora L. = Anemonastrum narcissiflorum
- subsp. *biarmiensis* (Juz.) Jalas = Anemonastrum biarmiense
- subsp. *crinita* (Juz.) Kitag. = Anemonastrum crinitum
- subsp. *sibirica* (L.) Hult. = Anemonastrum sibiricum
- subsp. *villosissima* (DC.) Hult. = Anemonastrum villosissimum
- subsp. *willdenowii* (Boiss.) P.H. Davis = Anemonastrum impexum
- var. *brevipedunculata* (Juz.) Tamura = Anemonastrum brevipedunculatum
- var. *crinita* (Juz.) Tamura = Anemonastrum crinitum
- var. *sibirica* (L.) Tamura = Anemonastrum sibiricum
- var. *willdenowii* Boiss. = Anemonastrum impexum
- f. *fasciculata* (L.) A. Nyarady = Anemonastrum fasciculatum

nemorosa L. = Anemonoides nemorosa
- var. *kamtschatica* Kom. = Anemonoides amurensis subsp. kamtschatica

obtusiloba D. Don - 6

ochotensis (Fisch. ex G. Pritz.) Juz. (*A. sylvestris* L. subsp. *ochotensis* (Fisch. ex G. Pritz.) Petrovsky) - 4, 5

oligotoma Juz. - 6

parviflora Michx. - 5

petiolulosa Juz. - 6

protracta (Ulbr.) Juz. = Anemonastrum protractum

raddeana Regel = Anemonoides raddeana

ranunculoides L. = Anemonoides ranunculoides

reflexa Steph. = Anemonoides reflexa

richardsonii Hook. = Anemonidium richardsonii

rossii S. Moore = Arsenjevia rossii

sachalinensis Juz. = Anemonastrum sachalinense

schrenkiana Juz. = Anemonastrum schrenkianum

sciaphila M. Pop. = Anemonoides sciaphila

serawschanica Kom. - 6

sibirica L. = Anemonastrum sibiricum

soyensis auct. = Anemonoides sciaphila

speciosa Adams ex G. Pritz. = Anemonastrum speciosum

sylvestris L. - 1, 2, 3, 4, 5, 6
- subsp. *ochotensis* (Fisch. ex G. Pritz.) Petrovsky = A. ochotensis

tamarae Charkev. - 5

tschernjaewii Regel - 6

udensis Trautv. & C.A. Mey. = Anemonoides udensis

umbrosa C.A. Mey. = Anemonoides umbrosa
- subsp. *extremiorientalis* Starodub. = Anemonoides extremiorientalis
- subsp. *sciaphila* (M. Pop.) Starodub. = Anemonoides sciaphila

uralensis Fisch. ex DC. = Anemonoides uralensis

verae Ovcz. & Scharipova - 6

villosissima (DC.) Juz. = Anemonastrum villosissimum

yezoensis (Miyabe) Koidz. p.p. = Anemonoides sciaphila

Anemonidium (Spach) Holub (*Jurtsevia* A. & D. Löve)

dichotomum (L.) Holub (*Anemone dichotoma* L., *Anemonidium dichotomum* (L.) A. & D. Löve, comb. superfl.) - 1, 3, 4, 5

richardsonii (Hook.) Starodub. (*Anemone richardsonii* Hook., *Jurtsevia richardsonii* (Hook.) A. & D. Löve) - 4, 5

Anemonoides Mill.

altaica (C.A. Mey.) Holub (*Anemone altaica* Fisch. ex C.A. Mey.) - 3, 4

amurensis (Korsh.) Holub (*Anemone amurensis* (Korsh.) Kom.) - 5

- subsp. **kamtschatica** (Kom.) Starodub. (*Anemone nemorosa* L. var. *kamtschatica* Kom., *A. amurensis* subsp. *kamtschatica* (Kom.) Starodub.) - 5

baikalensis (Turcz. ex Ledeb.) Holub = Arsenjevia baikalensis

blanda (Schott & Kotschy) Holub (*Anemone blanda* Schott & Kotschy) - 2

caerulea (DC.) Holub (*Anemone caerulea* DC., *A. fischeriana* DC., *Anemonoides fischeriana* (DC.) Holub) - 3, 4

caucasica (Rupr.) Holub (*Anemone caucasica* Willd. ex Rupr.) - 2

debilis (Turcz.) Holub (*Anemone debilis* Fisch. ex Turcz.) - 5

extremiorientalis (Starodub.) Starodub. (*Anemone umbrosa* C.A. Mey. subsp. *extremiorientalis* Starodub.) - 5

fischeriana (DC.) Holub = A. caerulea

flaccida (Fr. Schmidt) Holub = Arsenjevia flaccida

glabrata (Maxim.) Holub = Arsenjevia glabrata

jenisseensis (Korsh.) Holub (*Anemone jenisseensis* (Korsh.) Kryl.) - 3, 4

juzepczukii (Starodub.) Starodub. (*Anemone juzepczukii* Starodub.) - 5

x **korshinskyi** Saksonov & Rakov. - A. altaica (C.A. Mey.) Holub x A. ranunculoides (L.) Holub - 1

nemorosa (L.) Holub (*Anemone nemorosa* L.) - 1

raddeana (Regel) Holub (*Anemone raddeana* Regel) - 5

ranunculoides (L.) Holub (*Anemone ranunculoides* L.) - 1, 2

reflexa (Steph.) Holub (*Anemone reflexa* Steph.) - 1, 3, 4, 5

sciaphila (M.Pop.) Starodub. (*Anemone sciaphila* M. Pop., *A. umbrosa* C.A. Mey. subsp. *sciaphila* (M. Pop.) Starodub., *A. yezoensis* (Miyabe) Koidz. p.p. excl. typo, *A. soyensis* auct., *Anemonoides soyensis* auct.) - 5

soyensis auct. = A. sciaphila

udensis (Trautv. & C.A. Mey.) Holub (*Anemone udensis* Trautv. & C.A. Mey.) - 5

umbrosa (C.A. Mey.) Holub (*Anemone umbrosa* C.A. Mey.) - 3

uralensis (DC.) Holub (*Anemone uralensis* Fisch. ex DC.) - 1

Anetilla Galushko = Pulsatilla

aurea (Somm. & Levier) Galushko = Pulsatilla aurea

Aquilegia L.

amurensis Kom. - 4, 5

anemonoides Willd. = Paraquilegia anemonoides

atropurpurea Willd. - 4

atrovinosa M. Pop. ex Gamajun. - 6

borodinii Schischk. - 3, 4

brevicalcarata Kolokoln. ex Serg. - 4

buergeriana Siebold & Zucc. var. *oxysepala* (Trautv. & C.A. Mey.) Kitam. = A. oxysepala

buriatica Peschkova = A. tuvinica and A. viridiflora

caucasica (Ledeb.) Rupr. = A. olympica

coelestis Fed. - 6

colchica Kem.-Nath. - 3

darwasi Korsh. - 6

flabellata Siebold & Zucc. (*A. japonica* Nakai & Hara) - 5

gegica Jabr.-Kolak. - 2

glandulosa Fisch. ex Link - 3, 4, 6

japonica Nakai & Hara = A. flabellata

karatavica Mikesch. - 6

karelinii (Baker) O. & B. Fedtsch. - 6

kubanica I.M. Vassil. - 2

lactiflora Kar. & Kir. - 6

leptoceras Fisch. & C.A. Mey. = A. turczaninovii

longisepala Zimmeter = A. vulgaris

microphylla (Korsh.) Ikonn. (*A. moorcroftiana* Wall. ex Royle var. *microphylla* Korsh., *A. moorcroftiana* auct.) - 6

moorcroftiana auct. = A. microphylla

moorcroftiana Wall. ex Royle var. *microphylla* Korsh. = A. microphylla

nigricans Baumg. - 1

ochotensis Worosch. - 5

olympica Boiss. (*A. caucasica* (Ledeb.) Rupr., *A. vulgaris* L. var. *caucasica* Ledeb.) - 2

oxysepala Trautv. & C.A. Mey. (*A. buergeriana* Siebold & Zucc. var. *oxysepala* (Trautv. & C.A. Mey.) Kitam.) - 5

parviflora Ledeb. - 4, 5

sibirica Lam. - 3, 4, 6

tianschanica Butk. - 6

transsilvanica Schur - 1

turczaninovii R. Kam. & Gubanov (*A. leptoceras* Fisch. & C.A. Mey. 1837, non Nutt. 1834) - 4, 5

tuvinica I.M. Vassil. (*A. buriatica* Peshkova, p.p. excl.typo) - 4

vicaria Nevski - 6

viridiflora Pall. (*A. buriatica* Peschkova, p.p. quoad typum) - 4, 5

vitalii Gamajun. - 6

vulgaris L. (*A. longisepala* Zimmeter) - 1, 5(?)

- var. *caucasica* Ledeb. = A. olympica

Arsenjevia Starodub.

baikalensis (Turcz. ex Ledeb.) Starodub. (*Anemone baikalensis* Turcz. ex Ledeb., *Anemonoides baikalensis* (Turcz. ex Ledeb.) Holub) - 4

flaccida (Fr. Schmidt) Starodub. (*Anemone flaccida* Fr. Schmidt, *Anemonoides flaccida* (Fr. Schmidt) Holub) - 5

glabrata (Maxim.) Starodub. (*Anemone baikalensis* Turcz. ex Ledeb. subsp. *glabrata* (Maxim.) Kitag., *A. glabrata* (Maxim.) Juz., *Anemonoides glabrata* (Maxim.) Holub) - 5

rossii (S. Moore) Starodub. (*Anemone rossii* S. Moore, *A. baikalensis* Turcz. ex Ledeb. subsp. *rossii* (S. Moore) Starodub., *A. litoralis* (Litv.) Juz.) - 5

Atragene L.

alpina L. (*Clematis alpina* (L.) Mill.) - 1

- subsp. *sibirica* (L.) A. & D. Löve = A. sibirica

koreana (Kom.) Kom. (*Clematis koreana* Kom.) - 5

macropetala (Ledeb.) Ledeb. (*Clematis macropetala* Ledeb.) - 4, 5

ochotensis Pall. (*Clematis ochotensis* (Pall.) Poir., *C. sibirica* (L.) Mill. var. *ochotensis* (Pall.) S.H. Li & Y.H. Huang) - 4, 5

sibirica L. (*A. alpina* L. subsp. *sibirica* (L.) A. & D. Löve, *A. tianschanica* Pavl., *Clematis alpina* (L.) Mill. subsp. *sibirica* (L.) O. Kuntze) - 1, 3, 4, 6

tianschanica Pavl. = A. sibirica

Batrachium (DC.) S.F. Gray

aquatile (L.) Dumort. (*Ranunculus aquatilis* L., *Batrachium aquatile* subsp. *radicans* (Revel.) Soo, *B. carinatum* Schur, *B. diversifolium* (Gilib.) Min., *B. gilibertii* V. Krecz., *B. heterophyllum* (Web.) S.F. Gray, *B. mongolicum* (Kryl.) V. Krecz., *B. radicans* (Revel.) DesMoul., *B. triphyllum* (Wallr.) Dumort., *Ranunculus aquatilis* subsp. *radicans* (Revel.) Clapham, *R. diversifolius* Gilib., *R. heterophyllus* Web., *R. mongolicus* (Kryl.) Serg., *R. radicans* Revel.) - 1, 3, 4
- subsp. *radicans* (Revel.) Soo = B. aquatile
baudotii auct. = B. marinum
carinatum Schur = B. aquatile
circinatum (Sibth.) Spach (*Ranunculus circinatus* Sibth., *Batrachium foeniculaceum* (Gilib.) V. Krecz.) - 1, 3, 4, 5, 6
dichotomum Schmalh. = B. peltatum
divaricatum (Schrank) Wimm. p.p. = B. trichophyllum
diversifolium (Gilib.) Min. = B. aquatile
eradicatum (Laest.) Fries (*B. nipponicum* (Nakai) Czer., *B. trichophyllum* (Chaix) Bosch subsp. *eradicatum* (Laest.) A. Löve, *B. trichophyllum* subsp. *lutulentum* (Perrier & Song.) Janch. ex Petrovsky, *Ranunculus eradicatus* (Laest.) F. Johansen, *R. lutulentus* Perrier & Song., *R. nipponicus* Nakai, *R. trichophyllus* Chaix subsp. *eradicatus* (Laest.) C.D. Cook, *R. trichophyllus* subsp. *lutulentus* (Perrier & Song.) Gremli, *R. trichophyllus* subsp. *lutulentus* (Perrier & Song.) Vierh. comb. superfl.) - 1, 3, 4, 5
fluitans (Lam.) Wimm. (*Ranunculus fluitans* Lam.) - 1
foeniculaceum (Gilib.) V. Krecz. = B. circinatum
gilibertii V. Krecz. = B. aquatile
hederaceum (L.) S.F. Gray (*Ranunculus hederaceus* L.) - 1
heterophyllum (Web.) S.F. Gray = B. aquatile
kauffmannii (Clerc) V. Krecz. = B. trichophyllum
langei F. Schultz = B. peltatum
marinum Fries (*B. baudotii* auct., *Ranunculus baudotii* auct., *R. saniculifolius* auct. p.p.) - 1
mongolicum (Kryl.) V. Krecz. = B. aquatile
- subsp. *setosissimum* A. Khokhr. = B. setosissimum
nipponicum (Nakai) Czer. = B. eradicatum
pachycaulum Nevski - 6
peltatum (Schrank) Bercht. & J. Presl (*Ranunculus peltatus* Schrank, *Batrachium dichotomum* Schmalh., *B. langei* F. Schultz, *B. peltatum* (Schrank) Petrovsky, comb. superfl., *B. rhipiphyllum* (Bast. ex Boreau) Dumort., *Ranunculus dichotomus* (Schmalh.) Orlova, nom. invalid., *R. rhipiphyllus* Bast. ex Boreau) - 1
penicillatum Dumort. = B. trichophyllum
radicans (Revel.) DesMoul. = B. aquatile
rhipiphyllum (Bast. ex Boreau) Dumort. = B. peltatum
rionii (Lagger) Nym. (*Ranunculus rionii* Lagger, *Batrachium trichophyllum* (Chaix) Bosch subsp. *rionii* (Lagger) C.D.K. Cook, *Ranunculus trichophyllus* Chaix subsp. *rionii* (Lagger) Soo) - 1, 2, 3, 6
setosissimum (A. Khokhr.) A. Khokhr. (*B. mongolicum* (Kryl.) V. Krecz. subsp. *setosissimum* A. Khokhr.) - 4, 5
trichophyllum (Chaix) Bosch (*Ranunculus trichophyllus* Chaix, *Batrachium divaricatum* (Schrank) Wimm. p.p. quoad nomen, *B. kauffmannii* (Clerc) V. Krecz., *B. penicillatum* Dumort., *Ranunculus divaricatus* Schrank, *R. kauffmannii* Clerc, *R. penicillatus* (Dumort.) Bab., *R. pseudofluitans* (Syme) Newboult ex Baker & Foggitt) - 1, 2, 3, 4, 5, 6
- subsp. *eradicatum* (Laest.) A. Löve = B. eradicatum
- subsp. *lutulentum* (Perrier & Song.) Janch. ex Petrovsky =

B. eradicatum
- subsp. *rionii* (Lagger) C.D.K. Cook = B. rionii
tripartitum (DC.) S.F. Gray - 1
triphyllum (Wallr.) Dumort. = B. aquatile
yezoense (Nakai) Kitam. (*Ranunculus yezoensis* Nakai) - 5

Beckwithia Jeps.

chamissonis (Schlecht.) Tolm. (*Ranunculus chamissonis* Schlecht., *Beckwithia glacialis* (L.) A. & D. Löve subsp. *chamissonis* (Schlecht.) A. & D. Löve, *Oxygraphis chamissonis* (Schlecht.) Freyn, *Ranunculus glacialis* L. subsp. *chamissonis* (Schlecht.) Hult.) - 5
glacialis (L.) A. & D. Löve (*Ranunculus glacialis* L., *Oxygraphis gelida* (Hoffmgg.) O. Schwarz, *O. vulgaris* Freyn, *Ranunculus gelidus* Hoffmgg.) - 1
- subsp. *chamissonis* (Schlecht.) A. & D. Löve = B. chamissonis

Buschia Ovcz.

lateriflora (DC.) Ovcz. (*Ranunculus lateriflorus* DC.) - 1, 2, 3, 6

Callianthemum C.A. Mey.

alatavicum Freyn - 6
angustifolium Witas. - 3, 4, 6
isopyroides (DC.) Witas. - 4, 5
sachalinense Miyabe & Tatew. - 5
sajanense (Regel) Witas. - 3, 4

Caltha L.

See also Thacla.

arctica R. Br. (*C. minor* Mill. subsp. *arctica* (R. Br.) A. & D. Löve, *C. palustris* L. subsp. *arctica* (R. Br.) Hult.) - 1, 3, 4, 5
- subsp. *caespitosa* (Schipcz.) A. Khokhr. = C. caespitosa
- subsp. *membranacea* (Turcz.) A. Khokhr. = C. membranacea
- subsp. *sibirica* (Regel) Tolm. = C. sibirica
barthei (Hance) Koidz. = C. fistulosa
caespitosa Schipcz. (*C. arctica* R. Br. subsp. *caespitosa* (Schipcz.) A. Khokhr.) - 1, 3, 4, 5
cornuta Schott, Nym. & Kotschy = C. palustris subsp. cornuta
crenata Belaeva & Sipl. - 4
fistulosa Schipcz. (*C. barthei* (Hance) Koidz., *C. palustris* L. var. *barthei* Hance, *C. palustris* subsp. *barthei* (Hance) Kitam.) - 5
gorovoi Worosch. - 5
laeta Schott, Nym. & Kotschy = C. palustris subsp. laeta
membranacea (Turcz.) Schipcz. (*C. arctica* R. Br. subsp. *membranacea* (Turcz.) A. Khokhr., *C. palustris* L. subsp. *membranacea* (Turcz.) Hult., *C. palustris* var. *polysepala* Turcz., *C. palustris* subsp. *polysepala* (Turcz.) Worosch.) - 4, 5
minor Mill. (*C. palustris* L. subsp. *minor* (Mill.) Worosch. p.p. excl. pl.) - 1
- subsp. *arctica* (R. Br.) A. & D. Löve = C. arctica
natans Pall. ex Georgi = Thacla natans
palustris L. - 1, 2, 3, 4, 5, 6
- subsp. *arctica* (R. Br.) Hult. = C. arctica
- subsp. *barthei* (Hance) Kitam. = C. fistulosa
- subsp. **cornuta** (Schott, Nym. & Kotschy) Hegi (*C. cornuta* Schott, Nym. & Kotschy) - 1
- subsp. **laeta** (Schott, Nym. & Kotschy) Hegi (*C. laeta* Schott, Nym. & Kotschy) - 1

- subsp. *membranacea* (Turcz.) Hult. = C. membranacea
- subsp. *minor* (Mill.) Worosch. p.p. = C. minor
- subsp. **nymphaeifolia** Worosch. & Gorovoi (*C. pygmaea* auct.) - 5
- subsp. *polysepala* (Turcz.) Worosch. = C. membranacea
- subsp. **renifolia** (Tolm.) A. Luferov (*C. sibirica* (Regel) Makino var. *renifolia* Tolm.) - 5
- subsp. *sibirica* (Regel) Hult. = C. sibirica
- subsp. *violacea* (A. Khokhr.) A. Luferov = C. violacea
- var. *barthei* Hance = C. fistulosa
- var. *polysepala* Turcz. = C. membranacea
- var. *sibirica* Regel = C. sibirica
polypetala Hochst. - 2
pygmaea auct. = C. palustris subsp. nymphaeifolia
serotina Tolm. - 4
sibirica (Regel) Makino (*C. palustris* L. var. *sibirica* Regel, *C. arctica* R. Br. subsp. *sibirica* (Regel) Tolm., *C. palustris* L. subsp. *sibirica* (Regel) Hult., *C. sibirica* (Regel) Tolm. comb. superfl.) - 4, 5
- var. *renifolia* Tolm. = C. palustris subsp. renifolia
silvestris Worosch. - 5
violacea A. Khokhr. (*C. palustris* L. subsp. *violacea* (A. Khokhr.) A. Luferov) - 5

Ceratocephala Moench

falcata (L.) Pers. (*Ranunculus falcatus* L., *Ceratocephala glaberrima* Klok.) - 1, 2, 6
glaberrima Klok. = C. falcata
orthoceras DC. = C. testiculata
reflexa Stev. = C. testiculata
testiculata (Crantz) Bess. (*Ranunculus testiculatus* Crantz, *Ceratocephala orthoceras* DC., *C. reflexa* Stev.) - 1, 2, 3, 6

Chrysocyathus Falc. = Adonis

falconeri Chrtek & Slavikova = Adonis chrysocyathus

Cimicifuga Wernisch.

dahurica (Turcz.) Maxim. - 4, 5
europaea Schipcz. - 1
foetida L. - 3, 4
heracleifolia Kom. - 5
simplex (DC.) Wormsk. ex Turcz. - 4, 5

Clematis L. (*Viticella* auct.)

adrianowii Maxim. - 4
aethusifolia auct. = C. latisecta
alpina (L.) Mill. = Atragene alpina
- subsp. *sibirica* (L.) O. Kuntze = Atragene sibirica
asplenifolia Schrank - 6
brevicaudata DC. - 5
flammula L. - 2
fusca Turcz. - 5
- subsp. **violacea** (Maxim.) Kitag. (*Clematis fusca* var. *violacea* Maxim.) - 5
- var. *violacea* Maxim. = C. fusca subsp. violacea
glauca Willd. - 3, 4, 6
hexapetala Pall. - 4, 5
hilariae Kovalevsk. - 6
integrifolia L. - 1, 2, 3, 4, 6
ispahanica Boiss. - 6
koreana Kom. = Atragene koreana
lathyrifolia Bess. ex Reichenb. (*C. pseudoflammula* Schmalh. ex Lipsky) - 1, 2
latisecta (Maxim.) Prantl (*C. aethusifolia* auct.) - 4, 5

macropetala Ledeb. = Atragene macropetala
manschurica Rupr. - 5
ochotensis (Pall.) Poir. = Atragene ochotensis
orientalis L. (*Viticella orientalis* (L.) W.A. Weber) - 1, 2, 6
pseudoflammula Schmalh. ex Lipsky = C. lathyrifolia
recta L. - 1, 2
sarezica Ikonn. - 6
serratifolia Rehd. - 5
sibirica (L.) Mill. var. *ochotensis* (Pall.) S.H. Li & Y.H. Huang = Atragene ochotensis
sichotealinensis Ulanova - 5
songarica Bunge - 6
tangutica (Maxim.) Korsh. - 6
vitalba L. - 1, 2
viticella L. - 2

Consolida (DC.) S.F. Gray

ajacis (L.) Schur (*Delphinium ajacis* L., *Consolida ambigua* (L.) P.W. Ball & Heywood) - 1, 2, 6
ambigua (L.) P.W. Ball & Heywood = C. ajacis
barbata (Bunge) Schroding. = Aconitella barbata
camptocarpa (Fisch. & C.A. Mey.) Nevski (*Delphinium camptocarpum* Fisch. & C.A. Mey., *Consolida songorica* (Kar. & Kir.) Nevski, *Delphinium songoricum* (Kar. & Kir.) Nevski) - 6
divaricata (Ledeb.) Schroding. (*Delphinium divaricatum* Ledeb., *Consolida regalis* S.F. Gray subsp. *divaricata* (Ledeb.) Munz, *C. regalis* subsp. *paniculata* (Host) Soo var. *divaricata* (Ledeb.) P.H. Davis, *Delphinium consolida* L. subsp. *divaricatum* (Ledeb.) A. Nyarady) - 1, 2, 6
glandulosa (Boiss. & Huet) Bornm. (*Delphinium glandulosum* Boiss. & Huet) - 2
hohenackeri (Boiss.) Grossh. = Aconitella hohenackeri
leptocarpa Nevski (*Delphinium leptocarpum* (Nevski) Butk.) - 6
orientalis (J. Gay) Schroding. (*Delphinium orientale* J. Gay) - 1, 2, 6
paniculata (Host) Schur (*Delphinium paniculatum* Host, *Consolida regalis* S.F. Gray subsp. *paniculata* (Host) Soo) - 1, 2
paradoxa (Bunge) Nevski (*Delphinium paradoxum* Bunge) - 6
persica (Boiss.) Schroding. (*Delphinium persicum* Boiss.) - 2
regalis S.F. Gray (*Delphinium consolida* L.) - 1, 2, 3
- subsp. *divaricata* (Ledeb.) Munz = C. divaricata
- subsp. *paniculata* (Host) Soo = C. paniculata
- - var. *divaricata* (Ledeb.) P.H. Davis = C. divaricata
rugulosa (Boiss.) Schroding. (*Delphinium rugulosum* Boiss.) - 2, 6
songorica (Kar. & Kir.) Nevski = C. camptocarpa
stocksiana (Boiss.) Nevski (*Delphinium stocksianum* Boiss.) - 2, 6

Coptidium Beurl. = Ranunculus

lapponicum (L.) A. & D. Love = Ranunculus lapponicus
pallasii (Schlecht.) A. & D. Love = Ranunculus pallasii
spitzbergense (Hadac) Hadac = Ranunculus spitzbergensis

Coptis Salisb.

trifolia (L.) Salisb. - 4, 5

Delphinium L. (*Diedropetala* Galushko)

aemulans Nevski - 6

ajacis L. = Consolida ajacis

albiflorum DC. (*D. fissum* Waldst. & Kit. subsp. *albiflorum* (DC.) Greuter & Burdet, *D. ochroleucum* auct.) - 2

albomarginatum Simonova (*D. karataviense* Pavl.) - 6

alpinum Waldst. & Kit. = D. elatum

altaicum Nevski - 3

araraticum (N. Busch) Grossh. (*D. tomentellum* N. Busch var. *araraticum* N. Busch) - 2

arcuatum N. Busch - 2

atropurpureum Pall. = D. elatum

barbatum Bunge = Aconitella barbata

barlykense Lomonosova & V. Khanminchun - 4

batalinii Huth - 6

biternatum Huth - 6

blagovestschenskii Gubanov & Trusov = D. dictyocarpum

brachycentrum Ledeb. - 5

- subsp. **beringii** Jurtz. - 5

- subsp. **maydellianum** (Trautv.) Jurtz. (*D. maydellianum* Trautv.) - 5

bracteosum Somm. & Levier (*D. ruprechtii* Nevski) - 2

brunonianum Royle - 6

bucharicum M. Pop. (*D. karategini* Korsh. var. *bucharicum* (M. Pop.) Sosk. & Fachrieva) - 6

buschianum Grossh. - 2

camptocarpum Fisch. & C.A. Mey. = Consolida camptocarpa

caucasicum C.A. Mey. - 2

chamissonis G. Pritz. ex Walp. (*D. frigidum* Adams ex Nas., *D. pauciflorum* Reichenb. ex Ledeb. 1841, non D. Don, 1825) - 4, 5

charadzeae Kem.-Nath. & Gagnidze - 2

cheilanthum Fisch. - 4, 5

confusum M. Pop. - 6

connectens Pachom. - 6

consolida L. = Consolida regalis

- subsp. *divaricatum* (Ledeb.) A. Nyarady = Consolida divaricata

corymbosum auct. = D. turkestanicum

crassicaule Ledeb. - 4

crassifolium Schrad. ex Ledeb. (*D. korshinskyanum* Nevski) - 4, 5

crispulum Rupr. - 2

cryophilum Nevski (*D. elatum* L. subsp. *cryophilum* (Nevski) Jurtz.) - 1

cuneatum Stev. ex DC. (*D. litwinowii* Sambuk, *D. rossicum* Litv.) - 1

cyananthum Nevski - 3, 6

cyphoplectrum Boiss. (*D. laxiusculum* (Boiss.) Rouy) - 2

- var. *pallidiflorum* (Freyn) P.H. Davis = D. pallidiflorum

darginicum Dimitrova - 2

dasyanthum Kar. & Kir. - 6

dasycarpum Stev. ex DC. - 2

- var. *elisabethae* (N. Busch) Grossh. = D. elisabethae

decoloratum Ovcz. & Koczk. - 6

dictyocarpum DC. (*D. blagovestschenskii* Gubanov & Trusov) - 1, 3, 6

- subsp. *uralense* (Nevski) Pawl. = D. uralense

divaricatum Ledeb. = Consolida divaricata

duhmbergii Huth, p.p. = D. turkestanicum

duhmbergii sensu Munz, p.p. = D. iliense

dzavakhischwilii Kem.-Nath. - 2

elatum L. (*D. alpinum* Waldst. & Kit., ? *D. atropurpureum* Pall.) - 1, 3, 4, 6

- subsp. *cryophilum* (Nevski) Jurtz. = D. cryophilum

- subsp. *nacladense* (Zapal.) Holub = D. nacladense

elisabethae N. Busch (*D. dasycarpum* Stev. ex DC. var. *elisabethae* (N. Busch) Grossh.) - 2

fedorovii Dimitrova - 2

fissum Waldst. & Kit. (*D. fissum* subsp. *pallasii* (Nevski) Greuter & Burdet, *D. leiocarpum* Huth, *D. pallasii* Nevski) - 1, 2

- subsp. *albiflorum* (DC.) Greuter & Burdet = D. albiflorum

- subsp. *pallasii* (Nevski) Greuter & Burdet = D. fissum

flexuosum Bieb. - 2

foetidum Lomak. - 2

formosum Boiss. & Huet - 2

freynii Conrath (*D. schmalhausenii* Albov subsp. *freynii* (Conrath) Takht., *Diedropetala freynii* (Conrath) Galushko) - 2

frigidum Adams ex Nas. = D. chamissonis

gelmetzicum Dimitrova = D. prokhanovii

glandulosum Boiss. & Huet = Consolida glandulosa

grandiflorum L. - 3, 4, 5

hohenackeri Boiss. = Aconitella hohenackeri

iliense Huth (*D. duhmbergii* sensu Munz. p.p.) - 6

inconspicuum Serg. - 3, 4

inopinatum Nevski - 6

intermedium Soland. - 1

- subsp. *nacladense* (Zapal.) Jav. = D. nacladense

ironorum N. Busch - 2

karataviense Pavl. = D. albomarginatum

karategini Korsh. - 6

- var. *bucharicum* (M. Pop.) Sosk. & Fachrieva = D. bucharicum

keminense Pachom. - 6

knorringianum B. Fedtsch. - 6

kolymense A. Khokhr. - 5

korshinskyanum Nevski = D. crassifolium

lacostei Danguy - 6

laxiflorum DC. - 3

laxiusculum (Boiss.) Rouy = D. cyphoplectrum

leiocarpum Huth = D. fissum

leonidae Kem.-Nath. - 2

leptocarpum (Nevski) Butk. = Consolida leptocarpa

linearilobum (Trautv.) N. Busch - 2

lipskyi Korsh. - 6

litwinowii Sambuk = D. cuneatum

lomakinii Kem.-Nath. - 2

longipedunculatum Regel & Schmalh. - 6

maackianum Regel - 5

macropogon Prokh. (*Diedropetala macropogon* (Prokh.) Galushko) - 2

malyschevii Friesen - 4

mariae N. Busch - 2

maydellianum Trautv. = D. brachycentrum subsp. maydellianum

megalanthum Nevski - 2

middendorffii Trautv. - 1, 3, 4, 5

minjanense Rech. fil. - 6

mirabile Serg. - 3, 4

nachiczevanicum Tzvel. (*D. quercetorum* auct.) - 2

nacladense Zapal. (*D. elatum* L. subsp. *nacladense* (Zapal.) Holub, *D. intermedium* Soland. subsp. *nacladense* (Zapal.) Jav.) - 1

nevskii Zak. = D. oreophilum

nikitinae Pachom. - 6

ochotense Nevski - 4, 5

ochroleucum auct. = D. albiflorum

oreophilum Huth (*D. nevskii* Zak.) - 6

- subsp. *rotundifolium* (Afan.) Sosk. & Fachrieva = D. rotundifolium

- f. *rotundifolium* (Afan.) Ovcz. = D. rotundifolium

orientale J. Gay = Consolida orientalis

osseticum N. Busch (*D. speciosum* Bieb. var. *osseticum* (N. Busch) Grossh.) - 2

ovczinnikovii R. Kam. & Pissjauk. - 6

pallasii Nevski = D. fissum

pallidiflorum Freyn (*D. cyphoplectrum* Boiss. var. *pallidiflorum* (Freyn) P.H. Davis) - 2

paniculatum Host = Consolida paniculata

paradoxum Bunge = Consolida paradoxa

pauciflorum Reichenb. ex Ledeb. = D. chamissonis

pavlovii R. Kam. - 6

persicum Boiss. = Consolida persica

poltoratzkii Rupr. - 6

popovii Pachom. - 6

prokhanovii Dimitrova (*D. gelmetzicum* Dimitrova) - 2

propinquum Nevski - 6

puniceum Pall. (*Diedropetala punicea* (Pall.) Galushko) - 1, 2, 6

pyramidatum Albov - 2

quercetorum auct. = D. nachiczevanicum

raikovae Pachom. - 6

retropilosum (Huth) Sambuk - 3, 4

reverdattoanum Polozh. & Revyakina - 4

rossicum Litv. = D. cuneatum

rotundifolium Afan. (*D. oreophilum* Huth subsp. *rotundifolium* (Afan.) Sosk. & Fachrieva, *D. oreophilum* f. *rotundifolium* (Afan.) Ovcz.) - 6

rugulosum Boiss. = Consolida rugulosa

ruprechtii Nevski = D. bracteosum

sajanense Jurtz. - 4

sauricum Schischk. - 6

schmalhausenii Albov (*D. somcheticum* Conrath & Freyn, *Diedropetala schmalhausenii* (Albov) Galushko) - 1, 2

- subsp. *freynii* (Conrath) Takht. = D. freynii

semibarbatum Bien. ex Boiss. - 6

semiclavatum Nevski - 6

sergii Wissjul. - 1

somcheticum Conrath & Freyn = D. schmalhausenii

songoricum (Kar. & Kir.) Nevski = Consolida camptocarpa

speciosum Bieb. - 2

- var. *osseticum* (N. Busch) Grossh. = D. osseticum

stocksianum Boiss. = Consolida stocksiana

sylvaticum Turcz. = D. turczaninovii

szowitsianum Boiss. - 2

talyschense Tzvel. - 2

ternatum Huth - 6

thamarae Kem.-Nath. - 2

tianschanicum W.T. Wang - 6(?)

tomentellum N. Busch - 2

- var. *araraticum* N. Busch = D. araraticum

triste Fisch. - 4

turczaninovii Friesen (*D. sylvaticum* Turcz. 1842, nom. invalid., non Romel) - 4

turkestanicum Huth (*D. duhmbergii* Huth, p.p. excl. typo, *D. corymbosum* auct.) - 6

turkmenum Lipsky - 6

ukokense Serg. - 3, 4

uralense Nevski (*D. dictyocarpum* DC. subsp. *uralense* (Nevski) Pawl.) - 1

villosum Stev. - 1

vvedenskyi Pachom. - 6

Diedropetala Galushko = Delphinium

freynii (Conrath) Galushko = Delphinium freynii

macropogon (Prokh.) Galushko = Delphinium macropogon

punicea (Pall.) Galushko = Delphinium puniceum

schmalhausenii (Albov) Galushko = Delphinium schmalhausenii

Enemion Rafin.

raddeanum Regel (*Isopyrum raddeanum* (Regel) Maxim.) - 5

Eranthis auct. = Shibateranthis

longistipitata Regel = Shibateranthis longistipitata

sibirica DC. = Shibateranthis sibirica

stellata Maxim. = Shibateranthis stellata

uncinata Turcz. ex Ledeb. = Shibateranthis sibirica

Ficaria Guett.

calthifolia Reichenb. (*F. ledebourii* Grossh. & Schischk., *F. nudicaulis* A. Kerner, *F. popovii* A. Khokhr. subsp. *abchasica* A. Khokhr., *F. verna* Huds. subsp. *calthifolia* (Reichenb.) Velen., *F. verna* subsp. *ledebourii* (Grossh. & Schischk.) Soo, *Ranunculus ficaria* L. subsp. *calthifolius* (Reichb.) Arcang., *R. ficaria* subsp. *nudicaulis* (A. Kerner) Soo) - 1, 2

fascicularis C. Koch (*Ranunculus kochii* Ledeb.) - 2, 6

ficarioides (Bory & Chaub.) Halacsy (*Ranunculus ficarioides* Bory & Chaub.) - 2

glacialis Fisch. = Oxygraphis glacialis

grandiflora Robert (*F. popovii* A. Khokhr. p.p. excl. subsp. *abchasica* A. Khokhr., *Ranunculus ficaria* L. subsp. *ficariiformis* Rouy & Fouc.) - 1, 2

ledebourii Grossh. & Schischk. = F. calthifolia

nudicaulis A. Kerner = F. calthifolia

popovii A. Khokhr. = F. grandiflora

- subsp. *abchasica* A. Khokhr. = F. calthifolia

stepporum P. Smirn. - 1

varia Otschiauri - 2

verna Huds. (*Ranunculus ficaria* L.) - 1, 2, 3, 6

- subsp. *calthifolia* (Reichenb.) Velen. = F. calthifolia

- subsp. *ledebourii* (Grossh. & Schischk.) Soo = F. calthifolia

Garidella L.

nigellastrum L. (*Nigella garidella* Spenn., *N. nigellastrum* (L.) Willk.) - 1, 2, 6

Halerpestes Greene

ruthenica (Jacq.) Ovcz. = H. salsuginosa

salsuginosa (Pall. ex Georgi) Greene (*Ranunculus salsuginosus* Pall. ex Georgi, *Halerpestes ruthenica* (Jacq.) Ovcz., *Oxygraphis salsuginosa* (Pall. ex Georgi) Hyl., *Ranunculus ruthenicus* Jacq.) - 3, 4

salsuginosa sensu Ovcz. = H. sarmentosa

sarmentosa (Adams) Kom. (*Ranunculus sarmentosus* Adams, *Halerpestes subsimilis* (Printz) Tamura, *Ranunculus subsimilis* Printz, *Halerpestes salsuginosa* sensu Ovcz.) - 3, 4, 5, 6

subsimilis (Printz) Tamura = H. sarmentosa

Hegemone Bunge = Trollius

chartosepala (Schipcz.) A. Khokhr. = Trollius chartosepalus

lilacina (Bunge) Bunge = Trollius lilacinus

- var. *micrantha* C. Winkl. & Kom. = Trollius komarovii

micrantha (C. Winkl. & Kom.) Butk. = Trollius komarovii

Helleborus L.

abchasicus A. Br. (*H. polychromus* Kolak. p.p. nom. illegit., *H. orientalis* auct. p.p.) - 2

caucasicus A. Br. (*H. guttatus* A. Br. & Sauer, *H. polychro-*

mus Kolak. p.p. nom. illegit., *H. orientalis* auct. p.p.) - 2
*dumetorum Waldst. & Kit. - 1
guttatus A. Br. & Sauer = H. caucasicus
*niger L. - 1
orientalis auct. p.p. = H. abchasicus and H. caucasicus
polychromus Kolak. = H. abchasicus and H. caucasicus
purpurascens Waldst. & Kit. - 1
ranunculinus Smith = Trollius ranunculinus
*viridis L. - 1

Hepatica Hill

asiatica Nakai (*H. nobilis* Mill. var. *asiatica* (Nakai) Hara) - 5
falconeri (Thoms.) Steward (*H. falconeri* (Thoms.) Juz. comb. superfl.) - 6
nobilis Mill. - 1
- var. *asiatica* (Nakai) Hara = H. asiatica

Isopyrum L.

anemonoides Kar. & Kir. = Paropyrum anemonoides
manshuricum Kom. = Semiaquilegia manshurica
microphyllum Royle = Paraquilegia microphylla
raddeanum (Regel) Maxim. = Enemion raddeanum
thalictroides L. - 1

Jurtsevia A. & D. Löve = Anemonidium

richardsonii (Hook.) A. & D. Löve = Anemonidium richardsonii

Leptopyrum Reichenb. = Neoleptopyrum

fumarioides (L.) Reichenb. = Neoleptopyrum fumarioides

Lycoctonum (DC.) Fourr. = Aconitum

ajanense (Steinb.) Nakai = Aconitum ajanense
albo-violaceum (Kom.) Nakai = Aconitum albo-violaceum
barbatum (Pers.) Nakai = Aconitum barbatum
excelsum (Reichenb.) Nakai = Aconitum septentrionale
kirinense (Nakai) Nakai = Aconitum kirinense
ochranthum (C.A. Mey.) Nakai = Aconitum barbatum
ranunculoides (Turcz. ex Ledeb.) Nakai = Aconitum ranunculoides
sibiricum (Poir.) Nakai = Aconitum barbatum
umbrosum (Korsh.) Nakai = Aconitum umbrosum

Miyakea Miyabe & Tatew.

integrifolia Miyabe & Tatew. (*Pulsatilla integrifolia* (Miyabe & Tatew.) Worosch.) - 5

Myosurus L.

minimus L. - 1, 2, 3, 6

Neoleptopyrum Hutch. (*Leptopyrum* Reichenb. 1828, non Rafin.)

fumarioides (L.) ? (*Leptopyrum fumarioides* (L.) Reichenb.) - 3, 4, 5

Nigella L.

arvensis L. - 1, 2, 6
bucharica Schipcz. (*N. media* Pachom.) - 6
damascena L. - 1, 2, 5(alien)
garidella Spenn. = Garidella nigellastrum

glandulifera Freyn & Sint. - 6
integrifolia Regel - 6
latisecta P.H. Davis (*N. oxypetala* Boiss. subsp. *latisecta* (P.H. Davis) Takht.) - 2
media Pachom. = N. bucharica
nigellastrum (L.) Willk. = Garidella nigellastrum
orientalis L. - 2
oxypetala Boiss. (*N. persica* Boiss.) - 2
- subsp. *latisecta* (P.H. Davis) Takht. = N. latisecta
persica Boiss. = N. oxypetala
sativa L. - 1, 2
segetalis Bieb. - 1, 2

Oxygraphis Bunge

chamissonis (Schlecht.) Freyn = Beckwithia chamissonis
gelida (Hoffmgg.) O. Schwarz = Beckwithia glacialis
glacialis (Fisch.) Bunge (*Ficaria glacialis* Fisch., *Ranunculus kamtschaticus* DC.) - 1, 3, 4, 5, 6
salsuginosa (Pall. ex Georgi) Hyl. = Halerpestes salsuginosa
vulgaris Freyn = Beckwithia glacialis

Paraquilegia J. Drumm. & Hutch.

anemonoides (Kar. & Kir.) Schipcz. = Paropyrum anemonoides
anemonoides (Willd.) Ulbr. (*Aquilegia anemonoides* Willd., *Paraquilegia grandiflora* (Fisch. ex DC.) J. Drumm. & Hutch., *P. microphylla* auct. fl. As. Med. p.p.) - 3, 6
caespitosa (Boiss. & Hohen.) J. Drumm. & Hutch. - 6
grandiflora (Fisch. ex DC.) J. Drumm. & Hutch. = P. anemonoides
kareliniana Nevski = Paropyrum anemonoides
microphylla auct. fl. As. Med. = P. anemonoides and P. scabrifolia
microphylla (Royle) J. Drumm. & Hutch. (*Isopyrum microphyllum* Royle) - 3, 4, 5
scabrifolia Pachom. (*P. microphylla* auct. fl. As. Med. p.p.) - 6
uniflora (Aitch. & Hemsl.) J. Drumm. & Hutch. p.p. = Paropyrum anemonoides

Paropyrum Ulbr.

anemonoides (Kar. & Kir.) Ulbr. (*Isopyrum anemonoides* Kar. & Kir., *Paraquilegia anemonoides* (Kar. & Kir.) Schipcz. 1924, non Ulbr. 1922, *P. kareliniana* Nevski, *P. uniflora* (Aitch. & Hemsl.) J. Drumm. & Hutch. p.p. quoad pl.) - 6

Pulsatilla Hill (*Anetilla* Galushko)

ajanensis Regel & Til. - 5
- subsp. *tatewakii* (Kudo) Worosch. = P. tatewakii
alba Reichenb. - 1
albana (Stev.) Bercht. & J. Presl (*P. andina* (Rupr.) Woronow) - 2
- subsp. *armena* (Boiss.) Aichele & Schwegler = P. armena
ambigua (Turcz. ex Hayek) Juz. (*Anemone ambigua* Turcz. ex Hayek) - 3
andina (Rupr.) Woronow = P. albana
angustifolia subsp. *flavescens* (Zucc.) Holub = P. flavescens
armena (Boiss.) Rupr. (*Anemone albana* Stev. subsp. *armena* (Boiss.) Smirn., *Pulsatilla albana* (Stev.) Bercht. & J. Presl subsp. *armena* (Boiss.) Aichele & Schwegler) - 2
aurea (Somm. & Levier) Juz. (*Anetilla aurea* (Somm. & Levier) Galushko) - 2

bungeana C.A. Mey. - 3
campanella Fisch. ex Regel & Til. - 3, 6
cernua (Thunb.) Bercht. & Opiz - 5
chinensis (Bunge) Regel - 5
davurica (Fisch. ex DC.) Spreng. - 4, 5
donetzica Kotov = P. grandis
flavescens (Zucc.) Juz. 1937, non Boros, 1924 (*P. angustifo-lia* subsp. *flavescens* (Zucc.) Holub, *P. patens* (L.) Mill. subsp. *flavescens* (Zucc.) Zam.) - 1, 3, 4
georgica Rupr. = P. violacea
grandis Wend. (*P. donetzica* Kotov, *P. halleri* (All.) Willd. subsp. *grandis* (Wend.) Meikle, *P. vulgaris* Mill. subsp. *grandis* (Wend.) Zam.) - 1
- var. *turczaninovii* (Kryl. & Serg.) M. Pop. = P. turczani-novii
halleri auct. = P. taurica
halleri (All.) Willd. subsp. *grandis* (Wend.) Meikle = P. grandis
- subsp. *taurica* (Juz.) K. Krause = P. taurica
integrifolia (Miyabe & Tatew.) Worosch. = Miyakea integri-folia
x **intermedia** (Lasch) G. Don fil. - P. patens (L.) Mill. x P. vernalis (L.) Mill. - 1
kioviensis Wissjul. = P. patens
x **kissii** Mandl. - P. cernua (Thunb.) Bercht. & Opiz x P. chinensis (Bunge) Regel - 5
kostyczewii (Korsh.) Juz. - 6
latifolia Rupr. = P. patens
lithophila Kotov = P. taurica
magadanensis A. Khokhr. & Worosch. - 5
montana (Hoppe) Reichenb. - 1
- subsp. **dacica** J. Rummelspacher - 1
multiceps Greene = Anemone multiceps
multifida (G. Pritz.) Juz. (*P. nuttaliana* (DC.) Bercht. & J. Presl subsp. *multifida* (G. Pritz.) Aichele & Schwegler, *P. patens* (L.) Mill. var. *multifida* (G. Pritz.) Kitag. comb. superfl., *P. patens* var. *multifida* (G. Pritz.) S.H. Li & Y.H. Huang, *P. patens* subsp. *multifida* (G. Pritz.) Zam., *P. nuttaliana* auct.) - 1, 3, 4, 5
nigricans auct. = P. ucrainica
nuttaliana auct. = P. multifida
nuttaliana (DC.) Bercht. & J. Presl subsp. *multifida* (G. Pritz.) Aichele & Schwegler = P. multifida
ovczinnikovii Maximova = P. patens
patens (L.) Mill. (*P. kioviensis* Wissjul., *P. latifolia* Rupr., *P. ovczinnikovii* Maximova) - 1, 3, 4
- subsp. *flavescens* (Zucc.) Zam. = P. flavescens
- subsp. *multifida* (G. Pritz.) Zam. = P. multifida
- subsp. *teklae* (Zam.) Zam. = P. teklae
- var. *multifida* (G. Pritz.) Kitag. = P. multifida
- var. *multifida* (G. Pritz.) S.H. Li & Y.H. Huang = P. multifida
pratensis (L.) Mill. - 1
reverdattoi Polozh. & Maltzeva - 4
sachalinensis Hara - 5
sugawarae Miyabe & Tatew. - 5
sukaczewii Juz. (*P. tenuiloba* (Turcz.) Juz. var. *sukaczewii* (Juz.) M. Pop.) - 4
taraoi (Makino) Takeda ex Zam. & Paegle - 5
tatewakii Kudo (*P. ajanensis* Regel & Til. subsp. *tatewakii* (Kudo) Worosch.) - 5
taurica Juz. (*P. halleri* (All.) Wiild. subsp. *taurica* (Juz.) K. Krause, *P. lithophila* Kotov, *P. halleri* auct.) - 1
teklae Zam. (*P. patens* (L.) Mill. subsp. *teklae* (Zam.) Zam.) - 1
tenuiloba (Turcz.) Juz. - 4
- var. *sukaczewii* (Juz.) M. Pop. = P. sukaczewii

turczaninovii Kryl. & Serg. (*P. grandis* Wend. var. *turczani-novii* (Kryl. & Serg.) M. Pop.) - 3, 4, 5
ucrainica (Ugr.) Wissjul. (*P. nigricans* auct.) - 1
vernalis (L.) Mill. - 1
violacea Rupr. (*P. georgica* Rupr.) - 2
vulgaris Mill. - 1(?)
- subsp. *grandis* (Wend.) Zam. = P. grandis
x **wolfgangiana** (Bess.) Juz. - P. patens (L.) Mill. x P. pra-tensis (L.) Mill. - 1

Ranunculus L. (*Coptidium* Beurl. nom. invalid.)

abchasicus Freyn - 2
acer auct. = R. acris
aconitifolius auct. = R. platanifolius
aconitifolius L. subsp. *platanifolius* (L.) Rikli = R. platani-folius
- var. *platanifolius* (L.) Pacz. = R. platanifolius
acris L. (*R. acer* auct.) - 1, 2(?), 3, 4, 5
- subsp. *borealis* (Trautv.) Nym. = R. propinquus
- subsp. *friesianus* (Jord.) Rouy & Fouc. = R. friesianus
- subsp. *glabriusculus* (Rupr.) A. & D. Löve = R. glabrius-culus
- subsp. *japonicus* (Thunb.) Hult. = R. japonicus
- subsp. *novus* (Levl. & Vaniot) Worosch. = R. novus
- subsp. **pseudograndis** Worosch. - 5
- subsp. *scandinavicus* (Orlova) A. & D. Löve = R. frie-sianus
- subsp. **stevenii** Korsh. - 5
- subsp. *strigulosus* (Schur) Hyl. = R. stevenii
- subsp. *subcorymbosus* (Kom.) Toyokuni = R. subcorym-bosus
- subsp. *turneri* (Greene) Worosch. = R. turneri
- var. *austrokuriensis* Tatew. = R. novus
- var. *nipponicus* Hara = R. novus
- var. *subcorymbosus* (Kom.) Tatew. = R. subcorymbosus
acutidentatus Rupr. = R. oreophilus
acutilobus Ledeb. = R. oreophilus
- f. *araraticus* Rupr. = R. araraticus
affinis R. Br. (*R. pedatifidus* Smith subsp. *affinis* (R. Br.) Hult.) - 1, 3, 4, 5
afghanicus auct. = R. olgae
ageri Bertol.
    This species had been recorded for the investigated territory by mistake.
akkemensis Polozh. & Revyakina - 4
alajensis Ostenf. - 6
alberti Regel & Schmalh. - 6
aleae Willk. (*R. bulbosus* L. subsp. *aleae* (Willk.) Rouy & Fourc.) - 2
alexandri Kem.-Nath. = R. trachycarpus
alexeenkoi Grossh. = R. illyricus
allemannii Br.-Bl. = R. fallax
alpigenus Kom. - 6
altaicus Laxm. (*R. nivalis* L. subsp. *altaicus* (Laxm.) Worosch.) - 3, 4, 5, 6
ampelophyllus Somm. & Levier = R. cappadocicus
amurensis Kom. - 5
anadyriensis Ovcz. = R. monophyllus
anemonifolius DC. = R. grandiflorus
aquatilis L. = Batrachium aquatile
- subsp. *radicans* (Revel.) Clapham = Batrachium aquatile
arachnoideus C.A. Mey. - 2
aragazi Grossh. (*R. dissectus* Bieb. subsp. *aragazi* (Grossh.) Bulany & Derviz-Sokolova) - 2
araraticus (Rupr.) Grossh. ex Kem.-Nath. (*R. acutilobus* Ledeb. f. *araraticus* Rupr., *R. brachylobus* Boiss. & Hohen. subsp. *incisilobatus* P.H. Davis) - 2

arvensis L. - 1, 2, 6
astrantiifolius (Rupr.) Boiss. & Bal. = R. buhsei
astrantiifolius (Rupr.) Boiss. ex Trautv. = R. buhsei
aucheri Boiss. (*R. elbrusensis* Boiss.) - 2
aureopetalus Kom. (*R. sciatrophus* Ovcz.) - 6
auricomiformis Soo - 1
auricomus L. - 1, 2, 3
- subsp. **arctophilus** Markl. ex Fagerstrom - 1
- subsp. *cassubicus* (L.) Dostal = R. cassubicus
- subsp. **medians** Markl. ex Fagerstrom - 1
badachschanicus Ovcz. & Koczk. - 6
baidarae Rupr. (*R. dzhavakheticus* Ovcz., *R. ginkgolobus* Somm. & Levier) - 2
baldshuanicus Regel ex Kom. = R. sericeus
balkharicus N. Busch - 2
baudotii auct. = Batrachium marinum
borealis Trautv. = R. propinquus
botschantzevii Ovcz. - 6
brachylobus Boiss. & Hohen. (*R. buschii* Ovcz., *R. gymnadenus* Somm. & Levier, *R. svaneticus* Rupr., *R. tebulossicus* Prima) - 2
- subsp. *incisilobatus* P.H. Davis = R. araraticus
brevirostris Edgew. (*R. laetus* Wall. ex Royle, 1839, non Salisb. 1796, *R. laetus* subsp. *laetiformis* Ovcz., *R. laetus* subsp. *pseudolaetus* (Tamura) Ovcz., *R. pseudolaetus* Tamura) - 6
breyninus auct. = R. nemorosus
brotherusii Freyn - 6
brutius Ten. - 1, 2
- subsp. *crimaeus* (Juz.) A. Jelen. = R. crimaeus
budensis Soo - 1
buhsei Boiss. (*R. astrantiifolius* (Rupr.) Boiss. & Bal. 1888, comb. superfl., *R. astrantiifolius* (Rupr.) Boiss. ex Trautv. 1883, non Schur, 1853, *R. caucasicus* Bieb. var. *astrantiifolius* Rupr., *R. scherosii* Kem.-Nath., *R. trisectilis* Ovcz.) - 2
bulbosus L. - 1, 2
- subsp. *aleae* (Willk.) Rouy & Fourc. = R. aleae
buschii Ovcz. = R. brachylobus
cappadocicus Willd. (*R. ampelophyllus* Somm. & Levier) - 2
carpaticola Soo - 1
carpaticus Herbich - 1
cassubicus L. (*R. auricomus* L. subsp. *cassubicus* (L.) Dostal) - 1, 3
- subsp. (ap.) **archangeliensis** Fagerstrom - 1
- subsp. (ap.) **cajanderi** Fagerstrom - 1
- subsp. (ap.) **fedorovii** Fagerstrom - 1
- subsp. (ap.) **kemerovensis** K. Gustav. - 3
- subsp. (ap.) **tranzchellii** Fagerstrom - 1
caucasicus Bieb. (*R. transcaucasicus* Kem.-Nath.) - 1, 2
- subsp. **pavlii** A. Jelen. & Derviz-Sokolova - 2
- subsp. *subleiocarpus* (Somm. & Levier) P.H. Davis = R. raddeanus
- var. *astrantiifolius* Rupr. = R. buhsei
chaffanjonii Danguy - 6
chamissonis Schlecht. = Beckwithia chamissonis
chinensis Bunge - 4, 5, 6
chius DC. - 1, 2
chodzhamastonicus Ovcz. & Junussov - 6
cicutarius Schlecht. (*R. hyrcanus* Grossh.) - 2
circinatus Sibth. = Batrachium circinatum
constantinopolitanus (DC.) D'Urv. (*R. villosus* DC. subsp. *constantinopolitanus* (DC.) A. Jelen.) - 1, 2
convexiusculus Kovalevsk. - 6
cornutus DC. (*R. lomatocarpus* Fisch. & C.A. Mey.) - 2
crassifolius (Rupr.) Grossh. - 2

crenatus Waldst. & Kit. - 1
crimaeus Juz. (*R. brutius* Ten. subsp. *crimaeus* (Juz.) A. Jelen.) - 1
czimganicus Ovcz. - 6
dichotomus (Schmalh.) Orlova = Batrachium peltatum
dilatatus Ovcz. - 6
dissectus Bieb. - 1, 2
- subsp. *aragazi* (Grossh.) Bulany & Derviz-Sokolova = R. aragazi
- subsp. **glabrescens** (Boiss.) P.H. Davis (*R. huetii* Boiss. var. *glabrescens* Boiss.) - 2
- subsp. *huetii* (Boiss.) P.H. Davis = R. huetii
- subsp. *napellifolius* (DC.) P.H. Davis = R. napellifolius
- subsp. *szowitsianus* (Boiss.) A. Jelen. & Derviz-Sokolova = R. szowitsianus
divaricatus Schrank = Batrachium trichophyllum
diversifolius Gilib. = Batrachium aquatile
dolosus Fisch. & C.A. Mey. = R. sceleratus
dzhavakheticus Ovcz. = R. baidarae
elbrusensis Boiss. = R. aucheri
elegans C. Koch = R. grandiflorus
eradicatus (Laest.) F. Johansen = Batrachium eradicatum
eschscholtzii Schlecht. (*R. pauperculus* Ovcz.) - 5
falcatus L. = Ceratocephala falcata
fallax auct. p.p. = R. megacarpus
fallax (Wimm. & Graebn.) Sloboda (*R. allemannii* Br.-Bl., *R. fallax* (Wimm. & Graebn.) Schur, comb. superfl.) - 1
ficaria L. = Ficaria verna
- subsp. *calthifolius* (Reichenb.) Arcang. = Ficaria calthifolia
- subsp. *ficariiformis* Rouy & Fouc. = Ficaria grandiflora
- subsp. *nudicaulis* (A. Kerner) Soo = Ficaria calthifolia
ficarioides Bory & Chaub. = Ficaria ficarioides
flammula L. - 1, 3
fluitans Lam. = Batrachium fluitans
franchetii Boissieu (*R. ussuriensis* Kom.) - 5
fraternus Schrenk - 6
friesianus Jord. (*R. acris* L. subsp. *friesianus* (Jord.) Rouy & Fouc., *R. acris* subsp. *scandinavicus* (Orlova) A. & D. Löve, *R. scandinavicus* Orlova, *R. silvaticus* auct.) - 1
- subsp. *strigulosus* (Schur) A. Jelen. & Derviz-Sokolova = R. stevenii
gelidus Hoffmgg. = Beckwithia glacialis
gelidus Kar. & Kir. = R. karelinii
- subsp. *grayi* (Britt.) Hult. = R. grayi
georgicus Kem.-Nath. (*R. grandiflorus* L. subsp. *davisii* A. Jelen. & Derviz-Sokolova) - 2
ginkgolobus Somm. & Levier = R. baidarae
glabriusculus Rupr. (*R. acris* L. subsp. *glabriusculus* (Rupr.) A. & D. Löve) - 1, 3, 4
glacialis L. = Beckwithia glacialis
- subsp. *chamissonis* (Schlecht.) Hult. = Beckwithia chamissonis
gmelinii DC. - 1, 3, 4, 5
- subsp. **purshii** (Richards.) Hult. - 5
- subsp. *radicans* (C.A. Mey.) Hult. = R. radicans
grandiflorus L. (*R. anemonifolius* DC., *R. elegans* C. Koch) - 2
- subsp. *davisii* A. Jelen. & Derviz-Sokolova = R. georgicus
grandifolius C.A. Mey. - 3, 4, 6
grandis Honda = R. japonicus
- var. *austrokurilensis* (Tatew.) H. Hara = R. novus
- var. *transochotensis* (Hara) Hara = R. novus
grayi Britt. (*R. gelidus* Kar. & Kir. subsp. *grayi* (Britt.) Hult.) - 4, 5

grossheimii Kolak. = R. suukensis
gymnadenus Somm. & Levier = R. brachylobus
hederaceus L. = Batrachium hederaceum
helenae Albov - 2
heterophyllus Web. = Batrachium aquatile
hornschuchii Hoppe - 1
huetii Boiss. (*R. dissectus* Bieb. subsp. *huetii* (Boiss.) P.H. Davis) - 2(?)
- var. *glabrescens* Boiss. = R. dissectus subsp. glabrescens
hyperboreus Rottb. - 1, 3, 4, 5
- subsp. *arnellii* Scheutz = R. samojedorum
- subsp. *samojedorum* (Rupr.) Hult. = R. samojedorum
- subsp. *tricrenatus* (Rupr.) A. & D. Löve = R. tricrenatus
- subsp. *tricrenatus* (Rupr.) V. Sergienko = R. tricrenatus
- var. *samojedorum* (Rupr.) Perf. = R. samojedorum
- var. *tricrenatus* Rupr. = R. tricrenatus
hyrcanus Grossh. = R. cicutarius
illyricus L. (*R. alexeenkoi* Grossh., *meridionalis* Grossh.) - 1, 2
jacuticus Ocz. = R. turneri
japonicus Thunb. (*R. acris* L. subsp. *japonicus* (Thunb.) Hult., *R. grandis* Honda, *R. subcorymbosus* Kom. var. *grandis* (Honda) Kitag., *R. subcorymbosus* subsp. *grandis* (Honda) Tamura) - 5
- var. *smirnovii* (Ovcz.) L. Liou = R. propinquus
jazgulemicus Ovcz. - 6
jugentassicus N. Rubtz. = R. pedatifidus
kamtschaticus DC. = Oxygraphis glacialis
karelinii Czer. (*R. gelidus* Kar. & Kir. 1842, non Hoffmgg. 1832) - 6
kauffmannii Clerc = Batrachium trichophyllum
kitaibelii Soo - 1
kladnii auct. = R. malinovskii
kochii Ledeb. = Ficaria fascicularis
komarovii Freyn - 6
kopetdaghensis Litv. = R. trichocarpus
kotschyi Boiss. - 2, 6
krasnovii Ovcz. - 6
krylovii Ovcz. = R. monophyllus
laetus Wall. ex Royle = R. brevirostris
- subsp. *laetiformis* Ovcz. = R. brevirostris
- subsp. *pseudolaetus* (Tamura) Ovcz. = R. brevirostris
lanuginosiformis Selin ex N. Fellm. = R. propinquus
lanuginosus L. - 1
lapponicus L. (*Coptidium lapponicum* (L.) A. & D. Löve) - 1, 3, 4, 5
lasiocarpus C.A. Mey. - 3, 4
lateriflorus DC. = Buschia lateriflora
leptorrhynchus Aitch. & Hemsl. = R. sewerzowii
x **levenensis** Druce ex Gornall. - R. flammula L. x R. reptans L. - 1, 3
linearilobus Bunge - 6
- subsp. *mogoltavicus* M. Pop. = R. mogoltavicus
lingua L. - 1, 2, 3, 4, 6
lojkae Somm. & Levier - 2
lomatocarpus Fisch. & C.A. Mey. = R. cornutus
longicaulis C.A. Mey. - 3, 4, 6
- subsp. *pulchellus* (C.A. Mey.) Gubanov = R. longicaulis var. pulchellus
- var. **pseudohirculus** (Schrenk) Gubanov (*R. pseudohirculus* Schrenk) - 3, 4, 6
- var. **pulchellus** (C.A. Mey.) Gubanov (*R. pulchellus* C.A. Mey., *R. longicaulis* subsp. *pulchellus* (C.A. Mey.) Gubanov) - 3, 4, 6
longilobus Ovcz. - 6
lutulentus Perrier & Song. = Batrachium eradicatum
makaschwilii Kem.-Nath. = R. oreophilus

malinovskii A. Jelen. & Derviz-Sokolova (*R. kladnii* auct.) - 1
marginatus D'Urv. subsp. *trachycarpus* (Fisch. & C.A. Mey.) Hayek = R. trachycarpus
- subsp. *trachycarpus* (Fisch. & C.A. Mey.) A. Jelen. & Derviz-Sokolova = R. trachycarpus
- var. *trachycarpus* (Fisch. & C.A. Mey.) Aznav. = R. trachycarpus
marmarosensis Soo - 1
megacarpus Walo Koch (*R. fallax* auct. p.p.) - 1
meinshausenii Schrenk - 6
meridionalis Grossh. = R. illyricus
merovensis Grossh. = R. szowitsianus
meyerianus Rupr. (*R. polyanthemos* L. subsp. *meyerianus* (Rupr.) A. Jelen. & Derviz-Sokolova) - 1, 2, 6
michaelis Kovalevsk. - 6
migaricus Kem.-Nath. & Chinth. - 2
mindshelkensis B. Fedtsch. - 6
mogoltavicus (M. Pop.) Ovcz. (*R. linearilobus* Bunge subsp. *mogoltavicus* M. Pop.) - 6
mongolicus (Kryl.) Serg. = Batrachium aquatile
monophyllus Ovcz. (*R. anadyriensis* Ovcz., *R. krylovii* Ovcz.) - 1, 3, 4, 5, 6
- subsp.(ap.) **vytegrensis** Fagerstrom - 1
montanus Willd. - 1
muricatus L. - 1, 2, 6
napellifolius DC. (*R. dissectus* Bieb. subsp. *napellifolius* (DC.) P.H. Davis) - 2
natans C.A. Mey. - 3, 4, 5, 6
neapolitanus Ten. - 1
nemorosus DC. (*R. serpens* subsp. *nemorosus* (DC.) Gonzalez, *R. breyninus* auct.) - 1, 2
nipponicus Nakai = Batrachium eradicatum
nivalis L. - 1, 3, 4, 5
- subsp. *altaicus* (Laxm.) Worosch. = R. altaicus
- subsp. **intercedens** (Hult.) Worosch. (*R. sulphureus* C.J. Phipps var. *intercedens* Hult.) - 5
- subsp. *sulphureus* (C.J. Phipps) Worosch. = R. sulphureus
novus Levl. & Vaniot (*R. acris* L. var. *austrokurilensis* Tatew., *R. acris* var. *nipponicus* Hara, *R. acris* subsp. *novus* (Levl. & Vaniot) Worosch., *R. grandis* Hara var. *austrokurilensis* (Tatew.) Hara, *R. grandis* var. *transochotensis* (Hara) Hara, *R. subcorymbosus* Kom. subsp. *subcorymbosus* var. *austrokurilensis* (Tatew.) Tamura, *R. transochotensis* Hara) - 5
obesus Trautv. - 2
odessanus Klok. fil. - 1
olgae Regel (*R. afghanicus* auct.) - 6
oligophyllus Pissjauk. - 6
ophioglossifolius Vill. - 1, 2
oreophilus Bieb. (*R. acutidentatus* Rupr., *R. acutilobus* Ledeb., *R. makaschwilii* Kem.-Nath.) - 1, 2
▪ **orientalis** L.
osseticus Ovcz. - 2
ovczinnikovii Kovalevsk. (*R. stenopetalus* Ovcz. 1937, non Hook. 1844) - 6
oxyspermus Willd. - 1, 2, 6
pallasii Schlecht. (*Coptidium pallasii* (Schlecht.) A. & D. Löve) - 1, 3, 4, 5
- var. *minimus* Rupr. = R. spitzbergensis
pamiri Korsh. - 6
pannonicus Soo - 1(?)
paucidentatus Schrenk - 3, 6
pauperculus Ovcz. = R. eschscholtzii
pedatifidus Smith (*R. jugentassicus* N. Rubtz.) - 3, 4
- subsp. *affinis* (R. Br.) Hult. = R. affinis
pedatus Waldst. & Kit. - 1, 2, 3, 6

- subsp. *silvisteppaceus* (Dubovik) A. Jelen. & Derviz-Sokolova = R. silvisteppaceus
peltatus Schrank = Batrachium peltatum
penicillatus (Dumort.) Bab. = Batrachium trichophyllum
pinnatisectus M. Pop. - 6
platanifolius L. (*R. aconitifolius* L. subsp. *platanifolius* (L.) Rikli, *R. aconitifolius* var. *platanifolius* (L.) Pacz., *R. aconitifolius* auct.) - 1
platyspermus Fisch. ex DC. - 1, 3, 6
polyanthemos L. - 1, 3, 4, 5, 6
- subsp. *meyerianus* (Rupr.) A. Jelen. & Derviz-Sokolova = R. meyerianus
polyphyllus Waldst. & Kit. ex Willd. - 1, 3, 4
polyrhizos Steph. - 1, 2, 3, 6
popovii Ovcz. - 6
pronicus A. Skvorts. - 1
propinquus C.A. Mey. (*R. acris* L. subsp. *borealis* (Trautv.) Nym., *R. borealis* Trautv., *R. japonicus* var. *smirnovii* (Ovcz.) L. Liou, *R. lanuginosiformis* Selin ex N. Fellm., *R. smirnovii* Ovcz.) - 1, 3, 4, 5
- subsp. *subcorymbosus* (Kom.) A. Jelen & Derviz-Sokolova = R. subcorymbosus
- subsp. *turneri* (Greene) A. Jelen. & Derviz-Sokolova = R. turneri
pseudobulbosus Schur = R. sardous
pseudofluitans (Syme) Newboult ex Baker & Foggitt = Batrachium trichophyllum
pseudohirculus Schrenk = R. longicaulis var. pseudohirculus
pseudolaetus Tamura = R. brevirostris
pskemensis V. Pavl. - 6
pulchellus C.A. Mey. = R. longicaulis var. pulchellus
pulsatillifolius Litv. - 6
punctatus Jurtz. - 5
pygmaeus Wahlenb. - 1, 3, 4, 5
- subsp. *sabinii* (R. Br.) Hult. = R. sabinii
quelpaertensis (Levl.) Nakai (*R. repens* L. var. *quelpaertensis* Levl.) - 5
raddeanus Regel (*R. caucasicus* subsp. *subleiocarpus* (Somm. & Levier) P.H. Davis, *R. raddeanus* var. *subleiocarpus* Somm. & Levier, *R. raddeanus* subsp. *subleiocarpus* (Somm. & Levier) N. Busch, *R. sommieri* Albov) - 2
- subsp. *subleiocarpus* (Somm. & Levier) N. Busch = R. raddeanus
- var. *subleiocarpus* Somm. & Levier = R. raddeanus
radians Revel. = Batrachium aquatile
radicans C.A. Mey. (*R. gmelinii* DC. subsp. *radicans* (C.A. Mey.) Hult.) - 3, 4
recurvatus Poir. - 5
regelianus Ovcz. - 6
repens L. - 1, 2, 3, 4, 5
- var. *quelpaertensis* Levl. = R. quelpaertensis
reptabundus Rupr. (*R. sceleratus* L. subsp. *reptabundus* (Rupr.) Hult.) - 1, 3(?)
reptans L. - 1, 3, 4, 5
rhipiphyllus Bast. ex Boreau = Batrachium peltatum
rigescens Turcz. ex Trautv. - 4, 5
rionii Lagger = Batrachium rionii
rubrocalyx Regel ex Kom. - 6
rufosepalus Franch. - 6
ruthenicus Jacq. = Halerpestes salsuginosa
sabinii R. Br. (*R. pygmaeus* Wahlenb. subsp. *sabinii* (R.Br.) Hult.) - 4, 5
sajanensis M. Pop. - 4
salsuginosus Pall. ex Georgi = Halerpestes salsuginosa
samojedorum Rupr. (*R. hyperboreus* Rottb. subsp. *arnellii*

Scheutz, *R. hyperboreus* subsp. *samojedorum* (Rupr.) Hult., *R. hyperboreus* var. *samojedorum* (Rupr.) Perf.) - 1, 3, 4, 5
saniculifolius auct. p.p. = Batrachium marinum
sardous Crantz (*R. pseudobulbosus* Schur, *R. sardous* subsp. *laevis* (Schmalh.) N. Busch, *R. sardous* var. *pseudobulbosus* (Schur) Grossh.) - 1, 2
- subsp. *laevis* (Schmalh.) N. Busch = R. sardous
- var. *pseudobulbosus* (Schur) Grossh. = R. sardous
sarmentosus Adams = Halerpestes sarmentosa
sartorianus auct. = R. suukensis
scandinavicus Orlova = R. friesianus
sceleratus L. (*R. dolosus* Fisch. & C.A. Mey.) - 1, 2, 3, 4, 5, 6
- subsp. *reptabundus* (Rupr.) Hult. = R. reptabundus
schaftoanus (Aitch. & Hemsl.) Boiss. - 6
scherosii Kem.-Nath. = R. buhsei
sciatrophus Ovz. = R. aureopetalus
scythicus Klok. - 1
sericeus Banks & Soland. (*R. baldshuanicus* Regel ex Kom., *R. pseudolaetus* Tamura, p.p.) - 6
serpens subsp. *nemorosus* (DC.) Gonzalez = R. nemorosus
sewerzowii Regel (*R. leptorrhynchus* Aitch. & Hemsl., *R. walteri* Regel ex Kom.) - 6
silvaticus auct. = R. friesianus
silvisteppaceus Dubovik (*R. pedatus* Waldst. & Kit. subsp. *silvisteppaceus* (Dubovik) A. Jelen. & Derviz-Sokolova) - 1
smirnovii Ovcz. = R. propinquus
sommieri Albov = R. raddeanus
songaricus Schrenk - 6
sosnowskyi Kem.-Nath. - 2
spitzbergensis Hadac (*Coptidium spitzbergense* (Hadac) Hadac, *Ranunculus pallasii* Schlecht. var. *minimus* Rupr.) - 1, 3, 4, 5
stenopetalus Ovcz. = R. ovczinnikovii
stevenii Andrz. (*R. acris* L. subsp. *strigulosus* (Schur) Hyl., *R. friesianus* Jord. subsp. *strigulosus* (Schur) A. Jelen. & Derviz-Sokolova, *R. strigulosus* Schur) - 1
strigillosus Boiss. & Huet - 2
strigulosus Schur = R. stevenii
subcorymbosus Kom. (*R. acris* L. var. *subcorymbosus* (Kom.) Tatew., *R. acris* subsp. *subcorymbosus* (Kom.) Toyokuni, *R. propinquus* C.A. Mey. subsp. *subcorymbosus* (Kom.) A. Jelen. & Derviz-Sokolova) - 5
- subsp. *grandis* (Honda) Tamura = R. japonicus
- subsp. *subcorymbosus* var. *austrokurilensis* (Tatew.) Tamura = R. novus
- var. *grandis* (Honda) Kitag. = R. japonicus
submarginatus Ovcz. - 3, 4
subrigescens Ovcz. - 6
subsimilis Printz = Halerpestes sarmentosa
subtilis Trautv. - 2
sulphureus C.J. Phipps (*R. nivalis* L. subsp. *sulphureus* (C.J. Phipps) Worosch.) - 1, 3, 4, 5
- var. *intercedens* Hult. = R. nivalis subsp. intercedens
suukensis N. Busch (*R. grossheimii* Kolak., *R. sartorianus* auct.) - 2
svaneticus Rupr. = R. brachylobus
szowitsianus Boiss. (*R. dissectus* Bieb. subsp. *szowitsianus* (Boiss.) A. Jelen. & Derviz-Sokolova, *R. merovensis* Grossh.) - 2
tachiroei Franch. & Savat. - 5
tatrae Borb. = R. thora
tebulossicus Prima = R. brachylobus
tenuilobus Regel ex Kom. - 6
testiculatus Crantz = Ceratocephala testiculata

thora L. (*R. tatrae* Borb.) - 1

trachycarpus Fisch. & C.A. Mey. (*R. alexandri* Kem.-Nath., *R. marginatus* D'Urv. var. *trachycarpus* (Fisch. & C.A. Mey.) Aznav., *R. marginatus* subsp. *trachycarpus* (Fisch. & C.A. Mey.) Hayek, *R. marginatus* subsp. *trachycarpus* (Fisch. & C.A. Mey.) A. Jelen. & Derviz-Sokolova, comb. superfl.) - 1, 2

transalaicus Tzvel. - 6

transcaucasicus Kem.-Nath. = R. caucasicus

transiliensis M. Pop. ex Gamajun. - 6

transochotensis Hara = R. novus

trautvetterianus Regel ex Kom. - 6

trichocarpus Boiss. & Kotschy (*R. kopetdaghensis* Litv.) - 1, 2, 6

trichophyllus Chaix = Batrachium trichophyllum

- subsp. *eradicatus* (Laest.) C.D.K. Cook = Batrachium eradicatum

- subsp. *lutulentus* (Perrier & Song.) Gremli = Batrachium eradicatum

- subsp. *lutulentus* (Perrier & Song.) Vierh. = Batrachium eradicatum

- subsp. *rionii* (Lagger) Soo = Batrachium rionii

tricrenatus (Rupr.) Jurtz. & Petrovsky (*R. hyperboreus* Rottb. var. *tricrenatus* Rupr., *R. hyperboreus* subsp. *tricrenatus* (Rupr.) A. & D. Löve, *R. hyperboreus* subsp. *tricrenatus* (Rupr.) V. Sergienko, comb. superfl.) - 1, 3, 4, 5

trisectilis Ovcz. = R. buhsei

turkestanicus Franch. - 6

turneri Greene (*R. acris* L. subsp. *turneri* (Greene) Worosch. nom. invalid., *R. jacuticus* Ovcz., *R. propinquus* C.A. Mey. subsp. *turneri* (Greene) A. Jelen. & Derviz-Sokolova, *R. turneri* subsp. *jacuticus* (Ovcz.) Tolm.) - 4, 5

- subsp. *jacuticus* (Ovcz.) Tolm. = R. turneri

ussuriensis Kom. = R. franchetii

villosus DC. - 2

- subsp. *constantinopolitanus* (DC.) A. Jelen. = R. constantinopolitanus

vulgoramosum A. Khokhr. - 5

vvedenskyi Ovcz. - 6

walteri Regel ex Kom. = R. sewerzowii

yezoensis Nakai = Batrachium yezoense

zapalowiczii Pacz. - 1

Semiaquilegia Makino

manshurica Kom. (*Isopyrum manshuricum* Kom.) - 5

Shibateranthis Nakai (*Eranthis* auct.)

longistipitata (Regel) Nakai (*Eranthis longistipitata* Regel) - 6

sibirica (DC.) Nakai (*Eranthis sibirica* DC., *E. uncinata* Turcz. ex Ledeb., *Shibateranthis uncinata* (Turcz. ex Ledeb.) Nakai) - 3, 4

stellata (Maxim.) Nakai (*Eranthis stellata* Maxim.) - 5

uncinata (Turcz. ex Ledeb.) Nakai = S. sibirica

Thacla Spach

natans (Pall. ex Georgi) Deyl & Sojak (*Caltha natans* Pall. ex Georgi) - 3, 4, 5

Thalictrum L.

alpinum L. - 1, 2, 3, 4, 5, 6

- subsp. udocanicum Peschkova - 4

altaicum (Schischk.) Serg. (*T. simplex* L. var. *altaicum* Schischk.) - 3

amurense Maxim. - 5

angustifolium auct. = T. lucidum

appendiculatum C.A. Mey. (*T. minus* L. subsp. *appendiculatum* (C.A. Mey.) Gubanov) - 3, 4

aquilegifolium L. - 1

baikalense Turcz. ex Ledeb. - 4, 5

bauhinii Crantz (*T. simplex* L. var. *bauhinii* (Crantz) V. Osvacilova, *T. simplex* subsp. *bauhinii* (Crantz) Tutin) - 1

buschianum Kem.-Nath. - 2

bykovii Kotuchov - 3

collinum Wallr. = T. minus

contortum L. - 3, 4, 5

filamentosum Maxim. - 5

flavum L. - 1, 2, 3, 4, 5, 6

flexuosum Bernh. ex Reichenb. - 1, 2, 6

foetidum L. - 1, 2, 3, 4, 5, 6

friesii Rupr. = T. kemense

globiflorum Ledeb. (*T. minus* L. subsp. *globiflorum* (Ledeb.) Peschkova) - 4

hultenii Boivin = T. kemense

isopyroides C.A. Mey. - 2, 3, 6

kemense (Fries) Koch (*T. friesii* Rupr., *T. hultenii* Boivin, *T. leptophyllum* Nyl., *T. minus* L. subsp. *kemense* (Fries) Cajand., *T. minus* subsp. *kemense* (Fries) Hult. comb. superfl., *T. minus* subsp. *kemense* (Fries) Tutin, comb. superfl.) - 1, 3, 4, 5, 6

kochii auct. = T. minus

kuhistanicum Ovcz. & Koczk. - 6

leptophyllum Nyl. = T. kemense

lucidum L. (*T. angustifolium* auct.) - 1

macrophyllum V. Boczantzeva - 1, 2, 3, 4, 6

minus L. (*T. collinum* Wallr., *T. kochii* auct., *T. transsilvanicum* auct.) - 1, 2, 3, 4, 5, 6

- subsp. *appendiculatum* (C.A. Mey.) Gubanov = T. appendiculatum

- subsp. *globiflorum* (Ledeb.) Peschkova = T. globiflorum

- subsp. *kemense* (Fries) Cajand. = T. kemense

- subsp. *kemense* (Fries) Hult. = T. kemense

- subsp. *kemense* (Fries) Tutin = T. kemense

- subsp. *thunbergii* (DC.) Worosch. = T. thunbergii

neosachalinense Levl. = T. sachalinense

pavlovii Reverd. - 4

petaloideum L. - 3, 4, 5, 6

podolicum Lecoy. - 1

rariflorum Fries (*T. simplex* L. var. *boreale* Nyl., *T. simplex* subsp. *boreale* (Nyl.) A. & D. Löve, *T. simplex* subsp. *boreale* (Nyl.) Tutin, comb. superfl.) - 1

sachalinense Lecoy. (*T. neosachalinense* Levl.) - 5

saissanicum Kotuchov - 3

simplex L. (*T. simplex* subsp. *strictum* (Ledeb.) Worosch. p.p. quoad nomen, comb. invalid., *T. strictum* Ledeb., *T. trilobatum* Ovcz. & Koczk.) - 1, 2, 3, 4, 6

- subsp. *bauhinii* (Crantz) Tutin = T. bauhinii

- subsp. *boreale* (Nyl.) A. & D. Löve = T. rariflorum

- subsp. *boreale* (Nyl.) Tutin = T. rariflorum

- subsp. *strictum* (Ledeb.) Worosch. = T. simplex and T. ussuriense

- var. *altaicum* Schischk. = T. altaicum

- var. *bauhinii* (Crantz) V. Osvacilova = T. bauhinii

- var. *boreale* Nyl. = T. rariflorum

sparsiflorum Turcz. ex Fisch. & C.A. Mey. - 3, 4, 5

squarrosum Steph. - 4, 5

strictum Ledeb. = T. simplex

sultanabadense Stapf - 2, 6

thunbergii DC. (*T. minus* L. subsp. *thunbergii* (DC.) Worosch.) - 5

transsilvanicum auct. = T. minus
trilobatum Ovcz. & Koczk. = T. simplex
triternatum Rupr. - 2
tuberiferum Maxim. - 5
uncinatum Rehm. - 1
ussuriense A. Luferov (*T. simplex* L. subsp. *strictum* (Ledeb.) Worosch. p.p. quoad pl., comb. invalid.) - 5
yezoense Nakai - 5(?)

Trautvetteria Fisch. & C.A. Mey.

caroliniensis var. *japonica* (Siebold & Zucc.) T. Shimizu = T. japonica
japonica Siebold & Zucc. (*T. caroliniensis* var. *japonica* (Siebold & Zucc.) T. Shimizu) - 5

Trollius L. (*Hegemone* Bunge)

aldanensis Volotovsky - 4
altaicus C.A. Mey. - 3, 6
- var. *sajanensis* Malysch. = T. sajanensis
altissimus Crantz (*T. europaeus* L. subsp. *transsilvanicus* (Schur) Jav., *T. transsilvanicus* Schur) - 1
- subsp. **deylii** Chrtek - 1
apertus Perf. ex Igoschina - 1, 3
asiaticus L. (*T. asiaticus* var. *affinis* Regel, *T. asiaticus* subsp. *affinis* (Regel) Egor. & Sipl., *T. kytmanovii* Reverd.) - 3, 4, 6
- subsp. *affinis* (Regel) Egor. & Sipl. = T. asiaticus
- var. *affinis* Regel = T. asiaticus
- var. *stenopetalus* Regel = T. ircuticus
bargusinensis Sipl. = T. ircuticus
boreosibiricus Tolm. = T. sibiricus
chartosepalus Schipcz. (*Hegemone chartosepala* (Schipcz.) A. Khokhr.) - 5
chinensis auct. = T. macropetalus
chinensis Bunge subsp. *macropetalus* (Regel) A. Luferov = T. macropetalus
dschungaricus Regel - 6
europaeus L. - 1, 3
- subsp. *transsilvanicus* (Schur) Jav. = T. altissimus
ilmenensis Sipl. - 3
ircuticus Sipl. (*T. asiaticus* L. var. *stenopetalus* Regel, *T. bargusinensis* Sipl., *T. stenopetalus* (Regel) Egor. & Sipl. 1969, non Stapf, 1928) - 4
japonicus auct. = T. miyabei
komarovii Pachom. (*Hegemone lilacina* (Bunge) Bunge var. *micrantha* C. Winkl. & Kom., *H. micrantha* (C. Winkl. & Kom.) Butk., *Trollius micranthus* (C. Winkl. & Kom.) Pachom. 1972, non Hand.-Mazz. 1931) - 6
kurilensis Sipl. - 5
kytmanovii Reverd. = T. asiaticus
ledebourii Reichenb. - 4, 5
lilacinus Bunge (*Hegemone lilacina* (Bunge) Bunge) - 3, 4, 6
macropetalus (Regel) Fr. Schmidt (*T. chinensis* Bunge subsp. *macropetalus* (Regel) A. Luferov, *T. chinensis* auct.) - 5
membranostylis Hult. (*T. schipczinskyi* Miyabe, p.p., nom. superfl.) - 5
micranthus (C. Winkl. & Kom.) Pachom. = T. komarovii
miyabei Sipl. (*T. schipczinskyi* Miyabe, p.p., nom. superfl., *T. japonicus* auct.) - 5
patulus Salisb. = T. ranunculinus
pulcher Makino (*T. riederianus* Fisch. & C.A. Mey. var. *pulcher* (Makino) T. Shimizu) - 5
ranunculinus (Smith) Stearn (*Helleborus ranunculinus* Smith, *Trollius patulus* Salisb.) - 2

riederianus Fisch. & C.A. Mey. - 5
- subsp. *uncinatus* (Sipl.) A. Luferov = T. uncinatus
- var. *pulcher* (Makino) T. Shimizu = T. pulcher
sajanensis (Malysch.) Sipl. (*T. altaicus* C.A. Mey. var. *sajanensis* Malysch.) - 4
schipczinskyi Miyabe = T. membranostylis and T. miyabei
sibiricus Schipcz. (*T. boreosibiricus* Tolm. nom. superfl.) - 4, 5
stenopetalus (Regel) Egor. & Sipl. = T. ircuticus
transsilvanicus Schur = T. altissimus
uncinatus Sipl. (*T. riederianus* Fisch. & C.A. Mey. subsp. *uncinatus* (Sipl.) A. Luferov) - 4, 5
uniflorus Sipl. - 5
vicarius Sipl. - 4

Viticella auct. = Clematis

orientalis (L.) W.A. Weber = Clematis orientalis

RESEDACEAE S.F. Gray

Homalodiscus Bunge = Ochradenus

ochradeni (Boiss.) Boiss. = Ochradenus ochradeni

Ochradenus Delile (*Homalodiscus* Bunge)

ochradeni (Boiss.) Abdallah (*Homalodiscus ochradeni* (Boiss.) Boiss.) - 6

Reseda L.

alba L. - 1
aucheri Boiss. - 6
brevipedunculata (N. Busch) N. Busch = R. globulosa
bucharica Litv. (*R. hemithamnodes* Czerniak.) - 6
buhseana Muell.-Arg. var. *dshebeli* (Czerniak.) Abdallah & de Wit = R. dshebeli
dshebeli Czerniak. (*R. buhseana* Muell.-Arg. var. *dshebeli* (Czerniak.) Abdallah & de Wit) - 6
globulosa Fisch. & C.A. Mey. (*R. brevipedunculata* (N. Busch) N. Busch) - 2
hemithamnodes Czerniak. = R. bucharica
inodora Reichenb. - 1
lutea L. - 1, 2, 3, 6
luteola L. - 1, 2, 6
microcarpa Muell.-Arg. - 2
*****odorata** L. - 1, 2, 3, 4, 5, 6
phyteuma L. - 2

RHAMNACEAE Juss.
Frangula Hill

alnus Mill. - 1, 2, 3, 4, 6
grandiflora (Fisch. & C.A. Mey.) Grub. - 2
rupestris (Scop.) Schur - 1

Oreoherzogia W. Vent = Rhamnus

depressa (Grub.) W. Vent = Rhamnus depressa
imeretina (Booth) W. Vent = Rhamnus imeretina
microcarpa (Boiss.) W. Vent = Rhamnus microcarpa

Paliurus Hill

spina-christi Mill. - 1, 2, 6

Rhamnus L. (*Oreoherzogia* W. Vent)

alaternus L. - 1

awarica Sachok. = R. tortuosa
baldschuanica Grub. - 6
cathartica L. - 1, 2, 3, 6
cordata Medw. - 2
coriacea (Regel) Kom. - 6
davurica Pall. - 4, 5
depressa Grub. (*Oreoherzogia depressa* (Grub.) W. Vent) - 2
diamantiaca Nakai - 5
dolichophylla Gontsch. - 6
erythroxylon Pall. - 4
imeretina Booth (*Oreoherzogia imeretina* (Booth) W. Vent) - 2
medwedewii Sachok. = R. pallasii
microcarpa Boiss. (*Oreoherzogia microcarpa* (Boiss.) W. Vent) - 2
minuta Grub. - 6
pallasii Fisch. & C.A. Mey. (*R. medwedewii* Sachok.) - 2
- subsp. *sintenisii* (Rech. fil.) Browicz & J. Zielinski = R. sintenisii
parvifolia Bunge - 4
saxatilis Jacq. subsp. *tinctoria* (Waldst. & Kit.) Nym. = R. tinctoria
sintenisii Rech. fil. (*R. pallasii* Fisch. & C.A. Mey. subsp. *sintenisii* (Rech. fil.) Browicz & J. Zielinski) - 6
songorica Gontsch. - 6
spathulifolia Fisch. & C.A. Mey. - 2
thea Osbeck = Sageretia thea
tinctoria Waldst. & Kit. (*R. saxatilis* Jacq. subsp. *tinctoria* (Waldst. & Kit.) Nym.) - 1
tortuosa Somm. & Levier (*R. awarica* Sachok.) - 2
ussuriensis Ja. Vassil. - 5

Sageretia Brongn.

laetevirens (Kom.) Gontsch. = S. thea
thea (Osbeck) M.C. Johnst. (*Rhamnus thea* Osbeck, *Sageretia laetevirens* (Kom.) Gontsch.) - 6

Ziziphus Mill.

jujuba Mill. - 2, 6

ROSACEAE Juss.
Acomastylis Greene (*Novosieversia* F. Bolle)

glacialis (Adams) A. Khokhr. (*Novosieversia glacialis* (Adams) F. Bolle) - 3, 4, 5
rossii (R. Br.) Greene - 5

Actaea auct. = Aruncus

dioica Walt. = Aruncus dioicus

Aflatunia Vass. = Louiseania

ulmifolia (Franch.) Vass. = Louiseania ulmifolia

Agrimonia L.

agrimonoides L. = Aremonia agrimonoides
aitchisonii Schonbeck-Temesy - 6
asiatica Juz. (*A. eupatoria* L. subsp. *asiatica* (Juz.) Skalicky, *A. eupatoria* subsp. *asiatica* (Juz.) Schonbeck-Temesy, comb. superfl.) - 1, 2, 3, 6
coreana Nakai (*A. pilosa* Ledeb. var. *coreana* (Nakai) Liou & Cheng ex Liou & C.Y. Li, *A. velutina* Juz.) - 5
eupatoria L. - 1, 2
- subsp. *asiatica* (Juz.) Schonbeck-Temesy = A. asiatica

- subsp. *asiatica* (Juz.) Skalicky = A. asiatica
- subsp. *grandis* (Andrz. ex C.A. Mey.) Bornm. = A. grandis
gorovoii Rumjantsev - 5
grandis Andrz. ex C.A. Mey. (*A. eupatoria* L. subsp. *grandis* (Andrz. ex C.A. Mey.) Bornm.) - 1
granulosa Juz. (*A. japonica* (Miq.) Koidz. var. *granulosa* (Juz.) Worosch.) - 5
japonica (Miq.) Koidz. = A. viscidula
- var. *granulosa* (Juz.) Worosch. = A. granulosa
nipponica Koidz. - 2(alien)
odorata auct. = A. procera
pilosa Ledeb. - 1, 3, 4, 5
- subsp. *japonica* (Miq.) Hara = A. viscidula
- var. *coreana* (Nakai) Liou & Cheng ex Liou & C.Y. Li = A. coreana
procera Wallr. (*A. odorata* auct.) - 1
velutina Juz. = A. coreana
viscidula Bunge (*A. japonica* (Miq.) Koidz., *A. pilosa* Ledeb. subsp. *japonica* (Miq.) Hara) - 5

Alchemilla L.

abchasica Bus. - 2
acropsila Rothm. - 2
acutangula Bus. = A. acutiloba
acutidens Bus. subsp. *oxyodonta* Bus. = A. oxyodonta
acutiloba Opiz (*A. acutangula* Bus., *A. vulgaris* L. subsp. *acutangula* (Bus.) Palitz) - 1, 3
- var. *hirsutiflora* Bus. = A. hirsutiflora
- var. *mollis* Bus. = A. mollis
adelodictya Juz. - 2
aemula Juz. - 1
aenostipula Juz. - 2
alba Frohner - 2(?)
alexandri Juz. (*A. diversipes* Juz. f. *alexandri* (Juz.) Grossh.) - 2
alpestris F.W. Schmidt - 1
alpestris sensu Juz. = A. glabra
alpina L. - 1
altaica Juz. - 3
- f. *cryptocaula* (Juz.) Serg. = A. cryptocaula
amphipsila Juz. - 1
anisopoda Juz. - 3(?), 4
aperta Juz. - 4
appressipila Juz. - 4
arcuatiloba Juz. - 1
arguteserrata Lindb. fil. ex Juz. - 3
atrifolia Zam. - 1, 3
aurata Juz. - 2
auriculata Juz. - 1
austroaltaica V. Tichomirov (*A. pavlovii* Rothm. 1939, non Juz. 1929) - 3, 6
babiogorensis Pawl. - 1
bakurianica Sosn. - 2
baltica Sam. ex Juz. (*A. nebulosa* Sam.) - 1, 3
barbatiflora Juz. - 2
barbulata Juz. - 3
betuletorum Rothm. - 2
biquadrata Juz. - 3
biradiata Ovcz. - 6
bombycina Rothm. - 2
borealis Sam. ex Juz. - 1
brevidens Juz. - 1
breviloba Lindb. fil. - 1
brevituba Juz. - 1
bungei Juz. - 3, 6
buschii Juz. - 1
calvifolia Juz. - 4

calviformis Ovcz. - 6
calvipes Juz. - 1
camptopoda Juz. - 1
capillacea Juz. - 2
cartalinica Juz. - 2
cartilaginea Rothm. - 2
caucasica Bus. - 2
cheirochlora Juz. - 1
chionophila Juz. - 6
chlorosericea (Bus.) Juz. - 2
cinerascens Juz. (*A. lindbergiana* Juz. f. *cinerascens* (Juz.) Serg.) - 3
circassica Juz. - 2
circularis Juz. - 3
circumdentata Juz. - 3
commixta Juz. - 3
compactilis Juz. - 2
confertula Juz. - 1
conglobata Lindb. fil. (*A. juzepczukii* Alech.) - 1
connivens Bus. - 1(?)
- var. *wichurae* Bus. = A. wichurae
consobrina Juz. - 1
crassicaulis Juz. - 1
crebridens Juz. - 1
crinita Bus. - 1
cryptocaula Juz. (*A. altaica* Juz. f. *cryptocaula* (Juz.) Serg.) - 3
cunctatrix Juz. - 1
curaica Juz. - 3
curvidens Juz. - 3
- f. *hemicycla* (Juz.) Serg. = A. hemicycla
cymatophylla Juz. - 1
cyrtopleura Juz. - 3, 6
czywczynensis Pawl. - 1
daghestanica Juz. - 2
dasyclada Juz. - 3
dasycrater Juz. - 1
debilis Juz. - 2
decalvans Juz. - 1
denticulata Juz. - 3
- f. *rubricaulis* (Juz.) Serg. = A. rubricaulis
depexa Juz. - 2
devestiens Juz. - 1
deylii Plocek - 1
diglossa Juz. - 3
divaricans Bus. - 2
diversipes Juz. - 2
- f. *alexandri* (Juz.) Grossh. = A. alexandri
dombaica Juz. - 2
dura Bus. - 2
dzhavakhetica Juz. - 2
egens Juz. = A. exilis
elata Bus. - 2
elisabethae Juz. - 2
epidasys Rothm. (*A. pseudomollis* Juz.) - 2
epipsila Juz. - 2
erectilis Juz. - 2
erythropoda Juz. - 2
excentrica Zam. - 1
exculpta Juz. - 1
exilis Juz. (*A. egens* Juz.) - 1
exsanguis Juz. - 1
exuens Juz. - 1
exul Juz. - 1
filicaulis Bus. (*A. vulgaris* L. subsp. *filicaulis* (Bus.) Murb.) - 1
firma Bus. (*A. pyrenaica* auct.) - 1

fissa Gunth. & Schummel - 1
- agsp. *incisa* (Bus.) A. & D. Löve = A. incisa
flabellata Bus. - 1
flavescens Bus. - 4
flavescens Snarskis = A. snarskisii
fokinii Juz. - 1
fontinalis Juz. - 6
frondosa Juz. - 2
georgica Juz. - 2
gibberulosa Lindb. fil. - 1
glabra Neyg. (*A. suecica* Frohner, *A. vulgaris* L. subsp. *glabra* (Neyg.) O. Bolos & Vigo, *A. alpestris* sensu Juz.) - 1, 3
glabricaulis Lindb. fil. (*A. glabricaulis* f. *parcipila* (Juz.) Serg., *A. parcipila* Juz.) - 1
- f. *parcipila* (Juz.) Serg. = A. glabricaulis
glabriformis Juz. - 1
glaucescens Wallr. (*A. hybrida* (L.) Mill. subsp. *glaucescens* (Wallr.) O. Bolos & Vigo, *A. minor* auct.) - 1
glomerulans Bus. (*A. vulgaris* L. subsp. *glomerulans* (Bus.) Murb.) - 1
glyphodonta Juz. - 1
goloskokovii Juz. - 6
gorcensis Pawl. - 1
gorodkovii Juz. - 3
gortschakowskii Juz. - 1
gracilis Opiz (*A. micans* Bus., *A. vulgaris* L. subsp. *micans* (Bus.) Palitz) - 1, 3, 5(alien)
grandidens Juz. - 2
grossheimii Juz. - 2
haraldii Juz. - 1
hebescens Juz. - 1, 3, 4, 5
helenae Juz. - 1
hemicycla Juz. (*A. curvidens* Juz. f. *hemicycla* (Juz.) Serg.) - 3
heptagona Juz. - 1
heteroschista Juz. - 2
hians Juz. - 3
hirsuticaulis Lindb. fil. - 1, 3, 5(alien)
hirsutiflora (Bus.) Rothm. (*A. acutiloba* Stev. var. *hirsutiflora* Bus.) - 2
hirsutissima Juz. - 1
hirtipedicellata Juz. - 2
hissarica Ovcz. & Koczk. - 6
holotricha Juz. = A. orthotricha
homoeophylla Jus. - 1
hoverlensis M. Pawlus & O. Lovelius - 1
humilicaulis Juz. - 6
hybrida (L.) Mill. subsp. *glaucescens* (Wallr.) O. Bolos & Vigo = A. glaucescens
- subsp. *plicata* (Bus.) Palitz = A. plicata
hyperborea Juz. - 1
hypochlora Juz. - 2
hypotricha Juz. - 2
hyrcana (Bus.) Juz. - 2
imberbis Juz. - 1
impolita Juz. - 2
incisa Bus. (*A. fissa* Gunth. & Schummel agsp. *incisa* (Bus.) A. & D. Löve) - 1
indurata Juz. - 2
insignis Juz. - 2
integribasis Juz. - 3
inversa Juz. - 4
iremelica Juz. - 1
isfarensis Ovcz. & Koczk. - 6
jailae Juz. - 1
japonica Nakai & Hara - 5

jaroschenkoi Grossh. - 6
juzepczukii Alech. = A. conglobata
kolaensis Juz. - 1
kozlowskii Juz. - 2
krylovii Juz. - 3, 6
- f. *mastodonta* (Juz.) Serg. = A. mastodonta
kvarkushensis Juz. - 1
laeta Juz. - 2
laeticolor Juz. - 2
languescens Juz. - 1
languida Bus. - 2
ledebourii Juz. - 3
leiophylla Juz. - 1
lessingiana Juz. - 1
lindbergiana Juz. - 1
- f. *cinerascens* (Juz.) Serg. = A. cinerascens
lipschitzii Juz. - 3(?), 6
- f. *purpurascens* (Juz.) Serg. = A. purpurascens
lithophila Juz. - 1
litwinowii Juz. - 1
longipes Juz. - 1
lydiae Zam. - 3
macrescens Juz. - 1
macroclada Juz. - 1
malimontana Juz. - 1
mastodonta Juz. (*A. krylovii* Juz. f. *mastodonta* (Juz.) Serg.) - 3
micans Bus. = A. gracilis
michelsonii Juz. - 6
microdictya Juz. - 2
microdonta Juz. - 2
minor auct. = A. glaucescens
minusculiflora Bus. - 2
mollis (Bus.) Rothm. (*A. acutiloba* Opiz var. *mollis* Bus.) - 1
monticola Opiz (*A. pastoralis* Bus., *A. vulgaris* L. subsp. *pastoralis* (Bus.) Palitz) - 1, 3, 5(alien)
multiflora Bus. ex Rothm. = A. tytthantha
murbeckiana Bus. (*A. vulgaris* L. subsp. *murbeckiana* (Bus.) A. Löve) - 1, 3, 6
nebulosa Sam. = A. baltica
nemoralis Alech. - 1
neo-stevenii Juz. = A. stevenii
obconiciflora Juz. - 4
obtegens Juz. - 2
obtusa Bus. (*A. samuelssonii* Rothm. ex Frohner) - 1, 3, 6
obtusiformis Alech. - 1
oligantha Juz. - 1
oligotricha Juz. - 2
omalophylla Juz. - 3
ophioreina Juz. - 3
orbicans Juz. - 3, 4
orthotricha Rothm. (*A. holotricha* Juz.) - 2
oxyodonta (Bus.) C.G. Westerl. (*A. acutidens* Bus. subsp. *oxyodonta* Bus., *A. vulgaris* L. subsp. *oxyodonta* (Bus.) A. & D. Löve) - 1
oxysepala Juz. = A. persica
pachyphylla Juz. - 3
paeneglabra Juz. - 1
parcipila Juz. = A. glabricaulis
pascualis Juz. - 2
pastoralis Bus. = A. monticola
pavlovii Rothm. = A. austroaltaica
persica Rothm. (*A. oxysepala* Juz.) - 2
phalacropoda Juz. - 2
phegophila Juz. - 1
pilosiplica Juz. - 3

pinguis Juz. - 3
plicata Bus. (*A. hybrida* (L.) Mill. subsp. *plicata* (Bus.) Palitz) - 1
pogonophora Juz. - 2
polemochora Frohner - 3(?), 4(?)
porrectidens Juz. - 2
prasina Juz. - 1
propinqua Lindb. fil. ex Juz. - 1
pseudocalycina Juz. - 1
pseudocartalinica Juz. - 2
pseudoincisa Pawl. - 1
pseudomollis Juz. = A. epidasys
psilocaula Juz. - 2
psilomischa Rothm. (*A. woronowii* Juz.) - 2
psiloneura Juz. - 1
purpurascens Juz. (*A. lipschitzii* Juz. f. *purpurascens* (Juz.) Serg.) - 3
pycnantha Juz. - 1
pycnoloba Juz. - 1
pycnotricha Juz. - 2
pyrenaica auct. = A. firma
raddeana (Bus.) Juz. - 2
reniformis Bus. - 1
retinerviformis Juz. - 2
retinervis Bus. - 2
retropilosa Juz. - 6
rhiphaea Juz. - 1
rigescens Juz. - 1, 3
rigida Bus. - 2
rubens Bus. - 3, 6
rubricaulis Juz. (*A. denticulata* Juz. f. *rubricaulis* (Juz.) Serg.) - 3
samuelssonii Rothm. ex Frohner = A. obtusa
sanguinolenta Juz. - 3
sarmatica Juz. - 1
sauri Juz. - 6
scalaris Juz. - 6
schischkinii Juz. - 3
schistophylla Juz. - 1
sedelmeyeriana Juz. - 2
semilunaris Alech. - 1
semispoliata Juz. - 1
sergii V. Tichomirov - 1
sericata Reichenb. ex Bus. - 2
sericea Willd. - 2
sevangensis Juz. - 2
sibirica Zam. - 3, 6
smirnovii Juz. - 2
smyfniensis Pawl. - 1
snarskisii Czer. (*A. flavescens* Snarskis, 1971, non Bus. 1894) - 1
speciosa Bus. - 2
stellaris Juz. - 1
stellulata Juz. - 2
stenantha Juz. - 1
stevenii auct. = A. vinacea
stevenii Bus. (*A. neo-stevenii* Juz.) - 1
stichotricha Juz. - 1
stricta Rothm. (*A. undecimloba* Juz.) - 2
strictissima Juz. - 1
subconnivens Pawl. - 1
subcrenata Bus. (*A. vulgaris* L. subsp. *subcrenata* (Bus.) Murb., *A. vulgaris* subsp. *subcrenata* (Bus.) Palitz) - 1, 3, 5(alien)
subcrenatiformis Juz. - 2
subcrispata Juz. - 1
suberectipila Juz. - 2

subglobosa C.G. Westerl. - 1
sublessingiana Juz. - 1
submamillata Juz. - 1
subsplendens Bus. - 2
substrigosa Juz. - 1
suecica Frohner = A. glabra
sukaczevii V. Tichomirov - 1
supina Juz. - 1
szaferi Pawl. - 1
tamarae Juz. - 2
taurica Juz. - 1, 2
tephroserica (Bus.) Juz. - 2
tianschanica Juz. - 6
transcaucasica Rothm. - 2
transiliensis Juz. - 6
transpolaris Juz. (*A. vulgaris* L. subsp. *transpolaris* (Juz.) A. & D. Löve) - 1
tredecimloba Bus. - 2
trichocrater Juz. - 1
tubulosa Juz. - 1
turkulensis Pawl. - 1
turuchanica Juz. - 4
tytthantha Juz. (*A. multiflora* Bus. ex Rothm.) - 1
undecimloba Juz. = A. stricta
uralensis Galanin - 1
urceolata Juz. - 2
valdehirsuta Bus. - 2
venosa Juz. - 2
ventiana V. Tichomirov - 1
verae Ovcz. & Koczk. - 6
veronicae Juz. - 1
vinacea Juz. (*A. stevenii* auct.) - 1
viridifolia Snarskis - 1
vulgaris L. = A. xanthochlora
- subsp. *acutangula* (Bus.) Palitz = A. acutiloba
- subsp. *filicaulis* (Bus.) Murb. = A. filicaulis
- subsp. *glabra* (Neyg.) O. Bolos & Vigo = A. glabra
- subsp. *glomerulans* (Bus.) Murb. = A. glomerulans
- subsp. *micans* (Bus.) Palitz = A. gracilis
- subsp. *murbeckiana* (Bus.) A. Löve = A. murbeckiana
- subsp. *oxyodonta* (Bus.) A. & D. Löve = A. oxyodonta
- subsp. *pastoralis* (Bus.) Palitz = A. monticola
- subsp. *subcrenata* (Bus.) Murb. = A. subcrenata
- subsp. *subcrenata* (Bus.) Palitz = A. subcrenata
- subsp. *transpolaris* (Juz.) A. & D. Löve = A. transpolaris
- subsp. *wichurae* (Bus.) Gams = A. wichurae
- subsp. *xanthochlora* (Rothm.) O. Bolos & Vigo = A. xanthochloa
walasii Pawl. - 1
wichurae (Bus.) Stefanss. (*A. connivens* Bus. var. *wichurae* Bus., *A. vulgaris* L. subsp. *wichurae* (Bus.) Gams) - 1
wischniewskii Rothm. - 2
woronowii Juz. = A. psilomischa
xanthochlora Rothm. (*A. vulgaris* L. nom. ambig., *A. vulgaris* subsp. *xanthochlora* (Rothm.) A. Bolos & Vigo) - 1
zapalowiczii Pawl. - 1

Amelanchier Medik.

***canadensis** (L.) Medik. - 1
- subsp. *spicata* (Lam.) A. & D. Löve = A. spicata
■ **integrifolia** Boiss. & Hohen. (*A. ovalis* Medik. subsp. *integrifolia* (Boiss. & Hohen.) Bornm., *A. rotundifolia* (Lam.) Dum.-Cours. subsp. *integrifolia* (Boiss. & Hohen.) Browicz)
ovalis Medik. (*A. rotundifolia* (Lam.) Dum.-Cours. nom. illegit., *Crataegus rotundifolia* Lam. nom. illegit.) - 1, 2
- subsp. **embergeri** Favarger & Stearn - 1, 2

- subsp. *integrifolia* (Boiss. & Hohen.) Bornm. = A. integrifolia
rotundifolia (Lam.) Dum.-Cours. = A. ovalis
- subsp. *integrifolia* (Boiss. & Hohen.) Browicz = A. integrifolia
***spicata** (Lam.) C. Koch (*A. canadensis* (L.) Medik. subsp. *spicata* (Lam.) A. & D. Löve) - 1
turkestanica Litv. - 6

Amygdalus L.

bruhuica - 6
bucharica Korsh. (*Prunus bucharica* (Korsh.) Hand.-Mazz.) - 6
communis L. (*A. dulcis* Mill., *Prunus dulcis* (Mill.) D.A. Webb) - 2, 6
dulcis Mill. = A. communis
fenzliana (Fritsch) Lipsky - 2
georgica Desf. - 2
x **kalmykovii** O. Lincz. - A. communis L. x A. spinosissima Bunge - 6
ledebouriana Schlecht. - 3, 6
nairica Fed. & Takht. - 2
nana L. - 1, 2, 3, 6
pedunculata Pall. - 4
petunnikowii Litv. - 6
pseudopersica Tamamsch. - 2
x **saviczii** Pachom. - A. bucharica Korsh. x A. spinisissima Bunge - 6
scoparia Spach - 6
spinosissima Bunge - 6
- subsp. *turcomanica* (Lincz.) Browicz = A. turcomanica
susakensis Vass. - 6
turcomanica Lincz. (*A. spinosissima* Bunge subsp. *turcomanica* (Lincz.) Browicz, *Prunus turcomanica* (Lincz.) Kitam. 1960, non Gilli, 1966) - 6
ulmifolia (Franch.) M. Pop. = Louiseania ulmifolia
urartu Tamamsch. - 2
x **uzbekistanica** Sabirov (*A. vavilovii* M. Pop f. *uzbekistanica* (Sabirov) R. Kam.). - A. bucharica Korsh. x A. communis L. - 6
x **vavilovii** M. Pop. (*Prunus* x *vavilovii* (M. Pop.) E. Murr.). - A. communis L. x A. turcomanica Lincz. - 6
- f. *uzbekistanica* (Sabirov) R. Kam. = A. x uzbekistanica
zangezura Fed. & Takht. - 2

Aphanes L.

arvensis L. - 1, 2

Aremonia Neck. ex Nestl.

agrimonoides (L.) DC. (*Agrimonia agrimonoides* L.) - 1

Argentina Lam. = Potentilla

anserina (L.) Rydb. subsp. *egedii* (Wormsk.) A. Löve & Ritchie = Potentilla egedii
- subsp. *groenlandica* (Tratt.) A. Löve = Potentilla egedii

Armeniaca Hill

dasycarpa (Ehrh.) Borkh. = x Armeniaco-prunus dasycarpa
davidiana Carr. (*Prunus davidiana* (Carr.) Franch., *P. sibirica* L. f. *davidiana* (Carr.) Kitag.) - 5
kostiniae E. Lomakin - 6
mandshurica (Maxim.) Skvorts. - 5
sibirica (L.) Lam. - 4, 5
sogdiana Kudr. - 6

vulgaris Lam. - 6, cult. in 1, 2, 5, 6
- subsp. **pubescens** Kudr. - 6

*x **Armeniaco-prunus** Cinovskis

*****dasycarpa** (Ehrh.) Cinovskis (*Armeniaca dasycarpa* (Ehrh.) Borkh.). - Armeniaca vulgaris Lam. x Prunus divaricata Ledeb. - 1, 2, 6

Aruncus Hill (*Actaea* auct.)

americanus auct. = A. dioicus
asiaticus Pojark. = A. dioicus
dioicus (Walt.) Fern. (*Actaea dioica* Walt., *Aruncus asiaticus* Pojark., *A. dioicus* var. *asiaticus* (Pojark.) Kitag., *A. dioicus* var. *kamtschaticus* (Maxim.) Hara, *A. kamtschaticus* (Maxim.) Rydb., *A. sylvester* Kostel. ex Maxim., *A. americanus* auct.) - 4, 5
- var. *asiaticus* (Pojark.) Kitag. = A. dioicus
- var. *kamtschaticus* (Maxim.) Hara = A. dioicus
kamtschaticus (Maxim.) Rydb. = A. dioicus
parvulus Kom. - 5
sylvester Kostel. ex Maxim. = A. dioicus
vulgaris Rafin. - 2

x **Cerasolouiseania** E. Lomakin & Yushev

cziczanica E. Lomakin & Yushev. - Cerasus verrucosus (Franch.) Nevski x Louiseania ulmifolia (Spach) Pachom. - 6

Cerasus Hill (*Microcerasus* (Spach) M. Roem.)

alaica Pojark. (*Microcerasus prostrata* (Labill.) M. Roem. f. *alaica* (Pojark.) Eremin & Yushev, *Prunus alaica* (Pojark.) Gilli) - 6
amygdaliflora Nevski (*C. verrucosa* (Franch.) Nevski f. *amygdaliflora* (Nevski) V. Zapr., *Microcerasus prostrata* (Labill.) M. Roem. f. *amygdaliflora* (Nevski) Eremin & Yushev) - 6
angustifolia (Spach) Browicz, p.p. = C. araxina
araxina Pojark. (*C. angustifolia* (Spach) Browicz, p.p., nom. illegit., *Microcerasus incana* (Pall.) M. Roem. var. *araxina* (Pojark.) Eremin & Yushev) - 2
*****austera** (L.) Borkh. - 1, 2, 6
avium (L.) Moench - 1, 2, 6(cult.)
bifrons auct. p.p. = C. erythrocarpa
blinovskii Totschilina (*Microcerasus incana* (Pall.) M. Roem. var. *blinovskii* (Totschilina) Eremin & Yushev) - 6
x **chodshaatensis** Pjat. & Lincz. - C. erythrocarpa Nevski x Prunus divaricata Ledeb. - 6
*****collina** Lej. & Court.
erythrocarpa Nevski (*C. kulabensis* Sosk. & Junussov, *Prunus erythrocarpa* (Nevski) Gilli, *Cerasus bifrons* auct. p.p.) - 6
fruticosa Pall. (*C. fruticosa* (Pall.) Borkh. comb. superfl., *C. fruticosa* (Pall.) Woronow, comb. superfl., *Prunus fruticosa* Pall. nom. altern.) - 1, 2, 3, 6
glandulifolia (Rupr. & Maxim.) Kom. = Padus maackii
glandulosa (Thunb.) Loisel. - 5
griseola Pachom. (*Microcerasus prostrata* (Labill.) M. Roem. f. *griseola* (Pachom.) Eremin & Yushev) - 6
incana (Pall.) Spach (*Microcerasus incana* (Pall.) M. Roem.) - 2
jacquemontii auct. = C. tadshikistanica
karabastaviensis Vass. = C. tianschanica and C. verrucosa
klokovii Sobko - 1
kulabensis Sosk. & Junussov = C. erythrocarpa

kurilensis (Miyabe) Czer. (*C. kurilensis* (Miyabe) Kaban. ex Vorobiev, comb. invalid., *Prunus kurilensis* (Miyabe) Miyabe, *P. nipponica* Matsum. f. *kurilensis* (Miyabe) Hiroe) - 5
mahaleb (L.) Mill. = Padellus mahaleb
maximowiczii (Rupr.) Kom. (*Padus maximowiczii* (Rupr.) Sokolov) - 5
microcarpa (C.A. Mey.) Boiss. (*Microcerasus microcarpa* (C.A. Mey.) Eremin & Yushev) - 2, 6
petrovae S. Korov. = Prunus x ferganica
pseudoprostrata Pojark. (*Microcerasus prostrata* (Labill.) M. Roem. f. *pseudoprostrata* (Pojark.) Eremin & Yushev, *Prunus pseudoprostrata* (Pojark.) Rech. fil.) - 6
sachalinensis (Fr. Schmidt) Kom. (? *Prunus sargentii* Rehd.) - 5
schuebeleri Orlova = Padus borealis
tadshikistanica Vass. (*C. verrucosa* (Franch.) Nevski f. *tadshikistanica* (Vass.) V. Zapr., *Microcerasus prostrata* (Labill.) M. Roem. f. *tadshikistanica* (Vass.) Eremin & Yushev, *Cerasus jacquemontii* auct.) - 6
tianschanica Pojark. (*C. karabastaviensis* Vass. p.p. quoad paratypum, *Microcerasus prostrata* (Labill.) M. Roem. var. *tianschanica* (Pojark.) Eremin & Yushev) - 6
*****tomentosa** (Thunb.) Wall. (*Prunus tomentosa* Thunb., *Microcerasus tomentosa* (Thunb.) Eremin & Yushev) - 1, 2, 5
turcomanica Pojark. (*Microcerasus prostrata* (Labill.) M. Roem. f. *turcomanica* (Pojark.) Eremin & Yushev, *Prunus turcomanica* (Pojark.) Gilli, 1966, non Kitam. 1960) - 6
verrucosa (Franch.) Nevski (*C. karabastaviensis* Vass. p.p. excl. paratypo, *Microcerasus prostrata* (Labill.) M. Roem. f. *karabastaviensis* (Vass.) Eremin & Yushev, *M. prostrata* var. *verrucosa* (Franch.) Eremin & Yushev) - 6
- f. *amygdaliflora* (Nevski) V. Zapr. = C. amygdaliflora
- f. *tadshikistanica* (Vass.) V. Zapr. = C. tadshikistanica
*****vulgaris** Mill. - 1, 2

Chamaerhodos Bunge

altaica (Laxm.) Bunge - 3, 4,6
baicalensis M. Pop. = Ch. grandiflora
erecta (L.) Bunge - 3, 4, 5
grandiflora (Pall. ex Schult.) Bunge (*Ch. baicalensis* M. Pop.) - 4
sabulosa Bunge - 3, 6
songarica Juz. - 6
trifida Ledeb. - 4

Coluria R. Br.

geoides (Pall.) Ledeb. - 3, 4

Comarum L.(*Farinopsis* Chrtek & Sojak)

palustre L. - 1, 2, 3, 4
salesovianum (Steph.) Aschers. & Graebn. (*Farinopsis salesoviana* (Steph.) Chrtek & Sojak) - 3, 6

Cotoneaster Medik.

alatavicus M. Pop. - 6
alaunicus Golits. - 1
allochrous Pojark. - 6
x **antoninae** Juz. ex Orlova - 1
antoninae A. Vassil. = C. neo-antoninae
armenus Pojark. - 2

cinnabarinus Juz. - 1
discolor Pojark. - 6
goloskokovii Pojark. - 6
hissaricus Pojark. (*C. racemiflorus* (Desf.) Booth ex Bosse var. *hissaricus* (Pojark.) Kitam.) - 6
ignavus E. Wolf - 6
insignis Pojark. (*C. lindleyi* auct.) - 6
integerrimus Medik. - 1, 2
karatavicus Pojark. - 6
krasnovii Pojark. (*C. racemiflorus* auct. p.p.) - 6
lindleyi auct. = C. insignis
lucidus Schlecht. - 4
x **matrensis** Domokos. - C. interrimus Medik. x C. melanocarpus Fisch. ex Blytt - 1
megalocarpus M. Pop. - 3, 6
melanocarpus Fisch. ex Blytt (*C. niger* (Wahlenb.) Fries, *Mespilus cotoneaster* L. var. *nigra* Wahlenb.) - 1, 2, 3, 4, 5, 6
meyeri Pojark. - 2
mongolicus Pojark. - 4
morulus Pojark. - 2
multiflorus Bunge - 2, 3, 6
nefedovii Galushko - 2
neo-antoninae A. Vassil. (*C. antoninae* A. Vassil. 1961, non Juz. ex Orlova, 1959) - 6
neo-popovii Czer. (*C. popovii* Peschkova, 1979, non Pojark. 1975) - 4
niger (Wahlenb.) Fries = C. melanocarpus
nummularioides Pojark. (*C. racemiflorus* auct. p.p.) - 6
nummularius Fisch. & C.A. Mey. (*C. racemiflorus* auct. p.p.) - 6
obovatus Pojark. = C. transcaucasicus
oliganthus Pojark. - 6
ovatus Pojark. - 6
pojarkovae Zak. - 6
polyanthemus E. Wolf - 6
popovii Peschkova = C. neo-popovii
popovii Pojark. - 6
pseudomultiflorus M. Pop. = C. soongoricus
racemiflorus auct. = C. krasnovii, C. nummularioides, C. nummularius and C. suavis
racemiflorus (Desf.) Booth ex Bosse subsp. *suavis* (Pojark.) Fed. = C. suavis
- var. *hissaricus* (Pojark.) Kitam. = C. hissaricus
- var. *suavis* (Pojark.) Kitam. = C. suavis
roborowskii Pojark. - 6
saxatilis Pojark. - 2
soczavianus Pojark. - 2
soongoricus (Regel & Herd.) M. Pop. (*C. pseudomultiflorus* M. Pop.) - 6
suavis Pojark. (*C. racemiflorus* (Desf.) Booth ex Bosse subsp. *suavis* (Pojark.) Fed., *C. racemiflorus* var. *suavis* (Pojark.) Kitam., *C. racemiflorus* auct. p.p.) - 2, 6
subacutus Pojark. - 6
submultiflorus M. Pop. - 6
talgaricus M. Pop. - 6
tauricus Pojark. - 1
tjuliniae Pojark. ex Peschkova - 4
transcaucasicus Pojark. (*C. obovatus* Pojark. 1954, non Wall. ex Dunn, 1921) - 2
turcomanicus Pojark. - 6
tytthocarpus Pojark. - 6
uniflorus Bunge - 1, 3, 4, 6
zeravschanicus Pojark. - 6

Crataegus L.

*alemanniensis Cinovskis - 1
almaatensis Pojark. (*C. dshungarica* Zab. 1897, non C. songarica C. Koch, 1854) - 6
altaica (Loud.) Lange = C. chlorocarpa and C. korolkowii
alutacea Klok. - 1
ambigua C.A. Mey. ex A. Beck. - 1
androssovii Essenova & Kerimova - 6
aria L. = Sorbus aria
- var. *scandica* L. = Sorbus intermedia
armena Pojark. - 2
aronia (L.) Bosc. ex DC. var. *pontica* (C. Koch) Zohary & Danin = C. pontica
atrofusca (C. Koch) Kassumova (*Mespilus atrofusca* C. Koch) - 1, 2
atrosanguinea Pojark. - 2
azarella Griseb. (*C. monogyna* Jacq. subsp. *azarella* (Griseb.) Franco) - 1
azarolus L. subsp. *aronia* (L.) H. Riedl, p.p. = C. pontica
beckeriana Pojark. = C. pallasii
x **berolinensis** Cinovskis - 1
calycina auct. = C. lindmanii
calycina Peterm. - 1
- subsp. *curvisepala* (Lindm.) Franco = C. curvisepala
caucasica C. Koch - 2
ceratocarpa Kossych - 1
chlorocarpa Lenne & C. Koch (*C. altaica* (Loud.) Lange, p.p. incl. typo, *C. sanguinea* Pall. var. *sanguinea* f. *chlorocarpa* (Lenne & C. Koch) Cinovskis, *C. wattiana* auct. p.p.) - 1, 3, 4, 6
chlorosarca Maxim. (*C. jozana* Schneid.) - 5
cinovskisii Kassymova - 2
*coccinea L. - 1
colchica Grossh. = C. pentagyna
*crus-galli L. - 1
x **curonica** Cinovskis. - C. laevigata (Poir.) DC. x ? C. subborealis Cinovskis - 1
curvisepala Lindm. (*C. calycina* Peterm. subsp. *curvisepala* (Lindm.) Franco, *C. monogyna* Jacq. subsp. *curvisepala* (Lindm.) Soo, *C. oxyacantha* L. p.p. incl. lectotypo, nom. ambig., *C. pseudokyrtostyla* Klok., *C. kyrtostyla* auct.) - 1, 2
- subsp. *lindmanii* (Hrabetova-Uhrova) Byatt = C. lindmanii
dahurica Koehne & Schneid. - 4, 5
darvasica Pojark. = C. necopinata
dipyrena Pojark. - 1
dshungarica Zab. = C. almaatensis
x **dunensis** Cinovskis. - C. curvisepala Lindm. x C. lindmanii Hrabetova-Uhrova - 1
dzhairensis Vass. - 6
eriantha Pojark. - 2
x **estonica** Cinovskis. - C. laevigata (Poir.) DC. x ? C. sp. - 1
fallacina Klok. - Probably, C. curvisepala Lindm. x C. leiomonogyna Klok. - 1
ferganensis Pojark. - 6
fischeri Schneid. = C. songarica
x **gracilis** Cinovskis. - C. curvisepala Lindm. x C. orientobaltica Cinivskis - 1
helenae Grynj & Klok. = C. helenolae
helenolae Grynj & Klok. (*C. helenae* Grynj & Klok. 1952, non Sarg.) - 1
heterodonta Pojark. (*C. monogyna* Jacq. var. *heterodonta* (Pojark.) Gostynska-Jakuszewska) - 1
hissarica Pojark. - 6
insularis Cinovskis = C. osiliensis

isfajramensis Pachom. - 6
jozana Schneid. = C. chlorosarca
karadaghensis Pojark. - 1
klokovii Ivaschin = C. pentagyna
knorringiana Pojark. - 6
korolkowii L. Henry (*C. altaica* (Loud.) Lange, p.p., *C. russanovii* Cinovskis, *C. wattiana* auct. p.p.) - 6
x **kupfferi** Cinovskis. - C. curvisepala Lindm. x C. palmstruchii Lindm. - 1
kyrtostyla auct. = C. curvisepala
laciniata auct. p.p. = C. orientalis
laciniata Ucria subsp. *pojarkovae* (Kossych) Franco = C. pojarkoviae
laevigata (Poir.) DC. (*Mespilus laevigata* Poir., *Crataegus oxyacantha* sensu Pojark.) - 1
x **latvica** Cinovskis. - C. curvisepala Lindm. x C. subborealis Cinovskis - 1
leiomonogyna Klok. (*C. monogyna* Jacq. subsp. *leiomonogyna* (Klok.) Franco) - 1
lindmanii Hrabetova-Uhrova (*C. curvisepala* Lindm. subsp. *lindmanii* (Hrabetova-Uhrova) Byatt, *C. rosaeformis* subsp. *lindmanii* (Hrabetova-Uhrova) K.I. Christensen, *C. calycina* auct.) - 1
lipskyi Klok. - 1
x *maritima* Cinovskis = C. x viidumaegica
maximowiczii Schneid. - 4, 5
melanocarpa Bieb. = C. pentagyna
meyeri Pojark. - 2
microphylla C. Koch - 1, 2
monogyna Jacq. - 1, 2
- subsp. *azarella* (Griseb.) Franco = C. azarella
- subsp. *curvisepala* (Lindm.) Soo = C. curvisepala
- subsp. *leiomonogyna* (Klok.) Franco = C. leiomonogyna
- var. *heterodonta* (Pojark.) Gostynska-Jakuszewska = C. heterodonta
monticola Cinovskis - 6
necopinata Pojark. (*C. darvasica* Pojark.) - 6
nikitinii Essenova - 6
orientalis Pall. ex Bieb. (*C. laciniata* auct. p.p.) - 1, 2
- subsp. *pojarkoviae* (Kossych) Byatt = C. pojarkoviae
orientobaltica Cinovskis - 1
osiliensis Cinovskis (*C. insularis* Cinovskis) - 1
oxyacantha L. = C. curvisepala
- subsp. *palmstruchii* (Lindm.) Hrabetova-Uhrova = C. palmstruchii
oxyacantha sensu Pojark. = C. laevigata
pallasii Griseb. (*C. beckeriana* Pojark.) - 1, 2
palmstruchii Lindm. (*C. oxyacantha* L. subsp. *palmstruchii* (Lindm.) Hrabetova-Uhrova) - 1
pamiroalaica V. Zapr. - 6
pentagyna Waldst. & Kit. (*C. colchica* Grossh., *C. klokovii* Ivaschin, *C. melanocarpa* Bieb.) - 1, 2
pinnatifida Bunge - 5
plagiosepala Pojark. - 1
pojarkoviae Kossych (*C. laciniata* Ucria subsp. *pojarkovae* (Kossych) Franco, *C. orientalis* Pall. ex Bieb. subsp. *pojarkoviae* (Kossych) Byatt) - 1
poloniensis Cinovskis - 1
pontica C. Koch (*C. azarolus* L. subsp. *aronia* (L.) H. Riedl, p.p. quoad pl. kopetdagh., *C. aronia* (L.) Bosc. ex DC. var. *pontica* (C. Koch) Zohary & Danin) - 2, 6
popovii Chrshan. - 1
praearmata Klok. - 1
pseudoambigua Pojark. - 6
pseudoazarolus M. Pop. - 6
pseudoheterophylla Pojark. - 2
pseudokyrtostyla Klok. = C. curvisepala

pseudomelanocarpa M. Pop. ex Pojark. - 6
x **pseudooxyacantha** Cinovskis. - C. curvisepala Lindm. x C. laevigata (Poir.) DC. - 1
pseudosanguinea M. Pop. ex Pojark. - 6
remotilobata Raik. ex M. Pop. - 6
rosaeformis subsp. *lindmanii* (Hrabetova-Uhrova) K.I. Christensen = C. lindmanii
rotundifolia Lam. = Amelanchier ovalis
rubens Cinovskis - 6
russanovii Cinovskis = C. korolkowii
sanguinea Pall, - 1, 3, 4, 6
- var. *sanguinea* f. *chlorocarpa* (Lenne & C. Koch) Cinovskis = C. chlorocarpa
schraderiana Ledeb. = C. tournefortii
songarica C. Koch (*C. fischeri* Schneid.) - 6
sphaenophylla Pojark. - 1
stankovii Kossych - 1
stevenii Pojark. - 1
subborealis Cinovskis - 1
subrotunda Klok. - 1
szovitsii Pojark. - 2
tanaitica Klok. - 1
taurica Pojark. - 1
theodorii Essenova - 6
tianschanica Pojark. - 6
tournefortii Griseb. (*C. schraderiana* Ledeb.) - 1, 2
transcaspica Pojark. - 6
trilobata V. Tkaczenko - 6
turcomanica Pojark. - 6
turkestanica Pojark. - 6
ucrainica Pojark. - 1
x **uhrovae** Soo. - C. curvisepala Lindm. x C. laevigata (Poir.) DC. - 1
x **viidumaegica** Cinovskis (*C. x maritima* Cinovskis). - C. osiliensis Cinovskis x C. palmstruchii Lindm. - 1
volgensis Pojark. - 1
wattiana auct. p.p. = C. chlorocarpa and C. korolkowii
zangezura Pojark. - 2

Cydonia Hill

oblonga Mill. - 2, 6
- subsp. **integerrima** Lobachev - 2

Dasiphora Rafin. = Pentaphylloides

davurica (Nestl.) Kom. = Pentaphylloides davurica
dryadanthoides Juz. = Pentaphylloides dryadanthoides
fruticosa (L.) Rydb. = Pentaphylloides fruticosa
mandshurica (Maxim.) Juz. = Pentaphylloides mandshurica
parvifolia (Fisch. ex Lehm.) Juz. = Pentaphylloides parvifolia
phyllocalyx Juz. = Pentaphylloides phyllocalyx

Dryadanthe Endl. = Sibbaldia

tetrandra (Bunge) Juz. = Sibbaldia tetrandra

Dryas L.

ajanensis Juz. (*D. octopetala* L. subsp. *ajanensis* (Juz.) Hult., *D. octopetala* subsp. *tschonoskii* (Juz.) Hult. var. *ajanensis* (Juz.) Hult.) - 4, 5
- subsp. **beringensis** Jurtz. - 5
- subsp. **ochotensis** Jurtz. (*D. incanescens* Juz. nom. invalid.) - 5
- subsp. *tschonoskii* (Juz.) Jurtz. = D. tschonoskii
alaskensis A. Pors. = D. punctata subsp. alaskensis

caucasica Juz. (*D. octopetala* L. subsp. *caucasica* (Juz.) Hult.) - 2

chamissonis Spreng. ex Jurtz. (*D. integrifolia* Vahl subsp. *chamissonis* (Spreng.) H.J. Scoggan, nom. invalid.) - 5

crenulata Juz. (*D. integrifolia* Vahl subsp. *crenulata* (Juz.) Ju. Kozhevn. comb. superfl., *D. integrifolia* subsp. *crenulata* (Juz.) H.J. Scoggan) - 4

x **grandiformis** Jurtz. - D. grandis Juz. x D. punctata Juz. - 5

grandis Juz. - 4, 5

x **henricae** Juz. (*D. octopetala* L. subsp. *punctata* (Juz.) Hult. var. *henricae* (Juz.) Ju. Kozhevn. p.p. quoad nomen, nom. invalid., *D. punctata* Juz. subsp. *punctata* var. *henricae* (Juz.) A. Pors. p.p. quoad nomen). - D. incisa Juz. var. cana Jurtz. x D. punctata Juz. - 4

incanescens Juz. = D. ajanensis subsp. ochotensis

incisa Juz. (*D. octopetala* L. subsp. *incisa* (Juz.) Malysch.) - 1, 3, 4, 5

integrifolia Vahl - 5

- subsp. *chamissonis* (Spreng.) H.J. Scoggan = D. chamissonis
- subsp. *crenulata* (Juz.) Ju. Kozhevn. = D. crenulata
- subsp. *crenulata* (Juz.) H.J. Scoggan = D. crenulata

kamtschatica Juz. = D. punctata

octopetala L. subsp. **subincisa** Jurtz. (*D. octopetala* sensu Juz.) - 1, 3, 4, 5

- subsp. *ajanensis* (Juz.) Hult. = D. ajanensis
- subsp. *alaskensis* (A.Pors.) Hult. = D. punctata subsp. alaskensis
- subsp. *caucasica* (Juz.) Hult. = D. caucasica
- subsp. *incisa* (Juz.) Malysch. = D. incisa
- subsp. *octopetala* var. *kamtschatica* (Juz.) Hult. = D. punctata
- - var. *oxyodonta* (Juz.) Hult. = D. oxyodonta
- - var. *viscida* Hult. = D. punctata
- subsp. *oxyodonta* (Juz.) Hult. = D. oxyodonta
- subsp. *punctata* (Juz.) Hult. = D. punctata
- - var. *henricae* (Juz.) Ju. Kozhevn. p.p. = D. x henricae
- subsp. *tschonoskii* (Juz.) Hult. = D. tschonoskii
- - var. *ajanensis* (Juz.) Hult. = D. ajanensis
- subsp. *viscida* (Hult.) Ju. Kozhevn. = D. punctata
- subsp. *viscosa* (Juz.) Hult. = D. viscosa

octopetala sensu Juz. = D. octopetala subsp. subincisa

oxyodonta Juz. (*D. octopetala* L. subsp. *octopetala* var. *oxyodonta* (Juz.) Hult., *D. octopetala* subsp. *oxyodonta* (Juz.) Hult.) - 3, 4

punctata Juz. (*D. kamtschatica* Juz., *D. octopetala* L. subsp. *octopetala* var. *kamtschatica* (Juz.) Hult., *D. octopetala* subsp. *octopetala* var. *viscida* Hult., *D. octopetala* subsp. *punctata* (Juz.) Hult., *D. octopetala* subsp. *viscida* (Hult.) Ju. Kozhevn., *D. punctata* var. *kamtschatica* (Juz.) Ju. Kozhevn.) - 1, 3, 4, 5

- subsp. **alaskensis** (A. Pors.) Jurtz. (*D. alaskensis* A. Pors., *D. octopetala* L. subsp. *alaskensis* (A. Pors.) Hult.) - 5
- subsp. *punctata* var. *henricae* (Juz.) A. Pors. p.p. = D. x henricae
- var. *kamtschatica* (Juz.) Ju. Kozhevn. = D. punctata

sumneviczii Serg. - 4

tschonoskii Juz. (*D. ajanensis* Juz. subsp. *tschonoskii* (Juz.) Jurtz. nom. invalid., *D. octopetala* L. subsp. *tschonoskii* (Juz.) Hult.) - 5

x **vagans** Juz. - D. octopetala L. x D. punctata Juz. - 1, 3, 4

viscosa Juz. (*D. octopetala* L. subsp. *viscosa* (Juz.) Hult.) - 4

Duchesnea Smith

indica (Andr.) Focke - 2(alien), 5(alien)

*Eriobotrya Lindl.

*japonica (Thunb.) Lindl. - 1, 2

Exochorda Lindl.

alberti Regel = E. korolkowii

korolkowii Lavall. (*E. alberti* Regel) - 6

serratifolia S. Moore - 5

tianschanica Gontsch. - 6

Farinopsis Chrtek & Sojak = Comarum

salesoviana (Steph.) Chrtek & Sojak = Comarum salesovianum

Filipendula Mill.

angustiloba (Turcz.) Maxim. - 4, 5

camtschatica (Pall.) Maxim. (*F. kamtschatica* auct.) - 5

- var. *glaberrima* Nakai = F. glaberrima

denudata (J. & C. Presl) Fritsch (*F. ulmaria* (L.) Maxim. subsp. *denudata* (J. & C. Presl) Hayek) - 1, 2

glaberrima Nakai (*F. camtschatica* (Pall.) Maxim. var. *glaberrima* Nakai, nom. nud., *F. glabra* Nakai ex Kom., *F. koreana* Nakai, nom. nud., *F. multijuga* Maxim. var. *koreana* Nakai, *F. multijuga* subsp. *koreana* (Nakai) Worosch. comb. invalid., *F. yezoensis* Hara, *F. purpurea* auct.) - 5

glabra Nakai ex Kom. = F. glaberrima

hexapetala Gilib. = F. vulgaris

intermedia (Glehn) Juz. - 4, 5

kamtschatica auct. = F. camtschatica

koreana Nakai = F. glaberrima

megalocarpa Juz. (*F. ulmaria* (L.) Maxim. var. *ulmaria* f. *megalocarpa* (Juz.) T. Shimizu) - 2

multijuga Maxim. subsp. *koreana* (Nakai) Worosch. = F. glaberrima

- var. *koreana* Nakai = F. glaberrima

nuda Grub. = F. palmata

palmata (Pall.) Maxim. (*F. nuda* Grub.) - 4, 5

purpurea auct. = F. glaberrima

stepposa Juz. (*F. ulmaria* (L.) Maxim. subsp. *picbaueri* (Podp.) Smejkal, *F. ulmaria* var. *picbaueri* Podp.) - 1, 3

ulmaria (L.) Maxim. - 1, 2, 3, 4, 5, 6

- subsp. *denudata* (J. & C. Presl) Hayek = F. denudata
- subsp. *picbaueri* (Podp.) Smejkal = F. stepposa
- var. *picbaueri* Podp. = F. stepposa
- var. *ulmaria* f. *megalocarpa* (Juz.) T. Shimizu = F. megalocarpa

vulgaris Moench (*F. hexapetala* Gilib.) - 1, 2, 3, 4

yezoensis Hara = F. glaberrima

Fragaria L.

ananassa (Duch.) auct. comb. ? = F. magna

bucharica Losinsk. - 6

- subsp. **darvasica** V. Zapr. - 6

campestris Stev. (*F. viridis* (Duch.) Weston subsp. *campestris* (Stev.) Pawl.) - 1, 2, 4

chiloensis (L.) Mill.

iinumae Makino - 5

iturupensis Staudt - 5

magna Thuill. (*F. ananassa* (Duch.) auct. comb. ?, *F. vesca* L. "race" *ananassa* Duch.)

moschata (Duch.) Weston - 1

nipponica Makino var. *yezoensis* (Hara) Kitam. = F. yezoensis

nubicola (Lindl. ex Hook. fil.) Lacaita (*F. vesca* L. var.

nubicola Lindl. ex Hook. fil.) - 6
orientalis Losinsk. - 4, 5
sterilis L. = Potentilla sterilis
vesca L. - 1, 2, 3, 4, 6
- var. *nubicola* Lindl. ex Hook. fil. = F. nubicola
- "race" *ananassa* Duch. = F. magna
*****virginiana** (Duch.) Mill.
viridis (Duch.) Weston - 1, 2, 3, 4, 6
- subsp. *campestris* (Stev.) Pawl. = F. campestris
yezoensis Hara (*F. nipponica* Makino var. *yezoensis* (Hara) Kitam.) - 5

Geum L.

aleppicum Jacq. (*G. aleppicum* f. *glabricaule* (Juz.) Kitag., *G. aleppicum* var. *glabricaule* (Juz.) Worosch., *G. aleppicum* var. *strictum* (Ait.) Fern., *G. aleppicum* subsp. *strictum* (Ait.) Clausen, *G. glabricaule* Juz., *G. strictum* Ait.) - 1, 2, 3, 4, 5, 6
- subsp. *strictum* (Ait.) Clausen = G. aleppicum
- var. *glabricaule* (Juz.) Worosch. = G. aleppicum
- var. *sachalinense* (Koidz.) Ohwi = G. fauriei
- var. *strictum* (Ait.) Fern. = G. aleppicum
- f. *glabricaule* (Juz.) Kitag. = G. aleppicum
▪ coccineum Sibth. & Smith
fauriei Levl. (*G. aleppicum* Jacq. var. *sachalinense* (Koidz.) Ohwi, *G. japonicum* Thunb. var. *fauriei* (Levl.) Kudo, *G. japonicum* var. *sachalinense* Koidz., *G. macrophyllum* Willd. var. *sachalinense* (Koidz.) Hara) - 5
glabricaule Juz. = G. aleppicum
x **intermedium** Ehrh. - G. rivale L. x G. urbanum L. - 1, 2, 3, 6
japonicum Thunb. var. *fauriei* (Levl.) Kudo = G. fauriei
- var. *sachalinense* Koidz. = G. fauriei
latilobum Somm. & Levier - 2
macrophyllum Willd. - 1, 5
- var. *sachalinense* (Koidz.) Hara = G. fauriei
x **meinshausenii** Gams. - G. alppicum Jacq. x G. rivale L. - 1
montanum L. = Oreogeum montanum
potaninii Juz. - 5
rivale L. - 1, 2, 3, 4, 6
- subsp. *urbanum* (L.) A. & D. Löve = G. urbanum
x **spurium** Fisch. & C.A. Mey. - G. aleppicum Jacq. x G. urbanum L. - 1
strictum Ait. = G. aleppicum
urbanum L. (*G. rivale* L. subsp. *urbanum* (L.) A. & D. Löve) - 1, 2, 3, 5, 6

Hulthemia Dumort.

berberifolia (Pall.) Dumort. - 3, 6
persica (Michx. ex Juss.) Bornm. - 6

x Hulthemosa Juz.

guzarica Juz. - 6
kopetdaghensis (Meff.) Juz. (x *H. turcomanica* Blin. nom. invalid.) - 6
turcomanica Blin. = x H. kopetdaghensis

*Kerria DC.

*****japonica** (L.) DC. - 2

Laurocerasus Hill

officinalis M. Roem. - 2

Louiseania Carr. (*Aflatunia* Vass.)

triloba (Lindl.) Pachom. (*Prunus triloba* Lindl., *Persica triloba* (Lindl.) Drob.) - 6
ulmifolia (Franch.) Pachom. (*Aflatunia ulmifolia* (Franch.) Vass., *Amygdalus ulmifolia* (Franch.) M. Pop.) - 6

Malus Hill

anisophylla Sumn. = M. sieversii
*****baccata** (L.) Borkh. (*M. chamardabanica* V. Vartapetjan & L. Solovjeva, *M. pallasiana* Juz.) - 4, 5
- subsp. *mandshurica* (Maxim.) Likhonos = M. mandshurica
- subsp. *sachalinensis* (Juz.) Likhonos = M. mandshurica
- subsp. *zhukovskyi* (Ponomarenko) Likhonos = M. mandshurica
chamardabanica V. Vartapetjan & L. Solovjeva = M. baccata
*****domestica** Borkh. (*M. sylvestris* Mill. subsp. *mitis* (Wallr.) Mansf., *Pyrus malus* L. var. *mitis* Wallr.)
- subsp. **caucasica** Likhonos
- subsp. **cerasifera** (Spach) Likhonos
- subsp. **hybrida** Likhonos
- subsp. **intermedia** Likhonos
- subsp. **italo-taurica** Likhonos
- subsp. **macrocarpa** Likhonos
- subsp. **medio-asiatica** Likhonos
- subsp. **occidentali-europaea** Likhonos
- subsp. *prunifolia* (Willd.) Likhonos = M. prunifolia
- subsp. **rossica** Likhonos
heterophylla Sumn. = M. sieversii
hissarica Kudr. = M. sieversii
jarmolenkoi Poljak. = M. sieversii
juzepczukii Vass. = M. sieversii
kirghisorum Al. Theod. & Fed. = M. sieversii
kudrjaschevii Sumn. = M. sieversii
linczevskii Poljak. = M. sieversii
mandshurica (Maxim.) Kom. (*M. baccata* (L.) Borkh. subsp. *mandshurica* (Maxim.) Likhonos, *M. baccata* subsp. *sachalinensis* (Juz.) Likhonos, *M. baccata* subsp. *zhukovskyi* (Ponomarenko) Likhonos, *M. mandshurica* subsp. *sachalinensis* (Juz.) Ponomarenko, *M. mandshurica* var. *sachalinensis* (Juz.) Ponomarenko, *M. mandshurica* subsp. *zhukovskyi* Ponomarenko, *M. mandshurica* var. *zhukovskyi* (Ponomarenko) Ponomarenko, *M. sachalinensis* Juz.) - 5
- subsp. *sachalinensis* (Juz.) Ponomarenko = M. mandshurica
- subsp. *zhukovskyi* Ponomarenko = M. mandshurica
- var. *sachalinensis* (Juz.) Ponomarenko = M. mandshurica
- var. *zhukovskyi* (Ponomarenko) Ponomarenko = M. mandshurica
montana Uglitzk. = M. orientalis
niedzwetzkyana Dieck - 6
orientalis Uglitzk. (*M. montana* Uglitzk., *M. orientalis* subsp. *montana* (Uglitzk.) Likhonos, *M. sylvestris* Mill. subsp. *orientalis* (Uglitzk.) Soo, *M. sylvestris* subsp. *orientalis* (Uglitzk.) Browicz, comb. superfl.) - 1(?), 2
- subsp. *montana* (Uglitzk.) Likhonos = M. orientalis
pallasiana Juz. = M. baccata
persicifolia (M. Pop.) Sumn. = M. sieversii
*****praecox** (Pall.) Borkh. (*M. sylvestris* Mill. var. *praecox* (Pall.) Ponomarenko, *M. sylvestris* subsp. *praecox* (Pall.) Soo) - 1
*****prunifolia** (Willd.) Borkh. (*M. domestica* Borkh. subsp. *prunifolia* (Willd.) Likhonos)
pumila Mill. var. *persicifolia* M. Pop. = M. sieversii

sachalinensis Juz. = M. mandshurica
schischkinii Poljak. = M. sieversii
sieboldii (Regel) Rehd. = M. toringo
sieversii (Ledeb.) M. Roem. (*M. anisophylla* Sumn., *M. heterophylla* Sumn. nom. invalid., *M. hissarica* Kudr., *M. jarmolenkoi* Poljak., *M. juzepczukii* Vass. nom. invalid., *M. kirghisorum* Al. Theod. & Fed., *M. kudrjaschevii* Sumn., *M. linczevskii* Poljak., *M. persicifolia* (M. Pop.) Sumn., *M. pumila* Mill. var. *persicifolia* M. Pop., *M. schischkinii* Poljak., *M. sieversii* subsp. *hissarica* (Kudr.) Likhonos, *M. sieversii* subsp. *kirghisorum* (Al. Theod. & Fed.) Likhonos, *M. sieversii* subsp. *turkmenorum* (Juz. & M. Pop.) Likhonos, *M. sylvestris* Mill. subsp. *sieversii* (Ledeb.) Soo, *M. tianschanica* Sumn., *M. turkmenorum* Juz. & M. Pop.) - 6

> Sometimes many species listed in the synonymy are accepted as varieties.

- subsp. *hissarica* (Kudr.) Likhonos = M. sieversii
- subsp. *kirghisorum* (Al. Theod. & Fed.) Likhonos = M. sieversii
- subsp. *turkmenorum* (Juz. & M. Pop.) Likhonos = M. sieversii
sylvestris Mill. - 1
- subsp. *mitis* (Wallr.) Mansf. = M. domestica
- subsp. *orientalis* (Uglitzk.) Browicz = M. orientalis
- subsp. *orientalis* (Uglitzk.) Soo = M. orientalis
- subsp. *praecox* (Pall.) Soo = M. praecox
- subsp. *sieversii* (Ledeb.) Soo = M. sieversii
- var. *praecox* (Pall.) Ponomarenko = M. praecox
tianschanica Sumn. = M. sieversii
*toringo Siebold (*M. sieboldii* (Regel) Rehd., *Pyrus sieboldii* Regel) - 5
turkmenorum Juz. & M. Pop. = M. sieversii

Mespilus L.

atrofusca C. Koch = Crataegus atrofusca
cotoneaster L. var. *nigra* Wahlenb. = Cotoneaster melanocarpus
germanica L. - 1, 2, 6
laevigata Poir. = Crataegus laevigata

Microcerasus (Spach) M. Roem. = Cerasus

incana (Pall.) M. Roem. = Cerasus incana
- var. *araxina* (Pojark.) Eremin & Yushev = Cerasus araxina
- var. *blinovskii* (Totschilina) Eremin & Yushev = Cerasus blinovskii
microcarpa (C.A. Mey.) Eremin & Yushev = Cerasus microcarpa
prostrata (Labill.) M. Roem. var. *tianschanica* (Pojark.) Eremin & Yushev = Cerasus tianschanica
- var. *verrucosa* (Franch.) Eremin & Yushev = Cerasus verrucosa
- f. *alaica* (Pojark.) Eremin & Yushev = Cerasus alaica
- f. *amygdaliflora* (Nevski) Eremin & Yushev = Cerasus amygdaliflora
- f. *griseola* (Pachom.) Eremin & Yushev = Cerasus griseola
- f. *karabastaviensis* (Vass.) Eremin & Yushev = Cerasus verrucosa
- f. *pseudoprostrata* (Pojark.) Eremin & Yushev = Cerasus pseudoprostrata
- f. *tadshikistanica* (Vass.) Eremin & Yushev = Cerasus tadshikistanica
- f. *turcomanica* (Pojark.) Eremin & Yushev = Cerasus turcomanica

tomentosa (Thunb.) Eremin & Yushev = Cerasus tomentosa

Micromeles Decne. = Sorbus

alnifolia (Siebold & Zucc.) Koehne = Sorbus alnifolia

Novosieversia F. Bolle = Acomastylis

glacialis (Adams) F. Bolle = Acomastylis glacialis

Oreogeum (Ser.) E. Golubk.

montanum (L.) E.Golubk. (*Geum montanum* L., *Parageum montanum* (L.) Nakai & Hara, *Sieversia montana* (L.) R. Br.) - 1

Orthurus Juz.

heterocarpus (Boiss.) Juz. - 6
kokanicus (Regel & Schmalh.) Juz. - 6

Padellus Vass.

mahaleb (L.) Vass. (*Cerasus mahaleb* (L.) Mill.) - 1, 2, 6

Padus Hill

asiatica Kom. (*P. avium* Mill. subsp. *pubescens* (Regel & Til.) Browicz, *P. avium* var. *pubescens* (Regel & Til.) Cinovskis, comb. superfl., *P. avium* var. *pubescens* (Regel & Til.) Polozh., *Prunus padus* L. var. *pubescens* Regel & Til., *P. padus* f. *pubescens* (Regel & Til.) Kitag.) - 4, 5
avium Mill. (*Padus racemosa* (Lam.) Gilib.) - 1, 2, 3, 6
- subsp. *borealis* (Schubel.) Holub = P. borealis
- subsp. *pubescens* (Regel & Til.) Browicz = P. asiatica
- var. *pubescens* (Regel & Til.) Cinovskis = P. asiatica
- var. *pubescens* (Regel & Til.) Polozh. = P. asiatica
borealis Schubel. (*Cerasus schuebeleri* Orlova, *Padus avium* Mill. subsp. *borealis* (Schubel.) Holub, *P. borealis* (Schubel.) Belozor, comb. superfl., *Padus schuebeleri* (Orlova) Czer., *Prunus padus* L. var. *borealis* (Schubel.) Blytt, *P. padus* subsp. *borealis* (Schubel.) Nym.) - 1
maackii (Rupr.) Kom. (*Cerasus glandulifolia* (Rupr. & Maxim.) Kom.) - 5
maximowiczii (Rupr.) Sokolov = Cerasus maximowiczii
racemosa (Lam.) Gilib. = P. avium
schuebeleri (Orlova) Czer. = P. borealis
ssiori (Fr. Schmidt) Schneid. - 5
*virginiana (L.) Mill.

Parageum Nakai & Hara

calthifolium (Menz.) Nakai & Hara - 5
montanum (L.) Nakai & Hara = Oreogeum montanum

Pentaphylloides Hill (*Dasiphora* Rafin.)

davurica (Nestl.) Ikonn. (*Dasiphora davurica* (Nestl.) Kom., *Pentaphylloides glabrata* (Willd. ex Schlecht.) O. Schwarz, *Potentilla davurica* Nestl, *P. fruticosa* L. subsp. *glabrata* (Willd. ex Schlecht.) Worosch., *P. glabrata* Willd. ex Schlecht.) - 4, 5
dryadanthoides (Juz.) Sojak (*Dasiphora dryadanthoides* Juz., *Pentaphylloides dryadanthoides* (Juz.) Ikonn. comb. superfl., *Potentilla phyllocalyx* (Juz.) Schiman-Czeika subsp. *dryadanthoides* (Juz.) Podlech) - 6
floribunda (Pursh) A. Löve = P. fruticosa

fruticosa (L.) O. Schwarz (*Dasiphora fruticosa* (L.) Rydb., *Pentaphylloides floribunda* (Pursh) A. Löve, *Potentilla floribunda* Pursh, *P. fruticosa* L. subsp. *floribunda* (Pursh) Elkington) - 1, 2, 3, 4, 5, 6
glabrata (Willd. ex Schlecht.) O. Schwarz = P. davurica
mandshurica (Maxim.) Sojak (*Dasiphora mandshurica* (Maxim.) Juz., *Pentaphylloides mandshurica* (Maxim.) Ikonn. comb. superfl., *Potentilla mandshurica* (Maxim.) Ingwersen, *P. mandshurica* (Maxim.) Worosch. comb. superfl.) - 5
parvifolia (Fisch. ex Lehm.) Sojak (*Dasiphora parvifolia* (Fisch. ex Lehm.) Juz., *Pentaphylloides parvifolia* (Fisch. ex Lehm.) Ikonn. comb. superfl.) - 3, 4, 5
phyllocalyx (Juz.) Sojak (*Dasiphora phyllocalyx* (Juz.), *Pentaphylloides phyllocalyx* (Juz.) Ikonn. comb. superfl., *Potentilla phyllocalyx* (Juz.) Schiman-Czeika) - 6

*Persica Hill

*ferganensis (Kostina & Rjab.) Koval. & Kostina - 6
triloba (Lindl.) Drob. = Louiseania triloba
*vulgaris Mill. (*Prunus persica* (L.) Batsch) - 1, 2, 6

Physocarpus (Cambess.) Maxim.

amurensis (Maxim.) Maxim. - 5
*opulifolius (L.) Maxim.
ribesifolia Kom. - 5

Potentilla L. (*Agrentina* Lam., *Schistophyllidium* (Juz. ex Fed.) Ikonn.)

acaulis L. - 3, 4, 5(?)
acervata Sojak (*P. nudicaulis* sensu Juz.) - 4
adenophylla Boiss. & Hohen. - 2
adenotricha Vodopjanova - 4
adscendens Waldst. & Kit. ex Willd. = P. canescens
adscharica Somm. & Levier - 2
aemulans Juz. = P. rugulosa
agrimonioides Bieb. (*P. stanjukoviczii* Ovcz. & Koczk.) - 2, 3, 6
- var. **capitata** Sojak - 6
- var. **intercedens** Sojak - 6
- var. **malacotricha** (Juz.) Sojak (*P. malacotricha* Juz.) - 6
- var. **micans** Sojak - 3
- var. **strigosella** Sojak - 2
agrimonoides sensu Bunge = P. conferta
alba L. - 1
alexeenkoi Lipsky - 2
algida Sojak (*P. sericata* Th. Wolf, 1908, non Greene, 1887) - 6
altaica Bunge - 3, 4
amurensis Maxim. = P. heynii
anachoretica Sojak (*P. czukczorum* Jurtz. & Petrovsky, nom. invalid.) - 4, 5
anadyrensis Juz. - 5
ancistrifolia Bunge var. *rugulosa* (Kitag.) Liou & C.Y. Li = P. rugulosa
andrzejowskii Blocki = P. thyrsiflora
x **angarensis** M. Pop. - P. impolita Wahlenb. x P. tergemina Sojak - 4
anglica Laicharding - 1
angustifolia DC. - 1
x **angustiloba** Yu & C.Y. Li. - P. multifida L. x P. virgata Lehm. - 3, 4, 6
anjuica Petrovsky - 5
anserina L. - 1, 2, 3, 4, 5, 6
- subsp. *egedii* (Wormsk.) Hiit. = P. egedii

- [subsp.] *groenlandica* Tratt. = P. egedii
- subsp. *pacifica* (Howell) Rousi = P. pacifica
- var. *grandis* Torr. & Gray = P. pacifica
approximata Bunge - 3, 4, 5, 6
arctica Rouy (*P. lapponica* (Nyl.) Juz.) - 1
arenaria Borkh. (*P. glaucescens* Schlecht., *P. cinerea* auct. p.p.) - 1, 2, 3
arenosa (Turcz.) Juz. = P. nivea
argaea Boiss. & Bal. (*P. bertramii* Aznav., *P. raddeana* (Th. Wolf) Juz.) - 2
argentea L. - 1, 2, 3, 4, 5
- subsp. *impolita* (Wahlenb.) Arcang. = P. impolita
- subsp. *impolita* (Wahlenb.) Hiit. = P. impolita
- subsp. *impolita* (Wahlenb.) O. Schwarz = P. impolita
argenteiformis Kauffm. - 1
arnavatensis (Th. Wolf) Juz. - 6
asiae-mediae Ovcz. & Koczk. = P. multifida
asiatica (Th. Wolf) Juz. - 3(?), 4, 6
asperrima Turcz. - 4, 5
astracanica Jacq. (*P. bornmuelleri* Borb., *P. mollicrinis* (Borb.) Stank.) - 1, 2
astragalifolia Bunge (*P. transtuvinica* K. Sobol.) - 3, 4
aurea L. - 1
basanensis Serg. = P. nivea
beringensis Jurtz. - 5
beringii Jurtz. (*P. emarginata* Pursh subsp. *nana* (Willd. ex Schlecht.) Hult., *P. nana* auct.) - 5
bertramii Aznav. = P. argaea
biflora Willd. ex Schlecht. - 3, 4, 5, 6
bifurca L. (*Schistophyllidium bifurcum* (L.) Ikonn.) - 1, 3, 4, 5
- subsp. *orientalis* (Juz.) Sojak = P. orientalis
- subsp. *semiglabra* (Juz.) Worosch. = P. semiglabra
bimundorum Sojak - 4, 5
x **borealis** Sojak (*P. rubricaulis* auct. p.p.). - P. anachoretica Sojak x P. nivea L. - 5
borissii Ovcz. & Koczk. (*P. gelida* C.A. Mey. subsp. *borissii* (Ovcz. & Koczk.) Sojak) - 6
bornmuelleri Borb. = P. astracanica
botschantzeviana Adylov - 6
brachypetala Fisch. & C.A. Mey. ex Lehm. - 2
breviscapa Vest = P. micrantha
bungei Boiss. - 2
butkovii Botsch. - 6
callieri (Th. Wolf) Juz. (*P. taurica* Willd. ex Schlecht. var. *genuina* Th. Wolf f. *callieri* (Th. Wolf) Gusul.) - 1, 2
camillae Kolak. - 2
canescens Bess. (*P. adscendens* Waldst. & Kit. ex Willd., *P. inclinata* auct. p.p.) - 1, 2, 3, 4, 5, 6
caucasica Juz. - 2
centigrana Maxim. - 5
chalchorum Sojak - 3, 4
chamissonis Hult. = P. kuznetzowii
chinensis Ser. - 4, 5
x **chionea** Sojak (*P. x malyschevii* Peschkova). - P. crebridens Juz. x P. sericea L. - 4
chrysantha Trev. (*P. holopetala* Turcz.) - 1, 3, 4, 6
x **chulensis** Siegfr. & R. Keller. - P. argentea L. x P. canescens Bess. - 2
cinerea auct. p.p. = P. arenaria
collina Wib. (*P. wibeliana* Th. Wolf) - 1, 5(alien)
- subsp. *leucopolitana* (P.J. Muell.) Dostal = P. leucopolitana
- subsp. *wiemanniana* (Gunth. & Schummel) Dostal = P. wiemanniana
conferta Bunge (*P. martjanovii* Polozh., *P. agrimonoides* sensu Bunge) - 1, 3, 4, 5, 6
crantzii (Crantz) G. Beck ex Fritsch - 1, 2, 3, 4, 6

- var. *ternata* (A. Blytt) Pesmen = P. gelida
crassa Tausch (*P. recta* L. subsp. *crassa* (Tausch) Jav.) - 1
crebridens Juz. = P. matsuokana subsp. crebridens
- subsp. *hemicryophila* Jurtz. = P. matsuokana subsp. hemicryophilla
cryptophila Bornm. - 2
cryptotaeniae Maxim. - 5
czegitunica Jurtz. - 5
czerepninii Krasnob. - 4
czukczorum Jurtz. & Petrovsky = P. anachoretica
darvasica Juz. ex Botsch. - 6
davurica Nestl. = Pentaphylloides davurica
dealbata Bunge = P. virgata
depressa Willd. ex Schlecht. - 1
desertorum Bunge - 3, 4, 6
x **dezhnevii** Jurtz. - P. anachoretica Sojak x P. subvahliana Jurtz. - 5
dickinsii Franch. & Savat. - 5
discolor Bunge - 5
divina Albov - 2
doubjonneana Cambess. subsp. **doubjonneana** - 6
- subsp. **ossetica** Sojak - 2
x *drymeja* Sojak = P. prostrata
egedii Wormsk. (*Argentina anserina* (L.) Rydb. subsp. *egedii* (Wormsk.) A. Löve & Ritchie, *A. anserina* subsp. *groenlandica* (Tratt.) A. Löve, *Potentilla anserina* L. subsp. *egedii* (Wormsk.) Hitt., *P. anserina* [subsp.] *groenlandica* Tratt., *P. egedii* var. *groenlandica* (Tratt.) Polun.) - 1, 3, 4, 5
- subsp. *grandis* (Torr. & Gray) Hult. = P. pacifica
- subsp. *pacifica* (Howell) L. Sergienko = P. pacifica
- var. *grandis* (Torr. & Gray) Hara = P. pacifica
- var. *groenlandica* (Tratt.) Polun. = P. egedii
elatior Willd. ex Schlecht. - 2
elegans Cham. & Schlecht. - 4, 5
elegantissima Polozh. - 4
emarginata Pursh = P. hyparctica
- subsp. *nana* (Willd. ex Schlecht.) Hult. = P. beringii
erecta (L.) Raeusch. - 1, 2, 3
eremica Th. Wolf = P. turgaica
eversmanniana Fisch. ex Ledeb. - 1
evestita Th. Wolf (*P. regeliana* Th. Wolf, *P. reverdattoi* Polozh.) - 3, 4, 6
exigua auct. = P. multifida
exuta Sojak - 4, 6
fallacina Blocki (*P. obscura* Willd. var. *fallacina* (Blocki) Markova, *P. recta* L. subsp. *fallacina* (Blocki) Soo) - 1
fedtschenkoana Siegfr. ex Th.Wolf - 6
ferganensis Sojak - 6
filipendula Willd. ex Schlecht. = P. tanacetifolia
flabellata Regel & Schmalh. - 6
flabellifolia Hook. var. *emarginata* (Pursh) Boivin = P. hyparctica
flagellaris Willd. ex Schlecht. - 3, 4, 5
floribunda Pursh = Pentaphylloides fruticosa
foliosa Somm. & Levier ex R. Keller - 2
fragarioides L. - 3, 4, 5
- subsp. *togashii* (Ohwi) Ohwi & Satomi = P. togashii
fragiformis Willd. ex Schlecht. - 5
- subsp. *megalantha* (Takeda) Hult. = P. megalantha
freyniana Bornm. - 5
fruticosa L. subsp. *floribunda* (Pursh) Elkington = Pentaphylloides fruticosa
- subsp. *glabrata* (Willd. ex Schlecht.) Worosch. = Pentaphylloides davurica
gelida C.A. Mey. (*P. crantzii* (Crantz) G. Beck ex Fritsch var. *ternata* (A. Blytt) Pesmen, *P. hyparctica* Malte

subsp. *gelida* (C.A. Mey.) Worosch., *P. verna* L. var. *ternata* A. Blytt) - 1, 2, 3, 4, 5, 6
- subsp. **boreo-asiatica** Jurtz. & R. Kam. - 1, 3, 4, 5, 6
- subsp. **boreo-jacutica** Jurtz. - 4
- subsp. *borissii* (Ovcz. & Koczk.) Sojak = P. borissii
- subsp. **gelida** - 2
geoides Bieb. - 1, 2
ghalghana Juz. = P. oweriniana
glabrata Willd. ex Schlecht. = Pentaphylloides davurica
glaucescens Schlecht. = P. arenaria
goldbachii Rupr. (*P. thuringiaca* auct. p.p.) - 1, 3, 4
gordiaginii Juz. - 3
x **gorodkovii** Jurtz. (*P. nivea* L. subsp. *fallax* A. Pors.) - 4, 5
gracillima R. Kam. - 4
grisea Juz. - 6
heidenreichii Zimmeter - 1
heptaphylla L. - 1
heynii Roth (*P. amurensis* Maxim.) - 5
hirta L. subsp. *pedata* (Willd.) Prod. = P. pedata
hololeuca Boiss. ex Lehm. - 6
holopetala Turcz. = P. chrysantha
hookeriana auct. = P. nivea
hookeriana Lehm. subsp. *chamissonis* (Hult.) Hult. = P. kuznetzowii
humifusa Willd. ex Schlecht. - 1, 2, 3, 4, 6
hyparctica Malte (*P. emarginata* Pursh, 1814, non Desf. 1804, *P. flabellifolia* Hook. var. *emarginata* (Pursh) Boivin, *P. robbinsiana* Oakes subsp. *hyparctica* (Malte) A. & D. Löve) - 1, 3, 4, 5
- subsp. *gelida* (C.A. Mey.) Worosch. = P. gelida
- subsp. **nivicola** Jurtz. & Petrovsky - 4, 5
imbricata Kar. & Kir. (*P. porphyrii* K. Sobol.) - 6
imerethica Gagnidze & Sochadze - 2
impolita Wahlenb. (*P. argentea* L. subsp. *impolita* (Wahlenb.) Arcang., *P. argentea* subsp. *impolita* (Wahlenb.) Hiit. comb. superfl., *P. argentea* subsp. *impolita* (Wahlenb.) O. Schwarz, comb. superfl., *P. neglecta* auct. p.p.) - 1, 2, 3, 4, 6
- subsp. **magyarica** (Borb.) Soo - 1(?)
inclinata auct. p.p. = P. canescens
inquinans Turcz. - 3, 4, 5
intermedia L. - 1, 5(alien)
x **ivanoviae** Peschkova. - P. prostrata Rottb. subsp. floccosa Sojak x P. sericea L. - 4
jacutica Juz. - 1, 4, 5
jailae Juz. - 1
japonica auct. = P. stolonifera
jenissejensis Polozh. & W. Smirn. - 4
karatavica Juz. - 6
kasachstanica R. Kam. - 6
kemulariae Kapell. & A. Kuth. - 2
kleiniana Wight & Arn. - 5
komaroviana Th. Wolf = P. mollissima
kryloviana Th. Wolf - 3, 4
kulabensis Th. Wolf - 6
kuznetzowii (Govor.) Juz. (*P. nivea* L. var. *kuznetzowii* Govor., *P. chamissonis* Hult., *P. hookeriana* Lehm. subsp. *chamissonis* (Hult.) Hult., *P. nivea* var. *chamissonis* (Hult.) Lid, nom. invalid., *P. nivea* subsp. *hookeriana* (Lehm.) Hiit. var. *kuznetzowii* (Govor.) Hiit., *P. prostrata* Rottb. subsp. *chamissonis* (Hult.) Sojak) - 1, 3
laeta Reichenb. - 1
lapponica (Nyl.) Juz. = P. arctica
lazica Boiss. & Bal. - 2
ledebouriana A. Pors. = P. uniflora
leucophylla Pall. - 4, 5

leucopolitana P.J. Muell. (*P. collina* Wib. subsp. *leucopolitana* (P.J. Muell.) Dostal) - 1
leucotricha (Borb.) Borb. (*P. recta* L. subsp. *leucotricha* (Borb.) Jav., *P. recta* subsp. *leucotricha* (Borb.) Soo, comb. superfl.) - 1
lignosa Willd. ex Schlecht. = Tylosperma lignosa
lipskyana Th. Wolf = P. mollissima
lomakinii Grossh. - 2
longifolia Willd. ex Schlecht. (*P. tanacetifolia* Willd. ex Schlecht. var. *longifolia* (Willd. ex Schlecht.) Worosch., *P. viscosa* Donn ex Lehm.) - 1, 3, 4, 5, 6
longipes Ledeb. - 1, 3, 6
lydiae Kurbatsky - 3, 4
lyngei Jurtz. & Sojak - 1
macrantha Ledeb. - 4
macropoda Sojak - 3
malacotricha Juz. = P. agrimonioides var. malacotricha
x *malyschevii* Peschkova = P. x chionea
mandshurica (Maxim.) Ingwersen = Pentaphylloides mandshurica
mandshurica (Maxim.) Worosch. = Pentaphylloides mandshurica
martjanovii Polozh. = P. conferta
matsumurae Th. Wolf - 5
matsuokana Makino - 4, 5
- subsp. **crebridens** (Juz.) Sojak (*P. crebridens* Juz.) - 4
- subsp. **hemicryophilla** (Jurtz.) Sojak (*P. crebridens* subsp. *hemicryophila* Jurtz.) - 4, 5
megalantha Takeda (*P. fragiformis* Willd. ex Schlecht. subsp. *megalantha* (Takeda) Hult.) - 5
meyeri Boiss. - 2
micrantha Ramond ex DC. (*P. breviscapa* Vest) - 1, 2
mischkinii Juz. = P. prostrata
miyabei Makino - 5
mollicrinis (Borb.) Stank. = T. astracanica
mollissima Lehm. (*P. komaroviana* Th. Wolf, *P. lipskyana* Th. Wolf) - 6
moorcroftii Wall. ex Lehm. (*Schistophyllidium moorcroftii* (Wall. ex Lehm.) Ikonn.) - 6
mujensis Kurbatsky - 4
multifida L. (*P. asiae-mediae* Ovcz. & Koczk., *P. exigua* auct.) - 1, 2, 3, 4, 5, 6
- f. *ornithopoda* (Tausch) Kitag. = P. ornithopoda
multifidiformis Ikonn. - 6
nana auct. = P. beringii
neglecta auct. p.p. = P. imerethica
nervosa Juz. - 6
neumanniana Reichenb. (*P. tabernaemontani* Aschers.) - 1
nivea L. (*P. arenosa* (Turcz.) Juz., *P. basanensis* Serg., *P. nivea* subsp. *hookeriana* (Lehm.) Hiit. var. *arenosa* (Turcz.) Hiit., *P. hookeriana* auct.) - 1, 3, 4, 5
- subsp. *chamissonis* (Hult.) Hiit. = P. kuznetzowii
- subsp. *fallax* A. Pors. = P. x gorodkovii
- subsp. *hookeriana* (Lehm.) Hiit. var. *arenosa* (Turcz.) Hiit. = P. nivea
- - var. *kuznetzowii* (Govor.) Hiit. = P. kuznetzowii
- subsp. *mischkinii* (Juz.) Jurtz. = P. prostrata
- subsp. *subquinata* (Lange) Hult. p.p. = P. subquinata
- var. *chamissonis* (Hult.) Lid = P. kuznetzowii
- var. *kuznetzowii* Govor. = P. kuznetzowii
- var. *subquinata* Lange = P. subquinata
nivea sensu Juz. = P. prostrata subsp. floccosa
nordmanniana Ledeb. - 2
norvegica L. (*P. varians* Moench) - 1, 2(alien), 3, 4, 5
nudicaulis sensu Juz. = P. acervata
nudicaulis Willd. ex Schlecht. (*P. strigosa* Pall. ex Tratt., *P. tundrarum* Juz. ex Jurtz. nom. invalid.) - 4, 5

nurensis Boiss. & Hausskn. - 2
obscura Willd. (*P. recta* L. subsp. *obscura* (Willd.) Reichenb. ex Rothm.) - 1, 2
- var. *fallacina* (Blocki) Markova = P. fallacina
x **okensis** Petunn. - P. arenaria Borkh. x P. goldbachii Rupr. - 1
x **olchonensis** Peschkova - 4
        It is likely P. nivea L. (P. arenosa (Turcz.) Juz.) x P. sericea L.
omissa Sojak - 3, 4, 5
orbiculata Th. Wolf - 2
orientalis Juz. (*P. bifurca* L. subsp. *orientalis* (Juz.) Sojak, *Schistophyllidium orientale* (Juz.) Ikonn.) - 1, 2, 6
ornithopoda Tausch (*P. multifida* L. f. *ornithopoda* (Tausch) Kitag.) - 3, 4, 6
oweriniana Boiss. (*P. ghalghana* Juz.) - 2
ozjorensis Peschkova - 3
pacifica Howell (*P. anserina* L. var. *grandis* Torr. & Gray, *P. anserina* subsp. *pacifica* (Howell) Rousi, *P. egedii* Wormsk. subsp. *grandis* (Torr. & Gray) Hult., *P. egedii* var. *grandis* (Torr. & Gray) Hara, *P. egedii* subsp. *pacifica* (Howell) L. Sergienko) - 5
palczewskii Juz. = P. sprengeliana
pamirica Th. Wolf - 6
pamiroalaica Juz. - 6
paradoxa Nutt. ex Torr. & Gray = P. supina subsp. paradoxa
patula Waldst. & Kit. (*P. schurii* Fuss ex Zimmeter) - 1
pedata Willd. ex Hornem. (*P. hirta* L. subsp. *pedata* (Willd.) Prod., *P. transcaspia* Th. Wolf) - 1, 6
penniphylla Sojak - 6
pensylvanica L. (*P. strigosa* sensu Juz.) - 1, 3, 4, 5(alien), 6
x **petrovskyi** Sojak (*P. rubricaulis* auct. p.p.). - P. anachoretica Sojak x P. prostrata Rottb. subsp. floccosa Sojak (*P. nivea* sensu Juz.) - 5
phyllocalyx (Juz.) Schiman-Czeika = Pentaphylloides phyllocalyx
- subsp. *dryadanthoides* (Juz.) Podlech = Pentaphylloides dryadanthoides
pilosa Willd. (*P. recta* L. subsp. *pilosa* (Willd.) Jav.) - 1
pimpinelloides L. (*P. tanaitica* Zing.) - 1, 2
pindicola (Nym.) Hausskn. - 1
polyschista Boiss. - 2
porphyrantha Juz. - 2
porphyrii K. Sobol. = P. imbricata
prostrata Rottb. (*P. x drymeja* Sojak, *P. mischkinii* Juz., *P. nivea* L. subsp. *mischkinii* (Juz.) Jurtz., *P. x tomentulosa* Jurtz.) - 1, 2, 3, 4, 5, 6
- subsp. *chamissonis* (Hult.) Sojak = P. kuznetzowii
- subsp. **ciscaspica** Sojak - 2
- subsp. **floccosa** Sojak (*P. nivea* sensu Juz.) - 1, 3, 4, 5, 6
x **protea** Sojak. - P. crantzii (Crantz) G. Beck ex Fritsch x P. hyparctica Malte - 1
pulchella R. Br. - 1, 4, 5
- subsp. **gracilicaulis** (A. Pors.) Jurtz. (*P. pulchella* var. *gracilicaulis* A. Pors.) - 5
pulviniformis A. Khokhr. - 4, 5
x **pusilla** Host. (*P. arenaria* Borkh. x P. heptaphylla L. - 1
raddeana (Th. Wolf) Juz. = P. argaea
recta L. (*P. sulphurea* Lam.) - 1, 2, 3
- subsp. *crassa* (Tausch) Jav. = P. crassa
- subsp. *fallacina* (Blocki) Soo = P. fallacina
- subsp. *leucotricha* (Borb.) Jav. = P. leucotricha
- subsp. *leucotricha* (Borb.) Soo = P. leucotricha
- subsp. *obscura* (Willd.) Reichenb. ex Rothm. = P. obscura
- subsp. *pilosa* (Willd.) Jav. = P. pilosa
- subsp. *semilaciniosa* (Borb.) Jav. = P. semilaciniosa
regeliana Th. Wolf = P. evestita

reptans L. - 1, 2, 3, 5(alien), 6
reverdattoi Polozh. = P. evestita
rigidula Th. Wolf - 3
robbinsiana Oakes subsp. *hyparctica* (Malte) A. & D. Löve = P. hyparctica
rubella Sorens. - 4, 5
- var. **rubelloides** (Petrovsky) Jurtz. (*P. rubelloides* Petrovsky) - 4, 5
rubelloides Petrovsky = P. rubella var. rubelloides
rubricaulis auct. p.p. = P. x borealis, P. x petrovskyi and P. x tschaunensis
rugulosa Kitag. (*P. aemulans* Juz., *P. ancistrifolia* Bunge var. *rugulosa* (Kitag.) Liou & C.Y. Li, *P. tranzschelii* Juz., *P. ussuriensis* Juz.) - 5
rupestris L. - 1
rupifraga A. Khokhr. - 5
ruprechtii Boiss. - 2
sachalinensis Juz. = P. sprengeliana
x **safronoviae** Jurtz.& Sojak. - P. hyparctica Malte x P. pulchella R. Br. - 5
sajanensis Polozh. - 4
sanguisorba Willd. ex Schlecht. - 4
saposhnikovii Kurbatsky - 3, 4
schrenkiana Regel - 6
schugnanica Juz. ex Adylov - 6
schurii Fuss ex Zimmeter = P. patula
seidlitziana Bien. - 2
semiglabra Juz. (*P. bifurca* L. subsp. *semiglabra* (Juz.) Worosch.) - 4, 5
semilaciniosa Borb. (*P. recta* L. subsp. *semilaciniosa* (Borb.) Jav.) - 1, 2
sergievskajae Peschkova - 4
sericata Th. Wolf = P. algida
sericea L. - 1, 3, 4, 6
silesiaca Uechtr. - 1
sommieri Siegfr. & R. Keller - 2
soongarica Bunge - 3, 4, 6
sosnowskyi Kapell. - 2
sphenophylla Th. Wolf - 2
sprengeliana Lehm. (*P. palczewskii* Juz., *P. sachalinensis* Juz.) - 5
stanjukoviczii Ovcz. & Koczk. = P. agrimonioides
sterilis (L.) Garcke (*Fragaria sterilis* L.) - 2
stipularis L. - 1, 3, 4, 5
stolonifera Lehm. ex Ledeb. (*P. japonica* auct.) - 5
straussii Bornm. - 2(?)
strigosa Pall. ex Tratt. = P. nudicaulis
strigosa sensu Juz. = P. pensylvanica
x *subarenaria* Borb. ex Zimmeter = P. x subcinerea
x **subcinerea** Borb. (*P. x subarenaria* Borb. ex Zimmeter). - P. arenaria Borkh. x P. neumanniana Reichenb. - 1
subpalmata Ledeb. - 2
subquinata (Lange) Rydb. (*P. nivea* L. var. *subquinata* Lange, *P. nivea* subsp. *subquinata* (Lange) Hult. p.p.) - 1
x **subtrijuga** (Th. Wolf) Juz. - P. hololeuca Boiss. ex Lehm. x P. pamiroalaica Juz. - 6
subvahliana Jurtz. (*P. vahliana* Lehm. subsp. *subvahliana* Jurtz. nom. invalid., *P. vahliana* auct.) - 4, 5
sulphurea Lam. = P. recta
supina L. - 1, 2(?), 3, 4, 5, 6
- subsp. **costata** Sojak - 1, 2(?), 3, 4, 5
- subsp. **paradoxa** (Nutt. ex Torr. & Gray) Sojak (*P. paradoxa* Nutt. ex Torr. & Gray) - 3, 4, 6
- subsp. **supina** - 1
svanetica Siegfr. & R. Keller - 2
szovitsii Th. Wolf - 2

tabernaemontani Aschers. = P. neumanniana
tanacetifolia Willd. ex Schlecht. (*P. filipendula* Willd. ex Schlecht.) - 3, 4, 5
- var. *longifolia* (Willd. ex Schlecht.) Worosch. = P. longifolia
tanaitica Zing. = P. pimpinelloides
taurica Willd. ex Schlecht. - 1, 2
- var. *genuina* Th. Wolf f. *callieri* (Th. Wolf) Gusul. = P. callieri
tephroleuca (Th. Wolf) B. Fedtsch. - 6
tephroserica Juz. - 6
tergemina Sojak - 1(alien), 3, 4, 5(alien)
tericholicea K. Sobol. - 4
thuringiaca auct. p.p. = P. goldbachii
thyrsiflora Huels. ex Zimmeter (*P. andrzejowskii* Blocki, *P. thyrsiflora* subsp. *leucopolitanoides* (Blocki) Borhidi & Isepy) - 1
- subsp. *leucopolitanoides* (Blocki) Borhidi & Isepy = P. thyrsiflora
tianschanica Th. Wolf - 6
tikhomirovii Jurtz. - 1, 4, 5
tobolensis Th. Wolf ex Pavl. - 3
togashii Ohwi (*P. fragarioides* L. subsp. *togashii* (Ohwi) Ohwi & Satomi) - 4
tollii Trautv. - 4
x **tolmatchevii** Jurtz. & Sojak. - P. nivea L. (*A. arenosa* (Turcz.) Juz.) x P. pulchella R. Br. - 1, 4, 5
x *tomentulosa* Jurtz. = P. prostrata
transcaspia Th. Wolf = P. pedata
transtuvinica K. Sobol. = P. astragalifolia
tranzschelii Juz. = P. rugulosa
x **tschaunensis** Juz. ex Jurtz. (*P. rubricaulis* auct. p.p.). - P. anachoretica Sojak x P. prostrata subsp. floccosa Sojak (*P. nivea* sensu Juz.) - 4, 5
tschimganica Sojak - 6
tschukotica Jurtz. & Petrovsky - 4, 5
tundrarum Juz. ex Jurtz. = P. nudicaulis
turczaninowiana Stschegl. - 3, 4, 6
- subsp. **nephogena** Sojak - 3, 4, 6
- subsp. **turczaninowiana** - 6
turgaica Sojak (*P. eremica* Th. Wolf, 1908, non Cov. 1892) - 3, 6
x **tynieckii** Blocki. - P. argentea L. x P. collina Wib. - 1
umbrosa Stev. - 2
uniflora Ledeb. (*P. ledebouriana* A. Pors. nom. superfl.) - 4, 5
- subsp. *subvillosa* Jurtz. = P. villosula
uschakovii Jurtz. - 5
ussuriensis Juz. = P. rugulosa
vahliana auct. = P. subvahliana
vahliana Lehm. subsp. *subvahliana* Jurtz. = P. subvahliana
varians Moench = P. norvegica
verna L. var. *ternata* A. Blytt = P. gelida
verticillaris Steph. - 4, 5
villosa Pall. ex Pursh - 5
villosula Jurtz. (*P. uniflora* Ledeb. subsp. *subvillosa* Jurtz. nom. invalid.) - 5
virgata Lehm. (*P. dealbata* Bunge) - 3, 6
viscosa Donn ex Lehm. = P. longifolia
volgarica Juz. - 1
vorobievii T.I. Nechaeva & Sojak - 5
vulcanicola Juz. - 5
vvedenskyi Botsch. - 6
wibeliana Th. Wolf = P. collina
wiemanniana Gunth. & Schummel (*P. collina* Wib. subsp. *wiemanniana* (Gunth. & Schummel) Dostal) - 1
wrangelii Petrovsky - 5

Poterium L.

lasiocarpum Boiss. & Hausskn. (*Sanguisorba minor* Scop. subsp. *lasiocarpa* (Boiss. & Hausskn.) Nordborg) - 2, 6
polygamum Waldst. & Kit. (*P. sanguisorba* L. subsp. *polygamum* (Waldst. & Kit.) Simonk., *Sanguisorba minor* Scop. subsp. *muricata* (Spach) Briq., *S. minor* subsp. *polygama* (Waldst. & Kit.) Holub, *S. muricata* (Spach) Gremli) - 1, 2, 6
sanguisorba L. (*Sanguisorba minor* Scop.) - 1, 2, 3(?)
- subsp. *polygamum* (Waldst. & Kit.) Simonk. = P. polygamum

Prinsepia Royle

sinensis (Oliv.) Bean - 5

Prunus L.

alaica (Pojark.) Gilli = Cerasus alaica
bucharica (Korsh.) Hand.-Mazz. = Amygdalus bucharica
caspica Koval. & Ekim. = P. divaricata subsp. caspica
cerasifera auct. p.p. = P. divaricata
cerasifera Ehrh. - 2, 6
- subsp. **caspica** Luneva - 2
- subsp. *divaricata* (Ledeb.) Schneid. = P. divaricata
- subsp. *nachichevanica* Koval. = P. nachichevanica
- subsp. *sogdiana* (Vass.) Cinovskis = P. sogdiana
- subsp. **turcomanica** Luneva - 6
- var. *orientalis* M. Pop. = P. sogdiana
darvasica Temberg - 6
davidiana (Carr.) Franch. = Armeniaca davidiana
divaricata Ledeb. (*P. cerasifera* Ehrh. subsp. *divaricata* (Ledeb.) Schneid., *P. cerasifera* auct. p.p.) - 2, 6
- subsp. **caspica** Browicz (*P. caspica* Koval. & Ekim. nom. invalid.) - 2
- subsp. **pontica** Koval. & Ekim. - 2
***domestica** L.
dulcis (Mill.) D.A. Webb = Amygdalus communis
erythrocarpa (Nevski) Gilli = Cerasus erythrocarpa
x **ferganica** Lincz. (*Cerasus petrovae* S. Korov.). - Louiseania ulmifolia (Franch.) Pachom. x Prunus divaricata Ledeb. - 6
x **foveolata** Luneva - 6
fruticosa Pall. = Cerasus fruticosa
insititia L. - 1
kurilensis (Miyabe) Miyabe = Cerasus kurilensis
mirabilis Sumn. = P. sogdiana
moldavica Kotov - 1
***nachichevanica** Kudr. (*P. cerasifera* Ehrh. subsp. *nachichevanica* Koval. nom. invalid.) - 2
nipponica Matsum. f. *kurilensis* (Miyabe) Hiroe = Cerasus kurilensis
orientalis (M. Pop.) Kudr. = P. sogdiana
padus L. subsp. *borealis* (Schubel.) Nym. = Padus borealis
- var. *borealis* (Schubel.) Blytt = Padus borealis
- var. *pubescens* Regel & Til. = Padus asiatica
- f. *pubescens* (Regel & Til.) Kitag. = Padus asiatica
persica (L.) Batsch = Persica vulgaris
pseudoprostrata (Pojark.) Rech. fil. = Cerasus pseudoprostrata
***salicina** Lindl. - 5, 6
sargentii Rehd. = Cerasus sachalinensis
sibirica L. f. *davidiana* (Carr.) Kitag. = Armeniaca davidiana
***simonii** Carr. - 6
sogdiana Vass. (*P. cerasifera* Ehrh. subsp. *sogdiana* (Vass.) Cinovskis, *P. cerasifera* var. *orientalis* M. Pop., *P.*

mirabilis Sumn., *P. orientalis* (M. Pop.) Kudr.) - 6
spinosa L. - 1, 2, 6
stepposa Kotov - 1
tadzhikistanica V. Zapr. - 6
tomentosa Thunb. = Cerasus tomentosa
triloba Lindl. = Louiseania triloba
turcomanica (Lincz.) Kitam. = Amygdalus turcomanica
turcomanica (Pojark.) Gilli = Cerasus turcomanica
***ussuriensis** Koval. & Kostina - 4, 5
vachuschtii Bregadze - 2
x *vavilovii* (M. Pop.) E. Murr. = Amygdalus x vavilovii

Pyracantha M. Roem.

coccinea M. Roem. - 1, 2

Pyrus L.

acutiserrata Gladkova - 2
aria (L.) Ehrh. subsp. *rupicola* Syme = Sorbus rupicola
asiae-mediae (M. Pop.) Maleev (*P. sinensis* Lindl. subsp. *asiae-mediae* M. Pop.) - 6
aucuparia (L.) Gaertn. var. *glabrata* Wimm. & Grab. = Sorbus gorodkovii
balansae auct. = P. caucasica
boissieriana Buhse - 2, 6
browiczii Mulk. - 2
bucharica Litv. (*P. korshinskyi* Litv. subsp. *bucharica* (Litv.) Bondar.) - 6
- subsp. *daschtidshumica* V. Zapr. = P. korshinskyi
- subsp. *korshinskyi* (Litv.) V. Zapr. = P. korshinskyi
cajon V. Zapr. = P. lindleyi
caucasica Fed. (*P. communis* L. subsp. *caucasica* (Fed.) Browicz, *P. balansae* auct.) - 2
chosrovica Gladkova - 2
communis L. - 2, 6
- subsp. *caucasica* (Fed.) Browicz = P. caucasica
complexa Rubtz. - 2
costata Sumn. - 6
daralagezi Mulk. - 2
demetrii Kuth. - 2
elaeagrifolia Pall. - 1
- subsp. *kotschyana* (Boiss. ex Decne.) Browicz = P. kotschyana
elata Rubtz. - 2
eldarica Grossh. - 2
erythrocarpa Vass. = P. korshinskyi
fedorovii Kuth. - 2
ferganensis Vass. - 6
georgica Kuth. - 2
gergerana Gladkova - 2
grossheimii Fed. - 2
hajastana Mulk. - 2
hyrcana Fed. - 2
intermedia Ehrh. = Sorbus intermedia
ketzkhovelii Kuth. - 2
korshinskyi Litv. (*P. bucharica* Litv. subsp. *daschtidshumica* V. Zapr. nom. invalid., *P. bucharica* subsp. *korshinskyi* (Litv.) V. Zapr., *P. erythrocarpa* Vass., *P. korshinskyi* subsp. *daschtidshumica* (V. Zapr.) Bondar. nom. invalid.) - 6
- subsp. *bucharica* (Litv.) Bondar. = P. bucharica
- subsp. *daschtidshumica* (V. Zapr.) Bondar. = P. korshinskyi
▪ kotschyana Boiss. ex Decne. (*P. elaeagrifolia* Pall. subsp. *kotschyana* (Boiss. ex Decne.) Browicz, *P. taochia* Woronow)

lindleyi Rehd. (*P. cajon* V. Zapr., *P. sinensis* Lindl. 1826, non (Tonin) Poir. 1816) - 6
malus L. var. *mitis* Wallr. = Malus domestica
medvedevii Rubtz. - 2
megrica Gladkova - 2
nutans Rubtz. - 2
oxyprion Woronow - 2
pseudosyriaca Gladkova - 2
pyraster Burgsd. - 1
- var. *rossica* (Danil.) Tuz = P. rossica
raddeana Woronow - 2
regelii Rehd. - 6
rossica Danil. (*P. pyraster* Burgsd. var. *rossica* (Danil.) Tuz) - 1
rupicola (Syme) Bab. = Sorbus rupicola
sachokiana Kuth. - 2
salicifolia Pall. - 2
salviifolia DC. - 1(?)
sieboldii Regel = Malus toringo
sinensis Lindl. = P. lindleyi
- subsp. *asiae-mediae* M. Pop. = P. asiae-mediae
sogdiana Kudr. = P. ussuriensis
sosnovskyi Fed. - 2
syriaca Boiss. - 2
tadshikistanica V. Zapr. - 6
takhtadzhianii Fed. - 2
tamamschianae Fed. - 2
taochia Woronow = P. kotschyana
theodorovii Mulk. - 2
turcomanica Maleev - 2, 6
tuskaulensis Vass. - 6
ussuriensis Maxim. (*P. sogdiana* Kudr.) - 5, 6
x vavilovii M. Pop. - P. korshinskyi Litv. x P. turcomanica Maleev - 6
voronovii Rubtz. - 2
vsevolodii Heideman - 2
zangezura Maleev - 2

Rosa L.

abutalybovii Gadzhieva - 2
achburensis Chrshan. (*R. arnoldii* Sumn. ex V. Tkaczenko) - 6
acicularis Lindl. (*R. sichotealinensis* Kolesn., *R. suavis* auct.) - 1, 3, 4, 5, 6
adenodonta Dubovik = R. balsamica
adenophylla Galushko - 2
afzeliana Fries - 1
agrestis Savi (*R. rubiginosa* L. subsp. *agrestis* (Savi) Malagarriga) - 1
alabukensis V. Tkaczenko - 6
alaica Juz. = R. hissarica
*alba L.
alberti Regel - 3, 6
alexeenkoi Crep. ex Juz. - 2
alticola Bouleng. = R. korshinskiana
altidaghestanica Gussejnov - 2
amblyophylla Sumn. = R. kuhitangi
amblyotis C.A. Mey. - 5
- subsp. *jacutica* (Juz.) Worosch. = R. jacutica
andegavensis Bast. (*R. litvinovii* Chrshan. nom. invalid.) - 1
andrzejowskii Stev. (*R. scabriuscula* auct. p.p.) - 1
angreni Sumn. = R. nanothamnus
antonowii (Lonacz.) Dubovik (*R. klukii* Bess. var. *antonowii* Lonacz.) - 1
arensii Juz. & Galushko - 2
arnoldii Sumn. ex V. Tkaczenko = R. achburensis

*arvensis Huds. (*R. sempervirens* L. subsp. *arvensis* (Huds.) Malagarriga)
assyiensis Poljak. = R. platyacantha
atropatana Sosn. = R. orietalis
awarica Gussejnov - 2
azerbajdzhanica Novopokr. & Rzazade = R. pulverulenta
balcarica Galushko - 2
balsamica Bess. 1815, non Willd. ex Spreng. 1820 (*R. adenodonta* Dubovik, *R. fedoseevii* Chrshan., *R. klukii* Bess., *R. psammophila* Chrshan., *R. obtusifolia* auct. p.p.) - 1, 2
*banksiae R. Br. - 1, 2
baxanensis Galushko - 2
beggeriana Schrenk (*R. operta* Sumn.) - 6
bellicosa Nevski - 6
beschnauensis Sumn. = R. nanothamnus
biebersteiniana Tratt. - 1
*bifera (Poir.) Pers.
biserratifolia Sumn. = R. kuhitangi
blinovskyana Kult. - 6
boissieri Crep. (*R. dumalis* Bechst. subsp. *boissieri* (Crep.) O. Nilsson) - 2
bordzilowskii Chrshan. = R. rubiginosa
x boreykiana Bess. - R. crenatula Chrshan. x R. corymbifera Borkh. - 1
borissovae Chrshan. - 2
borysthenica Chrshan. - 1
botryoides Sumn. = R. nanothamnus
brotherorum Chrshan. - 2
bugensis Chrshan. - 1
bungeana Boiss. & Buhse = R. rapinii
buschiana Chrshan. - 2
caesia Smith (*R. coriifolia* Fries, *R. dumalis* Bechst. subsp. *coriifolia* (Fries) P. Fourn., *R. dumalis* subsp. *coriifolia* (Fries) A. Pedersen, comb. superfl.) - 1, 2
- subsp. *subcollina* (Christ) Soo = R. subcollina
calantha V. Tkaczenko - 6
calcarea Lipsch. & Sumn. - 6
canina L. (*R. ciliato-sepala* Blocki, *R. sosnovskyi* Chrshan.) - 1, 2, 6
- subsp. *dumetorum* (Thuill.) Parm. = R. corymbifera
- *vosagiaca* Desportes = R. dumalis
caraganifolia Sumn. = R. fedtschenkoana
caryophyllacea Bess. (*R. gypsicola* Blocki) - 1
*centifolia L.
chasmocarpa Juz. = R. hissarica
*chinensis Jacq.
chomutoviensis Chrshan. & Laseb. - 1
chrshanovskii Dubovik - 1
ciesielskii Blocki - 1
ciliato-petala Blocki = R. villosa
ciliato-sepala Blocki = R. canina
cinnamomea L. var. *pisiformis* Christ = R. pisiformis
cinnamomea sensu L. = R. majalis
coeruleifolia Sumn. = R. fedtschenkoana
coriifolia Fries = R. caesia
corymbifera Borkh. (*R. canina* L. subsp. *dumetorum* (Thuill.) Parm.) - 1, 2, 6
- subsp. *deseglisei* (Boreau) Stohr = R. deseglisei
crenata Sumn. = R. kuhitangi
crenatula Chrshan. - 1
curvata Sumn. = R. maracandica
czackiana Bess. - 1
cziragensis Gussejnov - 2
*damascena Mill.
darginica Gussejnov - 2
davurica Pall. - 4, 5

deseglisei Boreau (*R. corymbifera* Borkh. subsp. *deseglisei* (Boreau) Stohr) - 1
diacantha Chrshan. (*R. heteracantha* Chrshan. 1950, non Kar. & Kir. 1842) - 1
didoensis Boiss. - 2
diplodonta Dubovik - 1
divina Sumn. = R. kokanica
dolichocarpa Galushko - 2
doluchanovii Manden. - 2
donetzica Dubovik - 1
dryadophylla Sumn. = R. kuhitangi
dsharkenti Chrshan. - 6
dumalis Bechst. (*R. canina* L. *vosagiaca* Desportes, *R. glauca* Vill. ex Loisel. 1809, non Pourr. 1788, *R. vosagiaca* (Desportes) Schinz & R. Keller, *R. wilibaldii* Chrshan.) - 1, 2
- subsp. *boissieri* (Crep.) O. Nilsson = R. boissieri
- subsp. *coriifolia* (Fries) P. Fourn. = R. caesia
- subsp. *coriifolia* (Fries) A. Pedersen = R. caesia
- subsp. *subcanina* (Christ) Soo = R. subcanina
ecae Aitch. (*R. mogoltavica* Juz., *R. pedunculata* Kult.) - 6
echidna Sumn. = R. turkestanica
eglanteria L. = R. rubiginosa
elasmacantha Trautv. - 2
ellenae Chrshan. = R. hissarica
elliptica Tausch = R. inodora
elongata Galushko = R. pimpinellifolia
elymaitica Boiss. & Hausskn. - 2(?)
epipsila Sumn. = R. fedtschenkoana
ermanica Manden. - 2
fedorovii Sumn. ex Chrshan. = R. foetida
fedoseevii Chrshan. = R. balsamica
fedtschenkoana Regel (*R. caraganifolia* Sumn., *R. coeruleifolia* Sumn., *R. epipsila* Sumn., *R. lavrenkoi* Sumn., *R. lipschitzii* Sumn., *R. minusculifolia* Sumn., *R. oligosperma* Sumn., *R. webbiana* sensu Juz. p.p.) - 6
fertilis Kult. - 6
floribunda Stev. - 1, 2
foetida Herrm. (*R. fedorovii* Sumn. ex Chrshan.) - 2, 6
gadzhievii Chrshan. & Iskenderov - 2
gallica L. - 1, 2
galushkoi Demurova - 2
geninae Juz. - 6
glabrifolia C.A. Mey. ex Rupr. - 1, 3
glanduligera Kult. = R. korshinskiana
glanduloso-setosa Gadzhieva - 2
*glauca Pourr. (*R. rubrifolia* Vill.) - 1
glauca Vill. ex Loisel. = R. dumalis
glutinosa auct. = R. pulverulenta
gorinkensis Bess. - 1
gracilipes Chrshan. = R. pimpinellifolia
grossheimii Chrshan. - 1
gypsicola Blocki = R. caryophyllacea
haemisphaerica Herrm. - 2
hausi-mardonica Vass. = R. hissarica
heckeliana Tratt. subsp. *orientalis* (Dupont ex Ser.) Meikle = R. orietalis
- subsp. *vanheurckiana* (Crep.) O. Nilsson = R. orietalis
heteracantha Chrshan. = R. diacantha
heteracantha Kar. & Kir. = R. platyacantha
heterostyla Chrshan. - 1
hilariae Juz. = R. hissarica
hirtissima Lonacz. - 2
hissarica Slob. (*R. alaica* Juz., *R. chasmocarpa* Juz., *R. ellenae* Chrshan., *R. hausi-mardonica* Vass., *R. hilariae* Juz., *R. krylovii* Sumn., *R. reverdattoi* Sumn., *R. sergievskajae* Sumn., *R. tschimganica* Chrshan., *R. vvedenskyi*

Korotk.) - 6
holotricha Sumn. = R. maracandica
homoacantha Dubovik - 1
horrida Fisch. ex Crep. = R. turcica
hracziana Tamamsch. - 2
huntica Chrshan. - 6
iberica Stev. ex Bieb. - 2, 6
iliensis Chrshan. = R. silverhjelmii
iljinii Chrshan. ex Gadzhieva - 2
infracta Sumn. = R. kuhitangi
inodora Fries (*R. elliptica* Tausch) - 1
irinae Demurova - 2
irysthonica Manden. - 2
isaevii Gadzhieva & Iskenderova - 2
jacutica Juz. (*R. amblyotis* C.A. Mey. subsp. *jacutica* (Juz.) Worosch.) - 4, 5
jaroshenkoi Gadzhieva & Iskenderova - 2
jundzillii Bess. - 1, 2
juzepczukiana Vass. - 6
kalmiussica Chrshan. & Laseb. - 1
kamelinii Gussejnov - 5
x kamtschatica Vent. - R. amblyotis C.A. Mey. x R. rugosa Thunb. ? - 5
karaalmensis V. Tkaczenko - 6 .
karakalensis Kult. - 6
karataviensis Juz. = R. turkestanica
karjaginii Sosn. - 2
kazarjanii Sosn. - 2
khasautensis Galushko - 2
kirsanovae Sumn. = R. maracandica
klukii Bess. = R. balsamica
- var. *antonowii* Lonacz. = R. antonowii
kokanica (Regel) Juz. (*R. divina* Sumn.) - 6
kokijrimensis V. Tkaczenko - 6
komarovii Sosn. - 2
koreana Kom. (*R. koreana* f. *roseopetala* Gorovoi & Pankov, *R. ussuriensis* Juz.) - 5
- f. *roseopetala* Gorovoi & Pankov = R. koreana
korshinskiana Bouleng. (*R. alticola* Bouleng., *R. glanduligera* Kult., *R. micrantha* Sumn. 1846, non Smith, 1812, *R. oligophylla* Sumn.) - 6
koslowskii Chrshan. - 2
koso-poljanskii Chrshan. - 1
kossii Galushko - 2
krylovii Sumn. = R. hissarica
krynkensis Ostapko - 1
kudrjaschevii Sumn. = R. kuhitangi
kuhitangi Nevski (*R. amblyophylla* Sumn., *R. biserratifolia* Sumn., *R. crenata* Sumn., *R. dryadophylla* Sumn., *R. infracta* Sumn., *R. kudrjaschevii* Sumn., *R. opaca* Sumn., *R. paracantha* Sumn., *R. paucijuga* Sumn., *R. singularis* Sumn., *R. suberosa* Sumn.) - 6
kujmanica Golits. - 1
lacerans Boiss. & Buhse = R. lehmanniana
lapidosa Dubovik - 2
lavrenkoi Sumn. = R. fedtschenkoana
laxa Retz. - 3, 6
lazarenkoi Chrshan. - 1
lehmanniana Bunge (*R. lacerans* Boiss. & Buhse) - 6
leiophylla Sumn. = R. maracandica
leucantha Bieb. = R. marschalliana
lipschitzii Sumn. = R. fedtschenkoana
litvinovii Chrshan. = R. andegavensis
- var. *slobodjanii* Chrshan. = R. slobodjanii
livescens Bess. - 1
lonaczevskii Dubovik - 1
longipedicellata Sumn. = R. maracandica

longisepala Koczk. - 6
lupulina Dubovik - 1
luxurians Sumn. = R. turkestanica
maeotica Dubovik - 1
magnifica Sumn. = R. maracandica
majalis Herrm. (*R. cinnamomea* sensu L. 1759, non L. 1753) - 1, 3, 4
mandenovae Gadzhieva - 2
maracandica Bunge (*R. curvata* Sumn., *R. holotricha* Sumn., *R. kirsanovae* Sumn., *R. leiophylla* Sumn., *R. longipedicellata* Sumn. nom. invalid., *R. magnifica* Sumn., *R. nudiflora* Sumn., *R. nummularia* Sumn., *R. oligacantha* Sumn., *R. pycnantha* Sumn., *R. pyricarpa* Sumn., *R. rectinervis* Sumn., *R. rubens* Sumn., *R. scoparia* Sumn., *R. tytthantha* Sumn., *R. tytthotricha* Sumn.) - 6
marretii Levl. - 5
marschalliana Sosn. (*R. leucantha* Bieb. 1819, non Bast. 1809) - 2
maximowicziana Regel - 5
mediata Dubovik - 1
micrantha Smith - 1
micrantha Sumn. = R. korshinskiana
microcarpa Bess. = R. pimpinellifolia
minimalis Chrshan. - 1
minusculifolia Sumn. = R. fedtschenkoana
mogoltavica Juz. = R. ecae
mollis Smith (*R. villosa* L. subsp. *mollis* (Smith) R. Keller & Gams) - 1, 2
montana Chaix subsp. *woronowii* (Lonacz.) O. Nilsson = R. woronowii
mucatscheviensis Chrshan. - 1
***multiflora** Thunb. - 1, 2
myriacantha DC. (*R. pimpinellifolia* L. subsp. *myriacantha* (DC.) O. Bolos & Vigo) - 1, 2
nanothamnus Bouleng. (*R. angreni* Sumn., *R. beschnauensis* Sumn., *R. botryoides* Sumn., *R. vitaminifera* Sumn.) - 6
nisami Sosn. - 2
nitidula Bess. - 1
nudiflora Sumn. = R. maracandica
nummularia Sumn. = R. maracandica
obtegens Galushko - 2
obtusifolia auct. p.p. = R. balsamica
olgae Chrshan. & Barb. - 1
oligacantha Sumn. = R. maracandica
oligophylla Sumn. = R. korshinskiana
oligosperma Sumn. = R. fedtschenkoana
opaca Sumn. = R. kuhitangi
operta Sumn. = R. beggeriana
oplisthes Boiss. (*R. svanetica* Crep.) - 2
orietalis Dupont ex Ser. (*R. atropatana* Sosn., *R. heckeliana* Tratt. subsp. *orientalis* (Dupont ex Ser.) Meikle, *R. heckeliana* subsp. *vanheurckiana* (Crep.) O. Nilsson, *R. vanheurckiana* Crep.) - 2
ossethica Manden. - 2
ovczinnikovii Koczk. - 6
oxyacantha Bieb. - 3, 4
oxyodon Boiss. - 2
oxyodontoides Galushko - 2
paracantha Sumn. = R. kuhitangi
parviuscula Chrshan. & Laseb. - 1
paucijuga Sumn. = R. kuhitangi
pavlovii Chrshan. - 3
pedunculata Kult. = R. ecae
pendulina L. - 1
pimpinellifolia L. (*R. elongata* Galushko, *R. gracilipes* Chrshan., *R. microcarpa* Bess., *R. spinosissima* L. nom.

ambig., *R. spinosissima* L. subsp. *pimpinellifolia* (L.) Soo) - 1, 2, 3, 6
- subsp. *myriacantha* (DC.) O. Bolos & Vigo = R. myriacantha
x **piptocalyx** Juz. - R. beggeriana Schrenk x R. fedtschenkoana Regel - 6
pisiformis (Christ) Sosn. (*R. cinnamomea* L. var. *pisiformis* Christ) - 2(?)
platyacantha Schrenk (*R. assyiensis* Poljak., *R. heteracantha* Kar. & Kir. 1842, non Chrshan. 1950) - 6
- subsp. **crassipes** (Vass.) Buzun. (*R. platyacantha* var. *crassipes* Vass.) - 6
pohrebniakii Chrshan. & Laseb. - 1
pomifera Herrm. = R. villosa
popovii Chrshan. - 6
porrectidens Chrshan. & Laseb. (*R. tyraica* Chrshan. & Laseb. 1949, non Blocki) - 1
potentilliflora Chrshan. & M. Pop. - 6
praetermissa Galushko - 2
pratorum Sukacz. - 1
prilipkoana Sosn. - 2
prokhanovii Galushko - 2
prutensis Chrshan. - 2
psammophila Chrshan. = R. balsamica
pubicaulis Galushko - 2
pulverulenta Bieb. (*R. azerbajdzhanica* Novopokr. & Rzazade, *R. glutinosa* auct.) - 2
pycnantha Sumn. = R. maracandica
pygmaea Bieb. - 1
pyricarpa Sumn. = R. maracandica
rapinii Boiss. & Bal. (*R. bungeana* Boiss. & Buhse) - 2, 6
rectinervis Sumn. = R. maracandica
reuteri Godet f. *subcanica* Christ = R. subcanina
reverdattoi Sumn. = R. hissarica
roopae Lonacz. - 2
rubens Sumn. = R. maracandica
rubiginosa L. (*R. bordzilowskii* Chrshan., *R. eglanteria* L. nom. ambig., *R. volhynensis* Chrshan.) - 1, 2(?)
- subsp. *agrestis* (Savi) Malagarriga = R. agrestis
- var. *sassnowskyana* Regel = R. sassnowskyana
rubrifolia Vill. = R. glauca
rugosa Thunb. - 1(cult.), 5
russanovii V. Tkaczenko - 6
sachokiana P. Jarosch. - 2
sassnowskyana (Regel) Muscheg. (*R. rubiginosa* var. *sassnowskyana* Regel) - 3, 6
scabriuscula auct. p.p. = R. andrzejowskii
scabriuscula Smith - 1
x **schischkinii** Juz. - ? R. davurica Pall. x R. koreana Kom. - 5
schistosa Dubovik - 1
schmalhauseniana Chrshan. - 1
scoparia Sumn. = R. maracandica
sempervirens L. subsp. *arvensis* (Huds.) Malagarriga = R. arvensis
sergievskajae Sumn. = R. hissarica
sergievskiana Polozh. & Prozorova - 3
sherardii Davies - 1
sichotealinensis Kolesn. = R. acicularis
silverhjelmii Schrenk (*R. iliensis* Chrshan.) - 6
simplicidens Dubovik - 1
singularis Sumn. = R. kuhitangi
sjunikii P. Jarosch. - 2
slobodjanii (Chrshan.) Dubovik, nom. invalid. (*R. litvinovii* Chrshan. var. *slobodjanii* Chrshan. nom. invalid.) - 1
sogdiana V. Tkaczenko - 6

sosnovskyana Tamamsch. - 2
sosnovskyi Chrshan. = R. canina
spinosissima L. = R. pimpinellifolia
suavis auct. = R. acicularis
subafzeliana Chrshan. (*R. vosagiaca* auct. p.p.) - 1
subbuschiana Gussejnov - 2
subcanina (Christ) Dalla Torre & Sarnth. (*R. reuteri* Godet f. *subcanica* Christ, *R. dumalis* Bechst. subsp. *subcanina* (Christ) Soo) - 1
subcollina (Christ) Dalla Torre & Sarnth. (*R. caesia* Smith subsp. *subcollina* (Christ) Soo) - 1
suberosa Sumn. = R. kuhitangi
subpomifera Chrshan. - 1
subpygmaea Chrshan. - 1
x sumneviczii Korotk. - R. ecae Aitch. x R. ovczinnikovii Koczk. - 6
svanetica Crep. = R. oplisthes
- var. *tscherekensis* Galushko = R. tscherekensis
talijevii Dubovik - 1
tauriae Chrshan. - 1
tchegemensis Galushko - 2
teberdensis Chrshan. - 2
terscolensis Galushko - 2
tesquicola Dubovik - 1
tianschanica Juz. - 6
tlaratensis Gussejnov - 2
tomentosa Smith - 1, 2
transcaucasica Manden. - 2
transsilvanica Schur - 1
tschatyrdagi Chrshan. - 1
tscherekensis (Galushko) Galushko (*R. svanetica* Crep. var. *tscherekensis* Galushko) - 2
tschimganica Chrshan. = R. hissarica
*turbinata Ait.
turcica Rouy (*R. horrida* Fisch. ex Crep. 1872, non Spreng.) - 1, 2
turkestanica Regel (*R. echidna* Sumn., *R. karataviensis* Juz., *R. luxurians* Sumn.) - 6
tuschetica Boiss. - 2
tyraica Chrshan. & Laseb. = R. porrectidens
tytthantha Sumn. = R. maracandica
tytthotricha Sumn. = R. maracandica
ucrainica Chrshan. - 1
uncinella Bess. - 1
uniflora Galushko - 2
usischensis Gussejnov - 2
ussuriensis Juz. = R. koreana
uzbekistanica Sumn. = R. korshinskiana
vagiana Crep. - 1
valentinae Galushko - 2
vanheurckiana Crep. = R. orietalis
vassilczenkoi V. Tkaczenko - 6
villosa L. (*R. ciliato-petala* Blocki,*R. pomifera* Herrm.) - 1, 2
- subsp. *mollis* (Smith) R. Keller & Gams = R. mollis
vitaminifera Sumn. = R. nanothamnus
volhynensis Chrshan. = R. rubiginosa
vosagiaca auct. p.p. = R. subafzeliana
vosagiaca (Desportes) Schinz & R. Keller = R. dumalis
vvedenskyi Korotk. = R. hissarica
webbiana Wall. ex Royle - 6
wilibaldii Chrshan. = R. dumalis
woronowii Lonacz. (*R. montana* Chaix subsp. *woronowii* (Lonacz.) O. Nilsson) - 2
zakatalensis Gadzhieva - 2
zalana Wiesb. - 1
zangezura P. Jarosch. - 2
zaramagensis Demurova - 2

zuvandica Gadzhieva - 2

*Rubacer Rydb.

*odoratum (L.) Rydb.

Rubus L.

abchaziensis Sudre - 2
abnormis (Sudre) Sudre - 2(?)
acaulis auct. = R. arcticus
x adenocladus Juz. - R. hirtus Waldst. & Kit. x R. piceetorum Juz. - 2
adscharicus Sanadze - 2
aipetriensis Juz. - 1
almensis Juz. - 1
anatolicus (Focke) Focke ex Hausskn. (*R. sanguineus* auct.) - 1, 2, 6
apricus Wimm. - 1
arcticus L. (*R. acaulis* auct.) - 1, 3, 4, 5
- subsp. *stellatus* (Smith) Boivin = R. stellatus
■ x areschougii A. Blytt. - R. caesius L. x R. saxatilis L.
armeniacus Focke - 2
x baniskheviensis Juz. - R. caucasicus Focke x R. ochthodes Juz. ? - 2
bifrons Vest (*R. fruticosus* L. agsp. *bifrons* (Vest) A. & D. Löve) - 1
x borzhomicus Juz. - R. cyri Juz. x R. lloydianus Genev. ? - 2
buschii Grossh. ex Sinjkova (*R. vulgatus* Arrhen. subsp. *buschii* Rozan.) - 2
caesius L. - 1, 2, 3, 6
- var. *turkestanicus* Regel = R. turkestanicus
candicans Weihe (*R. fruticosus* L. agsp. *candicans* (Weihe) A. & D. Löve) - 1, 2
canescens DC. (*R. tomentosus* Borkh. nom. illegit.) - 1, 2
cartalinicus Juz. - 2
x castoreus Laest. - R. arcticus L. x R. saxatilis L. - 1
caucasicus Focke - 2
caucasigenus (Sudre) Juz. - 2
chamaemorus L. (*R. chamaemorus* var. *pseudochamaemorus* (Tolm.) Hult., *R. chamaemorus* var. *pseudochamaemorus* (Tolm.) Worosch. comb. superfl., *R. pseudochamaemorus* Tolm.) - 1, 3, 4, 5
- var. *pseudochamaemorus* (Tolm.) Hult. = R. chamaemorus
- var. *pseudochamaemorus* (Tolm.) Worosch. = R. chamaemorus
charadzeae Sanadze - 2
x collicolus Sudre. - R. candicans Weihe x R. lloydianus Genev. - 2
crataegifolius Bunge - 4(alien), 5
crimaeus Juz. - 1
cyri Juz. - 2
x deltoideus P.J. Muell. - R. caesius L. x R. lloydianus Genev. - 2
diffusus Sanadze - 2
■ x digeneus Lindb. fil. - R. idaeus L. x R. saxatilis L.
discernendus Sudre - 2
discolor Weihe & Nees (*R. fruticosus* L. agsp. *discolor* (Weihe & Nees) A. & D. Löve) - 1
x divergens P.J. Muell. - R. caesius L. x R. canescens DC. - 1
dolichocarpus Juz. - 2
x euroasiaticus Sinjkova. - R. idaeus L. x R. matsumuranus Levl. - 1, 3, 4
eurythyrsiger Juz. - 1

fruticosus L. agsp. *bifrons* (Vest) A. & D. Löve = R. bifrons
- agsp. *candicans* (Weihe) A. & D. Löve = R. candicans
- agsp. *discolor* (Weihe & Nees) A. & D. Löve = R. discolor
- agsp. *hirtus* (Waldst. & Kit.) A. & D. Löve = R. hirtus
- agsp. *suberectus* (G. Anders. ex Smith) A. & D. Löve = R. suberectus
- agsp. *sulcatus* (Vest) A. & D. Löve = R. sulcatus
- agsp. *villicaulis* (Koehler ex Weihe & Nees) A. & D. Löve = R. villicaulis
georgicus Focke - 2
glandulosus Bell. - 1
granulatus P.J. Muell. & Lefevre - 1
guenteri Weihe & Nees (*R. hirtus* Waldst. & Kit. subsp. *guenteri* (Weihe & Nees) Sudre) - 1
hirtimimus Juz. - 1
hirtus Waldst. & Kit. (*R. fruticosus* L. agsp. *hirtus* (Waldst. & Kit.) A. & D. Löve) - 1, 2
- subsp. *guenteri* (Weihe & Nees) Sudre = R. guenteri
- subsp. *nigricatus* (P.J. Muell. & Lefevre) Sudre = R. nigricatus
humifusus Weihe - 1
humilifolius C.A. Mey. - 1, 3, 4, 5
hyrcanus Juz. - 2
ibericus Juz. - 2
x **idaeoides** Ruthe. - R. caesius L. x R. idaeus L. - 1, 2, 3, 6
idaeus L. - 1, 2, 3, 4, 5, 6
- subsp. *komarovii* (Nakai) Worosch. = R. komarovii
- subsp. *melanolasius* Focke = R. matsumuranus
- subsp. *sibiricus* Kom. = R. sibiricus
juzepczukii Sanadze - 2
kachethicus Sanadze - 2
kalaidae Juz. - 1
kaltenbachii Metsch - 1(?)
x **karakalensis** Freyn. - R. anatolicus (Focke) Focke ex Hausskn. x R. caesius L. - 2, 6
ketzkhovelii Sanadze - 2
kinashii Levl. & Vaniot = R. mesogaeus
komarovii Nakai (*R. idaeus* L. subsp. *komarovii* (Nakai) Worosch., *R. sachalinensis* Levl. var. *komarovii* (Nakai) Bondar.) - 4, 5
kudagorensis Sanadze - 2
lanuginosus Stev. ex Ser. - 2
lepidulus (Sudre) Juz. - 2
leptostemon Juz. - 2
lloydianus Genev. (*R. tomentosus* Borkh. var. *lloydianus* (Genev.) E.I. Nyarady) - 1, 2
longipetiolatus Sanadze - 2
marschallianus Juz. - 2
matsumuranus Levl. & Vaniot (*R. idaeus* L. subsp. *melanolasius* Focke, *R. sachalinensis* Levl.) - 1, 3, 4, 5
mesogaeus Focke ex Diels (*R. kinashii* Levl. & Vaniot) - 5
miszczenkoi Juz. - 2
moestifrons Juz. - 1
moschus Juz. - 2
nakeralicus Sanadze - 2
nanitauricus Juz. - 1
x **neogardicus** Juz. - R. arcticus L. x R. chamaemorus L. - 1
nessensis W. Hall - 1
nigricatus P.J. Muell. & Lefevre (*R. hirtus* Waldst. & Kit. subsp. *nigricatus* (P.J. Muell. & Lefevre) Sudre) - 1(?)
ochthodes Juz. - 2
oenoxylon Juz. - 1
ossicus Juz. - 2
paratauricus Juz. - 1
parvifolius L. (*R. triphyllus* Thunb.) - 5

pedatus Smith - 5
persicus Boiss. - 2
peruncinatus (Sudre) Juz. - 2
piceetorum Juz. - 2
platyphylloides Sanadze - 2
platyphyllos C. Koch - 2
plicatus Weihe & Nees - 1
x **polyanthus** P.J. Muell. - R. candicans Weihe x R. lloydianus Genev. - 2
ponticus (Focke) Juz. - 2
pseudochamaemorus Tolm. = R. chamaemorus
pseudojaponicus Koidz. - 5
raddeanus Focke - 2
rivularis Wirtg. & P.J. Muell. - 1
rudis Weihe - 1
ruprechtii Juz. - 2
sachalinensis Levl. = R. matsumuranus
- subsp. *sibiricus* (Kom.) Sinjkova = R. sibiricus
- var. *komarovii* (Nakai) Bondar. = R. komarovii
sanguineus auct. = R. anatolicus
saxatilis L. - 1, 2, 3, 4, 5, 6
scenoreinus Juz. - 1
schleicheri Weihe ex Tratt. - 1
scissus W.C.R. Watson - 1
x **semicaucasicus** Sudre. - R. anatolicus (Focke) Focke ex Hausskn. x R. caucasicus Focke - 2
serpens Weihe ex Lej. & Court. - 2
sibiricus (Kom.) Sinjkova (*R. idaeus* L. subsp. *sibiricus* Kom., *R. sachalinensis* Levl. subsp. *sibiricus* (Kom.) Sinjkova) - 1, 3, 4, 5
stellatus Smith (*R. arcticus* L. subsp. *stellatus* (Smith) Boivin) - 5
stenophyllidium Juz. - 1
stevenii Juz. - 1
x **suberectiformis** Sudre. - R. caesius L. x R. nessensis W. Hall - 1
suberectus G. Anders. ex Smith (*R. fruticosus* L. agsp. *suberectus* (G. Anders. ex Smith) A. & D. Löve) - 1
subtauricus Juz. - 1
sulcatus Vest (*R. fruticosus* L. agsp. *sulcatus* (Vest) A. & D. Löve) - 1
svanensis Sanadze - 2
takhtadjanii Mulk. - 2
tauricus Schlecht. ex Juz. - 1
tereticaulis P.J. Muell. - 1
tomentosus Borkh. = R. canescens
- var. *lloydianus* (Genev.) E.I. Nyarady = R. lloydianus
x **tranzschelii** Juz. - R. chamaemorus L. x R. saxatilis L. - 1
triphyllus Thunb. = R. parvifolius
troitzkyi Juz. - 1
turkestanicus (Regel) Pavl. (*R. caesius* L. var. *turkestanicus* Regel) - 6
x **tzebeldensis** Sudre. - R. abchaziensis Sudre x R. anatolicus (Focke) Focke ex Hausskn. - 2
x **tzemiensis** Juz. - R. candicans Weihe x R. piceetorum Juz. - 2
undabundus Juz. - 1
utshansuensis Juz. - 1
villicaulis Koehler ex Weihe & Nees (*R. fruticosus* L. agsp. *villicaulis* (Koehler ex Weihe & Nees) A. & D. Löve) - 1
x **virgultorum** P.J. Muell. - R. caesius L. x R. candicans Weihe - 2
vulgatus Arrhen. subsp. *buschii* Rozan. = R. buschii
woronowii (Sudre) Sudre - 2
zangezurus Mulk. - 2

Sanguisorba L.

alpina Bunge - 3, 4, 6
azovtsevii Krasnob. & Pschen. - 3
glandulosa Kom. = S. officinalis
komaroviana Nedoluzhko - 5
magnifica I. Schischk. & Kom. - 5
minor Scop. = Poterium sanguisorba
- subsp. *lasiocarpa* (Boiss. & Hausskn.) Nordborg = Poterium lasiocarpum
- subsp. *muricata* (Spach) Briq. = Poterium polygamum
- subsp. *polygama* (Waldst. & Kit.) Holub = Poterium polygamum
muricata (Spach) Gremli = Poterium polygamum
officinalis L. (*S. glandulosa* Kom., *S. officinalis* var. *glandulosa* (Kom.) Worosch.) - 1, 3, 4, 5, 6
- var. *glandulosa* (Kom.) Worosch. = S. officinalis
- var. *polygama* (Nyl.) Serg. = S. polygama
parviflora (Maxim.) Takeda - 4, 5
polygama Nyl. (*S. officinalis* L. var. *polygama* (Nyl.) Serg.) - 1
riparia Juz. - 6
sitchensis C.A. Mey. = S. stipulata
stipulata Rafin. (*S. sitchensis* C.A. Mey.) - 5
taurica Juz. - 1
tenuifolia Fisch. ex Link - 4, 5

Schistophyllidium (Juz. ex Fed.) Ikonn. = Potentilla

bifurcum (L.) Ikonn. = Potentilla bifurca
moorcroftii (Wall. ex Lehm.) Ikonn. = Potentilla moorcroftii
orientale (Juz.) Ikonn. = Potentilla orientalis

Schizonotus Lindl. = Sorbaria

tomentosus Lindl. = Sorbaria tomentosa

Sibbaldia L. (*Dryadanthe* Endl.)

macrophylla Turcz. ex Juz. = S. procumbens
olgae Juz. & Ovcz. - 6
parviflora Willd. - 2
procumbens L. (*S. macrophylla* Turcz. ex Juz.) - 1, 3, 4, 5, 6
semiglabra C.A. Mey. - 2
tetrandra Bunge (*Dryadanthe tetrandra* (Bunge) Juz.) - 3, 4, 6

Sibbaldianthe Juz.

adpressa (Bunge) Juz. - 3, 4, 6

Sibiraea Maxim.

altaiensis (Laxm.) Schneid. = S. laevigata
laevigata (L.) Maxim. (*Spiraea laevigata* L. X 1771, *Sibiraea altaiensis* (Laxm.) Schneid., *Spiraea altaiensis* Laxm. XI 1771) - 3
tianschanica Pojark. - 6

Sieversia Willd.

montana (L.) R. Br. = Oreogeum montanum
pentapetala (L.) Greene - 5
pusilla (Gaertn.) Hult. - 5

Sorbaria (Ser. ex DC.) A. Br. (*Schizonotus* Lindl.)

olgae Zinserl. - 6

pallasii (G. Don fil.) Pojark. - 4, 5
- subsp. *rhoifolia* (Kom.) Worosch. = S. rhoifolia
rhoifolia Kom. (*S. pallasii* (G. Don fil.) Pojark. subsp. *rhoifolia* (Kom.) Worosch.) - 5
sorbifolia (L.) A. Br. - 3, 4, 5
stellipila (Maxim.) Schneid. - 5
tomentosa (Lindl.) Rehd. (*Schizonotus tomentosus* Lindl.) - 6

x Sorbocotoneaster Pojark.

pozdnjakovii Pojark. - Soorbus sibirica Hedl. x Cotoneaster melanocarpa Fisch. ex Blytt - 4

Sorbus L. (*Micromeles* Decne.)

adsharica Gatsch. = S. boissieri
albovii Zinserl. - 2
alnifolia (Siebold & Zucc.) C. Koch (*Micromeles alnifolia* (Siebold & Zucc.) Koehne) - 5
amurensis Koehne (*S. discolor* auct.) - 5
anadyrensis Kom. = S. sibirica
*aria (L.) Crantz (*Crataegus aria* L.)
- var. *concolor* Boiss. = S. concolor
- subsp. *luristanica* Bornm. = S. luristanica
armeniaca Hedl. - 2
aucuparia L. (*S. caucasigena* Kom. ex Gatsch.) - 1, 2
- subsp. *glabrata* (Wimm. & Grab.) Hayek = S. gorodkovii
- subsp. *gorodkovii* (Pojark.) Bondar. = S. gorodkovii
austriaca (G. Beck) Hedl. - 1(?)
bachmarensis Gatsch. = S. boissieri
baldaccii (Degen & Fritsch ex Schneid.) Zinserl. p.p. = S. graeca
boissieri Schneid. (*S. adsharica* Gatsch., *S. bachmarensis* Gatsch., *S. boissieri* var. *adsharica* (Gatsch.) Sosn., *S. boissieri* var. *bachmarensis* (Gatsch.) Sosn.) - 2
- var. *adsharica* (Gatsch.) Sosn. = S. boissieri
- var. *bachmarensis* (Gatsch.) Sosn. = S. boissieri
buschiana Zinserl. (*S. subtomentosa* (Albov) Zinserl.) - 2
caucasica Zinserl. (*S. woronowii* Zinserl.) - 2
caucasigena Kom. ex Gatsch. = S. aucuparia
colchica Zinserl. - 2
commixta Hedl. - 5
concolor (Boiss.) Schneid. (*S. aria* (L.) Crantz var. *concolor* Boiss.) - 2
cretica (Lindl.) Fritsch f. *danubialis* Jav. = S. danubialis
dacica Borb. - 1
danubialis (Jav.) Karpati (*S. cretica* (Lindl.) Fritsch f. *danubialis* Jav.) - 1(?)
discolor auct. = S. amurensis
domestica L. - 1, 2(cult.)
dualis Zinserl. = S. roopiana
fedorovii Zaikonn. - 2
fennica (Kalm) Fries = S. hybrida
glabrata (Wimm. & Grab.) Hedl. = S. gorodkovii
gorodkovii Pojark. (*S. aucuparia* L. subsp. *glabrata* (Wimm. & Grab.) Hayek, *S. aucuparia* subsp. *gorodkovii* (Pojark.) Bondar., *S. glabrata* (Wimm. & Grab.) Hedl. 1901, non Kirchn., *Pyrus aucuparia* (L.) Gaertn. var. *glabrata* Wimm. & Grab.) - 1
graeca (Spach) Lodd. ex Schauer (*S. baldaccii* (Degen & Fritsch ex Schneid.) Zinserl. p.p. quoad pl., *S. umbellata* var. *cretica* auct. p.p.) - 1, 2, 6
- var. *taurica* (Zinserl.) Gabr. = S. taurica
- var. *turcica* (Zinserl.) Gabr. = S. turcica
hajastana Gabr. - 2
*hybrida L. (*S. fennica* (Kalm) Fries)
*intermedia (Ehrh.) Pers. (*Pyrus intermedia* Ehrh., *Crataegus aria* L. var. *scandica* L., *Sorbus scandica* (L.) Fries)

kamtschatcensis Kom. - 5

kusnetzovii Zinserl. - 2

luristanica (Bornm.) Schonbeck-Temesy (*S. aria* (L.) Crantz subsp. *luristanica* Bornm., *S. persica* Hedl. subsp. *luristanica* (Bornm.) Gabr.) - 2(?)

migarica Zinserl. (*S. obtusidentata* Zinserl., *S. umbellata* (Desf.) Fritsch var. *orbiculata* Gabr. p.p., *S. umbellata* var. *cretica* auct. p.p.) - 2

obtusidentata Zinserl. = S. migarica

orientalis Schonbeck-Temesy = S. torminalis var. caucasica

persica Hedl. - 2, 6

- subsp. *luristanica* (Bornm.) Gabr. = S. luristanica

polaris Koehne = S. sibirica

pontica Zaikonn. - 2

pseudolatifolia K. Pop. = S. tauricola

roopiana Bordz. (*S. dualis* Zinserl.) - 2

rupicola (Syme) Hedl. (*Pyrus aria* (L.) Ehrh. subsp. *rupicola* Syme, *P. rupicola* (Syme) Bab.) - 1(?)

sambucifolia (Cham. & Schlecht.) M. Roem. (*S. schneideriana* Koehne) - 5

scandica (L.) Fries = S. intermedia

schemachensis Zinserl. (*S. umbellata* var. *cretica* auct. p.p.) - 2

schneideriana Koehne = S. sambucifolia

*semipinnata Borb. - 1

sibirica Hedl. (*S. anadyrensis* Kom., *S. polaris* Koehne) - 1, 3, 4, 5

stankovii Juz. - 1

subfusca (Ledeb.) Boiss. - 2

- subsp. zinserlingii Zaikonn. - 2

subtomentosa (Albov) Zinserl. = S. buschiana

takhtajanii Gabr. - 2

tamamschjanae Gabr. - 2

taurica Zinserl. (*S. graeca* (Spach) Lodd. ex Schauer var. *taurica* (Zinserl.) Gabr., *S. umbellata* (Desf.) Fritsch var. *taurica* (Zinserl.) Gabr.) - 1, 2

tauricola Zaikonn. (*S. pseudolatifolia* K. Pop. 1959, non Boros, 1937) - 1

*teodorii Liljefors - 1 (run wild)

tianschanica Rupr. - 6

torminalis (L.) Crantz - 1, 2

- var. caucasica Diapulis (*S. orientalis* Schonbeck-Temesy, *S. torminalis* var. *orientalis* (Schonbeck-Temesy) Gabr., *S. torminalis* f. *orientalis* (Schonbeck-Temesy) Browicz) - 2

- var. *orientalis* (Schonbeck-Temesy) Gabr. = S. torminalis var. caucasica

- f. *orientalis* (Schonbeck-Temesy) Browicz = S. torminalis var. caucasica

turcica Zinserl. (*S. graeca* (Spach) Lodd. ex Schauer var. *turcica* (Zinserl.) Gabr., *S. umbellata* auct. p.p.) - 1, 2

turkestanica (Franch.) Hedl. - 6

umbellata auct. p.p. = S. turcica

- var. *cretica* auct. p.p. = S. graeca, S. migarica and S. schemachensis

umbellata (Desf.) Fritsch var. *orbiculata* Gabr. p.p. = S. migarica

- var. *taurica* (Zinserl.) Gabr. = S. taurica

velutina (Albov) Schneid. - 2

woronowii Zinserl. = S. caucasica

Spiraea L.

aemiliana Schneid. = S. beauverdiana

- subsp. *stevenii* (Schneid.) Rydb. = S. beauverdiana

*alba DuRoi - 1

alpina Pall. - 3, 4

altaiensis Laxm. = Sibiraea laevigata

aquilegifolia Pall. - 4

- var. *vanhouttei* Briot. = S. vanhouttei

baldschuanica B. Fedtsch. - 6

beauverdiana Schneid. (*S. aemiliana* Schneid., *S. aemiliana* subsp. *stevenii* (Schneid.) Rydb., *S. betulifolia* Pall. subsp. *aemiliana* (Schneid.) Hara, *S. betulifolia* var. *aemiliana* (Schneid.) Koidz., *S. stevenii* (Schneid.) Rydb.) - 4, 5

betulifolia Pall. - 1(cult.), 4, 5

- subsp. *aemiliana* (Schneid.) Hara = S. beauverdiana

- var. *aemiliana* (Schneid.) Koidz. = S. beauverdiana

x billardii Dipp. - S. douglasii Hook. x S. salicifolia L. - 1

*cantoniensis Lour. - 1

chamaedrifolia L. (*S. chamaedrifolia* subsp. *ulmifolia* (Scop.) J. Duvigneaud, *S. ulmifolia* Scop.) - 1, 3, 4, 6

- subsp. *ulmifolia* (Scop.) J. Duvigneaud = S. chamaedrifolia

crenata L. - 1, 2, 3, 6

- subsp. *kotschyana* (Boiss.) Fed. = S. hypericifolia subsp. kotschyana

- var. *kotschyana* Boiss. = S. hypericifolia subsp. kotschyana

dahurica (Rupr.) Maxim. - 4, 5

*douglasii Hook. - 1

elegans Pojark. = S. flexuosa

ferganensis Pojark. = S. lasiocarpa

flexuosa Fisch. ex Cambess. (*S. elegans* Pojark.) - 3, 4, 5

humilis Pojark. = S. salicifolia

hypericifolia L. - 1, 2, 3, 4, 6

- subsp. *kotschyana* (Boiss.) Gladkova (*S. crenata* L. var. *kotschyana* Boiss., *S. crenata* subsp. *kotschyana* (Boiss.) Fed.) - 2

*japonica L. fil. - 1

laevigata L. = Sibiraea laevigata

lasiocarpa Kar. & Kir. (*S. ferganensis* Pojark.) - 6

*latifolia (Ait.) Borkh. (*S. salicifolia* L. var. *latifolia* Ait.) - 1

litwinowii Dobrocz. - 1, 2

media Franz Schmidt (*S. media* subsp. *polonica* (Blocki) Dostal, *S. media* subsp. *polonica* (Blocki) Pawl. comb. superfl., *S. polonica* Blocki) - 1, 3, 4, 5, 6

- subsp. *polonica* (Blocki) Dostal = S. media

- subsp. *polonica* (Blocki) Pawl. = S. media

pikoviensis Bess. - 1

pilosa Franch. - 6

polonica Blocki = S. media

*prunifolia Siebold & Zucc. - 1

pubescens Turcz. - 4, 5

salicifolia L. (*S. humilis* Pojark., *S. salicifolia* var. *humilis* (Pojark.) Hara) - 1(cult., run wild), 3, 4, 5

- var. *humilis* (Pojark.) Hara = S. salicifolia

- var. *latifolia* Ait. = S. latifolia

schlothgauerae Ignatov & Worosch. - 5

sericea Turcz. - 4, 5

stevenii (Schneid.) Rydb. = S. beauverdiana

*thunbergii Sieb. ex Blume - 1

tianschanica Pojark. - 6

*tomentosa L. - 1

trilobata L. - 1(cult.), 3, 4, 6

ulmifolia Scop. = S. chamaedrifolia

ussuriensis Pojark. - 5

*vanhouttei (Briot) Zab. (*S. aquilegifolia* Pall. var. *vanhouttei* Briot) - 1, 6

Spiraeanthus Maxim.

schrenkianus Maxim. - 6

Tylosperma Botsch.

lignosa (Willd. ex Schlecht.) Botsch. (*Potentilla lignosa* Willd. ex Schlecht.) - 6

Waldsteinia Willd.

geoides Willd. - 1, 2
ternata (Steph.) Fritsch - 4, 5
- subsp. **maximoveicziana** Teppner - 5

Woronowia Juz.

speciosa (Albov) Juz. - 2

RUBIACEAE Juss.
Asperula L.

abchasica V. Krecz. - 2
accrescens Klok. - 2
aemulans V. Krecz. ex Klok. (*A. cretacea* auct. p.p.) - 1
affinis Boiss. & Huet (*A. dolichophylla* Klok.) - 2
affrena Klok. = Galium ruthenicum
albiflora M. Pop. - 6
albovii Manden. - 2
alpina Bieb. - 2
aparine Bieb. = Galium pseudorivale
arvensis L. - 1, 2
attenuata Klok. - 1
azurea Jaub. & Spach = A. orientalis
badachschanica Pachom. - 6
balchanica Bobr. - 6
besseriana Klok. = Galium humifusum
bidentata Klok. = A. tenella
biebersteinii V. Krecz. - 2
botschantzevii Pachom. - 6
caespitans Juz. (*A. tranzshelii* Klok., *A. supina* auct. p.p.) - 1
campanulata (Vill.) Klok. = Galium campanulatum
caucasica Pobed. (*A. taurina* L. subsp. *caucasica* (Pobed.) Ehrend., *A. taurina* auct. p.p.) - 1(?), 2
ciliatula Pachom. - 6
cimmerica V. Krecz. ex Klok. (*A. cretacea* auct. p.p.) - 1
cincinnata Klok. = Galium humifusum
congesta Tscherneva (*A. popovii* Schischk. p.p.) - 6
cretacea auct. p.p. = A. aemulans, A. cimmerica, A. kotovii, A. praepilosa and A. praevestita
creticola Klok. = A. tephrocarpa
cristata (Somm. & Levier) V. Krecz. - 2
cynanchica L. (*A. semiamicta* Klok.) - 1
czukavinae Pachom. & Karim. - 6
danilewskiana Basin. (*A. laevissima* Klok.) - 1, 3, 6
dasyantha Klok. - 2
diminuta Klok. - 1, 2
dolichophylla Klok. = A. affinis
exasperata V. Krecz. ex Klok. - 1
fedtschenkoi Ovcz. & Tscherneva - 6
ferganica Pobed. - 6
galioides Bieb. = Galium biebersteinii
glabrata Tscherneva (*A. popovii* Schischk. p.p.) - 6
glomerata (Bieb.) Griseb. - 2
gracilis C.A. Mey. - 2
graniticola Klok. = A. montana
graveolens Bieb. ex Schult. & Schult. fil. (*A. leiograveolens* M. Pop. & Chrshan., *A. savranica* Klok.) - 1
gypsacea Pachom. - 6
hirsutiuscula Pobed. - 2
humifusa (Bieb.) Bess. = Galium humifusum
hypanica Klok. = A. montana

infracta Klok. = A. praevestita
insoluta Pachom. - 6
insuavis Pobed. = Galium insuave
intersita Klok. - 2
karataviensis Pavl. = Galium pseudorivale
karategini Pachom. & Karim. - 6
kemulariae Manden. - 2
kirghisorum Filat. - 6
kotovii Klok. (*A. cretacea* auct. p.p., *A. vestita* auct.) - 1
kovalevskiana Pachom. - 6
krylowiana Serg. = Galium krylowianum
laevis Schischk. - 6
laevissima Klok. = A. danilewskiana
laxiflora Boiss. - 2(?)
leiograveolens M. Pop. & Chrshan. = A. graveolens
lipskyana V. Krecz. - 2
markothensis Klok. - 2
maximowiczii Kom. = Galium maximowiczii
moldavica Dobrescu = Galium moldavicum
molluginoides (Bieb.) Reichenb. - 2
montana Waldst. & Kit. (*A. graniticola* Klok., *A. hypanica* Klok., *A. rumelica* Boiss.) - 1
neilreichii G. Beck - 1(?)
nuratensis Pachom. - 6
octonaria Klok. = Galium octonarium
odorata L. = Galium odoratum
oppositifolia Regel & Schmalh. (*A. popovii* Schischk. p.p. quoad nomen) - 6
orientalis Boiss. & Hohen. (*A. azurea* Jaub. & Spach) - 2
pamirica Pobed. - 6
paniculata Bunge = Galium paniculatum
pauciflora Tscherneva - 6
pedicellata Klok. - 2
pestolozzae auct. = A. woronowii
petraea V. Krecz. ex Klok. - 1, 3, 6
platygalium Maxim. = Galium platygalium
pontica Boiss. - 2
popovii Schischk. = A. congesta, A. glabrata and A. oppositifolia
praepilosa V. Krecz. ex Klok. (*A. cretacea* auct. p.p.) - 1
praevestita Klok. (*A. infracta* Klok., *A. cretacea* auct. p.p.) - 1
propinqua Pobed. (*A. taurina* auct. p.p.) - 1
prostrata (Adams) C. Koch - 2
pseudoglomerata Pachom. - 6
pugionifolia Tscherneva - 6
rivalis Sibth. & Smith = Galium rivale
- var. *schelkownikowiana* Bordz. = Galium schelkowniko-wianum
rumelica Boiss. = A. montana
savranica Klok. = A. graveolens
scabrella Tscherneva - 6
semiamicta Klok. = A. cynanchica
setosa Jaub. & Spach - 2, 6
setulosa Boiss. - 1
stevenii V. Krecz. = A. tenella
strishovae Pachom. & Karim. - 6
supina Bieb. - 1
taurica Pacz. - 1, 2
taurina auct. p.p. = A. caucasica and A. propinqua
taurina L. subsp. *caucasica* (Pobed.) Ehrend. = A. caucasica
tauro-scythica Klok. - 1
tenella Heuff. ex Degen (*A. bidentata* Klok., *A. stevenii* V. Krecz.) - 1, 2
tephrocarpa Czern. ex M. Pop. & Chrshan. (*A. creticola* Klok.) - 1

tinctoria L. = Galium tinctorium
tomentella Dubovik - 1
tranzshelii Klok. = A. caespitans
turcomanica Pobed. - 6
turkestanica Nevski = Galium humifusum
tyraica Bess. = Galium volhynicum
vestita auct. = A. kotovii
vestita V. Kreecz. - 2
woronowii V. Krecz. (*A. pestolozzae* auct.) - 2
xerotica Klok. = Galium xeroticum

Callipeltis Stev.

aperta Boiss. & Buhse - 6
cucullaria (L.) Stev. (*Valantia cucullaria* L., *Callipeltis cucullaris* (L.) Rothm.) - 2, 6
cucullaris (L.) Rothm. = C. cucullaria

Crucianella L.

angustifolia L. (*C. oxyloba* Janka) - 1, 2
baldshuanica Krasch. - 6
bucharica B. Fedtsch. - 6
catellata Klok. (*C. latifolia* auct.) - 1, 2(?)
chlorostachys Fisch. & C.A. Mey. - 2, 6
colchica Manden. (*C. gilanica* Trin. subsp. *pontica* (Ehrend.) Ehrend., *C. glauca* A. Rich. ex DC. subsp. *pontica* Ehrend.) - 2
divaricata Korov. - 6
exasperata Fisch. & C.A. Mey. - 2, 6
filifolia Regel & C. Winkl. - 6
gilanica Trin. - 2, 6
- subsp. *pontica* (Ehrend.) Ehrend. = C. colchica
- subsp. *transcaucasica* (Ehrend.) T.N. Pop. & Takht. (*C. glauca* A. Rich. ex DC. subsp. *transcaucasica* Ehrend.) - 2
glauca A. Rich. ex DC. subsp. *pontica* Ehrend. = C. colchica
- subsp. *transcaucasica* Ehrend. = C. gilanica subsp. transcaucasica
latifolia auct. = C. catellata
oxyloba Janka = C. angustifolia
sabulosa Korov. & Krasch. - 6
schischkinii Lincz. - 6
sintenisii Bornm. - 6
suaveolens C.A. Mey. - 2

Cruciata Hill

articulata (L.) Ehrend. (*Valantia articulata* L., *Galium cordatum* Roem. & Schult.) - 2
braunii (Zelen.) Pobed. (*Galium braunii* Zelen.) - 1
chersonensis (Willd.) Ehrend. = C. laevipes and C. tauruca
coronata (Sibth. & Smith) Ehrend. (*Galium coronatum* Sibth. & Smith) - 2
- subsp. *taurica* (Pall. ex Willd.) Ehrend. = C. taurica
elbrussica (Pobed.) Pobed. (*Galium elbrussicum* Pobed.) - 2
glabra (L.) Ehrend. (*Galium vernum* Scop.) - 1, 2, 3
kopetdaghensis (Pobed.) Pobed. (*Galium kopetdaghense* Pobed.) - 6
kryłovii (Iljin) Pobed. (*Galium krylovii* Iljin) - 3, 4
laevipes Opiz (*C. chersonensis* (Willd.) Ehrend. p.p. excl. typo, *Galium cruciata* (L.) Scop.) - 1, 2
neotaurica (Klok.) Pobed. (*Galium neotauricum* Klok.) - 1
pedemontana (Bell.) Ehrend. (*Valantia pedemontana* Bell., *Galium pedemontanum* (Bell.) All.) - 1, 2, 6
pseudopolycarpa (Somm. & Levier) Pobed. (*Galium pseudopolycarpon* Somm. & Levier) - 2
rugosa (Galushko) Galushko (*Galium rugosum* Galushko) - 2
schischkinii (Pobed.) Pobed. (*Galium schischkinii* Pobed.) - 2
sevanensis (Pobed.) Pobed. (*Galium sevanense* Pobed.) - 2
sosnowskyi (Manden.) Pobed. (*Galium sosnowskyi* Manden.) - 2
taurica (Pall. ex Willd.) Soo (*C. chersonensis* (Willd.) Ehrend. p.p. quoad typum, *C. coronata* (Sibth. & Smith) Ehrend. subsp. *taurica* (Pall. ex Willd.) Ehrend., *C. taurica* (Pall.) Ehrend. nom. invalid., *Galium chersonense* (Willd.) Roem. & Schult., *G. decoronatum* Klok., *G. tauricum* (Pall. ex Willd.) Roem. & Schult.) - 1, 2
valentinae (Galushko) Galushko (*Galium valentinae* Galushko) - 2

Gaillonia A. Rich. ex DC. = Neogaillonia

asperuliformis Lincz. = Neogaillonia asperuliformis
bruguierii A. Rich. ex DC. = Neogaillonia bruguierii
bucharica B. Fedtsch. & Desjat. = Neogaillonia bucharica
dubia Aitch. & Hemsl. = Neogaillonia dubia
inopinata Lincz. = Neogaillonia inopinata
szowitzii DC. = Neogaillonia szowitzii
trichophylla M. Pop. ex Tscherneva = Neogaillonia trichophylla

Galium L.

abaujense Borb. - 1
- subsp. *polonicum* (Blocki) Soo = G. x polonicum
achurense Grossh. (*G. subvelutinum* (DC.) C. Koch subsp. *problematicum* Ehrend. p.p.) - 2
album Mill. (*G. pseudomollugo* Klok., *G. erectum* sensu Huds. 1778, non Huds. 1762) - 1, 2, 3
- subsp. *prusense* (C. Koch) Ehrend. & Krendl, p.p. = G. calcareum and G. juzepczukii
- subsp. *pycnotrichum* (H. Br.) Krendl = G. pycnotrichum
amblyophyllum Schrenk - 3, 6
amurense Pobed. (*G. boreale* L. var. *amurense* (Pobed.) Kitag.) - 5
anfractum Somm. & Levier - 2
anisophyllon auct. p.p. = G. bellatulum
aparine L. - 1, 2, 3, 4, 5(alien), 6
- f. *spurium* (L.) Boivin = G. spurium
apsheronicum Pobed. - 2
articulatum Lam. = G. rubioides
atropatanum Grossh. - 2
attenuatum Klok. & Zaverucha - 1
auratum Klok. = G. juzepczukii
aureum Vis. - 2
baicalense Pobed. = G. brandegei
bellatulum Klok. (*G. anisophyllon* auct. p.p.) - 1
besseri Klok. - 1
biebersteinii Ehrend. (*Asperula galioides* Bieb. nom. illegit., *Galium galioides* (Bieb.) Soo) - 1, 2
boreale L. (*G. praeboreale* Klok., *G. septentrionale* sensu Pobed. p.p.) - 1, 2, 3, 4, 5
- subsp. *boreale* var. *boreale* f. *exoletum* (Klok.) Soo = G. exoletum
- subsp. *exoletum* (Klok.) Sojak = G. exoletum
- subsp. *mugodsharicum* (Pobed.) Sojak = G. mugodsharicum
- subsp. *ussuriense* (Pobed.) Worosch. = G. physocarpum
- var. *amurense* (Pobed.) Kitag. = G. amurense
borysthenicum Klok. - 1

brachyphyllum Roem. & Schult. - 1, 2

brandegei A. Gray (*G. baicalense* Pobed., *G. ruprechtii* Pobed. p.min.p., *G. subbiflorum* (Wieg.) Rydb., *G. trifidum* L. var. *subbiflorum* Wieg., *G. trifidum* subsp. *subbiflorum* (Wieg.) Puff) - 1, 3, 4, 5

braunii Zelen. = Cruciata braunii

bullatum Lipsky - 2

calcareum (Albov) Pobed. (*G. album* Mill. subsp. *prusense* (C. Koch) Ehrend. & Krendl, p.p., *G. fasciculatum* Klok., *G. zelenetzkii* Klok.) - 1, 2

campanulatum Vill. (*Asperula campanulata* (Vill.) Klok.) - 1

- subsp. *tyraicum* (Bess.) Sojak = G. volhynicum

carpaticum Klok. (*G. polonicum* sensu Pobed.) - 1

caspicum Stev. - 2

ceratopodum Boiss. - 6

chersonense (Willd.) Roem. & Schult. = Cruciata taurica

chloroleucum Fisch. & C.A. Mey. - 2

columbianum Rydb. = G. taquetii

congestum Klok. & Zaverucha - 1

consanguineum Boiss. (*G. majmechense* Bordz.) - 2

cordatum Roem. & Schult. = Cruciata articulata

coriaceum Bunge - 3, 4

coronatum Sibth. & Smith = Cruciata coronata

cruciata (L.) Scop. = Cruciata laevipes

czerepanovii Pobed. - 2

dasypodum Klok. = G. physocarpum and G. rubioides

davuricum Turcz. ex Ledeb. - 4, 5

debile Desf. (*G. krymense* Pobed., *G. elongatum* auct.) - 1, 2

decaisnei Boiss. (*G. setaceum* Lam. subsp. *decaisnei* (Boiss.) Ehrend.) - 2, 6

decoronatum Klok. = Cruciata taurica

densiflorum Ledeb. - 1, 3, 4, 5, 6

- var. **rosmarinifolium** (Bunge) Tzvel. (*G. verum* L. var. *rosmarinifolium* Bunge, *G. densiflorum* var. *saurense* (Litv.) Tzvel., *G. saurense* Litv.) - 3, 6

- var. *saurense* (Litv.) Tzvel. = G. densiflorum var. rosmarinifolium

elatum auct. = G. sphenophyllum

elbrussicum Pobed. = Cruciata elbrussica

eldaricum Grossh. - 2

elongatum auct. = G. debile

elongatum C. Presl (*G. maximum* G. Moris, *G. palustre* L. subsp. *elongatum* (C. Presl) Lange) - 1

erectum sensu Huds. = G. album

exoletum Klok. (*G. boreale* L. subsp. *boreale* var. *boreale* f. *exoletum* (Klok.) Soo, *G. boreale* subsp. *exoletum* (Klok.) Sojak) - 1

fagetorum Klok. = G. juzepczukii

fasciculatum Klok. = G. calcareum

fistulosum Somm. & Levier - 2

galioides (Bieb.) Soo = G. biebersteinii

ghilanicum Stapf (*G. transcaucasicum* Stapf) - 2, 6

glabratum Klok. - 1

glaucum L. subsp. *tyraicum* (Bess.) Sojak = G. volhynicum

grossheimii Pobed. = G. hyrcanicum

grusinum Trautv. (*G. incanum* Smith subsp. *elatius* (Boiss.) Ehrend., *G. orientale* Boiss. var. *elatius* Boiss.) - 2

hercynicum Weig. (*G. saxatile* auct.) - 1

humifusum Bieb. (*Asperula besseriana* Klok., *A. cincinnata* Klok., *A. humifusa* (Bieb.) Bess., *A. turkestanica* Nevski, *G. humifusum* subsp. *besserianum* (Klok.) Soo, *G. humifusum* subsp. *cincinnatum* (Klok.) Soo) - 1, 2, 6

- subsp. *besserianum* (Klok.) Soo = G. humifusum

- subsp. *cincinnatum* (Klok.) Soo = G. humifusum

hypanicum Klok. - 1

hyrcanicum C.A. Mey. (*G. grossheimii* Pobed., *G. subvelutinum* (DC.) C. Koch subsp. *hyrcanicum* (C.A. Mey.) Ehrend., *G. leiophyllum* sensu Pobed.) - 2

ibicinum Boiss. & Hausskn. (*G. linczevskyi* Pobed.) - 6

incanum Smith subsp. *elatius* (Boiss.) Ehrend. = G. grusinum

insuave (Pobed.) Pobed. (*Asperula insuavis* Pobed.) - 6

intermedium Schult. (*G. schultesii* Vest) - 1

irinae Pachom. - 6

japonicum Makino = G. triflorum

juzepczukii Pobed. (*G. album* Mill. subsp. *prusense* (C. Koch) Ehrend. & Krendl p.p., *G. auratum* Klok., *G. fagetorum* Klok.) - 1

kamtschaticum Stell. ex Schult. & Schult. fil. - 5

karakulense Pobed. - 6

karataviense (Pavl.) Pobed. = G. pseudorivale

kasachstanicum Pachom. - 6

kernerianum Klok. - 1

kiapazi Manden. = G. tianschanicum

kopetdaghense Pobed. = Cruciata kopetdaghensis

krylovii Iljin = Cruciata krylovii

krylowianum (Serg.) Pobed. (*Asperula krylowiana* Serg.) - 3

krymense Pobed. = G. debile

kutzingii Boiss. & Buhse - 2

lacteum (Maxim.) Pobed. - 5

leiophyllum Boiss. & Hohen. = G. psilophyllum

leiophyllum sensu Pobed. = G. hyrcanicum

linczevskyi Pobed. = G. ibicinum

lucidum All. - 1

macilentum Klok. & Zaverucha - 1

majmechense Bordz. = G. consanguineum

maximowiczii (Kom.) Pobed. (*Asperula maximowiczii* Kom.) - 5

maximum G. Moris = G. elongatum

mite Boiss. & Hohen. (*G. subvelutinum* (DC.) C. Koch subsp. *mite* (Boiss. & Hohen.) Ehrend.) - 2

moldavicum (Dobrescu) Franco (*Asperula moldavica* Dobrescu) - 1

mollugo L. (*G. tyrolense* Willd.) - 1, 2, 3, 5(alien)

- b. *pycnotrichum* H. Br. = G. pycnotrichum

mugodsharicum Pobed. (*G. boreale* L. subsp. *mugodsharicum* (Pobed.) Sojak) - 1, 3, 6

x **mutabile** Bess. - G. mollugo L. x G. ruthenicum Willd. - 1

neotauricum Klok. = Cruciata neotaurica

nupercreatum M. Pop. - 6

octonarium (Klok.) Soo (*Asperula octonaria* Klok., *Galium octonarium* (Klok.) Pobed. comb. superfl.) - 1, 2, 6

odessanum Klok. - 1

odoratum (L.) Scop. (*Asperula odorata* L.) - 1, 2, 3, 4, 5

olgae Klok. - 1

orientale Boiss. var. *elatius* Boiss. = G. grusinum

palustre L. - 1, 2, 3, 4, 6

- subsp. *elongatum* (C. Presl) Lange = G. elongatum

pamiro-alaicum Pobed. - 6

paniculatum (Bunge) Pobed. (*Asperula paniculata* Bunge) - 3, 4

paradoxum Maxim. - 1, 3, 4, 5

pedemontanum (Bell.) All. = Cruciata pedemontana

physocarpum Ledeb. (*G. boreale* L. subsp. *ussuriense* (Pobed.) Worosch., *G. dasypodum* Klok. p.p., *G. salicifolium* Klok., *G. ussuriense* Pobed., *G. volgense* Pobed.) - 1, 3, 4, 5, 6

platygalium (Maxim.) Pobed. (*Asperula platygalium* Maxim.) - 5

pojarkovae Pobed. - 6

x **polonicum** Blocki (*G. abaujense* Borb. subsp. *polonicum* (Blocki) Soo). - G. mollugo L. x G. verum L. - 1 *polonicum* sensu Pobed. = G. carpaticum

x **pomeranicum** Retz. - G. album Mill. x G. verum L. - 1

praeboreale Klok. = G. boreale

praemontanum T. Mardalejschvili - 2

pseudoaristatum Schur - 1(?)

pseudoasprellum Makino - 5

pseudoboreale Klok. = G. x pseudorubioides

pseudomollugo Klok. = G. album

pseudopolycarpon Somm. & Levier = Cruciata pseudopolycarpa

pseudorivale Tzvel. (*Asperula aparine* Bieb. non Galium aparine L., *A. karataviensis* Pavl., *Galium karataviense* (Pavl.) Pobed.) - 1, 2, 6

x **pseudorubioides** Klok. (*G. pseudoboreale* Klok., *G. rubioides* L. subsp. *pseudorubioides* (Schur) Soo & Draskovits var. *pseudorubioides* f. *pseudoboreale* (Klok.) Soo, *G. septentrionale* sensu Pobed. p.max.p.). - G. boreale L. x G. physocarpum Ledeb. - In places of the contact of parent species.

■ psilophyllum Ehrend. & Schonbeck-Temesy (*G. leiophyllum* Boiss. & Hohen. nom. illegit., *G. subvelutinum* (DC.) C. Koch subsp. *leiophyllum* Ehrend. p.p.)

pumilum Murr. - 1

pycnotrichum (H. Br.) Borb. (*G. mollugo* L. var. *pycnotrichum* H. Br., *G. album* Mill. subsp. *pycnotrichum* (H. Br.) Krendl) - 1

rivale (Sibth. & Smith) Griseb. (*Asperula rivalis* Sibth. & Smith) - 1, 3

rotundifolium L. (*G. scabrum* auct.) - 1, 2

rubioides L. (*G. articulatum* Lam., *G. dasypodum* Klok. p.p.) - 1, 2

- subsp. *pseudorubioides* (Schur) Soo & Draskovits var. *pseudorubioides* f. *pseudoboreale* (Klok.) Soo = G. x pseudorubioides

rugosum Galushko = Cruciata rugosa

ruprechtii Pobed. = G. brandegei and G. trifidum

ruthenicum Willd. (? *Asperula affrena* Klok., *G. verum* L. subsp. *ruthenicum* (Willd.) P. Fourn.) - 1, 2, 3, 4, 5, 6

Asperula affrena Klok. is probably a hybrid Galium octonarium (Klok.) Soo x G. verum L.

salicifolium Klok. = G. physocarpum

saturejifolium Trev. - 1, 2

saurense Litv. = G. densiflorum var. rosmarinifolium

saxatile auct. = G. hercynicum

scabrum auct. = G. rotundifolium

schelkownikowianum (Bordz.) Holub (*Asperula rivalis* Smith var. *schelkownikowiana* Bordz.) - 2

schischkinii Pobed. = Cruciata schischkinii

schultesii Vest = G. intermedium

semiamictum Klok. - 1

septentrionale sensu Pobed. = G. boreale and G. pseudorubioides

setaceum Lam. - 6

- subsp. *decaisnei* (Boiss.) Ehrend. = G. decaisnei

sevanense Pobed. = Cruciata sevanensis

songaricum Schrenk - 3, 6

sosnowskyi Manden. = Cruciata sosnowskyi

sphenophyllum Klok. (? *G. elatum* auct.) - 1

spurium L. (*G. aparine* L. f. *spurium* (L.) Boivin) - 1, 2, 3, 4, 5, 6

subbiflorum (Wieg.) Rydb. = G. brandegei

suberectum Klok. - 1

subnemorale Klok. & Zaverucha - 1

subuliferum Somm. & Levier - 2

subvelutinum (DC.) C. Koch subsp. *hyrcanicum* (C.A. Mey.) Ehrend. = G. hyrcanicum

- subsp. *leiophyllum* Ehrend. p.p. = G. psilophyllum

- subsp. *mite* (Boiss. & Hohen.) Ehrend. = G. mite

- subsp. *problematicum* Ehrend. = G. achurense and G. vartanii

taquetii Levl. (*G. columbianum* Rydb., *G. trifidum* L. subsp. *columbianum* (Rydb.) Hult., *G. trifidum* sensu Pobed.) - 5

tauricum (Pall. ex Will.) Roem. & Schult. = Cruciata taurica

tenderiense Klok. - 1

tenuissimum Bieb. - 1, 2, 6

- subsp. *trichophorum* (Kar. & Kir.) Ehrend. = G. trichophorum

tianschanicum M. Pop. (*G. kiapazi* Manden.) - 2, 6

tinctorium (L.) Scop. (*Asperula tinctoria* L.) - 1, 3

tomentellum Klok. - 1

transcarpaticum Stojko & Tasenkevitsch - 1

transcaucasicum Stapf = G. ghilanicum

trichophorum Kar. & Kir. (*G. tenuissimum* Bieb. subsp. *trichophorum* (Kar. & Kir.) Ehrend.) - 6

tricorne Stokes, p.p. = G. tricornutum

tricornutum Dandy (*G. tricorne* Stokes, p.p. excl. typo) - 1, 2, 6

trifidum L. (*G. ruprechtii* Pobed. p.max.p. incl. typo) - 1, 3, 4, 5, 6

- subsp. *columbianum* (Rydb.) Hult. = G. taquetii

- subsp. *subbiflorum* (Wieg.) Puff = G. brandegei

- var. *subbiflorum* Wieg. = G. brandegei

trifidum sensu Pobed. = G. taquetii

trifloriforme Kom. = G. triflorum

triflorum Michx. (*G. japonicum* Makino, *G. trifloriforme* Kom.) - 1, 3, 4, 5

turkestanicum Pobed. - 6

tyraicum Klok. - 1

tyrolense Willd. = G. mollugo

uliginosum L. - 1, 2, 3, 4, 5, 6

ussuriense Pobed. = G. physocarpum

vaillantii DC. - 1, 3, 4, 5, alien in all regions

valantioides Bieb. - 2

valdepilosum H. Br. - 1

valentinae Galushko = Cruciata valentinae

vartanii Grossh. (*G. subvelutinum* (DC.) C. Koch subsp. *problematicum* Ehrend. p.p.) - 2

vassilczenkoi Pobed. - 6

vernum Scop. = Cruciata glabra

verticillatum Danth. - 1, 2, 6

verum L. - 1, 2, 3, 4, 5, 6

- subsp. *ruthenicum* (Willd.) P. Fourn. = G. ruthenicum

- subsp. *wirtgenii* (F. Schultz) Oborny = G. wirtgenii

- var. *rosmarinifolium* Bunge = G. densiflorum var. rosmarinifolium

volgense Pobed. = G. physocarpum

volhynicum Pobed. (*Asperula tyraica* Bess., *Galium campanulatum* Vill. subsp. *tyraicum* (Bess.) Sojak, *G. glaucum* L. subsp. *tyraicum* (Bess.) Sojak) - 1

wirtgenii F. Schultz (*G. verum* L. subsp. *wirtgenii* (F. Schultz) Oborny) - 1, 3, 4, 5

xeroticum (Klok.) Soo (*Asperula xerotica* Klok., *Galium xeroticum* (Klok.) Pobed. comb. superfl.) - 1

xylorrhizum Boiss. & Huet - 2

zelenetzkii Klok. = G. calcareum

Jaubertia auct. = Neogaillonia

szowitzii (DC.) Takht. = Neogaillonia szowitzii

Karamyschewia Fisch. & C.A. Mey.

hedyotoides Fisch. & C.A. Mey. - 2

Leptunis Stev.
trichodes (J. Gay) Schischk. - 2, 6

Microphysa Schrenk
elongata (Schrenk) Pobed. - 6

Mitchella L.
undulata Siebold & Zucc. - 5

Neogaillonia Lincz. (*Gaillonia* A. Rich. ex DC. 1830, non *Gaillona* Bonnemaison, 1828, *Jaubertia* auct.)
asperuliformis (Lincz.) Lincz. (*Gaillonia asperuliformis* Lincz.) - 6
botschantzevii Lincz. - 6
bruguierii (A. Rich. ex DC.) Lincz. (*Gaillonia bruguierii* A. Rich. ex DC.) - 6
bucharica (B. Fedtsch. & Desjat.) Lincz. (*Gaillonia bucharica* B. Fedtsch. & Desjat.) - 6
dubia (Aitch. & Hemsl.) Lincz. (*Gaillonia dubia* Aitch. & Hemsl.) - 6
iljinii Lincz. - 6
inopinata (Lincz.) Lincz. (*Gaillonia inopinata* Lincz.) - 6
mestscherjakovii Lincz. - 6
szowitzii (DC.) Lincz. (*Gaillonia szowitzii* DC., *Jaubertia szowitzii* (DC.) Takht.) - 2
trichophylla (M. Pop. ex Tscherneva) Lincz. (*Gaillonia trichophylla* M. Pop. ex Tscherneva) - 6
vassilczenkoi Lincz. - 6

Phuopsis (Griseb.) Hook. fil.
stylosa (Trin.) Hook. fil. - 2

Rubia L.
alaica Pachom. - 6
albo-costata Ehrend. = R. rigidifolia
chinensis Regel & Maack - 5
chitralensis Ehrend. - 6
cordifolia L. (*R. cordifolia* var. *pratensis* Maxim., *R. cordifolia* f. *pratensis* (Maxim.) Kitag., *R. cordifolia* subsp. *pratensis* (Maxim.) Kitam.) - 4, 5
- subsp. *pratensis* (Maxim.) Kitam. = R. cordifolia
- var. *pratensis* Maxim. = R. cordifolia
- f. *pratensis* (Maxim.) Kitag. = R. cordifolia
cretacea Pojark. - 6
deserticola Pojark. - 6
dolichophylla Schrenk - 6
florida Boiss. - 6
iberica (Fisch. ex DC.) C. Koch = R. tinctorum
jesoensis (Miq.) Miyabe & Miyake - 5
komarovii Pojark. - 6
krascheninnikovii Pojark. - 3, 6
laevissima Tscherneva - 6
laxiflora Gontsch. - 6
pavlovii Bajt. & Myrzakulov - 6
rechingeri Ehrend. - 6
regelii Pojark. - 6
rezniczenkoana Litv. - 6
rigidifolia Pojark. (*R. albo-costata* Ehrend.) - 2
schugnanica B. Fedtsch. ex Pojark. - 6
tatarica (Trev.) Fr. Schmidt - 1, 2, 3, 6
tibetica Hook. fil. - 6
tinctorum L. (*R. iberica* (Fisch. ex DC.) C. Koch) - 1, 2, 6

transcaucasica Grossh. - 3

Sherardia L.
arvensis L. - 1, 2, 5(alien), 6

Valantia L.
articulata L. = Cruciata articulata
cucullaria L. = Callipeltis cucullaria
muralis L. - 2(?)
pedemontana Bell. = Cruciata pedemontana

RUPPIACEAE Hutch.
Buccaferrea Petagna = Ruppia
cirrhosa Petagna = Ruppia cirrhosa

Ruppia L. (*Buccaferrea* Petagna)
brachypus J. Gay (*R. maritima* L. subsp. *brachypus* (J. Gay) A. Löve, comb. superfl., *R. maritima* subsp. *brachypus* (J. Gay) Schlegel) - 1, 6
cirrhosa (Petagna) Grande (*Buccaferrea cirrhosa* Petagna) - 1, 2, 6
- subsp. *occidentalis* (S. Wats.) A. & D. Löve = R. occidentalis
drepanensis Tineo - 1, 2, 3, 6
maritima L. (*R. maritima* subsp. *rostellata* (Koch) Aschers. & Graebn., *R. rostellata* Koch) - 1, 2, 3, 4, 5, 6
- subsp. *brachypus* (J. Gay) A. Löve = R. brachypus
- subsp. *brachypus* (J. Gay) Schlegel = R. brachypus
- subsp. *rostellata* (Koch) Aschers. & Graebn. = R. maritima
occidentalis S. Wats. (*R. cirrhosa* (Petagna) Grande subsp. *occidentalis* (S. Wats.) A. & D. Löve) - 5
rostellata Koch = R. maritima
spiralis L. ex Dumort. - 1

RUSCACEAE Hutch.
Danae Medik.
racemosa (L.) Moench - 2

Ruscus L.
aculeatus L. (*R. ponticus* Woronow ex Grossh.) - 1, 2
colchicus P.F. Yeo (*Ruscus hypophyllum* auct.) - 2
hypoglossum L. - 1
hypophyllum auct. = R. colchicus
hyrcanus Woronow - 1, 2
ponticus Woronow ex Grossh. = R. aculeatus

RUTACEAE Juss.
*Citrus L.
*****limon** (L.) Burm. fil. - 2
*****unshiu** (Swingle) Marc. - 2

Dictamnus L.
albus L. - 1
- var. *dasycarpus* (Turcz.) T.N. Liou & Y.H. Chang = D. dasycarpus
angustifolius G. Don fil. ex Sweet - 3, 6
caucasicus (Fisch. & C.A. Mey.) Grossh. - 1, 2
dasycarpus Turcz. (*D. albus* L. var. *dasycarpus* (Turcz.) T.N. Liou & Y.H. Chang) - 4, 5

gymnostylis Stev. - 1, 2
tadshikorum Vved. - 6

*Fortunella Swingle
*japonica (Thunb.) Swingle - 2

Haplophyllum Adr. Juss.
acutifolium (DC.) G. Don fil. - 6
affine (Aitch. & Hemsl.) Korov. - 6
alberti-regelii Korov. - 6
armenum Spach (*H. bourgaei* Boiss., *H. ibericum* Kem.-
   Nath.) - 2
bourgaei Boiss. = H. armenum
bucharicum Litv. - 6
bungei Trautv. - 6
ciliatum Griseb. = H. suaveolens
ciscaucasicum (Rupr.) Grossh. & Vved. (*H. villosum*
   (Bieb.) G. Don fil. subsp. *ciscaucasicum* (Rupr.) C.C.
   Townsend) - 2
davuricum (L.) G. Don fil. - 3, 4
dshungaricum N. Rubtz. - 6
dubium Korov. - 6
eugenii-korovinii Pavl. - 6
ferganicum Vved. - 6
foliosum Vved. = H. griffithianum
griffithianum Boiss. (*H. foliosum* Vved., *H. leptomerum*
   Lincz. & Vved., *H. tenuisectum* Lincz. & Vved.) - 6
ibericum Kem.-Nath. = H. armenum
kowalenskyi Stschegl. - 2
latifolium Kar. & Kir. - 6
leptomerum Lincz. & Vved. = H. griffithianum
monadelphum Afan. - 6
multicaule Vved. - 6
obtusifolium (Ledeb.) Ledeb. - 6
pedicellatum Bunge - 6
perforatum Kar. & Kir. (*H. perforatum* (Bieb.) Kar. & Kir.
   ex Vved. nom. illegit., *H. sieversii* Fisch.) - 6
popovii Korov. - 6
ramosissimum (Pauls.) Vved. - 6
robustum Bunge - 6
schelkovnikovii Grossh. - 2
sieversii Fisch. = H. perforatum
suaveolens (DC.) G. Don fil. (*H. ciliatum* Griseb.) - 1
suaveolens sensu Vved. = H. thesioides
tauricum Spach = H. thesioides
tenue Boiss. - 2
tenuisectum Lincz. & Vved. = H. griffithianum
thesioides (Fisch. ex DC.) G. Don fil. (*H. tauricum* Spach,
   H. suaveolens sensu Vved.) - 1, 2
versicolor Fisch & C.A. Mey. (*H. versicolor* subsp. *australe*
   C.C. Townsend) - 6
- subsp. *australe* C.C. Townsend = H. versicolor
villosum (Bieb.) G. Don fil. - 2
- subsp. *ciscaucasicum* (Rupr.) C.C. Townsend = H. ciscaucasicum
vvedenskyi Nevski - 6

Phellodendron Rupr.
amurense Rupr. - 1(cult.), 2(cult.), 5, 6(cult.)
- var. *sachalinense* Fr. Schmidt = Ph. sachalinense
sachalinense (Fr. Schmidt) Sarg. (*Ph. amurense* Rupr. var.
   sachalinense Fr. Schmidt) - 5

*Poncirus Rafin.
*trifoliata (L.) Rafin. - 2

*Ptelea L.
*trifoliata L. - 1

Ruta L.
divaricata Ten. - 1
*graveolens L. (? *R. hortensis* Mill.) - 1
hortensis Mill. = R. graveolens

Skimmia Thunb.
japonica Thunb. var. *repens* (Nakai) Ohwi = S. repens
- f. *repens* (Nakai) Hara = S. repens
repens Nakai (*S. japonica* Thunb. f. *repens* (Nakai) Hara, *S.
   japonica* var. *repens* (Nakai) Ohwi) - 5

SALICACEAE Mirb.
Balsamiflua Griff. = Populus
diversifolia (Schrenk) Kimura = Populus diversifolia
euphratica (Olivier) Kimura = Populus euphratica
litwinowiana (Dode) Kimura = Populus diversifolia
pruinosa (Schrenk) Kimura = Populus pruinosa

Chosenia Nakai
arbutifolia (Pall.) A. Skvorts. (*Salix arbutifolia* Pall., *Chosenia macrolepis* (Turcz.) Kom.) - 4, 5
macrolepis (Turcz.) Kom. = Ch. arbutifolia

Populus L. (*Balsamiflua* Griff., *Turanga* (Bunge) Kimura)
afghanica (Aitch. & Hamsl.) Schneid. (*P. nigra* L. var.
   afghanica Aitch. & Hemsl., *P. afghanica* var. *tadshikistanica* (Kom.) C. Wang & C.Y. Yang, *P. iliensis* Drob.,
   P. kanjilaliana Dode subsp. *usbekistanica* (Kom.)
   Poljak., *P. tadshikistanica* Kom., *P. usbekistanica*
   Kom., *P. usbekistanica* subsp. *tadshikistaica* (Kom.)
   Bugula) - 6
- var. *tadshikistanica* (Kom.) C. Wang & C.Y. Yang = P.
   afghanica
alba L. (*P. alba* var. *nivea* Ait., *P. bachofenii* Wierzb. ex
   Rochel, *P. bolleana* Lauche, *P. nivea* (Ait.) Willd., *P.
   pseudonivea* Grossh.) - 1, 2, 3, 6
- subsp. **major** (Mill.) R. Kam. nom. invalid. (*P. major*
   Mill.) - 2
- var. *canescens* Ait. = P. x canescens
- var. *nivea* Ait. = P. alba
amurensis Kom. - 5
ariana Dode (*Turanga ariana* (Dode) Kimura) - 6
bachofenii Wierzb. ex Rochel = P. alba
baicalensis Kom. (*P. suaveolens* Fisch. subsp. *baicalensis*
   (Kom.) Egor. & Sipl.) - 4
*balsamifera L. - 1
berkarensis Poljak. - 6
*x berolinensis (C. Koch) Dipp. (*P. hybrida* Bieb. var. *berolinensis* C. Koch). - P. italica (DuRoi) Moench x P.
   laurifolia Ledeb. - 1
bolleana Lauche = P. alba
canadensis auct. = P. deltoides
x *canadensis* Moench var. *rasumowskiana* Schroed. ex
   Regel = P. x rasumowskiana
*candicans Ait. - 1, 6
x **canescens** (Ait.) Smith (*P. alba* L. var. *canescens* Ait., *P.
   hybrida* Bieb.). - P. alba L. x P. tremula L. - 1, 2, 3, 6
cataracti Kom. - 6

cathayana auct. = P. talassica
davidiana Dode (*P. tremula* L. subsp. *davidiana* (Dode) Hult.) - 5
***deltoides** Marsh. (*P. canadensis* auct.) - 1, 3, 6
densa Kom. = P. talassica
diversifolia Schrenk (*Balsamiflua diversifolia* (Schrenk) Kimura, *B. litwinowiana* (Dode) Kimura, *Populus litwinowiana* Dode, *Turanga diversifolia* (Schrenk) Kimura, *T. litwinowiana* (Dode) Kimura) - 6
euphratica Olivier (*Balsamiflua euphratica* (Olivier) Kimura, *Populus transcaucasica* Jarm. ex Grossh., *Turanga euphratica* (Olivier) Kimura) - 2
- f. *pruinosa* (Schrenk) Nevski = P. pruinosa
flexibilis Roz. (*P. nigra* L. subsp. *flexibilis* (Roz.) R. Kam.) - 2
gileadensis Rouleau - 1(?)
x **glaucicomans** Dode (*Turanga* x *glaucicomans* (Dode) Kimura). - P. ariana Dode x P. pruinosa Schrenk - 6
***gracilis** Grossh. - 2
hybrida Bieb. = P. x canescens
- var. *berolinensis* C. Koch = P. x berolinensis
hyrcana Grossh. - 2
iliensis Drob. = P. afghanica
***italica** (DuRoi) Moench (*P. nigra* L. var. *italica* DuRoi, *P. nigra* subsp. *pyramidalis* (Roz.) Celak., *P. pyramidalis* Roz.) - 1, 2, 6
jezoensis Nakai - 5
kanjilaliana Dode = P. talassica
- subsp. *usbekistanica* (Kom.) Poljak. = P. afghanica
komarovii Ja. Vassil. ex Worosch. = P. maximowiczii
koreana Rehd. - 5
x **krauseana** Dode. - P. nigra L. x P. pruinosa Schrenk ? - 6
laurifolia Ledeb. - 3, 4, 6
litwinowiana Dode = P. diversifolia
longifolia Fisch. - 1
macrocarpa (Schrenk) Pavl. & Lipsch. = P. pilosa
major Mill. = P. alba subsp. major
maximowiczii A. Henry (*P. komarovii* Ja. Vassil. ex Worosch., *P. suaveolens* Fisch. subsp. *maximowiczii* (A. Henry) Tatew., *P. ussuriensis* Kom.) - 5
***x moskoviensis** Schroed. - P. laurifolia Ledeb. x P. suaveolens Fisch. - 1
nigra L. (*P. nigra* var. *sosnowskyi* (Grossh.) Makaschvili, *P. sosnowskyi* Grossh.) - 1, 2, 3, 4, 6
- subsp. *flexibilis* (Roz.) R. Kam. = P. flexibilis
- subsp. *pyramidalis* (Roz.) Celak. = P. italica
- var. *afghanica* Aitch. & Hemsl. = P. afghanica
- var. *italica* DuRoi = P. italica
- var. *sosnowskyi* (Grossh.) Makaschvili = P. nigra
nivea (Ait.) Willd. = P. alba
pamirica Kom. - 6
***x petrowskiana** Schroed. - P. deltoides Marsh. x P. suaveolens Fisch. - 1
pilosa Rehd. (*P. macrocarpa* (Schrenk) Pavl. & Lipsch., *P. suaveolens* Fisch. var. *macrocarpa* Schrenk) - 6
pruinosa Schrenk (*Balsamiflua pruinosa* (Schrenk) Kimura, *Populus euphratica* Olivier f. *pruinosa* (Schrenk) Nevski, *Turanga pruinosa* (Schrenk) Kimura) - 6
pseudobalsamifera Fisch. = P. talassica
pseudonivea Grossh. = P. alba
pseudotremula N. Rubtz. = P. tremula
pyramidalis Roz. = P. italica
***x rasumowskiana** (Regel) Schneid. (*P. canadensis* Moench var. *rasumowskiana* Schroed. ex Regel.). - P. laurifolia Ledeb. x P. nigra L.
schischkinii Grossh. - 2
sieboldii Miq. - 5

***simonii** Carr. - 1, 2, 6
sosnowskyi Grossh. = P. nigra
suaveolens Fisch. - 4, 5
- subsp. *baicalensis* (Kom.) Egor. & Sipl. = P. baicalensis
- subsp. *maximowiczii* (A. Henry) Tatew. = P. maximowiczii
- var. *macrocarpa* Schrenk = P. pilosa
tadshikistanica Kom. = P. afghanica
talassica Kom. (*P. densa* Kom., ? *P. kanjilaliana* Dode, nom. ambig., *P. pseudobalsamifera* Fisch., *P. cathayana* auct.) - 6
x **tianschanica** V. Tkatschenko. - P. afganica (Aitch. & Hemsl.) Schneid. x P. talassica Kom. - 6
transcaucasica Jarm. ex Grossh. = P. euphratica
tremula L. (*P. pseudotremula* N. Rubtz.) - 1, 2, 3, 4, 5, 6
- subsp. *davidiana* (Dode) Hult. = P. davidiana
- subsp. **microtremula** (Grossh.) R. Kam. (*P. tremula* var. *microtremula* Grossh.) - 2
***tristis** Fisch. - 1
usbekistanica Kom. = P. afghanica
- subsp. *tadshikistaica* (Kom.) Bugula = P. afghanica
ussuriensis Kom. = P. maximowiczii
villosa Lang - 2

Salix L.

abscondita Laksch. (*S. enanderi* B. Floder., *S. floderusii* Nakai, p.p., *S. oleninii* Nas., *S. raddeana* Laksch. ex Nas., *S. sugawarana* Kumira) - 4, 5
acmophylla Boiss. (*S. daviesii* Boiss., *S. persica* Boiss.) - 6
acutifolia Willd. - 1, 2, 3, 4(cult.), 6
- subsp. **pomeranica** (Koch) Masan - 1
aegyptiaca L. (*S. phlomoides* Bieb. p.p.) - 2, 6
alatavica Kar. & Kir. ex Stschegl. - 3, 4, 6
alaxensis Cov. (*S. speciosa* Hook. & Arn. 1832, non Host, 1828) - 4, 5
alba L. - 1, 2, 3, 4, 6
- subsp. *micans* (Anderss.) Rech. fil. = S. micans
- var. *australior* (Anderss.) Poljak. = S. excelsa
alberti Regel = S. tenuijulis
alexii-skvortzovii A. Khokhr. - 4, 5
alifera Goerz = S. pseudomedemii
alpina Scop. (*S. jacquinii* Host, *S. myrsinites* L. subsp. *alpina* (Scop.) E. Murr.) - 1
altaica Lundstr. = S. torulosa
amygdalina L. = S. triandra
anadyrensis B. Floder. = S. pulchra
angrenica Drob. = S. olgae
anomala E. Wolf = S. rosmarinifolia
apoda Trautv. (*S. hastata* L. var. *apoda* (Trautv.) Laksch.) - 2
aquilonia Kimura = S. nakamurana
arbuscula L. - 1, 3
arbutifolia Pall. = Chosenia arbutifolia
arctica Pall. (*S. ehlei* B. Floder.) - 1, 3, 4, 5
- subsp. *crassijulis* (Trautv.) A. Skvorts. = S. crassijulis
- subsp. **jamutaridensis** Petrovsky - 4
- subsp *torulosa* (Trautv.) Hult. = S. torulosa
- subsp. *torulosa* (Trautv.) A. Skvorts. = S. torulosa
argyracea E. Wolf - 6
armena Schischk. = S. triandra
armeno-rossica A. Skvorts. - 2
aurita L. - 1, 3
australior Anderss. = S. excelsa
***babylonica** L. (*S. matsudana* Koidz.) - 1, 2, 5, 6
bakko Kimura = S. caprea
barclayi auct. = S. hastata
bebbiana Sarg. (*S. cinerascens* (Wahlenb.) B. Floder., *S.*

depressa L. var. *macropoda* (Stschegl.) Poljak. p.p., *S. floderusii* Nakai, p.p., *S. livida* Wahlenb. var. *cinerascens* Wahlenb., *S. macropoda* Stschegl., *S. orotchonorum* Kimura, *S. starkeana* Willd. subsp. *cinerascens* (Wahlenb.) Hult., *S. xerophila* B. Floder.) - 1, 3, 4, 5, 6

behringica Seemen - 5
    Dubious species.
berberifolia Pall. - 4, 5
- subsp. *brayi* (Ledeb.) A. Skvorts. = S. brayi
- subsp. *fimbriata* A. Skvorts. = S. fimbriata
- subsp. *kamtschatica* A. Skvorts. = S. kamtschatica
- subsp. *kimurana* (Miyabe & Tatew.) A. Skvorts. = S. kimurana
- subsp. *tschuktschorum* (A. Skvorts.) Worosch. = S. tschuktschorum
- var. *kimurana* Miyabe & Tatew. = S. kimurana
bicolor Ehrh. ex Willd. = S. rhaetica
- subsp. *rhaetica* (Anderss.) B. Floder. = S. rhaetica
blakii Goerz (*S. blakolgae* Drob., *S. linearifolia* E. Wolf, 1903, non Goeppert, 1855) - 6
blakolgae Drob. = S. blakii
*****blanda** Anderss. - 2
boganidensis Trautv. (*S. kolymensis* Seemen) - 4, 5
borealis (Fries) Nas. (*S. myrsinifolia* Salisb. subsp. *borealis* (Fries) A. Skvorts., *S. myrsinifolia* subsp. *borealis* (Fries) Hyl. comb. superfl.) - 1
bornmuelleri Hausskn. (*S. triandra* L. subsp. *bornmuelleri* (Hausskn.) A. Skvorts.) - 2
brachycarpa Nutt. subsp. **niphoclada** (Rydb.) Argus (*S. niphoclada* Rydb.) - 5
brachypoda (Trautv. & C.A. Mey.) Kom. (*S. repens* L. subsp. *brachypoda* (Trautv. & C.A. Mey.) Worosch., *S. rosmarinifolia* L. var. *brachypoda* (Trautv. & C.A. Mey.) Y.L. Chou) - 4, 5
brayi Ledeb. (*S. berberifolia* Pall. subsp. *brayi* (Ledeb.) A. Skvorts.) - 3, 4
brevijulis Turcz. = S. divaricata
burjatica Nas. = S. dasyclados
caesifolia Drob. = S. purpurea
caprea L. (*S. bakko* Kimura, *S. coaetanea* (C. Hartm.) B. Floder., *S. hultenii* B. Floder.) - 1, 2, 3, 4, 5, 6
capusii Franch. (*S. coerulea* E. Wolf, 1903, non Smith, 1812) - 6
cardiophylla Trautv. & C.A. Mey. - 4, 5
- subsp. *urbaniana* (Seemen) A. Skvorts. = S. urbaniana
caspica Pall. - 1, 2, 3, 6
- var. *michelsonii* (Goerz ex Nas.) Poljak. = S. michelsonii
caucasica Anderss. (*S. caucasica* var. *palibinii* (Goerz) Grossh., *S. caucasica* var. *paracaucasica* (Goerz) Grossh., *S. daghestanica* Goerz, *S. palibinii* Goerz, *S. paracaucasica* Goerz) - 2
- var. *palibinii* (Goerz) Grossh. = S. caucasica
- var. *paracaucasica* (Goerz) Grossh. = S. caucasica
chamissonis Anderss. (*S. pulchroides* Kimura) - 5
- subsp. **integerrima** Worosch. - 5
chlorostachya Turcz. = S. rhamnifolia
cinerascens (Wahlenb.) B. Floder. = S. bebbiana
cinerea L. (*S. deserticola* Goerz ex Pavl., *S. phlomoides* Bieb. p.p.) - 1, 2, 3, 6
coaetanea (C. Hartm.) B. Floder. = S. caprea
coerulangrenica Drob. = S. olgae
coerulea E. Wolf = S. capusii
coeruleiformis Drob. = S. niedzwieckii
coesia Vill. (*S. tuvinensis* Gudoschn.) - 3, 4, 6
- subsp. **coesia** - 6
- subsp. **tschujensis** N. Bolschakov - 3, 4
crassijulis Trautv. (*S. arctica* Pall. subsp. *crassijulis*

(Trautv.) A. Skvorts., *S. pallasii* Anderss.) - 5
cuneata Turcz. ex Ledeb. = S. sphenophylla
cyclophylla Rydb. (*S. ovalifolia* Trautv. subsp. *cyclophylla* (Rydb.) Jurtz. & Petrovsky) - 5
daghestanica Goerz = S. caucasica
dahurica Turcz. ex Laksch. = S. miyabeana
daphnoides Vill. - 1
darpirensis Jurtz. & A. Khokhr. - 4, 5
dasyclados Wimm. (*S. burjatica* Nas., *S. jacutica* Nas.) - 1, 3, 4, 6
daviesii Boiss. = S. acmophylla
depressa auct. p.p. = S. starkeana
depressa L. subsp. *iliensis* (Regel) Hiit. = S. iliensis
- var. *iliensis* (Regel) Poljak. = S. iliensis
- var. *macropoda* (Stschegl.) Poljak. = S. bebbiana and S. iliensis
deserticola Goerz ex Pavl. = S. cinerea
divaricata Pall. (*S. brevijulis* Turcz., *S. leptoclados* Anderss., *S. phylicifolia* L. subsp. *divaricata* (Pall.) N. Bolschakov) - 3, 4, 5
- subsp. *kalarica* (A. Skvorts.) A. Skvorts. = S. kalarica
- subsp. *parallelinervis* (B. Floder.) Worosch. = S. parallelinervis
- subsp. *pulchra* (Cham.) Worosch. = S. pulchra
dodgeana Rydb. (*S. rotundifolia* Trautv. subsp. *dodgeana* (Rydb.) Argus, *S. rotundifolia* var. *dodgeana* (Rydb.) E. Murr.) - 5
dolichostyla Seemen = S. pierotii
dshugdshurica A. Skvorts. - 4, 5
ehlei B. Floder. = S. arctica
elaeagnos Scop. (*S. incana* Schrank) - 1
elbursensis Boiss. (*S. roopi* (Goerz) Grossh.) - 2
elegans Bess. = S. myrtilloides
enanderi B. Floder. = S. abscondita
erythrocarpa Kom. - 5
euapiculata Nas. = S. excelsa
excelsa S.G. Gmel. (*S. alba* L. var. *australior* (Anderss.) Poljak., *S. australior* Anderss., *S. euapiculata* Nas., *S. litwinowii* Goerz ex Nas., *S. oxica* Dode) - 2, 6
fedtschenkoi Goerz - 6
ferganensis Nas. = S. pycnostachya
fimbriata (A. Skvorts.) Czer. (*S. berberifolia* Pall. subsp. *fimbriata* A. Skvorts.) - 4
flavellinervis A. Khokhr. - 5
floderusii Nakai = S. abscondita and S. bebbiana
fragilis L. - 1, 2, 3 (*S. alba* x *S. fragilis*), 6 (*S. alba* x *S. fragilis*)
fulcrata Anderss. = S. pulchra and S. udensis
fumosa Turcz. = S. saxatilis
fuscata Goerz = S. pseudomedemii
fuscescens Anderss. (? *S. fuscescens* Trautv. subsp. *poronaica* (Kimura) A. Skvorts., ? *S. poronaica* Kimura) - 4, 5
- subsp. *poronaica* (Kimura) A. Skvorts. = S. fuscescens
gilgiana Seemen - 5
glacialis Andress. (*S. ovalifolia* Trautv. var. *glacialis* (Anderss.) Argus, *S. ovalifolia* subsp. *glacialis* (Anderss.) Jurtz. & Petrovsky, *S. ovalifolia* subsp. *glacialis* (Anderss.) E. Murr. comb. superfl.) - 5
glandulifera B. Floder. = S. lanata
glauca L. (*S. glauca* subsp. *stipulifera* (B. Floder. ex Hayren) Hiit., *S. seemannii* Rydb., *S. stipulifera* B. Floder. ex Hayren) - 1, 3, 4, 5
- subsp. **callicarpaea** (Trautv.) Bocher - 5
- subsp. *stipulifera* (B. Floder. ex Hayren) Hiit. = S. glauca
gmelinii Pall. = S. viminalis
gordejevii Chang & Skvorts. - 4
gracilior (Siuzew) Nakai = S. miyabeana

gracilistyla Miq. - 5
gracilistyliformis Korkina - 5
hastata L. (*S. psiloides* (B. Floder.) Kom., *S. barclayi* auct.) - 1, 3, 4, 5
- var. *apoda* (Trautv.) Laksch. = S. apoda
helvetica Vill. subsp. *krylovii* (E. Wolf) B. Floder. = S. krylovii
herbacea L. - 1
heterandra Drob. = S. olgae
hidakamontana Hara (*S. subreniformis* Kimura) - 5
hidewoi Koidz. = S. tontomussirensis
holargyrea Goerz = S. pycnostachya
hultenii B. Floder. = S. caprea
hypericifolia Golosk. = S. songarica
iliensis Regel (*S. depressa* L. subsp. *iliensis* (Regel) Hiit., *S. depressa* var. *iliensis* (Regel) Poljak., *S. depressa* var. *macropoda* (Stschegl.) Poljak. p.p.) - 6
incana Schrank = S. elaeagnos
integra Thunb. - 5
issykiensis Goerz ex Nas. = S. kirilowiana
jacquinii Host = S. alpina
jacutica Nas. = S. dasyclados
jenisseensis (Fr. Schmidt) B. Floder. (*S. nigricans* Smith var. *jenisseensis* Fr. Schmidt, *S. borealis* (Fries) Nas. p.max.p. excl. typo, *S. viridula* (Anderss.) Nas. 1936, non Anderss. 1858) - 1, 3, 4, 5
jurtzevii A. Skvorts. - 5
kalarica (A. Skvorts.) Worosch. (*S. pulchra* Cham. subsp. *kalarica* A. Skvorts., *S. divaricata* Pall. subsp. *kalarica* (A. Skvorts.) A. Skvorts.) - 4, 5
kamtschatica (A. Skvorts.) Worosch. (*S. berberifolia* Pall. subsp. *kamtschatica* A. Skvorts., *S. tschuktschorum* A. Skvorts. subsp. *kamtschatica* (A. Skvorts.) Vorobiev) - 5
kangensis Nakai (*S. pierotii* sensu Nas.) - 5
karelinii Turcz. ex Stschegl. (*S. prunifolia* Kar. & Kir. 1842, non Smith, 1804) - 6
kazbekensis A. Skvorts. - 2
ketoiensis Kimura = S. nakamurana
khokhriakovii A. Skvorts. - 5
kikodseae Goerz - 5
kimurana (Miyabe & Tatew.) Miyabe & Tatew. (*S. berberifolia* Pall. var. *kimurana* Miyabe & Tatew., *S. berberifolia* subsp. *kimurana* (Miyabe & Tatew.) A. Skvorts.) - 5
kirilowiana Stschegl. (*S. issykiensis* Goerz ex Nas., *S. lipskyi* Nas.) - 6
kitaibeliana Willd. = S. retusa
kochiana Trautv. - 3, 4, 5
koidzumii Kimura - 5
kolaensis Schljak. = S. myrsinifolia
kolymensis Seemen = S. boganidensis
komarovii E. Wolf = S. pycnostachya
koreensis Anderss. = S. pierotii
***koriyanagi** Kimura ex Goerz - 5
korshinskyi Goerz = S. pycnostachya
krylovii E. Wolf (*S. helvetica* Vill. subsp. *krylovii* (E. Wolf) B. Floder.) - 3, 4, 5
x **kudoi** Kimura. - S. fuscescens Anderss. x S. udensis Trautv. & C.A. Mey. - 5
kurilensis Koidz. (*S. phanerodictya* Kimura) - 5
kuznetzowii Laksch. ex Goerz - 2
lanata L. (*S. glandulifera* B. Floder., *S. lanata* subsp. *glandulifera* (B. Floder.) Hiit.) - 1, 3, 4, 5
- subsp. *glandulifera* (B. Floder.) Hiit. = S. lanata
- subsp. *richardsonii* (Hook.) A. Skvorts. = S. richardsonii
lapponum L. - 1, 3, 4
ledebouriana Trautv. - 3, 4, 6

lenensis B. Floder. = S. myrtilloides
lepidostachya Seemen = S. miyabeana
leptoclados Anderss. = S. divaricata
liliputa Nas. = S. turczaninowii
linearifolia E. Wolf = S. blakii
lipskyi Nas. = S. kirilowiana
litwinowii Goerz ex Nas. = S. excelsa
livida sensu Nas. p.p. = S. taraikensis
livida Wahlenb. = S. starkeana
- var. *cinerascens* Wahlenb. = S. bebbiana
macilenta Anderss. - 5
Dubious species.
macropoda Stschegl. = S. bebbiana
macrostachya E. Wolf = S. pycnostachya
margaritifera E. Wolf = S. pycnostachya
matsudana Koidz. = S. babylonica
maximowiczii Kom. - 5
medwedewii Dode = S. triandra
mezereoides E. Wolf = S. udensis
micans Anderss. (*S. alba* L. subsp. *micans* (Anderss.) Rech. fil.) - 2
michelsonii Goerz ex Nas. (*S. caspica* Pall. var. *michelsonii* (Goerz ex Nas.) Poljak.) - 6
microstachya Turcz. ex Trautv. - 4
miyabeana Seemen (*S. dahurica* Turcz. ex Laksch., *S. gracilior* (Siuzew) Nakai, *S. lepidostachya* Seemen, *S. mongolica* Siuzew, *S. tenuifolia* Turcz. ex E. Wolf, 1903, non Smith, 1792) - 4, 5
mongolica Siuzew = S. miyabeana
myrsinifolia Salisb. (*S. kolaensis* Schljak., *S. nigricans* Smith) - 1, 3, 4
- subsp. *borealis* (Fries) Hyl. = S. borealis
- subsp. *borealis* (Fries) A. Skvorts. = S. borealis
myrsinites L. - 1
- subsp. *alpina* (Scop.) E. Murr. = S. alpina
myrtilloides L. (*S. elegans* Bess., *S. lenensis* B. Floder) - 1, 3, 4, 5
- subsp. *ustnerensis* N. Bolschakov - 4
nakamurana Koidz. (*S. aquilonia* Kimura, p.p., *S. ketoiensis* Kimura, *S. rashuwensis* Kimura) - 5
nasarovii A. Skvorts. - 4
niedzwieckii Goerz (*S. coeruleiformis* Drob.) - 6
nigricans Smith = S. myrsinifolia
- var. *jenisseensis* Fr. Schmidt = S. jenisseensis
niphoclada Rydb. = S. brachycarpa subsp. niphoclada
nipponica Franch. & Savat. (*S. triandra* L. subsp. *nipponica* (Franch. & Savat.) A. Skvorts.) - 4, 5
nummularia Anderss. (*S. nummularia* subsp *tundricola* (Schljak.) A. & D. Löve, *S. tundricola* Schljak.) - 1, 3, 4, 5
- subsp. *tundricola* (Schljak.) A. & D. Löve = S. nummularia
nyiwensis Kimura = S. saxatilis
oblongifolia Trautv. & C.A. Mey. = S. udensis
oleninii Nas. = S. abscondita
olgae Regel (*S. angrenica* Drob., *S. coerulangrenica* Drob., *S. heterandra* Drob., *S. olgangrenica* Drob., *S. pseudalba* E. Wolf) - 6
olgangrenica Drob. = S. olgae
opaca Anderss. ex Seemen = S. udensis
orbicularis Anderss. = S. reticulata
orotchonorum Kimura = S. bebbiana
ovalifolia Trautv. - 5
- subsp. *cyclophylla* (Rydb.) Jurtz. & Petrovsky = S. cyclophylla
- subsp. *glacialis* (Anderss.) Jurtz. & Petrovsky = S. glacialis
- subsp. *glacialis* (Anderss.) E. Murr. = S. glacialis

- var. *glacialis* (Anderss.) Argus = S. glacialis
oxica Dode = S. excelsa
oxycarpa auct. = S. pycnostachya
palibinii Goerz = S. caucasica
pallasii Anderss. = S. crassijulis
pamirica Drob. = S. pycnostachya
pantosericea Goerz - 2
paracaucasica Goerz = S. caucasica
parallelinervis B. Floder. (*S. divaricata* Pall. subsp. *parallelinervis* (B. Floder.) Worosch. comb. invalid., *S. pulchra* Cham. subsp. *parallelinervis* (B. Floder.) A. Skvorts.) - 5
paramushirensis Kudo = S. udensis
pedionoma Kimura - 5
pentandra L. - 1, 3, 4, 6
- subsp. *pseudopentandra* B. Floder. = S. pseudopentandra
pentandroides A. Skvorts. - 2
persica Boiss. = S. acmophylla
petsusu Kimura = S. schwerinii
phanerodictya Kimura = S. kurilensis
phlebophylla Anderss. - 5
phlomoides Bieb. = S. aegyptiaca and S. cinerea
phylicifolia L. - 1, 3, 4
- subsp. *bicolor* (Willd.) O. Bolos & Vigo = S. rhaetica
- subsp. *divaricata* (Pall.) N. Bolschakov = S. divaricata
- subsp. *pulchra* (Cham.) Hult. = S. pulchra
- subsp. *rhaetica* (Anderss.) A. Skvorts. = S. rhaetica
pierotii Miq. (*S. dolichostyla* Seemen, *S. koreensis* Anderss.) - 5
pierotii sensu Nas. = S. kangensis
planifolia Pursh ssp. *pulchra* (Cham.) Argus = S. pulchra
podophylla Andress. = S. rhamnifolia
polaris Wahlenb. (*S. polaris* subsp. *pseudopolaris* (B. Floder.) Hult., *S. pseudopolaris* B. Floder.) - 1, 3, 4, 5
- subsp. *pseudopolaris* (B. Floder.) Hult. = S. polaris
poronaica Kimura = S. fuscescens
prunifolia Kar. & Kir. = S. karelinii
przewalskii E. Wolf = S. tenuijulis
pseudalba E. Wolf = S. olgae
pseudodepressa A. Skvorts. - 2
pseudolinearis Nas. = S. schwerinii and S. viminalis
pseudomedemii E. Wolf (*S. alifera* Goerz, *S. fuscata* Goerz) - 2
pseudopentandra (B. Floder.) B. Floder. (*S. pentandra* L. subsp. *pseudopentandra* B. Floder.) - 4, 5
pseudopolaris B. Floder. = S. polaris
pseudotorulosa (A. Skvorts.) Czer. (*S. sphenophylla* A. Skvorts. subsp. *pseudotorulosa* A. Skvorts.) - 5
psiloides (B. Floder.) Kom. = S. hastata
pulchra Cham. (*S. anadyrensis* B. Floder., *S. divaricata* Pall. subsp. *pulchra* (Cham.) Worosch., *S. fulcrata* Anderss. p.p., *S. phylicifolia* L. subsp. *pulchra* (Cham.) Hult., *S. planifolia* Pursh subsp. *pulchra* (Cham.) Argus, *S. pulchra* var. *anadyrensis* (B. Floder.) A. Skvorts. comb. invalid.) - 1, 3, 4, 5
- subsp. *kalarica* A. Skvorts. = S. kalarica
- subsp. *parallelinervis* (B. Floder.) A. Skvorts. = S. parallelinervis
- var. *anadyrensis* (B. Floder.) A. Skvorts. = S. pulchra
pulchroides Kimura = S. chamissonis
purpurea L. (*S. caesifolia* Drob.) - 1, 6(cult.)
pycnostachya Anderss. (*S. ferganensis* Nas., *S. holargyrea* Goerz, *S. komarovii* E. Wolf, *S. korshinskyi* Goerz, *S. macrostachya* E. Wolf, *S. margaritifera* E. Wolf, *S. pamirica* Drob., *S. rubrobrunnea* Drob., *S. sarawschanica* Regel, *S. oxycarpa* auct.) - 6
pyrolifolia Ledeb. - 1, 3, 4, 5, 6

raddeana Laksch. ex Nas. = S. abscondita
rashuwensis Kimura = S. nakamurana
rectijulis Ledeb. ex Trautv. (*S. submyrsinites* B. Floder.) - 3, 4
recurvigemmis A. Skvorts. - 1, 4, 5
reinii Franch. & Savat. (*S. shikotanica* Kimura) - 5
repens L. - 1
- subsp. *brachypoda* (Trautv. & C.A. Mey.) Worosch. = S. brachypoda
- subsp. *rosmarinifolia* (L.) Celak. = S. rosmarinifolia
reptans Rupr. - 1, 3, 4, 5
reticulata L. (*S. orbicularis* Anderss.) - 1, 3, 4, 5
retusa L. (*S. kitaibeliana* Willd.) - 1
rhaetica Anderss. (*S. bicolor* Ehrh. ex Willd., *S. bicolor* subsp. *rhaetica* (Anderss.) B. Floder., *S. phylicifolia* L. subsp. *bicolor* (Willd.) O. Bolos & Vigo, *S. phylicifolia* subsp. *rhaetica* (Anderss.) A. Skvorts.) - 1
rhamnifolia Pall. (*S. chlorostachya* Turcz., *S. podophylla* Andress.) - 3, 4, 5
- subsp. *saposhnikovii* (A. Skvorts.) N. Bolschakov = S. saposhnikovii
richardsonii Hook. (*S. lanata* L. subsp. *richardsonii* (Hook.) A. Skvorts.) - 4, 5
roopii (Goerz) Grossh. = S. elbursensis
rorida Laksch. - 3, 4, 5
rosmarinifolia L. (? *S. anomala* E. Wolf, *S. repens* L. subsp. *rosmarinifolia* (L.) Celak., *S. sibirica* Pall.) - 1, 3, 4, 6
- subsp. *schugnanica* (Goerz) A. Skvorts. = S. schugnanica
- var. *brachypoda* (Trautv. & C.A. Mey.) Y.L. Chou = S. brachypoda
rossica Nas. = S. schwerinii and S. viminalis
rotundifolia Trautv. - 5
- subsp. *dodgeana* (Rydb.) Argus = S. dodgeana
- var. *dodgeana* (Rydb.) E. Murr. = S. dodgeana
rubrobrunnea Drob. = S. pycnostachya
rufescens (Turcz.) Nas. = S. viminalis
sachalinensis Fr. Schmidt = S. udensis
sajanensis Nas. - 3, 4
*****salomanii** Carr. - 1, 2
Probably a hybrid S. alba L. x S. babylonica L.
saposhnikovii A. Skvorts. (*S. rhamnifolia* Pall. subsp. *saposhnikovii* (A. Skvorts.) N. Bolschakov) - 3, 4
sarawschanica Regel = S. pycnostachya
saxatilis Turcz. ex Ledeb. (*S. fumosa* Turcz., *S. nyiwensis* Kimura, *S. saxatilis* subsp. *stoloniferoides* (Kimura) Worosch., *S. stoloniferoides* Kimura) - 4, 5
- subsp. *stoloniferoides* (Kimura) Worosch. = S. saxatilis
schugnanica Goerz (*S. rosmarinifolia* L. subsp. *schugnanica* (Goerz) A. Skvorts.) - 6
schwerinii E. Wolf (*S. petsusu* Kimura, *S. pseudolinearis* Nas. p.p., *S. rossica* Nas. p.p., *S. schwerinii* subsp. *yezoensis* (Schneid.) Worosch., *S. viminalis* L. var. *yezoensis* Schneid., *S. yezoensis* (Schneid.) Kimura) - 4, 5
- subsp. *yezoensis* (Schneid.) Worosch. = S. schwerinii
seemannii Rydb. = S. glauca
semiviminalis E. Wolf = S. viminalis
serotina Pall. = S. viminalis
serrulatifolia E. Wolf = S. tenuijulis
shikotanica Kimura = S. reinii
sibirica Pall. = S. rosmarinifolia
sichotensis Charkev. & Vyschin - 5
silesiaca Willd. - 1
siuzewii Seemen - 5
songarica Anderss. (*S. hypericifolia* Golosk.) - 6
speciosa Hook. & Arn. = S. alaxensis
sphenophylla A. Skvorts. (*S. cuneata* Turcz. ex Ledeb. 1850,

non Nutt. 1842) - 4, 5
- subsp. *pseudotorulosa* A. Skvorts. = S. pseudotorulosa
spinidens E. Wolf = S. tenuijulis
splendens (Turcz.) Nas. = S. viminalis
starkeana sensu Nas. = S. taraikensis
starkeana Willd. (*S. livida* Wahlenb., *S. depressa* auct. p.p.) - 1
- subsp. *cinerascens* (Wahlenb.) Hult. = S. bebbiana
stipulifera B. Floder. ex Hayren = S. glauca
stolonifera Cov. subsp. **carbonicola** Petrovsky - 5
stoloniferoides Kimura = S. saxatilis
strobilacea (E. Wolf) Nas. = S. viminalis
subfragilis auct. = S. triandra
submyrsinites B. Floder. = S. rectijulis
subreniformis Kimura = S. hidakamontana
sugawarana Kumira = S. abscondita
taraikensis Kimura (*S. livida* sensu Nas. p.p., *S. starkeana* sensu Nas.) - 3, 4, 5
tenuifolia Turcz. ex E. Wolf = S. miyabeana
tenuijulis Ledeb. (*S. alberti* Regel, *S. tenuijulis* var. *alberti* (Regel) Poljak., ? *S. przewalskii* E. Wolf, *S. serrulatifolia* E. Wolf, *S. spinidens* E. Wolf) - 3, 6
- var. *alberti* (Regel) Poljak. = S. tenuijulis
tianschanica Regel - 6
tontomussirensis Koidz. (*S. hidewoi* Koidz.) - 5
torulosa Trautv. (*S. altaica* Lundstr., *S. arctica* Pall. subsp. *torulosa* (Trautv.) Hult., *S. arctica* subsp. *torulosa* (Trautv.) A. Skvorts. comb. superfl.) - 3, 4, 6
triandra L. (*S. amygdalina* L., *S. armena* Schischk., *S. medwedewii* Dode, *S. triandra* subsp. *amygdalina* (L.) Masan, *S. subfragilis* auct.) - 1, 2, 3, 4, 5, 6
- subsp. *amygdalina* (L.) Masan = S. triandra
- subsp. *bommuelleri* (Hausskn.) A. Skvorts. = S. bornmuelleri
- subsp. *nipponica* (Franch. & Savat.) A. Skvorts. = S. nipponica
tschuktschorum A. Skvorts. (*S. berberifolia* subsp. *tschuktschorum* (A. Skvorts.) Worosch.) - 4, 5
- subsp. *kamtschatica* (A. Skvorts.) Vorobiev = S. kamtschatica
tundricola Schljak. = S. nummularia
turanica Nas. - 6
turczaninowii Laksch. (*S. liliputa* Nas.) - 3, 4, 5
tuvinensis Gudoschn. = S. coesia
udensis Trautv. & C.A. Mey. (*S. fulcrata* Anderss. p.p., *S. mezereoides* E. Wolf, *S. oblongifolia* Trautv. & C.A. Mey., *S. opaca* Anderss. ex Seemen, *S. paramushirensis* Kudo, *S. sachalinensis* Fr. Schmidt) - 4, 5
urbaniana Seemen (*S. cardiophylla* Trautv. & C.A. Mey. subsp. *urbaniana* (Seemen) A. Skvorts.) - 5
veriviminalis Nas. = S. viminalis
vestita Pursh - 3, 4
viminalis L. (*S. gmelinii* Pall., *S. pseudolinearis* Nas. p.p., *S. rossica* Nas. p.p., *S. rufescens* (Turcz.) Nas., *S. semiviminalis* E. Wolf, *S. serotina* Pall., *S. splendens* (Turcz.) Nas., *S. strobilacea* (E. Wolf) Nas., *S. veriviminalis* Nas., *S. viminalis* var. *semiviminalis* (E. Wolf) Poljak., *S. viminalis* subsp. *veriviminalis* (Nas.) Hyl.) - 1, 3, 4, 6
- subsp. *veriviminalis* (Nas.) Hyl. = S. viminalis
- var. *semiviminalis* (E. Wolf) Poljak. = S. viminalis
- var. *yezoensis* Schneid. = S. schwerinii
vinogradovii A. Skvorts. - 1, 3, 6
viridula (Anderss.) Nas. = S. jenisseensis
vorobievii Korkina - 5
vulpina Anderss. - 5
wilhelmsiana Bieb. - 2, 6
xerophila B. Floder. = S. bebbiana

yezoensis (Schneid.) Kimura = S. schwerinii

x **Toisochosenia** Kimura
tatewaki Kimura - 5

Turanga (Bunge) Kimura = Populus
ariana (Dode) Kimura = Populus ariana
diversifolia (Schrenk) Kimura = Populus diversifolia
euphratica (Olivier) Kimura = Populus euphratica
x *glaucicomans* (Dode) Kimura = Populus x glaucicomans
litwinowiana (Dode) Kimura = Populus diversifolia
pruinosa (Schrenk) Kimura = Populus pruinosa

SALVINIACEAE T. Lest.
Salvinia Seguier
natans (L.) All. - 1, 2, 3, 5, 6

SAMBUCACEAE Batsch ex Borkh.
Sambucus L.
coreana auct. = S. williamsii
ebulus L. - 1, 2, 6
kamtschatica E. Wolf - 5
latipinna Nakai = S. williamsii
manshurica Kitag. (*S. racemosa* L. subsp. *manshurica* (Kitag.) Worosch.) - 4, 5
miquelii (Nakai) Kom. (*S. racemosa* L. var. *miquelii* Nakai, *S. racemosa* subsp. *sieboldiana* auct. p.p., *S. sachalinensis* Pojark., *S. sibirica* Nakai subsp. *miquelii* (Nakai) Samut., *S. sieboldiana* sensu Pojark.) - 5
nigra L. - 1, 2
racemosa L. - 1
- subsp. *manshurica* (Kitag.) Worosch. = S. manshurica
- subsp. *miquelii* (Nakai) Samut. = S. miquelii
- subsp. *sibirica* (Nakai) Hara = S. sibirica
- subsp. *sieboldiana* auct. p.p. = S. miquelii
- var. *miquelii* Nakai = S. miquelii
sachalinensis Pojark. = S. miquelii
sibirica Nakai (*S. racemosa* L. subsp. *sibirica* (Nakai) Hara) - 1, 3, 4, 5
sieboldiana sensu Pojark. = S. miquelii
tigranii Troitzk. - 2
williamsii Hance (*S. latipinna* Nakai, *S. coreana* auct.) - 5

SANTALACEAE R.Br.
Thesium L.
alatavicum Kar. & Kir. - 6
alpinum L. - 1, 2
arvense Horvatovszky (*T. brevibracteatum* Sumn., *T. ramosum* Hayne) - 1, 2, 3, 6
bavarium Schrank = T. linophyllon
brachyphyllum Boiss. - 1, 2
brevibracteatum Sumn. = T. arvense
chinense Turcz. - 4, 5
compressum Boiss. & Heldr. - 2
dollineri Murb. subsp. **simplex** (Velen.) Stojan. & Stef. - 1
ebracteatum Hayne - 1, 3
ferganense Bobr. - 6
gontscharovii Bobr. - 6
kotschyanum Boiss. - 6
laxiflorum Trautv. - 2
linifolium Schrank = T. linophyllon

linophyllon L. (*T. bavarium* Schrank, *T. linifolium* Schrank) - 1
longifolium Turcz. ex Ledeb. - 4
maritimum C.A. Mey. - 2
minkwitzianum B. Fedtsch. - 6
multicaule Ledeb. - 3, 6
procumbens C.A. Mey. - 1, 2
ramosissimum Bobr. - 6
ramosum Hayne = T. arvense
refractum C.A. Mey. - 3, 4, 5, 6
- var. *saxatile* (Turcz. ex A. DC.) Worosch. = T. saxatile
repens Ledeb. - 3, 4
rupestre Ledeb. - 3
saxatile Turcz. ex A. DC. (*T. refractum* C.A. Mey. var. *saxatile* (Turcz. ex A. DC.) Worosch.) - 4, 5
szowitsii A. DC. - 2
tuvense Krasnob. - 4

SAURURACEAE E. Mey.
Houttuynia Thunb.
cordata Thunb. - 2

SAXIFRAGACEAE Juss.
Astilbe Buch.-Ham. ex D. Don
chinensis (Maxim.) Franch. & Savat. - 5
thunbergii (Siebold & Zucc.) Miq. - 5

Bergenia Moench
cordifolia (Haw.) Sternb. = B. crassifolia
crassifolia (L.) Fritsch (*B. cordifolia* (Haw.) Sternb., *B. crassifolia* var. *cordifolia* (Haw.) Boriss., *Saxifraga cordifolia* Haw.) - 3, 4
- var. *cordifolia* (Haw.) Boriss. = B. crassifolia
- var. *pacifica* (Kom.) Nekr. = B. pacifica
gorbunowii B. Fedtsch. = B. stracheyi
hissarica Boriss. - 6
pacifica Kom. (*B. crassifolia* (L.) Fritsch var. *pacifica* (Kom.) Nekr.) - 5
stracheyi (Hook. fil. & Thoms.) Engl. (*Saxifraga stracheyi* Hook. fil. & Thoms., *Bergenia gorbunowii* B. Fedtsch.) - 6
ugamica V. Pavl. - 6

Chrysosplenium L.
alberti Malysch. - 4
albowianum Kuth. - 2
alpinum Schur - 1
alternifolium L. - 1, 2, 3
- subsp. *arctomontanum* Petrovsky = Ch. arctomontanum
- subsp. *sibiricum* (Ser.) Hult. = Ch. sibiricum
- subsp. *tetrandrum* (Lund ex Malmgr.) Hult. = Ch. tetrandrum
- var. *sibiricum* Ser. = Ch. sibiricum
- var. *tetrandrum* Lund ex Malmgr. = Ch. tetrandrum
arctomontanum (Petrovsky) Charkev. (*Ch. alternifolium* L. subsp. *arctomontanum* Petrovsky) - 5
baicalense Maxim. - 4
beringianum Rose = Ch. wrightii
dezhnevii (Jurtz.) Charkev. (*Ch. rimosum* Kom. subsp. *dezhnevii* Jurtz.) - 5
dubium J. Gay ex Ser. - 2
filipes Kom. - 3, 4
flagelliferum Fr. Schmidt (*Ch. komarovii* Losinsk.) - 5

fulvum Terr. = Ch. pilosum
grayanum Maxim. - 5
kamtschaticum Fisch. - 5
komarovii Losinsk. = Ch. flagelliferum
nudicaule Bunge - 3, 4, 6
ovalifolium Bieb. ex Bunge - 3, 4
pacificum Hult. - 5
peltatum Turcz. - 4
pilosum Maxim. (*Ch. fulvum* Terr., *Ch. vidalii* Franch. & Savat.) - 5
- var. *villosum* (Franch.) Kitag. = Ch. villosum
pseudofauriei Levl. (*Ch. trachyspermum* auct.) - 5
ramosum Maxim. - 5
rimosum Kom. - 5
- subsp. *dezhnevii* Jurtz. = Ch. dezhnevii
rosendahlii Packer - 5
saxatile A. Khokhr. - 5
schagae Charkev. & Vyschin - 5
sedakowii Turcz. - 3, 4
sibiricum (Ser.) Charkev. (*Ch. alternifolium* L. var. *sibiricum* Ser., *Ch. alternifolium* subsp. *sibiricum* (Ser.) Hult.) - 4, 5
sinicum Maxim. - 5
tetrandrum (Lund ex Malmgr.) Th. Fries (*Ch. alternifolium* L. var. *tetrandrum* Lund ex Malmgr., *Ch. alternifolium* subsp. *tetrandrum* (Lund ex Malmgr.) Hult.) - 1, 3, 4, 5
thianschanicum Krasn. - 6
trachyspermum auct. = Ch. pseudofauriei
vidalii Franch. & Savat. = Ch. pilosum
villosum Franch. (*Ch. pilosum* Maxim. var. *villosum* (Franch.) Kitag.) - 5
woroschilovii Neczajeva - 5
wrightii Franch. & Savat. (*Ch. beringianum* Rose) - 5

Hirculus Haw. = Saxifraga
platysepalus (Trautv.) W.A. Weber = Saxifraga platysepala
properens (Fisch. ex Sternb.) A. & D. Löve = Saxifraga prorepens
serpyllifolius (Pursh) W.A. Weber = Saxifraga serpyllifolia

Leptarrhena R. Br. = Saxifraga
pyrolifolia (D.Don) R.Br. ex Ser. = Saxifraga pyrolifolia

Leptasea Haw. = Saxifraga
funstonii Small = Saxifraga funstonii

Micranthes Haw. = Saxifraga
sachalinensis (Fr. Schmidt) Hara = Saxifraga sachalinensis

Mitella L.
nuda L. - 3, 4, 5

Muscaria Haw. = Saxifraga
cespitosa (L.) Haw. subsp. *exaratoides* (Simm.) A. & D. Löve = Saxifraga cespitosa subsp. exaratoides
monticola Small = Saxifraga monticola

Saxifraga L. (*Hirculus* Haw., *Leptarrhena* R. Br., *Leptasea* Haw., *Micranthes* Haw., *Muscaria* Haw.)
abchasica Oetting. - 2
adenophora C. Koch - 2
adscendens L. - 2

aestivalis Fisch. & C.A. Mey. (*S. nelsoniana* D. Don subsp. *aestivalis* (Fisch. & C.A. Mey.) D.A. Webb, *S. punctata* sensu Losinsk. p.max.p.) - 1, 3, 4, 5, 6

aizoides L. - 1, 3

aizoon Jacq. = S. paniculata

ajanica Sipl. (*S. pseudoajanica* A. Khokhr.) - 5

x *akinfievii* Galushko & G. Kudrjaschova = S. x oettingenii

alberti Regel & Schmalh. - 6

algisii Egor. & Sipl. - 4, 5

anadyrensis Losinsk. (*S. bronchialis* L. subsp. *anadyrensis* (Losinsk.) Ju. Kozhevn.) - 4, 5

androsacea L. - 3, 4

arctolitoralis Jurtz. & Petrovsky - 5

ascoldica Sipl. (*S. cherlerioides* D. Don subsp. *ascoldica* (Sipl.) Worosch.) - 5

asiatica Hayek (*S. oppositifolia* L. subsp. *asiatica* (Hayek) Engl. & Irmsch.) - 3, 4, 6

aspera sensu Bieb. = S. flagellaris

astilbeoides Losinsk. (*S. punctata* L. subsp. *astilbeoides* (Losinsk.) Worosch.) - 5

biebersteinii Sipl. = S. pseudolaevis

brachypetala Malysch. - 4

bracteata D. Don - 5

bronchialis L. - 4, 5

- subsp. *anadyrensis* (Losinsk.) Ju. Kozhevn. = S. anadyrensis

- subsp. *cherlerioides* (D. Don) Hult. = S. cherlerioides

- subsp. *funstonii* (Small) Hult. = S. funstonii

- subsp. *spinulosa* (Adams) Hult. = S. spinulosa

- subsp. *stelleriana* (Merk ex Ser.) Malysch. = S. stelleriana

- var. *genuina* Trautv. f. *rebunshirensis* Engl. & Irmsch. = S. rebunshirensis

bryoides L. - 1

bulbifera L. - 1

calycina Sternb. (? *S. davurica* Willd. subsp. *grandipetala* (Engl. & Irmsch.) Hult. p.p. quoad nomen, ? *S. grandipetala* (Engl. & Irmsch.) Losinsk. p.p. quoad nomen) - 5

- subsp. *unalaschkensis* (Fisch. ex Sternb.) Hult. = S. pyrolifolia

carinata Oetting. - 2

carpathica Reichenb. - 1

cartilaginea Willd. (*S. paniculata* Mill. subsp. *cartilaginea* (Willd.) D.A. Webb) - 2

caspica Sipl. = S. meyeri

caucasica Somm. & Levier - 2

caulescens Sipl. - 4

cernua L. - 1, 3, 4, 5, 6

cespitosa L. - 1, 3, 4, 5

- subsp. *exaratoides* (Simm.) Engl. & Irmsch. (*S. groenlandica* L. subsp. *exaratoides* Simm., *Muscaria cespitosa* (L.) Haw. subsp. *exaratoides* (Simm.) A. & D. Löve) - 1

- subsp. *monticola* (Small) A. Pors. = S. monticola

- subsp. *sileniflora* (Sternb.) Hult. = S. sileniflora

charadzeae Otschiauri - 2

cherlerioides D. Don (*S. bronchialis* L. subsp. *cherlerioides* (D. Don) Hult., *S. pseudoburseriana* Fisch. ex Cham.) - 5

- subsp. *ascoldica* (Sipl.) Worosch. = S. ascoldica

- subsp. *rebunshirensis* (Engl. & Irmsch.) Hara = S. rebunshirensis

- subsp. *stelleriana* (Merk ex Ser.) Worosch. = S. stelleriana

cismagadanica Malysch. - 4

colchica Albov - 2

columnaris Schmalh. - 2

cordifolia Haw. = Bergenia crassifolia

coriifolia (Somm. & Levier) Grossh. = S. repanda

cortusifolia auct. = S. serotina

cymbalaria L. (*S. huetiana* auct.) - 2

cymosa Waldst. & Kit. (*S. pedemontana* All. subsp. *cymosa* (Waldst. & Kit.) Engl.) - 1

czekanowskii Sipl. - 4

davurica Willd. (*S. grandipetala* (Engl. & Irmsch.) Losinsk. p.p. quoad pl., *S. vicaria* Sipl.) - 4, 5

- subsp. *grandipetala* (Engl. & Irmsch.) Hult. p.p. = S. calycina

derbekii Sipl. (*S. spinulosa* Adams subsp. *derbekii* (Sipl.) Worosch.) - 5

- subsp. **xerophylla** A. Knokhr. - 5

desoulavyi Oetting. - 2

dinnikii Schmalh. - 2

engleri Dalla Torre (*S. stellaris* L. subsp. *alpigena* Temesy) - 1

eschscholtzii Sternb. - 5

exarata Vill. - 2

exilis Steph. ex Sternb. = S. radiata

firma Litv. ex Losinsk. = S. funstonii

flagellaris Willd. ex Sternb. (*S. flagellaris* subsp. *aspera* ("Bieb.") Tolm., *S. aspera* sensu Bieb. 1808, non L. 1753) - 2

- subsp. *aspera* ("Bieb.") Tolm. = S. flagellaris

- subsp. *flagellaris* var. *komarovii* (Losinsk.) Hara = S. komarovii

- subsp. *komarovii* (Losinsk.) Hult. = S. komarovii

- subsp *platysepala* (Trautv.) A. Pors. = S. platysepala

- - var. *macrocalyx* (Tolm.) Tolm. = S. macrocalyx

- subsp. *setigera* (Pursh) Tolm. = S. setigera

- subsp. *stenophylla* (Royle) Hult. = S. stenophylla

- var. *platysepala* Trautv. = S. platysepala

foliolosa R. Br. - 1, 3, 4, 5

fortunei Hook. fil. - 5

funstonii (Small) Fedde (*Leptasea funstonii* Small, *Saxifraga bronchialis* L. subsp *funstonii* (Small) Hult., *S. firma* Litv. ex Losinsk., *S. spinulosa* Adams subsp. *funstonii* (Small) Worosch.) - 5

fusca Maxim. (*S. ohwii* Tatew.) - 5

glutinosa Sipl. (*S. serpyllifolia* Pursh subsp. *glutinosa* (Sipl.) Ju. Kozhevn., *S. serpyllifolia* var. *viscosa* Trautv.) - 4

grandipetala (Engl. & Irmsch.) Losinsk. = S. calycina and S. davurica

granulata L. - 1

grisea Sipl. - 2

groenlandica L. subsp. *exaratoides* Simm. = S. cespitosa subsp. exaratoides

hieracifolia Waldst. & Kit. - 1, 3, 4, 5

- subsp. **czukczorum** Chrtek. & Sojak - 5

- subsp. *longifolia* (Engl. & Irmsch.) Jurtz. & Petrovsky (*S. hieracifolia* var. *typica* Engl. & Irmsch. f. *longifolia* Engl. & Irmsch.) - 5

hirculus L. (*S. hirculus* var. *alpina* Engl., *S. hirculus* subsp. *alpina* (Engl.) A. Löve) - 1, 2, 3, 4, 5

- subsp. *alpina* (Engl.) A. Löve = S. hirculus

- subsp. *propinqua* (R. Br.) A. & D. Löve = S. prorepens

- var. *alpina* Engl. = S. hirculus

huetiana auct. = S. cymbalaria

hyperborea R. Br. - 1, 3, 4, 5

idzuroei Franch. & Savat. - 5

insularis (Hult.) Sipl. (*S. punctata* L. subsp. *insularis* Hult., *S. nelsoniana* D. Don subsp. *insularis* (Hult.) Hult., *S. purpurascens* Kom. var. *insularis* (Hult.) Worosch.) - 5

irrigua Bieb. - 1

juniperifolia Adams - 2

kolenatiana Regel - 2

kolymensis A. Khokhr. = S. stelleriana

komarovii Losinsk. (*S. flagellaris* Willd. ex Sternb. subsp. *komarovii* (Losinsk.) Hult., *S. flagellaris* subsp. *flagellaris* var. *komarovii* (Losinsk.) Hara, *S. stenophylla* Royle subsp. *komarovii* (Losinsk.) Tolm., *S. stenophylla* subsp. *pamirica* Tolm. nom. invalid.) - 6

korshinskii Kom. - 5

■ **kotschyi** Boiss.

kruhsiana Fisch. ex Ser. (*S. microglobularis* A. Khokhr.) - 5

kusnezowiana Oetting. - 2

laciniata Nakai & Takeda - 5

lactea Turcz. - 5

laevis Bieb. = S. pseudolaevis

laevis sensu Losinsk. = S. meyeri

ledebouriana Holub = S. omolojensis

luteo-viridis Schott & Kotschy - 1

lyallii Engl. subsp. **hultenii** (Calder & Savile) Calder & Savile (*S. lyallii* var. *hultenii* Calder & Savile) - 4

macrocalyx Tolm. (*S. flagellaris* Willd. ex Sternb. subsp. *platysepala* (Trautv.) A. Pors. var. *macrocalyx* (Tolm.) Tolm., *S. setigera* sensu Losinsk. p.p.) - 3, 6

manchuriensis (Engl.) Kom. - 5

melaleuca Fisch. ex Spreng. - 3, 4

merkii Fisch. ex Sternb. - 4, 5

meyeri Manden. (*S. caspica* Sipl., *S. laevis* sensu Losinsk.) - 2

microglobularis A. Khokhr. = S. kruhsiana

mollis Smith = S. sibirica

monticola (Small) Fedde (*Muscaria monticola* Small, *Saxifraga cespitosa* L. subsp. *monticola* (Small) A. Pors., *S. monticola* (Small) A. & D. Löve, comb. superfl.) - 5

moschata Wulf. - 2

multiflora Ledeb. = S. omolojensis

nelsoniana D. Don (*S. punctata* L. subsp. *nelsoniana* (D. Don) Hult.) - 4, 5

- subsp. *aestivalis* (Fisch. & C.A. Mey.) D.A. Webb = S. aestivalis
- subsp. *insularis* (Hult.) Hult. = S. insularis
- subsp. *porsildiana* (Calder & Savile) Hult. = S. porsildiana
- subsp. *reniformis* (Ohwi) Hult. = S. reniformis

nivalis L. - 1, 3, 4, 5

nudicaulis D. Don - 5

- subsp. **soczavae** Rebr. - 5

nutans Adams = S. prorepens

oblongifolia Nakai - 5

x **oettingenii** Galushko & G. Kudrjaschova (*S.* x *akinfievii* Galushko & G. Kudrjaschova). - S. dinnikii Schmalh. x S. juniperifolia Adams - 2

ohwii Tatew. = S. fusca

omolojensis A. Khokhr. (*S. ledebouriana* Holub, *S. multiflora* Ledeb. 1815, non Schleicher, 1800) - 4, 5

oppositifolia L. - 1, 3, 4

- subsp. *asiatica* (Hayek) Engl. & Irmsch. = S. asiatica
- subsp. *smalliana* (Engl. & Irmsch.) Hult. = S. pulvinata
- var. *typica* Engl. & Irmsch. subvar. *smalliana* Engl. & Irmsch. = S. pulvinata

pallasiana Sternb. = S. serpyllifolia

paniculata Mill. (*S. aizoon* Jacq.) - 1

- subsp. *cartilaginea* (Willd.) D.A. Webb = S. cartilaginea

parnassioides Regel & Schmalh. - 6

pedemontana All. subsp. *cymosa* (Waldst. & Kit.) Engl. = S. cymosa

perdurans Kit. ex Kanitz = S. wahlenbergii

platysepala (Trautv.) Tolm. (*S. flagellaris* Willd. ex Sternb. var. *platysepala* Trautv., *Hirculus platysepalus* (Trautv.) W.A. Weber, *Saxifraga flagellaris* subsp. *platysepala* (Trautv.) A. Pors. p.p., *S. setigera* sensu Losinsk.

p.p.) - 1, 3, 4, 5

polytrichoides Sipl. - 2

pontica Albov - 2

porsildiana (Calder & Savile) Jurtz. & Petrovsky (*S. punctata* L. subsp. *porsildiana* Calder & Savile, *S. nelsoniana* D. Don subsp. *porsildiana* (Calder & Savile) Hult.) - 4, 5

propinqua R. Br. = S. prorepens

prorepens Fisch. ex Sternb. (*Hirculus prorepens* (Fisch. ex Sternb.) A. & D. Löve, *Saxifraga hirculus* L. subsp. *propinqua* (R. Br.) A. & D. Löve, *S. nutans* Adams, 1834, non D. Don, 1822, *S. propinqua* R. Br.) - 1, 3, 4, 5, 6

pseudoajanica A. Khokhr. = S. ajanica

pseudoburseriana Fisch. ex Cham. = S. cherlerioides

pseudolaevis Oetting. (*S. biebersteinii* Sipl., *S. laevis* Bieb. 1808, non Haw. 1803) - 2

pulvinaria H. Smith - 6

pulvinata Small (*S. opposotifolia* L. var. *typica* Engl. & Irmsch. subvar. *smalliana* Engl. & Irmsch., *S. oppositifolia* subsp. *smalliana* (Engl. & Irmsch.) Hult.) - 5

punctata L. (*S. punctata* subsp. *redowskyana* (Sternb.) Worosch., *S. redowskyana* Sternb.) - 4, 5

- subsp. *astilbeoides* (Losinsk.) Worosch. = S. astilbeoides
- subsp. *insularis* Hult. = S. insularis
- subsp. *nelsoniana* (D. Don) Hult. = S. nelsoniana
- subsp. *porsildiana* Calder & Savile = S. porsildiana
- subsp. *redowskyana* (Sternb.) Worosch. = S. punctata
- subsp. *reniformis* (Ohwi) Hara = S. reniformis

punctata sensu Losinsk. p.max.p. = S. aestivalis

purpurascens Kom. - 5

- var. *insularis* (Hult.) Worosch. = S. insularis

pyrolifolia D. Don, 3-9 XII 1822 (*Leptarrhena pyrolifolia* (D. Don) R. Br. ex Ser., *Saxifraga calycina* Sternb. subsp. *unalaschkensis* (Fisch. ex Sternb.) Hult., *S. unalaschkensis* Fisch. ex Sternb. 16 VI 1823) - 5

radiata Small (*S. exilis* Steph. ex Sternb. 1822, non Pall. 1816) - 4, 5

rebunshirensis (Engl. & Irmsch.) Sipl. (*S. bronchialis* L. var. *genuina* Trautv. f. *rebunshirensis* Engl. & Irmsch., *S. cherlerioides* D. Don subsp. *rebunshirensis* (Engl. & Irmsch.) Hara) - 5

redofskyi Adams - 4, 5

redowskyana Sternb. = S. punctata

reniformis Ohwi (*S. nelsoniana* D. Don subsp. *reniformis* (Ohwi) Hult., *S. punctata* L. subsp. *reniformis* (Ohwi) Hara) - 5

repanda Willd. ex Sternb. (*S. coriifolia* (Somm. & Levier) Grossh.) - 2

rivularis L. - 1

ruprechtiana Manden. - 2

sachalinensis Fr. Schmidt (*Micranthes sachalinensis* (Fr. Schmidt) Hara) - 5

scleropoda Somm. & Levier - 2

- var. *sommieri* Engl. & Irmsch. = S. sommieri

selemdzhensis Gorovoi & Worosch. - 5

serotina Sipl. (*S. cortusifolia* auct.) - 5

serpyllifolia Pursh (*Hirculus serpyllifolius* (Pursh) W.A. Weber, *Saxifraga pallasiana* Sternb.) - 4, 5

- subsp. *glutinosa* (Sipl.) Ju. Kozhevn. = S. glutinosa
- var. *viscosa* Trautv. = S. glutinosa

setigera Pursh (*S. flagellaris* Willd. ex Sternb. subsp. *setigera* (Pursh) Tolm., *S. sobolifera* Adams) - 3, 4, 5

setigera sensu Losinsk. = S. macrocalyx and S. platysepala

sibirica L. (*S. mollis* Smith, *S. sibirica* subsp. *mollis* (Smith) Matthews) - 1, 2, 3, 4, 5, 6

- subsp. *mollis* (Smith) Matthews = S. sibirica
sichotensis Gorovoi & N.S. Pavlova - 5
sieversiana Sternb. - 5
sileniflora Sternb. (*S. cespitosa* L. subsp. *sileniflora*
(Sternb.) Hult.) - 5
sobolifera Adams = S. setigera
sochondensis Maximova = S. tenuis
sommieri (Engl. & Irmsch.) Sipl. (*S. scleropoda* Somm. &
Levier var. *sommieri* Engl. & Irmsch.) - 2
sosnowskyi Manden. - 2
spinulosa Adams (*S. bronchialis* L. subsp. *spinulosa*
(Adams) Hult.) - 1, 3, 4, 5
- subsp. *derbekii* (Sipl.) Worosch. = S. derbekii
- subsp. *funstonii* (Small) Worosch. = S. funstonii
staminosa Schlothgauer & Worosch. - 5
stellaris L. - 1
- subsp. *alpigena* Temesy = S. engleri
stelleriana Merk ex Ser. (*S. bronchialis* L. subsp. *stelleriana*
(Merk ex Ser.) Malysch., *S. cherlerioides* D. Don
subsp. *stelleriana* (Merk ex Ser.) Worosch., *S. kolymensis* A. Khokhr.) - 4, 5
stenophylla Royle (*S. flagellaris* Willd. ex Sternb. subsp.
stenophylla (Royle) Hult.) - 6
- subsp. *komarovii* (Losinsk.) Tolm. = S. komarovii
- subsp. *pamirica* Tolm. = S. komarovii
stracheyi Hook. fil. & Thoms. = Bergenia stracheyi
subverticillata Boiss. - 2
svetlanae Worosch. - 5
tenuis (Wahlenb.) H. Smith (*S. sochondensis* Maximova) -
1, 3, 4, 5
terektensis Bunge - 3, 4, 6
tilingiana Regel & Til. - 5
trautvetteri Manden. - 2
tridactylites L. - 1, 2, 6
unalaschkensis Fisch. ex Sternb. = S. pyrolifolia
unifoveolata Sipl. - 2
ursina Sipl. - 1, 3, 4, 5
vaginalis Turcz.ex Ledeb. (*S. nudicaulis* D. Don subsp.
vaginalis (Turcz. ex Ledeb.) Rebr.) - 4, 5
verticillata Losinsk. - 2
vicaria Sipl. = S. davurica
voroschilovii Sipl. - 5
vulcanica Sipl. - 5
vvedenskyi Abdullaeva - 6
wahlenbergii Ball (*S. perdurans* Kit. ex Kanitz) - 1(?)
yoshimurae Miyabe & Tatew. - 5

SCHEUCHZERIACEAE Rudolphi
Scheuchzeria L.
palustris L. - 1, 2, 3, 4, 5

SCHISANDRACEAE Blume
SchisandraM ichx.
chinensis (Turcz.) Baill. - 5

SCROPHULARIACEAE Juss.
Alectorolophus Zinn = Rhinanthus
alpinus Walp. = Rhinanthus pulcher subsp. alpinus
erectus Sterneck = Rhinanthus pulcher subsp. erectus
grandiflorus Wallr. = Rhinanthus serotinus
lanceolatus Sterneck var. *subalpinus* Sterneck = Rhinanthus
glacialis subsp. subalpinus
major (L.) Reichenb. var. *bosnensis* Behrend. & Sterneck =

Rhinanthus bosnensis
pulcher (Gunther & Schummel ex Opiz) Wimm. var. *elatus*
Sterneck = Rhinanthus pulcher subsp. elatus
serotinus Schoenh. = Rhinanthus serotinus

Anagalloides Krock. = Lindernia
procumbens Krock. = Lindernia procumbens

*Antirrhinum L.
angustissimum Loisel. = Linaria angustissima
incarnatum Vent. = Linaria incarnata
majus L. - 1
minus L. = Chaenorhinum minus
orontium L. = Misopates orontium
pyramidatum Lam. = Linaria pyramidata
repens L. = Linaria repens

Bartsia L.
alpina L. - 1

Bellardia All.
trixago (L.) All. - 2

Bungea C.A. Mey.
trifida (Vahl) C.A. Mey. - 2
vesiculifera (Herd.) Pavl. & Lipsch. - 6

Castilleja Mutis ex L. fil.
arctica Kryl. & Serg. - 1, 3, 4
- subsp. **vorkutensis** Rebr. - 1
caudata (Pennell) Rebr. (*C. pallida* (L.) Spreng. subsp.
caudata Pennell) - 4, 5
chrymactis Pennell - 5
elegans Malte - 4, 5
hyparctica Rebr. (*C. pallida* (L.) Spreng. subsp. *hyparctica*
(Rebr.) A. & D. Löve) - 1, 3, 4, 5
hyperborea auct. p.p. = C. pseudohyperborea
lapponica Gand. (*C. pallida* (L.) Spreng. subsp. *lapponica*
(Gand.) A. & D. Löve) - 1
olgae A. Khokhr. - 5
pallida (L.) Spreng. - 1, 3, 4, 5
- subsp. *caudata* Pennell = C. caudata
- subsp. *dahurica* Pennell = C. rubra
- subsp. *hyparctica* (Rebr.) A. & D. Löve = C. hyparctica
- subsp. *lapponica* (Gand.) A. & D. Löve = C. lapponica
- subsp. *pavlovii* (Rebr.) A. & D. Löve = C. pavlovii
- var. *rubra* Drob. = C. rubra
pavlovii Rebr. (*C. pallida* (L.) Spreng. subsp. *pavlovii*
(Rebr.) A. & D. Löve) - 5
pseudohyperborea Rebr. (*C. hyperborea* auct. p.p. quoad pl.
asiat.) - 5
rubra (Drob.) Rebr. (*C. pallida* (L.) Spreng. var. *rubra*
Drob., *C. pallida* subsp. *dahurica* Pennell) - 3, 4, 5
schrenkii Rebr. - 1
tenella Rebr. - 4
unalaschcensis (Cham. & Schlecht.) Malte - 5
variocolorata A. Khokhr. - 5
yukonis Pennell - 5

Celsia L. = Verbascum
agrimoniifolia C. Koch = Verbascum agrimoniifolium
atroviolacea Somm. & Levier = Verbascum atroviolaceum
heterophylla Desf. = Verbascum agrimoniifolium

megrica Tzvel. = Verbascum megricum
nudicaulis (Wydl.) B. Fedtsch. = Verbascum nudicaule
orientalis L. = Verbascum orientale
patris Bordz. = Verbascum x patris
suworowiana C. Koch = Verbascum suworowianum

x *Celsioverbascum* Rech. fil. = Verbascum

gabrielianae Huber-Morath = Verbascum x gabrielianae

Chaenorhinum (DC.) Reichenb. (*Holzneria* F. Speta, *Hueblia* F. Speta, *Microrrhinum* (Endl.) Fourr.)

calycinum (Banks & Soland.) P.H. Davis, p.p. = Ch. persicum
gerense (Stapf) F. Speta (*Linaria gerensis* Stapf, *Chaenorhinum rubrifolium* auct. non (Robill. & Cast. ex DC.) Fourr.) - 2
klokovii Kotov (*Ch. minus* (L.) Lange subsp. *anatolicum* P.H. Davis, *Microrrhinum klokovii* (Kotov) F. Speta) - 1
minus (L.) Lange (*Antirrhinum minus* L., *Chaenorhinum viscidum* (Moench) Simonk., *Microrrhinum minus* (L.) Fourr.) - 1, 2, 5(alien)
- subsp. *anatolicum* P.H. Davis = Ch. klokovii
persicum (Chav.) O. & B. Fedtsch. (*Linaria persica* Chav., *Chaenorhinum calycinum* (Banks & Soland.) P.H. Davis, p.p., *Ch. rytidospermum* (Fisch. & C.A. Mey.) Kuprian., *Hueblia persica* (Chav.) F. Speta) - 2, 6
rubrifolium auct. = Ch. gerense
rytidospermum (Fisch. & C.A. Mey.) Kuprian. = Ch. persicum
spicatum Korov. (*Holzneria spicata* (Korov.) F. Speta) - 6
viscidum (Moench) Simonk. = Ch. minus

Cymbalaria Hill
muralis Gaertn., Mey. & Scherb. - 1, 2

Cymbaria L.
daurica L. - 4

Cymbochasma (Endl.) Klok. & Zoz
borysthenica (Pall. ex Schlecht.) Klok. & Zoz - 1

Deinostema Yamazaki
violacea (Maxim.) Yamazaki (*Gratiola violacea* Maxim.) - 5

Digitalis L.
ciliata Trautv. - 2
ferruginea L. - 2
- subsp. *schischkinii* (Ivanina) Werner = D. schischkinii
grandiflora Mill. - 1, 2, 3
lanata Ehrh. - 1
nervosa Steud. & Hochst. ex Benth. - 2
*****purpurea** L. - 1, 5(alien)
schischkinii Ivanina (*D. ferruginea* L. subsp. *schischkinii* (Ivanina) Werner) - 2

Dodartia L.
atro-coerulea Pavl. = D. orientalis
orientalis L. (*D. atro-coerulea* Pavl.) - 1, 2, 3, 6

Dopatrium Buch.-Ham. ex Benth.
junceum (Roxb.) Buch.-Ham. ex Benth. - 6

Euphrasia L.
adenocaulon Juz. - 2
ajanensis Worosch. - 5
alboffii Chabert - 2
altaica Serg. - 3
- subsp. *glabra* Kuvajev = E. putoranica
amblyodonta Juz. - 2
amurensis Freyn - 5
arctica Lange ex Rostrup subsp. **slovaca** P.F. Yeo - 1
- subsp. *tenuis* (Brenn.) P.F. Yeo, p.p. = E. vernalis
bajankolica Juz. - 6
bakurianica Juz. - 2
bicknellii Wettst. (*E. orae-australis* Juz.) - 1
brevipila Burn. & Gremli (*E. stricta* D. Wolff ex J.F. Lehm. var. *brevipila* (Burn. & Gremli) Hartl, *E. vernalis* List subsp. *brevipila* (Burn. & Gremli) O. Schwarz) - 1, 3, 4
carpatica Zapal. = E. salisburgensis
carthalinica Kem.-Nath. = E. sosnowskyi
caucasica Juz. - 2
chitrovoi Tzvel. - 1
coerulea Tausch ex Hoppe & Fuernrohr (*E. uechtritziana* Jung. & Engl.) - 1(?)
condensata Jord. = E. stricta
curta (Fries) Wettst. = E. parviflora
- subsp. *glabrescens* (Wettst.) Smejkal = E. parviflora
cyclophylla Juz. - 6
daghestanica Juz. - 2
drosophylla Juz. - 6
fedtschenkoana Wettst. ex Juz. - 6
fennica Kihlm. (*E. rostkoviana* Hayne var. *fennica* (Kihlm.) Jalas, *E. rostkoviana* subsp. *fennica* (Kihlm.) Karlsson) - 1
- subsp. *praecox* Ganesch. = E. onegensis
frigida Pugsl. - 1, 3
georgica Kem.-Nath. - 2
glabrescens (Wettst.) Wiinst. = E. parviflora
grossheimii Kem.-Nath. - 2
hirtella Jord. ex Reut. - 1, 2, 3, 4, 5(alien), 6
hyperborea Jorgens. (*E. saamica* Juz., *E. subpolaris* Juz.) - 1, 3, 4, 5
imbricans N. Vodopianova - 4
integriloba A. Dmitr. & N. Rubtz. ex Karmyscheva - 6
irenae Juz. (*E. liburnica* auct. p.p.) - 1
jacutica Juz. - 4, 5
juzepczukii Deniss. - 2
karataviensis Govor. ex Karmyscheva - 6
kemulariae Juz. - 2
kerneri Wettst. (*E. picta* Wimm. subsp. *kerneri* (Wettst.) P.F. Yeo) - 1
krassnovii Juz. - 3
lebardensis Kem.-Nath. - 1
liburnica auct. p.p. = E. irenae
macrocalyx Juz. - 6
macrodonta Juz. - 6
maximowiczii Wettst. - 5
micrantha Reichenb. - 1
minima Jacq. ex DC. subsp. *tatrae* (Wettst.) Hayek = E. tatrae
mollis (Ledeb.) Wettst. - 5
- subsp. *pseudomollis* (Juz.) Yamazuki = E. pseudomollis
montana Jord. - 1
x murbeckii Wettst. - E. brevipila Burn. & Gremli x E. parviflora Schag. - 1

nemorosa (Pers.) Wallr. (*E. officinalis* L. var. *nemorosa* Pers.) - 1
- subsp. *pectinata* (Ten.) Malagarriga & Barrau = E. pectinata
- subsp. *tatarica* (Fisch. ex Spreng.) Malagarriga & Barrau = E. pectinata
odontites L. var. *litoralis* Fries = Odontites litoralis
- var. *pratensis* Wirtg. = Odontites pratensis
officinalis L. var. *curta* Fries = E. parviflora
- var. *nemorosa* Pers. = E. nemorosa
onegensis Cajand. (*E. fennica* Kihlm. subsp. *praecox* Ganesch., *E. rostkoviana* Hayne var. *onegensis* (Cajand.) Jalas) - 1, 3
orae-australis Juz. = E. bicknellii
ossica Juz. - 2
parviflora Schag. (*E. curta* (Fries) Wettst., *E. curta* subsp. *glabrescens* (Wettst.) Smejkal, *E. glabrescens* (Wettst.) Wiinst., *E. officinalis* L. var. *curta* Fries, *E. parviflora* var. *glabrescens* (Wettst.) Tzvel., *E. stricta* D. Wolff ex J.F. Lehm. var. *curta* (Fries) Jalas) - 1
- var. *glabrescens* (Wettst.) Tzvel. = E. parviflora
pectinata Ten. (*E. nemorosa* (Pers.) Wallr. subsp. *pectinata* (Ten.) Malagarriga & Barrau, *E. nemorosa* subsp. *tatarica* (Fisch. ex Spreng.) Malagarriga & Barrau, *E. stricta* E. Wolff ex J.F. Lehm. var. *tatarica* (Fisch. ex Spreng.) Hartl, *E. tatarica* Fisch. ex Spreng.) - 1, 2, 3, 4, 6
peduncularis Juz. - 6
petiolaris Wettst. - 2
picta Wimm. - 1
- subsp. *kerneri* (Wettst.) P.F. Yeo = E. kerneri
pseudomollis Juz. (*E. mollis* (Ledeb.) Wettst. subsp. *pseudomollis* (Juz.) Yamazuki) - 5
putoranica N. Vodopianova (? *E. altaica* Serg. subsp. *glabra* Kuvajev) - 4
regelii Wettst. - 6
x **reuteri** Wettst. - E. parviflora Schag. x E. stricta D. Wolff ex J.F. Lehm. - 1
rostkoviana Hayne - 1
- subsp. *fennica* (Kihlm.) Karlsson = E. fennica
- var. *fennica* (Kihlm.) Jalas = E. fennica
- var. *onegensis* (Cajand.) Jalas = E. onegensis
rubra Baumg. = Odontites vulgaris
saamica Juz. = E. hyperborea
salisburgensis Funck (*E. carpatica* Zapal.) - 1
schischkinii Serg. - 3
schugnanica Juz. - 6
scottica Wettst. - 1
serotina Lam. = Odontites vulgaris
sevanensis Juz. - 2
sibirica Serg. - 3
sosnowskyi Kem.-Nath. (*E. carthalinica* Kem.-Nath.) - 2
stricta D. Wolff ex J.F. Lehm. (*E. condensata* Jord.) - 1
- var. *brevipila* (Burn. & Gremli) Hartl = E. brevipila
- var. *curta* (Fries) Jalas = E. parviflora
- var. *suecica* (Murb. & Wettst.) Karlsson = E. suecica
- var. *tatarica* (Fisch. ex Spreng.) Hartl = E. pectinata
- var. *tenuis* (Brenn.) Jalas = E. vernalis
subpolaris Juz. = E. hyperborea
suecica Murb. & Wettst. (*E. stricta* D. Wolff ex J.F. Lehm. var. *suecica* (Murb. & Wettst.) Karlsson) - 1
svanica Kem.-Nath. - 2
syreitschikovii Govor. - 3
tatarica Fisch. ex Spreng. = E. pectinata
tatrae Wettst. (*E. minima* Jacq. ex DC. subsp. *tatrae* (Wettst.) Hayek) - 1
taurica Ganesch. ex Popl. - 1

tenuis (Brenn.) Wettst. = E. vernalis
townsendiana Freyn ex Wettst. - 2
tranzschelii Juz. - 6
uechtritziana Jung. & Engl. = E. coerulea
ussuriensis Juz. - 5
vernalis List (*E. arctica* Lange ex subsp. *tenuis* (Brenn.) P.F. Yeo, p.p., *E. stricta* D. Wolff ex J.F. Lehm. var. *tenuis* (Brenn.) Jalas, *E. tenuis* (Brenn.) Wettst.) - 1, 3
- subsp. *brevipila* (Burn. & Gremli) O. Schwarz = E. brevipila
woronowii Juz. - 2
yesoensis Hara - 5

Gratiola L.
japonica Miq. - 5
officinalis L. - 1, 2, 3, 6
violacea Maxim. = Deinostema violacea

Gymnandra Pall. = Lagotis
globosa Kurz = Lagotis globosa

Holzneria F. Speta = Chaenorhinum
spicata (Korov.) F. Speta = Chaenorhinum spicatum

Hueblia F. Speta = Chaenorhinum
persica (Chav.) F. Speta = Chaenorhinum persicum

Kickxia Dumort.
caucasica (Muss.-Puschk. ex Spreng.) Kuprian. = K. elatine
elatine (L.) Dumort. (*K. caucasica* (Muss.-Puschk. ex Spreng.) Kuprian.) - 1, 2, 6(alien)
spuria (L.) Dumort. - 1

Lagotis Gaertn. (*Gymnandra* Pall.)
decumbens Rupr. - 6
glauca Gaertn. - 5
- subsp. *minor* (Willd.) Hult. = L. minor
globosa (Kurz) Hook. fil. (*Gymnandra globosa* Kurz) - 6
hultenii Polun. = L. minor
ikonnikovii Schischk. - 6
integrifolia (Willd.) Schischk. - 3, 4, 6
korolkowii (Regel & Schmalh.) Maxim. - 6
minor (Willd.) Standl. (*L. glauca* Gaertn. subsp. *minor* (Willd.) Hult., *L. hultenii* Polun.) - 1, 3, 4, 5
stolonifera (C. Koch) Maxim. - 2
uralensis Schischk. - 1

Lathraea L.
squamaria L. - 1, 2

Leptandra Nutt. = Veronicastrum
borissovae Czer. = Veronicastrum borissovae
cerasifolia (Monjuschko) Czer. = Veronicastrum cerasifolium
sibirica (L.) Nutt. ex G. Don fil. = Veronicastrum sibiricum
tubiflora (Fisch. & C.A. Mey.) Fisch. & C.A. Mey. = Veronicastrum tubiflorum

Leptorhabdos Schrenk
parviflora (Benth.) Benth. - 2, 6

Limosella L.

aquatica L. - 1, 2, 3, 4, 5, 6

Linaria Hill

acutiloba Fisch. ex Reichenb. (*L. vulgaris* Mill. subsp. *acutiloba* (Fisch. ex Reichenb.) D.Y. Hong, comb. superfl., *L. vulgaris* Mill. subsp. *acutiloba* (Fisch. ex Reichenb.) Hult.) - 1, 3, 4, 5

adzharica Kem.-Nath. - 2

alaica Junussov - 6

albifrons (Smith) Spreng. - 2

altaica Fisch. ex Kuprian. - 3, 4, 6

angustissima (Loisel.) Borb. (*Antirrhinum angustissimum* Loisel.) - 1(alien)

araratica Tzvel. = L. kurdica

armeniaca Chav. - 2

arvensis (L.) Desf. - 1

badachschanica Junussov - 6

bamianica auct. = L. sessilis

baxanensis Galushko - 2

bektauatensis Semiotr. - 6

bessarabica Kotov - 1

biebersteinii Bess. - 1, 2

- subsp. *maeotica* (Klok.) Ivanina = L. maeotica

- subsp. *ruthenica* (Blonski) Ivanina = L. ruthenica

- subsp. *strictissima* (Schur) Soo = L. ruthenica

bipartita auct. = L. incarnata

brachyceras (Bunge) Kuprian. - 3

bungei Kuprian. - 3

buriatica Turcz. ex Ledeb. - 4

*canadensis (L.) Dum.-Cours. - 1

caucasigena Kem.-Nath. = L. genistifolia

chalepensis (L.) Mill. - 2

concolor auct. = L. syspirensis

■ corifolia Desf.

corifolia sensu Kuprian. = L. corrugata

corrugata Karjag. (*L. corifolia* sensu Kuprian.) - 2

cretacea Fisch. ex Spreng. (*L. creticola* Kuprian.) - 1, 3, 6

creticola Kuprian. = L. cretacea

debilis Kuprian. - 3, 4

dmitrievae Semiotr. - 6

dolichocarpa Klok. - 3, 6

dolichoceras Kuprian. - 1, 6

dulcis Klok. (*L. odora* (Bieb.) Fisch. subsp. *dulcis* (Klok.) Ivanina) - 1

elymaitica (Boiss.) Kuprian. - 2

euxina Velen. = L. syspirensis

fedorovii R. Kam. - 6

genistifolia (L.) Mill. (*L. caucasigena* Kem.-Nath., *L. iberica* Kem.-Nath., *L. imerethica* Kem.-Nath., *L. kantschavelii* Kem.-Nath., *L. pontica* Kuprian.) - 1, 2, 3, 6

- subsp. *linifolia* (Boiss.) P.H. Davis = L. syspirensis

- var. *linifolia* Boiss. = L. syspirensis

gerensis Stapf = Chaenorhinum gerense

grandiflora Desf. - 2

grossheimii Kuprian. & Rzazade = L. schelkownikowii

hepatica Bunge - 3, 6

iberica Kem.-Nath. = L. genistifolia

imerethica Kem.-Nath. = L. genistifolia

*incarnata (Vent.) Spreng. (*Antirrhinum incarnatum* Vent., *Linaria bipartita* auct.) - 1, 3

incompleta Kuprian. - 1, 2, 3, 6

italica Trev. var. *strictissima* Schur = L. ruthenica

japonica Miq. - 5

jaxartica Levichev - 6

kantschavelii Kem.-Nath. = L. genistifolia

kokanica Regel - 6

kopetdaghensis Kuprian. = L. pyramidata

kulabensis B. Fedtsch. - 6

kurdica Boiss. & Hohen. (*L. araratica* Tzvel., *L. kurdica* subsp. *araratica* (Tzvel.) P.H. Davis, *L. lineolata* auct.) - 2

- subsp. *araratica* (Tzvel.) P.H. Davis = L. kurdica

lenkoranica Kuprian. = L. pyramidata

leptoceras Kuprian. - 6

lineolata auct. = L. kurdica

loeselii Schweigg. (*L. odora* (Bieb.) Fisch. subsp. *loeselii* (Schweigg.) Hartl) - 1

macrophylla Kuprian. - 6

macrostachya N. Vodopianova - 4

macroura (Bieb.) Bieb. - 1, 2

maeotica Klok. (*L. biebersteinii* Bess. subsp. *maeotica* (Klok.) Ivanina, *L. tesquicola* Klok.) - 1

megrica Tzvel. (*L. ordubadica* Tzvel.) - 2

melampyroides Kuprian. - 4, 5

meyeri Kuprian. - 2

micrantha (Cav.) Hoffmgg. & Link - 2, 6

monspessulana (L.) Mill. = L. repens

odora (Bieb.) Fisch. - 1, 3

- subsp. *dulcis* (Klok.) Ivanina = L. dulcis

- subsp. *loeselii* (Schweigg.) Hartl = L. loeselii

ordubadica Tzvel. = L. megrica

pedicellata Kuprian. - 6

pelisseriana (L.) Mill. - 2(?), 3

persica Chav. = Chaenorhinum persicum

petraea Stev. = L. syspirensis

pontica Kuprian. = L. genistifolia

popovii Kuprian. - 6

pyramidata (Lam.) Spreng. (*Antirrhinum pyramidatum* Lam., *Linaria kopetdaghensis* Kuprian., *L. lenkoranica* Kuprian.) - 2, 6

quasisessilis Levichev - 6

ramosa (Kar. & Kir.) Kuprian. - 6

reflexa (L.) Desf. - 2

repens (L.) Mill. (*Antirrhinum repens* L., *Linaria monspessulana* (L.) Mill.) - 1

ruthenica Blonski (*L. biebersteinii* Bess. subsp. *ruthenica* (Blonski) Ivanina, *L. biebersteinii* subsp. *strictissima* (Schur) Soo, *L. italica* Trev. var. *strictissima* Schur) - 1, 2, 3, 6

sabulosa Czern. ex Klok. - 1, 2

saposhnikovii E. Nikit. - 6

schelkownikowii Schischk. (*L. grossheimii* Kuprian. & Rzazade) - 2

schirvanica Fomin - 2

sessilis Kuprian. (*L. bamianica* auct.) - 6

simplex (Willd.) DC. (*L. turcomanica* Kuprian.) - 1, 2, 6

striatella Kuprian. - 6

syspirensis C. Koch (*L. euxina* Velen., *L. genistifolia* (L.) Mill. var. *linifolia* Boiss., *L. genistifolia* subsp. *linifolia* (Boiss.) P.H. Davis, *L. petraea* Stev., *L. concolor* auct.) - 1, 2

tesquicola Klok. = L. maeotica

tianschanica Semiotr. - 6

transiliensis Kuprian. - 6

turcomanica Kuprian. = L. simplex

vulgariformis E. Nikit. = L. vulgaris

vulgaris L. (*L. vulgariformis* E. Nikit.) - 1, 2, 3, 4, 5, 6

- subsp. *acutiloba* (Fisch. ex Reichenb.) D.Y. Hong = L. acutiloba

- subsp. *acutiloba* (Fisch. ex Reichenb.) Hult. = L. acutiloba

- subsp. arenosa Tzvel. - 1

zaissanica Semiotr. - 6

zangezura Grossh. - 2

Lindernia All. (*Anagalloides* Krock.)

procumbens (Krock.) Borb. (*Anagalloides procumbens* Krock., *Lindernia procumbens* (Krock.) Philcox, comb. superfl., *L. pyxidaria* L. p.p. nom. illegit.) - 1, 2, 3, 5, 6
pyxidaria L. p.p. = L. procumbens

Macrosyringion Rothm.

glutinosum (Bieb.) Rothm. (*Odontites glutinosa* (Bieb.) Benth.) - 1, 2

Mazus Lour.

japonicus (Thunb.) O. Kuntze - 5, 6
stachydifolius (Turcz.) Maxim. - 5

Melampyrum L.

alboffianum Beauverd - 2
alpestre Brugger = M. pratense subsp. alpestre
argyrocomum (Fisch. ex Ledeb.) K.-Pol. (*M. arvense* L. var. *argyrocomum* Fisch. ex Ledeb.) - 1, 3
arvense L. (*M. arvense* subsp. *aestivum* Govor., ? *M. arvense* var. *pseudobarbatum* Schur, ? *M. arvense* subsp. *pseudobarbatum* (Schur) Ronn., *M. arvense* subsp. *vernum* Govor., *M. pseudobarbatum* (Schur) Schur) - 1, 2, 3
- subsp. *aestivum* Govor. = M. arvense
- subsp. *pseudobarbatum* (Schur) Ronn. = M. arvense
- subsp. *vernum* Govor. = M. arvense
- var. *argyrocomum* Fisch. ex Ledeb. = M. argyrocomum
- var. *pseudobarbatum* Schur = M. arvense
carpathicum Schult. (*M. laricetorum* (A. Kerner) A. Kerner, *M. sylvaticum* L. subsp. *carpathicum* (Schult.) Soo, *M. sylvaticum* var. *laricetorum* A. Kerner, *M. sylvaticum* subsp. *laricetorum* (A. Kerner) Ronn.) - 1
caucasicum Bunge - 2
chlorostachyum Beauverd - 2
cristatum L. - 1, 2, 3, 4, 6
- subsp. **ronnigeri** (Poverl.) Ronn. - 1
- subsp. **solstitiale** (Ronn.) Ronn. (*M. solstitiale* Ronn.) - 1
elatius (Boiss.) Soo - 2
herbichii Woloszcz. - 1
- subsp. **woloszczakii** Jas. - 1
laciniatum Koshevn. & Zing. (*M. pratense* L. subsp. *laciniatum* (Koshevn. & Zing.) Tzvel.) - 1
laricetorum (A. Kerner) A. Kerner = M. carpathicum
moravicum H. Br. (*M. nemorosum* L. subsp. *moravicum* (H. Br.) Celak., *M. nemorosum* subsp. *moravicum* (H. Br.) Ronn. comb. superfl., *M. nemorosum* subsp. *zingeri* Ganesch.) - 1
mulkijanianii T.N. Pop. - 2
nemorosum L. (*M. nemorosum* subsp. *silesiacum* Ronn.) - 1, 4(alien)
- subsp. *moravicum* (H. Br.) Celak. = M. moravicum
- subsp. *moravicum* (H. Br.) Ronn. = M. moravicum
- subsp. *silesiacum* Ronn. = M. nemorosum
- subsp. *zingeri* Ganesch. = M. moravicum
polonicum (Beauverd) Soo (*M. subalpinum* auct.) - 1
- subsp. **hayekii** Soo - 1
pratense L. (*M. vulgatum* Pers. nom. illegit.) - 1, 3, 4
- subsp. **alpestre** Ronn. (*M. alpestre* Brugger, 1886, non Pers. 1807) - 1
- subsp. **divaricatum** (A. Kerner) Jas. - 1
- subsp. **hians** (Druce) Beauverd (*M. pratense* var. *hians* Druce) - 1
- subsp. *laciniatum* (Koshevn. & Zing.) Tzvel. = M. laciniatum
- subsp. **oligocladum** (Beauverd) Soo - 1

pseudobarbatum (Schur) Schur = M. arvense
roseum Maxim. - 5
saxosum Baumg. - 1
- subsp. **javorkae** Soo - 1(?)
setaceum (Maxim. ex Palib.) Nakai - 5
solstitiale Ronn. = M. cristatum subsp. solstitiale
stenophyllum Boiss. - 2
subalpinum auct. = M. polonicum
sylvaticum L. - 1
- subsp. *carpathicum* (Schult.) Soo = M. carpathicum
- subsp. *laricetorum* (A. Kerner) Ronn. = M. carpathicum
- var. *laricetorum* A. Kerner = M. carpathicum
vulgatum Pers. = M. pratense

Microrrhinum (Endl.) Fourr. = Chaenorhinum

klokovii (Kotov) F. Speta = Chaenorhinum klokovii
minus (L.) Fourr. = Chaenorhinum minus

Mimulus L. (*Torenia* auct.)

guttatus DC. - 1
inflatus (Miq.) Nakai (*Torenia inflata* Miq.) - 5
moschatus Dougl. ex Lindl. - 1
pilosiusculus H.B.K. - 1
sessilifolius Maxim. - 5
stolonifer Novopokr. - 5
tenellus Bunge - 5

Misopates Rafin.

orontium (L.) Rafin. (*Antirrhinum orontium* L.) - 1

Nathaliella B. Fedtsch.

alaica B. Fedtsch. - 6

Odicardis Rafin. = Veronica

crista-galli (Stev.) Rafin. = Veronica crista-galli

Odontites Lutw.

breviflora Regel - 1
    Critical species.
brevifolia Lindb. fil. = O. litoralis
fennica (Markl.) Tzvel. (*O. litoralis* (Fries) Fries subsp. *fennica* Markl.) - 1
glutinosa (Bieb.) Benth. = Macrosyringion glutinosum
litoralis (Fries) Fries (*Euphrasia odontites* L. var. *litoralis* Fries, *O. brevifolia* Lindb. fil., *O. rubra* (Baumg.) Opiz subsp. *litoralis* (Fries) A. & D. Löve, *O. verna* (Bell.) Dumort. subsp. *litoralis* (Fries) Nym.) - 1
- subsp. *fennica* Markl. = O. fennica
lutea (L.) Clairv. = Orthanthella lutea
pratensis (Wirtg.) Borb. (*Euphrasia odontites* L. var. *pratensis* Wirtg., *Odontites rubra* (Baumg.) Opiz subsp. *rothmaleri* U. Schneid., *O. vulgaris* Moench subsp. *rothmaleri* (U. Schneid.) Tzvel.) - 1
rubra (Baumg.) Opiz = O. vulgaris
- subsp. *litoralis* (Fries) A. & D. Löve = O. litoralis
- subsp. *pumila* (Nordst.) U. Schneid. = O. vulgaris subsp. pumila
- subsp. *rothmaleri* U. Schneid. = O. pratensis
rubra Gilib. = O. vulgaris
salina (Kotov) Kotov (*O. vulgaris* Moench subsp. *salina* (Kotov) Tzvel.) - 1
serotina (Lam.) Dumort. = O. vulgaris
- f. *pumila* Nordst. = O. vulgaris subsp. pumila
verna (Bell.) Dumort. - 1

- subsp. *litoralis* (Fries) Nym. = O. litoralis
- subsp. *pumila* (Nordst.) Pedersen = O. vulgaris subsp. pumila
- subsp. *serotina* (Dumort.) Corbiere = O. vulgaris
vulgaris Moench (*Euphrasia rubra* Baumg. nom. illegit., *E. serotina* Lam. nom. illegit., *Odontites rubra* (Baumg.) Opiz, nom. illegit., *O. rubra* Gilib. nom. invalid., *O. serotina* (Lam.) Dumort., *O. verna* (Bell.) Dumort. subsp. *serotina* (Dumort.) Corbiere) - 1, 2, 3, 4, 5, 6
- subsp. **pumila** (Nordst.) Tzvel. (*O. serotina* f. *pumila* Nordst., *O. rubra* subsp. *pumila* (Nordst.) U. Schneid., *O. verna* subsp. *pumila* (Nordst.) Pedersen) - 1
- subsp. *rothmaleri* (U. Schneid.) Tzvel. = O. pratensis
- subsp. *salina* (Kotov) Tzvel. = O. salina

Oligospermum D.Y. Hong = Veronica

crista-galli (Stev.) D.Y. Hong. = Veronica crista-galli

Omphalothrix Maxim.

longipes Maxim. - 4, 5

Orthantha (Benth.) A. Kerner = Orthanthella

aucheri (Boiss.) Wettst. = Orthanthella aucheri
lutea (L.) A. Kerner ex Wettst. = Orthanthella lutea

Orthanthella Rauschert (*Orthantha* (Benth.) A. Kerner, 1888, non *Orthanthe* Lem. 1856)

aucheri (Boiss.) Rauschert (*Orthantha aucheri* (Boiss.) Wettst.) - 2, 6
lutea (L.) Rauschert (*Odontites lutea* (L.) Clairv., *Orthantha lutea* (L.) A. Kerner ex Wettst.) - 1, 2

Paederotella (E. Wulf) Kem.-Nath.

daghestanica (Trautv.) Kem.-Nath. (*Veronica daghestanica* Trautv.) - 2
pontica (Rupr. ex Boiss.) Kem.-Nath. (*Veronica ruprechtii* Lipsky) - 2
teberdensis Kem.-Nath. (*Veronica teberdensis* (Kem.-Nath.) Boriss.) - 2

Parentucella Viv.

flaviflora (Boiss.) Nevski (*P. latifolia* (L.) Caruel subsp. *flaviflora* (Boiss.) Hand.-Mazz.) - 6
latifolia (L.) Caruel - 2
- subsp. *flaviflora* (Boiss.) Hand.-Mazz. = P. flaviflora
viscosa (L.) Caruel - 2

Pediculariopsis A. & D. Löve = Pedicularis

verticillata (L.) A. & D. Löve = Pedicularis verticillata

Pedicularis L. (*Pediculariopsis* A. & D. Löve)

abrotanifolia Bieb. ex Stev. - 3
achilleifolia Steph. - 3, 4, 6
acmodonta Boiss. - 2
adamsii Hult. = P. alopecuroides
adunca Bieb. ex Stev. - 4, 5
- subsp. *sachaliensis* (Miyabe & Miyake) Ivanina = P. rubinskii
alaica A. Li - 6
alatauica Stadlm. ex Vved. - 6
alberti Regel - 6
albolabiata (Hult.) Ju. Kozhevn. (*P. sudetica* Willd. subsp.

albolabiata Hult.) - 3, 4, 5
allorrhampha Vved. - 6
alopecuroides Stev. ex Spreng. (*P. adamsii* Hult., *P. kanei* Durand subsp. *adamsii* (Hult.) Hult., *P. lanata* Cham. & Schlecht. subsp. *adamsii* (Hult.) Hult.) - 3, 4, 5
altaica Steph. ex Stev. - 3, 6(?)
amoena Adams ex Stev. - 1, 3, 4, 5, 6
- var. *arguteserrata* (Vved.) Serg. = P. anthemifolia
- var. *elatior* Regel = P.macrochila
amoeniflora Vved. - 6
anthemifolia Fisch. ex Colla (*P. amoena* Adams ex Stev. var. *arguteserrata* (Vved.) Serg., *P. arguteserrata* Vved.) - 1, 3, 4
- subsp. *elatior* (Regel) Tsoong = P. macrochila
apodochila auct. = P. koidzumiana
apodochila Maxim. - 5
arctica R. Br. = P. langsdorfii
arguteserrata Vved. = P. anthemifolia
armena Boiss. & Huet - 2
atropurpurea Nordm. - 2
balkharica E. Busch - 2
brachystachys Bunge - 3, 4
breviflora Regel & C. Winkl. - 6
capitata Adams - 4, 5
caucasica Bieb. - 2
chamissonis Stev. - 5
- subsp. *japonica* (Miq.) Ivanina = P. japonica
cheilanthifolia Schrenk - 6
chroorrhyncha Vved. - 2
comosa L. var. *eriantha* Boiss. & Buhse = P. eriantha
- var. *venusta* Bunge = P.venusta
compacta Steph. - 1, 3, 4, 6
condensata Bieb. - 2
crassirostris Bunge - 2
czuiliensis Semiotr. - 6
daghestanica Bonati - 2
dasyantha Hadac (*P.kanei* Durand subsp. *dasyantha* (Hadac) Hult., *P. lanata* Cham. & Schlecht. subsp. *dasyantha* (Hadac) Hult.) - 1, 3
dasystachys Schrenk - 1, 3, 6
dolichorhiza Schrenk (*P. grandis* M. Pop., *P. jugentassica* Semiotr.) - 6
dubia B. Fedtsch. - 6
elata Willd. - 3, 4, 6
elisabethae T.N. Pop. - 2
eriantha (Boiss. & Buhse) T.N. Pop. (*P. comosa* L. var. *eriantha* Boiss. & Buhse, *P. sibthorpii* Boiss. var. *eriantha* (Boiss. & Buhse) T.N. Pop.) - 2
eriophora Turcz. - 5
exaltata Bess. - 1
fissa Turcz. - 4
flava Pall. - 4
grandiflora Fisch. - 4, 5
grandis M. Pop. = P. dolichorhiza
grigorjevii Ivanina - 6
gymnostachya (Trautv.) A. Khokhr. (*P. sudetica* Willd. var. *gymnostachya* Trautv., *P. sudetica* subsp. *gymnostachya* (Trautv.) Jurtz. & Petrovsky, *P. sudetica* subsp. *jacutica* Ju. Kozhevn.) - 4, 5
gypsicola Vved. - 6
hacquetii Graf - 1
hirsuta L. - 1, 3, 4, 5
hyperborea Vved. - 3
incarnata L. - 3, 4
inconspicua Vved. - 6
interioroides (Hult.) A. Khokhr. (*P. sudetica* Willd. subsp. *interioroides* Hult.) - 1, 3, 4, 5

interrupta Steph. - 6

japonica Miq. (*P. chamissonis* Stev. subsp. *japonica* (Miq.) Ivanina) - 5

jugentassica Semiotr. = P. dolichorhiza

kanei Durand = P. lanata

- subsp. *adamsii* (Hult.) Hult. = P. alopecuroides

'- subsp. *dasyantha* (Hadac) Hult. = P. dasyantha

- subsp. *pallasii* (Vved.) Hult. = P. pallasii

karatavica Pavl. - 6

karoi Freyn (*P. palustris* L. subsp. *karoi* (Freyn) Tsoong) - 1, 3, 4, 5, 6

kaufmannii Pinzg. - 1, 2, 3

koidzumiana Tatew. & Ohwi (*P. apodochila* auct. fl. sachal.) - 5

kokpakensis Semiotr. - 6

kolymensis A. Khokhr. - 5

korolkowii Regel - 6

krylovii Bonati - 6

kuljabensis Ivanina - 6

kungeica Bajt. - 6

kuznetzovii Kom. - 4(?), 5

labradorica Wirsing - 1, 3, 4, 5

lanata Cham. & Schlecht. (*P. kanei* Durand, *P. willdenowii* Vved.) - 5

- subsp. *adamsii* (Hult.) Hult. = P. alopecuroides

- subsp. *dasyantha* (Hadac) Hult. = P. dasyantha

- subsp. *pallasii* (Vved.) Hult. = P. pallasii

langsdorfii Fisch. ex Stev. (*P. arctica* R. Br., *P. langsdorfii* var. *arctica* (R. Br.) Ivanina, *P. langsdorfii* subsp. *arctica* (R. Br.) Pennell ex Hult., *P. langsdorfii* var. *purpurascens* (Cham. ex Spreng.) Ivanina, *P. purpurascens* Cham. ex Spreng.) - 4, 5

- subsp. *arctica* (R. Br.) Pennell ex Hult. = P. langsdorfii

- var. *arctica* (R. Br.) Ivanina = P. langsdorfii

- var. *purpurascens* (Cham. ex Spreng.) Ivanina = P. langsdorfii

lapponica L. - 1, 3, 4, 5

lasiostachys Bunge - 3

longiflora J. Rudolph - 3, 4

ludwigii Regel - 6

macrochila Vved. (*P. amoena* Adams ex Stev. var. *elatior* Regel, *P. anthemifolia* Fisch. ex Colla subsp. *elatior* (Regel) Tsoong) - 6

mandshurica Maxim. - 5

mariae Regel - 6

masalskyi Semiotr. - 6

maximowiczii Krasn. - 6

myriophylla Pall. - 3, 4

nasuta Bieb. ex Stev. - 5

nordmanniana Bunge - 2

novaiae-zemliae (Hult.) Ju. Kozhevn. (*P. sudetica* Willd. subsp. *novaiae-zemliae* Hult., *P. pseudoscopulorum* Yu. Kozhevn. & E. Tikhmenev) - 1, 3, 4, 5

ochotensis A. Khokhr. - 4, 5

ochrorrhyncha Galushko & T.N. Pop. (*P. sibthorpii* Boiss. var. *ochrorryncha* (Galushko & T.N. Pop.) T.N. Pop.) - 2

oederi Vahl - 1, 3, 4, 5, 6

olgae Regel - 6

opsiantha Ekman = P. palustris

pacifica (Hult.) Ju. Kozhevn. (*P. sudetica* Willd. subsp. *pacifica* Hult.) - 5

pallasii Vved. (*P. kanei* Durand subsp. *pallasii* (Vved.) Hult., *P. lanata* Cham. & Schlecht. subsp. *pallasii* (Vved.) Hult.) - 5

palustris L. (*P. opsiantha* Ekman, *P. palustris* subsp. *opsiantha* (Ekman) Almq., *P. palustris* var. *opsiantha* (Ekman) Hyl.) - 1, 2, 3

- subsp. **borealis** (I.W. Zett) Hyl. (*P. palustris* var. *borealis* I.W. Zett) - 1

- subsp. *karoi* (Freyn) Tsoong = P. karoi

- subsp. *opsiantha* (Ekman) Almq. = P. palustris

- var. *opsiantha* (Ekman) Hyl. = P. palustris

panjutinii E. Busch - 2

parviflora Smith subsp. *pennellii* (Hult.) Hult. = P. pennellii

peduncularis M. Pop. - 6

pennellii Hult. (*P. parviflora* Smith subsp. *pennellii* (Hult.) Hult.) - 3, 4, 5

physocalyx Bunge - 1, 3, 6

platyrhyncha Schrenk - 6

pontica Boiss. - 2

popovii Vved. - 6

proboscidea Stev. - 3, 6

pseudoscopulorum Yu. Kozhevn. & E. Tikhmenev = P. novaiae-zemliae

pubiflora Vved. - 6

pulchra Pauls. - 6

purpurascens Cham. ex Spreng. = P. langsdorfii

pycnantha Boiss. - 6

resupinata L. - 1, 3, 4, 5, 6

rhinanthoides Schrenk - 6

rubens Steph. - 4

rubinskii Kom. (*P. adunca* Bieb. ex Stev. subsp. *sachaliensis* (Miyabe & Miyake) Ivanina, *P. sachalinensis* Miyabe & Miyake) - 5

sachalinensis Miyabe & Miyake = P. rubinskii

sarawschanica Regel - 6

sceptrum-carolinum L. - 1, 3, 4, 5

- subsp. **pubescens** (Bunge) Tsoong (*P. sceptrum-carolinum* var. *pubescens* Bunge) - 4

schistostegia Vved. - 5

schugnana B. Fedtsch. - 6

semenowii Regel - 6

sibirica Vved. - 1, 3, 4

- subsp. *uralensis* (Vved.) Ivanina = P. uralensis

sibthorpii Boiss. - 1, 2

- var. *eriantha* (Boiss. & Buhse) T.N. Pop. = P. eriantha

- var. *ochrorryncha* (Galushko & T.N. Pop.) T.N. Pop. = P. ochrorrhyncha

songarica Schrenk - 6

spicata Pall. - 4, 5

striata Pall. - 4, 5

subrostrata C.A. Mey. - 2

sudetica Willd. subsp. *albolabiata* Hult. = P. albolabiata

- subsp. **arctoeuropaea** Hult. - 1, 3

- subsp. *gymnostachya* (Trautv.) Jurtz. & Petrovsky = P. gymnostachya

- subsp. *interioroides* Hult. = P. interioroides

- subsp. *jacutica* Ju. Kozhevn. = P. gymnostachya

- subsp. *novaiae-zemliae* Hult. = P. novaiae-zemliae

- subsp. *pacifica* Hult. = P. pacifica

- var. *gymnostachya* Trautv. = P. gymnostachya

sylvatica L. - 1

talassica Vved. - 6

tarbagataica Semiotr. - 6

tatianae Bordz. - 2

tianschanica Rupr. - 6

transversa Bajmuchambetova - 6

tristis L. - 3, 4, 5, 6

uliginosa Bunge - 3, 4, 6

uralensis Vved. (*P. sibirica* Vved. subsp. *uralensis* (Vved.) Ivanina) - 1, 3

venusta Schang. ex Bunge (*P. comosa* L. var. *venusta* Bunge, non basionymum) - 3, 4, 5

verae Vved. - 6
verticillata L. (*Pediculariopsis verticillata* (L.) A. & D. Löve) - 1
villosa Ledeb. ex Spreng. - 3, 4, 5
violascens Schrenk - 3(?), 6
waldheimii Bonati - 6
wilhelmsiana Fisch. ex Bieb. - 2
willdenowii Vved. = P. lanata
wlassowiana Stev. - 4
yezoensis Maxim. - 5(?)

Pennellianthus Crosswhite (*Penstemon* auct., *Pentastemon* auct.)

frutescens (Lamb.) Crosswhite (*Pentastemon frutescens* Lamb.) - 5

Penstemon auct. = Pennellianthus

Pentastemon auct. = Pennellianthus
frutescens Lamb. = Pennellianthus frutescens

Phtheirospermum Bunge

chinense Bunge - 5

Pseudolysimachion Opiz = Veronica

alatavicum (M. Pop.) Holub = Veronica alatavica
andrasovszkyi (Jav.) Holub = Veronica andrasovszkyi
arenosum (Serg.) Holub = Veronica laeta
barrelieri (Schott) Holub = Veronica barrelieri
- subsp. *andrasovszkyi* (Jav.) M. Fisch. = Veronica andrasovszkyi
dauricum (Stev.) Holub = Veronica daurica
dauricum (Stev.) Yamazaki = Veronica daurica
euxinum (Turrill) Holub = Veronica euxina
galactites (Hance) Holub = Veronica linariifolia
incanum (L.) Holub = Veronica incana
- subsp. *hololeucum* (Juz.) Holub = Veronica hololeuca
incanum (L.) Yamazaki = Veronica incana
komarovii (Monjuschko) Holub = Veronica komarovii
laetum (Kar. & Kir.) Holub = Veronica laeta
linariifolium (Pall. ex Link) Holub = Veronica linariifolia
linariifolium (Pall. ex Link) Yamazaki = Veronica linariifolia
longifolium (L.) Opiz = Veronica longifolia
- subsp. *maritimum* (L.) Hartl = Veronica longifolia
- subsp. *septentrionale* (Boriss.) Holub = Veronica longifolia
maritimum (L.) A. & D. Löve = Veronica longifolia
olgense (Kom.) Holub = Veronica olgensis
orchideum (Crantz) T. Wraber = Veronica orchidea
paniculatum (L.) Hartl = Veronica spuria
- subsp. *foliosum* (Waldst. & Kit.) Hartl = Veronica foliosa
pinnatum (L.) Holub = Veronica pinnata
porphyrianum (Pavl.) Holub = Veronica porphyriana
reverdattoi (Krasnob.) Holub = Veronica reverdattoi
rotundum (Nakai) Yamazaki var. *subintegrum* (Nakai) Yamazaki = Veronica komarovii
sachalinense (Yamazaki) Yamazaki = Veronica sachalinensis
sajanense (Printz) Holub = Veronica sajanensis
schmidtianum (Regel) Yamazaki = Veronica schmidtiana
septentrionale (Boriss.) A. & D. Löve = Veronica longifolia
sessiliflorum (Bunge) Holub = Veronica x sessiliflora
spicatum (L.) Opiz = Veronica spicata

- subsp. *hybridum* (L.) Holub = Veronica spicata
- subsp. *orchideum* (Crantz) Hartl = Veronica orchidea
- subsp. *transcaucasicum* (Bordz.) Gabr. = Veronica transcaucasica
spurium (L.) Rauschert = Veronica spuria
- subsp. *foliosum* (Waldst. & Kit.) Holub = Veronica foliosa
- subsp. *paniculatum* (L.) Dostal = Veronica spuria
transcaucasicum (Bordz.) Holub = Veronica transcaucasica

Rhamphicarpa Benth.

medwedewii Albov - 2

Rhinanthus L. (*Alectorolophus* Zinn)

aestivalis (N. Zing.) Schischk. & Serg. (*R. angustifolius* C.C. Gmel. var. *aestivalis* (N. Zing.) E. Mayer, *R. angustifolius* subsp. *aestivalis* (N. Zing.) Soo, *R. angustifolius* subsp. *grandiflorus* (Wallr.) D.A. Webb, p.p., *R. grandiflorus* (Wallr.) Soo subsp. *aestivalis* (N. Zing.) Soo, *R. serotinus* (Schoenh.) Oborny subsp. *aestivalis* (N. Zing.) Dostal, *R. vernalis* (N. Zing.) Schischk. & Serg. subsp. *aestivalis* (N. Zing.) Ivanina) - 1, 2, 3, 4, 5
alectorolophus (Scop.) Poll. (*R. alectorolophus* subsp. *patulus* (Sterneck) Soo, *R. major* L. nom. ambig., *R. patulus* (Sterneck) Thell. & Schinz) - 1
- subsp. *patulus* (Sterneck) Soo = R. alectorolophus
alpinus Baumg. = R. pulcher subsp. alpinus
angustifolius C.C. Gmel. = R. glacialis subsp. subalpinus
- subsp. *aestivalis* (N. Zing.) Soo = R. aestivalis
- subsp. *apterus* (Fries) Soo = R. apterus
- subsp. *cretaceus* (Vass.) Soo = R. cretaceus
- subsp. *grandiflorus* (Wallr.) D.A. Webb, p.p. = R. aestivalis, R. apterus and R. vernalis
- subsp. *vernalis* (N. Zing.) Soo = R. vernalis
- var. *aestivalis* (N. Zing.) E. Mayer = R. aestivalis
angustifolius sensu Vass. p.p. = R. glacialis subsp. aristatus and R. glacialis subsp. subalpinus
apterus (Fries) Ostenf. (*R. angustifolius* C.C. Gmel. subsp. *apterus* (Fries) Soo, *R. angustifolius* subsp. *grandiflorus* (Wallr.) D.A. Webb, p.p., *R. grandiflorus* (Wallr.) Soo subsp. *apterus* (Fries) Soo, *R. sachalinensis* Vass., *R. serotinus* (Schoenh.) Oborny subsp. *apterus* (Fries) Hyl., *R. serotinus* f. *apterus* (Fries) M. Mizianty) - 1, 3, 5(alien)
aristatus (Celak.) Hausskn. = R. glacialis subsp. aristatus
borbasii (Doerfl.) Soo subsp. *songaricus* (Sterneck) Soo = R. songaricus
borealis (Sterneck) Druce (*R. minor* L. subsp. *borealis* (Sterneck) P.D. Sell) - 5
bosnensis (Behrend. & Sterneck) Soo (*Alectorolophus major* (L.) Reichenb. var. *bosnensis* Behrend. & Sterneck) - 1(?)
colchicus Vass. - 2
cretaceus Vass. (*R. angustifolius* C.C. Gmel. subsp. *cretaceus* (Vass.) Soo, *R. grandiflorus* (Wallr.) Soo subsp. *cretaceus* (Vass.) Soo, *R. serotinus* (Schoenh.) Oborny subsp. *cretaceus* (Vass.) Soo) - 1
crista-galli L. Rasse *R. aristatus* Celak. = R. glacialis subsp. aristatus
erectus (Sterneck) O. Schwarz = R. pulcher subsp. erectus
x **fallax** (Wimm. & Grab.) Chabert. - R. minor L. x R. vernalis (N. Zing.) Schischk. & Serg. - In places of common occurrence of parent species.
ferganensis Vass. = R. songaricus
glacialis Personnat subsp. **aristatus** (Celak.) Rauschert (*R. crista-galli* L. Rasse *R. aristatus* Celak., *R. aristatus*

(Celak.) Hausskn., *R. angustifolius* sensu Vass. p.p.) - 1(?)
- subsp. **subalpinus** (Sterneck) Rauschert (*Alectorolophus lanceolatus* Sterneck var. *subalpinus* Sterneck, *Rhinanthus angustifolius* C.C. Gmel. nom. ambig., *R. glacialis* subsp. *subalpinus* (Sterneck) Dostal, comb. superfl., *R. serotinus* (Schoenh.) Oborny subsp. *angustifolius* (C.C. Gmel.) Dostal, *R. angustifolius* sensu Vass. p.p.) - 1
x **gracilis** Schur. - R. pulcher Gunther & Schummel ex Opiz x R. vernalis (N. Zing.) Schischk. & Serg. (or R. aestivalis (N. Zing.) Schischk. & Serg.) - 1
grandiflorus (Wallr.) Bluff & Fingerh. = R. serotinus
grandiflorus (Wallr.) Soo = R. serotinus
- subsp. *aestivalis* (N. Zing.) Soo = R. aestivalis
- subsp. *apterus* (Fries) Soo = R. apterus
- subsp. *cretaceus* (Vass.) Soo = R. cretaceus
- subsp. *vernalis* (N. Zing.) Soo = R. vernalis
groenlandicus (Ostenf.) Chabert (*R. minor* L. subsp. *groenlandicus* (Ostenf.) Neum.) - 1
major Ehrh. = R. serotinus
major L. = R. alectorolophus
mediterraneus auct. = R. vassilczenkoi
minor L. - 1, 2, 3, 5
- subsp. *borealis* (Sterneck) P.D. Sell = R. borealis
- subsp. *elatior* (Schur) O. Schwarz (*R. minor* var. *elatior* Schur) - 1
- subsp. *groenlandicus* (Ostenf.) Neum. = R. groenlandicus
- subsp. *rusticulus* (Chabert) O. Schwarz = R. rusticulus
- subsp. *stenophyllus* (Schur) O. Schwarz = R. nigricans
- d. *stenophyllus* Schur = R. nigricans
- var. *subulatus* Chabert = R. subulatus
montanus Saut. = R. serotinus
nigricans Meinsh. (*R. minor* L. d. *stenophyllus* Schur, *R. minor* subsp. *stenophyllus* (Schur) O. Schwarz) - 1, 2
osiliensis (Ronn. & Saars.) Vass. - 1
patulus (Sterneck) Thell. & Schinz = R. alectorolophus
pectinatus (Behrend.) Vass. - 1, 2
ponticus (Sterneck) Vass. - 2(?)
x **pseudomontanus** V. Krecz. ex Vass. - 1
  Probably hybrid R. serotinus (Schoenh.) Oborny x R. vernalis (N. Zing.) Schischk. & Serg.
x **pseudosongoricus** Vass. - 3, 4, 6
  Probably hybrid R. songaricus (Sterneck) B. Fedtsch. x R. vernalis (N. Zing.) Schischk. & Serg.
pulcher Gunther & Schummel ex Opiz subsp. **alpinus** (Walp.) Rauschert (*Alectorolophus alpinus* Walp., *Rhinanthus alpinus* Baumg. 1816, non *R. alpinus* (L.) Lam. 1778) - 1
- subsp. **elatus** (Sterneck) O. Schwarz (*Alectorolophus pulcher* (Gunther & Schummel ex Opiz) Wimm. var. *elatus* Sterneck) - 1
- subsp. **erectus** (Sterneck) O. Schwarz (*Alectorolophus erectus* Sterneck, *Rhinanthus erectus* (Sterneck) O. Schwarz) - 1
rusticulus (Chabert) Druce (*R. minor* L. subsp. *rusticulus* (Chabert) O. Schwarz) - 2
sachalinensis Vass. = R. apterus
schischkinii Vass. - 2
serotinus (Schoenh.) Oborny (*Alectorolophus serotinus* Schoenh., *A. grandiflorus* Wallr. nom. illegit., *Rhinanthus grandiflorus* (Wallr.) Bluff & Fingerh. nom. illegit., *R. grandiflorus* (Wallr.) Soo, nom. illegit., *R. major* Ehrh. 1791, non L. 1756, *R. montanus* Saut.) - 1, 3
- subsp. *aestivalis* (N. Zing.) Dostal = R. aestivalis
- subsp. *angustifolius* (C.C. Gmel.) Dostal = R. glacialis subsp. subalpinus
- subsp. *apterus* (Fries) Hyl. = R. apterus
- subsp. *cretaceus* (Vass.) Soo = R. cretaceus

- subsp. *vernalis* (N. Zing.) Hyl. = R. vernalis
- f. *apterus* (Fries) M. Mizianty = R. apterus
songaricus (Sterneck) B. Fedtsch. (*R. borbasii* (Doerfl.) Soo subsp. *songaricus* (Sterneck) Soo, *R. ferganensis* Vass.) - 1, 2, 3, 6
subulatus (Chabert) Soo (*R. minor* L. var. *subulatus* Chabert) - 1, 2
vassilczenkoi Ivanina & Karasjuk (*R. mediterraneus* auct.) - 1, 2
vernalis (N. Zing.) Schischk. & Serg. (*R. angustifolius* C.C. Gmel. subsp. *grandiflorus* (Wallr.) D.A. Webb, p.p., *R. angustifolius* subsp. *vernalis* (N. Zing.) Soo, *R. grandiflorus* (Wallr.) Soo subsp. *vernalis* (N. Zing.) Soo, *R. serotinus* (Schoenh.) Oborny subsp. *vernalis* (N. Zing.) Hyl.) - 1, 2, 3, 4, 5
- subsp. *aestivalis* (N. Zing.) Ivanina = R. aestivalis

Rhynchocorys Griseb.

elephas (L.) Griseb. (*R. intermedia* Albov) - 2
- subsp. **glabrescens** (Benth.) R.B. Burbidge & I.B.K. Richardson (*R. elephas* var. *glabrescens* Benth.) - 2
intermedia Albov = R. elephas
maxima K. Richt. - 2(?)
orientalis (L.) Benth. - 2
stricta (C. Koch) Albov - 2

Scrophularia L.

alata Gilib. = S. umbrosa
altaica Murr. - 3, 4
amgunensis Fr. Schmidt - 5
amplexicaulis Benth. - 2
armeniaca Bordz. = S. olgae
atropatana Grossh. - 2
azerbaijanica Grau (*S. decipiens* auct.) - 2
badghysi Botsch. - 6
benthamiana Boiss. (*S. syriaca* auct. p.p.) - 2
bicolor Smith (*S. canina* L. subsp. *bicolor* (Smith) Greuter, *S. canina* auct.) - 1
botschanzevii Turakulov - 6
buergeriana Miq. (*S. oldhamii* Oliv.) - 5
canescens Bong. - 3, 6
canina auct. = S. bicolor
canina L. subsp. *bicolor* (Smith) Greuter = S. bicolor
capusii Tzagolova = S. griffithii
charadzeae Kem.-Nath. - 2
chlorantha Kotschy & Boiss. - 2
chrysantha Jaub. & Spach (*S. lunariifolia* Boiss. & Bal.) - 2
cinerascens Boiss. (*S. grossheimii* Schischk., *S. variegata* Bieb. subsp. *cinerascens* (Boiss.) Grau) - 2
clausii Boiss. & Buhse (*S. hyrcana* (Grossh.) Grossh., *S. vernalis* L. subsp. *clausii* (Boiss. & Buhse) Grau) - 2
cretacea Fisch. ex Spreng. - 1
czapandaghi B. Fedtsch. - 6
czernjakowskiana B. Fedtsch. - 6
decipiens auct. = S. azerbaijanica
dissecta Gorschk. = S. rosulata
divaricata Ledeb. - 1, 2
donetzica Kotov - 1
dshungarica Golosk. & Tzagolova - 6
exilis Popl. - 1
fedtschenkoi Gorschk. - 6
frigida Boiss. - 6
glabella Botsch. & Junussov - 6
goldeana Juz. - 1
gontscharovii Gorschk. - 6
granitica Klok. & A. Krasnova - 1

grayana Maxim. ex Kom. - 5
griffithii Benth. (*S. capusii* Tzagolova, *S. zaravschanica* Gorschk. & Zak.) - 6
grossheimii Schischk. = S. cinerascens
haematantha Boiss. & Heldr. - 2
helenae Albov - 2
heucheriiflora Schrenk - 6
hyrcana (Grossh.) Grossh. = S. clausii
ilwensis C. Koch - 2
imerethica Kem.-Nath. - 2
incisa Weinm. - 3, 4, 6
- var. *pamirica* O. Fedtsch. = S. pamirica
integrifolia Pavl. - 2
kabadianensis B. Fedtsch. - 6
kakudensis Franch. - 5
kiriloviana Schischk. - 6
kjurendaghi Botsch. & Kurbanov - 6
kotschyana Benth. - 2
kurbanovii Botsch. - 6
lateriflora Trautv. - 2
leucoclada Bunge - 6
litwinowii B. Fedtsch. - 6
lucida auct. = S. rutifolia
lunariifolia Boiss. & Bal. = S. chrysantha
macrobotrys Ledeb. - 2
■ **mandshurica** Maxim.
maximowiczii Gorschk. - 5
minima Bieb. - 2
mollis Somm. & Levier - 2
multicaulis Turcz. - 4
nachitschevanica Grossh. - 2
nervosa Benth. - 2
nikitinii Gorschk. - 6
nodosa L. - 1, 2, 3, 4
nuraniae Tzagolova - 6
oldhamii Oliv. = S. buergeriana
olgae Grossh. (*S. armeniaca* Bordz.) - 2
olympica Boiss. - 1, 2
orientalis L. - 2
pamirica (O. Fedtsch.) Ivanina (*S. incisa* Weinm. var. *pamirica* O. Fedtsch.) - 6
pamiro-alaica Gorschk. - 6
peregrina L. - 1, 2
pruinosa auct. = S. rosulata
regelii Ivanina - 6
rostrata Boiss. & Buhse - 2
rosulata Stiefelh. (*S. dissecta* Gorschk., *S. pruinosa* auct.) - 6
rupestris Bieb. ex Willd. (*S. variegata* Bieb. subsp. *rupestris* (Bieb. ex Willd.) Grau) - 1, 2
ruprechtii Boiss. - 2
rutifolia Boiss. (*S. lucida* auct.) - 2
sangtodensis B. Fedtsch. - 6
sareptana Kleop. ex Ivanina - 1
scabiosifolia Benth. - 6
scoparia Pennell (*S. turcomanica* Bornm. & Sint. ex Rech. fil.) - 6
scopolii Hoppe ex Pers. - 1, 2
- subsp. **adenocalyx** (Somm. & Levier) Grau (*S. scopolii* var. *adenocalyx* Somm. & Levier) - 2
sosnowskyi Kem.-Nath. - 2
sprengeriana Somm. & Levier - 2
striata Boiss. - 6
strizhowiae Abduss. - 6
syriaca auct. = S. benthamiana
tadshicorum Gontsch. - 6
takhtajanii Gabr. - 2

talassica Tzagolova - 6
thesioides Boiss. & Buhse - 2
turcomanica Bornm. & Sint. ex Rech. fil. = S. scoparia
umbrosa Dumort. (*S. alata* Gilib. nom. invalid.) - 1, 2, 3, 4, 6
variegata Bieb. - 2
- subsp. *cinerascens* (Boiss.) Grau = S. cinerascens
- subsp. *rupestris* (Bieb. ex Willd.) Grau = S. rupestris
vernalis L. - 1
- subsp. *clausii* (Boiss. & Buhse) Grau = S. clausii
verticillata Gontsch. & Grig. - 6
vvedenskyi Bondar. & Filat. - 6
xanthoglossa Boiss. - 6
zaravschanica Gorschk. & Zak. = S. griffithii
zuvandica Grossh. - 2
zvartiana Gabr. - 2

Siphonostegia Benth.

chinensis Benth. - 5

Spirostegia Ivanina

bucharica (B. Fedtsch.) Ivanina - 6

Staurophragma Fisch. & C.A. Mey.

natolicum Fisch. & C.A. Mey. - 2

Torenia auct. = Mimulus

inflata Miq. = Mimulus inflatus

Tozzia L.

alpina L. subsp. *carpathica* (Woloszcz.) Dostal = T. carpathica
- subsp. *carpathica* (Woloszcz.) Pawl. & Jasiewicz = T. carpathica
carpathica Woloszcz. (*T. alpina* L. subsp. *carpathica* (Woloszcz.) Dostal, *T. alpina* subsp. *carpathica* (Woloszcz.) Pawl. & Jasiewicz, comb. superfl.) - 1

Vandellia L.

diffusa L. - 2

Verbascum L. (*Celsia* L., x *Celsioverbascum* Rech. fil.)

adzharicum Gritzenko - 2
agrimoniifolium (C. Koch) Huber-Morath (*Celsia agrimoniifolia* C. Koch, *C. heterophylla* Desf., *Verbascum heterophyllum* (Desf.) O. Kuntze, 1891, non Mill. 1760, nec Mor. 1822) - 2, 6
alpigenum C. Koch - 2
x **arpaczajicum** Bordz. - V. georgicum Benth. x V. speciosum Schrad. - 2
■ **artvinense** E. Wulf
atroviolaceum (Somm. & Levier) Murb. (*Celsia atroviolacea* Somm. & Levier) - 2
aureum (C. Koch) O. Kuntze = V. oreophilum
austriacum auct. = V. marschallianum
bactrianum Bunge = V. erianthum
banaticum Schrad. - 1
x **biebersteinii** Bess. - V. lychnitis L. x V. nigrum L. - 1
blattaria L. - 1, 2, 3, 6
■ **cedreti** Boiss.
chaixii Vill. subsp. *laxum* (Filar. & Jav.) Ivanina = V. laxum

- subsp. *orientale* (Bieb.) Hayek = V. marschallianum
cheiranthifolium Boiss. - 2, 6
x **clinantherum** Bordz. - V. georgicum Benth. x V. hajastanicum Bordz. - 2
x **collinum** Schrad. - V. nigrum L. x V. thapsus L. - In places
  of common occurrence of parent species.
compactum Bieb. = V. ovalifolium
densiflorum Bertol. (*V. thapsiforme* Schrad.) - 1, 2
drymophiloides Gritzenko (*V. drymophilum* auct. p.p.
  quoad pl. cauc.) - 2
drymophilum auct. p.p. = V. drymophiloides
drymophylum Huber-Morath - 2(?)
    Probably has been recorded in error, instead of V. dryomophi-
    loides Gritzenko.
erianthum Benth. (*V. bactrianum* Bunge) - 6
eriorrhabdon Boiss. - 2(?)
erivanicum E. Wulf - 2
flavidum (Boiss.) Freyn & Bornm. - 2
flexuosum E. Wulf = V. varians
formosum Fisch. ex Schrank - 2
x **gabrielianae** (Huber-Morath) Huber-Morath (x *Celsiov-
  erbascum gabrielianae* Huber-Morath). - V. suworowianum (C. Koch) O. Kuntze x V. szovitsianum Boiss.
  var. adenothyrsum Murb. - 2
georgicum Benth. - 2
glabratum Friv. - 1
glomeratum Boiss. - 2(?)
gnaphalodes Bieb. - 1, 2
gossypinum Bieb. - 2
hajastanicum Bordz. - 2
heterophyllum (Desf.) O. Kuntze = V. agrimoniifolium
x **horticolum** Huber-Morath. - V. georgicum Benth. x V.
  laxum Filar. & Jav. - 2
korovinii Gritzenko - 6
lanatum Schrad. - 1
laxum Filar. & Jav. (*V. chaixii* Vill. subsp. *laxum* (Filar. &
  Jav.) Ivanina) - 1, 2
lychnitis L. - 1, 2, 3
macrocarpum Boiss. - 2, 6
marschallianum Ivanina & Tzvel. (*V. chaixii* Vill. subsp.
  orientale (Bieb.) Hayek, *V. orientale* Bieb. 1808, non
  All. 1785, *V. austriacum* auct.) - 1, 2, 3, 4(?), 6
megaphlomos auct. = V. speciosum
megricum (Tzvel.) Huber-Morath (*Celsia megrica* Tzvel.) -
  2
nigrum L. - 1, 2(?), 3, 4
nudicaule (Wydl.) Takht. (*Celsia nudicaulis* (Wydl.) B.
  Fedtsch., *Verbascum nudicaule* (Wyld.) Huber-Morath, comb. superfl.) - 2
oreophilum C. Koch (*V. aureum* (C. Koch) O. Kuntze) - 2
orientale Bieb. = V. marschallianum
orientale (L.) All. (*Celsia orientalis* L.) - 1, 2
ovalifolium Donn ex Sims (*V. compactum* Bieb.) - 1, 2
paniculatum E. Wulf - 2
x **patris** (Bordz.) Bordz. (*Celsia patris* Bordz.) - 2
phlomoides L. - 1, 2, 3, 6(?)
phoeniceum L. - 1, 2, 3, 6
pinnatifidum Vahl - 1, 2
x **polyphyllo-pyramidatum** Bordz. - 2
x **pseudohajastanicum** Huber-Morath. - V. hajastanicum
  Bordz. x V. oreophilum C. Koch - 2
punalense Boiss. & Buhse - 2, 6
pyramidatum Bieb. - 1, 2
x **ramigerum** Link ex Schrad. - V. densiflorum Bertol. x V.
  lychnitis L. - 1
x **roopianeum** Bordz. - V. hajastanicum Bordz. x V. phoeniceum L. - 2

saccatum C. Koch - 2
schachdagense Gritzenko (*V. sevanense* Gritzenko, 1968,
  non *V.* x *sevanense* Huber-Morath, 1965) - 2
sessiliflorum Murb. - 2
sevanense Gritzenko = V. schachdagense
x **sevanense** Huber-Morath. - V. flavidum (Boiss.) Freyn &
  Bornm. x V. varians Freyn & Sint. - 2
sinuatum L. - 1, 2, 6
songaricum Schrenk - 2, 6
speciosum Schrad. (*V. megaphlomos* auct.) - 1, 2
spectabile Bieb. - 1, 2
stachydiforme Boiss. & Buhse - 2
suworowianum (C. Koch) O. Kuntze (*Celsia suworowiana*
  C. Koch) - 2
szovitsianum Boiss. - 2
x **tauricum** Hook. - V. phoeniceum L. x V. spectabile
  Bieb. - 1
thapsiforme Schrad. = V. densiflorum
thapsus L. - 1, 2, 3, 4, 5, 6
■ **transcaucasicum** E. Wulf
turcomanicum Murb. - 6
turkestanicum Franch. - 6
undulatum Lam. - 1
varians Freyn & Sint. (*V. flexuosum* E. Wulf) - 2
wilhelmsianum C. Koch - 2

Veronica L. (*Odicardis* Rafin., *Oligospermum* D.Y. Hong, *Pseudolysimachion* Opiz)

acinifolia L. - 1
- subsp. *reuteriana* (Boiss.) A. Jelen. = V. reuteriana
agrestis L. - 1, 6(?)
alatavica M. Pop. (*Pseudolysimachion alatavicum* (M. Pop.)
  Holub) - 6
albanica C. Koch = V. amoena
alpina L. - 1, 3, 4
- subsp. *pumila* (All.) Dostal (*V. pumila* All.) - 1(?)
americana (Rafin.) Schwein. ex Benth. (*V. beccabunga* L.
  var. *americana* Rafin.) - 5
amoena Bieb. (? *V. albanica* C. Koch) - 2, 6
- subsp. *transhyrcanica* Monjuschko ex A. Jelen. - 2
anagallidiformis auct. = V. connata
anagallidiformis Boreau = V. anagallis-aquatica
anagallis-aquatica L. (*V. anagallidiformis* Boreau, *V. anagallis-aquatica* subsp. *anagallidiformis* (Boreau) Soo) -
  1, 2, 3, 4, 5, 6
- subsp. *anagallidiformis* (Boreau) Soo = V. anagallis-
  aquatica
- subsp. *anagalloides* (Guss.) A. Jelen. = V. anagalloides
- subsp. *anagalloides* (Guss.) Rouy = V. anagalloides
- subsp. *lysimachioides* (Boiss.) M. Fisch. = V. lysimachioides
- subsp. *michauxii* (Lam.) A. Jelen. = V. michauxii
- subsp. *oxycarpa* (Boiss.) A. Jelen. = V. oxycarpa
anagalloides Guss. (*V. anagallis-aquatica* L. subsp. *anagalloides* (Guss.) A. Jelen. comb. superfl., *V. anagallis-
  aquatica* subsp. *anagalloides* (Guss.) Rouy, *V. poljensis*
  auct.) - 1, 2, 3, 5(alien), 6
- subsp. *heureka* M. Fisch. = V. heureka
andrasovszkyi Jav. (*Pseudolysimachion andrasovszkyi* (Jav.)
  Holub, *P. barrelieri* Schott subsp. *andrasovszkyi* (Jav.)
  M. Fisch.) - 1
aphylla L. - 1
arceuthobia Woronow = V. multifida
arenosa (Serg.) Boriss. = V. laeta
arguteserrata Regel & Schmalh. (*V. bornmuelleri* Hausskn.,
  V. karatavica Pavl. ex Nevski) - 2, 6

armena Boiss. & Huet - 2
arvensis L. - 1, 2, 5(alien), 6
austriaca auct. = V. dentata
austriaca L. subsp. *dentata* (F. W. Schmidt) Watzl = V. dentata
- subsp. *teucrium* (L.) D.A. Webb = V. teucrium
- var. *teucrium* (L.) O. Bolos & Vigo = V. teucrium
austriaca sensu Boriss. = V. jacquinii
bachofenii Heuff. - 1(?)
baranetzkii Bordz. = V. denudata
barrelieri Schott (*Pseudolysimachion barrelieri* (Schott) Holub, *Veronica spicata* L. subsp. *barrelieri* (Schott) Murb., *V. spicata* subsp. *barrelieri* (Schult.) A. Jelen. comb. superfl., *V. steppacea* Kotov) - 1, 2
bashkiriensis Klok. = V. spicata subsp. bashkiriensis
baumgartenii Roem. & Schult. - 1
beccabunga L. - 1, 2, 3, 6
- subsp. **abscondita** M. Fisch. - 2
- subsp. **muscosa** (Korsh.) A. Jelen. (*V. beccabunga* var. *muscosa* Korsh., *V. hjuleri* Pauls.) - 6
- var. *americana* Rafin. = V. americana
beccabungoides Bornm. = V. oxycarpa
bellidifolia Juz. = V. incana
bellidioides L. - 1
biloba Schreb. (*V. chantavica* Pavl., *V. nevskii* Boriss.) - 2, 3, 5(alien), 6
bobrovii Nevski = V. oxycarpa
bogosensis Tumadzhanov - 2
bordzilowskii Juz. (*V. taurica* Willd. subsp. *bordzilowskii* (Juz.) A. Jelen.) - 1
borissovae Holub (*V. glabrifolia* Boriss. 1955, non Kitag. 1941) - 2
bornmuelleri Hausskn. = V. arguteserrata
bozakmanii M. Fisch. - 2
bucharica B. Fedtsch. - 6
- subsp. *ramosissima* (Boriss.) A. Jelen. = V. ramosissima
callitrichoides Kom. - 5
campylopoda Boiss. - 2, 3, 6
capillipes Nevski (*V. pseudocapillipes* Golosk.) - 6
capsellicarpa Dubovik (*V. multifida* L. subsp. *capsellicarpa* (Dubovik) A. Jelen.) - 1, 3
cardiocarpa (Kar. & Kir.) Walp. - 6
- subsp. *nanella* (Vved.) M. Fisch. = V. nanella
catenata Pennell = V. connata
caucasica Bieb. - 2
cerasifolia Monjuschko = Veronicastrum cerasifolium
ceratocarpa C.A. Mey. - 2
chamaedrys L. - 1, 2, 3, 4, 5, 6
- subsp. *vindobonensis* M. Fisch. = V. vindobonensis
chantavica Pavl. = V. biloba
charadzeae Kem.-Nath. = V. gentianoides
ciliata Fisch. - 4, 6
colchica Kimeridze - 2
connata Rafin. (*V. catenata* Pennell, *V. anagallidiformis* auct.) - 1, 2, 3, 4, 6
crista-galli Stev. (*Odicardis crista-galli* (Stev.) Rafin., *Oligospermum crista-galli* (Stev.) D.Y. Hong) - 2
cymbalaria Bod. - 1(alien)
czerniakowskiana Monjuschko (*V. tripartita* Boriss.) - 6
daghestanica Trautv. = Paederotella daghestanica
daurica Stev. (*Pseudolysimachion dauricum* (Stev.) Holub, *P. dauricum* (Stev.) Yamazaki, comb. superfl.) - 4, 5
densiflora Ledeb. - 3, 4, 5, 6
dentata F. W. Schmidt (*V. austriaca* L. subsp. *dentata* (F.W. Schmidt) Watzl, *V. austriaca* auct.) - 1, 2, 5(alien)
denudata Albov (*V. baranetzkii* Bordz.) - 2
didyma auct. = V. polita

dillenii Crantz - 1, 2, 3, 6
euxina Turrill (*Pseudolysimachion euxinum* (Turrill) Holub, *Veronica gryniana* Klok.) - 1
exortiva Kitag. = V. longifolia subsp. exortiva
fedtschenkoi Boriss. - 6
ferganica M. Pop. (*V. rubrifolia* sensu Boriss.) - 3, 6
filifolia Lipsky - 2
filiformis Smith - 1(alien), 2
foliosa Waldt. & Kit. (*Pseudolysimachion paniculatum* (L.) Hartl subsp. *foliosum* (Waldst. & Kit.) Hartl, *P. spurium* (L.) Rauschert subsp. *foliosum* (Waldst. & Kit.) Holub, *Veronica paniculata* L. subsp. *foliosa* (Waldst. & Kit.) Skalicky, *V. spuria* L. subsp. *foliosa* (Waldst. & Kit.) Nym.) - 1
fruticans Jacq. - 1
fruticulosa L. - 1
galactites Hance = V. linariifolia
galathica Boiss. (*V. stenobotrys* auct.) - 2
gaubae Bornm. (*V. turkmenorum* B. Fedtsch. ex Boriss.) - 6
gentianoides Vahl (*V. charadzeae* Kem.-Nath., *V. kemulariae* Kuth.) - 1, 2
glabrifolia Boriss. = V. borissovae
glareosa Somm. & Levier (*V. telephiifolia* Vahl subsp. *glareosa* (Somm. & Levier) M. Fisch.) - 2
gorbunovii Gontsch. - 6
grandiflora Gaertn. - 5
gryniana Klok. = V. euxina
hederifolia L. - 1, 2, 6
heureka (M. Fisch.) Tzvel. (*V. anagalloides* Guss. subsp. *heureka* M. Fisch.) - 1, 2, 3, 5(alien), 6
hispidula Boiss. & Huet (*V. perpusilla* Boiss. ex A. DC. p.p. nom. illegit.) - 1, 2, 3, 6
hjuleri Pauls. = V. beccabunga subsp. muscosa
hololeuca Juz. (*Pseudolysimachion incanum* (L.) Holub subsp. *hololeucum* (Juz.) Holub, *Veronica incana* L. subsp. *hololeuca* (Juz.) A. Jelen.) - 1
humifusa Dicks. (*V. riederiana* Gand. ex Kom., *V. serpyllifolia* L. subsp. *humifusa* (Dicks.) Syme, *V. tenella* auct.) - 1, 2, 5, 6
hybrida L. = V. spicata
imerethica Kem.-Nath. - 2
incana L. (*Pseudolysimachion incanum* (L.) Holub, *P. incanum* (L.) Yamazaki, comb. superfl., *Veronica bellidifolia* Juz., *V. spicata* L. subsp. *incana* (L.) S.M. Walters) - 1, 3, 4, 5, 6
- subsp. *hololeuca* (Juz.) A. Jelen. = V. hololeuca
intercedens Bornm. - 2, 6
jacquinii Baumg. (*V. sclerophylla* Dubovik, *V. austriaca* sensu Boriss.) - 1, 2
karatavica Pavl. ex Nevski = V. arguteserrata
kemulariae Kuth. = V. gentianoides
khorassanica Czerniak. - 6
komarovii Monjuschko (*Pseudolysimachion komarovii* (Monjuschko) Holub, *P. rotundum* (Nakai) Yamazaki var. *subintegrum* (Nakai) Yamazaki, *Veronica rotunda* Nakai subsp. *subintegra* (Nakai) A. Jelen., *V. rotunda* var. *subintegra* (Nakai) Yamazaki, *V. spuria* L. var. *subintegra* Nakai) - 5
kopetdaghensis B. Fedtsch. ex Boriss. - 6
krylovii Schischk. (*V. teucrium* L. subsp. *altaica* Watzl) - 3, 4, 6
kuljabensis Ivanina - 6
kurdica auct. = V. orientalis
laeta Kar. & Kir. (*Pseudolysimachion arenosum* (Serg.) Holub, *P. laetum* (Kar. & Kir.) Holub, *Veronica arenosa* (Serg.) Boriss., *V. pinnata* L. subsp. *laeta* (Kar. & Kir.) A. Jelen.) - 3, 6

linariifolia Pall. ex Link (*Pseudolysimachion galactites* (Hance) Holub, *P. linariifolium* (Pall. ex Link) Holub, *P. linariifolium* (Pall. ex Link) Yamazaki, comb. superfl., *Veronica galactites* Hance) - 4, 5

liwanensis C. Koch (*V. telephiifolia* sensu Boriss.) - 1

longifolia L. (*Pseudolysimachion longifolium* (L.) Opiz, *P. longifolium* subsp. *maritimum* (L.) Hartl, *P. longifolium* subsp. *septentrionale* (Boriss.) Holub, *P. maritimum* (L.) A. & D. Löve, *P. septentrionale* (Boriss.) A. & D. Löve, *Veronica longifolia* L. var. *borealis* Trautv., *V. longifolia* subsp. *borealis* (Trautv.) Kuvajev, *V. maritima* L., *V. septentrionalis* Boriss.) - 1, 2, 3, 4, 5, 6

- subsp. *borealis* (Trautv.) Kuvajev = V. longifolia

- subsp. **exortiva** (Kitag.) Kitag. (*V. exortiva* Kitag., *V. pseudolongifolia* Printz) - 5

- var. *borealis* Trautv. = V. longifolia

luetkeana Rupr. (*V. macrostemon* Bunge subsp. *luetkeana* (Rupr.) A. Jelen., *V. macrostemonoides* Zak.) - 6

lysimachioides Boiss. (*V. anagallis-aquatica* L. subsp. *lysimachioides* (Boiss.) M. Fisch.) - 1, 2, 6

macrostemon Bunge (*V. serpylloides* Regel) - 3, 4, 6

- subsp. *luetkeana* (Rupr.) A. Jelen. = V. luetkeana

macrostemonoides Zak. = V. luetkeana

maeotica Klok. (*V. spicata* L. subsp. *maeotica* (Klok.) Tzvel.) - 1

magna M. Fisch. (*V. melissifolia* auct.) - 2

maritima L. = V. longifolia

maxima auct. = V. urticifolia

maximowicziana Worosch. (*V. peregrina* L. subsp. *asiatica* A. Jelen., *V. yedoensis* auct.) - 5

melissifolia auct. = V. magna

meskhetica Kem.-Nath. = V. persica

michauxii Lam. (*V. anagallis-aquatica* L. subsp. *michauxii* (Lam.) A. Jelen.) - 6

microcarpa Boiss. - 2

minima (C. Koch) C. Koch = V. pusilla

minuta C.A. Mey. (*V. telephiifolia* sensu A.Jelen.) - 2

montana L. - 1, 2

monticola Trautv. - 2

montioides Boiss. - 6

mthiuletica Kem.-Nath. = V. petraea

multifida L. (*V. arceuthobia* Woronow) - 2, 3, 6

- subsp. *capsellicarpa* (Dubovik) A. Jelen. = V. capsellicarpa

- subsp. *orientalis* (Mill.) A. Jelen. = V. orientalis

nanella Vved. (*V. cardiocarpa* (Kar. & Kir.) Walp. subsp. *nanella* (Vved.) M. Fisch.) - 6

nevskii Boriss. = V. biloba

nigricans C. Koch = V. peduncularis

officinalis L. - 1, 2, 5(alien)

olgensis Kom. (*Pseudolysimachion olgense* (Kom.) Holub) - 5

■ **oltensis** Woronow

opaca Fries - 1, 6(?)

orchidea Crantz (*Pseudolysimachion orchideum* (Crantz) T. Wraber, *P. spicatum* (L.) Opiz subsp. *orchideum* (Crantz) Hartl, *Veronica spicata* L. subsp. *orchidea* (Crantz) Hayek) - 1

orientalis Mill. (*V. multifida* L. subsp. *orientalis* (Mill.) A. Jelen., *V. kurdica* auct.) - 2

oxycarpa Boiss. (*V. anagallis-aquatica* L. subsp. *oxycarpa* (Boiss.) A. Jelen., *V. beccabungoides* Bornm., *V. bobrovii* Nevski) - 2

paczoskiana Klok. = V. spicata

paniculata L. = V. spuria

- subsp. *foliosa* (Waldst. & Kit.) Skalicky = V. foliosa

parvifolia Vahl - 2

peduncularis Bieb. (*V. nigricans* C. Koch) - 2

peregrina L. - 1, 4, 5(alien)

- subsp. *asiatica* A. Jelen. = V. maximowicziana

perpusilla Boiss. ex A. DC. = V. hispidula and V. pusilla

persica Poir. (*V. meskhetica* Kem.-Nath.) - 1, 2, 5(alien), 6

petraea (Bieb.) Stev. (*V. mthiuletica* Kem.-Nath.) - 2

pinnata L. (*Pseudolysimachion pinnatum* (L.) Holub) - 3, 4, 6

- subsp. *laeta* (Kar. & Kir.) A. Jelen. = V. laeta

polita Fries (*V. didyma* auct. vix Ten.) - 1, 2, 6

poljensis auct. = V. anagalloides

polozhiae Revusch. - 4

porphyriana Pavl. (*Pseudolysimachion porphyrianum* (Pavl.) Holub, *Veronica spicata* L. subsp. *porphyriana* (Pavl.) A. Jelen.) - 3, 6

praecox All. - 1, 2

propinqua Boriss. - 2

prostrata L. - 1, 2, 3

pseudocapillipes Golosk. = V. capillipes

pseudolongifolia Printz = V. longifolia subsp. exortiva

pseudoorchidea (Pacz.) Klok. (*V. spicata* L. var. *pseudoorchidea* Pacz.) - 1

pumila All. = V. alpina subsp. pumila

pusilla Kotschy (? *V. minima* (C. Koch) C. Koch, *V. perpusilla* Boiss. ex A. DC. p.p. nom. illegit.) - 2

ramosissima Boriss. (*V. bucharica* B. Fedtsch. subsp. *ramosissima* (Boriss.) A. Jelen.) - 6

reuteriana Boiss. (*V. acinifolia* L. subsp. *reuteriana* (Boiss.) A. Jelen.) - 2

reverdattoi Krasnob. (*Pseudolysimachion reverdattoi* (Krasnob.) Holub) - 4

riederiana Gand. ex Kom. = V. humifusa

rotunda Nakai subsp. *subintegra* (Nakai) A. Jelen = V. komarovii

- var. *subintegra* (Nakai) Yamazaki = V. komarovii

rubrifolia Boiss. subsp. **respectatissima** A. Jelen. - 6

rubrifolia sensu Boriss. = V. ferganica

ruprechtii Lipsky = Paederotella pontica

sachalinensis Boriss. = Veronicastrum borissovae

sachalinensis Yamazaki (*Pseudolysimachion sachalinense* (Yamazaki) Yamazaki, *Veronica yamazakii* A. Jelen., *V. subsessilis* auct.) - 5

sajanensis Printz (*Pseudolysimachion sajanense* (Printz) Holub) - 3, 4

scardica Griseb. - 1

schistosa E. Busch - 2

schmidtiana Regel (*Pseudolysimachion schmidtianum* (Regel) Yamazaki) - 5

sclerophylla Dubovik = V. jacquinii

scutellata L. - 1, 2, 3, 4, 5, 6

septentrionalis Boriss. = V. longifolia

serpyllifolia L. (*V. serpyllifolia* var. *nummularioides* Lecoq & Lamotte, *V. serpyllifolia* subsp. *nummularioides* (Lecoq & Lamotte) Dostal, *V. tenella* All.) - 1, 2, 3, 4, 5, 6

- subsp. *humifusa* (Dicks.) Syme = V. humifusa

- subsp. *nummularioides* (Lecoq & Lamotte) Dostal = V. serpyllifolia

- var. *nummularioides* Lecoq & Lamotte = V. serpyllifolia

serpylloides Regel = V. macrostemon

x **sessiliflora** Bunge (*Pseudolysimachion sessiliflorum* (Bunge) Holub). - V. pinnata L. x V. porphyriana Pavl. - 3

siaretensis Lehm. - 2

sibirica L. = Veronicastrum sibiricum

spicata L. (*Pseudolysimachion spicatum* (L.) Opiz, *P. spicatum* subsp. *hybridum* (L.) Holub, *Veronica hybrida* L., *V. paczoskiana* Klok.) - 1, 2, 3, 4, 5, 6

- subsp. *barrelieri* (Schott) Murb. = V. barrelieri
- subsp. *barrelieri* (Schult.) A. Jelen. = V. barrelieri
- subsp. **bashkiriensis** Klok. ex Tzvel. (*V. bashkiriensis* Klok. nom. invalid.) - 1, 3
- subsp. *incana* (L.) S.M. Walters = V. incana
- subsp. **klokovii** Tzvel. - 2
- subsp. *maeotica* (Klok.) Tzvel. = V. maeotica
- subsp. *orchidea* (Crantz) Hayek = V. orchidea
- subsp. **petschorica** Tzvel. - 1
- subsp. *porphyriana* (Pavl.) A. Jelen. = V. porphyriana
- subsp. *transcaucasica* Bordz. = V. transcaucasica
- var. *pseudoorchidea* Pacz. = V. pseudoorchidea
spuria L. (*Pseudolysimachion paniculatum* (L.) Hartl, *P. spurium* (L.) Rauschert, *P. spurium* subsp. *paniculatum* (L.) Dostal, *Veronica paniculata* L.) - 1, 2, 3, 6
- subsp. *foliosa* (Waldst. & Kit.) Nym. = V. foliosa
- var. *subintegra* Nakai = V. komarovii
stelleri Pall. ex Link - 5
stenobotrys auct. = V. galathica
steppacea Kotov = V. barrelieri
stylophora M. Pop. - 6
sublobata M. Fisch. - 1
subsessilis auct. = V. sachalinensis
taurica Willd. - 1
- subsp. *bordzilowskii* (Juz.) A. Jelen. = V. bordzilowskii
teberdensis (Kem.-Nath.) Boriss. = Paederotella teberdensis
telephiifolia sensu Boriss. = V. liwanensis
telephiifolia sensu A. Jelen. = V. minuta
telephiifolia Vahl subsp. *glareosa* (Somm. & Levier) M. Fisch. = V. glareosa
tenella All. = V. serpyllifolia
tenella auct. = V. humifusa
tenuissima Boriss. - 6
teucrium L. (*V. austriaca* L. var. *teucrium* (L.) O. Bolos & Vigo, *V. austriaca* subsp. *teucrium* (L.) D.A. Webb) - 1, 2, 3
- subsp. *altaica* Watzl = V. krylovii
tianschanica Lincz. - 6
transcaucasica (Bordz.) Grossh. (*V. spicata* L. subsp. *transcaucasica* Bordz., *Pseudolysimachion spicatum* (L.) Opiz subsp. *transcaucasicum* (Bordz.) Gabr., *P. transcaucasicum* (Bordz.) Holub) - 2
triloba Opiz - 1(alien)
tripartita Boriss. = V. czerniakowskiana
triphyllos L. - 1, 2, 4(alien)
tubiflora Fisch. & C.A. Mey. = Veronicastrum tubiflorum
tumadzhanovii T. Mardalejschvili - 2
turkmenorum B. Fedtsch. ex Boriss. = V. gaubae
umbrosa Bieb. - 1, 2
urticifolia Jacq. (*V. maxima* auct.) - 1, 4
verna L. - 1, 2, 3, 6
vindobonensis (M. Fisch.) M. Fisch. (*V. chamaedrys* L. subsp. *vindobonensis* M. Fisch.) - 1
viscosula Klok. - 1
yamazakii A. Jelen. = V. sachalinensis
yedoensis auct. = V. maximowicziana

Veronicastrum Heist. ex Fabr. (*Leptandra* Nutt.)

borissovae (Czer.) Sojak (*Leptandra borissovae* Czer., *Veronica sachalinensis* Boriss. 1955, non Yamazaki, 1952, *Veronicastrum sachalinense* (Boriss.) Yamazaki) - 5
cerasifolium (Monjuschko) Yamazaki (*Veronica cerasifolia* Monjuschko, *Leptandra cerasifolia* (Monjuschko) Czer.) - 5
sachalinense (Boriss.) Yamazaki = V. borissovae

sibiricum (L.) Pennell (*Veronica sibirica* L., *Leptandra sibirica* (L.) Nutt. ex G. Don fil., *Veronicastrum sibiricum* (L.) Hara, comb. superfl., *V. virginicum* (L.) Farm. var. *sibiricum* (L.) B. Boivin) - 4, 5
tubiflorum (Fisch. & C.A. Mey.) Sojak (*Veronica tubiflora* Fisch. & C.A. Mey., *Leptandra tubiflora* (Fisch. & C.A. Mey.) Fisch. & C.A. Mey.) - 4, 5
virginicum (L.) Farw. var. *sibiricum* (L.) B. Boivin = V. sibiricum

SELAGINELLACEAE Willk.
Lycopodioides Boehm. = Selaginella
helvetica (L.) O. Kuntze = Selaginella helvetica

Selaginella Beauv. (*Lycopodioides* Boehm., *Stachygynandrum* Beauv. ex Mirb.)

aitchisonii Hieron.
This species had been recorded for the investigated territory by mistake.
borealis (Kaulf.) Rupr. - 4, 5
helvetica (L.) Spring (*Lycopodioides helvetica* (L.) O. Kuntze) - 1, 2, 4, 5
involvens (Sw.) Spring, p.p. = S. tamariscina
rossii (Baker) Warb. - 5
rupestris (L.) Spring (*Lycopodium rupestre* L., *Selaginella sibirica* (Milde) Hieron.) - 3, 4, 5
- var. *shakotanensis* Franch. ex Takeda = S. shakotanensis
sanguinolenta (L.) Spring - 3, 4, 5
selaginoides (L.) C. Mart. - 1, 2, 3, 4, 5
shakotanensis (Franch. ex Takeda) Miyabe & Kudo (*S. rupestris* (L.) Spring var. *shakotanensis* Franch. ex Takeda) - 5
sibirica (Milde) Hieron. = S. rupestris
tamariscina (Beauv.) Spring (*Stachygynandrum tamariscinum* Beauv., *Selaginella involvens* (Sw.) Spring, p.p. excl. basionymo) - 5

Stachygynandrum Beauv. ex Mirb. = Selaginella
tamariscinum Beauv. = Selaginella tamariscina

*SIMAROUBACEAE DC.
*Ailanthus Desf.
altissima (Mill.) Swingle - 1, 2, 6

SINOPTERIDACEAE Koidz. (*CHEILAN-THACEAE* B.K. Nayar)
Aleuritopteris Fee

argentea (S.F. Gmel.) Fee (*Cheilanthes argentea* (S.G. Gmel.) G. Kunze) - 3, 4, 5
kuhnii (Milde) Ching (*Cheilanthes kuhnii* Milde) - 5

Cheilanthes Sw.

acrosticha (Balb.) Tod. (*Ch. fragrans* auct., *Ch. pteridioides* auct.) - 2, 6
The name Ch. pteridioides (Reichard) C. Chr. is related to the diploid species Ch. maderensis Lowe.
argentea (S.G. Gmel.) G. Kunze = Aleuritopteris argentea
fragrans auct. = Ch. acrosticha
kuhnii Milde = Aleuritopteris kuhnii
marantae (L.) Domin = Notholaena marantae

persica (Bory) Mett. ex Kuhn - 1, 2, 6
pteridioides auct. = Ch. acrosticha

Notholaena R. Br.

marantae (L.) Desv. (*Cheilanthes marantae* (L.) Domin) - 1, 2

SMILACACEAE Vent.
Smilax L.

excelsa L. (*S. panduriformis* D. Aliev) - 2
- var. *ussuriensis* Regel = S. maximowiczii
maximowiczii Koidz. (*S. excelsa* L. var. *ussuriensis* Regel, *S. riparia* A. DC. var. *ussuriensis* (Regel) Hara, *S. riparia* subsp. *ussuriensis* (Regel) Kitag., *S. oldhamii* auct., *S. riparia* auct.) - 5
oldhamii auct. = S. maximowiczii
panduriformis D. Aliev = S. excelsa
riparia auct. = S. maximowiczii
riparia A. DC. subsp. *ussuriensis* (Regel) Kitag. = S. maximowiczii
- var. *ussuriensis* (Regel) Hara = S. maximowiczii

SOLANACEAE Juss.
Atropa L.

bella-donna L. - 1
- subsp. *caucasica* (Kreyer) V. Avet. = A. caucasica
caucasica Kreyer (*A. bella-donna* L. subsp. *caucasica* (Kreyer) V. Avet.) - 2
komarovii Blin. & Shal. - 6

*Capsicum L.

***annuum** L. - 1, 2, 3, 5, 6

Datura L.

ferox L. - 5(alien)
***inoxia** Mill. - 2, 6
***metel** L. - 2, 6
meteloides DC. - 1
stramonium L. - 1, 2, 3, 5, 6
tatula L. - 1, 2, 5, 6

Hyoscyamus L.

albus L. - 1
bohemicus F.W. Schmidt = H. niger
camerarii Fisch. & C.A. Mey. = H. reticulatus
kopetdaghi Pojark. - 6
kurdicus Bornm. - 2(?)
niger L. (*H. bohemicus* F.W. Schmidt, *H. pallidus* Waldst. & Kit. ex Willd.) - 1, 2, 3, 4, 5, 6
pallidus Waldst. & Kit. ex Willd. = H. niger
pusillus L. - 1, 2, 3, 6
reticulatus L. (*H. camerarii* Fisch. & C.A. Mey.) - 2
turcomanicus Pojark. - 6

Lycium L.

anatolicum A. Baytop & R. Mill - 2
barbarum L. - 1, 2, 6
***chinense** Mill. - 5(runs wild)
dasystemum Pojark. - 6
depressum Stocks, VI 1852 (*L. turcomanicum* Fisch. & C.A. Mey. ex Bunge, 7 XI 1852, *L. turcomanicum* Turcz. ex

Miers, 1854) - 2, 6
flexicaule Pojark. - 6
foliosum Stocks = L. ruthenicum
kopetdaghi Pojark. - 6
ruthenicum Murr. (*L. foliosum* Stocks) - 1, 2, 6
turcomanicum Fisch. & C.A. Mey. ex Bunge = L. depressum
turcomanicum Turcz. ex Miers = L. depressum

*Lycopersicon Hill

***esculentum** Mill.
***x humboldtii** (Willd.) Dun. - L. esculentum Mill. x L. pimpinellifolium (L.) Mill.
***peruvianum** (L.) Mill.
***pimpinellifolium** (L.) Mill.

Mandragora L.

turcomanica Mizg. - 6

Nicandra Adans.

physalodes (L.) Gaertn. - 1, 2, 3, 5, 6

*Nicotiana L.

***alata** Link & Otto - 1
***rustica** L. - 1, 2, 5, 6
***tabacum** L. - 1, 2, 5, 6

Physaliastrum Makino

echinatum (Yatabe) Makino = Ph. japonicum
japonicum (Franch. & Savat.) Honda (*Ph. echinatum* (Yatabe) Makino) - 5

Physalis L.

alkekengi L. - 1, 2, 6
- var. *glabripes* (Pojark.) Grub. = Ph. franchetii
***angulata** L. - 1, 2, 5, 6
franchetii Mast. (*Ph. alkekengi* L. var. *glabripes* (Pojark.) Grub., *Ph. glabripes* Pojark.) - 5
glabripes Pojark. = Ph. franchetii
hermannii Dun. - 6
***ixocarpa** Brot. ex Hornem. - 1, 2, 5
***peruviana** L. - 1, 2
praetermissa Pojark. - 6
***pubescens** L. - 1, 2, 5
viscosa L. - 6(alien)

Physochlaina G. Don fil.

alaica Korotk. - 6
orientalis (Bieb.) G. Don fil. - 2, 6
physaloides (L.) G. Don fil. - 3, 4, 5, 6
semenowii Regel - 6

Salpichroa Miers

rhomboidea Miers - 2

Scopolia Jacq.

carniolica Jacq. (*S. tubiflora* Kreyer) - 1
caucasica Kolesn. ex Kreyer - 2
tubiflora Kreyer = S. carniolica

Solanum L.

alatum Moench (*S. luteum* Mill. subsp. *alatum* (Moench) Dostal, *S. nigrum* L. subsp. *puniceum* Kirschl., *S. puniceum* C.C.Gmel. nom. superfl., *S. villosum* Mill. subsp. *alatum* (Moench) J.M. Edmonds, *S. villosum* subsp. *alatum* (Moench) Dostal, comb. superfl., *S. villosum* subsp. *puniceum* (Kirschl.) J.M. Edmonds) - 1

asiae-mediae Pojark. - 6

aviculare Forst. - 2(alien)

capsicastrum Link - 2(alien)

carolinense L. - 2(alien), 5(alien)

cornutum Lam. (*S. rostratum* Dun.) - 1, 2, 5

decipiens Opiz = S. schultesii

depilatum Kitag. = S. kitagawae

dulcamara L. - 1, 2, 3, 6(alien)

flavescens Andrz. = S. nigrum

heterodoxum Dun. - 1

humile Bernh. ex Willd. - 1

judaicum (L.) Bess. - 1

kieseritzkii C.A. Mey. - 2

kitagawae Schonbeck-Temesy (*S. depilatum* Kitag. 1939, non Bitt. 1913) - 1, 3, 4, 5, 6

littorale Raab = S. ulugurense

luteum Mill. = S. villosum

- subsp. *alatum* (Moench) Dostal = S. alatum

marinum (Bab.) Pojark. - 1

megacarpum Koidz. - 5

*****melongena** L. - 1, 2, 5, 6

nigrum L. (*S. flavescens* Andrz. 1862, non Dun. 1813) - 1, 2, 3, 4, 5, 6

- subsp. *puniceum* Kirschl. = S. alatum

- subsp. *schultesii* (Opiz) Wessely = S. schultesii

nitidibaccatum Bitt. - 1

olgae Pojark. (*S. pseudoflavum* Pojark.) - 6

persicum Willd. ex Roem. & Schult. - 1, 2, 6

- subsp. *pseudopersicum* (Pojark.) Schonbeck-Temesy = S. pseudopersicum

pseudoflavum Pojark. = S. olgae

pseudopersicum Pojark. (*S. persicum* Willd. ex Roem. & Schult. subsp. *pseudopersicum* (Pojark.) Schonbeck-Temesy) - 2

puniceum C.C. Gmel. = S. alatum

rostratum Dun. = S. cornutum

schultesii Opiz (*S. decipiens* Opiz, *S. nigrum* L. subsp. *schultesii* (Opiz) Wessely) - 1, 2

septemlobum Bunge - 4

sisymbriifolium Lam. - 1, 2

transcaucasicum Pojark. - 2

triflorum Nutt. - 5

*****tuberosum** L.

ulugurense Holub (*S. littorale* Raab, nom. illegit.) - 1

villosum Mill. (*S. luteum* Mill.) - 1, 2, 6

- subsp. *alatum* (Moench) J.M. Edmonds = S. alatum

- subsp. *alatum* (Moench) Dostal = S. alatum

- subsp. *puniceum* (Kirschl.) J.M. Edmonds = S. alatum

woronowii Pojark. - 2

zelenetzkii Pojark. - 1

SPARGANIACEAE Rudolphi

Sparganium L.

affine Schnizl. = S. angustifolium

angustifolium Michx. (*S. affine* Schnizl., *S. emersum* Rehm. var. *angustifolium* (Michx.) R.L. Taylor & B. MacBryde) - 1, 5

coreanum Levl. - 5

emersum Rehm. (*S. emersum* subsp. *simplex* (Huds.) Soo, *S. simplex* Huds. nom. illegit.) - 1, 2, 3, 4, 5, 6

- subsp. *simplex* (Huds.) Soo = S. emersum

- var. *angustifolium* (Michx.) R.L. Taylor & B. MacBryde = S. angustifolium

erectum L. (*S. erectum* subsp. *polyedrum* (Aschers. & Graebn.) Schinz & Thell., *S. polyedrum* (Aschers. & Graebn.) Juz., *S. ramosum* Huds. nom. illegit., *S. ramosum* subsp. *polyedrum* Aschers. & Graebn.) - 1, 2, 3, 4

- subsp. *microcarpum* (Neum.) Domin = S. microcarpum

- subsp. *microcarpum* (Neum.) Hyl. = S. microcarpum

- subsp. *neglectum* (Beeby) K. Richt. = S. neglectum

- subsp. *neglectum* (Beeby) Schinz & Thell. = S. neglectum

- subsp. *oocarpum* (Celak.) Domin = S. oocarpum

- subsp. *polyedrum* (Aschers. & Graebn.) Schinz & Thell. = S. erectum

- subsp. *stoloniferum* (Graebn.) Hara = S. stoloniferum

- var. *glomeratum* Laest. = S. glomeratum

friesii Beurl. = S. gramineum

glehnii Meinsh. = S. glomeratum

glomeratum (Laest.) L. Neum. (*S. erectum* L. var. *glomeratum* Laest., *S. glehnii* Meinsh., *S. glomeratum* Laest. ex Beurl. nom. nud.) - 1, 4, 5

gramineum Georgi (*S. friesii* Beurl.) - 1, 4, 5

hyperboreum Laest. 1853 (*S. hyperboreum* Laest. ex Beurl. 1852, nom. nud.) - 1, 4, 5

japonicum Rothert - 5

kawakamii Hara - 5

x **longifolium** Turcz. ex Ledeb. - S. emersum Rehm. x S. gramineum Georgi - 1, 4

microcarpum (Neum.) Raunk. (*S. erectum* L. subsp. *microcarpum* (Neum.) Domin, *S. erectum* subsp. *microcarpum* (Neum.) Hyl. comb. superfl., *S. ramosum* Huds. subsp. *microcarpum* (Neum.) Hyl.) - 1, 2, 3, 6

minimum Wallr. - 1, 2, 3, 4, 5

neglectum Beeby (*S. erectum* L. subsp. *neglectum* (Beeby) K. Richt., *S. erectum* subsp. *neglectum* (Beeby) Schinz & Thell. comb. superfl., *S. ramosum* Huds. subsp. *neglectum* (Beeby) Neum., *S. ramosum* subsp. *neglectum* (Beeby) Nym. comb. superfl.) - 1, 2, 6

- var. *oocarpum* Celak. = S. oocarpum

x **oligocarpum** Angstr. - E. emersum Rehm. x S. minimum Wallr. - 1, 3, 4

oocarpum (Celak.) Fritsch (*S. neglectum* Beeby var. *oocarpum* Celak., *S. erectum* L. subsp. *oocarpum* (Celak.) Domin) - 2

polyedrum (Aschers. & Graebn.) Juz. = S. erectum

probatovae Tzvel. - 5

ramosum Huds. = S. erectum

- subsp. *microcarpum* (Neum.) Hyl. = S. microcarpum

- subsp. *neglectum* (Beeby) Neum. = S. neglectum

- subsp. *neglectum* (Beeby) Nym. = S. neglectum

- subsp. *polyedrum* Aschers. & Graebn. = S. erectum

rothertii Tzvel. - 4, 5

simplex Huds. = S. emersum

x **speirocephalum** Neum. - S. angustifolium Michx. x S. gramineum Georgi - 1

x **splendens** Meinsh. - S. angustifolium Michx. x S. emersum Rehm. - 1

stenophyllum Maxim. ex Meinsh. - 5

stoloniferum (Graebn.) Buch.-Ham. ex Juz. (*S. erectum* L. subsp. *stoloniferum* (Graebn.) Hara) - 3, 4, 6

*SPHENOCLEACEAE DC.
*Sphenoclea Gaertn.
*zeylanica Gaertn. - 6

STAPHYLEACEAE Lindl.
Staphylea L.
colchica Stev. - 2
pinnata L. - 1, 2
*trifolia L. - 1

TAMARICACEAE Link
Myricaria Desv. (*Myrtama* Ovcz. & Kinzikaeva, *Tamaricaria* Qaiser & Ali)

alopecuroides Schrenk = M. bracteata
bracteata Royle (*M. alopecuroides* Schrenk, *M. germanica* (L.) Desv. subsp. *alopecuroides* (Schrenk) Kitam.) - 1, 2, 3, 4, 6
dahurica (Willd.) Ehrenb. - 3, 4
elegans Royle (*Myrtama elegans* (Royle) Ovcz. & Kinzikaeva, *Tamaricaria elegans* (Royle) Qaiser & Ali) - 6
germanica (L.) Desv. - 1
- subsp. *alopecuroides* (Schrenk) Kitam. = M. bracteata
longifolia (Willd.) Ehrenb. - 3, 4
prostrata Hook. fil. & Thoms. ex Benth. & Hook. fil. - 6
scharti Vass. - 6
squamosa Desv. - 1, 2, 3, 4, 6

Myrtama Ovcz. & Kinzikaeva = Myricaria
elegans (Royle) Ovcz. & Kinzikaeva = Myricaria elegans

Reaumuria L.
alternifolia auct. p.p. = R. cistoides, R. kuznetzovii, R. reflexa, R. sogdiana and R. turkestanica
■ **atreki** Botsch. & Zucker.
babataghi Botsch. - 6
badhysi Korov. - 6
botschantzevii Zucker. & Kurbanov - 6
cistoides Adams (*R. alternifolia* auct. p.p.) - 2, 6
fruticosa Bunge - 6
kaschgarica Rupr. - 6
korovinii Botsch. & Lincz. - 6
kuznetzovii Sosn. & Manden. (*R. alternifolia* auct. p.p.) - 2
oxiana (Ledeb.) Boiss. - 6
persica (Boiss.) Boiss. - 2
reflexa Lipsky (*R. alternifolia* auct. p.p.) - 6
- subsp. **odonta** Botsch. & Zucker. (*R. zakirovii* Gorschk.) - 6
sogdiana Kom. (*R. alternifolia* auct. p.p.) - 6
songarica (Pall.) Maxim. - 3, 6
squarrosa auct. p.p. = R. turkestanica
tatarica Jaub. & Spach - 6
turkestanica Gorschk. (*R. alternifolia* auct. p.p., *R. squarrosa* auct. p.p.) - 6
- subsp. **dentata** Botsch. & Zucker. - 6
zakirovii Gorschk. = R. reflexa subsp. odonta

Tamaricaria Qaiser & Ali = Myricaria
elegans (Royle) Qaiser & Ali = Myricaria elegans

Tamarix L.
affinis Bunge = T. gracilis
altaica Niedenzu = T. ramosissima
androssowii Litv. - 6
x aralensis Bunge. - T. gracilis Willd. x T. hispida Willd. - 6
araratica (Bunge) Gorschk. = T. kotschyi
arceuthoides Bunge - 6
aschabadensis Freyn & Sint. = T. florida
aucheriana (Decne. ex Walp.) Baum = T. passerinoides
brachystachys Bunge - 2, 6
bungei Boiss. - 6
elongata Ledeb. - 3, 6
x ewersmannii C. Presl ex Bunge. - T. leptostachya Bunge x T. ramosissima Ledeb. - 6
florida Bunge (*T. aschabadensis* Freyn & Sint., *T. karakalensis* Freyn & Sint.) - 6
gracilis Willd. (*T. affinis* Bunge, *T. polystachya* Ledeb.) - 1, 2, 3, 6
hispida Willd. - 1, 6
- var. *karelinii* (Bunge) Baum = T. x karelinii
hohenackeri Bunge (*T. rosea* Bunge, *T. smyrnensis* auct. p.p.) - 1, 2, 6
karakalensis Freyn & Sint. = T. florida
x karelinii Bunge (*T. hispida* Willd. var. *karelinii* (Bunge) Baum). - T. hispida Willd. x T. ramosissima Ledeb. - 6
kasakhorum Gorschk. (*T. leptostachys* Bunge var. *kasakhorum* (Gorschk.) Baum) - 6
komarovii Gorschk. - 6
korolkowii Regel & Schmalh. - 6
kotschyi Bunge (*T. araratica* (Bunge) Gorschk., *T. leptopetala* Bunge) - 2, 6
laxa Willd. - 1, 2, 3, 6
leptopetala Bunge = T. kotschyi
leptostachys Bunge - 6
- var. *kasakhorum* (Gorschk.) Baum = T. kasakhorum
litwinowii Gorschk. - 6
macrocarpa (Ehrenb.) Bunge = T. passerinoides
meyeri Boiss. - 1, 2, 6
octandra Bunge - 2
odessana Stev. ex Bunge = T. ramosissima
passerinoides Delile ex Desv. (*T. aucheriana* (Decne. ex Walp.) Baum, *T. macrocarpa* (Ehrenb.) Bunge) - 6
pentandra Pall. = T. ramosissima
polystachya Ledeb. = T. gracilis
ramosissima Ledeb. (*T. altaica* Niedenzu, *T. odessana* Stev. ex Bunge, *T. pentandra* Pall. p.p. nom. illegit.) - 1, 2, 6
rosea Bunge = T. hohenackeri
smyrnensis auct. p.p. = T. hohenackeri
szowitsiana Bunge - 6
tetrandra Pall. ex Bieb. - 1, 2

TAXACEAE S.F. Gray
Taxus L.
baccata L. - 1, 2
*canadensis Marsh. - 1
cuspidata Siebold & Zucc. ex Endl. - 1(cult.), 5

TETRADICLIDACEAE (Engl.) Takht.
Tetradiclis Stev.
corniculata Chalkuziev - 6
tenella (Ehrenb.) Litv. - 1, 2, 6

*THEACEAE D. Don
*Thea L.
*sinensis L. - 2

THELIGONACEAE Dumort.
Theligonum L.
cynocrambe L. - 1

THELYPTERIDACEAE Pichi Sermolli
Crenitis auct. = Oreopteris
quelpaertensis (Christ) H. Ito = Oreopteris quelpaertensis

Lastrea Bory = Oreopteris and Thelypteris
limbosperma (All.) Ching = Oreopteris limbosperma
limbosperma (All.) Heywood = Oreopteris limbosperma
limbosperma (All.) Holub & Pouzar = Oreopteris limbosperma
nipponica (Franch. & Savat.) Copel. = Parathelypteris nipponica
phegopteris (L.) Bory = Phegopteris connectilis
quelpaertensis (Christ) Copel. = Oreopteris quelpaertensis
thelypteris (L.) Bory = Thelypteris palustris

Nephrodium auct. = Thelypteris
thelypteroides Michx. = Thelypteris thelypteroides

Oreopteris Holub (*Lastrea* Bory, p.p. nom. ambig., *Crenitis* auct.)
limbosperma (All.) Holub (*Polypodium limbospermum* All., *Dryopteris limbosperma* (All.) Becherer, *D. oreopteris* (Ehrh.) Maxon, *Lastrea limbosperma* (All.) Ching, comb. superfl., *L. limbosperma* (All.) Heywood, *L. limbosperma* (All.) Holub & Pouzar, comb. superfl., *Polypodium oreopteris* Ehrh., *Thelypteris limbosperma* (All.) H.P. Fuchs, *T. oreopteris* (Ehrh.) Sloss.) - 1, 2, 4
quelpaertensis (Christ) Holub (*Dryopteris quelpaertensis* Christ, *Athyrium quelpaertensis* (Christ) Ching, *Ctenitis quelpaertensis* (Christ) H. Ito, *Dryopteris kamtschatica* Kom., *Lastrea quelpaertensis* (Christ) Copel., *Thelypteris quelpaertensis* (Christ) Ching) - 5

Parathelypteris (H. Ito) Ching (*Wagneriopteris* A. & D. Löve)
nipponica (Franch. & Savat.) Ching (*Aspidium nipponicum* Franch. & Savat., *Dryopteris nipponica* (Franch. & Savat.) C. Chr., *Lastrea nipponica* (Franch. & Savat.) Copel., *Thelypteris nipponica* (Franch. & Savat.) Ching, *Wagneriopteris nipponica* (Franch. & Savat.) A. & D. Löve) - 5

Phegopteris (C. Presl) Fee
connectilis (Michx.) Watt (*Polypodium connectile* Michx., *Dryopteris phegopteris* (L.) C. Chr., *Lastrea phegopteris* (L.) Bory, *Phegopteris polypodioides* Fee, *Thelypteris phegopteris* (L.) Sloss.) - 1, 2, 3, 4, 5, 6
polypodioides Fee = Ph. connectilis

Thelypteris Schmidel (*Aspidium* Sw. p.p. nom. illegit., *Lastrea* Bory, p.p., *Nephrodium* auct.)
limbosperma (All.) H.P. Fuchs = Oreopteris limbosperma
nipponica (Franch. & Savat.) Ching = Parathelypteris nipponica
oreopteris (Ehrh.) Sloss. = Oreopteris limbosperma
palustris Schott (*Polypodium palustre* Salisb. nom. illegit., *Dryopteris thelypteris* (L.) A. Gray, *Lastrea thelypteris* (L.) Bory, *Thelypteris thelypteroides* (Michx.) Holub subsp. *glabra* Holub) - 1, 2
phegopteris (L.) Sloss. = Phegopteris connectilis
quelpaertensis (Christ) Ching = Oreopteris quelpaertensis
thelypteroides (Michx.) Holub (*Nephrodium thelypteroides* Michx.) - 1, 3, 4, 5, 6
- subsp. *glabra* Holub = T. palustris

Wagneriopteris A. & D. Löve = Parathelypteris
nipponica (Franch. & Savat.) A. & D. Löve = Parathelypteris nipponica

THYMELAEACEAE Juss.
Daphne L.
albowiana Woronow ex Pobed. - 2
altaica Pall. - 3, 6
angustifolia C. Koch = D. mucronata
axilliflora (Keissl.) Pobed. - 2
baksanica Pobed. - 2
*blagayana Freyer
caucasica Pall. - 2
circassica Woronow ex Pobed. - 2
cneorum L. (*D. julia* K.-Pol.) - 1
glomerata Lam. - 2
jezoensis Maxim. (*D. kamtschatica* Maxim. var. *jezoensis* (Maxim.) Ohwi, *D. kamtschatica* subsp. *yezoensis* (Maxim.) Worosch.) - 5
julia K.-Pol. = D. cneorum
kamtschatica Maxim. - 5
- subsp. *yezoensis* (Maxim.) Worosch. = D. jezoensis
- var. *jezoensis* (Maxim.) Ohwi = D. jezoensis
kurdica (Bornm.) Bornm. (*D. oleoides* Schreb. var. *kurdica* Bornm., *D. oleoides* subsp. *kurdica* (Bornm.) Bornm., *D. transcaucasica* Pobed., *D. oleoides* auct.) - 2
laureola L. - 1
mezereum L. - 1, 2, 3
mucronata Royle (*D. angustifolia* C. Koch) - 2
oleoides auct. = D. kurdica
oleoides Schreb. subsp. *kurdica* (Bornm.) Bornm. = D. kurdica
- var. *kurdica* Bornm. = D. kurdica
pontica L. - 2
pseudosericea Pobed. - 2
sophia Kolen. - 1
taurica Kotov - 1
transcaucasica Pobed. = D. kurdica
woronowii Kolak. - 2

Dendrostellera (C.A. Mey.) Van Tiegh.
ammodendron (Kar. & Kir.) Botsch. (*D. stachyoides* (Schrenk) Van Tiegh., *Diarthron stachyoides* (Schrenk) Kit Tan) - 6
arenaria Pobed. (*Diarthron arenarium* (Pobed.) Kit Tan) - 6
linearifolia Pobed. (*Diarthron linearifolium* (Pobed.) Kit Tan) - 6

macrorhachis Pobed. (*Diarthron macrorhachis* (Pobed.) Kit Tan) - 6
olgae Pobed. (*Diarthron olgae* (Pobed.) Kit Tan) - 6
stachyoides (Schrenk) Van Tiegh. = D. ammodendron
turkmenorum Pobed. (*Diarthron turkmenorum* (Pobed.) Kit Tan) - 6

Diarthron Turcz.

altaicum (Thieb.) Kit Tan = Stelleropsis altaica
antoninae (Pobed.) Kit Tan = Stelleropsis antoninae
arenarium (Pobed.) Kit Tan = Dendrostellera arenaria
caucasicum (Pobed.) Kit Tan = Stelleropsis caucasica
iranicum (Pobed.) Kit Tan = Stelleropsis iranica
issykkulense (Pobed.) Kit Tan = Stelleropsis issykkulensis
linearifolium (Pobed.) Kit Tan = Dendrostellera linearifolia
linifolium Turcz. - 4, 5
macrorhachis (Pobed.) Kit Tan = Dendrostellera macrorhachis
magakjanii (Sosn.) Kit Tan = Stelleropsis magakjanii
olgae (Pobed.) Kit Tan = Dendrostellera olgae
stachyoides (Schrenk) Kit Tan = Dendrostellera ammodendron
tarbagataicum (Pobed.) Kit Tan = Stelleropsis tarbagataica
tianschanicum (Pobed.) Kit Tan = Stelleropsis tianschanica
turcomanicum (Czerniak.) Kit Tan = Stelleropsis turcomanica
turkmenorum (Pobed.) Kit Tan = Dendrostellera turkmenorum
vesiculosum (Fisch. & C.A. Mey. ex Kar. & Kir.) C.A. Mey. - 1, 2, 6

Restella Pobed.

alberti (Regel) Pobed. - 6

Stellera L.

chamaejasme L. - 4, 5

Stelleropsis Pobed.

altaica (Thieb.) Pobed. (*Diarthron altaicum* (Thieb.) Kit Tan) - 3, 6
antoninae Pobed. (*Diarthron antoninae* (Pobed.) Kit Tan) - 6
caucasica Pobed. (*Diarthron caucasicum* (Pobed.) Kit Tan) - 2
■ **iranica** Pobed. (*Diarthron iranicum* (Pobed.) Kit Tan)
issykkulensis Pobed. (*Diarthron issykkulense* (Pobed.) Kit Tan) - 6
magakjanii (Sosn.) Pobed. (*Diarthron magakjanii* (Sosn.) Kit Tan) - 2
tarbagataica Pobed. (*Diarthron tarbagataicum* (Pobed.) Kit Tan) - 6
tianschanica Pobed. (*Diarthron tianschanicum* (Pobed.) Kit Tan) - 6
turcomanica (Czerniak.) Pobed. (*Diarthron turcomanicum* (Czerniak.) Kit Tan) - 6

Thymelaea Mill.

passerina (L.) Coss. & Germ. - 1, 2, 3, 6

TILIACEAE Juss.
*Corchorus L.

***capsularis** L. - 2, 6

***olitorius** L. - 2, 6

Tilia L.

amurensis Rupr. (*T. amurensis* var. *koreana* (Nakai) Worosch., *T. divaricata* Ig. Vassil. p.p., *T. komarovii* Ig. Vassil., *T. koreana* Nakai) - 5
- var. *koreana* (Nakai) Worosch. = T. amurensis
- var. *rufa* Nakai = T. taquetii
argentea Desf. ex DC. = T. tomentosa
begoniifolia Stev. (*T. caucasica* Rupr., *T. caucasica* f. *begoniifolia* (Stev.) Ig. Vassil., *T. officinarum* Crantz subsp. *caucasica* (Rupr.) Andonovski, *T. platyphyllos* Scop. subsp. *caucasica* (Rupr.) Loria, *T. prilipkoana* Grossh. & J. Wagner) - 1, 2
caucasica Rupr. = T. begoniifolia
- f. *begoniifolia* (Stev.) Ig. Vassil. = T. begoniifolia
cordata Mill. - 1, 2, 3
- f. *vulgaris* (Hayne) Ig. Vassil. = T. x vulgaris
cordifolia Bess. = T. europaea
dasystyla Stev. - 1
divaricata Ig. Vassil. p.p. = T. amurensis and T. taquetii
euchlora C. Koch - 1
europaea L. (*T. cordifolia* Bess., *T. platyphyllos* Scop. subsp. *cordifolia* (Bess.) Schneid., *T. platyphyllos* subsp. *grandiflora* (Ehrh.) Hayek var. *cordifolia* (Bess.) Beldie) - 1, 2(?)
komarovii Ig. Vassil. = T. amurensis
koreana Nakai = T. amurensis
ledebourii Borb. (*T. multiflora* Ledeb. 1842, non Vent.) - 2
mandshurica Rupr. - 5
maximowicziana Shirasawa - 5
mongolica Maxim. - 5
multiflora Ledeb. = T. ledebourii
officinarum Crantz subsp. *caucasica* (Rupr.) Andonovski = T. begoniifolia
pekinensis Rupr. - 5
petiolaris DC. (*T. tomentosa* Moench subsp. *petiolaris* (DC.) Soo) - 1
platyphyllos Scop. - 1
- subsp. *caucasica* (Rupr.) Loria = T. begoniifolia
- subsp. *cordifolia* (Bess.) Schneid. = T. europaea
- subsp. *grandiflora* (Ehrh.) Hayek var. *cordifolia* (Bess.) Beldie = T. europaea
prilipkoana Grossh. & J. Wagner = T. begoniifolia
sibirica Bayer - 3
taquetii Schneid. (*T. amurensis* Rupr. var. *rufa* Nakai, *T. divaricata* Ig. Vassil. p.p.) - 5
tomentosa Moench (*T. argentea* Desf. ex DC.) - 1
- subsp. *petiolaris* (DC.) Soo = T. petiolaris
x **vulgaris** Hayne (*T. cordata* Mill. f. *vulgaris* (Hayne) Ig. Vassil.). - T. cordata Mill. x T. platyphyllos Scop. - 1

TRAPACEAE Dumort. (*Hydrocaryaceae* Raimann)

Trapa L.

Many species of this genus must be carefully investigated to establish their real taxonomic rank.

alatyrica Spryg. ex V. Vassil. - 1
amurensis Fler. - 4, 5
astrachanica (Fler.) N. Wint. - 1, 3
borysthenica V. Vassil. - 1
carinthiaca (G. Beck) V. Vassil. - 1
caspica V. Vassil. - 1
colchica Albov - 1, 2
conocarpa (Aresch.) Fler. - 1

- subsp. *laevigata* (Nath.) Tacik var. *laevigata* f. *flerovii*
(Dobrocz.) Tacik = T. flerovii
cruciata (Gluck) V. Vassil. - 1
danubialis Dobrocz. - 1
europaea Fler. (*T. muzzanensis* (Jaggi) Szaf., Kulcz & Pawl.
var. *europaea* (Fler.) Tacik) - 1
flerovii Dobrocz. (*T. conocarpa* (Aresch.) Fler. subsp. *laevigata* (Nath.) Tacik var. *laevigata* f. *flerovii* (Dobrocz.)
Tacik) - 1
hyrcana Woronow - 2
incisa Siebold & Zucc. - 5
japonica Fler. (*T. korshinskyi* V. Vassil. p.p., *T. litwinowii* V.
Vassil. p.p.) - 5
kasachstanica V. Vassil. - 3
kazakorum V. Vassil. - 1
komarovii V. Vassil. = T.pseudoincisa
korshinskyi V. Vassil. = T. japonica and T. pseudoincisa
litwinowii V. Vassil. p.p. = T. japonica
longicornis V. Vassil. - 4
macrorhiza Dobrocz. - 1
maeotica Woronow - 1
maleevii V. Vassil. - 2
manshurica Fler. (*T. manshurica* subsp. *tranzschelii* (V.
Vassil.) Kitag., *T. tranzschelii* V. Vassil., *T. tuberculifera* V. Vassil.) - 5
- subsp. *tranzschelii* (V. Vassil.) Kitag. = T. manshurica
maximowiczii Korsh. - 5
media (Gluck) V. Vassil. = T.rossica
metschorica V. Vassil. - 1
muzzanensis (Jaggi) Szaf., Kulcz. & Pawl. var. *europaea*
(Fler.) Tacik = T. europaea
natans L. - 1
okensis V. Vassil. - 1
pectinata V. Vassil. - 1
potaninii V. Vassil. - 5
pseudocolchica V. Vassil. - 1
pseudoincisa Nakai (*T. komarovii* V. Vassil., *T. korshinskyi*
V. Vassil. p.p.) - 5
pseudorossica V. Vassil. - 1
pyramidalis V. Vassil. - 1
rossica V. Vassil. (*T. media* (Gluck) V. Vassil.) - 1
saissanica (Fler.) V. Vassil. (*T. sibirica* Fler. var. *saissanica*
Fler.) - 6
sajanensis V. Vassil. - 4
septentrionalis V. Vassil. - 1
sibirica Fler. - 3, 4, 5
- var. *saissanica* Fler. = T. saissanica
spryginii V. Vassil. - 1, 3, 6
tranzschelii V. Vassil. = T. manshurica
tuberculifera V. Vassil. = T. manshurica
turbinata V. Vassil. - 1
ucrainica V. Vassil. - 1
uralensis V. Vassil. - 1
wolgensis V. Vassil. - 1

TRAPELLACEAE Honda & Sakisaka
Trapella Oliv.
sinensis Oliv. - 5

TRICHOMANACEAE Burmeist. = HYME-
NOPHYLLACEAE

TRILLIACEAE Lindl.
Paris L.
hexaphylla Cham. = P. verticillata
- var. *manshurica* (Kom.) Worosch. = P. manshurica
incompleta Bieb. - 2
manshurica Kom. (*P. hexaphylla* Cham. var. *manshurica*
(Kom.) Worosch., *P. verticillata* Bieb. subsp. *manshurica* (Kom.) Kitag., *P. verticillata* var. *manshurica*
(Kom.) Hara) - 5
quadrifolia L. - 1, 2, 3, 4
- var. *setchuenensis* Franch. = P. setchuenensis
setchuenensis (Franch.) Barkalov (*P. quadrifolia* L. var.
setchuenensis Franch.) - 5
tetraphylla auct. = P. verticillata
verticillata Bieb. (*P. hexaphylla* Cham., *P. tetraphylla*
auct.) - 4, 5
- subsp. *manshurica* (Kom.) Kitag. = P. manshurica
- var. *manshurica* (Kom.) Hara = P. manshurica

Trillium L.
apetalon Makino - 5
camschatcense Ker-Gawl. (*T. kamtschaticum* Pall. ex
Pursh) - 5
- var. *tschonoskii* (Maxim.) Worosch. = T. tschonoskii
x **hagae** Miyabe & Tatew. - T. camschatcense Ker-Gawl. x
T. tschonoskii Maxim. - 5
kamtschaticum Pall. ex Pursh = T. camschatcense
rhombifolium Kom. - 5
smallii Maxim. - 5
tschonoskii Maxim. (*T. camschatcense* Ker-Gawl. var.
tschonoskii (Maxim.) Worosch.) - 5

*TROPAEOLACEAE DC.
*Tropaeolum L.
***majus** L.
***minus** L.

TYPHACEAE Juss.
Typha L.
angustata Bory & Chaub. = T. domingensis
angustifolia L. (*T. foveolata* Pobed., *T. pontica* Klok. fil. &
A. Krasnova) - 1, 2, 3, 4, 5, 6
- var. *angustata* (Bory & Chaub.) Jordanov = T. domingensis
australis Schum. & Thonn. = T. domingensis
bethulona auct. = T. shuttleworthii
caspica Pobed. - 1, 2
domingensis Pers. (*T. angustata* Bory & Chaub., *T. angustifolia* L. var. *angustata* (Bory & Chaub.) Jordanov, *T.
australis* Schum. & Thonn., *T. salgirica* A. Krasnova) -
1, 5, 6
elephantina Roxb. - 6
foveolata Pobed. = T. angustifolia
grossheimii Pobed. - 1, 2, 6
latifolia L. - 1, 2, 3, 4, 5, 6
- subsp. *shuttleworthii* (Koch & Sond.) Stojan. & Stef. = T.
shuttleworthii
laxmannii Lepech. (*T. veresczaginii* Kryl. & Schischk., *T.
zerovii* Klok. fil & A. Krasnova) - 1, 2, 3, 4, 5, 6
minima Funck (*T. pallida* Pobed.) - 1, 2, 3, 4, 6
orientalis C. Presl - 5
pallida Pobed. = T. minima

pontica Klok. fil. & A. Krasnova = T. angustifolia
przewalskii Skvorts. - 5
salgirica A. Krasnova = T. domingensis
shuttleworthii Koch & Sond. (*T. latifolia* L. subsp. *shuttle-worthii* (Koch & Sond.) Stojan. & Stef., *T. bethulona* auct.) - 1
turcomanica Pobed. - 6
vereszaginii Kryl. & Schischk. = T. laxmannii
zerovii Klok. fil & A. Krasnova = T. laxmannii

ULMACEAE Mirb.

Corchorus auct. = Zelkova

serratus Thunb. = Zelkova serrata

Ulmus L.

*androssowii Litv. - 6
araxina Takht. = U. minor
campestris L. = U. minor
carpinifolia Rupp. ex Suckow. = U. minor
celtidea (Rogow.) Litv. = U. laevis
densa Litv. = U. minor
elliptica C. Koch = U. glabra
foliacea Gilib. = U. minor
georgica Schchian = U. minor
glabra Huds. (*U. elliptica* C. Koch, *U. podolica* (Wilcz.) Klok., *U. sukaczevii* Andron.) - 1, 2
- subsp. *scabra* (Mill.) Dostal = U. scabra
grossheimii Takht. = U. minor
japonica (Rehd.) Sarg. (*U. propinqua* Koidz.) - 4, 5
laciniata (Trautv.) Mayr - 5
laevis Pall. (*U. celtidea* (Rogow.) Litv., *U. simplicidens* E. Wolf) - 1, 2
macrocarpa Hance - 4, 5
minor Mill. (*U. araxina* Takht., *U. campestris* L. nom. ambig., ? *U. carpinifolia* Rupp. ex Suckow., *U. densa* Litv., *U. foliacea* Gilib. nom. invalid., *U. georgica* Schchian, *U. grossheimii* Takht., *U. minor* var. *suberosa* (Moench) Dostal, *U. suberosa* Moench, *U. uzbekistanica* Drob., *U. wyssotzkyi* Kotov) - 1, 2, 6
- var. *suberosa* (Moench) Dostal = U. minor
pinnato-ramosa Dieck ex Koehne = U. pumila
podolica (Wilcz.) Klok. = U. glabra
propinqua Koidz. = U. japonica
pumila L. (*U. pinnato-ramosa* Dieck ex Koehne) - 4, 5, 6
scabra Mill. (*U. glabra* Huds. subsp. *scabra* (Mill.) Dostal) - 1, 2
simplicidens E. Wolf = U. laevis
suberosa Moench = U. minor
sukaczevii Andron. = U. glabra
uzbekistanica Drob. = U. minor
wyssotzkyi Kotov = U. minor

Zelkova Spach (*Corchorus* auct.)

carpinifolia (Pall.) C.Koch (*Zelkova hyrcana* Grossh. & Jarm.) - 2
hyrcana Grossh. & Jarm. = Z. carpinifolia
serrata (Thunb.) Makino (*Corchorus serratus* Thunb.) - 5

UMBELLIFERAE Juss. = APIACEAE

UROSTACHYACEAE Rothm. = HUPER-ZIACEAE

URTICACEAE Juss.

Achudemia Blume

japonica Maxim. (*Pilea japonica* (Maxim.) Hand.-Mazz.) - 5

Boehmeria Jacq.

*nivea (L.) Gaudich. - 2, 6
platyphylla var. *tricuspis* Hance = B. tricuspis
tricuspis (Hance) Makino (*B. platyphylla* var. *tricuspis* Hance) - 5

Dubreulia Gaudich. = Pilea

peploides Gaudich. = Pilea peploides

Freirea Gaudich. = Parietaria

chersonensis (Lang & Szov.) Jarm. = Parietaria serbica
micrantha (Ledeb.) Jarm. = Parietaria micrantha
serbica (Panc.) Jarm. = Parietaria serbica

Girardinia Gaudich.

cuspidata auct. = G. septentrionalis
septentrionalis Grudz. (*G. cuspidata* auct.) - 5

Laportea Gaudich.

bulbifera (Siebold & Zucc.) Wedd. - 5

Parietaria L. (*Freirea* Gaudich.)

alsinifolia Delile - 6
caespitosa Jarm. ex Schchian - 2
chersonensis (Lang & Szov.) Doerfl. = P. serbica
cryptorum C. Koch - 2
debilis auct. = P. micrantha
debilis Forst. fil. var. *micrantha* (Ledeb.) Wedd. = P. micrantha
diffusa Mert. & Koch (*P. ramiflora* Moench, nom. illegit., *P. officinalis* sensu Jarm.) - 1, 2
erecta Mert. & Koch = P. officinalis
jaxartica Pavl. = P. judaica
judaica L. (*P. jaxartica* Pavl.) - 2, 6
- subsp. *persica* (Stapf) Chrtek = P. persica
kemulariae Schchian - 2
littoralis Schchian - 2
lusitanica auct. = P. serbica
lusitanica L. subsp. *chersonensis* (Lang & Szov.) Chrtek = P. serbica
- - var. *micrantha* (Ledeb.) Chrtek = P. micrantha
- subsp. *serbica* (Panc.) P.W. Ball = P. serbica
micrantha Ledeb. (*Freirea micrantha* (Ledeb.) Jarm., *Parietaria debilis* Forst. fil. var. *micrantha* (Ledeb.) Wedd., *P. lusitanica* L. subsp. *chersonensis* (Lang & Szov.) Chrtek var. *micrantha* (Ledeb.) Chrtek, *P. debilis* auct.) - 1, 2, 3, 4, 5, 6
nitens C. Koch - 2
officinalis L. (*P. erecta* Mert. & Koch) - 1, 2, 6
officinalis sensu Jarm. = P. diffusa
persica Stapf (*P. judaica* L. subsp. *persica* (Stapf) Chrtek) - 6
ramiflora Moench = P. diffusa
ruschanica Jarm. ex Ikonn. - 6
serbica Panc. (*Freirea chersonensis* (Lang & Szov.) Jarm., *F. serbica* (Panc.) Jarm., *Parietaria chersonensis* (Lang & Szov.) Doerfl., *P. lusitanica* L. subsp. *chersonensis*

(Lang & Szov.) Chrtek, p.p., *P. lusitanica* subsp. *serbica* (Panc.) P.W. Ball, *P. lusitanica* auct.) - 1, 2, 6

Pilea Lindl. (*Dubreulia* Gaudich.)

hamaoi Makino - 5
japonica (Maxim.) Hand.-Mazz. = Achudemia japonica
mongolica Wedd. (*P. viridissima* Makino) - 4, 5
peploides (Gaudich.) Hook. & Arn. (*Dubreulia peploides* Gaudich.) - 5
viridissima Makino = P. mongolica

Urtica L.

angustifolia Fisch. ex Hornem. (*U. dioica* L. subsp. *angustifolia* (Fisch. ex Hornem.) Domin) - 3, 4, 5
cannabina L. - 1(alien), 3, 4, 5(alien), 6
cyanescens Kom. = U. laetevirens
dioica L. - 1, 2, 3, 4, 5, 6
- subsp. *angustifolia* (Fisch. ex Hornem.) Domin = U. angustifolia
- subsp. *cyanescens* (Kom.) Domin = U. laetevirens
- subsp. *galeopsifolia* (Wierzb. ex Opiz) Chrtek = U. galeopsifolia
- subsp. *kioviensis* (Rogow.) Buia = U. kioviensis
- subsp. *kioviensis* (Rogow.) Domin = U. kioviensis
- subsp. *laetevirens* (Maxim.) Domin = U. laetevirens
- subsp. *platyphylla* (Wedd.) Domin = U. platyphylla
- subsp. *platyphylla* (Wedd.) P. Medvedev = U. platyphylla
- subsp. *pubescens* (Ledeb.) Domin = U. pubescens
- subsp. *sondenii* (Simm.) Hyl. = U. sondenii
- var. *sondenii* Simm. = U. sondenii
galeopsifolia Wierzb. ex Opiz (*U. dioica* L. subsp. *galeopsifolia* (Wierzb. ex Opiz) Chrtek) - 1, 2, 3, 4
gracilis Ait. subsp. *sondenii* (Simm.) A. & D. Löve = U. sondenii
kioviensis Rogow. (*U. dioica* L. subsp. *kioviensis* (Rogow.) Buia, comb. superfl., *U. dioica* subsp. *kioviensis* (Rogow.) Domin) - 1
laetevirens Maxim. (*U. cyanescens* Kom., *U. dioica* L. subsp. *cyanescens* (Kom.) Domin, *U. dioica* subsp. *laetevirens* (Maxim.) Domin) - 5
pilulifera L. - 1, 2
platyphylla Wedd. (*U. dioica* L. subsp. *platyphylla* (Wedd.) Domin, comb. superfl., *U. dioica* subsp. *platyphylla* (Wedd.) P. Medvedev, *U. takedana* Ohwi) - 5
pubescens Ledeb. (*U. dioica* L. subsp. *pubescens* (Ledeb.) Domin) - 1, 2
sondenii (Simm.) Avror. ex Geltm. (*U. dioica* L. var. *sondenii* Simm., *U. dioica* subsp. *sondenii* (Simm.) Hyl., *U. gracilis* Ait. subsp. *sondenii* (Simm.) A. & D. Löve) - 1, 3, 4, 6
takedana Ohwi = U. platyphylla
urens L. - 1, 2, 3, 4, 5, 6

VACCINIACEAE S.F. Gray = ERICACEAE

VALERIANACEAE Batsch

Centranthus DC. (*Kentranthus* Neck.nom. invalid.)

calcitrapa (L.) Dufr. - 1
longifolius Stev. - 2

ruber (L.) DC. - 1

Kentranthus Neck. = Centranthus

Patrinia Juss.

gibbosa Maxim. - 5
intermedia (Hornem.) Roem. & Schult. - 3, 6
rupestris (Pall.) Dufr. - 4, 5
scabiosifolia Fisch. ex Link - 4, 5
sibirica (L.) Juss. - 1, 3, 4, 5, 6

Pseudobetckea (Hoeck) Lincz.

caucasica (Boiss.) Lincz. - 2

Valeriana L.

ajanensis (Regel & Til.) Kom. - 5
alliariifolia Adams - 2
- var. *tiliifolia* (Troitzk.) V. Avet. = V. tiliifolia
alpestris Stev. - 2
altaica Sumn. (*V. turczaninovii* Grub.) - 3, 4
alternifolia Ledeb. (*V. alternifolia* subsp. *stubendorfiana* (Sumn.) Worosch., *V. jacutica* Sumn., *V. stubendorfiana* Sumn., *V. stubendorfii* Kreyer ex Kom.) - 3, 4, 5
- subsp. *stubendorfiana* (Sumn.) Worosch. = V. alternifolia
amurensis P. Smirn. ex Kom. (*V. sambucifolia* Mikan fil. subsp. *amurensis* (P. Smirn. ex Kom.) Hara) - 5
angustifolia Tausch = V. collina
- subsp. *stolonifera* (Czern.) Rostanski = V. collina
armena P. Smirn. - 2
- subsp. *grossheimii* (Worosch.) Worosch. = V. grossheimii
baltica Pleijel = V. officinalis
capitata Pall. ex Link - 1, 3, 4, 5
cardamines Bieb. - 2
chionophila M. Pop. & Kult. - 6
colchica Utkin - 2
collina Wallr. (*V. angustifolia* Tausch, 1823, non Mill. 1768, *V. angustifolia* subsp. *stolonifera* (Czern.) Rostanski, *V. collina* subsp. *pratensis* (Dierb. ex Walter) A. & D. Löve, *V. officinalis* L. subsp. *collina* (Wallr.) Nym., *V. pratensis* Dierb. ex Walter, *V. stolonifera* Czern., *V. wallrothii* Kreyer, *V. wallrothii* subsp. *pratensis* (Dierb. ex Walter) Dostal) - 1
- subsp. *nitida* (Kreyer) A. & D. Löve = V. wolgensis
- subsp. *pratensis* (Dierb. ex Walter) A. & D. Löve = V. collina
coreana Briq. (*V. fauriei* Briq.) - 5
daghestanica Rupr. ex Boiss. - 2
dioica L. - 1
- subsp. *simplicifolia* (Reichenb.) Nym. = V. simplicifolia
dubia Bunge (*V. proximata* Sumn., *V. turkestanica* Sumn.) - 3, 4, 6
- subsp. *rossica* (P. Smirn.) Worosch. = V. rossica
eriophylla (Ledeb.) Utkin - 2
exaltata Mikan fil. = V. officinalis
excelsa auct. = V. sambucifolia
fasciculata Worosch. & Gorovoi - 5
fauriei Briq. = V. coreana
fedtschenkoi Coincy - 3, 6
ficariifolia Boiss. - 6
gotvanskyi Worosch. & Schlothgauer - 5
grossheimii Worosch. (*V. armena* P. Smirn. subsp. *gros-*

sheimii (Worosch.) Worosch.) - 1, 2
jacutica Sumn. = V. alternifolia
jelenevskyi P. Smirn. - 2
kamelinii B. Scharipova - 6
- subsp. **albiflora** B. Scharipova - 6
kassarica Charadze & Kapell. - 2
leucophaea DC. - 2
martjanovii Kryl. - 3, 6
minuta Wendelbo - 6
murmanica Orlova = V. sambucifolia
nitida Kreyer = V. wolgensis
officinalis L. (*V. baltica* Pleijel, *V. exaltata* Mikan fil., *V. officinalis* subsp. *baltica* (Pleijel) A. & D. Löve, *V. officinalis* subsp. *exaltata* (Mikan fil.) Soo, *V. palustris* Kreyer) - 1
- subsp. *baltica* (Pleijel) A. & D. Löve = V. officinalis
- subsp. *collina* (Wallr.) Nym. = V. collina
- subsp. *exaltata* (Mikan fil.) Soo = V. officinalis
- subsp. *nitida* (Kreyer) Soo = V. wolgensis
- subsp. *sambucifolia* (Mikan fil.) Celak. = V. sambucifolia
- var. *nitida* (Kreyer) Rostanski = V. wolgensis
- var. **simplicifolia** Ledeb. - 1
palustris Kreyer = V. officinalis
paucijuga Sumn. - 4
petrophila Bunge - 3, 4
pleijelii Kreyer = V. sambucifolia
pratensis Dierb. ex Walter = V. collina
procurrens Wallr. subsp. *salina* (Pleijel) A. & D. Löve = V. salina
proximata Sumn. = V. dubia
pseudodubia Sumn. = V. rossica
pseudoumbrosa Worosch. = V. transjenisensis
rossica P. Smirn. (*V. dubia* Bunge subsp. *rossica* (P. Smirn.) Worosch., *V. pseudodubia* Sumn., *V. spryginii* Sumn., *V. sumneviczii* Worosch.) - 1, 3, 4
salina Pleijel (*V. procurrens* Wallr. subsp. *salina* (Pleijel) A. & D. Löve) - 1
sambucifolia Mikan fil. (*V. murmanica* Orlova, *V. officinalis* L. subsp. *sambucifolia* (Mikan fil.) Celak., *V. pleijelii* Kreyer, *V. excelsa* auct.) - 1
- subsp. *amurensis* (P. Smirn. ex Kom.) Hara = V. amurensis
saxicola C.A. Mey. - 2
schachristanica R. Kam. & B. Scharipova - 6
simplicifolia (Reichenb.) Kabath (*V. dioica* L. subsp. *simplicifolia* (Reichenb.) Nym.) - 1
sisymbriifolia Vahl - 2, 6
spryginii Sumn. = V. rossica
stolonifera Czern. = V. collina
stubendorfiana Sumn. = V. alternifolia
stubendorfii Kreyer ex Kom. = V. alternifolia
sumneviczii Worosch. = V. rossica
tanaitica Worosch. = V. wolgensis
tiliifolia Troitzk. (*V. alliariifolia* Adams var. *tiliifolia* (Troitzk.) V. Avet.) - 2
transjenisensis Kreyer (*V. pseudoumbrosa* Worosch., *V. umbrosa* Sumn.) - 3, 4
transsilvanica Schur = V. tripteris
tripteris L. (*V. transsilvanica* Schur) - 1
tuberosa L. - 1, 2, 3, 6
turczaninovii Grub. = V. altaica
turkestanica Sumn. = V. dubia
umbrosa Sumn. = V. transjenisensis

wallrothii Kreyer = V. collina
- subsp. *nitida* (Kreyer) Dostal = V. wolgensis
- subsp. *pratensis* (Dierb. ex Walter) Dostal = V. collina
wolgensis Kazak. (*V. collina* Wallr. subsp. *nitida* (Kreyer) A. & D. Löve, *V. nitida* Kreyer, *V. officinalis* L. subsp. *nitida* (Kreyer) Soo, *V. officinalis* var. *nitida* (Kreyer) Rostanski, *V. tanaitica* Worosch., *V. wallrothii* Kreyer subsp. *nitida* (Kreyer) Dostal, *V. wolgensis* var. *nitida* (Kazak.) Worosch.) - 1, 3
- var. *nitida* (Kazak.) Worosch. = V. wolgensis

Valerianella Hill

amblyotis Fisch. & C.A. Mey. - 2
anodon Lincz. - 6
brachystephana (Ten.) Bertol. - 1
carinata Loisel. - 1, 2
corniculata C.A. Mey. - 2
coronata (L.) DC. - 1, 2, 6
costata (Stev.) Betcke - 1
cymbocarpa C.A. Mey. - 2, 6
dactylophylla Boiss. & Hohen. - 6
dentata (L.) Poll. - 1, 2, 6
diodon Boiss. - 2, 6
dufresnia Bunge - 2, 6
echinata (L.) DC. - 1
falconida N. Schvetsch. - 1
kotschyi Boiss. - 1, 2, 6
kulabensis Lipsky ex Lincz. - 6
lasiocarpa (Stev.) Betcke - 1, 2
leiocarpa (C. Koch) O. Kuntze - 2, 6
lipskyi Lincz. - 6
locusta (L.) Laterrade - 1, 2
mixta (L.) Dufr. - 1
muricata (Stev. ex Bieb.) J.W. Loud. - 1, 2, 6
ovczinnikovii B. Scharipova - 6
oxyrrhyncha Fisch. & C.A. Mey. - 2, 6
plagiostephana Fisch. & C.A. Mey. - 2, 6
platycarpa Trautv. - 2, 6
pontica Lipsky - 1, 2
pumila (L.) DC. - 1, 2, 6
rimosa Bast. - 1, 2
sclerocarpa Fisch. & C.A. Mey. - 2, 6
szovitsiana Fisch. & C.A. Mey. - 2, 6
triplaris Boiss. & Buhse - 6
tuberculata Boiss. - 6
turgida (Stev.) Betcke - 1, 2
turkestanica Regel & Schmalh. - 6
uncinata (Bieb.) Dufr. - 1, 2, 6
varzobica B. Scharipova - 6
vesicaria (L.) Moench - 6
vvedenskyi Lincz. - 6

VERBENACEAE J. St.-Hil.
Caryopteris Bunge
mongholica Bunge - 4

Clerodendrum L.
bungei Steud. (*C. foetidum* Bunge, 1833, non D. Don, 1825) - 2
foetidum Bunge = C. bungei

Lippia L.
nodiflora (L.) Michx. - 2, 6

Verbena L.
bracteata auct. = V. bracteosa
bracteosa Michx. (*V. bracteata* auct.) - 5
hastata L. - 2
officinalis L. - 1, 2, 6
supina L. - 1, 2, 6
urticifolia L. - 1
venosa Gill. & Hook. - 2(run wild)

Vitex L.
agnus-castus L. - 1, 2, 6
- var. *pseudonegundo* Hausskn. ex Bornm. = V. pseudonegundo
pseudonegundo (Hausskn. ex Bornm.) Hand.-Mazz. (*V. agnus-castus* L. var. *pseudonegundo* Hausskn. ex Bornm.) - 6

VIBURNACEAE Rafin.
Viburnum L.
burejaeticum Regel & Herd. - 5
edule (Michx.) Rafin. (*V. opulus* L.var. *edule* Michx.) - 5
furcatum Blume ex Maxim. - 5
lantana L. - 1, 2
mongolicum (Pall.) Rehd. - 4
opulus L. - 1, 2, 3, 4, 6
- var. *edule* Michx. = V. edule
orientale Pall. - 2
sargentii Koehne - 4, 5
wrightii Miq. - 5

VIOLACEAE Batsch
Viola L.
accrescens Klok. - 1
acuminata Ledeb. (*V. acuminata* var. *brevistipulata* (W. Beck.) Kitag., *V. turczaninowii* Juz.) - 4, 5
- var. *austro-ussuriensis* (W. Beck.) Kitag. = V. austro-ussuriensis
- var. *brevistipulata* (W. Beck.) Kitag. = V. acuminata
acutifolia (Kar. & Kir.) W. Beck. - 6
alaica Vved. (*V. oxycentra* Juz.) - 6
alba Bess. - 1, 2
- subsp. *scotophylla* (Jord.) Nym. = V. scotophylla
albida Palib. (*V. extremiorientalis* Worosch. & N.S. Pavlo-va) - 5
- var. *chaerophylloides* (Regel) F. Maek. = V. chaerophyll-oides
alexandrowiana (W. Beck.) Juz. - 4
alexejana R. Kam. & Junussov - 6
alisoviana Kiss = V. yedoensis
allochroa Botsch. - 6
altaica Ker-Gawl. (*V. monochroa* Klok.) - 3, 4, 6
ambigua Waldst. & Kit. - 1, 2
amurica W. Beck. - 5
arenaria DC. (*V. rupestris* F.W. Schmidt subsp. *arenaria* (DC.) Tzvel.) - 1, 2, 3, 4, 5
armena Boiss. & Huet - 2

arvensis Murr. - 1, 2, 3, 4, 5, 6
atroviolacea W. Beck. = V. disjuncta
austro-ussuriensis (W. Beck.) Kom. (*V. acuminata* Ledeb. var. *austro-ussuriensis* (W. Beck.) Kitag.) - 5
avatschensis W. Beck. & Hult. (*V. biflora* L. subsp. *avat-schensis* (W. Beck. & Hult.) Tzvel.) - 5
x **baltica** W. Beck. - V. canina L. x V. riviniana Reichenb. - 1
bezdelevae Worosch. - 5
biflora L. - 1, 3, 4, 5, 6
- subsp. *avatschensis* (W. Beck. & Hult.) Tzvel. = V. avat-schensis
blandiformis auct. = V. hultenii
brachyceras Turcz. - 3, 4, 5
brachysepala Maxim. (*V. mirabilis* L. subsp. *brachysepala* (Maxim.) Worosch.) - 4, 5
canina L. - 1, 2, 4
- subsp. *montana* (L.) C. Hartm. = V. montana
- var. *montana* (L.) Fries = V. montana
carnosula W. Beck. = V. selkirkii
caucasica Kolenati - 2
chaerophylloides (Regel) W. Beck. (*V. albida* Palib. var. *chaerophylloides* (Regel) F. Maek.) - 5
chassanica Korkischko - 5
collina Bess. (*V. teshioensis* Miyabe & Tatew.) - 1, 3, 4, 5, 6
commutata Waisb. - 1
crassa (Makino) Makino - 5
crassicornis W. Beck. & Hult. = V. selkirkii
cretacea Klok. - 1
dacica Borb. - 1
dactyloides Schult. - 4, 5
declinata Waldst. & Kit. - 1
disjuncta W. Beck. (*V. atroviolacea* W. Beck.) - 3
dissecta Ledeb. - 3, 4, 5, 6
dolichocentra Botsch. - 6
donetzkiensis Klok. - 1
elatior Fries = V. montana
- subsp. *jordanii* (Hanry) Soo = V. jordanii
elisabethae Klok. - 1
epipsila Ledeb. - 1, 3, 4
- subsp. *palustroides* W. Beck. = V. palustroides
epipsiloides A. & D. Löve (*V. repens* Turcz. ex Trautv. & C.A. Mey. 1856, non Schwein. 1822) - 1, 3, 4, 5
extremiorientalis Worosch. & N.S. Pavlova = V. albida
fedtschenkoana W. Beck. - 6
fischeri W. Beck. - 3
glaberrima (Murb.) C. Serg. = V. rupestris
gmeliniana Schult. - 4, 5
grypoceras A. Gray - 5
hirta L. - 1, 2, 3, 4, 6
hirtipes S. Moore - 5
hissarica Juz. - 6
hultenii W. Beck. (*V. blandiformis* auct.) - 5
ignobilis Rupr. - 2
incisa Turcz. - 3, 4
ircutiana Turcz. (*V. variegata* Fisch. ex Link f. *ircutiana* (Turcz.) F. Maek.) - 4, 5
- subsp. *tenuicornis* (W. Beck.) Worosch. = V. tenuicornis
irinae N. Zolot. - 3
isopetala Juz. - 6
jagellonica Zapal. - 1
jooi Janka - 1
jordanii Hanry (*V. elatior* Fries subsp. *jordanii* (Hanry) Soo) - 1

kamtschadalorum W. Beck. & Hult. (*V. langsdorfii* Fisch. ex Ging. subsp. *kamtschadalorum* (W. Beck. & Hult.) Tzvel.) - 5

karakalensis Klok. - 6

kitaibeliana Schult. - 1, 2

x kozo-poljanskii Grosset. - V. odorata L. x V. suavis Bieb. - 1

kunawarensis auct. = V. tianschanica

kupfferi Klok. - 2

kusanoana Makino (*V. silvestriformis* W. Beck.) - 5

kusnezowiana W. Beck. - 5

langsdorfii Fisch. ex Ging. - 5

- subsp. *kamtschadalorum* (W. Beck. & Hult.) Tzvel. = V. kamtschadalorum

lavrenkoana Klok. - 1

littoralis Spreng. - 1

macroceras Bunge - 3, 6

majchurensis Pissjauk. - 6

mandshurica W. Beck. - 5

maritima (Schweigg. ex Clausen) Tzvel. (*V. tricolor* L. subsp. *maritima* Schweigg. ex Clausen) - 1

matutina Klok. (*V. tricolor* L. subsp. *matutina* (Klok.) Valentine) - 1

mauritii Tepl. - 1, 3, 4, 5

meyeriana (Rupr.) Klok. - 2

minuta Bieb. - 2

mirabilis L. - 1, 2, 3, 4, 5, 6

- subsp. *brachysepala* (Maxim.) Worosch. = V. brachysepala

modestula Klok. - 6

monochroa Klok. = V. altaica

montana L. (*V. canina* L. subsp. *montana* (L.) C. Hartm., *V. canina* var. *montana* (L.) Fries, *V. elatior* Fries) - 1, 2, 3, 6

montana sensu Juz. = V. ruppii

muehldorfii Kiss - 5

nemausensis Jord. - 1, 2

occulta Lehm. - 1, 2, 6

odorata L. - 1, 2

oreades Bieb. - 1, 2

orientalis (Maxim.) W. Beck. - 5

orthoceras Ledeb. - 2

oxycentra Juz. = V. alaica

pacifica Juz. - 5

palustris L. - 1

palustroides (W. Beck.) Tzvel. (*V. epipsila* Ledeb. subsp. *palustroides* W. Beck.) - 5

parvula Tineo (*V. sosnowskyi* Kapell.) - 2

patrinii DC. - 4, 5

persicifolia Schreb. (*V. stagnina* Kit.) - 1, 3, 4

phalacrocarpa Maxim. - 5

pobedimovae C. Serg. = V. ruppii

pontica W. Beck. = V. suavis

primorskajensis (W. Beck.) Worosch. (*V. tenuicornis* W. Beck. subsp. *primorskajensis* W. Beck., *V. variegata* Fisch. ex Link var. *primorskajensis* (W. Beck.) Worosch.) - 5

prionantha Bunge - 5

pumila Chaix - 1, 3, 4, 6

pyrenaica Ramond ex DC. - 2

raddeana Regel - 5

reichenbachiana Jord. ex Boreau (*V. sylvestris* Lam. p.p. nom. illegit.) - 1, 2

repens Turcz. ex Trautv. & C.A. Mey. = V. epipsiloides

riviniana Reichenb. - 1

rossii Hemsl. - 5

rupestris F.W. Schmidt (*V. glaberrima* (Murb.) C. Serg., *V. rupestris* var. *glaberrima* Murb.) - 1, 4

- subsp. *arenaria* (DC.) Tzvel. = V. arenaria

- subsp. *sacchalinensis* (Boissieu) Worosch. = V. sacchalinensis

- var. *glaberrima* Murb. = V. rupestris

ruppii All. (*V. pobedimovae* C. Serg., *V. montana* sensu Juz.) - 1, 2, 3, 4, 5

x ruprechtiana Borb. - V. epipsila Ledeb. x V. palustris L. - 1

sacchalinensis Boissieu (*V. rupestris* F.W. Schmidt subsp. *sacchalinensis* (Boissieu) Worosch.) - 3, 4, 5

saxatilis F.W. Schmidt - 1

schachimardanica Chalkuziev - 6

scotophylla Jord. (*V. alba* Bess. subsp. *scotophylla* (Jord.) Nym.) - 1

selkirkii Pursh ex Goldie (*V. carnosula* W. Beck., *V. crassicornis* W. Beck. & Hult.) - 1, 2, 3, 4, 5

semilunaris auct. = V. verecunda

sieheana W. Beck. - 1, 2

silvestriformis W. Beck. = V. kusanoana

sintenisii W. Beck. - 6

somchetica C. Koch - 2

sosnowskyi Kapell. = V. parvula

stagnina Kit. = V. persicifolia

suavis Bieb. (*V. pontica* W. Beck.) - 1, 2, 6

sylvestris Lam. p.p. = V. reichenbachiana

tanaitica Grosset - 1

tarbagataica Klok. - 6

tenuicornis W. Beck. (*V. ircutiana* Turcz. subsp. *tenuicornis* (W. Beck.) Worosch., *V. trichosepala* (W. Beck.) Juz.) - 5

- subsp. *primorskajensis* W. Beck. = V. primorskajensis

teplouchovii Juz. - 4

teshioensis Miyabe & Tatew. = V. collina

tianschanica Maxim. (*V. kunawarensis* auct.) - 6

trichosepala (W. Beck.) Juz. = V. tenuicornis

tricolor L. - 1, 3, 4, 5

- subsp. *maritima* Schweigg. ex Clausen = V. maritima

- subsp. *matutina* (Klok.) Valentine = V. matutina

turczaninowii Juz. = V. acuminata

turkestanica Regel & Schmalh. - 6

uliginosa Bess. - 1

uniflora L. - 3, 4, 5

ursina Kom. - 5

variegata Fisch. ex Link - 4, 5

- var. *primorskajensis* (W. Beck.) Worosch. = V. primorskajensis

- f. *ircutiana* (Turcz.) F. Maek. = V. ircutiana

verecunda A. Gray (*V. semilunaris* auct.) - 5

vespertina Klok. - 2

wiedemannii Boiss. - 2

yamatsutae Ishidoya - 3

yedoensis Makino (*V. alisoviana* Kiss) - 5

VISCACEAE Batsch

Arceuthobium Bieb. (*Razoumofskya* Hoffm.)

oxycedri (DC.) Bieb. (*Razoumofskya oxycedri* (DC.) F. Schultz) - 1, 2, 6

Razoumofskya Hoffm. = Arceuthobium

oxycerdi (DC.) F. Schultz = Arceuthobium oxycedri

Viscum L.

abietis (Wiesb.) Fritsch (*V. austriacum* Wiesb. var. *abietis* Wiesb., *V. album* L. subsb. *abietis* (Wiesb.) Abrom., *V. laxum* Boiss. & Reut. subsp. *abietis* (Wiesb.) O. Schwartz, nom. illegit.) - 1
album L. - 1, 2
- subsb. *abietis* (Wiesb.) Abrom. = V. abietis
- subsp. *austriacum* (Wiesb.) Vollm. = V. austriacum
- var. *coloratum* (Kom.) Ohwi = V. coloratum
austriacum Wiesb. (*V. album* L. subsp. *austriacum* (Wiesb.) Vollm.) - 1
- var. *abietis* Wiesb. = V. abietis
coloratum (Kom.) Nakai (*V. album* L. var. *coloratum* (Kom.) Ohwi) - 5
laxum Boiss. & Reut. subsp. *abietis* (Wiesb.) O. Schwarz = V. abietis

VITACEAE Juss.
Ampelopsis Michx.

aegirophylla (Bunge) Planch. (*A. vitifolia* auct.) - 6
- var. *chondisensis* Vass. & L. Vassil. = A. chondisensis
brevipedunculata (Maxim.) Trautv. (*A. brevipedunculata* var. *maximowiczii* (Regel) Rehd., *A. glandulosa* (Wall.) Y. Momiyama var. *brevipedunculata* (Maxim.) Y. Momiyama, *A. glandulosa* var. *heterophylla* (Thunb.) Y. Momiyama, *A. heterophylla* (Thunb.) Siebold & Zucc. 1846, non Blume, 1825) - 5
- var. *maximowiczii* (Regel) Rehd. = A. brevipedunculata
chondisensis (Vass.& L.Vassil.) Tuljaganova (*A. aegirophylla* (Bunge) Planch. var. *chondisensis* Vass. & L. Vassil.) - 6
glandulosa (Wall.) Y. Momiayama var. *brevipedunculata* (Maxim.) Y. Momiyama = A. brevipedunculata
- var. *heterophylla* (Thunb.) Y. Momiyama = A. brevipedunculata
heterophylla (Thunb.) Siebold & Zucc. = A. brevipedunculata
japonica (Thunb.) Makino - 6
tadshikistanica V. Zapr. - 6
vitifolia auct. = A. aegirophylla

Parthenocissus Planch.

*inserta (A. Kerner) Fritsch
*quinquefolia (L.) Planch.
tricuspidata (Siebold & Zucc.) Planch. - 5

Vitis L.

amurensis Rupr. - 5
*berlandieri Planch.
bosturgaiensis Vass. = V. vinifera
coignetiae Pulliat ex Planch. (*Vitis kaempferi* auct.) - 5
hissarica Vass. = V. vinifera
hyrcanica Vass. = V. vinifera
kaempferi auct. = V. coignetiae
*labrusca L.
*rupestris Scheele
sylvestris C.C. Gmel. - 1, 2, 6

thunbergii Siebold & Zucc.
> This species had been recorded for the investigated territory by mistake.

usunachmatica Vass. = V. vinifera
*vinifera L. (*V. bosturgaiensis* Vass., *V. hissarica* Vass., *V. hyrcanica* Vass., *V. usunachmatica* Vass.) - 1, 2, 6; ofter runs wild
*vulpina L.

WOODSIACEAE (Diels) Herter
Acrostichum L. p.p. = Woodsia

ilvense L. = Woodsia ilvense

Hymenocystis C.A. Mey. = Woodsia

fragilis (Trev.) A. Askerov = Woodsia fragilis

Protowoodsia Ching

manchuriensis (Hook.) Ching (*Woodsia manchuriensis* Hook.) - 5

Woodsia R. Br. (*Acrostichum* L. p.p., *Hymenocystis* C.A. Mey.)

acuminata (Fomin) Sipl. = W. ilvensis
alpina (Bolt.) S.F. Gray (*W. ilvensis* (L.) R. Br. subsp. *alpina* (Bolt.) Aschers., *W. intermedia* Rupr. 1845, non Tagawa, 1936) - 1, 2, 3, 4, 5
commixta Ching = W. subcordata
fragilis (Trev.) Moore (*Hymenocystis fragilis* (Trev.) A. Askerov) - 2
glabella R. Br. (*W. pulchella* auct. p.p.) - 1, 2, 3, 4, 5
gracillima C. Chr. = W. hancockii
■ hancockii Baker (*W. gracillima* C. Chr.)
ilvensis (L.) R. Br. (*Acrostichum ilvense* L., *Woodsia acuminata* (Fomin) Sipl., *W. ilvensis* var. *acuminata* Fomin, *W. uralensis* Gand.) - 1, 2, 3, 4, 5, 6
- subsp. *alpina* (Bolt.) Aschers. = W. alpina
- var. *acuminata* Fomin = W. ilvensis
intermedia Rupr. = W. alpina
intermedia Tagawa = W. subintermedia
longifolia Tagawa = W. subcordata var. longifolia
macrochlaena Mett. ex Kuhn (*W. sinuata* Makino, 1897, non (Hook.) Christ, 1902) - 5
manchuriensis Hook. = Protowoodsia manchuriensis
polystichoides D. Eat. - 5
pulchella auct. p.p. = W. glabella
sinuata (Hook.) Christ = W. subcordata
sinuata Makino = W. macrochlaena
subcordata Turcz. (*W. commixta* Ching, *W. sinuata* (Hook.) Christ, 1902, non Makino, 1897) - 5
- var. **longifolia** (Tagawa) Tzvel. (*W. longifolia* Tagawa) - 5
subintermedia Tzvel. (*W. intermedia* Tagawa, 1936, non Rupr. 1845) - 5
uralensis Gand. = W. ilvensis

ZANNICHELLIACEAE Dumort.
Althenia F. Petit

filiformis F. Petit subsp. **betpakdalensis** Tzvel.- 6
- subsp. **orientalis** Tzvel. - 1, 3, 6

Zannichellia L.

clausii Tzvel. - 1
dentata Willd. = Z. pedunculata
komarovii Tzvel. - 5
major Boenn. (*Z. palustris* L. var. *major* C. Hartm., *Z. palustris* subsp. *major* (C. Hartm.) Van Ooststroom & Reichgelt) - 1, 2, 6
palustris L. (*Z. polycarpa* auct.) - 1, 3
- subsp. *dentata* (Willd.) K. Richt. = Z. pedunculata
- subsp. *major* (C. Hartm.) Van Ooststroom & Reichgelt = Z. major
- subsp. *pedicellata* (Rosen & Wahlenb.) Hegi = Z. pedunculata
- subsp. *repens* (Boenn.) Uotila = Z. repens
- subsp. *repens* (Boenn.) Schuebl. & Martens = Z. repens
- var. *major* C. Hartm. = Z. major
- var. *pedicellata* Rosen & Wahlenb. = Z. pedunculata
palustris sensu Juz. = Z. repens
pedicellata (Rosen & Wahlenb.) Fries = Z. pedunculata
pedunculata Reichenb. (? *Z. dentata* Willd., ? *Z. palustris* L. subsp. *dentata* (Willd.) K. Richt., *Z. palustris* var. *pedicellata* Rosen & Wahlenb., *Z. palustris* subsp. *pedicellata* (Rosen & Wahlenb.) Hegi, *Z. pedicellata* (Rosen & Wahlenb.) Fries) - 1, 2, 3, 4, 5, 6
repens Boenn. (*Z. palustris* L. subsp. *repens* (Boenn.) Uotila, comb. superfl., *Z. palustris* subsp. *repens* (Boenn.) Schuebl. & Martens, *Z. palustris* sensu Juz. p.max.p.) - 1, 2, 3, 4, 5, 6

ZOSTERACEAE Dumort.
Phyllospadix Hook.

iwatensis Makino (*P. scouleri* auct.) - 5
juzepczukii Tzvel. - 5
scouleri auct. = Ph. iwatensis

Zostera L.

angustifolia (Hornem.) Reichenb. (*Z. marina* L. var. *angustifolia* Hornem., *Z. hornemanniana* Tutin, *Z. marina* L. subsp. *hornemanniana* (Tutin) Lemke, nom. invalid.) - 1, 5
asiatica Miki (*Z. pacifica* auct.) - 5
caespitosa Miki - 5
caulescens Miki - 5
hornemanniana Tutin = Z. angustifolia
japonica Aschers. & Graebn. - 5
marina L. - 1, 2, 5
- subsp. *hornemanniana* (Tutin) Lemke = Z. angustifolia
- var. *angustifolia* Hornem. = Z. angustifolia
minor (Cavol.) Nolte ex Reichenb. = Z. noltii
nana Roth = Z. noltii
noltii Hornem. (*Z. minor* (Cavol.) Nolte ex Reichenb., *Z. nana* Roth, nom. illegit.) - 1, 2, 6
pacifica auct. = Z. asiatica

ZYGOPHYLLACEAE R. Br.
Halimiphyllum (Engl.) Boriss. = Zygophyllum

atriplicoides (Fisch. & C.A. Mey.) Boriss. = Zygophyllum atriplicoides

darvasicum (Boriss.) Boriss. = Zygophyllum darvasicum
eurypterum (Boiss. & Buhse) Boriss. = Zygophyllum eurypterum
gontscharovii (Boriss.) Boriss. = Zygophyllum gontscharovii
megacarpum (Boriss.) Boriss. = Zygophyllum megacarpum

Tribulus L.

longipetalus Viv. subsp. *macropterus* (Boiss.) Maire ex Ozenda & Quezel = T. macropterus
- var. *macropterus* (Boiss.) Zohary = T. macropterus
macropterus Boiss. (*T. longipetalus* Viv. subsp. *macropterus* (Boiss.) Maire ex Ozenda & Quezel, *T. longipetalus* var. *macropterus* (Boiss.) Zohary, *T. pentandrus* var. *macropterus* (Boiss.) P. Singh & V. Singh) - 6
pentandrus var. *macropterus* (Boiss.) P. Singh & V. Singh = T. macropterus
terrestris L. - 1, 2, 3, 4, 5(alien), 6

Zygophyllum L. (*Halimiphyllum* (Engl.) Boriss.)

atriplicoides Fisch. & C.A. Mey. (*Halimiphyllum atriplicoides* (Fisch. & C.A. Mey.) Boriss.) - 2, 6
balchaschense Boriss. - 6
betpakdalense Golosk. & Semiotr. - 6
brachypterum Kar. & Kir. (*Z. fabago* L. subsp. *brachypterum* (Kar. & Kir.) M. Pop.) - 3, 6
bucharicum B. Fedtsch. - 6
budunense Semiotr. = Z. oxycarpum
cuspidatum Boriss. - 6
darvasicum Boriss. (*Halimiphyllum darvasicum* (Boriss.) Boriss.) - 6
dielsianum (M. Pop.) M. Pop. = Z. jaxarticum
eichwaldii C.A. Mey. - 6
eurypterum Boiss. & Buhse (*Halimiphyllum eurypterum* (Boiss. & Buhse) Boriss.) - 6
- subsp. *gontscharovii* (Boriss.) Hadidi = Z. gontscharovii
fabago L. - 1, 2, 6
- subsp. *brachypterum* (Kar. & Kir.) M. Pop. = Z. brachypterum
- subsp. *dolichocarpum* M. Pop. ex Hadidi - 6
- subsp. *orientale* Boriss. ex Hadidi - 6
- var. *oxianum* (Boriss.) Kitam. = Z. oxianum
fabagoides M. Pop. - 6
ferganense (Drob.) Boriss. - 6
furcatum C.A. Mey. - 3, 6
gontscharovii Boriss. (*Halimiphyllum gontscharovii* (Boriss.) Boriss., *Zygophyllum eurypterum* Boiss. & Buhse subsp. *gontscharovii* (Boriss.) Hadidi) - 6
iliense M. Pop. - 6
jaxarticum M. Pop. (*Z. dielsianum* (M. Pop.) M. Pop. 1928, non Schlechter) - 6
karatavicum Boriss. - 6
kaschgaricum Boriss. - 6
kegense Boriss. - 6
kopalense Boriss. - 6
latifolium Schrenk - 6
lehmannianum Bunge - 6
macrophyllum Regel & Schmalh. (*Z. portulacoides* Cham. 1830, non Forssk. 1775) - 6
macropodum Boriss. - 6
macropterum C.A. Mey. = Z. pinnatum
megacarpum Boriss. (*Halimiphyllum megacarpum* (Boriss.) Boriss.) - 6
melongena Bunge - 3
microcarpum Boriss. - 6
miniatum Cham. & Schlecht. - 6

obliquum M. Pop. - 6
ovigerum Fisch. & C.A. Mey. ex Bunge - 1, 6
oxianum Boriss. (*Z. fabago* L. var. *oxianum* (Boriss.)
    Kitam.) - 6
oxycarpum M. Pop. (*Z. budunense* Semiotr.) - 6
pinnatum Cham. (*Z. macropterum* C.A. Mey.) - 1, 3, 6
portulacoides Cham. = Z. macrophyllum
potaninii Maxim. - 6
pterocarpum Bunge - 3, 6
rosowii Bunge - 6
stenopterum Schrenk - 6
subtrijugum C.A. Mey. - 3, 6
taldy-kurganicum Boriss. - 6
turcomanicum Fisch. ex Bunge - 6

INDEX TO FAMILY AND GENERIC NAMES